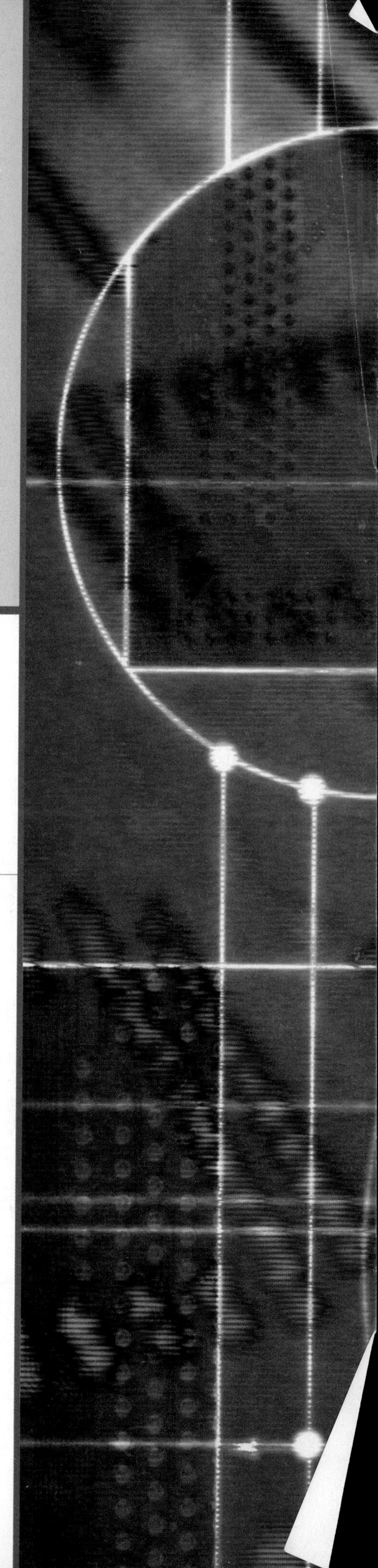

INTRODUCTORY CIRCUIT ANALYSIS

Robert L. Boylestad

Eleventh Edition

Upper Saddle River, New Jersey
Columbus, Ohio

Library of Congress Cataloging-in-Publication Data

Boylestad, Robert L.
Introductory circuit analysis / Robert L. Boylestad.—11th ed.
p. cm.
Includes index.
ISBN 0-13-173044-4
1. Electric circuits—Textbooks. 2. Electric circuit analysis—Data processing—Textbooks. 3. PSpice. I. Title.
TK454.B68. 2007
621.319'2—dc.22 2006044772

Editor in Chief: Vern Anthony
Production Editor: Rex Davidson
Design Coordinator: Diane Ernsberger
Editorial Assistant: Lara Dimmick
Cover Designer: Candace Rowley
Cover art: Getty One
Illustrations: Rolin Graphics
Production Manager: Matt Ottenweller
Senior Marketing Manager: Ben Leonard
Marketing Assistant: Les Roberts
Senior Marketing Coordinator: Liz Farrell

This book was set in Times Roman by Carlisle Publishing Services and was printed and bound by R. R. Donnelley & Sons Company. The cover was printed by Phoenix Color Corp.

Pearson Education Ltd.
Pearson Education Singapore Pte. Ltd.
Pearson Education Canada, Ltd.
Pearson Education—Japan
Pearson Education Australia Pty. Limited
Pearson Education North Asia Ltd.
Pearson Educación de Mexico, S.A. de C.V.
Pearson Education Malaysia Pte. Ltd.

10 9 8 7 6 5 4 3 2
ISBN: 0-13-173044-4

To Else Marie

PREFACE

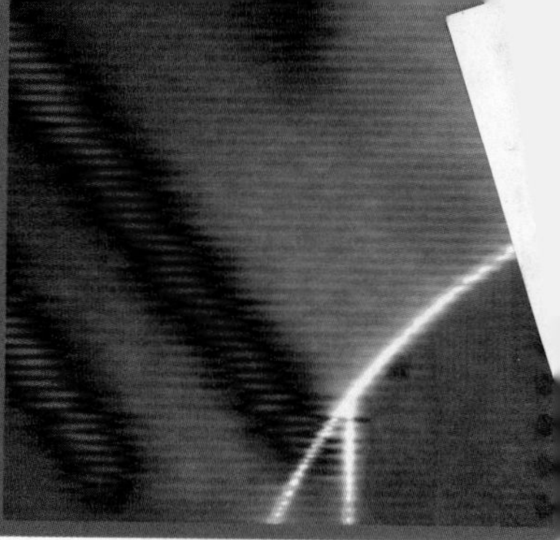

The primary goal for this extensive revision of *Introductory Circuit Analysis* is to incorporate the comments of current users and reviewers, include current trends in curriculum content, and expand on specific areas of coverage.

To accomplish this, I rearranged the order in which the information is presented so that similar topics are covered sequentially, added new material, removed less-used material, and updated content where possible. Extensive editing has ensured that the fundamental laws covered in the first half of the text are clearly understood, establishing a firm foundation for the ac section and concepts to follow in other courses in the curriculum.

The following are some of the improvements that will aid both students and teachers in using this text:

- Objectives are added to the beginning of each chapter to reveal on a broad scale what the student should learn from the chapter.
- The chapter on Inductors now follows directly after the chapter on Capacitors. The more natural flow from coverage of capacitors to inductors makes use of the similarities in circuit analysis for the two elements.
- The chapter on Magnetic Circuits now follows the chapter on Inductors. This chapter contains excellent material for practical applications that students can use as a resource later, but many teachers rarely had time to cover the detail presented in the chapter.
- The Filters chapter now follows the Resonance chapter, which again permits a continuing discussion of similar material.
- The chapter on System Analysis was deleted to permit expanded coverage of some of the important topics in the first half of the text and because so few institutions cover the material in an introductory course.
- Coverage is increased in the earlier chapters for some very important topics, such as series and parallel circuits, mesh and nodal analysis, and the transient response of capacitors and inductors.
- Problems are added in almost every chapter of the text. A review of the Solutions Manual prompted me to revise many problems so that the results better reinforce the concept under investigation and further challenge the student.

Every technique, approach, analogy, and maneuver I could imagine to clarify some of the most important concepts in this field has been adapted for this text to improve the presentation. Some of these elements include:

- A broader use of instrumentation throughout the text
- Additional examples for the most important topics
- Expanded comments to teach students how to avoid some common pitfalls
- Additional problems appropriate to introduce in this text

Throughout the text, every effort was made to ensure that the content is current. For instance, film resistors are now emphasized instead of carbon. All photos of devices and instrumentation in the text are up to date. Protoboards, surface mount resistors, fuel cells, and improved solar cells are now covered. Outdated applications have been replaced. Because the TI-86 calculator is no longer manufactured, examples now include instructions for the TI-89 calculator. However, since so many students still have the TI-86, an appendix was added so that instructions for that calculator are still available.

All of the software programs covered in this text have been upgraded. This edition uses PSpice Release 10, Multisim Version 8, and Mathcad Version 12. Users of earlier versions of these programs will find that most of the changes are minimal.

The depth of coverage for software packages always generates some discussion with each edition. I am particularly pleased that at least five editions ago *Introductory Circuit Analysis* was the first to cover software packages in the detail, depth, and range that enabled students to feel comfortable with the applications. This level of detail is not offered in other publications or in the information packages provided by the software distributors; therefore, I have decided to maintain this depth of coverage. Because it is a constant source of discussion with the publisher and some users, I deeply appreciate any comments you would like to offer on the subject. Hopefully, it is a positive aspect of the text, but coverage may need to be modified in some areas.

QBASIC and C++ have been removed from this edition due to the growing sophistication of the software packages and the need for some additional coverage in other critical areas.

For review purposes, additional summary tables are introduced to emphasize the duality that exists between various facets of the material and to review the important conclusions associated with each type of analysis. In general, these tables should help the student in developing a broad sense of understanding about how various laws can be applied to different configurations and how best to analyze a system. For some topics such as mesh and nodal analysis, power in the ac domain, and parallel ac networks, additional methods of analysis and comments have been included that should help clarify the approach and remove the confusion surrounding some unique situations.

The Laboratory Manual associated with the text is also extensively revised. One major change, based on the comments of Professor James Fiore, is the increased use of tables throughout to display the results better and permit immediate comparisons between results obtained using different approaches. Only after revising a few experiments did I realize what a wonderful improvement to the Laboratory Manual this change made. Computer exercises have been introduced into a number of the other laboratory experiments, so the computer labs have been removed.

SUPPLEMENTS

To enhance the learning process, a full supplements package accompanies this text and is available to students and instructors using the text for a course.

Student Resources

- **Laboratory Manual,** ISBN 0-13-219615-8
- **Companion Website** (student study guide) at www.prenhall.com/boylestad
- **CD-ROM.** Packaged with this textbook, this CD contains a set of Multisim circuit files and a set of PSpice circuit files. These files (also available on the Companion Website) are provided at no extra cost to the consumer and are for use by anyone who chooses to use Multisim and/or PSpice software. Multisim and PSpice are widely regarded as excellent simulation tools for classroom and laboratory learning. However, successful use of this textbook is not dependent upon the use of the circuit files.

If you do not currently have access to Multisim software on your computer or in your school lab and you wish to purchase it in order to use the circuits created for this text, visit www.prenhall.com/ewb or request information via e-mail from phewb@prenhall.com. The PSpice software can be purchased by visiting www.orcad.com.

At the time of publication, the CD provided with this text contains files appropriate for use with Multisim 2001, Multisim 7, Multisim 8, and PSpice Release 10. However, circuits created in subsequent versions of software also may be available at www.prenhall.com/boylestad whenever later versions of the software are developed by their respective manufacturers.

Instructor Resources

To access supplementary materials online, instructors need to request an instructor access code. Go to www.prenhall.com, click the **Instructor Resource Center** link, and then click **Register Today** for an instructor access code. Within 48 hours after registering you will receive a confirming e-mail including an instructor access code. Once you have received your code, go to the site and log on for full instructions on downloading the materials you wish to use.

- **Instructor's Resource Manual,** containing text solutions and test item file. Print version (ISBN 0-13-219616-6) and online version (ISBN 0-13-221446-6).
- **PowerPoint® Lecture Notes.** Available on CD (ISBN 0-13-188761-0) and online (ISBN 0-13-173557-8).
- **TestGen®,** a computerized test bank. Available on CD (ISBN 0-13-188850-1) and online (ISBN 0-13-198670-8)

ACKNOWLEDGMENTS

As I approach the completion of each edition, I always hope that the material has now been presented to the best of my ability. However, I always end up with a folder full of suggestions to consider for the next edition. This edition is no different, although I believe the revisions made have improved the content and presentation so that future changes will be minimal.

As with every edition, I have found that certain individuals in the academic world have made a significant contribution to the revision. For all the hours he spent working with me on the revision of the software printouts, I must thank my good friend Professor Louis Nashelsky. For questions of a practical nature, Jerry Sitbon's input was simply invaluable. The vast improvement in the Laboratory Manual is due primarily to the constructive criticisms of Professor James Fiore.

A number of reviewers and users provided outstanding input—I am deeply grateful for the many hours of effort on their part. Their reviews helped me to define which changes should be made and which topics should be left as is. Many comments included requests that I not lower or raise the level of the text significantly. Therefore, this text continues to provide the coverage appropriate for today's technology student without removing the challenge associated with some advanced topics.

I thank the reviewers of this edition: Sami Antoun, DeVry University; Don Barrett, DeVry University; David Barth, Edison State Community College; Jim Fiore, Mohawk Valley Community College; George Flantinis, New Hampshire Technical Institute; Curtis Johnson, University of Houston; Angela Lemons, North Carolina A&T University; Richard McKinney, Nashville State Community College, and Paul Svatik, Owens Community College.

I also thank the following individuals for their help: Leslie Bondaryk, MathSoft Corporation; Delphine Gerard, Pearson Education (Norway); Marc Herniter, Rose-Hulman Institute of Technology; Lenda Hill and Tracee Larson, Texas Instruments; Erica Kaleda, Edison Electric Institute, and Nell Mathot, Cadence Design Systems.

I have been fortunate over the years to have Rex Davidson as my production editor. He helps to ensure that the text is a quality publication. Lara Dimmick at Prentice Hall has to be one of the most capable individuals I have ever worked with. Ask her anything about anything, and she will get the answer for you—not tomorrow, but today. Thank you, Lara.

With each edition, I am always amazed at the abilities of the copy editors to detect and correct mistakes. Their suggestions and comments are valuable–they are a special group of people. For this edition, I was blessed to have Colleen Brosnan as copy editor, who transformed my input into something on another plane entirely. There are, of course, numerous other individuals in the production, sales, and marketing departments that I want to thank for this edition. The production of any publication of this magnitude is not an individual effort but one that requires teamwork and dedication. Thank you.

To all of you, both students and faculty, you have my best wishes for a healthy, productive, and pleasant school year.

Robert Boylestad

Brief Contents

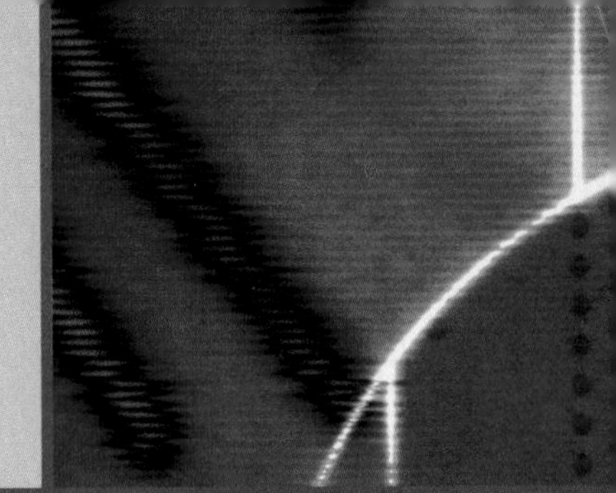

Contents

5

Series dc Circuits 131

6

Parallel dc Circuits 185

7

Series-Parallel Circuits 243

8

Methods of Analysis and Selected Topics (dc) 283

9

Network Theorems 345

10

Capacitors 397

11

Inductors 461

12

Magnetic Circuits 513

13

Sinusoidal Alternating Waveforms 539

14

The Basic Elements and Phasors 587

1 INTRODUCTION

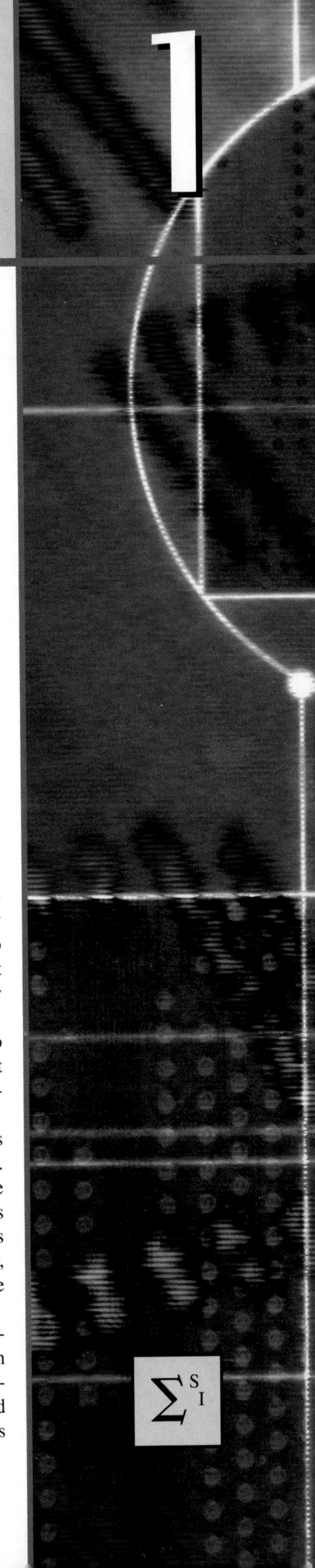

Objectives

- *Become aware of the rapid growth of the electrical/electronics industry over the past century.*
- *Understand the importance of applying a unit of measurement to a result or measurement and to ensuring that the numerical values substituted into an equation are consistent with the unit of measurement of the various quantities.*
- *Become familiar with the SI system of units used throughout the electrical/electronics industry.*
- *Understand the importance of powers of ten and how to work with them in any numerical calculation.*
- *Be able to convert any quantity, in any system of units, to another system with confidence.*

1.1 THE ELECTRICAL/ELECTRONICS INDUSTRY

Over the past few decades, technology has been changing at an ever-increasing rate. The pressure to develop new products, improve the performance of existing systems, and create new markets will only accelerate that rate. This pressure, however, is also what makes the field so exciting. New ways of storing information, constructing integrated circuits, or hardware that contains software components that can "think" on their own based on data input are only a few possibilities.

Change has always been part of the human experience, but it used to be gradual. This is no longer true. Just think, for example, that it was only a few years ago that TVs with wide, flat screens were introduced. Already, these have been eclipsed by high-definition TVs with images so crystal clear that they seem almost three-dimensional.

Miniaturization has also made possible huge advances in electronic systems. Cell phones that originally were the size of notebooks are now about the size of a deck of playing cards. Plus, these new versions record short videos, transmit photos, send text messages, and have calendars, reminders, calculators, games, and lists of frequently called numbers. Boom boxes playing audio cassettes have been replaced by pocket-sized iPods® that can store 15,000 songs or 25,000 photos. Hearing aids with higher power levels that are almost invisible in the ear, TVs with 1-inch screens—the list of new or improved products continues to expand because significantly smaller electronic systems have been developed.

This reduction in size of electronic systems is due primarily to an important innovation introduced less than 50 years ago (1958)—the **integrated circuit (IC).** An integrated circuit can now contain features less than 100 nanometers across. The fact that measurements are now being made in nanometers has resulted in the terminology **nanotechnology** producing integrated circuits called *nanochips.* To understand nanometers, consider drawing one hundred lines

Σ^{S}_{I}

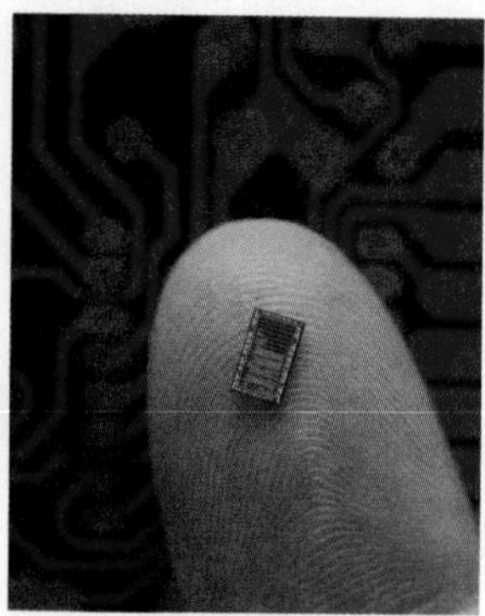

FIG. 1.1
Computer chip.
(© Roger Du Buisson/Corbis)

within the boundaries of one inch. Then attempt drawing 1000 lines within the same length. Cutting 100 nanometer features would require drawing over 250,000 lines in one inch. The integrated circuit shown in Fig. 1.1 is the Intel® Pentium® 4 processor, which has 42 million transistors in an area measuring 0.34 square inches. Intel Corporation recently presented a technical paper describing 20 nanometer (0.02 micrometers) transistors, developed in its silicon research laboratory. These very small, ultrafast transistors will permit placing nearly 1 billion transistors on a sliver of silicon smaller than a stamp—an incredible accomplishment.

However, before a decision is made on such dramatic reductions in size, the system must be designed and tested to determine if it is worth constructing as an integrated circuit. That design process requires engineers who know the characteristics of each device used in the system, including undesirable characteristics that are part of any electronic element. In other words, there are *no ideal (perfect) elements* in an electronic design. Considering the limitations of each component is necessary to ensure a reliable response under all conditions of temperature, vibration, and effects of the surrounding environment. To develop this awareness requires time and must begin with understanding the basic characteristics of the device, as covered in this text. One of the objectives of this text is to explain how ideal components work and their function in a network. Another is to explain conditions in which components may not be ideal.

One of the very positive aspects of the learning process associated with electric and electronic circuits is that once a concept or procedure is clearly and correctly understood, it will be useful throughout the career of the individual at any level in the industry. Once a law or equation is understood, it will not be replaced by another equation as the material becomes more advanced and complicated. For instance, one of the first laws to be introduced is Ohm's law. This law provides a relationship between forces and components that will always be true, no matter how complicated the system becomes. In fact, it is an equation that will be applied in various forms throughout the design of the entire system. The use of the basic laws may change, but the laws will not change and will always be applicable.

It is vitally important to understand that the learning process for circuit analysis is sequential. That is, the first few chapters establish the foundation for the remaining chapters. Failure to properly understand the opening chapters will only lead to difficulties understanding the material in the chapters to follow. This first chapter provides a brief history of the field followed by a review of mathematical concepts necessary to understand the rest of the material.

1.2 A BRIEF HISTORY

In the sciences, once a hypothesis is proven and accepted, it becomes one of the building blocks of that area of study, permitting additional investigation and development. Naturally, the more pieces of a puzzle available, the more obvious the avenue toward a possible solution. In fact, history demonstrates that a single development may provide the key that will result in a mushroom effect that brings the science to a new plateau of understanding and impact.

If the opportunity presents itself, read one of the many publications reviewing the history of this field. Space requirements are such that only a brief review can be provided here. There are many more contributors than could be listed, and their efforts have often provided important keys to the solution of some very important concepts.

Throughout history, some periods were characterized by what appeared to be an explosion of interest and development in particular areas. As you will see from the discussion of the late 1700s and the early 1800s, inventions, discoveries, and theories came fast and furiously. Each new concept has broadened the possible areas of application until it becomes almost impossible to trace developments without picking a particular area of interest and following it through. In the review, as you read about the development of the radio, television, and computer, keep in mind that similar progressive steps were occurring in the areas of the telegraph, the telephone, power generation, the phonograph, appliances, and so on.

There is a tendency when reading about the great scientists, inventors, and innovators to believe that their contribution was a totally individual effort. In many instances, this was not the case. In fact, many of the great contributors were friends or associates who provided support and encouragement in their efforts to investigate various theories. At the very least, they were aware of one another's efforts to the degree possible in the days when a letter was often the best form of communication. In particular, note the closeness of the dates during periods of rapid development. One contributor seemed to spur on the efforts of the others or possibly provided the key needed to continue with the area of interest.

In the early stages, the contributors were not electrical, electronic, or computer engineers as we know them today. In most cases, they were physicists, chemists, mathematicians, or even philosophers. In addition, they were not from one or two communities of the Old World. The home country of many of the major contributors introduced in the paragraphs to follow is provided to show that almost every established community had some impact on the development of the fundamental laws of electrical circuits.

As you proceed through the remaining chapters of the text, you will find that a number of the units of measurement bear the name of major contributors in those areas—*volt* after Count Alessandro Volta, *ampere* after André Ampère, *ohm* after Georg Ohm, and so forth—fitting recognition for their important contributions to the birth of a major field of study.

Time charts indicating a limited number of major developments are provided in Fig. 1.2, primarily to identify specific periods of rapid development and to reveal how far we have come in the last few decades. In essence, the current state of the art is a result of efforts that began in earnest some 250 years ago, with progress in the last 100 years almost exponential.

As you read through the following brief review, try to sense the growing interest in the field and the enthusiasm and excitement that must have accompanied each new revelation. Although you may find some of the terms used in the review new and essentially meaningless, the remaining chapters will explain them thoroughly.

The Beginning

The phenomenon of **static electricity** has intrigued scholars throughout history. The Greeks called the fossil resin substance so often used to demonstrate the effects of static electricity *elektron,* but no extensive study was made of the subject until William Gilbert researched the event in 1600. In the years to follow, there was a continuing investigation of electrostatic charge by many individuals such as Otto von Guericke, who

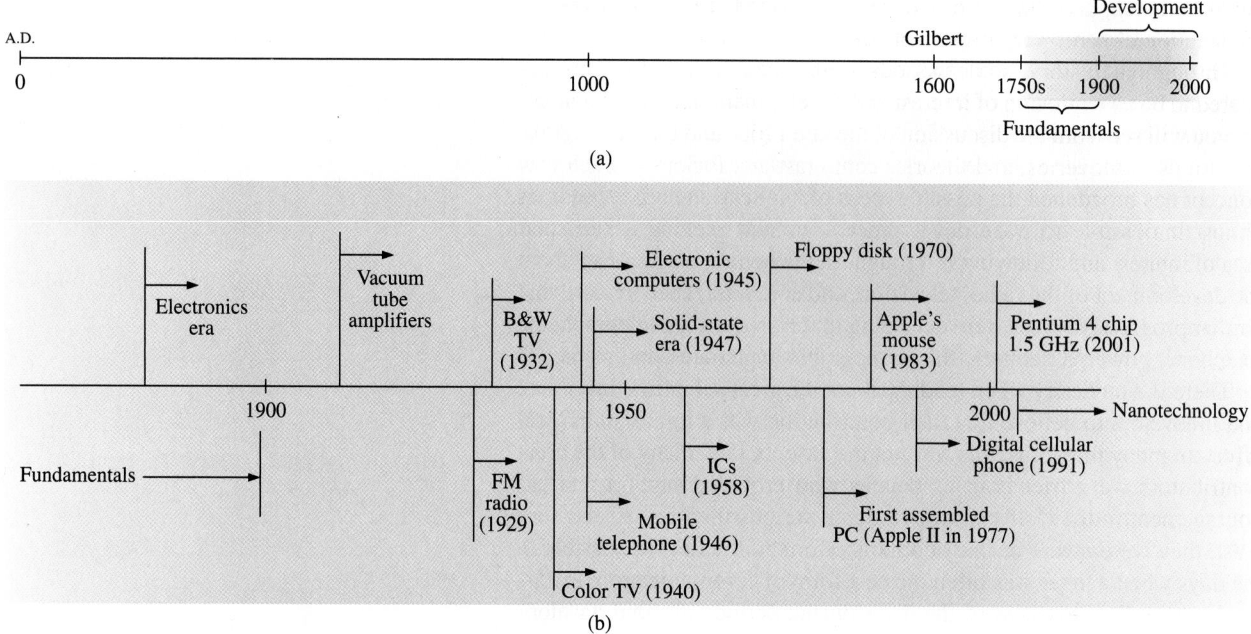

FIG. 1.2
Time charts: (a) long-range; (b) expanded.

developed the first machine to generate large amounts of charge, and Stephen Gray, who was able to transmit electrical charge over long distances on silk threads. Charles DuFay demonstrated that charges either attract or repel each other, leading him to believe that there were two types of charge—a theory we subscribe to today with our defined positive and negative charges.

There are many who believe that the true beginnings of the electrical era lie with the efforts of Pieter van Musschenbroek and Benjamin Franklin. In 1745, van Musschenbroek introduced the **Leyden jar** for the storage of electrical charge (the first capacitor) and demonstrated electrical shock (and therefore the power of this new form of energy). Franklin used the Leyden jar some seven years later to establish that lightning is simply an electrical discharge, and he expanded on a number of other important theories including the definition of the two types of charge as *positive* and *negative.* From this point on, new discoveries and theories seemed to occur at an increasing rate as the number of individuals performing research in the area grew.

In 1784, Charles Coulomb demonstrated in Paris that the force between charges is inversely related to the square of the distance between the charges. In 1791, Luigi Galvani, professor of anatomy at the University of Bologna, Italy, performed experiments on the effects of electricity on animal nerves and muscles. The first **voltaic cell,** with its ability to produce electricity through the chemical action of a metal dissolving in an acid, was developed by another Italian, Alessandro Volta, in 1799.

The fever pitch continued into the early 1800s with Hans Christian Oersted, a Danish professor of physics, announcing in 1820 a relationship between magnetism and electricity that serves as the foundation for the theory of **electromagnetism** as we know it today. In the same year, a French physicist, André Ampère, demonstrated that there are magnetic effects around every current-carrying conductor and that current-carrying conductors can attract and repel each other just like magnets. In the period 1826 to 1827, a German physicist, Georg Ohm, introduced an important relationship between potential, current, and resistance which we now refer to as *Ohm's law.* In 1831, an English physicist, Michael Faraday, demonstrated his theory of *electromagnetic induction,* whereby a changing current in one coil can induce a changing current in another coil, even though the two coils are not directly connected. Professor Faraday also did extensive work on a storage device he called the condenser, which we refer to today as a *capacitor.* He introduced the idea of adding a dielectric between the plates of a capacitor to increase the storage capacity (Chapter 10). James Clerk Maxwell, a Scottish professor of natural philosophy, performed extensive mathematical analyses to develop what are currently called *Maxwell's equations,* which support the efforts of Faraday linking electric and magnetic effects. Maxwell also developed the *electromagnetic theory of light* in 1862, which, among other things, revealed that electromagnetic waves travel through air at the velocity of light (186,000 miles per second or 3×10^8 meters per second). In 1888, a German physicist, Heinrich Rudolph Hertz, through experimentation with lower-frequency electromagnetic waves (microwaves), substantiated Maxwell's predictions and equations. In the mid-1800s, Professor Gustav Robert Kirchhoff introduced a series of laws of voltages and currents that find application at every level and area of this field (Chapters 5 and 6). In 1895, another German physicist, Wilhelm Röntgen, discovered electromagnetic waves of high frequency, commonly called *X-rays* today.

By the end of the 1800s, a significant number of the fundamental equations, laws, and relationships had been established, and various fields of study, including electricity, electronics, power generation and distribution, and communication systems, started to develop in earnest.

The Age of Electronics

Radio The true beginning of the electronics era is open to debate and is sometimes attributed to efforts by early scientists in applying potentials across evacuated glass envelopes. However, many trace the beginning to Thomas Edison, who added a metallic electrode to the vacuum of the tube and discovered that a current was established between the metal electrode and the filament when a positive voltage was applied to the metal electrode. The phenomenon, demonstrated in 1883, was referred to as the **Edison effect.** In the period to follow, the transmission of radio waves and the development of the radio received widespread attention. In 1887, Heinrich Hertz, in his efforts to verify Maxwell's equations, transmitted radio waves for the first time in his laboratory. In 1896, an Italian scientist, Guglielmo Marconi (often called the father of the radio), demonstrated that telegraph signals could be sent through the air over long distances (2.5 kilometers) using a grounded antenna. In the same year, Aleksandr Popov sent what might have been the first radio message some 300 yards. The message was the name "*Heinrich Hertz*" in respect for Hertz's earlier contributions. In 1901, Marconi established radio communication across the Atlantic.

In 1904, John Ambrose Fleming expanded on the efforts of Edison to develop the first diode, commonly called **Fleming's valve**—actually the first of the *electronic devices.* The device had a profound impact on the design of detectors in the receiving section of radios. In 1906, Lee De Forest added a third element to the vacuum structure and created the first amplifier, the triode. Shortly thereafter, in 1912, Edwin Armstrong built the first regenerative circuit to improve receiver capabilities and then used the same contribution to develop the first nonmechanical oscillator. By 1915, radio signals were being transmitted across the United States, and in 1918 Armstrong applied for a patent for the superheterodyne circuit employed in virtually every television and radio to permit amplification at one frequency rather than at the full range of incoming signals. The major components of the modern-day radio were now in place, and sales in radios grew from a few million dollars in the early 1920s to over $1 billion by the 1930s. The 1930s were truly the golden years of radio, with a wide range of productions for the listening audience.

Television The 1930s were also the true beginnings of the television era, although development on the picture tube began in earlier years with Paul Nipkow and his *electrical telescope* in 1884 and John Baird and his long list of successes, including the transmission of television pictures over telephone lines in 1927 and over radio waves in 1928, and simultaneous transmission of pictures and sound in 1930. In 1932, NBC installed the first commercial television antenna on top of the Empire State Building in New York City, and RCA began regular broadcasting in 1939. The war slowed development and sales, but in the mid-1940s the number of sets grew from a few thousand to a few million. Color television became popular in the early 1960s.

Computers The earliest computer system can be traced back to Blaise Pascal in 1642 with his mechanical machine for adding and subtracting numbers. In 1673, Gottfried Wilhelm von Leibniz used the *Leibniz wheel* to add multiplication and division to the range of operations, and in 1823 Charles Babbage developed the **difference engine** to add the mathematical operations of sine, cosine, logs, and several others. In the years to follow, improvements were made, but the system remained primarily mechanical until the 1930s when electromechanical systems using components such as relays were introduced. It was not until the 1940s that totally electronic systems became the new wave. It is interesting to note that, even though IBM was formed in 1924, it did not enter the computer industry until 1937. An entirely electronic system known as **ENIAC** was dedicated at the University of Pennsylvania in 1946. It contained 18,000 tubes and weighed 30 tons but was several times faster than most electromechanical systems. Although other vacuum tube systems were built, it was not until the birth of the solid-state era that computer systems experienced a major change in size, speed, and capability.

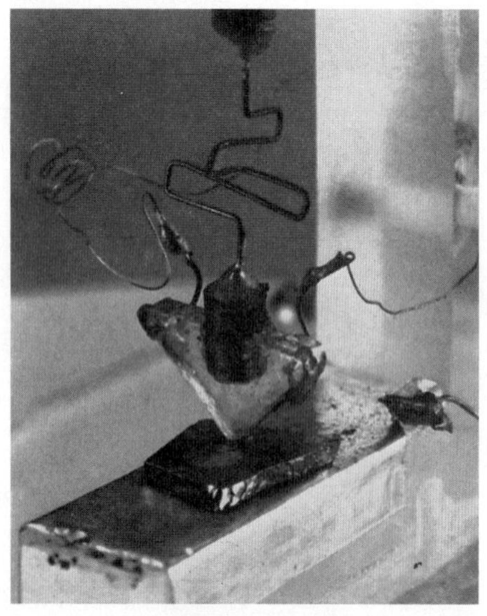

FIG. 1.3
The first transistor.
(Used with permission of Lucent Technologies Inc./ Bell Labs.)

The Solid-State Era

In 1947, physicists William Shockley, John Bardeen, and Walter H. Brattain of Bell Telephone Laboratories demonstrated the point-contact **transistor** (Fig. 1.3), an amplifier constructed entirely of solid-state materials with no requirement for a vacuum, glass envelope, or heater voltage for the filament. Although reluctant at first due to the vast amount of material available on the design, analysis, and synthesis of tube networks, the industry eventually accepted this new technology as the wave of the future. In 1958, the first **integrated circuit (IC)** was developed at Texas

Instruments, and in 1961 the first commercial integrated circuit was manufactured by the Fairchild Corporation.

It is impossible to review properly the entire history of the electrical/electronics field in a few pages. The effort here, both through the discussion and the time graphs in Fig. 1.2, was to reveal the amazing progress of this field in the last 50 years. The growth appears to be truly exponential since the early 1900s, raising the interesting question, Where do we go from here? The time chart suggests that the next few decades will probably contain many important innovative contributions that may cause an even faster growth curve than we are now experiencing.

1.3 UNITS OF MEASUREMENT

One of the most important rules to remember and apply when working in any field of technology is to use the correct units when substituting numbers into an equation. Too often we are so intent on obtaining a numerical solution that we overlook checking the units associated with the numbers being substituted into an equation. Results obtained, therefore, are often meaningless. Consider, for example, the following very fundamental physics equation:

$$v = \frac{d}{t} \qquad \begin{aligned} v &= \text{velocity} \\ d &= \text{distance} \\ t &= \text{time} \end{aligned} \tag{1.1}$$

Assume, for the moment, that the following data are obtained for a moving object:

$$d = 4000 \text{ ft}$$
$$t = 1 \text{ min}$$

and v is desired in miles per hour. Often, without a second thought or consideration, the numerical values are simply substituted into the equation, with the result here that

$$v = \frac{d}{t} = \frac{4000 \text{ ft}}{1 \text{ min}} = \cancel{4000 \text{ mph}}$$

As indicated above, the solution is totally incorrect. If the result is desired in *miles per hour,* the unit of measurement for distance must be *miles,* and that for time, *hours.* In a moment, when the problem is analyzed properly, the extent of the error will demonstrate the importance of ensuring that

the numerical value substituted into an equation must have the unit of measurement specified by the equation.

The next question is normally, How do I convert the distance and time to the proper unit of measurement? A method is presented in Section 1.9 of this chapter, but for now it is given that

$$1 \text{ mi} = 5280 \text{ ft}$$
$$4000 \text{ ft} = 0.76 \text{ mi}$$
$$1 \text{ min} = \tfrac{1}{60} \text{ h} = 0.017 \text{ h}$$

Substituting into Eq. (1.1), we have

$$v = \frac{d}{t} = \frac{0.76 \text{ mi}}{0.017 \text{ h}} = \mathbf{44.71\ mph}$$

which is significantly different from the result obtained before.

To complicate the matter further, suppose the distance is given in kilometers, as is now the case on many road signs. First, we must realize that the prefix *kilo* stands for a multiplier of 1000 (to be introduced in Section 1.5), and then we must find the conversion factor between kilometers and miles. If this conversion factor is not readily available, we must be able to make the conversion between units using the conversion factors between meters and feet or inches, as described in Section 1.9.

Before substituting numerical values into an equation, try to mentally establish a reasonable range of solutions for comparison purposes. For instance, if a car travels 4000 ft in 1 min, does it seem reasonable that the speed would be 4000 mph? Obviously not! This self-checking procedure is particularly important in this day of the hand-held calculator, when ridiculous results may be accepted simply because they appear on the digital display of the instrument.

Finally,

if a unit of measurement is applicable to a result or piece of data, then it must be applied to the numerical value.

To state that $v = 44.71$ without including the unit of measurement *mph* is meaningless.

Eq. (1.1) is not a difficult one. A simple algebraic manipulation will result in the solution for any one of the three variables. However, in light of the number of questions arising from this equation, the reader may wonder if the difficulty associated with an equation will increase at the same rate as the number of terms in the equation. In the broad sense, this will not be the case. There is, of course, more room for a mathematical error with a more complex equation, but once the proper system of units is chosen and each term properly found in that system, there should be very little added difficulty associated with an equation requiring an increased number of mathematical calculations.

In review, before substituting numerical values into an equation, be absolutely sure of the following:

1. ***Each quantity has the proper unit of measurement as defined by the equation.***
2. ***The proper magnitude of each quantity as determined by the defining equation is substituted.***
3. ***Each quantity is in the same system of units (or as defined by the equation).***
4. ***The magnitude of the result is of a reasonable nature when compared to the level of the substituted quantities.***
5. ***The proper unit of measurement is applied to the result.***

1.4 SYSTEMS OF UNITS

In the past, the *systems of units* most commonly used were the English and metric, as outlined in Table 1.1. Note that while the English system is based on a single standard, the metric is subdivided into two interrelated standards: the **MKS** and the **CGS.** Fundamental quantities of these systems are compared in Table 1.1 along with their abbreviations. The MKS and CGS systems draw their names from the units of measurement used with each system; the MKS system uses *M*eters, *K*ilograms, and *S*econds, while the CGS system uses *C*entimeters, *G*rams, and *S*econds.

Understandably, the use of more than one system of units in a world that finds itself continually shrinking in size, due to advanced technical

TABLE 1.1
Comparison of the English and metric systems of units.

English	Metric		
	MKS	CGS	SI
Length: Yard (yd) (0.914 m)	Meter (m) (39.37 in.) (100 cm)	Centimeter (cm) (2.54 cm = 1 in.)	**Meter (m)**
Mass: **Slug** (14.6 kg)	Kilogram (kg) (1000 g)	Gram (g)	**Kilogram (kg)**
Force: **Pound** (lb) (4.45 N)	Newton (N) (100,000 dynes)	Dyne	**Newton (N)**
Temperature: Fahrenheit (°F) $\left(=\frac{9}{5}°C + 32\right)$	Celsius or Centigrade (°C) $\left(=\frac{5}{9}(°F - 32)\right)$	Centigrade (°C)	**Kelvin (K)** K = 273.15 + °C
Energy: Foot-pound (ft-lb) (1.356 joules)	Newton-meter (N•m) or joule (J) (0.7376 ft-lb)	Dyne-centimeter or erg (1 joule = 10^7 ergs)	**Joule (J)**
Time: Second (s)	Second (s)	Second (s)	**Second (s)**

developments in communications and transportation, would introduce unnecessary complications to the basic understanding of any technical data. The need for a standard set of units to be adopted by all nations has become increasingly obvious. The International Bureau of Weights and Measures located at Sèvres, France, has been the host for the General Conference of Weights and Measures, attended by representatives from all nations of the world. In 1960, the General Conference adopted a system called Le Système International d'Unités (International System of Units), which has the international abbreviation **SI.** Since then, it has been adopted by the Institute of Electrical and Electronic Engineers, Inc. (IEEE) in 1965 and by the United States of America Standards Institute in 1967 as a standard for all scientific and engineering literature.

For comparison, the SI units of measurement and their abbreviations appear in Table 1.1. These abbreviations are those usually applied to each unit of measurement, and they were carefully chosen to be the most effective. Therefore, it is important that they be used whenever applicable to ensure universal understanding. Note the similarities of the SI system to the MKS system. This text uses, whenever possible and practical, all of the major units and abbreviations of the SI system in an effort to support the need for a universal system. Those readers requiring additional information on the SI system should contact the information office of the American Society for Engineering Education (ASEE).*

Figure 1.4 should help you develop some feeling for the relative magnitudes of the units of measurement of each system of units. Note

*American Society for Engineering Education (ASEE), 1818 N Street N.W., Suite 600, Washington, D.C. 20036-2479; (202) 331-3500; http://www.asee.org/.

Length:

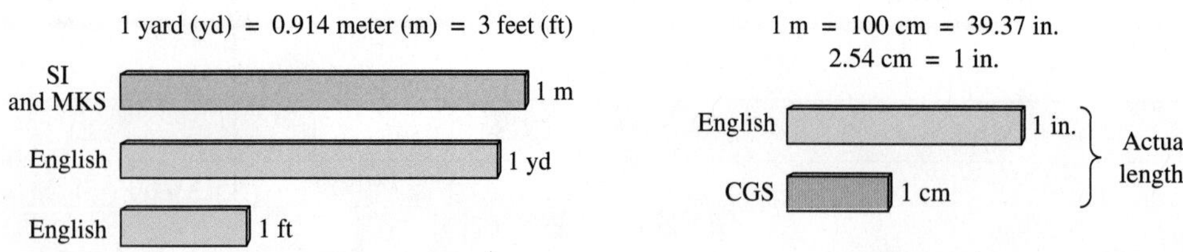

1 ft

Mass:

1 slug = 14.6 kilograms

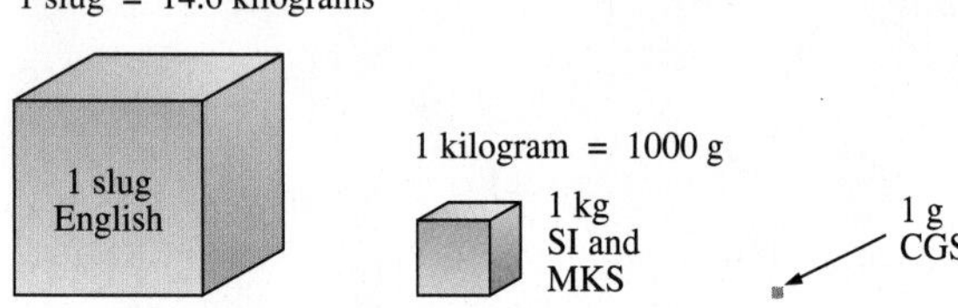

Force:

English
1 pound (lb)

SI and
MKS
1 newton (N)

1 dyne (CGS)

1 pound (lb) = 4.45 newtons (N)
1 newton = 100,000 dynes (dyn)

Temperature:

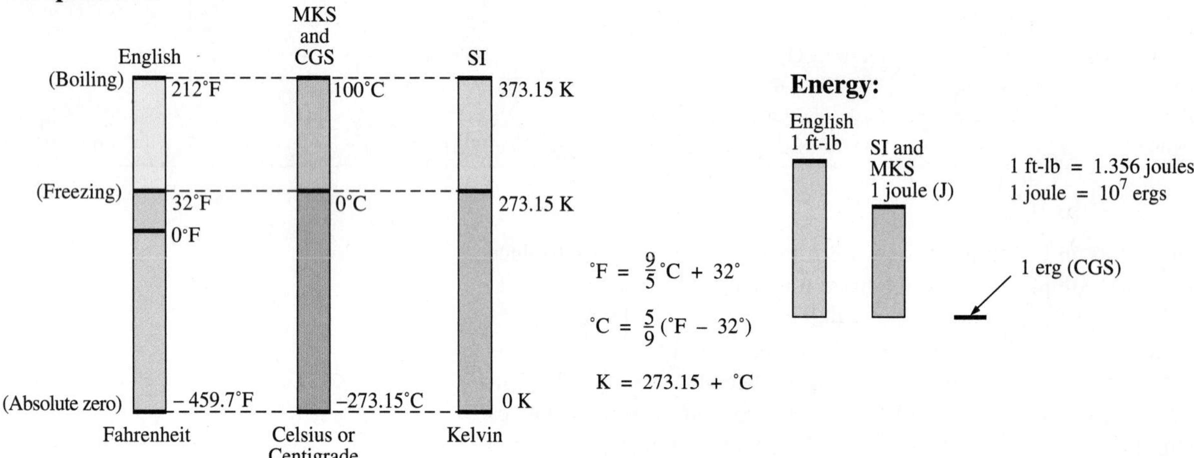

$$°F = \frac{9}{5}°C + 32°$$

$$°C = \frac{5}{9}(°F - 32°)$$

$$K = 273.15 + °C$$

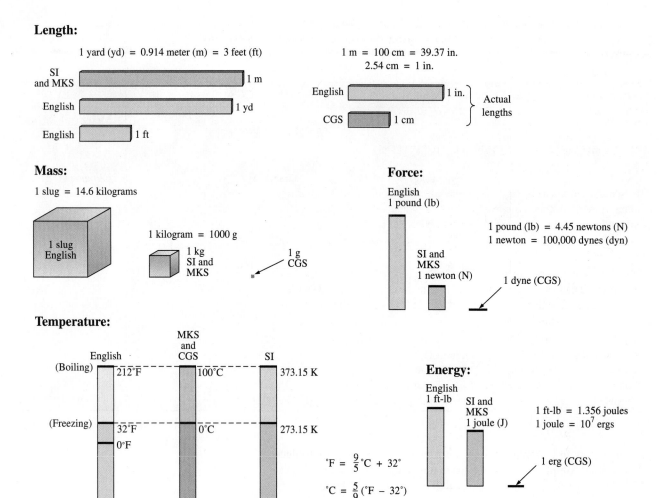

FIG. 1.4
Comparison of units of the various systems of units.

in the figure the relatively small magnitude of the units of measurement for the CGS system.

A standard exists for each unit of measurement of each system. The standards of some units are quite interesting.

The **meter** was originally defined in 1790 to be 1/10,000,000 the distance between the equator and either pole at sea level, a length preserved on a platinum-iridium bar at the International Bureau of Weights and Measures at Sèvres, France.

The meter is now defined with reference to the speed of light in a vacuum, which is 299,792,458 m/s.

The kilogram is defined as a mass equal to 1000 times the mass of one cubic centimeter of pure water at 4°C.

This standard is preserved in the form of a platinum-iridium cylinder in Sèvres.

The **second** was originally defined as 1/86,400 of the mean solar day. However, since Earth's rotation is slowing down by almost 1 second every 10 years,

the second was redefined in 1967 as 9,192,631,770 periods of the electromagnetic radiation emitted by a particular transition of cesium atom.

1.5 SIGNIFICANT FIGURES, ACCURACY, AND ROUNDING OFF

This section emphasizes the importance of knowing the source of a piece of data, how a number appears, and how it should be treated. Too often we write numbers in various forms with little concern for the format used, the number of digits that should be included, and the unit of measurement to be applied.

For instance, measurements of 22.1″ and 22.10″ imply different levels of accuracy. The first suggests that the measurement was made by an instrument accurate only to the tenths place; the latter was obtained with instrumentation capable of reading to the hundredths place. The use of zeros in a number, therefore, must be treated with care and the implications must be understood.

In general, there are two types of numbers: *exact* and *approximate.* Exact numbers are precise to the exact number of digits presented, just as we know that there are 12 apples in a dozen and not 12.1. Throughout the text, the numbers that appear in the descriptions, diagrams, and examples are considered *exact,* so that a battery of 100 V can be written as 100.0 V, 100.00 V, and so on, since it is 100 V at any level of precision. The additional zeros were not included for purposes of clarity. However, in the laboratory environment, where measurements are continually being taken and the level of accuracy can vary from one instrument to another, it is important to understand how to work with the results. Any reading obtained in the laboratory should be considered *approximate.* The analog scales with their pointers may be difficult to read, and even though the digital meter provides only specific digits on its display, it is limited to the number of digits it can provide, leaving us to wonder about the less significant digits not appearing on the display.

The precision of a reading can be determined by the number of *significant figures (digits)* present. Significant digits are those integers (0 to 9) that can be assumed to be accurate for the measurement being made. The result is that all nonzero numbers are considered significant, with zeros being significant in only some cases. For instance, the zeros in 1005 are considered significant because they define the size of the number and are surrounded by nonzero digits. For the number 0.4020, the zero to the left of the decimal point is not significant, but the other two zeros are because they define the magnitude of the number and the fourth-place accuracy of the reading.

When adding approximate numbers, it is important to be sure that the accuracy of the readings is consistent throughout. To add a quantity accurate only to the tenths place to a number accurate to the thousandths place will result in a total having accuracy only to the tenths place. One cannot expect the reading with the higher level of accuracy to improve the reading with only tenths-place accuracy.

In the addition or subtraction of approximate numbers, the entry with the lowest level of accuracy determines the format of the solution.

For the multiplication and division of approximate numbers, the result has the same number of significant figures as the number with the least number of significant figures.

For approximate numbers (and exact, for that matter), there is often a need to *round off* the result; that is, you must decide on the appropriate level of accuracy and alter the result accordingly. The accepted procedure is simply to note the digit following the last to appear in the rounded-off form, and add a 1 to the last digit if it is greater than or equal to 5, and leave it alone if it is less than 5. For example, $3.186 \cong 3.19 \cong 3.2$, depending on the level of precision desired. The symbol $\cong$ appearing means *approximately equal to.*

EXAMPLE 1.1 Perform the indicated operations with the following approximate numbers and round off to the appropriate level of accuracy.

a. $532.6 + 4.02 + 0.036 = 536.656 \cong$ **536.7** (as determined by 532.6)
b. $0.04 + 0.003 + 0.0064 = 0.0494 \cong$ **0.05** (as determined by 0.04)

EXAMPLE 1.2 Round off the following numbers to the hundredths place.

a. 32.419 = **32.42**
b. 0.05328 = **0.05**

EXAMPLE 1.3 Round off the following numbers to the thousandths place.

a. 0.00738 = **0.007**
b. 46.23550 = **46.236**

1.6 POWERS OF TEN

It should be apparent from the relative magnitude of the various units of measurement that very large and very small numbers are frequently encountered in the sciences. To ease the difficulty of mathematical operations with numbers of such varying size, *powers of ten* are usually employed. This notation takes full advantage of the mathematical properties of powers of ten. The notation used to represent numbers that are integer powers of ten is as follows:

$$1 = 10^0 \qquad 1/10 = 0.1 = 10^{-1}$$
$$10 = 10^1 \qquad 1/100 = 0.01 = 10^{-2}$$
$$100 = 10^2 \qquad 1/1000 = 0.001 = 10^{-3}$$
$$1000 = 10^3 \qquad 1/10{,}000 = 0.0001 = 10^{-4}$$

In particular, note that $10^0 = 1$, and, in fact, any quantity to the zero power is 1 ($x^0 = 1$, $1000^0 = 1$, and so on). Numbers in the list *greater than 1 are associated with positive powers of ten,* and numbers in the list *less than 1 are associated with negative powers of ten.*

A quick method of determining the proper power of ten is to place a caret mark to the right of the numeral 1 wherever it may occur; then count from this point to the number of places to the right or left before arriving at the decimal point. Moving to the right indicates a positive power of ten, whereas moving to the left indicates a negative power. For example,

$$10{,}000.0 = 1\underbrace{0{,}000}_{1\ 2\ 3\ 4}. = 10^{+4}$$

$$0.00001 = 0.\underbrace{00001}_{5\ 4\ 3\ 2\ 1} = 10^{-5}$$

Some important mathematical equations and relationships pertaining to powers of ten are listed below, along with a few examples. In each case, n and m can be any positive or negative real number.

$$\frac{1}{10^n} = 10^{-n} \qquad \frac{1}{10^{-n}} = 10^n \tag{1.2}$$

Eq. (1.2) clearly reveals that shifting a power of ten from the denominator to the numerator, or the reverse, requires simply changing the sign of the power.

EXAMPLE 1.4

a. $\dfrac{1}{1000} = \dfrac{1}{10^{+3}} = \mathbf{10^{-3}}$

b. $\dfrac{1}{0.00001} = \dfrac{1}{10^{-5}} = \mathbf{10^{+5}}$

The product of powers of ten:

$$(10^n)(10^m) = (10)^{(n+m)} \tag{1.3}$$

EXAMPLE 1.5

a. $(1000)(10{,}000) = (10^3)(10^4) = 10^{(3+4)} = \mathbf{10^7}$
b. $(0.00001)(100) = (10^{-5})(10^2) = 10^{(-5+2)} = \mathbf{10^{-3}}$

The division of powers of ten:

$$\frac{10^n}{10^m} = 10^{(n-m)} \tag{1.4}$$

EXAMPLE 1.6

a. $\dfrac{100{,}000}{100} = \dfrac{10^5}{10^2} = 10^{(5-2)} = \mathbf{10^3}$

b. $\dfrac{1000}{0.0001} = \dfrac{10^3}{10^{-4}} = 10^{(3-(-4))} = 10^{(3+4)} = \mathbf{10^7}$

Note the use of parentheses in part (b) to ensure that the proper sign is established between operators.

The power of powers of ten:

$$(10^n)^m = 10^{nm} \tag{1.5}$$

EXAMPLE 1.7

a. $(100)^4 = (10^2)^4 = 10^{(2)(4)} = \mathbf{10^8}$
b. $(1000)^{-2} = (10^3)^{-2} = 10^{(3)(-2)} = \mathbf{10^{-6}}$
c. $(0.01)^{-3} = (10^{-2})^{-3} = 10^{(-2)(-3)} = \mathbf{10^6}$

Basic Arithmetic Operations

Let us now examine the use of powers of ten to perform some basic arithmetic operations using numbers that are not just powers of ten. The number 5000 can be written as $5 \times 1000 = 5 \times 10^3$, and the number 0.0004 can be written as $4 \times 0.0001 = 4 \times 10^{-4}$. Of course, 10^5 can also be written as 1×10^5 if it clarifies the operation to be performed.

Addition and Subtraction To perform addition or subtraction using powers of ten, the power of ten *must be the same for each term;* that is,

$$A \times 10^n \pm B \times 10^n = (A \pm B) \times 10^n \tag{1.6}$$

Eq. (1.6) covers all possibilities, but students often prefer to remember a verbal description of how to perform the operation.

Eq. (1.6) states

when adding or subtracting numbers in a power-of-ten format, be sure that the power of ten is the same for each number. Then separate the multipliers, perform the required operation, and apply the same power of ten to the result.

EXAMPLE 1.8

a.
$$\begin{aligned} 6300 + 75{,}000 &= (6.3)(1000) + (75)(1000) \\ &= 6.3 \times 10^3 + 75 \times 10^3 \\ &= (6.3 + 75) \times 10^3 \\ &= \mathbf{81.3 \times 10^3} \end{aligned}$$

b.
$$\begin{aligned} 0.00096 - 0.000086 &= (96)(0.00001) - (8.6)(0.00001) \\ &= 96 \times 10^{-5} - 8.6 \times 10^{-5} \\ &= (96 - 8.6) \times 10^{-5} \\ &= \mathbf{87.4 \times 10^{-5}} \end{aligned}$$

Multiplication In general,

$$(A \times 10^n)(B \times 10^m) = (A)(B) \times 10^{n+m} \tag{1.7}$$

revealing that the *operations with the power of ten can be separated from the operation with the multipliers.*

Eq. (1.7) states

when multiplying numbers in the power-of-ten format, first find the product of the multipliers and then determine the power of ten for the result by adding the power-of-ten exponents.

EXAMPLE 1.9

a. $(0.0002)(0.000007) = [(2)(0.0001)][(7)(0.000001)]$
$= (2 \times 10^{-4})(7 \times 10^{-6})$
$= (2)(7) \times (10^{-4})(10^{-6})$
$= \mathbf{14 \times 10^{-10}}$

b. $(340{,}000)(0.00061) = (3.4 \times 10^{5})(61 \times 10^{-5})$
$= (3.4)(61) \times (10^{5})(10^{-5})$
$= 207.4 \times 10^{0}$
$= \mathbf{207.4}$

Division In general,

$$\frac{A \times 10^{n}}{B \times 10^{m}} = \frac{A}{B} \times 10^{n-m} \qquad \textbf{(1.8)}$$

revealing again that the *operations with the power of ten can be separated from the same operation with the multipliers.*

Eq. (1.8) states

when dividing numbers in the power-of-ten format, first find the result of dividing the mutipliers. Then determine the associated power for the result by subtracting the power of ten of the denominator from the power of ten of the numerator.

EXAMPLE 1.10

a. $$\frac{0.00047}{0.002} = \frac{47 \times 10^{-5}}{2 \times 10^{-3}} = \left(\frac{47}{2}\right) \times \left(\frac{10^{-5}}{10^{-3}}\right) = \mathbf{23.5 \times 10^{-2}}$$

b. $$\frac{690{,}000}{0.00000013} = \frac{69 \times 10^{4}}{13 \times 10^{-8}} = \left(\frac{69}{13}\right) \times \left(\frac{10^{4}}{10^{-8}}\right) = \mathbf{5.31 \times 10^{12}}$$

Powers In general,

$$(A \times 10^{n})^{m} = A^{m} \times 10^{nm} \qquad \textbf{(1.9)}$$

which again permits the separation of the *operation with the power of ten from the multiplier.*

Eq. (1.9) states

when finding the power of a number in the power-of-ten format, first separate the multiplier from the power of ten and determine each separately. Determine the power-of-ten component by multiplying the power of ten by the power to be determined.

EXAMPLE 1.11

a. $(0.00003)^3 = (3 \times 10^{-5})^3 = (3)^3 \times (10^{-5})^3$
$= \mathbf{27 \times 10^{-15}}$

b. $(90{,}800{,}000)^2 = (9.08 \times 10^7)^2 = (9.08)^2 \times (10^7)^2$
$= \mathbf{82.45 \times 10^{14}}$

In particular, remember that the following operations are not the same. One is the product of two numbers in the power-of-ten format, while the other is a number in the power-of-ten format taken to a power. As noted below, the results of each are quite different:

$$(10^3)(10^3) \neq (10^3)^3$$
$$(10^3)(10^3) = 10^6 = 1{,}000{,}000$$
$$(10^3)^3 = (10^3)(10^3)(10^3) = 10^9 = 1{,}000{,}000{,}000$$

1.7 FIXED-POINT, FLOATING-POINT, SCIENTIFIC, AND ENGINEERING NOTATION

When you are using a computer or a calculator, numbers generally appear in one of four ways. If powers of ten are not employed, numbers are written in the **fixed-point** or **floating-point notation.**

The fixed-point format requires that the decimal point appear in the same place each time. In the floating-point format, the decimal point appears in a location defined by the number to be displayed.

Most computers and calculators permit a choice of fixed- or floating-point notation. In the fixed format, the user can choose the level of accuracy for the output as tenths place, hundredths place, thousandths place, and so on. Every output will then fix the decimal point to one location, such as the following examples using thousandths-place accuracy:

$$\frac{1}{3} = \mathbf{0.333} \qquad \frac{1}{16} = \mathbf{0.063} \qquad \frac{2300}{2} = \mathbf{1150.000}$$

If left in the floating-point format, the results will appear as follows for the above operations:

$$\frac{1}{3} = \mathbf{0.333333333333} \qquad \frac{1}{16} = \mathbf{0.0625} \qquad \frac{2300}{2} = \mathbf{1150}$$

Powers of ten will creep into the fixed- or floating-point notation if the number is too small or too large to be displayed properly.

Scientific (also called *standard*) **notation** and **engineering notation** make use of powers of ten, with restrictions on the mantissa (multiplier) or scale factor (power of ten).

Scientific notation requires that the decimal point appear directly after the first digit greater than or equal to 1 but less than 10.

A power of ten will then appear with the number (usually following the power notation E), even if it has to be to the zero power. A few examples:

$$\frac{1}{3} = \mathbf{3.33333333333E{-}1} \qquad \frac{1}{16} = \mathbf{6.25E{-}2} \qquad \frac{2300}{2} = \mathbf{1.15E3}$$

Within scientific notation, the fixed- or floating-point format can be chosen. In the above examples, floating was employed. If fixed is chosen and set at the hundredths-point accuracy, the following will result for the above operations:

$$\frac{1}{3} = \mathbf{3.33E{-}1} \quad \frac{1}{16} = \mathbf{6.25E{-}2} \quad \frac{2300}{2} = \mathbf{1.15E3}$$

Engineering notation specifies that

all powers of ten must be multiples of 3, and the mantissa must be greater than or equal to 1 but less than 1000.

This restriction on the powers of ten is because specific powers of ten have been assigned prefixes that are introduced in the next few paragraphs. Using scientific notation in the floating-point mode results in the following for the above operations:

$$\frac{1}{3} = \mathbf{333.333333333E{-}3} \quad \frac{1}{16} = \mathbf{62.5E{-}3} \quad \frac{2300}{2} = \mathbf{1.15E3}$$

Using engineering notation with two-place accuracy will result in the following:

$$\frac{1}{3} = \mathbf{333.33E{-}3} \quad \frac{1}{16} = \mathbf{62.50E{-}3} \quad \frac{2300}{2} = \mathbf{1.15E3}$$

Prefixes

Specific powers of ten in engineering notation have been assigned prefixes and symbols, as appearing in Table 1.2. They permit easy recognition of the power of ten and an improved channel of communication between technologists.

TABLE 1.2

Multiplication Factors	SI Prefix	SI Symbol
1 000 000 000 000 000 000 = 10^{18}	exa	E
1 000 000 000 000 000 = 10^{15}	peta	P
1 000 000 000 000 = 10^{12}	tera	T
1 000 000 000 = 10^{9}	giga	G
1 000 000 = 10^{6}	mega	M
1 000 = 10^{3}	kilo	k
0.001 = 10^{-3}	milli	m
0.000 001 = 10^{-6}	micro	μ
0.000 000 001 = 10^{-9}	nano	n
0.000 000 000 001 = 10^{-12}	pico	p
0.000 000 000 000 001 = 10^{-15}	femto	f
0.000 000 000 000 000 001 = 10^{-18}	atto	a

EXAMPLE 1.12

a. 1,000,000 ohms $= 1 \times 10^6$ ohms
$= \mathbf{1\ megohm} = \mathbf{1\ M\Omega}$

b. 100,000 meters $= 100 \times 10^3$ meters
$= \mathbf{100\ kilometers} = \mathbf{100\ km}$

c. 0.0001 second $= 0.1 \times 10^{-3}$ second
$= \mathbf{0.1\ millisecond} = \mathbf{0.1\ ms}$

d. 0.000001 farad $= 1 \times 10^{-6}$ farad
$= \mathbf{1\ microfarad} = \mathbf{1\ \mu F}$

Here are a few examples with numbers that are not strictly powers of ten.

EXAMPLE 1.13

a. 41,200 m is equivalent to 41.2×10^3 m = 41.2 kilometers = **41.2 km.**
b. 0.00956 J is equivalent to 9.56×10^{-3} J = 9.56 millijoules = **9.56 mJ.**
c. 0.000768 s is equivalent to 768×10^{-6} s = 768 microseconds = **768 μs.**
d. $$\frac{8400\text{ m}}{0.06} = \frac{8.4 \times 10^3\text{m}}{6 \times 10^{-2}} = \left(\frac{8.4}{6}\right) \times \left(\frac{10^3}{10^{-2}}\right)\text{ m}$$
$$= 1.4 \times 10^5\text{ m} = 140 \times 10^3\text{ m} = 140\text{ kilometers} = \mathbf{140\ km}$$
e. $(0.0003)^4\text{ s} = (3 \times 10^{-4})^4\text{ s} = 81 \times 10^{-16}\text{ s}$
$= 0.0081 \times 10^{-12}\text{ s} = 0.0081\text{ picosecond} = \mathbf{0.0081\ ps}$

1.8 CONVERSION BETWEEN LEVELS OF POWERS OF TEN

It is often necessary to convert from one power of ten to another. For instance, if a meter measures kilohertz (kHz—a unit of measurement for the frequency of an ac waveform), it may be necessary to find the corresponding level in megahertz (MHz). If time is measured in milliseconds (ms), it may be necessary to find the corresponding time in microseconds (μs) for a graphical plot. The process is not difficult if we simply keep in mind that an increase or a decrease in the power of ten must be associated with the opposite effect on the multiplying factor. The procedure is best described by the following steps:

1. ***Replace the prefix by its corresponding power of ten.***
2. ***Rewrite the expression, and set it equal to an unknown multiplier and the new power of ten.***
3. ***Note the change in power of ten from the original to the new format. If it is an increase, move the decimal point of the original multiplier to the left (smaller value) by the same number. If it is a decrease, move the decimal point of the original multiplier to the right (larger value) by the same number.***

EXAMPLE 1.14 Convert 20 kHz to megahertz.

Solution: In the power-of-ten format:

$$20\text{ kHz} = 20 \times 10^3\text{ Hz}$$

The conversion requires that we find the multiplying factor to appear in the space below:

Increase by 3

$$20 \times 10^3\text{ Hz} \Rightarrow \_\_ \times 10^6\text{ Hz}$$

Decrease by 3

Since the power of ten will be *increased* by a factor of *three*, the multiplying factor must be *decreased* by moving the decimal point *three* places to the left, as shown below:

$$\underbrace{020.}_{3} = 0.02$$

and $$20 \times 10^3 \text{ Hz} = 0.02 \times 10^6 \text{ Hz} = \mathbf{0.02\ MHz}$$

EXAMPLE 1.15 Convert 0.01 ms to microseconds.

Solution: In the power-of-ten format:

$$0.01 \text{ ms} = 0.01 \times 10^{-3} \text{ s}$$

Decrease by 3

and $$0.01 \times 10^{-3} \text{ s} \Rightarrow \_\_ \times 10^{-6} \text{ s}$$

Increase by 3

Since the power of ten will be *reduced* by a factor of three, the multiplying factor must be *increased* by moving the decimal point three places to the right, as follows:

$$0.\underbrace{010}_{3} = 10$$

and $$0.01 \times 10^{-3} \text{ s} = 10 \times 10^{-6} \text{ s} = \mathbf{10\ \mu s}$$

There is a tendency when comparing -3 to -6 to think that the power of ten has increased, but keep in mind when making your judgment about increasing or decreasing the magnitude of the multiplier that 10^{-6} is a great deal smaller than 10^{-3}.

EXAMPLE 1.16 Convert 0.002 km to millimeters.

Solution:

Decrease by 6

$$0.002 \times 10^3 \text{ m} \Rightarrow \_\_ \times 10^{-3} \text{ m}$$

Increase by 6

In this example we have to be very careful because the difference between $+3$ and -3 is a factor of 6, requiring that the multiplying factor be modified as follows:

$$0.\underbrace{002000}_{6} = 2000$$

and $$0.002 \times 10^3 \text{ m} = 2000 \times 10^{-3} \text{ m} = \mathbf{2000\ mm}$$

1.9 CONVERSION WITHIN AND BETWEEN SYSTEMS OF UNITS

The conversion within and between systems of units is a process that cannot be avoided in the study of any technical field. It is an operation, however, that is performed incorrectly so often that this section was included to provide one approach that, if applied properly, will lead to the correct result.

There is more than one method of performing the conversion process. In fact, some people prefer to determine mentally whether the conversion factor is multiplied or divided. This approach is acceptable for some elementary conversions, but it is risky with more complex operations.

The procedure to be described here is best introduced by examining a relatively simple problem such as converting inches to meters. Specifically, let us convert 48 in. (4 ft) to meters.

If we multiply the 48 in. by a factor of **1,** the magnitude of the quantity remains the same:

$$\boxed{48\text{ in.} = 48\text{ in.}(\mathbf{1})} \tag{1.10}$$

Let us now look at the conversion factor for this example:

$$1\text{ m} = 39.37\text{ in.}$$

Dividing both sides of the conversion factor by 39.37 in. results in the following format:

$$\frac{1\text{ m}}{39.37\text{ in.}} = (\mathbf{1})$$

Note that the end result is that the ratio 1 m/39.37 in. equals 1, as it should since they are equal quantities. If we now substitute this factor (**1**) into Eq. (1.10), we obtain

$$48\text{ in.}(\mathbf{1}) = 48\,\cancel{\text{in.}}\left(\frac{1\text{ m}}{39.37\,\cancel{\text{in.}}}\right)$$

which results in the cancellation of inches as a unit of measure and leaves meters as the unit of measure. In addition, since the 39.37 is in the denominator, it must be divided into the 48 to complete the operation:

$$\frac{48}{39.37}\text{ m} = \mathbf{1.219\text{ m}}$$

Let us now review the method:

1. ***Set up the conversion factor to form a numerical value of (1) with the unit of measurement to be removed from the original quantity in the denominator.***
2. ***Perform the required mathematics to obtain the proper magnitude for the remaining unit of measurement.***

EXAMPLE 1.17 Convert 6.8 min to seconds.

Solution: The conversion factor is

$$1\text{ min} = 60\text{ s}$$

Since the minute is to be removed as the unit of measurement, it must appear in the denominator of the (**1**) factor, as follows:

Step 1.
$$\left(\frac{60\text{ s}}{1\text{ min}}\right) = (\mathbf{1})$$

Step 2.
$$6.8\text{ min}(\mathbf{1}) = 6.8\,\cancel{\text{min}}\left(\frac{60\text{ s}}{1\,\cancel{\text{min}}}\right) = (6.8)(60)\text{ s}$$
$$= \mathbf{408\text{ s}}$$

EXAMPLE 1.18 Convert 0.24 m to centimeters.

Solution: The conversion factor is

$$1 \text{ m} = 100 \text{ cm}$$

Since the meter is to be removed as the unit of measurement, it must appear in the denominator of the (**1**) factor as follows:

Step 1. $$\left(\frac{100 \text{ cm}}{1 \text{ m}}\right) = \mathbf{1}$$

Step 2. $$0.24 \text{ m}(\mathbf{1}) = 0.24 \cancel{\text{m}}\left(\frac{100 \text{ cm}}{1 \cancel{\text{m}}}\right) = (0.24)(100) \text{ cm}$$
$$= \mathbf{24\ cm}$$

The products (**1**)(**1**) and (**1**)(**1**)(**1**) are still **1.** Using this fact, we can perform a series of conversions in the same operation.

EXAMPLE 1.19 Determine the number of minutes in half a day.

Solution: Working our way through from days to hours to minutes, always ensuring that the unit of measurement to be removed is in the denominator, results in the following sequence:

$$0.5 \cancel{\text{day}}\left(\frac{3 \cancel{\text{h}}}{1 \cancel{\text{day}}}\right)\left(\frac{60 \text{ min}}{1 \cancel{\text{h}}}\right) = (0.5)(24)(60) \text{ min}$$
$$= \mathbf{720\ min}$$

EXAMPLE 1.20 Convert 2.2 yards to meters.

Solution: Working our way through from yards to feet to inches to meters results in the following:

$$2.2 \cancel{\text{yards}}\left(\frac{3 \cancel{\text{ft}}}{1 \cancel{\text{yard}}}\right)\left(\frac{12 \cancel{\text{in.}}}{1 \cancel{\text{ft}}}\right)\left(\frac{1 \text{ m}}{39.37 \cancel{\text{in.}}}\right) = \frac{(2.2)(3)(12)}{39.37} \text{ m}$$
$$= \mathbf{2.012\ m}$$

The following examples are variations of the above in practical situations.

EXAMPLE 1.21 In Europe, Canada, and many countries, the speed limit is posted in kilometers per hour. How fast in miles per hour is 100 km/h?

Solution:

$$\left(\frac{100 \text{ km}}{\text{h}}\right)(\mathbf{1})(\mathbf{1})(\mathbf{1})(\mathbf{1})$$
$$= \left(\frac{100 \cancel{\text{km}}}{\text{h}}\right)\left(\frac{1000 \cancel{\text{m}}}{1 \cancel{\text{km}}}\right)\left(\frac{39.37 \cancel{\text{in.}}}{1 \cancel{\text{m}}}\right)\left(\frac{1 \cancel{\text{ft}}}{12 \cancel{\text{in.}}}\right)\left(\frac{1 \text{ mi}}{5280 \cancel{\text{ft}}}\right)$$
$$= \frac{(100)(1000)(39.37)}{(12)(5280)} \frac{\text{mi}}{\text{h}}$$
$$= \mathbf{62.14\ mph}$$

TABLE 1.3

Symbol	Meaning
$\neq$	Not equal to $6.12 \neq 6.13$
$>$	Greater than $4.78 > 4.20$
$>>$	Much greater than $840 >> 16$
$<$	Less than $430 < 540$
$<<$	Much less than $0.002 << 46$
$\geq$	Greater than or equal to $x \geq y$ is satisfied for $y = 3$ and $x > 3$ or $x = 3$
$\leq$	Less than or equal to $x \leq y$ is satisfied for $y = 3$ and $x < 3$ or $x = 3$
$\cong$	Appoximately equal to $3.14159 \cong 3.14$
Σ	Sum of $\Sigma\,(4 + 6 + 8) = 18$
$\vert\,\vert$	Absolute magnitude of $\vert a\vert = 4$, where $a = -4$ or $+4$
$\therefore$	Therefore $x = \sqrt{4}\ \therefore x = \pm 2$
$\equiv$	By definition Establishes a relationship between two or more quantities
$a{:}b$	Ratio defined by $\frac{a}{b}$
$a{:}b = c{:}d$	Proportion defined by $\frac{a}{b} = \frac{c}{d}$

Many travelers use 0.6 as a conversion factor to simplify the math involved; that is,

$$(100\text{ km/h})(0.6) \cong 60\text{ mph}$$

and

$$(60\text{ km/h})(0.6) \cong 36\text{ mph}$$

EXAMPLE 1.22 Determine the speed in miles per hour of a competitor who can run a 4-min mile.

Solution: Inverting the factor 4 min/1 mi to 1 mi/4 min, we can proceed as follows:

$$\left(\frac{1\text{ mi}}{4\text{ }\cancel{\text{min}}}\right)\left(\frac{60\text{ }\cancel{\text{min}}}{\text{h}}\right) = \frac{60}{4}\text{ mi/h} = \mathbf{15\ mph}$$

1.10 SYMBOLS

Throughout the text, various symbols will be employed that you may not have had occasion to use. Some are defined in Table 1.3, and others will be defined in the text as the need arises.

1.11 CONVERSION TABLES

Conversion tables such as those appearing in Appendix A can be very useful when time does not permit the application of methods described in this chapter. However, even though such tables appear easy to use, frequent errors occur because the operations appearing at the head of the table are not performed properly. In any case, when using such tables, try to establish mentally some order of magnitude for the quantity to be determined compared to the magnitude of the quantity in its original set of units. This simple operation should prevent several impossible results that may occur if the conversion operation is improperly applied.

For example, consider the following from such a conversion table:

To convert from	To	Multiply by
Miles	Meters	1.609×10^3

A conversion of 2.5 mi to meters would require that we multiply 2.5 by the conversion factor; that is,

$$2.5\text{ mi}(1.609 \times 10^3) = \mathbf{4.0225 \times 10^3\ m}$$

A conversion from 4000 m to miles would require a division process:

$$\frac{4000\text{ m}}{1.609 \times 10^3} = 2486.02 \times 10^{-3} = \mathbf{2.486\ mi}$$

In each of the above, there should have been little difficulty realizing that 2.5 mi would convert to a few thousand meters and 4000 m would be only a few miles. As indicated above, this kind of anticipatory thinking will eliminate the possibility of ridiculous conversion results.

1.12 CALCULATORS

In most texts, the calculator is not discussed in detail. Instead, students are left with the general exercise of choosing an appropriate calculator and learning to use it properly on their own. However, some discussion about the use of the calculator is needed to eliminate some of the impossible results obtained (and often strongly defended by the user—because the calculator says so) through a correct understanding of the process by which a calculator performs the various tasks. Time and space do not permit a detailed explanation of all the possible operations, but the following discussion explains why it is important to understand how a calculator proceeds with a calculation and that the unit cannot accept data in any form and still generate the correct answer.

When choosing a scientific calculator, be absolutely certain that it can operate on complex numbers and determinants, needed for the concepts introduced in this text. The simplest way to determine this is to look up the terms in the index of the operator's manual. Next, be aware that some calculators perform the required operations in a minimum number of steps while others require a lengthy and complex series of steps. Speak to your instructor if you are unsure about your purchase.

The calculator examples in this text have used the Texas Instruments TI-89 (see Fig.1.5). (For those students who still have the Texas Instruments TI-86, Appendix B was developed to provide the same coverage as offered in the text for the TI-89 calculator.)

When using any calculator for the first time, the unit must be set up to provide the answers in a desired format. Following are the steps needed to use a calculator correctly.

FIG. 1.5
Texas Instruments TI-89 calculator.
(Courtesy of Texas Instruments, Inc.)

Initial Settings

In the following sequences, the arrows within the square indicate the direction of the scrolling required to reach the desired location. The shape of the keys is a relatively close match of the actual keys on the TI-89.

Notation

The first sequence sets the **engineering notation** as the choice for all answers. It is particularly important to take note that you must select the ENTER key twice to ensure the process is set in memory.

Accuracy Level Next, the accuracy level can be set to two places as follows:

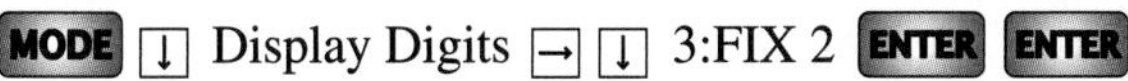

Approximate Mode For all solutions, the solution should be in decimal form to second-place accuracy. If this is not set, some answers will appear in fractional form to ensure the answer is EXACT (another option). This selection is made with the following sequence:

MODE **F2** ↓ Exact/Approx → ↓ 3: APPROXIMATE **ENTER** **ENTER**

Clear Screen To clear the screen of all entries and results, use the following sequence:

F1 ↓ 8: Clear Home **ENTER**

Clear Current Entries To delete the sequence of current entries at the bottom of the screen, select the **CLEAR** key.

Order of Operations

Although setting the correct format and accurate input is important, improper results occur primarily because users fail to realize that no matter how simple or complex an equation, the calculator performs the required operations in a specific order.

For instance, the operation

$$\frac{8}{3+1}$$

is often entered as

8 **÷** **3** **+** **1** **ENTER** $= \frac{8}{3} + 1 = 2.67 + 1 = \mathbf{3.67}$

which is totally incorrect (2 is the answer).

The calculator *will not* perform the addition first and then the division. In fact, addition and subtraction are the last operations to be performed in any equation. It is therefore very important that you carefully study and thoroughly understand the next few paragraphs in order to use the calculator properly.

1. ***The first operations to be performed by a calculator can be set using parentheses ( ). It does not matter which operations are within the parentheses. The parentheses simply dictate that this part of the equation is to be determined first. There is no limit to the number of parentheses in each equation—all operations within parentheses will be performed first. For instance, for the example above, if parentheses are added as shown below, the addition will be performed first and the correct answer obtained:***

$\frac{\mathbf{8}}{\mathbf{(3+1)}} =$ **8** **÷** **(** **3** **+** **1** **)** **ENTER** $= \mathbf{2.00}$

2. ***Next, powers and roots are performed, such as x^2, $\sqrt{x}$, and so on.***
3. ***Negation (applying a negative sign to a quantity) and single-key operations such as sin, $\tan^{-1}$, and so on, are performed.***
4. ***Multiplication and division are then performed.***
5. ***Addition and subtraction are performed last.***

It may take a few moments and some repetition to remember the order, but at least you are now aware that there is an order to the operations and that ignoring them can result in meaningless results.

EXAMPLE 1.23 Determine

$$\sqrt{\frac{9}{3}}$$

Solution:

2ND √ 9 ÷ 3) ENTER = **1.73**

In this case, the left bracket is automatically entered after the square root sign. The right bracket must then be entered before performing the calculation.

For all calculator operations, the number of right brackets must always equal the number of left brackets.

EXAMPLE 1.24 Find

$$\frac{3 + 9}{4}$$

Solution: If the problem is entered as it appears, the *incorrect answer* of 5.25 will result.

3 + 9 ÷ 4 ENTER $= 3 + \frac{9}{4} = 5.25$

Using brackets to ensure that the addition takes place before the division will result in the correct answer as shown below:

(3 + 9) ÷ 4 ENTER

$$= \frac{(3 + 9)}{4} = \frac{12}{4} = \mathbf{3.00}$$

EXAMPLE 1.25 Determine

$$\frac{1}{4} + \frac{1}{6} + \frac{2}{3}$$

Solution: Since the division will occur first, the correct result will be obtained by simply performing the operations as indicated. That is,

1 ÷ 4 + 1 ÷ 6 + 2 ÷

3 ENTER $= \frac{1}{4} + \frac{1}{6} + \frac{2}{3} = \mathbf{1.08}$

Powers of Ten

The EE key is used to set the power of ten of a number. Setting up the number $2200 = 2.2 \times 10^3$ requires the following keypad selections:

2 • 2 EE 3 ENTER = **2.20E3**

Setting up the number 8.2×10^{-6} requires the negative sign (−) from the *numerical keypad. Do not* use the negative sign from the mathematical listing of ÷, ×, −, and +. That is,

8 • 2 EE (−) 6 ENTER = **8.20E−6**

EXAMPLE 1.26 Perform the addition $6.3 \times 10^3 + 75 \times 10^3$ and compare your answer with the longhand solution of Example 1.8(a).

Solution:

6 • 3 EE 3 + 7 5 EE 3

ENTER = **81.30E3**

which confirms the results of Example 1.8(a).

EXAMPLE 1.27 Perform the division $(69 \times 10^4)/(13 \times 10^{-8})$ and compare your answer with the longhand solution of Example 1.10(b).

Solution:

6 9 EE 4 ÷ 1 3 EE (−) 8

ENTER = **5.31E12**

which confirms the results of Example 1.10(b).

EXAMPLE 1.28 Using the provided format of each number, perform the following calculation in one series of keypad entries:

$$\frac{(0.004)(6 \times 10^{-4})}{(2 \times 10^{-3})^2} = ?$$

Solution:

((0 • 0 0 4) × (

6 EE (−) 4)) ÷ 2 EE (−)

3 ∧ 2 ENTER = **600.00E−3 = 0.6**

Brackets were used to ensure that the calculations were performed in the correct order. Note also that the number of left brackets equals the number of right brackets.

1.13 COMPUTER ANALYSIS

The use of computers in the educational process has grown exponentially in the past decade. Very few texts at this introductory level fail to include some discussion of current popular computer techniques. In fact, the very accreditation of a technology program may be a function of the depth to which computer methods are incorporated in the program.

There is no question that a basic knowledge of computer methods is something that the graduating student must learn in a two-year or four-year program. Industry now requires students to be proficient in the use of a computer.

Two general directions can be taken to develop the necessary computer skills: the study of computer languages or the use of software packages.

Languages

There are several languages that provide a direct line of communication with the computer and the operations it can perform. A **language** is a set of symbols, letters, words, or statements that the user can enter into the computer. The computer system will "understand" these entries and will perform them in the order established by a series of commands called a **program.** The program tells the computer what to do on a sequential, line-by-line basis in the same order a student would perform the calculations in longhand. The computer can respond only to the commands entered by the user. This requires that the programmer understand fully the sequence of operations and calculations required to obtain a particular solution. A lengthy analysis can result in a program having hundreds or thousands of lines. Once written, the program must be checked carefully to ensure that the results have meaning and are valid for an expected range of input variables. Some of the popular languages applied in the electrical-electronics field today include C++, QBASIC, Pascal, and FORTRAN. Each has its own set of commands and statements to communicate with the computer, but each can be used to perform the same type of analysis.

Software Packages

The second approach to computer analysis—**software packages**—avoids the need to know a particular language; in fact, the user may not be aware of which language was used to write the programs within the package. All that is required is a knowledge of how to input the network parameters, define the operations to be performed, and extract the results; the package will do the rest. However, there is a problem with using packaged programs without understanding the basic steps the program uses. You can obtain solution without the faintest idea of either how the solution was obtained or whether the results are valid or way off base. It is imperative that you realize that the computer should be used as a tool to assist the user—it must not be allowed to control the scope and potential of the user! Therefore, as we progress through the chapters of the text, be sure that you clearly understand the concepts before turning to the computer for support and efficiency.

Each software package has a **menu,** which defines the range of application of the package. Once the software is entered into the computer, the system will perform all the functions appearing in the menu, as it was preprogrammed to do. Be aware, however, that if a particular type of analysis is requested that is not on the menu, the software package cannot provide the desired results. The package is limited solely to those maneuvers developed by the team of programmers who developed the software package. In such situations the user must turn to another software package or write a program using one of the languages listed above.

In broad terms, if a software package is available to perform a particular analysis, then that package should be used rather than developing new routines. Most popular software packages are the result of many hours of effort by teams of programmers with years of experience. However, if the results are not in the desired format, or if the software package does not provide all the desired results, then the user's innovative talents should be put to use to develop a software package. As noted above, any program the user writes that passes the tests of range and accuracy can be considered a software package of his or her authorship for future use.

Three software packages are used throughout this text: Cadence's OrCAD's PSpice Release 10, Electronics Workbench Multisim Version 8, and MathSoft's Mathcad 12. Although both PSpice and Multisim are designed to analyze electric circuits, there are sufficient differences between the two to warrant covering each approach. The growing use of some form of mathematical support in the educational and industrial environments supports the introduction and use of Mathcad throughout the text. However, you are not required to obtain all three programs in order to proceed with the content of this text. The primary reason for including these programs is simply to introduce each and demonstrate how each can support the learning process. In most cases, sufficient detail has been provided to actually use the software package to perform the examples provided, although it would certainly be helpful to have someone to turn to if questions arise. In addition, the literature supporting all three packages has improved dramatically in recent years and should be available through your bookstore or a major publisher. Appendix C lists all the system requirements, including how to get in touch with each company.

PROBLEMS

Note: More difficult problems are denoted by an asterisk (*) throughout the text.

SECTION 1.2 A Brief History

1. Visit your local library (at school or home) and describe the extent to which it provides literature and computer support for the technologies—in particular, electricity, electronics, electromagnetics, and computers.

2. Choose an area of particular interest in this field and write a very brief report on the history of the subject.

3. Choose an individual of particular importance in this field and write a very brief review of his or her life and important contributions.

SECTION 1.3 Units of Measurement

4. What is the velocity of a rocket in mph if it travels 20,000 ft in 10 s?

5. In a recent Tour de France time trial, Lance Armstrong traveled 31 miles in a time trial in 1 hour and 4 minutes. What was his average speed in mph?

* 6. A pitcher has the ability to throw a baseball at 95 mph.
   a. How fast is the speed in ft/s?
   b. How long does the hitter have to make a decision about swinging at the ball if the plate and the mound are separated by 60 feet?
   c. If the batter wanted a full second to make a decision, what would the speed in mph have to be?

SECTION 1.4 Systems of Units

7. Are there any relative advantages associated with the metric system compared to the English system with respect to length, mass, force, and temperature? If so, explain.

8. Which of the four systems of units appearing in Table 1.1 has the smallest units for length, mass, and force? When would this system be used most effectively?

* 9. Which system of Table 1.1 is closest in definition to the SI system? How are the two systems different? Why do you think the units of measurement for the SI system were chosen as listed in Table 1.1? Give the best reasons you can without referencing additional literature.

10. What is room temperature (68°F) in the MKS, CGS, and SI systems?

11. How many foot-pounds of energy are associated with 1000 J?

12. How many centimeters are there in 1/2 yd?

SECTIONS 1.6 and 1.7 Powers of Ten and Notation

13. Express the following numbers as powers of ten:
   a. 10,000 b. 1,000,000
   c. 1000 d. 0.001
   e. 1 f. 0.1

14. Using only those powers of ten listed in Table 1.2, express the following numbers in what seems to you the most logical form for future calculations:
   a. 15,000 b. 0.030
   c. 2,400,000 d. 150,000
   e. 0.00040200 f. 0.0000000002

15. Perform the following operations and express your answer as a power of ten using scientific notation:
   a. $4200 + 48{,}000$
   b. $9 \times 10^4 + 3.6 \times 10^5$
   c. $0.5 \times 10^{-3} - 6 \times 10^{-5}$
   d. $1.2 \times 10^3 + 50{,}000 \times 10^{-3} - 0.6 \times 10^3$

16. Perform the following operations and express your answer as a power of ten using engineering notation:
   a. $(100)(1000)$ b. $(0.01)(1000)$
   c. $(10^3)(10^6)$ d. $(100)(0.00001)$
   e. $(10^{-6})(10{,}000{,}000)$ f. $(10{,}000)(10^{-8})(10^{28})$

17. Perform the following operations and express your answer in scientific notation:
   a. $(50{,}000)(0.0003)$
   b. 2200×0.002
   c. $(0.000082)(2{,}800{,}000)$
   d. $(30 \times 10^{-4})(0.004)(7 \times 10^8)$

18. Perform the following operations and express your answer in engineering notation:
   a. $\dfrac{100}{10{,}000}$ b. $\dfrac{0.010}{1000}$
   c. $\dfrac{10{,}000}{0.001}$ d. $\dfrac{0.0000001}{100}$
   e. $\dfrac{10^{38}}{0.000100}$ f. $\dfrac{(100)^{1/2}}{0.01}$

19. Perform the following operations and express your answer in scientific notation:
   a. $\dfrac{2000}{0.00008}$ b. $\dfrac{0.004}{60{,}000}$
   c. $\dfrac{0.000220}{0.00005}$ d. $\dfrac{78 \times 10^{18}}{4 \times 10^{-6}}$

20. Perform the following operations and express your answer in engineering notation:
   a. $(100)^3$ b. $(0.0001)^{1/2}$
   c. $(10{,}000)^8$ d. $(0.00000010)^9$

21. Perform the following operations and express your answer in scientific notation:
   a. $(400)^2$
   b. $(0.006)^3$
   c. $(0.004)(6 \times 10^2)^2$
   d. $((2 \times 10^{-3})(0.8 \times 10^4)(0.003 \times 10^5))^3$

22. Perform the following operations and express your answer in scientific notation:
   a. $(-0.001)^2$ b. $\dfrac{(100)(10^{-4})}{1000}$
   c. $\dfrac{(0.001)^2(100)}{10{,}000}$ d. $\dfrac{(10^3)(10{,}000)}{1 \times 10^{-4}}$
   e. $\dfrac{(0.0001)^3(100)}{1 \times 10^6}$ *f. $\dfrac{[(100)(0.01)]^{-3}}{[(100)^2][0.001]}$

23. Perform the following operations and express your answer in engineering notation:
   a. $\dfrac{(300)^2(100)}{3 \times 10^4}$ b. $[(40{,}000)^2][(20)^{-3}]$
   c. $\dfrac{(60{,}000)^2}{(0.02)^2}$ d. $\dfrac{(0.000027)^{1/3}}{200{,}000}$
   e. $\dfrac{[(4000)^2][300]}{2 \times 10^{-4}}$
   f. $[(0.000016)^{1/2}][(100{,}000)^5][0.02]$
   *g. $\dfrac{[(0.003)^3][0.00007]^{-2}[(160)^2]}{[(200)(0.0008)]^{-1/2}}$ (a challenge)

SECTION 1.8 Conversion between Levels of Powers of Ten

24. Fill in the blanks of the following conversions:
   a. $6 \times 10^3 =$ _____ $\times 10^6$
   b. $4 \times 10^{-3} =$ _____ $\times 10^{-6}$

c. $50 \times 10^5 =$ _____ $\times 10^3 =$ _____ $\times 10^6$
$=$ _____ $\times 10^9$

d. $30 \times 10^{-8} =$ _____ $\times 10^{-3} =$ _____ $\times 10^{-6}$
$=$ _____ $\times 10^{-9}$

25. Perform the following conversions:
a. 0.05 s to milliseconds
b. 2000 μs to milliseconds
c. 0.04 ms to microseconds
d. 8400 ps to microseconds
e. 4×10^{-3} km to millimeters
f. 260×10^3 mm to kilometers

SECTION 1.9 Conversion within and between Systems of Units

26. Perform the following conversions:
a. 1.5 min to seconds
b. 0.04 h to seconds
c. 0.05 s to microseconds
d. 0.16 m to millimeters
e. 0.00000012 s to nanoseconds
f. 3,620,000 s to days

27. Perform the following conversions:
a. 0.1 μF to picofarads
b. 80 mm to centimeters
c. 60 cm to kilometers
d. 3.2 h to milliseconds
e. 0.016 mm to micrometers
f. 60 sq cm (cm^2) to square meters (m^2)

28. Perform the following conversions:
a. 100 in. to meters
b. 4 ft to meters
c. 6 lb to newtons
d. 60,000 dyn to pounds
e. 150,000 cm to feet
f. 0.002 mi to meters (5280 ft = 1 mi)

29. What is a mile in feet, yards, meters, and kilometers?

30. Calculate the speed of light in miles per hour using the speed defined in Section 1.4.

31. How long in seconds will it take a car traveling at 60 mph to travel the length of a football field (100 yd)?

32. Convert 30 mph to meters per second.

33. If an athlete can row at a rate of 50 yd/min, how many days would it take to cross the Atlantic ($\cong$ 3000 mi)?

34. How long would it take a runner to complete a 10 km race if a pace of 6.5 min/mi were maintained?

35. Quarters are about 1 in. in diameter. How many would be required to stretch from one end of a football field to the other (100 yd)?

36. Compare the total time required to drive 100 miles at an average speed of 60 mph versus an average speed of 75 mph. Is the time saved for such a long trip worth the added risk of the higher speed?

***37.** Find the distance in meters that a mass traveling at 600 cm/s will cover in 0.016 h.

***38.** Each spring there is a race up 86 floors of the 102 story Empire State Building in New York City. If you were able to climb 2 steps/second, how long would it take in minutes to reach the 86th floor if each floor is 14 ft high and each step is about 9 in.?

***39.** The record for the race in Problem 38 is 10 minutes, 47 seconds. What was the racer's speed in min/mi for the race?

***40.** If the race of Problem 38 were a horizontal distance, how long would it take a runner who can run 5 min miles to cover the distance? Compare this with the record speed of Problem 39. Gravity is certainly a factor to be reckoned with!

SECTION 1.11 Conversion Tables

41. Using Appendix A, determine the number of
a. Btu in 5 J of energy.
b. cubic meters in 24 oz of a liquid.
c. seconds in 1.4 days.
d. pints in 1 m^3 of a liquid.

SECTION 1.12 Calculators

Perform the following operations using a single sequence of calculator keys:

42. $6\,(4 + 8) =$

43. $\dfrac{20 + 32}{4} =$

44. $\sqrt{8^2 + 12^2} =$

45. $\cos 50° =$

46. $\tan^{-1}\dfrac{3}{4} =$

47. $\sqrt{\dfrac{400}{6^2 + 10}} =$

48. $\dfrac{8.2 \times 10^{-3}}{0.04 \times 10^3}$ (in engineering notation) $=$

***49.** $\dfrac{(0.06 \times 10^5)(20 \times 10^3)}{(0.01)^2}$ (in engineering notation) $=$

***50.** $\dfrac{4 \times 10^4}{2 \times 10^{-3} + 400 \times 10^{-5}} + \dfrac{1}{2 \times 10^{-6}}$
(in engineering notation) $=$

SECTION 1.13 Computer Analysis

51. Investigate the availability of computer courses and computer time in your curriculum. Which languages are commonly used, and which software packages are popular?

52. Develop a list of three popular computer languages, including a few characteristics of each. Why do you think some languages are better for the analysis of electric circuits than others?

GLOSSARY

Cathode-ray tube (CRT) A glass enclosure with a relatively flat face (screen) and a vacuum inside that will display the light generated from the bombardment of the screen by electrons.

CGS system The system of units employing the *C*entimeter, *G*ram, and *S*econd as its fundamental units of measure.

Difference engine One of the first mechanical calculators.

Edison effect Establishing a flow of charge between two elements in an evacuated tube.

Electromagnetism The relationship between magnetic and electrical effects.

Engineering notation A method of notation that specifies that all powers of ten used to define a number be multiples of 3 with a mantissa greater than or equal to 1 but less than 1000.

ENIAC The first totally electronic computer.

Fixed-point notation Notation using a decimal point in a particular location to define the magnitude of a number.

Fleming's valve The first of the electronic devices, the diode.

Floating-point notation Notation that allows the magnitude of a number to define where the decimal point should be placed.

Integrated circuit (IC) A subminiature structure containing a vast number of electronic devices designed to perform a particular set of functions.

Joule (J) A unit of measurement for energy in the SI or MKS system. Equal to 0.7378 foot-pound in the English system and 10^7 ergs in the CGS system.

Kelvin (K) A unit of measurement for temperature in the SI system. Equal to 273.15 + °C in the MKS and CGS systems.

Kilogram (kg) A unit of measure for mass in the SI and MKS systems. Equal to 1000 grams in the CGS system.

Language A communication link between user and computer to define the operations to be performed and the results to be displayed or printed.

Leyden jar One of the first charge-storage devices.

Menu A computer-generated list of choices for the user to determine the next operation to be performed.

Meter (m) A unit of measure for length in the SI and MKS systems. Equal to 1.094 yards in the English system and 100 centimeters in the CGS system.

MKS system The system of units employing the *M*eter, *K*ilogram, and *S*econd as its fundamental units of measure.

Nanotechnology The production of integrated circuits in which the nanometer is the typical unit of measurement.

Newton (N) A unit of measurement for force in the SI and MKS systems. Equal to 100,000 dynes in the CGS system.

Pound (lb) A unit of measurement for force in the English system. Equal to 4.45 newtons in the SI or MKS system.

Program A sequential list of commands, instructions, and so on, to perform a specified task using a computer.

Scientific notation A method for describing very large and very small numbers through the use of powers of ten, which requires that the multiplier be a number between 1 and 10.

Second (s) A unit of measurement for time in the SI, MKS, English, and CGS systems.

SI system The system of units adopted by the IEEE in 1965 and the USASI in 1967 as the International System of Units (*S*ystème *I*nternational d'Unités).

Slug A unit of measure for mass in the English system. Equal to 14.6 kilograms in the SI or MKS system.

Software package A computer program designed to perform specific analysis and design operations or generate results in a particular format.

Static electricity Stationary charge in a state of equilibrium.

Transistor The first semiconductor amplifier.

Voltaic cell A storage device that converts chemical to electrical energy.

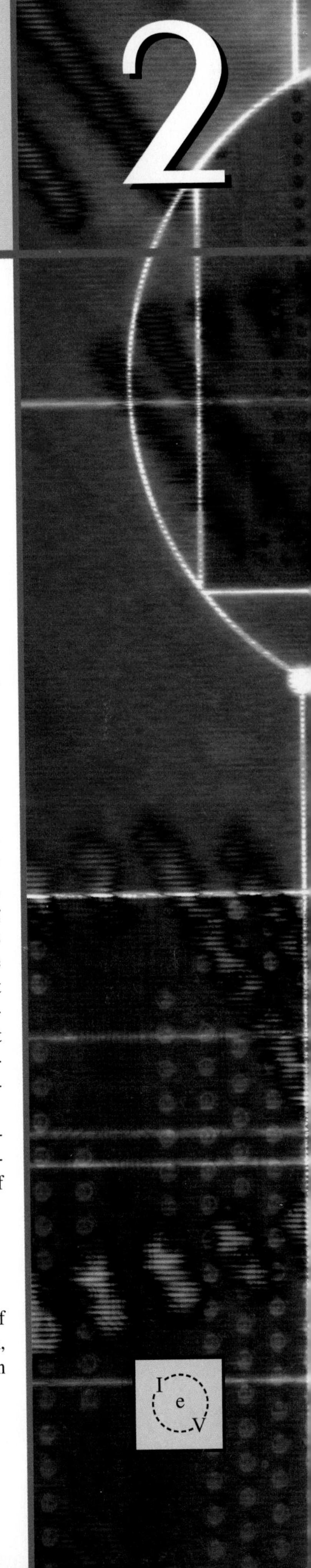

Voltage and Current

Objectives

- ***Become aware of the basic atomic structure of conductors such as copper and aluminum and understand why they are used so extensively in the field.***
- ***Understand how the terminal voltage of a battery or any dc supply is established and how it creates a flow of charge in the system.***
- ***Understand how current is established in a circuit and how its magnitude is affected by the charge flowing in the system and the time involved.***
- ***Become familiar with the factors that affect the terminal voltage of a battery and how long a battery will remain effective.***
- ***Be able to apply a voltmeter and ammeter correctly to measure the voltage and current of a network.***

2.1 INTRODUCTION

Now that the foundation for the study of electricity/electronics has been established, the concepts of voltage and current can be investigated. The term **voltage** is encountered practically every day. We have all replaced batteries in our flashlights, answering machines, calculators, automobiles, and so on, that had specific voltage ratings. We are aware that most outlets in our homes are 120 volts. Although **current** may be a less familiar term, we know what happens when we place too many appliances on the same outlet—the circuit breaker opens due to the excessive current that results. It is fairly common knowledge that current is something that moves through the wires and causes sparks and possibly fire if there is a "short circuit." Current heats up the coils of an electric heater or the range of an electric stove; it generates light when passing through the filament of a bulb; it causes twists and kinks in the wire of an electric iron over time, and so on. All in all, the terms *voltage* and *current* are part of the vocabulary of most individuals.

In this chapter, the basic impact of current and voltage and the properties of each are introduced and discussed in some detail. Hopefully, any mysteries surrounding the general characteristics of each will be eliminated, and you will gain a clear understanding of the impact of each on an electric/electronics circuit.

2.2 ATOMS AND THEIR STRUCTURE

A basic understanding of the fundamental concepts of current and voltage requires a degree of familiarity with the atom and its structure. The simplest of all atoms is the hydrogen atom, made up of two basic particles, the **proton** and the **electron,** in the relative positions shown in Fig. 2.1(a). The **nucleus** of the hydrogen atom is the proton, a positively charged particle.

The orbiting electron carries a negative charge equal in magnitude to the positive charge of the proton.

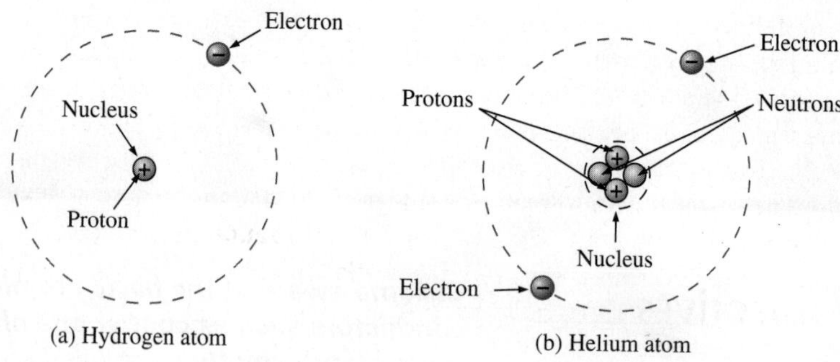

FIG. 2.1
Hydrogen and helium atoms.

In all other elements, the nucleus also contains **neutrons,** which are slightly heavier than protons and have *no electrical charge.* The helium atom, for example, has two neutrons in addition to two electrons and two protons, as shown in Fig. 2.1(b). In general,

the atomic structure of any stable atom has an equal number of electrons and protons.

Different atoms have various numbers of electrons in concentric orbits called *shells* around the nucleus. The first shell, which is closest to the nucleus, can contain only two electrons. If an atom has three electrons, the extra electron must be placed in the next shell. The number of electrons in each succeeding shell is determined by $2n^2$ where n is the shell number. Each shell is then broken down into subshells where the number of electrons is limited to 2, 6, 10, and 14 in that order as you move away from the nucleus.

Copper is the most commonly used metal in the electrical/electronics industry. An examination of its atomic structure will reveal why it has such widespread application. As shown in Fig. 2.2, it has 29 electrons in orbits around the nucleus, with the 29th electron appearing all by itself in the 4th shell. Note that the number of electrons in each shell and subshell

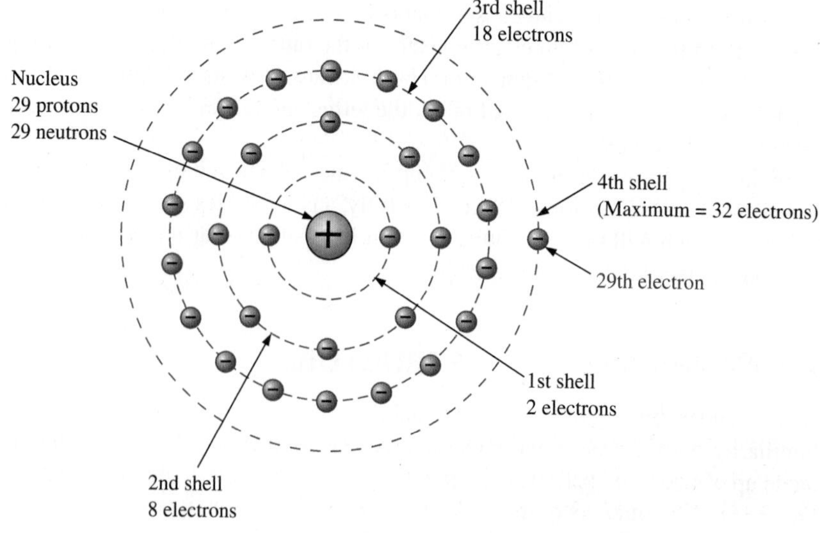

FIG. 2.2
The atomic structure of copper.

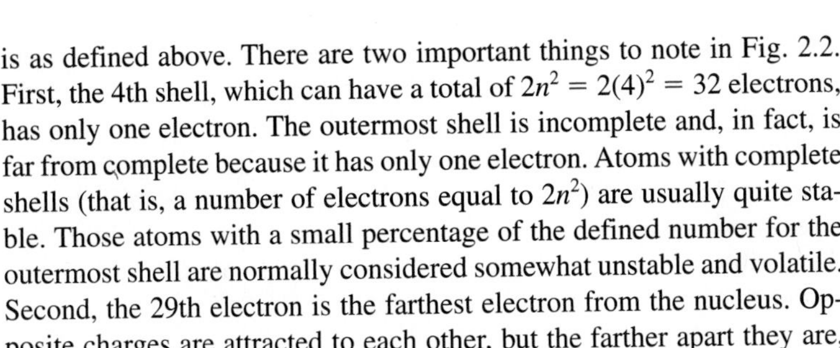

is as defined above. There are two important things to note in Fig. 2.2. First, the 4th shell, which can have a total of $2n^2 = 2(4)^2 = 32$ electrons, has only one electron. The outermost shell is incomplete and, in fact, is far from complete because it has only one electron. Atoms with complete shells (that is, a number of electrons equal to $2n^2$) are usually quite stable. Those atoms with a small percentage of the defined number for the outermost shell are normally considered somewhat unstable and volatile. Second, the 29th electron is the farthest electron from the nucleus. Opposite charges are attracted to each other, but the farther apart they are, the less the attraction. In fact, the force of attraction between the nucleus and the 29th electron of copper can be determined by **Coulomb's law** developed by Charles Augustin Coulomb (Fig. 2.3) in the late 18th century:

FIG. 2.3
Charles Augustin Coulomb.
Courtesy of the Smithsonian Institution, Photo No. 52,597

French (Angoulème, Paris)
(1736–1806) Scientist and Inventor
Military Engineer, West Indies

Attended the engineering school at Mezieres, the first such school of its kind. Formulated *Coulomb's law,* which defines the force between two electrical charges and is, in fact, one of the principal forces in atomic reactions. Performed extensive research on the friction encountered in machinery and windmills and the elasticity of metal and silk fibers.

$$F = k\frac{Q_1Q_2}{r^2} \quad \text{(newtons, N)} \qquad \textbf{(2.1)}$$

where F is in newtons (N), k = a constant = 9.0×10^9 N·m²/C², Q_1 and Q_2 are the charges in coulombs (a unit of measure discussed in the next section), and r is the distance between the two charges in meters.

At this point, the most important thing to note is that the distance between the charges appears as a squared term in the denominator. First, the fact that this term is in the denominator clearly reveals that as it increases, the force will decrease. However, since it is a squared term, the force will drop dramatically with distance. For instance, if the distance is doubled, the force will drop to 1/4 because $(2)^2 = 4$. If the distance is increased by a factor of 4, it will drop by 1/16, and so on. The result, therefore, is that the force of attraction between the 29th electron and the nucleus is significantly less than that between an electron in the first shell and the nucleus. The result is that the 29th electron is loosely bound to the atomic structure and with a little bit of pressure from outside sources could be encouraged to leave the parent atom.

If this 29th electron gains sufficient energy from the surrounding medium to leave the parent atom, it is called a **free electron.** In 1 cubic in. of copper at room temperature, there are approximately 1.4×10^{24} free electrons. Expanded, that is 1,400,000,000,000,000,000,000,000 free electrons in a 1 in. square cube. The point is that we are dealing with enormous numbers of electrons when we talk about the number of free electrons in a copper wire—not just a few that you could leisurely count. Further, the numbers involved are clear evidence of the need to become proficient in the use of powers of ten to represent numbers and use them in mathematical calculations.

Other metals that exhibit the same properties as copper, but to a different degree, are silver, gold, and aluminum, and some rarer metals such as tungsten. Additional comments on the characteristics of conductors are in the following sections.

2.3 VOLTAGE

If we separate the 29th electron in Fig. 2.2 from the rest of the atomic structure of copper by a dashed line as shown in Fig. 2.4(a), we create regions that have a net positive and negative charge as shown in Fig. 2.4(b) and (c). For the region inside the dashed boundary, the number of protons in the nucleus exceeds the number of orbiting electrons by 1, so the net charge is positive as shown in both figures. This positive region created by separating the free electron from the basic atomic structure is called a

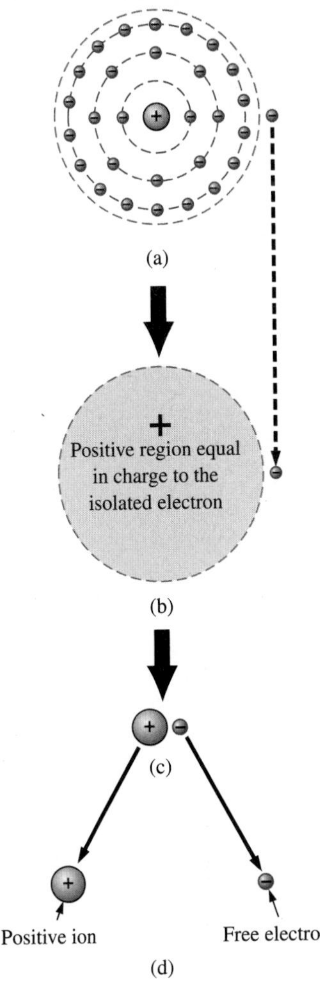

FIG. 2.4
Defining the positive ion.

positive ion. If the free electron then leaves the vicinity of the parent atom as shown in Fig. 2.4(d), regions of positive and negative charge have been established.

This separation of charge to establish regions of positive and negative charge is the action that occurs in every battery. Through chemical action, a heavy concentration of positive charge (positive ions) is established at the positive terminal, with an equally heavy concentration of negative charge (electrons) at the negative terminal.

In general,

every source of voltage is established by simply creating a separation of positive and negative charges.

It is that simple: If you want to create a voltage level of any magnitude, simply establish a region of positive and negative charge. The more the required voltage, the greater the quantity of positive and negative charge.

In Fig. 2.5(a), for example, a region of positive charge has been established by a packaged number of positive ions, and a region of negative charge by a similar number of electrons, both separated by a distance *r*. Since it would be inconsequential to talk about the voltage established by the separation of a single electron, a package of electrons called a **coulomb (C)** of charge was defined as follows:

One coulomb of charge is the total charge associated with 6.242×10^{18} electrons.

A coulomb of positive charge would have the same magnitude but opposite polarity.

In Fig. 2.5(b), if we take a coulomb of negative charge near the surface of the positive charge and move it toward the negative charge, energy must be expended to overcome the repulsive forces of the larger negative charge and the attractive forces of the positive charge. In the process of moving the charge from point *a* to point *b* in Fig. 2.5(b):

if a total of 1 joule (J) of energy is used to move the negative charge of 1 coulomb (C), there is a difference of 1 volt (V) between the two points.

The defining equation is

$$V = \frac{W}{Q} \qquad \begin{aligned} V &= \text{volts (V)} \\ W &= \text{joules (J)} \\ Q &= \text{coulombs (C)} \end{aligned} \tag{2.2}$$

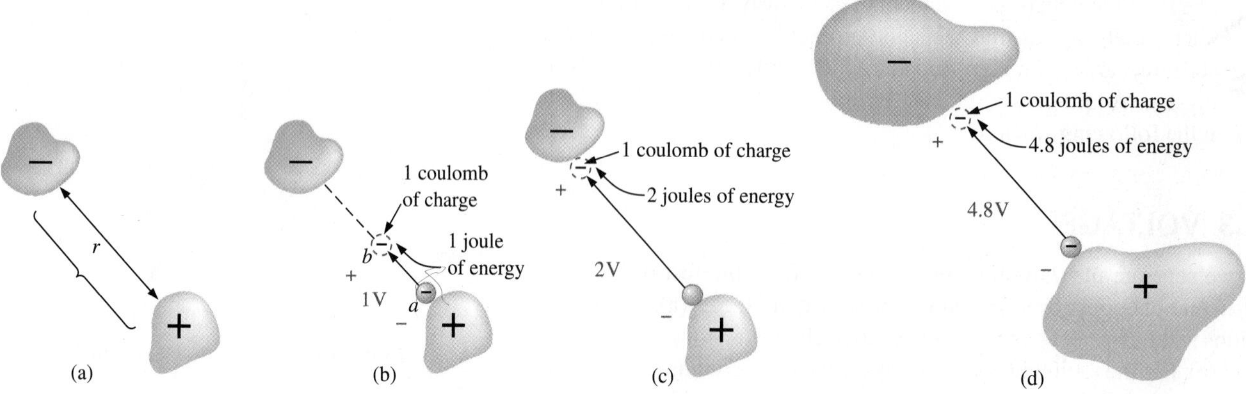

FIG. 2.5
Defining the voltage between two points.

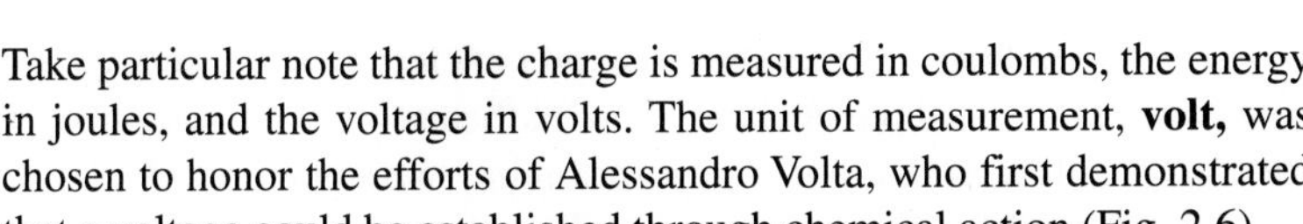

Take particular note that the charge is measured in coulombs, the energy in joules, and the voltage in volts. The unit of measurement, **volt,** was chosen to honor the efforts of Alessandro Volta, who first demonstrated that a voltage could be established through chemical action (Fig. 2.6).

If the charge is now moved all the way to the surface of the larger negative charge as shown in Fig. 2.5(c), using 2 joules of energy for the whole trip, there are 2 volts between the two charged bodies. If the package of positive and negative charge is larger, as shown in Fig. 2.5(d), more energy will have to be expended to overcome the larger repulsive forces of the large negative charge and attractive forces of the large positive charge. As shown in Fig. 2.5(d), 4.8 joules of energy were expended, resulting in a voltage of 4.8 V between the two points. We can therefore conclude that it would take 12 joules of energy to move 1 coulomb of negative charge from the positive terminal to the negative terminal of a 12 V car battery.

Through algebraic manipulations, we can define an equation to determine the energy required to move charge through a difference in voltage:

$$\boxed{W = QV} \quad \text{(joules, J)} \tag{2.3}$$

Finally, if we want to know how much charge was involved:

$$\boxed{Q = \frac{W}{V}} \quad \text{(coulombs, C)} \tag{2.4}$$

FIG. 2.6
Count Alessandro Volta.
Courtesy of the Smithsonian Institution, Photo No. 55,393

Italian (Como, Pavia)
(1745–1827)
Physicist
Professor of Physics,
Pavia, Italy

Began electrical experiments at the age of 18 working with other European investigators. Major contribution was the development of an electrical energy source from chemical action in 1800. For the first time, electrical energy was available on a continuous basis and could be used for practical purposes. Developed the first *condenser* known today as the *capacitor.* Was invited to Paris to demonstrate the *voltaic cell* to Napoleon. The International Electrical Congress meeting in Paris in 1881 honored his efforts by choosing the *volt* as the unit of measure for electromotive force.

EXAMPLE 2.1 Find the voltage between two points if 60 J of energy are required to move a charge of 20 C between the two points.

Solution: Eq. (2.2): $V = \dfrac{W}{Q} = \dfrac{60\text{ J}}{20\text{ C}} = \mathbf{3\ V}$

EXAMPLE 2.2 Determine the energy expended moving a charge of 50 μC between two points if the voltage between the points is 6 V.

Solution: Eq. (2.3):

$$W = QV = (50 \times 10^{-6}\text{ C})(6\text{ V}) = 300 \times 10^{-6}\text{ J} = \mathbf{300\ \mu J}$$

There are a variety of ways to separate charge to establish the desired voltage. The most common is the chemical action used in car batteries, flashlight batteries, and, in fact, all portable batteries. Other sources use mechanical methods such as car generators and steam power plants or alternative sources such as solar cells and windmills. In total, however, the sole purpose of the system is to create a separation of charge. In the future, therefore, when you see a positive and a negative terminal on any type of battery, you can think of it as a point where a large concentration of charge has gathered to create a voltage between the two points. More important is to recognize that a voltage exists between two points—for a battery between the positive and negative terminals. Hooking up just the positive or the negative terminal of a battery and not the other would be meaningless.

Both terminals must be connected to define the applied voltage.

As we moved the 1 coulomb of charge in Fig. 2.5(b), the energy expended would depend on where we were in the crossing. The *position* of the charge is therefore a factor in determining the voltage level at each point in the crossing. Since the **potential energy** associated with a body is defined by its position, the term *potential* is often applied to define voltage levels. For example, the difference in potential is 4 V between the two points, or the **potential difference** between a point and ground is 12 V, and so on.

2.4 CURRENT

The question, "Which came first—the chicken or the egg?" can be applied here also because the layperson has a tendency to use the terms *current* and *voltage* interchangeably as if both were sources of energy. It is time to set things straight:

The applied voltage is the starting mechanism—the current is a reaction to the applied voltage.

In Fig. 2.7(a), a copper wire sits isolated on a laboratory bench. If we cut the wire with an imaginary perpendicular plane, producing the circular cross section shown in Fig. 2.7(b), we would be amazed to find that there are free electrons crossing the surface in both directions. Those free electrons generated at room temperature are in constant motion in random directions. However, at any instant of time, the number of electrons crossing the imaginary plane in one direction is exactly equal to that crossing in the opposite direction, so the *net flow in any one direction is zero.* Even though the wire seems dead to the world sitting by itself on the bench, internally, it is quite active. The same would be true for any other good conductor.

Now, to make this electron flow do work for us, we need to give it a direction and be able to control its magnitude. This is accomplished by simply applying a voltage across the wire to force the electrons to move toward the positive terminal of the battery, as shown in Fig. 2.8. The instant the wire is placed across the terminals, the free electrons in the wire drift toward the positive terminal. The positive ions in the copper wire simply oscillate in a mean fixed position. As the electrons pass through the wire, the negative terminal of the battery acts as a supply of additional

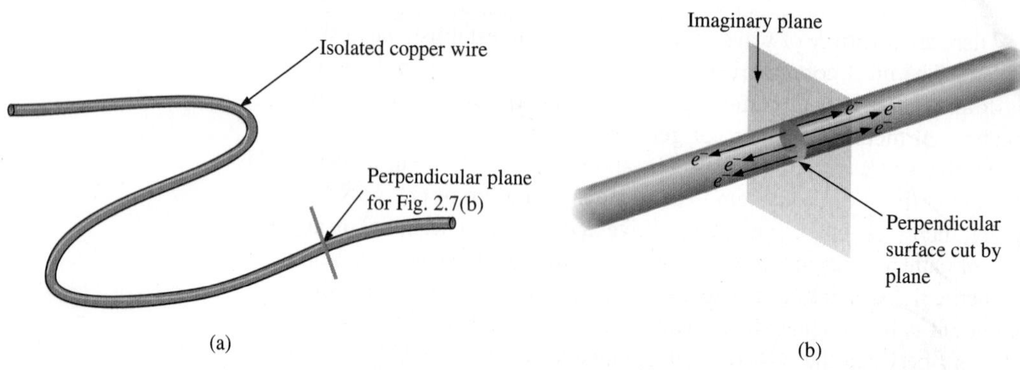

FIG. 2.7
There is motion of free carriers in an isolated piece of copper wire, but the flow of charge fails to have a particular direction.

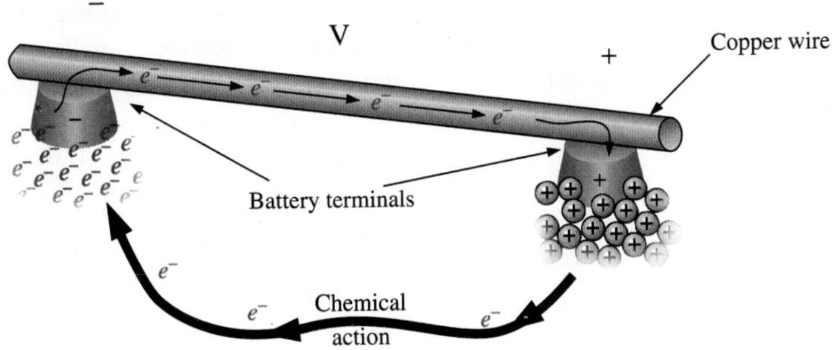

FIG. 2.8
Motion of negatively charged electrons in a copper wire when placed across battery terminals with a difference in potential of volts (V).

electrons to keep the process moving. The electrons arriving at the positive terminal are absorbed, and through the chemical action of the battery, additional electrons are deposited at the negative terminal to make up for those that left.

To take the process a step further, consider the configuration in Fig. 2.9, where a copper wire has been used to connect a light bulb to a battery to create the simplest of electric circuits. The instant the final connection is made, the free electrons of negative charge drift toward the positive terminal, while the positive ions left behind in the copper wire simply oscillate in a mean fixed position. The flow of charge (the electrons) through the bulb heats up the filament of the bulb through friction to the point that it glows red-hot and emits the desired light.

In total, therefore, the applied voltage has established a flow of electrons in a particular direction. In fact, by definition,

if 6.242×10^{18} electrons (1 coulomb) pass through the imaginary plane in Fig. 2.9 in 1 second, the flow of charge, or current, is said to be 1 ampere (A).

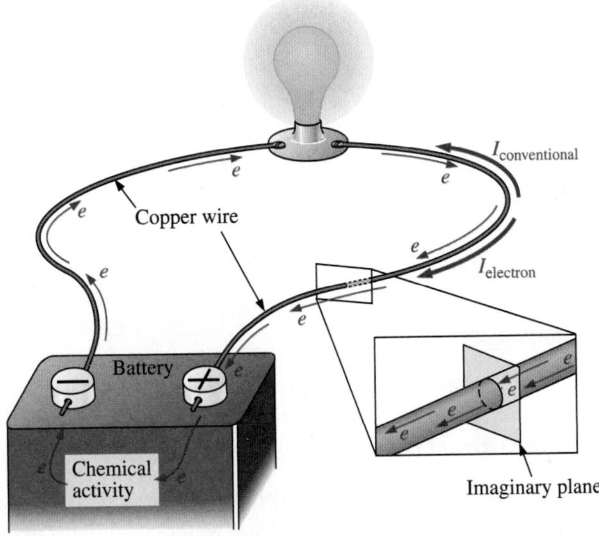

FIG. 2.9
Basic electric circuit.

FIG. 2.10
André Marie Ampère.
Courtesy of the Smithsonian Institution, Photo No. 76,524

French (Lyon, Paris)
(1775–1836)
Mathematician and Physicist
Professor of Mathematics,
École, Polytechnique in Paris

On September 18, 1820, introduced a new field of study, electrodynamics, devoted to the effect of electricity in motion, including the interaction between currents in adjoining conductors and the interplay of the surrounding magnetic fields. Constructed the first *solenoid* and demonstrated how it could behave like a magnet (the first *electromagnet*). Suggested the name *galvanometer* for an instrument designed to measure current levels.

The unit of current measurement, **ampere,** was chosen to honor the efforts of André Ampère in the study of electricity in motion (Fig. 2.10).

Using the coulomb as the unit of charge, the current in amperes can be determined using the following equation:

$$\boxed{I = \frac{Q}{t}} \qquad \begin{aligned} I &= \text{amperes (A)} \\ Q &= \text{coulombs (C)} \\ t &= \text{time (s)} \end{aligned} \qquad \textbf{(2.5)}$$

The capital letter I was chosen from the French word for current, *intensité.* The SI abbreviation for each quantity in Eq. (2.5) is provided to the right of the equation. The equation clearly reveals that for equal time intervals, the more charge that flows through the wire, the larger the resulting current.

Through algebraic manipulations, the other two quantities can be determined as follows:

$$\boxed{Q = It} \qquad \text{(coulombs, C)} \qquad \textbf{(2.6)}$$

and

$$\boxed{t = \frac{Q}{I}} \qquad \text{(seconds, s)} \qquad \textbf{(2.7)}$$

EXAMPLE 2.3 The charge flowing through the imaginary surface in Fig. 2.9 is 0.16 C every 64 ms. Determine the current in amperes.

Solution: Eq. (2.5): $I = \dfrac{Q}{t} = \dfrac{0.16\text{ C}}{64 \times 10^{-3}\text{s}} = \dfrac{160 \times 10^{-3}\text{ C}}{64 \times 10^{-3}\text{ s}} = \mathbf{2.50 A}$

EXAMPLE 2.4 Determine how long it will take 4×10^{16} electrons to pass through the imaginary surface in Fig. 2.9 if the current is 5 mA.

Solution: Determine the charge in coulombs:

$$4 \times 10^{16}\ \cancel{\text{electrons}} \left(\frac{1\text{ C}}{6.242 \times 10^{18}\ \cancel{\text{electrons}}}\right) = 0.641 \times 10^{-2}\text{C}$$

$$= 6.41\text{ mC}$$

$$\text{Eq. (2.7): } t = \frac{Q}{I} = \frac{6.41 \times 10^{-3}\text{C}}{5 \times 10^{-3}\text{ A}} = \mathbf{1.28\ s}$$

In summary, therefore,

the applied voltage (or potential difference) in an electrical/electronics system is the "pressure" to set the system in motion, and the current is the reaction to that pressure.

A mechanical analogy often used to explain the above is the simple garden hose. In the absence of any pressure, the water sits quietly in the hose with no general direction, just as electrons do not have a net direction in the absence of an applied voltage. However, release the spigot, and the applied pressure forces the water to flow through the hose. Similarly, apply a voltage to the circuit, and a flow of charge or current results.

A second glance at Fig. 2.9 reveals that two directions of charge flow have been indicated. One is called *conventional flow,* and the other is called *electron flow.* This text discusses only conventional flow for a variety of reasons; namely, it is the most widely used at educational institutions and in industry, it is employed in the design of all electronic device symbols, and it is the popular choice for all major computer software packages. The flow controversy is a result of an assumption made at the time electricity was discovered that the positive charge was the moving particle in metallic conductors. Be assured that the choice of conventional flow will not create great difficulty and confusion in the chapters to follow. Once the direction of I is established, the issue is dropped and the analysis can continue without confusion.

Safety Considerations

It is important to realize that even small levels of current through the human body can cause serious, dangerous side effects. Experimental results reveal that the human body begins to react to currents of only a few milliamperes. Although most individuals can withstand currents up to perhaps 10 mA for very short periods of time without serious side effects, any current over 10 mA should be considered dangerous. In fact, currents of 50 mA can cause severe shock, and currents of over 100 mA can be fatal. In most cases, the skin resistance of the body when dry is sufficiently high to limit the current through the body to relatively safe levels for voltage levels typically found in the home. However, if the skin is wet due to perspiration, bathing, and so on, or if the skin barrier is broken due to an injury, the skin resistance drops dramatically, and current levels could rise to dangerous levels for the same voltage shock. In general, therefore, simply remember that *water and electricity don't mix.* Granted, there are safety devices in the home today [such as the ground fault circuit interrupt (GFCI) breaker, discussed in Chapter 4] that are designed specifically for use in wet areas such as the bathroom and kitchen, but accidents happen. Treat electricity with respect—not fear.

2.5 VOLTAGE SOURCES

The term **dc,** used throughout this text, is an abbreviation for **direct current,** which encompasses all systems where there is a unidirectional (one direction) flow of charge. This section reviews dc voltage supplies that apply a fixed voltage to electrical/electronics systems.

The graphic symbol for all dc voltage sources is shown in Fig. 2.11. Note that the relative length of the bars at each end define the polarity of the supply. The long bar represents the positive side; the short bar, the negative. Note also the use of the letter E to denote *voltage source.* It comes from the fact that

an electromotive force (emf) is a force that establishes the flow of charge (or current) in a system due to the application of a difference in potential.

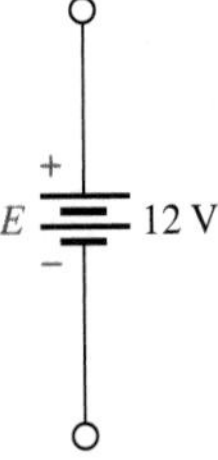

FIG. 2.11
Standard symbol for a dc voltage source.

In general, dc voltage sources can be divided into three basic types: (1) batteries (chemical action or solar energy), (2) generators (electromechanical), and (3) power supplies (rectification—a conversion process to be described in your electronics courses).

Batteries

General Information For the layperson, the battery is the most common of the dc sources. By definition, a battery (derived from the expression "battery of cells") consists of a combination of two or more similar **cells,** a cell being the fundamental source of electrical energy developed through the conversion of chemical or solar energy. All cells can be divided into the **primary** or **secondary** types. The secondary is rechargeable, whereas the primary is not. That is, the chemical reaction of the secondary cell can be reversed to restore its capacity. The two most common rechargeable batteries are the lead-acid unit (used primarily in automobiles) and the nickel-metal hydride (NiMH) battery (used in calculators, tools, photoflash units, shavers, and so on). The obvious advantages of rechargeable units are the savings in time and money of not continually replacing discharged primary cells.

All the cells discussed in this chapter (except the **solar cell,** which absorbs energy from incident light in the form of photons) establish a potential difference at the expense of chemical energy. In addition, each has a positive and a negative *electrode* and an **electrolyte** to complete the circuit between electrodes within the battery. The electrolyte is the contact element and the source of ions for conduction between the terminals.

Primary Cells The popular alkaline primary battery uses a powdered zinc anode (+); a potassium (alkali metal) hydroxide electrolyte; and a manganese dioxide, carbon cathode (−) as shown in Fig. 2.12(a). In Fig. 2.12(b), note that for the cylindrical types (AAA, AA, C, and D), the voltage is the same for each, but the ampere-hour (Ah) rating increases significantly with size. The ampere-hour rating is an indication of the level of current that the battery can provide for a specified period of time (to be discussed in detail in Section 2.6). In particular, note that for the large, lantern-type battery, the voltage is only 4 times that of the AAA

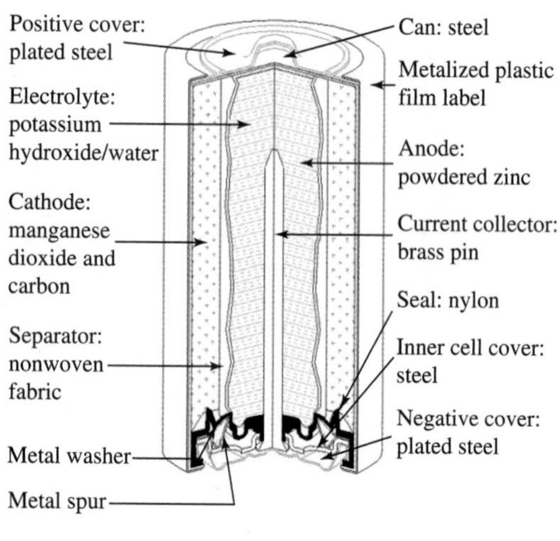

(a)

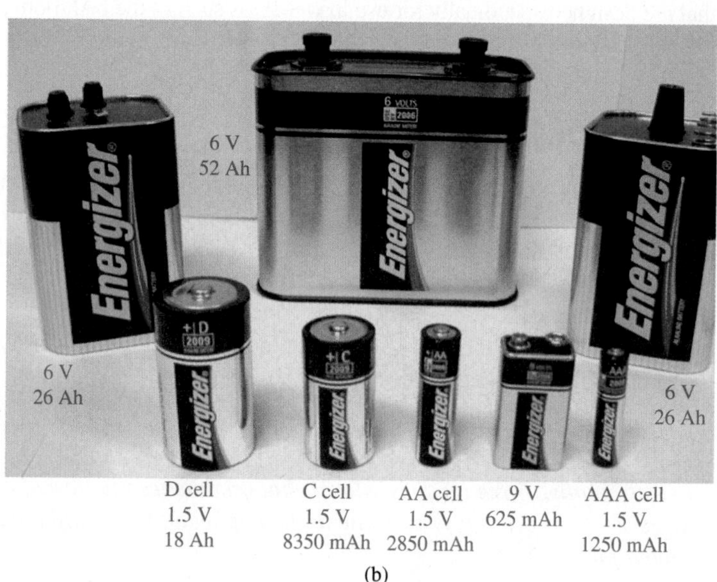

(b)

FIG. 2.12
Alkaline primary cell: (a) Cutaway of cylindrical Energizer® cell; (b) various types of Eveready Energizer® primary cells.
(© Eveready Battery Company, Inc., St. Louis Missouri)

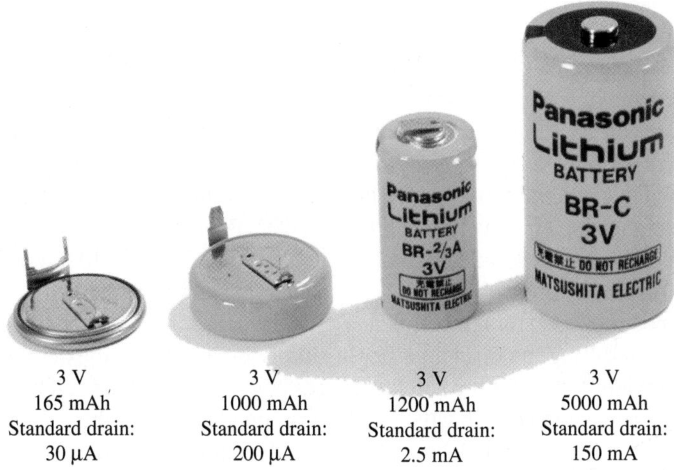

FIG. 2.13
Lithium primary batteries.

battery, but the ampere-hour rating of 52 Ah is almost 42 times that of the AAA battery.

Another type of popular primary cell is the lithium battery, shown in Fig. 2.13. Again, note that the voltage is the same for each, but the size increases substantially with the ampere-hour rating and the rated drain current.

In general, therefore,

for batteries of the same type, the size is dictated primarily by the standard drain current or ampere-hour rating, not by the terminal voltage rating.

Lead-Acid Secondary Cell For the secondary lead-acid unit shown in Fig. 2.14, the electrolyte is sulfuric acid, and the electrodes are spongy lead (Pb) and lead peroxide (PbO_2). When a load is applied to the battery terminals, there is a transfer of electrons from the spongy lead electrode

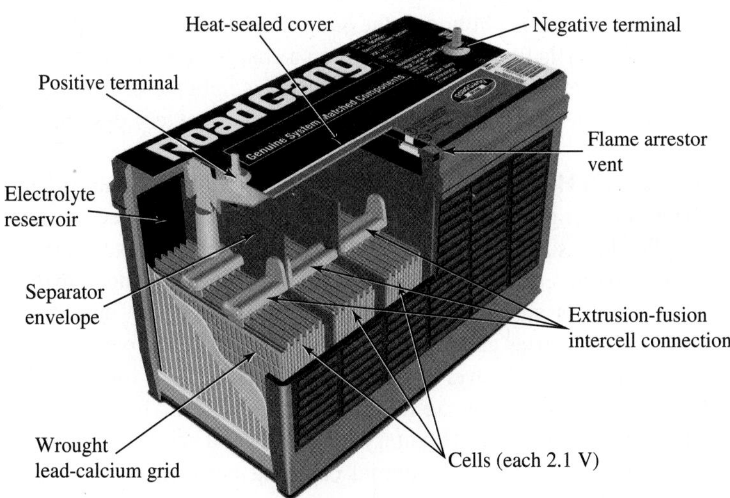

FIG. 2.14
Maintenance-free 12 V (actually 12.6 V) lead-acid battery.
(Courtesy of Remy International, Inc.)

to the lead peroxide electrode through the load. This transfer of electrons will continue until the battery is completely discharged. The discharge time is determined by how diluted the acid has become and how heavy the coating of lead sulfate is on each plate. The state of discharge of a lead storage cell can be determined by measuring the **specific gravity** of the electrolyte with a hydrometer. The specific gravity of a substance is defined to be the ratio of the weight of a given volume of the substance to the weight of an equal volume of water at 4°C. For fully charged batteries, the specific gravity should be somewhere between 1.28 and 1.30. When the specific gravity drops to about 1.1, the battery should be recharged.

Since the lead storage cell is a secondary cell, it can be recharged at any point during the discharge phase simply by applying an external **dc current source** across the cell that passes current through the cell in a direction opposite to that in which the cell supplied current to the load. This removes the lead sulfate from the plates and restores the concentration of sulfuric acid.

The output of a lead storage cell over most of the discharge phase is about 2.1 V. In the commercial lead storage batteries used in automobiles, 12.6 V can be produced by six cells in series, as shown in Fig. 2.14. In general, lead-acid storage batteries are used in situations where a high current is required for relatively short periods of time. At one time, all lead-acid batteries were vented. Gases created during the discharge cycle could escape, and the vent plugs provided access to replace the water or electrolyte and to check the acid level with a hydrometer. The use of a grid made from a wrought lead-calcium alloy strip, rather than the lead-antimony cast grid commonly used, has resulted in maintenance-free batteries, shown in Fig. 2.14. The lead-antimony structure was susceptible to corrosion, overcharge, gasing, water usage, and self-discharge. Improved design with the lead-calcium grid has either eliminated or substantially reduced most of these problems.

It would seem with all the years of technology surrounding batteries that smaller, more powerful units would now be available. However, when it comes to the electric car, which is slowly gaining interest and popularity throughout the world, the lead-acid battery is still the primary source of power. A "station car," manufactured in Norway and used on a test basis in San Francisco for typical commuter runs, has a total weight of 1650 pounds, with 550 pounds (a third of its weight) for the lead-acid rechargeable batteries. Although the station car will travel at speeds of 55 mph, its range is limited to 65 miles on a charge. It would appear that long-distance travel with significantly reduced weight factors for the batteries will depend on a new, innovative approach to battery design.

Nickel–Metal Hydride Secondary Cells The rechargeable battery has been receiving enormous interest and development in recent years. For applications such as flashlights, shavers, portable televisions, power drills, and so on, rechargeable batteries such as the nickel–metal hydride (NiMH) batteries shown in Fig. 2.15 are the secondary batteries of choice. These batteries are so well made that they can survive over 1000 charge/discharge cycles over a period of time and can last for years.

It is important to recognize that if an appliance calls for a rechargeable battery such as a NiMH battery, a primary cell should not be used. The appliance may have an internal charging network that would be dysfunctional with a primary cell. In addition, note that NiMH batteries are about 1.2 V per cell, whereas the common primary cells are typically 1.5 V per cell.

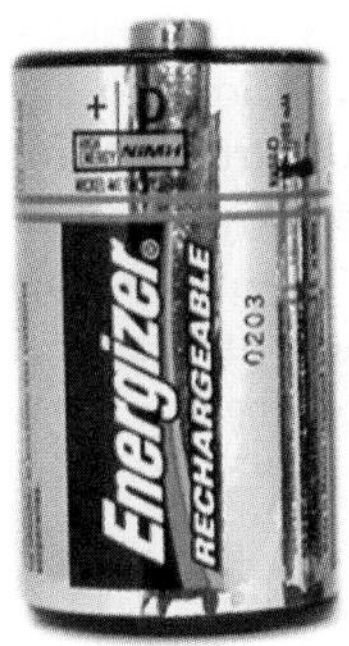

D cell	C cell	AA cell	AAA cell	9 V (7.2 V nominal)
1.2 V	1.2 V	1.2 V	1.2 V	150 mAh
2200 mAh	2200 mAh	1850 mAh	750 mAh	@ 30 mA
@ 440 mA	@ 440 mA	@ 370 mA	@ 150 mA	

FIG. 2.15
Nickel–metal hydride (NiMH) rechargeable batteries.
(© Eveready Battery Company, Inc., St. Louis, Missouri)

There is some ambiguity about how often a secondary cell should be recharged. Generally, the battery can be used until there is some indication that the energy level is low, such as a dimming light from a flashlight, less power from a drill, or a low-battery indicator. Keep in mind that secondary cells do have some "memory." If they are recharged continuously after being used for a short period of time, they may begin to believe they are short-term units and actually fail to hold the charge for the rated period of time. In any event, always try to avoid a "hard" discharge, which results when every bit of energy is drained from a cell. Too many hard-discharge cycles will reduce the cycle life of the battery. Finally, be aware that the charging mechanism for nickel-cadmium cells is quite different from that for lead-acid batteries. The nickel-cadmium battery is charged by a constant current source, with the terminal voltage staying fairly steady through the entire charging cycle. The lead-acid battery is charged by a constant voltage source, permitting the current to vary as determined by the state of the battery. The capacity of the NiMH battery increases almost linearly throughout most of the charging cycle. Ni-Cad batteries become relatively warm when charging. The lower the capacity level of the battery when charging, the higher the temperature of the cell. As the battery approaches rated capacity, the temperature of the cell approaches room temperature.

Other types of rechargeable batteries include the nickel-cadmium (Ni-Cad) and nickel hydrogen (Ni-H) batteries. In reality, however, the NiMH battery is a hybrid of the nickel-cadmium and nickel-hydrogen cells, combining the positive characteristics of each to create a product with a high power level in a small package that has a long life. Another type of rechargeable battery is the lithium-ion variety shown in Fig. 2.16, used in the IBM laptop computer.

10.8 V, 10.8 Ah
Charge time:
System operational: 6 h max.
Power off: 2.5 h max.

FIG. 2.16
IBM ThinkPad T-20 lithium-ion rechargeable battery.
(Courtesy of IBM.)

Solar Cell The SX 20 and SX 30 solar modules (a combination of connected cells) shown in Fig. 2.17 provide 20 W and 30 W of electrical power, respectively. The size and orientation of such units are important because the maximum available wattage on an average bright, sunlit day is 100 mW/cm^2. Since conversion efficiencies are currently only at 10% to 14%, the maximum available power per square centimeter from most

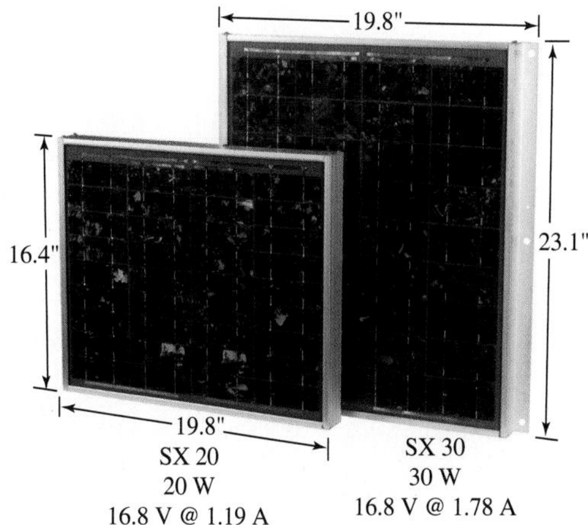

FIG. 2.17
Photovoltaic solar module.
(Photograph courtesy of BP Solar.)

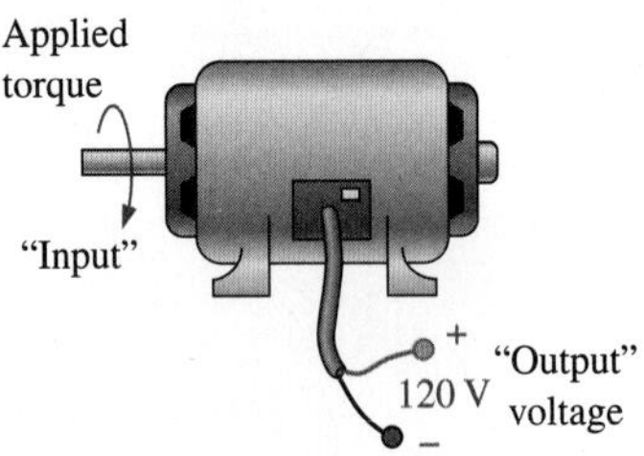

FIG. 2.18
dc generator.

FIG. 2.19
dc laboratory supply (30 V, 3 A).
(Image compliments of Leader Instruments Corporation.)

commercial units is between 10 mW and 14 mW. For a square meter, however, the return would be 100 W to 140 W. The units shown in Fig. 2.17 are typically used for remote telemetry, isolated instrumentation systems, security sensors, signal sources, and land-based navigation aids. A more detailed description of the **solar cell** will appear in your electronics courses, but for now it is important to realize that a fairly steady source of electrical dc power can be obtained from the sun.

Generators

The **dc generator** is quite different from the battery, both in construction (Fig. 2.18) and in mode of operation. When the shaft of the generator is rotating at the nameplate speed due to the applied torque of some external source of mechanical power, a voltage of rated value appears across the external terminals. The terminal voltage and power-handling capabilities of the dc generator are typically higher than those of most batteries, and its lifetime is determined only by its construction. Commercially used dc generators are typically 120 V or 240 V. For the purposes of this text, the same symbols are used for a battery and a generator.

Power Supplies

The dc supply encountered most frequently in the laboratory uses the **rectification** and *filtering* processes as its means toward obtaining a steady dc voltage. Both processes will be covered in detail in your basic electronics courses. In total, a time-varying voltage (such as ac voltage available from a home outlet) is converted to one of a fixed magnitude. A dc laboratory supply of this type is shown in Fig. 2.19.

Most dc laboratory supplies have a regulated, adjustable voltage output with three available terminals, as indicated horizontally at the bottom of Fig 2.19 and vertically in Fig 2.20(a). The symbol for ground or zero potential (the reference) is also shown in Fig. 2.20(a). If 10 V above ground potential are required, the connections are made as shown in Fig. 2.20(b). If 15 V below ground potential are required, the connections are made as shown in Fig. 2.20(c). If connections are as shown in Fig. 2.20(d), we say we have a "floating" voltage of 5 V since the reference level is not included. Seldom is the configuration in Fig. 2.20(d) used since it fails to

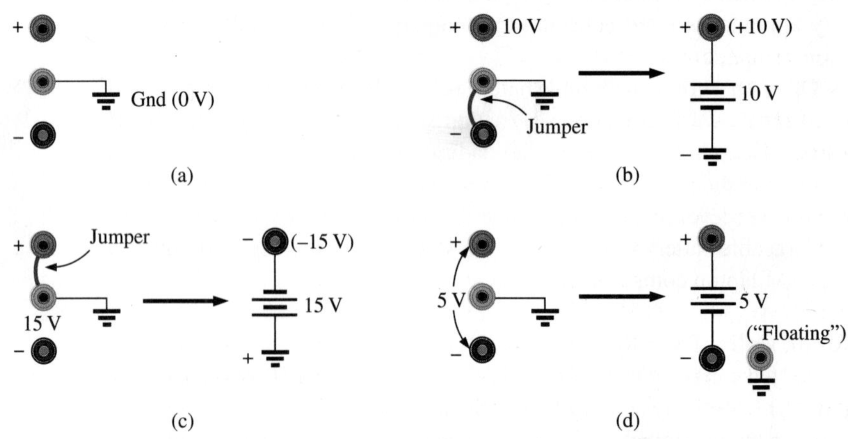

FIG. 2.20
dc laboratory supply: (a) available terminals; (b) positive voltage with respect to (w.r.t.) ground; (c) negative voltage w.r.t. ground; (d) floating supply.

protect the operator by providing a direct low-resistance path to ground and to establish a common ground for the system. In any case, *the positive and negative terminals must be part of any circuit configuration.*

Fuel Cells

One of the most exciting developments in recent years has been the steadily rising interest in **fuel cells** as an alternative energy source. Fuel cells are now being used in the small stationary power plants, transportation (buses), and a wide variety of applications where portability is a major factor, such as the space shuttle. Millions are now being spent by major automobile manufacturers to build affordable fuel-cell vehicles.

Fuel cells have the distinct advantage of operating at efficiencies of 70% to 80% rather than the typical 20% to 25% efficiency of current internal combustion engine of today's automobiles. They also have no moving parts, produce little or no pollution, generate very little noise, and use fuels such as hydrogen and oxygen that are readily available. Fuel cells are considered primary cells (of the continuous-feed variety) because they cannot be recharged. They hold their characteristics as long as the fuel (hydrogen) and oxygen are supplied to the cell. The only byproducts of the conversion process are small amounts of heat (which is often used elsewhere in the system design), water (which may also be reused), and negligible levels of some oxides, depending on the components of the process. Overall, fuel cells are environmentally friendly.

The operation of the fuel cell is essentially opposite to that of the chemical process of electrolysis. Electrolysis is the process whereby electric current is passed through an electrolyte to break it down into its fundamental components. An electrolyte is any solution that will permit conduction through the movement of ions between adjoining electrodes. For instance, passing current through water results in a hydrogen gas by the cathode (negative terminal) and oxygen gas at the anode (positive terminal). In 1839, Sir William Grove believed this process could be reversed and demonstrated that the proper application of the hydrogen gas and oxygen results in a current through an applied load connected to the electrodes of the system. The first commercial unit was used in a tractor in 1959, followed by an energy pack in the 1965 Gemini program. In 1996, the first small power plant was designed, and today it is an important component of the shuttle program.

The basic components of a fuel cell are depicted in Fig. 2.21(a) with details of the construction in Fig. 2.21(b). Hydrogen gas (the fuel) is supplied to the system at a rate proportional to the current required by the load. At the opposite end of the cell, oxygen is supplied as needed. The net result is a flow of electrons through the load and a discharge of water with a release of some heat developed in the process. The amount of heat is minimal although it can also be used as a component in the design to improve the efficiency of the cell. The water (very clean) can simply be discharged or used for other applications such as cooling in the overall application. If the source of hydrogen or oxygen is removed, the system breaks down. The flow diagram of the system is relatively simple, as shown in Fig. 2.21(a). In an actual cell, shown in Fig. 2.21(b), the hydrogen gas is applied to a porous electrode called the *anode* that is coated with a platinum catalyst. The catalyst on the anode serves to speed up the process of breaking down the hydrogen atom into positive hydrogen ions and free electrons. The electrolyte between the electrodes is a solution or membrane that permits the passage of positive hydrogen ions, but not electrons. Facing this wall, the

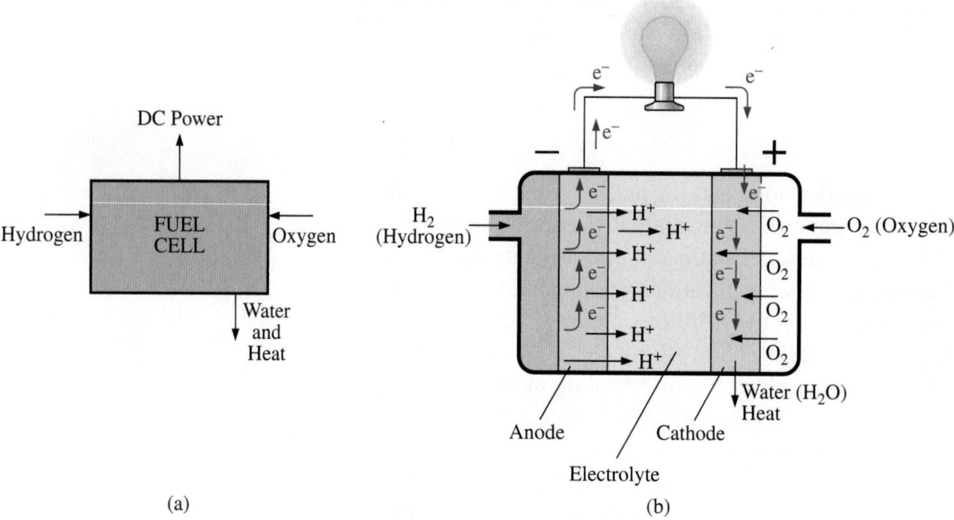

FIG. 2.21
Fuel cell (a) components; (b) basic construction.

electrons choose to pass through the load and light up the bulb, while the positive hydrogen ions migrate toward the cathode. At the porous cathode (also coated with the catalyst), the incoming oxygen atoms combine with the arriving hydrogen ions and the electrons from the circuit to create water (H_2O) and heat. The circuit is, therefore, complete. The electrons are generated and then absorbed. If the hydrogen supply is cut off, the source of electrons is shut down, and the system is no longer an operating fuel cell.

In some fuel cells, either a liquid or molten electrolyte membrane is used. Depending on which the system uses, the chemical reactions will change slightly, but not dramatically from that described above. The phosphoric acid fuel cell is a popular cell using a liquid electrolyte, while the PEM uses a polymer electrolyte membrane. The liquid or molten type is typically used in stationary power plants, while the membrane type is favored for vehicular use.

The output from a single fuel cell is a low voltage, high current dc output. Stacking the cells in series or parallel increases the output voltage or current level.

Fuel cells are receiving an enormous amount of attention and development effort today. It is certainly possible that fuel cells may some day replace batteries in the vast majority of applications requiring a portable energy source. Fig. 2.22 shows the components of a hydrogen fuel-cell automobile.

2.6 AMPERE-HOUR RATING

The most important piece of data for any battery (other than its voltage rating) is its **ampere-hour (Ah) rating.** You have probably noted in the photographs of batteries in this chapter that both the voltage and the ampere-hour rating have been provided for each battery.

The ampere-hour (Ah) rating provides an indication of how long a battery of fixed voltage will be able to supply a particular current.

A battery with an ampere-hour rating of 100 will theoretically provide a current of 1 A for 100 hours, 10 A for 10 hours, or 100 A for 1 hour. Quite

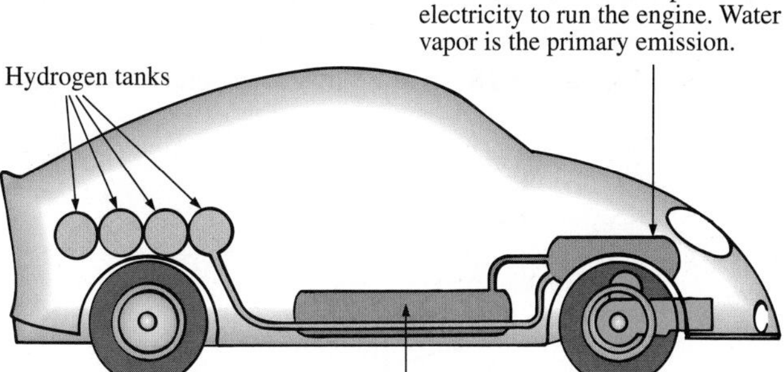
Fuel cells use hydrogen and oxygen to create a reaction that produces electricity to run the engine. Water vapor is the primary emission.
Hydrogen tanks
Air compressor pumps air into the fuel cell.

FIG. 2.22
Hydrogen fuel-cell automobile.

obviously, the greater the current, the shorter the time. An equation for determining the length of time a battery will supply a particular current is the following:

$$\text{Life (hours)} = \frac{\text{ampere-hour (Ah) rating}}{\text{amperes drawn (A)}} \quad \textbf{(2.8)}$$

EXAMPLE 2.5 How long will a 9 V transistor battery with an ampere-hour rating of 520 mAh provide a current of 20 mA?

Solution: Eq. (2.8): $\text{Life} = \dfrac{520 \text{ mAh}}{20 \text{ mA}} = \dfrac{520}{20} \text{ h} = \mathbf{26\ h}$

EXAMPLE 2.6 How long can a 1.5 V flashlight battery provide a current of 250 mA to light the bulb if the ampere-hour rating is 16 Ah?

Solution: Eq. (2.8): $\text{Life} = \dfrac{16 \text{ Ah}}{250 \text{ mA}} = \dfrac{16}{250 \times 10^{-3}} \text{ h} = \mathbf{64\ h}$

2.7 BATTERY LIFE FACTORS

The previous section made it clear that the life of a battery is directly related to the magnitude of the current drawn from the supply. However, there are factors that affect the given ampere-hour rating of a battery, so we may find that a battery with an ampere-hour rating of 100 can supply a current of 10 A for 10 hours but can supply a current of 100 A for only 20 minutes rather than the full 1 hour calculated using Eq. (2.8). In other words,

the capacity of a battery (in ampere-hours) will change with change in current demand.

This is not to say that Eq. (2.8) is totally invalid. It can always be used to gain some insight into how long a battery can supply a particular current. However, be aware that there are factors that affect the ampere-hour rating. Just as with most systems, including the human body, the more we demand, the shorter the time that the output level can be maintained. This is clearly verified by the curves in Fig. 2.23 for the Eveready Energizer D cell. As the constant current drain increased, the ampere-hour rating decreased from about 18 Ah at 25 mA to around 12 Ah at 300 mA.

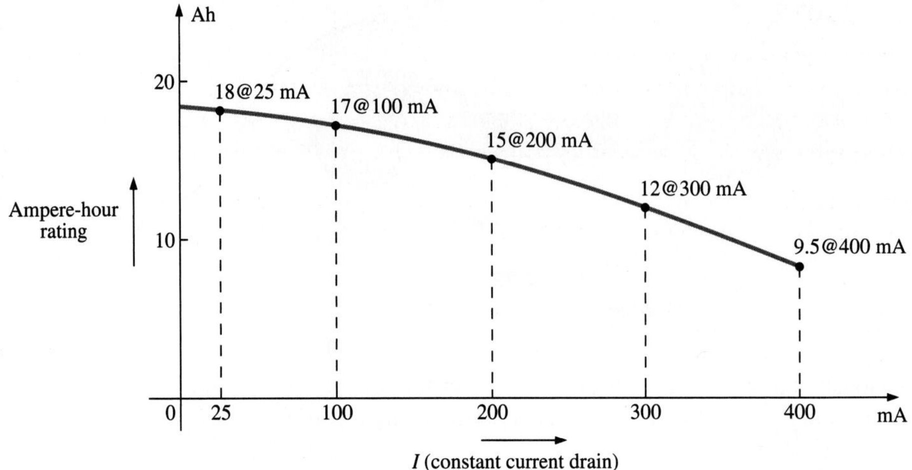

FIG. 2.23
Ampere-hour rating (capacity) versus drain current for an Energizer® D cell.

Another factor that affects the ampere-hour rating is the temperature of the unit and the surrounding medium. In Fig. 2.24, the capacity of the same battery plotted in Fig. 2.23 shows a peak value near the common room temperature of 68°F. At very cold temperatures and very warm temperatures, the capacity drops. Clearly, the ampere-hour rating will be provided at or near room temperature to give it a maximum value, but be aware that it will drop off with an increase or decrease in temperature. Most of us have noted that the battery in a car, radio, two-way radio, flashlight, or whatever seems to have less power in really cold weather. It would seem, then, that the battery capacity would increase with higher temperatures—apparently not the case. In general, therefore,

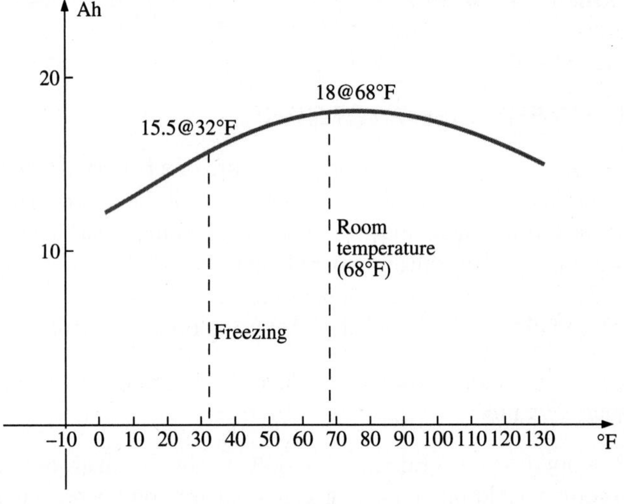

FIG. 2.24
Ampere-hour rating (capacity) versus temperature for an Energizer® D cell.

the ampere-hour rating of a battery will decrease from the room-temperature level with very cold and very warm temperatures.

Another interesting factor that affects the performance of a battery is how long it is asked to supply a particular voltage at a continuous drain current. Note the curves in Fig. 2.25, where the terminal voltage dropped at each level of drain current as the time period increased. The lower the current drain, the longer it could supply the desired current. At 100 mA, it was limited to about 100 hours near the rated voltage, but at 25 mA, it did not drop below 1.2 V until about 500 hours had passed. That is an increase in time of 5 : 1, which is significant. The result is that

the terminal voltage of a battery will eventually drop (at any level of current drain) if the time period of continuous discharge is too long.

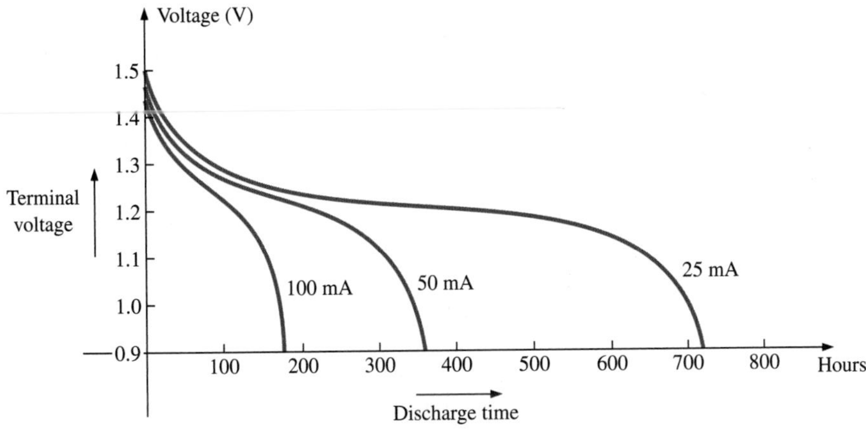

FIG. 2.25
Terminal voltage versus discharge time for specific drain currents for an Energizer® D cell.

2.8 CONDUCTORS AND INSULATORS

Different wires placed across the same two battery terminals allow different amounts of charge to flow between the terminals. Many factors, such as the density, mobility, and stability characteristics of a material, account for these variations in charge flow. In general, however,

conductors are those materials that permit a generous flow of electrons with very little external force (voltage) applied.

In addition,

good conductors typically have only one electron in the valence (most distant from the nucleus) ring.

Since **copper** is used most frequently, it serves as the standard of comparison for the relative conductivity in Table 2.1. Note that aluminum, which has seen some commercial use, has only 61% of the conductivity level of copper. The choice of material must be weighed against the cost and weight factors, however.

TABLE 2.1
Relative conductivity of various materials.

Metal	Relative Conductivity (%)
Silver	105
Copper	**100**
Gold	70.5
Aluminum	61
Tungsten	31.2
Nickel	22.1
Iron	14
Constantan	3.52
Nichrome	1.73
Calorite	1.44

Insulators are those materials that have very few free electrons and require a large applied potential (voltage) to establish a measurable current level.

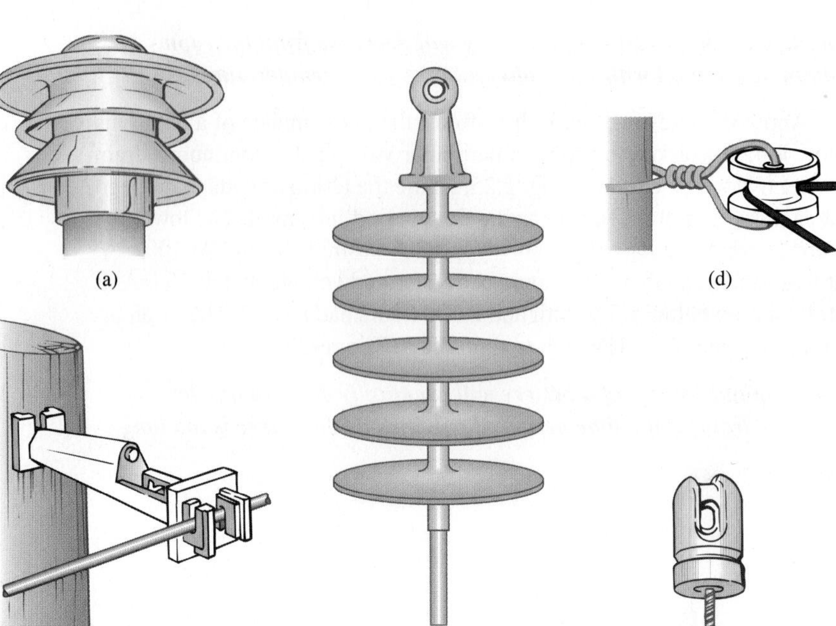

FIG. 2.26
Various types of insulators and their applications: (a) Corning Glass Works Pyrex™ power line insulator; (b) Fi-Shock extender insulator; (c) Lapp power line insulator; (d) Fi-Shock corner insulator; (e) Fi-Shock screw-in post insulator.

TABLE 2.2
Breakdown strength of some common insulators.

Material	Average Breakdown Strength (kV/cm)
Air	30
Porcelain	70
Oils	140
Bakelite®	150
Rubber	270
Paper (paraffin-coated)	500
Teflon®	600
Glass	900
Mica	2000

A common use of insulating material is for covering current-carrying wire, which, if uninsulated, could cause dangerous side effects. Power line workers wear rubber gloves and stand on rubber mats as safety measures when working on high voltage transmission lines. A number of different types of insulators and their applications appear in Fig. 2.26.

Be aware, however, that even the best insulator will break down (permit charge to flow through it) if a sufficiently large potential is applied across it. The breakdown strengths of some common insulators are listed in Table 2.2. According to this table, for insulators with the same geometric shape, it would require $270/30 = 9$ times as much potential to pass current through rubber compared to air and approximately 67 times as much voltage to pass current through mica as through air.

2.9 SEMICONDUCTORS

Semiconductors are a specific group of elements that exhibit characteristics between those of insulators and those of conductors.

The prefix *semi,* included in the terminology, has the dictionary definition of *half, partial,* or *between,* as defined by its use. The entire electronics industry is dependent on this class of materials since the electronic devices and integrated circuits (ICs) are constructed of semiconductor materials. Although *silicon* (Si) is the most extensively employed material, *germanium* (Ge) and *gallium arsenide* (GaAs) are also used in many important devices.

Semiconductor materials typically have four electrons in the outermost valence ring.

Semiconductors are further characterized as being photoconductive and having a negative temperature coefficient. Photoconductivity is a phenomenon where the photons (small packages of energy) from incident light can increase the carrier density in the material and thereby the charge flow level. A negative temperature coefficient reveals that the resistance (a characteristic to be described in detail in the next chapter) decreases with an increase in temperature (opposite to that of most conductors). A great deal more will be said about semiconductors in the chapters to follow and in your basic electronics courses.

2.10 AMMETERS AND VOLTMETERS

It is important to be able to measure the current and voltage levels of an operating electrical system to check its operation, isolate malfunctions, and investigate effects impossible to predict on paper. As the names imply, **ammeters** are used to measure current levels; **voltmeters,** the potential difference between two points. If the current levels are usually of the order of milliamperes, the instrument will typically be referred to as a *milliammeter,* and if the current levels are in the microampere range, as a *microammeter.* Similar statements can be made for voltage levels. Throughout the industry, voltage levels are measured more frequently than current levels, primarily because measurement of the former does not require that the network connections be disturbed.

The potential difference between two points can be measured by simply connecting the leads of the meter *across the two points,* as indicated in Fig. 2.27. An up-scale reading is obtained by placing the positive lead of the meter to the point of higher potential of the network and the common or negative lead to the point of lower potential. The reverse connection results in a negative reading or a below-zero indication.

Ammeters are connected as shown in Fig. 2.28. Since ammeters measure the rate of flow of charge, the meter must be placed in the network such that the charge flows through the meter. The only way this can be accomplished is to open the path in which the current is to be measured and place the meter between the two resulting terminals. For the configuration in Fig. 2.28, the voltage source lead (+) must be disconnected

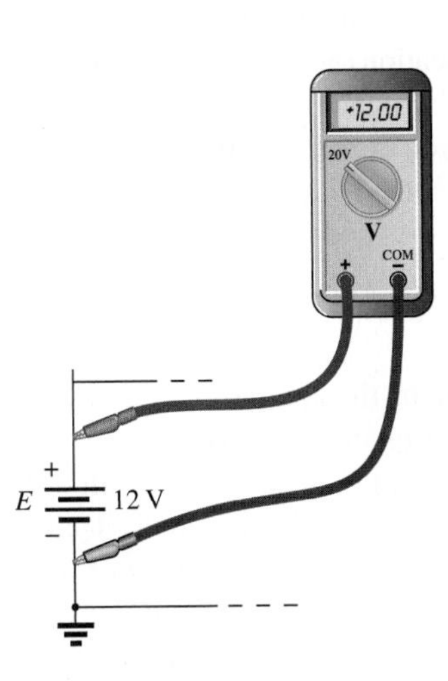

FIG. 2.27
Voltmeter connection for an up-scale (+) reading.

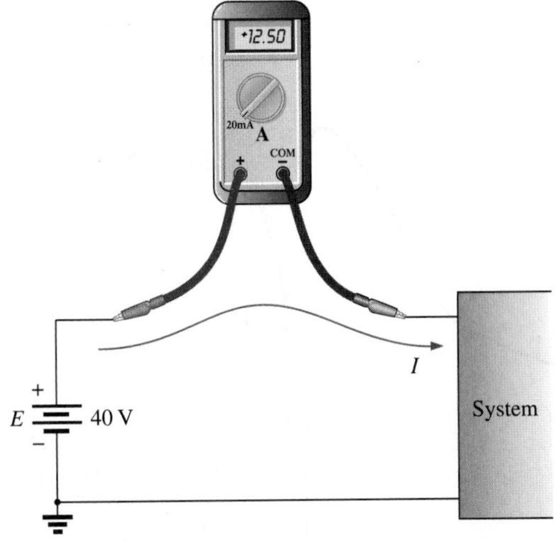

FIG. 2.28
Ammeter connection for an up-scale (+) reading.

FIG. 2.29
Volt-ohm-milliammeter (VOM) analog meter.
(Courtesy of Simpson Electric Co.)

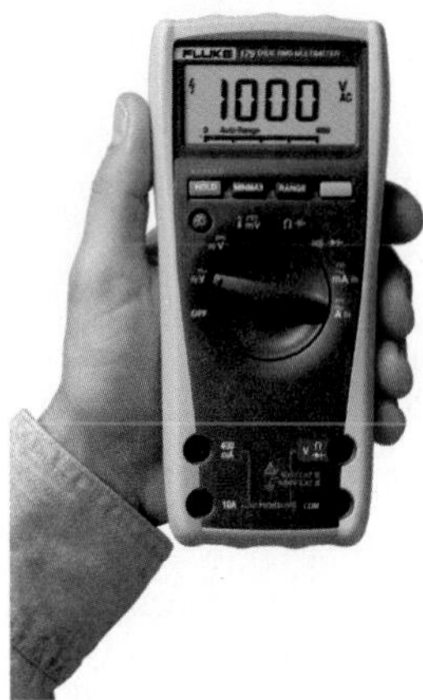

FIG. 2.30
Digital multimeter (DMM).
(Courtesy of Fluke Corporation. Reproduced with permission.)

from the system, and the ammeter inserted as shown. An up-scale reading will be obtained if the polarities on the terminals of the ammeter are such that the current of the system enters the positive terminal.

The introduction of any meter into an electrical/electronic system raises a concern about whether the meter will affect the behavior of the system. This question and others will be examined in Chapters 5 and 6 after additional terms and concepts have been introduced. For the moment, let it be said that since voltmeters and ammeters do not have internal components, they will affect the network when introduced for measurement purposes. The design of each, however, is such that the impact is minimized.

There are instruments designed to measure just current or just voltage levels. However, the most common laboratory meters include the *volt-ohm-milliammeter* (VOM) and the *digital multimeter* (DMM) in Figs. 2.29 and 2.30, respectively. Both instruments measure voltage and current and a third quantity, resistance (introduced in the next chapter). The VOM uses an analog scale, which requires interpreting the position of a pointer on a continuous scale, while the DMM provides a display of numbers with decimal-point accuracy determined by the chosen scale. Comments on the characteristics and use of various meters will be made throughout the text. However, the major study of meters will be left for the laboratory sessions.

2.11 APPLICATIONS

Throughout the text, Applications sections such as this one have been included to permit a further investigation of terms, quantities, or systems introduced in the chapter. The primary purpose of these Applications is to establish a link between the theoretical concepts of the text and the real, practical world. Although the majority of components that appear in a system may not have been introduced (and, in fact, some components will not be examined until more advanced studies), the topics were chosen very carefully and should be quite interesting to a new student of the subject matter. Sufficient comment is included to provide a surface understanding of the role of each part of the system, with the understanding that the details will come at a later date. Since exercises on the subject matter of the Applications do not appear at the end of the chapter, the content is designed not to challenge the student but rather to stimulate his or her interest and answer some basic questions such as how the system looks inside, what role specific elements play in the system, and, of course, how the system works. In essence, therefore, each Applications section provides an opportunity to begin to establish a practical background beyond simply the content of the chapter. Do not be concerned if you do not understand every detail of each application. Understanding will come with time and experience. For now, take what you can from the examples and then proceed with the material.

Flashlight

Although the flashlight uses one of the simplest of electrical circuits, a few fundamentals about its operation do carry over to more sophisticated systems. First, and quite obviously, it is a dc system with a lifetime totally dependent on the state of the batteries and bulb. Unless it is the rechargeable type, each time you use it, you take some of the life out of it. For many hours, the brightness will not diminish noticeably. Then, however, as it reaches the end of its ampere-hour capacity, the light becomes dimmer at an increasingly rapid rate (almost exponentially). The standard two-battery flashlight is shown in Fig. 2.31(a) with its electrical

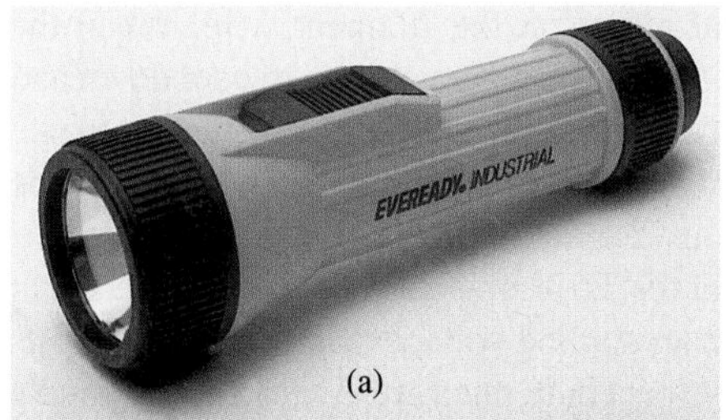

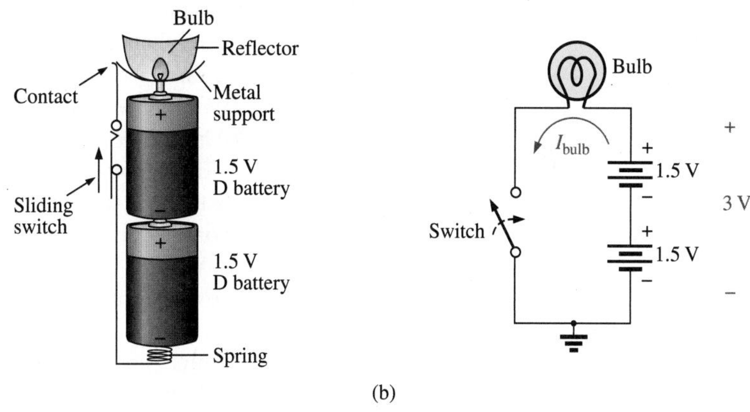

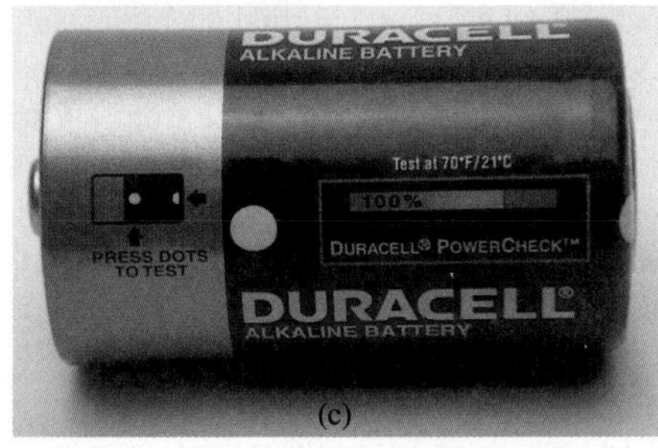

FIG. 2.31
(a) Eveready® D cell flashlight; (b) electrical schematic of flashlight of part (a); (c) Duracell® Powercheck™ D cell battery.

schematic in Fig. 2.31(b). Each 1.5 V battery has an ampere-hour rating of about 18 as indicated in Fig. 2.12. The single-contact miniature flange-base bulb is rated at 2.5 V and 300 mA with good brightness and a lifetime of about 30 hours. Thirty hours may not seem like a long lifetime, but you have to consider how long you usually use a flashlight on each occasion. If we assume a 300 mA drain from the battery for the bulb when in use, the lifetime of the battery, by Eq. (2.8), is about 60 hours. Comparing the 60 hour lifetime of the battery to the 30 hour life expectancy of the bulb suggests that we normally have to replace bulbs more frequently than batteries.

However, most of us have experienced the opposite effect. We can change batteries two or three times before we need to replace the bulb. This is simply one example of the fact that one cannot be guided solely by the specifications of each component of an electrical design. The operating conditions, terminal characteristics, and details about the actual response of the system for short and long periods of time must be considered. As mentioned earlier, the battery loses some of its power each time it is used. Although the terminal voltage may not change much at first, its ability to provide the same level of current drops with each usage. Further, batteries slowly discharge due to "leakage currents" even if the switch is not on. The air surrounding the battery is not "clean" in the sense that moisture and other elements in the air can provide a conduction path for leakage currents through the air through the surface of the battery itself, or through other nearby surfaces, and the battery eventually discharges. How often have we left a flashlight with new batteries in a car for a long period of time only to find the light very dim or the batteries dead when we need the flashlight the most? An additional problem is acid leaks that appear as brown stains or corrosion on the casing of the battery. These leaks also affect the life of the battery. Further, when the flashlight is turned on, there is an initial surge in current that drains the battery more than continuous use for a period of time. In other words, continually turning the flashlight on and off has a very detrimental effect on its life. We must also realize that the 30 hour rating of the bulb is for continuous use, that is, 300 mA flowing through the bulb for a continuous 30 hours. Certainly, the filament in the bulb and the bulb itself will get hotter with time, and this heat has a detrimental effect on the filament wire. When the flashlight is turned on and off, it gives the bulb a chance to cool down and regain its normal characteristics, thereby avoiding any real damage. Therefore, with normal use we can expect the bulb to last longer than the 30 hours specified for continuous use.

Even though the bulb is rated for 2.5 V operation, it would appear that the two batteries would result in an applied voltage of 3 V which suggests poor operating conditions. However, a bulb rated at 2.5 V can easily handle 2.5 V to 3 V. In addition, as was pointed out in this chapter, the terminal voltage drops with the current demand and usage. Under normal operating conditions, a 1.5 V battery is considered to be in good condition if the loaded terminal voltage is 1.3 V to 1.5 V. When it drops to the range from 1 V to 1.1 V, it is weak, and when it drops to the range from 0.8 V to 0.9 V, it has lost its effectiveness. The levels can be related directly to the test band now appearing on Duracell® batteries, such as on the one shown in Fig. 2.31(c). In the test band on this battery, the upper voltage area (green) is near 1.5 V (labeled 100%); the lighter area to the right, from about 1.3 V down to 1 V; and the replace area (red) on the far right, below 1 V.

Be aware that the total supplied voltage of 3 V will be obtained only if the batteries are connected as shown in Fig. 2.31(b). Accidentally placing the two positive terminals together will result in a total voltage of 0 V, and the bulb will not light at all. *For the vast majority of systems with more than one battery, the positive terminal of one battery will always be connected to the negative terminal of another. For all low-voltage batteries, the end with the nipple is the positive terminal, and the end with the flat end is the negative terminal. In addition, the flat or negative end of a battery is always connected to the battery casing with the helical coil to keep the batteries in place. The positive end of the battery is always connected to a flat spring connection or the element to be operated.* If you look carefully at the bulb, you will find that the nipple connected to the positive end of the battery is insulated from the jacket around the base of the bulb. The jacket is the second terminal of the battery used to complete the circuit through the on/off switch.

If a flashlight fails to operate properly, the first thing to check is the state of the batteries. It is best to replace both batteries at once. A system with one good battery and one nearing the end of its life will result in pressure on the good battery to supply the current demand, and, in fact, the bad battery will actually be a drain on the good battery. Next, check the condition of the bulb by checking the filament to see whether it has opened at some point because a long-term, continuous current level occurred or because the flashlight was dropped. If the battery and bulb seem to be in good shape, the next area of concern is the contacts between the positive terminal and the bulb and the switch. Cleaning both with emery cloth often eliminates this problem.

12 V Car Battery Charger

Battery chargers are a common household piece of equipment used to charge everything from small flashlight batteries to heavy-duty, marine, lead-acid batteries. Since all are plugged into a 120 V ac outlet such as found in the home, the basic construction of each is quite similar. In every charging system, a *transformer* (Chapter 22) must be included to cut the ac voltage to a level appropriate for the dc level to be established. A *diode* (also called *rectifier*) arrangement must be included to convert the ac voltage which varies with time to a fixed dc level such as described in this chapter. Diodes and/or rectifiers will be discussed in detail in your first electronics course. Some dc chargers also include a *regulator* to provide

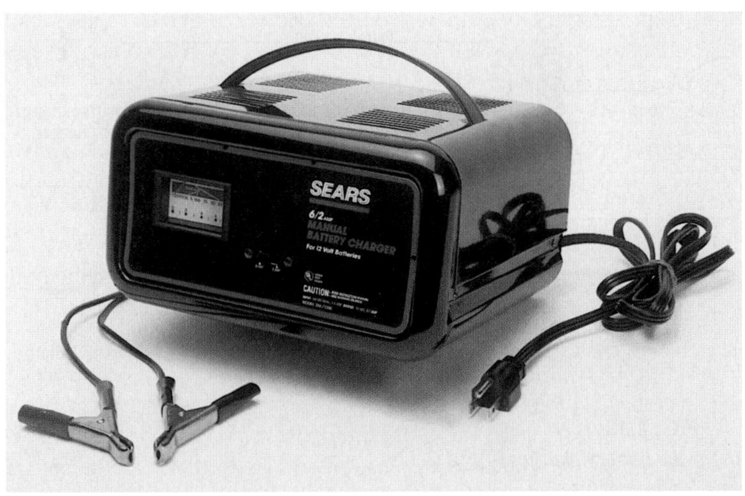

(a)

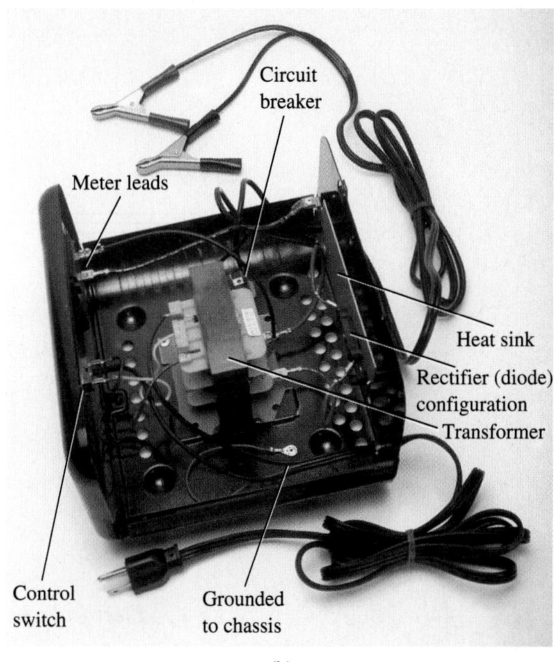

(b)

FIG. 2.32
Battery charger: (a) external appearance; (b) internal construction.

an improved dc level (one that varies less with time or load). The car battery charger, one of the most common, is described here.

The outside appearance and the internal construction of a Sears 6/2 AMP Manual Battery Charger are provided in Fig. 2.32. Note in Fig. 2.32(b) that the transformer (as in most chargers) takes up most of the internal space. The additional air space and the holes in the casing are there to ensure an outlet for the heat that will develop due to the resulting current levels.

The schematic in Fig. 2.33 includes all the basic components of the charger. Note first that the 120 V from the outlet are applied directly across the primary of the transformer. The charging rate of 6 A or 2 A is determined by the switch, which simply controls how many windings of the primary will be in the circuit for the chosen charging rate. If the battery is charging at the 2 A level, the full primary will be in the circuit, and the ratio of the turns in the primary to the turns in the secondary will be a maximum. If it is charging at the 6 A level, fewer turns of the primary are in the circuit, and the ratio drops. When you study transformers, you will find that the voltage at the primary and secondary is directly related to the *turns ratio.* If the ratio from primary to secondary drops, the voltage drops also. The reverse effect occurs if the turns on the secondary exceed those on the primary.

The general appearance of the waveforms appears in Fig. 2.33 for the 6 A charging level. Note that so far, the ac voltage has the same wave shape across the primary and secondary. The only difference is in the peak value of the waveforms. Now the diodes take over and convert the ac waveform which has zero average value (the waveform above equals the waveform below) to one that has an average value (all above the axis) as shown in the same figure. For the moment simply recognize that diodes are semiconductor electronic devices that permit only conventional current to flow through them in the direction indicated by the arrow in the symbol. Even though the waveform resulting from the diode

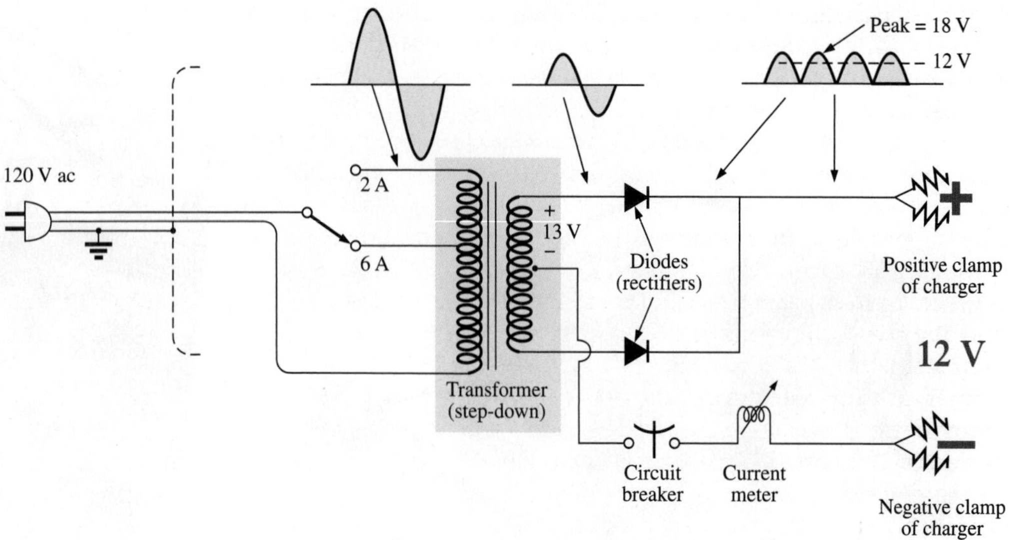

FIG. 2.33
Electrical schematic for the battery charger of Fig. 2.32.

action has a pulsing appearance with a peak value of about 18 V, it charges the 12 V battery whenever its voltage is greater than that of the battery, as shown by the shaded area. Below the 12 V level, the battery cannot discharge back into the charging network because the diodes permit current flow in only one direction.

In particular, note in Fig. 2.32(b) the large plate that carries the current from the rectifier (diode) configuration to the positive terminal of the battery. Its primary purpose is to provide a *heat sink* (a place for the heat to be distributed to the surrounding air) for the diode configuration. Otherwise, the diodes would eventually melt down and self-destruct due to the resulting current levels. Each component of Fig. 2.33 has been carefully labeled in Fig. 2.32(b) for reference.

When current is first applied to a battery at the 6 A charge rate, the current demand as indicated by the meter on the face of the instrument may rise to 7 A or almost 8 A. However, the level of current decreases as the battery charges until it drops to a level of 2 A or 3 A. For units such as this that do not have an automatic shutoff, it is important to disconnect the charger when the current drops to the fully charged level; otherwise, the battery becomes overcharged and may be damaged. A battery that is at its 50% level can take as long as 10 hours to charge, so don't expect it to be a 10-minute operation. In addition, if a battery is in very bad shape with a lower than normal voltage, the initial charging current may be too high for the design. To protect against such situations, the circuit breaker opens and stops the charging process. Because of the high current levels, it is important that the directions provided with the charger be carefully read and applied.

FIG. 2.34
Answering machine/phone 9 V dc supply.

Answering Machines/Phones dc Supply

A wide variety of systems in the home and office receive their dc operating voltage from an ac/dc conversion system plugged right into a 120 V ac outlet. Laptop computers, answering machines/phones, radios, clocks, cellular phones, CD players, and so on, all receive their dc power from a packaged system such as shown in Fig. 2.34. The conversion from ac to dc occurs within the unit which is plugged directly into the outlet. The dc voltage is available at the end of the long wire which is designed to be plugged into the operating unit. As small as the unit may be, it contains basically the same components as in the battery charger in Fig. 2.32.

In Fig. 2.35, you can see the transformer used to cut the voltage down to appropriate levels (again the largest component of the system). Note that two diodes establish a dc level, and a capacitive filter (Chapter 10) is added to smooth out the dc as shown. The system can be relatively small because the operating current levels are quite small, permitting the use of thin wires to construct the transformer and limit its size. The lower currents also reduce the concerns about heating effects, permitting a small housing structure. The unit in Fig. 2.35, rated at 9 V at 200 mA, is commonly used to provide power to answering machines/phones. Further smoothing of the dc voltage is accomplished by a regulator built into the receiving unit. The regulator is normally a small IC chip placed in the receiving unit to separate the heat that it generates from the heat generated by the transformer, thereby reducing the net heat at the outlet close to the wall. In addition, its placement in the receiving unit reduces the possibility of picking up noise and oscillations along the long wire from the conversion unit to the operating unit, and it ensures that the full rated voltage is available at the unit itself, not a lesser value due to losses along the line.

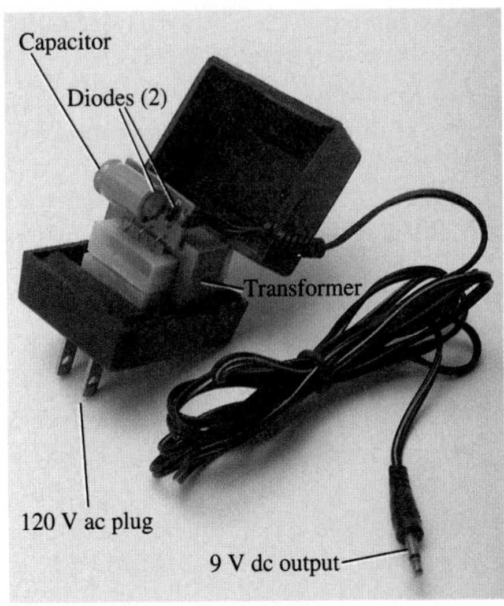

FIG. 2.35
Internal construction of the 9 V dc supply in Fig. 2.34.

2.12 COMPUTER ANALYSIS

In some texts, the procedure for choosing a dc voltage source and placing it on the schematic using computer methods is introduced at this point. This approach, however, requires students to turn back to this chapter for the procedure when the first complete network is installed and examined. Therefore, the procedure is introduced in Chapter 4 when the first complete network is examined, thereby localizing the material and removing the need to reread this chapter and Chapter 3.

PROBLEMS

SECTION 2.2 Atoms and Their Structure

1. The numbers of orbiting electrons in aluminum and silver are 13 and 47, respectively. Draw the electronic configuration for each, and discuss briefly why each is a good conductor.

2. Find the force of attraction in newtons between the charges Q_1 and Q_2 in Fig. 2.36 when
   a. $r = 1$ m
   b. $r = 3$ m
   c. $r = 10$ m
   d. Did the force drop off quickly with an increase in distance?

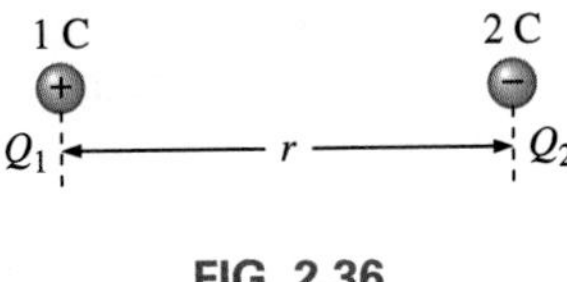

FIG. 2.36
Problem 2.

*3. Find the force of repulsion in newtons between Q_1 and Q_2 in Fig. 2.37 when
   a. $r = 1$ mi
   b. $r = 10$ ft
   c. $r = 1/16$ in.

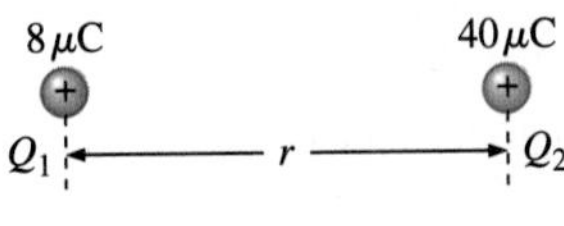

FIG. 2.37
Problem 3.

*4. Plot the force of attraction (in newtons) versus separation (in meters) for two charges of 2 C and 8 C. Set r to 0.5 m and 1 m, followed by 1 m intervals to 10 m. Comment on the shape of the curve. Is it linear or nonlinear? What does it tell you about the force of attraction between charges as they are separated? What does it tell you about any function plotted against a squared term in the denominator?

5. Determine the distance between two charges of 20 μC if the force between the two charges is 3.6×10^4 N.

*6. Two charged bodies, Q_1 and Q_2, when separated by a distance of 2 m, experience a force of repulsion equal to 1.8 N.
   a. What will the force of repulsion be when they are 10 m apart?
   b. If the ratio $Q_1/Q_2 = 1/2$, find Q_1 and Q_2 ($r = 10$ m).

SECTION 2.3 Voltage

7. What is the voltage between two points if 1.2 J of energy are required to move 0.4 mC between the two points?

8. If the potential difference between two points is 60 V, how much energy is expended to bring 8 mC from one point to the other?

9. Find the charge Q that requires 96 J of energy to be moved through a potential difference of 16 V.

10. How much charge passes through a radio battery of 9 V if the energy expended is 72 J?

SECTION 2.4 Current

11. Find the current in amperes if 12 mC of charge pass through a wire in 2.8 s.

12. If 312 C of charge pass through a wire in 2 min, find the current in amperes.

13. If a current of 40 mA exists for 0.8 min, how many coulombs of charge have passed through the wire?

14. How many coulombs of charge pass through a lamp in 1.2 min if the current is constant at 250 mA?

15. If the current in a conductor is constant at 2 mA, how much time is required for 6 mC to pass through the conductor?

16. If $21.847 \times 10^{+18}$ electrons pass through a wire in 12 s, find the current.

17. How many electrons pass through a conductor in 1 min and 30 s if the current is 4 mA?

18. Will a fuse rated at 1 A "blow" if 86 C pass through it in 1.2 min?

*19. If $0.84 \times 10^{+16}$ electrons pass through a wire in 60 ms, find the current.

*20. Which would you prefer?
   a. A penny for every electron that passes through a wire in 0.01 μs at a current of 2 mA, or
   b. A dollar for every electron that passes through a wire in 1.5 ns if the current is 100 μA.

*21. If a conductor with a current of 200 mA passing through it converts 40 J of electrical energy into heat in 30 s, what is the potential drop across the conductor?

*22. Charge is flowing through a conductor at the rate of 420 C/min. If 742 J of electrical energy are converted to heat in 30 s, what is the potential drop across the conductor?

*23. The potential difference between two points in an electric circuit is 24 V. If 0.4 J of energy were dissipated in a period of 5 ms, what would the current be between the two points?

SECTION 2.6 Ampere-Hour Rating

24. What current will a battery with an Ah rating of 200 theoretically provide for 40 h?

25. What is the Ah rating of a battery that can provide 0.8 A for 75 h?

26. For how many hours will a battery with an Ah rating of 32 theoretically provide a current of 1.28 A?

27. A standard 12 V car battery has an ampere-hour rating of 40 Ah, whereas a heavy-duty battery has a rating of 60 Ah. How would you compare the energy levels of each and the available current for starting purposes?

*28. A portable television using a 12 V, 3 Ah rechargeable battery can operate for a period of about 6 h. What is the average current drawn during this period? What is the energy expended by the battery in joules?

SECTION 2.8 Conductors and Insulators

29. Discuss two properties of the atomic structure of copper that make it a good conductor.

30. Explain the terms *insulator* and *breakdown strength.*

31. List three uses of insulators not mentioned in Section 2.8.

SECTION 2.9 Semiconductors

32. What is a semiconductor? How does it compare with a conductor and an insulator?

33. Consult a semiconductor electronics text and note the extensive use of germanium and silicon semiconductor materials. Review the characteristics of each material.

SECTION 2.10 Ammeters and Voltmeters

34. What are the significant differences in the way ammeters and voltmeters are connected?

35. If an ammeter reads 2.5 A for a period of 4 min, determine the charge that has passed through the meter.

36. Between two points in an electric circuit, a voltmeter reads 12.5 V for a period of 20 s. If the current measured by an ammeter is 10 mA, determine the energy expended and the charge that flowed between the two points.

GLOSSARY

Ammeter An instrument designed to read the current through elements in series with the meter.

Ampere (A) The SI unit of measurement applied to the flow of charge through a conductor.

Ampere-hour (Ah) rating The rating applied to a source of energy that will reveal how long a particular level of current can be drawn from that source.

Cell A fundamental source of electrical energy developed through the conversion of chemical or solar energy.

Conductors Materials that permit a generous flow of electrons with very little voltage applied.

Copper A material possessing physical properties that make it particularly useful as a conductor of electricity.

Coulomb (C) The fundamental SI unit of measure for charge. It is equal to the charge carried by 6.242×10^{18} electrons.

Coulomb's law An equation defining the force of attraction or repulsion between two charges.

Current The flow of charge resulting from the application of a difference in potential between two points in an electrical system.

dc current source A source that will provide a fixed current level even though the load to which it is applied may cause its terminal voltage to change.

dc generator A source of dc voltage available through the turning of the shaft of the device by some external means.

Direct current (dc) Current having a single direction (unidirectional) and a fixed magnitude over time.

Electrolysis The process of passing a current through an electrolyte to break it down into its fundamental components.

Electrolytes The contact element and the source of ions between the electrodes of the battery.

Electron The particle with negative polarity that orbits the nucleus of an atom.

Free electron An electron unassociated with any particular atom, relatively free to move through a crystal lattice structure under the influence of external forces.

Fuel cell A nonpolluting source of energy that can establish current through a load by simply applying the correct levels of hydrogen and oxygen.

Insulators Materials in which a very high voltage must be applied to produce any measurable current flow.

Neutron The particle having no electrical charge, found in the nucleus of the atom.

Nucleus The structural center of an atom that contains both protons and neutrons.

Positive ion An atom having a net positive charge due to the loss of one of its negatively charged electrons.

Potential difference The algebraic difference in potential (or voltage) between two points in an electrical system.

Potential energy The energy that a mass possesses by virtue of its position.

Primary cell Sources of voltage that cannot be recharged.

Proton The particle of positive polarity found in the nucleus of an atom.

Rectification The process by which an ac signal is converted to one that has an average dc level.

Secondary cell Sources of voltage that can be recharged.

Semiconductor A material having a conductance value between that of an insulator and that of a conductor. Of significant importance in the manufacture of semiconductor electronic devices.

Solar cell Sources of voltage available through the conversion of light energy (photons) into electrical energy.

Specific gravity The ratio of the weight of a given volume of a substance to the weight of an equal volume of water at 4°C.

Volt (V) The unit of measurement applied to the difference in potential between two points. If 1 joule of energy is required to move 1 coulomb of charge between two points, the difference in potential is said to be 1 volt.

Voltage The term applied to the difference in potential between two points as established by a separation of opposite charges.

Voltmeter An instrument designed to read the voltage across an element or between any two points in a network.

RESISTANCE

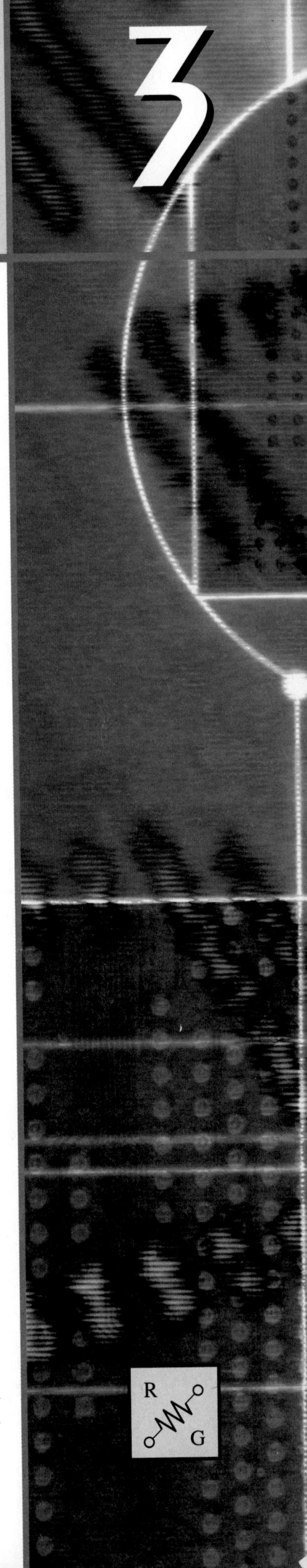

Objectives

- *Become familiar with the parameters that determine the resistance of an element and be able to calculate the resistance from the given dimensions and material characteristics.*
- *Understand the effects of temperature on the resistance of a material and how to calculate the change in resistance with temperature.*
- *Develop some understanding of superconductors and how they can affect future development in the industry.*
- *Become familiar with the broad range of commercially available resistors available today and how to read the value of each from the color code or labeling.*
- *Become aware of a variety of elements such as thermistors, photoconductive cells, and varistors and how their terminal resistance is controlled.*

3.1 INTRODUCTION

In the previous chapter, we found that placing a voltage across a wire or simple circuit results in a flow of charge or current through the wire or circuit. The question remains, however, What determines the level of current that results when a particular voltage is applied? Why is the current heavier in some circuits than in others? The answers lie in the fact that there is an opposition to the flow of charge in the system that depends on the components of the circuit. This opposition to the flow of charge through an electrical circuit, called **resistance,** has the units of **ohms** and uses the Greek letter *omega* (Ω) as its symbol. The graphic symbol for resistance, which resembles the cutting edge of a saw, is provided in Fig. 3.1.

FIG. 3.1
Resistance symbol and notation.

This opposition, due primarily to collisions and friction between the free electrons and other electrons, ions, and atoms in the path of motion, converts the supplied electrical energy into **heat** that raises the temperature of the electrical component and surrounding medium. The heat you feel from an electrical heater is simply due to passing current through a high-resistance material.

Each material with its unique atomic structure reacts differently to pressures to establish current through its core. Conductors that permit a generous flow of charge with little external pressure have low resistance levels, while insulators have high resistance characteristics.

3.2 RESISTANCE: CIRCULAR WIRES

The resistance of any material is due primarily to four factors:

1. *Material*
2. *Length*
3. *Cross-sectional area*
4. *Temperature of the material*

As noted in Section 3.1, the atomic structure determines how easily a free electron will pass through a material. The longer the path through which the free electron must pass, the greater the resistance factor. Free electrons pass more easily through conductors with larger cross-sectional areas. In addition, the higher the temperature of the conductive materials, the greater the internal vibration and motion of the components that make up the atomic structure of the wire, and the more difficult it is for the free electrons to find a path through the material.

The first three elements are related by the following basic equation for resistance:

$$R = \rho \frac{l}{A} \qquad \begin{aligned} \rho &= \text{CM-}\Omega\text{/ft at } T = 20°\text{C} \\ l &= \text{feet} \\ A &= \text{area in circular mils (CM)} \end{aligned} \tag{3.1}$$

with each component of the equation defined by Fig. 3.2.

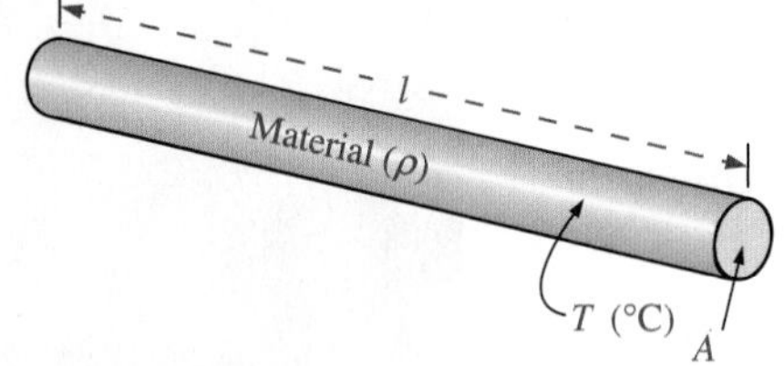

FIG. 3.2
Factors affecting the resistance of a conductor.

The material is identified by a factor called the **resistivity,** which uses the Greek letter *rho* (ρ) as its symbol and is measured in CM-Ω/ft. Its value at a temperature of 20°C (room temperature = 68°F) is provided in Table 3.1 for a variety of common materials. Since the larger the resistivity, the greater the resistance to setting up a flow of charge, it appears as a multiplying factor in Eq. (3.1); that is, it appears in the numerator of the equation. It is important to realize at this point that since the resistivity is provided at a particular temperature, *Eq. (3.1) is applicable only at room temperature.* The effect of higher and lower temperatures is considered in Section 3.4.

Since the resistivity is in the numerator of Eq. (3.1),

the higher the resistivity, the greater the resistance of a conductor

as shown for two conductors of the same length in Fig. 3.3(a).

Further,

the longer the conductor, the greater the resistance

since the length also appears in the numerator of Eq. (3.1). Note Fig. 3.3(b). Finally,

the greater the area of a conductor, the less the resistance

because the area appears in the denominator of Eq. (3.1). Note Fig. 3.3(c).

TABLE 3.1
Resistivity (ρ) of various materials.

Material	**ρ (CM · Ω/ft)@20°C**
Silver	9.9
Copper	**10.37**
Gold	14.7
Aluminum	17.0
Tungsten	33.0
Nickel	47.0
Iron	74.0
Constantan	295.0
Nichrome	600.0
Calorite	720.0
Carbon	21,000.0

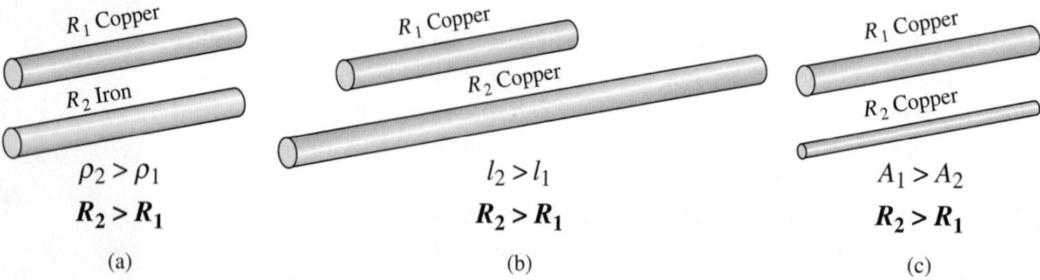

FIG. 3.3
Cases in which $R_2 > R_1$. For each case, all remaining parameters that control the resistance level are the same.

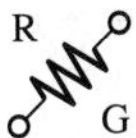

Circular Mils (CM)

In Eq. (3.1), the area is measured in a quantity called **circular mils** (CM). It is the quantity used in most commercial wire tables, and thus it needs to be carefully defined. The *mil* is a unit of measurement for length and is related to the inch by

$$\mathbf{1\ mil} = \frac{\mathbf{1}}{\mathbf{1000}}\ \mathbf{in.}$$

or

$$\mathbf{1000\ mils = 1\ in.}$$

In general, therefore, the mil is a very small unit of measurement for length. There are 1000 mils in an inch, or 1 mil is only 1/1000 of an inch. It is a length that is not visible with the naked eye although it can be measured with special instrumentation. The phrase *milling* used in steel factories is derived from the fact that a few mils of material are often removed by heavy machinery such as a lathe, and the thickness of steel is usually measured in mils.

By definition,

a wire with a diameter of 1 mil has an area of 1 CM.

as shown in Fig. 3.4.

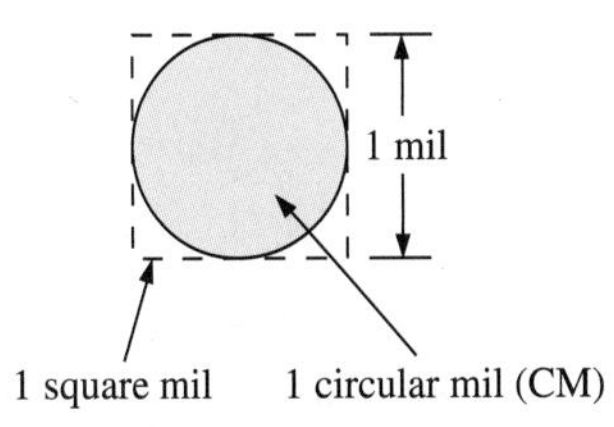

FIG. 3.4
Defining the circular mil (CM).

An interesting result of such a definition is that the area of a circular wire in circular mils can be defined by the following equation:

$$\boxed{A_{CM} = (d_{\text{mils}})^2} \qquad \textbf{(3.2)}$$

Verification of this equation appears in Fig. 3.5 which shows that a wire with a diameter of 2 mils has a total area of 4 CM and a wire with a diameter of 3 mils has a total area of 9 CM.

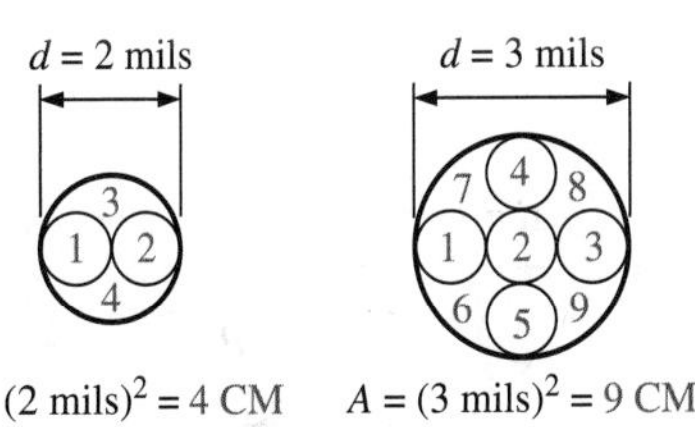

FIG. 3.5
Verification of Eq. (3.2): $A_{CM} = (d_{mils})^2$.

Remember, to compute the area of a wire in circular mils when the diameter is given in inches, first convert the diameter to mils by simply writing the diameter in decimal form and moving the decimal point three places to the right. For example,

$$\frac{1}{8}\text{ in.} = \underset{\text{3 places}}{0.125\text{ in.}} = 125\text{ mils}$$

Then the area is determined by

$$A_{CM} = (d_{\text{mils}})^2 = (125\text{ mils})^2 = \mathbf{15{,}625\ CM}$$

Sometimes when you are working with conductors that are not circular, you will need to convert square mils to circular mils, and vice versa. Applying the basic equation for the area of a circle and substituting a diameter of 1 mil results in

$$A = \frac{\pi d^2}{4} = \frac{\pi}{4}(1\text{ mil})^2 = \frac{\pi}{4}\text{ sq mils} \overset{\text{by definition}}{\equiv} 1\text{ CM}$$

from which we can conclude the following:

$$\boxed{1\text{ CM} = \frac{\pi}{4}\text{ sq mils}} \qquad \textbf{(3.3)}$$

or

$$\boxed{1\text{ sq mil} = \frac{4}{\pi}\text{ CM}} \qquad \textbf{(3.4)}$$

EXAMPLE 3.1 What is the resistance of a 100 ft length of copper wire with a diameter of 0.020 in. at 20°C?

Solution:

$$\rho = 10.37 \frac{\text{CM-}\Omega}{\text{ft}} \qquad 0.020 \text{ in.} = 20 \text{ mils}$$

$$A_{CM} = (d_{\text{mils}})^2 = (20 \text{ mils})^2 = 400 \text{ CM}$$

$$R = \rho\frac{l}{A} = \frac{(10.37 \text{ CM-}\Omega/\text{ft})(100 \text{ ft})}{400 \text{ CM}}$$

$$R = \mathbf{2.59\ \Omega}$$

EXAMPLE 3.2 An undetermined number of feet of wire have been used from the carton in Fig. 3.6. Find the length of the remaining copper wire if it has a diameter of 1/16 in. and a resistance of 0.5 Ω.

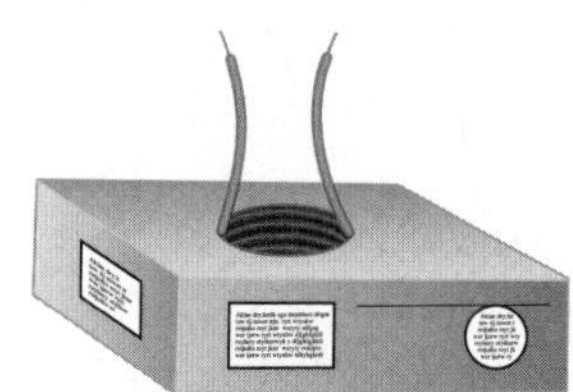

FIG. 3.6
Example 3.2.

Solution:

$$\rho = 10.37 \text{ CM-}\Omega/\text{ft} \qquad \frac{1}{16} \text{ in.} = 0.0625 \text{ in.} = 62.5 \text{ mils}$$

$$A_{\text{CM}} = (d_{\text{mils}})^2 = (62.5 \text{ mils})^2 = 3906.25 \text{ CM}$$

$$R = \rho\frac{l}{A} \Rightarrow l = \frac{RA}{\rho} = \frac{(0.5\ \Omega)(3906.25 \text{ CM})}{10.37 \frac{\text{CM-}\Omega}{\text{ft}}} = \frac{1953.125}{10.37}$$

$$l = \mathbf{188.34\ ft}$$

EXAMPLE 3.3 What is the resistance of a copper bus-bar, as used in the power distribution panel of a high-rise office building, with the dimensions indicated in Fig. 3.7?

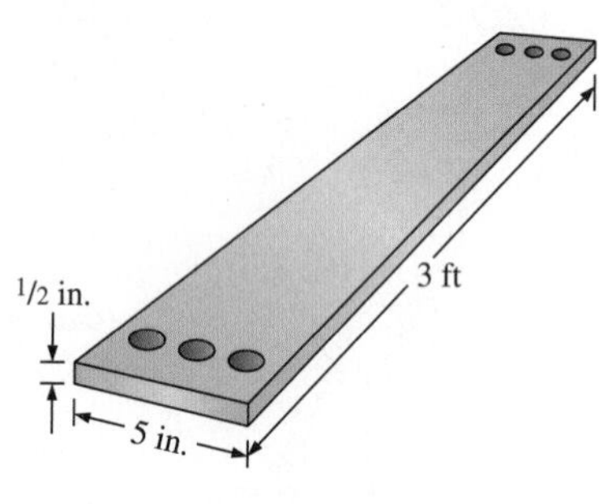

FIG. 3.7
Example 3.3.

Solution:

$$A_{\text{CM}}\begin{cases} 5.0 \text{ in.} = 5000 \text{ mils} \\ \frac{1}{2} \text{ in.} = 500 \text{ mils} \\ A = (5000 \text{ mils})(500 \text{ mils}) = 2.5 \times 10^6 \text{ sq mils} \\ \quad = 2.5 \times 10^6 \cancel{\text{sq mils}} \left(\frac{4/\pi \text{ CM}}{1 \cancel{\text{sq mil}}}\right) \\ A = 3.183 \times 10^6 \text{ CM} \end{cases}$$

$$R = \rho\frac{l}{A} = \frac{(10.37 \text{ CM-}\Omega/\text{ft})(3 \text{ ft})}{3.183 \times 10^6 \text{ CM}} = \frac{31.11}{3.183 \times 10^6}$$

$$R = \mathbf{9.774 \times 10^{-6}\ \Omega}$$

(quite small, 0.000009774 Ω≅0 Ω)

You will learn in the following chapters that the less the resistance of a conductor, the lower the losses in conduction from the source to the load. Similarly, since resistivity is a major factor in determining the resistance of

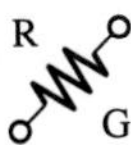

a conductor, the lower the resistivity, the lower the resistance for the same size conductor. It would appear from Table 3.1 that silver, copper, gold, and aluminum would be the best conductors and the most common. In general, there are other factors, however, such as **malleability** (ability of a material to be shaped), **ductility** (ability of a material to be drawn into long, thin wires), temperature sensitivity, resistance to abuse, and, of course, cost, that must all be weighed when choosing a conductor for a particular application.

In general, copper is the most widely used material because it is quite malleable, ductile, and available; has good thermal characteristics; and is less expensive than silver or gold. It is certainly not cheap, however. Contractors always ensure that the copper wiring has been removed before leveling a building because of its salvage value. Aluminum was once used for general wiring because it is cheaper than copper, but its thermal characteristics created some difficulties. The heating due to current flow and the cooling that occurred when the circuit was turned off resulted in expansion and contraction of the aluminum wire to the point where connections eventually loosened, and resulting in dangerous side effects. Aluminum is still used today, however, in areas such as integrated circuit manufacturing and in situations where the connections can be made secure. Silver and gold are, of course, much more expensive than copper or aluminum, but the cost is justified for certain applications. Silver has excellent plating characteristics for surface preparations, and gold is used quite extensively in integrated circuits. Tungsten has a resistivity three times that of copper, but there are occasions when its physical characteristics (durability, hardness) are the overriding considerations.

3.3 WIRE TABLES

The wire table was designed primarily to standardize the size of wire produced by manufacturers. As a result, the manufacturer has a larger market, and the consumer knows that standard wire sizes will always be available. The table was designed to assist the user in every way possible; it usually includes data such as the cross-sectional area in circular mils, diameter in mils, ohms per 1000 feet at 20°C, and weight per 1000 feet.

The American Wire Gage (AWG) sizes are given in Table 3.2 for solid round copper wire. A column indicating the maximum allowable current in amperes, as determined by the National Fire Protection Association, has also been included.

The chosen sizes have an interesting relationship: For every drop in 3 gage numbers, the area is doubled; and for every drop in 10 gage numbers, the area increases by a factor of 10.

Examining Eq. (3.1), we note also that *doubling the area cuts the resistance in half, and increasing the area by a factor of 10 decreases the resistance of 1/10 the original*, everything else kept constant.

The actual sizes of some of the gage wires listed in Table 3.2 are shown in Fig. 3.8 with a few of their areas of application. A few examples using Table 3.2 follow.

EXAMPLE 3.4 Find the resistance of 650 ft of #8 copper wire ($T =$ 20°C).

Solution: For #8 copper wire (solid), Ω/1000 ft at 20°C = 0.6282 Ω, and

$$650\ \cancel{\text{ft}}\left(\frac{0.6282\ \Omega}{1000\ \cancel{\text{ft}}}\right) = \mathbf{0.408\ \Omega}$$

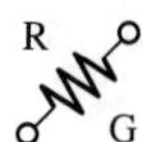

TABLE 3.2
American Wire Gage (AWG) sizes.

	AWG #	Area (CM)	Ω/1000 ft at 20°C	Maximum Allowable Current for RHW Insulation (A)*
(4/0)	**0000**	211,600	0.0490	**230**
(3/0)	**000**	167,810	0.0618	**200**
(2/0)	**00**	133,080	0.0780	**175**
(1/0)	**0**	105,530	0.0983	**150**
	1	83,694	0.1240	**130**
	2	66,373	0.1563	**115**
	3	52,634	0.1970	**100**
	4	41,742	0.2485	**85**
	5	33,102	0.3133	—
	6	26,250	0.3951	**65**
	7	20,816	0.4982	—
	8	16,509	0.6282	**50**
	9	13,094	0.7921	—
	10	10,381	0.9989	**30**
	11	8,234.0	1.260	—
	12	6,529.9	1.588	**20**
	13	5,178.4	2.003	—
	14	4,106.8	2.525	**15**
	15	3,256.7	3.184	
	16	2,582.9	4.016	
	17	2,048.2	5.064	
	18	1,624.3	6.385	
	19	1,288.1	8.051	
	20	1,021.5	10.15	
	21	810.10	12.80	
	22	642.40	16.14	
	23	509.45	20.36	
	24	404.01	25.67	
	25	320.40	32.37	
	26	254.10	40.81	
	27	201.50	51.47	
	28	159.79	64.90	
	29	126.72	81.83	
	30	100.50	103.2	
	31	79.70	130.1	
	32	63.21	164.1	
	33	50.13	206.9	
	34	39.75	260.9	
	35	31.52	329.0	
	36	25.00	414.8	
	37	19.83	523.1	
	38	15.72	659.6	
	39	12.47	831.8	
	40	9.89	1049.0	

*Not more than three conductors in raceway, cable, or direct burial.

Source: Reprinted by permission from NFPA No. SPP-6C, National Electrical Code®, copyright © 1996, National Fire Protection Association, Quincy, MA 02269. This reprinted material is not the complete and official position of the NFPA on the referenced subject which is represented only by the standard in its entirety, *National Electrical Code* is a registered trademark of the National Fire Protection Association, Inc., Quincy, MA for a triennial electrical publication. The term *National Electrical Code,* as used herein, means the triennial publication constituting the National Electrical Code and is used with permission of the National Fire Protection Association.

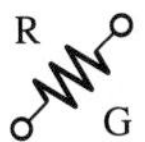

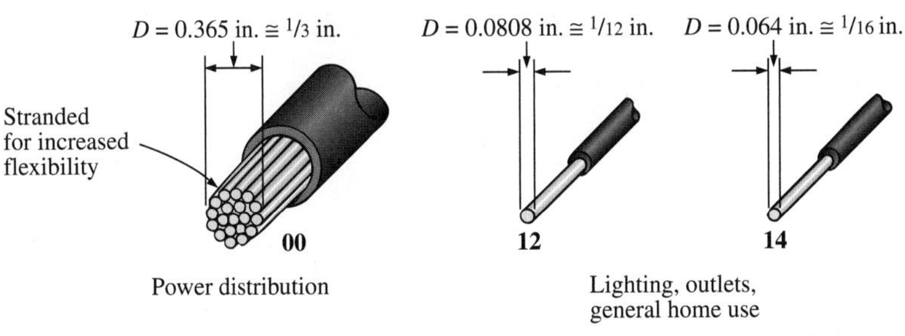

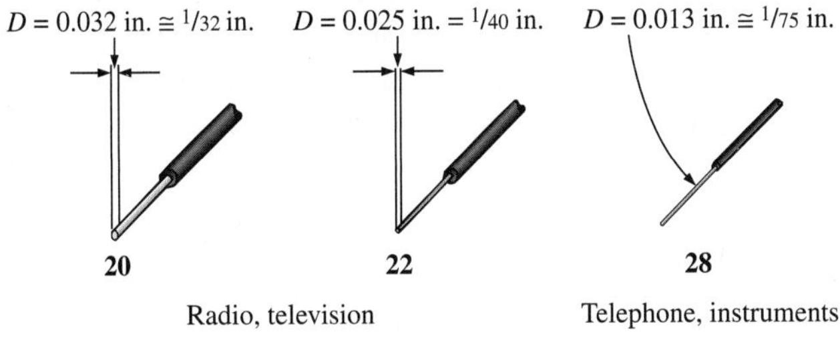

FIG. 3.8
Popular wire sizes and some of their areas of application.

EXAMPLE 3.5 What is the diameter, in inches, of a #12 copper wire?

Solution: For #12 copper wire (solid), $A = 6529.9$ CM, and

$$d_{\text{mils}} = \sqrt{A_{\text{CM}}} = \sqrt{6529.9 \text{ CM}} \cong 80.81 \text{ mils}$$

$$d = \mathbf{0.0808 \text{ in.}} \text{ (or close to } 1/12 \text{ in.)}$$

EXAMPLE 3.6 For the system in Fig. 3.9, the total resistance of *each* power line cannot exceed 0.025 Ω, and the maximum current to be drawn by the load is 95 A. What gage wire should be used?

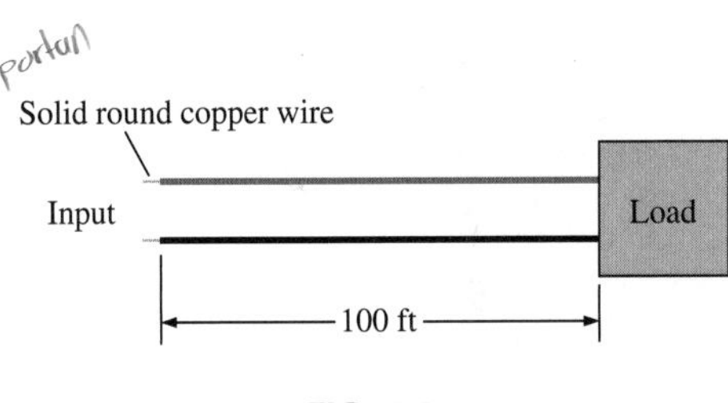

FIG. 3.9
Example 3.6.

Solution:

$$R = \rho\frac{l}{A} \Rightarrow A = \rho\frac{l}{R} = \frac{(10.37 \text{ CM-}\Omega/\text{ft})(100 \text{ ft})}{0.025\ \Omega} = \mathbf{41{,}480 \text{ CM}}$$

Using the wire table, we choose the wire with the next largest area, which is #4, to satisfy the resistance requirement. We note, however, that 95 A must flow through the line. This specification requires that **#3 wire** be used since the #4 wire can carry a maximum current of only 85 A.

3.4 RESISTANCE: METRIC UNITS

The design of resistive elements for various areas of application, including thin-film resistors and integrated circuits, uses metric units for the quantities of Eq. (3.1). In SI units, the resistivity would be measured in ohm-meters, the area in square meters, and the length in meters. However, the meter is generally too large a unit of measure for most applications, and so the centimeter is usually employed. The resulting dimensions for Eq. (3.1) are therefore

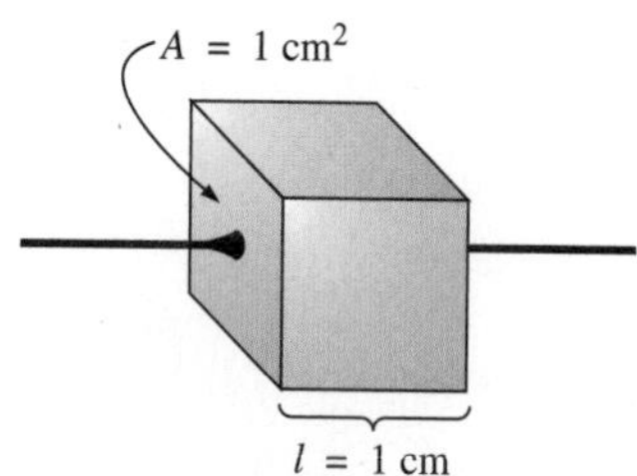

FIG. 3.10
Defining ρ in ohm-centimeters.

TABLE 3.3
Resistivity (ρ) of various materials.

Material	Ω-cm
Silver	1.645×10^{-6}
Copper	$\mathbf{1.723 \times 10^{-6}}$
Gold	2.443×10^{-6}
Aluminum	2.825×10^{-6}
Tungsten	5.485×10^{-6}
Nickel	7.811×10^{-6}
Iron	12.299×10^{-6}
Tantalum	15.54×10^{-6}
Nichrome	99.72×10^{-6}
Tin oxide	250×10^{-6}
Carbon	3500×10^{-6}

ρ:	ohm-centimeters
l:	centimeters
A:	square centimeters

The units for ρ can be derived from

$$\rho = \frac{RA}{l} = \frac{\Omega\text{-cm}^2}{\text{cm}} = \Omega\text{-cm}$$

The resistivity of a material is actually the resistance of a sample such as that appearing in Fig. 3.10. Table 3.3 provides a list of values of ρ in ohm-centimeters. Note that the area now is expressed in square centimeters, which can be determined using the basic equation $A = \pi d^2/4$, eliminating the need to work with circular mils, the special unit of measure associated with circular wires.

EXAMPLE 3.7 Determine the resistance of 100 ft of #28 copper telephone wire if the diameter is 0.0126 in.

Solution: Unit conversions:

$$l = 100\ \cancel{\text{ft}}\left(\frac{12\ \cancel{\text{in.}}}{1\ \cancel{\text{ft}}}\right)\left(\frac{2.54\ \text{cm}}{1\ \cancel{\text{in.}}}\right) = 3048\ \text{cm}$$

$$d = 0.0126\ \text{in.}\left(\frac{2.54\ \text{cm}}{1\ \text{in.}}\right) = 0.032\ \text{cm}$$

Therefore,

$$A = \frac{\pi d^2}{4} = \frac{(3.1416)(0.032\ \text{cm})^2}{4} = 8.04 \times 10^{-4}\ \text{cm}^2$$

$$R = \rho\frac{l}{A} = \frac{(1.723 \times 10^{-6}\Omega\text{-cm})(3048\ \text{cm})}{8.04 \times 10^{-4}\ \text{cm}^2} \cong \mathbf{6.5\ \Omega}$$

Using the units for circular wires and Table 3.2 for the area of a #28 wire, we find

$$R = \rho\frac{l}{A} = \frac{(10.37\ \text{CM-}\Omega/\text{ft})(100\ \text{ft})}{159.79\ \text{CM}} \cong \mathbf{6.5\ \Omega}$$

EXAMPLE 3.8 Determine the resistance of the thin-film resistor in Fig. 3.11 if the **sheet resistance** R_s (defined by $R_s = \rho/d$) is 100 Ω.

Solution: For deposited materials of the same thickness, the sheet resistance factor is usually employed in the design of thin-film resistors.

Eq. (3.1) can be written

$$R = \rho\frac{l}{A} = \rho\frac{l}{dw} = \left(\frac{\rho}{d}\right)\left(\frac{l}{w}\right) = R_s\frac{l}{w}$$

where l is the length of the sample and w is the width. Substituting into the above equation yields

$$R = R_s\frac{l}{w} = \frac{(100\ \Omega)(0.6\ \text{cm})}{0.3\ \text{cm}} = \mathbf{200\ \Omega}$$

as one might expect since $l = 2w$.

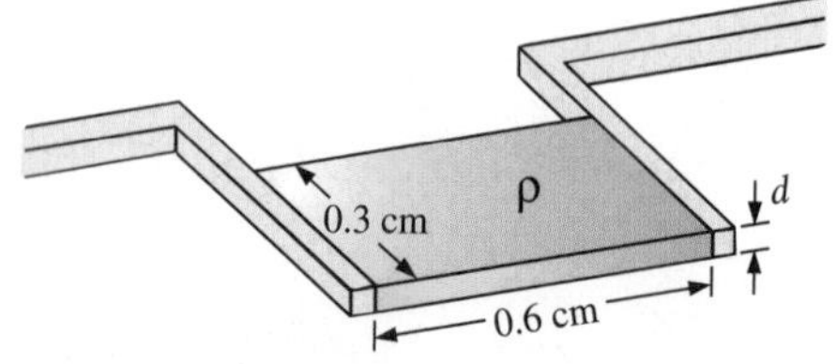

FIG. 3.11
Thin-film resistor (note Fig. 3.25).

The conversion factor between resistivity in circular mil-ohms per foot and ohm-centimeters is the following:

$$\rho\,(\Omega\text{-cm}) = (1.662 \times 10^{-7}) \times (\text{value in CM-}\Omega/\text{ft})$$

For example, for copper, $\rho = 10.37$ CM-Ω/ft:

$$\begin{aligned}\rho\,(\Omega\text{-cm}) &= 1.662 \times 10^{-7}\,(10.37\ \text{CM-}\Omega/\text{ft})\\ &= 1.723 \times 10^{-6}\ \Omega\text{-cm}\end{aligned}$$

as indicated in Table 3.3.

The resistivity in integrated circuit design is typically in ohm-centimeter units, although tables often provide ρ in ohm-meters or microhm-centimeters. Using the conversion technique of Chapter 1, we find that the conversion factor between ohm-centimeters and ohm-meters is the following:

$$1.723 \times 10^{-6}\Omega\text{-}\cancel{\text{cm}}\left[\frac{1\ \text{m}}{100\ \cancel{\text{cm}}}\right] = \frac{1}{100}[1.723 \times 10^{-6}]\Omega\text{-m}$$

or the value in ohm-meters is 1/100 the value in ohm-centimeters, and

$$\rho(\Omega\text{-m}) = \left(\frac{1}{100}\right) \times (\text{value in }\Omega\text{-cm}) \qquad \textbf{(3.5)}$$

Similarly,

$$\rho\,(\mu\Omega\text{-cm}) = (10^{6}) \times (\text{value in }\Omega\text{-cm}) \qquad \textbf{(3.6)}$$

For comparison purposes, typical values of ρ in ohm-centimeters for conductors, semiconductors, and insulators are provided in Table 3.4.

TABLE 3.4
Comparing levels of ρ in Ω-cm.

Conductor (Ω-cm)	Semiconductor (Ω-cm)	Insulator (Ω-cm)
Copper 1.723×10^{-6}	Ge 50	In general: 10^{15}
	Si 200×10^{3}	
	GaAs 70×10^{6}	

In particular, note the power-of-ten difference between conductors and insulators (10^{21})—a difference of huge proportions. There is a significant difference in levels of ρ for the list of semiconductors, but the power-of-ten difference between the conductor and insulator levels is at least 10^{6} for each of the semiconductors listed.

3.5 TEMPERATURE EFFECTS

Temperature has a significant effect on the resistance of conductors, semiconductors, and insulators.

Conductors

Conductors have a generous number of free electrons, and any introduction of thermal energy will have little impact on the total number of free

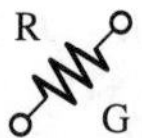

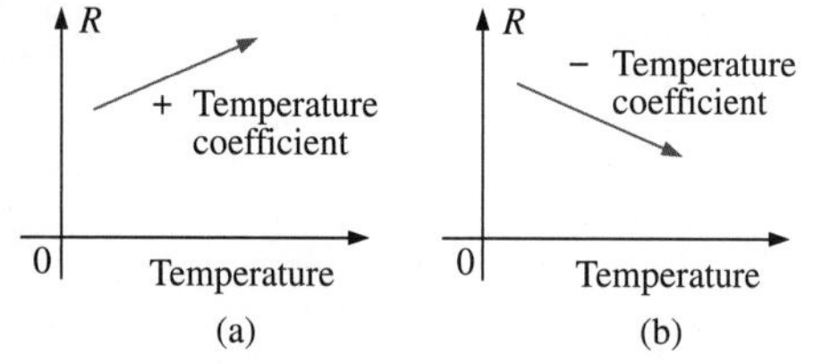

FIG. 3.12
Demonstrating the effect of a positive and a negative temperature coefficient on the resistance of a conductor.

carriers. In fact, the thermal energy only increases the intensity of the random motion of the particles within the material and makes it increasingly difficult for a general drift of electrons in any one direction to be established. The result is that

for good conductors, an increase in temperature results in an increase in the resistance level. Consequently, conductors have a positive temperature coefficient.

The plot in Fig. 3.12(a) has a positive temperature coefficient.

Semiconductors

In semiconductors, an increase in temperature imparts a measure of thermal energy to the system that results in an increase in the number of free carriers in the material for conduction. The result is that

for semiconductor materials, an increase in temperature results in a decrease in the resistance level. Consequently, semiconductors have negative temperature coefficients.

The thermistor and photoconductive cell discussed in Sections 3.11 and 3.12 are excellent examples of semiconductor devices with negative temperature coefficients. The plot in Fig. 3.12(b) has a negative temperature coefficient.

Insulators

As with semiconductors, an increase in temperature results in a decrease in the resistance of an insulator. The result is a negative temperature coefficient.

Inferred Absolute Temperature

Fig. 3.13 reveals that for copper (and most other metallic conductors), the resistance increases almost linearly (in a straight-line relationship) with an increase in temperature. Since temperature can have such a pronounced effect on the resistance of a conductor, it is important that we have some method of determining the resistance at any temperature within operating limits. An equation for this purpose can be obtained by approximating the curve in Fig. 3.13 by the straight dashed line that intersects the temperature scale at −234.5°C. Although the actual curve extends to **absolute zero** (−273.15°C, or 0 K), the straight-line approximation is quite accurate for

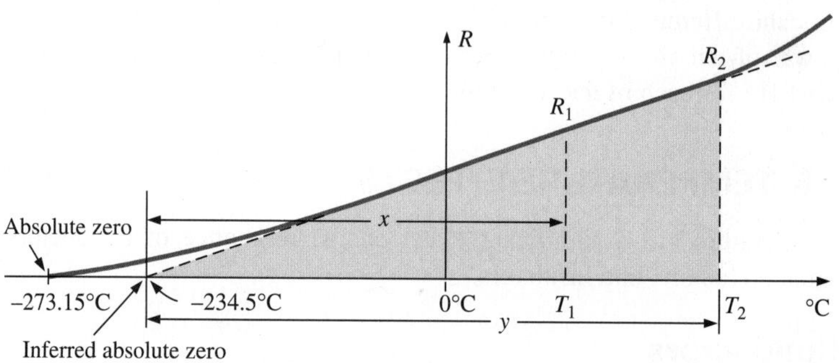

FIG. 3.13
Effect of temperature on the resistance of copper.

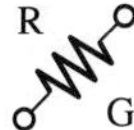

the normal operating temperature range. At two different temperatures, T_1 and T_2, the resistance of copper is R_1 and R_2, as indicated on the curve. Using a property of similar triangles, we may develop a mathematical relationship between these values of resistances at different temperatures. Let x equal the distance from $-234.5°C$ to T_1 and y the distance from $-234.5°C$ to T_2, as shown in Fig. 3.13. From similar triangles,

$$\frac{x}{R_1} = \frac{y}{R_2}$$

or

$$\boxed{\frac{234.5 + T_1}{R_1} = \frac{234.5 + T_2}{R_2}} \qquad \textbf{(3.7)}$$

The temperature of $-234.5°C$ is called the **inferred absolute temperature** of copper. For different conducting materials, the intersection of the straight-line approximation occurs at different temperatures. A few typical values are listed in Table 3.5.

The minus sign does not appear with the inferred absolute temperature on either side of Eq. (3.7) because x and y are the *distances* from $-234.5°C$ to T_1 and T_2, respectively, and therefore are simply magnitudes. For T_1 and T_2 less than zero, x and y are less than $-234.5°C$, and the distances are the differences between the inferred absolute temperature and the temperature of interest.

Eq. (3.7) can easily be adapted to any material by inserting the proper inferred absolute temperature. It may therefore be written as follows:

$$\boxed{\frac{|T_1| + T_1}{R_1} = \frac{|T_1| + T_2}{R_2}} \qquad \textbf{(3.8)}$$

where $|T_1|$ indicates that the inferred absolute temperature of the material involved is inserted as a positive value in the equation. In general, therefore, associate the sign only with T_1 and T_2.

TABLE 3.5

Inferred absolute temperatures (T_i).

Material	°C
Silver	−243
Copper	−234.5
Gold	−274
Aluminum	−236
Tungsten	−204
Nickel	−147
Iron	−162
Nichrome	−2,250
Constantan	−125,000

EXAMPLE 3.9 If the resistance of a copper wire is 50 Ω at 20°C, what is its resistance at 100°C (boiling point of water)?

Solution: Eq. (3.7):

$$\frac{234.5°\text{C} + 20°\text{C}}{50\ \Omega} = \frac{234.5°\text{C} + 100°\text{C}}{R_2}$$

$$R_2 = \frac{(50\ \Omega)(334.5°\text{C})}{254.5°\text{C}} = \mathbf{65.72\ \Omega}$$

EXAMPLE 3.10 If the resistance of a copper wire at freezing (0°C) is 30 Ω, what is its resistance at −40°C?

Solution: Eq. (3.7):

$$\frac{234.5°\text{C} + 0}{30\ \Omega} = \frac{234.5°\text{C} - 40°\text{C}}{R_2}$$

$$R_2 = \frac{(30\ \Omega)(194.5°\text{C})}{234.5°\text{C}} = \mathbf{24.88\ \Omega}$$

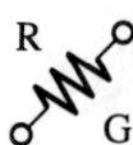

EXAMPLE 3.11 If the resistance of an aluminum wire at room temperature (20°C) is 100 mΩ (measured by a milliohmmeter), at what temperature will its resistance increase to 120 mΩ?

Solution: Eq. (3.7):

$$\frac{236°\text{C} + 20°\text{C}}{100 \text{ m}\Omega} = \frac{236°\text{C} + T_2}{120 \text{ m}\Omega}$$

and

$$T_2 = 120 \text{ m}\Omega\left(\frac{256°\text{C}}{100 \text{ m}\Omega}\right) - 236°\text{C}$$

$$T_2 = \mathbf{71.2°C}$$

Temperature Coefficient of Resistance

There is a second popular equation for calculating the resistance of a conductor at different temperatures. Defining

$$\alpha_{20} = \frac{1}{|T_1| + 20°\text{C}} \qquad (\Omega/°\text{C}/\Omega) \tag{3.9}$$

as the **temperature coefficient of resistance** at a temperature of 20°C, and R_{20} as the resistance of the sample at 20°C, the resistance R_1 at a temperature T_1 is determined by

$$R_1 = R_{20}[1 + \alpha_{20}(T_1 - 20°\text{C})] \tag{3.10}$$

TABLE 3.6
Temperature coefficient of resistance for various conductors at 20°C.

Material	Temperature Coefficient (α_{20})
Silver	0.0038
Copper	0.00393
Gold	0.0034
Aluminum	0.00391
Tungsten	0.005
Nickel	0.006
Iron	0.0055
Constantan	0.000008
Nichrome	0.00044

The values of α_{20} for different materials have been evaluated, and a few are listed in Table 3.6.

Eq. (3.10) can be written in the following form:

$$\alpha_{20} = \frac{\left(\dfrac{R_1 - R_{20}}{T_1 - 20°\text{C}}\right)}{R_{20}} = \frac{\dfrac{\Delta R}{\Delta T}}{R_{20}}$$

from which the units of Ω/°C/Ω for α_{20} are defined.

Since $\Delta R/\Delta T$ is the slope of the curve in Fig. 3.13, we can conclude that

the higher the temperature coefficient of resistance for a material, the more sensitive the resistance level to changes in temperature.

Referring to Table 3.5, we find that copper is more sensitive to temperature variations than is silver, gold, or aluminum, although the differences are quite small. The slope defined by α_{20} for constantan is so small that the curve is almost horizontal.

Since R_{20} of Eq. (3.10) is the resistance of the conductor at 20°C and $T_1 - 20°\text{C}$ is the change in temperature from 20°C, Eq. (3.10) can be written in the following form:

$$R = \rho\frac{l}{A}[1 + \alpha_{20}\,\Delta T] \tag{3.11}$$

providing an equation for resistance in terms of all the controlling parameters.

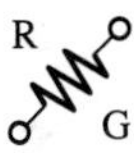

PPM/°C

For resistors, as for conductors, resistance changes with a change in temperature. The specification is normally provided in parts per million per degree Celsius (**PPM/°C**), providing an immediate indication of the sensitivity level of the resistor to temperature. For resistors, a 5000 PPM level is considered high, whereas 20 PPM is quite low. A 1000 PPM/°C characteristic reveals that a 1° change in temperature results in a change in resistance equal to 1000 PPM, or 1000/1,000,000 = 1/1000 of its nameplate value—not a significant change for most applications. However, a 10° change results in a change equal to 1/100 (1%) of its nameplate value, which is becoming significant. The concern, therefore, lies not only with the PPM level but with the range of expected temperature variation.

In equation form, the change in resistance is given by

$$\boxed{\Delta R = \frac{R_{\text{nominal}}}{10^6}(\text{PPM})(\Delta T)} \qquad \textbf{(3.12)}$$

where R_{nominal} is the nameplate value of the resistor at room temperature and ΔT is the change in temperature from the reference level of 20°C.

EXAMPLE 3.12 For a 1 kΩ carbon composition resistor with a PPM of 2500, determine the resistance at 60°C.

Solution:

$$\Delta R = \frac{1000\ \Omega}{10^6}(2500)(60°\text{C} - 20°\text{C})$$
$$= 100\ \Omega$$

and

$$R = R_{\text{nominal}} + \Delta R = 1000\ \Omega + 100\ \Omega$$
$$= \mathbf{1100\ \Omega}$$

3.6 SUPERCONDUCTORS

The field of electricity/electronics is one of the most exciting of our time. New developments appear almost weekly from extensive research and development activities. The research drive to develop a room-temperature **superconductor** is generating even more excitement. This advancement rivals the introduction of semiconductor devices such as the transistor (to replace tubes), wireless communication, or the electric light.

What are superconductors? Why is their development so important? In a nutshell,

superconductors are conductors of electric charge that, for all practical purposes, have zero resistance.

In a conventional conductor, electrons travel at average speeds of about 1000 mi/s (they can cross the United States in about 3 seconds), even though Einstein's theory of relativity suggests that the maximum speed of information transmission is the speed of light, or 186,000 mi/s. The relatively slow speed of conventional conduction is due to collisions with atoms in the material, repulsive forces between electrons (like charges repel), thermal agitation that results in indirect paths due to the

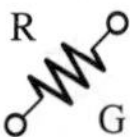

increased motion of the neighboring atoms, impurities in the conductor, and so on. In the superconductive state, there is a pairing of electrons, denoted by the **Cooper effect,** in which electrons travel in pairs and help each other maintain a significantly higher velocity through the medium. In some ways this is like "drafting" by competitive cyclists or runners. There is an oscillation of energy between partners or even "new" partners (as the need arises) to ensure passage through the conductor at the highest possible velocity with the least total expenditure of energy.

Even though the concept of superconductivity first surfaced in 1911, it was not until 1986 that the possibility of superconductivity at room temperature became a renewed goal of the research community. For over 70 years, superconductivity could be established only at temperatures colder than 23 K. (Kelvin temperature is universally accepted as the unit of measurement for temperature for superconductive effects. Recall that K = 273.15° + °C, so a temperature of 23 K is −250°C, or −418°F.) In 1986, however, physicists Alex Muller and George Bednorz of the IBM Zurich Research Center found a ceramic material—lanthanum barium copper oxide—that exhibited superconductivity at 30 K. This discovery introduced a new direction to the research effort and spurred others to improve on the new standard. (In 1987, both scientists received the Nobel prize for their contribution to an important area of development.)

In just a few short months, Professors Paul Chu of the University of Houston and Man Kven Wu of the University of Alabama raised the temperature to 95 K using a superconductor of yttrium barium copper oxide. The result was a level of excitement in the scientific community that brought research in the area to a new level of effort and investment. The major impact of this discovery was that liquid nitrogen (boiling point of 77 K) rather than liquid helium, (boiling point of 4 K) could now be used to bring the material down to the required temperature. The result is a tremendous saving in the cooling expense since liquid nitrogen is at least ten times less expensive than liquid helium. Pursuing the same direction, some success has been achieved at 125 K and 162 K using a thallium compound (unfortunately, however, thallium is a very poisonous substance).

Fig. 3.14 illustrates that the discovery in 1986 of using a ceramic material in superconductors led to rapid developments in the field. Un-

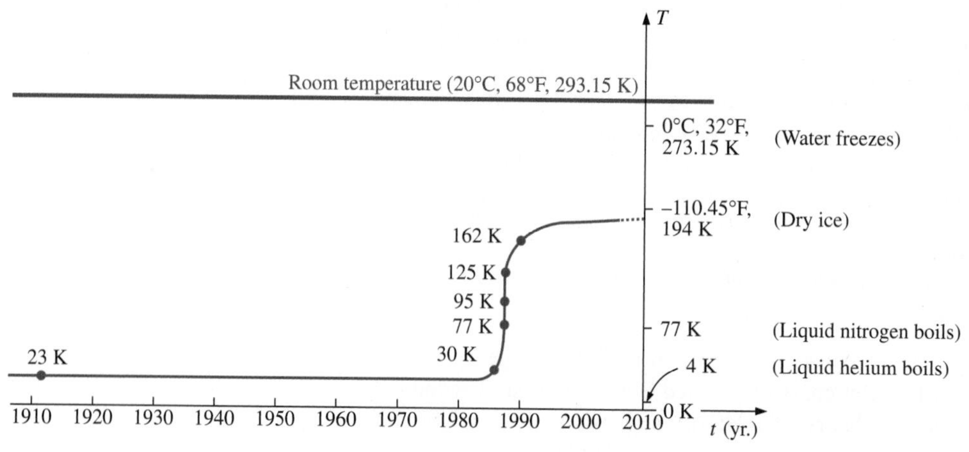

FIG. 3.14
Rising temperatures of superconductors.

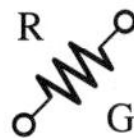

fortunately, however, the pace has slowed in recent years. The effort continues and is receiving an increasing level of financing and worldwide attention. Now, increasing numbers of corporations are trying to capitalize on the success already attained, as will be discussed later in this section.

The fact that ceramics provided the recent breakthrough in superconductivity is probably a surprise when you consider that they are also an important class of insulators. However, the ceramics that exhibit the characteristics of superconductivity are compounds that include copper, oxygen, and rare earth elements such as yttrium, lanthanum, and thallium. There are also indicators that the current compounds may be limited to a maximum temperature of 200 K (about 100 K short of room temperature), leaving the door wide open to innovative approaches to compound selection. The temperature at which a superconductor reverts back to the characteristics of a conventional conductor is called the *critical temperature,* denoted by T_c. Note in Fig. 3.15 that the resistivity level changes abruptly at T_c. The sharpness of the transition region is a function of the purity of the sample. Long listings of critical temperatures for a variety of tested compounds can be found in reference materials providing tables of a wide variety to support research in physics, chemistry, geology, and related fields. Two such publications include the CRC (The Chemical Rubber Co.) *Handbook of Tables for Applied Engineering Science* and the CRC *Handbook of Chemistry and Physics.*

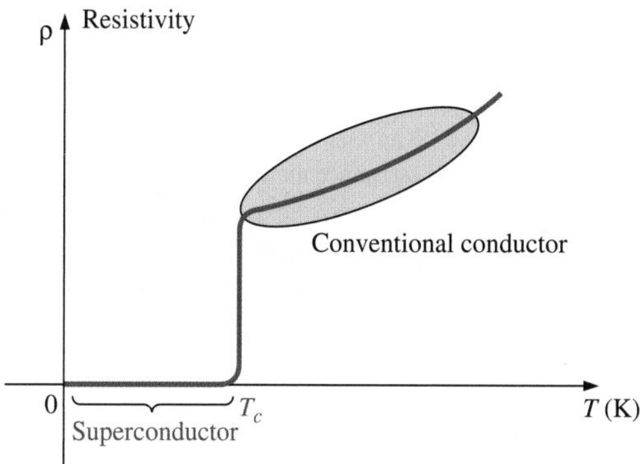

FIG. 3.15
Defining the critical temperature T_c.

Even though ceramic compounds have established higher transition temperatures, there is concern about their brittleness and current density limitations. In the area of integrated circuit manufacturing, current density levels must equal or exceed 1 MA/cm^2, or 1 million amperes through a cross-sectional area about one-half the size of a dime. IBM has attained a level of 4 MA/cm^2 at 77 K, permitting the use of superconductors in the design of some new-generation, high-speed computers.

Although room-temperature success has not been attained, numerous applications for some of the superconductors have been developed. It is simply a matter of balancing the additional cost against the results obtained or deciding whether any results at all can be obtained without the use of this zero-resistance state. Some research efforts require high-energy accelerators or strong magnets attainable only with superconductive materials. Superconductivity is currently applied in the design of Maglev trains (trains that ride on a cushion of air established by opposite magnetic poles) that exceed 300 mi/h, in powerful motors and generators, in nuclear magnetic resonance imaging systems to obtain cross-sectional images of the brain (and other parts of the body), in the design of computers with operating speeds four times that of conventional systems, and in improved power distribution systems.

The range of future uses for superconductors is a function of how much success physicists have in raising the operating temperature and how well they can utilize the successes obtained thus far. However, it may be only a matter of time before magnetically levitated trains increase in number, improved medical diagnostic equipment is available, computers operate at much higher speeds, high-efficiency power and storage systems are available, and transmission systems operate at very high efficiency levels due to this area of developing interest. Only time will reveal the impact that this new direction will have on the quality of life.

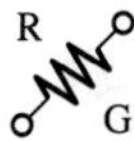

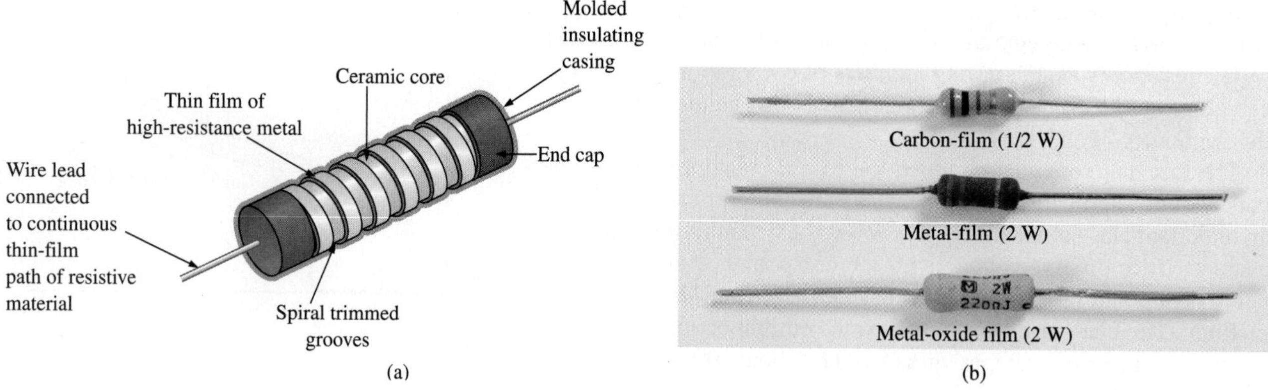

FIG. 3.16
Film resistors: (a) construction; (b) types.

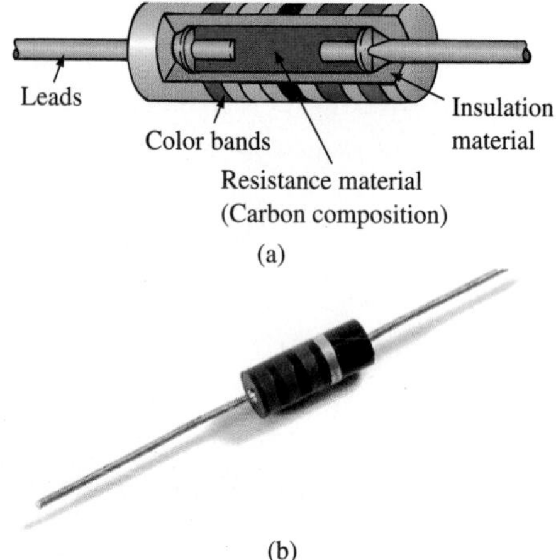

FIG. 3.17
Fixed composition resistors: (a) construction; (b) appearance.

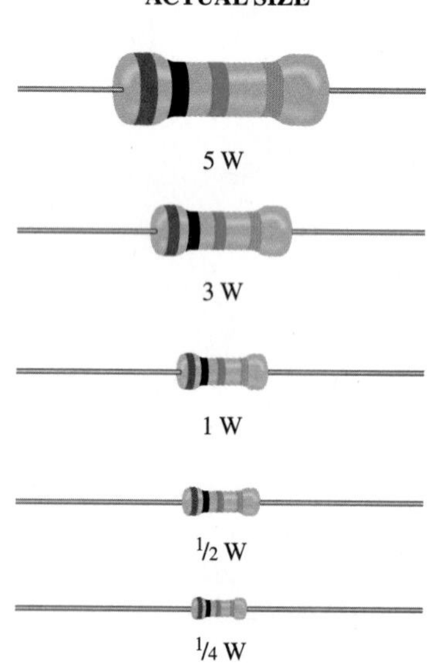

FIG. 3.18
Fixed metal-oxide resistors of different wattage ratings.

3.7 TYPES OF RESISTORS

Fixed Resistors

Resistors are made in many forms, but all belong in either of two groups: fixed or variable. The most common of the low-wattage, fixed-type resistors is the film resistor shown in Fig. 3.16. It is constructed by depositing a thin layer of resistive material (typically carbon, metal, or metal oxide) on a ceramic rod. The desired resistance is then obtained by cutting away some of the resistive material in a helical manner to establish a long, continuous band of high-resistance material from one end of the resistor to the other. In general, carbon-film resistors have a beige body and a lower wattage rating. The metal-film resistor is typically a stronger color, such as brick red or dark green, with higher wattage ratings. The metal-oxide resistor is usually a softer pastel color, such as rating powder blue shown in Fig. 3.16(b), and has the highest wattage rating of the three.

When you search through most electronics catalogs or visit a local electronics dealer such as Radio Shack to purchase resistors, you will find that the most common resistor is the film resistor. In years past, the carbon composition resistor in Fig. 3.17 was the most common, but fewer and fewer companies are manufacturing this variety, with its range of applications reduced to applications in which very high temperatures and inductive effects (Chapter 11) can be a problem. Its resistance is determined by the carbon composition material molded directly to each end of the resistor. The high resistivity characteristics of carbon ($\rho = 21{,}000$ CM-Ω/ft) provide a high-resistance path for the current through the element.

For a particular style and manufacturer, the size of a resistor increases with the power or wattage rating.

The concept of power is covered in detail in Chapter 4, but for the moment recognize that increased power ratings are normally associated with the ability to handle higher current and temperature levels. Fig. 3.18 depicts the actual size of thin-film, metal-oxide resistors in the 1/4 W to 5 W rating range. All the resistors in Fig. 3.18 are 1 MΩ, revealing that

the size of a resistor does not define its resistance level.

A variety of other fixed resistors are depicted in Fig. 3.19. The wire-wound resistors of Fig. 3.19(a) are formed by winding a high-resistance wire around a ceramic core. The entire structure is then baked in a ceramic cement to provide a protective covering. Wire-wound resistors are

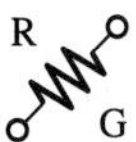

470 Ω, 35 W
Thick-film power resistor
(b)

1 kΩ, 25 W

Aluminum-housed, chassis-mount resistor–precision wire-mount
(c)

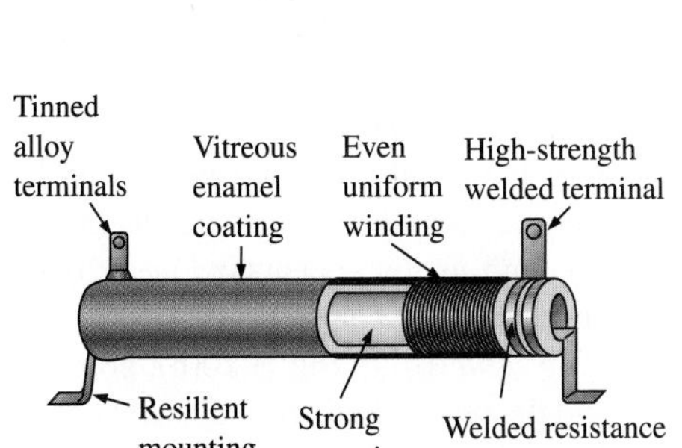

Wire-wound resistors
(a)

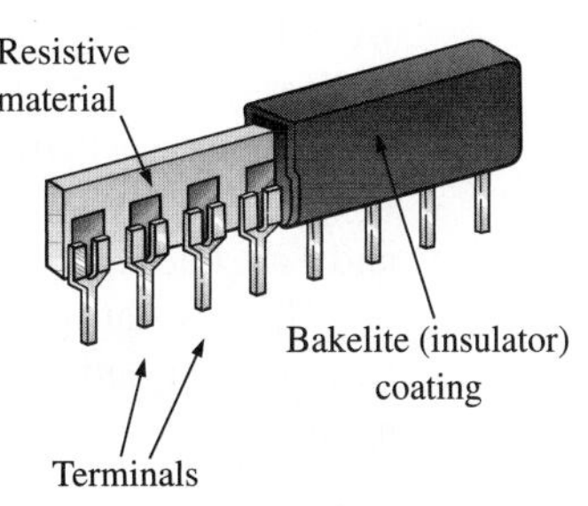

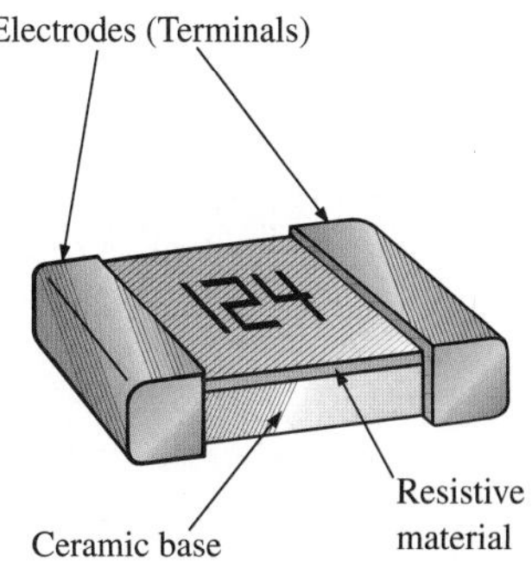

100 MΩ, 0.75 W
Precision power film resistor
(d)

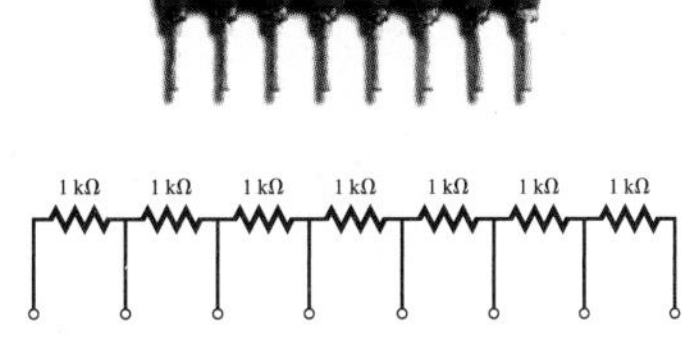

1 kΩ bussed (all connected on one side) single in-line resistor network
(e)

22 kΩ, 1W
Surface mount thick-film chip resistors with gold electrodes
(f)

25 kΩ, 5 W
Silicon-coated, wire-wound resistor
(g)

FIG. 3.19
Various types of fixed resistors.

typically used for larger power applications, although they are also available with very small wattage ratings and very high accuracy.

Fig. 3.19(c) and (g) are special types of wire-wound resistors with a low percent tolerance. Note, in particular, the high power ratings for the wire-wound resistors for their relatively small size. Figs. 3.19(b), (d), and (f) are power film resistors that use a thicker layer of film material than used in the variety shown in Fig. 3.16. The chip resistors in Fig. 3.19(f) are used where space is a priority, such as on the surface of circuit board. Units of this type can be less than 1/16 in. in length or width, with thickness as small as 1/30 in., yet they can still handle 0.5 W of power with resistance levels as

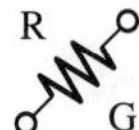

high as 1000 MΩ—clear evidence that size does not determine the resistance level. The fixed resistor in Fig. 3.19(e) has terminals applied to a layer of resistor material, with the resistance between the terminals a function of the dimensions of the resistive material and the placement of the terminal pads.

Variable Resistors

Variable resistors, as the name implies, have a terminal resistance that can be varied by turning a dial, knob, screw, or whatever seems appropriate for the application. They can have two or three terminals, but most have three terminals. If the two- or three-terminal device is used as a variable resistor, it is usually referred to as a **rheostat.** If the three-terminal device is used for controlling potential levels, it is then commonly called a **potentiometer.** Even though a three-terminal device can be used as a rheostat or a potentiometer (depending on how it is connected), it is typically called a *potentiometer* when listed in trade magazines or requested for a particular application.

The symbol for a three-terminal potentiometer appears in Fig. 3.20(a). When used as a variable resistor (or rheostat), it can be hooked up in one of two ways, as shown in Figs. 3.20(b) and (c). In Fig. 3.20(b), points a and b are hooked up to the circuit, and the remaining terminal is left hanging. The resistance introduced is determined by that portion of the resistive element between points a and b. In Fig. 3.20(c), the resistance is again between points a and b, but now the remaining resistance is "shorted-out" (effect removed) by the connection from b to c. The universally accepted symbol for a rheostat appears in Fig. 3.20(d).

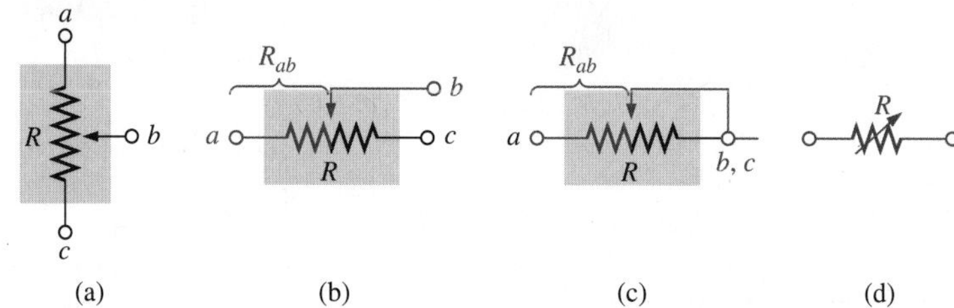

FIG. 3.20
Potentiometer: (a) symbol; (b) and (c) rheostat connections; (d) rheostat symbol.

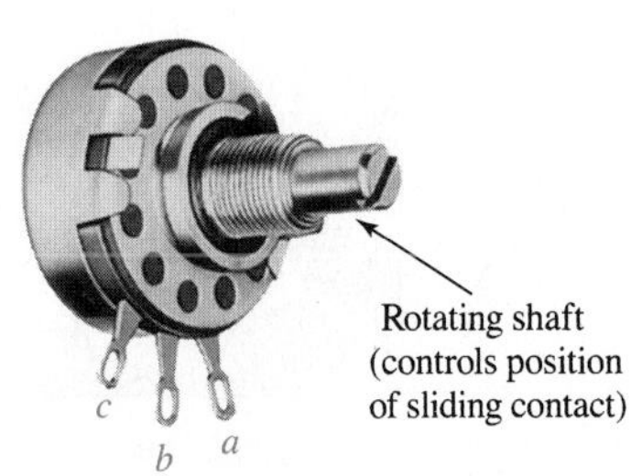

(a) External view

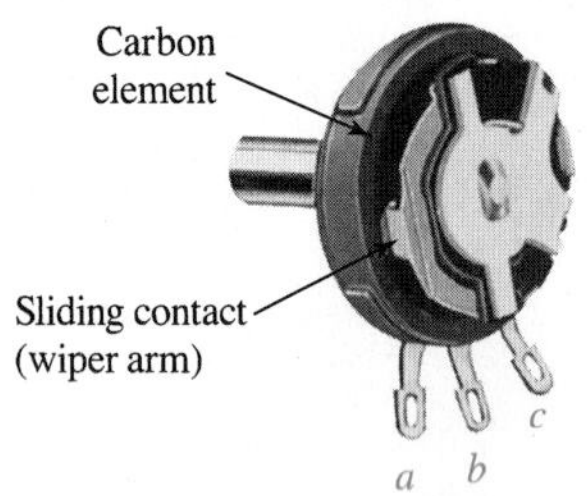

(b) Internal view

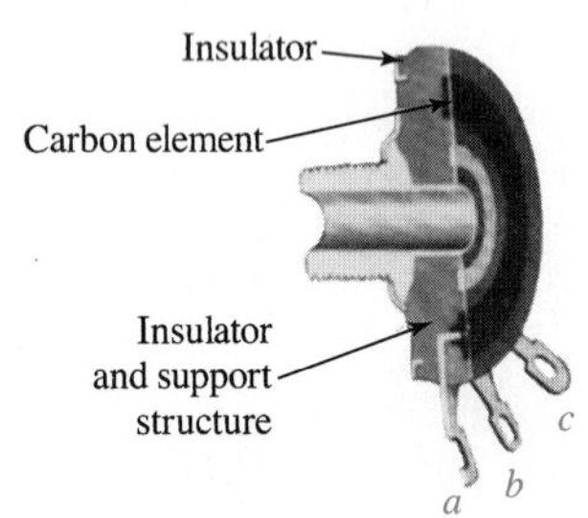

(c) Carbon element

FIG. 3.21
Molded composition-type potentiometer.
(Courtesy of Allen-Bradley Co.)

Most potentiometers have three terminals in the relative positions shown in Fig. 3.21. The knob, dial, or screw in the center of the housing controls the motion of a contact that can move along the resistive element connected between the outer two terminals. The contact is connected to the center terminal, establishing a resistance from movable contact to each outer terminal.

The resistance between the outside terminals a and c in Fig. 3.22(a) (and Fig. 3.21) is always fixed at the full rated value of the potentiometer, regardless of the position of the wiper arm b.

In other words, the resistance between terminals a and c in Fig. 3.22(a) for a 1 MΩ potentiometer will always be 1 MΩ, no matter how we turn the control element and move the contact. In Fig. 3.22(a), the center contact is not part of the network configuration.

The resistance between the wiper arm and either outside terminal can be varied from a minimum of 0 Ω to a maximum value equal to the full rated value of the potentiometer.

In Fig. 3.22(b), the wiper arm has been placed 1/4 of the way down from point a to point c. The resulting resistance between points a and b will

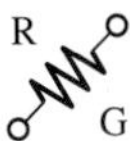

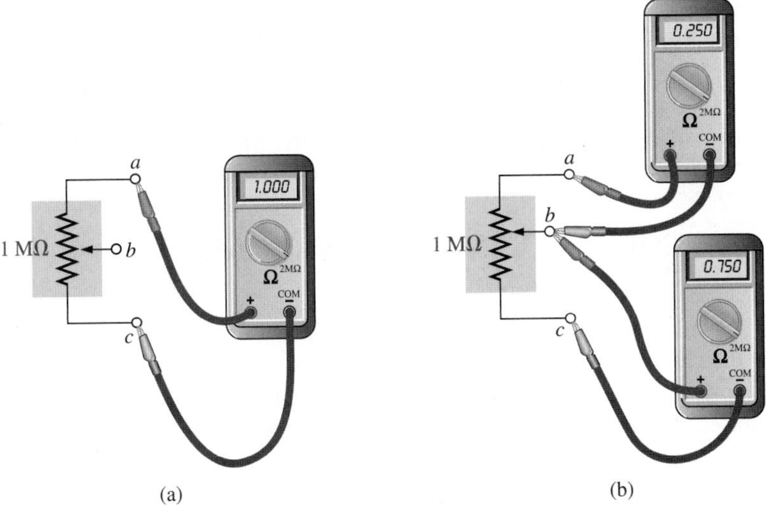

FIG. 3.22
Resistance components of a potentiometer: (a) between outside terminals; (b) between wiper arm and each outside terminal.

therefore be 1/4 of the total, or 250 kΩ (for a 1 MΩ potentiometer), and the resistance between b and c will be 3/4 of the total, or 750 kΩ.

The sum of the resistances between the wiper arm and each outside terminal equals the full rated resistance of the potentiometer.

This was demonstrated in Fig. 3.22(b), where 250 kΩ + 750 kΩ = 1 MΩ. Specifically,

$$\boxed{R_{ac} = R_{ab} + R_{bc}} \qquad \textbf{(3.13)}$$

Therefore, as the resistance from the wiper arm to one outside contact increases, the resistance between the wiper arm and the other outside terminal must decrease accordingly. For example, if R_{ab} of a 1 kΩ potentiometer is 200 Ω, then the resistance R_{bc} must be 800 Ω. If R_{ab} is further decreased to 50 Ω, then R_{bc} must increase to 950 Ω, and so on.

The molded carbon composition potentiometer is typically applied in networks with smaller power demands, and it ranges in size from 20 Ω to 22 MΩ (maximum values). A miniature trimmer (less than 1/4 in. in diameter) appears in Fig. 3.23(a), and a variety of potentiometers that use a cermet resistive material appear in Fig. 3.23(b). The contact point of the

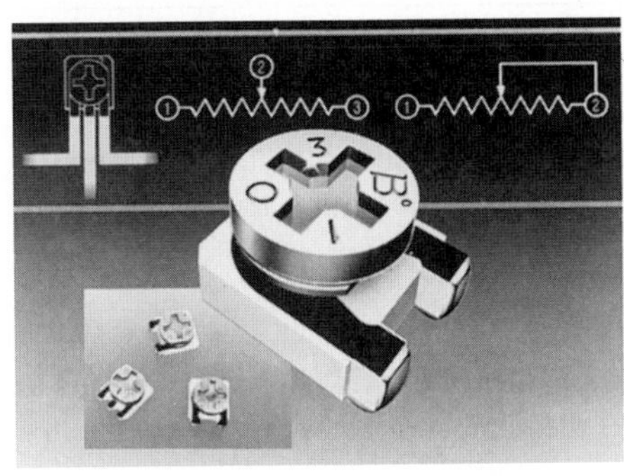

(a)

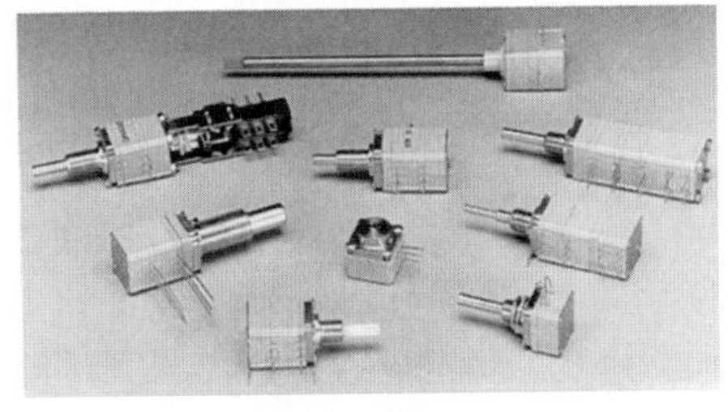

(b)

(c)

FIG. 3.23
Variable resistors: (a) 4 mm (≈5/32″) trimmer (courtesy of Bourns, Inc.); (b) conductive plastic and cermet elements (courtesy of Honeywell Clarostat); (c) three-point wire-wound resistor.

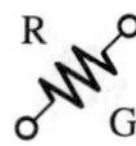

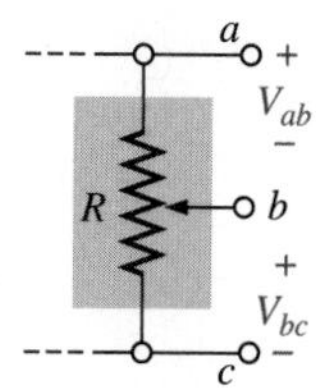

FIG. 3.24
Potentiometer control of voltage levels.

FIG. 3.25
Color coding for fixed resistors.

Number	Color
0	Black
1	Brown
2	Red
3	Orange
4	Yellow
5	Green
6	Blue
7	Violet
8	Gray
9	White
±5% (0.1 multiplier if 3rd band)	Gold
±10% (0.01 multiplier if 3rd band)	Silver

FIG. 3.26
Color coding.

FIG. 3.27
Example 3.13.

three-point wire-wound resistor in Fig. 3.23(c) can be moved to set the resistance between the three terminals.

When the device is used as a potentiometer, the connections are as shown in Fig. 3.24. It can be used to control the level of V_{ab}, V_{bc}, or both, depending on the application. Additional discussion of the potentiometer in a loaded situation can be found in later chapters.

3.8 COLOR CODING AND STANDARD RESISTOR VALUES

A wide variety of resistors, fixed or variable, are large enough to have their resistance in ohms printed on the casing. Some, however, are too small to have numbers printed on them, so a system of **color coding** is used. For the thin-film resistor, four, five, or six bands may be used. The four-band scheme is described. Later in this section the purpose of the fifth and sixth bands will be described.

For the four-band scheme, the bands are *always read from the end that has a band closest to it,* as shown in Fig. 3.25. The bands are numbered as shown for reference in the discussion to follow.

The first two bands represent the first and second digits, respectively.

They are the actual first two numbers that define the numerical value of the resistor.

The third band determines the power-of-ten multiplier for the first two digits (actually the number of zeros that follow the second digit for resistors greater than 10 Ω).

The fourth band is the manufacturer's tolerance, which is an indication of the precision by which the resistor was made.

If the fourth band is omitted, the tolerance is assumed to be ±20%.

The number corresponding to each color is defined in Fig. 3.26. The fourth band will be either ±5% or ±10% as defined by gold and silver, respectively. To remember which color goes with which percent, simply remember that ±5% resistors cost more and gold is more valuable than silver.

Remembering which color goes with each digit takes a bit of practice. In general, the colors start with the very dark shades and move toward the lighter shades. The best way to memorize is to simply repeat over and over that red is 2, yellow is 4, and so on. Simply practice with a friend or a fellow student, and you will learn most of the colors in short order.

EXAMPLE 3.13 Find the value of the resistor in Fig. 3.27.

Solution: Reading from the band closest to the left edge, we find that the first two colors of brown and red represent the numbers 1 and 2, respectively. The third band is orange, representing the number 3 for the power of the multiplier as follows:

$$12 \times 10^3\ \Omega$$

resulting in a value of 12 kΩ. As indicated above, if 12 kΩ is written as 12,000 Ω, the third band reveals the number of zeros that follow the first two digits.

Now for the fourth band of gold, representing a tolerance of ±5%: To find the range into which the manufacturer has guaranteed the resistor will fall, first convert the 5% to a decimal number by moving the decimal point two places to the left:

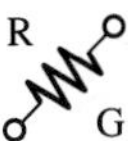

$$5\% \Rightarrow 0.05$$

Then multiply the resistor value by this decimal number:

$$0.05(12\text{ k}\Omega) = 600\ \Omega$$

Finally, add the resulting number to the resistor value to determine the maximum value, and subtract the number to find the minimum value. That is,

$$\text{Maximum} = 12{,}000\ \Omega + 600\ \Omega = 12.6\text{ k}\Omega$$
$$\text{Minimum} = 12{,}000\ \Omega - 600\ \Omega = 11.4\text{ k}\Omega$$
$$\text{Range} = \mathbf{11.4\ k\Omega\ to\ 12.6\ k\Omega}$$

The result is that the manufacturer has guaranteed with the 5% gold band that the resistor will fall in the range just determined. In other words, the manufacturer does not guarantee that the resistor will be exactly 12 kΩ but rather that it will fall in a range as defined above.

Using the above procedure, the smallest resistor that can be labeled with the color code is 10 Ω. However,

the range can be extended to include resistors from 0.1 Ω to 10 Ω by simply using gold as a multiplier color (third band) to represent 0.1 and using silver to represent 0.01.

This is demonstrated in the next example.

EXAMPLE 3.14 Find the value of the resistor in Fig. 3.28.

FIG. 3.28
Example 3.14.

Solution: The first two colors are gray and red, representing the numbers 8 and 2. The third color is gold, representing a multiplier of 0.1. Using the multiplier, we obtain a resistance of

$$(0.1)(82\ \Omega) = 8.2\ \Omega$$

The fourth band is silver, representing a tolerance of ±10%. Converting to a decimal number and multiplying through yields

$$10\% = 0.10 \quad \text{and} \quad (0.1)(8.2\ \Omega) = 0.82\ \Omega$$

$$\text{Maximum} = 8.2\ \Omega + 0.82\ \Omega = 9.02\ \Omega$$
$$\text{Minimum} = 8.2\ \Omega - 0.82\ \Omega = 7.38\ \Omega$$

so that

$$\text{Range} = \mathbf{7.38\ \Omega\ to\ 9.02\ \Omega}$$

Although it will take some time to learn the numbers associated with each color, it is certainly encouraging to become aware that

the same color scheme to represent numbers is used for all the important elements of electrical circuits.

Later on, you will find that the numerical value associated with each color is the same for capacitors and inductors. Therefore, once learned, the scheme has repeated areas of application.

Some manufacturers prefer to use a **five-band color code.** In such cases, as shown in the top portion of Fig. 3.29, three digits are provided before the multiplier. The fifth band remains the tolerance indicator. If the manufacturer decides to include the temperature coefficient, a sixth band will appear as shown in the lower portion of Fig. 3.29, with the color indicating the PPM level.

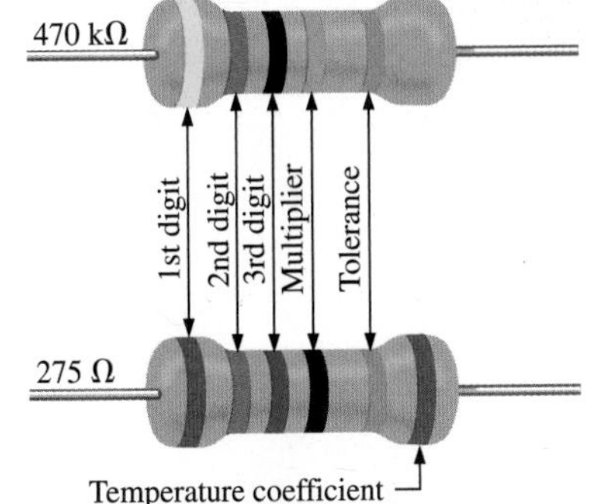

FIG. 3.29
Five-band color coding for fixed resistors.

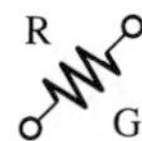

For four, five, or six bands, if the tolerance is less than 5%, the following colors are used to reflect the % tolerances:

brown* = ±*1%, red* = ±*2%, green* = ±*0.5%, blue* = ±*0.25%, and violet* = ±*0.1%.

You might expect that resistors would be available for a full range of values such as 10 Ω, 20 Ω, 30 Ω, 40 Ω, 50 Ω, and so on. However, this is not the case with some typical commercial values, such as 27 Ω, 56 Ω, and 68 Ω. There is a reason for the chosen values, which is best demonstrated by examining the list of standard values of commercially available resistors in Table 3.7. The values in boldface blue are available with 5%, 10%, and 20% tolerances, making them the most common of the commercial variety. The values in boldface black are typically available with 5% and 10% tolerances, and those in normal print are available only in the 5% variety.

TABLE 3.7
Standard values of commercially available resistors.

Ohms (Ω)					Kilohms (kΩ)		Megohms (MΩ)	
0.10	**1.0**	**10**	**100**	**1000**	**10**	**100**	**1.0**	**10.0**
0.11	1.1	11	110	1100	11	110	1.1	11.0
0.12	**1.2**	**12**	**120**	**1200**	**12**	**120**	**1.2**	**12.0**
0.13	1.3	13	130	1300	13	130	1.3	13.0
0.15	**1.5**	**15**	**150**	**1500**	**15**	**150**	**1.5**	**15.0**
0.16	1.6	16	160	1600	16	160	1.6	16.0
0.18	**1.8**	**18**	**180**	**1800**	**18**	**180**	**1.8**	**18.0**
0.20	2.0	20	200	2000	20	200	2.0	20.0
0.22	**2.2**	**22**	**220**	**2200**	**22**	**220**	**2.2**	**22.0**
0.24	2.4	24	240	2400	24	240	2.4	
0.27	**2.7**	**27**	**270**	**2700**	**27**	**270**	**2.7**	
0.30	3.0	30	300	3000	30	300	3.0	
0.33	**3.3**	**33**	**330**	**3300**	**33**	**330**	**3.3**	
0.36	3.6	36	360	3600	36	360	3.6	
0.39	**3.9**	**39**	**390**	**3900**	**39**	**390**	**3.9**	
0.43	4.3	43	430	4300	43	430	4.3	
0.47	**4.7**	**47**	**470**	**4700**	**47**	**470**	**4.7**	
0.51	5.1	51	510	5100	51	510	5.1	
0.56	**5.6**	**56**	**560**	**5600**	**56**	**560**	**5.6**	
0.62	6.2	62	620	6200	62	620	6.2	
0.68	**6.8**	**68**	**680**	**6800**	**68**	**680**	**6.8**	
0.75	7.5	75	750	7500	75	750	7.5	
0.82	**8.2**	**82**	**820**	**8200**	**82**	**820**	**8.2**	
0.91	9.1	91	910	9100	91	910	9.1	

Examining the impact of the tolerance level will help explain the choice of numbers for the commercial values. Take the sequence 47 Ω–68 Ω–100 Ω, which are all available with 20% tolerances. In Fig. 3.30(a), the tolerance band for each has been determined and plotted on a single axis. Note that with this tolerance (which is all that the manufacturer will guarantee), the full range of resistor values is available from 37.6 Ω to 120 Ω. In other words, the manufacturer is guaranteeing

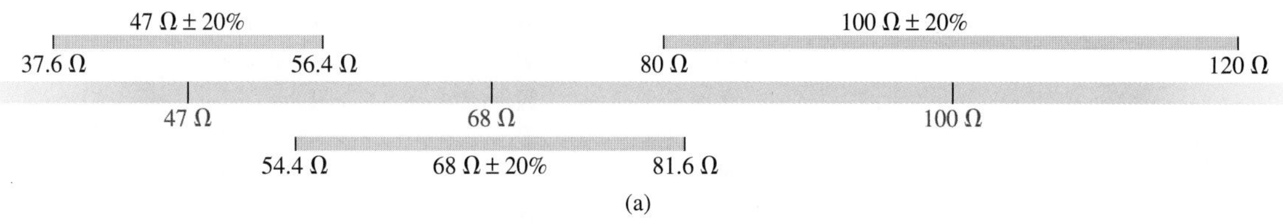

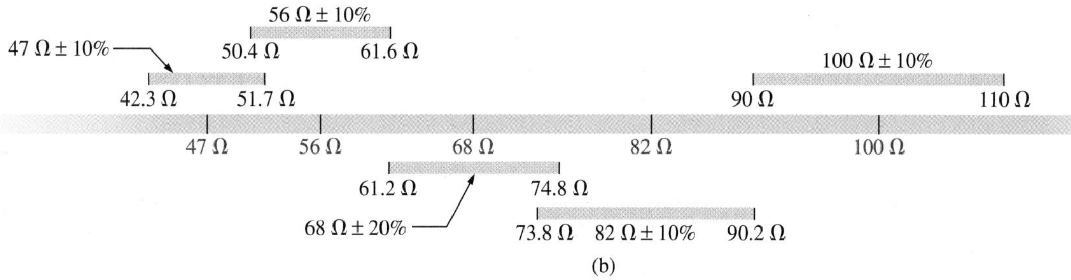

FIG. 3.30
Guaranteeing the full range of resistor values for the given tolerance: (a) 20%; (b) 10%.

the full range, using the tolerances to fill in the gaps. Dropping to the 10% level introduces the 56 Ω and 82 Ω resistors to fill in the gaps, as shown in Fig. 3.30(b). Dropping to the 5% level would require additional resistor values to fill in the gaps. In total, therefore, the resistor values were chosen to ensure that the full range was covered, as determined by the tolerances employed. Of course, if a specific value is desired but is not one of the standard values, combinations of standard values often result in a total resistance very close to the desired level. If this approach is still not satisfactory, a potentiometer can be set to the exact value and then inserted in the network.

Throughout the text, you will find that many of the resistor values are not standard values. This was done to reduce the mathematical complexity, which might make it more difficult to understand the procedure or analysis technique being introduced. In the problem sections, however, standard values are frequently used to ensure that you start to become familiar with the commercial values available.

Surface Mount Resistors

In general, surface mount resistors are marked in three ways: color coding, three symbols, and two symbols.

The **color coding** is the same as just described earlier in this section for through-hole resistors.

The **three-symbol** approach uses three digits. The first two define the first two digits of the value; the last digit, the power of the power-of-ten multiplier.

For instance:

$$820 \text{ is } 82 \times 10^0\ \Omega = \mathbf{82\ \Omega}$$
$$222 \text{ is } 22 \times 10^2\ \Omega = 2200\ \Omega = \mathbf{2.2\ k\Omega}$$
$$010 \text{ is } 1 \times 10^0\ \Omega = \mathbf{1\ \Omega}$$

The **two-symbol** marking uses a letter followed by a number. The letter defines the value as listed below. Note that all the numbers of the commercially available list of Table 3.7 are included.

A = 1.0	**B** = 1.1	**C** = 1.2	**D** = 1.3
E = 1.5	**F** = 1.6	**G** = 1.8	**H** = 2
J = 2.2	**K** = 2.4	**L** = 2.7	**M** = 3
N = 3.3	**P** = 3.6	**Q** = 3.9	**R** = 4.3
S = 4.7	**T** = 5.1	**U** = 5.6	**V** = 6.2
W = 6.8	**X** = 7.5	**Y** = 8.2	**Z** = 9.1

The second symbol is the power of the power-of-ten multiplier. For example:

$$\text{C3} = 1.2 \times 10^3\,\Omega = \mathbf{1.2\ k\Omega}$$
$$\text{T0} = 5.1 \times 10^0\,\Omega = \mathbf{5.1\ \Omega}$$
$$\text{Z1} = 9.1 \times 10^1\,\Omega = \mathbf{91\ \Omega}$$

Additional symbols may precede or follow the codes above and may differ depending on the manufacturer. These may provide information on the internal resistance structure, power rating, surface material, tapping, and tolerance.

3.9 CONDUCTANCE

By finding the reciprocal of the resistance of a material, we have a measure of how well the material conducts electricity. The quantity is called **conductance,** has the symbol G, and is measured in *siemens* (S) (note Fig. 3.31). In equation form, conductance is

$$\boxed{G = \frac{1}{R}} \qquad \text{(siemens, S)} \qquad \textbf{(3.14)}$$

FIG. 3.31
Werner von Siemens.
© Bettmann/Corbis

German (Lenthe, Berlin)
(1816–92)
Electrical Engineer
Telegraph Manufacturer,
Siemens & Halske AG

Developed an *electroplating process* during a brief stay in prison for acting as a second in a duel between fellow officers of the Prussian army. Inspired by the electronic telegraph invented by Sir Charles Wheatstone in 1817, he improved on the design and proceeded to lay cable with the help of his brother Carl across the Mediterranean and from Europe to India. His inventions included the first *self-excited generator,* which depended on the *residual* magnetism of its electromagnet rather than an inefficient permanent magnet. In 1888 he was raised to the rank of nobility with the addition of *von* to his name. The current firm of Siemens AG has manufacturing outlets in some 35 countries with sales offices in some 125 countries.

A resistance of 1 MΩ is equivalent to a conductance of 10^{-6} S, and a resistance of 10 Ω is equivalent to a conductance of 10^{-1} S. The larger the conductance, therefore, the less the resistance and the greater the conductivity.

In equation form, the conductance is determined by

$$\boxed{G = \frac{A}{\rho l}} \qquad \text{(S)} \qquad \textbf{(3.15)}$$

indicating that increasing the area or decreasing either the length or the resistivity increases the conductance.

EXAMPLE 3.15

a. Determine the conductance of a 1 Ω, 50 kΩ, and 10 MΩ resistor.
b. How does the conductance level change with increase in resistance?

Solution: Eq. (3.14):

a. $1\Omega\text{: } G = \dfrac{1}{R} = \dfrac{1}{1\,\Omega} = \mathbf{1\ S}$

$$50\text{ k}\Omega\text{: } G = \frac{1}{R} = \frac{1}{50\text{ k}\Omega} = \frac{1}{50 \times 10^3\,\Omega} = 0.02 \times 10^{-3}\text{ S} = \mathbf{0.02\ mS}$$

$$10\text{ M}\Omega\text{: } G = \frac{1}{R} = \frac{1}{10\text{ M}\Omega} = \frac{1}{10 \times 10^6\,\Omega} = 0.1 \times 10^{-6}\text{ S} = \mathbf{0.1\ \mu S}$$

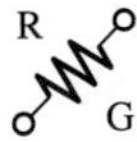

b. The conductance level decreases rapidly with significant increase in resistance levels.

EXAMPLE 3.16 What is the relative increase or decrease in conductivity of a conductor if the area is reduced by 30% and the length is increased by 40%? The resistivity is fixed.

Solution: Eq. (3.15):

$$G_i = \frac{1}{R_i} = \frac{1}{\frac{\rho_i l_i}{A_i}} = \frac{A_i}{\rho_i l_i}$$

with the subscript i for the initial value. Using the subscript n for new value:

$$G_n = \frac{A_n}{\rho_n l_n} = \frac{0.70A_i}{\rho_i(1.4l_i)} = \frac{0.70}{1.4}\frac{A_i}{\rho_i l_i} = \frac{0.70G_i}{1.4}$$

and $\qquad G_n = \mathbf{0.5}\boldsymbol{G_i}$

3.10 OHMMETERS

The **ohmmeter** is an instrument used to perform the following tasks and several other useful functions:

1. ***Measure the resistance of individual or combined elements.***
2. ***Detect open-circuit (high-resistance) and short-circuit (low-resistance) situations.***
3. ***Check the continuity of network connections and identify wires of a multilead cable.***
4. ***Test some semiconductor (electronic) devices.***

For most applications, the ohmmeters used most frequently are the ohmmeter section of a VOM or DMM. The details of the internal circuitry and the method of using the meter will be left primarily for a laboratory exercise. In general, however, the resistance of a resistor can be measured by simply connecting the two leads of the meter across the resistor, as shown in Fig. 3.32. There is no need to be concerned about which lead goes on which end; the result is the same in either case since resistors offer the same resistance to the flow of charge (current) in either direction. If the VOM is used, a switch must be set to the proper resistance range, and a nonlinear scale (usually the top scale of the meter) must be properly read to obtain the resistance value. The DMM also requires choosing the best scale setting for the resistance to be measured, but the result appears as a numerical display, with the proper placement of the decimal point determined by the chosen scale. When measuring the resistance of a single resistor, it is usually best to remove the resistor from the network before making the measurement. If this is difficult or impossible, at least one end of the resistor must not be connected to the network, or the reading may include the effects of the other elements of the system.

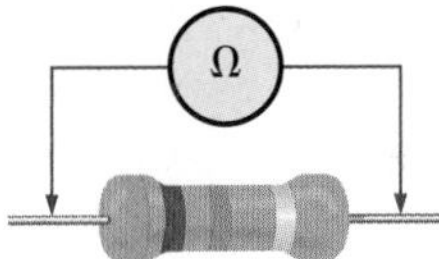

FIG. 3.32
Measuring the resistance of a single element.

If the two leads of the meter are touching in the ohmmeter mode, the resulting resistance is zero. A connection can be checked as shown in Fig. 3.33 by simply hooking up the meter to either side of the connection. If the resistance is zero, the connection is secure. If it is other than zero, the connection could be weak; if it is infinite, there is no connection at all.

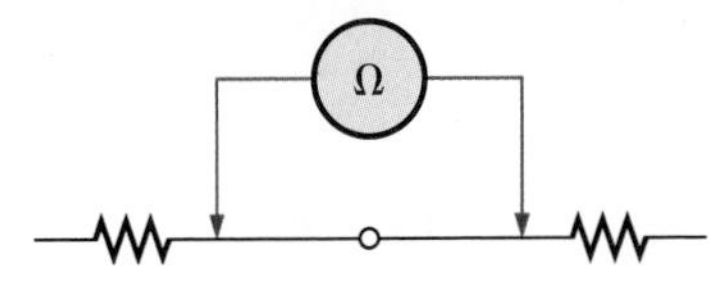

FIG. 3.33
Checking the continuity of a connection.

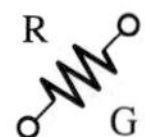

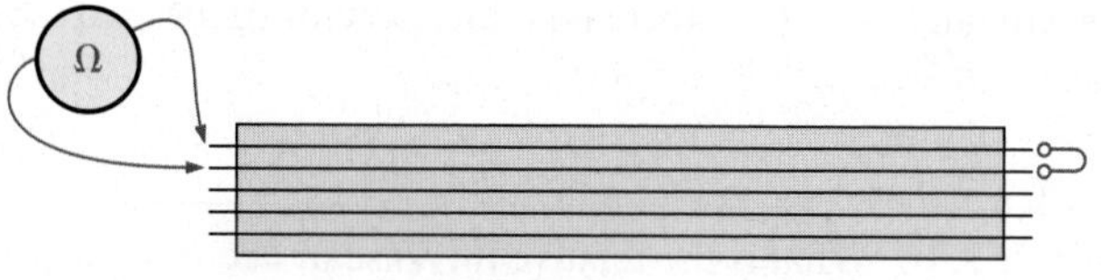

FIG. 3.34
Identifying the leads of a multilead cable.

If one wire of a harness is known, a second can be found as shown in Fig. 3.34. Simply connect the end of the known lead to the end of any other lead. When the ohmmeter indicates zero ohms (or very low resistance), the second lead has been identified. The above procedure can also be used to determine the first known lead by simply connecting the meter to any wire at one end and then touching all the leads at the other end until a zero ohm indication is obtained.

Preliminary measurements of the condition of some electronic devices such as the diode and the transistor can be made using the ohmmeter. The meter can also be used to identify the terminals of such devices.

One important note about the use of any ohmmeter:

Never hook up an ohmmeter to a live circuit!

The reading will be meaningless, and you may damage the instrument. The ohmmeter section of any meter is designed to pass a small sensing current through the resistance to be measured. A large external current could damage the movement and would certainly throw off the calibration of the instrument. In addition:

Never store a VOM or a DMM in the resistance mode.

If the two leads of the meter touch, the small sensing current could drain the internal battery. VOMs should be stored with the selector switch on the highest voltage range, and the selector switch of DMMs should be in the off position.

3.11 THERMISTORS

The **thermistor** is a two-terminal semiconductor device whose resistance, as the name suggests, is temperature sensitive. A representative characteristic appears in Fig. 3.35 with the graphic symbol for the device. Note the nonlinearity of the curve and the drop in resistance from about 5000 Ω to 100 Ω for an increase in temperature from 20°C to 100°C. The decrease in resistance with an increase in temperature indicates a negative temperature coefficient.

The temperature of the device can be changed internally or externally. An increase in current through the device raises its temperature, causing a drop in its terminal resistance. Any externally applied heat source results in an increase in its body temperature and a drop in resistance. This type of action (internal or external) lends itself well to control mechanisms. Many different types of thermistors are shown in Fig. 3.36. Materials used in the manufacture of thermistors include oxides of cobalt, nickel, strontium, and manganese.

Note the use of a log scale (to be discussed in Chapter 23) in Fig. 3.35 for the vertical axis. The log scale permits the display of a wider range of

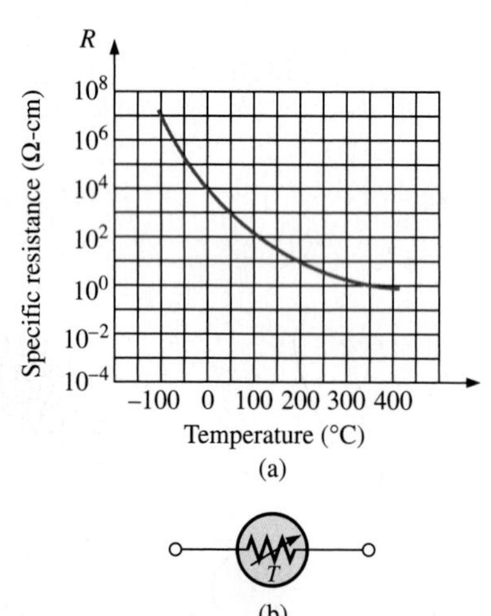

FIG. 3.35
Thermistor: (a) characteristics; (b) symbol.

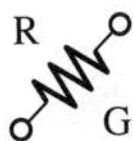

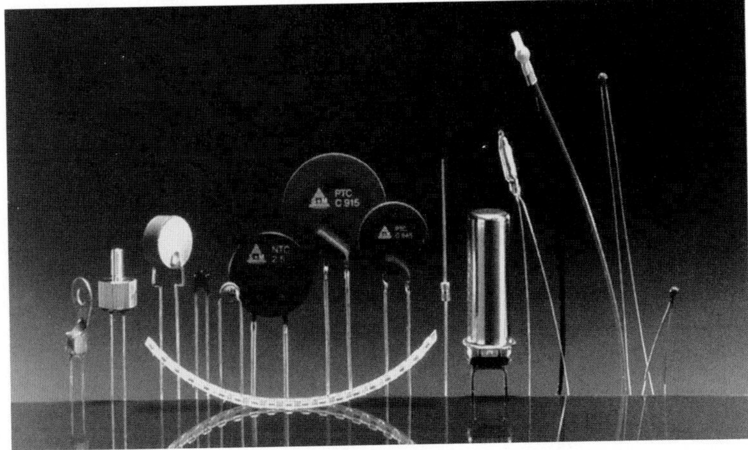

FIG. 3.36
NTC (negative temperature coefficient) and PTC (positive temperature coefficient) thermistors.
(Courtesy of Siemens Components, Inc.)

specific resistance levels than a linear scale such as the horizontal axis. Note that it extends from 0.0001 Ω-cm to 100,000,000 Ω-cm over a very short interval. The log scale is used for both the vertical and the horizontal axis in Fig. 3.37.

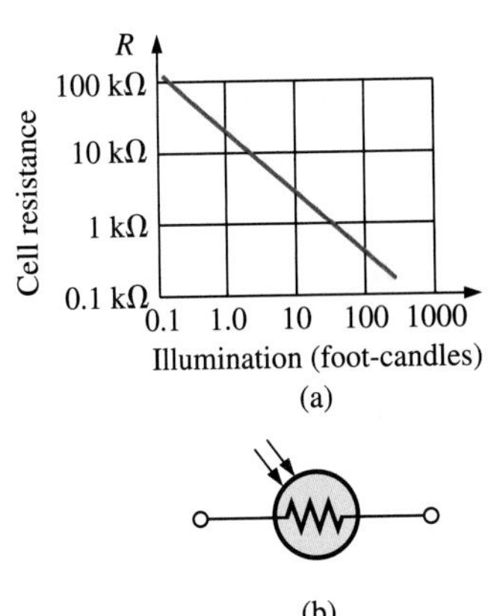

FIG. 3.37
Photoconductive cell: (a) characteristics. (b) symbol.

3.12 PHOTOCONDUCTIVE CELL

The **photoconductive cell** is a two-terminal semiconductor device whose terminal resistance is determined by the intensity of the incident light on its exposed surface. As the applied illumination increases in intensity, the energy state of the surface electrons and atoms increases, with a resultant increase in the number of "free carriers" and a corresponding drop in resistance. A typical set of characteristics and the photoconductive cell's graphic symbol appear in Fig. 3.37. Note the negative illumination coefficient. Several cadmium sulfide photoconductive cells appear in Fig. 3.38.

FIG. 3.38
Photoconductive cells.
(Courtesy of EG&G VACTEC, Inc.)

3.13 VARISTORS

Varistors are voltage-dependent, nonlinear resistors used to suppress high-voltage transients; that is, their characteristics enable them to limit the voltage that can appear across the terminals of a sensitive device or system. A typical set of characteristics appears in Fig. 3.39(a), along with a linear resistance characteristic for comparison purposes. Note that at a particular "firing voltage," the current rises rapidly, but the voltage is limited to a level just above this firing potential. In other words, the magnitude of the voltage that can appear across this device cannot exceed that level defined by its characteristics. Through proper design techniques, this device can therefore limit the voltage appearing across sensitive regions of a network. The current is simply limited by the network to which it is connected. A photograph of a number of commercial units appears in Fig. 3.39(b).

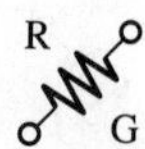

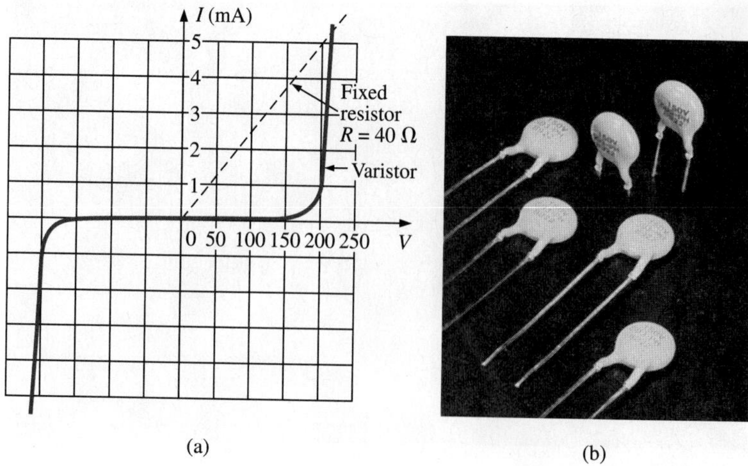

FIG. 3.39
Varistors available with maximum dc voltage ratings between 18 V and 615 V.
(Courtesy of Philips Components, Inc.)

3.14 APPLICATIONS

The following are examples of how resistance can be used to perform a variety of tasks, from heating to measuring the stress or strain on a supporting member of a structure. In general, resistance is a component of every electrical or electronic application.

Electric Baseboard Heating Element

One of the most common applications of resistance is in household fixtures such as toasters and baseboard heating where the heat generated by current passing through a resistive element is employed to perform a useful function.

Recently, as we remodeled our house, the local electrician informed us that we were limited to 16 ft of electric baseboard on a single circuit. That naturally had me wondering about the wattage per foot, the resulting current level, and whether the 16-ft limitation was a national standard. Reading the label on the 2-ft section appearing in Fig. 3.40(a). I found VOLTS AC 240/208, WATTS 750/575 (the power rating is described in Chapter 4), AMPS 3.2/2.8. Since my panel is rated 240 V (as are those in most residential homes), the wattage rating per foot is 575 W/2 or 287.5 W at a current of 2.8 A. The total wattage for the 16 ft is therefore 16×287.5 W or 4600 W.

In Chapter 4, you will find that the power to a resistive load is related to the current and applied voltage by the equation $P = VI$. The total resulting current can then be determined using this equation in the following manner: $I = P/V = 4600\text{ W}/240\text{ V} = 19.17\text{ A}$. The result was that we needed a circuit breaker larger than 19.17 A; otherwise, the circuit breaker would trip every time we turned the heat on. In my case, the electrician used a 30 A breaker to meet the National Fire Code requirement that does not permit exceeding 80% of the rated current for a conductor or breaker. In most panels, a 30 A breaker takes two slots of the panel, whereas the more common 20 A breaker takes only one slot. If you have

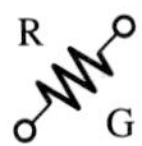

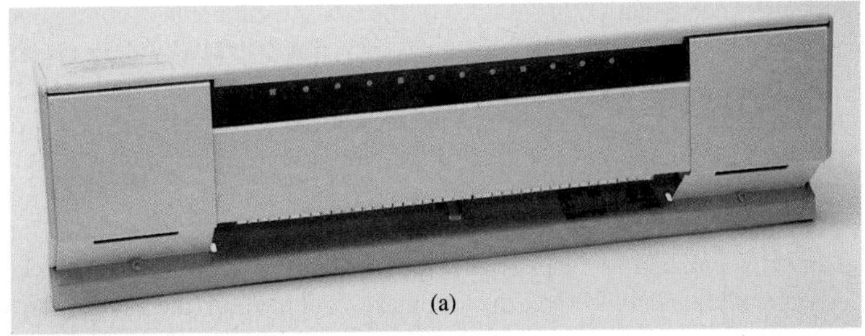

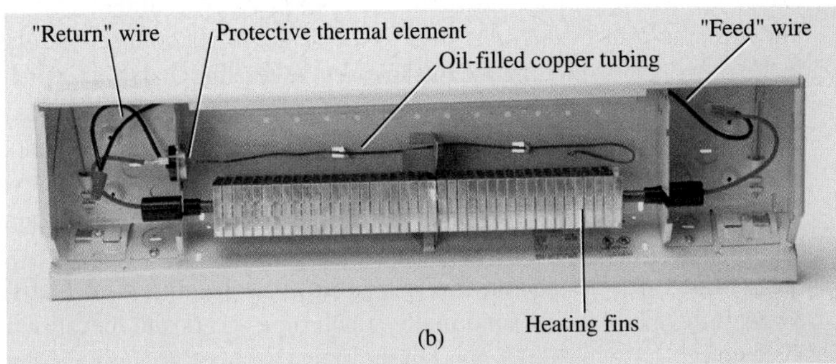

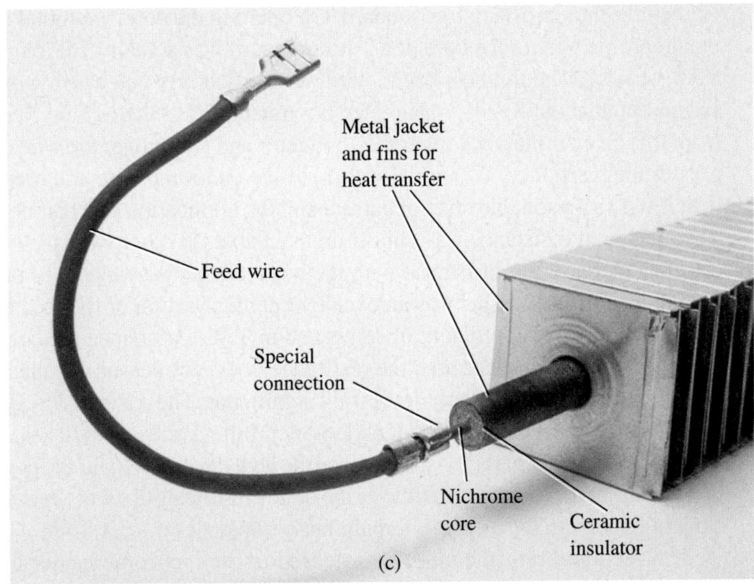

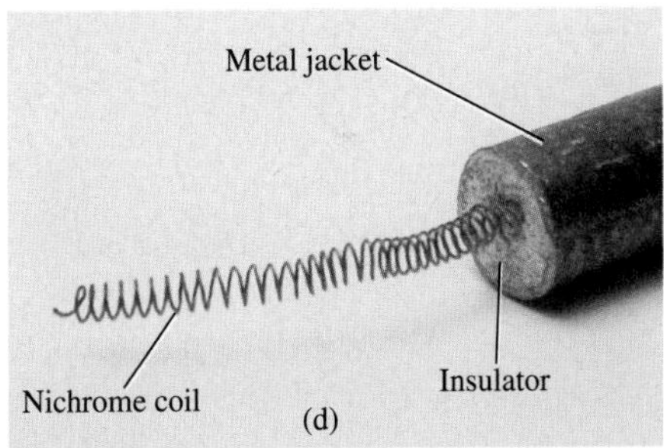

FIG. 3.40
Electric baseboard: (a) 2-ft section; (b) interior; (c) heating element; (d) nichrome coil.

a moment, take a look in your own panel and note the rating of the breakers used for various circuits of your home.

Going back to Table 3.2, we find that the #12 wire commonly used for most circuits in the home has a maximum rating of 20 A and would not be suitable for the electric baseboard. Since #11 is usually not commercially available, a #10 wire with a maximum rating of 30 A was used. You might wonder why the current drawn from the supply is 19.17 A while that required for one unit was only 2.8 A. This difference is due to the

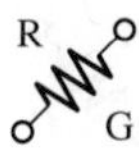

parallel combination of sections of the heating elements, a configuration that will be described in Chapter 6. It is now clear why they specify a 16-ft limitation on a single circuit. Additional elements would raise the current to a level that would exceed the code level for #10 wire and would approach the maximum rating of the circuit breaker.

Fig. 3.40(b) shows a photo of the interior construction of the heating element. The red feed wire on the right is connected to the core of the heating element, and the black wire at the other end passes through a protective heater element and back to the terminal box of the unit (the place where the exterior wires are brought in and connected). If you look carefully at the end of the heating unit as shown in Fig. 3.40(c), you will find that the heating wire that runs through the core of the heater is not connected directly to the round jacket holding the fins in place. A ceramic material (insulator) separates the heating wire from the fins to remove any possibility of conduction between the current passing through the bare heating element and the outer fin structure. Ceramic materials are used because they are excellent conductors of heat. They also have a high retentivity for heat so the surrounding area remains heated for a period of time even after the current has been turned off. As shown in Fig. 3.40(d), the heating wire that runs through the metal jacket is normally a nichrome composite (because pure nichrome is quite brittle) wound in the shape of a coil to compensate for expansion and contraction with heating and also to permit a longer heating element in standard-length baseboard. On opening the core, we found that the nichrome wire in the core of a 2-ft baseboard was actually 7 ft long, or a 3.5 : 1 ratio. The thinness of the wire was particularly noteworthy, measuring out at about 8 mils in diameter, not much thicker than a hair. Recall from this chapter that the longer the conductor and the thinner the wire, the greater the resistance. We took a section of the nichrome wire and tried to heat it with a reasonable level of current and the application of a hair dryer. The change in resistance was almost unnoticeable. In other words, all our effort to increase the resistance with the basic elements available to us in the lab was fruitless. This was an excellent demonstration of the meaning of the temperature coefficient of resistance in Table 3.6. Since the coefficient is so small for nichrome, the resistance does not measurably change unless the change in temperature is truly significant. The curve in Fig. 3.13 would therefore be close to horizontal for nichrome. For baseboard heaters, this is an excellent characteristic because the heat developed, and the power dissipated, will not vary with time as the conductor heats up with time. The flow of heat from the unit will remain fairly constant.

The feed and return cannot be soldered to the nichrome heater wire for two reasons. First, you cannot solder nichrome wires to each other or to other types of wire. Second, if you could, there might be a problem because the heat of the unit could rise above 880°F at the point where the wires are connected, the solder could melt, and the connection could be broken. Nichrome must be spot welded or crimped onto the copper wires of the unit. Using Eq. (3.1) and the 8-mil measured diameter, and assuming pure nichrome for the moment, the resistance of the 7-ft length is

$$R = \frac{\rho l}{A}$$

$$= \frac{(600)(7')}{(8 \text{ mils})^2} = \frac{4200}{64}$$

$$R = \mathbf{65.6\ \Omega}$$

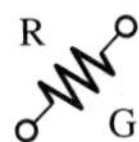

In Chapter 4, a power equation will be introduced in detail relating power, current, and resistance in the following manner: $P = I^2R$. Using the above data and solving for the resistance, we obtain

$$R = \frac{P}{I^2}$$
$$= \frac{575\text{ W}}{(2.8\text{ A})^2}$$
$$R = \mathbf{73.34\ \Omega}$$

which is very close to the value calculated above from the geometric shape since we cannot be absolutely sure about the resistivity value for the composite.

During normal operation, the wire heats up and passes that heat on to the fins, which in turn heat the room via the air flowing through them. The flow of air through the unit is enhanced by the fact that hot air rises, so when the heated air leaves the top of the unit, it draws cold air from the bottom to contribute to the convection effect. Closing off the top or bottom of the unit would effectively eliminate the convection effect, and the room would not heat up. A condition could occur in which the inside of the heater became too hot, causing the metal casing also to get too hot. This concern is the primary reason for the thermal protective element introduced above and appearing in Fig. 3.40(b). The long, thin copper tubing in Fig. 3.40 is actually filled with an oil-type fluid that expands when heated. If too hot, it expands, depresses a switch in the housing, and turns off the heater by cutting off the current to the heater wire.

Dimmer Control in an Automobile

A two-point rheostat is the primary element in the control of the light intensity on the dashboard and accessories of a car. The basic network appears in Fig. 3.41 with typical voltage and current levels. When the light switch is closed (usually by pulling the light control knob out from the dashboard), current is established through the 50 Ω rheostat and then to the various lights on the dashboard (including the panel lights, ashtray light, radio display, and glove compartment light). As the knob of the control switch is turned, it controls the amount of resistance between points a and b of the rheostat. The more resistance between points a and

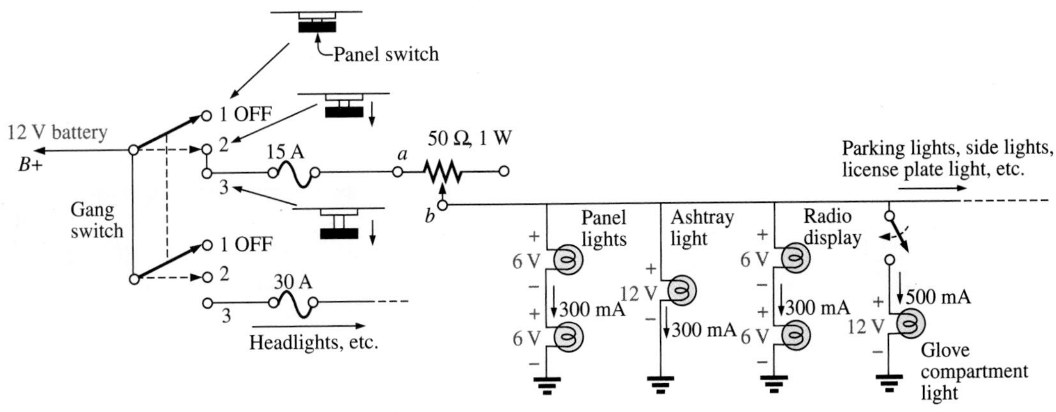

FIG. 3.41
Dashboard dimmer control in an automobile.

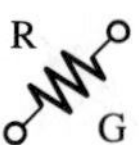

b, the less the current and the less the brightness of the various lights. Note the additional switch in the glove compartment light which is activated by the opening of the door of the compartment. Aside from the glove compartment light, all the lights in Fig. 3.41 will be on at the same time when the light switch is activated. The first branch after the rheostat contains two bulbs of 6 V rating rather than the 12 V bulbs appearing in the other branches. The smaller bulbs of this branch produce a softer, more even light for specific areas of the panel. Note that the sum of the two bulbs (in series) is 12 V to match that across the other branches. The division of voltage in any network is covered in detail in Chapters 5 and 6.

Typical current levels for the various branches have also been provided in Fig. 3.41. You will learn in Chapter 6 that the current drain from the battery and through the fuse and rheostat approximately equals the sum of the currents in the branches of the network. The result is that the fuse must be able to handle current in amperes, so a 15 A fuse was used (even though the bulbs appear in Fig. 3.41 as 12 V bulbs to match the battery).

Whenever the operating voltage and current levels of a component are known, the internal "hot" resistance of the unit can be determined using Ohm's law, introduced in detail in Chapter 4. Basically this law relates voltage, current, and resistance by $I = V/R$. For the 12 V bulb at a rated current of 300 mA, the resistance is $R = V/I = 12\text{ V}/300\text{ mA} = 40\ \Omega$. For the 6 V bulbs, it is $6\text{ V}/300\text{ mA} = 20\ \Omega$. Additional information regarding the power levels and resistance levels is discussed in later chapters.

The preceding description assumed an ideal level of 12 V for the battery. In actuality, 6.3 V and 14 V bulbs are used to match the charging level of most automobiles.

Strain Gauges

Any change in the shape of a structure can be detected using strain gauges whose resistance changes with applied stress or flex. An example of a strain gauge is shown in Fig. 3.42. Metallic strain gauges are constructed of a fine wire or thin metallic foil in a grid pattern. The terminal resistance of the strain gauge will change when exposed to compression or extension. One simple example of the use of resistive strain gauges is to monitor earthquake activity. When the gauge is placed across an area of suspected earthquake activity, the slightest separation in the earth changes the terminal resistance, and the processor displays a result sen-

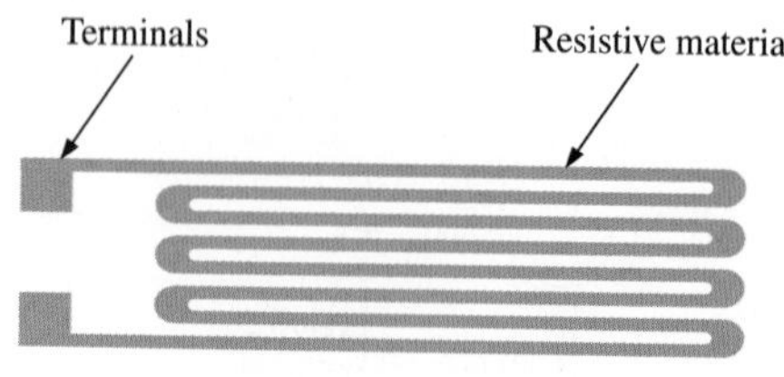

(a) Typical strain gauge configuration.

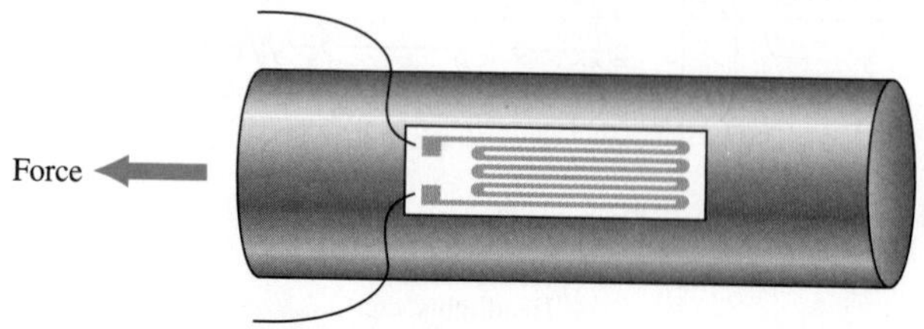

(b) The strain gauge is bonded to the surface to be measured along the line of force. When the surface lengthens, the strain gauge stretches.

FIG. 3.42
Resistive strain gauge.

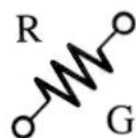

sitive to the amount of separation. Another example is in alarm systems where the slightest change in the shape of a supporting beam when someone walks overhead results in a change in terminal resistance, and an alarm sounds. Other examples include placing strain gauges on bridges to maintain an awareness of their rigidity and on very large generators to check whether various moving components are beginning to separate because of a wearing of the bearings or spacers. The small mouse control within a computer keyboard can be a series of strain gauges that reveal the direction of compression or extension applied to the controlling element on the keyboard. Movement in one direction can extend or compress a resistance gauge which can monitor and control the motion of the mouse on the screen.

3.15 MATHCAD

Throughout the text the mathematical software package Mathcad 12 is used to introduce a variety of operations that a math software package can perform. There is no need to obtain a copy of the software package to continue with the material covered in this text. The coverage is at a very introductory level simply to introduce the scope and power of the package. All the exercises appearing at the end of each chapter can be done without Mathcad.

Once the package is installed, all operations begin with the basic screen in Fig. 3.43. The operations must be performed in the sequence appearing in Fig. 3.44; that is, from left to right and then from top to bottom. For example, if an equation on the second line is to operate on a specific variable, the variable must be defined to the left of or above the equation.

To perform any mathematical calculation, simply click on the screen at any convenient point to establish a crosshair on the display (the location of the first entry). Then type in the mathematical operation such as

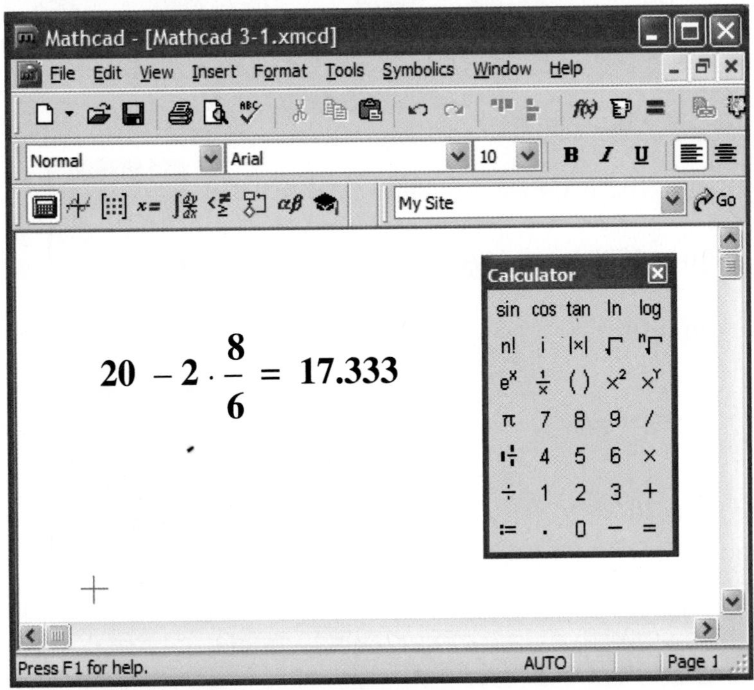

FIG. 3.43
Using Mathcad to perform a basic mathematical operation.

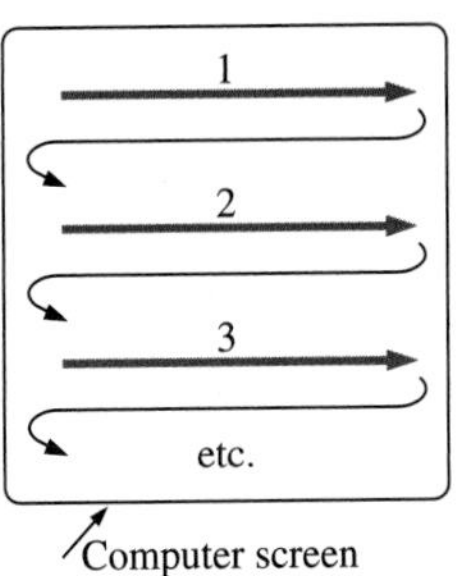

FIG. 3.44
Defining the order of mathematical operations for Mathcad.

20 − 2 · 8/6 as shown in Fig. 3.43; the instant the equal sign is selected, the result, **17.333,** will appear as shown in Fig. 3.43. The multiplication is obtained using the asterisk (*) appearing at the top of the number 8 key (under the SHIFT CONTROL key). The division is set by the / key at the bottom right of the keyboard. The equal sign can be selected from the top right corner of the keyboard. Another option is to apply the sequence **View-Toolbars-Calculator** to obtain the calculator in Fig. 3.43. Then use the calculator to enter the entire expression and the result will appear as soon as the equal sign is selected.

As an example in which variables must be defined, the resistance of a 200-ft length of copper wire with a diameter of 0.01 in. will be determined. First, as shown in Fig. 3.45, the variables for resistivity, length, and diameter must be defined. This is accomplished by first calling for the **Greek** palette through **View-Toolbars-Greek** and selecting the Greek letter *rho* (ρ) followed by a combined **Shift-colon** operation. A colon and an equal sign will appear, after which **10.37** is entered. For all the calculations to follow, the value of ρ has been defined. A left click on the screen removes the rectangular enclosure and places the variable and its value in memory. Proceed in the same way to define the length l and the diameter d. Next, the diameter in mils is defined by multiplying the diameter in inches by 1000, and the area is defined by the diameter in mils squared. Note that m had to be defined to the left of the expression for the area, and the variable d was defined in the line above. The power of 2 was obtained by first selecting the superscript symbol (^) at the top of the number 6 on the keyboard and then entering the number 2 in the Mathcad bracket. Or you can simply type the letter m and choose x^2 from the **Calculator** palette. In fact, all the operations of multiplication,

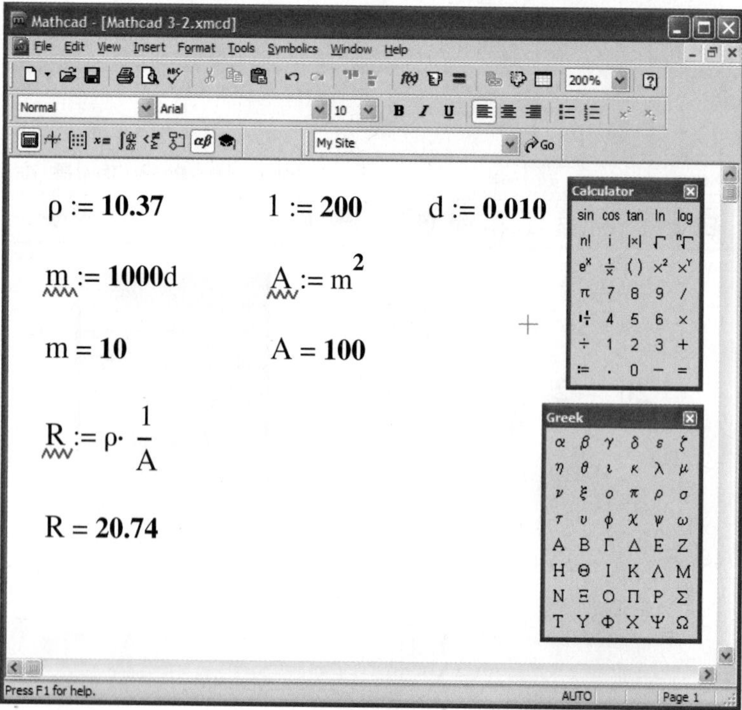

FIG. 3.45
Using Mathcad to calculate the resistance of a copper conductor.

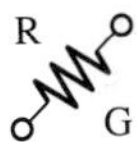

division, etc., required to determine the resistance R can be lifted from the **Calculator** palette.

On the next line in Fig. 3.45, the values of m and A were calculated by simply typing in m followed by the keyboard equal sign. Finally, the equation for the resistance R is defined in terms of the variables, and the result is obtained. The true value of developing mathematical equations in the above sequence is that you can place the program in memory and, when the need arises, call it up and change a variable or two—the result will appear immediately. There is no need to reenter all the definitions—just change the numerical value.

The following chapters have additional examples of how using Mathcad to perform calculation can save time and ensure accuracy.

PROBLEMS

SECTION 3.2 Resistance: Circular Wires

1. Convert the following to mils:
   a. 0.5 in. b. 0.02 in. c. 1/4 in.
   d. 1 in. e. 0.02 ft f. 0.1 cm

2. Calculate the area in circular mils (CM) of wires having the following diameters:
   a. 30 mils b. 0.016 in. c. 1/8 in.
   d. 1 cm e. 0.02 ft f. 0.0042 m

3. The area in circular mils is
   a. 1600 CM b. 820 CM c. 40,000 CM
   d. 625 CM e. 6.25 CM f. 100 CM

   What is the diameter of each wire in inches?

4. What is the resistance of a copper wire 200 ft long and 0.01 in. in diameter ($T = 20°C$)?

5. Find the resistance of a silver wire 50 yd long and 4 mils in diameter ($T = 20°C$).

6. a. What is the area in circular mils of an aluminum conductor that is 80 ft long with a resistance of 2.5 Ω?
   b. What is its diameter in inches?

7. A 2.2 Ω resistor is to be made of nichrome wire. If the available wire is 1/32 in. in diameter, how much wire is required?

8. a. What is the diameter in inches of a copper wire that has a resistance of 3.3 Ω and is as long as a football field (100 yd) ($T = 20°C$)?
   b. Without working out the numerical solution, determine whether the area of an aluminum wire will be smaller or larger than that of the copper wire. Explain.
   c. Repeat (b) for a silver wire.

9. In Fig. 3.46, three conductors of different materials are presented.
   a. Without working out the numerical solution, which do you think has the most resistance between its ends? Explain.
   b. Find the resistance of each section and compare your answer with the result of (a) ($T = 20°C$).

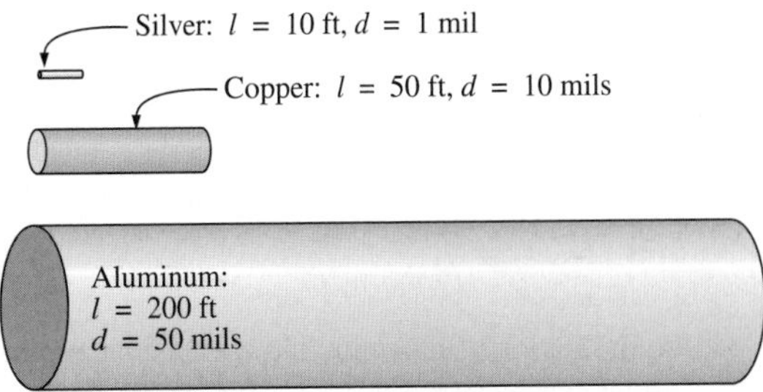

FIG. 3.46
Problem 9.

10. A wire 1000 ft long has a resistance of 0.5 kΩ and an area of 94 CM. Of what material is the wire made ($T = 20°C$)?

*11. a. What is the resistance of a copper bus-bar for a high-rise building with the dimensions shown ($T = 20°C$) in Fig. 3.47?
   b. Repeat (a) for aluminum and compare the results.

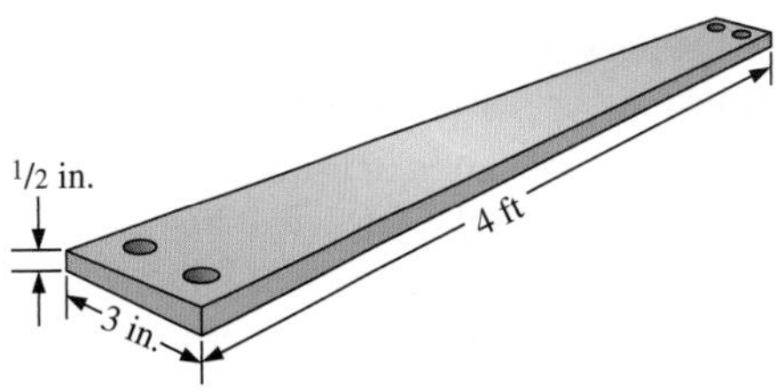

FIG. 3.47
Problem 11.

12. Determine the increase in resistance of a copper conductor if the area is reduced by a factor of 4 and the length is doubled. The original resistance was 0.2 Ω. The temperature remains fixed.

*13. What is the new resistance level of a copper wire if the length is changed from 200 ft to 100 yd, the area is changed from 40,000 CM to 0.04 in.^{2}, and the original resistance was 800 mΩ?

SECTION 3.3 Wire Tables

14. a. Using Table 3.2, find the resistance of 450 ft of #11 and #14 AWG wires.
    b. Compare the resistances of the two wires.
    c. Compare the areas of the two wires.

15. a. Using Table 3.2, find the resistance of 1800 ft of #8 and #18 AWG wires.
    b. Compare the resistances of the two wires.
    c. Compare the areas of the two wires.

16. a. For the system in Fig. 3.48, the resistance of each line cannot exceed 0.006 Ω, and the maximum current drawn by the load is 110 A. What minimum size gage wire should be used?
    b. Repeat (a) for a maximum resistance of 0.003 Ω, $d = 30$ ft, and a maximum current of 110 A.

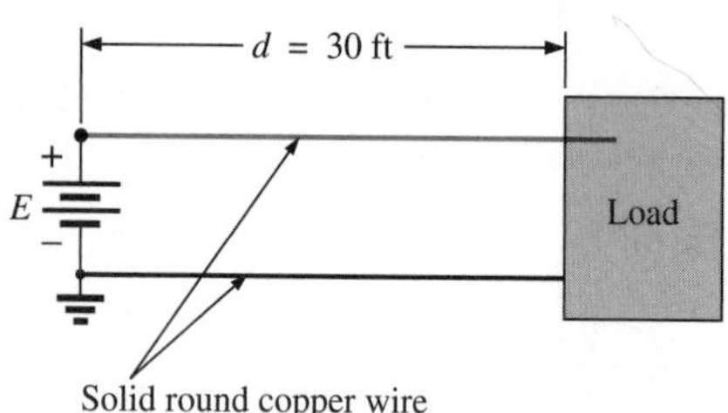

FIG. 3.48
Problem 16.

*17. a. From Table 3.2, determine the maximum permissible current density (A/CM) for an AWG #0000 wire.
    b. Convert the result of (a) to A/in.^{2}
    c. Using the result of (b), determine the cross-sectional area required to carry a current of 5000 A.

SECTION 3.4 Resistance: Metric Units

18. Using metric units, determine the length of a copper wire that has a resistance of 0.2 Ω and a diameter of 1/12 in.

19. Repeat Problem 11 using metric units; that is, convert the given dimensions to metric units before determining the resistance.

20. If the sheet resistance of a tin oxide sample is 100 Ω, what is the thickness of the oxide layer?

21. Determine the width of a carbon resistor having a sheet resistance of 15 0 Ω if the length is 1/2 in. and the resistance is 500 Ω.

*22. Derive the conversion factor between ρ (CM-Ω/ft) and ρ (Ω-cm) by
    a. Solving for ρ for the wire in Fig. 3.49 in CM-Ω/ft.
    b. Solving for ρ for the same wire in Fig. 3.49 in Ω-cm by making the necessary conversions.
    c. Use the equation $\rho_2 = k\rho_1$ to determine the conversion factor k if ρ_1 is the solution of part (a) and ρ_2 the solution of part (b).

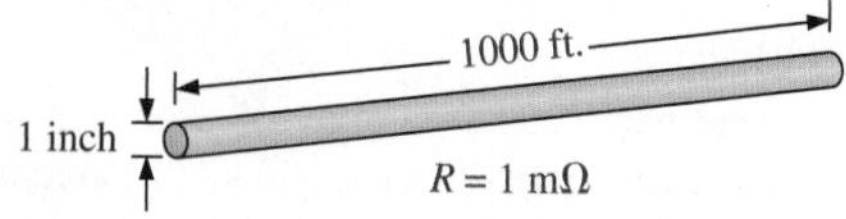

FIG. 3.49
Problem 22.

SECTION 3.5 Temperature Effects

23. The resistance of a copper wire is 2 Ω at 10°C. What is its resistance at 80°C?

24. The resistance of an aluminum bus-bar is 0.02 Ω at 0°C. What is its resistance at 100°C?

25. The resistance of a copper wire is 4 Ω at 70°F. What is its resistance at 32°F?

26. The resistance of a copper wire is 0.76 Ω at 30°C. What is its resistance at −40°C?

27. If the resistance of a silver wire is 0.04 Ω at −30°C, what is its resistance at 32°F?

*28. a. The resistance of a copper wire is 0.002 Ω at room temperature (68°F). What is its resistance at 32°F (freezing) and 212°F (boiling)?
    b. For part (a), determine the change in resistance for each 10° change in temperature between room temperature and 212°F.

29. a. The resistance of a copper wire is 1 Ω at 4°C. At what temperature (°C) will it be 1.1 Ω?
    b. At what temperature will it be 0.1 Ω?

*30. a. If the resistance of a 1000-ft length of copper wire is 10 Ω at room temperature (20°C), what will its resistance be at 50 K (Kelvin units) using Eq. (3.8)?
    b. Repeat part (a) for a temperature of 38.65 K. Comment on the results obtained by reviewing the curve of Fig. 3.13.
    c. What is the temperature of absolute zero in Fahrenheit units?

31. a. Verify the value of α_{20} for copper in Table 3.6 by substituting the inferred absolute temperature into Eq. (3.9).
    b. Using Eq. (3.10) find the temperature at which the resistance of a copper conductor will increase to 1 Ω from a level of 0.8 Ω at 20°C.

32. Using Eq. (3.10), find the resistance of a copper wire at 16°C if its resistance at 20°C is 0.4 Ω.

*33. Determine the resistance of a 1000-ft coil of #12 copper wire sitting in the desert at a temperature of 115°F.

34. A 22 Ω wire-wound resistor is rated at +200 PPM for a temperature range of −10°C to +75°C. Determine its resistance at 65°C.

35. A 100 Ω wire-wound resistor is rated at +100 PPM for a temperature range of 0°C to +100°C. Determine its resistance at 50°C.

SECTION 3.6 Superconductors

36. Visit your local library and find a table listing the critical temperatures for a variety of materials. List at least five materials with critical temperatures that are not mentioned in this text. Choose a few materials that have relatively high critical temperatures.

37. Find at least one article on the application of superconductivity in the commercial sector, and write a short summary, including all interesting facts and figures.

*38. Using the required 1 MA/cm^2 density level for integrated circuit manufacturing, determine what the resulting current would be through a #12 house wire. Compare the result obtained with the allowable limit of Table 3.2.

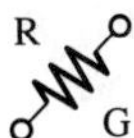

*39. Research the SQUID magnetic field detector and review its basic mode of operation and an application or two.

SECTION 3.7 Types of Resistors

40. a. What is the approximate increase in size from a 1 W to a 2 W carbon resistor?
    b. What is the approximate increase in size from a 1/2 W to a 2 W carbon resistor?
    c. In general, can we conclude that for the same type of resistor, an increase in wattage rating requires an increase in size (volume)? Is it almost a linear relationship? That is, does twice the wattage require an increase in size of 2:1?

41. If the resistance between the outside terminals of a linear potentiometer is 10 kΩ, what is its resistance between the wiper (movable) arm and an outside terminal if the resistance between the wiper arm and the other outside terminal is 3.5 kΩ?

42. If the wiper arm of a linear potentiometer is one-quarter the way around the contact surface, what is the resistance between the wiper arm and each terminal if the total resistance is 2.5 kΩ?

*43. Show the connections required to establish 4 kΩ between the wiper arm and one outside terminal of a 10 kΩ potentiometer while having only zero ohms between the other outside terminal and the wiper arm.

SECTION 3.8 Color Coding and Standard Resistor Values

44. Find the range in which a resistor having the following color bands must exist to satisfy the manufacturer's tolerance:

	1st band	2nd band	3rd band	4th band
a.	green	blue	yellow	gold
b.	red	red	brown	silver
c.	brown	black	brown	—

45. Find the color code for the following 10% resistors:
    a. 120 Ω b. 0.2 Ω
    c. 68 kΩ d. 3.3 MΩ

46. Is there an overlap in coverage between 20% resistors? That is, determine the tolerance range for a 10 Ω 20% resistor and a 15 Ω 20% resistor, and note whether their tolerance ranges overlap.

47. Repeat Problem 46 for 10% resistors of the same value.

48. Find the value of the following surface mount resistors:
    a. 621 b. 333
    c. Q2 d. C6

SECTION 3.9 Conductance

49. Find the conductance of each of the following resistances:
    a. 120 Ω
    b. 4 kΩ
    c. 2.2 MΩ
    d. Compare the three results.

50. Find the conductance of 1000 ft of #12 AWG wire made of
    a. copper
    b. aluminum
    c. iron

*51. The conductance of a wire is 100 S. If the area of the wire is increased by 2/3 and the length is reduced by the same amount, find the new conductance of the wire if the temperature remains fixed.

SECTION 3.10 Ohmmeters

52. How would you check the status of a fuse with an ohmmeter?

53. How would you determine the on and off states of a switch using an ohmmeter?

54. How would you use an ohmmeter to check the status of a light bulb?

SECTION 3.11 Thermistors

*55. a. Find the resistance of the thermistor having the characteristics of Fig. 3.35 at −50°C, 50°C, and 200°C. Note that it is a log scale. If necessary, consult a reference with an expanded log scale.
    b. Does the thermistor have a positive or a negative temperature coefficient?
    c. Is the coefficient a fixed value for the range −100°C to 400°C? Why?
    d. What is the approximate rate of change of ρ with temperature at 100°C?

SECTION 3.12 Photoconductive Cell

56. a. Using the characteristics of Fig. 3.37, determine the resistance of the photoconductive cell at 10 and 100 foot-candles of illumination. As in Problem 55, note that it is a log scale.
    b. Does the cell have a positive or a negative illumination coefficient?
    c. Is the coefficient a fixed value for the range 0.1 to 1000 foot-candles? Why?
    d. What is the approximate rate of change of R with illumination at 10 foot-candles?

SECTION 3.13 Varistors

57. a. Referring to Fig. 3.39(a), find the terminal voltage of the device at 0.5 mA, 1 mA, 3 mA, and 5 mA.
    b. What is the total change in voltage for the indicated range of current levels?
    c. Compare the ratio of maximum to minimum current levels above to the corresponding ratio of voltage levels.

SECTION 3.15 Mathcad

58. Verify the results of Example 3.2 using Mathcad.

59. Verify the results of Example 3.10 using Mathcad.

GLOSSARY

Absolute zero The temperature at which all molecular motion ceases; −273.15°C.

Circular mil (CM) The cross-sectional area of a wire having a diameter of one mil.

Color coding A technique using bands of color to indicate the resistance levels and tolerance of resistors.

Conductance (G) An indication of the relative ease with which current can be established in a material. It is measured in siemens (S).

Cooper effect The "pairing" of electrons as they travel through a medium.

Ductility The property of a material that allows it to be drawn into long, thin wires.

Inferred absolute temperature The temperature through which a straight-line approximation for the actual resistance-versus-temperature curve intersects the temperature axis.

Malleability The property of a material that allows it to be worked into many different shapes.

Negative temperature coefficient of resistance The value revealing that the resistance of a material will decrease with an increase in temperature.

Ohm (Ω) The unit of measurement applied to resistance.

Ohmmeter An instrument for measuring resistance levels.

Photoconductive cell A two-terminal semiconductor device whose terminal resistance is determined by the intensity of the incident light on its exposed surface.

Positive temperature coefficient of resistance The value revealing that the resistance of a material will increase with an increase in temperature.

Potentiometer A three-terminal device through which potential levels can be varied in a linear or nonlinear manner.

PPM/°C Temperature sensitivity of a resistor in parts per million per degree Celsius.

Resistance A measure of the opposition to the flow of charge through a material.

Resistivity (ρ) A constant of proportionality between the resistance of a material and its physical dimensions.

Rheostat An element whose terminal resistance can be varied in a linear or nonlinear manner.

Sheet resistance Defined by ρ/d for thin-film and integrated circuit design.

Superconductor Conductors of electric charge that have for all practical purposes zero ohms.

Thermistor A two-terminal semiconductor device whose resistance is temperature sensitive.

Varistor A voltage-dependent, nonlinear resistor used to suppress high-voltage transients.

Ohm's Law, Power, and Energy

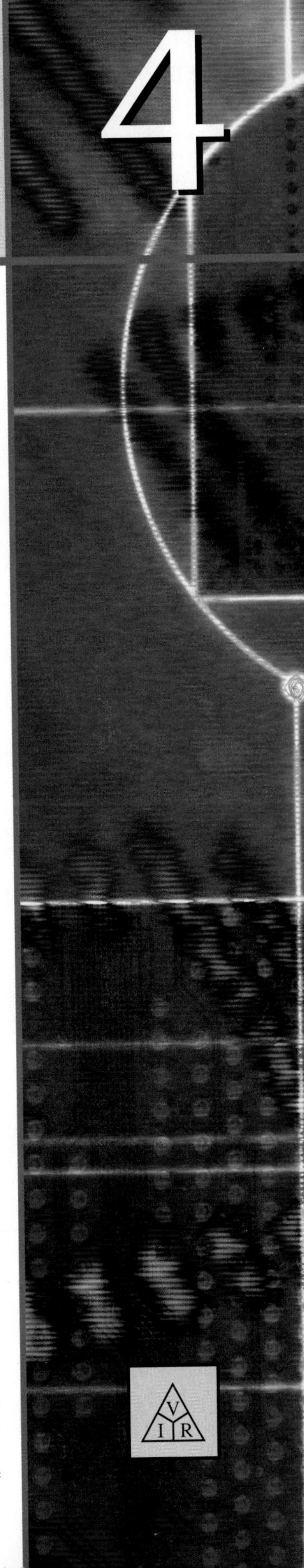

Objectives

- *Understand the importance of Ohm's law and how to apply it to a variety of situations.*
- *Be able to plot Ohm's law and understand how to "read" a graphical plot of voltage versus current.*
- *Become aware of the differences between power and energy levels and how to solve for each.*
- *Understand the power and energy flow of a system, including how the flow impacts on the efficiency of operation.*
- *Become aware of the operation of a variety of fuses and circuit breakers and where each is employed.*

4.1 INTRODUCTION

Now that the three important quantities of an electric circuit have been introduced, this chapter reveals how they are interrelated. The most important equation in the study of electric circuits is introduced, and various other equations that allow us to find power and energy levels are discussed in detail. It is the first chapter where we tie things together and develop a feeling for the way an electric circuit behaves and what affects its response. For the first time, the data provided on the labels of household appliances and the manner in which your electric bill is calculated will have some meaning. It is indeed a chapter that should open your eyes to a wide array of past experiences with electrical systems.

4.2 OHM'S LAW

As mentioned above, the first equation to be described is without question one of the most important to be learned in this field. It is not particularly difficult mathematically, but it is very powerful because it can be applied to any network in any time frame. That is, it is applicable to dc circuits, ac circuits, digital and microwave circuits, and, in fact, any type of applied signal. In addition, it can be applied over a period of time or for instantaneous responses. The equation can be derived directly from the following basic equation for all physical systems:

$$\text{Effect} = \frac{\text{cause}}{\text{opposition}} \tag{4.1}$$

Every conversion of energy from one form to another can be related to this equation. In electric circuits, the *effect* we are trying to establish is the flow of charge, or *current.* The *potential difference,* or voltage, between two points is the *cause* ("pressure"), and the opposition is the *resistance* encountered.

An excellent analogy for the simplest of electrical circuits is the water in a hose connected to a pressure valve, as discussed in Chapter 2. Think of the electrons in the copper wire as the water in the hose, the pressure valve as the applied voltage, and the size of the hose as the factor that determines the resistance. If the pressure valve is closed, the water simply sits in the

hose without a general direction, much like the oscillating electrons in a conductor without an applied voltage. When we open the pressure valve, water will flow through the hose much like the electrons in a copper wire when the voltage is applied. In other words, the absence of the "pressure" in one case and the voltage in the other simply results in a system without direction or reaction. The rate at which the water will flow in the hose is a function of the size of the hose. A hose with a very small diameter will limit the rate at which water can flow through the hose, just as a copper wire with a small diameter will have a high resistance and will limit the current.

In summary, therefore, the absence of an applied "pressure" such as voltage in an electric circuit will result in no reaction in the system and no current in the electric circuit. Current is a reaction to the applied voltage and not the factor that gets the system in motion. To continue with the analogy, the more the pressure at the spigot, the greater the rate of water flow through the hose, just as applying a higher voltage to the same circuit results in a higher current.

Substituting the terms introduced above into Eq. (4.1) results in

$$\text{Current} = \frac{\text{potential difference}}{\text{resistance}}$$

and

$$I = \frac{E}{R} \quad \text{(amperes, A)} \qquad \textbf{(4.2)}$$

FIG. 4.1
George Simon Ohm.
Courtesy of the Smithsonian Institution. Photo No. 51,145.

German (Erlangen, Cologne)
(1789–1854)
Physicist and Mathematician
Professor of Physics, University of Cologne

In 1827, developed one of the most important laws of electric circuits: *Ohm's law.* When the law was first introduced, the supporting documentation was considered lacking and foolish, causing him to lose his teaching position and search for a living doing odd jobs and some tutoring. It took some 22 years for his work to be recognized as a major contribution to the field. He was then awarded a chair at the University of Munich and received the Copley Medal of the Royal Society in 1841. His research also extended into the areas of molecular physics, acoustics, and telegraphic communication.

Eq. (4.2) is known as **Ohm's law** in honor of Georg Simon Ohm (Fig. 4.1). The law states that for a fixed resistance, the greater the voltage (or pressure) across a resistor, the more the current, and the more the resistance for the same voltage, the less the current. In other words, the current is proportional to the applied voltage and inversely proportional to the resistance.

By simple mathematical manipulations, the voltage and resistance can be found in terms of the other two quantities:

$$E = IR \quad \text{(volts, V)} \qquad \textbf{(4.3)}$$

and

$$R = \frac{E}{I} \quad \text{(ohms, } \Omega\text{)} \qquad \textbf{(4.4)}$$

All the quantities of Eq. (4.2) appear in the simple electrical circuit in Fig. 4.2. A resistor has been connected directly across a battery to establish a current through the resistor and supply. Note that

the symbol E is applied to all sources of voltage

and

the symbol V is applied to all voltage drops across components of the network.

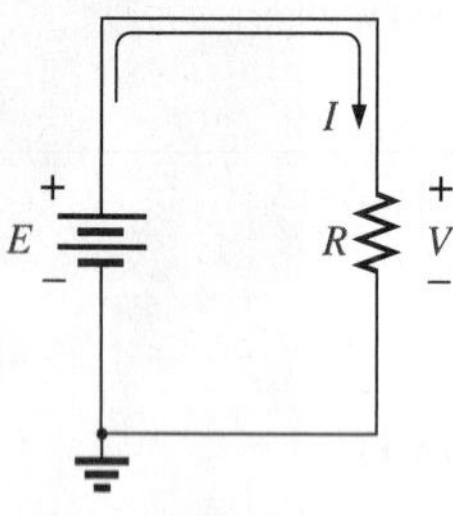

FIG. 4.2
Basic circuit.

Both are measured in volts and can be applied interchangeably in Eqs. (4.2) through (4.4).

Since the battery in Fig. 4.2 is connected directly across the resistor, the voltage V_R across the resistor must be equal to that of the supply. Applying Ohm's law:

$$I = \frac{V_R}{R} = \frac{E}{R}$$

Note in Fig. 4.2 that the voltage source "pressures" current (conventional current) in a direction that leaves the positive terminal of the supply and

returns to the negative terminal of the battery. *This will always be the case for single-source networks.* (The effect of more than one source in the same network is investigated in a later chapter.) Note also that the current enters the positive terminal and leaves the negative terminal for the load resistor R.

For any resistor, in any network, the direction of current through a resistor will define the polarity of the voltage drop across the resistor

as shown in Fig. 4.3 for two directions of current. Polarities as established by current direction become increasingly important in the analyses to follow.

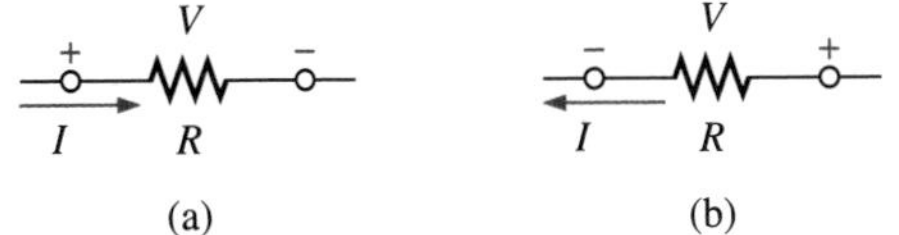

FIG. 4.3
Defining polarities.

EXAMPLE 4.1 Determine the current resulting from the application of a 9 V battery across a network with a resistance of 2.2 Ω.

Solution: Eq. (4.2):

$$I = \frac{V_R}{R} = \frac{E}{R} = \frac{9\text{ V}}{2.2\ \Omega} = \mathbf{4.09\ A}$$

EXAMPLE 4.2 Calculate the resistance of a 60 W bulb if a current of 500 mA results from an applied voltage of 120 V.

Solution: Eq. (4.4):

$$R = \frac{V_R}{I} = \frac{E}{I} = \frac{120\text{ V}}{500 \times 10^{-3}\text{ A}} = \mathbf{240\ \Omega}$$

EXAMPLE 4.3 Calculate the current through the 2 kΩ resistor in Fig. 4.4 if the voltage drop across it is 16 V.

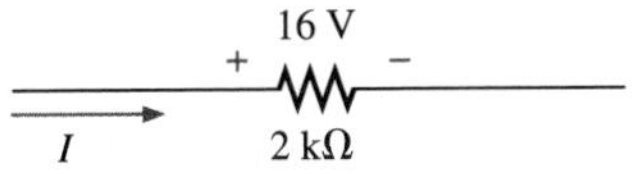

Solution:

$$I = \frac{V}{R} = \frac{16\text{ V}}{2 \times 10^3\ \Omega} = \mathbf{8\ mA}$$

16 V
\+ −
I
2 kΩ

FIG. 4.4
Example 4.3.

EXAMPLE 4.4 Calculate the voltage that must be applied across the soldering iron in Fig. 4.5 to establish a current of 1.5 A through the iron if its internal resistance is 80 Ω.

Solution:

$$E = V_R = IR = (1.5\text{ A})(80\ \Omega) = \mathbf{120\ V}$$

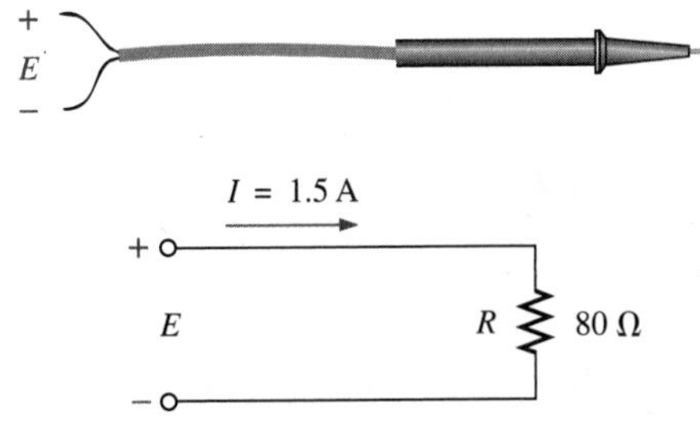

FIG. 4.5
Example 4.4.

In a number of the examples in this chapter, such as Example 4.4, the voltage applied is actually that obtained from an ac outlet in the home, office, or laboratory. This approach was used to provide an opportunity for the student to relate to real-world situations as soon as possible and to demonstrate that a number of the equations derived in this chapter are applicable to ac networks also. Chapter 13 will provide a direct relationship between ac and dc voltages that permits the mathematical substitutions used in this chapter. In other words, don't be concerned that some of the voltages and currents appearing in the examples of this chapter are actually ac voltages, because the equations for dc networks have exactly the same format, and all the solutions will be correct.

4.3 PLOTTING OHM'S LAW

Graphs, characteristics, plots, and the like, play an important role in every technical field as a mode through which the broad picture of the behavior or response of a system can be conveniently displayed. It is therefore critical to develop the skills necessary both to read data and to plot them in such a manner that they can be interpreted easily.

For most sets of characteristics of electronic devices, the current is represented by the vertical axis (ordinate), and the voltage by the horizontal axis (abscissa), as shown in Fig. 4.6. First note that the vertical axis is in amperes and the horizontal axis is in volts. For some plots, I may be in milliamperes (mA), microamperes (μA), or whatever is appropriate for the range of interest. The same is true for the levels of voltage on the horizontal axis. Note also that the chosen parameters require that the spacing between numerical values of the vertical axis be different from that of the horizontal axis. The linear (straight-line) graph reveals that the resistance is not changing with current or voltage level; rather, it is a fixed quantity throughout. The current direction and the voltage polarity appearing at the top of Fig. 4.6 are the defined direction and polarity for the provided plot. If the current direction is opposite to the defined direction, the region below the horizontal axis is the region of interest for the current I. If the voltage polarity is opposite to that defined, the region to the left of the current axis is the region of interest. For the standard fixed resistor, the first quadrant, or region, of Fig. 4.6 is the only region of interest. However, you will encounter many devices in your electronics courses that use the other quadrants of a graph.

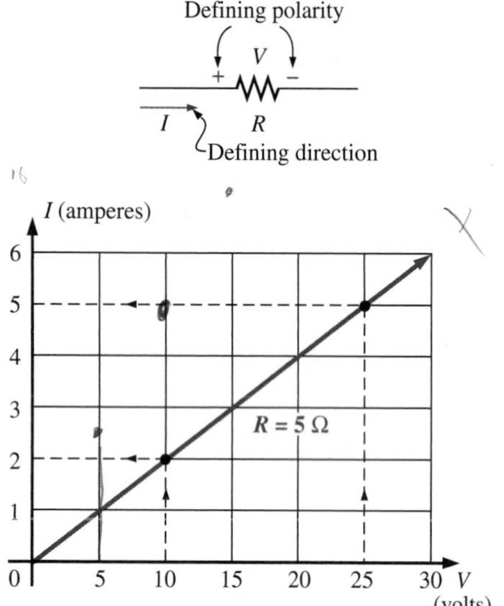

FIG. 4.6
Plotting Ohm's law.

Once a graph such as Fig. 4.6 is developed, the current or voltage at any level can be found from the other quantity by simply using the resulting plot. For instance, at $V = 25$ V, if a vertical line is drawn on Fig. 4.6 to the curve as shown, the resulting current can be found by drawing a horizontal line over to the current axis, where a result of 5 A is obtained. Similarly, at $V = 10$ V, a vertical line to the plot and a horizontal line to the current axis results in a current of 2 A, as determined by Ohm's law.

If the resistance of a plot is unknown, it can be determined at any point on the plot since a straight line indicates a fixed resistance. At any point on the plot, find the resulting current and voltage, and simply substitute into the following equation:

$$R_{dc} = \frac{V}{I} \tag{4.5}$$

To test Eq. (4.5), consider a point on the plot where $V = 20$ V and $I = 4$ A. The resulting resistance is $R_{dc} = 20\text{ V}/I = 20\text{ V}/4\text{ A} = 5\ \Omega$. For comparison purposes, a 1 Ω and a 10 Ω resistor were plotted on the graph in Fig. 4.7. Note that the less the resistance, the steeper the slope (closer to the vertical axis) of the curve.

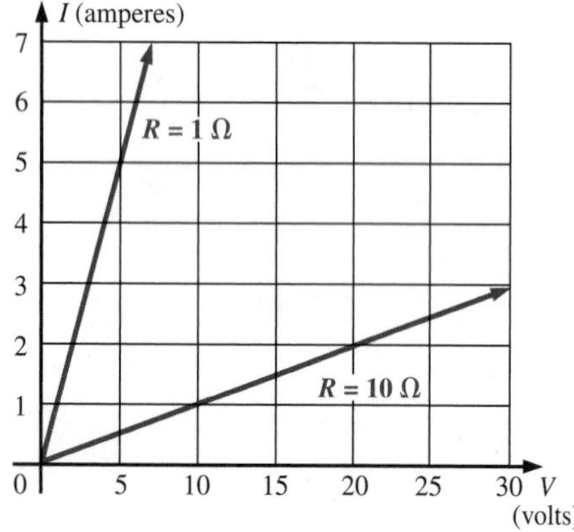

FIG. 4.7
Demonstrating on an I-V plot that the less the resistance, the steeper the slope.

If we write Ohm's law in the following manner and relate it to the basic straight-line equation

$$\begin{array}{ccccccc} I & = & \dfrac{1}{R} & \cdot & E & + & 0 \\ \downarrow & & \downarrow & & \downarrow & & \downarrow \\ y & = & m & \cdot & x & + & b \end{array}$$

we find that the slope is equal to 1 divided by the resistance value, as indicated by the following:

$$m = \text{slope} = \frac{\Delta y}{\Delta x} = \frac{\Delta I}{\Delta V} = \frac{1}{R} \tag{4.6}$$

where Δ signifies a small, finite change in the variable.

Eq. (4.6) reveals that the greater the resistance, the less the slope. If written in the following form, Eq. (4.6) can be used to determine the resistance from the linear curve:

$$R = \frac{\Delta V}{\Delta I} \quad \text{(ohms)} \tag{4.7}$$

The equation states that by choosing a particular ΔV (or ΔI), you can obtain the corresponding ΔI (or ΔV, respectively) from the graph, as shown in Fig. 4.8, and then determine the resistance. If the plot is a straight line, Eq. (4.7) will provide the same result no matter where the equation is applied. However, if the plot curves at all, the resistance will change.

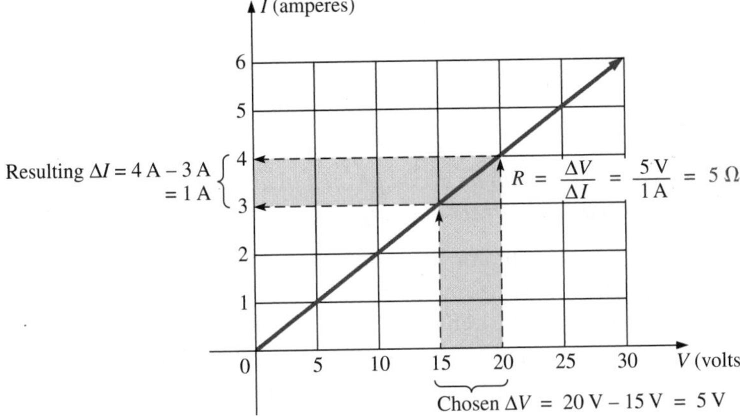

FIG. 4.8
Applying Eq. (4.7)

EXAMPLE 4.5 Determine the resistance associated with the curve in Fig. 4.9 using Eqs. (4.5) and (4.7), and compare results.

Solution: At $V = 6$ V, $I = 3$ mA, and

$$R_{dc} = \frac{V}{I} = \frac{6\text{ V}}{3\text{ mA}} = \mathbf{2\ k\Omega}$$

For the interval between 6 V and 8 V,

$$R = \frac{\Delta V}{\Delta I} = \frac{2\text{ V}}{1\text{ mA}} = \mathbf{2\ k\Omega}$$

The results are equivalent.

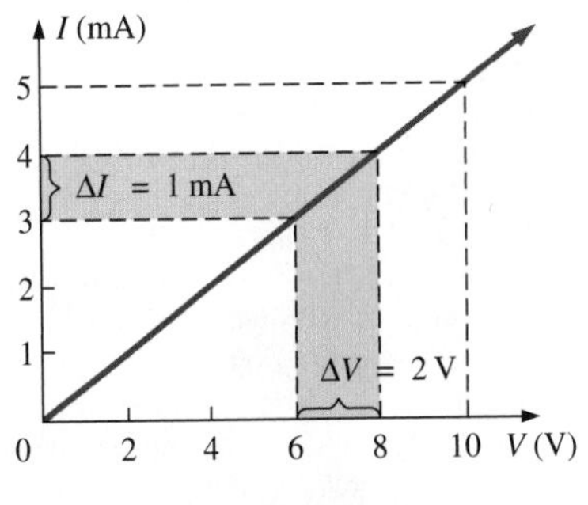

FIG. 4.9
Example 4.5.

Before leaving the subject, let us first investigate the characteristics of a very important semiconductor device called the **diode,** which will be examined in detail in basic electronics courses. This device ideally acts

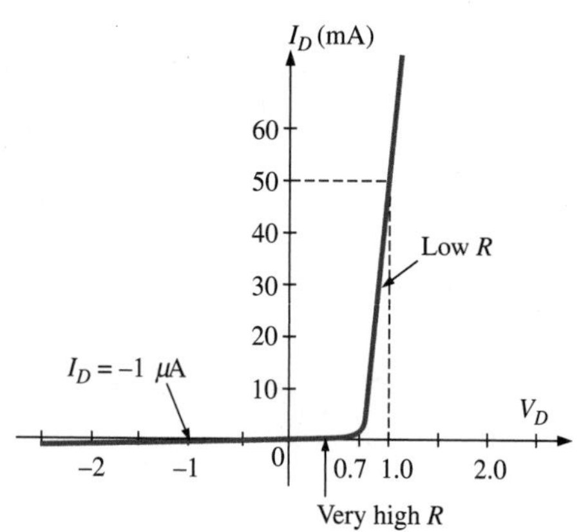

FIG. 4.10
Semiconductor diode characteristics.

as a low-resistance path to current in one direction and a high-resistance path to current in the reverse direction, much like a switch that passes current in only one direction. A typical set of characteristics appears in Fig. 4.10. Without any mathematical calculations, the closeness of the characteristic to the voltage axis for negative values of applied voltage indicates that this is the low-conductance (high resistance, switch opened) region. Note that this region extends to approximately 0.7 V positive. However, for values of applied voltage greater than 0.7 V, the vertical rise in the characteristics indicates a high-conductivity (low resistance, switch closed) region. Application of Ohm's law will now verify the above conclusions.

At $V_D = +1$ V,

$$R_{\text{diode}} = \frac{V_D}{I_D} = \frac{1\text{ V}}{50\text{ mA}} = \frac{1\text{ V}}{50 \times 10^{-3}\text{ A}} = 20\ \Omega$$

(a relatively low value for most applications)

At $V_D = -1$ V,

$$R_{\text{diode}} = \frac{V_D}{I_D} = \frac{1\text{V}}{1\ \mu\text{A}} = 1\text{ M}\Omega$$

(which is often represented by an open-circuit equivalent)

4.4 POWER

In general,

the term power is applied to provide an indication of how much work (energy conversion) can be accomplished in a specified amount of time; that is, power is a rate of doing work.

For instance, a large motor has more power than a smaller motor because it has the ability to convert more electrical energy into mechanical energy in the same period of time. Since energy is measured in joules (J) and time in seconds (s), power is measured in joules/second (J/s). The electrical unit of measurement for power is the watt (W) defined by

$$\boxed{1\text{ watt (W)} = 1\text{ joule/second (J/s)}} \qquad \textbf{(4.8)}$$

In equation form, power is determined by

$$\boxed{P = \frac{W}{t}} \qquad \text{(watts, W, or joules/second, J/s)} \qquad \textbf{(4.9)}$$

with the **energy** (W) measured in joules and the time t in seconds.

The unit of measurement—the watt—is derived from the surname of James Watt (Fig. 4.11), who was instrumental in establishing the standards for power measurements. He introduced the **horsepower** (hp) as a measure of the average power of a strong dray horse over a full working day. It is approximately 50% more than can be expected from the average horse. The horsepower and watt are related in the following manner:

$$\boxed{1\text{ horsepower} \cong 746\text{ watts}}$$

FIG. 4.11
James Watt.
Courtesy of the Smithsonian Institution. Photo No. 30,391.

Scottish (Greenock, Birmingham)
(1736–1819)
Instrument Maker and Inventor
Elected Fellow of the Royal Society of London in 1785

In 1757, at the age of 21, used his innovative talents to design mathematical instruments such as the *quadrant, compass,* and various *scales.* In 1765, introduced the use of a separate *condenser* to increase the efficiency of steam engines. In the following years, he received a number of important patents on improved engine design, including a rotary motion for the steam engine (versus the reciprocating action) and a double-action engine, in which the piston pulled as well as pushed in its cyclic motion. Introduced the term **horsepower** as the average power of a strong dray (small cart) horse over a full working day.

The power delivered to, or absorbed by, an electrical device or system can be found in terms of the current and voltage by first substituting Eq. (2.5) into Eq. (4.9):

$$P = \frac{W}{t} = \frac{QV}{t} = V\frac{Q}{t}$$

But

$$I = \frac{Q}{t}$$

so that

$$\boxed{P = VI} \qquad \text{(watts, W)} \tag{4.10}$$

By direct substitution of Ohm's law, the equation for power can be obtained in two other forms:

$$P = VI = V\left(\frac{V}{R}\right)$$

and

$$\boxed{P = \frac{V^2}{R}} \qquad \text{(watts, W)} \tag{4.11}$$

or

$$P = VI = (IR)I$$

and

$$\boxed{P = I^2R} \qquad \text{(watts, W)} \tag{4.12}$$

The result is that the power absorbed by the resistor in Fig. 4.12 can be found directly, depending on the information available. In other words, if the current and resistance are known, it pays to use Eq. (4.12) directly, and if V and I are known, use of Eq. (4.10) is appropriate. It saves having to apply Ohm's law before determining the power.

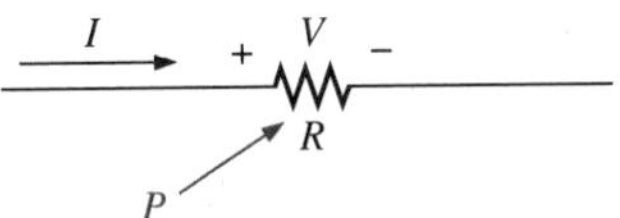

FIG. 4.12
Defining the power to a resistive element.

The power supplied by a battery can be determined by simply inserting the supply voltage into Eq. (4.10) to produce

$$\boxed{P = EI} \qquad \text{(watts, W)} \tag{4.13}$$

The importance of Eq. (4.13) cannot be overstated. It clearly states the following:

The power associated with any supply is not simply a function of the supply voltage. It is determined by the product of the supply voltage and its maximum current rating.

The simplest example is the car battery—a battery that is large, difficult to handle, and relatively heavy. It is only 12 V, a voltage level that could be supplied by a battery slightly larger than the small 9 V portable radio battery. However, to provide the **power** necessary to start a car, the battery must be able to supply the high surge current required at starting—a component that requires size and mass. In total, therefore, it is not the voltage or current rating of a supply that determines its power capabilities; it is the product of the two.

Throughout the text, the abbreviation for energy (W) can be distinguished from that for the watt (W) because the one for energy is in italics while the one for watt is in roman. In fact, all variables in the dc section appear in italics while the units appear in roman.

EXAMPLE 4.6 Find the power delivered to the dc motor of Fig. 4.13.

Solution: $P = EI = (120\text{ V})(5\text{ A}) = 600\text{ W} = \mathbf{0.6\ kW}$

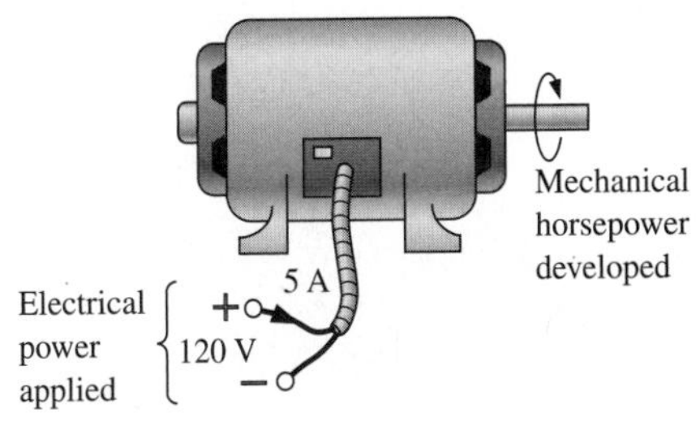

FIG. 4.13
Example 4.6.

EXAMPLE 4.7 What is the power dissipated by a 5 Ω resistor if the current is 4 A?

Solution:

$$P = I^2R = (4\text{ A})^2(5\ \Omega) = \mathbf{80\ W}$$

EXAMPLE 4.8 The *I-V* characteristics of a light bulb are provided in Fig. 4.14. Note the nonlinearity of the curve, indicating a wide range in resistance of the bulb with applied voltage. If the rated voltage is 120 V, find the wattage rating of the bulb. Also calculate the resistance of the bulb under rated conditions.

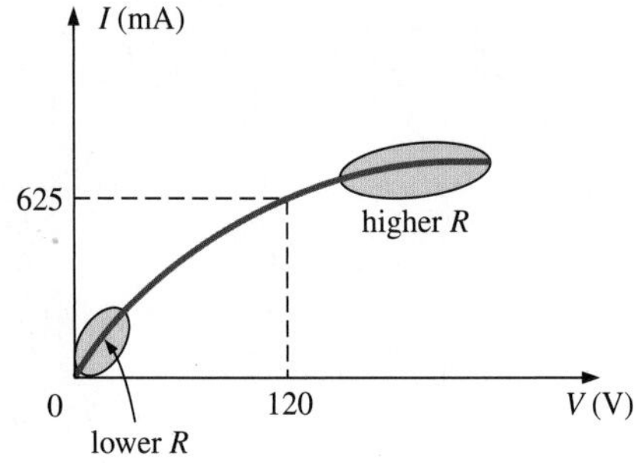

FIG. 4.14
The nonlinear I-V characteristics of a 75 W light bulb (Example 4.8).

Solution: At 120 V,

$$I = 0.625\text{ A}$$

and

$$P = VI = (120\text{ V})(0.625\text{ A}) = \mathbf{75\ W}$$

At 120 V,

$$R = \frac{V}{I} = \frac{120\text{ V}}{0.625\text{ A}} = \mathbf{192\ \Omega}$$

Sometimes the power is given and the current or voltage must be determined. Through algebraic manipulations, an equation for each variable is derived as follows:

$$P = I^2R \Rightarrow I^2 = \frac{P}{R}$$

and

$$\boxed{I = \sqrt{\frac{P}{R}}} \qquad \text{(amperes, A)} \qquad \mathbf{(4.14)}$$

$$P = \frac{V^2}{R} \Rightarrow V^2 = PR$$

and

$$\boxed{V = \sqrt{PR}} \qquad \text{(volts, V)} \qquad \mathbf{(4.15)}$$

EXAMPLE 4.9 Determine the current through a 5 kΩ resistor when the power dissipated by the element is 20 mW.

Solution: Eq. (4.14):

$$I = \sqrt{\frac{P}{R}} = \sqrt{\frac{20 \times 10^{-3}\text{ W}}{5 \times 10^3\ \Omega}} = \sqrt{4 \times 10^{-6}} = 2 \times 10^{-3}\text{ A}$$
$$= \mathbf{2\ mA}$$

4.5 ENERGY

For power, which is the rate of doing work, to produce an energy conversion of any form, it must be *used over a period of time.* For example, a motor may have the horsepower to run a heavy load, but unless the motor is *used* over a period of time, there will be no energy conversion. In addition, the longer the motor is used to drive the load, the greater will be the energy expended.

The **energy** (W) lost or gained by any system is therefore determined by

$$\boxed{W = Pt} \qquad \text{(wattseconds, Ws, or joules)} \qquad \textbf{(4.16)}$$

Since power is measured in watts (or joules per second) and time in seconds, the unit of energy is the *wattsecond* or *joule* (note Fig. 4.15). The wattsecond, however, is too small a quantity for most practical purposes, so the *watthour* (Wh) and the *kilowatthour* (kWh) are defined, as follows:

$$\boxed{\text{Energy (Wh)} = \text{power (W)} \times \text{time (h)}} \qquad \textbf{(4.17)}$$

$$\boxed{\text{Energy (kWh)} = \frac{\text{power (W)} \times \text{time (h)}}{1000}} \qquad \textbf{(4.18)}$$

Note that the energy in kilowatthours is simply the energy in watthours divided by 1000. To develop some sense for the kilowatthour energy level, consider that *1 kWh is the energy dissipated by a 100 W bulb in 10 h.*

The **kilowatthour meter** is an instrument for measuring the energy supplied to the residential or commercial user of electricity. It is normally connected directly to the lines at a point just prior to entering the power distribution panel of the building. A typical set of dials is shown in Fig. 4.16, along with a photograph of an analog kilowatthour meter. As indicated, each power of ten below a dial is in kilowatthours. The more rapidly the aluminum disc rotates, the greater the energy demand. The dials are connected through a set of gears to the rotation of this disc. A solid-state digital meter with an extended range of capabilities also appears in Fig. 4.16.

FIG. 4.15
James Prescott Joule.

British (Salford, Manchester)
(1818–89)
Physicist
Honorary Doctorates from the Universities of Dublin and Oxford

Contributed to the important fundamental *law of conservation of energy* by establishing that various forms of energy, whether electrical, mechanical, or heat, are in the same family and can be exchanged from one form to another. In 1841 introduced *Joule's law,* which stated that the heat developed by electric current in a wire is proportional to the product of the current squared and the resistance of the wire (I^2R). He further determined that the heat emitted was equivalent to the power absorbed and therefore heat is a form of energy.

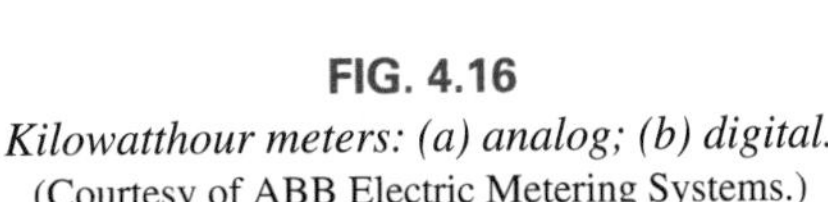

(a)

(b)

FIG. 4.16
Kilowatthour meters: (a) analog; (b) digital.
(Courtesy of ABB Electric Metering Systems.)

EXAMPLE 4.10 For the dial positions in Fig. 4.16(a), calculate the electricity bill if the previous reading was 4650 kWh and the average cost in your area is 9¢ per kilowatthour.

Solution:

$$5360 \text{ kWh} - 4650 \text{ kWh} = 710 \text{ kWh used}$$

$$710 \cancel{\text{kWh}}\left(\frac{9¢}{\cancel{\text{kWh}}}\right) = \textbf{\$63.90}$$

EXAMPLE 4.11 How much energy (in kilowatthours) is required to light a 60 W bulb continuously for 1 year (365 days)?

Solution:

$$W = \frac{Pt}{1000} = \frac{(60\text{ W})(24\text{ h/day})(365\text{ days})}{1000} = \frac{525{,}600\text{ Wh}}{1000}$$
$$= \mathbf{525.60\ kWh}$$

EXAMPLE 4.12 How long can a 205 W television set be on before using more than 4 kWh of energy?

Solution:

$$W = \frac{Pt}{1000} \Rightarrow t\,(\text{hours}) = \frac{(W)(1000)}{P} = \frac{(4\text{ kWh})(1000)}{205\text{ W}} = \mathbf{19.51\ h}$$

EXAMPLE 4.13 What is the cost of using a 5 hp motor for 2 h if the rate is 9¢ per kilowatthour?

Solution:

$$W\,(\text{kilowatthours}) = \frac{Pt}{1000} = \frac{(5\text{ hp} \times 746\text{ W/hp})(2\text{ h})}{1000} = 7.46\text{ kWh}$$
$$\text{Cost} = (7.46\text{ kWh})(9¢/\text{kWh}) = \mathbf{67.14¢}$$

EXAMPLE 4.14 What is the total cost of using all of the following at 9¢ per kilowatthour?

A 1200 W toaster for 30 min
Six 50 W bulbs for 4 h
A 400 W washing machine for 45 min
A 4800 W electric clothes dryer for 20 min

Solution:

$$W = \frac{(1200\text{ W})(\tfrac{1}{2}\text{ h}) + (6)(50\text{ W})(4\text{ h}) + (400\text{ W})(\tfrac{3}{4}\text{ h}) + (4800\text{ W})(\tfrac{1}{3}\text{ h})}{1000}$$

$$= \frac{600\text{ Wh} + 1200\text{ Wh} + 300\text{ Wh} + 1600\text{ Wh}}{1000} = \frac{3700\text{ Wh}}{1000}$$

$$W = 3.7\text{ kWh}$$
$$\text{Cost} = (3.7\text{ kWh})(9¢/\text{kWh}) = \mathbf{33.3¢}$$

The chart in Fig. 4.17 shows the national average cost per kilowatthour compared to the kilowatthours used per customer. Note that the cost today is just above the level of 1926, but the average customer uses more than 20 times as much electrical energy in a year. Keep in mind that the chart in Fig. 4.17 is the average cost across the nation. Some states have average rates close to 6¢ per kilowatthour, whereas others approach 18¢ per kilowatthour.

Table 4.1 lists some common household appliances with their typical wattage ratings. You might find it interesting to calculate the cost of

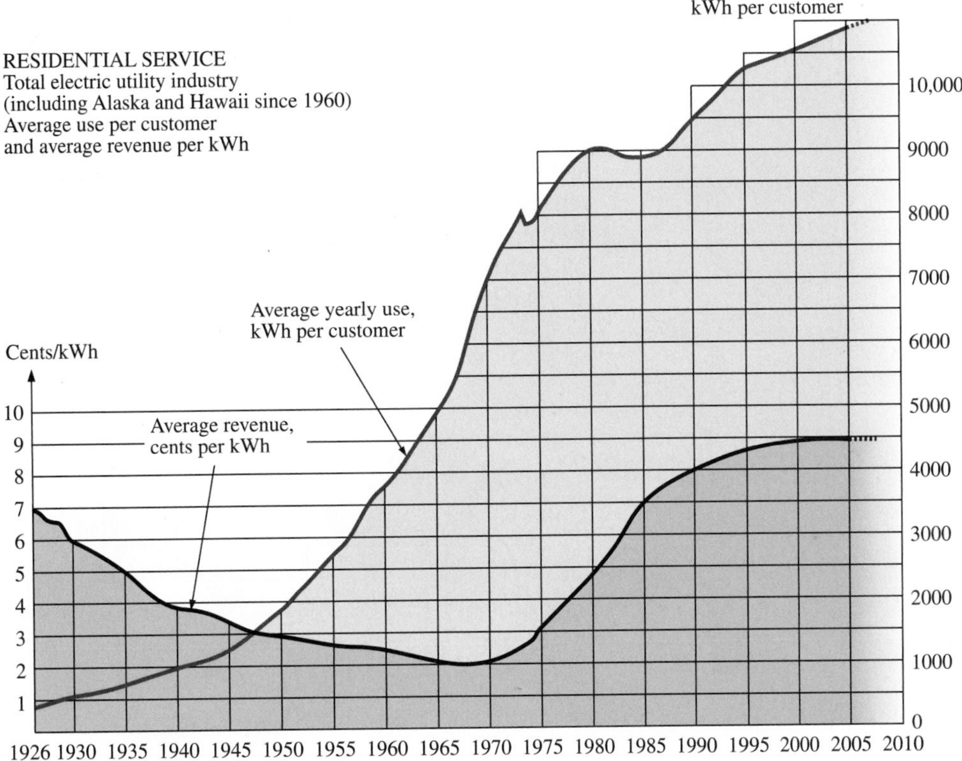

FIG. 4.17

Cost per kWh and average kWh per customer versus time.

(Based on data from Edison Electric Institute.)

TABLE 4.1

Typical wattage ratings of some common household items.

Appliance	Wattage Rating	Appliance	Wattage Rating
Air conditioner	860	Laptop computer:	
Blow dryer	1300	Sleep	<1 W (Typically 0.3 W to 0.5 W)
Cassette player/recorder	5	Normal	10–20 W
Cellular phone:		High	25–35 W
Standby	≅35 mW	Microwave oven	1200
Talk	≅4.3 W	Pager	1–2 mW
Clock	2	Radio	70
Clothes dryer (electric)	4800	Range (self-cleaning)	12,200
Coffee maker	900	Refrigerator (automatic defrost)	1800
Dishwasher	1200	Shaver	15
Fan:		Stereo equipment	110
Portable	90	Sun lamp	280
Window	200	Toaster	1200
Heater	1322	Trash compactor	400
Heating equipment:		TV (color)	200
Furnace fan	320	Videocassette recorder	110
Oil-burner motor	230	Washing machine	500
Iron, dry or steam	1100	Water heater	4500

operating some of these appliances over a period of time using the chart in Fig. 4.17 to find the cost per kilowatthour.

4.6 EFFICIENCY

A flowchart for the energy levels associated with any system that converts energy from one form to another is provided in Fig. 4.18. Note that the output energy level must always be less than the applied energy due to losses and storage within the system. The best one can hope for is that W_{out} and W_{in} are relatively close in magnitude.

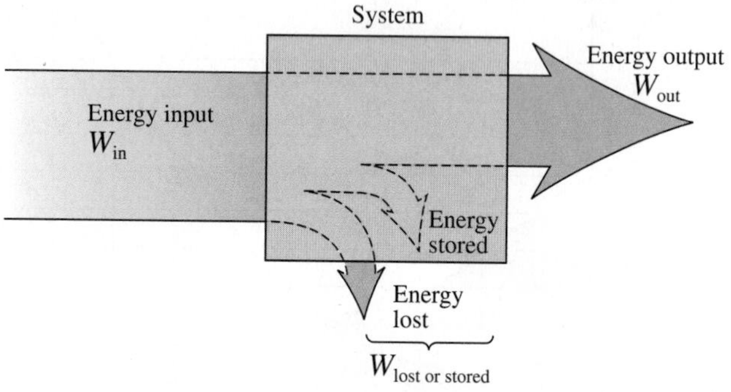

FIG. 4.18
Energy flow through a system.

Conservation of energy requires that

Energy input = energy output + energy lost or stored by the system

Dividing both sides of the relationship by t gives

$$\frac{W_{in}}{t} = \frac{W_{out}}{t} + \frac{W_{\text{lost or stored by the system}}}{t}$$

Since $P = W/t$, we have the following:

$$\boxed{P_i = P_o + P_{\text{lost or stored}}} \qquad \text{(W)} \qquad \textbf{(4.19)}$$

The **efficiency** (η) of the system is then determined by the following equation:

$$\text{Efficiency} = \frac{\text{power output}}{\text{power input}}$$

and

$$\boxed{\eta = \frac{P_o}{P_i}} \qquad \text{(decimal number)} \qquad \textbf{(4.20)}$$

where η (the lowercase Greek letter *eta*) is a decimal number. Expressed as a percentage,

$$\boxed{\eta\% = \frac{P_o}{P_i} \times 100\%} \qquad \text{(percent)} \qquad \textbf{(4.21)}$$

In terms of the input and output energy, the efficiency in percent is given by

$$\eta\% = \frac{W_o}{W_i} \times 100\% \qquad \text{(percent)} \qquad \textbf{(4.22)}$$

The maximum possible efficiency is 100%, which occurs when $P_o = P_i$, or when the power lost or stored in the system is zero. Obviously, the greater the internal losses of the system in generating the necessary output power or energy, the lower the net efficiency.

EXAMPLE 4.15 A 2 hp motor operates at an efficiency of 75%. What is the power input in watts? If the applied voltage is 220 V, what is the input current?

Solution:

$$\eta\% = \frac{P_o}{P_i} \times 100\%$$

$$0.75 = \frac{(2\text{ hp})(746\text{ W/hp})}{P_i}$$

and $$P_i = \frac{1492\text{ W}}{0.75} = \mathbf{1989.33\ W}$$

$$P_i = EI \quad \text{or} \quad I = \frac{P_i}{E} = \frac{1989.33\text{ W}}{220\text{ V}} = \mathbf{9.04\ A}$$

EXAMPLE 4.16 What is the output in horsepower of a motor with an efficiency of 80% and an input current of 8 A at 120 V?

Solution:

$$\eta\% = \frac{P_o}{P_i} \times 100\%$$

$$0.80 = \frac{P_o}{(120\text{ V})(8\text{ A})}$$

and $$P_o = (0.80)(120\text{ V})(8\text{ A}) = 768\text{ W}$$

with $$768\text{ W}\left(\frac{1\text{ hp}}{746\text{ W}}\right) = \mathbf{1.03\ hp}$$

EXAMPLE 4.17 If $\eta = 0.85$, determine the output energy level if the applied energy is 50 J.

Solution:

$$\eta = \frac{W_o}{W_i} \Rightarrow W_o = \eta W_i = (0.85)(50\text{ J}) = \mathbf{42.5\ J}$$

The very basic components of a generating (voltage) system are depicted in Fig. 4.19. The source of mechanical power is a structure such as a paddlewheel that is turned by water rushing over the dam. The gear train ensures that the rotating member of the generator is turning at rated speed. The output voltage must then be fed through a transmission system to the load. For each component of the system, an input and

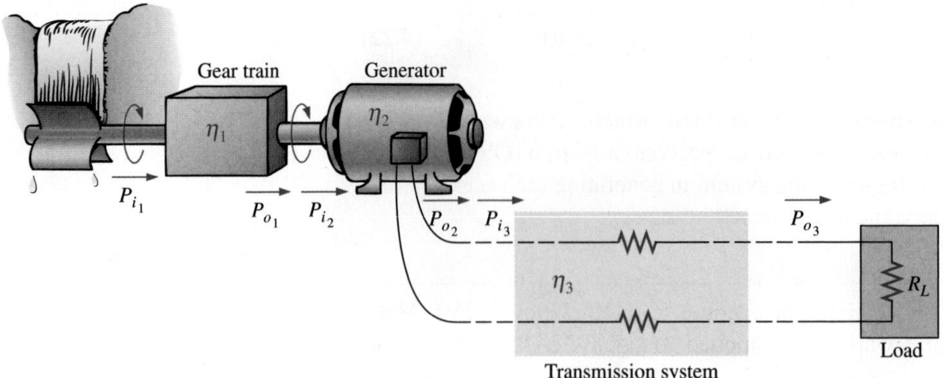

FIG. 4.19
Basic components of a generating system.

output power have been indicated. The efficiency of each system is given by

$$\eta_1 = \frac{P_{o_1}}{P_{i_1}} \qquad \eta_2 = \frac{P_{o_2}}{P_{i_2}} \qquad \eta_3 = \frac{P_{o_3}}{P_{i_3}}$$

If we form the product of these three efficiencies,

$$\eta_1 \cdot \eta_2 \cdot \eta_3 = \frac{P_{o_1}}{P_{i_1}} \cdot \frac{P_{o_2}}{P_{i_2}} \cdot \frac{P_{o_3}}{P_{i_3}} = \frac{P_{o_3}}{P_{i_1}}$$

and substitute the fact that $P_{i_2} = P_{o_1}$ and $P_{i_3} = P_{o_2}$, we find that the quantities indicated above will cancel, resulting in P_{o_3}/P_{i_1}, which is a measure of the efficiency of the entire system.

In general, for the representative cascaded system in Fig. 4.20,

$$\eta_{\text{total}} = \eta_1 \cdot \eta_2 \cdot \eta_3 \cdot \cdot \cdot \eta_n \tag{4.23}$$

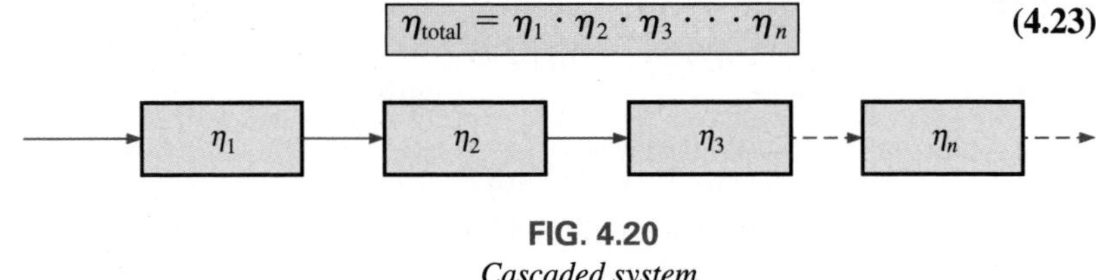

FIG. 4.20
Cascaded system.

EXAMPLE 4.18 Find the overall efficiency of the system in Fig. 4.19 if $\eta_1 = 90\%$, $\eta_2 = 85\%$, and $\eta_3 = 95\%$.

Solution:

$$\eta_T = \eta_1 \cdot \eta_2 \cdot \eta_3 = (0.90)(0.85)(0.95) = 0.727, \text{ or } \mathbf{72.7\%}$$

EXAMPLE 4.19 If the efficiency η_1 drops to 40%, find the new overall efficiency and compare the result with that obtained in Example 4.18.

Solution:

$$\eta_T = \eta_1 \cdot \eta_2 \cdot \eta_3 = (0.40)(0.85)(0.95) = 0.323, \text{ or } \mathbf{32.3\%}$$

Certainly 32.3% is noticeably less than 72.7%. The total efficiency of a cascaded system is therefore determined primarily by the lowest efficiency (weakest link) and is less than (or equal to if the remaining efficiencies are 100%) the least efficient link of the system.

4.7 CIRCUIT BREAKERS, GFCIs, AND FUSES

The incoming power to any large industrial plant, heavy equipment, simple circuit in the home, or to meters used in the laboratory must be limited to ensure that the current through the lines is not above the rated value. Otherwise, the conductors or the electrical or electronic equipment may be damaged, and dangerous side effects such as fire or smoke may result.

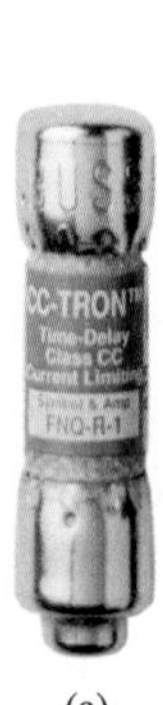

(a)

(b)

(c)

FIG. 4.21
Fuses: (a) CC-TRON® (0–10 A); (b) Semitron (0–600 A); (c) subminiature surface-mount chip fuses.
(Courtesy of Cooper Bussmann.)

To limit the current level, **fuses** or **circuit breakers** are installed where the power enters the installation, such as in the panel in the basement of most homes at the point where the outside feeder lines enter the dwelling. The fuses in Fig. 4.21 have an internal metallic conductor through which the current passes; a fuse begins to melt if the current through the system exceeds the rated value printed on the casing. Of course, if the fuse melts through, the current path is broken and the load in its path is protected.

In homes built in recent years, fuses have been replaced by circuit breakers such as those appearing in Fig. 4.22. When the current exceeds rated conditions, an electromagnet in the device will have sufficient strength to draw the connecting metallic link in the breaker out of the circuit and open the current path. When conditions have been corrected, the breaker can be reset and used again.

FIG. 4.22
Circuit breakers.
(Courtesy of Potter and Brumfield Division, AMF, Inc.)

The most recent National Electrical Code requires that outlets in the bathroom and other sensitive areas be of the ground fault circuit interrupt (GFCI) variety; GFCIs (often abbreviated GFI) are designed to trip more quickly than the standard circuit breaker. The commercial unit in Fig. 4.23 trips in 5 ms. It has been determined that 6 mA is the maximum level that most individuals can be exposed to for a short period of time without the risk of serious injury. A current higher than 11 mA can cause involuntary muscle contractions that could prevent a person from letting go of the conductor and possibly cause him or her to enter a state of shock. Higher currents lasting more than a second can cause the heart to go into fibrillation and possibly cause death in a few minutes. The GFCI is able to react as quickly as it does by sensing the difference between the input and output currents to the outlet; the currents should be the same if everything is working properly. An errant path, such as through an individual, establishes a difference in the two current levels and causes the breaker to trip and disconnect the power source.

FIG. 4.23
Ground fault circuit interrupter (GFCI): 125 V ac, 60 Hz, 15 A outlet.
(Reprinted with permission of the Leviton Manufacturing Company. Leviton SmartLock™ GFCI.)

4.8 APPLICATIONS

Microwave Oven

It is probably safe to say that most homes today have a microwave oven (see Fig. 4.24). Most users are not concerned with its operating efficiency. However, it is interesting to learn how the units operate and apply some of the theory presented in this chapter.

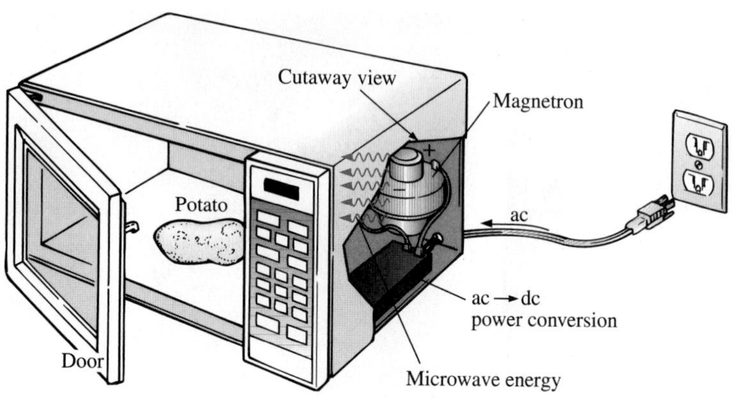

FIG. 4.24
Microwave oven.

First, some general comments. Most microwaves are rated at 500 W to 1200 W at a frequency of 2.45 GHz (almost 2.5 billion cycles per second compared to the 60 cycles per second for the ac voltage at the typical home outlet—details in Chapter 13). The heating occurs because the water molecules in the food are vibrated at such a high frequency that the friction with neighboring molecules causes the heating effect. Since it is the high frequency of vibration that heats the food, there is no need for the material to be a conductor of electricity. However, any metal placed in the microwave can act as an antenna (especially if it has any points or sharp edges) that will attract the microwave energy and reach very high temperatures. In fact, a browning skillet is now made for microwaves that has some metal embedded in the bottom and sides to attract the microwave energy and raise the temperature at the surface between the food and skillet to give the food a brown color and a crisp texture. Even if the metal did not act as an antenna, it is a good conductor of heat and could get quite hot as it draws heat from the food.

Any container with low moisture content can be used to heat foods in a microwave. Because of this requirement, manufacturers have developed a whole line of microwave cookware that is very low in moisture content. Theoretically, glass and plastic have very little moisture content, but even so when heated in the oven for a minute or so, they do get warm. It could be the moisture in the air that clings to the surface of each or perhaps the lead used in good crystal. In any case, microwaves should be used only to prepare food. They were not designed to be dryers or evaporators.

The instructions with every microwave specify that the oven should not be turned on when empty. Even though the oven may be empty, microwave energy will be generated and will make every effort to find a channel for absorption. If the oven is empty, the energy might be attracted to the oven itself and could do some damage. To demonstrate that a dry empty glass or plastic container will not attract a significant amount of microwave energy, place two glasses in an oven, one with water and the other empty. After one minute, you will find the glass with the water quite

warm due to the heating effect of the hot water while the other is close to its original temperature. In other words, the water created a heat sink for the majority of the microwave energy, leaving the empty glass as a less attractive path for heat conduction. Dry paper towels and plastic wrap can be used in the oven to cover dishes since they initially have low water molecule content, and paper and plastic are not good conductors of heat. However, it would be very unsafe to place a paper towel in an oven alone because, as said above, the microwave energy will look for an absorbing medium and could set the paper on fire.

The cooking of food by a conventional oven is from the outside in. The same is true for microwave ovens, but they have the additional advantage of being able to penetrate the outside few centimeters of the food, reducing the cooking time substantially. The cooking time with a microwave oven is related to the amount of food in the oven. Two cups of water will take longer to heat than one cup, although it is not a linear relationship so it won't take twice as long—perhaps 75% to 90% longer. Eventually, if you place enough food in the microwave oven and compare the longer cooking time to that with a conventional oven, you will reach a crossover point where it would be just as wise to use a conventional oven and get the texture in the food you might prefer.

The basic construction of the microwave is depicted in Fig. 4.24. It uses a 120 V ac supply which is then converted through a high-voltage transformer to one having peak values approaching 5000 V (at substantial current levels)—sufficient warning to leave microwave repair to the local service location. Through the rectifying process briefly described in Chapter 2, a high dc voltage of a few thousand volts is generated that appears across a magnetron. The magnetron, through its very special design (currently the same design as in WW II when it was invented by the British for their high-power radar units), generates the required 2.45 GHz signal for the oven. It should be pointed out also that the magnetron has a specific power level of operation that cannot be controlled—once it's on, it's on at a set power level. One may then wonder how the cooking temperature and duration can be controlled. This is accomplished through a controlling network that determines the amount of off and on time during the input cycle of the 120 V supply. Higher temperatures are achieved by setting a high ratio of on to off time, while low temperatures are set by the reverse action.

One unfortunate characteristic of the magnetron is that in the conversion process, it generates a great deal of heat that does not go toward the heating of the food and that must be absorbed by heat sinks or dispersed by a cooling fan. Typical conversion efficiencies are between 55% and 75%. Considering other losses inherent in any operating system, it is reasonable to assume that most microwaves are between 50% and 60% efficient. However, the conventional oven with its continually operating exhaust fan and heating of the oven, cookware, surrounding air, and so on, also has significant losses, even if it is less sensitive to the amount of food to be cooked. All in all, the convenience factor is probably the other factor that weighs the heaviest in this discussion. It also leaves the question of how our time is figured into the efficiency equation.

For specific numbers, let us consider the energy associated with baking a 5-oz potato in a 1200 W microwave oven for 5 min if the conversion efficiency is an average value of 55%. First, it is important to realize that when a unit is rated as 1200 W, that is the rated power drawn from the line during the cooking process. If the microwave is plugged into a 120 V outlet, the current drawn is

$$I = P/V = 1200\text{ W}/120\text{ V} = 10.0\text{ A}$$

which is a significant level of current. Next, we can determine the amount of power dedicated solely to the cooking process by using the efficiency level. That is,

$$P_o = \eta P_i = (0.55)(1200 \text{ W}) = 600 \text{ W}$$

The energy transferred to the potato over a period of 5 min can then be determined from

$$W = Pt = (660 \text{ W})(5 \text{ min})(60 \text{ s}/1 \text{ min}) = 198 \text{ kJ}$$

which is about half of the energy (nutritional value) derived from eating a 5-oz potato. The number of kilowatthours drawn by the unit is determined from

$$W = Pt/1000 = (1200 \text{ W})(5/60 \text{ h})/1000 = 0.1 \text{ kWh}$$

At a rate of 10¢/kWh we find that we can cook the potato for 1 penny—relatively speaking, pretty cheap. A typical 1550 W toaster oven would take an hour to heat the same potato, using 1.55 kWh and costing 15.5 cents—a significant increase in cost.

Household Wiring

A number of facets of household wiring can be discussed without examining the manner in which they are physically connected. In the following chapters, additional coverage is provided to ensure that you develop a solid fundamental understanding of the overall household wiring system. At the very least you will establish a background that will permit you to answer questions that you should be able to answer as a student of this field.

The one specification that defines the overall system is the maximum current that can be drawn from the power lines since the voltage is fixed at 120 V or 240 V (sometimes 208 V). For most older homes with a heating system other than electric, a 100 A service is the norm. Today, with all the electronic systems becoming commonplace in the home, many people are opting for the 200 A service even if they don't have electric heat. A 100 A service specifies that the maximum current that can be drawn through the power lines into your home is 100 A. Using the line-to-line rated voltage and the full-service current (and assuming all resistive-type loads), we can determine the maximum power that can be delivered using the basic power equation:

$$P = EI = (240 \text{ V})(100 \text{ A}) = 24{,}000 \text{ W} = 24 \text{ kW}$$

This rating reveals that the total rating of all the units turned on in the home cannot exceed 24 kW at any one time. If it did, we could expect the main breaker at the top of the power panel to open. Initially, 24 kW may seem like quite a large rating, but when you consider that a self-cleaning electric oven may draw 12.2 kW, a dryer 4.8 kW, a water heater 4.5 kW, and a dishwasher 1.2 kW, we are already at 22.7 kW (if all the units are operating at peak demand), and we haven't turned the lights or TV on yet. Obviously, the use of an electric oven alone may strongly suggest considering a 200 A service. However, be aware that seldom are all the burners of a stove used at once, and the oven incorporates a thermostat to control the temperature so that it is not on all the time. The same is true for the water heater and dishwasher, so the chances of all the units in a home demanding full service at the same time is very slim. Certainly, a typical home with electric heat that may draw 16 kW just for heating in cold weather must consider a 200 A service. You must also understand that there is some leeway in maximum ratings for safety purposes. In

other words, a system designed for a maximum load of 100 A can accept a slightly higher current for short periods of time without significant damage. For the long term, however, it should not be exceeded.

Changing the service to 200 A is not simply a matter of changing the panel in the basement—a new, heavier line must be run from the road to the house. In some areas feeder cables are aluminum because of the reduced cost and weight. In other areas, aluminum is not permitted because of its temperature sensitivity (expansion and contraction), and copper must be used. In any event, when aluminum is used, the contractor must be absolutely sure that the connections at both ends are very secure. The National Electric Code specifies that 100 A service must use a #4 AWG copper conductor or #2 aluminum conductor. For 200 A service, a 2/0 copper wire or a 4/0 aluminum conductor must be used as shown in Fig. 4.25(a). A 100 A or 200 A service must have two lines and a service neutral as shown in Fig. 4.25(b). Note in Fig. 4.25(b) that the lines are coated and insulated from each other, and the service neutral is spread around the inside of the wire coating. At the terminal point, all the strands of the service neutral are gathered together and securely attached to the panel. It is fairly obvious that the cables of Fig. 4.25(a) are stranded for added flexibility.

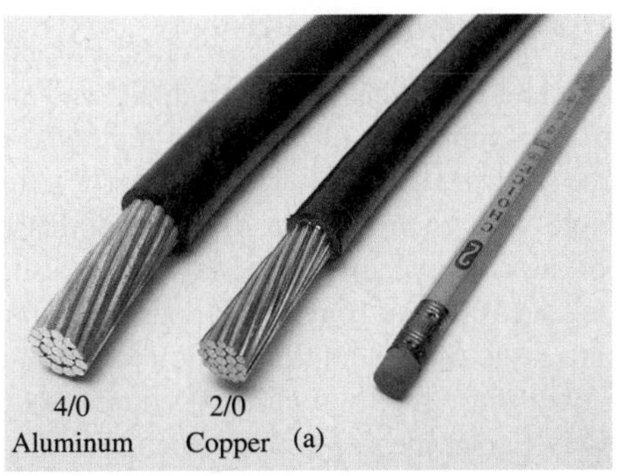

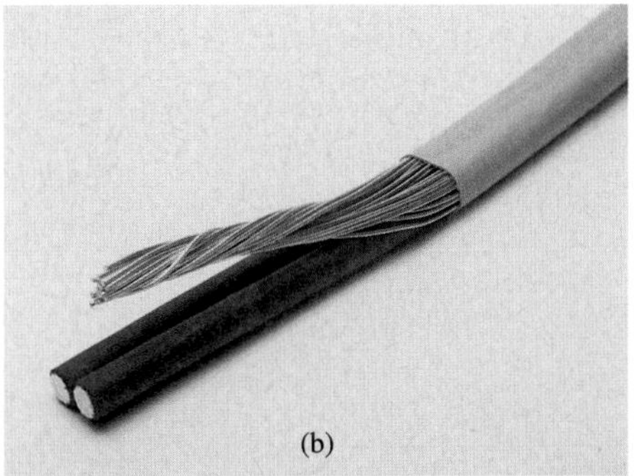

FIG. 4.25
200 A service conductors: (a) 4/0 aluminum and 2/0 copper; (b) three-wire 4/0 aluminum service.

Within the system, the incoming power is broken down into a number of circuits with lower current ratings utilizing 15 A, 20 A, 30 A, and 40 A protective breakers. Since the load on each breaker should not exceed 80% of its rating, in a 15 A breaker the maximum current should be limited to 80% of 15 A, or 12 A, with 16 A for a 20 A breaker, 24 A for a 30 A breaker, and 32 A for a 40 A breaker. The result is that a home with 200 A service can theoretically have a maximum of 12 circuits (200 A/16 A = 12.5) utilizing the 16 A maximum current ratings associated with 20 A breakers. However, if aware of the loads on each circuit, electricians can install as many circuits as they feel appropriate. The code further specifies that a #14 wire should not carry a current in excess of 15 A, a #12 in excess of 20 A, and a #10 in excess of 30 A. Thus, #12 wire is now the most common for general home wiring to ensure that it can handle any excursions beyond 15 A on the 20 A breaker (the most common breaker size). The #14 wire is often used in conjunction with the #12 wire in areas where it is known that the current levels are limited. The #10 wire is typically used for high-demand appliances such as dryers and ovens.

The circuits themselves are usually broken down into those that provide lighting, outlets, and so on. Some circuits (such as ovens and dryers)

require a higher voltage of 240 V, obtained by using two power lines and the neutral. The higher voltage reduces the current requirement for the same power rating, with the net result that the appliance can usually be smaller. For example, the size of an air conditioner with the same cooling ability is measurably smaller when designed for a 240 V line than when designed for 120 V. Most 240 V lines, however, demand a current level that requires 30 A or 40 A breakers and special outlets to ensure that appliances rated at 120 V are not connected to the same outlet. If time permits, check the panel in your home and take note of the number of circuits—in particular, the rating of each breaker and the number of 240 V lines indicated by breakers requiring two slots of the panel. Total the current ratings of all the breakers in your panel, and explain, using the above information, why the total exceeds your feed level.

For safety sake, grounding is a very important part of the electrical system in your home. The National Electric Code requires that the neutral wire of a system be grounded to an earth-driven rod, a metallic water piping system of 10 ft or more, or a buried metal plate. That ground is then passed on through the electrical circuits of the home for further protection. In a later chapter, the details of the connections and grounding methods are discussed.

4.9 COMPUTER ANALYSIS

Now that a complete circuit has been introduced and examined in detail, we can begin the application of computer methods. As mentioned in Chapter 1, three software packages will be introduced to demonstrate the options available with each and the differences that exist. All have a broad range of support in the educational and industrial communities. The student version of PSpice (OrCAD Release 10 from Cadence Design Systems) receives the most attention, followed by Multisim. Each approach has its own characteristics with procedures that must be followed exactly; otherwise, error messages will appear. Do not assume that you can "force" the system to respond the way you would prefer—every step is well defined, and one error on the input side can result in results of a meaningless nature. At times you may believe that the system is in error because you are absolutely sure you followed every step correctly. In such cases, accept the fact that something was entered incorrectly, and review all your work very carefully. All it takes is a comma instead of a period or a decimal point to generate incorrect results.

Be patient with the learning process; keep notes of specific maneuvers that you learn; and don't be afraid to ask for help when you need it. For each approach, there is always the initial concern about how to start and proceed through the first phases of the analysis. However, be assured that with time and exposure you will work through the required maneuvers at a speed you never would have expected. In time you will be absolutely delighted with the results you can obtain using computer methods.

In this section, Ohm's law is investigated using the software packages PSpice and Multisim to analyze the circuit in Fig. 4.26. Both require that the circuit first be "drawn" on the computer screen and then analyzed (simulated) to obtain the desired results. As mentioned above, the analysis program cannot be changed by the user. The proficient user is one who can draw the most out of a computer software package.

Although the author feels that there is sufficient material in the text to carry a new student of the material through the programs provided, be aware that this is not a computer text. Rather, it is one whose primary pur-

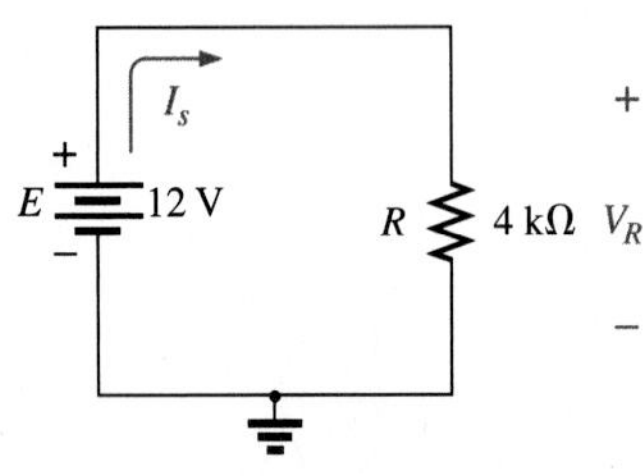

FIG. 4.26
Circuit to be analyzed using PSpice and Multisim.

pose is simply to introduce the different approaches and how they can be applied effectively. Today, some excellent texts and manuals are available that cover the material in a great deal more detail and perhaps at a slower pace. In fact, the quality of the available literature has improved dramatically in recent years.

PSpice

Readers who were familiar with older versions of PSpice such as Versions 8 and 9.2 will find that the changes in Version 10 are primarily in the front end and the simulation process. After executing a few programs, you will find that most of the procedures you learned from older versions will be applicable here also—at least the sequential process has a number of strong similarities.

Once OrCAD Version 10 has been installed, you must first open a **Folder** in the **C:** drive for storage of the circuit files that result from the analysis. Be aware, however, that

once the folder has been defined, it does not have to be defined for each new project unless you choose to do so. If you are satisfied with one location (folder) for all your projects, this is a one-time operation that does not have to be repeated with each network.

To establish the **Folder,** simply right-click the mouse on **Start** at the bottom left of the screen to obtain a listing that includes **Explore.** Select **Explore** to obtain the **Start Menu** dialog box and then use the sequence **File-New Folder** to open a new folder. Type in **PSpice** (the author's choice) and left-click to install. Then exit (using the **X** at the top right of the screen). The folder **PSpice** will be used for all the projects you plan to work on in this text.

Our first project can now be initiated by double-clicking on the **OrCAD 10.0 Demo** icon on the screen, or you can use the sequence **Start All Programs-CAPTURE CIS DEMO.** The resulting screen has only a few active keys on the top toolbar. The first keypad at the top left is the **Create document** key (or you can use the sequence **File-New Project**). A **New Project** dialog box opens in which you must enter the **Name** of the project. For our purposes we will choose **PSpice 4-1** as shown in the **SCHEMATIC** listing in Fig. 4.27 and select **Analog or Mixed A/D** (to be used for all the analyses of this text). Note at the bottom of the dialog box that the **Location** appears as **C:\PSpice** as set above. Click **OK,** and another dialog box appears titled **Create PSpice Project.** Select **Create a blank project** (again, for all the analyses to be performed in this text). Click **OK,** and a third toolbar appears at the top of the screen with some of the keys enabled. An additional toolbar can be established along the right edge of the screen by left-clicking anywhere on the **SCHEMATIC 1: PAGE 1** screen. A **Project Manager Window** appears with **PSpice 4-1** next to an icon and an associated + sign in a small square. Clicking on the + sign will take the listing a step further to **SCHEMATIC1.** Click + again, and **PAGE1** appears; clicking on a − sign reverses the process. Double-clicking on **PAGE1** creates a working window titled **SCHEMATIC1: PAGE1,** revealing that a project can have more than one schematic file and more than one associated page. The width and height of the window can be adjusted by grabbing an edge until you see a double-headed arrow and dragging the border to the desired location. Either window on the screen can be moved by clicking on the top heading to make it dark blue and then dragging it to any location.

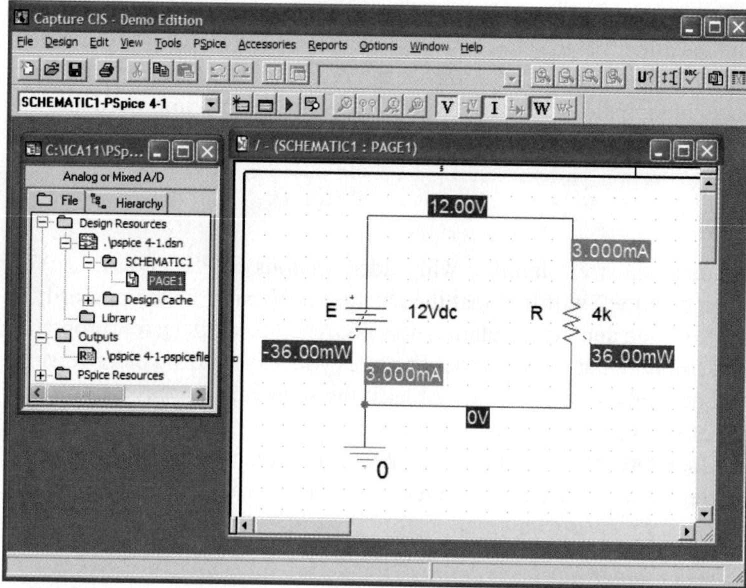

FIG. 4.27
Using PSpice to determine the voltage, current, and power levels for the circuit in Fig. 4.26.

Now you are ready to build the simple circuit in Fig. 4.26. Select the **Place a part** key (the second key from the top of the toolbar on the right) to obtain the **Place Part** dialog box. Since this is the first circuit to be constructed, you must ensure that the parts appear in the list of active libraries. Select **Add Library-Browse File,** and select **analog.olb,** and when it appears under the **File name** heading, select **Open.** It will now appear in the **Libraries** listing at the bottom left of the dialog box. Repeat for the **eval.olb** and **source.olb** libraries. All three files are required to build the networks appearing in this text. However, it is important to realize that

once the library files have been selected, they will appear in the active listing for each new project without having to add them each time—a step, such as the Folder step above, that does not have to be repeated with each similar project.

Click **OK,** and you can now place components on the screen. For the dc voltage source, first select the **Place a part** key and then select **SOURCE** in the library listing. Under **Part List,** a list of available sources appears; select **VDC** for this project. Once **VDC** has been selected, its symbol, label, and value appears on the picture window at the bottom right of the dialog box. Click **OK,** and the **VDC** source follows the cursor across the screen. Move it to a convenient location, left-click the mouse, and it will be set in place as shown in Fig. 4.27. Since only one source is required, right-clicking results in a list of options, in which **End Mode** appears at the top. Choosing this option ends the procedure, leaving the source in a red dashed box. If it is red, it is an active mode and can be operated on. Left-clicking puts the source in place and removes the red active status.

One of the most important steps in the procedure is to ensure that a 0 V ground potential is defined for the network so that voltages at any point in the network have a reference point. *The result is a requirement that every network must have a ground defined.* For our purposes, the **0/SOURCE**

option will be our choice when the **GND** key is selected. It ensures that one side of the source is defined as 0 V. It is obtained by selecting the ground symbol from the toolbar at the right edge of the screen. A **Place Ground** dialog box appears under which **0/SOURCE** can be selected. Finally, you need to add a resistor to the network by selecting the **Place a part** key again and then selecting the **ANALOG** library. Scrolling the options, note that **R** appears and should be selected. Click **OK,** and the resistor appears next to the cursor on the screen. Move it to the desired location and click it in place. Then right-click and **End Mode,** and the resistor has been entered into the schematic's memory. Unfortunately, the resistor ended up in the horizontal position, and the circuit of Fig. 4.26 has the resistor in the vertical position. No problem: Simply select the resistor again to make it red, and right-click. A listing appears in which **Rotate** is an option. It turns the resistor 90° in the counterclockwise direction. It can also be rotated 90° by simultaneously selecting **Crtl-R.**

All the required elements are on the screen, but they need to be connected. To accomplish this, select the **Place a wire** key that looks like a step in the right toolbar. The result is a crosshair with the center that should be placed at the point to be connected. Place the crosshair at the top of the voltage source, and left-click it once to connect it to that point. Then draw a line to the end of the next element, and click again when the crosshair is at the correct point. A red line results with a square at each end to confirm that the connection has been made. Then move the crosshair to the other elements, and build the circuit. Once everything is connected, right-clicking provides the **End Mode** option. Don't forget to connect the source to ground as shown in Fig. 4.27.

Now you have all the elements in place, but their labels and values are wrong. To change any parameter, simply double-click on the parameter (the label or the value) to obtain the **Display Properties** dialog box. Type in the correct label or value, click **OK,** and the quantity is changed on the screen. The labels and values can be moved by simply clicking on the center of the parameter until it is closely surrounded by the four small squares and then dragging it to the new location. Left-clicking again deposits it in its new location.

Finally, you can initiate the analysis process, called **Simulation,** by selecting the **Create a new simulation profile** key at the far left of the second toolbar down—it resembles a data page with a star in the top left corner. **A New Simulation** dialog box opens that first asks for the **Name** of the simulation. The New Simulation dialog box can also be obtained by using the sequence **PSpice-New Simulation Profile. Bias Point** is entered for a dc solution, and **none** is left in the **Inherit From** request. Then select **Create,** and a **Simulation Settings** dialog box appears in which **Analysis-Analysis Type-Bias Point** is sequentially selected. Click **OK,** and select the **Run** key (which looks like an isolated blue arrowhead) or choose **PSpice-Run** from the menu bar. An **Output Window** opens that appears to be somewhat inactive. It will not be used in the current analysis, so close (**X**) the window, and the circuit in Fig. 4.27 appears with the voltage, current, and power levels of the network. The voltage, current, or power levels can be removed (or replaced) from the display by simply selecting the **V, I,** or **W** in the third toolbar from the top. Individual values can be removed by simply selecting the value and pressing the **Delete** key or the scissors key in the top menu bar. Resulting values can be moved by simply left-clicking the value and dragging it to the desired location.

Note in Fig. 4.27 that the current is 3 mA (as expected) at each point in the network, and the power delivered by the source and dissipated by

the resistor is the same, or 36 mW. There are also 12 V across the resistor as required by the configuration.

There is no question that this procedure seems long for such a simple circuit. However, keep in mind that we needed to introduce many new facets of using PSpice that are not discussed again. By the time you finish analyzing your third or fourth network, the procedure will be routine and easy to do.

Multisim

For comparison purposes, Multisim is also used to analyze the circuit in Fig. 4.26. Although there are differences between PSpice and Multisim, such as initiating the process, constructing the networks, making the measurements, and setting up the simulation procedure, there are sufficient similarities between the two approaches to make it easier to learn one if you are already familiar with the other. The similarities will be obvious only if you make an attempt to learn both. One of the major differences between the two is the option to use actual instruments in Multisim to make the measurements—a positive trait in preparation for the laboratory experience. However, in Multisim, you may not find the extensive list of options available with PSpice. In general, however, both software packages are well prepared to take us through the types of analyses to be encountered in this text.

When the **Multisim** icon is selected from the opening window, a screen appears with the heading **Circuit 1-Multisim.** A menu bar appears across the top of the screen, with a **Standard toolbar** directly below as shown in Fig. 4.28. Initially, the third level includes the

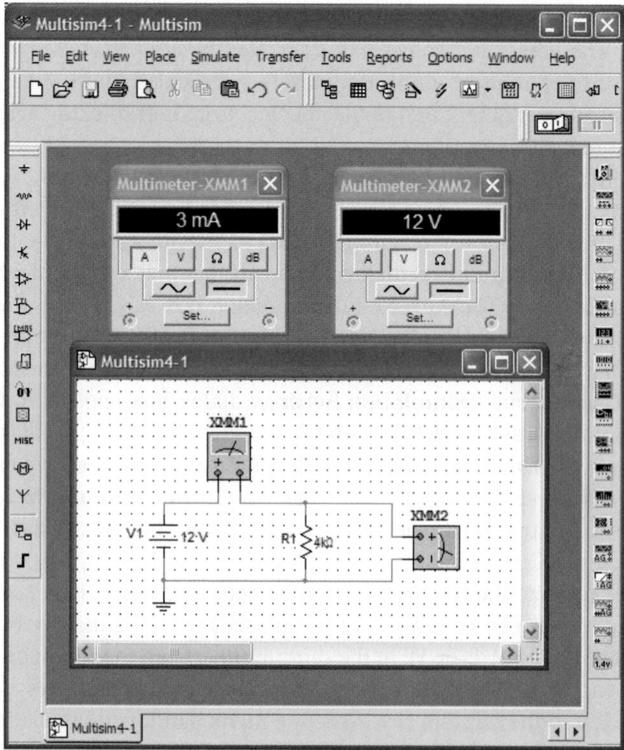

FIG. 4.28

Using Multisim to determine the voltage and current levels for the circuit of Fig. 4.26.

Component toolbar, Ladder diagram toolbar and **Virtual toolbar.** For our purposes, the **Ladder diagram toolbar** and **Virtual toolbar** were deleted. You do this by selecting **View** from the top menu bar and then **Toolbars.** By removing the check next to **Ladder diagram** and **Virtual,** both toolbars are deleted. The remaining **Component toolbar** can be moved to the left edge by grabbing an edge of the toolbar and moving to the left edge as shown in Fig. 4.28. Some users of Multisim prefer to leave the **Virtual toolbar** and use it more than the **Component toolbar.** Either approach is simply a matter of preference. The **Virtual toolbar** provides a list of components for which the value must be set. The **Components list** includes the same list under **Basic** but also includes a list of standard values. The heading can be changed to **Multisim 4-1** by selecting **File-Save As** to open the **Save As** dialog box. Enter **Multisim 4-1** as the **File name** to obtain the listing of Fig. 4.28.

For the placement of components, **View-Show Grid** was selected so that a grid would appear on the screen. As you place an element, it will automatically be placed in a relationship specific to the grid structure.

To build the circuit in Fig. 4.26, first take the cursor and place it on the battery symbol at the top of the component toolbar at the left of the screen. Left-click and a list of sources will appear. Under **Component,** select **DC-POWER.** The symbol appears in the adjoining box area. Click **OK.** The battery symbol appears on the screen next to the location of the cursor. Move the cursor to the desired location, and left-click to set the battery symbol in place. The operation is complete. If you want to delete the source, simply left-click on the symbol again to create a dashed rectangle around the source. These rectangles indicate that the source is in the active mode and can be operated on. If you want to delete it, click on the **Delete** key or select the scissor keypad on the top toolbar. If you want to modify the source, right-click *outside* the rectangle, and you get one list. Right-click *within* the rectangle, and you have a different set of options. At any time, if you want to remove the active state, left-click anywhere on the screen. If you want to move the source, click on the source symbol to create the rectangle, but do not release the mouse. Hold it down and drag the source to the preferred location. When the source is in place, release the mouse. Click again to remove the active state. *From now on, whenever possible, the word click means a left-click.* The need for a right click will continue to be spelled out.

For the simple circuit in Fig. 4.28, you need to place a resistor across the source. Select the second keypad down that looks like a resistor symbol. A dialog box opens with a **Family** listing. Selecting **RESISTOR** results in a list of standard values that can be quickly selected for the deposited resistor. However, in this case, you want to use a 4 kΩ resistor, which is not a standard value, so you must choose the **BASIC_VIRTUAL** option. Recall that the term *virtual* indicates that we can set the parameters of the component. When **BASIC_VIRTUAL** is selected, a **Component** list appears under which **RESISTOR_VIRTUAL** is chosen resulting in the resistor symbol in the symbol box. Select **OK** to place the resistor on the grid structure which can then be deposited as described for the supply. Because it is the first resistor to be placed, the resistor is labeled **R1,** has a value of 1 kΩ as its initial value, and appears in the horizontal position.

In Fig. 4.26, the resistor is in the vertical position, so a rotation must be made. Click on the resistor to obtain the active state and then right-click within the rectangle. A number of options appear, including **Cut, Delete, Flip, Rotate, Font,** and **Color.** To rotate 90° counterclockwise, select that option, and the resistor is automatically rotated 90°. Now, as with the battery, to place the resistor in position, click on the resistor symbol to create the rectangle and then, holding it down, drag the resistor

to the desired position. When the resistor is in place, release the mouse, and click again to remove the rectangle—the resistor is in place.

Finally, you need a ground for all networks. Going back to the **Sources** parts bin, find **GROUND,** which is the fourth option down under **Component.** Select **GROUND** and place it on the screen below the voltage source as shown in Fig. 4.28. Now, before connecting the components together, move the labels and the value of each component to the relative positions shown in Fig. 4.28. Do this by clicking on the label or value to create a small set of squares around the element and then dragging the element to the desired location. Release the mouse, and then click again to set the element in place. To change the label or value, double-click on the label (such as **V1**) to open a **POWER_SOURCES** dialog box. Select **Label** and enter **E** as the **Reference Designation (Ref Des).** Then, before leaving the dialog box, go to **Value** and change the value if necessary. It is very important to realize that you cannot type in the units where the **V** now appears to the right of the value. The prefix is controlled by the scroll keys at the left of the unit of measure. For practice, try the scroll keys, and you will find that you can go from **pV** to **TV.** For now, leave it as **V.** Click **OK,** and both have been changed on the screen. The same process can be applied to the resistive element to obtain the label and value appearing in Fig. 4.28.

Next, you need to tell the system which results should be generated and how they should be displayed. For this example, we use a multimeter to measure both the current and the voltage of the circuit. The **Multimeter** is the first option in the list of instruments appearing in the toolbar to the right of the screen. When selected, it appears on the screen and can be placed anywhere using the same procedure defined for the components above. The voltmeter was turned clockwise using the procedure described above for the elements. Double-click on the meter symbol, and a **Multimeter** dialog box opens in which the function of the meter must be defined. Since the meter **XMM1** will be used as an ammeter, select the letter **A** and the horizontal line to indicate dc level. There is no need to select **Set** for the default values since they have been chosen for the broad range of applications. The dialog meters can be moved to any location by clicking on their heading bar to make it dark blue and then dragging the meter to the preferred position. For the voltmeter, **V** and the horizontal bar were selected as shown in Fig. 4.28.

Finally, the elements need to be connected. To do this, bring the cursor to one end of an element, say, the top of the voltage source. A small dot and a crosshair appears at the top end of the element. Click once, follow the path you want, and place the crosshair over the positive terminal of the ammeter. Click again, and the wire appears in place.

At this point, you should be aware that the software package has its preferences about how it wants the elements to be connected. That is, you may try to draw it one way, but the computer generation may be a different path. Eventually, you will learn these preferences and can set up the network to your liking. Now continue making the connections appearing in Fig. 4.28, moving elements or adjusting lines as necessary. Be sure that the small dot appears at any point where you want a connection. Its absence suggests that the connection has not been made and the software program has not accepted the entry.

You are now ready to run the program and view the solution. The analysis can be initiated in a number of ways. One option is to select **Simulate** from the top toolbar, followed by **RUN.** Another is to select the **Simulate** key in the design bar grouping in the top toolbar. It appears as a sharp, jagged lightning bolt on a black background. The last option, and

the one we use the most, requires an **OFF/ON, 0/1** switch on the screen. It is obtained through **VIEW-Show Simulate Switch** and appears as shown in the top right corner of Fig. 4.28. Using this last option, the analysis (called **Simulation**) is initiated by placing the cursor on the 1 of the switch and left-clicking. The analysis is performed, and the current and voltage appear on the meter as shown in Fig. 4.28. Note that both provide the expected results.

One of the most important things to learn about applying Multisim:

Always stop or end the simulation (clicking on 0 or choosing OFF) before making any changes in the network. When the simulation is initiated, it stays in that mode until turned off.

There was obviously a great deal of material to learn in this first exercise using Multisim. Be assured, however, that as we continue with more examples, you will find the procedure quite straightforward and actually enjoyable to apply.

PROBLEMS

SECTION 4.2 Ohm's Law

1. What is the voltage across a 47 Ω resistor if the current through it is 2.5 A?
2. What is the current through a 6.8 Ω resistor if the voltage drop across it is 12 V?
3. How much resistance is required to limit the current to 1.5 mA if the potential drop across the resistor is 6 V?
4. At starting, what is the current drain on a 12 V car battery if the resistance of the starting motor is 40 MΩ?
5. If the current through a 0.02 MΩ resistor is 3.6 μA, what is the voltage drop across the resistor?
6. If a voltmeter has an internal resistance of 15 kΩ, find the current through the meter when it reads 62 V.
7. If a refrigerator draws 2.2 A at 120 V, what is its resistance?
8. If a clock has an internal resistance of 7.5 kΩ, find the current through the clock if it is plugged into a 120 V outlet.
9. A washing machine is rated at 4.2 A at 120 V. What is its internal resistance?
10. A CD player draws 125 mA when 4.5 V is applied. What is the internal resistance?
11. The input current to a transistor is 20 μA. If the applied (input) voltage is 24 mV, determine the input resistance of the transistor.
12. The internal resistance of a dc generator is 0.5 Ω. Determine the loss in terminal voltage across this internal resistance if the current is 15 A.

*13. a. If an electric heater draws 9.5 A when connected to a 120 V supply, what is the internal resistance of the heater?
 b. Using the basic relationships of Chapter 2, how much energy is converted in 1 h?

14. In a TV camera, a current of 2.4 μA passes through a resistor of 3.3 MΩ. What is the voltage drop across the resistor?

SECTION 4.3 Plotting Ohm's Law

15. a. Plot the curve of I (vertical axis) versus V (horizontal axis) for a 120 Ω resistor. Use a horizontal scale of 0 to 100 V and a vertical scale of 0 to 1 A.
 b. Using the graph of part (a), find the current at a voltage of 20 V and 50 V.
16. a. Plot the I-V curve for a 5 Ω and a 20 Ω resistor on the same graph. Use a horizontal scale of 0 to 40 V and a vertical scale of 0 to 2 A.
 b. Which is the steeper curve? Can you offer any general conclusions based on results?
 c. If the horizontal and vertical scales were interchanged, which would be the steeper curve?
17. a. Plot the I-V characteristics of a 1 Ω, 100 Ω, and 1000 Ω resistor on the same graph. Use a horizontal axis of 0 to 100 V and a vertical axis of 0 to 100 A.
 b. Comment on the steepness of a curve with increasing levels of resistance.

*18. Sketch the internal resistance characteristics of a device that has an internal resistance of 20 Ω from 0 to 10 V, an internal resistance of 4 Ω from 10 V to 15 V, and an internal resistance of 1 Ω for any voltage greater than 15 V. Use a horizontal scale that extends from 0 to 20 V and a vertical scale that permits plotting the current for all values of voltage from 0 to 20 V.

*19. a. Plot the I-V characteristics of a 2 kΩ, 1 MΩ, and a 100 Ω resistor on the same graph. Use a horizontal axis of 0 to 20 V and a vertical axis of 0 to 10 mA.
 b. Comment on the steepness of the curve with decreasing levels of resistance.
 c. Are the curves linear or nonlinear? Why?

SECTION 4.4 Power

20. If 420 J of energy are absorbed by a resistor in 4 min, what is the power to the resistor?
21. The power to a device is 40 joules per second (J/s). How long will it take to deliver 640 J?

22. **a.** How many joules of energy does a 2 W nightlight dissipate in 8 h?
b. How many kilowatthours does it dissipate?

23. A resistor of 10 Ω has charge flowing through it at the rate of 300 coulombs per minute (C/min). How much power is dissipated?

24. How long must a steady current of 1.4 A exist in a resistor that has 3 V across it to dissipate 12 J of energy?

25. What is the power delivered by a 6 V battery if the charge flows at the rate of 48 C/min?

26. The current through a 4 kΩ resistor is 7.2 mA. What is the power delivered to the resistor?

27. The power consumed by a 2.2 kΩ resistor is 240 mW. What is the current level through the resistor?

28. What is the maximum permissible current in a 120 Ω, 2 W resistor? What is the maximum voltage that can be applied across the resistor?

29. The voltage drop across a transistor network is 12 V. If the total resistance is 5.6 kΩ, what is the current level? What is the power delivered? How much energy is dissipated in 1 h?

30. If the power applied to a system is 324 W, what is the voltage across the line if the current is 2.7 A?

31. A 1 W resistor has a resistance of 4.7 MΩ. What is the maximum current level for the resistor? If the wattage rating is increased to 2 W, will the current rating double?

32. A 2.2 kΩ resistor in a stereo system dissipates 42 mW of power. What is the voltage across the resistor?

33. A dc battery can deliver 45 mA at 9 V. What is the power rating?

34. What are the "hot" resistance level and current rating of a 120 V, 100 W bulb?

35. What are the internal resistance and voltage rating of a 450 W automatic washer that draws 3.75 A?

36. A calculator with an internal 3 V battery draws 0.4 mW when fully functional.
a. What is the current demand from the supply?
b. If the calculator is rated to operate 500 h on the same battery, what is the ampere-hour rating of the battery?

37. A 20 kΩ resistor has a rating of 100 W. What are the maximum current and the maximum voltage that can be applied to the resistor?

SECTION 4.5 Energy

38. A 10 Ω resistor is connected across a 12 V battery.
a. How many joules of energy will it dissipate in 1 min?
b. If the resistor is left connected for 2 min instead of 1 min, will the energy used increase? Will the power dissipation level increase?

39. How much energy in kilowatthours is required to keep a 230 W oil-burner motor running 12 h a week for 5 months? (Use 4½ weeks = 1 month.)

40. How long can a 1500 W heater be on before using more than 12 kWh of energy?

41. How much does it cost to use a 24 W radio for 3 h at 9¢ per kilowatthour?

42. **a.** In 10 h an electrical system converts 1200 kWh of electrical energy into heat. What is the power level of the system?
b. If the applied voltage is 208 V, what is the current drawn from the supply?
c. If the efficiency of the system is 82%, how much energy is lost or stored in 10 h?

43. At 9¢ per kilowatthour, how long can you play a 250 W color television for $1?

44. A 60 W bulb is on for one hour. Find the energy converted in
a. watthours
b. wattseconds
c. joules
d. kilowatthours

*45. A small, portable black-and-white television draws 0.455 A at 9 V.
a. What is the power rating of the television?
b. What is the internal resistance of the television?
c. What is the energy converted in 6 h of typical battery life?

*46. **a.** If a house is supplied with 120 V, 100 A service, find the maximum power capability.
b. Can the homeowner safely operate the following loads at the same time?
5 hp motor
3000 W clothes dryer
2400 W electric range
1000 W steam iron
c. If all the appliances are used for 2 hours, how much energy is converted in kWh?

*47. What is the total cost of using the following at 9¢ per kilowatthour?
a. 860 W air conditioner for 6 h
b. 4800 W clothes dryer for 30 min
c. 900 W coffee maker for 20 min
d. 110 W stereo for 3.5 h

*48. What is the total cost of using the following at 9¢ per kilowatthour?
a. 200 W fan for 4 h
b. 1200 W dryer for 20 min
c. 70 W radio for 1.5 h
d. 150 W color television set for 2 h 10 min

SECTION 4.6 Efficiency

49. What is the efficiency of a motor that has an output of 0.5 hp with an input of 395 W?

50. The motor of a power saw is rated 68.5% efficient. If 1.8 hp are required to cut a particular piece of lumber, what is the current drawn from a 120 V supply?

51. What is the efficiency of a dryer motor that delivers 1 hp when the input current and voltage are 4 A and 220 V, respectively?

52. A stereo system draws 2.4 A at 120 V. The audio output power is 50 W.
a. How much power is lost in the form of heat in the system?
b. What is the efficiency of the system?

53. If an electric motor having an efficiency of 76% and operating off a 220 V line delivers 3.6 hp, what input current does the motor draw?

54. A motor is rated to deliver 2 hp.
    a. If it runs on 110 V and is 90% efficient, how many watts does it draw from the power line?
    b. What is the input current?
    c. What is the input current if the motor is only 70% efficient?

55. An electric motor used in an elevator system has an efficiency of 90%. If the input voltage is 220 V, what is the input current when the motor is delivering 15 hp?

56. The motor used on a conveyor belt is 85% efficient. If the overall efficiency is 75%, what is the efficiency of the conveyor belt assembly?

57. A 2 hp motor drives a sanding belt. If the efficiency of the motor is 87% and that of the sanding belt is 75% due to slippage, what is the overall efficiency of the system?

58. If two systems in cascade each have an efficiency of 80% and the input energy is 60 J, what is the output energy?

59. The overall efficiency of two systems in cascade is 72%. If the efficiency of one is 0.9, what is the efficiency in percent of the other?

60. a. What is the total efficiency of three systems in cascade with efficiencies of 98%, 87%, and 21%?
    b. If the system with the least efficiency (21%) were removed and replaced by one with an efficiency of 90%, what would be the percentage increase in total efficiency?

*61. If the total input and output power of two systems in cascade are 400 W and 128 W, respectively, what is the efficiency of each system if one has twice the efficiency of the other?

SECTION 4.9 Computer Analysis

62. Using PSpice or Multisim, repeat the analysis of the circuit in Fig. 4.26 with $E = 400$ mV and $R = 0.04$ MΩ.

63. Using PSpice or Multisim, repeat the analysis of the circuit in Fig. 4.26, but reverse the polarity of the battery and use $E = 0.02$ V and $R = 240$ Ω.

GLOSSARY

Circuit breaker A two-terminal device designed to ensure that current levels do not exceed safe levels. If "tripped," it can be reset with a switch or a reset button.

Diode A semiconductor device whose behavior is much like that of a simple switch; that is, it will pass current ideally in only one direction when operating within specified limits.

Efficiency (η) A ratio of output to input power that provides immediate information about the energy-converting characteristics of a system.

Energy (W) A quantity whose change in state is determined by the product of the rate of conversion (P) and the period involved (t). It is measured in joules (J) or wattseconds (Ws).

Fuse A two-terminal device whose sole purpose is to ensure that current levels in a circuit do not exceed safe levels.

Horsepower (hp) Equivalent to 746 watts in the electrical system.

Kilowatthour meter An instrument for measuring kilowatthours of energy supplied to a residential or commercial user of electricity.

Ohm's law An equation that establishes a relationship among the current, voltage, and resistance of an electrical system.

Power An indication of how much work can be done in a specified amount of time; a *rate* of doing work. It is measured in joules/second (J/s) or watts (W).

Series dc Circuits

Objectives

- *Become familiar with the characteristics of a series circuit and how to solve for the voltage, current, and power to each of the elements.*
- *Develop a clear understanding of Kirchhoff's voltage law and how important it is to the analysis of electric circuits.*
- *Become aware of how an applied voltage will divide among series components and how to properly apply the voltage divider rule.*
- *Understand the use of single- and double-subscript notation to define the voltage levels of a network.*
- *Learn how to use a voltmeter, ammeter, and ohmmeter to measure the important quantities of a network.*

5.1 INTRODUCTION

Two types of current are readily available to the consumer today. One is *direct current* (dc), in which ideally the flow of charge (current) does not change in magnitude (or direction) with time. The other is *sinusoidal alternating current* (ac), in which the flow of charge is continually changing in magnitude (and direction) with time. The next few chapters are an introduction to circuit analysis purely from a dc approach. The methods and concepts are discussed in detail for direct current; when possible, a short discussion suffices to cover any variations we may encounter when we consider ac in the later chapters.

The battery in Fig. 5.1, by virtue of the potential difference between its terminals, has the ability to cause (or "pressure") charge to flow through the simple circuit. The positive terminal attracts the electrons through the wire at the same rate at which electrons are supplied by the negative terminal. As long as the battery is connected in the circuit and maintains its terminal characteristics, the current (dc) through the circuit will not change in magnitude or direction.

If we consider the wire to be an ideal conductor (that is, having no opposition to flow), the potential difference V across the resistor equals the applied voltage of the battery: V (volts) = E (volts).

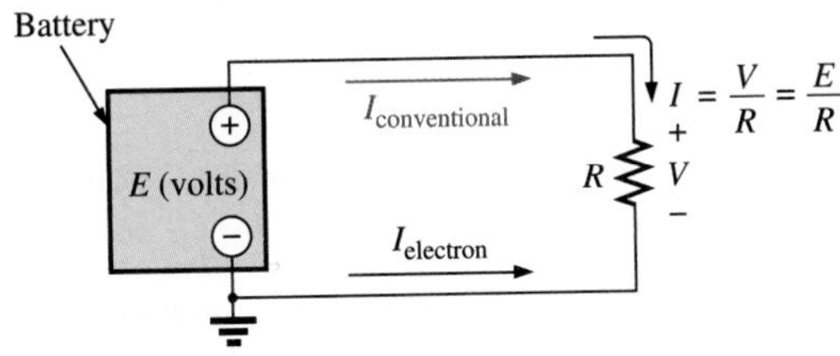

FIG. 5.1
Introducing the basic components of an electric circuit.

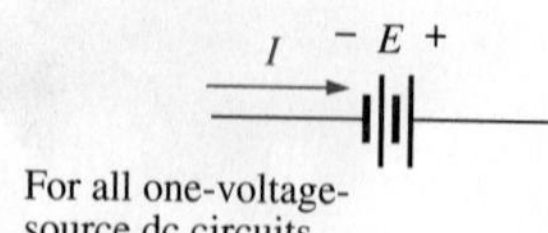

FIG. 5.2
Defining the direction of conventional flow for single-source dc circuits.

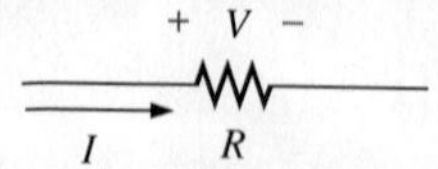

For any combination of voltage sources in the same dc circuit

FIG. 5.3
Defining the polarity resulting from a conventional current I through a resistive element.

The current is limited only by the resistor R. The higher the resistance, the less the current, and conversely, as determined by Ohm's law.

By convention (as discussed in Chapter 2), the direction of **conventional current flow** ($I_{\text{conventional}}$) as shown in Fig. 5.1 is opposite to that of **electron flow** (I_{electron}). Also, the uniform flow of charge dictates that the direct current I be the same everywhere in the circuit. By following the direction of conventional flow, we notice that there is a rise in potential across the battery ($-$ to $+$), and a drop in potential across the resistor ($+$ to $-$). For single-voltage-source dc circuits, conventional flow always passes from a low potential to a high potential when passing through a voltage source, as shown in Fig. 5.2. However, conventional flow always passes from a high to a low potential when passing through a resistor for any number of voltage sources in the same circuit, as shown in Fig. 5.3.

The circuit in Fig. 5.1 is the simplest possible configuration. This chapter and the following chapters add elements to the system in a very specific manner to introduce a range of concepts that will form a major part of the foundation required to analyze the most complex system. Be aware that the laws, rules, and so on, introduced in Chapters 5 and 6 will be used throughout your studies of electrical, electronic, or computer systems. They are not replaced by a more advanced set as you progress to more sophisticated material. It is therefore critical that you understand the concepts thoroughly and are able to apply the various procedures and methods with confidence.

5.2 SERIES RESISTORS

Before the series connection is described, first recognize that every fixed resistor has only two terminals to connect in a configuration—it is therefore referred to as a **two-terminal device.** In Fig. 5.4, one terminal of resistor R_2 is connected to resistor R_1 on one side, and the remaining terminal is connected to resistor R_3 on the other side, resulting in one, and only one, connection between adjoining resistors. When connected in this manner, the resistors have established a series connection. If three elements were connected to the same point, as shown in Fig. 5.5, there would not be a series connection between resistors R_1 and R_2.

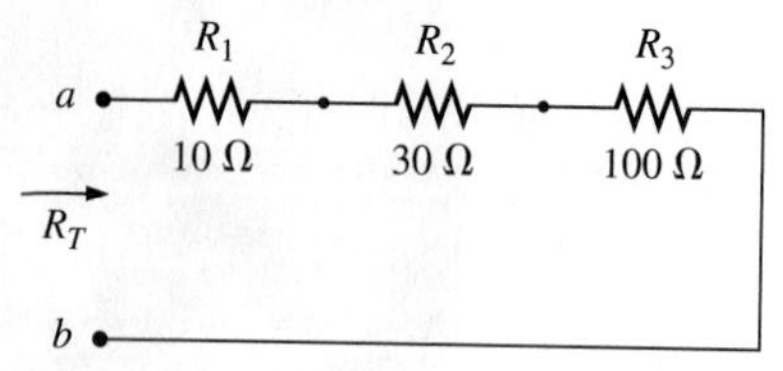

FIG. 5.4
Series connection of resistors.

For resistors in series,

the total resistance of a series configuration is the sum of the resistance levels.

In equation form for any number (N) of resistors,

$$R_T = R_1 + R_2 + R_3 + R_4 + \cdots + R_N \qquad \textbf{(5.1)}$$

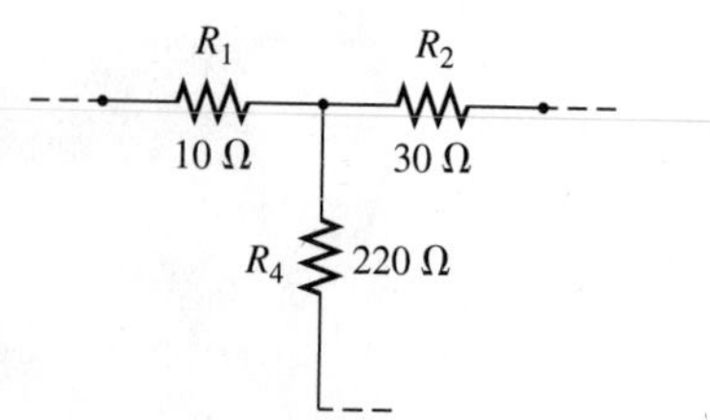

FIG. 5.5
Configuration in which none of the resistors are in series.

A result of Eq. (5.1) is that

the more resistors we add in series, the greater the resistance, no matter what their value.

Further,

the largest resistor in a series combination will have the most impact on the total resistance.

For the configuration in Fig. 5.4, the total resistance would be

$$R_T = R_1 + R_2 + R_3$$
$$= 10\ \Omega + 30\ \Omega + 100\ \Omega$$

and

$$R_T = \mathbf{140\ \Omega}$$

EXAMPLE 5.1 Determine the total resistance of the series connection in Fig. 5.6. Note that all the resistors appearing in this network are standard values.

Solution: Note in Fig. 5.6 that even though resistor R_3 is on the vertical and resistor R_4 returns at the bottom to terminal b, all the resistors are in series since there are only two resistor leads at each connection point.

Applying Eq. (5.1):

$$R_T = R_1 + R_2 + R_3 + R_4$$
$$R_T = 20\ \Omega + 220\ \Omega + 1.2\ \text{k}\Omega + 5.6\ \text{k}\Omega$$

and $$R_T = 7040\ \Omega = \mathbf{7.04\ k\Omega}$$

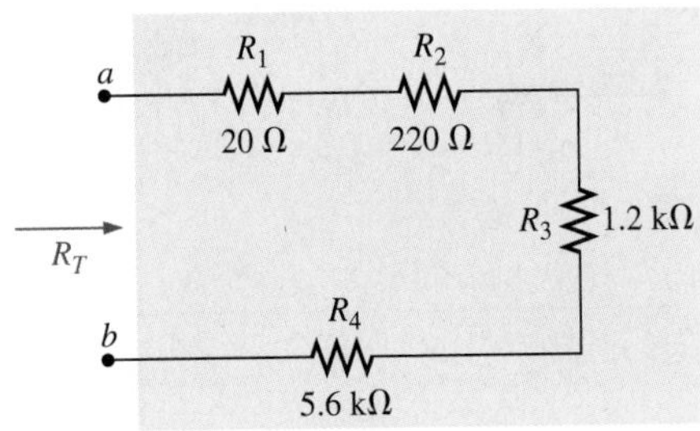

FIG. 5.6
Series connection of resistors for Example 5.1.

For the special case where resistors are the *same value,* Eq. (5.1) can be modified as follows:

$$\boxed{R_T = NR} \tag{5.2}$$

where N is the number of resistors in series of value R.

EXAMPLE 5.2 Find the total resistance of the series resistors in Fig. 5.7. Again, recognize 3.3 kΩ as a standard value.

Solution: Again, don't be concerned about the change in configuration. Neighboring resistors are connected only at one point, satisfying the definition of series elements.

Eq. (5.2): $$R_T = NR$$
$$= (4)(3.3\ \text{k}\Omega) = \mathbf{13.2\ k\Omega}$$

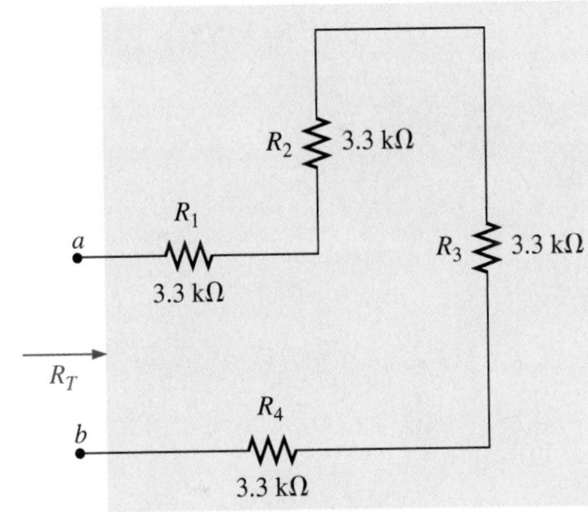

FIG. 5.7
Series connection of four resistors of the same value (Example 5.2).

It is important to realize that since the parameters of Eq. (5.1) can be put in any order,

the total resistance of resistors in series is unaffected by the order in which they are connected.

The result is that the total resistance in Fig. 5.8(a) and (b) are both the same. Again, note that all the resistors are standard values.

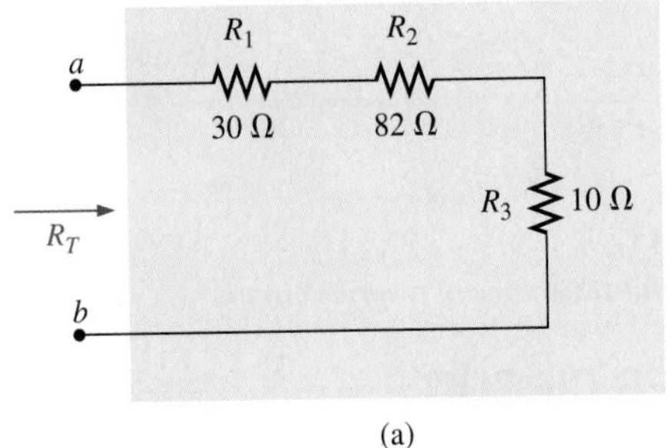

(a)

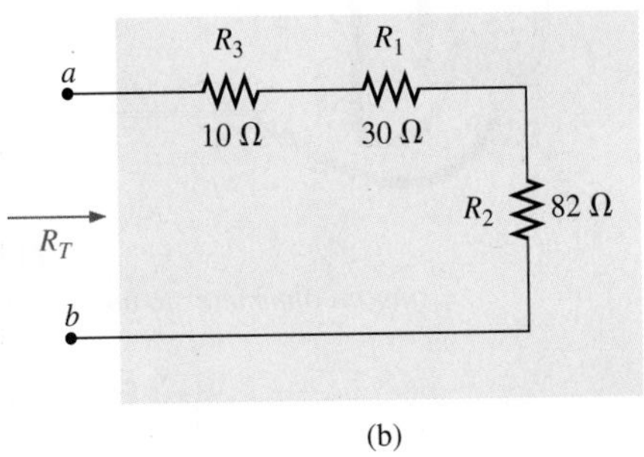

(b)

FIG. 5.8
Two series combinations of the same elements with the same total resistance.

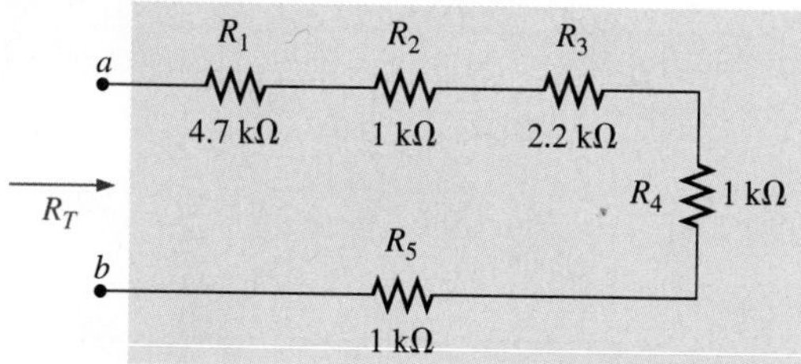

FIG. 5.9
Series combination of resistors for Example 5.3.

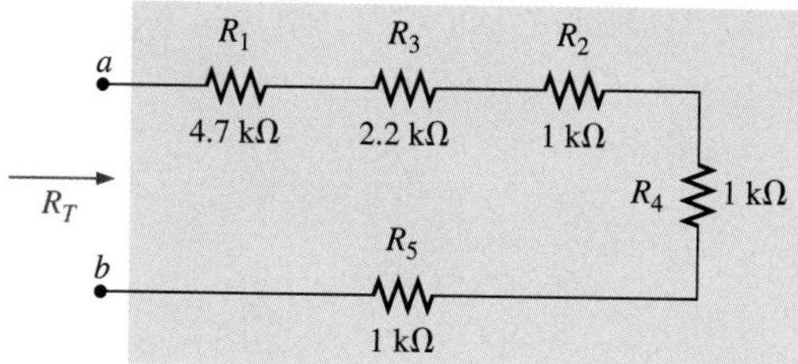

FIG. 5.10
Series circuit of Fig. 5.9 redrawn to permit the use of Eq. (5.2): $R_T = NR$.

EXAMPLE 5.3 Determine the total resistance for the series resistors (standard values) in Fig. 5.9.

Solution: First, the order of the resistors is changed as shown in Fig. 5.10 to permit the use of Eq. (5.2). The total resistance is then

$$R_T = R_1 + R_3 + NR_2$$
$$= 4.7\text{ k}\Omega + 2.2\text{ k}\Omega + (3)(1\text{ k}\Omega) = \mathbf{9.9\text{ k}\Omega}$$

Analogies

Throughout the text, analogies are used to help explain some of the important fundamental relationships in electrical circuits. An analogy is simply a combination of elements of a different type that are helpful in explaining a particular concept, relationship, or equation.

Two analogies that work well for the series combination of elements are connecting different lengths of rope together to make the rope longer. Adjoining pieces of rope are connected at only one point, satisfying the definition of series elements. Connecting a third rope to the common point would mean that the sections of rope are no longer in a series.

Another analogy is connecting hoses together to form a longer hose. Again, there is still only one connection point between adjoining sections, resulting in a series connection.

Instrumentation

The total resistance of any configuration can be measured by simply connecting an ohmmeter across the access terminals as shown in Fig. 5.11 for the circuit in Fig. 5.4. *Since there is no polarity associated with resistance,* either lead can be connected to point *a,* with the other lead connected to point *b.* Choose a scale that will exceed the total resistance of the circuit, and remember when you read the response on the meter, if a kilohm scale was selected, the result will be in kilohms. For Fig. 5.11, the 200 Ω scale of our chosen multimeter was used because the total resistance is 140 Ω. If the 2 kΩ scale of our meter were selected, the digital display would read 0.140, and you must recognize that the result is in kilohms.

In the next section, another method for determining the total resistance of a circuit is introduced using Ohm's law.

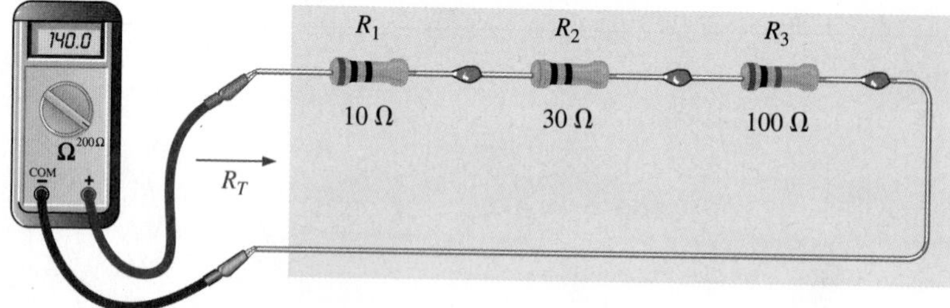

FIG. 5.11
Using an ohmmeter to measure the total resistance of a series circuit.

5.3 SERIES CIRCUITS

If we now take an 8.4 V dc supply and connect it in series with the series resistors in Fig. 5.4, we have the **series circuit** in Fig. 5.12.

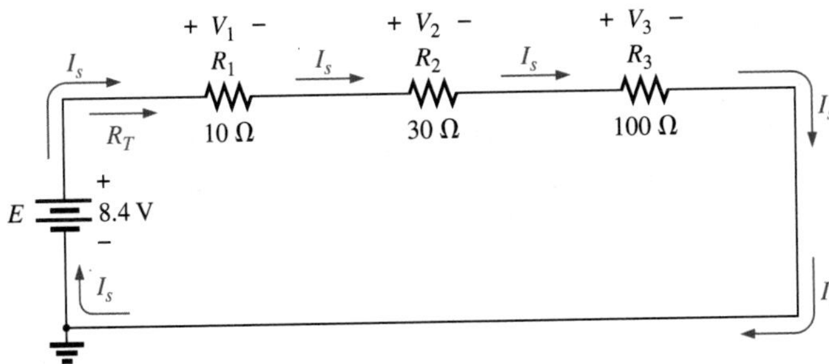

FIG. 5.12
Schematic representation for a dc series circuit.

A circuit is any combination of elements that will result in a continuous flow of charge, or current, through the configuration.

First, recognize that the *dc supply is also a two-terminal device* with two points to be connected. If we simply ensure that there is only one connection made at each end of the supply to the series combination of resistors, we can be sure that we have established a series circuit.

The manner in which the supply is connected determines the direction of the resulting conventional current. For series dc circuits:

the direction of conventional current in a series dc circuit is such that it leaves the positive terminal of the supply and returns to the negative terminal, as shown in Fig. 5.12.

One of the most important concepts to remember when analyzing series circuits and defining elements that are in series is:

The current is the same at every point in a series circuit.

For the circuit in Fig. 5.12, the above statement dictates that the current is the same through the three resistors and the voltage source. In addition, if you are ever concerned about whether two elements are in series, simply check whether the current is the same through each element.

In any configuration, if two elements are in series, the current must be the same. However, if the current is the same for two adjoining elements, the elements may or may not be in series.

The need for this constraint in the last sentence will be demonstrated in the chapters to follow.

Now that we have a complete circuit and current has been established, the level of current and the voltage across each resistor should be determined. To do this, return to Ohm's law and replace the resistance in the equation by the total resistance of the circuit. That is,

$$I_s = \frac{E}{R_T} \tag{5.3}$$

with the subscript s used to indicate source current.

It is important to realize that when a dc supply is connected, it does not "see" the individual connection of elements but simply the total resistance "seen" at the connection terminals, as shown in Fig. 5.13(a). In other words, it reduces the entire configuration to one such as in Fig. 5.13(b) to which Ohm's law can easily be applied.

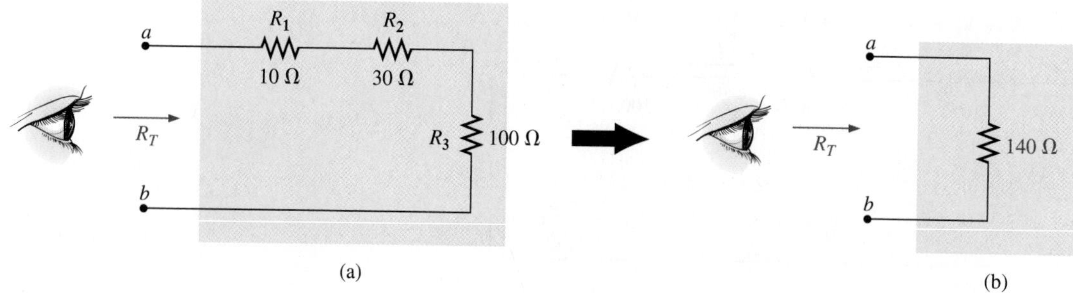

FIG. 5.13
Resistance "seen" at the terminals of a series circuit.

For the configuration in Fig. 5.12, using the total resistance calculated in the last section, the resulting current is

$$I_s = \frac{E}{R_T} = \frac{8.4\text{ V}}{140\ \Omega} = 0.06\text{ A} = \mathbf{60\ mA}$$

Note that the current I_s at every point or corner of the network is the same. Furthermore, note that the current is also indicated on the current display of the power supply.

Now that we have the current level, we can calculate the voltage across each resistor. First recognize that

the polarity of the voltage across a resistor is determined by the direction of the current.

Current entering a resistor creates a drop in voltage with the polarity indicated in Fig. 5.14(a). Reverse the direction of the current, and the polarity will reverse as shown in Fig. 5.14(b). Change the orientation of the resistor, and the same rules apply as shown in Fig. 5.14(c). Applying the above to the circuit in Fig. 5.12 will result in the polarities appearing in that figure.

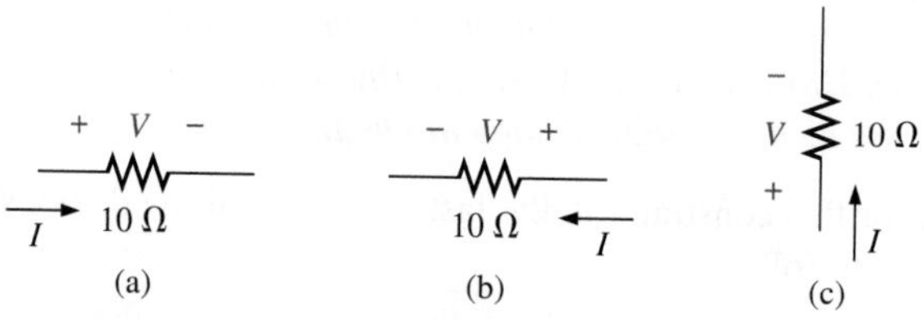

FIG. 5.14
Inserting the polarities across a resistor as determined by the direction of the current.

The magnitude of the voltage drop across each resistor can then be found by applying Ohm's law using only the resistance of each resistor. That is,

$$\boxed{\begin{aligned} V_1 &= I_1R_1 \\ V_2 &= I_2R_2 \\ V_3 &= I_3R_3 \end{aligned}} \qquad \textbf{(5.4)}$$

which for Fig. 5.12 results in

$$V_1 = I_1R_1 = I_sR_1 = (60\text{ mA})(10\ \Omega) = \mathbf{0.6\ V}$$
$$V_2 = I_2R_2 = I_sR_2 = (60\text{ mA})(30\ \Omega) = \mathbf{1.8\ V}$$
$$V_3 = I_3R_3 = I_sR_3 = (60\text{ mA})(100\ \Omega) = \mathbf{6.0\ V}$$

Note that in all the numerical calculations appearing in the text thus far, a unit of measurement has been applied to each calculated quantity. Always remember that a quantity without a unit of measurement is often meaningless.

EXAMPLE 5.4 For the series circuit in Fig. 5.15:

a. Find the total resistance R_T.
b. Calculate the resulting source current I_s.
c. Determine the voltage across each resistor.

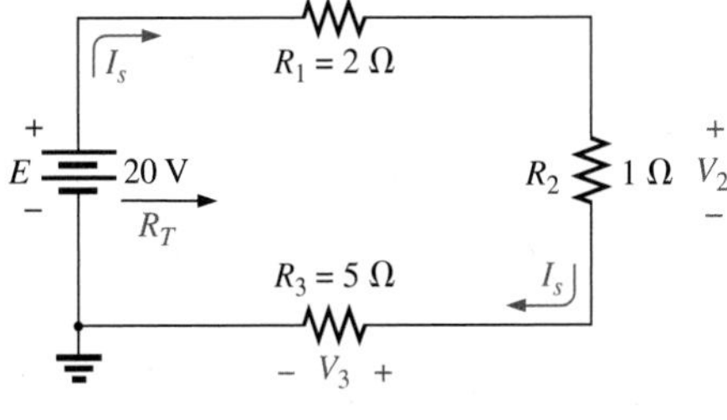

FIG. 5.15
Series circuit to be investigated in Example 5.4.

Solutions:

a. $R_T = R_1 + R_2 + R_3$
$= 2\ \Omega + 1\ \Omega + 5\ \Omega$
$R_T = \mathbf{8\ \Omega}$

b. $I_s = \dfrac{E}{R_T} = \dfrac{20\text{ V}}{8\ \Omega} = \mathbf{2.5\ A}$

c. $V_1 = I_1R_1 = I_sR_1 = (2.5\text{ A})(2\ \Omega) = \mathbf{5\ V}$
$V_2 = I_2R_2 = I_sR_2 = (2.5\text{ A})(1\ \Omega) = \mathbf{2.5\ V}$
$V_3 = I_3R_3 = I_sR_3 = (2.5\text{ A})(5\ \Omega) = \mathbf{12.5\ V}$

EXAMPLE 5.5 For the series circuit in Fig. 5.16:

a. Find the total resistance R_T.
b. Determine the source current I_s and indicate its direction on the circuit.
c. Find the voltage across resistor R_2 and indicate its polarity on the circuit.

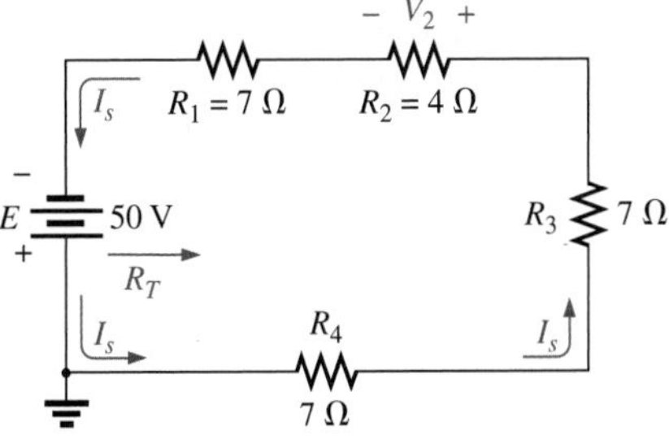

FIG. 5.16
Series circuit to be analyzed in Example 5.5.

Solutions:

a. The elements of the circuit are rearranged as shown in Fig. 5.17.

$$R_T = R_2 + NR$$
$$= 4\ \Omega + (3)(7\ \Omega)$$
$$= 4\ \Omega + 21\ \Omega$$
$$R_T = \mathbf{25\ \Omega}$$

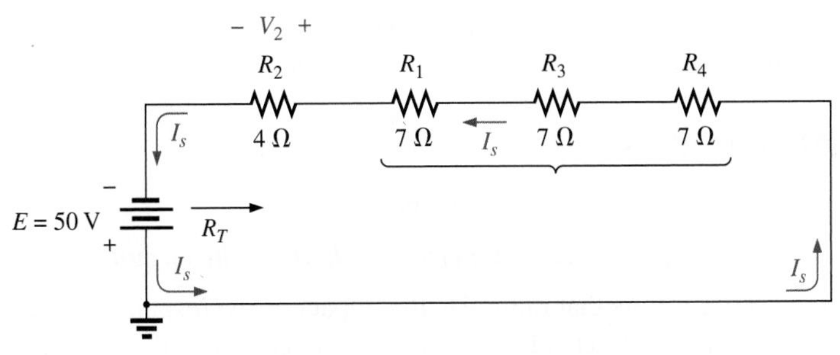

FIG. 5.17
Circuit in Fig. 5.16 redrawn to permit the use of Eq. (5.2).

b. Note that because of the manner in which the dc supply was connected, the current now has a counterclockwise direction as shown in Fig. 5.17.

$$I_s = \frac{E}{R_T} = \frac{50\text{ V}}{25\ \Omega} = \mathbf{2\ A}$$

c. The direction of the current will define the polarity for V_2 appearing in Fig. 5.17.

$$V_2 = I_2R_2 = I_sR_2 = (2\text{ A})(4\ \Omega) = \mathbf{8\ V}$$

Examples 5.4 and 5.5 are straightforward, substitution-type problems that are relatively easy to solve with some practice. Example 5.6, however, is another type of problem that requires both a firm grasp of the fundamental laws and equations and an ability to identify which quantity should be determined first. The best preparation for this type of exercise is to work through as many problems of this kind as possible.

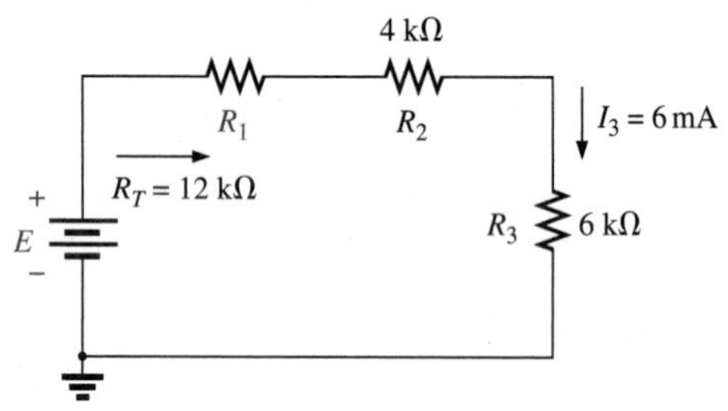

FIG. 5.18
Series circuit to be analyzed in Example 5.6.

EXAMPLE 5.6 Given R_T and I_3, calculate R_1 and E for the circuit in Fig. 5.18.

Solution: Since we are given the total resistance, it seems natural to first write the equation for the total resistance and then insert what we know.

$$R_T = R_1 + R_2 + R_3$$

We find that there is only one unknown, and it can be determined with some simple mathematical manipulations. That is,

$$12\text{ k}\Omega = R_1 + 4\text{ k}\Omega + 6\text{ k}\Omega = R_1 + 10\text{ k}\Omega$$

and $\qquad 12\text{ k}\Omega - 10\text{ k}\Omega = R_1$

so that $\qquad R_1 = \mathbf{2\ k\Omega}$

The dc voltage can be determined directly from Ohm's law.

$$E = I_sR_T = I_3R_T = (6\text{ mA})(12\text{ k}\Omega) = \mathbf{72\ V}$$

Analogies

The analogies used earlier to define the series connection are also excellent for the current of a series circuit. For instance, for the series-connected ropes, the stress on each rope **is the same** as they try to hold the heavy weight. For the water analogy, the flow of water **is the same** through each section of hose as the water is carried to its destination.

Instrumentation

Another important concept to remember is:

The insertion of any meter in a circuit will affect the circuit.

You must use meters that minimize the impact on the response of the circuit. The loading effects of meters are discussed in detail in a later section of this chapter. For now, we will assume that the meters are ideal and do not affect the networks to which they are applied so that we can concentrate on their proper usage.

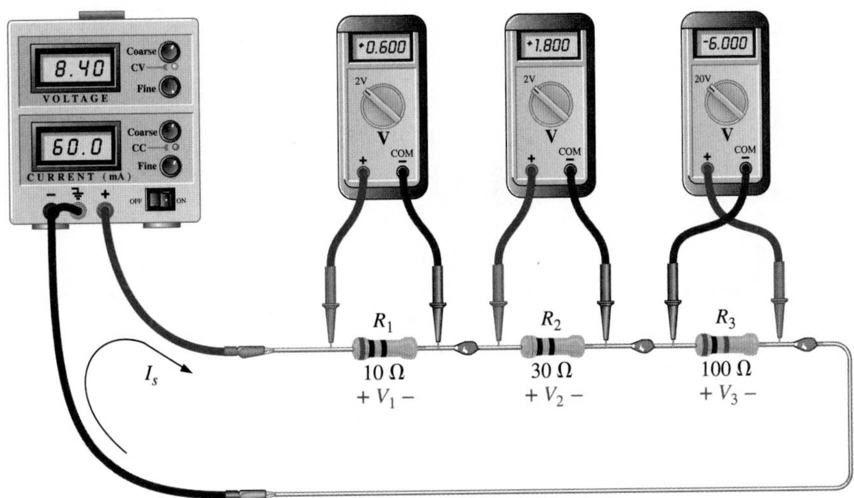

FIG. 5.19

Using voltmeters to measure the voltages across the resistors in Fig. 5.12.

Further, it is particularly helpful in the laboratory to realize that

the voltages of a circuit can be measured without disturbing (breaking the connections in) the circuit.

In Fig. 5.19, all the voltages of the circuit in Fig. 5.12 are being measured by voltmeters that were connected without disturbing the original configuration. Note that all the voltmeters are placed **across** the resistive elements. In addition, note that the positive (normally red) lead of the voltmeter is connected to the point of higher potential (positive sign), with the negative (normally black) lead of the voltmeter connected to the point of lower potential (negative sign) for V_1 and V_2. The result is a positive reading on the display. If the leads were reversed, the magnitude would remain the same, but a negative sign would appear as shown for V_3.

Take special note that the 20 V scale of our meter was used to measure the −6 V level, while the 2 V scale of our meter was used to measure the 0.6 V and 1.8 V levels. The maximum value of the chosen scale must always exceed the maximum value to be measured. In general,

when using a voltmeter, start with a scale that will ensure that the reading is less than the maximum value of the scale. Then work your way down in scales until the reading with the highest level of precision is obtained.

Turning our attention to the current of the circuit, we find that

using an ammeter to measure the current of a circuit requires that the circuit be broken at some point and the meter inserted in series with the branch in which the current is to be determined.

For instance, to measure the current leaving the positive terminal of the supply, the connection to the positive terminal must be removed to create an open circuit between the supply and resistor R_1. The ammeter is then inserted between these two points to form a bridge between the supply and the first resistor, as shown in Fig. 5.20. The ammeter is now in series with the supply and the other elements of the circuit. If each meter is to provide a positive reading, the connection must be made such that conventional current enters the positive terminal of the meter and leaves the

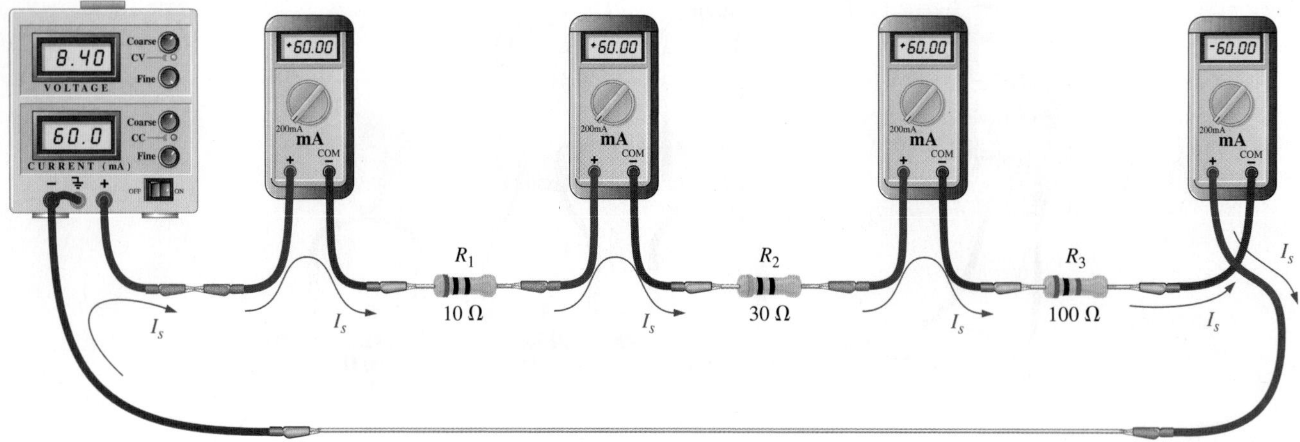

FIG. 5.20
Measuring the current throughout the series circuit in Fig. 5.12.

negative terminal. This was done for three of the ammeters, with the ammeter to the right of R_3 connected in the reverse manner. The result is a negative sign for the current. However, also note that the current has the correct magnitude. Since the current is 60 mA, the 200 mA scale of our meter was used for each meter.

As expected, the current at each point in the series circuit is the same using our ideal ammeters.

5.4 POWER DISTRIBUTION IN A SERIES CIRCUIT

In any electrical system, the power applied will equal the power dissipated or absorbed. For any series circuit, such as that in Fig. 5.21,

the power applied by the dc supply must equal that dissipated by the resistive elements.

In equation form,

$$\boxed{P_E = P_{R_1} + P_{R_2} + P_{R_3}} \qquad (5.5)$$

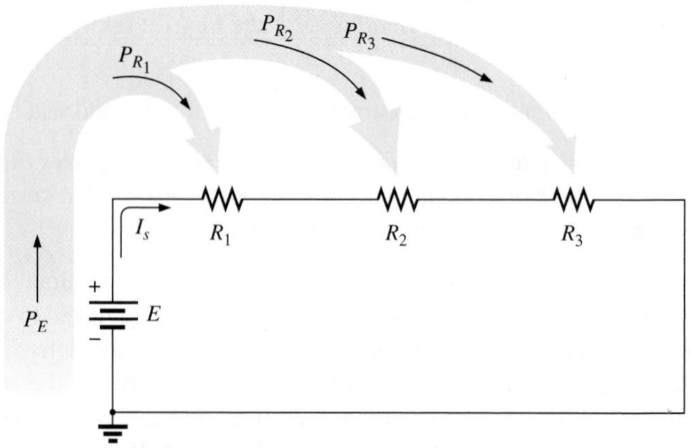

FIG. 5.21
Power distribution in a series circuit.

The power delivered by the supply can be determined using

$$P_E = EI_s \qquad \text{(watts, W)} \qquad \textbf{(5.6)}$$

The power dissipated by the resistive elements can be determined by any of the following forms (shown for resistor R_1 only):

$$P_1 = V_1 I_1 = I_1^2 R_1 = \frac{V_1^2}{R_1} \qquad \text{(watts, W)} \qquad \textbf{(5.7)}$$

Since the current is the same through series elements, you will find in the following examples that

in a series configuration, maximum power is delivered to the largest resistor.

EXAMPLE 5.7 For the series circuit in Fig. 5.22 (all standard values):

a. Determine the total resistance R_T.
b. Calculate the current I_s.
c. Determine the voltage across each resistor.
d. Find the power supplied by the battery.
e. Determine the power dissipated by each resistor.
f. Comment on whether the total power supplied equals the total power dissipated.

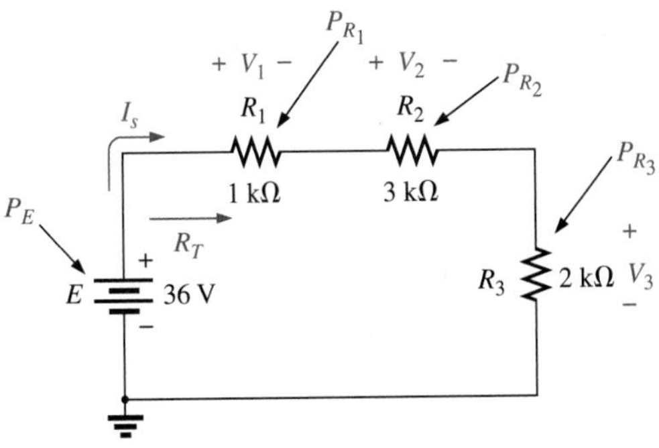

FIG. 5.22
Series circuit to be investigated in Example 5.7.

Solutions:

a. $R_T = R_1 + R_2 + R_3$
$= 1\text{ k}\Omega + 3\text{ k}\Omega + 2\text{ k}\Omega$
$R_T = \mathbf{6\text{ k}\Omega}$

b. $I_s = \dfrac{E}{R_T} = \dfrac{36\text{ V}}{6\text{ k}\Omega} = \mathbf{6\text{ mA}}$

c. $V_1 = I_1R_1 = I_sR_1 = (6\text{ mA})(1\text{ k}\Omega) = \mathbf{6\text{ V}}$
$V_2 = I_2R_2 = I_sR_2 = (6\text{ mA})(3\text{ k}\Omega) = \mathbf{18\text{ V}}$
$V_3 = I_3R_3 = I_sR_3 = (6\text{ mA})(2\text{ k}\Omega) = \mathbf{12\text{ V}}$

d. $P_E = EI_s = (36\text{ V})(6\text{ mA}) = \mathbf{216\text{ mW}}$

e. $P_1 = V_1I_1 = (6\text{ V})(6\text{ mA}) = \mathbf{36\text{ mW}}$
$P_2 = I_2^2R_2 = (6\text{ mA})^2(3\text{ k}\Omega) = \mathbf{108\text{ mW}}$
$P_3 = \dfrac{V_3^2}{R_3} = \dfrac{(12\text{ V})^2}{2\text{ k}\Omega} = \mathbf{72\text{ mW}}$

f. $P_E = P_{R_1} + P_{R_2} + P_{R_3}$
$216\text{ mW} = 36\text{ mW} + 108\text{ mW} + 72\text{ mW} = \mathbf{216\text{ mW}}$ (checks)

5.5 VOLTAGE SOURCES IN SERIES

Voltage sources can be connected in series, as shown in Fig. 5.23, to increase or decrease the total voltage applied to a system. The net voltage is determined by summing the sources with the same polarity and subtracting the total of the sources with the opposite polarity. The net polarity is the polarity of the larger sum.

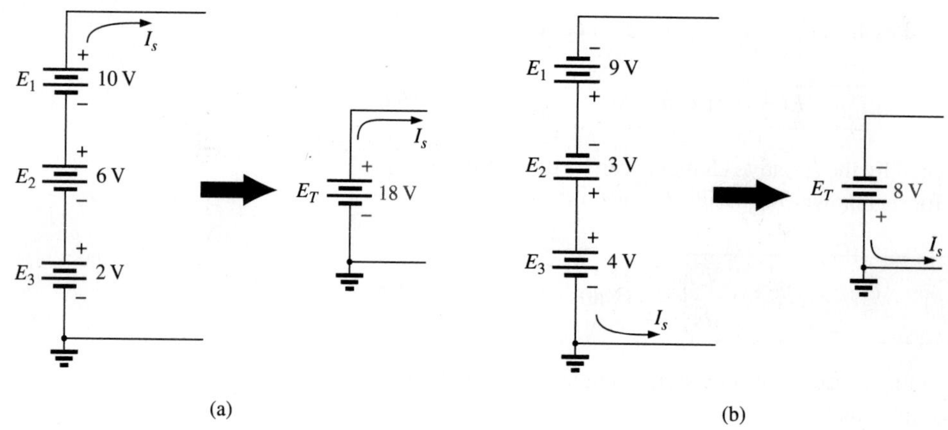

FIG. 5.23
Reducing series dc voltage sources to a single source.

In Fig. 5.23(a), for example, the sources are all "pressuring" current to follow a clockwise path, so the net voltage is

$$E_T = E_1 + E_2 + E_3 = 10\text{ V} + 6\text{ V} + 2\text{ V} = \mathbf{18\text{ V}}$$

as shown in the figure. In Fig. 5.23(b), however, the 4 V source is "pressuring" current in the clockwise direction while the other two are trying to establish current in the counterclockwise direction. In this case, the applied voltage for a counterclockwise direction is greater than that for the clockwise direction. The result is the counterclockwise direction for the current as shown in Fig. 5.23(b). The net effect can be determined by finding the difference in applied voltage between those supplies "pressuring" current in one direction and the total in the other direction. In this case,

$$E_T = E_1 + E_2 - E_3 = 9\text{ V} + 3\text{ V} - 4\text{ V} = \mathbf{8\text{ V}}$$

with the polarity shown in the figure.

Instrumentation

The connection of batteries in series to obtain a higher voltage is common in much of today's portable electronic equipment. For example, in Fig. 5.24(a), four 1.5 V AAA batteries have been connected in series to obtain a source voltage of 6 V. Although the voltage has increased, keep in mind that the maximum current for each AAA battery and for the 6 V supply is still the same. However, the power available has increased by a factor of 4 due to the increase in terminal voltage. Note also, as mentioned in Chapter 2, that the negative end of each battery is connected to the spring, and the positive end to the solid contact. In addition, note how the connection is made between batteries using the horizontal connecting tabs.

In general, supplies with only two terminals (+ and −) can be connected as shown for the batteries. A problem arises, however, if the supply has an optional or fixed internal ground connection. In Fig. 5.24(b), two laboratory supplies have been connected in series with both grounds connected. The result is a shorting out of the lower source E_1 (which may damage the supply if the protective fuse does not activate quickly enough) because both grounds are at zero potential. In such cases, the supply E_2 must be left ungrounded (floating), as shown in Fig. 5.24(c), to provide the 60 V terminal voltage. If the laboratory supplies have an internal connection from the negative terminal to ground as a protective

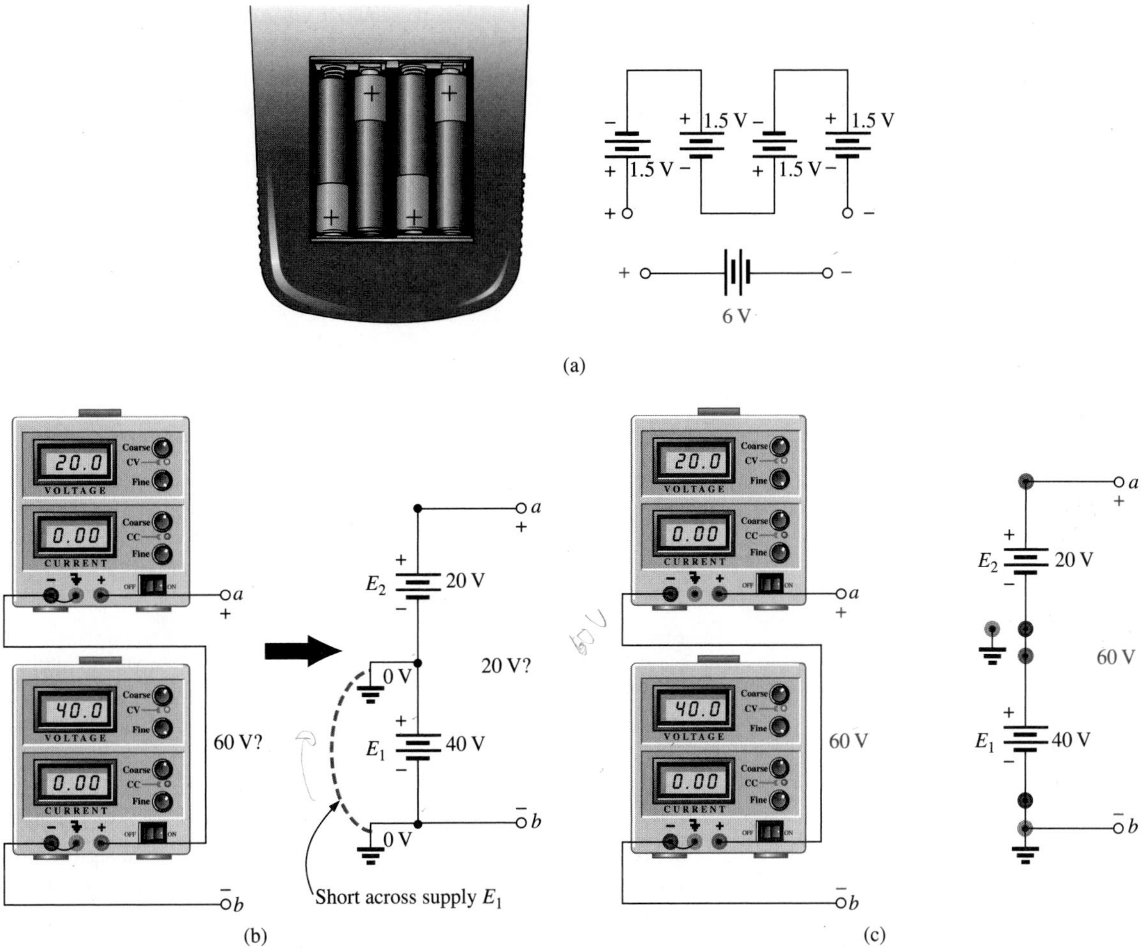

FIG. 5.24
Series connection of dc supplies: (a) four 1.5 V batteries in series to establish a terminal voltage of 6 V; (b) incorrect connections for two series dc supplies; (c) correct connection of two series supplies to establish 60 V at the output terminals.

feature for the users, a series connection of supplies cannot be made. Be aware of this fact, because some educational institutions add an internal ground to the supplies as a protective feature even though the panel still displays the ground connection as an optional feature.

5.6 KIRCHHOFF'S VOLTAGE LAW

The law to be described in this section is one of the most important in this field. It has application not only to dc circuits but also to any type of signal—whether it be ac, digital, and so on. This law is far-reaching and can be very helpful in working out solutions to networks that sometimes leave us lost for a direction of investigation.

The law, called **Kirchhoff's voltage law (KVL),** was developed by Gustav Kirchhoff (Fig. 5.25) in the mid-1800s. It is a cornerstone of the entire field and, in fact, will never be outdated or replaced.

The application of the law requires that we define a closed path of investigation, permitting us to start at one point in the network, travel through the network, and find our way back to the original starting point. The path does not have to be circular, square, or any other defined shape; it must simply provide a way to leave a point and get back to it without leaving the

FIG. 5.25
Gustav Robert Kirchhoff.
Courtesy of the Smithsonian Institution, Photo No. 58,283.

German (Königsberg, Berlin)
(1824–87),
Physicist
Professor of Physics, University of Heidelberg

Although a contributor to a number of areas in the physics domain, he is best known for his work in the electrical area with his definition of the relationships between the currents and voltages of a network in 1847. Did extensive research with German chemist Robert Bunsen (developed the *Bunsen burner*), resulting in the discovery of the important elements of *cesium and rubidium.*

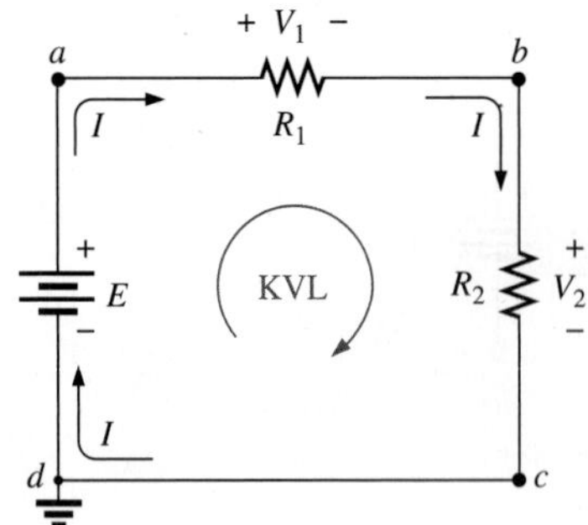

FIG. 5.26
Applying Kirchhoff's voltage law to a series dc circuit.

network. In Fig. 5.26, if we leave point a and follow the current, we will end up at point b. Continuing, we can pass through points c and d and eventually return through the voltage source to point a, our starting point. The path *abcda* is therefore a closed path, or **closed loop.** The law specifies that

the algebraic sum of the potential rises and drops around a closed path (or closed loop) is zero.

In symbolic form it can be written as

$$\Sigma_{\circlearrowright} V = 0 \qquad \text{(Kirchhoff's voltage law in symbolic form)} \quad \mathbf{(5.8)}$$

where Σ represents summation, $\circlearrowright$ the closed loop, and V the potential drops and rises. The term *algebraic* simply means paying attention to the signs that result in the equations as we add and subtract terms.

The first question that often arises is, Which way should I go around the closed path? Should I always follow the direction of the current? To simplify matters, this text will always try to move in a clockwise direction. By selecting a direction, you eliminate the need to think about which way would be more appropriate. Any direction will work as long as you get back to the starting point.

Another question is, How do I apply a sign to the various voltages as I proceed in a clockwise direction? For a particular voltage, we will assign a positive sign when proceeding from the negative to positive potential—a positive experience such as moving from a negative checking balance to a positive one. The opposite change in potential level results in a negative sign. In Fig. 5.26, as we proceed from point d to point a across the voltage source, we move from a negative potential (the negative sign) to a positive potential (the positive sign), so a positive sign is given to the source voltage E. As we proceed from point a to point b, we encounter a positive sign followed by a negative sign, so a drop in potential has occurred, and a negative sign is applied. Continuing from b to c, we encounter another drop in potential, so another negative sign is applied. We then arrive back at the starting point d, and the resulting sum is set equal to zero as defined by Eq. (5.8).

Writing out the sequence with the voltages and the signs results in the following:

$$+E - V_1 - V_2 = 0$$

which can be rewritten as $E = V_1 + V_2$

The result is particularly interesting because it tells us that

the applied voltage of a series dc circuit will equal the sum of the voltage drops of the circuit.

Kirchhoff's voltage law can also be written in the following form:

$$\Sigma_{\circlearrowright} V_{\text{rises}} = \Sigma_{\circlearrowright} V_{\text{drops}} \qquad \mathbf{(5.9)}$$

revealing that

the sum of the voltage rises around a closed path will always equal the sum of the voltage drops.

To demonstrate that the direction that you take around the loop has no effect on the results, let's take the counterclockwise path and compare results. The resulting sequence appears as

$$-E + V_2 + V_1 = 0$$

yielding the same result of $E = V_1 + V_2$

EXAMPLE 5.8 Use Kirchhoff's voltage law to determine the unknown voltage for the circuit in Fig. 5.27.

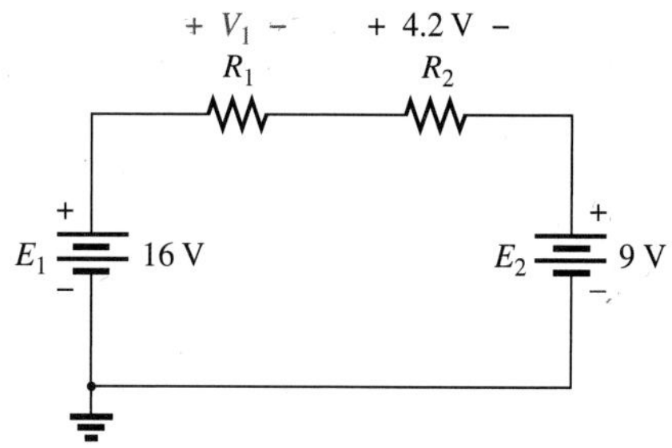

FIG. 5.27
Series circuit to be examined in Example 5.8.

Solution: When applying Kirchhoff's voltage law, be sure to concentrate on the polarities of the voltage rise or drop rather than on the type of element. In other words, do not treat a voltage drop across a resistive element differently from a voltage rise (or drop) across a source. If the polarity dictates that a drop has occurred, that is the important fact, not whether it is a resistive element or source.

Application of Kirchhoff's voltage law to the circuit in Fig. 5.27 in the clockwise direction results in

$$+E_1 - V_1 - V_2 - E_2 = 0$$

and
$$V_1 = E_1 - V_2 - E_2$$
$$= 16\text{ V} - 4.2\text{ V} - 9\text{ V}$$

so
$$V_1 = \mathbf{2.8\text{ V}}$$

The result clearly indicates that you do not need to know the values of the resistors or the current to determine the unknown voltage. Sufficient information was carried by the other voltage levels to determine the unknown.

EXAMPLE 5.9 Determine the unknown voltage for the circuit in Fig. 5.28.

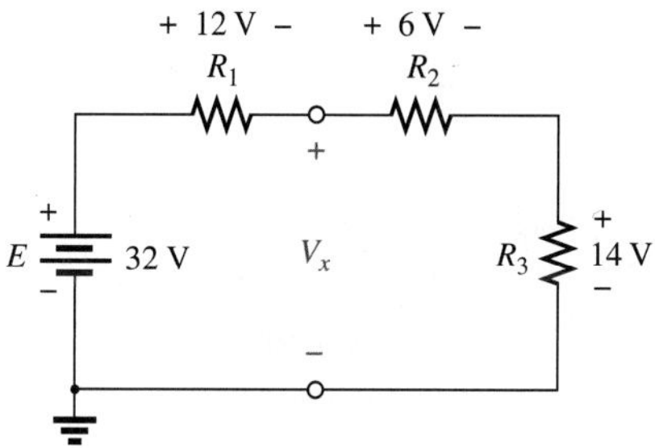

FIG. 5.28
Series dc circuit to be analyzed in Example 5.9.

Solution: In this case, the unknown voltage is not across a single resistive element but between two arbitrary points in the circuit. Simply apply Kirchhoff's voltage law around a path, including the source or resistor R_3. For the clockwise path, including the source, the resulting equation is the following:

$$+E - V_1 - V_x = 0$$

and
$$V_x = E - V_1 = 32\text{ V} - 12\text{ V} = \mathbf{20\text{ V}}$$

For the clockwise path, including resistor R_3, the following results:

$$+V_x - V_2 - V_3 = 0$$

and
$$V_x = V_2 + V_3$$
$$= 6\text{ V} + 14\text{ V}$$

with
$$V_x = \mathbf{20\text{ V}}$$

providing exactly the same solution.

There is no requirement that the followed path have charge flow or current. In Example 5.10, the current is zero everywhere, but Kirchhoff's voltage law can still be applied to determine the voltage between the points of interest. Also, there will be situations where the actual polarity will not be provided. In such cases, simply assume a polarity. If the answer is negative, the magnitude of the result is correct, but the polarity should be reversed.

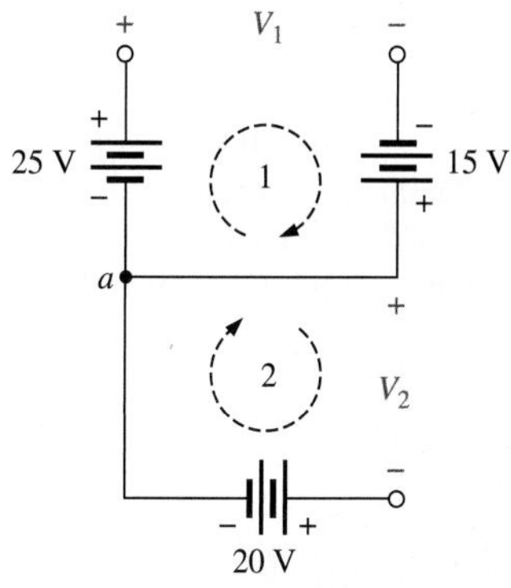

FIG. 5.29
Combination of voltage sources to be examined in Example 5.10.

EXAMPLE 5.10 Using Kirchhoff's voltage law, determine voltages V_1 and V_2 for the network in Fig. 5.29.

Solution: For path 1, starting at point a in a clockwise direction,

$$+25\text{ V} - V_1 + 15\text{ V} = 0$$

and
$$V_1 = \mathbf{40\text{ V}}$$

For path 2, starting at point a in a clockwise direction,

$$-V_2 - 20\text{ V} = 0$$

and $V_2 = \mathbf{-20\text{ V}}$

The minus sign in the solution simply indicates that the actual polarities are different from those assumed.

The next example demonstrates that you do not need to know what elements are inside a container when applying Kirchhoff's voltage law. They could all be voltage sources or a mix of sources and resistors. It doesn't matter—simply pay strict attention to the polarities encountered.

Try to find the unknown quantities in the next examples without looking at the solutions. It will help define where you may be having trouble.

Example 5.11 emphasizes the fact that when you are applying Kirchhoff's voltage law, the polarities of the voltage rise or drop are the important parameters, not the type of element involved.

EXAMPLE 5.11 Using Kirchhoff's voltage law, determine the unknown voltage for the circuit in Fig. 5.30.

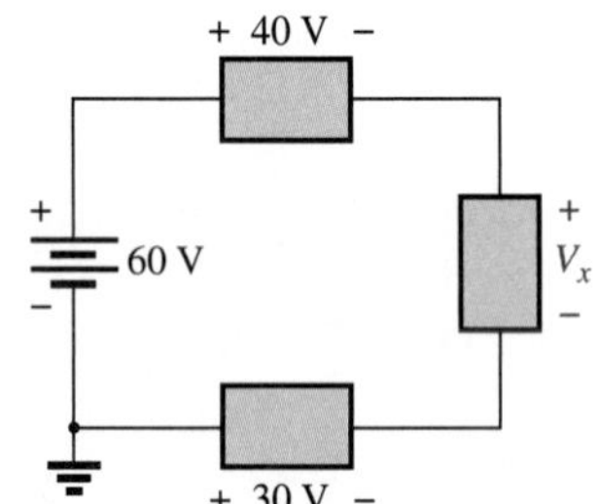

FIG. 5.30
Series configuration to be examined in Example 5.11.

Solution: Note that in this circuit, there are various polarities across the unknown elements since they can contain any mixture of components. Applying Kirchhoff's voltage law in the clockwise direction results in

$$+60\text{ V} - 40\text{ V} - V_x + 30\text{ V} = 0$$

and $V_x = 60\text{ V} + 30\text{ V} - 40\text{ V} = 90\text{ V} - 40\text{ V}$

with $V_x = \mathbf{50\text{ V}}$

EXAMPLE 5.12 Determine the voltage V_x for the circuit in Fig. 5.31. Note that the polarity of V_x was not provided.

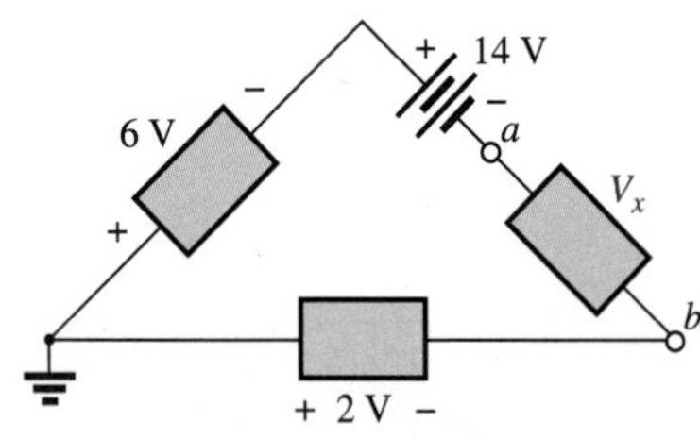

FIG. 5.31
Applying Kirchhoff's voltage law to a circuit in which the polarities have not been provided for one of the voltages (Example 5.12).

Solution: For cases where the polarity is not included, simply make an assumption about the polarity, and apply Kirchhoff's voltage law as before. If the result has a positive sign, the assumed polarity was correct. If the result has a minus sign, the **magnitude is correct,** but the assumed polarity must be reversed. In this case, if we assume point a to be positive and point b to be negative, an application of Kirchhoff's voltage law in the clockwise direction results in

$$-6\text{ V} - 14\text{ V} - V_x + 2\text{ V} = 0$$

and $V_x = -20\text{ V} + 2\text{ V}$

so that $V_x = \mathbf{-18\text{ V}}$

Since the result is negative, we know that point a should be negative and point b should be positive, but the magnitude of 18 V is correct.

EXAMPLE 5.13 For the series circuit in Fig. 5.32.

a. Determine V_2 using Kirchhoff's voltage law.
b. Determine current I_2.
c. Find R_1 and R_3.

Solution:

a. Applying Kirchhoff's voltage law in the clockwise direction starting at the negative terminal of the supply results in

$$-E + V_3 + V_2 + V_1 = 0$$

and $E = V_1 + V_2 + V_3$ (as expected)

so that $V_2 = E - V_1 - V_3 = 54\text{ V} - 18\text{ V} - 15\text{ V}$

and $V_2 = \mathbf{21\text{ V}}$

b. $I_2 = \dfrac{V_2}{R_2} = \dfrac{21\text{ V}}{7\,\Omega}$

$I_2 = \mathbf{3A}$

c. $R_1 = \dfrac{V_1}{I_1} = \dfrac{18\text{ V}}{3\text{ A}} = \mathbf{6\,\Omega}$

with $R_3 = \dfrac{V_3}{I_3} = \dfrac{15\text{ V}}{3\text{ A}} = \mathbf{5\,\Omega}$

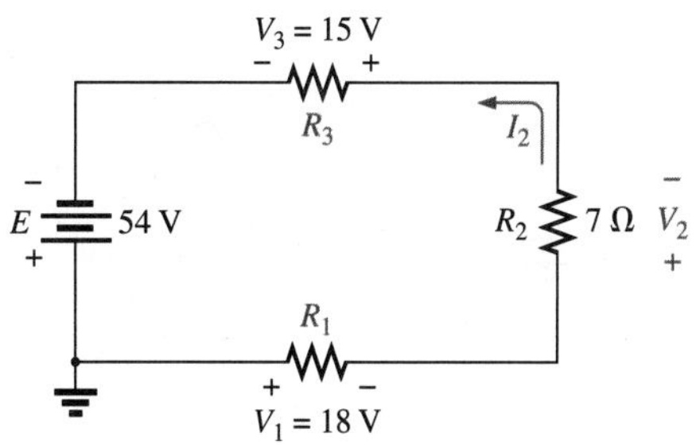

FIG. 5.32
Series configuration to be examined in Example 5.13.

EXAMPLE 5.14 Using Kirchhoff's voltage law and Fig. 5.12, verify Eq. (5.1).

Solution: Applying Kirchhoff's voltage law around the closed path:

$$E = V_1 + V_2 + V_3$$

Substituting Ohm's law:

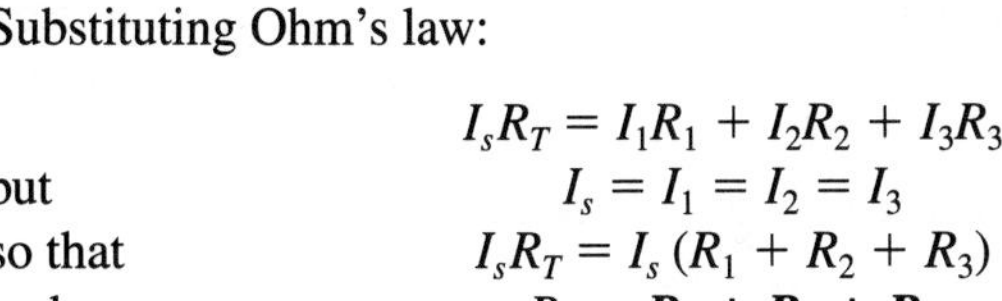

$$I_sR_T = I_1R_1 + I_2R_2 + I_3R_3$$

but $I_s = I_1 = I_2 = I_3$

so that $I_sR_T = I_s(R_1 + R_2 + R_3)$

and $R_T = \mathbf{R_1 + R_2 + R_3}$

which is Eq. (5.1).

5.7 VOLTAGE DIVISION IN A SERIES CIRCUIT

The previous section demonstrated that the sum of the voltages across the resistors of a series circuit will always equal the applied voltage. It cannot be more or less than that value. The next question is, How will a resistor's value affect the voltage across the resistor? It turns out that

the voltage across series resistive elements will divide as the magnitude of the resistance levels.

In other words,

in a series resistive circuit, the larger the resistance, the more of the applied voltage it will capture.

In addition,

the ratio of the voltages across series resistors will be the same as the ratio of their resistance levels.

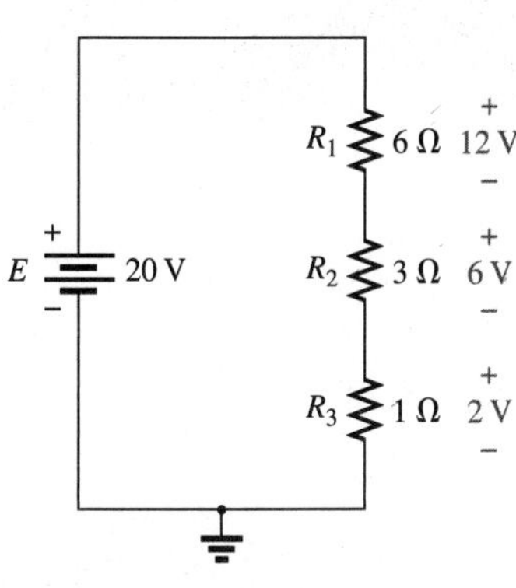

FIG. 5.33
Revealing how the voltage will divide across series resistive elements.

All of the above statements can best be described by a few simple examples. In Fig. 5.33, all the voltages across the resistive elements are provided. The largest resistor of 6 Ω captures the bulk of the applied voltage, while the smallest resistor, R_3, has the least. In addition, note that since the resistance level of R_1 is six times that of R_3, the voltage across R_1 is six times that of R_3. The fact that the resistance level of R_2 is three times that of R_1

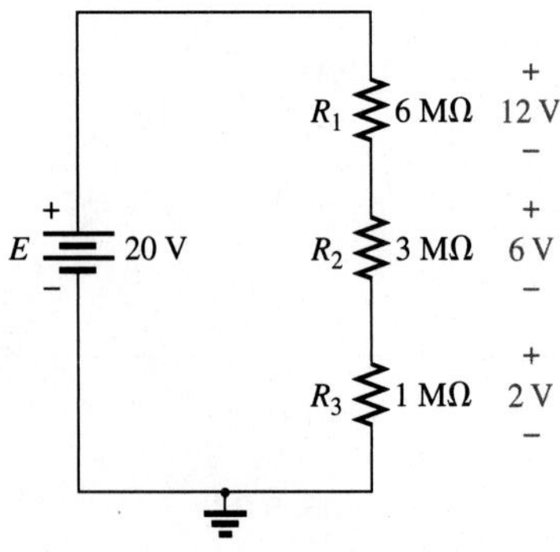

FIG. 5.34
The ratio of the resistive values determines the voltage division of a series dc circuit.

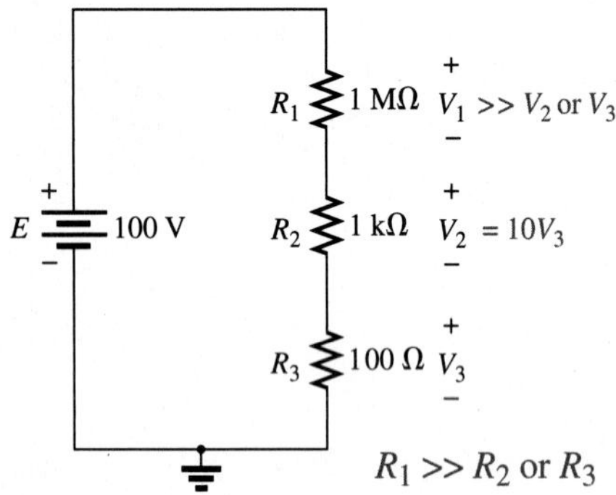

FIG. 5.35
The largest of the series resistive elements will capture the major share of the applied voltage.

results in three times the voltage across R_2. Finally, since R_1 is twice R_2, the voltage across R_1 is twice that of R_2. In general, therefore, the voltage across series resistors will have the same ratio as their resistance levels.

Note that if the resistance levels of all the resistors in Fig. 5.33 are increased by the same amount, as shown in Fig. 5.34, the voltage levels all remain the same. In other words, even though the resistance levels were increased by a factor of 1 million, the voltage ratios remained the same. Clearly, therefore, it is the ratio of resistor values that counts when it comes to voltage division, not the magnitude of the resistors. The current level of the network will be severely affected by this change in resistance level, but the voltage levels remain unaffected.

Based on the above, it should now be clear that when you first encounter a circuit such as that in Fig. 5.35, you will expect that the voltage across the 1 MΩ resistor will be much greater than that across the 1 kΩ or the 100 Ω resistors. In addition, based on a statement above, the voltage across the 1 kΩ resistor will be 10 times as great as that across the 100 Ω resistor since the resistance level is 10 times as much. Certainly, you would expect that very little voltage will be left for the 100 Ω resistor. Note that the current was never mentioned in the above analysis. The distribution of the applied voltage is determined solely by the ratio of the resistance levels. Of course, the magnitude of the resistors will determine the resulting current level.

To continue with the above, since 1 MΩ is 1000 times larger than 1 kΩ, voltage V_1 will be 1000 times larger than V_2. In addition, voltage V_2 will be 10 times larger than V_3. Finally, the voltage across the largest resistor of 1 MΩ will be (10)(1000) = 10,000 times larger than V_3.

Now for some details. The total resistance is

$$\begin{aligned} R_T &= R_1 + R_2 + R_3 \\ &= 1\text{ M}\Omega + 1\text{ k}\Omega + 100\ \Omega \\ R_T &= \mathbf{1{,}001{,}100\ \Omega} \end{aligned}$$

The current is

$$I_s = \frac{E}{R_T} = \frac{100\text{ V}}{1{,}001{,}100\ \Omega} \cong 99.89\ \mu\text{A} \qquad \text{(about } 100\ \mu\text{A)}$$

with

$$\begin{aligned} V_1 &= I_1R_1 = I_sR_1 = (99.89\ \mu\text{A})(1\text{ M}\Omega) = \mathbf{99.89\ V} \quad \text{(almost the full 100 V)} \\ V_2 &= I_2R_2 = I_sR_2 = (99.89\ \mu\text{A})(1\text{ k}\Omega) = \mathbf{99.89\ mV} \quad \text{(about 100 mV)} \\ V_3 &= I_3R_3 = I_sR_3 = (99.89\ \mu\text{A})(100\ \Omega) = \mathbf{9.989\ mV} \quad \text{(about 10 mV)} \end{aligned}$$

As illustrated above, the major part of the applied voltage is across the 1 MΩ resistor. The current is in the microampere due primarily to the large 1 MΩ resistor. Voltage V_2 is about 0.1 V compared to almost 100 V for V_1. The voltage across R_3 is only about 10 mV, or 0.010 V.

Before making any detailed, lengthy calculations, you should first examine the resistance levels of the series resistors to develop some idea of how the applied voltage will be divided through the circuit. It will reveal, with a minumum amount of effort, what you should expect when performing the calculations (a checking mechanism). It also allows you to speak intelligently about the response of the circuit without having to resort to any calculations.

Voltage Divider Rule (VDR)

The **voltage divider rule (VDR)** permits the determination of the voltage across a series resistor without first having to determine the current of the circuit. The rule itself can be derived by analyzing the simple series circuit in Fig. 5.36.

First, determine the total resistance as follows:

$$R_T = R_1 + R_2$$

Then
$$I_s = I_1 = I_2 = \frac{E}{R_T}$$

Apply Ohm's law to each resistor:

$$V_1 = I_1R_1 = \left(\frac{E}{R_T}\right)R_1 = R_1\frac{E}{R_T}$$

$$V_2 = I_2R_2 = \left(\frac{E}{R_T}\right)R_2 = R_2\frac{E}{R_T}$$

The resulting format for V_1 and V_2 is

$$V_x = R_x\frac{E}{R_T} \qquad \text{(voltage divider rule)} \qquad \textbf{(5.10)}$$

where V_x is the voltage across the resistor R_x, E is the impressed voltage across the series elements, and R_T is the total resistance of the series circuit.

The voltage divider rule states that

the voltage across a resistor in a series circuit is equal to the value of that resistor times the total applied voltage divided by the total resistance of the series configuration.

Although Eq. (5.10) was derived using a series circuit of only two elements, it can be used for series circuits with any number of series resistors.

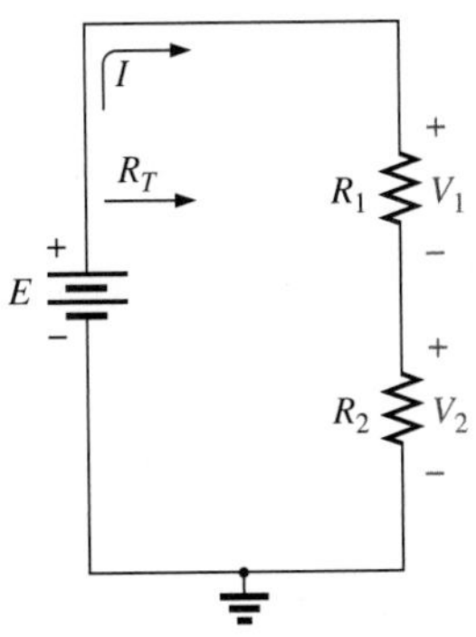

FIG. 5.36
Developing the voltage divider rule.

EXAMPLE 5.15 For the series circuit in Fig. 5.37.

a. Without making any calculations, how much larger would you expect the voltage across R_2 to be compared to that across R_1?
b. Find the voltage V_1 using only the voltage divider rule.
c. Using the conclusion of part (a), determine the voltage across R_2.
d. Use the voltage divider rule to determine the voltage across R_2, and compare your answer to your conclusion in part (c).
e. How does the sum of V_1 and V_2 compare to the applied voltage?

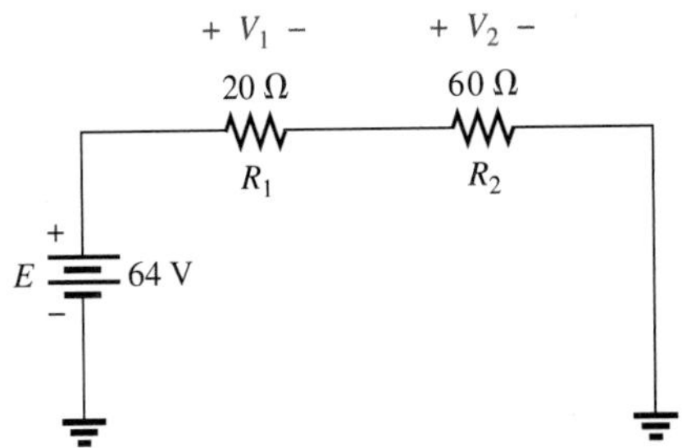

FIG. 5.37
Series circuit to be examined using the voltage divider rule in Example 5.15.

Solutions:

a. Since resistor R_2 is three times R_1, it is expected that $V_2 = 3V_1$.

b. $V_1 = R_1\frac{E}{R_T} = 20\ \Omega\left(\frac{64\text{ V}}{20\ \Omega + 60\ \Omega}\right) = 20\ \Omega\left(\frac{64\text{ V}}{80\ \Omega}\right) = \mathbf{16\ V}$

c. $V_2 = 3V_1 = 3(16\text{ V}) = \mathbf{48\ V}$

d. $V_2 = R_2\frac{E}{R_T} = (60\ \Omega)\left(\frac{64\text{ V}}{80\ \Omega}\right) = \mathbf{48\ V}$

   The results are an exact match.

e. $E = V_1 + V_2$
   $64\text{ V} = 16\text{ V} + 48\text{ V} = \mathbf{64\ V}$ (checks)

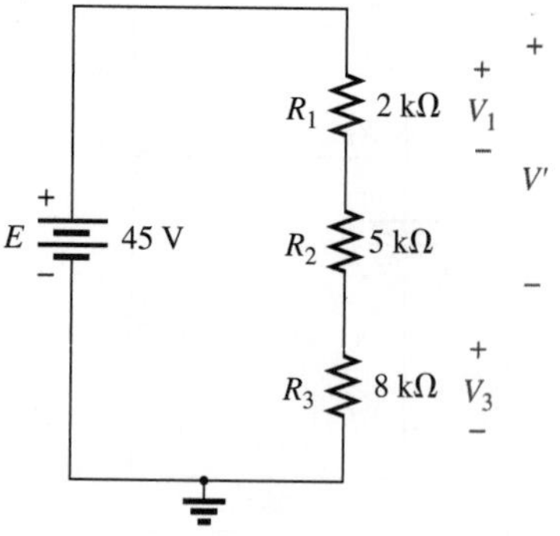

FIG. 5.38
Series circuit to be investigated in Examples 5.16 and 5.17.

EXAMPLE 5.16 Using the voltage divider rule, determine voltages V_1 and V_3 for the series circuit in Fig. 5.38.

Solution:

$$R_T = R_1 + R_2 + R_3$$
$$= 2\ \text{k}\Omega + 5\ \text{k}\Omega + 8\ \text{k}\Omega$$
$$R_T = 15\text{k}\Omega$$

$$V_1 = R_1\frac{E}{R_T} = 2\ \text{k}\Omega\left(\frac{45\ \text{V}}{15\text{k}\Omega}\right) = \mathbf{6\ V}$$

and

$$V_3 = R_3\frac{E}{R_T} = 8\ \text{k}\Omega\left(\frac{45\ \text{V}}{15\ \Omega}\right) = \mathbf{24\ V}$$

The voltage divider rule can be extended to the voltage across two or more series elements if the resistance in the numerator of Eq. (5.10) is expanded to include the total resistance of the series resistors across which the voltage is to be found (R'). That is,

$$V' = R'\frac{E}{R_T} \qquad \textbf{(5.11)}$$

EXAMPLE 5.17 Determine the voltage (denoted V') across the series combination of resistors R_1 and R_2 in Fig. 5.38.

Solution: Since the voltage desired is across both R_1 and R_2, the sum of R_1 and R_2 will be substituted as R' in Eq. (5.11). The result is

$$R' = R_1 + R_2 = 2\ \text{k}\Omega + 5\ \text{k}\Omega = 7\ \text{k}\Omega$$

and

$$V' = R'\frac{E}{R_T} = 7\ \text{k}\Omega\left(\frac{45\ \text{V}}{15\ \text{k}\Omega}\right) = \mathbf{21\ V}$$

In the next example you are presented with a problem of the other kind: Given the voltage division, you must determine the required resistor values. In most cases, problems of this kind simply require that you are able to use the basic equations introduced thus far in the text.

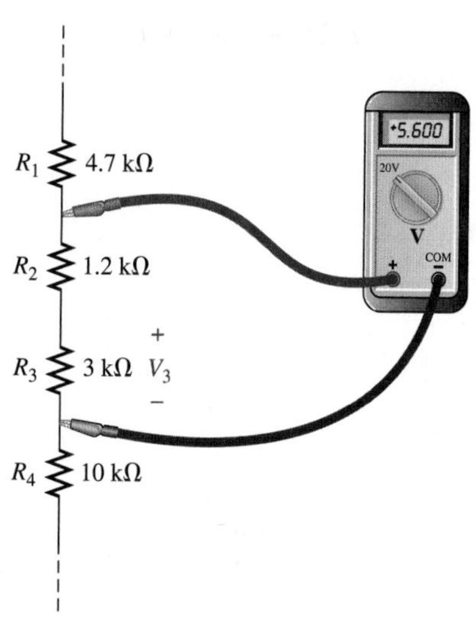

FIG. 5.39
Voltage divider action for Example 5.18.

EXAMPLE 5.18 Given the voltmeter reading in Fig. 5.39, find voltage V_3.

Solution: Even though the rest of the network is not shown and the current level has not been determined, the voltage divider rule can be applied by using the voltmeter reading as the full voltage across the series combination of resistors. That is,

$$V_3 = R_3\frac{(V_{\text{meter}})}{R_3 + R_2} = \frac{3\ \text{k}\Omega(5.6\ \text{V})}{3\ \text{k}\Omega + 1.2\ \text{k}\Omega}$$
$$V_3 = \mathbf{4\ V}$$

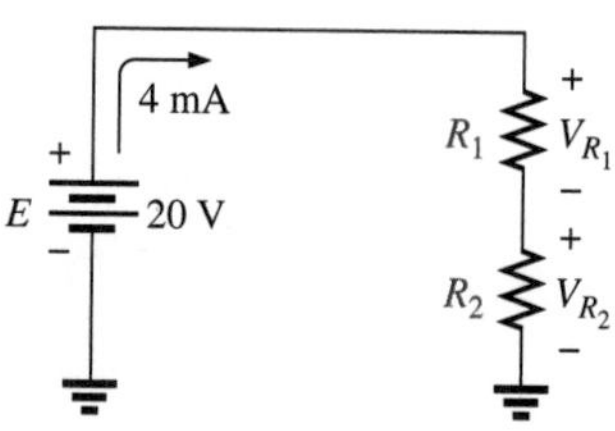

FIG. 5.40
Designing a voltage divider circuit (Example 5.19).

EXAMPLE 5.19 Design the voltage divider circuit in Fig. 5.40 such that the voltage across R_1 will be four times the voltage across R_2; that is, $V_{R_1} = 4V_{R_2}$.

Solution: The total resistance is defined by

$$R_T = R_1 + R_2$$

However, if $V_{R_1} = 4V_{R_2}$

then $R_1 = 4R_2$

so that $R_T = R_1 + R_2 = 4R_2 + R_2 = 5R_2$

Applying Ohm's law, we can determine the total resistance of the circuit:

$$R_T = \frac{E}{I_s} = \frac{20 \text{ V}}{4 \text{ mA}} = 5 \text{ k}\Omega$$

so $R_T = 5R_2 = 5 \text{ k}\Omega$

and $R_2 = \frac{5 \text{ k}\Omega}{5} = \mathbf{1\ k\Omega}$

Then $R_1 = 4R_2 = 4(1 \text{ k}\Omega) = \mathbf{4\ k\Omega}$

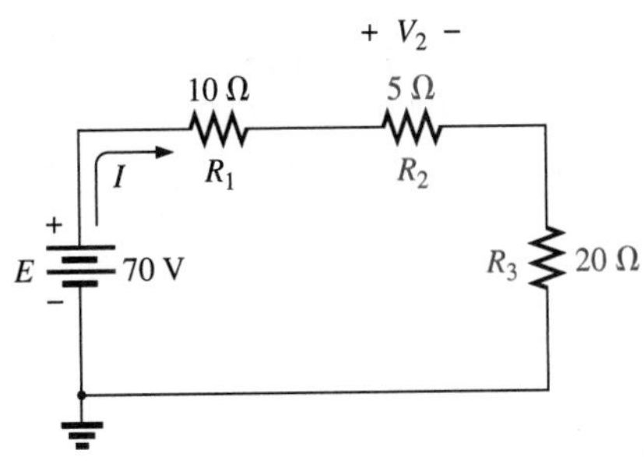

FIG. 5.41
Series dc circuit with elements to be interchanged.

5.8 INTERCHANGING SERIES ELEMENTS

The elements of a series circuit can be interchanged without affecting the total resistance, current, or power to each element. For instance, the network in Fig. 5.41 can be redrawn as shown in Fig. 5.42 without affecting I or V_2. The total resistance R_T is 35 Ω in both cases, and $I = 70 \text{ V}/35\ \Omega = 2$ A. The voltage $V_2 = IR_2 = (2 \text{ A})(5\ \Omega) = 10$ V for both configurations.

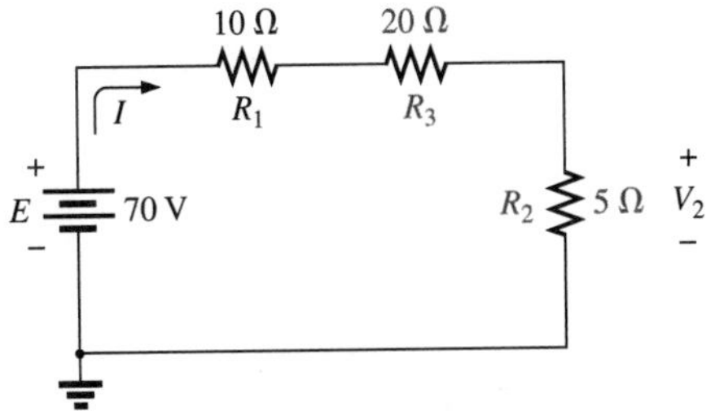

FIG. 5.42
Circuit in Fig. 5.41 with R_2 and R_3 interchanged.

EXAMPLE 5.20 Determine I and the voltage across the 7 Ω resistor for the network in Fig. 5.43.

Solution: The network is redrawn in Fig. 5.44.

$$R_T = (2)(4\ \Omega) + 7\ \Omega = 15\ \Omega$$

$$I = \frac{E}{R_T} = \frac{37.5 \text{ V}}{15\ \Omega} = \mathbf{2.5\ A}$$

$$V_{7\Omega} = IR = (2.5 \text{ A})(7\ \Omega) = \mathbf{17.5\ V}$$

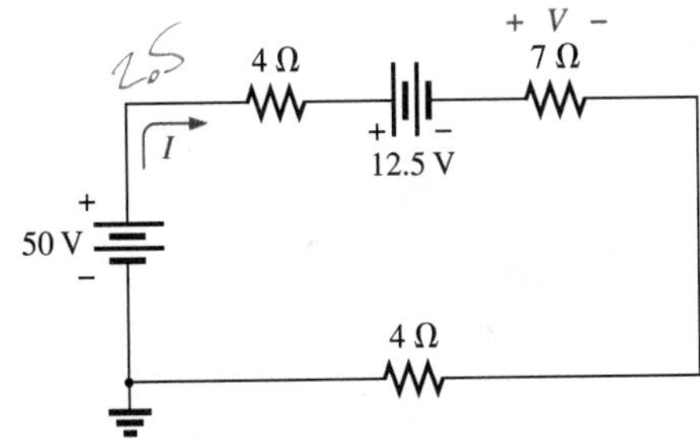

FIG. 5.43
Example 5.20.

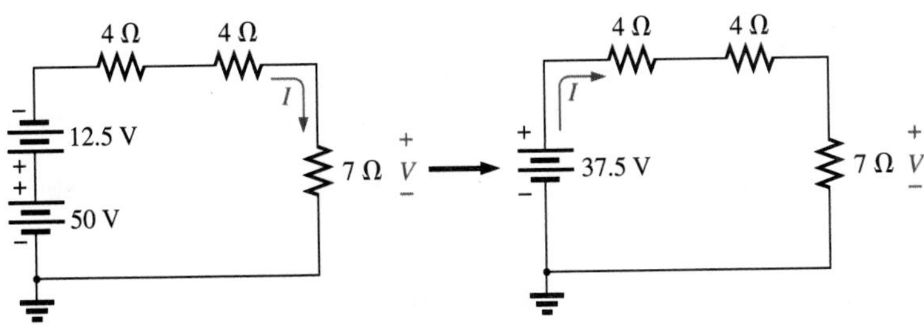

FIG. 5.44
Redrawing the circuit in Fig. 5.43.

5.9 NOTATION

Notation plays an increasingly important role in the analysis to follow. It is important, therefore, that we begin to examine the notation used throughout the industry.

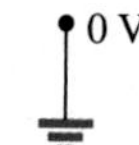

FIG. 5.45
Ground potential.

Voltage Sources and Ground

Except for a few special cases, electrical and electronic systems are grounded for reference and safety purposes. The symbol for the ground connection appears in Fig. 5.45 with its defined potential level—zero volts. A grounded circuit may appear as shown in Fig. 5.46(a), (b), or (c). In any case, it is understood that the negative terminal of the battery and the bottom of the resistor R_2 are at ground potential. Although Fig. 5.46(c) shows no connection between the two grounds, it is recognized that such a connection exists for the continuous flow of charge. If $E = 12$ V, then point a is 12 V positive with respect to ground potential, and 12 V exist across the series combination of resistors R_1 and R_2. If a voltmeter placed from point b to ground reads 4 V, then the voltage across R_2 is 4 V, with the higher potential at point b.

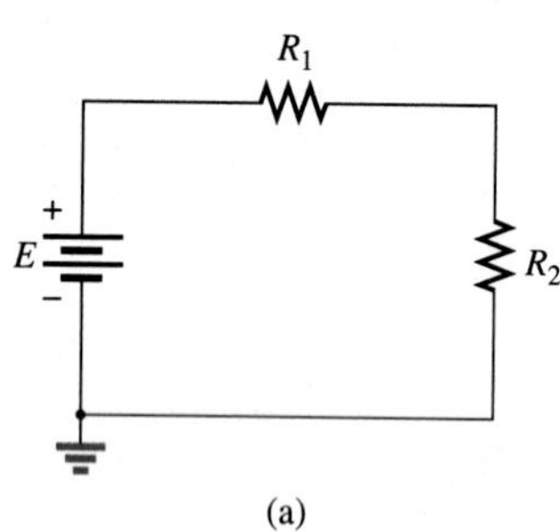

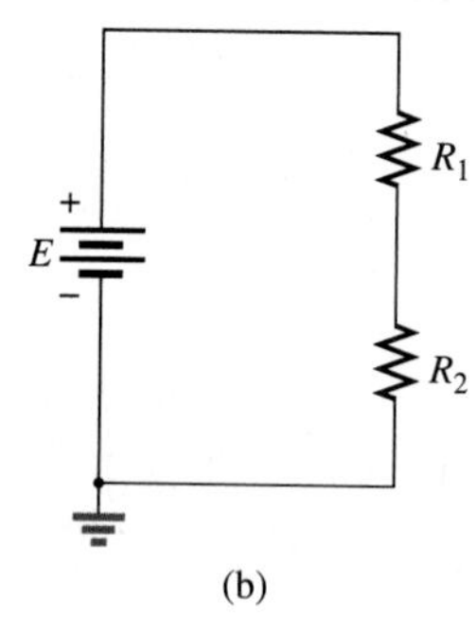

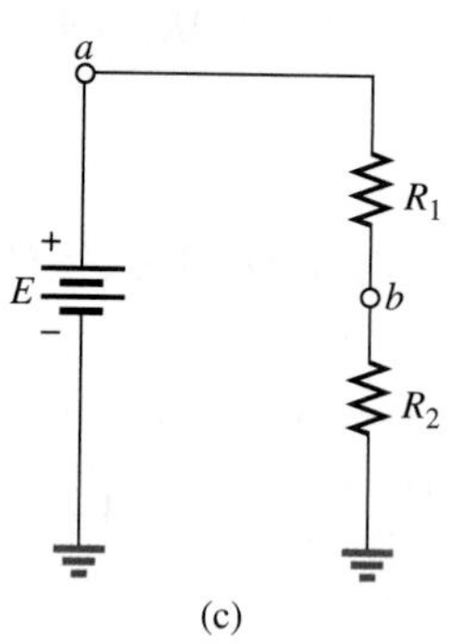

FIG. 5.46
Three ways to sketch the same series dc circuit.

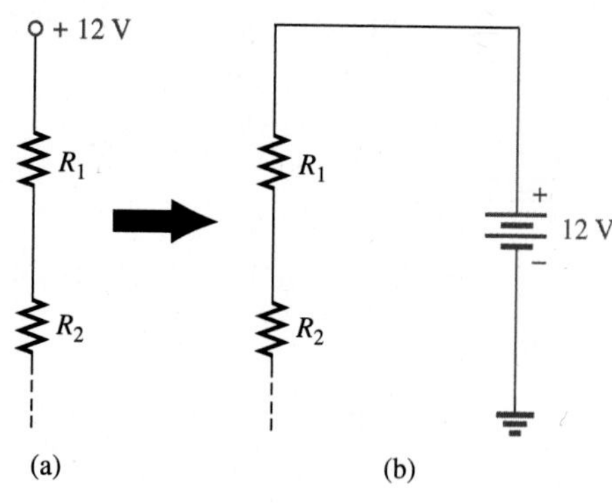

FIG. 5.47
Replacing the special notation for a dc voltage source with the standard symbol.

On large schematics where space is at a premium and clarity is important, voltage sources may be indicated as shown in Figs. 5.47(a) and 5.48(a) rather than as illustrated in Figs. 5.47(b) and 5.48(b). In addition, potential levels may be indicated as in Fig. 5.49, to permit a rapid check of the potential levels at various points in a network with respect to ground to ensure that the system is operating properly.

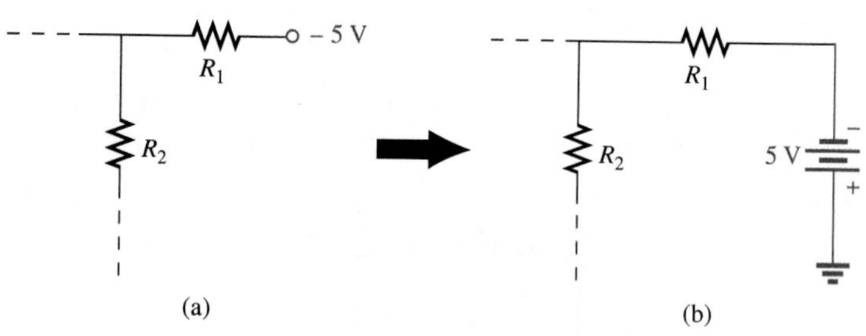

FIG. 5.48
Replacing the notation for a negative dc supply with the standard notation.

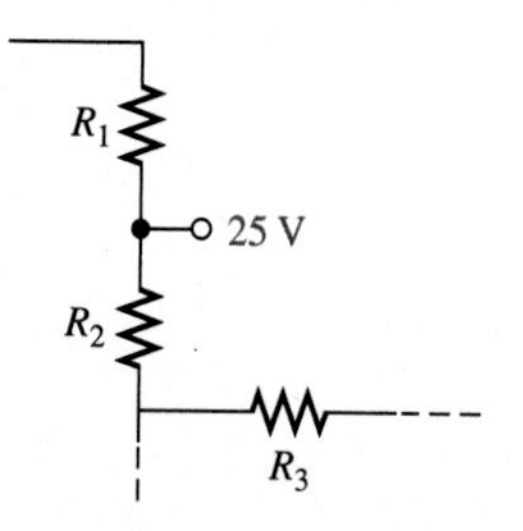

FIG. 5.49
The expected voltage level at a particular point in a network if the system is functioning properly.

Double-Subscript Notation

The fact that voltage is an *across* variable and exists between two points has resulted in a double-subscript notation that defines the first subscript as the higher potential. In Fig. 5.50(a), the two points that define the voltage across the resistor R are denoted by a and b. Since a is the first subscript for V_{ab}, point a must have a higher potential than point b if V_{ab} is to have a positive value. If, in fact, point b is at a higher poten-

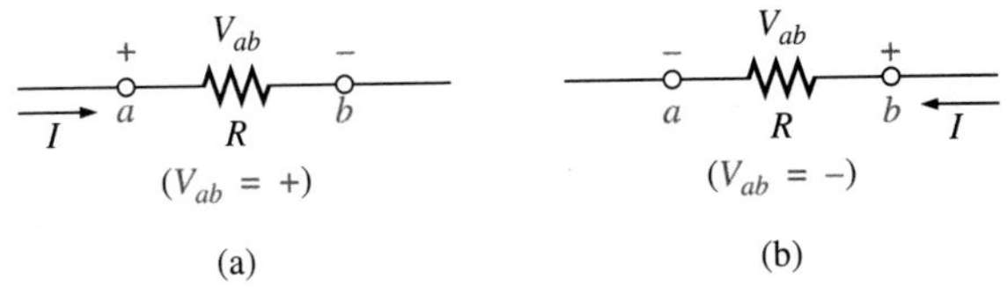

FIG. 5.50
Defining the sign for double-subscript notation.

tial than point *a*, V_{ab} will have a negative value, as indicated in Fig. 5.50(b).

In summary:

The double-subscript notation V_{ab} specifies point a as the higher potential. If this is not the case, a negative sign must be associated with the magnitude of V_{ab}.

In other words,

the voltage V_{ab} is the voltage at point a with respect to (w.r.t.) point b.

Single-Subscript Notation

If point *b* of the notation V_{ab} is specified as ground potential (zero volts), then a single-subscript notation can be used that provides the voltage at a point with respect to ground.

In Fig. 5.51, V_a is the voltage from point *a* to ground. In this case, it is obviously 10 V since it is right across the source voltage *E*. The voltage V_b is the voltage from point *b* to ground. Because it is directly across the 4 Ω resistor, $V_b = 4$ V.

In summary:

The single-subscript notation V_a specifies the voltage at point a with respect to ground (zero volts). If the voltage is less than zero volts, a negative sign must be associated with the magnitude of V_a.

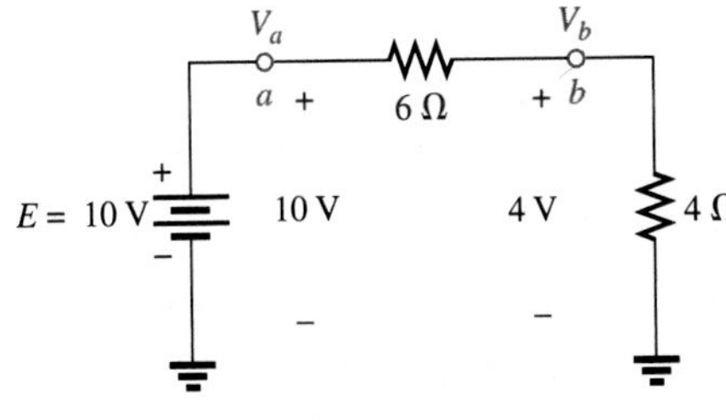

FIG. 5.51
Defining the use of single-subscript notation for voltage levels.

General Comments

A particularly useful relationship can now be established that has extensive applications in the analysis of electronic circuits. For the above notational standards, the following relationship exists:

$$\boxed{V_{ab} = V_a - V_b} \tag{5.12}$$

In other words, if the voltage at points *a* and *b* is known with respect to ground, then the voltage V_{ab} can be determined using Eq. (5.12). In Fig. 5.51, for example,

$$V_{ab} = V_a - V_b = 10\text{ V} - 4\text{ V}$$
$$= 6\text{ V}$$

EXAMPLE 5.21 Find the voltage V_{ab} for the conditions in Fig. 5.52.

Solution: Applying Eq. (5.12):

$$V_{ab} = V_a - V_b = 16\text{ V} - 20\text{ V}$$
$$= \mathbf{-4\text{ V}}$$

V_a = +16 V V_b = +20 V
a R b

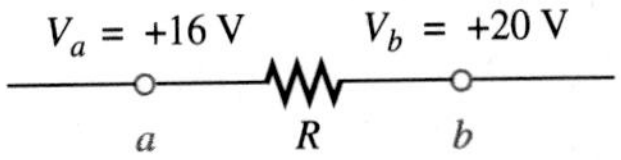

FIG. 5.52
Example 5.21.

Note the negative sign to reflect the fact that point b is at a higher potential than point a.

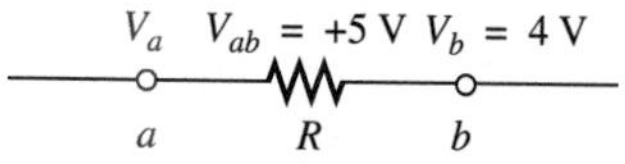

V_a V_{ab} = +5 V V_b = 4 V
a R b

FIG. 5.53
Example 5.22.

EXAMPLE 5.22 Find the voltage V_a for the configuration in Fig. 5.53.

Solution: Applying Eq. (5.12):

$$V_{ab} = V_a - V_b$$

and

$$V_a = V_{ab} + V_b = 5\text{ V} + 4\text{ V} = \mathbf{9\text{ V}}$$

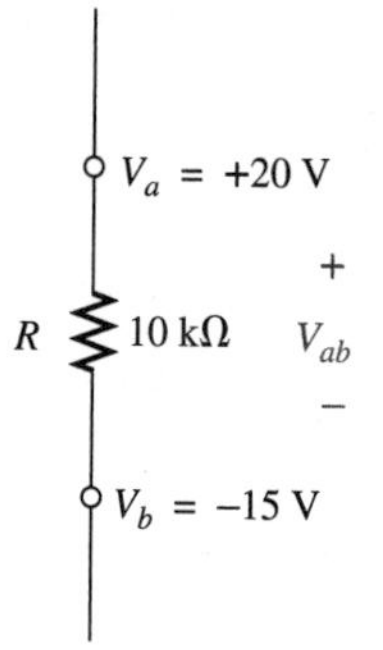

V_a = +20 V
R 10 kΩ + V_{ab} −
V_b = −15 V

FIG. 5.54
Example 5.23.

EXAMPLE 5.23 Find the voltage V_{ab} for the configuration in Fig. 5.54.

Solution: Applying Eq. (5.12):

$$V_{ab} = V_a - V_b = 20\text{ V} - (-15\text{ V}) = 20\text{ V} + 15\text{ V} = \mathbf{35\text{ V}}$$

Note in Example 5.23 you must be careful with the signs when applying the equation. The voltage is dropping from a high level of +20 V to a negative voltage of −15 V. As shown in Fig. 5.55, this represents a drop in voltage of 35 V. In some ways it's like going from a positive checking balance of $20 to owing $15; the total expenditure is $35.

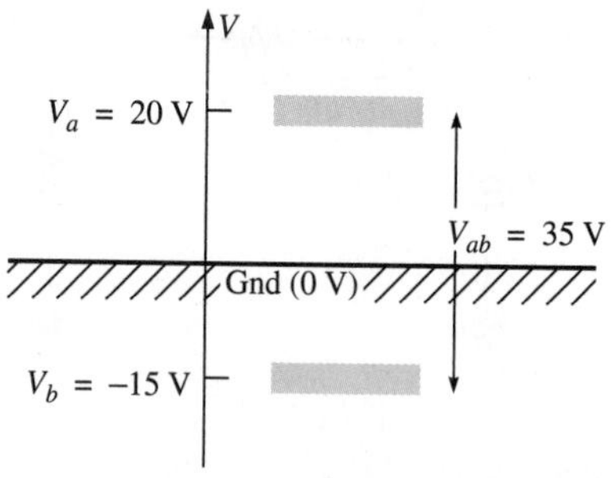

FIG. 5.55
The impact of positive and negative voltages on the total voltage drop.

EXAMPLE 5.24 Find the voltages V_b, V_c, and V_{ac} for the network in Fig. 5.56.

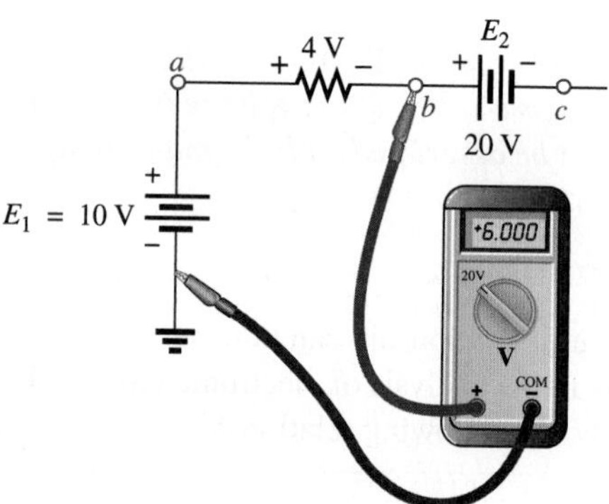

FIG. 5.56
Example 5.24.

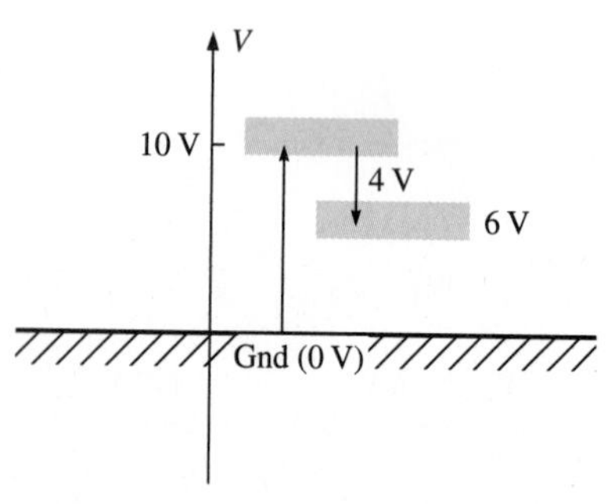

FIG. 5.57
Determining V_b using the defined voltage levels.

Solution: Starting at ground potential (zero volts), we proceed through a rise of 10 V to reach point a and then pass through a drop in potential of 4 V to point b. The result is that the meter reads

$$V_b = +10\text{ V} - 4\text{ V} = \mathbf{6\text{ V}}$$

as clearly demonstrated by Fig. 5.57.

If we then proceed to point c, there is an additional drop of 20 V, resulting in

$$V_c = V_b - 20\text{ V} = 6\text{ V} - 20\text{ V} = \mathbf{-14\text{ V}}$$

as shown in Fig. 5.58.

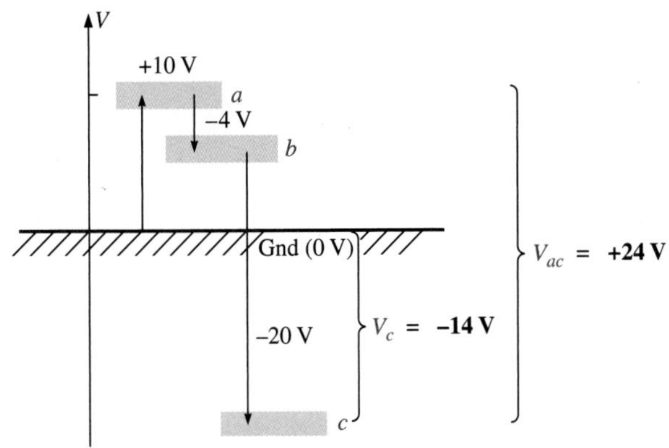

FIG. 5.58
Review of the potential levels for the circuit in Fig. 5.56.

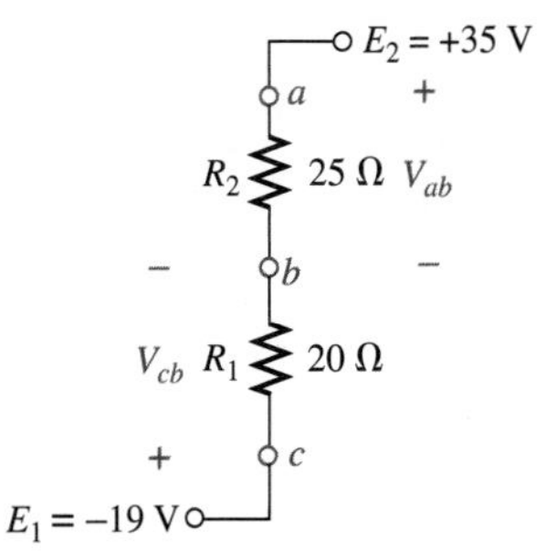

FIG. 5.59
Example 5.25.

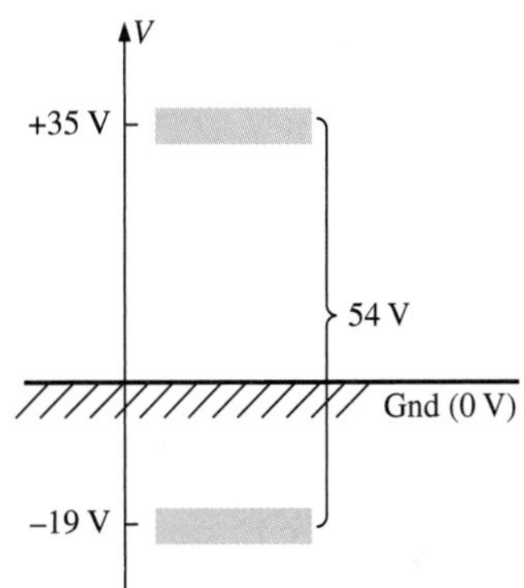

FIG. 5.60
Determining the total voltage drop across the resistive elements in Fig 5.59.

The voltage V_{ac} can be obtainted using Eq. (5.12) or by simply referring to Fig. 5.58.

$$V_{ac} = V_a - V_c = 10\text{ V} - (-14\text{ V}) = \mathbf{24\text{ V}}$$

EXAMPLE 5.25 Determine V_{ab}, V_{cb}, and V_c for the network in Fig. 5.59.

Solution: There are two ways to approach this problem. The first is to sketch the diagram in Fig. 5.60 and note that there is a 54 V drop across the series resistors R_1 and R_2. The current can then be determined using Ohm's law and the voltage levels as follows:

$$I = \frac{54\text{ V}}{45\ \Omega} = 1.2\text{ A}$$

$$V_{ab} = IR_2 = (1.2\text{ A})(25\ \Omega) = \mathbf{30\text{ V}}$$

$$V_{cb} = -IR_1 = -(1.2\text{ A})(20\ \Omega) = \mathbf{-24\text{ V}}$$

$$V_c = E_1 = \mathbf{-19\text{ V}}$$

The other approach is to redraw the network as shown in Fig. 5.61 to clearly establish the aiding effect of E_1 and E_2 and then solve the resulting series circuit.

$$I = \frac{E_1 + E_2}{R_T} = \frac{19\text{ V} + 35\text{ V}}{45\ \Omega} = \frac{54\text{ V}}{45\ \Omega} = 1.2\text{ A}$$

and $\quad V_{ab} = \mathbf{30\text{ V}} \quad V_{cb} = \mathbf{-24\text{ V}} \quad V_c = \mathbf{-19\text{ V}}$

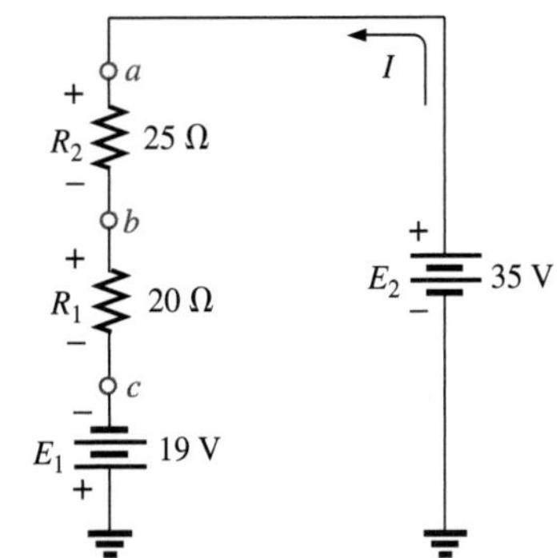

FIG. 5.61
Redrawing the circuit in Fig. 5.59 using standard dc voltage supply symbols.

EXAMPLE 5.26 Using the voltage divider rule, determine the voltages V_1 and V_2 of Fig. 5.62.

Solution: Redrawing the network with the standard battery symbol results in the network in Fig. 5.63. Applying the voltage divider rule,

$$V_1 = \frac{R_1 E}{R_1 + R_2} = \frac{(4\ \Omega)(24\text{ V})}{4\ \Omega + 2\ \Omega} = \mathbf{16\text{ V}}$$

$$V_2 = \frac{R_2 E}{R_1 + R_2} = \frac{(2\ \Omega)(24\text{ V})}{4\ \Omega + 2\ \Omega} = \mathbf{8\text{ V}}$$

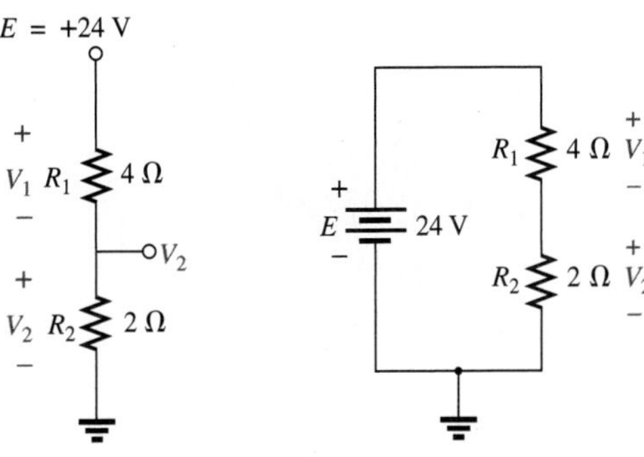

FIG. 5.62
Example 5.26.

FIG. 5.63
Circuit of Fig. 5.62 redrawn.

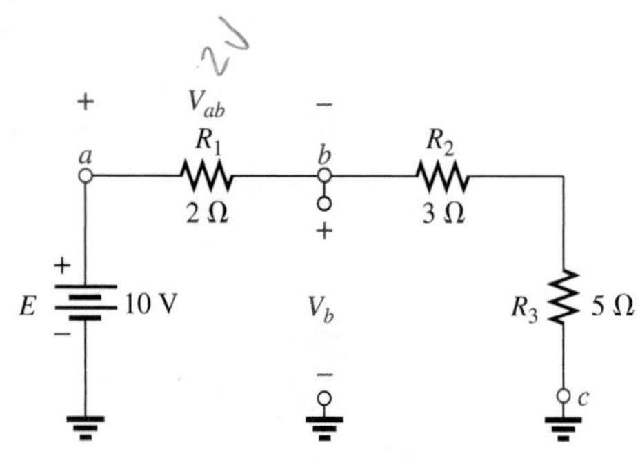

FIG. 5.64
Example 5.27.

EXAMPLE 5.27 For the network in Fig. 5.64.

a. Calculate V_{ab}.
b. Determine V_b.
c. Calculate V_c.

Solutions:

a. Voltage divider rule:

$$V_{ab} = \frac{R_1E}{R_T} = \frac{(2\ \Omega)(10\ \text{V})}{2\ \Omega + 3\Omega + 5\ \Omega} = \mathbf{+2\ V}$$

b. Voltage divider rule:

$$V_b = V_{R_2} + V_{R_3} = \frac{(R_2 + R_3)E}{R_T} = \frac{(3\ \Omega + 5\ \Omega)(10\ \text{V})}{10\ \Omega} = \mathbf{8\ V}$$

or $\quad V_b = V_a - V_{ab} = E - V_{ab} = 10\ \text{V} - 2\ \text{V} = \mathbf{8\ V}$

c. V_c = ground potential = **0 V**

5.10 VOLTAGE REGULATION AND THE INTERNAL RESISTANCE OF VOLTAGE SOURCES

When you use a dc supply such as the generator, battery or supply in Fig. 5.65, you initially assume that it will provide the desired voltage for any resistive load you may hook up to the supply. In other words, if the battery is labeled 1.5 V or the supply is set at 20 V, you assume that they will provide that voltage no matter what load we may apply. Unfortunately, this is not always the case. For instance, if we apply a 1 kΩ resistor to a dc laboratory supply, it is fairly easy to set the voltage across the resistor to 20 V. However, if we remove the 1 kΩ resistor and replace it with a 100 Ω resistor and don't touch the controls on the supply at all, we may find that the voltage has dropped to 19.14 V. Change the load to a 68 Ω resistor, and the terminal voltage drops to 18.72 V. We discover that the load applied affects the terminal voltage of the supply. In fact, this example points out that

a network should always be connected to a supply before the level of supply voltage is set.

The reason the terminal voltage drops with changes in load (current demand) is that

every practical (real-world) supply has an internal resistance in series with the idealized voltage source

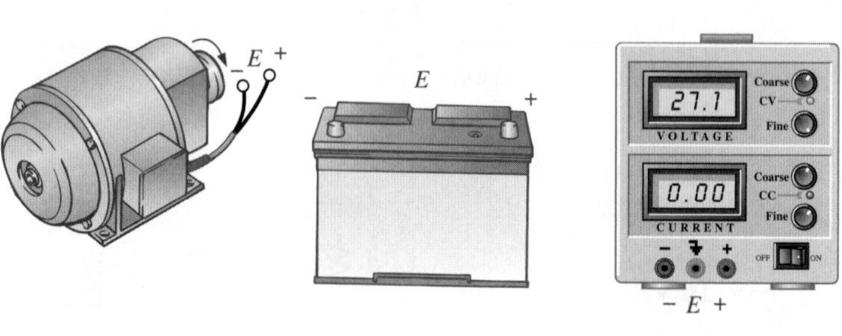

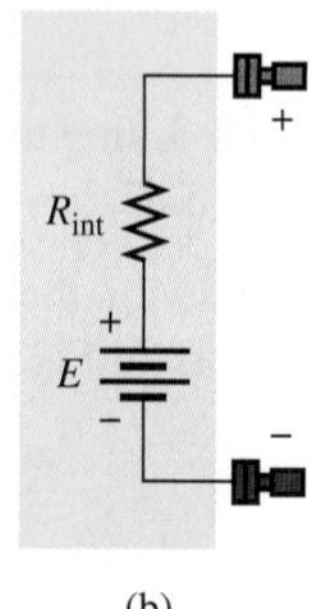

FIG. 5.65
(a) Sources of dc voltage; (b) equivalent circuit.

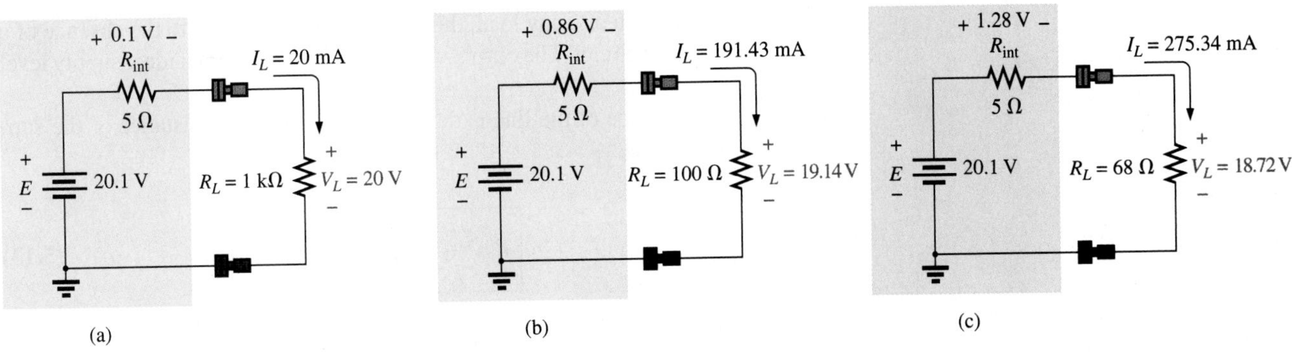

FIG. 5.66
Demonstrating the effect of changing a load on the terminal voltage of a supply.

as shown in Fig. 5.65(b). The resistance level depends on the type of supply, but it is always present. Every year new supplies come out that are less sensitive to the load applied, but even so, some sensitivity still remains.

The supply in Fig. 5.66 helps explain the action that occurred above as we changed the load resistor. Due to the **internal resistance** of the supply, the ideal internal supply must be set to 20.1 V in Fig. 5.66(a) if 20 V are to appear across the 1 kΩ resistor. The internal resistance will capture 0.1 V of the applied voltage. The current in the circuit is determined by simply looking at the load and using Ohm's law; that is, $I_L = V_L/R_L = 20\text{ V}/1\text{ k}\Omega = 20\text{ mA}$, which is a relatively low current.

In Fig. 5.66(b), all the settings of the supply are left untouched, but the 1 kΩ load is replaced by a 100 Ω resistor. The resulting current is now $I_L = E/R_T = 20.1\text{ V}/105\ \Omega = 191.43\text{ mA}$, and the output voltage is $V_L = I_LR = (191.43\text{ mA})(100\ \Omega) = 19.14\text{ V}$, a drop of 0.86 V. In Fig. 5.66(c), a 68 Ω load is applied, and the current increases substantially to 275.34 mA with a terminal voltage of only 18.72 V. This is a drop of 1.28 V from the expected level. Quite obviously, therefore, as the current drawn from the supply increases, the terminal voltage continues to drop.

If we plot the terminal voltage versus current demand from 0 A to 275.34 mA, we obtain the plot in Fig. 5.67. Interestingly enough, it turns out to be a straight line that continues to drop with an increase in current demand. Note, in particular, that the curve begins at a current level of 0 A. Under no-load conditions, where the output terminals of the supply are

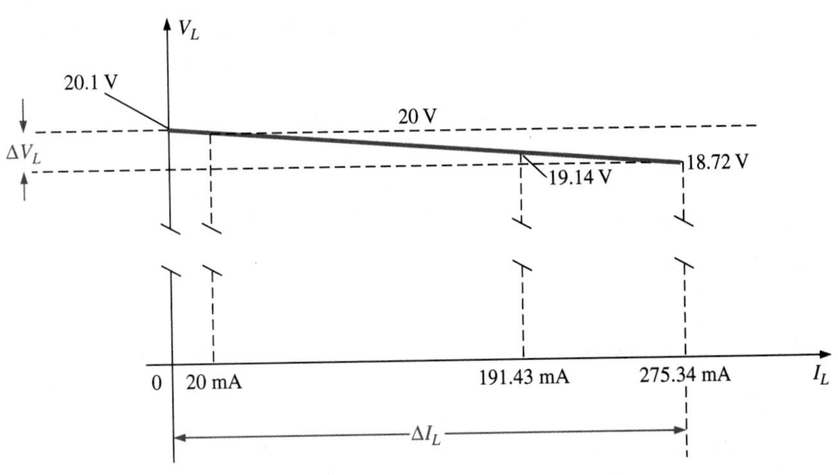

FIG. 5.67
Plotting V_L versus I_L for the supply in Fig. 5.66.

not connected to any load, the current will be 0 A due to the absence of a complete circuit. The output voltage will be the internal ideal supply level of 20.1 V.

The slope of the line is defined by the internal resistance of the supply. That is,

$$R_{int} = \frac{\Delta V_L}{\Delta I_L} \quad \text{(ohms, } \Omega) \tag{5.13}$$

which for the plot in Fig. 5.67 results in

$$R_{int} = \frac{\Delta V_L}{\Delta I_L} = \frac{20.1\text{ V} - 18.72\text{ V}}{275.34\text{ mA} - 0\text{ mA}} = \frac{1.38\text{ V}}{275.34\text{ mA}} = \mathbf{5\ \Omega}$$

For supplies of any kind, the plot of particular importance is the output voltage versus current drawn from the supply, as shown in Fig. 5.68(a). Note that the maximum value is achieved under no-load (NL) conditions as defined by Fig. 5.68(b) and the description above. Full-load (FL) conditions are defined by the maximum current the supply can provide on a continuous basis, as shown in Fig. 5.68(c).

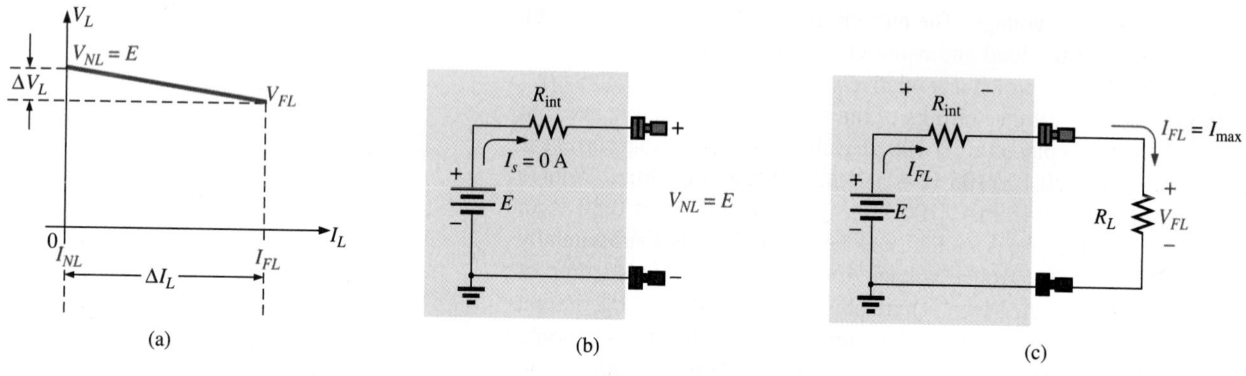

FIG. 5.68
Defining the properties of importance for a power supply.

As a basis for comparison, an ideal power supply and its response curve are provided in Fig. 5.69. Note the absence of the internal resistance and the fact that the plot is a horizontal line (no variation at all with load demand)—an impossible response curve. When we compare the

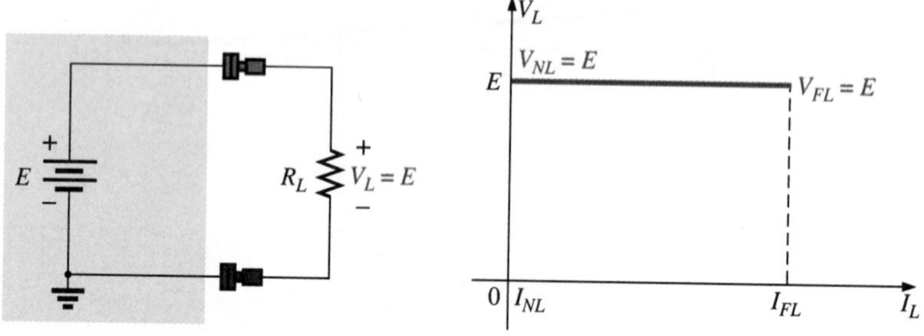

FIG. 5.69
Ideal supply and its terminal characteristics.

curve in Fig. 5.69 with that in Fig. 5.68(a), however, we now realize that the *steeper the slope,* the more sensitive the supply is to the change in load and therefore the *less desirable* it is for many laboratory procedures. In fact,

the larger the internal resistance, the steeper the drop in voltage with an increase in load demand (current).

To help us anticipate the expected response of a supply, a defining quantity called **voltage regulation** (abbreviated *VR;* often called *load regulation* on specification sheets) was established. The basic equation in terms of the quantities in Fig. 5.68(a) is the following:

$$VR = \frac{V_{NL} - V_{FL}}{V_{FL}} \times 100\% \quad \textbf{(5.14)}$$

The examples to follow demonstrate that

the smaller the voltage or load regulation of a supply, the less the terminal voltage will change with increasing levels of current demand.

For the supply above with a no-load voltage of 20.1 V and a full-load voltage of 18.72 V, at 275.34 mA the voltage regulation is

$$VR = \frac{V_{NL} - V_{FL}}{V_{FL}} \times 100\% = \frac{20.1\text{ V} - 18.72\text{ V}}{18.72\text{ V}} \times 100\% \cong \mathbf{7.37\%}$$

which is quite high, revealing that we have a very sensitive supply. Most modern commercial supplies have regulation factors less than 1%, with 0.01% being very typical.

EXAMPLE 5.28

a. Given the characteristics in Fig. 5.70, determine the voltage regulation of the supply.
b. Determine the internal resistance of the supply.
c. Sketch the equivalent circuit for the supply.

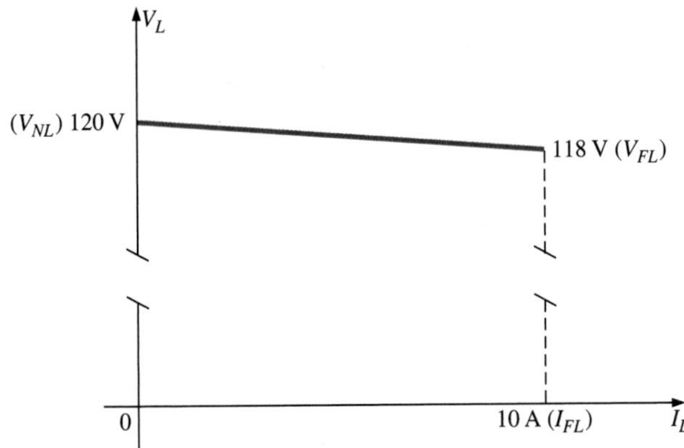

FIG. 5.70
Terminal characteristics for the supply of Example 5.28.

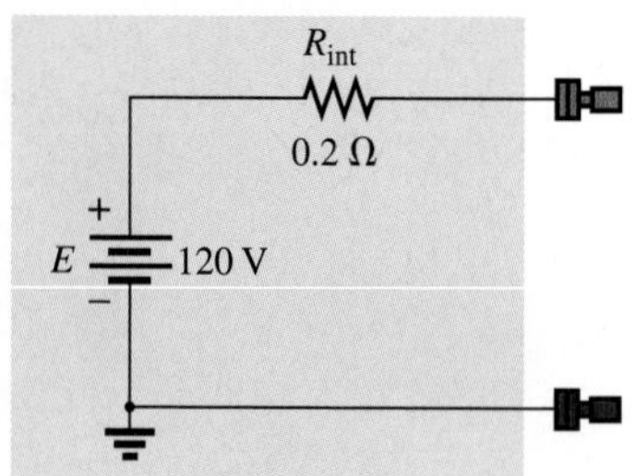

FIG. 5.71
dc supply with the terminal characteristics of Fig. 5.70.

Solutions:

a. $$VR = \frac{V_{NL} - V_{FL}}{V_{FL}} \times 100\%$$

$$= \frac{120\text{ V} - 118\text{ V}}{118\text{ V}} \times 100\% = \frac{2}{118} \times 100\%$$

$$VR \cong \mathbf{1.7\%}$$

b. $$R_{int} = \frac{\Delta V_L}{\Delta I_L} = \frac{120\text{ V} - 118\text{ V}}{10\text{ A} - 0\text{ A}} = \frac{2\text{ V}}{10\text{ A}} = \mathbf{0.2\ \Omega}$$

c. See Fig. 5.71.

EXAMPLE 5.29 Given a 60 V supply with a voltage regulation of 2%:

a. Determine the terminal voltage of the supply under full-load conditions.
b. If the half-load current is 5 A, determine the internal resistance of the supply.
c. Sketch the curve of the terminal voltage versus load demand and the equivalent circuit for the supply.

Solutions:

a. $$VR = \frac{V_{NL} - V_{FL}}{V_{FL}} \times 100\%$$

$$2\% = \frac{60\text{ V} - V_{FL}}{V_{FL}} \times 100\%$$

$$\frac{2\%}{100\%} = \frac{60\text{ V} - V_{FL}}{V_{FL}}$$

$$0.02V_{FL} = 60\text{ V} - V_{FL}$$

$$1.02V_{FL} = 60\text{ V}$$

$$V_{FL} = \frac{60\text{ V}}{1.02} = \mathbf{58.82\ V}$$

b. $I_{FL} = 10\text{ A}$

$$R_{int} = \frac{\Delta V_L}{\Delta I_L} = \frac{60\text{ V} - 58.82\text{ V}}{10\text{ A} - 0\text{ A}} = \frac{1.18\text{ V}}{10\text{ A}} \cong \mathbf{0.12\ \Omega}$$

c. See Fig. 5.72.

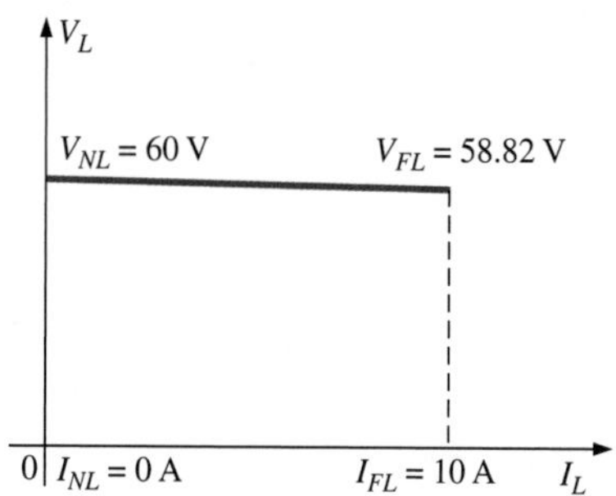

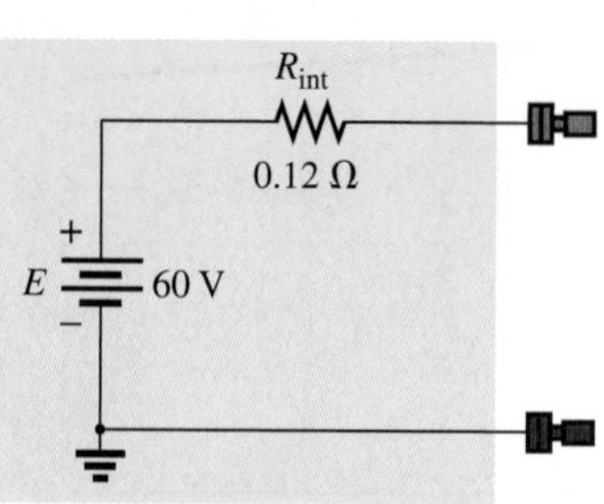

FIG. 5.72
Characteristics and equivalent circuit for the supply of Example 5.29.

5.11 LOADING EFFECTS OF INSTRUMENTS

In the previous section, we learned that power supplies are not the ideal instruments we may have thought they were. The applied load can have an effect on the terminal voltage. Fortunately, since today's supplies have such small load regulation factors, the change in terminal voltage with load can usually be ignored for most applications. If we now turn our attention to the various meters we use in the lab, we again find that they are not totally ideal:

Whenever you apply a meter to a circuit, you change the circuit and the response of the system. Fortunately, however, for most applications, considering the meters to be ideal is a valid approximation as long as certain factors are considered.

For instance,

any ammeter connected in a series circuit will introduce resistance to the series combination that will affect the current and voltages of the configuration.

The resistance between the terminals of an ammeter is determined by the chosen scale of the ammeter. In general,

for ammeters, the higher the maximum value of the current for a particular scale, the smaller the internal resistance will be.

For example, it is not uncommon for the resistance between the terminals of an ammeter to be 250 Ω for a 2 mA scale but only 1.5 Ω for the 2 A scale, as shown in Fig. 5.73(a) and (b). If analyzing a circuit in detail, you can include the internal resistance as shown in Fig. 5.73 as a resistor between the two terminals of the meter.

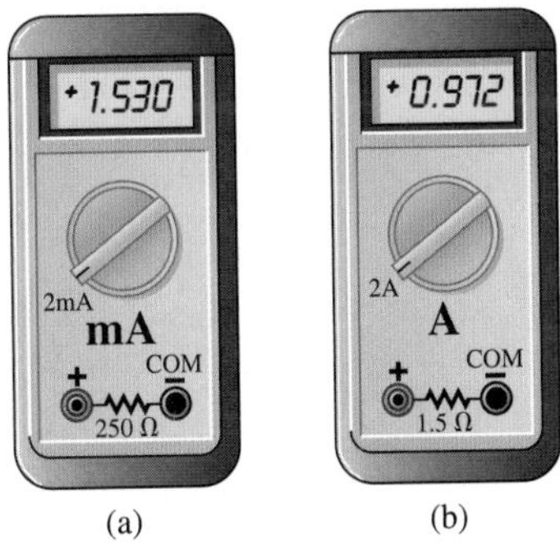

FIG. 5.73
Including the effects of the internal resistance of an ammeter: (a) 2 mA scale; (b) 2 A scale.

At first reading, such resistance levels at low currents give the impression that ammeters are far from ideal, that they should be used only to obtain a general idea of the current and should not be expected to provide a true reading. Fortunately, however, when you are reading currents below the 2 mA range, the resistors in series with the ammeter are typically in the kilohm range. For example, in Fig. 5.74(a), using an ideal ammeter, the current displayed is 0.6 mA as determined from $I_s = E/R_T = 12\text{ V}/20\text{ k}\Omega = 0.6\text{ mA}$. If we now insert a meter with an internal resistance of 250 Ω as shown in Fig. 5.74(b), the additional resistance in the circuit will drop the current to 0.593 mA as determined from $I_s = E/R_T = 12\text{ V}/20.25\text{ k}\Omega = 0.593\text{ mA}$. Now, certainly the current has dropped from the ideal level, but the difference in results is only about 1%—nothing major, and the measurement can be used for most purposes. If the series resistors were in the same range as the 250 Ω resistors, we would have a different problem, and we would have to look at the results very carefully.

Let us go back to Fig. 5.20 and determine the actual current if each meter on the 2 A scale has an internal resistance of 1.5 Ω. The fact that there are four meters will result in an additional resistance of (4)(1.5 Ω) = 6 Ω in the circuit, and the current will be $I_s = E/R_T = 8.4\text{ V}/146\ \Omega \cong 58\text{ mA}$, rather than the 60 mA under ideal conditions. This value is still close enough to be considered a helpful reading. However, keep in mind that if we were measuring the current in the circuit, we would use only one ammeter, and the current would be $I_s = E/R_T = 8.4\text{ V}/141.5\ \Omega \cong 59\text{ mA}$, which can certainly be approximated as 60 mA.

In general, therefore, be aware that this internal resistance must be factored in, but for the reasons just described, most readings can be used as an excellent first approximation to the actual current.

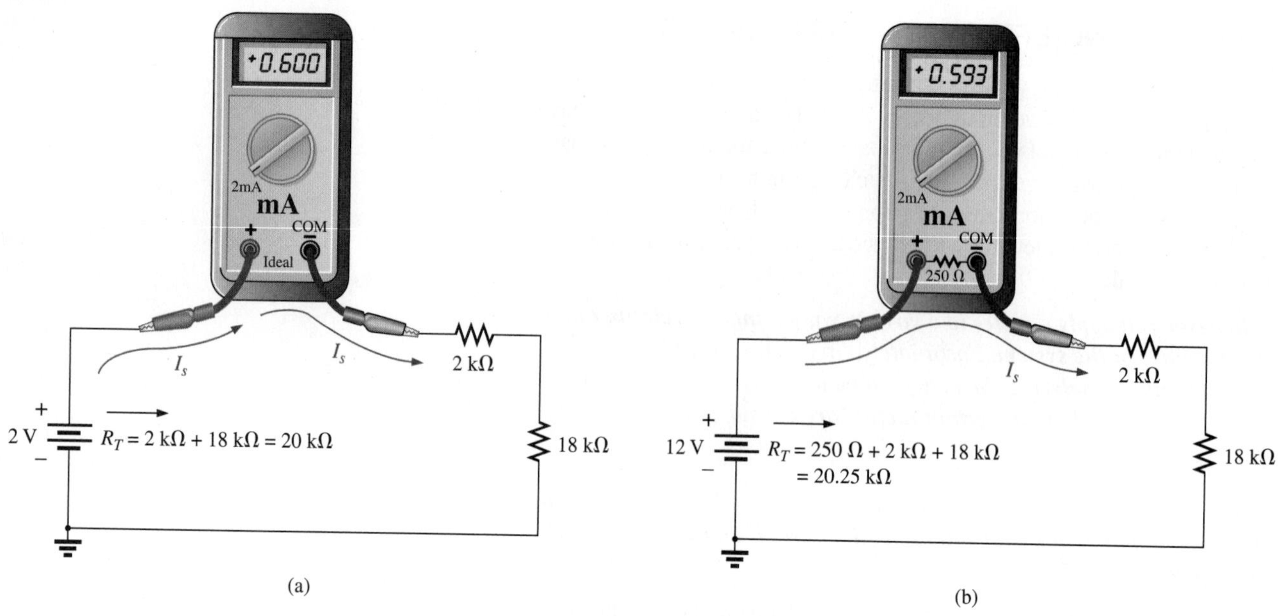

FIG. 5.74
Applying an ammeter, set on the 2 mA scale, to a circuit with resistors in the kilohm range: (a) ideal; (b) practical.

It should be added that because of this *insertion problem* with ammeters, and because of the very important fact that the *circuit must be disturbed* to measure a current, ammeters are not used as much as you might initially expect. Rather than break a circuit to insert a meter, the voltage across a resistor is often measured and the current then calculated using Ohm's law. This eliminates the need to worry about the level of the meter resistance and having to disturb the circuit. Another option is to use the clamp-type ammeters introduced in Chapter 2, removing the concerns about insertion loss and disturbing the circuit. Of course, for many practical applications (such as on power supplies), it is convenient to have an ammeter permanently installed so that the current can quickly be read from the display. In such cases, however, the design is such as to compensate for the insertion loss.

In summary, therefore, keep in mind that the insertion of an ammeter will add resistance to the branch and will affect the current and voltage levels. However, in most cases the effect is minimal, and the reading will provide a good first approximation to the actual level.

The loading effect of voltmeters are discussed in detail in the next chapter because loading is not a series effect. In general, however, the results will be similar in many ways to those of the ammeter, but the major difference is that the circuit does not have to be disturbed to apply the meter.

5.12 PROTOBOARDS (BREADBOARDS)

At some point in the design of any electrical/electronic system, a prototype must be built and tested. One of the most effective ways to build a testing model is to use the **protoboard** (in the past most commonly called a **breadboard**) in Fig. 5.75. It permits a direct connection of the power supply and provides a convenient method to hold and connect the components. There isn't a great deal to learn about the protoboard, but it is important to point out some of its characteristics, including the way the elements are typically connected.

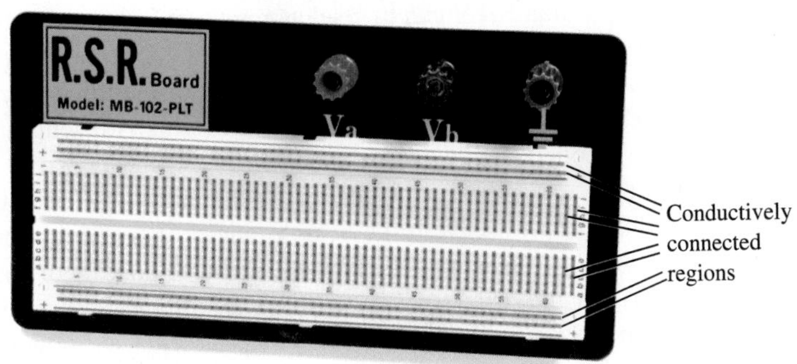

FIG. 5.75
Protoboard with areas of conductivity defined using two different approaches.

The red terminal V_a is connected directly to the positive terminal of the dc power supply, with the black lead V_b connected to the negative terminal and the green terminal used for the ground connection. Under the hole pattern, there are continuous horizontal copper strips under the top and bottom rows, as shown by the copper bands in Fig. 5.75. In the center region, the conductive strips are vertical but do not extend beyond the deep notch running the horizontal length of the board. That's all there is to it, although it will take some practice to make the most effective use of the conductive patterns.

As examples, the network in Fig. 5.12 is connected on the protoboard in the photo in Fig. 5.76 using *two different approaches.* After the dc power supply has been hooked up, a lead is brought down from the positive red terminal to the top conductive strip marked "+." Keep in mind that now the entire strip is connected to the positive terminal of the supply. The negative terminal is connected to the bottom strip marked with

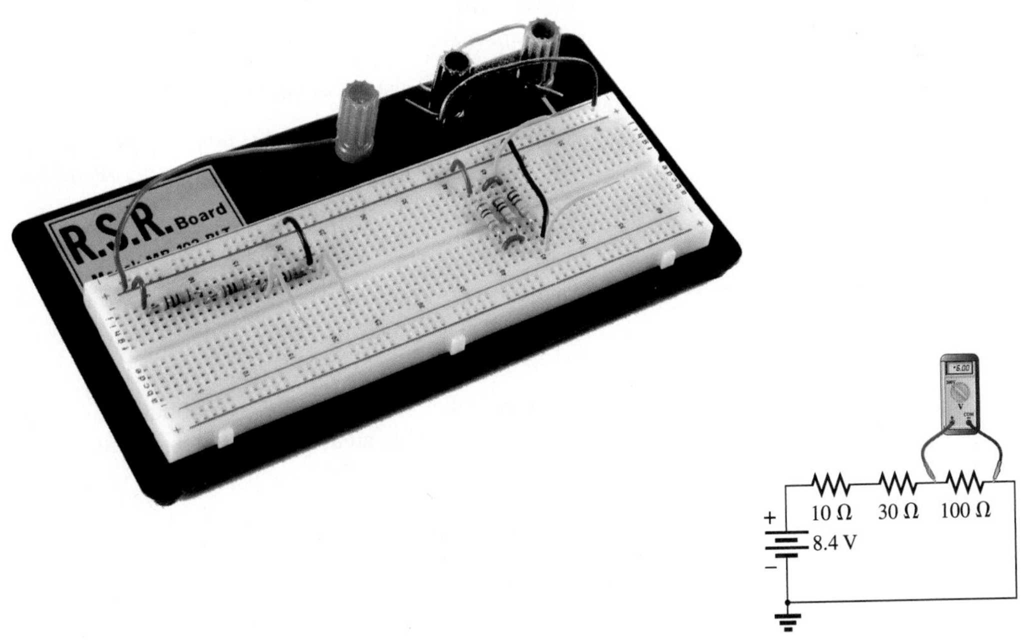

FIG. 5.76
Two setups for the network in Fig. 5.12 on a protoboard with yellow leads added to each configuration to measure voltage V_3 with a voltmeter.

a minus sign (−), so that 8.4 V can be read at any point between the top positive strip and the bottom negative strip. A ground connection to the negative terminal of the battery was made at the site of the three terminals. For the user's convenience, kits are available in which the length of the wires is color coded. Otherwise, a spool of 24 gage wire is cut to length and the ends are stripped. In general, feel free to use the extra length—everything doesn't have to be at right angles. For most protoboards, 1/4 through 1 W resistors will insert nicely in the board. For clarity, 1/2 W resistors are used in Fig. 5.76. The voltage across any component can be easily read by inserting additional leads as shown in the figure (yellow leads) for the voltage V_3 of each configuration (the yellow wires) and attaching the meter. For any network, the components can be wired in a variety of ways. Note in the configuration on the right that the horizontal break through the center of the board was used to isolate the two terminals of each resistor. Even though there are no set standards, it is important that the arrangement can *easily be understood* by someone else.

Additional setups using the protoboard are in the chapters to follow so that you can become accustomed to the manner in which it is used most effectively. You will probably see the protoboard quite frequently in your laboratory sessions or in an industrial setting.

5.13 APPLICATIONS

Before looking at a few applications, we need to consider a few general characteristics of the series configuration that you should always keep in mind when designing a system. First, and probably the most important, is that

if one element of a series combination of elements should fail, it will disrupt the response of all the series elements. If an open circuit occurs, the current will be zero. If a short circuit results, the voltage will increase across the other elements, and the current will increase in magnitude.

Second, and just a thought you should always keep in mind, is that

for the same source voltage, the more elements you place is series, the less the current and the less the voltage across all the elements of the series combination.

Last, and a result discussed in detail in this chapter, is that

the current is the same for each element of a series combination, but the voltage across each element is a function of its terminal resistance.

There are other characteristics of importance that you will learn as you investigate possible areas of application, but the above are the most important.

Series Control

One common use of the series configuration is in setting up a system that ensures that everything is in place before full power is applied. In Fig. 5.77, various sensing mechanisms can be tied to series switches, preventing power to the load until all the switches are in the closed or on position. For instance, as shown in Fig. 5.77, one component may test the environment for dangers such as gases, high temperatures, and so on. The

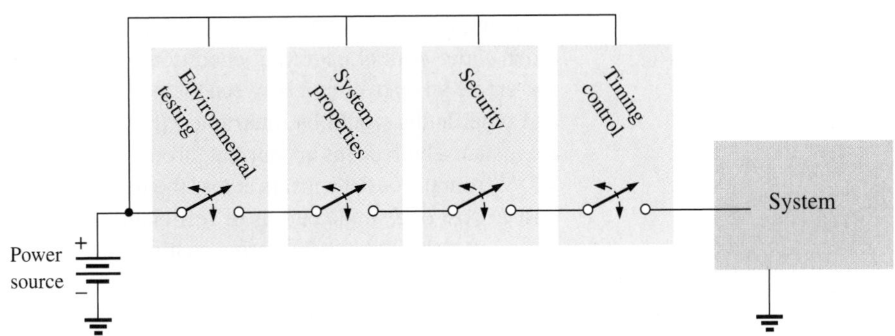

FIG. 5.77
Series control over an operating system.

next component may be sensitive to the properties of the system to be energized to be sure all components are working. Security is another factor in the series sequence, and finally a timing mechanism may be present to ensure limited hours of operation or to restrict operating periods. The list is endless, but the fact remains that "all systems must be go" before power reaches the operating system.

Holiday Lights

In recent years, the small blinking holiday lights with 50 to 100 bulbs on a string have become very popular [see Fig. 5.78(a)]. Although holiday lights can be connected in series or parallel (to be described in the next chapter), the smaller blinking light sets are normally connected in series. It is relatively easy to determine if the lights are connected in series. If one wire enters and leaves the bulb casing, they are in series. If two wires enter and leave, they are probably in parallel. *Normally, when bulbs are connected in series, if one burns out (the filament breaks and the circuit opens), all the bulbs go out since the current path has been interrupted.* However, the bulbs in Fig. 5.78(a) are specially designed, as shown in Fig. 5.78(b), to permit current to conitnue to flow to the other bulbs when the filament burns out. At the base of each bulb, there is a fuse link wrapped around the two posts holding the filament. The fuse link of a soft conducting metal appears to be touching the two vertical posts, but in fact

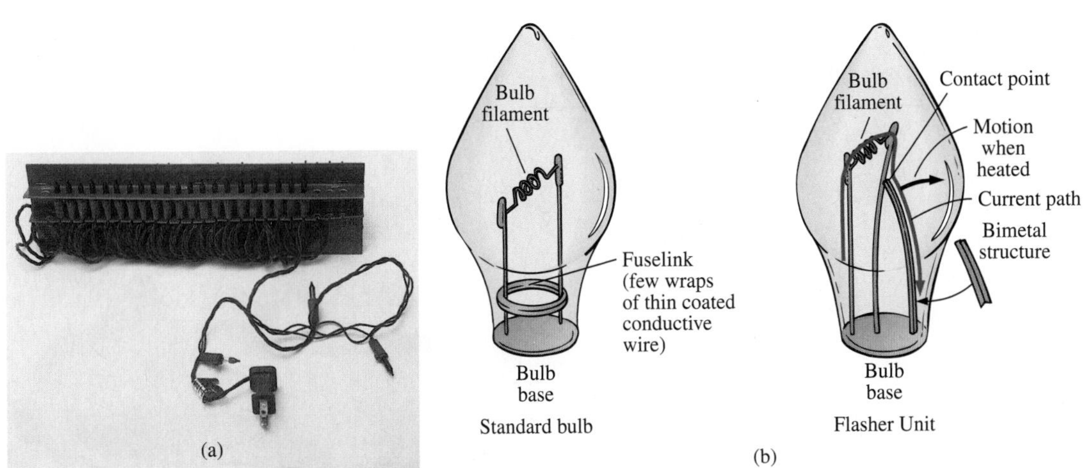

FIG. 5.78
Holiday lights: (a) 50-unit set; (b) bulb construction.

a coating on the posts or fuse link prevents conduction from one to the other under normal operating conditions. If a filament should burn out and create an open circuit between the posts, the current through the bulb and other bulbs would be interrupted if it were not for the fuse link. At the instant a bulb opens up, current through the circuit is zero, and the full 120 V from the outlet appears across the bad bulb. This high voltage from post to post of a single bulb is of sufficient potential difference to establish current through the insulating coatings and spot-weld the fuse link to the two posts. The circuit is again complete, and all the bulbs light except the one with the activated fuse link. Keep in mind, however, that each time a bulb burns out, there is more voltage across the other bulbs of the circuit, making them burn brighter. Eventually, if too many bulbs burn out, the voltage reaches a point where the other bulbs burn out in rapid succession. To prevent this, you must replace burned-out bulbs at the earliest opportunity.

The bulbs in Fig. 5.78(b) are rated 2.5 V at 0.2 A or 200 mA. Since there are 50 bulbs in series, the total voltage across the bulbs will be 50×2.5 V or 125 V, which matches the voltage available at the typical home outlet. Since the bulbs are in series, the current through each bulb will be 200 mA. The power rating of each bulb is therefore $P = VI = (2.5 \text{ V})(0.2 \text{ A}) = 0.5$ W with a total wattage demand of $50 \times 0.5 \text{ W} = 25$ W.

A schematic representation for the set of Fig. 5.78(a) is provided in Fig. 5.79(a). Note that only one flasher unit is required. Since the bulbs are in

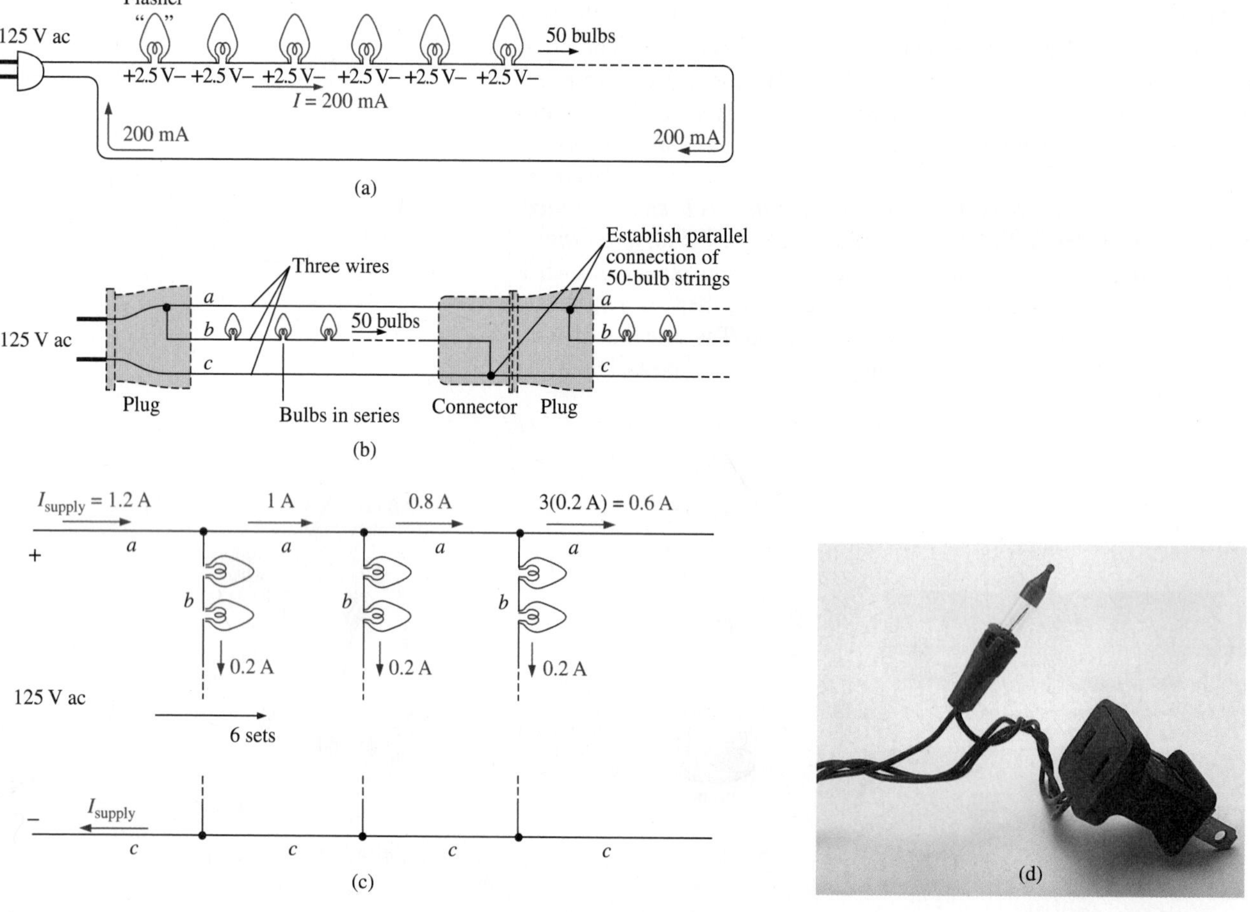

FIG. 5.79

(a) Single-set wiring diagram; (b) special wiring arrangement; (c) redrawn schematic; (d) special plug and flasher unit.

series, when the flasher unit interrupts the current flow, it turns off all the bulbs. As shown in Fig. 5.78(b), the flasher unit incorporates a bimetal thermal switch that opens when heated to a preset level by the current. As soon as it opens, it begins to cool down and closes again so that current can return to the bulbs. It then heats up again, opens up, and repeats the entire process. The result is an on-and-off action that creates the flashing pattern we are so familiar with. Naturally, in a colder climate (for example, outside in the snow and ice), it initially takes longer to heat up, so the flashing pattern is slow at first, but as the bulbs warm up, the frequency increases.

The manufacturer specifies that no more than six sets should be connected together. How can you connect the sets together, end to end, without reducing the voltage across each bulb and making all the lights dimmer? If you look closely at the wiring, you will find that since the bulbs are connected in series, there is one wire to each bulb with additional wires from plug to plug. Why would they need two additional wires if the bulbs are connected in series? Because when each set is connected together, they are actually in a parallel arrangement (to be discussed in the next chapter). This unique wiring arrangement is shown in Fig. 5.79(b) and redrawn in Fig. 5.79(c). Note that the top line is the hot line to all the connected sets, and the bottom line is the return, neutral, or ground line for all the sets. Inside the plug in Fig. 5.79(d), the hot line and return are connected to each set, with the connections to the metal spades of the plug as shown in Fig. 5.79(b). We will find in the next chapter that the current drawn from the wall outlet for parallel loads is the sum of the current to each branch. The result, as shown in Fig. 5.79(c), is that the current drawn from the supply is 6×200 mA $= 1.2$ A, and the total wattage for all six sets is the product of the applied voltage and the source current or (120 V)(1.2 A) $=$ 144 W with 144 W/6 $=$ 24 W per set.

Microwave Oven

Series circuits can be very effective in the design of safety equipment. Although we all recognize the usefulness of the microwave oven, it can be

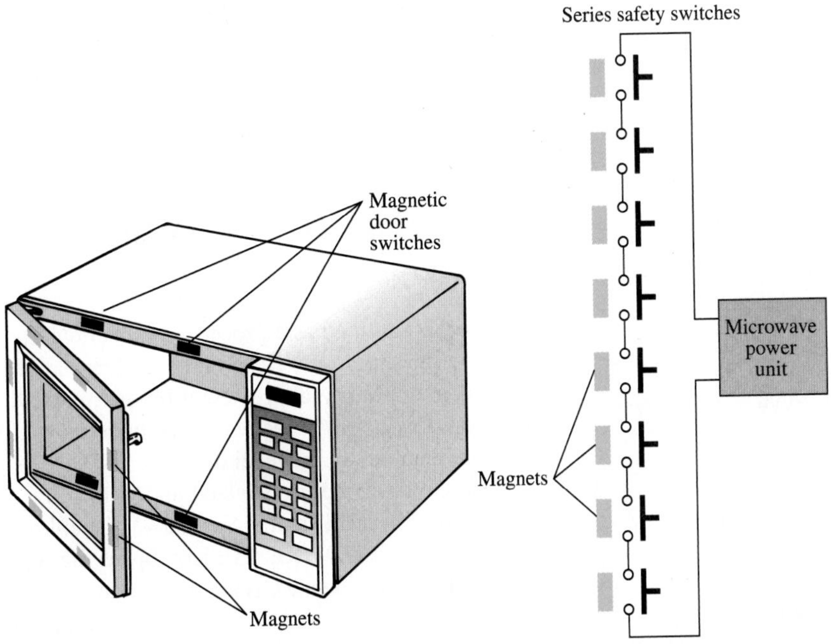

FIG. 5.80
Series safety switches in a microwave oven.

quite dangerous if the door is not closed or sealed properly. It is not enough to test the closure at only one point around the door because the door may be bent or distorted from continual use, and leakage can result at some point distant from the test point. One common safety arrangement appears in Fig. 5.80. Note that magnetic switches are located all around the door, with the magnet in the door itself and the magnetic door switch in the main frame. Magnetic switches are simply switches where the magnet draws a magnetic conducting bar between two contacts to complete the circuit—somewhat revealed by the symbol for the device in the circuit diagram in Fig. 5.80. Since the magnetic switches are all in series, they must all be closed to complete the circuit and turn on the power unit. If the door is sufficiently out of shape to prevent a single magnet from getting close enough to the switching mechanism, the circuit will not be complete, and the power cannot be turned on. Within the control unit of the power supply, either the series circuit completes a circuit for operation or a sensing current is established and monitored that controls the system operation.

Series Alarm Circuit

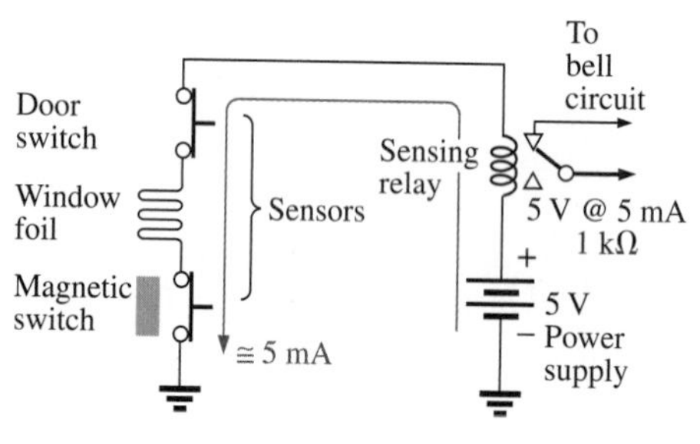

FIG. 5.81
Series alarm circuit.

The circuit in Fig. 5.81 is a simple alarm circuit. Note that every element of the design is in a series configuration. The power supply is a 5 V dc supply that can be provided through a design similar to that in Fig. 2.33, a dc battery, or a combination of an ac and a dc supply that ensures that the battery will always be at full charge. If all the sensors are closed, a current of 5 mA results because of the terminal load of the relay of about 1 kΩ. That current energizes the relay and maintains an off position for the alarm. However, if any of the sensors and opened, the current will be interrupted, the relay will let go, and the alarm circuit will be energized. With relatively short wires and a few sensors, the system should work well since the voltage drop across each is minimal. However, since the alarm wire is usually relatively thin, resulting in a measurable resistance level, if the wire to the sensors is too long, a sufficient voltage drop could occur across the line, reducing the voltage across the relay to a point where the alarm fails to operate properly. Thus, wire length is a factor that must be considered if a series configuration is used. Proper sensitivity to the length of the line should remove any concerns about its operation. An improved design is described in Chapter 8.

5.14 COMPUTER ANALYSIS

PSpice

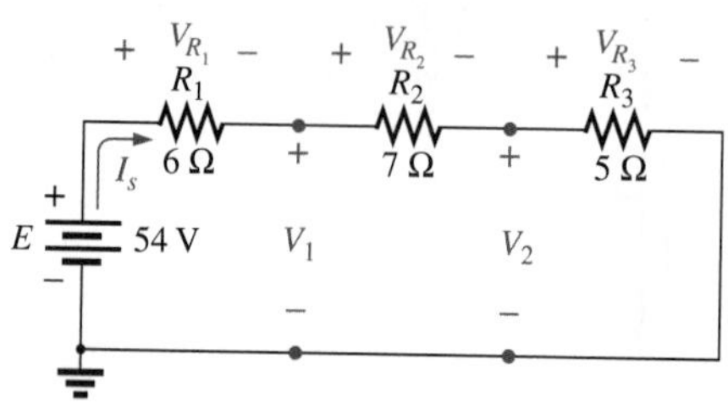

FIG. 5.82
Series dc network to be analyzed using PSpice.

In Section 4.9, the basic procedure for setting up the PSpice folder and running the program were presented. Because of the detail provided in that section, you should review it before proceeding with this example. Because this is only the second example using PSpice, some detail is provided, but not at the level of Section 4.9.

The circuit to be investigated appears in Fig. 5.82. You will use the **PSpice** folder established in Section 4.9. Double-clicking on the **OrCAD 10.0 DEMO/CAPTURE CIS** icon opens the window. A new project is initiated by selecting the **Create document** key at the top left of the screen (it looks like a page with a star in the upper left corner). The result is the **New Project** dialog box in which **PSpice 5–1** is entered as the **Name.** The

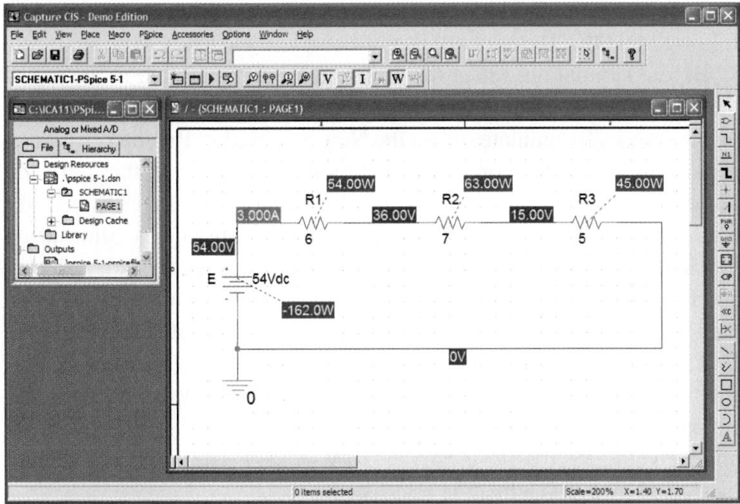

FIG. 5.83
Applying PSpice to a series dc circuit.

Analog or Mixed A/D is already selected, and **PSpice** appears as the **Location.** Click **OK,** and the **Create PSpice Project** dialog box appears. Select **Create a blank project,** click **OK,** and the working windows appear. Grab the left edge of the **SCHEMATIC1:PAGE1** window to move it to the right so that you can see both screens. Clicking the + sign in the **Project Manager** window allows you to set the sequence down to **PAGE1.** You can change the name of the **SCHEMATIC1** by selecting it and right-clicking. Choose **Rename** from the list. Enter **PSpice 5–1** in the **Rename Schematic** dialog box. In Fig. 5.83 it was left as **SCHEMATIC1.**

This next step is important. If the toolbar on the right does not appear, left-click anywhere on the **SCHEMATIC1:PAGE1** screen. To start building the circuit, select **Place a part** key (the second one down) to open the **Place Part** dialog box. Note that the **SOURCE** library is already in place in the **Library** list (from the efforts of Chapter 4). Selecting **SOURCE** results in the list of sources under **Part List,** and **VDC** can be selected. Click **OK,** and the cursor can put it in place with a single left click. Right-click and select **End Mode** to end the process since the network has only one source. One more left click, and the source is in place. Select the **Place a Part** key again, followed by **ANALOG** library to find the resistor **R.** Once the resistor has been selected, click **OK** to place it next to the cursor on the screen. This time, since three resistors need to be placed, there is no need to go to **End Mode** between depositing each. Simply click one in place, then the next, and finally the third. Then right-click to end the process with **End Mode.** Finally, add a **GND** by selecting the appropriate key from the right toolbar and selecting **0/SOURCE** in the **Place Ground** dialog box. Click **OK,** and place the ground as shown in Fig. 5.83.

Connect the elements by using the **Place a wire** key to obtain the crosshair on the screen. Start at the top of the voltage source with a left click, and draw the wire, left-clicking it at every 90° turn. When a wire is connected from one element to another, move on to the next connection to be made—there is no need to go **End Mode** between connections. Now set the labels and values by double-clicking on each parameter to obtain a **Display Properties** dialog box. Since the dialog box appears with the quantity of interest in a blue background, type in the desired label or value, followed by **OK.** The network is now complete and ready to be analyzed.

Before simulation, select the **V, I,** and **W** in the toolbar at the top of the window to ensure that the voltages, currents, and power are displayed on the screen. To simulate, select the **New Simulation Profile** key (which appears as a data sheet on the second toolbar down with a star in the top left corner) to obtain the **New Simulation** dialog box. Enter **Bias Point** for a dc solution under **Name,** and hit the **Create** key. A **Simulation Settings-Bias Point** dialog box appears in which **Analysis** is selected and **Bias Point** is found under the **Analysis type** heading. Click **OK,** and then select the **Run PSpice** key (the blue arrow) to initiate the simulation. Exit the resulting screen. The resulting display (Fig. 5.83) shows the current is 3 A for the circuit with 15 V across R_3, and 36 V from a point between R_1 and R_2 to ground. The voltage across R_2 is 36 V − 15 V = 21 V, and the voltage across R_1 is 54 V − 36 V = 18 V. The power supplied or dissipated by each element is also listed.

Multisim

The construction of the network in Fig 5.84 using Multisim is simply an extension of the procedure outlined in Chapter 4. For each resistive element or meter, the process is repeated. The label for each increases by one as additional resistors or meters are added. Remember from the discussion of Chapter 4 that you should add the meters before connecting the elements together because the meters take space and must be properly oriented. The current is determined by the **XMM1** ammeter and the voltages by **XMM2** through **XMM5.** Of particular importance, note that

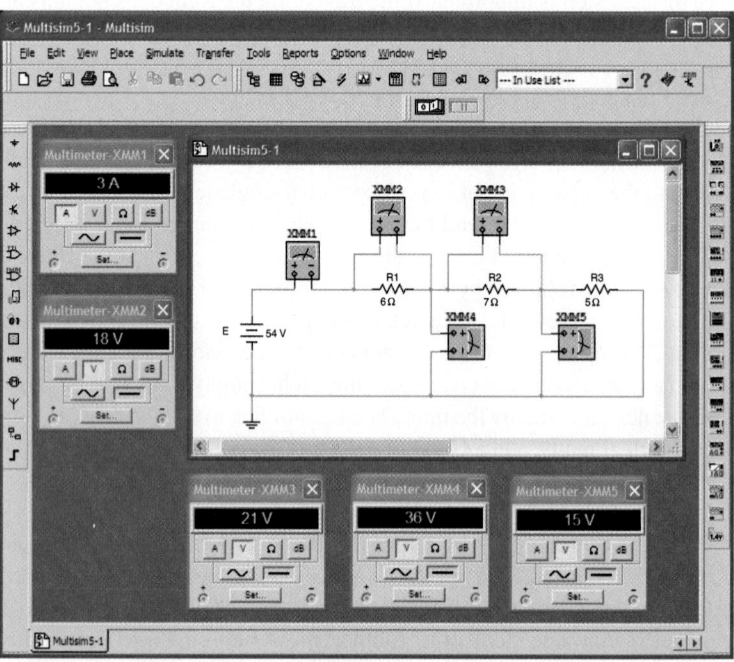

FIG. 5.84
Applying Multisim to a series dc circuit.

in Multisim the meters are connected in exactly the same way they would be placed in an active circuit in the laboratory. Ammeters are in series with the branch in which the current is to be determined, and voltmeters are connected between the two points of interest (across resistors). In addition, for positive readings, ammeters are connected so that conventional current enters the positive terminal, and voltmeters are connected so that the point of higher potential is connected to the positive terminal.

The meter settings are made by double-clicking on the meter symbol on the schematic. In each case, **V** or **I** had to be chosen, but the horizontal line for dc analysis is the same for each. Again, you can select the **Set** key to see what it controls, but the default values of meter input resistance levels are fine for all the analyses described in this text. Leave the meters on the screen so that the various voltages and the current level will be displayed after the simulation.

Recall from Chapter 4 that elements can be moved by simply clicking on each schematic symbol and dragging it to the desired location. The same is true for labels and values. Labels and values are set by double-clicking on the label or value and entering your preference. Click **OK,** and they are changed on the schematic. You do not have to first select a special key to connect the elements. Simply bring the cursor to the starting point to generate the small circle and crosshair. Click on the starting point, and follow the desired path to the next connection path. When in the correct location, click again, and the line appears. All connecting lines can make 90° turns. However, you cannot follow a diagonal path from one point to another. To remove any element, label, or line, click on the quantity to obtain the four-square active status, and select the **Delete** key or the scissors key on the top menu bar.

Recall from Chapter 4 that you can initiate simulation through the sequence **Simulate-Run** selecting the **Lightning Key,** or switching the **Simulate Switch** to the **1** position. You can stop the simulation by selecting the same **Lightning** key or switching the **Simulate Switch** to the **0** position.

Note from the results that the sum of the voltages measured by XMM2 and XMM4 equals the applied voltage. All the meters are considered ideal so there is no voltage drop across the XMM1 ammeter. In addition, they do not affect the value of the current measured by XMM1. All the voltmeters have essentially infinite internal resistance while the ammeters all have zero internal resistance. Of course, the meters can be entered as anything but ideal using the **Set** option. Note also that the sum of the voltages measured by XMM3 and XMM5 equals that measured by XMM4 as required by Kirchhoff's voltage law.

PROBLEMS

SECTION 5.2 Series Resistors

1. For each configuration in Fig. 5.85, find the individual (not combinations of) elements (voltage sources and/or resistors) that are in series. If necessary, use the fact that elements in series have the same current. Simply list those that satisfy the conditions for a series relationship. We will learn more about other combinations later.

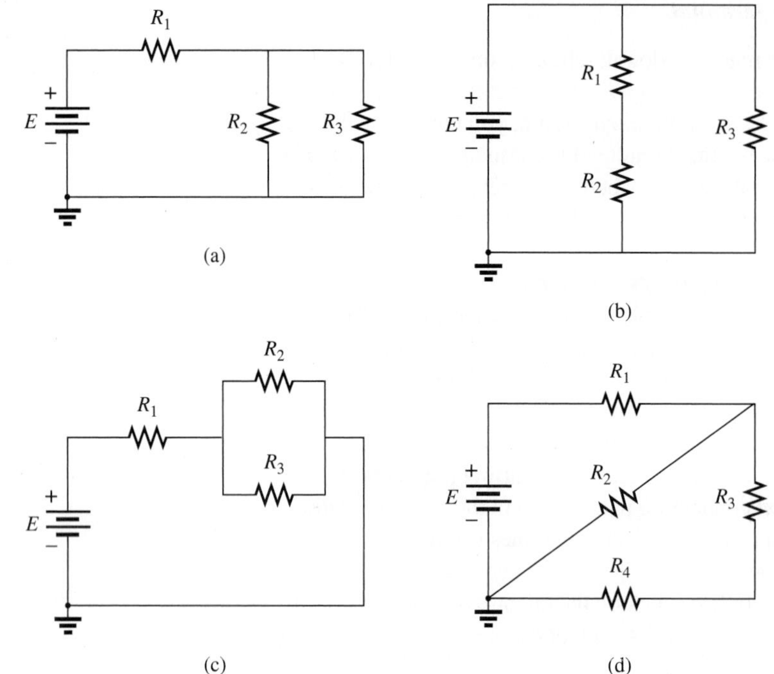

FIG. 5.85
Problem 1.

2. Find the total resistance R_T for each configuration in Fig. 5.86. Note that only standard resistor values were used.

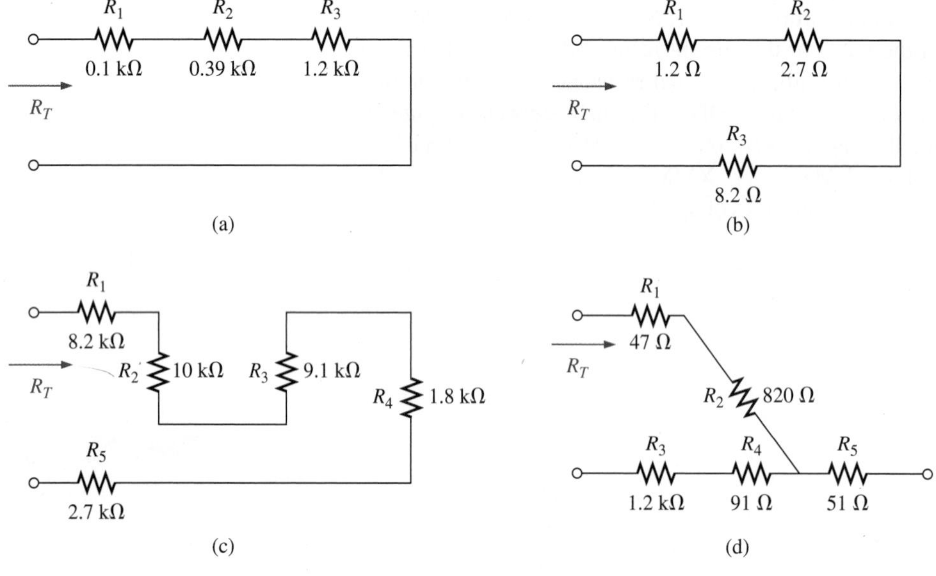

FIG. 5.86
Problem 2.

3. For each circuit board in Fig. 5.87, find the total resistance between connection tabs 1 and 2.

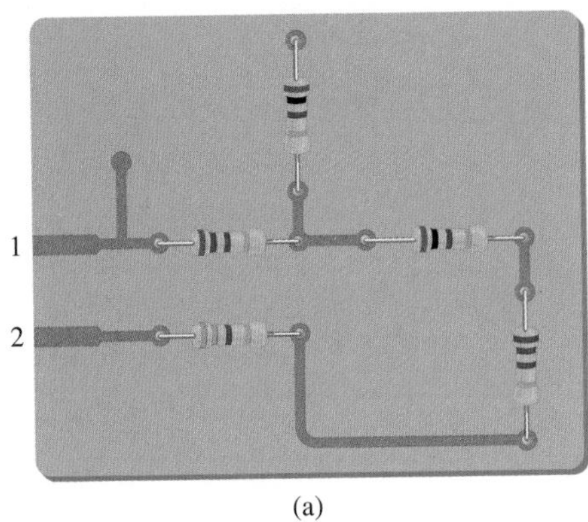

(a)

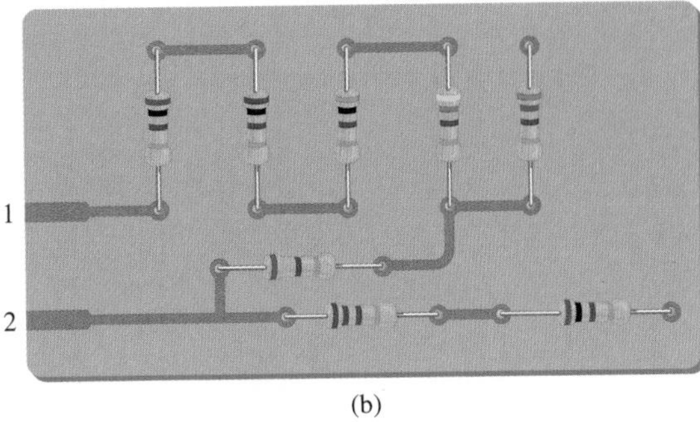

(b)

FIG. 5.87
Problem 3.

4. For the circuit in Fig. 5.88, composed of standard values:
   a. Which resistor will have the most impact on the total resistance?
   b. On an approximate basis, which resistors can be ignored when determining the total resistance?
   c. Find the total resistance, and comment on your results for parts (a) and (b).

5. For each configuration in Fig. 5.89, find the unknown resistors using the ohmmeter reading.

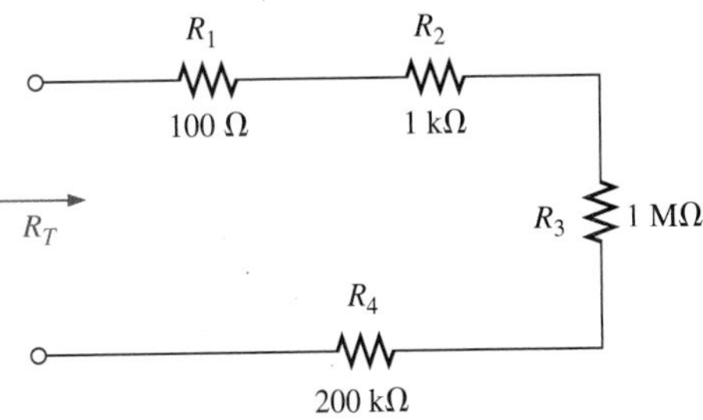

FIG. 5.88
Problem 4.

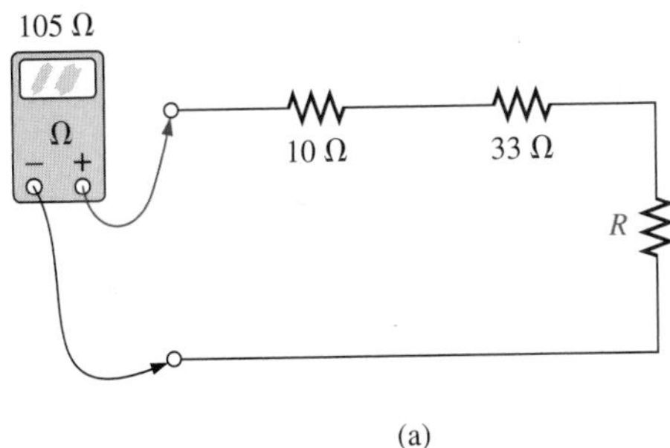

(a)

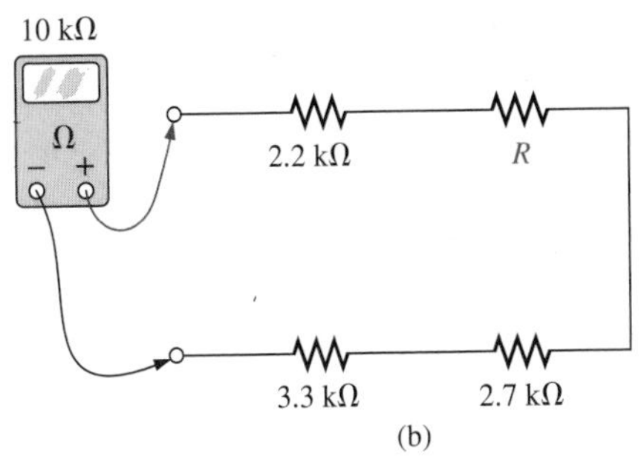

(b)

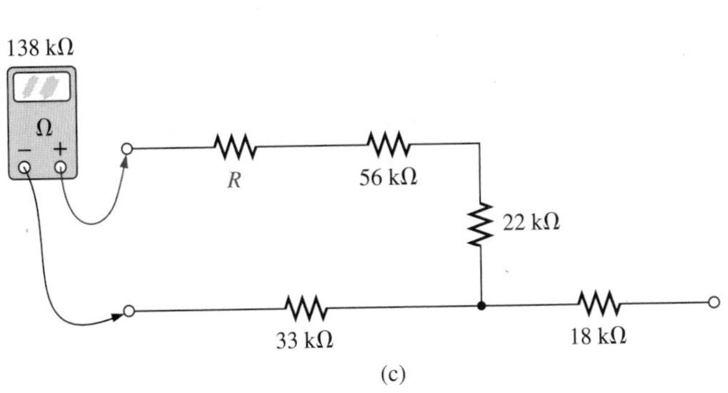

(c)

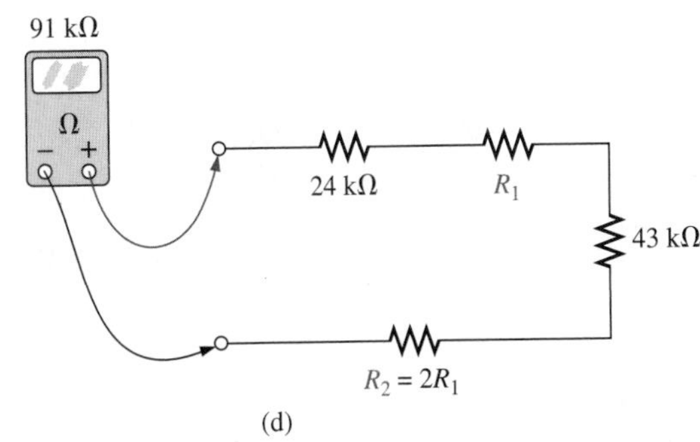

(d)

FIG. 5.89
Problem 5.

6. What is the ohmmeter reading for each configuration in Fig. 5.90?

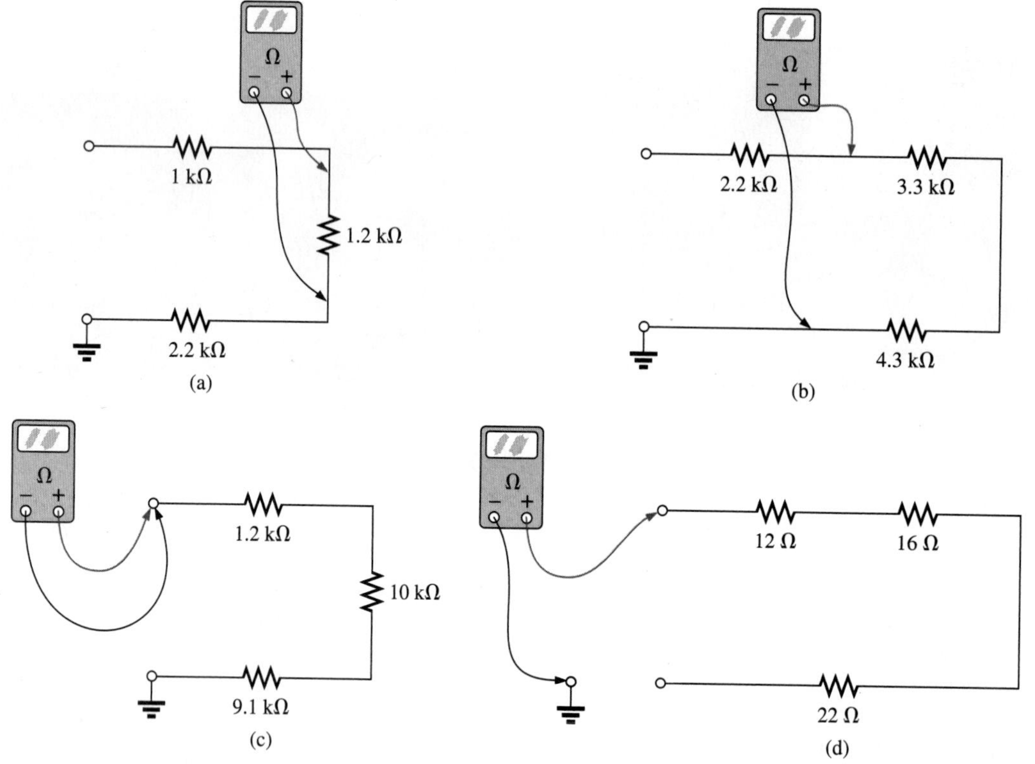

FIG. 5.90
Problem 6.

SECTION 5.3 Series Circuits

7. For the series configuration in Fig. 5.91, constructed of standard values:
   a. Find the total resistance.
   b. Calculate the current.
   c. Find the voltage across each resistive element.
8. For the series configuration in Fig. 5.92, constructed using standard value resistors:
   a. Without making a single calculation, which resistive element will have the most voltage across it? Which will have the least?
   b. Which resistor will have the most impact on the total resistance and the resulting current? Find the total resistance and the current.
   c. Find the voltage across each element and review your response to part (a).
9. Find the applied voltage necessary to develop the current specified in each circuit in Fig. 5.93.
10. For each network in Fig. 5.94, constructed of standard values, determine:
    a. The current *I*.
    b. The source voltage *E*.
    c. The unknown resistance.
    d. The voltage across each element.
11. For each configuration in Fig. 5.95, what are the readings of the ammeter and the voltmeter?

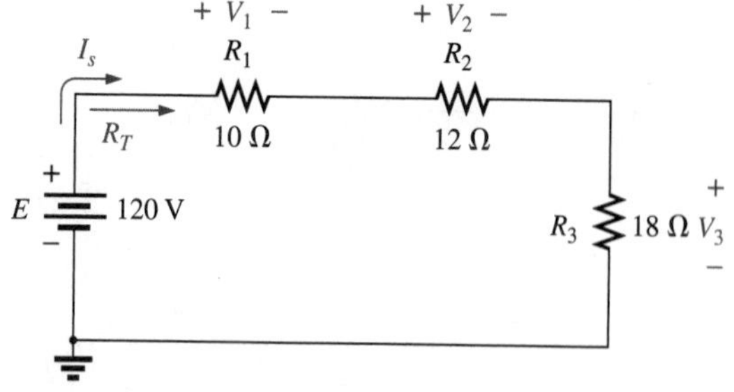

FIG. 5.91
Problem 7.

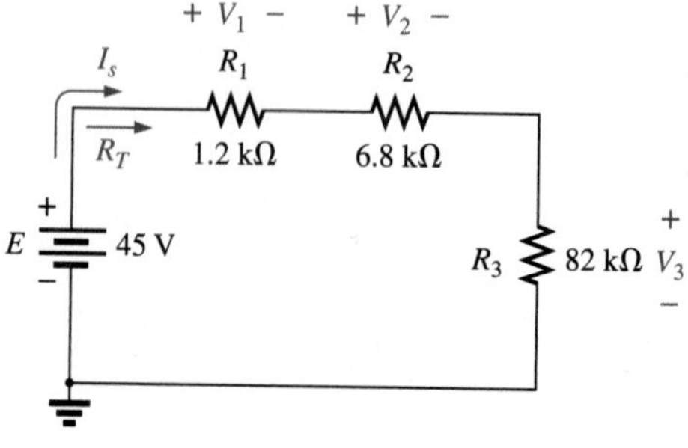

FIG. 5.92
Problem 8.

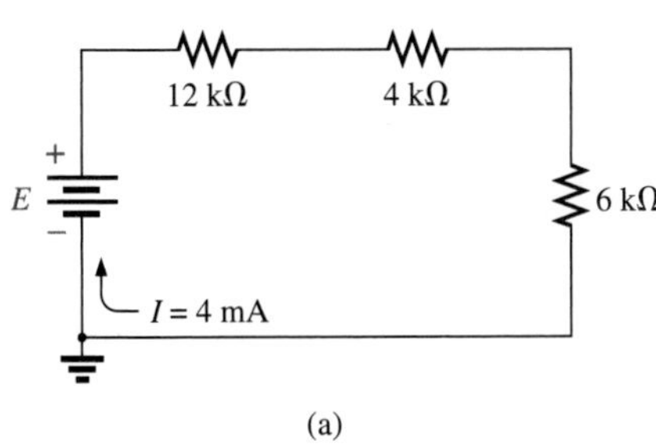

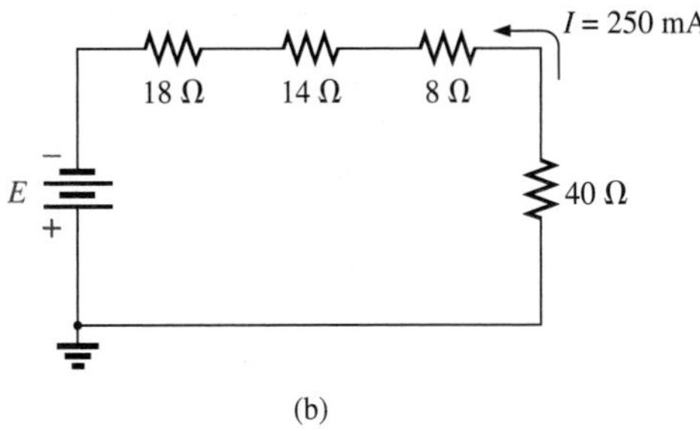

FIG. 5.93
Problem 9.

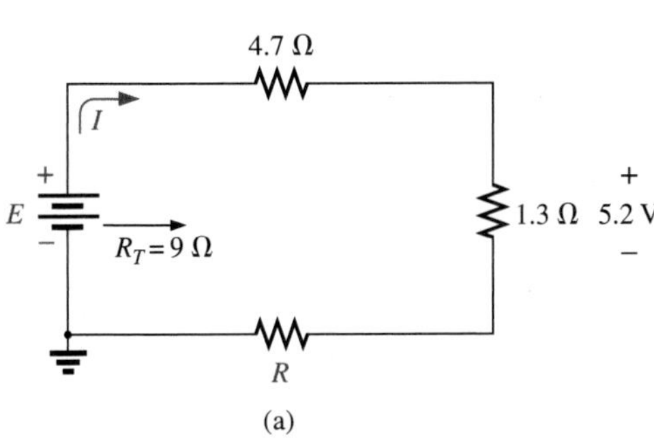

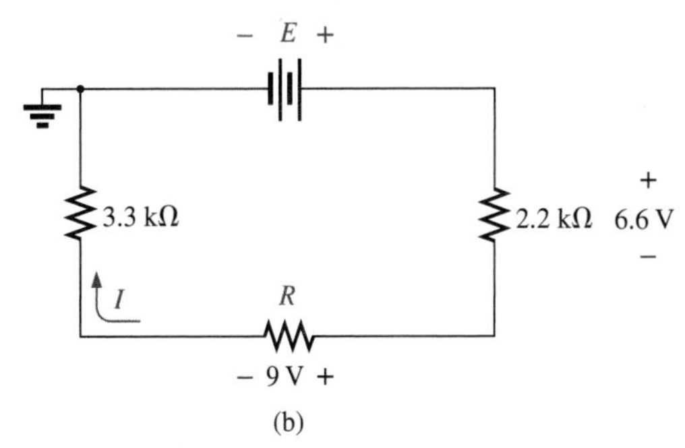

FIG. 5.94
Problem 10.

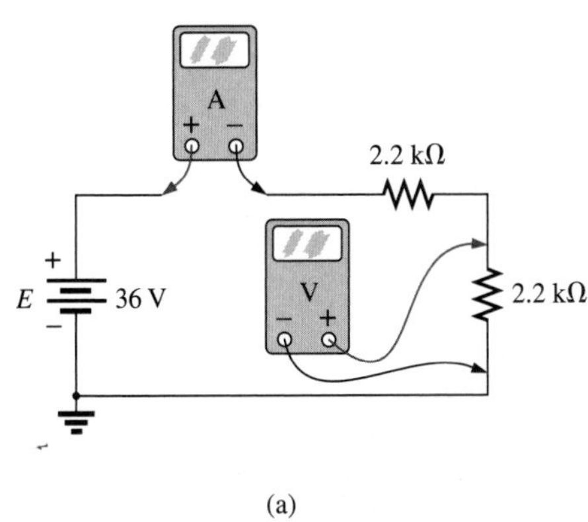

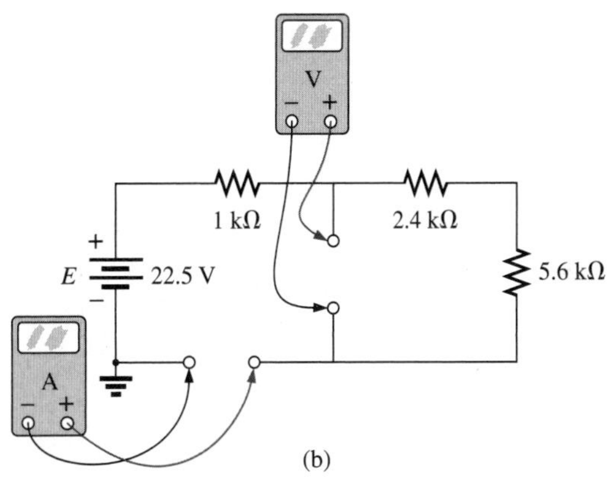

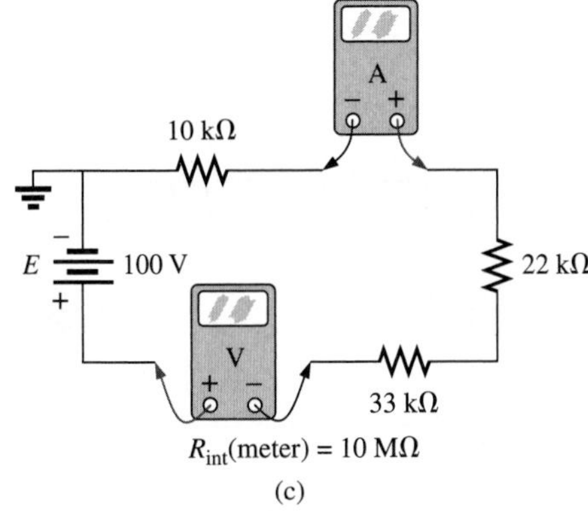

FIG. 5.95
Problem 11.

SECTION 5.4 Power Distribution in a Series Circuit

12. For the circuit in Fig. 5.96, constructed of standard value resistors:

a. Find the total resistance, current, and voltage across each element.

b. Find the power delivered to each resistor.

c. Calculate the total power delivered to all the resistors.

d. Find the power delivered by the source.

e. How does the power delivered by the source compare to that delivered to all the resistors?

f. Which resistor received the most power? Why?

g. What happened to all the power delivered to the resistors?

h. If the resistors are available with wattage ratings of 1/2 W, 1 W, 2 W, and 5 W, what minimum wattage rating can be used for each resistor?

13. Repeat Problem 12 for the circuit in Fig. 5.97.

14. Find the unknown quantities for the circuits in Fig. 5.98 using the information provided.

15. Eight holiday lights are connected in series as shown in Fig. 5.99.

a. If the set is connected to a 120 V source, what is the current through the bulbs if each bulb has an internal resistance of 28⅛ Ω?

b. Determine the power delivered to each bulb.

c. Calculate the voltage drop across each bulb.

d. If one bulb burns out (that is, the filament opens), what is the effect on the remaining bulbs? Why?

16. For the conditions specified in Fig. 5.100, determine the unknown resistance.

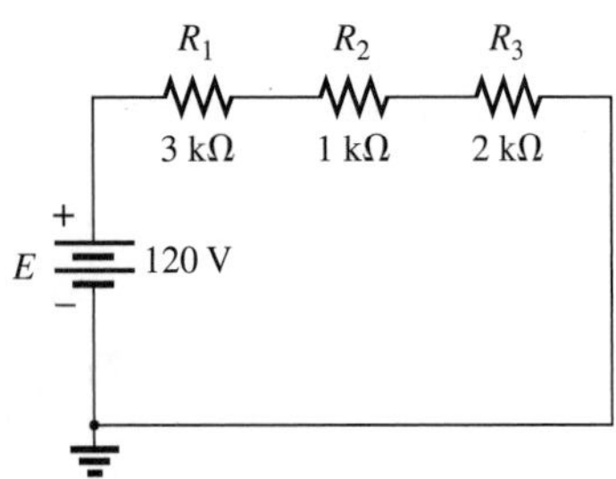

FIG. 5.96
Problem 12.

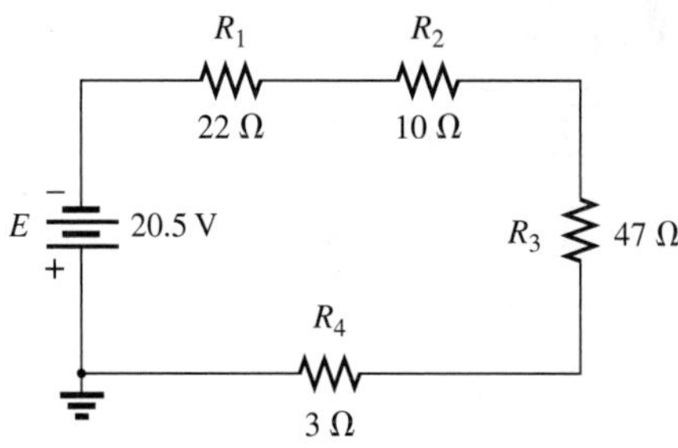

FIG. 5.97
Problem 13.

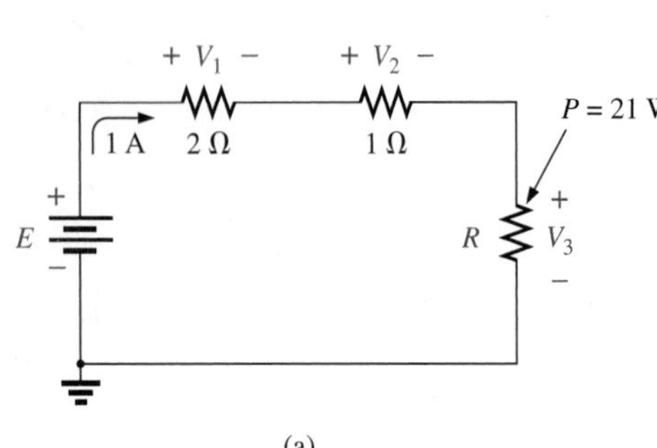

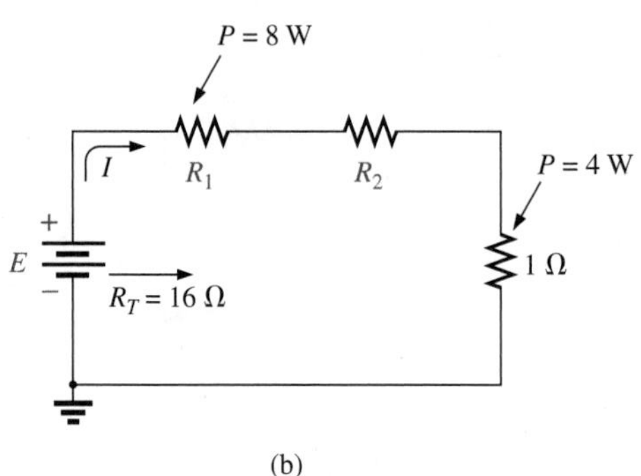

FIG. 5.98
Problem 14.

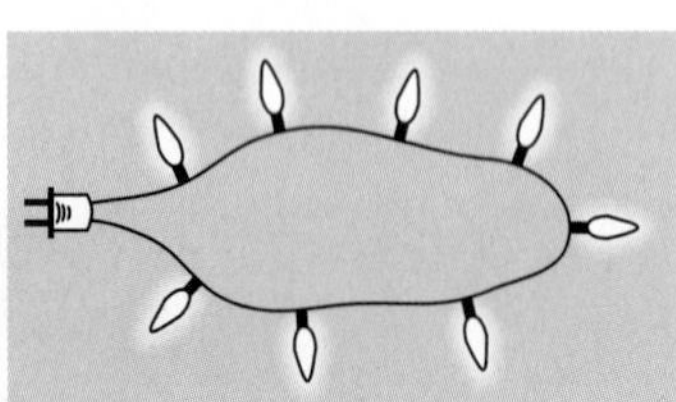

FIG. 5.99
Problem 15.

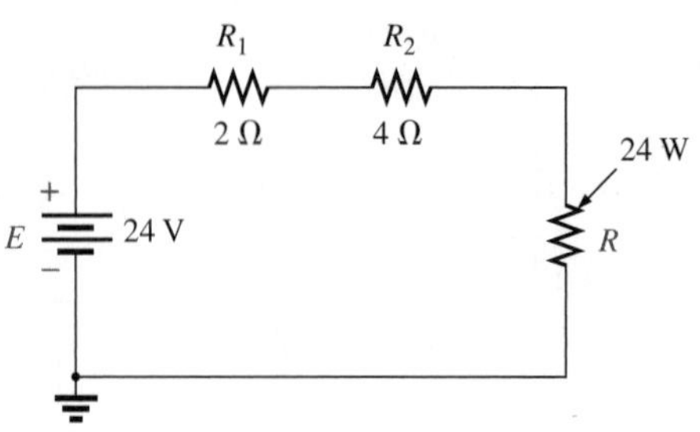

FIG. 5.100
Problem 16.

SECTION 5.5 Voltage Sources in Series

17. Combine the series voltage sources in Fig. 5.101 into a single voltage source between points *a* and *b*.

18. Determine the current *I* and its direction for each network in Fig. 5.102. Before solving for *I*, redraw each network with a single voltage source.

19. Find the unknown voltage source and resistor for the networks in Fig. 5.103. First combine the series voltage sources into a single source. Indicate the direction of the resulting current.

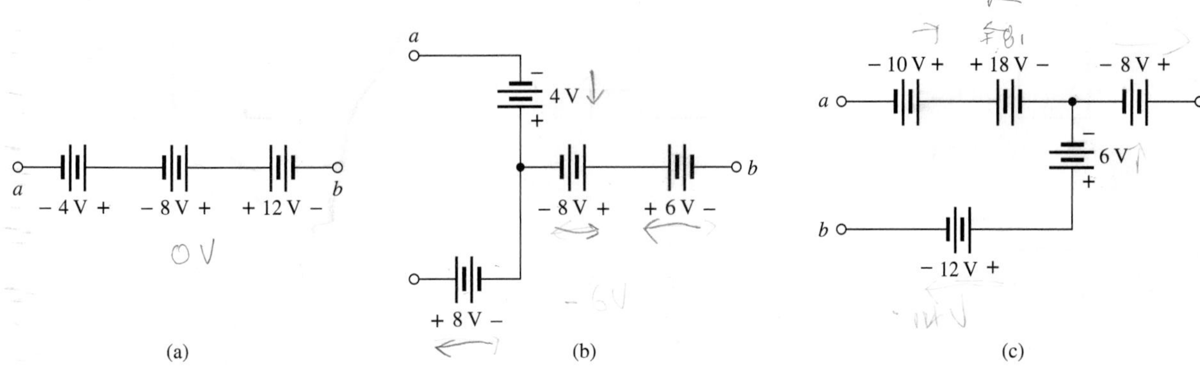

FIG. 5.101
Problem 17.

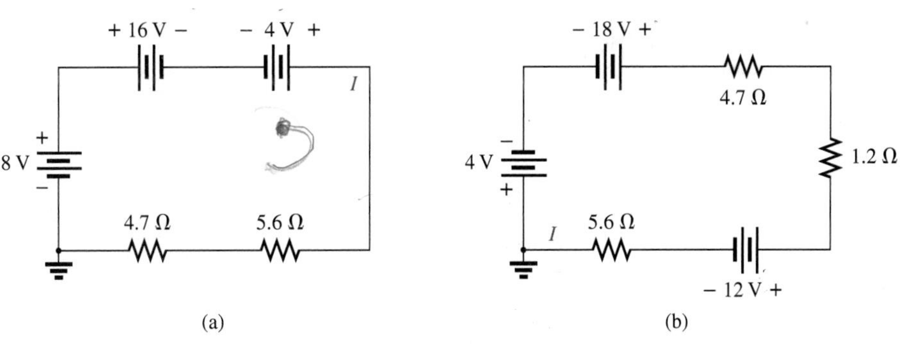

FIG. 5.102
Problem 18.

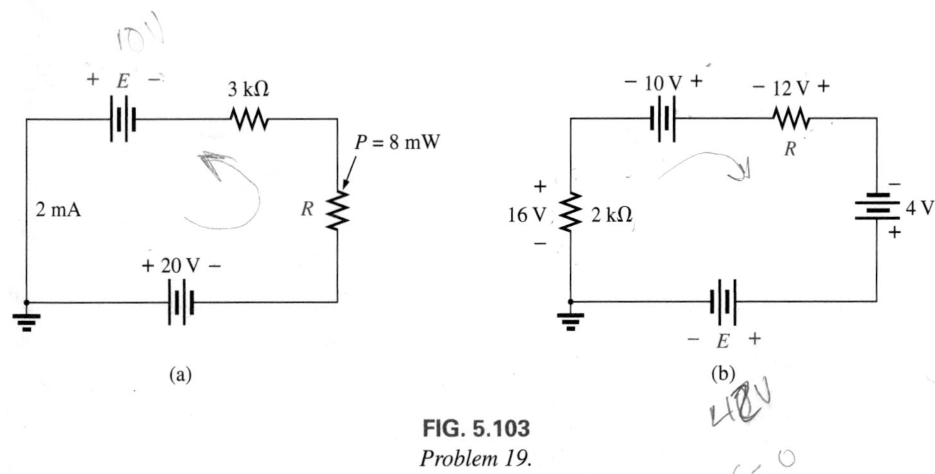

FIG. 5.103
Problem 19.

SECTION 5.6 Kirchhoff's Voltage Law

20. Using Kirchhoff's voltage law, find the unknown voltages for the circuits in Fig. 5.104.

21. Using Kirchhoff's voltage law, determine the unknown voltages for the configurations in Fig. 5.105.

22. Using Kirchhoff's voltage law, determine the unknown voltages for the series circuits in Fig. 5.106.

23. Using Kirchhoff's voltage law, find the unknown voltages for the configurations in Fig. 5.107.

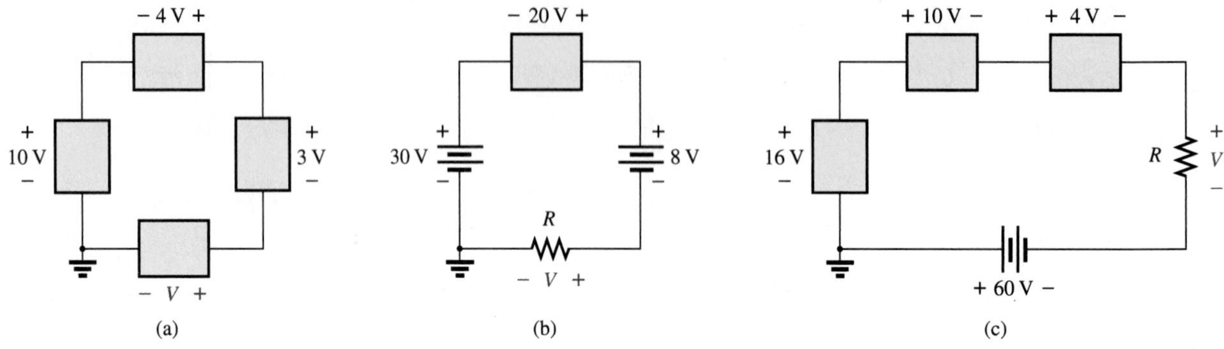

FIG. 5.104
Problem 20.

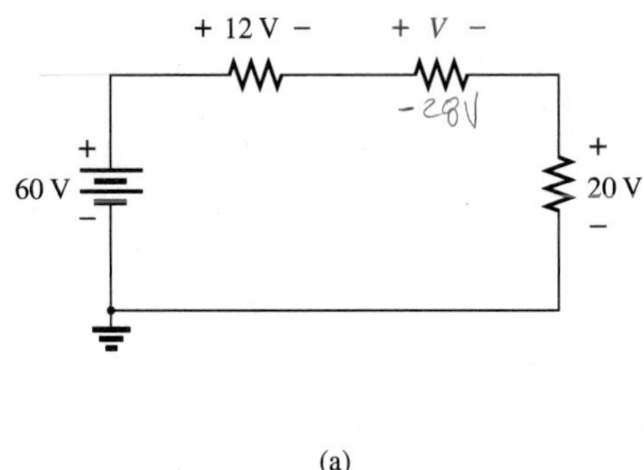

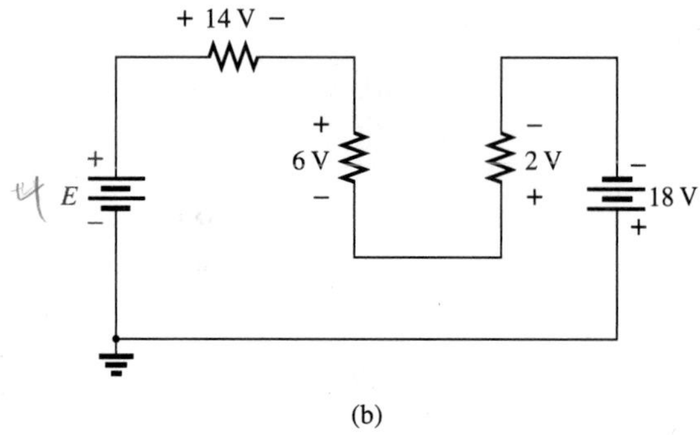

FIG. 5.105
Problem 21.

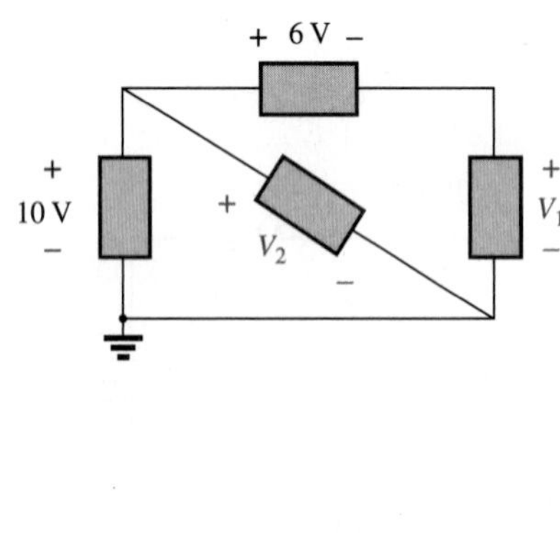

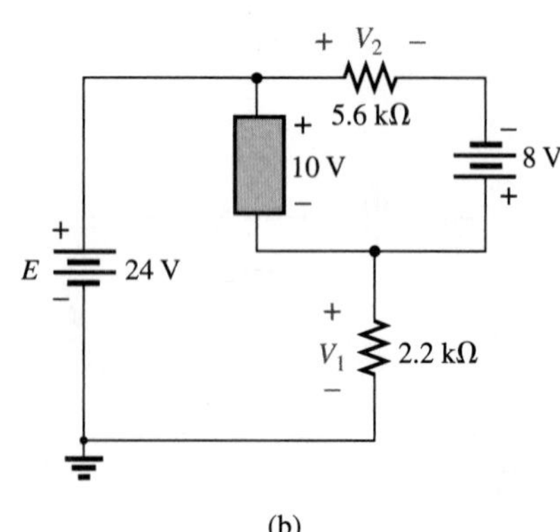

FIG. 5.106
Problem 22.

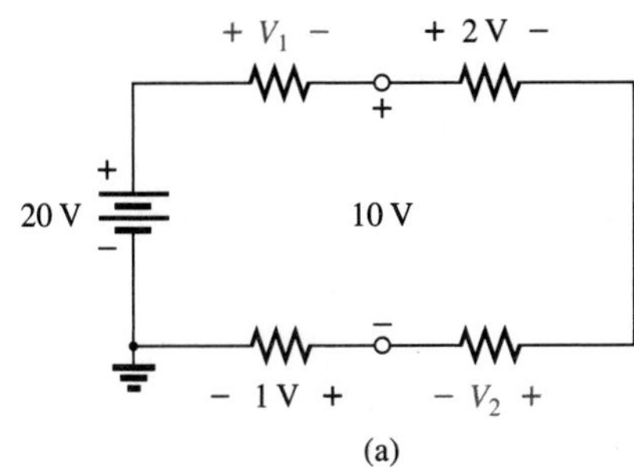

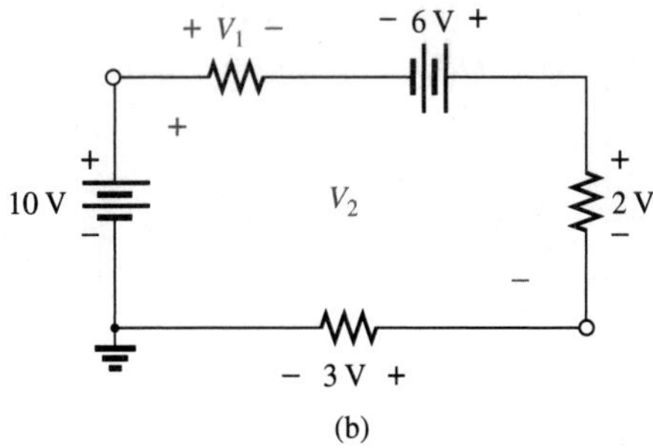

FIG. 5.107
Problem 23.

SECTION 5.7 Voltage Division in a Series Circuit

24. Determine the values of the unknown resistors in Fig. 5.108 using the provided voltage levels.

25. For the configuration in Fig. 5.109, with standard resistor values:

a. By inspection, which resistor will receive the largest share of the applied voltage? Why?

b. How much larger will voltage V_3 be compared to V_2 and V_1?

c. Find the voltage across the largest resistor using the voltage divider rule.

d. Find the voltage across the series combination of resistors R_2 and R_3.

26. Using the voltage divider rule, find the indicated voltages in Fig. 5.110.

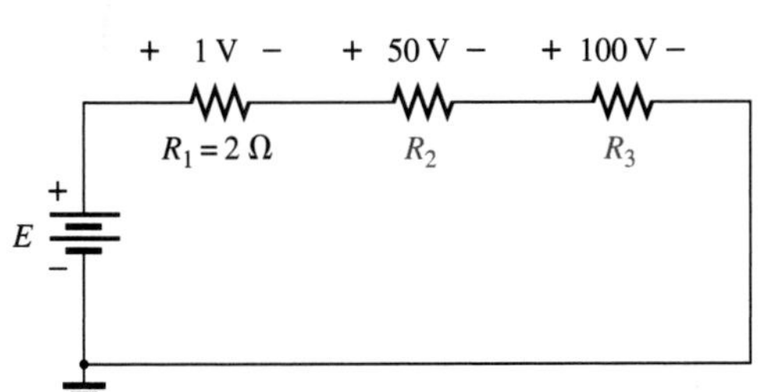

FIG. 5.108
Problem 24.

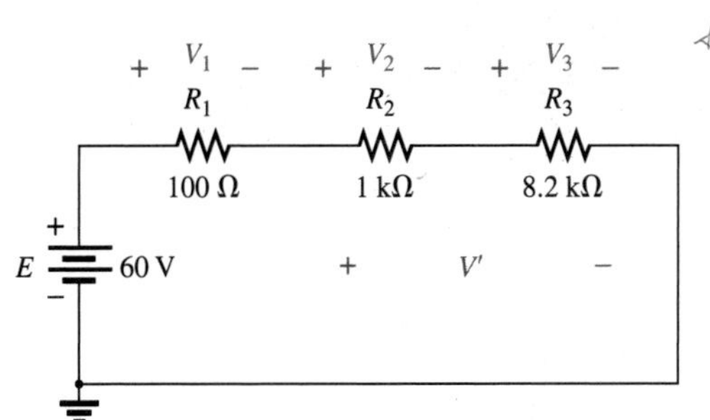

FIG. 5.109
Problem 25.

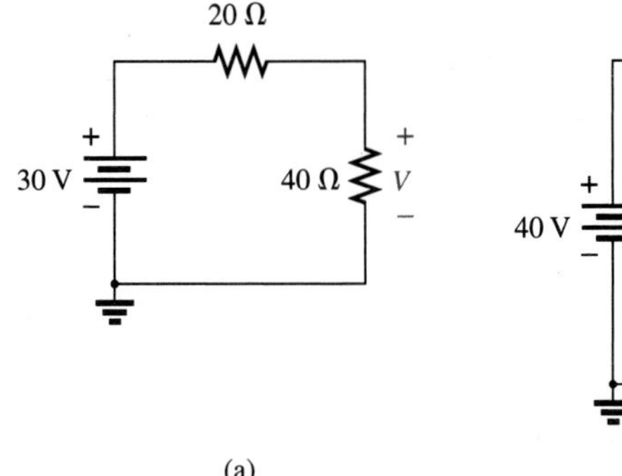

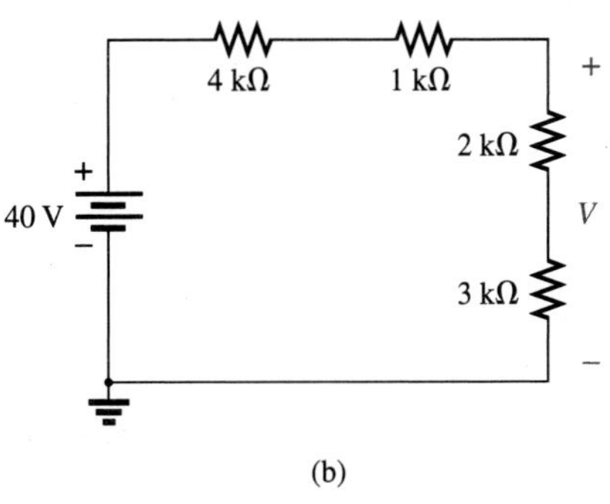

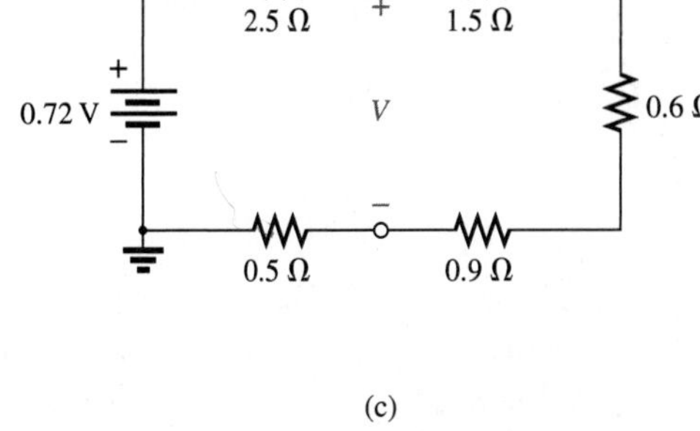

FIG. 5.110
Problem 26.

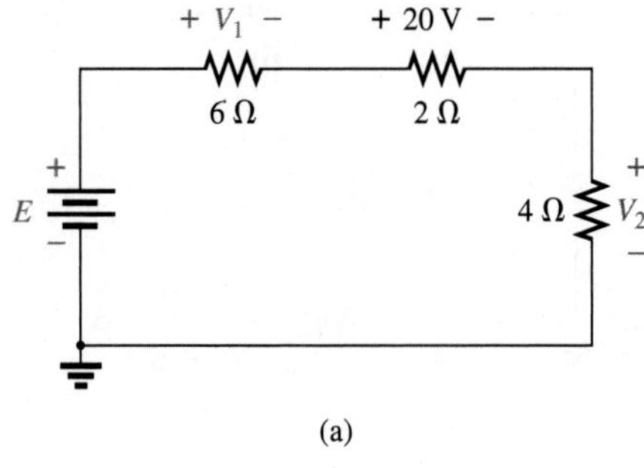

(a)

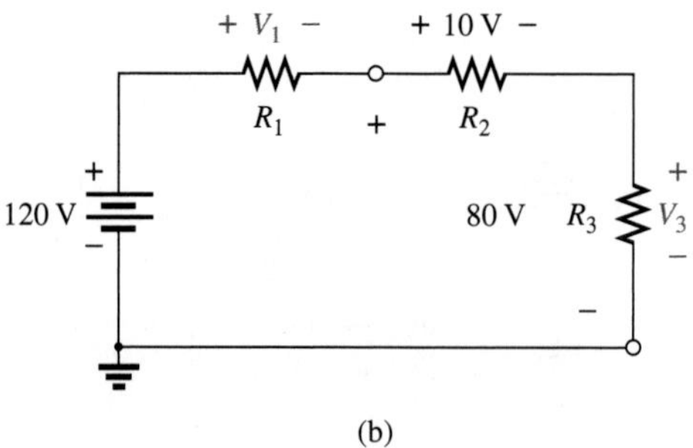

(b)

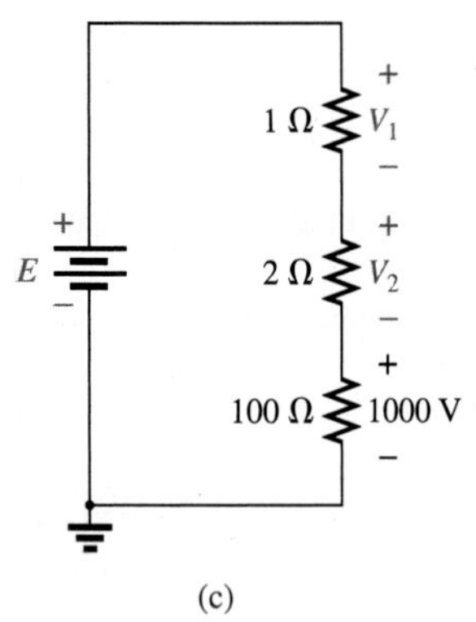

(c)

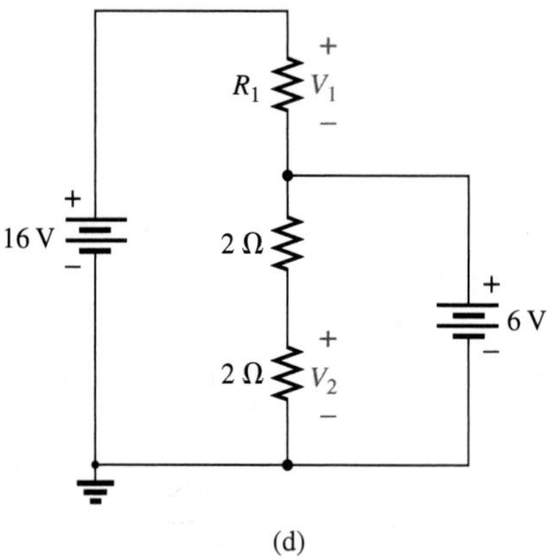

(d)

FIG. 5.111
Problem 27.

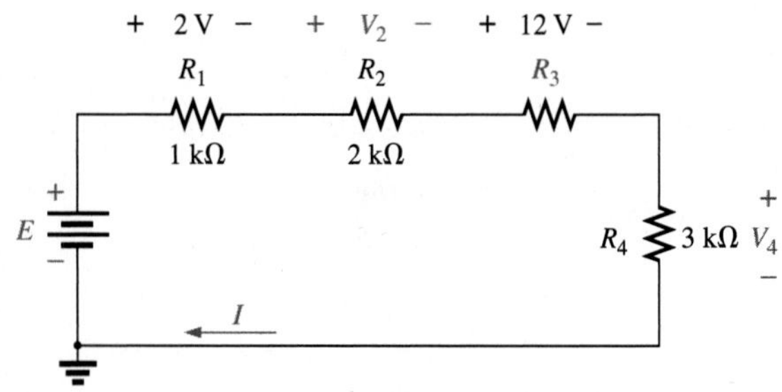

FIG. 5.112
Problem 28.

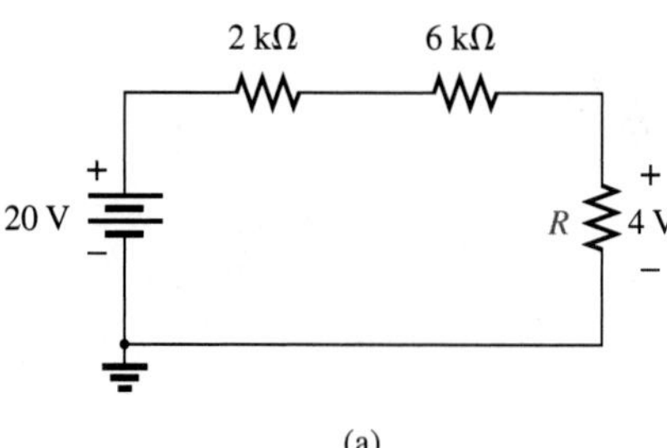

(a)

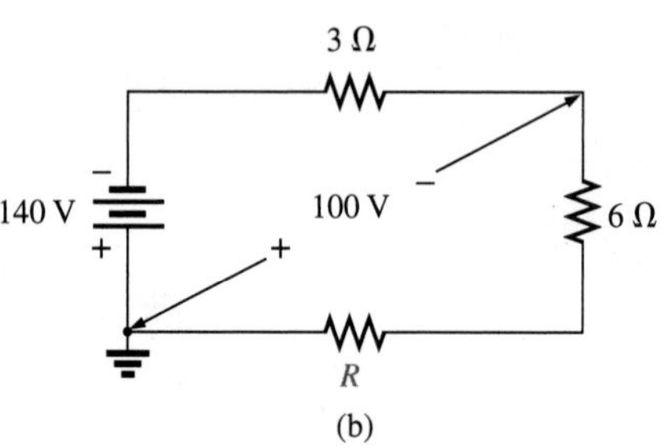

(b)

FIG. 5.113
Problem 29.

27. Using the voltage divider rule or Kirchhoff's voltage law, determine the unknown voltages for the configurations in Fig. 5.111.

28. Using the information provided, find the unknown quantities of Fig. 5.112.

***29.** Using the voltage divider rule, find the unknown resistance for the configurations in Fig. 5.113.

30. Referring to Fig. 5.114.
  a. Determine V_2.
  b. Calculate V_3.
  c. Determine R_3.

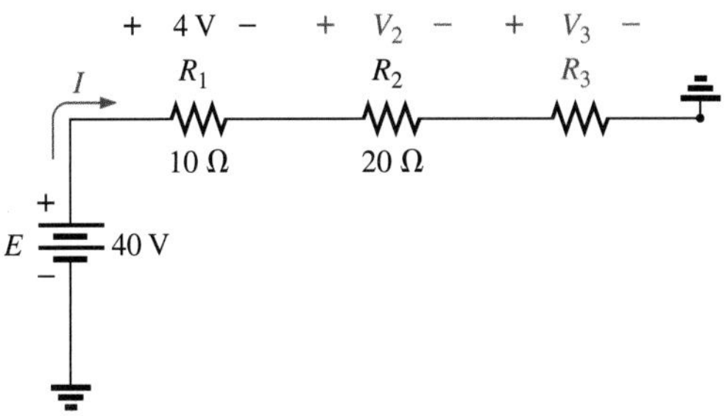

FIG. 5.114
Problem 30.

31. **a.** Design a voltage divider circuit that will permit the use of an 8 V, 50 mA bulb in an automobile with a 12 V electrical system.
  b. What is the minimum wattage rating of the chosen resistor if 1/4 W, 1/2 W, and 1 W resistors are available?

32. Design the voltage divider in Fig. 5.115 such that $V_{R_1} = 1/5V_{R_1}$. That is, find R_1 and R_2.

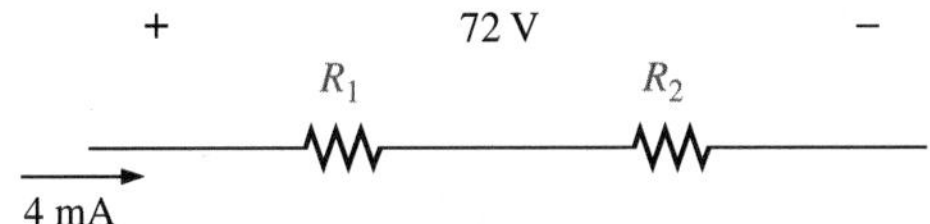

FIG. 5.115
Problem 32.

33. Find the voltage across each resistor in Fig. 5.116 if $R_1 = 2R_3$ and $R_2 = 7R_3$.

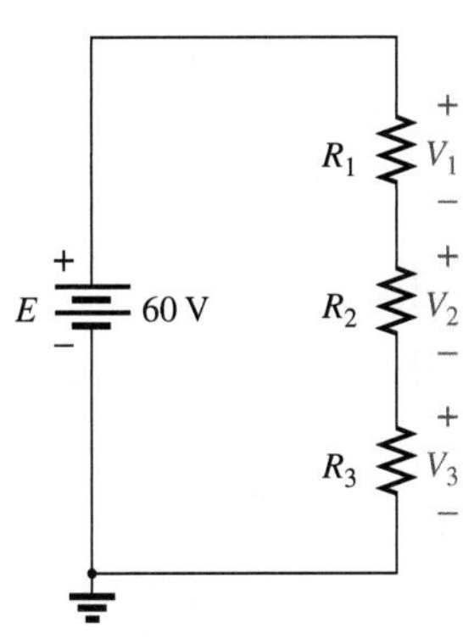

FIG. 5.116
Problem 33.

***34.** **a.** Design the circuit in Fig. 5.117 such that $V_{R_2} = 3V_{R_1}$ and $V_{R_3} = 4V_{R_2}$.
  b. If the current is reduced to 10 μA, what are the new values of R_1, R_2, and R_3? How do they compare to the results of part (a)?

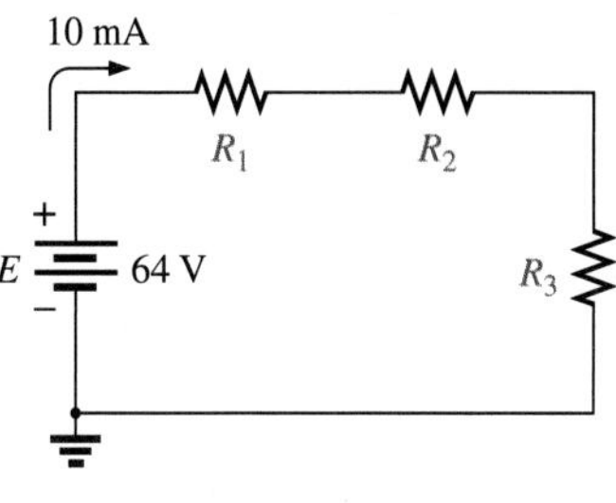

FIG. 5.117
Problem 34.

SECTION 5.9 Notation

35. Determine the voltages V_a, V_b, and V_{ab} for the networks in Fig. 5.118.

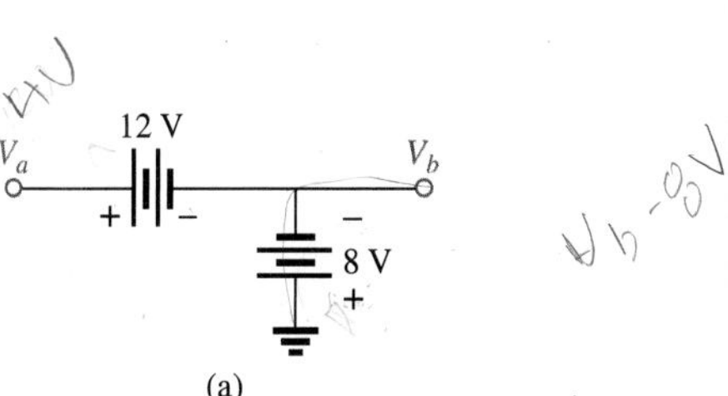

FIG. 5.118
Problem 35.

36. Determine the current I (with direction) and the voltage V (with polarity) for the networks in Fig. 5.119.

37. Determine the voltages V_a and V_1 for the networks in Fig. 5.120.

38. For the network in Fig. 5.121 determine the voltages:
 a. V_a, V_b, V_c, V_d, V_e
 b. V_{ab}, V_{dc}, V_{cb}
 c. V_{ac}, V_{db}

39. Given the information appearing in Fig. 5.122, find the level of resistance for R_1 and R_3.

40. Determine the values of R_1, R_2, R_3, and R_4 for the voltage divider of Fig. 5.123 if the source current is 16 mA.

41. For the network in Fig. 5.124, determine the voltages:
 a. V_a, V_b, V_c, V_d
 b. V_{ab}, V_{cb}, V_{cd}
 c. V_{ad}, V_{ca}

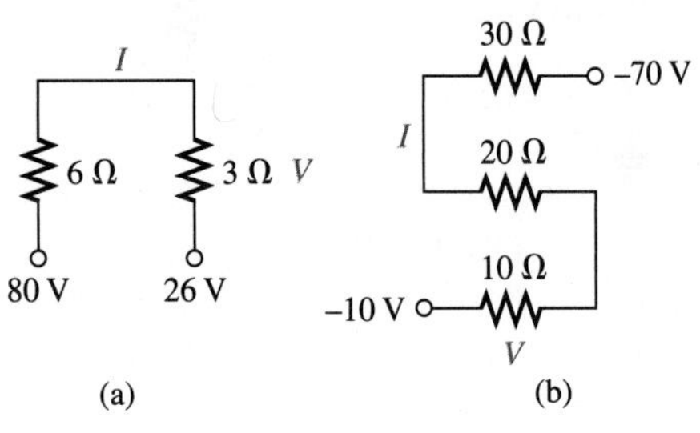

FIG. 5.119
Problem 36.

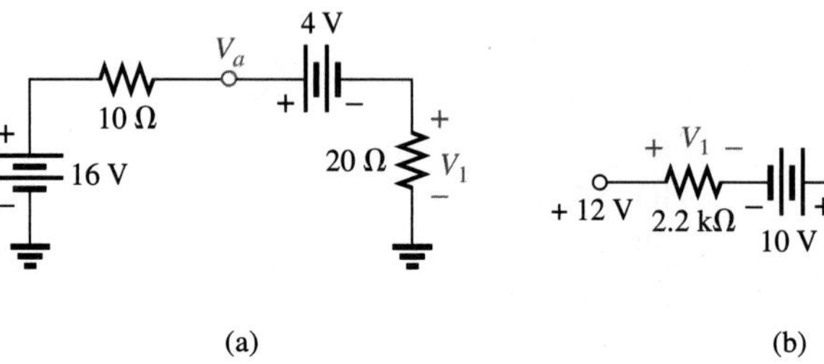

FIG. 5.120
Problem 37.

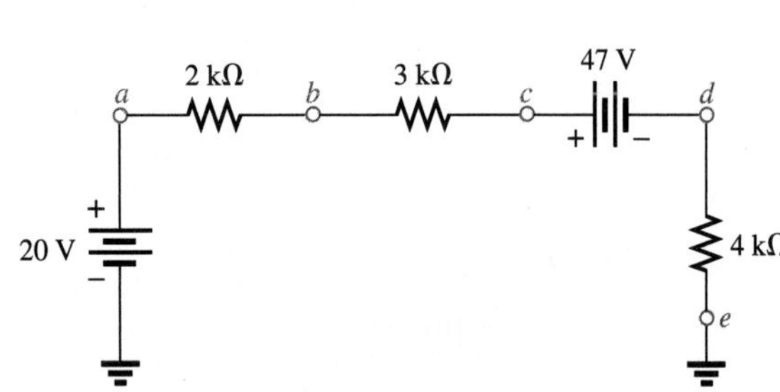

FIG. 5.121
Problem 38.

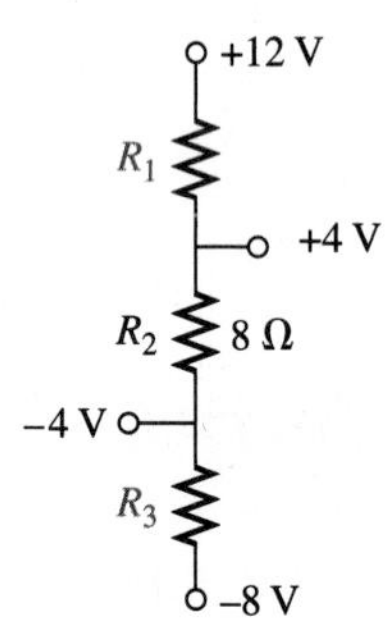

FIG. 5.122
Problem 39.

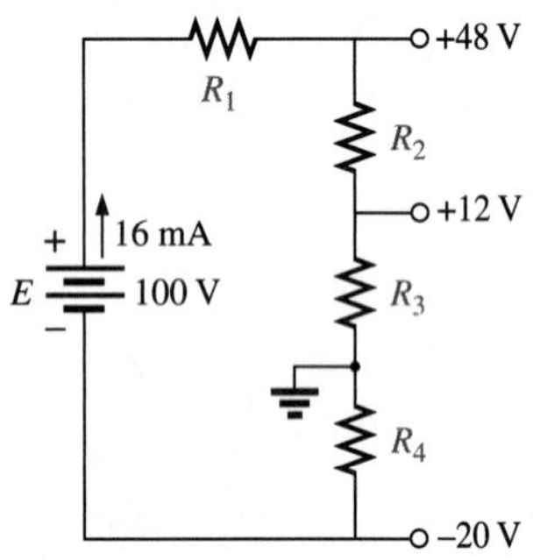

FIG. 5.123
Problem 40.

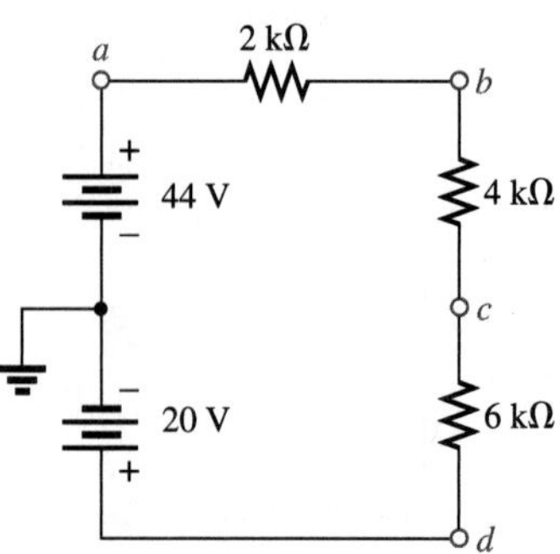

FIG. 5.124
Problem 41.

*42. For the integrated circuit in Fig. 5.125, determine V_0, V_4, V_7, V_{10}, V_{23}, V_{30}, V_{67}, V_{56}, and I (magnitude and direction.)

*43. For the integrated circuit in Fig. 5.126, determine V_0, V_{03}, V_2, V_{23}, V_{12}, and I_i.

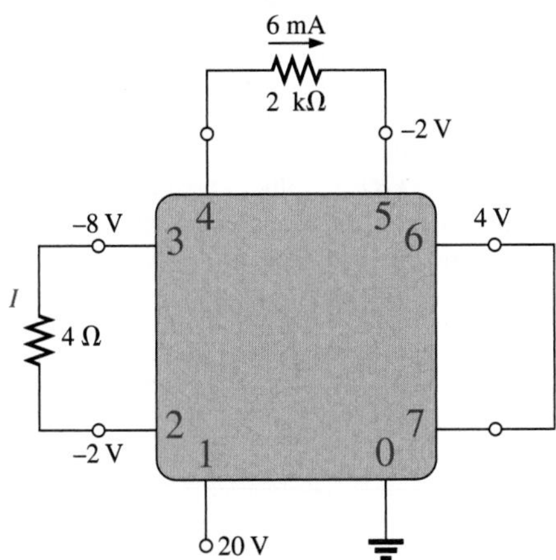

FIG. 5.125
Problem 42.

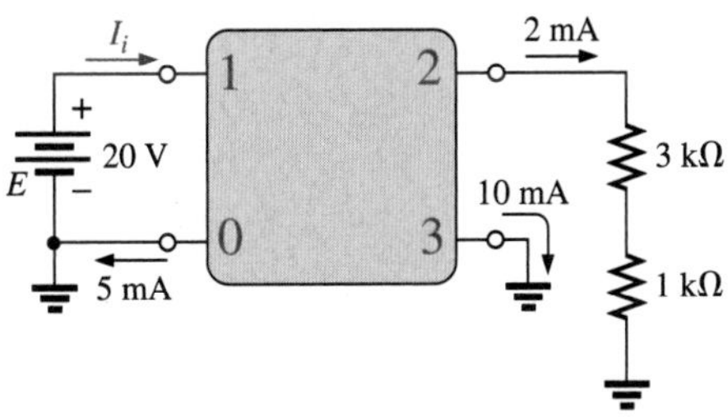

FIG. 5.126
Problem 43.

SECTION 5.10 Voltage Regulation and the Internal Resistance of Voltage Sources

44. a. Find the internal resistance of a battery that has a no-load output of 60 V and that supplies a full-load current of 2 A to a load of 28 Ω.
    b. Find the voltage regulation of the supply.

45. a. Find the voltage to the load (full-load conditions) for the supply in Fig. 5.127.
    b. Find the voltage regulation of the supply.
    c. How much power is supplied by the source and lost to the internal resistance under full-load conditions?

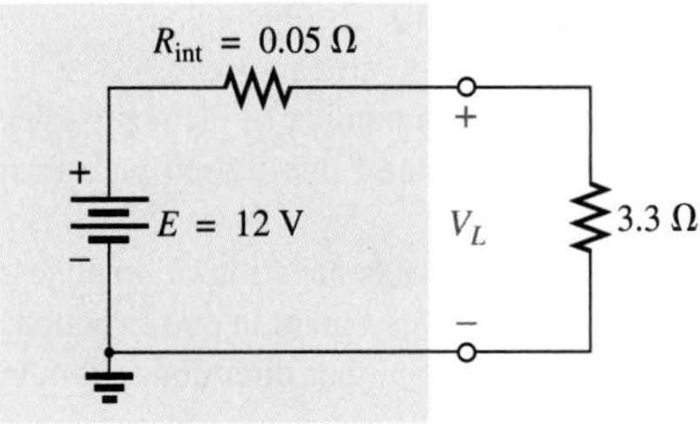

FIG. 5.127
Problem 45.

SECTION 5.11 Loading Effects of Instruments

46. a. Determine the current through the circuit in Fig. 5.128.
    b. If an ammeter with an internal resistance of 250 Ω is inserted into the circuit in Fig. 5.128, what effect will it have on the current level?
    c. Is the difference in current level a major concern for most applications?

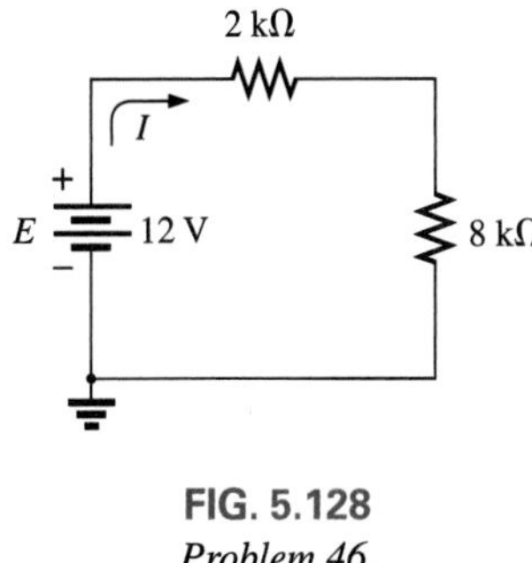

FIG. 5.128
Problem 46.

SECTION 5.14 Computer Analysis

47. Use the computer to verify the results of Example 5.4.
48. Use the computer to verify the results of Example 5.5.
49. Use the computer to verify the results of Example 5.15.

GLOSSARY

Circuit A combination of a number of elements joined at terminal points providing at least one closed path through which charge can flow.

Closed loop Any continuous connection of branches that allows tracing of a path that leaves a point in one direction and returns to that same point from another direction without leaving the circuit.

Internal resistance The inherent resistance found internal to any source of energy.

Kirchhoff's voltage law (KVL) The algebraic sum of the potential rises and drops around a closed loop (or path) is zero.

Protoboard (breadboard) A flat board with a set pattern of conductively connected holes designed to accept 24-gage wire and components with leads of about the same diameter.

Series circuit A circuit configuration in which the elements have only one point in common and each terminal is not connected to a third, current-carrying element.

Two-terminal device Any element or component with two external terminals for connection to a network configuration.

Voltage divider rule (VDR) A method by which a voltage in a series circuit can be determined without first calculating the current in the circuit.

Voltage regulation (*VR*) A value, given as a percent, that provides an indication of the change in terminal voltage of a supply with a change in load demand.

Parallel dc Circuits

6

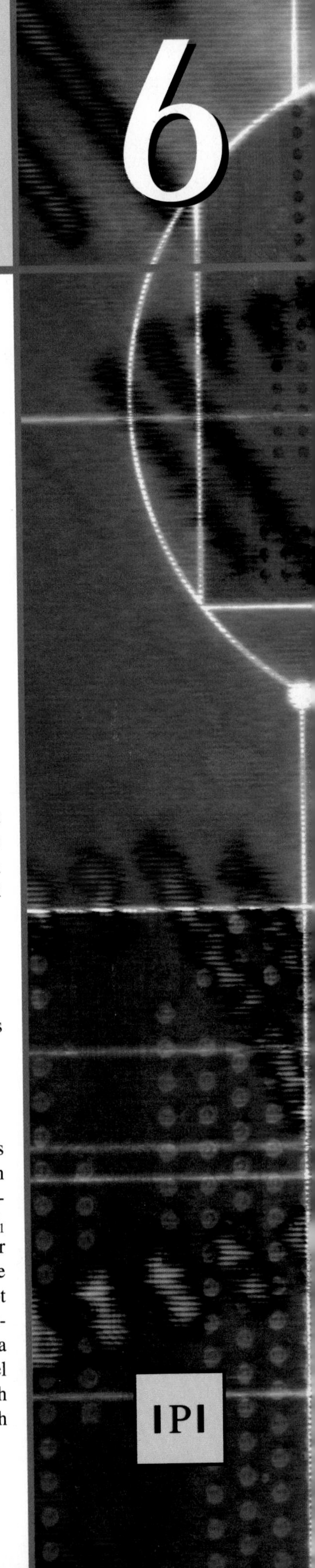

Objectives

- *Become familiar with the characteristics of a parallel network and how to solve for the voltage, current, and power to each element.*
- *Develop a clear understanding of Kirchhoff's current law and its importance to the analysis of electric circuits.*
- *Become aware of how the source current will split between parallel elements and how to properly apply the current divider rule.*
- *Clearly understand the impact of open and short circuits on the behavior of a network.*
- *Learn how to use an ohmmeter, voltmeter, and ammeter to measure the important parameters of a parallel network.*

6.1 INTRODUCTION

Two network configurations, series and parallel, form the framework for some of the most complex network structures. A clear understanding of each will pay enormous dividends as more complex methods and networks are examined. The series connection was discussed in detail in the last chapter. We will now examine the **parallel circuit** and all the methods and laws associated with this important configuration.

6.2 PARALLEL RESISTORS

The term *parallel* is used so often to describe a physical arrangement between two elements that most individuals are aware of its general characteristics.

In general,

two elements, branches, or circuits are in parallel if they have two points in common.

For instance, in Fig. 6.1(a), the two resistors are in parallel because they are connected at points *a* and *b*. If both ends were *not* connected as shown, the resistors would not be in parallel. In Fig. 6.1(b), resistors R_1 and R_2 are in parallel because they again have points *a* and *b* in common. R_1 is not in parallel with R_3 because they are connected at only one point (*b*). Further, R_1 and R_3 are not in series because a third connection appears at point *b*. The same can be said for resistors R_2 and R_3. In Fig. 6.1(c), resistors R_1 and R_2 are in series because they have only one point in common that is not connected elsewhere in the network. Resistors R_1 and R_3 are not in parallel because they have only point *a* in common. In addition, they are not in series because of the third connection to point *a*. The same can be said for resistors R_2 and R_3. In a broader context, it can be said that the series combination of resistors R_1 and R_2 is in parallel with resistor R_3 (more will be said about this option in Chapter 7). Furthermore, even though the discussion above was only for resistors, it can be applied to any two-terminal elements such as voltage sources and meters.

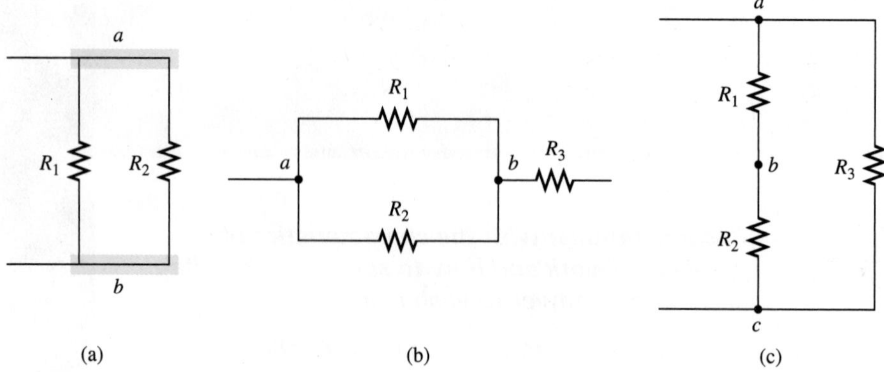

FIG. 6.1
(a) Parallel resistors; (b) R_1 and R_2 are in parallel; (c) R_3 is in parallel with the series combination of R_1 and R_2.

On schematics, the parallel combination can appear in a number of ways, as shown in Fig. 6.2. In each case, the three resistors are in parallel. They all have points *a* and *b* in common.

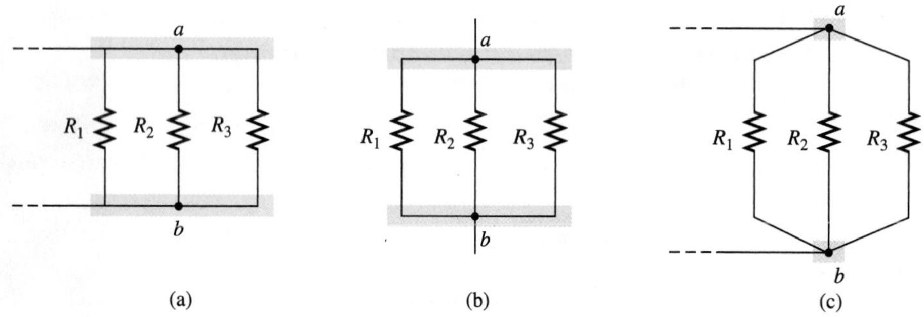

FIG. 6.2
Schematic representations of three parallel resistors.

For resistors in parallel as shown in Fig. 6.3, the total resistance is determined from the following equation:

$$\frac{1}{R_T} = \frac{1}{R_1} + \frac{1}{R_2} + \frac{1}{R_3} + \cdots + \frac{1}{R_N} \qquad \textbf{(6.1)}$$

Since $G = 1/R$, the equation can also be written in terms of conductance levels as follows:

$$G_T = G_1 + G_2 + G_3 + \cdots + G_N \qquad \text{(siemens, S)} \qquad \textbf{(6.2)}$$

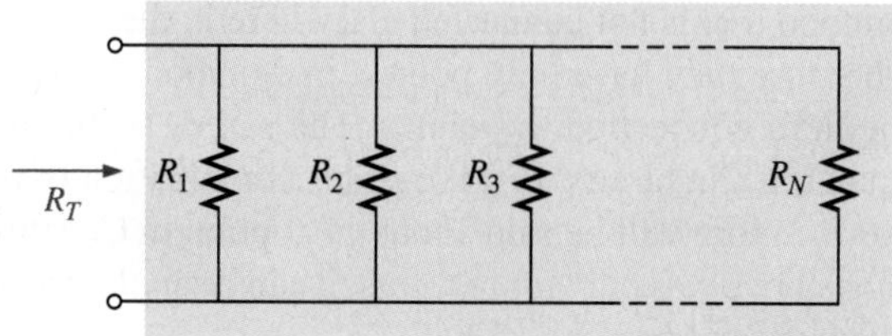

FIG. 6.3
Parallel combination of resistors.

which is an exact match in format with the equation for the total resistance of resistors in series: $R_T = R_1 + R_2 + R_3 + \cdots + R_N$. The result of this duality is that you can go from one equation to the other simply by interchanging R and G.

In general, however, when the total resistance is desired, the following format is applied:

$$R_T = \frac{1}{\dfrac{1}{R_1} + \dfrac{1}{R_2} + \dfrac{1}{R_3} + \cdots + \dfrac{1}{R_N}} \tag{6.3}$$

Quite obviously, Eq. (6.3) is not as "clean" as the equation for the total resistance of series resistors. You must be careful when dealing with all the divisions into 1. The great feature about the equation, however, is that it can be applied to any number of resistors in parallel.

EXAMPLE 6.1

a. Find the total conductance of the parallel network in Fig. 6.4.
b. Find the total resistance of the same network using the results of part (a) and using Eq. (6.3).

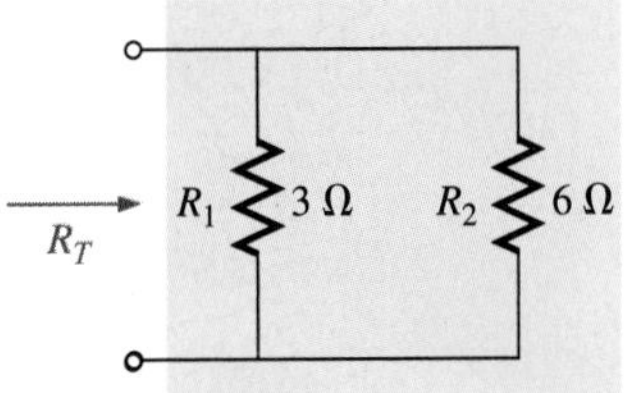

FIG. 6.4
Parallel resistors for Example 6.1.

Solutions:

a. $G_{1\pi} = \dfrac{1}{R_1} = \dfrac{1}{3\,\Omega} = 0.333\text{ S}, \quad G_2 = \dfrac{1}{R_2} = \dfrac{1}{6\,\Omega} = 0.167\text{ S}$

and $G_T = G_1 + G_2 = 0.333\text{ S} + 0.167\text{ S} = \mathbf{0.5\text{ S}}$

b. $R_T = \dfrac{1}{G_T} = \dfrac{1}{0.5\text{ S}} = \mathbf{2\,\Omega}$

Applying Eq. (6.3)

$$R_T = \frac{1}{\dfrac{1}{R_1} + \dfrac{1}{R_2}} = \frac{1}{\dfrac{1}{3\,\Omega} + \dfrac{1}{6\,\Omega}}$$

$$= \frac{1}{0.333\text{ S} + 0.167\text{ S}} = \frac{1}{0.5\text{ S}} = \mathbf{2\,\Omega}$$

EXAMPLE 6.2

a. By inspection, which parallel element in Fig. 6.5 has the least conductance? Determine the total conductance of the network and note whether your conclusion was verified.

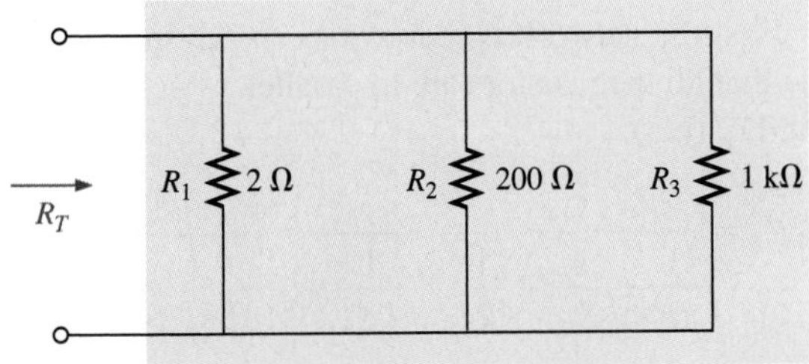

FIG. 6.5
Parallel resistors for Example 6.2.

b. Determine the total resistance from the results of part (a) and by applying Eq. (6.3).

Solutions:

a. Since the 1 kΩ resistor has the largest resistance and therefore the largest opposition to the flow of charge (level of conductivity), it will have the least level of conductance.

$$G_1 = \frac{1}{R_1} = \frac{1}{2\ \Omega} = 0.5\ \text{S},\ G_2 = \frac{1}{R_2} + \frac{1}{200\ \Omega} = 0.005\ \text{S} = 5\ \text{mS}$$

$$G_3 = \frac{1}{R_3} = \frac{1}{1\ \text{k}\Omega} = \frac{1}{1000\ \Omega} = 0.001\ \text{S} = 1\ \text{mS}$$

$$G_T = G_1 + G_2 + G_3 = 0.5\ \text{S} + 5\ \text{mS} + 1\ \text{mS}$$
$$= \mathbf{506\ mS}$$

Note the difference in conductance level between the 2 Ω (500 mS) and the 1 kΩ (1 mS) resistor.

b. $R_T = \dfrac{1}{G_T} = \dfrac{1}{506\ \text{mS}} = \mathbf{1.976\ \Omega}$

Applying Eq. (6.3):

$$R_T = \frac{1}{\dfrac{1}{R_1} + \dfrac{1}{R_2} + \dfrac{1}{R_3}} = \frac{1}{\dfrac{1}{2\ \Omega} + \dfrac{1}{200\ \Omega} + \dfrac{1}{1\ \text{k}\Omega}}$$

$$= \frac{1}{0.5\ \text{S} + 0.005\ \text{S} + 0.001\ \text{S}} = \frac{1}{0.506\ \text{S}} = \mathbf{1.98\ \Omega}$$

EXAMPLE 6.3 Find the total resistance of the configuration in Fig. 6.6.

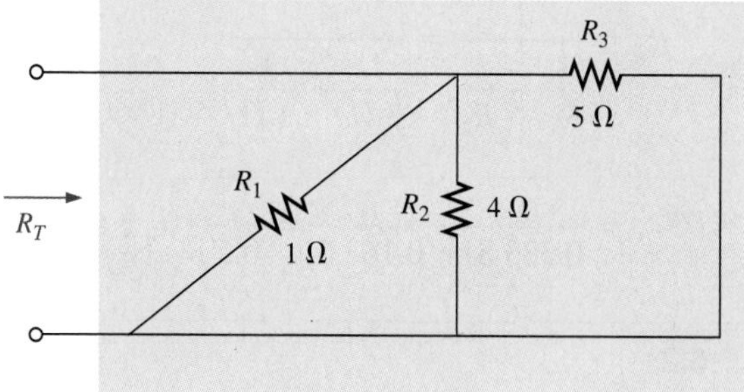

FIG. 6.6
Network to be investigated in Example 6.3.

Solution: First the network is redrawn as shown in Fig. 6.7 to clearly demonstrate that all the resistors are in parallel.

Applying Eq. (6.3):

$$R_T = \frac{1}{\dfrac{1}{R_1} + \dfrac{1}{R_2} + \dfrac{1}{R_3}} = \frac{1}{\dfrac{1}{1\ \Omega} + \dfrac{1}{4\ \Omega} + \dfrac{1}{5\ \Omega}}$$

$$= \frac{1}{1\ \text{S} + 0.25\ \text{S} + 0.2\ \text{S}} = \frac{1}{1.45\ \text{S}} \cong \mathbf{0.69\ \Omega}$$

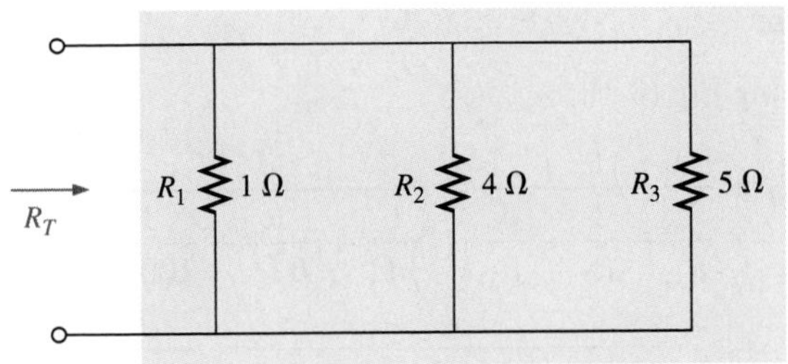

FIG. 6.7
Network in Fig. 6.6 redrawn.

If you review the examples above, you will find that the total resistance is less than the smallest parallel resistor. That is, in Example 6.1, 2 Ω is less than 3 Ω or 6 Ω. In Example 6.2, 1.976 Ω is less than 2 Ω, 100 Ω, or 1 kΩ; and in Example 6.3, 0.69 Ω is less than 1 Ω, 4 Ω, or 5 Ω. In general, therefore,

the total resistance of parallel resistors is always less than the value of the smallest resistor.

This is particularly important when you want a quick estimate of the total resistance of a parallel combination. Simply find the smallest value, and you know that the total resistance will be less than that value. It is also a great check on your calculations. In addition, you will find that

if the smallest resistor of a parallel combination is much smaller than the other parallel resistors, the total resistance will be very close to the smallest resistor value.

This fact is obvious in Example 6.2 where the total resistance of 1.976 Ω is very close to the smallest resistor of 2 Ω.

Another interesting characteristic of parallel resistors is demonstrated in Example 6.4.

EXAMPLE 6.4

a. What is the effect of adding another resistor of 100 Ω in parallel with the parallel resistors of Example 6.1 as shown in Fig. 6.8?
b. What is the effect of adding a parallel 1 Ω resistor to the configuration in Fig. 6.8?

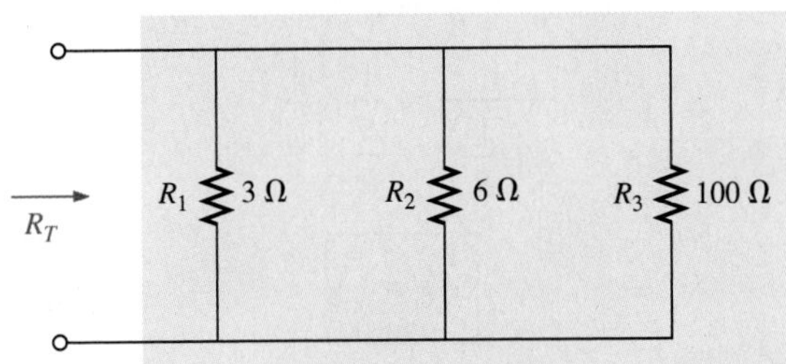

FIG. 6.8
Adding a parallel 100 Ω resistor to the network in Fig. 6.4.

Solutions:

a. Applying Eq. (6.3):

$$R_T = \frac{1}{\dfrac{1}{R_1} + \dfrac{1}{R_2} + \dfrac{1}{R_3}} = \frac{1}{\dfrac{1}{3\ \Omega} + \dfrac{1}{6\ \Omega} + \dfrac{1}{100\ \Omega}}$$

$$= \frac{1}{0.333\ \text{S} + 0.167\ \text{S} + 0.010\ \text{S}} = \frac{1}{0.510\ \text{S}} = \mathbf{1.96\ \Omega}$$

The parallel combination of the 3 Ω and 6 Ω resistors resulted in a total resistance of 2 Ω in Example 6.1. The effect of adding a resistor in parallel of 100 Ω had little effect on the total resistance because its resistance level is significantly higher (and conductance level significantly less) than the other two resistors. The total change in resistance was less than 2%. However, do note that the total resistance dropped with the addition of the 100 Ω resistor.

b. Applying Eq. (6.3):

$$R_T = \frac{1}{\dfrac{1}{R_1} + \dfrac{1}{R_2} + \dfrac{1}{R_3} + \dfrac{1}{R_4}} = \frac{1}{\dfrac{1}{3\ \Omega} + \dfrac{1}{6\ \Omega} + \dfrac{1}{100\ \Omega} + \dfrac{1}{1\ \Omega}}$$

$$= \frac{1}{0.333\ \text{S} + 0.167\ \text{S} + 0.010\ \text{S} + 1\ \text{S}} = \frac{1}{0.51\ \text{S}} = \mathbf{0.66\ \Omega}$$

The introduction of the 1 Ω resistor reduced the total resistance from 2 Ω to only 0.66 Ω—a decrease of almost 67%. The fact that the added resistor has a resistance level less than the other parallel elements and one-third that of the smallest contributed to the significant drop in resistance level.

In part (a) of Example 6.4, the total resistance dropped from 2 Ω to 1.96 Ω. In part (b), it dropped to 0.66 Ω. The results clearly reveal that

the total resistance of parallel resistors will always drop as new resistors are added in parallel, irrespective of their value.

Recall that this is the opposite of series resistors, where additional resistors of any value increase the total resistance.

For equal resistors in parallel, the equation for the total resistance becomes significantly easier to apply. For N equal resistors in parallel, Eq. (6.3) becomes

$$R_T = \frac{1}{\dfrac{1}{R} + \dfrac{1}{R} + \dfrac{1}{R} + \cdots + \dfrac{1}{R_N}}$$

$$= \frac{1}{N\left(\dfrac{1}{R}\right)} = \frac{1}{\dfrac{N}{R}}$$

and

$$\boxed{R_T = \frac{R}{N}} \qquad \textbf{(6.4)}$$

In other words,

the total resistance of N parallel resistors of equal value is the resistance of one resistor divided by the number (N) of parallel resistors.

EXAMPLE 6.5 Find the total resistance of the parallel resistors in Fig. 6.9.

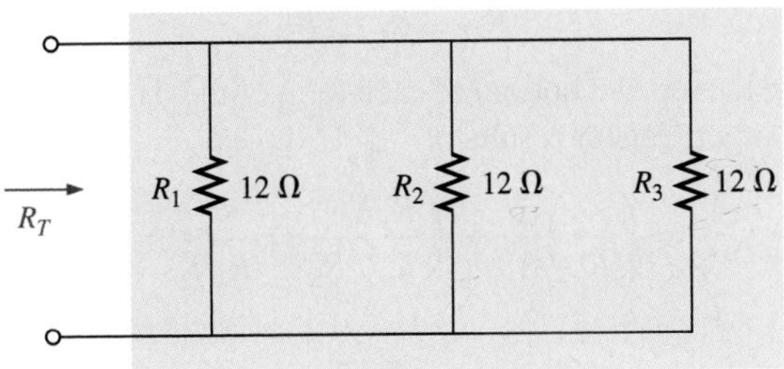

FIG. 6.9
Three equal parallel resistors to be investigated in Example 6.5.

Solution: Applying Eq. (6.4):

$$R_T = \frac{R}{N} = \frac{12\ \Omega}{3} = \mathbf{4\ \Omega}$$

EXAMPLE 6.6 Find the total resistance for the configuration in Fig. 6.10.

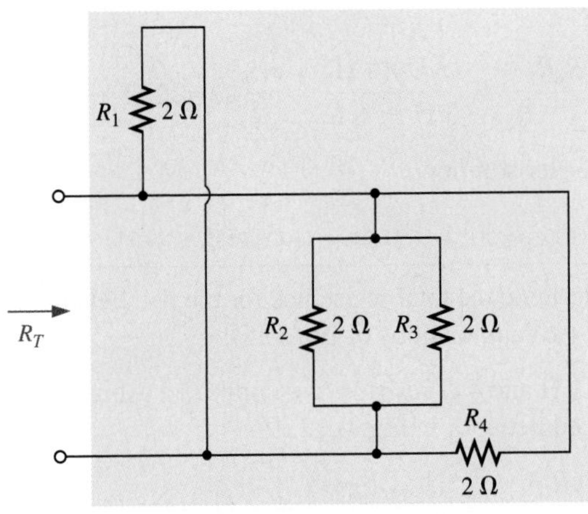

FIG. 6.10
Parallel configuration for Example 6.6.

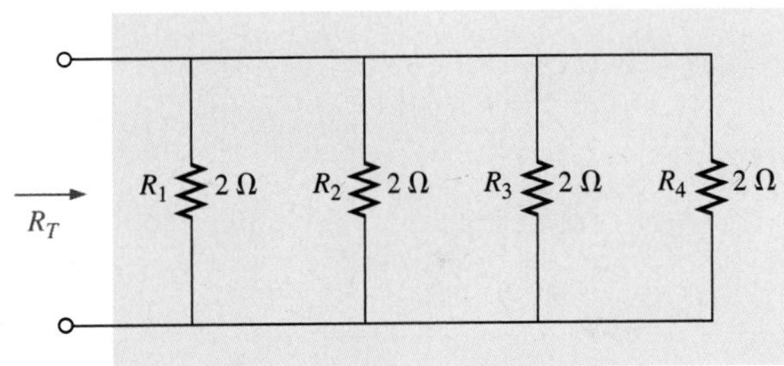

FIG. 6.11
Network in Fig. 6.10 redrawn.

Solution: Redrawing the network results in the parallel network in Fig. 6.11.

Applying Eq. (6.4):

$$R_T = \frac{R}{N} = \frac{2\ \Omega}{4} = \mathbf{0.5\ \Omega}$$

Special Case: Two Parallel Resistors

In the vast majority of cases, only two or three parallel resistors will have to be combined. With this in mind, an equation has been derived for two parallel resistors that is easy to apply and removes the need to continually worry about dividing into 1 and possibly misplacing a decimal point. For three parallel resistors, the equation to be derived here can be applied twice, or Eq. (6.3) can be used.

For two parallel resistors, the total resistance is determined by Eq. (6.1):

$$\frac{1}{R_T} = \frac{1}{R_1} + \frac{1}{R_2}$$

Multiplying the top and bottom of each term of the right side of the equation by the other resistor results in

$$\frac{1}{R_T} = \left(\frac{R_2}{R_2}\right)\frac{1}{R_1} + \left(\frac{R_1}{R_1}\right)\frac{1}{R_2} = \frac{R_2}{R_1R_2} + \frac{R_1}{R_1R_2}$$

$$\frac{1}{R_T} = \frac{R_2 + R_1}{R_1R_2}$$

and

$$\boxed{R_T = \frac{R_1R_2}{R_1 + R_2}} \qquad \textbf{(6.5)}$$

In words, the equation states that

the total resistance of two parallel resistors is simply the product of their values divided by their sum.

EXAMPLE 6.7 Repeat Example 6.1 using Eq. (6.5).

Solution: Eq. (6.5):

$$R_T = \frac{R_1R_2}{R_1 + R_2} = \frac{(3\ \Omega)(6\ \Omega)}{3\ \Omega + 6\ \Omega} = \frac{18}{9}\ \Omega = \mathbf{2\ \Omega}$$

which matches the earlier solution.

EXAMPLE 6.8 Determine the total resistance for the parallel combination in Fig. 6.7 using two applications of Eq. (6.5).

Solution: First the 1 Ω and 4 Ω resistors are combined using Eq. (6.5), resulting in the reduced network in Fig. 6.12.

$$\text{Eq. (6.4): } R'_T = \frac{R_1R_2}{R_1 + R_2} = \frac{(1\ \Omega)(4\ \Omega)}{1\ \Omega + 4\ \Omega} = \frac{4}{5}\ \Omega = 0.8\ \Omega$$

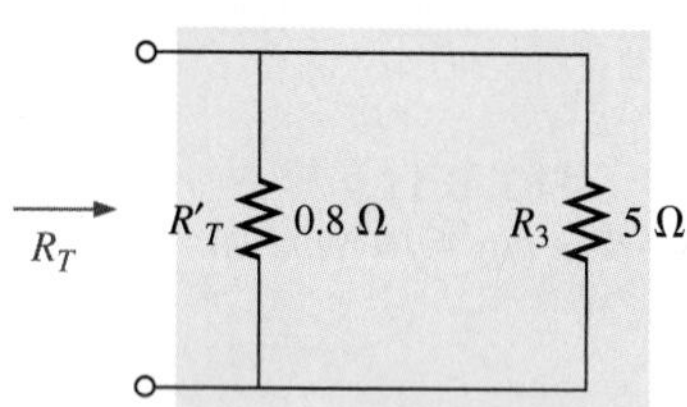

FIG. 6.12
Reduced equivalent in Fig. 6.7.

Then Eq. (6.5) is applied again using the equivalent value:

$$R_T = \frac{R'_T R_3}{R'_T + R_3} = \frac{(0.8\ \Omega)(5\ \Omega)}{0.8\ \Omega + 5\ \Omega} = \frac{4}{5.8}\ \Omega = \mathbf{0.69\ \Omega}$$

The result matches that obtained in Example 6.3.

Recall that series elements can be interchanged without affecting the magnitude of the total resistance. In parallel networks,

parallel resistors can be interchanged without affecting the total resistance.

The next example demonstrates this and reveals how redrawing a network can often define which operations or equations should be applied.

EXAMPLE 6.9 Determine the total resistance of the parallel elements in Fig. 6.13.

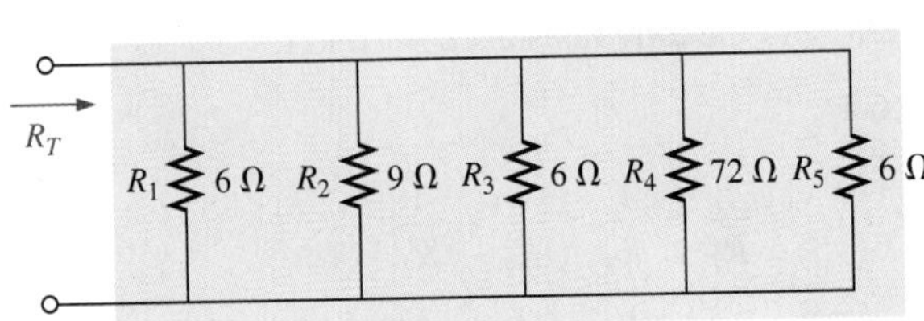

FIG. 6.13
Parallel network for Example 6.9.

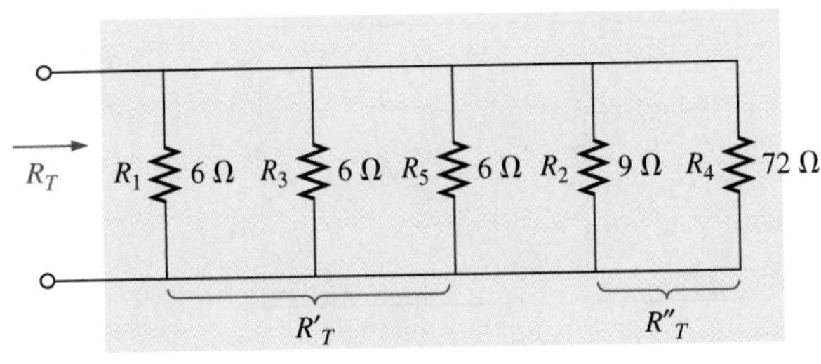

FIG. 6.14
Redrawn network in Fig. 6.13 (Example 6.9).

Solution: The network is redrawn in Fig. 6.14.

Eq. (6.4):
$$R'_T = \frac{R}{N} = \frac{6\ \Omega}{3} = 2\ \Omega$$

Eq. (6.5):
$$R''_T = \frac{R_2R_4}{R_2 + R_4} = \frac{(9\ \Omega)(72\ \Omega)}{9\ \Omega + 72\ \Omega} = \frac{648}{81}\ \Omega = 8\ \Omega$$

Eq. (6.5):
$$R_T = \frac{R'_TR''_T}{R'_T + R''_T} = \frac{(2\ \Omega)(8\ \Omega)}{2\ \Omega + 8\ \Omega} = \frac{16}{10}\ \Omega = \mathbf{1.6\ \Omega}$$

The preceding examples involve direct substitution; that is, once the proper equation has been defined, it is only a matter of plugging in the numbers and performing the required algebraic manipulations. The next two examples have a design orientation, in which specific network parameters are defined and the circuit elements must be determined.

EXAMPLE 6.10 Determine the value of R_2 in Fig. 6.15 to establish a total resistance of 9 kΩ.

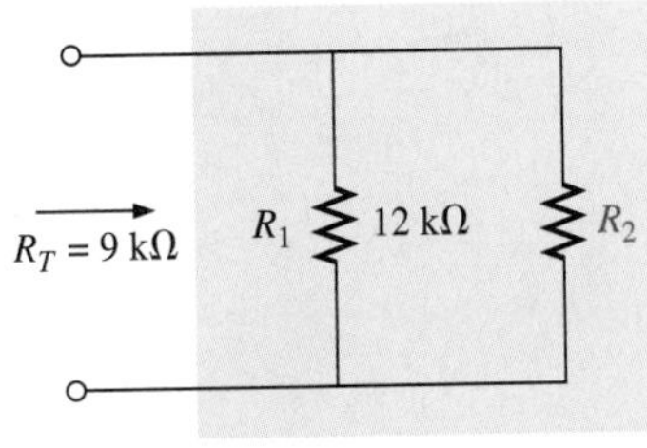

FIG. 6.15
Parallel network for Example 6.10.

Solution:

$$R_T = \frac{R_1R_2}{R_1 + R_2}$$
$$R_T(R_1 + R_2) = R_1R_2$$
$$R_TR_1 + R_TR_2 = R_1R_2$$
$$R_TR_1 = R_1R_2 - R_TR_2$$
$$R_TR_1 = (R_1 - R_T)R_2$$

and
$$R_2 = \frac{R_TR_1}{R_1 - R_T}$$

Substituting values:

$$R_2 = \frac{(9\ \text{k}\Omega)(12\ \text{k}\Omega)}{12\ \text{k}\Omega - 9\ \text{k}\Omega} = \frac{108}{3}\ \text{k}\Omega = \mathbf{36\ k\Omega}$$

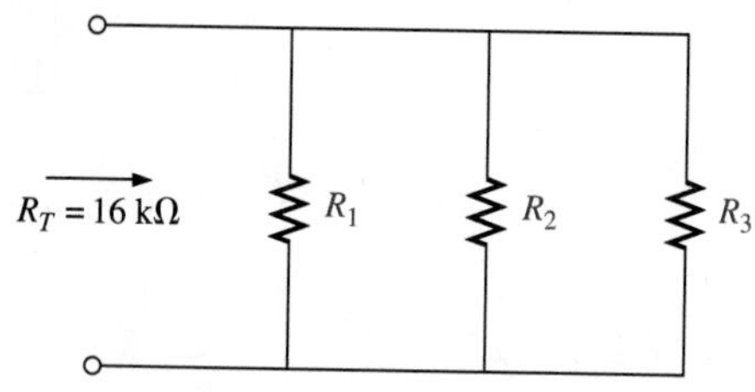

FIG. 6.16
Parallel network for Example 6.11.

EXAMPLE 6.11 Determine the values of R_1, R_2 and R_3 in Fig. 6.16 if $R_2 = 2R_1$, $R_3 = 2R_2$, and the total resistance is 16 kΩ.

Solution: Eq. (6.1):

$$\frac{1}{R_T} = \frac{1}{R_1} + \frac{1}{R_2} + \frac{1}{R_3}$$

However, $\quad R_2 = 2R_1 \quad \text{and} \quad R_3 = 2R_2 = 2(2R_1) = 4R_1$

so that

$$\frac{1}{16\text{ k}\Omega} = \frac{1}{R_1} + \frac{1}{2R_1} + \frac{1}{4R_1}$$

and

$$\frac{1}{16\text{ k}\Omega} = \frac{1}{R_1} + \frac{1}{2}\left(\frac{1}{R_1}\right) + \frac{1}{4}\left(\frac{1}{R_1}\right)$$

or

$$\frac{1}{16\text{ k}\Omega} = 1.75\left(\frac{1}{R_1}\right)$$

resulting in $\quad R_1 = 1.75(16\text{ k}\Omega) = \mathbf{28\text{ k}\Omega}$

so that $\quad R_2 = 2R_1 = 2(28\text{ k}\Omega) = \mathbf{56\text{ k}\Omega}$

and $\quad R_3 = 2R_2 = 2(56\text{ k}\Omega) = \mathbf{112\text{ k}\Omega}$

Analogies

Analogies were effectively used to introduce the concept of parallel elements. They can also be used to help define *parallel configuration.* On a ladder, the rungs of the ladder form a parallel configuration. When ropes are tied between a grappling hook and a load, they effectively absorb the stress in a parallel configuration. The cables of a suspended roadway form a parallel configuration. There are numerous other analogies that demonstrate how connections between the same two points permit a distribution of stress between the parallel elements.

Instrumentation

As shown in Fig. 6.17, the total resistance of a parallel combination of resistive elements can be found by simply applying an ohmmeter. There is no polarity to resistance, so either lead of the ohmmeter can be connected to either side of the network. Although there are no supplies in Fig. 6.17, always keep in mind that ohmmeters can never be applied to a "live" circuit. It is not enough to set the supply to 0 V or to turn it off. It may still

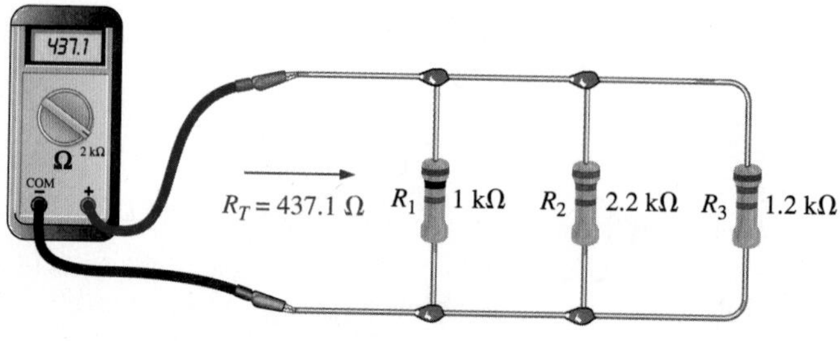

FIG. 6.17
Using an ohmmeter to measure the total resistance of a parallel network.

load down (change the network configuration of) the circuit and change the reading. It is best to remove the supply and apply the ohmmeter to the two resulting terminals. Since all the resistors are in the kilohm range, the 20 kΩ scale was chosen first. We then moved down to the 2 kΩ scale for increased precision. Moving down to the 200 Ω scale resulted in an "OL" indication since we were below the measured resistance value.

6.3 PARALLEL CIRCUITS

A **parallel circuit** can now be established by connecting a supply across a set of parallel resistors as shown in Fig. 6.18. The positive terminal of the supply is directly connected to the top of each resistor, while the negative terminal is connected to the bottom of each resistor. Therefore, it should be quite clear that the applied voltage is the same across each resistor. In general,

the voltage is always the same across parallel elements.

Therefore, remember that

if two elements are in parallel, the voltage across them must be the same. However, if the voltage across two neighboring elements is the same, the two elements may or may not be in parallel.

The reason for this qualifying comment in the above statement is discussed in detail in Chapter 7.

For the voltages of the circuit in Fig. 6.18, the result is that

$$V_1 = V_2 = E \tag{6.6}$$

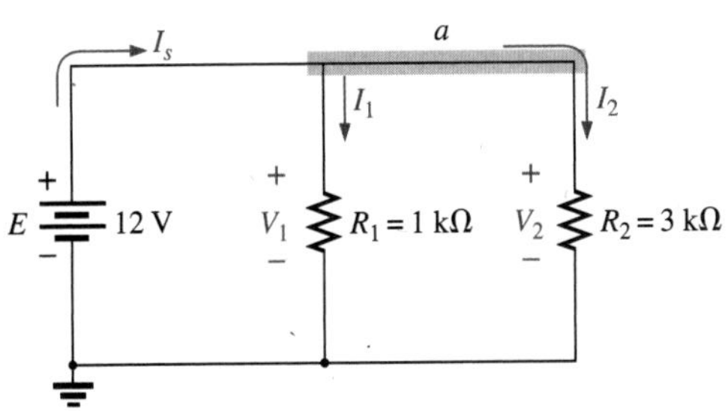

FIG. 6.18
Parallel network.

Once the supply has been connected, a source current is established through the supply that passes through the parallel resistors. The current that results is a direct function of the total resistance of the parallel circuit. The smaller the total resistance, the more the current, as occurred for series circuits also.

Recall from series circuits that the source does not "see" the parallel combination of elements. It reacts only to the total resistance of the circuit, as shown in Fig. 6.19. The source current can then be determined using Ohm's law:

$$I_s = \frac{E}{R_T} \tag{6.7}$$

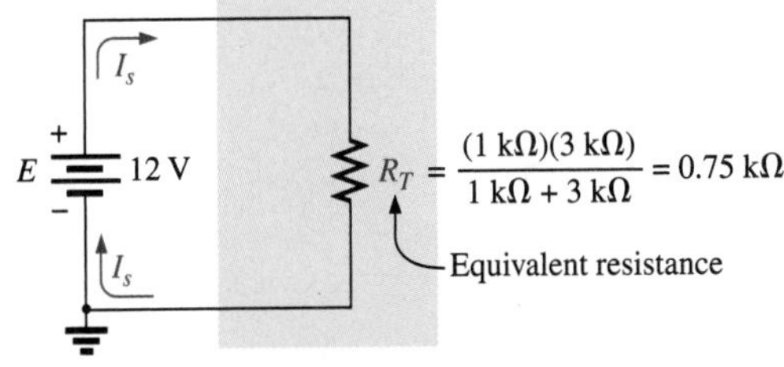

FIG. 6.19
Replacing the parallel resistors in Fig. 6.18 with the equivalent total resistance.

Since the voltage is the same across parallel elements, the current through each resistor can also be determined using Ohm's law. That is,

$$I_1 = \frac{V_1}{R_1} = \frac{E}{R_1} \quad \text{and} \quad I_2 = \frac{V_2}{R_2} = \frac{E}{R_2} \tag{6.8}$$

The direction for the currents is dictated by the polarity of the voltage across the resistors. Recall that for a resistor, current enters the positive side of a potential drop and leaves the negative. The result, as shown in Fig. 6.18, is that the source current enters point *a*, and currents I_1 and I_2 leave the same point. An excellent analogy for describing the flow of charge through the network of Fig. 6.18 is the flow of water through the parallel pipes of Fig. 6.20. The larger pipe with less "resistance" to the

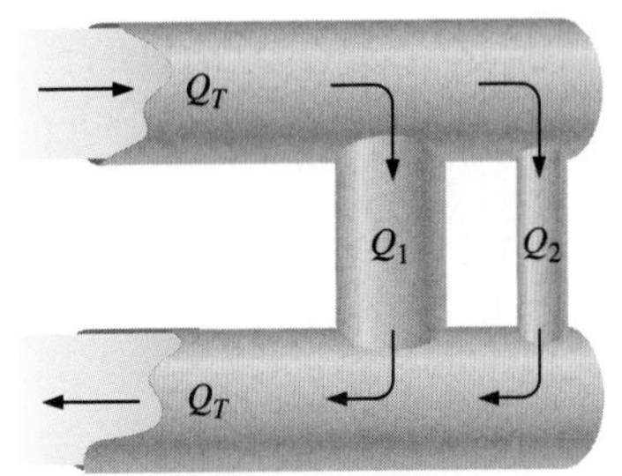

FIG. 6.20
Mechanical analogy for Fig. 6.18.

flow of water will have a larger flow of water through it. The thinner pipe with its increased "resistance" level will have less water through it. In any case, the total water entering the pipes at the top Q_T must equal that leaving at the bottom, with $Q_T = Q_1 + Q_2$.

The relationship between the source current and the parallel resistor currents can be derived by simply taking the equation for the total resistance in Eq. (6.1):

$$\frac{1}{R_T} = \frac{1}{R_1} + \frac{1}{R_2}$$

Multiplying both sides by the applied voltage:

$$E\left(\frac{1}{R_T}\right) = E\left(\frac{1}{R_1} + \frac{1}{R_2}\right)$$

resulting in

$$\frac{E}{R_T} = \frac{E}{R_1} + \frac{E}{R_2}$$

Then note that $E/R_1 = I_1$ and $E/R_2 = I_2$ to obtain

$$\boxed{I_s = I_1 + I_2} \qquad \textbf{(6.9)}$$

The result reveals a very important property of parallel circuits:

For single-source parallel networks, the source current (I_s) is always equal to the sum of the individual branch currents.

The duality that exists between series and parallel circuits continues to surface as we proceed through the basic equations for electric circuits. This is fortunate because it provides a way of remembering the characteristics of one using the results of another. For instance, in Fig. 6.21(a), we have a parallel circuit where it is clear that $I_T = I_1 + I_2$. By simply replacing the currents of the equation in Fig. 6.21(a) by a voltage level, as shown in Fig. 6.21(b), we have Kirchhoff's voltage law for a series circuit: $E = V_1 + V_2$. In other words,

for a parallel circuit, the source current equals the sum of the branch currents, while for a series circuit, the applied voltage equals the sum of the voltage drops.

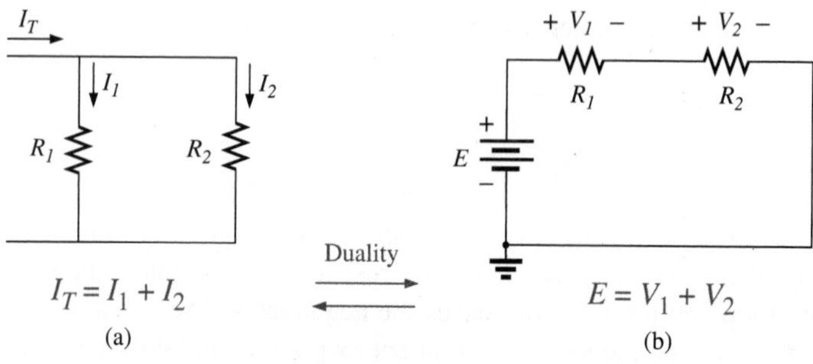

FIG. 6.21
Demonstrating the duality that exists between series and parallel circuits.

EXAMPLE 6.12 For the parallel network in Fig. 6.22:

a. Find the total resistance.
b. Calculate the source current.
c. Determine the current through each parallel branch.
d. Show that Eq. (6.9) is satisfied.

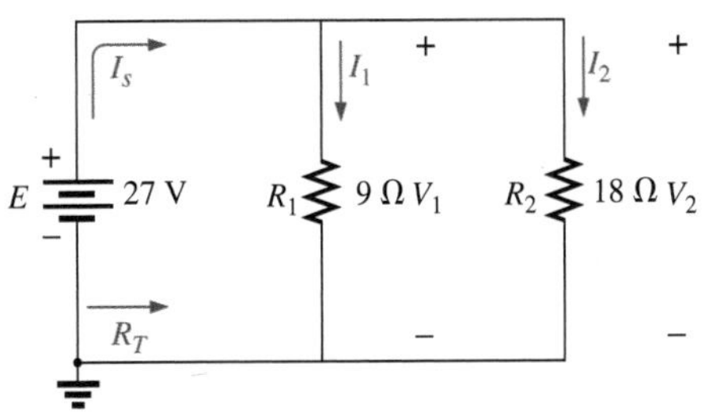

FIG. 6.22
Parallel network for Example 6.12.

Solutions:

a. Using Eq. (6.5):

$$R_T = \frac{R_1R_2}{R_1 + R_2} = \frac{(9\ \Omega)(18\ \Omega)}{9\ \Omega + 18\ \Omega} = \frac{162}{27}\ \Omega = \mathbf{6\ \Omega}$$

b. Applying Ohm's law:

$$I_s = \frac{E}{R_T} = \frac{27\text{ V}}{6\ \Omega} = \mathbf{4.5\ A}$$

c. Applying Ohm's law:

$$I_1 = \frac{V_1}{R_1} = \frac{E}{R_1} = \frac{27\text{ V}}{9\ \Omega} = \mathbf{3\ A}$$

$$I_2 = \frac{V_2}{R_2} = \frac{E}{R_2} = \frac{27\text{ V}}{18\ \Omega} = \mathbf{1.5\ A}$$

d. Substituting values from parts (b) and (c):

$$I_s = \mathbf{4.5\ A} = I_1 + I_2 = 3\text{ A} + 1.5\text{ A} = \mathbf{4.5\ A} \quad \text{(checks)}$$

EXAMPLE 6.13 For the parallel network in Fig. 6.23.

a. Find the total resistance.
b. Calculate the source current.
c. Determine the current through each branch.

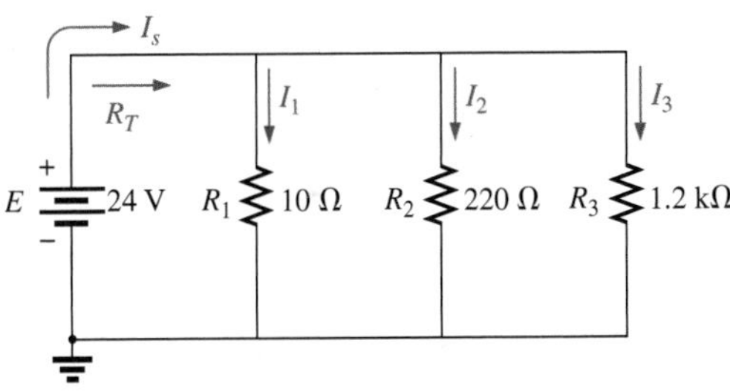

FIG. 6.23
Parallel network for Example 6.13.

Solutions:

a. Applying Eq. (6.3):

$$R_T = \frac{1}{\dfrac{1}{R_1} + \dfrac{1}{R_2} + \dfrac{1}{R_3}} = \frac{1}{\dfrac{1}{10\ \Omega} + \dfrac{1}{220\ \Omega} + \dfrac{1}{1.2\text{ k}\Omega}}$$

$$= \frac{1}{100 \times 10^{-3} + 4.545 \times 10^{-3} + 0.833 \times 10^{-3}} = \frac{1}{105.38 \times 10^{-3}}$$

$$R_T = \mathbf{9.49\ \Omega}$$

Note that the total resistance is less than the smallest parallel resistor, and the magnitude is very close to the smallest resistor because the other resistors are larger by a factor greater than 10 : 1.

b. Using Ohm's law:

$$I_s = \frac{E}{R_T} = \frac{24\text{ V}}{9.49\ \Omega} = \mathbf{2.53\ A}$$

c. Applying Ohm's law:

$$I_1 = \frac{V_1}{R_1} = \frac{E}{R_1} = \frac{24\text{ V}}{10\ \Omega} = \mathbf{2.4\ A}$$

$$I_2 = \frac{V_2}{R_2} = \frac{E}{R_2} = \frac{24\text{ V}}{220\ \Omega} = \mathbf{0.11\ A}$$

$$I_3 = \frac{V_3}{R_3} = \frac{E}{R_3} = \frac{24\text{ V}}{1.2\text{ k}\Omega} = \mathbf{0.02\ A}$$

A careful examination of the results of Example 6.13 reveals that the larger the parallel resistor, the smaller the branch current. In general, therefore,

for parallel resistors, the greatest current will exist in the branch with the least resistance.

A more powerful statement is that

current always seeks the path of least resistance.

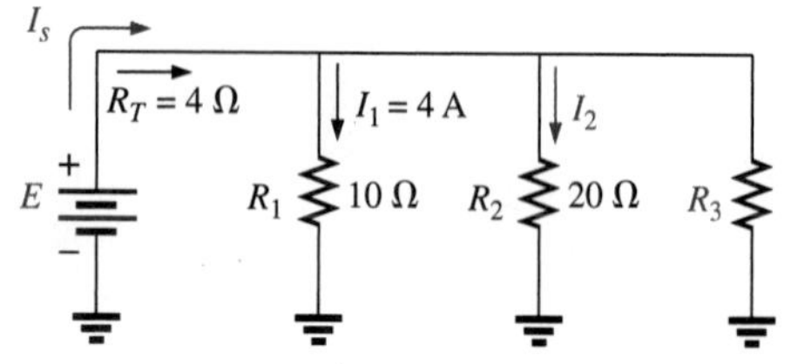

FIG. 6.24
Parallel network for Example 6.14.

EXAMPLE 6.14 Given the information provided in Fig. 6.24.

a. Determine R_3.
b. Find the applied voltage E.
c. Find the source current I_s.
d. Find I_2.

Solutions:

a. Applying Eq. (6.1):

$$\frac{1}{R_T} = \frac{1}{R_1} + \frac{1}{R_2} + \frac{1}{R_3}$$

Substituting:
$$\frac{1}{4\ \Omega} = \frac{1}{10\ \Omega} + \frac{1}{20\ \Omega} + \frac{1}{R_3}$$

so that
$$0.25\text{ S} = 0.1\text{ S} + 0.05\text{ S} + \frac{1}{R_3}$$

and
$$0.25\text{ S} = 0.15\text{ S} + \frac{1}{R_3}$$

with
$$\frac{1}{R_3} = 0.1\text{ S}$$

and
$$R_3 = \frac{1}{0.1\text{ S}} = \mathbf{10\ \Omega}$$

b. Using Ohm's law:

$$E = V_1 = I_1R_1 = (4\text{ A})(10\ \Omega) = \mathbf{40\ V}$$

c.
$$I_s = \frac{E}{R_T} = \frac{40\text{ V}}{4\ \Omega} = \mathbf{10\ A}$$

d. Applying Ohm's law:

$$I_2 = \frac{V_2}{R_2} = \frac{E}{R_2} = \frac{40\text{ V}}{20\ \Omega} = \mathbf{2\ A}$$

Mathcad Solution: This example provides an excellent opportunity to practice our skills using Mathcad. As shown in Fig. 6.25, the known parameters and quantities of the network are entered first, followed by an equation for the unknown resistor R_3. Note that after the first division operator is selected, a left bracket appears (to be followed eventually by a right enclosure bracket) to tell the computer that the mathematical operations in the denominator must be carried out first before the division into 1. In addition, each individual division into 1 is separated by brackets to ensure that the division operation is performed before each quantity is added to the neighboring factor. Finally, keep in mind that the Mathcad bracket must encompass each individual expression of the denominator before you place the right bracket in place.

This example shows that when units are added to all the parameters, the results appear with the proper unit of measurement. The units for each are entered by entering the magnitude of the quantity followed by the multiplication operation. Select the **Insert** option from the menu at the top of the screen and then select **Unit.** The **Unit** dialog box appears. For the resistor RT, select **Resistance** under the **Dimension** scroll list followed by Ohm (Ω) under the Unit listing. Click **OK,** and the unit of measurement appears beside the magnitude of the quantity.

After you enter the equation for R3, you can obtain the value of R3 by typing **R3** followed by the equal sign. As soon as you enter the equal sign,

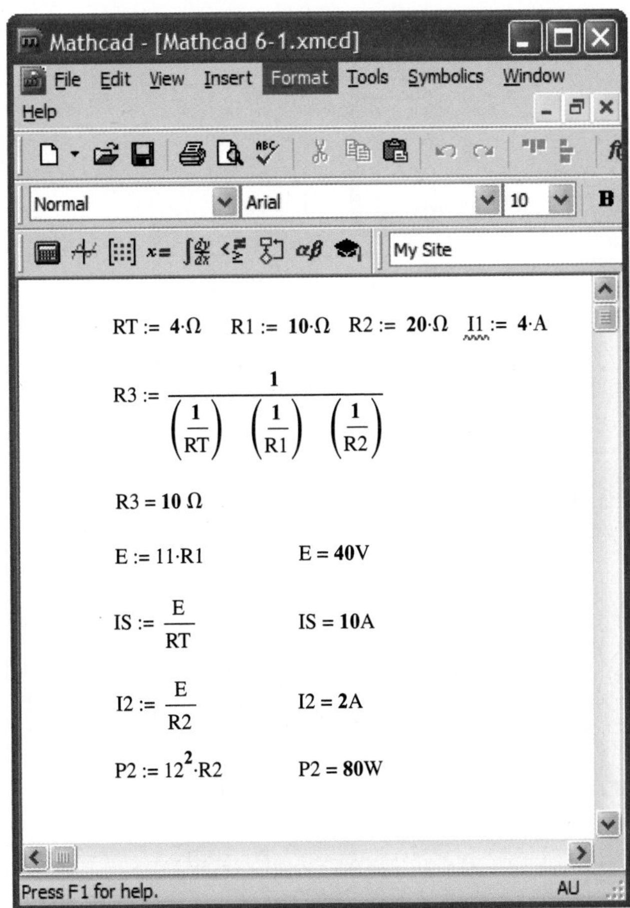

FIG. 6.25
Using Mathcad to confirm the results of Example 6.14.

the result of 10 Ω appears. Enter the equation for *E*. When you retype **E** followed by the equal sign, the result of 40 V appears. The remaining unknowns can then be found in the same manner.

One final note. You will find when you enter the variable **I1** and the equal sign, the word *function* appears. This is because the capital letter *I* is reserved for a defined Bessel function in Mathcad 12. However, you can override this result by using the sequence SHIFT-colon to enter the value of I1.

In each case, the quantity of interest was entered below the defining equation to obtain the numerical result by selecting an equal sign. As expected, all the results match the longhand solution.

Instrumentation

In Fig. 6.26, voltmeters have been connected to verify that the voltage across parallel elements is the same. Note that the positive or red lead of each voltmeter is connected to the high (positive) side of the voltage across each resistor to obtain a positive reading. The 20 V scale was used because the applied voltage exceeded the range of the 2 V scale.

In Fig. 6.27, an ammeter has been hooked up to measure the source current. First, the connection to the supply had to be broken at the positive terminal and the meter inserted as shown. Be sure to use ammeter terminals on your meter for such measurements. The red or positive lead of

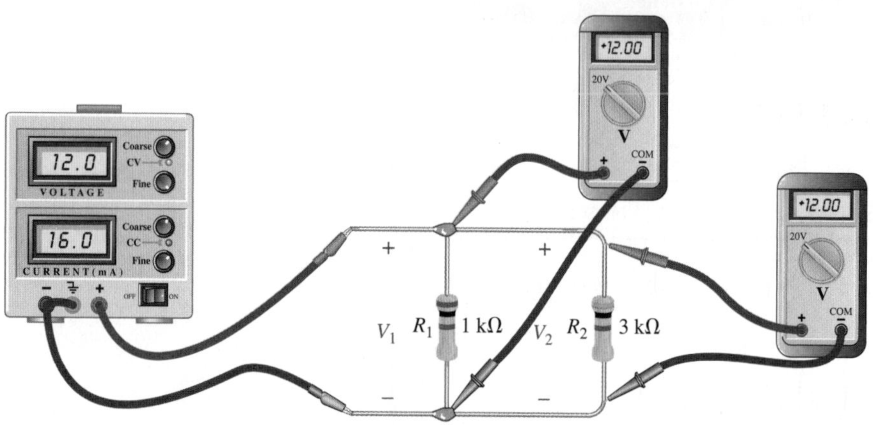

FIG. 6.26
Measuring the voltages of a parallel dc network.

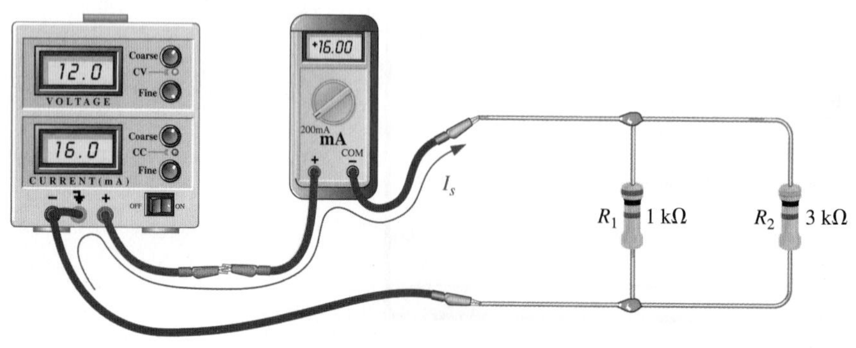

FIG. 6.27
Measuring the source current of a parallel network.

the meter is connected so that the source current enters that lead and leaves the negative or black lead to ensure a positive reading. The 200 mA scale was used because the source current exceeded the maximum value of the 2 mA scale. For the moment, we assume that the internal resistance of the meter can be ignored. Since the internal resistance of an ammeter on the 200 mA scale is typically only a few ohms, compared to the parallel resistors in the kilohm range, it is an excellent assumption.

A more difficult measurement is for the current through resistor R_1. This measurement often gives trouble in the laboratory session. First, as shown in Fig. 6.28(a), resistor R_1 must be disconnected from the upper connection point to establish an open circuit. The ammeter is then inserted between the resulting terminals so that the current enters the positive or red terminal, as shown in Fig. 6.28(b). Always remember: When using an ammeter, first establish an open circuit in the branch in which the current is to be measured, and then insert the meter.

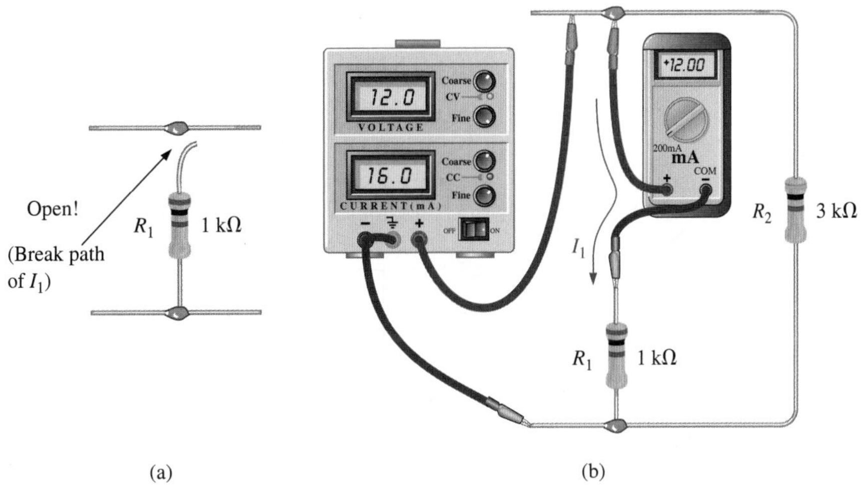

FIG. 6.28
Measuring the current through resistor R_1.

The easiest measurement is for the current through resistor R_2. Break the connection to R_2 above or below the resistor, and insert the ammeter with the current entering the positive or red lead to obtain a positive reading.

6.4 POWER DISTRIBUTION IN A PARALLEL CIRCUIT

Recall from the discussion of series circuits that the power applied to a series resistive circuit equals the power dissipated by the resistive elements. The same is true for parallel resistive networks. In fact,

for any network composed of resistive elements, the power applied by the battery will equal that dissipated by the resistive elements.

For the parallel circuit in Fig. 6.29:

$$P_E = P_{R_1} + P_{R_2} + P_{R_3} \tag{6.10}$$

which is exactly the same as obtained for the series combination.

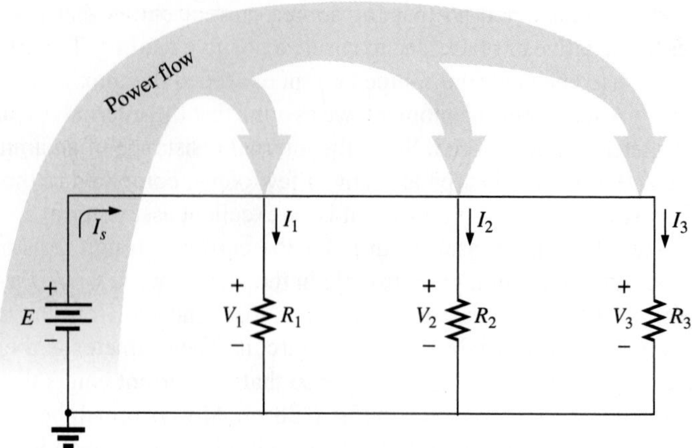

FIG. 6.29
Power flow in a dc parallel network.

The power delivered by the source in the same:

$$P_E = EI_s \qquad \text{(watts, W)} \qquad \textbf{(6.11)}$$

as is the equation for the power to each resistor (shown for R_1 only):

$$P_1 = V_1I_1 = I_1^2R_1 = \frac{V_1^2}{R_1} \qquad \text{(watts, W)} \qquad \textbf{(6.12)}$$

In the equation $P = V^2/R$, the voltage across each resistor in a parallel circuit will be the same. The only factor that changes is the resistance in the denominator of the equation. The result is that

in a parallel resistive network, the larger the resistor, the less the power absorbed.

EXAMPLE 6.15 For the parallel network in Fig. 6.30 (all standard values):

a. Determine the total resistance R_T.
b. Find the source current and the current through each resistor.
c. Calculate the power delivered by the source.

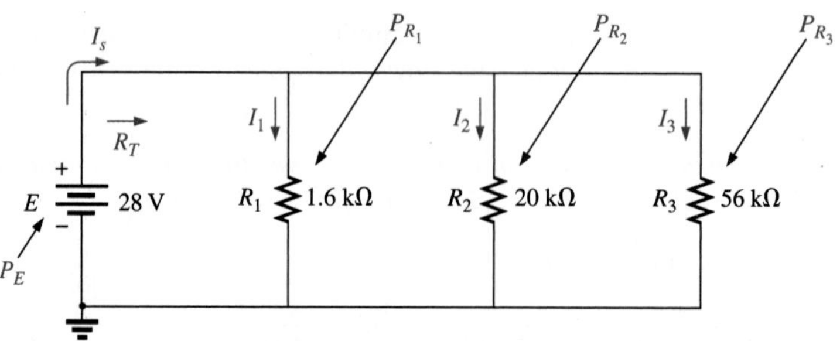

FIG. 6.30
Parallel network for Example 6.15.

d. Determine the power absorbed by each parallel resistor.
e. Verify Eq. (6.10).

Solutions:

a. Without making a single calculation, it should now be apparent from previous examples that the total resistance is less than 1.6 kΩ and very close to this value because of the magnitude of the other resistance levels.

$$R_T = \frac{1}{\dfrac{1}{R_1} + \dfrac{1}{R_2} + \dfrac{1}{R_3}} = \frac{1}{\dfrac{1}{1.6\text{ k}\Omega} + \dfrac{1}{20\text{ k}\Omega} + \dfrac{1}{56\text{ k}\Omega}}$$

$$= \frac{1}{625 \times 10^{-6} + 50 \times 10^{-6} + 17.867 \times 10^{-6}} = \frac{1}{692.867 \times 10^{-6}}$$

and $R_T = \mathbf{1.44\text{ k}\Omega}$

b. Applying Ohm's law:

$$I_s = \frac{E}{R_T} = \frac{28\text{ V}}{1.44\text{ k}\Omega} = \mathbf{19.44\text{ mA}}$$

Recalling that current always seeks the path of least resistance immediately tells us that the current through the 1.6 kΩ resistor will be the largest and the current through the 56 kΩ resistor the smallest.

Applying Ohm's law again:

$$I_1 = \frac{V_1}{R_1} = \frac{E}{R_1} = \frac{28\text{ V}}{1.6\text{ k}\Omega} = \mathbf{17.5\text{ mA}}$$

$$I_2 = \frac{V_2}{R_2} = \frac{E}{R_2} = \frac{28\text{ V}}{20\text{ k}\Omega} = \mathbf{1.4\text{ mA}}$$

$$I_3 = \frac{V_3}{R_3} = \frac{E}{R_3} = \frac{28\text{ V}}{56\text{ k}\Omega} = \mathbf{0.5\text{ mA}}$$

c. Applying Eq. (6.11):

$$P_E = EI_s = (28\text{ V})(19.4\text{ mA}) = \mathbf{543.2\text{ mW}}$$

d. Applying each form of the power equation:

$$P_1 = V_1I_1 = EI_1 = (28\text{ V})(17.5\text{ mA}) = \mathbf{490\text{ mW}}$$

$$P_2 = I_2^2R_2 = (1.4\text{ mA})^2(20\text{ k}\Omega) = \mathbf{39.2\text{ mW}}$$

$$P_3 = \frac{V_3^2}{R_3} = \frac{E^2}{R_3} = \frac{(28\text{ V})^2}{56\text{ k}\Omega} = \mathbf{14\text{ mW}}$$

A review of the results clearly substantiates the fact that the larger the resistor, the less the power absorbed.

e. $$P_E = P_{R_1} + P_{R_2} + P_{R_3}$$

$$\mathbf{543.2\text{ mW}} = 490\text{ mW} + 39.2\text{ mW} + 14\text{ mW} = \mathbf{543.2\text{ mW}} \quad \text{(checks)}$$

6.5 KIRCHHOFF'S CURRENT LAW

In the previous chapter, Kirchhoff's voltage law was introduced, providing a very important relationship between the voltages of a closed path.

Professor Gustav Kirchhoff is also credited with developing the following equally important relationship between the currents of a network, called **Kirchhoff's current law (KCL):**

The algebraic sum of the currents entering and leaving a junction (or region) of a network is zero.

The law can also be stated in the following way:

The sum of the currents entering a junction (or region) of a network must equal the sum of the currents leaving the same junction (or region).

In equation form, the above statement can be written as follows:

$$\Sigma I_i = \Sigma I_o \tag{6.13}$$

with I_i representing the current entering, or "in," and I_o representing the current leaving, or "out."

In Fig. 6.31, for example, the shaded area can enclose an entire system or a complex network, or it can simply provide a connection point (junction) for the displayed currents. In each case, the current entering must equal that leaving, as required by Eq. (6.13):

$$\Sigma I_i = \Sigma I_o$$
$$I_1 + I_4 = I_2 + I_3$$
$$4\text{ A} + 8\text{ A} = 2\text{ A} + 10\text{ A}$$
$$\mathbf{12\text{ A} = 12\text{ A}} \quad \text{(checks)}$$

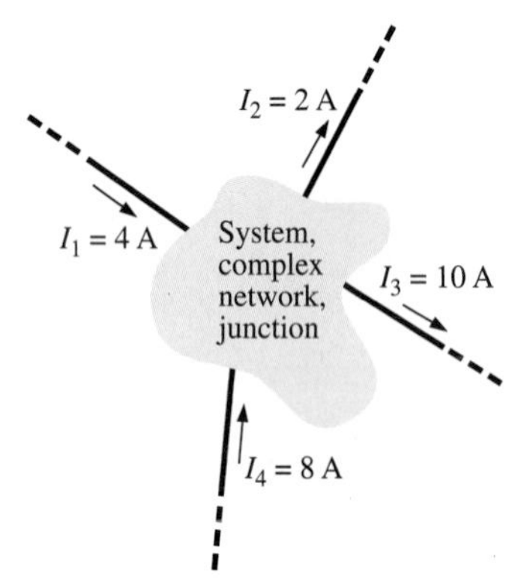

FIG. 6.31
Introducing Kirchhoff's current law.

The most common application of the law will be at a junction of two or more current paths, as shown in Fig. 6.32(a). Some students have difficulty initially determining whether a current is entering or leaving a junction. One approach that may help is to use the water analog in Fig. 6.32(b) where the junction in Fig. 6.32(a) is the small bridge across the stream. Simply relate the current of I_1 to the fluid flow of Q_1, the smaller branch current I_2 to the water flow Q_2, and the larger branch current I_3 to the flow Q_3. The water arriving at the bridge must equal the sum of that leaving the bridge so that $Q_1 = Q_2 + Q_3$. Since the current I_1 is pointing *at* the junction and the fluid flow Q_1 is *toward* the person on the bridge, both quantities are seen as approaching the junction and can be considered *entering* the junction. The currents I_2 and I_3 are both leaving the junction just as Q_2 and Q_3 are leaving the fork in the river. The quantities I_2, I_3, Q_2, and Q_3 are therefore all *leaving* the junction.

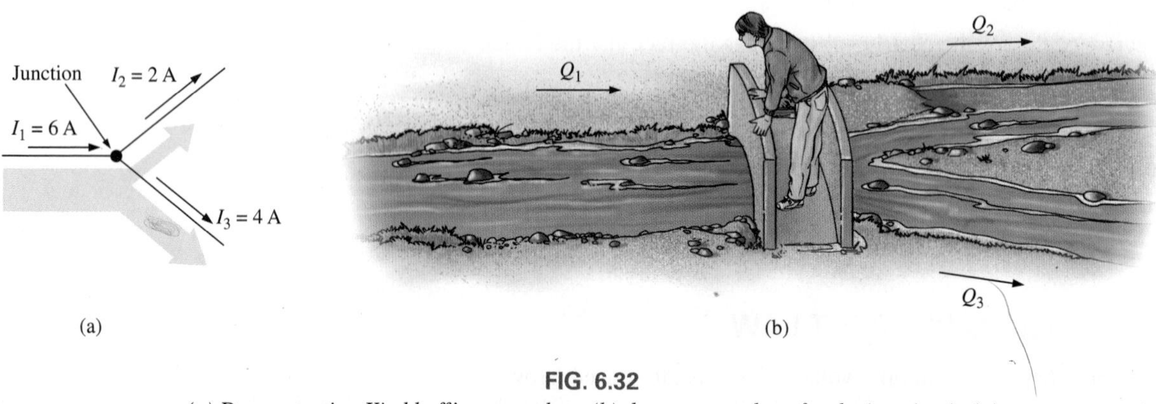

FIG. 6.32
(a) Demonstrating Kirchhoff's current law; (b) the water analogy for the junction in (a).

In the next few examples, unknown currents can be determined by applying Kirchhoff's current law. Remember to place all current levels entering the junction to the left of the equals sign and the sum of all currents leaving the junction to the right of the equals sign.

In technology, the term **node** is commonly used to refer to a junction of two or more branches. Therefore, this term is used frequently in the analyses to follow.

EXAMPLE 6.16 Determine currents I_3 and I_4 in Fig. 6.33 using Kirchhoff's current law.

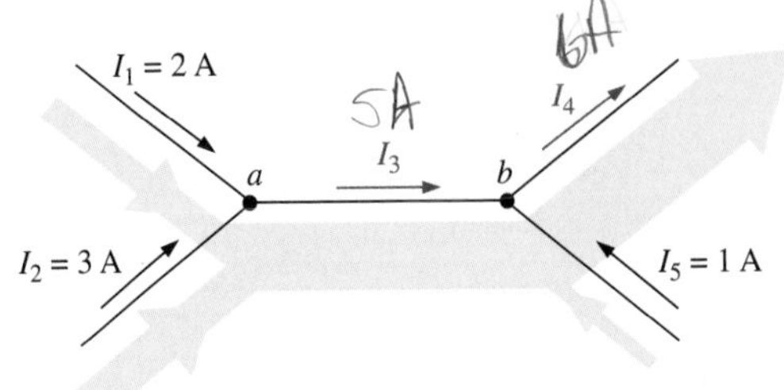

FIG. 6.33
Two-node configuration for Example 6.16.

Solution: There are two junctions or nodes in Fig. 6.33. Node *a* has only one unknown, while node *b* has two unknowns. Since a single equation can be used to solve for only one unknown, we must apply Kirchhoff's current law to node *a* first.

At node *a:*

$$\Sigma I_i = \Sigma I_o$$
$$I_1 + I_2 = I_3$$
$$2\text{ A} + 3\text{ A} = I_3 = \mathbf{5\text{ A}}$$

At node *b,* using the result just obtained:

$$\Sigma I_i = \Sigma I_o$$
$$I_3 + I_5 = I_4$$
$$5\text{ A} + 1\text{ A} = I_4 = \mathbf{6\text{ A}}$$

Note that in Fig. 6.33, the width of the blue shaded regions matches the magnitude of the current in that region.

EXAMPLE 6.17 Determine currents I_1, I_3, I_4, and I_5 for the network in Fig. 6.34.

Solution: In this configuration, four nodes are defined. Nodes *a* and *c* have only one unknown current at the junction, so Kirchhoff's current law can be applied at either junction.

At node *a:*

$$\Sigma I_i = \Sigma I_o$$
$$I = I_1 + I_2$$
$$5\text{ A} = I_1 + 4\text{ A}$$

and

$$I_1 = 5\text{ A} - 4\text{ A} = \mathbf{1\text{ A}}$$

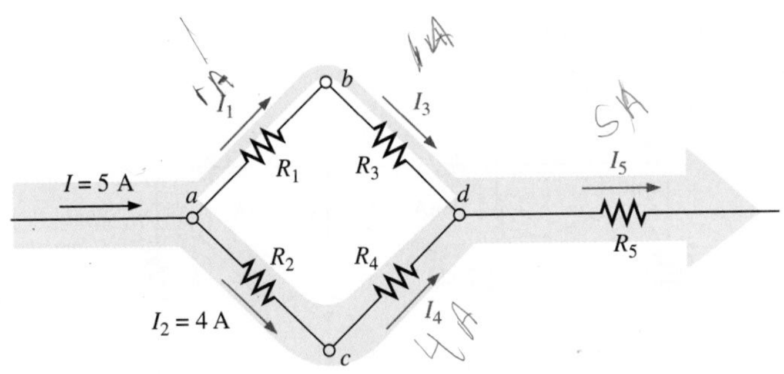

FIG. 6.34
Four-node configuration for Example 6.17.

At node *c:*

$$\Sigma I_i = \Sigma I_o$$
$$I_2 = I_4$$

and $$I_4 = I_2 = \mathbf{4\,A}$$

Using the above results at the other junctions results in the following.

At node *b:*

$$\Sigma I_i = \Sigma I_o$$
$$I_1 = I_3$$

and $$I_3 = I_1 = \mathbf{1\,A}$$

At node *d:*

$$\Sigma I_i = \Sigma I_o$$
$$I_3 + I_4 = I_5$$
$$1\text{ A} + 4\text{ A} = I_5 = \mathbf{5\,A}$$

If we enclose the entire network, we find that the current entering from the far left is $I = 5$ A, while the current leaving from the far right is $I_5 = 5$ A. The two must be equal since the net current entering any system must equal the net current leaving.

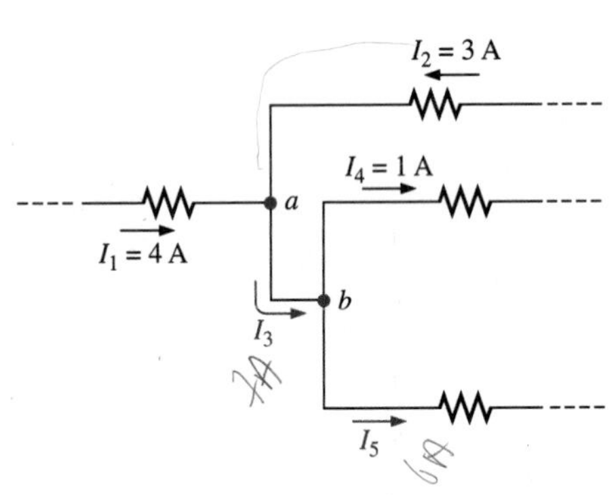

FIG. 6.35
Network for Example 6.18.

EXAMPLE 6.18 Determine currents I_3 and I_5 in Fig. 6.35 through applications of Kirchhoff's current law.

Solution: Note first that since node *b* has two unknown quantities (I_3 and I_5), and node *a* has only one, Kirchhoff's current law must first be applied to node *a*. The result is then applied to node *b*.

At node *a:*

$$\Sigma I_i = \Sigma I_o$$
$$I_1 + I_2 = I_3$$
$$4\text{ A} + 3\text{ A} = I_3 = \mathbf{7\,A}$$

At node *b:*

$$\Sigma I_i = \Sigma I_o$$
$$I_3 = I_4 + I_5$$
$$7\text{ A} = 1\text{ A} + I_5$$

and $$I_5 = 7\text{ A} - 1\text{ A} = \mathbf{6\,A}$$

EXAMPLE 6.19 For the parallel dc network in Fig. 6.36.

a. Determine the source current I_s.
b. Find the source voltage E.

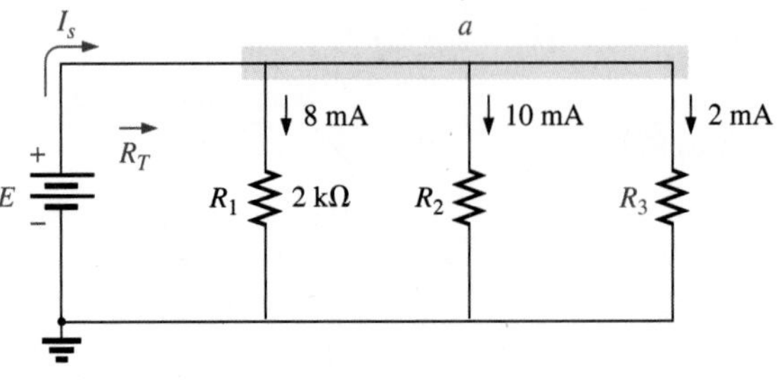

FIG. 6.36
Parallel network for Example 6.19.

c. Determine R_3.
d. Calculate R_T.

Solutions:

a. First apply Eq. (6.13) at node *a*. Although node *a* in Fig. 6.36 may not initially appear as a single junction, it can be redrawn as shown in Fig. 6.37, where it is clearly a common point for all the branches. The result is

$$\Sigma I_i = \Sigma I_o$$
$$I_s = I_1 + I_2 + I_3$$

Substituting values: $I_s = 8\text{ mA} + 10\text{ mA} + 2\text{ mA} = \mathbf{20\text{ mA}}$

Note in this solution that you do not need to know the resistor values or the voltage applied. The solution is determined solely by the current levels.

b. Applying Ohm's law:

$$E = V_1 = I_1R_1 = (8\text{ mA})(2\text{ k}\Omega) = \mathbf{16\text{ V}}$$

c. Applying Ohm's law in a different form:

$$R_3 = \frac{V_3}{I_3} = \frac{E}{I_3} = \frac{16\text{ V}}{2\text{ mA}} = \mathbf{8\text{ k}\Omega}$$

d. Applying Ohm's law again:

$$R_T = \frac{E}{I_s} = \frac{16\text{ V}}{20\text{ mA}} = \mathbf{0.8\text{ k}\Omega}$$

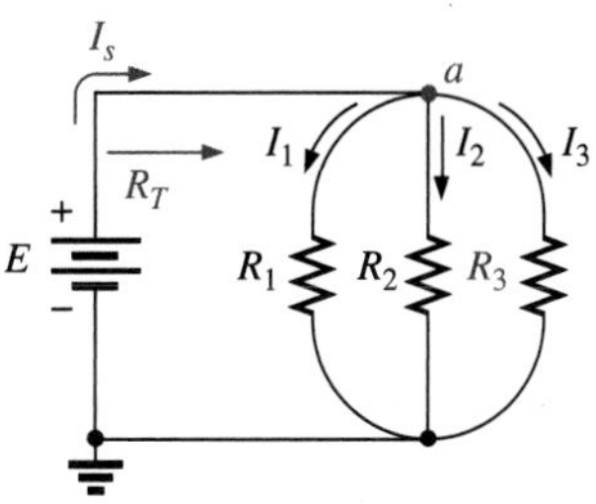

FIG. 6.37
Redrawn network in Fig. 6.36.

The application of Kirchhoff's current law is not limited to networks where all the internal connections are known or visible. For instance, all the currents of the integrated circuit in Fig. 6.38 are known except I_1. By treating the entire system (which could contain over a million elements) as a single node, we can apply Kirchhoff's current law as shown in Example 6.20.

Before looking at Example 6.20 in detail, note that the direction of the unknown current I_1 is not provided in Fig. 6.38. On many occasions, this will be true. With so many currents entering or leaving the system, it is difficult to know by inspection which direction should be assigned to I_1. *In such cases, simply make an assumption about the direction and then check out the result. If the result is negative, the wrong direction was assumed. If the result is positive, the correct direction was assumed. In either case, the magnitude of the current will be correct.*

EXAMPLE 6.20 Determine I_1 for the integrated circuit in Fig. 6.38.

Solution: Assuming that the current I_1 entering the chip results in the following when Kirchhoff's current law is applied:

$$\Sigma I_i = \Sigma I_o$$
$$I_1 + 10\text{ mA} + 4\text{ mA} + 8\text{ mA} = 5\text{ mA} + 4\text{ mA} + 2\text{ mA} + 6\text{ mA}$$
$$I_1 + 22\text{ mA} = 17\text{ mA}$$
$$I_1 = 17\text{ mA} - 22\text{ mA} = \mathbf{-5\text{ mA}}$$

We find that the direction for I_1 is *leaving* the IC, although the magnitude of 5 mA is correct.

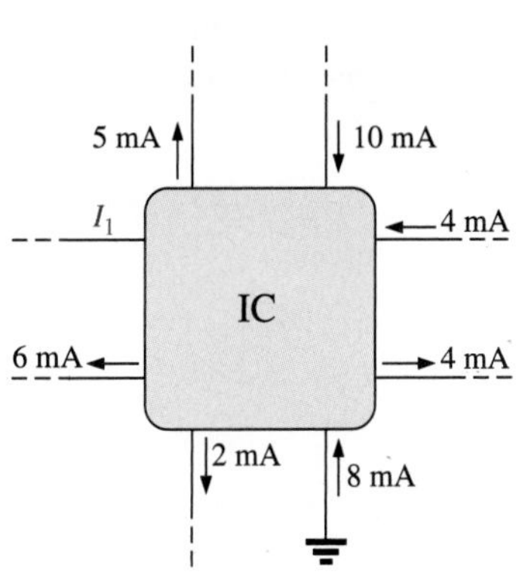

FIG. 6.38
Integrated circuit for Example 6.20.

As we leave this important section, be aware that Kirchhoff's current law will be applied in one form or another throughout the text. *Kirchhoff's laws are unquestionably two of the most important in this field because they are applicable to the most complex configurations in existence today.* They will not be replaced by a more important law or dropped for a more sophisticated approach.

6.6 CURRENT DIVIDER RULE

For series circuits we have the powerful voltage divider rule for finding the voltage across a resistor in a series circuit. We now introduce the equally powerful **current divider rule (CDR)** for finding the current through a resistor in a parallel circuit.

In Section 6.4, it was pointed out that current will always seek the path of least resistance. In Fig. 6.39, for example, the current of 9 A is faced with splitting between the three parallel resistors. Based on the previous sections, it should now be clear without a single calculation that the majority of the current will pass through the smallest resistor of 10 Ω, and the least current will pass through the 1 kΩ resistor. In fact, the current through the 100 Ω resistor will also exceed that through the 1 kΩ resistor. We can take it one step further by recognizing that the resistance of the 100 Ω resistor is 10 times that of the 10 Ω resistor. The result is a current through the 10 Ω resistor that is 10 times that of the 100 Ω resistor. Similarly, the current through the 100 Ω resistor is 10 times that through the 1 kΩ resistor.

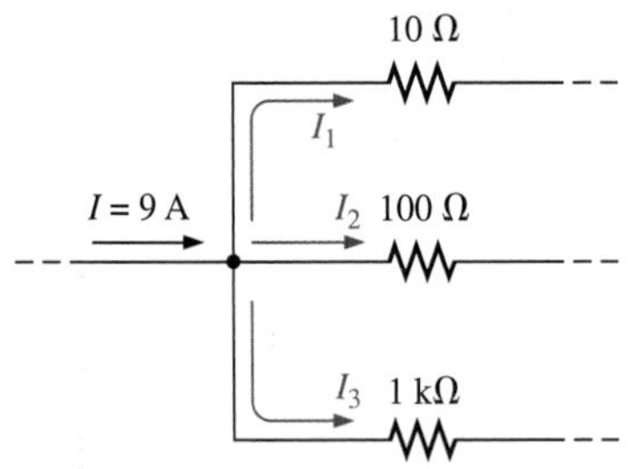

FIG. 6.39
Discussing the manner in which the current will split between three parallel branches of different resistive value.

In general,

For two parallel elements of equal value, the current will divide equally.

For parallel elements with different values, the smaller the resistance, the greater the share of input current.

For parallel elements of different values, the current will split with a ratio equal to the inverse of their resistor values.

EXAMPLE 6.21

a. Determine currents I_1 and I_3 for the network in Fig. 6.40.
b. Find the source current I_s.

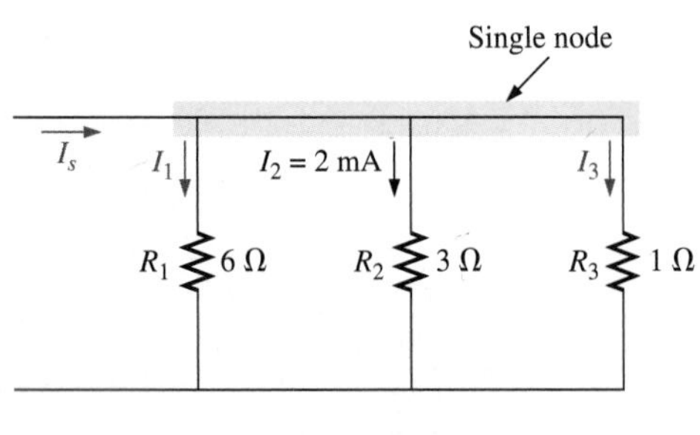

FIG. 6.40
Parallel network for Example 6.21.

Solutions:

a. Since R_1 is twice R_2, the current I_1 must be one-half I_2, and

$$I_1 = \frac{I_2}{2} = \frac{2\text{ mA}}{2} = \mathbf{1\text{ mA}}$$

Since R_2 is three times R_3, the current I_3 must be three times I_2, and

$$I_3 = 3I_2 = 3(2\text{ mA}) = \mathbf{6\text{ mA}}$$

b. Applying Kirchhoff's current law:

$$\Sigma I_i = \Sigma I_o$$
$$I_s = I_1 + I_2 + I_3$$
$$I_s = 1\text{ mA} + 2\text{ mA} + 6\text{ mA} = \mathbf{9\text{ mA}}$$

Although the above discussions and examples allowed us to determine the relative magnitude of a current based on a known level, they do not provide the magnitude of a current through a branch of a parallel network

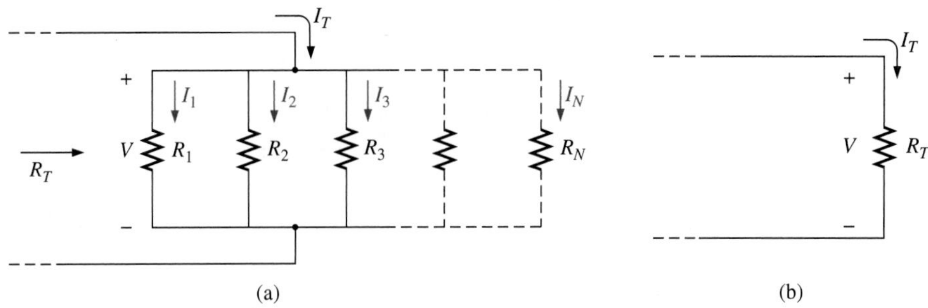

FIG. 6.41
Deriving the current divider rule: (a) parallel network of N parallel resistors; (b) reduced equivalent of part (a).

if only the total entering current is known. The result is a need for the current divider rule which will be derived using the parallel configuration in Fig. 6.41(a). The current I_T (using the subscript T to indicate the total entering current) splits between the N parallel resistors and then gathers itself together again at the bottom of the configuration. In Fig. 6.41(b), the parallel combination of resistors has been replaced by a single resistor equal to the total resistance of the parallel combination as determined in the previous sections.

The current I_T can then be determined using Ohm's law:

$$I_T = \frac{V}{R_T}$$

Since the voltage V is the same across parallel elements, the following is true:

$$V = I_1R_1 = I_2R_2 = I_3R_3 = \cdots = I_xR_x$$

where the product I_xR_x refers to any combination in the series.

Substituting for V in the above equation for I_T, we have

$$I_T = \frac{I_xR_x}{R_T}$$

Solving for I_x, the final result is the **current divider rule:**

$$I_x = \frac{R_T}{R_x}I_T \qquad \textbf{(6.14)}$$

which states that

the current through any branch of a parallel resistive network is equal to the total resistance of the parallel network divided by the resistor of interest and multiplied by the total current entering the parallel configuration.

Since R_T and I_T are constants, for a particular configuration the larger the value of R_x (in the denominator), the smaller the value of I_x for that branch, confirming the fact that current always seeks the path of least resistance.

EXAMPLE 6.22 For the parallel network in Fig. 6.42, determine current I_1 using Eq. (6.14).

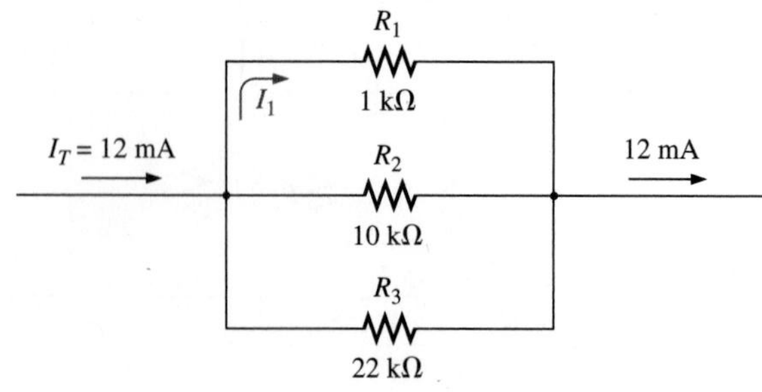

FIG. 6.42
Using the current divider rule to calculate current I_1 in Example 6.22.

Solution: Eq. (6.3):

$$R_T = \frac{1}{\dfrac{1}{R_1} + \dfrac{1}{R_2} + \dfrac{1}{R_3}}$$

$$= \frac{1}{\dfrac{1}{1\text{ k}\Omega} + \dfrac{1}{10\text{ k}\Omega} + \dfrac{1}{22\text{ k}\Omega}}$$

$$= \frac{1}{1 \times 10^{-3} + 100 \times 10^{-6} + 45.46 \times 10^{-6}}$$

$$= \frac{1}{1.145 \times 10^{-3}} = \mathbf{873.01\ \Omega}$$

Eq. (6.14): $$I_1 = \frac{R_T}{R_1} I_T$$

$$= \frac{(873.01\ \Omega)}{1\text{ k}\Omega}(12\text{ mA}) = (0.873)(12\text{ mA}) = \mathbf{10.48\ mA}$$

and the smallest parallel resistor receives the majority of the current.

Note also that

for a parallel network, the current through the smallest resistor will be very close to the total entering current if the other parallel elements of the configuration are much larger in magnitude.

In Example 6.22, the current through R_1 is very close to the total current because R_1 is 10 times less than the next smallest resistor.

Special Case: Two Parallel Resistors

For the case of two parallel resistors as shown in Fig 6.43, the total resistance is determined by

$$R_T = \frac{R_1 R_2}{R_1 + R_2}$$

Substituting R_T into Eq. (6.14) for current I_1 results in

$$I_1 = \frac{R_T}{R_1} I_T = \frac{\left(\dfrac{R_1 R_2}{R_1 + R_2}\right)}{R_1} I_T$$

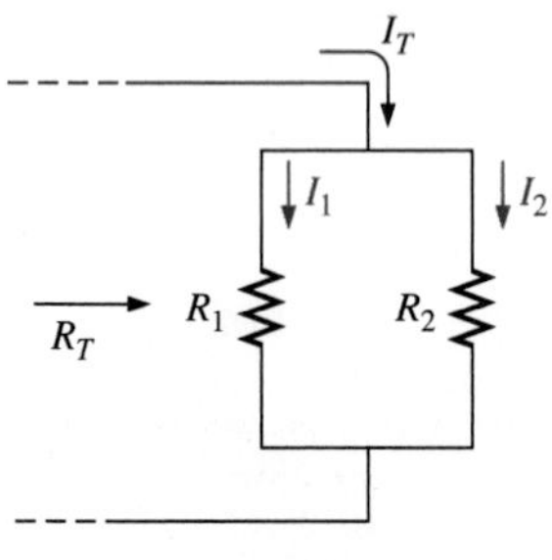

FIG. 6.43
Deriving the current divider rule for the special case of only two parallel resistors.

IPI

and

$$I_1 = \left(\frac{R_2}{R_1 + R_2}\right) I_T \qquad \textbf{(6.15a)}$$

Similarly, for I_2,

$$I_2 = \left(\frac{R_1}{R_1 + R_2}\right) I_T \qquad \textbf{(6.15b)}$$

Eq. (6.15) states that

for two parallel resistors, the current through one is equal to the other resistor times the total entering current divided by the sum of the two resistors.

Since the combination of two parallel resistors is probably the most common parallel configuration, the simplicity of the format for Eq. (6.15) suggests that it is worth memorizing. Take particular note, however, that the denominator of the equation is simply the *sum,* not the total resistance, of the combination.

EXAMPLE 6.23 Determine current I_2 for the network in Fig. 6.44 using the current divider rule.

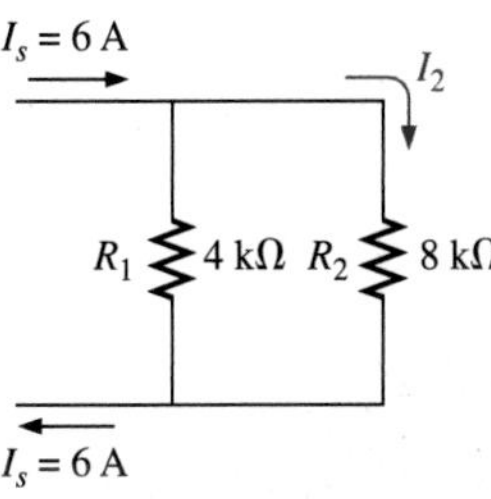

FIG. 6.44
Using the current divider rule to determine current I_2 in Example 6.23.

Solution: Using Eq. (6.15b):

$$I_2 = \left(\frac{R_1}{R_1 + R_2}\right) I_T$$

$$= \left(\frac{4\text{ k}\Omega}{4\text{ k}\Omega + 8\text{ k}\Omega}\right) 6\text{ A} = (0.333)(6\text{ A}) = \mathbf{2\text{ A}}$$

Using Eq. (6.14):

$$I_2 = \frac{R_T}{R_2} I_T$$

with

$$R_T = 4\text{ k}\Omega \parallel 8\text{ k}\Omega = \frac{(4\text{ k}\Omega)(8\text{ k}\Omega)}{4\text{ k}\Omega + 8\text{ k}\Omega} = 2.667\text{ k}\Omega$$

and

$$I_2 = \left(\frac{2.667\text{ k}\Omega}{8\text{ k}\Omega}\right) 6\text{ A} = (0.333)(6\text{ A}) = \mathbf{2\text{ A}}$$

matching the above solution.

It would appear that the solution with Eq. 6.15(b) is more direct in Example 6.23. However, keep in mind that Eq. (6.14) is applicable to any parallel configuration, removing the necessity to remember two equations.

Now we present a design-type problem.

EXAMPLE 6.24 Determine resistor R_1 in Fig. 6.45 to implement the division of current shown.

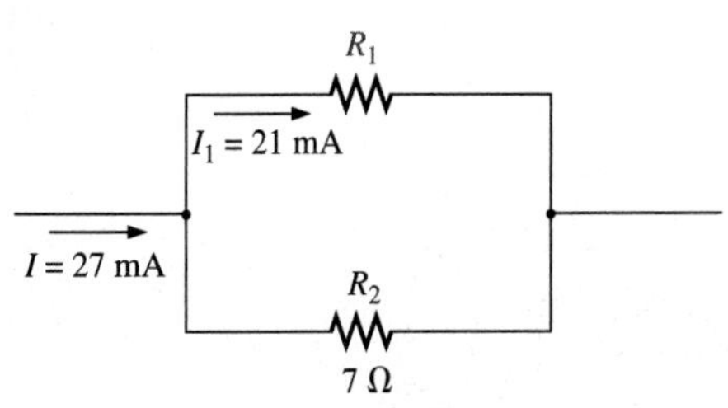

FIG. 6.45
A design-type problem for two parallel resistors (Example 6.24).

Solution: There are essentially two approaches to this type of problem. One involves the direct substitution of known values into the current divider rule equation followed by a mathematical analysis. The other is the sequential application of the basic laws of electric circuits. First we will use the latter approach.

Applying Kirchhoff's current law:

$$\Sigma I_i = \Sigma I_o$$

$$I = I_1 + I_2$$

$$27 \text{ mA} = 21 \text{ mA} + I_2$$

and $$I_2 = 27 \text{ mA} - 21 \text{ mA} = 6 \text{ mA}$$

The voltage V_2: $$V_2 = I_2R_2 = (6 \text{ mA})(7\ \Omega) = 42 \text{ mV}$$

so that $$V_1 = V_2 = 42 \text{ mV}$$

Finally, $$R_1 = \frac{V_1}{I_1} = \frac{42 \text{ mV}}{21 \text{ mA}} = \mathbf{2\ \Omega}$$

Now for the other approach using the current divider rule:

$$I_1 = \frac{R_2}{R_1 + R_2} I_T$$

$$21 \text{ mA} = \left(\frac{7\ \Omega}{R_1 + 7\ \Omega}\right) 27 \text{ mA}$$

$$(R_1 + 7\ \Omega)(21 \text{ mA}) = (7\ \Omega)(27 \text{ mA})$$

$$(21 \text{ mA})R_1 + 147 \text{ mV} = 189 \text{ mV}$$

$$(21 \text{ mA})R_1 = 189 \text{ mV} - 147 \text{ mV} = 42 \text{ mV}$$

and $$R_1 = \frac{42 \text{ mV}}{21 \text{ mA}} = \mathbf{2\ \Omega}$$

In summary, therefore, remember that current always seeks the path of least resistance, and the ratio of the resistor values is the inverse of the resulting current levels, as shown in Fig 6.46. The thickness of the blue bands in Fig. 6.46 reflects the relative magnitude of the current in each branch.

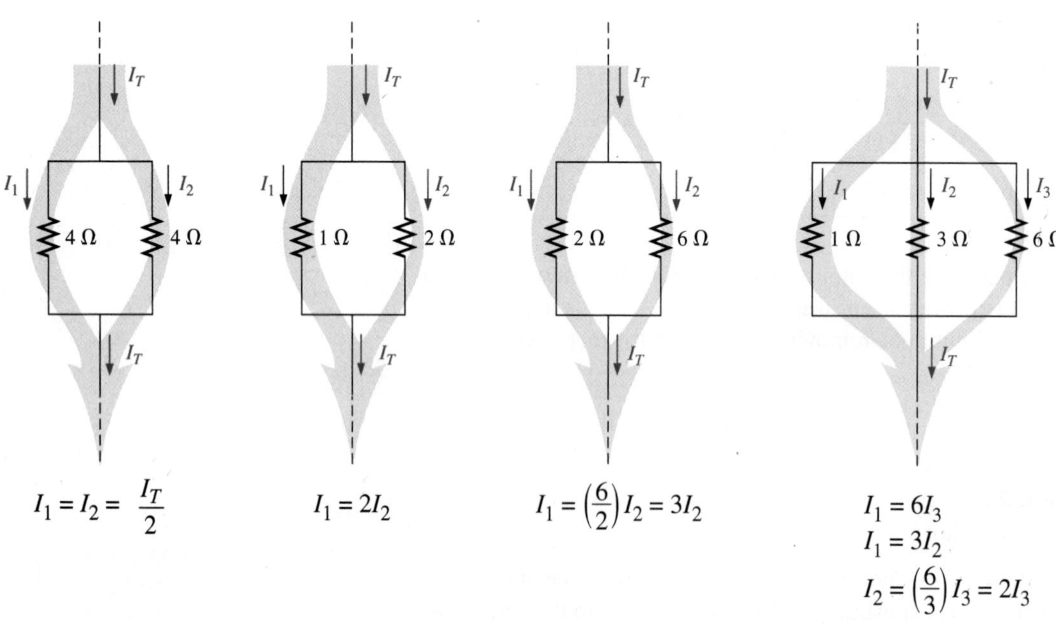

$I_1 = I_2 = \frac{I_T}{2}$ $\qquad I_1 = 2I_2$ $\qquad I_1 = \left(\frac{6}{2}\right)I_2 = 3I_2$ $\qquad I_1 = 6I_3$, $I_1 = 3I_2$, $I_2 = \left(\frac{6}{3}\right)I_3 = 2I_3$

FIG. 6.46
Demonstrating how current divides through equal and unequal parallel resistors.

6.7 VOLTAGE SOURCES IN PARALLEL

Because the voltage is the same across parallel elements,

voltage sources can be placed in parallel only if they have the same voltage.

The primary reason for placing two or more batteries or supplies in parallel is to increase the current rating above that of a single supply. For example, in Fig 6.47, two ideal batteries of 12 V have been placed in parallel. The total source current using Kirchhoff's current law is now the sum of the rated currents of each supply. The resulting power available will be twice that of a single supply if the rated supply current of each is the same. That is,

with $\quad I_1 = I_2 = I$

then $\quad P_T = E(I_1 + I_2) = E(I + I) = E(2I) = 2(EI) = 2P_{(\text{one supply})}$

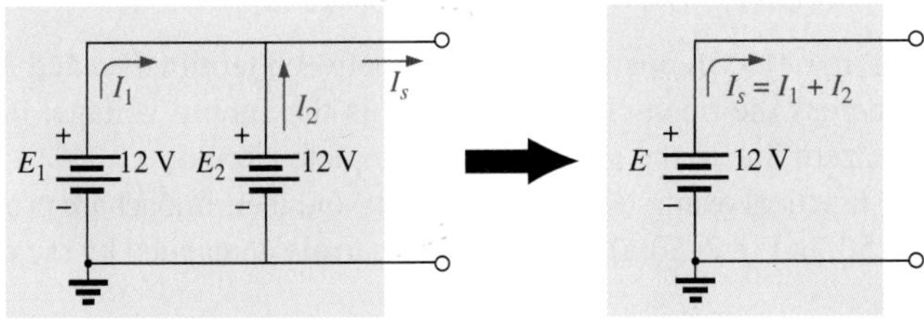

FIG. 6.47
Demonstrating the effect of placing two ideal supplies of the same voltage in parallel.

If for some reason two batteries of different voltages are placed in parallel, both will become ineffective or damaged because the battery with the larger voltage rapidly discharges through the battery with the smaller terminal voltage. For example, consider two lead-acid batteries of different terminal voltages placed in parallel as shown in Fig 6.48. It makes no sense to talk about placing an ideal 12 V battery in parallel with a 6 V battery, because Kirchhoff's voltage law would be violated. However, we can examine the effects if we include the internal resistance levels as shown in Fig. 6.48.

The only current-limiting resistors in the network are the internal resistances, resulting in a very high discharge current for the battery with the larger supply voltage. The resulting current for the case in Fig. 6.48 would be

$$I = \frac{E_1 - E_2}{R_{\text{int}_1} + R_{\text{int}_2}} = \frac{12\text{ V} - 6\text{ V}}{0.03\ \Omega + 0.02\ \Omega} = \frac{6\text{ V}}{0.05\ \Omega} = 120\text{ A}$$

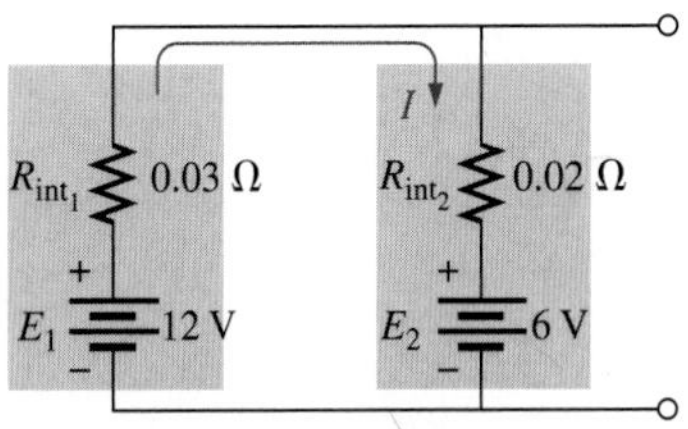

FIG. 6.48
Examining the impact of placing two lead-acid batteries of different terminal voltages in parallel.

This value far exceeds the rated drain current of the 12 V battery, resulting in rapid discharge of E_1 and a destructive impact on the smaller supply due to the excessive currents. This type of situation did arise on occasion when some cars still had 6 V batteries. Some people thought, "If I have a 6 V battery, a 12 V battery will work twice as well"—not true!

In general,

it is always recommended that when you are replacing batteries in series or parallel, all the batteries be replaced.

A fresh battery placed in parallel with an older battery probably has a higher terminal voltage and immediately starts discharging through the older battery. In addition, the available current is less for the older battery, resulting in a higher-than-rated current drain from the newer battery when a load is applied.

6.8 OPEN AND SHORT CIRCUITS

Open circuits and short circuits can often cause more confusion and difficulty in the analysis of a system than standard series or parallel configurations. This will become more obvious in the chapters to follow when we apply some of the methods and theorems.

An **open circuit** is two isolated terminals not connected by an element of any kind, as shown in Fig. 6.49(a). Since a path for conduction does not exist, the current associated with an open circuit must always be zero. The voltage across the open circuit, however, can be any value, as determined by the system it is connected to. In summary, therefore,

an open circuit can have a potential difference (voltage) across its terminals, but the current is always zero amperes.

In Fig. 6.49(b), an open circuit exists between terminals *a* and *b*. The voltage across the open-circuit terminals is the supply voltage, but the current is zero due to the absence of a complete circuit.

Some practical examples of open circuits and their impact are provided in Fig. 6.50. In Fig. 6.50(a), the excessive current demanded by the circuit

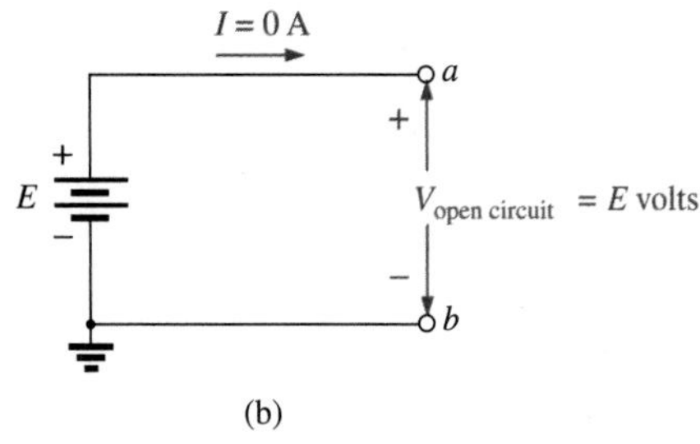

FIG. 6.49
Defining an open circuit.

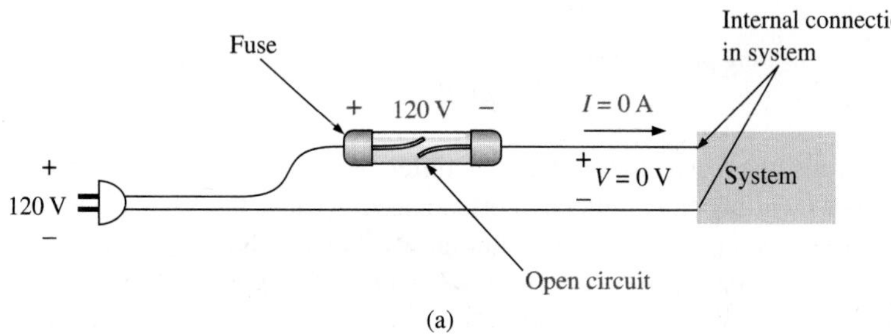

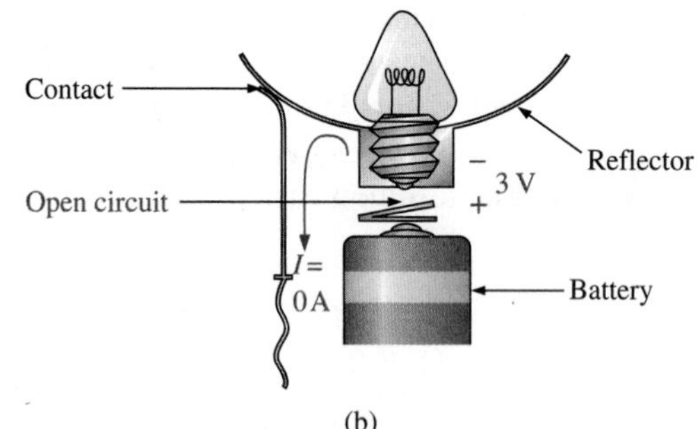

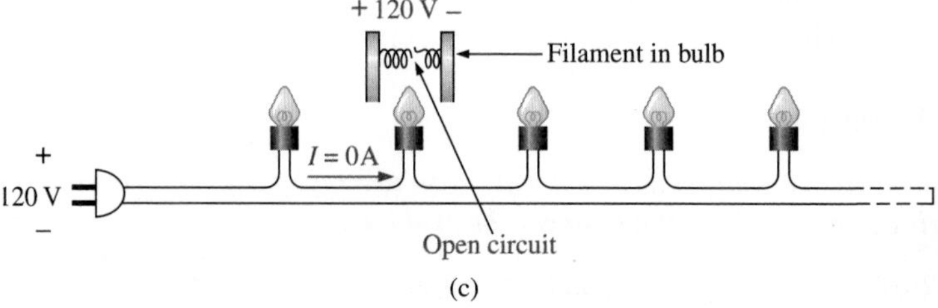

FIG. 6.50
Examples of open circuits.

caused a fuse to fail, creating an open circuit that reduced the current to zero amperes. However, it is important to note that *the full applied voltage is now across the open circuit,* so you must be careful when changing the fuse. If there is a main breaker ahead of the fuse, throw it first to remove the possibility of getting a shock. This situation clearly reveals the benefit of circuit breakers: You can reset the breaker without having to get near the hot wires.

In Fig. 6.50(b), the pressure plate at the bottom of the bulb cavity in a flashlight was bent when the flashlight was dropped. An open circuit now exists between the contact point of the bulb and the plate connected to the batteries. The current has dropped to zero amperes, but the 3 V provided by the series batteries appears across the open circuit. The situation can be corrected by placing a flat-edge screwdriver under the plate and bending it toward the bulb.

Finally, in Fig. 6.50(c), the filament in a bulb in a series connection has opened due to excessive current or old age, creating an open circuit that knocks out all the bulbs in the series configuration. Again, the current has dropped to zero amperes, but the full 120 V will appear across the contact points of the bad bulb. For situations such as this, *you should remove the plug from the wall before changing the bulb.*

A **short circuit** is a very low resistance, direct connection between two terminals of a network, as shown in Fig. 6.51. The current through the short circuit can be any value, as determined by the system it is connected to, but the voltage across the short circuit is always zero volts because the resistance of the short circuit is assumed to be essentially zero ohms and $V = IR = I(0\ \Omega) = 0\ \text{V}$.

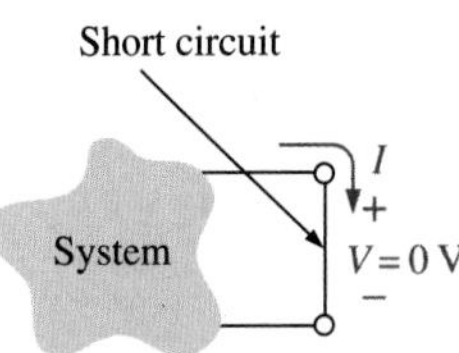

FIG. 6.51
Defining a short circuit.

In summary, therefore,

a short circuit can carry a current of a level determined by the external circuit, but the potential difference (voltage) across its terminals is always zero volts.

In Fig. 6.52(a), the current through the 2 Ω resistor is 5 A. If a short circuit should develop across the 2 Ω resistor, the total resistance of the parallel combination of the 2 Ω resistor and the short (of essentially zero ohms) will be

$$2\ \Omega \parallel 0\ \Omega = \frac{(2\ \Omega)(0\ \Omega)}{2\ \Omega + 0\ \Omega} = 0\ \Omega$$

as indicated in Fig. 6.52(b), and the current will rise to very high levels, as determined by Ohm's law:

$$I = \frac{E}{R} = \frac{10\ \text{V}}{0\ \Omega} \rightarrow \infty\ \text{A}$$

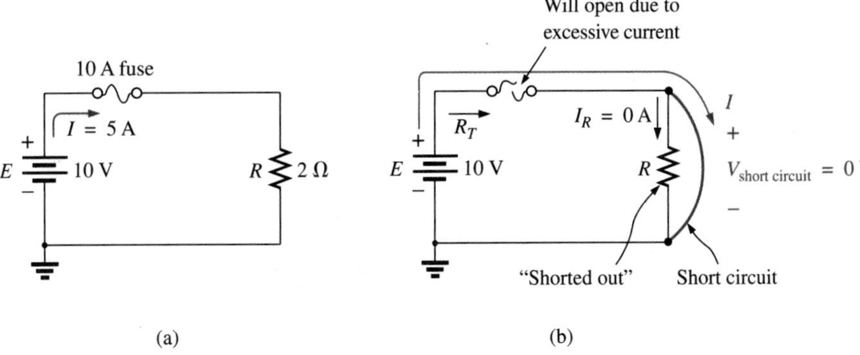

FIG. 6.52
Demonstrating the effect of a short circuit on current levels.

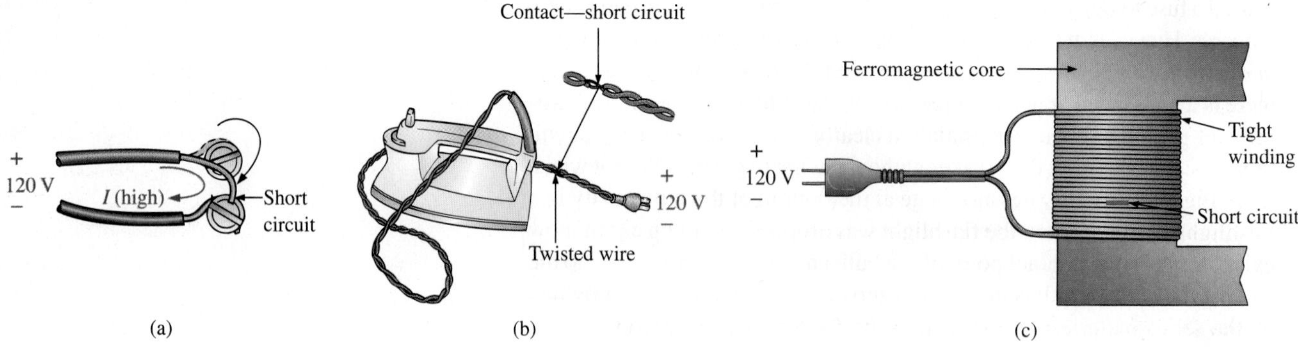

FIG. 6.53
Examples of short circuits.

The effect of the 2 Ω resistor has effectively been "shorted out" by the low-resistance connection. The maximum current is now limited only by the circuit breaker or fuse in series with the source.

Some practical examples of short circuits and their impact are provided in Fig. 6.53. In Fig. 6.53(a), a hot (the feed) wire wrapped around a screw became loose and is touching the return connection. A short-circuit connection between the two terminals has been established that could result in a very heavy current and a possible fire hazard. Hopefully, the breaker will "pop," and the circuit will be deactivated. Problems such as this is one of the reasons aluminum wires (cheaper and lighter than copper) are not permitted in residential or industrial wiring. Aluminum is more sensitive to temperature than copper and will expand and contract due to the heat developed by the current passing through the wire. Eventually, this expansion and contraction can loosen the screw, and a wire under some torsional stress from the installation can move and make contact as shown in Fig. 6.53(a). Aluminum is still used in large panels as a bus-bar connection, but it is bolted down.

In Fig. 6.53(b), the wires of an iron have started to twist and crack due to excessive currents or long-term use of the iron. Once the insulation breaks down, the twisting can cause the two wires to touch and establish a short circuit. Hopefully, a circuit breaker or fuse will quickly disconnect the circuit. Often, it is not the wire of the iron that causes the problem but a cheap extension cord with the wrong gage wire. Be aware that you cannot tell the capacity of an extension cord by its outside jacket. It may have a thick orange covering but have a very thin wire inside. Check the gage on the wire the next time you buy an extension cord, and be sure that it is at least #14 gage, with #12 being the better choice for high-current appliances.

Finally, in Fig. 6.53(c), the windings in a transformer or motor for residential or industrial use are illustrated. The windings are wound so tightly together with such a very thin coating of insulation that it is possible with age and use for the insulation to break down and short out the windings. In many cases, shorts can develop, but a short will simply reduce the number of effective windings in the unit. The tool or appliance may still work, but with less strength or rotational speed. If you notice such a change in the response, you should check the windings because a short can lead to a dangerous situation. In many cases, the state of the windings can be checked with a simple ohmmeter reading. If a short has occurred, the length of usable wire in the winding has been reduced, and the resistance drops. If you know what the resistance normally is, you can compare and make a judgment.

For the layperson, the terminology *short circuit* or *open circuit* is usually associated with dire situations such as power loss, smoke, or fire. However, in network analysis, both can play an integral role in determining specific parameters about a system. Most often, however, if a short-circuit condition is to be established, it is accomplished with a *jumper*—a lead of negligible resistance to be connected between the points of interest. Establishing an open circuit just requires making sure that the terminals of interest are isolated from each other.

EXAMPLE 6.25 Determine voltage V_{ab} for the network in Fig. 6.54.

Solution: The open circuit requires that I be zero amperes. The voltage drop across both resistors is therefore zero volts since $V = IR = (0)R = 0$ V. Applying Kirchhoff's voltage law around the closed loop,

$$V_{ab} = E = \mathbf{20\ V}$$

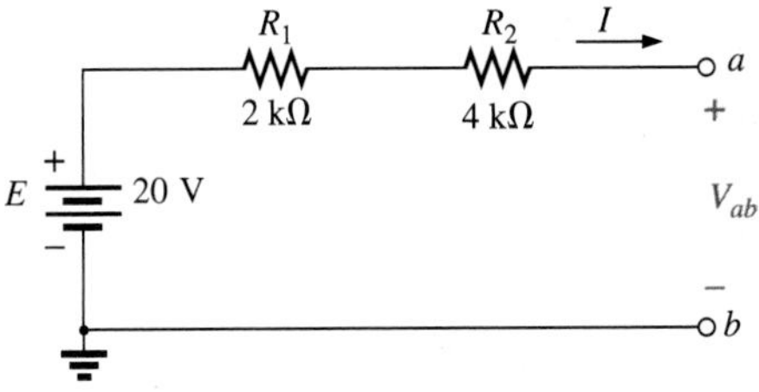

FIG. 6.54
Network for Example 6.25.

EXAMPLE 6.26 Determine voltages V_{ab} and V_{cd} for the network in Fig. 6.55.

Solution: The current through the system is zero amperes due to the open circuit, resulting in a 0 V drop across each resistor. Both resistors can therefore be replaced by short circuits, as shown in Fig. 6.56. Voltage V_{ab} is then directly across the 10 V battery, and

$$V_{ab} = E_1 = \mathbf{10\ V}$$

Voltage V_{cd} requires an application of Kirchhoff's voltage law:

$$+E_1 - E_2 - V_{cd} = 0$$

or

$$V_{cd} = E_1 - E_2 = 10\text{ V} - 30\text{ V} = \mathbf{-20\ V}$$

The negative sign in the solution indicates that the actual voltage V_{cd} has the opposite polarity of that appearing in Fig. 6.55.

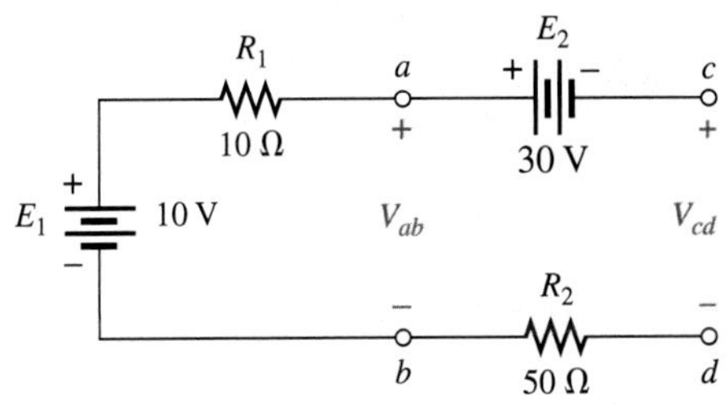

FIG. 6.55
Network for Example 6.26.

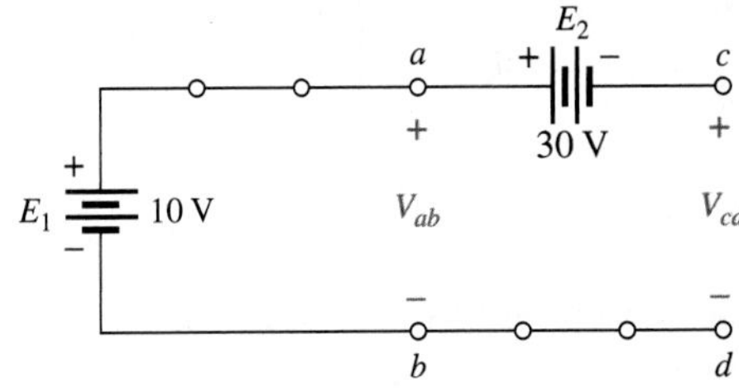

FIG. 6.56
Circuit in Fig. 6.55 redrawn.

EXAMPLE 6.27 Determine the unknown voltage and current for each network in Fig. 6.57.

Solution: For the network in Fig. 6.57(a), the current I_T will take the path of least resistance, and since the short-circuit condition at the end of the network is the least-resistance path, all the current will pass through the short circuit. This conclusion can be verified using the current divider rule. The voltage across the network is the same as that across the short circuit and will be zero volts, as shown in Fig. 6.58(a).

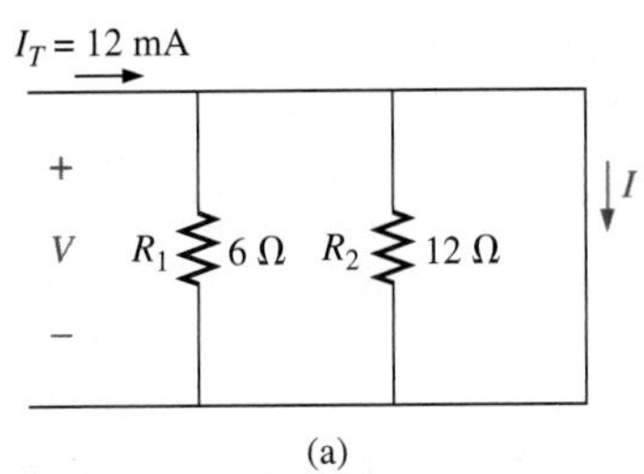

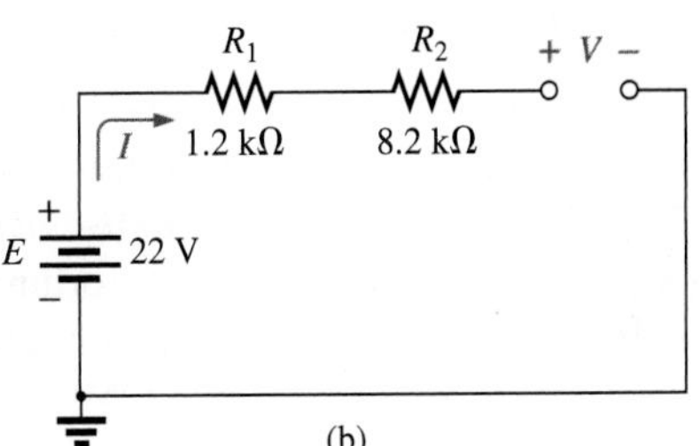

FIG. 6.57
Networks for Example 6.27.

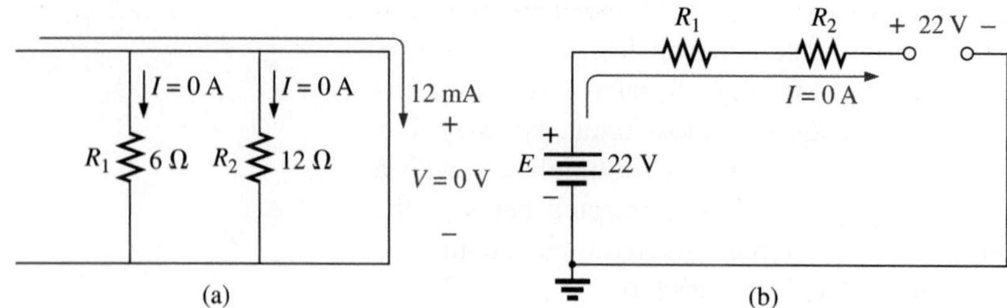

FIG. 6.58
Solutions to Example 6.27.

For the network in Fig. 6.57(b), the open-circuit condition requires that the current be zero amperes. The voltage drops across the resistors must therefore be zero volts, as determined by Ohm's law [$V_R = IR = (0)R = 0\text{ V}$], with the resistors acting as a connection from the supply to the open circuit. The result is that the open-circuit voltage is $E = 22\text{ V}$, as shown in Fig. 6.58(b).

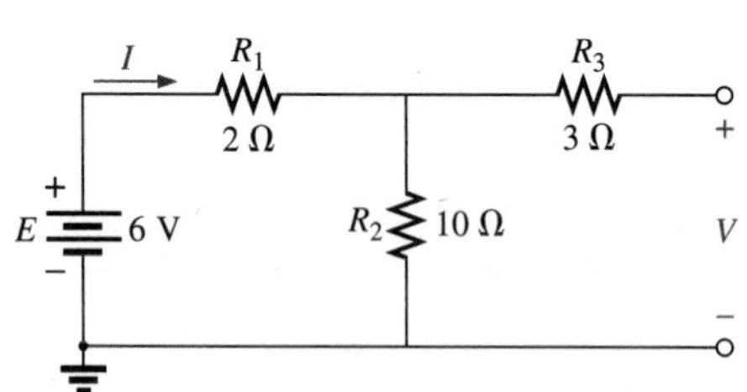

FIG. 6.59
Network for Example 6.28.

EXAMPLE 6.28 Determine V and I for the network in Fig. 6.59 if resistor R_2 is shorted out.

Solution: The redrawn network appears in Fig. 6.60. The current through the 3 Ω resistor is zero due to the open circuit, causing all the current I to pass through the jumper. Since $V_{3\Omega} = IR = (0)R = 0\text{ V}$, the voltage V is directly across the short, and

$$V = \mathbf{0\ V}$$

with
$$I = \frac{E}{R_1} = \frac{6\text{ V}}{2\ \Omega} = \mathbf{3\ A}$$

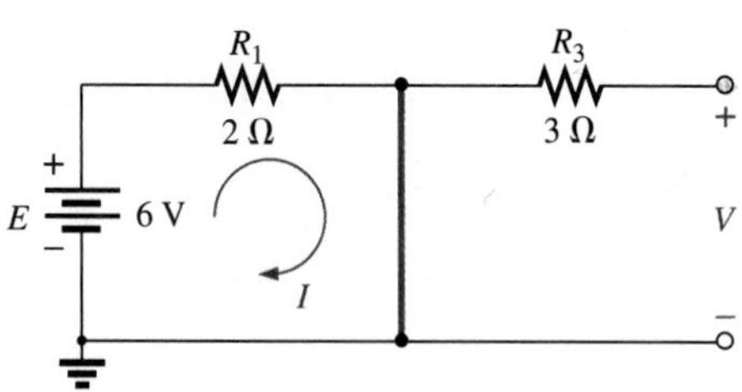

FIG. 6.60
Network in Fig. 6.59 with R_2 replaced by a jumper.

6.9 VOLTMETER LOADING EFFECTS

In previous chapters, we learned that ammeters are not ideal instruments. When you insert an ammeter, you actually introduce an additional resistance in series with the branch in which you are measuring the current. Generally, this is not a serious problem, but it can have a troubling effect on your readings, so it is important to be aware of it.

Voltmeters also have an internal resistance that appears between the two terminals of interest when a measurement is being made. While an ammeter places an additional resistance in series with the branch of interest, a voltmeter places an additional resistance *across* the element, as shown in Fig. 6.61. Since it appears in parallel with the element of interest, *the ideal level for the internal resistance of a voltmeter would be infinite ohms, just as zero ohms would be ideal for an ammeter.* Unfortunately, the internal resistance of any voltmeter is not infinite and changes from one type of meter to another.

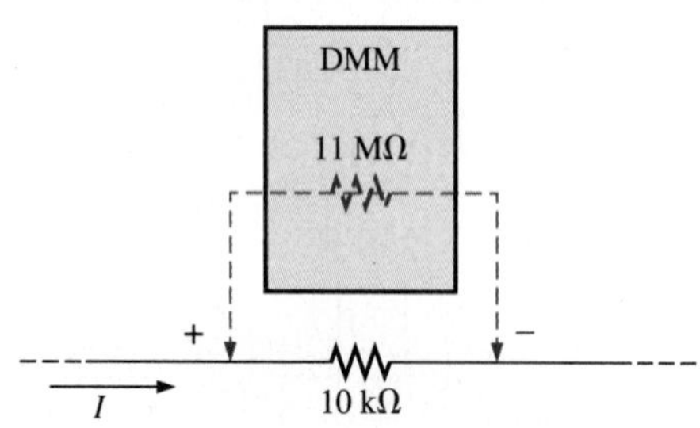

FIG. 6.61
Voltmeter loading.

Most digital meters have a fixed internal resistance level in the megohm range that remains the same *for all its scales.* For example, the meter in Fig. 6.61 has the typical level of 11 MΩ for its internal resistance, no matter which voltage scale is used. When the meter is placed across the 10 kΩ resistor, the total resistance of the combination is

$$R_T = 10\text{ k}\Omega \parallel 11\text{ M}\Omega = \frac{(10^4\,\Omega)(11 \times 10^6\,\Omega)}{10^4\,\Omega + (11 \times 10^6)} = 9.99\text{ k}\Omega$$

and the behavior of the network is not seriously affected. The result, therefore, is that

most digital voltmeters can be used in circuits with resistances up to the high-kilohm range without concern for the effect of the internal resistance on the reading.

However, if the resistances are in the megohm range, you should investigate the effect of the internal resistance.

An analog VOM is a different matter, however, because the internal resistance levels are much lower and the internal resistance levels are a function of the scale used. If a VOM on the 2.5 V scale were placed across the 10 kΩ resistor in Fig. 6.61, the internal resistance might be 50 kΩ, resulting in a combined resistance of

$$R_T = 10\text{ k}\Omega \parallel 50\text{ k}\Omega = \frac{(10^4\,\Omega)(50 \times 10^3\,\Omega)}{10^4\,\Omega + (50 \times 10^3\,\Omega)} = 8.33\text{ k}\Omega$$

and the behavior of the network would be affected because the 10 kΩ resistor would appear as an 8.33 kΩ resistor.

To determine the resistance R_m of any scale of a VOM, simply multiply the **maximum voltage** of the chosen scale by the **ohm/volt (Ω/V) rating** normally appearing at the bottom of the face of the meter. That is,

$$R_m\ (\text{VOM}) = (\text{scale})(\Omega/\text{V rating})$$

For a typical Ω/V rating of 20,000, the 2.5 V scale would have an internal resistance of

$$(2.5\text{ V})(20{,}000\ \Omega/\text{V}) = \mathbf{50\ k\Omega}$$

whereas for the 100 V scale, the internal resistance of the VOM would be

$$(100\text{ V})(20{,}000\ \Omega/\text{V}) = \mathbf{2\ M\Omega}$$

and for the 250 V scale,

$$(250\text{ V})(20{,}000\ \Omega/\text{V}) = \mathbf{5\ M\Omega}$$

EXAMPLE 6.29 For the relatively simple circuit in Fig. 6.62(a):

a. What is the open-circuit voltage V_{ab}?
b. What will a DMM indicate if it has an internal resistance of 11 MΩ? Compare your answer to that of part (a).
c. Repeat part (b) for a VOM with an Ω/V rating of 20,000 on the 100 V scale.

Solutions:

a. Due to the open circuit, the current is zero, and the voltage drop across the 1 MΩ resistor is zero volts. The result is that the entire source voltage appears between points *a* and *b*, and

$$V_{ab} = \mathbf{20\ V}$$

b. When the meter is connected as shown in Fig. 6.62(b), a complete circuit has been established, and current can pass through the circuit.

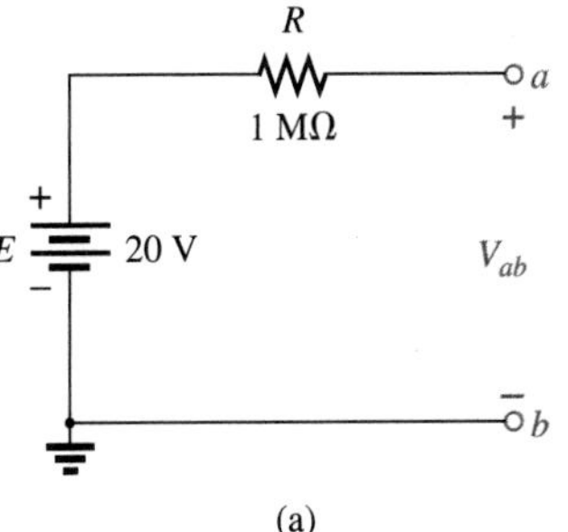

(a)

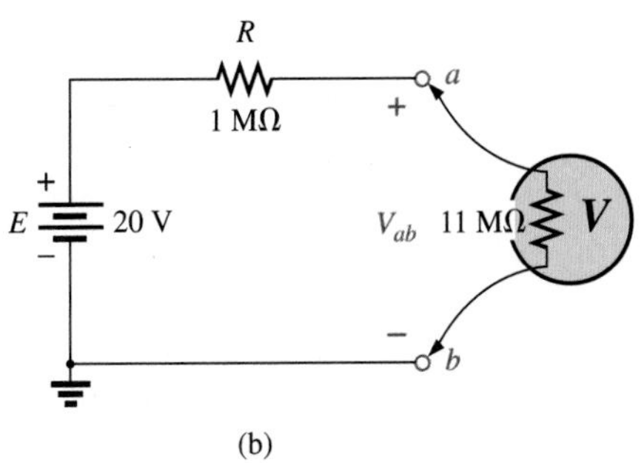

(b)

FIG. 6.62
(a) Measuring an open-circuit voltage with a voltmeter; (b) determining the effect of using a digital voltmeter with an internal resistance of 11 MΩ on measuring an open-circuit voltage (Example 6.29).

The voltmeter reading can be determined using the voltage divider rule as follows:

$$V_{ab} = \frac{(11\ \text{M}\Omega)(20\ \text{V})}{11\ \text{M}\Omega + 1\ \text{M}\Omega)} = \mathbf{18.33\ V}$$

and the reading is affected somewhat.

c. For the VOM, the internal resistance of the meter is

$$R_m = (100\ \text{V})\ (20{,}000\ \Omega/\text{V}) = 2\ \text{M}\Omega$$

and
$$V_{ab} = \frac{(2\ \text{M}\Omega)(20\ \text{V})}{2\ \text{M}\Omega + 1\ \text{M}\Omega)} = \mathbf{13.33\ V}$$

which is considerably below the desired level of 20 V.

6.10 SUMMARY TABLE

Now that the series and parallel configurations have been covered in detail, we will review the salient equations and characteristics of each. The equations for the two configurations have a number of similarities. In fact, the equations for one can often be obtained directly from the other by simply applying the **duality** principle. Duality between equations means that the format for an equation can be applied to two different situations by just changing the variable of interest. For instance, the equation for the total resistance of a series circuit is the sum of the resistances. By changing the resistance parameters to conductance parameters, you can obtain the equation for the total conductance of a parallel network—an easy way to remember the two equations. Similarly, by starting with the total conductance equation, you can easily write the total resistance equation for series circuits by replacing the conductance parameters by resistance parameters. Series and parallel networks share two important dual relationships: (1) between resistance of series circuits and conductance of parallel circuits and (2) between the voltage or current of a series circuit and the current or voltage, respectively, of a parallel circuit. Table 6.1 summarizes this duality.

The format for the total resistance for a series circuit has the same format as the total conductance of a parallel network as shown in Table 6.1. All that is required to move back and forth between the series and parallel headings is to interchange the letters R and G. For the special case of two elements, the equations have the same format, but the equation applied for the total resistance of the parallel configuration has changed. In the series configuration, the total resistance increases with each added resistor. For parallel networks, the total conductance increases with each additional conductance. The result is that the total conductance of a series circuit drops with added resistive elements while the total resistance of parallel networks decreases with added elements.

In a series circuit, the current is the same everywhere. In a parallel network, the voltage is the same across each element. The result is a duality between voltage and current for the two configurations. What is true for one in one configuration is true for the other in the other configuration. In a series circuit, the applied voltage divides between the series elements. In a parallel network, the current divides between parallel elements. For series circuits, the largest resistor captures the largest share of the applied voltage. For parallel networks, the branch with the highest conductance captures the greater share of the incoming current. In addition, for series circuits, the applied voltage equals the sum of the voltage drops across the

TABLE 6.1
Summary table.

Series and Parallel Circuits		
Series	**Duality**	**Parallel**
$R_T = R_1 + R_2 + R_3 + \cdots + R_N$	$R \rightleftarrows G$	$G_T = G_1 + G_2 + G_3 + \cdots + G_N$
R_T increases (G_T decreases) if additional resistors are added in series	$R \rightleftarrows G$	G_T increases (R_T decreases) if additional resistors are added in parallel
Special case: two elements $R_T = R_1 + R_2$	$R \rightleftarrows G$	$G_T = G_1 + G_2$ and $R_T = \dfrac{1}{\dfrac{1}{R_1} + \dfrac{1}{R_2}} = \dfrac{R_1R_2}{R_1 + R_2}$
I the same through series elements	$I \rightleftarrows V$	V the same across parallel elements
$E = V_1 + V_2 + V_3$	$E, V \rightleftarrows I$	$I_T = I_1 + I_2 + I_3$
Largest V across largest R	$V \rightleftarrows I$ and $R \rightleftarrows G$	Greatest I through largest G (smallest R)
$V_x = \dfrac{R_xE}{R_T}$	$E, V \rightleftarrows I$ and $R \rightleftarrows G$	$I_x = \dfrac{G_xI_T}{G_T} = \dfrac{R_TI_T}{R_x}$ with $I_1 = \dfrac{R_2 I_T}{R_1 + R_2}$ and $I_2 = \dfrac{R_1 I_T}{R_1 + R_2}$
$P = EI_T$	$E \rightleftarrows I$ and $I \rightleftarrows E$	$P = I_T E$
$P = I^2R$	$I \rightleftarrows V$ and $R \rightleftarrows G$	$P = V^2G = V^2/R$
$P = V^2/R$	$V \rightleftarrows I$ and $R \rightleftarrows G$	$P = I^2/G = I^2R$

series elements of the circuit, while the source current for parallel branches equals the sum of the currents through all the parallel branches.

The total power delivered to a series or parallel network is determined by the product of the applied voltage and resulting source current. The power delivered to each element is also the same for each configuration. Duality can be applied again, but the equation $P = EI$ results in the same result as $P = IE$. Also, $P = I^2R$ can be replaced by $P = V^2G$ for parallel elements, but essentially each can be used for each configuration. The duality principle can be very helpful in the learning process. Remember this as you progress through the next few chapters. You will find in the later chapters that this duality can also be applied between two important elements—inductors and capacitors.

6.11 TROUBLESHOOTING TECHNIQUES

The art of *troubleshooting* is not limited solely to electrical or electronic systems. In the broad sense,

troubleshooting is a process by which acquired knowledge and experience are used to localize a problem and offer or implement a solution.

There are many reasons why the simplest electrical circuit does not operate correctly. A connection may be open; the measuring instruments may need calibration; the power supply may not be on or may have been connected incorrectly to the circuit; an element may not be performing correctly due to earlier damage or poor manufacturing; a fuse may have blown; and so on. Unfortunately, a defined sequence of steps does not

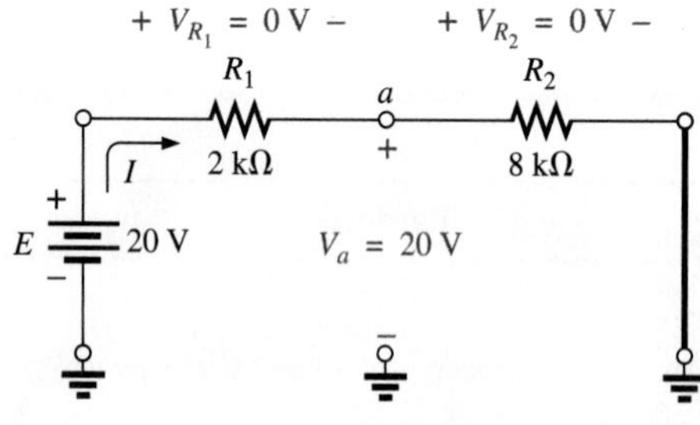

FIG. 6.63
A malfunctioning network.

exist for identifying the wide range of problems that can surface in an electrical system. It is only through experience and a clear understanding of the basic laws of electric circuits that you can become proficient at quickly locating the cause of an erroneous output.

It should be fairly obvious, however, that the first step in checking a network or identifying a problem area is to have some idea of the expected voltage and current levels. For instance, the circuit in Fig. 6.63 should have a current in the low milliampere range, with the majority of the supply voltage across the 8 kΩ resistor. However, as indicated in Fig. 6.63, $V_{R_1} = V_{R_2} = 0$ V and $V_a = 20$ V. Since $V = IR$, the results immediately suggest that $I = 0$ A and an open circuit exists in the circuit. The fact that $V_a = 20$ V immediately tells us that the connections are true from the ground of the supply to point *a*. The open circuit must therefore exist between R_1 and R_2 or at the ground connection of R_2. An open circuit at either point results in $I = 0$ A and the readings obtained previously. Keep in mind that, even though $I = 0$ A, R_1 does form a connection between the supply and point *a*. That is, if $I = 0$ A, $V_{R_1} = IR_2 = (0)R_2 = 0$ V, as obtained for a short circuit.

In Fig. 6.63, if $V_{R_1} \cong 20$ V and V_{R_2} is quite small ($\cong 0.08$ V), it first suggests that the circuit is complete, a current does exist, and a problem surrounds the resistor R_2. R_2 is not shorted out since such a condition would result in $V_{R_2} = 0$ V. A careful check of the inserted resistor reveals that an 8 Ω resistor was used rather than the 8 kΩ resistor specified—an incorrect reading of the color code. To avoid this, an ohmmeter should be used to check a resistor to validate the color-code reading or to ensure that its value is still in the prescribed range set by the color code.

Occasionally, the problem may be difficult to diagnose. You've checked all the elements, and all the connections appear tight. The supply is on and set at the proper level; the meters appear to be functioning correctly. In situations such as this, experience becomes a key factor. Perhaps you can recall when a recent check of a resistor revealed that the internal connection (not externally visible) was a "make or break" situation or that the resistor was damaged earlier by excessive current levels, so its actual resistance was much lower than called for by the color code. Recheck the supply! Perhaps the terminal voltage was set correctly, but the current control knob was left in the zero or minimum position. Is the ground connection stable? The questions that arise may seem endless. However, as you gain experience, you will be able to localize problems more rapidly. Of course, the more complicated the system, the longer the list of possibilities, but it is often possible to identify a particular area of the system that is behaving improperly before checking individual elements.

6.12 PROTOBOARDS (BREADBOARDS)

In Section 5.12, the protoboard was introduced with the connections for a simple series circuit. To continue the development, the network in Fig. 6.17 was set up on the board in Fig. 6.64(a) using two different techniques. The possibilities are endless, but these two solutions use a fairly straightforward approach.

First, note that the supply lines and ground are established across the length of the board using the horizontal conduction zones at the top and bottom of the board through the connections to the terminals. The network to the left on the board was used to set up the circuit in much the same manner as it appears in the schematic of Fig. 6.64(b). This approach required that the resistors be connected between two vertical conducting

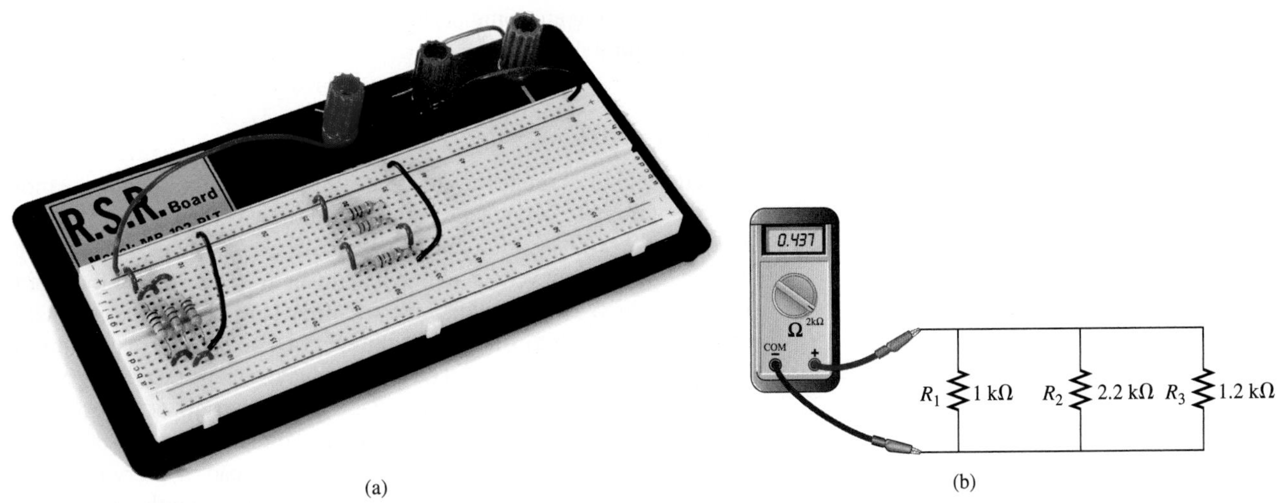

FIG. 6.64
Using a protoboard to set up the circuit in Fig. 6.17.

strips. If placed perfectly vertical in a single conducting strip, the resistors would have shorted out. Often, setting the network up in a manner that best copies the original can make it easier to check and make measurements. The network to the right in part (a) used the vertical conducting strips to connect the resistors together at each end. Since there wasn't enough room for all three, a connection had to be added from the upper vertical set to the lower set. The resistors are in order R_1, R_2, and R_3 from the top-down. For both configurations, the ohmmeter can be connected to the positive lead of the supply terminal and the negative or ground terminal.

Take a moment to review the connections and think of other possibilities. Improvements can often be made, and it can be satisfying to find the most effective setup with the least number of connecting wires.

6.13 APPLICATIONS

One of the most important advantages of the parallel configuration is that

if one branch of the configuration should fail (open circuit), the remaining branches will still have full operating power.

In a home, the parallel connection is used throughout to ensure that if one circuit has a problem and opens the circuit breaker, the remaining circuits still have the full 120 V. The same is true in automobiles, computer systems, industrial plants, and wherever it would be disastrous for one circuit to control the total power distribution.

Another important advantage is that

branches can be added at any time without affecting the behavior of those already in place.

In other words, unlike the series connection where an additional component reduces the current level and perhaps affects the response of some of the existing components, an additional parallel branch will not affect the current level in the other branches. Of course, the current demand from the supply increases as determined by Kirchhoff's current law, so you must be aware of the limitations of the supply.

The following are some of the most common applications of the parallel configuration.

Car System

As you begin to examine the electrical system of an automobile, the most important thing to understand is that the entire electrical system of a car is run as a *dc system.* Although the generator produces a varying ac signal, rectification converts it to one having an average dc level for charging the battery. In particular, note the filter capacitor in the alternator branch in Fig. 6.65 to smooth out the rectified ac waveform and to provide an improved dc supply. The charged battery must therefore provide the required direct current for the entire electrical system of the car. Thus, the power demand on the battery at any instant is the product of the terminal voltage and the current drain of the total load of every operating system of the car. This certainly places an enormous burden on the battery and its internal chemical reaction and warrants all the battery care we can provide.

Since the electrical system of a car is essentially a parallel system, the total current drain on the battery is the sum of the currents to all the parallel branches of the car connected directly to the battery. In Fig. 6.65, a few branches of the wiring diagram for a car have been sketched to provide some background information on basic wiring, current levels, and fuse configurations. Every automobile has fuse links and fuses, and some also have circuit breakers, to protect the various components of the car and to ensure that a dangerous fire situation does not develop. Except for a few branches that may have series elements, the operating voltage for most components of a car is the terminal voltage of the battery which we will designate as 12 V even though it will typically vary between 12 V and the charging level of 14.6 V. In other words, each component is connected to the battery at one end and to the ground or chassis of the car at the other end.

Referring to Fig. 6.65, note that the alternator or charging branch of the system is connected directly across the battery to provide the charging current as indicated. Once the car is started, the rotor of the alternator turns,

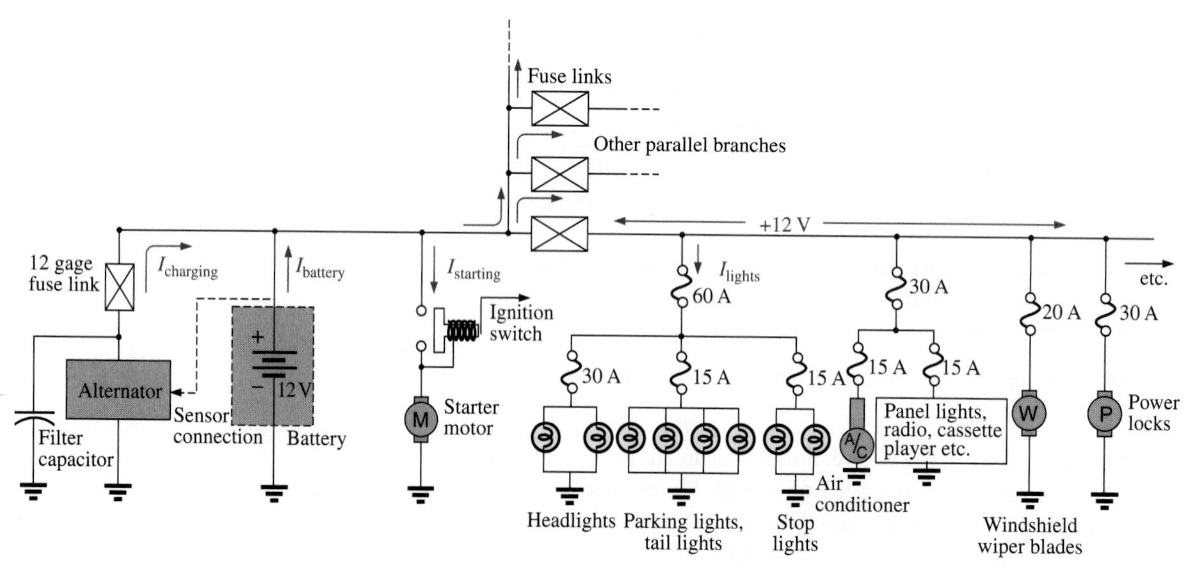

FIG. 6.65
Expanded view of an automobile's electrical system.

generating an ac varying voltage which then passes through a rectifier network and filter to provide the dc charging voltage for the battery. Charging occurs only when the sensor, connected directly to the battery, signals that the terminal voltage of the battery is too low. Just to the right of the battery the starter branch was included to demonstrate that there is no fusing action between the battery and starter when the ignition switch is activated. The lack of fusing action is provided because enormous starting currents (hundreds of amperes) flow through the starter to start a car that has not been used for days and/or has been sitting in a cold climate—and high friction occurs between components until the oil starts flowing. The starting level can vary so much that it would be difficult to find the right fuse level, and frequent high currents may damage the fuse link and cause a failure at expected levels of current. When the ignition switch is activated, the starting relay completes the circuit between the battery and starter, and hopefully the car starts. If a car fails to start, the first thing to check is the connections at the battery, starting relay, and starter to be sure that they are not providing an unexpected open circuit due to vibration, corrosion, or moisture.

Once the car has started, the starting relay opens, and the battery begins to activate the operating components of the car. Although the diagram in Fig. 6.65 does not display the switching mechanism, the entire electrical network of the car, except for the important external lights, is usually disengaged so that the full strength of the battery can be dedicated to the starting process. The lights are included for situations where turning the lights off, even for short periods of time, could create a dangerous situation. If the car is in a safe environment, it is best to leave the lights off when starting to save the battery an additional 30 A of drain. If the lights are on, they dim because of the starter drain, which may exceed 500 A. Today, batteries are typically rated in cranking (starting) current rather than ampere-hours. Batteries rated with cold cranking ampere ratings between 700 A and 1000 A are typical today.

Separating the alternator from the battery and the battery from the numerous networks of the car are fuse links such as shown in Fig. 6.66. Fuse links are actually wires of a specific gage designed to open at fairly high current levels of 100 A or more. They are included to protect against those situations where there is an unexpected current drawn from the many circuits to which they are connected. That heavy drain can, of course, be from a short circuit in one of the branches, but in such cases the fuse in that branch will probably release. The fuse link is an additional protection for the line if the total current drawn by the parallel-connected branches begins to exceed safe levels. The fuses following the fuse link have the appearance shown in Fig. 6.66(b), where a gap between the legs of the fuse

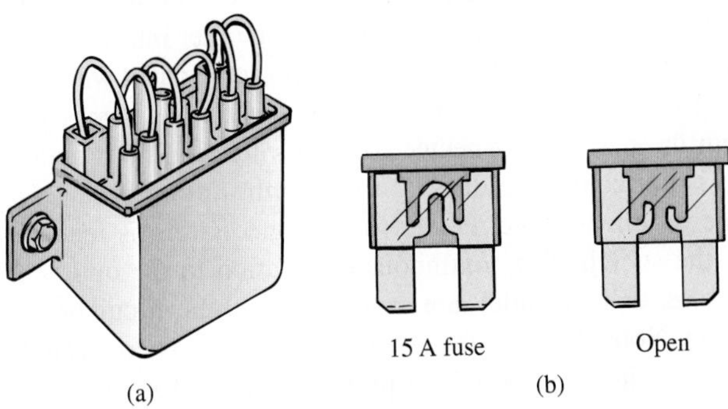

FIG. 6.66
Car fuses: (a) fuse link; (b) plug-in.

indicates a blown fuse. As shown in Fig. 6.65, the 60 A fuse (often called a *power distribution fuse*) for the lights is a second-tier fuse sensitive to the total drain from the three light circuits. Finally, the third fuse level is for the individual units of a car such as the lights, air conditioner, and power locks. In each case, the fuse rating exceeds the normal load (current level) of the operating component, but the level of each fuse does give some indication of the demand to be expected under normal operating conditions. For instance, headlights typically draw more than 10 A, tail lights more than 5 A, air conditioner about 10 A (when the clutch engages), and power windows 10 A to 20 A depending on how many are operated at once.

Some details for only one section of the total car network are provided in Fig. 6.65. In the same figure, additional parallel paths with their respective fuses have been provided to further reveal the parallel arrangement of all the circuits.

In most vehicles the return path to the battery through the ground connection is through the chassis of the car. That is, there is only one wire to each electrical load, with the other end simply grounded to the chassis. The return to the battery (chassis to negative terminal) is therefore a heavy-gage wire matching that connected to the positive terminal. In some cars constructed of a mixture of materials such as metal, plastic, and rubber, the return path through the metallic chassis may be lost, and two wires must be connected to each electrical load of the car.

House Wiring

In Chapter 4, the basic power levels of importance were discussed for various services to the home. We are now ready to take the next step and examine the actual connection of elements in the home.

First, it is important to realize that except for some very special circumstances, the basic wiring is done in a parallel configuration. Each parallel branch, however, can have a combination of parallel and series elements. Every full branch of the circuit receives the full 120 V or 208 V, with the current determined by the applied load. Figure 6.67(a) provides the detailed wiring of a single circuit having a light bulb and two outlets. Figure 6.67(b) shows the schematic representation. Note that although each load is in parallel with the supply, switches are always connected in series with the load. The power is transmitted to the lamp only when the switch is closed and the full 120 V appears across the bulb. The connection point for the two outlets is in the ceiling box holding the light bulb. Since a switch is not present, both outlets are always "hot" unless the circuit breaker in the main panel is opened. This is important to understand in case you are tempted to change the light fixture by simply turning off the wall switch. True, if you're very careful, you can work with one line at a time (being sure that you don't touch the other line at any time), but it is much safer to throw the circuit breaker on the panel whenever working on a circuit. Note in Fig. 6.67(a) that the *feed* wire (black) into the fixture from the panel is connected to the switch and both outlets at one point. It is not connected directly to the light fixture because the lamp would be on all the time. Power to the light fixture is made available through the switch. The continuous connection to the outlets from the panel ensures that the outlets are "hot" whenever the circuit breaker in the panel is on. Note also how the *return* wire (white) is connected directly to the light switch and outlets to provide a return for each component. There is no need for the white wire to go through the switch since an applied voltage is a two-point connection and the black wire is controlled by the switch.

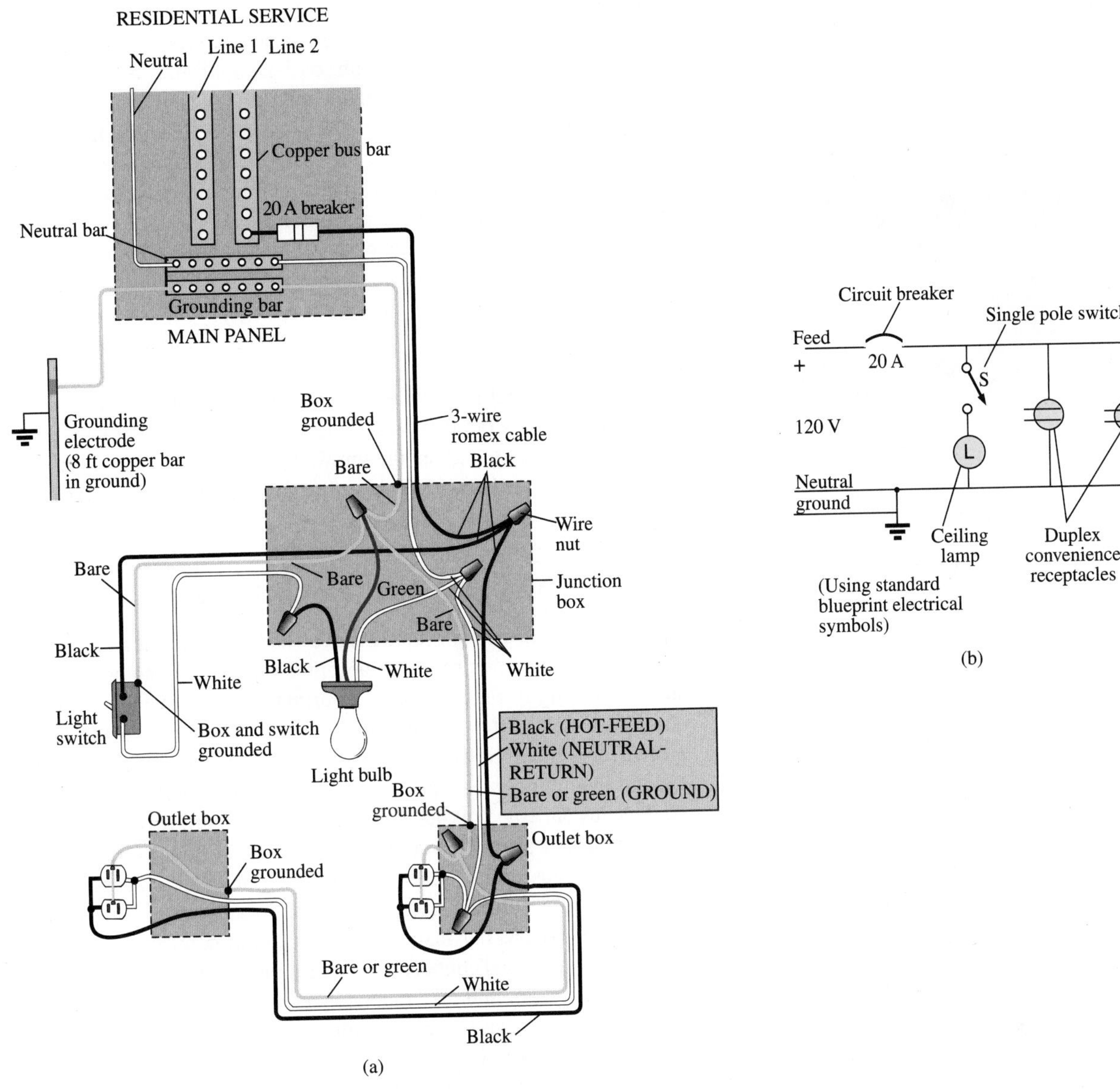

FIG. 6.67
Single phase of house wiring: (a) physical details; (b) schematic representation.

Proper grounding of the system in total and of the individual loads is one of the most important facets in the installation of any system. There is a tendency at times to be satisfied that the system is working and to pay less attention to proper grounding technique. Always keep in mind that a properly grounded system has a direct path to ground if an undesirable situation should develop. The absence of a direct ground causes the system to determine its own path to ground, and you could be that path if you happened to touch the wrong wire, metal box, metal pipe, and so on. In Fig. 6.67(a), the connections for the ground wires have been included. For the romex (plastic-coated wire) used in Fig. 6.67(a), the ground wire is provided as a bare copper wire. Note that it is connected to the panel which in turn is directly connected to the grounded 8-ft copper rod. In addition, note that the ground connection is carried through the entire circuit, including the switch, light fixture, and outlets. It is one continuous connection. If the outlet box, switch box, and housing for the light fixture

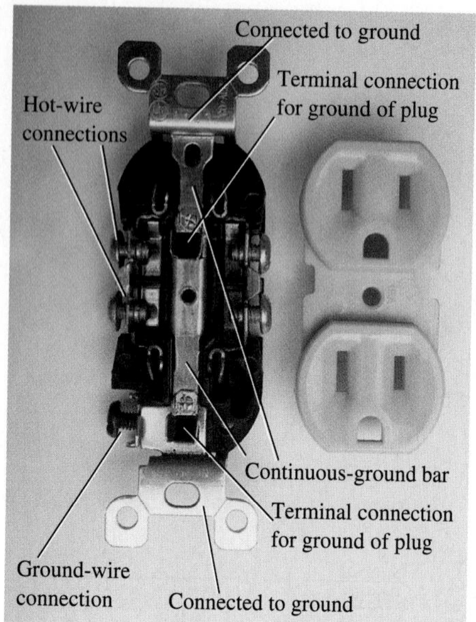

FIG. 6.68
Continuous ground connection in a duplex outlet.

are made of a conductive material such as metal, the ground will be connected to each. If each is plastic, there is no need for the ground connection. However, the switch, both outlets, and the fixture itself are connected to ground. For the switch and outlets, there is usually a green screw for the ground wire which is connected to the entire framework of the switch or outlet as shown in Fig. 6.68, including the ground connection of the outlet. For both the switch and the outlet, even the screw or screws used to hold the outside plate in place are grounded since they are screwed into the metal housing of the switch or outlet. When screwed into a metal box, the ground connection can be made by the screws that hold the switch or outlet in the box as shown in Fig. 6.68. *Always pay strict attention to the grounding process whenever installing any electrical equipment.*

On the practical side, whenever hooking up a wire to a screw-type terminal, always wrap the wire around the screw in the clockwise manner so that when you tighten the screw, it grabs the wire and turns it in the same direction. An expanded view of a typical house-wiring arrangement appears in Chapter 15.

Parallel Computer Bus Connections

The internal construction (hardware) of large mainframe computers and personal computers is set up to accept a variety of adapter cards in the slots appearing in Fig. 6.69(a). The primary board (usually the largest), commonly called the *motherboard,* contains most of the functions required for full computer operation. Adapter cards are normally added to expand the memory, set up a network, add peripheral equipment, and so on. For instance, if you decide to add another hard drive to your computer, you can simply insert the card into the proper channel of Fig. 6.69(a). The bus connectors are connected in parallel with common connections to the power supply, address and data buses, control signals, ground, and so on. For instance, if the bottom connection of each bus connector is a ground connection, that ground connection carries through each bus connector and is immediately connected to any adapter card installed. Each card has a slot connector that fits directly into the bus connector without the need for any soldering or construction. The pins of the adapter card are then designed to provide a path between the motherboard and its components to support the desired function. Note in Fig. 6.69(b), which is a back view of the region identified in Fig. 6.69(a), that if you follow the path of the second pin from the top on the far left, you will see that it is connected to the same pin on the other three bus connectors.

Most small laptop computers today have all the options already installed, thereby bypassing the need for bus connectors. Additional memory and other upgrades are added as direct inserts into the motherboard.

6.14 COMPUTER ANALYSIS

PSpice

Parallel dc Network The computer analysis coverage for parallel dc circuits is very similar to that for series dc circuits. However, in this case the voltage is the same across all the parallel elements, and the current through each branch changes with the resistance value. The parallel network to be analyzed will have a wide range of resistor values to demonstrate the effect on the resulting current. The following is

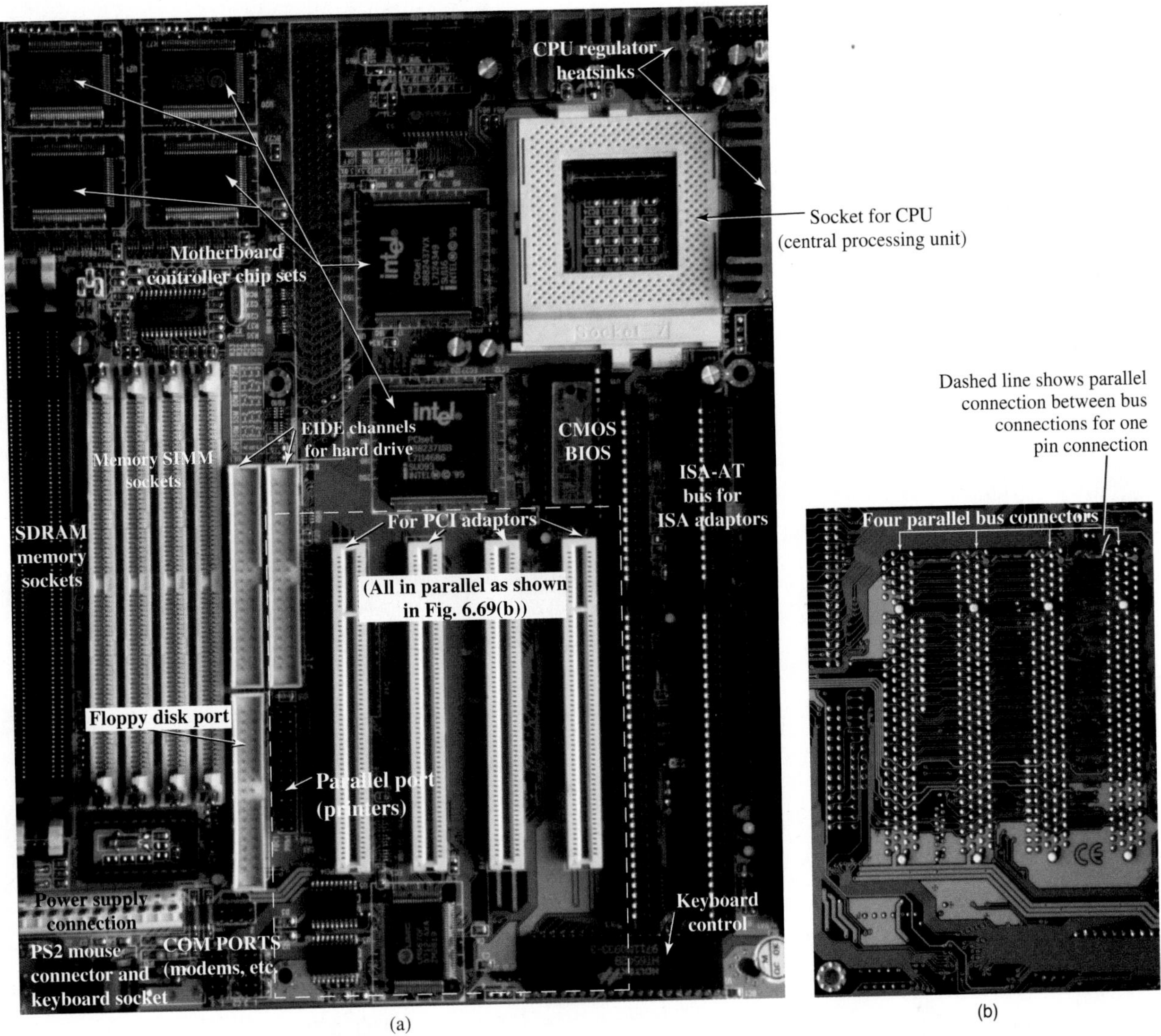

FIG. 6.69
(a) Motherboard for a desktop computer; (b) the printed circuit board connections for the region indicated in part (a).

a list of abbreviations for any parameter of a network when using PSpice:

$$\mathbf{f} = 10^{-15}$$
$$\mathbf{p} = 10^{-12}$$
$$\mathbf{n} = 10^{-9}$$
$$\mathbf{u} = 10^{-6}$$
$$\mathbf{m} = 10^{-3}$$
$$\mathbf{k} = 10^{+3}$$
$$\mathbf{MEG} = 10^{+6}$$
$$\mathbf{G} = 10^{+9}$$
$$\mathbf{T} = 10^{+12}$$

In particular, note that **m** (or **M**) is used for "milli," and **MEG** for "megohms." Also, PSpice does not distinguish between upper- and lowercase units, but certain parameters typically use either the upper- or lowercase abbreviation as shown above.

IPI

Since the details of setting up a network and going through the simulation process were covered in detail in Sections 4.9 and 5.14 for dc circuits, the coverage here is limited solely to the various steps required. These steps will help you learn how to "draw" a circuit and then run a simulation fairly quickly and easily.

After selecting the **Create document** key (the top left of the screen), the following sequence opens the **Schematic** window: **PSpice 6-1-OK-Create a blank project-OK-PAGE1** (if necessary).

Add the voltage source and resistors as described in detail in earlier sections, but now you need to turn the resistors 90°. You do this by right-clicking before setting a resistor in place. Choose **Rotate** from the list of options, which turns the resistor counterclockwise 90°. It can also be rotated by simultaneously selecting **Ctrl-R.** The resistor can then be placed in position by a left click. An additional benefit of this maneuver is that the remaining resistors to be placed will already be in the vertical position. The values selected for the voltage source and resistors appear in Fig. 6.70.

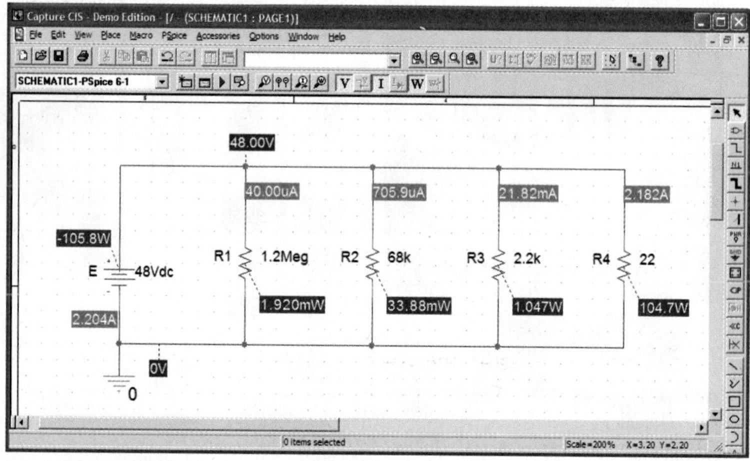

FIG. 6.70
Applying PSpice to a parallel network.

Once the network is complete, you can obtain the simulation and the results through the following sequence: **Select New Simulation Profile** key-**Bias Point-Create-Analysis-Bias Point-OK-Run PSpice** key-**Exit(X).**

The result, shown in Fig. 6.70, reveals that the voltage is the same across all the parallel elements and the current increases significantly with decrease in resistance. The range in resistor values suggests, by inspection, that the total resistance is just less than the smallest resistance of 22 Ω. Using Ohm's law and the source current of 2.204 A results in a total resistance of $R_T = E/I_s = 48\text{ V}/2.204\text{ A} = 21.78\ \Omega$, confirming the above conclusion.

Multisim

Parallel dc Network For comparison purposes with the PSpice approach, the same parallel network in Fig. 6.70 is now analyzed using Multisim. The source and ground are selected and placed as shown in Fig. 6.71 using the procedure defined in previous chapters. For the resistors, choose **VIRTUAL_RESISTOR,** but you must rotate it 90° to match the configuration of Fig. 6.70. You do this by first clicking on the resistor symbol to place it in the active state. (Be sure that the resulting small black squares surround the symbol, label, and value; otherwise, you may have activated only the label or value.) Then right-click inside the rectangle. Select **90° Clockwise,** and the resistor is turned automatically. Unfortunately, there is no continuum here, so the next resistor has to be turned using the same procedure. The values of each resistor are set by double-clicking on the resistor symbol to obtain the **BASIC-VIRTUAL** dialog box. Remember that the unit of measurement is controlled by the scrolls at the right of the unit of measurement. For Multisim, unlike PSpice, megohm uses capital **M** and milliohm uses lowercase **m.**

This time, rather than using meters to make the measurements, you will use indicators. The **Indicators** key pad is the tenth down on the left toolbar. It has the appearance of an LCD display with the number 8. Select it, and eight possible indicators appear. For this example, the **AMMETER** indicator, representing an ammeter, is used since we are interested only in the current levels. When you select **AMMETER,** a **Component** listing appears with four choices, with each option referring to a position for the ammeter. The **H** means "horizontal" as shown in the picture window when the dialog box is first opened. The **HR** means "horizontal," but with the polarity reversed. The **V** is for a vertical configuration with the positive sign at the top, and the **VR** is the vertical position with the positive sign at the bottom. Select the one you want followed by an **OK,** and your choice appears in that position on the screen. Click it into position, and you can return for the next indicator. Once all the elements are in place and their values set, initiate simulation with the sequence **Simulate-Run.** The results shown in Fig. 6.71 appear.

Note that all the results appear with the indicator boxes. All are positive results because the ammeters were all entered with a configuration that would result in conventional current entering the positive current. Also note that as was true for inserting the meters, the indicators are placed in series with the branch in which the current is to be measured.

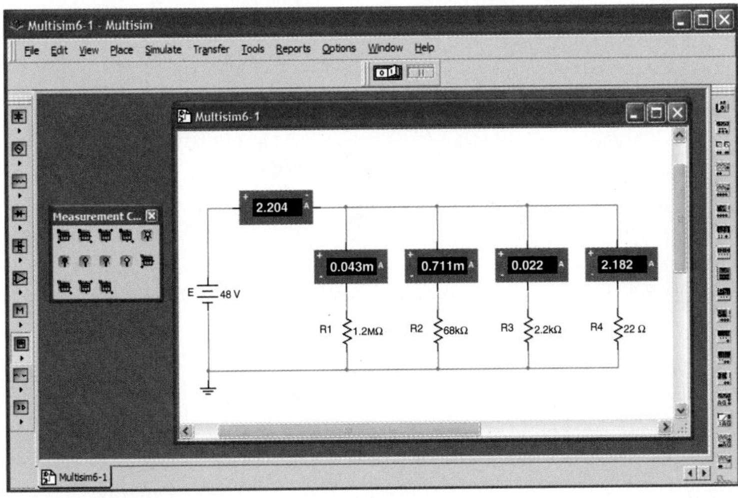

FIG. 6.71
Using the indicators of Multisim to display the currents of a parallel dc network.

IPI

PROBLEMS

SECTION 6.2 Parallel Resistors

1. For each configuration in Fig. 6.72, find the voltage sources and/or resistors elements (individual elements, not combinations of elements) that are in parallel. Remember that elements in parallel have the same voltage.

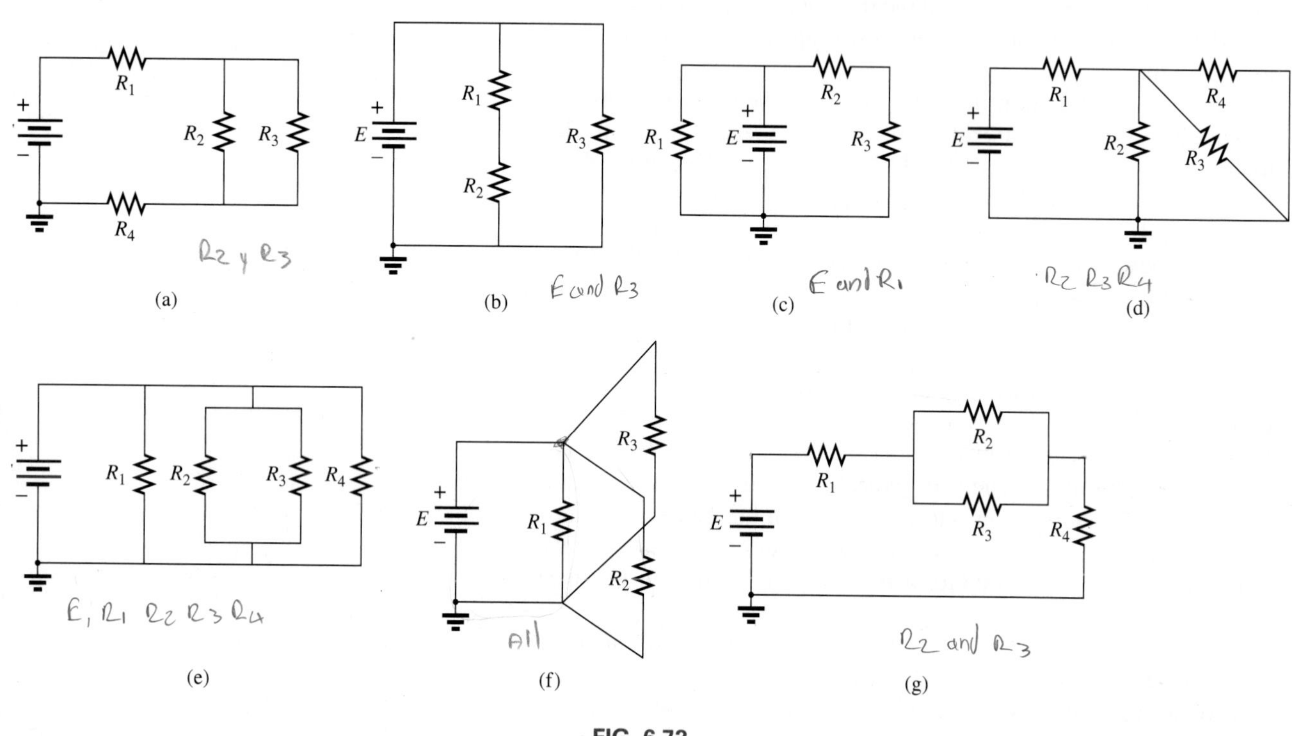

FIG. 6.72
Problem 1.

2. For the network in Fig. 6.73:
   a. Find the elements (voltage sources and/or resistors) that are in parallel.
   b. Find the elements (voltage sources and/or resistors) that are in series.
3. Find the total resistance for each configuration in Fig. 6.74. Note that only standard value resistors were used.
4. For each circuit board in Fig. 6.75, find the total resistance between connection tabs 1 and 2.
5. The total resistance of each of the configurations in Fig. 6.76 is specified. Find the unknown resistance.

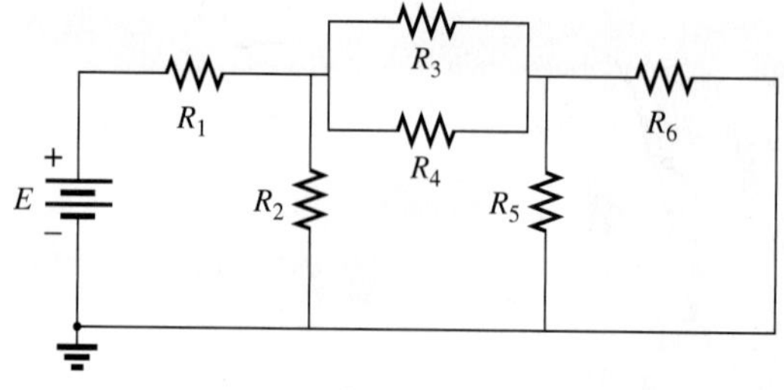

FIG. 6.73
Problem 2.

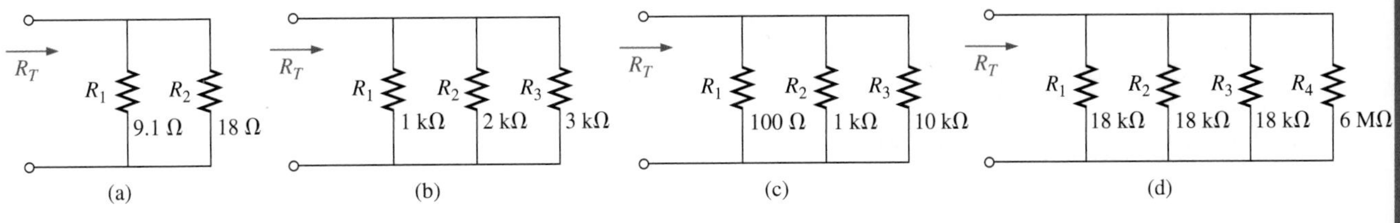

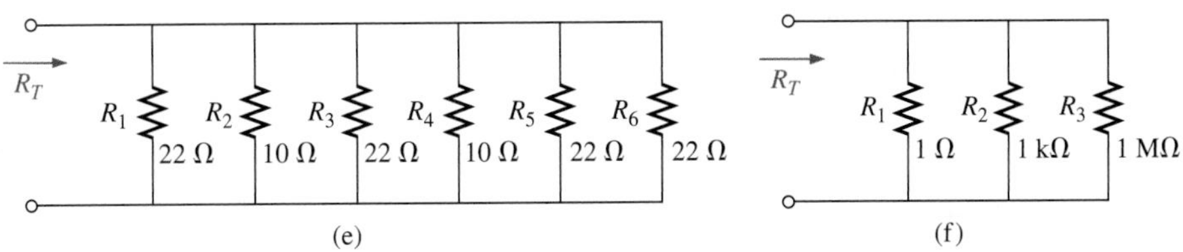

FIG. 6.74
Problem 3.

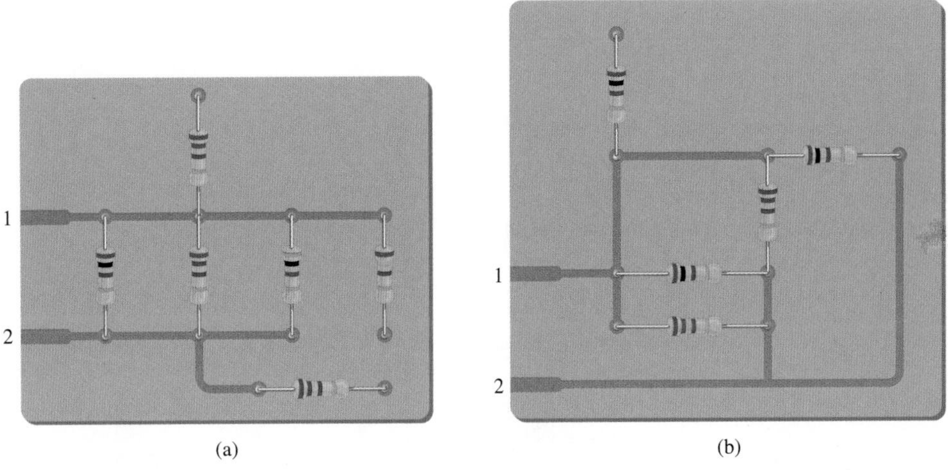

FIG. 6.75
Problem 4.

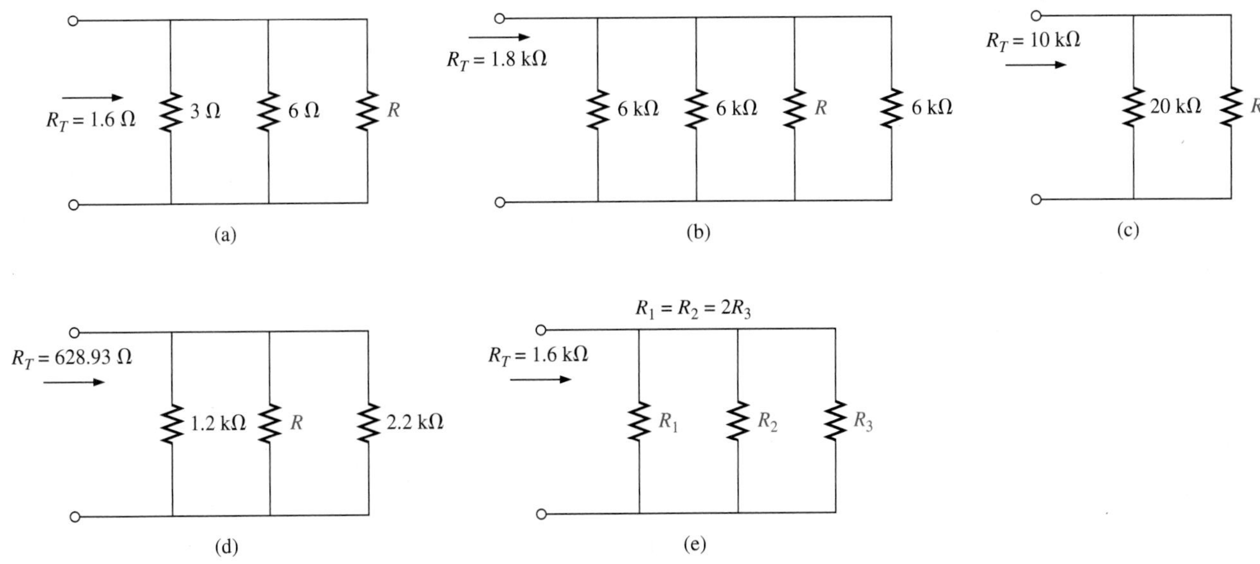

FIG. 6.76
Problem 5.

6. For the parallel network in Fig. 6.77, composed of standard values:
   a. Which resistor has the most impact on the total resistance?
   b. Without making a single calculation, what is an approximate value for the total resistance?
   c. Calculate the total resistance and comment on your response to part (b).
   d. On an approximate basis, which resistors can be ignored when determining the total reistance?
   e. If we add another parallel resistor of any value to the network, what is the impact on the total resistance?

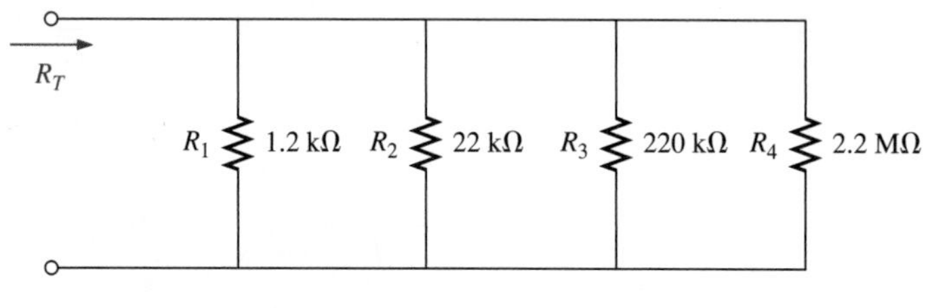

FIG. 6.77
Problem 6.

7. What is the ohmmeter reading for each configuration in Fig. 6.78?

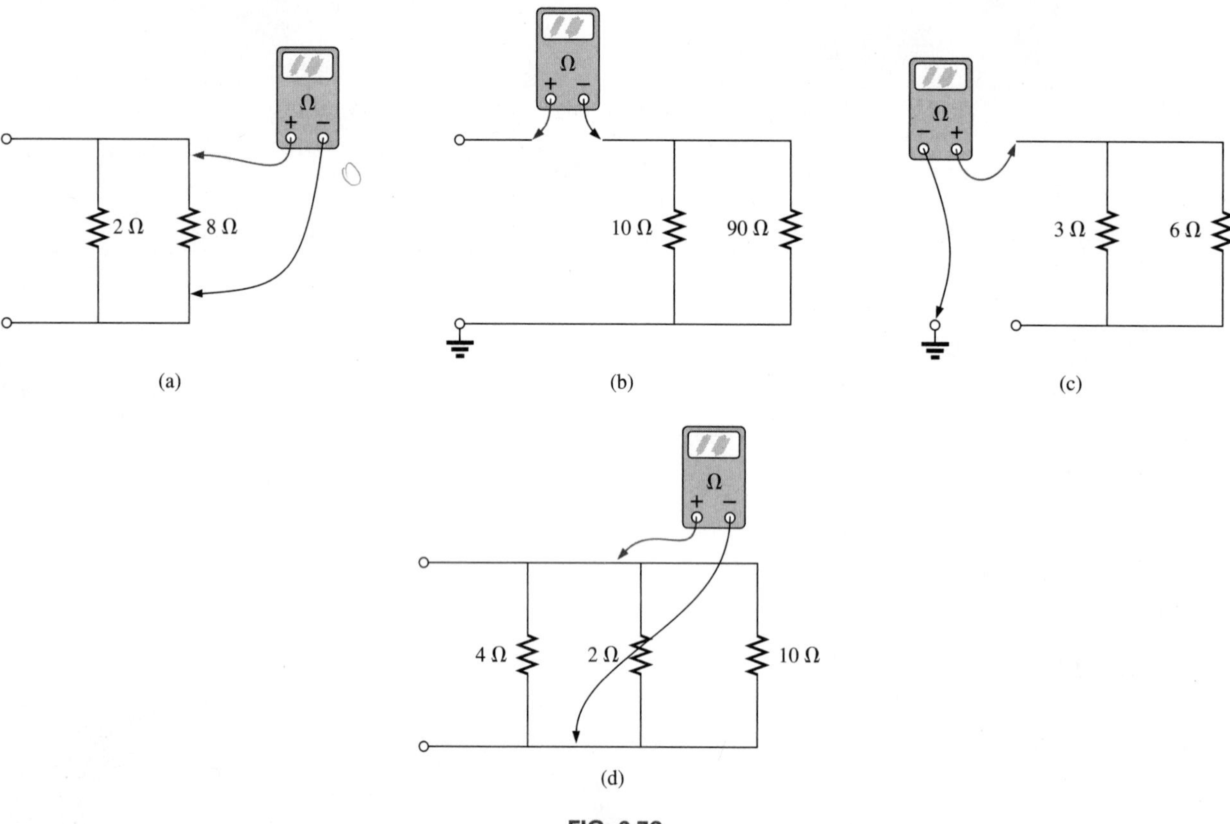

FIG. 6.78
Problem 7.

*8. Determine the unknown resistors in Fig. 6.79 given the fact that $R_2 = 5R_1$ and $R_3 = (1/2)R_1$.

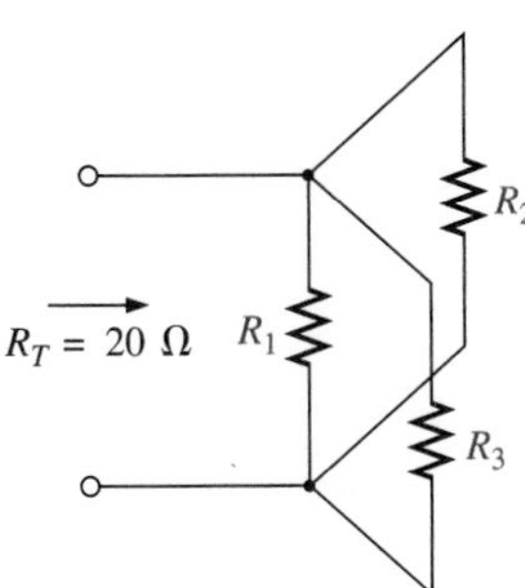

FIG. 6.79
Problem 8.

*9. Determine R_1 for the network in Fig. 6.80.

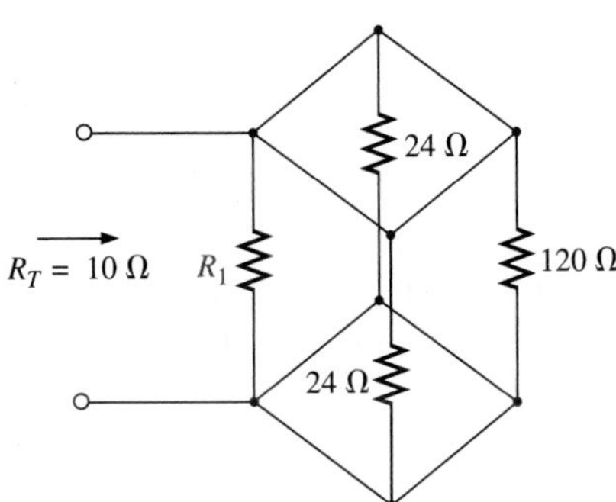

FIG. 6.80
Problem 9.

SECTION 6.3 Parallel Circuits

10. For the parallel network in Fig. 6.81:
   a. Find the total resistance.
   b. What is the voltage across each branch?
   c. Determine the source current and the current through each branch.
   d. Verify that the source current equals the sum of the branch currents.

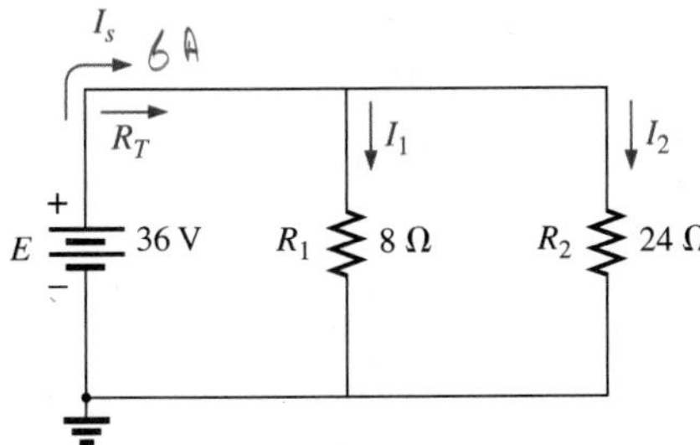

FIG. 6.81
Problem 10.

11. Repeat the analysis of Problem 10 for the network in Fig. 6.82.

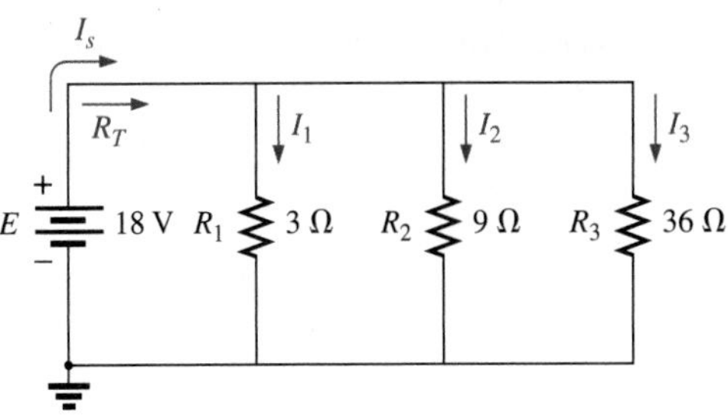

FIG. 6.82
Problems 11 and 14.

12. Repeat the analysis of Problem 10 for the network in Fig. 6.83, constructed of standard value resistors.

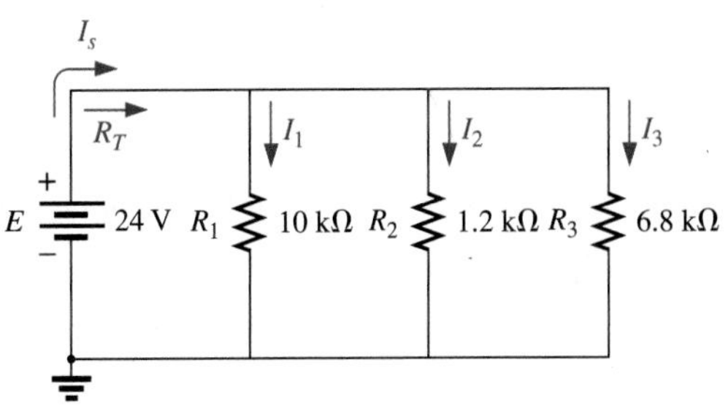

FIG. 6.83
Problem 12.

13. For the parallel network in Fig. 6.84:
   a. Without making a single calculation, make a guess on the total resistance.
   b. Calculate the total resistance and compare it to your guess in part (a).
   c. Without making a single calculation, which branch will have the most current? Which will have the least?
   d. Calculate the current through each branch, and compare your results to the assumptions of part (c).
   e. Find the source current and test whether it equals the sum of the branch currents.
   f. How does the magnitude of the source current compare to that of the branch currents?

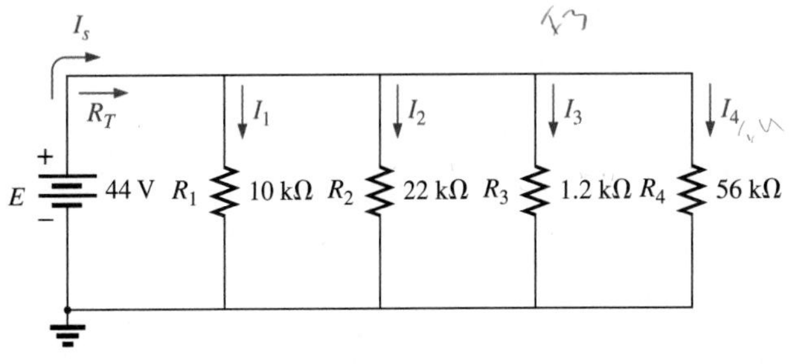

FIG. 6.84
Problem 13.

14. For the network in Fig. 6.82:
    a. Redraw the network and insert ammeters to measure the source current and the current through each branch.
    b. Connect a voltmeter to measure the source voltage and the voltage across resistor, R_3. Is there any difference in the connections? Why?

15. What is the response of the voltmeter and ammeters connected in Fig. 6.85?

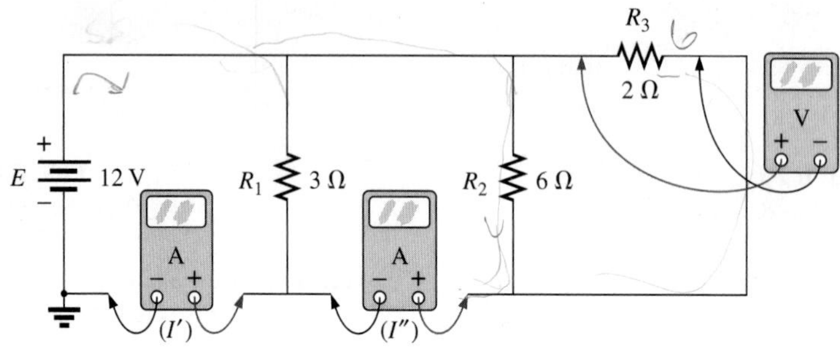

FIG. 6.85
Problem 15.

16. Given the information provided in Fig. 6.86, find the unknown quantities: E, R_1, and I_3.

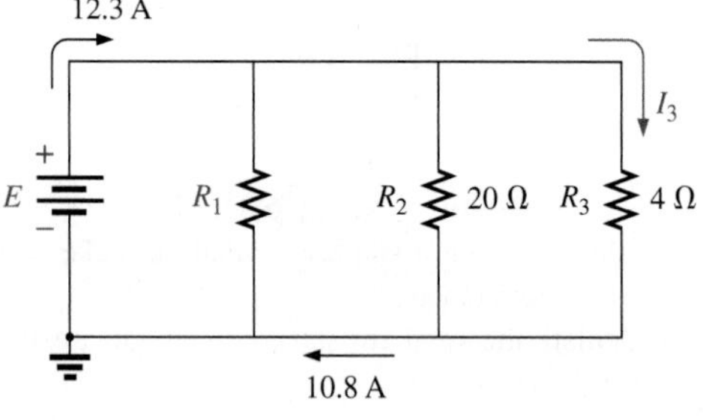

FIG. 6.86
Problem 16.

17. Determine the currents I_1 and I_s for the networks in Fig. 6.87.

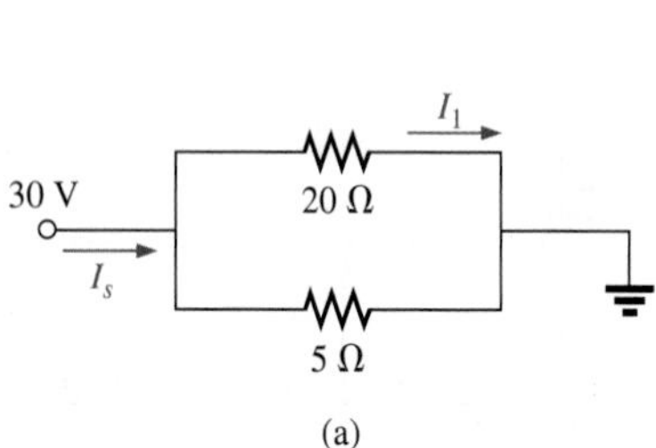

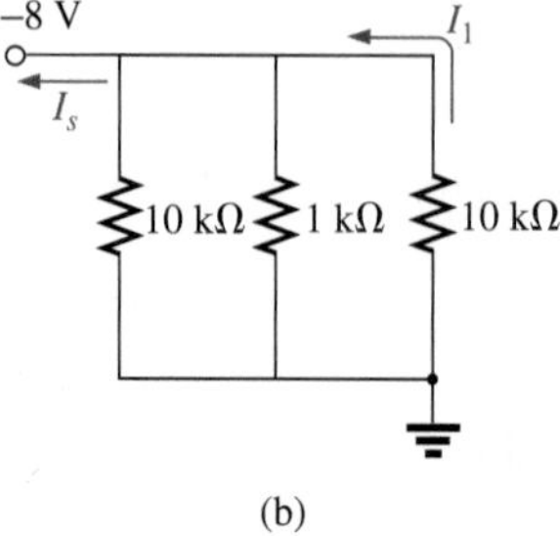

FIG. 6.87
Problem 17.

*18. For the network in Fig. 6.88:
    a. Find the current I.
    b. Determine the voltage V.
    c. Calculate the source current I_s.

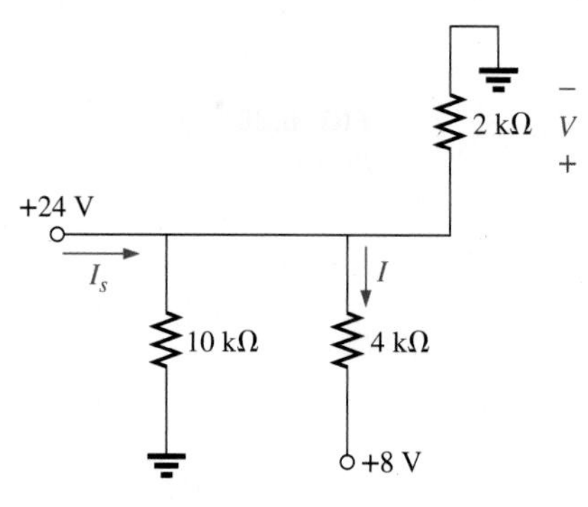

FIG. 6.88
Problem 18.

SECTION 6.4 Power Distribution in a Parallel Circuit

19. For the configuration in Fig. 6.89:
  a. Find the total resistance and the current through each branch.
  b. Find the power delivered to each resistor.
  c. Calculate the power delivered by the source.
  d. Compare the power delivered by the source to the sum of the powers delivered to the resistors.
  e. Which resistor received the most power? Why?

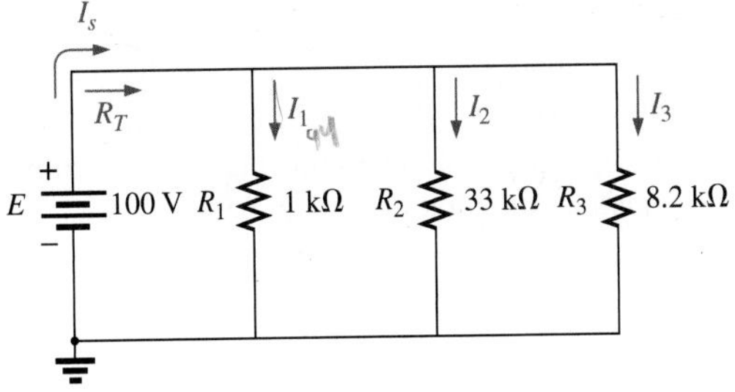

FIG. 6.89
Problem 19.

20. Eight holiday lights are connected in parallel as shown in Fig. 6.90.
  a. If the set is connected to a 120 V source, what is the current through each bulb if each bulb has an internal resistance of 1.8 kΩ?
  b. Determine the total resistance of the network.
  c. Find the current drain from the supply.
  d. What is the power delivered to each bulb?
  e. Using the results of part (d), what is the power delivered by the source?
  f. If one bulb burns out (that is, the filament opens up), what is the effect on the remaining bulbs? What is the effect on the source current? Why?

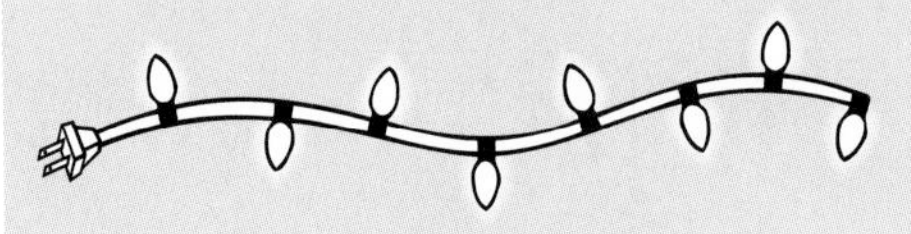

FIG. 6.90
Problem 20.

21. Determine the power delivered by the dc battery in Fig. 6.91.

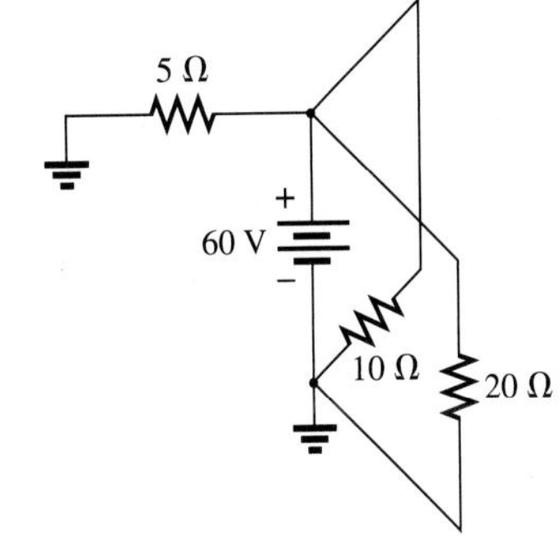

FIG. 6.91
Problem 21.

22. A portion of a residential service to a home is depicted in Fig. 6.92.
  a. Determine the current through each parallel branch of the system.
  b. Calculate the current drawn from the 120 V source. Will the 20 A breaker trip?
  c. What is the total resistance of the network?
  d. Determine the power delivered by the source. How does it compare to the sum of the wattage ratings appearing in Fig. 6.92?

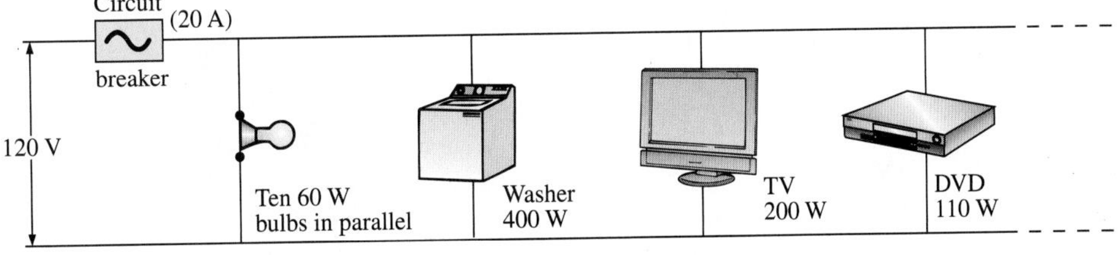

FIG. 6.92
Problem 22.

***23.** For the network in Fig. 6.93:
  a. Find the current I_1.
  b. Calculate the power dissipated by the 4 Ω resistor.
  c. Find the current I_2.

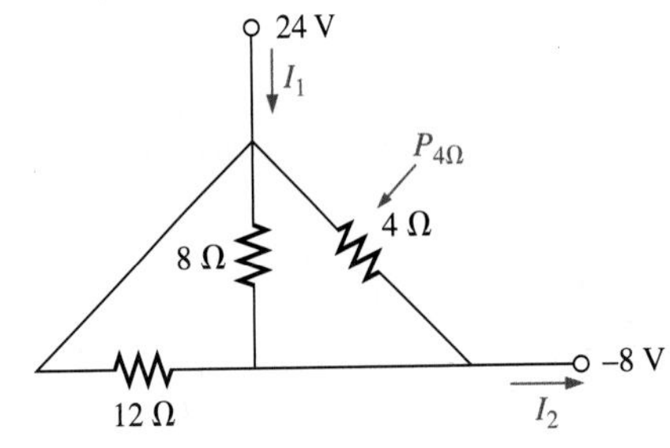

FIG. 6.93
Problem 23.

SECTION 6.5 Kirchhoff's Current Law

24. Using Kirchhoff's current law, determine the unknown currents for the parallel network in Fig. 6.94.

25. Using Kirchoff's current law, find the unknown currents for the complex configurations in Fig. 6.95.

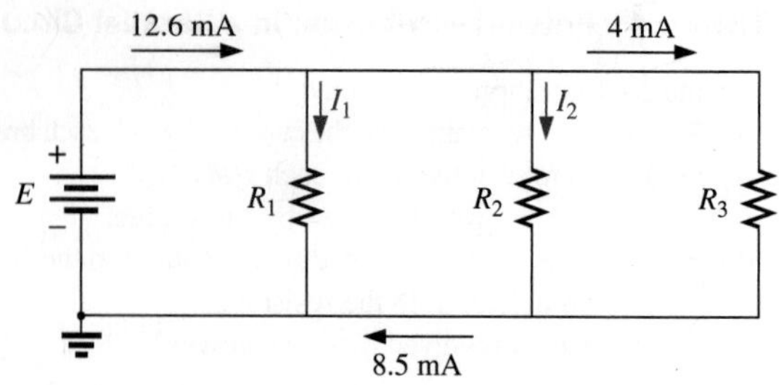

FIG. 6.94
Problem 24.

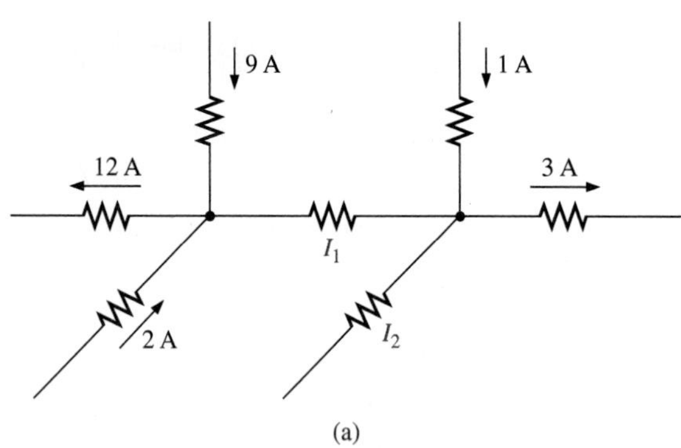

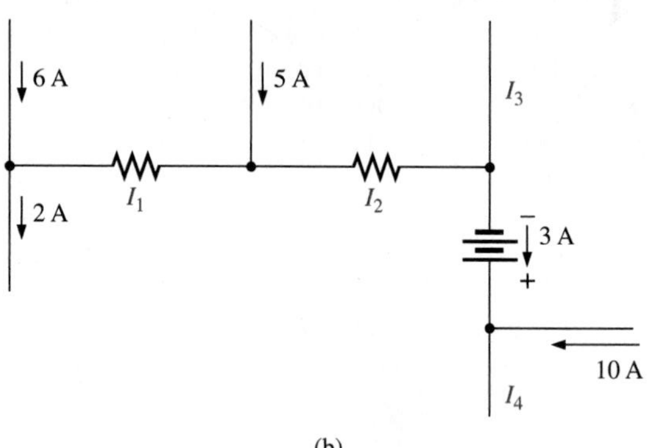

FIG. 6.95
Problem 25.

26. Using Kirchhoff's current law, determine the unknown currents for the networks in Fig. 6.96.

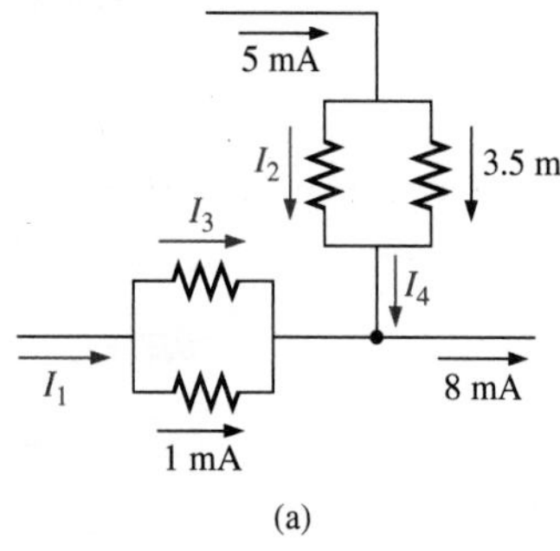

FIG. 6.96
Problem 26.

27. Using the information provided in Fig. 6.97, find the branch resistors R_1 and R_3, the total resistance R_T, and the voltage source E.

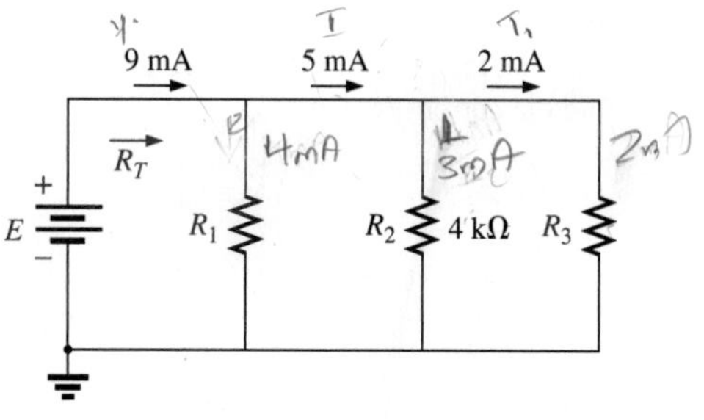

FIG. 6.97
Problem 27.

28. Find the unknown quantities for the networks in Fig. 6.98 using the information provided.

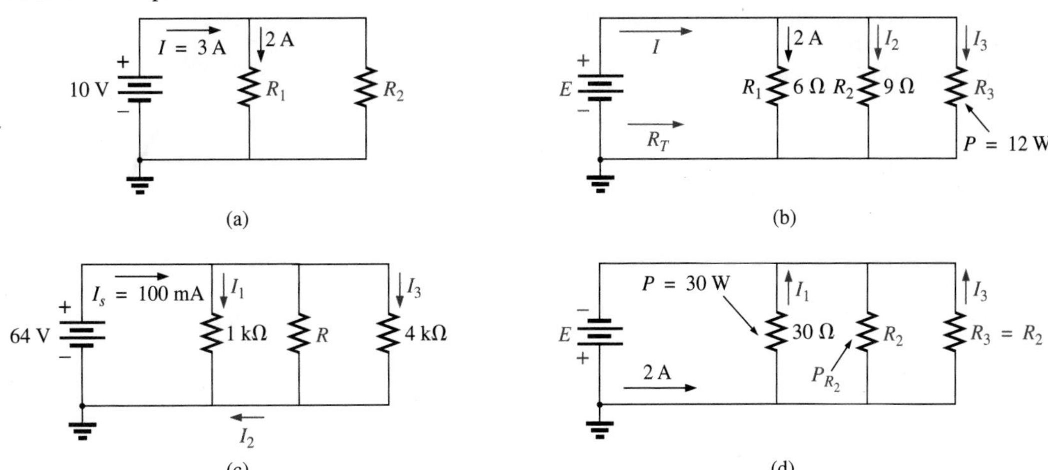

FIG. 6.98
Problem 28.

SECTION 6.6 Current Divider Rule

29. Based solely on the resistor values, determine all the currents for the configuration in Fig. 6.99. Do not use Ohm's law.

30. Determine the currents for the configurations in Fig. 6.100.

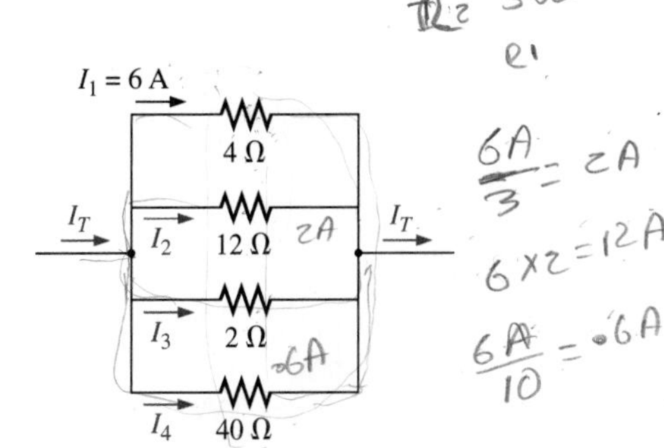

FIG. 6.99
Problem 29.

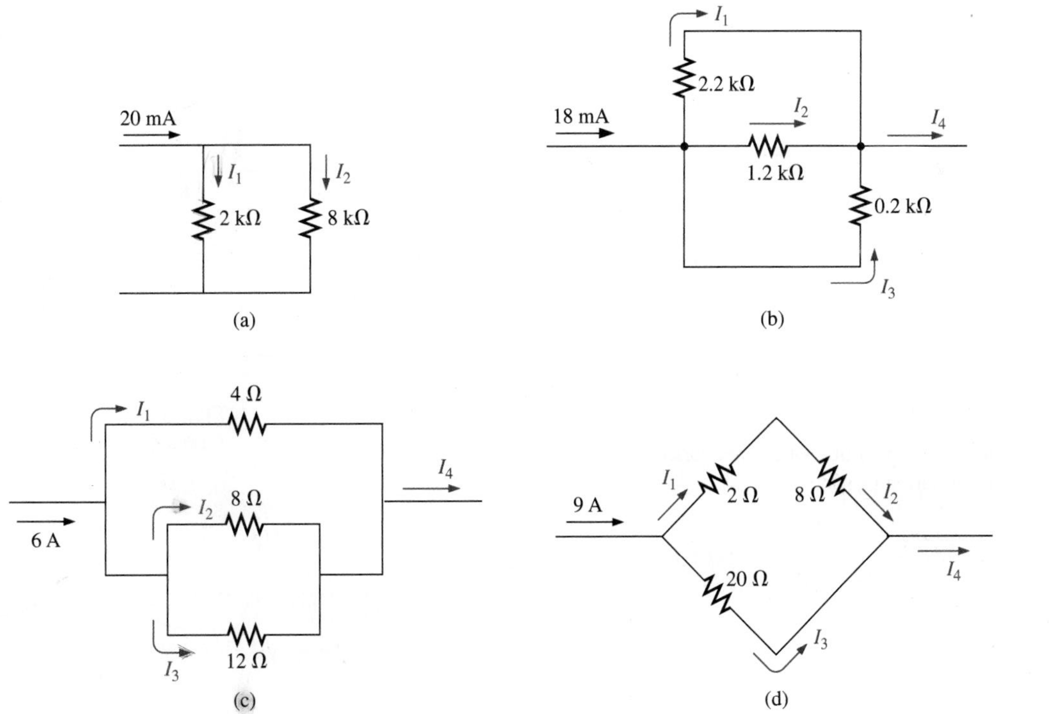

FIG. 6.100
Problem 30.

31. Parts (a) through (e) of this problem should be done by inspection—that is, mentally. The intent is to obtain an approximate solution without a lengthy series of calculations. For the network in Fig. 6.101:

a. What is the approximate value of I_1 considering the magnitude of the parallel elements?

b. What is the ratio I_1/I_2? Using the result of part (a), what is an approximate value of I_2?

c. What is the ratio I_1/I_3? Using the result, what is an approximate value of I_3?

d. What is the ratio I_1/I_4? Using the result, what is an approximate value of I_4?

e. What is the effect of the parallel 100 kΩ resistor on the above calculations? How much smaller will the current I_4 be than the current I_1?

f. Calculate the current through the 1 Ω resistor using the current divider rule. How does it compare to the result of part (a)?

g. Calculate the current through the 10 Ω resistor. How does it compare to the result of part (b)?

h. Calculate the current through the 1 kΩ resistor. How does it compare to the result of part (c)?

i. Calculate the current through the 100 kΩ resistor. How does it compare to the solutions to part (e)?

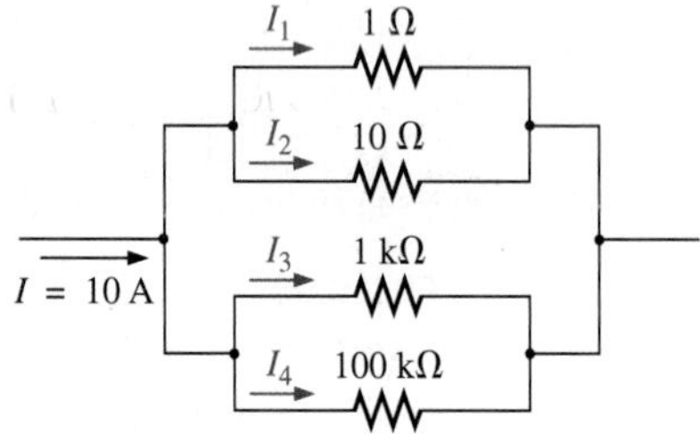

FIG. 6.101
Problem 31.

32. Find the unknown quantities for the networks in Fig. 6.102 using the information provided.

33. a. Find resistor R for the network in Fig. 6.103 that will ensure that $I_1 = 3I_2$.

b. Find I_1 and I_2.

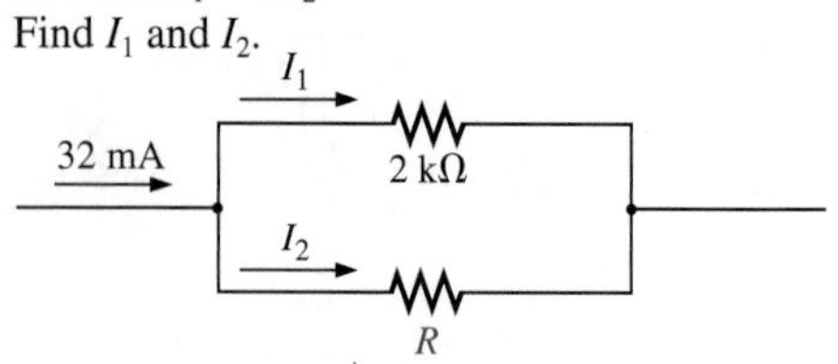

FIG. 6.103
Problem 33.

34. Design the network in Fig. 6.104 such that $I_2 = 2I_1$ and $I_3 = 2I_2$.

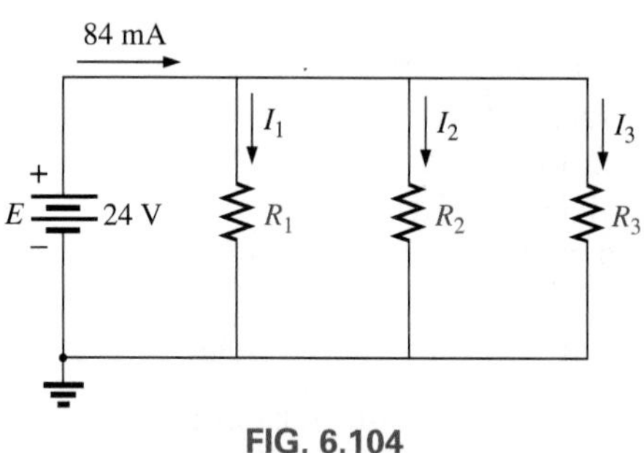

FIG. 6.104
Problem 34.

SECTION 6.7 Voltage Source in Parallel

35. Assuming identical supplies in Fig. 6.105.

a. Find the indicated currents.

b. Find the power delivered by each source.

c. Find the total power delivered by both sources, and compare it to the power delivered to the load R_L.

d. If only source current were available, what would the current drain be to supply the same power to the load? How does the current level compare to the calculated level of part (a)?

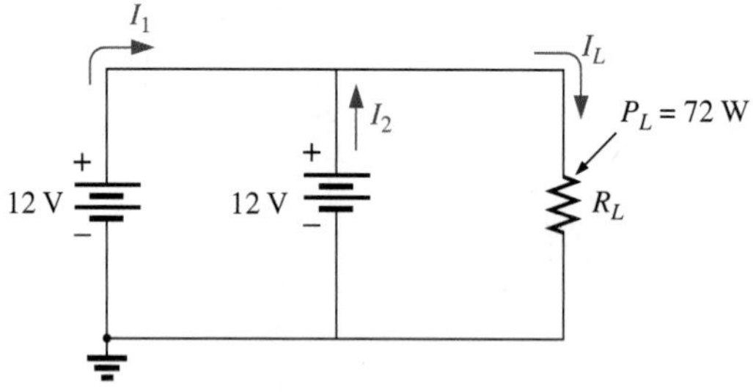

FIG. 6.105
Problem 35.

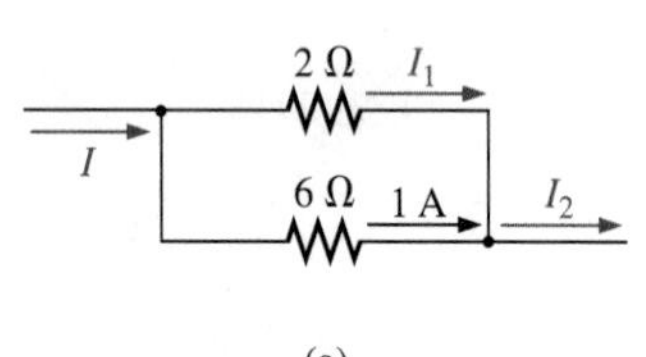

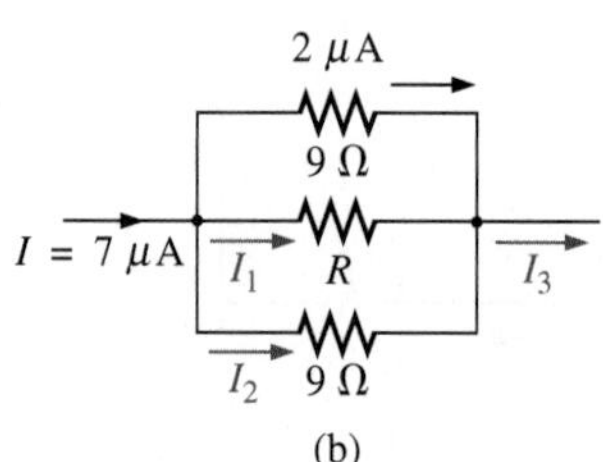

FIG. 6.102
Problem 32.

36. Assuming identical supplies, determine currents I_1, I_2, and I_3 for the configuration in Fig. 6.106.

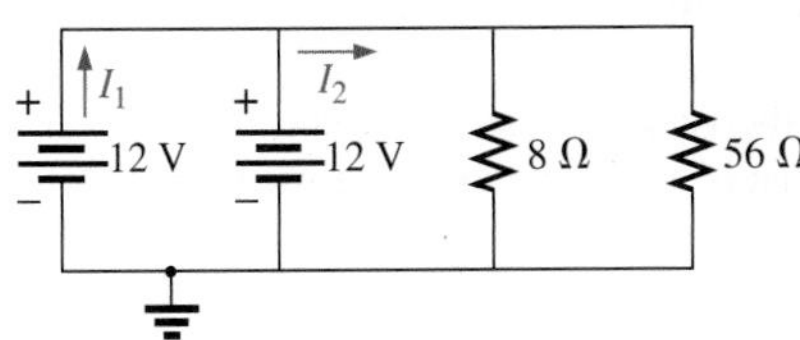

FIG. 6.106
Problem 36.

37. Assuming identical supplies, determine the current I and resistance R for the parallel network in Fig. 6.107.

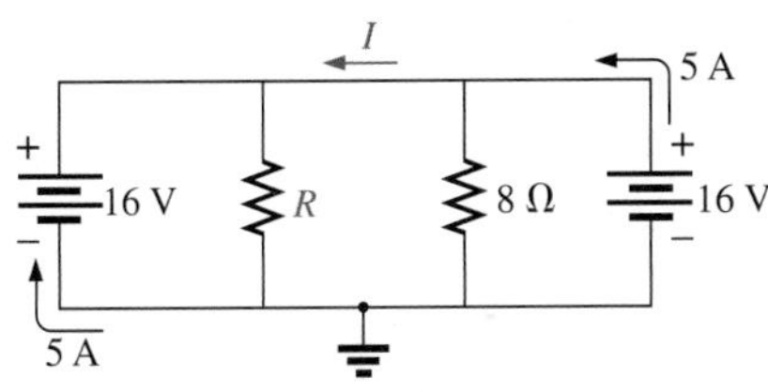

FIG. 6.107
Problem 37.

SECTION 6.8 Open and Short Circuits

38. For the network in Fig. 6.108:

a. Determine I_s and V_L.
b. Determine I_s if R_L is shorted out.
c. Determine V_L if R_L is replaced by an open circuit.

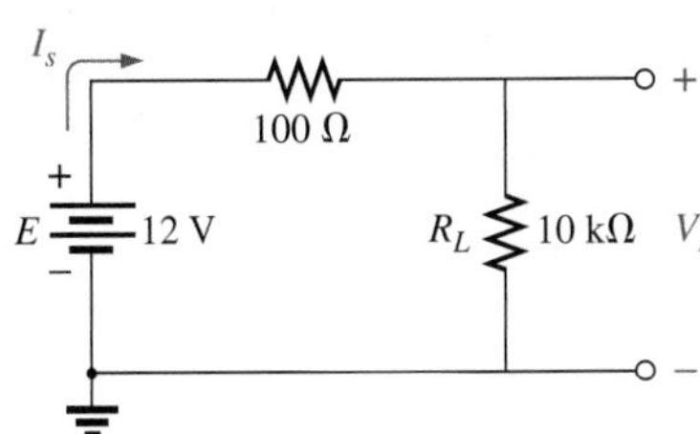

FIG. 6.108
Problem 38.

39. For the network in Fig. 6.109:

a. Determine the open-circuit voltage V_L.
b. If the 2.2 kΩ resistor is short circuited, what is the new value of V_L?
c. Determine V_L if the 4.7 kΩ resistor is replaced by an open circuit.

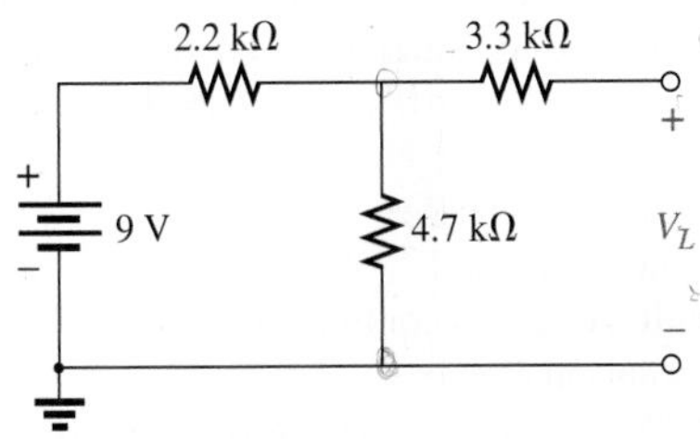

FIG. 6.109
Problem 39.

***40.** For the network in Fig. 6.110, determine

a. the short-circuit currents I_1 and I_2.
b. the voltages V_1 and V_2.
c. the source current I_s.

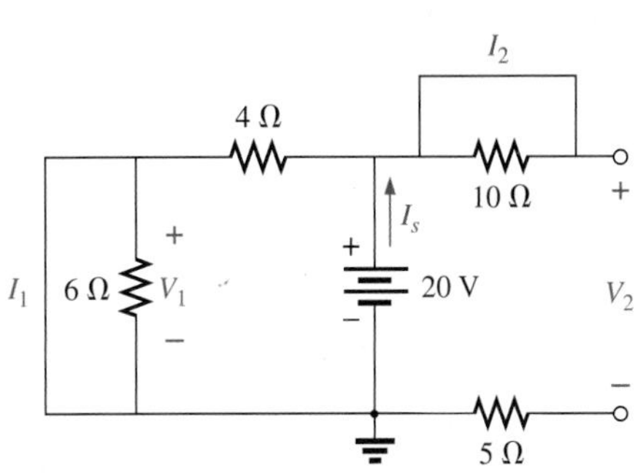

FIG. 6.110
Problem 40.

SECTION 6.9 Voltmeter Loading Effects

41. For the simple series configuration in Fig. 6.111:

a. Determine voltage V_2.
b. Determine the reading of a DMM having an internal resistance of 11 MΩ when used to measure V_2.
c. Repeat part (b) with a VOM having an Ω/V rating of 20,000 using the 20 V scale. Compare the results of parts (b) and (c). Explain any differences.
d. Repeat parts (a) through (c) with R_1 = 100 kΩ and R_2 = 200 kΩ.
e. Based on the above, what general conclusions can you make about the use of a DMM or a VOM in the voltmeter mode?

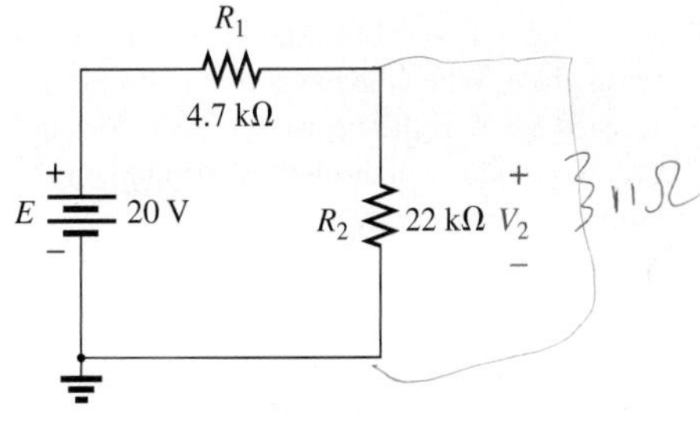

FIG. 6.111
Problem 41.

42. Given the configuration in Fig. 6.112:
    a. What is the voltage between points *a* and *b*?
    b. What will the reading of a DMM be when placed across terminals *a* and *b* if the internal resistance of the meter is 11 MΩ?
    c. Repeat part (b) if a VOM having an Ω/V rating of 20,000 using the 200 V scale is used. What is the reading using the 20 V scale? Is there a difference? Why?

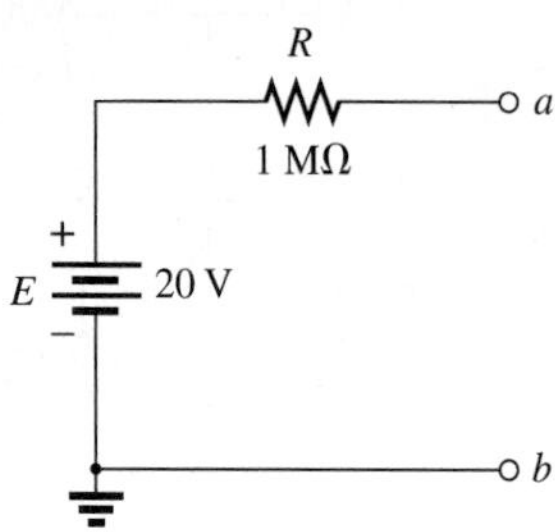

FIG. 6.112
Problem 42.

SECTION 6.10 Troubleshooting Techniques

43. Based on the measurements of Fig. 6.113, determine whether the network is operating correctly. If not, try to determine why.

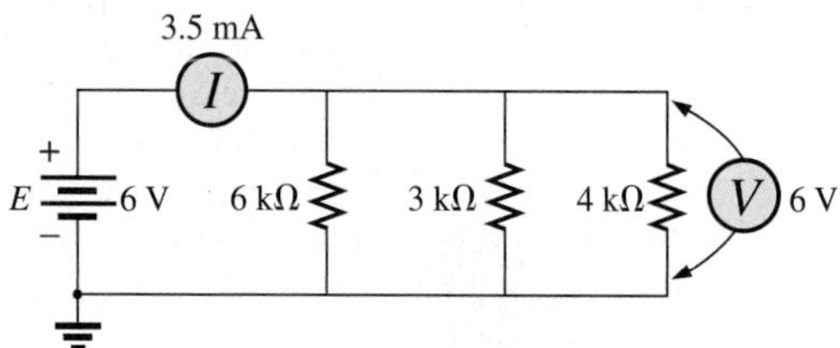

FIG. 6.113
Problem 43.

44. Referring to Fig. 6.114, find the voltage V_{ab} without the meter in place. When the meter is applied to the active network, it reads 8.8 V. If the measured value does not equal the theoretical value, which element or elements may have been connected incorrectly?

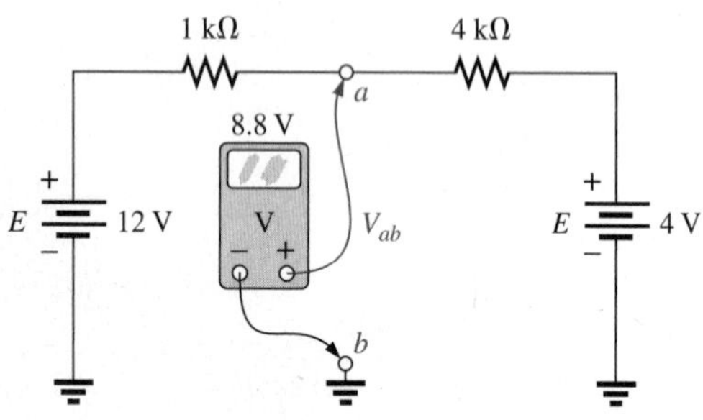

FIG. 6.114
Problem 44.

45. a. The voltage V_a for the network in Fig. 6.115 is −1 V. If it suddenly jumped to 20 V, what could have happened to the circuit structure? Localize the problem area.
    b. If the voltage V_a is 6 V rather than −1 V, try to explain what is wrong about the network construction.

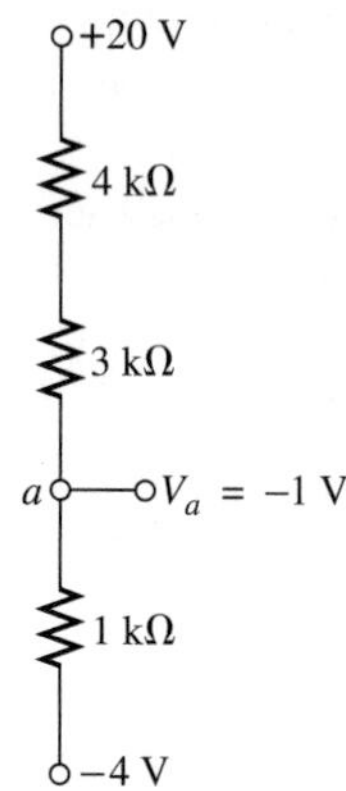

FIG. 6.115
Problem 45.

SECTION 6.14 Computer Analysis

46. Using PSpice or Multisim, verify the results of Example 6.13.
47. Using PSpice or Multisim, determine the solution to Problem 10, and compare your answer to the longhand solution.
48. Using PSpice or Multisim, determine the solution to Problem 12, and compare your answer to the longhand solution.

GLOSSARY

Current divider rule (CDR) A method by which the current through parallel elements can be determined without first finding the voltage across those parallel elements.

Kirchhoff's current law (KCL) The algebraic sum of the currents entering and leaving a node is zero.

Node A junction of two or more branches.

Ohm/volt (Ω/V) rating A rating used to determine both the current sensitivity of the movement and the internal resistance of the meter.

Open circuit The absence of a direct connection between two points in a network.

Parallel circuit A circuit configuration in which the elements have two points in common.

Short circuit A direct connection of low resistive value that can significantly alter the behavior of an element or system.

Series-Parallel Circuits

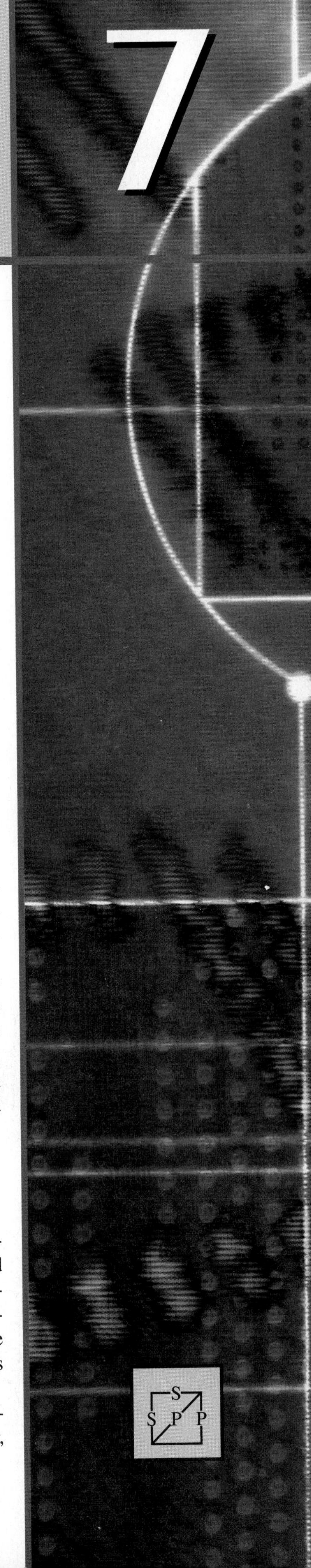

Objectives

- *Learn about the unique characteristics of series-parallel configurations and how to solve for the voltage, current, or power to any individual element or combination of elements.*
- *Become familiar with the voltage divider supply and the conditions needed to use it effectively.*
- *Learn how to use a potentiometer to control the voltage across any given load.*

7.1 INTRODUCTION

Chapters 5 and 6 were dedicated to the fundamentals of series and parallel circuits. In some ways, these chapters may be the most important ones in the text, because they form a foundation for all the material to follow. The remaining network configurations cannot be defined by a strict list of conditions because of the variety of configurations that exists. In broad terms, we can look upon the remaining possibilities as either **series-parallel** or **complex**.

A series-parallel configuration is one that is formed by a combination of series and parallel elements.

A complex configuration is one in which none of the elements are in series or parallel.

In this chapter, we examine the series-parallel combination using the basic laws introduced for series and parallel circuits. There are no new laws or rules to learn—simply an approach that permits the analysis of such structures. In the next chapter, we consider complex networks using methods of analysis that allow us to analyze any type of network.

The possibilities for series-parallel configurations are infinite. Therefore, you need to examine each network as a separate entity and define the approach that provides the best path to determining the unknown quantities. In time, you will find similarities between configurations that make it easier to define the best route to a solution, but this occurs only with exposure, practice, and patience. The best preparation for the analysis of series-parallel networks is a firm understanding of the concepts introduced for series and parallel networks. All the rules and laws to be applied in this chapter have already been introduced in the previous two chapters.

7.2 SERIES-PARALLEL NETWORKS

The network in Fig. 7.1 is a series-parallel network. At first, you must be very careful to determine which elements are in series and which are in parallel. For instance, resistors R_1 and R_2 are *not* in series due to resistor R_3 connected to the common point b between R_1 and R_2. Resistors R_2 and R_4 are *not* in parallel because they are not connected at both ends. They are separated at one end by resistor R_3. The need to be absolutely sure of your definitions from the last two chapters now becomes obvious. In fact, it may be a good idea to refer to those rules as we progress through this chapter.

If we look carefully enough at Fig. 7.1, we do find that the two resistors R_3 and R_4 are in series because they share only point c, and no other element is connected to that point. Further,

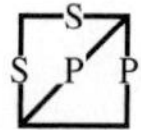

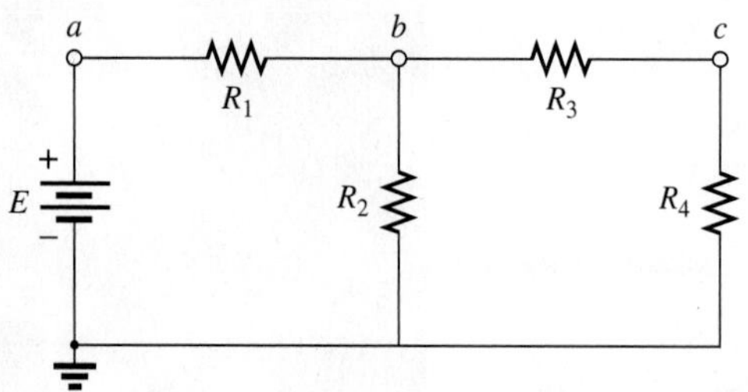

FIG. 7.1
Series-parallel dc network.

the voltage source E and resistor R_1 are in series because they share point a, with no other elements connected to the same point. In the entire configuration, there are no two elements in parallel.

How do we analyze such configurations? The approach is one that requires us to first identify elements that can be combined. Since there are no parallel elements, we must turn to the possibilities with series elements. The voltage source and the series resistor cannot be combined because they are different types of elements. However, resistors R_3 and R_4 can be combined to form a single resistor. The total resistance of the two is their sum as defined by series circuits. The resulting resistance is then in parallel with resistor R_2, and they can be combined using the laws for parallel elements. The process has begun: We are slowly reducing the network to one that will be represented by a single resistor equal to the total resistance "seen" by the source.

The source current can now be determined using Ohm's law, and we can work back through the network to find all the other currents and voltages. The ability to define the first step in the analysis can sometimes be difficult. However, combinations can be made only by using the rules for series or parallel elements, so naturally the first step may simply be to define which elements are in series or parallel. You must then define how to find such things as the total resistance and the source current and proceed with the analysis. In general, the following steps will provide some guidance for the wide variety of possible combinations that you might encounter.

General Approach:

1. Take a moment to study the problem "in total" and make a brief mental sketch of the overall approach you plan to use. The result may be time- and energy-saving shortcuts.
2. Examine each region of the network independently before tying them together in series-parallel combinations. This usually simplifies the network and possibly reveals a direct approach toward obtaining one or more desired unknowns. It also eliminates many of the errors that may result due to the lack of a systematic approach.
3. Redraw the network as often as possible with the reduced branches and undisturbed unknown quantities to maintain clarity and provide the reduced networks for the trip back to unknown quantities from the source.
4. When you have a solution, check that it is reasonable by considering the magnitudes of the energy source and the elements in the network. If it does not seem reasonable, either solve the circuit using another approach or review your calculations.

7.3 REDUCE AND RETURN APPROACH

The network of Fig. 7.1 is redrawn as Fig. 7.2(a). For this discussion, let us assume that voltage V_4 is desired. As described in Section 7.2, first combine the series resistors R_3 and R_4 to form an equivalent resistor R' as shown in Fig. 7.2(b). Resistors R_2 and R' are then in parallel and can be combined to establish an equivalent resistor R'_T as shown in Fig. 7.2(c). Resistors R_1 and R'_T are then in series and can be combined to establish the total resistance of the network as shown in Fig. 7.2(d). The **reduction phase** of the analysis is now complete. The network cannot be put in a simpler form.

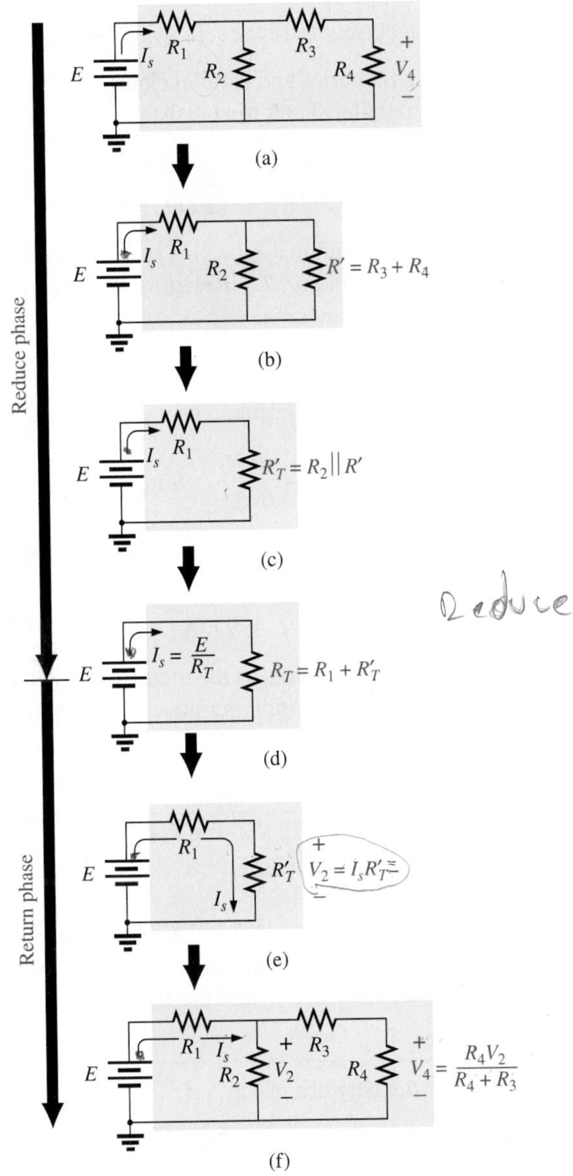

FIG. 7.2
Introducing the reduce and return approach.

We can now proceed with the **return phase** whereby we work our way back to the desired voltage V_4. Due to the resulting series configuration, the source current is also the current through R_1 and R'_T. The voltage across R'_T (and therefore across R_2) can be determined using Ohm's law as shown in Fig. 7.2(e). Finally, the desired voltage V_4 can be determined by an application of the voltage divider rule as shown in Fig. 7.2(f).

The *reduce and return approach* has now been introduced. This process enables you to reduce the network to its simplest form across the source and then determine the source current. In the return phase, you use the resulting source current to work back to the desired unknown. For most single-source series-parallel networks, the above approach provides a viable option toward the solution. In some cases, shortcuts can be applied that save some time and energy. Now for a few examples.

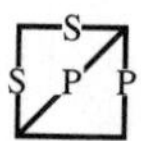

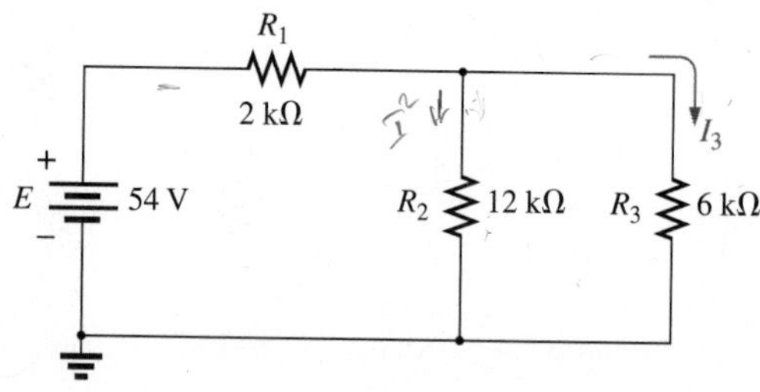

FIG. 7.3
Series-parallel network for Example 7.1.

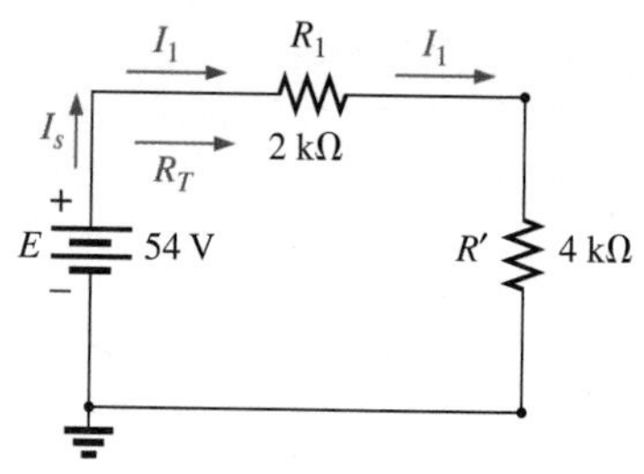

FIG. 7.4
Substituting the parallel equivalent resistance for resistors R_2 and R_3 in Fig. 7.3.

EXAMPLE 7.1 Find current I_3 for the series-parallel network in Fig. 7.3.

Solution: Checking for series and parallel elements, we find that resistors R_2 and R_3 are in parallel. Their total resistance is

$$R' = R_2 \| R_3 = \frac{R_2R_3}{R_2 + R_3} = \frac{(12\text{ k}\Omega)(6\text{ k}\Omega)}{12\text{ k}\Omega + 6\text{ k}\Omega} = 4\text{ k}\Omega$$

Replacing the parallel combination with a single equivalent resistance results in the configuration in Fig. 7.4. Resistors R_1 and R' are then in series, resulting in a total resistance of

$$R_T = R_1 + R' = 2\text{ k}\Omega + 4\text{ k}\Omega = 6\text{ k}\Omega$$

The source current is then determined using Ohm's law:

$$I_s = \frac{E}{R_T} = \frac{54\text{ V}}{6\text{ k}\Omega} = 9\text{ mA}$$

In Fig. 7.4, since R_1 and R' are in series, they have the same current I_s. The result is

$$I_1 = I_s = 9\text{ mA}$$

Returning to Fig. 7.3, we find that I_1 is the total current entering the parallel combination of R_2 and R_3. Applying the current divider rule results in the desired current:

$$I_3 = \left(\frac{R_2}{R_2 + R_3}\right)I_1 = \left(\frac{12\text{ k}\Omega}{12\text{ k}\Omega + 6\text{ k}\Omega}\right)9\text{ mA} = \mathbf{6\text{ mA}}$$

Note in the solution for Example 7.1 that all of the equations used were introduced in the last two chapters—nothing new was introduced except how to approach the problem and use the equations properly.

EXAMPLE 7.2 For the network in Fig. 7.5:

a. Determine currents I_4 and I_s and voltage V_2.
b. Insert the meters to measure current I_4 and voltage V_2.

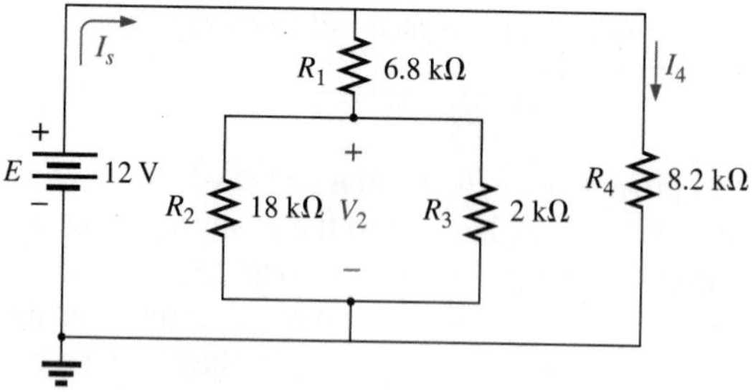

FIG. 7.5
Series-parallel network for Example 7.2.

Solutions:

a. Checking out the network, we find that there are no two resistors in series and the only parallel combination is resistors R_2 and R_3. Combining the two parallel resistors results in a total resistance of

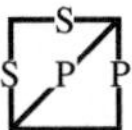

$$R' = R_2 \| R_3 = \frac{R_2R_3}{R_2 + R_3} = \frac{(18\text{ k}\Omega)(2\text{ k}\Omega)}{18\text{ k}\Omega + 2\text{ k}\Omega} = 1.8\text{ k}\Omega$$

Redrawing the network with resistance R' inserted results in the configuration in Fig. 7.6.

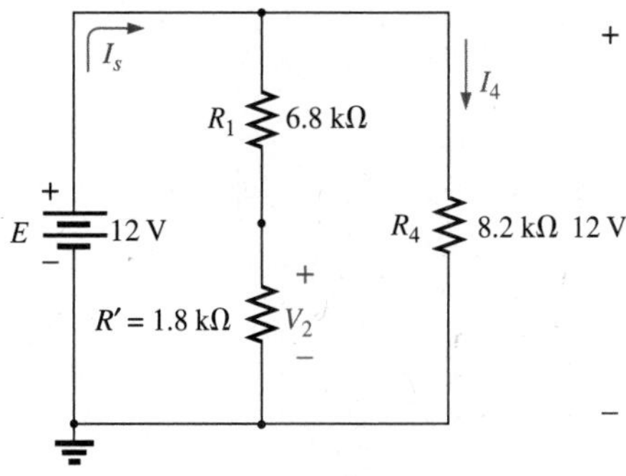

FIG. 7.6
Schematic representation of the network in Fig. 7.5 after substituting the equivalent resistance R' for the parallel combination of R_2 and R_3.

You may now be tempted to combine the series resistors R_1 and R' and redraw the network. However, a careful examination of Fig. 7.6 reveals that since the two resistive branches are in parallel, the voltage is the same across each branch. That is, the voltage across the series combination of R_1 and R' is 12 V and that across resistor R_4 is 12 V. The result is that I_4 can be determined directly using Ohm's law as follows:

$$I_4 = \frac{V_4}{R_4} = \frac{E}{R_4} = \frac{12\text{ V}}{8.2\text{ k}\Omega} = \mathbf{1.46\text{ mA}}$$

In fact, for the same reason, I_4 could have been determined directly from Fig. 7.5. Because the total voltage across the series combination of R_1 and R'_T is 12 V, the voltage divider rule can be applied to determine voltage V_2 as follows:

$$V_2 = \left(\frac{R'}{R' + R_1}\right)E = \left(\frac{1.8\text{ k}\Omega}{1.8\text{ k}\Omega + 6.8\text{ k}\Omega}\right)12\text{ V} = \mathbf{2.51\text{ V}}$$

The current I_s can be found one of two ways. Find the total resistance and use Ohm's law or find the current through the other parallel branch and apply Kirchhoff's current law. Since we already have the current I_4, the latter approach will be applied:

$$I_1 = \frac{E}{R_1 + R'} = \frac{12\text{ V}}{6.8\text{ k}\Omega + 1.8\text{ k}\Omega} = 1.40\text{ mA}$$

and $\quad I_s = I_1 + I_4 = 1.40\text{ mA} + 1.46\text{ mA} = \mathbf{2.86\text{ mA}}$

b. The meters have been properly inserted in Fig. 7.7. Note that the voltmeter is across both resistors since the voltage across parallel

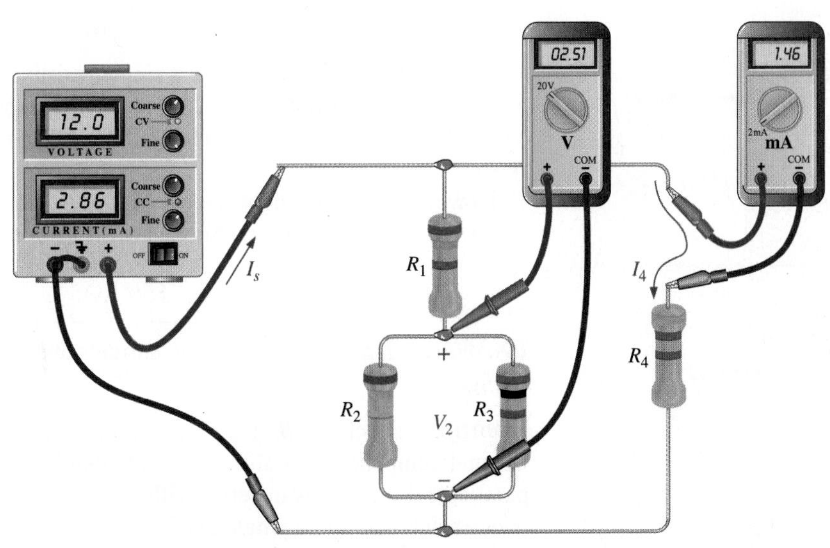

FIG. 7.7
Inserting an ammeter and a voltmeter to measure I_4 and V_2, respectively.

elements is the same. In addition, note that the ammeter is in series with resistor R_4, forcing the current through the meter to be the same as that through the series resistor. The power supply is displaying the source current.

Clearly, Example 7.2 revealed how a careful study of a network can eliminate unnecessary steps toward the desired solution. It is often worth the extra time to sit back and carefully examine a network before trying every equation that seems appropriate.

7.4 BLOCK DIAGRAM APPROACH

In the previous example, we used the reduce and return approach to find the desired unknowns. The direction seemed fairly obvious and the solution relatively easy to understand. However, occasionally the approach is not as obvious, and you may need to look at groups of elements rather than the individual components. Once the grouping of elements reveals the most direct approach, you can examine the impact of the individual components in each group. This grouping of elements is called the *block diagram approach* and is used in the following examples.

In Fig. 7.8, blocks B and C are in parallel (points b and c in common), and the voltage source E is in series with block A (point a in common). The parallel combination of B and C is also in series with A and the voltage source E due to the common points b and c, respectively.

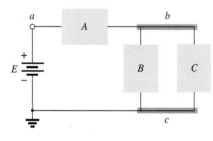

FIG. 7.8
Introducing the block diagram approach.

To ensure that the analysis to follow is as clear and uncluttered as possible, the following notation is used for series and parallel combinations of elements. For series resistors R_1 and R_2, a comma is inserted between their subscript notations, as shown here:

$$R_{1,2} = R_1 + R_2$$

For parallel resistors R_1 and R_2, the parallel symbol is inserted between their subscripted notations, as follows:

$$R_{1\|2} = R_1 \| R_2 = \frac{R_1 R_2}{R_1 + R_2}$$

If each block in Fig. 7.8 were a single resistive element, the network in Fig. 7.9 would result. Note that it is an exact replica of Fig. 7.3 in Example 7.1. Blocks B and C are in parallel, and their combination is in series with block A.

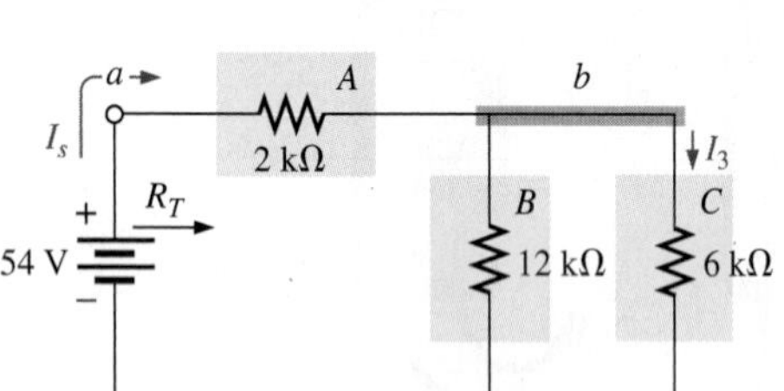

FIG. 7.9
Block diagram format of Fig. 7.3.

However, as shown in the next example, the same block configuration can result in a totally different network.

EXAMPLE 7.3 Determine all the currents and voltages of the network in Fig. 7.10.

Solution: Blocks A, B, and C have the same relative position, but the internal components are different. Note that blocks B and C are still in parallel and block A is in series with the parallel combination. First, reduce each block into a single element and proceed as described for Example 7.1.

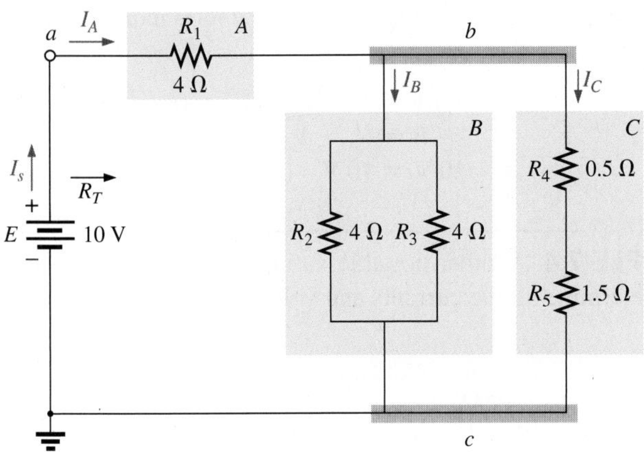

FIG. 7.10
Example 7.3.

In this case:

$$A: R_A = 4\ \Omega$$

$$B: R_B = R_2 \parallel R_3 = R_{2\parallel 3} = \frac{R}{N} = \frac{4\ \Omega}{2} = 2\ \Omega$$

$$C: R_C = R_4 + R_5 = R_{4,5} = 0.5\ \Omega + 1.5\ \Omega = 2\ \Omega$$

Blocks B and C are still in parallel, and

$$R_{B\parallel C} = \frac{R}{N} = \frac{2\ \Omega}{2} = 1\ \Omega$$

with

$$R_T = R_A + R_{B\parallel C}$$
$$= 4\ \Omega + 1\ \Omega = \mathbf{5\ \Omega}$$

(Note the similarity between this equation and that obtained for Example 7.1.)

and $$I_s = \frac{E}{R_T} = \frac{10\text{ V}}{5\ \Omega} = \mathbf{2\ A}$$

We can find the currents I_A, I_B, and I_C using the reduction of the network in Fig. 7.10 (recall Step 3) as found in Fig. 7.11. Note that I_A, I_B, and I_C are the same in Figs. 7.10 and Fig. 7.11 and therefore also appear in Fig. 7.11. In other words, the currents I_A, I_B, and I_C in Fig. 7.11 have the same magnitude as the same currents in Fig. 7.10.

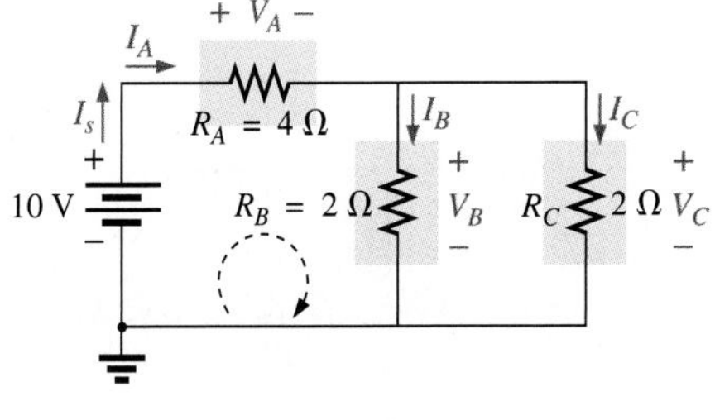

FIG. 7.11
Reduced equivalent of Fig. 7.10.

$$I_A = I_s = \mathbf{2\ A}$$

and $$I_B = I_C = \frac{I_A}{2} = \frac{I_s}{2} = \frac{2\text{ A}}{2} = \mathbf{1\ A}$$

Returning to the network in Fig. 7.10, we have

$$I_{R_2} = I_{R_3} = \frac{I_B}{2} = \mathbf{0.5\ A}$$

The voltages V_A, V_B, and V_C from either figure are

$$V_A = I_A R_A = (2\text{ A})(4\ \Omega) = \mathbf{8\ V}$$
$$V_B = I_B R_B = (1\text{ A})(2\ \Omega) = \mathbf{2\ V}$$
$$V_C = V_B = \mathbf{2\ V}$$

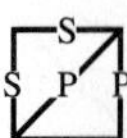

Applying Kirchhoff's voltage law for the loop indicated in Fig. 7.11, we obtain

$$\Sigma_{\circlearrowright} V = E - V_A - V_B = 0$$

$$E = V_A + V_B = 8\text{ V} + 2\text{ V}$$

or

$$10\text{ V} = 10\text{ V} \quad \text{(checks)}$$

EXAMPLE 7.4 Another possible variation of Fig. 7.8 appears in Fig. 7.12. Determine all the currents and voltages.

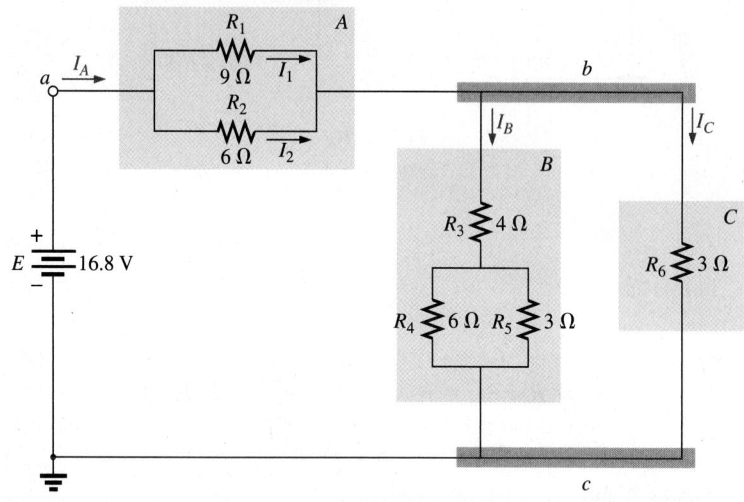

FIG. 7.12
Example 7.4.

Solution:

$$R_A = R_{1\|2} = \frac{(9\,\Omega)(6\,\Omega)}{9\,\Omega + 6\,\Omega} = \frac{54\,\Omega}{15} = 3.6\,\Omega$$

$$R_B = R_3 + R_{4\|5} = 4\,\Omega + \frac{(6\,\Omega)(3\,\Omega)}{6\,\Omega + 3\,\Omega} = 4\,\Omega + 2\,\Omega = 6\,\Omega$$

$$R_C = 3\,\Omega$$

The network in Fig. 7.12 can then be redrawn in reduced form, as shown in Fig. 7.13. Note the similarities between this circuit and the circuits in Figs. 7.9 and 7.11.

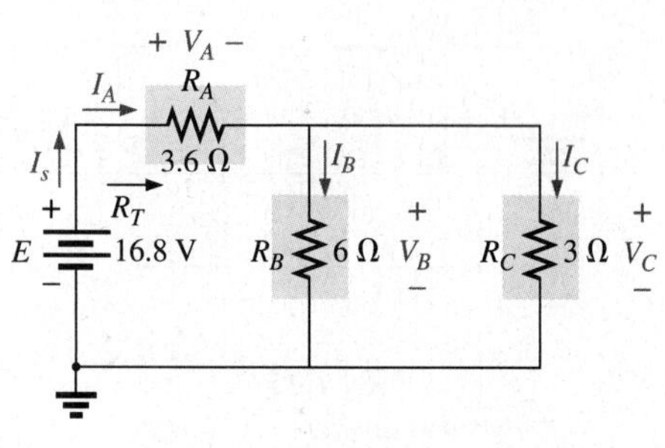

FIG. 7.13
Reduced equivalent of Fig. 7.12.

$$R_T = R_A + R_{B\|C} = 3.6\,\Omega + \frac{(6\,\Omega)(3\,\Omega)}{6\,\Omega + 3\,\Omega}$$

$$= 3.6\,\Omega + 2\,\Omega = \mathbf{5.6\,\Omega}$$

$$I_s = \frac{E}{R_T} = \frac{16.8\text{ V}}{5.6\,\Omega} = \mathbf{3\text{ A}}$$

$$I_A = I_s = \mathbf{3\text{ A}}$$

Applying the current divider rule yields

$$I_B = \frac{R_C I_A}{R_C + R_B} = \frac{(3\,\Omega)(3\text{ A})}{3\,\Omega + 6\,\Omega} = \frac{9\text{ A}}{9} = \mathbf{1\text{ A}}$$

By Kirchhoff's current law,

$$I_C = I_A - I_B = 3\text{ A} - 1\text{ A} = \mathbf{2\text{ A}}$$

By Ohm's law,

$$V_A = I_A R_A = (3\text{ A})(3.6\ \Omega) = \mathbf{10.8\ V}$$
$$V_B = I_B R_B = V_C = I_C R_C = (2\text{ A})(3\ \Omega) = \mathbf{6\ V}$$

Returning to the original network (Fig. 7.12) and applying the current divider rule,

$$I_1 = \frac{R_2 I_A}{R_2 + R_1} = \frac{(6\ \Omega)(3\text{ A})}{6\ \Omega + 9\ \Omega} = \frac{18\text{ A}}{15} = \mathbf{1.2\ A}$$

By Kirchhoff's current law,

$$I_2 = I_A - I_1 = 3\text{ A} - 1.2\text{ A} = \mathbf{1.8\ A}$$

Figs. 7.9, 7.10, and 7.12 are only a few of the infinite variety of configurations that the network can assume starting with the basic arrangement in Fig. 7.8. They were included in our discussion to emphasize the importance of considering each region of the network independently before finding the solution for the network as a whole.

The blocks in Fig. 7.8 can be arranged in a variety of ways. In fact, there is no limit on the number of series-parallel configurations that can appear within a given network. In reverse, the block diagram approach can be used effectively to reduce the apparent complexity of a system by identifying the major series and parallel components of the network. This approach is demonstrated in the next few examples.

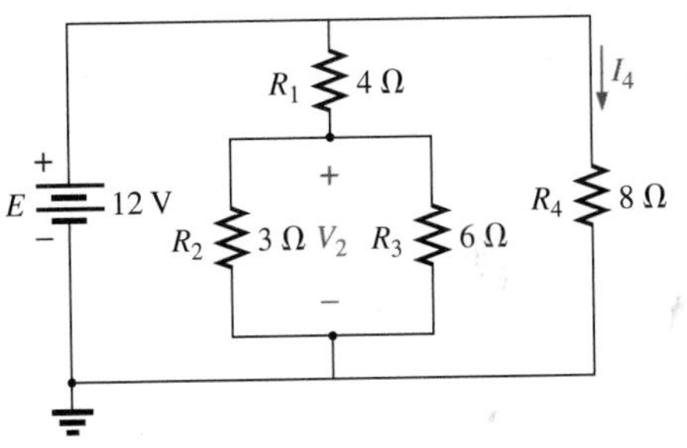

FIG. 7.14
Example 7.5.

7.5 DESCRIPTIVE EXAMPLES

EXAMPLE 7.5 Find the current I_4 and the voltage V_2 for the network in Fig. 7.14 using the block diagram approach.

Solution: Note the similarities with the network in Fig. 7.5. In this case, particular unknowns are requested instead of a complete solution. It would, therefore, be a waste of time to find all the currents and voltages of the network. The method used should concentrate on obtaining only the unknowns requested. With the block diagram approach, the network has the basic structure in Fig. 7.15, clearly indicating that the three branches are in parallel and the voltage across A and B is the supply voltage. The current I_4 is now immediately obvious as simply the supply voltage divided by the resultant resistance for B. If desired, block A can be broken down further, as shown in Fig. 7.16, to identify C and D as series elements, with the voltage V_2 capable of being determined using the voltage divider rule once the resistance of C and D is reduced to a single value. This is an example of how making a mental sketch of the approach before applying laws, rules, and so on, can help avoid dead ends and frustration.

Applying Ohm's law,

$$I_4 = \frac{E}{R_B} = \frac{E}{R_4} = \frac{12\text{ V}}{8\ \Omega} = \mathbf{1.5\ A}$$

Combining the resistors R_2 and R_3 in Fig. 7.14 results in

$$R_D = R_2 \| R_3 = 3\ \Omega \| 6\ \Omega = \frac{(3\ \Omega)(6\ \Omega)}{3\ \Omega + 6\ \Omega} = \frac{18\ \Omega}{9} = 2\ \Omega$$

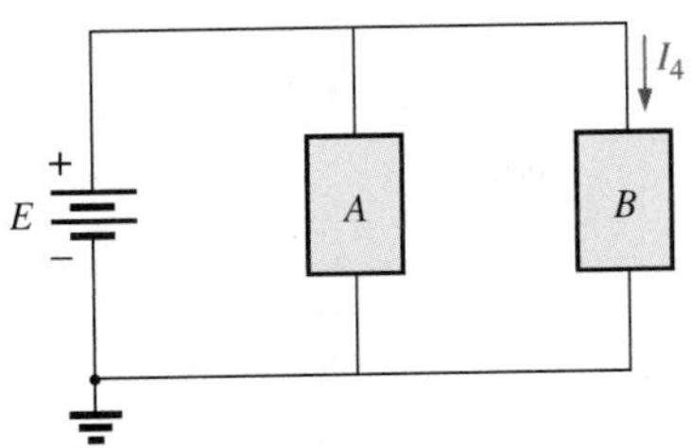

FIG. 7.15
Block diagram of Fig. 7.14.

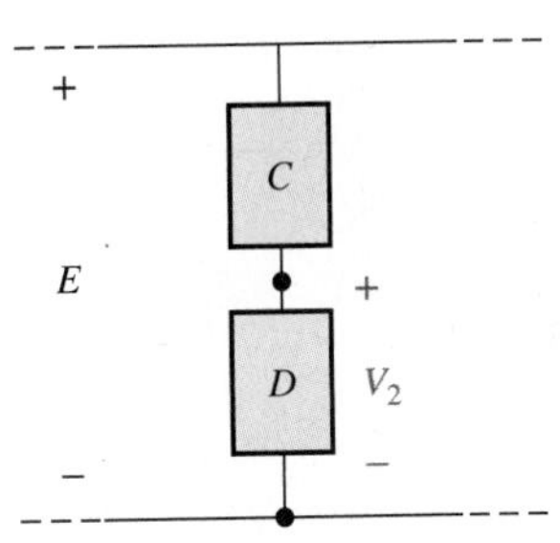

FIG. 7.16
Alternative block diagram for the first parallel branch in Fig. 7.14.

and, applying the voltage divider rule,

$$V_2 = \frac{R_D E}{R_D + R_C} = \frac{(2\ \Omega)(12\ \text{V})}{2\ \Omega + 4\ \Omega} = \frac{24\ \text{V}}{6} = \mathbf{4\ V}$$

EXAMPLE 7.6 Find the indicated currents and voltages for the network in Fig. 7.17.

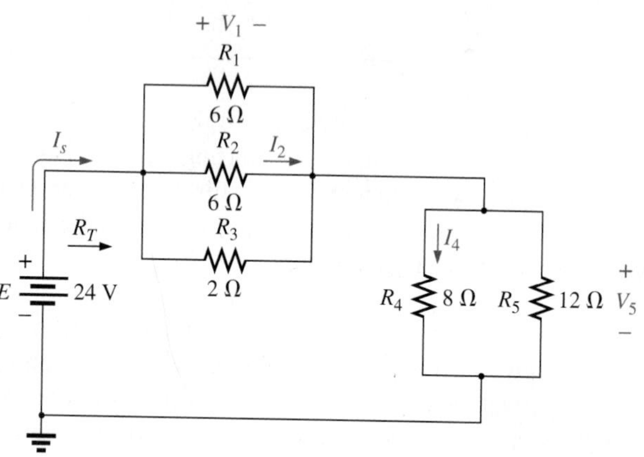

FIG. 7.17
Example 7.6.

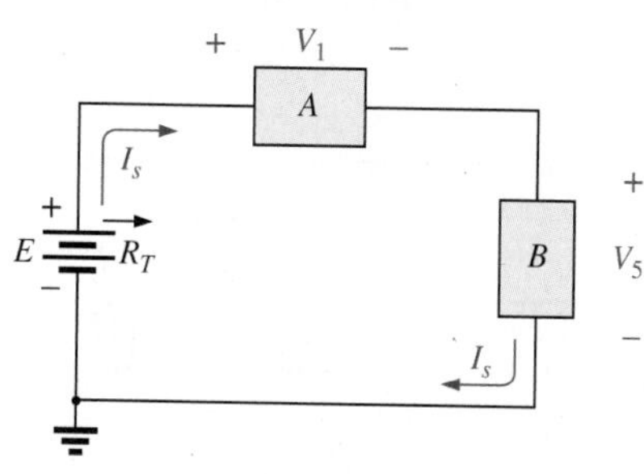

FIG. 7.18
Block diagram for Fig. 7.17.

Solution: Again, only specific unknowns are requested. When the network is redrawn, be sure to note which unknowns are preserved and which have to be determined using the original configuration. The block diagram of the network may appear as shown in Fig. 7.18, clearly revealing that *A* and *B* are in series. Note in this form the number of unknowns that have been preserved. The voltage V_1 is the same across the three parallel branches in Fig. 7.17, and V_5 is the same across R_4 and R_5. The unknown currents I_2 and I_4 are lost since they represent the currents through only one of the parallel branches. However, once V_1 and V_5 are known, you can find the required currents using Ohm's law.

$$R_{1\|2} = \frac{R}{N} = \frac{6\ \Omega}{2} = 3\ \Omega$$

$$R_A = R_{1\|2\|3} = \frac{(3\ \Omega)(2\ \Omega)}{3\ \Omega + 2\ \Omega} = \frac{6\ \Omega}{5} = 1.2\ \Omega$$

$$R_B = R_{4\|5} = \frac{(8\ \Omega)(12\ \Omega)}{8\ \Omega + 12\ \Omega} = \frac{96\ \Omega}{20} = 4.8\ \Omega$$

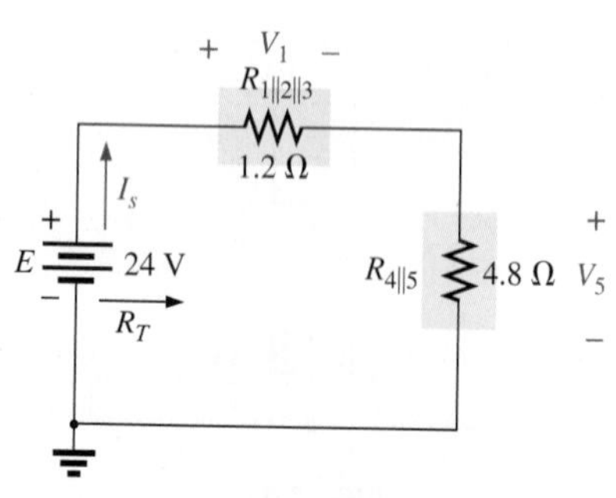

FIG. 7.19
Reduced form of Fig. 7.17.

The reduced form of Fig. 7.17 then appears as shown in Fig. 7.19, and

$$R_T = R_{1\|2\|3} + R_{4\|5} = 1.2\ \Omega + 4.8\ \Omega = \mathbf{6\ \Omega}$$

$$I_s = \frac{E}{R_T} = \frac{24\ \text{V}}{6\ \Omega} = \mathbf{4\ A}$$

with

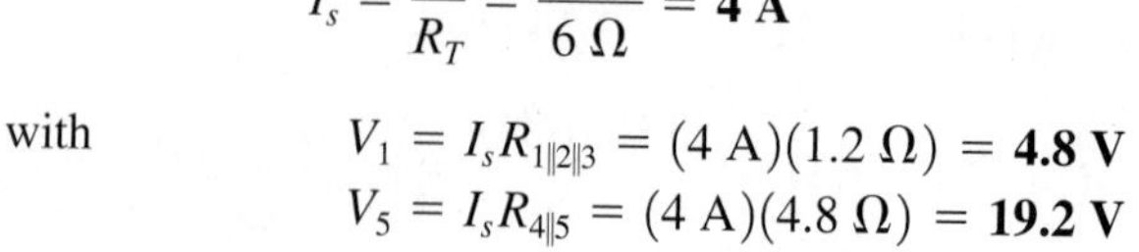

$$V_1 = I_s R_{1\|2\|3} = (4\ \text{A})(1.2\ \Omega) = \mathbf{4.8\ V}$$

$$V_5 = I_s R_{4\|5} = (4\ \text{A})(4.8\ \Omega) = \mathbf{19.2\ V}$$

Applying Ohm's law,

$$I_4 = \frac{V_5}{R_4} = \frac{19.2 \text{ V}}{8 \ \Omega} = \mathbf{2.4 \text{ A}}$$

$$I_2 = \frac{V_2}{R_2} = \frac{V_1}{R_2} = \frac{4.8 \text{ V}}{6 \ \Omega} = \mathbf{0.8 \text{ A}}$$

The next example demonstrates that unknown voltages do not have to be across elements but can exist between any two points in a network. In addition, the importance of redrawing the network in a more familiar form is clearly revealed by the analysis to follow.

EXAMPLE 7.7

a. Find the voltages V_1, V_3, and V_{ab} for the network in Fig. 7.20.
b. Calculate the source current I_s.

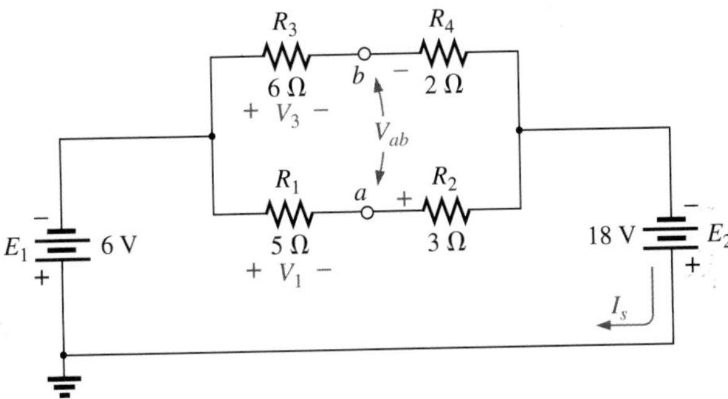

FIG. 7.20
Example 7.7.

Solutions: This is one of those situations where it may be best to redraw the network before beginning the analysis. Since combining both sources will not affect the unknowns, the network is redrawn as shown in Fig. 7.21, establishing a parallel network with the total source voltage across each parallel branch. The net source voltage is the difference between the two with the polarity of the larger.

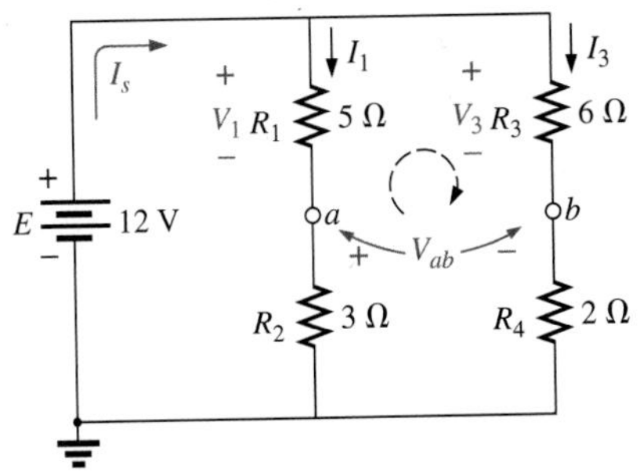

FIG. 7.21
Network in Fig. 7.20 redrawn.

a. Note the similarities with Fig. 7.16, permitting the use of the voltage divider rule to determine V_1 and V_3:

$$V_1 = \frac{R_1 E}{R_1 + R_2} = \frac{(5 \ \Omega)(12 \text{ V})}{5 \ \Omega + 3 \ \Omega} = \frac{60 \text{ V}}{8} = \mathbf{7.5 \text{ V}}$$

$$V_3 = \frac{R_3 E}{R_3 + R_4} = \frac{(6 \ \Omega)(12 \text{ V})}{6 \ \Omega + 2 \ \Omega} = \frac{72 \text{ V}}{8} = \mathbf{9 \text{ V}}$$

The open-circuit voltage V_{ab} is determined by applying Kirchhoff's voltage law around the indicated loop in Fig. 7.21 in the clockwise direction starting at terminal a.

$$+V_1 - V_3 + V_{ab} = 0$$

and $\quad V_{ab} = V_3 - V_1 = 9 \text{ V} - 7.5 \text{ V} = \mathbf{1.5 \text{ V}}$

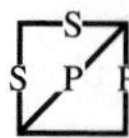

b. By Ohm's law,

$$I_1 = \frac{V_1}{R_1} = \frac{7.5\text{ V}}{5\ \Omega} = 1.5\text{ A}$$
$$I_3 = \frac{V_3}{R_3} = \frac{9\text{ V}}{6\ \Omega} = 1.5\text{ A}$$

Applying Kirchhoff's current law,

$$I_s = I_1 + I_3 = 1.5\text{ A} + 1.5\text{ A} = \mathbf{3\ A}$$

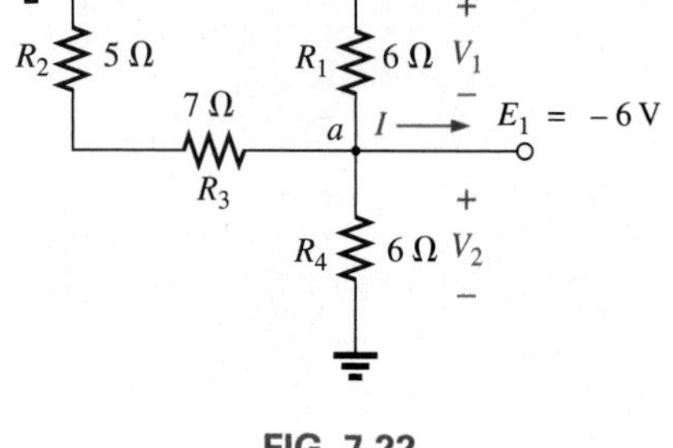

FIG. 7.22
Example 7.8.

EXAMPLE 7.8 For the network in Fig. 7.22, determine the voltages V_1 and V_2 and the current I.

Solution: It would indeed be difficult to analyze the network in the form in Fig. 7.22 with the symbolic notation for the sources and the reference or ground connection in the upper left corner of the diagram. However, when the network is redrawn as shown in Fig. 7.23, the unknowns and the relationship between branches become significantly clearer. Note the common connection of the grounds and the replacing of the terminal notation by actual supplies.

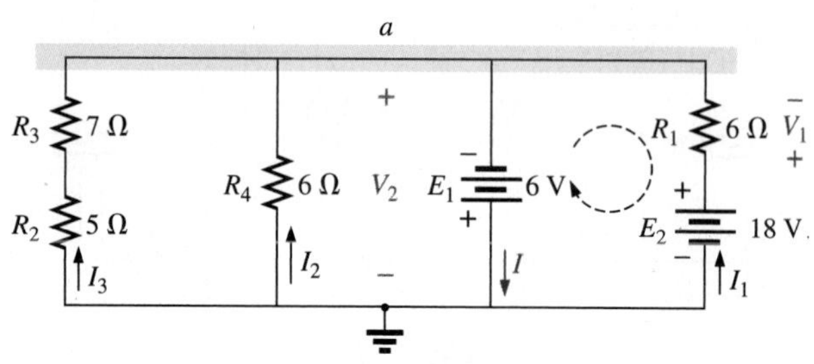

FIG. 7.23
Network in Fig. 7.22 redrawn.

It is now obvious that

$$V_2 = -E_1 = \mathbf{-6\ V}$$

The minus sign simply indicates that the chosen polarity for V_2 in Fig. 7.18 is opposite to that of the actual voltage. Applying Kirchhoff's voltage law to the loop indicated, we obtain

$$-E_1 + V_1 - E_2 = 0$$

and

$$V_1 = E_2 + E_1 = 18\text{ V} + 6\text{ V} = \mathbf{24\ V}$$

Applying Kirchhoff's current law to note a yields

$$\begin{aligned}I &= I_1 + I_2 + I_3\\ &= \frac{V_1}{R_1} + \frac{E_1}{R_4} + \frac{E_1}{R_2 + R_3}\\ &= \frac{24\text{ V}}{6\ \Omega} + \frac{6\text{ V}}{6\ \Omega} + \frac{6\text{ V}}{12\ \Omega}\\ &= 4\text{ A} + 1\text{ A} + 0.5\text{ A}\\ I &= \mathbf{5.5\ A}\end{aligned}$$

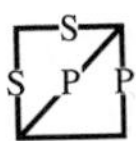

The next example is clear evidence that techniques learned in the current chapters will have far-reaching applications and will not be dropped for improved methods. Even though we have not studied the **transistor** yet, the dc levels of a transistor network can be examined using the basic rules and laws introduced in earlier chapters.

EXAMPLE 7.9 For the transistor configuration in Fig. 7.24, in which V_B and V_{BE} have been provided:

a. Determine the voltage V_E and the current I_E.
b. Calculate V_1.
c. Determine V_{BC} using the fact that the approximation $I_C = I_E$ is often applied to transistor networks.
d. Calculate V_{CE} using the information obtained in parts (a) through (c).

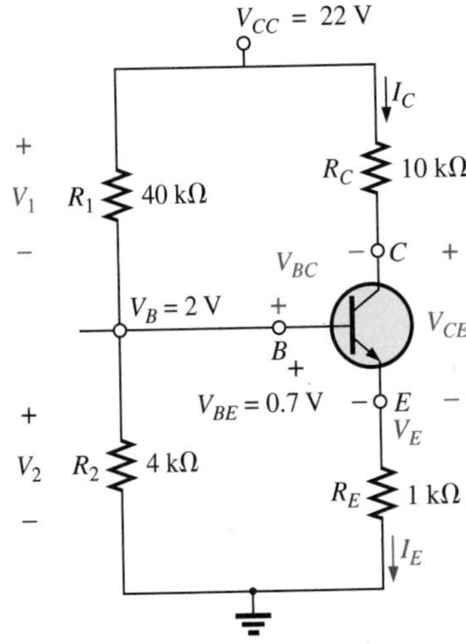

FIG. 7.24
Example 7.9.

Solutions:

a. From Fig. 7.24, we find

$$V_2 = V_B = 2\text{ V}$$

Writing Kirchhoff's voltage law around the lower loop yields

$$V_2 - V_{BE} - V_E = 0$$

or $$V_E = V_2 - V_{BE} = 2\text{ V} - 0.7\text{ V} = \mathbf{1.3\text{ V}}$$

and $$I_E = \frac{V_E}{R_E} = \frac{1.3\text{ V}}{1\text{ k}\Omega} = \mathbf{1.3\text{ mA}}$$

b. Applying Kirchhoff's voltage law to the input side (left region of the network) results in

$$V_2 + V_1 - V_{CC} = 0$$

and $$V_1 = V_{CC} - V_2$$

but $$V_2 = V_B$$

and $$V_1 = V_{CC} - V_2 = 22\text{ V} - 2\text{ V} = \mathbf{20\text{ V}}$$

c. Redrawing the section of the network of immediate interest results in Fig. 7.25, where Kirchhoff's voltage law yields

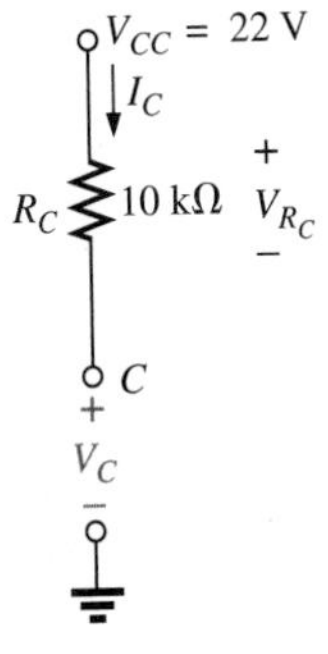

$$V_C + V_{R_C} - V_{CC} = 0$$

and $$V_C = V_{CC} - V_{R_C} = V_{CC} - I_C R_C$$

but $$I_C = I_E$$

and $$V_C = V_{CC} - I_E R_C = 22\text{ V} - (1.3\text{ mA})(10\text{ k}\Omega) = 9\text{ V}$$

Then $$V_{BC} = V_B - V_C = 2\text{ V} - 9\text{ V} = \mathbf{-7\text{ V}}$$

d. $$V_{CE} = V_C - V_E = 9\text{ V} - 1.3\text{ V} = \mathbf{7.7\text{ V}}$$

FIG. 7.25
Determining V_C for the network in Fig. 7.24.

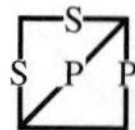

EXAMPLE 7.10 Calculate the indicated currents and voltage in Fig. 7.26.

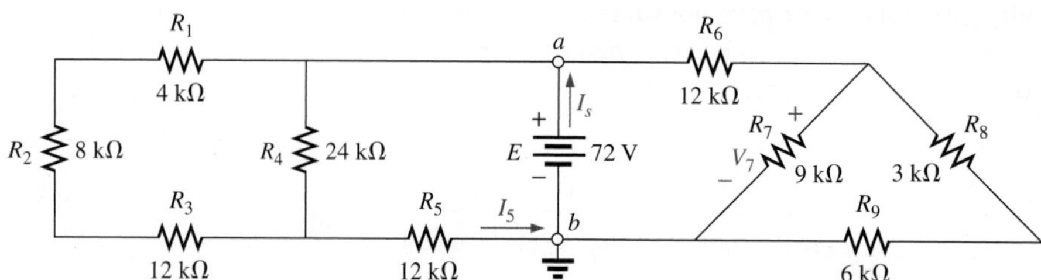

FIG. 7.26
Example 7.10.

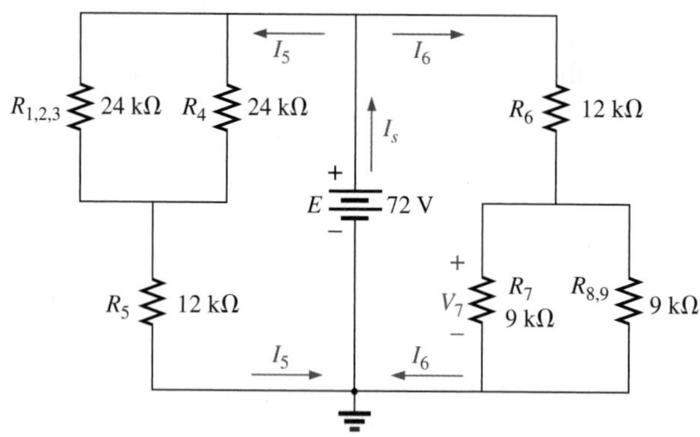

FIG. 7.27
Network in Fig. 7.26 redrawn.

Solution: Redrawing the network after combining series elements yields Fig. 7.27, and

$$I_5 = \frac{E}{R_{(1,2,3)\|4} + R_5} = \frac{72\text{ V}}{12\text{ k}\Omega + 12\text{ k}\Omega} = \frac{72\text{ V}}{24\text{ k}\Omega} = \mathbf{3\text{ mA}}$$

with

$$V_7 = \frac{R_{7\|(8,9)}E}{R_{7\|(8,9)} + R_6} = \frac{(4.5\text{ k}\Omega)(72\text{ V})}{4.5\text{ k}\Omega + 12\text{ k}\Omega} = \frac{324\text{ V}}{16.5} = \mathbf{19.6\text{ V}}$$

$$I_6 = \frac{V_7}{R_{7\|(8,9)}} = \frac{19.6\text{ V}}{4.5\text{ k}\Omega} = \mathbf{4.35\text{ mA}}$$

and $$I_s = I_5 + I_6 = 3\text{ mA} + 4.35\text{ mA} = \mathbf{7.35\text{ mA}}$$

Since the potential difference between points a and b in Fig. 7.26 is fixed at E volts, the circuit to the right or left is unaffected if the network is reconstructed as shown in Fig. 7.28.

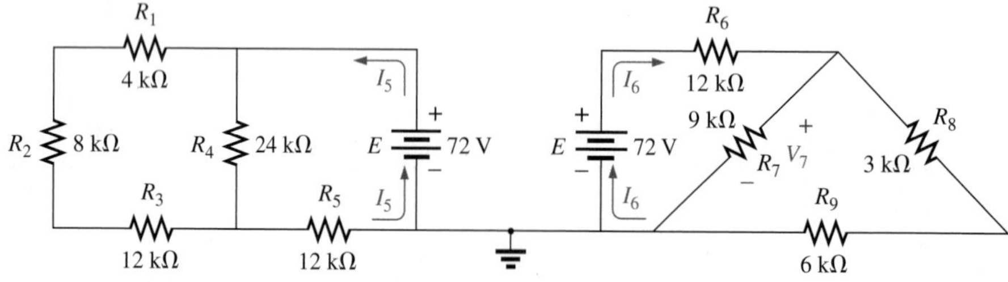

FIG. 7.28
An alternative approach to Example 7.10.

We can find each quantity required, except I_s, by analyzing each circuit independently. To find I_s, we must find the source current for each circuit and add it as in the above solution; that is, $I_s = I_5 + I_6$.

EXAMPLE 7.11 For the network in Fig. 7.29:

a. Determine voltages V_a, V_b, and V_c.
b. Find voltages V_{ac} and V_{bc}.
c. Find current I_2.
d. Find the source current I_{s_3}.
e. Insert voltmeters to measure voltages V_a and V_{bc} and current I_{s_3}.

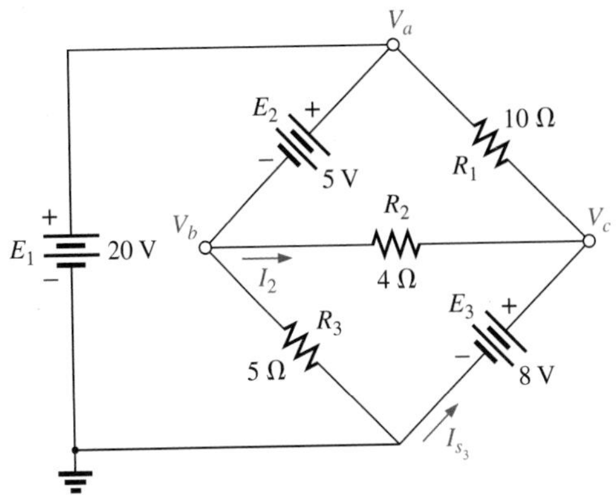

FIG. 7.29
Example 7.11.

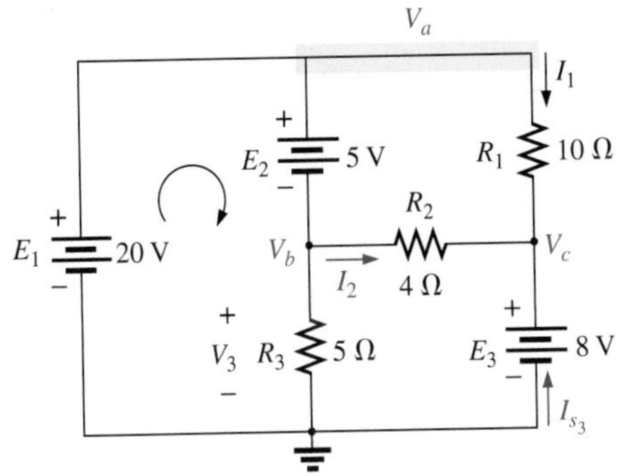

FIG. 7.30
Network in Fig. 7.29 redrawn to better define a path toward the desired unknowns.

Solutions:

a. The network is redrawn in Fig. 7.30 to clearly indicate the arrangement between elements.

First, note that voltage V_a is directly across voltage source E_1. Therefore,

$$V_a = E_1 = \mathbf{20\ V}$$

The same is true for voltage V_c, which is directly across the voltage source E_3. Therefore,

$$V_c = E_3 = \mathbf{8\ V}$$

To find voltage V_b, which is actually the voltage across R_3, we must apply Kirchhoff's voltage law around loop 1 as follows:

$$+E_1 - E_2 - V_3 = 0$$

and $\quad V_3 = E_1 - E_2 = 20\ \text{V} - 5\ \text{V} = 15\ \text{V}$

and $\quad V_b = V_3 = \mathbf{15\ V}$

b. Voltage V_{ac}, which is actually the voltage across resistor R_1, can then be determined as follows:

$$V_{ac} = V_a - V_c = 20\ \text{V} - 8\ \text{V} = \mathbf{12\ V}$$

Similarly, voltage V_{bc}, which is actually the voltage across resistor R_2, can then be determined as follows:

$$V_{bc} = V_b - V_c = 15\ \text{V} - 8\ \text{V} = \mathbf{7\ V}$$

c. Current I_2 can be determined using Ohm's law:

$$I_2 = \frac{V_2}{R_2} = \frac{V_{bc}}{R_2} = \frac{7\ \text{V}}{4\ \Omega} = \mathbf{1.75\ A}$$

d. The source current I_{s_3} can be determined using Kirchhoff's current law at note c:

$$\Sigma I_i = \Sigma I_o$$
$$I_1 + I_2 + I_{s_3} = 0$$

and $\quad I_{s_3} = -I_1 - I_2 = -\dfrac{V_1}{R_1} - I_2$

with $\quad V_1 = V_{ac} = V_a - V_c = 20\ \text{V} - 8\ \text{V} = 12\ \text{V}$

so that

$$I_{s_3} = -\frac{12\ \text{V}}{10\ \Omega} - 1.75\ \text{A} = -1.2\ \text{A} - 1.75\ \text{A} = \mathbf{-2.95\ A}$$

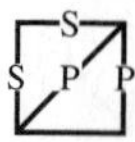

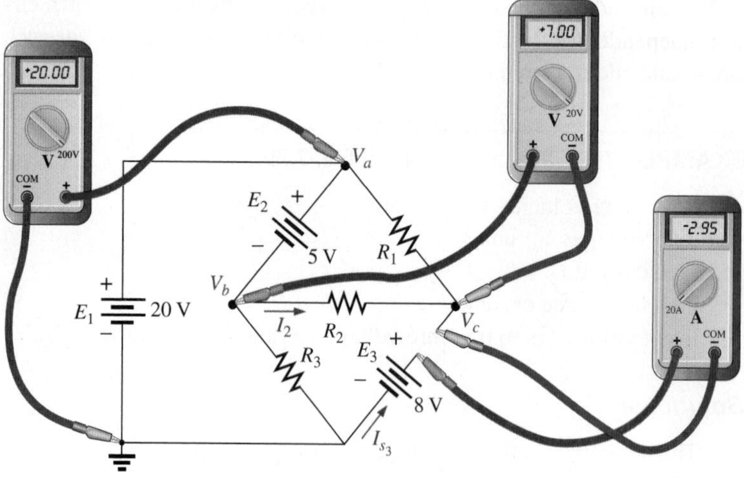

FIG. 7.31
Complex network for Example 7.11.

revealing that current is actually being forced through source E_3 in a direction opposite to that shown in Fig. 7.29.

e. Both voltmeters have a positive reading as shown in Fig. 7.31 while the ammeter has a negative reading.

7.6 LADDER NETWORKS

A three-section **ladder network** appears in Fig. 7.32. The reason for the terminology is quite obvious for the repetitive structure. Basically two approaches are used to solve networks of this type.

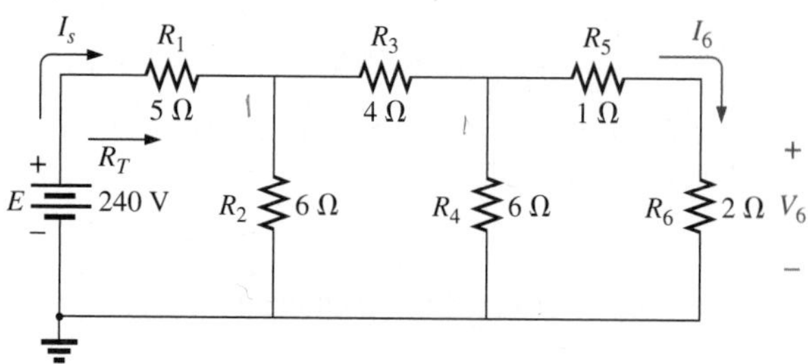

FIG. 7.32
Ladder network.

Method 1

Calculate the total resistance and resulting source current, and then work back through the ladder until the desired current or voltage is obtained. This method is now employed to determine V_6 in Fig. 7.32.

Combining parallel and series elements as shown in Fig. 7.33 results in the reduced network in Fig. 7.34, and

$$R_T = 5\ \Omega + 3\ \Omega = 8\ \Omega$$

$$I_s = \frac{E}{R_T} = \frac{240\ \text{V}}{8\ \Omega} = 30\ \text{A}$$

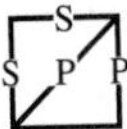

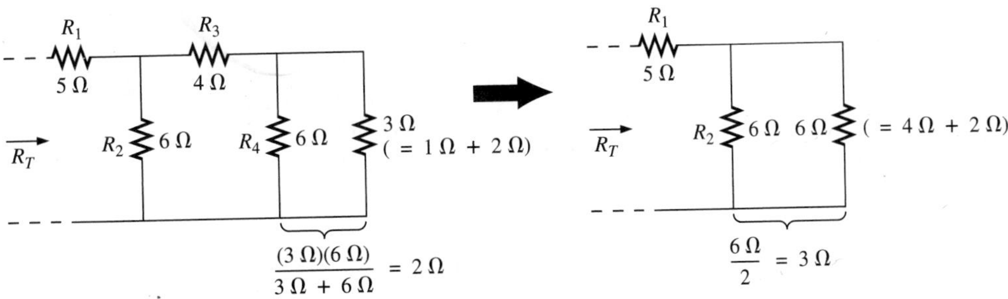

FIG. 7.33
Working back to the source to determine R_T for the network in Fig. 7.32.

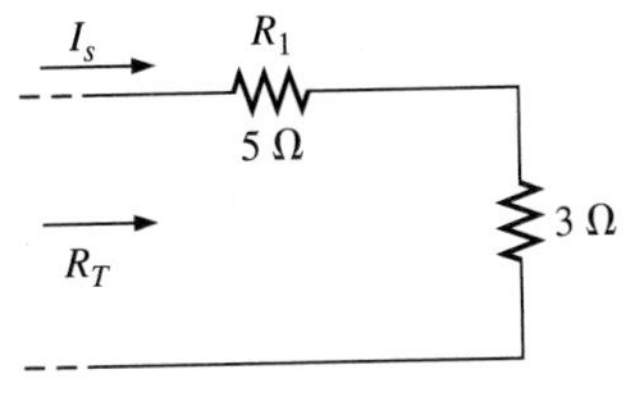

FIG. 7.34
Calculating R_T and I_s.

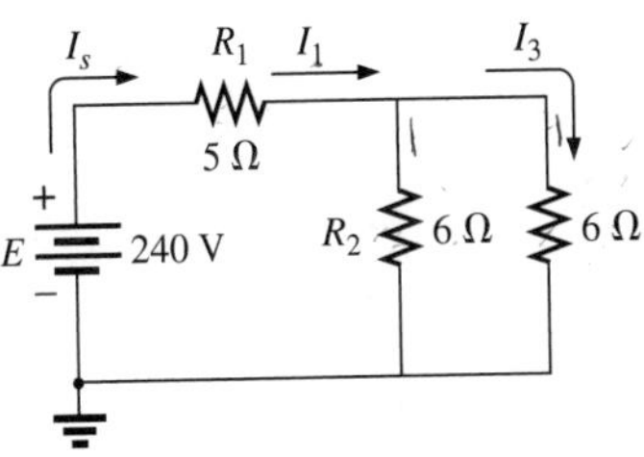

FIG. 7.35
Working back toward I_6.

Working our way back to I_6 (Fig. 7.35), we find that

$$I_1 = I_s$$

and

$$I_3 = \frac{I_s}{2} = \frac{30\text{ A}}{2} = 15\text{ A}$$

and, finally (Fig. 7.36),

$$I_6 = \frac{(6\ \Omega)I_3}{6\ \Omega + 3\ \Omega} = \frac{6}{9}(15\text{ A}) = 10\text{ A}$$

and

$$V_6 = I_6R_6 = (10\text{ A})(2\ \Omega) = \mathbf{20\text{ V}}$$

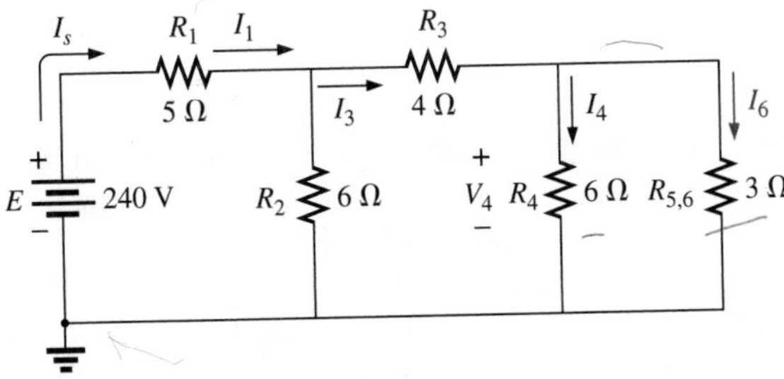

FIG. 7.36
Calculating I_6.

Method 2

Assign a letter symbol to the last branch current and work back through the network to the source, maintaining this assigned current or other current of interest. The desired current can then be found directly. This

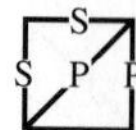

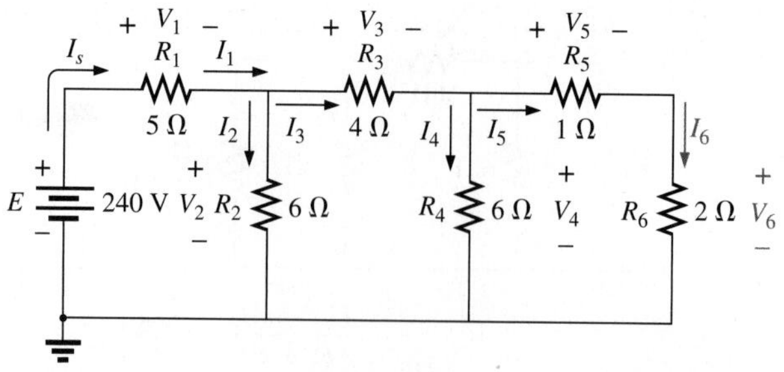

FIG. 7.37
An alternative approach for ladder networks.

method can best be described through the analysis of the same network considered in Fig. 7.32, redrawn in Fig. 7.37.

The assigned notation for the current through the final branch is I_6:

$$I_6 = \frac{V_4}{R_5 + R_6} = \frac{V_4}{1\ \Omega + 2\ \Omega} = \frac{V_4}{3\ \Omega}$$

or
$$V_4 = (3\ \Omega)I_6$$

so that
$$I_4 = \frac{V_4}{R_4} = \frac{(3\ \Omega)I_6}{6\ \Omega} = 0.5I_6$$

and
$$I_3 = I_4 + I_6 = 0.5I_6 + I_6 = 1.5I_6$$
$$V_3 = I_3R_3 = (1.5I_6)(4\ \Omega) = (6\ \Omega)I_6$$

Also,
$$V_2 = V_3 + V_4 = (6\ \Omega)I_6 + (3\ \Omega)I_6 = (9\ \Omega)I_6$$

so that
$$I_2 = \frac{V_2}{R_2} = \frac{(9\ \Omega)I_6}{6\ \Omega} = 1.5I_6$$

and
$$I_s = I_2 + I_3 = 1.5I_6 + 1.5I_6 = 3I_6$$

with
$$V_1 = I_1R_1 = I_sR_1 = (5\ \Omega)I_s$$

so that
$$E = V_1 + V_2 = (5\ \Omega)I_s + (9\ \Omega)I_6$$
$$= (5\ \Omega)(3I_6) + (9\ \Omega)I_6 = (24\ \Omega)I_6$$

and
$$I_6 = \frac{E}{24\ \Omega} = \frac{240\text{ V}}{24\ \Omega} = 10\text{ A}$$

with
$$V_6 = I_6R_6 = (10\text{ A})(2\ \Omega) = \mathbf{20\ V}$$

as was obtained using method 1.

Mathcad

We will now use Mathcad to analyze the ladder network in Fig. 7.32 using method 1. It provides an excellent opportunity to practice the basic maneuvers introduced in Sections 3.15 and 6.3.

First, as shown in Fig. 7.38, we define all the parameters of the network. Then we follow the same sequence as included in the text material. For Mathcad, however, we must be sure that the defining sequence for each new variable flows from left to right, as shown in Fig. 7.38, until R_{10} is defined. We are then ready to write the equation for the total resistance and display the result. All the remaining parameters are then defined and displayed as shown. The results are an exact match with the longhand solution.

The wonderful thing about Mathcad is that we can save this sequence in memory and use it as needed for different networks. Simply redefine the parameters of the network, and all the new values for the important parameters of the network will be displayed immediately.

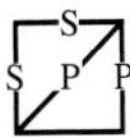

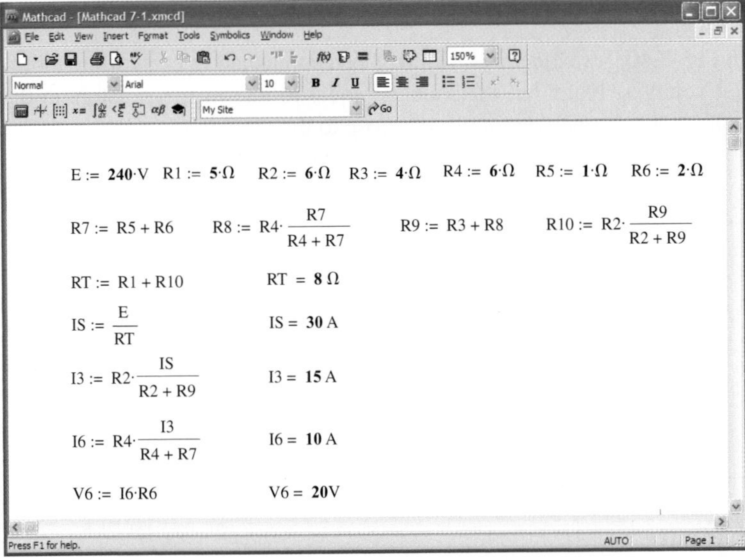

FIG. 7.38

Using Mathcad to analyze the ladder network in Fig. 7.32.

7.7 VOLTAGE DIVIDER SUPPLY (UNLOADED AND LOADED)

When the term *loaded* is used to describe voltage divider supply, it refers to the application of an element, network, or system to a supply that draws current from the supply. In other words,

the loading down of a system is the process of introducing elements that will draw current from the system. The heavier the current, the greater the loading effect.

Recall from Section 5.10 that the application of a load can affect the terminal voltage of a supply due to the internal resistance.

No-Load Conditions

Through a voltage divider network such as that in Fig. 7.39, a number of different terminal voltages can be made available from a single supply. Instead of having a single supply of 120 V, we now have terminal voltages of 100 V and 60 V available—a wonderful result for such a simple network. However, there can be disadvantages. One is that the applied resistive loads can have values too close to those making up the voltage divider network.

In general,

for a voltage divider supply to be effective, the applied resistive loads should be significantly larger than the resistors appearing in the voltage divider network.

To demonstrate the validity of the above statement, let us now examine the effect of applying resistors with values very close to those of the voltage divider network.

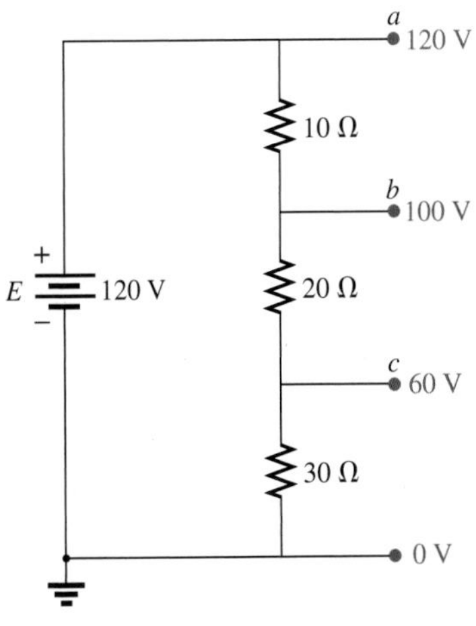

FIG. 7.39

Voltage divider supply.

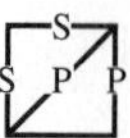

Loaded Conditions

In Fig. 7.40, resistors of 20 Ω have been connected to each of the terminal voltages. Note that this value is equal to one of the resistors in the voltage divider network and very close to the other two.

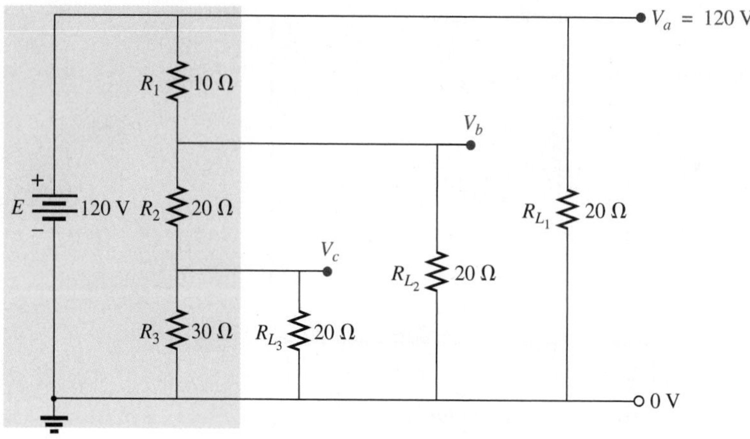

FIG. 7.40
Voltage divider supply with loads equal to the average value of the resistive elements that make up the supply.

Voltage V_a is unaffected by the load R_{L_1} since the load is in parallel with the supply voltage E. The result is $V_a = 120$ V, which is the same as the no-load level. To determine V_b, we must first note that R_3 and R_{L_3} are in parallel and $R'_3 = R_3 \parallel R_{L_3} = 30\ \Omega \parallel 20\ \Omega = 12\ \Omega$. The parallel combination

$$R'_2 = (R_2 + R'_3) \parallel R_{L_2} = (20\ \Omega + 12\ \Omega) \parallel 20\ \Omega$$
$$= 32\ \Omega \parallel 20\ \Omega = 12.31\ \Omega$$

Applying the voltage divider rule gives

$$V_b = \frac{(12.31\ \Omega)(120\ \text{V})}{12.31\ \Omega + 10\ \Omega} = \mathbf{66.21\ V}$$

versus 100 V under no-load conditions.

Voltage V_c is

$$V_c = \frac{(12\ \Omega)(66.21\ \text{V})}{12\ \Omega + 20\ \Omega} = \mathbf{24.83\ V}$$

versus 60 V under no-load conditions.

The effect of load resistors close in value to the resistor employed in the voltage divider network is, therefore, to decrease significantly some of the terminal voltages.

If the load resistors are changed to the 1 kΩ level, the terminal voltages will all be relatively close to the no-load values. The analysis is similar to the above, with the following results:

$$V_a = \mathbf{120\ V} \quad V_b = \mathbf{98.88\ V} \quad V_c = \mathbf{58.63\ V}$$

If we compare current drains established by the applied loads, we find for the network in Fig. 7.40 that

$$I_{L_2} = \frac{V_{L_2}}{R_{L_2}} = \frac{66.21\ \text{V}}{20\ \Omega} = 3.31\ \text{A}$$

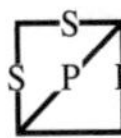

and for the 1 kΩ level,

$$I_{L_2} = \frac{98.88\text{ V}}{1\text{ k}\Omega} = 98.88\text{ mA} < 0.1\text{ A}$$

As demonstrated above, the greater the current drain, the greater the change in terminal voltage with the application of the load. This is certainly verified by the fact that I_{L_2} is about 33.5 times larger with the 20 Ω loads.

The next example is a design exercise. The voltage and current ratings of each load are provided, along with the terminal ratings of the supply. The required voltage divider resistors must be found.

EXAMPLE 7.12 Determine R_1, R_2, and R_3 for the voltage divider supply in Fig. 7.41. Can 2 W resistors be used in the design?

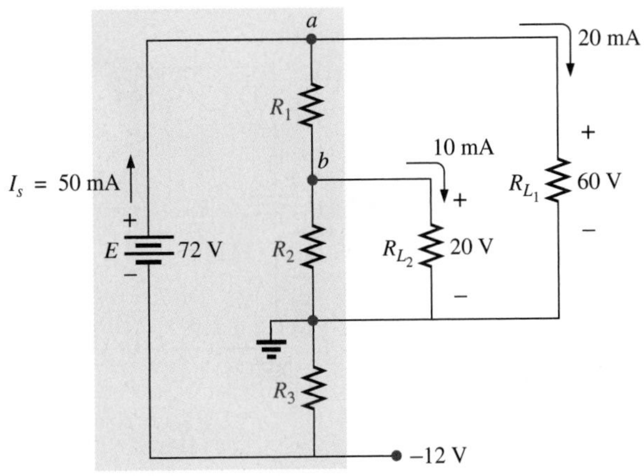

FIG. 7.41
Voltage divider supply for Example 7.12.

Solution: R_3:

$$R_3 = \frac{V_{R_3}}{I_{R_3}} = \frac{V_{R_3}}{I_s} = \frac{12\text{ V}}{50\text{ mA}} = \mathbf{240\ \Omega}$$

$$P_{R_3} = (I_{R_3})^2R_3 = (50\text{ mA})^2\,240\ \Omega = 0.6\text{ W} < 2\text{ W}$$

R_1: Applying Kirchhoff's current law to node a:

$$I_s - I_{R_1} - I_{L_1} = 0$$

and $I_{R_1} = I_s - I_{L_1} = 50\text{ mA} - 20\text{ mA} = 30\text{ mA}$

$$R_1 = \frac{V_{R_1}}{I_{R_1}} = \frac{V_{L_1} - V_{L_2}}{I_{R_1}} = \frac{60\text{ V} - 20\text{ V}}{30\text{ mA}} = \frac{40\text{ V}}{30\text{ mA}} = \mathbf{1.33\text{ k}\Omega}$$

$$P_{R_1} = (I_{R_1})^2R_1 = (30\text{ mA})^2\,1.33\text{ k}\Omega = 1.197\text{ W} < 2\text{ W}$$

R_2: Applying Kirchhoff's current law at node b:

$$I_{R_1} - I_{R_2} - I_{L_2} = 0$$

and $$I_{R_2} = I_{R_1} - I_{L_2} = 30\text{ mA} - 10\text{ mA} = 20\text{ mA}$$

$$R_2 = \frac{V_{R_2}}{I_{R_2}} = \frac{20\text{ V}}{20\text{ mA}} = \mathbf{1\text{ k}\Omega}$$

$$P_{R_2} = (I_{R_2})^2R_2 = (20\text{ mA})^2\,1\text{ k}\Omega = 0.4\text{ W} < 2\text{ W}$$

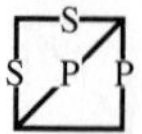

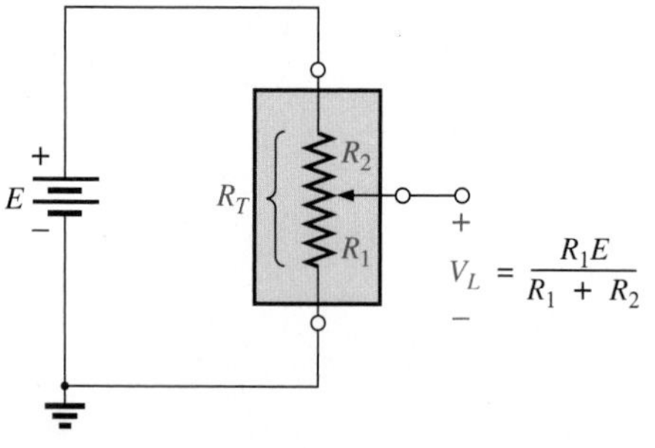

FIG. 7.42
Unloaded potentiometer.

Since P_{R_1}, P_{R_2}, and P_{R_3} are less than 2 W, 2 W resistors can be used for the design.

7.8 POTENTIOMETER LOADING

For the unloaded potentiometer in Fig. 7.42, the output voltage is determined by the voltage divider rule, with R_T in the figure representing the total resistance of the potentiometer. Too often it is assumed that the voltage across a load connected to the wiper arm is determined solely by the potentiometer and the effect of the load can be ignored. This is definitely not the case, as is demonstrated here.

When a load is applied as shown in Fig. 7.43, the output voltage V_L is now a function of the magnitude of the load applied since R_1 is not as shown in Fig. 7.42 but is instead the parallel combination of R_1 and R_L.

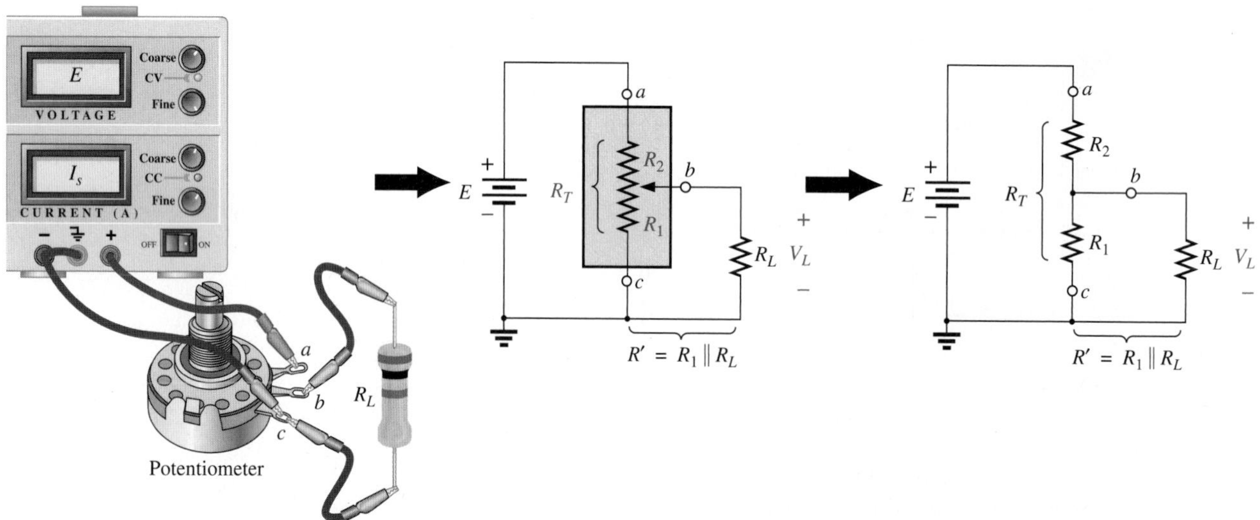

FIG. 7.43
Loaded potentiometer.

The output voltage is now

$$V_L = \frac{R'E}{R' + R_2} \quad \text{with } R' = R_1 \| R_L \tag{7.1}$$

If you want to have good control of the output voltage V_L through the controlling dial, knob, screw, or whatever, you must choose a load or potentiometer that satisfies the following relationship:

$$R_L \gg R_T \tag{7.2}$$

In general,

when hooking up a load to a potentiometer, be sure that the load resistance far exceeds the maximum terminal resistance of the potentiometer if good control of the output voltage is desired.

For example, let's disregard Eq. (7.2) and choose a 1 MΩ potentiometer with a 100 Ω load and set the wiper arm to 1/10 the total resistance, as shown in Fig. 7.44. Then

$$R' = 100\text{ k}\Omega \parallel 100\ \Omega = 99.9\ \Omega$$

and
$$V_L = \frac{99.9\ \Omega(10\text{ V})}{99.9\ \Omega + 900\text{ k}\Omega} \cong 0.001\text{ V} = 1\text{ mV}$$

which is extremely small compared to the expected level of 1 V.

In fact, if we move the wiper arm to the midpoint,

$$R' = 500\text{ k}\Omega \parallel 100\ \Omega = 99.98\ \Omega$$

and
$$V_L = \frac{(99.98\ \Omega)(10\text{ V})}{99.98\ \Omega + 500\text{ k}\Omega} \cong 0.002\text{ V} = 2\text{ mV}$$

which is negligible compared to the expected level of 5 V. Even at $R_1 = 900\text{ k}\Omega$, V_L is only 0.01 V, or 1/1000 of the available voltage.

Using the reverse situation of $R_T = 100\ \Omega$ and $R_L = 1\text{ M}\Omega$ and the wiper arm at the 1/10 position, as in Fig. 7.45, we find

$$R' = 10\ \Omega \parallel 1\text{ M}\Omega \cong 10\ \Omega$$

and
$$V_L = \frac{10\ \Omega(10\text{ V})}{10\ \Omega + 90\ \Omega} = 1\text{ V}$$

as desired.

For the lower limit (worst-case design) of $R_L = R_T = 100\ \Omega$, as defined by Eq. (7.2) and the halfway position of Fig. 7.43,

$$R' = 50\ \Omega \parallel 100\ \Omega = 33.33\ \Omega$$

and
$$V_L = \frac{33.33\ \Omega(10\text{ V})}{33.33\ \Omega + 50\ \Omega} \cong 4\text{ V}$$

It may not be the ideal level of 5 V, but at least 40% of the voltage E has been achieved at the halfway position rather than the 0.02% obtained with $R_L = 100\ \Omega$ and $R_T = 1\text{ M}\Omega$.

In general, therefore, try to establish a situation for potentiometer control in which Eq. (7.2) is satisfied to the highest degree possible.

Someone might suggest that we make R_T as small as possible to bring the percent result as close to the ideal as possible. Keep in mind, however, that the potentiometer has a power rating, and for networks such as Fig. 7.45, $P_{max} \cong E^2/R_T = (10\text{ V})^2/100\ \Omega = 1\text{ W}$. If R_T is reduced to 10 Ω, $P_{max} = (10\text{ V})^2/10\ \Omega = 10\text{ W}$, which would require a *much larger* unit.

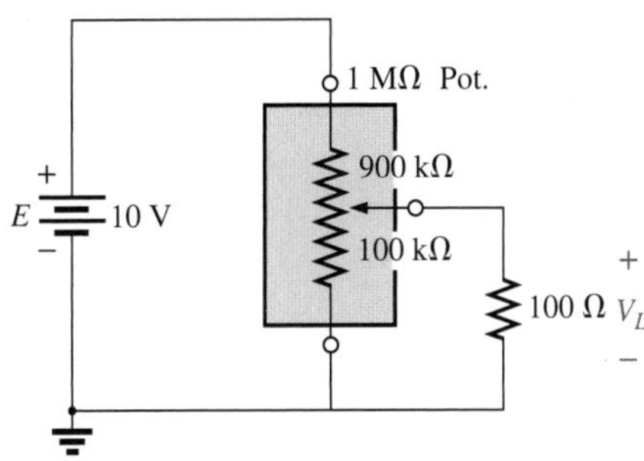

FIG. 7.44
Loaded potentiometer with $R_L \ll R_T$.

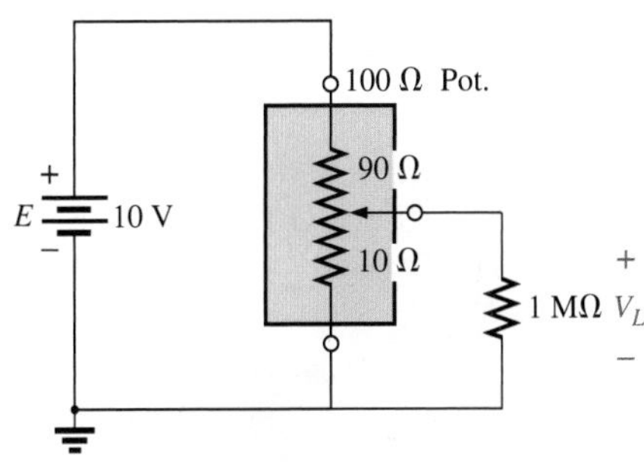

FIG. 7.45
Loaded potentiometer with $R_L \gg R_T$.

EXAMPLE 7.13 Find voltages V_1 and V_2 for the loaded potentiometer of Fig. 7.46.

Solution: Ideal (no load):

$$V_1 = \frac{4\text{ k}\Omega(120\text{ V})}{10\text{ k}\Omega} = \mathbf{48\text{ V}}$$

$$V_2 = \frac{6\text{ k}\Omega(120\text{ V})}{10\text{ k}\Omega} = \mathbf{72\text{ V}}$$

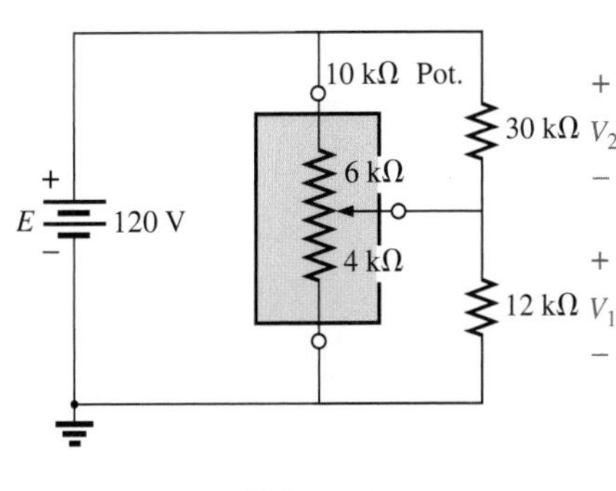

FIG. 7.46
Example 7.13.

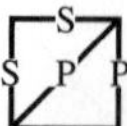

Loaded:

$$R' = 4\text{ k}\Omega \parallel 12\text{ k}\Omega = 3\text{ k}\Omega$$
$$R'' = 6\text{ k}\Omega \parallel 30\text{ k}\Omega = 5\text{ k}\Omega$$
$$V_1 = \frac{3\text{ k}\Omega(120\text{ V})}{8\text{ k}\Omega} = \mathbf{45\text{ V}}$$
$$V_2 = \frac{5\text{ k}\Omega(120\text{ V})}{8\text{ k}\Omega} = \mathbf{75\text{ V}}$$

The ideal and loaded voltage levels are so close that the design can be considered a good one for the applied loads. A slight variation in the position of the wiper arm will establish the ideal voltage levels across the two loads.

7.9 AMMETER, VOLTMETER, AND OHMMETER DESIGN

Now that the fundamentals of series, parallel, and series-parallel networks have been introduced, we are prepared to investigate the fundamental design of an ammeter, voltmeter, and ohmmeter. Our design of each uses the **d'Arsonval analog movement** of Fig. 7.47. The movement consists basically of an iron-core coil mounted on jewel bearings between a permanent magnet. The helical springs limit the turning motion of the coil and provide a path for the current to reach the coil. When a current is passed through the movable coil, the fluxes of the coil and permanent magnet interact to develop a torque on the coil that cause it to rotate on its bearings. The movement is adjusted to indicate zero deflection on a meter scale when the current through the coil is zero. The direction of current through the coil then determines whether the pointer displays an up-scale or below-zero indication. For this reason, ammeters and voltmeters have an assigned polarity on their terminals to ensure an up-scale reading.

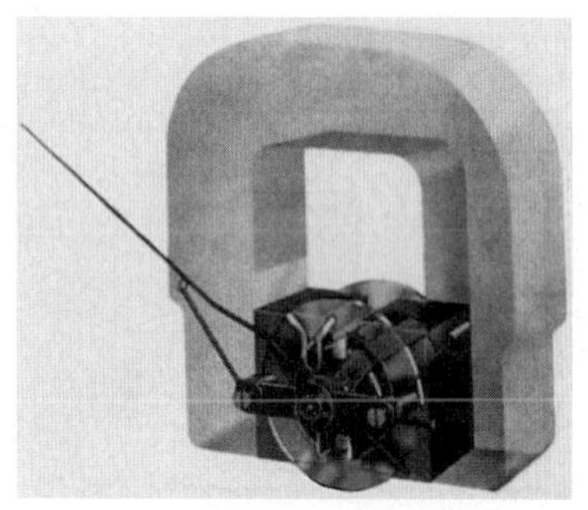

FIG. 7.47
d'Arsonval analog movement.
(Courtesy of Weston Instruments, Inc.)

D'Arsonval movements are usually rated by current and resistance. The specifications of a typical movement may be 1 mA, 50 Ω. The 1 mA is the *current sensitivity (CS)* of the movement, which is the current required for a full-scale deflection. It is denoted by the symbol I_{CS}. The 50 Ω represents the internal resistance (R_m) of the movement. A common notation for the movement and its specifications is provided in Fig. 7.48.

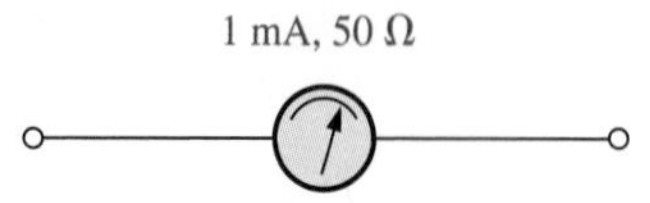

FIG. 7.48
Movement notation.

The Ammeter

The maximum current that the d'Arsonval movement can read independently is equal to the current sensitivity of the movement. However, higher currents can be measured if additional circuitry is introduced. This additional circuitry, as shown in Fig. 7.49, results in the basic construction of an ammeter.

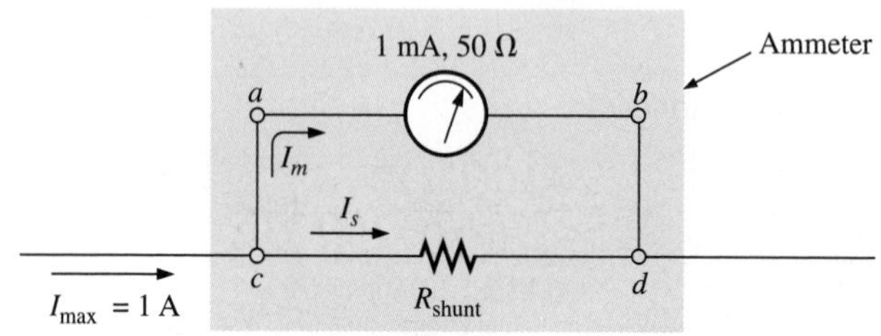

FIG. 7.49
Basic ammeter.

The resistance R_{shunt} is chosen for the ammeter in Fig. 7.49 to allow 1 mA to flow through the movement when a maximum current of 1 A enters the ammeter. If less than 1 A flows through the ammeter, the movement will have less than 1 mA flowing through it and will indicate less than full-scale deflection.

Since the voltage across parallel elements must be the same, the potential drop across *a-b* in Fig. 7.49 must equal that across *c-d*; that is,

$$(1\text{ mA})(50\ \Omega) = R_{shunt}I_s$$

Also, I_s must equal 1 A − 1 mA = 999 mA if the current is to be limited to 1 mA through the movement (Kirchhoff's current law). Therefore,

$$(1\text{ mA})(50\ \Omega) = R_{shunt}(999\text{ mA})$$

$$R_{shunt} = \frac{(1\text{ mA})(50\ \Omega)}{999\text{ mA}}$$

$$\cong 0.05\ \Omega$$

In general,

$$R_{shunt} = \frac{R_m I_{CS}}{I_{max} - I_{CS}} \qquad \textbf{(7.4)}$$

One method of constructing a multirange ammeter is shown in Fig. 7.50, where the rotary switch determines the R_{shunt} to be used for the maximum current indicated on the face of the meter. Most meters use the same scale for various values of maximum current. If you read 375 on the 0–5 mA scale with the switch on the 5 setting, the current is 3.75 mA; on the 50 setting, the current is 37.5 mA; and so on.

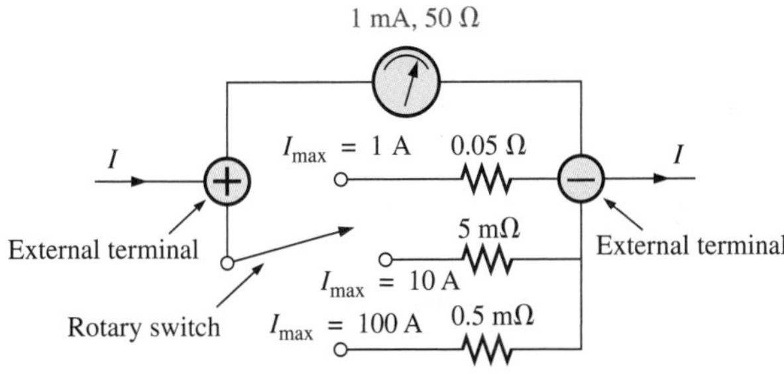

FIG. 7.50
Multirange ammeter.

The Voltmeter

A variation in the additional circuitry permits the use of the d'Arsonval movement in the design of a voltmeter. The 1 mA, 50 Ω movement can also be rated as a 50 mV (1 mA × 50 Ω), 50 Ω movement, indicating that the maximum voltage that the movement can measure independently is 50 mV. The millivolt rating is sometimes referred to as the *voltage sensitivity (VS)*. The basic construction of the voltmeter is shown in Fig. 7.51.

The R_{series} is adjusted to limit the current through the movement to 1 mA when the maximum voltage is applied across the voltmeter. A lesser voltage simply reduces the current in the circuit and thereby the deflection of the movement.

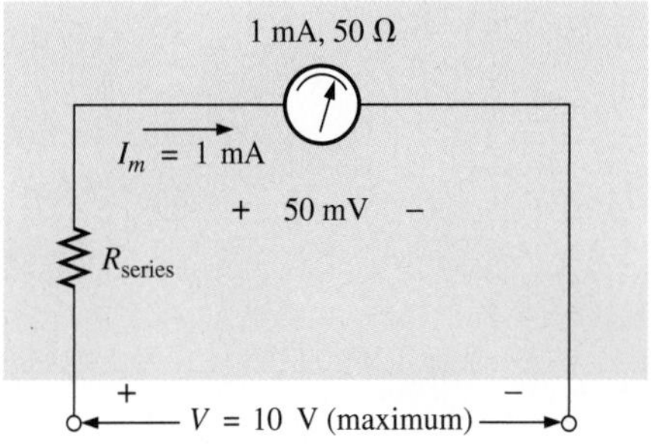

FIG. 7.51
Basic voltmeter.

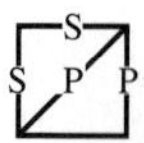

Applying Kirchhoff's voltage law around the closed loop of Fig. 7.51, we obtain

$$[10\text{ V} - (1\text{ mA})(R_{\text{series}})] - 50\text{ mV} = 0$$

or

$$R_{\text{series}} = \frac{10\text{ V} - (50\text{ mV})}{1\text{ mA}} = 9950\ \Omega$$

In general,

$$R_{\text{series}} = \frac{V_{\max} - V_{VS}}{I_{CS}} \tag{7.5}$$

One method of constructing a multirange voltmeter is shown in Fig. 7.52. If the rotary switch is at 10 V, $R_{\text{series}} = 9.95\text{ k}\Omega$; at 50 V, $R_{\text{series}} = 40\text{ k}\Omega + 9.95\text{ k}\Omega = 49.95\text{ k}\Omega$; and at 100 V, $R_{\text{series}} = 50\text{ k}\Omega + 40\text{ k}\Omega + 9.95\text{ k}\Omega = 99.95\text{ k}\Omega$.

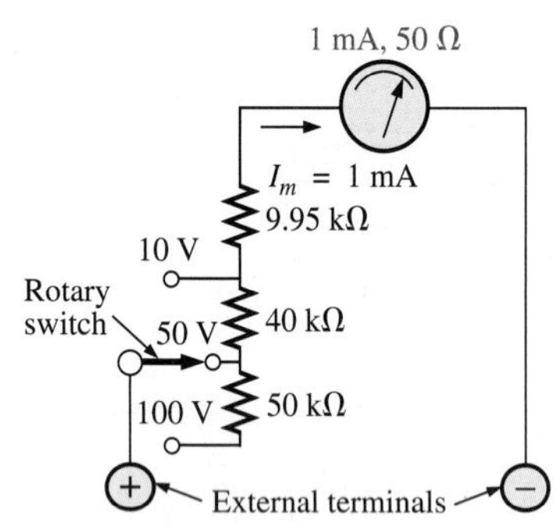

FIG. 7.52
Multirange voltmeter.

The Ohmmeter

In general, ohmmeters are designed to measure resistance in the low, mid-, or high range. The most common is the **series ohmmeter**, designed to read resistance levels in the midrange. It uses the series configuration in Fig. 7.53. The design is quite different from that of the ammeter or voltmeter because it shows a full-scale deflection for zero ohms and no deflection for infinite resistance.

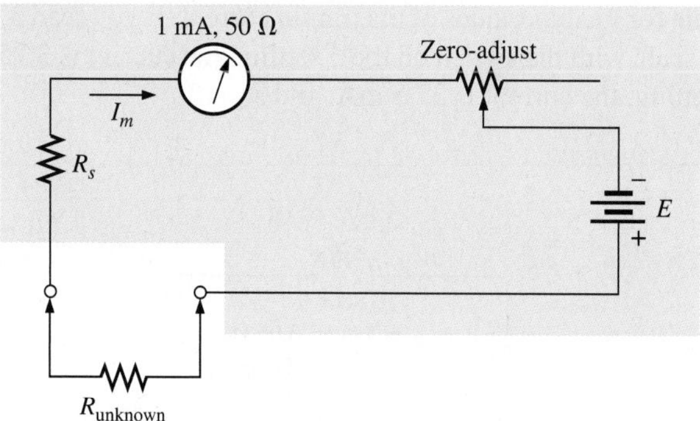

FIG. 7.53
Series ohmmeter.

To determine the series resistance R_s, the external terminals are shorted (a direct connection of zero ohms between the two) to simulate zero ohms, and the zero-adjust is set to half its maximum value. The resistance R_s is then adjusted to allow a current equal to the current sensitivity of the movement (1 mA) to flow in the circuit. The zero-adjust is set to half its value so that any variation in the components of the meter that may produce a current more or less than the current sensitivity can be compensated for. The current I_m is

$$I_m(\text{full scale}) = I_{CS} = \frac{E}{R_s + R_m + \dfrac{\text{zero} - \text{adjust}}{2}} \tag{7.6}$$

and

$$R_s = \frac{E}{I_{CS}} - R_m - \frac{\text{zero-adjust}}{2} \tag{7.7}$$

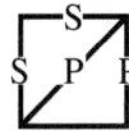

If an unknown resistance is then placed between the external terminals, the current is reduced, causing a deflection less than full scale. If the terminals are left open, simulating infinite resistance, the pointer does not deflect since the current through the circuit is zero.

An instrument designed to read very low values of resistance appears in Fig. 7.54. Because of its low-range capability, the network design must be a great deal more sophisticated than described above. It uses electronic components that eliminate the inaccuracies introduced by lead and contact resistances. It is similar to the above system in the sense that it is completely portable and does require a dc battery to establish measurement conditions. Special leads are used to limit any introduced resistance levels. The maximum scale setting can be set as low as 0.00352 (3.52 mΩ).

The **megohmmeter** (often called a *megger*) is an instrument for measuring very high resistance values. Its primary function is to test the insulation found in power transmission systems, electrical machinery, transformers, and so on. To measure the high-resistance values, a high dc voltage is established by a hand-driven generator. If the shaft is rotated above some set value, the output of the generator is fixed at one selectable voltage, typically 250 V, 500 V, or 1000 V. A photograph of a commercially available tester is shown in Fig. 7.55. For this instrument, the range is zero to 5000 MΩ.

FIG. 7.54
Milliohmmeter.
(Courtesy of Keithley Instruments, Inc.)

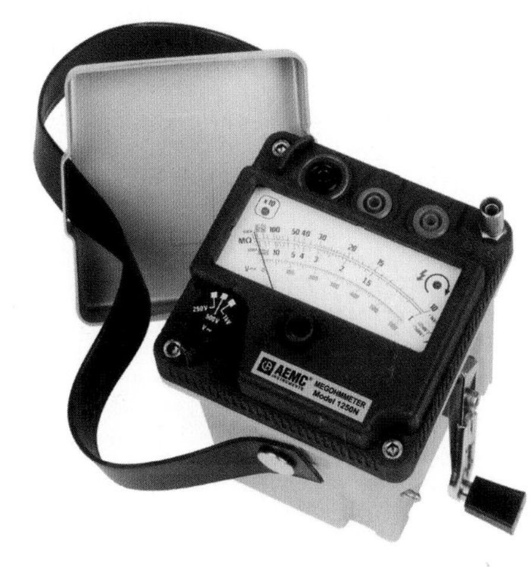

FIG. 7.55
Megohmmeter.
(Courtesy of AEMC® Instruments, Foxborough, MA.)

7.10 APPLICATIONS

Boosting a Car Battery

Although boosting a car battery may initially appear to be a simple application of parallel networks, it is really a series-parallel operation that is worthy of some investigation. As indicated in Chapter 2, every dc supply has some internal resistance. For the typical 12 V lead-acid car battery, the resistance is quite small—in the milliohm range. In most cases, the low internal resistance ensures that most of the voltage (or power) is delivered to the load and not lost on the internal resistance. In Fig. 7.56, battery #2 has discharged because the lights were left on for three hours during a movie. Fortunately, a friend who made sure his own lights were off has a fully charged battery #1 and a good set of 16-ft cables with #6 gage stranded wire and well-designed clips. The investment in a good set of cables with sufficient length and heavy wire is a wise one, particularly if you live in a cold climate. Flexibility, as provided by stranded wire, is also a very desirable characteristic under some conditions. Be sure to check the gage of the

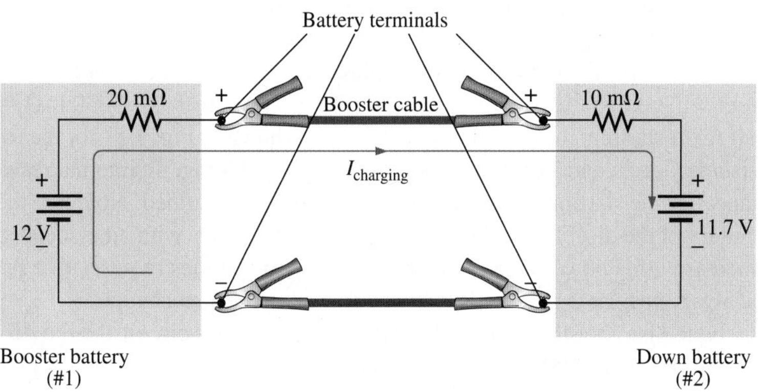

FIG. 7.56
Boosting a car battery.

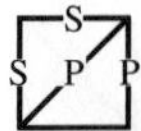

wire and not just the thickness of the insulating jacket. You get what you pay for, and the copper is the most expensive part of the cables. Too often the label says "heavy-duty," but the gage number of the wire is too high.

The proper sequence of events in boosting a car is often a function of to whom you speak or what information you read. For safety sake, some people recommend that the car with the good battery be turned off when making the connections. This, however, can create an immediate problem if the "dead" battery is in such a bad state that when it is hooked up to the good battery, it immediately drains the good battery to the point that neither car will start. With this in mind, it does make some sense to leave the car running to ensure that the charging process continues until the starting of the disabled car is initiated. *Because accidents do happen, it is strongly recommended that the person making the connections wear the proper type of protective eye equipment. Take sufficient time to be sure that you know which are the positive and negative terminals for both cars.* If it's not immediately obvious, keep in mind that the negative or ground side is usually connected to the chassis of the car with a relatively short, heavy wire.

When you are sure of which are the positive and negative terminals, first connect one of the red wire clamps of the booster cables to the positive terminal of the discharged battery—all the while being sure that the other red clamp *is not touching the battery or car.* Then connect the other end of the red wire to the positive terminal of the fully charged battery. Next, connect one end of the black cable of the booster cables to the negative terminal of the booster battery, and finally connect the other end of the black cable to the engine block of the stalled vehicle (not the negative post of the dead battery) away from the carburetor, fuel lines, or moving parts of the car. Lastly, have someone maintain a constant idle speed in the car with the good battery as you start the car with the bad battery. After the vehicle starts, remove the cables in the *reverse order* starting with the cable connected to the engine block. Always be careful to ensure that clamps don't touch the battery or chassis of the car or get near any moving parts.

Some people feel that the car with the good battery should charge the bad battery for 5 to 10 minutes before starting the disabled car so the disabled car will be essentially using its own battery in the starting process. Keep in mind that the instant the booster cables are connected, the booster car is making a concerted effort to charge both its own battery and the drained battery. At starting, the good battery is asked to supply a heavy current to start the other car. It's a pretty heavy load to put on a single battery. For the situation in Fig. 7.56, the voltage of battery #2 is less than that of battery #1, and the charging current will flow as shown. The resistance in series with the boosting battery is more because of the long length of the booster cable to the other car. The current is limited only by the series milliohm resistors of the batteries, but the voltage difference is so small that the starting current will be in safe range for the cables involved. The initial charging current will be $I = (12\text{ V} - 11.7\text{ V})/(20\text{ m}\Omega + 10\text{ m}\Omega) = 0.3\text{ V}/30\text{ m}\Omega = 10\text{ A}$. At starting, the current levels will be as shown in Fig. 7.57 for the resistance levels and battery voltages assumed. At starting, an internal resistance for the starting circuit of $0.1\ \Omega = 100\text{ m}\Omega$ is assumed. Note that the battery of the disabled car has now charged up to 11.8 V with an associated increase in its power level. The presence of two batteries requires that the analysis wait for the methods to be introduced in the next chapter.

Note also that the current drawn from the starting circuit for the disabled car is over 100 A and that the majority of the starting current is provided by the battery being charged. In essence, therefore, the majority of the starting current is coming from the disabled car. The good battery has provided an initial charge to the bad battery and has provided the addi-

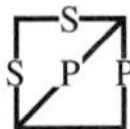

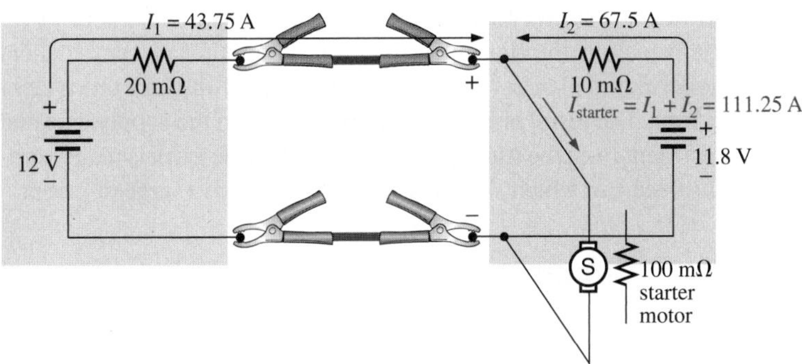

FIG. 7.57
Current levels at starting.

tional current necessary to start the car. But, in total, it is the battery of the disabled car that is the primary source of the starting current. For this very reason, the charging action should continue for 5 or 10 minutes before starting the car. If the disabled car is in really bad shape with a voltage level of only 11 V, the resulting levels of current will reverse, with the good battery providing 68.75 A and the bad battery only 37.5 A. Quite obviously, therefore, the worse the condition of the dead battery, the heavier the drain on the good battery. A point can also be reached where the bad battery is in such bad shape that it cannot accept a good charge or provide its share of the starting current. The result can be continuous cranking of the disabled car without starting and possible damage to the battery of the running car due to the enormous current drain. Once the car is started and the booster cables are removed, the car with the discharged battery will continue to run because the alternator will carry the load (charging the battery and providing the necessary dc voltage) after ignition.

The above discussion was all rather straightforward, but let's investigate what may happen if it is a dark and rainy night, you are rushed, and you hook up the cables incorrectly as shown in Fig. 7.58. The result is two series-aiding batteries and a very low resistance path. The resulting current can then theoretically be extremely high [$I = (12\text{ V} + 11.7\text{ V})/30\text{ m}\Omega = 23.7\text{ V}/30\text{ m}\Omega = 790\text{ A}$], perhaps permanently damaging the electrical system of both cars and, worst of all, causing an explosion that may seriously injure someone. It is therefore very important that you treat the process of boosting a car with great care. Find that flashlight, double-check the connections, and be sure that everyone is clear when you start that car.

Before leaving the subject, we should point out that getting a boost from a tow truck results in a somewhat different situation: The connections to the

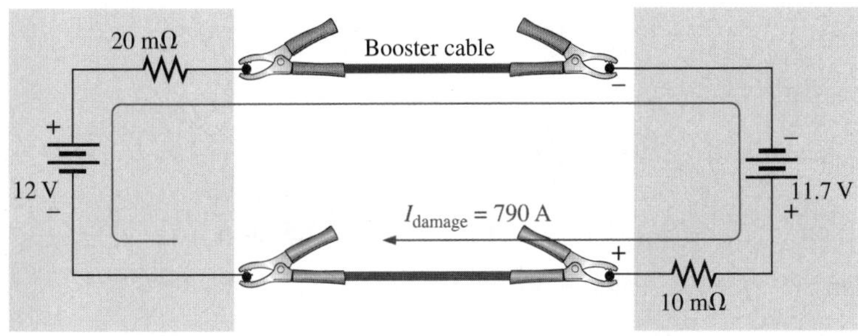

FIG. 7.58
Current levels if the booster battery is improperly connected.

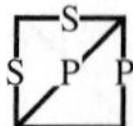

battery in the truck are very secure; the cable from the truck is a heavy wire with thick insulation; the clamps are also quite large and make an excellent connection with your battery; and the battery is heavy-duty for this type of expected load. The result is less internal resistance on the supply side and a heavier current from the truck battery. In this case, the truck is really starting the disabled car, which simply reacts to the provided surge of power.

Electronic Circuits

The operation of most electronic systems requires a distribution of dc voltages throughout the design. Although a full explanation of why the dc level is required (since it is an ac signal to be amplified) will have to wait for the introductory courses in electronic circuits, the dc analysis will proceed in much the same manner as described in this chapter. In other words, this chapter and the preceding chapters are sufficient background to perform the dc analysis of the majority of electronic networks you will encounter if given the dc terminal characteristics of the electronic elements. For example, the network in Fig. 7.59 using a transistor will be covered in detail in any introductory electronics course. The dc voltage between the base (B) of the transistor and the emitter (E) is about 0.7 V under normal operating conditions, and the collector (C) is related to the base current by $I_C = \beta I_B = 50I_B$ (β varies from transistor to transistor). Using these facts will enable us to determine all the dc currents and voltages of the network using the laws introduced in this chapter. In general, therefore, be encouraged that you will use the content of this chapter in numerous applications in the courses to follow.

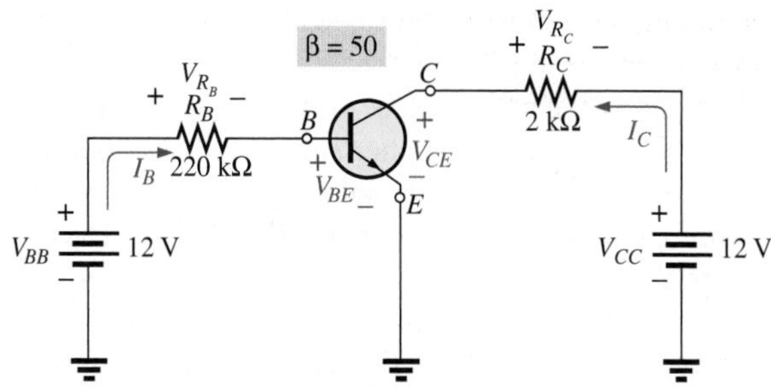

FIG. 7.59
dc bias levels of a transistor amplifier.

For the network in Fig. 7.59 we begin our analysis by applying Kirchhoff's voltage law to the base circuit:

$$+V_{BB} - V_{R_B} - V_{BE} = 0 \quad \text{or} \quad V_{BB} = V_{R_B} + V_{BE}$$

and $$V_{R_B} = V_{BB} - V_{BE} = 12\text{ V} - 0.7\text{ V} = 11.3\text{ V}$$

so that $$V_{R_B} = I_B R_B = 11.3\text{ V}$$

and $$I_B = \frac{V_{R_B}}{R_B} = \frac{11.3\text{ V}}{220\text{ k}\Omega} = \mathbf{51.4\ \mu A}$$

Then $$I_C = \beta I_B = 50I_B = 50(51.4\ \mu\text{A}) = \mathbf{2.57\ mA}$$

and $$+V_{CE} + V_{R_C} - V_{CC} = 0 \quad \text{or} \quad V_{CC} = V_{R_C} + V_{CE}$$

with $$V_{CE} = V_{CC} - V_{R_C} = V_{CC} - I_C R_C = 12\text{ V} - (2.57\text{ mA})(2\text{ k}\Omega)$$
$$= 12\text{ V} - 5.14\text{ V} = \mathbf{6.86\ V}$$

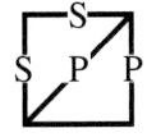

For a typical dc analysis of a transistor, all the currents and voltages of interest are now known: I_B, V_{BE}, I_C, and V_{CE}. All the remaining voltage, current, and power levels for the other elements of the network can now be found using the basic laws applied in this chapter.

The above example is typical of the type of exercise you will be asked to perform in your first electronics course. For now you only need to be exposed to the device and to understand the reason for the relationships between the various currents and voltages of the device.

7.11 COMPUTER ANALYSIS

PSpice

Voltage Divider Supply We will now use PSpice to verify the results of Example 7.12. The calculated resistor values will be substituted and the voltage and current levels checked to see if they match the handwritten solution.

As shown in Fig. 7.60, the network is drawn as in earlier chapters using only the tools described thus far—in one way, a practice exercise for everything learned about the **Capture CIS Edition.** Note in this case that rotating the first resistor sets everything up for the remaining resistors. Further, it is a nice advantage that you can place one resistor after another without going to the **End Mode** option. Be especially careful with the placement of the ground, and be sure that **0/SOURCE** is used. Note also that resistor R_1 in Fig. 7.60 was entered as 1.333 kΩ rather than 1.33 kΩ as in Example 7.12. When running the program, we found that the computer solutions were not a perfect match to the longhand solution to the level of accuracy desired unless this change was made.

Since all the voltages are to ground, the voltage across R_{L_1} is 60 V; across R_{L_2}, 20 V; and across R_3, −12 V. The currents are also an excellent

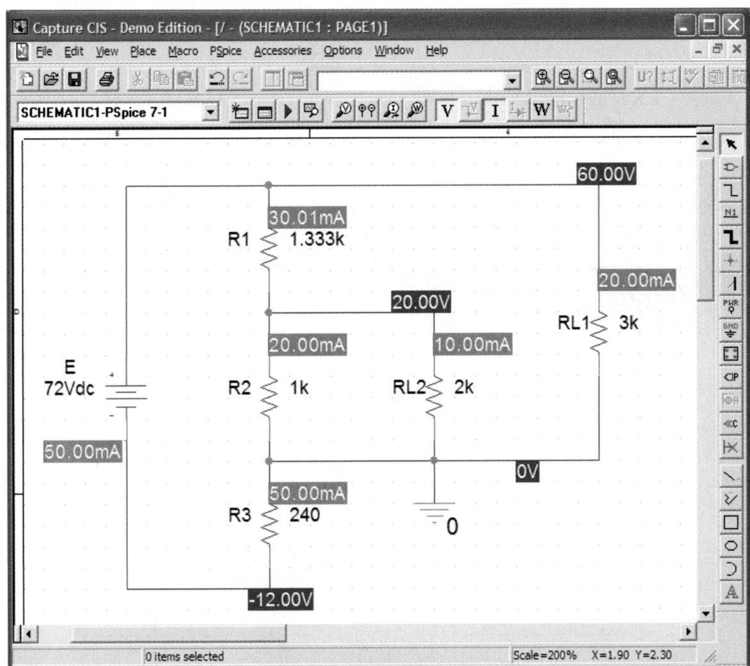

FIG. 7.60
Using PSpice to verify the results of Example 7.12.

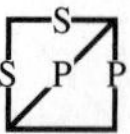

match with the handwritten solution, with $I_E = 50$ mA, $I_{R_1} = 30$ mA, $I_{R_2} = 20$ mA, $I_{R_3} = 50$ mA, $I_{R_{L2}} = 10$ mA, and $I_{R_{L1}} = 20$ mA. For the display in Fig. 7.60, the **W** option was disabled to permit concentrating on the voltage and current levels. This time, there is an exact match with the longhand solution.

PROBLEMS

SECTIONS 7.2–7.5 Series-Parallel Networks

1. Which elements (individual elements, not combinations of elements) of the networks in Fig. 7.61 are in series? Which are in parallel? As a check on your assumptions, be sure that the elements in series have the same current and that the elements in parallel have the same voltage. Restrict your decisions to single elements, not combinations of elements.

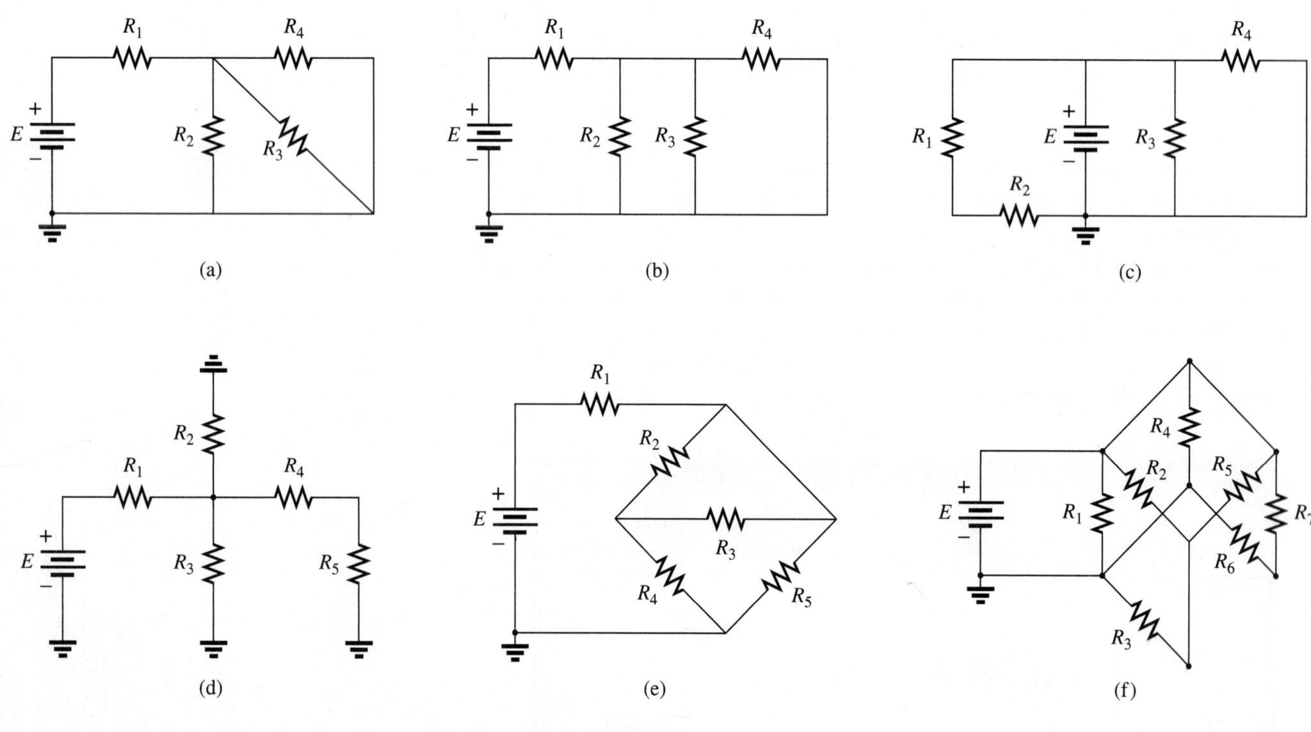

FIG. 7.61
Problem 1.

2. Determine R_T for the networks in Fig. 7.62.

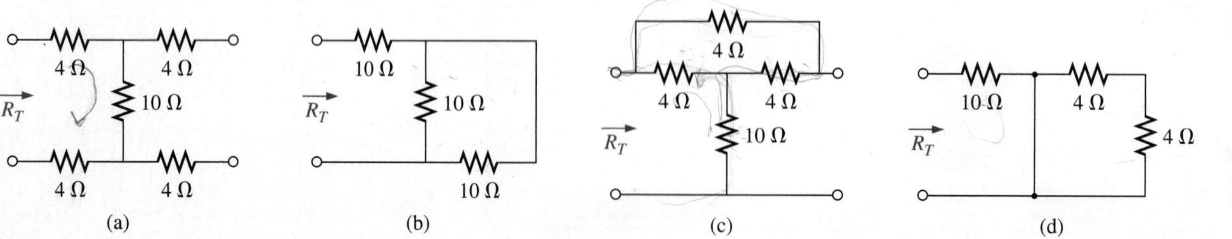

FIG. 7.62
Problem 2.

3. For the network in Fig. 7.63:
   a. Does $I_s = I_5 = I_6$? Explain.
   b. If $I_s = 10$ A and $I_1 = 4$ A, find I_2.
   c. Does $I_1 + I_2 = I_3 + I_4$? Explain.
   d. If $V_2 = 8$ V and $E = 14$ V, find V_3.
   e. If $R_1 = 4\ \Omega$, $R_2 = 2\ \Omega$, $R_3 = 4\ \Omega$, and $R_4 = 6\ \Omega$, what is R_T?
   f. If all the resistors of the configuration are 20 Ω, what is the source current if the applied voltage is 20 V?
   g. Using the values of part (f), find the power delivered by the battery and the power absorbed by the total resistance R_T.

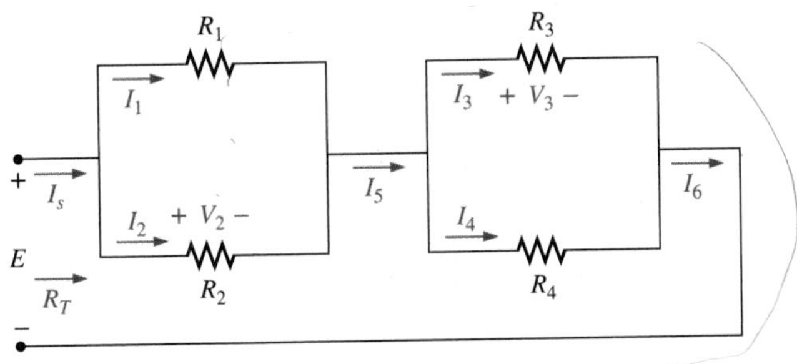

FIG. 7.63
Problem 3.

4. For the network in Fig. 7.64:
   a. Find the total resistance R_T.
   b. Find the source current I_s and currents I_2 and I_3.
   c. Find current I_5.
   d. Find voltages V_2 and V_4.

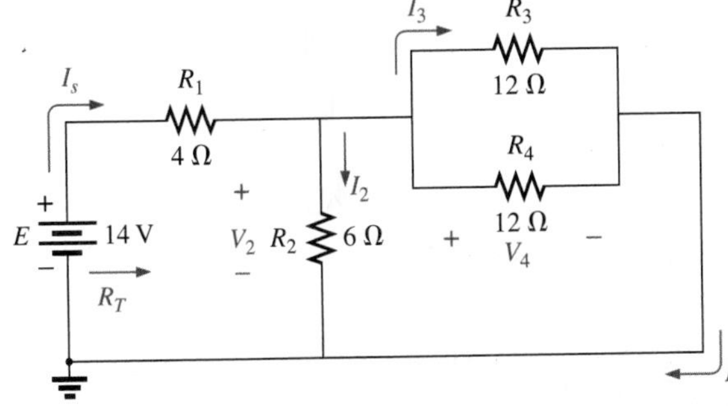

FIG. 7.64
Problem 4.

5. For the network in Fig. 7.65:
   a. Determine R_T.
   b. Find I_s, I_1, and I_2.
   c. Find voltage V_4.

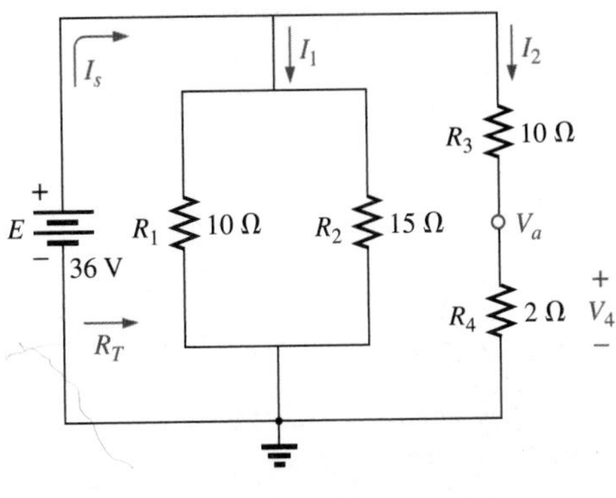

FIG. 7.65
Problem 5.

6. For the circuit board in Fig. 7.66:
   a. Find the total resistance R_T of the configuration.
   b. Find the current drawn from the supply if the applied voltage is 48 V.
   c. Find the reading of the applied voltmeter.

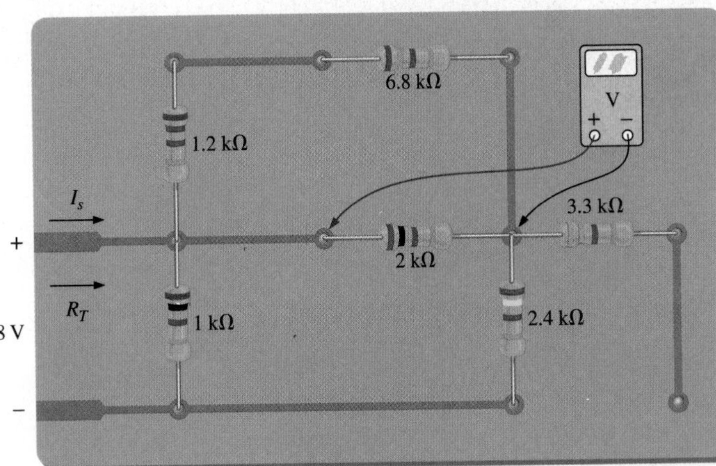

FIG. 7.66
Problem 6.

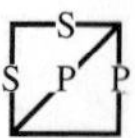

*7. For the network in Fig. 7.67:
   a. Find currents I_s, I_2, and I_6.
   b. Find voltages V_1 and V_5.
   c. Find the power delivered to the 3 kΩ resistor.

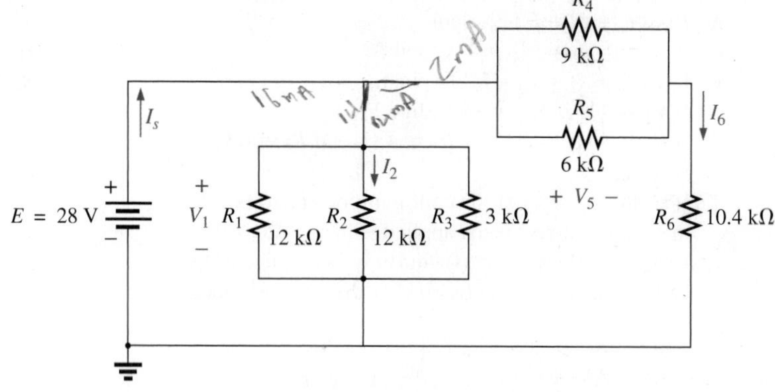

FIG. 7.67
Problem 7.

*8. For the series-parallel configuration in Fig. 7.68:
   a. Find the source current I_s.
   b. Find currents I_3 and I_9.
   c. Find current I_8.
   d. Find voltage V_x.

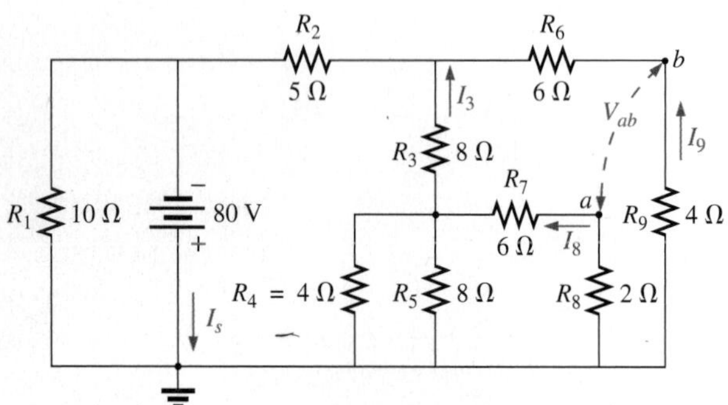

FIG. 7.68
Problem 8.

9. Determine the currents I_1 and I_2 for the network in Fig. 7.69.

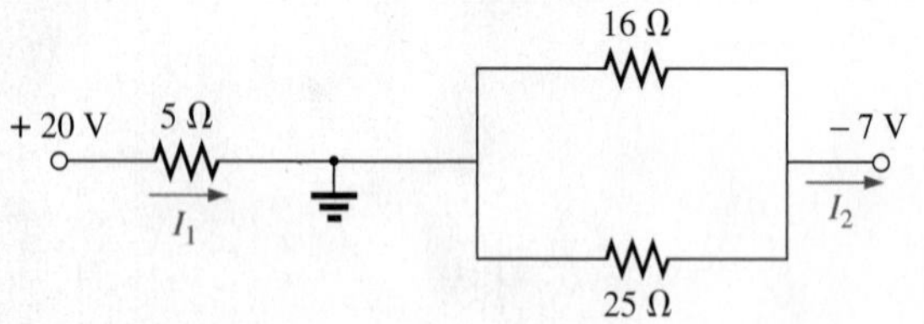

FIG. 7.69
Problem 9.

10. a. Find the magnitude and direction of the currents I, I_1, I_2, and I_3 for the network in Fig. 7.70.
   b. Indicate their direction on Fig. 7.70.

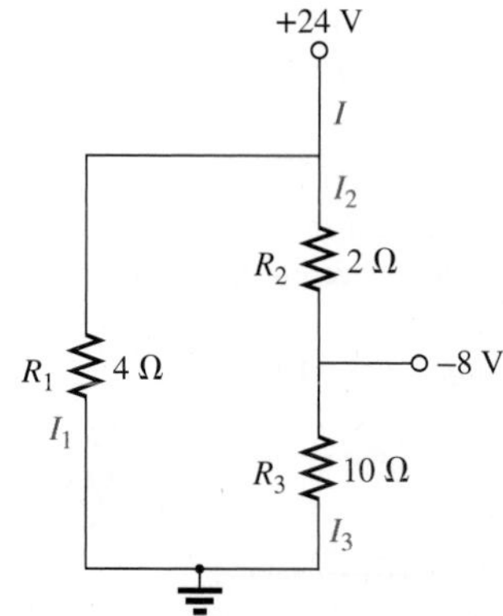

FIG. 7.70
Problem 10.

*11. For the network in Fig. 7.71:
   a. Determine the currents I_s, I_1, I_3, and I_4.
   b. Calculate V_a and V_{bc}.

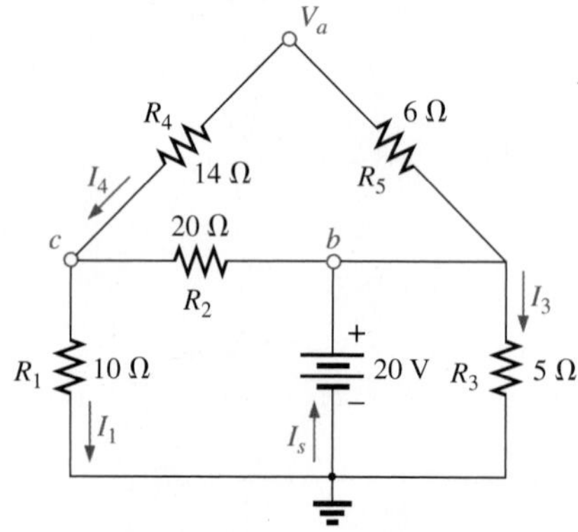

FIG. 7.71
Problem 11.

12. For the network in Fig. 7.72:

a. Determine the current I_1.

b. Calculate the currents I_2 and I_3.

c. Determine the voltage levels V_a and V_b.

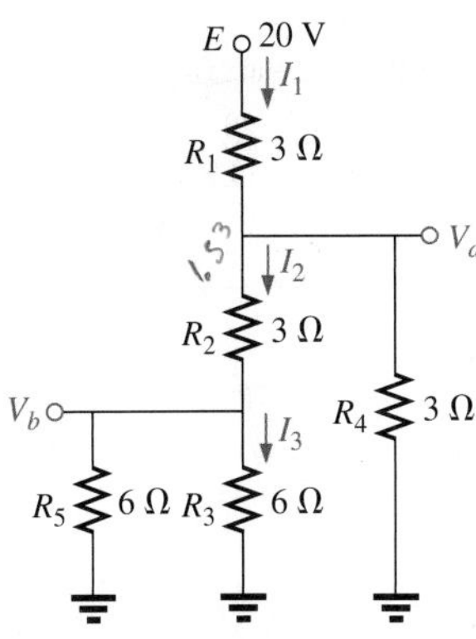

FIG. 7.72
Problem 12.

***13.** Determine the dc levels for the transistor network in Fig. 7.73 using the fact that $V_{BE} = 0.7$ V, $V_E = 2$ V, and $I_C = I_E$. That is:

a. Determine I_E and I_C.

b. Calculate I_B.

c. Determine V_B and V_C.

d. Find V_{CE} and V_{BC}.

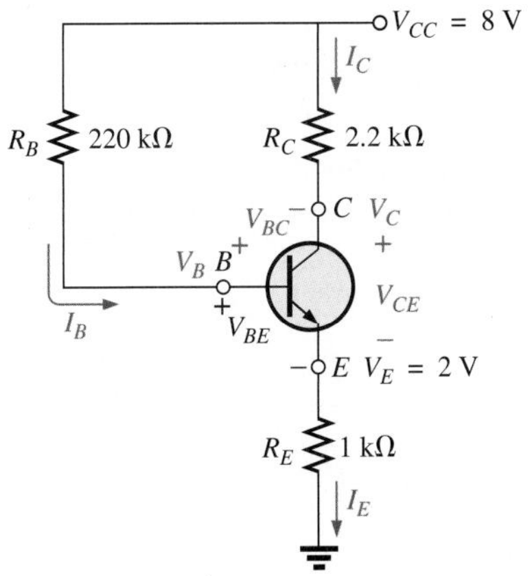

FIG. 7.73
Problem 13.

***14.** The network in Fig. 7.74 is the basic biasing arrangement for the *field-effect transistor* (FET), a device of increasing importance in electronic design. (*Biasing* simply means the application of dc levels to establish a particular set of operating conditions.) Even though you may be unfamiliar with the FET, you can perform the following analysis using only the basic laws introduced in this chapter and the information provided on the diagram.

a. Determine the voltages V_G and V_S.

b. Find the currents I_1, I_2, I_D, and I_S.

c. Determine V_{DS}.

d. Calculate V_{DG}.

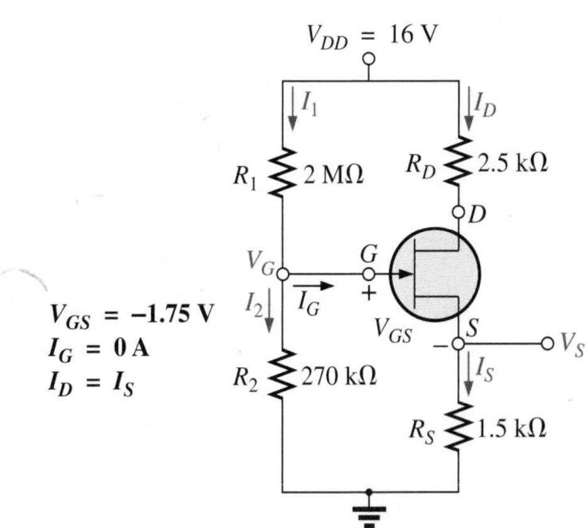

FIG. 7.74
Problem 14.

***15.** For the network in Fig. 7.75:

a. Determine R_T.

b. Calculate V_a.

c. Find V_1.

d. Calculate V_2.

e. Determine I (with direction).

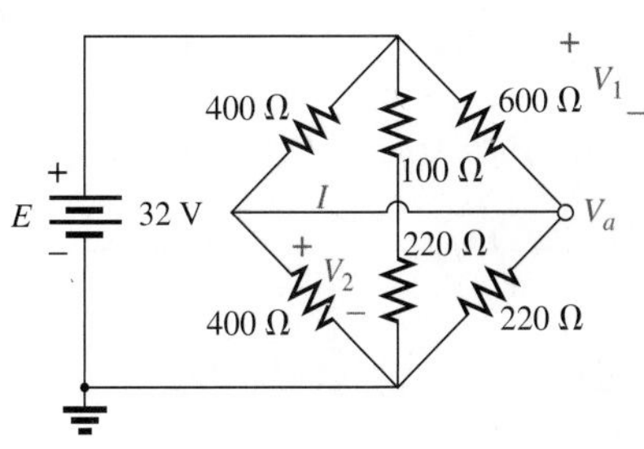

FIG. 7.75
Problem 15.

16. For the network in Fig. 7.76:

a. Determine the current I.

b. Find V_1.

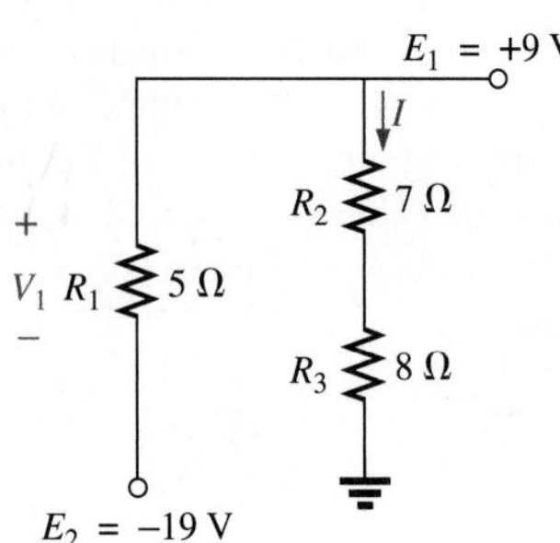

FIG. 7.76
Problem 16.

17. For the configuration in Fig. 7.77:

a. Find the currents I_2, I_6, and I_8.

b. Find the voltages V_4 and V_8.

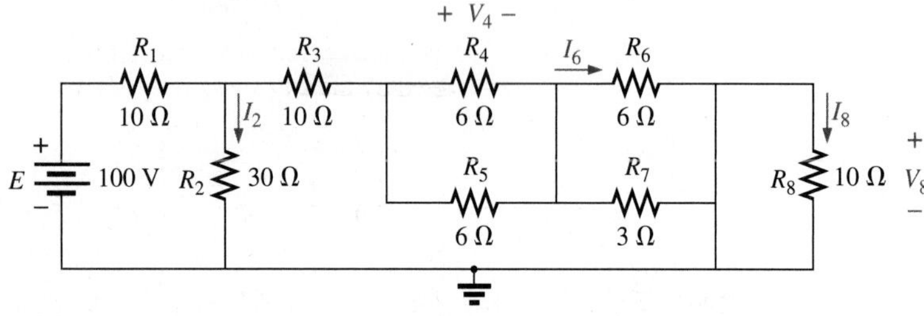

FIG. 7.77
Problem 17.

18. Determine the voltage V and the current I for the network in Fig. 7.78.

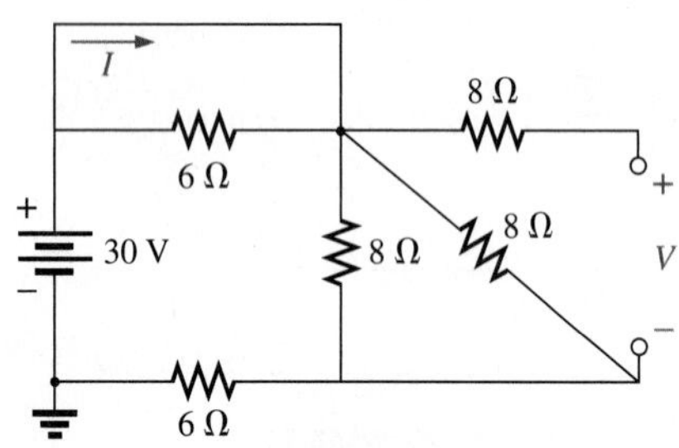

FIG. 7.78
Problem 18.

***19.** For the network in Fig. 7.79:

a. Determine R_T by combining resistive elements.

b. Find V_1 and V_4.

c. Calculate I_3 (with direction).

d. Determine I_s by finding the current through each element and then applying Kirchhoff's current law. Then calculate R_T from $R_T = E/I_s$, and compare the answer with the solution of part (a).

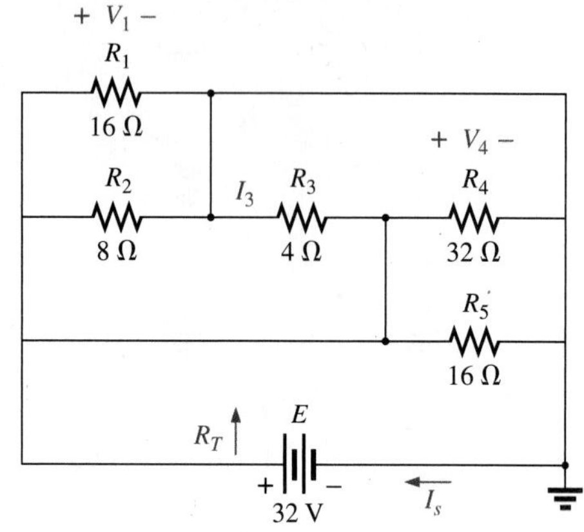

FIG. 7.79
Problem 19.

20. For the network in Fig. 7.80:

a. Determine the voltage V_{ab}. (*Hint:* Just use Kirchhoff's voltage law.)

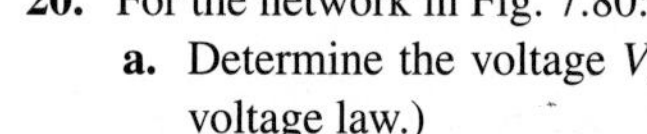

b. Calculate the current I.

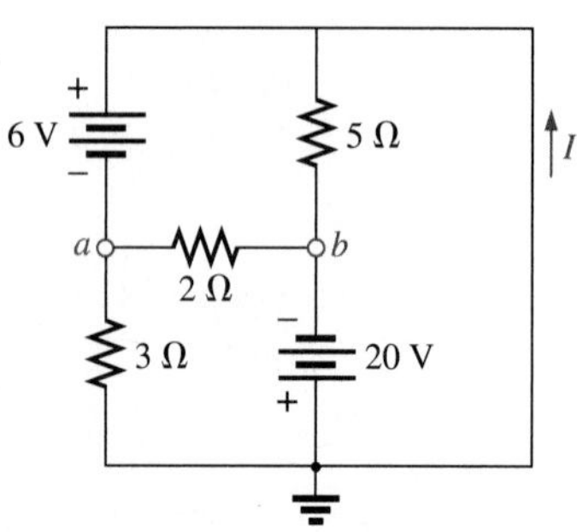

FIG. 7.80
Problem 20.

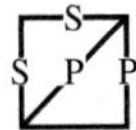

*21. For the network in Fig. 7.81:
   a. Determine the current I.
   b. Calculate the open-circuit voltage V.

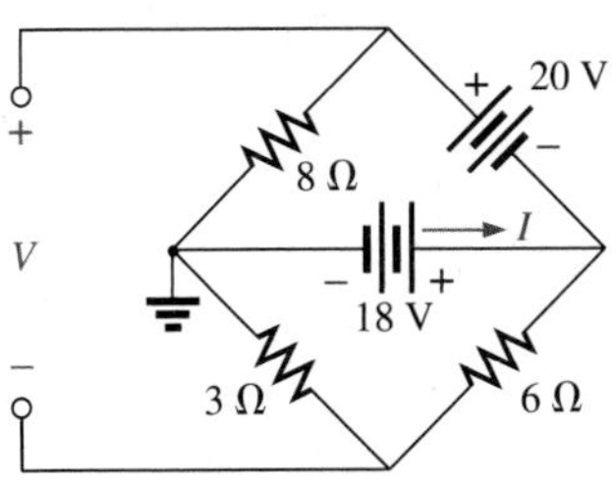

FIG. 7.81
Problem 21.

*22. For the network in Fig. 7.82, find the resistance R_3 if the current through it is 2 A.

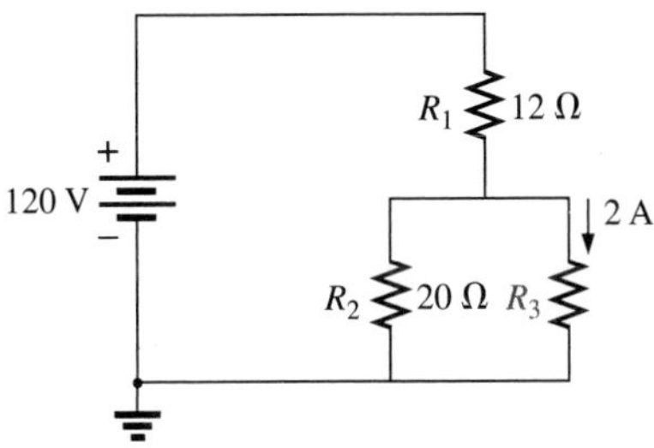

FIG. 7.82
Problem 22.

*23. If all the resistors of the cube in Fig. 7.83 are 10 Ω, what is the total resistance? (*Hint:* Make some basic assumptions about current division through the cube.)

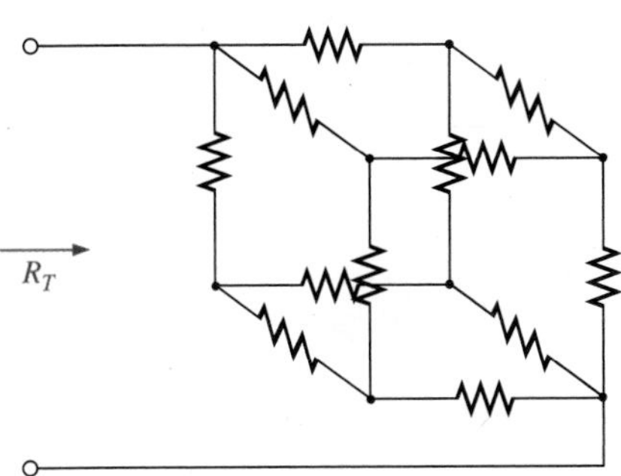

FIG. 7.83
Problem 23.

*24. Given the voltmeter reading $V = 27$ V in Fig. 7.84:
   a. Is the network operating properly?
   b. If not, what could be the cause of the incorrect reading?

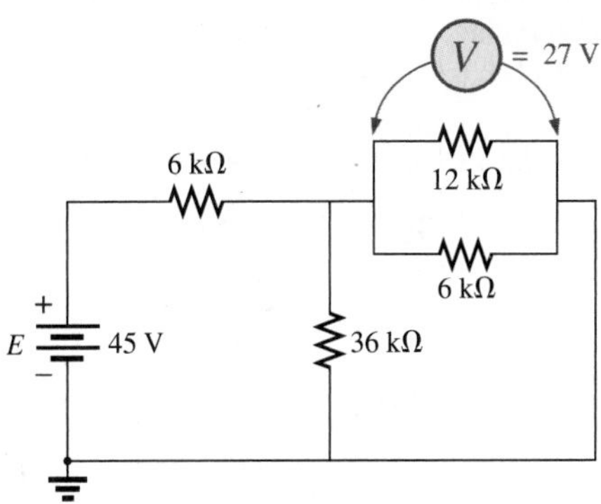

FIG. 7.84
Problem 24.

SECTION 7.6 Ladder Networks

25. For the ladder network in Fig. 7.85:
   a. Find the current I.
   b. Find the current I_7.
   c. Determine the voltages V_3, V_5, and V_7.
   d. Calculate the power delivered to R_7, and compare it to the power delivered by the 240 V supply.

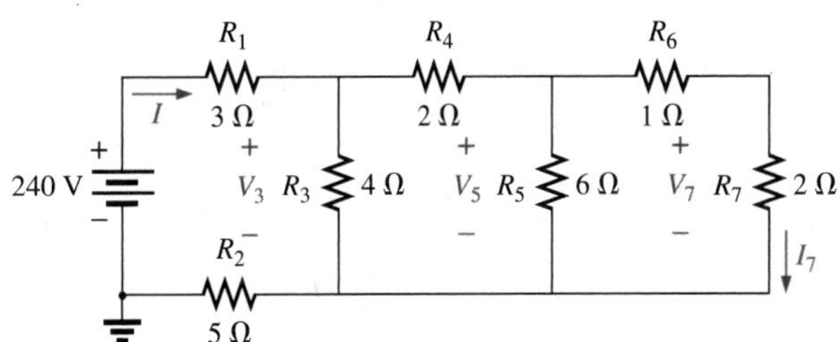

FIG. 7.85
Problem 25.

26. For the ladder network in Fig. 7.86:
   a. Determine R_T.
   b. Calculate I.
   c. Find I_8.

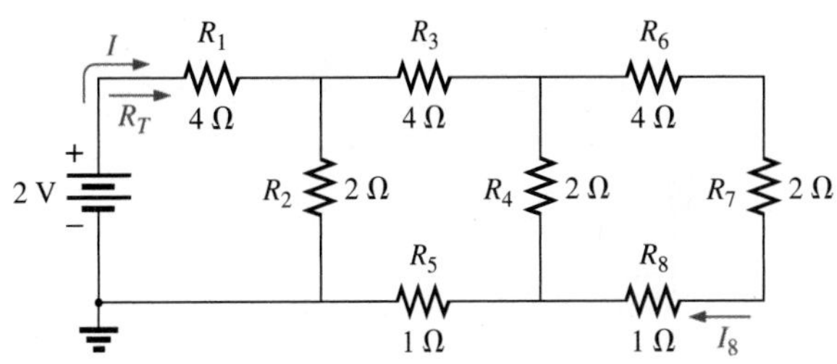

FIG. 7.86
Problem 26.

*27. Determine the power delivered to the 10 Ω load in Fig. 7.87.

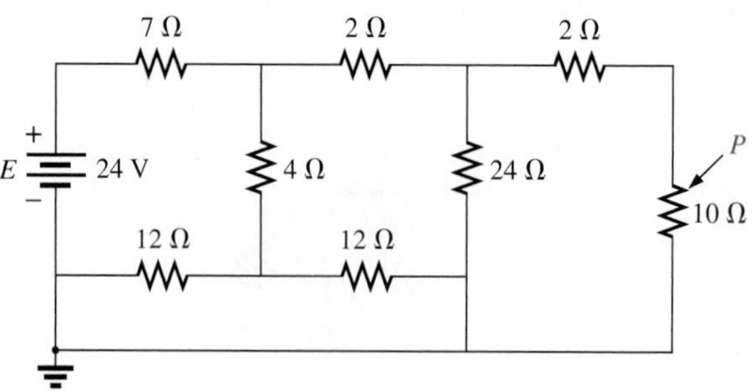

FIG. 7.87
Problem 27.

28. For the multiple ladder configuration in Fig. 7.88:
   a. Determine I.
   b. Calculate I_4.
   c. Find I_6.
   d. Find I_{10}.

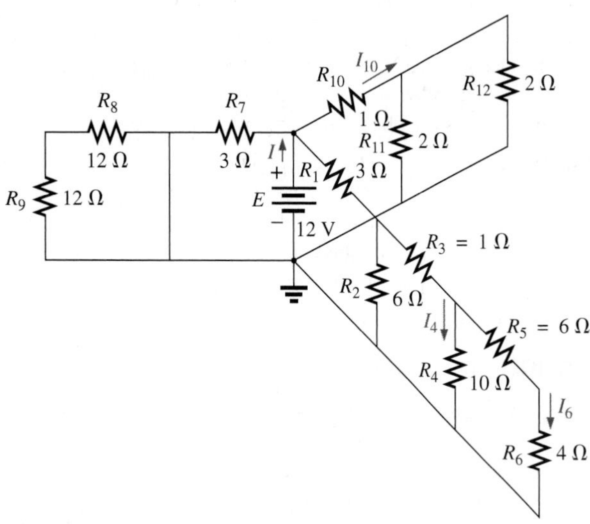

FIG. 7.88
Problem 28.

SECTION 7.7 Voltage Divider Supply (Unloaded and Loaded)

29. Given the voltage divider supply in Fig. 7.89:
   a. Determine the supply voltage E.
   b. Find the load resistors R_{L_2} and R_{L_3}.
   c. Determine the voltage divider resistors R_1, R_2, and R_3.

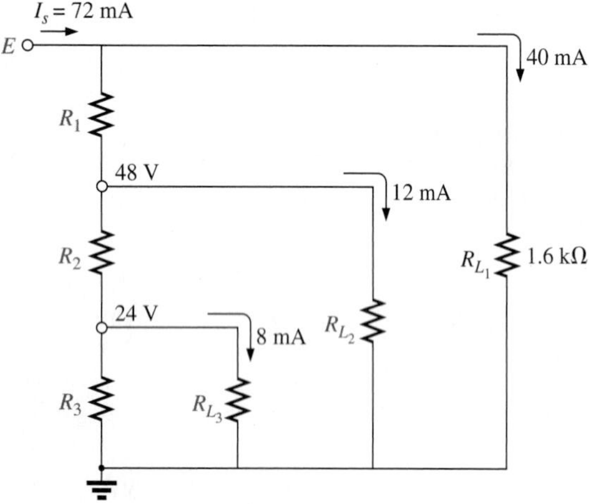

FIG. 7.89
Problem 29.

*30. Determine the voltage divider supply resistors for the configuration in Fig. 7.90. Also determine the required wattage rating for each resistor, and compare their levels.

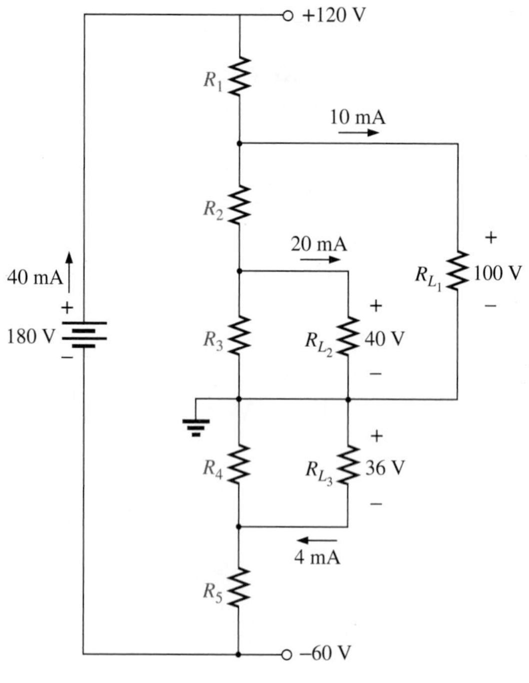

FIG. 7.90
Problem 30.

SECTION 7.8 Potentiometer Loading

*31. For the system in Fig. 7.91:

a. At first exposure, does the design appear to be a good one?

b. In the absence of the 10 kΩ load, what are the values of R_1 and R_2 to establish 3 V across R_2?

c. Determine the values of R_1 and R_2 to establish $V_{R_L} = 3$ V when the load is applied, and compare them to the results of part (b).

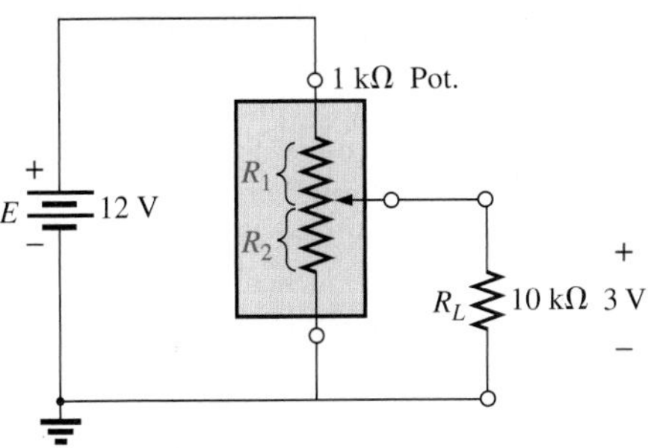

FIG. 7.91
Problem 31.

*32. For the potentiometer in Fig. 7.92:

a. What are the voltages V_{ab} and V_{bc} with no load applied ($R_{L_1} = R_{L_2} = \infty\ \Omega$)?

b. What are the voltages V_{ab} and V_{bc} with the indicated loads applied?

c. What is the power dissipated by the potentiometer under the loaded conditions in Fig. 7.92?

d. What is the power dissipated by the potentiometer with no loads applied? Compare it to the results of part (c).

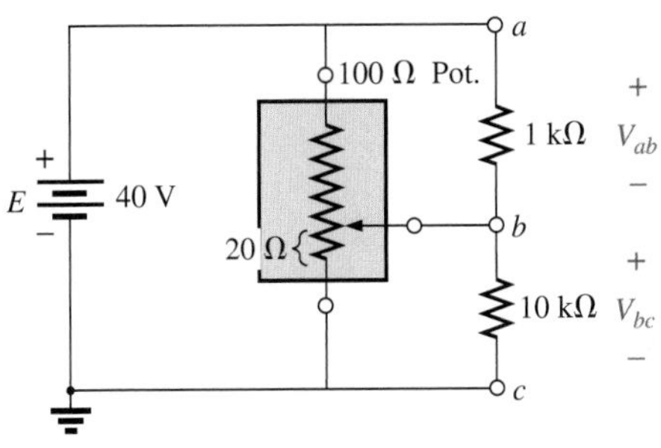

FIG. 7.92
Problem 32.

SECTION 7.9 Ammeter, Voltmeter, and Ohmmeter Design

33. A d'Arsonval movement is rated 1 mA, 100 Ω.

a. What is the current sensitivity?

b. Design a 20 A ammeter using the above movement. Show the circuit and component values.

34. Using a 50 μA, 1000 Ω d'Arsonval movement, design a multirange milliammeter having scales of 25 mA, 50 mA, and 100 mA. Show the circuit and component values.

35. A d'Arsonval movement is rated 50 μA, 1000 Ω.

a. Design a 15 V dc voltmeter. Show the circuit and component values.

b. What is the ohm/volt rating of the voltmeter?

36. Using a 1 mA, 100 Ω d'Arsonval movement, design a multirange voltmeter having scales of 5 V, 50 V, and 500 V. Show the circuit and component values.

37. A digital meter has an internal resistance of 10 MΩ on its 0.5 V range. If you had to build a voltmeter with a d'Arsonval movement, what current sensitivity would you need if the meter were to have the same internal resistance on the same voltage scale?

*38. a. Design a series ohmmeter using a 100 μA, 1000 Ω movement; a zero-adjust with a maximum value of 2 kΩ; a battery of 3 V; and a series resistor whose value is to be determined.

b. Find the resistance required for full-scale, 3/4-scale, 1/2-scale, and 1/4-scale deflection.

c. Using the results of part (b), draw the scale to be used with the ohmmeter.

39. Describe the basic construction and operation of the megohmmeter.

*40. Determine the reading of the ohmmeter for the configuration in Fig. 7.93.

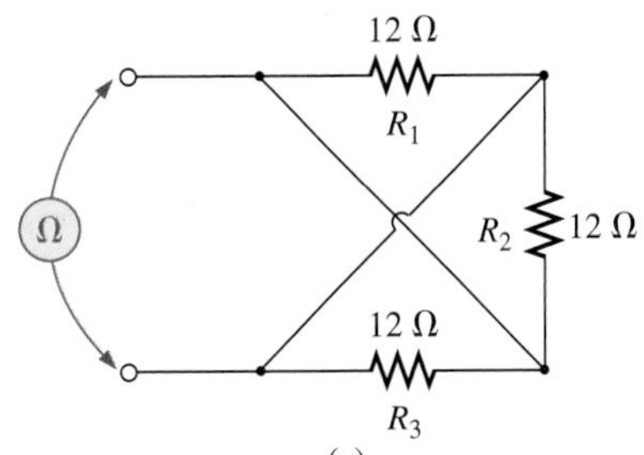

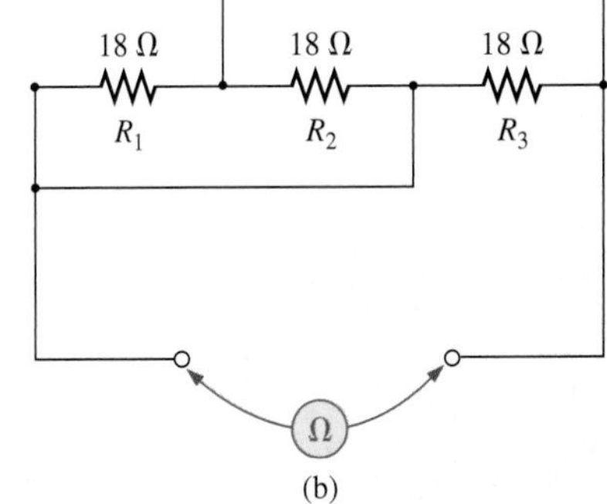

FIG. 7.93
Problem 40.

SECTION 7.11 Computer Analysis

41. Using PSpice or Multisim, verify the results of Example 7.2.
42. Using PSpice or Multisim, confirm the solutions of Example 7.5.
43. Using PSpice or Multisim, verify the results of Example 7.10.
44. Using PSpice or Multisim, find voltage V_6 of Fig. 7.32.
45. Using PSpice or Multisim, find voltages V_b and V_c of Fig. 7.40.

GLOSSARY

Complex configuration A network in which none of the elements are in series or parallel.

d'Arsonval movement An iron-core coil mounted on bearings between a permanent magnet. A pointer connected to the movable core indicates the strength of the current passing through the coil.

Ladder network A network that consists of a cascaded set of series-parallel combinations and has the appearance of a ladder.

Megohmmeter An instrument for measuring very high resistance levels, such as in the megohm range.

Series ohmmeter A resistance-measuring instrument in which the movement is placed in series with the unknown resistance.

Series-parallel network A network consisting of a combination of both series and parallel branches.

Transistor A three-terminal semiconductor electronic device that can be used for amplification and switching purposes.

Voltage divider supply A series network that can provide a range of voltage levels for an application.

Methods of Analysis and Selected Topics (dc)

8

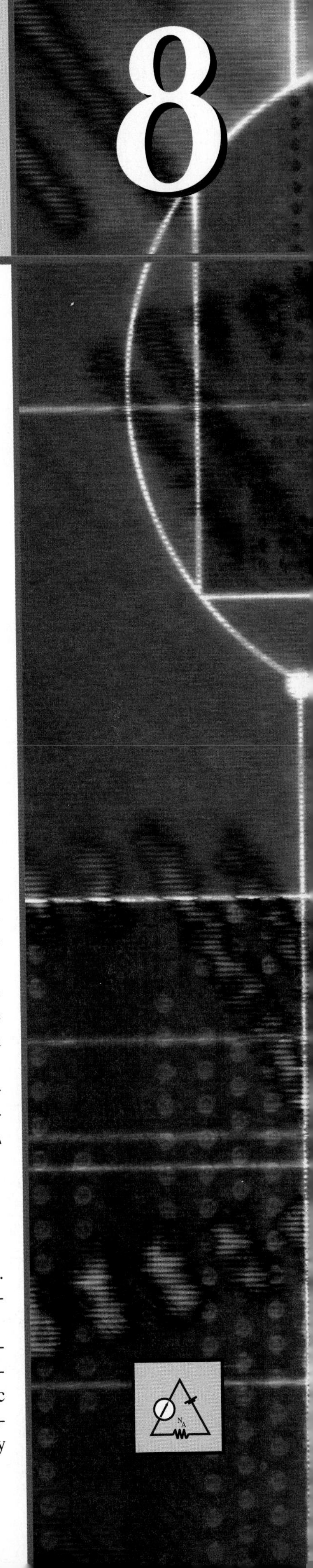

Objectives

- *Become familiar with the terminal characteristics of a current source and how to solve for the voltages and currents of a network using current sources and/or current sources and voltage sources.*
- *Be able to apply branch-current analysis and mesh analysis to find the currents of network with one or more independent paths.*
- *Be able to apply nodal analysis to find all the terminal voltages of any series-parallel network with one or more independent sources.*
- *Become familiar with bridge network configurations and how to perform Δ–Y or Y–Δ conversions.*

8.1 INTRODUCTION

The circuits described in previous chapters had only one source or two or more sources in series or parallel. The step-by-step procedures outlined in those chapters can be applied only if the sources are in series or parallel. There will be an interaction of sources that will not permit the reduction techniques used to find quantities such as the total resistance and the source current.

For such situations, methods of analysis have been developed that allow us to approach, in a systematic manner, networks with any number of sources in any arrangement. To our benefit, the methods to be introduced can also be applied to networks with only *one source* or to networks in which sources are in *series or parallel.*

The methods to be introduced in this chapter include branch-current analysis, mesh analysis, and nodal analysis. Each can be applied to the same network, although usually one is more appropriate than the other. The "best" method cannot be defined by a strict set of rules but can be determined only after developing an understanding of the relative advantages of each.

Before considering the first of the methods, we will examine current sources in detail because they appear throughout the analyses to follow. The chapter concludes with an investigation of a complex network called the *bridge configuration,* followed by the use of Δ-Y and Y-Δ conversions to analyze such configurations.

8.2 CURRENT SOURCES

In previous chapters, the voltage source was the only source appearing in the circuit analysis. This was primarily because voltage sources such as the battery and supply are the most common in our daily lives and in the laboratory environment.

We now turn our attention to a second type of source called the **current source** which appears throughout the analyses in this chapter. Although current sources are available as laboratory supplies (introduced in Chapter 2), they appear extensively in the modeling of electronic devices such as the transistor. Their characteristics and their impact on the currents and voltages of a network must therefore be clearly understood if electronic systems are to be properly investigated.

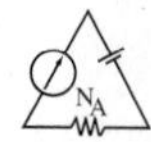

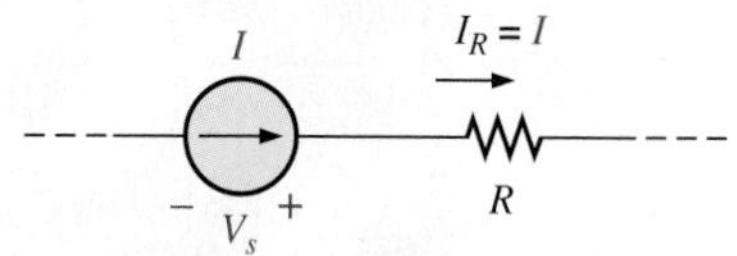

(a)

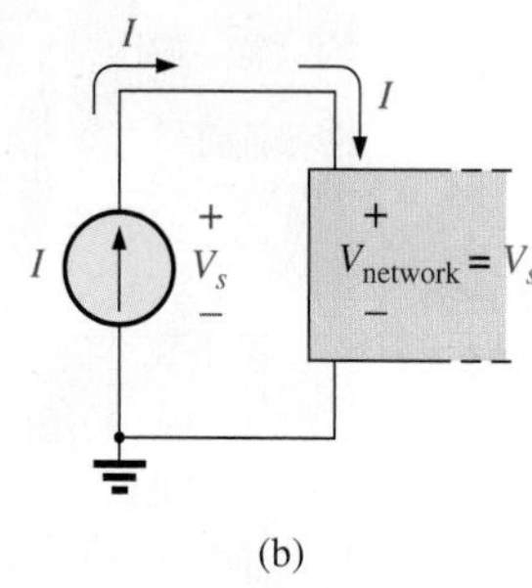

(b)

FIG. 8.1
Introducing the current source symbol.

The current source is often described as the *dual* of the voltage source. Just as a battery provides a fixed voltage to a network, a current source establishes a fixed current in the branch where it is located. Further, the current through a battery is a function of the network to which it is applied, just as the voltage across a current source is a function of the connected network. The term *dual* is applied to any two elements in which the traits of one variable can be interchanged with the traits of another. This is certainly true for the current and voltage of the two types of sources.

The symbol for a current source appears in Fig. 8.1(a). The arrow indicates the direction in which it is supplying current to the branch where it is located. The result is a current equal to the source current through the series resistor. In Fig. 8.1(b), we find that the voltage across a current source is determined by the polarity of the voltage drop caused by the current source. For single-source networks, it always has the polarity of Fig. 8.1(b), but for multisource networks it can have either polarity.

In general, therefore,

a current source determines the direction and magnitude of the current in the branch where it is located.

Furthermore,

the magnitude and the polarity of the voltage across a current source are each a function of the network to which the voltage is applied.

A few examples will demonstrate the similarities between solving for the source current of a voltage source and the terminal voltage of a current source. All the rules and laws developed in the previous chapter still apply, so we just have to remember what we are looking for and properly understand the characteristics of each source.

The simplest possible configuration with a current source appears in Example 8.1.

EXAMPLE 8.1 Find the source voltage, the voltage V_1, and current I_1 for the circuit in Fig. 8.2.

Solution: Since the current source establishes the current in the branch in which it is located, the current I_1 must equal I, and

$$I_1 = I = \mathbf{10\ mA}$$

The voltage across R_1 is then determined by Ohm's law:

$$V_1 = I_1R_1 = (10\ \text{mA})(20\ \Omega) = \mathbf{200\ V}$$

Since resistor R_1 and the current source are in parallel, the voltage across each must be the same, and

$$V_s = V_1 = \mathbf{200\ V}$$

with the polarity shown.

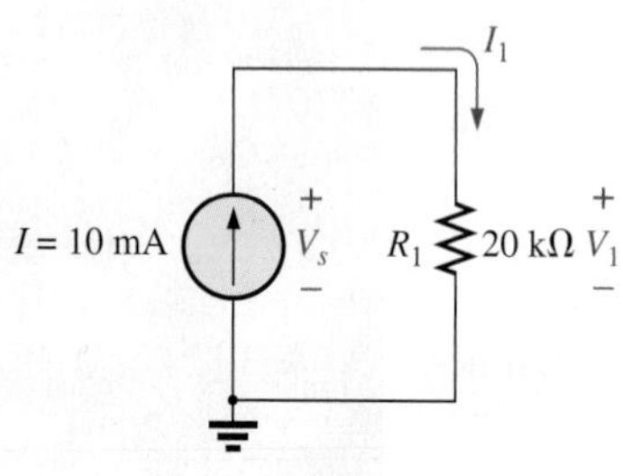

FIG. 8.2
Circuit for Example 8.1.

EXAMPLE 8.2 Find the voltage V_s and currents I_1 and I_2 for the network in Fig. 8.3.

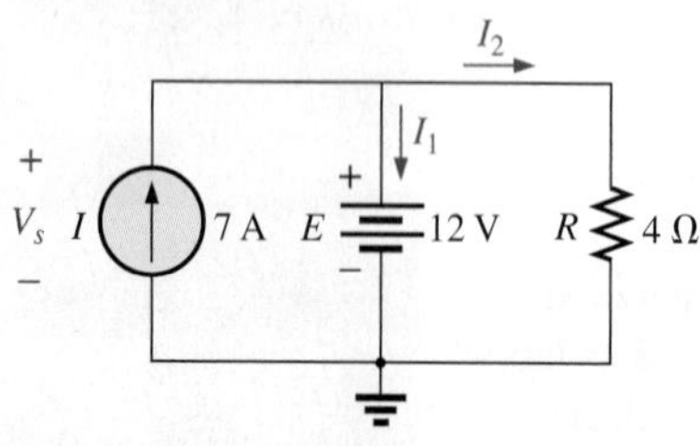

FIG. 8.3
Network for Example 8.2.

Solution: This is an interesting problem because it has both a current source and a voltage source. For each source, the dependent (a function

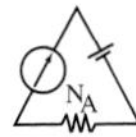

of something else) variable will be determined. That is, for the current source, V_s must be determined, and for the voltage source, I_s must be determined.

Since the current source and voltage source are in parallel,

$$V_s = E = \mathbf{12\ V}$$

Further, since the voltage source and resistor R are in parallel,

$$V_R = E = 12\ \text{V}$$

and $$I_2 = \frac{V_R}{R} = \frac{12\ \text{V}}{4\ \Omega} = \mathbf{3\ A}$$

The current I_1 of the voltage source can then be determined by applying Kirchhoff's current law at the top of the network as follows:

$$\Sigma I_i = \Sigma I_o$$
$$I = I_1 + I_2$$

and $$I_1 = I - I_2 = 7\ \text{A} - 3\ \text{A} = \mathbf{4\ A}$$

EXAMPLE 8.3 Determine the current I_1 and the voltage V_s for the network in Fig. 8.4.

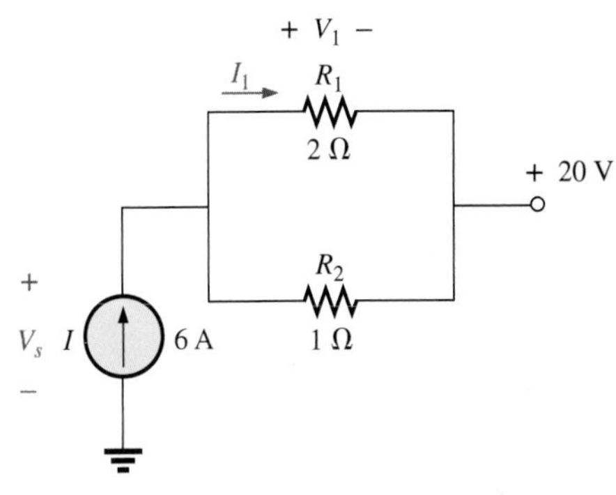

FIG. 8.4
Example 8.3.

Solution: First note that the current in the branch with the current source must be 6 A, no matter what the magnitude of the voltage source to the right. In other words, the currents of the network are defined by I, R_1, and R_2. However, the voltage across the current source is directly affected by the magnitude and polarity of the applied source.

Using the current divider rule:

$$I_1 = \frac{R_2 I}{R_2 + R_1} = \frac{(1\ \Omega)(6\ \text{A})}{1\ \Omega + 2\ \Omega} = \frac{1}{3}(6\ \text{A}) = \mathbf{2\ A}$$

The voltage V_1:

$$V_1 = I_1 R_1 = (2\ \text{A})(2\ \Omega) = 4\ \text{V}$$

Applying Kirchhoff's voltage rule to determine V_s:

$$+V_s - V_1 - 20\ \text{V} = 0$$

and $$V_s = V_1 + 20\ \text{V} = 4\ \text{V} + 20\ \text{V} = \mathbf{24\ V}$$

In particular, note the polarity of the voltage V_s as determined by the network.

8.3 SOURCE CONVERSIONS

The current source appearing is the previous section is called an *ideal source* due to the absence of any internal resistance. In reality, all sources—whether they are voltage sources or current sources—have some internal resistance in the relative positions shown in Fig. 8.5. For the voltage source, if $R_s = 0\ \Omega$, or if it is so small compared to any series resistors that it can be ignored, then we have an "ideal" voltage source for all practical purposes. For the current source, since the resistor R_P is in parallel, if $R_P = \infty\ \Omega$, or if it is large enough compared to any parallel resistive elements that it can be ignored, then we have an "ideal" current source.

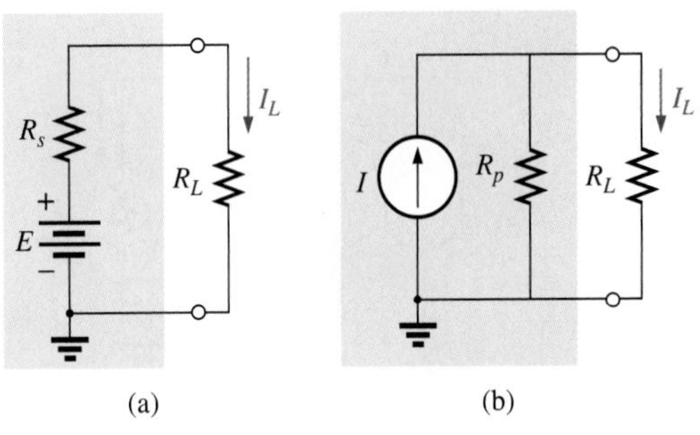

FIG. 8.5
Practical sources: (a) voltage; (b) current.

Unfortunately, however, ideal sources *cannot be converted* from one type to another. That is, a voltage source cannot be converted to a current

source, and vice versa—*the internal resistance must be present.* If the voltage source in Fig. 8.5(a) is to be equivalent to the source in Fig. 8.5(b), any load connected to the sources such as R_L should receive the same current, voltage, and power from each configuration. In other words, if the source were enclosed in a container, the load R_L would not know which source it was connected to.

This type of equivalence is established using the equations appearing in Fig. 8.6. First note that the resistance is the same in each configuration—a nice advantage. For the voltage source equivalent, the voltage is determined by a simple application of Ohm's law to the current source: $E = IR_P$. For the current source equivalent, the current is again determined by applying Ohm's law to the voltage source: $I = E/R_s$. At first glance, it all seems too simple, but Example 8.4 verifies the results.

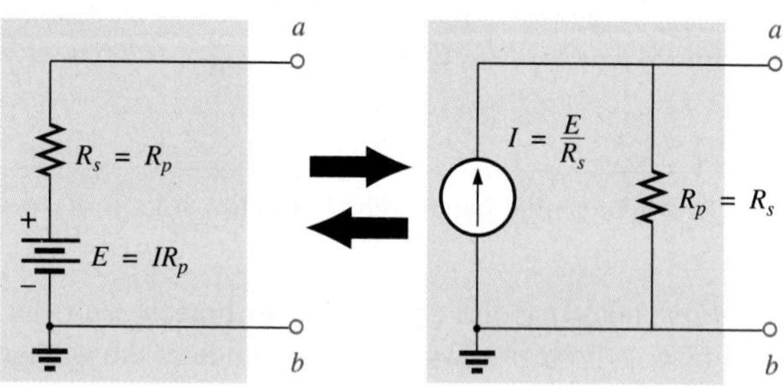

FIG. 8.6
Source conversion.

It is important to realize, however, that

the equivalence between a current source and a voltage source exists only at their external terminals.

The internal characteristics of each are quite different.

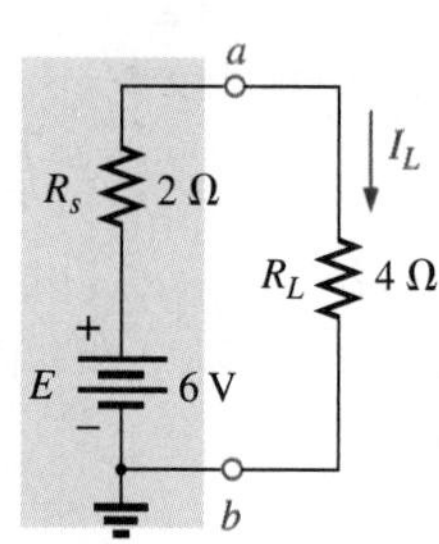

FIG. 8.7
Practical voltage source and load for Example 8.4.

EXAMPLE 8.4 For the circuit in Fig. 8.7:

a. Determine the current I_L.
b. Convert the voltage source to a current source.
c. Using the resulting current source of part (b), calculate the current through the load resistor, and compare your answer to the result of part (a).

Solutions:

a. Applying Ohm's law:

$$I_L = \frac{E}{R_s + R_L} = \frac{6\text{ V}}{2\ \Omega + 4\ \Omega} = \frac{6\text{ V}}{6\ \Omega} = \mathbf{1\ A}$$

b. Using Ohm's law again:

$$I = \frac{E}{R_s} = \frac{6\text{ V}}{2\ \Omega} = \mathbf{3\ A}$$

and the equivalent source appears in Fig. 8.8 with the load reapplied.

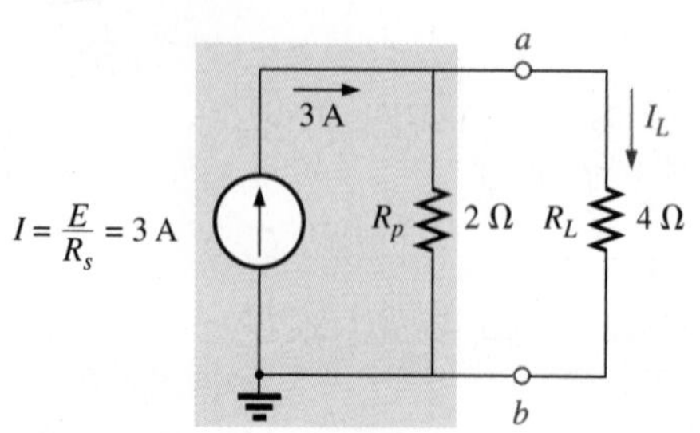

FIG. 8.8
Equivalent current source and load for the voltage source in Fig. 8.7.

c. Using the current divider rule:

$$I_L = \frac{R_p I}{R_p + R_L} = \frac{(2\ \Omega)(3\ \text{A})}{2\ \Omega + 4\ \Omega} = \frac{1}{3}(3\ \text{A}) = \mathbf{1\ A}$$

We find that the current I_L is the same for the voltage source as it was for the equivalent current source—the sources are therefore equivalent.

As demonstrated in Fig. 8.5 and in Example 8.4, note that

a source and its equivalent will establish current in the same direction through the applied load.

In Example 8.4, note that both sources pressure or establish current up through the circuit to establish the same direction for the load current I_L and the same polarity for the voltage V_L.

EXAMPLE 8.5 Determine current I_2 for the network in Fig. 8.9.

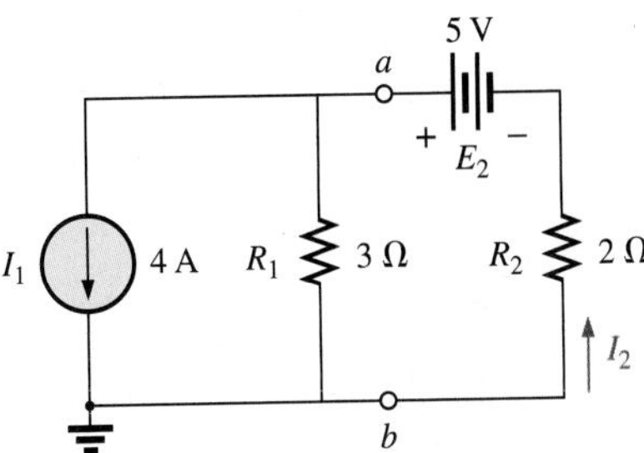

FIG. 8.9
Two-source network for Example 8.5.

Solution: Although it may appear that the network cannot be solved using methods introduced thus far, one source conversion, as shown in Fig. 8.10, results in a simple series circuit. It does not make sense to convert the voltage source to a current source because you would lose the current I_2 in the redrawn network. Note the polarity for the equivalent voltage source as determined by the current source.

For the source conversion:

$$E_1 = I_1 R_1 = (4\ \text{A})(3\ \Omega) = 12\ \text{V}$$

and

$$I_2 = \frac{E_1 + E_2}{R_1 + R_2} = \frac{12\ \text{V} + 5\ \text{V}}{3\ \Omega + 2\ \Omega} = \frac{17\ \text{V}}{5\ \Omega} = \mathbf{3.4\ A}$$

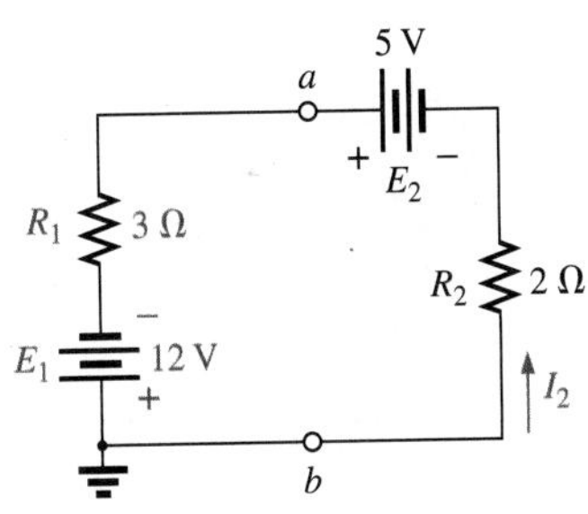

FIG. 8.10
Network in Fig. 8.9 following the conversion of the current source to a voltage source.

8.4 CURRENT SOURCES IN PARALLEL

We found that voltage sources of different terminal voltages cannot be placed in parallel because of a violation of Kirchhoff's voltage law. Similarly,

current sources of different values cannot be placed in series due to a violation of Kirchhoff's current law.

However, current sources can be placed in parallel just as voltage sources can be placed in series. In general,

two or more current sources in parallel can be replaced by a single current source having a magnitude determined by the difference of the sum of the currents in one direction and the sum in the opposite direction. The new parallel internal resistance is the total resistance of the resulting parallel resistive elements.

Consider the following examples.

EXAMPLE 8.6 Reduce the parallel current sources in Fig. 8.11 to a single current source.

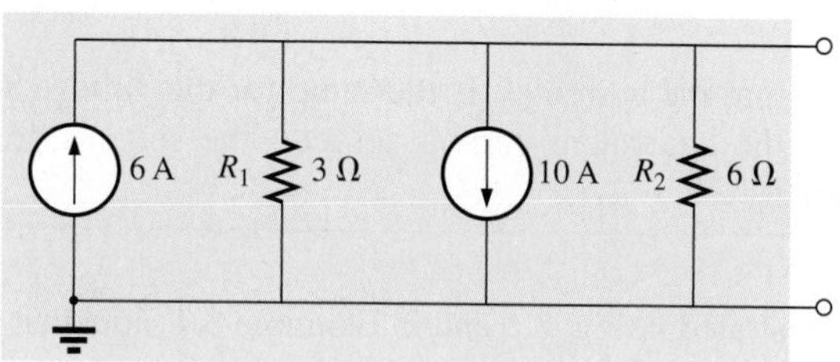

FIG. 8.11
Parallel current sources for Example 8.6.

Solution: The net source current is

$$I = 10\text{ A} - 6\text{ A} = \mathbf{4\text{ A}}$$

with the direction of the larger.

The net internal resistance is the parallel combination of resistors, R_1 and R_2:

$$R_p = 3\ \Omega \| 6\ \Omega = \mathbf{2\ \Omega}$$

The reduced equivalent appears in Fig. 8.12.

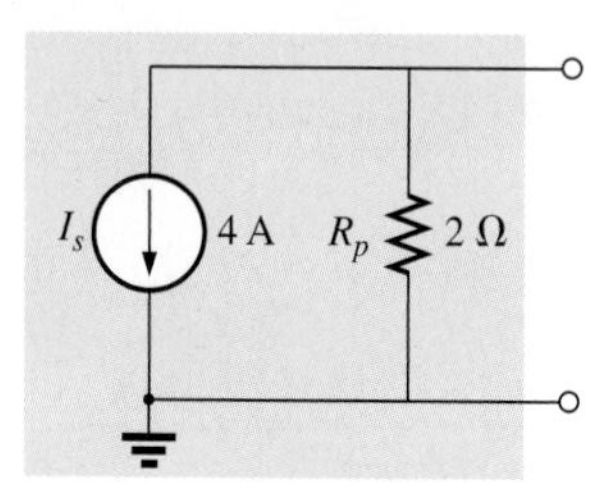

FIG. 8.12
Reduced equivalent for the configuration of Fig. 8.11.

EXAMPLE 8.7 Reduce the parallel current sources in Fig. 8.13 to a single current source.

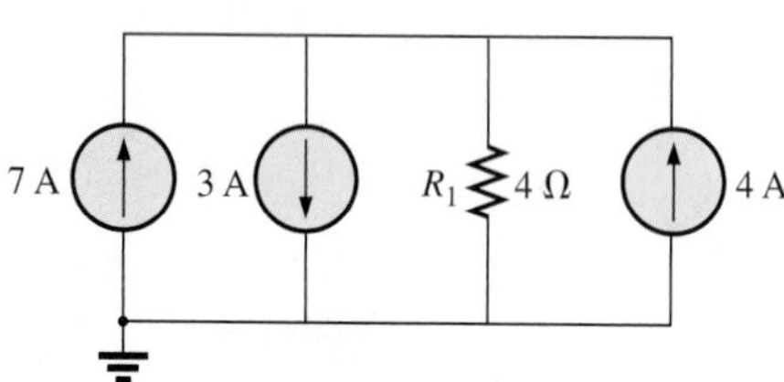

FIG. 8.13
Parallel current sources for Example 8.7.

Solution: The net current is

$$I = 7\text{ A} + 4\text{ A} - 3\text{ A} = \mathbf{8\text{ A}}$$

with the direction shown in Fig. 8.14. The net internal resistance remains the same.

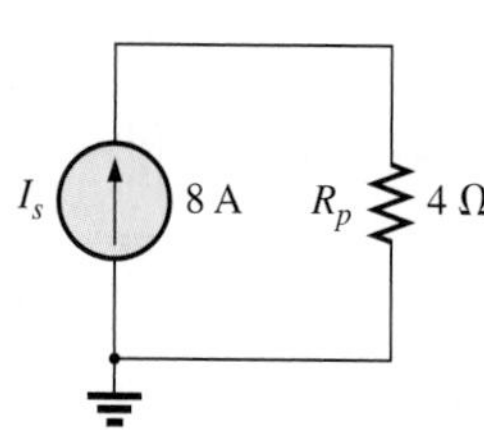

FIG. 8.14
Reduced equivalent for Fig. 8.13.

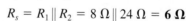

EXAMPLE 8.8 Reduce the network in Fig. 8.15 to a single current source, and calculate the current through R_L.

Solution: In this example, the voltage source will first be converted to a current source as shown in Fig. 8.16. Combining current sources,

$$I_s = I_1 + I_2 = 4\text{ A} + 6\text{ A} = \mathbf{10\text{ A}}$$

and $$R_s = R_1 \| R_2 = 8\ \Omega \| 24\ \Omega = \mathbf{6\ \Omega}$$

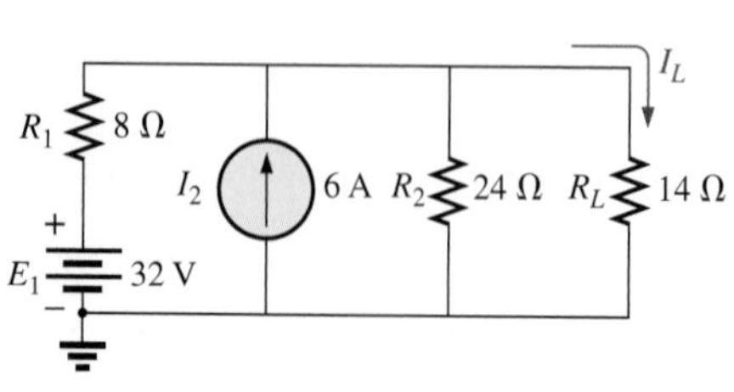

FIG. 8.15
Example 8.8.

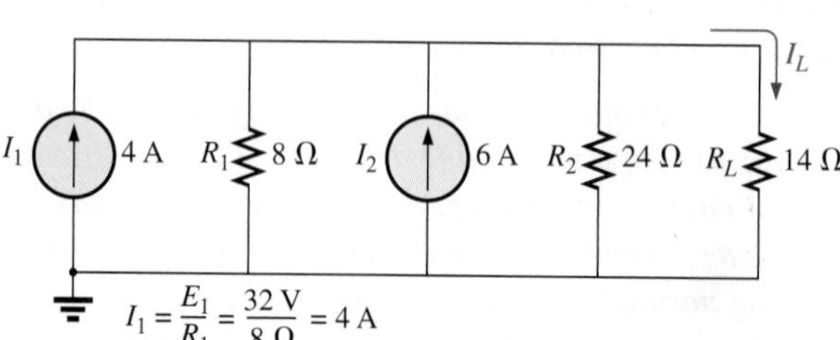

FIG. 8.16
Network in Fig. 8.15 following the conversion of the voltage source to a current source.

Applying the current divider rule to the resulting network in Fig. 8.17,

$$I_L = \frac{R_p I_s}{R_p + R_L} = \frac{(6\ \Omega)(10\ \text{A})}{6\ \Omega + 14\ \Omega} = \frac{60\ \text{A}}{20} = \mathbf{3\ A}$$

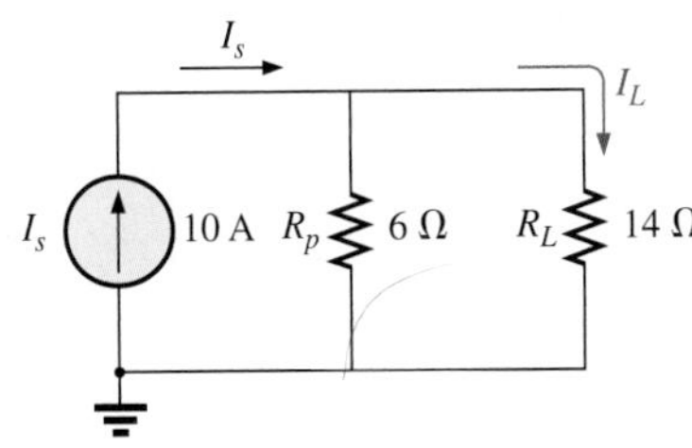

FIG. 8.17
Network in Fig. 8.16 reduced to its simplest form.

8.5 CURRENT SOURCES IN SERIES

The current through any branch of a network can be only single-valued. For the situation indicated at point *a* in Fig. 8.18, we find by application of Kirchhoff's current law that the current leaving that point is greater than that entering—an impossible situation. Therefore,

current sources of different current ratings are not connected in series,

just as voltage sources of different voltage ratings are not connected in parallel.

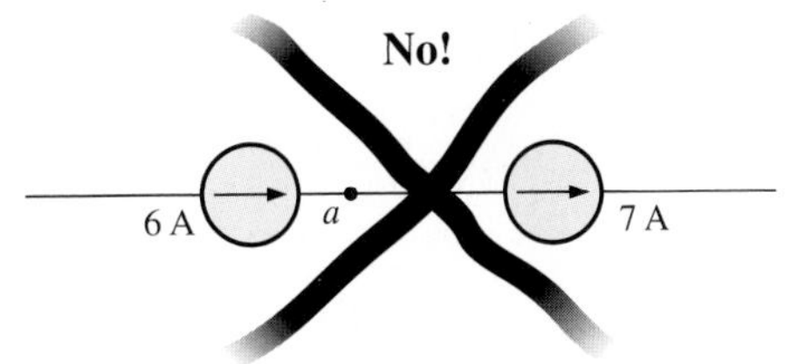

FIG. 8.18
Invalid situation.

8.6 BRANCH-CURRENT ANALYSIS

Before examining the details of the first important method of analysis, let us examine the network in Fig. 8.19 to be sure that you understand the need for these special methods.

Initially, it may appear that we can use the reduce and return approach to work our way back to the source E_1 and calculate the source current I_{s_1}. Unfortunately, however, the series elements R_3 and E_2 cannot be combined because they are different types of elements. A further examination of the network reveals that there are no two like elements that are in series or parallel. No combination of elements can be performed, and it is clear that another approach must be defined.

The first approach to be introduced is called the **branch-current method** because we will define and solve for the currents of each branch of the network. The best way to introduce this method and understand its application is to follow a series of steps, as listed below. Each step is carefully defined in the examples to follow.

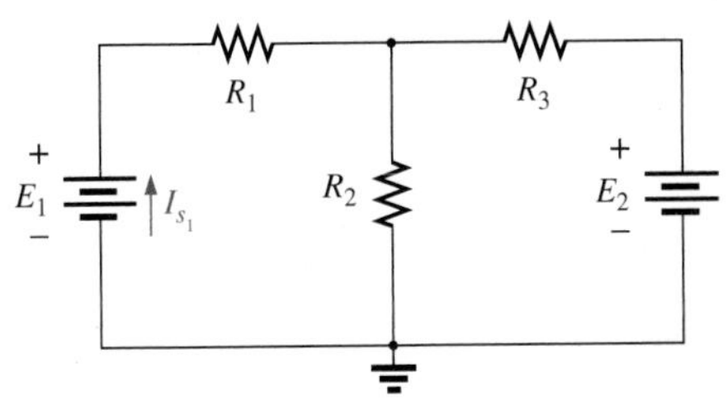

FIG. 8.19
Demonstrating the need for an approach such as branch-current analysis.

Branch-Current Analysis Procedure

1. ***Assign a distinct current of arbitrary direction to each branch of the network.***
2. ***Indicate the polarities for each resistor as determined by the assumed current direction.***
3. ***Apply Kirchhoff's voltage law around each closed, independent loop of the network.***

The best way to determine how many times Kirchhoff's voltage law has to be applied is to determine the number of "windows" in the network. The network in Example 8.9 has a definite similarity to the two-window configuration in Fig. 8.20(a). The result is a need to apply Kirchhoff's voltage law twice. For networks with three windows, as shown in Fig. 8.20(b), three applications of Kirchhoff's voltage law are required, and so on.

4. ***Apply Kirchhoff's current law at the minimum number of nodes that will include all the branch currents of the network.***

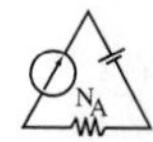

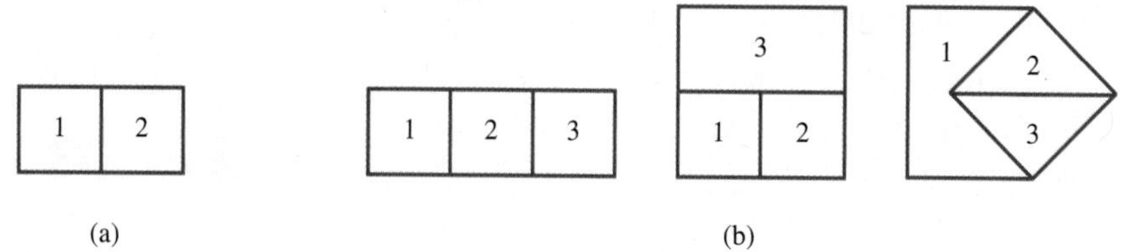

FIG. 8.20
Determining the number of independent closed loops.

The minimum number is one less than the number of independent nodes of the network. For the purposes of this analysis, a **node** is a junction of two or more branches, where a branch is any combination of series elements. Fig. 8.21 defines the number of applications of Kirchhoff's current law for each configuration in Fig. 8.20.

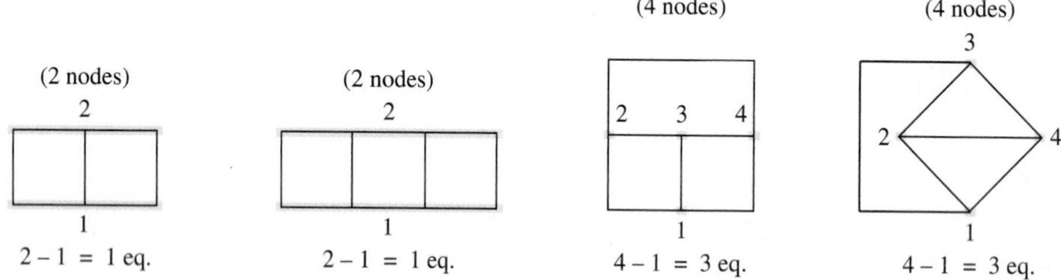

FIG. 8.21
Determining the number of applications of Kirchhoff's current law required.

5. ***Solve the resulting simultaneous linear equations for assumed branch currents.***

It is assumed that the use of the **determinants method** to solve for the currents I_1, I_2, and I_3 is understood and is a part of the student's mathematical background. If not, a detailed explanation of the procedure is provided in Appendix D. Calculators and computer software packages such as Mathcad can find the solutions quickly and accurately.

EXAMPLE 8.9 Apply the branch-current method to the network in Fig. 8.22.

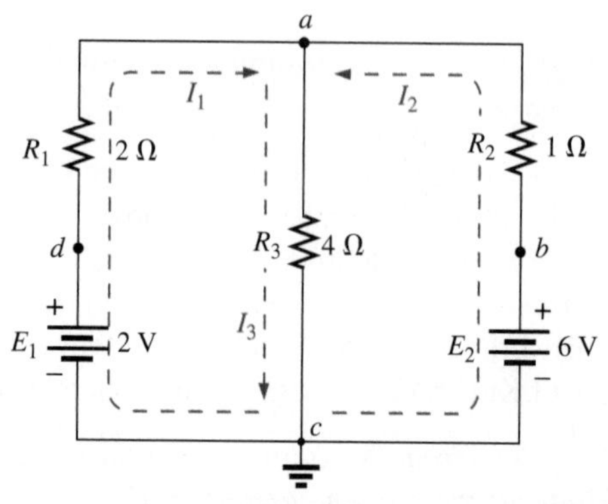

FIG. 8.22
Example 8.9.

Solution 1:

Step 1: Since there are three distinct branches (*cda*, *cba*, *ca*), three currents of arbitrary directions (I_1, I_2, I_3) are chosen, as indicated in Fig. 8.22. The current directions for I_1 and I_2 were chosen to match the "pressure" applied by sources E_1 and E_2, respectively. Since both I_1 and I_2 enter node *a*, I_3 is leaving.

Step 2: Polarities for each resistor are drawn to agree with assumed current directions, as indicated in Fig. 8.23.

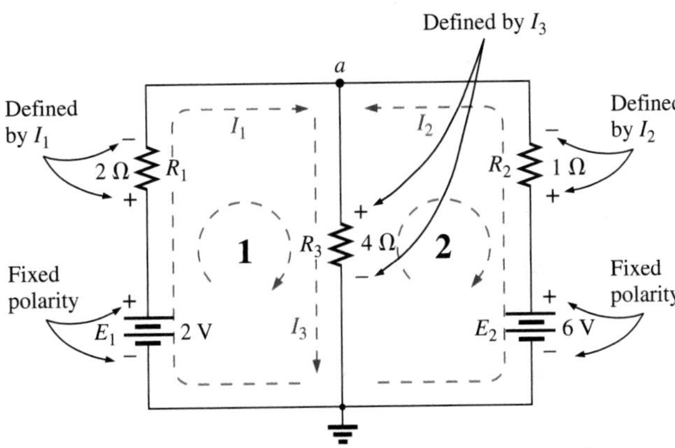

FIG. 8.23
Inserting the polarities across the resistive elements as defined by the chosen branch currents.

Step 3: Kirchhoff's voltage law is applied around each closed loop (1 and 2) in the clockwise direction:

$$\text{loop 1:} \quad \Sigma_{\circlearrowright} V = \underset{\text{Rise in potential}}{+E_1} - V_{R_1} - V_{R_3} = 0 \quad (V_{R_1}, V_{R_3}: \text{Drop in potential})$$

$$\text{loop 2:} \quad \Sigma_{\circlearrowright} V = +V_{R_3} + V_{R_2} - E_2 = 0 \quad (V_{R_3}, V_{R_2}: \text{Rise in potential};\ E_2: \text{Drop in potential})$$

and

$$\text{loop 1:} \quad \Sigma_{\circlearrowright} V = \underbrace{+2\text{ V}}_{\substack{\text{Battery}\\\text{potential}}} - \underbrace{(2\ \Omega)I_1}_{\substack{\text{Voltage drop}\\\text{across 2 }\Omega\\\text{resistor}}} - \underbrace{(4\ \Omega)I_3}_{\substack{\text{Voltage drop}\\\text{across 4 }\Omega\\\text{resistor}}} = 0$$

$$\text{loop 2:} \quad \Sigma_{\circlearrowright} V = (4\ \Omega)I_3 + (1\ \Omega)I_2 - 6\text{ V} = 0$$

Step 4: Applying Kirchhoff's current law at node *a* (in a two-node network, the law is applied at only one node),

$$I_1 + I_2 = I_3$$

Step 5: There are three equations and three unknowns (units removed for clarity):

$$\begin{aligned} 2 - 2I_1 - 4I_3 &= 0 \\ 4I_3 + 1I_2 - 6 &= 0 \\ I_1 + I_2 &= I_3 \end{aligned} \qquad \text{Rewritten:} \quad \begin{aligned} 2I_1 + 0 + 4I_3 &= 2 \\ 0 + I_2 + 4I_3 &= 6 \\ I_1 + I_2 - \ I_3 &= 0 \end{aligned}$$

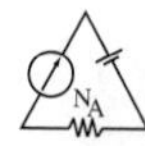

Using third-order determinants (Appendix D), we have

$$I_1 = \frac{\begin{vmatrix} 2 & 0 & 4 \\ 6 & 1 & 4 \\ 0 & 1 & -1 \end{vmatrix}}{D = \begin{vmatrix} 2 & 0 & 4 \\ 0 & 1 & 4 \\ 1 & 1 & -1 \end{vmatrix}} = -1\text{ A}$$

A negative sign in front of a branch current indicates only that the actual current is in the direction opposite to that assumed.

$$I_2 = \frac{\begin{vmatrix} 2 & 2 & 4 \\ 0 & 6 & 4 \\ 1 & 0 & -1 \end{vmatrix}}{D} = 2\text{ A}$$

$$I_3 = \frac{\begin{vmatrix} 2 & 0 & 2 \\ 0 & 1 & 6 \\ 1 & 1 & 0 \end{vmatrix}}{D} = 1\text{ A}$$

Mathcad Solution: Once you understand the procedure for entering the parameters, you can use Mathcad to solve determinants such as appearing in Solution 1 in a very short time frame. The numerator is defined by n with the sequence **n =**. Then you apply the sequence **View-Toolbars-Matrix** to obtain the **Matrix** toolbar appearing in Fig. 8.24. Selecting the top left option called **Matrix** results in the **Insert Matrix** dialog box in which **3 × 3** is selected. The 3 × 3 matrix appears with a bracket to signal which parameter should be entered. Enter that parameter, and then left-click to select the next parameter you want to enter. When you have finished, move on to define the denominator d in the same manner. Then define the current of interest, select **Determinant (1 × 1)** from the **Matrix** toolbar, and insert the numerator variable n. Follow with a division sign, and enter the **Determinant** of the denominator as shown in Fig. 8.24. Retype **I1** and select the equal sign; the correct result of **−1** will appear.

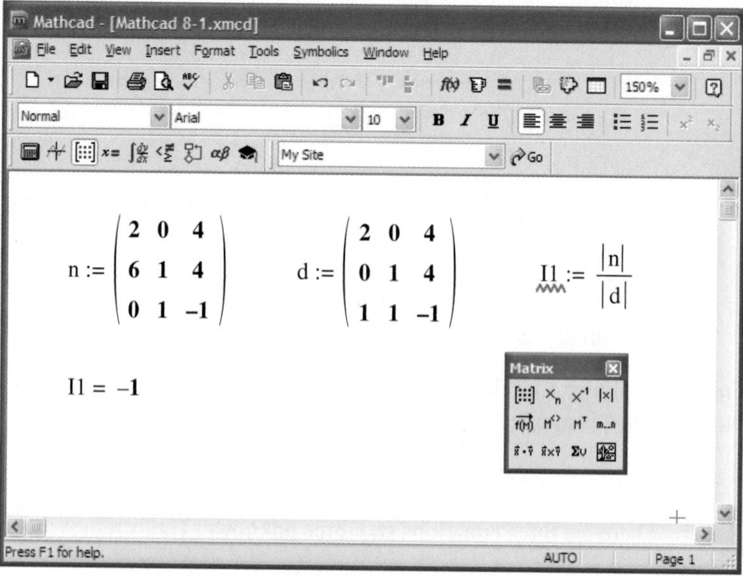

FIG. 8.24
Using Mathcad to verify the numerical calculations of Example 8.9.

Once you have mastered the rather simple and direct process just described, you will deeply appreciate being able to use Mathcad as a checking tool or solving mechanism.

Solution 2: Instead of using third-order determinants as in Solution 1, we can reduce the three equations to two by substituting the third equation in the first and second equations:

$$\left.\begin{aligned} 2 - 2I_1 - 4\overbrace{(I_1 + I_2)}^{I_3} &= 0 \\ 4\overbrace{(I_1 + I_2)}^{I_3} + I_2 - 6 &= 0 \end{aligned}\right\} \begin{aligned} 2 - 2I_1 - 4I_1 - 4I_2 &= 0 \\ 4I_1 + 4I_2 + I_2 - 6 &= 0 \end{aligned}$$

or

$$\begin{aligned} -6I_1 - 4I_2 &= -2 \\ +4I_1 + 5I_2 &= +6 \end{aligned}$$

Multiplying through by -1 in the top equation yields

$$\begin{aligned} 6I_1 - 4I_2 &= +2 \\ 4I_1 + 5I_2 &= +6 \end{aligned}$$

and using determinants,

$$I_1 = \frac{\begin{vmatrix} 2 & 4 \\ 6 & 5 \end{vmatrix}}{\begin{vmatrix} 6 & 4 \\ 4 & 5 \end{vmatrix}} = \frac{10 - 24}{30 - 16} = \frac{-14}{14} = \mathbf{-1\ A}$$

TI-89 Solution: The procedure for determining the determinant in Example 8.9 requires some scrolling to obtain the desired math functions, but in time that procedure can be performed quite rapidly. As with any computer or calculator system, it is paramount that you enter all parameters correctly. One error in the sequence negates the entire process. For the T1-89, the entries are shown in Fig. 8.25.

Home 2ND MATH ↓ Matrix ENTER ↓ det ENTER 2ND [2 , 4 2ND ; 6 , 5 2ND]) ÷ 2ND MATH ↓ Matrix ENTER ↓ det ENTER 2ND [6 , 4 2ND ; 4 , 5 2ND]) ENTER

FIG. 8.25

After you select the last ENTER key, the screen shown in Fig. 8.26 appears.

$$\frac{\det\left(\begin{bmatrix} 2 & 4 \\ 6 & 5 \end{bmatrix}\right)}{\det\left(\begin{bmatrix} 6 & 4 \\ 4 & 5 \end{bmatrix}\right)} \qquad -1.00\text{E}0$$

FIG. 8.26

$$I_2 = \frac{\begin{vmatrix} 6 & 2 \\ 4 & 6 \end{vmatrix}}{14} = \frac{36 - 8}{14} = \frac{28}{14} = \mathbf{2\ A}$$

$$I_3 = I_1 + I_2 = -1 + 2 = \mathbf{1\ A}$$

It is now important that the impact of the results obtained be understood. The currents I_1, I_2, and I_3 are the actual currents in the branches in

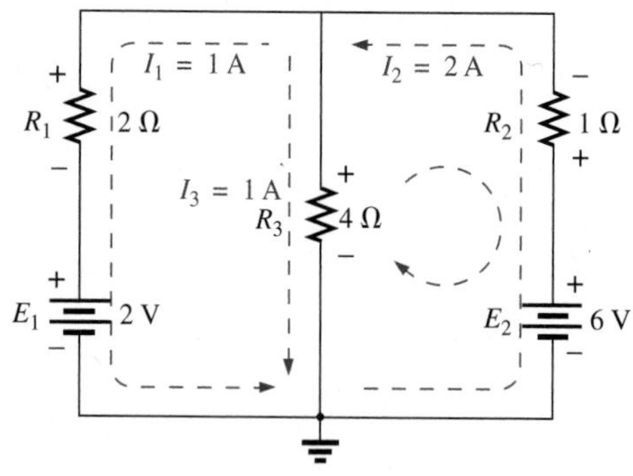

FIG. 8.27
Reviewing the results of the analysis of the network in Fig. 8.22.

which they were defined. A negative sign in the solution means that the actual current has the opposite direction than initially defined—the magnitude is correct. Once the actual current directions and their magnitudes are inserted in the original network, the various voltages and power levels can be determined. For this example, the actual current directions and their magnitudes have been entered on the original network in Fig. 8.27. Note that the current through the series elements R_1 and E_1 is 1 A; the current through R_3, 1 A; and the current through the series elements R_2 and E_2, 2 A. Due to the minus sign in the solution, the direction of I_1 is opposite to that shown in Fig. 8.22. The voltage across any resistor can now be found using Ohm's law, and the power delivered by either source or to any one of the three resistors can be found using the appropriate power equation.

Applying Kirchhoff's voltage law around the loop indicated in Fig. 8.27,

$$\Sigma_{\text{C}} V = +(4\ \Omega)I_3 + (1\ \Omega)I_2 - 6\text{ V} = 0$$

or

$$(4\ \Omega)I_3 + (1\ \Omega)I_2 = 6\text{ V}$$

and

$$(4\ \Omega)(1\text{ A}) + (1\ \Omega)(2\text{ A}) = 6\text{ V}$$
$$4\text{ V} + 2\text{ V} = 6\text{ V}$$
$$6\text{ V} = 6\text{ V} \quad \text{(checks)}$$

EXAMPLE 8.10 Apply branch-current analysis to the network in Fig. 8.28.

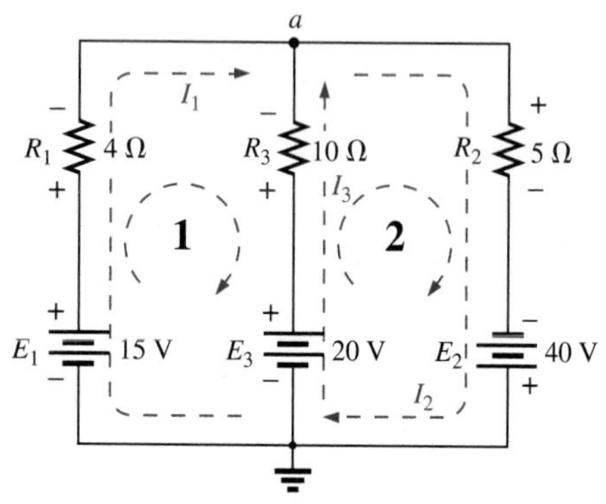

FIG. 8.28
Example 8.10.

Solution: Again, the current directions were chosen to match the "pressure" of each battery. The polarities are then added, and Kirchhoff's voltage law is applied around each closed loop in the clockwise direction. The result is as follows:

$$\text{loop 1:}\quad +15\text{ V} - (4\ \Omega)I_1 + (10\ \Omega)I_3 - 20\text{ V} = 0$$
$$\text{loop 2:}\quad +20\text{ V} - (10\ \Omega)I_3 - (5\ \Omega)I_2 + 40\text{ V} = 0$$

Applying Kirchhoff's current law at node *a*,

$$I_1 + I_3 = I_2$$

Substituting the third equation into the other two yields (with units removed for clarity)

$$\left.\begin{aligned} 15 - 4I_1 + 10I_3 - 20 &= 0 \\ 20 - 10I_3 - 5(I_1 + I_3) + 40 &= 0 \end{aligned}\right\} \text{Substituting for } I_2 \text{ (since it occurs only once in the two equations)}$$

or

$$-4I_1 + 10I_3 = 5$$
$$-5I_1 - 15I_3 = -60$$

Multiplying the lower equation by -1, we have

$$-4I_1 + 10I_3 = 5$$
$$5I_1 + 15I_3 = 60$$

$$I_1 = \frac{\begin{vmatrix} 5 & 10 \\ 60 & 15 \end{vmatrix}}{\begin{vmatrix} -4 & 10 \\ 5 & 15 \end{vmatrix}} = \frac{75 - 600}{-60 - 50} = \frac{-525}{-110} = \mathbf{4.77\ A}$$

$$I_3 = \frac{\begin{vmatrix} -4 & 5 \\ 5 & 60 \end{vmatrix}}{-110} = \frac{-240 - 25}{-110} = \frac{-265}{-110} = \mathbf{2.41\ A}$$

$$I_2 = I_1 + I_3 = 4.77\text{ A} + 2.41\text{ A} = \mathbf{7.18\ A}$$

revealing that the assumed directions were the actual directions, with I_2 equal to the sum of I_1 and I_3.

8.7 MESH ANALYSIS (GENERAL APPROACH)

The next method to be described—**mesh analysis**—is actually an extension of the branch-current analysis approach just introduced. By defining a unique array of currents to the network, the information provided by the application of Kirchhoff's current law is already included when we apply Kirchhoff's voltage law. In other words, there is no need to apply step 4 of the branch-current method.

The currents to be defined are called **mesh** or **loop currents.** The two terms are used interchangeably. In Fig. 8.29(a), a network with two "windows" has had two mesh currents defined. Note that each forms a closed "loop" around the inside of each window; these loops are similar to the loops defined in the wire mesh fence in Fig. 8.29(b)—hence the use of the term *mesh* for the loop currents. We will find that

the number of mesh currents required to analyze a network will equal the number of "windows" of the configuration.

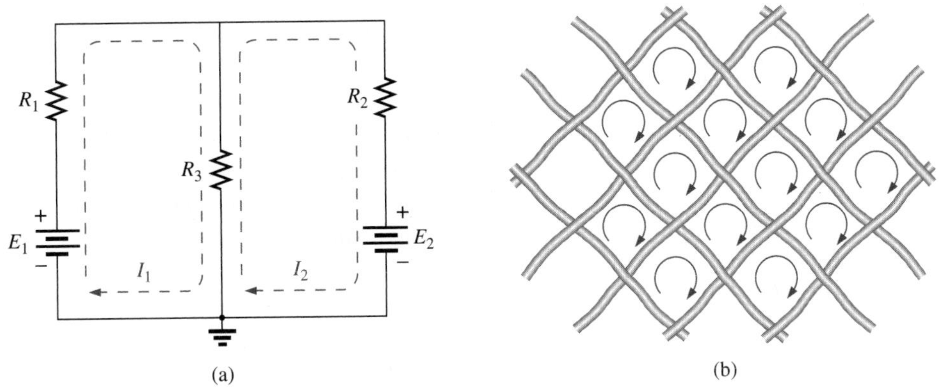

FIG. 8.29
Defining the mesh (loop) current: (a) "two-window" network: (b) wire mesh fence analogy.

The defined mesh currents can initially be a little confusing because it appears that two currents have been defined for resistor R_3. There is no problem with E_1 and R_1, which have only current I_1, or with E_2 and R_2, which have only current I_2. However, defining the current through R_3 may seem a little troublesome. Actually, it is quite straightforward. The current through R_3 is simply the difference between I_1 and I_2, with the direction of the larger. This is demonstrated in the examples to follow.

Because mesh currents can result in more than one current through an element, branch-current analysis was introduced first. Branch-current analysis is the straightforward application of the basic laws of electric circuits. Mesh analysis employs a maneuver ("trick," if you prefer) that removes the need to apply Kirchhoff's current law.

Mesh Analysis Procedure

1. ***Assign a distinct current in the clockwise direction to each independent, closed loop of the network. It is not absolutely necessary to choose the clockwise direction for each loop current. In fact, any direction can be chosen for each loop current with no loss in accuracy, as long as the remaining steps are followed properly. However, by choosing the clockwise direction as a standard, we can develop a shorthand method (Section 8.8) for writing the required equations that will save time and possibly prevent some common errors.***

This first step is accomplished most effectively by placing a loop current *within* each "window" of the network, as demonstrated in the previous section, to ensure that they are all independent. A variety of other loop currents can be assigned. In each case, however, be sure, that the information carried by any one loop equation is not included in a combination of the other network equations. This is the crux of the terminology: *independent.* No matter how you choose your loop currents, the number of loop currents required is always equal to the number of windows of a planar (no-crossovers) network. On occasion, a network may appear to be nonplanar. However, a redrawing of the network may reveal that it is, in fact, planar. This may be true for one or two problems at the end of the chapter.

Before continuing to the next step, let us ensure that the concept of a loop current is clear. For the network in Fig. 8.30, the loop current I_1 is the branch current of the branch containing the 2 Ω resistor and 2 V battery. The current through the 4 Ω resistor is not I_1, however, since there is also a loop current I_2 through it. Since they have opposite directions, $I_{4\Omega}$ equals the difference between the two, $I_1 - I_2$ or $I_2 - I_1$, depending on which you choose to be the defining direction. In other words, *a loop current is a branch current only when it is the only loop current assigned to that branch.*

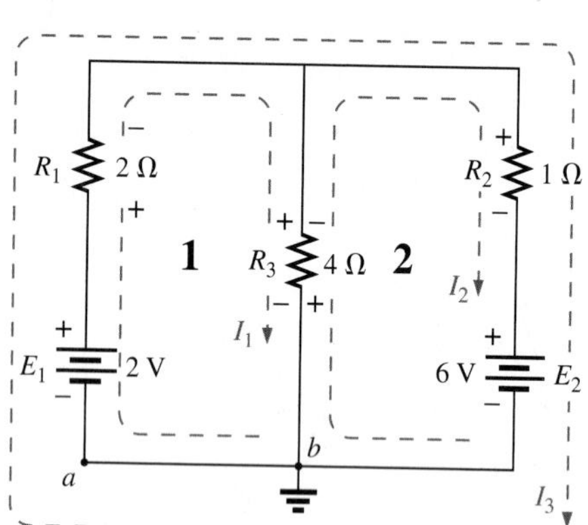

FIG. 8.30
Defining the mesh currents for a "two-window" network.

2. ***Indicate the polarities within each loop for each resistor as determined by the assumed direction of loop current for that loop. Note the requirement that the polarities be placed within each loop. This requires, as shown in Fig. 8.30, that the 4 Ω resistor have two sets of polarities across it.***
3. ***Apply Kirchhoff's voltage law around each closed loop in the clockwise direction. Again, the clockwise direction was chosen to establish uniformity and prepare us for the method to be introduced in the next section.***
   a. ***If a resistor has two or more assumed currents through it, the total current through the resistor is the assumed current of the loop in which Kirchhoff's voltage law is being applied, plus the assumed currents of the other loops passing through in the same direction, minus the assumed currents through in the opposite direction.***
   b. ***The polarity of a voltage source is unaffected by the direction of the assigned loop currents.***
4. ***Solve the resulting simultaneous linear equations for the assumed loop currents.***

EXAMPLE 8.11 Consider the same basic network as in Example 8.9, now appearing as Fig. 8.30.

Solution:

Step 1: Two loop currents (I_1 and I_2) are assigned in the clockwise direction in the windows of the network. A third loop (I_3) could have been included around the entire network, but the information carried by this loop is already included in the other two.

Step 2: Polarities are drawn within each window to agree with assumed current directions. Note that for this case, the polarities across the 4 Ω resistor are the opposite for each loop current.

Step 3: Kirchhoff's voltage law is applied around each loop in the clockwise direction. Keep in mind as this step is performed that the law is con-

cerned only with the magnitude and polarity of the voltages around the closed loop and not with whether a voltage rise or drop is due to a battery or a resistive element. The voltage across each resistor is determined by $V = IR$. For a resistor with more than one current through it, the current is the loop current of the loop being examined plus or minus the other loop currents as determined by their directions. If clockwise applications of Kirchhoff's voltage law are always chosen, the other loop currents are always subtracted from the loop current of the loop being analyzed.

loop 1: $+E_1 - V_1 - V_3 = 0$ (clockwise starting at point a)

$$+2\text{ V} - (2\ \Omega)I_1 - \overbrace{(4\ \Omega)\underbrace{(I_1 - I_2)}_{\substack{\text{Total current}\\ \text{through}\\ 4\ \Omega\text{ resistor}}}}^{\substack{\text{Voltage drop across}\\ 4\ \Omega\text{ resistor}}} = 0$$

Subtracted since I_2 is opposite in direction to I_1.

loop 2: $-V_3 - V_2 - E_2 = 0$ (clockwise starting at point b)

$$-(4\ \Omega)(I_2 - I_1) - (1\ \Omega)I_2 - 6\text{ V} = 0$$

Step 4: The equations are then rewritten as follows (without units for clarity):

$$\begin{aligned} \text{loop 1:}\quad & +2 - 2I_1 - 4I_1 + 4I_2 = 0 \\ \text{loop 2:}\quad & -4I_2 + 4I_1 - 1I_2 - 6 = 0 \end{aligned}$$

and

$$\begin{aligned} \text{loop 1:}\quad & +2 - 6I_1 + 4I_2 = 0 \\ \text{loop 2:}\quad & -5I_2 + 4I_1 - 6 = 0 \end{aligned}$$

or

$$\begin{aligned} \text{loop 1:}\quad & -6I_1 + 4I_2 = -2 \\ \text{loop 2:}\quad & +4I_1 - 5I_2 = +6 \end{aligned}$$

Applying determinants results in

$$I_1 = \mathbf{-1\ A} \quad \text{and} \quad I_2 = \mathbf{-2\ A}$$

The minus signs indicate that the currents have a direction opposite to that indicated by the assumed loop current.

The actual current through the 2 V source and 2 Ω resistor is therefore 1 A in the other direction, and the current through the 6 V source and 1 Ω resistor is 2 A in the opposite direction indicated on the circuit. The current through the 4 Ω resistor is determined by the following equation from the original network:

$$\begin{aligned} \text{loop 1:}\quad I_{4\Omega} &= I_1 - I_2 = -1\text{ A} - (-2\text{ A}) = -1\text{ A} + 2\text{ A} \\ &= \mathbf{1\ A} \quad (\text{in the direction of } I_1) \end{aligned}$$

The outer loop (I_3) and *one* inner loop (either I_1 or I_2) would also have produced the correct results. This approach, however, often leads to errors since the loop equations may be more difficult to write. The best method of picking these loop currents is the window approach.

EXAMPLE 8.12 Find the current through each branch of the network in Fig. 8.31.

Solution:

Steps 1 and 2: These are as indicated in the circuit. Note that the polarities of the 6 Ω resistor are different for each loop current.

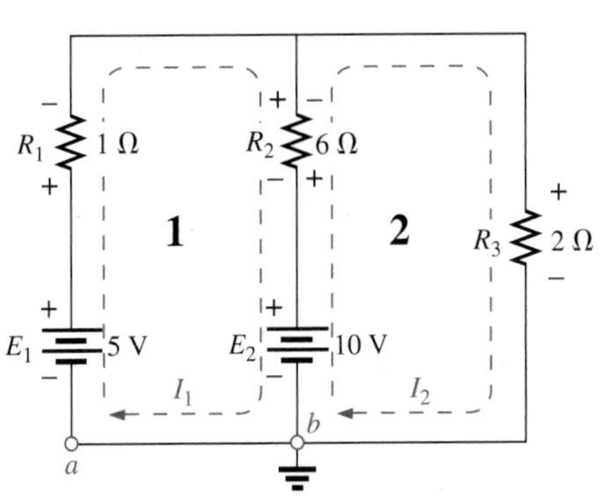

FIG. 8.31
Example 8.12.

Step 3: Kirchhoff's voltage law is applied around each closed loop in the clockwise direction:

loop 1: $+E_1 - V_1 - V_2 - E_2 = 0$ (clockwise starting at point a)

$$+5\text{ V} - (1\ \Omega)I_1 - (6\ \Omega)(I_1 - I_2) - 10\text{ V} = 0$$

I_2 flows through the 6 Ω resistor in the direction opposite to I_1.

loop 2: $E_2 - V_2 - V_3 = 0$ (clockwise starting at point b)

$$+10\text{ V} - (6\ \Omega)(I_2 - I_1) - (2\ \Omega)I_2 = 0$$

The equations are rewritten as

$$\left.\begin{aligned} 5 - I_1 - 6I_1 + 6I_2 - 10 &= 0 \\ 10 - 6I_2 + 6I_1 - 2I_2 &= 0 \end{aligned}\right\} \begin{aligned} -7I_1 + 6I_2 &= 5 \\ +6I_1 - 8I_2 &= -10 \end{aligned}$$

Step 4:

$$I_1 = \frac{\begin{vmatrix} 5 & 6 \\ -10 & -8 \end{vmatrix}}{\begin{vmatrix} -7 & 6 \\ 6 & -8 \end{vmatrix}} = \frac{-40 + 60}{56 - 36} = \frac{20}{20} = \mathbf{1\ A}$$

$$I_2 = \frac{\begin{vmatrix} -7 & 5 \\ 6 & -10 \end{vmatrix}}{20} = \frac{70 - 30}{20} = \frac{40}{20} = \mathbf{2\ A}$$

Since I_1 and I_2 are positive and flow in opposite directions through the 6 Ω resistor and 10 V source, the total current in this branch is equal to the difference of the two currents in the direction of the larger:

$$I_2 > I_1 \quad (2\text{ A} > 1\text{ A})$$

Therefore,

$$I_{R_2} = I_2 - I_1 = 2\text{ A} - 1\text{ A} = \mathbf{1\ A} \quad \text{in the direction of } I_2$$

It is sometimes impractical to draw all the branches of a circuit at right angles to one another. The next example demonstrates how a portion of a network may appear due to various constraints. The method of analysis is no different with this change in configuration.

EXAMPLE 8.13 Find the branch currents of the networks in Fig. 8.32.

Solution:

Steps 1 and 2: These are as indicated in the circuit.

Step 3: Kirchhoff's voltage law is applied around each closed loop:

loop 1: $-E_1 - I_1R_1 - E_2 - V_2 = 0$ (clockwise from point a)

$$-6\text{ V} - (2\ \Omega)I_1 - 4\text{ V} - (4\ \Omega)(I_1 - I_2) = 0$$

loop 2: $-V_2 + E_2 - V_3 - E_3 = 0$ (clockwise from point b)

$$-(4\ \Omega)(I_2 - I_1) + 4\text{ V} - (6\ \Omega)(I_2) - 3\text{ V} = 0$$

which are rewritten as

$$\left.\begin{aligned} -10 - 4I_1 - 2I_1 + 4I_2 &= 0 \\ +1 + 4I_1 + 4I_2 - 6I_2 &= 0 \end{aligned}\right\} \begin{aligned} -6I_1 + 4I_2 &= +10 \\ +4I_1 - 10I_2 &= -1 \end{aligned}$$

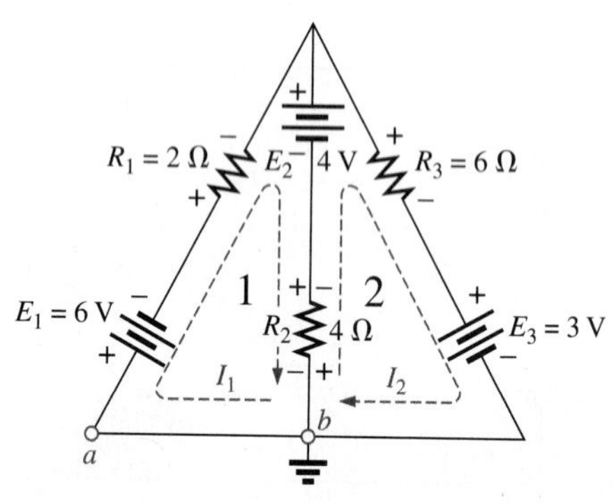

FIG. 8.32
Example 8.13.

or, by multiplying the top equation by -1, we obtain

$$\begin{aligned} 6I_1 - 4I_2 &= -10 \\ 4I_1 - 10I_2 &= -1 \end{aligned}$$

Step 4: $$I_1 = \frac{\begin{vmatrix} -10 & -4 \\ -1 & -10 \end{vmatrix}}{\begin{vmatrix} 6 & -4 \\ 4 & -10 \end{vmatrix}} = \frac{100 - 4}{-60 + 16} = \frac{96}{-44} = \mathbf{-2.18\ A}$$

$$I_2 = \frac{\begin{vmatrix} 6 & -10 \\ 4 & -1 \end{vmatrix}}{-44} = \frac{-6 + 40}{-44} = \frac{34}{-44} = \mathbf{-0.77\ A}$$

The current in the 4 Ω resistor and 4 V source for loop 1 is

$$\begin{aligned} I_1 - I_2 &= -2.18\text{ A} - (-0.77\text{ A}) \\ &= -2.18\text{ A} + 0.77\text{ A} \\ &= \mathbf{-1.41\ A} \end{aligned}$$

revealing that it is 1.41 A in a direction opposite (due to the minus sign) to I_1 in loop 1.

Supermesh Currents

Occasionally, you will find current sources in a network without a parallel resistance. This removes the possibility of converting the source to a voltage source as required by the given procedure. In such cases, you have a choice of two approaches.

The simplest and most direct approach is to place a resistor in parallel with the current source that has a much higher value than the other resistors of the network. For instance, if most of the resistors of the network are in the 1 to 10 Ω range, choosing a resistor of 100 Ω or higher would provide one level of accuracy for the answer. However, choosing a resistor of 1000 Ω or higher would increase the accuracy of the answer. You will never get the exact answer because the network has been modified by this introduced element. However for most applications, the answer will be sufficiently accurate.

The other choice is to use the **Supermesh approach** described in the following steps. Although this approach will provide the exact solution, it does require some practice to become proficient in its use. The procedure is as follows.

Start as before and assign a mesh current to each independent loop, including the current sources, as if they were resistors or voltage sources. Then mentally (redraw the network if necessary) remove the current sources (replace with open-circuit equivalents), and apply Kirchhoff's voltage law to all the remaining independent paths of the network using the mesh currents just defined. Any resulting path, including two or more mesh currents, is said to be the path of a **supermesh current.** Then relate the chosen mesh currents of the network to the independent current sources of the network, and solve for the mesh currents. The next example clarifies the definition of supermesh current and the procedure.

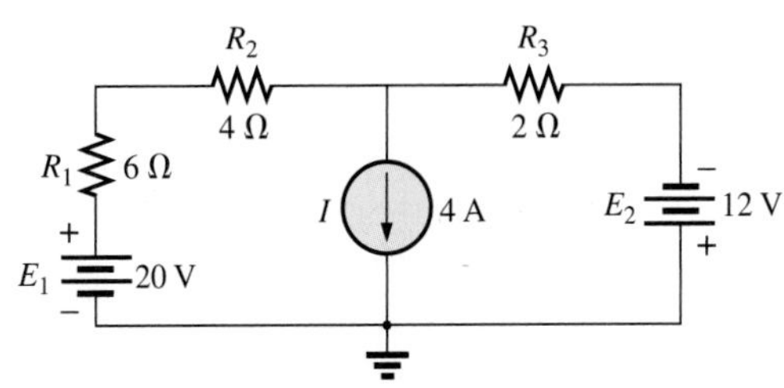

FIG. 8.33
Example 8.14.

EXAMPLE 8.14 Using mesh analysis, determine the currents of the network in Fig. 8.33.

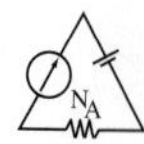

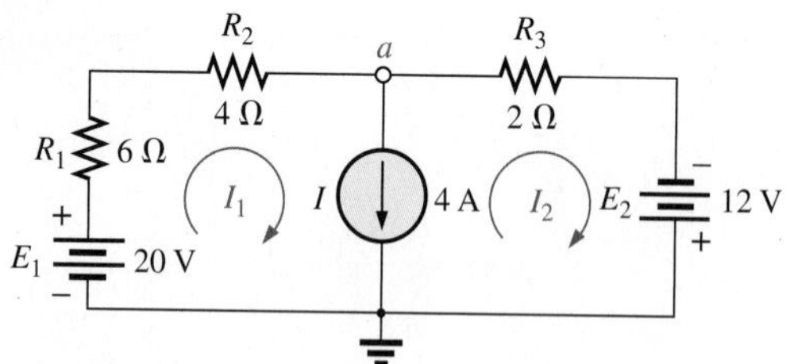

FIG. 8.34
Defining the mesh currents for the network in Fig. 8.33.

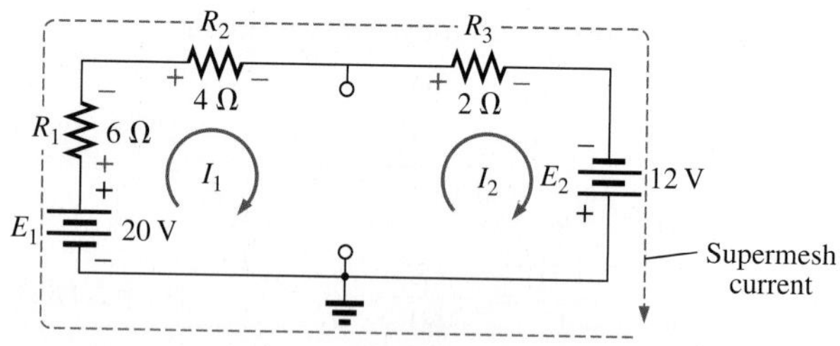

FIG. 8.35
Defining the supermesh current.

Solution: First, the mesh currents for the network are defined, as shown in Fig. 8.34. Then the current source is mentally removed, as shown in Fig. 8.35, and Kirchhoff's voltage law is applied to the resulting network. The single path now including the effects of two mesh currents is referred to as the path of a *supermesh current.*

Applying Kirchhoff's law:

$$20\text{ V} - I_1(6\ \Omega) - I_1(4\ \Omega) - I_2(2\ \Omega) + 12\text{ V} = 0$$

or

$$10I_1 + 2I_2 = 32$$

Node *a* is then used to relate the mesh currents and the current source using Kirchhoff's current law:

$$I_1 = I + I_2$$

The result is two equations and two unknowns:

$$\begin{aligned} 10I_1 + 2I_2 &= 32 \\ I_1 - \ \ I_2 &= 4 \end{aligned}$$

Applying determinants:

$$I_1 = \frac{\begin{vmatrix} 32 & 2 \\ 4 & -1 \end{vmatrix}}{\begin{vmatrix} 10 & 2 \\ 1 & -1 \end{vmatrix}} = \frac{(32)(-1) - (2)(4)}{(10)(-1) - (2)(1)} = \frac{40}{12} = \mathbf{3.33\ A}$$

and

$$I_2 = I_1 - I = 3.33\text{ A} - 4\text{ A} = \mathbf{-0.67\ A}$$

In the above analysis, it may appear that when the current source was removed, $I_1 = I_2$. However, the supermesh approach requires that we stick with the original definition of each mesh current and not alter those definitions when current sources are removed.

EXAMPLE 8.15 Using mesh analysis, determine the currents for the network in Fig. 8.36.

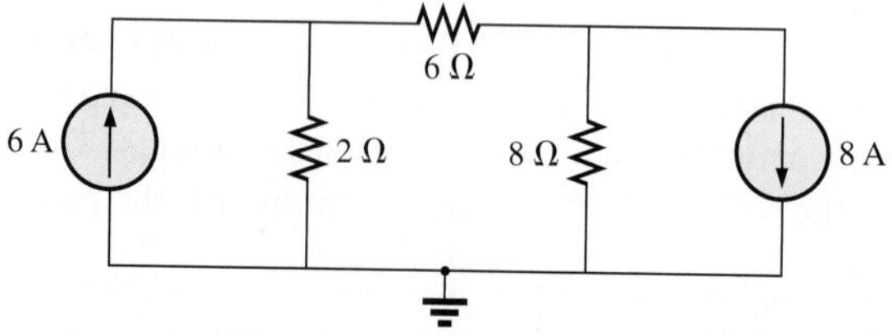

FIG. 8.36
Example 8.15.

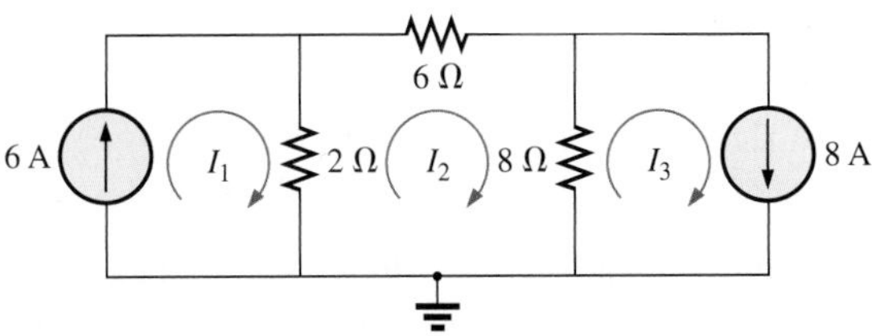

FIG. 8.37
Defining the mesh currents for the network in Fig. 8.36.

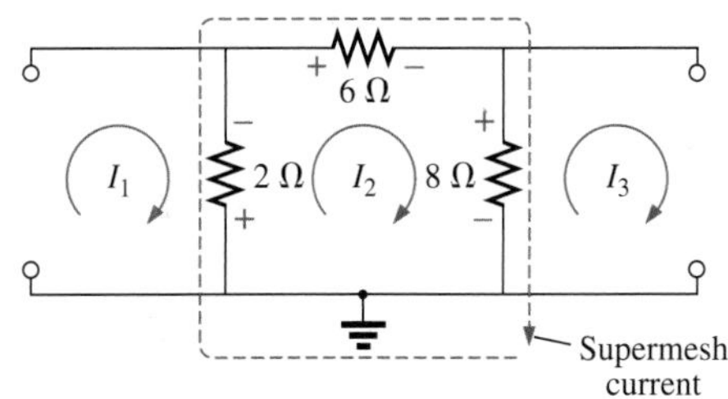

FIG. 8.38
Defining the supermesh current for the network in Fig. 8.36.

Solution: The mesh currents are defined in Fig. 8.37. The current sources are removed, and the single supermesh path is defined in Fig. 8.38.

Applying Kirchhoff's voltage law around the supermesh path:

$$-V_{2\Omega} - V_{6\Omega} - V_{8\Omega} = 0$$
$$-(I_2 - I_1)2\ \Omega - I_2(6\ \Omega) - (I_2 - I_3)8\ \Omega = 0$$
$$-2I_2 + 2I_1 - 6I_2 - 8I_2 + 8I_3 = 0$$
$$2I_1 - 16I_2 + 8I_3 = 0$$

Introducing the relationship between the mesh currents and the current sources:

$$I_1 = 6\text{ A}$$
$$I_3 = 8\text{ A}$$

results in the following solutions:

$$2I_1 - 16I_2 + 8I_3 = 0$$
$$2(6\text{ A}) - 16I_2 + 8(8\text{ A}) = 0$$

and $$I_2 = \frac{76\text{ A}}{16} = \mathbf{4.75\ A}$$

Then $$I_{2\Omega}\downarrow = I_1 - I_2 = 6\text{ A} - 4.75\text{ A} = \mathbf{1.25\ A}$$

and $$I_{8\Omega}\uparrow = I_3 - I_2 = 8\text{ A} - 4.75\text{ A} = \mathbf{3.25\ A}$$

Again, note that you must stick with your original definitions of the various mesh currents when applying Kirchhoff's voltage law around the resulting supermesh paths.

8.8 MESH ANALYSIS (FORMAT APPROACH)

Now that the basis for the mesh-analysis approach has been established, we now examine a technique for writing the mesh equations more rapidly and usually with fewer errors. As an aid in introducing the procedure, the network in Example 8.12 (Fig. 8.31) has been redrawn in Fig. 8.39 with the assigned loop currents. (Note that each loop current has a clockwise direction.)

The equations obtained are

$$-7I_1 + 6I_2 = 5$$
$$6I_1 - 8I_2 = -10$$

which can also be written as

$$7I_1 - 6I_2 = -5$$
$$8I_2 - 6I_1 = 10$$

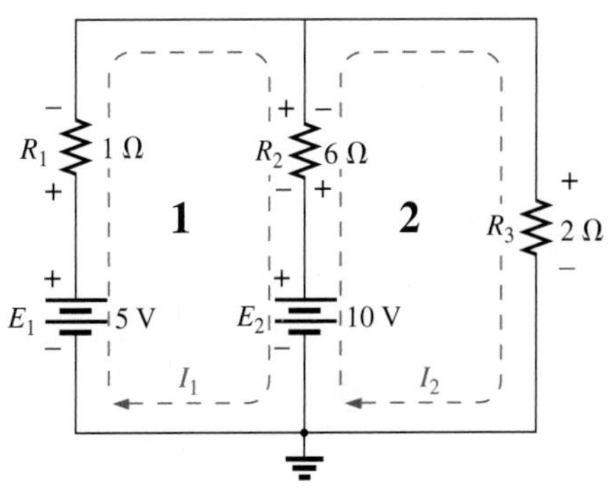

FIG. 8.39
Network in Fig. 8.31 redrawn with assigned loop currents.

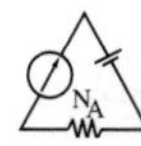

and expanded as

Col. 1	Col. 2	Col. 3
$(1 + 6)I_1 -$	$6I_2$	$= (5 - 10)$
$(2 + 6)I_2 -$	$6I_1$	$= 10$

Note in the above equations that column 1 is composed of a loop current times the sum of the resistors through which that loop current passes. Column 2 is the product of the resistors common to another loop current times that other loop current. Note that in each equation, this column is subtracted from column 1. Column 3 is the *algebraic* sum of the voltage sources through which the loop current of interest passes. A source is assigned a positive sign if the loop current passes from the negative to the positive terminal, and a negative value is assigned if the polarities are reversed. The comments above are correct only for a standard direction of loop current in each window, the one chosen being the clockwise direction.

The above statements can be extended to develop the following *format approach* to mesh analysis.

Mesh Analysis Procedure

1. *Assign a loop current to each independent, closed loop (as in the previous section) in a clockwise direction.*
2. *The number of required equations is equal to the number of chosen independent, closed loops. Column 1 of each equation is formed by summing the resistance values of those resistors through which the loop current of interest passes and multiplying the result by that loop current.*
3. *We must now consider the mutual terms, which, as noted in the examples above, are always subtracted from the first column. A* **mutual term** *is simply any resistive element having an additional loop current passing through it. It is possible to have more than one mutual term if the loop current of interest has an element in common with more than one other loop current. This will be demonstrated in an example to follow. Each term is the product of the mutual resistor and the other loop current passing through the same element.*
4. *The column to the right of the equality sign is the algebraic sum of the voltage sources through which the loop current of interest passes. Positive signs are assigned to those sources of voltage having a polarity such that the loop current passes from the negative to the positive terminal. A negative sign is assigned to those potentials for which the reverse is true.*
5. *Solve the resulting simultaneous equations for the desired loop currents.*

Before considering a few examples, be aware that since the column to the right of the equals sign is the algebraic sum of the voltage sources in that loop, *the format approach can be applied only to networks in which all current sources have been converted to their equivalent voltage source.*

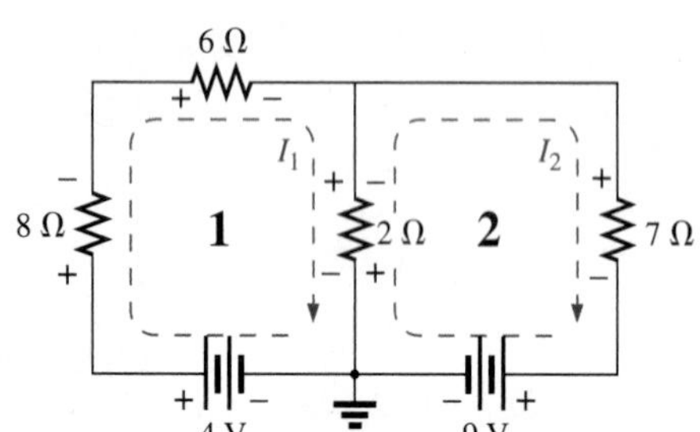

FIG. 8.40
Example 8.16.

EXAMPLE 8.16 Write the mesh equations for the network in Fig. 8.40, and find the current through the 7 Ω resistor.

Solution:

Step 1: As indicated in Fig. 8.40, each assigned loop current has a clockwise direction.

Steps 2 to 4:

$$I_1: \quad (8\ \Omega + 6\ \Omega + 2\ \Omega)I_1 - (2\ \Omega)I_2 = 4\text{ V}$$
$$I_2: \quad (7\ \Omega + 2\ \Omega)I_2 - (2\ \Omega)I_1 = -9\text{ V}$$

and

$$16I_1 - 2I_2 = 4$$
$$9I_2 - 2I_1 = -9$$

which, for determinants, are

$$16I_1 - 2I_2 = 4$$
$$-2I_1 + 9I_2 = -9$$

and

$$I_2 = I_{7\Omega} = \frac{\begin{vmatrix} 16 & 4 \\ -2 & -9 \end{vmatrix}}{\begin{vmatrix} 16 & -2 \\ -2 & 9 \end{vmatrix}} = \frac{-144 + 8}{144 - 4} = \frac{-136}{140}$$
$$= \mathbf{-0.97\ A}$$

EXAMPLE 8.17 Write the mesh equations for the network in Fig. 8.41.

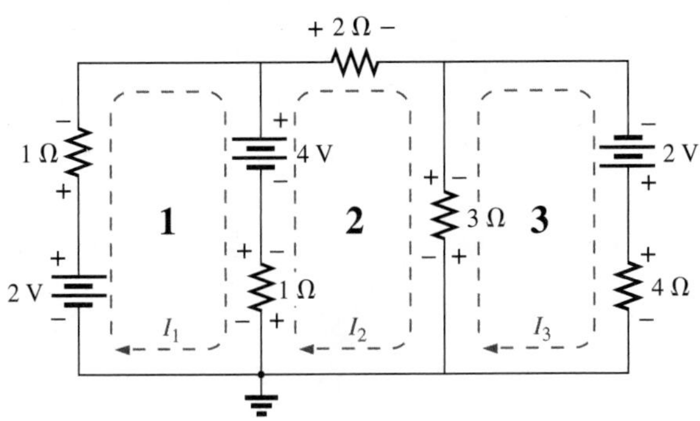

FIG. 8.41
Example 8.17.

Solution: Each window is assigned a loop current in the clockwise direction:

I_1 does not pass through an element mutual with I_3.

$$I_1: \quad (1\ \Omega + 1\ \Omega)I_1 - (1\ \Omega)I_2 + 0 = 2\text{ V} - 4\text{ V}$$
$$I_2: \quad (1\ \Omega + 2\ \Omega + 3\ \Omega)I_2 - (1\ \Omega)I_1 - (3\ \Omega)I_3 = 4\text{ V}$$
$$I_3: \quad (3\ \Omega + 4\ \Omega)I_3 - (3\ \Omega)I_2 + 0 = 2\text{ V}$$

I_3 does not pass through an element mutual with I_1.

Summing terms yields

$$2I_1 - I_2 + 0 = -2$$
$$6I_2 - I_1 - 3I_3 = 4$$
$$7I_3 - 3I_2 + 0 = 2$$

which are rewritten for determinants as

$$\begin{aligned} 2I_1 - I_2 + 0 &= -2 \\ -I_1 + 6I_2 - 3I_3 &= 4 \\ 0 - 3I_2 + 7I_3 &= 2 \end{aligned}$$

Note that the coefficients of the a and b diagonals are equal. This *symmetry* about the c-axis will always be true for equations written using the format approach. It is a check on whether the equations were obtained correctly.

We now consider a network with only one source of voltage to point out that mesh analysis can be used to advantage in other than multisource networks.

EXAMPLE 8.18 Find the current through the 10 Ω resistor of the network in Fig. 8.42.

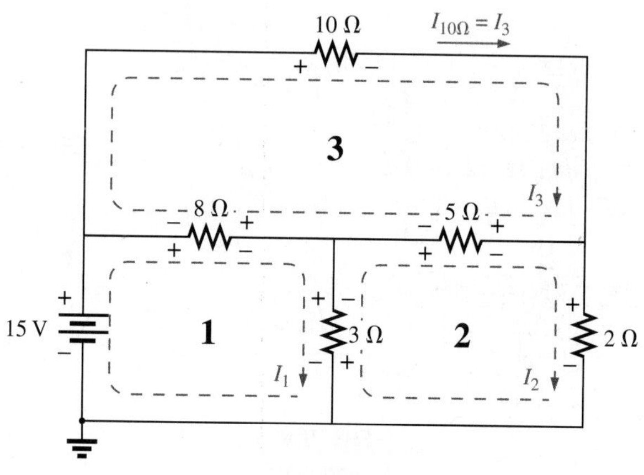

FIG. 8.42
Example 8.18.

Solution:

$$\begin{aligned} I_1: \quad & (8\ \Omega + 3\ \Omega)I_1 - (8\ \Omega)I_3 - (3\ \Omega)I_2 = 15\ \text{V} \\ I_2: \quad & (3\ \Omega + 5\ \Omega + 2\ \Omega)I_2 - (3\ \Omega)I_1 - (5\ \Omega)I_3 = 0 \\ I_3: \quad & (8\ \Omega + 10\ \Omega + 5\ \Omega)I_3 - (8\ \Omega)I_1 - (5\ \Omega)I_2 = 0 \end{aligned}$$

$$\begin{aligned} 11I_1 - 8I_3 - 3I_2 &= 15\ \text{V} \\ 10I_2 - 3I_1 - 5I_3 &= 0 \\ 23I_3 - 8I_1 - 5I_2 &= 0 \end{aligned}$$

or

$$\begin{aligned} 11I_1 - 3I_2 - 8I_3 &= 15\ \text{V} \\ -3I_1 + 10I_2 - 5I_3 &= 0 \\ -8I_1 - 5I_2 + 23I_3 &= 0 \end{aligned}$$

and
$$I_3 = I_{10\Omega} = \frac{\begin{vmatrix} 11 & -3 & 15 \\ -3 & 10 & 0 \\ -8 & -5 & 0 \end{vmatrix}}{\begin{vmatrix} 11 & -3 & -8 \\ -3 & 10 & -5 \\ -8 & -5 & 23 \end{vmatrix}} = \mathbf{1.22\ A}$$

Mathcad Solution: For this example, rather than take the time to develop the determinant form for each variable, we apply Mathcad directly to the resulting equations. As shown in Fig. 8.43, a **Guess** value for each variable must first be defined. Such guessing helps the computer begin its iteration process as it searches for the solution. By providing a rough estimate of 1, the computer recognizes that the result is probably a number with a magnitude less than 100 rather than have to worry about solutions that extend into the thousands or tens of thousands—the search has been narrowed considerably.

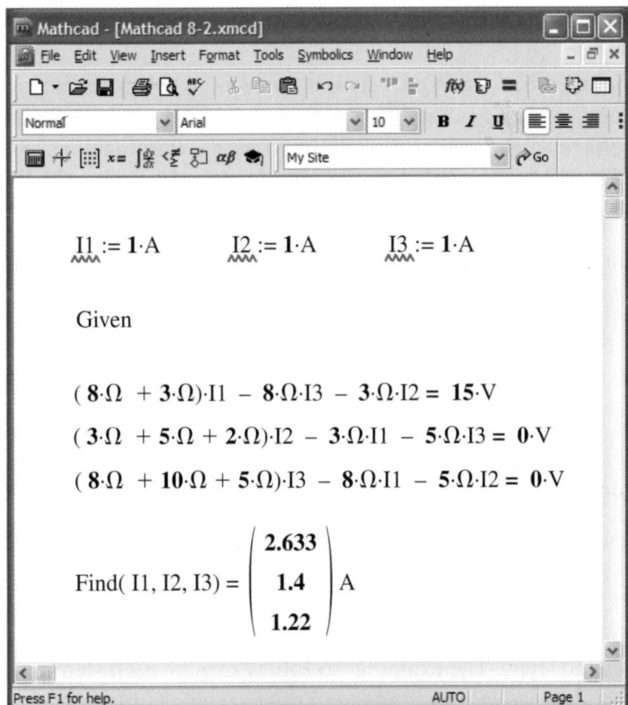

FIG. 8.43

Using Mathcad to verify the numerical calculations of Example 8.18.

Next, as shown, enter the word **Given** to tell the computer that the defining equations will follow. Finally, carefully enter each equation and set each one equal to the constant on the right using the **Ctrl=** operation.

The results are then obtained with the **Find (I1,I2,I3)** expression and an equal sign. As shown, the results are available with an acceptable degree of accuracy even though entering the equations and performing the analysis took only a minute or two (with practice).

TI-89 Calculator Solution: Using the TI-89 calculator, the sequence in Fig. 8.44 results. The intermediary 2ND and scrolling steps were not

det([11,−3,15;−3,10,0;−8,−5,0])/det([11,−3,−8;−3,10,−5;−8,−5,23]) (ENTER)	1.22

FIG. 8.44
Using the TI-89 calculator to solve for the current I_3.

$$\frac{\det\left(\begin{bmatrix} 11 & -3 & 15 \\ -3 & 10 & 0 \\ -8 & -5 & 0 \end{bmatrix}\right)}{\det\left(\begin{bmatrix} 11 & -3 & -8 \\ -3 & 10 & -5 \\ -8 & -5 & 23 \end{bmatrix}\right)} \quad 1.22\text{E}0$$

FIG. 8.45
The resulting display after properly entering the data for the current I_3.

included. This sequence certainly requires some care in entering the data in the required format, but it is still a rather neat, compact format.

The resulting display in Fig. 8.45 confirms our solution.

8.9 NODAL ANALYSIS (GENERAL APPROACH)

The methods introduced thus far have all been to find the currents of the network. We now turn our attention to **nodal analysis**—a method that provides the nodal voltages of a network, that is, the voltage from the various **nodes** (junction points) of the network to ground. The method is developed through the use of Kirchhoff's current law in much the same manner as Kirchhoff's voltage law was used to develop the mesh analysis approach.

Although it is not a requirement, we make it a policy to make ground our reference node and assign it a potential level of zero volts. All the other voltage levels are then found with respect to this reference level. For a network of N nodes, by assigning one as our reference node, we have $(N - 1)$ nodes for which the voltage must be determined. In other words,

the number of nodes for which the voltage must be determined using nodal analysis is 1 less than the total number of nodes.

The result of the above is $(N - 1)$ nodal voltages that need to be determined, requiring that $(N - 1)$ independent equations be written to find the nodal voltages. In other words,

the number of equations required to solve for all the nodal voltages of a network is 1 less than the total number of independent nodes.

Since each equation is the result of an application of Kirchhoff's current law, Kirchhoff's current law must be applied $(N - 1)$ times for each network.

Nodal analysis, like mesh analysis, can be applied by a series of carefully defined steps. The examples to follow explain each step in detail.

Nodal Analysis Procedure

1. ***Determine the number of nodes within the network.***
2. ***Pick a reference node, and label each remaining node with a subscripted value of voltage: V_1, V_2, and so on.***
3. ***Apply Kirchhoff's current law at each node except the reference. Assume that all unknown currents leave the node for each application of Kirchhoff's current law. In other words, for each node, don't be influenced by the direction that an unknown current for another node may have had. Each node is to be treated as a separate entity, independent of the application of Kirchhoff's current law to the other nodes.***
4. ***Solve the resulting equations for the nodal voltages.***

A few examples clarify the procedure defined by step 3. It initially takes some practice writing the equations for Kirchhoff's current law correctly, but in time the advantage of assuming that all the currents leave a node rather than identifying a specific direction for each branch become obvious. (The same type of advantage is associated with assuming that all the mesh currents are clockwise when applying mesh analysis.)

EXAMPLE 8.19 Apply nodal analysis to the network in Fig. 8.46.

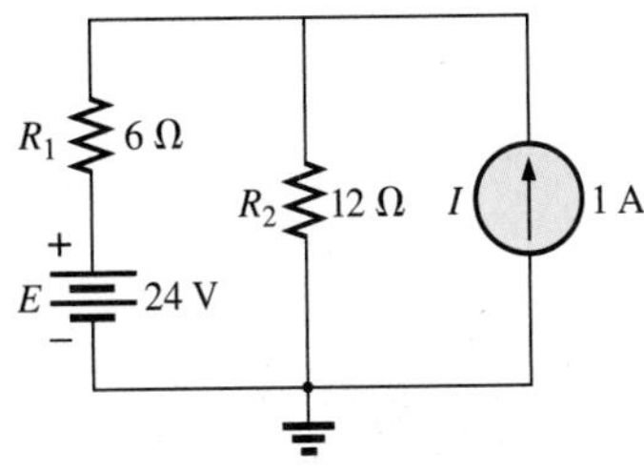

FIG. 8.46
Example 8.19.

Solution:

Steps 1 and 2: The network has two nodes, as shown in Fig. 8.47. The lower node is defined as the reference node at ground potential (zero volts), and the other node as V_1, the voltage from node 1 to ground.

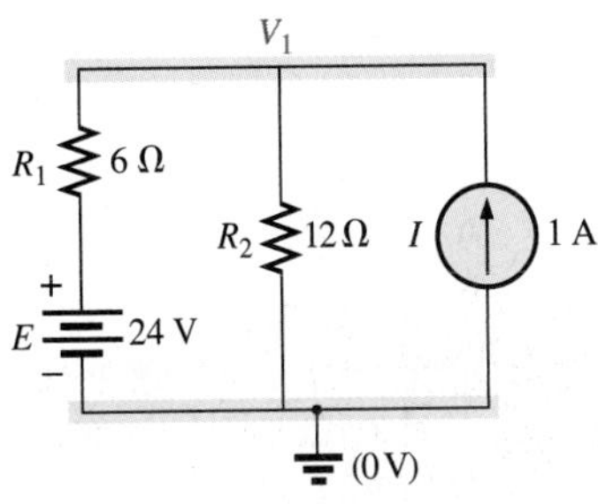

FIG. 8.47
Network in Fig. 8.46 with assigned nodes.

Step 3: I_1 and I_2 are defined as leaving the node in Fig. 8.48, and Kirchhoff's current law is applied as follows:

$$I = I_1 + I_2$$

The current I_2 is related to the nodal voltage V_1 by Ohm's law:

$$I_2 = \frac{V_{R_2}}{R_2} = \frac{V_1}{R_2}$$

The current I_1 is also determined by Ohm's law as follows:

$$I_1 = \frac{V_{R_1}}{R_1}$$

with

$$V_{R_1} = V_1 - E$$

Substituting into the Kirchhoff's current law equation:

$$I = \frac{V_1 - E}{R_1} + \frac{V_1}{R_2}$$

and rearranging, we have

$$I = \frac{V_1}{R_1} - \frac{E}{R_1} + \frac{V_1}{R_2} = V_1\left(\frac{1}{R_1} + \frac{1}{R_2}\right) - \frac{E}{R_1}$$

or

$$V_1\left(\frac{1}{R_1} + \frac{1}{R_2}\right) = \frac{E}{R_1} + 1$$

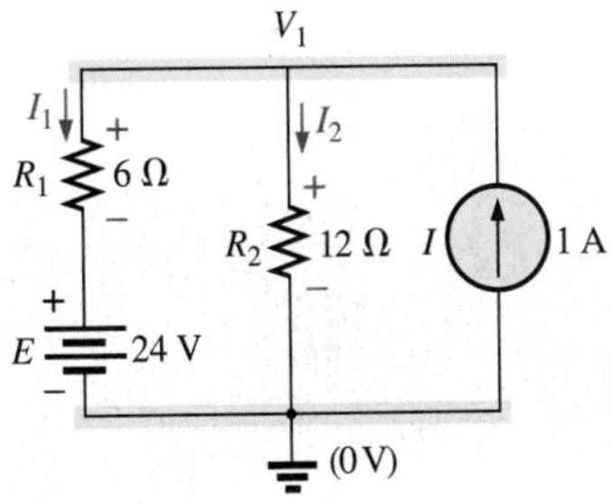

FIG. 8.48
Applying Kirchhoff's current law to the node V_1.

Substituting numerical values, we obtain

$$V_1\left(\frac{1}{6\ \Omega} + \frac{1}{12\ \Omega}\right) = \frac{24\ \text{V}}{6\ \Omega} + 1\ \text{A} = 4\ \text{A} + 1\ \text{A}$$

$$V_1\left(\frac{1}{4\ \Omega}\right) = 5\ \text{A}$$

$$V_1 = \mathbf{20\ V}$$

The currents I_1 and I_2 can then be determined using the preceding equations:

$$I_1 = \frac{V_1 - E}{R_1} = \frac{20\ \text{V} - 24\ \text{V}}{6\ \Omega} = \frac{-4\ \text{V}}{6\ \Omega}$$

$$= \mathbf{-0.67\ A}$$

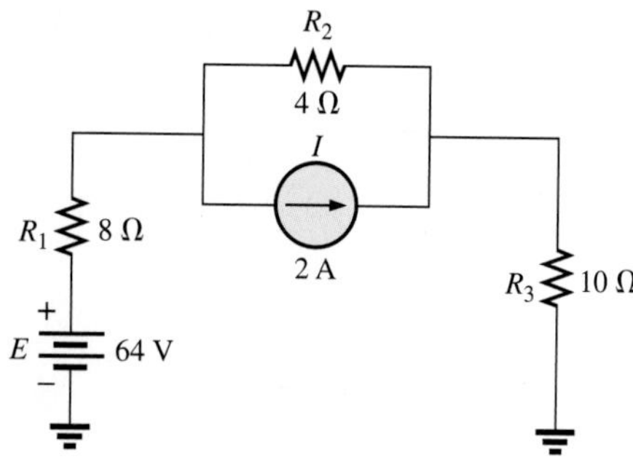

FIG. 8.49
Example 8.20.

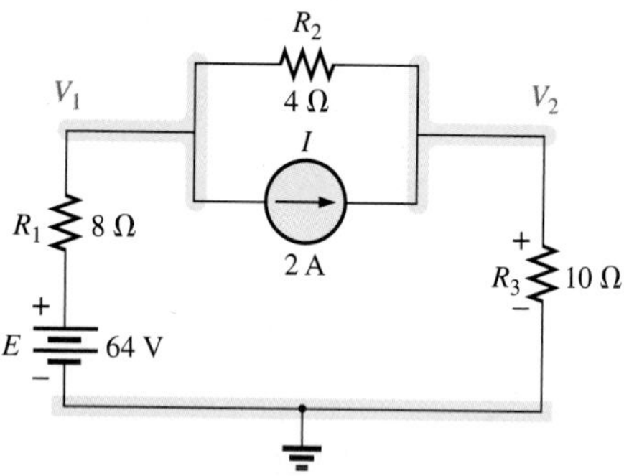

FIG. 8.50
Defining the nodes for the network in Fig. 8.49.

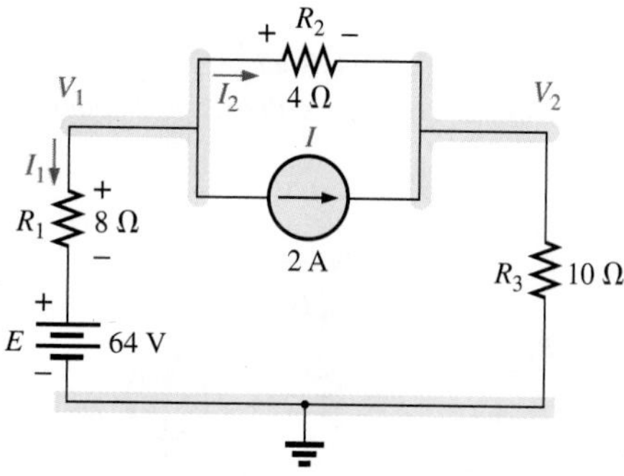

FIG. 8.51
Applying Kirchhoff's current law to node V_1.

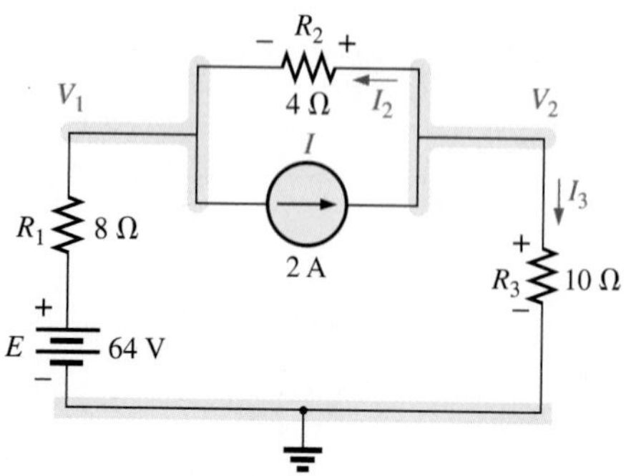

FIG. 8.52
Applying Kirchhoff's current law to node V_2.

The minus sign indicates that the current I_1 has a direction opposite to that appearing in Fig. 8.48.

$$I_2 = \frac{V_1}{R_2} = \frac{20\text{ V}}{12\ \Omega} = \mathbf{1.67\ A}$$

EXAMPLE 8.20 Apply nodal analysis to the network in Fig. 8.49.

Solution:

Steps 1 and 2: The network has three nodes, as defined in Fig. 8.50, with the bottom node again defined as the reference node (at ground potential, or zero volts), and the other nodes as V_1 and V_2.

Step 3: For node V_1, the currents are defined as shown in Fig. 8.51 and Kirchhoff's current law is applied:

$$0 = I_1 + I_2 + I$$

with

$$I_1 = \frac{V_1 - E}{R_1}$$

and

$$I_2 = \frac{V_{R_2}}{R_2} = \frac{V_1 - V_2}{R_2}$$

so that

$$\frac{V_1 - E}{R_1} + \frac{V_1 - V_2}{R_2} + I = 0$$

or

$$\frac{V_1}{R_1} - \frac{E}{R_1} + \frac{V_1}{R_2} - \frac{V_2}{R_2} + I = 0$$

and

$$V_1\left(\frac{1}{R_1} + \frac{1}{R_2}\right) - V_2\left(\frac{1}{R_2}\right) = -I + \frac{E}{R_1}$$

Substituting values:

$$V_1\left(\frac{1}{8\ \Omega} + \frac{1}{4\ \Omega}\right) - V_2\left(\frac{1}{4\ \Omega}\right) = -2\text{ A} + \frac{64\text{ V}}{8\ \Omega} = 6\text{ A}$$

For node V_2 the currents are defined as shown in Fig. 8.52, and Kirchhoff's current law is applied:

$$I = I_2 + I_3$$

with

$$I = \frac{V_2 - V_1}{R_2} + \frac{V_2}{R_3}$$

or

$$I = \frac{V_2}{R_2} - \frac{V_1}{R_2} + \frac{V_2}{R_3}$$

and

$$V_2\left(\frac{1}{R_2} + \frac{1}{R_3}\right) - V_1\left(\frac{1}{R_2}\right) = I$$

Substituting values:

$$V_2\left(\frac{1}{4\ \Omega} + \frac{1}{10\ \Omega}\right) - V_1\left(\frac{1}{4\ \Omega}\right) = 2\text{ A}$$

Step 4: The result is two equations and two unknowns:

$$V_1\left(\frac{1}{8\ \Omega} + \frac{1}{4\ \Omega}\right) - V_2\left(\frac{1}{4\ \Omega}\right) = 6\text{ A}$$

$$-V_1\left(\frac{1}{4\ \Omega}\right) + V_2\left(\frac{1}{4\ \Omega} + \frac{1}{10\ \Omega}\right) = 2\text{ A}$$

which become

$$\begin{aligned} 0.375V_1 - 0.25V_2 &= 6 \\ -0.25V_1 + 0.35V_2 &= 2 \end{aligned}$$

Using determinants,

$$V_1 = \mathbf{37.82\ V}$$
$$V_2 = \mathbf{32.73\ V}$$

Since E is greater than V_1, the current I_1 flows from ground to V_1 and is equal to

$$I_{R_1} = \frac{E - V_1}{R_1} = \frac{64\text{ V} - 37.82\text{ V}}{8\ \Omega} = \mathbf{3.27\ A}$$

The positive value for V_2 results in a current I_{R_3} from node V_2 to ground equal to

$$I_{R_3} = \frac{V_{R_3}}{R_3} = \frac{V_2}{R_3} = \frac{32.73\text{ V}}{10\ \Omega} = \mathbf{3.27\ A}$$

Since V_1 is greater than V_2, the current I_{R_2} flows from V_1 to V_2 and is equal to

$$I_{R_2} = \frac{V_1 - V_2}{R_2} = \frac{37.82\text{ V} - 32.73\text{ V}}{4\ \Omega} = \mathbf{1.27\ A}$$

Mathcad Solution: For this example, we will use Mathcad to work directly with the Kirchhoff's current law equations rather than solving it mathematically. Simply define everything correctly, provide the **Guess** values, and insert **Given** where required. The process should be quite straightforward.

Note in Fig. 8.53 that the first equation comes from the fact that $I_1 + I_2 + I = 0$ while the second equation comes from $I_2 + I_3 = I$. Note that the first equation is defined by Fig. 8.51 and the second by Fig. 8.52 because the direction of I_2 is different for each.

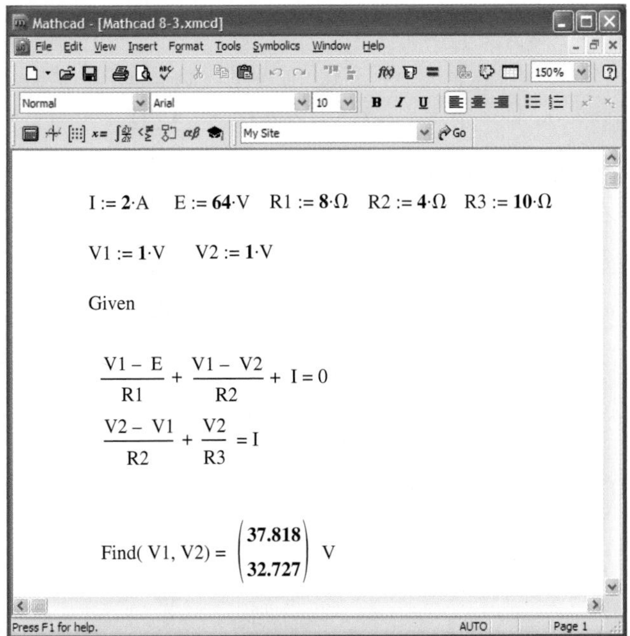

FIG. 8.53
Using Mathcad to verify the mathematical calculations of Example 8.20.

The results of $V_1 = 37.82$ V and $V_2 = 32.73$ V confirm the theoretical solution.

EXAMPLE 8.21 Determine the nodal voltages for the network in Fig. 8.54.

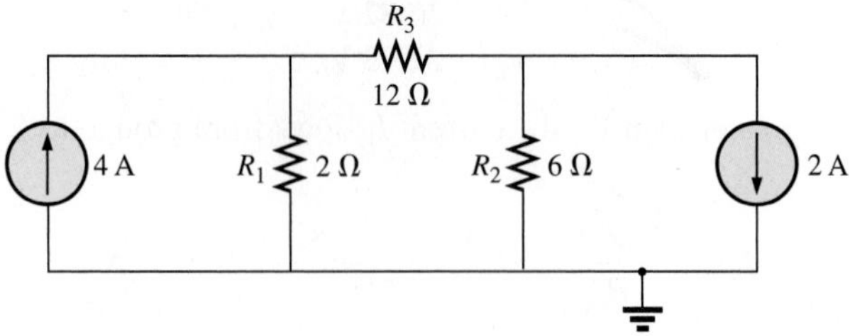

FIG. 8.54
Example 8.21.

Solution:

Steps 1 and 2: As indicated in Fig. 8.55:

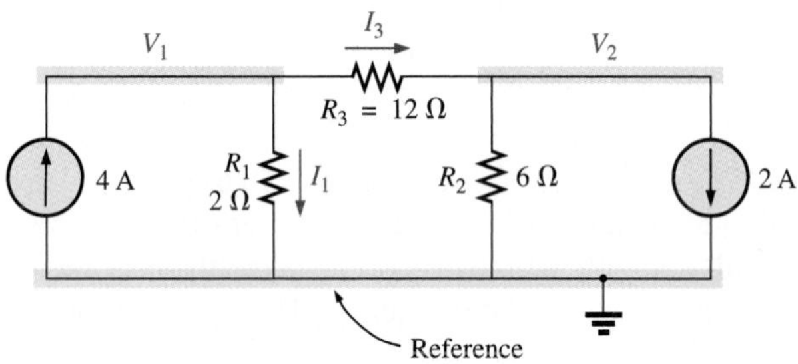

FIG. 8.55
Defining the nodes and applying Kirchhoff's current law to the node V_1.

Step 3: Included in Fig. 8.55 for the node V_1. Applying Kirchhoff's current law:

$$4\text{ A} = I_1 + I_3$$

and
$$4\text{ A} = \frac{V_1}{R_1} + \frac{V_1 - V_2}{R_3} = \frac{V_1}{2\ \Omega} + \frac{V_1 - V_2}{12\ \Omega}$$

Expanding and rearranging:

$$V_1\left(\frac{1}{2\ \Omega} + \frac{1}{12\ \Omega}\right) - V_2\left(\frac{1}{12\ \Omega}\right) = 4\text{ A}$$

For node V_2, the currents are defined as in Fig. 8.56.

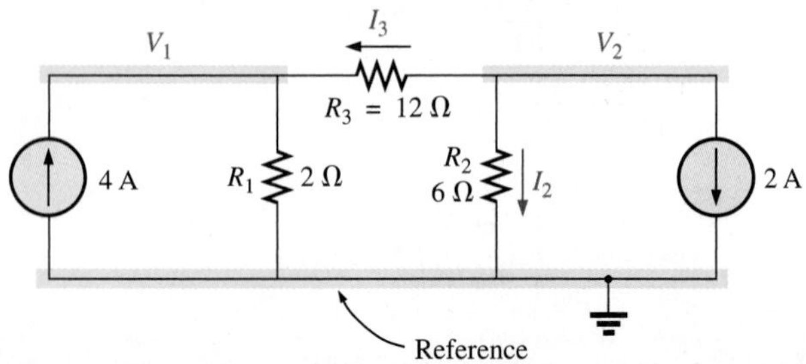

FIG. 8.56
Applying Kirchhoff's current law to the node V_2.

Applying Kirchhoff's current law:

$$0 = I_3 + I_2 + 2\text{ A}$$

and $$\frac{V_2 - V_1}{R_3} + \frac{V_2}{R_2} + 2\text{ A} = 0 \longrightarrow \frac{V_2 - V_1}{12\,\Omega} + \frac{V_2}{2\,\Omega} + 2\text{ A} = 0$$

Expanding and rearranging:

$$V_2\left(\frac{1}{12\,\Omega} + \frac{1}{6\,\Omega}\right) - V_1\left(\frac{1}{12\,\Omega}\right) = -2\text{ A}$$

resulting in Eq. (8.1), which consists of two equations and two unknowns:

$$\left.\begin{aligned} V_1\left(\frac{1}{2\,\Omega} + \frac{1}{12\,\Omega}\right) - V_2\left(\frac{1}{12\,\Omega}\right) &= +4\text{ A} \\ V_2\left(\frac{1}{12\,\Omega} + \frac{1}{6\,\Omega}\right) - V_1\left(\frac{1}{12\,\Omega}\right) &= -2\text{ A} \end{aligned}\right\} \quad \textbf{(8.1)}$$

producing

$$\left.\begin{aligned} \frac{7}{12}V_1 - \frac{1}{12}V_2 &= +4 \\ -\frac{1}{12}V_1 + \frac{3}{12}V_2 &= -2 \end{aligned}\right\} \quad \begin{aligned} 7V_1 - V_2 &= 48 \\ -1V_1 + 3V_2 &= -24 \end{aligned}$$

and $$V_1 = \frac{\begin{vmatrix} 48 & -1 \\ -24 & 3 \end{vmatrix}}{\begin{vmatrix} 7 & -1 \\ -1 & 3 \end{vmatrix}} = \frac{120}{20} = \mathbf{+6\text{ V}}$$

$$V_2 = \frac{\begin{vmatrix} 7 & 48 \\ -1 & -24 \end{vmatrix}}{20} = \frac{-120}{20} = \mathbf{-6\text{ V}}$$

Since V_1 is greater than V_2, the current through R_3 passes from V_1 to V_2. Its value is

$$I_{R_3} = \frac{V_1 - V_2}{R_3} = \frac{6\text{ V} - (-6\text{ V})}{12\,\Omega} = \frac{12\text{ V}}{12\,\Omega} = \mathbf{1\text{ A}}$$

The fact that V_1 is positive results in a current I_{R_1} from V_1 to ground equal to

$$I_{R_1} = \frac{V_{R_1}}{R_1} = \frac{V_1}{R_1} = \frac{6\text{ V}}{2\,\Omega} = \mathbf{3\text{ A}}$$

Finally, since V_2 is negative, the current I_{R_2} flows from ground to V_2 and is equal to

$$I_{R_2} = \frac{V_{R_2}}{R_2} = \frac{V_2}{R_2} = \frac{6\text{ V}}{6\,\Omega} = \mathbf{1\text{ A}}$$

Supernode

Occasionally, you may encounter voltage sources in a network that do not have a series internal resistance that would permit a conversion to a current source. In such cases, you have two options.

The simplest and most direct approach is to *place a resistor in series with the source of a very small value compared to the other resistive elements of*

the network. For instance, if most of the resistors are 10 Ω or larger, placing a 1 Ω resistor in series with a voltage source provides one level of accuracy for your answer. However, choosing a resistor of 0.1 Ω or less increases the accuracy of your answer. You will never get an exact answer because the network has been modified by the introduced element. However, for most applications, the accuracy will be sufficiently high.

The other approach is to use the **Supernode approach** described below. This approach provides an exact solution but requires some practice to become proficient.

Start as usual and assign a nodal voltage to each independent node of the network, including each independent voltage source as if it were a resistor or current source. Then mentally replace the independent voltage sources with short-circuit equivalents, and apply Kirchhoff's current law to the defined nodes of the network. Any node including the effect of elements tied only to *other* nodes is referred to as a *supernode* (since it has an additional number of terms). Finally, relate the defined nodes to the independent voltage sources of the network, and solve for the nodal voltages. The next example clarifies the definition of *supernode.*

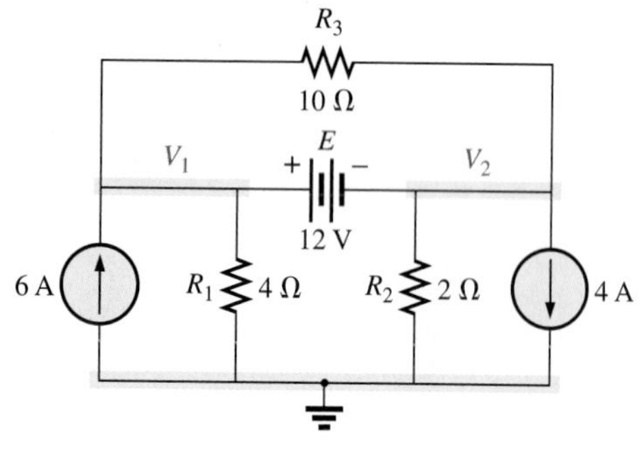

FIG. 8.57
Example 8.22.

EXAMPLE 8.22 Determine the nodal voltages V_1 and V_2 in Fig. 8.57 using the concept of a supernode.

Solution: Replacing the independent voltage source of 12 V with a short-circuit equivalent results in the network in Fig. 8.58. Even though the mental application of a short-circuit equivalent is discussed above, it would be wise in the early stage of development to redraw the network as shown in Fig. 8.58. The result is a single supernode for which Kirchhoff's current law must be applied. Be sure to leave the other defined nodes in place and use them to define the currents from that region of the network. In particular, note that the current I_3 leaves the supernode at V_1 and then enters the same supernode at V_2. It must therefore appear twice when applying Kirchhoff's current law, as shown below:

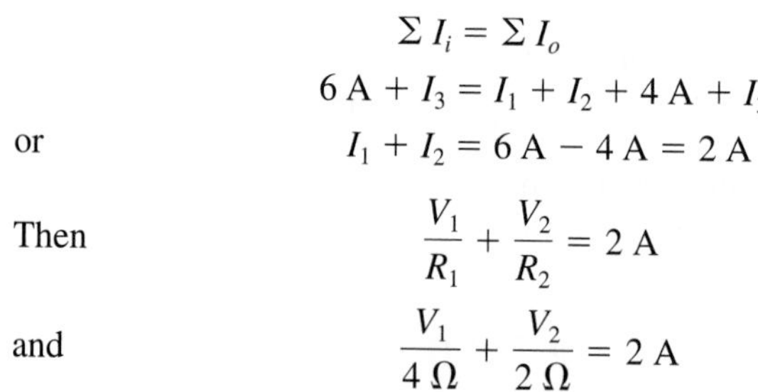

$$\Sigma I_i = \Sigma I_o$$

$$6\text{ A} + I_3 = I_1 + I_2 + 4\text{ A} + I_3$$

or

$$I_1 + I_2 = 6\text{ A} - 4\text{ A} = 2\text{ A}$$

Then

$$\frac{V_1}{R_1} + \frac{V_2}{R_2} = 2\text{ A}$$

and

$$\frac{V_1}{4\ \Omega} + \frac{V_2}{2\ \Omega} = 2\text{ A}$$

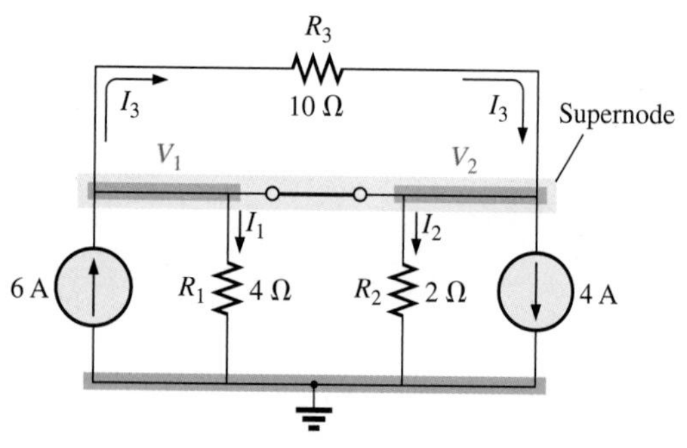

FIG. 8.58
Defining the supernode for the network in Fig. 8.57.

Relating the defined nodal voltages to the independent voltage source, we have

$$V_1 - V_2 = E = 12\text{ V}$$

which results in two equations and two unknowns:

$$\begin{aligned} 0.25V_1 + 0.5V_2 &= 2 \\ V_1 - \ 1V_2 &= 12 \end{aligned}$$

Substituting:

$$V_1 = V_2 + 12$$

$$0.25(V_2 + 12) + 0.5V_2 = 2$$

and

$$0.75V_2 = 2 - 3 = -1$$

so that $$V_2 = \frac{-1}{0.75} = -\mathbf{1.33\ V}$$

and $$V_1 = V_2 + 12\text{ V} = -1.33\text{ V} + 12\text{ V} = \mathbf{+10.67\ V}$$

The current of the network can then be determined as follows:

$$I_1\downarrow = \frac{V}{R_1} = \frac{10.67\text{ V}}{4\ \Omega} = \mathbf{2.67\ A}$$

$$I_2\uparrow = \frac{V_2}{R_2} = \frac{1.33\text{ V}}{2\ \Omega} = \mathbf{0.67\ A}$$

$$\underset{\rightarrow}{I_3} = \frac{V_1 - V_2}{10\ \Omega} = \frac{10.67\text{ V} - (-1.33\text{ V})}{10\ \Omega} = \frac{12\ \Omega}{10\ \Omega} = \mathbf{1.2\ A}$$

A careful examination of the network at the beginning of the analysis would have revealed that the voltage across the resistor R_3 must be 12 V and I_3 must be equal to 1.2 A.

8.10 NODAL ANALYSIS (FORMAT APPROACH)

A close examination of Eq. (8.1) appearing in Example 8.21 reveals that the subscripted voltage at the node in which Kirchhoff's current law is applied is multiplied by the sum of the conductances attached to that node. Note also that the other nodal voltages within the same equation are multiplied by the negative of the conductance between the two nodes. The current sources are represented to the right of the equals sign with a positive sign if they supply current to the node and with a negative sign if they draw current from the node.

These conclusions can be expanded to include networks with any number of nodes. This allows us to write nodal equations rapidly and in a form that is convenient for the use of determinants. A major requirement, however, is that *all voltage sources must first be converted to current sources before the procedure is applied.* Note the parallelism between the following four steps of application and those required for mesh analysis in Section 8.8.

Nodal Analysis Procedure

1. *Choose a reference node and assign a subscripted voltage label to the (N − 1) remaining nodes of the network.*
2. *The number of equations required for a complete solution is equal to the number of subscripted voltages (N − 1). Column 1 of each equation is formed by summing the conductances tied to the node of interest and multiplying the result by that subscripted nodal voltage.*
3. *We must now consider the mutual terms that, as noted in the preceding example, are always subtracted from the first column. It is possible to have more than one mutual term if the nodal voltage of current interest has an element in common with more than one other nodal voltage. This is demonstrated in an example to follow. Each mutual term is the product of the mutual conductance and the other nodal voltage tied to that conductance.*
4. *The column to the right of the equality sign is the algebraic sum of the current sources tied to the node of interest. A current source is*

assigned a positive sign if it supplies current to a node and a negative sign if it draws current from the node.

5. *Solve the resulting simultaneous equations for the desired voltages.*

Let us now consider a few examples.

EXAMPLE 8.23 Write the nodal equations for the network in Fig. 8.59.

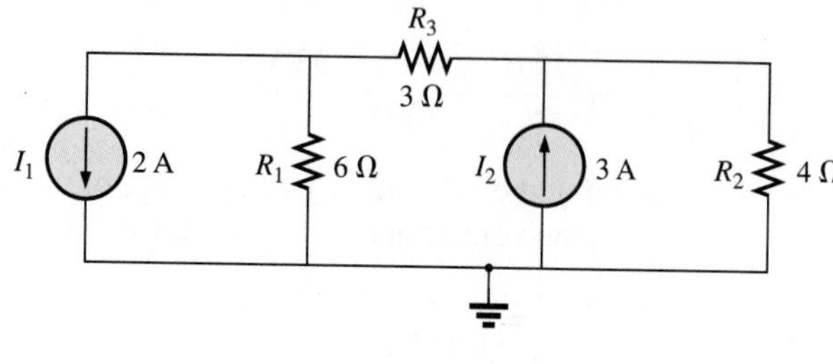

FIG. 8.59
Example 8.23.

Solution:

Step 1: Redraw the figure with assigned subscripted voltages in Fig. 8.60.

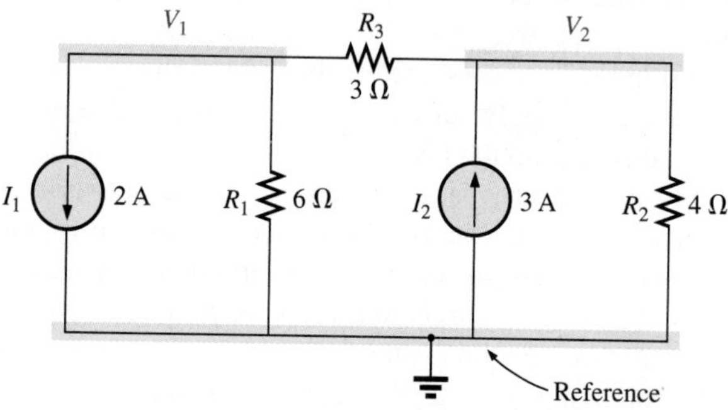

FIG. 8.60
Defining the nodes for the network in Fig. 8.59.

Steps 2 to 4:

Drawing current from node 1

$$V_1\text{:}\quad \underbrace{\left(\frac{1}{6\ \Omega} + \frac{1}{3\ \Omega}\right)}_{\substack{\text{Sum of}\\ \text{conductances}\\ \text{connected}\\ \text{to node 1}}} V_1 - \underbrace{\left(\frac{1}{3\ \Omega}\right)}_{\substack{\text{Mutual}\\ \text{conductance}}} V_2 = -2\text{ A}$$

Supplying current to node 2

$$V_2\text{:}\quad \underbrace{\left(\frac{1}{4\ \Omega} + \frac{1}{3\ \Omega}\right)}_{\substack{\text{Sum of}\\ \text{conductances}\\ \text{connected}\\ \text{to node 2}}} V_2 - \underbrace{\left(\frac{1}{3\ \Omega}\right)}_{\substack{\text{Mutual}\\ \text{conductance}}} V_1 = +3\text{ A}$$

and
$$\frac{1}{2}V_1 - \frac{1}{3}V_2 = -2$$
$$-\frac{1}{3}V_1 + \frac{7}{12}V_2 = 3$$

EXAMPLE 8.24 Find the voltage across the 3 Ω resistor in Fig. 8.61 by nodal analysis.

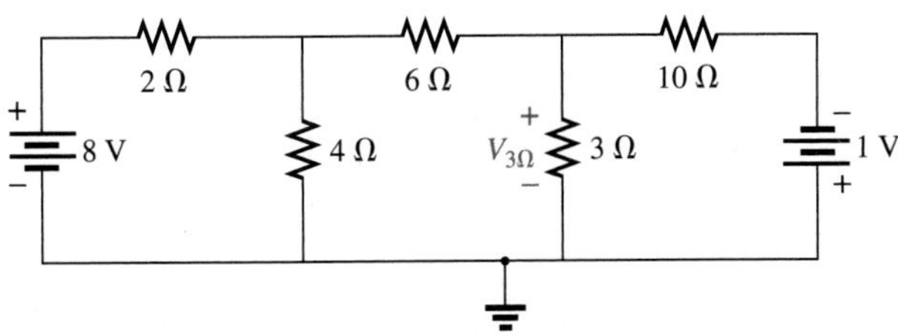

FIG. 8.61
Example 8.24.

Solution: Converting sources and choosing nodes (Fig. 8.62), we have

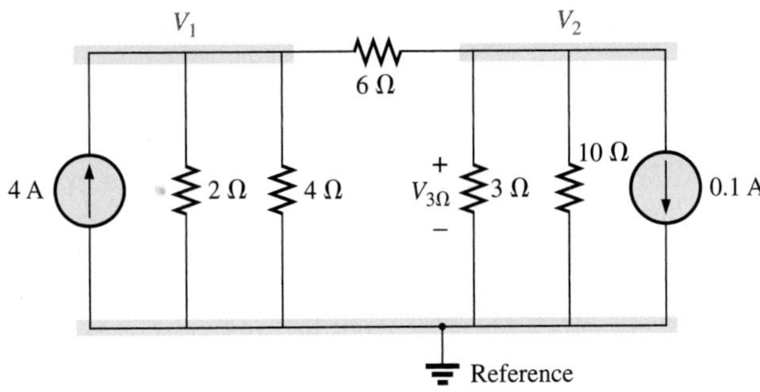

FIG. 8.62
Defining the nodes for the network in Fig. 8.61.

$$\left.\begin{aligned}\left(\frac{1}{2\,\Omega} + \frac{1}{4\,\Omega} + \frac{1}{6\,\Omega}\right)V_1 - \left(\frac{1}{6\,\Omega}\right)V_2 &= +4\text{ A}\\ \left(\frac{1}{10\,\Omega} + \frac{1}{3\,\Omega} + \frac{1}{6\,\Omega}\right)V_2 - \left(\frac{1}{6\,\Omega}\right)V_1 &= -0.1\text{ A}\end{aligned}\right\}$$

$$\frac{11}{12}V_1 - \frac{1}{6}V_2 = 4$$
$$-\frac{1}{6}V_1 + \frac{3}{5}V_2 = -0.1$$

resulting in

$$11V_1 - 2V_2 = +48$$
$$-5V_1 + 18V_2 = -3$$

and

$$V_2 = V_{3\Omega} = \frac{\begin{vmatrix} 11 & 48 \\ -5 & -3 \end{vmatrix}}{\begin{vmatrix} 11 & -2 \\ -5 & 18 \end{vmatrix}} = \frac{-33 + 240}{198 - 10} = \frac{207}{188} = \mathbf{1.10\ V}$$

As demonstrated for mesh analysis, nodal analysis can also be a very useful technique for solving networks with only one source.

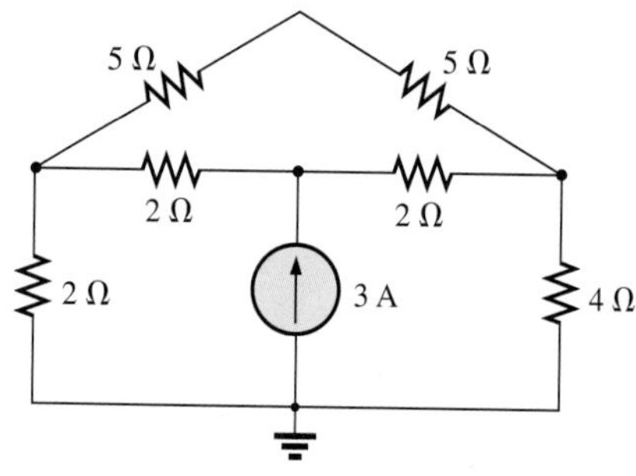

FIG. 8.63
Example 8.25.

EXAMPLE 8.25 Using nodal analysis, determine the potential across the 4 Ω resistor in Fig. 8.63.

Solution: The reference and four subscripted voltage levels were chosen as shown in Fig. 8.64. Remember that for any difference in potential between V_1 and V_3, the current through and the potential drop across each 5 Ω resistor is the same. Therefore, V_4 is simply a mid-voltage level between V_1 and V_3 and is known if V_1 and V_3 are available. We will therefore not include it in a nodal voltage and will redraw the network as shown in Fig. 8.65. Understand, however, that V_4 can be included if desired, although four nodal voltages will result rather than three as in the solution of this problem.

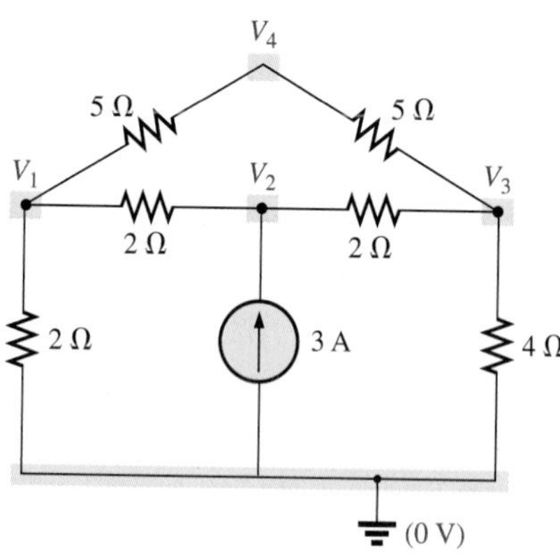

FIG. 8.64
Defining the nodes for the network in Fig. 8.63.

$$V_1:\ \left(\frac{1}{2\ \Omega} + \frac{1}{2\ \Omega} + \frac{1}{10\ \Omega}\right)V_1 - \left(\frac{1}{2\ \Omega}\right)V_2 - \left(\frac{1}{10\ \Omega}\right)V_3 = 0$$

$$V_2:\ \left(\frac{1}{2\ \Omega} + \frac{1}{2\ \Omega}\right)V_2 - \left(\frac{1}{2\ \Omega}\right)V_1 - \left(\frac{1}{2\ \Omega}\right)V_3 = 3\ \text{A}$$

$$V_3:\ \left(\frac{1}{10\ \Omega} + \frac{1}{2\ \Omega} + \frac{1}{4\ \Omega}\right)V_3 - \left(\frac{1}{2\ \Omega}\right)V_2 - \left(\frac{1}{10\ \Omega}\right)V_1 = 0$$

which are rewritten as

$$\begin{aligned} 1.1V_1 - 0.5V_2 - 0.1V_3 &= 0 \\ V_2 - 0.5V_1 - 0.5V_3 &= 3 \\ 0.85V_3 - 0.5V_2 - 0.1V_1 &= 0 \end{aligned}$$

For determinants,

$$\begin{aligned} 1.1V_1 - 0.5V_2 - 0.1V_3 &= 0 \\ -0.5V_1 + 1V_2 - 0.5V_3 &= 3 \\ -0.1V_1 - 0.5V_2 + 0.85V_3 &= 0 \end{aligned}$$

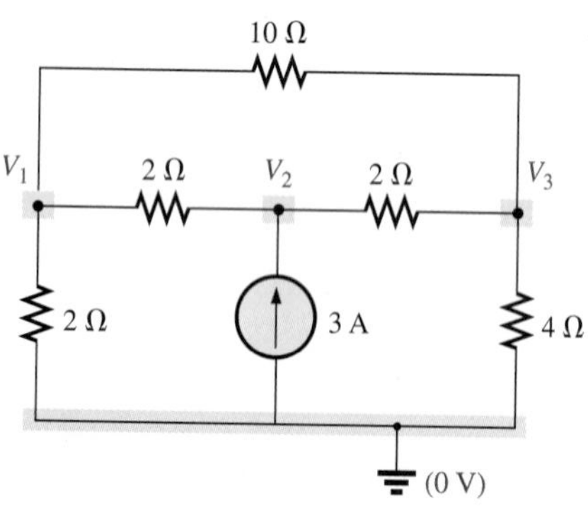

FIG. 8.65
Reducing the number of nodes for the network in Fig. 8.63 by combining the two 5 Ω resistors.

Before continuing, note the symmetry about the major diagonal in the equation above. Recall a similar result for mesh analysis. Examples 8.23 and 8.24 also exhibit this property in the resulting equations. Keep this in mind as a check on future applications of nodal analysis.

$$V_3 = V_{4\Omega} = \frac{\begin{vmatrix} 1.1 & -0.5 & 0 \\ -0.5 & +1 & 3 \\ -0.1 & -0.5 & 0 \end{vmatrix}}{\begin{vmatrix} 1.1 & -0.5 & -0.1 \\ -0.5 & +1 & -0.5 \\ -0.1 & -0.5 & +0.85 \end{vmatrix}} = \mathbf{4.65\ V}$$

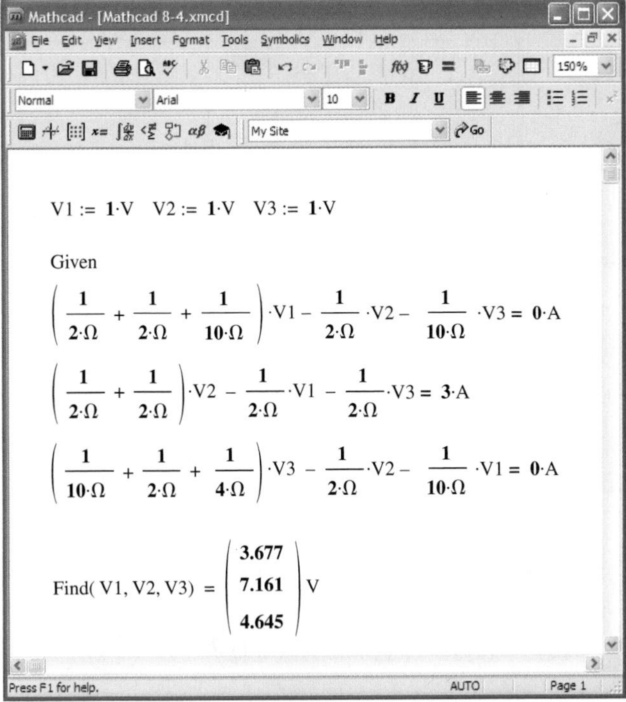

FIG. 8.66
Using Mathcad to verify the mathematical calculations of Example 8.25.

Mathcad Solution: By now, the sequence of steps necessary to solve a series of equations using Mathcad should be more familiar and easier to remember. For this example, all the parameters are entered in the three simultaneous equations, avoiding the need to define each parameter of the network. Simply provide a **Guess** at the three nodal voltages, apply the word **Given,** and enter the three equations properly as shown in Fig. 8.66. It does take some practice to remember to move the bracket to the proper location before making an entry, but you will soon understand how the rules are needed to maintain control of the operations to be performed. Finally, request the desired nodal voltages using the correct format. The numerical results appear, again confirming our theoretical solutions.

The next example has only one source applied to a ladder network.

EXAMPLE 8.26 Write the nodal equations and find the voltage across the 2 Ω resistor for the network in Fig. 8.67.

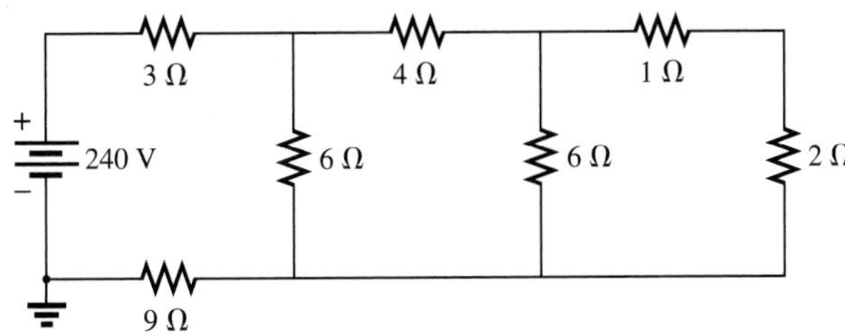

FIG. 8.67
Example 8.26.

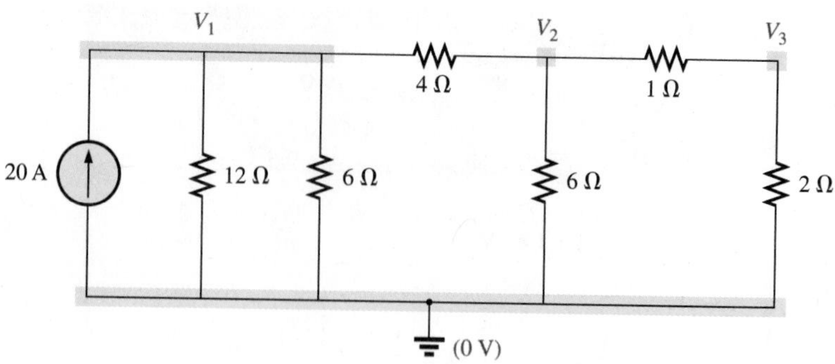

FIG. 8.68
Converting the voltage source to a current source and defining the nodes for the network in Fig. 8.67.

Solution: The nodal voltages are chosen as shown in Fig. 8.68.

$$V_1: \left(\frac{1}{12\ \Omega} + \frac{1}{6\ \Omega} + \frac{1}{4\ \Omega}\right)V_1 - \left(\frac{1}{4\ \Omega}\right)V_2 + 0 = 20\ \text{A}$$

$$V_2: \left(\frac{1}{4\ \Omega} + \frac{1}{6\ \Omega} + \frac{1}{1\ \Omega}\right)V_2 - \left(\frac{1}{4\ \Omega}\right)V_1 - \left(\frac{1}{1\ \Omega}\right)V_3 = 0$$

$$V_3: \left(\frac{1}{1\ \Omega} + \frac{1}{2\ \Omega}\right)V_3 - \left(\frac{1}{1\ \Omega}\right)V_2 + 0 = 0$$

and

$$\begin{aligned} 0.5V_1 - 0.25V_2 + \quad 0 &= 20 \\ -0.25V_1 + \frac{17}{12}V_2 - 1V_3 &= 0 \\ 0 - 1V_2 + 1.5V_3 &= 0 \end{aligned}$$

Note the symmetry present about the major axis. Application of determinants reveals that

$$V_3 = V_{2\Omega} = \mathbf{10.67\ V}$$

8.11 BRIDGE NETWORKS

This section introduces the **bridge network,** a configuration that has a multitude of applications. In the following chapters, this type of network is used in both dc and ac meters. Electronics courses introduce these in the discussion of rectifying circuits used in converting a varying signal to one of a steady nature (such as dc). A number of other areas of application also require some knowledge of ac networks; these areas are discussed later.

The bridge network may appear in one of the three forms as indicated in Fig. 8.69. The network in Fig. 8.69(c) is also called a *symmetrical lat-*

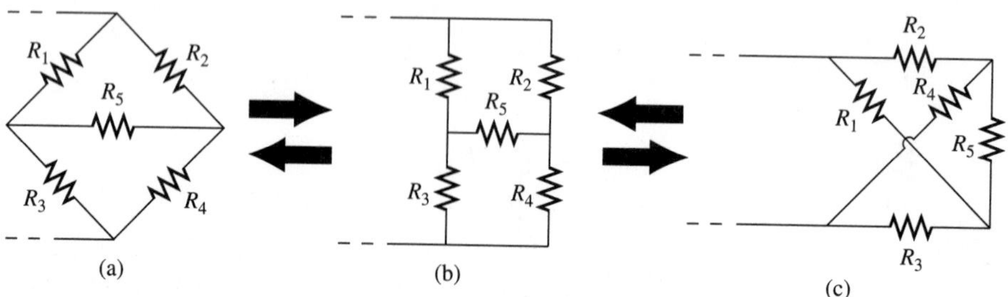

FIG. 8.69
Various formats for a bridge network.

tice network if $R_2 = R_3$ and $R_1 = R_4$. Fig. 8.69(c) is an excellent example of how a planar network can be made to appear nonplanar. For the purposes of investigation, let us examine the network in Fig. 8.70 using mesh and nodal analysis.

Mesh analysis (Fig 8.71) yields

$$(3\ \Omega + 4\ \Omega + 2\ \Omega)I_1 - (4\ \Omega)I_2 - (2\ \Omega)I_3 = 20\ \text{V}$$
$$(4\ \Omega + 5\ \Omega + 2\ \Omega)I_2 - (4\ \Omega)I_1 - (5\ \Omega)I_3 = 0$$
$$(2\ \Omega + 5\ \Omega + 1\ \Omega)I_3 - (2\ \Omega)I_1 - (5\ \Omega)I_2 = 0$$

and

$$9I_1 - 4I_2 - 2I_3 = 20$$
$$-4I_1 + 11I_2 - 5I_3 = 0$$
$$-2I_1 - 5I_2 + 8I_3 = 0$$

with the result that

$$I_1 = \mathbf{4\ A}$$
$$I_2 = \mathbf{2.67\ A}$$
$$I_3 = \mathbf{2.67\ A}$$

The net current through the 5 Ω resistor is

$$I_{5\Omega} = I_2 - I_3 = 2.67\ \text{A} - 2.67\ \text{A} = 0\ \text{A}$$

Nodal analysis (Fig. 8.72) yields

$$\left(\frac{1}{3\ \Omega} + \frac{1}{4\ \Omega} + \frac{1}{2\ \Omega}\right)V_1 - \left(\frac{1}{4\ \Omega}\right)V_2 - \left(\frac{1}{2\ \Omega}\right)V_3 = \frac{20}{3}\ \text{A}$$
$$\left(\frac{1}{4\ \Omega} + \frac{1}{2\ \Omega} + \frac{1}{5\ \Omega}\right)V_2 - \left(\frac{1}{4\ \Omega}\right)V_1 - \left(\frac{1}{5\ \Omega}\right)V_3 = 0$$
$$\left(\frac{1}{5\ \Omega} + \frac{1}{2\ \Omega} + \frac{1}{1\ \Omega}\right)V_3 - \left(\frac{1}{2\ \Omega}\right)V_1 - \left(\frac{1}{5\ \Omega}\right)V_2 = 0$$

and

$$\left(\frac{1}{3\ \Omega} + \frac{1}{4\ \Omega} + \frac{1}{2\ \Omega}\right)V_1 - \left(\frac{1}{4\ \Omega}\right)V_2 - \left(\frac{1}{2\ \Omega}\right)V_3 = 6.67\ \text{A}$$
$$-\left(\frac{1}{4\ \Omega}\right)V_1 + \left(\frac{1}{4\ \Omega} + \frac{1}{2\ \Omega} + \frac{1}{5\ \Omega}\right)V_2 - \left(\frac{1}{5\ \Omega}\right)V_3 = 0$$
$$-\left(\frac{1}{2\ \Omega}\right)V_1 - \left(\frac{1}{5\ \Omega}\right)V_2 + \left(\frac{1}{5\ \Omega} + \frac{1}{2\ \Omega} + \frac{1}{1\ \Omega}\right)V_3 = 0$$

Note the symmetry of the solution.

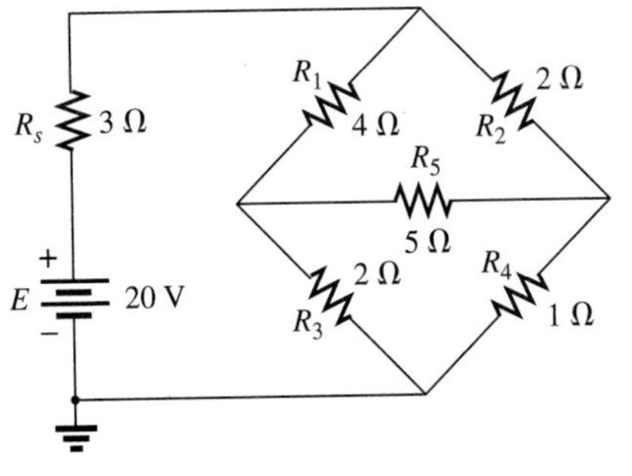

FIG. 8.70
Standard bridge configuration.

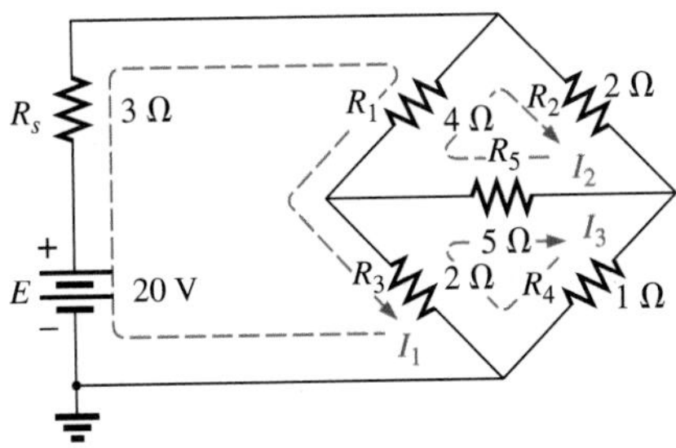

FIG. 8.71
Assigning the mesh currents to the network in Fig. 8.70.

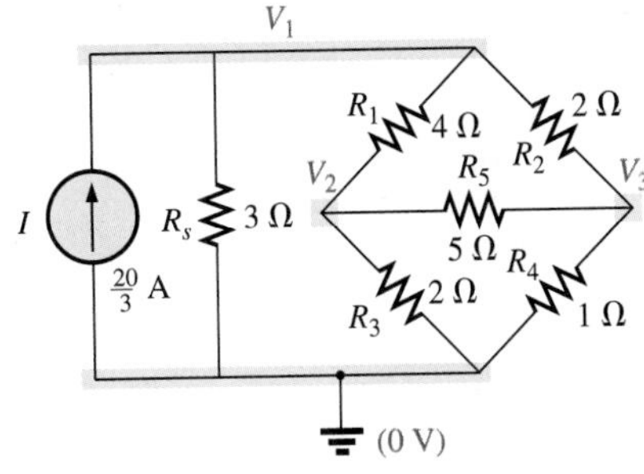

FIG. 8.72
Defining the nodal voltages for the network in Fig. 8.70.

TI-89 Calculator Solution

With the TI-89 calculator, the top part of the determinant is determined by the sequence in Fig. 8.73 (take note of the calculations within parentheses):

det([6.67,−1/4,−1/2;0,(1/4+1/2+1/5),−1/5;0,−1/5,(1/5+1/2+1/1)]) (ENTER)	10.51E0

FIG. 8.73

with the bottom of the determinant determined by the sequence in Fig. 8.74.

det([(1/3+1/4+1/2),−1/4,−1/2;−1/4,(1/4+1/2+1/5),−1/5;−1/2,−1/5,(1/5+1/2+1/1)]) (ENTER)	1.31E0

FIG. 8.74

Finally, the simple division in Fig. 8.75 provides the desired result.

and $$V_1 = \mathbf{8.02\ V}$$

FIG. 8.75

Similarly, $$V_2 = \mathbf{2.67\ V} \quad \text{and} \quad V_3 = \mathbf{2.67\ V}$$

and the voltage across the 5 Ω resistor is

$$V_{5\Omega} = V_2 - V_3 = 2.67\ \text{V} - 2.67\ \text{V} = \mathbf{0\ V}$$

Since $V_{5\Omega} = 0$ V, we can insert a short in place of the bridge arm without affecting the network behavior. (Certainly $V = IR = I\cdot(0) = 0$ V.) In Fig. 8.76, a short circuit has replaced the resistor R_5, and the voltage across R_4 is to be determined. The network is redrawn in Fig. 8.77, and

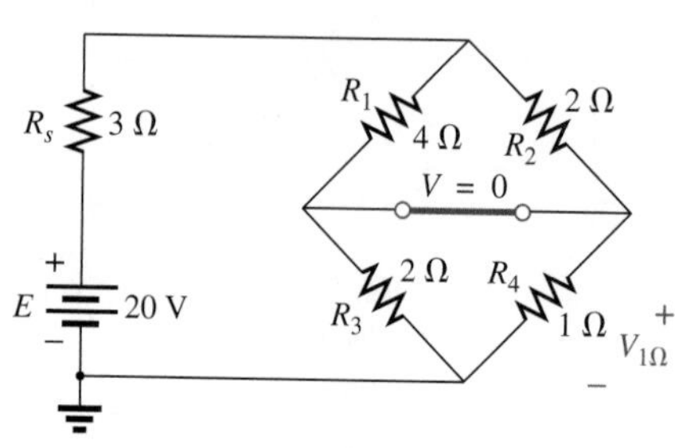

FIG. 8.76
Substituting the short-circuit equivalent for the balance arm of a balanced bridge.

$$V_{1\Omega} = \frac{(2\ \Omega \parallel 1\ \Omega)20\ \text{V}}{(2\ \Omega \parallel 1\ \Omega) + (4\ \Omega \parallel 2\ \Omega) + 3\ \Omega} \quad \text{(voltage divider rule)}$$

$$= \frac{\frac{2}{3}(20\ \text{V})}{\frac{2}{3} + \frac{8}{6} + 3} = \frac{\frac{2}{3}(20\ \text{V})}{\frac{2}{3} + \frac{4}{3} + \frac{9}{3}}$$

$$= \frac{2(20\ \text{V})}{2 + 4 + 9} = \frac{40\ \text{V}}{15} = \mathbf{2.67\ V}$$

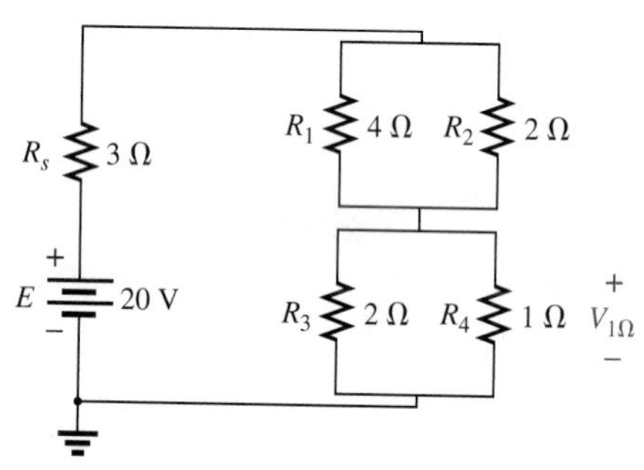

FIG. 8.77
Redrawing the network in Fig. 8.76.

as obtained earlier.

We found through mesh analysis that $I_{5\Omega} = 0$ A, which has as its equivalent an open circuit as shown in Fig. 8.78(a). (Certainly $I = V/R = 0/(\infty\ \Omega) = 0$ A.) The voltage across the resistor R_4 is again determined and compared with the result above.

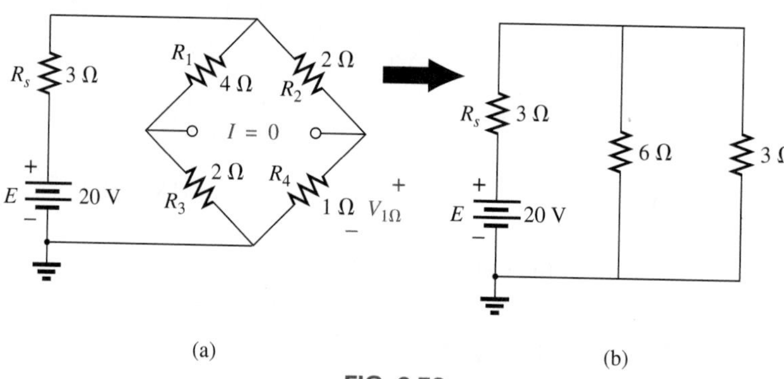

FIG. 8.78
Substituting the open-circuit equivalent for the balance arm of a balanced bridge.

The network is redrawn after combining series elements, as shown in Fig. 8.78(b), and

$$V_{3\Omega} = \frac{(6\ \Omega \parallel 3\ \Omega)(20\ \text{V})}{6\ \Omega \parallel 3\ \Omega + 3\ \Omega} = \frac{2\ \Omega(20\ \text{V})}{2\ \Omega + 3\ \Omega} = 8\ \text{V}$$

and $$V_{1\Omega} = \frac{1\ \Omega(8\ \text{V})}{1\ \Omega + 2\ \Omega} = \frac{8\ \text{V}}{3} = \mathbf{2.67\ V}$$

as above.

The condition $V_{5\Omega} = 0$ V or $I_{5\Omega} = 0$ A exists only for a particular relationship between the resistors of the network. Let us now derive this re-

lationship using the network in Fig. 8.79, in which it is indicated that $I = 0$ A and $V = 0$ V. Note that resistor R_s of the network in Fig. 8.78 does not appear in the following analysis.

The bridge network is said to be *balanced* when the condition of $I = 0$ A or $V = 0$ V exists.

If $V = 0$ V (short circuit between a and b), then

$$V_1 = V_2$$

and

$$I_1R_1 = I_2R_2$$

or

$$I_1 = \frac{I_2R_2}{R_1}$$

In addition, when $V = 0$ V,

$$V_3 = V_4$$

and

$$I_3R_3 = I_4R_4$$

If we set $I = 0$ A, then $I_3 = I_1$ and $I_4 = I_2$, with the result that the above equation becomes

$$I_1R_3 = I_2R_4$$

Substituting for I_1 from above yields

$$\left(\frac{I_2R_2}{R_1}\right)R_3 = I_2R_4$$

or, rearranging, we have

$$\boxed{\frac{R_1}{R_3} = \frac{R_2}{R_4}} \qquad \textbf{(8.2)}$$

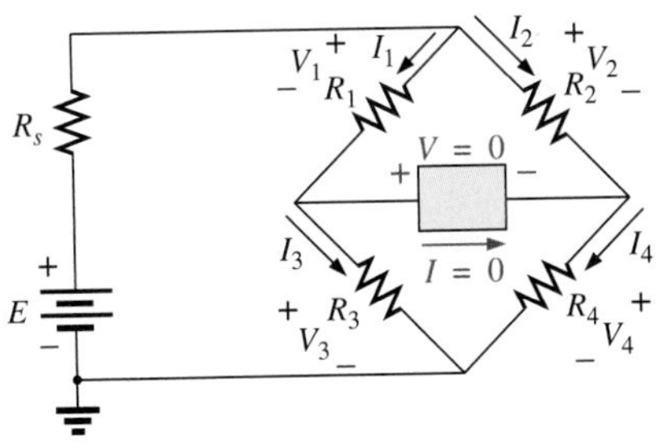

FIG. 8.79
Establishing the balance criteria for a bridge network.

This conclusion states that if the ratio of R_1 to R_3 is equal to that of R_2 to R_4, the bridge is balanced, and $I = 0$ A or $V = 0$ V. A method of memorizing this form is indicated in Fig. 8.80.

For the example above, $R_1 = 4\ \Omega$, $R_2 = 2\ \Omega$, $R_3 = 2\ \Omega$, $R_4 = 1\ \Omega$, and

$$\frac{R_1}{R_3} = \frac{R_2}{R_4} \rightarrow \frac{4\ \Omega}{2\ \Omega} = \frac{2\ \Omega}{1\ \Omega} = 2$$

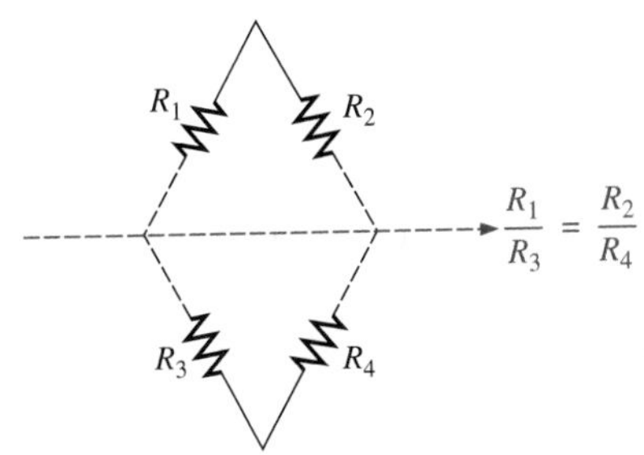

FIG. 8.80
A visual approach to remembering the balance condition.

The emphasis in this section has been on the balanced situation. Understand that if the ratio is not satisfied, there will be a potential drop across the balance arm and a current through it. The methods just described (mesh and nodal analysis) will yield any and all potentials or currents desired, just as they did for the balanced situation.

8.12 Y-Δ (T-π) AND Δ-Y (π-T) CONVERSIONS

Circuit configurations are often encountered in which the resistors do not appear to be in series or parallel. Under these conditions, it may be necessary to convert the circuit from one form to another to solve for any unknown quantities if mesh or nodal analysis is not applied. Two circuit configurations that often account for these difficulties are the **wye (Y)** and **delta (Δ) configurations** depicted in Fig. 8.81(a). They are also referred to as the **tee (T)** and **pi (π),** respectively, as indicated in Fig. 8.81(b). Note that the pi is actually an inverted delta.

The purpose of this section is to develop the equations for converting from Δ to Y, or vice versa. This type of conversion normally leads to a

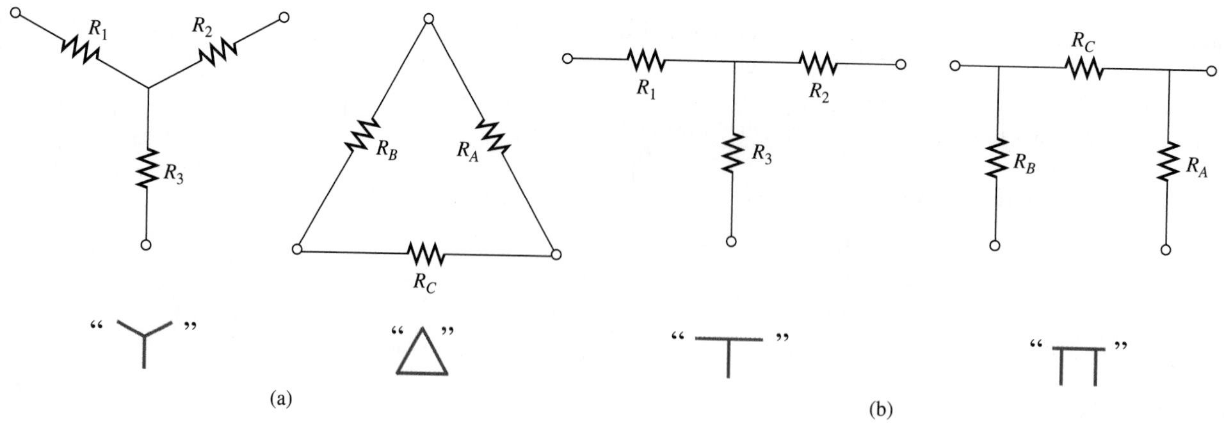

FIG. 8.81
The Y(T) and Δ (π) configurations.

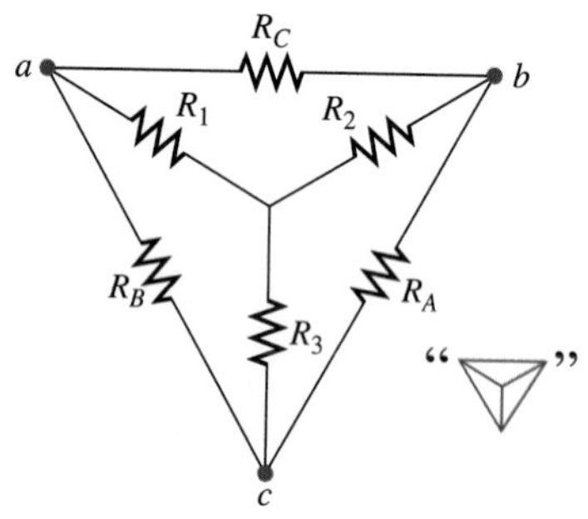

FIG. 8.82
Introducing the concept of Δ-Y or Y-Δ conversions.

network that can be solved using techniques such as those described in Chapter 7. In other words, in Fig. 8.82, with terminals *a, b,* and *c* held fast, if the wye (Y) configuration were desired *instead of* the inverted delta (Δ) configuration, all that would be necessary is a direct application of the equations to be derived. The phrase *instead of* is emphasized to ensure that it is understood that only one of these configurations is to appear at one time between the indicated terminals.

It is our purpose (referring to Fig. 8.82) to find some expression for R_1, R_2, and R_3 in terms of R_A, R_B, and R_C, and vice versa, that will ensure that the resistance between any two terminals of the Y configuration will be the same with the Δ configuration inserted in place of the Y configuration (and vice versa). If the two circuits are to be equivalent, the total resistance between any two terminals must be the same. Consider terminals *a-c* in the Δ-Y configurations in Fig. 8.83.

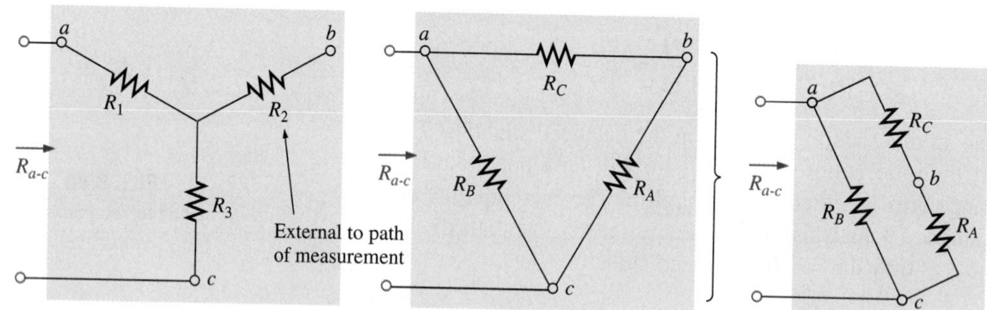

FIG. 8.83
Finding the resistance $R_{a\text{-}c}$ for the Y and Δ configurations.

Let us first assume that we want to convert the Δ (R_A, R_B, R_C) to the Y (R_1, R_2, R_3). This requires that we have a relationship for R_1, R_2, and R_3 in terms of R_A, R_B, and R_C. If the resistance is to be the same between terminals *a-c* for both the Δ and the Y, the following must be true:

$$R_{a\text{-}c}\,(\text{Y}) = R_{a\text{-}c}\,(\Delta)$$

so that

$$R_{a\text{-}c} = R_1 + R_3 = \frac{R_B(R_A + R_C)}{R_B + (R_A + R_C)} \qquad \textbf{(8.3a)}$$

Using the same approach for a-b and b-c, we obtain the following relationships:

$$R_{a\text{-}b} = R_1 + R_2 = \frac{R_C(R_A + R_B)}{R_C + (R_A + R_B)} \qquad \textbf{(8.3b)}$$

and

$$R_{b\text{-}c} = R_2 + R_3 = \frac{R_A(R_B + R_C)}{R_A + (R_B + R_C)} \qquad \textbf{(8.3c)}$$

Subtracting Eq. (8.3a) from Eq. (8.3b), we have

$$(R_1 + R_2) - (R_1 + R_3) = \left(\frac{R_C R_B + R_C R_A}{R_A + R_B + R_C}\right) - \left(\frac{R_B R_A + R_B R_A}{R_A + R_B + R_C}\right)$$

so that

$$R_2 - R_3 = \frac{R_A R_C - R_B R_A}{R_A + R_B + R_C} \qquad \textbf{(8.3d)}$$

Subtracting Eq. (8.3d) from Eq. (8.3c) yields

$$(R_2 + R_3) - (R_2 - R_3) = \left(\frac{R_A R_B + R_A R_C}{R_A + R_B + R_C}\right) - \left(\frac{R_A R_C - R_B R_A}{R_A + R_B + R_C}\right)$$

so that

$$2R_3 = \frac{2R_B R_A}{R_A + R_B + R_C}$$

resulting in the following expression for R_3 in terms of R_A, R_B, and R_C:

$$R_3 = \frac{R_A R_B}{R_A + R_B + R_C} \qquad \textbf{(8.4a)}$$

Following the same procedure for R_1 and R_2, we have

$$R_1 = \frac{R_B R_C}{R_A + R_B + R_C} \qquad \textbf{(8.4b)}$$

and

$$R_2 = \frac{R_A R_C}{R_A + R_B + R_C} \qquad \textbf{(8.4c)}$$

Note that each resistor of the Y is equal to the product of the resistors in the two closest branches of the Δ divided by the sum of the resistors in the Δ.

To obtain the relationships necessary to convert from a Y to a Δ, first divide Eq. (8.4a) by Eq. (8.4b):

$$\frac{R_3}{R_1} = \frac{(R_A R_B)/(R_A + R_B + R_C)}{(R_B R_C)/(R_A + R_B + R_C)} = \frac{R_A}{R_C}$$

or

$$R_A = \frac{R_C R_3}{R_1}$$

Then divide Eq. (8.4a) by Eq. (8.4c)

$$\frac{R_3}{R_2} = \frac{(R_A R_B)/(R_A + R_B + R_C)}{(R_A R_C)/(R_A + R_B + R_C)} = \frac{R_B}{R_C}$$

or
$$R_B = \frac{R_3 R_C}{R_2}$$

Substituting for R_A and R_B in Eq. (8.4c) yields

$$R_2 = \frac{(R_C R_3/R_1)R_C}{(R_3 R_C/R_2) + (R_C R_3/R_1) + R_C}$$
$$= \frac{(R_3/R_1)R_C}{(R_3/R_2) + (R_3/R_1) + 1}$$

Placing these over a common denominator, we obtain

$$R_2 = \frac{(R_3 R_C/R_1)}{(R_1 R_2 + R_1 R_3 + R_2 R_3)/(R_1 R_2)}$$
$$= \frac{R_2 R_3 R_C}{R_1 R_2 + R_1 R_3 + R_2 R_3}$$

and
$$R_C = \frac{R_1 R_2 + R_1 R_3 + R_2 R_3}{R_3} \qquad \textbf{(8.5a)}$$

We follow the same procedure for R_B and R_A:

$$R_A = \frac{R_1 R_2 + R_1 R_3 + R_2 R_3}{R_1} \qquad \textbf{(8.5b)}$$

and
$$R_B = \frac{R_1 R_2 + R_1 R_3 + R_2 R_3}{R_2} \qquad \textbf{(8.5c)}$$

Note that the value of each resistor of the Δ is equal to the sum of the possible product combinations of the resistances of the Y divided by the resistance of the Y farthest from the resistor to be determined.

Let us consider what would occur if all the values of a Δ or Y were the same. If $R_A = R_B = R_C$, Eq. (8.4a) would become (using R_A only) the following:

$$R_3 = \frac{R_A R_B}{R_A + R_B + R_C} = \frac{R_A R_A}{R_A + R_A + R_A} = \frac{R_A^2}{3R_A} = \frac{R_A}{3}$$

and, following the same procedure,

$$R_1 = \frac{R_A}{3} \qquad R_2 = \frac{R_A}{3}$$

In general, therefore,

$$R_Y = \frac{R_\Delta}{3} \qquad \textbf{(8.6a)}$$

or
$$R_\Delta = 3R_Y \qquad \textbf{(8.6b)}$$

which indicates that *for a* Y *of three equal resistors, the value of each resistor of the* Δ *is equal to three times the value of any resistor of the* Y. If only two elements of a Y or a Δ are the same, the corresponding Δ or Y of each will also have two equal elements. The converting of equations is left as an exercise for you.

The Y and the Δ often appear as shown in Fig. 8.84. They are then referred to as a **tee (T)** and a **pi (π)** network, respectively. The equations used to convert from one form to the other are exactly the same as those developed for the Y and Δ transformation.

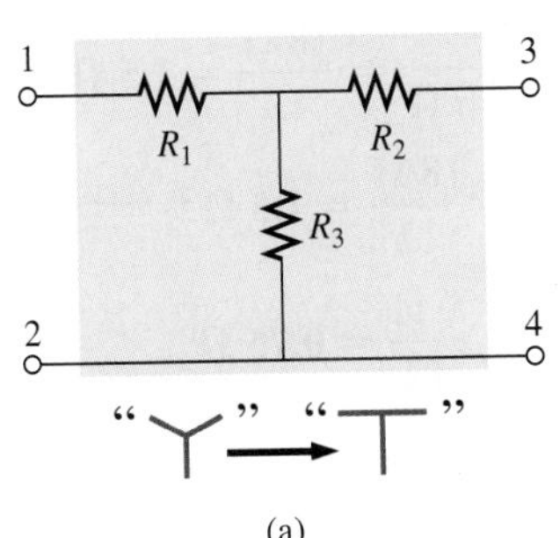

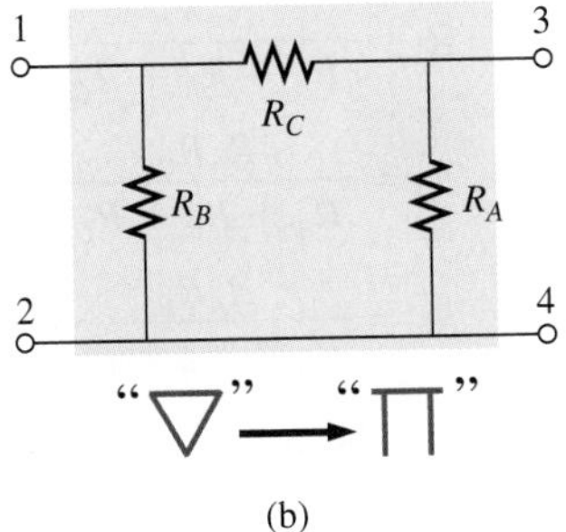

FIG. 8.84
The relationship between the Y and T configurations and the Δ and π configurations.

EXAMPLE 8.27 Convert the Δ in Fig. 8.85 to a Y.

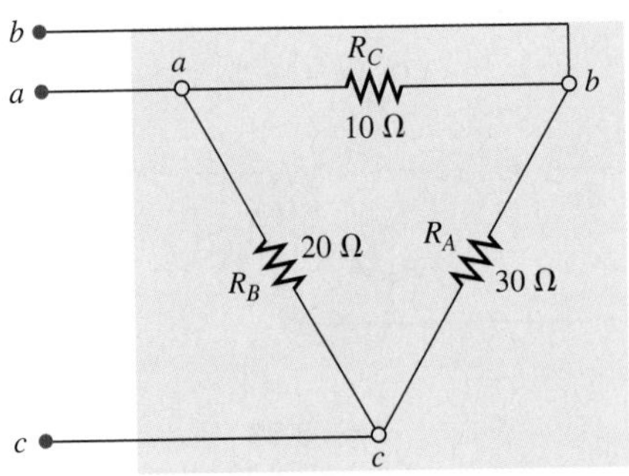

FIG. 8.85
Example 8.27.

Solution:

$$R_1 = \frac{R_B R_C}{R_A + R_B + R_C} = \frac{(20\ \Omega)(10\ \Omega)}{30\ \Omega + 20\ \Omega + 10\ \Omega} = \frac{200\ \Omega}{60} = \mathbf{3\tfrac{1}{3}\ \Omega}$$

$$R_2 = \frac{R_A R_C}{R_A + R_B + R_C} = \frac{(30\ \Omega)(10\ \Omega)}{60\ \Omega} = \frac{300\ \Omega}{60} = \mathbf{5\ \Omega}$$

$$R_3 = \frac{R_A R_B}{R_A + R_B + R_C} = \frac{(20\ \Omega)(30\ \Omega)}{60\ \Omega} = \frac{600\ \Omega}{60} = \mathbf{10\ \Omega}$$

The equivalent network is shown in Fig. 8.86.

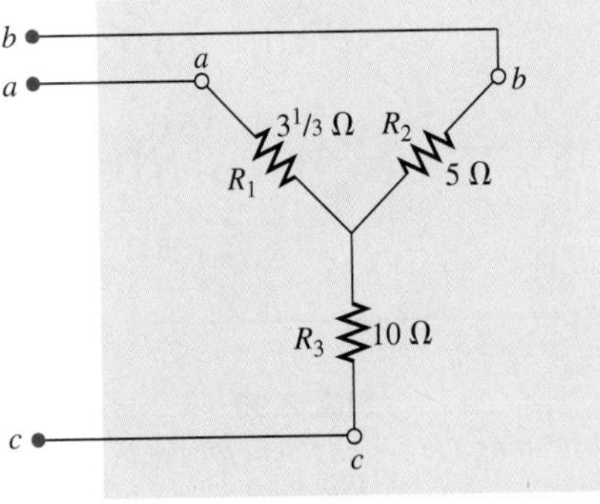

FIG. 8.86
The Y equivalent for the Δ in Fig. 8.85.

EXAMPLE 8.28 Convert the Y in Fig. 8.87 to a Δ.

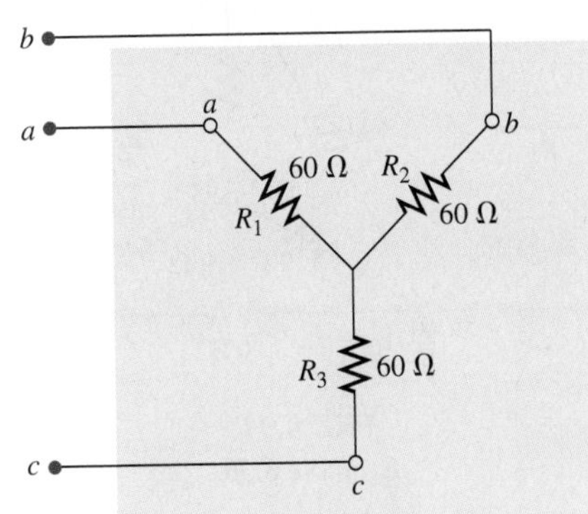

FIG. 8.87
Example 8.28.

Solution:

$$R_A = \frac{R_1R_2 + R_1R_3 + R_2R_3}{R_1}$$
$$= \frac{(60\ \Omega)(60\ \Omega) + (60\ \Omega)(60\ \Omega) + (60\ \Omega)(60\ \Omega)}{60\ \Omega}$$
$$= \frac{3600\ \Omega + 3600\ \Omega + 3600\ \Omega}{60} = \frac{10{,}800\ \Omega}{60}$$

$$R_A = \mathbf{180\ \Omega}$$

However, the three resistors for the Y are equal, permitting the use of Eq. (8.6) and yielding

$$R_\Delta = 3R_Y = 3(60\ \Omega) = 180\ \Omega$$

and $$R_B = R_C = \mathbf{180\ \Omega}$$

The equivalent network is shown in Fig. 8.88.

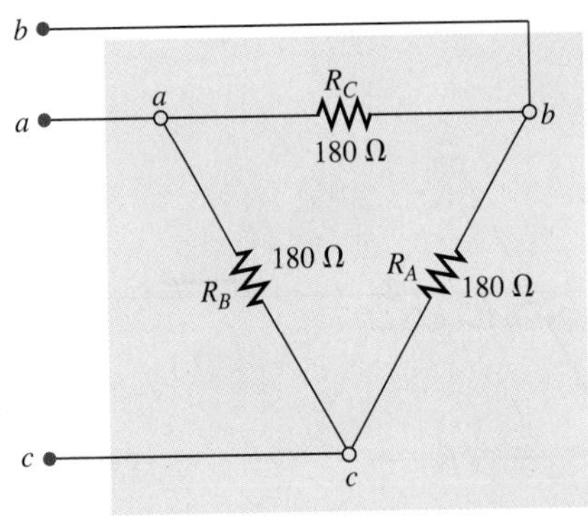

FIG. 8.88
The Δ equivalent for the Y in Fig. 8.87.

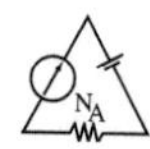

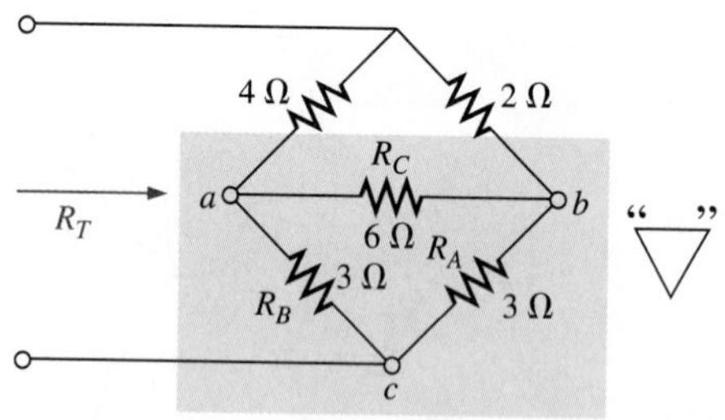

FIG. 8.89
Example 8.29.

EXAMPLE 8.29 Find the total resistance of the network in Fig. 8.89, where $R_A = 3\ \Omega$, $R_B = 3\ \Omega$, and $R_C = 6\ \Omega$.

Solution:

Two resistors of the Δ were equal; therefore, two resistors of the Y will be equal.

$$R_1 = \frac{R_B R_C}{R_A + R_B + R_C} = \frac{(3\ \Omega)(6\ \Omega)}{3\ \Omega + 3\ \Omega + 6\ \Omega} = \frac{18\ \Omega}{12} = \mathbf{1.5\ \Omega}$$

$$R_2 = \frac{R_A R_C}{R_A + R_B + R_C} = \frac{(3\ \Omega)(6\ \Omega)}{12\ \Omega} = \frac{18\ \Omega}{12} = \mathbf{1.5\ \Omega}$$

$$R_3 = \frac{R_A R_B}{R_A + R_B + R_C} = \frac{(3\ \Omega)(3\ \Omega)}{12\ \Omega} = \frac{9\ \Omega}{12} = \mathbf{0.75\ \Omega}$$

Replacing the Δ by the Y, as shown in Fig. 8.90, yields

$$R_T = 0.75\ \Omega + \frac{(4\ \Omega + 1.5\ \Omega)(2\ \Omega + 1.5\ \Omega)}{(4\ \Omega + 1.5\ \Omega) + (2\ \Omega + 1.5\ \Omega)}$$

$$= 0.75\ \Omega + \frac{(5.5\ \Omega)(3.5\ \Omega)}{5.5\ \Omega + 3.5\ \Omega}$$

$$= 0.75\ \Omega + 2.139\ \Omega$$

$$R_T = \mathbf{2.89\ \Omega}$$

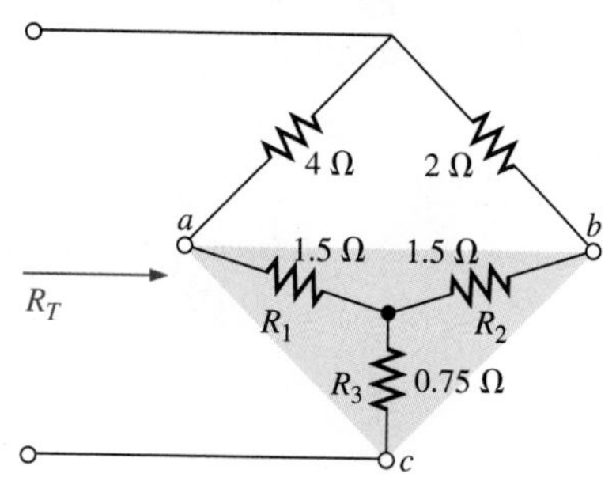

FIG. 8.90
Substituting the Y equivalent for the bottom Δ in Fig. 8.89.

EXAMPLE 8.30 Find the total resistance of the network in Fig. 8.91.

Solutions: Since all the resistors of the Δ or Y are the same, Eqs. (8.6a) and (8.6b) can be used to convert either form to the other.

a. *Converting the Δ to a Y:* Note: When this is done, the resulting d' of the new Y will be the same as the point d shown in the original figure, only because both systems are "balanced." That is, the resistance in each branch of each system has the same value:

$$R_Y = \frac{R_\Delta}{3} = \frac{6\ \Omega}{3} = 2\ \Omega \qquad \text{(Fig. 8.92)}$$

The network then appears as shown in Fig. 8.93.

$$R_T = 2\left[\frac{(2\ \Omega)(9\ \Omega)}{2\ \Omega + 9\ \Omega}\right] = \mathbf{3.27\ \Omega}$$

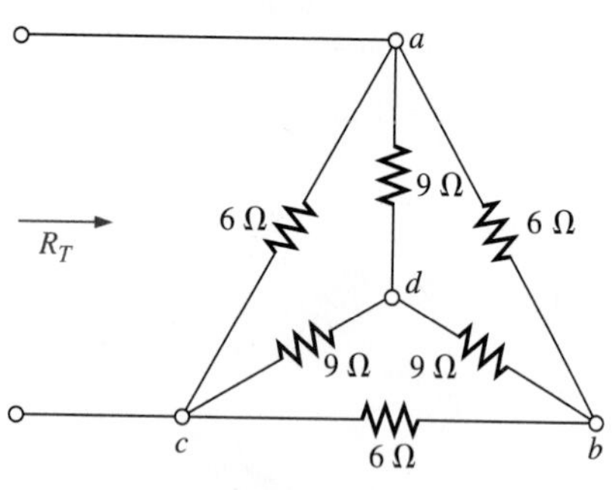

FIG. 8.91
Example 8.30.

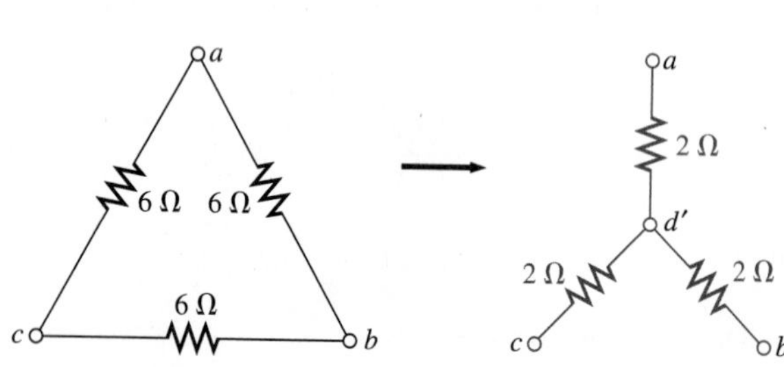

FIG. 8.92
Converting the Δ configuration of Fig. 8.91 to a Y configuration.

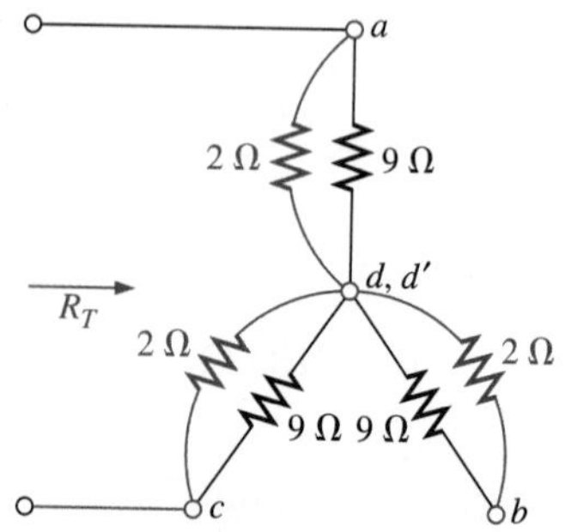

FIG. 8.93
Substituting the Y configuration for the converted Δ into the network in Fig. 8.91.

b. *Converting the Y to a Δ:*

$$R_\Delta = 3R_Y = (3)(9\ \Omega) = 27\ \Omega \qquad \text{(Fig. 8.94)}$$

$$R'_T = \frac{(6\ \Omega)(27\ \Omega)}{6\ \Omega + 27\ \Omega} = \frac{162\ \Omega}{33} = 4.91\ \Omega$$

$$R_T = \frac{R'_T(R'_T + R'_T)}{R'_T + (R'_T + R'_T)} = \frac{R'_T 2R'_T}{3R'_T} = \frac{2R'_T}{3}$$

$$= \frac{2(4.91\ \Omega)}{3} = \mathbf{3.27\ \Omega}$$

which checks with the previous solution.

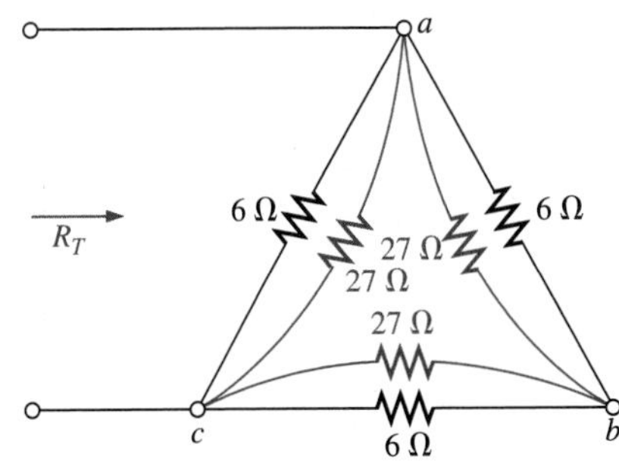

FIG. 8.94
Substituting the converted Y configuration into the network in Fig. 8.91.

8.13 APPLICATIONS

This section discusses the constant current characteristic in the design of security systems, the bridge circuit in a common residential smoke detector, and the nodal voltages of a digital logic probe.

Constant Current Alarm Systems

The basic components of an alarm system using a constant current supply are provided in Fig. 8.95. This design is improved over that provided in Chapter 5 in the sense that it is less sensitive to changes in resistance in the circuit due to heating, humidity, changes in the length of the line to the sensors, and so on. The 1.5 kΩ rheostat (total resistance between points *a* and *b*) is adjusted to ensure a current of 5 mA through the single-series security circuit. The adjustable rheostat is necessary to compensate for variations in the total resistance of the circuit introduced by the resistance of the wire, sensors, sensing relay, and milliammeter. The milliammeter is included to set the rheostat and ensure a current of 5 mA.

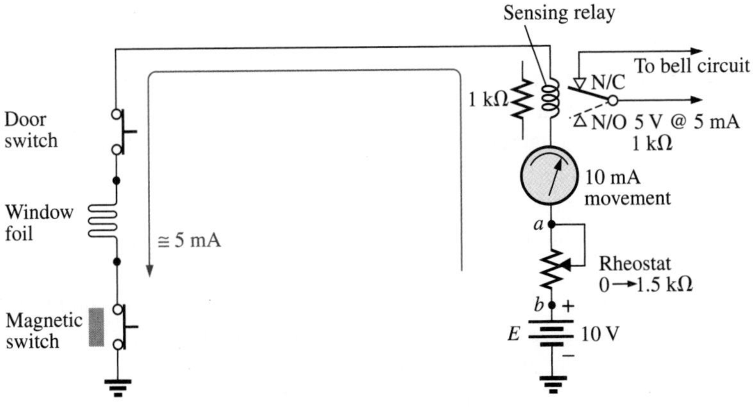

FIG. 8.95
Constant current alarm system.

If any of the sensors open, the current through the entire circuit drops to zero, the coil of the relay releases the plunger, and contact is made with the N/C position of the relay. This action completes the circuit for the bell circuit, and the alarm sounds. For the future, keep in mind that

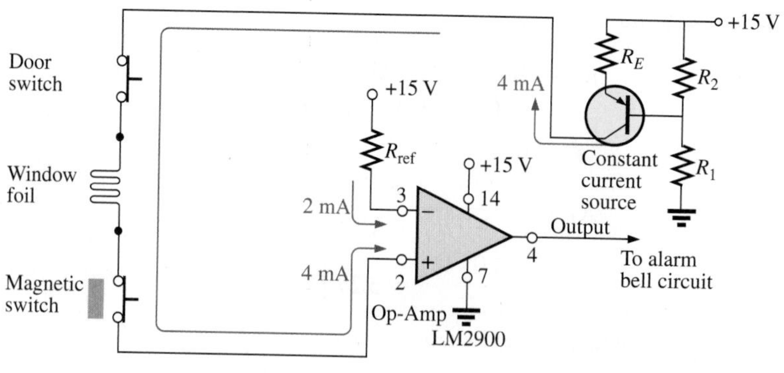

FIG. 8.96
Constant current alarm system with electronic components.

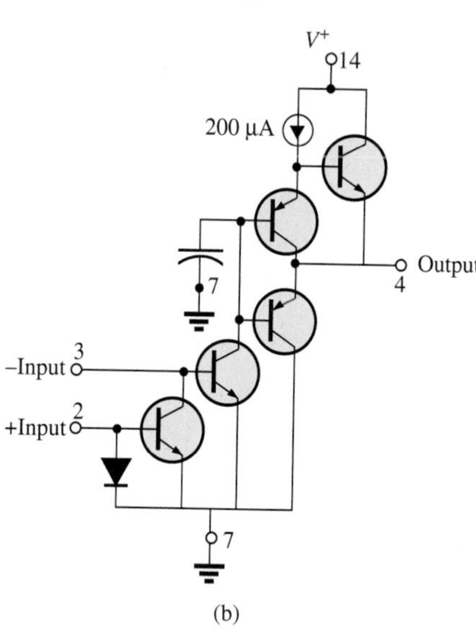

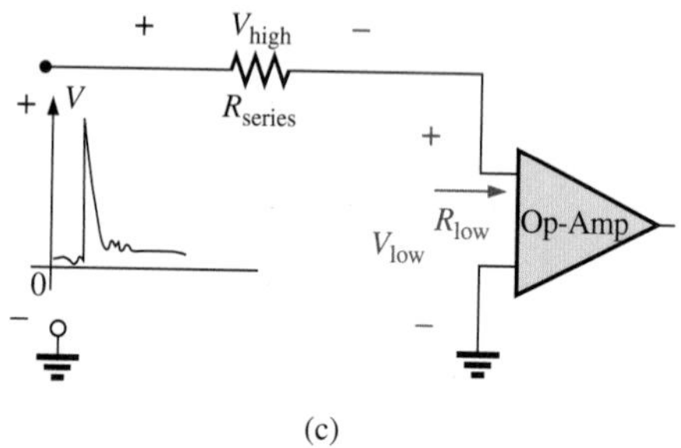

FIG. 8.97
LM2900 operational amplifier: (a) dual-in-line package (DIP); (b) components; (c) impact of low-input impedance.

switch positions for a relay are always shown with no power to the network, resulting in the N/C position in Fig. 8.95. When power is applied, the switch will have the position indicated by the dashed line. That is, various factors, such as a change in resistance of any of the elements due to heating, humidity, and so on, cause the applied voltage to redistribute itself and create a sensitive situation. With an adjusted 5 mA, the loading can change, but the current will always be 5 mA and the chance of tripping reduced. Note that the relay is rated as 5 V at 5 mA, indicating that in the on state the voltage across the relay is 5 V and the current through the relay is 5 mA. Its internal resistance is therefore 5 V/5 mA = 1 kΩ in this state.

A more advanced alarm system using a constant current is illustrated in Fig. 8.96. In this case, an electronic system using a single transistor, biasing resistors, and a dc battery are establishing a current of 4 mA through the series sensor circuit connected to the positive side of an operational amplifier (op-amp). Transistors and op-amp devices may be new to you (these are discussed in detail in electronics courses), but for now you just need to know that the transistor in this application is being used not as an amplifier but as part of a design to establish a constant current through the circuit. The op-amp is a very useful component of numerous electronic systems, and it has important terminal characteristics established by a variety of components internal to its design. The LM2900 operational amplifier in Fig. 8.96 is one of four found in the dual-in-line integrated circuit package appearing in Fig. 8.97(a). Pins 2, 3, 4, 7, and 14 were used for the design in Fig. 8.96. Note in Fig. 8.97(b) the number of elements required to establish the desired terminal characteristics—the details of which will be investigated in your electronics courses.

In Fig. 8.96, the designed 15 V dc supply, biasing resistors, and transistor in the upper right corner of the schematic establish a constant 4 mA current through the circuit. It is referred to as a *constant current source* because the current remains fairly constant at 4 mA even though there may be moderate variations in the total resistance of the series sensor circuit connected to the transistor. Following the 4 mA through the circuit, we find that it enters terminal 2 (positive side of the input) of the op-amp. A second current of 2 mA, called the *reference current,* is established by the 15 V source and resistance R and enters terminal 3 (negative side of the input) of the op-amp. The reference current of 2 mA is necessary to establish a current for the 4 mA current of the network to be compared against. As long as the 4 mA current exists, the operational

amplifier provides a "high" output voltage that exceeds 13.5 V, with a typical level of 14.2 V (according to the specification sheet for the op-amp). However, if the sensor current drops from 4 mA to a level below the reference level of 2 mA, the op-amp responds with a "low" output voltage that is typically about 0.1 V. The output of the operational amplifier then signals the alarm circuit about the disturbance. Note from the above that it is not necessary for the sensor current to drop to 0 mA to signal the alarm circuit—just a variation around the reference level that appears unusual.

One very important characteristic of this particular op-amp is that the input impedance to the op-amp is relatively low. This feature is important because you don't want alarm circuits reacting to every voltage spike or turbulence that comes down the line because of external switching action or outside forces such as lightning. In Fig. 8.97(c), for instance, if a high voltage should appear at the input to the series configuration, most of the voltage will be absorbed by the series resistance of the sensor circuit rather than traveling across the input terminals of the operational amplifier—thus preventing a false output and an activation of the alarm.

Wheatstone Bridge Smoke Detector

The Wheatstone bridge is a popular network configuration whenever detection of small changes in a quantity is required. In Fig. 8.98(a), the dc bridge configuration uses a photoelectric device to detect the presence of smoke and to sound the alarm. A photograph of an actual photoelectric smoke detector appears in Fig. 8.98(b), and the internal construction of the unit is shown in Fig. 8.98(c). First, note that air vents are provided to permit the smoke to enter the chamber below the clear plastic. The clear plastic prevents the smoke from entering the upper chamber but permits the light from the bulb in the upper chamber to bounce off the lower reflector to the semiconductor light sensor (a cadmium photocell) at the left side of the chamber. The clear plastic separation ensures that the light hitting the light sensor in the upper chamber is not affected by the entering smoke. It establishes a reference level to compare against the chamber with the entering smoke. If no smoke is present, the difference in response between the sensor cells will be registered as the normal situation. Of course, if both cells were exactly identical, and if the clear plastic did not cut down on the light, both sensors would establish the same reference level, and their difference would be zero. However, this is seldom the case, so a reference difference is recognized as the sign that smoke is not present. However, once smoke is present, there will be a sharp difference in the sensor reaction from the norm, and the alarm should sound.

In Fig. 8.98(a), we find that the two sensors are located on opposite arms of the bridge. With no smoke present, the balance-adjust rheostat is used to ensure that the voltage V between points a and b is zero volts and the resulting current through the primary of the sensitive relay is zero amperes. Taking a look at the relay, we find that the absence of a voltage from a to b leaves the relay coil unenergized and the switch in the N/O position (recall that the position of a relay switch is always drawn in the unenergized state). An unbalanced situation results in a voltage across the coil and activation of the relay, and the switch moves to the N/C position to complete the alarm circuit and activate the alarm. Relays with two contacts and one movable arm are called *single-pole–double-throw* (SPDT) relays. The dc power is required to set up the balanced situation, energize

(a)

(b)

(c)

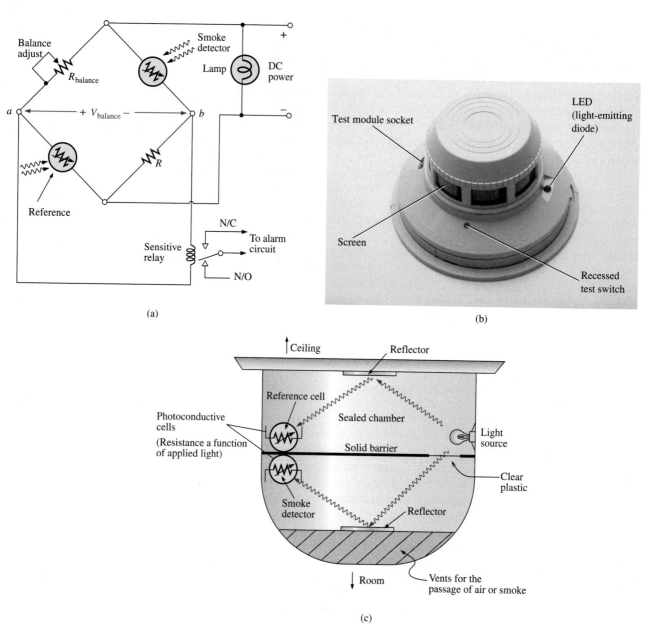

FIG. 8.98
Wheatstone bridge smoke detector: (a) dc bridge configuration; (b) outside appearance; (c) internal construction.

the parallel bulb so we know that the system is on, and provide the voltage from a to b if an unbalanced situation should develop.

Why do you suppose only one sensor isn't used since its resistance would be sensitive to the presence of smoke? The answer is that the smoke detector may generate a false readout if the supply voltage or output light intensity of the bulb should vary. Smoke detectors of the type just described must be used in gas stations, kitchens, dentist offices, etc., where the range of gas fumes present may set off an ionizing type smoke detector.

Schematic with Nodal Voltages

When an investigator is presented with a system that is down or not operating properly, one of the first options is to check the system's specified voltages on the schematic. These specified voltage levels are actually the

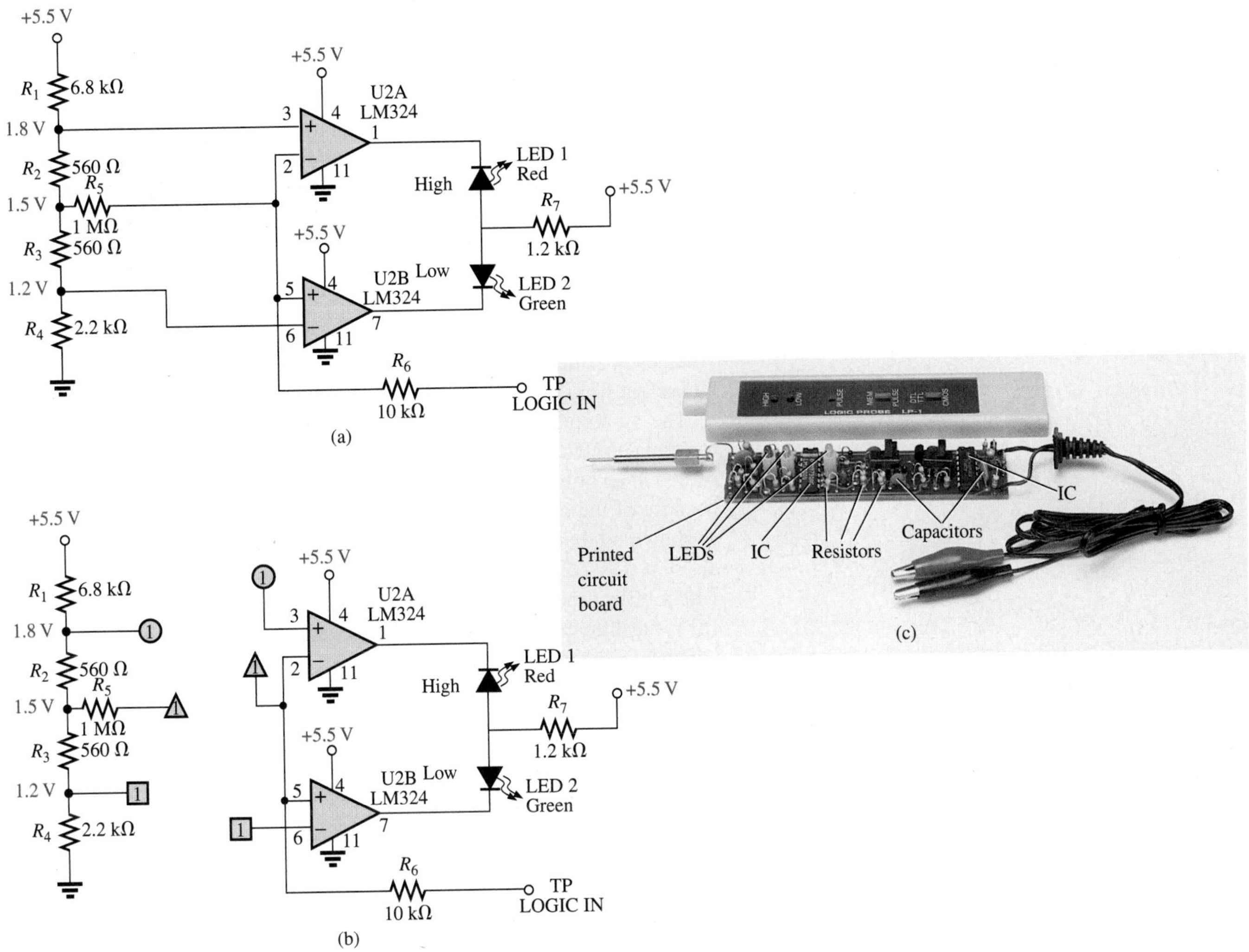

FIG. 8.99
Logic probe: (a) schematic with nodal voltages; (b) network with global connections; (c) photograph of commercially available unit.

nodal voltages determined in this chapter. *Nodal voltage* is simply a special term for a voltage measured from that point to ground. The technician attaches the negative or lower-potential lead to the ground of the network (often the chassis) and then places the positive or higher-potential lead on the specified points of the network to check the nodal voltages. If they match, it means that section of the system is operating properly. If one or more fail to match the given values, the problem area can usually be identified. Be aware that a reading of −15.87 V is significantly different from an expected reading of +16 V if the leads have been properly attached. Although the actual numbers seem close, the difference is actually more than 30 V. You must expect some deviation from the given value as shown, but always be very sensitive to the resulting sign of the reading.

The schematic in Fig. 8.99(a) includes the nodal voltages for a logic probe used to measure the input and output states of integrated circuit logic chips. In other words, the probe determines whether the measured voltage is one of two states: high or low (often referred to as "on" or "off" or 1 or 0). If the LOGIC IN terminal of the probe is placed on a chip at a location where the voltage is between 0 and 1.2 V, the voltage

is considered to be a low level, and the green LED lights. (LEDs are light-emitting semiconductor diodes that emit light when current is passed through them.) If the measured voltage is between 1.8 V and 5 V, the reading is considered high, and the red LED lights. Any voltage between 1.2 V and 1.8 V is considered a "floating level" and is an indication that the system being measured is not operating correctly. Note that the reference levels mentioned above are established by the voltage divider network to the left of the schematic. The op-amps used are of such high input impedance that their loading on the voltage divider network can be ignored and the voltage divider network considered a network unto itself. Even though three 5.5 V dc supply voltages are indicated on the diagram, be aware that all three points are connected to the same supply. The other voltages provided (the nodal voltages) are the voltage levels that should be present from that point to ground if the system is working properly.

The op-amps are used to sense the difference between the reference at points 3 and 6 and the voltage picked up in LOGIC IN. Any difference results in an output that lights either the green or the red LED. Be aware, because of the direct connection, that the voltage at point 3 is the same as shown by the nodal voltage to the left, or 1.8 V. Likewise, the voltage at point 6 is 1.2 V for comparison with the voltages at points 5 and 2, which reflect the measured voltage. If the input voltage happened to be 1.0 V, the difference between the voltages at points 5 and 6 would be 0.2 V, which ideally would appear at point 7. This low potential at point 7 would result in a current flowing from the much higher 5.5 V dc supply through the green LED, causing it to light and indicating a low condition. By the way, LEDs, like diodes, permit current through them only in the direction of the arrow in the symbol. Also note that the voltage at point 6 must be higher than that at point 5 for the output to turn on the LED. The same is true for point 2 over point 3, which reveals why the red LED does not light when the 1.0 V level is measured.

Often it is impractical to draw the full network as shown in Fig. 8.99(b) because there are space limitations or because the same voltage divider network is used to supply other parts of the system. In such cases, you should recognize that points having the same shape are connected, and the number in the figure reveals how many connections are made to that point.

A photograph of the outside and inside of a commercially available logic probe is shown in Fig. 8.99(c). Note the increased complexity of system because of the variety of functions that the probe can perform.

8.14 COMPUTER ANALYSIS

PSpice

We will now analyze the bridge network in Fig. 8.72 using PSpice to ensure that it is in the balanced state. The only component that has not been introduced in earlier chapters is the dc current source. To obtain it, first select the **Place a part** key and then the **SOURCE** library. Scrolling the **Part List** results in the option **IDC.** A left click of **IDC** followed by **OK** results in a dc current source whose direction is toward the bottom of the screen. One left click (to make it red, or active) followed by a right click results in a listing having a **Mirror Vertically** option. Selecting that option flips the source and gives it the direction in Fig. 8.72.

The remaining parts of the PSpice analysis are pretty straightforward, with the results in Fig. 8.100 matching those obtained in the analysis of Fig. 8.72. The voltage across the current source is 8 V positive to ground, and the voltage at either end of the bridge arm is 2.667 V. The voltage

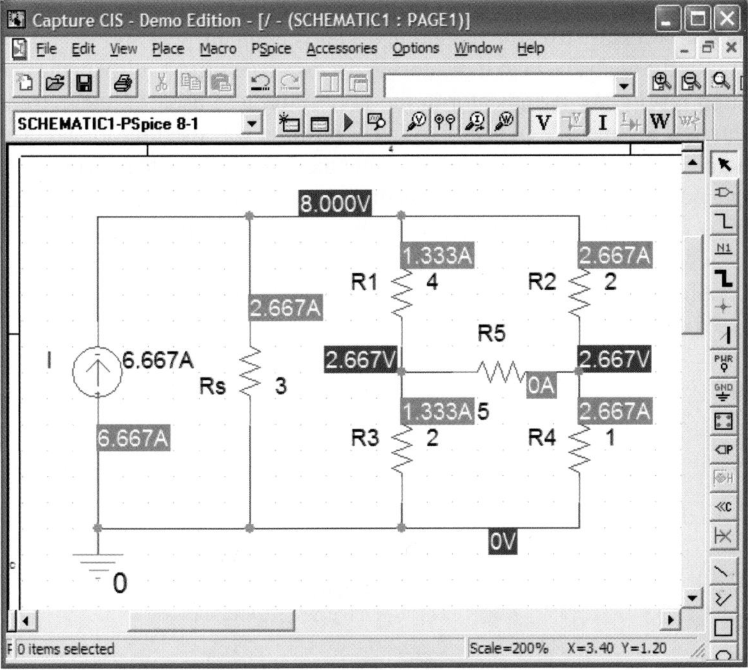

FIG. 8.100
Applying PSpice to the bridge network in Fig. 8.72.

across R_5 is obviously 0 V for the level of accuracy displayed, and the current is such a small magnitude compared to the other current levels of the network that it can essentially be considered 0 A. Note also for the balanced bridge that the current through R_1 equals that of R_3, and the current through R_2 equals that of R_4.

Multisim

We will now use Multisim to verify the results in Example 8.18. All the elements of creating the schematic in Fig. 8.101 have been presented in earlier

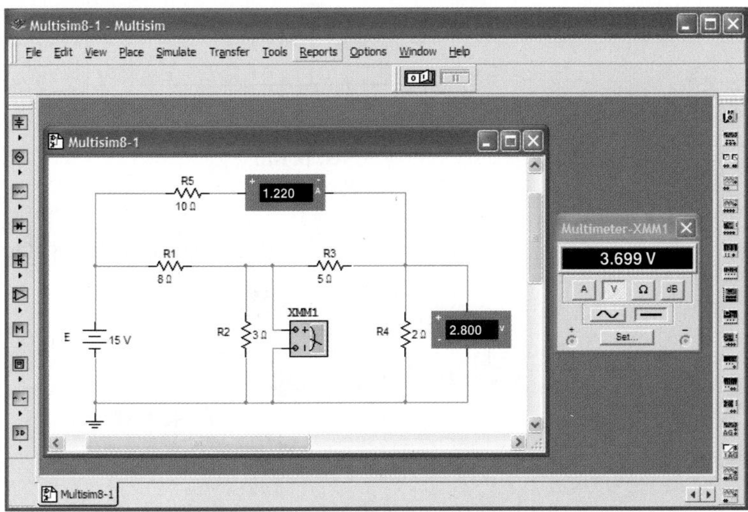

FIG. 8.101
Using Multisim to verify the results in Example 8.18.

chapters; they are not repeated here to demonstrate how little documentation is now necessary to carry you through a fairly complex network.

For the analysis, both indicators and a meter are used to display the desired results. An **A** indicator in the **H** position was used for the current through R_5, and a **V** indicator in the **V** position was used for the voltage across R_2. A multimeter in the voltmeter mode was placed to read the voltage across R_4. The ammeter is reading the mesh or loop current for that branch, and the two voltmeters are displaying the nodal voltages of the network.

After simulation, the results displayed are an exact match with those in Example 8.18.

PROBLEMS

SECTION 8.2 Current Sources

1. For the network in Fig. 8.102:
   - **a.** Determine currents I_2 and I_3.
   - **b.** Find voltage V_1.
   - **c.** Find the voltage across the source V_s.

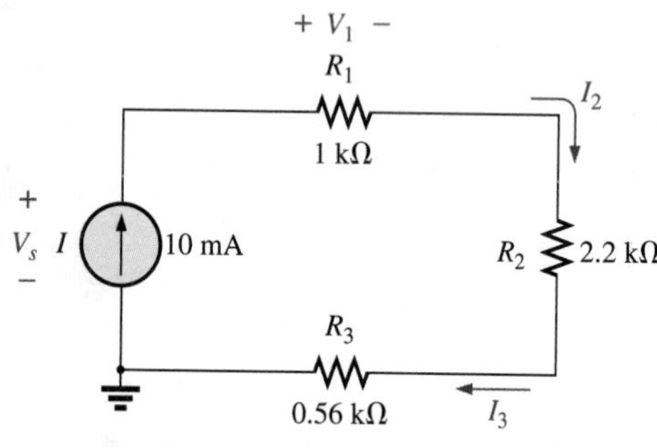

FIG. 8.102
Problem 1.

2. For the network in Fig. 8.103:
   - **a.** Find current I_2. Comment on the impact of $R_p \gg R_1$ or R_2.
   - **b.** Calculate voltage V_2.
   - **c.** Find the source voltage V_s.

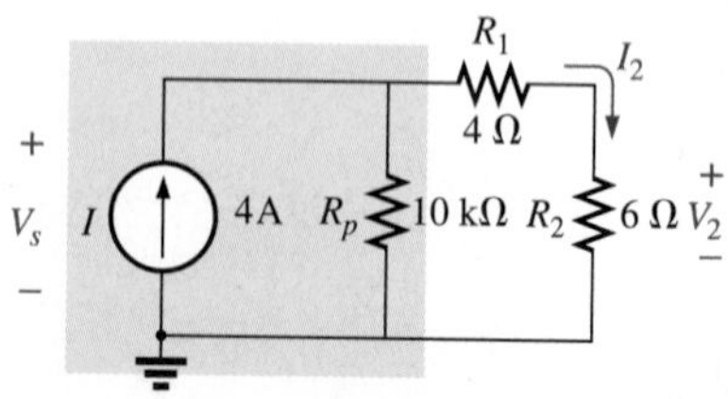

FIG. 8.103
Problem 2.

3. Find voltage V_s (with polarity) across the ideal current source in Fig. 8.104.

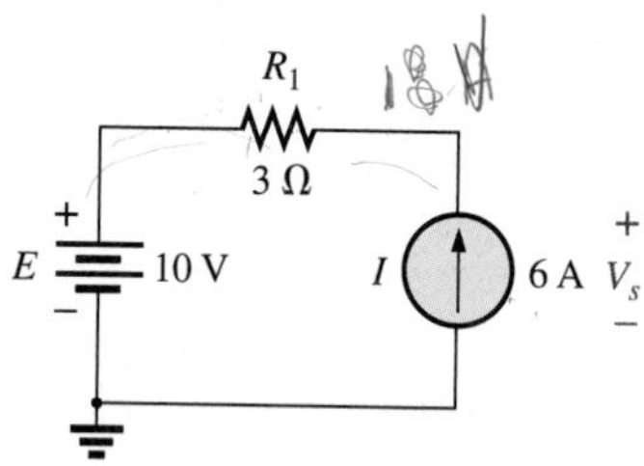

FIG. 8.104
Problem 3.

4. For the network in Fig. 8.105:
   - **a.** Find voltage V_s.
   - **b.** Calculate current I_2.
   - **c.** Find the source current I_s.

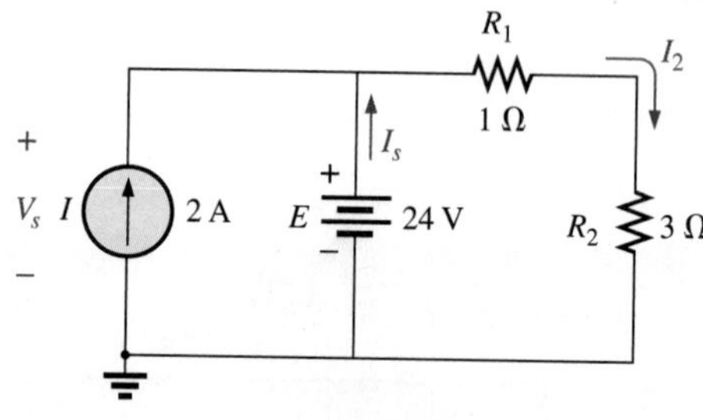

FIG. 8.105
Problem 4.

5. Find voltage V_3 and current I_2 for the network in Fig. 8.106.

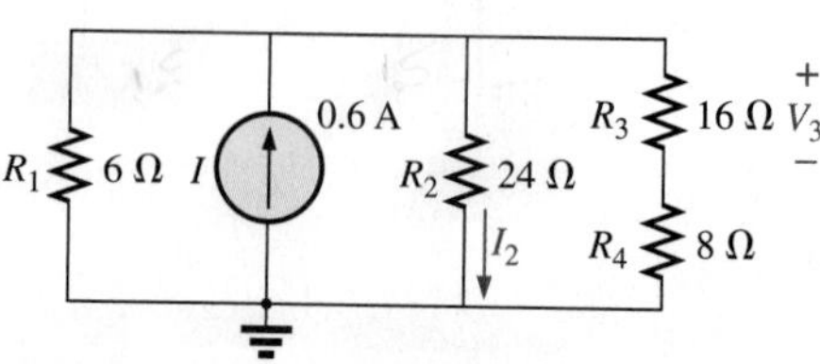

FIG. 8.106
Problem 5.

6. For the network in Fig. 8.107:
   a. Find the currents I_1 and I_s.
   b. Find the voltages V_s and V_3.

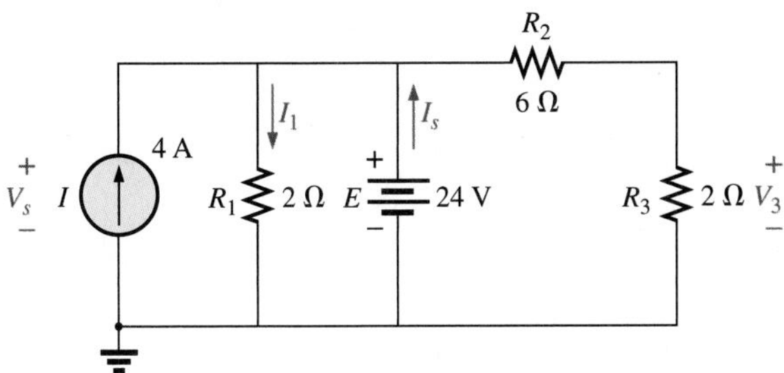

FIG. 8.107
Problem 6.

SECTION 8.3 Source Conversions

7. Convert the voltage sources in Fig. 8.108 to current sources.

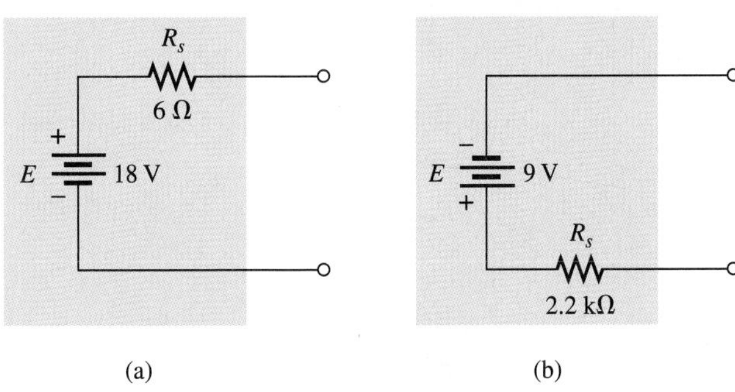

FIG. 8.108
Problem 7.

8. Convert the current sources in Fig. 8.109 to voltage sources.

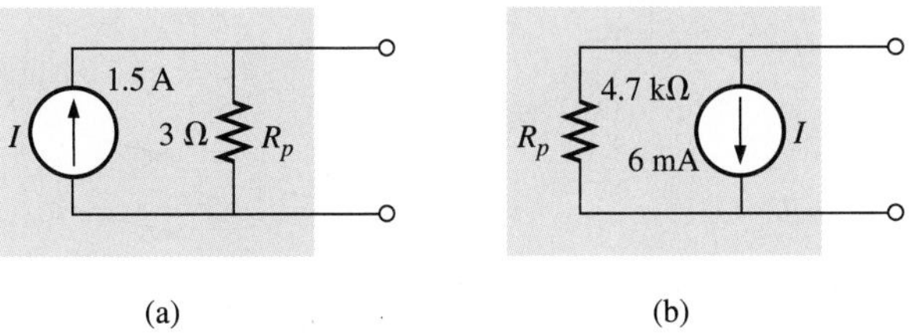

FIG. 8.109
Problem 8.

9. For the network in Fig. 8.110:

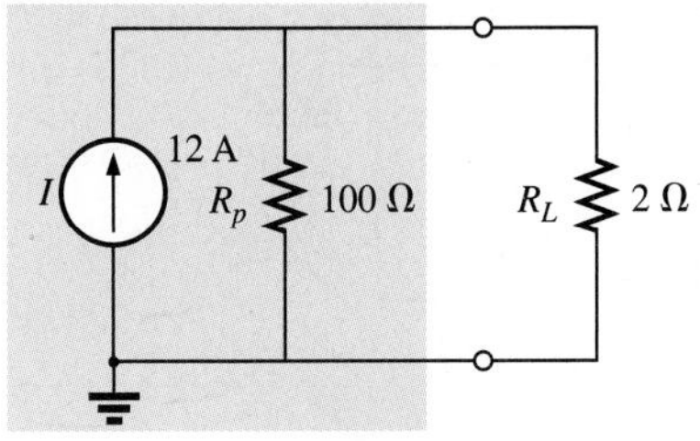

FIG. 8.110
Problem 9.

   a. Find the current through the 2 Ω resistor. Comment on the impact of $R_p = 50\ R_L$.
   b. Convert the current source and 100 Ω resistor to a voltage source, and again solve for the current in the 2 Ω resistor. Compare the results.

10. For the configuration in Fig. 8.111:
    a. Convert the current source and 6.8 Ω resistor to a voltage source.
    b. Find the magnitude and direction of the current I_1.
    c. Find the voltage V_{ab} and the polarity of points a and b.

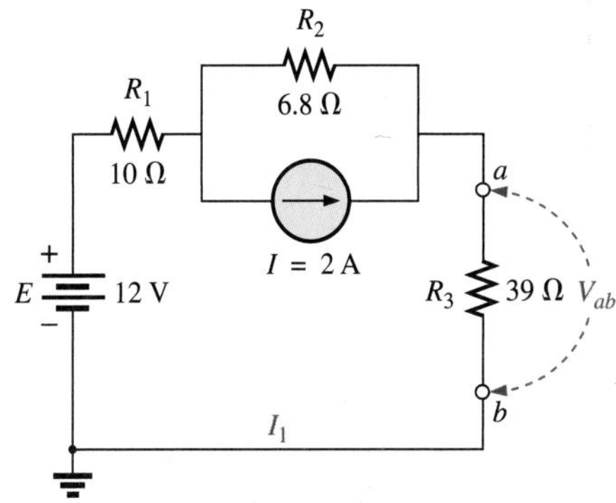

FIG. 8.111
Problem 10.

SECTION 8.4 Current Sources in Parallel

11. For the network in Fig. 8.112:
    a. Replace all the current sources by a single current source.
    b. Find the source voltage V_s.

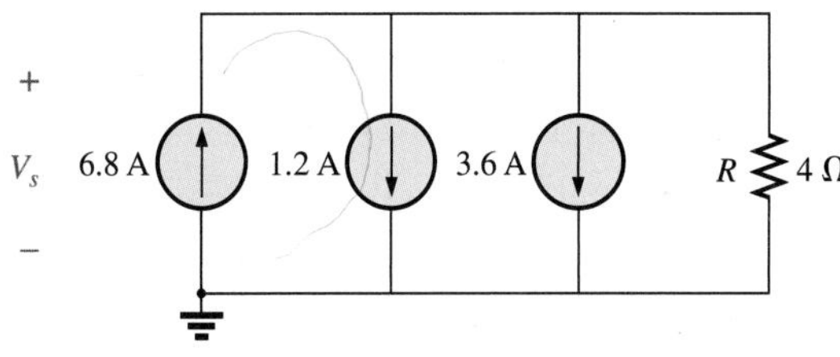

FIG. 8.112
Problem 11.

12. Find the voltage V_2 and the current I_1 for the network in Fig. 8.113.

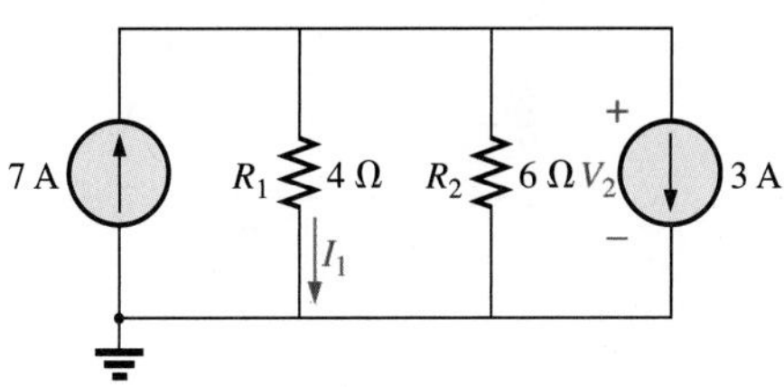

FIG. 8.113
Problem 12.

13. Convert the voltage sources in Fig. 8.114 to current sources.
   a. Find the voltage V_{ab} and the polarity of points a and b.
   b. Find the magnitude and direction of the current I_3.

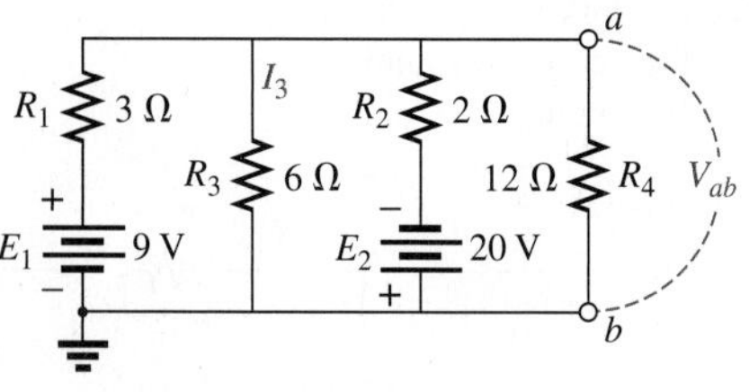

FIG. 8.114
Problem 13.

14. For the network in Fig. 8.115:
   a. Convert the voltage source to a current source.
   b. Reduce the network to a single current source, and determine the voltage V_1.
   c. Using the results of part (b), determine V_2.
   d. Calculate the current I_2.

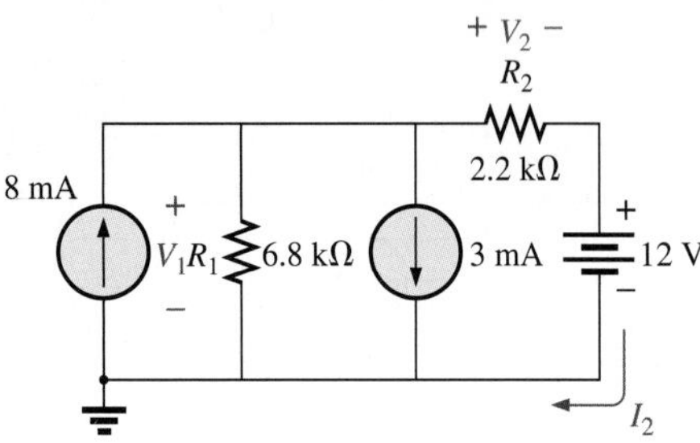

FIG. 8.115
Problem 14.

SECTION 8.6 Branch-Current Analysis

15. Using branch-current analysis, find the magnitude and direction of the current through each resistor for the networks in Fig. 8.116.

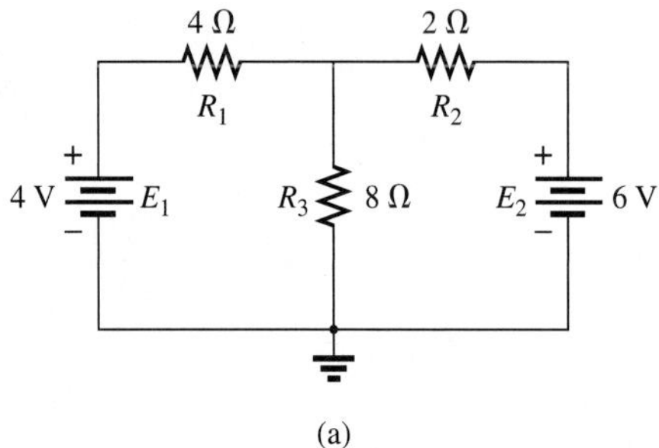

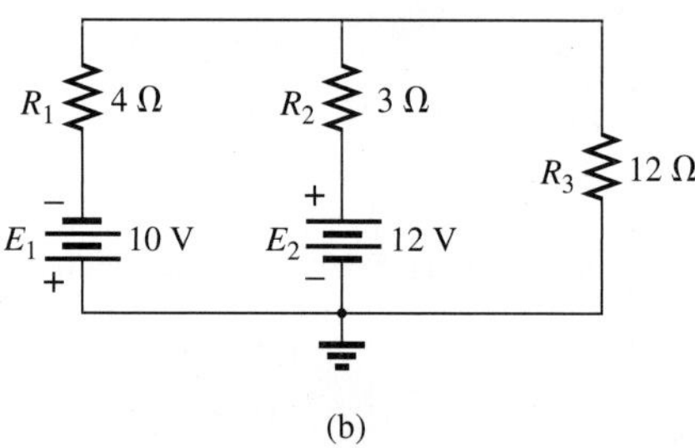

FIG. 8.116
Problems 15, 20, 28, and 57.

*16. Using branch-current analysis, find the current through each resistor for the networks in Fig. 8.117. The resistors are all standard values.

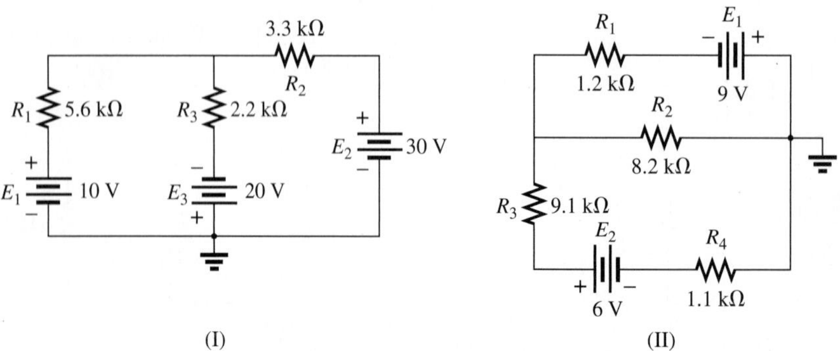

FIG. 8.117
Problems 16, 21, and 29.

***17.** For the networks in Fig. 8.118, determine the current I_2 using branch-current analysis, and then find the voltage V_{ab}.

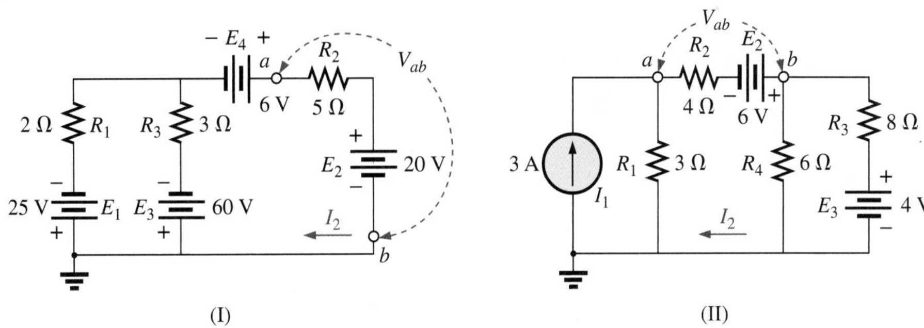

FIG. 8.118
Problems 17, 22, and 30.

***18.** For the network in Fig. 8.119:

a. Write the equations necessary to solve for the branch currents.

b. By substitution of Kirchhoff's current law, reduce the set to three equations.

c. Rewrite the equations in a format that can be solved using third-order determinants.

d. Solve for the branch current through the resistor R_3.

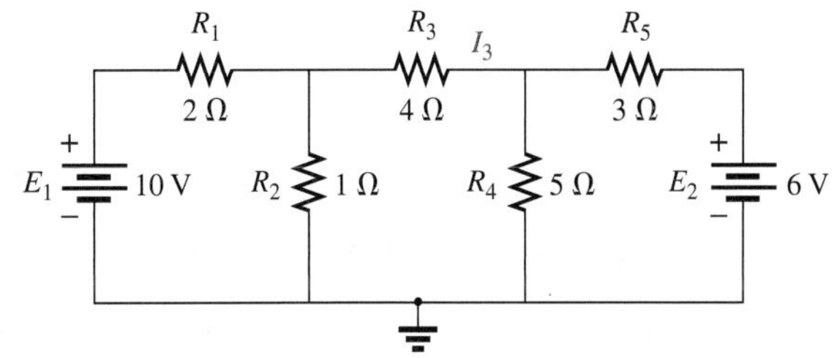

FIG. 8.119
Problems 18, 23, and 31.

***19.** For the transistor configuration in Fig. 8.120:

a. Solve for the currents I_B, I_C, and I_E using the fact that $V_{BE} = 0.7$ V and $V_{CE} = 8$ V.

b. Find the voltages V_B, V_C, and V_E with respect to ground.

c. What is the ratio of output current I_C to input current I_B? [*Note:* In transistor analysis, this ratio is referred to as the *dc beta* of the transistor (β_{dc}).]

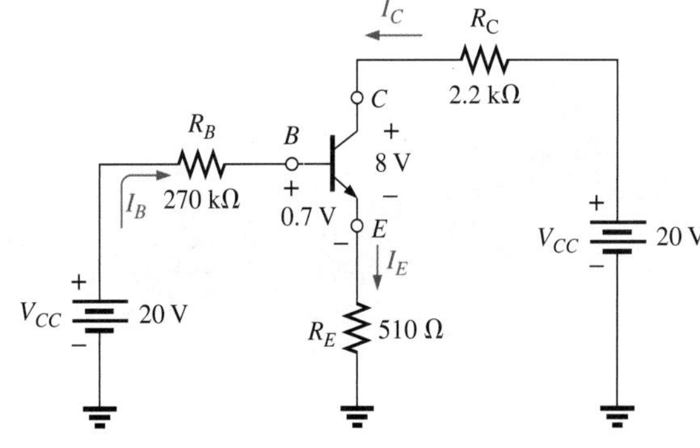

FIG. 8.120
Problem 19.

SECTION 8.7 Mesh Analysis (General Approach)

20. Find the current through each resistor for the networks in Fig. 8.116.

21. Find the current through each resistor for the networks in Fig. 8.117.

22. Find the mesh currents and the voltage V_{ab} for each network in Fig. 8.118. Use clockwise mesh currents.

23. a. Find the current I_3 for the network in Fig. 8.119 using mesh analysis.

b. Based on the results of part (a), how would you compare the application of mesh analysis to the branch-current method?

*__24.__ Using mesh analysis, determine the current through the 5 Ω resistor for each network in Fig. 8.121. Then determine the voltage V_a.

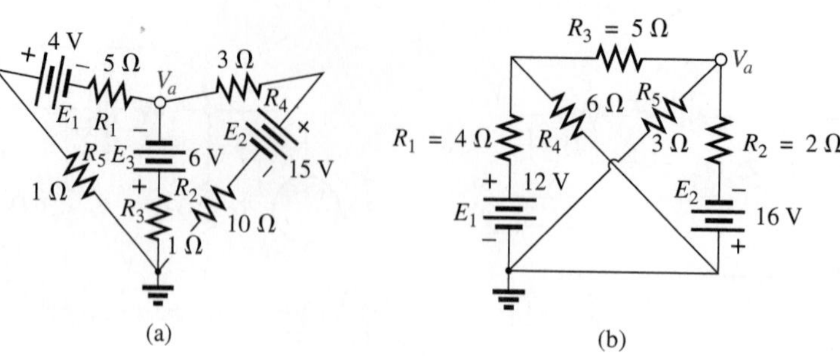

FIG. 8.121
Problems 24 and 32.

*__25.__ Write the mesh equations for each of the networks in Fig. 8.122. Using determinants, solve for the loop currents in each network. Use clockwise mesh currents.

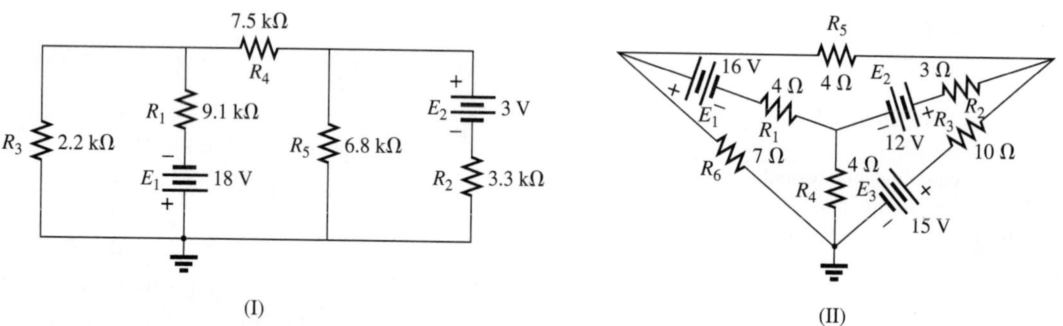

FIG. 8.122
Problems 25, 33 and 37.

*__26.__ Write the mesh equations for each of the networks in Fig. 8.123. Using determinants, solve for the loop currents in each network.

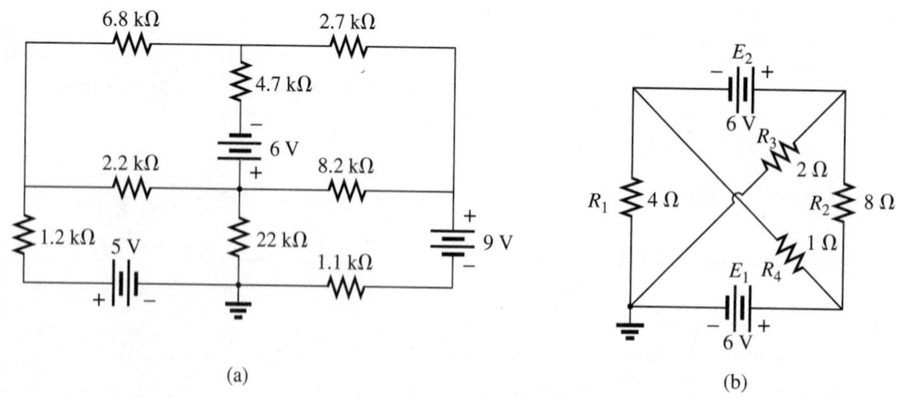

FIG. 8.123
Problems 26, 34, and 58.

*27. Using the supermesh approach, find the current through each element of the networks in Fig. 8.124.

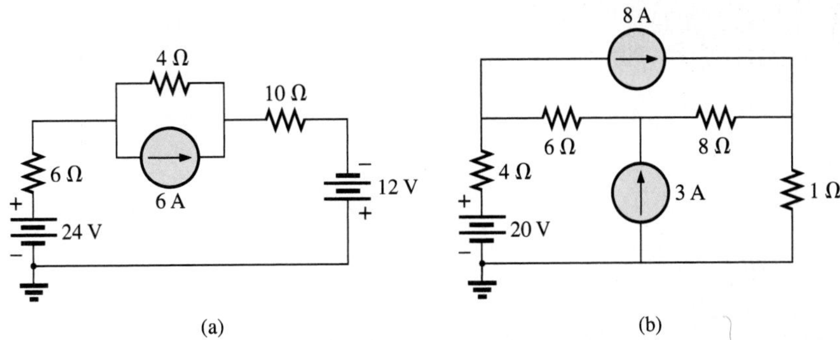

FIG. 8.124
Problem 27.

SECTION 8.8 Mesh Analysis (Format Approach)

28. Using the format approach, write the mesh equations for the networks in Fig. 8.116. Is symmetry present? Using determinants, solve for the mesh currents.

29. a. Using the format approach, write the mesh equations for the networks in Fig. 8.117.
   b. Using determinants, solve for the mesh currents.
   c. Determine the magnitude and direction of the current through each resistor.

30. a. Using the format approach, write the mesh equations for the networks in Fig. 8.118.
   b. Using determinants, solve for the mesh currents.
   c. Determine the magnitude and direction of the current through each resistor.

31. Using mesh analysis, determine the current I_3 for the network in Fig. 8.119. Compare your answer to the solution of Problem 18.

32. Using mesh analysis, determine $I_{5\Omega}$ and V_a for the network in Fig. 8.121(b).

33. Using mesh analysis, determine the mesh currents for the networks in Fig. 8.122.

34. Using mesh analysis, determine the mesh currents for the networks in Fig. 8.123.

SECTION 8.9 Nodal Analysis (General Approach)

35. Write the nodal equations for the networks in Fig. 8.125. Using determinants, solve for the nodal voltages. Is symmetry present?

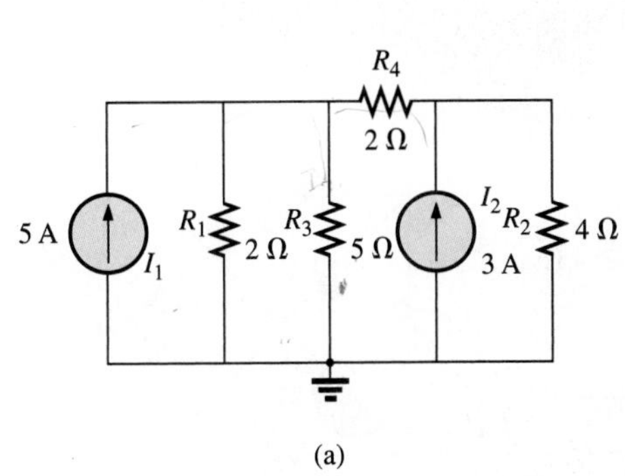

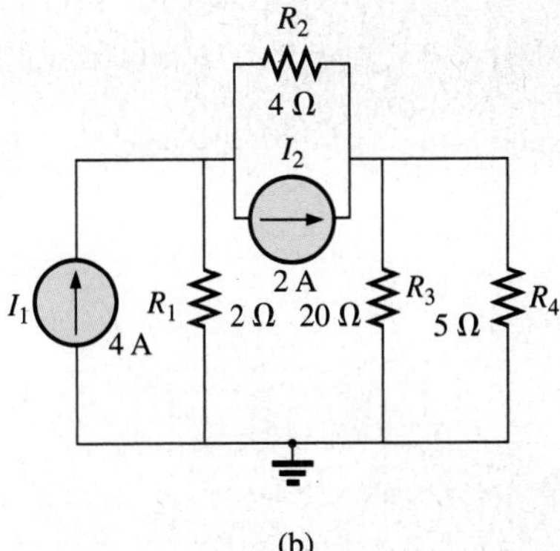

FIG. 8.125
Problems 35 and 41.

36. **a.** Write the nodal equations for the networks in Fig. 8.126.
b. Using determinants, solve for the nodal voltages.
c. Determine the magnitude and polarity of the voltage across each resistor.

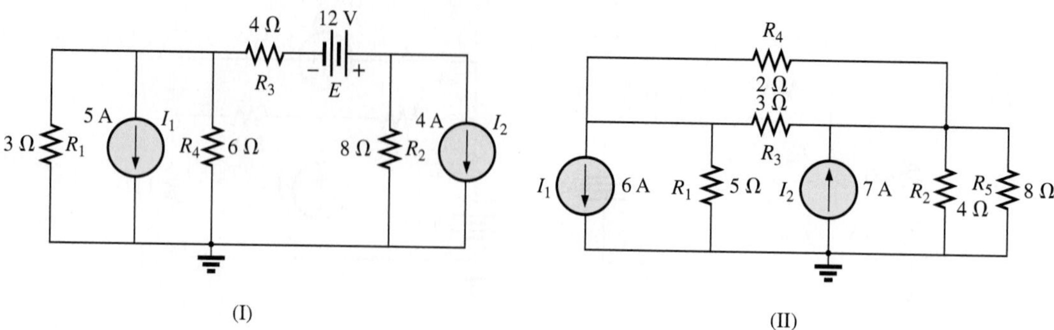

FIG. 8.126
Problems 36 and 42.

37. **a.** Write the nodal equations for the networks in Fig. 8.122.
b. Using determinants, solve for the nodal voltages.
c. Determine the magnitude and polarity of the voltage across each resistor.

***38.** For the networks in Fig. 8.127, write the nodal equations and solve for the nodal voltages.

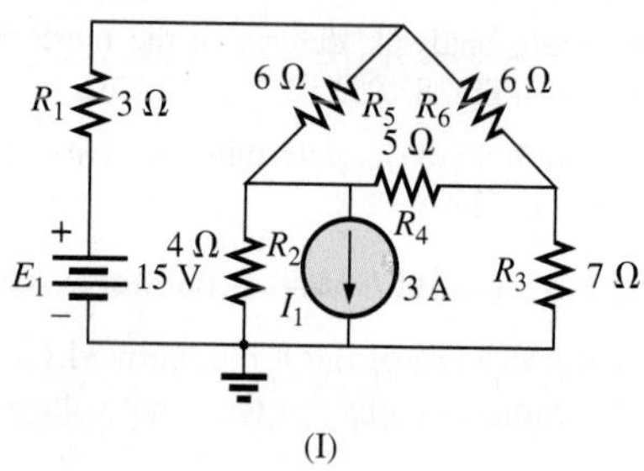

FIG. 8.127
Problems 38 and 43.

39. **a.** Determine the nodal voltages for the networks in Fig. 8.128.
b. Find the voltage across each current source.

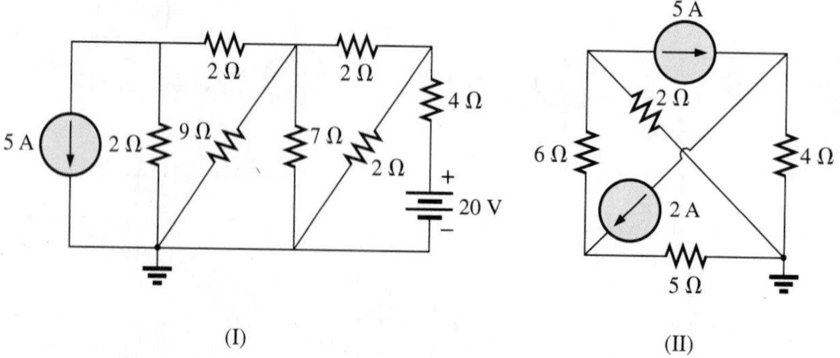

FIG. 8.128
Problems 39 and 44.

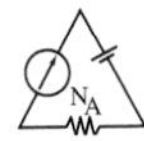

*40. Using the supernode approach, determine the nodal voltages for the networks in Fig. 8.129.

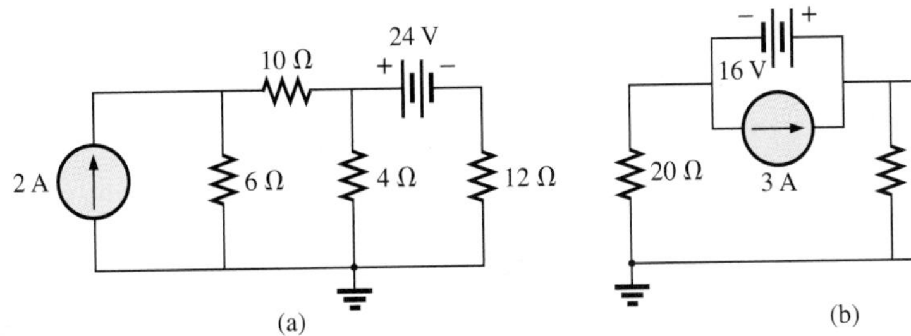

FIG. 8.129
Problems 40 and 59.

SECTION 8.10 Nodal Analysis (Format Approach)

41. Using the format approach, write the nodal equations for the networks in Fig. 8.125. Is symmetry present? Using determinants, solve for the nodal voltages.

42. **a.** Write the nodal equations for the networks in Fig. 8.126.
  b. Solve for the nodal voltages.
  c. Find the magnitude and polarity of the voltage across each resistor.

43. **a.** Write the nodal equations for the networks in Fig. 8.127.
  b. Solve for the nodal voltages.
  c. Find the magnitude and polarity of the voltage across each resistor.

44. Determine the nodal voltages for the networks in Fig. 8.128. Then determine the voltage across each current source.

SECTION 8.11 Bridge Networks

45. For the bridge network in Fig. 8.130:
  a. Write the mesh equations using the format approach.
  b. Determine the current through R_5.
  c. Is the bridge balanced?
  d. Is Eq. (8.2) satisfied?

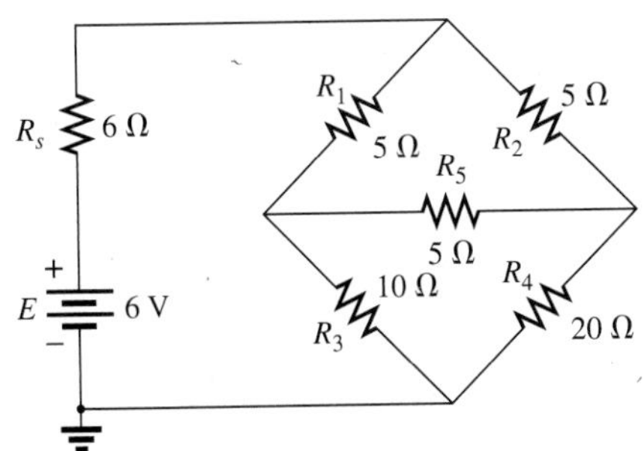

FIG. 8.130
Problems 45 and 46.

46. For the network in Fig. 8.130:
  a. Write the nodal equations using the format approach.
  b. Determine the voltage across R_5.
  c. Is the bridge balanced?
  d. Is Eq. (8.2) satisfied?

47. For the bridge in Fig. 8.131:
  a. Write the mesh equations using the format approach.
  b. Determine the current through R_5.
  c. Is the bridge balanced?
  d. Is Eq. (8.2) satisfied?

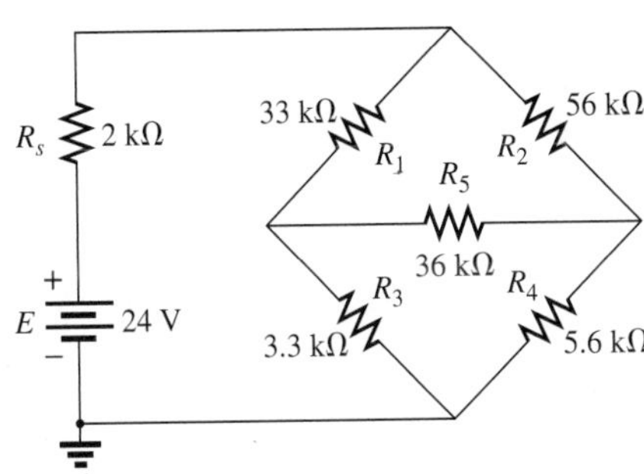

FIG. 8.131
Problems 47 and 48.

48. For the bridge network in Fig. 8.131:
  a. Write the nodal equations using the format approach.
  b. Determine the current across R_5.
  c. Is the bridge balanced?
  d. Is Eq. (8.2) satisfied?

49. Write the nodal equations for the bridge configuration in Fig. 8.132. Use the format approach.

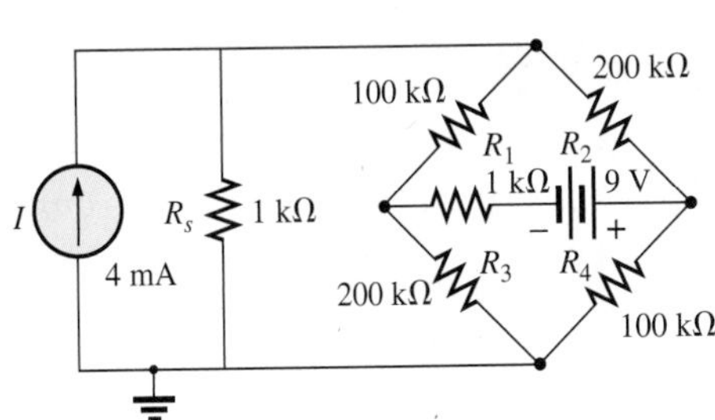

FIG. 8.132
Problem 49.

***50.** Determine the current through the source resistor R_s of each network in Fig. 8.133 using either mesh or nodal analysis. Explain why you chose one method over the other.

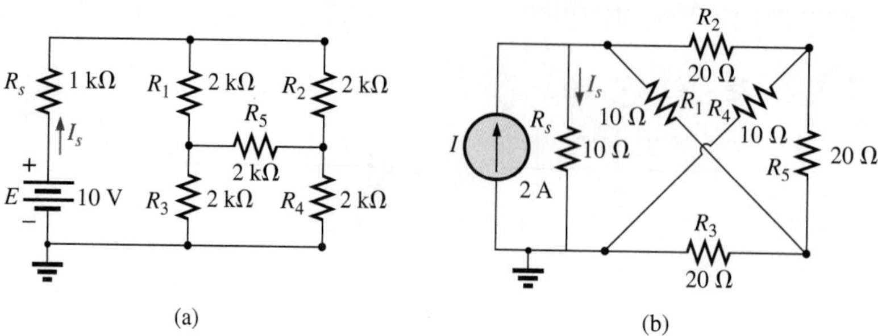

FIG. 8.133
Problem 50.

SECTION 8.12 Y-Δ (T-π) and Δ-Y (π-T) Conversions

51. Using a Δ-Y or Y-Δ conversion, find the current I in each of the networks in Fig. 8.134.

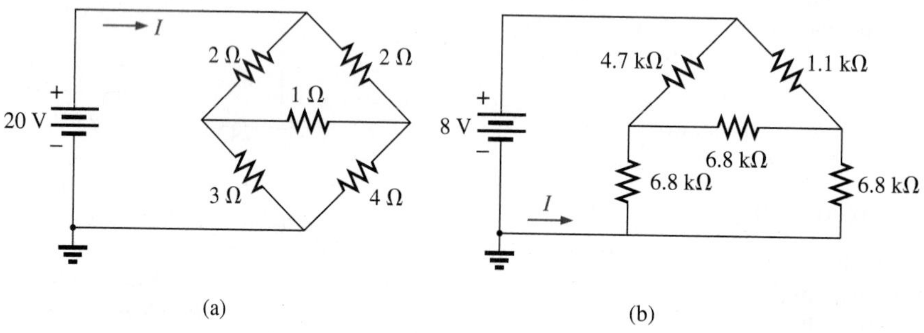

FIG. 8.134
Problem 51.

***52.** Repeat Problem 51 for the networks in Fig. 8.135.

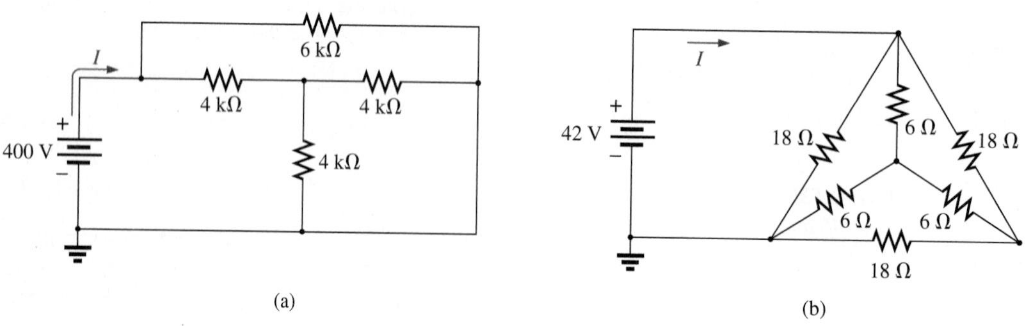

FIG. 8.135
Problem 52.

*53. Determine the current I for the network in Fig. 8.136.

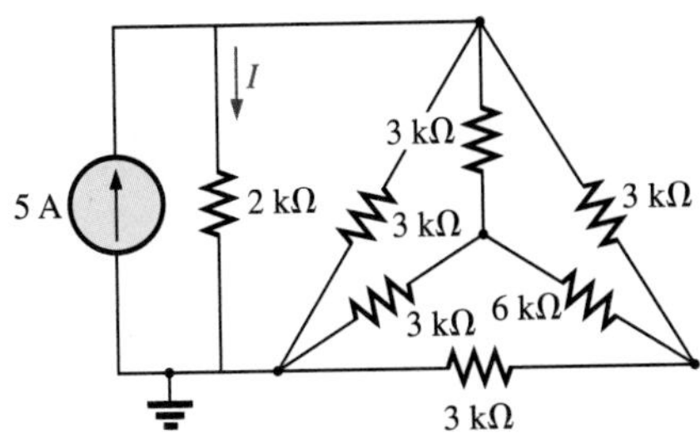

FIG. 8.136
Problem 53.

*54. **a.** Replace the T configuration in Fig. 8.137 (composed of 6 kΩ resistors) with a π configuration.

b. Solve for the source current I_{s_1}.

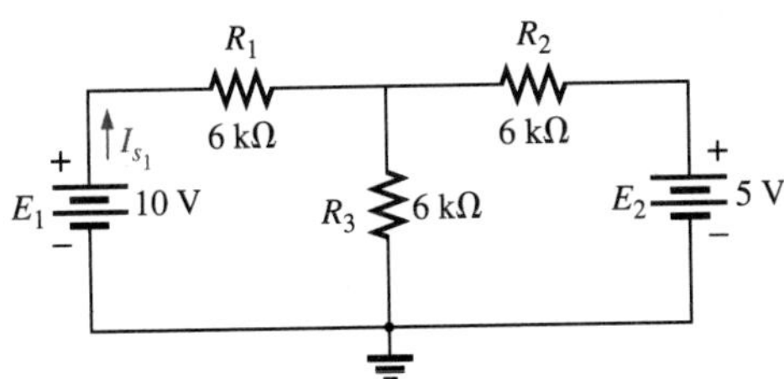

FIG. 8.137
Problem 54.

*55. **a.** Replace the π configuration in Fig. 8.138 (composed of 3 kΩ resistors) with a T configuration.

b. Solve for the source current I_s.

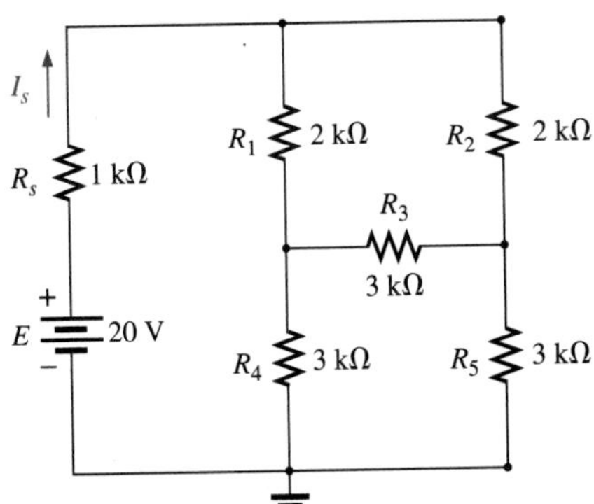

FIG. 8.138
Problem 55.

*56. Using Y-Δ or Δ-Y conversions, determine the total resistance of the network in Fig. 8.139.

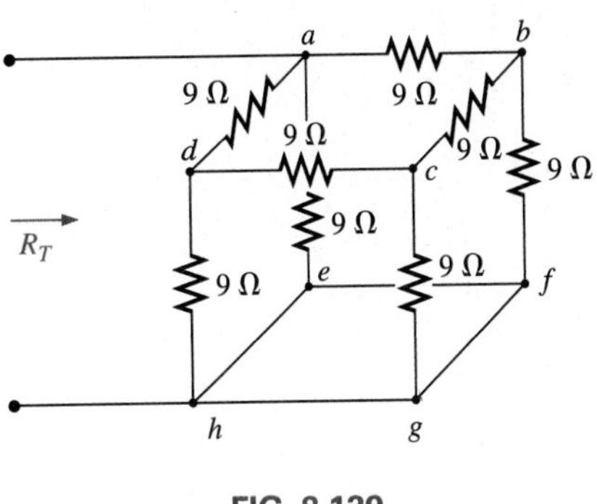

FIG. 8.139
Problem 56.

SECTION 8.14 Computer Analysis

PSpice or Multisim

57. Using schematics, find the current through each element in Fig. 8.116.

*58. Using schematics, find the mesh currents for the network in Fig. 8.123(a).

*59. Using schematics, determine the nodal voltages for the network in Fig. 8.129(II).

GLOSSARY

Branch-current method A technique for determining the branch currents of a multiloop network.

Bridge network A network configuration typically having a diamond appearance in which no two elements are in series or parallel.

Current sources Sources that supply a fixed current to a network and have a terminal voltage dependent on the network to which they are applied.

Delta (Δ), pi (π) configuration A network structure that consists of three branches and has the appearance of the Greek letter delta (Δ) or pi (π).

Determinants method A mathematical technique for finding the unknown variables of two or more simultaneous linear equations.

Mesh analysis A technique for determining the mesh (loop) currents of a network that results in a reduced set of equations compared to the branch-current method.

Mesh (loop) current A labeled current assigned to each distinct closed loop of a network that can, individually or in combination with other mesh currents, define all of the branch currents of a network.

Nodal analysis A technique for determining the nodal voltages of a network.

Node A junction of two or more branches in a network.

Supermesh current A current defined in a network with ideal current sources that permits the use of mesh analysis.

Supernode A node defined in a network with ideal voltage sources that permits the use of nodal analysis.

Wye (Y), tee (T) configuration A network structure that consists of three branches and has the appearance of the capital letter Y or T.

Network Theorems

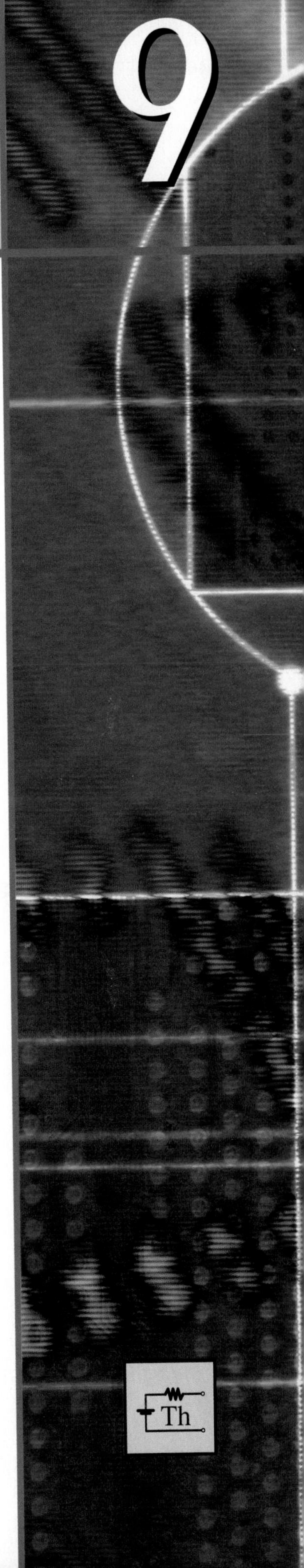

Objectives

- *Become familiar with the superposition theorem and its unique ability to separate the impact of each source on the quantity of interest.*
- *Be able to apply Thévenin's theorem to reduce any two-terminal, series-parallel network with any number of sources to a single voltage source and series resistor.*
- *Become familiar with Norton's theorem and how it can be used to reduce any two-terminal, series-parallel network with any number of sources to a single current source and a parallel resistor.*
- *Understand how to apply the maximum power transfer theorem to determine the maximum power to a load and to choose a load that will receive maximum power.*
- *Become aware of the reduction powers of Millman's theorem and the powerful implications of the substitution and reciprocity theorems.*

9.1 INTRODUCTION

This chapter introduces a number of theorems that have application throughout the field of electricity and electronics. Not only can they be used to solve networks such as encountered in the previous chapter, but they also provide an opportunity to determine the impact of a particular source or element on the response of the entire system. In most cases, the network to be analyzed and the mathematics required to find the solution are simplified. All of the theorems appear again in the analysis of ac networks. In fact, the application of each theorem to ac networks is very similar in content to that found in this chapter.

The first theorem to be introduced is the superposition theorem, followed by Thévenin's theorem, Norton's theorem, and the maximum power transfer theorem. The chapter concludes with a brief introduction to Millman's theorem and the substitution and reciprocity theorems.

9.2 SUPERPOSITION THEOREM

The **superposition theorem** is unquestionably one of the most powerful in this field. It has such widespread application that people often apply it without recognizing that their maneuvers are valid only because of this theorem.

In general, the theorem can be used to do the following:

- *Analyze networks such as introduced in the last chapter that have two or more sources that are not in series or parallel.*
- *Reveal the effect of each source on a particular quantity of interest.*
- *For sources of different types (such as dc and ac which affect the parameters of the network in a different manner), apply a separate analysis for each type, with the total result simply the algebraic sum of the results.*

The first two areas of application are described in detail in this section. The last are covered in the discussion of the superposition theorem in the ac portion of the text.

The superposition theorem states the following:

The current through, or voltage across, any element of a network is equal to the algebraic sum of the currents or voltages produced independently by each source.

In other words, this theorem allows us to find a solution for a current or voltage using *only one source at a time.* Once we have the solution for each source, we can combine the results to obtain the total solution. The term *algebraic* appears in the above theorem statement because the currents resulting from the sources of the network can have different directions, just as the resulting voltages can have opposite polarities.

If we are to consider the effects of each source, the other sources obviously must be removed. Setting a voltage source to zero volts is like placing a short circuit across its terminals. Therefore,

when removing a voltage source from a network schematic, replace it with a direct connection (short circuit) of zero ohms. Any internal resistance associated with the source must remain in the network.

Setting a current source to zero amperes is like replacing it with an open circuit. Therefore,

when removing a current source from a network schematic, replace it by an open circuit of infinite ohms. Any internal resistance associated with the source must remain in the network.

The above statements are illustrated in Fig. 9.1.

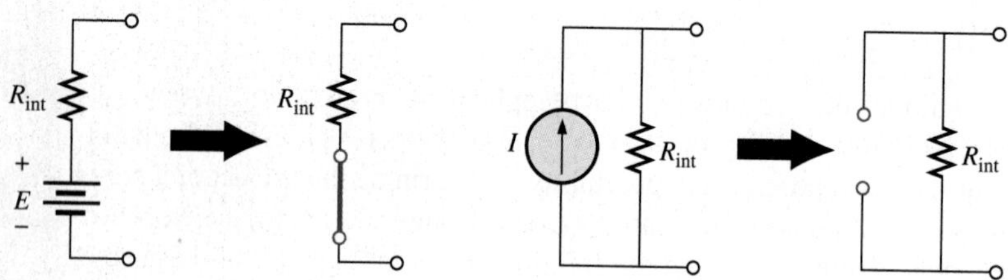

FIG. 9.1
Removing a voltage source and a current source to permit the application of the superposition theorem.

Since the effect of each source will be determined independently, the number of networks to be analyzed will equal the number of sources.

If a particular current of a network is to be determined, the contribution to that current must be determined for *each source.* When the effect of each source has been determined, those currents in the same direction are added, and those having the opposite direction are subtracted; the algebraic sum is being determined. The total result is the direction of the larger sum and the magnitude of the difference.

Similarly, if a particular voltage of a network is to be determined, the contribution to that voltage must be determined for each source. When the effect of each source has been determined, those voltages with the same polarity are added, and those with the opposite polarity are subtracted; the algebraic sum is being determined. The total result has the polarity of the larger sum and the magnitude of the difference.

Superposition cannot be applied to power effects because the power is related to the square of the voltage across a resistor or the current through a resistor. The squared term results in a nonlinear (a curve, not a straight line) relationship between the power and the determining current or voltage. For example, doubling the current through a resistor does not double the power to the resistor (as defined by a linear relationship) but, in fact, increases it by a factor of 4 (due to the squared term). Tripling the current increases the power level by a factor of 9. Example 9.3 demonstrates the differences between a linear and a nonlinear relationship.

A few examples clarify how sources are removed and total solutions obtained.

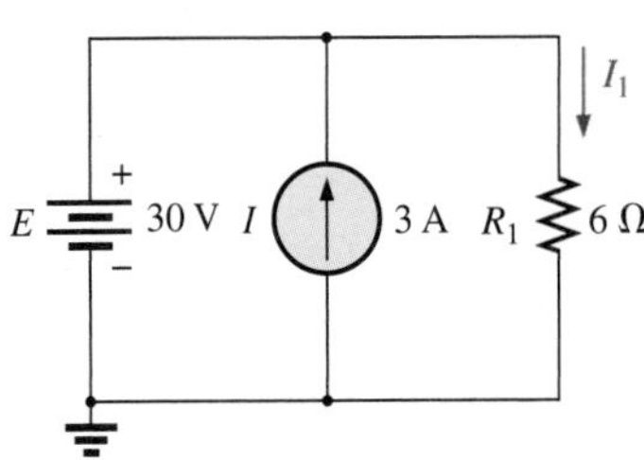

FIG. 9.2
Two-source network to be analyzed using the superposition theorem in Example 9.1.

EXAMPLE 9.1 Using the superposition theorem, determine current I_1 for the network in Fig. 9.2.

Solution: Since two sources are present, there are two networks to be analyzed. First let us determine the effects of the voltage source by setting the current source to zero amperes as shown in Fig. 9.3. Note that the resulting current is defined as I'_1 because it is the current through resistor R_1 due to the voltage source only.

Due to the open circuit, resistor R_1 is in series (and, in fact, in parallel) with the voltage source E. The voltage across the resistor is the applied voltage, and current I'_1 is determined by

$$I'_1 = \frac{V_1}{R_1} = \frac{E}{R_1} = \frac{30\text{ V}}{6\ \Omega} = 5\text{ A}$$

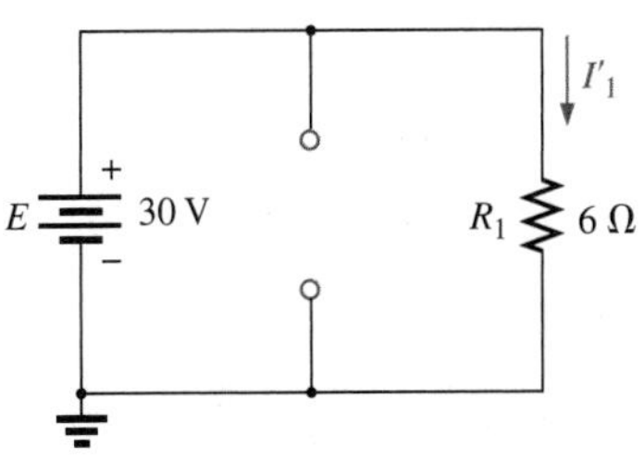

FIG. 9.3
Determining the effect of the 30 V supply on the current I_1 in Fig. 9.2.

Now for the contribution due to the current source. Setting the voltage source to zero volts results in the network in Fig. 9.4, which presents us with an interesting situation. The current source has been replaced with a short-circuit equivalent that is directly across the current source and resistor R_1. Since the source current takes the path of least resistance, it chooses the zero ohm path of the inserted short-circuit equivalent, and the current through R_1 is zero amperes. This is clearly demonstrated by an application of the current divider rule as follows:

$$I''_1 = \frac{R_{sc}I}{R_{sc} + R_1} = \frac{(0\ \Omega)I}{0\ \Omega + 6\ \Omega} = 0\text{ A}$$

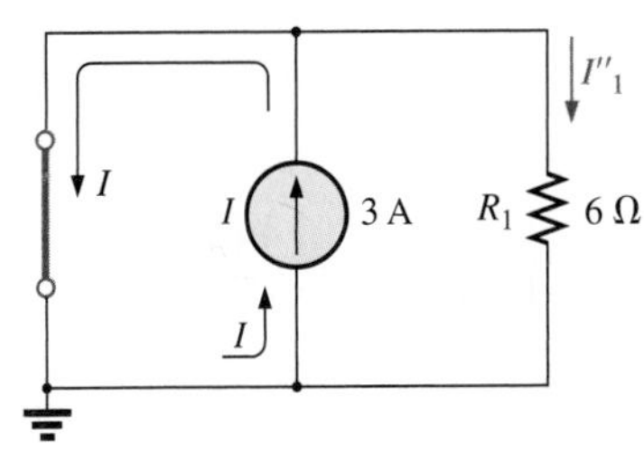

FIG. 9.4
Determining the effect of the 3 A current source on the current I_1 in Fig. 9.2.

Since I'_1 and I''_1 have the same defined direction in Figs. 9.3 and 9.4, the total current is defined by

$$I_1 = I'_1 + I''_1 = 5\text{ A} + 0\text{ A} = \mathbf{5\ A}$$

Although this has been an excellent introduction to the application of the superposition theorem, it should be immediately clear in Fig. 9.2 that the voltage source is in parallel with the current source and load resistor R_1, so the voltage across each must be 30 V. The result is that I_1 must be determined solely by

$$I_1 = \frac{V_1}{R_1} = \frac{E}{R_1} = \frac{30\text{ V}}{6\ \Omega} = \mathbf{5\ A}$$

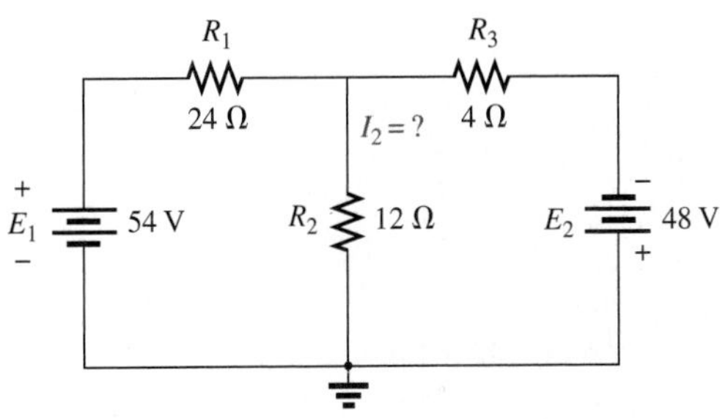

FIG. 9.5
Using the superposition theorem to determine the current through the 12 Ω resistor (Example 9.2).

EXAMPLE 9.2 Using the superposition theorem, determine the current through the 12 Ω resistor in Fig. 9.5. Note that this is a two-source network of the type examined in the previous chapter when we applied branch-current analysis and mesh analysis.

Solution: Considering the effects of the 54 V source requires replacing the 48 V source by a short-circuit equivalent as shown in Fig. 9.6. The result is that the 12 Ω and 4 Ω resistors are in parallel.

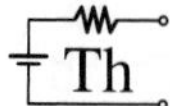

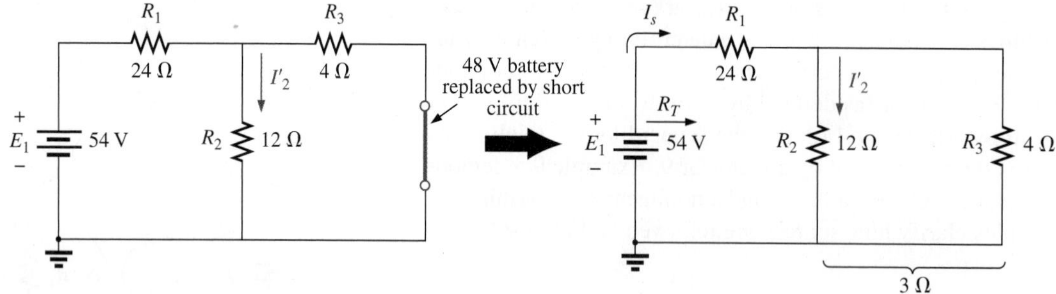

FIG. 9.6
Using the superposition theorem to determine the effect of the 54 V voltage source on current I_2 in Fig. 9.5.

The total resistance seen by the source is therefore

$$R_T = R_1 + R_2 \| R_3 = 24\ \Omega + 12\ \Omega \| 4\ \Omega = 24\ \Omega + 3\ \Omega = 27\ \Omega$$

and the source current is

$$I_s = \frac{E_1}{R_T} = \frac{54\text{ V}}{27\ \Omega} = 2\text{ A}$$

Using the current divider rule results in the contribution to I_2 due to the 54 V source:

$$I'_2 = \frac{R_3 I_s}{R_3 + R_2} = \frac{(4\ \Omega)(2\text{ A})}{4\ \Omega + 12\ \Omega} = 0.5\text{ A}$$

If we now replace the 54 V source by a short-circuit equivalent, the network in Fig. 9.7 results. The result is a parallel connection for the 12 Ω and 24 Ω resistors.

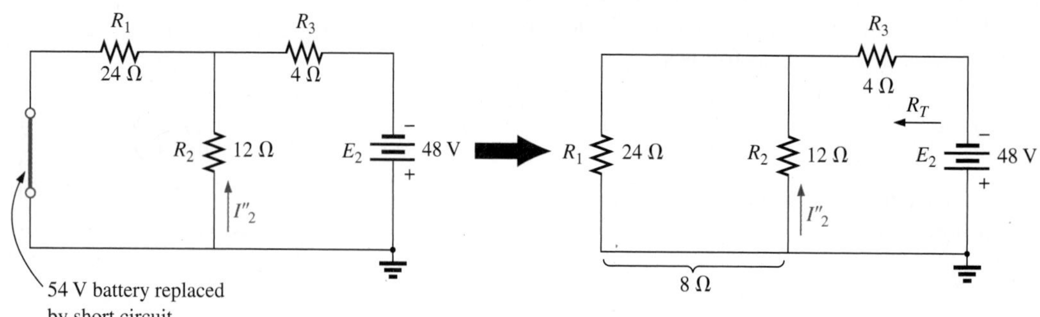

FIG. 9.7
Using the superposition theorem to determine the effect of the 48 V voltage source on current I_2 in Fig. 9.5.

Therefore, the total resistance seen by the 48 V source is

$$R_T = R_3 + R_2 \| R_1 = 4\ \Omega + 12\ \Omega \| 24\ \Omega = 4\ \Omega + 8\ \Omega = 12\ \Omega$$

and the source current is

$$I_s = \frac{E_2}{R_T} = \frac{48\text{ V}}{12\ \Omega} = 4\text{ A}$$

Applying the current divider rule results in

$$I''_2 = \frac{R_1(I_s)}{R_1 + R_2} = \frac{(24\ \Omega)(4\text{ A})}{24\ \Omega + 12\ \Omega} = 2.67\text{ A}$$

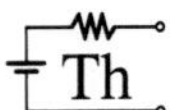

It is now important to realize that current I_2 due to each source has a different direction, as shown in Fig. 9.8. The net current therefore is the difference of the two and the direction of the larger as follows:

$$I_2 = I''_2 - I'_2 = 2.67\text{ A} - 0.5\text{ A} = \mathbf{2.17\text{ A}}$$

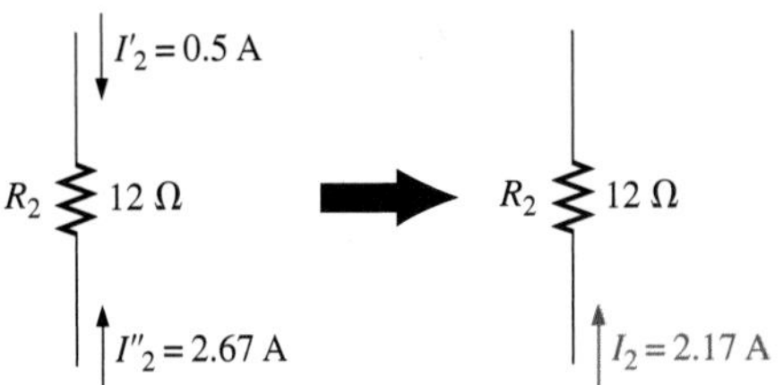

FIG. 9.8
Using the results of Figs. 9.6 and 9.7 to determine current I_2 for the network in Fig. 9.5.

Using Figs. 9.6 and 9.7 in Example 9.2, we can determine the other currents of the network with little added effort. That is, we can determine all the branch currents of the network, matching an application of the branch-current analysis or mesh analysis approach. In general, therefore, not only can the superposition theorem provide a complete solution for the network, but it also reveals the effect of each source on the desired quantity.

EXAMPLE 9.3

a. Using the superposition theorem, determine the current through resistor R_2 for the network in Fig. 9.9.
b. Demonstrate that the superposition theorem is not applicable to power levels.

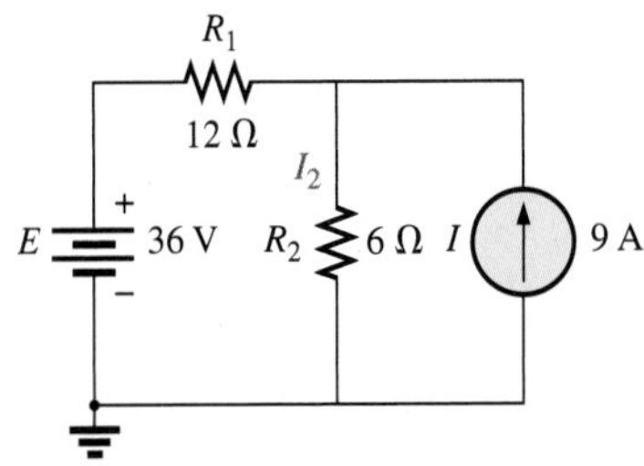

FIG. 9.9
Network to be analyzed in Example 9.3 using the superposition theorem.

Solutions:

a. In order to determine the effect of the 36 V voltage source, the current source must be replaced by an open-circuit equivalent as shown in Fig. 9.10. The result is a simple series circuit with a current equal to

$$I'_2 = \frac{E}{R_T} = \frac{E}{R_1 + R_2} = \frac{36\text{ V}}{12\ \Omega + 6\ \Omega} = \frac{36\text{ V}}{18\ \Omega} = 2\text{ A}$$

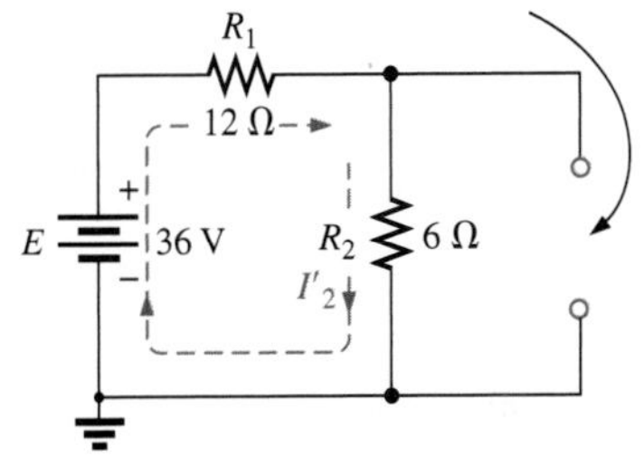

FIG. 9.10
Replacing the 9 A current source in Fig. 9.9 by an open circuit to determine the effect of the 36 V voltage source on current I_2.

Examining the effect of the 9 A current source requires replacing the 36 V voltage source by a short-circuit equivalent as shown in Fig. 9.11. The result is a parallel combination of resistors R_1 and R_2. Applying the current divider rule results in

$$I''_2 = \frac{R_1(I)}{R_1 + R_2} = \frac{(12\ \Omega)(9\text{ A})}{12\ \Omega + 6\ \Omega} = 6\text{ A}$$

Since the contribution to current I_2 has the same direction for each source, as shown in Fig. 9.12, the total solution for current I_2 is the sum of the currents established by the two sources. That is,

$$I_2 = I'_2 + I''_2 = 2\text{ A} + 6\text{ A} = \mathbf{8\text{ A}}$$

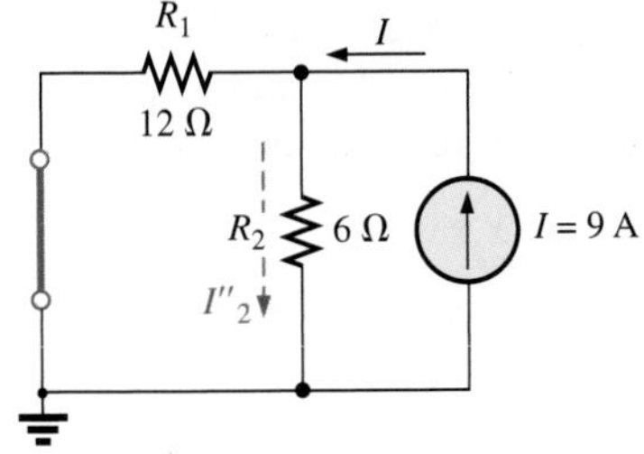

FIG. 9.11
Replacing the 36 V voltage source by a short-circuit equivalent to determine the effect of the 9 A current source on current I_2.

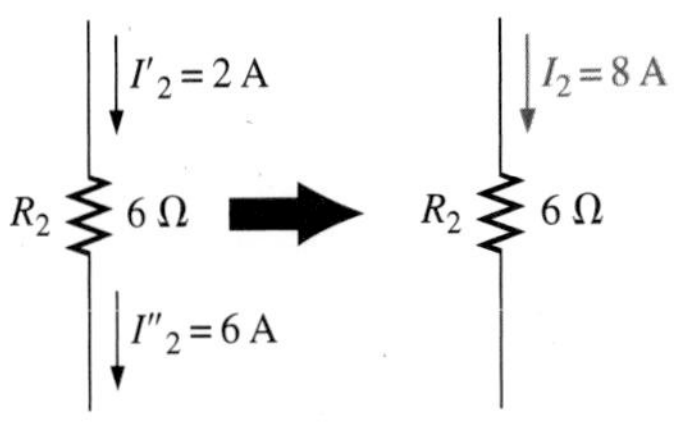

FIG. 9.12
Using the results of Figs. 9.10 and 9.11 to determine current I_2 for the network in Fig. 9.9.

b. Using Fig. 9.10 and the results obtained, the power delivered to the 6 Ω resistor is

$$P_1 = (I'_2)^2(R_2) = (2\text{ A})^2(6\ \Omega) = \mathbf{24\ W}$$

Using Fig. 9.11 and the results obtained, the power delivered to the 6 Ω resistor is

$$P_2 = (I''_2)^2(R_2) = (6\text{ A})^2(6\ \Omega) = \mathbf{216\ W}$$

Using the total results of Fig. 9.12, the power delivered to the 6 Ω resistor is

$$P_T = I_2^2R_2 = (8\text{ A})^2(6\ \Omega) = \mathbf{384\ W}$$

It is now quite clear that the power delivered to the 6 Ω resistor using the total current of 8 A is not equal to the sum of the power levels due to each source independently. That is,

$$P_1 + P_2 = 24\text{ W} + 216\text{ W} = 240\text{ W} \neq P_T = 348\text{ W}$$

To expand on the above conclusion and further demonstrate what is meant by a *nonlinear relationship,* the power to the 6 Ω resistor versus current through the 6 Ω resistor is plotted in Fig. 9.13. Note that the curve is not a straight line but one whose rise gets steeper with increase in current level.

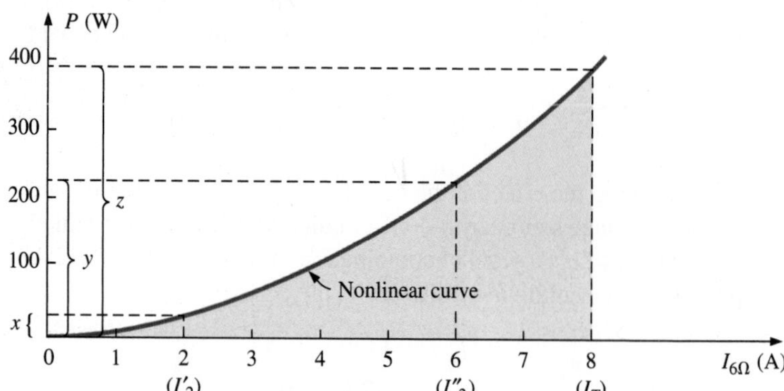

FIG. 9.13
Plotting power delivered to the 6 Ω resistor versus current through the resistor.

Recall from Fig. 9.11 that the power level was 24 W for a current of 2 A developed by the 36 V voltage source, shown in Fig. 9.13. From Fig. 9.12, we found that the current level was 6 A for a power level of 216 W, shown in Fig. 9.13. Using the total current of 8 A, we find that the power level in 384 W, shown in Fig. 9.13. Quite clearly, the sum of power levels due to the 2 A and 6 A current levels does not equal that due to the 8 A level. That is,

$$x + y \neq z$$

Now, the relationship between the voltage across a resistor and the current through a resistor is a linear (straight line) one as shown in Fig. 9.14, with

$$c = a + b$$

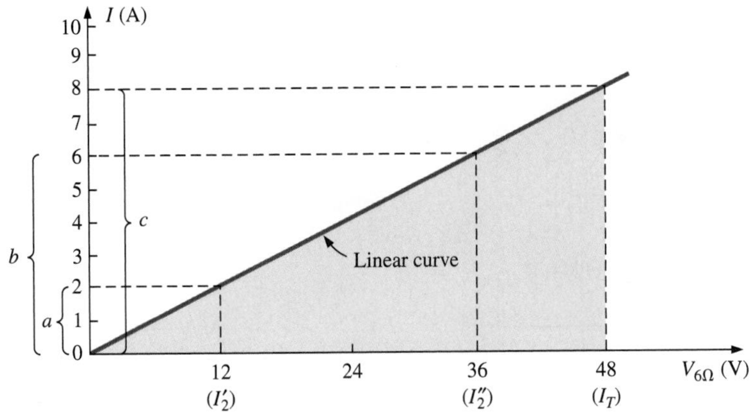

FIG. 9.14
Plotting I versus V for the 6 Ω resistor.

EXAMPLE 9.4 Using the principle of superposition, find the current I_2 through the 12 kΩ resistor in Fig. 9.15.

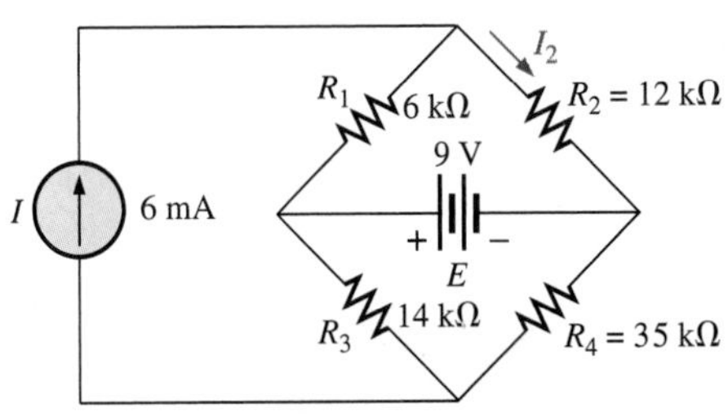

FIG. 9.15
Example 9.4.

Solution: Considering the effect of the 6 mA current source (Fig. 9.16):

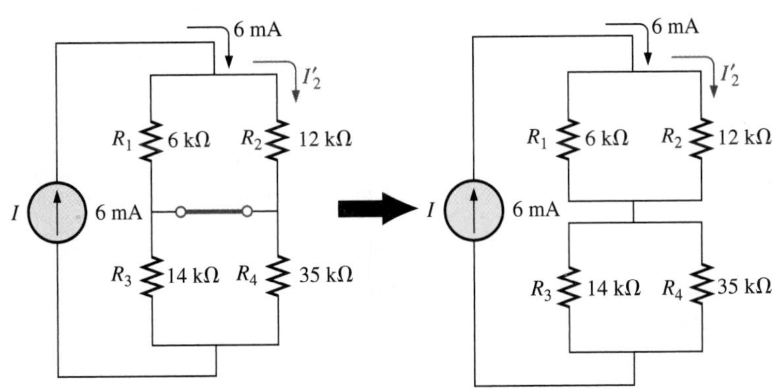

FIG. 9.16
The effect of the current source I on the current I_2.

Current divider rule:

$$I'_2 = \frac{R_1 I}{R_1 + R_2} = \frac{(6\text{ k}\Omega)(6\text{ mA})}{6\text{ k}\Omega + 12\text{ k}\Omega} = 2\text{ mA}$$

Considering the effect of the 9 V voltage source (Fig 9.17):

$$I''_2 = \frac{E}{R_1 + R_2} = \frac{9\text{ V}}{6\text{ k}\Omega + 12\text{ k}\Omega} = 0.5\text{ mA}$$

Since I'_2 and I''_2 have the same direction through R_2, the desired current is the sum of the two:

$$\begin{aligned} I_2 &= I'_2 + I''_2 \\ &= 2\text{ mA} + 0.5\text{ mA} \\ &= \mathbf{2.5\text{ mA}} \end{aligned}$$

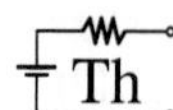

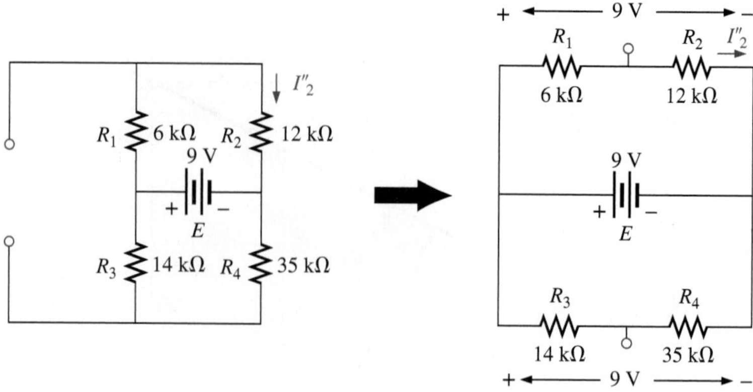

FIG. 9.17
The effect of the voltage source E on the current I_2.

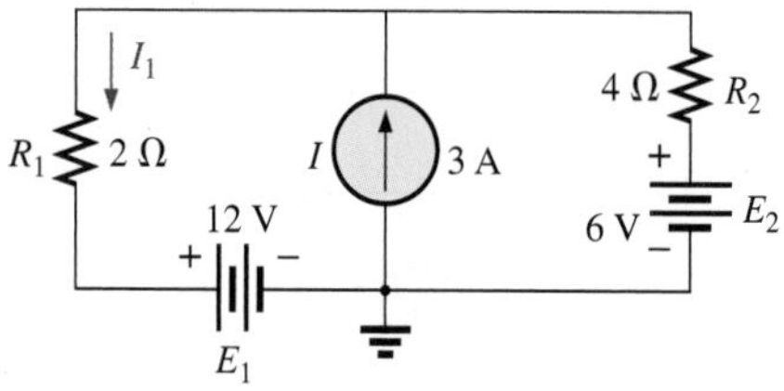

FIG. 9.18
Example 9.5.

EXAMPLE 9.5 Find the current through the 2 Ω resistor of the network in Fig. 9.18. The presence of three sources results in three different networks to be analyzed.

Solution: Considering the effect of the 12 V source (Fig. 9.19):

$$I'_1 = \frac{E_1}{R_1 + R_2} = \frac{12\text{ V}}{2\ \Omega + 4\ \Omega} = \frac{12\text{ V}}{6\ \Omega} = 2\text{ A}$$

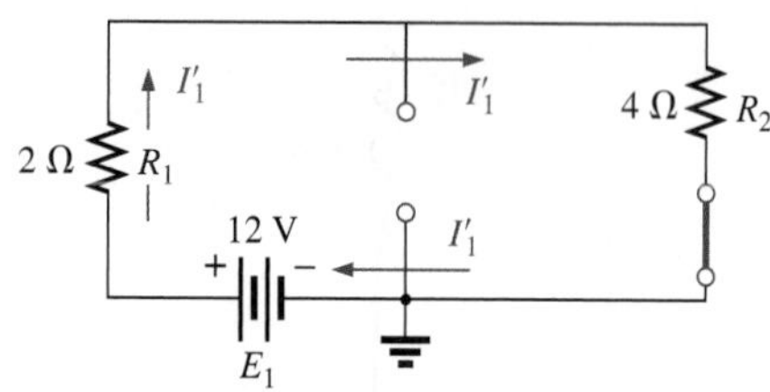

FIG. 9.19
The effect of E_1 on the current I.

Considering the effect of the 6 V source (Fig. 9.20):

$$I''_1 = \frac{E_2}{R_1 + R_2} = \frac{6\text{ V}}{2\ \Omega + 4\ \Omega} = \frac{6\text{ V}}{6\ \Omega} = 1\text{ A}$$

Considering the effect of the 3 A source (Fig. 9.21):
Applying the current divider rule,

$$I'''_1 = \frac{R_2 I}{R_1 + R_2} = \frac{(4\ \Omega)(3\text{ A})}{2\ \Omega + 4\ \Omega} = \frac{12\text{ A}}{6} = 2\text{ A}$$

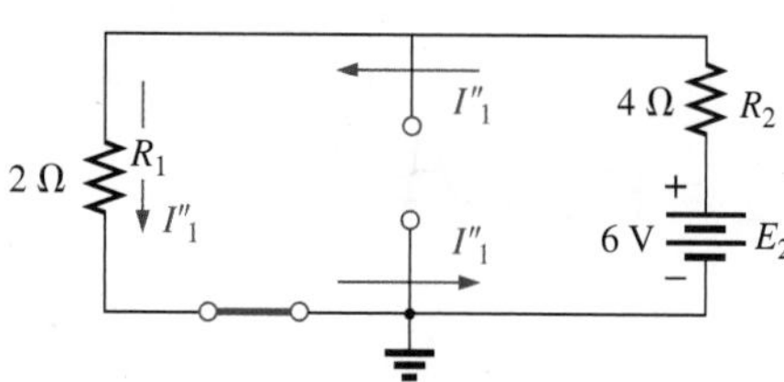

FIG. 9.20
The effect of E_2 on the current I_1.

The total current through the 2 Ω resistor appears in Fig. 9.22 and

Same direction as I_1 in Fig. 9.18 — Opposite direction to I_1 in Fig. 9.18

$$I_1 = \underbrace{I''_1 + I'''_1}_{} - I'_1$$
$$= 1\text{ A} + 2\text{ A} - 2\text{ A} = \mathbf{1\ A}$$

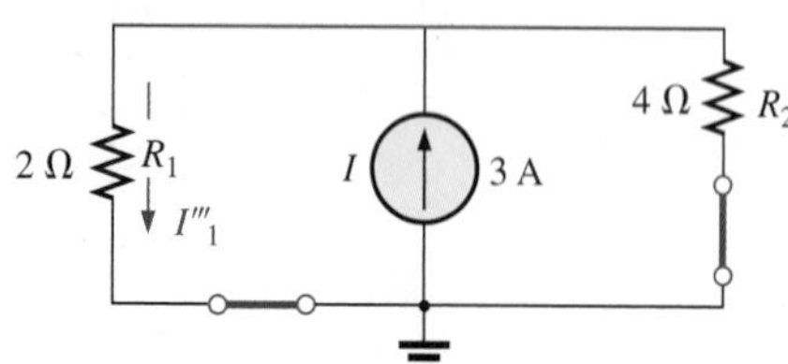

FIG. 9.21
The effect of I on the current I_1.

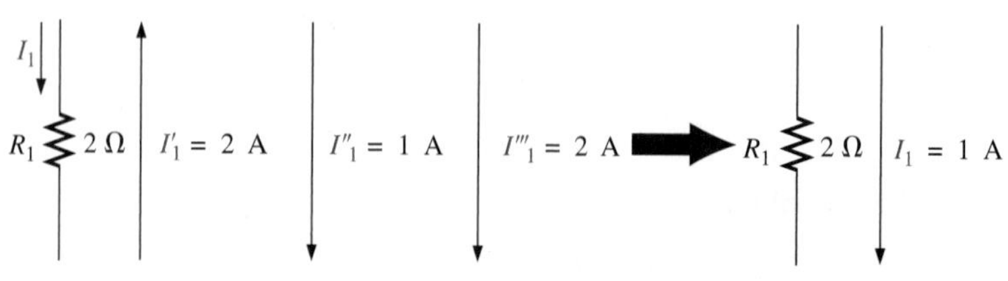

FIG. 9.22
The resultant current I_1.

9.3 THÉVENIN'S THEOREM

The next theorem to be introduced, **Thévenin's theorem,** is probably one of the most interesting in that it permits the reduction of complex networks to a simpler form for analysis and design.

In general, the theorem can be used to do the following:

- ***Analyze networks with sources that are not in series or parallel.***
- ***Reduce the number of components required to establish the same characteristics at the output terminals.***
- ***Investigate the effect of changing a particular component on the behavior of a network without having to analyze the entire network after each change.***

All three areas of application are demonstrated in the examples to follow.

Thévenin's theorem states the following:

Any two-terminal dc network can be replaced by an equivalent circuit consisting solely of a voltage source and a series resistor as shown in Fig. 9.23.

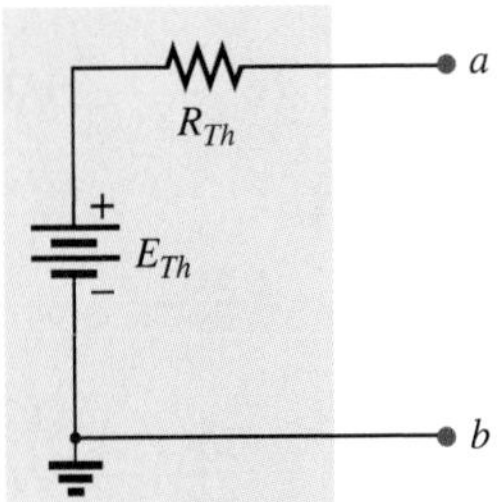

FIG. 9.23
Thévenin equivalent circuit.

The theorem was developed by Commandant Leon-Charles Thévenin in 1883 as described in Fig. 9.24.

To demonstrate the power of the theorem, consider the fairly complex network of Fig. 9.25(a) with its two sources and series-parallel connections. The theorem states that the entire network inside the blue shaded area can be replaced by one voltage source and one resistor as shown in Fig. 9.25(b). If the replacement is done properly, the voltage across, and the current through, the resistor R_L will be the same for each network. The value of R_L can be changed to any value, and the voltage, current, or power to the load resistor is the same for each configuration. Now, this is a very powerful statement—one that is verified in the examples to follow.

The question then is, How can you determine the proper value of Thévenin voltage and resistance? In general, finding the Thévenin *resistance* value is quite straightforward. Finding the Thévenin *voltage* can be more of a challenge and, in fact, may require using the superposition theorem or one of the methods described in Chapter 8.

Fortunately, there are a series of steps that will lead to the proper value of each parameter. Although a few of the steps may seem trivial at first, they can become quite important when the network becomes complex.

Thévenin's Theorem Procedure

Preliminary:

1. ***Remove that portion of the network where the Thévenin equivalent circuit is found. In Fig. 9.25(a), this requires that the load resistor R_L be temporarily removed from the network.***
2. ***Mark the terminals of the remaining two-terminal network. (The importance of this step will become obvious as we progress through some complex networks.)***

R_{Th}:

3. ***Calculate R_{Th} by first setting all sources to zero (voltage sources are replaced by short circuits, and current sources by open circuits) and then finding the resultant resistance between the two marked terminals. (If the internal resistance of the voltage and/or***

FIG. 9.24
Leon-Charles Thévenin.
Courtesy of the Bibliothèque École Polytechnique, Paris, France.

French (Meaux, Paris)
(1857–1927)
Telegraph Engineer, Commandant and Educator
École Polytechnique and École Supérieure de Télégraphie

Although active in the study and design of telegraphic systems (including underground transmission), cylindrical condensers (capacitors), and electromagnetism, he is best known for a theorem first presented in the French *Journal of Physics—Theory and Applications* in 1883. It appeared under the heading of "Sur un nouveau théorème d'électricité dynamique" ("On a new theorem of dynamic electricity") and was originally referred to as the *equivalent generator theorem.* There is some evidence that a similar theorem was introduced by Hermann von Helmholtz in 1853. However, Professor Helmholtz applied the theorem to animal physiology and not to communication or generator systems, and therefore he has not received the credit in this field that he might deserve. In the early 1920s AT&T did some pioneering work using the equivalent circuit and may have initiated the reference to the theorem as simply Thévenin's theorem. In fact, Edward L. Norton, an engineer at AT&T at the time, introduced a current source equivalent of the Thévenin equivalent currently referred to as the Norton equivalent circuit. As an aside, Commandant Thévenin was an avid skier and in fact was commissioner of an international ski competition in Chamonix. France, in 1912.

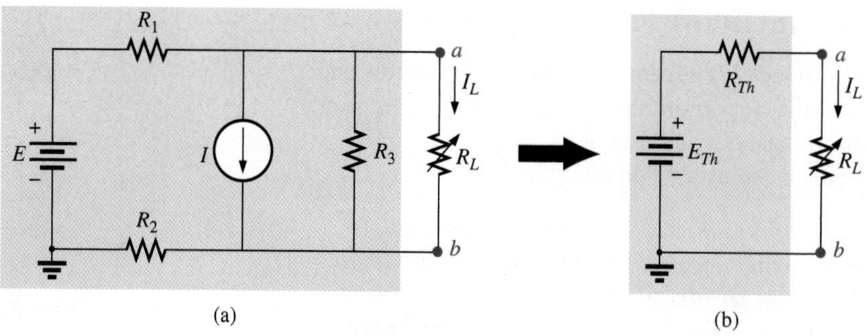

FIG. 9.25
Substituting the Thévenin equivalent circuit for a complex network.

current sources is included in the original network, it must remain when the sources are set to zero.)

E_{Th}:

4. ***Calculate E_{Th} by first returning all sources to their original position and finding the open-circuit voltage between the marked terminals. (This step is invariably the one that causes most confusion and errors. In all cases, keep in mind that it is the open-circuit potential between the two terminals marked in step 2.)***

Conclusion:

5. ***Draw the Thévenin equivalent circuit with the portion of the circuit previously removed replaced between the terminals of the equivalent circuit. This step is indicated by the placement of the resistor R_L between the terminals of the Thévenin equivalent circuit as shown in Fig. 9.25(b).***

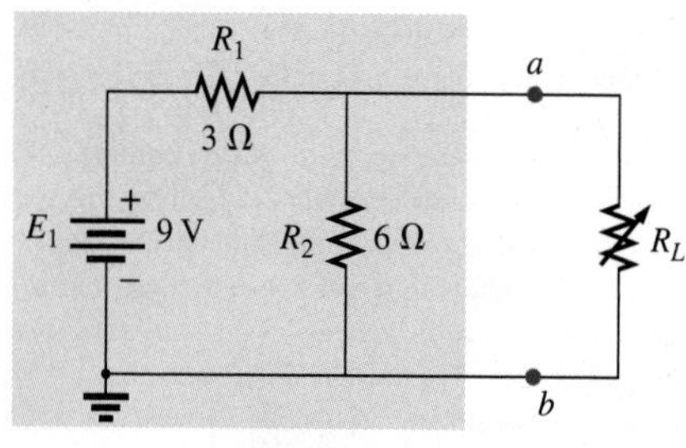

FIG. 9.26
Example 9.6.

EXAMPLE 9.6 Find the Thévenin equivalent circuit for the network in the shaded area of the network in Fig. 9.26. Then find the current through R_L for values of 2 Ω, 10 Ω, and 100 Ω.

Solution:

Steps 1 and 2: These produce the network in Fig. 9.27. Note that the load resistor R_L has been removed and the two "holding" terminals have been defined as *a* and *b*.

Steps 3: Replacing the voltage source E_1 with a short-circuit equivalent yields the network in Fig. 9.28(a), where

$$R_{Th} = R_1 \| R_2 = \frac{(3\ \Omega)(6\ \Omega)}{3\ \Omega + 6\ \Omega} = \mathbf{2\ \Omega}$$

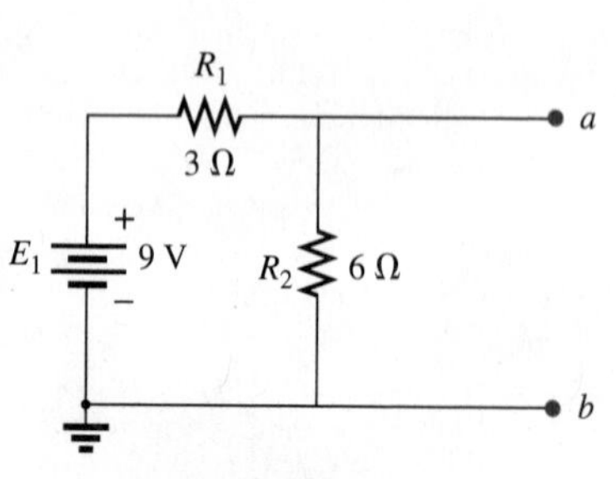

FIG. 9.27
Identifying the terminals of particular importance when applying Thévenin's theorem.

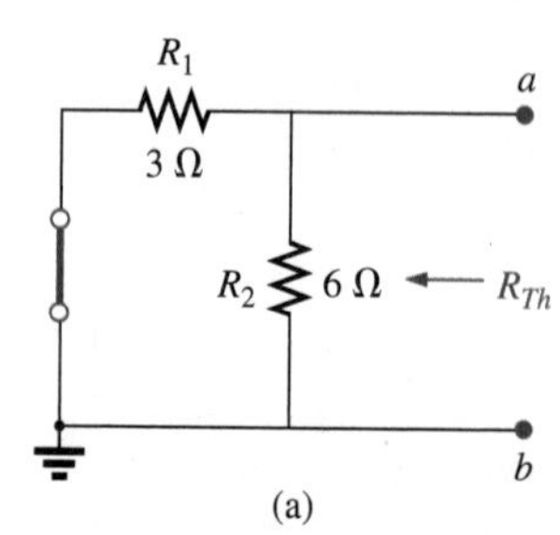

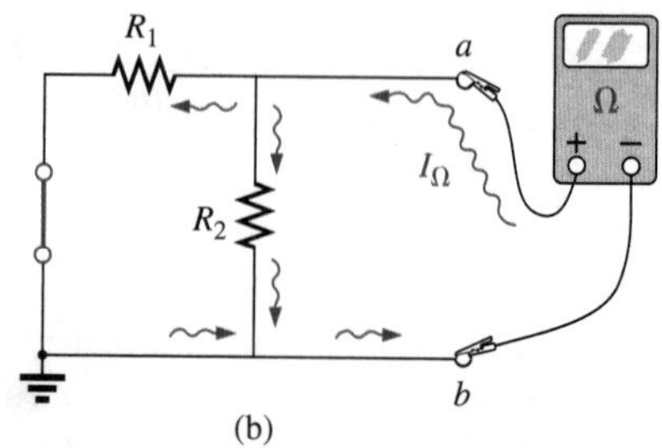

FIG. 9.28
Determining R_{Th} for the network in Fig. 9.27.

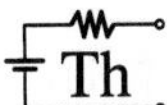

The importance of the two marked terminals now begins to surface. They are the two terminals across which the Thévenin resistance is measured. It is no longer the total resistance as seen by the source, as determined in the majority of problems of Chapter 7. If some difficulty develops when determining R_{Th} with regard to whether the resistive elements are in series or parallel, consider recalling that the ohmmeter sends out a trickle current into a resistive combination and senses the level of the resulting voltage to establish the measured resistance level. In Fig. 9.28(b), the trickle current of the ohmmeter approaches the network through terminal *a*, and when it reaches the junction of R_1 and R_2, it splits as shown. The fact that the trickle current splits and then recombines at the lower node reveals that the resistors are in parallel as far as the ohmmeter reading is concerned. In essence, the path of the sensing current of the ohmmeter has revealed how the resistors are connected to the two terminals of interest and how the Thévenin resistance should be determined. Remember this as you work through the various examples in this section.

Step 4: Replace the voltage source (Fig. 9.29). For this case, the open-circuit voltage E_{Th} is the same as the voltage drop across the 6 Ω resistor. Applying the voltage divider rule,

$$E_{Th} = \frac{R_2E_1}{R_2 + R_1} = \frac{(6\,\Omega)(9\,\text{V})}{6\,\Omega + 3\,\Omega} = \frac{54\,\text{V}}{9} = \mathbf{6\,V}$$

It is particularly important to recognize that E_{Th} is the open-circuit potential between points *a* and *b*. Remember that an open circuit can have any voltage across it, but the current must be zero. In fact, the current through any element in series with the open circuit must be zero also. The use of a voltmeter to measure E_{Th} appears in Fig. 9.30. Note that it is placed directly across the resistor R_2 since E_{Th} and V_{R_2} are in parallel.

Step 5 (Fig. 9.31):

$$I_L = \frac{E_{Th}}{R_{Th} + R_L}$$

$$R_L = 2\,\Omega: \quad I_L = \frac{6\,\text{V}}{2\,\Omega + 2\,\Omega} = \mathbf{1.5\,A}$$

$$R_L = 10\,\Omega: \quad I_L = \frac{6\,\text{V}}{2\,\Omega + 10\,\Omega} = \mathbf{0.5\,A}$$

$$R_L = 100\,\Omega: \quad I_L = \frac{6\,\text{V}}{2\,\Omega + 100\,\Omega} = \mathbf{0.06\,A}$$

If Thévenin's theorem were unavailable, each change in R_L would require that the entire network in Fig. 9.26 be reexamined to find the new value of R_L.

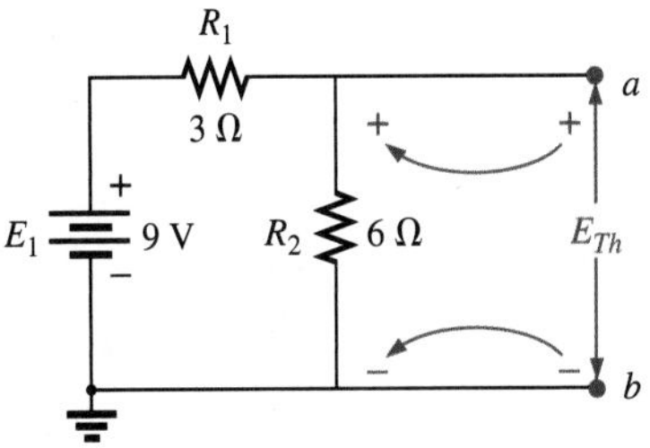

FIG. 9.29
Determining E_{Th} for the network in Fig. 9.27.

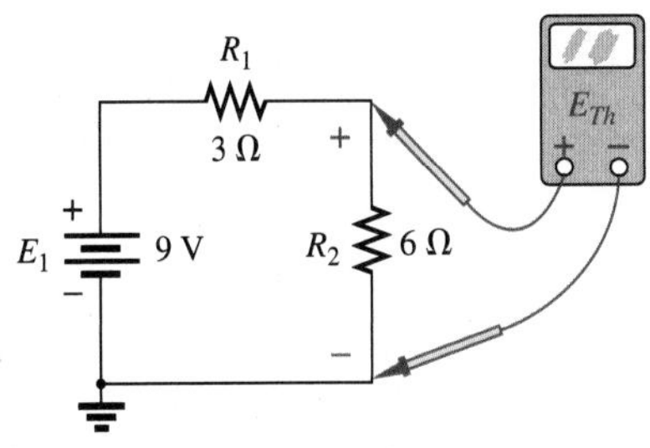

FIG. 9.30
Measuring E_{Th} for the network in Fig. 9.27.

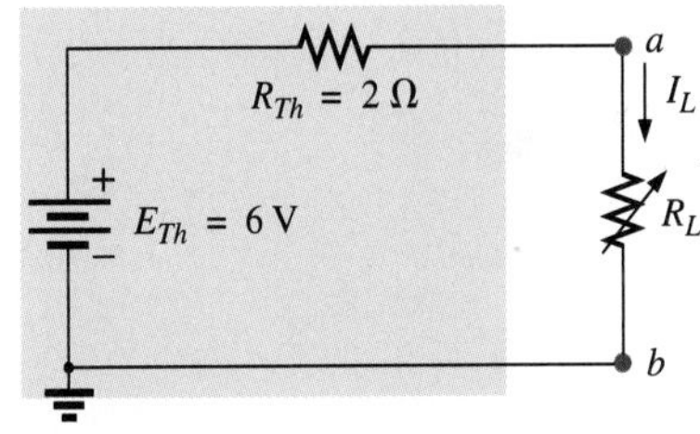

FIG. 9.31
Substituting the Thévenin equivalent circuit for the network external to R_L in Fig. 9.26.

EXAMPLE 9.7 Find the Thévenin equivalent circuit for the network in the shaded area of the network in Fig. 9.32.

Solution:

Steps 1 and 2: See Fig. 9.33.

Step 3: See Fig. 9.34. The current source has been replaced with an open-circuit equivalent, and the resistance determined between terminals *a* and *b*.

In this case, an ohmmeter connected between terminals *a* and *b* sends out a sensing current that flows directly through R_1 and R_2 (at the

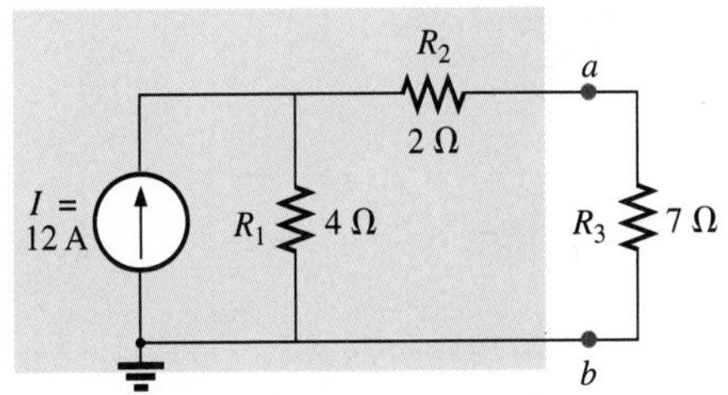

FIG. 9.32
Example 9.7.

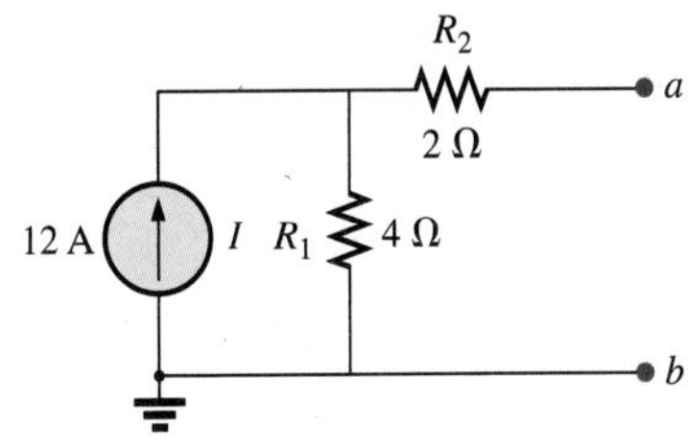

FIG. 9.33
Establishing the terminals of particular interest for the network in Fig. 9.32.

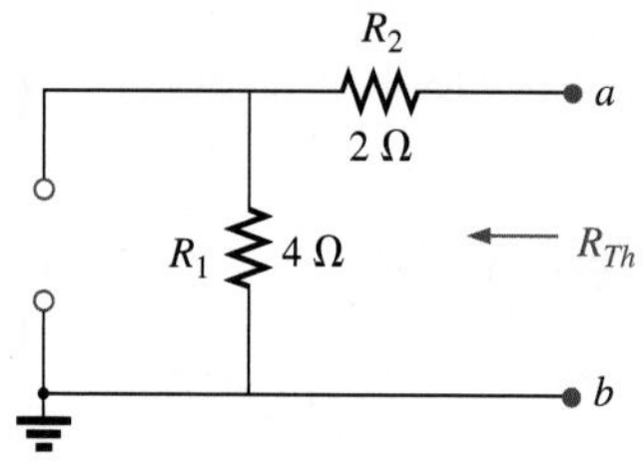

FIG. 9.34
Determining R_{Th} for the network in Fig. 9.33.

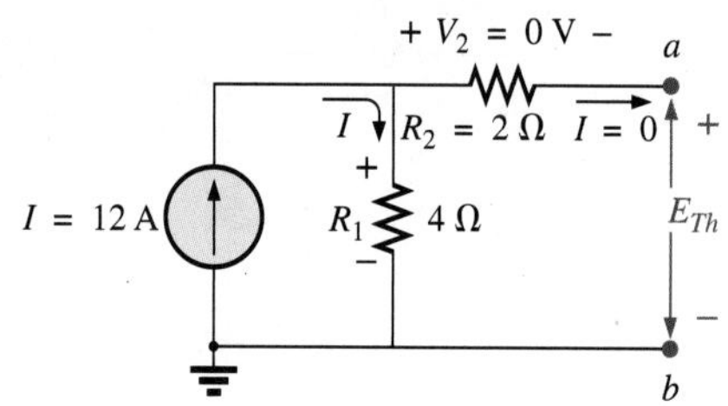

FIG. 9.35
Determining E_{Th} for the network in Fig. 9.33.

same level). The result is that R_1 and R_2 are in series and the Thévenin resistance is the sum of the two.

$$R_{Th} = R_1 + R_2 = 4\,\Omega + 2\,\Omega = \mathbf{6\,\Omega}$$

Step 4: See Fig. 9.35. In this case, since an open circuit exists between the two marked terminals, the current is zero between these terminals and through the 2 Ω resistor. The voltage drop across R_2 is, therefore,

$$V_2 = I_2R_2 = (0)R_2 = 0\text{ V}$$

and $$E_{Th} = V_1 = I_1R_1 = IR_1 = (12\text{ A})(4\,\Omega) = \mathbf{48\text{ V}}$$

Step 5: See Fig. 9.36.

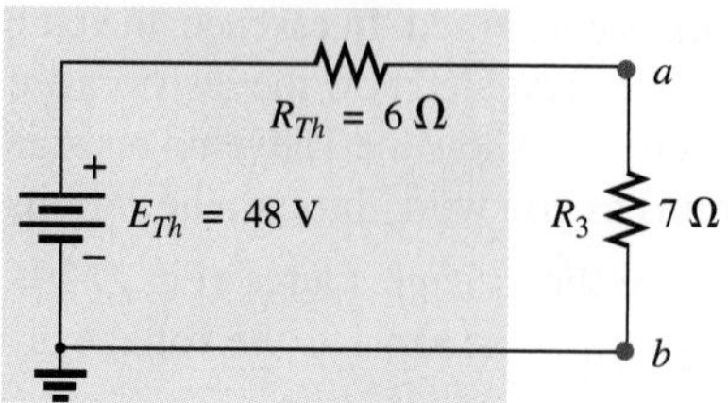

FIG. 9.36
Substituting the Thévenin equivalent circuit in the network external to the resistor R_3 in Fig. 9.32.

EXAMPLE 9.8 Find the Thévenin equivalent circuit for the network in the shaded area of the network in Fig. 9.37. Note in this example that there is no need for the section of the network to be preserved to be at the "end" of the configuration.

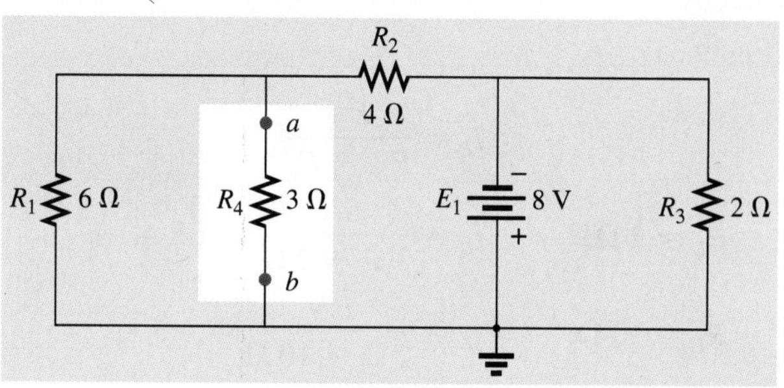

FIG. 9.37
Example 9.8.

Solution:

Steps 1 and 2: See Fig. 9.38.

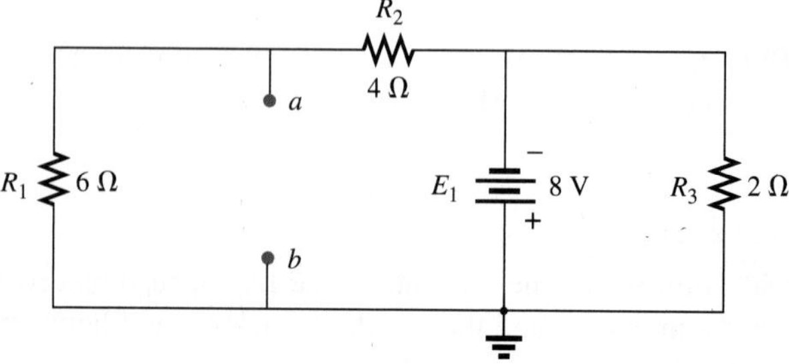

FIG. 9.38
Identifying the terminals of particular interest for the network in Fig. 9.37.

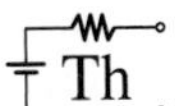

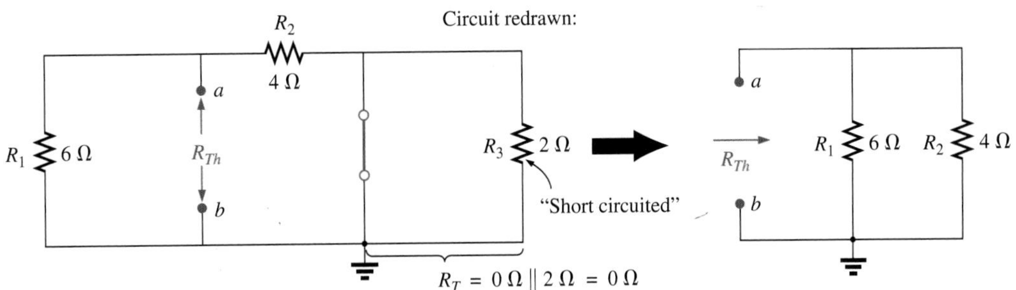

FIG. 9.39
Determining R_{Th} for the network in Fig. 9.38.

Step 3: See Fig. 9.39. Steps 1 and 2 are relatively easy to apply, but now we must be careful to "hold" onto the terminals a and b as the Thévenin resistance and voltage are determined. In Fig. 9.39, all the remaining elements turn out to be in parallel, and the network can be redrawn as shown.

$$R_{Th} = R_1 \| R_2 = \frac{(6\ \Omega)(4\ \Omega)}{6\ \Omega + 4\ \Omega} = \frac{24\ \Omega}{10} = \mathbf{2.4\ \Omega}$$

Step 4: See Fig. 9.40. In this case, the network can be redrawn as shown in Fig. 9.41. Since the voltage is the same across parallel elements, the voltage across the series resistors R_1 and R_2 is E_1, or 8 V. Applying the voltage divider rule,

$$E_{Th} = \frac{R_1 E_1}{R_1 + R_2} = \frac{(6\ \Omega)(8\ \text{V})}{6\ \Omega + 4\ \Omega} = \frac{48\ \text{V}}{10} = \mathbf{4.8\ V}$$

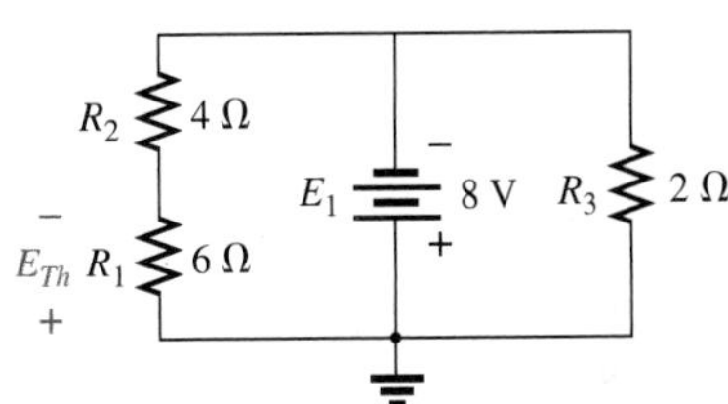

FIG. 9.41
Network of Fig. 9.40 redrawn.

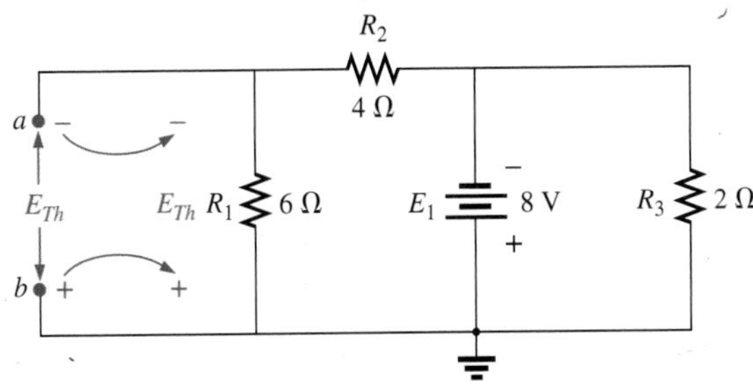

FIG. 9.40
Determining E_{Th} for the network in Fig. 9.38.

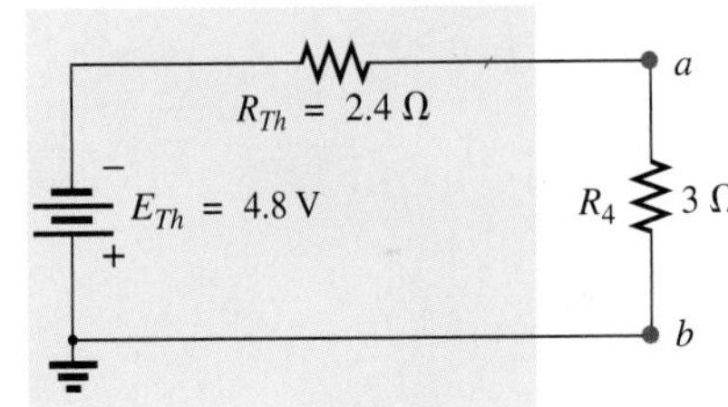

FIG. 9.42
Substituting the Thévenin equivalent circuit for the network external to the resistor R_4 in Fig. 9.37.

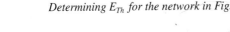

Step 5: See Fig. 9.42.

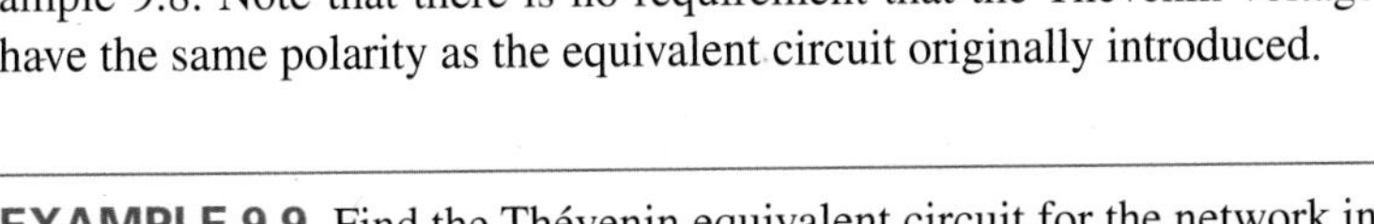

The importance of marking the terminals should be obvious from Example 9.8. Note that there is no requirement that the Thévenin voltage have the same polarity as the equivalent circuit originally introduced.

EXAMPLE 9.9 Find the Thévenin equivalent circuit for the network in the shaded area of the bridge network in Fig. 9.43.

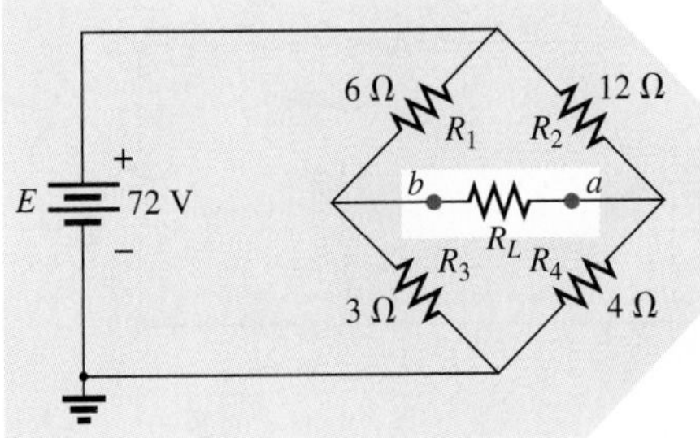

FIG. 9.43
Example 9.9.

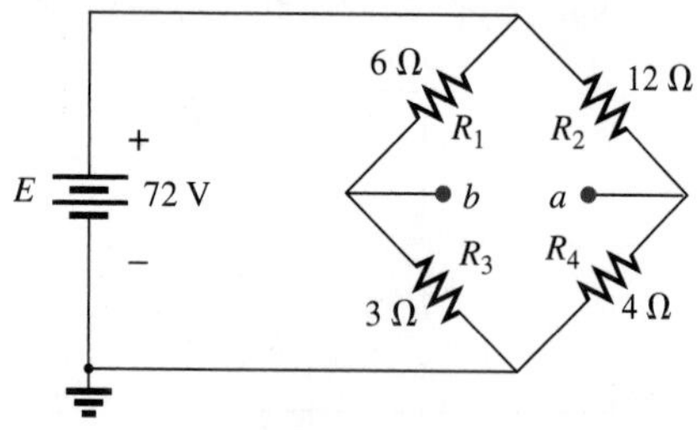

FIG. 9.44
Identifying the terminals of particular interest for the network in Fig. 9.43.

Solution:

Steps 1 and 2: See Fig. 9.44.

Step 3: See Fig. 9.45. In this case, the short-circuit replacement of the voltage source E provides a direct connection between c and c' in Fig. 9.45(a), permitting a "folding" of the network around the horizontal line of a-b to produce the configuration in Fig. 9.45(b).

$$R_{Th} = R_{a-b} = R_1 \parallel R_3 + R_2 \parallel R_4$$
$$= 6\ \Omega \parallel 3\ \Omega + 4\ \Omega \parallel 12\ \Omega$$
$$= 2\ \Omega + 3\ \Omega = \mathbf{5\ \Omega}$$

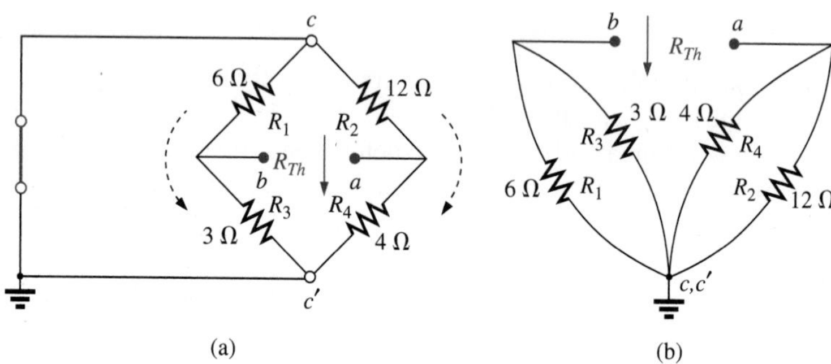

FIG. 9.45
Solving for R_{Th} for the network in Fig. 9.44.

Step 4: The circuit is redrawn in Fig. 9.46. The absence of a direct connection between a and b results in a network with three parallel branches. The voltages V_1 and V_2 can therefore be determined using the voltage divider rule:

$$V_1 = \frac{R_1 E}{R_1 + R_3} = \frac{(6\ \Omega)(72\ \text{V})}{6\ \Omega + 3\ \Omega} = \frac{432\ \text{V}}{9} = 48\ \text{V}$$

$$V_2 = \frac{R_2 E}{R_2 + R_4} = \frac{(12\ \Omega)(72\ \text{V})}{12\ \Omega + 4\ \Omega} = \frac{864\ \text{V}}{16} = 54\ \text{V}$$

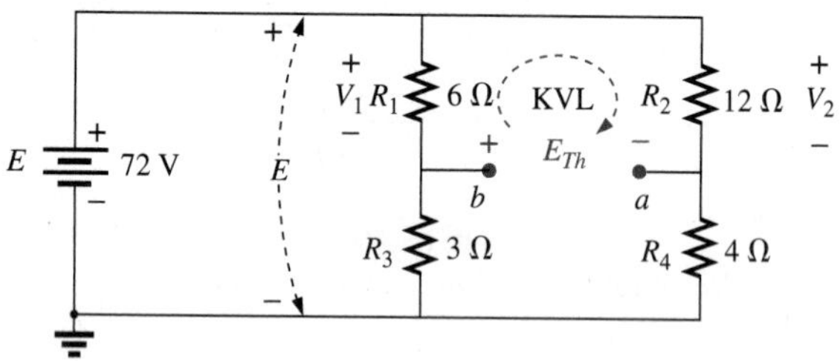

FIG. 9.46
Determining E_{Th} for the network in Fig. 9.44.

Assuming the polarity shown for E_{Th} and applying Kirchhoff's voltage law to the top loop in the clockwise direction results in

$$\Sigma_{\circlearrowright} V = +E_{Th} + V_1 - V_2 = 0$$

and
$$E_{Th} = V_2 - V_1 = 54\ \text{V} - 48\ \text{V} = \mathbf{6\ V}$$

Step 5: See Fig. 9.47.

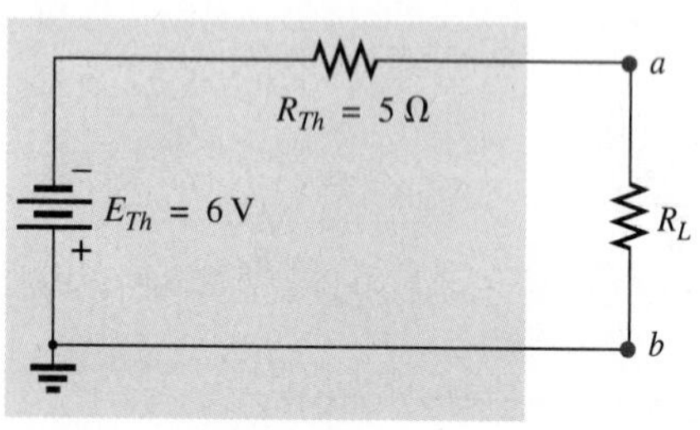

FIG. 9.47
Substituting the Thévenin equivalent circuit for the network external to the resistor R_L in Fig. 9.43.

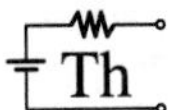

Thévenin's theorem is not restricted to a single passive element, as shown in the preceding examples, but can be applied across sources, whole branches, portions of networks, or any circuit configuration as shown in the following example. It is also possible that you may have to use one of the methods previously described, such as mesh analysis or superposition, to find the Thévenin equivalent circuit.

EXAMPLE 9.10 (Two sources) Find the Thévenin circuit for the network within the shaded area of Fig. 9.48.

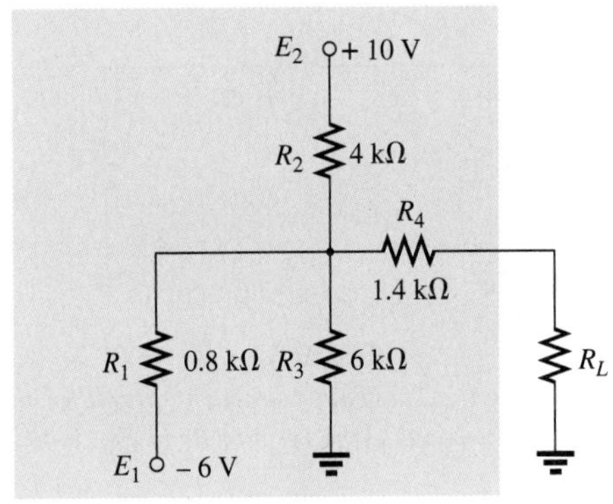

FIG. 9.48
Example 9.10.

Solution:

Steps 1 and 2: See Fig. 9.49. The network is redrawn.

Step 3: See Fig. 9.50.

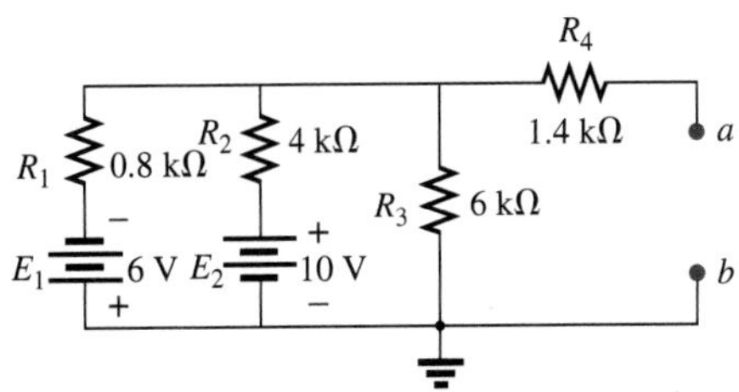

FIG. 9.49
Identifying the terminals of particular interest for the network in Fig. 9.48.

$$
\begin{aligned}
R_{Th} &= R_4 + R_1 \| R_2 \| R_3 \\
&= 1.4\ \text{k}\Omega + 0.8\ \text{k}\Omega \| 4\ \text{k}\Omega \| 6\ \text{k}\Omega \\
&= 1.4\ \text{k}\Omega + 0.8\ \text{k}\Omega \| 2.4\ \text{k}\Omega \\
&= 1.4\ \text{k}\Omega + 0.6\ \text{k}\Omega \\
&= \mathbf{2\ k\Omega}
\end{aligned}
$$

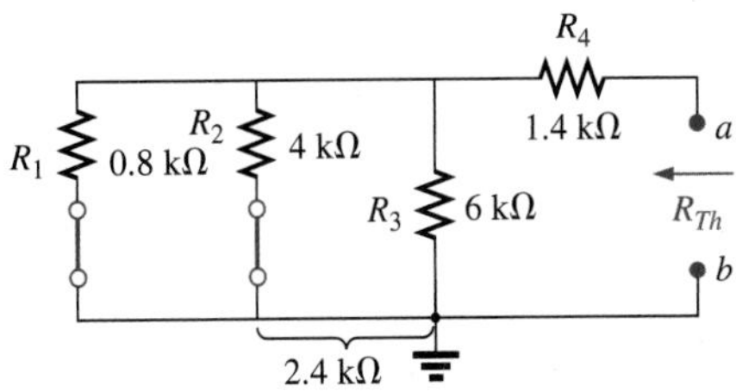

FIG. 9.50
Determining R_{Th} for the network in Fig. 9.49.

Step 4: Applying superposition, we will consider the effects of the voltage source E_1 first. Note Fig. 9.51. The open circuit requires that $V_4 = I_4R_4 = (0)R_4 = 0$ V, and

$$E'_{Th} = V_3$$

$$R'_T = R_2 \| R_3 = 4\ \text{k}\Omega \| 6\ \text{k}\Omega = 2.4\ \text{k}\Omega$$

Applying the voltage divider rule,

$$V_3 = \frac{R'_T E_1}{R'_T} = \frac{(2.4\ \text{k}\Omega)(6\ \text{V})}{2.4\ \text{k}\Omega + 0.8\ \text{k}\Omega} = \frac{14.4\ \text{V}}{3.2} = 4.5\ \text{V}$$

$$E'_{Th} = V_3 = 4.5\ \text{V}$$

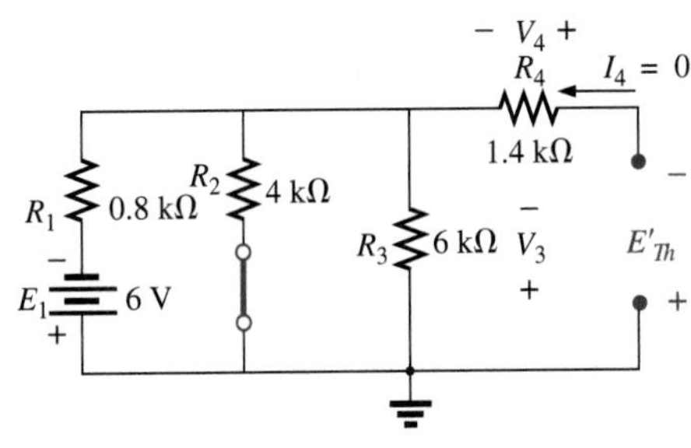

FIG. 9.51
Determining the contribution to E_{Th} from the source E_1 for the network in Fig. 9.49.

For the source E_2, the network in Fig. 9.52 results. Again, $V_4 = I_4R_4 = (0)R_4 = 0$ V, and

$$E''_{Th} = V_3$$

$$R'_T = R_1 \| R_3 = 0.8\ \text{k}\Omega \| 6\ \text{k}\Omega = 0.706\ \text{k}\Omega$$

and $$V_3 = \frac{R'_T E_2}{R'_T + R_2} = \frac{(0.706\ \text{k}\Omega)(10\ \text{V})}{0.706\ \text{k}\Omega + 4\ \text{k}\Omega} = \frac{7.06\ \text{V}}{4.706} = 1.5\ \text{V}$$

$$E''_{Th} = V_3 = 1.5\ \text{V}$$

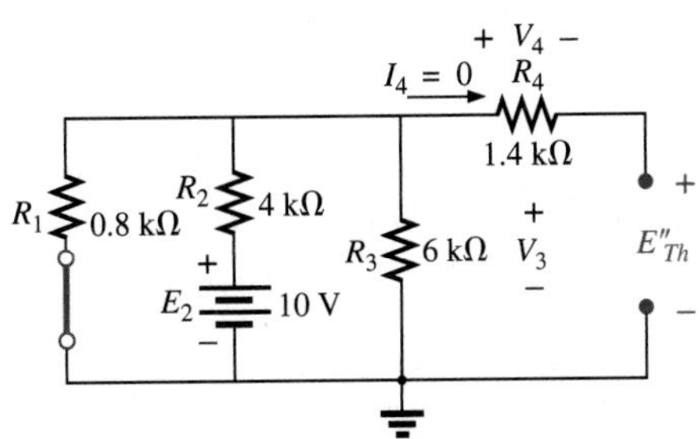

FIG. 9.52
Determining the contribution to E_{Th} from the source E_2 for the network in Fig. 9.49.

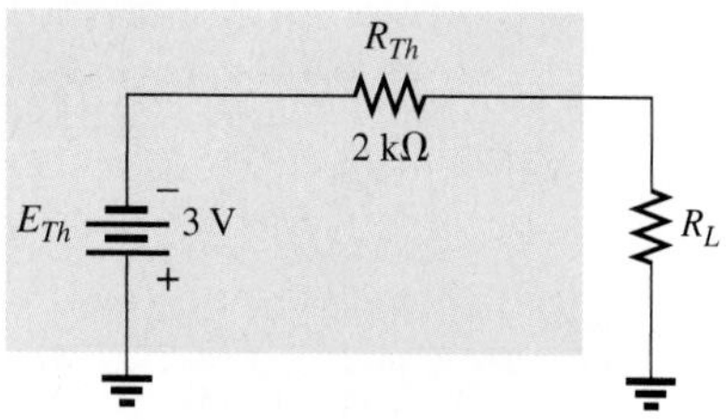

FIG. 9.53
Substituting the Thévenin equivalent circuit for the network external to the resistor R_L in Fig. 9.48.

Since E'_{Th} and E''_{Th} have opposite polarities,

$$\begin{aligned} E_{Th} &= E'_{Th} - E''_{Th} \\ &= 4.5 \text{ V} - 1.5 \text{ V} \\ &= \mathbf{3\ V} \qquad \text{(polarity of } E'_{Th}) \end{aligned}$$

Step 5: See Fig. 9.53.

Experimental Procedures

Now that the analytical procedure has been described in detail and a sense for the Thévenin impedance and voltage established, it is time to investigate how both quantities can be determined using an experimental procedure.

Even though the Thévenin resistance is usually the easiest to determine analytically, the Thévenin voltage is often the easiest to determine experimentally, and therefore it will be examined first.

Measuring E_{Th} The network of Fig. 9.54(a) has the equivalent Thévenin circuit appearing in Fig. 9.54(b). The open-circuit Thévenin voltage can be determined by simply placing a voltmeter on the output terminals in Fig. 9.54(a) as shown. This is due to the fact that the open circuit in Fig. 9.54(b) dictates that the current through and the voltage across the Thévenin resistance must be zero. The result for Fig. 9.54(b) is that

$$V_{oc} = E_{Th} = \mathbf{4.5\ V}$$

In general, therefore,

the Thévenin voltage is determined by connecting a voltmeter to the output terminals of the network. Be sure the internal resistance of the voltmeter is significantly more than the expected level of R_{Th}.

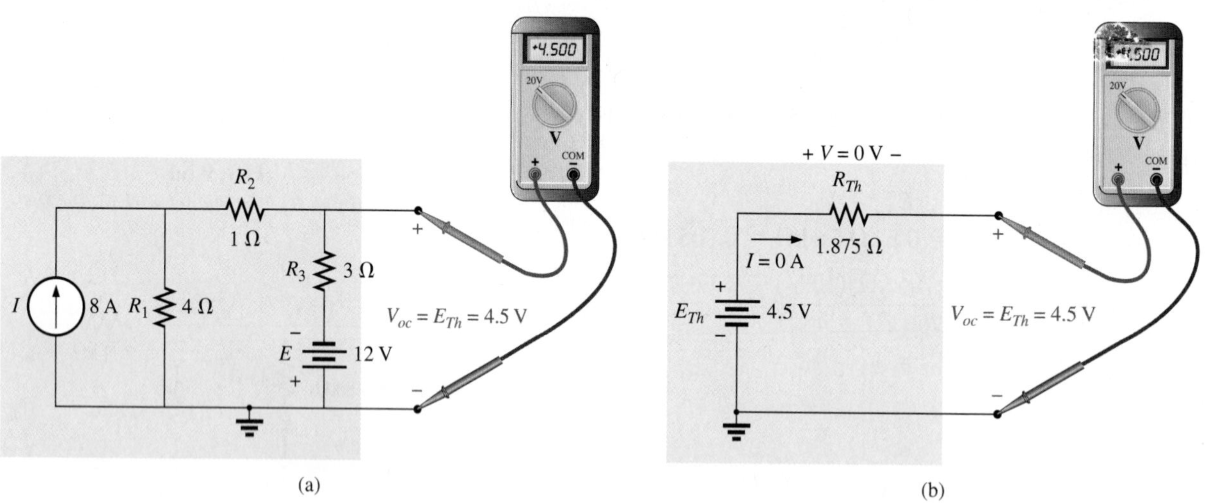

FIG. 9.54
Measuring the Thévenin voltage with a voltmeter: (a) actual network; (b) Thévenin equivalent.

Measuring R_{Th}

USING AN OHMMETER:

In Fig. 9.55, the sources in Fig. 9.54(a) have been set to zero, and an ohmmeter has been applied to measure the Thévenin resistance. In Fig. 9.54(b), it is clear that if the Thévenin voltage is set to zero volts, the ohmmeter will read the Thévenin resistance directly.

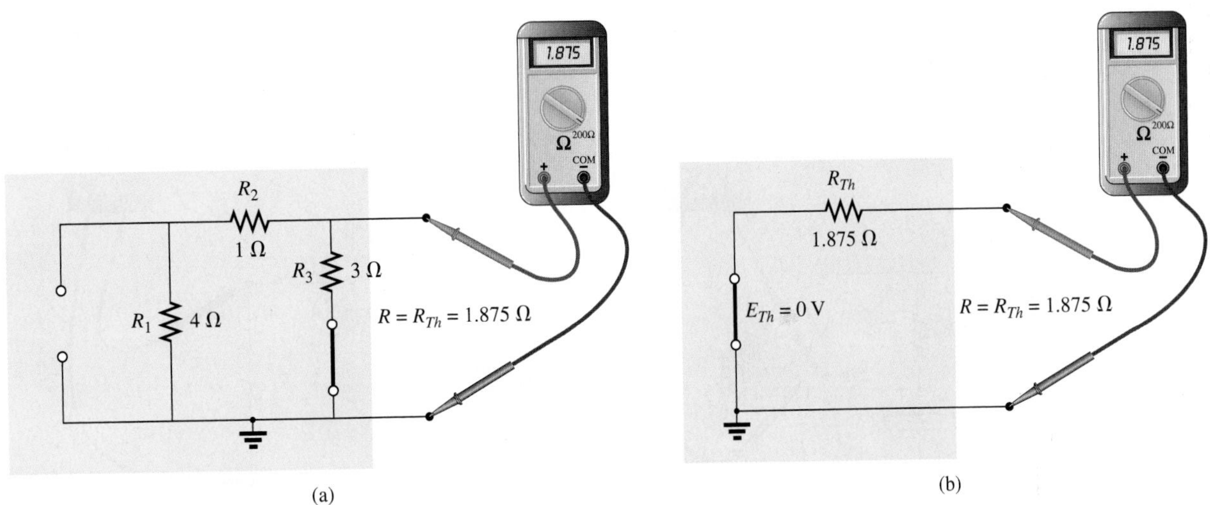

FIG. 9.55
Measuring R_{Th} with an ohmmeter: (a) actual network; (b) Thévenin equivalent.

In general, therefore,

the Thévenin resistance can be measured by setting all the sources to zero and measuring the resistance at the output terminals.

It is important to remember, however, that ohmmeters cannot be used on live circuits, and you cannot set a voltage source by putting a short circuit across it—it causes instant damage. The source must either be set to zero or removed entirely and then replaced by a direct connection. For the current source, the open-circuit condition must be clearly established; otherwise, the measured resistance will be incorrect. For most situations, it is usually best to remove the sources and replace them by the appropriate equivalent.

USING A POTENTIOMETER:

If we use a potentiometer to measure the Thévenin resistance, the sources can be left as is. For this reason alone, this approach is one of the more popular. In Fig. 9.56(a), a potentiometer has been connected across the output terminals of the network to establish the condition appearing in Fig. 9.56(b) for the Thévenin equivalent. If the resistance of the potentiometer is now adjusted so that the voltage across the potentiometer is one-half the measured Thévenin voltage, the Thévenin resistance must match that of the potentiometer. Recall that for a series circuit, the applied voltage will divide equally across two equal series resistors.

If the potentiometer is then disconnected and the resistance measured with an ohmmeter as shown in Fig. 9.56(c), the ohmmeter displays the Thévenin resistance of the network. In general, therefore,

the Thévenin resistance can be measured by applying a potentiometer to the output terminals and varying the resistance until the output voltage is one-half the measured Thévenin voltage. The resistance of the potentiometer is the Thévenin resistance for the network.

USING THE SHORT-CIRCUIT CURRENT:

The Thévenin resistance can also be determined by placing a short circuit across the output terminals and finding the current through the short circuit. Since ammeters ideally have zero internal ohms between their terminals, hooking up an ammeter as shown in Fig. 9.57(a) has the effect of both hooking up a short circuit across the terminals and measuring the resulting

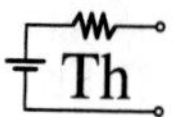

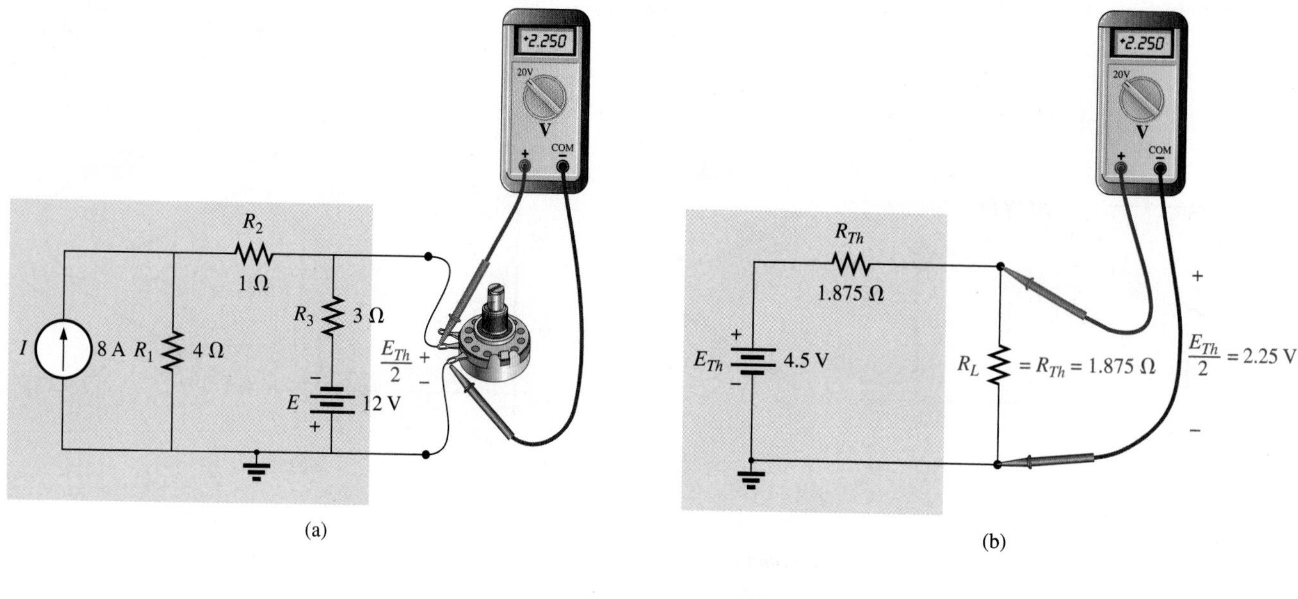

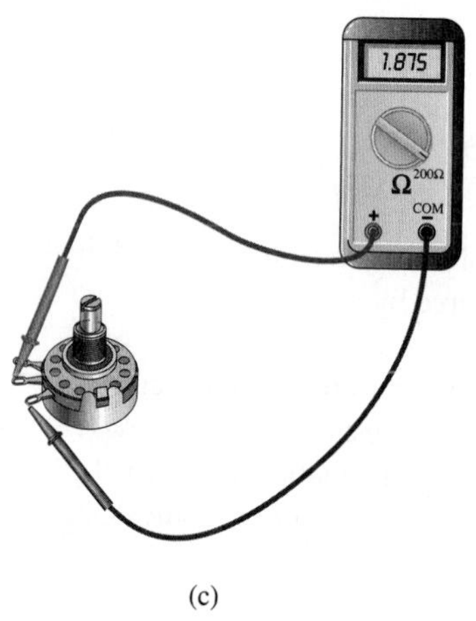

FIG. 9.56
Using a potentiometer to determine R_{Th}: (a) actual network; (b) Thévenin equivalent; (c) measuring R_{Th}.

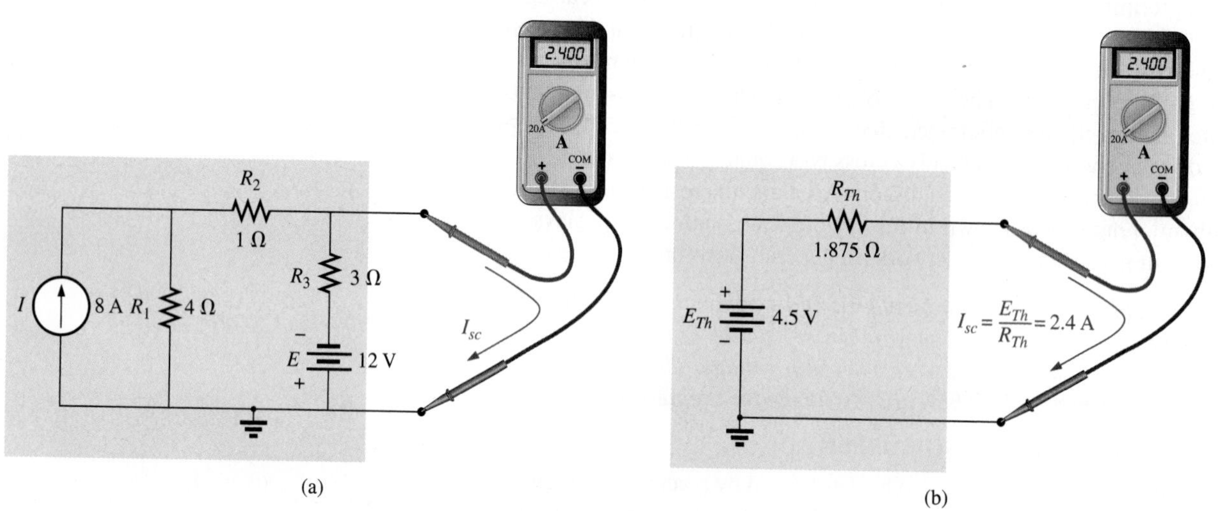

FIG. 9.57
Determining R_{Th} using the short-circuit current: (a) actual network; (b) Thévenin equivalent.

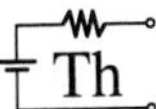

current. The same ammeter was connected across the Thévenin equivalent circuit in Fig. 9.57(b).

On a practical level, it is assumed, of course, that the internal resistance of the ammeter is approximately zero ohms in comparison to the other resistors of the network. It is also important to be sure that the resulting current does not exceed the maximum current for the chosen ammeter scale.

In Fig. 9.57(b), since the short-circuit current is

$$I_{sc} = \frac{E_{Th}}{R_{Th}}$$

the Thévenin resistance can be determined by

$$R_{Th} = \frac{E_{Th}}{I_{sc}}$$

In general, therefore,

the Thévenin resistance can be determined by hooking up an ammeter across the output terminals to measure the short-circuit current and then using the open-circuit voltage to calculate the Thévenin resistance in the following manner:

$$\boxed{R_{Th} = \frac{V_{oc}}{I_{sc}}} \qquad \textbf{(9.1)}$$

As a result, we have three ways to measure the Thévenin resistance of a configuration. Because of the concern about setting the sources to zero in the first procedure and the concern about current levels in the last, the second method is often chosen.

FIG. 9.58
Edward L. Norton.
Courtesy of AT&T Archives.

American (Rockland, Maine; Summit, New Jersey)
1898–1983
Electrical Engineer, Scientist, Inventor
Department Head: Bell Laboratories
Fellow: Acoustical Society and Institute of Radio Engineers

Although interested primarily in communications circuit theory and the transmission of data at high speeds over telephone lines, Edward L. Norton is best remembered for development of the dual of Thévenin equivalent circuit, currently referred to as *Norton's equivalent circuit.* In fact, Norton and his associates at AT&T in the early 1920s are recognized as some of the first to perform pioneering work applying Thévenin's equivalent circuit and who referred to this concept simply as Thévenin's theorem. In 1926, he proposed the equivalent circuit using a current source and parallel resistor to assist in the design of recording instrumentation that was primarily current driven. He began his telephone career in 1922 with the Western Electric Company's Engineering Department, which later became Bell Laboratories. His areas of active research included network theory, acoustical systems, electromagnetic apparatus, and data transmission. A graduate of MIT and Columbia University, he held nineteen patents on his work.

9.4 NORTON'S THEOREM

In Section 8.3, we learned that every voltage source with a series internal resistance has a current source equivalent. The current source equivalent can be determined by **Norton's theorem** (Fig. 9.58). It can also be found through the conversions of Section 8.3.

The theorem states the following:

Any two-terminal linear bilateral dc network can be replaced by an equivalent circuit consisting of a current source and a parallel resistor, as shown in Fig. 9.59.

The discussion of Thévenin's theorem with respect to the equivalent circuit can also be applied to the Norton equivalent circuit. The steps leading to the proper values of I_N and R_N are now listed.

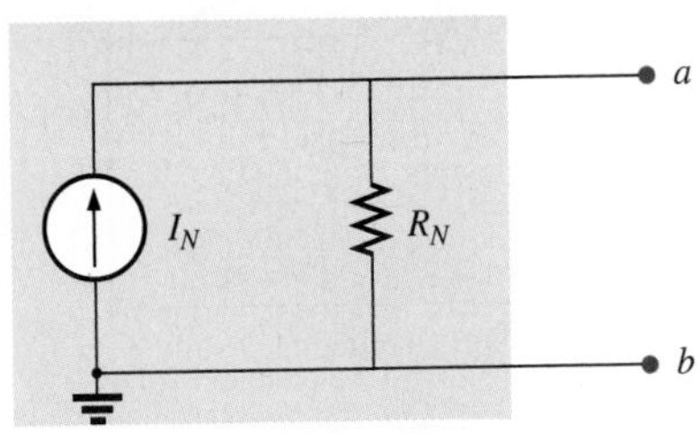

FIG. 9.59
Norton equivalent circuit.

Norton's Theorem Procedure

Preliminary:

1. ***Remove that portion of the network across which the Norton equivalent circuit is found.***
2. ***Mark the terminals of the remaining two-terminal network.***

R_N:

3. ***Calculate R_N by first setting all sources to zero (voltage sources are replaced with short circuits, and current sources with open circuits) and then finding the resultant resistance between the two***

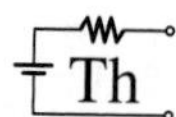

marked terminals. (If the internal resistance of the voltage and/or current sources is included in the original network, it must remain when the sources are set to zero.) Since $R_N = R_{Th}$, the procedure and value obtained using the approach described for Thévenin's theorem will determine the proper value of R_N.

I_N:

4. ***Calculate I_N by first returning all sources to their original position and then finding the short-circuit current between the marked terminals. It is the same current that would be measured by an ammeter placed between the marked terminals.***

Conclusion:

5. ***Draw the Norton equivalent circuit with the portion of the circuit previously removed replaced between the terminals of the equivalent circuit.***

The Norton and Thévenin equivalent circuits can also be found from each other by using the source transformation discussed earlier in this chapter and reproduced in Fig. 9.60.

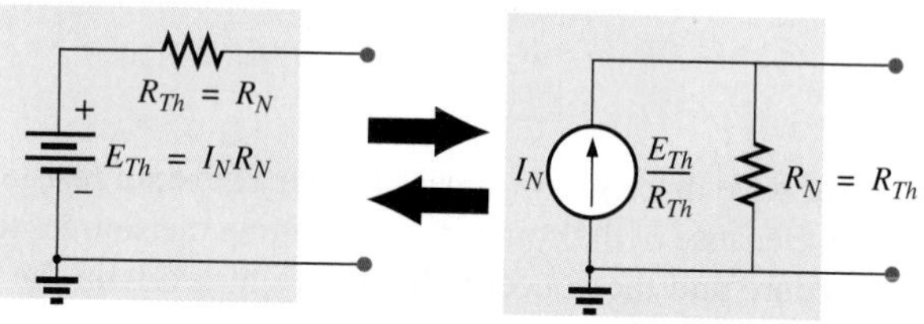

FIG. 9.60
Converting between Thévenin and Norton equivalent circuits.

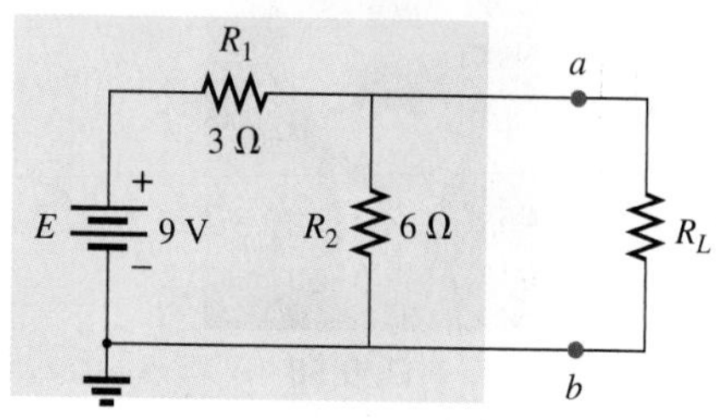

FIG. 9.61
Example 9.11.

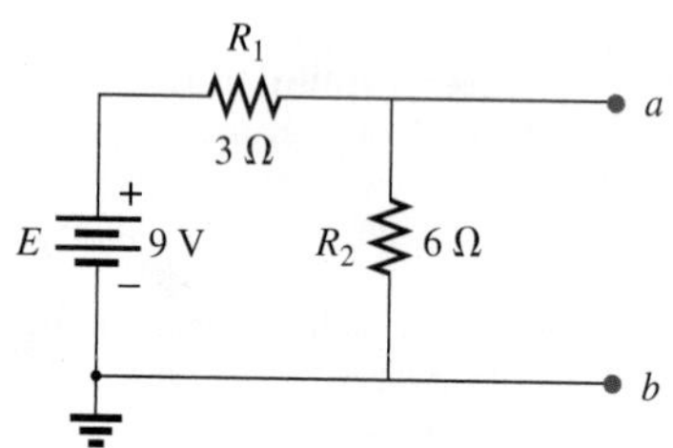

FIG. 9.62
Identifying the terminals of particular interest for the network in Fig. 9.61.

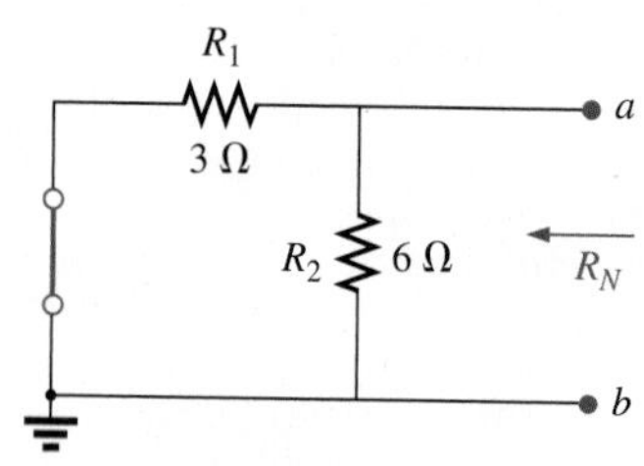

FIG. 9.63
Determining R_N for the network in Fig. 9.62.

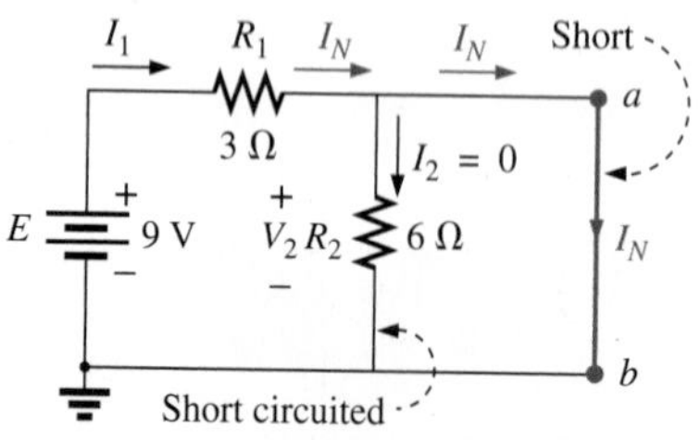

FIG. 9.64
Determining I_N for the network in Fig. 9.62.

EXAMPLE 9.11 Find the Norton equivalent circuit for the network in the shaded area in Fig. 9.61.

Solution:

Steps 1 and 2: See Fig. 9.62.

Step 3: See Fig. 9.63, and

$$R_N = R_1 \parallel R_2 = 3\ \Omega \parallel 6\ \Omega = \frac{(3\ \Omega)(6\ \Omega)}{3\ \Omega + 6\ \Omega} = \frac{18\ \Omega}{9} = \mathbf{2\ \Omega}$$

Step 4: See Fig. 9.64, which clearly indicates that the short-circuit connection between terminals *a* and *b* is in parallel with R_2 and eliminates its effect. I_N is therefore the same as through R_1, and the full battery voltage appears across R_1 since

$$V_2 = I_2R_2 = (0)6\ \Omega = 0\ \text{V}$$

Therefore,

$$I_N = \frac{E}{R_1} = \frac{9\ \text{V}}{3\ \Omega} = \mathbf{3\ A}$$

Step 5: See Fig. 9.65. This circuit is the same as the first one considered in the development of Thévenin's theorem. A simple conversion indicates that the Thévenin circuits are, in fact, the same (Fig. 9.66).

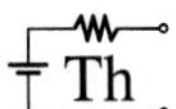

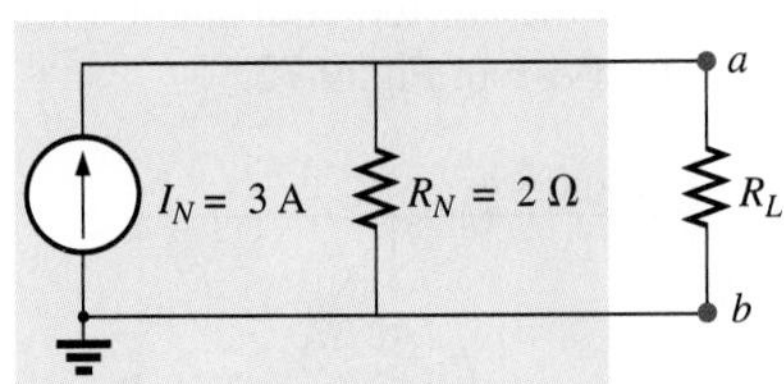

FIG. 9.65
Substituting the Norton equivalent circuit for the network external to the resistor R_L in Fig. 9.61.

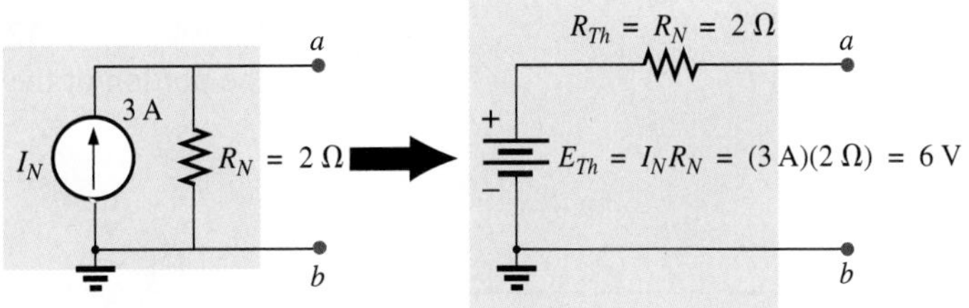

FIG. 9.66
Converting the Norton equivalent circuit in Fig. 9.65 to a Thévenin equivalent circuit.

EXAMPLE 9.12 Find the Norton equivalent circuit for the network external to the 9 Ω resistor in Fig. 9.67.

Solution:

Steps 1 and 2: See Fig. 9.68.

Step 3: See Fig. 9.69, and

$$R_N = R_1 + R_2 = 5\ \Omega + 4\ \Omega = \mathbf{9\ \Omega}$$

Step 4: As shown in Fig. 9.70, the Norton current is the same as the current through the 4 Ω resistor. Applying the current divider rule,

$$I_N = \frac{R_1 I}{R_1 + R_2} = \frac{(5\ \Omega)(10\ \text{A})}{5\ \Omega + 4\ \Omega} = \frac{50\ \text{A}}{9} = \mathbf{5.56\ A}$$

Step 5: See Fig. 9.71.

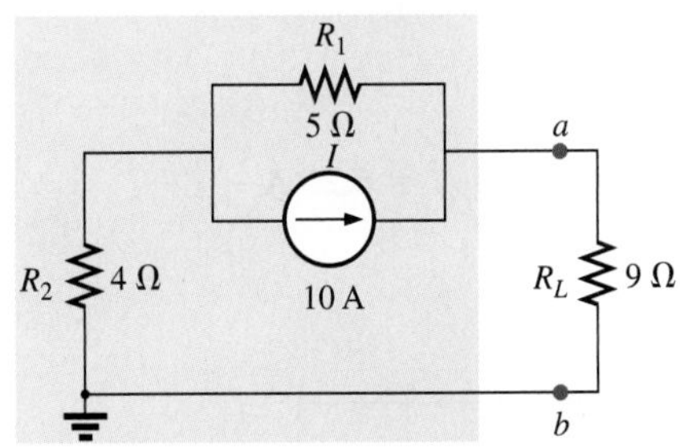

FIG. 9.67
Example 9.12.

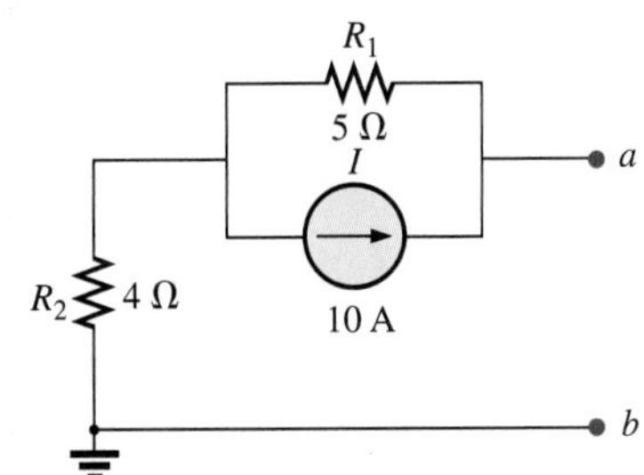

FIG. 9.68
Identifying the terminals of particular interest for the network in Fig. 9.67.

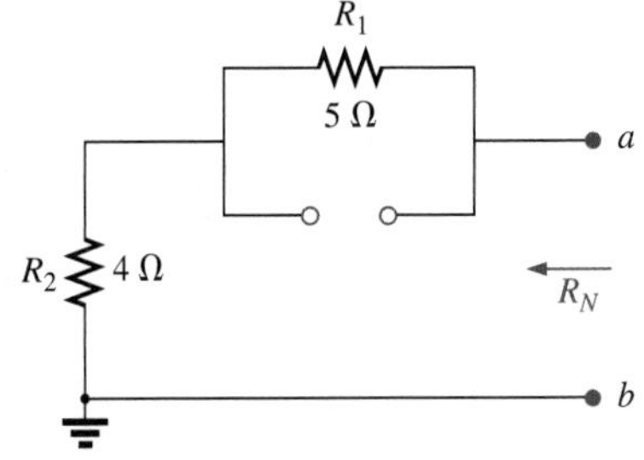

FIG. 9.69
Determining R_N for the network in Fig. 9.68.

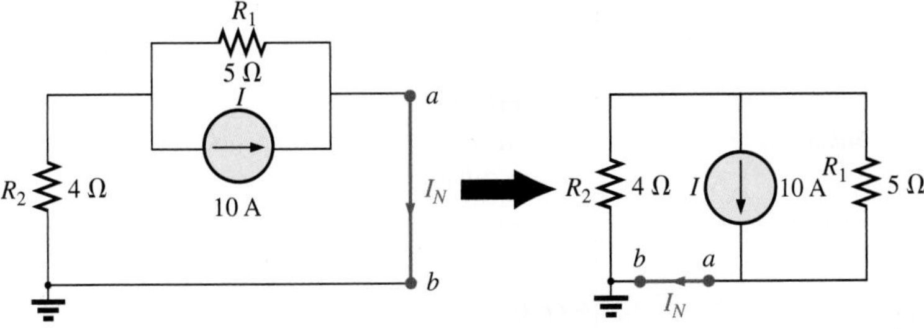

FIG. 9.70
Determining I_N for the network in Fig. 9.68.

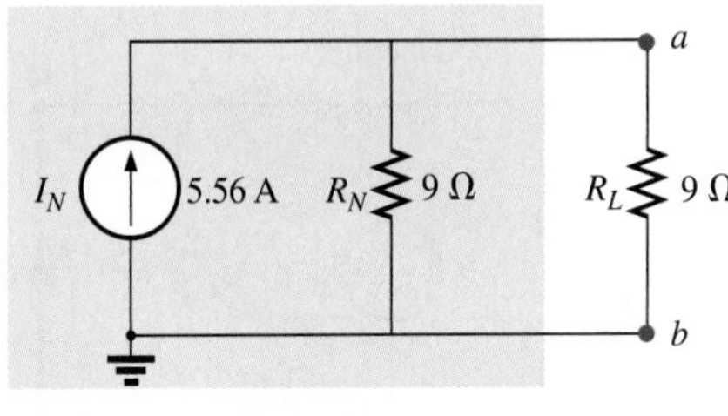

FIG. 9.71
Substituting the Norton equivalent circuit for the network external to the resistor R_L in Fig. 9.67.

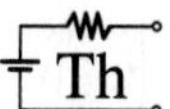

EXAMPLE 9.13 (Two sources) Find the Norton equivalent circuit for the portion of the network to the left of *a-b* in Fig. 9.72.

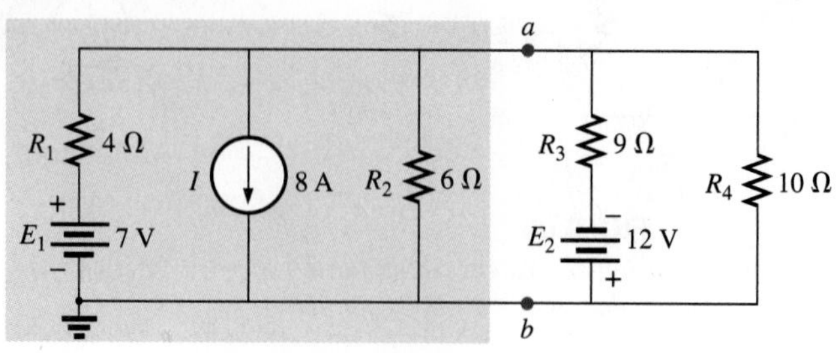

FIG. 9.72
Example 9.13.

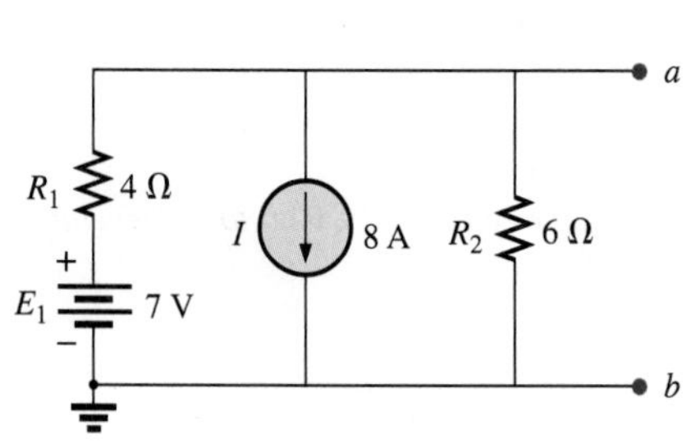

FIG. 9.73
Identifying the terminals of particular interest for the network in Fig. 9.72.

Solution:

Steps 1 and 2: See Fig. 9.73.

Step 3: See Fig. 9.74, and

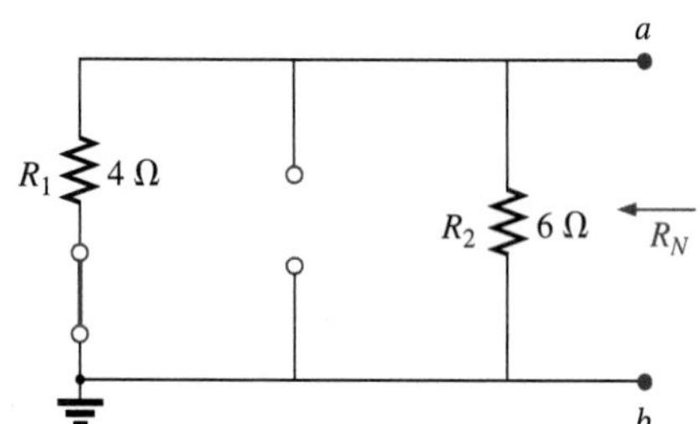

FIG. 9.74
Determining R_N for the network in Fig. 9.73.

$$R_N = R_1 \| R_2 = 4\ \Omega \| 6\ \Omega = \frac{(4\ \Omega)(6\ \Omega)}{4\ \Omega + 6\ \Omega} = \frac{24\ \Omega}{10} = \mathbf{2.4\ \Omega}$$

Step 4: (Using superposition) For the 7 V battery (Fig. 9.75),

$$I'_N = \frac{E_1}{R_1} = \frac{7\text{ V}}{4\ \Omega} = 1.75\text{ A}$$

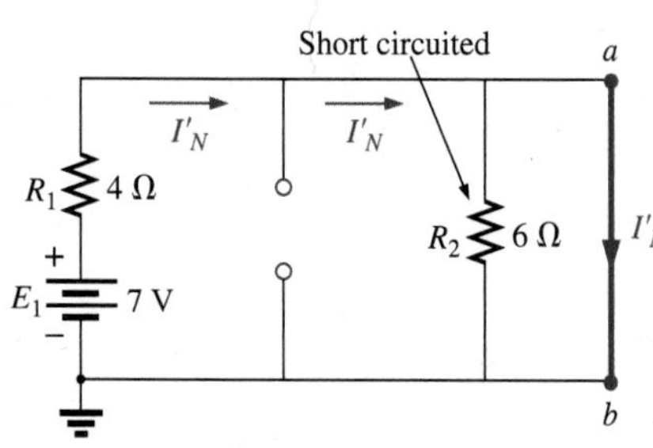

FIG. 9.75
Determining the contribution to I_N from the voltage source E_1.

For the 8 A source (Fig. 9.76), we find that both R_1 and R_2 have been "short circuited" by the direct connection between *a* and *b*, and

$$I''_N = I = 8\text{ A}$$

The result is

$$I_N = I''_N - I'_N = 8\text{ A} - 1.75\text{ A} = \mathbf{6.25\ A}$$

Step 5: See Fig. 9.77.

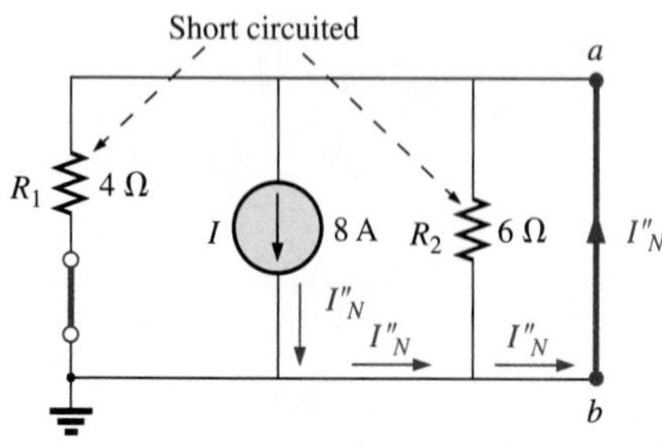

FIG. 9.76
Determining the contribution to I_N from the current source I.

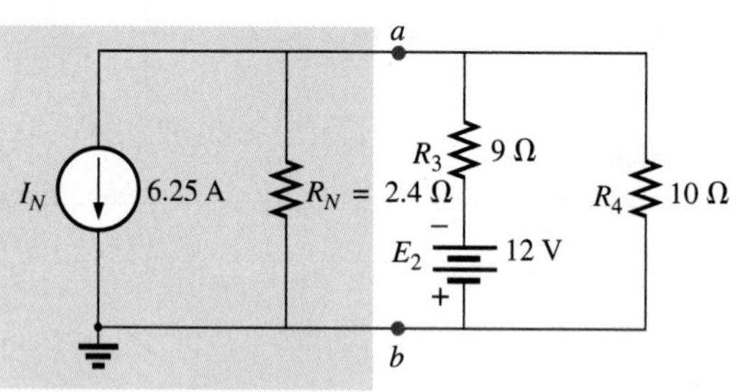

FIG. 9.77
Substituting the Norton equivalent circuit for the network to the left of terminals a-b in Fig. 9.72.

Experimental Procedure

The Norton current is measured in the same way as described for the short-circuit current (I_{sc}) for the Thévenin network. Since the Norton and Thévenin resistances are the same, the same procedures can be followed as described for the Thévenin network.

9.5 MAXIMUM POWER TRANSFER THEOREM

When designing a circuit, it is often important to be able to answer one of the following questions:

What load should be applied to a system to ensure that the load is receiving maximum power from the system?

and, conversely:

For a particular load, what conditions should be imposed on the source to ensure that it will deliver the maximum power available?

Even if a load cannot be set at the value that would result in maximum power transfer, it is often helpful to have some idea of the value that will draw maximum power so that you can compare it to the load at hand. For instance, if a design calls for a load of 100 Ω, to ensure that the load receives maximum power, using a resistor of 1 Ω or 1 kΩ results in a power transfer that is much less than the maximum possible. However, using a load of 82 Ω or 120 Ω probably results in a fairly good level of power transfer.

Fortunately, the process of finding the load that will receive maximum power from a particular system is quite straightforward due to the **maximum power transfer theorem,** which states the following:

A load will receive maximum power from a network when its resistance is exactly equal to the Thévenin resistance of the network applied to the load. That is,

$$\boxed{R_L = R_{Th}} \qquad \textbf{(9.2)}$$

In other words, for the Thévenin equivalent circuit in Fig. 9.78, when the load is set equal to the Thévenin resistance, the load will receive maximum power from the network.

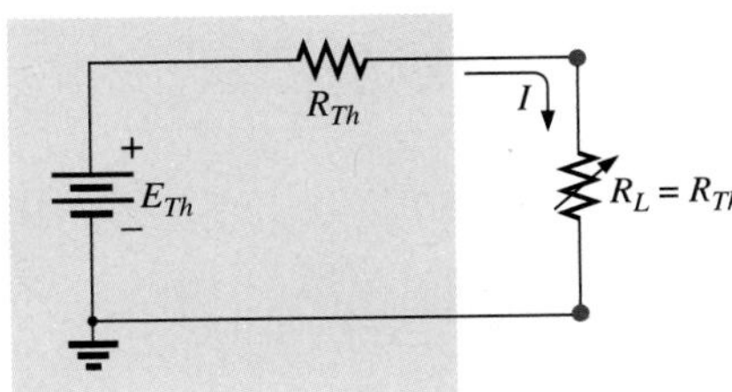

FIG. 9.78
Defining the conditions for maximum power to a load using the Thévenin equivalent circuit.

Using Fig. 9.78 with $R_L = R_{Th}$, the maximum power delivered to the load can be determined by first finding the current:

$$I_L = \frac{E_{Th}}{R_{Th} + R_L} = \frac{E_{Th}}{R_{Th} + R_{Th}} = \frac{E_{Th}}{2R_{Th}}$$

Then substitute into the power equation:

$$P_L = I_L^2 R_L = \left(\frac{E_{Th}}{2R_{Th}}\right)^2 (R_{Th}) = \frac{E_{Th}^2 R_{Th}}{4R_{Th}^2}$$

and

$$\boxed{P_{L_{max}} = \frac{E_{Th}^2}{4R_{Th}}} \qquad \textbf{(9.3)}$$

To demonstrate that maximum power is indeed transferred to the load under the conditions defined above, consider the Thévenin equivalent circuit in Fig. 9.79.

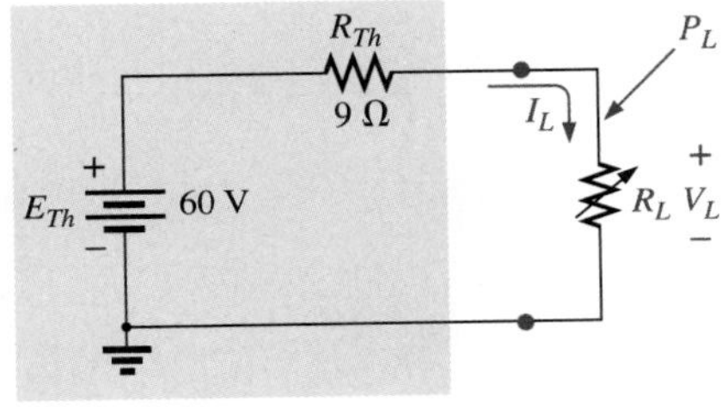

FIG. 9.79
Thévenin equivalent network to be used to validate the maximum power transfer theorem.

Before getting into detail, however, if you were to guess what value of R_L would result in maximum power transfer to R_L, you may think that the smaller the value of R_L, the better, because the current reaches a maximum when it is squared in the power equation. The problem is, however, that in the equation $P_L = I_L^2 R_L$, the load resistance is a multiplier. As it gets smaller, it forms a smaller product. Then again, you may suggest larger values of R_L, because the output voltage increases and power is determined by $P_L = V_L^2/R_L$. This time, however, the load resistance is in the denominator of the equation and causes the resulting power to decrease. A balance must obviously be made between the load resistance and the resulting current or voltage. The following discussion shows that

maximum power transfer occurs when the load voltage and current are one-half of their maximum possible values.

For the circuit in Fig. 9.79, the current through the load is determined by

$$I_L = \frac{E_{Th}}{R_{Th} + R_L} = \frac{60\text{ V}}{9\ \Omega + R_L}$$

The voltage is determined by

$$V_L = \frac{R_L E_{Th}}{R_L + R_{Th}} = \frac{R_L(60\text{ V})}{R_L + R_{Th}}$$

and the power by

$$P_L = I_L^2 R_L = \left(\frac{60\text{ V}}{9\ \Omega + R_L}\right)^2 (R_L) = \frac{3600 R_L}{(9\ \Omega + R_L)^2}$$

If we tabulate the three quantities versus a range of values for R_L from 0.1 Ω to 30 Ω, we obtain the results appearing in Table 9.1. Note in particular that when R_L is equal to the Thévenin resistance of 9 Ω, the power has a maximum value of 100 W, the current is 3.33 A or one-half its max-

TABLE 9.1

R_L (Ω)	P_L (W)		I_L (A)		V_L (V)	
0.1	4.35		6.60		0.66	
0.2	8.51		6.52		1.30	
0.5	19.94		6.32		3.16	
1	36.00		6.00		6.00	
2	59.50		5.46		10.91	
3	75.00		5.00		15.00	
4	85.21		4.62		18.46	
5	91.84		4.29		21.43	
6	96.00		4.00		24.00	
7	98.44	Increase	3.75	Decrease	26.25	Increase
8	99.65 ↓		3.53 ↓		28.23 ↓	
9 (R_{Th})	100.00 (Maximum)		3.33 ($I_{max}/2$)		30.00 ($E_{Th}/2$)	
10	99.72		3.16		31.58	
11	99.00		3.00		33.00	
12	97.96		2.86		34.29	
13	96.69		2.73		35.46	
14	95.27		2.61		36.52	
15	93.75		2.50		37.50	
16	92.16		2.40		38.40	
17	90.53		2.31		39.23	
18	88.89		2.22		40.00	
19	87.24		2.14		40.71	
20	85.61		2.07		41.38	
25	77.86		1.77		44.12	
30	71.00		1.54		46.15	
40	59.98		1.22		48.98	
100	30.30		0.55		55.05	
500	6.95	Decrease	0.12	Decrease	58.94	Increase
1000	3.54 ↓		0.06 ↓		59.47 ↓	

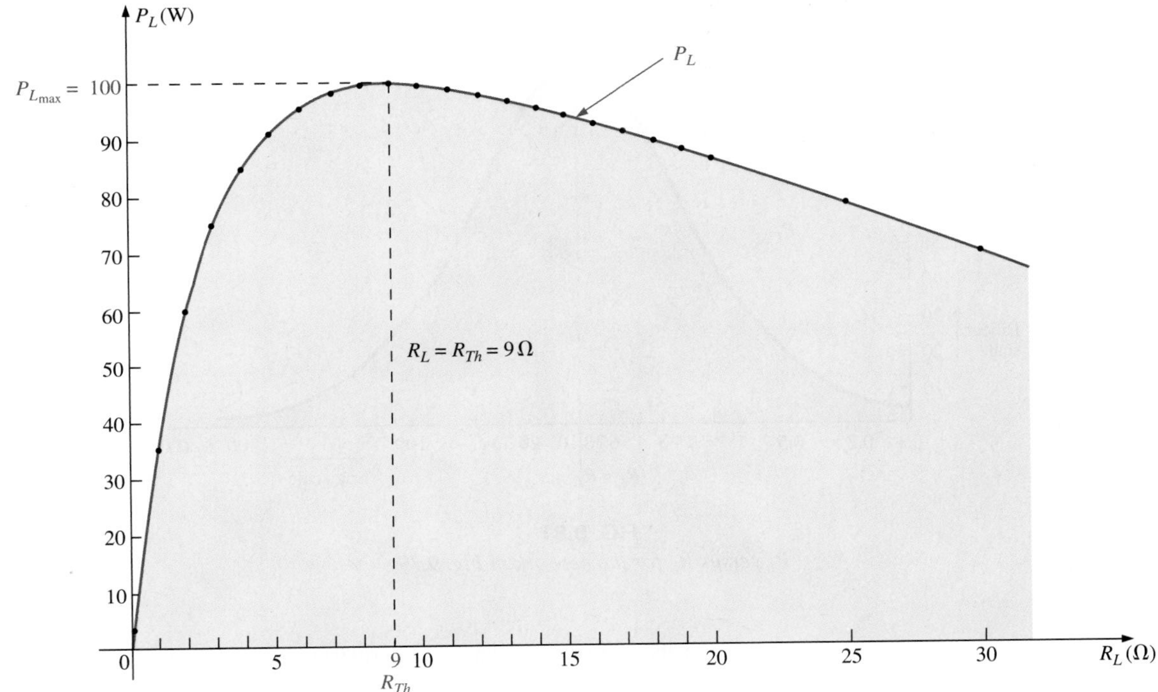

FIG. 9.80
P_L versus R_L for the network in Fig. 9.79.

imum value of 6.60 A (as would result with a short circuit across the output terminals), and the voltage across the load is 30 V or one-half its maximum value of 60 V (as would result with an open circuit across its output terminals). As you can see, there is no question that maximum power is transferred to the load when the load equals the Thévenin value.

The power to the load versus the range of resistor values is provided in Fig. 9.80. Note in particular that for values of load resistance less than the Thévenin value, the change is dramatic as it approaches the peak value. However, for values greater than the Thévenin value, the drop is a great deal more gradual. This is important because it tells you the following:

If the load applied is less than the Thévenin resistance, the power to the load will drop off rapidly as it gets smaller. However, if the applied load is greater than the Thévenin resistance, the power to the load will not drop off as rapidly as it increases.

For instance, the power to the load is at least 90 W for the range of about 4.5 Ω to 9 Ω below the peak value, but it is at least the same level for a range of about 9 Ω to 18 Ω above the peak value. The range below the peak is 4.5 Ω, while the range above the peak is almost twice as much at 9 Ω. As mentioned above, if maximum transfer conditions cannot be established, at least we now know from Fig. 9.80 that any resistance relatively close to the Thévenin value results in a strong transfer of power. More distant values such as 1 Ω or 100 Ω result in much lower levels.

It is particularly interesting to plot the power to the load versus load resistance using a log scale, as shown in Fig. 9.81. Logarithms will be discussed in detail in Chapter 21, but for now notice that the spacing between values of R_L is not linear, but the distance between powers of ten (such as 0.1 and 1, 1 and 10, and 10 and 100) are all equal. The advantage

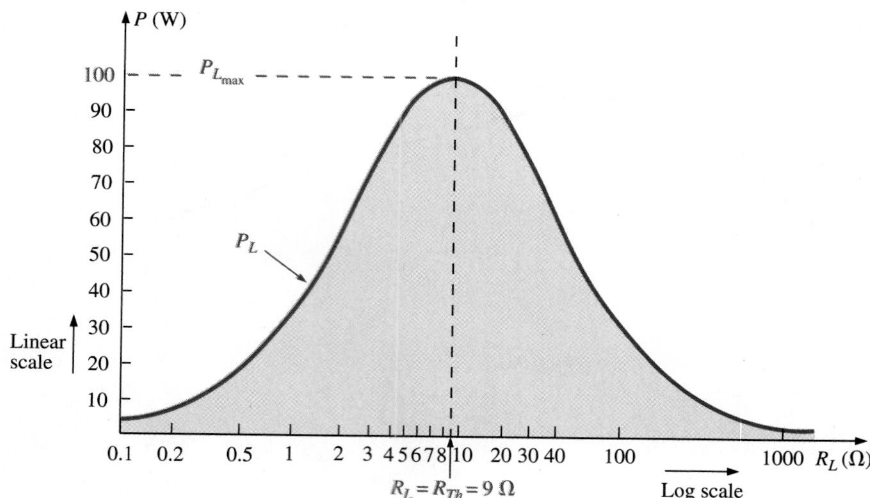

FIG. 9.81
P_L versus R_L for the network in Fig. 9.79.

of the log scale is that a wide resistance range can be plotted on a relatively small graph.

Note in Fig. 9.81 that a smooth, bell-shaped curve results that is symmetrical about the Thévenin resistance of 9 Ω. At 0.1 Ω, the power has dropped to about the same level as that at 1000 Ω, and at 1 Ω and 100 Ω, the power has dropped to the neighborhood of 30 W.

Although all of the above discussion centers on the power to the load, it is important to remember the following:

The total power delivered by a supply such as E_{Th} is absorbed by both the Thévenin equivalent resistance and the load resistance. Any power delivered by the source that does not get to the load is lost to the Thévenin resistance.

Under maximum power conditions, only half the power delivered by the source gets to the load. Now, that sounds disastrous, but remember that we are starting out with a fixed Thévenin voltage and resistance, and the above simply tells us that we must make the two resistance levels equal if we want maximum power to the load. On an efficiency basis, we are working at only a 50% level, but we are content because *we are getting maximum power out of our system.*

The dc operating efficiency is defined as the ratio of the power delivered to the load (P_L) to the power delivered by the source (P_s). That is,

$$\eta\% = \frac{P_L}{P_s} \times 100\% \tag{9.4}$$

For the situation where $R_L = R_{Th}$,

$$\eta\% = \frac{I_L^2 R_L}{I_L^2 R_T} \times 100\% = \frac{R_L}{R_T} \times 100\% = \frac{R_{Th}}{R_{Th} + R_{Th}} \times 100\%$$

$$= \frac{R_{Th}}{2R_{Th}} \times 100\% = \frac{1}{2} \times 100\% = \mathbf{50\%}$$

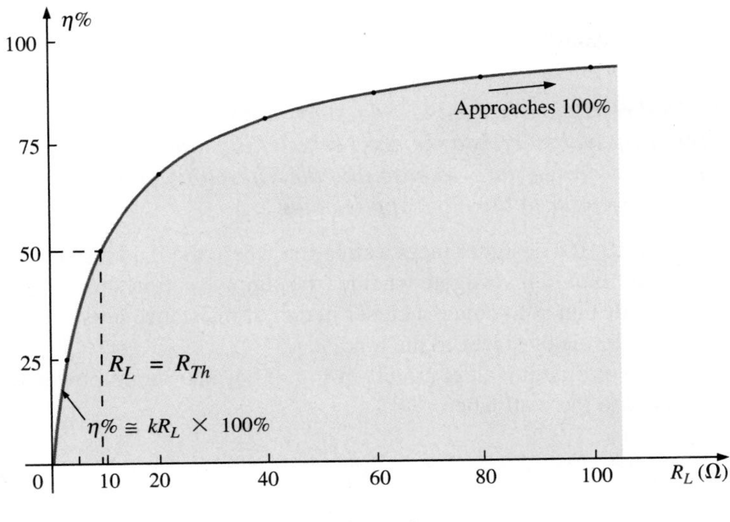

FIG. 9.82
Efficiency of operation versus increasing values of R_L.

For the circuit in Fig. 9.79, if we plot the efficiency of operation versus load resistance, we obtain the plot in Fig. 9.82, which clearly shows that the efficiency continues to rise to a 100% level as R_L gets larger. Note in particular that the efficiency is 50% when $R_L = R_{Th}$.

To ensure that you completely understand the effect of the maximum power transfer theorem and the efficiency criteria, consider the circuit in Fig. 9.83 where the load resistance is set at 100 Ω and the power to the Thévenin resistance and to the load are calculated as follows:

$$I_L = \frac{E_{Th}}{R_{Th} + R_L} = \frac{60\text{ V}}{9\ \Omega + 100\ \Omega} = \frac{60\text{ V}}{109\ \Omega} = 550.5\text{ mA}$$

with $\qquad P_{R_{Th}} = I_L^2 R_{Th} = (550.5\text{ mA})^2(9\ \Omega) \cong \mathbf{2.73\ W}$

and $\qquad P_L = I_L^2 R_L = (550.5\text{ mA})^2(100\ \Omega) \cong \mathbf{30.3\ W}$

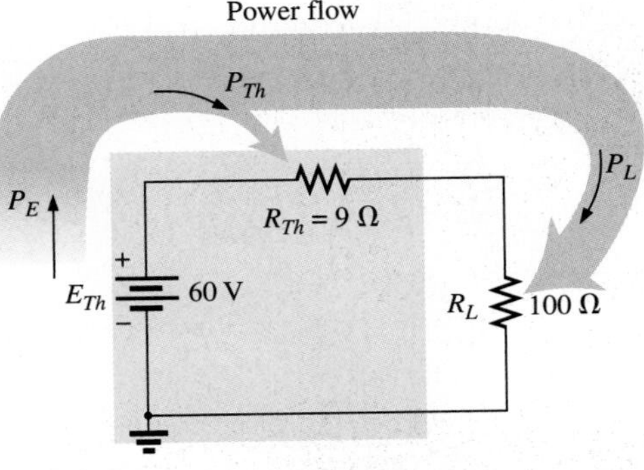

FIG. 9.83
Examining a circuit with high efficiency but a relatively low level of power to the load.

The results clearly show that most of the power supplied by the battery is getting to the load—a desirable attribute on an efficiency basis. However, the power getting to the load is only 30.3 W compared to the 100 W obtained under maximum power conditions. In general, therefore, the following guidelines apply:

If efficiency is the overriding factor, then the load should be much larger than the internal resistance of the supply. If maximum power transfer is desired and efficiency less of a concern, then the conditions dictated by the maximum power transfer theorem should be applied.

A relatively low efficiency of 50% can be tolerated in situations where power levels are relatively low, such as in a wide variety of electronic systems, where maximum power transfer for the given system is usually more important. However, when large power levels are involved, such as at generating plants, efficiencies of 50% cannot be tolerated. In fact, a great deal of expense and research is dedicated to raising power generating and transmission efficiencies a few percentage points. Raising an efficiency level of a 10 MkW power plant from 94% to 95% (a 1% increase) can save 0.1 MkW, or 100 million watts, of power—an enormous saving.

In all of the above discussions, the effect of changing the load was discussed for a fixed Thévenin resistance. Looking at the situation from a different viewpoint,

if the load resistance is fixed and does not match the applied Thévenin equivalent resistance, then some effort should be made (if possible) to redesign the system so that the Thévenin equivalent resistance is closer to the fixed applied load.

In other words, if a designer faces a situation where the load resistance is fixed, he/she should investigate whether the supply section should be replaced or redesigned to create a closer match of resistance levels to produce higher levels of power to the load.

For the Norton equivalent circuit in Fig. 9.84, maximum power will be delivered to the load when

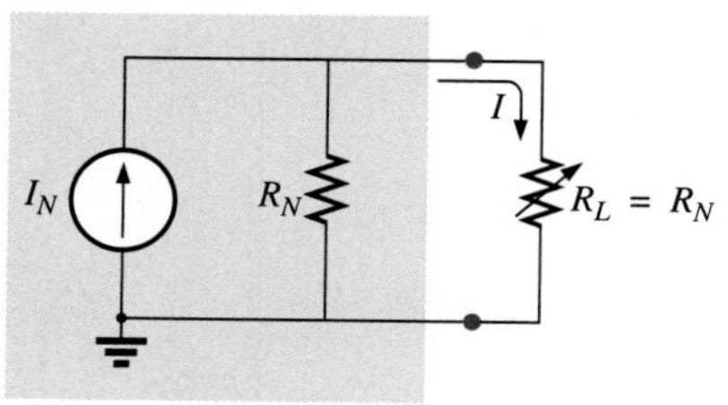

FIG. 9.84
Defining the conditions for maximum power to a load using the Norton equivalent circuit.

$$R_L = R_N \quad \textbf{(9.5)}$$

This result [Eq. (9.5)] will be used to its fullest advantage in the analysis of transistor networks, where the most frequently applied transistor circuit model uses a current source rather than a voltage source.

For the Norton circuit in Fig. 9.84,

$$P_{L_{max}} = \frac{I_N^2 R_N}{4} \quad \text{(W)} \quad \textbf{(9.6)}$$

EXAMPLE 9.14 A dc generator, battery, and laboratory supply are connected to resistive load R_L in Fig. 9.85.

a. For each, determine the value of R_L for maximum power transfer to R_L.
b. Under maximum power conditions, what are the current level and the power to the load for each configuration?
c. What is the efficiency of operation for each supply in part (b)?
d. If a load of 1 kΩ were applied to the laboratory supply, what would the power delivered to the load be? Compare your answer to the level of part (b). What is the level of efficiency?
e. For each supply, determine the value of R_L for 75% efficiency.

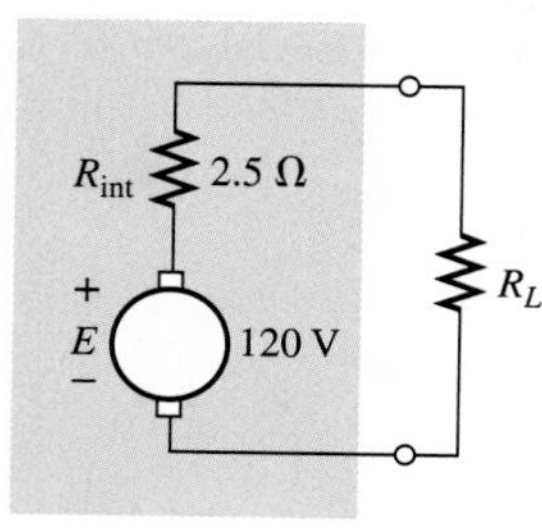

(a) dc generator

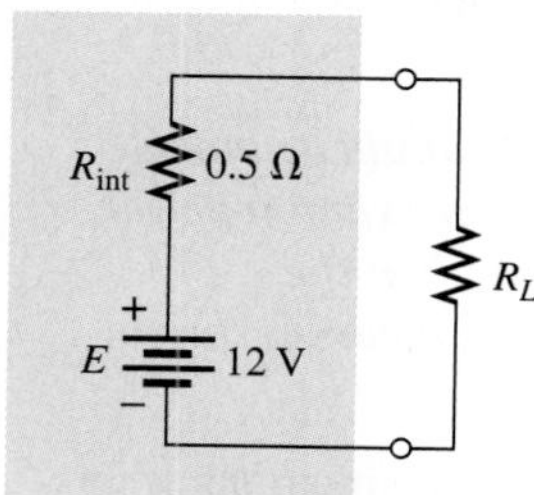

(b) Battery

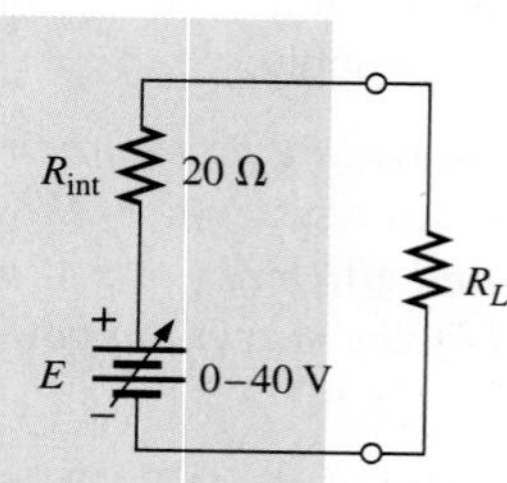

(c) Laboratory supply

FIG. 9.85
Example 9.14.

Solutions:

a. For the dc generator,

$$R_L = R_{Th} = R_{int} = \mathbf{2.5\ \Omega}$$

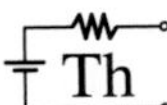

For the 12 V car battery,

$$R_L = R_{Th} = R_{\text{int}} = \mathbf{0.05\ \Omega}$$

For the dc laboratory supply,

$$R_L = R_{Th} = R_{\text{int}} = \mathbf{20\ \Omega}$$

b. For the dc generator,

$$P_{L_{\max}} = \frac{E_{Th}^2}{4R_{Th}} = \frac{E^2}{4R_{\text{int}}} = \frac{(120\text{ V})^2}{4(2.5\ \Omega)} = \mathbf{1.44\ kW}$$

For the 12 V car battery,

$$P_{L_{\max}} = \frac{E_{Th}^2}{4R_{Th}} = \frac{E^2}{4R_{\text{int}}} = \frac{(12\text{ V})^2}{4(0.05\ \Omega)} = \mathbf{720\ W}$$

For the dc laboratory supply,

$$P_{L_{\max}} = \frac{E_{Th}^2}{4R_{Th}} = \frac{E^2}{4R_{\text{int}}} = \frac{(40\text{ V})^2}{4(20\ \Omega)} = \mathbf{20\ W}$$

c. They are all operating under a 50% efficiency level because $R_L = R_{Th}$.

d. The power to the load is determined as follows:

$$I_L = \frac{E}{R_{\text{int}} + R_L} = \frac{40\text{ V}}{20\ \Omega + 1000\ \Omega} = \frac{40\text{ V}}{1020\ \Omega} = 39.22\text{ mA}$$

and $\quad P_L = I_L^2 R_L = (39.22\text{ mA})^2(1000\ \Omega) = \mathbf{1.54\ W}$

The power level is significantly less than the 20 W achieved in part (b). The efficiency level is

$$\eta\% = \frac{P_L}{P_s} \times 100\% = \frac{1.54\text{ W}}{EI_s} \times 100\% = \frac{1.54\text{ W}}{(40\text{ V})(39.22\text{ mA})} \times 100\%$$

$$= \frac{1.54\text{ W}}{1.57\text{ W}} \times 100\% = \mathbf{98.09\%}$$

which is markedly higher than achieved under maximum power conditions—albeit at the expense of the power level.

e. For the dc generator,

$$\eta = \frac{P_o}{P_s} = \frac{R_L}{R_{Th} + R_L} \qquad (\eta \text{ in decimal form})$$

and

$$\eta = \frac{R_L}{R_{Th} + R_L}$$

$$\eta(R_{Th} + R_L) = R_L$$

$$\eta R_{Th} + \eta R_L = R_L$$

$$R_L(1 - \eta) = \eta R_{Th}$$

and

$$\boxed{R_L = \frac{\eta R_{Th}}{1 - \eta}} \qquad \textbf{(9.7)}$$

$$R_L = \frac{0.75(2.5\ \Omega)}{1 - 0.75} = \mathbf{7.5\ \Omega}$$

For the battery,

$$R_L = \frac{0.75(0.05\ \Omega)}{1 - 0.75} = \mathbf{0.15\ \Omega}$$

For the laboratory supply,

$$R_L = \frac{0.75(20\ \Omega)}{1 - 0.75} = \mathbf{60\ \Omega}$$

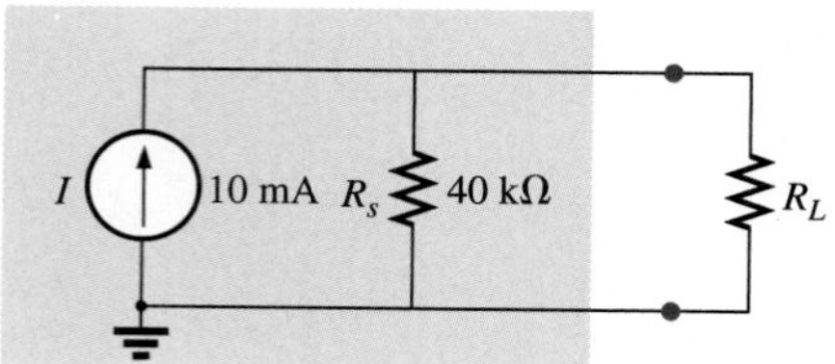

FIG. 9.86
Example 9.15.

EXAMPLE 9.15 The analysis of a transistor network resulted in the reduced equivalent in Fig. 9.86.

a. Find the load resistance that will result in maximum power transfer to the load, and find the maximum power delivered.
b. If the load were changed to 68 kΩ, would you expect a fairly high level of power transfer to the load based on the results of part (a)? What would the new power level be? Is your initial assumption verified?
c. If the load were changed to 8.2 kΩ, would you expect a fairly high level of power transfer to the load based on the results of part (a)? What would the new power level be? Is your initial assumption verified?

Solutions:

a. Replacing the current source by an open-circuit equivalent results in

$$R_{Th} = R_s = 40\ \text{k}\Omega$$

Restoring the current source and finding the open-circuit voltage at the output terminals results in

$$E_{Th} = V_{oc} = IR_s = (10\ \text{mA})(40\ \text{k}\Omega) = 400\ \text{V}$$

For maximum power transfer to the load,

$$R_L = R_{Th} = \mathbf{40\ k\Omega}$$

with a maximum power level of

$$P_{L_{\text{max}}} = \frac{E_{Th}^2}{4R_{Th}} = \frac{(400\ \text{V})^2}{4(40\ \text{k}\Omega)} = \mathbf{1\ W}$$

b. Yes, because the 68 kΩ load is greater (note Fig. 9.80) than the 40 kΩ load, but relatively close in magnitude.

$$I_L = \frac{E_{Th}}{R_{Th} + R_L} = \frac{400\ \text{V}}{40\ \text{k}\Omega + 68\ \text{k}\Omega} = \frac{400}{108\ \text{k}\Omega} \cong 3.7\ \text{mA}$$

$$P_L = I_L^2 R_L = (3.7\ \text{mA})^2(68\ \text{k}\Omega \cong \mathbf{0.93\ W}$$

Yes, the power level of 0.93 W compared to the 1 W level of part (a) verifies the assumption.

c. No, 8.2 kΩ is quite a bit less (note Fig. 9.80) than the 40 kΩ value.

$$I_L = \frac{E_{Th}}{R_{Th} + R_L} = \frac{400\ \text{V}}{40\ \text{k}\Omega + 8.2\ \text{k}\Omega} = \frac{400\ \text{V}}{48.2\ \text{k}\Omega} \cong 8.3\ \text{mA}$$

$$P_L = I_L^2 R_L = (8.3\ \text{mA})^2(8.2\ \text{k}\Omega) \cong \mathbf{0.57\ W}$$

Yes, the power level of 0.57 W compared to the 1 W level of part (a) verifies the assumption.

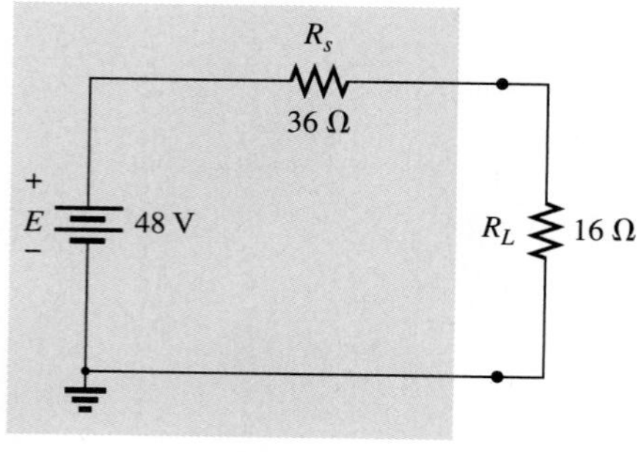

FIG. 9.87
dc supply with a fixed 16 Ω load (Example 9.16).

EXAMPLE 9.16 In Fig. 9.87, a fixed load of 16 Ω is applied to a 48 V supply with an internal resistance of 36 Ω.

a. For the conditions in Fig. 9.87, what is the power delivered to the load and lost to the internal resistance of the supply?
b. If the designer has some control over the internal resistance level of the supply, what value should he/she make it for maximum power to the load? What is the maximum power to the load? How does it compare to the level obtained in part (a)?
c. Without making a single calculation, if the designer could change the internal resistance to 22 Ω or 8.2 Ω, which value would result in more power to the load? Verify your conclusion by calculating the power to the load for each value.

Solutions:

a. $$I_L = \frac{E}{R_s + R_L} = \frac{48 \text{ V}}{36\ \Omega + 16\ \Omega} = \frac{48 \text{ V}}{52\ \Omega} = 923.1 \text{ mA}$$

$$P_{R_s} = I_L^2 R_s = (923.1 \text{ mA})^2(36\ \Omega) = \mathbf{30.68\ W}$$
$$P_L = I_L^2 R_L = (923.1 \text{ mA})^2(16\ \Omega) = \mathbf{13.63\ W}$$

b. Be careful here. The quick response is to make the source resistance R_s equal to the load resistance to satisfy the criteria of the maximum power transfer theorem. However, this is a totally different type of problem from what was examined earlier in this section. If the load is fixed, the smaller the source resistance R_s, the more applied voltage will reach the load and the less will be lost in the internal series resistor. In fact, the source resistance should be made as small as possible. If zero ohms were possible for R_s, the voltage across the load would be the full supply voltage, and the power delivered to the load would equal

$$P_L = \frac{V_L^2}{R_L} = \frac{(48 \text{ V})^2}{16\ \Omega} = \mathbf{144\ W}$$

which is more than 10 times the value with a source resistance of 36 Ω.

c. Again, forget the impact in Fig. 9.80: The smaller the source resistance, the greater the power to the fixed 16 Ω load. Therefore, the 8.2 Ω resistance level results in a higher power transfer to the load than the 22 Ω resistor.

For $R_s = 8.2\ \Omega$:

$$I_L = \frac{E}{R_s + R_L} = \frac{48 \text{ V}}{8.2\ \Omega + 16\ \Omega} = \frac{48 \text{ V}}{24.2\ \Omega} = 1.983 \text{ A}$$

and $$P_L = I_L^2 R_L = (1.983 \text{ A})^2(16\ \Omega) \cong \mathbf{62.92\ W}$$

For $R_s = 22\ \Omega$:

$$I_L = \frac{E}{R_s + R_L} = \frac{48 \text{ V}}{22\ \Omega + 16\ \Omega} = \frac{48 \text{ V}}{38\ \Omega} = 1.263 \text{ A}$$

and $$P_L = I_L^2 R_L = (1.263 \text{ A})^2(16\ \Omega) \cong \mathbf{25.52\ W}$$

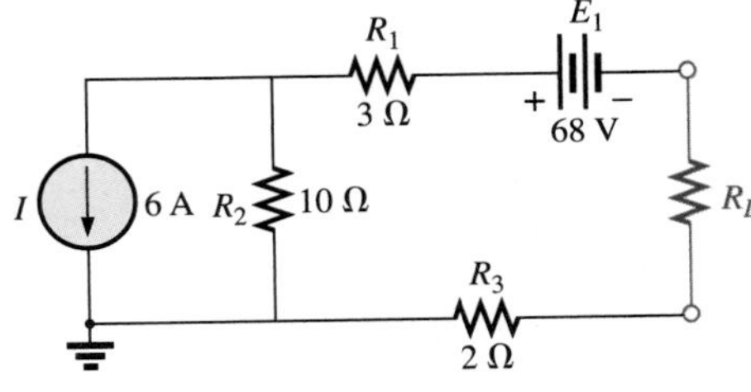

FIG. 9.88
Example 9.17.

EXAMPLE 9.17 Given the network in Fig. 9.88, find the value of R_L for maximum power to the load, and find the maximum power to the load.

Solution: The Thévenin resistance is determined from Fig. 9.89.

$$R_{Th} = R_1 + R_2 + R_3 = 3\ \Omega + 10\ \Omega + 2\ \Omega = 15\ \Omega$$

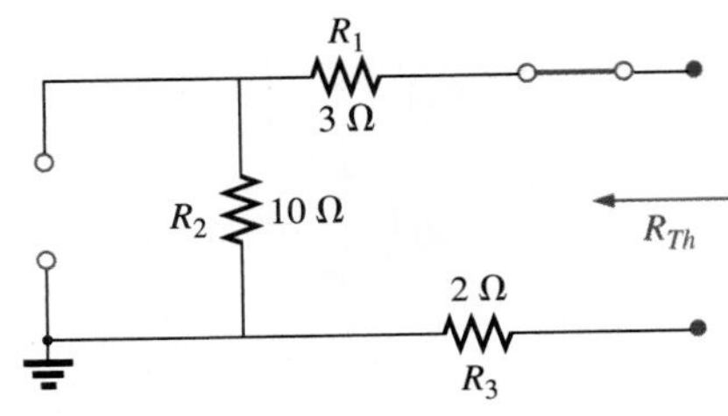

FIG. 9.89
Determining R_{Th} for the network external to resistor R_L in Fig. 9.88.

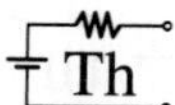

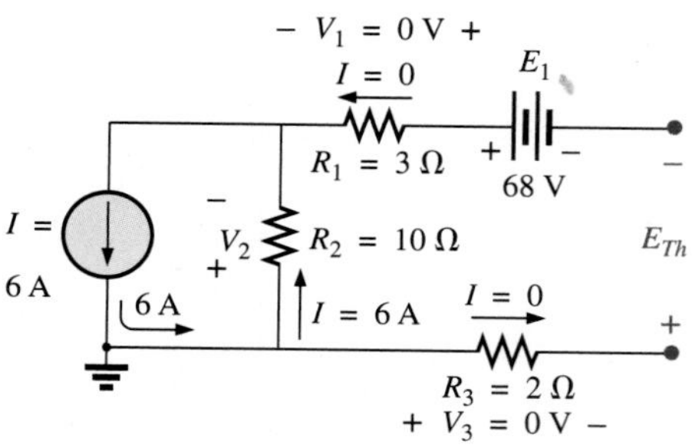

FIG. 9.90
Determining E_{Th} for the network external to resistor R_L in Fig. 9.88.

so that
$$R_L = R_{Th} = \mathbf{15\ \Omega}$$

The Thévenin voltage is determined using Fig. 9.90, where

$$V_1 = V_3 = 0\text{ V} \quad \text{and} \quad V_2 = I_2R_2 = IR_2 = (6\text{ A})(10\ \Omega) = 60\text{ V}$$

Applying Kirchhoff's voltage law:

$$-V_2 - E + E_{Th} = 0$$

and
$$E_{Th} = V_2 + E = 60\text{ V} + 68\text{ V} = \mathbf{128\ V}$$

with the maximum power equal to

$$P_{L_{\max}} = \frac{E_{Th}^2}{4R_{Th}} = \frac{(128\text{ V})^2}{4(15\text{ k}\Omega)} = \mathbf{273.07\ W}$$

9.6 MILLMAN'S THEOREM

Through the application of **Millman's theorem,** any number of parallel voltage sources can be reduced to one. In Fig. 9.91, for example, the three voltage sources can be reduced to one. This permits finding the current through or voltage across R_L without having to apply a method such as mesh analysis, nodal analysis, superposition, and so on. The theorem can best be described by applying it to the network in Fig. 9.91. Basically, three steps are included in its application.

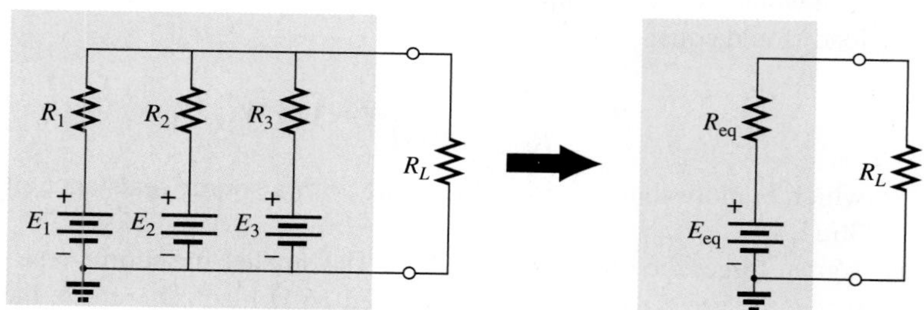

FIG. 9.91
Demonstrating the effect of applying Millman's theorem.

Step 1: Convert all voltage sources to current sources as outlined in Section 8.3. This is performed in Fig. 9.92 for the network in Fig. 9.91.

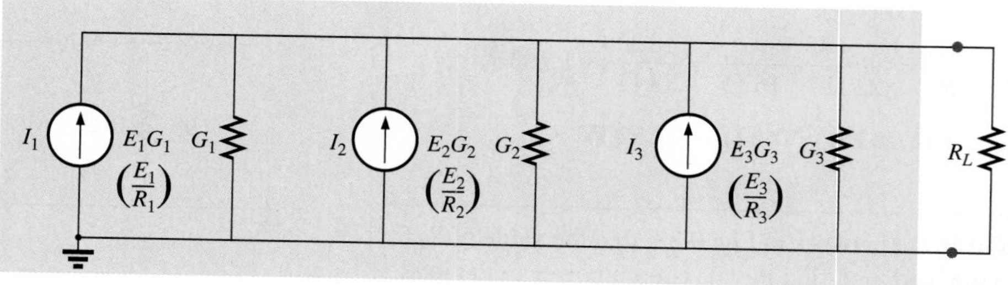

FIG. 9.92
Converting all the sources in Fig. 9.91 to current sources.

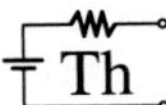

Step 2: Combine parallel current sources as described in Section 8.4. The resulting network is shown in Fig. 9.93, where

$$I_T = I_1 + I_2 + I_3 \quad \text{and} \quad G_T = G_1 + G_2 + G_3$$

Step 3: Convert the resulting current source to a voltage source, and the desired single-source network is obtained, as shown in Fig. 9.94.

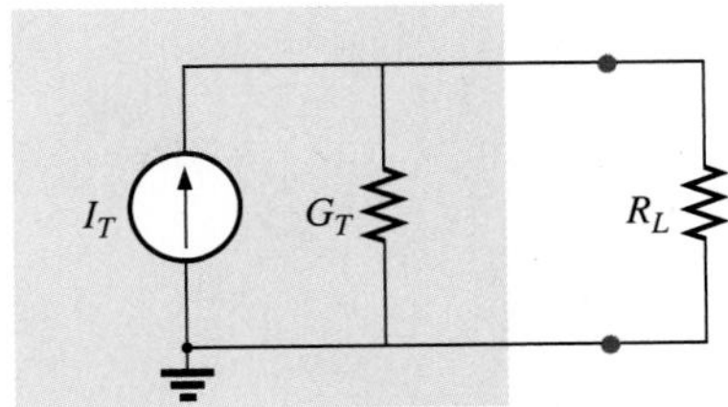

FIG. 9.93
Reducing all the current sources in Fig. 9.92 to a single current source.

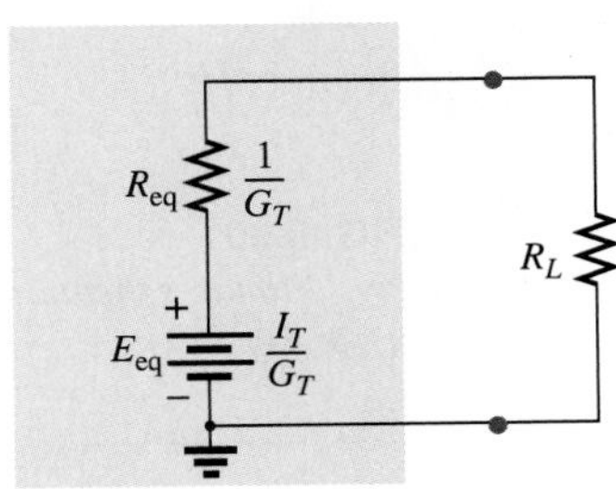

FIG. 9.94
Converting the current source in Fig. 9.93 to a voltage source.

In general, Millman's theorem states that for any number of parallel voltage sources,

$$E_{eq} = \frac{I_T}{G_T} = \frac{\pm I_1 \pm I_2 \pm I_3 \pm \cdots \pm I_N}{G_1 + G_2 + G_3 + \cdots + G_N}$$

or

$$E_{eq} = \frac{\pm E_1G_1 \pm E_2G_2 \pm E_3G_3 \pm \cdots \pm E_NG_N}{G_1 + G_2 + G_3 + \cdots + G_N} \tag{9.8}$$

The plus-and-minus signs appear in Eq. (9.8) to include those cases where the sources may not be supplying energy in the same direction. (Note Example 9.18.)

The equivalent resistance is

$$R_{eq} = \frac{1}{G_T} = \frac{1}{G_1 + G_2 + G_3 + \cdots + G_N} \tag{9.9}$$

In terms of the resistance values,

$$E_{eq} = \frac{\pm\dfrac{E_1}{R_1} \pm \dfrac{E_2}{R_2} \pm \dfrac{E_3}{R_3} \pm \cdots \pm \dfrac{E_N}{R_N}}{\dfrac{1}{R_1} + \dfrac{1}{R_2} + \dfrac{1}{R_3} + \cdots + \dfrac{1}{R_N}} \tag{9.10}$$

and

$$R_{eq} = \frac{1}{\dfrac{1}{R_1} + \dfrac{1}{R_2} + \dfrac{1}{R_3} + \cdots + \dfrac{1}{R_N}} \tag{9.11}$$

Because of the relatively few direct steps required, you may find it easier to apply each step rather than memorizing and employing Eqs. (9.8) through (9.11).

EXAMPLE 9.18 Using Millman's theorem, find the current through and voltage across the resistor R_L in Fig. 9.95.

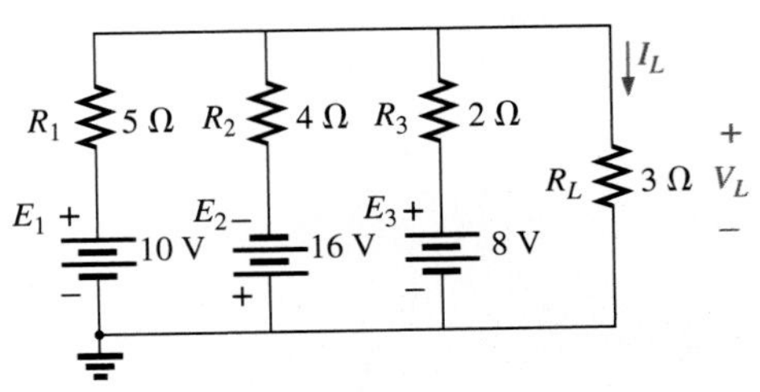

FIG. 9.95
Example 9.18.

Solution: By Eq. (9.10),

$$E_{eq} = \frac{+\dfrac{E_1}{R_1} - \dfrac{E_2}{R_2} + \dfrac{E_3}{R_3}}{\dfrac{1}{R_1} + \dfrac{1}{R_2} + \dfrac{1}{R_3}}$$

The minus sign is used for E_2/R_2 because that supply has the opposite polarity of the other two. The chosen reference direction is therefore that of E_1 and E_3. The total conductance is unaffected by the direction, and

$$E_{eq} = \frac{+\dfrac{10\text{ V}}{5\ \Omega} - \dfrac{16\text{ V}}{4\ \Omega} + \dfrac{8\text{ V}}{2\ \Omega}}{\dfrac{1}{5\ \Omega} + \dfrac{1}{4\ \Omega} + \dfrac{1}{2\ \Omega}} = \frac{2\text{ A} - 4\text{ A} + 4\text{ A}}{0.2\text{ S} + 0.25\text{ S} + 0.5\text{ S}}$$

$$= \frac{2\text{ A}}{0.95\text{ S}} = \mathbf{2.11\ V}$$

with
$$R_{eq} = \frac{1}{\dfrac{1}{5\ \Omega} + \dfrac{1}{4\ \Omega} + \dfrac{1}{2\ \Omega}} = \frac{1}{0.95\text{ S}} = \mathbf{1.05\ \Omega}$$

The resultant source is shown in Fig. 9.96, and

$$I_L = \frac{2.11\text{ V}}{1.05\ \Omega + 3\ \Omega} = \frac{2.11\text{ V}}{4.05\ \Omega} = \mathbf{0.52\ A}$$

with
$$V_L = I_L R_L = (0.52\text{ A})(3\ \Omega) = \mathbf{1.56\ V}$$

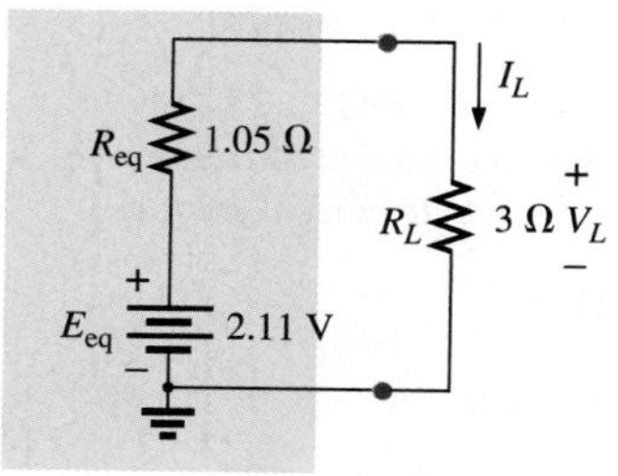

FIG. 9.96
The result of applying Millman's theorem to the network in Fig. 9.95.

EXAMPLE 9.19 Let us now consider the type of problem encountered in the introduction to mesh and nodal analysis in Chapter 8. Mesh analysis was applied to the network of Fig. 9.97 (Example 8.12). Let us now use Millman's theorem to find the current through the 2 Ω resistor and compare the results.

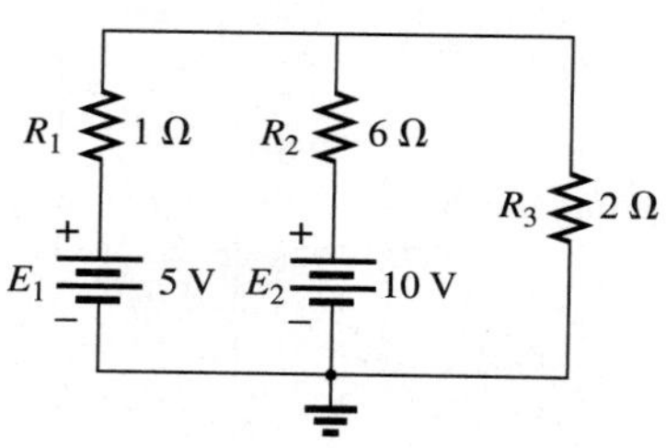

FIG. 9.97
Example 9.19.

Solutions:

a. Let us first apply each step and, in the (b) solution, Eq. (9.10). Converting sources yields Fig. 9.98. Combining sources and parallel conductance branches (Fig. 9.99) yields

$$I_T = I_1 + I_2 = 5\text{ A} + \frac{5}{3}\text{ A} = \frac{15}{3}\text{ A} + \frac{5}{3}\text{ A} = \frac{20}{3}\text{ A}$$

$$G_T = G_1 + G_2 = 1\text{ S} + \frac{1}{6}\text{ S} = \frac{6}{6}\text{ S} + \frac{1}{6}\text{ S} = \frac{7}{6}\text{ S}$$

Converting the current source to a voltage source (Fig. 9.100), we obtain

$$E_{eq} = \frac{I_T}{G_T} = \frac{\dfrac{20}{3}\text{ A}}{\dfrac{7}{6}\text{ S}} = \frac{(6)(20)}{(3)(7)}\text{ V} = \mathbf{\frac{40}{7}\ V}$$

and
$$R_{eq} = \frac{1}{G_T} = \frac{1}{\dfrac{7}{6}\text{ S}} = \mathbf{\frac{6}{7}\ \Omega}$$

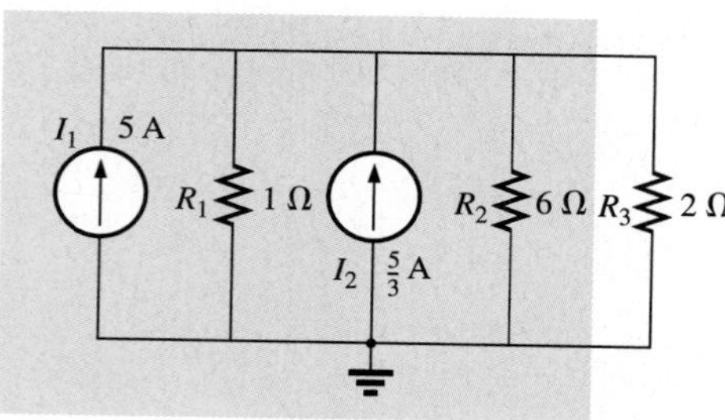

FIG. 9.98
Converting the sources in Fig. 9.97 to current sources.

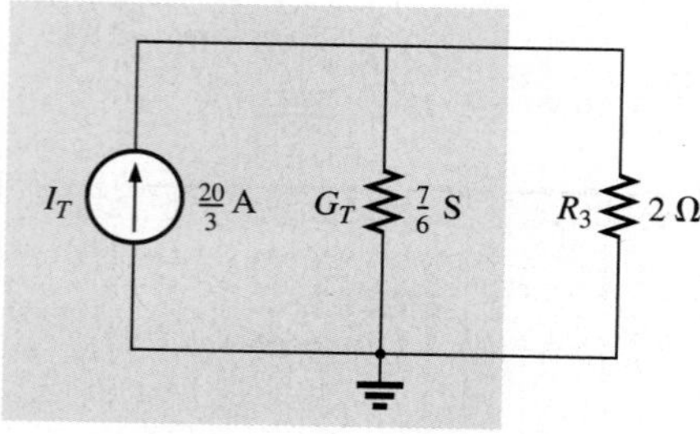

FIG. 9.99
Reducing the current sources in Fig. 9.98 to a single source.

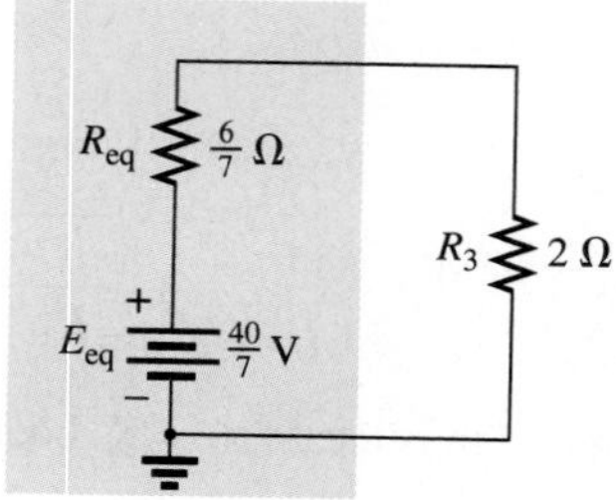

FIG. 9.100
Converting the current source in Fig. 9.99 to a voltage source.

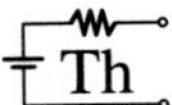

so that

$$I_{2\,\Omega} = \frac{E_{eq}}{R_{eq} + R_3} = \frac{\frac{40}{7}\,\text{V}}{\frac{6}{7}\,\Omega + 2\,\Omega} = \frac{\frac{40}{7}\,\text{V}}{\frac{6}{7}\,\Omega + \frac{14}{7}\,\Omega} = \frac{40\,\text{V}}{20\,\Omega} = \mathbf{2\,A}$$

which agrees with the result obtained in Example 8.18.

b. Let us now simply apply the proper equation, Eq. (9.10):

$$E_{eq} = \frac{+\frac{5\,\text{V}}{1\,\Omega} + \frac{10\,\text{V}}{6\,\Omega}}{\frac{1}{1\,\Omega} + \frac{1}{6\,\Omega}} = \frac{\frac{30\,\text{V}}{6\,\Omega} + \frac{10\,\text{V}}{6\,\Omega}}{\frac{6}{6\,\Omega} + \frac{1}{6\,\Omega}} = \mathbf{\frac{40}{7}\,V}$$

and

$$R_{eq} = \frac{1}{\frac{1}{1\,\Omega} + \frac{1}{6\,\Omega}} = \frac{1}{\frac{6}{6\,\Omega} + \frac{1}{6\,\Omega}} = \frac{1}{\frac{7}{6}\,\text{S}} = \mathbf{\frac{6}{7}\,\Omega}$$

which are the same values obtained above.

The dual of Millman's theorem (Fig. 9.91) appears in Fig. 9.101. It can be shown that I_{eq} and R_{eq}, as in Fig. 9.101, are given by

$$I_{eq} = \frac{\pm I_1R_1 \pm I_2R_2 \pm I_3R_3}{R_1 + R_2 + R_3} \tag{9.12}$$

and

$$R_{eq} = R_1 + R_2 + R_3 \tag{9.13}$$

The derivation appears as a problem at the end of the chapter.

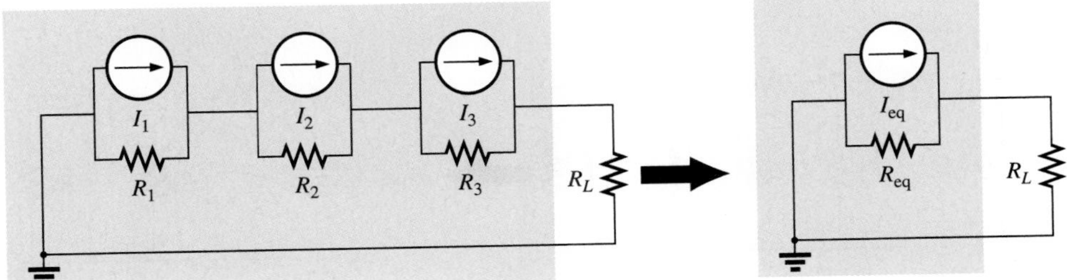

FIG. 9.101
The dual effect of Millman's theorem.

9.7 SUBSTITUTION THEOREM

The **substitution theorem** states the following:

If the voltage across and the current through any branch of a dc bilateral network are known, this branch can be replaced by any combination of elements that will maintain the same voltage across and current through the chosen branch.

More simply, the theorem states that for branch equivalence, the terminal voltage and current must be the same. Consider the circuit in Fig. 9.102,

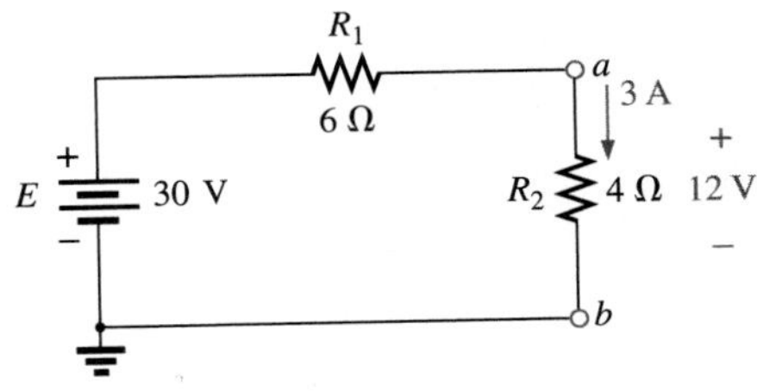

FIG. 9.102
Demonstrating the effect of the substitution theorem.

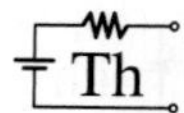

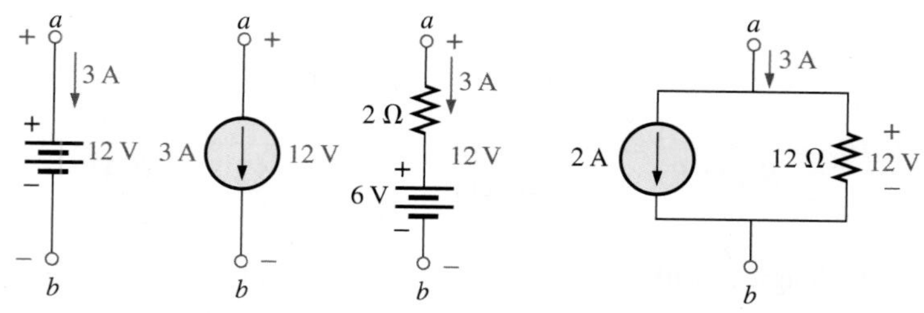

FIG. 9.103
Equivalent branches for the branch a-b in Fig. 9.102.

in which the voltage across and current through the branch *a-b* are determined. Through the use of the substitution theorem, a number of equivalent *a-a′* branches are shown in Fig. 9.103.

Note that for each equivalent, the terminal voltage and current are the same. Also consider that the response of the remainder of the circuit in Fig. 9.102 is unchanged by substituting any one of the equivalent branches. As demonstrated by the single-source equivalents in Fig. 9.103, *a known potential difference and current in a network can be replaced by an ideal voltage source and current source, respectively.*

Understand that this theorem cannot be used to *solve* networks with two or more sources that are not in series or parallel. For it to be applied, a potential difference or current value must be known or found using one of the techniques discussed earlier. One application of the theorem is shown in Fig. 9.104. Note that in the figure the known potential difference V was replaced by a voltage source, permitting the isolation of the portion of the network including R_3, R_4, and R_5. Recall that this was basically the approach used in the analysis of the ladder network as we worked our way back toward the terminal resistance R_5.

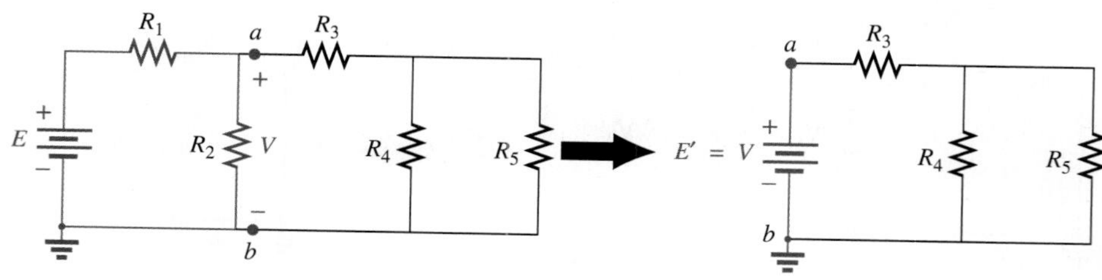

FIG. 9.104
Demonstrating the effect of knowing a voltage at some point in a complex network.

The current source equivalence of the above is shown in Fig. 9.105, where a known current is replaced by an ideal current source, permitting the isolation of R_4 and R_5.

Recall from the discussion of bridge networks that $V = 0$ and $I = 0$ were replaced by a short circuit and an open circuit, respectively. This substitution is a very specific application of the substitution theorem.

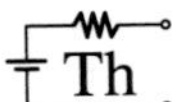

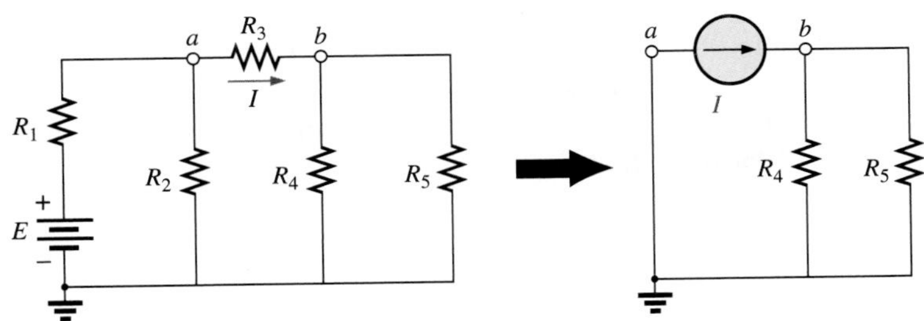

FIG. 9.105
Demonstrating the effect of knowing a current at some point in a complex network.

9.8 RECIPROCITY THEOREM

The **reciprocity theorem** is applicable only to single-source networks. It is, therefore, not a theorem used in the analysis of multisource networks described thus far. The theorem states the following:

The current I in any branch of a network, due to a single voltage source E anywhere else in the network, will equal the current through the branch in which the source was originally located if the source is placed in the branch in which the current I was originally measured.

In other words, the location of the voltage source and the resulting current may be interchanged without a change in current. The theorem requires that the polarity of the voltage source have the same correspondence with the direction of the branch current in each position.

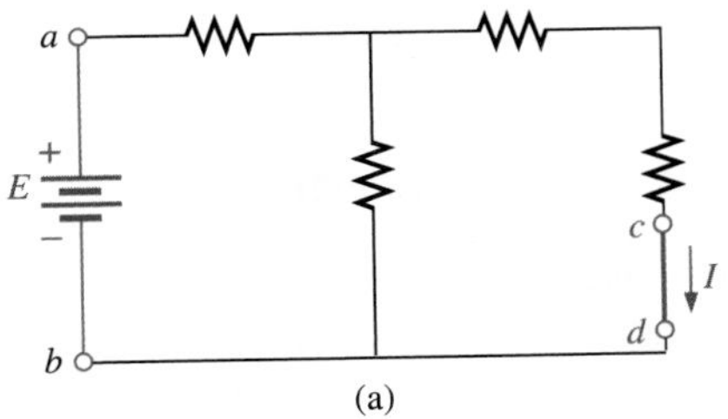

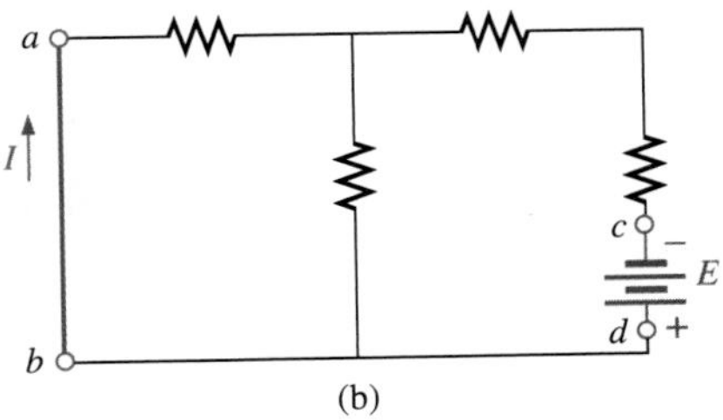

FIG. 9.106
Demonstrating the impact of the reciprocity theorem.

In the representative network in Fig. 9.106(a), the current I due to the voltage source E was determined. If the position of each is interchanged as shown in Fig. 9.106(b), the current I will be the same value as indicated. To demonstrate the validity of this statement and the theorem, consider the network in Fig. 9.107, in which values for the elements of Fig. 9.106(a) have been assigned.

The total resistance is

$$R_T = R_1 + R_2 \| (R_3 + R_4) = 12\ \Omega + 6\ \Omega \| (2\ \Omega + 4\ \Omega)$$
$$= 12\ \Omega + 6\ \Omega \| 6\ \Omega = 12\ \Omega + 3\ \Omega = 15\ \Omega$$

and

$$I_s = \frac{E}{R_T} = \frac{45\ \text{V}}{15\ \Omega} = 3\ \text{A}$$

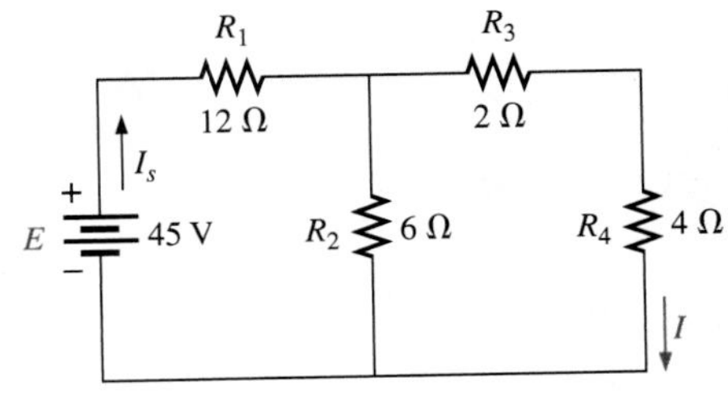

FIG. 9.107
Finding the current I due to a source E.

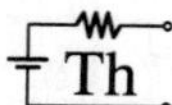

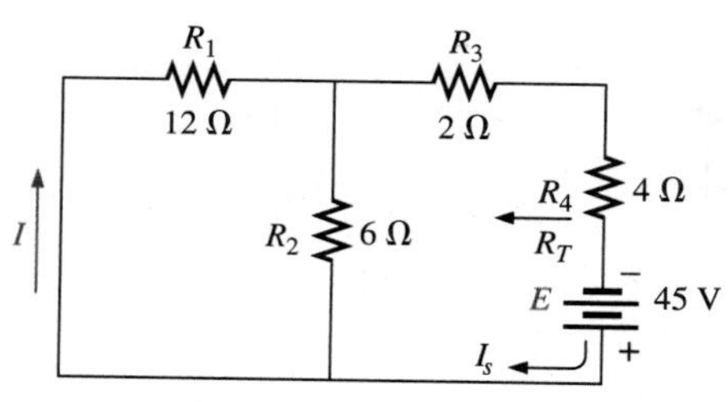

FIG. 9.108
Interchanging the location of E and I of Fig. 9.107 to demonstrate the validity of the reciprocity theorem.

with $$I = \frac{3\text{ A}}{2} = \mathbf{1.5\ A}$$

For the network in Fig. 9.108, which corresponds to that in Fig. 9.106(b), we find

$$R_T = R_4 + R_3 + R_1 \parallel R_2$$
$$= 4\ \Omega + 2\ \Omega + 12\ \Omega \parallel 6\ \Omega = 10\ \Omega$$

and $$I_s = \frac{E}{R_T} = \frac{45\text{ V}}{10\ \Omega} = 4.5\text{ A}$$

so that $$I = \frac{(6\ \Omega)(4.5\text{ A})}{12\ \Omega + 6\ \Omega} = \frac{4.5\text{ A}}{3} = \mathbf{1.5\ A}$$

which agrees with the above.

The uniqueness and power of this theorem can best be demonstrated by considering a complex, single-source network such as the one shown in Fig. 9.109.

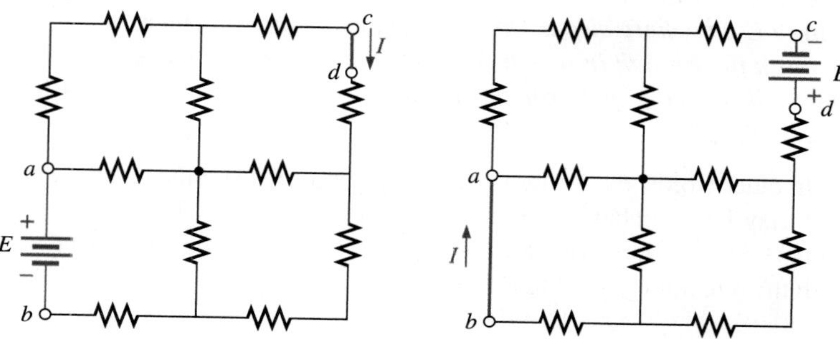

FIG. 9.109
Demonstrating the power and uniqueness of the reciprocity theorem.

9.9 COMPUTER ANALYSIS

Once you understand the mechanics of applying a software package or language, the opportunity to be creative and innovative presents itself. Through years of exposure and trial-and-error experiences, professional programmers develop a catalog of innovative techniques that are not only functional but very interesting and truly artistic in nature. Now that some of the basic operations associated with PSpice have been introduced, a few innovative maneuvers will be made in the examples to follow.

PSpice

Thévenin's Theorem The application of Thévenin's theorem requires an interesting maneuver to determine the Thévenin resistance. It is a maneuver, however, that has application beyond Thévenin's theorem whenever a resistance level is required. The network to be analyzed appears in Fig. 9.110 and is the same one analyzed in Example 9.10 (Fig. 9.48).

Since PSpice is not set up to measure resistance levels directly, a 1 A current source can be applied as shown in Fig. 9.111, and Ohm's law can be used to determine the magnitude of the Thévenin resistance in the following manner:

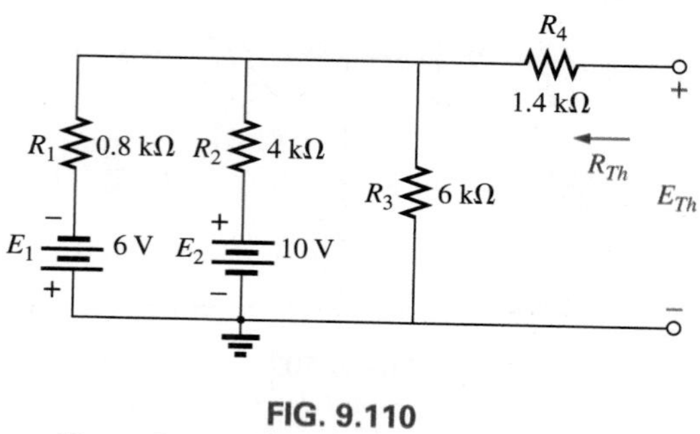

$$|R_{Th}| = \left|\frac{V_s}{I_s}\right| = \left|\frac{V_s}{1\text{ A}}\right| = |V_s| \qquad \textbf{(9.14)}$$

FIG. 9.110
Network to which PSpice is to be applied to determine E_{Th} and R_{Th}.

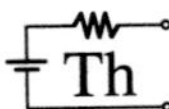

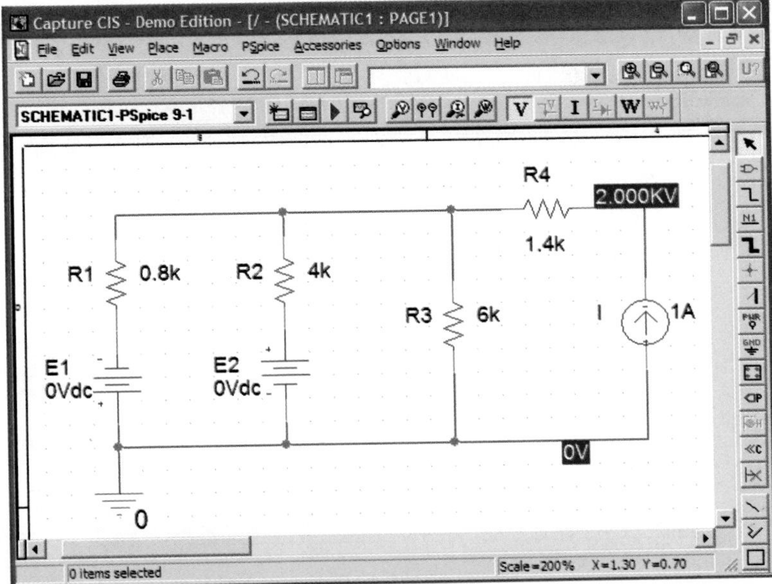

FIG. 9.111

Using PSpice to determine the Thévenin resistance of a network through the application of a 1 A current source.

In Eq. (9.14), since $I_s = 1$ A, the magnitude of R_{Th} in ohms is the same as the magnitude of the voltage V_s (in volts) across the current source. The result is that when the voltage across the current source is displayed, it can be read as ohms rather than volts.

When PSpice is applied, the network appears as shown in Fig. 9.111. Flip the voltage source E_1 and the current source by right-clicking on the source and choosing the **Mirror Vertically** option. Set both voltage sources to zero through the **Display Properties** dialog box obtained by double-clicking on the source symbol. The result of the **Bias Point** simulation is 2 kV across the current source. The Thévenin resistance is therefore 2 kΩ between the two terminals of the network to the left of the current source (to match the results of Example 9.10). In total, by setting the voltage source to 0 V, we have dictated that the voltage is the same at both ends of the voltage source, replicating the effect of a short-circuit connection between the two points.

For the open-circuit Thévenin voltage between the terminals of interest, the network must be constructed as shown in Fig. 9.112. The resistance of 1 T (= 1 million MΩ) is considered large enough to represent an open circuit to permit an analysis of the network using PSpice. PSpice does not recognize floating nodes and generates an error signal if a connection is not made from the top right node to ground. Both voltage sources are now set on their prescribed values, and a simulation results in 3 V across the 1 T resistor. The open-circuit Thévenin voltage is therefore 3 V, which agrees with the solution in Example 9.10.

Maximum Power Transfer The procedure for plotting a quantity versus a parameter of the network is now introduced. In this case, the output power versus values of load resistance is used to verify that maximum power is delivered to the load when its value equals the series Thévenin resistance. A number of new steps are introduced, but keep in mind that the method has broad application beyond Thévenin's theorem and is therefore well worth the learning process.

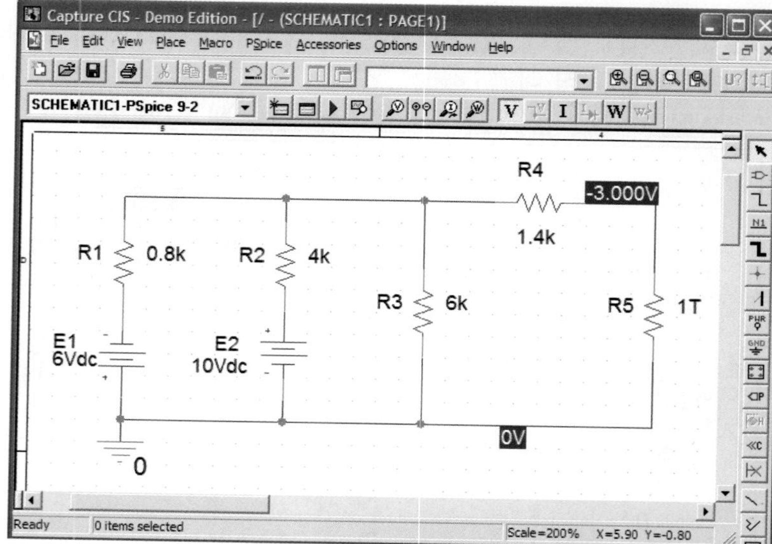

FIG. 9.112
Using PSpice to determine the Thévenin voltage for a network using a very large resistance value to represent the open-circuit condition between the terminals of interest.

The circuit to be analyzed appears in Fig. 9.113. The circuit is constructed in exactly the same manner as described earlier except for the value of the load resistance. Begin the process by starting a **New Project** labeled **PSpice 9-3,** and build the circuit in Fig. 9.113. For the moment, do not set the value of the load resistance.

The first step is to establish the value of the load resistance as a variable since it will not be assigned a fixed value. Double-click on the value of **RL** to obtain the **Display Properties** dialog box. For **Value,** type in **{Rval}** and click in place. The brackets (*not* parentheses) are required, but the variable does not have to be called **Rval**—it is the choice of the user.

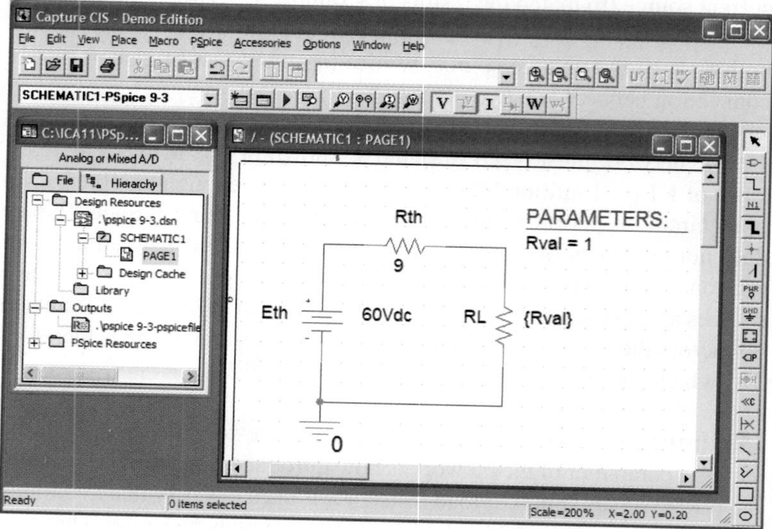

FIG. 9.113
Using PSpice to plot the power to R_L for a range of values for R_L.

Next select the **Place part** key to obtain the **Place Part** dialog box. If you are not already in the **Libraries** list, choose **Add Library** and add **SPECIAL** to the list. Select the **SPECIAL** library and scroll the **Part List** until **PARAM** appears. Select it; then click **OK** to obtain a rectangular box next to the cursor on the screen. Select a spot near **Rval,** and deposit the rectangle. The result is **PARAMETERS:** as shown in Fig. 9.113.

Next double-click on **PARAMETERS:** to obtain a **Property Editor** dialog box which should have **SCHEMATIC1:PAGE1** in the second column from the left. Now select the **New Column** option from the top list of choices to obtain the **Add New Column** dialog box. Under **Name,** enter **Rval** and under **Value,** enter **1** followed by an **OK** to leave the dialog box. The result is a return to the **Property Editor** dialog box but with **Rval** and its value (below **Rval**) added to the horizontal list. Now select **Rval/1** by clicking on **Rval** to surround **Rval** by a dashed line and add a black background around the **1.** Choose **Display** to produce the **Display Properties** dialog box, and select **Name and Value** followed by **OK.** Then exit the **Property Editor** dialog box (**X**) to display the screen in Fig. 9.113. Note that now the first value (1 Ω) of **Rval** is displayed.

We are now ready to set up the simulation process. Under **PSpice,** select the **New Simulation Profile** key to open the **New Simulation** dialog box. Enter **DC Sweep** under **Name** followed by **Create.** The **Simulation Settings-DC Sweep** dialog box appears. After selecting **Analysis,** select **DC Sweep** under the **Analysis type** heading. Then leave the **Primary Sweep** under the **Options** heading, and select **Global parameter** under the **Sweep variable.** The **Parameter name** should then be entered as **Rval.** For the **Sweep type,** the **Start value** should be 1 Ω; but if we use 1 Ω, the curve to be generated will start at 1 Ω, leaving a blank from 0 to 1 Ω. The curve will look incomplete. To solve this problem, select 0.001 Ω as the **Start value** (very close to 0 Ω) with an **Increment** of 1 Ω. Enter the **End value** as 30.001 Ω to ensure a calculation at $R_L = 30\ \Omega$. If we used 30 Ω as the end value, the last calculation would be at 29.001 Ω since 29.001 Ω + 1 Ω = 30.001 Ω, which is beyond the range of 30 Ω. The values of **RL** will therefore be 0.001 Ω, 1.001 Ω, 2.001 Ω, . . . 29.001 Ω, 30.001 Ω, and so on, although the plot will look as if the values were 0 Ω, 1 Ω, 2 Ω, 29 Ω, 30 Ω, and so on. Click **OK,** and select **Run** under **PSpice** to obtain the display in Fig. 9.114.

Note that there are no plots on the graph and that the graph extends to 35 Ω rather than 30 Ω as desired. It did not respond with a plot of power versus **RL** because we have not defined the plot of interest for the computer. To do this, select the **Add Trace** key (the key in the middle of the lower toolbar that looks like a red sawtooth waveform) or **Trace-Add Trace** from the top menu bar. Either choice results in the **Add Traces** dialog box. The most important region of this dialog box is the **Trace Expression** listing at the bottom. The desired trace can be typed in directly, or the quantities of interest can be chosen from the list of **Simulation Output Variables** and deposited in the **Trace Expression** listing. To find the power to **RL** for the chosen range of values for **RL,** select **W(RL)** in the listing; it then appears as the **Trace Expression.** Click **OK,** and the plot in Fig. 9.115 appears. Originally, the plot extended from 0 Ω to 35 Ω. We reduced the range to 0 Ω to 30 Ω by selecting **Plot-Axis Settings-X Axis-User Defined 0 to 30-OK.**

Select the **Toggle cursor** key (which looks like a red curve passing through the origin of a graph), and then left-click. A vertical line and a horizontal line appears, with the vertical line controlled by the position

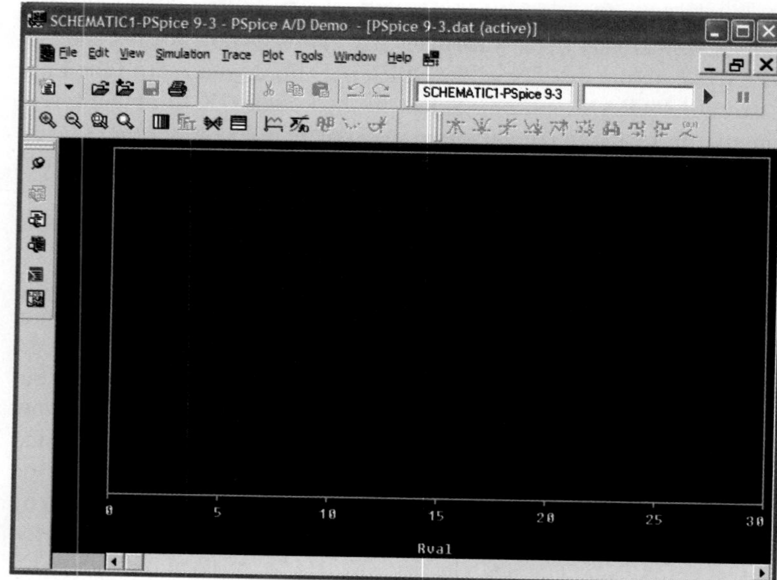

FIG. 9.114

Plot resulting from the dc sweep of R_L for the network in Fig. 9.113 before defining the parameters to be displayed.

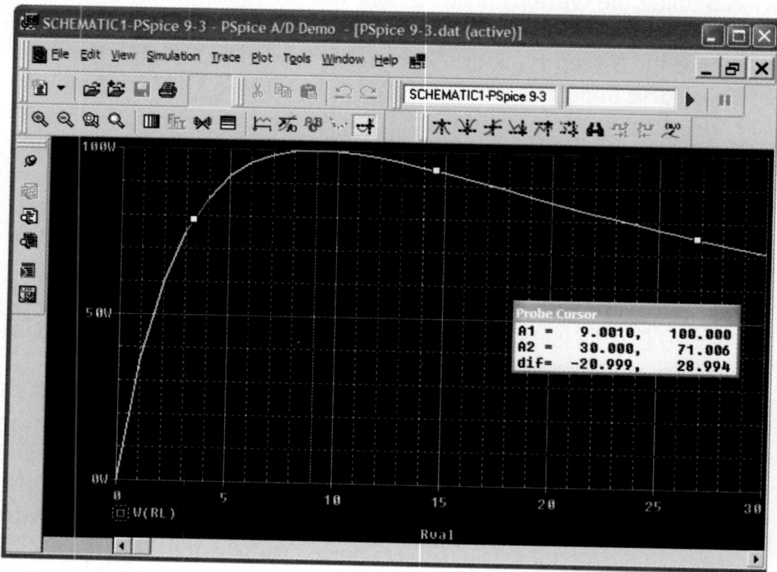

FIG. 9.115

A plot of the power delivered to R_L in Fig. 9.113 for a range of values for R_L extending from 0 Ω to 30 Ω.

of the cursor. Moving the cursor to the peak value results in **A1** = 9.0010 as the x value and 100.000 W as the y value, shown in the **Probe Cursor** box at the right of the screen. A second cursor can be generated by a right click, which was set at **RL** = 30.001 Ω to result in a power of 71.005 W. Notice also that the plot generated appears as a listing at the bottom left of the screen as **W(RL).**

Note that the power to **RL** can be determined in more ways than one from the **Add Traces** dialog box. For example, first enter a minus sign because of the resulting current direction through the resistor, and then select **V2 (RL)** followed by the multiplication of **I(RL)** using the multiplication operation under the **Functions** or **Macros** heading. The following expression appears in the **Trace Expression** box: **− V2(RL)*I(RL),** which is an expression having the basic power format of $P = V*I$. Click **OK,** and the same power curve in Fig. 9.115 appears. Other quantities, such as the voltage across the load and the current through the load, can be plotted against **RL** by following the sequence **Trace-Delete All Traces-Trace-Add Trace-V1(RL)** or **I(RL).**

Multisim

Superposition Let us now apply superposition to the network in Fig. 9.116, which appeared earlier as Fig. 9.9 in Example 9.3, to permit a comparison of resulting solutions. The current through R_2 is to be determined. Using methods described in earlier chapters for the application of Multisim, the network in Fig. 9.117 results to determine the effect of the 36 V voltage source. Note in Fig. 9.117 that both the voltage source and current source are present even though we are finding the contribution due solely to the voltage source. Obtain the voltage source by selecting the **Place Source** option at the top of the left toolbar to open the **Select a Component** dialog box. Then select **POWER_SOURCES** followed by **DC_POWER** as described in earlier chapters. You can also obtain the current source from the same dialog box by selecting **SIGNAL_CURRENT** under **Family** followed by **DC_CURRENT** under **Component.** The current source can be flipped vertically by right-clicking the source and selecting **Flip Vertical.** Set the current source to zero by left-clicking the

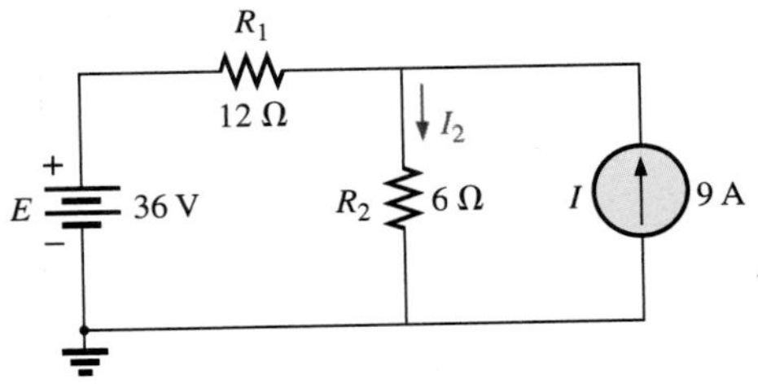

FIG. 9.116
Applying Multisim to determine the current I_2 using superposition.

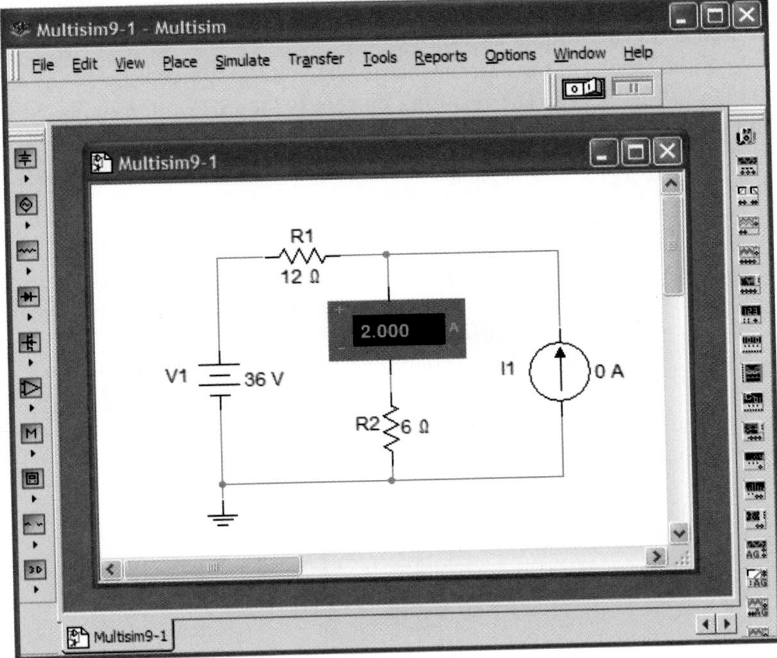

FIG. 9.117
Using Multisim to determine the contribution of the 36 V voltage source to the current through R_2.

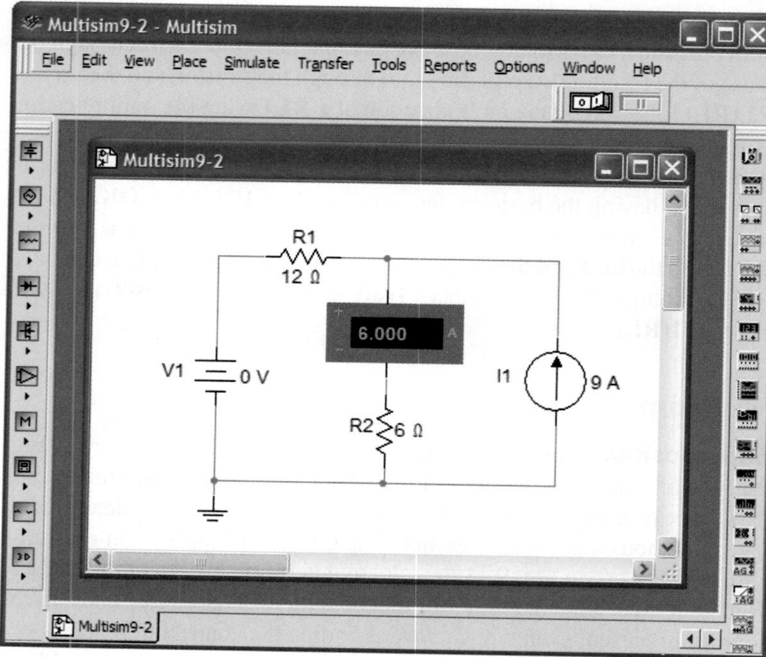

FIG. 9.118
Using Multisim to determine the contribution of the 9 A current source to the current through R_2.

source twice to obtain the **SIGNAL_CURRENT_SOURCES** dialog box. After choosing **Value,** set **Current(I)** to 0 A.

Following simulation, the results appear as in Fig. 9.117. The current through the 6 Ω resistor is 2 A due solely to the 36 V voltage source. The positive value for the 2 A reading reveals that the current due to the 36 V source is down through resistor R_2.

For the effects of the current source, the voltage source is set to 0 V as shown in Fig. 9.118. The resulting current is then 6 A through R_2, with the same direction as the contribution due to the voltage source.

The resulting current for the resistor R_2 is the sum of the two currents: $I_T = 2\text{ A} + 6\text{ A} = 8\text{ A}$, as determined in Example 9.3.

PROBLEMS

SECTION 9.2 Superposition Theorem

1. a. Using superposition, find the current through each resistor of the network in Fig. 9.119.

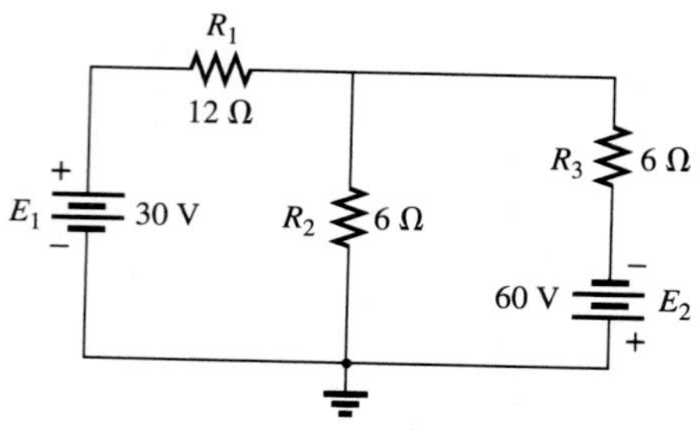

FIG. 9.119
Problem 1.

   b. Find the power delivered to R_1 for each source.
   c. Find the power delivered to R_1 using the total current through R_1.
   d. Does superposition apply to power effects? Explain.

2. Using superposition, find the current I through the 10 Ω resistor for the network in Fig. 9.120.

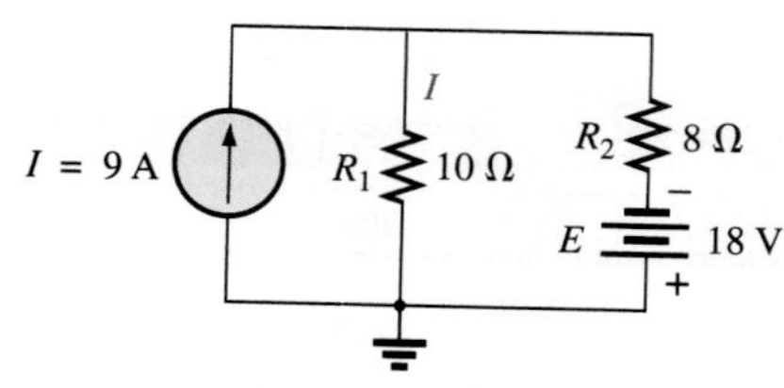

FIG. 9.120
Problem 2.

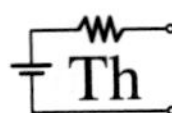

3. Using superposition, find the current I through the 24 V source in Fig. 9.121.

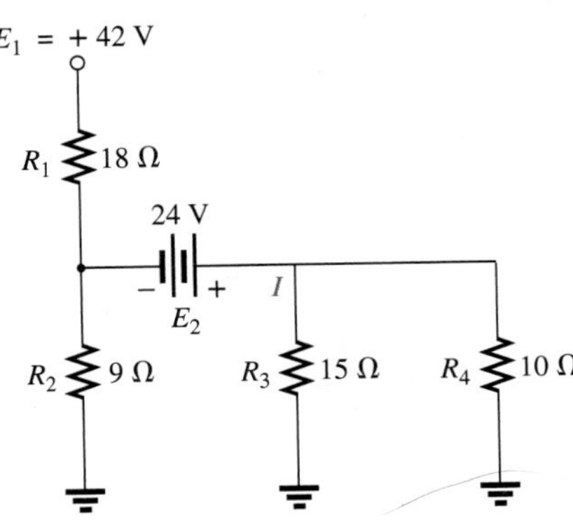

FIG. 9.121
Problem 3.

*4. Using superposition, find the current through R_1 for the network in Fig. 9.122.

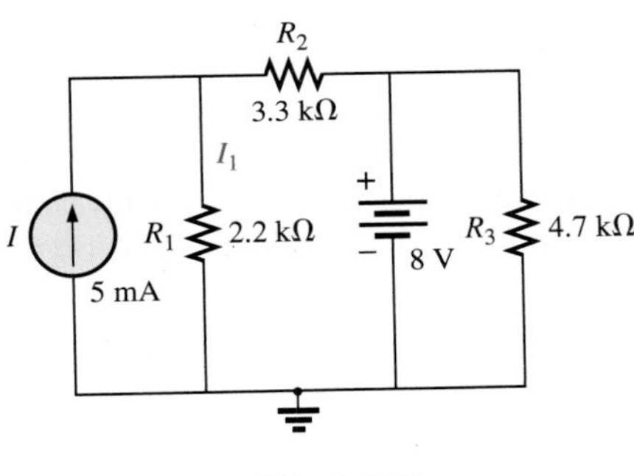

FIG. 9.122
Problem 4.

*5. Using superposition, find the voltage across the 6 A source in Fig. 9.123.

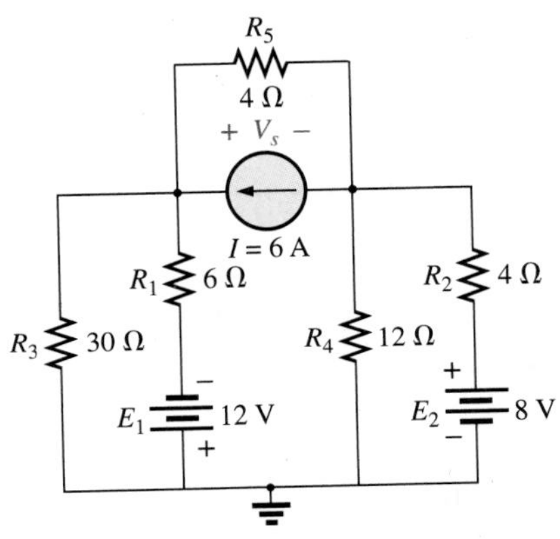

FIG. 9.123
Problem 5.

6. Using superposition, find the voltage V_2 for the network in Fig. 9.124.

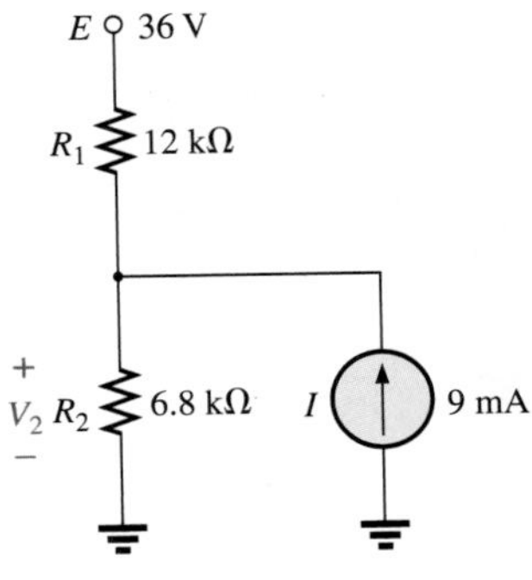

FIG. 9.124
Problems 6 and 42.

SECTION 9.3 Thévenin's Theorem

7. **a.** Find the Thévenin equivalent circuit for the network external to the resistor R in Fig. 9.125.
   b. Find the current through R when R is 2 Ω, 30 Ω, and 100 Ω.

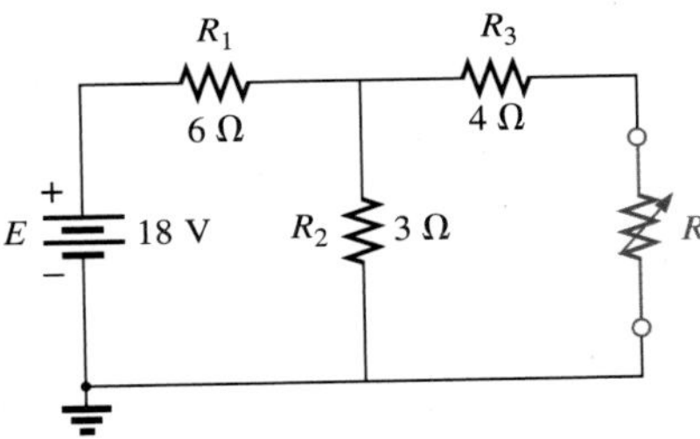

FIG. 9.125
Problem 7.

8. **a.** Find the Thévenin equivalent circuit for the network external to the resistor R for the network in Fig. 9.126.
   b. Find the power delivered to R when R is 2 Ω and 100 Ω.

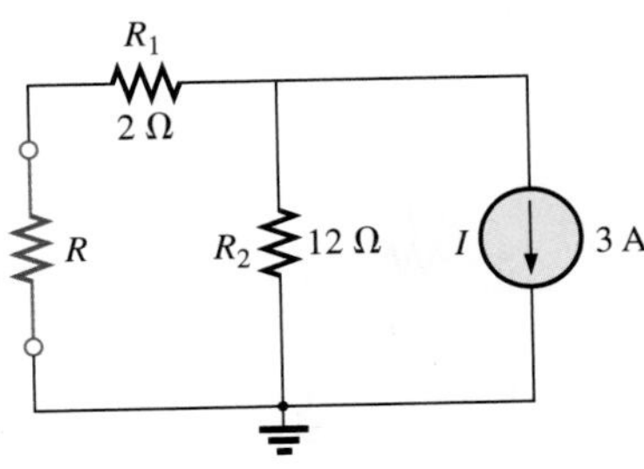

FIG. 9.126
Problems 8 and 18.

9. **a.** Find the Thévenin equivalent circuit for the network external to the resistor R for the network in Fig. 9.127.
   b. Find the power delivered to R when R is 2 Ω and 100 Ω.

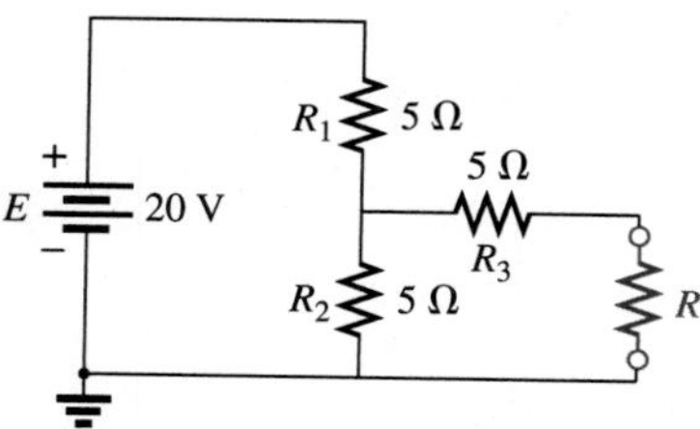

FIG. 9.127
Problems 9 and 19.

10. Find the Thévenin equivalent circuit for the network external to the resistor R for the network in Fig. 9.128.

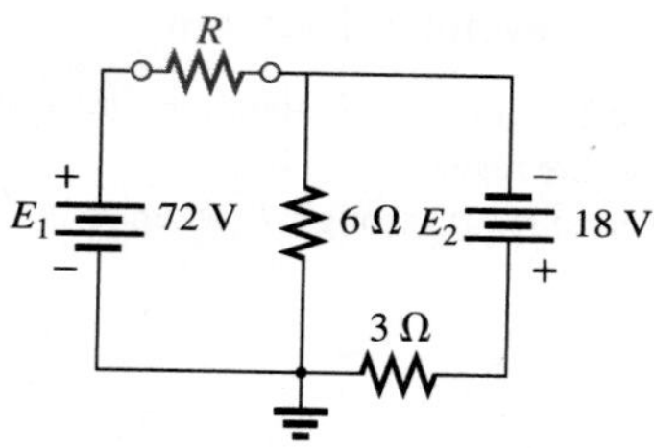

FIG. 9.128
Problem 10.

11. Find the Thévenin equivalent circuit for the network external to the resistor R for the network in Fig. 9.129.

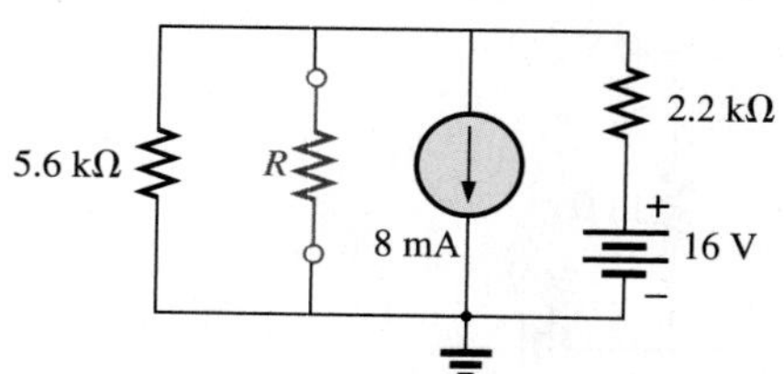

FIG. 9.129
Problems 11 and 20.

*12. Find the Thévenin equivalent circuit for the network external to the resistor R in each of the networks in Fig. 9.130.

*13. Find the Thévenin equivalent circuit for the portions of the networks in Fig. 9.131 external to points a and b.

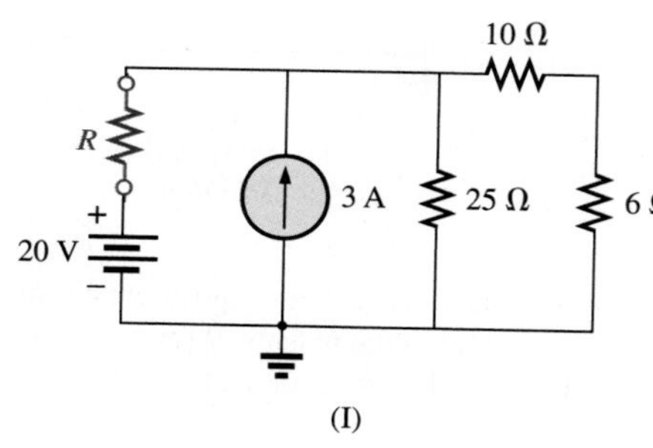

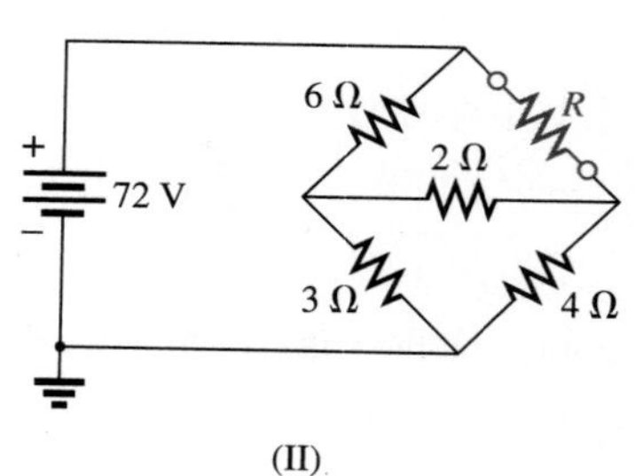

FIG. 9.130
Problems 12, 21, 24, 43, and 44.

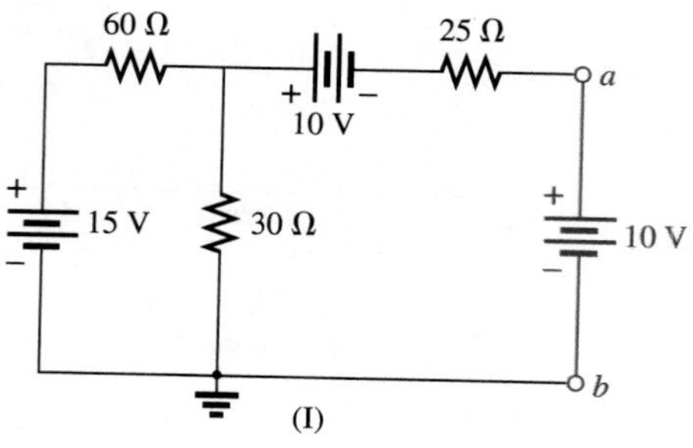

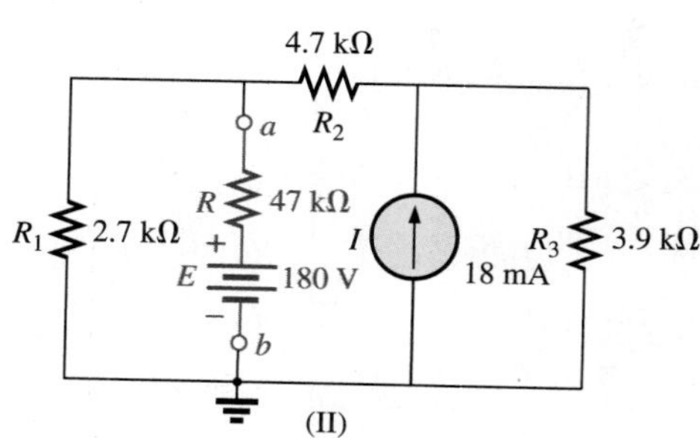

FIG. 9.131
Problem 13.

*14. Determine the Thévenin equivalent circuit for the network external to the resistor R in both networks in Fig. 9.132.

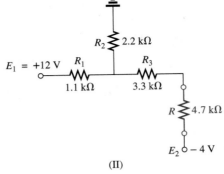

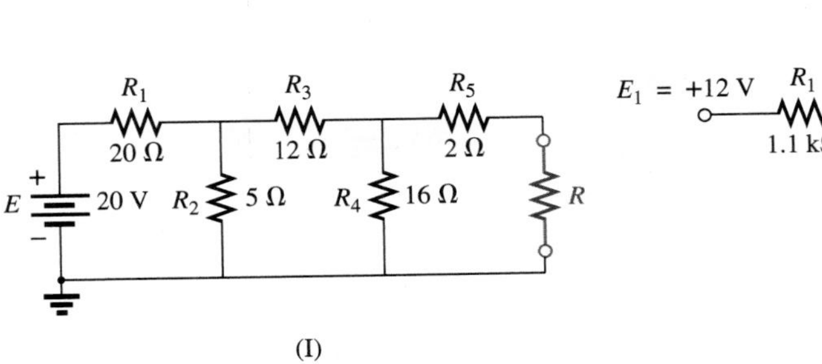

FIG. 9.132
Problems 14, 22, and 25.

*15. For the network in Fig. 9.133, find the Thévenin equivalent circuit for the network external to the load resistor R_L.

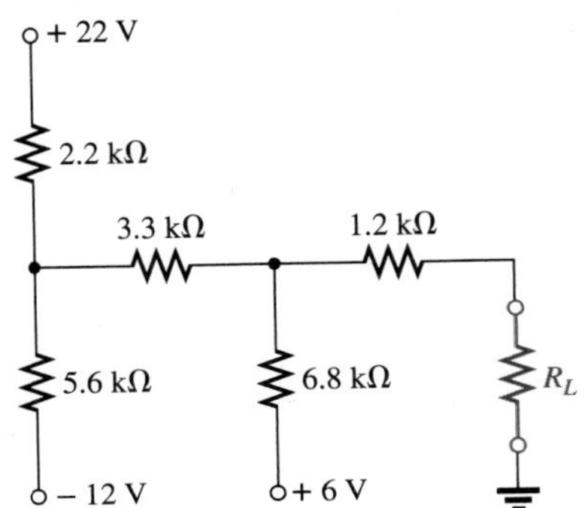

FIG. 9.133
Problem 15.

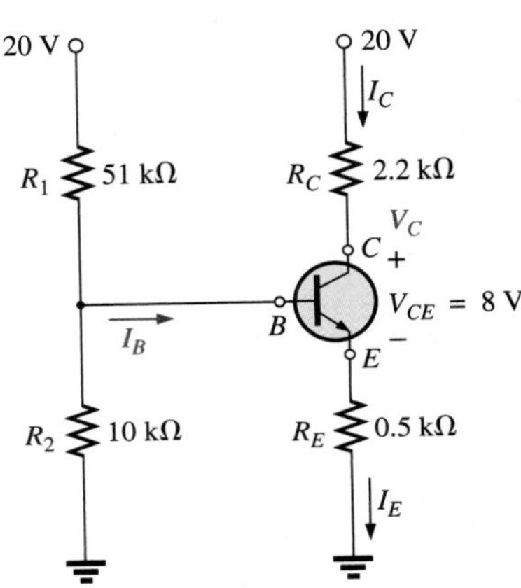

FIG. 9.134
Problem 16.

*16. For the transistor network in Fig. 9.134:

a. Find the Thévenin equivalent circuit for that portion of the network to the left of the base (B) terminal.

b. Using the fact that $I_C = I_E$ and $V_{CE} = 8$ V, determine the magnitude of I_E.

c. Using the results of parts (a) and (b), calculate the base current I_B if $V_{BE} = 0.7$ V.

d. What is the voltage V_C?

17. For each vertical set of measurements appearing in Fig. 9.135, determine the Thévenin equivalent circuit.

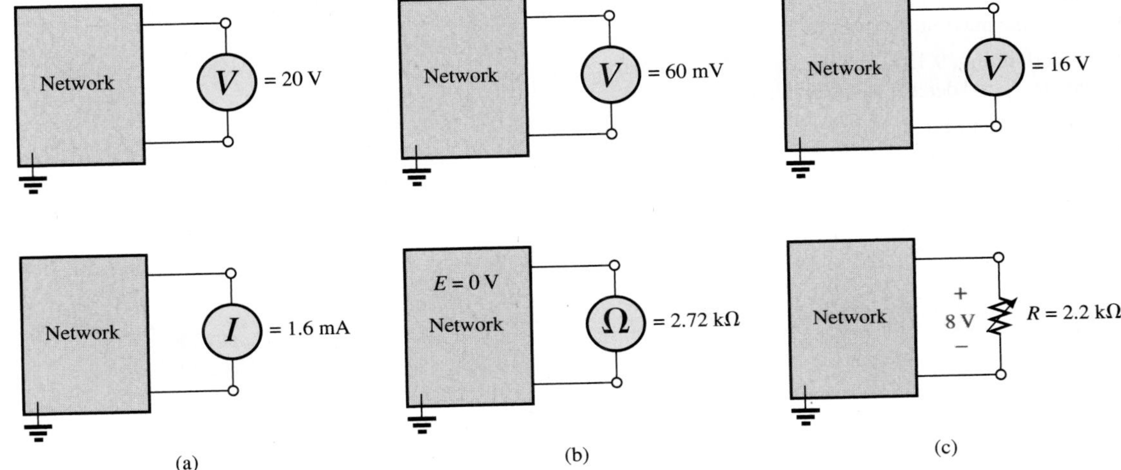

FIG. 9.135
Problem 17.

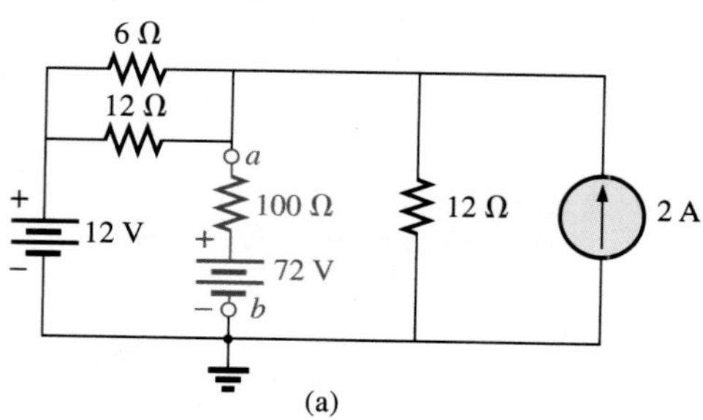

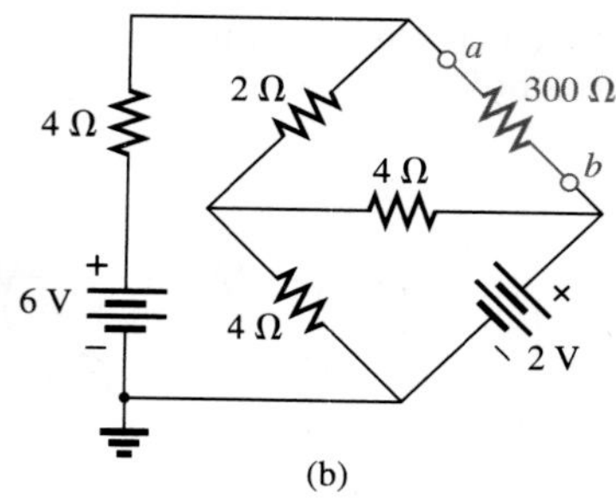

FIG. 9.136
Problems 23 and 45.

SECTION 9.4 Norton's Theorem

18. Find the Norton equivalent circuit for the network external to the resistor R for the network in Fig. 9.126.

19. a. Find the Norton equivalent circuit for the network external to the resistor R for the network in Fig. 9.127.
b. Convert to the Thévenin equivalent circuit, and compare your solution for E_{Th} and R_{Th} to that appearing in the solution for Problem 9.

20. Find the Norton equivalent circuit for the network external to the resistor R for each network in Fig. 9.129.

21. a. Find the Norton equivalent circuit for the network external to the resistor R for each network in Fig. 9.130.
b. Convert to the Thévenin equivalent circuit, and compare your solution for E_{Th} and R_{Th} to that appearing in the solutions for Problem 12.

22. Find the Norton equivalent circuit for the network external to resistor R for each network in Fig. 9.132.

23. Find the Norton equivalent circuit for the portions of the networks in Fig. 9.136 external to branch a-b.

SECTION 9.5 Maximum Power Transfer Theorem

24. a. For each network in Fig. 9.130, find the value of R for maximum power to R.
b. Determine the maximum power to R for each network.

25. a. For each network in Fig. 9.132, find the value of R for maximum power to R.
b. Determine the maximum power to R for each network.

26. For the network in Fig. 9.133, find the value of R_L for maximum power to R_L and determine the maximum power to R_L.

27. a. For the network in Fig. 9.137, determine the value of R for maximum power to R.

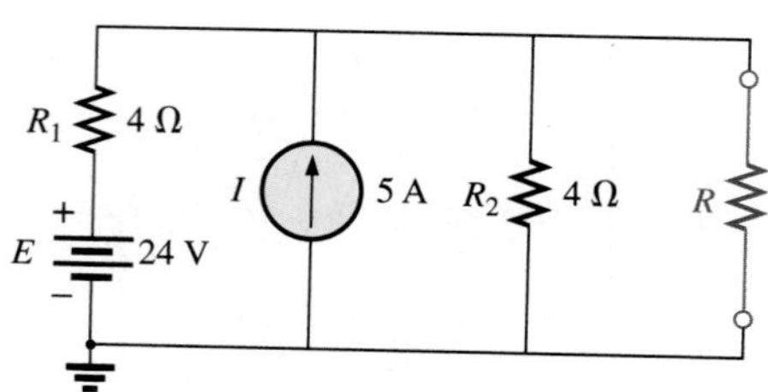

FIG. 9.137
Problem 27.

b. Determine the maximum power to R.
c. Plot a curve of power to R versus R for R equal to $\frac{1}{4}, \frac{1}{2}, \frac{3}{4}, 1, 1\frac{1}{4}, 1\frac{1}{2}, 1\frac{3}{4}$, and 2 times the value obtained in part (a).

***28.** Find the resistance R_1 in Fig. 9.138 such that the resistor R_4 will receive maximum power. Think!

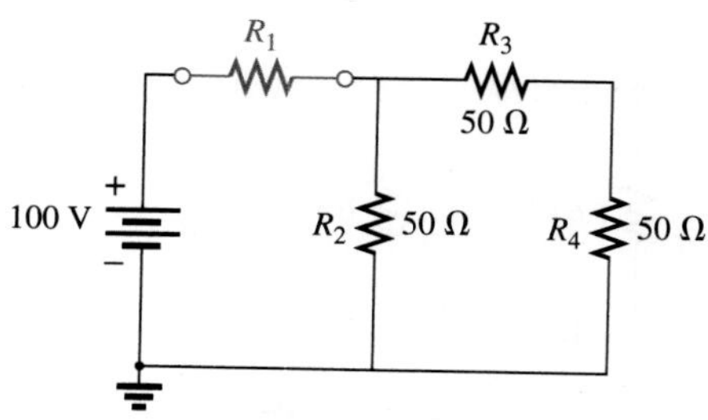

FIG. 9.138
Problem 28.

***29. a.** For the network in Fig. 9.139, determine the value of R_2 for maximum power to R_4.
b. Is there a general statement that can be made about situations such as those presented here and in Problem 28?

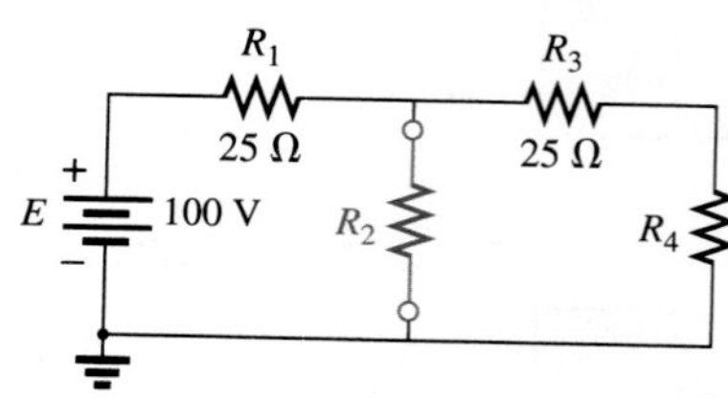

FIG. 9.139
Problem 29.

***30.** For the network in Fig. 9.140, determine the level of R that will ensure maximum power to the 100 Ω resistor.

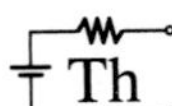

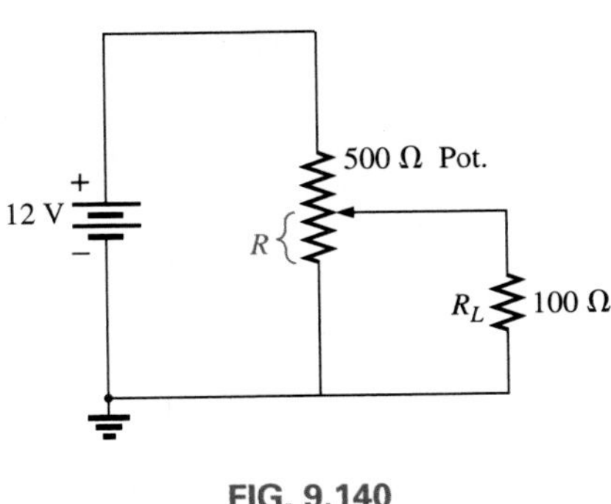

FIG. 9.140
Problem 30.

SECTION 9.6 Millman's Theorem

31. Using Millman's theorem, find the current through and voltage across the resistor R_L in Fig. 9.141.

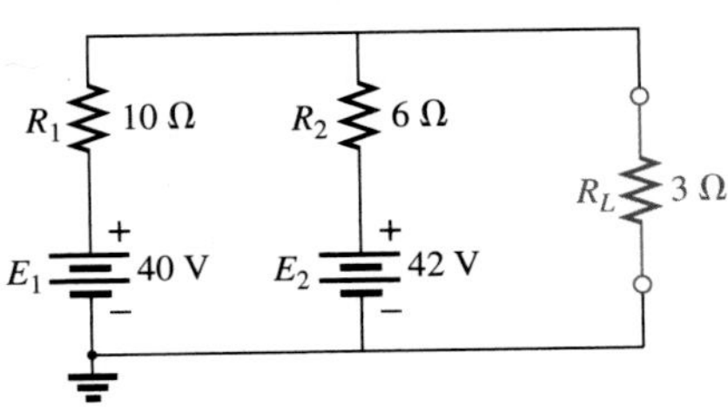

FIG. 9.141
Problem 31.

32. Repeat Problem 31 for the network in Fig. 9.142.

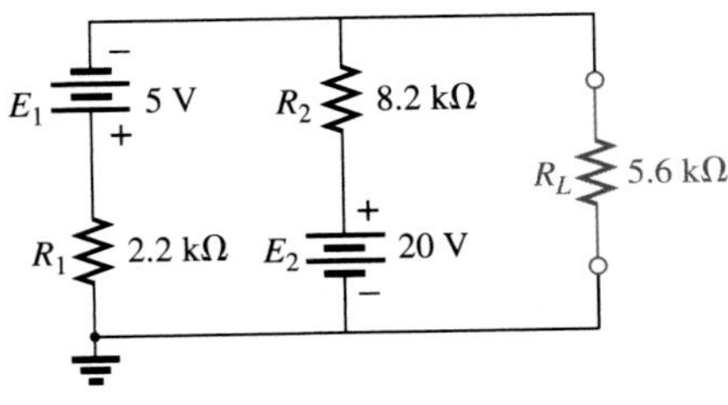

FIG. 9.142
Problem 32.

33. Repeat Problem 31 for the network in Fig. 9.143.

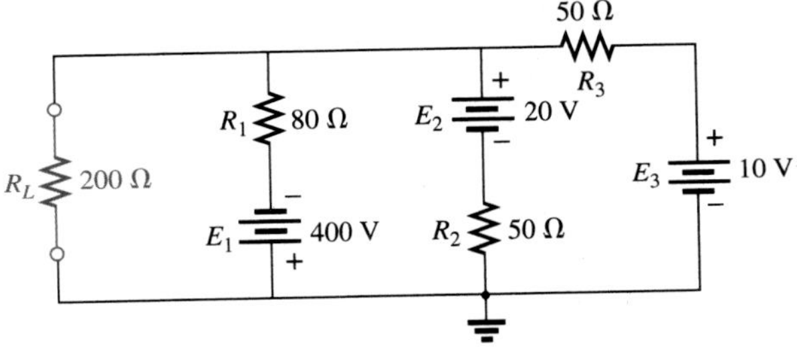

FIG. 9.143
Problem 33.

34. Using the dual of Millman's theorem, find the current through and voltage across the resistor R_L in Fig. 9.144.

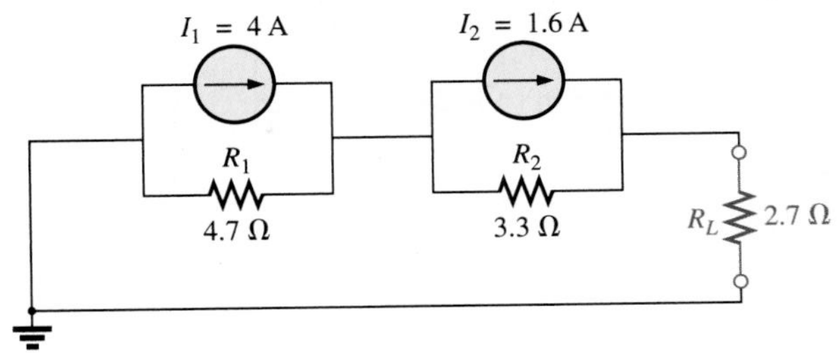

FIG. 9.144
Problem 34.

***35.** Repeat Problem 34 for the network in Fig. 9.145.

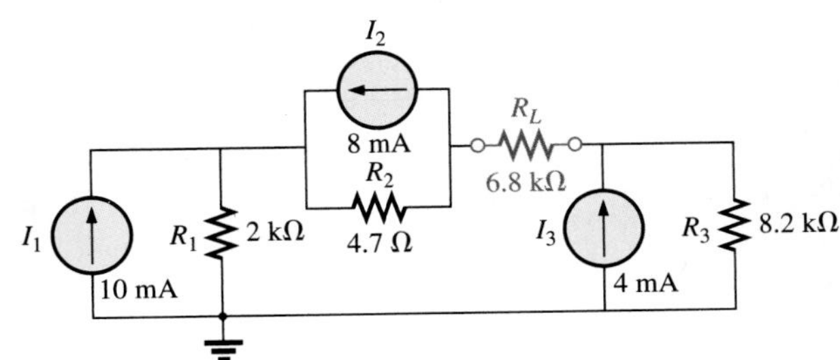

FIG. 9.145
Problem 35.

SECTION 9.7 Substitution Theorem

36. Using the substitution theorem, draw three equivalent branches for the branch *a-b* of the network in Fig. 9.146.

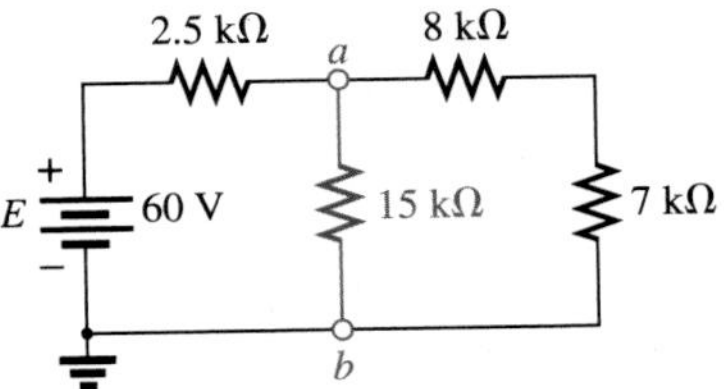

FIG. 9.146
Problem 36.

37. Repeat Problem 36 for the network in Fig. 9.147.

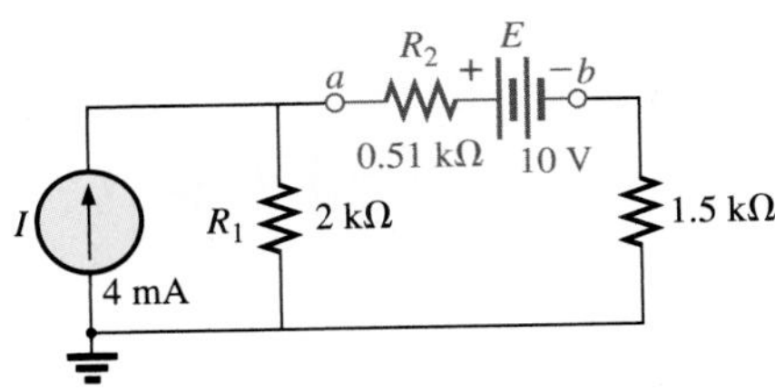

FIG. 9.147
Problem 37.

*38. Repeat Problem 36 for the network in Fig. 9.148. Be careful!

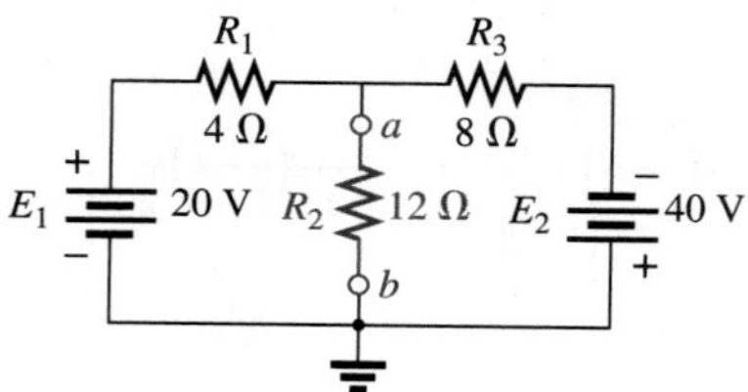

FIG. 9.148
Problem 38.

SECTION 9.8 Reciprocity Theorem

39. **a.** For the network in Fig. 9.149(a), determine the current I.
b. Repeat part (a) for the network in Fig. 9.149(b).
c. Is the reciprocity theorem satisfied?

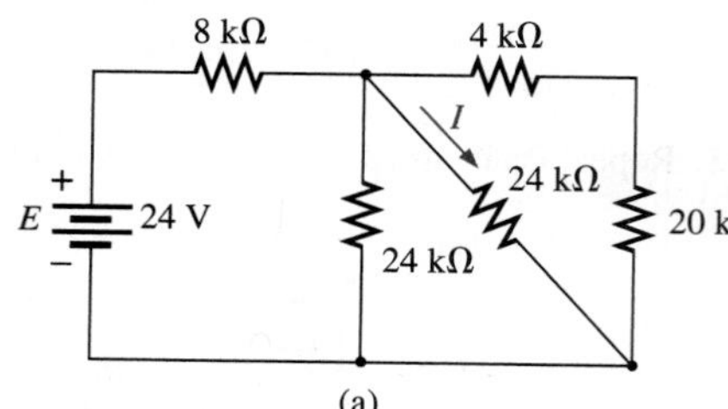

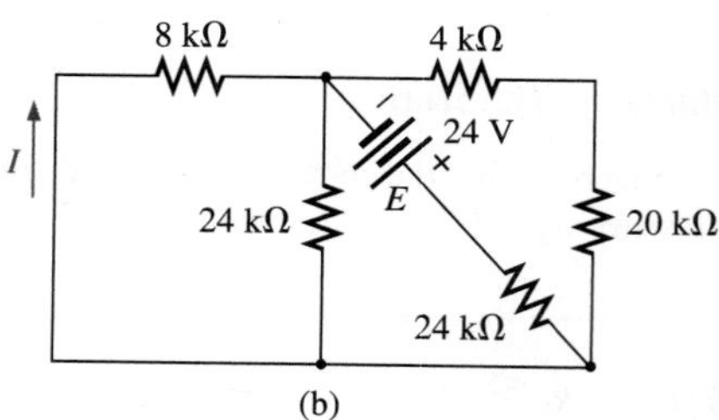

FIG. 9.149
Problem 39.

40. Repeat Problem 39 for the networks in Fig. 9.150.

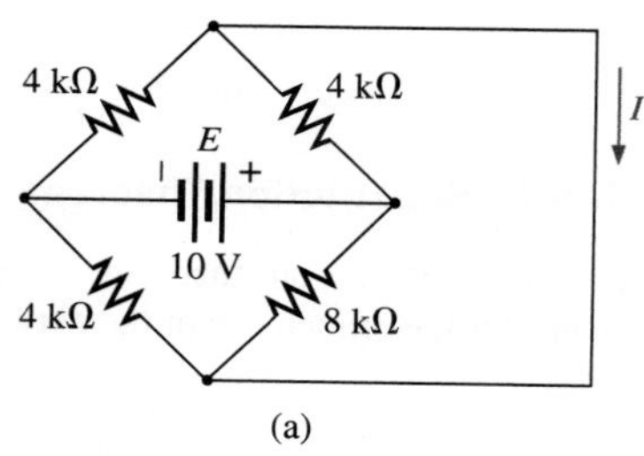

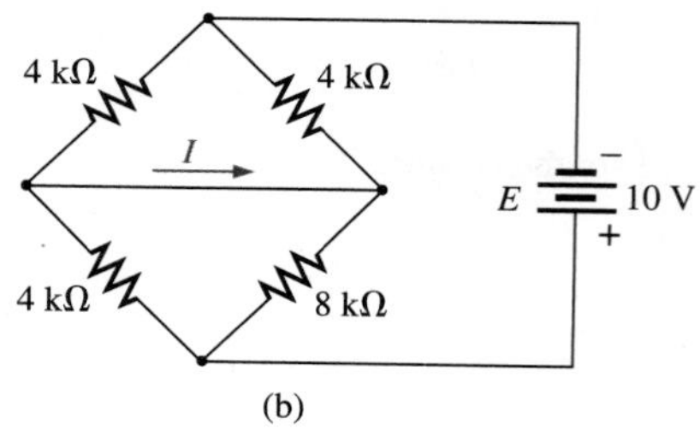

FIG. 9.150
Problem 40.

41. **a.** Determine the voltage V for the network in Fig. 9.151(a).
b. Repeat part (a) for the network in Fig. 9.151(b).
c. Is the dual of the reciprocity theorem satisfied?

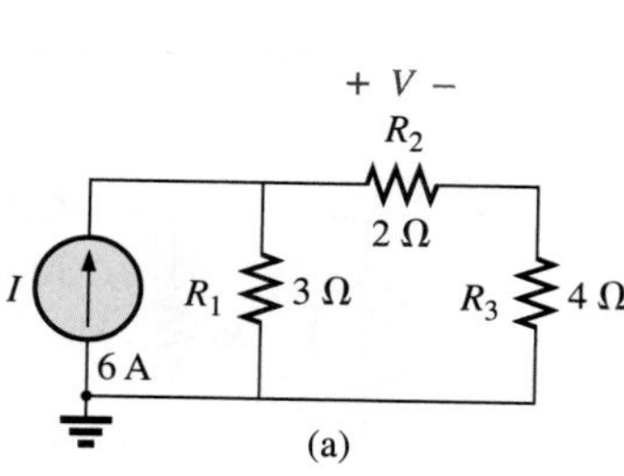

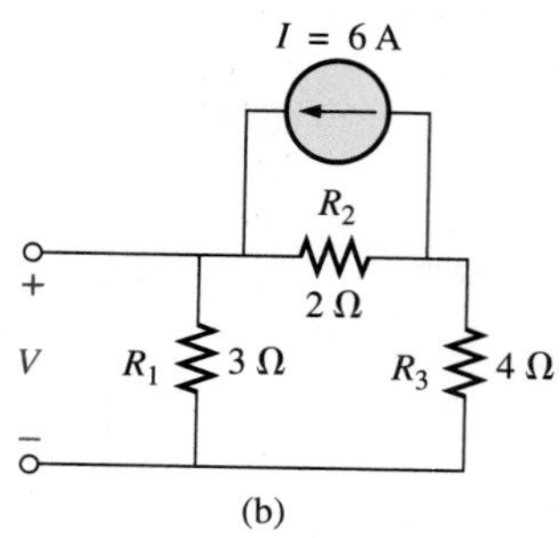

FIG. 9.151
Problem 41.

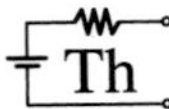

SECTION 9.9 Computer Analysis

42. Using PSpice or Multisim, determine the voltage V_2 and its components for the network in Fig. 9.124.

43. Using PSpice or Multisim, determine the Thévenin equivalent circuit for the network in Fig. 9.130(b).

***44.** **a.** Using PSpice, plot the power delivered to the resistor R in Fig. 9.130(a) for R having values from 1 Ω to 50 Ω.
b. From the plot, determine the value of R resulting in maximum power to R and the maximum power to R.
c. Compare the results of part (a) to the numerical solution.
d. Plot V_R and I_R versus R, and find the value of each under maximum power conditions.

***45.** Change the 300 Ω resistor in Fig. 9.136(b) to a variable resistor, and using PSpice plot the power delivered to the resistor versus values of the resistor. Determine the range of resistance by trial and error rather than first performing a longhand calculation. Determine the Norton equivalent circuit from the results. The Norton current can be determined from the maximum power level.

GLOSSARY

Maximum power transfer theorem A theorem used to determine the load resistance necessary to ensure maximum power transfer to the load.

Millman's theorem A method using source conversions that will permit the determination of unknown variables in a multiloop network.

Norton's theorem A theorem that permits the reduction of any two-terminal linear dc network to one having a single current source and parallel resistor.

Reciprocity theorem A theorem that states that for single-source networks, the current in any branch of a network, due to a single voltage source in the network, will equal the current through the branch in which the source was originally located if the source is placed in the branch in which the current was originally measured.

Substitution theorem A theorem that states that if the voltage across and current through any branch of a dc bilateral network are known, the branch can be replaced by any combination of elements that will maintain the same voltage across and current through the chosen branch.

Superposition theorem A network theorem that permits considering the effects of each source independently. The resulting current and/or voltage is the algebraic sum of the currents and/or voltages developed by each source independently.

Thévenin's theorem A theorem that permits the reduction of any two-terminal, linear dc network to one having a single voltage source and series resistor.

Capacitors

Objectives

- ***Become familiar with the basic construction of a capacitor and the factors that affect its ability to store charge on its plates.***
- ***Be able to determine the transient (time-varying) response of a capacitive network and plot the resulting voltages and currents.***
- ***Understand the impact of combining capacitors in series or parallel and how to read the nameplate data.***
- ***Develop some familiarity with the use of computer methods to analyze networks with capacitive elements.***

10.1 INTRODUCTION

Thus far, the resistor has been the only network component appearing in our analyses. In this chapter, we introduce the **capacitor,** which has a significant impact on the types of networks that you will be able to design and analyze. Like the resistor, it is a two-terminal device, but its characteristics are totally different from those of a resistor. In fact, *the capacitor displays its true characteristics only when a change in the voltage or current is made in the network.* All the power delivered to a resistor is dissipated in the form of heat. An ideal capacitor, however, stores the energy delivered to it in a form that can be returned to the system.

Although the basic construction of capacitors is actually quite simple, it is a component that opens the door to all types of practical applications, extending from touch pads to sophisticated control systems. A few applications are introduced and discussed in detail later in this chapter.

10.2 THE ELECTRIC FIELD

Recall from Chapter 2 that a force of attraction or repulsion exists between two charged bodies. We now examine this phenomenon in greater detail by considering the electric field that exists in the region around any charged body. This electric field is represented by **electric flux lines,** which are drawn to indicate the strength of the electric field at any point around the charged body. The denser the lines of flux, the stronger the electric field. In Fig. 10.1, for example, the electric field strength is stronger in region a than region b because the flux lines are denser in region a than b. That is, the same number of flux lines pass through each region, but the area A_1 is much smaller than area A_2. The symbol for electric flux is the Greek letter ψ (*psi*). The flux per unit area (flux density) is represented by the capital letter D and is determined by

$$D = \frac{\psi}{A} \quad \text{(flux/unit area)} \tag{10.1}$$

The larger the charge Q in coulombs, the greater the number of flux lines extending or terminating per unit area, independent of the surrounding medium. Twice the charge produces twice the flux per unit area. The two can therefore be equated:

$$\psi \equiv Q \quad \text{(coulombs, C)} \tag{10.2}$$

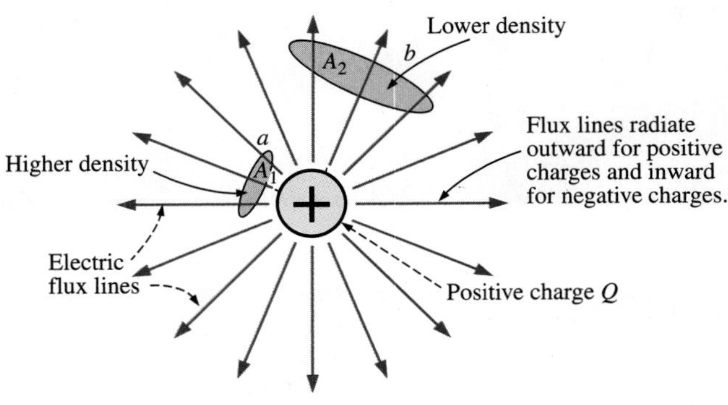

FIG. 10.1
Flux distribution from an isolated positive charge.

By definition, the **electric field strength** (designated by the capital script letter $\mathscr{E}$) at a point is the force acting on a unit positive charge at that point; that is,

$$\boxed{\mathscr{E} = \frac{F}{Q}} \quad \text{(newtons/coulomb, N/C)} \tag{10.3}$$

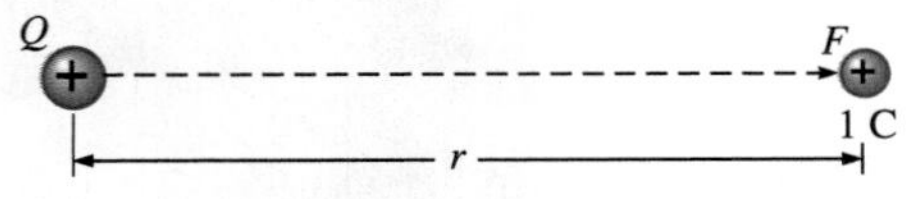

FIG. 10.2
Determining the force on a unit charge r meters from a charge Q of similar polarity.

In Fig. 10.2, the force exerted on a unit (1 coulomb) positive charge by a charge Q, r meters away, can be determined using **Coulomb's law** (Eq. 2.1) as follows:

$$F = k\frac{Q_1Q_2}{r^2} = k\frac{Q(1\text{ C})}{r^2} = \frac{kQ}{r^2} \quad (k = 9 \times 10^9\ \text{N}\cdot\text{m}^2/\text{C}^2)$$

Substituting the result into Eq. 10.3 for a unit positive charge results in

$$\mathscr{E} = \frac{F}{Q} = \frac{kQ/r^2}{1/\text{C}}$$

and

$$\boxed{\mathscr{E} = \frac{kQ}{r^2}} \quad \text{(N/C)} \tag{10.4}$$

The result clearly reveals that the electric field strength is directly related to the size of the charge Q. The greater the charge Q, the greater the electric field intensity on a unit charge at any point in the neighborhood. However, the distance is a squared term in the denominator. The result is that the greater the distance from the charge Q, the less the electric field strength, and dramatically so because of the squared term. In Fig. 10.1, the electric field strength at region A_2 is therefore significantly less than at region A_1.

For two charges of similar and opposite polarities, the flux distribution appears as shown in Fig. 10.3. In general,

electric flux lines always extend from a positively charged body to a negatively charged body, always extend or terminate perpendicular to the charged surfaces, and never intersect.

Note in Fig. 10.3(a) that the electric flux lines establish the most direct pattern possible from the positive to negative charge. They are evenly distributed and have the shortest distance on the horizontal between the two charges. This pattern is a direct result of the fact that electric flux

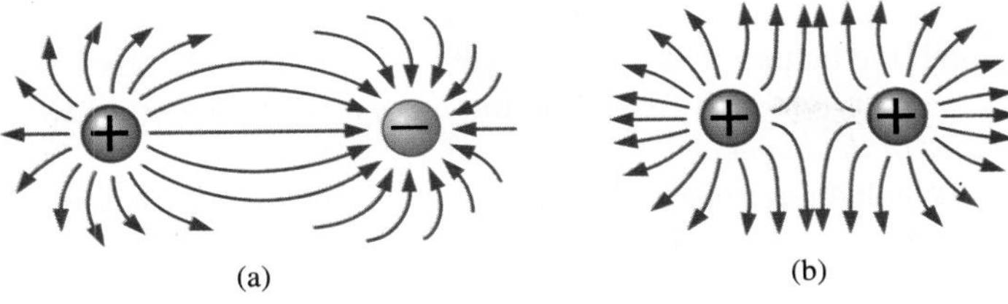

FIG. 10.3
Electric flux distributions: (a) opposite charges; (b) like charges.

lines strive to establish the shortest path from one charged body to another. The result is a natural pressure to be as close as possible. If two bodies of the same polarity are in the same vicinity, as shown in Fig. 10.3(b), the result is the direct opposite. The flux lines tend to establish a buffer action between the two with a repulsive action that grows as the two charges are brought closer to one another.

10.3 CAPACITANCE

Thus far, we have examined only isolated positive and negative spherical charges, but the description can be extended to charged surfaces of any shape and size. In Fig. 10.4, for example, two parallel plates of a material such as aluminum (the most commonly used metal in the construction of capacitors) have been connected through a switch and a resistor to a battery. If the parallel plates are initially uncharged and the switch is left open, no net positive or negative charge exists on either plate. The instant the switch is closed, however, electrons are drawn from the upper plate through the resistor to the positive terminal of the battery. There will be a surge of current at first, limited in magnitude by the resistance present. The level of flow then declines, as will be demonstrated in the sections to follow. This action creates a net positive charge on the top plate. Electrons are being repelled by the negative terminal through the lower conductor to the bottom plate at the same rate they are being drawn to the positive terminal. This transfer of electrons continues until the potential difference across the parallel plates is exactly equal to the battery voltage. The final result is a net positive charge on the top plate and a negative charge on the bottom plate, very similar in many respects to the two isolated charges in Fig. 10.3(a).

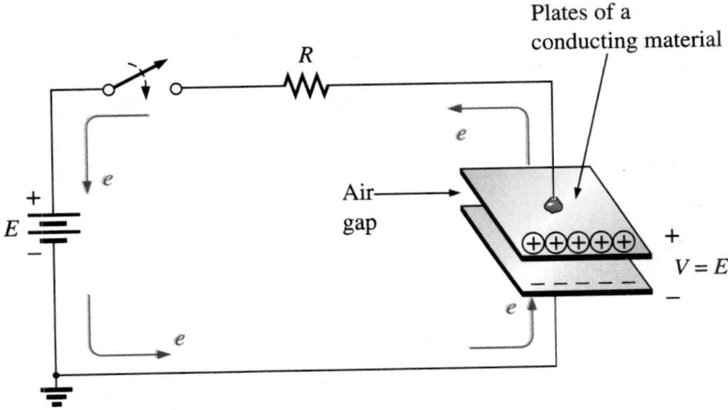

FIG. 10.4
Fundamental charging circuit.

FIG. 10.5
Michael Faraday.
Courtesy of the Smithsonian Institution
Photo No. 51,147

English (London)
(1791–1867)
Chemist and Electrical Experimenter
Honorary Doctorate, Oxford University, 1832

An experimenter with no formal education, he began his research career at the Royal Institute in London as a laboratory assistant. Intrigued by the interaction between electrical and magnetic effects, he discovered *electromagnetic induction,* demonstrating that electrical effects can be generated from a magnetic field (the birth of the generator as we know it today). He also discovered *self-induced currents* and introduced the concept of *lines and fields of magnetic force.* Having received over one hundred academic and scientific honors, he became a Fellow of the Royal Society in 1824 at the young age of 32.

Before continuing, it is important to note that the entire flow of charge is through the battery and resistor—not through the region between the plates. In every sense of the definition, *there is an open circuit between the plates of the capacitor.*

This element, constructed simply of two conducting surfaces separated by the air gap, is called a **capacitor.**

Capacitance is a measure of a capacitor's ability to store charge on its plates—in other words, its storage capacity.

In addition,

the higher the capacitance of a capacitor, the greater the amount of charge stored on the plates for the same applied voltage.

The unit of measure applied to capacitors is the farad (F), named after an English scientist, Michael Faraday, who did extensive research in the field (Fig. 10.5). In particular,

a capacitor has a capacitance of 1 F if 1 C of charge (6.242×10^{18} electrons) is deposited on the plates by a potential difference of 1 V across its plates.

The farad, however, is generally too large a measure of capacitance for most practical applications, so the microfarad (10^{-6}) or picofarad(10^{-12}) are more commonly encountered.

The relationship between the applied voltage, the charge on the plates, and the capacitance level is defined by the following equation:

$$\boxed{C = \frac{Q}{V}} \qquad \begin{aligned} C &= \text{farads (F)} \\ Q &= \text{coulombs (C)} \\ V &= \text{volts (V)} \end{aligned} \qquad \textbf{(10.5)}$$

Eq. (10.5) reveals that for the same voltage (V), the greater the charge (Q) on the plates (in the numerator of the equation), the higher the capacitance level (C).

If we write the equation in the following form:

$$\boxed{Q = CV} \qquad \text{(coulombs, C)} \qquad \textbf{(10.6)}$$

it becomes obvious through the product relationship that the higher the capacitance (C) or applied voltage (V), the greater the charge on the plates.

EXAMPLE 10.1

a. If 82.4×10^{14} electrons are deposited on the negative plate of a capacitor by an applied voltage of 60 V, find the capacitance of the capacitor.
b. If 40 V are applied across a 470 μF capacitor, find the charge on the plates.

Solutions:

a. First find the number of coulombs of charge as follows:

$$82.4 \times 10^{14} \cancel{\text{electrons}} \left(\frac{1\text{ C}}{6.242 \times 10^{18} \cancel{\text{electrons}}} \right) = 1.32\text{ mC}$$

and then

$$C = \frac{Q}{V} = \frac{1.32\ \text{mC}}{60\ \text{V}} = \mathbf{22\ \mu F} \qquad \text{(a standard value)}$$

b. Applying Eq. (10.6):

$$Q = CV = (470\ \mu\text{F})\,(40\ \text{V}) = \mathbf{18.8\ mC}$$

A cross-sectional view of the parallel plates in Fig. 10.4 is provided in Fig. 10.6(a). Note the **fringing** that occurs at the edges as the flux lines originating from the points farthest away from the negative plate strive to complete the connection. This fringing, which has the effect of reducing the net capacitance somewhat, can be ignored for most applications. Ideally, and the way we will assume the distribution to be in this text, the electric flux distribution appears as shown in Fig. 10.6(b), where all the flux lines are equally distributed and "fringing" does not occur.

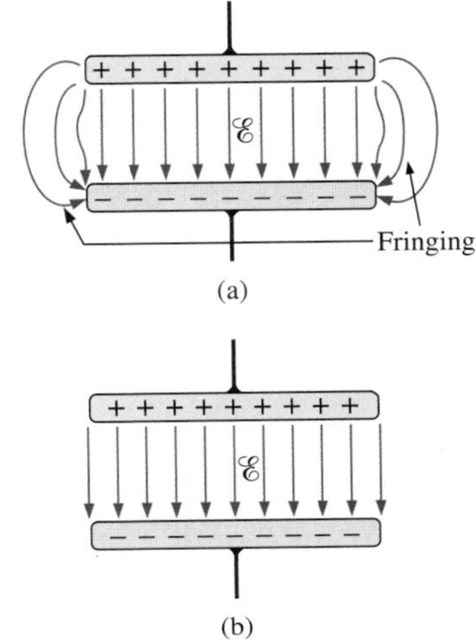

FIG. 10.6
Electric flux distribution between the plates of a capacitor: (a) including fringing; (b) ideal.

The **electric field strength** between the plates is determined by the voltage across the plates and the distance between the plates as follows:

$$\boxed{\mathscr{E} = \frac{V}{d}} \qquad \begin{aligned} \mathscr{E} &= \text{volts/m (V/m)} \\ V &= \text{volts (V)} \\ d &= \text{meters (m)} \end{aligned} \qquad \mathbf{(10.7)}$$

Note that the distance between the plates is measured in meters, not centimeters or inches.

The equation for the electric field strength is determined by two factors only: *the applied voltage and the distance between the plates.* The charge on the plates does not appear in the equation, nor does the size of the capacitor or the plate material.

Many values of capacitance can be obtained for the same set of parallel plates by the addition of certain insulating materials between the plates. In Fig. 10.7, an insulating material has been placed between a set of parallel plates having a potential difference of V volts across them.

Since the material is an insulator, the electrons within the insulator are unable to leave the parent atom and travel to the positive plate. The positive components (protons) and negative components (electrons) of each atom do shift, however [as shown in Fig. 10.7(a)], to form *dipoles.*

When the dipoles align themselves as shown in Fig. 10.7(a), the material is *polarized.* A close examination within this polarized material reveals that the positive and negative components of adjoining dipoles are

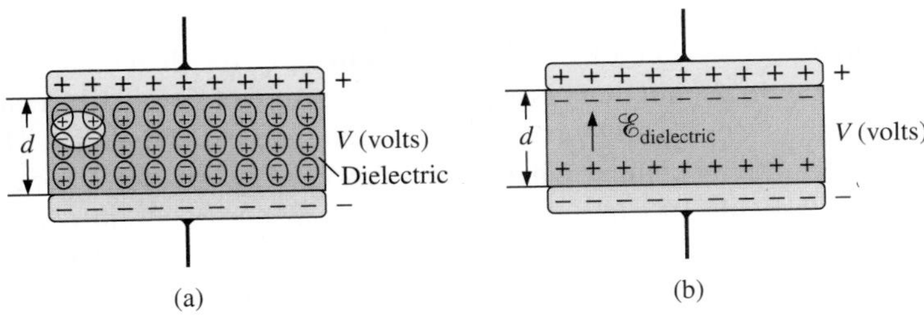

FIG. 10.7
Effect of a dielectric on the field distribution between the plates of a capacitor: (a) alignment of dipoles in the dielectric; (b) electric field components between the plates of a capacitor with a dielectric present.

neutralizing the effects of each other [note the dashed area in Fig. 10.7(a)]. The layer of positive charge on one surface and the negative charge on the other are not neutralized, however, resulting in the establishment of an electric field within the insulator [$\mathscr{E}_{\text{dielectric}}$; Fig. 10.7(b)].

In Fig. 10.8(a), two plates are separated by an air gap and have layers of charge on the plates as established by the applied voltage and the distance between the plates. The electric field strength is $\mathscr{E}_1$ as defined by Eq. (10.7). In Fig. 10.8(b), a slice of mica is introduced which, through an alignment of cells within the dielectric, establishes an electric field $\mathscr{E}_2$ that will oppose electric field $\mathscr{E}_1$. The effect is to try to reduce the electric field strength between the plates. However, Eq. (10.7) states that the electric field strength *must be* the value established by the applied voltage and the distance between the plates. This condition is maintained by placing more charge on the plates, thereby increasing the electric field strength between the plates to a level that cancels out the opposing electric field introduced by the mica sheet. The net result is an increase in charge on the plates and an increase in the capacitance level as established by Eq. (10.5).

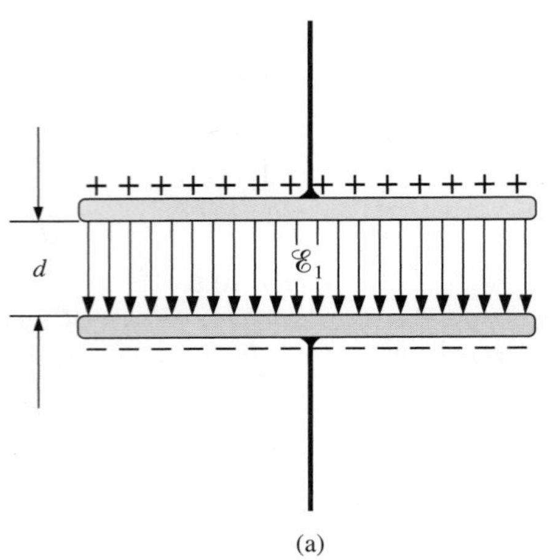

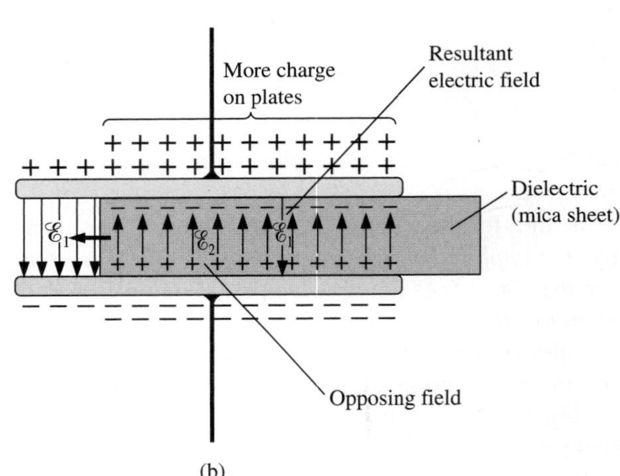

FIG. 10.8
Demonstrating the effect of inserting a dielectric between the plates of a capacitor: (a) air capacitor; (b) dielectric being inserted.

Different materials placed between the plates establish different amounts of additional charge on the plates. All, however, must be insulators and must have the ability to set up an electric field within the structure. A list of common materials appears in Table 10.1 using air as the reference level of 1.* All of these materials are referred to as **dielectrics,** the "di" for *opposing,* and the "electric" from *electric field.* The symbol ϵ_r in Table 10.1 is called the **relative permittivity** (or **dielectric constant**). The term **permittivity** is applied as a measure of how easily a material "permits" the establishment of an electric field in the material. The relative permittivity compares the permittivity of a material to that of air. For instance, Table 10.1 reveals that mica, with a relative permittivity of 5, "permits" the establishment of an opposing electric field in the material five times better than in air. Note the ceramic material at the bottom of the chart with a relative permittivity of 7500—a relative permittivity that makes it a very special dielectric in the manufacture of capacitors.

*Although there is a difference in dielectric characteristics between air and a vacuum, the difference is so small that air is commonly used as the reference level.

TABLE 10.1
Relative permittivity (dielectric constant) ϵ_r of various dielectrics.

Dielectric	ϵ_r (Average Values)
Vacuum	1.0
Air	1.0006
Teflon®	2.0
Paper, paraffined	2.5
Rubber	3.0
Polystyrene	3.0
Oil	4.0
Mica	5.0
Porcelain	6.0
Bakelite®	7.0
Aluminum oxide	7
Glass	7.5
Tantalum oxide	30
Ceramics	20–7500
Barium-strontium titanite (ceramic)	7500.0

Defining ϵ_o as the permittivity of air, the relative permittivity of a material with a permittivity ϵ is defined by Eq. (10.8):

$$\epsilon_r = \frac{\epsilon}{\epsilon_o} \quad \text{(dimensionless)} \quad \textbf{(10.8)}$$

Note that ϵ_r, which (as mentioned previously) is often called the **dielectric constant,** is a dimensionless quantity because it is a ratio of similar quantities. However, permittivity does have the units of farads/meter (F/m) and is 8.85×10^{-12} F/m for air. Although the relative permittivity for the air we breathe is listed as 1.006, a value of 1 is normally used for the relative permittivity of air.

For every dielectric there is a potential that, if applied across the dielectric, will break down the bonds within it and cause current to flow through it. The voltage required per unit length is an indication of its **dielectric strength** and is called the **breakdown voltage.** When breakdown occurs, the capacitor has characteristics very similar to those of a conductor. A typical example of dielectric breakdown is lightning, which occurs when the potential between the clouds and the earth is so high that charge can pass from one to the other through the atmosphere (the dielectric). The average dielectric strengths for various dielectrics are tabulated in volts/mil in Table 10.2 (1 mil = 1/1000 inch).

One of the important parameters of a capacitor is the **maximum working voltage.** It defines the maximum voltage that can be placed across the capacitor on a continuous basis without damaging it or changing its characteristics. For most capacitors, it is the dielectric strength that defines the maximum working voltage.

TABLE 10.2
Dielectric strength of some dielectric materials.

Dielectric	Dielectric Strength (Average Value) in Volts/Mil
Air	75
Barium-strontium titanite (ceramic)	75
Ceramics	75–1000
Porcelain	200
Oil	400
Bakelite®	400
Rubber	700
Paper paraffined	1300
Teflon®	1500
Glass	3000
Mica	5000

10.4 CAPACITORS

Capacitor Construction

We are now aware of the basic components of a capacitor: conductive plates, separation, and dielectric. However, the question remains, How do

all these factors interact to determine the capacitance of a capacitor? *Larger plates* permit an increased area for the storage of charge, so the area of the plates should be in the numerator of the defining equation. *The smaller the distance between the plates,* the larger the capacitance so this factor should appear in the numerator of the equation. Finally, since *higher levels of permittivity* result in higher levels of capacitance, the factor ϵ should appear in the numerator of the defining equation.

The result is the following general equation for capacitance:

$$\boxed{C = \epsilon\frac{A}{d}} \qquad \begin{aligned} C &= \text{farads (F)} \\ \epsilon &= \text{permittivity (F/m)} \\ A &= \text{m}^2 \\ d &= \text{m} \end{aligned} \qquad \textbf{(10.9)}$$

If we substitute Eq. (10.8) for the permittivity of the material, we obtain the following equation for the capacitance:

$$\boxed{C = \epsilon_o\epsilon_r\frac{A}{d}} \qquad \text{(farads, F)} \qquad \textbf{(10.10)}$$

or if we substitute the known value for the permittivity of air, we obtain the following useful equation:

$$\boxed{C = 8.85 \times 10^{-12}\,\epsilon_r\frac{A}{d}} \qquad \text{(farads, F)} \qquad \textbf{(10.11)}$$

It is important to note in Eq. (10.11) that the area of the plates (actually the area of only one plate) is in meters squared (m^2); the distance between the plates is measured in meters; and the numerical value of ϵ_r is simply taken from Table 10.1.

You should also be aware that most capacitors are in the μF, nF, or pF range, not the 1 F or greater range. A 1 F capacitor can be as large as a typical flashlight requiring that the housing for the system be quite large. Most capacitors in electronic systems are the size of a thumbnail or smaller.

If we form the ratio of the equation for the capacitance of a capacitor with a specific dielectric to that of the same capacitor with air as the dielectric, the following results:

$$\frac{C = \varepsilon\dfrac{A}{d}}{C_o = \varepsilon_o\dfrac{A}{d}} \Rightarrow \frac{C}{C_o} = \frac{\varepsilon}{\varepsilon_o} = \varepsilon_r$$

and

$$\boxed{C = \epsilon_r C_o} \qquad \textbf{(10.12)}$$

The result is that

the capacitance of a capacitor with a dielectric having a relative permittivity of ϵ_r is ϵ_r times the capacitance using air as the dielectric.

The next few examples review the concepts and equations just presented.

EXAMPLE 10.2 In Fig. 10.9, if each air capacitor in the left column is changed to the type appearing in the right column, find the new capacitance level. For each change, the other factors remain the same.

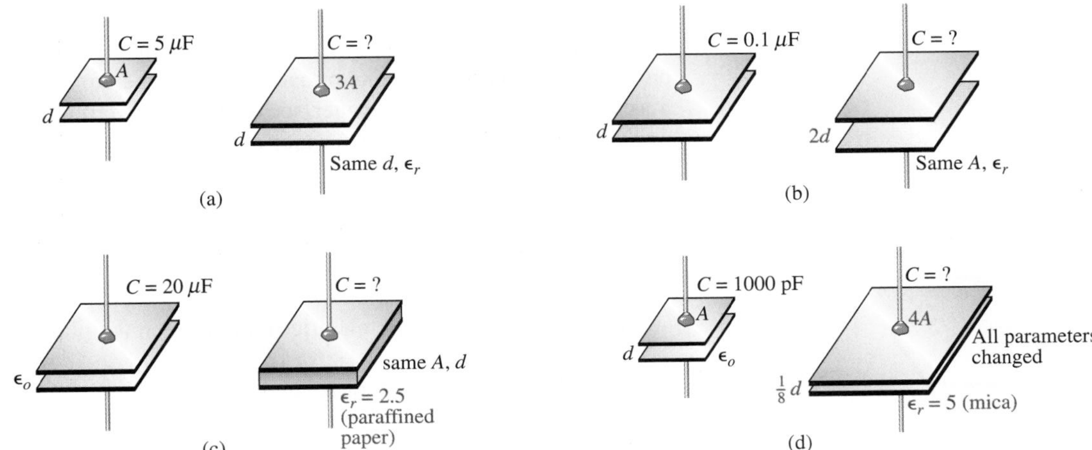

FIG. 10.9
Example 10.2.

Solutions:

a. In Fig. 10.9(a), the area has increased by a factor of three, providing more space for the storage of charge on each plate. Since the area appears in the numerator of the capacitance equation, the capacitance increases by a factor of three. That is,

$$C = 3(C_o) = 3(5\ \mu\text{F}) = \mathbf{15\ \mu F}$$

b. In Fig. 10.9(b), the area stayed the same, but the distance between the plates was increased by a factor of two. Increasing the distance reduces the capacitance level, so the resulting capacitance is one-half of what it was before. That is,

$$C = \frac{1}{2}(0.1\ \mu\text{F}) = \mathbf{0.05\ \mu F}$$

c. In Fig. 10.9(c), the area and the distance between the plates were maintained, but a dielectric of paraffined (waxed) paper was added between the plates. Since the permittivity appears in the numerator of the capacitance equation, the capacitance increases by a factor determined by the relative permittivity. That is,

$$C = \epsilon_r C_o = 2.5(20\ \mu\text{F}) = \mathbf{50\ \mu F}$$

d. In Fig. 10.9(d), a multitude of changes are happening at the same time. However, solving the problem is simply a matter of determining whether the change increases or decreases the capacitance and then placing the multiplying factor in the numerator or denominator of the equation. The increase in area by a factor of four produces a multiplier of four in the numerator, as shown in the equation below. Reducing the distance by a factor of 1/8 will increase the capacitance by its inverse, or a factor of eight. Inserting the mica dielectric increases the capacitance by a factor of five. The result is

$$C = (5)\frac{4}{(1/8)}(C_o) = 160(1000\ \text{pF}) = \mathbf{0.16\ \mu F}$$

In the next example, the dimensions of an air capacitor are provided and the capacitance is to be determined. The example emphasizes the importance of knowing the units of each factor of the equation. Failing to make a conversion to the proper set of units will probably produce a meaningless result, even if the proper equation were used and the mathematics properly executed.

FIG. 10.10
Air capacitor for Example 10.3.

EXAMPLE 10.3 For the capacitor in Fig. 10.10:

a. Find the capacitance.
b. Find the strength of the electric field between the plates if 48 V are applied across the plates.
c. Find the charge on each plate.

Solutions:

a. First, the area and the distance between the plates must be converted to the SI system as required by Eq. (10.11):

$$d = \frac{1}{32}\,\text{in.}\left(\frac{1\text{ m}}{39.37\text{ in.}}\right) = 0.794\text{ mm}$$

and $$A = (2\text{ in.})(2\text{ in.})\left(\frac{1\text{ m}}{39.37\text{ in.}}\right)\left(\frac{1\text{ m}}{39.37\text{ in.}}\right) = 2.581 \times 10^{-3}\text{ m}^2$$

Eq. (10.11):

$$C = 8.85 \times 10^{-12}\,\epsilon_r\frac{A}{d} = 8.85 \times 10^{-12}(1)\frac{(2.581 \times 10^{-3}\text{ m}^2)}{0.794\text{ mm}} = \mathbf{28.8\text{ pF}}$$

b. The electric field between the plates is determined by Eq. (10.7):

$$\mathscr{E} = \frac{V}{d} = \frac{48\text{ V}}{0.794\text{ mm}} = \mathbf{60.5\text{ kV/m}}$$

c. The charge on the plates is determined by Eq. (10.6):

$$Q = CV = (28.8\text{ pF})(48\text{ V}) = \mathbf{1.38\text{ nC}}$$

In the next example, we will insert a ceramic dielectric between the plates of the air capacitor in Fig. 10.10 and see how it affects the capacitance level, electric field, and charge on the plates.

EXAMPLE 10.4

a. Insert a ceramic dielectric with an ϵ_r of 250 between the plates of the capacitor in Fig. 10.10. Then determine the new level of capacitance. Compare your results to the solution in Example 10.3.
b. Find the resulting electric field strength between the plates, and compare your answer to the result in Example 10.3.
c. Determine the charge on each of the plates, and compare your answer to the result in Example 10.3.

Solutions:

a. Using Eq. (10.12), the new capacitance level is

$$C = \epsilon_r C_o = (250)(28.8\text{ pF}) = \mathbf{7200\text{ pF} = 7.2\text{ nF} = 0.0072\ \mu F}$$

which is *significantly higher* than the level in Example 10.3.

b. $\mathscr{E} = \dfrac{V}{d} = \dfrac{48\text{ V}}{0.794\text{ mm}} = \mathbf{60.5\ kV/m}$

Since the applied voltage and the distance between the plates did not change, *the electric field between the plates remains the same.*

c. $Q = CV = (7200\text{ pF})(48\text{ V}) = \mathbf{345.6\ nC = 0.35\ \mu C}$

We now know that the insertion of a dielectric between the plates increases the amount of charge stored on the plates. In Example 10.4, since the relative permittivity increased by a factor of 250, the charge on the plates *increased by the same amount.*

EXAMPLE 10.5 Find the maximum voltage that can be applied across the capacitor in Example 10.4 if the dielectric strength is 80 V/mil.

Solution:

$$d = \frac{1}{32}\text{ in.}\left(\frac{1000\text{ mils}}{1\text{ in.}}\right) = 31.25\text{ mils}$$

and

$$V_{max} = 31.25\text{ mils}\left(\frac{80\text{ V}}{\text{mil}}\right) = \mathbf{2.5\ kV}$$

although the provided working voltage may be only 2 kV to provide a margin of safety.

Types of Capacitors

Capacitors, like resistors, can be listed under two general headings: **fixed** and **variable.** The symbol for the fixed capacitor appears in Fig. 10.11(a). Note that the curved side is normally connected to ground or to the point of lower dc potential. The symbol for variable capacitors appears in Fig. 10.11(b).

FIG. 10.11
Symbols for the capacitor: (a) fixed; (b) variable.

Fixed Capacitors Fixed-type capacitors come in all shapes and sizes. However,

in general, for the same type of construction and dielectric, the larger the required capacitance, the larger the physical size of the capacitor.

In Fig. 10.12(a), the 10,000 μF electrolytic capacitor is significantly larger than the 1 μF capacitor. However, it is certainly not 10,000 times larger. For the polyester-film type of Fig. 10.12(b), the 2.2 μF capacitor is significantly larger than the 0.01 μF capacitor, but again it is not 220 times larger. The 22 μF tantalum capacitor of Fig. 10.12(c) is about 6 times larger than the 1.5 μF capacitor, even though the capacitance level is about 15 times higher. It is particularly interesting to note that due to the difference in dielectric and construction, the 22 μF tantalum capacitor is significantly smaller than the 2.2 μF polyester-film capacitor, and much smaller than 1/5 the size of the 100 μF electrolytic capacitor. The relatively large 10,000 μF electrolytic capacitor is normally used for high-power applications, such as in power supplies and high-output speaker systems. All the others may appear in any commercial electronic system.

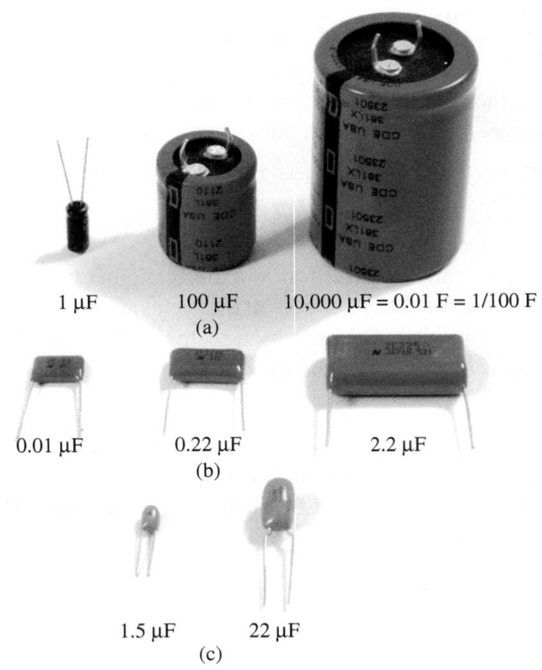

FIG. 10.12
Demonstrating that, in general, for each type of construction, the size of a capacitor increases with the capacitance value: (a) electrolytic; (b) polyester-film; (c) tantalum.

The increase in size is due primarily to the effect of area and thickness of the dielectric on the capacitance level. There are a number of ways to increase the area without making the capacitor too large. One is to lay out the plates and the dielectric in long, narrow strips and then roll them all together, as shown in Fig. 10.13(a). The dielectric (remember that it has the characteristics of an insulator) between the conducting strips ensures the strips never touch. Of course, the dielectric must be the type that can be rolled without breaking up. Depending on how the materials are wrapped, the capacitor can be either a cylindrical or a rectangular, box-type shape.

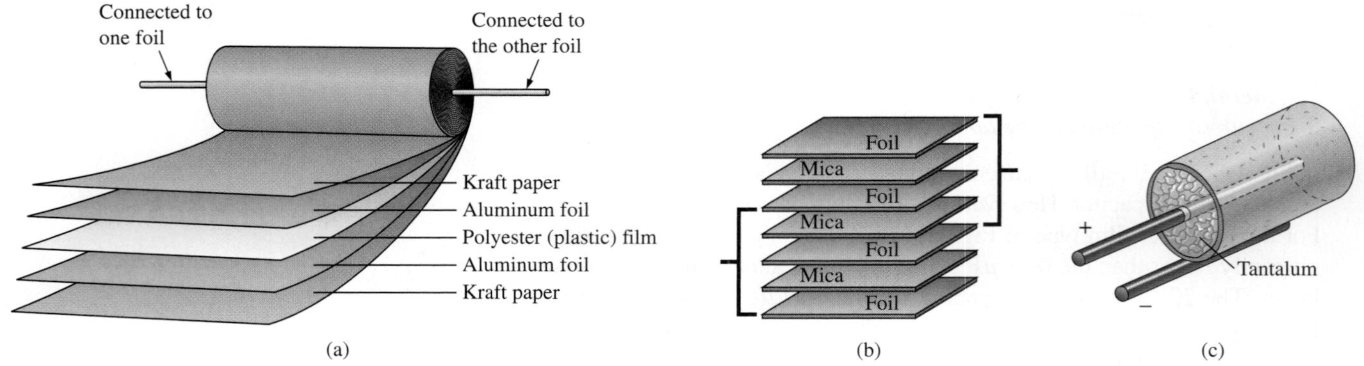

FIG. 10.13
Three ways to increase the area of a capacitor: (a) rolling; (b) stacking; (c) insertion.

A second popular method is to stack the plates and the dielectrics, as shown in Fig. 10.14(b). The area is now a multiple of the number of dielectric layers. This construction is very popular for smaller capacitors.

A third method is to use the dielectric to establish the body shape [a cylinder in Fig. 10.13(c)]. Then simply insert a rod for the positive plate, and coat the surface of the cylinder to form the negative plate, as shown in Fig. 10.13(c). Although the resulting "plates" are not the same in construction or surface area, the effect is to provide a large surface area for storage (the density of electric field lines will be different on the two "plates"), although the resulting distance factor may be larger than desired. Using a dielectric with a high ϵ_r, however, compensates for the increased distance between the plates.

There are other variations of the above to increase the area factor, but the three depicted in Fig. 10.13 are the most popular.

The next controllable factor is the distance between the plates. This factor, however, is very sensitive to how thin the dielectric can be made, with natural concerns because the working voltage (the breakdown voltage) drops as the gap decreases. Some of the thinnest dielectrics are just oxide coatings on one of the conducting surfaces (plates). A very thin polyester material, such as Mylar®, Teflon®, or even paper with a paraffin coating, provides a thin sheet of material than can easily be wrapped for increased areas. Materials such as mica and some ceramic materials can be made only so thin before crumbling or breaking down under stress.

The last factor is the dielectric, for which there is a wide range of possibilities. However, the following factors greatly influence which dielectric is used:

The level of capacitance desired
The resulting size
The possibilities for rolling, stacking and so on
Temperature sensitivity
Working voltage

The range of relative permittivities is enormous, as shown in Table 10.2, but all the factors listed above must be considered in the construction process.

In general, the most common fixed capacitors are the electrolytic, film, polyester, foil, ceramic, mica, dipped, and oil.

The **electrolytic capacitors** in Fig. 10.14 are usually easy to identify by their shape and the fact that they usually have a polarity marking on the body (although special-application electrolytics are available that are not polarized). Few capacitors have a polarity marking, but those that do must be connected with the negative terminal connected to ground or to the point of lower potential. The markings often used to denote the positive terminal or plate include +, □, and Δ. In general, electrolytic

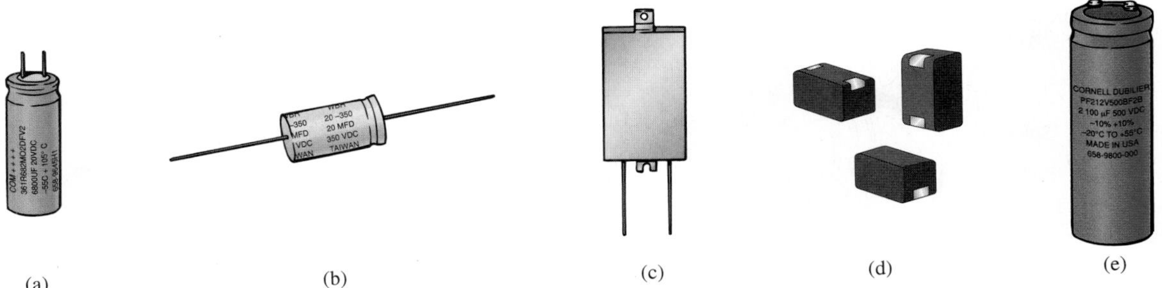

FIG. 10.14
Various types of electrolytic capacitors: (a) miniature radial leads; (b) axial leads; (c) flatpack; (d) surface-mount; (e) screw-in terminals. (Courtesy of Cornell-Dubilier.)

capacitors offer some of the highest capacitance values available, although their working voltage levels are limited. Typical values range from 0.1 μF to 15,000 μF, with working voltages from 5 V to 450 V. The basic construction uses the rolling process in Fig. 10.13(a) in which a roll of aluminum foil is coated on one side with aluminum oxide—the aluminum being the positive plate, and the oxide the dielectric. A layer of paper or gauze saturated with an electrolyte (a solution or paste that forms the conducting medium between the electrodes of the capacitor) is placed over the aluminum-oxide coating of the positive plate. Another layer of aluminum without the oxide coating is then placed over this layer to assume the role of the negative plate. In most cases, the negative plate is connected directly to the aluminum container, which then serves as the negative terminal for external connections. Because of the size of the roll of aluminum foil, the overall size of the electrolytic capacitor is greater than most.

Film, polyester, foil, polypropylene or **Teflon® capacitors** use a rolling or stacking process to increase the surface area, as shown in Fig. 10.15. The resulting shape can be either round or rectangular, with radial or axial leads. The typical range for such capacitors is 100 pF to 10 μF, with units available up to 100 μF. The name of the unit defines the type of dielectric employed. Working voltages can extend from a few volts to 2000 V, depending on the type of unit.

FIG. 10.15
(a) Film/foil polyester radial lead; (b) metalized polyester-film axial lead; (c) surface-mount polyester-film; (d) polypropylene-film, radial lead.

Ceramic capacitors (often called **disc capacitors**) use a ceramic dielectric, as shown in Fig. 10.16(a), to utilize the excellent ϵ_r values and high working voltages associated with a number of ceramic materials.

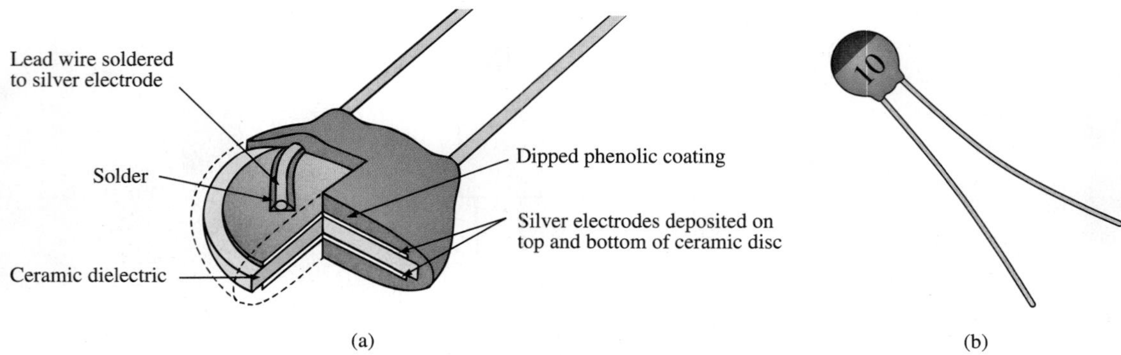

FIG. 10.16
Ceramic (disc) capacitor: (a) construction; (b) appearance.

(a)

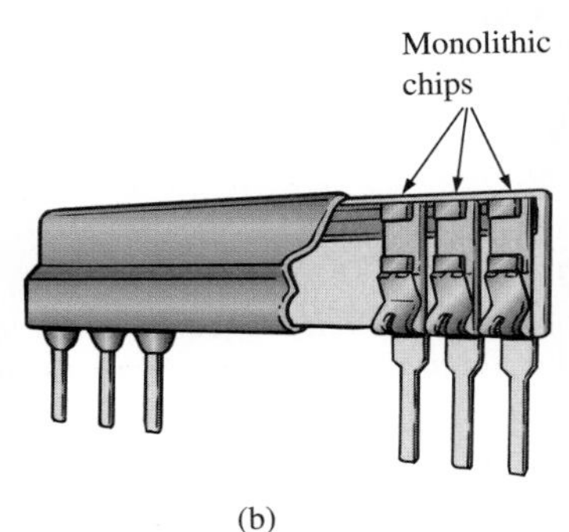

(b)

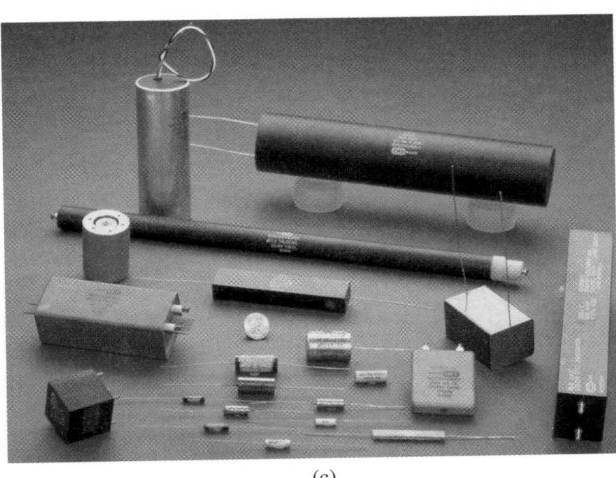

(c)

FIG. 10.17
Mica capacitors: (a) and (b) surface-mount monolithic chips; (c) high-voltage/temperature mica paper capacitors.
[(a) and (b) courtesy of Vishay Intertechnology, Inc.; (c) courtesy of Custom Electronics, Inc.]

Stacking can also be applied to increase the surface area. An example of the disc variety appears in Fig. 10.16(b). Ceramic capacitors typically range in value from 10 pF to 0.047 μF, with high working voltages that can reach as high as 10 kV.

Mica capacitors use a mica dielectric that can be monolithic (single chip) or stacked. The relatively small size of monolithic mica chip capacitors is demonstrated in Fig. 10.17(a), with their placement shown in Fig. 10.17(b). A variety of high-voltage mica paper capacitors are displayed in Fig. 10.17(c). Mica capacitors typically range in value from 2 pF to several microfarads, with working voltages up to 20 kV.

Dipped capacitors are made by dipping the dielectric (tantalum or mica) into a conductor in a molten state to form a thin, conductive sheet on the dielectric. Due to the presence of an electrolyte in the manufacturing process, dipped tantalum capacitors require a polarity marking to ensure that the positive plate is always at a higher potential than the negative plate, as shown in Fig. 10.18(a). A series of small positive signs is typically applied to the casing near the positive lead. A group of nonpolarized, mica dipped capacitors are shown in Fig. 10.18(b). They typically range in value from 0.1 μF to 680 μF, but with lower working voltages ranging from 6 V to 50 V.

Most **oil capacitors** such as appearing in Fig. 10.19 are used for industrial applications such as welding, high-voltage power supplies, surge protection, and power-factor correction (Chapter 19). They can provide capacitance levels extending from 0.001 μF all the way up to 10,000 μF, with working voltages up to 150 kV. Internally, there are a number of parallel plates sitting in a bath of oil or oil-impregnated material (the dielectric).

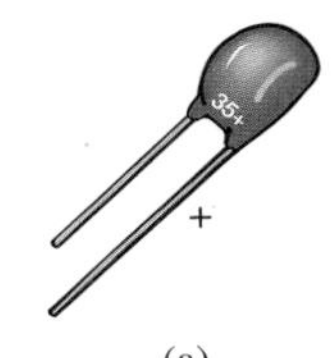

(a)

(b)

FIG. 10.18
Dipped capacitors: (a) polarized tantalum; (b) nonpolarized mica.

FIG. 10.19
Oil-filled, metallic oval case snubber capacitor (the snubber removes unwanted voltage spikes).
(Courtesy of Cornell-Dubilier.)

Variable Capacitors All the parameters in Eq. (10.11) can be changed to some degree to create a **variable capacitor.** For example, in Fig. 10.20(a), the capacitance of the variable air capacitor is changed by turning the shaft at the end of the unit. By turning the shaft, you control the amount of common area between the plates: The less common area there is, the less capacitance. In Fig. 10.20(b), we have a much smaller **air trimmer capacitor.** It works under the same principle, but the rotating blades are totally hidden inside the structure. In Fig. 10.20(c), the

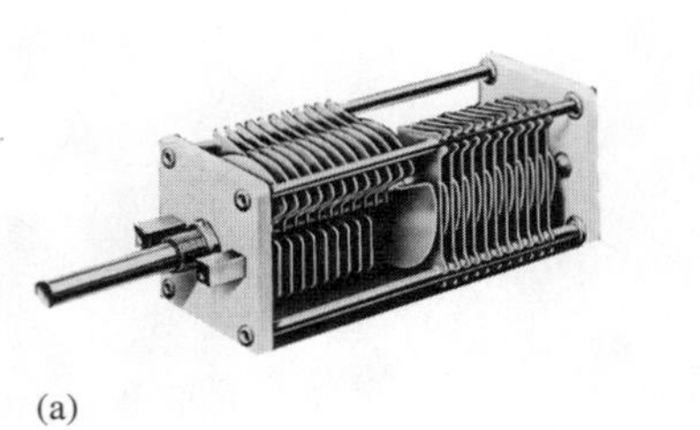

(a)

(b)

(c)

FIG. 10.20
Variable capacitors: (a) air; (b) air trimmer; (c) ceramic dielectric compression trimmer.
[(a) courtesy of James Millen Manufacturing Co.; (c) courtesy of Sprague-Goodman, Inc.]

ceramic trimmer capacitor permits varying the capacitance by changing the common area as above or by applying pressure to the ceramic plate to reduce the distance between the plates.

Leakage Current

Although we would like to think of capacitors as ideal elements, unfortunately, this is not the case. Up to this point, we have assumed that the insulating characteristics of dielectrics prevent any flow of charge between the plates unless the breakdown voltage is exceeded. In reality, however, dielectrics are not perfect insulators, and they do carry a few free electrons in their atomic structure.

When a voltage is applied across a capacitor, a **leakage current** is established between the plates. This current is usually so small that it can be ignored for the application under investigation. The availability of free electrons to support current flow is represented by a large parallel resistor in the equivalent circuit for a capacitor as shown in Fig. 10.21(a). If we apply 10 V across a capacitor with an internal resistance of 1000 MΩ, the current will be 0.01 μA—a level that can be ignored for most applications.

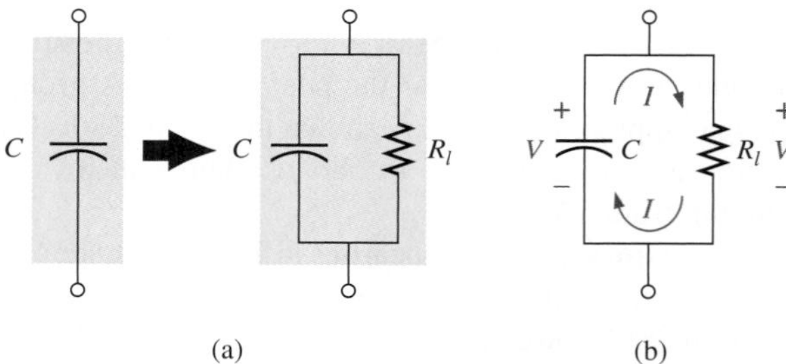

FIG. 10.21
Leakage current: (a) including the leakage resistance in the equivalent model for a capacitor; (b) internal discharge of a capacitor due to the leakage current.

The real problem associated with leakage currents is not evident until you ask the capacitors to sit in a charged state for long periods of time. As shown in Fig. 10.21(b), the voltage ($V = Q/C$) across a charged capacitor also appears across the parallel leakage resistance and establishes a discharge current through the resistor. In time, the capacitor is totally discharged. Capacitors such as the electrolytic that have high leakage currents (a leakage resistance of 0.5 MΩ is typical) usually have

a limited shelf life due to this internal discharge characteristic. Ceramic, tantalum, and mica capacitors typically have unlimited shelf life due to leakage resistances in excess of 1000 MΩ. Thin-film capacitors have lower levels of leakage resistances that result in some concern about shelf life.

Temperature Effects: ppm

Every capacitor is temperature sensitive, with the nameplate capacitance level specified at room temperature. Depending on the type of dielectric, increasing or decreasing temperatures can cause either a drop or a rise in capacitance. If temperature is a concern for a particular application, the manufacturer will provide a temperature plot, such as shown in Fig. 10.22, or a **ppm/°C** (parts per million per degree Celsius) rating for the capacitor. Note in Fig. 10.20 the 0% variation from the nominal (nameplate) value at 25°C (room temperature). At 0°C (freezing), it has dropped 20%, while at 100°C (the boiling point of water), it has dropped 70%—a factor to consider for some applications.

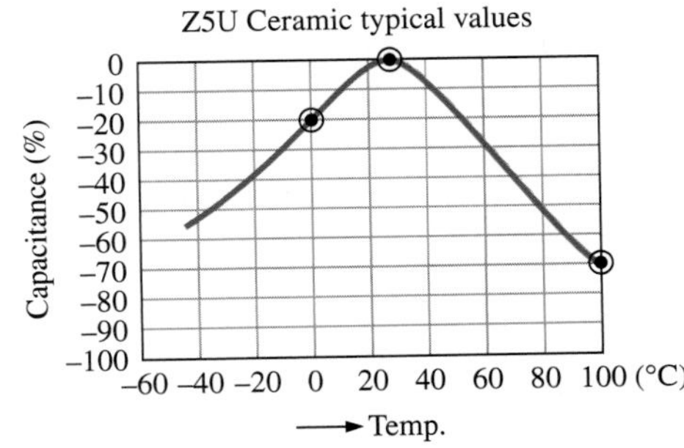

FIG. 10.22
Variation of capacitor value with temperature.

As an example using the ppm level, consider a 100 μF capacitor with a **temperature coefficient** or **ppm** of −150 ppm/°C. It is important to note the negative sign in front of the ppm value, because it reveals that the capacitance will drop with increase in temperature. It takes a moment to fully appreciate a term such as *parts per million.* In equation form, 150 parts per million can be written as

$$-\frac{150}{1{,}000{,}000} \times$$

If we then multiply this term by the capacitor value, we can obtain the change in capacitance for each 1°C change in temperature. That is,

$$-\frac{150}{1{,}000{,}000}(100\ \mu\text{F})/°\text{C} = \mathbf{-0.015\ \mu F/°C} = \mathbf{-15{,}000\ pF/°C}$$

If the temperature should rise 25°C, the capacitance decreases by

$$-\frac{15{,}000\ \text{pF}}{°\cancel{\text{C}}}(25°\cancel{\text{C}}) = \mathbf{-0.38\ \mu F}$$

changing the capacitance level to

$$100\ \mu\text{F} - 0.38\ \mu\text{F} = \mathbf{99.62\ \mu F}$$

Capacitor Labeling

Due to the small size of some capacitors, various marking schemes have been adopted to provide the capacitance level, tolerance, and, if possible, working voltage. In general, however, as pointed out above, *the size of the capacitor is the first indicator of its value.* In fact, most marking schemes do not indicate whether it is in μF or pF. It is assumed that you can make that judgment purely from the size. The smaller units are typically in pF, and the larger units in μF. Unless indicated by an **n** or **N,** most units are not provided in nF. On larger μF units, the value can often be printed on the jacket with the tolerance and working voltage. However, smaller units need to use some form of abbreviation as shown in Fig. 10.23. For very small units such as those in Fig. 10.23(a) with only two numbers, the value is recognized immediately as being in pF with the **K** an indicator of a

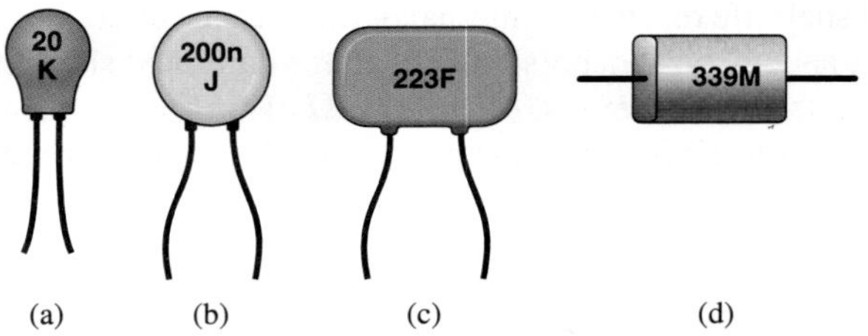

FIG. 10.23
Various marking schemes for small capacitors.

±10% tolerance level. Too often the K is read as a multiplier of 10^3, and the capacitance is read as 20,000 pF or 20 nF rather than the actual 20 pF.

For the unit in Fig. 10.23(b), there was room for a lowercase **n** to represent a multiplier of 10^{-9}, resulting in a value of 200 nF. To avoid unnecessary confusion, the letters used for tolerance do not include **N** or **U** or **P,** so the presence of any of these letters in upper- or lowercase normally refers to the multiplier level. The **J** appearing on the unit in Fig. 10.23(b) represents a ±5% tolerance level. For the capacitor in Fig. 10.23(c), the first two numbers are the numerical value of the capacitor, while the third number is the power of the multiplier (or number of zeros to be added to the first two numbers). The question then remains whether the units are μF or pF. With the 223 representing a number of 22,000, the units are certainly not μF, because the unit is too small for such a large capacitance. It is a 22,000 pF = 22 nF capacitor. The **F** represents a ±1% tolerance level. Multipliers of 0.01 use an 8 for the third digit, while multipliers of 0.1 use a 9. The capacitor in Fig. 10.23(d) is a $33 \times 0.1 = 3.3\ \mu$F capacitor with a tolerance of ±20% as defined by the capital letter **M.** The capacitance is not 3.3 pF because the unit is too large; again, the factor of size is very helpful in making a judgment about the capacitance level. It should also be noted that **MFD** is sometimes used to signify microfarads.

FIG. 10.24
Digital reading capacitance meter.
(Image provided by B & K Precision Corporation.)

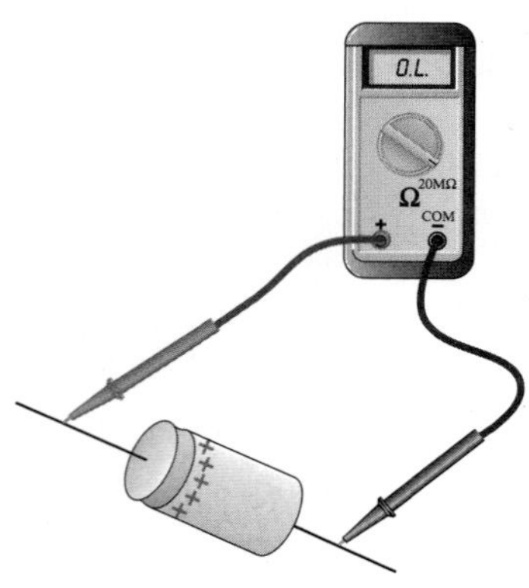

FIG. 10.25
Checking the dielectric of an electrolytic capacitor.

Measurement and Testing of Capacitors

The capacitance of a capacitor can be read directly using a meter such as the Universal LCR Meter in Fig. 10.24. If you set the meter on **C** for *capacitance,* it will automatically choose the most appropriate unit of measurement for the element, that is, F, μF, nF, or pF. Note the polarity markings on the meter for capacitors that have a specified polarity.

The best check is to use a meter such as the one in Fig. 10.24. However, if it is unavailable, an ohmmeter can be used to determine whether the dielectric is still in good working order or whether it has deteriorated due to age or use (especially for paper and electrolytics). As the dielectric breaks down, the insulating qualities of the material decrease to the point where the resistance between the plates drops to a relatively low level. To use an ohmmeter, be sure that the capacitor is fully discharged by placing a lead directly across its terminals. Then hook up the meter (paying attention to the polarities if the unit is polarized) as shown in Fig. 10.25, and note whether the resistance has dropped to a relatively low value (0 to a few kilohms). If so, the capacitor should be discarded. You may find that the reading changes when the meter is first connected. This change is due to the charging of the capacitor by the internal supply of the ohmmeter. In time the capacitor becomes stable, and the correct read-

ing can be observed. Typically, it should pin at the highest level on the megohm scales or indicate OL on a digital meter.

The above ohmmeter test is not all-inclusive, since some capacitors exhibit the breakdown characteristics only when a large voltage is applied. The test, however, does help isolate capacitors in which the dielectric has deteriorated.

Standard Capacitor Values

The most common capacitors use the same numerical multipliers encountered for resistors.

The vast majority are available with 5%, 10%, or 20% tolerances. There are capacitors available, however, with tolerances of 1%, 2%, or 3%, if you are willing to pay the price. Typical values include 0.1 μF, 0.15 μF, 0.22 μF, 0.33 μF, 0.47 μF, 0.68 μF; or 1 μF, 1.5 μF, 2.2 μF, 3.3 μF, 4.7 μF, 6.8 μF; and 10 pF, 22 pF, 33 pF, 100 pF; and so on.

10.5 TRANSIENTS IN CAPACITIVE NETWORKS: THE CHARGING PHASE

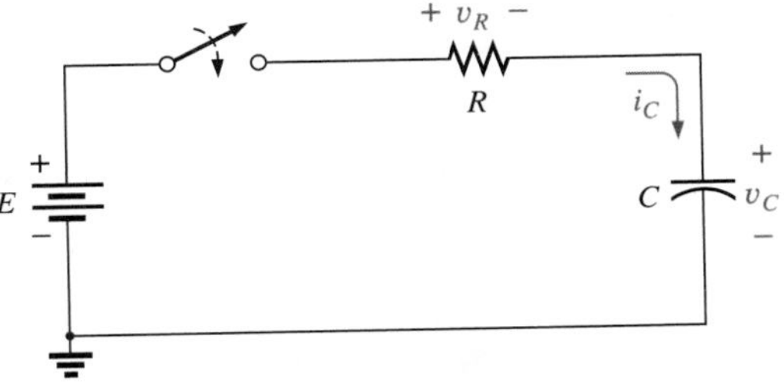

FIG. 10.26
Basic R-C charging network.

The placement of charge on the plates of a capacitor does not occur instantaneously. Instead, it occurs over a period of time determined by the components of the network. The charging phase, the phase during which charge is deposited on the plates, can be described by reviewing the response of the simple series circuit in Fig. 10.4. The circuit has been redrawn in Fig. 10.26 with the symbol for a fixed capacitor.

Recall that the instant the switch is closed, electrons are drawn from the top plate and deposited on the bottom plate by the battery, resulting in a net positive charge on the top plate and a negative charge on the bottom plate. The transfer of electrons is very rapid at first, slowing down as the potential across the plates approaches the applied voltage of the battery. Eventually, when the voltage across the capacitor equals the applied voltage, the transfer of electrons ceases, and the plates have a net charge determined by $Q = CV_C = CE$. This period of time during which charge is being deposited on the plates is called the **transient period**—a period of time where the voltage or current changes from one steady-state level to another.

Since the voltage across the plates is directly related to the charge on the plates by $V = Q/C$, a plot of the voltage across the capacitor will have the same shape as a plot of the charge on the plates over time. As shown in Fig. 10.27, the voltage across the capacitor is zero volts when the switch is closed ($t = 0$ s). It then builds up very quickly at first since charge is being deposited at a very high rate of speed. As time passes, the charge is deposited at a slower rate, and the change in voltage drops off. The voltage continues to grow, but at a much slower rate. Eventually, as the voltage across the plates approaches the applied voltage, the charging rate is very slow, until finally the voltage across the plates is equal to the applied voltage—the transient phase has passed.

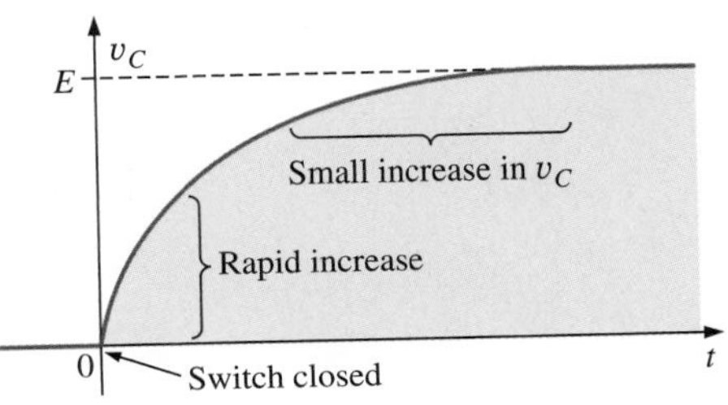

FIG. 10.27
v_C during the charging phase.

Fortunately, the waveform in Fig. 10.27 from beginning to end can be described using the mathematical function e^{-x}. It is an exponential function that decreases with time, as shown in Fig. 10.28. If we substitute zero for x, we obtain e^{-0} which by definition is 1, as shown in

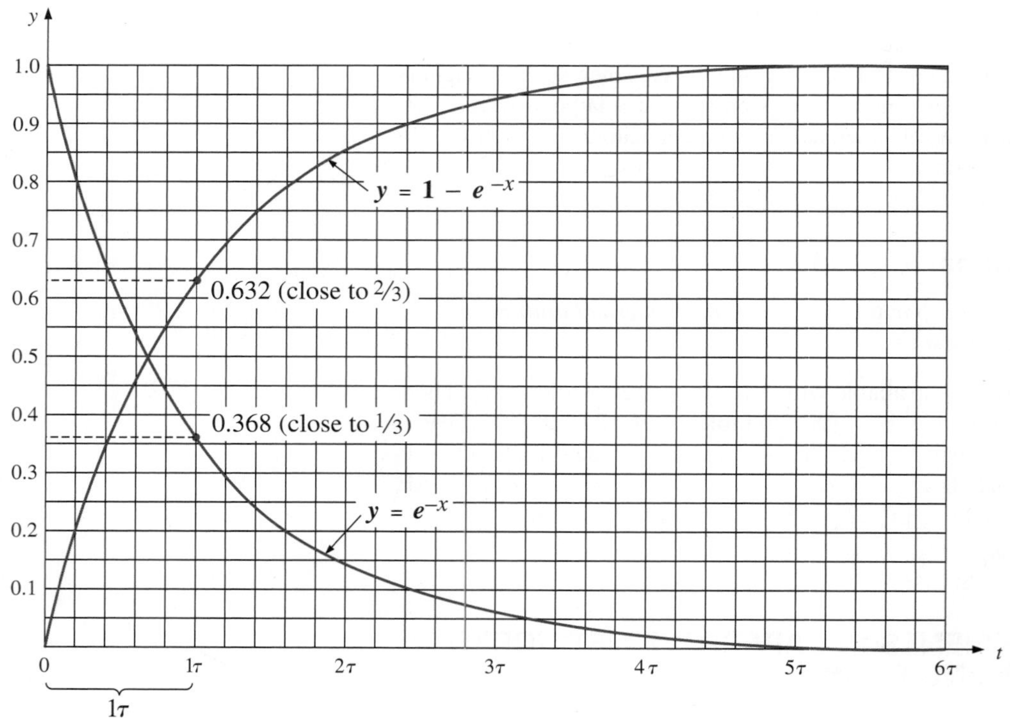

FIG. 10.28
Universal time constant chart.

TABLE 10.3
Selected values of e^{-x}.

$x = 0$	$e^{-x} = e^{-0} = \frac{1}{e^0} = \frac{1}{1} = 1$
$x = 1$	$e^{-1} = \frac{1}{e} = \frac{1}{2.71828\ldots} = 0.3679$
$x = 2$	$e^{-2} = \frac{1}{e^2} = 0.1353$
$x = 5$	$e^{-5} = \frac{1}{e^5} = 0.00674$
$x = 10$	$e^{-10} = \frac{1}{e^{10}} = 0.0000454$
$x = 100$	$e^{-100} = \frac{1}{e^{100}} = 3.72 \times 10^{-44}$

Table 10.3 and on the plot in Fig. 10.28. Table 10.3 reveals that as x increases, the function e^{-x} decreases in magnitude until it is very close to zero after $x = 5$. As noted in Table 10.3, the exponential factor $e^1 = e = 2.71828$.

A plot of $1 - e^{-x}$ is also provided in Fig. 10.28 since it is a component of the voltage v_C in Fig. 10.27. When e^{-x} is 1, $1 - e^{-x}$ is zero as shown in Fig. 10.28 and when e^{-x} decreases in magnitude, $1 - e^{-x}$ approaches 1 as shown in the same figure.

You may wonder how this function can help us if it decreases with time and the curve for the voltage across the capacitor increases with time. We simply place the exponential in the proper mathematical form as shown in Eq. (10.13):

$$\boxed{v_C = E(1 - e^{-t/\tau})}_{\text{charging}} \qquad \text{(volts, V)} \qquad \textbf{(10.13)}$$

First note in Eq. (10.13) that the voltage v_C is written in *lowercase (not capital) italic* to point out that it is a function that will change with time—it is not a constant. The exponent of the exponential function is no longer just x, but now is time (t) divided by a constant τ, the Greek letter *tau*. The quantity τ is defined by

$$\boxed{\tau = RC} \qquad \text{(time, s)} \qquad \textbf{(10.14)}$$

The factor τ, called the **time constant** of the network, has the units of time as shown below using some of the basic equations introduced earlier in this text:

$$\tau = RC = \left(\frac{V}{I}\right)\left(\frac{Q}{V}\right) = \left(\frac{\not{V}}{\not{Q}/t}\right)\left(\frac{\not{Q}}{\not{V}}\right) = t \text{ (seconds)}$$

A plot of Eq. (10.13) results in the curve in Fig. 10.29 whose shape is an exact match with that in Fig. 10.27.

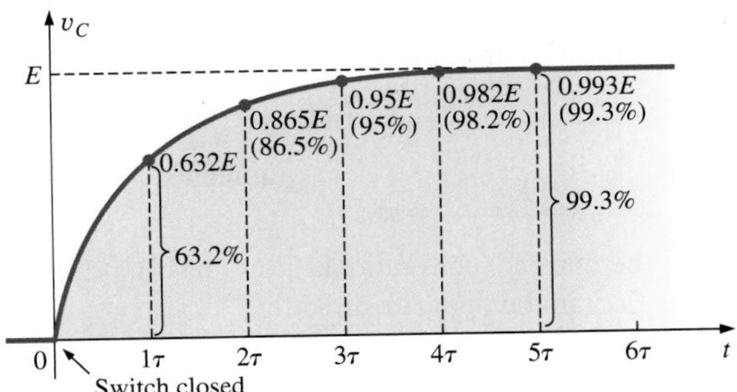

FIG. 10.29
Plotting the equation $v_C = E(1 - e^{-t/\tau})$ versus time (t).

In Eq. (10.13), if we substitute $t = 0$ s, we find that

$$e^{-t/\tau} = e^{-0/\tau} = e^{-0} = \frac{1}{e^0} = \frac{1}{1} = 1$$

and $$v_C = E(1 - e^{-t/\tau}) = E(1 - 1) = \mathbf{0\ V}$$

as appearing in the plot in Fig. 10.29.

It is important to realize at this point that the plot in Fig. 10.29 is not against simply time but against τ, the time constant of the network. If we want to know the voltage across the plates after one time constant, we simply plug $t = 1\tau$ into Eq. (10.13). The result is

$$e^{-t/\tau} = e^{-\tau/\tau} = e^{-1} \cong 0.368$$

and $$v_C = E(1 - e^{-t/\tau}) = E(1 - 0.368) = \mathbf{0.632E}$$

as shown in Fig. 10.29.

At $t = 2\tau$:

$$e^{-t/\tau} = e^{-2\tau/\tau} = e^{-2} \cong 0.135$$

and $$v_C = E(1 - e^{-t/\tau}) = E(1 - 0.135) = \mathbf{0.865E}$$

as shown in Fig. 10.29.

As the number of time constants increases, the voltage across the capacitor does indeed approach the applied voltage.

At $t = 5\tau$:

$$e^{-t/\tau} = e^{-5\tau/\tau} = e^{-5} \cong 0.007$$

and $$v_C = E(1 - e^{-t/\tau}) = E(1 - 0.007) = \mathbf{0.993E} \cong E$$

In fact, we can conclude from the results just obtained that

the voltage across a capacitor in a dc network is essentially equal to the applied voltage after five time constants of the charging phase have passed.

Or, in more general terms,

the transient or charging phase of a capacitor has essentially ended after five time constants.

It is indeed fortunate that the same exponential function can be used to plot the current of the capacitor versus time. When the switch is first closed, the flow of charge or current jumps very quickly to a value limited by the applied voltage and the circuit resistance, as shown in Fig. 10.30. The rate of deposit, and hence the current, then decreases quite rapidly, until eventually charge is not being deposited on the plates and the current drops to zero amperes.

The equation for the current is:

$$\boxed{i_C = \frac{E}{R}e^{-t/\tau}}_{\text{charging}} \qquad \text{(amperes, A)} \qquad \textbf{(10.15)}$$

In Fig. 10.26, the current (conventional flow) has the direction shown since electrons flow in the opposite direction.

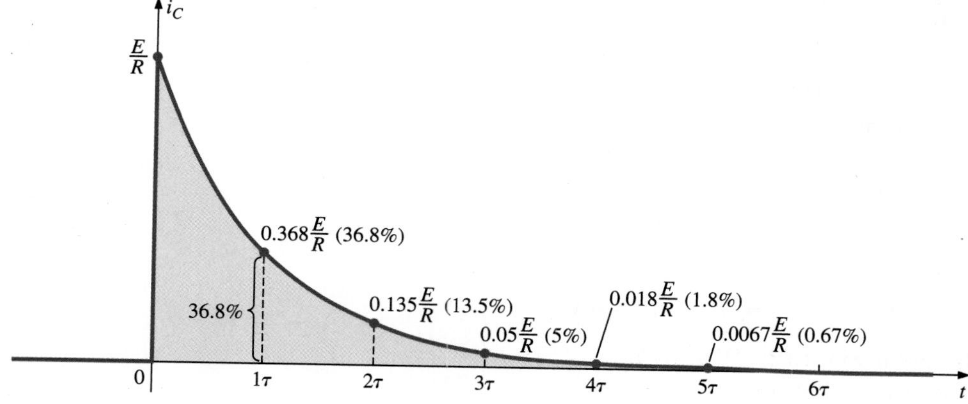

FIG. 10.30

Plotting the equation $i_C = \frac{E}{R}e^{-t/\tau}$ versus time (t).

At $t = 0$ s:

$$e^{-t/\tau} = e^{-0} = 1$$

and
$$i_C = \frac{E}{R}e^{-t/\tau} = \frac{E}{R}(1) = \mathbf{\frac{E}{R}}$$

At $t = 1\tau$:

$$e^{-t/\tau} = e^{-\tau/\tau} = e^{-1} \cong 0.368$$

and
$$i_C = \frac{E}{R}e^{-t/\tau} = \frac{E}{R}(0.368) = \mathbf{0.368\frac{E}{R}}$$

In general, Figure 10.30 clearly reveals that

the current of a capacitive dc network is essentially zero amperes after five time constants of the charging phase have passed.

It is also important to recognize that

during the charging phase, the major change in voltage and current occurs during the first time constant.

The voltage across the capacitor reaches about 63.2% (about 2/3) of its final value, whereas the current drops to 36.8% (about 1/3) of its peak value. During the next time constant, the voltage increases only about 23.3%, whereas the current drops to 13.5%. The first time constant is

therefore a very dramatic time for the changing parameters. Between the fourth and fifth time constants, the voltage increases only about 1.2%, whereas the current drops to less than 1% of its peak value.

Returning to Figs. 10.29 and 10.30, note that when the voltage across the capacitor reaches the applied voltage *E*, the current drops to zero amperes, as reviewed in Fig. 10.31. These conditions match those of an open circuit, permitting the following conclusion:

A capacitor can be replaced by an open-circuit equivalent once the charging phase in a dc network has passed.

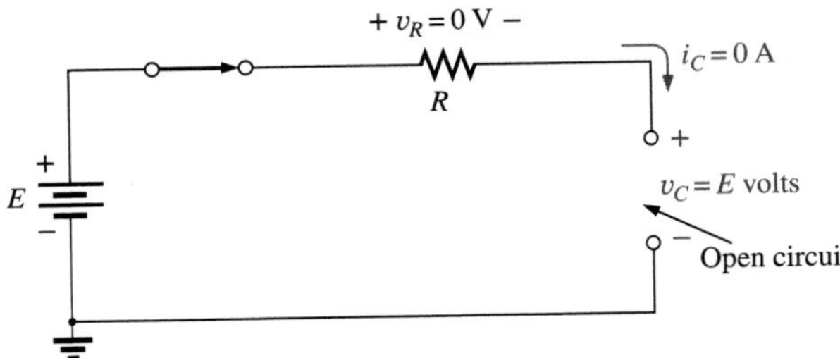

FIG. 10.31
Demonstrating that a capacitor has the characteristics of an open circuit after the charging phase has passed.

This conclusion will be particularly useful when analyzing dc networks that have been on for a long period of time or have passed the transient phase that normally occurs when a system is first turned on.

A similar conclusion can be reached if we consider the instant the switch is closed in the circuit in Fig. 10.26. Referring to Figs. 10.29 and 10.30 again, we find that the current is a peak value at $t = 0$ s, whereas the voltage across the capacitor is 0 V, as shown in the equivalent circuit in Fig. 10.32. The result is that

a capacitor has the characteristics of a short-circuit equivalent at the instant the switch is closed in an uncharged series R-C circuit.

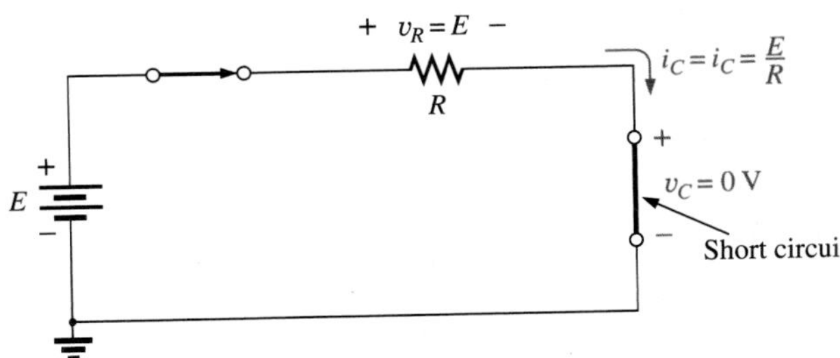

FIG. 10.32
Revealing the short-circuit equivalent for the capacitor that occurs when the switch is first closed.

In Eq. (10.13), the time constant τ will always have some value because some resistance is always present in a capacitive network. In some cases, the value of τ may be very small, but five times that value of τ, no matter how small, must therefore always exist; it cannot be zero. The result is the following very important conclusion:

The voltage across a capacitor cannot change instantaneously.

In fact, we can take this statement a step further by saying that the capacitance of a network is a measure of how much it will oppose a change in voltage in a network. The larger the capacitance, the larger the time constant, and the longer it will take the voltage across the capacitor to reach the applied value. This can prove very helpful when lightning arresters and surge suppressors are designed to protect equipment from unexpected high surges in voltage.

Since the resistor and the capacitor in Fig. 10.26 are in series, the current through the resistor is the same as that associated with the capacitor. The voltage across the resistor can be determined by using Ohm's law in the following manner:

$$\upsilon_R = i_R R = i_C R$$

so that

$$\upsilon_R = \left(\frac{E}{R}e^{-t/\tau}\right)R$$

and

$$\boxed{\upsilon_R = Ee^{-t/\tau}}_{\text{charging}} \qquad \text{(volts, V)} \qquad \textbf{(10.16)}$$

A plot of the voltage as shown in Fig. 10.33 has the same shape as that for the current because they are related by the constant R. Note, however, that the voltage across the resistor starts at a level of E volts because the voltage across the capacitor is zero volts and Kirchhoff's voltage law must always be satisfied. When the capacitor has reached the applied voltage, the voltage across the resistor must drop to zero volts for the same reason. Always remember that

Kirchhoff's voltage law is applicable at any instant of time for any type of voltage in any type of network.

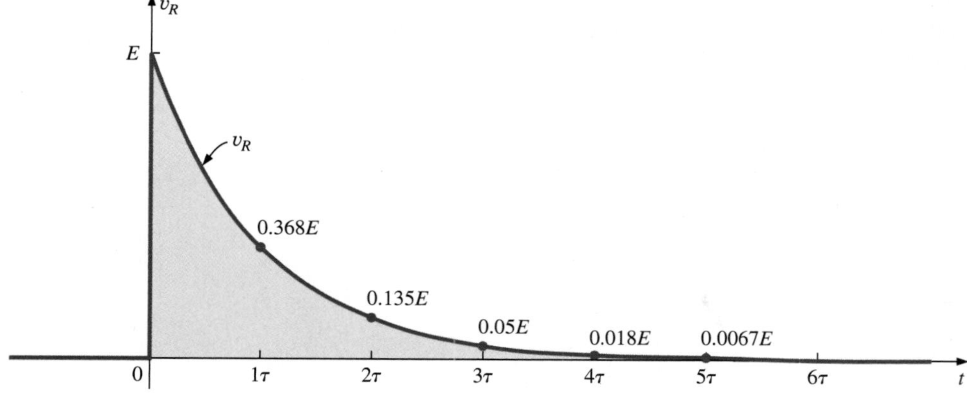

FIG. 10.33
Plotting the equation $\upsilon_R = Ee^{-t/\tau}$ versus time (t).

Using the Calculator to Solve Exponential Functions

Before looking at an example, we will first discuss the use of the TI-89 calculator with exponential functions. The process is actually quite simple for a number such as $e^{-1.2}$. Just select the 2nd function (diamond) key, followed by the function e^x. Then insert the (−) sign from the numerical keyboard (not the mathematical functions), and insert the number 1.2 followed by ENTER to obtain the result of 0.301, as shown in Fig. 10.34. The use of the computer software program Mathcad is demonstrated in a later example.

◆ e^x (–) 1 • 2) ENTER **301.2E-3**

FIG. 10.34
Calculator key strokes to determine $e^{-1.2}$.

EXAMPLE 10.6 For the circuit in Fig. 10.35:

a. Find the mathematical expression for the transient behavior of v_C, i_C, and v_R if the switch is closed at $t = 0$ s.
b. Plot the waveform of v_C versus the time constant of the network.
c. Plot the waveform of v_C versus time.
d. Plot the waveforms of i_C and v_R versus the time constant of the network.
e. What is the value of v_C at $t = 20$ ms?
f. On a practical basis, how much time must pass before we can assume that the charging phase has passed?
g. When the charging phase has passed, how much charge is sitting on the plates?
h. If the capacitor has a leakage resistance of 10,000 MΩ, what is the initial leakage current? Once the capacitor is separated from the circuit, how long will it take to totally discharge, assuming a linear (unchanging) discharge rate?

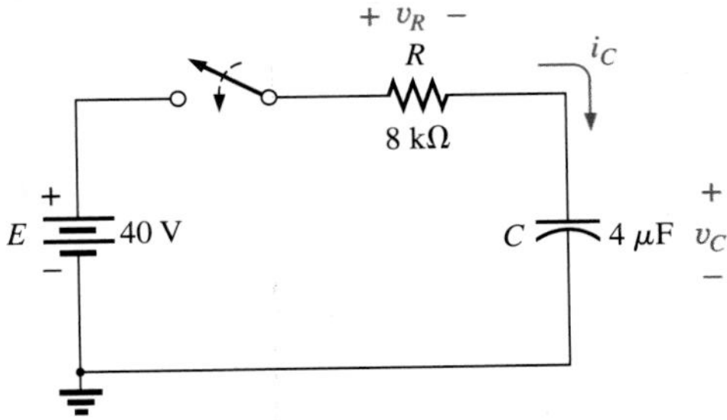

FIG. 10.35
Transient network for Example 10.6.

Solutions:

a. The time constant of the network is

$$\tau = RC = (8\text{ k}\Omega)(4\ \mu\text{F}) = 32\text{ ms}$$

resulting in the following mathematical equations:

$$v_C = E(1 - e^{-t/\tau}) = \mathbf{40\ V(1 - e^{-t/32ms})}$$

$$i_C = \frac{E}{R}e^{-t/\tau} = \frac{40\text{ V}}{8\text{ k}\Omega}e^{-t/32\text{ms}} = \mathbf{5\ mAe^{-t/32ms}}$$

$$v_R = Ee^{-t/\tau} = \mathbf{40\ Ve^{-t/32ms}}$$

b. The resulting plot appears in Fig. 10.36.
c. The horizontal scale will now be against time rather than time constants, as shown in Fig. 10.37. The plot points in Fig. 10.37 were taken from Fig. 10.36.

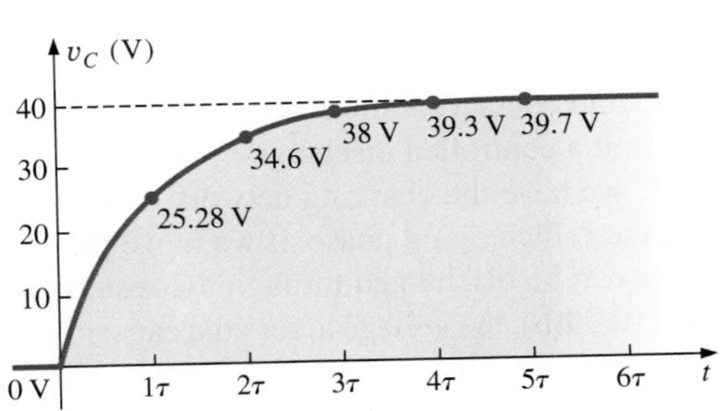

FIG. 10.36
v_C versus time for the charging network in Fig. 10.35.

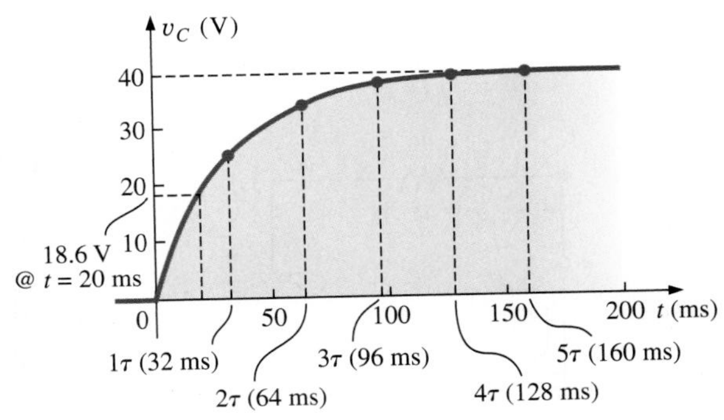

FIG. 10.37
Plotting the waveform in Fig. 10.36 versus time (t).

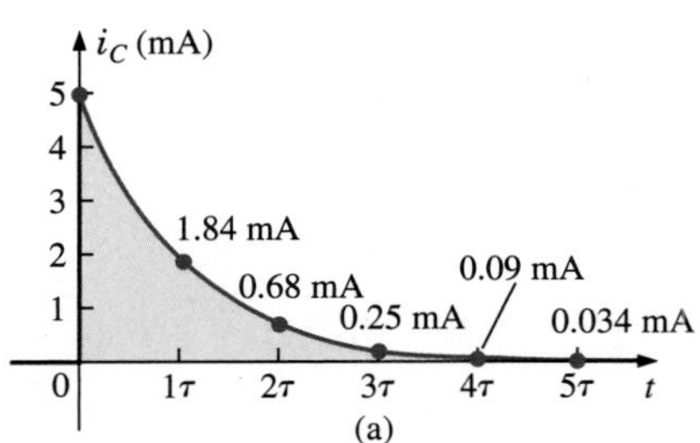

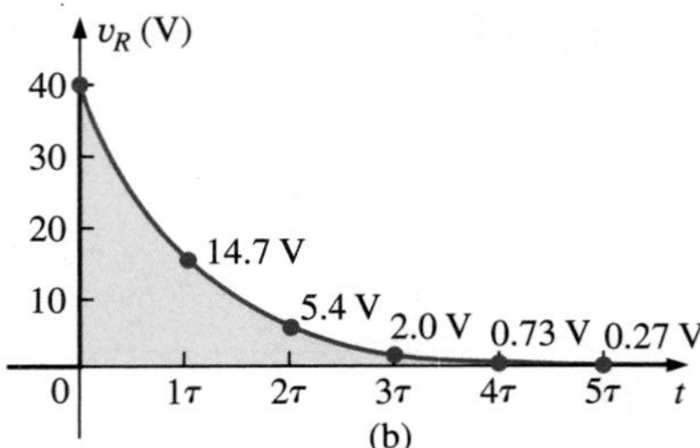

FIG. 10.38
i_C and v_R for the charging network in Fig. 10.36.

d. Both plots appear in Fig. 10.38.
e. Substituting the time $t = 20$ ms results in the following for the exponential part of the equation:

$$e^{-t/\tau} = e^{-20\text{ms}/32\text{ms}} = e^{-0.625} = 0.535 \text{ (using a calculator)}$$

so that
$$v_C = 40\text{ V}(1 - e^{-t/32\text{ms}}) = 40\text{ V}(1 - 0.535)$$
$$= (40\text{ V})(0.465) = \mathbf{18.6\ V} \text{ (as verified by Fig. 10.37)}$$

f. Assuming a full charge in five time constants results in

$$5\tau = 5(32\text{ ms}) = \mathbf{160\ ms} = \mathbf{0.16\ s}$$

g. Using Eq. (10.6):

$$Q = CV = (4\ \mu\text{F})(40\text{ V}) = \mathbf{160\ \mu C}$$

h. Using Ohm's law:

$$I_{\text{leakage}} = \frac{40\text{ V}}{10{,}000\text{ M}\Omega} = 4\text{ nA}$$

Finally, the basic equation $I = Q/t$ results in

$$t = \frac{Q}{I} = \frac{160\ \mu\text{C}}{4\text{ nA}} = (40{,}000\text{ s})\left(\frac{1\text{ min}}{60\text{ s}}\right)\left(\frac{1\text{ h}}{60\text{ min}}\right) = \mathbf{11.11\ h}$$

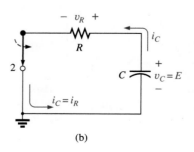

10.6 TRANSIENTS IN CAPACITIVE NETWORKS: THE DISCHARGING PHASE

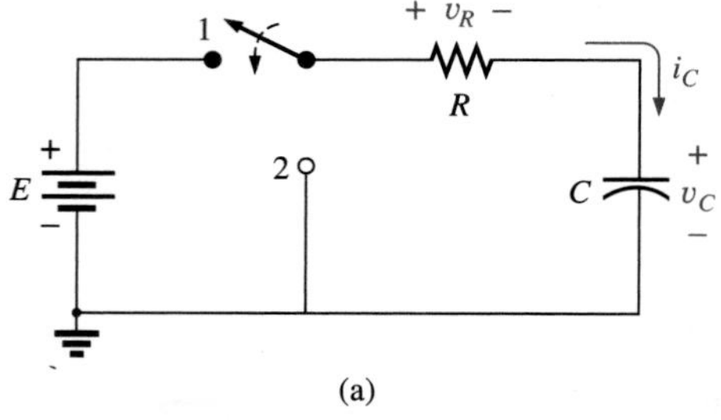

FIG. 10.39
(a) Charging network; (b) discharging configuration.

We now investigate how to discharge a capacitor while exerting some control on how long the discharge time will be. You can, of course, place a lead directly across a capacitor to discharge it very quickly—and possibly cause a visible spark. For larger capacitors such those in TV sets, this procedure should not be attempted because of the high voltages involved—unless, of course, you are trained in the maneuver.

In Fig. 10.39(a), a second contact for the switch was added to the circuit in Fig. 10.26 to permit a controlled discharge of the capacitor. With the switch in position 1, we have the charging network described in the last section. Following the full charging phase, if we move the switch to position 1, the capacitor can be discharged through the resulting circuit in Fig. 10.39(b). In Fig. 10.39(b), the voltage across the capacitor appears directly across the resistor to establish a discharge current. Initially, the current jumps to a relatively high value; then it begins to drop. It drops with time because charge is leaving the plates of the capacitor, which in turn reduces the voltage across the capacitor and thereby the voltage across the resistor and the resulting current.

Before looking at the wave shapes for each quantity of interest, note that current i_C has now reversed direction as shown in Fig. 10.39(b). As shown in parts (a) and (b) in Fig. 10.39, the voltage across the capacitor does not reverse polarity, but the current reverses direction. We will show the reversals on the resulting plots by sketching the waveforms in the negative regions of the graph. In all the waveforms, note that all the mathematical expressions use the same e^{-x} factor appearing during the charging phase.

For the voltage across the capacitor that is decreasing with time, the mathematical expression is:

$$\boxed{\upsilon_C = Ee^{-t/\tau}}_{\text{discharging}} \tag{10.17}$$

For this circuit, the time constant τ is defined by the same equation as used for the charging phase. That is,

$$\boxed{\tau = RC}_{\text{discharging}} \tag{10.18}$$

Since the current decreases with time, it will have a similar format:

$$\boxed{i_C = \frac{E}{R}e^{-t/\tau}}_{\text{discharging}} \tag{10.19}$$

For the configuration in Fig. 10.39(b), since $\upsilon_R = \upsilon_C$ (in parallel), the equation for the voltage υ_R has the same format:

$$\boxed{\upsilon_R = Ee^{-t/\tau}}_{\text{discharging}} \tag{10.20}$$

The complete discharge will occur, for all practical purposes, in five time constants. If the switch is moved between terminals 1 and 2 every five time constants, the wave shapes in Fig. 10.40 will result for υ_C, i_C,

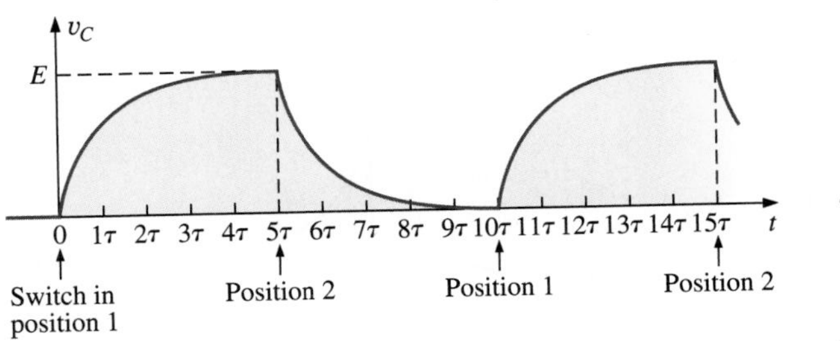

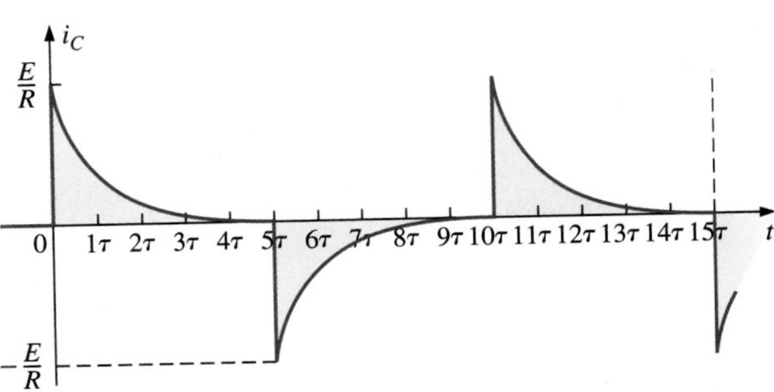

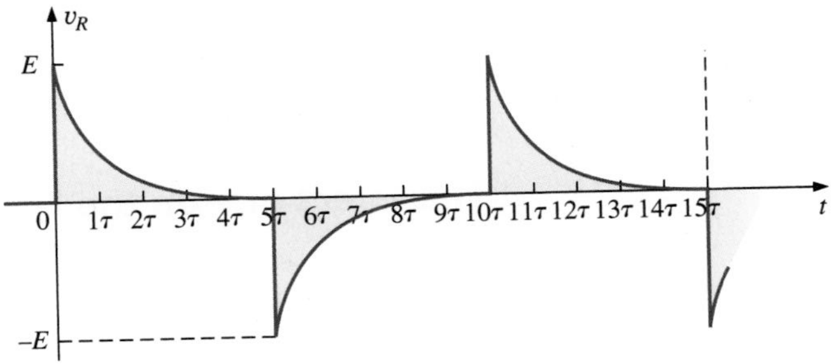

FIG. 10.40
υ_C, i_C, and υ_R for 5τ switching between contacts in Fig. 10.39(a).

and v_R. For each curve, the current directions and voltage polarities are as defined by the configurations in Fig. 10.39. Note, as pointed out above, that the current reverses direction during the discharge phase.

The discharge rate does not have to equal the charging rate if a different switching arrangement is used. In fact, Example 10.8 will demonstrate how to change the discharge rate.

EXAMPLE 10.7 Using the values in Example 10.6, plot the waveforms for v_C and i_C resulting from switching between contacts 1 and 2 in Fig. 10.39 every five time constants.

Solution: The time constant is the same for the charging and discharging phases. That is,

$$\tau = RC = (8\text{ k}\Omega)(4\ \mu\text{F}) = 32\text{ ms}$$

For the discharge phase, the equations are

$$v_C = Ee^{-t/\tau} = \mathbf{40\ Ve^{-t/32ms}}$$

$$i_C = -\frac{E}{R}e^{-t/\tau} = \frac{40\text{ V}}{8\text{ k}\Omega}e^{-t/32\text{ms}} = \mathbf{-5\ mAe^{-t/32ms}}$$

$$v_R = v_C = \mathbf{40\ Ve^{-t/32ms}}$$

A continuous plot for the charging and discharging phases appears in Fig. 10.41.

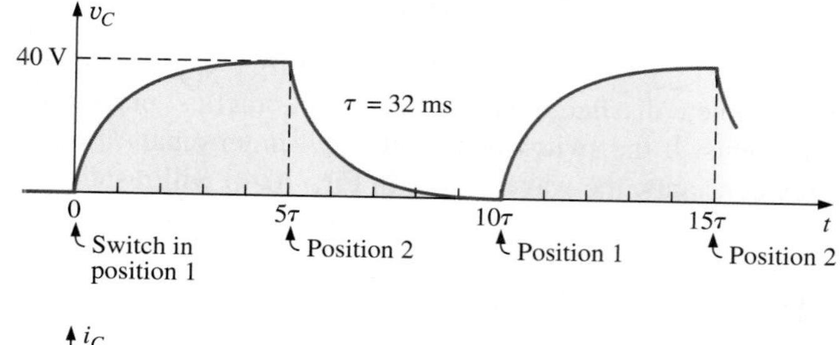

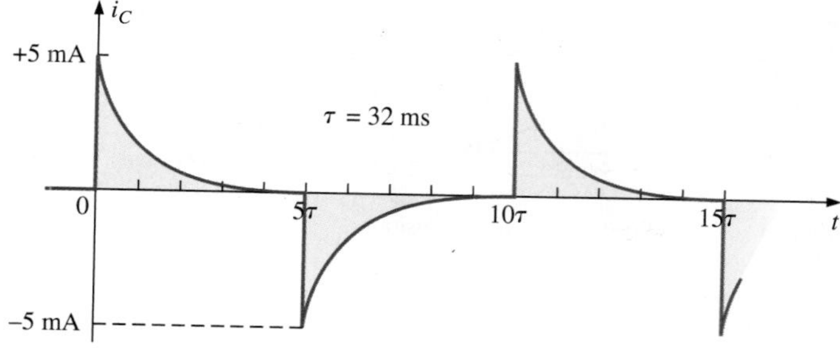

FIG. 10.41
v_C and i_C for the network in Fig. 10.39(a) with the values in Example 10.6.

The Effect of τ on the Response

In Example 10.7, if the value of τ were changed by changing the resistance, or the capacitor, or both, *the resulting waveforms would appear the same because they were plotted against the time constant of the network.* If they were plotted against time, there could be a dramatic change in the appearance of the resulting plots. In fact, on an oscilloscope, an instru-

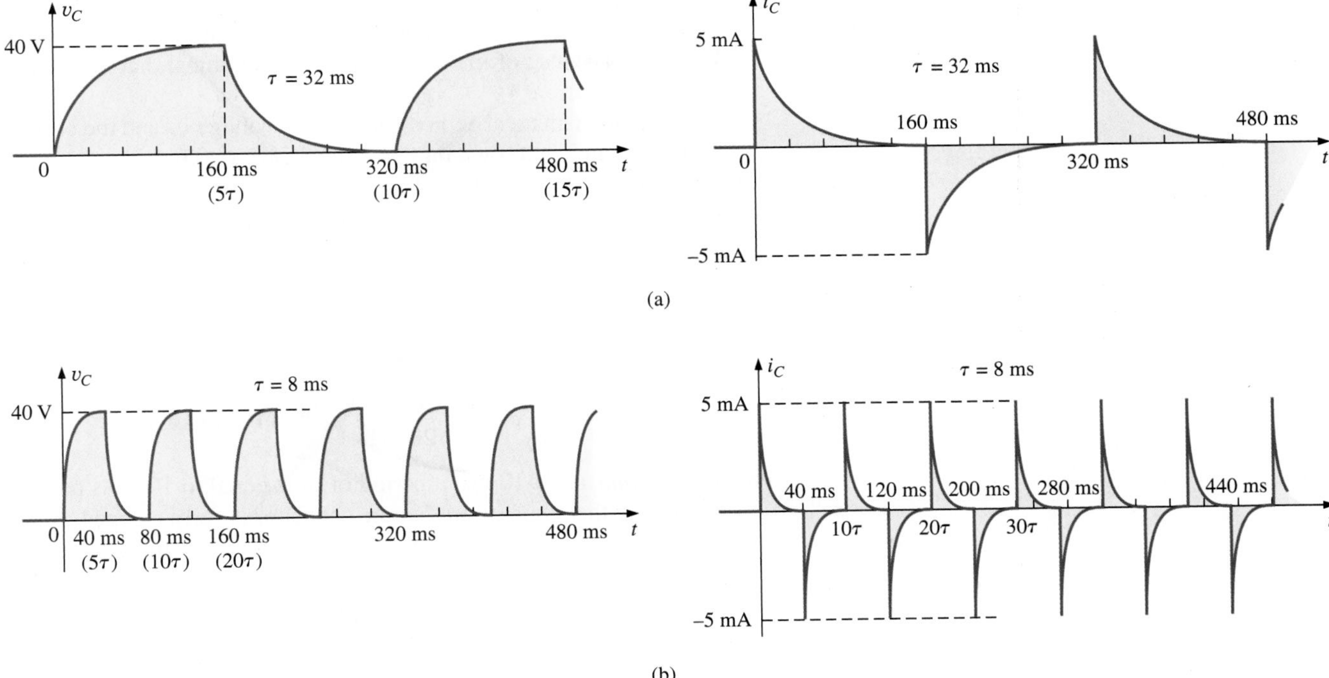

FIG. 10.42
Plotting v_C and i_C versus time in ms: (a) τ = 32 ms; (b) τ = 8 ms.

ment designed to display such waveforms, the plots are against time, and the change will be immediately apparent. In Fig. 10.42(a), the waveforms in Fig. 10.41 for v_C and i_C were plotted against time. In Fig. 10.42(b), the capacitance was decreased to 1 μF which reduces the time constant to 8 ms. Note the dramatic effect on the appearance of the waveform.

For a fixed-resistance network, the effect of increasing the capacitance is clearly demonstrated in Fig. 10.43. The larger the capacitance, and hence the time constant, the longer it takes the capacitor to charge up—there is more charge to be stored. The same effect can be created by holding the capacitance constant and increasing the resistance, but now the longer time is due to the lower currents that are a result of the higher resistance.

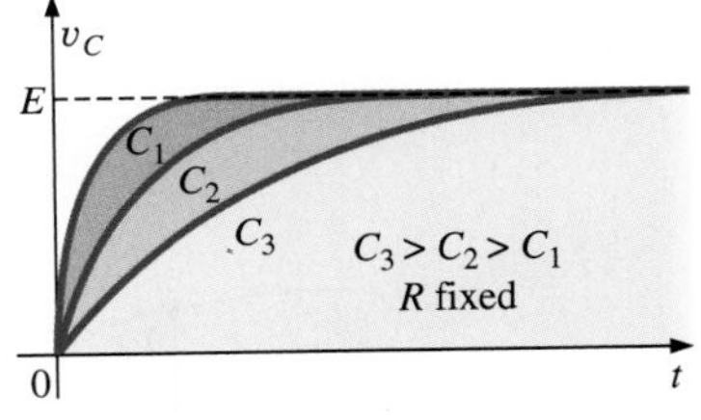

FIG. 10.43
Effect of increasing values of C (with R constant) on the charging curve for v_C.

EXAMPLE 10.8 For the circuit in Fig. 10.44:

a. Find the mathematical expressions for the transient behavior of the voltage v_C and the current i_C if the capacitor was initially uncharged and the switch is thrown into position 1 at $t = 0$ s.

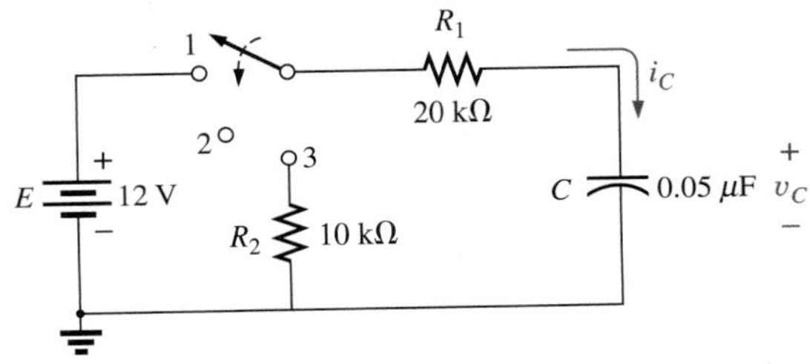

FIG. 10.44
Network to be analyzed in Example 10.8.

b. Find the mathematical expressions for the voltage v_C and the current i_C if the switch is moved to position 2 at $t = 10$ ms. (Assume that the leakage resistance of the capacitor is infinite ohms; that is, there is no leakage current.)
c. Find the mathematical expressions for the voltage v_C and the current i_C if the switch is thrown into position 3 at $t = 20$ ms.
d. Plot the waveforms obtained in parts (a)–(c) on the same time axis using the defined polarities in Fig. 10.44.

Solutions:

a. *Charging phase:*

$$\tau = R_1C = (20\text{ k}\Omega)(0.05\ \mu\text{F}) = 1\text{ ms}$$
$$v_C = E(1 - e^{-t/\tau}) = \mathbf{12\ V(1 - e^{-t/1ms})}$$
$$i_C = \frac{E}{R_1}e^{-t/\tau} = \frac{12\text{ V}}{20\text{ k}\Omega}e^{-t/1\text{ms}} = \mathbf{0.6\ mAe^{-t/1ms}}$$

b. *Storage phase:* At 10 ms, a period of time equal to 10τ has passed, permitting the assumption that the capacitor is fully charged. Since $R_{\text{leakage}} = \infty\ \Omega$, the capacitor will hold its charge indefinitely. The result is that both v_C and i_C will remain at a fixed value of

$$v_C = \mathbf{12\ V}$$
$$i_C = \mathbf{0\ A}$$

c. *Discharge phase* (using 20 ms as the new $t = 0$ s for the equations): The new time constant is

$$\tau' = RC = (R_1 + R_2)C = (20\text{ k}\Omega + 10\text{ k}\Omega)(0.05\ \mu\text{F}) = 1.5\text{ ms}$$
$$v_C = Ee^{-t/\tau'} = \mathbf{12\ Ve^{-t/1.5ms}}$$
$$i_C = -\frac{E}{R}e^{-t/\tau'} = -\frac{E}{R_1 + R_2}e^{-t/\tau'}$$
$$= -\frac{12\text{ V}}{20\text{ k}\Omega + 10\text{ k}\Omega}e^{-t/1.5\text{ms}} = \mathbf{-0.4\ mAe^{-t/1.5ms}}$$

d. See Fig. 10.45.

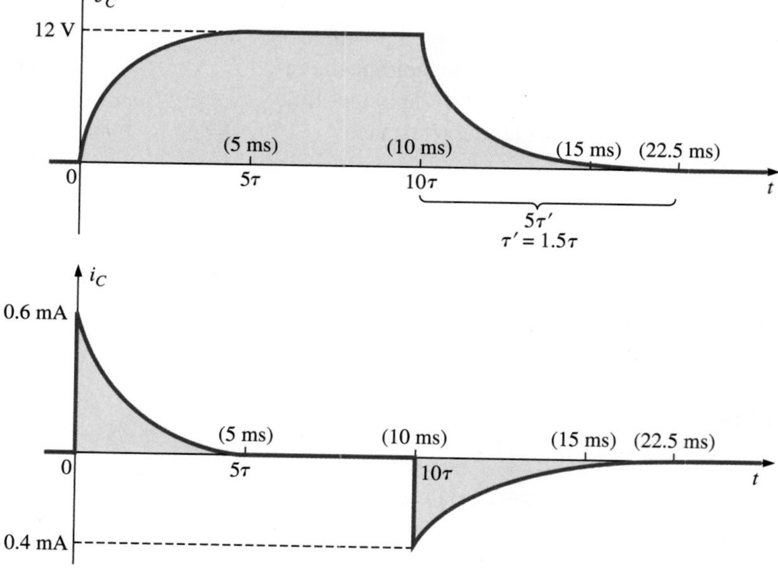

FIG. 10.45
v_C and i_C for the network in Fig. 10.44.

EXAMPLE 10.9 For the network in Fig. 10.46:

a. Find the mathematical expression for the transient behavior of the voltage across the capacitor if the switch is thrown into position 1 at $t = 0$ s.
b. Find the mathematical expression for the transient behavior of the voltage across the capacitor if the switch is moved to position 2 at $t = 1\tau$.
c. Plot the resulting waveform for the voltage v_C as determined by parts (a) and (b).
d. Repeat parts (a)–(c) for the current i_C.

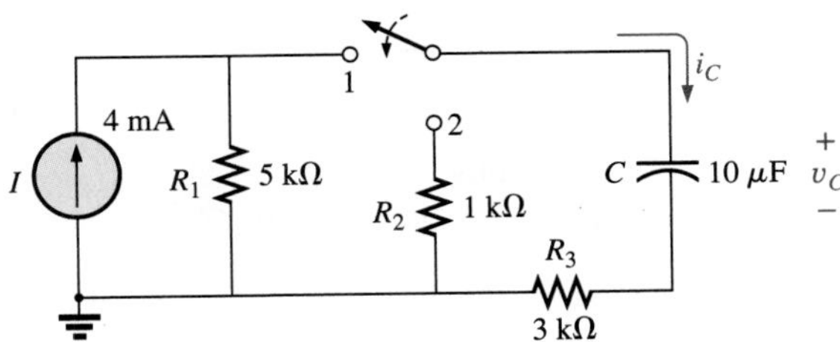

FIG. 10.46
Network to be analyzed in Example 10.9.

Solutions:

a. Converting the current source to a voltage source results in the configuration in Fig. 10.47 for the charging phase.

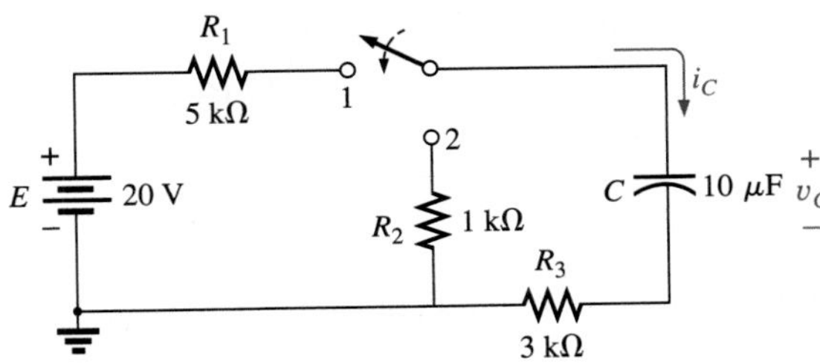

FIG. 10.47
The charging phase for the network in Fig. 10.46.

For the source conversion: $E = IR = (4\text{ mA})\,(5\text{ k}\Omega) = 20\text{ V}$

and $R_s = R_p = 5\text{ k}\Omega$

$$\tau = RC = (R_1 + R_3)C = (5\text{ k}\Omega + 3\text{ k}\Omega)(10\ \mu\text{F}) = 80\text{ ms}$$

$$v_C = E(1 - e^{-t/\tau}) = \mathbf{20\ V\ (1 - e^{-t/80ms})}$$

b. With the switch in position 2, the network appears as shown in Fig. 10.48. The voltage at 1τ can be found by using the fact that the voltage is 63.2% of its final value of 20 V, so that 0.632(20 V) = 12.64 V. Or you can substitute into the derived equation as follows:

$$e^{-t/\tau} = e^{-\tau/\tau} = e^{-1} = 0.368$$

and
$$v_C = 20\text{ V}(1 - e^{-t/80\text{ms}}) = 20\text{ V}(1 - 0.368)$$
$$= (20\text{ V})(0.632) = 12.64\text{ V}$$

Using this voltage as the starting point and substituting into the discharge equation results in

$$\tau' = RC = (R_2 + R_3)C = (1\text{ k}\Omega + 3\text{ k}\Omega)(10\ \mu\text{F}) = 40\text{ ms}$$
$$v_C = Ee^{-t/\tau'} = \mathbf{12.64\ Ve^{-t/40ms}}$$

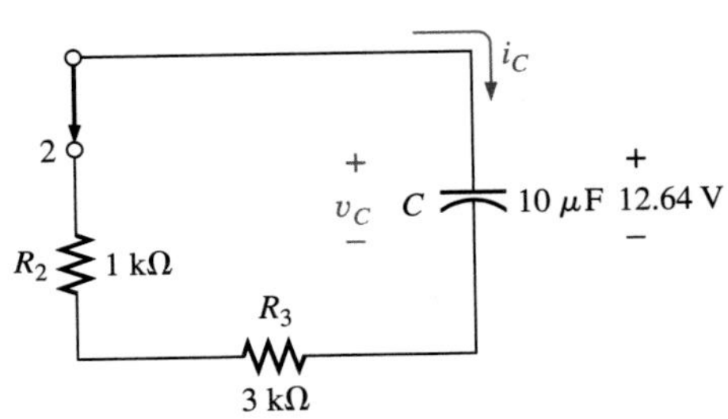

FIG. 10.48
Network in Fig. 10.47 when the switch is moved to position 2 at $t = 1\tau_1$.

c. See Fig. 10.49.

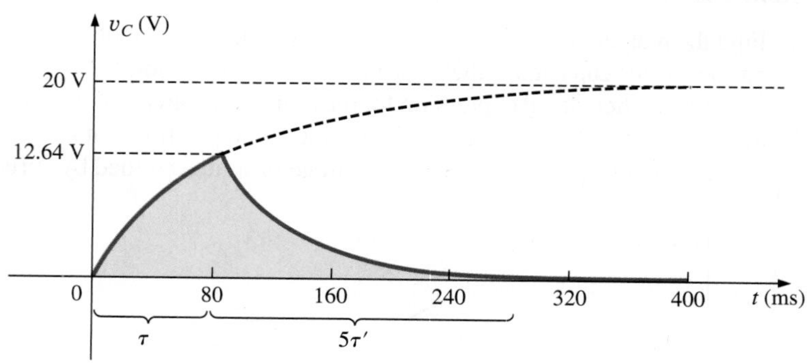

FIG. 10.49
v_C for the network in Fig. 10.47.

d. The charging equation for the current is

$$i_C = \frac{E}{R}e^{-t/\tau} = \frac{E}{R_1 + R_3}e^{-t/\tau} = \frac{20\text{ V}}{8\text{ k}\Omega}e^{-t/80\text{ms}} = \mathbf{2.5\ mA}\boldsymbol{e^{-t/80\text{ms}}}$$

which, at $t = 80$ ms, results in

$$i_C = 2.5\text{ mA}e^{-80\text{ms}/80\text{ms}} = 2.5\text{ mA}e^{-1} = (2.5\text{ mA})(0.368) = 0.92\text{ mA}$$

When the switch is moved to position 2, the 12.64 V across the capacitor appears across the resistor to establish a current of 12.64 V/4 kΩ = 3.16 mA. Substituting into the discharge equation with $V_i = 12.64$ V and $\tau' = 40$ ms yields

$$i_C = -\frac{V_i}{R_2 + R_3}e^{-t/\tau'} = -\frac{12.64\text{ V}}{1\text{ k}\Omega + 3\text{ k}\Omega}e^{-t/40\text{ms}}$$
$$= -\frac{12.64\text{ V}}{4\text{ k}\Omega}e^{-t/40\text{ms}} = \mathbf{-3.16\ mA}\boldsymbol{e^{-t/40\text{ms}}}$$

The equation has a minus sign because the direction of the discharge current is opposite to that defined for the current in Fig. 10.48. The resulting plot appears in Fig. 10.50.

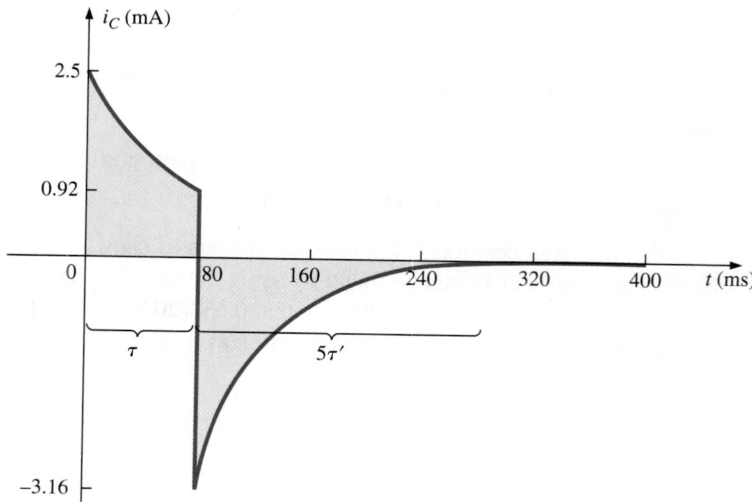

FIG. 10.50
i_c for the network in Fig. 10.47.

10.7 INITIAL CONDITIONS

In all the examples in the previous sections, the capacitor was uncharged before the switch was thrown. We now examine the effect of a charge, and therefore a voltage ($V = Q/C$), on the plates at the instant the switching action takes place. The voltage across the capacitor at this instant is called the **initial value,** as shown for the general waveform in Fig. 10.51.

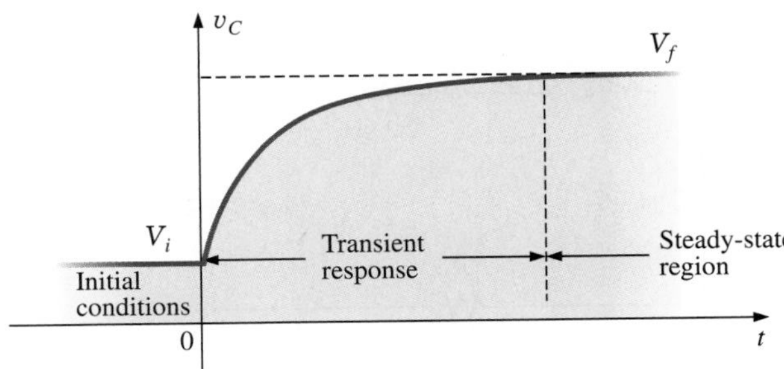

FIG. 10.51
Defining the regions associated with a transient response.

Once the switch is thrown, the transient phase commences until a leveling off occurs after five time constants. This region of relatively fixed value that follows the transient response is called the **steady-state region,** and the resulting value is called the **steady-state** or **final value.** The steady-state value is found by substituting the open-circuit equivalent for the capacitor and finding the voltage across the plates. Using the transient equation developed in the previous section, an equation for the voltage v_C can be written for the entire time interval in Fig. 10.51. That is, for the transient period, the voltage rises from V_i (previously 0 V) to a final value of V_f. Therefore,

$$v_C = E(1 - e^{-t/\tau}) = (V_f - V_i)(1 - e^{-t/\tau})$$

Adding the starting value of V_i to the equation results in

$$v_C = V_i + (V_f - V_i)(1 - e^{-t/\tau})$$

However, by multiplying through and rearranging terms:

$$\begin{aligned} v_C &= V_i + V_f - V_f e^{-t/\tau} - V_i + V_i e^{-t/\tau} \\ &= V_f - V_f e^{-t/\tau} + V_i e^{-t/\tau} \end{aligned}$$

we find

$$\boxed{v_C = V_f + (V_i - V_f)e^{-t/\tau}} \qquad \textbf{(10.21)}$$

Now that the equation has been developed, it is important to recognize that

Eq. (10.21) is a universal equation for the transient response of a capacitor.

That is, it can be used whether or not the capacitor has an initial value. If the initial value is 0 V as it was in all the previous examples, simply set V_i equal to zero in the equation, and the desired equation results. The final value is the voltage across the capacitor when the open-circuit equivalent is substituted.

EXAMPLE 10.10 The capacitor in Fig. 10.52 has an initial voltage of 4 V.

a. Find the mathematical expression for the voltage across the capacitor once the switch is closed.
b. Find the mathematical expression for the current during the transient period.
c. Sketch the waveform for each from initial value to final value.

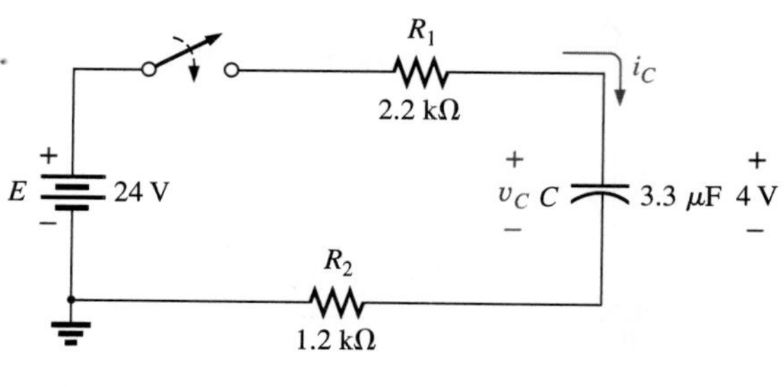

FIG. 10.52
Example 10.10.

Solutions:

a. Substituting the open-circuit equivalent for the capacitor results in a final or steady-state voltage v_C of 24 V.

The time constant is determined by

$$\tau = (R_1 + R_2)C$$
$$= (2.2\text{ k}\Omega + 1.2\text{ k}\Omega)(3.3\ \mu\text{F}) = 11.22\text{ ms}$$

with $\quad 5\tau = 56.1\text{ ms}$

Applying Eq. (10.21):

$$v_C = V_f + (V_i - V_f)e^{-t/\tau} = 24\text{ V} + (4\text{ V} - 24\text{ V})e^{-t/11.22\text{ms}}$$

and $v_C = \mathbf{24\ V - 20\ Ve^{-t/11.22ms}}$

b. Since the voltage across the capacitor is constant at 4 V prior to the closing of the switch, the current (whose level is sensitive only to changes in voltage across the capacitor) must have an initial value of 0 mA. At the instant the switch is closed, the voltage across the capacitor cannot change instantaneously, so the voltage across the resistive elements at this instant is the applied voltage less the initial voltage across the capacitor. The resulting peak current is

$$I_m = \frac{E - V_C}{R_1 + R_2} = \frac{24\text{ V} - 4\text{ V}}{2.2\text{ k}\Omega + 1.2\text{ k}\Omega} = \frac{20\text{ V}}{3.4\text{ k}\Omega} = 5.88\text{ mA}$$

The current then decays (with the same time constant as the voltage v_C) to zero because the capacitor is approaching its open-circuit equivalence.

The equation for i_C is therefore:

$$i_C = \mathbf{5.88\ mAe^{-t/11.22ms}}$$

c. See Fig. 10.53. The initial and final values of the voltage were drawn first, and then the transient response was included between these levels. For the current, the waveform begins and ends at zero, with the

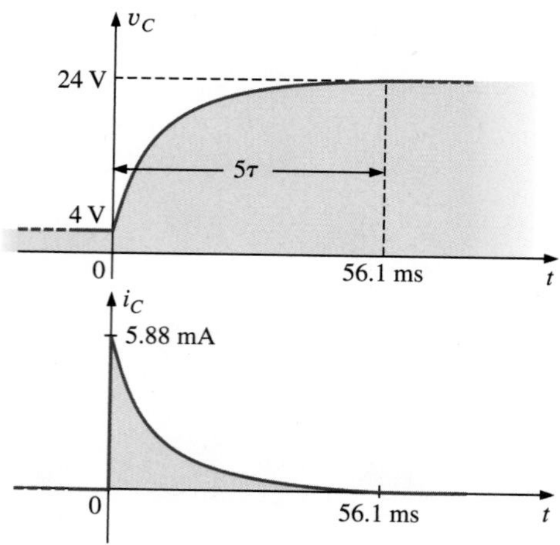

FIG. 10.53
v_C and i_C for the network in Fig. 10.52.

peak value having a sign sensitive to the defined direction of i_C in Fig. 10.52.

Let us now test the validity of the equation for v_C by substituting $t = 0$ s to reflect the instant the switch is closed.

$$e^{-t/\tau} = e^{-0} = 1$$

and $$v_C = 24\text{ V} - 20\text{ V}e^{-t/\tau} = 24\text{ V} - 20\text{ V} = 4\text{ V}$$

When $t > 5\tau$,

$$e^{-t/\tau} \cong 0$$

and $$v_C = 24\text{ V} - 20\text{ V}e^{-t/\tau} = 24\text{ V} - 0\text{ V} = 24\text{ V}$$

Eq. (10.21) can also be applied to the discharge phase by applying the correct levels of V_i and V_f.

For the discharge pattern in Fig. 10.54, $V_f = 0$ V, and Eq. (10.21) becomes

$$v_C = V_f + (V_i - V_f)e^{-t/\tau} = 0\text{ V} + (V_i - 0\text{ V})e^{-t/\tau}$$

and $$\boxed{v_C = V_i e^{-t/\tau}}_{\text{discharging}} \qquad \textbf{(10.22)}$$

Substituting $V_i = E$ volts results in Eq. (10.17).

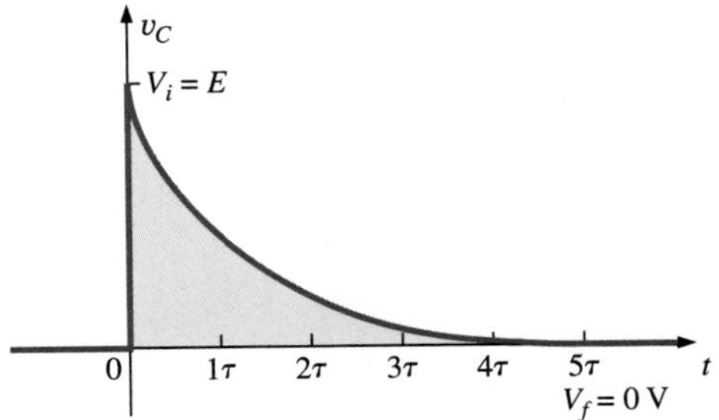

FIG. 10.54
Defining the parameters in Eq. (10.21) for the discharge phase.

10.8 INSTANTANEOUS VALUES

Occasionally, you may need to determine the voltage or current at a particular instant of time that is not an integral multiple of τ, as in the previous sections. For example, if

$$v_C = 20\text{ V}(1 - e^{(-t/2\text{ms})}$$

the voltage v_C may be required at $t = 5$ ms, which does not correspond to a particular value of τ. Fig. 10.28 reveals that $(1 - e^{-t/\tau})$ is approximately 0.93 at $t = 5\text{ ms} = 2.5\tau$, resulting in $v_C = 20(0.93) = 18.6$ V. Additional accuracy can be obtained by substituting $v = 5$ ms into the equation and solving for v_C using a calculator or table to determine $e^{-2.5}$. Thus,

$$v_C = 20\text{ V}(1 - e^{-5\text{ms}/2\text{ms}}) = (20\text{ V})(1 - e^{-2.5}) = (20\text{ V})(1 - 0.082)$$
$$= (20\text{ V})(0.918) = \mathbf{18.36\text{ V}}$$

The results are close, but accuracy beyond the tenths place is suspect using Fig. 10.29. The above procedure can also be applied to any other equation introduced in this chapter for currents or other voltages.

Occasionally, you may need to determine the time required to reach a particular voltage or current. The procedure is complicated somewhat by the use of natural logs ($\log_e$, or ln), but today's calculators are equipped to handle the operation with ease.

For example, solving for t in the equation

$$v_C = V_f + (V_i - V_f)\,e^{-t/\tau}$$

results in

$$\boxed{t = \tau(\log_e)\frac{(V_i - V_f)}{(v_C - V_f)}} \qquad \textbf{(10.23)}$$

For example, suppose that

$$v_C = 20\text{ V}(1 - e^{-t/2\text{ms}})$$

and the time t to reach 10 V is desired. Since $V_i = 0$ V, and $V_f = 20$ V, we have

$$\begin{aligned} t &= \tau(\log_e)\frac{(V_i - V_f)}{(v_C - V_f)} = (2\text{ ms})(\log_e)\frac{(0\text{ V} - 20\text{ V})}{(10\text{ V} - 20\text{ V})} \\ &= (2\text{ ms})\left[\log_e\left(\frac{-20\text{ V}}{-10\text{ V}}\right)\right] = (2\text{ ms})(\log_e 2) = (2\text{ ms})(0.693) \\ &= \mathbf{1.386\text{ ms}} \end{aligned}$$

The TI-89 calculator key strokes appear in Fig. 10.55.

FIG. 10.55
Key strokes to determine (2 ms)($\log_e 2$) using the TI-89 calculator.

For the discharge equation,

$$v_C = Ee^{-t/\tau} = V_i\,(e^{-t/\tau}) \quad \text{with } V_f = 0\text{ V}$$

Using Eq. (10.23):

$$t = \tau(\log_e)\frac{(V_i - V_f)}{(v_C - V_f)} = \tau(\log_e)\frac{(V_i - 0\text{ V})}{(v_C - 0\text{ V})}$$

and

$$\boxed{t = \tau \log_e \frac{V_i}{v_C}} \qquad \textbf{(10.24)}$$

For the current equation,

$$i_C = \frac{E}{R}e^{-t/\tau} \qquad I_i = \frac{E}{R} \qquad I_f = 0\text{ A}$$

and

$$\boxed{t = \log_e \frac{I_i}{i_C}} \qquad \textbf{(10.25)}$$

Using Mathcad to Perform Transient Analysis

Mathcad will now be applied to Eq. (10.13), the equation for the voltage across a capacitor as it changes to the supply voltage of the series circuit. The value of t must be defined before the expression is written, or the value can simply be inserted in the equation. The former approach is often the better choice, because changing the defined value of t results in an immediate change in the result. In other words, the value can be used for further calculations. In Fig. 10.56, the value of t was defined as 5 ms. The equation was then entered using the e function from the **Calculator** palette obtained from **View-Toolbars-Caculator.** Be sure to insert a multiplication operator between the initial 20 and the main left bracket. Also, be careful that the control bracket is in the correct place before placing the right bracket to enclose the equation. It takes some practice to ensure

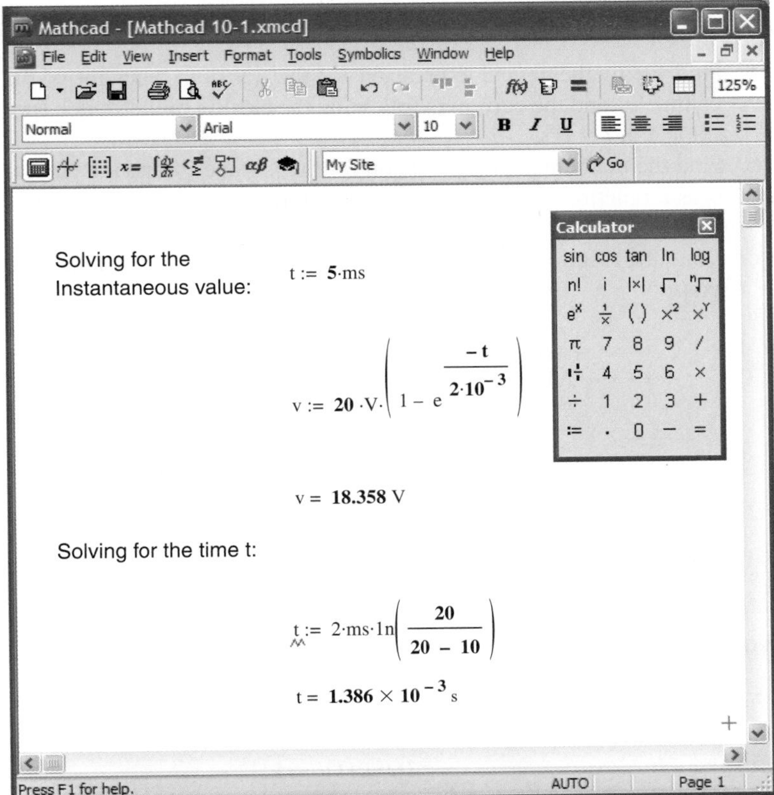

FIG. 10.56
Applying Mathcad to the transient R-C equations.

that the insertion bracket is in the correct place before entering a parameter, but in time you will find that it is a fairly direct procedure. The **−3** is placed using the shift operator over the number **6** on the standard keyboard. The result is displayed by entering υ again, followed by an equals sign. The result for $t = 1$ ms can now be obtained by changing the defined value for t. The result of 7.869 V appears immediately.

For the example above, the equation for t can be entered directly as shown in the bottom of Fig. 10.56. The **ln** from the **Calculator** is for a base e calculation, whereas **log** is for a base 10 calculation. The result appears the instant the equals sign is placed after the t on the bottom line.

The text you see on the screen to define each operation is obtained from **Insert-Text Region.** Then simply type in the text material. Change to boldface by clicking on the text material and swiping the text to establish a black background. Then select **B** from the toolbar, and the boldface appears.

10.9 THÉVENIN EQUIVALENT: $\tau = R_{Th}C$

You may encounter instances in which the network does not have the simple series form in Fig. 10.26. You then need to find the Thévenin equivalent circuit for the network external to the capacitive element. E_{Th} will be the source voltage E in Eqs. (10.13) through (10.25), and R_{Th} will be the resistance R. The time constant is then $\tau = R_{Th}C$.

EXAMPLE 10.11 For the network in Fig. 10.57:

a. Find the mathematical expression for the transient behavior of the voltage υ_C and the current i_C following the closing of the switch (position 1 at $t = 0$ s).
b. Find the mathematical expression for the voltage υ_C and the current i_C as a function of time if the switch is thrown into position 2 at $t = 9$ ms.
c. Draw the resultant waveforms of parts (a) and (b) on the same time axis.

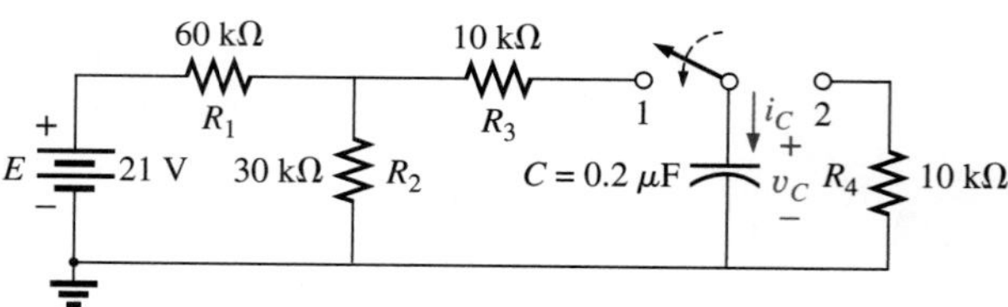

FIG. 10.57
Example 10.11.

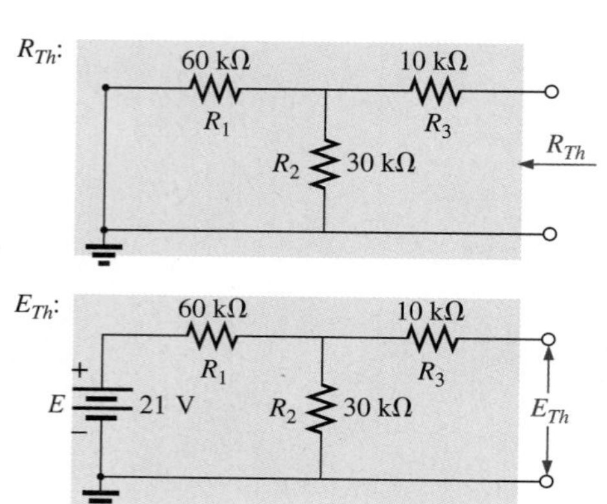

FIG. 10.58
Applying Thévenin's theorem to the network in Fig. 10.57.

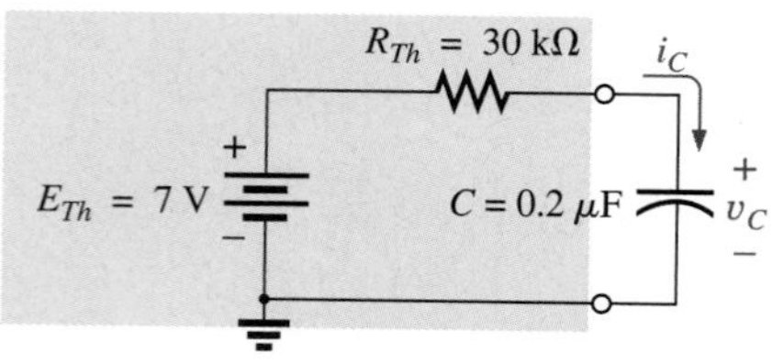

FIG. 10.59
Substituting the Thévenin equivalent for the network in Fig. 10.57.

Solutions:

a. Applying Thévenin's theorem to the 0.2 μF capacitor, we obtain Fig. 10.58.

$$R_{Th} = R_1 \parallel R_2 + R_3 = \frac{(60\text{ k}\Omega)(30\text{ k}\Omega)}{90\text{ k}\Omega} + 10\text{ k}\Omega$$
$$= 20\text{ k}\Omega + 10\text{ k}\Omega = 30\text{ k}\Omega$$
$$E_{Th} = \frac{R_2 E}{R_2 + R_1} = \frac{(30\text{ k}\Omega)(21\text{ V})}{30\text{ k}\Omega + 60\text{ k}\Omega} = \frac{1}{3}(21\text{ V}) = 7\text{ V}$$

The resultant Thévenin equivalent circuit with the capacitor replaced is shown in Fig. 10.59.

Using Eq. (10.21) with $V_f = E_{Th}$ and $V_i = 0$ V, we find that

$$\upsilon_C = V_f + (V_i - V_f)e^{-t/\tau}$$

becomes
$$\upsilon_C = E_{Th} + (0\text{ V} - E_{Th})e^{-t/\tau}$$

or
$$\upsilon_C = E_{Th}(1 - e^{-t/\tau})$$

with
$$\tau = RC = (30\text{ k}\Omega)(0.2\ \mu\text{F}) = 6\text{ ms}$$

Therefore,
$$\boldsymbol{\upsilon_C = 7\textbf{ V}(1 - e^{-t/6\textbf{ms}})}$$

For the current i_C:

$$i_C = \frac{E_{Th}}{R}e^{-t/RC} = \frac{7\text{ V}}{30\text{ k}\Omega}e^{-t/6\text{ms}}$$
$$= \boldsymbol{0.23\textbf{ mA}e^{-t/6\textbf{ms}}}$$

b. At $t = 9$ ms,

$$\upsilon_C = E_{Th}(1 - e^{-t/\tau}) = 7\text{ V}(1 - e^{-(9\text{ms}/6\text{ms})})$$
$$= (7\text{ V})(1 - e^{-1.5}) = (7\text{ V})(1 - 0.223)$$
$$= (7\text{ V})(0.777) = 5.44\text{ V}$$

and
$$i_C = \frac{E_{Th}}{R}e^{-t/\tau} = 0.23\text{ mA}e^{-1.5}$$
$$= (0.23 \times 10^{-3})(0.233) = 0.052 \times 10^{-3} = 0.05\text{ mA}$$

Using Eq. (10.21) with $V_f = 0$ V and $V_i = 5.44$ V, we find that

$$v_C = V_f + (V_i - V_f)e^{-t/\tau'}$$

becomes
$$v_C = 0\text{ V} + (5.44\text{ V} - 0\text{ V})e^{-t/\tau'}$$
$$= 5.44\text{ V}e^{-t/\tau'}$$

with
$$\tau' = R_4C = (10\text{ k}\Omega)(0.2\ \mu\text{F}) = 2\text{ ms}$$

and
$$v_C = \mathbf{5.44\ Ve^{-t/2ms}}$$

By Eq. (10.19):

$$I_i = \frac{5.44\text{ V}}{10\text{ k}\Omega} = 0.54\text{ mA}$$

and
$$i_C = I_i e^{-t/\tau} = \mathbf{-0.54\ mAe^{-t/2ms}}$$

c. See Fig. 10.60.

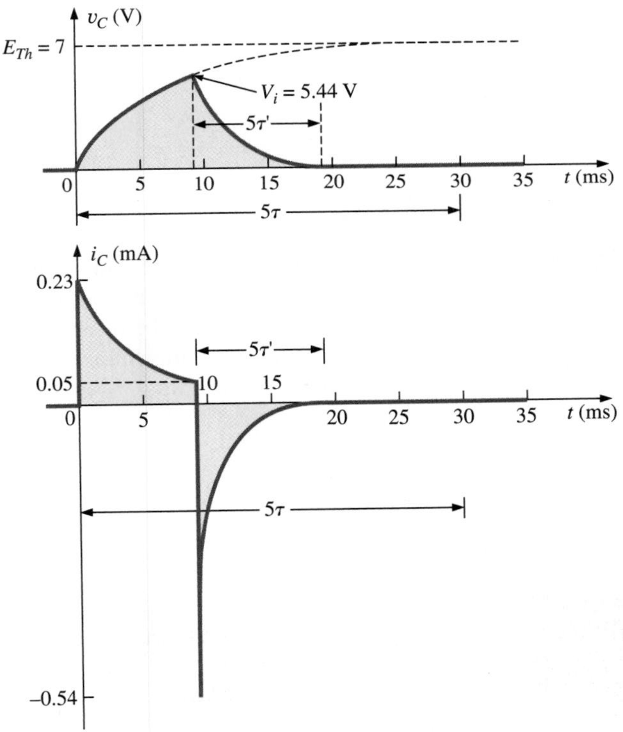

FIG. 10.60
The resulting waveforms for the network in Fig. 10.57.

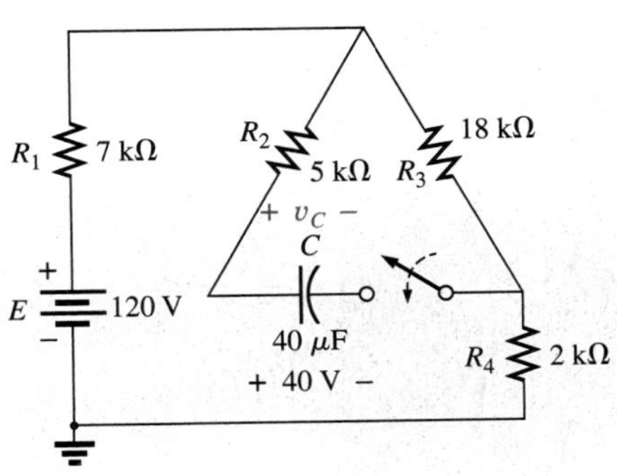

FIG. 10.61
Example 10.12.

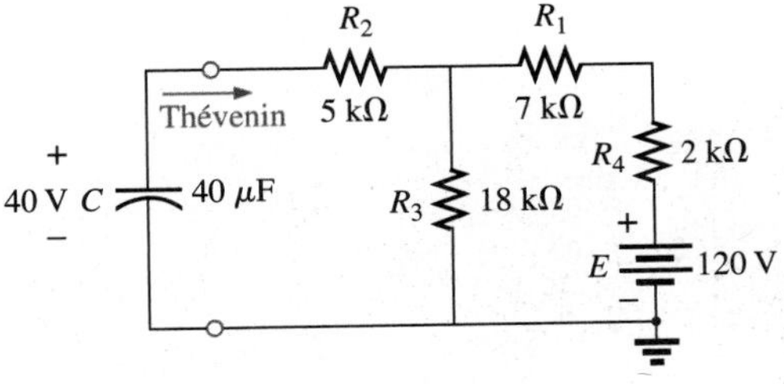

FIG. 10.62
Network in Fig. 10.61 redrawn.

EXAMPLE 10.12 The capacitor in Fig. 10.61 is initially charged to 40 V. Find the mathematical expression for v_C after the closing of the switch. Plot the waveform for v_C.

Solution: The network is redrawn in Fig. 10.62.

E_{Th}:

$$E_{Th} = \frac{R_3E}{R_3 + R_1 + R_4} = \frac{(18\text{ k}\Omega)(120\text{ V})}{18\text{ k}\Omega + 7\text{ k}\Omega + 2\text{ k}\Omega} = 80\text{ V}$$

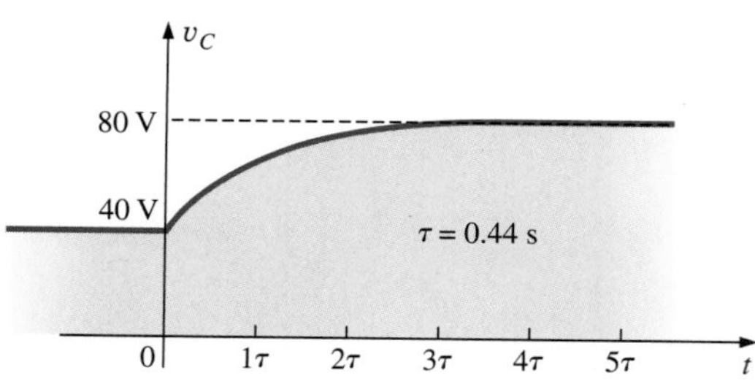

FIG. 10.63
v_C for the network in Fig. 10.61.

R_{Th}:

$$R_{Th} = 5\text{ k}\Omega + (18\text{ k}\Omega) \parallel (7\text{ k}\Omega + 2\text{ k}\Omega)$$
$$= 5\text{ k}\Omega + 6\text{ k}\Omega = 11\text{ k}\Omega$$

Therefore, $V_i = 40\text{ V}$ and $V_f = 80\text{ V}$

and $\tau = R_{Th}C = (11\text{ k}\Omega)(40\ \mu\text{F}) = 0.44\text{ s}$

Eq. (10.21): $v_C = V_f + (V_i - V_f)e^{-t/\tau}$

$$= 80\text{ V} + (40\text{ V} - 80\text{ V})e^{-t/0.44\text{s}}$$

and $v_C = \mathbf{80\ V - 40\ V}e^{-t/0.44s}$

The waveform appears as in Fig. 10.63.

EXAMPLE 10.13 For the network in Fig. 10.64, find the mathematical expression for the voltage v_C after the closing of the switch (at $t = 0$).

Solution:

$$R_{Th} = R_1 + R_2 = 6\ \Omega + 10\ \Omega = 16\ \Omega$$
$$E_{Th} = V_1 + V_2 = IR_1 + 0$$
$$= (20 \times 10^{-3}\text{ A})(6\ \Omega) = 120 \times 10^{-3}\text{ V} = 0.12\text{ V}$$

and $\tau = R_{Th}C = (16\ \Omega)(500 \times 10^{-6}\text{ F}) = 8\text{ ms}$

so that $v_C = \mathbf{0.12\ V(1 - e^{-t/8ms})}$

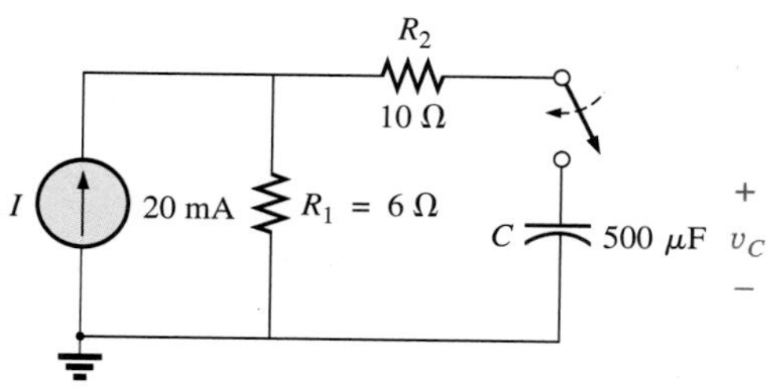

FIG. 10.64
Example 10.13.

10.10 THE CURRENT i_C

There is a very special relationship between the current of a capacitor and the voltage across it. For the resistor, it is defined by Ohm's law: $i_R = v_R/R$. The current through and the voltage across the resistor are related by a constant R—a very simple direct linear relationship. For the capacitor, it is the more complex relationship defined by

$$i_C = C\frac{dv_C}{dt} \tag{10.26}$$

The factor C reveals that the higher the capacitance, the greater the resulting current. Intuitively, this relationship makes sense, because higher capacitance levels result in increased levels of stored charge, providing a source for increased current levels. The second term, dv_C/dt, is sensitive to the *rate of change* of v_C with time. The function dv_C/dt is called the **derivative** (calculus) of the voltage v_C with respect to time t. The faster the voltage v_C changes with time, the larger the factor dv_C/dt will be and the larger the resulting current i_C will be. That is why the current jumps to its maximum of E/R in a charging circuit the instant the switch is closed. At that instant, if you look at the charging curve for v_C, the voltage is *changing* at its greatest rate. As it approaches its final value, the rate of change decreases, and, as confirmed by Eq. (10.26), the level of current decreases.

Take special note of the following:

The capacitive current is directly related to the rate of change of the voltage across the capacitor, not the levels of voltage involved.

For example, the current of a capacitor will be *greater* when the voltage changes from 1 V to 10 V in 1 ms than when it changes from 10 V to 100 V in 1 s; in fact, it will be 100 times more.

If the voltage fails to change over time, then

$$\frac{dv_C}{dt} = 0$$

and
$$i_C = C\frac{dv_C}{dt} = C(0) = 0 \text{ A}$$

In an effort to develop a clearer understanding of Eq. (10.26), let us calculate the **average current** associated with a capacitor for various voltages impressed across the capacitor. The average current is defined by the equation

$$i_{C_{av}} = C\frac{\Delta v_C}{\Delta t} \qquad \textbf{(10.27)}$$

where Δ indicates a finite (measurable) change in voltage or time. The instantaneous current can be derived from Eq. (10.27) by letting Δt become vanishingly small; that is,

$$i_{C_{inst}} = \lim_{\Delta t \to 0} C\frac{\Delta v_C}{\Delta t} = C\frac{dv_C}{dt}$$

In the following example, the change in voltage Δv_C will be considered for each slope of the voltage waveform. If the voltage increases with time, the average current is the change in voltage divided by the change in time, with a positive sign. If the voltage decreases with time, the average current is again the change in voltage divided by the change in time, but with a negative sign.

EXAMPLE 10.14 Find the waveform for the average current if the voltage across a 2 μF capacitor is as shown in Fig. 10.65.

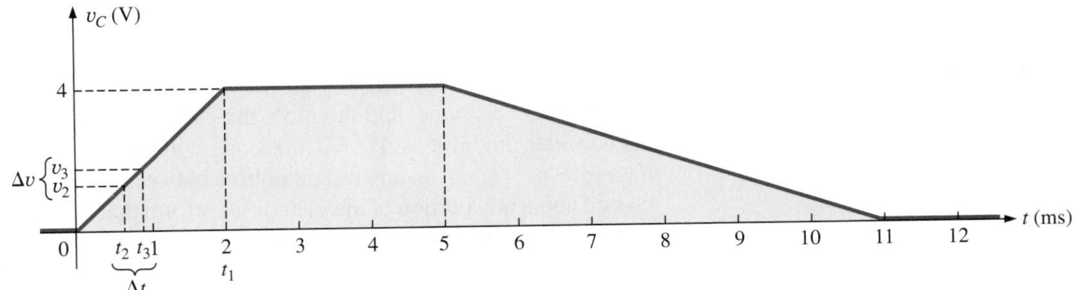

FIG. 10.65
v_C for Example 10.14.

Solutions:

a. From 0 ms to 2 ms, the voltage increases linearly from 0 V to 4 V; the change in voltage $\Delta v = 4 \text{ V} - 0 = 4 \text{ V}$ (with a positive sign since the voltage increases with time). The change in time $\Delta t = 2 \text{ ms} - 0 = 2 \text{ ms}$, and

$$i_{C_{av}} = C\frac{\Delta v_C}{\Delta t} = (2 \times 10^{-6} \text{ F})\left(\frac{4 \text{ V}}{2 \times 10^{-3} \text{ s}}\right)$$
$$= 4 \times 10^{-3} \text{ A} = \mathbf{4 \text{ mA}}$$

b. From 2 ms to 5 ms, the voltage remains constant at 4 V; the change in voltage $\Delta v = 0$. The change in time $\Delta t = 3$ ms, and

$$i_{C_{av}} = C\frac{\Delta v_C}{\Delta t} = C\frac{0}{\Delta t} = \mathbf{0\ mA}$$

c. From 5 ms to 11 ms, the voltage decreases from 4 V to 0 V. The change in voltage Δv is, therefore, 4 V − 0 = 4 V (with a negative sign since the voltage is decreasing with time). The change in time Δt = 11 ms − 5 ms = 6 ms, and

$$i_{C_{av}} = C\frac{\Delta v_C}{\Delta t} = -(2 \times 10^{-6}\ \text{F})\left(\frac{4\ \text{V}}{6 \times 10^{-3}\ \text{s}}\right)$$
$$= -1.33 \times 10^{-3}\ \text{A} = \mathbf{-1.33\ mA}$$

d. From 11 ms on, the voltage remains constant at 0 and $\Delta v = 0$, so $i_{C_{av}} = 0$ mA. The waveform for the average current for the impressed voltage is as shown in Fig. 10.66.

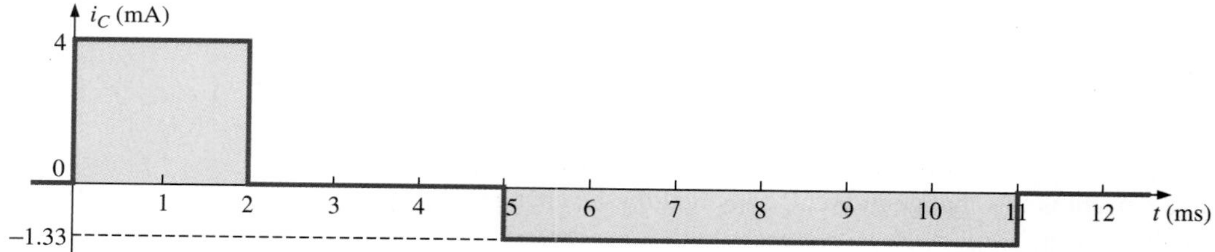

FIG. 10.66
The resulting current i_C for the applied voltage in Fig. 10.65.

Note in Example 10.14 that, in general, the steeper the slope, the greater the current, and when the voltage fails to change, the current is zero. In addition, the average value is the same as the instantaneous value at any point along the slope over which the average value was found. For example, if the interval Δt is reduced from $0 \rightarrow t_1$ to $t_2 - t_3$, as noted in Fig. 10.65, $\Delta v/\Delta t$ is still the same. In fact, no matter how small the interval Δt, the slope will be the same, and therefore the current $i_{C_{av}}$ will be the same. If we consider the limit as $\Delta t \rightarrow 0$, the slope will still remain the same, and therefore $i_{C_{av}} = i_{C_{inst}}$ at any instant of time between 0 and t_1. The same can be said about any portion of the voltage waveform that has a constant slope.

An important point to be gained from this discussion is that it is not the magnitude of the voltage across a capacitor that determines the current but rather how quickly the voltage *changes* across the capacitor. An applied steady dc voltage of 10,000 V would (ideally) not create any flow of charge (current), but a change in voltage of 1 V in a very brief period of time could create a significant current.

The method described above is only for waveforms with straight-line (linear) segments. For nonlinear (curved) waveforms, a method of calculus (differentiation) must be used.

10.11 CAPACITORS IN SERIES AND IN PARALLEL

Capacitors, like resistors, can be placed in series and in parallel. Increasing levels of capacitance can be obtained by placing capacitors in parallel, while decreasing levels can be obtained by placing capacitors in series.

For capacitors in series, the charge is the same on each capacitor (Fig. 10.67):

$$Q_T = Q_1 = Q_2 = Q_3 \tag{10.28}$$

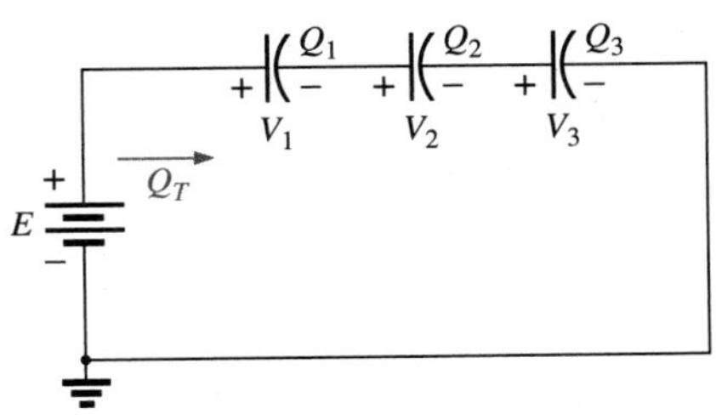

FIG. 10.67
Series capacitors.

Applying Kirchhoff's voltage law around the closed loop gives

$$E = V_1 + V_2 + V_3$$

However,
$$V = \frac{Q}{C}$$

so that
$$\frac{Q_T}{C_T} = \frac{Q_1}{C_1} + \frac{Q_2}{C_2} + \frac{Q_3}{C_3}$$

Using Eq. (10.28) and dividing both sides by Q yields

$$\frac{1}{C_T} = \frac{1}{C_1} + \frac{1}{C_2} + \frac{1}{C_3} \tag{10.29}$$

which is similar to the manner in which we found the total resistance of a parallel resistive circuit. The total capacitance of two capacitors in series is

$$C_T = \frac{C_1C_2}{C_1 + C_2} \tag{10.30}$$

The voltage across each capacitor in Fig. 10.67 can be found by first recognizing that

$$Q_T = Q_1$$

or
$$C_TE = C_1V_1$$

Solving for V_1:
$$V_1 = \frac{C_TE}{C_1}$$

and substituting for C_T:

$$V_1 = \left(\frac{1/C_1}{1/C_1 + 1/C_2 + 1/C_3}\right)E \tag{10.31}$$

A similar equation results for each capacitor of the network.

For capacitors in parallel, as shown in Fig. 10.68, the voltage is the same across each capacitor, and the total charge is the sum of that on each capacitor:

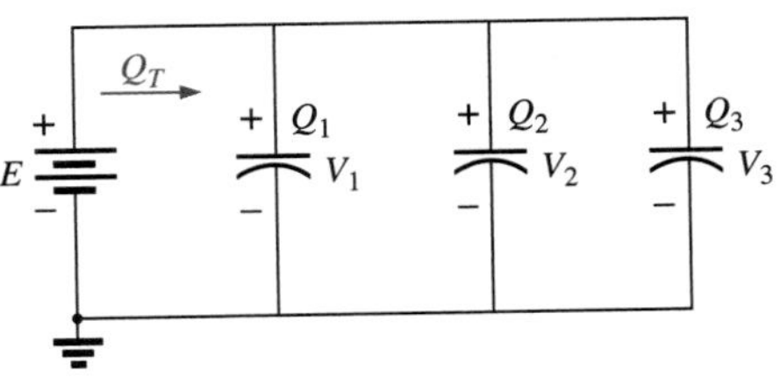

FIG. 10.68
Parallel capacitors.

$$Q_T = Q_1 + Q_2 + Q_3 \tag{10.32}$$

However,
$$Q = CV$$

Therefore,
$$C_TE = C_1V_1 = C_2V_2 = C_3V_3$$

but
$$E = V_1 = V_2 = V_3$$

Thus,
$$C_T = C_1 + C_2 + C_3 \tag{10.33}$$

which is similar to the manner in which the total resistance of a series circuit is found.

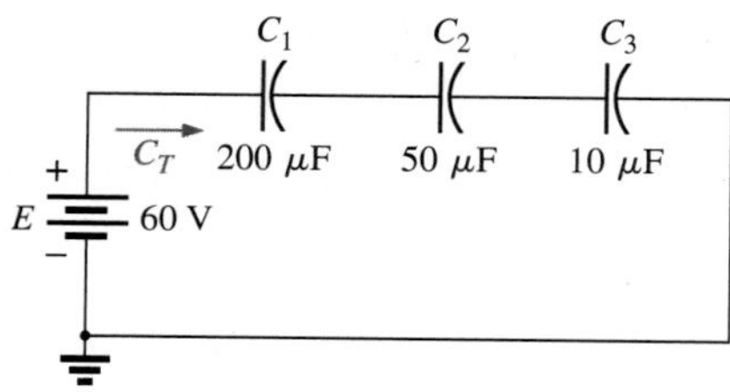

FIG. 10.69
Example 10.15.

EXAMPLE 10.15 For the circuit in Fig. 10.69:

a. Find the total capacitance.
b. Determine the charge on each plate.
c. Find the voltage across each capacitor.

Solutions:

a. $$\frac{1}{C_T} = \frac{1}{C_1} + \frac{1}{C_2} + \frac{1}{C_3}$$

$$= \frac{1}{200 \times 10^{-6}\text{ F}} + \frac{1}{50 \times 10^{-6}\text{ F}} + \frac{1}{10 \times 10^{-6}\text{ F}}$$
$$= 0.005 \times 10^6 + 0.02 \times 10^6 + 0.1 \times 10^6$$
$$= 0.125 \times 10^6$$

and $$C_T = \frac{1}{0.125 \times 10^6} = \mathbf{8\ \mu F}$$

b. $$Q_T = Q_1 = Q_2 = Q_3$$
$$= C_T E = (8 \times 10^{-6}\text{ F})(60\text{ V}) = \mathbf{480\ \mu C}$$

c. $$V_1 = \frac{Q_1}{C_1} = \frac{480 \times 10^{-6}\text{ C}}{200 \times 10^{-6}\text{ F}} = \mathbf{2.4\ V}$$

$$V_2 = \frac{Q_2}{C_2} = \frac{480 \times 10^{-6}\text{ C}}{50 \times 10^{-6}\text{ F}} = \mathbf{9.6\ V}$$

$$V_3 = \frac{Q_3}{C_3} = \frac{480 \times 10^{-6}\text{ C}}{10 \times 10^{-6}\text{ F}} = \mathbf{48.0\ V}$$

and $$E = V_1 + V_2 + V_3 = 2.4\text{ V} + 9.6\text{ V} + 48\text{ V} = \mathbf{60\ V} \quad \text{(checks)}$$

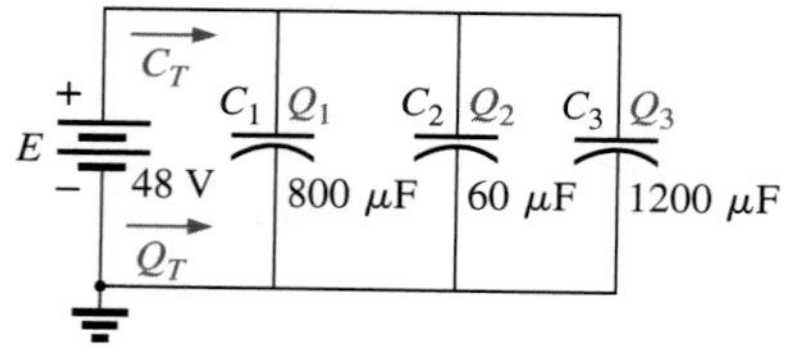

FIG. 10.70
Example 10.16.

EXAMPLE 10.16 For the network in Fig. 10.70:

a. Find the total capacitance.
b. Determine the charge on each plate.
c. Find the total charge.

Solutions:

a. $C_T = C_1 + C_2 + C_3 = 800\ \mu\text{F} + 60\ \mu\text{F} + 1200\ \mu\text{F} = \mathbf{2060\ \mu F}$

b. $Q_1 = C_1 E = (800 \times 10^{-6}\text{ F})(48\text{ V}) = \mathbf{38.4\ mC}$
$Q_2 = C_2 E = (60 \times 10^{-6}\text{ F})(48\text{ V}) = \mathbf{2.88\ mC}$
$Q_3 = C_3 E = (1200 \times 10^{-6}\text{ F})(48\text{ V}) = \mathbf{57.6\ mC}$

c. $Q_T = Q_1 + Q_2 + Q_3 = 38.4\text{ mC} + 2.88\text{ mC} + 57.6\text{ mC} = \mathbf{98.88\ mC}$

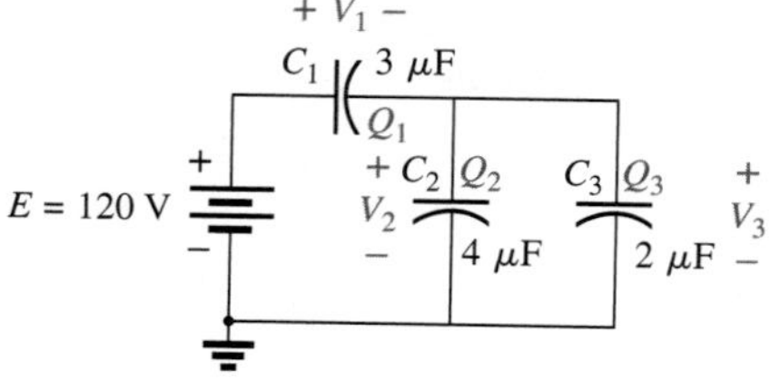

FIG. 10.71
Example 10.17.

EXAMPLE 10.17 Find the voltage across and the charge on each capacitor for the network in Fig. 10.71.

Solution:

$$C'_T = C_2 + C_3 = 4\ \mu\text{F} + 2\ \mu\text{F} = 6\ \mu\text{F}$$

$$C_T = \frac{C_1 C'_T}{C_1 + C'_T} = \frac{(3\ \mu\text{F})(6\ \mu\text{F})}{3\ \mu\text{F} + 6\ \mu\text{F}} = 2\ \mu\text{F}$$

$$Q_T = C_T E = (2 \times 10^{-6}\text{ F})(120\text{ V}) = \mathbf{240\ \mu C}$$

An equivalent circuit (Fig. 10.72) has

$$Q_T = Q_1 = Q'_T$$

and, therefore,

$$Q_1 = \mathbf{240\ \mu C}$$

and

$$V_1 = \frac{Q_1}{C_1} = \frac{240 \times 10^{-6}\ \text{C}}{3 \times 10^{-6}\ \text{F}} = \mathbf{80\ V}$$

$$Q'_T = 240\ \mu\text{C}$$

Therefore,

$$V'_T = \frac{Q'_T}{C'_T} = \frac{240 \times 10^{-6}\ \text{C}}{6 \times 10^{-6}\ \text{F}} = \mathbf{40\ V}$$

and

$$Q_2 = C_2V'_T = (4 \times 10^{-6}\ \text{F})(40\ \text{V}) = \mathbf{160\ \mu C}$$

$$Q_3 = C_3V'_T = (2 \times 10^{-6}\ \text{F})(40\ \text{V}) = \mathbf{80\ \mu C}$$

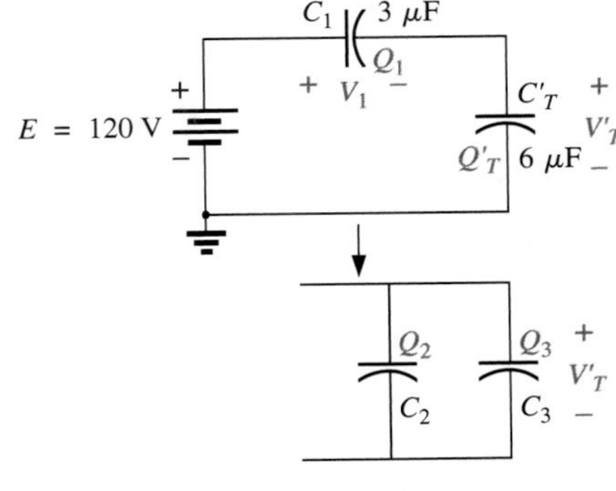

FIG. 10.72
Reduced equivalent for the network in Fig. 10.71.

EXAMPLE 10.18 Find the voltage across and the charge on capacitor C_1 in Fig. 10.73 after it has charged up to its final value.

Solution: As previously discussed, the capacitor is effectively an open circuit for dc after charging up to its final value (Fig. 10.74).

Therefore,

$$V_C = \frac{(8\ \Omega)(24\ \text{V})}{4\ \Omega + 8\ \Omega} = \mathbf{16\ V}$$

$$Q_1 = C_1V_C = (20 \times 10^{-6}\ \text{F})(16\ \text{V}) = \mathbf{320\ \mu C}$$

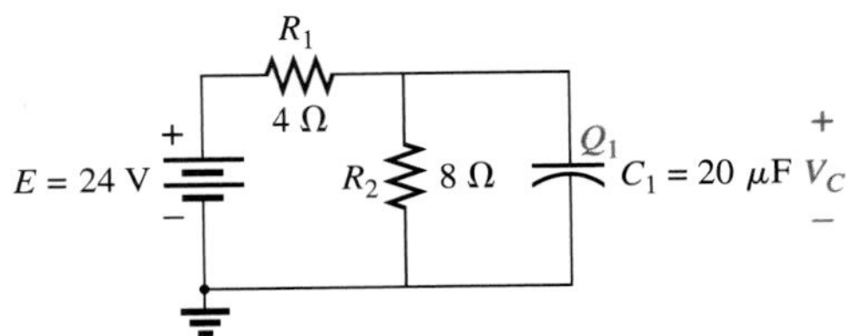

FIG. 10.73
Example 10.18.

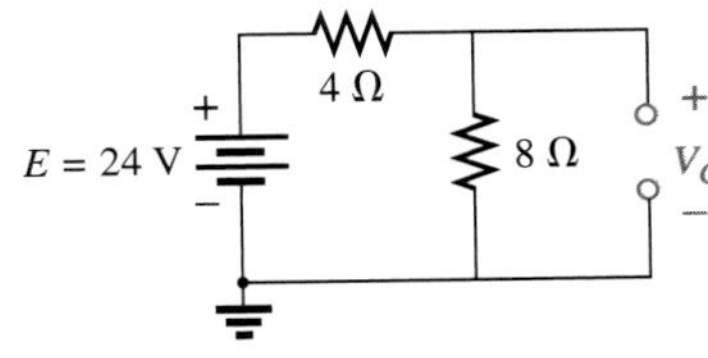

FIG. 10.74
Determining the final (steady-state) value for v_C.

EXAMPLE 10.19 Find the voltage across and the charge on each capacitor of the network in Fig. 10.75(a) after each has charged up to its final value.

Solution: See Fig. 10.75(b).

$$V_{C_2} = \frac{(7\ \Omega)(72\ \text{V})}{7\ \Omega + 2\ \Omega} = \mathbf{56\ V}$$

$$V_{C_1} = \frac{(2\ \Omega)(72\ \text{V})}{2\ \Omega + 7\ \Omega} = \mathbf{16\ V}$$

$$Q_1 = C_1V_{C_1} = (2 \times 10^{-6}\ \text{F})(16\ \text{V}) = \mathbf{32\ \mu C}$$

$$Q_2 = C_2V_{C_2} = (3 \times 10^{-6}\ \text{F})(56\ \text{V}) = \mathbf{168\ \mu C}$$

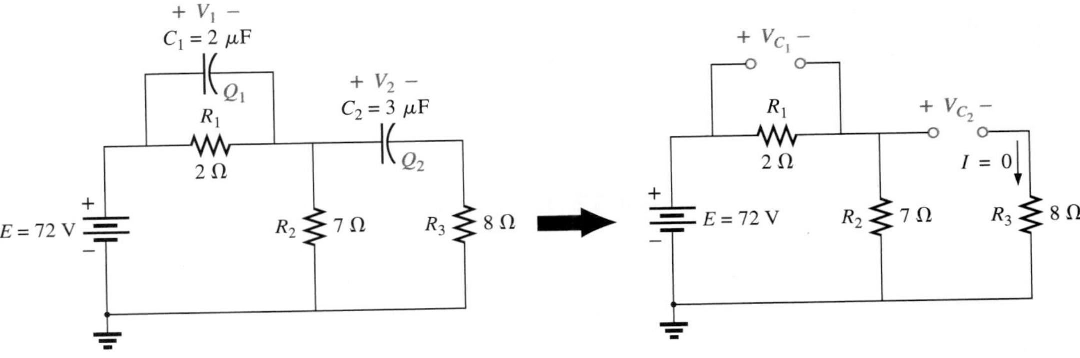

FIG. 10.75
Example 10.19.

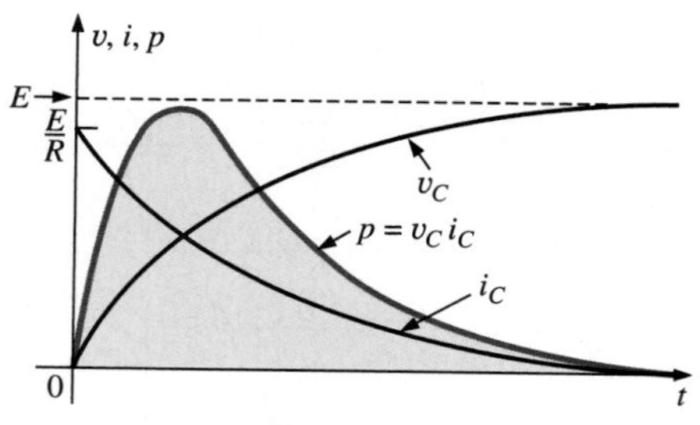

FIG. 10.76
Plotting the power to a capacitive element during the transient phase.

10.12 ENERGY STORED BY A CAPACITOR

An ideal capacitor does not dissipate any of the energy supplied to it. It stores the energy in the form of an electric field between the conducting surfaces. A plot of the voltage, current, and power to a capacitor during the charging phase is shown in Fig. 10.76. The power curve can be obtained by finding the product of the voltage and current at selected instants of time and connecting the points obtained. *The energy stored is represented by the shaded area under the power curve.* Using calculus, we can determine the area under the curve:

$$W_C = \frac{1}{2}CE^2$$

In general,

$$\boxed{W_C = \frac{1}{2}CV^2} \quad \text{(J)} \qquad \textbf{(10.34)}$$

where V is the steady-state voltage across the capacitor. In terms of Q and C,

$$W_C = \frac{1}{2}C\left(\frac{Q}{C}\right)^2$$

or

$$\boxed{W_C = \frac{Q^2}{2C}} \quad \text{(J)} \qquad \textbf{(10.35)}$$

EXAMPLE 10.20 For the network in Fig. 10.75(a), determine the energy stored by each capacitor.

Solution: For C_1:

$$W_C = \frac{1}{2}CV^2$$
$$= \frac{1}{2}(2 \times 10^{-6}\,\text{F})(16\,\text{V})^2 = (1 \times 10^{-6})(256) = \mathbf{256\ \mu J}$$

For C_2:

$$W_C = \frac{1}{2}CV^2$$
$$= \frac{1}{2}(3 \times 10^{-6}\,\text{F})(56\,\text{V})^2 = (1.5 \times 10^{-6})(3136) = \mathbf{4704\ \mu J}$$

Due to the squared term, the energy stored increases rapidly with increasing voltages.

10.13 STRAY CAPACITANCES

In addition to the capacitors discussed so far in this chapter, there are **stray capacitances** that exist not through design but simply because two conducting surfaces are relatively close to each other. Two conducting wires in the same network have a capacitive effect between them, as shown in Fig. 10.77(a). In electronic circuits, capacitance levels exist between conducting surfaces of the transistor, as shown in Fig. 10.77(b). In Chapter 11,

Conductors

(a)

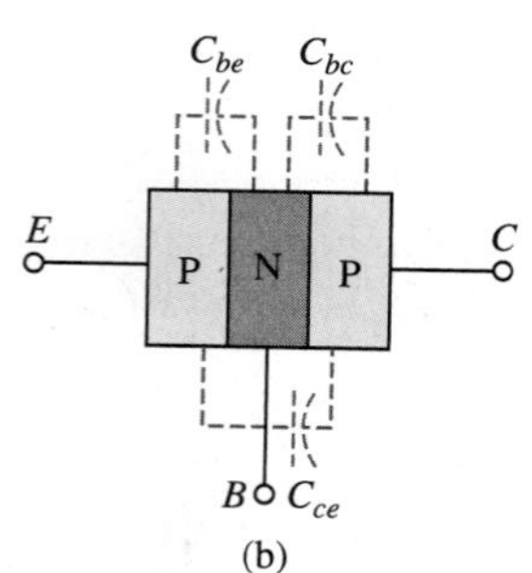

(b)

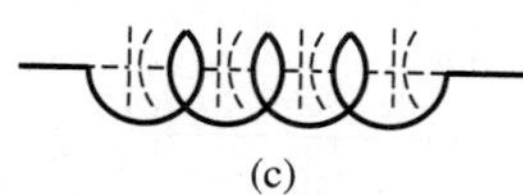

(c)

FIG. 10.77
Examples of stray capacitance.

we will discuss another element called the *inductor,* which has capacitive effects between the windings [Fig. 10.77(c)]. Stray capacitances can often lead to serious errors in system design if they are not considered carefully.

10.14 APPLICATIONS

This section includes both a description of the operation of one of the less expensive, throwaway cameras that have become so popular today and a discussion of the use of capacitors in the line conditioners (surge protectors) that are used in many homes and throughout the business world. Additional examples of the use of capacitors appear in Chapter 11.

Flash Lamp

The basic circuitry for the flash lamp of the popular, inexpensive, throwaway camera in Fig. 10.78(a) is provided in Fig. 10.78(b). The physical circuitry is in Fig. 10.78(c) on the next page. The labels added to Fig. 10.78(c) identify broad areas of the design and some individual components. The major components of the electronic circuitry include a large 160 μF, 330 V, polarized electrolytic capacitor to store the necessary charge for the flash lamp, a flash lamp to generate the required light, a dc battery

(a)

FIG. 10.78(a)
Flash camera: general appearance.

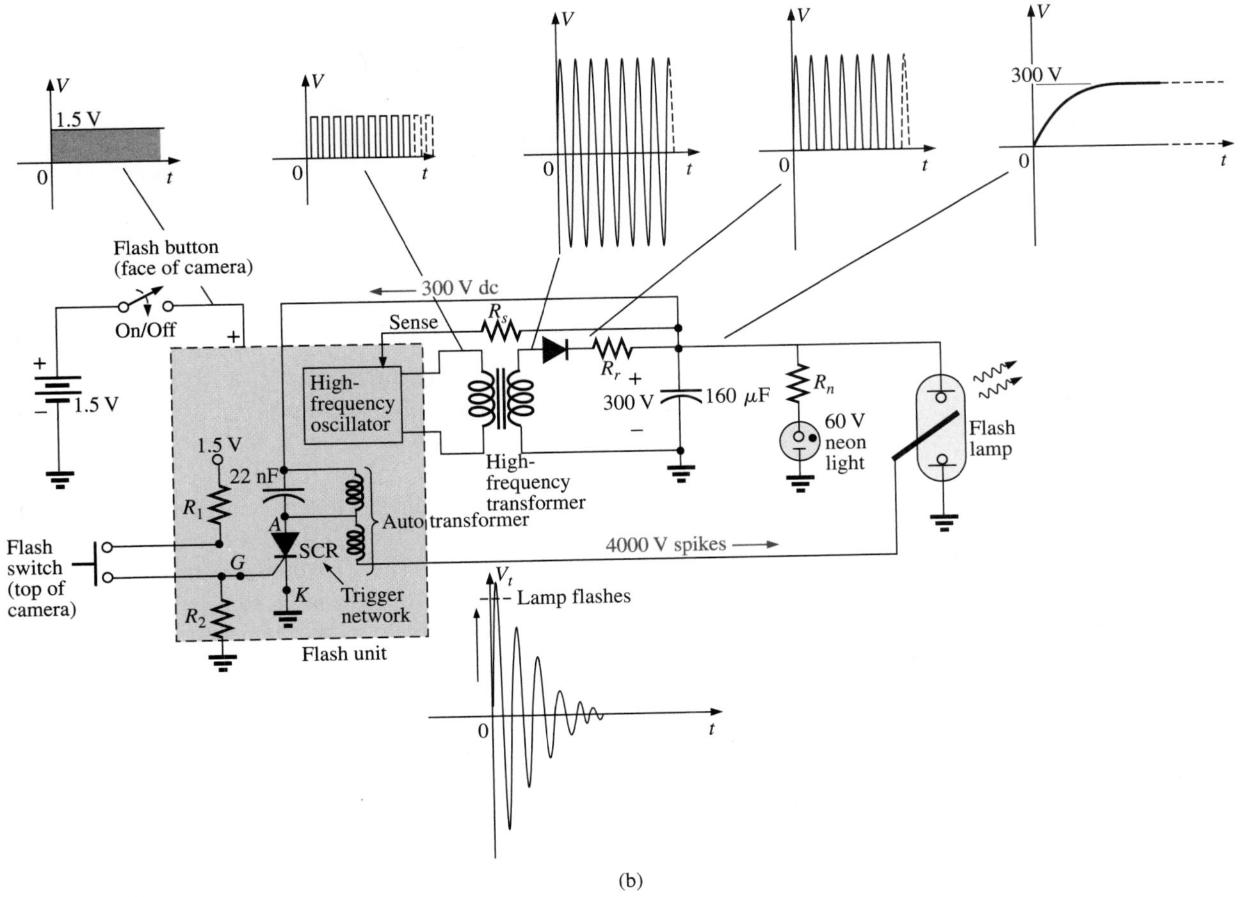

(b)

FIG. 10.78(b)
Flash camera: basic circuitry.

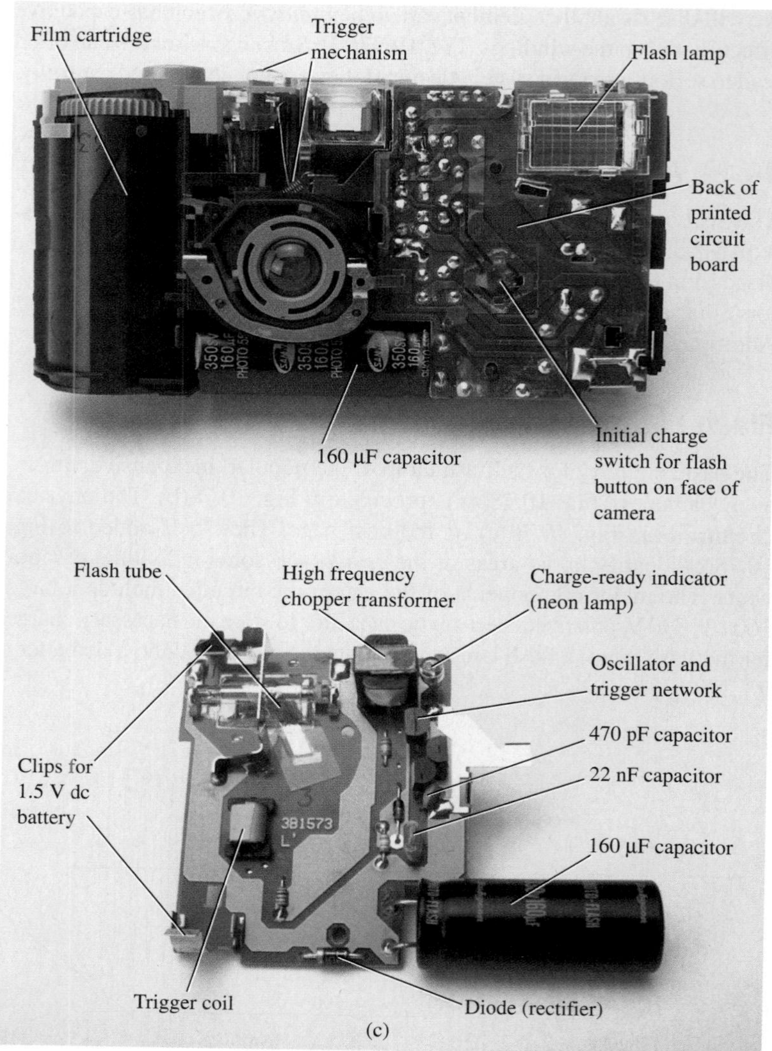

FIG. 10.78(c)
Flash camera: internal construction.

of 1.5 V, a chopper network to generate a dc voltage in excess of 300 V, and a trigger network to establish a few thousand volts for a very short period of time to fire the flash lamp. There are both a 22 nF capacitor in the trigger network as shown in Fig. 10.78(b) and (c) and a third capacitor of 470 pF in the high-frequency oscillator of the chopper network. In particular, note that the size of each capacitor is directly related to its capacitance level. It should certainly be of some interest that a single source of energy of only 1.5 V dc can be converted to one of a few thousand volts (albeit for a very short period of time) to fire the flash lamp. In fact, that single, small battery has sufficient power for the entire run of film through the camera. Always keep in mind that energy is related to power and time by $W = Pt = (VI)t$. That is, a high level of voltage can be generated for a defined energy level as long as the factors I and t are sufficiently small.

When you first use the camera, you are directed to press the flash button on the face of the camera and wait for the flash-ready light to come on. As soon as the flash button is depressed, the full 1.5 V of the dc bat-

tery are applied to an electronic network (a variety of networks can perform the same function) that generates an oscillating waveform of very high frequency (with a high repetitive rate) as shown in Fig. 10.78(c). The high-frequency transformer then significantly increases the magnitude of the generated voltage and passes it on to a half-wave rectification system (introduced in earlier chapters), resulting in a dc voltage of about 300 V across the 160 μF capacitor to charge the capacitor (as determined by $Q = CV$). Once the 300 V level is reached, the lead marked "sense" in Fig. 10.78(b) feeds the information back to the oscillator and turns it off until the output dc voltage drops to a low threshold level. When the capacitor is fully charged, the neon light in parallel with the capacitor turns on (labeled "flash-ready lamp" on the camera) to let you know that the camera is ready to use. The entire network from the 1.5 V dc level to the final 300 V level is called a *dc-dc converter.* The terminology *chopper network* comes from the fact that the applied dc voltage of 1.5 V was chopped up into one that changes level at a very high frequency so that the transformer can perform its function.

Even though the camera may use a 60 V neon light, the neon light and series resistor R_n must have a full 300 V across the branch before the neon light turns on. Neon lights are simply bulbs with a neon gas that support conduction when the voltage across the terminals reaches a sufficiently high level. There is no filament, or hot wire as in a light bulb, but simply conduction through the gaseous medium. For new cameras, the first charging sequence may take 12 s to 15 s. Succeeding charging cycles may only take some 7 s or 8 s because the capacitor still has some residual charge on its plates. If the flash unit is not used, the neon light begins to drain the 300 V dc supply with a drain current in microamperes. As the terminal voltage drops, the neon light eventually turns off. For the unit in Fig. 10.78, it takes about 15 min before the light turns off. Once off, the neon light no longer drains the capacitor, and the terminal voltage of the capacitor remains fairly constant. Eventually, however, the capacitor discharges due to its own leakage current, and the terminal voltage drops to very low levels. The discharge process is very rapid when the flash unit is used, causing the terminal voltage to drop very quickly ($V = Q/C$) and, through the feedback-sense connection signal, causing the oscillator to start up again and recharge the capacitor. You may have noticed when using a camera of this type that once the camera has its initial charge, you do not need to press the charge button between pictures—it is done automatically. However, if the camera sits for a long period of time, you must depress the charge button, but the charge time is only 3 s or 4 s due to the residual charge on the plates of the capacitor.

The 300 V across the capacitor are insufficient to fire the flash lamp. Additional circuitry, called the *trigger network,* must be incorporated to generate the few thousand volts necessary to fire the flash lamp. The resulting high voltage is one reason that there is a CAUTION note on each camera regarding the high internal voltages generated and the possibility of electrical shock if the camera is opened.

The thousands of volts required to fire the flash lamp require a discussion that introduces elements and concepts beyond the current level of the text. This description is simply a first exposure to some of the interesting possibilities available from the right mix of elements. When the flash switch at the bottom left of Fig. 10.78(b) is closed, it establishes a connection between the resistors R_1 and R_2. Through a voltage divider action, a dc voltage appears at the gate (G) terminal of the SCR (silicon-controlled rectifier—a device whose state is controlled by the voltage at

the gate terminal). This dc voltage turns the SCR "on" and establishes a very low resistance path (like a short circuit) between its anode (A) and cathode (K) terminals. At this point the trigger capacitor, which is connected directly to the 300 V sitting across the capacitor, rapidly charges to 300 V because it now has a direct, low-resistance path to ground through the SCR. Once it reaches 300 V, the charging current in this part of the network drops to 0 A, and the SCR opens up again since it is a device that needs a steady current in the anode circuit to stay on. The capacitor then sits across the parallel coil (with no connection to ground through the SCR) with its full 300 V and begins to quickly discharge through the coil because the only resistance in the circuit affecting the time constant is the resistance of the parallel coil. As a result, a rapidly changing current through the coil generates a high voltage across the coil for reasons to be introduced in Chapter 11.

When the capacitor decays to zero volts, the current through the coil will be zero amperes, but a strong magnetic field has been established around the coil. This strong magnetic field then quickly collapses, establishing a current in the parallel network that recharges the capacitor again. This continual exchange between the two storage elements continues for a period of time, depending on the resistance in the circuit. The more the resistance, the shorter the "ringing" of the voltage at the output. This action of the energy "flying back" to the other element is the basis for the "flyback" effect that is frequently used to generate high dc voltages such as needed in TVs. In Fig. 10.78(b), you will find that the trigger coil is connected directly to a second coil to form an autotransformer (a transformer with one end connected). Through transformer action, the high voltage generated across the trigger coil increases further, resulting in the 4000 V necessary to fire the flash lamp. Note in Fig. 10.78(c) that the 4000 V are applied to a grid that actually lies on the surface of the glass tube of the flash lamp (not internally connected or in contact with the gases). When the trigger voltage is applied, it excites the gases in the lamp, causing a very high current to develop in the bulb for a very short period of time and producing the desired bright light. The current in the lamp is supported by the charge on the 160 μF capacitor which is dissipated very quickly. The capacitor voltage drops very quickly, the photo lamp shuts down, and the charging process begins again. If the entire process didn't occur as quickly as it does, the lamp would burn out after a single use.

Surge Protector (Line Conditioner)

In recent years we have all become familiar with the surge protector as a safety measure for our computers, TVs, DVD players, and other sensitive instrumentation. In addition to protecting equipment from unexpected surges in voltage and current, most quality units also filter out (remove) electromagnetic interference (EMI) and radio-frequency interference (RFI). EMI encompasses any unwanted disturbances down the power line established by any combination of electromagnetic effects such as those generated by motors on the line, power equipment in the area emitting signals picked up by the power line acting as an antenna, and so on. RFI includes all signals in the air in the audio range and beyond which may also be picked up by power lines inside or outside the house.

The unit in Fig. 10.79 has all the design features expected in a good line conditioner. Figure 10.79(a) reveals that it can handle the power drawn by six outlets and that it is set up for FAX/MODEM protection. Also note that it has both LED (light-emitting diode) displays which

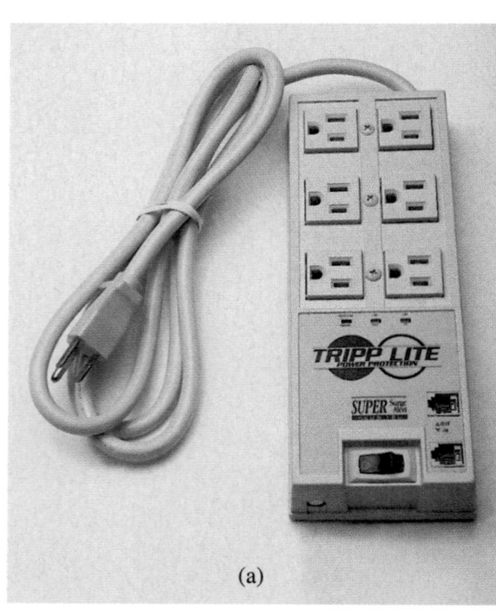

(a)

FIG. 10.79(a)
Surge protector: general appearance.

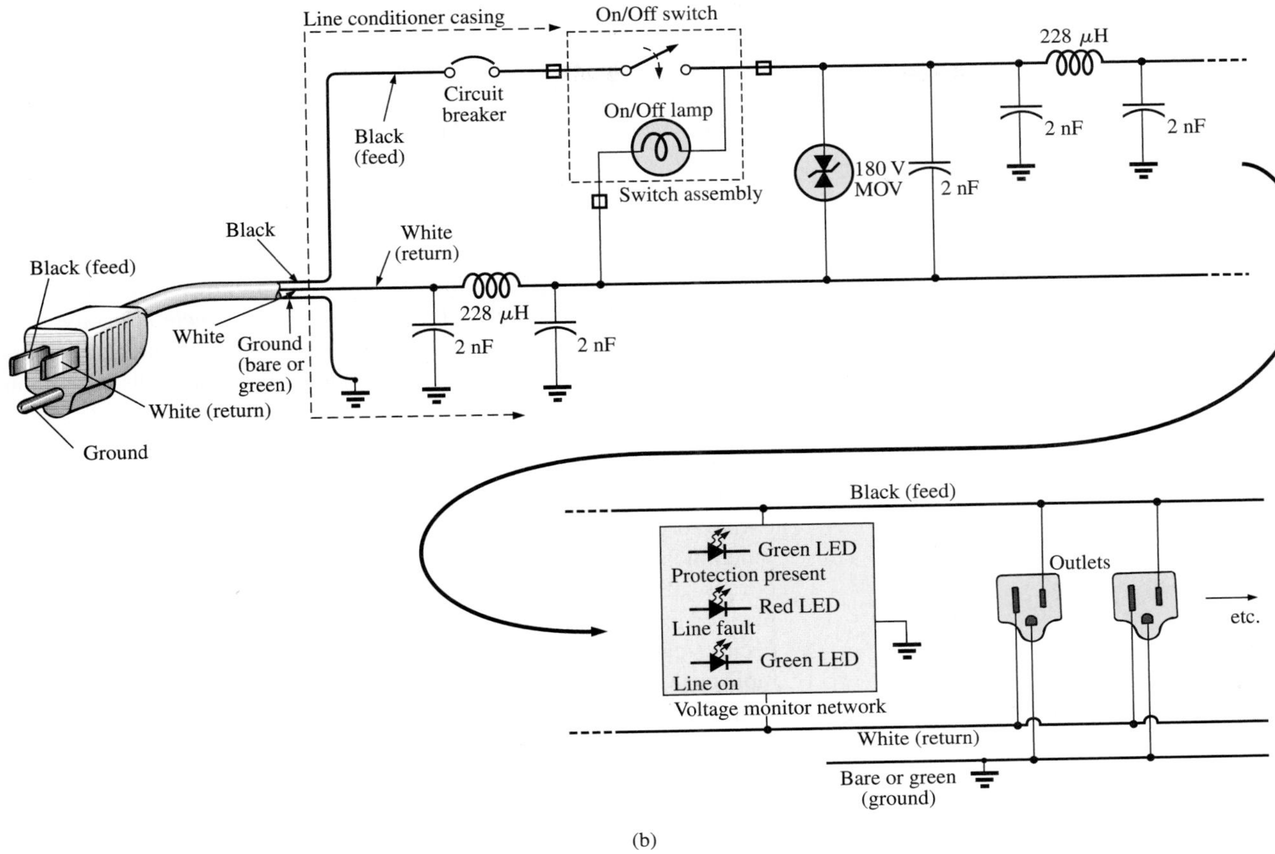

(b)

FIG. 10.79(b)
Electrical schematic.

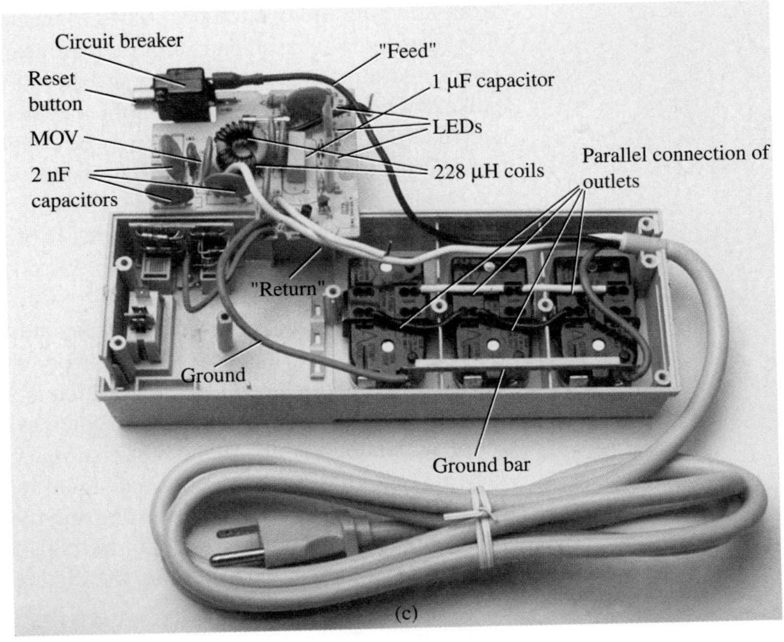

FIG. 10.79(c)
Internal construction.

reveal whether there is fault on the line or whether the line is OK and an external circuit breaker to reset the system. In addition, when the surge protector is on, a red light is visible at the power switch.

The schematic in Fig. 10.79(b) does not include all the details of the design, but it does include the major components that appear in most good line conditioners. First note in the photograph in Fig. 10.79(c) that the outlets are all connected in parallel, with a ground bar used to establish a ground connection for each outlet. The circuit board had to be flipped over to show the components, so it will take some adjustment to relate the position of the elements on the board to the casing. The *feed line* or *hot lead wire* (black in the actual unit) is connected directly from the line to the circuit breaker. The other end of the circuit breaker is connected to the other side of the circuit board. All the large discs that you see are 2 nF capacitors [not all have been included in Fig 10.79(b) for clarity]. There are quite a few capacitors to handle all the possibilities. For instance, there are capacitors from line to return (black wire to white wire), from line to ground (black to green), and from return to ground (white to ground). Each has two functions. The first and most obvious function is to prevent any spikes in voltage that may come down the line because of external effects such as lightning from reaching the equipment plugged into the unit. Recall from this chapter that the voltage across capacitors cannot change instantaneously and, in fact, acts to squelch any rapid change in voltage across its terminals. The capacitor, therefore, prevents the line to neutral voltage from changing too quickly, and any spike that tries to come down the line has to find another point in the feed circuit to fall across. In this way, the appliances plugged into the surge protector are well protected.

The second function requires some knowledge of the reaction of capacitors to different frequencies and is discussed in more detail in later chapters. For the moment, let it suffice to say that the capacitor has a different impedance to different frequencies, thereby preventing undesired frequencies, such as those associated with EMI and RFI disturbances, from affecting the operation of units connected to the line conditioner. The rectangular-shaped capacitor of 1 μF near the center of the board is connected directly across the line to take the brunt of a strong voltage spike down the line. Its larger size is clear evidence that it is designed to absorb a fairly high energy level that may be established by a large voltage—significant current over a period of time that may exceed a few milliseconds.

The large, toroidal-shaped structure in the center of the circuit board in Fig. 10.79(c) has two coils (Chapter 11) of 228 μH that appear in the line and neutral in Fig. 10.79(b). Their purpose, like that of the capacitors, is twofold: to block spikes in current from coming down the line and to block unwanted EMI and RFI frequencies from getting to the connected systems. In the next chapter you will find that coils act as "chokes" to quick changes in current; that is, the current through a coil cannot change instantaneously. For increasing frequencies, such as those associated with EMI and RFI disturbances, the reactance of a coil increases and absorbs the undesired signal rather than let it pass down the line. Using a choke in both the line and the neutral makes the conditioner network balanced to ground. In total, capacitors in a line conditioner have the effect of *bypassing* the disturbances, whereas inductors *block* the disturbance.

The smaller disc (blue) between two capacitors and near the circuit breaker is an MOV (metal-oxide varistor) which is the heart of most line conditioners. It is an electronic device whose terminal characteristics change with the voltage applied across its terminals. For the normal range

of voltages down the line, its terminal resistance is sufficiently large to be considered an open circuit, and its presence can be ignored. However, if the voltage is too large, its terminal characteristics change from a very large resistance to a very small resistance that can essentially be considered a short circuit. This variation in resistance with applied voltage is the reason for the name *varistor.* For MOVs in North America where the line voltage is 120 V, the MOVs are 180 V or more. The reason for the 60 V difference is that the 120 V rating is an effective value related to dc voltage levels, whereas the waveform for the voltage at any 120 V outlet has a peak value of about 170 V. A great deal more will be said about this topic in Chapter 13.

Taking a look at the symbol for an MOV in Fig. 10.79(b), note that it has an arrow in each direction, revealing that the MOV is bidirectional and blocks voltages with either polarity. In general, therefore, for normal operating conditions, the presence of the MOV can be ignored; but, if a large spike should appear down the line, exceeding the MOV rating, it acts as a short across the line to protect the connected circuitry. It is a significant improvement to simply putting a fuse in the line because it is voltage sensitive, can react much quicker than a fuse, and displays its low-resistance characteristics for only a short period of time. When the spike has passed, it returns to its normal open-circuit characteristic. If you're wondering where the spike goes if the load is protected by a short circuit, remember that all sources of disturbance, such as lightning, generators, inductive motors (such as in air conditioners, dishwashers, power saws, and so on), have their own "source resistance," and there is always some resistance down the line to absorb the disturbance.

Most line conditioners, as part of their advertising, mention their energy absorption level. The rating of the unit in Fig. 10.79 is 1200 J which is actually higher than most. Remembering that $W = Pt = EIt$ from the earlier discussion of cameras, we now realize that if a 5000 V spike occurred, we would be left with the product $It = W/E = 1200\text{ J}/5000\text{ V} = 240\text{ mAs}$. Assuming a linear relationship between all quantities, the rated energy level reveals that a current of 100 A could be sustained for $t = 240\text{ mAs}/100\text{ A} = 2.4\text{ ms}$, a current of 1000 A for 240 μs, and a current of 10,000 A for 24 μs. Obviously, the higher the power product of E and I, the less the time element.

The technical specifications of the unit in Fig. 10.79 include an instantaneous response time in the order of ps, with a phone line protection of 5 ns. The unit is rated to dissipate surges up to 6000 V and current spikes up to 96,000 A. It has a very high noise suppression ratio (80 dB; see Chapter 21) at frequencies from 50 kHz to 1000 MHz, and (a credit to the company) it has a lifetime warranty.

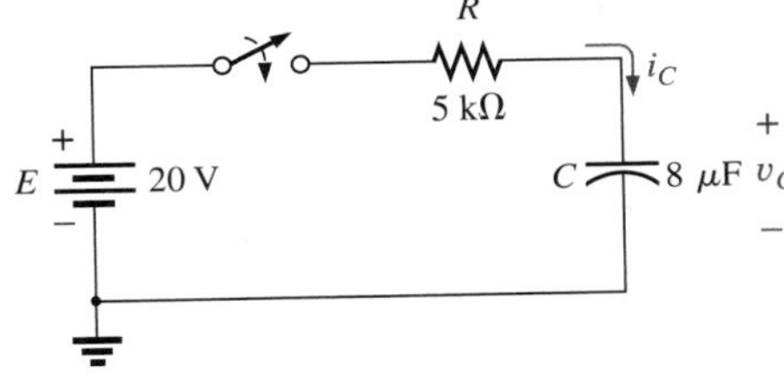

FIG. 10.80
Circuit to be analyzed using PSpice.

10.15 COMPUTER ANALYSIS

PSpice

Transient *RC* Response We now use PSpice to investigate the transient response for the voltage across the capacitor in Fig. 10.80. In all the examples in the text involving a transient response, a switch appeared in series with the source as shown in Fig. 10.81(a). When applying PSpice, we establish this instantaneous change in voltage level by applying a pulse waveform as shown in Fig. 10.81(b) with a pulse width *(PW)* longer than the period (5τ) of interest for the network.

To obtain a pulse source, start with the sequence **Place part** key-**Libraries-SOURCE-VPULSE-OK.** Once in place, set the label and all the parameters by double-clicking on each to obtain the **Display**

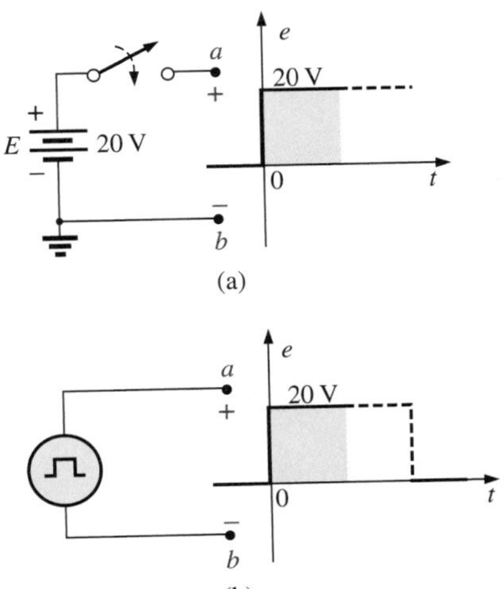

FIG. 10.81
Establishing a switching dc voltage level: (a) series dc voltage-switch combination; (b) PSpice pulse option.

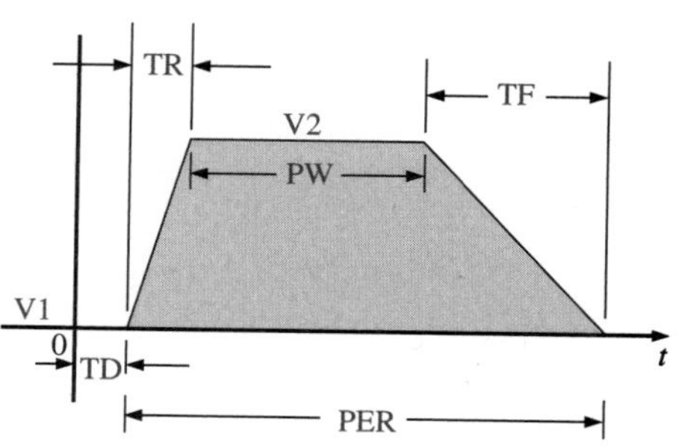

FIG. 10.82
The defining parameters of PSpice ***VPULSE.***

Properties dialog box. As you scroll down the list of attributes, you will see the following parameters defined by Fig. 10.82:

V1 is the initial value.
V2 is the pulse level.
TD is the delay time.
TR is the rise time.
TF is the fall time.
PW is the pulse width at the V_2 level.
PER is the period of the waveform.

All the parameters have been set as shown on the schematic in Fig. 10.83 for the network in Fig. 10.80. Since a rise and fall time of 0 s is unrealistic from a practical standpoint, 0.1 ms was chosen for each in this example. Further, since $\tau = RC = (5\text{ k}\Omega) \times (8\ \mu\text{F}) = 20$ ms and $5\tau = 200$ ms, a pulse width of 500 ms was selected. The period was simply chosen as twice the pulse width.

Now for the simulation process. First select the **New Simulation Profile** key to obtain the **New Simulation** dialog box in which **PSpice 10-1** is inserted for the **Name** and **Create** is chosen to leave the dialog box. The **Simulation Settings-PSpice 10-1** dialog box results, and under **Analysis,** choose the **Time Domain (Transient)** option under **Analysis type.** Set the **Run to time** at 200 ms so that only the first five time constants will be plotted. Set the **Start saving data after** option at 0 s to ensure that the data are collected immediately. The **Maximum step size** is 1 ms to provide sufficient data points for a good plot. Click **OK,** and you are ready to select the **Run PSpice** key. The result is a graph without a plot (since it has not been defined yet) and an *x*-axis that extends from 0 s to 200 ms as defined above. To obtain a plot of the voltage across the capacitor versus time, apply the following sequence: **Add Trace** key-**Add Traces** dialog box-**V1(C)-OK.** The plot in Fig. 10.84 results. The color and thickness of the plot and the axis can be changed by placing the cursor on the plot line and right-clicking. Select **Properties** from the list that appears. A **Trace Properties** dialog box appears in which you can change the color and thickness of the line. Since the plot is against a black background, a better printout occurred when yellow was selected and the line was made thicker as shown in Fig. 10.84. For comparison, plot the

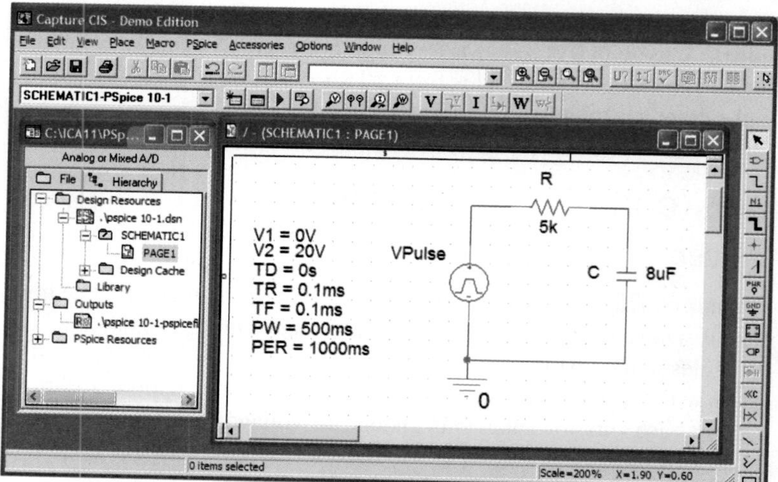

FIG. 10.83
Using PSpice to investigate the transient response of the series R-C circuit in Fig. 10.80.

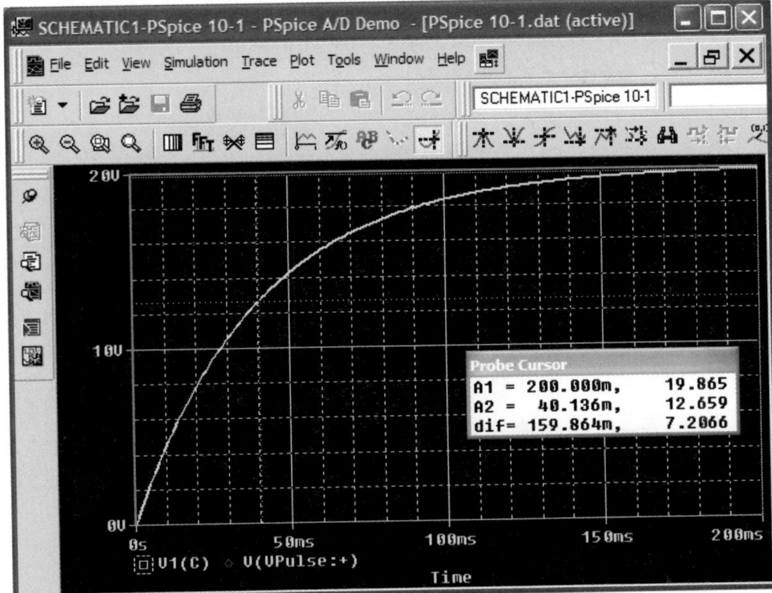

FIG. 10.84
Transient response for the voltage across the capacitor in Fig. 10.80 when ***VPulse*** *is applied.*

applied pulse signal also. This is accomplished by going back to **Trace** and selecting **Add Trace** followed by **V(Vpulse:+)** and **OK.** Now both waveforms appear on the same screen as shown in Fig. 10.84. In this case, the plot has a reddish tint so it can be distinguished from the axis and the other plot. Note that it follows the left axis to the top and travels across the screen at 20 V.

If you want the magnitude of either plot at any instant, simply select the **Toggle cursor** key. Then click on **V1(C)** at the bottom left of the screen. A box appears around **V1(C)** that reveals the spacing between the dots of the cursor on the screen. This is important when more than one cursor is used. By moving the cursor to 200 ms, you find that the magnitude **(A1)** is 19.865 V (in the **Probe Cursor** dialog box), clearly showing how close it is to the final value of 20 V. A second cursor can be placed on the screen with a right click and then a click on the same **V1(C)** on the bottom of the screen. The box around **V1(C)** cannot show two boxes, but the spacing and the width of the lines of the box have definitely changed. There is no box around the **Pulse** symbol since it was not selected—although it could have been selected by either cursor. If you now move the second cursor to one time constant of 40 ms, you find that the voltage is 12.659 V as shown in the **Probe Cursor** dialog box. This confirms that the voltage should be 63.2% of its final value of 20 V in one time constant ($0.632 \times 20\text{ V} = 12.4\text{ V}$). Two separate plots could have been obtained by going to **Plot-Add Plot to Window** and then using the trace sequence again.

Average Capacitive Current As an exercise in using the pulse source and to verify our analysis of the average current for a purely capacitive network, the description to follow verifies the results of Example 10.14. For the pulse waveform in Fig. 10.65, the parameters of the pulse supply appear in Fig. 10.85. Note that the rise time is now 2 ms, starting at 0 s, and the fall time is 6 ms. The period was set at 15 ms to permit monitoring the current after the pulse had passed.

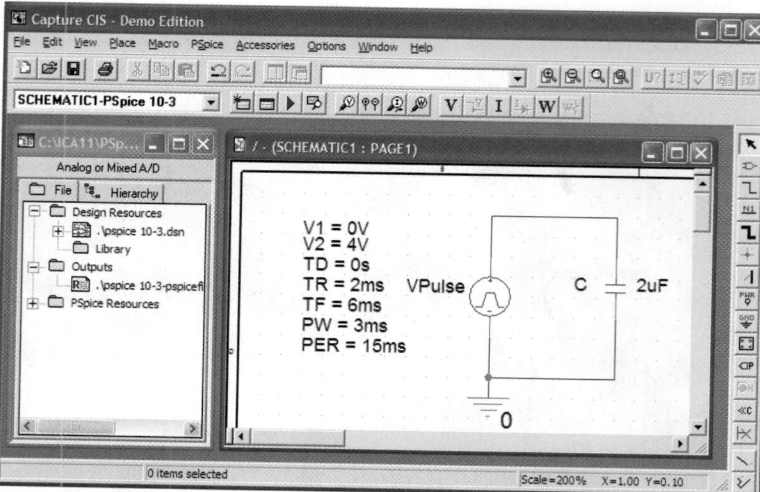

FIG. 10.85
Using PSpice to verify the results in Example 10.14.

Initiate simulation by first selecting the **New Simulation Profile** key to obtain the **New Simulation** dialog box in which **AverageIC** is entered as the **Name.** Choose **Create** to obtain the **Simulation Settings-AverageIC** dialog box. Select **Analysis** and choose **Time Domain(Transient)** under the **Analysis type** options. Set the **Run to time** to 15 ms to encompass the period of interest, and set the **Start saving data after** at 0 s to ensure data points starting at $t = 0$ s. Select the **Maximum step size** from 15 ms/1000 = 15 μs to ensure 1000 data points for the plot. Click **OK,** and select the **Run PSpice** key. A window appears with a horizontal scale that extends from 0 to 15 ms as defined above. Then select the **Add Trace** key, and choose **I(C)** to appear in the **Trace Expression** below. Click **OK,** and the plot of **I(C)** appears in the bottom of Fig. 10.86. This time it would be

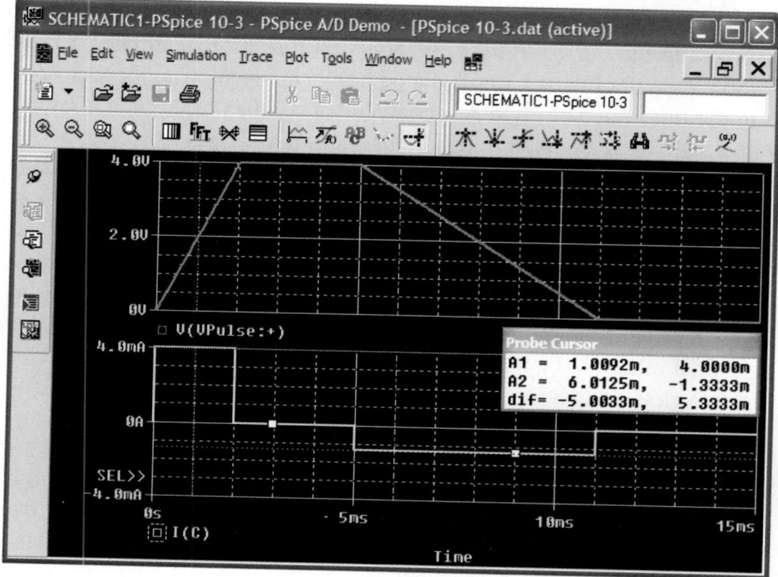

FIG. 10.86
The applied pulse and resulting current for the 2 μF capacitor in Fig. 10.85.

nice to see the pulse waveform in the same window but as a separate plot. Therefore, continue with **Plot-Add Plot to Window-Trace-Add Trace-V(Vpulse:+)-OK,** and both plots appear as shown in Fig. 10.86.

Now use the cursors to measure the resulting average current levels. First, select the **I(C)** plot to move the **SEL>>** notation to the lower plot. The **SEL>>** defines which plot for multiplot screens is active. Then select the **Toggle cursor** key, and left-click on the **I(C)** plot to establish the crosshairs of the cursor. Set the value at 1 ms, and the magnitude **A1** is displayed as 4 mA. Right-click on the same plot, and a second cursor results that can be placed at 6 ms to get a response of −1.33 mA (**A2**) as expected from Example 10.14. The plot for **I(C)** was set in the yellow color with a wider line by right-clicking on the curve and choosing **Properties.**

PROBLEMS

SECTION 10.2 The Electric Field

1. **a.** Find the electric field strength at a point 2 m from a charge of 4 μC.
   b. Find the electric field strength at a point 1 mm from the same charge as part (a) and compare results.

2. The electric field strength is 72 newtons/coulomb (N/C) at a point *r* meters from a charge of 2 μC. Find the distance *r.*

SECTIONS 10.3 AND 10.4 Capacitance and Capacitors

3. Find the capacitance of a parallel plate capacitor if 1200 μC of charge are deposited on its plates when 10 V are applied across the plates.

4. How much charge is deposited on the plates of a 0.15 μF capacitor if 45 V are applied across the capacitor?

5. Find the electric field strength between the plates of a parallel plate capacitor if 100 mV are applied across the plates and the plates are 2 mm apart.

6. Repeat Problem 5 if the plates are separated by 10 mils.

7. A 4 μF parallel plate capacitor has 160 μC of charge on its plates. If the plates are 5 mm apart, find the electric field strength between the plates.

8. Find the capacitance of a parallel plate capacitor if the area of each plate is 0.1 m^2 and the distance between the plates is 2 mm. The dielectric is air.

9. Repeat Problem 8 if the dielectric is paraffin-coated paper.

10. Find the distance in mils between the plates of a 2 μF capacitor if the area of each plate is 0.15 m^2 and the dielectric is transformer oil.

11. The capacitance of a capacitor with a dielectric of air is 1200 pF. When a dielectric is inserted between the plates, the capacitance increases to 6 nF. Of what material is the dielectric made?

12. The plates of a parallel plate air capacitor are 0.2 mm apart and have an area of 0.08 m^2, and 200 V are applied across the plates.
    a. Determine the capacitance.
    b. Find the electric field intensity between the plates.
    c. Find the charge on each plate if the dielectric is air.

13. A sheet of Bakelite 0.2 mm thick having an area of 0.08 m^2 is inserted between the plates of Problem 12.
    a. Find the electric field strength between the plates.
    b. Determine the charge on each plate.
    c. Determine the capacitance.

14. A parallel plate air capacitor has a capacitance of 5 μF. Find the new capacitance if:
    a. The distance between the plates is doubled (everything else remains the same).
    b. The area of the plates is doubled (everything else remains the same as for the 5 μF level).
    c. A dielectric with a relative permittivity of 20 is inserted between the plates (everything else remains the same as for the 5 μF level).
    d. A dielectric is inserted with a relative permittivity of 4, and the area is reduced to 1/3 and the distance to 1/4 of their original dimensions.

*15. Find the maximum voltage that can be applied across a parallel plate capacitor of 6 nF if the area of one plate is 0.02 m^2 and the dielectric is mica. Assume a linear relationship between the dielectric strength and the thickness of the dielectric.

*16. Find the distance in micrometers between the plates of a parallel plate mica capacitor if the maximum voltage that can be applied across the capacitor is 1200 V. Assume a linear relationship between the breakdown strength and the thickness of the dielectric.

17. A 22 μF capacitor has −200 ppm/°C at room temperature of 20°C. What is the capacitance if the temperature increases to 100°C, the boiling point of water?

18. What is the capacitance of a small teardrop capacitor labeled 40 J? What is the range of expected values as established by the tolerance?

19. A large, flat, mica capacitor is labeled 220M. What are the capacitance and the expected range of values guaranteed by the manufacturer?

20. A small flat disc ceramic capacitor is labeled 333K. What are the capacitance level and the expected range of values?

SECTION 10.5 Transients in Capacitive Networks: The Charging Phase

21. For the circuit in Fig. 10.87, composed of standard values:
    a. Determine the time constant of the circuit.
    b. Write the mathematical equation for the voltage v_C following the closing of the switch.
    c. Determine the voltage v_C after one, three, and five time constants.
    d. Write the equations for the current i_C and the voltage v_R.
    e. Sketch the waveforms for v_C and i_C.

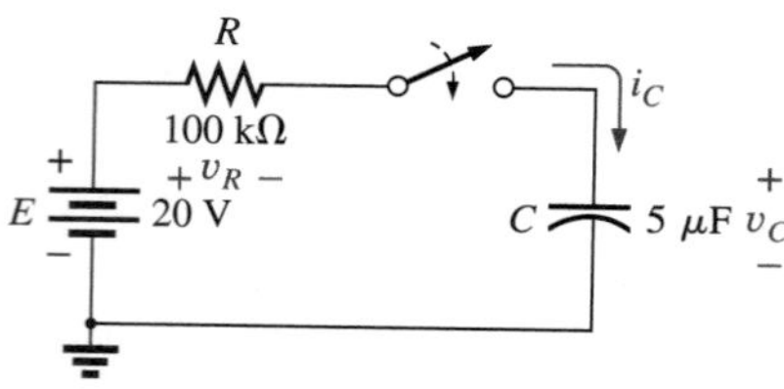

FIG. 10.87
Problems 21 and 22.

22. Repeat Problem 21 for $R = 1\ \text{M}\Omega$, and compare the results.
23. For the circuit in Fig. 10.88, composed of standard values:
    a. Determine the time constant of the circuit.
    b. Write the mathematical equation for the voltage v_C following the closing of the switch.
    c. Determine v_C after one, three, and five time constants.
    d. Write the equations for the current i_C and the voltage v_R.
    e. Sketch the waveforms for v_C and i_C.

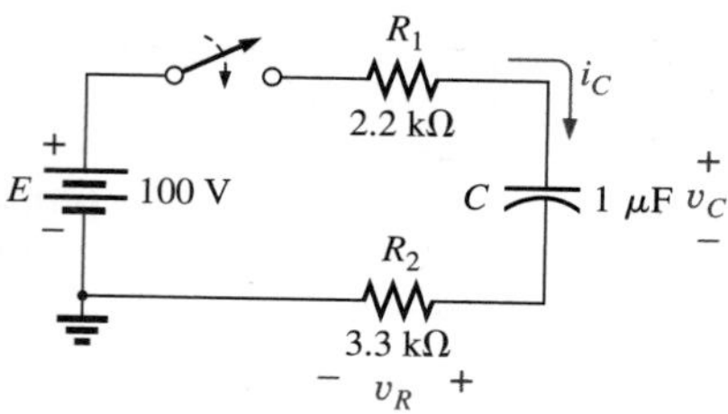

FIG. 10.88
Problem 23.

24. For the circuit in Fig. 10.89, composed of standard values:
    a. Determine the time constant of the circuit.
    b. Write the mathematical equation for the voltage v_C following the closing of the switch.
    c. Write the mathematical expression for the current i_C following the closing of the switch.
    d. Sketch the waveforms of v_C and i_C.

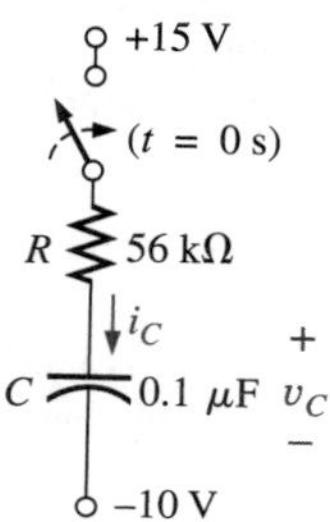

FIG. 10.89
Problem 24.

25. Given the voltage $v_C = 60\ \text{mV}(1 - e^{-t/5\text{ms}})$:
    a. What is the time constant?
    b. What is the voltage at $t = 2$ ms?
    c. What is the voltage at $t = 100$ ms?
26. The voltage across a 10 μF capacitor in a series *R-C* circuit is $v_C = 12\ \text{V}(1 - e^{-t/40\text{ms}})$.
    a. On a practical basis, how much time must pass before the charging phase has passed?
    b. What is the resistance of the circuit?
    c. What is the voltage at $t = 20$ ms?
    d. What is the voltage at 10 time constants?
    e. Under steady-state conditions, how much charge is on the plates?
    f. If the leakage resistance is 1000 MΩ, how long will it take (in hours) for the capacitor to discharge if we assume that the discharge rate is constant throughout the discharge period?

SECTION 10.6 Transients in Capacitive Networks: The Discharging Phase

27. For the *R-C* circuit in Fig. 10.90, composed of standard values:
    a. Determine the time constant of the circuit when the switch is thrown into position 1.
    b. Find the mathematical expression for the voltage across the capacitor and the current after the switch is thrown into position 1.
    c. Determine the magnitude of the voltage v_C and the current i_C the instant the switch is thrown into position 2 at $t = 1$ s.
    d. Determine the mathematical expression for the voltage v_C and the current i_C for the discharge phase.
    e. Plot the waveforms of v_C and i_C for a period of time extending from 0 to 2 s from when the switch was thrown into position 1.

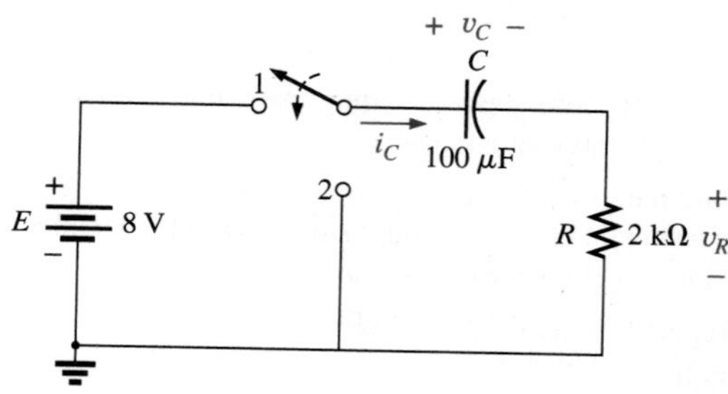

FIG. 10.90
Problem 27.

28. For the network in Fig. 10.91, composed of standard values:
    a. Write the mathematical expressions for the voltages v_C, and v_{R_1} and the current i_C after the switch is thrown into position 1.
    b. Find the values of v_C, v_{R_1}, and i_C when the switch is moved to position 2 at $t = 100$ ms.

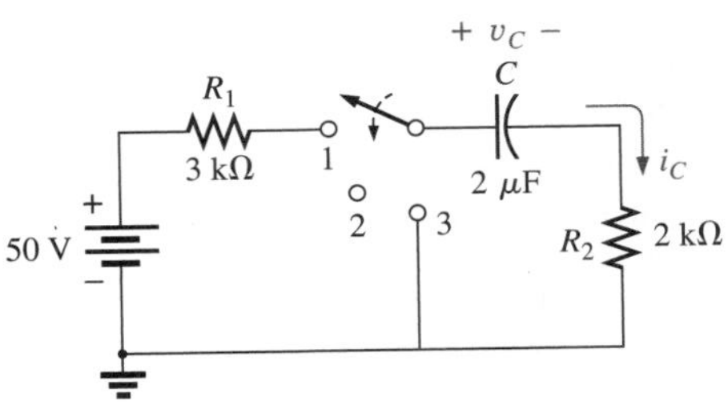

FIG. 10.91
Problems 28 and 29.

c. Write the mathematical expressions for the voltages v_C and v_{R_2} and the current i_C if the switch is moved to position 3 at $t = 200$ ms.
d. Plot the waveforms of v_C, v_{R_2}, and i_C for the time period extending from 0 to 300 ms.

29. Repeat Problem 28 for a capacitance of 20 μF assuming that the leakage resistance of the capacitor is ∞ Ω.

30. For the network in Fig. 10.92, composed of standard values:
a. Find the mathematical expressions for the voltage v_C and the current i_C when the switch is thrown into position 1.
b. Find the mathematical expressions for the voltage v_C and the current i_C if the switch is thrown into position 2 at a time equal to five time constants of the charging circuit.
c. Plot the waveforms of v_C and i_C for a period of time extending from 0 to 30 μs.

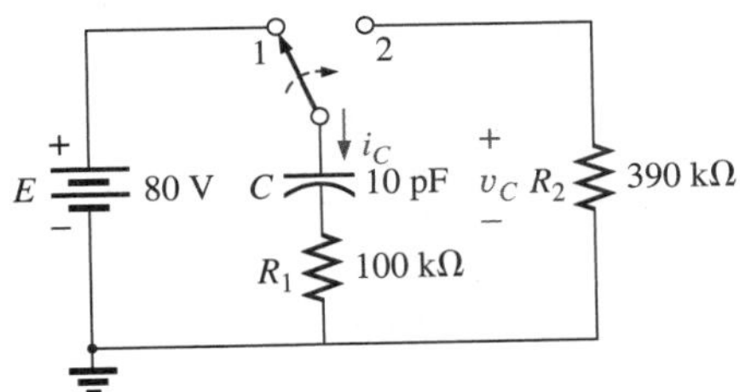

FIG. 10.92
Problem 30.

31. The 1000 μF capacitor in Fig. 10.93 is charged to 6 V. To discharge the capacitor before further use, a wire with a resistance of 2 mΩ is placed across the capacitor.
a. How long will it take to discharge the capacitor?
b. What is the peak value of the current?
c. Based on the answer to part (b), is a spark expected when contact is made with both ends of the capacitor?

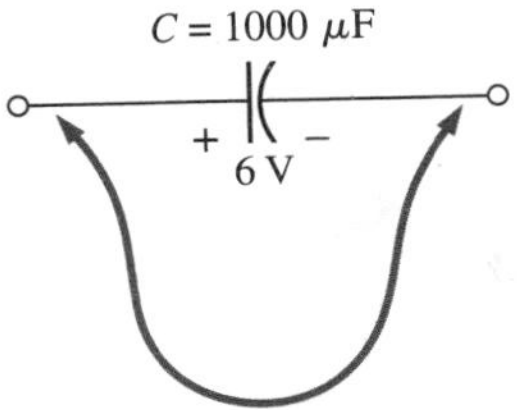

FIG. 10.93
Problem 31.

SECTION 10.7 Initial Conditions

32. The capacitor in Fig. 10.94 is initially charged to 10 V with the polarity shown.
a. Write the expression for the voltage v_C after the switch is closed.
b. Write the expression for the current i_C after the switch is closed.
c. Plot the results of parts (a) and (b).

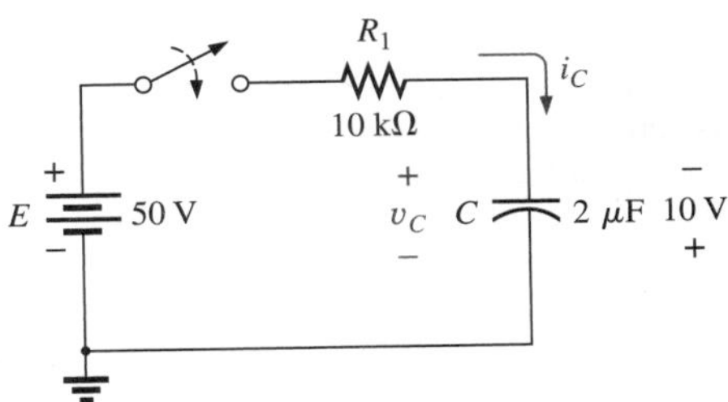

FIG. 10.94
Problem 32.

33. The capacitor in Fig. 10.95 is initially charged to 40 V before the switch is closed. Write the expressions for the voltages v_C and v_R and the current i_C following the closing of the switch. Plot the resulting waveforms.

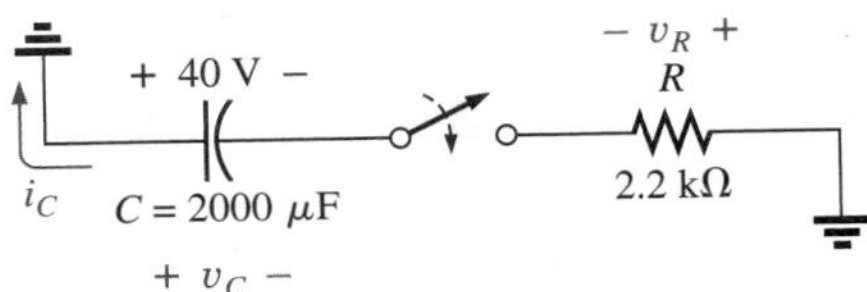

FIG. 10.95
Problem 33.

***34.** The capacitor in Fig. 10.96 is initially charged to 20 V with the polarity shown. Write the expressions for the voltage v_C and the current i_C following the closing of the switch. Plot the resulting waveforms.

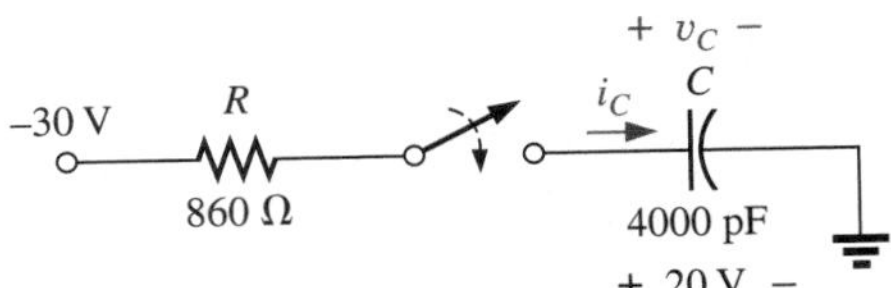

FIG. 10.96
Problem 34.

***35.** The capacitor in Fig. 10.97 is initially charged to 12 V with the polarity shown.

$+ v_C -$
40 V R_1 10 kΩ R_2 8.2 kΩ i_C C 6.8 μF −12 V
+ 12 V −

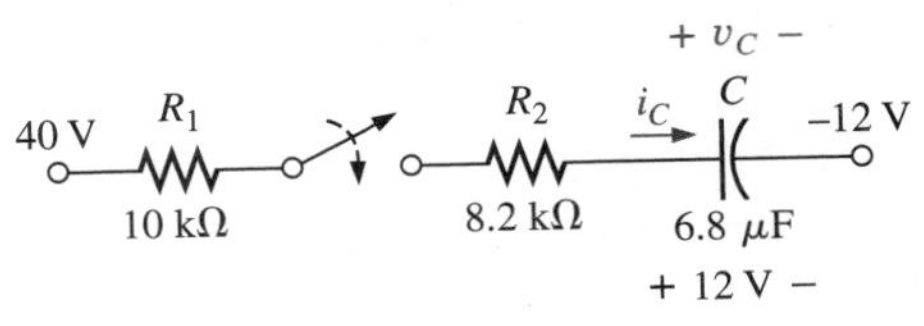

FIG. 10.97
Problem 35.

a. Find the mathematical expressions for the voltage v_C and the current i_C when the switch is closed.
b. Sketch the waveforms of v_C and i_C.

SECTION 10.8 Instantaneous Values

36. Given the expression $v_C = 12\text{ V}(1 - e^{-t/20\mu s})$:
   a. Determine v_C at $t = 10\ \mu s$.
   b. Determine v_C at $t = 10\tau$.
   c. Find the time t for v_C to reach 6 V.
   d. Find the time t for v_C to reach 11.98 V.

37. For the network in Fig. 10.98, V_L must be 8 V before the system is activated. If the switch is closed at $t = 0$ s, how long will it take for the system to be activated?

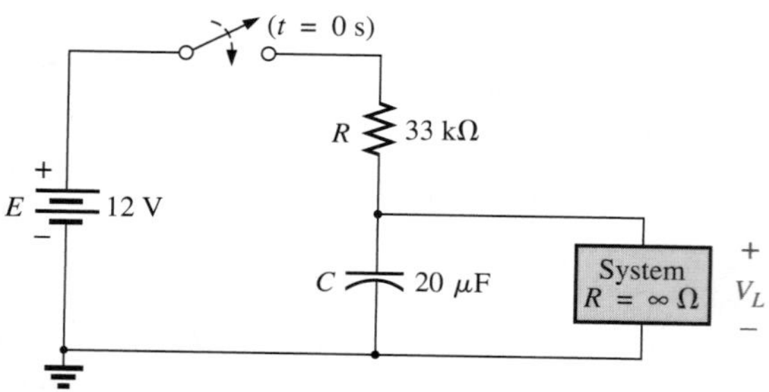

FIG. 10.98
Problem 37.

*38. Design the network in Fig. 10.99 such that the system turns on 10 s after the switch is closed.

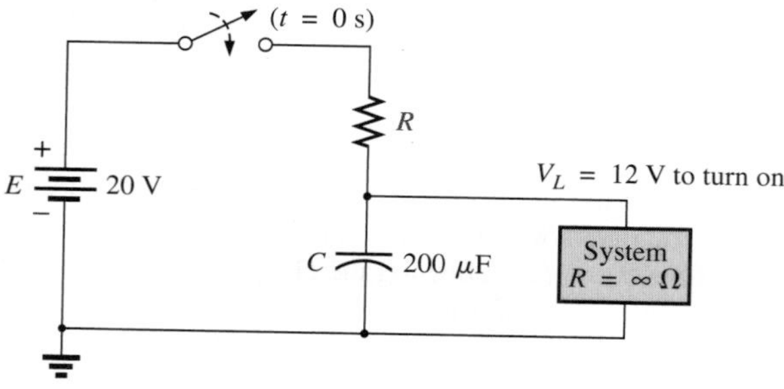

FIG. 10.99
Problem 38.

39. For the circuit in Fig. 10.100:
   a. Find the time required for v_C to reach 60 V following the closing of the switch.
   b. Calculate the current i_C at the instant $v_C = 60$ V.
   c. Determine the power delivered by the source at the instant $t = 2\tau$.

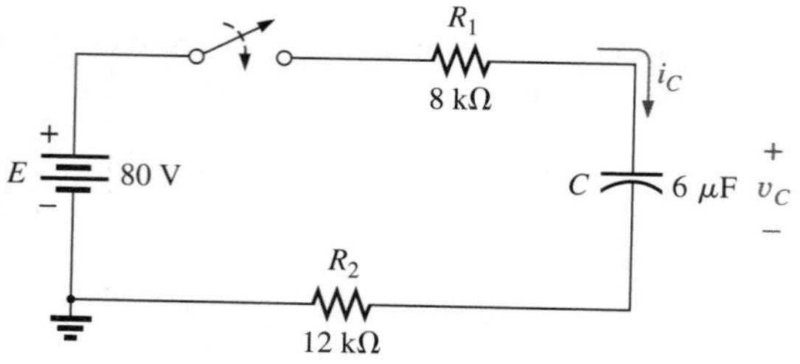

FIG. 10.100
Problem 39.

*40. For the network in Fig. 10.101:
   a. Calculate v_C, i_C, and v_{R_1} at 0.5 s and 1 s after the switch makes contact with position 1.
   b. The network sits in position 1 10 min before the switch is moved to position 2. How long after making contact with position 2 will it take for the current i_C to drop to 8 μA? How much *longer* will it take for v_C to drop to 10 V?

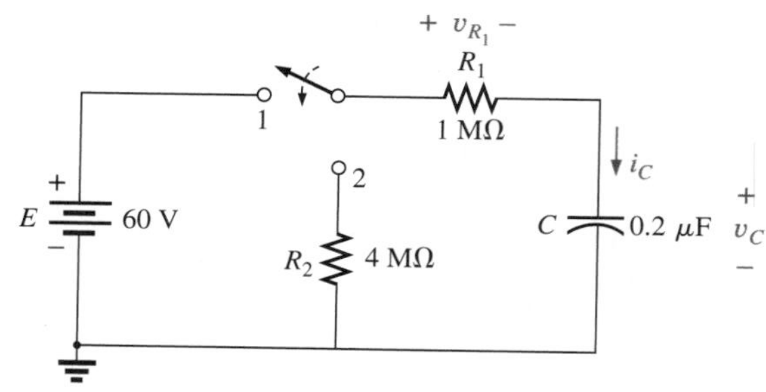

FIG. 10.101
Problem 40.

41. For the system in Fig. 10.102, using a DMM with a 10 MΩ internal resistance in the voltmeter mode:
   a. Determine the voltmeter reading one time constant after the switch is closed.
   b. Find the current i_C two time constants after the switch is closed.
   c. Calculate the time that must pass after the closing of the switch for the voltage v_C to be 50 V.

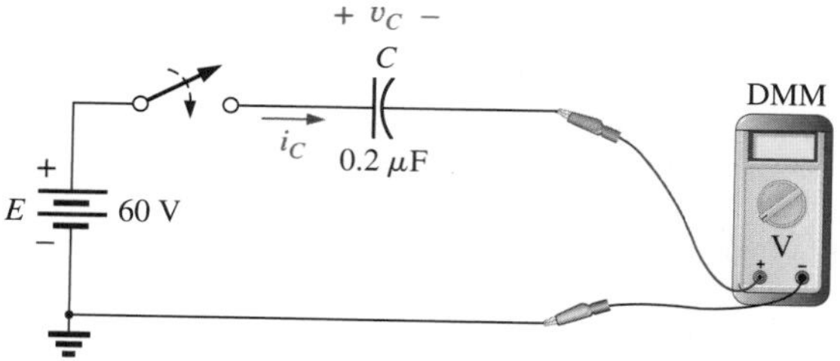

FIG. 10.102
Problem 41.

SECTION 10.9 Thévenin Equivalent: $\tau = R_{Th}C$

42. For the circuit in Fig. 10.103:

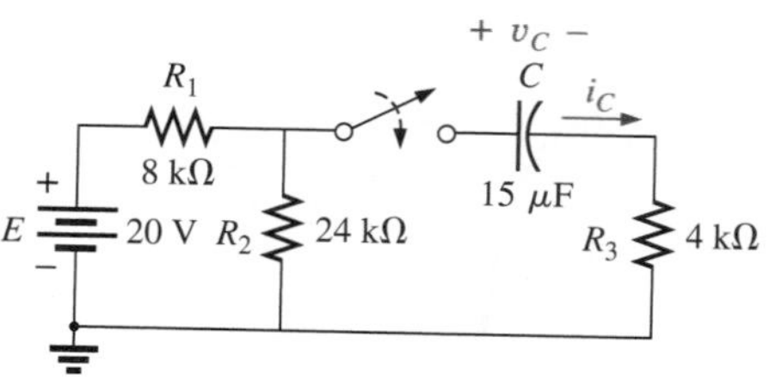

FIG. 10.103
Problem 42.

a. Find the mathematical expressions for the transient behavior of the voltage v_C and the current i_C following the closing of the switch.
b. Sketch the waveforms of v_C and i_C.

43. The capacitor in Fig. 10.104 is initially charged to 2 V with the polarity shown.
a. Write the mathematical expressions for the voltage v_C and the current i_C when the switch is closed.
b. Sketch the waveforms of v_C and i_C.

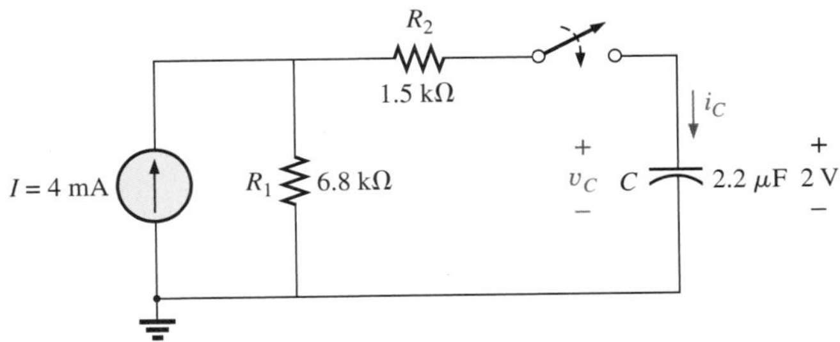

FIG. 10.104
Problem 43.

44. The capacitor in Fig. 10.105 is initially charged to 4 V with the polarity shown.
a. Write the mathematical expressions for the voltage v_C and the current i_C when the switch is closed.
b. Sketch the waveforms of v_C and i_C.

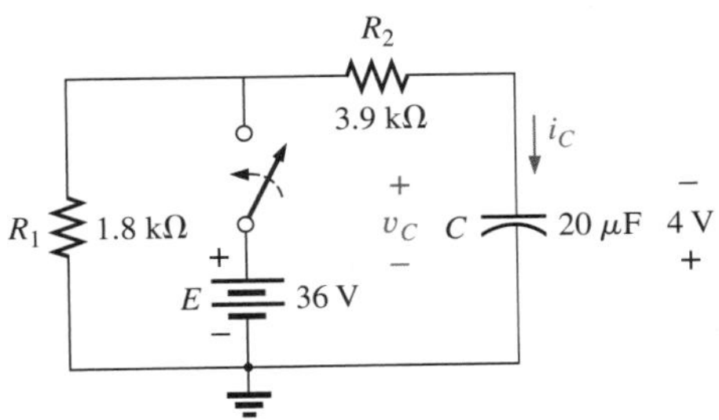

FIG. 10.105
Problem 44.

45. For the circuit in Fig. 10.106:
a. Find the mathematical expressions for the transient behavior of the voltage v_C and the current i_C following the closing of the switch.
b. Sketch the waveforms of v_C and i_C.

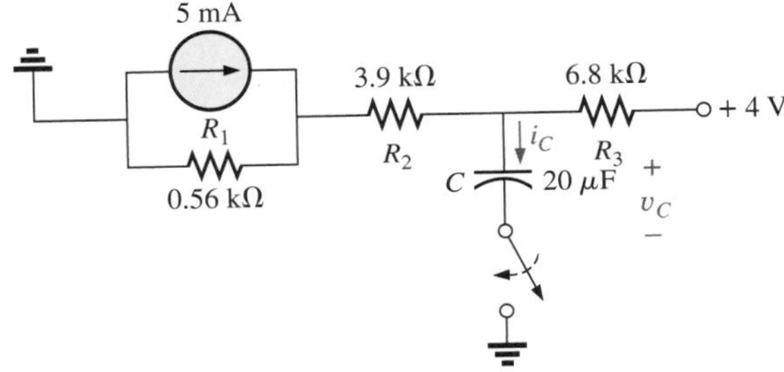

FIG. 10.106
Problem 45.

*46. The capacitor in Fig. 10.107 is initially charged to 3 V with the polarity shown.
a. Write the mathematical expressions for the voltage v_C and the current i_C when the switch is closed.
b. Sketch the waveforms of v_C and i_C.

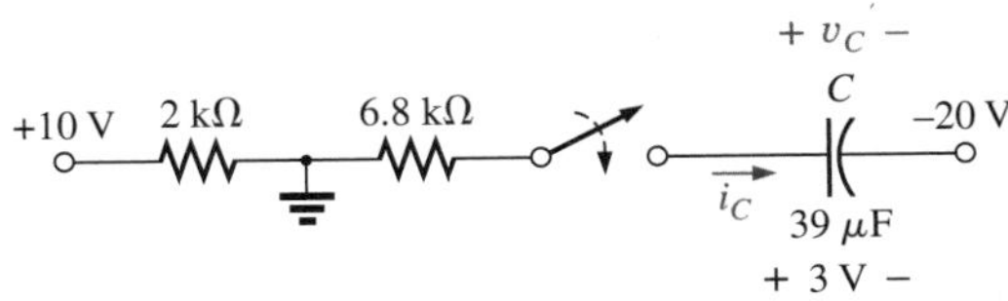

FIG. 10.107
Problem 46.

47. For the system in Fig. 10.108, using a DMM with a 10 MΩ internal resistance in the voltmeter mode:
a. Determine the voltmeter reading four time constants after the switch is closed.
b. Find the time that must pass before i_C drops to 3 μA.
c. Find the time that must pass after the closing of the switch for the voltage across the meter to reach 10 V.

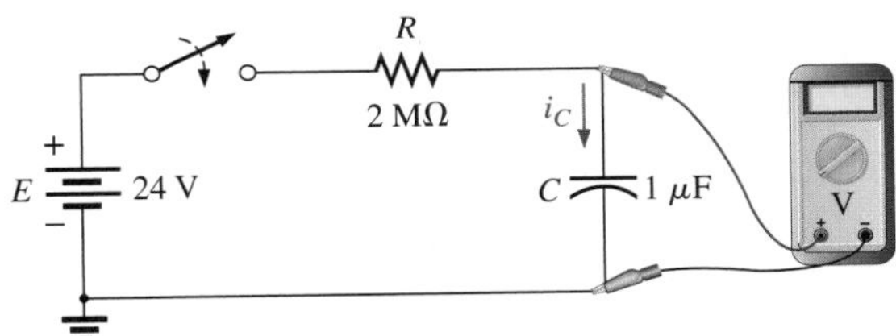

FIG. 10.108
Problem 47.

SECTION 10.10 The Current i_C

48. Find the waveform for the average current if the voltage across the 2 μF capacitor is as shown in Fig. 10.109.

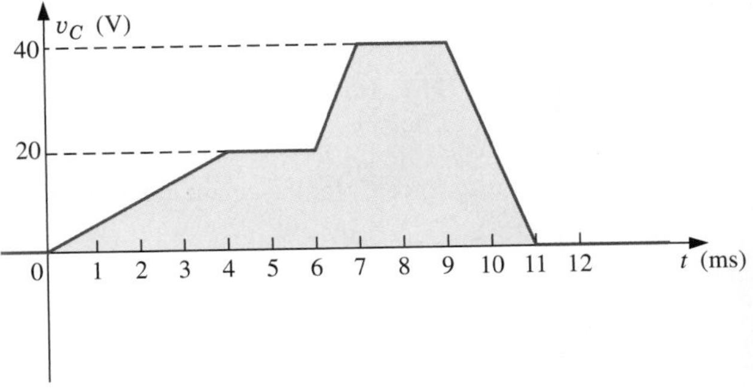

FIG. 10.109
Problem 48.

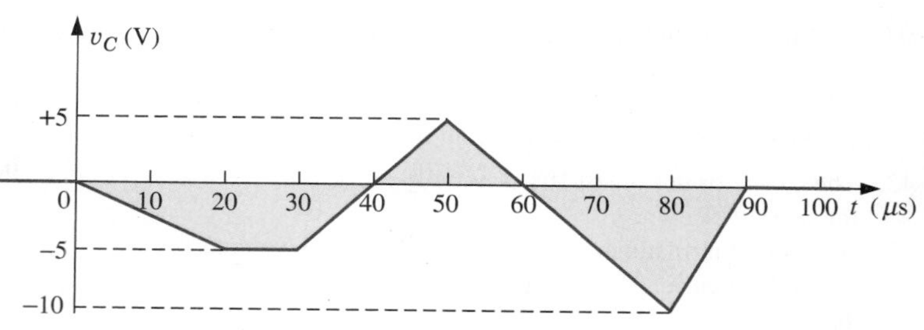

FIG. 10.110
Problem 49.

49. Find the waveform for the average current if the voltage across the 4.7 μF capacitor is as shown in Fig. 10.110.

50. Given the waveform in Fig. 10.111 for the current of a 20 μF capacitor, sketch the waveform of the voltage v_C across the capacitor if $v_C = 0$ V at $t = 0$ s.

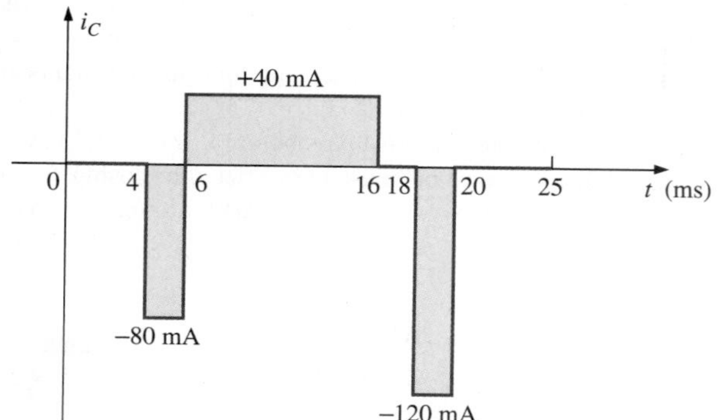

FIG. 10.111
Problem 50.

SECTION 10.11 Capacitors in Series and in Parallel

51. Find the total capacitance C_T for the circuit in Fig. 10.112.

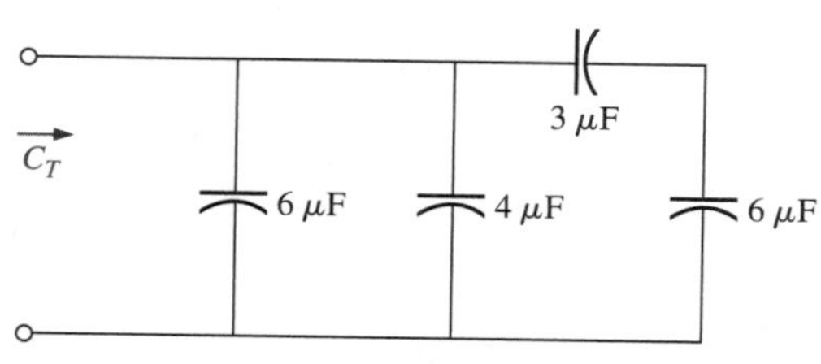

FIG. 10.112
Problem 51.

52. Find the total capacitance C_T for the circuit in Fig. 10.113.

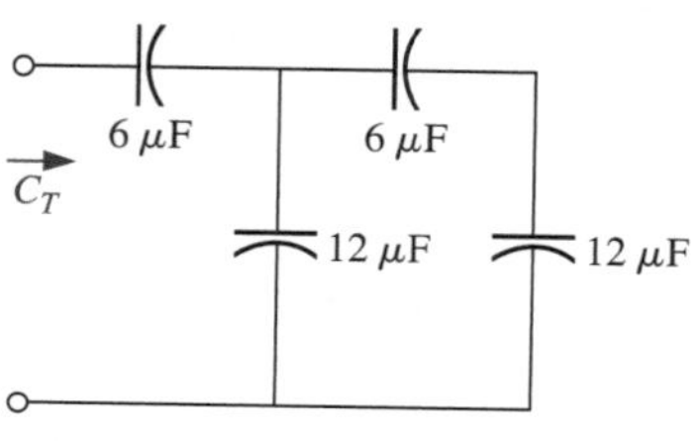

FIG. 10.113
Problem 52.

53. Find the voltage across and the charge on each capacitor for the circuit in Fig. 10.114.

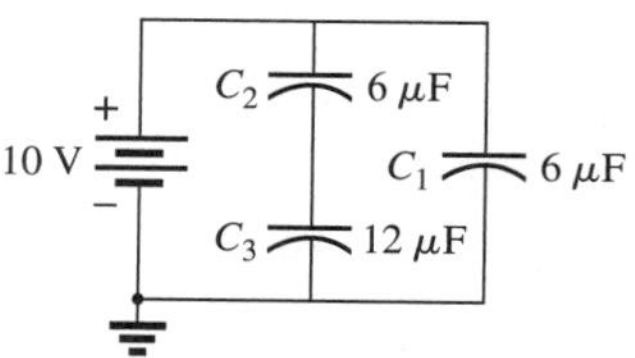

FIG. 10.114
Problem 53.

54. Find the voltage across and the charge on each capacitor for the circuit in Fig. 10.115.

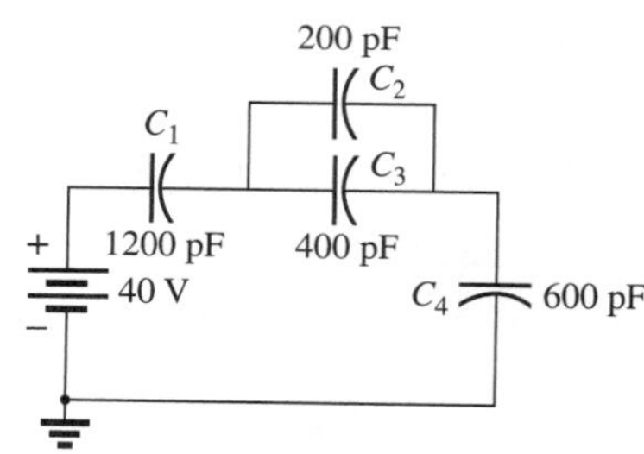

FIG. 10.115
Problem 54.

55. For the configuration in Fig. 10.116, determine the voltage across each capcitor and the charge on each capacitor.

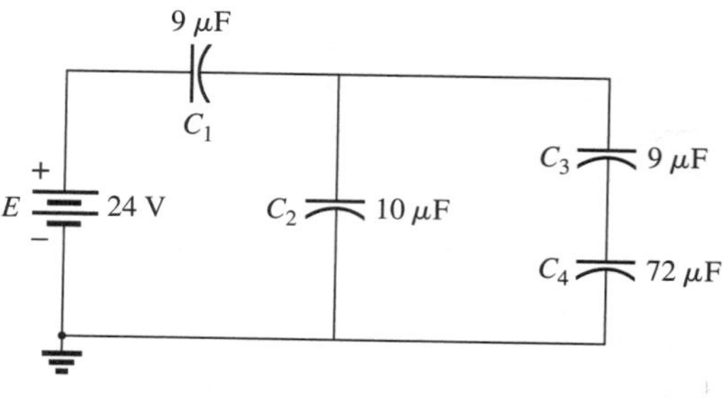

FIG. 10.116
Problem 55.

56. For the configuration in Fig. 10.117, determine the voltage across each capacitor and the charge on each capacitor.

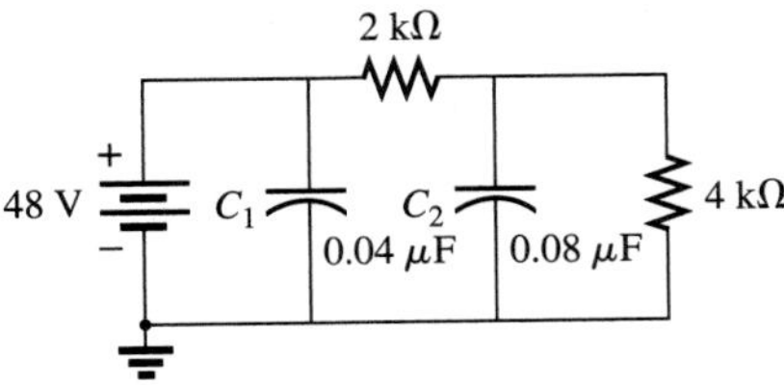

FIG. 10.117
Problem 56.

SECTION 10.12 Energy Stored by a Capacitor

57. Find the energy stored by a 120 pF capacitor with 12 V across its plates.

58. If the energy stored by a 6 μF capacitor is 1200 J, find the charge Q on each plate of the capacitor.

59. For the network in Fig. 10.118:
  a. Determine the energy stored by each capacitor under steady-state conditions.
  b. Repeat part (a) if the capacitors are in series.

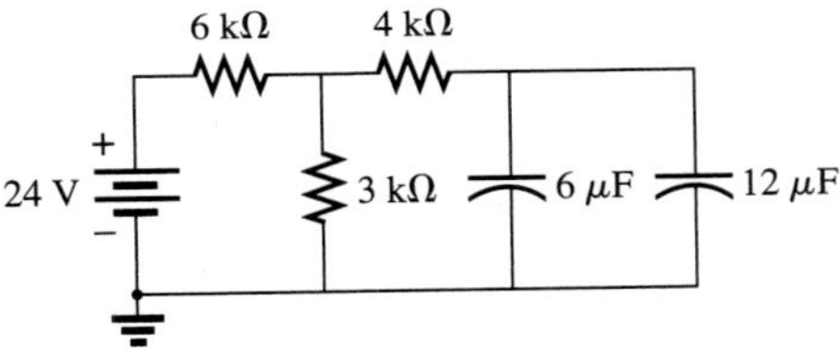

FIG. 10.118
Problem 59.

***60.** An electronic flashgun has a 1000 μF capacitor that is charged to 100 V.
  a. How much energy is stored by the capacitor?
  b. What is the charge on the capacitor?
  c. When the photographer takes a picture, the flash fires for 1/2000 s. What is the average current through the flashtube?
  d. Find the power delivered to the flashtube.
  e. After a picture is taken, the capacitor has to be recharged by a power supply that delivers a maximum current of 10 mA. How long will it take to charge the capacitor?

SECTION 10.15 Computer Analysis

61. Using PSpice or Multisim, verify the results in Example 10.6.

62. Using the initial condition operator, verify the results in Example 10.8 for the charging phase using PSpice or Multisim.

63. Using PSpice or Multisim, verify the results for v_C during the charging phase in Example 10.11.

64. Using PSpice or Multisim, verify the results in Problem 48.

GLOSSARY

Average current The current defined by a linear (straight line) change in voltage across a capacitor for a specific period of time.

Breakdown voltage Another term for *dielectric strength*, listed below.

Capacitance A measure of a capacitor's ability to store charge; measured in farads (F).

Capacitor A fundamental electrical element having two conducting surfaces separated by an insulating material and having the capacity to store charge on its plates.

Coulomb's law An equation relating the force between two like or unlike charges.

Derivative The instantaneous change in a quantity at a particular instant in time.

Dielectic The insulating material between the plates of a capacitor that can have a pronounced effect on the charge stored on the plates of a capacitor.

Dielectric constant Another term for *relative permittivity*, listed below.

Dielectric strength An indication of the voltage required for unit length to establish conduction in a dielectric.

Electric field strength The force acting on a unit positive charge in the region of interest.

Electric flux lines Lines drawn to indicate the strength and direction of an electric field in a particular region.

Fringing An effect established by flux lines that do not pass directly from one conducting surface to another.

Initial value The steady-state voltage across a capacitor before a transient period begins.

Leakage current The current that results in the total discharge of a capacitor if the capacitor is disconnected from the charging network for a sufficient length of time.

Maximum working voltage That voltage level at which a capacitor can perform its function without concern about breakdown or change in characteristics.

Permittivity A measure of how well a dielectric *permits* the establishment of flux lines within the dielectric.

Relative permittivity The permittivity of a material compared to that of air.

Steady-state region A period of time defined by the fact that the voltage across a capacitor has reached a level that, for all practical purposes, remains constant.

Stray capacitance Capacitances that exist not through design but simply because two conducting surfaces are relatively close to each other.

Temperature coefficient An indication of how much the capacitance value of a capacitor will change with change in temperature.

Time constant A period of time defined by the parameters of the network that defines how long the transient behavior of the voltage or current of a capacitor will last.

Transient period That period of time where the voltage across a capacitor or the current of a capacitor will change in value at a rate determined by the time constant of the network.

Inductors

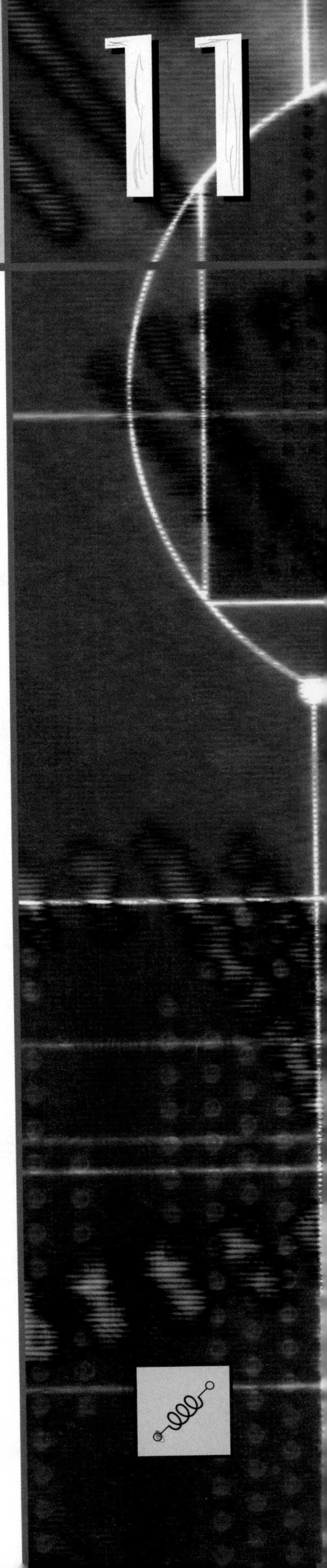

Objectives

- *Become familiar with the basic construction of an inductor, the factors that affect the strength of the magnetic field established by the element, and how to read the nameplate data.*
- *Be able to determine the transient (time-varying) response of an inductive network and plot the resulting voltages and currents.*
- *Understand the impact of combining inductors in series or parallel.*
- *Develop some familiarity with the use of PSpice or Multisim to analyze networks with inductive elements.*

11.1 INTRODUCTION

Three basic components appear in the majority of electrical/electronic systems in use today. They include the *resistor* and the *capacitor,* which have already been introduced, and the **inductor,** to be examined in detail in this chapter. In many ways, the inductor is the dual of the capacitor; that is, the voltage of one is applicable to the current of the other, and vice versa. In fact, some sections in this chapter parallel those in Chapter 10 on the capacitor. Like the capacitor, *the inductor exhibits its true characteristics only when a change in voltage or current is made in the network.*

Recall from Chapter 10 that a capacitor can be replaced by an open-circuit equivalent under steady-state conditions. You will see in this chapter that an inductor can be replaced by a short-circuit equivalent under steady-state conditions. Finally, you will learn that while resistors dissipate the power delivered to them in the form of heat, ideal capacitors store the energy delivered to them in the form of an electric field. Inductors, in the ideal sense, are like capacitors in that they also store the energy delivered to them—but in the form of a magnetic field.

11.2 MAGNETIC FIELD

Magnetism plays an integral part in almost every electrical device used today in industry, research, or the home. Generators, motors, transformers, circuit breakers, televisions, computers, tape recorders, and telephones all employ magnetic effects to perform a variety of important tasks.

The compass, used by Chinese sailors as early as the second century A.D., relies on a **permanent magnet** for indicating direction. A permanent magnet is made of a material, such as steel or iron, that remains magnetized for long periods of time without the need for an external source of energy.

In 1820, the Danish physicist Hans Christian Oersted discovered that the needle of a compass deflects if brought near a current-carrying conductor. This was the first demonstration that electricity and magnetism were related. In the same year, the French physicist André-Marie Ampère performed experiments in this area and developed what is presently known as **Ampère's circuital law.** In subsequent years, others such as Michael Faraday, Karl Friedrich

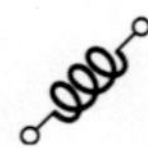

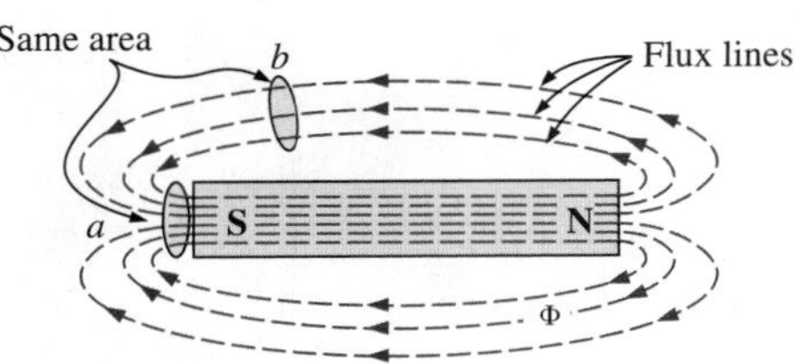

FIG. 11.1
Flux distribution for a permanent magnet.

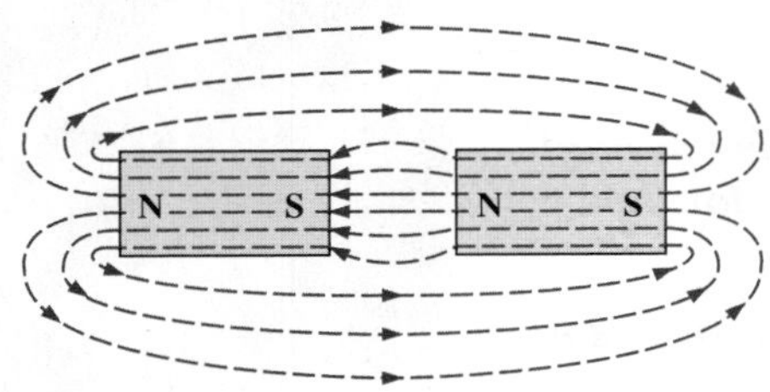

FIG. 11.2
Flux distribution for two adjacent, opposite poles.

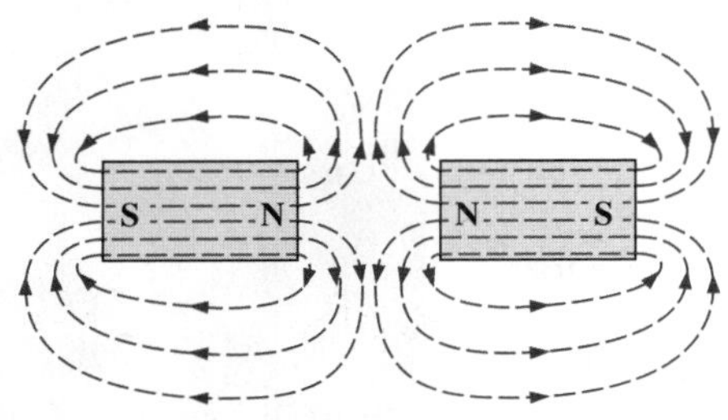

FIG. 11.3
Flux distribution for two adjacent, like poles.

Gauss, and James Clerk Maxwell continued to experiment in this area and developed many of the basic concepts of **electromagnetism**—magnetic effects induced by the flow of charge, or current.

A magnetic field exists in the region surrounding a permanent magnet, which can be represented by **magnetic flux lines** similar to electric flux lines. Magnetic flux lines, however, do not have origins or terminating points as do electric flux lines but exist in *continuous loops,* as shown in Fig. 11.1.

The magnetic flux lines radiate from the north pole to the south pole, returning to the north pole through the metallic bar. Note the equal spacing between the flux lines within the core and the symmetric distribution outside the magnetic material. These are additional properties of magnetic flux lines in homogeneous materials (that is, materials having uniform structure or composition throughout). It is also important to realize that the continuous magnetic flux line will strive to occupy as small an area as possible. This results in magnetic flux lines of minimum length between the unlike poles, as shown in Fig. 11.2. The strength of a magnetic field in a particular region is directly related to the density of flux lines in that region. In Fig. 11.1, for example, the magnetic field strength at point *a* is twice that at point *b* since twice as many magnetic flux lines are associated with the perpendicular plane at point *a* than at point *b.* Recall from childhood experiments that the strength of permanent magnets is always stronger near the poles.

If unlike poles of two permanent magnets are brought together, the magnets attract, and the flux distribution is as shown in Fig. 11.2. If like poles are brought together, the magnets repel, and the flux distribution is as shown in Fig. 11.3.

If a nonmagnetic material, such as glass or copper, is placed in the flux paths surrounding a permanent magnet, an almost unnoticeable change occurs in the flux distribution (Fig. 11.4). However, if a magnetic material, such as soft iron, is placed in the flux path, the flux lines pass through the soft iron rather than the surrounding air because flux lines pass with greater ease through magnetic materials than through air. This principle is used in shielding sensitive electrical elements and instruments that can be affected by stray magnetic fields (Fig. 11.5).

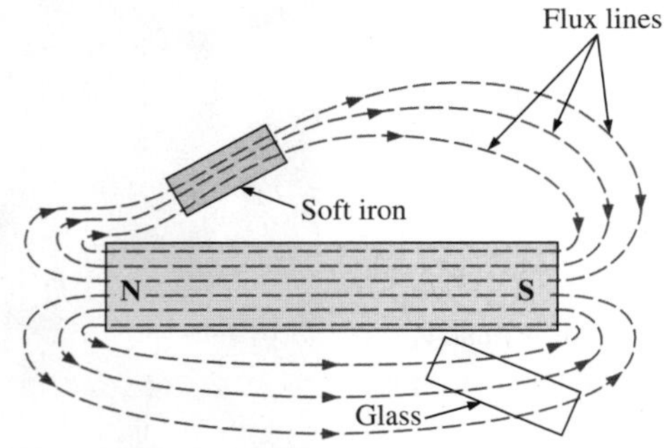

FIG. 11.4
Effect of a ferromagnetic sample on the flux distribution of a permanent magnet.

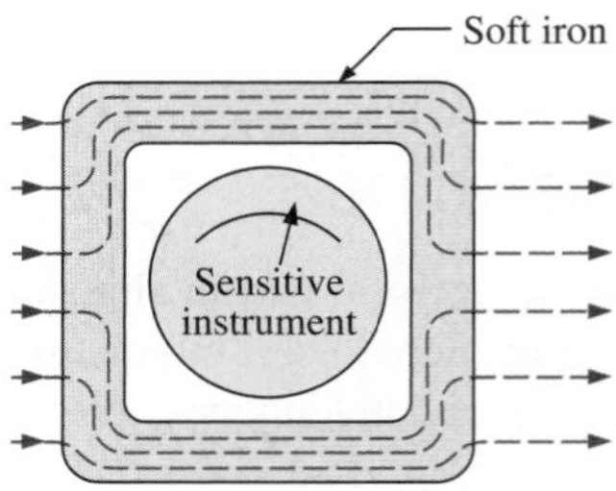

FIG. 11.5
Effect of a magnetic shield on the flux distribution.

A magnetic field (represented by concentric magnetic flux lines, as in Fig. 11.6) is present around every wire that carries an electric current. The direction of the magnetic flux lines can be found simply by placing the thumb of the *right* hand in the direction of *conventional* current flow and noting the direction of the fingers. (This method is commonly called the *right-hand rule.*) If the conductor is wound in a

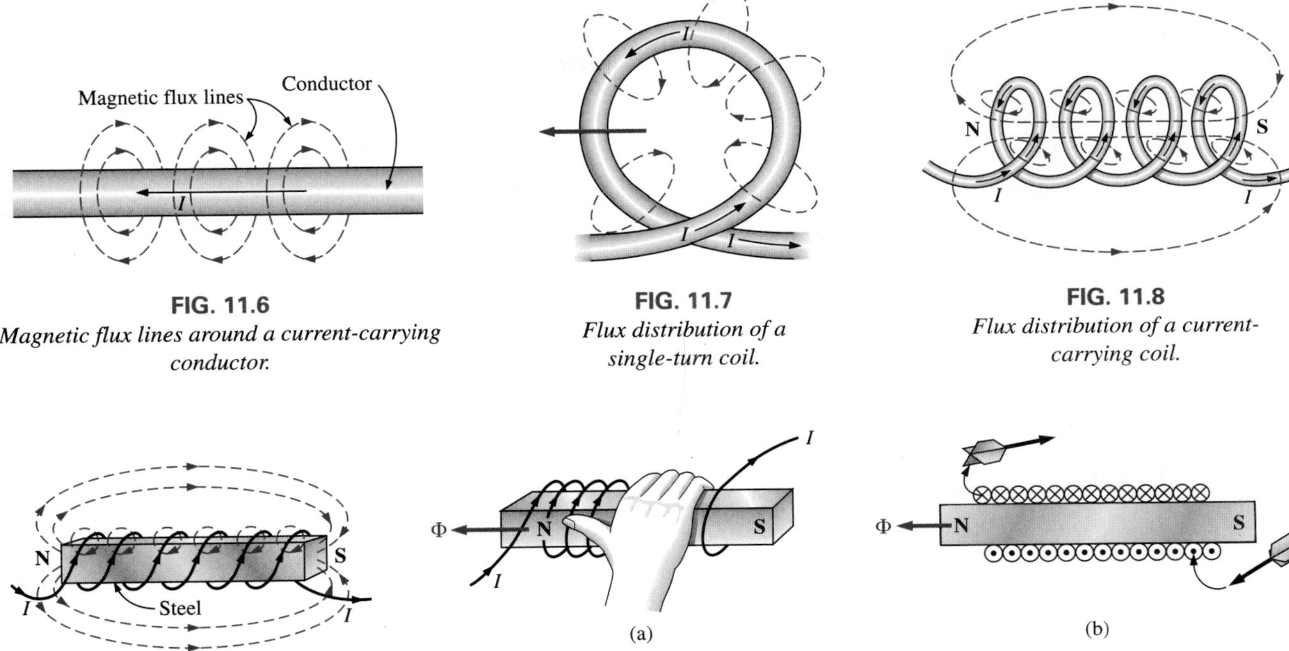

FIG. 11.6
Magnetic flux lines around a current-carrying conductor.

FIG. 11.7
Flux distribution of a single-turn coil.

FIG. 11.8
Flux distribution of a current-carrying coil.

FIG. 11.9
Electromagnet.

FIG. 11.10
Determining the direction of flux for an electromagnet: (a) method; (b) notation.

single-turn coil (Fig. 11.7), the resulting flux flows in a common direction through the center of the coil. A coil of more than one turn produces a magnetic field that exists in a continuous path through and around the coil (Fig. 11.8).

The flux distribution of the coil is quite similar to that of the permanent magnet. The flux lines leaving the coil from the left and entering to the right simulate a north and a south pole, respectively. The principal difference between the two flux distributions is that the flux lines are more concentrated for the permanent magnet than for the coil. Also, since *the strength of a magnetic field is determined by the density of the flux lines,* the coil has a weaker field strength. The field strength of the coil can be effectively increased by placing certain materials, such as iron, steel, or cobalt, within the coil to increase the flux density within the coil. By increasing the field strength with the addition of the core, we have devised an *electromagnet* (Fig. 11.9) that not only has all the properties of a permanent magnet but also has a field strength that can be varied by changing one of the component values (current, turns, and so on). Of course, current must pass through the coil of the electromagnet for magnetic flux to be developed, whereas there is no need for the coil or current in the permanent magnet. The direction of flux lines can be determined for the electromagnet (or in any core with a wrapping of turns) by placing the fingers of your right hand in the direction of current flow around the core. Your thumb then points in the direction of the north pole of the induced magnetic flux, as demonstrated in Fig. 11.10(a). A cross section of the same electromagnet is in Fig. 11.10(b) to introduce the convention for directions perpendicular to the page. The cross and the dot refer to the tail and the head of the arrow, respectively.

In the SI system of units, magnetic flux is measured in **webers (Wb)** as derived from the surname of Wilhelm Eduard Weber (Fig. 11.11). The

FIG. 11.11
Wilhelm Eduard Weber.
Courtesy of the Smithsonian Institution, Photo No. 52,604.

German (Wittenberg, Göttingen)
(1804–91)
Physicist
Professor of Physics, University of Göttingen

An important contributor to the establishment of a system of *absolute units* for the electrical sciences, which was beginning to become a very active area of research and development. Established a definition of electric current in an electromagnetic system based on the magnetic field produced by the current. He was politically active and, in fact, was dismissed from the faculty of the University of Göttingen for protesting the suppression of the constitution by the King of Hanover in 1837. However, he found other faculty positions and eventually returned to Göttingen as director of the astronomical observatory. He received honors from England, France, and Germany, including the Copley Medal of the Royal Society of London.

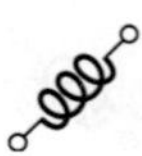

FIG. 11.12
Nikola Tesla.
Courtesy of the Smithsonian Institution, Photo No. 52,223.

Croatian-American (Smiljan, Paris, Colorado Springs, New York City)
(1856–1943)
Electrical Engineer and Inventor Recipient of the Edison Medal in 1917

Often regarded as one of the most innovative and inventive individuals in the history of the sciences. He was the first to introduce the *alternating-current machine,* removing the need for commutator bars of dc machines. After emigrating to the United States in 1884, he sold a number of his patents on *ac machines, transformers,* and *induction coils* (including the Tesla coil as we know it today) to the Westinghouse Electric Company. Some say that his most important discovery was made at his laboratory in Colorado Springs, where in 1900 he discovered *terrestrial stationary waves.* The range of his discoveries and inventions is too extensive to list here but extends from lighting systems to *polyphase power systems* to a *wireless world broadcasting system.*

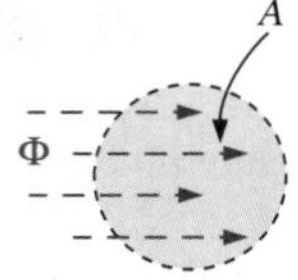

FIG. 11.13
Defining the flux density B.

applied symbol is the capital Greek letter *phi,* Φ. The number of flux lines per unit area, called the **flux density,** is denoted by the capital letter B and is measured in **teslas (T)** to honor the efforts of Nikola Tesla, a scientist of the late 1800s (Fig. 11.12).

In equation form:

$$\boxed{B = \frac{\Phi}{A}} \qquad \begin{aligned} B &= \text{Wb/m}^2 = \text{teslas (T)} \\ \Phi &= \text{webers (Wb)} \\ A &= \text{m}^2 \end{aligned} \tag{11.1}$$

where Φ is the number of flux lines passing through area A in Fig. 11.13. The flux density at point a in Fig. 11.1 is twice that at point b because twice as many flux lines pass through the same area.

In Eq. (11.1), the equivalence is given by

$$\boxed{1 \text{ tesla} = 1 \text{ T} = 1 \text{ Wb/m}^2} \tag{11.2}$$

which states in words that if 1 weber of magnetic flux passes through an area of 1 square meter, the flux density is 1 tesla.

For the CGS system, magnetic flux is measured in maxwells and the flux density in gauss. For the English system, magnetic flux is measured in lines and the flux density in lines per square inch. The relationship between such systems is defined in Appendix F.

The flux density of an electromagnet is directly related to the number of turns of, and current through, the coil. The product of the two, called the **magnetomotive force,** is measured in **ampere-turns (At)** as defined by

$$\boxed{\mathscr{F} = NI} \qquad \text{(ampere-turns, At)} \tag{11.3}$$

In other words, if you increase the number of turns around a core and/or increase the current through the coil, the magnetic field strength also increases. In many ways, the magnetomotive force for magnetic circuits is similar to the applied voltage in an electric circuit. Increasing either one results in an increase in the desired effect: magnetic flux for magnetic circuits and current for electric circuits.

For the CGS system, the magnetomotive force is measured in gilberts, while for the English system, it is measured in ampere-turns.

Another factor that affects the magnetic field strength is the type of core used. Materials in which magnetic flux lines can readily be set up are said to be **magnetic** and to have a high **permeability.** Again, note the similarity with the word "permit" used to describe permittivity for the dielectrics of capacitors. Similarly, the permeability (represented by the Greek letter *mu,* μ) of a material is a measure of the ease with which magnetic flux lines can be established in the material.

Just as there is a specific value for the permittivity of air, there is a specific number associated with the permeability of air:

$$\boxed{\mu_o = 4\pi \times 10^{-7} \text{ Wb/A}\cdot\text{m}} \tag{11.4}$$

Practically speaking, the permeability of all nonmagnetic materials, such as copper, aluminum, wood, glass, and air, is the same as that for free space. Materials that have permeabilities slightly less than that of free space are said to be **diamagnetic,** and those with permeabilities slightly greater than that of free space are said to be **paramagnetic.** Magnetic materials,

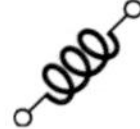

such as iron, nickel, steel, cobalt, and alloys of these metals, have permeabilities hundreds and even thousands of times that of free space. Materials with these very high permeabilities are referred to as **ferromagnetic.**

The ratio of the permeability of a material to that of free space is called its **relative permeability;** that is,

$$\boxed{\mu_r = \frac{\mu}{\mu_o}} \tag{11.5}$$

In general, for ferromagnetic materials, $\mu_r \geq 100$, and for nonmagnetic materials, $\mu_r = 1$.

A table of values for μ to match the provided table for permittivity levels of specific dielectrics would be helpful. Unfortunately, such a table cannot be provided because *relative permeability is a function of the operating conditions.* If you change the magnetomotive force applied, the level of μ can vary between extreme limits. At one level of magnetomotive force, the permeability of a material can be 10 times that at another level.

An instrument designed to measure flux density in gauss (CGS system) appears in Fig. 11.14. Appendix F reveals that 1 T = 10^4 gauss. The magnitude of the reading appearing on the face of the meter in Fig. 11.14 is therefore

$$20.2159 \cancel{\text{gauss}}\left(\frac{1\text{ T}}{10^4 \cancel{\text{gauss}}}\right) = 2.02 \times 10^{-3}\text{ T}$$

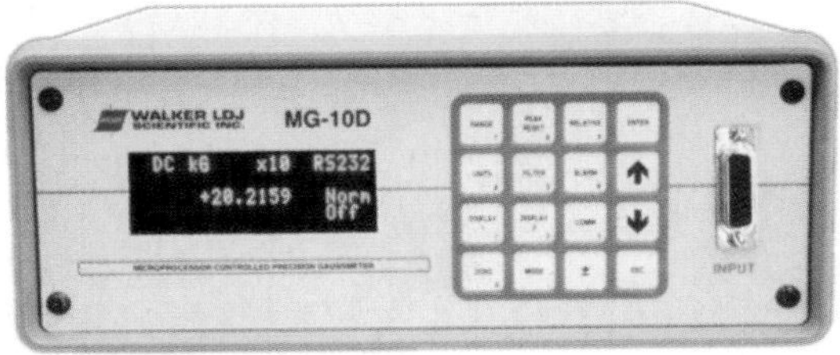

FIG. 11.14
Digital display gaussmeter.
(Courtesy of Walker LDJ Scientific Inc.
www.walkerldjscientific.com)

Although our emphasis in this chapter is to introduce the parameters that affect the nameplate data of an inductor, the use of magnetics has widespread application in the electrical/electronics industry, as shown by a few areas of application in Fig. 11.15.

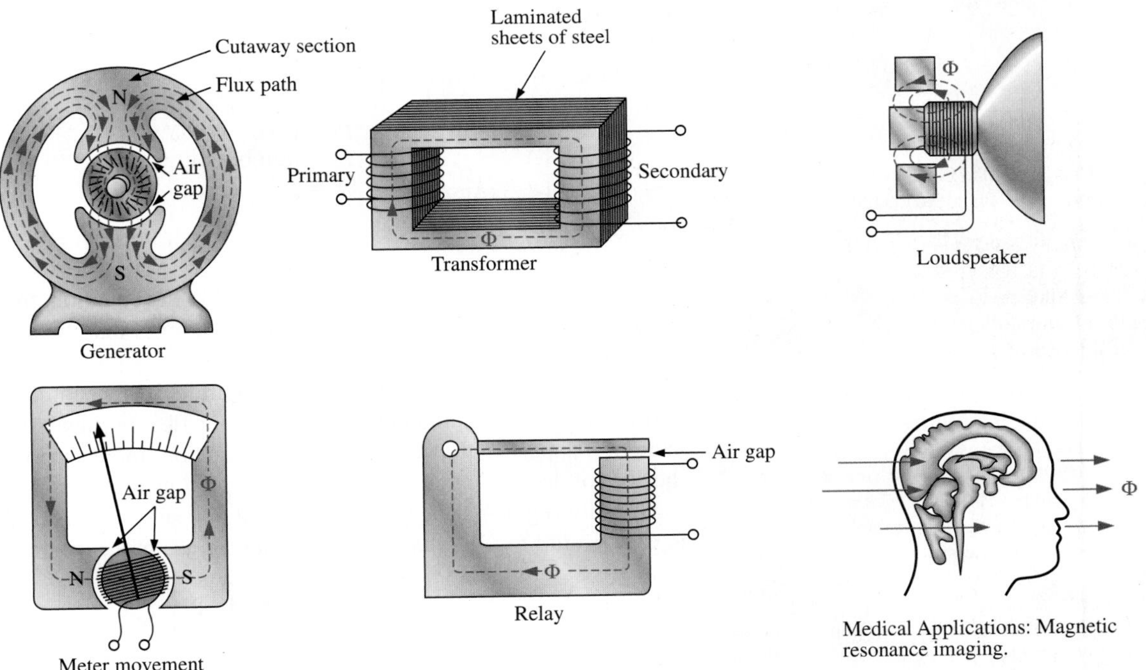

FIG. 11.15
Some areas of application of magnetic effects.

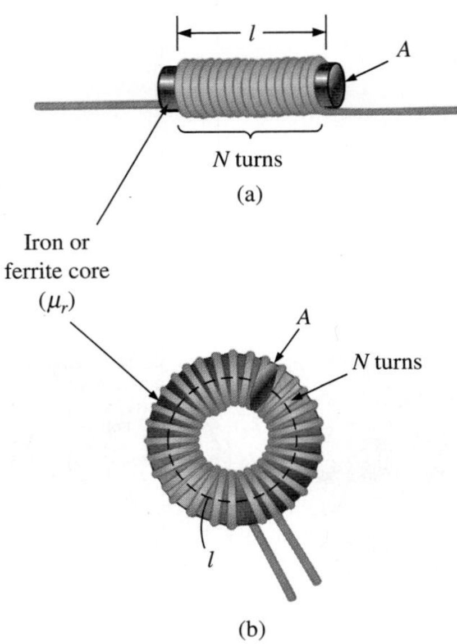

FIG. 11.16
Defining the parameters for Eq. (11.6).

FIG. 11.17
Joseph Henry.
Courtesy of the Smithsonian Institutions, Photo No. 59,054.

American (Albany, NY; Princeton, NJ)
(1797–1878)
Physicist and Mathematician
Professor of Natural Philosophy,
Princeton University

In the early 1800s the title Professor of Natural Philosophy was applied to educators in the sciences. As a student and teacher at the Albany Academy, Henry performed extensive research in the area of electromagnetism. He improved the design of *electromagnets* by insulating the coil wire to permit a tighter wrap on the core. One of his earlier designs was capable of lifting 3600 pounds. In 1832 he discovered and delivered a paper on *self-induction.* This was followed by the construction of an effective *electric telegraph transmitter and receiver* and extensive research on the oscillatory nature of lightning and discharges from a *Leyden jar.* In 1845 he was appointed the first Secretary of the Smithsonian.

11.3 INDUCTANCE

In the previous section, we learned that sending a current through a coil of wire, with or without a core, establishes a magnetic field through and surrounding the unit. This component, of rather simple construction (see Fig. 11.16), is called an **inductor** (often referred to as a **coil**). Its **inductance** level determines the strength of the magnetic field around the coil due to an applied current. The higher the inductance level, the greater the strength of the magnetic field. In total, therefore,

inductors are designed to set up a strong magnetic field linking the unit, whereas capacitors are designed to set up a strong electric field between the plates.

Inductance is measured in **henries (H),** after the American physicist Joseph Henry (Fig. 11.17). However, just as the farad is too large a unit for most applications, most inductors are of the millihenry (mH) or microhenry (μH) range.

In Chapter 10, 1 farad was defined as a capacitance level that would result in 1 coulomb of charge on the plates due to the application of 1 volt across the plates. For inductors,

1 henry is the inductance level that will establish a voltage of 1 volt across the coil due to a change in current of 1 A/s through the coil.

Inductor Construction

In Chapter 10, we found that capacitance is sensitive to the area of the plates, the distance between the plates, and the dielectic employed. The level of inductance has similar construction sensitivities in that it is dependent on the area within the coil, the length of the unit, and the permeability of the core material. It is also sensitive to the number of turns of wire in the coil as dictated by Eq. (11.6) and defined in Fig. 11.16 for two of the most popular shapes:

$$L = \frac{\mu N^2 A}{l} \qquad \begin{aligned} \mu &= \text{permeability (Wb/A}\cdot\text{m)} \\ N &= \text{number of turns (t)} \\ A &= \text{m}^2 \\ l &= \text{m} \\ L &= \text{henries (H)} \end{aligned} \tag{11.6}$$

First note that since the turns are squared in the equation, the number of turns is a big factor. However, also keep in mind that the more turns, the bigger the unit. If the wire is made too thin to get more windings on the core, the rated current of the inductor is limited. Since higher levels of permeability result in higher levels of magnetic flux, permeability should, and does, appear in the numerator of the equation. Increasing the area of the core or decreasing the length also increases the inductance level.

Substituting $\mu = \mu_r \mu_o$ for the permeability results in Eq. (11.7), which is very similar to the equation for the capacitance of a capacitor:

$$L = \frac{\mu_r \mu_o N^2 A}{l}$$

or

$$L = 4\pi \times 10^{-7} \frac{\mu_r N^2 A}{l} \qquad \text{(henries, H)} \tag{11.7}$$

If we break out the relative permeability as follows:

$$L = \mu_r\left(\frac{\mu_o N^2 A}{l}\right)$$

we obtain the following useful equation:

$$\boxed{L = \mu_r L_o} \qquad \textbf{(11.8)}$$

which is very similar to the equation $C = \epsilon_r C_o$. Eq. (11.8) states the following:

The inductance of an inductor with a ferromagnetic core is μ_r times the inductance obtained with an air core.

Although Eq. (11.6) is approximate at best, the equations for the inductance of a wide variety of coils can be found in reference handbooks. Most of the equations are mathematically more complex than Eq. (11.6), but the impact of each factor is the same in each equation.

EXAMPLE 11.1 For the air-core coil in Fig. 11.18:

a. Find the inductance.
b. Find the inductance if a metallic core with $\mu_r = 2000$ is inserted in the coil.

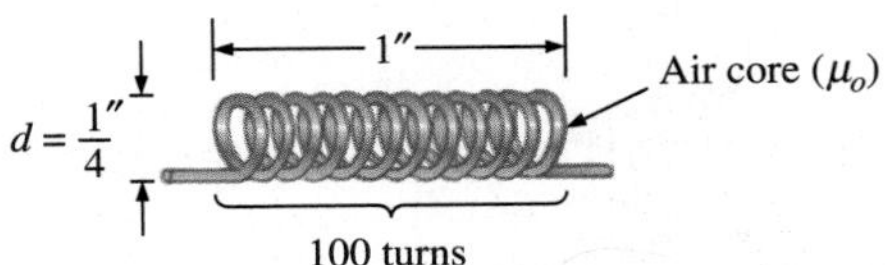

FIG. 11.18
Air-core coil for Example 11.1.

Solutions:

a. $d = \frac{1}{4}\text{in.}\left(\frac{1\text{ m}}{39.37\text{ in.}}\right) = 6.35\text{ mm}$

$$A = \frac{\pi d^2}{4} = \frac{\pi(6.35\text{ mm})^2}{4} = 31.7\ \mu\text{m}^2$$

$$l = 1\text{ in.}\left(\frac{1\text{ m}}{39.37\text{ in.}}\right) = 25.4\text{ mm}$$

$$L = 4\pi \times 10^{-7}\frac{\mu_r N^2 A}{l}$$

$$= 4\pi \times 10^{-7}\frac{(1)(100\text{ t})^2(31.7\ \mu\text{m}^2)}{25.4\text{ mm}} = \mathbf{15.68\ \mu H}$$

b. Eq. (11.8): $\quad L = \mu_r L_o = (2000)(15.68\ \mu\text{H}) = \mathbf{31.36\ mH}$

EXAMPLE 11.2 In Fig. 11.19, if each inductor in the left column is changed to the type appearing in the right column, find the new inductance level. For each change, assume that the other factors remain the same.

Solutions:

a. The only change was the number of turns, but it is a squared factor, resulting in

$$L = (2)^2 L_o = (4)(20\ \mu\text{H}) = \mathbf{80\ \mu H}$$

b. In this case, the area is three times the original size, and the number of turns is 1/2. Since the area is in the numerator, it increases

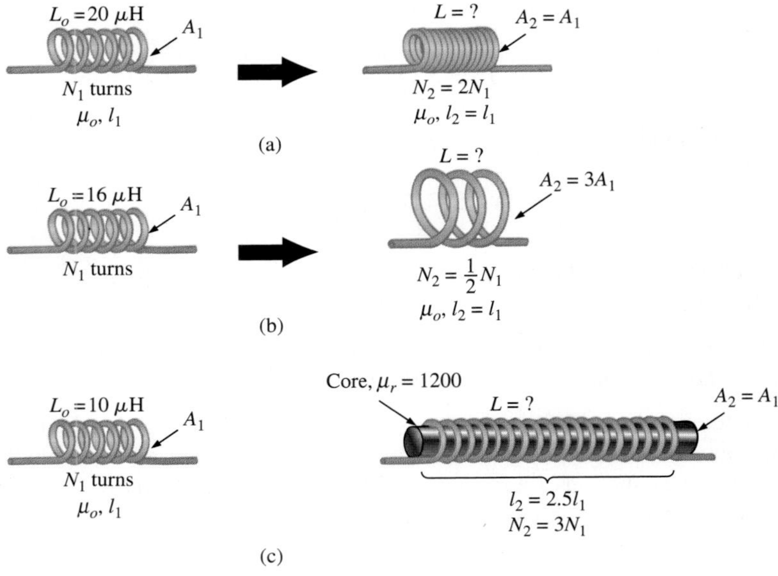

FIG. 11.19
Inductors for Example 11.2.

the inductance by a factor of three. The drop in the number of turns reduces the inductance by a factor of $(1/2)^2 = 1/4$. Therefore,

$$L = (3)\left(\frac{1}{4}\right)L_o = \frac{3}{4}(16\ \mu H) = \mathbf{12\ \mu H}$$

c. Both μ and the number of turns have increased, although the increase in the number of turns is squared. The increased length reduces the inductance. Therefore,

$$L = \frac{(3)^2(1200)}{2.5}L_o = (4.32 \times 10^3)(10\ \mu H) = \mathbf{43.2\ mH}$$

Types of Inductors

Inductors, like capacitors and resistors, can be categorized under the general headings **fixed** or **variable.** The symbol for a fixed air-core inductor is provided in Fig. 11.20(a), for an inductor with a ferromagnetic core in Fig. 11.20(b), for a tapped coil in Fig. 11.20(c), and for a variable inductor in Fig. 11.20(d).

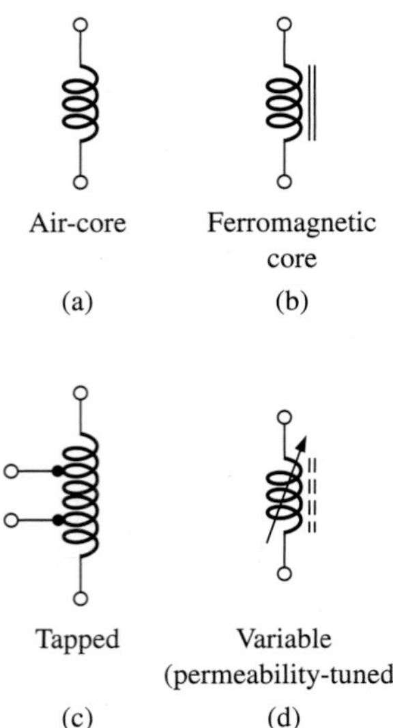

FIG. 11.20
Inductor (coil) symbols.

Fixed Fixed-type inductors come in all shapes and sizes. However,

in general, the size of an inductor is determined primarily by the type of construction, the core used, or the current rating.

In Fig. 11.21(a), the 10 μH and 1 mH coils are about the same size because a thinner wire was used for the 1 mH coil to permit more turns in the same space. The result, however, is a drop in rated current from 10 A to only 1.3 A. If the wire of the 10 μH coil had been used to make the 1 mH coil, the resulting coil would have been many times the size of the

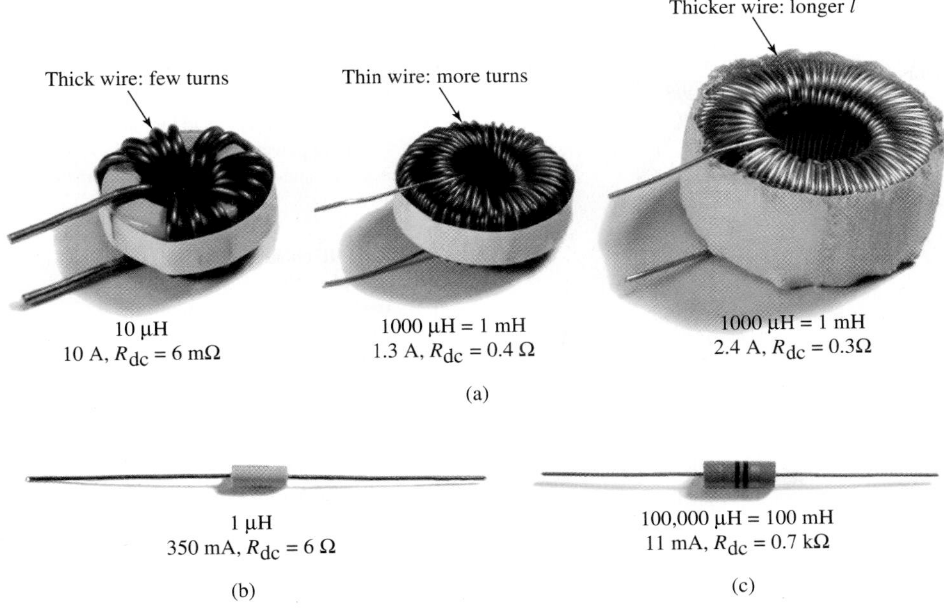

FIG. 11.21
Relative sizes of different types of inductors: (a) toroid, high-current; (b) phenolic (resin or plastic core); (c) ferrite core.

10 μH coil. The impact of the wire thickness is clearly revealed by the 1 mH coil at the far right in Fig. 11.21(a), where a thicker wire was used to raise the rated current level from 1.3 A to 2.4 A. Even though the inductance level is the same, the size of the toroid is 4 or 5 times greater.

The phenolic inductor (using a nonferromagnetic core of resin or plastic) in Fig. 11.21(b) is quite small for its level of inductance. We must assume that it has a high number of turns of very thin wire. Note, however, that the use of a very thin wire has resulted in a relatively low current rating of only 350 mA (0.35 A). The use of a ferrite (ferromagnetic) core in the inductor in Fig. 11.21(c) has resulted in an amazingly high level of inductance for its size. However, the wire is so thin that the current rating is only 11 mA $=$ 0.011 A. Note that for all the inductors, the dc resistance of the inductor increases with a decrease in the thickness of the wire. The 10 μH toroid has a dc resistance of only 6 mΩ, whereas the dc resistance of the 100 mH ferrite inductor is 700 Ω—a price to be paid for the smaller size and high inductance level.

Different types of fixed inductive elements are displayed in Fig. 11.22 on the next page, including their typical range of values and common areas of application. Based on the earlier discussion of inductor construction, it is fairly easy to identify an inductive element. The shape of a molded film resistor is similar to that of an inductor. However, careful examination of the typical shapes of each reveals some differences, such as the ridges at each end of a resistor that do not appear on most inductors.

Variable A number of variable inductors are depicted in Fig. 11.23 on the next page. In each case, the inductance is changed by turning the slot at the end of the core to move it in and out of the unit. The farther in the core is, the more the ferromagnetic material is part of the magnetic circuit, and the higher the magnetic field strength and the inductance level.

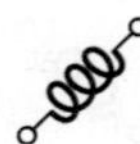

Type: Air-core inductors (1–32 turns)
Typical values: 2.5 nH–1 μH
Applications: High-frequency applications

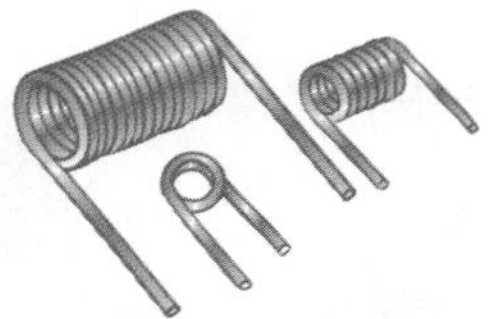

Type: Toroid coil
Typical values: 10μH–30 mH
Applications: Used as a choke in ac power line circuits to filter transient and reduce EMI interference. This coil is found in many electronic appliances.

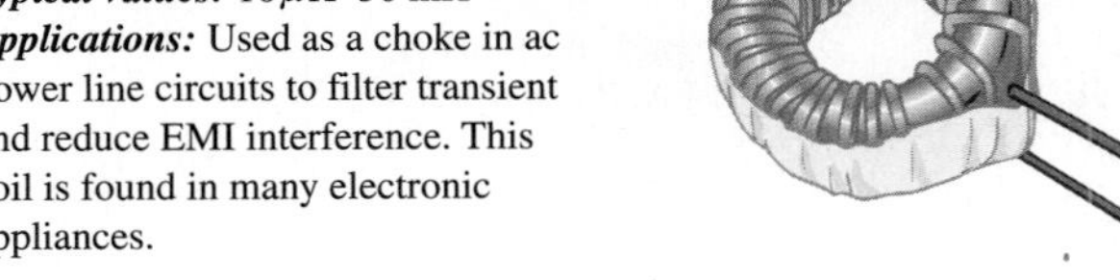

Type: Hash choke coil
Typical values: 3μH–1 mH
Applications: Used in ac supply lines that deliver high currents.

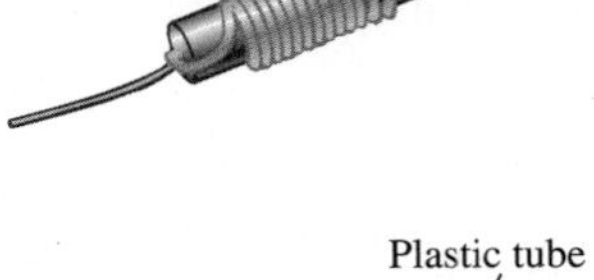

Type: Delay line coil
Typical values: 10μH–50μH
Applications: Used in color televisions to correct for timing differences between the color signal and the black-and-white signal.

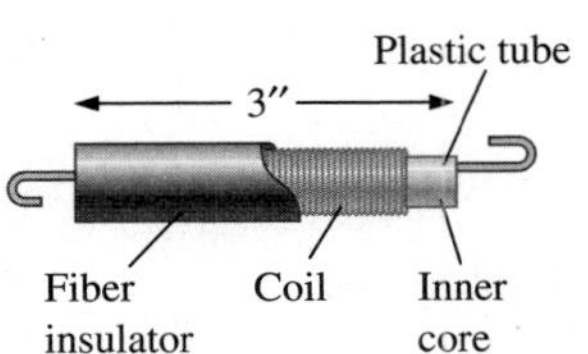

Type: Common-mode choke coil
Typical values: 0.6 mH–50 mH
Applications: Used in ac line filters, switching power supplies, battery chargers, and other electronic equipment.

Type: RF chokes
Typical values: 10μH–470 mH
Applications: Used in radio, television, and communication circuits. Found in AM, FM, and UHF circuits.

Type: Molded coils
Typical values: 0.1 μH–100 mH
Applications: Used in a wide variety of circuits such as oscillators, filters, pass-band filters, and others.

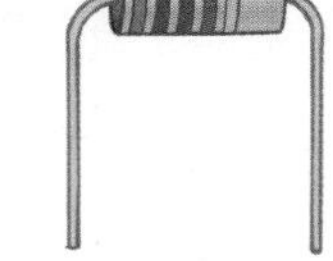

Type: Surface-mount inductors
Typical values: 0.01 μH–250 μH
Applications: Found in many electronic circuits that require miniature components on multilayered PCBs (printed circuit boards).

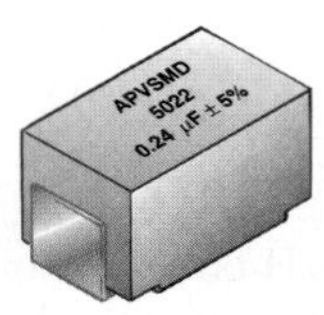

FIG. 11.22
Typical areas of application for inductive elements.

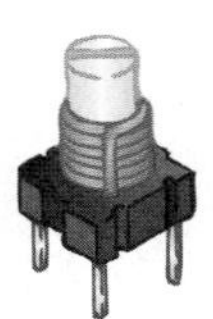

FIG. 11.23
Variable inductors with a typical range of values from 1 μH to 100 μH; commonly used in oscillators and various RF circuits such as CB transceivers, televisions, and radios.

Practical Equivalent Inductors

Inductors, like capacitors, are not ideal. Associated with every inductor is a resistance determined by the resistance of the turns of wire (the thinner the wire, the greater the resistance for the same material) and by the core losses (radiation and skin effect, eddy current and hysteresis losses—all discussed in more advanced texts). There is also some stray capacitance due to the capacitance between the current-carrying turns of wire of the coil. Recall that capacitance appears whenever there are two conducting surfaces separated by an insulator, such as air, and when those wrappings are fairly tight and are parallel. Both elements are included in the equivalent circuit in Fig. 11.24. For most applications in this text, the capacitance can be ignored, resulting in the equivalent model in Fig. 11.25. The resistance R_l plays an im-

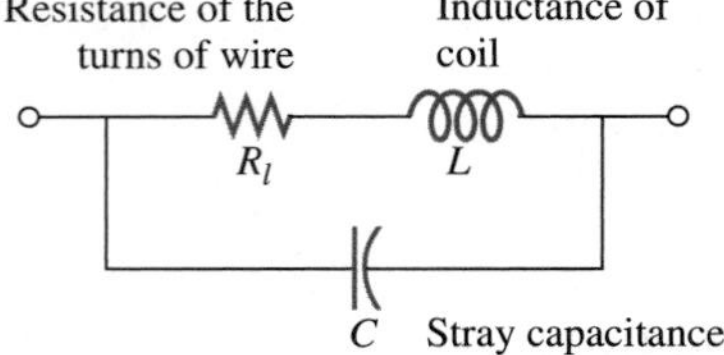

FIG. 11.24
Complete equivalent model for an inductor.

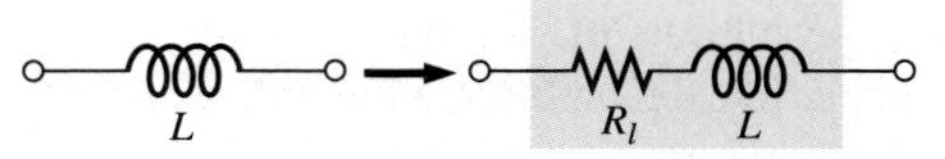

FIG. 11.25
Practical equivalent model for an inductor.

portant part in some areas (such as resonance, discussed in Chapter 20), because the resistance can extend from a few ohms to a few hundred ohms depending on the construction. For this chapter, the inductor is considered an ideal element and the series resistance is dropped from Fig. 11.25.

Inductor Labeling

Because some inductors are larger in size, their nameplate value can often be printed on the body of the element. However, on smaller units, there may not be enough room to print the actual value, so an abbreviation is used that is fairly easy to understand. First, realize that the *microhenry* (μH) is the fundamental unit of measurement for this marking. Most manuals list the inductance value in μH even if the value must be reported as 470,000 μH rather than as 470 mH. If the label reads 223K, the third number (3) is the power to be applied to the first two. The K is not from *kilo,* representing a power of three, but is used to denote a tolerance of ±10% as described for capacitors. The resulting number of 22,000 is, therefore, in μH so the 223K unit is a 22,000 μH or 22 mH inductor. The letters J and M indicate a tolerance of ±5% and ±20%, respectively.

For molded inductors, a color-coding system very similar to that used for resistors is used. The major difference is that *the resulting value is always in μH,* and a wide band at the beginning of the labeling is an MIL ("meets military standards") indicator. Always read the colors in sequence, starting with the band closest to one end as shown in Fig. 11.26.

The standard values for inductors employ the same numerical values and multipliers used with resistors and capacitors. In general, therefore,

Color Code Table

Color[1]	Significant Figure	Multiplier[2]	Inductance Tolerance (%)
Black	0	1	
Brown	1	10	
Red	2	100	
Orange	3	1000	
Yellow	4		
Green	5		
Blue	6		
Violet	7		
Gray	8		
White	9		
None[2]			±20
Silver			±10
Gold	Decimal point		±5

[1] Indicates body color.
[2] The multiplier is the factor by which the two significant figures are multiplied to yield the nominal inductance value.

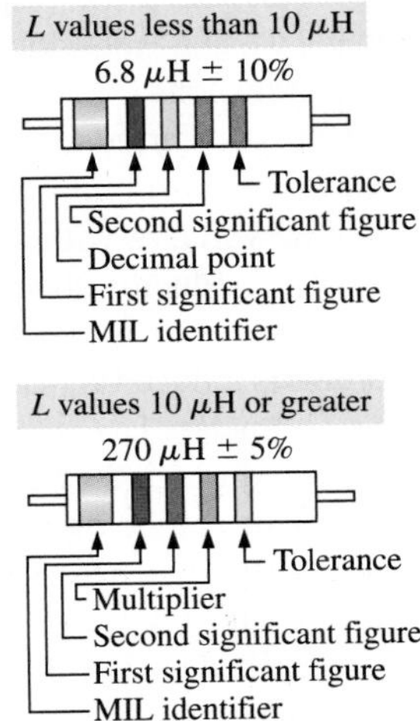

Cylindrical molded choke coils are marked with five colored bands. A wide silver band, located at one end of the coil, identifies military radio-frequency coils. The next three bands indicate the inductance in microhenries, and the fourth band is the tolerance.

Color coding is in accordance with the color code table, shown on the left. If the first or second band is gold, it represents the decimal point for inductance values less than 10. Then the following two bands are significant figures. For inductance values of 10 or more, the first two bands represent significant figures, and the third is the multiplier.

FIG. 11.26
Molded inductor color coding.

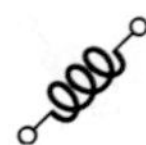

expect to find inductors with the following multipliers: 1 μH, 1.5 μH, 2.2 μH, 3.3 μH, 4.7 μH, 6.8 μH, 10 μH, and so on.

Measurement and Testing of Inductors

The inductance of an inductor can be read directly using a meter such as the Universal LCR Meter (Fig. 11.27), also discussed in Chapter 10 on capacitors. Set the meter to L for inductance, and the meter automatically chooses the most appropriate unit of measurement for the element, that is, H, mH, μH, or pH.

An inductance meter is the best choice, but an ohmmeter can also be used to check whether a short has developed between the windings or whether an open circuit has developed. The open-circuit possibility is easy to check because a reading of infinite ohms or very high resistance results. The short-circuit condition is harder to check because the resistance of many good inductors is relatively small, and the shorting of a few windings may not adversely affect the total resistance. Of course, if you are aware of the typical resistance of the coil, you can compare it to the measured value. A short between the windings and the core can be checked by simply placing one lead of the ohmmeter on one wire (perhaps a terminal) and the other on the core itself. A reading of zero ohms reveals a short between the two that may be due to a breakdown in the insulation jacket around the wire resulting from excessive currents, environmental conditions, or simply old age and cracking.

FIG. 11.27
Digital reading inductance meter.
(Image provided by B & K Precision Corporation.)

11.4 INDUCED VOLTAGE v_L

Before analyzing the response of inductive elements to an applied dc voltage, we must introduce a number of laws and equations that affect the transient response.

The first, referred to as **Faraday's law of electromagnetic induction,** is one of the most important in this field because it enables us to establish ac and dc voltages with a generator. If we move a conductor (any material with conductor characteristics as defined in Chapter 2) through a magnetic field so that it cuts magnetic lines of flux as shown in Fig. 11.28, a voltage is induced across the conductor that can be measured with a sensitive voltmeter. That's all it takes, and, in fact, the faster you move the conductor through the magnetic flux, the greater the induced voltage. The same effect can be produced if you hold the conductor still and move the magnetic field across the conductor. Note that the direction in which you move the conductor through the field determines the polarity of the induced voltage. Also, if you move the conductor through the field at right angles to the magnetic flux, you generate the maximum induced voltage. Moving the conductor parallel with the magnetic flux lines results in an induced voltage of zero volts since magnetic lines of flux are not crossed.

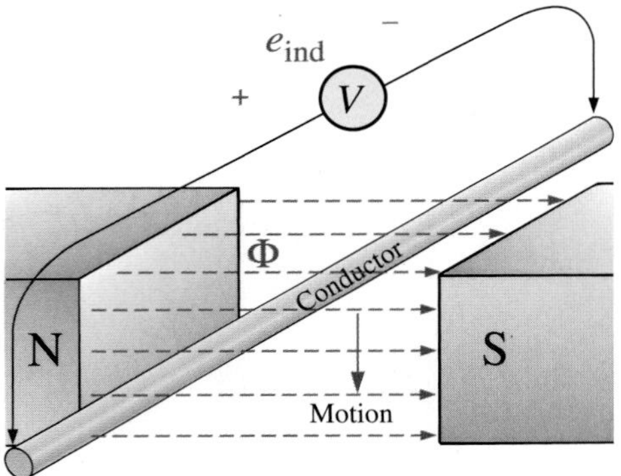

FIG. 11.28
Generating an induced voltage by moving a conductor through a magnetic field.

If we now go a step further and move a coil of N turns through the magnetic field as shown in Fig. 11.29, a voltage will be induced across the coil as determined by **Faraday's law:**

$$e = N\frac{d\phi}{dt} \quad \text{(volts, V)} \qquad \textbf{(11.9)}$$

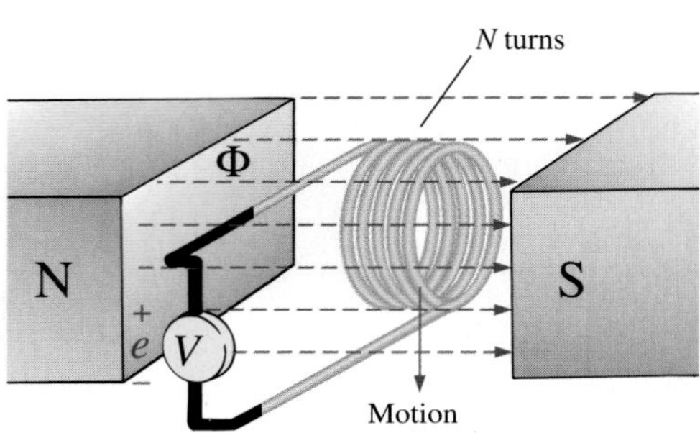

FIG. 11.29
Demonstrating Faraday's law.

The greater the number of turns or the faster the coil is moved through the magnetic flux pattern, the greater the induced voltage. The term $d\phi/dt$

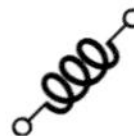

is the differential change in magnetic flux through the coil at a particular instant in time. If the magnetic flux passing through a coil remains constant—no matter how strong the magnetic field—the term will be zero, and the induced voltage zero volts. It doesn't matter whether the changing flux is due to moving the magnetic field or moving the coil in the vicinity of a magnetic field: The only requirement is that the flux linking (passing through) the coil changes with time. Before the coil passes through the magnetic poles, the induced voltage is zero because there are no magnetic flux lines passing through the coil. As the coil enters the flux pattern, the number of flux lines cut per instant of time increases until it peaks at the center of the poles. The induced voltage then decreases with time as it leaves the magnetic field.

This important phenomenon can now be applied to the inductor in Fig. 11.30, which is simply an extended version of the coil in Fig. 11.29. In Section 11.2, we found that the magnetic flux linking the coil of N turns with a current I has the distribution shown in Fig. 11.30. If the current through the coil increases in magnitude, the flux linking the coil also increases. We just learned through Faraday's law, however, that a coil in the vicinity of a changing magnetic flux will have a voltage induced across it. The result is that a voltage is induced across the coil in Fig. 11.30 due to the *change in current through the coil.*

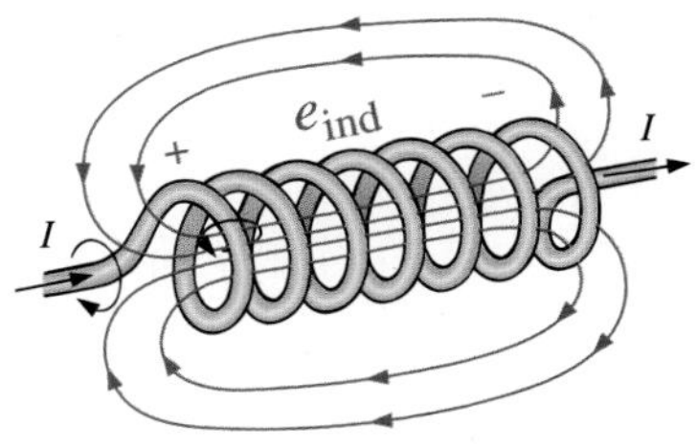

FIG. 11.30
Demonstrating the effect of Lenz's law.

It is very important to note in Fig. 11.30 that the polarity of the induced voltage across the coil is such that it opposes the increasing level of current in the coil. In other words, the changing current through the coil induces a voltage across the coil that is opposing the applied voltage that establishes the increase in current in the first place. The quicker the change in current through the coil, the greater the opposing induced voltage to squelch the attempt of the current to increase in magnitude. The "choking" action of the coil is the reason inductors or coils are often referred to as **chokes.** This effect is a result of an important law referred to as **Lenz's law,** which states that

an induced effect is always such as to oppose the cause that produced it.

The inductance of a coil is also a measure of the change in flux linking the coil due to a change in current through the coil. That is,

$$L = N\frac{d\phi}{di_L} \quad \text{(henries, H)} \qquad \textbf{(11.10)}$$

The equation reveals that the greater the number of turns or the greater the change in flux linking the coil due to a particular change in current, the greater the level of inductance. In other words, coils with smaller levels of inductance generate smaller changes in flux linking the coil for the same change in current through the coil. If the inductance level is very small, there will be almost no change in flux linking the coil, and the induced voltage across the coil will be very small. In fact, if we now write Eq. (11.9) in the following form:

$$e = N\frac{d\phi}{dt} = \left(N\frac{d\phi}{di_L}\right)\left(\frac{di_L}{dt}\right)$$

and substitute Eq. (11.10), we obtain

$$e_L = L\frac{di_L}{dt} \quad \text{(volts, V)} \qquad \textbf{(11.11)}$$

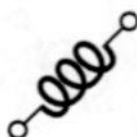

which relates the voltage across a coil to the number of turns of the coil and the change in current through the coil.

When induced effects are used in the generation of voltages such as those from dc or ac generators, the symbol e is applied to the induced voltage. However, in network analysis, the voltage induced across an inductor will always have a polarity that opposes the applied voltage (like the voltage across a resistor). Therefore, the following notation is used for the induced voltage across an inductor:

$$\boxed{v_L = L\frac{di_L}{dt}} \qquad \text{(volts, V)} \qquad \mathbf{(11.12)}$$

The equation clearly states that

the larger the inductance and/or the more rapid the change in current through a coil, the larger will be the induced voltage across the coil.

If the current through the coil fails to change with time, the induced voltage across the coil will be zero. We will find in the next section that for dc applications, when the transient phase has passed, $di_L/dt = 0$, and the induced voltage across the coil is

$$v_L = L\frac{di_L}{dt} = L(0) = 0\text{ V}$$

The duality that exists between inductive and capacitive elements is now abundantly clear. Simply interchange the voltages and currents of Eq. (11.12), and interchange the inductance and capacitance. The following equation for the current of a capacitor results:

$$v_L = L\frac{di_L}{dt}$$

$$\downarrow \quad \downarrow$$

$$i_C = C\frac{dv_C}{dt}$$

We are now at a point where we have all the background relationships necessary to investigate the transient behavior of inductive elements.

11.5 *R-L* TRANSIENTS: THE STORAGE PHASE

A great number of similarities exist between the analyses of inductive and capacitive networks. That is, what is true for the voltage of a capacitor is also true for the current of an inductor, and what is true for the current of a capacitor can be matched in many ways by the voltage of an inductor. The storage waveforms have the same shape, and time constants are defined for each configuration. Because these concepts are so similar (refer to Section 10.5 on the charging of a capacitor), you have an opportunity to reinforce concepts introduced earlier and still learn more about the behavior of inductive elements.

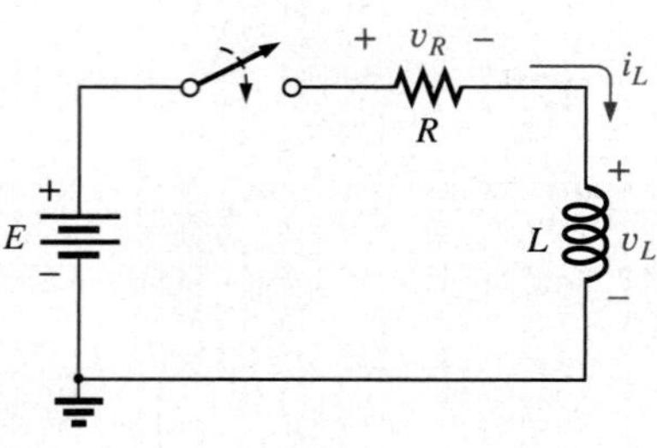

FIG. 11.31
Basic R-L transient network.

The circuit in Fig. 11.31 is used to describe the storage phase. Note that it is the same circuit used to describe the charging phase of capacitors, with a simple replacement of the capacitor by an ideal inductor. Throughout the analysis, it is important to remember that energy is stored in the form of an electric field between the plates of a capacitor. For inductors, on the other hand, energy is stored in the form of a magnetic field linking the coil.

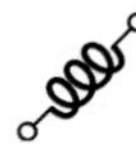

At the instant the switch is closed, the choking action of the coil prevents an instantaneous change in current through the coil, resulting in $i_L = 0$ A as shown in Fig. 11.32(a). The absence of a current through the coil and circuit at the instant the switch is closed results in zero volts across the resistor as determined by $v_R = i_R R = i_L R = (0 \text{ A})R = 0$ V, as shown in Fig. 11.32(c). Applying Kirchhoff's voltage law around the closed loop results in E volts across the coil at the instant the switch is closed, as shown in Fig. 11.32(b).

Initially, the current increases very rapidly as shown in Fig. 11.32(a) and then at a much slower rate as it approaches its steady-state value determined by the parameters of the network (E/R). The voltage across the resistor rises at the same rate because $v_R = i_R R = i_L R$. Since the voltage across the coil is sensitive to the rate of change of current through the coil, the voltage will be at or near its maximum value early in the storage phase. Finally, when the current reaches its steady-state value of E/R amperes, the current through the coil ceases to change, and the voltage across the coil drops to zero volts. At any instant of time, the voltage

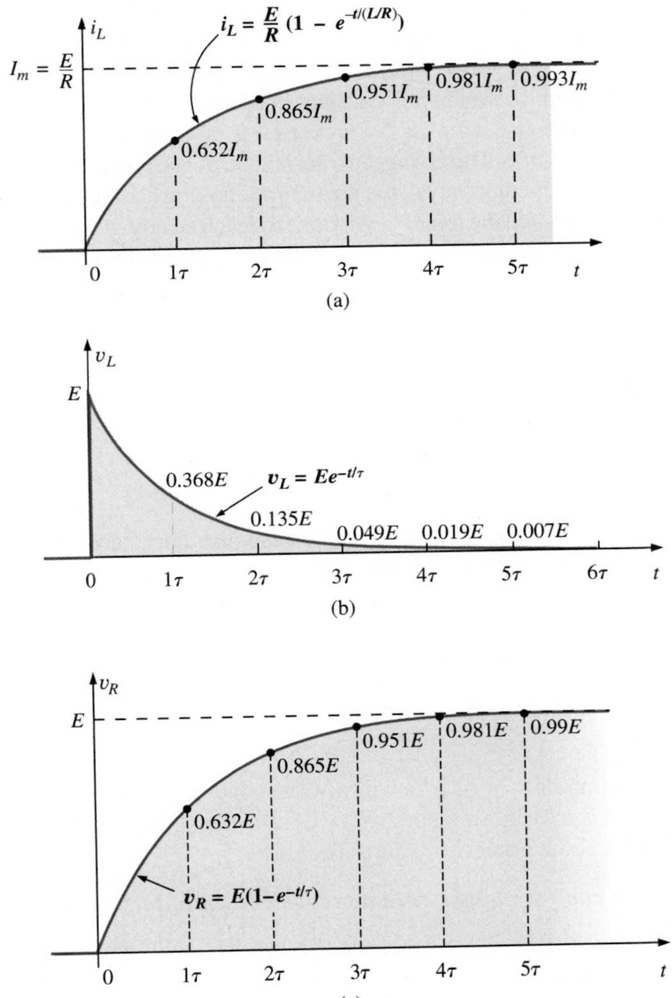

FIG. 11.32
i_L, v_L, and v_R for the circuit in Fig. 11.31 following the closing of the switch.

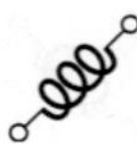

across the coil can be determined using Kirchhoff's voltage law in the following manner: $v_L = E - v_R$.

Because the waveforms for the inductor have the same shape as obtained for capacitive networks, we are familiar with the mathematical format and can feel comfortable calculating the quantities of interest using a calculator or computer.

The equation for the transient response of the current through an inductor is the following:

$$i_L = \frac{E}{R}(1 - e^{-t/\tau}) \quad \text{(amperes, A)} \qquad \textbf{(11.13)}$$

with the time constant now defined by

$$\tau = \frac{L}{R} \quad \text{(seconds, s)} \qquad \textbf{(11.14)}$$

Note that Eq. (11.14) is a ratio of parameters rather than a product as used for capacitive networks, yet the units used are still seconds (for time).

Our experience with the factor $(1 - e^{-t/\tau})$ verifies the level of 63.2% for the inductor current after one time constant, 86.5% after two time constants, and so on. If we keep R constant and increase L, the ratio L/R increases, and the rise time of 5τ increases as shown in Fig. 11.33 for increasing levels of L. The change in transient response is expected because the higher the inductance level, the greater the choking action on the changing current level, and the longer it will take to reach steady-state conditions.

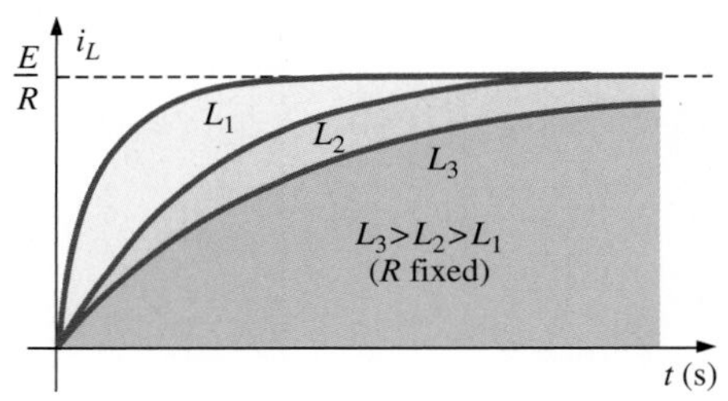

FIG. 11.33
Effect of L on the shape of the i_L storage waveform.

The equation for the voltage across the coil is the following:

$$v_L = Ee^{-t/\tau} \quad \text{(volts, V)} \qquad \textbf{(11.15)}$$

and for the voltage across the resistor:

$$v_R = E(1 - e^{-t/\tau}) \quad \text{(volts, V)} \qquad \textbf{(11.16)}$$

As mentioned earlier, the shape of the response curve for the voltage across the resistor must match that of the current i_L since $v_R = i_R R = i_L R$.

Since the waveforms are similar to those obtained for capacitive networks, we will assume that

the storage phase has passed and steady-state conditions have been established once a period of time equal to five time constants has occurred.

In addition, since $\tau = L/R$ will always have some numerical value, even though it may be very small at times, the transient period of 5τ will always have some numerical value. Therefore,

the current cannot change instantaneously in an inductive network.

If we examine the conditions that exist at the instant the switch is closed, we find that the voltage across the coil is E volts, although the current is zero amperes as shown in Fig. 11.34. In essence, therefore,

the inductor takes on the characteristics of an open circuit at the instant the switch is closed.

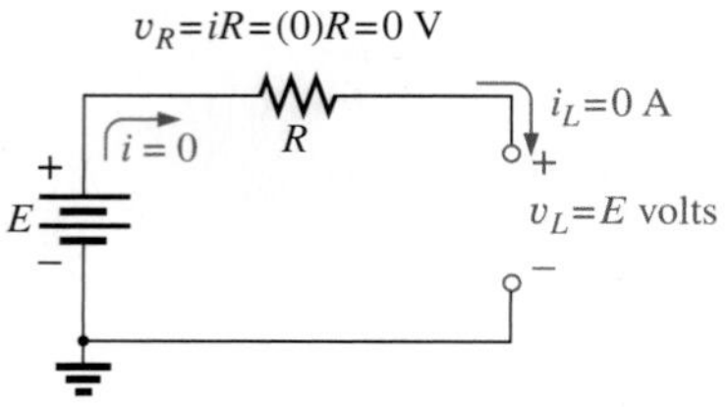

FIG. 11.34
Circuit in Figure 11.31 the instant the switch is closed.

However, if we consider the conditions that exist when steady-state conditions have been established, we find that the voltage across the coil is zero volts and the current is a maximum value of E/R amperes as shown in Fig. 11.35. In essence, therefore,

the inductor takes on the characteristics of a short circuit when steady-state conditions have been established.

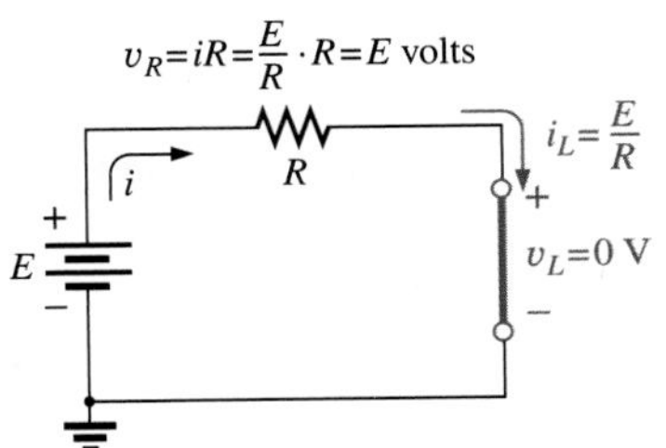

FIG. 11.35
Circuit in Fig. 11.31 under steady-state conditions.

EXAMPLE 11.3 Find the mathematical expressions for the transient behavior of i_L and v_L for the circuit in Fig. 11.36 if the switch is closed at $t = 0$ s. Sketch the resulting curves.

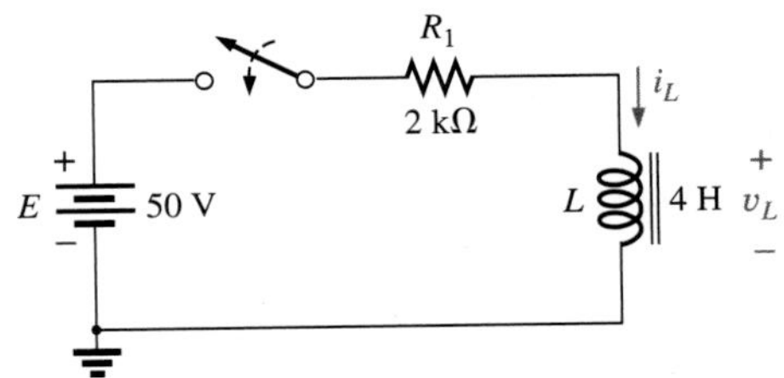

FIG. 11.36
Series R-L circuit for Example 11.3.

Solution: First, the time constant is determined:

$$\tau = \frac{L}{R_1} = \frac{4\text{ H}}{2\text{ k}\Omega} = 2\text{ ms}$$

Then the maximum or steady-state current is

$$I_m = \frac{E}{R_1} = \frac{50\text{ V}}{2\text{ k}\Omega} = 25 \times 10^{-3}\text{A} = 25\text{ mA}$$

Substituting into Eq. (11.13):

$$i_L = \mathbf{25\ mA\ (1 - e^{-t/2ms})}$$

Using Eq. (11.15):

$$v_L = \mathbf{50\ Ve^{-t/2ms}}$$

The resulting waveforms appear in Fig. 11.37.

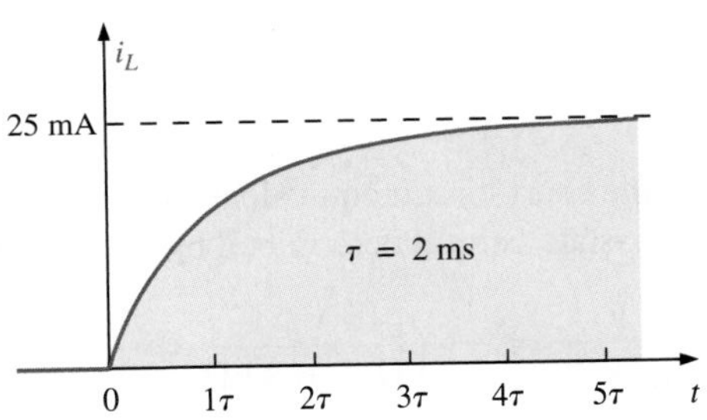

FIG. 11.37
i_L and v_L for the network in Fig. 11.36.

11.6 INITIAL CONDITIONS

This section parallels Section 10.7 ("Initial Conditions") on the effect of initial values on the transient phase. Since the current through a coil cannot change instantaneously, the current through a coil begins the transient phase at the initial value established by the network (note Fig. 11.38) before the switch was closed. It then passes through the transient phase until it reaches the steady-state (or final) level after about five time constants. The steady-state level of the inductor current can be found by substituting its short-circuit equivalent (or R_l for the practical equivalent) and finding the resulting current through the element.

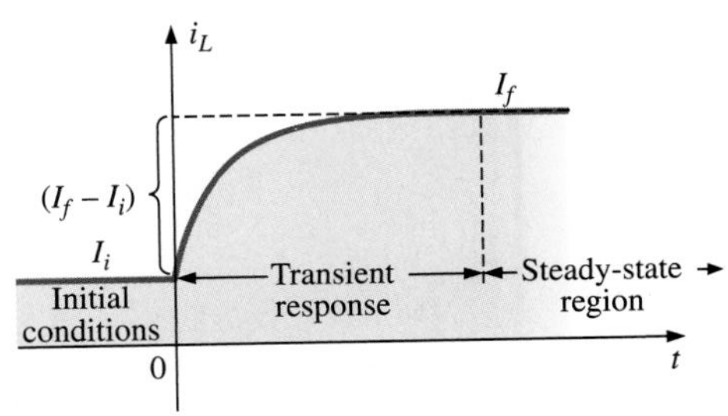

FIG. 11.38
Defining the three phases of a transient waveform.

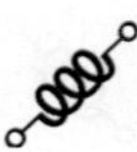

Using the transient equation developed in the previous section, an equation for the current i_L can be written for the entire time interval in Fig. 11.38; that is,

$$i_L = I_i + (I_f - I_i)(1 - e^{-t/\tau})$$

with $(I_f - I_i)$ representing the total change during the transient phase. However, by multiplying through and rearranging terms:

$$\begin{aligned} i_L &= I_i + I_f - I_f e^{-t/\tau} - I_i + I_i e^{-t/\tau} \\ &= I_f - I_f e^{-t/\tau} + I_i e^{-t/\tau} \end{aligned}$$

we find

$$\boxed{i_L = I_f + (I_i - I_f)e^{-t/\tau}} \qquad \textbf{(11.17)}$$

If you are required to draw the waveform for the current i_L from initial value to final value, start by drawing a line at the initial value and steady-state levels, and then add the transient response (sensitive to the time constant) between the two levels. The following example will clarify the procedure.

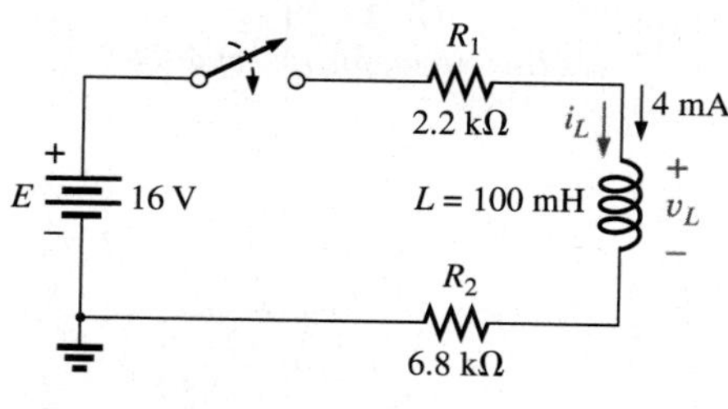

FIG. 11.39
Example 11.4.

EXAMPLE 11.4 The inductor in Fig. 11.39 has an initial current level of 4 mA in the direction shown. (Specific methods to establish the initial current are presented in the sections and problems to follow.)

a. Find the mathematical expression for the current through the coil once the switch is closed.
b. Find the mathematical expression for the voltage across the coil during the same transient period.
c. Sketch the waveform for each from initial value to final value.

Solutions:

a. Substituting the short-circuit equivalent for the inductor results in a final or steady-state current determined by Ohm's law:

$$I_f = \frac{E}{R_1 + R_2} = \frac{16\text{ V}}{2.2\text{ k}\Omega + 6.8\text{ k}\Omega} = \frac{16\text{ V}}{9\text{ k}\Omega} = 1.78\text{ mA}$$

The time constant is determined by

$$\tau = \frac{L}{R_T} = \frac{100\text{ mH}}{2.2\text{ k}\Omega + 6.8\text{ k}\Omega} = \frac{100\text{ mH}}{9\text{ k}\Omega} = 11.11\ \mu\text{s}$$

Applying Eq. (11.17):

$$\begin{aligned} i_L &= I_f + (I_i - I_f)\,e^{-t/\tau} = 1.78\text{ mA} + (4\text{ mA} - 1.78\text{ mA})e^{-t/11.11\mu\text{s}} \\ &= \mathbf{1.78\ mA + 2.22\ mAe^{-t/11.11\mu s}} \end{aligned}$$

b. Since the current through the inductor is constant at 4 mA prior to the closing of the switch, the voltage (whose level is sensitive only to changes in current through the coil) must have an initial value of 0 V. At the instant the switch is closed, the current through the coil cannot change instantaneously, so the current through the resistive elements is 4 mA. The resulting peak voltage at $t = 0$ s can then be found using Kirchhoff's voltage law as follows:

$$\begin{aligned} V_m &= E - V_{R_1} - V_{R_2} = 16\text{ V} - (4\text{ mA})(2.2\text{ k}\Omega) - (4\text{ mA})(6.8\text{ k}\Omega) \\ &= 16\text{ V} - 8.8\text{ V} - 27.2\text{ V} = 16\text{ V} - 36\text{ V} = -20\text{ V} \end{aligned}$$

Note the minus sign to indicate that the polarity of the voltage v_L is opposite to the defined polarity of Fig. 11.39.

The voltage then decays (with the same time constant as the current i_L) to zero because the inductor is approaching its short-circuit equivalence.

The equation for v_L is therefore

$$v_L = \mathbf{-20Ve^{-t/11.11\mu s}}$$

c. See Fig. 11.40. The initial and final values of the current were drawn first, and then the transient response was included between these levels. For the voltage, the waveform begins and ends at zero, with the peak value having a sign sensitive to the defined polarity of v_L in Fig. 11.39.

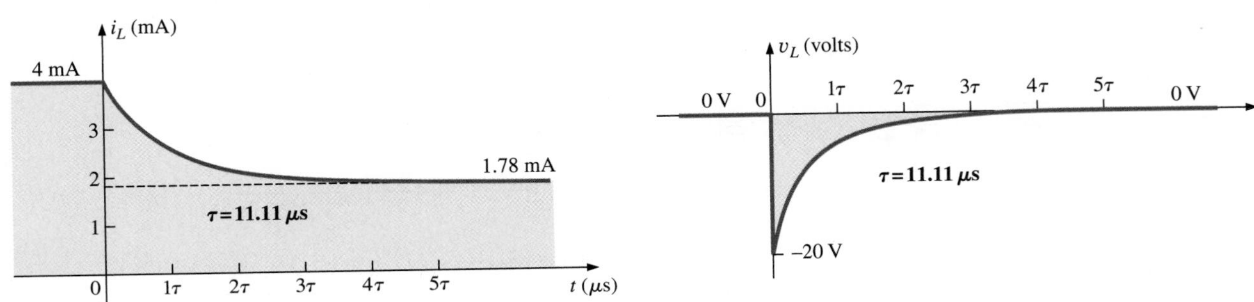

FIG. 11.40
i_L and v_L for the network in Fig. 11.39.

Let us now test the validity of the equation for i_L by substituting $t = 0$ s to reflect the instant the switch is closed.

$$e^{-t/\tau} = e^{-0} = 1$$

and $\quad i_L = 1.78 \text{ mA} + 2.22 \text{ mA}e^{-t/\tau} = 1.78 \text{ mA} + 2.22 \text{ mA} = 4 \text{ mA}$

When $t > 5\tau$, $\quad e^{-t/\tau} \cong 0$

and $\quad i_L = 1.78 \text{ mA} + 2.22 \text{ mA}e^{-t/\tau} = 1.78 \text{ mA}$

11.7 *R-L* TRANSIENTS: THE RELEASE PHASE

In the analysis of *R-C* circuits, we found that the capacitor could hold its charge and store energy in the form of an electric field for a period of time determined by the leakage factors. In *R-L* circuits, the energy is stored in the form of a magnetic field established by the current through the coil. Unlike the capacitor, however, an isolated inductor cannot continue to store energy, because the absence of a closed path causes the current to drop to zero, releasing the energy stored in the form of a magnetic field. If the series *R-L* circuit in Fig. 11.41 reaches steady-state conditions and the switch is quickly opened, a spark will occur across the contacts due to the rapid change in current from a maximum of *E/R* to zero amperes. The change in current *di/dt* of the equation $v_L = L(di/dt)$ establishes a high voltage v_L across the coil that, in conjunction with the applied voltage *E*, appears across the points of the switch. This is the same mechanism used in the ignition system of a car to ignite the fuel in the cylinder. Some 25,000 V are

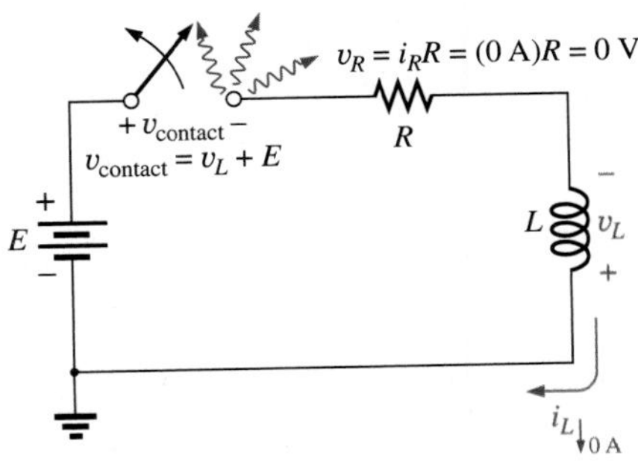

FIG. 11.41
Demonstrating the effect of opening a switch in series with an inductor with a steady-state current.

generated by the rapid decrease in ignition coil current that occurs when the switch in the system is opened. (In older systems, the "points" in the distributor served as the switch.) This inductive reaction is significant when you consider that the only independent source in a car is a 12 V battery.

If opening the switch to move it to another position causes such a rapid discharge in stored energy, how can the decay phase of an *R-L* circuit be analyzed in much the same manner as for the *R-C* circuit? The solution is to use a network like that in Fig. 11.42(a). When the switch is closed, the voltage across resistor R_2 is E volts, and the *R-L* branch responds in the same manner as described above, with the same waveforms and levels. A Thévenin network of E in parallel with R_2 results in the source as shown in Fig. 11.42(b), since R_2 will be shorted out by the short-circuit replacement of the voltage source E when the Thévenin resistance is determined.

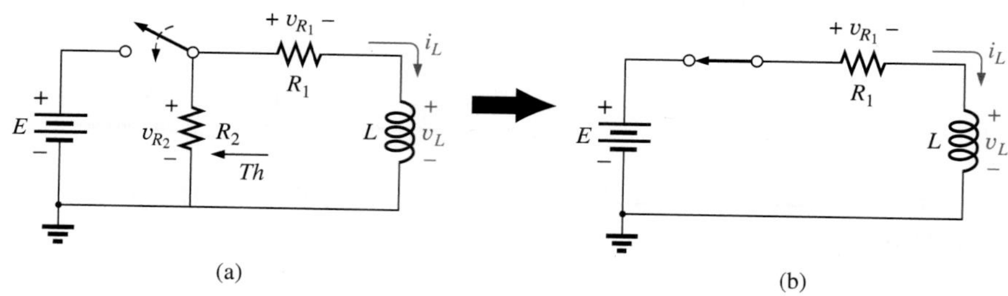

FIG. 11.42
Initiating the storage phase for an inductor by closing the switch.

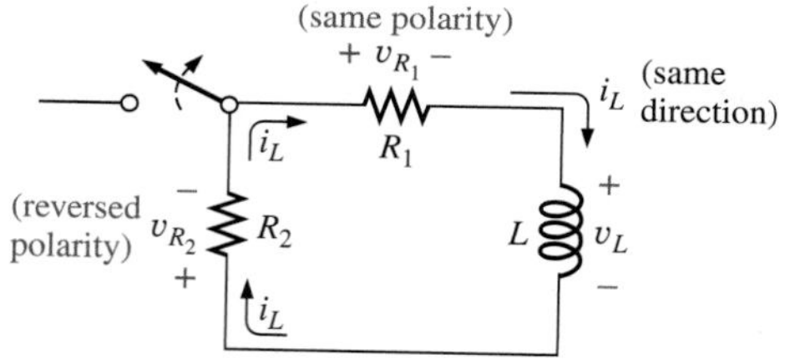

FIG. 11.43
Network in Fig. 11.42 the instant the switch is opened.

After the storage phase has passed and steady-state conditions are established, the switch can be opened without the sparking effect or rapid discharge due to resistor R_2, which provides a complete path for the current i_L. In fact, for clarity the discharge path is isolated in Fig. 11.43. The voltage v_L across the inductor reverses polarity and has a magnitude determined by

$$v_L = -(v_{R_1} + v_{R_2}) \tag{11.18}$$

Recall that the voltage across an inductor can change instantaneously but the current cannot. The result is that the current i_L must maintain the same direction and magnitude as shown in Fig. 11.43. Therefore, the instant after the switch is opened, i_L is still $I_m = E/R_1$, and

$$v_L = -(v_{R_1} + v_{R_2}) = -(i_1R_1 + i_2R_2)$$
$$= -i_L(R_1 + R_2) = -\frac{E}{R_1}(R_1 + R_2) = -\left(\frac{R_1}{R_1} + \frac{R_2}{R_1}\right)E$$

and

$$v_L = -\left(1 + \frac{R_2}{R_1}\right)E \qquad \text{switch opened} \tag{11.19}$$

which is bigger than E volts by the ratio R_2/R_1. In other words, when the switch is opened, the voltage across the inductor reverses polarity and drops instantaneously from E to $-[1 + (R_2/R_1)]E$ volts.

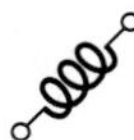

As an inductor releases its stored energy, the voltage across the coil decays to zero in the following manner:

$$\boxed{v_L = -V_i e^{-t/\tau'}} \tag{11.20}$$

with $$V_i = \left(1 + \frac{R_2}{R_1}\right)E$$

and $$\tau' = \frac{L}{R_T} = \frac{L}{R_1 + R_2}$$

The current decays from a maximum of $I_m = E/R_1$ to zero.

Using Eq. (11.17):

$$I_i = \frac{E}{R_1} \quad \text{and} \quad I_f = 0\text{ A}$$

so that $$i_L = I_f + (I_i - I_f)e^{-t/\tau'} = 0\text{ A} + \left(\frac{E}{R_1} - 0\text{ A}\right)e^{-t/\tau'}$$

and $$\boxed{i_L = \frac{E}{R_1}e^{-t/\tau'}} \tag{11.21}$$

with $$\tau' = \frac{L}{R_1 + R_2}$$

The mathematical expression for the voltage across either resistor can then be determined using Ohm's law:

$$v_{R_1} = i_{R_1}R_1 = i_L R_1 = \frac{E}{R_1}R_1 e^{-t/\tau'}$$

and $$\boxed{v_{R_1} = Ee^{-t/\tau'}} \tag{11.22}$$

The voltage v_{R_1} has the same polarity as during the storage phase since the current i_L has the same direction. The voltage v_{R_2} is expressed as follows using the defined polarity of Fig. 11.42:

$$v_{R_2} = -i_{R_2}R_2 = -i_L R_2 = -\frac{E}{R_1}R_2 e^{-t/\tau'}$$

and $$\boxed{v_{R_2} = -\frac{R_2}{R_1}Ee^{-t/\tau'}} \tag{11.23}$$

EXAMPLE 11.5 Resistor R_2 was added to the network in Fig. 11.36 as shown in Fig. 11.44.

a. Find the mathematical expressions for i_L, v_L, v_{R_1}, and v_{R_2} for five time constants of the storage phase.
b. Find the mathematical expressions for i_L, v_L, v_{R_1}, and v_{R_2} if the switch is opened after five time constants of the storage phase.
c. Sketch the waveforms for each voltage and current for both phases covered by this example. Use the defined polarities in Fig. 11.43.

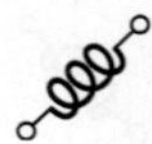

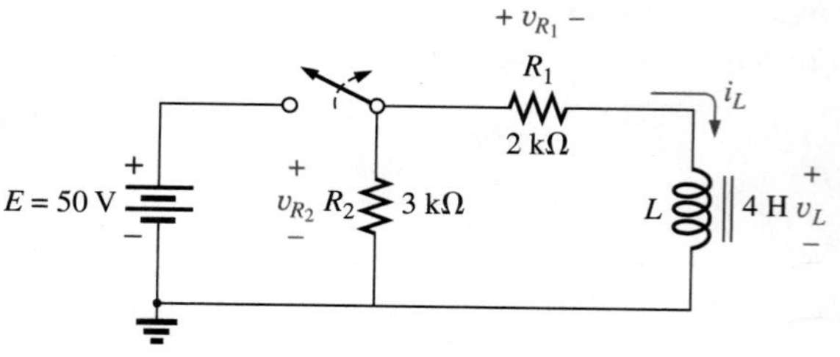

FIG. 11.44
Defined polarities for v_{R_1}, v_{R_2}, v_L, and current direction for i_L for Example 11.5.

Solutions:

a. From Example 11.3:

$$i_L = \mathbf{25\ mA(1 - e^{-t/2ms})}$$
$$v_L = \mathbf{50\ Ve^{-t/2ms}}$$
$$v_{R_1} = i_{R_1}R_1 = i_L R_1$$
$$= \left[\frac{E}{R_1}(1 - e^{-t/\tau})\right]R_1$$
$$= E(1 - e^{-t/\tau})$$

and
$$v_{R_1} = \mathbf{50\ V(1 - e^{-t/2ms})}$$
$$v_{R_2} = E = \mathbf{50\ V}$$

b. $$\tau' = \frac{L}{R_1 + R_2} = \frac{4\ H}{2\ k\Omega + 3\ k\Omega} = \frac{4\ H}{5 \times 10^3\ \Omega}$$
$$= 0.8 \times 10^{-3}\ s = 0.8\ ms$$

By Eqs. (11.19) and (11.20):

$$V_i = \left(1 + \frac{R_2}{R_1}\right)E = \left(1 + \frac{3\ k\Omega}{2\ k\Omega}\right)(50\ V) = 125\ V$$

and
$$v_L = -V_i e^{-t/\tau'} = \mathbf{-125\ Ve^{-t/-0.8ms}}$$

By Eq. (11.21):

$$I_m = \frac{E}{R_1} = \frac{50\ V}{2\ k\Omega} = 25\ mA$$

and
$$i_L = I_m e^{-t/\tau'} = \mathbf{25\ mAe^{-t/0.8ms}}$$

By Eq. (11.22):

$$v_{R_1} = Ee^{-t/\tau'} = \mathbf{50\ Ve^{-t/0.8ms}}$$

By Eq. (11.23):

$$v_{R_2} = -\frac{R_2}{R_1}Ee^{-t/\tau'} = -\frac{3\ k\Omega}{2\ k\Omega}(50\ V)e^{-t/\tau'} = \mathbf{-75\ Ve^{-t/0.8\ ms}}$$

c. See Fig. 11.45:

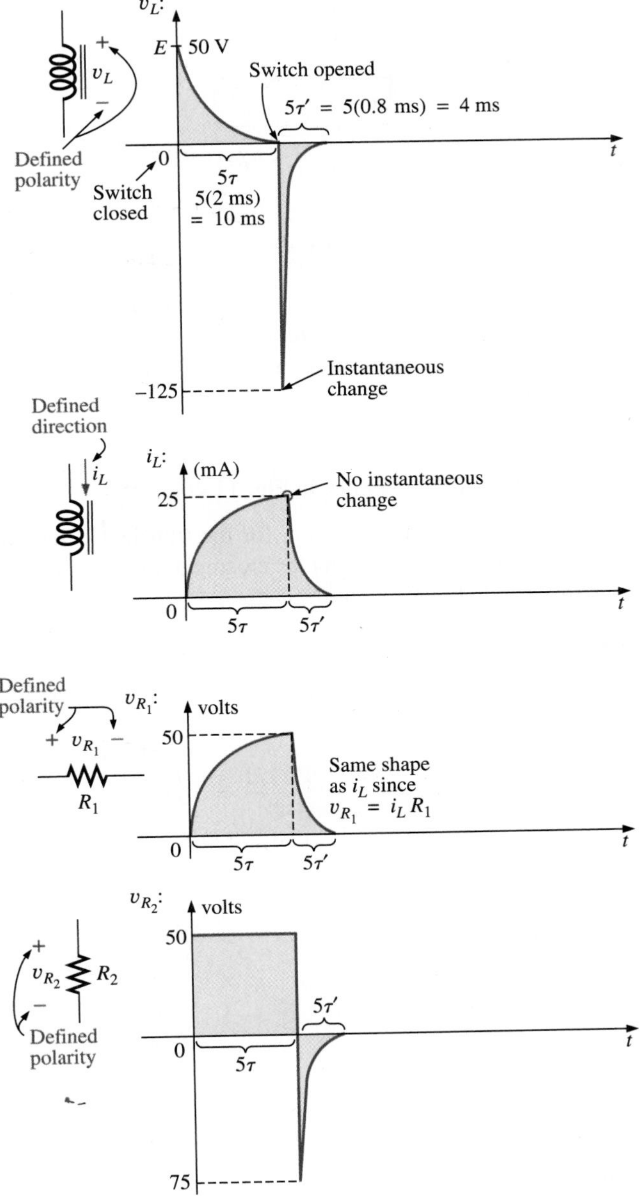

FIG. 11.45

The various voltages and the current for the network in Fig. 11.44.

In the preceding analysis, it was assumed that steady-state conditions were established during the charging phase and $I_m = E/R_1$, with $v_L = 0$ V. However, if the switch in Fig. 11.42 is opened before i_L reaches its maximum value, the equation for the decaying current of Fig. 11.42 must change to

$$\boxed{i_L = I_i e^{-t/\tau'}} \qquad \textbf{(11.24)}$$

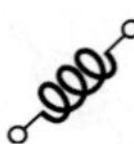

where I_i is the starting or initial current. The voltage across the coil is defined by the following:

$$v_L = -V_i e^{-t/\tau'} \tag{11.25}$$

with

$$V_i = I_i(R_1 + R_2)$$

11.8 THÉVENIN EQUIVALENT: $\tau = L/R_{Th}$

In Chapter 10 on capacitors, we found that a circuit does not always have the basic form in Fig. 11.31. The solution is to find the Thévenin equivalent circuit before proceeding in the manner described in this chapter. Consider the following example.

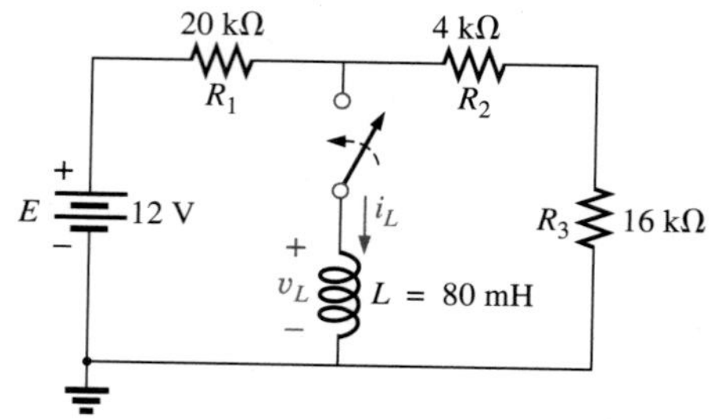

FIG. 11.46
Example 11.6.

EXAMPLE 11.6 For the network in Fig. 11.46.

a. Find the mathematical expression for the transient behavior of the current i_L and the voltage v_L after the closing of the switch ($I_i = 0$ mA).
b. Draw the resultant waveform for each.

Solutions:

a. Applying Thévenin's theorem to the 80 mH inductor (Fig. 11.47) yields

$$R_{Th} = \frac{R}{N} = \frac{20\text{ k}\Omega}{2} = 10\text{ k}\Omega$$

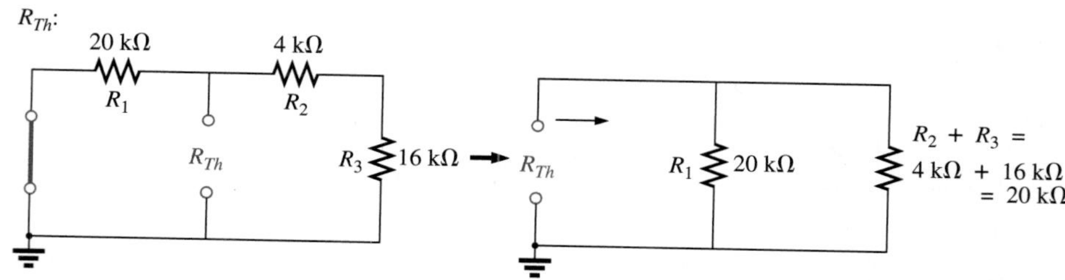

FIG. 11.47
Determining R_{Th} for the network in Fig. 11.46.

Applying the voltage divider rule (Fig. 11.48), we obtain

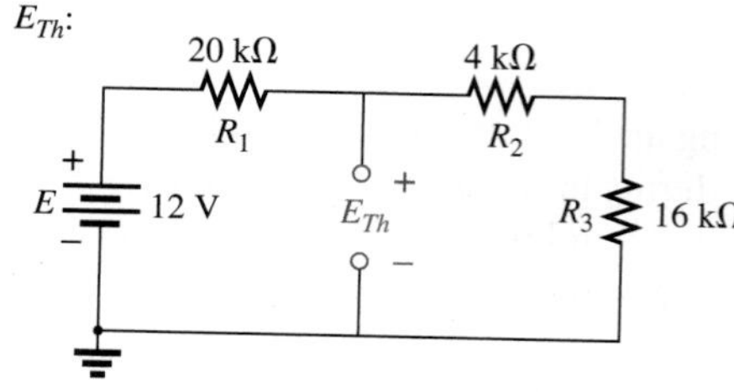

FIG. 11.48
Determining E_{Th} for the network in Fig. 11.46.

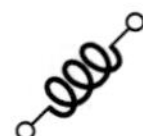

$$E_{Th} = \frac{(R_2 + R_3)E}{R_1 + R_2 + R_3}$$
$$= \frac{(4\ \text{k}\Omega + 16\ \text{k}\Omega)(12\ \text{V})}{20\ \text{k}\Omega + 4\ \text{k}\Omega + 16\ \text{k}\Omega} = \frac{(20\ \text{k}\Omega)(12\ \text{V})}{40\ \text{k}\Omega} = 6\ \text{V}$$

The Thévenin equivalent circuit is shown in Fig. 11.49. Using Eq. (11.13):

$$i_L = \frac{E_{Th}}{R}(1 - e^{-t/\tau})$$
$$\tau = \frac{L}{R_{Th}} = \frac{80 \times 10^{-3}\ \text{H}}{10 \times 10^3\ \Omega} = 8 \times 10^{-6}\ \text{s} = 8\ \mu\text{s}$$
$$I_m = \frac{E_{Th}}{R_{Th}} = \frac{6\ \text{V}}{10 \times 10^3\ \Omega} = 0.6 \times 10^{-3}\ \text{A} = 0.6\ \text{mA}$$

and $\quad i_L = \mathbf{0.6\ mA\ (1 - e^{-t/8\mu s})}$

Using Eq. (11.15):

$$v_L = E_{Th}e^{-t/\tau}$$

so that $\quad v_L = \mathbf{6\ V e^{-t/8\mu s}}$

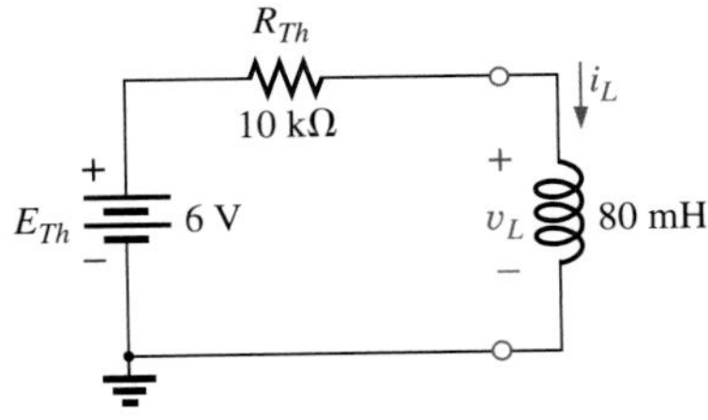

FIG. 11.49
The resulting Thévenin equivalent circuit for the network in Fig. 11.46.

b. See Fig. 11.50.

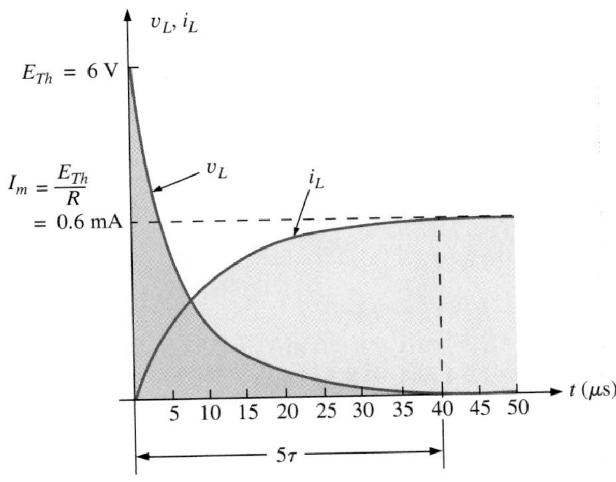

FIG. 11.50
The resulting waveforms for i_L and v_L for the network in Fig. 11.46.

EXAMPLE 11.7 Switch S_1 in Fig. 11.51 has been closed for a long time. At $t = 0$ s, S_1 is opened at the same instant that S_2 is closed to avoid an interruption in current through the coil.

a. Find the initial current through the coil. Pay particular attention to its direction.
b. Find the mathematical expression for the current i_L following the closing of switch S_2.
c. Sketch the waveform for i_L.

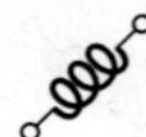

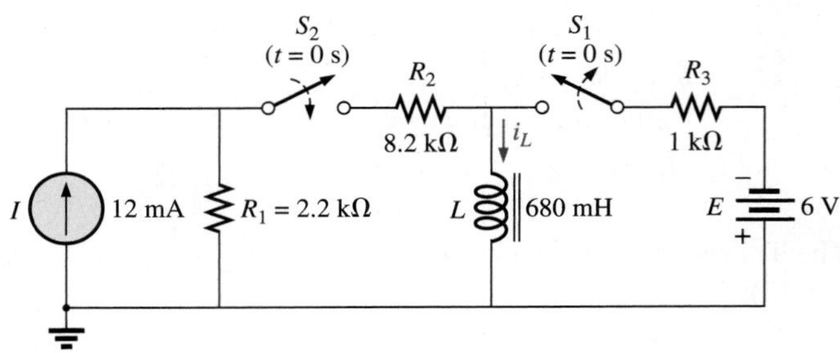

FIG. 11.51
Example 11.7.

Solutions:

a. Using Ohm's law, the initial current through the coil is determined by

$$I_i = -\frac{E}{R_3} = -\frac{6\text{ V}}{1\text{ k}\Omega} = -6\text{ mA}$$

b. Applying Thévenin's theorem:

$$R_{Th} = R_1 + R_2 = 2.2\text{ k}\Omega + 8.2\text{ k}\Omega = 10.4\text{ k}\Omega$$
$$E_{Th} = IR_1 = (12\text{ mA})(2.2\text{ k}\Omega) = 26.4\text{ V}$$

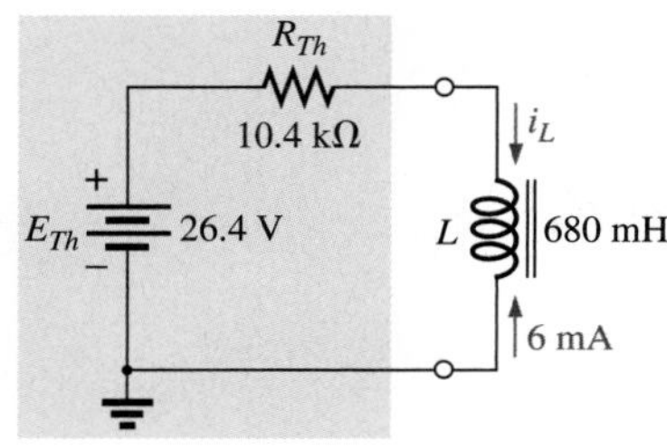

FIG. 11.52
Thévenin equivalent circuit for the network in Fig. 11.51 for t ≥ 0 s.

The Thévenin equivalent network appears in Fig. 11.52.

The steady-state current can then be determined by substituting the short-circuit equivalent for the inductor:

$$I_f = \frac{E}{R_{Th}} = \frac{26.4\text{ V}}{10.4\text{ k}\Omega} = 2.54\text{ mA}$$

The time constant is

$$\tau = \frac{L}{R_{Th}} = \frac{680\text{ mH}}{10.4\text{ k}\Omega} = 65.39\ \mu\text{s}$$

Applying Eq. (11.17):

$$i_L = I_f + (I_i - I_f)e^{-t/\tau}$$
$$= 2.54\text{ mA} + (-6\text{ mA} - 2.54\text{ mA})e^{-t/65.39\mu\text{s}}$$
$$= \mathbf{2.54\text{ mA} - 8.54\text{ mA}e^{-t/65.39\mu s}}$$

c. Note Fig. 11.53.

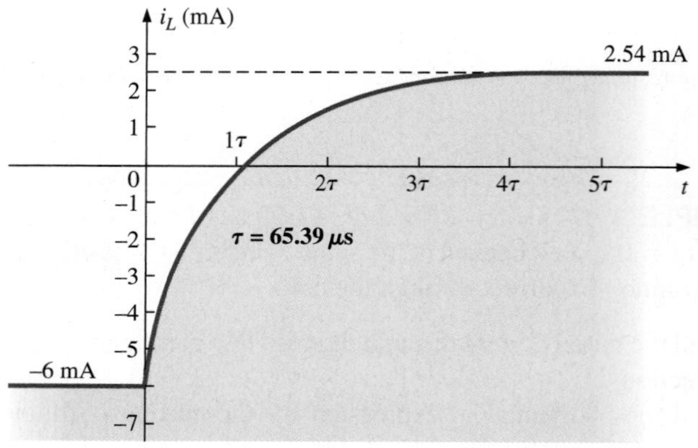

FIG. 11.53
The current i_L for the network in Fig. 11.51.

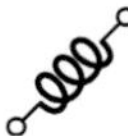

11.9 INSTANTANEOUS VALUES

The development presented in Section 10.8 for capacitive networks can also be applied to *R-L* networks to determine instantaneous voltages, currents, and time. The instantaneous values of any voltage or current can be determined by simply inserting *t* into the equation and using a calculator or table to determine the magnitude of the exponential term.

The similarity between the equations

$$v_C = V_f + (V_i - V_f)e^{-t/\tau}$$

and

$$i_L = I_f + (I_i - I_f)e^{-t/\tau}$$

results in a derivation of the following for *t* that is identical to that used to obtain Eq. (10.22):

$$t = \tau \log_e \frac{(I_i - I_f)}{(i_L - I_f)} \quad \text{(seconds, s)} \qquad \textbf{(11.26)}$$

For the other form, the equation $v_C = Ee^{-t/\tau}$ is a close match with $v_L = Ee^{-t/\tau} = V_i e^{-t/\tau}$, permitting a derivation similar to that employed for Eq. (10.23):

$$t = \tau \log_e \frac{V_i}{v_L} \quad \text{(seconds, s)} \qquad \textbf{(11.27)}$$

For the voltage v_R, $V_i = 0$ V and $V_f = E$ V since $v_R = E(1 - e^{-t/\tau})$. Solving for *t* yields

$$t = \tau \log_e \left(\frac{E}{E - v_R} \right)$$

or

$$t = \tau \log_e \left(\frac{V_f}{V_f - v_R} \right) \quad \text{(seconds, s)} \qquad \textbf{(11.28)}$$

11.10 AVERAGE INDUCED VOLTAGE: $v_{L_{av}}$

In an effort to develop some feeling for the impact of the derivative in an equation, the average value was defined for capacitors in Section 10.10, and a number of plots for the current were developed for an applied voltage. For inductors, a similar relationship exists between the induced voltage across a coil and the current through the coil. For inductors, the average induced voltage is defined by

$$v_{L_{av}} = L \frac{\Delta i_L}{\Delta t} \quad \text{(volts, V)} \qquad \textbf{(11.29)}$$

where Δ indicates a finite (measurable) change in current or time. Eq. (11.12) for the instantaneous voltage across a coil can be derived from Eq. (11.29) by letting V_L become vanishingly small. That is,

$$v_{L_{inst}} = \lim_{\Delta t \to 0} L \frac{\Delta i_L}{\Delta t} = L \frac{di_L}{dt}$$

In the following example, the change in current Δi_L is considered for each slope of the current waveform. *If the current increases with time, the average current is the change in current divided by the change in time,*

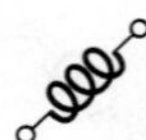

with a positive sign. If the current decreases with time, a negative sign is applied. Note in the example that the faster the current changes with time, the greater the induced voltage across the coil. When making the necessary calculations, do not forget to multiply by the inductance of the coil. Larger inductances result in increased levels of induced voltage for the same change in current through the coil.

EXAMPLE 11.8 Find the waveform for the average voltage across the coil if the current through a 4 mH coil is as shown in Fig. 11.54.

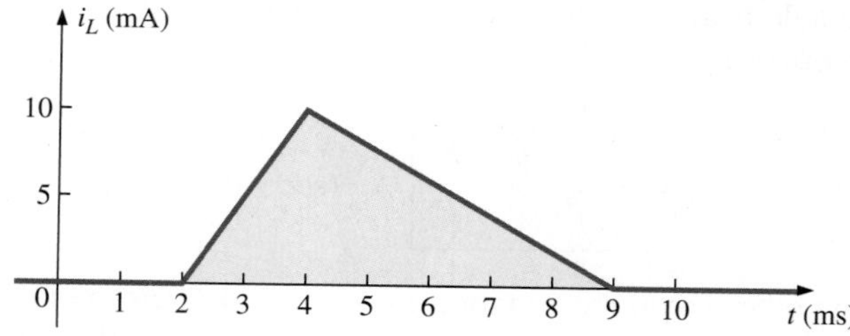

FIG. 11.54
Current i_L to be applied to a 4 mH coil in Example 11.8.

Solutions:

a. *0 to 2 ms:* Since there is no change in current through the coil, there is no voltage induced across the coil. That is,

$$v_L = L\frac{\Delta i}{\Delta t} = L\frac{0}{\Delta t} = \mathbf{0\ V}$$

b. *2 ms to 4 ms:*

$$v_L = L\frac{\Delta i}{\Delta t} = (4 \times 10^{-3}\,\mathrm{H})\left(\frac{10 \times 10^{-3}\,\mathrm{A}}{2 \times 10^{-3}\,\mathrm{s}}\right) = 20 \times 10^{-3}\,\mathrm{V} = \mathbf{20\ mV}$$

c. *4 ms to 9 ms:*

$$v_L = L\frac{\Delta i}{\Delta t} = (-4 \times 10^{-3}\,\mathrm{H})\left(\frac{10 \times 10^{-3}\,\mathrm{A}}{5 \times 10^{-3}\,\mathrm{s}}\right) = -8 \times 10^{-3}\,\mathrm{V} = \mathbf{-8\ mV}$$

d. *9 ms to ∞:*

$$v_L = L\frac{\Delta i}{\Delta t} = L\frac{0}{\Delta t} = \mathbf{0\ V}$$

The waveform for the average voltage across the coil is shown in Fig. 11.55. Note from the curve that

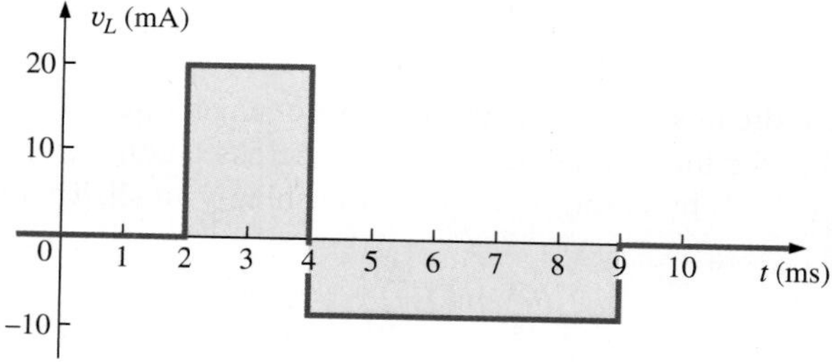

FIG. 11.55
Voltage across a 4 mH coil due to the current in Fig. 11.54.

the voltage across the coil is not determined solely by the magnitude of the change in current through the coil (Δi), but by the rate of change of current through the coil ($\Delta i/\Delta t$).

A similar statement was made for the current of a capacitor due to change in voltage across the capacitor.

A careful examination of Fig. 11.55 also reveals that the area under the positive pulse from 2 ms to 4 ms equals the area under the negative pulse from 4 ms to 9 ms. In Section 11.13, we will find that the area under the curves represents the energy stored or released by the inductor. From 2 ms to 4 ms, the inductor is storing energy, whereas from 4 ms to 9 ms, the inductor is releasing the energy stored. For the full period zero to 10 ms, energy has been stored and released; there has been no dissipation as experienced for the resistive elements. Over a full cycle, both the ideal capacitor and inductor do not consume energy but store and release it in their respective forms.

11.11 INDUCTORS IN SERIES AND IN PARALLEL

Inductors, like resistors and capacitors, can be placed in series or in parallel. Increasing levels of inductance can be obtained by placing inductors in series, while decreasing levels can be obtained by placing inductors in parallel.

For inductors in series, the total inductance is found in the same manner as the total resistance of resistors in series (Fig. 11.56):

$$L_T = L_1 + L_2 + L_3 + \cdots + L_N \tag{11.30}$$

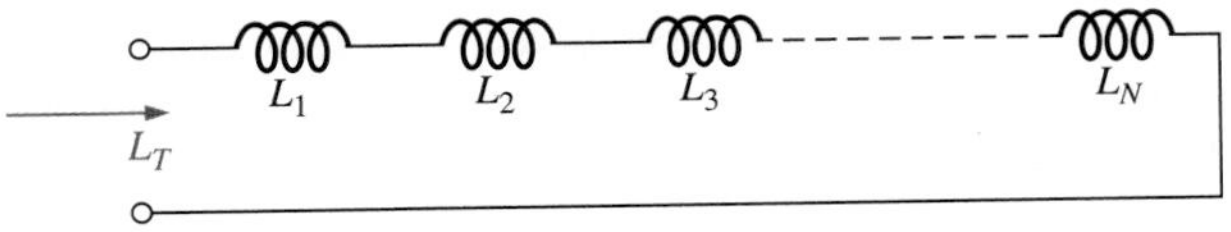

FIG. 11.56
Inductors in series.

For inductors in parallel, the total inductance is found in the same manner as the total resistance of resistors in parallel (Fig. 11.57):

$$\frac{1}{L_T} = \frac{1}{L_1} + \frac{1}{L_2} + \frac{1}{L_3} + \cdots + \frac{1}{L_N} \tag{11.31}$$

For two inductors in parallel,

$$L_T = \frac{L_1 L_2}{L_1 + L_2} \tag{11.32}$$

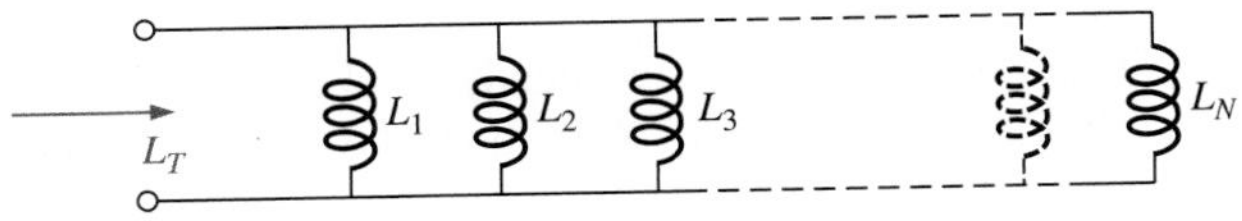

FIG. 11.57
Inductors in parallel.

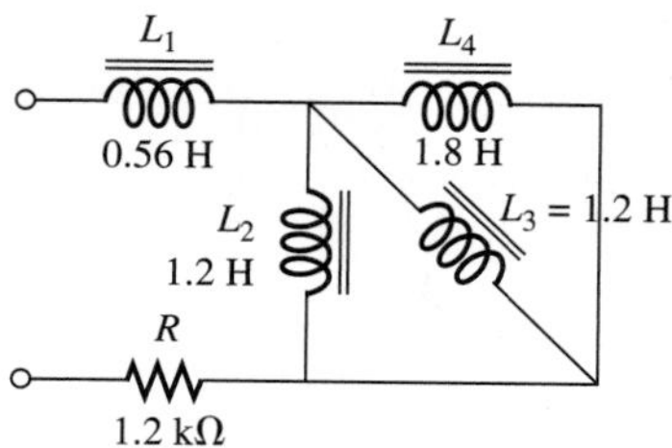

FIG. 11.58
Example 11.9.

EXAMPLE 11.9 Reduce the network in Fig. 11.58 to its simplest form.

Solution: Inductors L_2 and L_3 are equal in value and they are in parallel, resulting in an equivalent parallel value of

$$L'_T = \frac{L}{N} = \frac{1.2\text{ H}}{2} = 0.6\text{ H}$$

The resulting 0.6 H is then in parallel with the 1.8 H inductor, and

$$L''_T = \frac{(L'_T)(L_4)}{L'_T + L_4} = \frac{(0.6\text{ H})(1.8\text{ H})}{0.6\text{ H} + 1.8\text{ H}} = 0.45\text{ H}$$

Inductor L_1 is then in series with the equivalent parallel value, and

$$L_T = L_1 + L''_T = 0.56\text{ H} + 0.45\text{ H} = \mathbf{1.01\text{ H}}$$

The reduced equivalent network appears in Fig. 11.59.

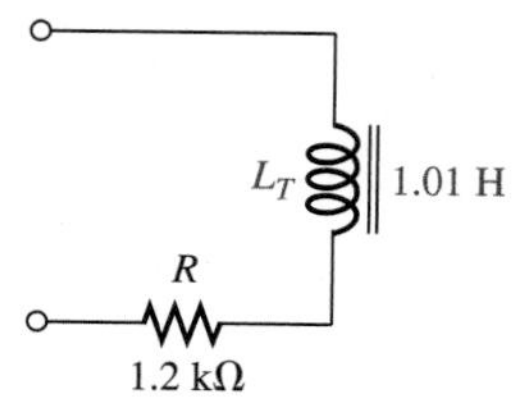

FIG. 11.59
Terminal equivalent of the network in Fig. 11.58.

11.12 STEADY-STATE CONDITIONS

We found in Section 11.5 that, for all practical purposes, an ideal (ignoring internal resistance and stray capacitances) inductor can be replaced by a short-circuit equivalent once steady-state conditions have been established. Recall that the term *steady state* implies that the voltage and current levels have reached their final resting value and will no longer change unless a change is made in the applied voltage or circuit configuration. For all practical purposes, our assumption is that steady-state conditions have been established after five time constants of the storage or release phase have passed.

For the circuit in Fig. 11.60(a), for example, if we assume that steady-state conditions have been established, the inductor can be removed and replaced by a short-circuit equivalent as shown in Fig. 11.60(b). The short-circuit equivalent shorts out the 3 Ω resistor, and current I_1 is determined by

$$I_1 = \frac{E}{R_1} = \frac{10\text{ V}}{2\ \Omega} = \mathbf{5\text{ A}}$$

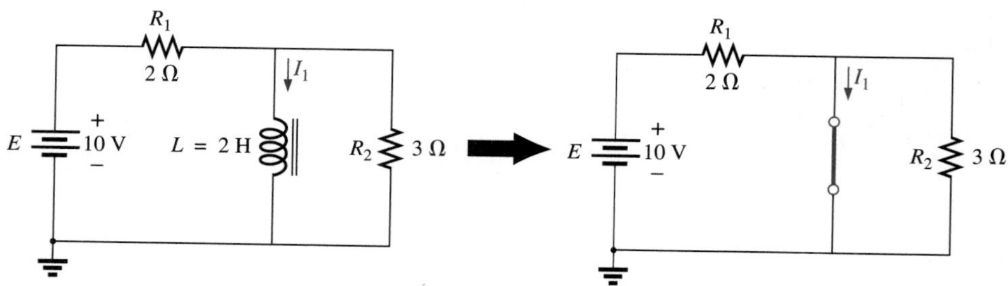

FIG. 11.60
Substituting the short-circuit equivalent for the inductor for t > 5τ.

For the circuit in Fig. 11.61(a), the steady-state equivalent is as shown in Fig. 11.61(b). This time, resistor R_1 is shorted out, and resistors R_2 and R_3 now appear in parallel. The result is

$$I = \frac{E}{R_2 \parallel R_3} = \frac{21\text{ V}}{2\ \Omega} = \mathbf{10.5\text{ A}}$$

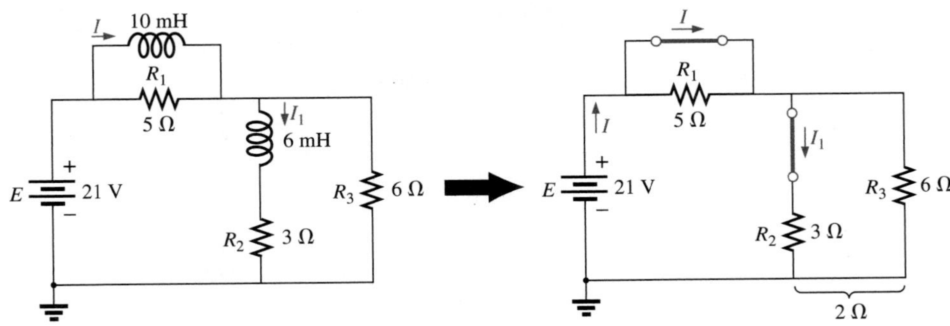

FIG. 11.61
Establishing the equivalent network for $t > 5\tau$.

Applying the current divider rule yields

$$I_1 = \frac{R_3 I}{R_3 + R_2} = \frac{(6\ \Omega)(10.5\ \text{A})}{6\ \Omega + 3\ \Omega} = \frac{63}{9}\text{A} = \mathbf{7\ A}$$

In the examples to follow, it is assumed that steady-state conditions have been established.

EXAMPLE 11.10 Find the current I_L and the voltage V_C for the network in Fig. 11.62.

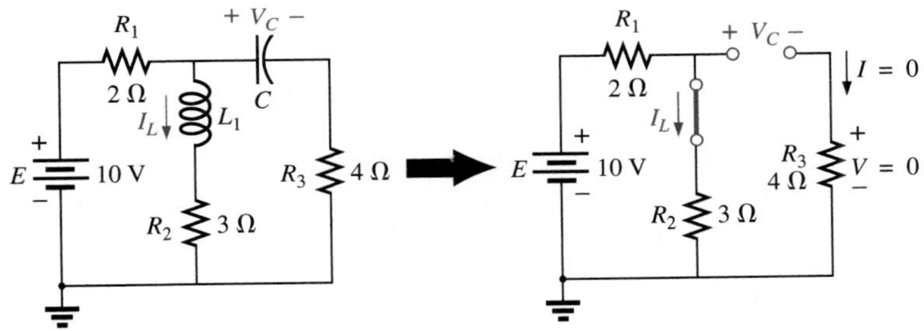

FIG. 11.62
Example 11.10.

Solution:

$$I_L = \frac{E}{R_1 + R_2} = \frac{10\ \text{V}}{5\ \Omega} = \mathbf{2\ A}$$

$$V_C = \frac{R_2 E}{R_2 + R_1} = \frac{(3\ \Omega)(10\ \text{V})}{3\ \Omega + 2\ \Omega} = \mathbf{6\ V}$$

EXAMPLE 11.11 Find currents I_1 and I_2 and voltages V_1 and V_2 for the network in Fig. 11.63.

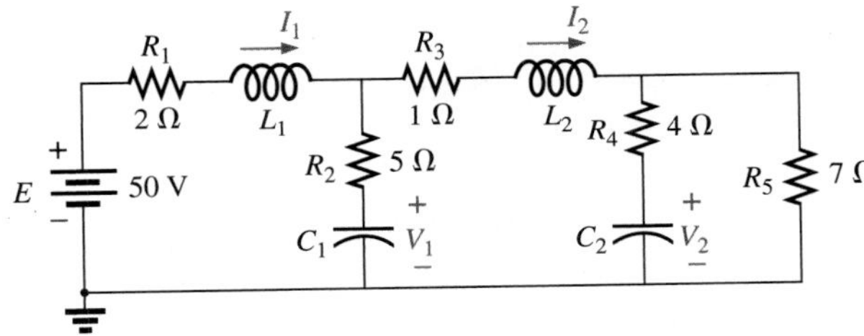

FIG. 11.63
Example 11.11.

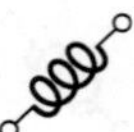

Solution: Note Fig. 11.64.

$$I_1 = I_2$$

$$= \frac{E}{R_1 + R_3 + R_5} = \frac{50\text{ V}}{2\ \Omega + 1\ \Omega + 7\ \Omega} = \frac{50\text{ V}}{10\ \Omega} = \mathbf{5\ A}$$

$$V_2 = I_2R_5 = (5\text{ A})(7\ \Omega) = \mathbf{35\ V}$$

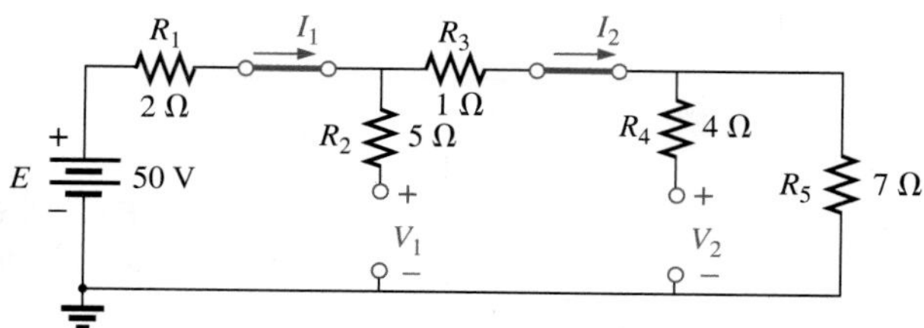

FIG. 11.64
Substituting the short-circuit equivalents for the inductors and the open-circuit equivalents for the capacitor for $t > 5\tau$.

Applying the voltage divider rule yields

$$V_1 = \frac{(R_3 + R_5)E}{R_1 + R_3 + R_5} = \frac{(1\ \Omega + 7\ \Omega)(50\text{ V})}{2\ \Omega + 1\ \Omega + 7\ \Omega} = \frac{(8\ \Omega)(50\text{ V})}{10\ \Omega} = \mathbf{40\ V}$$

11.13 ENERGY STORED BY AN INDUCTOR

The ideal inductor, like the ideal capacitor, does not dissipate the electrical energy supplied to it. It stores the energy in the form of a magnetic field. A plot of the voltage, current, and power to an inductor is shown in Fig. 11.65 during the buildup of the magnetic field surrounding the inductor. The energy stored is represented by the shaded area under the power curve. Using calculus, we can show that the evaluation of the area under the curve yields

$$\boxed{W_{\text{stored}} = \frac{1}{2}LI_m^2} \qquad \text{(joules, J)} \qquad \textbf{(11.33)}$$

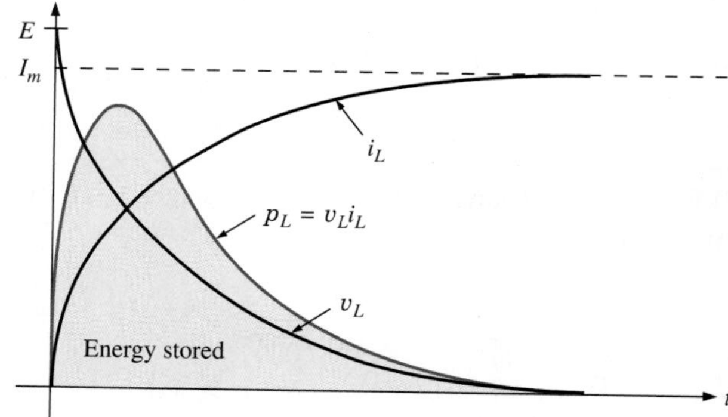

FIG. 11.65
The power curve for an inductive element under transient conditions.

EXAMPLE 11.12 Find the energy stored by the inductor in the circuit in Fig. 11.66 when the current through it has reached its final value.

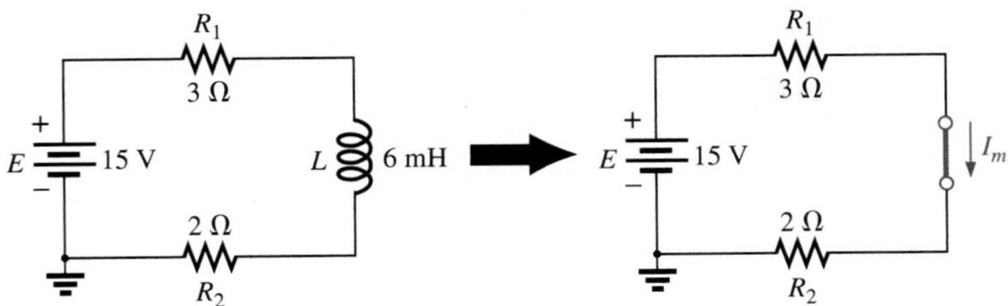

FIG. 11.66
Example 11.12.

Solution:

$$I_m = \frac{E}{R_1 + R_2} = \frac{15\text{ V}}{3\ \Omega + 2\ \Omega} = \frac{15\text{ V}}{5\ \Omega} = 3\text{ A}$$

$$W_{\text{stored}} = \frac{1}{2}LI_m^2 = \frac{1}{2}(6 \times 10^{-3}\text{ H})(3\text{ A})^2 = \frac{54}{2} \times 10^{-3}\text{ J} = \mathbf{27\ mJ}$$

11.14 APPLICATIONS

Camera Flash Lamp

The inductor played an important role in the camera flash lamp circuitry described in the Application section of Chapter 10 on capacitors. For the camera, the inductor was the important component that resulted in the high spike voltage across the trigger coil, which was then magnified by the autotransformer action of the secondary to generate the 4000 V necessary to ignite the flash lamp. Recall that the capacitor in parallel with the trigger coil charged up to 300 V using the low-resistance path provided by the SCR (silicon-controlled rectifier). However, once the capacitor was fully charged, the short-circuit path to ground provided by the SCR was removed, and the capacitor immediately started to discharge through the trigger coil. Since the only resistance in the time constant for the inductive network is the relatively low resistance of the coil itself, the current through the coil grew at a very rapid rate. A significant voltage was then developed across the coil as defined by Eq. (11.12): $v_L = L\,(di_L/dt)$. This voltage was in turn increased by transformer action to the secondary coil of the autotransformer, and the flash lamp was ignited. That high voltage generated across the trigger coil also appears directly across the capacitor of the trigger network. The result is that it begins to charge up again until the generated voltage across the coil drops to zero volts. However, when it does drop, the capacitor again discharges through the coil, establishes another charging current through the coil, and again develops a voltage across the coil. The high-frequency exchange of energy between the coil and capacitor is called *flyback* because of the "flying back" of energy from one storage element to the other. It begins to decay with time because of the resistive elements in the loop. The more resistance, the more quickly it dies out. If the capacitor-inductor pairing is isolated and "tickled" along the way with the application of a dc voltage, the high frequency-generated voltage across the coil can be maintained and put to good use. In fact, it is this flyback effect that is used to generate a steady dc voltage (using rectification to convert the oscillating waveform to one of a steady dc nature) that is commonly used in TVs.

Household Dimmer Switch

Inductors can be found in a wide variety of common electronic circuits in the home. The typical household dimmer uses an inductor to protect the other components and the applied load from "rush" currents—currents that increase at very high rates and often to excessively high levels. This feature is particularly important for dimmers since they are most commonly used to control the light intensity of an incandescent lamp. When a lamp is turned on, the resistance is typically very low, and relatively high currents may flow for short periods of time until the filament of the bulb heats up. The inductor is also effective in blocking high-frequency noise (RFI) generated by the switching action of the triac in the dimmer. A capacitor is also normally included from line to neutral to prevent any voltage spikes from affecting the operation of the dimmer and the applied load (lamp, etc.) and to assist with the suppression of RFI disturbances.

A photograph of one of the most common dimmers is provided in Fig. 11.67(a), with an internal view shown in Fig. 11.67(b). The basic compo-

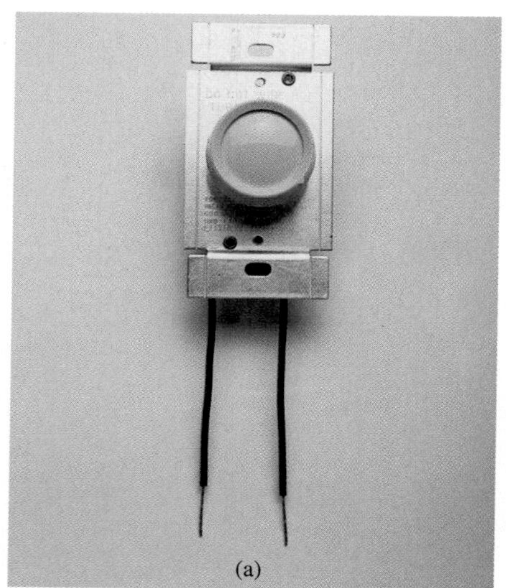

(a)

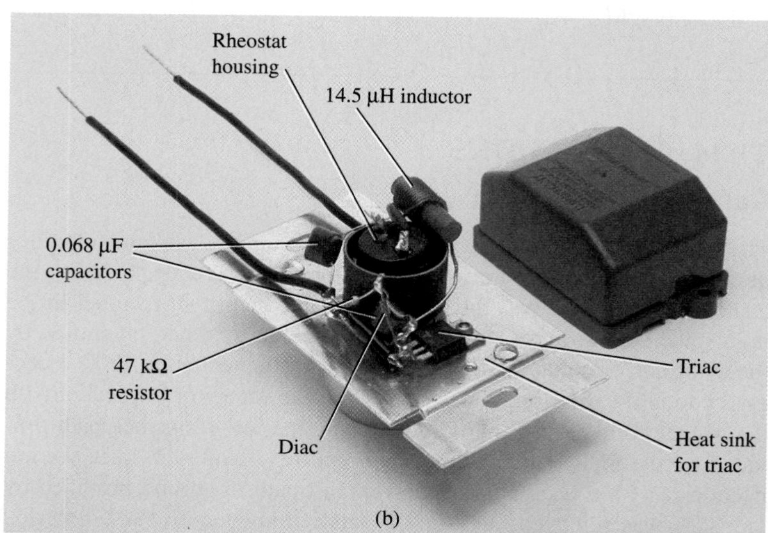

(b)

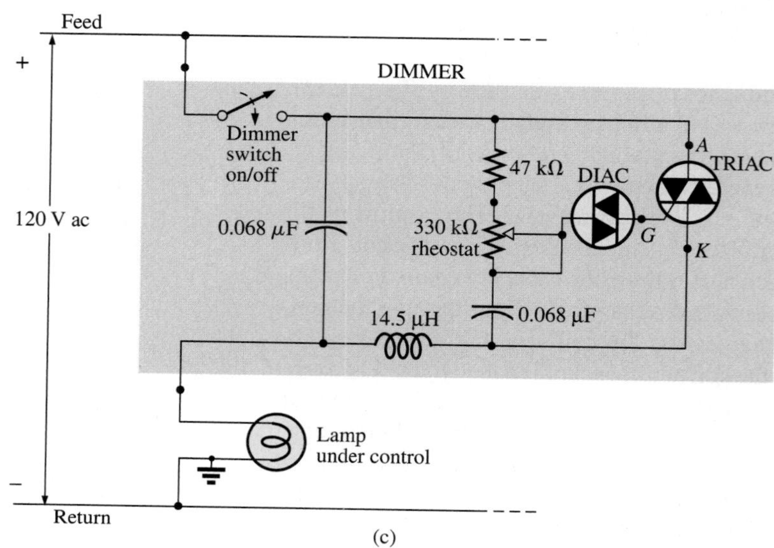

(c)

FIG. 11.67

Dimmer control: (a) external appearance; (b) internal construction; (c) schematic.

nents of most commercially available dimmers appear in the schematic in Fig. 11.67(c). In this design, a 14.5 μH inductor is used in the choking capacity described above, with a 0.068 μF capacitor for the "bypass" operation. Note the size of the inductor with its heavy wire and large ferromagnetic core and the relatively large size of the two 0.068 μF capacitors. Both suggest that they are designed to absorb high-energy disturbances.

The general operation of a dimmer is shown in Fig. 11.68. The controlling network is in series with the lamp and essentially acts as an impedance (like resistance—to be introduced in Chapter 15) that can vary between very low and very high levels. Very low impedance levels resemble a short circuit so that the majority of the applied voltage appears across the lamp [Fig. 11.68(a)] and very high impedances approach an open circuit where very little voltage appears across the lamp [Fig. 11.68(b)]. Intermediate levels of impedance control the terminal voltage of the bulb accordingly. For instance, if the controlling network has a very high impedance (open-circuit equivalent) through half the cycle, as shown in Fig. 11.68(c), the brightness of the bulb will be less than full voltage but not 50% due to the nonlinear relationship between the brightness of a bulb and the applied voltage. A lagging effect is also present in the actual operation of the dimmer, which we will learn about when leading and lagging networks are examined in the ac chapters.

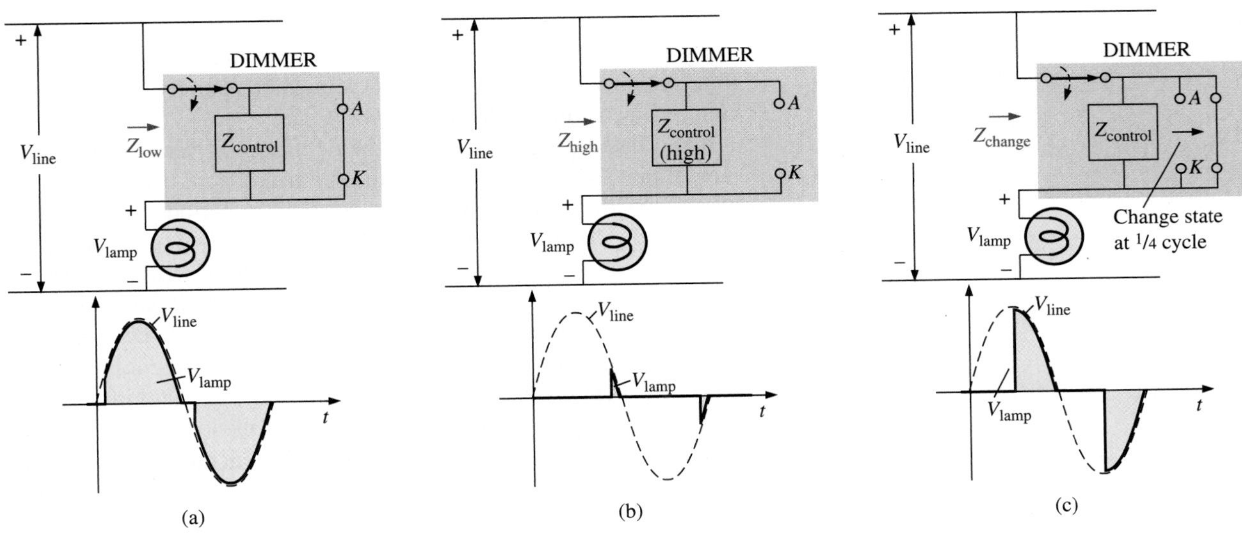

FIG. 11.68

Basic operation of the dimmer in Fig. 11.67: (a) full voltage to the lamp; (b) approaching the cutoff point for the bulb; (c) reduced illumination of the lamp.

The controlling knob, slide, or whatever other method is used on the face of the switch to control the light intensity is connected directly to the rheostat in the branch parallel to the triac. Its setting determines when the voltage across the capacitor reaches a sufficiently high level to turn on the diac (a bidirectional diode) and establish a voltage at the gate (G) of the triac to turn it on. When it does, it establishes a very low resistance path from the anode (A) to the cathode (K), and the applied voltage appears directly across the lamp. When the SCR is off, its terminal resistance between anode and cathode is very high and can be approximated by an open circuit. During this period, the applied voltage does not reach the load (lamp). At this time, the impedance of the parallel branch containing the rheostat, fixed resistor, and capacitor is sufficiently high compared to the load that

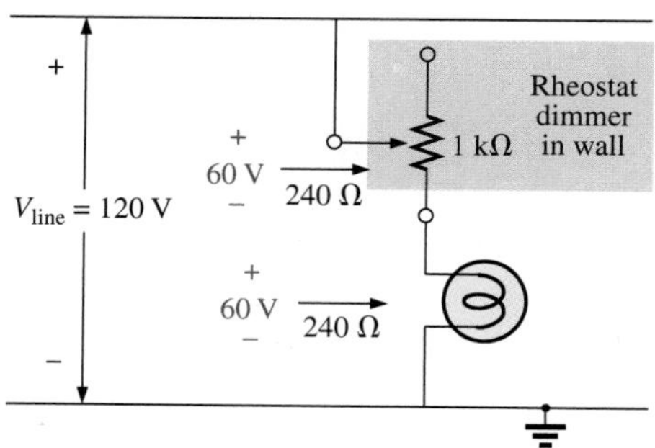

FIG. 11.69
Direct rheostat control of the brightness of a 60 W bulb.

it can also be ignored, completing the open-circuit equivalent in series with the load. Note the placement of the elements in the photograph in Fig. 11.67(b) and that the metal plate to which the triac is connected is actually a heat sink for the device. The on/off switch is in the same housing as the rheostat. The total design is certainly well planned to maintain a relatively small size for the dimmer.

Since the effort here is to control the amount of power getting to the load, the question is often asked, Why don't we just use a rheostat in series with the lamp? The question is best answered by examining Fig. 11.69, which shows a rather simple network with a rheostat in series with the lamp. At full wattage, a 60 W bulb on a 120 V line theoretically has an internal resistance of $R = V^2/P$ (from the equation $P = V^2/R$) $=$ $(120\text{ V})^2/60\text{ W} = 240\ \Omega$. Although the resistance is sensitive to the applied voltage, we will assume this level for the following calculations.

If we consider the case where the rheostat is set for the same level as the bulb, as shown in Fig. 11.69, there will be 60 V across the rheostat and the bulb. The power to each element is then $P = V^2/R = (60\text{ V})^2/240\ \Omega =$ 15 W. The bulb is certainly quite dim, but the rheostat inside the dimmer switch is dissipating 15 W of power on a continuous basis. When you consider the size of a 2 W potentiometer in your laboratory, you can imagine the size rheostat you would need for 15 W, not to mention the purchase cost, although the biggest concern would probably be all the heat developed in the walls of the house. You would be paying for electric power that was not performing a useful function. Also, if you had four dimmers set at the same level, you would actually be wasting sufficient power to fully light another 60 W bulb.

On occasion, especially when the lights are set very low by the dimmer, a faint "singing" can sometimes be heard from the light bulb. This effect sometimes occurs when the conduction period of the dimmer is very small. The short, repetitive voltage pulse applied to the bulb sets the bulb into a condition similar to a resonance state (Chapter 20). The short pulses are just enough to heat up the filament and its supporting structures, and then the pulses are removed to allow the filament to cool down again for a longer period of time. This repetitive heating and cooling cycle can set the filament in motion, and the "singing" can be heard in a quiet environment. Incidentally, the longer the filament, the louder the "singing." A further condition for this effect is that the filament must be in the shape of a coil and not a straight wire so that the "slinky" effect can develop.

TV or PC Monitor Yolk

Inductors and capacitors play a multitude of roles in the operation of a TV or PC monitor. However, the most obvious use of the coil is in the yolk assembly wrapped around the neck of the tube as shown in Fig. 11.70. The tube itself, in addition to providing the screen for viewing, is actually a large capacitor which plays an integral part in establishing the high dc voltage for the proper operation of the monitor.

A photograph of the yolk assembly of a cathode-ray tube is provided in Fig. 11.71(a). It is constructed of four 28 mH coils with two sets of two coils connected at one point [Fig 11.71(b)] so that they share the same current and establish the same magnetic field. The purpose of the yolk assembly is to control the direction of the electron beam from the cathode to the screen of the tube. When the cathode is heated to a very temperature by a filament internal to the structure, electrons are emitted into the surrounding media. The placement of a very high positive potential (10 kV to 25 kV

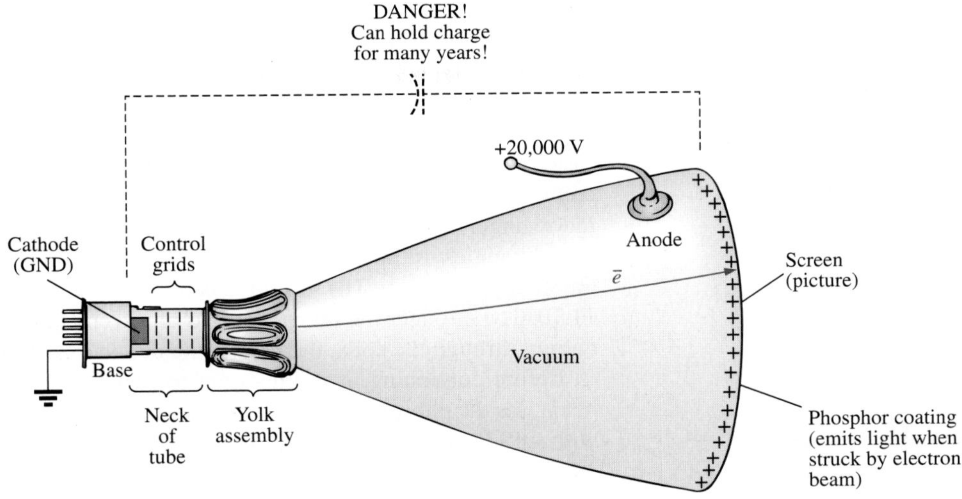

FIG. 11.70
Yolk assembly for a TV or PC tube.

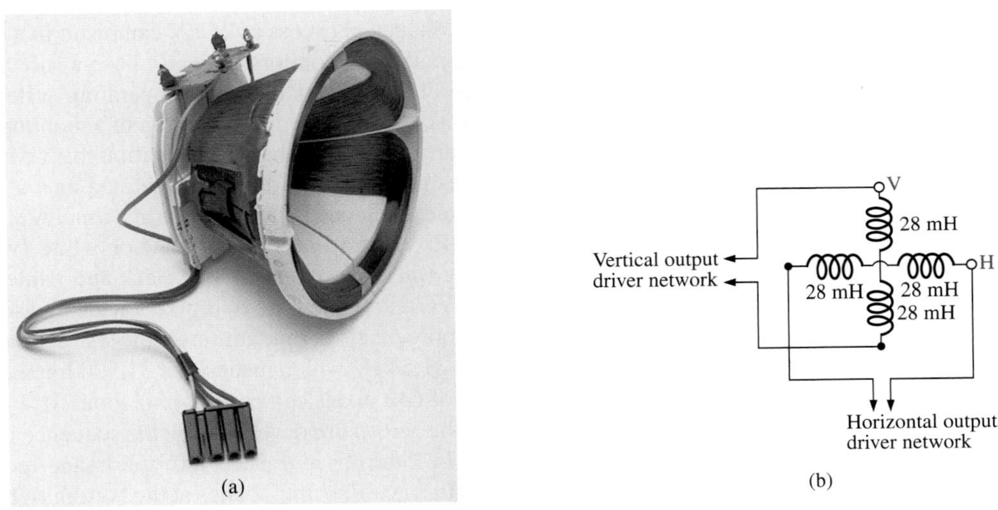

FIG. 11.71
(a) Black-and-white TV yolk assembly; (b) schematic representation.

dc) to the conductive coating on the face of the tube attracts the emitted electrons at a very high speed and therefore at a high level of kinetic energy. When the electrons hit the phosphorescent coating (usually white, green, or amber) on the screen, light is emitted which can be seen by someone facing the monitor. The beam characteristics (such as intensity, focus, and shape) are controlled by a series of grids placed relatively close to the cathode in the neck of the tube. The grid is such that the negatively charged electrons can easily pass through, but the number and speed with which they pass can be controlled by a negative potential applied to the grid. The grids cannot have a positive potential because the negatively charged electrons would be attracted to the grid structure and would eventually disintegrate from the high rate of conduction. Negative potentials on the grids control the flow of electrons by repulsion and by masking the attraction for the large positive potential applied to the face of the tube.

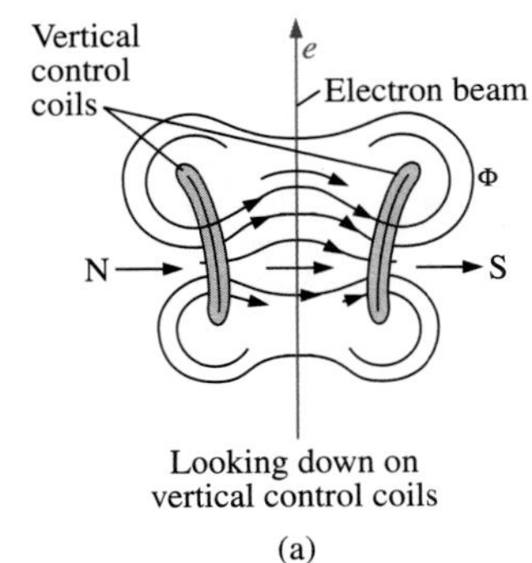

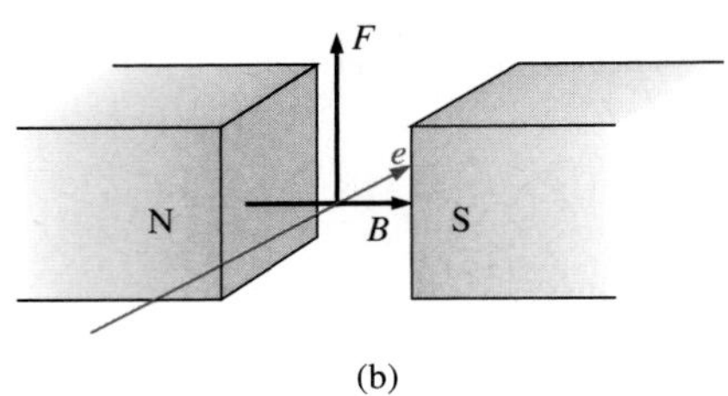

FIG. 11.72
Deflection coils: (a) vertical control; (b) right-hand-rule (RHR) for electron flow.

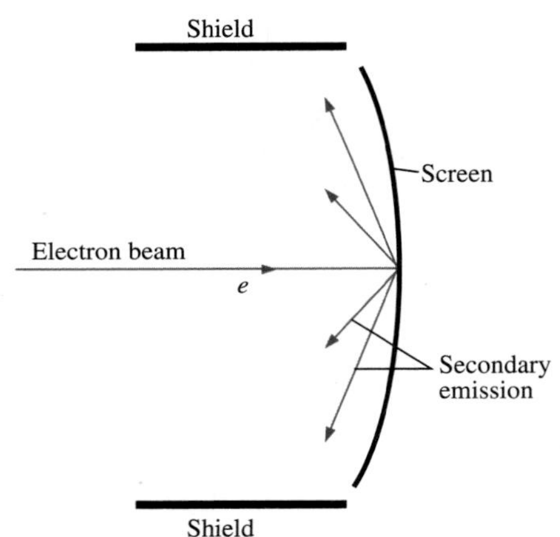

FIG. 11.73
Secondary emission from and protective measures for a TV or PC monitor.

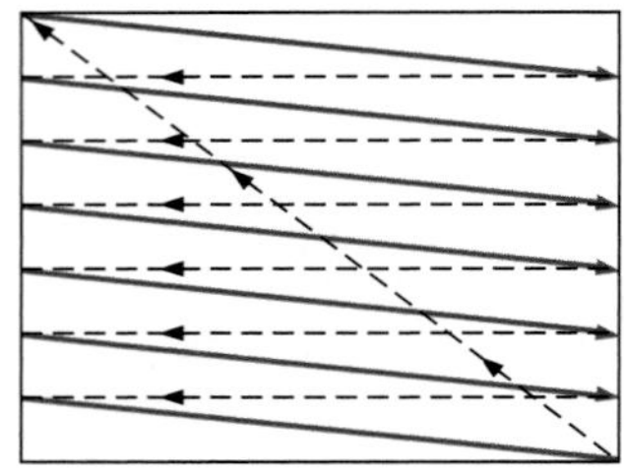

FIG. 11.74
Pattern generation.

Once the beam has been established with the desired intensity and shape, it must be directed to a particular location on the screen using the yolk assembly. For vertical control, the two coils on the side establish a magnetic flux pattern as shown in Fig. 11.72(a). The resulting direction of the magnetic field is from left to right as shown in Figs. 11.72(a) and 11.72(b). Using your right hand, with the index finger pointing in the direction of the magnetic field and the middle finger (at right angles to the index finger) in the direction of electron flow, results in the thumb (also at right angles to the index finger) pointing in the direction of the force on the electron beam. The result is a bending of the beam as shown in Fig. 11.70. The stronger the magnetic field of the coils as determined by the current through the coils, the greater the deflection of the beam.

Before continuing, it is important to realize that when the electron beam hits the phosphorescent screen as shown in Fig. 11.73, it is moving with sufficient velocity to cause a secondary emission of X-rays that scatter to all sides of the monitor. Even though the X-rays die off exponentially with distance from the source, there is some concern about safety, and monitors today have shields all around the outside surface of the tube as shown in Fig. 11.73. Actually, it is not direct viewing that is of some concern but rather viewing by individuals to the side, above, or below the screen. Monitors are currently limited to 25 kV at the anode because the application of voltages in excess of 25 kV can result in a direct emission of X-rays. Internally, all monitors currently have a safety shutoff to ensure that this level is never attained in the operating system.

A detailed discussion of the full operation of a monitor is not possible here, but there are some facts about its operation that reveal the sophistication of the design. When an image is generated on a screen, it is done one *pixel* at a time along one horizontal line at a time. A pixel is one point on the screen. Pixels are black (no signal) or white (with signal) for black-and-white (monochromic) TVs or black and white or some color for color TVs. VGA monitors are 640 pixels wide and 480 pixels high. Obviously the more pixels in the same area, the sharper the image. A typical scan rate is 31.5 kHz which means that 31,500 lines can be drawn in 1 s, or one line of 640 pixels can be drawn in about 31.7 μs.

Patterns on the screen are developed by the sequence of lines appearing in Fig. 11.74. Starting at the top left, the image moves across the screen down to the next line until it ends at the bottom right of the screen, at which point there is a rapid retrace (invisible) back to the starting point. Typical scanning rates (full image generated) extend from 60 frames per second to 80 frames per second. The slower the rate, the higher the possibility of flickering in the images. At 60 frames per second, one entire frame is generated every $1/60 = 16.67$ ms $= 0.017$ s.

Color monitors are particularly interesting because all colors on the screen are generated by the colors red, blue, and green. The reason is that the human eye responds to the wavelengths and energy levels of the various colors. The absence of any color is black, and the result of full energy to each of the three colors is white. The color yellow is a combination of red and green with no blue, and pink is primarily red energy with smaller amounts of blue and green. An in-depth description of this "additive" type of color generation must be left for another course.

The fact that three colors define the resulting color requires that there be three cathodes in a color monitor to generate three electron beams. However, the three beams must sweep the screen in the same relative positions. Each pixel is now made up of three color dots in the same relative position

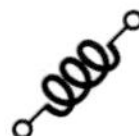

for each pixel, as shown in Fig. 11.75. Each dot has a phosphorescent material that generates the desired color when hit with an electron beam. For situations where the desired color has no green, the electron beam associated with the color green is turned off. In fact, between each pixel, each beam is shut down to provide definition between the color pixels. The dots within the pixel are so close that the human eye cannot pick up the individual colors but simply the color that results from the "additive" process.

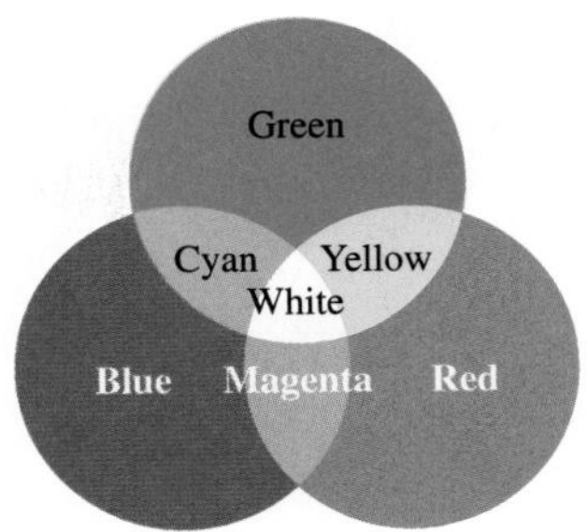

FIG. 11.75
Color television pixels.

During the entire "on" time of a monitor, a full 10 kV to 25 kV are applied to the conductor on the screen to attract electrons. Over time there is naturally an accumulation of negative charge on the screen which remains after the power is turned off—a typical capacitive storage charge. For a brief period of time, it sits with 25 kV across the plates which drop as the "capacitor" begins to discharge. However, the lack of a low-resistance path often results in a storage of the charge for a fairly long period of time. This stored charge and the associated voltage across the plates are sufficiently high to cause severe damage. It is therefore paramount that TVs and monitors be repaired or investigated only by someone who is well versed in how to discharge the tube. One commonly applied procedure is to attach a long lead from the metal shaft of a flat-edge screwdriver to a good ground connection. Then leave the anode connection to the tube in place, and insert the screwdriver under the cap until it touches the metal clip of the cap. You will probably hear a loud snap when discharge occurs. Because of the enormous amount of residual charge, it is recommended that the above procedure be repeated two or three times. Even then, treat the tube with a great deal of respect. In short, until you become familiar with the discharge procedure, leave the investigation of TVs and monitors to someone with the necessary experience. Very high pulse voltages are also generated in an operating system. Be aware that they are of a magnitude that could destroy standard test equipment.

The capacitive effect of the tube is an integral part of developing the high dc anode potential. Its filtering action smooths out the repetitive, high-voltage pulses generated by the flyback action of the TV. Otherwise, the screen would simply be a flickering pattern as the anode potential switched on and off with the pulsating signal.

11.15 COMPUTER ANALYSIS

PSpice

Transient *RL* Response The computer analysis begins with a transient analysis of the network of parallel inductive elements in Fig. 11.76. The inductors are picked up from the **ANALOG** library in the **Place Part** dialog box. As noted in Fig. 11.76, the inductor is displayed with its terminal identification which is helpful for identifying nodes when calling for specific output plots and values. In general, when an element is first placed on a schematic, the number 1 is assigned to the left end on a horizontal display and to the top on a vertical display. Similarly, the number 2 is assigned to the right end of an element in the horizontal position and to the bottom in the vertical position. Be aware, however, that the option **Rotate** rotates the element in the CCW direction, so taking a horizontal resistor to the vertical position requires three rotations to get the number 1 to the top again. In previous chapters, you may have noted that a number of the outputs were taken off terminal 2 because a single rotation placed this terminal at the top of the vertical display.

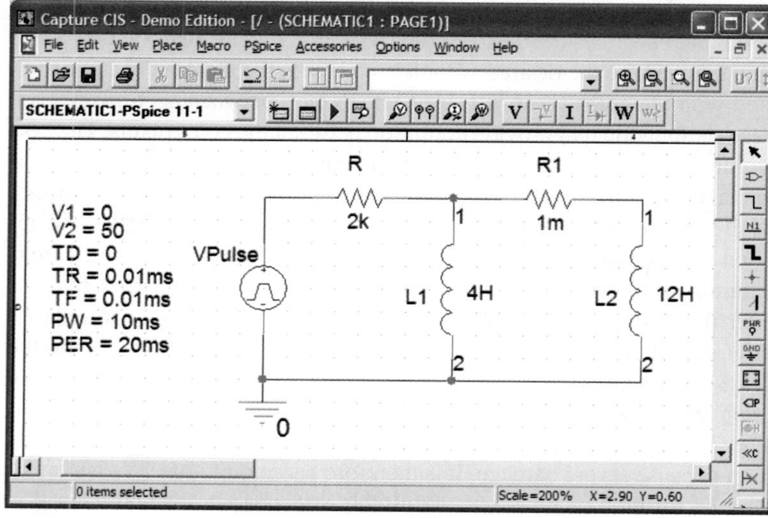

FIG. 11.76
Using PSpice to obtain the transient response of a parallel inductive network due to an applied pulse of 50 V.

Also note in Fig. 11.76 the need for a series resistor R_1 within the parallel loop of inductors. In PSpice, inductors must have a series resistor to reflect real-world conditions. The chosen value of 1 mΩ is so small, however, that it will not affect the response of the system. For **VPulse** (obtained from the SOURCE Library), the rise and fall times were selected as 0.01 ms, and the pulse width was chosen as 10 ms because the time constant of the network is $\tau = L_T/R = (4\text{ H} \parallel 12\text{ H})/2\text{ k}\Omega = 1.5\text{ ms}$ and $5\tau = 7.5\text{ ms}$.

The simulation is the same as applied when obtaining the transient response of capacitive networks. In condensed form, the sequence to obtain a plot of the voltage across the coils versus time is as follows: **New SimulationProfile** key-**PSpice 11-1-Create-TimeDomain(Transient)-Run to time:**10ms-**Start saving data after:**0s and **Maximum step size:**5μs-**OK-Run PSpice** key-**Add Trace** key-**V1(L2)-OK.** The resulting trace appears in the bottom of Fig. 11.77. A maximum step size of 5 μs was chosen to ensure that it was less than the rise or fall times of 10 μs. Note that the voltage across the coil jumps to the 50 V level almost immediately; then it decays to 0 V in about 8 ms. A plot of the total current through the parallel coils can be obtained through **Plot-Plot to Window-Add Trace** key-**I(R)-OK,** resulting in the trace appearing at the top of Fig. 11.77. When the trace first appeared, the vertical scale extended from 0 A to 40 mA even though the maximum value of i_R was 25 mA. To bring the maximum value to the top of the graph, **Plot** was selected followed by **Axis Settings-Y Axis-User Defined-0A to 25mA-OK.**

For values, the voltage plot was selected, **SEL>>,** followed by the **Toggle cursor** key and a click on the screen to establish the crosshairs. The left-click cursor was set on one time constant to reveal a value of 18.461 V for **A1** (about 36.8% of the maximum as defined by the exponential waveform). The right-click cursor was set at 7.5 ms or five time constants, resulting in a relatively low 0.338 V for **A2.**

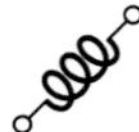

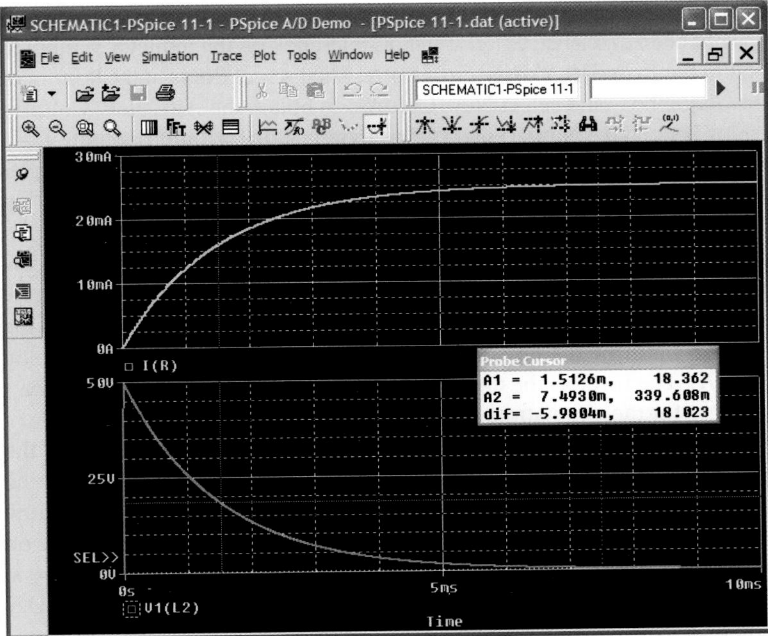

FIG. 11.77

The transient response of v_L and i_R for the network in Fig. 11.76.

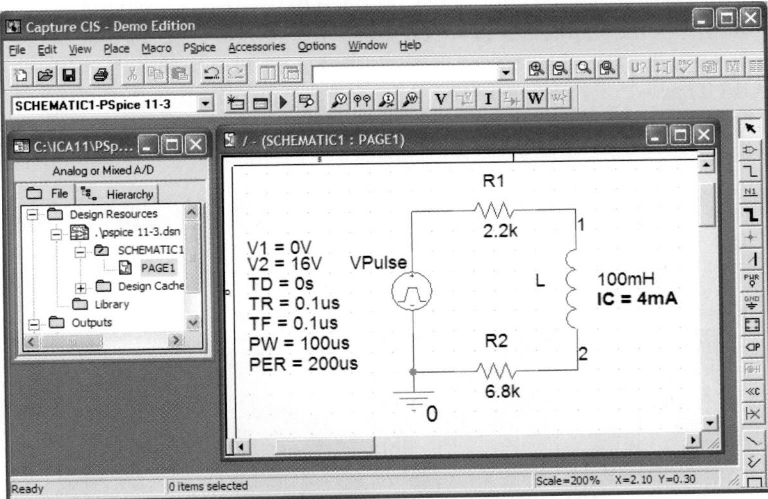

FIG. 11.78

Using PSpice to determine the transient response for a circuit in which the inductive element has an initial condition.

Transient Response with Initial Conditions The next application verifies the results of Example 11.4 which has an initial condition associated with the inductive element. **VPULSE** is again employed with the parameters appearing in Fig. 11.78. Since $\tau = L/R = 100 \text{ mH}/(2.2 \text{ k}\Omega + 6.8 \text{ k}\Omega) = 100 \text{ mH}/9 \text{ k}\Omega = 11.11 \ \mu\text{s}$ and $5\tau = 55.55 \ \mu\text{s}$, the pulse width (**PW**) was set to 100 μs. The rise and fall times were set at $100 \ \mu\text{s}/1000 = 0.1 \ \mu\text{s}$. Note again that the labels 1 and 2 appear with the inductive element.

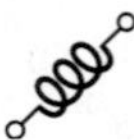

Setting the initial conditions for the inductor requires a procedure that has not been described as yet. First double-click on the inductor symbol to obtain the **Property Editor** dialog box. Then select **Parts** at the bottom of the dialog box, and select **New Column** to obtain the **Add New Column** dialog box. Under **Name,** enter **IC** (an abbreviation for "initial condition"—not "capacitive current") followed by the initial condition of 4 mA under **Value;** then click **OK.** The **Property Editor** dialog box appears again, but now the initial condition appears as a **New Column** in the horizontal listing dedicated to the inductive element. Now select **Display** to obtain the **Display Properties** dialog box, and under **Display Format** choose **Name and Value** so that both **IC** and **4mA** appear. Click **OK** to return to the **Property Editor** dialog box. Finally, click on **Apply** and exit the dialog box (**X**). The result is the display in Fig. 11.78 for the inductive element.

Now for the simulation. First select the **New Simulation Profile** key, insert the name **PSpice 11-3,** and follow up with **Create.** Then in the **Simulation Settings** dialog box, select **Time Domain(Transient)** for the **Analysis type** and **General Settings** for the **Options.** The **Run to time** should be 200 μs so that you can see the full effect of the pulse source on the transient response. The **Start saving data after** should remain at 0 s, and the **Maximum step size** should be 200 μs/1000 = 200 ns. Click **OK** and then select the **Run PSpice** key. The result is a screen with an x-axis extending from 0 to 200 μs. Selecting **Trace** to get to the **Add Traces** dialog box and then selecting **I(L)** followed by **OK** results in the display in Fig. 11.79. The plot for **I(L)** clearly starts at the initial value of 4 mA and then decays to 1.78 mA as defined by the left-click cursor. The right-click cursor reveals that the current has dropped to 0.222 μA (essentially 0 A) after the pulse source has dropped to 0 V for 100 μs. The **VPulse** source was placed in the same figure through **Plot-Add Plot to Window-Trace-Add Trace-V(VPulse:+)-OK** to permit a comparison between the applied voltage and the resulting inductor current.

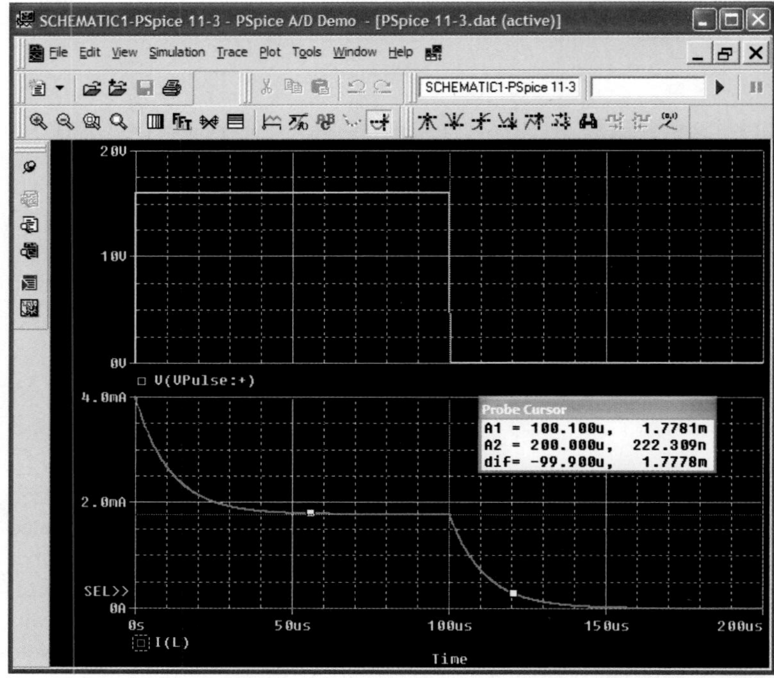

FIG. 11.79
A plot of the applied pulse and resulting current for the circuit in Fig. 11.78.

Multisim

The transient response of an R-L network can also be obtained using Multisim. The circuit to be examined appears in Fig. 11.80 with a pulse voltage source to simulate the closing of a switch at $t = 0$ s. The source, **PULSE_VOLTAGE,** is found under **SIGNAL_VOLTAGE** Source Family. When selected, it appears with a label, an initial voltage, a step voltage, and the time period for each level. All can be changed by double-clicking on the source symbol to obtain the dialog box. As shown in Fig. 11.80, the **Pulsed Value** will be set at 20 V, and the **Delay Time** to 0 s. The **Rise Time** and **Fall Time** will both remain at the default levels of 1 ns. For our analysis we want a **Pulse Width** that is at least twice the 5τ transient period of the circuit. For the chosen values of R and L, $\tau = L/R = 10\text{ mH}/100\ \Omega = 0.1\text{ ms} = 100\ \mu\text{s}$. The transient period of 5τ is therefore 500 μs or 0.5 ms. Thus, a **Pulse Width** of 1 ms would seem appropriate with a **Period** of 2 ms. The result is a frequency of $f = I/T = 1/2\text{ ms} = 500\text{ Hz}$. When all have been set and selected, the parameters of the pulse source appear as shown in Fig. 11.80. Next the resistor, inductor, and ground are placed on the screen to complete the circuit.

The simulation process is initiated by the following sequence: **Simulate-Analyses-Transient Analysis.** The result is the **Transient Analysis** dialog box in which **Analysis Parameters** is chosen first. Under **Parameters,** use 0 s as the **Start time** and 4 ms as the **End time** so that we get two full cycles of the applied voltage. After enabling the **Maximum time step settings(TMAX),** set the **Minimum number of time points** at 1000 to get a reasonably good plot during the rapidly changing transient period. Next, select the **Output variables** section and tell the program which voltage and current levels you are interested in. On the left side of the dialog box is a list of **Variables** that have been defined for the circuit. On the right is a list of **Selected variables for analysis.** In between you see **Add** or **Remove.** To move a variable from the left to the right column, select it in the left column and choose **Add.** It then appears in the right column. To plot both the applied voltage and the voltage across the coil, move **1** and **2** to the right column. Then select **Simulate.** A window titled **Grapher View** appears with the selected plots as shown in Fig. 11.80. Click on the **Show/Hide Grid** key (a red grid on a black axis), and the grid lines appear. Selecting the **Show/Hide**

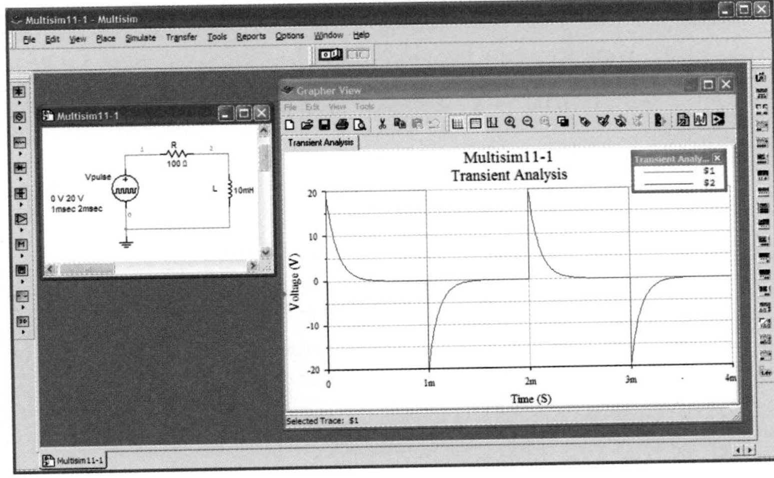

FIG. 11.80
Using Multisim to obtain the transient response for an inductive circuit.

Legend key on the immediate right results in the small **Transient Analysis** dialog box that identifies the color that goes with each nodal voltage. In this case, red is the color of the applied voltage, and blue is the color of the voltage across the coil.

The source voltage appears as expected with its transition to 20 V, 50% duty cycle, and the period of 2 ms. The voltage across the coil jumped immediately to the 20 V level and then began its decay to 0 V in about 0.5 ms as predicted. When the source voltage dropped to zero, the voltage across the coil reversed polarity to maintain the same direction of current in the inductive circuit. Remember that for a coil, the voltage can change instantaneously, but the inductor "chokes" any instantaneous change in current. By reversing its polarity, the voltage across the coil ensures the same polarity of voltage across the resistor and therefore the same direction of current through the coil and circuit.

PROBLEMS

SECTION 11.2 Magnetic Field

1. For the electromagnet in Fig. 11.81:
   a. Find the flux density in Wb/m^2.
   b. What is the flux density in teslas?
   c. What is the applied magnetomotive force?
   d. What would the reading of the meter in Fig. 11.14 read in gauss?

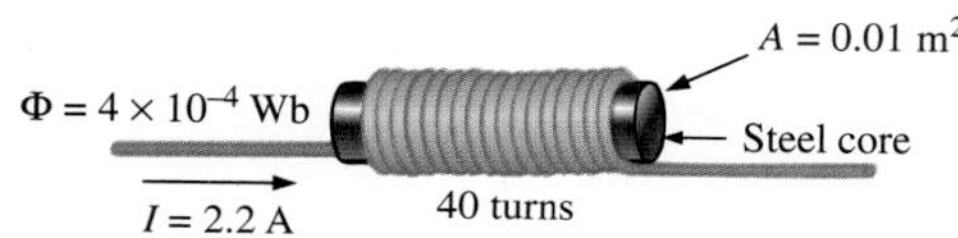

FIG. 11.81
Problem 1.

SECTION 11.3 Inductance

2. For the inductor in Fig. 11.82, find the inductance L in henries.

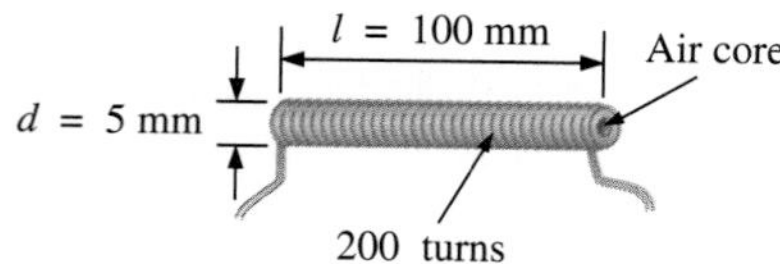

FIG. 11.82
Problems 2 and 3.

3. Repeat Problem 2 with l = 1.6 in., d = 0.2 in., and a ferromagnetic core with μ_r = 500.
4. For the inductor in Fig. 11.83, find the inductance L in henries.
5. An air-core inductor has a total inductance of 5 mH.
   a. What is the inductance if the only change is to increase the number of turns by a factor of three?

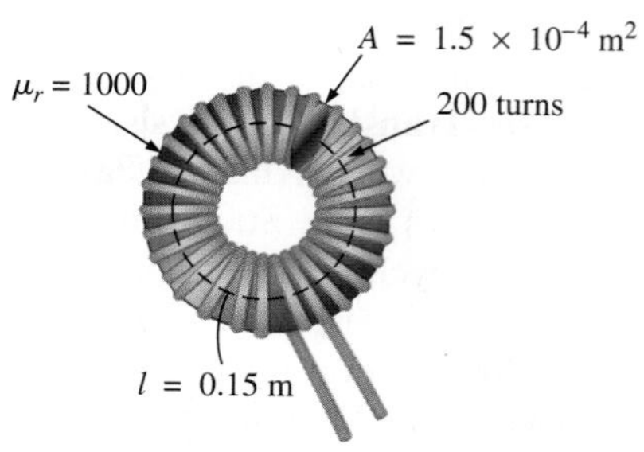

FIG. 11.83
Problem 4.

   b. What is the inductance if the only change is to increase the length by a factor of three?
   c. What is the inductance if the area is doubled, the length cut in half, and the number of turns doubled?
   d. What is the inductance if the area, length, and number of turns are cut in half and a ferromagnetic core with a μ_r of 1500 is inserted?
6. What are the inductance and the range of expected values for an inductor with the following label?
   a. 123J
   b. 47K

SECTION 11.4 Induced Voltage v_L

7. If the flux linking a coil of 50 turns changes at a rate of 120 mW/s, what is the induced voltage across the coil?
8. Determine the rate of change of flux linking a coil if 20 V are induced across a coil of 200 turns.
9. How many turns does a coil have if 42 mV are induced across the coil by a change in flux of 3 mW/s?
10. Find the voltage induced across a coil of 5 H if the rate of change of current through the coil is:
    a. 1 A/s
    b. 60 mA/s
    c. 0.5 A/ms
11. Find the induced voltage across a 50 mH inductor if the current through the coil changes at a rate of 0.1 mA/μs.

SECTION 11.5 *R-L* Transients: The Storage Phase

12. For the circuit in Fig. 11.84:
- **a.** Determine the time constant.
- **b.** Write the mathematical expression for the current i_L after the switch is closed.
- **c.** Repeat part (b) for v_L and v_R.
- **d.** Determine i_L and v_L at one, three, and five time constants.
- **e.** Sketch the waveforms of i_L, v_L, and v_R.

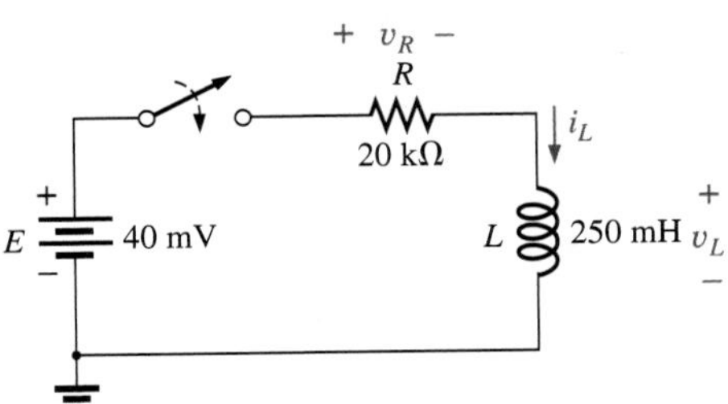

FIG. 11.84
Problem 12.

13. For the circuit in Fig. 11.85:
- **a.** Determine τ.
- **b.** Write the mathematical expression for the current i_L after the switch is closed at $t = 0$ s.
- **c.** Write the mathematical expression for v_L and v_R after the switch is closed at $t = 0$ s.
- **d.** Determine i_L and v_L at $t = 1\tau$, 3τ, and 5τ.
- **e.** Sketch the waveforms of i_L, v_L, and v_R for the storage phase.

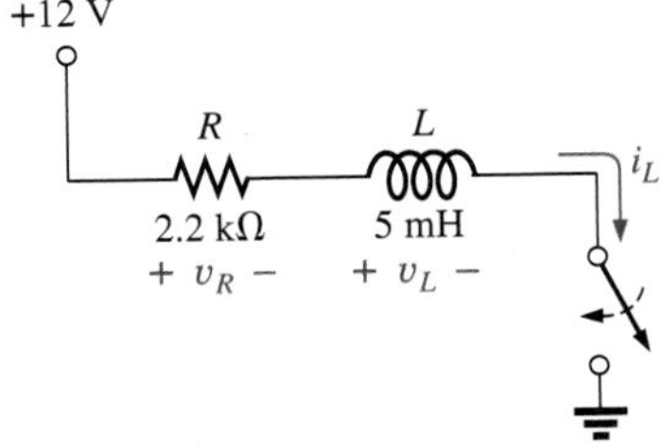

FIG. 11.85
Problem 13.

***14.** For the circuit in Fig. 11.86:
- **a.** Determine the time constant.
- **b.** Write the mathematical expression for the voltage v_L and the current i_L using the defined polarities and direction.
- **c.** Sketch the waveforms of v_L and i_L.

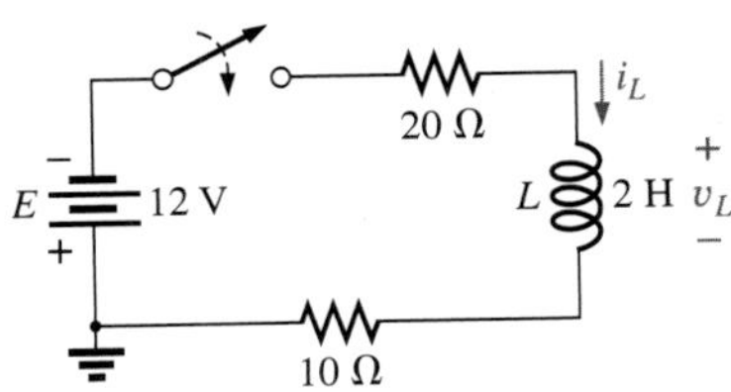

FIG. 11.86
Problem 14.

SECTION 11.6 Initial Conditions

15. For the circuit in Fig. 11.87:
- **a.** Write the mathematical expressions for the current i_L and the voltage v_L following the closing of the switch. Note the magnitude and the direction of the initial current.
- **b.** Sketch the waveform of i_L and v_L for the entire period from initial value to steady-state level.

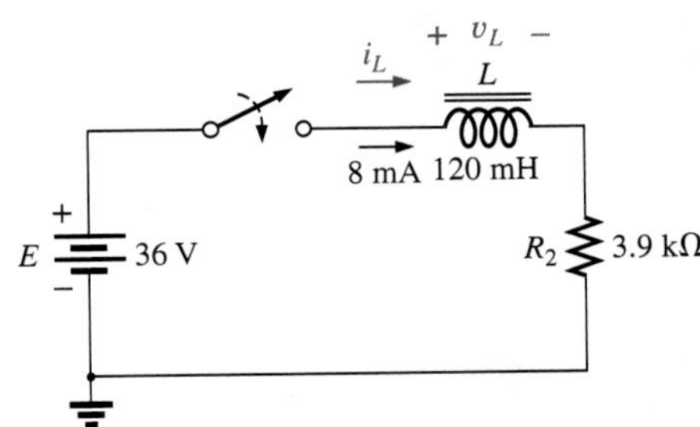

FIG. 11.87
Problems 15 and 47.

16. In this problem, the effect of reversing the initial current is investigated. The circuit in Fig. 11.88 is the same as that appearing in Fig. 11.87, with the only change being the direction of the initial current.
- **a.** Write the mathematical expressions for the current i_L and the voltage v_L following the closing of the switch. Take careful note of the defined polarity for v_L and the direction for i_L.
- **b.** Sketch the waveform of i_L and v_L for the entire period from initial value to steady-state level.
- **c.** Compare the results with those of Problem 15.

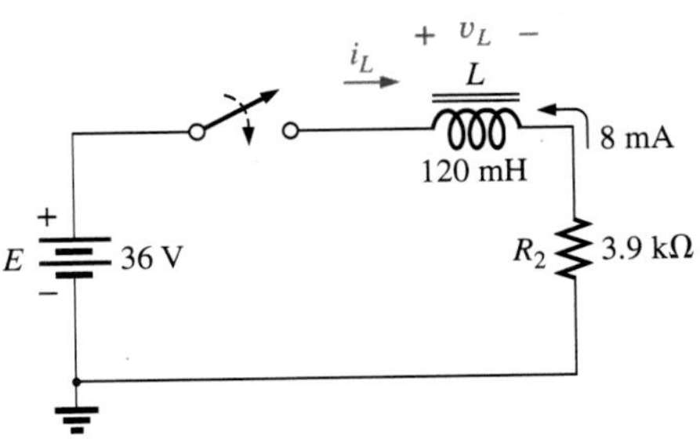

FIG. 11.88
Problem 16.

17. For the network in Fig. 11.89:

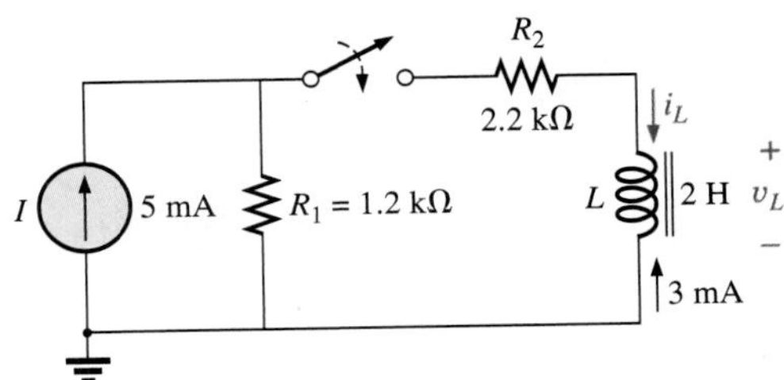

FIG. 11.89
Problem 17.

a. Write the mathematical expressions for the current i_L and the voltage v_L following the closing of the switch. Note the magnitude and the direction of the initial current.
b. Sketch the waveform of i_L and v_L for the entire period from initial value to steady-state level.

***18.** For the network in Fig. 11.90:
a. Write the mathematical expressions for the current i_L and the voltage v_L following the closing of the switch. Note the magnitude and direction of the initial current.
b. Sketch the waveform of i_L and v_L for the entire period from initial value to steady-state level.

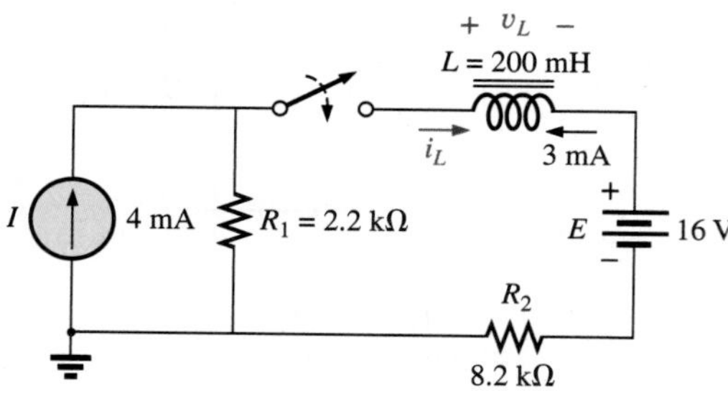

FIG. 11.90
Problem 18.

SECTION 11.7 *R-L* Transients: The Release Phase

19. For the network in Fig. 11.91:
a. Determine the mathematical expressions for the current i_L and the voltage v_L when the switch is closed.
b. Repeat part (a) if the switch is opened after a period of five time constants has passed.
c. Sketch the waveforms of parts (a) and (b) on the same set of axes.

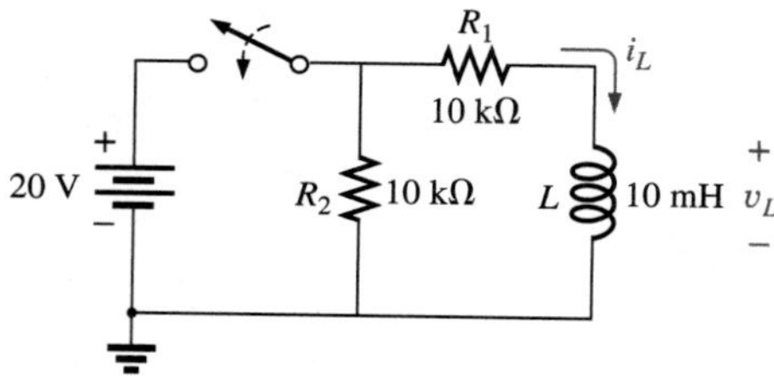

FIG. 11.91
Problem 19.

***20.** For the network in Fig. 11.92:
a. Determine the mathematical expressions for the current i_L and the voltage v_L following the closing of the switch.

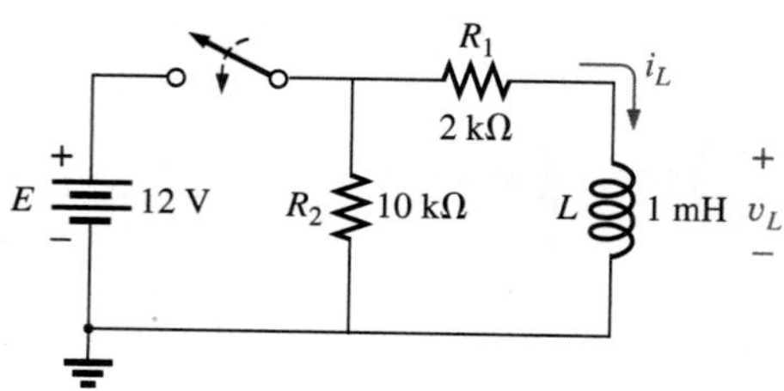

FIG. 11.92
Problem 20.

b. Repeat part (a) if the switch is opened at $t = 1\ \mu s$.
c. Sketch the waveforms of parts (a) and (b) on the same set of axes.

***21.** For the network in Fig. 11.93:
a. Write the mathematical expression for the current i_L and the voltage v_L following the closing of the switch.
b. Determine the mathematical expressions for i_L and v_L if the switch is opened after a period of five time constants has passed.
c. Sketch the waveforms of i_L and v_L for the time periods defined by parts (a) and (b).
d. Sketch the waveform for the voltage across R_2 for the same period of time encompassed by i_L and v_L. Take careful note of the defined polarities and directions in Fig. 11.93.

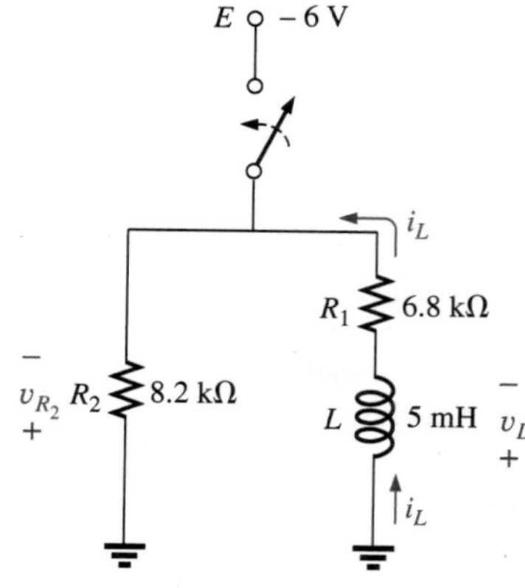

FIG. 11.93
Problem 21.

SECTION 11.8 Thévenin Equivalent: $\tau = L/R_{Th}$

22. For Fig. 11.94:
a. Determine the mathematical expressions for i_L and v_L following the closing of the switch.
b. Determine i_L and v_L after one time constant.

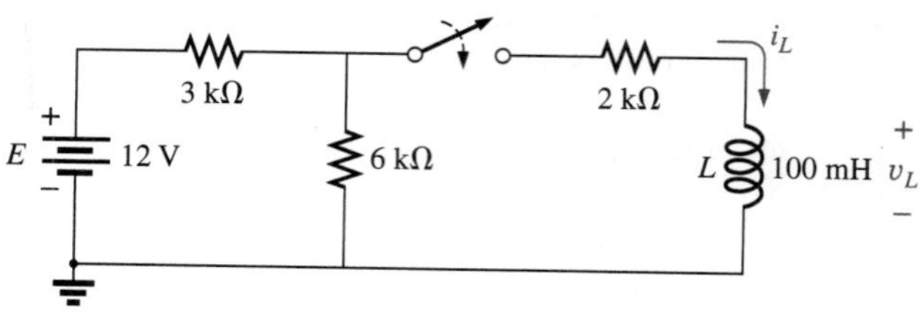

FIG. 11.94
Problems 22 and 48.

23. For Fig. 11.95:
a. Determine the mathematical expressions for i_L and v_L following the closing of the switch.
b. Determine i_L and v_L at $t = 100$ ns.

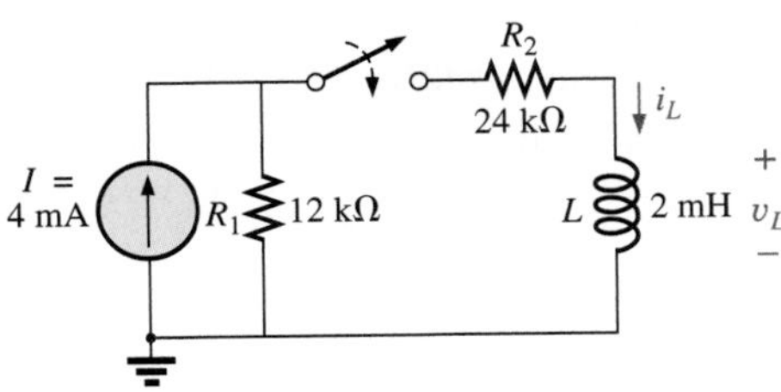

FIG. 11.95
Problem 23.

***24.** For Fig. 11.96:

a. Determine the mathematical expressions for i_L and v_L following the closing of the switch.
b. Calculate i_L and v_L at $t = 10\ \mu s$.
c. Write the mathematical expressions for the current i_L and the voltage v_L if the switch is opened at $t = 10\ \mu s$.
d. Sketch the waveforms of i_L and v_L for parts (a) and (c).

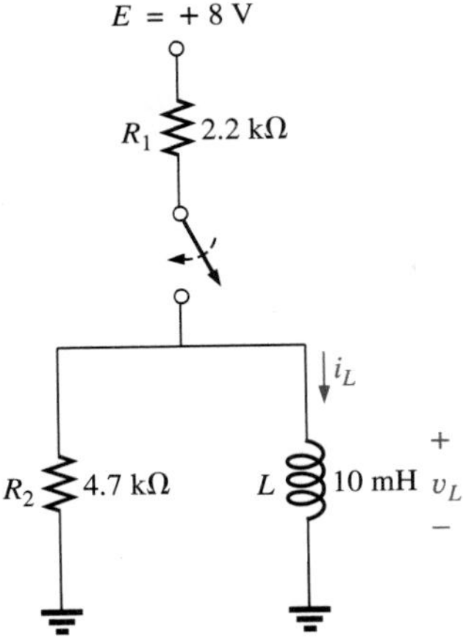

FIG. 11.96
Problem 24.

***25.** For the network in Fig. 11.97, the switch is closed at $t = 0$ s.

a. Determine v_L at $t = 25$ ms.
b. Find v_L at $t = 1$ ms.
c. Calculate v_{R_1} at $t = 1\tau$.
d. Find the time required for the current i_L to reach 100 mA.

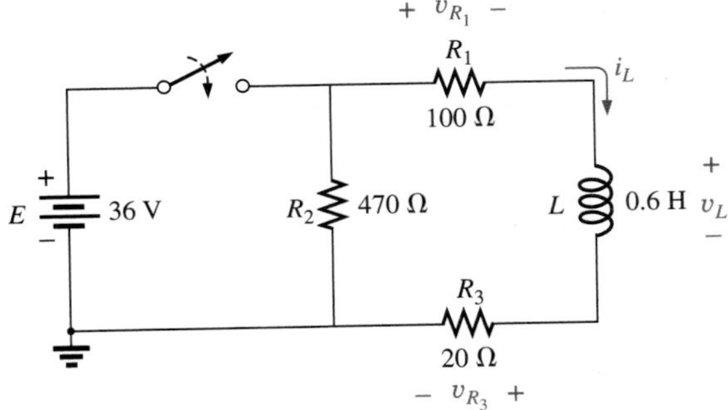

FIG. 11.97
Problem 25.

***26.** The switch in Fig. 11.98 has been open for a long time. It is then closed at $t = 0$ s.

a. Write the mathematical expression for the current i_L and the voltage v_L after the switch is closed.
b. Sketch the waveform of i_L and v_L from the initial value to the steady-state level.

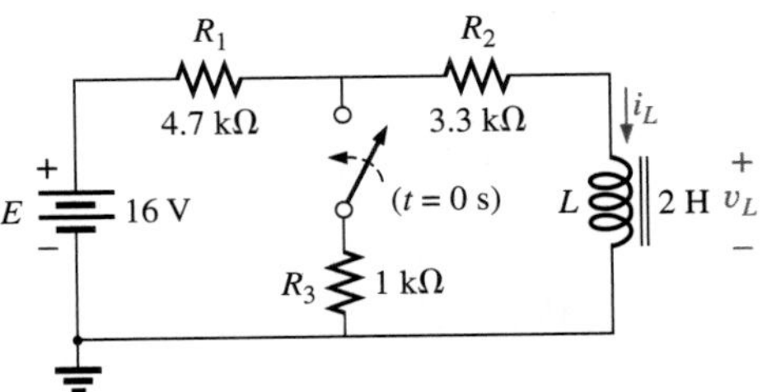

FIG. 11.98
Problem 26.

***27. a.** Determine the mathematical expressions for i_L and v_L following the closing of the switch in Fig. 11.99.
b. Determine i_L and v_L after two time constants of the storage phase.
c. Write the mathematical expressions for the current i_L and the voltage v_L if the switch is opened at the instant defined by part (b).
d. Sketch the waveforms of i_L and v_L for parts (a) and (c).

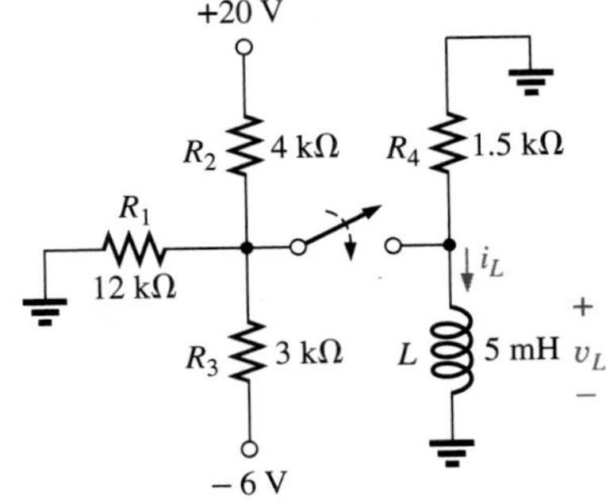

FIG. 11.99
Problem 27.

***28.** The switch for the network in Fig. 11.100 has been closed for about 1 h. It is then opened at the time defined as $t = 0$ s.

a. Determine the time required for the current i_R to drop to 1 mA.

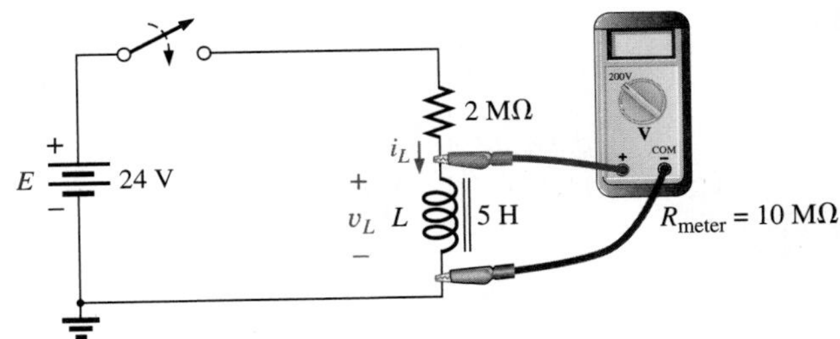

FIG. 11.100
Problem 28.

b. Find the voltage v_L at $t = 1$ ms.
c. Calculate v_R at $t = 5\tau$.

***29.** The switch in Fig. 11.101 has been closed for a long time. It is then opened at $t = 0$ s.
a. Write the mathematical expression for the current i_L and the voltage v_L after the switch is opened.
b. Sketch the waveform of i_L and v_L from initial value to the steady-state level.

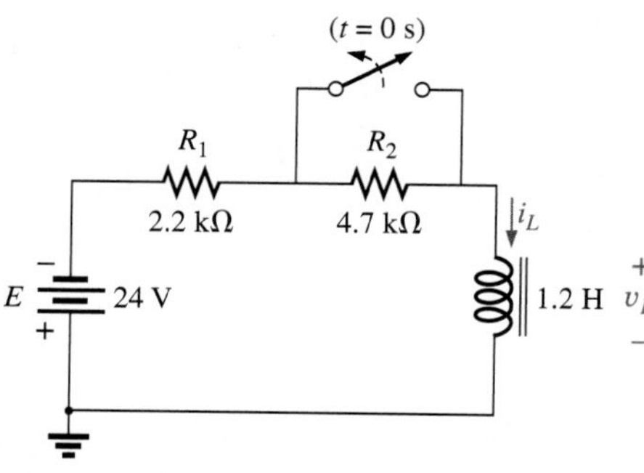

FIG. 11.101
Problem 29.

SECTION 11.9 Instantaneous Values

30. Given $i_L = 100\text{ mA}\,(1 - e^{-t/20\text{ms}})$:
a. Determine i_L at $t = 1$ ms.
b. Determine i_L at $t = 100$ ms.
c. Find the time t when i_L will equal 50 mA.
d. Find the time t when i_L will equal 99 mA.

31. The network in Fig. 11.102 employs a DMM with an internal resistance of 10 MΩ in the voltmeter mode. The switch is closed at $t = 0$ s.
a. Find the voltage across the coil the instant after the switch is closed.
b. What is the final value of the current i_L?
c. How much time must pass before i_L reaches 10 μA?
d. What is the voltmeter reading at $t = 12$ μs?

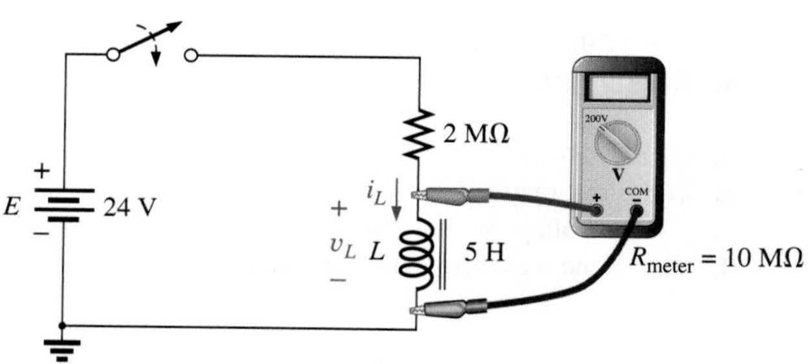

FIG. 11.102
Problem 31.

SECTION 11.10 Average Induced Voltage: $v_{L_{av}}$

32. Find the waveform for the voltage induced across a 200 mH coil if the current through the coil is as shown in Fig. 11.103.

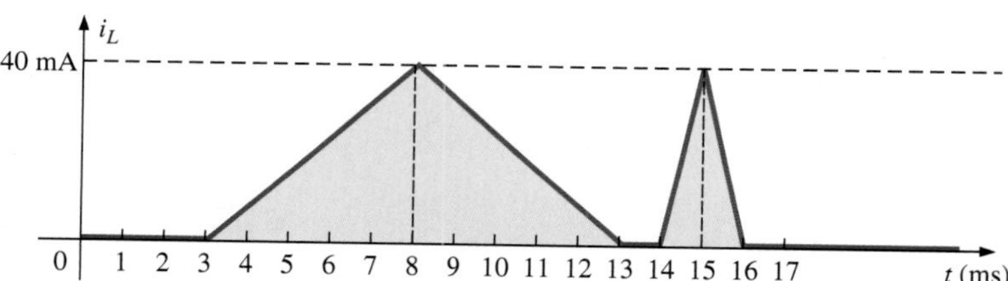

FIG. 11.103
Problem 32.

33. Find the waveform for the voltage induced across a 5 mH coil if the current through the coil is as shown in Fig. 11.104.

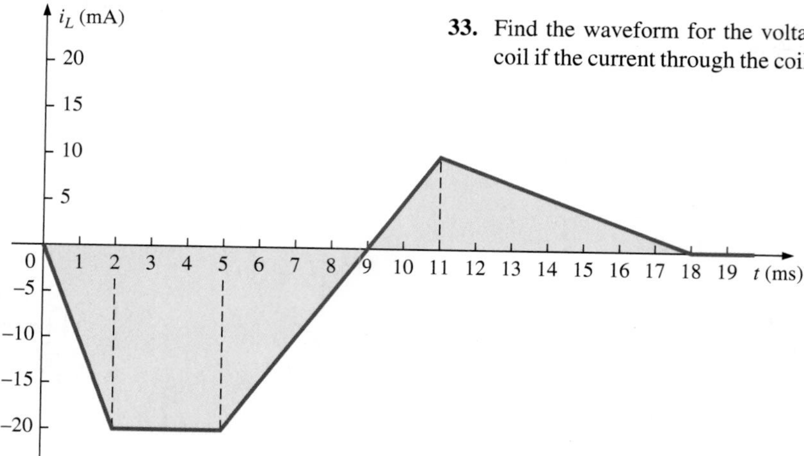

FIG. 11.104
Problem 33.

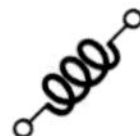

*34. Find the waveform for the current of a 10 mH coil if the voltage across the coil follows the pattern in Fig. 11.105. The current i_L is 4 mA at $t = 0$ s.

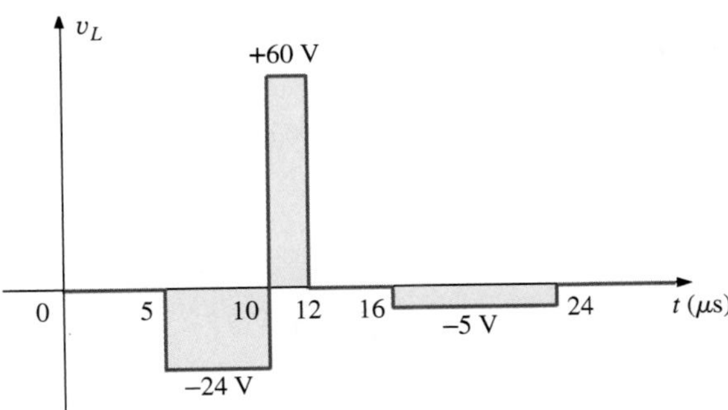

FIG. 11.105
Problem 34.

SECTION 11.11 Inductors in Series and in Parallel

35. Find the total inductance of the circuits in Fig. 11.106.

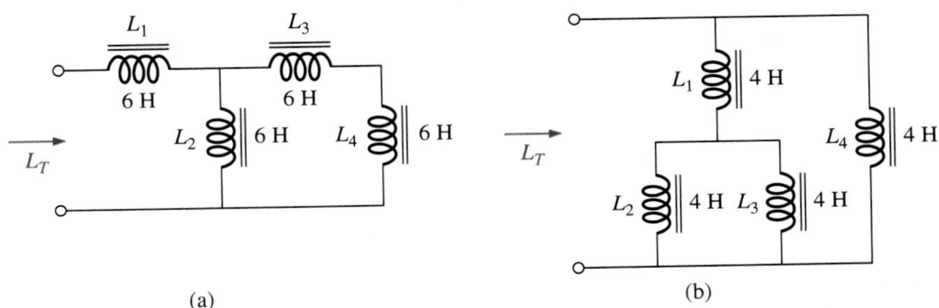

FIG. 11.106
Problem 35.

36. Reduce the network in Fig. 11.107 to the fewest number of components.

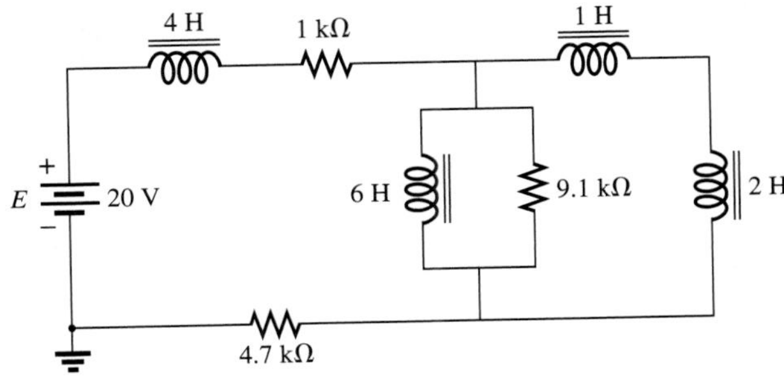

FIG. 11.107
Problem 36.

37. Reduce the network in Fig. 11.108 to the fewest elements.

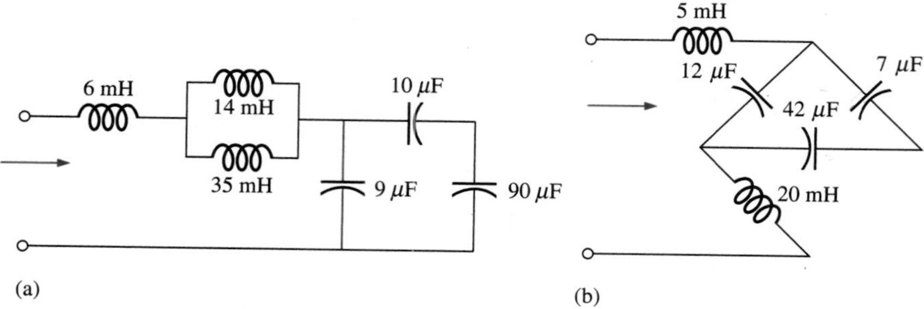

FIG. 11.108
Problem 37.

***38.** For the network in Fig. 11.109:

a. Write the mathematical expressions for the voltages v_L and v_R and the current i_L if the switch is closed at $t = 0$ s.

b. Sketch the waveforms of v_L, v_R, and i_L.

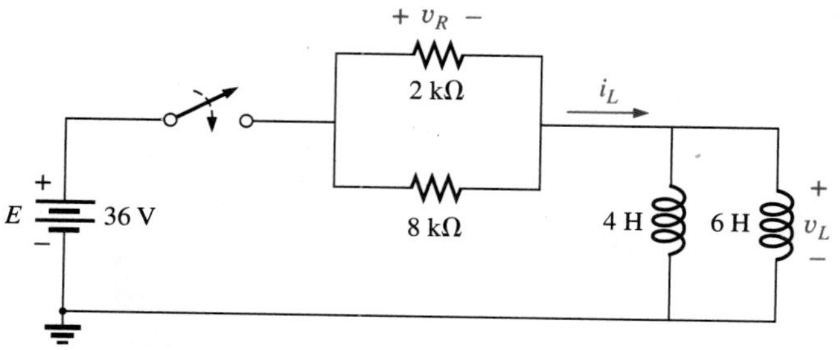

FIG. 11.109
Problem 38.

***39.** For the network in Fig. 11.110:

a. Write the mathematical expressions for the voltage v_L and the current i_L if the switch is closed at $t = 0$ s.

b. Sketch the waveforms of v_L and i_L.

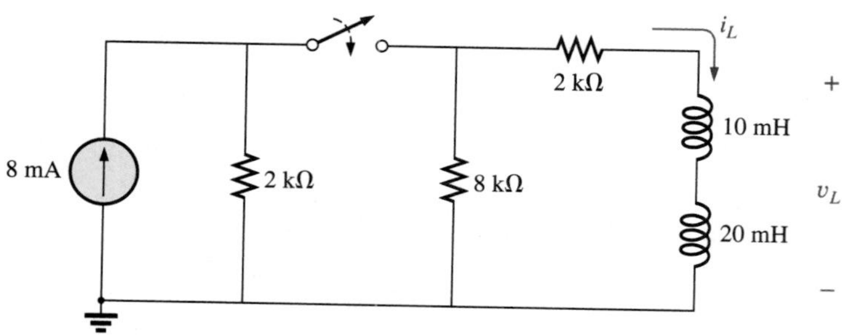

FIG. 11.110
Problem 39.

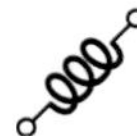

*40. For the network in Fig. 11.111:
  a. Find the mathematical expressions for the voltage v_L and the current i_L following the closing of the switch.
  b. Sketch the waveforms of v_L and i_L obtained in part (a).
  c. Determine the mathematical expression for the voltage v_{L_3} following the closing of the switch, and sketch the waveform.

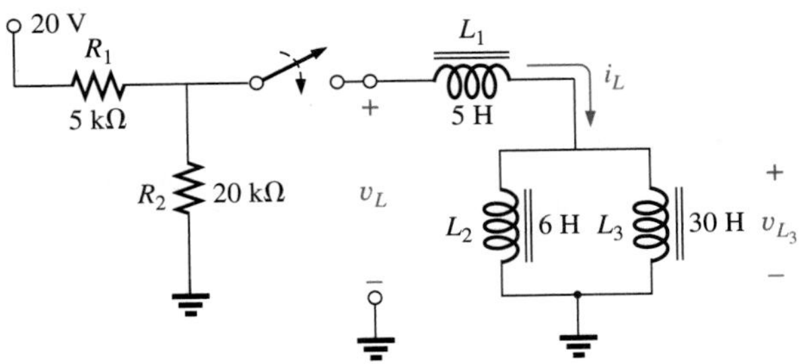

FIG. 11.111
Problem 40.

SECTION 11.12 Steady-State Conditions

41. Find the steady-state currents I_1 and I_2 for the network in Fig. 11.112.

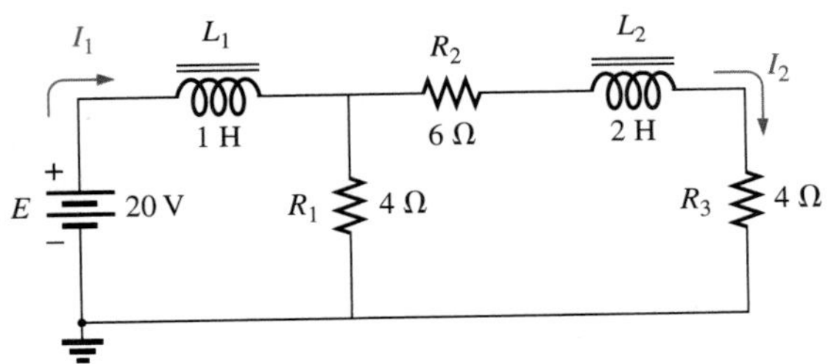

FIG. 11.112
Problem 41.

42. Find the steady-state currents and voltages for the network in Fig. 11.113.

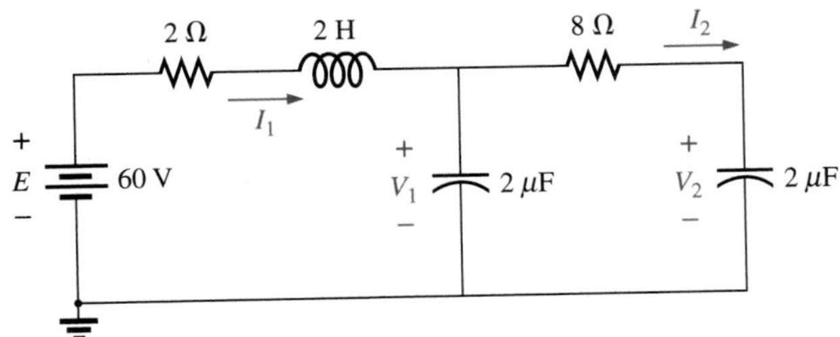

FIG. 11.113
Problem 42.

43. Find the steady-state currents and voltages for the network in Fig. 11.114.

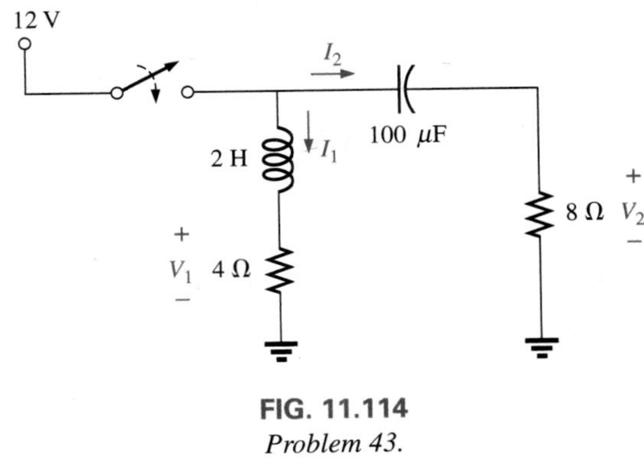

FIG. 11.114
Problem 43.

44. Find the steady-state currents and voltages for the network in Fig. 11.115.

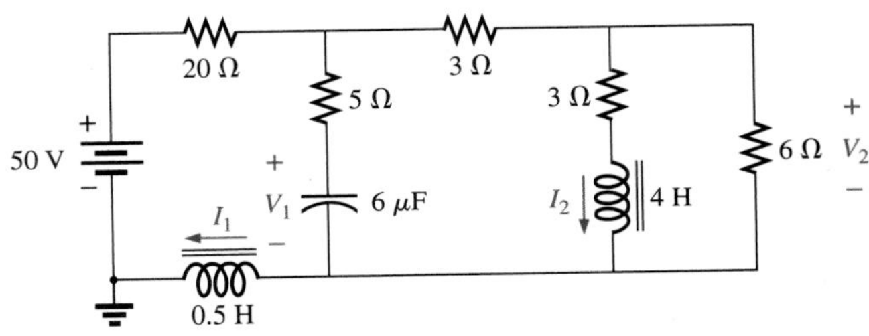

FIG. 11.115
Problem 44.

SECTION 11.15 Computer Analysis

45. Using PSpice or Multisim, verify the results of Example 11.3.
46. Using PSpice or Multisim, verify the results of Example 11.4.
47. Using PSpice or Multisim, find the solution to Problem 15.
48. Using PSpice or Multisim, find the solution to Problem 22.
49. Using PSpice or Multisim, verify the results of Example 11.8.

GLOSSARY

Ampère's circuital law A law establishing the fact that the algebraic sum of the rises and drops of the magnetomotive force (mmf) around a closed loop of a magnetic circuit is equal to zero.

Choke A term often applied to an inductor, due to the ability of an inductor to resist a change in current through it.

Diamagnetic materials Materials that have permeabilities slightly less than that of free space.

Electromagnetism Magnetic effects introduced by the flow of charge, or current.

Faraday's law A law stating the relationship between the voltage induced across a coil and the number of turns in the coil and the rate at which the flux linking the coil is changing.

Ferromagnetic materials Materials having permeabilities hundreds and thousands of times greater than that of free space.

Flux density (*B*) A measure of the flux per unit area perpendicular to a magnetic flux path. It is measured in teslas (T) or webers per square meter (Wb/m^2).

Inductance (*L*) A measure of the ability of a coil to oppose any change in current through the coil and to store energy in the form of a magnetic field in the region surrounding the coil.

Inductor (coil) A fundamental element of electrical systems constructed of numerous turns of wire around a ferromagnetic core or an air core.

Lenz's law A law stating that an induced effect is always such as to oppose the cause that produced it.

Magnetic flux lines Lines of a continuous nature that reveal the strength and direction of a magnetic field.

Magnetomotive force (mmf) ($\mathscr{F}$) The "pressure" required to establish magnetic flux in a ferromagnetic material. It is measured in ampere-turns (At).

Paramagnetic materials Materials that have permeabilities slightly greater than that of free space.

Permanent magnet A material such as steel or iron that will remain magnetized for long periods of time without the aid of external means.

Permeability (μ) A measure of the ease with which magnetic flux can be established in a material. It is measured in Wb/A · m.

Relative permeability (μ_r) The ratio of the permeability of a material to that of free space.

MAGNETIC CIRCUITS

12

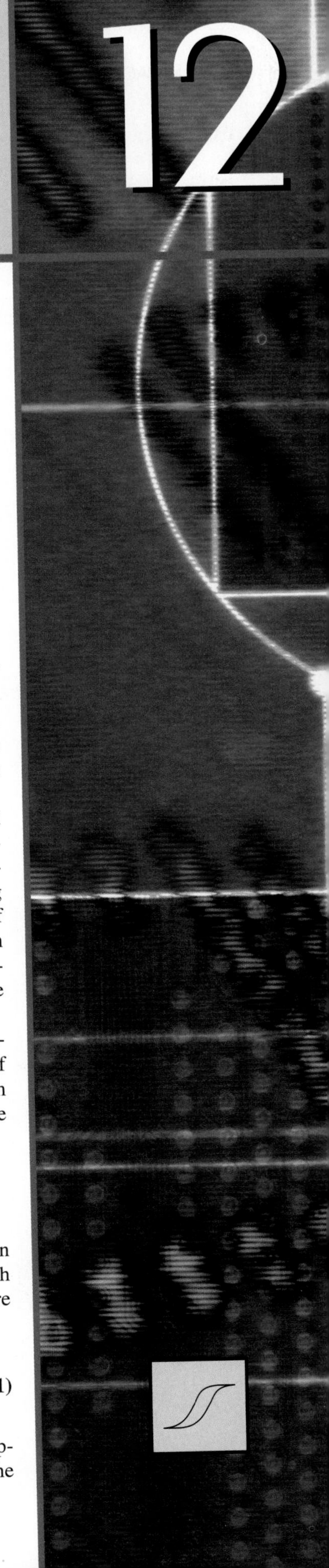

Objectives

- ***Become aware of the similarities between the analysis of magnetic circuits and electric circuits.***
- ***Develop a clear understanding of the important parameters of a magnetic circuit and how to find each quantity for a variety of magnetic circuit configurations.***
- ***Begin to appreciate why a clear understanding of magnetic circuit parameters is an important component in the design of electrical/electronic systems.***

12.1 INTRODUCTION

Magnetic and electromagnetic effects play an important role in the design of a wide variety of electrical/electronic systems in use today. Motors, generators, transformers, loudspeakers, relays, medical equipment and movements of all kinds depend on magnetic effects to function properly. The response and characteristics of each have an impact on the current and voltage levels of the system, the efficiency of the design, the resulting size, and many other important considerations.

Fortunately, there is a great deal of similarity between the analyses of electric circuits and magnetic circuits. The magnetic flux of magnetic circuits has properties very similar to the current of electric circuits. As shown in Fig. 11.15, it has a direction and a closed path. The magnitude of the established flux is a direct function of the applied **magnetomotive force** resulting in a duality with electric circuits where the resulting current is a function of the magnitude of the applied voltage. The flux established is also inversely related to the structural opposition of the magnetic path in the same way the current in a network is inversely related to the resistance of the network. All of these similarities are used throughout the analysis to clarify the approach.

One of the difficulties associated with studying magnetic circuits is that three different systems of units are commonly used in the industry. The manufacturer, application, and type of component all have an impact on which system is used. To the extent practical, the SI system is applied throughout the chapter. References to the CGS and English systems require the use of Appendix F.

12.2 MAGNETIC FIELD

The magnetic field distribution around a permanent magnet or **electromagnet** was covered in detail in Chapter 11. Recall that flux lines strive to be as short as possible and take the path with the highest permeability. The **flux density** is defined by Eq. 12.1 (Eq. 11.1 repeated here for convenience).

$$\boxed{B = \frac{\Phi}{A}} \qquad \begin{aligned} B &= \text{Wb/m}^2 = \text{teslas (T)} \\ \Phi &= \text{webers (Wb)} \\ A &= \text{m}^2 \end{aligned} \tag{12.1}$$

The "pressure" on the system to establish magnetic lines of force is determined by the applied magnetomotive force which is directly related to the number of turns and current of the

magnetizing coil as appearing in Eq. 12.2 (Eq. 11.3 repeated here for convenience).

$$\boxed{\mathscr{F} = NI} \qquad \begin{aligned} \mathscr{F} &= \text{ampere-turns (At)} \\ N &= \text{turns (t)} \\ I &= \text{amperes (A)} \end{aligned} \tag{12.2}$$

The level of magnetic flux established in a ferromagnetic core is a direction function of the permeability of the material. **Ferromagnetic materials** have a very high level of **permeability** while non-magnetic materials such as air and wood have very low levels. The ratio of the permeability of the material to that of air is called the **relative permeability** and is defined by Eq. 12.3 (Eq. 11.5 repeated here for convenience).

$$\boxed{\mu_r = \frac{\mu}{\mu_o}} \qquad \mu_o = 4\pi \times 10^{-7}\text{Wb/A}\cdot\text{m} \tag{12.3}$$

As mentioned in Chapter 11, the values of μ_r are not provided in a table format because the value is determined by the other quantities of the magnetic circuit. Change the magnetomotive force, and the relative permeability changes.

12.3 RELUCTANCE

The resistance of a material to the flow of charge (current) is determined for electric circuits by the equation

$$R = \rho\frac{l}{A} \qquad (\text{ohms}, \Omega)$$

The **reluctance** of a material to the setting up of magnetic flux lines in the material is determined by the following equation:

$$\boxed{\mathscr{R} = \frac{l}{\mu A}} \qquad (\text{rels, or At/Wb}) \tag{12.4}$$

where $\mathscr{R}$ is the reluctance, l is the length of the magnetic path, and A is the cross-sectional area. The t in the units At/Wb is the number of turns of the applied winding. More is said about ampere-turns (At) in the next section. Note that the resistance and reluctance are inversely proportional to the area, indicating that an increase in area results in a reduction in each and an *increase* in the desired result: current and flux. For an increase in length, the opposite is true, and the desired effect is reduced. The reluctance, however, is inversely proportional to the permeability, while the resistance is directly proportional to the resistivity. The larger the μ or the smaller the ρ, the smaller the reluctance and resistance, respectively. Obviously, therefore, materials with high permeability, such as the ferromagnetics, have very small reluctances and result in an increased measure of flux through the core. There is no widely accepted unit for reluctance, although the *rel* and the At/Wb are usually applied.

12.4 OHM'S LAW FOR MAGNETIC CIRCUITS

Recall the equation

$$\text{Effect} = \frac{\text{cause}}{\text{opposition}}$$

appearing in Chapter 4 to introduce Ohm's law for electric circuits. For magnetic circuits, the effect desired is the flux Φ. The cause is the **magnetomotive force (mmf)** $\mathscr{F}$, which is the external force (or "pressure") required to set up the **magnetic flux lines** within the magnetic material. The opposition to the setting up of the flux Φ is the reluctance $\mathscr{R}$.

Substituting, we have

$$\Phi = \frac{\mathscr{F}}{\mathscr{R}} \qquad (12.5)$$

Since $\mathscr{F} = NI$, Eq. 12.5 clearly reveals that an increase in the number of turns or the current through the wire in Fig. 12.1 results in an increased "pressure" on the system to establish the flux lines through the core.

Although there is a great deal of similarity between electric and magnetic circuits, you must understand that the flux Φ is not a "flow" variable such as current in an electric circuit. Magnetic flux is established in the core through the alteration of the atomic structure of the core due to external pressure and is not a measure of the flow of some charged particles through the core.

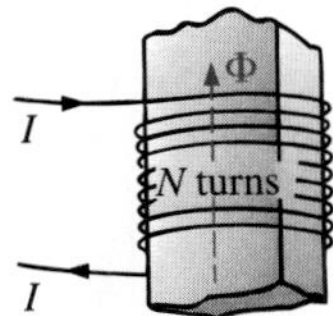

FIG. 12.1
Defining the components of a magnetomotive force.

12.5 MAGNETIZING FORCE

The magnetomotive force per unit length is called the **magnetizing force** (H). In equation form,

$$H = \frac{\mathscr{F}}{l} \quad \text{(At/m)} \qquad (12.6)$$

Substituting for the magnetomotive force results in

$$H = \frac{NI}{l} \quad \text{(At/m)} \qquad (12.7)$$

For the magnetic circuit in Fig. 12.2, if $NI = 40$ At and $l = 0.2$ m, then

$$H = \frac{NI}{l} = \frac{40\text{ At}}{0.2\text{ m}} = 200\text{ At/m}$$

In words, the result indicates that there are 200 At of "pressure" per meter to establish flux in the core.

Note in Fig. 12.2 that the direction of the flux Φ can be determined by placing the fingers of your right hand in the direction of current around the core and noting the direction of the thumb. It is interesting to realize that *the magnetizing force is independent of the type of core material*—it is determined solely by the number of turns, the current, and the length of the core.

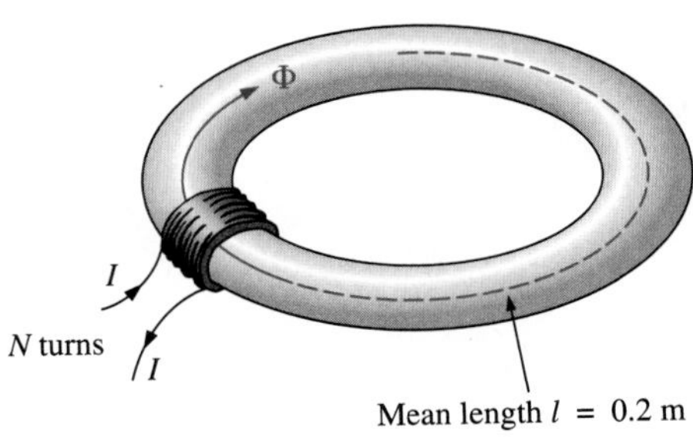

FIG. 12.2
Defining the magnetizing force of a magnetic circuit.

The applied magnetizing force has a pronounced effect on the resulting permeability of a magnetic material. As the magnetizing force increases, the permeability rises to a maximum and then drops to a minimum, as shown in Fig. 12.3 for three commonly employed magnetic materials.

The flux density and the magnetizing force are related by the following equation:

$$B = \mu H \qquad (12.8)$$

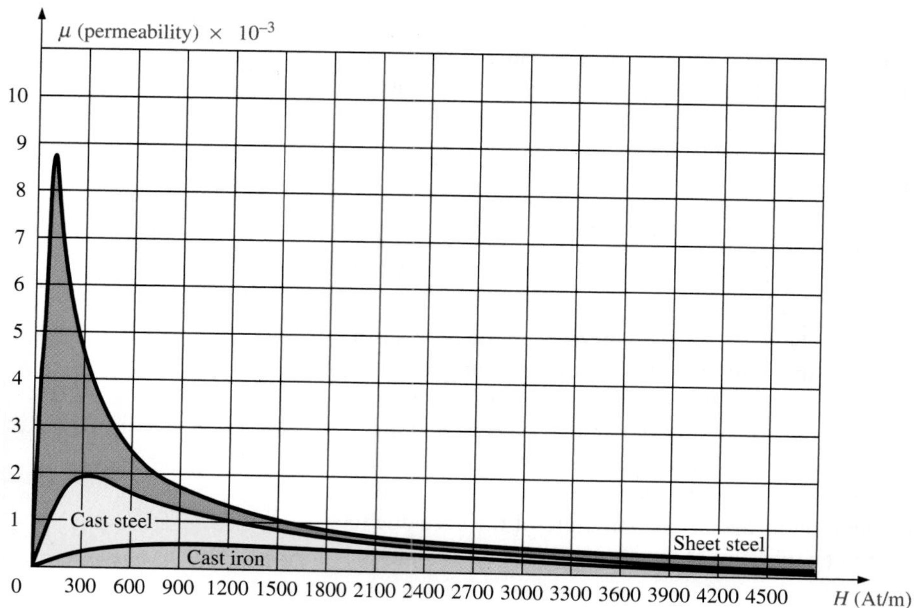

FIG. 12.3
Variation of μ with the magnetizing force.

This equation indicates that for a particular magnetizing force, the greater the permeability, the greater the induced flux density.

Since henries (H) and the magnetizing force (H) use the same capital letter, it must be pointed out that all units of measurement in the text, such as henries, use roman letters, such as H, whereas variables such as the magnetizing force use italic letters, such as H.

12.6 HYSTERESIS

A curve of the flux density B versus the magnetizing force H of a material is of particular importance to the engineer. Curves of this type can usually be found in manuals, descriptive pamphlets, and brochures published by manufacturers of magnetic materials. A typical B-H curve for a ferromagnetic material such as steel can be derived using the setup in Fig. 12.4.

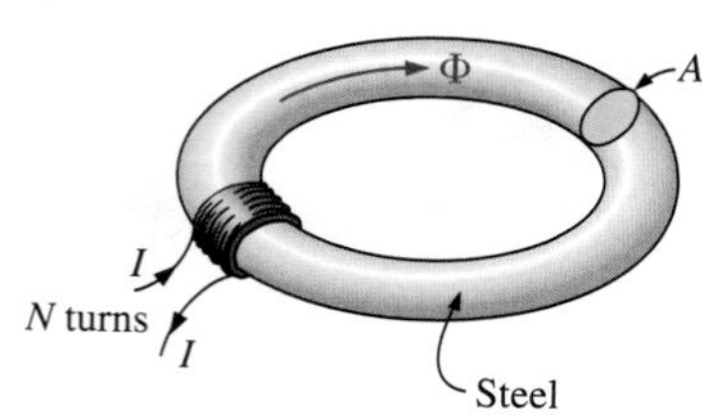

FIG. 12.4
Series magnetic circuit used to define the hysteresis curve.

The core is initially unmagnetized, and the current $I = 0$. If the current I is increased to some value above zero, the magnetizing force H increases to a value determined by

$$H\uparrow = \frac{NI\uparrow}{l}$$

The flux Φ and the flux density B ($B = \Phi/A$) also increase with the current I (or H). If the material has no residual magnetism, and the magnetizing force H is increased from zero to some value H_a, the B-H curve follows the path shown in Fig. 12.5 between o and a. If the magnetizing force H is increased until saturation (H_s) occurs, the curve continues as shown in the figure to point b. When saturation occurs, the flux density has, *for all practical purposes,* reached its maximum value. Any further increase in current through the coil increasing $H = NI/l$ results in a very small increase in flux density B.

If the magnetizing force is reduced to zero by letting I decrease to zero, the curve follows the path of the curve between b and c. The flux

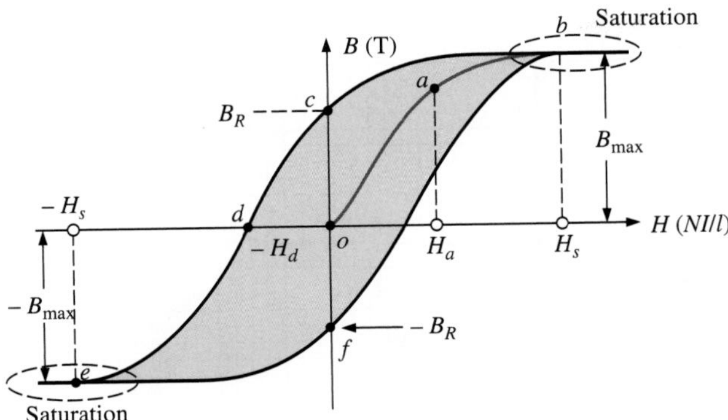

FIG. 12.5
Hysteresis curve.

density B_R, which remains when the magnetizing force is zero, is called the *residual flux density.* It is this residual flux density that makes it possible to create permanent magnets. If the coil is now removed from the core in Fig. 12.4, the core will still have the magnetic properties determined by the residual flux density, a measure of its "retentivity." If the current I is reversed, developing a magnetizing force, $-H$, the flux density B decreases with an increase in I. Eventually, the flux density will be zero when $-H_d$ (the portion of curve from c to d) is reached. The magnetizing force $-H_d$ required to "coerce" the flux density to reduce its level to zero is called the *coercive force,* a measure of the coercivity of the magnetic sample. As the force $-H$ is increased until saturation again occurs and is then reversed and brought back to zero, the path *def* results. If the magnetizing force is increased in the positive direction ($+H$), the curve traces the path shown from f to b. The entire curve represented by *bcdefb* is called the **hysteresis** curve for the ferromagnetic material, from the Greek *hysterein,* meaning "to lag behind." The flux density B *lagged* behind the magnetizing force H during the entire plotting of the curve. When H was zero at c, B was not zero but had only begun to decline. Long after H had passed through zero and had become equal to $-H_d$ did the flux density B finally become equal to zero.

If the entire cycle is repeated, the curve obtained for the same core will be determined by the maximum H applied. Three hysteresis loops for the same material for maximum values of H less than the saturation value are shown in Fig. 12.6. In addition, the saturation curve is repeated for comparison purposes.

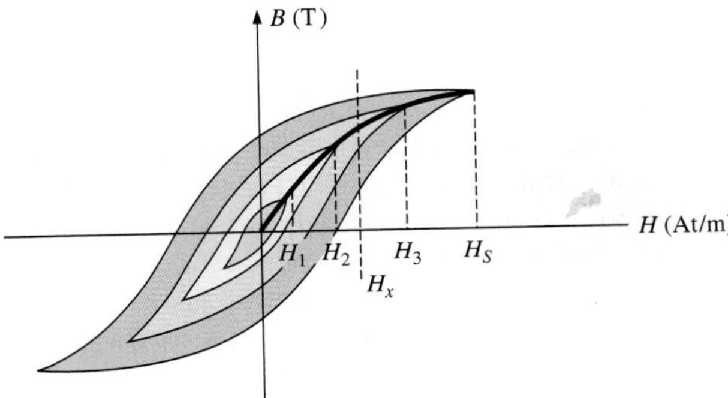

FIG. 12.6
Defining the normal magnetization curve.

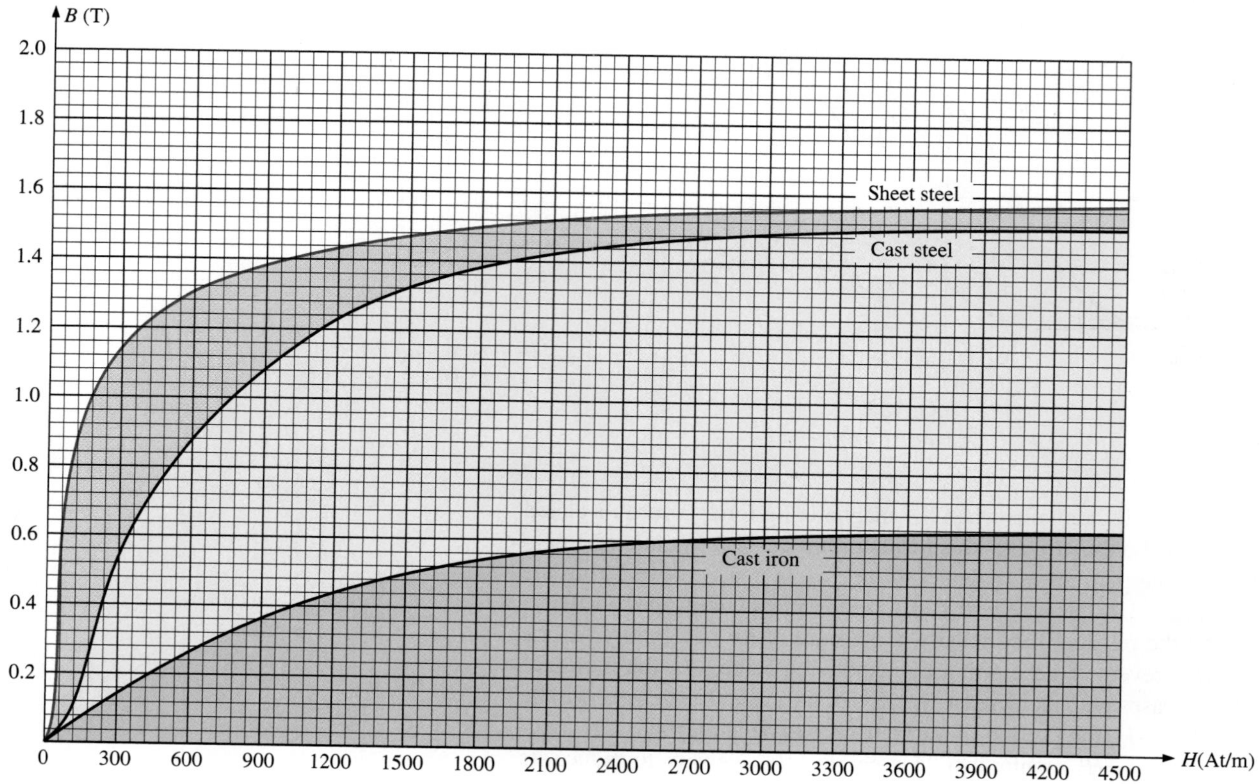

FIG. 12.7
Normal magnetization curve for three ferromagnetic materials.

Note from the various curves that for a particular value of *H*, say, H_x, the value of *B* can vary widely, as determined by the history of the core. In an effort to assign a particular value of *B* to each value of *H*, we compromise by connecting the tips of the hysteresis loops. The resulting curve, shown by the heavy, solid line in Fig. 12.6 and for various materials in Fig. 12.7, is called the *normal magnetization curve.* An expanded view of one region appears in Fig. 12.8.

A comparison of Figs. 12.3 and 12.7 shows that for the same value of *H*, the value of *B* is higher in Fig. 12.7 for the materials with the higher μ in Fig. 12.3. This is particularly obvious for low values of *H*. This correspondence between the two figures must exist since $B = \mu H$. In fact, if in Fig. 12.7 we find μ for each value of *H* using the equation $\mu = B/H$, we obtain the curves in Fig. 12.3.

It is interesting to note that the hysteresis curves in Fig. 12.6 have a *point symmetry* about the origin; that is, the inverted pattern to the left of the vertical axis is the same as that appearing to the right of the vertical axis. In addition, you will find that a further application of the same magnetizing forces to the sample results in the same plot. For a current *I* in $H = NI/l$ that moves between positive and negative maximums at a fixed rate, the same *B-H* curve results during each cycle. Such will be the case when we examine ac (sinusoidal) networks in the later chapters. The reversal of the field (Φ) due to the changing current direction results in a loss of energy that can best be described by first introducing the *domain theory of magnetism.*

Within each atom, the orbiting electrons (described in Chapter 2) are also spinning as they revolve around the nucleus. The atom, due to its

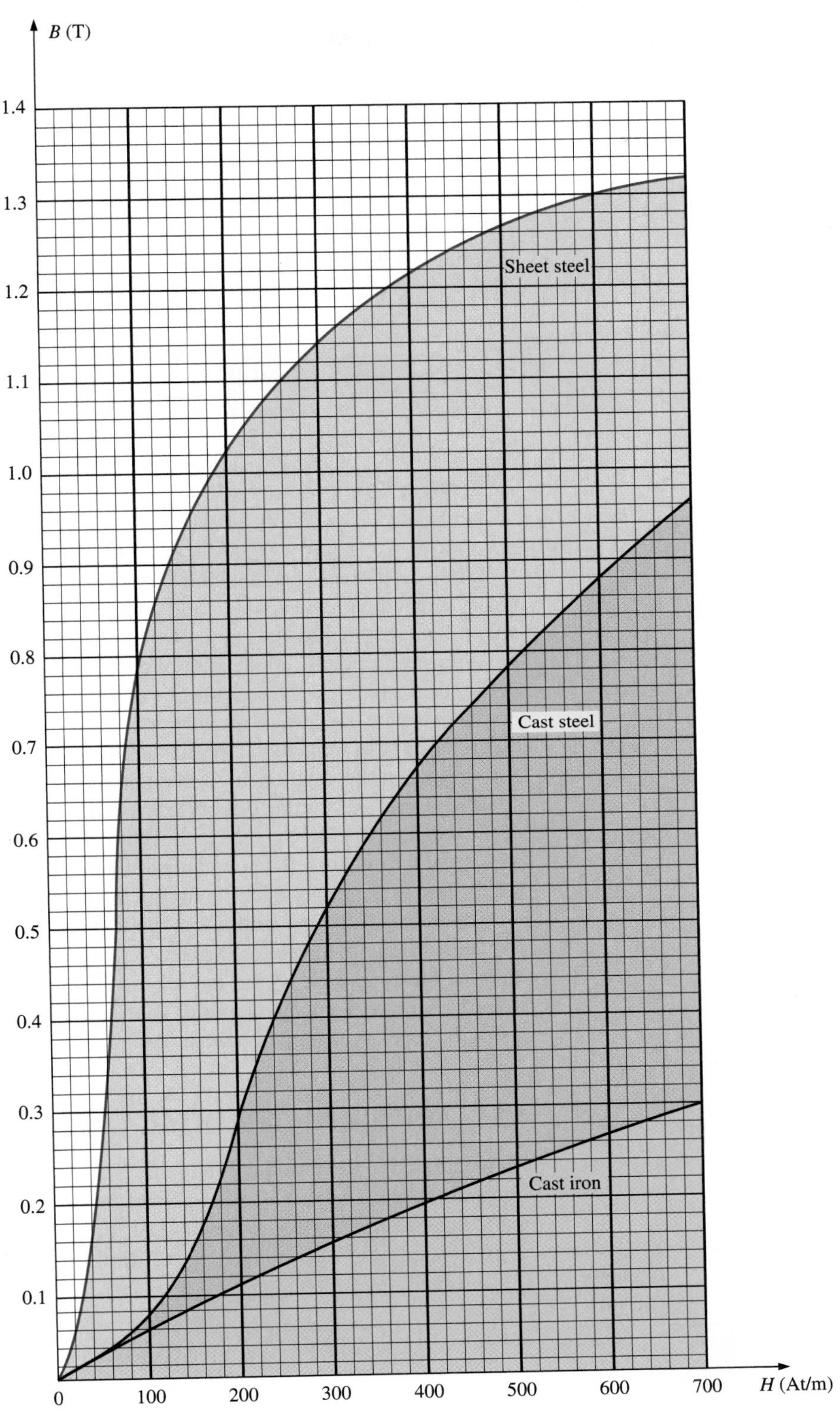

FIG. 12.8
Expanded view of Fig. 12.7 for the low magnetizing force region.

spinning electrons, has a magnetic field associated with it. In nonmagnetic materials, the net magnetic field is effectively zero since the magnetic fields due to the atoms of the material oppose each other. In magnetic materials such as iron and steel, however, the magnetic fields of groups of atoms numbering in the order of 10^{12} are aligned, forming very small bar magnets. This group of magnetically aligned atoms is called a **domain.** Each domain is a separate entity; that is, each domain is independent of the surrounding domains. For an unmagnetized sample of magnetic material, these domains appear in a random manner, such as shown in Fig. 12.9(a). The net magnetic field in any one direction is zero.

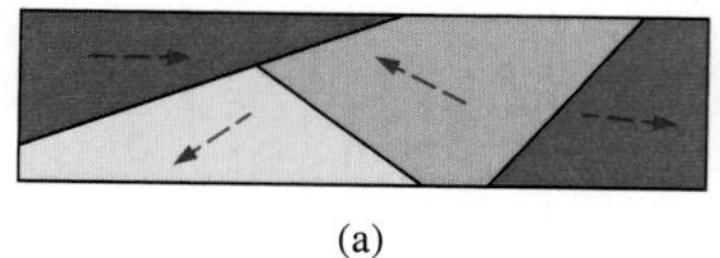

(a)

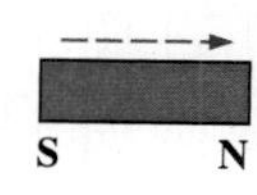

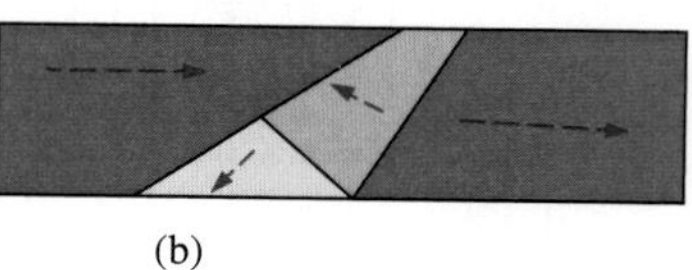

(b)

FIG. 12.9
Demonstrating the domain theory of magnetism.

When an external magnetizing force is applied, the domains that are nearly aligned with the applied field grow at the expense of the less favorably oriented domains, such as shown in Fig. 12.9(b). Eventually, if a sufficiently strong field is applied, all of the domains have the orientation of the applied magnetizing force, and any further increase in external field will not increase the strength of the magnetic flux through the core—a condition referred to as *saturation.* The elasticity of the above is evidenced by the fact that when the magnetizing force is removed, the alignment is lost to some measure, and the flux density drops to B_R. In other words, the removal of the magnetizing force results in the return of a number of misaligned domains within the core. The continued alignment of a number of the domains, however, accounts for our ability to create **permanent magnets.**

At a point just before saturation, the opposing unaligned domains are reduced to small cylinders of various shapes referred to as *bubbles.* These bubbles can be moved within the magnetic sample through the application of a *controlling* magnetic field. These magnetic bubbles form the basis of the recently designed bubble memory system for computers.

12.7 AMPÈRE'S CIRCUITAL LAW

As mentioned in the introduction to this chapter, there is a broad similarity between the analyses of electric and magnetic circuits. This has already been demonstrated to some extent for the quantities in Table 12.1.

TABLE 12.1

	Electric Circuits	Magnetic Circuits
Cause	E	$\mathscr{F}$
Effect	I	Φ
Opposition	R	$\mathscr{R}$

If we apply the "cause" analogy to Kirchhoff's voltage law ($\Sigma_{\circlearrowright} V = 0$), we obtain the following:

$$\boxed{\Sigma_{\circlearrowright} \mathscr{F} = 0} \qquad \text{(for magnetic circuits)} \qquad \textbf{(12.9)}$$

which, in words, states that the algebraic sum of the rises and drops of the mmf around a closed loop of a magnetic circuit is equal to zero; that is, the sum of the rises in mmf equals the sum of the drops in mmf around a closed loop.

Eq. (12.9) is referred to as **Ampère's circuital law.** When it is applied to magnetic circuits, sources of mmf are expressed by the equation

$$\boxed{\mathscr{F} = NI} \qquad \text{(At)} \qquad \textbf{(12.10)}$$

The equation for the mmf drop across a portion of a magnetic circuit can be found by applying the relationships listed in Table 12.1; that is, for electric circuits,

$$V = IR$$

resulting in the following for magnetic circuits:

$$\boxed{\mathcal{F} = \Phi\mathcal{R}} \qquad \text{(At)} \qquad \textbf{(12.11)}$$

where Φ is the flux passing through a section of the magnetic circuit and $\mathcal{R}$ is the reluctance of that section. The reluctance, however, is seldom calculated in the analysis of magnetic circuits. A more practical equation for the mmf drop is

$$\boxed{\mathcal{F} = Hl} \qquad \text{(At)} \qquad \textbf{(12.12)}$$

as derived from Eq. (12.6), where H is the magnetizing force on a section of a magnetic circuit and l is the length of the section.

As an example of Eq. (12.9), consider the magnetic circuit appearing in Fig. 12.10 constructed of three different ferromagnetic materials.

Applying Ampère's circuital law, we have

$$\Sigma_{\circlearrowright}\, \mathcal{F} = 0$$

$$\underbrace{+NI}_{\text{Rise}} - \underbrace{H_{ab}l_{ab}}_{\text{Drop}} - \underbrace{H_{bc}l_{bc}}_{\text{Drop}} - \underbrace{H_{ca}l_{ca}}_{\text{Drop}} = 0$$

$$\underbrace{NI}_{\substack{\text{Impressed}\\ \text{mmf}}} = \underbrace{H_{ab}l_{ab} + H_{bc}l_{bc} + H_{ca}l_{ca}}_{\text{mmf drops}}$$

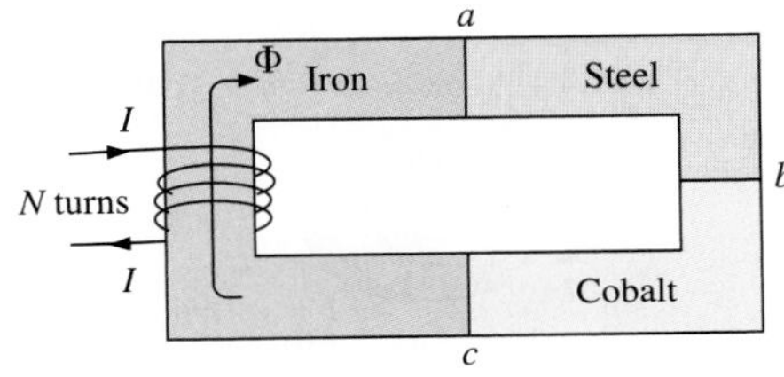

FIG. 12.10
Series magnetic circuit of three different materials.

All the terms of the equation are known except the magnetizing force for each portion of the magnetic circuit, which can be found by using the *B-H* curve if the flux density *B* is known.

12.8 FLUX Φ

If we continue to apply the relationships described in the previous section to Kirchhoff's current law, we find that the sum of the fluxes entering a junction is equal to the sum of the fluxes leaving a junction; that is, for the circuit in Fig. 12.11,

$$\Phi_a = \Phi_b + \Phi_c \qquad \text{(at junction } a\text{)}$$

or

$$\Phi_b + \Phi_c = \Phi_a \qquad \text{(at junction } b\text{)}$$

both of which are equivalent.

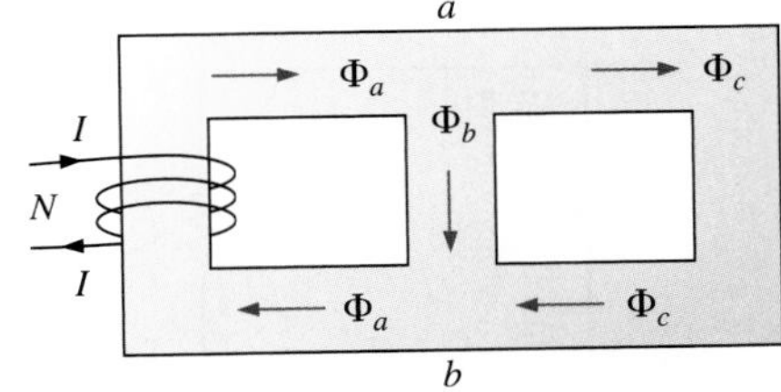

FIG. 12.11
Flux distribution of a series-parallel magnetic network.

12.9 SERIES MAGNETIC CIRCUITS: DETERMINING *NI*

We are now in a position to solve a few magnetic circuit problems, which are basically of two types. *In one type, Φ is given, and the impressed mmf NI must be computed.* This is the type of problem encountered in the design of motors, generators, and transformers. *In the other type, NI is given, and the flux Φ of the magnetic circuit must be found.* This type of problem is encountered primarily in the design of magnetic amplifiers and is more difficult since the approach is "hit or miss."

As indicated in earlier discussions, the value of μ varies from point to point along the magnetization curve. This eliminates the possibility of finding the reluctance of each "branch" or the "total reluctance" of a network, as was done for electric circuits where ρ had a fixed value for any applied current or voltage. If the total reluctance can be determined, Φ can then be determined using the Ohm's law analogy for magnetic circuits.

For magnetic circuits, the level of B or H is determined from the other using the B-H curve, and μ is seldom calculated unless asked for.

An approach frequently used in the analysis of magnetic circuits is the *table* method. Before a problem is analyzed in detail, a table is prepared listing in the far left column the various sections of the magnetic circuit (see Table 12.2). The columns on the right are reserved for the quantities to be found for each section. In this way, when you are solving a problem, you can keep track of what the next step should be and what is required to complete the problem. After a few examples, the usefulness of this method should become clear.

This section considers only *series* magnetic circuits in which the flux Φ is the same throughout. In each example, the magnitude of the magnetomotive force is to be determined.

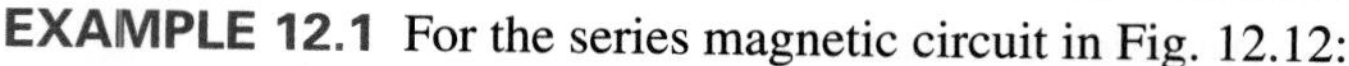

EXAMPLE 12.1 For the series magnetic circuit in Fig. 12.12:

a. Find the value of I required to develop a magnetic flux of $\Phi = 4 \times 10^{-4}$ Wb.
b. Determine μ and μ_r for the material under these conditions.

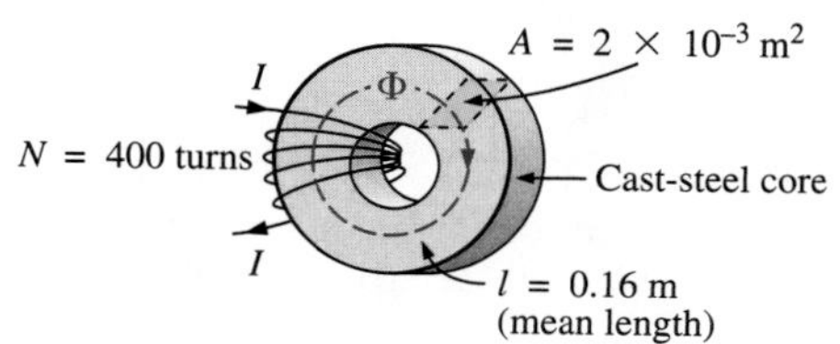

FIG. 12.12
Example 12.1.

Solutions: The magnetic circuit can be represented by the system shown in Fig. 12.13(a). The electric circuit analogy is shown in Fig. 12.13(b). Analogies of this type can be very helpful in the solution of magnetic circuits. Table 12.2 is for part (a) of this problem. The table is fairly trivial for this example, but it does define the quantities to be found.

a. The flux density B is

$$B = \frac{\Phi}{A} = \frac{4 \times 10^{-4}\text{ Wb}}{2 \times 10^{-3}\text{ m}^2} = 2 \times 10^{-1}\text{ T} = 0.2\text{ T}$$

Using the B-H curves in Fig. 12.8, we can determine the magnetizing force H:

$$H\text{ (cast steel)} = 170\text{ At/m}$$

Applying Ampère's circuital law yields

$$NI = Hl$$

and $$I = \frac{Hl}{N} = \frac{(170\text{ At/m})(0.16\text{ m})}{400\text{ t}} = \mathbf{68\text{ mA}}$$

(Recall that t represents turns.)

b. The permeability of the material can be found using Eq. (12.8):

$$\mu = \frac{B}{H} = \frac{0.2\text{ T}}{170\text{ At/m}} = \mathbf{1.176 \times 10^{-3}\text{ Wb/A}\cdot\text{m}}$$

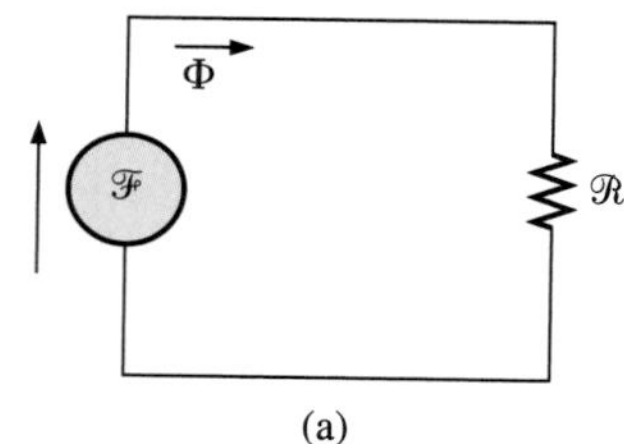

(a)

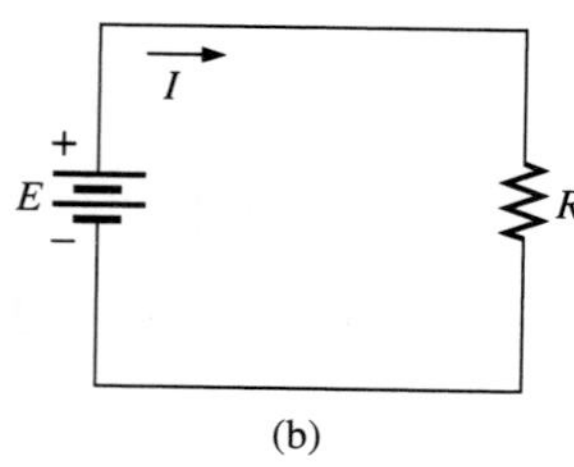

(b)

FIG. 12.13
(a) Magnetic circuit equivalent and (b) electric circuit analogy.

TABLE 12.2

Section	Φ (Wb)	A (m²)	B (T)	H (At/m)	l (m)	Hl (At)
One continuous section	4×10^{-4}	2×10^{-3}			0.16	

and the relative permeability is

$$\mu_r = \frac{\mu}{\mu_o} = \frac{1.176 \times 10^{-3}}{4\pi \times 10^{-7}} = \mathbf{935.83}$$

EXAMPLE 12.2 The electromagnet in Fig. 12.14 has picked up a section of cast iron. Determine the current I required to establish the indicated flux in the core.

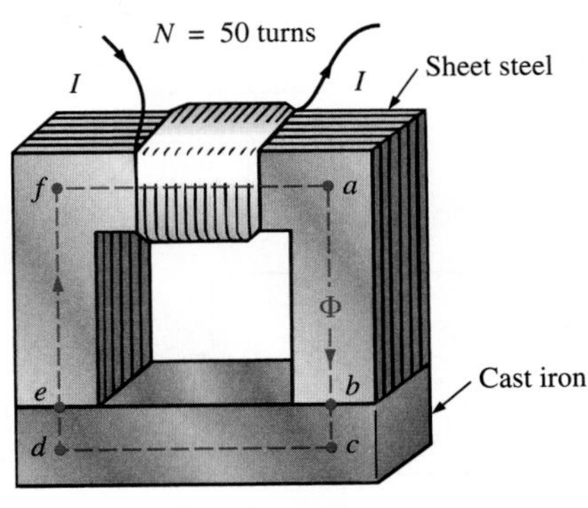

$l_{ab} = l_{cd} = l_{ef} = l_{fa} = 4$ in.
$l_{bc} = l_{de} = 0.5$ in.
Area (throughout) = 1 in.^{2}
$\Phi = 3.5 \times 10^{-4}$ Wb

FIG. 12.14
Electromagnet for Example 12.2.

Solution: To be able to use Figs. 12.7 and 12.8, we must first convert to the metric system. However, since the area is the same throughout, we can determine the length for each material rather than work with the individual sections:

$$l_{efab} = 4 \text{ in.} + 4 \text{ in.} + 4 \text{ in.} = 12 \text{ in.}$$
$$l_{bcde} = 0.5 \text{ in.} + 4 \text{ in.} + 0.5 \text{ in.} = 5 \text{ in.}$$
$$12 \text{ in.}\left(\frac{1 \text{ m}}{39.37 \text{ in.}}\right) = 304.8 \times 10^{-3} \text{ m}$$
$$5 \text{ in.}\left(\frac{1 \text{ m}}{39.37 \text{ in.}}\right) = 127 \times 10^{-3} \text{ m}$$
$$1 \text{ in.}^2\left(\frac{1 \text{ m}}{39.37 \text{ in.}}\right)\left(\frac{1 \text{ m}}{39.37 \text{ in.}}\right) = 6.452 \times 10^{-4} \text{ m}^2$$

The information available from the *efab* and *bcde* specifications of the problem has been inserted in Table 12.3. When the problem has been completed, each space will contain some information. Sufficient data to complete the problem can be found if we fill in each column from left to right. As the various quantities are calculated, they will be placed in a similar table found at the end of the example.

TABLE 12.3

Section	Φ (Wb)	A (m²)	B (T)	H (At/m)	l (m)	Hl (At)
efab	3.5×10^{-4}	6.452×10^{-4}			304.8×10^{-3}	
bcde	3.5×10^{-4}	6.452×10^{-4}			127×10^{-3}	

The flux density for each section is

$$B = \frac{\Phi}{A} = \frac{3.5 \times 10^{-4} \text{ Wb}}{6.452 \times 10^{-4} \text{ m}^2} = 0.542 \text{ T}$$

and the magnetizing force is

$$H \text{ (sheet steel, Fig. 12.8)} \cong 70 \text{ At/m}$$
$$H \text{ (cast iron, Fig. 12.7)} \cong 1600 \text{ At/m}$$

Note the extreme difference in magnetizing force for each material for the required flux density. In fact, when we apply Ampère's circuital law, we find that the sheet steel section can be ignored with a minimal error in the solution.

Determining Hl for each section yields

$$H_{efab}l_{efab} = (70 \text{ At/m})(304.8 \times 10^{-3} \text{ m}) = 21.34 \text{ At}$$
$$H_{bcde}l_{bcde} = (1600 \text{ At/m})(127 \times 10^{-3} \text{ m}) = 203.2 \text{ At}$$

Inserting the above data in Table 12.3 results in Table 12.4.

TABLE 12.4

Section	Φ (Wb)	A (m²)	B (T)	H (At/m)	l (m)	Hl (At)
efab	3.5×10^{-4}	6.452×10^{-4}	0.542	70	304.8×10^{-3}	21.34
bcde	3.5×10^{-4}	6.452×10^{-4}	0.542	1600	127×10^{-3}	203.2

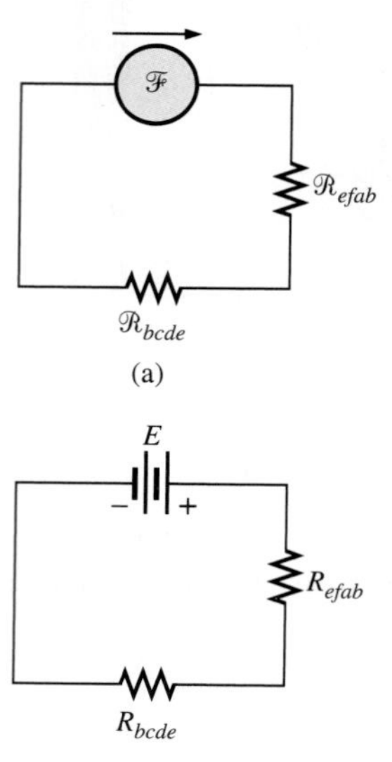

FIG. 12.15
(a) Magnetic circuit equivalent and (b) electric circuit analogy for the electromagnet in Fig. 12.14.

The magnetic circuit equivalent and the electric circuit analogy for the system in Fig. 12.14 appear in Fig. 12.15.

Applying Ampère's circuital law,

$$NI = H_{efab}l_{efab} + H_{bcde}l_{bcde}$$
$$= 21.34 \text{ At} + 203.2 \text{ At} = 224.54 \text{ At}$$

and
$$(50 \text{ t})I = 224.54 \text{ At}$$

so that
$$I = \frac{224.54 \text{ At}}{50 \text{ t}} = \mathbf{4.49\ A}$$

EXAMPLE 12.3 Determine the secondary current I_2 for the transformer in Fig. 12.16 if the resultant clockwise flux in the core is 1.5×10^{-5} Wb.

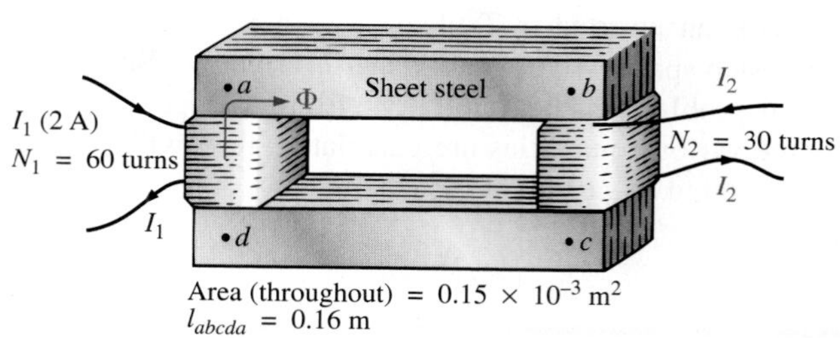

FIG. 12.16
Transformer for Example 12.3.

Solution: This is the first example with two magnetizing forces to consider. In the analogies in Fig. 12.17, note that the resulting flux of each is opposing, just as the two sources of voltage are opposing in the electric circuit analogy.

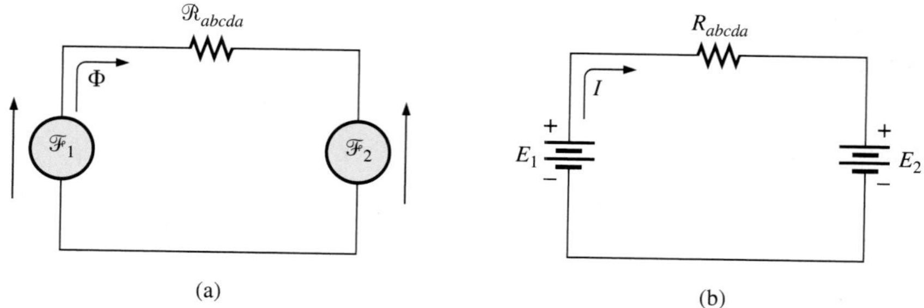

FIG. 12.17
(a) Magnetic circuit equivalent and (b) electric circuit analogy for the transformer in Fig. 12.16.

The *abcda* structural data appear in Table 12.5.

TABLE 12.5

Section	Φ (Wb)	A (m²)	B (T)	H (At/m)	l (m)	Hl (At)
abcda	1.5×10^{-5}	0.15×10^{-3}			0.16	

The flux density throughout is

$$B = \frac{\Phi}{A} = \frac{1.5 \times 10^{-5}\ \text{Wb}}{0.15 \times 10^{-3}\ \text{m}^2} = 10 \times 10^{-2}\ \text{T} = 0.10\ \text{T}$$

and

$$H\ (\text{from Fig. 12.8}) \cong \frac{1}{5}(100\ \text{At/m}) = 20\ \text{At/m}$$

Applying Ampère's circuital law,

$$N_1I_1 - N_2I_2 = H_{abcda}l_{abcda}$$
$$(60\ \text{t})(2\ \text{A}) - (30\ \text{t})(I_2) = (20\ \text{At/m})(0.16\text{m})$$
$$120\ \text{At} - (30\ \text{t})I_2 = 3.2\ \text{At}$$

and

$$(30\ \text{t})I_2 = 120\ \text{At} - 3.2\ \text{At}$$

or

$$I_2 = \frac{116.8\ \text{At}}{30\ \text{t}} = \mathbf{3.89\ A}$$

For the analysis of most transformer systems, the equation $N_1I_1 = N_2I_2$ is used. This results in 4 A versus 3.89 A above. This difference is normally ignored, however, and the equation $N_1I_1 = N_2I_2$ considered exact.

Because of the nonlinearity of the *B-H* curve, *it is not possible to apply superposition to magnetic circuits;* that is, in Example 12.3, we cannot consider the effects of each source independently and then find the total effects by using superposition.

12.10 AIR GAPS

Before continuing with the illustrative examples, let us consider the effects that an air gap has on a magnetic circuit. Note the presence of air gaps in the magnetic circuits of the motor and meter in Fig. 11.15. The spreading of the flux lines outside the common area of the core for the air gap in Fig. 12.18(a) is known as *fringing.* For our purposes, we shall ignore this effect and assume the flux distribution to be as in Fig. 12.18(b).

The flux density of the air gap in Fig. 12.18(b) is given by

$$B_g = \frac{\Phi_g}{A_g} \tag{12.13}$$

where, for our purposes,

$$\Phi_g = \Phi_{\text{core}}$$

and

$$A_g = A_{\text{core}}$$

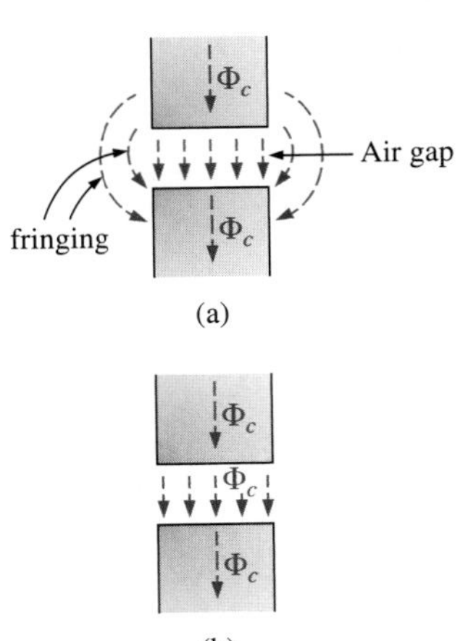

FIG. 12.18
Air gaps: (a) with fringing; (b) ideal.

For most practical applications, the permeability of air is taken to be equal to that of free space. The magnetizing force of the air gap is then determined by

$$H_g = \frac{B_g}{\mu_o} \tag{12.14}$$

and the mmf drop across the air gap is equal to $H_g L_g$. An equation for H_g is as follows:

$$H_g = \frac{B_g}{\mu_o} = \frac{B_g}{4\pi \times 10^{-7}}$$

and

$$H_g = (7.96 \times 10^5)B_g \quad \text{(At/m)} \tag{12.15}$$

EXAMPLE 12.4 Find the value of I required to establish a magnetic flux of $\Phi = 0.75 \times 10^{-4}$ Wb in the series magnetic circuit in Fig. 12.19.

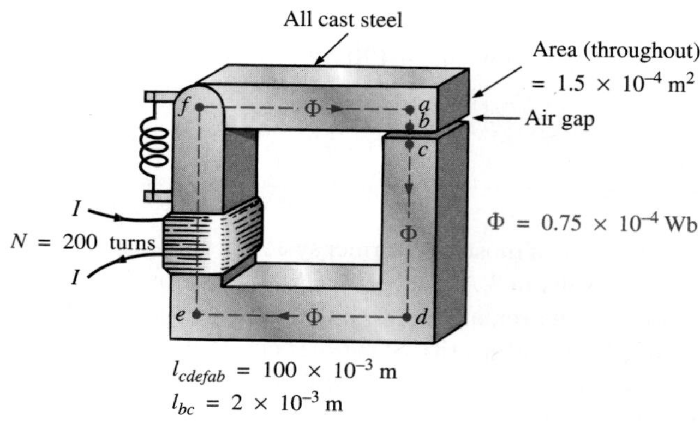

FIG. 12.19
Relay for Example 12.4.

Solution: An equivalent magnetic circuit and its electric circuit analogy are shown in Fig. 12.20.

The flux density for each section is

$$B = \frac{\Phi}{A} = \frac{0.75 \times 10^{-4}\text{ Wb}}{1.5 \times 10^{-4}\text{ m}^2} = 0.5\text{ T}$$

From the *B-H* curves in Fig. 12.8,

$$H\text{ (cast steel)} \cong 280\text{ At/m}$$

Applying Eq. (12.15),

$$H_g = (7.96 \times 10^5)B_g = (7.96 \times 10^5)(0.5\text{ T}) = 3.98 \times 10^5\text{ At/m}$$

The mmf drops are

$$H_{core}l_{core} = (280\text{ At/m})(100 \times 10^{-3}\text{ m}) = 28\text{ At}$$
$$H_g l_g = (3.98 \times 10^5\text{ At/m})(2 \times 10^{-3}\text{ m}) = 796\text{ At}$$

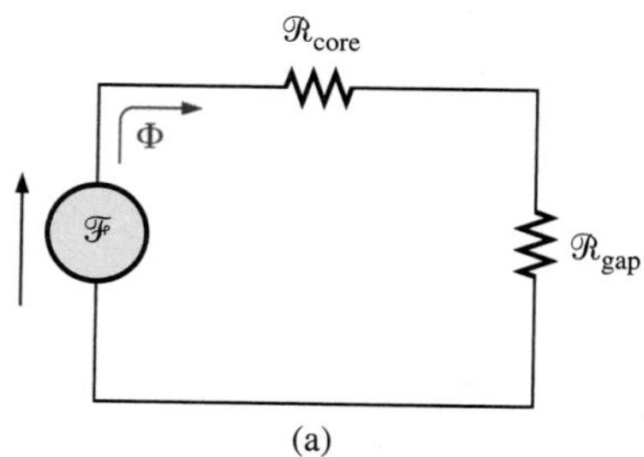

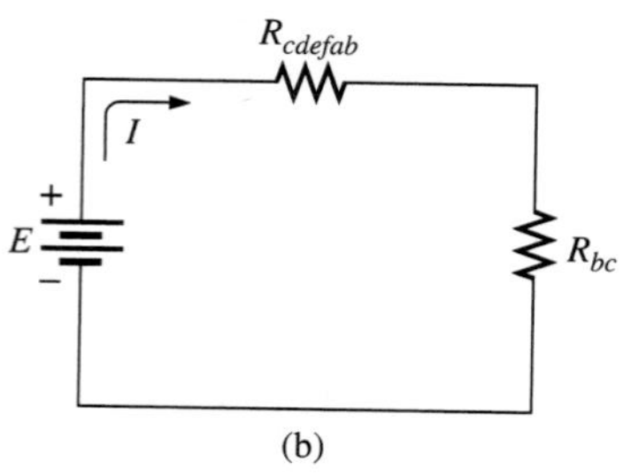

FIG. 12.20
(a) Magnetic circuit equivalent and (b) electric circuit analogy for the relay in Fig. 12.19.

Applying Ampère's circuital law,

$$\begin{aligned} NI &= H_{core}l_{core} + H_g l_g \\ &= 28 \text{ At} + 796 \text{ At} \\ (200 \text{ t})I &= 824 \text{ At} \\ I &= \mathbf{4.12 \text{ A}} \end{aligned}$$

Note from the above that the air gap requires the biggest share (by far) of the impressed NI because air is nonmagnetic.

12.11 SERIES-PARALLEL MAGNETIC CIRCUITS

As one might expect, the close analogies between electric and magnetic circuits eventually lead to series-parallel magnetic circuits similar in many respects to those encountered in Chapter 7. In fact, the electric circuit analogy will prove helpful in defining the procedure to follow toward a solution.

EXAMPLE 12.5 Determine the current I required to establish a flux of 1.5×10^{-4} Wb in the section of the core indicated in Fig. 12.21.

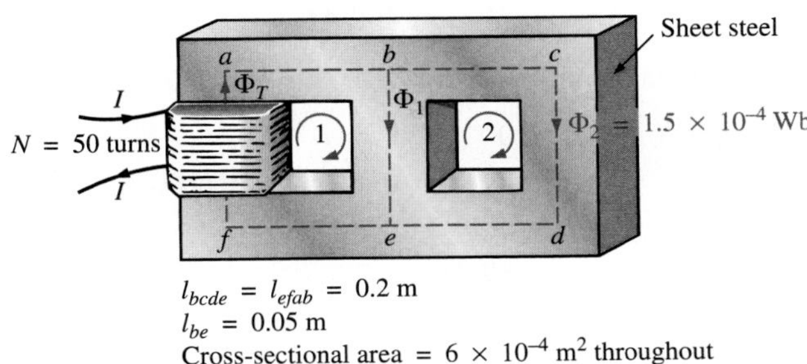

$l_{bcde} = l_{efab} = 0.2$ m
$l_{be} = 0.05$ m
Cross-sectional area = 6×10^{-4} m^2 throughout

FIG. 12.21
Example 12.5.

Solution: The equivalent magnetic circuit and the electric circuit analogy appear in Fig. 12.22. We have

$$B_2 = \frac{\Phi_2}{A} = \frac{1.5 \times 10^{-4} \text{ Wb}}{6 \times 10^{-4} \text{ m}^2} = 0.25 \text{ T}$$

From Fig. 12.8,

$$H_{bcde} \cong 40 \text{ At/m}$$

Applying Ampère's circuital law around loop 2 in Figs. 12.21 and 12.22,

$$\begin{aligned} \Sigma_{\circlearrowleft} \mathcal{F} &= 0 \\ H_{be}l_{be} - H_{bcde}l_{bcde} &= 0 \\ H_{be}(0.05 \text{ m}) - (40 \text{ At/m})(0.2\text{m}) &= 0 \\ H_{be} = \frac{8 \text{ At}}{0.05 \text{ m}} &= 160 \text{ At/m} \end{aligned}$$

From Fig. 12.8,

$$B_1 \cong 0.97 \text{ T}$$

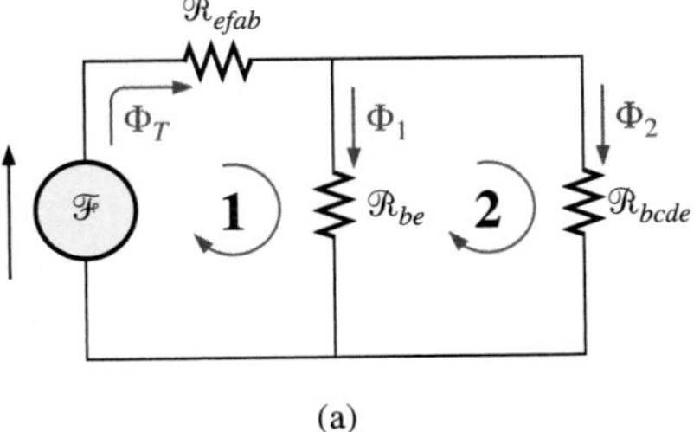

(a)

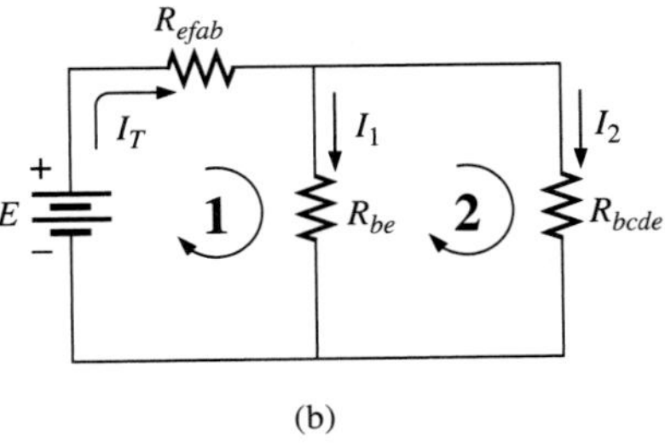

(b)

FIG. 12.22
(a) Magnetic circuit equivalent and (b) electric circuit analogy for the series-parallel system in Fig. 12.21.

and

$$\Phi_1 = B_1 A = (0.97\text{ T})(6 \times 10^{-4}\text{ m}^2) = 5.82 \times 10^{-4}\text{ Wb}$$

The results for *bcde, be,* and *efab* are entered in Table 12.6.

TABLE 12.6

Section	Φ (Wb)	A (m²)	B (T)	H (At/m)	l (m)	Hl (At)
bcde	1.5×10^{-4}	6×10^{-4}	0.25	40	0.2	8
be	5.82×10^{-4}	6×10^{-4}	0.97	160	0.05	8
efab		6×10^{-4}			0.2	

Table 12.6 reveals that we must now turn our attention to section *efab:*

$$\Phi_T = \Phi_1 + \Phi_2 = 5.82 \times 10^{-4}\text{ Wb} + 1.5 \times 10^{-4}\text{ Wb} = 7.32 \times 10^{-4}\text{ Wb}$$

$$B = \frac{\Phi_T}{A} = \frac{7.32 \times 10^{-4}\text{ Wb}}{6 \times 10^{-4}\text{ m}^2} = 1.22\text{ T}$$

From Fig. 12.7,

$$H_{efab} \cong 400\text{ At}$$

Applying Ampère's circuital law,

$$+NI - H_{efab}l_{efab} - H_{be}l_{be} = 0$$

$$NI = (400\text{ At/m})(0.2\text{ m}) + (160\text{ At/m})(0.05\text{ m})$$

$$(50\text{ t})I = 80\text{ At} + 8\text{ At}$$

$$I = \frac{88\text{ At}}{50\text{ t}} = \mathbf{1.76\text{ A}}$$

To demonstrate that μ is sensitive to the magnetizing force H, the permeability of each section is determined as follows. For section *bcde,*

$$\mu = \frac{B}{H} = \frac{0.25\text{ T}}{40\text{ At/m}} = 6.25 \times 10^{-3}$$

and

$$\mu_r = \frac{\mu}{\mu_o} = \frac{6.25 \times 10^{-3}}{12.57 \times 10^{-7}} = \mathbf{4972.2}$$

For section *be,*

$$\mu = \frac{B}{H} = \frac{0.97\text{ T}}{160\text{ At/m}} = 6.06 \times 10^{-3}$$

and

$$\mu_r = \frac{\mu}{\mu_o} = \frac{6.06 \times 10^{-3}}{12.57 \times 10^{-7}} = \mathbf{4821}$$

For section *efab,*

$$\mu = \frac{B}{H} = \frac{1.22\text{ T}}{400\text{ At/m}} = 3.05 \times 10^{-3}$$

and

$$\mu_r = \frac{\mu}{\mu_o} = \frac{3.05 \times 10^{-3}}{12.57 \times 10^{-7}} = \mathbf{2426.41}$$

12.12 DETERMINING Φ

The examples of this section are of the second type, where NI is given and the flux Φ must be found. This is a relatively straightforward problem if only one magnetic section is involved. Then

$$H = \frac{NI}{l} \qquad H \rightarrow B \qquad (B\text{-}H \text{ curve})$$

and

$$\Phi = BA$$

For magnetic circuits with more than one section, there is no set order of steps that lead to an exact solution for every problem on the first attempt. In general, however, we proceed as follows. We must find the impressed mmf for a *calculated guess* of the flux Φ and then compare this with the specified value of mmf. We can then make adjustments to our guess to bring it closer to the actual value. For most applications, a value within ±5% of the actual Φ or specified NI is acceptable.

We can make a reasonable guess at the value of Φ if we realize that the maximum mmf drop appears across the material with the smallest permeability if the length and area of each material are the same. As shown in Example 12.4, if there is an air gap in the magnetic circuit, there will be a considerable drop in mmf across the gap. As a starting point for problems of this type, therefore, we shall assume that the total mmf (NI) is across the section with the lowest μ or greatest $\mathcal{R}$ (if the other physical dimensions are relatively similar). This assumption gives a value of Φ that will produce a calculated NI greater than the specified value. Then, after considering the results of our original assumption very carefully, we shall *cut* Φ and NI by introducing the effects (reluctance) of the other portions of the magnetic circuit and *try* the new solution. For obvious reasons, this approach is frequently called the *cut and try* method.

EXAMPLE 12.6 Calculate the magnetic flux Φ for the magnetic circuit in Fig. 12.23.

Solution: By Ampère's circuital law,

$$NI = H_{abcda}l_{abcda}$$

or

$$H_{abcda} = \frac{NI}{l_{abcda}} = \frac{(60\text{ t})(5\text{ A})}{0.3\text{ m}}$$

$$= \frac{300\text{ At}}{0.3\text{ m}} = 1000\text{ At/m}$$

and

$$B_{abcda}\text{ (from Fig. 12.7)} \cong 0.39\text{ T}$$

Since $B = \Phi/A$, we have

$$\Phi = BA = (0.39\text{ T})(2 \times 10^{-4}\text{ m}^2) = \mathbf{0.78 \times 10^{-4}\ Wb}$$

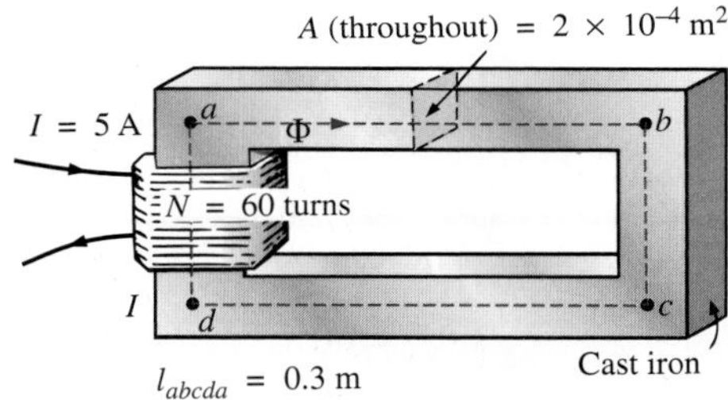

FIG. 12.23
Example 12.6.

EXAMPLE 12.7 Find the magnetic flux Φ for the series magnetic circuit in Fig. 12.24 for the specified impressed mmf.

Solution: Assuming that the total impressed mmf NI is across the air gap,

$$NI = H_g l_g$$

or

$$H_g = \frac{NI}{l_g} = \frac{400\text{ At}}{0.001\text{ m}} = 4 \times 10^5\text{ At/m}$$

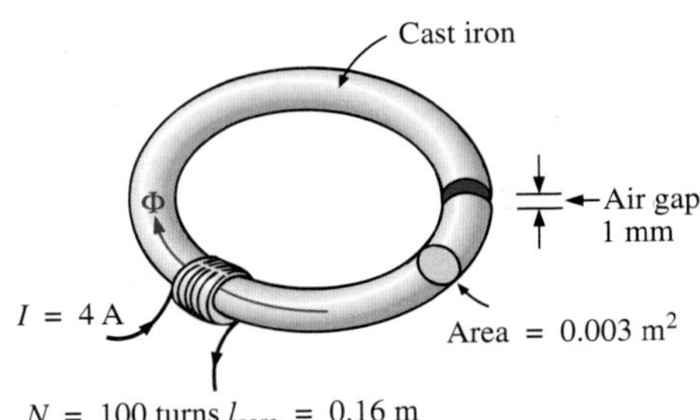

FIG. 12.24
Example 12.7.

and $$B_g = \mu_o H_g = (4\pi \times 10^{-7})(4 \times 10^5 \text{ At/m}) \\ = 0.503 \text{ T}$$

The flux

$$\Phi_g = \Phi_{core} = B_g A \\ = (0.503 \text{ T})(0.003 \text{ m}^2) \\ \Phi_{core} = 1.51 \times 10^{-3} \text{ Wb}$$

Using this value of Φ, we can find NI. The core and gap data are inserted in Table 12.7.

TABLE 12.7

Section	Φ (Wb)	A (m²)	B (T)	H (At/m)	l (m)	Hl (At)
Core	1.51×10^{-3}	0.003	0.503	1500 (B-H curve)	0.16	
Gap	1.51×10^{-3}	0.003	0.503	4×10^5	0.001	400

$$H_{core} l_{core} = (1500 \text{ At/m})(0.16 \text{ m}) = 240 \text{ At}$$

Applying Ampère's circuital law results in

$$NI = H_{core} l_{core} + H_g l_g \\ = 240 \text{ At} + 400 \text{ At} \\ 400 \text{ At} \neq 640 \text{ At}$$

Since we neglected the reluctance of all the magnetic paths but the air gap, the calculated value is greater than the specified value. We must therefore reduce this value by including the effect of these reluctances. Since approximately (640 At − 400 At)/640 At = 240 At/640 At ≅ 37.5% of our calculated value is above the desired value, let us reduce Φ by 30% and see how close we come to the impressed mmf of 400 At:

$$\Phi = (1 - 0.3)(1.51 \times 10^{-3} \text{ Wb}) \\ = 1.057 \times 10^{-3} \text{ Wb}$$

See Table 12.8.

TABLE 12.8

Section	Φ (Wb)	A (m²)	B (T)	H (At/m)	l (m)	Hl (At)
Core	1.057×10^{-3}	0.003			0.16	
Gap	1.057×10^{-3}	0.003			0.001	

$$B = \frac{\Phi}{A} = \frac{1.057 \times 10^{-3} \text{ Wb}}{0.003 \text{ m}^3} \cong 0.352 \text{ T}$$

$$H_g l_g = (7.96 \times 10^5) B_g l_g \\ = (7.96 \times 10^5)(0.352 \text{ T})(0.001 \text{ m}) \\ \cong 280.19 \text{ At}$$

From the B-H curves,

$$H_{core} \cong 850 \text{ At/m} \\ H_{core} l_{core} = (850 \text{ At/m})(0.16 \text{ m}) = 136 \text{ At}$$

Applying Ampère's circuital law yields

$$NI = H_{core} l_{core} + H_g l_g \\ = 136 \text{ At} + 280.19 \text{ At}$$

$400 \text{ At} = \mathbf{416.19\ At}$ (but within ±5% and therefore acceptable)

The solution is, therefore,

$$\Phi \cong \mathbf{1.057 \times 10^{-3}\ Wb}$$

12.13 APPLICATIONS

Speakers and Microphones

Electromagnetic effects are the moving force in the design of speakers such as the one shown in Fig. 12.25. The shape of the pulsating waveform of the input current is determined by the sound to be reproduced by the speaker at a high audio level. As the current peaks and returns to the valleys of the sound pattern, the strength of the electromagnet varies in exactly the same manner. This causes the cone of the speaker to vibrate at a frequency directly proportional to the pulsating input. The higher the pitch of the sound pattern, the higher the oscillating frequency between the peaks and valleys and the higher the frequency of vibration of the cone.

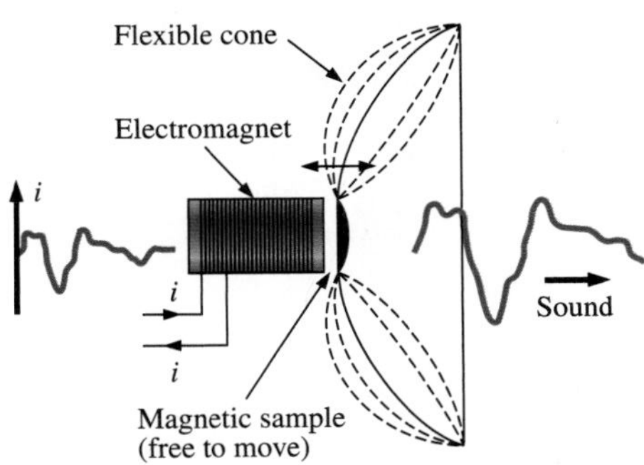

FIG. 12.25
Speaker.

A second design used more frequently in more expensive speaker systems appears in Fig. 12.26. In this case, the permanent magnet is fixed, and the input is applied to a movable core within the magnet, as shown in the figure. High peaking currents at the input produce a strong flux pattern in the voice coil, causing it to be drawn well into the flux pattern of the permanent magnet. As occurred for the speaker in Fig. 12.25, the core then vibrates at a rate determined by the input and provides the audible sound.

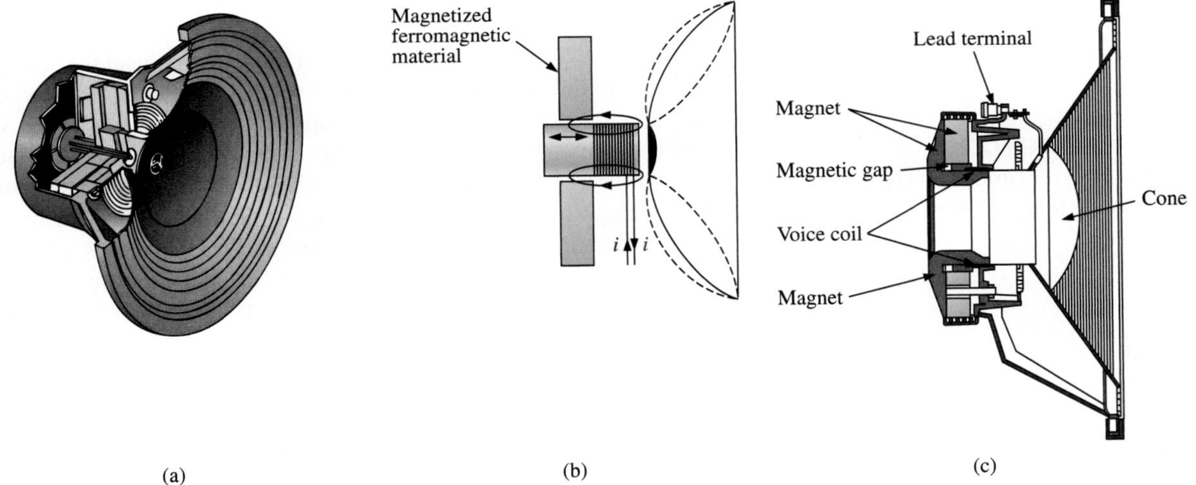

FIG. 12.26
Coaxial high-fidelity loudspeaker: (a) construction: (b) basic operation; (c) cross section of actual unit.
(Courtesy of Electro-Voice, Inc.)

Microphones also employ electromagnetic effects. The incoming sound causes the core and attached moving coil to move within the magnetic field of the permanent magnet. Through Faraday's law ($e = N\,d\phi/dt$), a voltage is induced across the movable coil proportional to the speed with which it is moving through the magnetic field. The resulting induced voltage pattern can then be amplified and reproduced at a much higher audio level through the use of speakers, as described earlier. Microphones of this type are the most frequently employed, although other

types that use capacitive, carbon granular, and piezoelectric* effects are available. This particular design is commercially referred to as a *dynamic* microphone.

Hall Effect Sensor

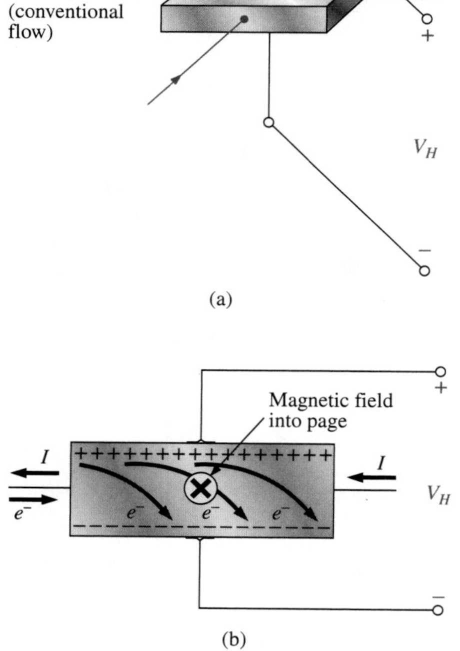

FIG. 12.27
Hall effect sensor: (a) orientation of controlling parameters; (b) effect on electron flow.

The Hall effect sensor is a semiconductor device that generates an output voltage when exposed to a magnetic field. The basic construction consists of a slab of semiconductor material through which a current is passed, as shown in Fig. 12.27(a). If a magnetic field is applied as shown in the figure perpendicular to the direction of the current, a voltage V_H is generated between the two terminals, as indicated in Fig. 12.27(a). The difference in potential is due to the separation of charge established by the Lorentz force first studied by Professor Hendrick Lorentz in the late 1800s. He found that electrons in a magnetic field are subjected to a force proportional to the velocity of the electrons through the field and the strength of the magnetic field. The direction of the force is determined by the left-hand rule. Simply place the index finger of your left hand in the direction of the magnetic field, with the second finger at right angles to the index finger in the direction of conventional current through the semiconductor material, as shown in Fig. 12.27(b). The thumb, if placed at right angles to the index finger, will indicate the direction of the force on the electrons. In Fig. 12.27(b), the force causes the electrons to accumulate in the bottom region of the semiconductor (connected to the negative terminal of the voltage V_H), leaving a net positive charge in the upper region of the material (connected to the positive terminal of V_H). The stronger the current or strength of the magnetic field, the greater the induced voltage V_H.

In essence, therefore, the Hall effect sensor can reveal the strength of a magnetic field or the level of current through a device if the other determining factor is held fixed. Two applications of the sensor are therefore apparent—to measure the strength of a magnetic field in the vicinity of a sensor (for an applied fixed current) and to measure the level of current through a sensor (with knowledge of the strength of the magnetic field linking the sensor). The gaussmeter in Fig. 11.14 uses a Hall effect sensor. Internal to the meter, a fixed current is passed through the sensor with the voltage V_H indicating the relative strength of the field. Through amplification, calibration, and proper scaling, the meter can display the relative strength in gauss.

The Hall effect sensor has a broad range of applications that are often quite interesting and innovative. The most widespread is as a trigger for an alarm system in large department stores, where theft is often a difficult problem. A magnetic strip attached to the merchandise sounds an alarm when a customer passes through the exit gates without paying for the product. The sensor, control current, and monitoring system are housed in the exit fence and react to the presence of the magnetic field as the product leaves the store. When the product is paid for, the cashier removes the strip or demagnetizes the strip by applying a magnetizing force that reduces the residual magnetism in the strip to essentially zero.

The Hall effect sensor is also used to indicate the speed of a bicycle on a digital display conveniently mounted on the handlebars. As shown in Fig. 12.28(a), the sensor is mounted on the frame of the bike, and a small

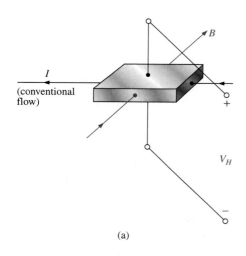

*Piezoelectricity is the generation of a small voltage by exerting pressure across certain crystals.

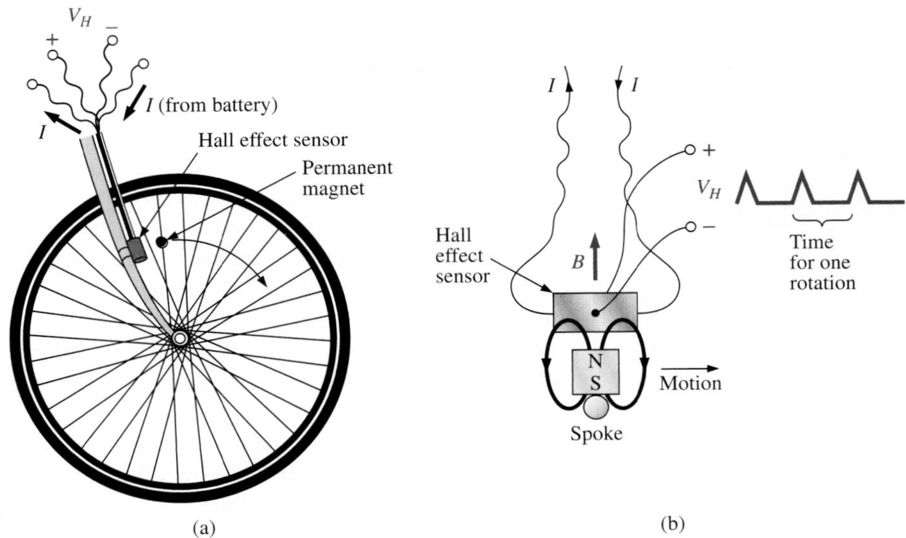

FIG. 12.28
Obtaining a speed indication for a bicycle using a Hall effect sensor: (a) mounting the components; (b) Hall effect response.

permanent magnet is mounted on a spoke of the front wheel. The magnet must be carefully mounted to be sure that it passes over the proper region of the sensor. When the magnet passes over the sensor, the flux pattern in Fig. 12.28(b) results, and a voltage with a sharp peak is developed by the sensor. Assuming a bicycle with a 26-in.-diameter wheel, the circumference will be about 82 in. Over 1 mi, the number of rotations is

$$5280\ \text{ft}\left(\frac{12\ \text{in.}}{1\ \text{ft}}\right)\left(\frac{1\ \text{rotation}}{82\ \text{in.}}\right) \cong 773\ \text{rotations}$$

If the bicycle is traveling at 20 mph, an output pulse occurs at a rate of 4.29 per second. It is interesting to note that at a speed of 20 mph, the wheel is rotating at more than 4 revolutions per second, and the total number of rotations over 20 mi is 15,460.

Magnetic Reed Switch

One of the most frequently employed switches in alarm systems is the *magnetic reed switch* shown in Fig. 12.29. As shown by the figure, there are two components of the reed switch—a permanent magnet embedded in one unit that is normally connected to the movable element (door, window, and so on) and a reed switch in the other unit that is connected to the electrical control circuit. The reed switch is constructed of two iron-alloy (ferromagnetic) reeds in a hermetically sealed capsule. The cantilevered ends of the two reeds do not touch but are in very close proximity to one another. In the absence of a magnetic field, the reeds remain separated. However, if a magnetic field is introduced, the reeds are drawn to each other because flux lines seek the path of least reluctance and, if possible, exercise every alternative to establish the path of least reluctance. It is similar to placing a ferromagnetic bar close to the ends of a U-shaped magnet. The bar is drawn to the poles of the magnet, establishing a magnetic flux path without air gaps and with minimum reluctance. In the open-circuit state, the resistance between reeds is in excess of 100 MΩ, while in the on state it drops to less than 1 Ω.

In Fig. 12.30 a reed switch has been placed on the fixed frame of a window and a magnet on the movable window unit. When the window is

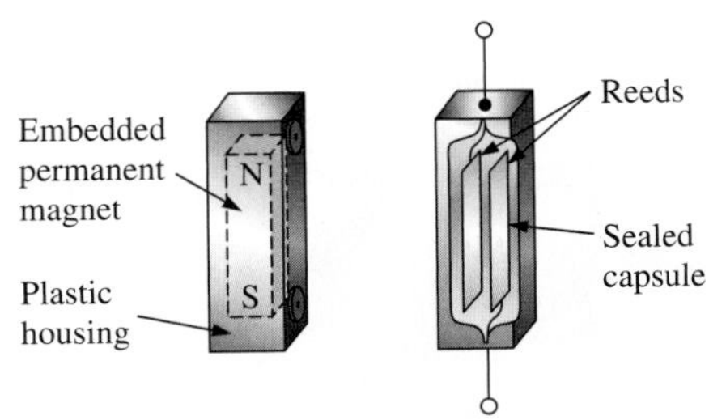

FIG. 12.29
Magnetic reed switch.

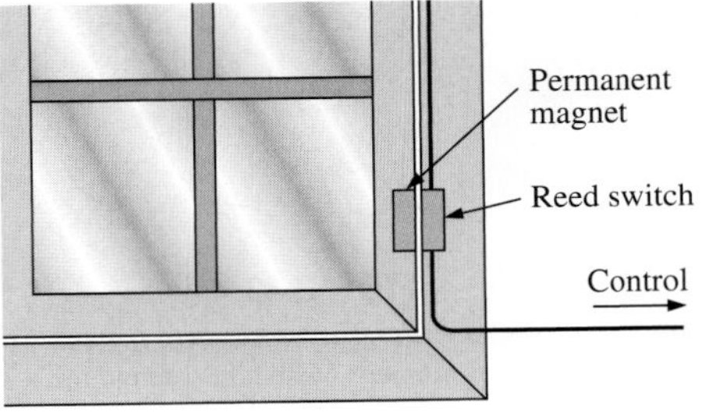

FIG. 12.30
Using a magnetic reed switch to monitor the state of a window.

closed as shown in Fig. 12.30, the magnet and reed switch are sufficiently close to establish contact between the reeds, and a current is established through the reed switch to the control panel. In the armed state, the alarm system accepts the resulting current flow as a normal secure response. If the window is opened, the magnet leaves the vicinity of the reed switch, and the switch opens. The current through the switch is interrupted, and the alarm reacts appropriately.

One of the distinct advantages of the magnetic reed switch is that the proper operation of any switch can be checked with a portable magnetic element. Simply bring the magnet to the switch and note the output response. There is no need to continually open and close windows and doors. In addition, the reed switch is hermetically enclosed so that oxidation and foreign objects cannot damage it, and the result is a unit that can last indefinitely. Magnetic reed switches are also available in other shapes and sizes, allowing them to be concealed from obvious view. One is a circular variety that can be set into the edge of a door and door jam, resulting in only two small visible disks when the door is open.

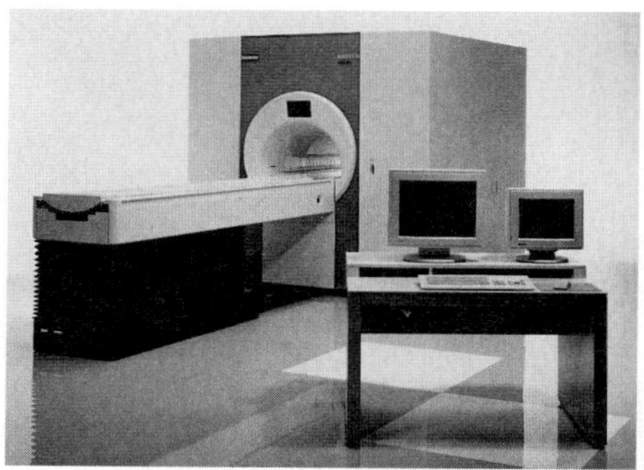

FIG. 12.31
Magnetic resonance imaging equipment.
(Courtesy of Siemens Medical Systems, Inc.)

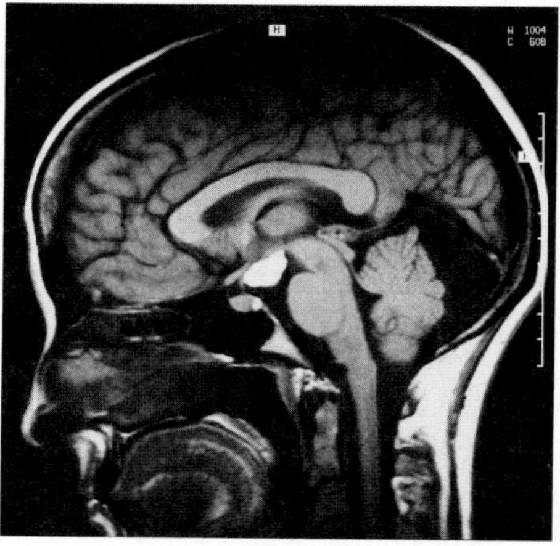

FIG. 12.32
Magnetic resonance image.
(Courtesy of Siemens Medical Systems, Inc.)

Magnetic Resonance Imaging

Magnetic resonance imaging (MRI) provides quality cross-sectional images of the body for medical diagnosis and treatment. MRI does not expose the patient to potentially hazardous X-rays or injected contrast materials such as those used to obtain computerized axial tomography (CAT) scans.

The three major components of an MRI system are a strong magnet, a table for transporting the patient into the circular hole in the magnet, and a control center, as shown in Fig. 12.31. The image is obtained by placing the patient in the tube to a precise depth depending on the cross section to be obtained and applying a strong magnetic field that causes the nuclei of certain atoms in the body to line up. Radio waves of different frequencies are then applied to the patient in the region of interest, and if the frequency of the wave matches the natural frequency of the atom, the nuclei is set into a state of resonance and absorbs energy from the applied signal. When the signal is removed, the nuclei release the acquired energy in the form of weak but detectable signals. The strength and duration of the energy emission vary from one tissue of the body to another. The weak signals are then amplified, digitized, and translated to provide a cross-sectional image such as the one shown in Fig. 12.32. For some patients the claustrophobic feeling they experience while in the circular tube is difficult to contend with. Today, however, a more open unit has been developed, as shown in Fig. 12.33, that has removed most of this discomfort.

Patients who have metallic implants or pacemakers or those who have worked in industrial environments where minute ferromagnetic particles may have become lodged in open, sensitive areas such as the eyes, nose, and so on, may have to use a CAT scan system because it does not employ magnetic effects. The attending physician is well trained in such areas of concern and will remove any unfounded fears or suggest alternative methods.

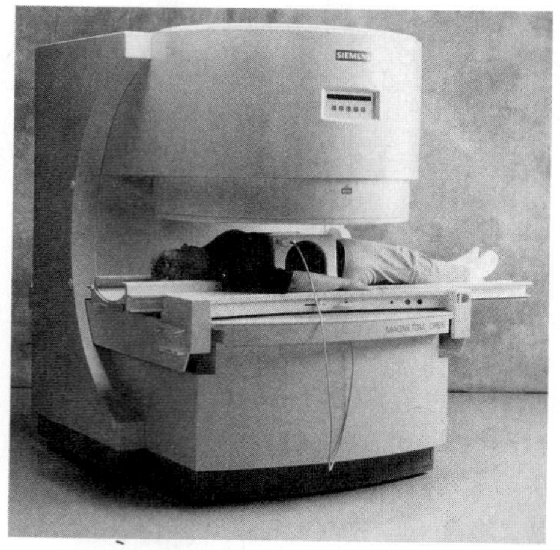

FIG. 12.33
Magnetic resonance imaging equipment (open variety).
(Courtesy of Siemens Medical Systems, Inc.)

PROBLEMS

SECTION 12.2 Magnetic Field

1. Using Appendix F, fill in the blanks in the following table. Indicate the units for each quantity.

	Φ	B
SI	5×10^{-4} Wb	8×10^{-4} T
CGS	________	________
English	________	________

2. Repeat Problem 1 for the following table if area = 2 in.^{2}:

	Φ	B
SI	________	________
CGS	60,000 maxwells	________
English	________	________

3. For the electromagnet in Fig. 12.34:
   a. Find the flux density in the core.
   b. Sketch the magnetic flux lines and indicate their direction.
   c. Indicate the north and south poles of the magnet.

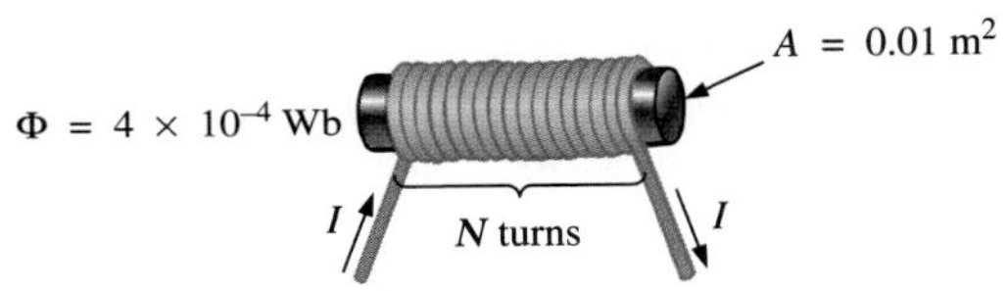

FIG. 12.34
Problem 3.

SECTION 12.3 Reluctance

4. Which section of Fig. 12.35—(a), (b), or (c)—has the largest reluctance to the setting up of flux lines through its longest dimension?

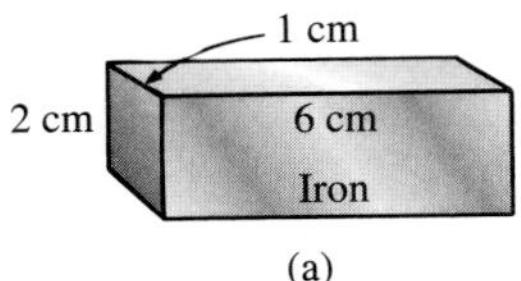

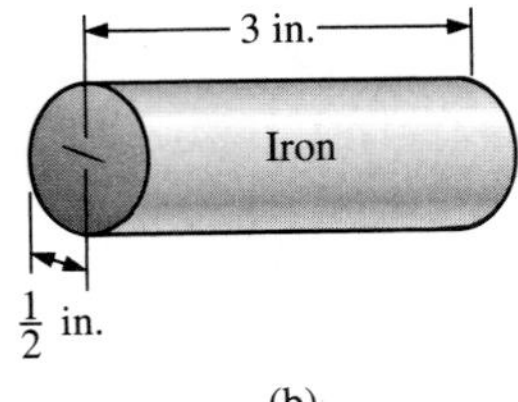

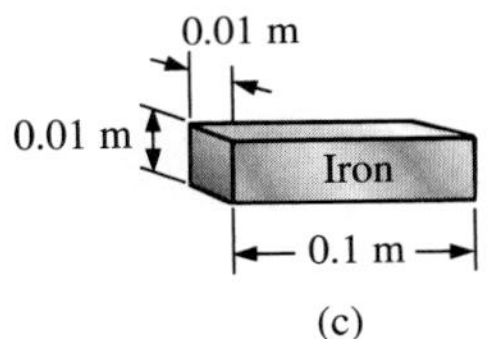

FIG. 12.35
Problem 4.

SECTION 12.4 Ohm's Law for Magnetic Circuits

5. Find the reluctance of a magnetic circuit if a magnetic flux $\Phi = 4.2 \times 10^{-4}$ Wb is established by an impressed mmf of 400 At.

6. Repeat Problem 5 for Φ = 72,000 maxwells and an impressed mmf of 120 gilberts.

SECTION 12.5 Magnetizing Force

7. Find the magnetizing force H for Problem 5 in SI units if the magnetic circuit is 6 in. long.

8. If a magnetizing force H of 600 At/m is applied to a magnetic circuit, a flux density B of 1200×10^{-4} Wb/m^2 is established. Find the permeability μ of a material that will produce twice the original flux density for the same magnetizing force.

SECTIONS 12.6–12.9 Hysteresis through Series Magnetic Circuits

9. For the series magnetic circuit in Fig. 12.36, determine the current I necessary to establish the indicated flux.

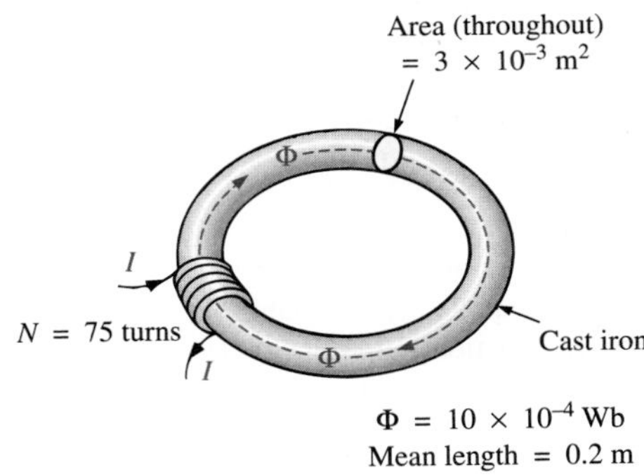

FIG. 12.36
Problem 9.

10. Find the current necessary to establish a flux of $\Phi = 3 \times 10^{-4}$ Wb in the series magnetic circuit in Fig 12.37.

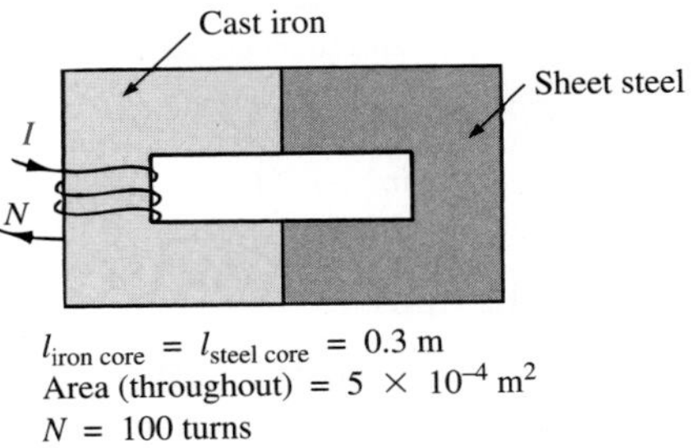

FIG. 12.37
Problem 10.

11. **a.** Find the number of turns N_1 required to establish a flux $\Phi = 12 \times 10^{-4}$ Wb in the magnetic circuit in Fig. 12.38.
    b. Find the permeability μ of the material.

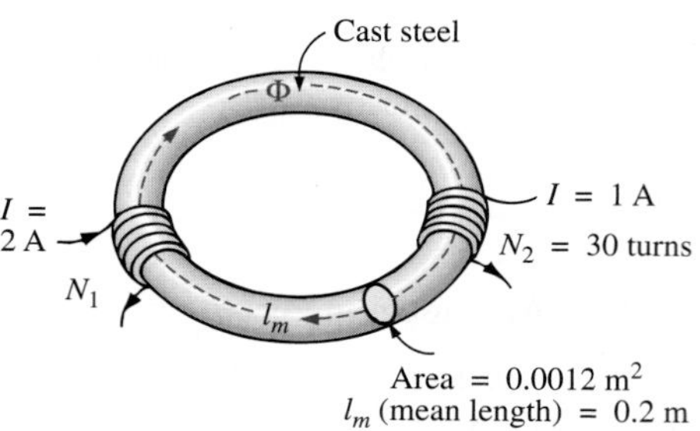

FIG. 12.38
Problem 11.

12. **a.** Find the mmf (NI) required to establish a flux Φ = 80,000 lines in the magnetic circuit in Fig. 12.39.
    b. Find the permeability of each material.

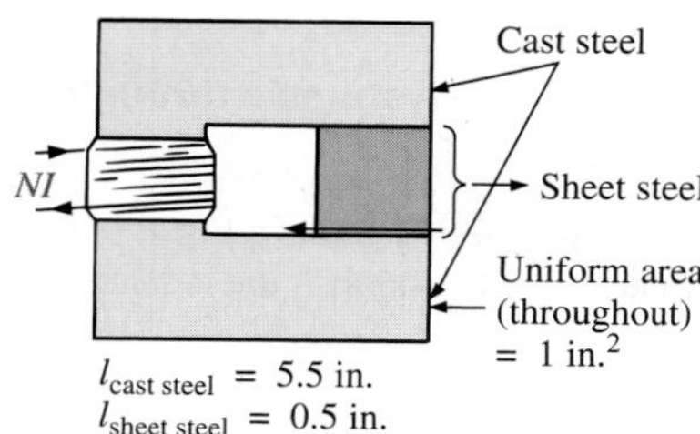

FIG. 12.39
Problem 12.

*13. For the series magnetic circuit in Fig. 12.40 with two impressed sources of magnetic "pressure," determine the current I. Each applied mmf establishes a flux pattern in the clockwise direction.

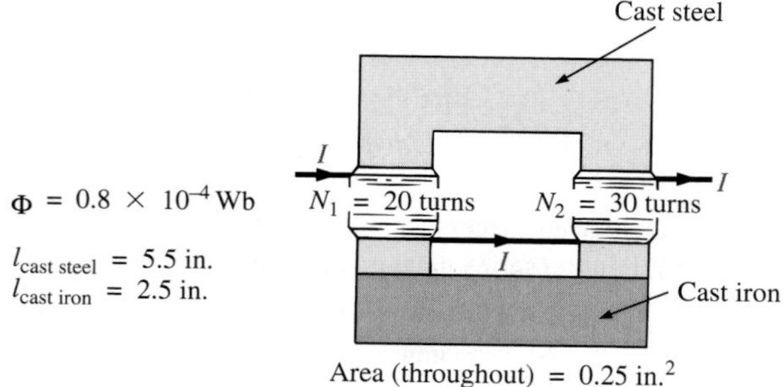

FIG. 12.40
Problem 13.

SECTION 12.10 Air Gaps

14. **a.** Find the current I required to establish a flux $\Phi = 2.4 \times 10^{-4}$ Wb in the magnetic circuit in Fig. 12.41.
    b. Compare the mmf drop across the air gap to that across the rest of the magnetic circuit. Discuss your results using the value of μ for each material.

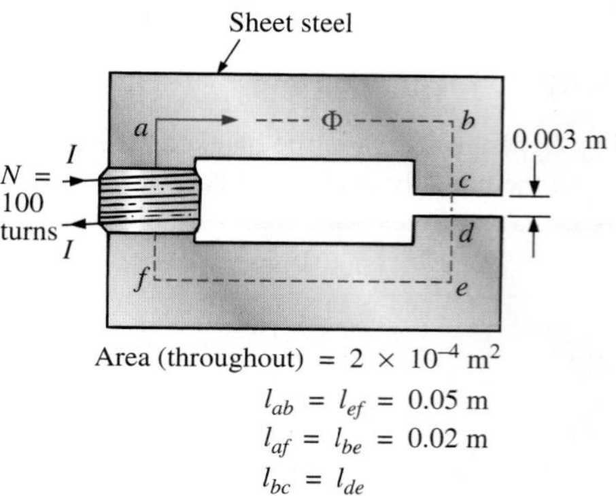

FIG. 12.41
Problem 14.

*15. The force carried by the plunger of the door chime in Fig. 12.42 is determined by

$$f = \frac{1}{2} NI \frac{d\phi}{dx} \qquad \text{(newtons)}$$

where $d\phi/dx$ is the rate of change of flux linking the coil as the core is drawn into the coil. The greatest rate of change of flux occurs when the core is ¼ to ¾ the way through. In this region, if Φ changes from 0.5×10^{-4} Wb to 8×10^{-4} Wb, what is the force carried by the plunger?

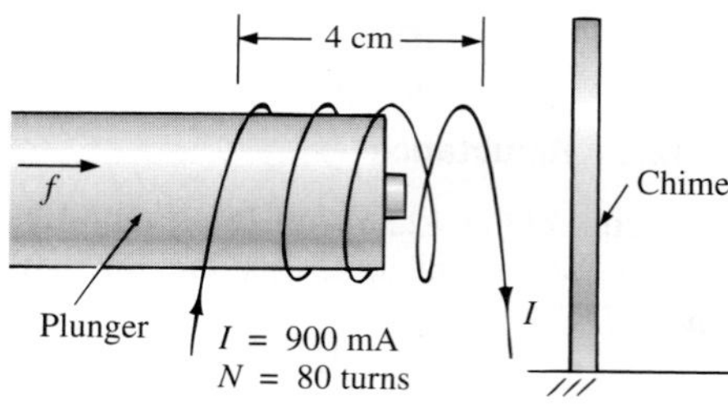

FIG. 12.42
Door chime for Problem 15.

16. Determine the current I_1 required to establish a flux of $\Phi = 2 \times 10^{-4}$ Wb in the magnetic circuit in Fig. 12.43.

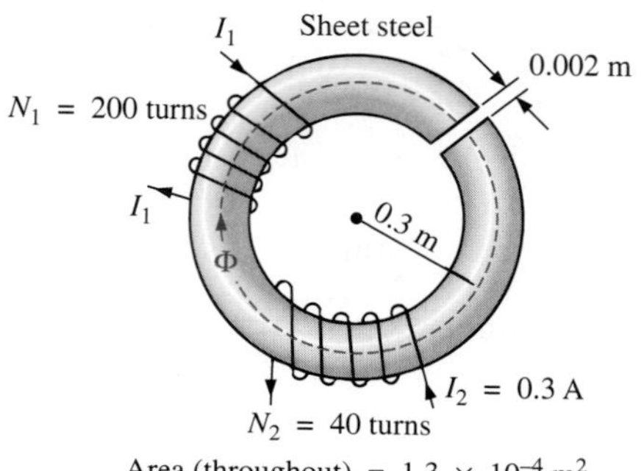

FIG. 12.43
Problem 16.

*17. **a.** A flux of 0.2×10^{-4} Wb will establish sufficient attractive force for the armature of the relay in Fig. 12.44 to close the contacts. Determine the required current to establish this flux level if we assume that the total mmf drop is across the air gap.

b. The force exerted on the armature is determined by the equation

$$F(\text{newtons}) = \frac{1}{2} \cdot \frac{B_g^2 A}{\mu_o}$$

where B_g is the flux density within the air gap and A is the common area of the air gap. Find the force in newtons exerted when the flux Φ specified in part (a) is established.

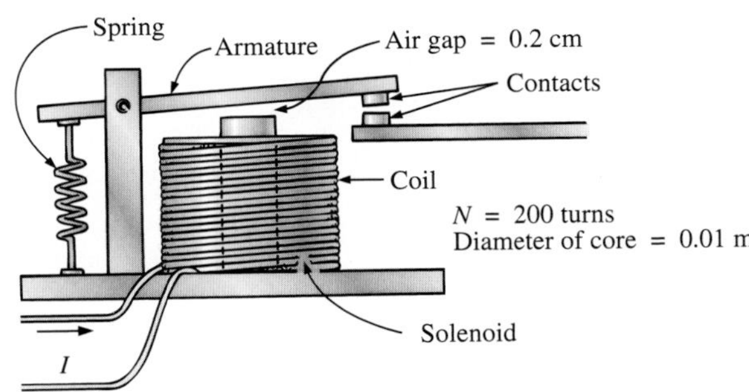

FIG. 12.44
Relay for Problem 17.

SECTION 12.11 Series-Parallel Magnetic Circuits

*18. For the series-parallel magnetic circuit in Fig. 12.45, find the value of I required to establish a flux in the gap of $\Phi_g = 2 \times 10^{-4}$ Wb.

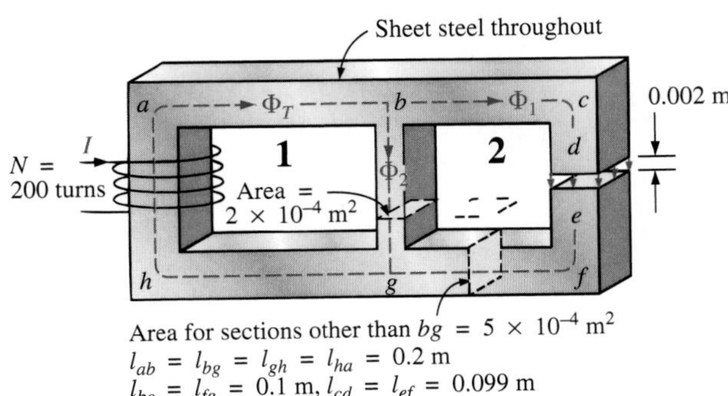

FIG. 12.45
Problem 18.

SECTION 12.12 Determining Φ

19. Find the magnetic flux Φ established in the series magnetic circuit in Fig. 12.46.

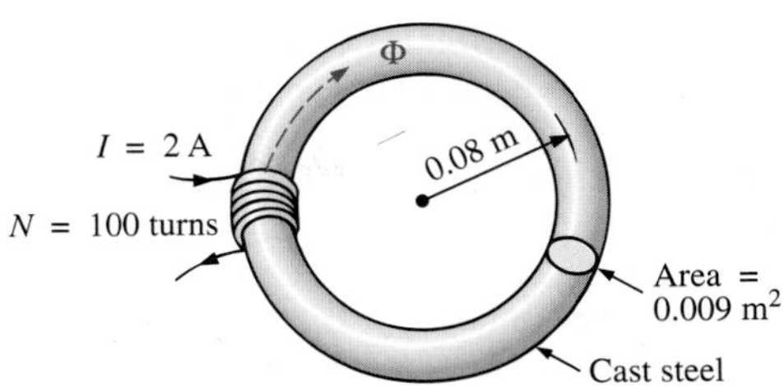

FIG. 12.46
Problem 19.

*20. Determine the magnetic flux Φ established in the series magnetic circuit in Fig. 12.47.

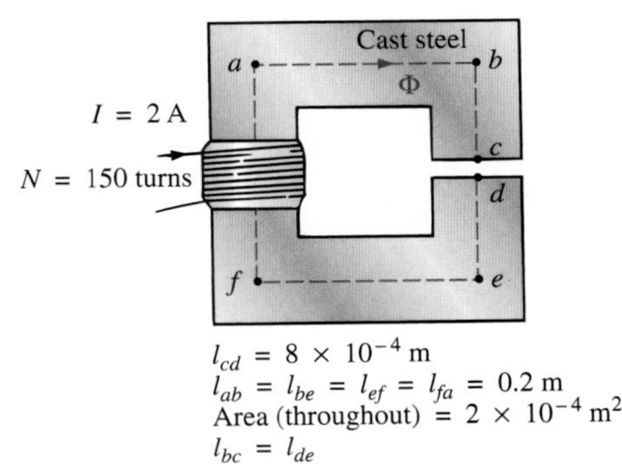

FIG. 12.47
Problem 20.

*21. Note how closely the B-H curve of cast steel in Fig. 12.7 matches the curve for the voltage across a capacitor as it charges from zero volts to its final value.

a. Using the equation for the charging voltage as a guide, write an equation for B as a function of H [$B = f(H)$] for cast steel.

b. Test the resulting equation at $H = 900$ At/m, 1800 At/m, and 2700 At/m.

c. Using the equation of part (a), derive an equation for H in terms of B [$H = f(B)$].

d. Test the resulting equation at $B = 1$ T and $B = 1.4$ T.

e. Using the result of part (c), perform the analysis of Example 12.1, and compare the results for the current I.

GLOSSARY

Ampère's circuital law A law establishing the fact that the algebraic sum of the rises and drops of the mmf around a closed loop of a magnetic circuit is equal to zero.

Domain A group of magnetically aligned atoms.

Electromagnetism Magnetic effects introduced by the flow of charge or current.

Ferromagnetic materials Materials having permeabilities hundreds and thousands of times greater than that of free space.

Flux density (*B*) A measure of the flux per unit area perpendicular to a magnetic flux path. It is measured in teslas (T) or webers per square meter (Wb/m²).

Hysteresis The lagging effect between the flux density of a material and the magnetizing force applied.

Magnetic flux lines Lines of a continuous nature that reveal the strength and direction of a magnetic field.

Magnetizing force (H) A measure of the magnetomotive force per unit length of a magnetic circuit.

Magnetomotive force (mmf) ($\mathscr{F}$) The "pressure" required to establish magnetic flux in a ferromagnetic material. It is measured in ampere-turns (At).

Permanent magnet A material such as steel or iron that will remain magnetized for long periods of time without the aid of external means.

Permeability (μ) A measure of the ease with which magnetic flux can be established in a material. It is measured in Wb/Am.

Relative permeability (μ_r) The ratio of the permeability of a material to that of free space.

Reluctance ($\mathscr{R}$) A quantity determined by the physical characteristics of a material that will provide an indication of the "reluctance" of that material to the setting up of magnetic flux lines in the material. It is measured in rels or At/Wb.

Sinusoidal Alternating Waveforms

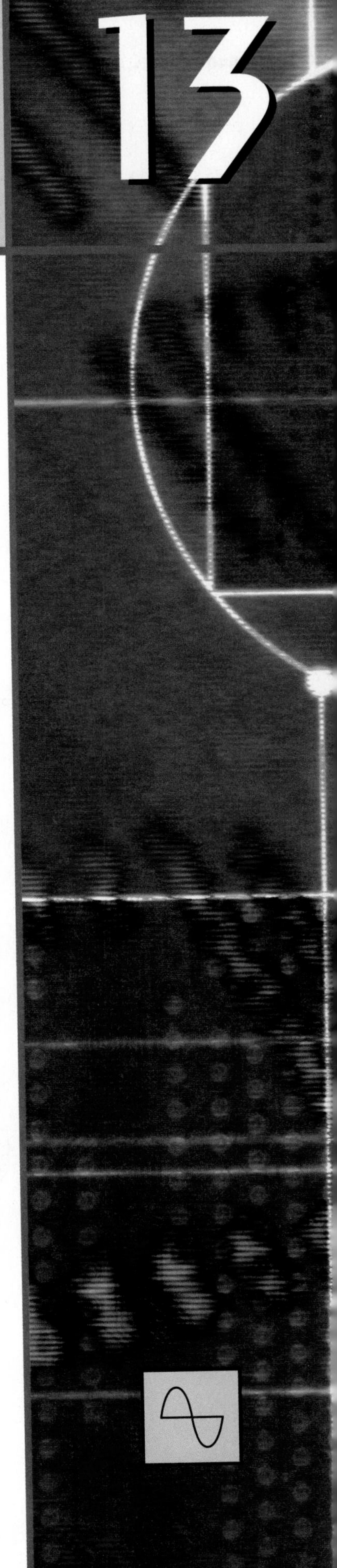

Objectives

- *Become familiar with the characteristics of a sinusoidal waveform including its general format, average value, and effective value.*
- *Be able to determine the phase relationship between two sinusoidal waveforms of the same frequency.*
- *Understand how to calculate the average and effective values of any waveform.*
- *Become familiar with the use of instruments designed to measure ac quantities.*

13.1 INTRODUCTION

The analysis thus far has been limited to dc networks, networks in which the currents or voltages are fixed in magnitude except for transient effects. We now turn our attention to the analysis of networks in which the magnitude of the source varies in a set manner. Of particular interest is the time-varying voltage that is commercially available in large quantities and is commonly called the *ac voltage.* (The letters *ac* are an abbreviation for *alternating current.*) To be absolutely rigorous, the terminology *ac voltage* or *ac current* is not sufficient to describe the type of signal we will be analyzing. Each waveform in Fig. 13.1 is an **alternating waveform** available from commercial supplies. The term *alternating* indicates only that the waveform alternates between two prescribed levels in a set time sequence. To be absolutely correct, the term *sinusoidal, square-wave,* or *triangular* must also be applied.

The pattern of particular interest is the **sinusoidal ac voltage** in Fig. 13.1. Since this type of signal is encountered in the vast majority of instances, the abbreviated phrases *ac voltage* and *ac current* are commonly applied without confusion. For the other patterns in Fig. 13.1, the descriptive term is always present, but frequently the *ac* abbreviation is dropped, resulting in the designation *square-wave* or *triangular* waveforms.

One of the important reasons for concentrating on the sinusoidal ac voltage is that it is the voltage generated by utilities throughout the world. Other reasons include its application throughout electrical, electronic, communication, and industrial systems. In addition, the chapters to follow will reveal that the waveform itself has a number of characteristics that result in a unique response when it is applied to basic electrical elements. The wide range of theorems and methods introduced for dc networks will also be applied to sinusoidal ac systems. Although the application of sinusoidal signals raise the required math level, once the notation

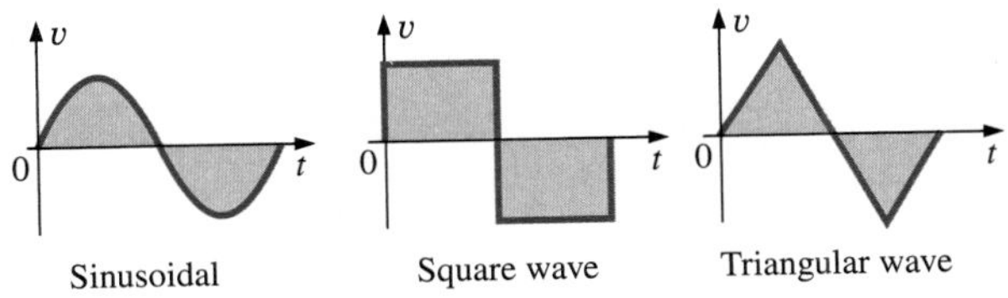

FIG. 13.1
Alternating waveforms.

given in Chapter 14 is understood, most of the concepts introduced in the dc chapters can be applied to ac networks with a minimum of added difficulty.

13.2 SINUSOIDAL ac VOLTAGE CHARACTERISTICS AND DEFINITIONS

Generation

Sinusoidal ac voltages are available from a variety of sources. The most common source is the typical home outlet, which provides an ac voltage that originates at a power plant. Most power plants are fueled by water power, oil, gas, or nuclear fusion. In each case, an **ac generator** (also called an *alternator*), as shown in Fig. 13.2(a), is the primary component in the energy-conversion process. The power to the shaft developed by one of the energy sources listed turns a *rotor* (constructed of alternating magnetic poles) inside a set of windings housed in the *stator* (the stationary part of the dynamo) and induces a voltage across the windings of the stator, as defined by Faraday's law:

$$e = N\frac{d\phi}{dt}$$

Through proper design of the generator, a sinusoidal ac voltage is developed that can be transformed to higher levels for distribution through the power lines to the consumer. For isolated locations where power lines have not been installed, portable ac generators [Fig. 13.2(b)] are available that run on gasoline. As in the larger power plants, however, an ac generator is an integral part of the design.

In an effort to conserve our natural resources and reduce pollution, wind power, solar energy, and fuel cells are receiving increasing interest from various districts of the world that have such energy sources available in level and duration that make the conversion process viable. The turning propellers of the wind-power station [Fig. 13.2(c)] are connected directly to the shaft of an ac generator to provide the ac voltage described above. Through light energy absorbed in the form of *photons,* solar cells

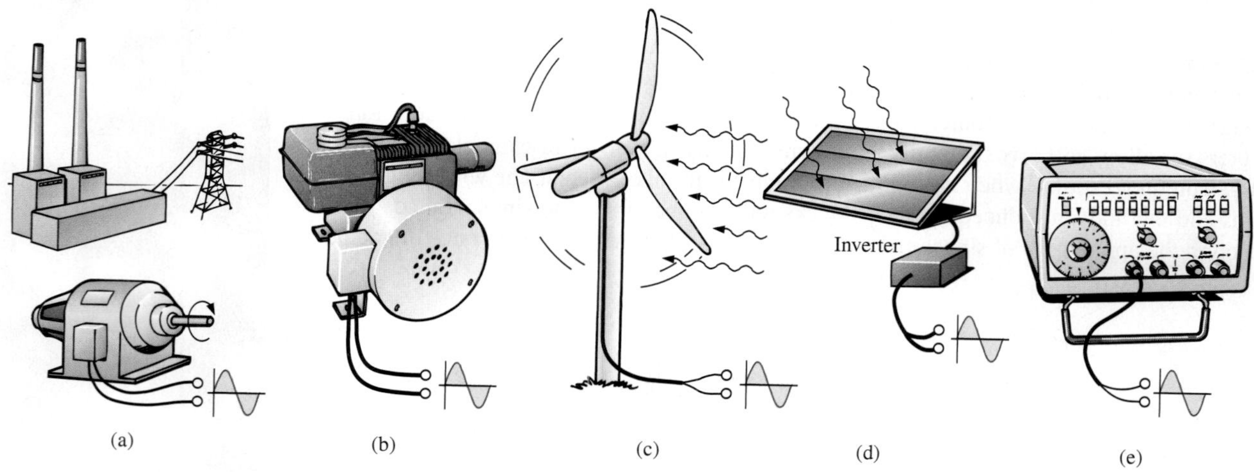

FIG. 13.2
Various sources of ac power: (a) generating plant; (b) portable ac generator; (c) wind-power station; (d) solar panel; (e) function generator.

[Fig. 13.2(d)] can generate dc voltages. Through an electronic package called an *inverter*, the dc voltage can be converted to one of a sinusoidal nature. Boats, recreational vehicles (RVs), and so on, make frequent use of the inversion process in isolated areas.

Sinusoidal ac voltages with characteristics that can be controlled by the user are available from **function generators,** such as the one in Fig. 13.2(e). By setting the various switches and controlling the position of the knobs on the face of the instrument, you can make available sinusoidal voltages of different peak values and different repetition rates. The function generator plays an integral role in the investigation of the variety of theorems, methods of analysis, and topics to be introduced in the chapters that follow.

Definitions

The sinusoidal waveform in Fig. 13.3 with its additional notation will now be used as a model in defining a few basic terms. These terms, however, can be applied to any alternating waveform. It is important to remember, as you proceed through the various definitions, that the vertical scaling is in volts or amperes and the horizontal scaling is in units of time.

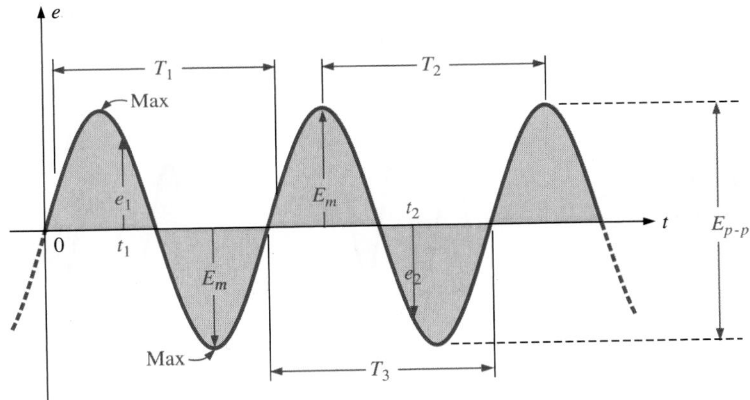

FIG. 13.3
Important parameters for a sinusoidal voltage.

Waveform: The path traced by a quantity, such as the voltage in Fig. 13.3, plotted as a function of some variable such as time (as above), position, degrees, radians, temperature, and so on.
Instantaneous value: The magnitude of a waveform at any instant of time; denoted by lowercase letters (e_1, e_2 in Fig. 13.3).
Peak amplitude: The maximum value of a waveform as measured from its *average,* or *mean,* value, denoted by uppercase letters [such as E_m (Fig. 13.3) for sources of voltage and V_m for the voltage drop across a load]. For the waveform in Fig. 13.3, the average value is zero volts, and E_m is as defined by the figure.
Peak value: The maximum instantaneous value of a function as measured from the zero volt level. For the waveform in Fig. 13.3, the peak amplitude and peak value are the same, since the average value of the function is zero volts.
Peak-to-peak value: Denoted by $E_{p\text{-}p}$ or $V_{p\text{-}p}$ (as shown in Fig. 13.3), the full voltage between positive and negative peaks of the waveform, that is, the sum of the magnitude of the positive and negative peaks.

Periodic waveform: A waveform that continually repeats itself after the same time interval. The waveform in Fig. 13.3 is a periodic waveform.

Period (*T*): The time of a periodic waveform.

Cycle: The portion of a waveform contained in one period of time. The cycles within T_1, T_2, and T_3 in Fig. 13.3 may appear different in Fig. 13.4, but they are all bounded by one period of time and therefore satisfy the definition of a cycle.

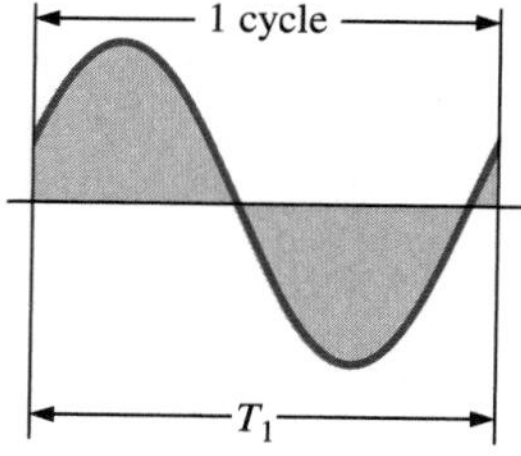

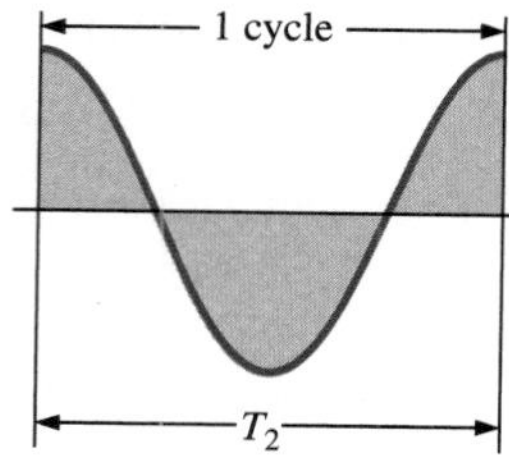

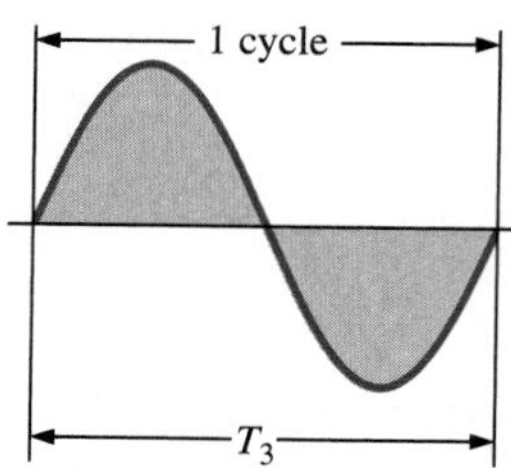

FIG. 13.4
Defining the cycle and period of a sinusoidal waveform.

Frequency (*f*): The number of cycles that occur in 1 s. The frequency of the waveform in Fig. 13.5(a) is 1 cycle per second, and for Fig. 13.5(b), 2½ cycles per second. If a waveform of similar shape had a period of 0.5 s [Fig. 13.5(c)], the frequency would be 2 cycles per second.

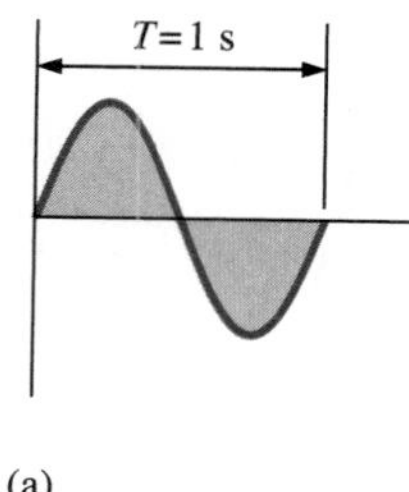

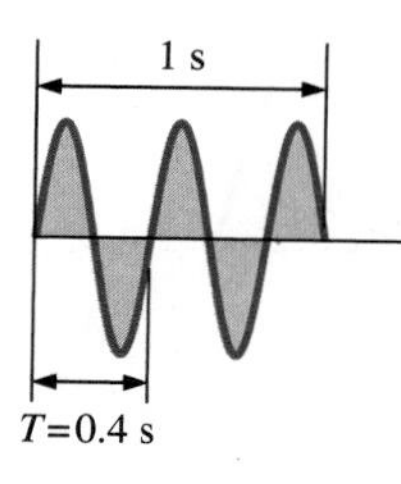

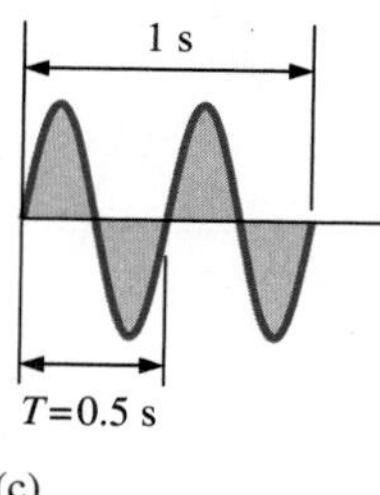

FIG. 13.5
Demonstrating the effect of a changing frequency on the period of a sinusoidal waveform.

The unit of measure for frequency is the *hertz* (Hz), where

$$\boxed{1 \text{ hertz (Hz)} = 1 \text{ cycle per second (cps)}} \qquad \textbf{(13.1)}$$

The unit hertz is derived from the surname of Heinrich Rudolph Hertz (Fig. 13.6), who did original research in the area of alternating currents and voltages and their effect on the basic *R, L,* and *C* elements. The frequency standard for North America is 60 Hz, whereas for Europe it is predominantly 50 Hz.

As with all standards, any variation from the norm will cause difficulties. In 1993, Berlin, Germany, received all its power from eastern plants, whose output frequency was varying between 50.03 Hz and 51 Hz. The result was that clocks were gaining as much as 4 minutes a day. Alarms went off too soon, VCRs clicked off before the end of the program, and so on, requiring that clocks be continually reset. In 1994, however, when power was linked with the rest of Europe, the precise standard of 50 Hz was reestablished and everyone was on time again.

FIG. 13.6
Heinrich Rudolph Hertz.
Courtesy of the Smithsonian Institution, Photo No. 66,606.

German (Hamburg, Berlin, Karlsruhe)
(1857–94)
Physicist
Professor of Physics, Karlsruhe Polytechnic and University of Bonn

Spurred on by the earlier predictions of the English physicist James Clerk Maxwell, Heinrich Hertz produced *electromagnetic waves* in his laboratory at the Karlsruhe Polytechnic while in his early 30s. The rudimentary *transmitter* and *receiver* were in essence the first to broadcast and receive radio waves. He was able to measure the *wavelength* of the electromagnetic waves and confirmed that the *velocity of propagation* is in the same order of magnitude as light. In addition, he demonstrated that the *reflective* and *refractive* properties of electromagnetic waves are the same as those for heat and light waves. It was indeed unfortunate that such an ingenious, industrious individual should pass away at the very early age of 37 due to a bone disease.

EXAMPLE 13.1 For the sinusoidal waveform in Fig. 13.7.

a. What is the peak value?
b. What is the instantaneous value at 0.3 s and 0.6 s?
c. What is the peak-to-peak value of the waveform?
d. What is the period of the waveform?
e. How many cycles are shown?
f. What is the frequency of the waveform?

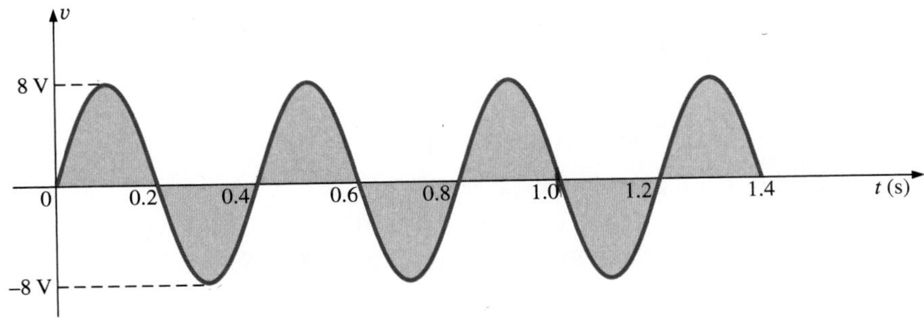

FIG. 13.7
Example 13.1.

Solutions:

a. **8 V.**
b. At 0.3 s, **−8 V;** at 0.6 s, **0 V.**
c. **16 V.**
d. **0.4 s.**
e. **3.5 cycles.**
f. **2.5 cps,** or **2.5 Hz.**

13.3 FREQUENCY SPECTRUM

Using a log scale (described in detail in Chapter 20), a frequency spectrum from 1 Hz to 1000 GHz can be scaled off on the same axis, as shown in Fig. 13.8. A number of terms in the various spectrums are probably familiar to you from everyday experiences. Note that the audio range (human ear) extends from only 15 Hz to 20 kHz, but the transmission of radio signals can occur between 3 kHz and 300 GHz. The uniform process of defining the intervals of the radio-frequency spectrum from VLF to EHF is quite evident from the length of the bars in the figure (although keep in mind that it is a log scale, so the frequencies encompassed within each segment are quite different). Other frequencies of particular interest (TV, CB, microwave, and so on) are also included for reference purposes. Although it is numerically easy to talk about frequencies in the megahertz and gigahertz range, keep in mind that a frequency of 100 MHz, for instance, represents a sinusoidal waveform that passes through 100,000,000 cycles in only 1 s—an incredible number when we compare it to the 60 Hz of our conventional power sources. The Intel® Pentium® 4 chip manufactured by Intel can run at speeds over 2 GHz. Imagine a product able to handle 2 billion instructions per second—an incredible achievement.

exito

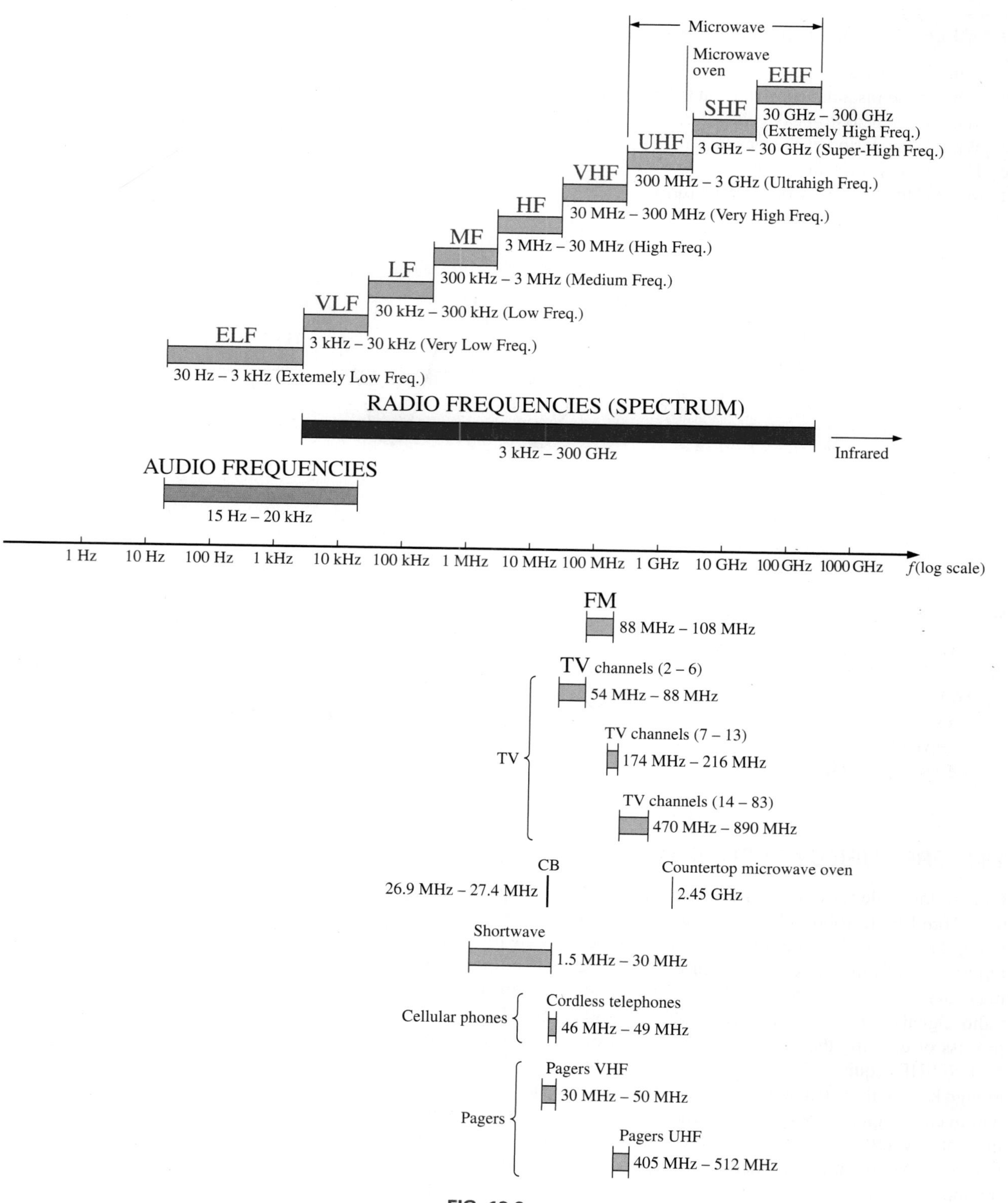

FIG. 13.8
Areas of application for specific frequency bands.

Since the frequency is inversely related to the period—that is, as one increases, the other decreases by an equal amount—the two can be related by the following equation:

$$\boxed{f = \frac{1}{T}} \qquad \begin{aligned} f &= \text{Hz} \\ T &= \text{seconds (s)} \end{aligned} \qquad \textbf{(13.2)}$$

or

$$\boxed{T = \frac{1}{f}} \qquad \textbf{(13.3)}$$

EXAMPLE 13.2 Find the period of periodic waveform with a frequency of

a. 60 Hz.
b. 1000 Hz.

Solutions:

a. $T = \dfrac{1}{f} = \dfrac{1}{60\text{ Hz}} \cong 0.01667\text{ s or } \mathbf{16.67\text{ ms}}$

(a recurring value since 60 Hz is so prevalent)

b. $T = \dfrac{1}{f} = \dfrac{1}{1000\text{ Hz}} = 10^{-3}\text{ s } = \mathbf{1\text{ ms}}$

EXAMPLE 13.3 Determine the frequency of the waveform in Fig. 13.9.

Solution: From the figure, $T = (25\text{ ms} - 5\text{ ms})$ or $(35\text{ ms} - 15\text{ ms}) = 20\text{ ms}$, and

$$f = \frac{1}{T} = \frac{1}{20 \times 10^{-3}\text{ s}} = \mathbf{50\text{ Hz}}$$

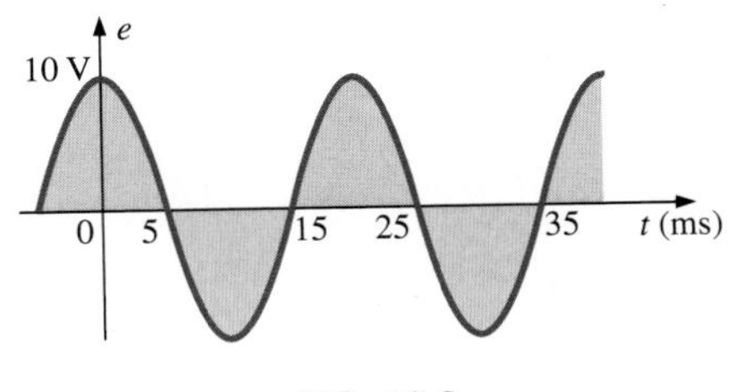

FIG. 13.9
Example 13.3.

In Fig. 13.10, the seismogram resulting from a seismometer near an earthquake is displayed. Prior to the disturbance, the waveform has a relatively steady level, but as the event is about to occur, the frequency begins

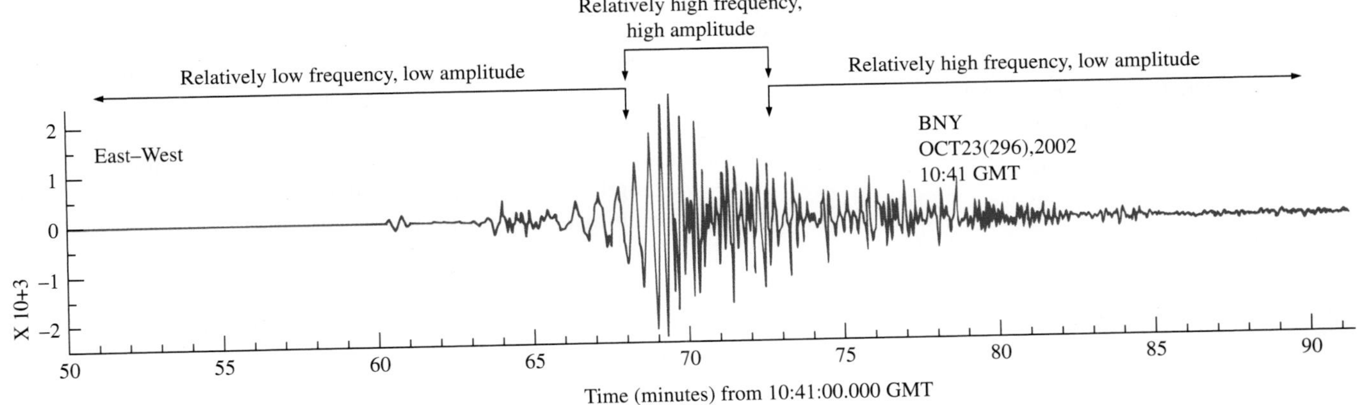

FIG. 13.10
Seismogram from station BNY (Binghamton University) in New York due to magnitude 6.7 earthquake in Central Alaska that occurred at 63.62°N, 148.04°W, with a depth of 10 km, on Wednesday, October 23, 2002.

to increase along with the amplitude. Finally, the earthquake occurs, and the frequency and the amplitude increase dramatically. In other words, the relative frequencies can be determined simply by looking at the tightness of the waveform and the associated period. The change in amplitude is immediately obvious from the resulting waveform. The fact that the earthquake lasts for only a few minutes is clear from the horizontal scale.

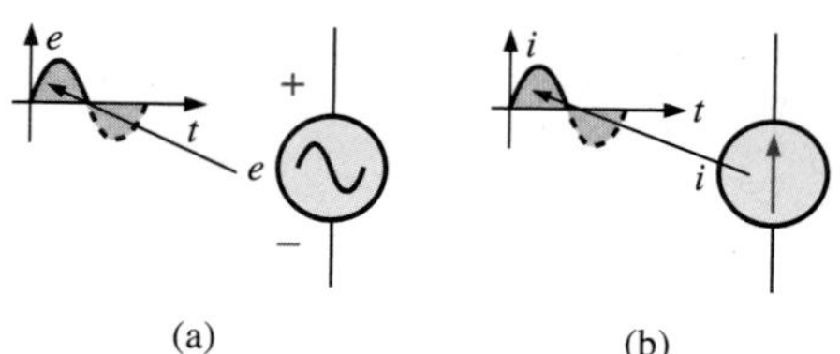

FIG. 13.11
(a) Sinusoidal ac voltage sources; (b) sinusoidal current sources.

Defined Polarities and Direction

You may be wondering how a polarity for a voltage or a direction for a current can be established if the waveform moves back and forth from the positive to the negative region. For a period of time, a voltage has one polarity, while for the next equal period it reverses. To take care of this problem, a positive sign is applied if the voltage is above the axis, as shown in Fig. 13.11(a). For a current source, the direction in the symbol corresponds with the positive region of the waveform, as shown in Fig. 13.11(b).

For any quantity that will not change with time, an uppercase letter such as V or I is used. For expressions that are time dependent or that represent a particular instant of time, a lowercase letter such as e or i is used.

The need for defining polarities and current direction becomes quite obvious when we consider multisource ac networks. Note in the last sentence the absence of the term *sinusoidal* before the phrase *ac networks.* This phrase will be used to an increasing degree as we progress; *sinusoidal* is to be understood unless otherwise indicated.

13.4 THE SINUSOIDAL WAVEFORM

The terms defined in the previous section can be applied to any type of periodic waveform, whether smooth or discontinuous. The sinusoidal waveform is of particular importance, however, since it lends itself readily to the mathematics and the physical phenomena associated with electric circuits. Consider the power of the following statement:

The sinusoidal waveform is the only alternating waveform whose shape is unaffected by the response characteristics of R, L, and C elements.

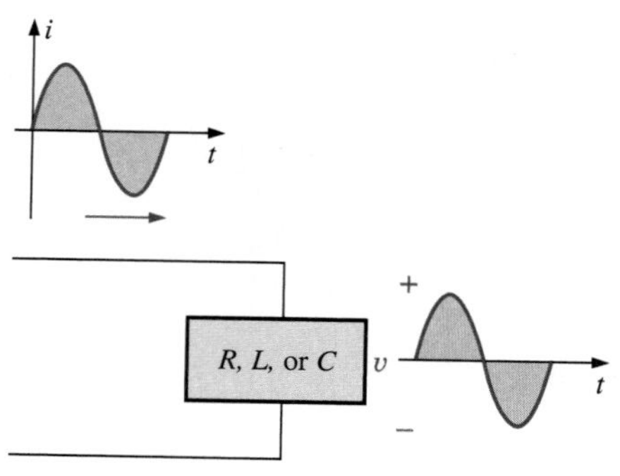

FIG. 13.12
The sine wave is the only alternating waveform whose shape is not altered by the response characteristics of a pure resistor, inductor, or capacitor.

In other words, if the voltage across (or current through) a resistor, inductor, or capacitor is sinusoidal in nature, the resulting current (or voltage, respectively) for each will also have sinusoidal characteristics, as shown in Fig. 13.12. If any other alternating waveform such as a square wave or a triangular wave were applied, such would not be the case.

The unit of measurement for the horizontal axis can be **time** (as appearing in the figures thus far), **degrees**, or **radians**. The term **radian** can be defined as follows: If we mark off a portion of the circumference of a circle by a length equal to the radius of the circle, as shown in Fig. 13.13, the angle resulting is called *1 radian.* The result is

$$\boxed{1 \text{ rad} = 57.296° \cong 57.3°} \qquad (13.4)$$

where 57.3° is the usual approximation applied.

One full circle has 2π radians, as shown in Fig. 13.14. That is,

$$\boxed{2\pi \text{ rad} = 360°} \qquad (13.5)$$

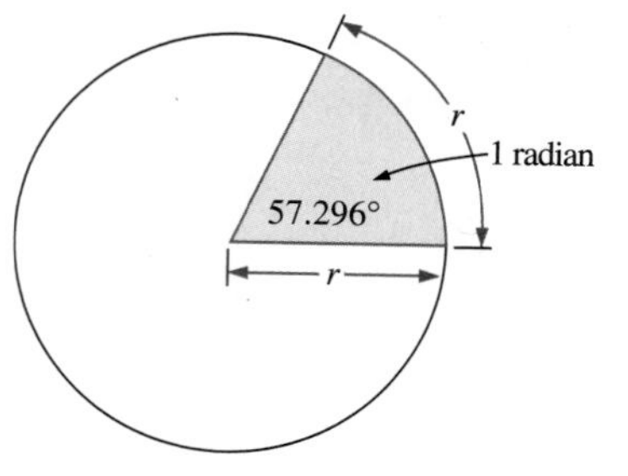

FIG. 13.13
Defining the radian.

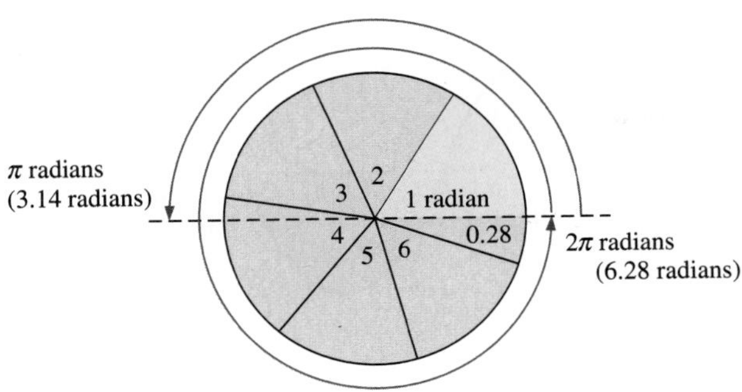

FIG. 13.14
There are 2π radians in one full circle of 360°.

so that $$2\pi = 2(3.142) = 6.28$$

and $$2\pi(57.3°) = 6.28(57.3°) = 359.84° \cong 360°$$

A number of electrical formulas contain a multiplier of π. For this reason, it is sometimes preferable to measure angles in radians rather than in degrees.

The quantity π is the ratio of the circumference of a circle to its diameter.

π has been determined to an extended number of places, primarily in an attempt to see if a repetitive sequence of numbers appears. It does not. A sampling of the effort appears below:

$$\pi = 3.14159\ 26535\ 89793\ 23846\ 26433\ \ldots$$

Although the approximation $\pi \cong 3.14$ is often applied, all the calculations in the text use the π function as provided on all scientific calculators.

For 180° and 360°, the two units of measurement are related as shown in Fig. 13.14. The conversions equations between the two are the following:

$$\text{Radians} = \left(\frac{\pi}{180°}\right) \times (\text{degrees}) \tag{13.6}$$

$$\text{Degrees} = \left(\frac{180°}{\pi}\right) \times (\text{radians}) \tag{13.7}$$

Applying these equations, we find

$$\mathbf{90°:}\ \text{Radians} = \frac{\pi}{180°}(90°) = \frac{\pi}{2}\ \textbf{rad}$$

$$\mathbf{30°:}\ \text{Radians} = \frac{\pi}{180°}(30°) = \frac{\pi}{6}\ \textbf{rad}$$

$$\frac{\pi}{3}\ \textbf{rad:}\ \text{Degrees} = \frac{180°}{\pi}\left(\frac{\pi}{3}\right) = \mathbf{60°}$$

$$\frac{3\pi}{2}\ \textbf{rad:}\ \text{Degrees} = \frac{180°}{\pi}\left(\frac{3\pi}{2}\right) = \mathbf{270°}$$

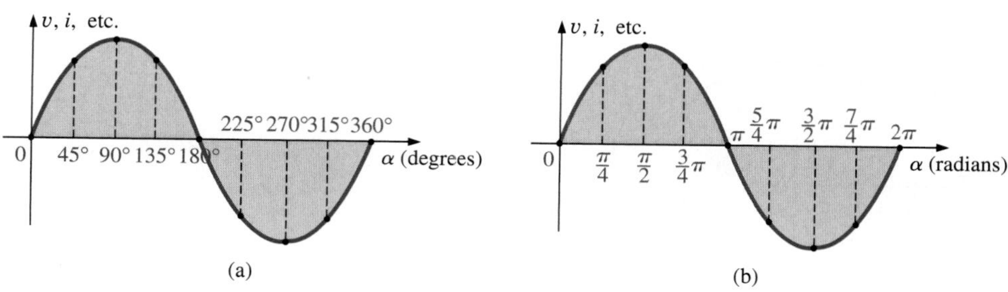

FIG. 13.15
Plotting a sine wave versus (a) degrees and (b) radians.

For comparison purposes, two sinusoidal voltages are plotted in Fig. 13.15 using degrees and radians as the units of measurement for the horizontal axis.

It is of particular interest that the sinusoidal waveform can be derived from the length of the *vertical projection* of a radius vector rotating in a uniform circular motion about a fixed point. Starting as shown in Fig. 13.16(a) and plotting the amplitude (above and below zero) on the coordinates drawn to the right [Figs. 13.16(b) through (i)], we will trace a complete sinusoidal waveform after the radius vector has completed a 360° rotation about the center.

The velocity with which the radius vector rotates about the center, called the **angular velocity,** can be determined from the following equation:

$$\text{Angular velocity} = \frac{\text{distance (degrees or radians)}}{\text{time (seconds)}} \tag{13.8}$$

Substituting into Eq. (13.8) and assigning the lowercase Greek letter *omega* (ω) to the angular velocity, we have

$$\omega = \frac{\alpha}{t} \tag{13.9}$$

and

$$\alpha = \omega t \tag{13.10}$$

Since ω is typically provided in radians per second, the angle α obtained using Eq. (13.10) is usually in radians. If α is required in degrees, Eq. (13.7) must be applied. The importance of remembering the above will become obvious in the examples to follow.

In Fig. 13.16, the time required to complete one revolution is equal to the period (T) of the sinusoidal waveform in Fig. 13.16(i). The radians subtended in this time interval are 2π. Substituting, we have

$$\omega = \frac{2\pi}{T} \qquad \text{(rad/s)} \tag{13.11}$$

In words, this equation states that the smaller the period of the sinusoidal waveform of Fig. 13.16(i), or the smaller the time interval before one complete cycle is generated, the greater must be the angular velocity of the rotating radius vector. Certainly this statement agrees with what we have learned thus far. We can now go one step further and apply the fact

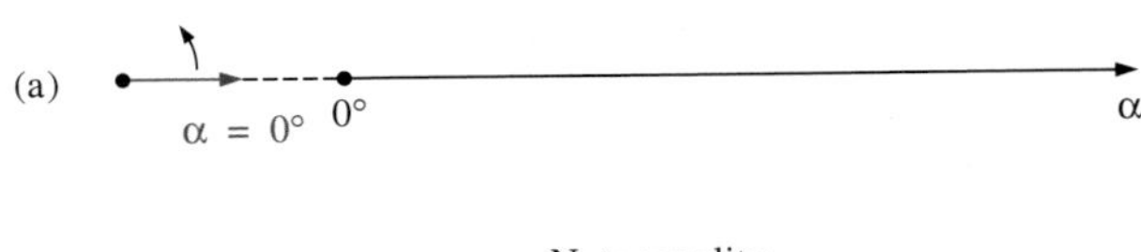

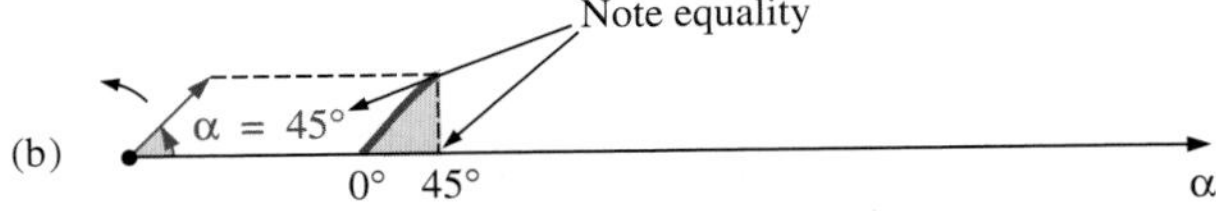

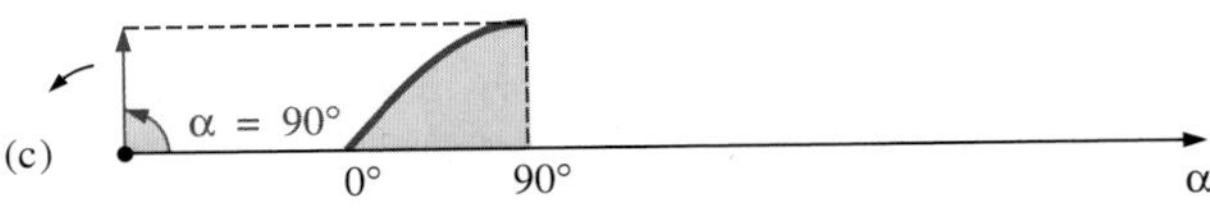

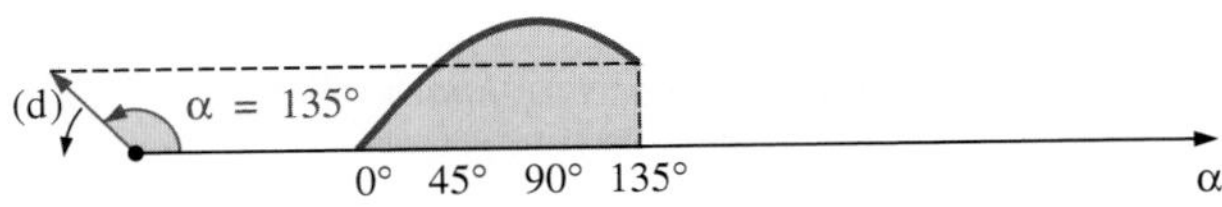

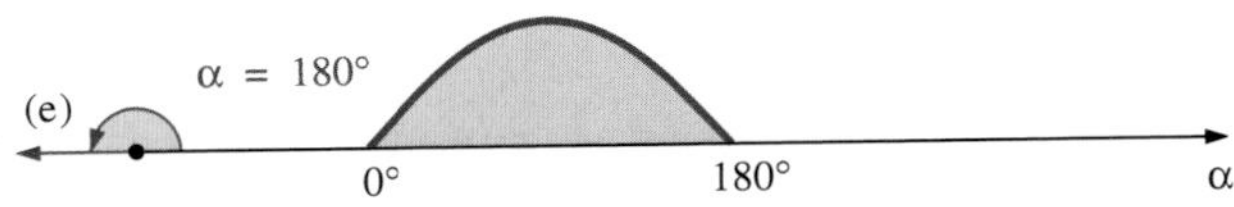

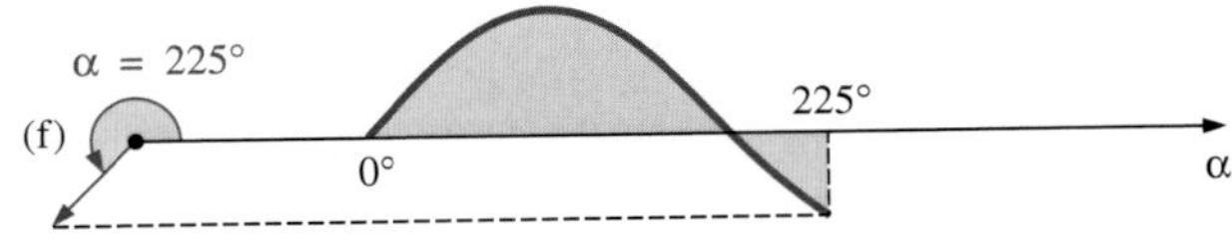

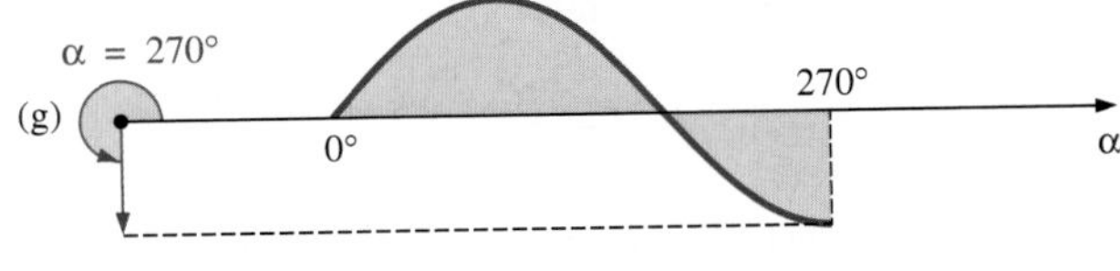

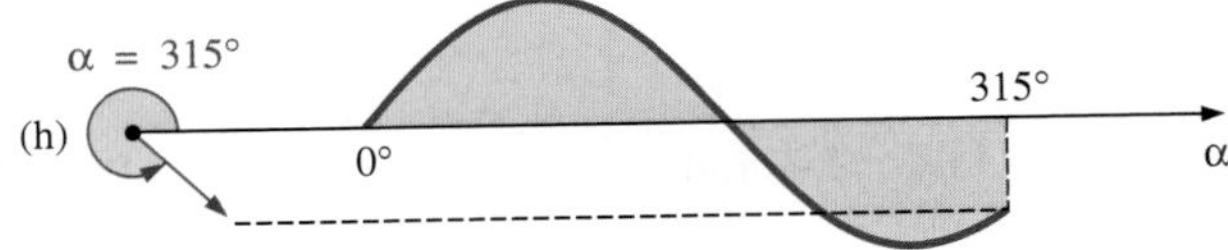

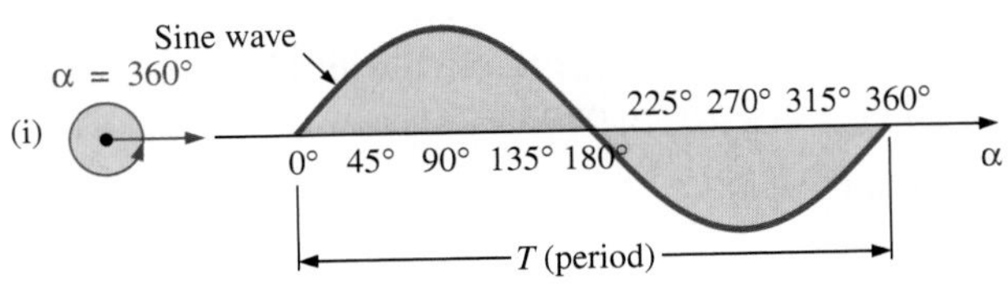

FIG. 13.16

Generating a sinusoidal waveform through the vertical projection of a rotating vector.

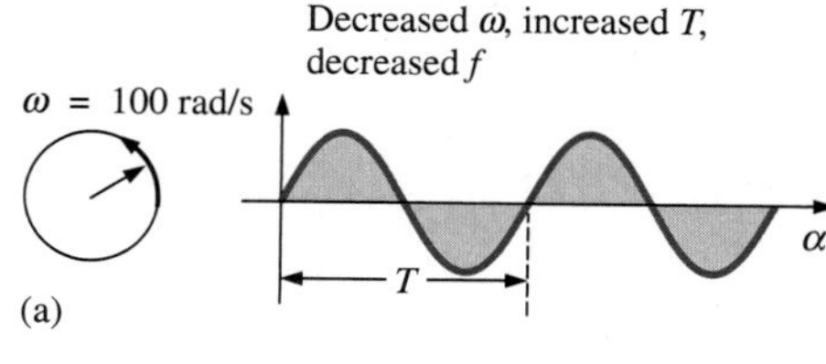

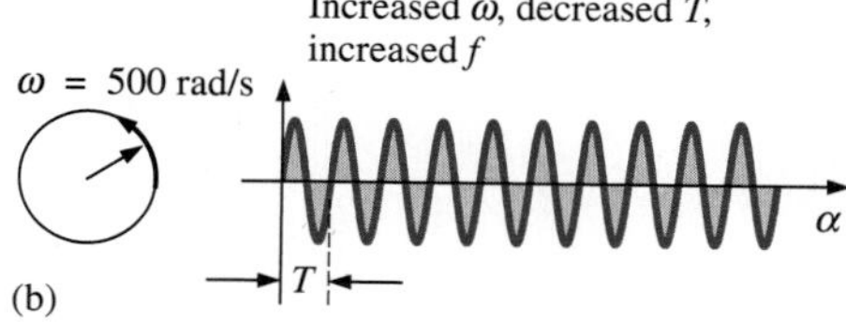

FIG. 13.17
Demonstrating the effect of ω on the frequency and period.

that the frequency of the generated waveform is inversely related to the period of the waveform; that is, $f = 1/T$. Thus,

$$\boxed{\omega = 2\pi f} \qquad \text{(rad/s)} \qquad \textbf{(13.12)}$$

This equation states that the higher the frequency of the generated sinusoidal waveform, the higher must be the angular velocity. Eqs. (13.11) and (13.12) are verified somewhat by Fig. 13.17, where for the same radius vector, $\omega = 100$ rad/s and 500 rad/s.

EXAMPLE 13.4 Determine the angular velocity of a sine wave having a frequency of 60 Hz.

Solution:

$$\omega = 2\pi f = (2\pi)(60 \text{ Hz}) \cong \mathbf{377\ rad/s}$$

(a recurring value due to 60 Hz predominance)

EXAMPLE 13.5 Determine the frequency and period of the sine wave in Fig. 13.17(b).

Solution: Since $\omega = 2\pi/T$,

$$T = \frac{2\pi}{\omega} = \frac{2\pi \text{ rad}}{500 \text{ rad/s}} = \frac{2\pi \text{ rad}}{500 \text{ rad/s}} = \mathbf{12.57\ ms}$$

and

$$f = \frac{1}{T} = \frac{1}{12.57 \times 10^{-3} \text{ s}} = \mathbf{79.58\ Hz}$$

EXAMPLE 13.6 Given $\omega = 200$ rad/s, determine how long it will take the sinusoidal waveform to pass through an angle of 90°.

Solution: Eq. (13.10): $\alpha = \omega t$, and

$$t = \frac{\alpha}{\omega}$$

However, α must be substituted as $\pi/2$ (= 90°) since ω is in radians per second:

$$t = \frac{\alpha}{\omega} = \frac{\pi/2 \text{ rad}}{200 \text{ rad/s}} = \frac{\pi}{400} \text{ s} = \mathbf{7.85\ ms}$$

EXAMPLE 13.7 Find the angle through which a sinusoidal waveform of 60 Hz will pass in a period of 5 ms.

Solution: Eq. (13.11): $\alpha = \omega t$, or

$$\alpha = 2\pi f t = (2\pi)(60 \text{ Hz})(5 \times 10^{-3} \text{ s}) = \mathbf{1.885\ rad}$$

If not careful, you might be tempted to interpret the answer as 1.885°. However,

$$\alpha\,(°) = \frac{180°}{\pi \cancel{\text{rad}}}(1.885 \cancel{\text{rad}}) = \mathbf{108°}$$

13.5 GENERAL FORMAT FOR THE SINUSOIDAL VOLTAGE OR CURRENT

The basic mathematical format for the sinusoidal waveform is

$$\boxed{A_m \sin \alpha} \tag{13.13}$$

where A_m is the peak value of the waveform and α is the unit of measure for the horizontal axis, as shown in Fig. 13.18.

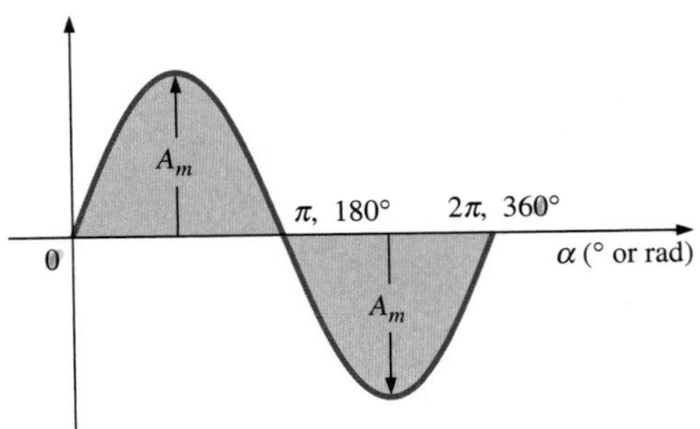

FIG. 13.18
Basic sinusoidal function.

The equation $\alpha = \omega t$ states that the angle α through which the rotating vector in Fig. 13.16 will pass is determined by the angular velocity of the rotating vector and the length of time the vector rotates. For example, for a particular angular velocity (fixed ω), the longer the radius vector is permitted to rotate (that is, the greater the value of t), the greater the number of degrees or radians through which the vector will pass. Relating this statement to the sinusoidal waveform, for a particular angular velocity, the longer the time, the greater the number of cycles shown. For a fixed time interval, the greater the angular velocity, the greater the number of cycles generated.

Due to Eq. (13.10), the general format of a sine wave can also be written

$$\boxed{A_m \sin \omega t} \tag{13.14}$$

with ωt as the horizontal unit of measure.

For electrical quantities such as current and voltage, the general format is

$$i = I_m \sin \omega t = I_m \sin \alpha$$
$$e = E_m \sin \omega t = E_m \sin \alpha$$

where the capital letters with the subscript m represent the amplitude, and the lowercase letters i and e represent the instantaneous value of current and voltage, respectively, at any time t. This format is particularly important because it presents the sinusoidal voltage or current as a function of time, which is the horizontal scale for the oscilloscope. Recall that the horizontal sensitivity of a scope is in time per division, not degrees per centimeter.

EXAMPLE 13.8 Given $e = 5 \sin \alpha$, determine e at $\alpha = 40°$ and $\alpha = 0.8\pi$.

Solution: For $\alpha = 40°$,

$$e = 5 \sin 40° = 5(0.6428) = \mathbf{3.21\ V}$$

For $\alpha = 0.8\pi$,

$$\alpha\,(°) = \frac{180°}{\pi}(0.8\pi) = 144°$$

and
$$e = 5 \sin 144° = 5(0.5878) = \mathbf{2.94\ V}$$

The angle at which a particular voltage level is attained can be determined by rearranging the equation

$$e = E_m \sin \alpha$$

in the following manner:

$$\sin \alpha = \frac{e}{E_m}$$

which can be written

$$\alpha = \sin^{-1} \frac{e}{E_m} \quad (13.15)$$

Similarly, for a particular current level,

$$\alpha = \sin^{-1} \frac{i}{I_m} \quad (13.16)$$

EXAMPLE 13.9

a. Determine the angle at which the magnitude of the sinusoidal function $v = 10 \sin 377t$ is 4 V.
b. Determine the time at which the magnitude is attained.

Solutions:

a. Eq. (13.15):

$$\alpha_1 = \sin^{-1} \frac{v}{E_m} = \sin^{-1} \frac{4 \text{ V}}{10 \text{ V}} = \sin^{-1} 0.4 = \mathbf{23.58^\circ}$$

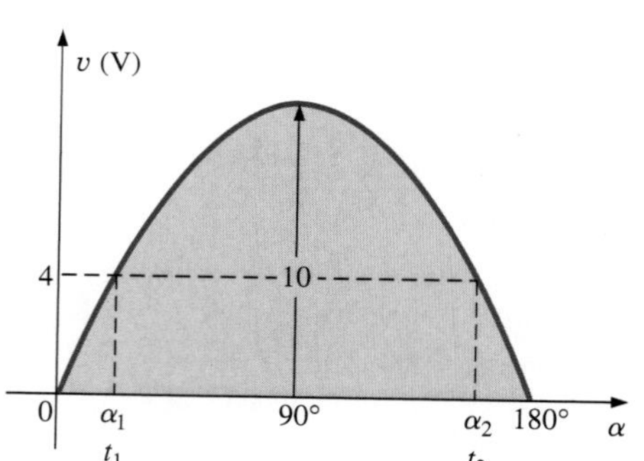

FIG. 13.19
Example 13.9.

However, Fig. 13.19 reveals that the magnitude of 4 V (positive) will be attained at two points between 0° and 180°. The second intersection is determined by

$$\alpha_2 = 180^\circ - 23.578^\circ = \mathbf{156.42^\circ}$$

In general, therefore, keep in mind that Eqs. (13.15) and (13.16) will provide an angle with a magnitude between 0° and 90°.

b. Eq. (13.10): $\alpha = \omega t$, and so $t = \alpha/\omega$. However, α must be in radians. Thus,

$$\alpha \text{ (rad)} = \frac{\pi}{180^\circ}(23.578^\circ) = 0.412 \text{ rad}$$

and

$$t_1 = \frac{\alpha}{\omega} = \frac{0.412 \text{ rad}}{377 \text{ rad/s}} = \mathbf{1.09 \text{ ms}}$$

For the second intersection,

$$\alpha \text{ (rad)} = \frac{\pi}{180^\circ}(156.422^\circ) = 2.73 \text{ rad}$$

$$t_2 = \frac{\alpha}{\omega} = \frac{2.73 \text{ rad}}{377 \text{ rad/s}} = \mathbf{7.24 \text{ ms}}$$

Calculator Operations

Both sin and $\sin^{-1}$ are available on all scientific calculators. You can also use them to work with the angle in degrees or radians without having to convert from one form to the other. That is, if the angle is in radians and the mode setting is for radians, you can enter the radian measure directly.

To set the DEGREE mode, proceed as outlined in Fig. 13.20(a) using the TI-89 calculator. The magnitude of the voltage e at 40° can then be found using the sequence in Fig. 13.20(b).

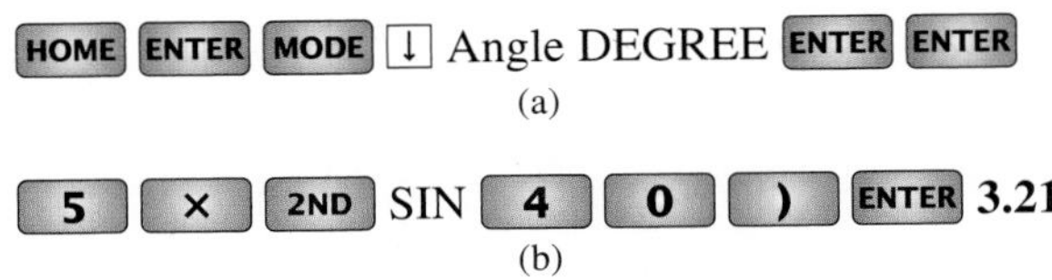

FIG. 13.20
(a) Setting the DEGREE mode; (b) evaluating 5 sin 40°.

After establishing the RADIAN mode, the sequence in Fig. 13.21 determines the voltage at 0.8π.

FIG. 13.21
Finding $e = 5 \sin 0.8\pi$ using the calculator in the RADIAN mode.

Finally, the angle in degrees for α_1 in part (a) of Example 13.9 can be determined by the sequence in Fig. 13.22 with the mode set in degrees, whereas the angle in radians for part (a) of Example 13.9 can be determined by the sequence in Fig. 13.23 with the mode set in radians.

FIG. 13.22
Finding $\alpha_1 = \sin^{-1}(4/10)$ using the calculator in the DEGREE mode.

FIG. 13.23
Finding $\alpha_1 = \sin^{-1}(4/10)$ using the calculator in the RADIAN mode.

The sinusoidal waveform can also be plotted against *time* on the horizontal axis. The time period for each interval can be determined from $t = \alpha/\omega$, but the most direct route is simply to find the period T from $T = 1/f$ and break it up into the required intervals. This latter technique is demonstrated in Example 13.10.

Before reviewing the example, take special note of the relative simplicity of the mathematical equation that can represent a sinusoidal waveform. Any alternating waveform whose characteristics differ from those of the sine wave cannot be represented by a single term, but may require two, four, six, or perhaps an infinite number of terms to be represented accurately.

EXAMPLE 13.10 Sketch $e = 10 \sin 314t$ with the abscissa

a. angle (α) in degrees.
b. angle (α) in radians.
c. time (t) in seconds.

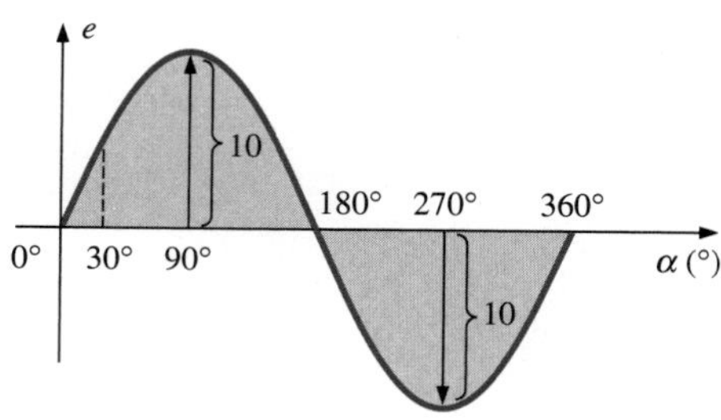

FIG. 13.24
Example 13.10, horizontal axis in degrees.

Solutions:

a. See Fig. 13.24. (Note that no calculations are required.)
b. See Fig. 13.25. (Once the relationship between degrees and radians is understood, no calculations are required.)
c. See Fig 13.26.

$$360°: \quad T = \frac{2\pi}{\omega} = \frac{2\pi}{314} = 20 \text{ ms}$$
$$180°: \quad \frac{T}{2} = \frac{20 \text{ ms}}{2} = 10 \text{ ms}$$
$$90°: \quad \frac{T}{4} = \frac{20 \text{ ms}}{4} = 5 \text{ ms}$$
$$30°: \quad \frac{T}{12} = \frac{20 \text{ ms}}{12} = 1.67 \text{ ms}$$

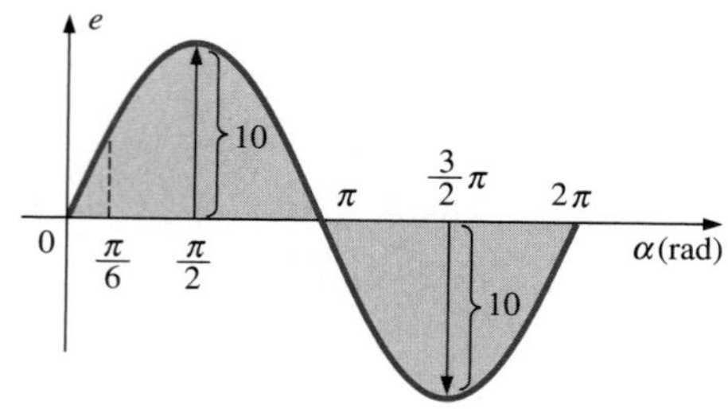

FIG. 13.25
Example 13.10, horizontal axis in radians.

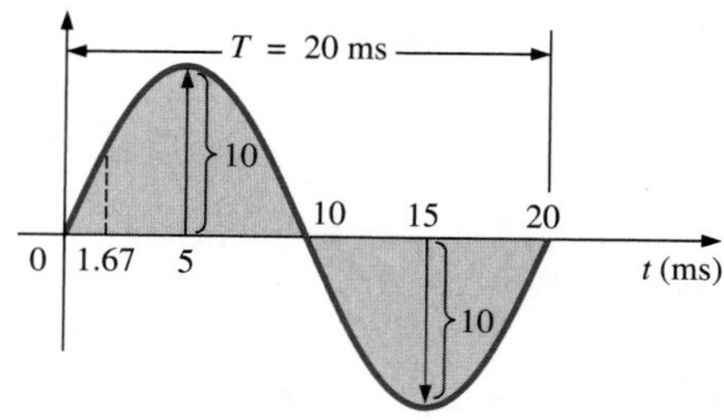

FIG. 13.26
Example 13.10, horizontal axis in milliseconds.

EXAMPLE 13.11 Given $i = 6 \times 10^{-3} \sin 1000t$, determine i at $t =$ 2 ms.

Solution:

$$\alpha = \omega t = 1000t = (1000 \text{ rad/s})(2 \times 10^{-3} \text{ s}) = 2 \text{ rad}$$
$$\alpha\,(°) = \frac{180°}{\pi \text{ rad}}(2 \text{ rad}) = 114.59°$$
$$i = (6 \times 10^{-3})(\sin 114.59°) = (6 \text{ mA})(0.9093) = \mathbf{5.46 \text{ mA}}$$

13.6 PHASE RELATIONS

Thus far, we have considered only sine waves that have maxima at $\pi/2$ and $3\pi/2$, with a zero value at 0, π, and 2π, as shown in Fig. 13.25. If the waveform is shifted to the right or left of 0°, the expression becomes

$$\boxed{A_m \sin(\omega t \pm \theta)} \qquad \textbf{(13.17)}$$

where θ is the angle in degrees or radians that the waveform has been shifted.

If the waveform passes through the horizontal axis with a *positive-going* (increasing with time) slope *before* 0°, as shown in Fig. 13.27, the expression is

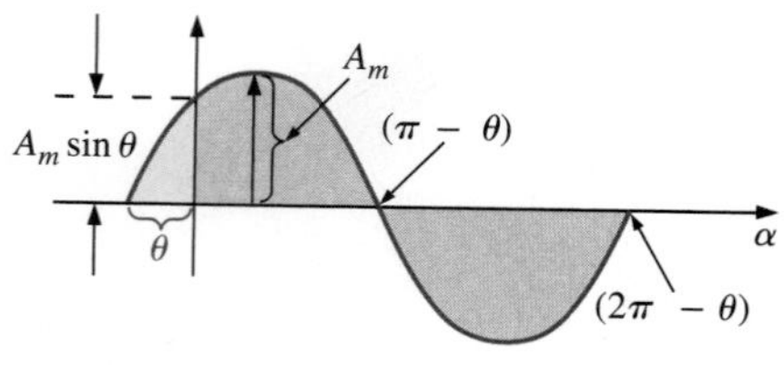

FIG. 13.27
Defining the phase shift for a sinusoidal function that crosses the horizontal axis with a positive slope before 0°.

$$\boxed{A_m \sin(\omega t + \theta)} \qquad \textbf{(13.18)}$$

At $\omega t = \alpha = 0°$, the magnitude is determined by $A_m \sin\theta$. If the waveform passes through the horizontal axis with a positive-going slope *after* 0°, as shown in Fig. 13.28, the expression is

$$\boxed{A_m \sin(\omega t - \theta)} \tag{13.19}$$

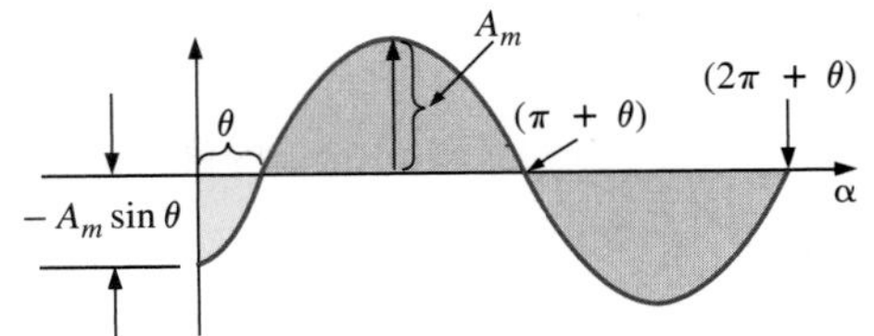

FIG. 13.28
Defining the phase shift for a sinusoidal function that crosses the horizontal axis with a positive slope after 0°.

Finally, at $\omega t = \alpha = 0°$, the magnitude is $A_m \sin(-\theta)$, which, by a trigonometric identity, is $-A_m \sin\theta$.

If the waveform crosses the horizontal axis with a positive-going slope 90° ($\pi/2$) sooner, as shown in Fig. 13.29, it is called a *cosine wave;* that is,

$$\boxed{\sin(\omega t + 90°) = \sin\left(\omega t + \frac{\pi}{2}\right) = \cos\omega t} \tag{13.20}$$

or

$$\boxed{\sin\omega t = \cos(\omega t - 90°) = \cos\left(\omega t - \frac{\pi}{2}\right)} \tag{13.21}$$

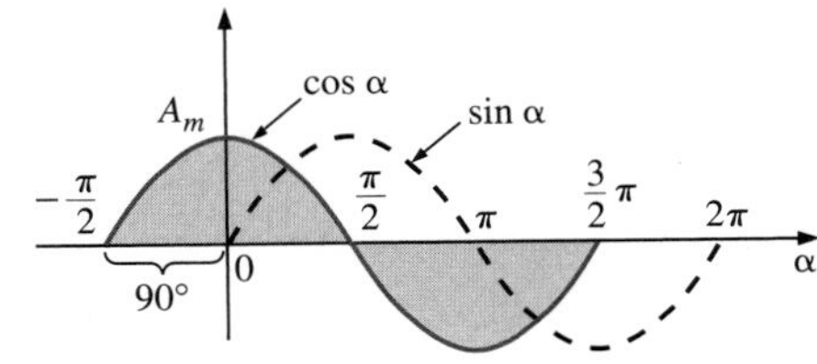

FIG. 13.29
Phase relationship between a sine wave and a cosine wave.

The terms **leading** and **lagging** are used to indicate the relationship between two sinusoidal waveforms of the *same frequency* plotted on the same set of axes. In Fig. 13.29, the cosine curve is said to *lead* the sine curve by 90°, and the sine curve is said to *lag* the cosine curve by 90°. The 90° is referred to as the phase angle between the two waveforms. In language commonly applied, the waveforms are *out of phase* by 90°. Note that the phase angle between the two waveforms is measured between those two points on the horizontal axis through which each passes with the *same slope.* If both waveforms cross the axis at the same point with the same slope, they are *in phase.*

The geometric relationship between various forms of the sine and cosine functions can be derived from Fig. 13.30. For instance, starting at the $+\sin\alpha$ position, we find that $+\cos\alpha$ is an additional 90° in the counterclockwise direction. Therefore, $\cos\alpha = \sin(\alpha + 90°)$. For $-\sin\alpha$ we must travel 180° in the counterclockwise (or clockwise) direction so that $-\sin\alpha = \sin(\alpha \pm 180°)$, and so on, as listed below:

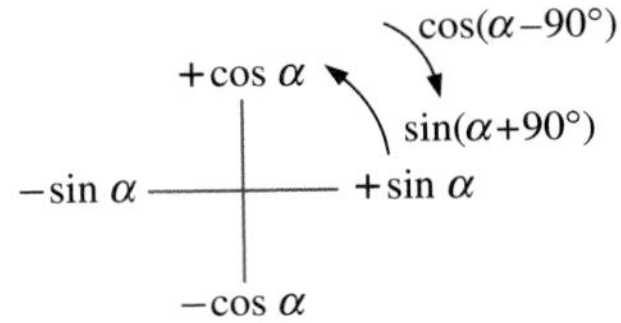

FIG. 13.30
Graphic tool for finding the relationship between specific sine and cosine functions.

$$\boxed{\begin{aligned} \cos\alpha &= \sin(\alpha + 90°) \\ \sin\alpha &= \cos(\alpha - 90°) \\ -\sin\alpha &= \sin(\alpha \pm 180°) \\ -\cos\alpha &= \sin(\alpha + 270°) = \sin(\alpha - 90°) \\ &\text{etc.} \end{aligned}} \tag{13.22}$$

In addition, note that

$$\boxed{\begin{aligned} \sin(-\alpha) &= -\sin\alpha \\ \cos(-\alpha) &= \cos\alpha \end{aligned}} \tag{13.23}$$

If a sinusoidal expression appears as

$$e = -E_m \sin\omega t$$

the negative sign is associated with the sine portion of the expression, not the peak value E_m. In other words, the expression, if not for convenience, would be written

$$e = E_m(-\sin\omega t)$$

Since $$-\sin \omega t = \sin(\omega t \pm 180°)$$

the expression can also be written

$$e = E_m \sin(\omega t \pm 180°)$$

revealing that a negative sign can be replaced by a 180° change in phase angle (+ or −); that is,

$$e = -E_m \sin \omega t = E_m \sin(\omega t + 180°) = E_m \sin(\omega t - 180°)$$

A plot of each will clearly show their equivalence. There are, therefore, two correct mathematical representations for the functions.

The **phase relationship** between two waveforms indicates which one leads or lags the other, and by how many degrees or radians.

EXAMPLE 13.12 What is the phase relationship between the sinusoidal waveforms of each of the following sets?

a. $v = 10 \sin(\omega t + 30°)$
   $i = 5 \sin(\omega t + 70°)$
b. $i = 15 \sin(\omega t + 60°)$
   $v = 10 \sin(\omega t - 20°)$
c. $i = 2 \cos(\omega t + 10°)$
   $v = 3 \sin(\omega t - 10°)$
d. $i = -\sin(\omega t + 30°)$
   $v = 2 \sin(\omega t + 10°)$
e. $i = -2 \cos(\omega t - 60°)$
   $v = 3 \sin(\omega t - 150°)$

Solutions:

a. See Fig. 13.31.
   ***i* leads *v* by 40°, or *v* lags *i* by 40°.**

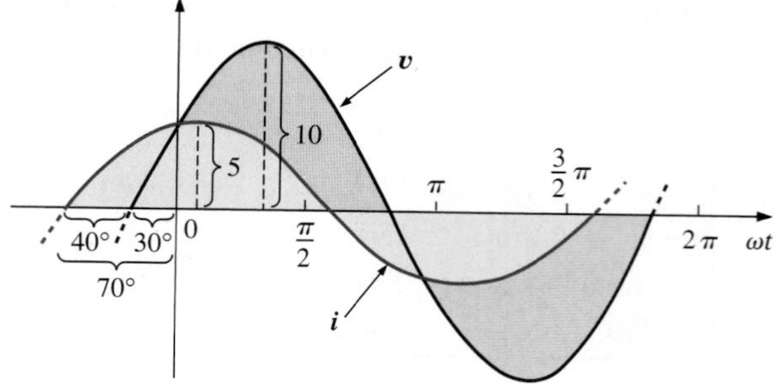

FIG. 13.31
Example 13.12(a): i leads v by 40°.

b. See Fig. 13.32.
   ***i* leads *v* by 80°, or *v* lags *i* by 80°.**

c. See Fig. 13.33.

$$i = 2 \cos(\omega t + 10°) = 2 \sin(\omega t + 10° + 90°)$$
$$= 2 \sin(\omega t + 100°)$$

***i* leads *v* by 110°, or *v* lags *i* by 110°.**

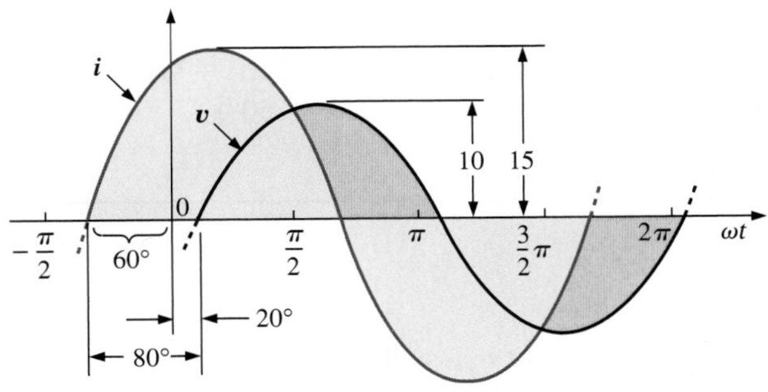

FIG. 13.32
Example 13.12(b): i leads v by 80°.

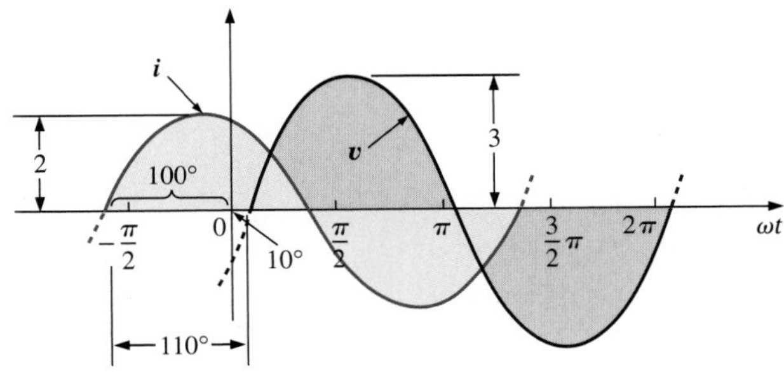

FIG. 13.33
Example 13.12(c): i leads v by 110°.

d. See Fig. 13.34.

$$-\sin(\omega t + 30°) = \sin(\omega t + 30° \overset{\text{Note}}{-} 180°)$$
$$= \sin(\omega t - 150°)$$

***v* leads *i* by 160°, or *i* lags *v* by 160°.**

Or using

$$-\sin(\omega t + 30°) = \sin(\omega t + 30° \overset{\text{Note}}{+} 180°)$$
$$= \sin(\omega t + 210°)$$

***i* leads *v* by 200°, or *v* lags *i* by 200°.**

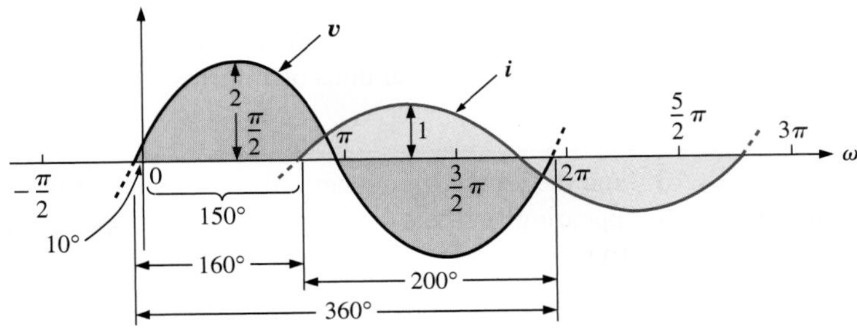

FIG. 13.34
Example 13.12(d): v leads i by 160°.

e. See Fig. 13.35.

$$i = -2\cos(\omega t - 60°) = 2\cos(\omega t - 60° \overset{\text{By choice}}{-} 180°)$$
$$= 2\cos(\omega t - 240°)$$

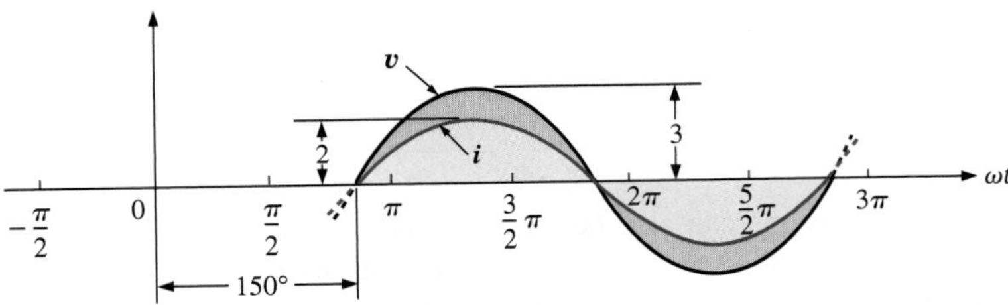

FIG. 13.35
Example 13.12(e): v and i are in phase.

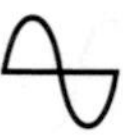

However, $\cos \alpha = \sin(\alpha + 90°)$

so that

$$2 \cos(\omega t - 240°) = 2 \sin(\omega t - 240° + 90°)$$
$$= 2 \sin(\omega t - 150°)$$

***v* and *i* are in phase.**

The Oscilloscope

The **oscilloscope** is an instrument that will display the sinusoidal alternating waveform in a way that will permit the reviewing of all of the waveform's characteristics. In some ways, the screen and the dials give an oscilloscope the appearance of a small TV, but remember that *it can display only what you feed into it.* You can't turn it on and ask for a sine wave, a square wave, and so on; it must be connected to a source or an active circuit to pick up the desired waveform.

The screen has a standard appearance, with 10 horizontal divisions and 8 vertical divisions. The distance between divisions is 1 cm on the vertical and horizontal scales, providing you with an excellent opportunity to become aware of the length of 1 cm. *The vertical scale is set to display voltage levels, whereas the horizontal scale is always in units of time.* The vertical sensitivity control sets the voltage level for each division, whereas the horizontal sensitivity control sets the time associated with each division. In other words, if the vertical sensitivity is set at 1 V/div., each division displays a 1 V swing, so that a total vertical swing of 8 divisions represents 8 V peak-to-peak. If the horizontal control is set on 10 μs/div., 4 divisions equal a time period of 40 μs. Remember, the oscilloscope display presents a sinusoidal voltage versus time, not degrees or radians. Further, the vertical scale is always a voltage sensitivity, never units of amperes.

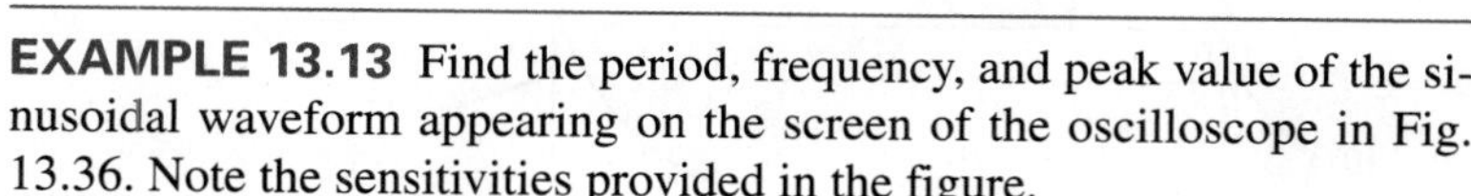

EXAMPLE 13.13 Find the period, frequency, and peak value of the sinusoidal waveform appearing on the screen of the oscilloscope in Fig. 13.36. Note the sensitivities provided in the figure.

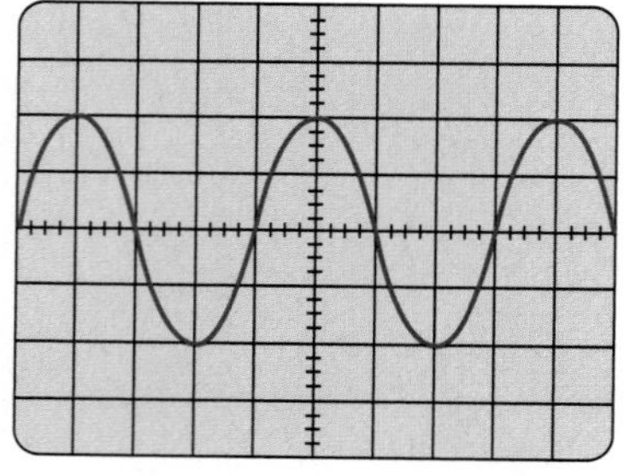

Vertical sensitivity = 0.1 V/div.
Horizontal sensitivity = 50 μs/div.

FIG. 13.36
Example 13.13.

Solution: One cycle spans 4 divisions. Therefore, the period is

$$T = 4\ \cancel{\text{div.}} \left(\frac{50\ \mu\text{s}}{\cancel{\text{div.}}}\right) = \mathbf{200\ \mu s}$$

and the frequency is

$$f = \frac{1}{T} = \frac{1}{200 \times 10^{-6}\ \text{s}} = \mathbf{5\ kHz}$$

The vertical height above the horizontal axis encompasses 2 divisions. Therefore,

$$V_m = 2\ \cancel{\text{div.}} \left(\frac{0.1\ \text{V}}{\cancel{\text{div.}}}\right) = \mathbf{0.2\ V}$$

An oscilloscope can also be used to make phase measurements between two sinusoidal waveforms. Virtually all laboratory oscilloscopes today have the dual-trace option, that is, the ability to show two waveforms at the same time. It is important to remember, however, that both waveforms will and must have the same frequency. The hookup procedure for using an oscilloscope to measure phase angles is covered in de-

tail in Section 15.13. However, the equation for determining the phase angle can be introduced using Fig. 13.37.

First, note that each sinusoidal function *has the same frequency,* permitting the use of either waveform to determine the period. For the waveform chosen in Fig. 13.37, the period encompasses 5 divisions at 0.2 ms/div. The phase shift between the waveforms (irrespective of which is leading or lagging) is 2 divisions. Since the full period represents a cycle of 360°, the following ratio [from which Eq. (13.24) can be derived] can be formed:

$$\frac{360^\circ}{T\,(\text{no. of div.})} = \frac{\theta}{\text{phase shift (no. of div.)}}$$

and

$$\theta = \frac{\text{phase shift (no. of div.)}}{T\,(\text{no. of div.})} \times 360^\circ \qquad \textbf{(13.24)}$$

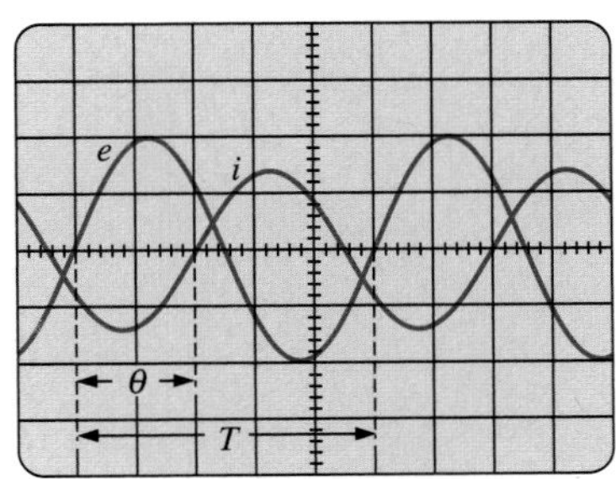

Vertical sensitivity = 2 V/div.
Horizontal sensitivity = 0.2 ms/div.

FIG. 13.37
Finding the phase angle between waveforms using a dual-trace oscilloscope.

Substituting into Eq. (13.24) results in

$$\theta = \frac{(2\text{ div.})}{(5\text{ div.})} \times 360^\circ = \mathbf{144^\circ}$$

and e leads i by 144°.

13.7 AVERAGE VALUE

Even though the concept of the **average value** is an important one in most technical fields, its true meaning is often misunderstood. In Fig. 13.38(a), for example, the average height of the sand may be required to determine the volume of sand available. The average height of the sand is that height obtained if the distance from one end to the other is maintained while the sand is leveled off, as shown in Fig. 13.38(b). The area under the mound in Fig. 13.38(a) then equals the area under the rectangular shape in Fig. 13.38(b) as determined by $A = b \times h$. Of course, the depth (into the page) of the sand must be the same for Fig. 13.38(a) and (b) for the preceding conclusions to have any meaning.

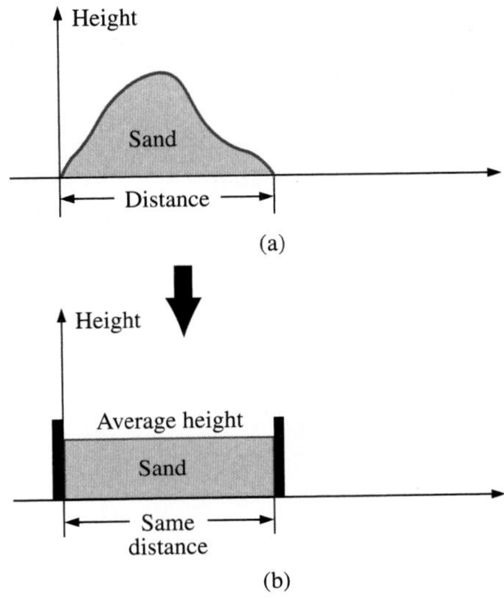

FIG. 13.38
Defining average value.

In Fig. 13.38, the distance was measured from one end to the other. In Fig. 13.39(a), the distance extends beyond the end of the original pile in Fig. 13.38. The situation could be one where a landscaper wants to know the average height of the sand if spread out over a distance such as defined in Fig. 13.39(a). The result of an increased distance is shown in Fig. 13.39(b). The average height has decreased compared to Fig. 13.38. Quite obviously, therefore, the longer the distance, the lower the average value.

If the distance parameter includes a depression, as shown in Fig. 13.40(a), some of the sand will be used to fill the depression, resulting in an even lower average value for the landscaper, as shown in Fig. 13.40(b). For a sinusoidal waveform, the depression would have the same shape as the mound of sand (over one full cycle), resulting in an average value at ground level (or zero volts for a sinusoidal voltage over one full period).

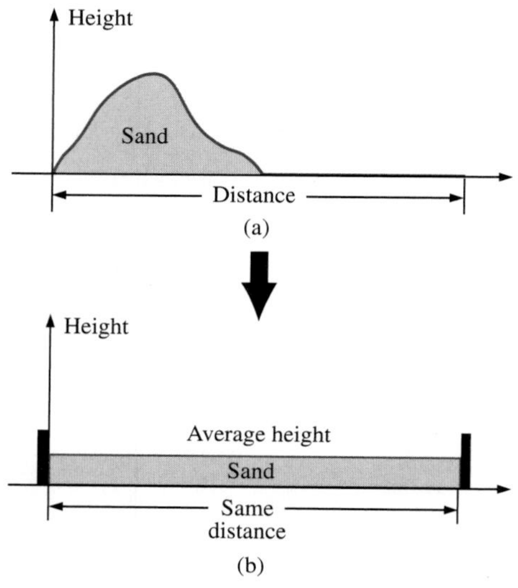

FIG. 13.39
Effect of distance (length) on average value.

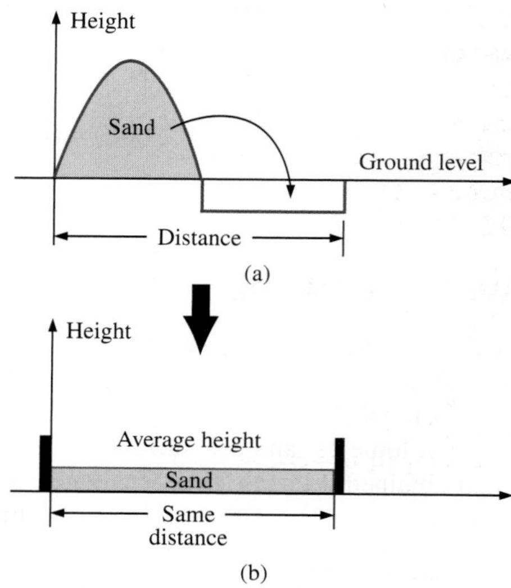

FIG. 13.40
Effect of depressions (negative excursions) on average value.

After traveling a considerable distance by car, some drivers like to calculate their average speed for the entire trip. This is usually done by dividing the miles traveled by the hours required to drive that distance. For example, if a person traveled 225 mi in 5 h, the average speed was 225 mi/5 h, or 45 mi/h. This same distance may have been traveled at various speeds for various intervals of time, as shown in Fig. 13.41.

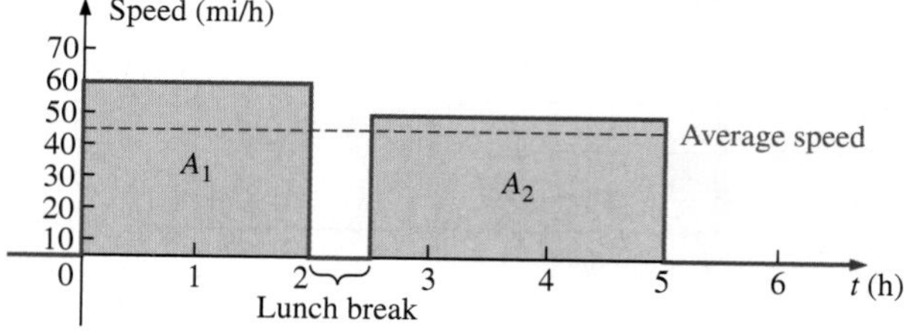

FIG. 13.41
Plotting speed versus time for an automobile excursion.

By finding the total area under the curve for the 5 h and then dividing the area by 5 h (the total time for the trip), we obtain the same result of 45 mi/h; that is,

$$\text{Average speed} = \frac{\text{area under curve}}{\text{length of curve}} \tag{13.25}$$

$$\begin{aligned}\text{Average speed} &= \frac{A_1 + A_2}{5\text{ h}} = \frac{(60\text{ mi/h})(2\text{ h}) + (50\text{ mi/h})(2.5\text{ h})}{5\text{ h}}\\ &= \frac{225}{5}\text{ mi/h} = \mathbf{45\ mi/h}\end{aligned}$$

Eq. (13.25) can be extended to include any variable quantity, such as current or voltage, if we let G denote the average value, as follows:

$$G\text{ (average value)} = \frac{\text{algebraic sum of areas}}{\text{length of curve}} \tag{13.26}$$

The *algebraic* sum of the areas must be determined, since some area contributions are from below the horizontal axis. Areas above the axis are assigned a positive sign, and those below, a negative sign. A positive average value is then above the axis, and a negative value, below.

The average value of *any* current or voltage is the value indicated on a dc meter. In other words, over a complete cycle, the average value is the equivalent dc value. In the analysis of electronic circuits to be considered in a later course, both dc and ac sources of voltage will be applied to the same network. You will then need to know or determine the dc (or average value) and ac components of the voltage or current in various parts of the system.

EXAMPLE 13.14 Determine the average value of the waveforms in Fig. 13.42.

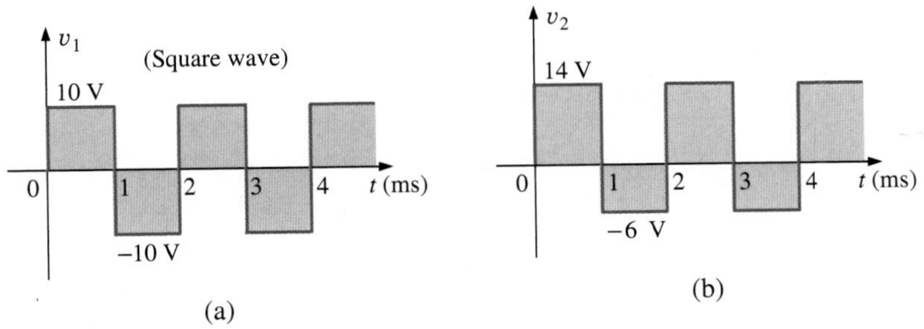

FIG. 13.42
Example 13.14.

Solutions:

a. By inspection, the area above the axis equals the area below over one cycle, resulting in an average value of zero volts. Using Eq. (13.26):

$$G = \frac{(10\text{ V})(1\text{ ms}) - (10\text{ V})(1\text{ ms})}{2\text{ ms}} = \frac{0}{2\text{ ms}} = \mathbf{0\ V}$$

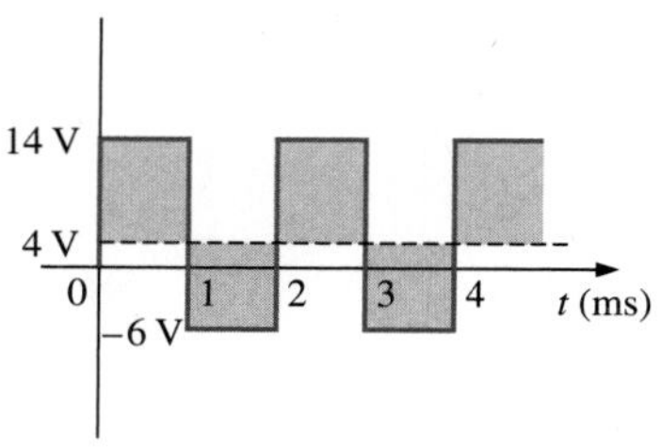

FIG. 13.43
Defining the average value for the waveform in Fig. 13.42(b).

b. Using Eq. (13.26):

$$G = \frac{(14\text{ V})(1\text{ ms}) - (6\text{ V})(1\text{ ms})}{2\text{ ms}} = \frac{14\text{ V} - 6\text{ V}}{2} = \frac{8\text{ V}}{2} = \mathbf{4\text{ V}}$$

as shown in Fig. 13.43.

In reality, the waveform in Fig. 13.42(b) is simply the square wave in Fig. 13.42(a) with a dc shift of 4 V; that is,

$$v_2 = v_1 + 4\text{ V}$$

EXAMPLE 13.15 Find the average values of the following waveforms over one full cycle:

a. Fig. 13.44.
b. Fig. 13.45.

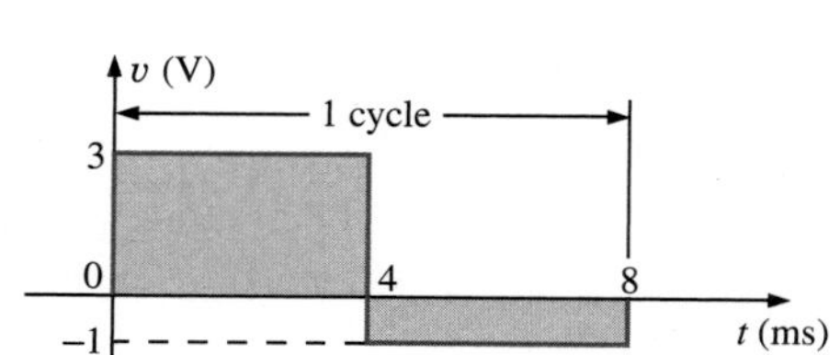

FIG. 13.44
Example 13.15(a).

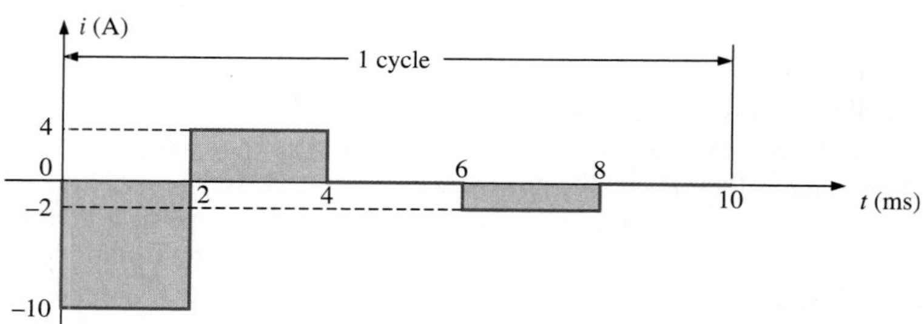

FIG. 13.45
Example 13.15(b).

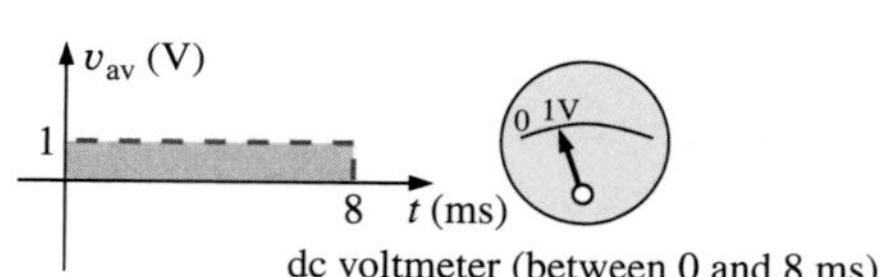

FIG. 13.46
The response of a dc meter to the waveform in Fig. 13.44.

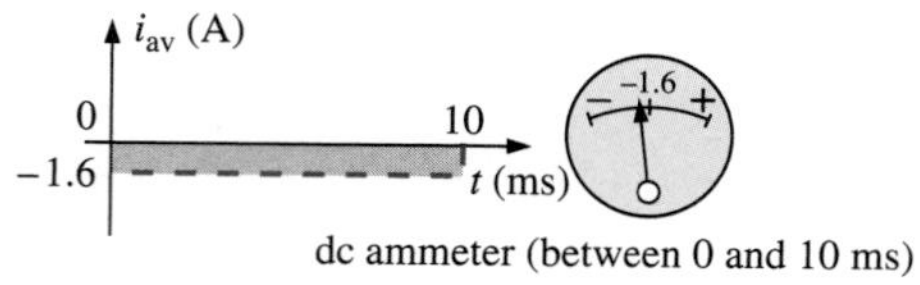

FIG. 13.47
The response of a dc meter to the waveform in Fig. 13.45.

Solutions:

a. $$G = \frac{+(3\text{ V})(4\text{ ms}) - (1\text{ V})(4\text{ ms})}{8\text{ ms}} = \frac{12\text{ V} - 4\text{ V}}{8} = \mathbf{1\text{ V}}$$

Note Fig. 13.46.

b. $$G = \frac{-(10\text{ V})(2\text{ ms}) + (4\text{ V})(2\text{ ms}) - (2\text{ V})(2\text{ ms})}{10\text{ ms}}$$

$$= \frac{-20\text{ V} + 8\text{ V} - 4\text{ V}}{10} = -\frac{16\text{ V}}{10} = \mathbf{-1.6\text{ V}}$$

Note Fig. 13.47.

We found the areas under the curves in Example 13.15 by using a simple geometric formula. If we should encounter a sine wave or any other unusual shape, however, we must find the area by some other means. We can obtain a good approximation of the area by attempting to reproduce the original wave shape using a number of small rectangles or other familiar shapes, the area of which we already know through simple geometric formulas. For example,

the area of the positive (or negative) pulse of a sine wave is $2A_m$.

Approximating this waveform by two triangles (Fig. 13.48), we obtain (using *area = 1/2 base × height* for the area of a triangle) a rough idea of the actual area:

$$\text{Area shaded} = 2\left(\frac{1}{2}bh\right) = 2\left[\left(\frac{1}{2}\right)\overbrace{\left(\frac{\pi}{2}\right)}^{b}\overbrace{(A_m)}^{h}\right] = \frac{\pi}{2}A_m \cong 1.58A_m$$

A closer approximation may be a rectangle with two similar triangles (Fig. 13.49):

$$\text{Area} = A_m\frac{\pi}{3} + 2\left(\frac{1}{2}bh\right) = A_m\frac{\pi}{3} + \frac{\pi}{3}A_m = \frac{2}{3}\pi A_m = 2.094A_m$$

which is certainly close to the actual area. If an infinite number of forms is used, an exact answer of $2A_m$ can be obtained. For irregular waveforms, this method can be especially useful if data such as the average value are desired.

The procedure of calculus that gives the exact solution $2A_m$ is known as *integration.* Integration is presented here only to make the method recognizable to you; it is not necessary to be proficient in its use to continue with this text. It is a useful mathematical tool, however, and should be learned. Finding the area under the positive pulse of a sine wave using integration, we have

$$\text{Area} = \int_0^{\pi} A_m \sin\alpha \, d\alpha$$

where $\int$ is the sign of integration, 0 and π are the limits of integration, $A_m \sin\alpha$ is the function to be integrated, and $d\alpha$ indicates that we are integrating with respect to α.

Integrating, we obtain

$$\begin{aligned}\text{Area} &= A_m[-\cos\alpha]_0^{\pi}\\ &= -A_m(\cos\pi - \cos 0°)\\ &= -A_m[-1 - (+1)] = -A_m(-2)\end{aligned}$$

$$\boxed{\text{Area} = 2A_m} \qquad \textbf{(13.27)}$$

Since we know the area under the positive (or negative) pulse, we can easily determine the average value of the positive (or negative) region of a sine wave pulse by applying Eq. (13.26):

$$G = \frac{2A_m}{\pi}$$

and

$$\boxed{G = \frac{2A_m}{\pi} = 0.637A_m} \qquad \textbf{(13.28)}$$

For the waveform in Fig. 13.50,

$$G = \frac{(2A_m/2)}{\pi/2} = \frac{2A_m}{\pi} \quad \text{(The average is the same as for a full pulse.)}$$

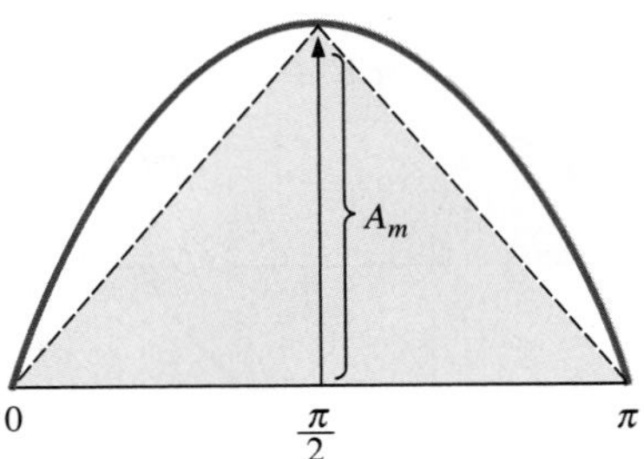

FIG. 13.48
Approximating the shape of the positive pulse of a sinusoidal waveform with two right triangles.

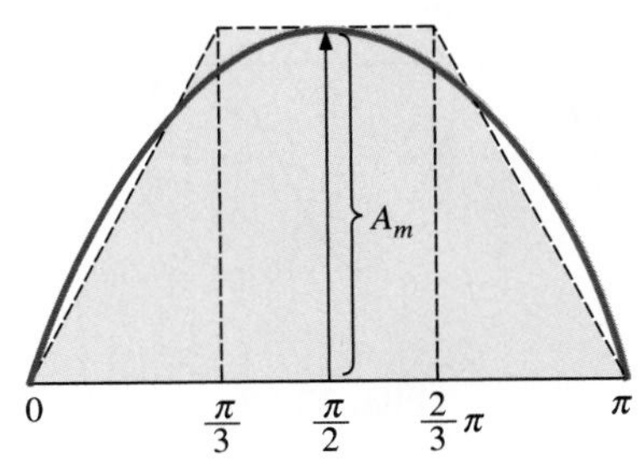

FIG. 13.49
A better approximation for the shape of the positive pulse of a sinusoidal waveform.

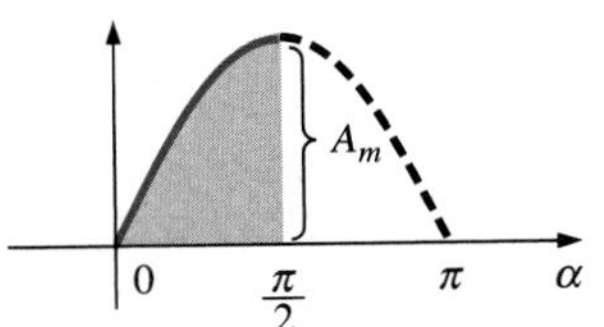

FIG. 13.50
Finding the average value of one-half the positive pulse of a sinusoidal waveform.

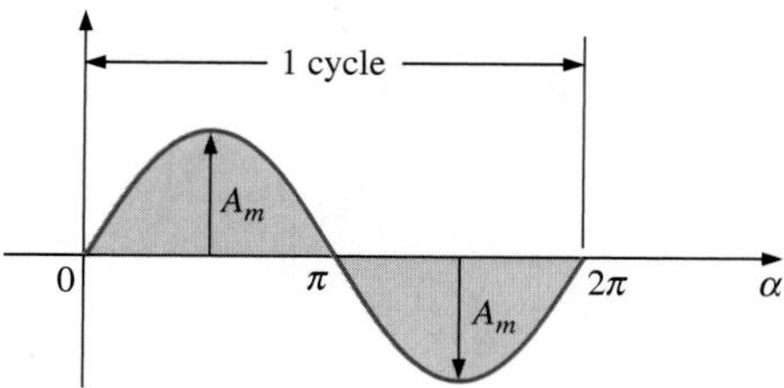

FIG. 13.51
Example 13.16.

EXAMPLE 13.16 Determine the average value of the sinusoidal waveform in Fig. 13.51.

Solution: By inspection it is fairly obvious that

the average value of a pure sinusoidal waveform over one full cycle is zero.

Eq. (13.26):

$$G = \frac{+2A_m - 2A_m}{2\pi} = \mathbf{0\ V}$$

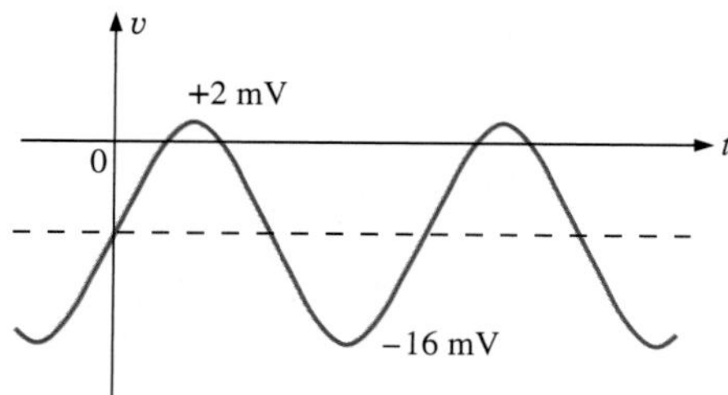

FIG. 13.52
Example 13.17.

EXAMPLE 13.17 Determine the average value of the waveform in Fig. 13.52.

Solution: The peak-to-peak value of the sinusoidal function is 16 mV + 2 mV = 18 mV. The peak amplitude of the sinusoidal waveform is, therefore, 18 mV/2 = 9 mV. Counting down 9 mV from 2 mV (or 9 mV up from −16 mV) results in an average or dc level of **−7 mV,** as noted by the dashed line in Fig. 13.52.

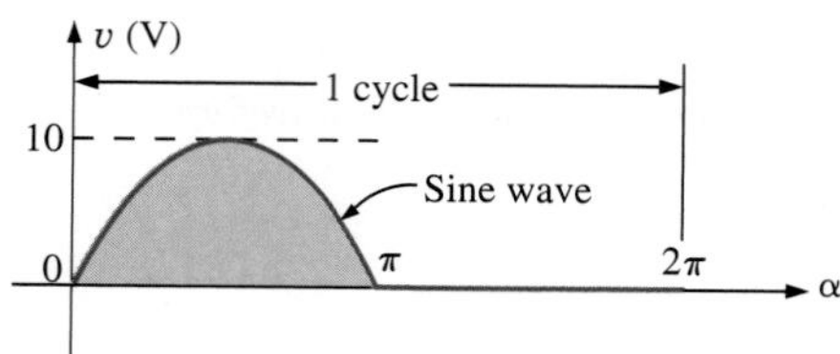

FIG. 13.53
Example 13.18.

EXAMPLE 13.18 Determine the average value of the waveform in Fig. 13.53.

Solution:

$$G = \frac{2A_m + 0}{2\pi} = \frac{2(10\ \text{V})}{2\pi} \cong \mathbf{3.18\ V}$$

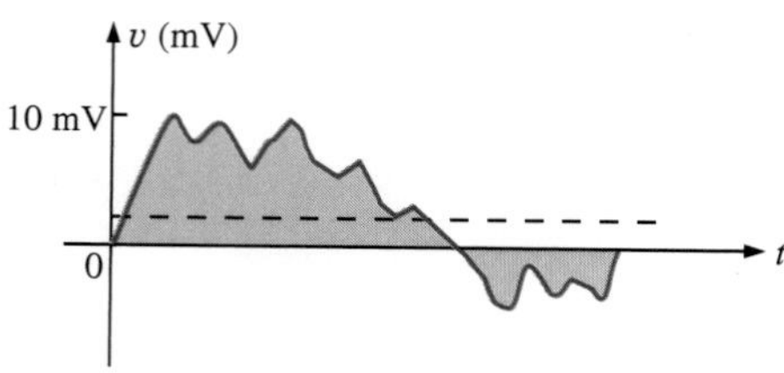

FIG. 13.54
Example 13.19.

EXAMPLE 13.19 For the waveform in Fig. 13.54, determine whether the average value is positive or negative, and determine its approximate value.

Solution: From the appearance of the waveform, the average value is positive and in the vicinity of 2 mV. Occasionally, judgments of this type will have to be made.

Instrumentation

The dc level or average value of any waveform can be found using a digital multimeter (DMM) or an **oscilloscope.** For purely dc circuits, set the DMM on dc, and read the voltage or current levels. Oscilloscopes are limited to voltage levels using the sequence of steps listed below:

1. First choose GND from the DC-GND-AC option list associated with each vertical channel. The GND option blocks any signal to which the oscilloscope probe may be connected from entering the oscilloscope and responds with just a horizontal line. Set the resulting line in the middle of the vertical axis on the horizontal axis, as shown in Fig. 13.55(a).
2. Apply the oscilloscope probe to the voltage to be measured (if not already connected), and switch to the DC option. If a dc voltage is present, the horizontal line shifts up or down, as demonstrated in Fig. 13.55(b). Multiplying the shift by the vertical sensitivity

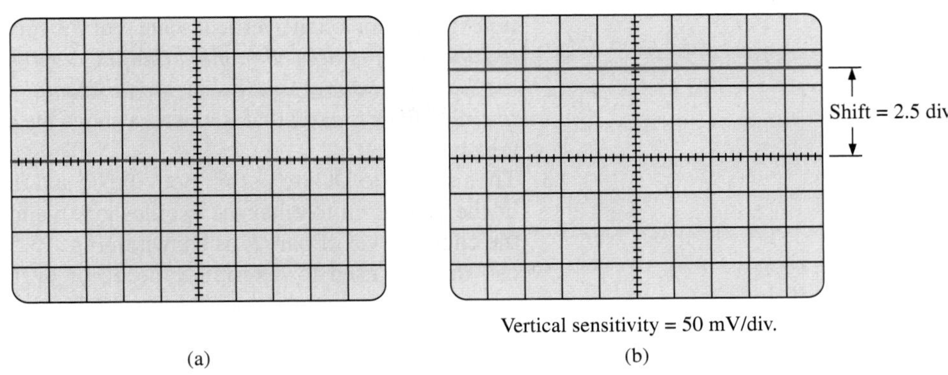

FIG. 13.55
Using the oscilloscope to measure dc voltages; (a) setting the GND condition; (b) the vertical shift resulting from a dc voltage when shifted to the DC option.

results in the dc voltage. An upward shift is a positive voltage (higher potential at the red or positive lead of the oscilloscope), while a downward shift is a negative voltage (lower potential at the red or positive lead of the oscilloscope).

In general,

$$V_{dc} = (\text{vertical shift in div.}) \times (\text{vertical sensitivity in V/div.}) \quad \textbf{(13.29)}$$

For the waveform in Fig. 13.55(b),

$$V_{dc} = (2.5 \text{ div.})(50 \text{ mV/div.}) = \mathbf{125\ mV}$$

The oscilloscope can also be used to measure the dc or average level of any waveform using the following sequence:

1. Using the GND option, reset the horizontal line to the middle of the screen.
2. Switch to AC (all dc components of the signal to which the probe is connected will be blocked from entering the oscilloscope—only the alternating, or changing, components are displayed). Note the location of some definitive point on the waveform, such as the bottom of the half-wave rectified waveform of Fig. 13.56(a); that is,

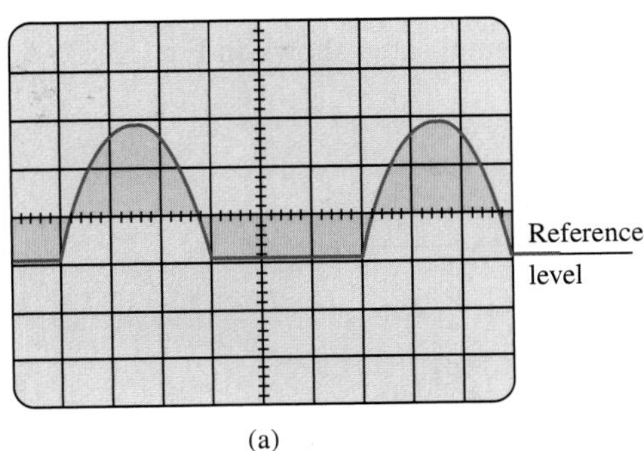

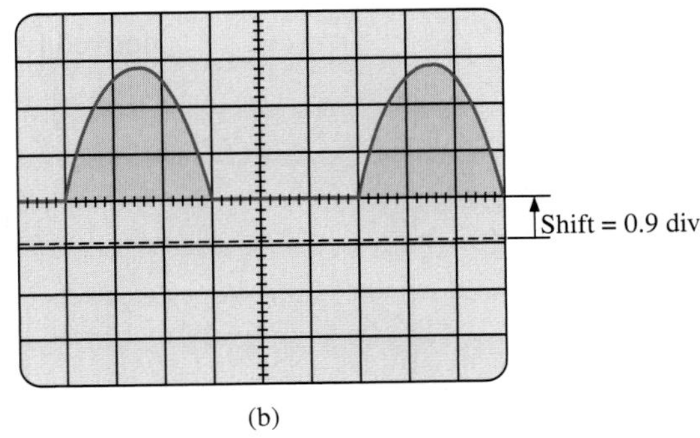

FIG. 13.56
Determining the average value of a nonsinusoidal waveform using the oscilloscope: (a) vertical channel on the ac mode; (b) vertical channel on the dc mode.

note its position on the vertical scale. For the future, *whenever you use the AC option, keep in mind that the computer will distribute the waveform above and below the horizontal axis such that the average value is zero;* that is, the area above the axis will equal the area below.

3. Then switch to DC (to permit both the dc and the ac components of the waveform to enter the oscilloscope), and note the shift in the chosen level of part 2, as shown in Fig. 13.56(b). Eq. (13.29) can then be used to determine the dc or average value of the waveform. For the waveform in Fig. 13.56(b), the average value is about

$$V_{av} = V_{dc} = (0.9 \text{ div.})(5 \text{ V/div.}) = \mathbf{4.5 \text{ V}}$$

The procedure outlined above can be applied to any alternating waveform such as the one in Fig. 13.54. In some cases the average value may require moving the starting position of the waveform under the AC option to a different region of the screen or choosing a higher voltage scale. By choosing the appropriate scale, you can enable DMMs to read the average or dc level of any waveform.

13.8 EFFECTIVE (rms) VALUES

This section begins to relate dc and ac quantities with respect to the power delivered to a load. It will help us determine the amplitude of a sinusoidal ac current required to deliver the same power as a particular dc current. The question frequently arises, How is it possible for a sinusoidal ac quantity to deliver a net power if, over a full cycle, the net current in any one direction is zero (average value = 0)? It would almost appear that the power delivered during the positive portion of the sinusoidal waveform is withdrawn during the negative portion, and since the two are equal in magnitude, the net power delivered is zero. However, understand that regardless of *direction,* current of any magnitude through a resistor delivers power *to that resistor.* In other words, during the positive or negative portions of a sinusoidal ac current, power is being delivered at *each instant of time* to the resistor. The power delivered at each instant, of course, varies with the magnitude of the sinusoidal ac current, but there will be a net flow during either the positive or the negative pulses with a net flow over the full cycle. The net power flow equals twice that delivered by either the positive or the negative regions of sinusoidal quantity.

A fixed relationship between ac and dc voltages and currents can be derived from the experimental setup shown in Fig. 13.57. A resistor in a

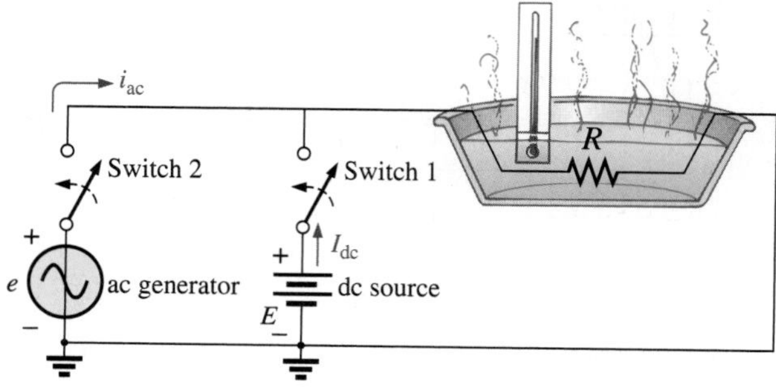

FIG. 13.57
An experimental setup to establish a relationship between dc and ac quantities.

water bath is connected by switches to a dc and an ac supply. If switch 1 is closed, a dc current I, determined by the resistance R and battery voltage E, is established through the resistor R. The temperature reached by the water is determined by the dc power dissipated in the form of heat by the resistor.

If switch 2 is closed and switch 1 left open, the ac current through the resistor has a peak value of I_m. The temperature reached by the water is now determined by the ac power dissipated in the form of heat by the resistor. The ac input is varied until the temperature is the same as that reached with the dc input. When this is accomplished, the average electrical power delivered to the resistor R by the ac source is the same as that delivered by the dc source.

The power delivered by the ac supply at any instant of time is

$$P_{ac} = (i_{ac})^2 R = (I_m \sin \omega t)^2 R = (I_m^2 \sin^2 \omega t) R$$

However,

$$\sin^2 \omega t = \frac{1}{2}(1 - \cos 2\omega t) \qquad \text{(trigonometric identity)}$$

Therefore,

$$P_{ac} = I_m^2 \left[\frac{1}{2}(1 - \cos 2\omega t) \right] R$$

and

$$P_{ac} = \frac{I_m^2 R}{2} - \frac{I_m^2 R}{2} \cos 2\omega t \qquad \textbf{(13.30)}$$

The *average power* delivered by the ac source is just the first term, since the average value of a cosine wave is zero even though the wave may have twice the frequency of the original input current waveform. Equating the average power delivered by the ac generator to that delivered by the dc source,

$$P_{av(ac)} = P_{dc}$$

$$\frac{I_m^2 R}{2} = I_{dc}^2 R$$

and

$$I_{dc} = \frac{I_m}{\sqrt{2}} = 0.707 I_m$$

which, in words, states that

the equivalent dc value of a sinusoidal current or voltage is $1/\sqrt{2}$ or 0.707 of its peak value.

The equivalent dc value is called the **rms** or **effective value** of the sinusoidal quantity.

As a simple numerical example, it requires an ac current with a peak value of $\sqrt{2}(10) = 14.14$ A to deliver the same power to the resistor in Fig. 13.57 as a dc current of 10 A. The effective value of any quantity plotted as a function of time can be found by using the following equation derived from the experiment just described.

Calculus format:

$$I_{rms} = \sqrt{\frac{\int_0^T i^2(t)\, dt}{T}} \qquad \textbf{(13.31)}$$

which means:

$$I_{\text{rms}} = \sqrt{\frac{\text{area } (i^2(t))}{T}} \tag{13.32}$$

In words, Eqs. (13.31) and (13.32) state that to find the rms value, the function $i(t)$ must first be squared. After $i(t)$ is squared, the area under the curve is found by integration. It is then divided by T, the length of the cycle or the period of the waveform, to obtain the average or *mean* value of the squared waveform. The final step is to take the *square root* of the mean value. This procedure is the source for the other designation for the effective value, the **root-mean-square (rms) value.** In fact, since *rms* is the most commonly used term in the educational and industrial communities, it is used throughout this text.

The relationship between the peak value and the rms value is the same for voltages, resulting in the following set of relationships for the examples and text material to follow:

$$\begin{aligned} I_{\text{rms}} &= \frac{1}{\sqrt{2}} I_m = 0.707 I_m \\ E_{\text{rms}} &= \frac{1}{\sqrt{2}} E_m = 0.707 E_m \end{aligned} \tag{13.33}$$

Similarly,

$$\begin{aligned} I_m &= \sqrt{2} I_{\text{rms}} = 1.414 I_{\text{rms}} \\ E_m &= \sqrt{2} E_{\text{rms}} = 1.414 E_{\text{rms}} \end{aligned} \tag{13.34}$$

EXAMPLE 13.20 Find the rms values of the sinusoidal waveform in each part in Fig. 13.58.

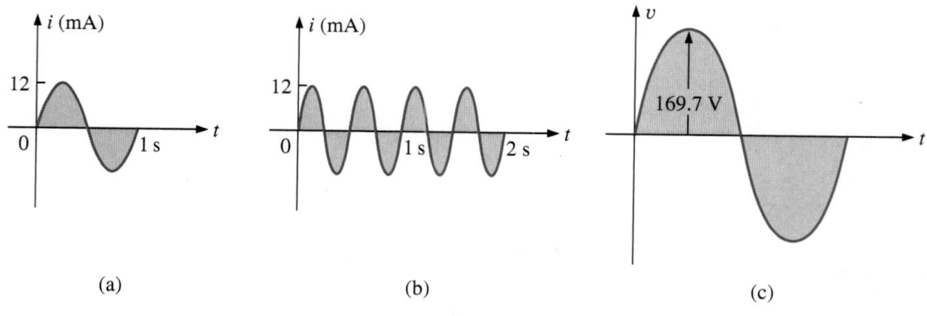

FIG. 13.58
Example 13.20.

Solution: For part (a), $I_{\text{rms}} = 0.707(12 \times 10^{-3} \text{ A}) =$ **8.48 mA.** For part (b), again $I_{\text{rms}} =$ **8.48 mA.** Note that frequency did not change the effective value in (b) compared to (a). For part (c), $V_{\text{rms}} = 0.707(169.73 \text{ V}) \cong$ **120 V,** the same as available from a home outlet.

EXAMPLE 13.21 The 120 V dc source in Fig. 13.59(a) delivers 3.6 W to the load. Determine the peak value of the applied voltage (E_m) and the current (I_m) if the ac source [Fig. 13.59(b)] is to deliver the same power to the load.

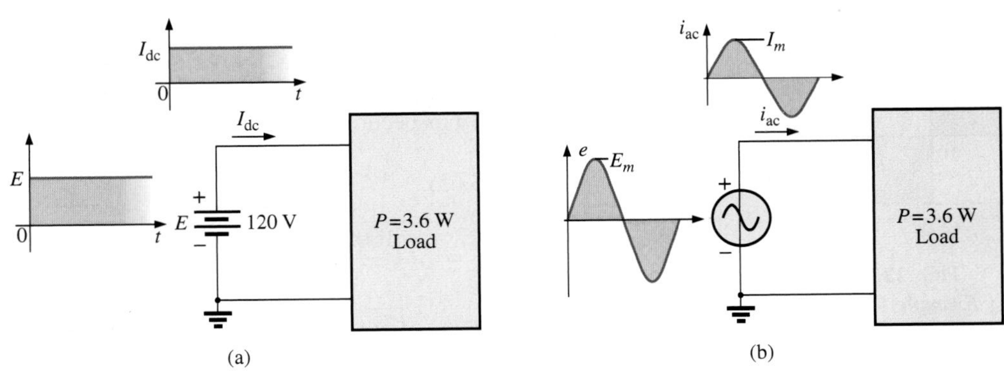

FIG. 13.59
Example 13.21.

Solution:

$$P_{dc} = V_{dc}I_{dc}$$

and
$$I_{dc} = \frac{P_{dc}}{V_{dc}} = \frac{3.6 \text{ W}}{120 \text{ V}} = 30 \text{ mA}$$

$$I_m = \sqrt{2}I_{dc} = (1.414)(30 \text{ mA}) = \mathbf{42.42 \text{ mA}}$$

$$E_m = \sqrt{2}E_{dc} = (1.414)(120 \text{ V}) = \mathbf{169.68 \text{ V}}$$

EXAMPLE 13.22 Find the rms value of the waveform in Fig. 13.60.

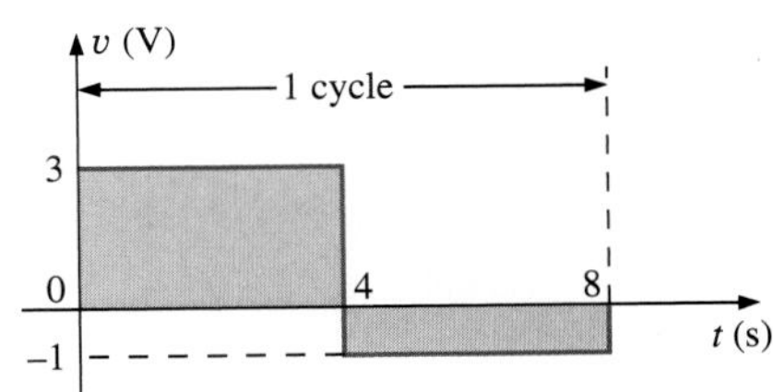

FIG. 13.60
Example 13.22.

Solution: v^2 (Fig. 13.61):

$$V_{rms} = \sqrt{\frac{(9)(4) + (1)(4)}{8}} = \sqrt{\frac{40}{8}} = \mathbf{2.24 \text{ V}}$$

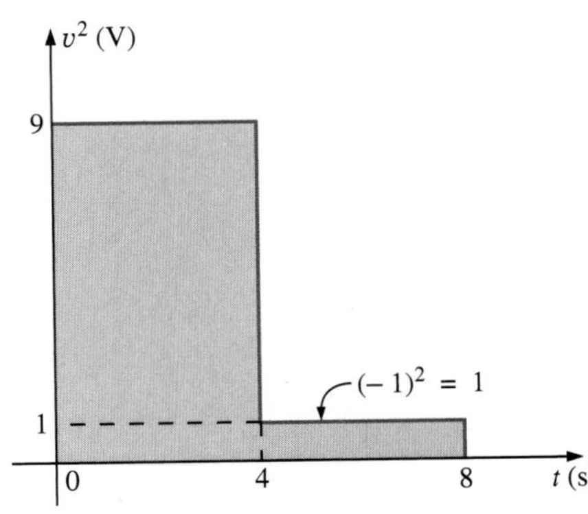

FIG. 13.61
The squared waveform of Fig. 13.60.

EXAMPLE 13.23 Calculate the rms value of the voltage in Fig. 13.62.

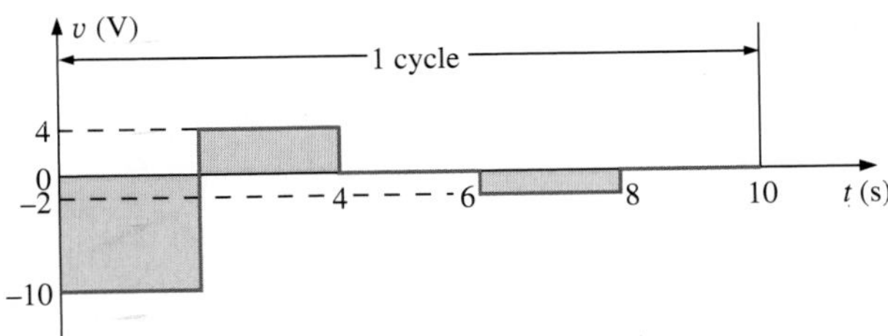

FIG. 13.62
Example 13.23.

Solution: v^2 (Fig. 13.63):

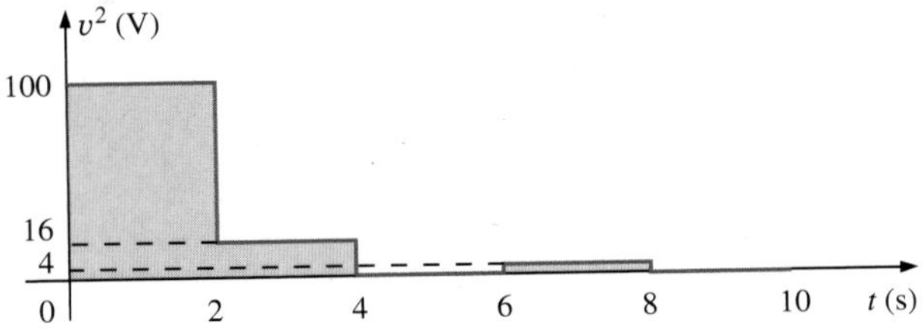

FIG. 13.63
The squared waveform of Fig. 13.62.

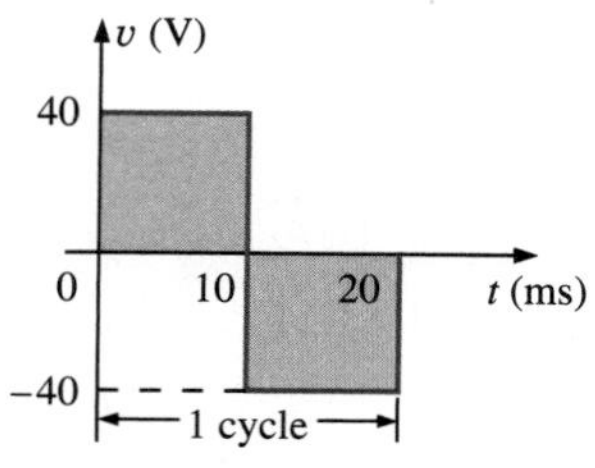

FIG. 13.64
Example 13.24.

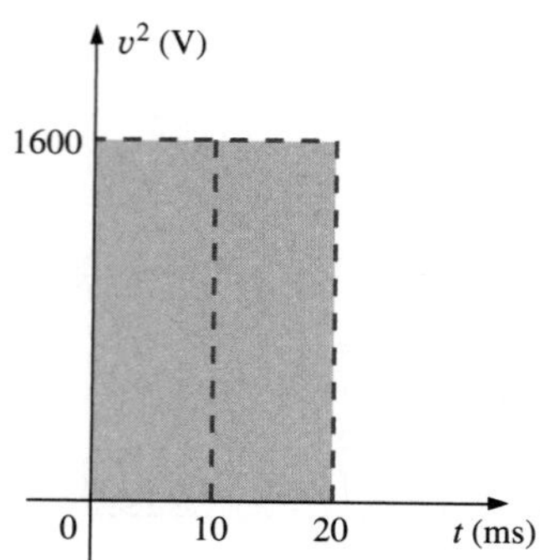

FIG. 13.65
The squared waveform of Fig. 13.64.

EXAMPLE 13.24 Determine the average and rms values of the square wave in Fig. 13.64.

Solution: By inspection, the average value is zero.

v^2 (Fig. 13.65):

$$V_{\text{rms}} = \sqrt{\frac{(1600)(10 \times 10^{-3}) + (1600)(10 \times 10^{-3})}{20 \times 10^{-3}}}$$

$$= \sqrt{\frac{(32{,}000 \times 10^{-3})}{20 \times 10^{-3}}} = \sqrt{1600} = \mathbf{40\ V}$$

(the maximum value of the waveform in Fig. 13.64).

The waveforms appearing in these examples are the same as those used in the examples on the average value. It may prove interesting to compare the rms and average values of these waveforms.

The rms values of sinusoidal quantities such as voltage or current are represented by E and I. These symbols are the same as those used for dc voltages and currents. To avoid confusion, the peak value of a waveform always has a subscript m associated with it: $I_m \sin \omega t$. *Caution:* When finding the rms value of the positive pulse of a sine wave, note that the squared area is *not* simply $(2A_m)^2 = 4A_m^2$; it must be found by a completely new integration. This is always true for any waveform that is not rectangular.

A unique situation arises if a waveform has both a dc and an ac component that may be due to a source such as the one in Fig. 13.66. The combination appears frequently in the analysis of electronic networks where both dc and ac levels are present in the same system.

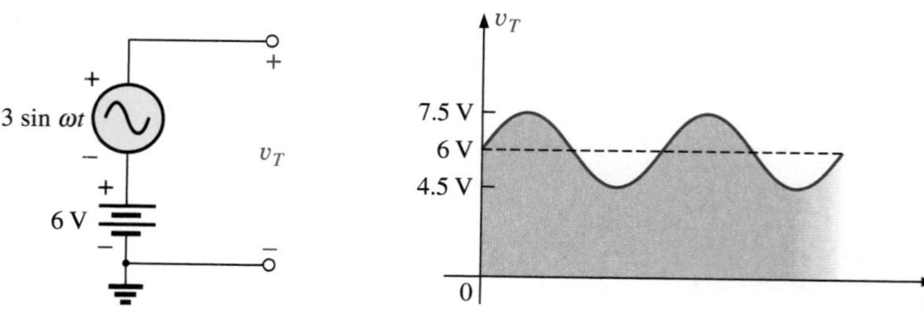

FIG. 13.66
Generation and display of a waveform having a dc and an ac component.

The question arises, What is the rms value of the voltage v_T? You may be tempted to assume that it is the sum of the rms values of each component of the waveform; that is, $V_{T_{\text{rms}}} = 0.7071(1.5\ \text{V}) + 6\ \text{V} = 1.06\ \text{V} + 6\ \text{V} = 7.06\ \text{V}$. However, the rms value is actually determined by

$$\boxed{V_{\text{rms}} = \sqrt{V_{\text{dc}}^2 + V_{\text{ac(rms)}}^2}} \tag{13.35}$$

which for the waveform in Fig. 13.66 is

$$V_{\text{rms}} = \sqrt{(6\ \text{V})^2 + (1.06\ \text{V})^2} = \sqrt{37.124}\ \text{V} \cong \mathbf{6.1\ V}$$

This result is noticeably less than the solution of 7.06 V.

13.9 ac METERS AND INSTRUMENTS

It is important to note whether the DMM in use is a *true rms* meter or simply a meter where the average value is calibrated (as described in the next section) to indicate the rms level. *A true rms meter reads the effective value of any waveform (such as Figs. 13.54 and 13.66) and is not limited to only sinusoidal waveforms.* Since the label *true rms* is normally not placed on the face of the meter, it is prudent to check the manual if waveforms other than purely sinusoidal are to be encountered. For any type of rms meter, be sure to check the manual for its frequency range of application. For most, it is less than 1 kHz.

If an average reading movement such as the d'Arsonval movement used in the **VOM** of Fig. 2.29 is used to measure an ac current or voltage, the level indicated by the movement must be multiplied by a **calibration factor.** In other words, if the movement of any voltmeter or ammeter is reading the average value, that level must be multiplied by a specific constant, or calibration factor, to indicate the rms level. For ac waveforms, the signal must first be converted to one having an average value over the time period. Recall that it is zero over a full period for a sinusoidal waveform. This is usually accomplished for sinusoidal waveforms using a bridge rectifier such as in Fig. 13.67. The conversion process, involving four diodes in a bridge configuration, is well documented in most electronic texts.

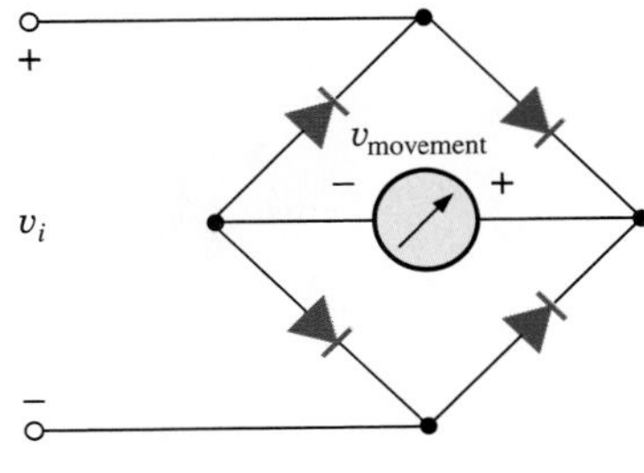

FIG. 13.67
Full-wave bridge rectifier.

Fundamentally, conduction is permitted through the diodes in such a manner as to convert the sinusoidal input of Fig. 13.68(a) to one having the appearance of Fig. 13.68(b). The negative portion of the input has been effectively "flipped over" by the bridge configuration. The resulting waveform in Fig. 13.68(b) is called a *full-wave rectified waveform.*

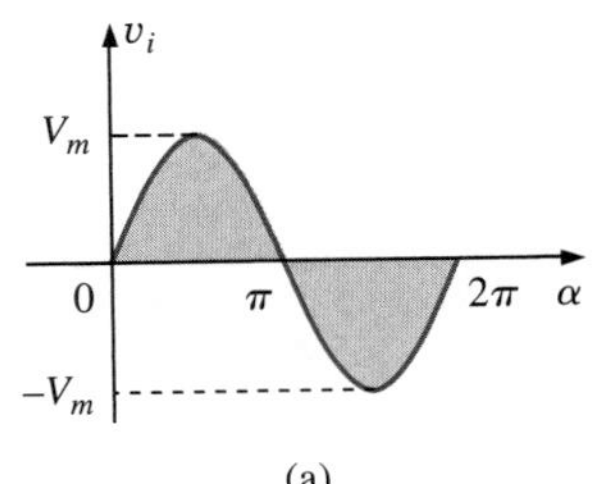

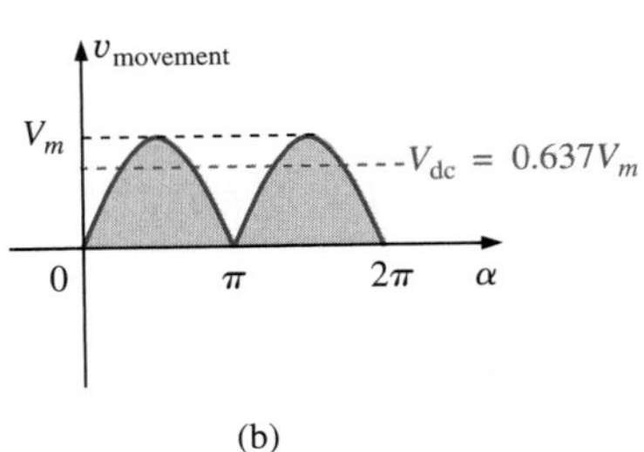

FIG. 13.68
(a) Sinusoidal input; (b) full-wave rectified signal.

The zero average value in Fig. 13.68(a) has been replaced by a pattern having an average value determined by

$$G = \frac{2V_m + 2V_m}{2\pi} = \frac{4V_m}{2\pi} = \frac{2V_m}{\pi} = 0.637V_m$$

The movement of the pointer is therefore directly related to the peak value of the signal by the factor 0.637.

Forming the ratio between the rms and dc levels results in

$$\frac{V_{rms}}{V_{dc}} = \frac{0.707V_m}{0.637V_m} \cong 1.11$$

revealing that the scale indication is 1.11 times the dc level measured by the movement; that is,

$$\boxed{\text{Meter indication} = 1.11\ (\text{dc or average value})}\ \text{full-wave} \qquad \textbf{(13.36)}$$

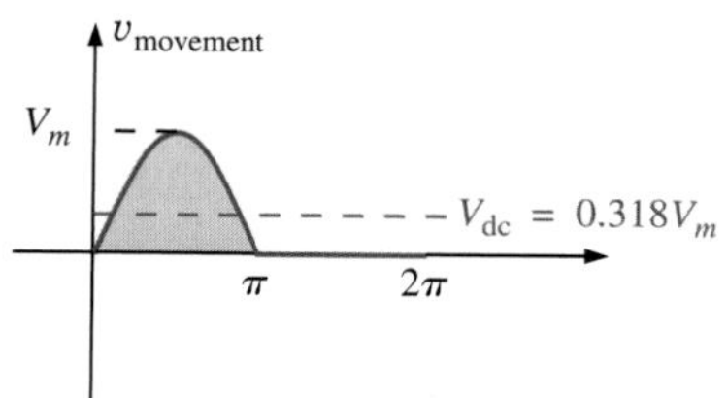

FIG. 13.69
Half-wave rectified signal.

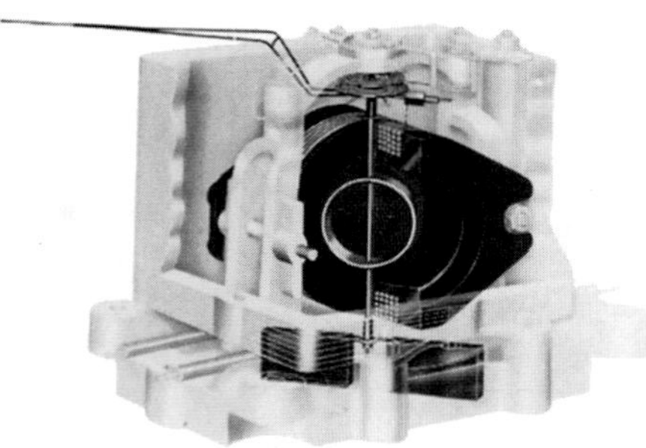

FIG. 13.70
Electrodynamometer movement.
(Courtesy of Schlumberger Technology Corp.)

Some ac meters use a half-wave rectifier arrangement that results in the waveform in Fig. 13.69, which has half the average value in Fig. 13.68(b) over one full cycle. The result is

$$\boxed{\text{Meter indication} = 2.22\text{ (dc or average value)}} \text{ half-wave} \quad \textbf{(13.37)}$$

A second movement, called the **electrodynamometer movement** (Fig. 13.70), can measure both ac and dc quantities without a change in internal circuitry. The movement can, in fact, read the effective value of any periodic or nonperiodic waveform because a reversal in current direction reverses the fields of both the stationary and the movable coils, so the deflection of the pointer is always up-scale.

EXAMPLE 13.25 Determine the reading of each meter for each situation in Fig. 13.71(a) and (b).

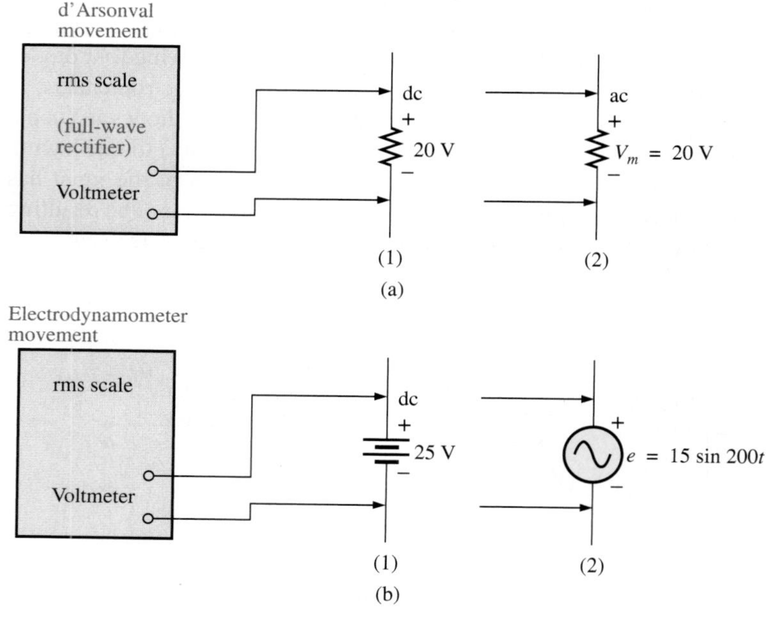

FIG. 13.71
Example 13.25.

Solution: For Fig. 13.71(a), situation (1): By Eq. (13.36),

$$\text{Meter indication} = 1.11(20\text{ V}) = \mathbf{22.2\ V}$$

For Fig. 13.71(a), situation (2):

$$V_{rms} = 0.707V_m = 0.707(20\text{ V}) = \mathbf{14.14\ V}$$

For Fig. 13.71(b), situation (1):

$$V_{rms} = V_{dc} = \mathbf{25\ V}$$

For Fig. 13.71(b), situation (2):

$$V_{rms} = 0.707V_m = 0.707(15\text{ V}) \cong \mathbf{10.6\ V}$$

Most DMMs employ a full-wave rectification system to convert the input ac signal to one with an average value. In fact, for the VOM in Fig. 2.29, the same scale factor of Eq. (13.36) is employed; that is, the average value is scaled up by a factor of 1.11 to obtain the rms value. In digital meters, however, there are no moving parts such as in the d'Arsonval movement to display the signal level. Rather, the average value is sensed by a multiprocessor integrated circuit (IC), which in turn determines which digits should appear on the digital display.

Digital meters can also be used to measure nonsinusoidal signals, but the scale factor of each input waveform must first be known (normally provided by the manufacturer in the operator's manual.) For instance, the scale factor for an average responding DMM on the ac rms scale produces an indication for a square-wave input that is 1.11 times the peak value. For a triangular input, the response is 0.555 times the peak value. Obviously, for a sine wave input, the response is 0.707 times the peak value.

For any instrument, it is always good practice to read the operator's manual if you will use the instrument on a regular basis.

For frequency measurements, the **frequency counter** in Fig. 13.72 provides a digital readout of sine, square, and triangular waves from 1 Hz to 1.3 GHz. Note the relative simplicity of the panel and the high degree of accuracy available. The temperature-compensated, crystal-controlled time base is stable to ± 1 part per million per year.

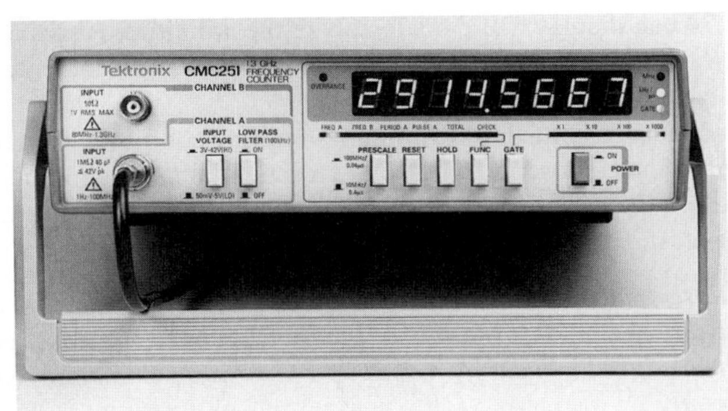

FIG. 13.72
Frequency counter. Tektronix CMC251 1.3 GHz multifunction counter.
(Photo courtesy of Tektronix, Inc.)

The AEMC® **Clamp Meter** in Fig. 13.73 is an instrument that can measure alternating current in the ampere range without having to open the circuit. The loop is opened by squeezing the "trigger"; then it is placed around the current-carrying conductor. Through transformer action, the level of current in rms units appears on the appropriate scale. The Model 501 is auto-ranging (that is, each scale changes automatically) and can measure dc or ac currents up to 400 mA. Through the use of additional leads, it can also be used as a voltmeter (up to 400 V, dc or ac) and an ohmmeter (from zero to 400 Ω).

One of the most versatile and important instruments in the electronics industry is the **oscilloscope,** which has already been introduced in this chapter. It provides a display of the waveform on a cathode-ray tube to permit the detection of irregularities and the determination of quantities such as magnitude, frequency, period, dc component, and so on. The

FIG. 13.73
Clamp-on ammeter and voltmeter.
(Courtesy of AEMC® Instruments, Foxborough, MA.)

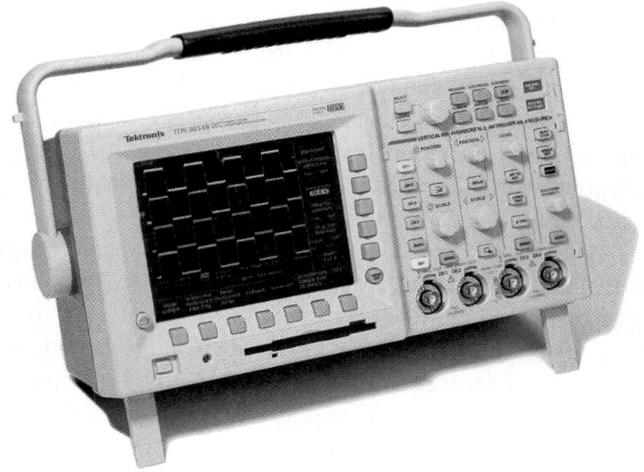

FIG. 13.74
Four-channel digital phosphor oscilloscope. Tektronix TDS3000B series oscilloscope.
(Photo courtesy of Tektronix, Inc.)

digital oscilloscope in Fig. 13.74 can display four waveforms at the same time. You use menu buttons to set the vertical and horizontal scales by choosing from selections appearing on the screen. The TDS model in Fig. 13.74 can display, store, and analyze the amplitude, time, and distribution of amplitude over time. It is also completely portable due to its battery-capable design.

A student accustomed to watching TV may be confused when first introduced to an oscilloscope. There is, at least initially, an assumption that the oscilloscope is generating the waveform on the screen—much like a TV broadcast. However, it is important to clearly understand that

an oscilloscope displays only those signals generated elsewhere and connected to the input terminals of the oscilloscope. The absence of an external signal will simply result in a horizontal line on the screen of the scope.

On most oscilloscopes today, there is a switch or knob with the choice DC/GND/AC, as shown in Fig. 13.75(a), that is often ignored or treated too lightly in the early stages of scope utilization. The effect of each position is fundamentally as shown in Fig. 13.75(b). In the DC mode, the dc and ac components of the input signal can pass directly to the display. In the AC mode, the dc input is blocked by the capacitor, but the ac portion of the signal can pass through to the screen. In the GND position, the input signal is prevented from reaching the scope display by a direct ground connection, which reduces the scope display to a single horizontal line.

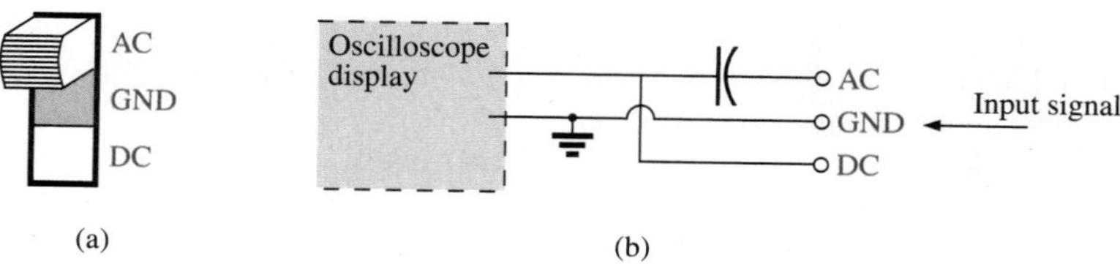

FIG. 13.75
AC-GND-DC switch for the vertical channel of an oscilloscope.

Before we leave the subject of ac meters and instrumentation, you should understand that

an ohmmeter cannot be used to measure the ac reactance or impedance of an element or system even though reactance and impedance are measured in ohms.

Recall that ohmmeters cannot be used on energized networks—the power must be shut off or disconnected. For an inductor, if the ac power is removed, the reactance of the coil is simply the dc resistance of the windings because the applicable frequency will be 0 Hz. For a capacitor, if the ac power is removed, the reactance of the capacitor is simply the leakage resistance of the capacitor. In general, therefore, always keep in mind that *ohmmeters can read only the dc resistance of an element or network, and only after the applied power has been removed.*

13.10 APPLICATIONS

(120 V at 60 Hz) versus (220 V at 50 Hz)

In North and South America, the most common available ac supply is 120 V at 60 Hz; in Europe and the Eastern countries, it is 220 V at 50 Hz. The choices of rms value and frequency were obviously made carefully because they have such an important impact on the design and operation of so many systems.

The fact that the frequency difference is only 10 Hz reveals that there was agreement on the general frequency range that should be used for power generation and distribution. History suggests that the question of frequency selection originally focused on the frequency that would not exhibit *flicker in the incandescent lamps* available in those days. Technically, however, there really wouldn't be a noticeable difference between 50 and 60 cycles per second based on this criterion. Another important factor in the early design stages was the effect of frequency on the size of transformers, which play a major role in power generation and distribution. Working through the fundamental equations for transformer design, you will find that *the size of a transformer is inversely proportional to frequency.* The result is that transformers operating at 50 Hz must be larger (on a purely mathematical basis about 17% larger) than those operating at 60 Hz. You will therefore find that transformers designed for the international market where they can operate on 50 Hz or 60 Hz are designed around the 50 Hz frequency. On the other side of the coin, however, higher frequencies result in increased concerns about arcing, increased losses in the transformer core due to eddy current and hysteresis losses, and skin effect phenomena. Somewhere in the discussion we may wonder about the fact that 60 Hz is an exact multiple of 60 seconds in a minute and 60 minutes in an hour. On the other side of the coin, however, a 60 Hz signal has a period of 16.67 ms (an awkward number), but the period of a 50 Hz signal is exactly 20 ms. Since accurate timing is such a critical part of our technological design, was this a significant motive in the final choice? There is also the question about whether the 50 Hz is a result of the close affinity of this value to the metric system. Keep in mind that powers of ten are all powerful in the metric system, with 100 cm in a meter, 100°C the boiling point of water, and so on. Note that 50 Hz is exactly half of this special number. All in all, it would seem that both sides have an argument that is worth defending. However, in the final analysis, we must also wonder whether the difference is simply political in nature.

The difference in voltage between the Americas and Europe is a different matter entirely in the sense that the difference is close to 100%. Again, however, there are valid arguments for both sides. There is no question that larger voltages such as 220 V *raise safety issues* beyond those raised by voltages of 120 V. However, when higher voltages are supplied, there is less current in the wire for the same power demand, permitting the use of smaller conductors—a real money saver. In addition, motors and some appliances *can be smaller in size.* Higher voltages, however, also bring back the concern about arcing effects, insulation requirements, and, due to real safety concerns, higher installation costs. In general, however, international travelers are prepared for most situations if they have a transformer that can convert from their home level to that of the country they plan to visit. Most equipment (not clocks, of course) can run quite well on 50 Hz or 60 Hz for most travel periods. For any unit not operating at its design frequency, it simply has to "work a little harder" to perform the given task. The major problem for the traveler is not the transformer itself but the wide variety of plugs used from one country to another. Each country has its own design for the "female" plug in the wall. For a three-week tour, this could mean as many as 6 to 10 different plugs of the type shown in Fig. 13.76. For a 120 V, 60 Hz supply, the plug is quite standard in appearance with its two spade leads (and possible ground connection).

FIG. 13.76
Variety of plugs for a 220 V, 50 Hz connection.

In any event, both the 120 V at 60 Hz and the 220 V at 50 Hz are obviously meeting the needs of the consumer. It is a debate that could go on at length without an ultimate victor.

Safety Concerns (High Voltages and dc versus ac)

Be aware that any "live" network should be treated with a calculated level of respect. Electricity in its various forms is not to be feared but used with some awareness of its potentially dangerous side effects. It is common knowledge that electricity and water do not mix (never use extension cords or plug in TVs or radios in the bathroom) because a full 120 V in a layer of water of any height (from a shallow puddle to a full bath) can be *lethal.* However, other effects of dc and ac voltages are less known. In general, as the voltage and current increase, your concern about safety should increase exponentially. For instance, under dry conditions, most human beings can survive a 120 V ac shock such as obtained when changing a light bulb, turning on a switch, and so on. Most electricians have experienced such a jolt many times in their careers. However, ask an electrician to relate how it feels to hit 220 V, and the response (if he or she has been unfortunate to have had such an experience) will be totally different. How often have you heard of a back-hoe operator hitting a 220 V line and having a fatal heart attack? Remember, the operator is sitting in a metal container on a damp ground which provides an excellent path for the resulting current to flow from the line to ground. If only for a short period of time, with the best environment (rubber-sole shoes, and so on), in a situation where you can quickly escape the situation, most human beings can also survive a 220 V shock. However, as mentioned above, it is one you will not quickly forget. For voltages beyond 220 V rms, the chances of survival go down exponentially with increase in voltage. It takes only about 10 mA of steady current through the heart to put it in defibrillation. In general, therefore, always be sure that the power is disconnected when working on the repair of electrical equipment. Don't assume that throwing a wall switch will disconnect the power. Throw the main circuit breaker and test the lines with a voltmeter before working on

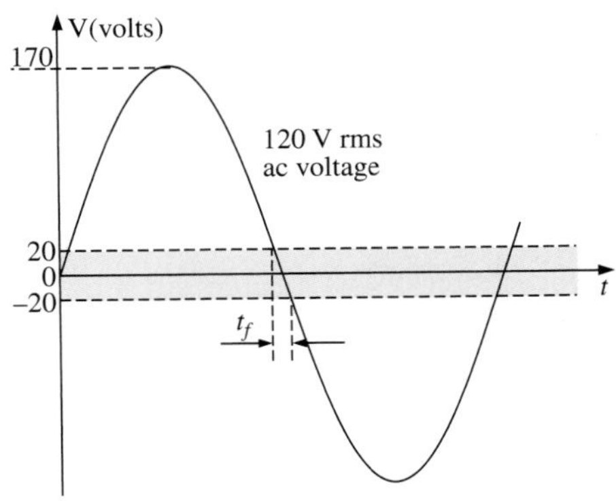

FIG. 13.77
Interval of time when sinusoidal voltage is near zero volts.

the system. Since voltage is a two-point phenomenon, be sure to work with only one line at at time—accidents happen!

You should also be aware that the reaction to dc voltages is quite different from that to ac voltages. You have probably seen in movies or comic strips that people are often unable to let go of a *hot* wire. This is evidence of the most important difference between the two types of voltages. As mentioned above, if you happen to touch a "hot" 120 V ac line, you will probably get a good sting, but *you can let go.* If it happens to be a "hot" 120 V dc line, you will probably not be able to let go, and you could die. Time plays an important role when this happens, because the longer you are subjected to the dc voltage, the more the resistance in the body decreases until a fatal current can be established. The reason that we can let go of an ac line is best demonstrated by carefully examining the 120 V rms, 60 Hz voltage in Fig. 13.77. Since the voltage is oscillating, there is a period when the voltage is near zero or less than, say, 20 V, and is reversing in direction. Although this time interval is very short, it appears every 8.3 ms and provides a window for you to *let go.*

Now that we are aware of the additional dangers of dc voltages, it is important to mention that under the wrong conditions, dc voltages as low as 12 V such as from a car battery can be quite dangerous. If you happen to be working on a car under wet conditions, or if you are sweating badly for some reason or, worse yet, wearing a wedding ring that may have moisture and body salt underneath, touching the positive terminal may initiate the process whereby the body resistance begins to drop and serious injury could take place. It is one of the reasons you seldom see a professional electrician wearing any rings or jewelry—it is just not worth the risk.

Before leaving this topic of safety concerns, you should also be aware of the dangers of high-frequency supplies. We are all aware of what 2.45 GHz at 120 V can do to a meat product in a microwave oven, and it is therefore very important that the seal around the oven be as tight as possible. However, don't ever assume that anything is absolutely perfect in design—so don't make it a habit to view the cooking process in the microwave 6 in. from the door on a continuing basis. Find something else to do, and check the food only when the cooking process is complete. If you ever visit the Empire State Building, you will notice that you are unable to get close to the antenna on the dome due to the high-frequency signals being emitted with a great deal of power. Also note the large KEEP OUT signs near radio transmission towers for local radio stations. Standing within 10 ft of an AM transmitter working at 540 kHz would bring on disaster. Simply holding (do not try!) a fluorescent bulb near the tower could make it light up due to the excitation of the molecules inside the bulb.

In total, therefore, treat any situation with high ac voltages or currents, high-energy dc levels, and high frequencies with added care.

13.11 COMPUTER ANALYSIS

PSpice

OrCAD Capture offers a variety of ac voltage and current sources. However, for the purposes of this text, the voltage source **VSIN** and the current source **ISIN** are the most appropriate because they have a list of attributes that cover current areas of interest. Under the library **SOURCE,** a number of others are listed, but they don't have the full range of the above, or they are dedicated to only one type of analysis. On

occasion, **ISRC** is used because it has an arrow symbol like that appearing in the text, and it can be used for dc, ac, and some transient analyses. The symbol for **ISIN** is a sine wave that utilizes the plus-and-minus sign (±) to indicate direction. The sources **VAC, IAC, VSRC,** and **ISRC** are fine if the magnitude and the phase of a specific quantity are desired or if a transient plot against frequency is desired. However, they will not provide a transient response against time even if the frequency and the transient information are provided for the simulation.

For all of the sinusoidal sources, the magnitude (**VAMPL**) is the peak value of the waveform, not the rms value. This becomes clear when a plot of a quantity is desired and the magnitude calculated by PSpice is the peak value of the transient response. However, for a purely steady-state ac response, the magnitude provided can be the rms value, and the output read as the rms value. Only when a plot is desired will it be clear that PSpice is accepting every ac magnitude as the peak value of the waveform. Of course, the phase angle is the same whether the magnitude is the peak or the rms value.

Before examining the mechanics of getting the various sources, remember that

Transient Analysis provides an ac or a dc output versus time, while AC Sweep is used to obtain a plot versus frequency.

To obtain any of the sources listed above, apply the following sequence: **Place part** key-**Place Part** dialog box-**Source-**(enter type of source). Once you select the source, the ac source **VSIN** appears on the schematic with **OFF, VAMPL,** and **FREQ.** Always specify **VOFF** as 0 V (unless a specific value is part of the analysis), and provide a value for the amplitude and frequency. Enter the remaining quantities of **PHASE, AC, DC, DF,** and **TD** by double-clicking on the source symbol to obtain the **Property Editor,** although **PHASE, DF** (damping factor), and **TD** (time delay) do have a default of 0 s. To add a phase angle, click on **PHASE,** enter the phase angle in the box below, and then select **Apply.** If you want to display a factor such as a phase angle of 60°, click on **PHASE** followed by **Display** to obtain the **Display Properties** dialog box. Then choose **Name and Value** followed by **OK** and **Apply,** and leave the **Properties Editor** dialog box (**X**) to see **PHASE=60** next to the **VSIN** source. The next chapter includes the use of the ac source in a simple circuit.

Multisim

For Multisim, the ac voltage source is available from two sources—the **Sources** parts bin and the **Function Generator.** The major difference between the two is that the phase angle can be set when using the **Sources** parts bin, whereas it cannot be set using the **Function Generator.**

Under **Sources,** select **SIGNAL_VOLTAGE_SOURCES** group under the **Family** heading. When selected and placed, it displays the default values for the amplitude, frequency, and phase. All the parameters of the source can be changed by double-clicking on the source symbol to obtain the dialog box. The listing clearly indicates that the set voltage is the peak value. Note that the unit of measurement is controlled by the scrolls to the right of the default label and cannot be set by typing in the desired unit of measurement. The label can be changed by switching the **Label** heading and inserting the desired label. After all the changes have been made in the dialog box, click **OK,** and all the changes appear next to the

FIG. 13.78
*Using the oscilloscope to display the sinusoidal ac voltage source available in the Multisim **Sources** tool bin.*

ac voltage source symbol. In Fig. 13.78, the label was changed to **Vs** and the amplitude to 10 V while the freqency and phase angle were left with their default values. It is particularly important to realize that

for any frequency analysis (that is, where the frequency will change), the AC Magnitude of the ac source must be set under Analysis Setup in the SIGNAL_VOLTAGE_SOURCES dialog box. Failure to do so will create results linked to the default values rather than the value set under the Value heading.

To view the sinusoidal voltage set in Fig. 13.78, select an oscilloscope from the **Instrument** toolbar at the right of the screen. It is the fourth option down and has the appearance shown in Fig. 13.78 when selected. Note that it is a dual-channel oscilloscope with an **A** channel and a **B** channel. It has a ground (**G**) connection and a trigger (**T**) connection. The connections for viewing the ac voltage source on the **A** channel are provided in Fig. 13.78. Note that the trigger control is also connected to the **A** channel for sync control. The screen appearing in Fig. 13.78 can be displayed by double-clicking on the oscilloscope symbol on the screen. It has all the major controls of a typical laboratory oscilloscope. When you select **Simulate-Run** or select **1** on the **Simulate Switch,** the ac voltage appears on the screen. Changing the **Time base** to 100 μs/div. results in the display of Fig. 13.78 since there are 10 divisions across the screen and $10(100\ \mu\text{s}) = 1$ ms (the period of the applied signal). Changes in the **Time base** are made by clicking on the default value to obtain the scrolls in the same box. It is important to remember, however, that

changes in the oscilloscope setting or any network should not be made until the simulation is ended by disabling the Simulate-Run option or placing the Simulate switch in the 0 mode.

The options within the time base are set by the scroll bars and cannot be changed—again they match those typically available on a laboratory oscilloscope. The vertical sensitivity of the **A** channel was automatically set by the program at 5 V/div. to result in two vertical boxes for the peak

value as shown in Fig. 13.78. Note the **AC** and **DC** keypads below Channel **A.** Since there is no dc component in the applied signal, either one results in the same display. The **Trigger** control is set on the positive transition at a level of 0 V. The **T1** and **T2** refer to the cursor positions on the horizontal time axis. By clicking on the small red triangle at the top of the red line at the far left edge of the screen and dragging the triangle, you can move the vertical red line to any position along the axis. In Fig. 13.78, it was moved to the peak value of the waveform at one-quarter of the total period or 0.25 ms = 250 μs. Note the value of **T1** (250 μs) and the corresponding value of **VA1** (9.995 V $\cong$ 10.0 V). By moving the other cursor with a blue triangle at the top to one-half the total period or 0.5 ms = 500 μs, we find that the value at **T2** (500 μs) is 0.008 pV (**VA2**), which is essentially 0 V for a waveform with a peak value of 10 V. The accuracy is controlled by the number of data points called for in the simulation setup. The more data points, the higher the likelihood of a higher degree of accuracy for the desired quantity. However, an increased number of data points also extends the running time of the simulation. The third line provides the difference between **T2** and **T1** as 250 μs and difference between their magnitudes (**VA2–VA1**) as -9.995 V, with the negative sign appearing because **VA1** is greater than **VA2.**

As mentioned above, you can also obtain an ac voltage from the **Function Generator** appearing as the second option down on the **Instrument** toolbar. Its symbol appears in Fig. 13.79 with positive, negative, and ground connections. Double-click on the generator graphic symbol, and the **Function Generator** dialog box appears in which selections can be made. For this example, the sinusoidal waveform is chosen. The **Frequency** is set at 1 kHz, the **Amplitude** is set at 10 V, and the **Offset** is left at 0 V. Note that there is no option to set the phase angle as was possible for the source above. Double-clicking on the oscilloscope generates the **Oscilloscope-XSCI** dialog box in which a **Timebase** of 100 μs/div. can be set again with a vertical sensitivity of 5 V/div. Select **1** on the **Simulate** switch, and the waveform of Fig. 13.79 appears. Choosing **Singular** under **Trigger** results in a fixed display. Set the **Simulate** switch on **0** to end the simulation. Placing the cursors in the same position shows that the waveforms for Figs. 13.78 and 13.79 are the same.

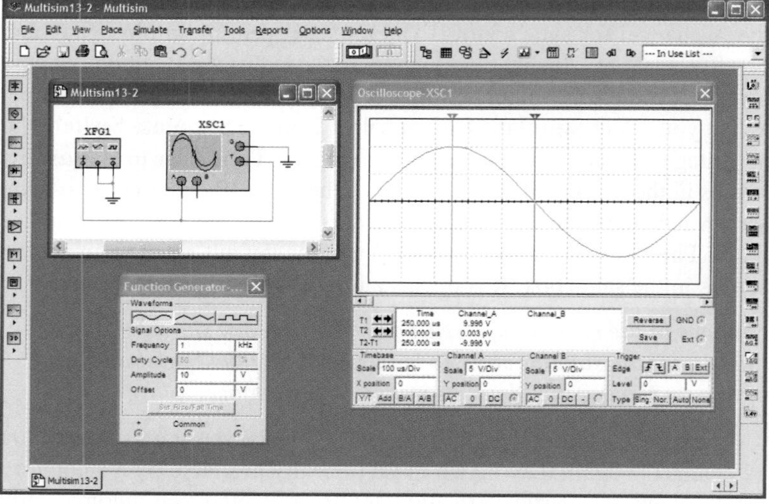

FIG. 13.79
Using the function generator to place a sinusoidal ac voltage waveform on the screen of the oscilloscope.

For most of the Multisim analyses to appear in this text, the **AC_VOLTAGE_SOURCE** under **Sources** will be employed. However, with such a limited introduction to Multisim, it seemed appropriate to introduce the use of the **Function Generator** because of its close linkage to the laboratory experience.

PROBLEMS

SECTION 13.2 Sinusoidal ac Voltage Characteristics and Definitions

1. For the sinusoidal waveform in Fig. 13.80:
   a. What is the peak value?
   b. What is the instantaneous value at 15 ms and at 20 ms?
   c. What is the peak-to-peak value of the waveform?
   d. What is the period of the waveform?
   e. How many cycles are shown?

2. For the square-wave signal in Fig. 13.81:
   a. What is the peak value?
   b. What is the instantaneous value at 5 μs and at 11 μs?
   c. What is the peak-to-peak value of the waveform?
   d. What is the period of the waveform?
   e. How many cycles are shown?

3. For the periodic waveform in Fig. 13.82:
   a. What is the peak value?
   b. What is the instantaneous value at 3 μs and at 9 μs?
   c. What is the peak-to-peak value of the waveform?
   d. What is the period of the waveform?
   e. How many cycles are shown?

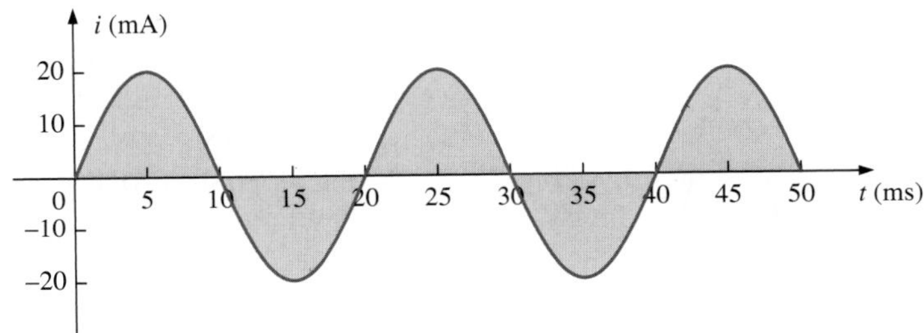

FIG. 13.80
Problem 1.

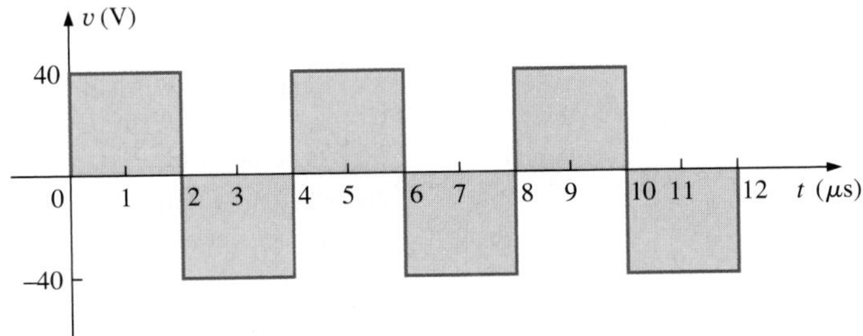

FIG. 13.81
Problem 2.

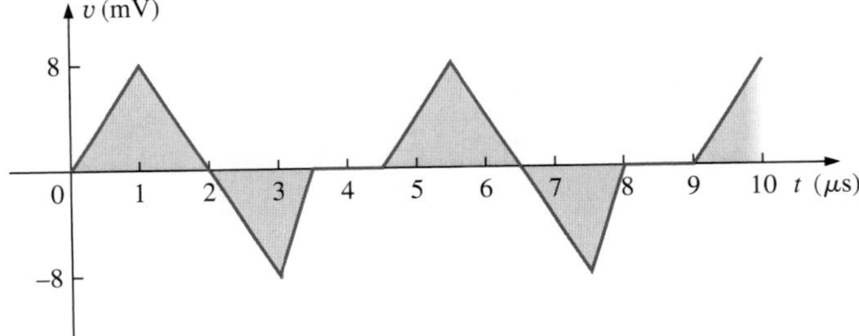

FIG. 13.82
Problem 3.

SECTION 13.3 Frequency Spectrum

4. Find the period of a periodic waveform whose frequency is
a. 25 Hz. **b.** 40 MHz.
c. 25 kHz. **d.** 1 Hz.

5. Find the frequency of a repeating waveform whose period is
a. 1/60 s. **b.** 0.01 s.
c. 40 ms. **d.** 25 μs.

6. If a periodic waveform has a frequency of 20 Hz, how long (in seconds) will it take to complete five cycles?

7. Find the period of a sinusoidal waveform that completes 80 cycles in 24 ms.

8. What is the frequency of a periodic waveform that completes 42 cycles in 6 s?

9. For the oscilloscope pattern of Fig. 13.83:
a. Determine the peak amplitude.
b. Find the period.
c. Calculate the frequency.
Redraw the oscilloscope pattern if a +25 mV dc level were added to the input waveform.

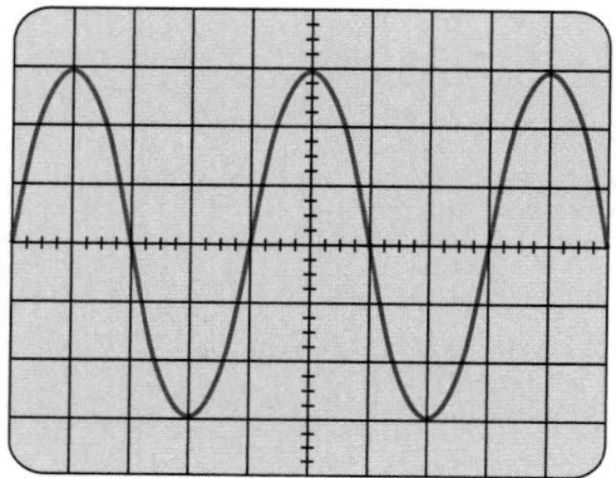

Vertical sensitivity = 50 mV/div.
Horizontal sensitivity = 10 μs/div.

FIG. 13.83
Problem 9.

SECTION 13.4 The Sinusoidal Waveform

10. Convert the following degrees to radians:
a. 45° **b.** 60° **c.** 270° **d.** 170°

11. Convert the following radians to degrees:
a. $\pi/4$ **b.** $\pi/6$ **c.** $\frac{1}{10}\pi$ **d.** 0.6π

12. Find the angular velocity of a waveform with a period of
a. 2 s. **b.** 0.3 ms. **c.** 4 μs. **d.** $\frac{1}{26}$ s.

13. Find the angular velocity of a waveform with a frequency of
a. 50 Hz. **b.** 600 Hz.
c. 2 kHz. **d.** 0.004 MHz.

14. Find the frequency and period of sine waves having an angular velocity of
a. 754 rad/s. **b.** 8.4 rad/s.
c. 6000 rad/s. **d.** $\frac{1}{16}$ rad/s.

***15.** Given $f = 60$ Hz, determine how long it will take the sinusoidal waveform to pass through an angle of 45°.

***16.** If a sinusoidal waveform passes through an angle of 30° in 5 ms, determine the angular velocity of the waveform.

SECTION 13.5 General Format for the Sinusoidal Voltage or Current

17. Find the amplitude and frequency of the following waves:
a. $20 \sin 377t$ **b.** $5 \sin 754t$
c. $10^6 \sin 10{,}000t$ **d.** $-6.4 \sin 942t$

18. Sketch $5 \sin 754t$ with the abscissa
a. angle in degrees. **b.** angle in radians.
c. time in seconds.

***19.** Sketch $-7.6 \sin 43.6t$ with the abscissa
a. angle in degrees. **b.** angle in radians.
c. time in seconds.

20. If $e = 300 \sin 157t$, how long (in seconds) does it take this waveform to complete 1/2 cycle?

21. Given $i = 0.5 \sin \alpha$, determine i at $\alpha = 72°$.

22. Given $v = 20 \sin \alpha$, determine v at $\alpha = 1.2\pi$.

***23.** Given $v = 30 \times 10^{-3} \sin \alpha$, determine the angles at which v will be 6 mV.

***24.** If $v = 40$ V at $\alpha = 30$ and $t = 1$ ms, determine the mathematical expression for the sinusoidal voltage.

SECTION 13.6 Phase Relations

25. Sketch $\sin(377t + 60°)$ with the abscissa
a. angle in degrees. **b.** angle in radians.
c. time in seconds.

26. Sketch the following waveforms:
a. $50 \sin(\omega t + 0°)$ **b.** $5 \sin(\omega t + 60°)$
c. $2 \cos(\omega t + 10°)$ **d.** $-20 \sin(\omega t + 2°)$

27. Write the analytical expression for the waveforms in Fig. 13.84 with the phase angle in degrees.

v (V)
25
f = 60 Hz
0
ωt
30°

(a)

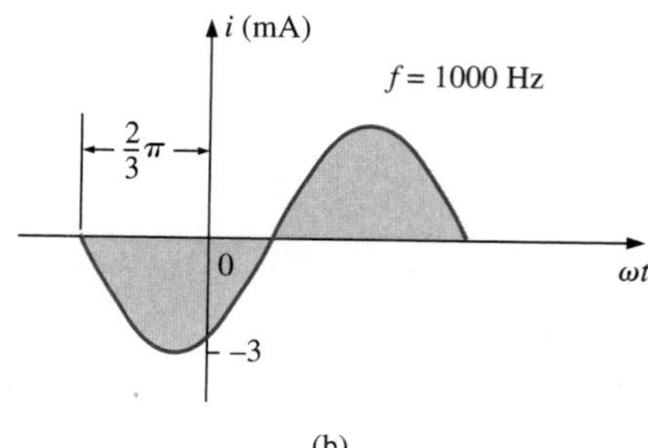

(b)

FIG. 13.84
Problem 27.

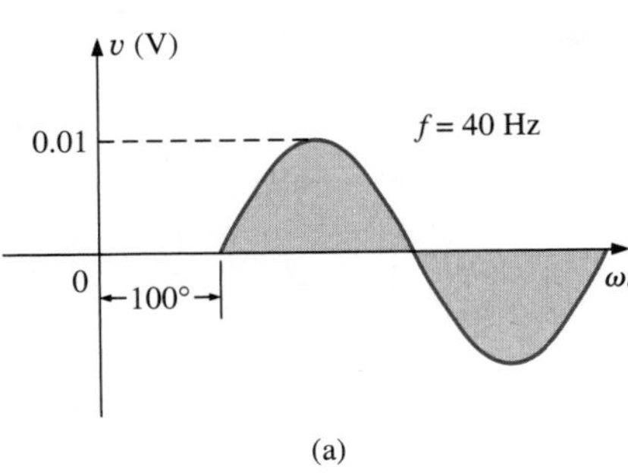

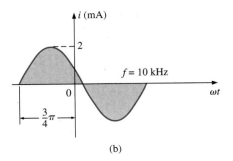

FIG. 13.85
Problem 28.

28. Write the analytical expression for the waveforms in Fig. 13.85 with the phase angle in degrees.

29. Find the phase relationship between the following waveforms:

$$v = 4\sin(\omega t + 50^\circ)$$
$$i = 6\sin(\omega t + 40^\circ)$$

30. Find the phase relationship between the following waveforms:

$$v = 25\sin(\omega t - 80^\circ)$$
$$i = 5 \times 10^{-3}\sin(\omega t - 10^\circ)$$

31. Find the phase relationship between the following waveforms:

$$v = 0.2\sin(\omega t - 60^\circ)$$
$$i = 0.1\sin(\omega t + 20^\circ)$$

***32.** Find the phase relationship between the following waveforms:

$$v = 2\cos(\omega t - 30^\circ)$$
$$i = 5\sin(\omega t + 60^\circ)$$

***33.** Find the phase relationship between the following waveforms:

$$v = -4\cos(\omega t + 90^\circ)$$
$$i = -2\sin(\omega t + 10^\circ)$$

***34.** The sinusoidal voltage $v = 200\sin(2\pi 1000t + 60^\circ)$ is plotted in Fig. 13.86. Determine the time t_1 when the waveform crosses the axis.

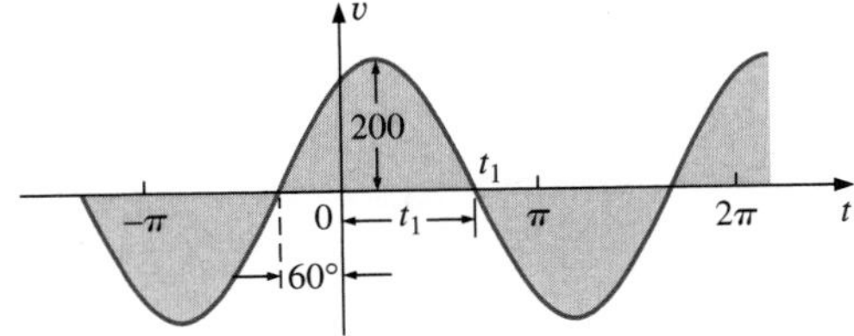

FIG. 13.86
Problem 34.

***35.** The sinusoidal current $i = 4\sin(50{,}000t - 40^\circ)$ is plotted in Fig. 13.87. Determine the time t_1 when the waveform crosses the axis.

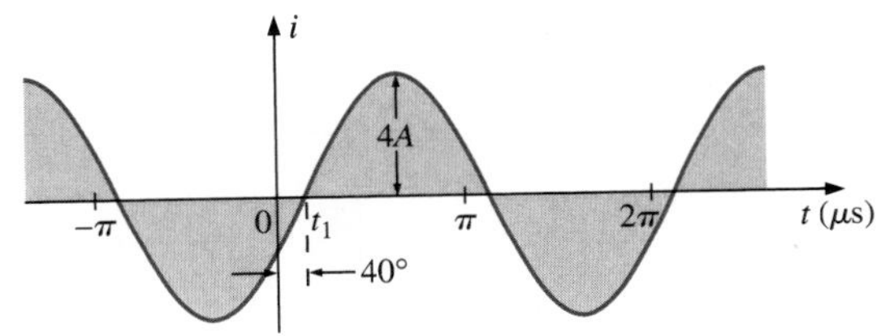

FIG. 13.87
Problem 35.

36. For the oscilloscope display in Fig. 13.88:
- **a.** Determine the period of the waveform.
- **b.** Determine the frequency of each waveform.
- **c.** Find the rms value of each waveform.
- **d.** Determine the phase shift between the two waveforms and determine which leads and which lags.

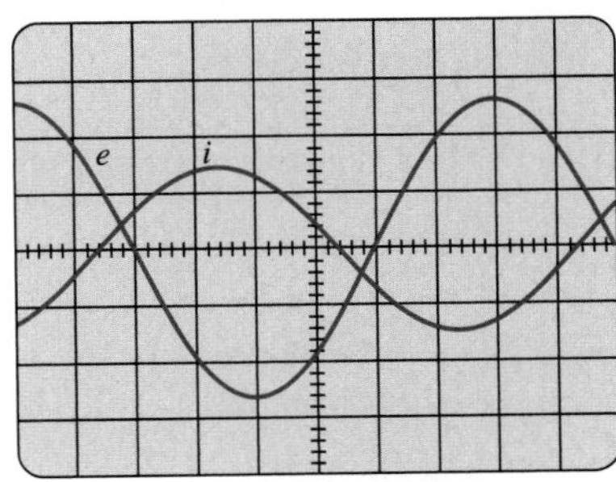

FIG. 13.88
Problem 36.

SECTION 13.7 Average Value

37. Find the average value of the periodic waveform in Fig. 13.89.

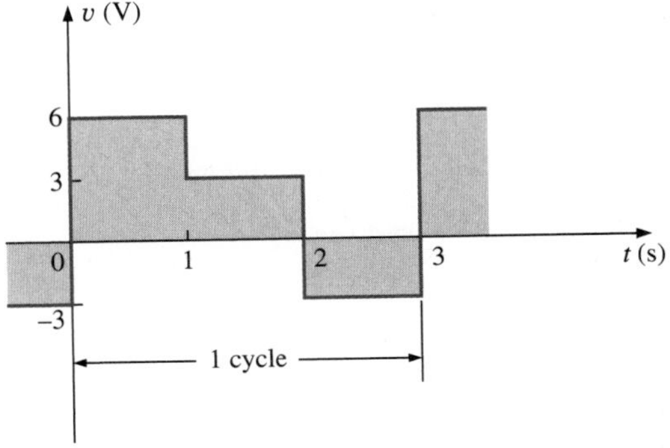

FIG. 13.89
Problem 37.

38. Find the average value of the periodic waveform in Fig. 13.90.

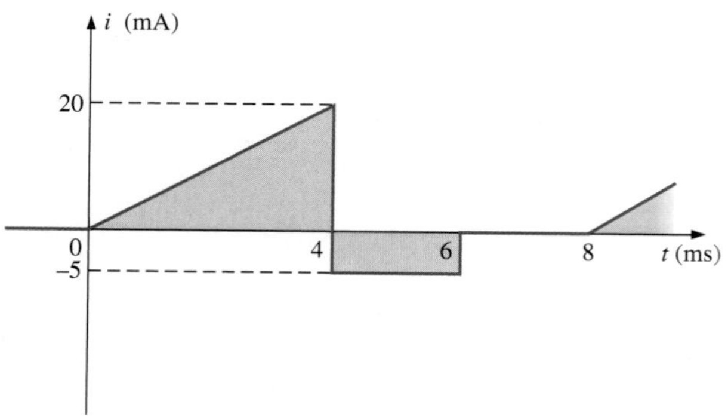

FIG. 13.90
Problem 38.

39. Find the average value of the periodic waveform in Fig. 13.91.

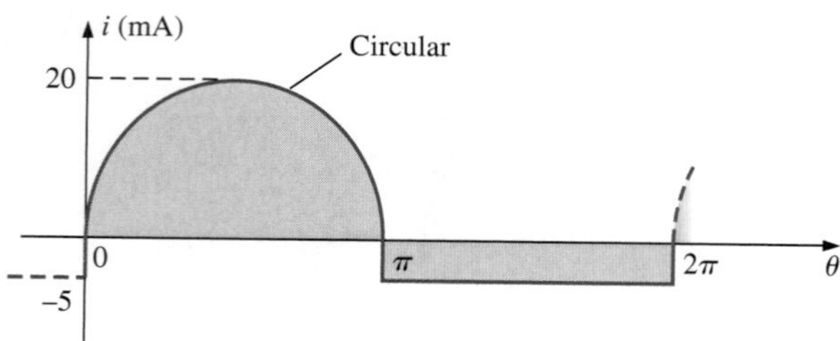

FIG. 13.91
Problem 39.

40. For the waveform in Fig. 13.92:
a. Determine the period.
b. Find the frequency.
c. Determine the average value.
d. Sketch the resulting oscilloscope display if the vertical channel is switched from DC to AC.

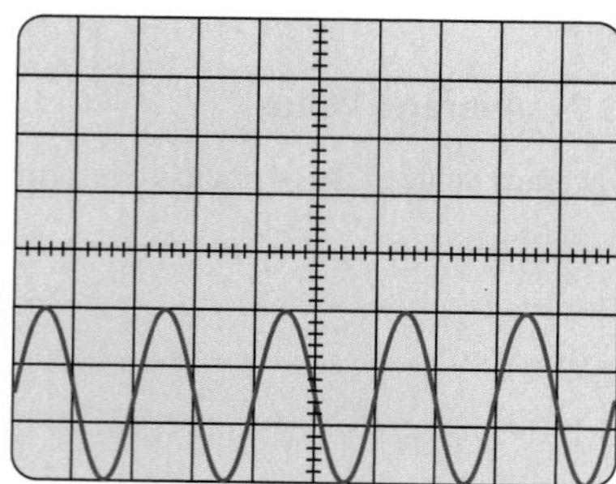

Vertical sensitivity = 10 mV/div.
Horizontal sensitivity = 0.2 ms/div.

FIG. 13.92
Problem 40.

*__41.__ For the waveform in Fig. 13.93:
a. Determine the period.
b. Find the frequency.
c. Determine the average value.
d. Sketch the resulting oscilloscope display if the vertical channel is switched from DC to AC.

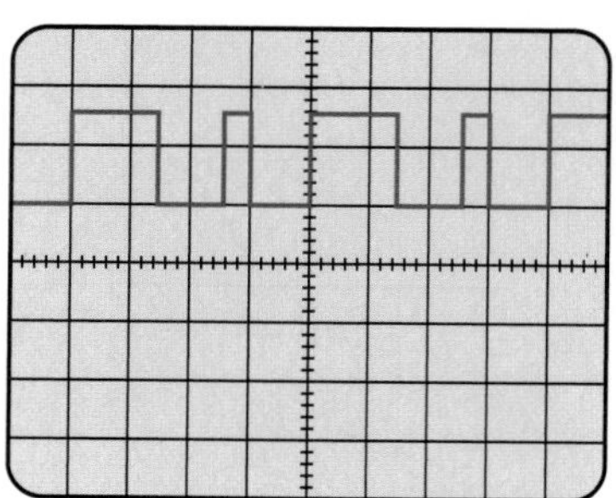

Vertical sensitivity = 10 mV/div.
Horizontal sensitivity = 10 μs/div.

FIG. 13.93
Problem 41.

SECTION 13.8 Effective (rms) Values

42. Find the rms values of the following sinusoidal waveforms:
a. $v = 140 \sin(377t + 60°)$
b. $i = 6 \times 10^{-3} \sin(2\pi\, 1000t)$
c. $v = 40 \times 10^{-6} (\sin(2\pi\, 5000t + 30°)$

43. Write the sinusoidal expressions for voltages and currents having the following rms values at a frequency of 60 Hz with zero phase shift:
a. 10 V
b. 50 mA
c. 2 kV

44. Find the rms value of the periodic waveform in Fig. 13.94 over one full cycle.

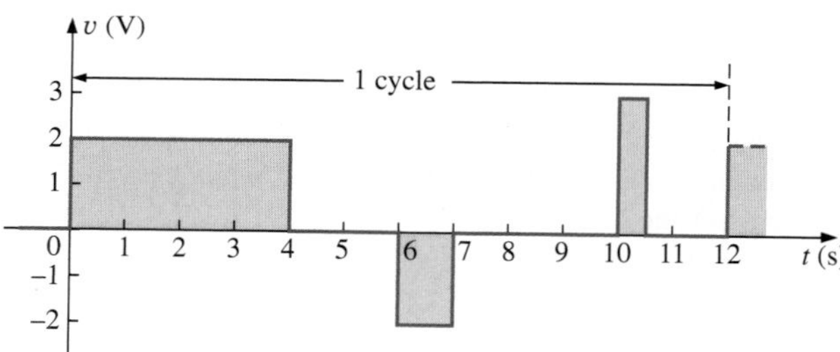

FIG. 13.94
Problem 44.

45. Find the rms value of the periodic waveform in Fig. 13.95 over one full cycle.

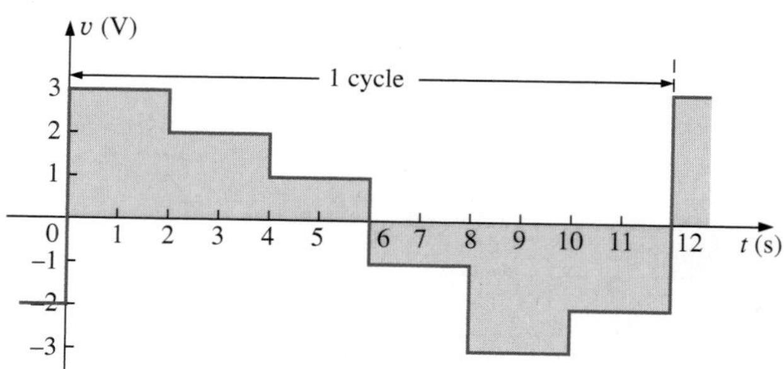

FIG. 13.95
Problem 45.

46. What are the average and rms values of the square wave in Fig. 13.96?

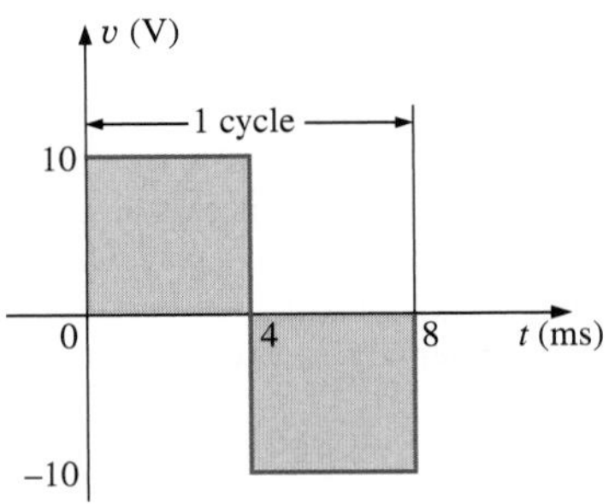

FIG. 13.96
Problem 46.

***47.** For each waveform in Fig. 13.97, determine the period, frequency, average value, and rms value.

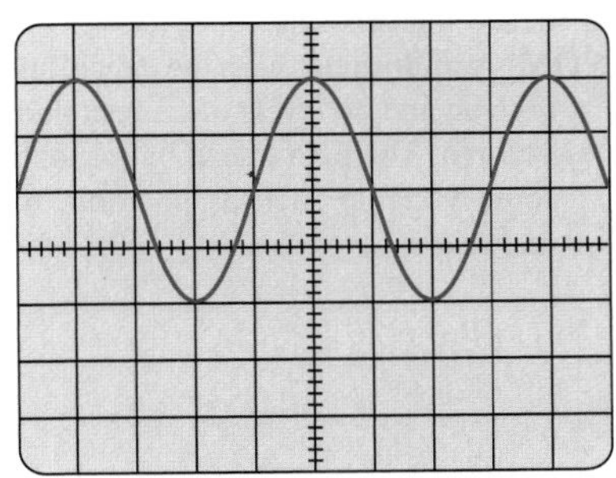
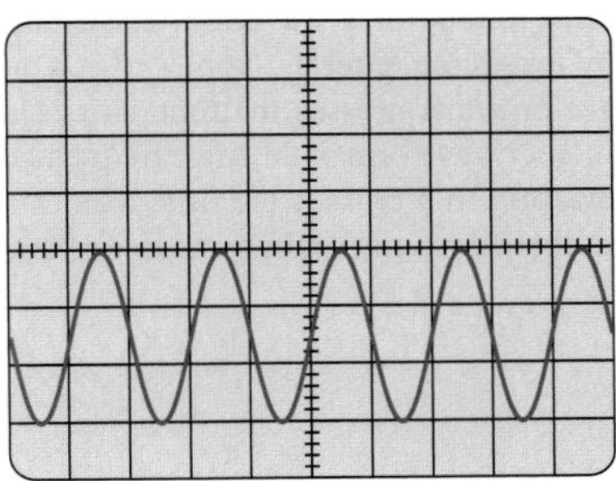

Vertical sensitivity = 20 mV/div.
Horizontal sensitivity = 10 μs/div.
(a)

Vertical sensitivity = 0.2 V/div.
Horizontal sensitivity = 50 μs/div.
(b)

FIG. 13.97
Problem 47.

SECTION 13.9 ac Meters and Instruments

48. Determine the reading of the meter for each situation in Fig. 13.98.

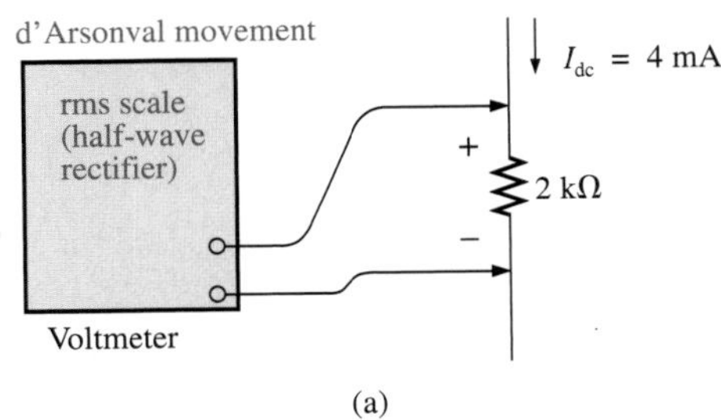

(a)

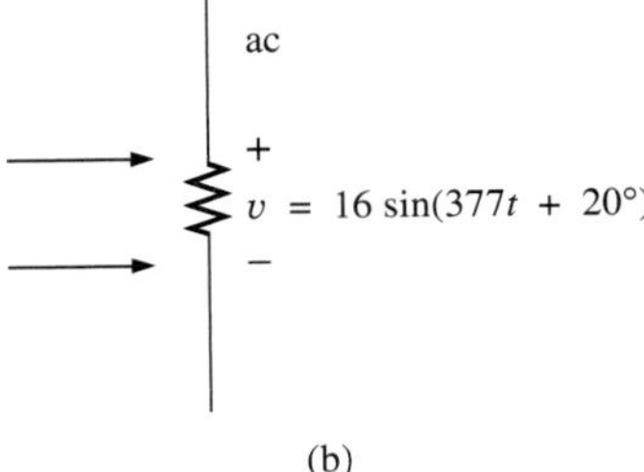

(b)

FIG. 13.98
Problem 48.

GLOSSARY

Alternating waveform A waveform that oscillates above and below a defined reference level.

Angular velocity The velocity with which a radius vector projecting a sinusoidal function rotates about its center.

Average value The level of a waveform defined by the condition that the area enclosed by the curve above this level is exactly equal to the area enclosed by the curve below this level.

Calibration factor A multiplying factor used to convert from one meter indication to another.

Clamp Meter® A clamp-type instrument that will permit noninvasive current measurements and that can be used as a conventional voltmeter or ohmmeter.

Cycle A portion of a waveform contained in one period of time.

Effective value The equivalent dc value of any alternating voltage or current.

Electrodynamometer meters Instruments that can measure both ac and dc quantities without a change in internal circuitry.

Frequency (*f*) The number of cycles of a periodic waveform that occur in 1 second.

Frequency counter An instrument that will provide a digital display of the frequency or period of a periodic time-varying signal.

Instantaneous value The magnitude of a waveform at any instant of time, denoted by lowercase letters.

Lagging waveform A waveform that crosses the time axis at a point in time later than another waveform of the same frequency.

Leading waveform A waveform that crosses the time axis at a point in time ahead of another waveform of the same frequency.

Oscilloscope An instrument that will display, through the use of a cathode-ray tube, the characteristics of a time-varying signal.

Peak amplitude The maximum value of a waveform as measured from its average, or mean, value, denoted by uppercase letters.

Peak-to-peak value The magnitude of the total swing of a signal from positive to negative peaks. The sum of the absolute values of the positive and negative peak values.

Peak value The maximum value of a waveform, denoted by uppercase letters.

Period (*T*) The time interval necessary for one cycle of a periodic waveform.

Periodic waveform A waveform that continually repeats itself after a defined time interval.

Phase relationship An indication of which of two waveforms leads or lags the other, and by how many degrees or radians.

Radian (rad) A unit of measure used to define a particular segment of a circle. One radian is approximately equal to 57.3°; 2π rad are equal to 360°.

Root-mean-square (rms) value The root-mean-square or effective value of a waveform.

Sinusoidal ac waveform An alternating waveform of unique characteristics that oscillates with equal amplitude above and below a given axis.

VOM A multimeter with the capability to measure resistance and both ac and dc levels of current and voltage.

Waveform The path traced by a quantity, plotted as a function of some variable such as position, time, degrees, temperature, and so on.

The Basic Elements and Phasors

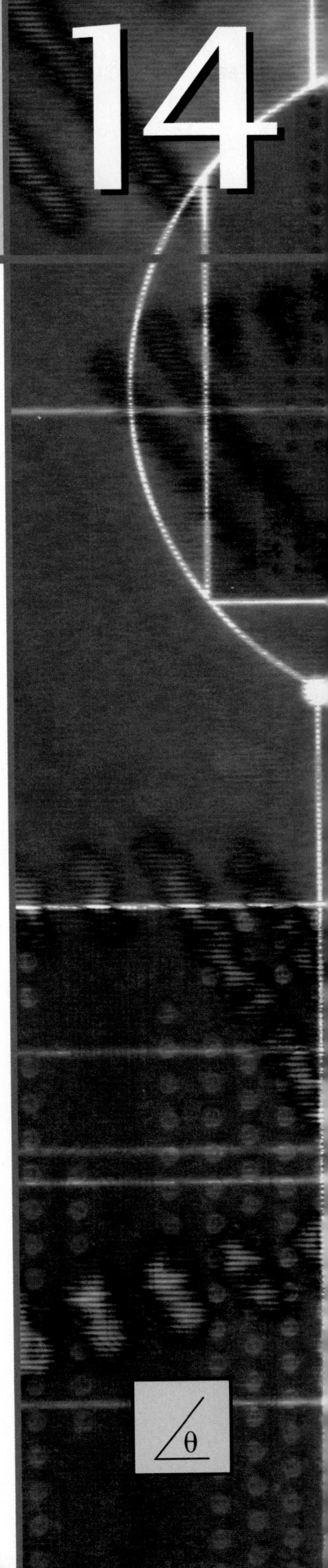

Objectives

- *Become familiar with the response of a resistor, inductor, and capacitor to the application of a sinusoidal voltage or current.*
- *Learn how to apply the phasor format to add and subtract sinusoidal waveforms.*
- *Understand how to calculate the real power to resistive elements and the reactive power to inductive and capacitive elements.*
- *Become aware of the differences between the frequency response of ideal and practical elements.*
- *Become proficient in the use of a calculator or Mathcad to work with complex numbers.*

14.1 INTRODUCTION

The response of the basic *R*, *L*, and *C* elements to a sinusoidal voltage and current are examined in this chapter, with special note of how frequency affects the "opposing" characteristic of each element. Phasor notation is then introduced to establish a method of analysis that permits a direct correspondence with a number of the methods, theorems, and concepts introduced in the dc chapters.

14.2 DERIVATIVE

To understand the response of the basic *R*, *L*, and *C* elements to a sinusoidal signal, you need to examine the concept of the **derivative** in some detail. You do not have to become proficient in the mathematical technique but simply understand the impact of a relationship defined by a derivative.

Recall from Section 10.10 that the derivative dx/dt is defined as the rate of change of x with respect to time. If x fails to change at a particular instant, $dx = 0$, and the derivative is zero. For the sinusoidal waveform, dx/dt is zero only at the positive and negative peaks ($\omega t = \pi/2$ and $\frac{3}{2}\pi$ in Fig. 14.1), since x fails to change at these instants of time. The derivative dx/dt is actually the slope of the graph at any instant of time.

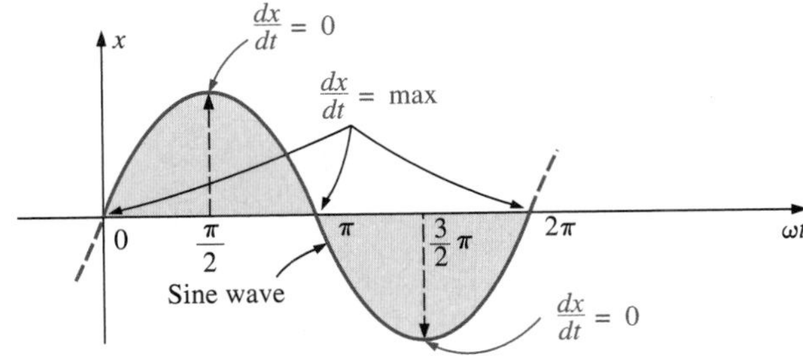

FIG. 14.1
Defining those points in a sinusoidal waveform that have maximum and minimum derivatives.

A close examination of the sinusoidal waveform will also indicate that the greatest change in x occurs at the instants $\omega t = 0, \pi$, and 2π. The derivative is therefore a maximum at these points. At 0 and 2π, x increases at its greatest rate, and the derivative is given a positive sign since x increases with time. At π, dx/dt decreases at the same rate as it increases at 0 and 2π, but the derivative is given a negative sign since x decreases with time. Since the rate of change at 0, π, and 2π is the same, the magnitude of the derivative at these points is the same also. For various values of ωt between these maxima and minima, the derivative will exist and have values from the minimum to the maximum inclusive. A plot of the derivative in Fig. 14.2 shows that

the derivative of a sine wave is a cosine wave.

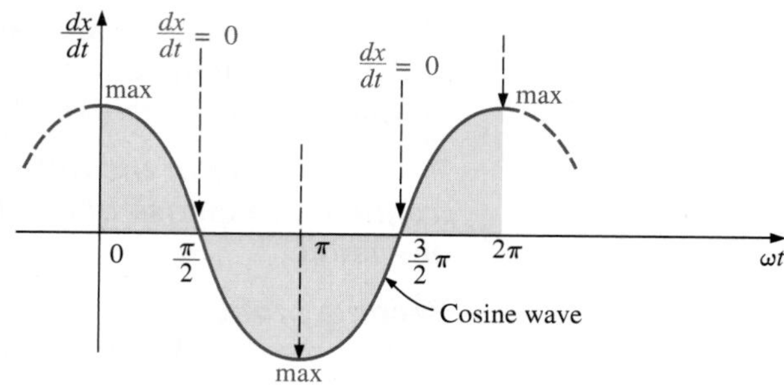

FIG. 14.2
Derivative of the sine wave of Fig. 14.1.

The peak value of the cosine wave is directly related to the frequency of the original waveform. The higher the frequency, the steeper the slope at the horizontal axis and the greater the value of dx/dt, as shown in Fig. 14.3 for two different frequencies.

Note in Fig. 14.3 that even though both waveforms (x_1 and x_2) have the same peak value, the sinusoidal function with the higher frequency produces the larger peak value for the derivative. In addition, note that

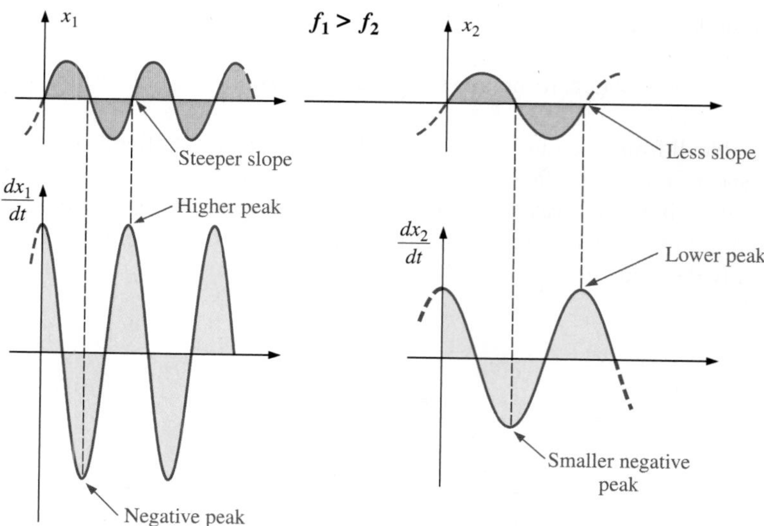

FIG. 14.3
Effect of frequency on the peak value of the derivative.

the derivative of a sine wave has the same period and frequency as the original sinusoidal waveform.

For the sinusoidal voltage

$$e(t) = E_m \sin(\omega t \pm \theta)$$

the derivative can be found directly by differentiation (calculus) to produce the following:

$$\begin{aligned}\frac{d}{dt}e(t) &= \omega E_m \cos(\omega t \pm \theta)\\ &= 2\pi f E_m \cos(\omega t \pm \theta)\end{aligned} \qquad \textbf{(14.1)}$$

The mechanics of the differentiation process are not discussed or investigated here; nor are they required to continue with the text. Note, however, that the peak value of the derivative, $2\pi f E_m$, is a function of the frequency of $e(t)$, and the derivative of a sine wave is a cosine wave.

14.3 RESPONSE OF BASIC *R, L,* AND *C* ELEMENTS TO A SINUSOIDAL VOLTAGE OR CURRENT

Now that we are familiar with the characteristics of the derivative of a sinusoidal function, we can investigate the response of the basic elements *R, L,* and *C* to a sinusoidal voltage or current.

Resistor

For power-line frequencies and frequencies up to a few hundred kilohertz, resistance is, for all practical purposes, unaffected by the frequency of the applied sinusoidal voltage or current. For this frequency region, the resistor *R* in Fig. 14.4 can be treated as a constant, and Ohm's law can be applied as follows. For $v = V_m \sin \omega t$,

$$i = \frac{v}{R} = \frac{V_m \sin \omega t}{R} = \frac{V_m}{R}\sin \omega t = I_m \sin \omega t$$

where

$$I_m = \frac{V_m}{R} \qquad \textbf{(14.2)}$$

In addition, for a given *i*,

$$v = iR = (I_m \sin \omega t)R = I_m R \sin \omega t = V_m \sin \omega t$$

where

$$V_m = I_m R \qquad \textbf{(14.3)}$$

A plot of *v* and *i* in Fig. 14.5 reveals that

for a purely resistive element, the voltage across and the current through the element are in phase, with their peak values related by Ohm's law.

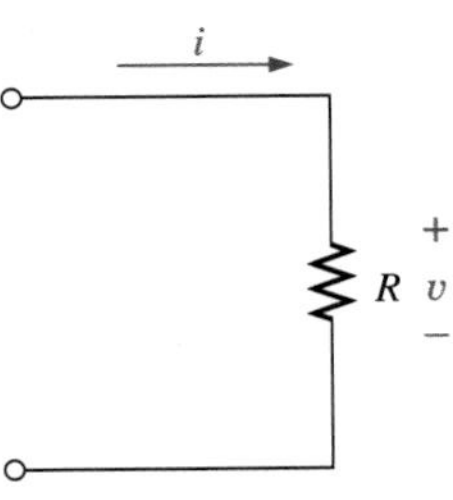

FIG. 14.4
Determining the sinusoidal response for a resistive element.

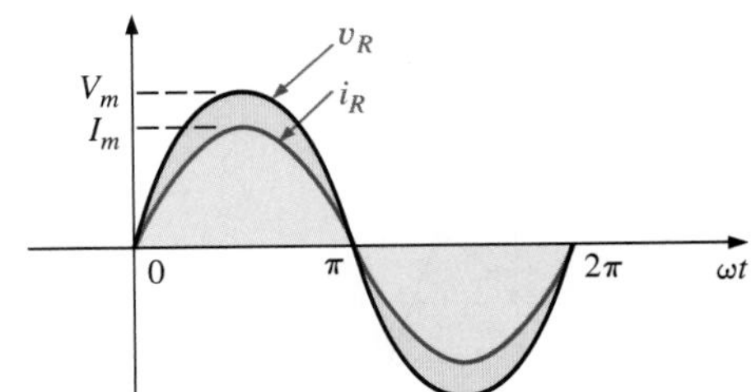

FIG. 14.5
The voltage and current of a resistive element are in phase.

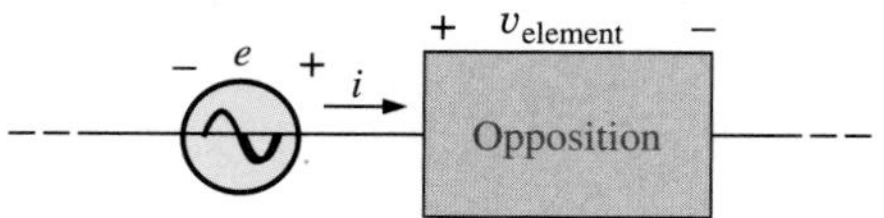

FIG. 14.6
Defining the opposition of an element to the flow of charge through the element.

Inductor

For the series configuration in Fig. 14.6, the voltage $v_{element}$ of the boxed-in element opposes the source e and thereby reduces the magnitude of the current i. The magnitude of the voltage across the element is determined by the opposition of the element to the flow of charge, or current i. For a resistive element, we have found that the opposition is its resistance and that $v_{element}$ and i are determined by $v_{element} = iR$.

We found in Chapter 11 that the voltage across an inductor is directly related to the rate of change of current through the coil. Consequently, the higher the frequency, the greater the rate of change of current through the coil, and the greater the magnitude of the voltage. In addition, we found in the same chapter that the inductance of a coil determines the rate of change of the flux linking a coil for a particular change in current through the coil. The higher the inductance, the greater the rate of change of the flux linkages, and the greater the resulting voltage across the coil.

The inductive voltage, therefore, is directly related to the frequency (or, more specifically, the angular velocity of the sinusoidal ac current through the coil) and the inductance of the coil. For increasing values of f and L in Fig. 14.7, the magnitude of v_L increases as described above.

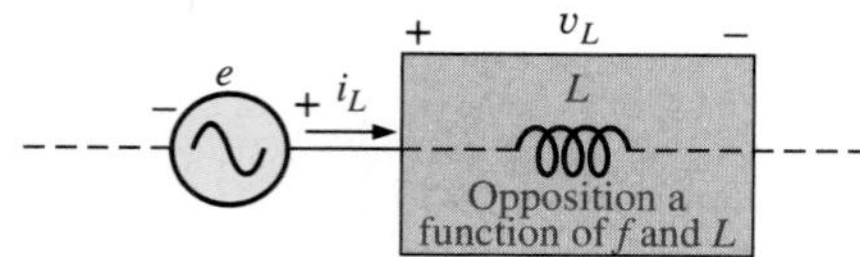

FIG. 14.7
Defining the parameters that determine the opposition of an inductive element to the flow of charge.

Using the similarities between Figs. 14.6 and 14.7, we find that increasing levels of v_L are directly related to increasing levels of opposition in Fig. 14.6. Since v_L increases with both ω (= $2\pi f$) and L, the opposition of an inductive element is as defined in Fig. 14.7.

We will now verify some of the preceding conclusions using a more mathematical approach and then define a few important quantities to be used in the sections and chapters to follow.

For the inductor in Fig. 14.8, we recall from Chapter 11 that

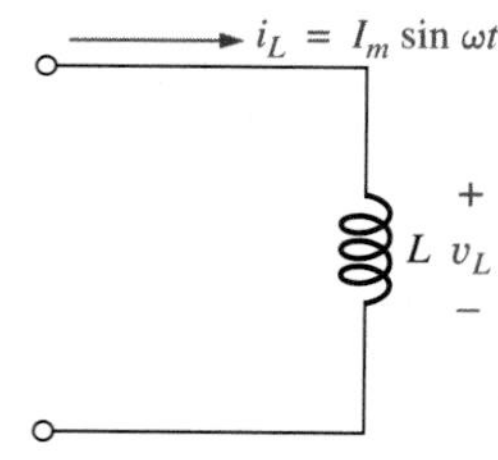

FIG. 14.8
Investigating the sinusoidal response of an inductive element.

$$v_L = L\frac{di_L}{dt}$$

and, applying differentiation,

$$\frac{di_L}{dt} = \frac{d}{dt}(I_m \sin \omega t) = \omega I_m \cos \omega t$$

Therefore, $$v_L = L\frac{di_L}{dt} = L(\omega I_m \cos \omega t) = \omega L I_m \cos \omega t$$

or $$v_L = V_m \sin(\omega t + 90°)$$

where $$V_m = \omega L I_m$$

Note that the peak value of v_L is directly related to ω (= $2\pi f$) and L as predicted in the discussion above.

A plot of v_L and i_L in Fig. 14.9 reveals that

for an inductor, v_L leads i_L by 90°, or i_L lags v_L by 90°.

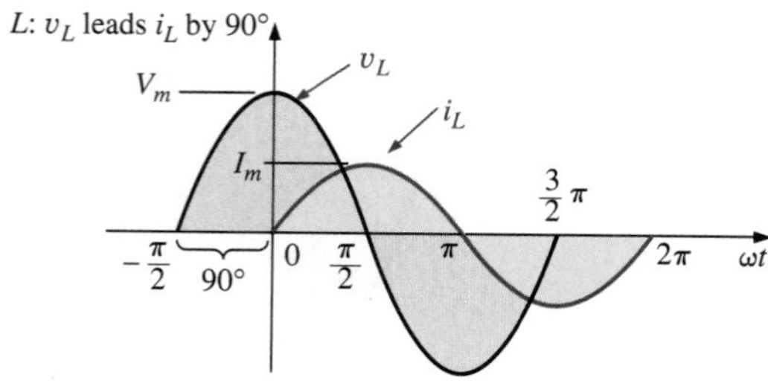

FIG. 14.9
For a pure inductor, the voltage across the coil leads the current through the coil by 90°.

If a phase angle is included in the sinusoidal expression for i_L, such as

$$i_L = I_m \sin(\omega t \pm \theta)$$

then $$v_L = \omega L I_m \sin(\omega t \pm \theta + 90°)$$

The opposition established by an inductor in a sinusoidal ac network can now be found by applying Eq. (4.1):

$$\text{Effect} = \frac{\text{cause}}{\text{opposition}}$$

which, for our purposes, can be written

$$\text{Opposition} = \frac{\text{cause}}{\text{effect}}$$

Substituting values, we have

$$\text{Opposition} = \frac{V_m}{I_m} = \frac{\omega L I_m}{I_m} = \omega L$$

revealing that the opposition established by an inductor in an ac sinusoidal network is directly related to the product of the angular velocity ($\omega = 2\pi f$) and the inductance, verifying our earlier conclusions.

The quantity ωL, called the **reactance** (from the word *reaction*) of an inductor, is symbolically represented by X_L and is measured in ohms; that is,

$$\boxed{X_L = \omega L} \qquad \text{(ohms, } \Omega) \qquad \textbf{(14.4)}$$

In an Ohm's law format, its magnitude can be determined from

$$\boxed{X_L = \frac{V_m}{I_m}} \qquad \text{(ohms, } \Omega) \qquad \textbf{(14.5)}$$

Inductive reactance is the opposition to the flow of current, which results in the continual interchange of energy between the source and the magnetic field of the inductor. In other words, inductive reactance, unlike resistance (which dissipates energy in the form of heat), does not dissipate electrical energy (ignoring the effects of the internal resistance of the inductor.)

Capacitor

Let us now return to the series configuration in Fig. 14.6 and insert the capacitor as the element of interest. For the capacitor, however, we will determine i for a particular voltage across the element. When this approach reaches its conclusion, we will know the relationship between the voltage and current and can determine the opposing voltage (v_{element}) for any sinusoidal current i.

Our investigation of the inductor revealed that the inductive voltage across a coil opposes the instantaneous change in current through the coil. For capacitive networks, the voltage across the capacitor is limited by the rate at which charge can be deposited on, or released by, the plates of the capacitor during the charging and discharging phases, respectively. In other words, an instantaneous change in voltage across a capacitor is opposed by the fact that there is an element of time required to deposit charge on (or release charge from) the plates of a capacitor, and $V = Q/C$.

Since capacitance is a measure of the rate at which a capacitor will store charge on its plates,

for a particular change in voltage across the capacitor, the greater the value of capacitance, the greater the resulting capacitive current.

In addition, the fundamental equation relating the voltage across a capacitor to the current of a capacitor [$i = C(dv/dt)$] indicates that

for a particular capacitance, the greater the rate of change of voltage across the capacitor, the greater the capacitive current.

Certainly, an increase in frequency corresponds to an increase in the rate of change of voltage across the capacitor and to an increase in the current of the capacitor.

The current of a capacitor is therefore directly related to the frequency (or, again more specifically, the angular velocity) and the capacitance of the capacitor. An increase in either quantity results in an increase in the current of the capacitor. For the basic configuration in Fig. 14.10, however, we are interested in determining the opposition of the capacitor as related to the resistance of a resistor and ωL for the inductor. Since an increase in current corresponds to a decrease in opposition, and i_C is proportional to ω and C, the opposition of a capacitor is inversely related to ω $(= 2\pi f)$ and C.

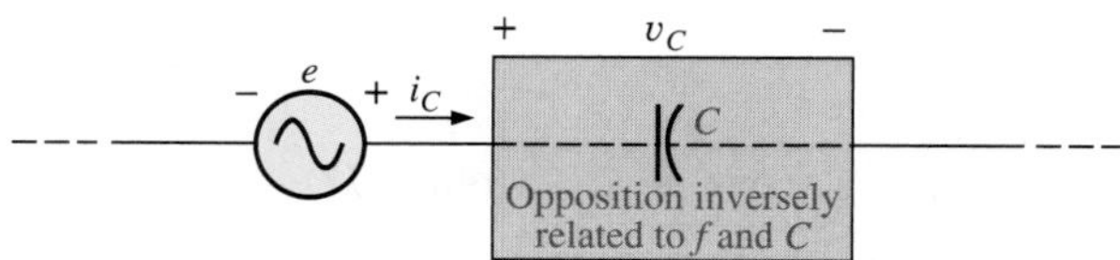

FIG. 14.10
Defining the parameters that determine the opposition of a capacitive element to the flow of charge.

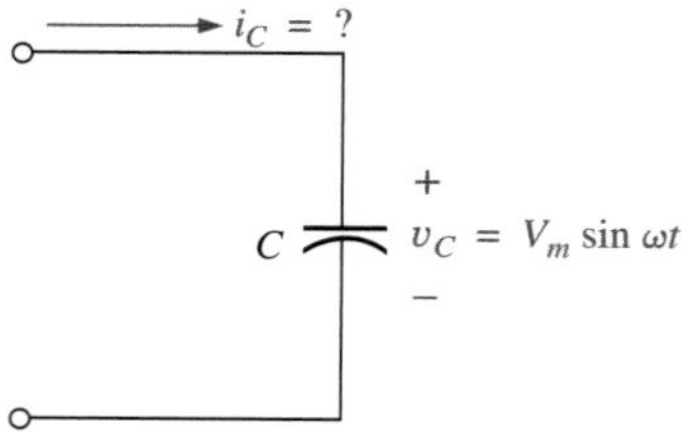

FIG. 14.11
Investigating the sinusoidal response of a capacitive element.

We will now verify, as we did for the inductor, some of the above conclusions using a more mathematical approach.

For the capacitor of Fig. 14.11, we recall from Chapter 11 that

$$i_C = C\frac{dv_C}{dt}$$

and, applying differentiation,

$$\frac{dv_C}{dt} = \frac{d}{dt}(V_m \sin \omega t) = \omega V_m \cos \omega t$$

Therefore,

$$i_C = C\frac{dv_C}{dt} = C(\omega V_m \cos \omega t) = \omega C V_m \cos \omega t$$

or

$$i_C = I_m \sin(\omega t + 90°)$$

where

$$I_m = \omega C V_m$$

Note that the peak value of i_C is directly related to ω $(= 2\pi f)$ and C, as predicted in the discussion above.

A plot of v_C and i_C in Fig. 14.12 reveals that

for a capacitor, i_C leads v_C by 90°, or v_C lags i_C by 90°.*

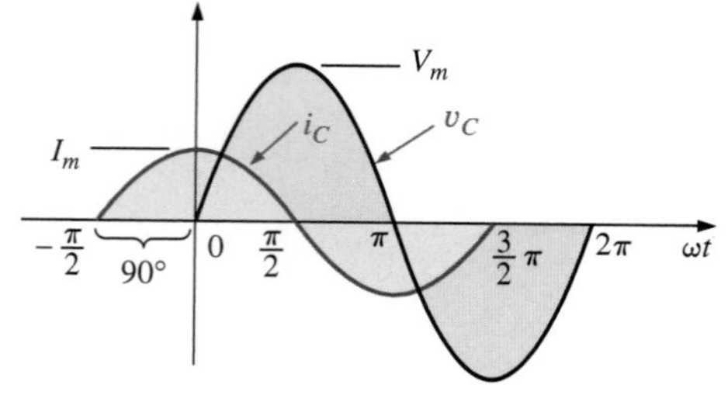

FIG. 14.12
The current of a purely capacitive element leads the voltage across the element by 90°.

If a phase angle is included in the sinusoidal expression for v_C, such as

$$v_C = V_m \sin(\omega t \pm \theta)$$

then

$$i_C = \omega C V_m \sin(\omega t \pm \theta + 90°)$$

---

*A mnemonic phrase sometimes used to remember the phase relationship between the voltage and current of a coil and capacitor is "*ELI* the *ICE* man." Note that the L (inductor) has the E before the I (e leads i by 90°), and the C (capacitor) has the I before the E (i leads e by 90°).

Applying

$$\text{Opposition} = \frac{\text{cause}}{\text{effect}}$$

and substituting values, we obtain

$$\text{Opposition} = \frac{V_m}{I_m} = \frac{V_m}{\omega C V_m} = \frac{1}{\omega C}$$

which agrees with the results obtained above.

The quantity $1/\omega C$, called the **reactance** of a capacitor, is symbolically represented by X_C and is measured in ohms; that is,

$$X_C = \frac{1}{\omega C} \qquad \text{(ohms, } \Omega\text{)} \qquad \textbf{(14.6)}$$

In an Ohm's law format, its magnitude can be determined from

$$X_C = \frac{V_m}{I_m} \qquad \text{(ohms, } \Omega\text{)} \qquad \textbf{(14.7)}$$

Capacitive reactance is the opposition to the flow of charge, which results in the continual interchange of energy between the source and the electric field of the capacitor. Like the inductor, the capacitor does *not* dissipate energy in any form (ignoring the effects of the leakage resistance).

In the circuits just considered, the current was given in the inductive circuit, and the voltage in the capacitive circuit. This was done to avoid the use of integration in finding the unknown quantities. In the inductive circuit,

$$v_L = L\frac{di_L}{dt}$$

but

$$i_L = \frac{1}{L}\int v_L\,dt \qquad \textbf{(14.8)}$$

In the capacitive circuit,

$$i_C = C\frac{dv_C}{dt}$$

but

$$v_C = \frac{1}{C}\int i_C\,dt \qquad \textbf{(14.9)}$$

Shortly, we shall consider a method of analyzing ac circuits that will permit us to solve for an unknown quantity with sinusoidal input without having to use direct integration or differentiation.

It is possible to determine whether a network with one or more elements is predominantly capacitive or inductive by noting the phase relationship between the input voltage and current.

If the source current leads the applied voltage, the network is predominantly capacitive, and if the applied voltage leads the source current, it is predominantly inductive.

Since we now have an equation for the reactance of an inductor or capacitor, we do not need to use derivatives or integration in the examples to be considered. Simply applying Ohm's law, $I_m = E_m/X_L$ (or X_C), and keeping in mind the phase relationship between the voltage and current for each element, will be sufficient to complete the examples.

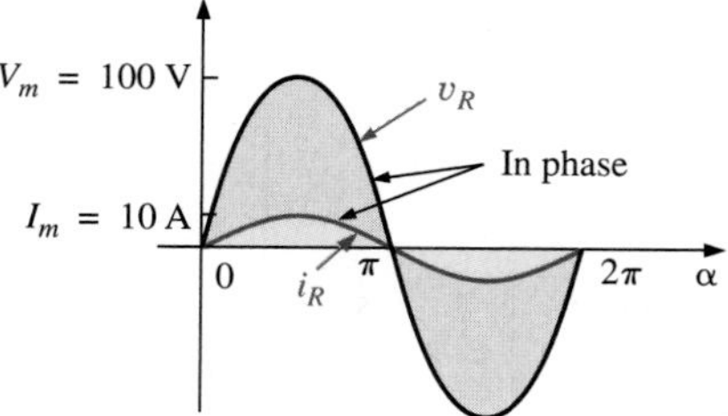

FIG. 14.13
Example 14.1(a).

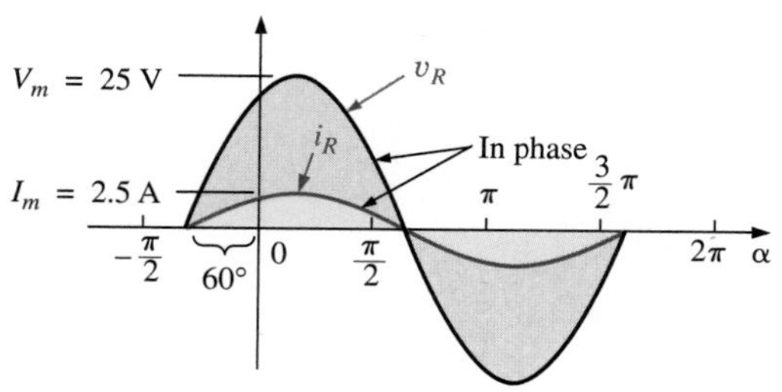

FIG. 14.14
Example 14.1(b).

EXAMPLE 14.1 The voltage across a resistor is indicated. Find the sinusoidal expression for the current if the resistor is 10 Ω. Sketch the curves for v and i.

a. $v = 100 \sin 377t$
b. $v = 25 \sin(377t + 60°)$

Solutions:

a. Eq. (14.2): $I_m = \dfrac{V_m}{R} = \dfrac{100\text{ V}}{10\ \Omega} = 10\text{ A}$

(v and i are in phase), resulting in

$$i = \mathbf{10 \sin 377}t$$

The curves are sketched in Fig. 14.13.

b. Eq. (14.2): $I_m = \dfrac{V_m}{R} = \dfrac{25\text{ V}}{10\ \Omega} = 2.5\text{ A}$

(v and i are in phase), resulting in

$$i = \mathbf{2.5 \sin(377}t\mathbf{ + 60°)}$$

The curves are sketched in Fig. 14.14.

EXAMPLE 14.2 The current through a 5 Ω resistor is given. Find the sinusoidal expression for the voltage across the resistor for $i = 40 \sin(377t + 30°)$.

Solution: Eq. (14.3): $V_m = I_m R = (40\text{ A})(5\ \Omega) = 200\text{ V}$

(v and i are in phase), resulting in

$$v = \mathbf{200 \sin(377}t\mathbf{ + 30°)}$$

EXAMPLE 14.3 The current through a 0.1 H coil is provided. Find the sinusoidal expression for the voltage across the coil. Sketch the v and i curves.

a. $i = 10 \sin 377t$
b. $i = 7 \sin(377t - 70°)$

Solutions:

a. Eq. (14.4): $X_L = \omega L = (377\text{ rad/s})(0.1\text{ H}) = 37.7\ \Omega$
Eq. (14.5): $V_m = I_m X_L = (10\text{ A})(37.7\ \Omega) = 377\text{ V}$

and we know that for a coil v leads i by 90°. Therefore,

$$v = \mathbf{377 \sin(377}t\mathbf{ + 90°)}$$

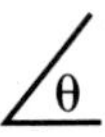

The curves are sketched in Fig. 14.15.

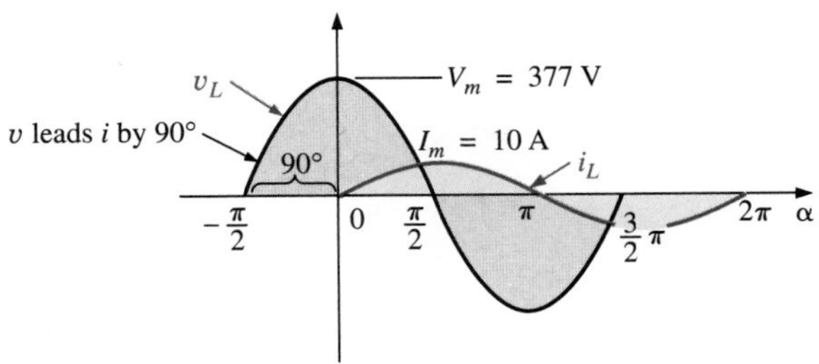

FIG. 14.15
Example 14.3(a).

b. X_L remains at 37.7 Ω.

$V_m = I_m X_L = (7\text{ A})(37.7\ \Omega) = 263.9\text{ V}$

and we know that for a coil v leads i by 90°. Therefore,

$$v = 263.9 \sin(377t - 70° + 90°)$$

and

$$\mathbf{v = 263.9 \sin(377t + 20°)}$$

The curves are sketched in Fig. 14.16.

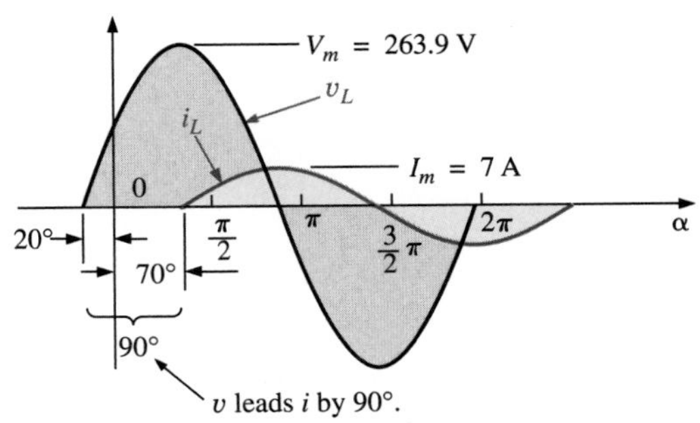

FIG. 14.16
Example 14.3(b).

EXAMPLE 14.4 The voltage across a 0.5 H coil is provided below. What is the sinusoidal expression for the current?

$$v = 100 \sin 20t$$

Solution:

$$X_L = \omega L = (20\text{ rad/s})(0.5\text{ H}) = 10\ \Omega$$

$$I_m = \frac{V_m}{X_L} = \frac{100\text{ V}}{10\ \Omega} = 10\text{ A}$$

and we know the i lags v by 90°. Therefore,

$$\mathbf{i = 10 \sin(20t - 90°)}$$

EXAMPLE 14.5 The voltage across a 1 μF capacitor is provided below. What is the sinusoidal expression for the current? Sketch the v and i curves.

$$v = 30 \sin 400t$$

Solution:

$$\text{Eq. (14.6):}\quad X_C = \frac{1}{\omega C} = \frac{1}{(400\text{ rad/s})(1 \times 10^{-6}\text{ F})} = \frac{10^6\ \Omega}{400} = 2500\ \Omega$$

$$\text{Eq. (14.7):}\quad I_m = \frac{V_m}{X_C} = \frac{30\text{ V}}{2500\ \Omega} = 0.0120\text{ A} = 12\text{ mA}$$

and we know that for a capacitor i leads v by 90°. Therefore,

$$\mathbf{i = 12 \times 10^{-3} \sin(400t + 90°)}$$

The curves are sketched in Fig. 14.17.

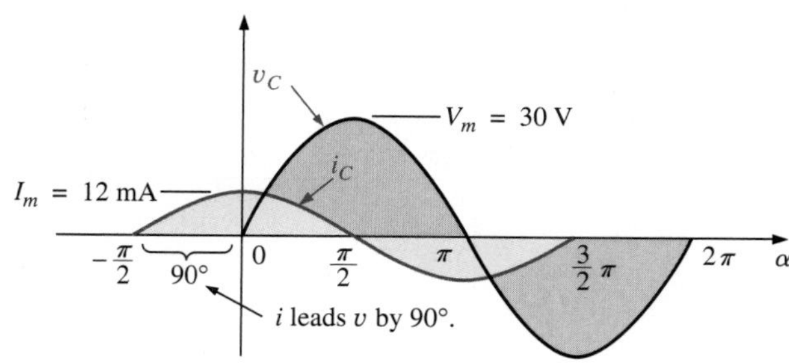

FIG. 14.17
Example 14.5.

EXAMPLE 14.6 The current through a 100 μF capacitor is given. Find the sinusoidal expression for the voltage across the capacitor.

$$i = 40 \sin(500t + 60°)$$

Solution:

$$X_C = \frac{1}{\omega C} = \frac{1}{(500 \text{ rad/s})(100 \times 10^{-6} \text{ F})} = \frac{10^6 \ \Omega}{5 \times 10^4} = \frac{10^2 \ \Omega}{5} = 20 \ \Omega$$

$$V_M = I_M X_C = (40 \text{ A})(20 \ \Omega) = 800 \text{ V}$$

and we know that for a capacitor, v lags i by 90°. Therefore,

$$v = 800 \sin(500t + 60° - 90°)$$

and $$v = \mathbf{800 \sin(500}\boldsymbol{t} \mathbf{- 30°)}$$

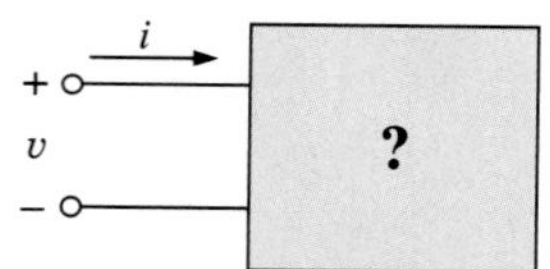

FIG. 14.18
Example 14.7.

EXAMPLE 14.7 For the following pairs of voltages and currents, determine whether the element involved is a capacitor, an inductor, or a resistor. Determine the value of C, L, or R if sufficient data are provided (Fig. 14.18):

a. $v = 100 \sin(\omega t + 40°)$
   $i = 20 \sin(\omega t + 40°)$
b. $v = 1000 \sin(377t + 10°)$
   $i = 5 \sin(377t - 80°)$
c. $v = 500 \sin(157t + 30°)$
   $i = 1 \sin(157t + 120°)$
d. $v = 50 \cos(\omega t + 20°)$
   $i = 5 \sin(\omega t + 110°)$

Solutions:

a. Since v and i are *in phase,* the element is a *resistor,* and

$$R = \frac{V_m}{I_m} = \frac{100 \text{ V}}{20 \text{ A}} = \mathbf{5 \ \Omega}$$

b. Since v *leads* i by 90°, the element is an *inductor,* and

$$X_L = \frac{V_m}{I_m} = \frac{1000 \text{ V}}{5 \text{ A}} = 200 \ \Omega$$

so that $X_L = \omega L = 200 \ \Omega$ or

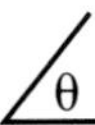

$$L = \frac{200\ \Omega}{\omega} = \frac{200\ \Omega}{377\ \text{rad/s}} = \mathbf{0.53\ H}$$

c. Since i *leads* v by 90°, the element is a *capacitor,* and

$$X_C = \frac{V_m}{I_m} = \frac{500\ \text{V}}{1\ \text{A}} = 500\ \Omega$$

so that $X_C = \dfrac{1}{\omega C} = 500\ \Omega$ or

$$C = \frac{1}{\omega 500\ \Omega} = \frac{1}{(157\ \text{rad/s})(500\ \Omega)} = \mathbf{12.74\ \mu F}$$

d. $v = 50\cos(\omega t + 20°) = 50\sin(\omega t + 20° + 90°)$
$= 50\sin(\omega t + 110°)$

Since v and i are *in phase,* the element is a *resistor,* and

$$R = \frac{V_m}{I_m} = \frac{50\ \text{V}}{5\ \text{A}} = \mathbf{10\ \Omega}$$

14.4 FREQUENCY RESPONSE OF THE BASIC ELEMENTS

Thus far, each description has been for a set frequency, resulting in a fixed level of impedance for each of the basic elements. We must now investigate how a change in frequency affects the impedance level of the basic elements. It is an important consideration because most signals other than those provided by a power plant contain a variety of frequency levels. The last section made it quite clear that the reactance of an inductor or a capacitor is sensitive to the applied frequency. However, the question is, How will these reactance levels change if we steadily increase the frequency from a very low level to a much higher level?

Although we would like to think of every element as ideal, it is important to realize that every commercial element available today *will not respond in an ideal fashion for the full range of possible frequencies.* That is, each element is such that for a particular range of frequencies, it performs in an essentially ideal manner. However, there is always a range of frequencies in which the performance varies from the ideal. Fortunately, the designer is aware of these limitations and will take them into account in the design.

The discussion begins with a look at the response of the *ideal elements*—a response that will be assumed for the remaining chapters of this text and one that can be assumed for any initial investigation of a network. This discussion is followed by a look at the factors that cause an element to deviate from an ideal response as frequency levels become too low or high.

Ideal Response

Resistor *R* For an ideal resistor, you can assume that *frequency will have absolutely no effect on the impedance level,* as shown by the response in Fig. 14.19. Note that at 5 kHz or 20 kHz, the resistance of the resistor remains at 22 Ω; there is no change whatsoever. For the rest of

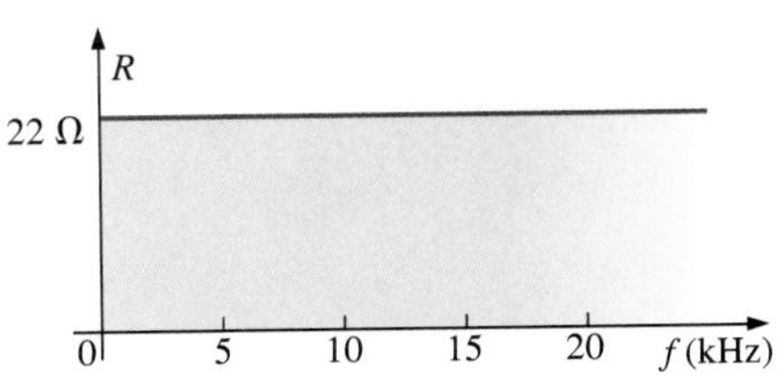

FIG. 14.19
R versus f for the range of interest.

the analyses in this text, the resistance level remains as the nameplate value, no matter what frequency is applied.

Inductor *L* For the ideal inductor, the equation for the reactance can be written as follows to isolate the frequency term in the equation. The result is a constant times the frequency variable that changes as we move down the horizontal axis of a plot:

$$X_L = \omega L = 2\pi f L = (2\pi L)f = kf \quad \text{with } k = 2\pi L$$

The resulting equation can be compared directly with the equation for a straight line:

$$y = mx + b = kf + 0 = kf$$

where $b = 0$ and the slope is k or $2\pi L$. X_L is the y variable, and f is the x variable, as shown in Fig. 14.20. Since the inductance determines the slope of the curve, the higher the inductance, the steeper the straight-line plot as shown in Fig. 14.20 for two levels of inductance.

In particular, note that at $f = 0$ Hz, the reactance of each plot is zero ohms as determined by substituting $f = 0$ Hz into the basic equation for the reactance of an inductor:

$$X_L = 2\pi f L = 2\pi(0 \text{ Hz})L = 0\ \Omega$$

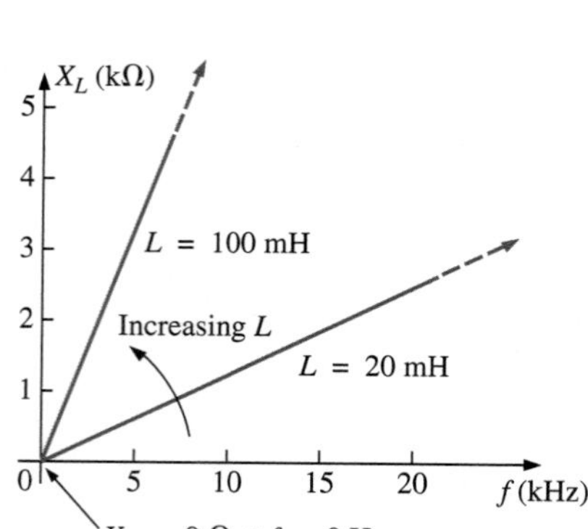

FIG. 14.20
X_L versus frequency.

Since a reactance of zero ohms corresponds with the characteristics of a short circuit, we can conclude that

at a frequency of 0 Hz, an inductor takes on the characteristics of a short circuit, as shown in Fig. 14.21.

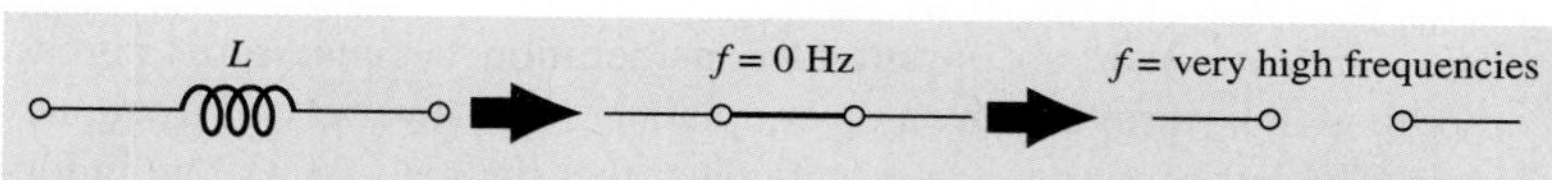

FIG. 14.21
Effect of low and high frequencies on the circuit model of an inductor.

As shown in Fig. 14.21, as the frequency increases, the reactance increases, until it reaches an extremely high level at very high frequencies. The result is that

at very high frequencies, the characteristics of an inductor approach those of an open circuit, as shown in Fig. 14.21.

The inductor, therefore, is capable of handling impedance levels that cover the entire range, from zero ohms to infinite ohms, changing at a *steady rate* determined by the inductance level. The higher the inductance, the faster it approaches the open-circuit equivalent.

Capacitor *C* For the capacitor, the equation for the reactance

$$X_C = \frac{1}{2\pi f C}$$

can be written as

$$X_C f = \frac{1}{2\pi C} = k \qquad \text{(a constant)}$$

which matches the basic format for a hyberbola:

$$yx = k$$

where X_C is the y variable, f the x variable, and k a constant equal to $1/(2\pi C)$.

Hyberbolas have the shape appearing in Fig. 14.22 for two levels of capacitance. Note that the higher the capacitance, the closer the curve approaches the vertical and horizontal axes at low and high frequencies.

At or near 0 Hz, the reactance of any capacitor is extremely high, as determined by the basic equation for capacitance:

$$X_C = \frac{1}{2\pi fC} = \frac{1}{2\pi(0\text{ Hz})C} \Rightarrow \infty\ \Omega$$

The result is that

at or near 0 Hz, the characteristics of a capacitor approach those of an open circuit, as shown in Fig. 14.23.

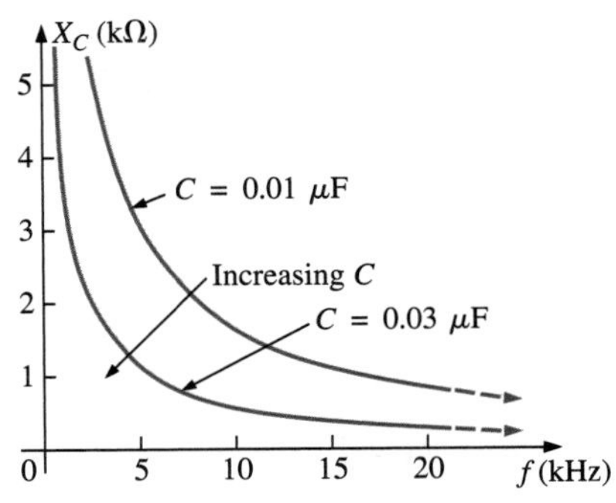

FIG. 14.22
X_C versus frequency.

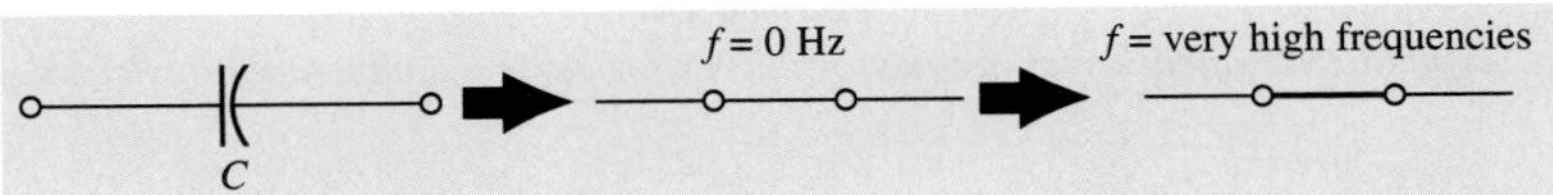

FIG. 14.23
Effect of low and high frequencies on the circuit model of a capacitor.

As the frequency increases, the reactance approaches a value of zero ohms. The result is that

at very high frequencies, a capacitor takes on the characteristics of a short circuit, as shown in Fig. 14.23.

It is important to note in Fig. 14.22 that the reactance drops very rapidly as the frequency increases. It is not a gradual drop as encountered for the rise in inductive reactance. In addition, the reactance sits at a fairly low level for a broad range of frequencies. In general, therefore, recognize that for capacitive elements, the change in reactance level can be dramatic with a relatively small change in frequency level.

Finally, recognize the following:

As frequency increases, the reactance of an inductive element increases while that of a capacitor decreases, with one approaching an open-circuit equivalent as the other approaches a short-circuit equivalent.

Practical Response

Resistor *R* In the manufacturing process, every resistive element inherits some stray capacitance levels and lead inductances. For most applications, the levels are so low that their effects can be ignored. However, as the frequency extends beyond a few megahertz, it may be necessary to be aware of their effects. For instance, a number of carbon composition resistors have the frequency response appearing in Fig. 14.24. The 100 Ω resistor is essentially stable up to about 300 MHz, whereas the 100 kΩ resistor starts to drop off at about 15 MHz. In general, therefore, this type of carbon composition resistor has the ideal characteristics of Fig. 14.19 for frequencies up to about 15 MHz. For frequencies of 100 Hz, 1 kHz, 150 kHz, and so on, the resistor can be considered ideal.

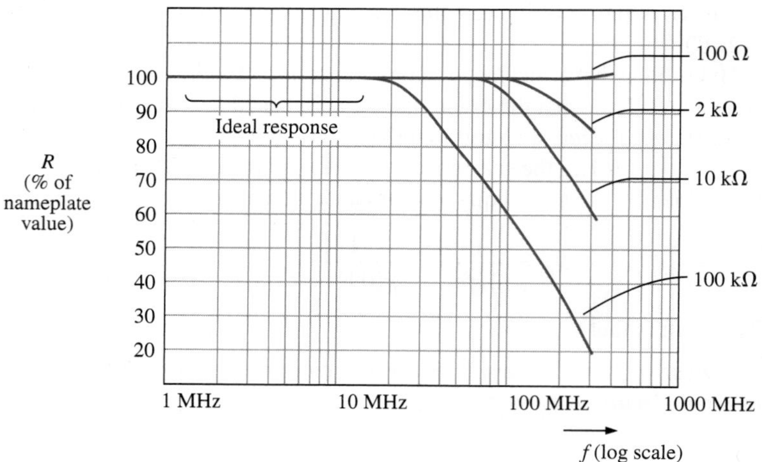

FIG. 14.24
Typical resistance-versus-frequency curves for carbon composition resistors.

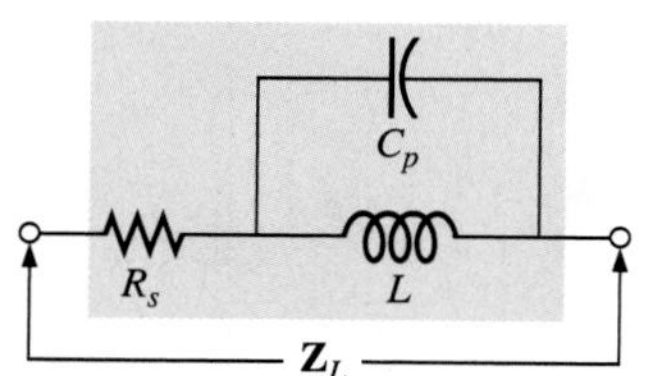

FIG. 14.25
Practical equivalent for an inductor.

Inductor *L* In reality, inductance can be affected by frequency, temperature, and current. A true equivalent for an inductor appears in Fig. 14.25. The series resistance R_s represents the copper losses (resistance of the many turns of thin copper wire); the eddy current losses (losses due to small circular currents in the core when an ac voltage is applied); and the hysteresis losses (losses due to core losses created by the rapidly reversing field in the core). The capacitance C_p is the stray capacitance that exists between the windings of the inductor.

For most inductors, the construction is usually such that the larger the inductance, the lower the frequency at which the parasitic elements become important. That is, for inductors in the millihenry range (which is very typical), frequencies approaching 100 kHz can have an effect on the ideal characteristics of the element. For inductors in the microhenry range, a frequency of 1 MHz may introduce negative effects. This is not to suggest that the inductors lose their effect at these frequencies but rather that they can no longer be considered ideal (purely inductive elements).

Fig. 14.26 is a plot of the magnitude of the impedance Z_L of Fig. 14.25 versus frequency. Note that up to about 2 MHz, the impedance increases almost linearly with frequency, clearly suggesting that the 100 μH inductor is essentially ideal. However, above 2 MHz, all the factors contributing to R_s start to increase, while the reactance due to the capacitive element C_p

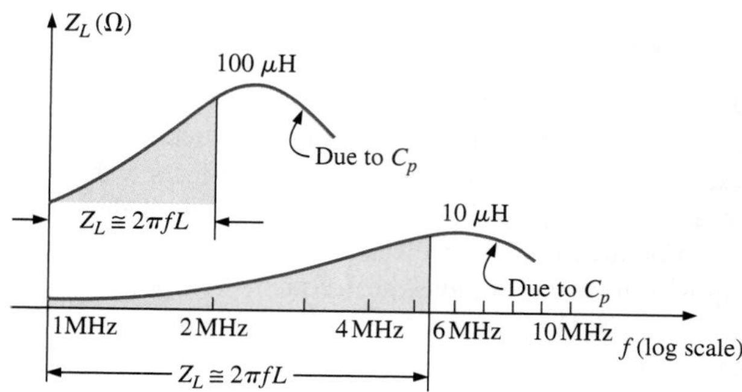

FIG. 14.26
Z_L versus frequency for the practical inductor equivalent of Fig. 14.25.

is more pronounced. The dropping level of capacitive reactance begins to have a shorting effect across the windings of the inductor and reduces the overall inductive effect. Eventually, if the frequency continues to increase, the capacitive effects overcome the inductive effects, and the element actually begins to behave in a capacitive fashion. Note the similarities of this region with the curves in Fig. 14.22. Also, note that decreasing levels of inductance (available with fewer turns and therefore lower levels of C_p) do not demonstrate the degrading effect until higher frequencies are applied.

In general, therefore, the frequency of application for a coil becomes important at increasing frequencies. Inductors lose their ideal characteristics and, in fact, begin to act as capacitive elements with increasing losses at very high frequencies.

Capacitor *C* The capacitor, like the inductor, is not ideal at higher frequencies. In fact, a transition point can be defined where the characteristics of the capacitor will actually be inductive. The complete equivalent model for a capacitor is provided in Fig. 14.27. The resistance R_s, defined by the resistivity of the dielectric (typically $10^{12}\ \Omega \cdot \text{m}$ or better) and the case resistance, determines the level of leakage current to expect during the discharge cycle. In other words, a charged capacitor can discharge both through the case and through the dielectric at a rate determined by the resistance of each path. Depending on the capacitor, the discharge time can extend from a few seconds for some electrolytic capacitors to hours (paper) or perhaps days (polystyrene). Inversely, therefore, electrolytics obviously have much lower levels of R_s than paper or polystyrene.

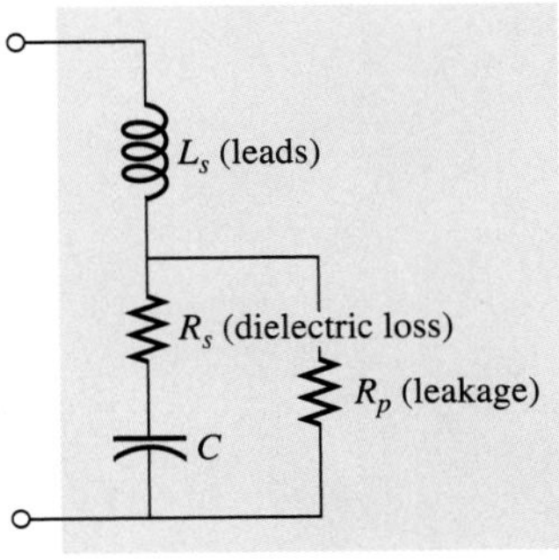

FIG. 14.27
Practical equivalent for a capacitor.

The resistance R_p reflects the energy lost as the atoms continually realign themselves in the dielectric due to the applied alternating ac voltage. Molecular friction is present due to the motion of the atoms as they respond to the alternating applied electric field. Interestingly enough, however, the relative permittivity decreases with increasing frequencies but eventually takes a complete turnaround and begins to increase at very high frequencies. The inductance L_s includes the inductance of the capacitor leads and any inductive effects introduced by the design of the capacitor. Be aware that the inductance of the leads is about 0.05 μH per centimeter or 0.2 μH for a capacitor with two 2 cm leads—a level that can be important at high frequencies. As for the inductor, the capacitor behaves quite ideally for the low- and mid-frequency range, as shown by the plot in Fig. 14.28 for a 0.01 μF metalized film capacitor with 2 cm

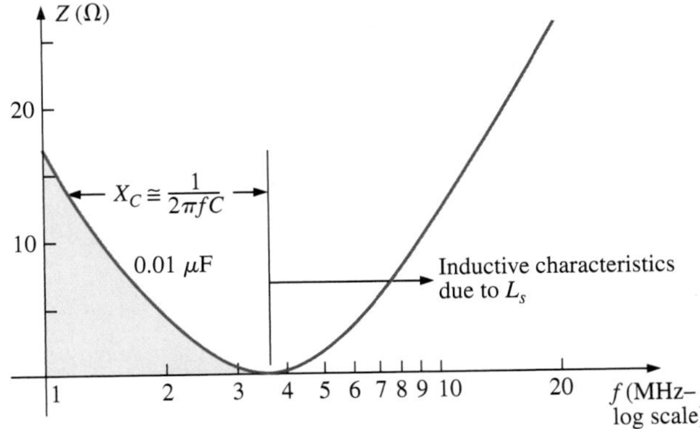

FIG. 14.28
Impedance characteristics of a 0.01 μF metalized film capacitor versus frequency.

leads. As the frequency increases, however, and the reactance X_s becomes larger, a frequency is eventually reached where the reactance of the coil equals that of the capacitor (a resonant condition to be described in Chapter 20). Any additional increase in frequency results in X_s being greater than X_C, and the element behaves like an inductor.

In general, therefore, the frequency of application is important for capacitive elements because when the frequency increases to a certain level, the element takes on inductive characteristics. Also, the frequency of application defines the type of capacitor (or inductor) that is applied: Electrolytics are limited to frequencies to perhaps 10 kHz, while ceramic or mica can handle frequencies higher than 10 MHz.

The expected temperature range of operation can have an important impact on the type of capacitor chosen for a particular application. Electrolytics, tantalum, and some high-*k* ceramic capacitors are very sensitive to colder temperatures. In fact, most electrolytics lose 20% of their room-temperature capacitance at 0°C (freezing). Higher temperatures (up to 100°C or 212°F) seem to have less impact in general than colder temperatures, but high-*k* ceramics can lose up to 30% of their capacitance level at 100°C compared to room temperature. With experience, you will learn the type of capacitor to use for each application and only be concerned when you encounter very high frequencies, extreme temperatures, or very high currents or voltages.

EXAMPLE 14.8 At what frequency will the reactance of a 200 mH inductor match the resistance level of a 5 kΩ resistor?

Solution: The resistance remains constant at 5 kΩ for the frequency range of the inductor. Therefore,

$$R = 5000\ \Omega = X_L = 2\pi f L = 2\pi L f$$
$$= 2\pi(200 \times 10^{-3}\ \text{H})f = 1.257f$$

and

$$f = \frac{5000\ \text{Hz}}{1.257} \cong \mathbf{3.98\ kHz}$$

EXAMPLE 14.9 At what frequency will an inductor of 5 mH have the same reactance as a capacitor of 0.1 μF?

Solution:

$$X_L = X_C$$
$$2\pi f L = \frac{1}{2\pi f C}$$
$$f^2 = \frac{1}{4\pi^2 LC}$$

and

$$f = \frac{1}{2\pi\sqrt{LC}} = \frac{1}{2\pi\sqrt{(5 \times 10^{-3}\ \text{H})(0.1 \times 10^{-6}\ \text{F})}}$$
$$= \frac{1}{2\pi\sqrt{5 \times 10^{-10}}} = \frac{1}{(2\pi)(2.236 \times 10^{-5})} = \frac{10^5\ \text{Hz}}{14.05} \cong \mathbf{7.12\ kHz}$$

14.5 AVERAGE POWER AND POWER FACTOR

A common question is, How can a sinusoidal voltage or current deliver power to a load if it seems to be delivering power during one part of its cycle and taking it back during the negative part of the sinusoidal cycle? The equal oscillations above and below the axis seem to suggest that over one full cycle there is no net transfer of power or energy. However, as mentioned in the last chapter, there is a net transfer of power over one full cycle because power is delivered to the load *at each instant* of the applied voltage or current (except when either is crossing the axis) no matter what the direction is of the current or polarity of the voltage.

To demonstrate this, consider the relatively simple configuration in Fig. 14.29 where an 8 V peak sinusoidal voltage is applied across a 2 Ω resistor. When the voltage is at its positive peak, the power delivered at that instant is 32 W as shown in the figure. At the midpoint of 4 V, the instantaneous power delivered drops to 8 W; when the voltage crosses the axis, it drops to 0 W. Note, however, that when the applied voltage is at its negative peak, the current may reverse but, at that instant, 32 W is still being delivered to the resistor.

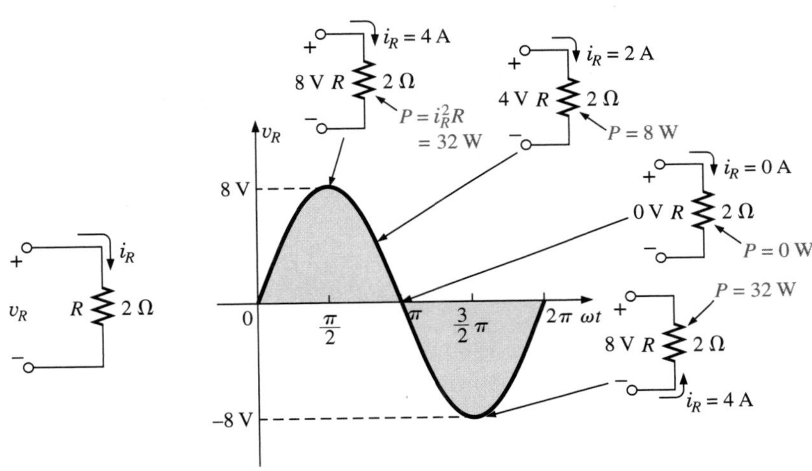

FIG. 14.29
Demonstrating that power is delivered at every instant of a sinusoidal voltage waveform (except $v_R = 0$ V).

In total, therefore,

even though the current through and the voltage across reverse direction and polarity, respectively, power is delivered to the resistive load at each instant of time.

If we plot the power delivered over a full cycle, the curve in Fig. 14.30 results. Note that the applied voltage and resulting current are in phase and have twice the frequency of the power curve. For one full cycle of the applied voltage having a period *T*, the power level peaks for each pulse of the sinusoidal waveform.

The fact that the power curve is always above the horizontal axis reveals that power is being delivered to the load at each instant of time of the applied sinusoidal voltage.

Any portion of the power curve below the axis reveals that power is being returned to the source. The average value of the power curve occurs at a level equal to $V_m I_m/2$ as shown in Fig. 14.30. This power level

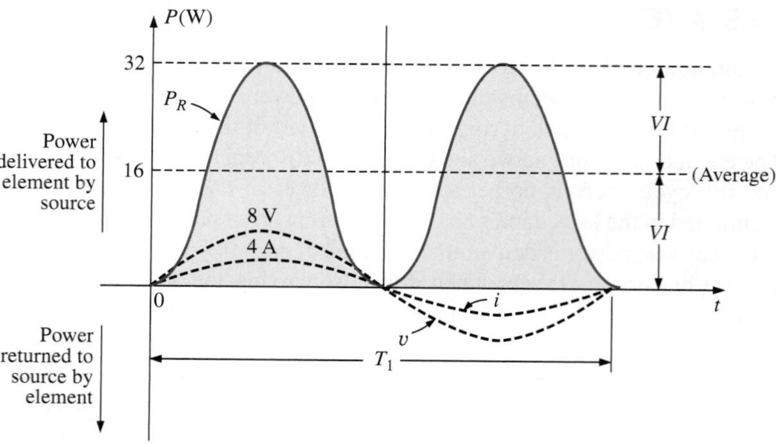

FIG. 14.30
Power versus time for a purely resistive load.

is called the **average** or **real power** level. It establishes a particular level of power transfer for the full cycle, so that we do not have to determine the level of power to apply to a quantity that varies in a sinusoidal nature.

If we substitute the equation for the peak value in terms of the rms value as follows:

$$P_{av} = \frac{V_m I_m}{2} = \frac{(\sqrt{2}\,V_{rms})(\sqrt{2}\,I_{rms})}{2} = \frac{2\,V_{rms} I_{rms}}{2}$$

we find that the average or real power delivered to a resistor takes on the following very convenient form:

$$\boxed{P_{av} = V_{rms}\,I_{rms}} \qquad \textbf{(14.10)}$$

Note that the power equation is exactly the same when applied to dc networks as long as we work with rms values.

The above analysis was for a purely resistive load. If the sinusoidal voltage is applied to a network with a combination of *R, L,* and *C* components, the instantaneous equation for the power levels is more complex. However, if we are careful in developing the general equation and examine the results, we find some general conclusions that will be very helpful in the analysis to follow.

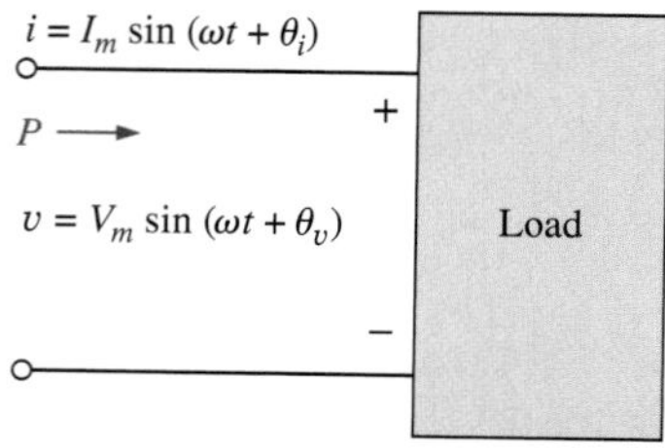

FIG. 14.31
Determining the power delivered in a sinusoidal ac network.

In Fig. 14.31, a voltage with an initial phase angle is applied to a network with any combination of elements that results in a current with the indicated phase angle.

The power delivered at each instant of time is then defined by

$$\begin{aligned} p = vi &= V_m \sin(\omega t + \theta_v) I_m \sin(\omega t + \theta_i) \\ &= V_m I_m \sin(\omega t + \theta_v) \sin(\omega t + \theta_i) \end{aligned}$$

Using the trigonometric identity

$$\sin A \sin B = \frac{\cos(A - B) - \cos(A + B)}{2}$$

the function $\sin(\omega t + \theta_v)\sin(\omega t + \theta_i)$ becomes

$$\sin(\omega t + \theta_v)\sin(\omega t + \theta_i)$$
$$= \frac{\cos[(\omega t + \theta_v) - (\omega t + \theta_i)] - \cos[(\omega t + \theta_v) + (\omega t + \theta_i)]}{2}$$
$$= \frac{\cos(\theta_v - \theta_i) - \cos(2\omega t + \theta_v + \theta_i)}{2}$$

so that

$$p = \overbrace{\left[\frac{V_m I_m}{2}\cos(\theta_v - \theta_i)\right]}^{\text{Fixed value}} - \overbrace{\left[\frac{V_m I_m}{2}\cos(2\omega t + \theta_v + \theta_i)\right]}^{\text{Time-varying (function of } t)}$$

A plot of v, i, and p on the same set of axes is shown in Fig. 14.32.

Note that the second factor in the preceding equation is a cosine wave with an amplitude of $V_m I_m/2$ and with a frequency twice that of the voltage or current. The average value of this term is zero over one cycle, producing no net transfer of energy in any one direction.

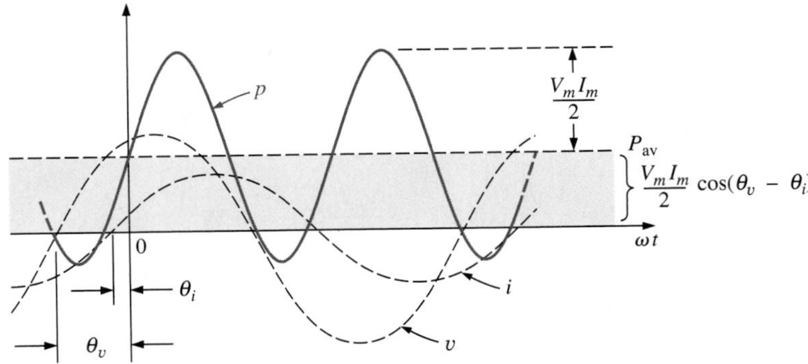

FIG. 14.32
Defining the average power for a sinusoidal ac network.

The first term in the preceding equation, however, has a constant magnitude (no time dependence) and therefore provides some net transfer of energy. This term is referred to as the **average power** or **real power** as introduced earlier. The angle $(\theta_v - \theta_i)$ is the phase angle between v and i. Since $\cos(-\alpha) = \cos\alpha$,

the magnitude of average power delivered is independent of whether v leads i or i leads v.

Defining θ as equal to $|\theta_v - \theta_i|$, where $|\ |$ indicates that only the magnitude is important and the sign is immaterial, we have

$$\boxed{P = \frac{V_m I_m}{2}\cos\theta} \qquad \text{(watts, W)} \qquad \textbf{(14.11)}$$

where P is the average power in watts. This equation can also be written

$$P = \left(\frac{V_m}{\sqrt{2}}\right)\left(\frac{I_m}{\sqrt{2}}\right)\cos\theta$$

or, since $\quad V_{\text{eff}} = \dfrac{V_m}{\sqrt{2}} \quad \text{and} \quad I_{\text{eff}} = \dfrac{I_m}{\sqrt{2}}$

Eq. (14.11) becomes

$$\boxed{P = V_{\text{rms}} I_{\text{rms}} \cos\theta} \qquad \textbf{(14.12)}$$

Let us now apply Eqs. (14.11) and (14.12) to the basic R, L, and C elements.

Resistor

In a purely resistive circuit, since v and i are in phase, $|\theta_v - \theta_i| = \theta = 0°$, and $\cos\theta = \cos 0° = 1$, so that

$$\boxed{P = \frac{V_m I_m}{2} = V_{rms} I_{rms}} \quad \text{(W)} \qquad \textbf{(14.13)}$$

Or, since

$$I_{rms} = \frac{V_{rms}}{R}$$

then

$$\boxed{P = \frac{V_{rms}^2}{R} = I_{rms}^2 R} \quad \text{(W)} \qquad \textbf{(14.14)}$$

Inductor

In a purely inductive circuit, since v leads i by 90°, $|\theta_v - \theta_i| = \theta = |-90°| = 90°$. Therefore,

$$P = \frac{V_m I_m}{2}\cos 90° = \frac{V_m I_m}{2}(0) = \mathbf{0\ W}$$

The average power or power dissipated by the ideal inductor (no associated resistance) is zero watts.

Capacitor

In a purely capacitive circuit, since i leads v by 90°, $|\theta_v - \theta_i| = \theta = |-90°| = 90°$. Therefore,

$$P = \frac{V_m I_m}{2}\cos(90°) = \frac{V_m I_m}{2}(0) = \mathbf{0\ W}$$

The average power or power dissipated by the ideal capacitor (no associated resistance) is zero watts.

EXAMPLE 14.10 Find the average power dissipated in a network whose input current and voltage are the following:

$$i = 5\sin(\omega t + 40°)$$
$$v = 10\sin(\omega t + 40°)$$

Solution: Since v and i are in phase, the circuit appears to be purely resistive at the input terminals. Therefore,

$$P = \frac{V_m I_m}{2} = \frac{(10\text{ V})(5\text{ A})}{2} = \mathbf{25\ W}$$

or

$$R = \frac{V_m}{I_m} = \frac{10\text{ V}}{5\text{ A}} = 2\ \Omega$$

and

$$P = \frac{V_{rms}^2}{R} = \frac{[(0.707)(10\text{ V})]^2}{2} = \mathbf{25\ W}$$

or

$$P = I_{rms}^2 R = [(0.707)(5\text{ A})]^2(2) = \mathbf{25\ W}$$

For the following example, the circuit consists of a combination of resistances and reactances producing phase angles between the input current and voltage different from 0° or 90°.

EXAMPLE 14.11 Determine the average power delivered to networks having the following input voltage and current:

a. $v = 100 \sin(\omega t + 40°)$
   $i = 20 \sin(\omega t + 70°)$

b. $v = 150 \sin(\omega t - 70°)$
   $i = 3 \sin(\omega t - 50°)$

Solutions:

a. $V_m = 100, \quad \theta_v = 40°$
   $I_m = 20\text{ A}, \quad \theta_i = 70°$
   $\theta = |\theta_v - \theta_i| = |40° - 70°| = |-30°| = 30°$

and

$$P = \frac{V_m I_m}{2}\cos\theta = \frac{(100\text{ V})(20\text{ A})}{2}\cos(30°) = (1000\text{ W})(0.866)$$
$$= \mathbf{866\ W}$$

b. $V_m = 150\text{ V}, \quad \theta_v = -70°$
   $I_m = 3\text{ A}, \quad \theta_i = -50°$
   $\theta = |\theta_v - \theta_i| = |-70° - (-50°)|$
   $= |-70° + 50°| = |-20°| = 20°$

and

$$P = \frac{V_m I_m}{2}\cos\theta = \frac{(150\text{ V})(3\text{ A})}{2}\cos(20°) = (225\text{ W})(0.9397)$$
$$= \mathbf{211.43\ W}$$

Power Factor

In the equation $P = (V_m I_m/2)\cos\theta$, the factor that has significant control over the delivered power level is the $\cos\theta$. No matter how large the voltage or current, if $\cos\theta = 0$, the power is zero; if $\cos\theta = 1$, the power delivered is a maximum. Since it has such control, the expression was given the name **power factor** and is defined by

$$\boxed{\text{Power factor} = F_p = \cos\theta} \qquad \textbf{(14.15)}$$

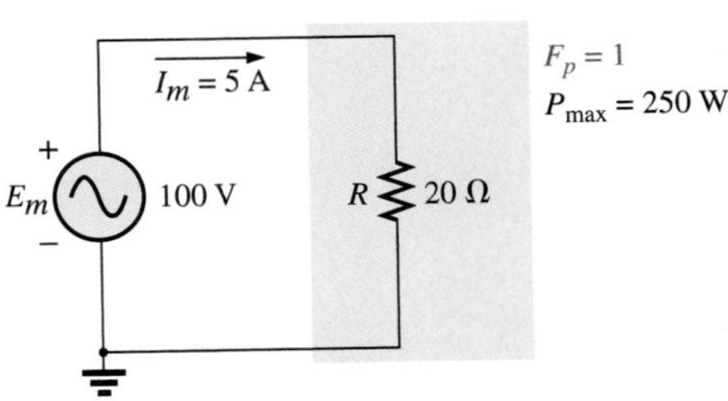

FIG. 14.33
Purely resistive load with $F_p = 1$.

For a purely resistive load such as the one shown in Fig. 14.33, the phase angle between v and i is 0° and $F_p = \cos\theta = \cos 0° = 1$. The power delivered is a maximum of $(V_m I_m/2)\cos\theta = ((100\text{ V})(5\text{ A})/2)(1) = 250$ W.

For a purely reactive load (inductive or capactitive) such as the one shown in Fig. 14.34, the phase angle between v and i is 90° and $F_p = \cos\theta = \cos 90° = 0$. The power delivered is then the minimum value of zero watts, *even though the current has the same peak value* as that encountered in Fig. 14.33.

For situations where the load is a combination of resistive and reactive elements, the power factor varies between 0 and 1. The more resistive the total impedance, the closer the power factor is to 1; the more reactive the total impedance, the closer the power factor is to 0.

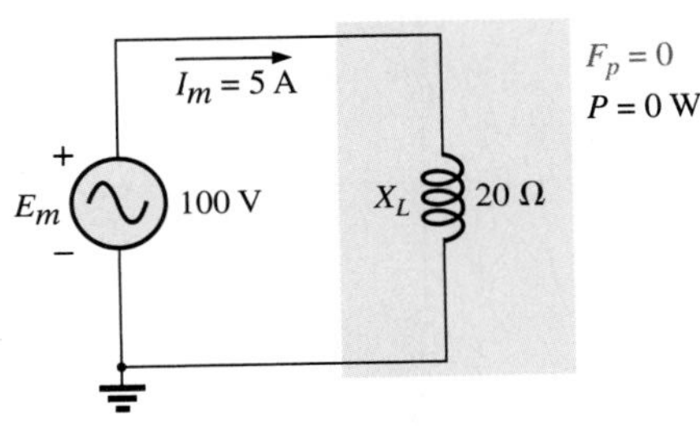

FIG. 14.34
Purely inductive load with $F_p = 0$.

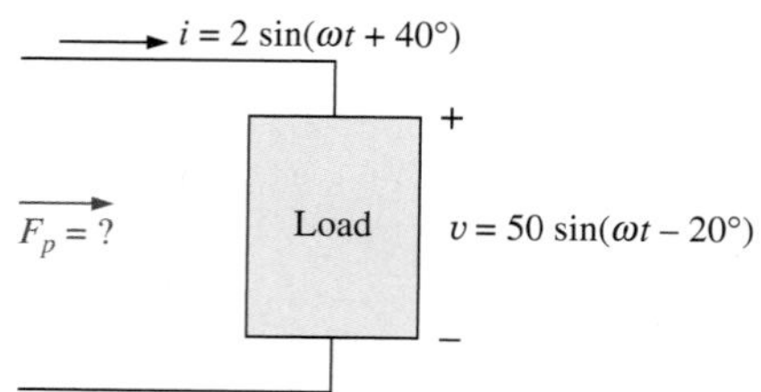

FIG. 14.35
Example 14.12(a).

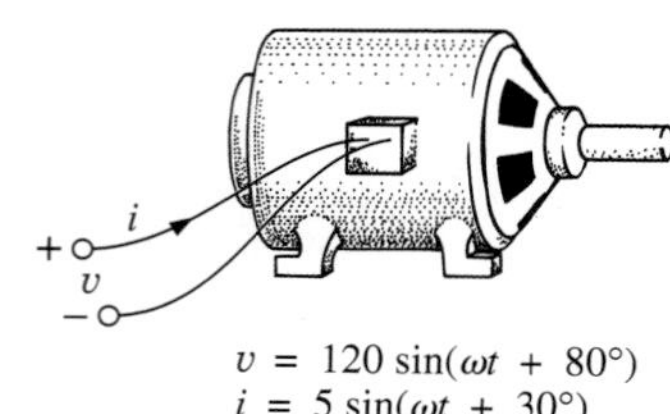

FIG. 14.36
Example 14.12(b).

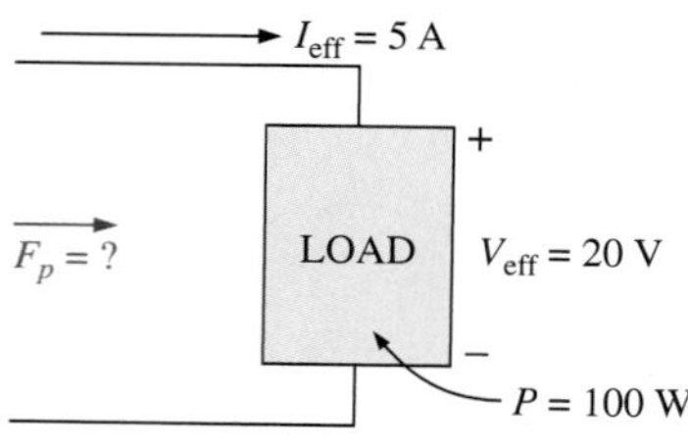

FIG. 14.37
Example 14.12(c).

In terms of the average power and the terminal voltage and current,

$$F_p = \cos\theta = \frac{P}{V_{rms}I_{rms}} \tag{14.16}$$

The terms *leading* and *lagging* are often written in conjunction with the power factor. *They are defined by the current through the load.* If the current leads the voltage across a load, the load has a **leading power factor.** If the current lags the voltage across the load, the load has a **lagging power factor.** In other words,

capacitive networks have leading power factors, and inductive networks have lagging power factors.

The importance of the power factor to power distribution systems is examined in Chapter 19. In fact, one section is devoted to power-factor correction.

EXAMPLE 14.12 Determine the power factors of the following loads, and indicate whether they are leading or lagging:

a. Fig. 14.35
b. Fig. 14.36
c. Fig. 14.37

Solutions:

a. $F_p = \cos\theta = \cos|40° - (-20°)| = \cos 60° = \mathbf{0.5\ leading}$
b. $F_p = \cos\theta\ |80° - 30°| = \cos 50° = \mathbf{0.64\ lagging}$
c. $F_p = \cos\theta = \dfrac{P}{V_{eff}I_{eff}} = \dfrac{100\text{ W}}{(20\text{ V})(5\text{ A})} = \dfrac{100\text{ W}}{100\text{ W}} = \mathbf{1}$

The load is resistive, and F_p is neither leading nor lagging.

14.6 COMPLEX NUMBERS

In our analysis of dc networks, we found it necessary to determine the algebraic sum of voltages and currents. Since the same will also be true for ac networks, the question arises, How do we determine the algebraic sum of two or more voltages (or currents) that are varying sinusoidally? Although one solution would be to find the algebraic sum on a point-to-point basis (as shown in Section 14.12), this would be a long and tedious process in which accuracy would be directly related to the scale used.

It is the purpose of this chapter to introduce a system of **complex numbers** that, when related to the sinusoidal ac waveform, results in a technique for finding the algebraic sum of sinusoidal waveforms that is quick, direct, and accurate. In the following chapters, the technique is extended to permit the analysis of sinusoidal ac networks in a manner very similar to that applied to dc networks. The methods and theorems as described for dc networks can then be applied to sinusoidal ac networks with little difficulty.

A **complex number** represents a point in a two-dimensional plane located with reference to two distinct axes. This point can also determine a radius vector drawn from the origin to the point. The horizontal axis is called the *real* axis, while the vertical axis is called the *imaginary* axis. Both are labeled in Fig. 14.38. Every number from zero to ±∞ can be

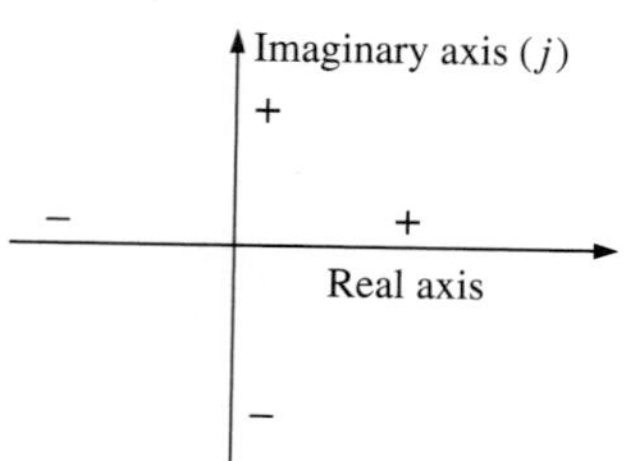

FIG. 14.38
Defining the real and imaginary axes of a complex plane.

represented by some point along the real axis. Prior to the development of this system of complex numbers, it was believed that any number not on the real axis did not exist—hence the term *imaginary* for the vertical axis.

In the complex plane, the horizontal or real axis represents all positive numbers to the right of the imaginary axis and all negative numbers to the left of the imaginary axis. All positive imaginary numbers are represented above the real axis, and all negative imaginary numbers, below the real axis. The symbol j (or sometimes i) is used to denote the imaginary component.

Two forms are used to represent a complex number: **rectangular** and **polar.** Each can represent a point in the plane or a radius vector drawn from the origin to that point.

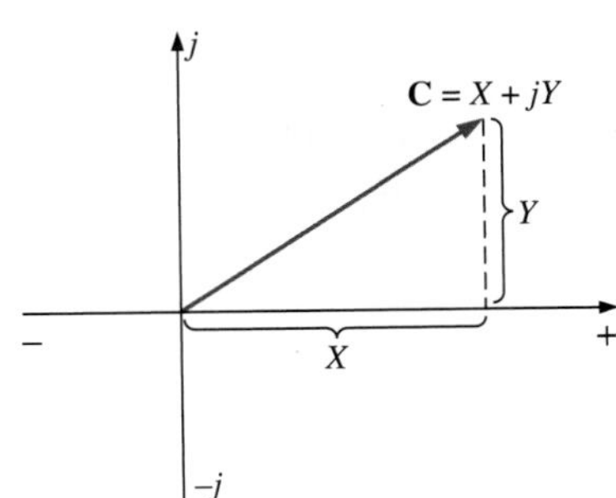

FIG. 14.39
Defining the rectangular form.

14.7 RECTANGULAR FORM

The format for the **rectangular form** is

$$\mathbf{C} = X + jY \tag{14.17}$$

as shown in Fig. 14.39. The letter **C** was chosen from the word "complex." The **boldface** notation is for any number with magnitude and direction. The *italic* is for magnitude only.

EXAMPLE 14.13 Sketch the following complex numbers in the complex plane:

a. $\mathbf{C} = 3 + j4$
b. $\mathbf{C} = 0 - j6$
c. $\mathbf{C} = -10 - j20$

Solutions:

a. See Fig. 14.40.
b. See Fig. 14.41.
c. See Fig. 14.42.

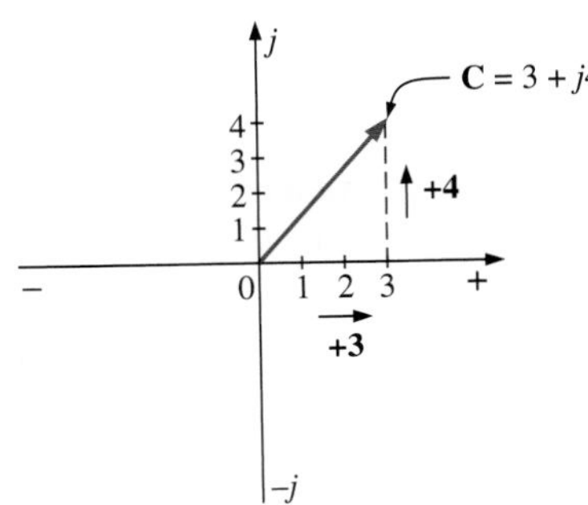

FIG. 14.40
Example 14.13(a).

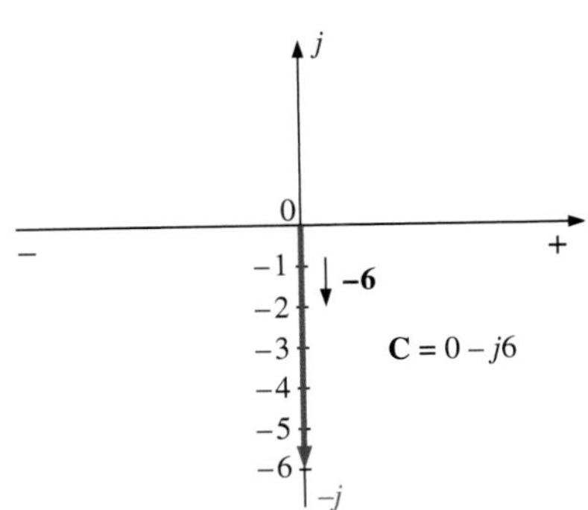

FIG. 14.41
Example 14.13(b).

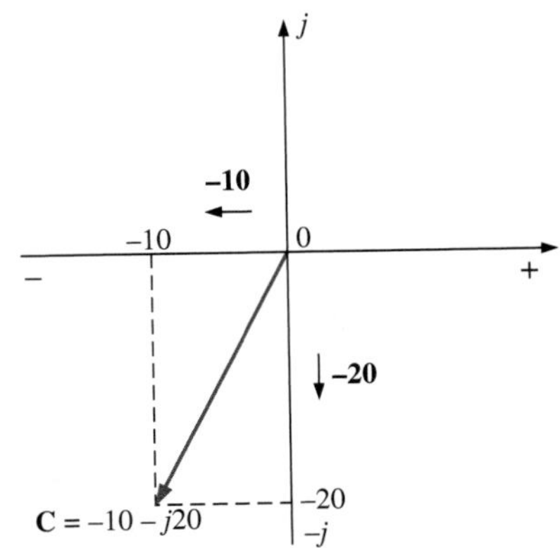

FIG. 14.42
Example 14.13(c).

14.8 POLAR FORM

The format for the **polar form** is

$$\mathbf{C} = Z \angle \theta \tag{14.18}$$

with the letter Z chosen from the sequence X, Y, Z.

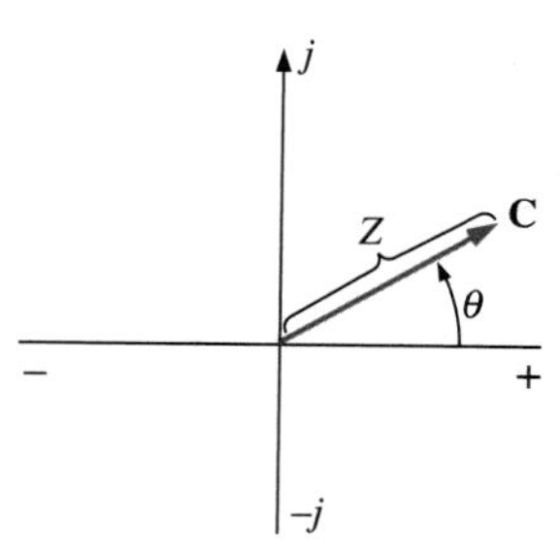

FIG. 14.43
Defining the polar form.

Z indicates magnitude only and θ is *always measured counterclockwise (CCW) from the positive real axis,* as shown in Fig. 14.43. Angles measured in the clockwise direction from the positive real axis must have a negative sign associated with them.

A negative sign in front of the polar form has the effect shown in Fig. 14.44. Note that it results in a complex number directly opposite the complex number with a positive sign.

$$-\mathbf{C} = -Z\angle\theta = Z\angle\theta \pm 180° \quad (14.19)$$

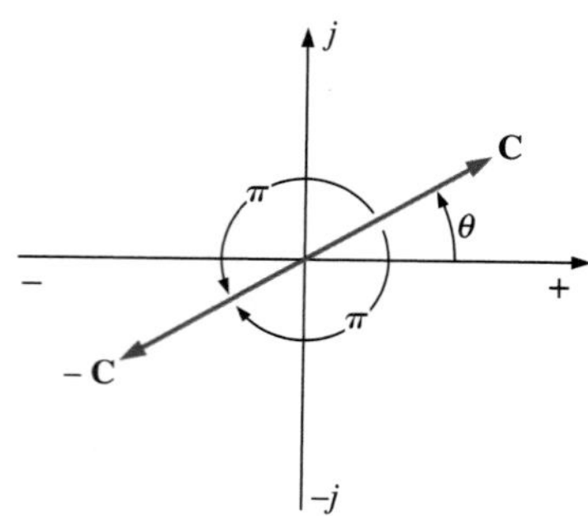

FIG. 14.44
Demonstrating the effect of a negative sign on the polar form.

EXAMPLE 14.14 Sketch the following complex numbers in the complex plane:

a. $\mathbf{C} = 5\angle 30°$
b. $\mathbf{C} = 7\angle -120°$
c. $\mathbf{C} = -4.2\angle 60°$

Solutions:

a. See Fig. 14.45.
b. See Fig. 14.46.
c. See Fig. 14.47.

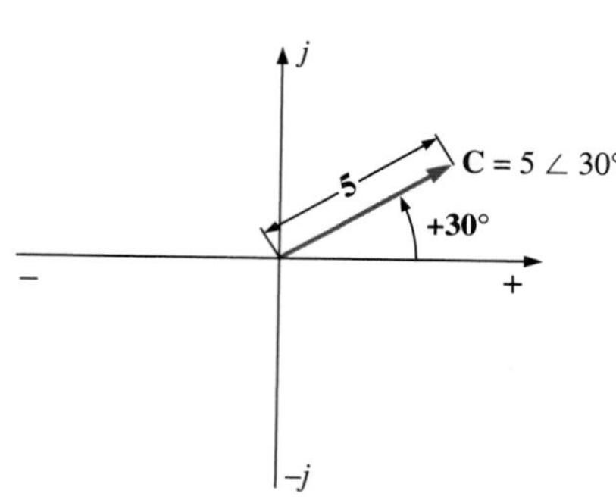

FIG. 14.45
Example 14.14(a).

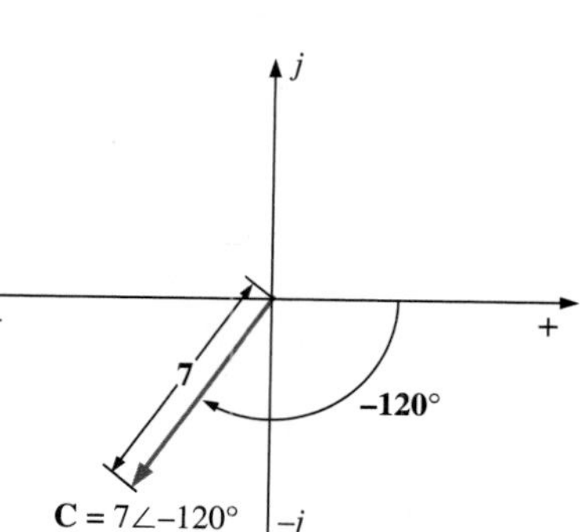

FIG. 14.46
Example 14.14(b).

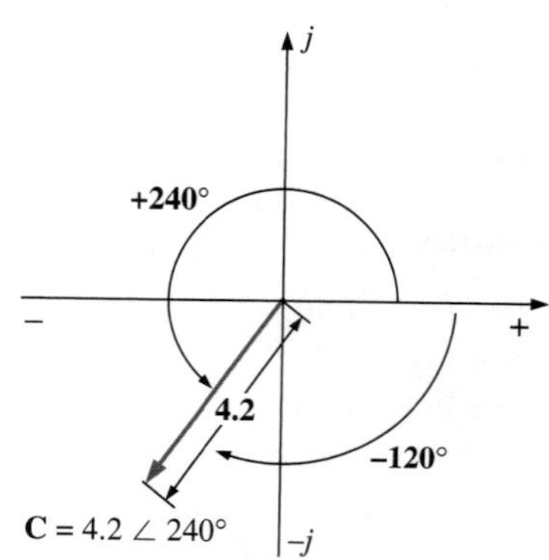

FIG. 14.47
Example 14.14(c).

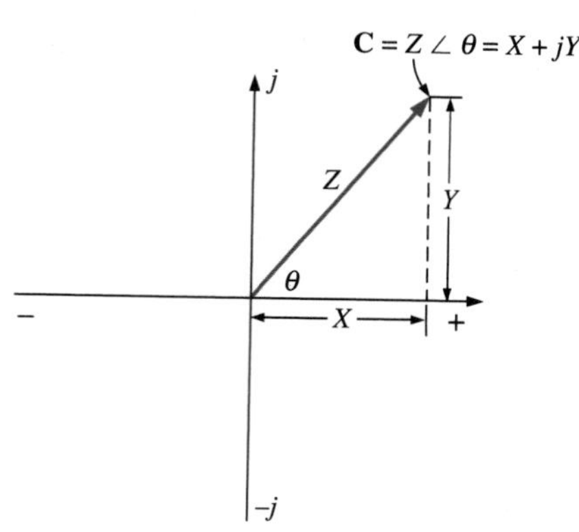

FIG. 14.48
Conversion between forms.

14.9 CONVERSION BETWEEN FORMS

The two forms are related by the following equations, as illustrated in Fig. 14.48.

Rectangular to Polar

$$Z = \sqrt{X^2 + Y^2} \quad (14.20)$$

$$\theta = \tan^{-1}\frac{Y}{X} \quad (14.21)$$

Polar to Rectangular

$$\boxed{X = Z \cos \theta} \qquad \textbf{(14.22)}$$

$$\boxed{Y = Z \sin \theta} \qquad \textbf{(14.23)}$$

EXAMPLE 14.15 Convert the following from rectangular to polar form:

$$\mathbf{C} = 3 + j4 \qquad \text{(Fig. 14.49)}$$

Solution:

$$Z = \sqrt{(3)^2 + (4)^2} = \sqrt{25} = 5$$

$$\theta = \tan^{-1}\left(\frac{4}{3}\right) = 53.13°$$

and

$$\mathbf{C = 5\angle 53.13°}$$

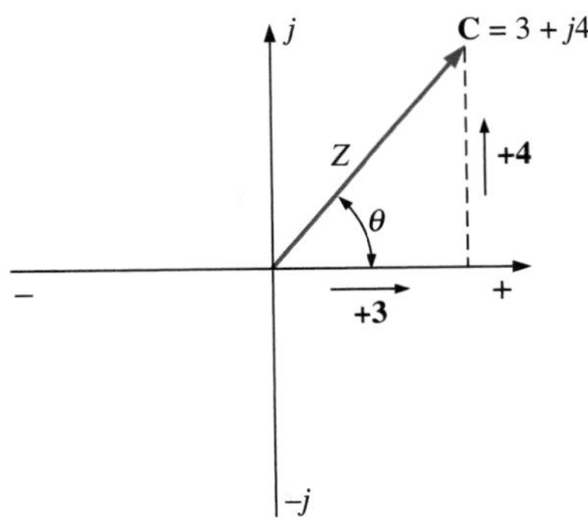

FIG. 14.49
Example 14.15.

EXAMPLE 14.16 Convert the following from polar to rectangular form:

$$\mathbf{C} = 10\,\angle 45° \qquad \text{(Fig. 14.50)}$$

Solution:

$$X = 10 \cos 45° = (10)(0.707) = 7.07$$
$$Y = 10 \sin 45° = (10)(0.707) = 7.07$$

and

$$\mathbf{C = 7.07 + \mathit{j}7.07}$$

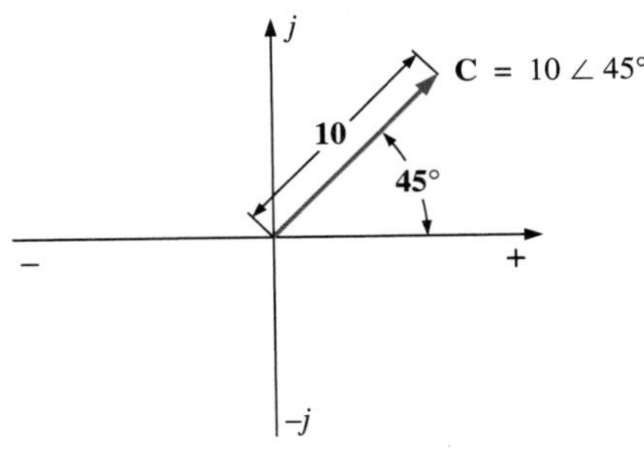

FIG. 14.50
Example 14.16.

If the complex number should appear in the second, third, or fourth quadrant, simply convert it in that quadrant, and carefully determine the proper angle to be associated with the magnitude of the vector.

EXAMPLE 14.17 Convert the following from rectangular to polar form:

$$\mathbf{C} = -6 + j3 \qquad \text{(Fig. 14.51)}$$

Solution:

$$Z = \sqrt{(6)^2 + (3)^2} = \sqrt{45} = 6.71$$

$$\beta = \tan^{-1}\left(\frac{3}{6}\right) = 26.57°$$

$$\theta = 180° - 26.57° = 153.43°$$

and

$$\mathbf{C = 6.71\,\angle 153.43°}$$

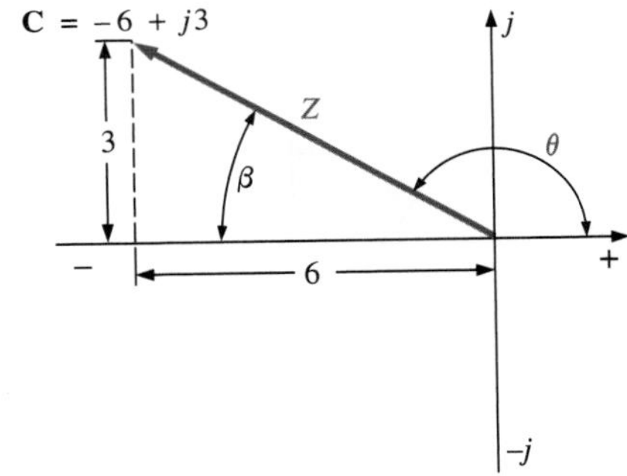

FIG. 14.51
Example 14.17.

EXAMPLE 14.18 Convert the following from polar to rectangular form:

$$\mathbf{C} = 10\,\angle 230° \qquad \text{(Fig. 14.52)}$$

Solution:

$$X = Z \cos \beta = 10 \cos(230° - 180°) = 10 \cos 50°$$
$$= (10)(0.6428) = 6.428$$
$$Y = Z \sin \beta = 10 \sin 50° = (10)(0.7660) = 7.660$$

and

$$\mathbf{C = -6.43 - \mathit{j}7.66}$$

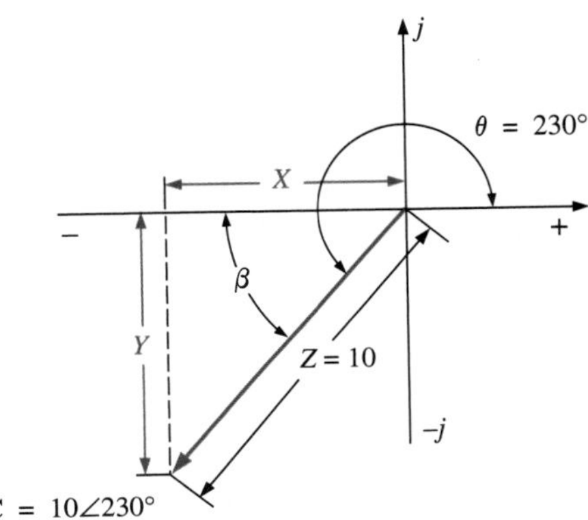

FIG. 14.52
Example 14.18.

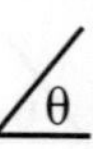

14.10 MATHEMATICAL OPERATIONS WITH COMPLEX NUMBERS

Complex numbers lend themselves readily to the basic mathematical operations of addition, subtraction, multiplication, and division. A few basic rules and definitions must be understood before considering these operations.

Let us first examine the symbol j associated with imaginary numbers. By definition,

$$j = \sqrt{-1} \tag{14.24}$$

Thus,

$$j^2 = -1 \tag{14.25}$$

and

$$j^3 = j^2 j = -1j = -j$$

with

$$j^4 = j^2 j^2 = (-1)(-1) = +1$$
$$j^5 = j$$

and so on. Further,

$$\frac{1}{j} = (1)\left(\frac{1}{j}\right) = \left(\frac{j}{j}\right)\left(\frac{1}{j}\right) = \frac{j}{j^2} = \frac{j}{-1}$$

and

$$\frac{1}{j} = -j \tag{14.26}$$

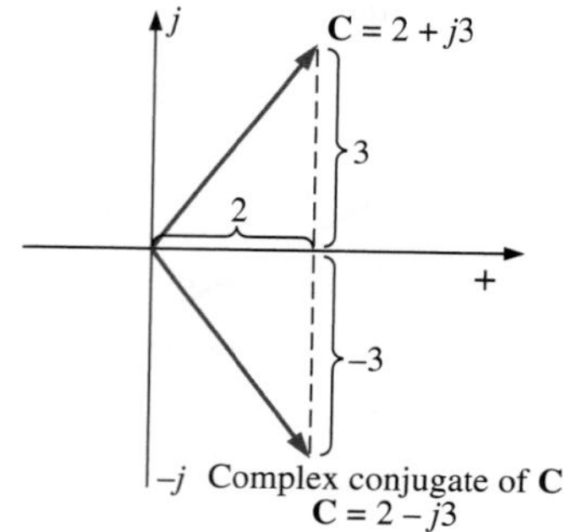

FIG. 14.53
Defining the complex conjugate of a complex number in rectangular form.

Complex Conjugate

The **conjugate** or **complex conjugate** of a complex number can be found by simply changing the sign of the imaginary part in the rectangular form or by using the negative of the angle of the polar form. For example, the conjugate of

$$\mathbf{C} = 2 + j3$$

is

$$2 - j3$$

as shown in Fig. 14.53. The conjugate of

$$\mathbf{C} = 2 \angle 30°$$

is

$$2 \angle -30°$$

as shown in Fig. 14.54.

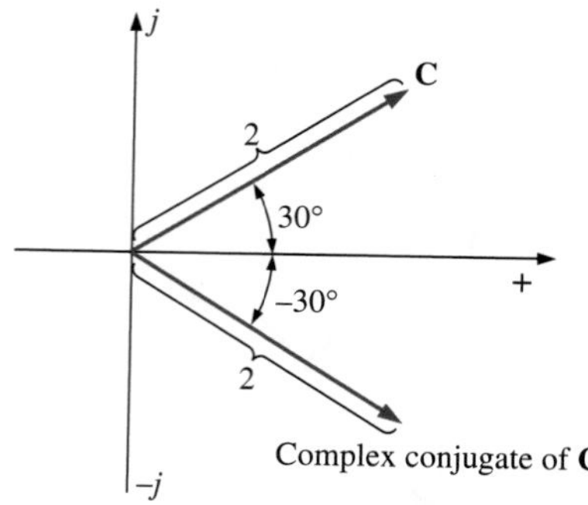

FIG. 14.54
Defining the complex conjugate of a complex number in polar form.

Reciprocal

The **reciprocal** of a complex number is 1 divided by the complex number. For example, the reciprocal of

$$\mathbf{C} = X + jY$$

is

$$\frac{1}{X + jY}$$

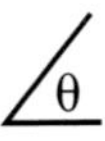

and of $Z \angle\theta$,

$$\frac{1}{Z \angle\theta}$$

We are now prepared to consider the four basic operations of *addition, subtraction, multiplication,* and *division* with complex numbers.

Addition

To add two or more complex numbers, add the real and imaginary parts separately. For example, if

$$\mathbf{C}_1 = \pm X_1 \pm jY_1 \quad \text{and} \quad \mathbf{C}_2 = \pm X_2 \pm jY_2$$

then

$$\boxed{\mathbf{C}_1 + \mathbf{C}_2 = (\pm X_1 \pm X_2) + j(\pm Y_1 \pm Y_2)} \qquad \textbf{(14.27)}$$

There is really no need to memorize the equation. Simply set one above the other and consider the real and imaginary parts separately, as shown in Example 14.19.

EXAMPLE 14.19

a. Add $\mathbf{C}_1 = 2 + j4$ and $\mathbf{C}_2 = 3 + j1$.
b. Add $\mathbf{C}_1 = 3 + j6$ and $\mathbf{C}_2 = -6 + j3$.

Solutions:

a. By Eq. (14.27),

$$\mathbf{C}_1 + \mathbf{C}_2 = (2 + 3) + j(4 + 1) = \mathbf{5 + j5}$$

Note Fig. 14.55. An alternative method is

$$\begin{array}{c} 2 + j4 \\ \underline{3 + j1} \\ \downarrow \quad \downarrow \\ \mathbf{5 + j5} \end{array}$$

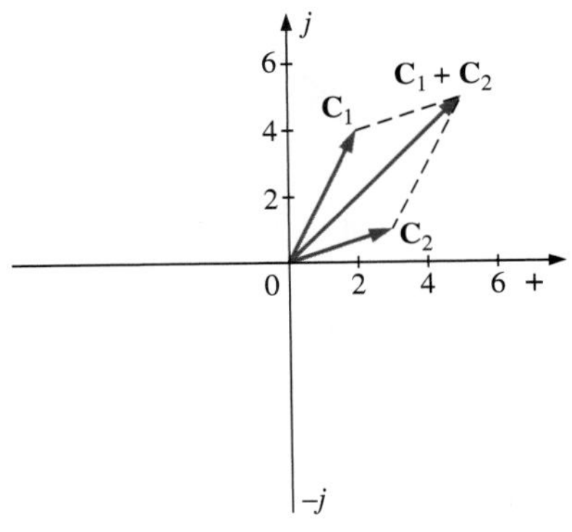

FIG. 14.55
Example 14.19(a).

b. By Eq. (14.27),

$$\mathbf{C}_1 + \mathbf{C}_2 = (3 - 6) + j(6 + 3) = \mathbf{-3 + j9}$$

Note Fig. 14.56. An alternative method is

$$\begin{array}{c} 3 + j6 \\ \underline{-6 + j3} \\ \downarrow \quad \downarrow \\ \mathbf{-3 + j9} \end{array}$$

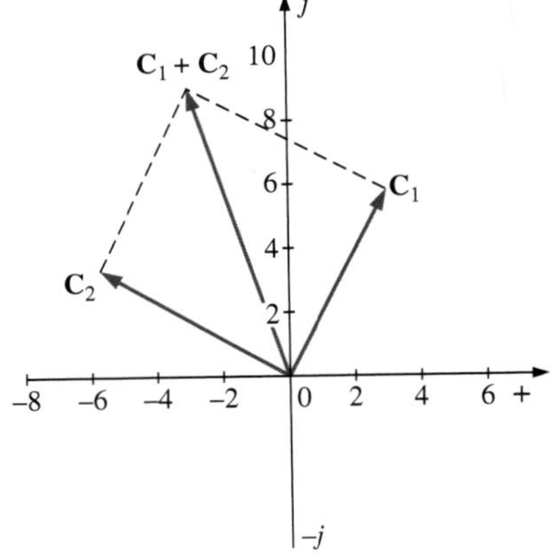

FIG. 14.56
Example 14.19(b).

Subtraction

In subtraction, the real and imaginary parts are again considered separately. For example, if

$$\mathbf{C}_1 = \pm X_1 \pm jY_1 \quad \text{and} \quad \mathbf{C}_2 = \pm X_2 \pm jY_2$$

then

$$\boxed{\mathbf{C}_1 - \mathbf{C}_2 = [\pm X_1 - (\pm X_2)] + j[\pm Y_1 - (\pm Y_2)]} \qquad \textbf{(14.28)}$$

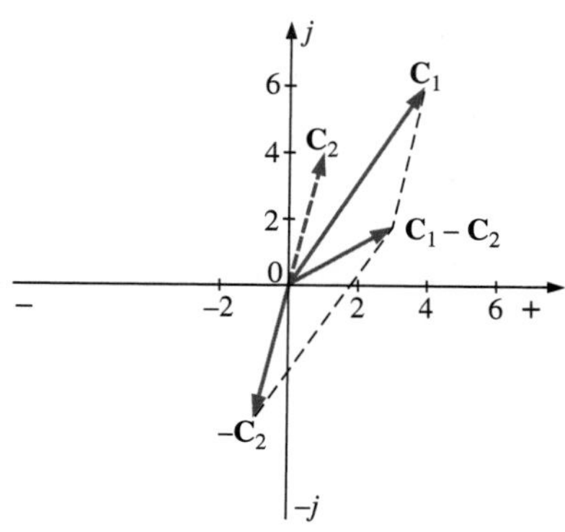

FIG. 14.57
Example 14.20(a).

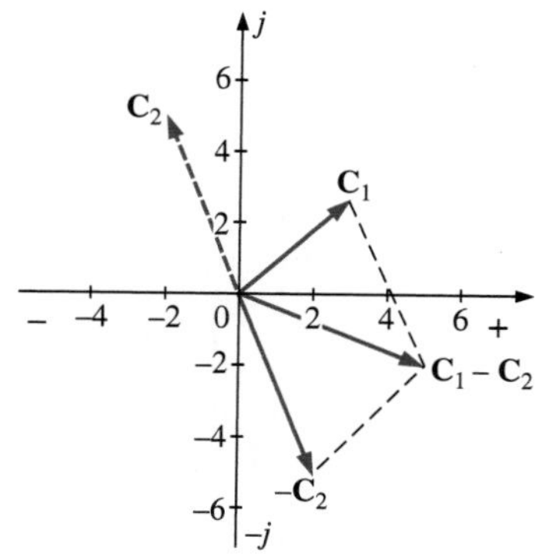

FIG. 14.58
Example 14.20(b).

Again, there is no need to memorize the equation if the alternative method of Example 14.20 is used.

EXAMPLE 14.20

a. Subtract $\mathbf{C}_2 = 1 + j4$ from $\mathbf{C}_1 = 4 + j6$.
b. Subtract $\mathbf{C}_2 = -2 + j5$ from $\mathbf{C}_1 = +3 + j3$.

Solutions:

a. By Eq. (14.28),

$$\mathbf{C}_1 - \mathbf{C}_2 = (4 - 1) + j(6 - 4) = \mathbf{3 + j2}$$

Note Fig. 14.57. An alternative method is

$$\begin{array}{r} 4 + j6 \\ \underline{-(1 + j4)} \\ \downarrow \quad \downarrow \\ \mathbf{3 + j2} \end{array}$$

b. By Eq. (14.28),

$$\mathbf{C}_1 - \mathbf{C}_2 = [3 - (-2)] + j(3 - 5) = \mathbf{5 - j2}$$

Note Fig. 14.58. An alternative method is

$$\begin{array}{r} 3 + j3 \\ \underline{-(-2 + j5)} \\ \downarrow \quad \downarrow \\ \mathbf{5 - j2} \end{array}$$

Addition or subtraction cannot be performed in polar form unless the complex numbers have the same angle θ or unless they differ only by multiples of 180°.

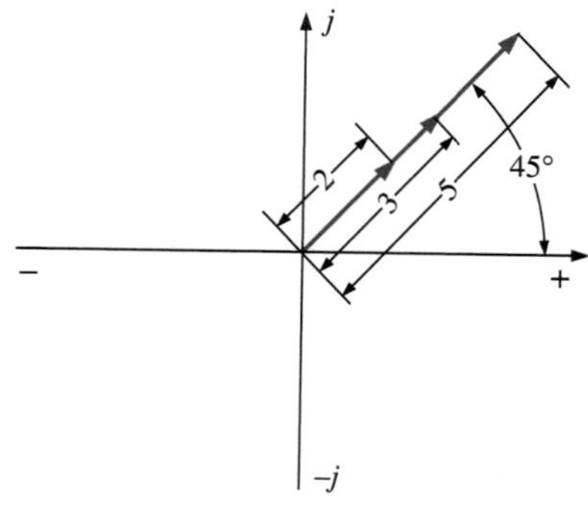

FIG. 14.59
Example 14.21(a).

EXAMPLE 14.21

a. $2 \angle 45° + 3 \angle 45° = \mathbf{5 \angle 45°}$. Note Fig. 14.59.

b. $2 \angle 0° - 4 \angle 180° = \mathbf{6 \angle 0°}$. Note Fig. 14.60.

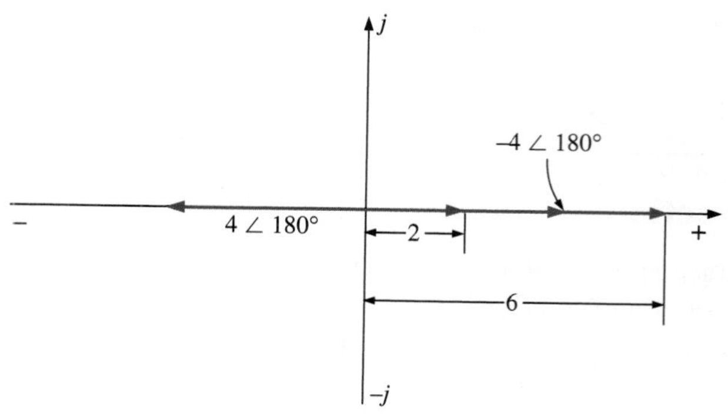

FIG. 14.60
Example 14.21(b).

Multiplication

To multiply two complex numbers in *rectangular* form, multiply the real and imaginary parts of one in turn by the real and imaginary parts of the other. For example, if

$$\mathbf{C}_1 = X_1 + jY_1 \quad \text{and} \quad \mathbf{C}_2 = X_2 + jY_2$$

then $\mathbf{C}_1 \cdot \mathbf{C}_2$:

$$\begin{array}{r} X_1 + jY_1 \\ \underline{X_2 + jY_2} \\ X_1X_2 + jY_1X_2 \\ \underline{+ jX_1Y_2 + j^2Y_1Y_2} \\ X_1X_2 + j(Y_1X_2 + X_1Y_2) + Y_1Y_2(-1) \end{array}$$

and

$$\boxed{\mathbf{C}_1 \cdot \mathbf{C}_2 = (X_1X_2 - Y_1Y_2) + j(Y_1X_2 + X_1Y_2)} \tag{14.29}$$

In Example 14.22(b), we obtain a solution without resorting to memorizing Eq. (14.29). Simply carry along the j factor when multiplying each part of one vector with the real and imaginary parts of the other.

EXAMPLE 14.22

a. Find $\mathbf{C}_1 \cdot \mathbf{C}_2$ if

$$\mathbf{C}_1 = 2 + j3 \quad \text{and} \quad \mathbf{C}_2 = 5 + j10$$

b. Find $\mathbf{C}_1 \cdot \mathbf{C}_2$ if

$$\mathbf{C}_1 = -2 - j3 \quad \text{and} \quad \mathbf{C}_2 = +4 - j6$$

Solutions:

a. Using the format above, we have

$$\begin{aligned} \mathbf{C}_1 \cdot \mathbf{C}_2 &= [(2)(5) - (3)(10)] + j[(3)(5) + (2)(10)] \\ &= \mathbf{-20 + j35} \end{aligned}$$

b. Without using the format, we obtain

$$\begin{array}{l} -2 - j3 \\ \underline{+4 - j6} \\ -8 - j12 \\ \underline{\quad + j12 + j^2 18} \\ -8 + j(-12 + 12) - 18 \end{array}$$

and

$$\mathbf{C}_1 \cdot \mathbf{C}_2 = \mathbf{-26 = 26 \angle 180^\circ}$$

In *polar* form, the magnitudes are multiplied and the angles added algebraically. For example, for

$$\mathbf{C}_1 = Z_1 \angle \theta_1 \quad \text{and} \quad \mathbf{C}_2 = Z_2 \angle \theta_2$$

we write

$$\boxed{\mathbf{C}_1 \cdot \mathbf{C}_2 = Z_1Z_2 \angle \theta_1 + \theta_2} \tag{14.30}$$

EXAMPLE 14.23

a. Find $\mathbf{C}_1 \cdot \mathbf{C}_2$ if

$$\mathbf{C}_1 = 5 \angle 20^\circ \quad \text{and} \quad \mathbf{C}_2 = 10 \angle 30^\circ$$

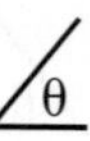

b. Find $\mathbf{C}_1 \cdot \mathbf{C}_2$ if

$$\mathbf{C}_1 = 2 \angle -40^\circ \quad \text{and} \quad \mathbf{C}_2 = 7 \angle +120^\circ$$

Solutions:

a. $\mathbf{C}_1 \cdot \mathbf{C}_2 = (5 \angle 20^\circ)(10 \angle 30^\circ) = (5)(10) \underline{/20^\circ + 30^\circ} = \mathbf{50 \angle 50^\circ}$

b. $\mathbf{C}_1 \cdot \mathbf{C}_2 = (2 \angle -40^\circ)(7 \angle +120^\circ) = (2)(7) \underline{/-40^\circ + 120^\circ}$
$= \mathbf{14 \angle +80^\circ}$

To multiply a complex number in rectangular form by a real number requires that both the real part and the imaginary part be multiplied by the real number. For example,

$$(10)(2 + j3) = 20 + j30$$

and

$$50 \angle 0^\circ \, (0 + j6) = j300 = 300 \angle 90^\circ$$

Division

To divide two complex numbers in *rectangular* form, multiply the numerator and denominator by the conjugate of the denominator and the resulting real and imaginary parts collected. That is, if

$$\mathbf{C}_1 = X_1 + jY_1 \quad \text{and} \quad \mathbf{C}_2 = X_2 + jY_2$$

then

$$\frac{\mathbf{C}_1}{\mathbf{C}_2} = \frac{(X_1 + jY_1)(X_2 - jY_2)}{(X_2 + jY_2)(X_2 - jY_2)}$$
$$= \frac{(X_1X_2 + Y_1Y_2) + j(X_2Y_1 - X_1Y_2)}{X_2^2 + Y_2^2}$$

and

$$\boxed{\frac{\mathbf{C}_1}{\mathbf{C}_2} = \frac{X_1X_2 + Y_1Y_2}{X_2^2 + Y_2^2} + j\frac{X_2Y_1 - X_1Y_2}{X_2^2 + Y_2^2}} \qquad \textbf{(14.31)}$$

The equation does not have to be memorized if the steps above used to obtain it are employed. That is, first multiply the numerator by the complex conjugate of the denominator and separate the real and imaginary terms. Then divide each term by the sum of each term of the denominator squared.

EXAMPLE 14.24

a. Find $\mathbf{C}_1/\mathbf{C}_2$ if $\mathbf{C}_1 = 1 + j4$ and $\mathbf{C}_2 = 4 + j5$.
b. Find $\mathbf{C}_1/\mathbf{C}_2$ if $\mathbf{C}_1 = -4 - j8$ and $\mathbf{C}_2 = +6 - j1$.

Solutions:

a. By Eq. (14.31),

$$\frac{\mathbf{C}_1}{\mathbf{C}_2} = \frac{(1)(4) + (4)(5)}{4^2 + 5^2} + j\frac{(4)(4) - (1)(5)}{4^2 + 5^2}$$
$$= \frac{24}{41} + \frac{j11}{41} \cong \mathbf{0.59 + j0.27}$$

b. Using an alternative method, we obtain

$$\begin{array}{l} -4 - j8 \\ \underline{+6 + j1} \\ -24 - j48 \\ \underline{\quad - j4 - j^2 8} \\ -24 - j52 + 8 = -16 - j52 \end{array}$$

$$\begin{array}{l} +6 - j1 \\ \underline{+6 + j1} \\ 36 + j6 \\ \underline{\quad - j6 - j^2 1} \\ 36 + \ 0 + 1 \ = 37 \end{array}$$

and
$$\frac{\mathbf{C}_1}{\mathbf{C}_2} = \frac{-16}{37} - \frac{j52}{37} = \mathbf{-0.43 - j1.41}$$

To divide a complex number in rectangular form by a real number, both the real part and the imaginary part must be divided by the real number. For example,

$$\frac{8 + j10}{2} = 4 + j5$$

and
$$\frac{6.8 - j0}{2} = 3.4 - j0 = 3.4\angle 0°$$

In *polar* form, division is accomplished by dividing the magnitude of the numerator by the magnitude of the denominator and subtracting the angle of the denominator from that of the numerator. That is, for

$$\mathbf{C}_1 = Z_1 \angle\theta_1 \quad \text{and} \quad \mathbf{C}_2 = Z_2 \angle\theta_2$$

we write

$$\boxed{\frac{\mathbf{C}_1}{\mathbf{C}_2} = \frac{Z_1}{Z_2} \underline{/\theta_1 - \theta_2}} \tag{14.32}$$

EXAMPLE 14.25

a. Find $\mathbf{C}_1/\mathbf{C}_2$ if $\mathbf{C}_1 = 15 \angle 10°$ and $\mathbf{C}_2 = 2 \angle 7°$.
b. Find $\mathbf{C}_1/\mathbf{C}_2$ if $\mathbf{C}_1 = 8 \angle 120°$ and $\mathbf{C}_2 = 16 \angle -50°$.

Solutions:

a. $$\frac{\mathbf{C}_1}{\mathbf{C}_2} = \frac{15 \angle 10°}{2 \angle 7°} = \frac{15}{2} \underline{/10° - 7°} = \mathbf{7.5 \angle 3°}$$

b. $$\frac{\mathbf{C}_1}{\mathbf{C}_2} = \frac{8 \angle 120°}{16 \angle -50°} = \frac{8}{16} \underline{/120° - (-50°)} = \mathbf{0.5 \angle 170°}$$

We obtain the *reciprocal* in the rectangular form by multiplying the numerator and denominator by the complex conjugate of the denominator:

$$\frac{1}{X + jY} = \left(\frac{1}{X + jY}\right)\left(\frac{X - jY}{X - jY}\right) = \frac{X - jY}{X^2 + Y^2}$$

and

$$\frac{1}{X + jY} = \frac{X}{X^2 + Y^2} - j\frac{Y}{X^2 + Y^2} \tag{14.33}$$

In polar form, the reciprocal is

$$\frac{1}{Z \angle \theta} = \frac{1}{Z} \angle -\theta \tag{14.34}$$

A concluding example using the four basic operations follows.

EXAMPLE 14.26 Perform the following operations, leaving the answer in polar or rectangular form:

a.
$$\frac{(2 + j3) + (4 + j6)}{(7 + j7) - (3 - j3)} = \frac{(2 + 4) + j(3 + 6)}{(7 - 3) + j(7 + 3)}$$
$$= \frac{(6 + j9)(4 - j10)}{(4 + j10)(4 - j10)}$$
$$= \frac{[(6)(4) + (9)(10)] + j[(4)(9) - (6)(10)]}{4^2 + 10^2}$$
$$= \frac{114 - j24}{116} = \mathbf{0.98 - j0.21}$$

b.
$$\frac{(50 \angle 30^\circ)(5 + j5)}{10 \angle -20^\circ} = \frac{(50 \angle 30^\circ)(7.07 \angle 45^\circ)}{10 \angle -20^\circ} = \frac{353.5 \angle 75^\circ}{10 \angle -20^\circ}$$
$$= 35.35 \angle 75^\circ - (-20^\circ) = \mathbf{35.35 \angle 95^\circ}$$

c.
$$\frac{(2 \angle 20^\circ)^2(3 + j4)}{8 - j6} = \frac{(2 \angle 20^\circ)(2 \angle 20^\circ)(5 \angle 53.13^\circ)}{10 \angle -36.87^\circ}$$
$$= \frac{(4 \angle 40^\circ)(5 \angle 53.13^\circ)}{10 \angle -36.87^\circ} = \frac{20 \angle 93.13^\circ}{10 \angle -36.87^\circ}$$
$$= 2 \angle 93.13^\circ - (-36.87^\circ) = \mathbf{2.0 \angle 130^\circ}$$

d.
$$3 \angle 27^\circ - 6 \angle -40^\circ = (2.673 + j1.362) - (4.596 - j3.857)$$
$$= (2.673 - 4.596) + j(1.362 + 3.857)$$
$$= \mathbf{-1.92 + j5.22}$$

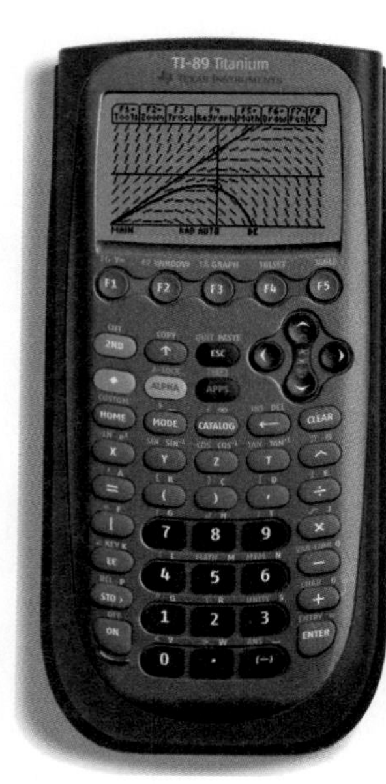

FIG. 14.61
TI-89 scientific calculator.
(Courtesy of Texas Instruments, Inc.)

14.11 CALCULATOR AND COMPUTER METHODS WITH COMPLEX NUMBERS

The process of converting from one form to another or working through lengthy operations with complex numbers can be time-consuming and often frustrating if one lost minus sign or decimal point invalidates the solution. Fortunately, technologists of today have calculators and computer methods that make the process measurably easier with higher degrees of reliability and accuracy.

Calculators

The TI-89 calculator in Fig. 14.61 is only one of numerous calculators that can convert from one form to another and perform lengthy calcula-

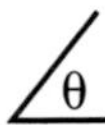

tions with complex numbers in a concise, neat form. Not all of the details of using a specific calculator are included here because each has its own format and sequence of steps. However, the basic operations with the TI-89 is included primarily to demonstrate the ease with which the conversions can be made and the format for more complex operations. If you have a TI-86 calculator, Appendix B provides details for using that calculator to perform these operations.

There are different routes to perform the conversions and operations below, but these instructions give you one approach that is fairly direct and straightforward. Since most operations are in the DEGREE rather than RADIAN mode, the sequence in Fig. 14.62 shows how to set the DEGREE mode for the operations to follow. A similar sequence sets the RADIAN mode if required.

FIG. 14.62
Setting the DEGREE mode on the TI-89 calculator.

Rectangular to Polar Conversion The sequence in Fig. 14.63 provides a detailed listing of the steps needed to convert from rectangular to polar form. In the examples to follow, the scrolling steps are not listed to simplify the sequence.

In the sequence in Fig. 14.63, an up scroll is chosen after Matrix because that is a more direct path to Vector ops. A down scroll generates the same result, but it requires going through the whole listing. The sequence seems quite long for such a simple conversion, but with practice you will be able to perform the scrolling steps quite rapidly. Always be sure the input data is entered correctly, such as including the i after the y component. Any incorrect entry will result in an error listing.

FIG. 14.63
Converting $3 + j5$ to the polar form using the TI-89 calculator.

Polar to Rectangular Conversion The sequence in Fig. 14.64 is a detailed listing of the steps needed to convert from polar to rectangular form. Note in the format that the brackets must surround the polar form. Also, the degree sign must be included with the angle to perform the calculation. The answer is displayed in the engineering notation selected.

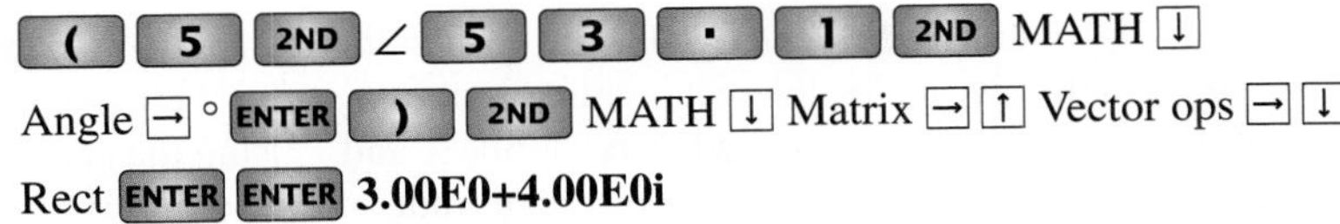

FIG. 14.64
Converting $5\angle 53.1°$ to the rectangular form using the TI-89 calculator.

Mathematical Operations Mathematical operations are performed in the natural order of operations, but you must remember to select the format for the solution. For instance, if the sequence in Fig. 14.65 did not include the polar designation, the answer would be in rectangular form

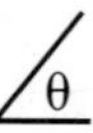

(1 0 ∠ 5 0 °) × (2 ∠ 2 0 °)

▶ Polar ENTER 20.00E0 ∠ 70.00E0

FIG. 14.65
Performing the operation (10∠50°)(2∠20°).

even though both quantities in the calculation are in polar form. In the rest of the examples, the scrolling required to obtain mathematical functions is not included to minimize the length of the sequence.

For the product of mixed complex numbers, the sequence of Fig. 14.66 results. Again, the polar form was selected for the solution.

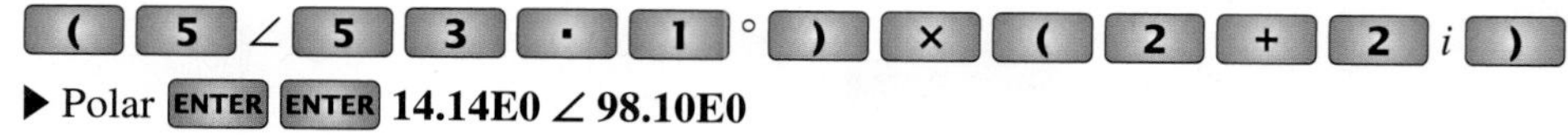

FIG. 14.66
Performing the operation (5∠53.1°)(2 + j2).

Finally, Example 14.26(c) is entered as shown by the sequence in Fig. 14.67. Note that the results exactly match those obtained earlier.

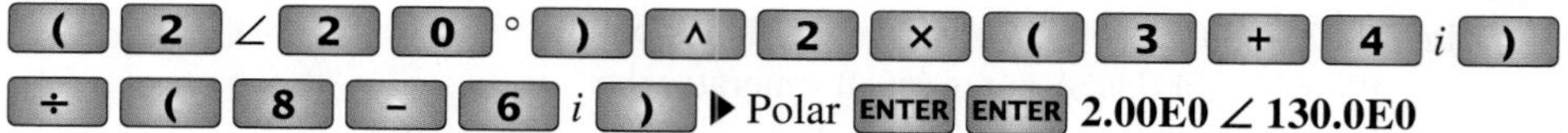

FIG. 14.67
Verifying the results of Example 14.26(c).

Mathcad

The Mathcad format for complex numbers is now introduced in preparation for the chapters to follow. We continue to use j when we define a complex number in rectangular form even though the Mathcad result always appears with the letter i. You can change this by going to the **Format** menu, but for this presentation we decided to use the default operators as much as possible.

When entering j to define the imaginary component of a complex number, be sure to enter it as $1j$, but do not put a multiplication operator between the 1 and the j. Just type 1 and then j. In addition, place the j after the constant rather than before as in the text material.

When Mathcad operates on an angle, it assumes that the angle is in radians and not degrees. Further, all results appear in radians rather than degrees.

The first operation to be developed is the conversion from rectangular to polar form. In Fig. 14.68, the rectangular number $4 + j3$ is being converted to polar form using Mathcad. First define X and Y using the colon operator. Next, write the equation for the magnitude of the polar form in terms of the two variables just defined. The magnitude of the polar form is then revealed by writing the variable again and using the equal sign. It takes some practice, but be careful when writing the equation for Z; you must pay particular attention to the location of the bracket before performing the next operation. The resulting magnitude of 5 is as expected.

For the angle, the sequence **View-Toolbars-Greek** is first applied to obtain the **Greek** toolbar appearing in Fig. 14.68. It can be moved to any location by clicking on the blue at the top of the toolbar and dragging it

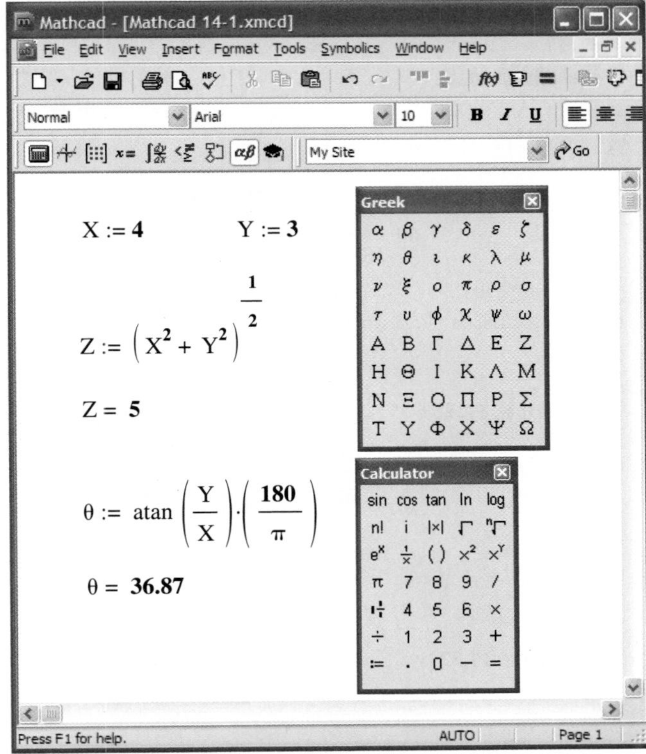

FIG. 14.68
Using Mathcad to convert from rectangular to polar form.

to the preferred location. Then select θ from the toolbar as the variable to be defined. Obtain the $\tan^{-1}\theta$ through the sequence **Insert-*f*(*x*)-Insert Function** dialog box**-trigonometric-atan-OK** in which *Y/X* is inserted. Then bring the controlling bracket to the outside of the entire expression, and multiply by the ratio of $180/\pi$ with π selected from the **Calculator** toolbar (available from the same sequence used to obtain the **Greek** toolbar). The multiplication by the last factor of the equation ensures that the angle is in degrees. Selecting θ again followed by an equal sign results in the correct angle of 36.87° as shown in Fig. 14.68.

We now look at two forms for the polar form of a complex number. The first is defined by the basic equations introduced in this chapter, while the second uses a special format. For all the Mathcad analyses to be provided in this text, the latter format is used. First define the magnitude of the polar form followed by the conversion of the angle of 60° to radians by multiplying by the factor $\pi/180$ as shown in Fig. 14.69. In this example, the resulting angular measure is $\pi/3$ radians. Next define the rectangular format by a real part $X = Z\cos\theta$ and by an imaginary part $Y = Z\sin\theta$. Both the cos and the sin are obtained by the sequence **Insert-*f(x)*-trigonometric-cos**(or **sin**)**-OK.** Note the multiplication by j which was actually entered as $1j$ without the multiplication operator between the 1 and the j. Entering C again followed by an equal sign results in the correct conversion shown in Fig. 14.69.

The next format is based on the mathematical relationship that $e^{j\theta} = \cos\theta + j\sin\theta$. Both Z and θ are as defined above, but now the complex number is written as shown in Fig. 14.69 using the notation just introduced. Note that both Z and θ are part of this defining form. The e^x is obtained directly from the **Calculator** toolbar. Remember to enter the j as

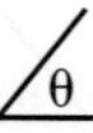

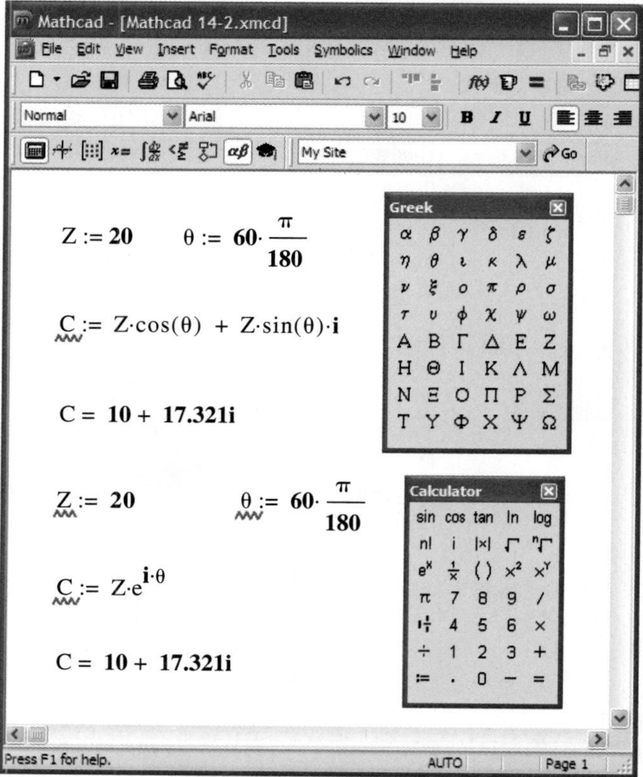

FIG. 14.69
Using Mathcad to convert from polar to rectangular form.

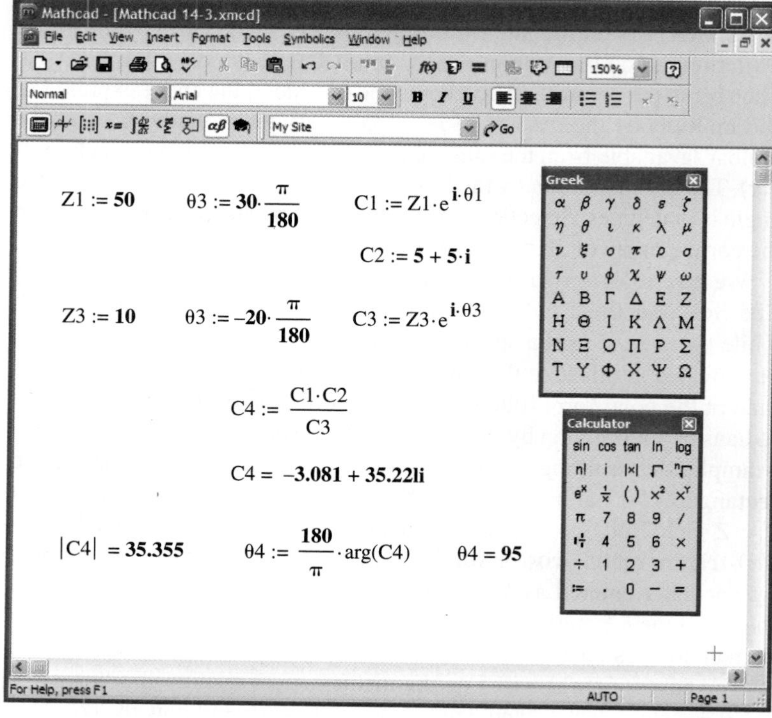

FIG. 14.70
Using Mathcad to confirm the results of Example 14.26(b).

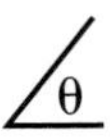

1*j* without a multiplication sign between the 1 and the *j*. However, there is a multiplication operator placed between the *j* and θ. When entered again followed by an equal sign, the rectangular form appears to match the above results. As mentioned above, it is this latter format that will be used throughout the text due to its cleaner form and more direct entering path.

The last example using Mathcad is a confirmation of the results of Example 14.26(b) as shown in Fig. 14.70. First define the three complex numbers as shown. Then enter the equation for the desired result using C_4, and the results are displayed. Note the relative simplicity of the equation for C_4 now that all the other variables have been defined. As shown, however, the immediate result is in the rectangular form. The components of the polar form can be obtained using the |x| function from the **Calculator** toolbar and the **arg** function from **Insert-*f(x)*-Complex Numbers-arg.** There are many other examples in the chapters to follow on the use of Mathcad with complex numbers.

14.12 PHASORS

As noted earlier in this chapter, the addition of sinusoidal voltages and currents is frequently required in the analysis of ac circuits. One lengthy but valid method of performing this operation is to place both sinusoidal waveforms on the same set of axes and add algebraically the magnitudes of each at every point along the abscissa, as shown for $c = a + b$ in Fig. 14.71. This, however, can be a long and tedious process with limited accuracy. A shorter method uses the rotating radius vector first appearing in Fig. 13.16. This *radius vector,* having a *constant magnitude* (length) with *one end fixed at the origin,* is called a **phasor** when applied to electric circuits. During its rotational development of the sine wave, the phasor will, at the instant $t = 0$, have the positions shown in Fig. 14.72(a) for each waveform in Fig. 14.72(b).

Note in Fig. 14.72(b) that v_2 passes through the horizontal axis at $t = 0$ s, requiring that the radius vector in Fig. 14.72(a) be on the horizontal axis to ensure a vertical projection of zero volts at $t = 0$ s. Its length in Fig. 14.72(a) is equal to the peak value of the sinusoid as required by the radius vector in Fig. 13.16. The other sinusoid has passed through 90° of its rotation by the time $t = 0$ s is reached and therefore has its maximum

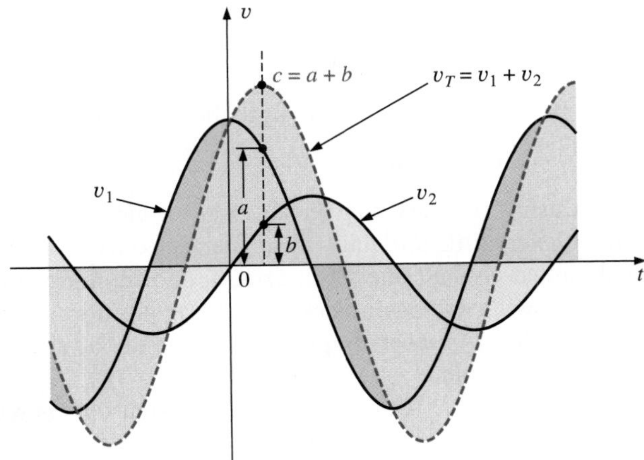

FIG. 14.71
Adding two sinusoidal waveforms on a point-by-point basis.

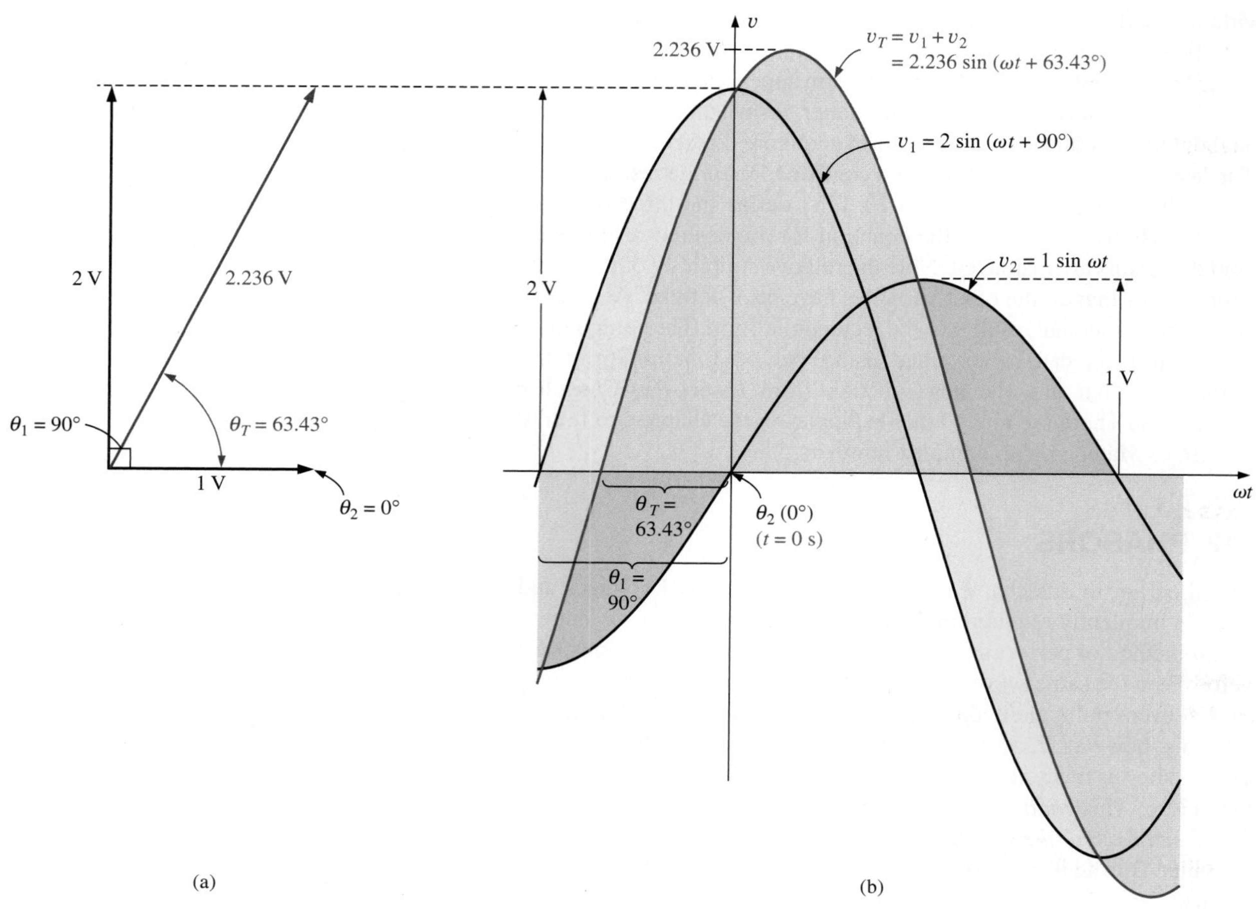

FIG. 14.72
(a) The phasor representation of the sinusoidal waveforms of Fig. 14.72(b); (b) finding the sum of two sinusoidal waveforms of v_1 and v_2.

vertical projection as shown in Fig. 14.72(a). Since the vertical projection is a maximum, the peak value of the sinusoid that it generates is also attained at $t = 0$ s, as shown in Fig. 14.72(b). Note also that $v_T = v_1$ at $t = 0$ s since $v_2 = 0$ V at this instant.

It can be shown [see Fig. 14.72(a)] using the vector algebra described in Section 14.10 that

$$1\text{ V} \angle 0° + 2\text{ V} \angle 90° = 2.236\text{ V} \angle 63.43°$$

In other words, if we convert v_1 and v_2 to the phasor form using

$$v = V_m \sin(\omega t \pm \theta) \Rightarrow V_m \angle \pm\theta$$

and add them using complex number algebra, we can find the phasor form for v_T with very little difficulty. It can then be converted to the time domain and plotted on the same set of axes, as shown in Fig. 14.72(b). Fig. 14.72(a), showing the magnitudes and relative positions of the various phasors, is called a **phasor diagram.** It is actually a "snapshot" of the rotating radius vectors at $t = 0$ s.

In the future, therefore, if the addition of two sinusoids is required, you should first convert them to the phasor domain and find the sum using complex algebra. You can then convert the result to the time domain.

The case of two sinusoidal functions having phase angles different from 0° and 90° appears in Fig. 14.73. Note again that the vertical height

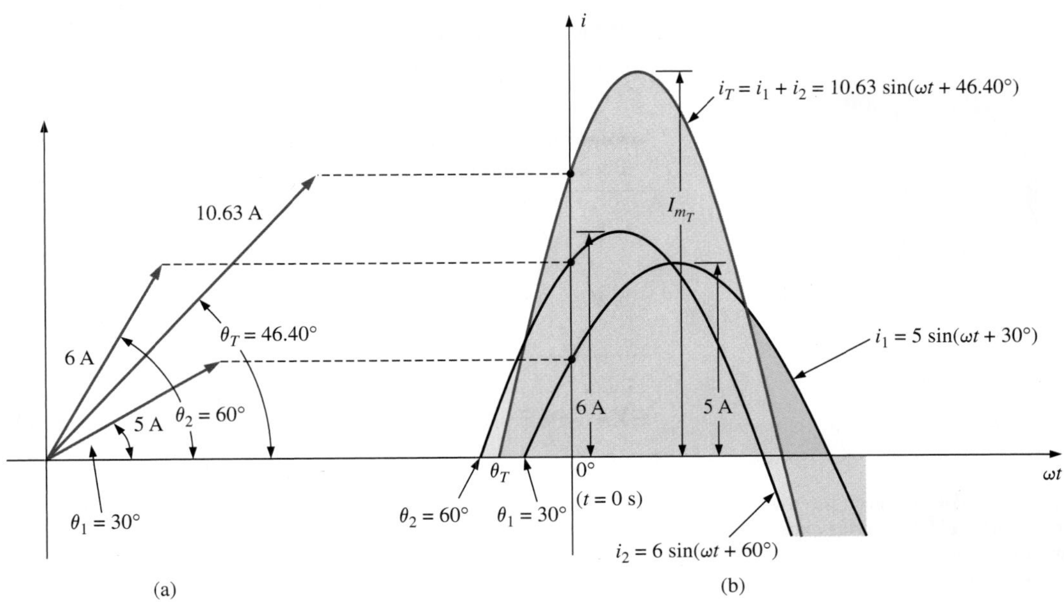

FIG. 14.73
Adding two sinusoidal currents with phase angles other than 90°.

of the functions in Fig. 14.73(b) at $t = 0$ s is determined by the rotational positions of the radius vectors in Fig. 14.73(a).

Since the rms, rather than the peak, values are used almost exclusively in the analysis of ac circuits, the phasor will now be redefined for the purposes of practicality and uniformity as having a magnitude equal to the *rms value* of the sine wave it represents. The angle associated with the phasor will remain as previously described—the phase angle.

In general, for all of the analyses to follow, the phasor form of a sinusoidal voltage or current will be

$$\mathbf{V} = V \angle\theta \quad \text{and} \quad \mathbf{I} = I \angle\theta$$

where V and I are rms values and θ is the phase angle. It should be pointed out that in phasor notation, the sine wave is always the reference, and the frequency is not represented.

Phasor algebra for sinusoidal quantities is applicable only for waveforms having the same frequency.

The use of phasor notation in the analysis of ac networks was first introduced by Professor Charles Proteus Steinmetz in 1897 (Fig. 14.74).

EXAMPLE 14.27 Convert the following from the time to the phasor domain:

Time Domain	Phasor Domain
a. $\sqrt{2}(50) \sin \omega t$	$\mathbf{50 \angle 0°}$
b. $69.6 \sin(\omega t + 72°)$	$(0.707)(69.6) \angle 72° = \mathbf{49.21 \angle 72°}$
c. $45 \cos \omega t$	$(0.707)(45) \angle 90° = \mathbf{31.82 \angle 90°}$

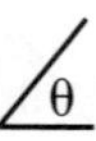

FIG. 14.74
Charles Proteus Steinmetz.
Courtesy of the Hall of History Foundaton, Schenectady, New York

German-American (Breslau, Germany; Yonkers and Schenectady, NY, USA)
(1865–1923)
Mathematician, Scientist, Engineer, Inventor, Professor of Electrical Engineering and Electrophysics, Union College
Department Head, General Electric Co.

Although the holder of some 200 patents and recognized worldwide for his contributions to the study of hysteresis losses and electrical transients. Charles Proteus Steinmetz is best recognized for his contribution to the study of ac networks. His "Symbolic Method of Alternating-current Calculations" provided an approach to the analysis of ac networks that removed a great deal of the confusion and frustration experienced by engineers of that day as they made the transition from dc to ac systems. His approach (from which the phasor notation of this text is premised) permitted a direct analysis of ac systems using many of the theorems and methods of analysis developed for dc systems. In 1897 he authored the epic work *Theory and Calculation of Alternating Current Phenomena,* which became the authoritative guide for practicing engineers. Dr. Steinmetz was fondly referred to as "The Doctor" at General Electric Company where he worked for some 30 years in a number of important capacities. His recognition as a "multigifted genius" is supported by the fact that he maintained active friendships with such individuals as Albert Einstein, Guglielmo Marconi (radio), and Thomas A. Edison, to name just a few. He was President of the American Institute of Electrical Engineers (AIEE) and the National Association of Corporation Schools and actively supported his local community (Schenectady) as president of the Board of Education and the Commission on Parks and City Planning.

EXAMPLE 14.28 Write the sinusoidal expression for the following phasors if the frequency is 60 Hz:

Phasor Domain	Time Domain
a. $\mathbf{I} = 10\ \angle 30^\circ$	$i = \sqrt{2}(10)\sin(2\pi 60t + 30^\circ)$ and $i =$ $\mathbf{14.14\sin(377t + 30^\circ)}$
b. $\mathbf{V} = 115\ \angle -70^\circ$	$v = \sqrt{2}(115)\sin(377t - 70^\circ)$ and $v =$ $\mathbf{162.6\sin(377t - 70^\circ)}$

EXAMPLE 14.29 Find the input voltage of the circuit in Fig. 14.75 if

$$\left.\begin{aligned} v_a &= 50\sin(377t + 30^\circ) \\ v_b &= 30\sin(377t + 60^\circ) \end{aligned}\right\} f = 60\text{ Hz}$$

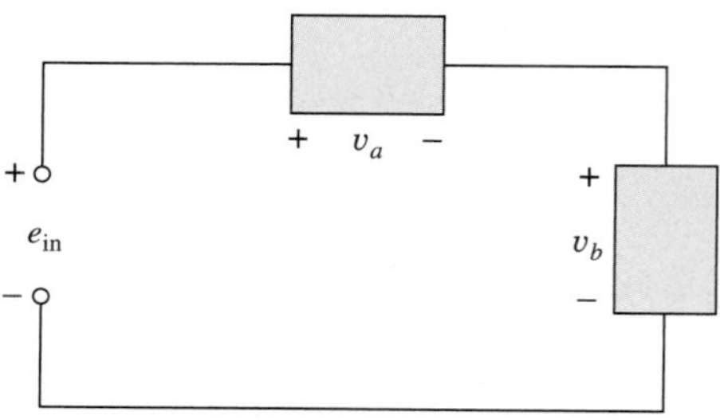

FIG. 14.75
Example 14.31.

Solution: Applying Kirchhoff's voltage law, we have

$$e_{in} = v_a + v_b$$

Converting from the time to the phasor domain yields

$$v_a = 50\sin(377t + 30^\circ) \Rightarrow \mathbf{V}_a = 35.35\text{ V}\ \angle 30^\circ$$
$$v_b = 30\sin(377t + 60^\circ) \Rightarrow \mathbf{V}_b = 21.21\text{ V}\ \angle 60^\circ$$

Converting from polar to rectangular form for addition yields

$$\mathbf{V}_a = 35.35\text{ V}\ \angle 30^\circ = 30.61\text{ V} + j17.68\text{ V}$$
$$\mathbf{V}_b = 21.21\text{ V}\ \angle 60^\circ = 10.61\text{ V} + j18.37\text{ V}$$

Then

$$\mathbf{E}_{in} = \mathbf{V}_a + \mathbf{V}_b = (30.61\text{ V} + j17.68\text{ V}) + (10.61\text{ V} + j18.37\text{ V}) = 41.22\text{ V} + j36.05\text{ V}$$

Converting from rectangular to polar form, we have

$$\mathbf{E}_{in} = 41.22\text{ V} + j36.05\text{ V} = 54.76\text{ V}\ \angle 41.17^\circ$$

Converting from the phasor to the time domain, we obtain

$$\mathbf{E}_{in} = 54.76\text{ V}\ \angle 41.17^\circ \Rightarrow e_{in} = \sqrt{2}(54.76)\sin(377t + 41.17^\circ)$$

and $$e_{in} = \mathbf{77.43\sin(377t + 41.17^\circ)}$$

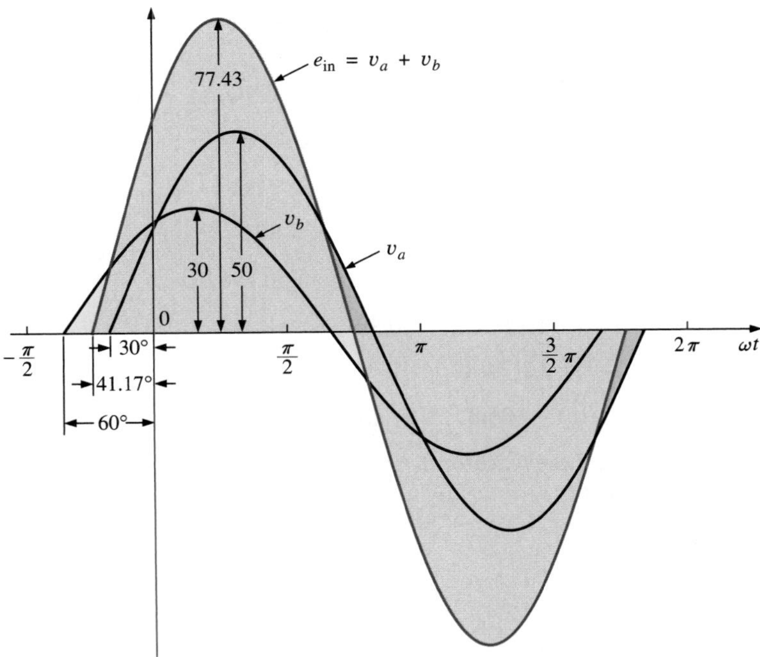

FIG. 14.76
Solution to Example 14.29.

A plot of the three waveforms is shown in Fig. 14.76. Note that at each instant of time, the sum of the two waveforms does in fact add up to e_{in}. At $t = 0$ ($\omega t = 0$), e_{in} is the sum of the two positive values, while at a value of ωt, almost midway between $\pi/2$ and π, the sum of the positive value of v_a and the negative value of v_b results in $e_{in} = 0$.

EXAMPLE 14.30 Determine the current i_2 for the network in Fig. 14.77.

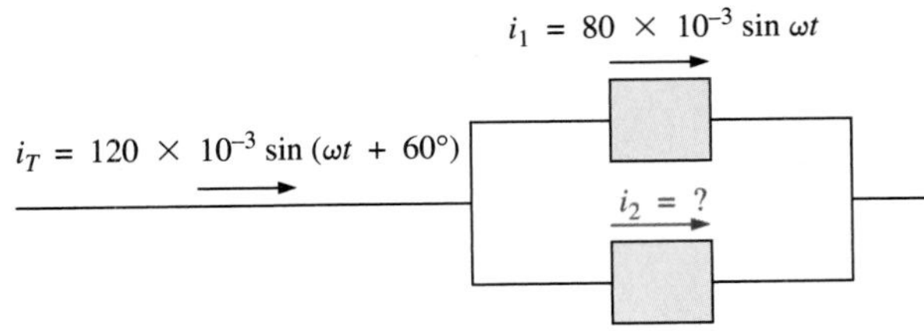

FIG. 14.77
Example 14.30.

Solution: Applying Kirchhoff's current law, we obtain

$$i_T = i_1 + i_2 \quad \text{or} \quad i_2 = i_T - i_1$$

Converting from the time to the phasor domain yields

$$i_T = 120 \times 10^{-3} \sin(\omega t + 60°) \Rightarrow 84.84 \text{ mA } \angle 60°$$
$$i_1 = 80 \times 10^{-3} \sin \omega t \Rightarrow 56.56 \text{ mA } \angle 0°$$

Converting from polar to rectangular form for subtraction yields

$$\mathbf{I}_T = 84.84 \text{ mA } \angle 60° = 42.42 \text{ mA} + j73.47 \text{ mA}$$
$$\mathbf{I}_1 = 56.56 \text{ mA } \angle 0° = 56.56 \text{ mA} + j0$$

Then

$$\mathbf{I}_2 = \mathbf{I}_T - \mathbf{I}_1$$
$$= (42.42 \text{ mA} + j73.47 \text{ mA}) - (56.56 \text{ mA} + j0)$$

and
$$\mathbf{I}_2 = -14.14 \text{ mA} + j73.47 \text{ mA}$$

Converting from rectangular to polar form, we have

$$\mathbf{I}_2 = 74.82 \text{ mA} \angle 100.89°$$

Converting from the phasor to the time domain, we have

$$\mathbf{I}_2 = 74.82 \text{ mA} \angle 100.89° \Rightarrow$$
$$i_2 = \sqrt{2}(74.82 \times 10^{-3}) \sin(\omega t + 100.89°)$$

and
$$i_2 = \mathbf{105.8 \times 10^{-3} \sin(\omega t + 100.89°)}$$

A plot of the three waveforms appears in Fig. 14.78. The waveforms clearly indicate that $i_T = i_1 + i_2$.

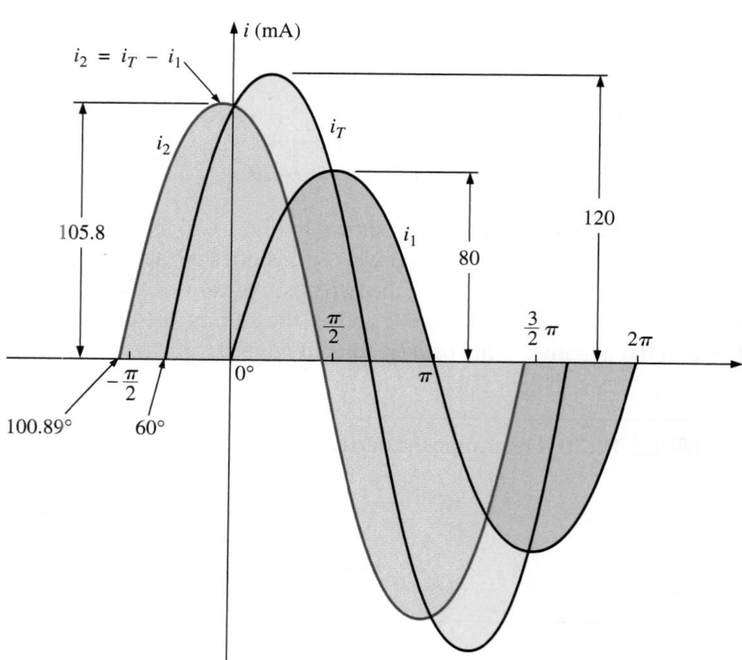

FIG. 14.78
Solution to Example 14.30.

14.13 COMPUTER ANALYSIS

PSpice

Capacitors and the ac Response The simplest of ac capacitive circuits is now analyzed to introduce the process of setting up an ac source and running an ac transient simulation. The ac source in Fig. 14.79 is obtained through **Place part** key-**SOURCE-VSIN-OK.** Change the name or value of any parameter by double-clicking on the parameter on the display or by double-clicking on the source symbol to get the **Property Editor** dialog box. Within the dialog box, set the values appearing in Fig. 14.79, and under **Display,** select **Name and Value.** After

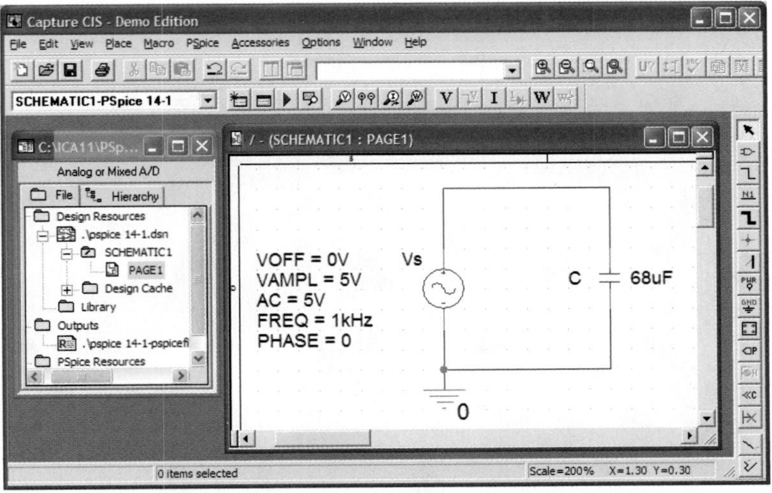

FIG. 14.79
Using PSpice to analyze the response of a capacitor to a sinusoidal ac signal.

you select **Apply** and exit the dialog box, the parameters appear as shown in Fig. 14.79.

The simulation process is initiated by selecting the **New Simulation Profile.** Under **New Simulation,** enter **PSpice 14-1** for the **Name** followed by **Create.** In the **Simulation Settings** dialog box, select **Analysis** and choose **Time Domain(Transient)** under **Analysis type.** Set the **Run to time** at 3 ms to permit a display of three cycles of the sinusoidal waveforms ($T = 1/f = 1/1000$ Hz $= 1$ ms). Leave the **Start saving data after** at 0 s, and set the **Maximum step size** at 3 ms/1000 = 3 μs. Clicking **OK** and then selecting the **Run PSpice** icon results in a plot having a horizontal axis that extends from 0 to 3 ms.

Now you must tell the computer which waveforms you are interested in. First, take a look at the applied ac source by selecting **Trace-Add Trace-V(Vs:+)** followed by **OK.** The result is the sweeping ac voltage in the bottom region of the screen in Fig. 14.80. Note that it has a peak value of 5 V, and three cycles appear in the 3 ms time frame. The current for the capacitor can be added by selecting **Trace-Add Trace** and choosing **I(C)** followed by **OK.** The resulting waveform for **I(C)** appears at a 90° phase shift from the applied voltage, with the current leading the voltage (the current has already peaked as the voltage crosses the 0 V axis). Since the peak value of each plot is in the same magnitude range, the 5 appearing on the vertical scale can be used for both. A theoretical analysis results in $X_C = 2.34\ \Omega$, and the peak value of $I_C = E/X_C = 5\text{ V}/2.34\ \Omega = 2.136$ A, as shown in Fig. 14.80.

For practice, let us obtain the curve for the power delivered to the capacitor over the same time period. First select **Plot-Add Plot to Window-Trace-Add Trace** to obtain the **Add Traces** dialog box. Then choose **V(Vs:+)**, follow it with a * for multiplication, and finish by selecting **I(C).** The result is the expression **V(Vs:+)*I(C)** of the power format: $p = vi$. Click **OK,** and the power plot at the top of Fig. 14.80 appears. Note that over the full three cycles, the area above the axis equals the area below—there is no net transfer of power over the 3 ms period. Note also that the power curve is sinusoidal (which is quite interesting) with a frequency twice that of the applied signal. Using the cursor control, we can determine that the maximum power (peak value of the sinusoidal waveform)

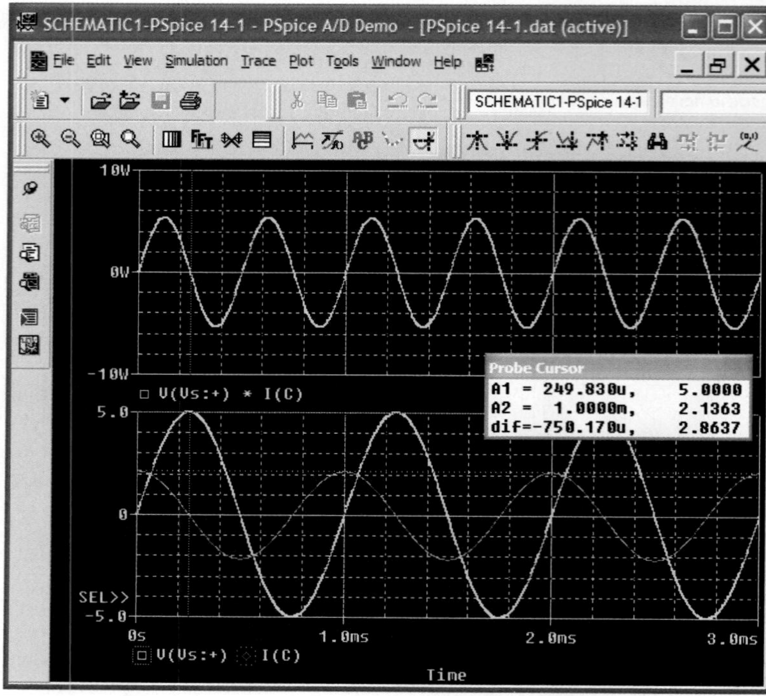

FIG. 14.80

A plot of the voltage, current, and power for the capacitor in Fig. 14.79.

is 5.34 W. The cursors, in fact, have been added to the lower curves to show the peak value of the applied sinusoid and the resulting current.

After selecting the **Toggle cursor** icon, left-click to surround the **V(Vs:+)** at the bottom of the plot with a dashed line to show that the cursor is providing the levels of that quantity. When placed at ¼ of the total period of 250 μs **(A1),** the peak value is exactly 5 V as shown in the **Probe Cursor** dialog box. Placing the cursor over the symbol next to **I(C)** at the bottom of the plot and right-clicking assigns the right cursor to the current. Placing it at exactly 1 ms (**A2**) results in a peak value of 2.136 A to match the solution above. To further distinguish between the voltage and current waveforms, the color and the width of the lines of the traces were changed. Place the cursor right on the plot line and right-click. The **Properties** option appears. When **Properties** is selected, a **Trace Properties** dialog box appears in which the yellow color can be selected and the width widened to improve the visibility on the black background. Note that yellow was chosen for **Vs** and green for **I(C).** Note also that the axis and the grid have been changed to a more visible color using the same procedure.

Multisim

Since PSpice reviewed the response of a capacitive element to an ac voltage, Multisim repeats the analysis for an inductive element. The ac voltage source was derived from the **Sources** parts bin as described in Chapter 13 with the values appearing in Fig. 14.81 set in the **AC Voltage** dialog box. Since the transient response of Multisim is limited to a plot of voltage versus time, a plot of the current of the circuit requires the addition of a resistor of 1 Ω in series with the inductive element. The magnitude of the current through the resistor and, of course, the series inductor is then determined by

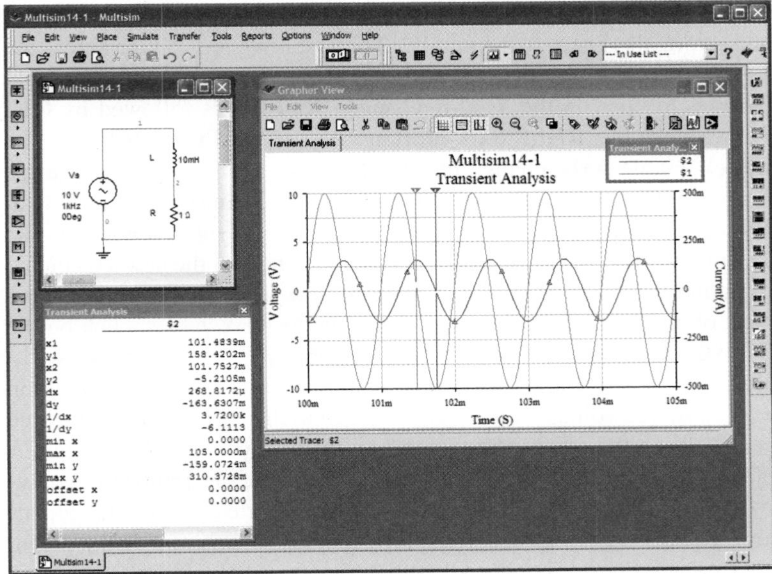

FIG. 14.81
Using Multisim to review the response of an inductive element to a sinusoidal ac signal.

$$|\,i_R\,| = \left|\frac{v_R}{R}\right| = \left|\frac{v_R}{1\,\Omega}\right| = |\,v_R\,| = |\,i_L\,|$$

revealing that the current has the same peak value as the voltage across the resistor due to the division by 1. When viewed on the graph, it can simply be considered a plot of the current. In actuality, all inductors require a series resistance, so the 1 Ω resistor serves an important dual purpose. The 1 Ω resistance is also so small compared to the reactance of the coil at the 1 kHz frequency that its effect on the total impedance or voltage across the coil can be ignored.

Once the circuit has been constructed, the sequence **Simulate-Analyses-Transient Analysis** results in a **Transient Analysis** dialog box in which the **Start time** is set at 0 s and the **End time** at 105 ms. The 105 ms was set as the **End time** to give the network 100 ms to settle down in its steady-state mode and 5 ms for five cycles in the output display. The **Minimum number of time points** was set at 10,000 to ensure a good display for the rapidly changing waveforms.

Next the **Output variables** heading was chosen within the dialog box, and nodes **1** and **2** were moved from the **Variables in Circuit** to **Selected variables for analysis** using the **Add** option. Choosing **Simulate** results in a waveform that extends from 0 s to 105 ms. Even though we plan to save only the response that occurs after 100 ms, the computer is unaware of our interest, and it plots the response for the entire period. This is corrected by selecting the **Properties** keypad in the toolbar at the top of the graph (it looks like a tag and pencil) to obtain the **Graph Properties** dialog box. Selecting **Bottom Axis** permits setting the **Range** from a **Minimum of 0.100s=100ms** to a **Maximum of 0.105s=105ms.** Click **OK,** and the time period of Fig. 14.81 is displayed. The grid structure is added by selecting the **Show/Hide Grid** keypad, and the color associated with each nodal voltage is displayed if we choose the **Show/Hide Legend** key next to it.

The scale for the plot of i_L can be improved by first going to **Traces** and setting the **Trace** to the number **2** representing the voltage across the 1 Ω resistor. When **2** is selected, the **Color** displayed automatically changes to blue. In the **Y Range,** select **Right Axis** followed by **OK.** Then select the **Right Axis** heading, and enter **Current(A)** for the **Label,** enable **Axis,** change the **Pen Size** to 1, and change the **Range** from −500 mA to +500 mA. Finally, set the **Total Ticks** at 8 with **Minor Ticks** at 2 to match the **Left Axis,** and leave the box with an **OK.** The plot in Fig. 14.81 results. Take immediate note of the new axis on the right and the **Current(A)** label. We can now see that the current has a peak of about 160 mA. For more detail on the peak values, click on the **Show/Hide Cursors** keypad on the top toolbar. A **Transient Analysis** dialog box appears with a **1** and a red line to indicate that it is working on the full source voltage at node **1.** To switch to the current curve (the blue curve), bring the cursor to any point on the blue curve and left-click. A blue line and the number **2** appear at the heading of the **Transient Analysis** dialog box. Clicking on the **1** in the small inverted arrow at the top allows you to drag the vertical red line to any horizontal point on the graph. As shown in Fig. 14.81, when the cursor is set on 101.5 ms (**x1**), the peak value of the current curve is 159.05 mA (**y1**). A second cursor appears in blue with a number **2** in the inverted arrowhead that can also be moved with a left click on the number **2** at the top of the line. If set at 101.75 ms (**x2**), it has a minimum value of −5.18 mA (**y2**), the smallest value available for the calculated data points. Note that the difference between horizontal time values **dx** = 252 μs = 0.25 ms which is ¼ of the period of the wave (at 1 ms).

PROBLEMS

SECTION 14.2 Derivative

1. Plot the following waveform versus time showing one clear, complete cycle. Then determine the derivative of the waveform using Eq. (14.1), and sketch one complete cycle of the derivative directly under the original waveform. Compare the magnitude of the derivative at various points versus the slope of the original sinusoidal function.

$$v = 1 \sin 3.14t$$

2. Repeat Problem 1 for the following sinusoidal function, and compare results. In particular, determine the frequency of the waveforms of Problems 1 and 2, and compare the magnitude of the derivative.

$$v = 1 \sin 15.71t$$

3. What is the derivative of each of the following sinusoidal expressions?
   a. $10 \sin 377t$ **b.** $0.6 \sin(754t + 20°)$
   c. $\sqrt{2}\, 20 \sin(157t - 20°)$ **d.** $-200 \sin(t + 180°)$

SECTION 14.3 Response of Basic *R, L,* and *C* Elements to a Sinusoidal Voltage or Current

4. The voltage across a 5 Ω resistor is as indicated. Find the sinusoidal expression for the current. In addition, sketch the v and i sinusoidal waveforms on the same axis.
   a. $150 \sin 200t$ **b.** $30 \sin(377t + 20°)$
   c. $40 \cos(\omega t + 10°)$ **d.** $-80 \sin(\omega t + 40°)$

5. The current through a 7 kΩ resistor is as indicated. Find the sinusoidal expression for the voltage. In addition, sketch the v and i sinusoidal waveforms on the same axis.
   a. $0.1 \sin 1000t$
   b. $2 \times 10^{-3} \sin(400t - 120°)$
   c. $6 \times 10^{-6} \cos(\omega t - 2°)$
   d. $-0.004 \cos(\omega t + 90°)$

6. Determine the inductive reactance (in ohms) of a 2 H coil for
   a. dc
   and for the following frequencies:
   b. 10 Hz **c.** 60 Hz
   d. 2000 Hz **e.** 100,000 Hz

7. Determine the inductance of a coil that has a reactance of
   a. 20 Ω at f = 2 Hz.
   b. 1000 Ω at f = 60 Hz.
   c. 5280 Ω at f = 500 Hz.

8. Determine the frequency at which a 10 H inductance has the following inductive reactances:
   a. 100 Ω **b.** 3770 Ω
   c. 15.7 kΩ **d.** 243 Ω

9. The current through a 20 Ω inductive reactance is given. What is the sinusoidal expression for the voltage? Sketch the v and i sinusoidal waveforms on the same axis.
   a. $i = 5 \sin \omega t$
   b. $i = 40 \times 10^{-3} \sin(\omega t + 60°)$
   c. $i = -6 \sin(\omega t - 30°)$
   d. $i = 3 \cos(\omega t + 10°)$

10. The current through a 0.1 H coil is given. What is the sinusoidal expression for the voltage?
    a. $10 \sin 100t$
    b. $6 \times 10^{-3} \sin 377t$
    c. $5 \times 10^{-6} \sin(400t + 20°)$
    d. $-4 \cos(20t - 70°)$

11. The voltage across a 50 Ω inductive reactance is given. What is the sinusoidal expression for the current? Sketch the v and i sinusoidal waveforms on the same set of axes.
    a. $120 \sin \omega t$
    b. $30 \sin(\omega t + 20°)$
    c. $40 \cos(\omega t + 10°)$
    d. $-80 \sin(377t + 40°)$

12. The voltage across a 0.2 H coil is given. What is the sinusoidal expression for the current?
    a. $1.5 \sin 60t$
    b. $16 \times 10^{-3} \sin(10t + 2°)$
    c. $-4.8 \sin(0.05t + 50°)$
    d. $9 \times 10^{-3} \cos(377t + 360°)$

13. Determine the capacitive reactance (in ohms) of a 5 μF capacitor for
    a. dc
    and for the following frequencies:
    b. 60 Hz
    c. 120 Hz
    d. 2 kHz
    e. 2 MHz

14. Determine the capacitance in microfarads if a capacitor has a reactance of
    a. 250 Ω at f = 60 Hz.
    b. 55 Ω at f = 312 Hz.
    c. 10 Ω at f = 25 Hz.

15. Determine the frequency at which a 50 μF capacitor has the following capacitive reactances:
    a. 100 Ω
    b. 684 Ω
    c. 342 Ω
    d. 2000 Ω

16. The voltage across a 2.5 Ω capacitive reactance is given. What is the sinusoidal expression for the current? Sketch the v and i sinusoidal waveforms on the same set of axes.
    a. $120 \sin \omega t$
    b. $0.4 \sin(\omega t + 20°)$
    c. $8 \cos(\omega t + 10°)$
    d. $-70 \sin(\omega t + 40°)$

17. The voltage across a 1 μF capacitor is given. What is the sinusoidal expression for the current?
    a. $30 \sin 200t$
    b. $60 \times 10^{-3} \sin 377t$
    c. $-120 \sin(374t + 30°)$
    d. $70 \cos(800t - 20°)$

18. The current through a 10 Ω capacitive reactance is given. Write the sinusoidal expression for the voltage. Sketch the v and i sinusoidal waveforms on the same set of axes.
    a. $i = 50 \times 10^{-3} \sin \omega t$
    b. $i = 2 \times 10^{-6} \sin(\omega t + 60°)$
    c. $i = -6 \sin(\omega t - 30°)$
    d. $i = 3 \cos(\omega t + 10°)$

19. The current through a 0.5 μF capacitor is given. What is the sinusoidal expression for the voltage?
    a. $0.20 \sin 300t$
    b. $8 \times 10^{-3} \sin 377t$
    c. $60 \times 10^{-3} \cos 754t$
    d. $0.08 \sin(1600t - 80°)$

*20. For the following pairs of voltages and currents, indicate whether the element involved is a capacitor, an inductor, or a resistor, and the value of C, L or R if sufficient data are given:
    a. $v = 550 \sin(377t + 50°)$
       $i = 11 \sin(377t - 40°)$
    b. $v = 36 \sin(754t - 80°)$
       $i = 4 \sin(754t - 170°)$
    c. $v = 10.5 \sin(\omega t - 13°)$
       $i = 1.5 \sin(\omega t - 13°)$

*21. Repeat Problem 20 for the following pairs of voltages and currents:
    a. $v = 2000 \sin \omega t$
       $i = 5 \cos \omega t$
    b. $v = 80 \sin(157t + 150°)$
       $i = 2 \sin(157t + 60°)$
    c. $v = 35 \sin(\omega t - 20°)$
       $i = 7 \cos(\omega t - 110°)$

SECTION 14.4 Frequency Response of the Basic Elements

22. Plot X_L versus frequency for a 5 mH coil using a frequency range of zero to 100 kHz on a linear scale.

23. Plot X_C versus frequency for a 1 μF capacitor using a frequency range of zero to 10 kHz on a linear scale.

24. At what frequency will the reactance of a 1 μF capacitor equal the resistance of a 2 kΩ resistor?

25. The reactance of a coil equals the resistance of a 10 kΩ resistor at a frequency of 5 kHz. Determine the inductance of the coil.

26. Determine the frequency at which a 1 μF capacitor and a 10 mH inductor will have the same reactance.

27. Determine the capacitance required to establish a capacitive reactance that will match that of a 2 mH coil at a frequency of 50 kHz.

SECTION 14.5 Average Power and Power Factor

28. Find the average power loss in watts for each set in Problem 20.

29. Find the average power loss in watts for each set in Problem 21.

*30. Find the average power loss and power factor for each of the circuits whose input current and voltage are as follows:
    a. $v = 60 \sin(\omega t + 30°)$
       $i = 15 \sin(\omega t + 60°)$
    b. $v = -50 \sin(\omega t - 20°)$
       $i = -2 \sin(\omega t - 20°)$
    c. $v = 50 \sin(\omega t + 80°)$
       $i = 3 \cos(\omega t - 20°)$
    d. $v = 75 \sin(\omega t - 5°)$
       $i = 0.08 \sin(\omega t + 35°)$

31. If the current through and voltage across an element are $i = 8 \sin(\omega t + 40°)$ and $v = 48 \sin(\omega t + 40°)$, respectively, compute the power by I^2R, $(V_mI_m/2) \cos \theta$, and $VI \cos \theta$, and compare answers.

32. A circuit dissipates 100 W (average power) at 150 V (effective input voltage) and 2 A (effective input current). What is the power factor? Repeat if the power is 0 W; 300 W.

*33. The power factor of a circuit is 0.5 lagging. The power delivered in watts is 500. If the input voltage is $50 \sin(\omega t + 10°)$, find the sinusoidal expression for the input current.

34. In Fig. 14.82, $e = 30 \sin(377t + 20°)$.
   a. What is the sinusoidal expression for the current?
   b. Find the power loss in the circuit.
   c. How long (in seconds) does it take the current to complete six cycles?

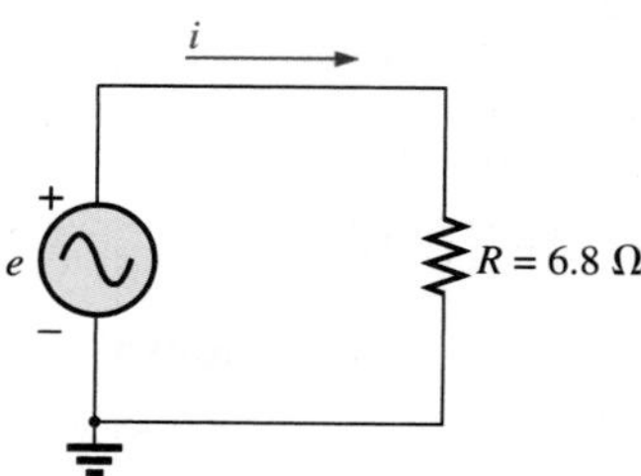

FIG. 14.82
Problem 34.

35. In Fig. 14.83, $e = 100 \sin(314t + 60°)$.
   a. Find the sinusoidal expression for i.
   b. Find the value of the inductance L.
   c. Find the average power loss by the inductor.

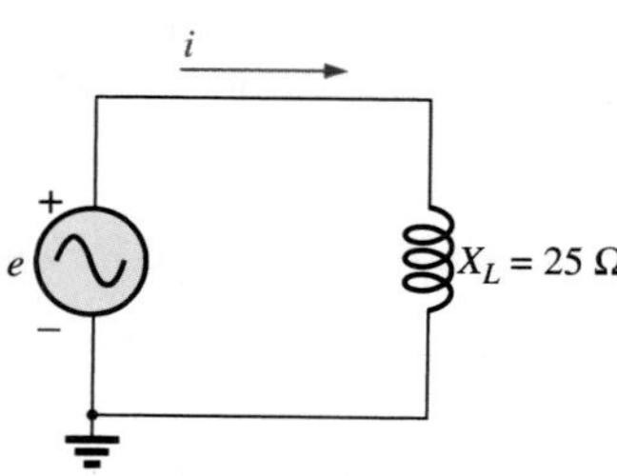

FIG. 14.83
Problem 35.

36. In Fig. 14.84, $i = 30 \times 10^{-3} \sin(377t - 20°)$.
   a. Find the sinusoidal expression for e.
   b. Find the value of the capacitance C in microfarads.
   c. Find the average power loss in the capacitor.

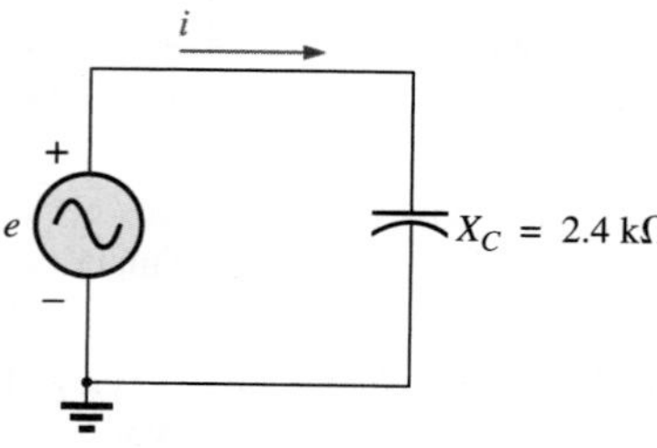

FIG. 14.84
Problem 36.

*37. For the network in Fig. 14.85 and the applied signal:
   a. Determine i_1 and i_2.
   b. Find i_s.

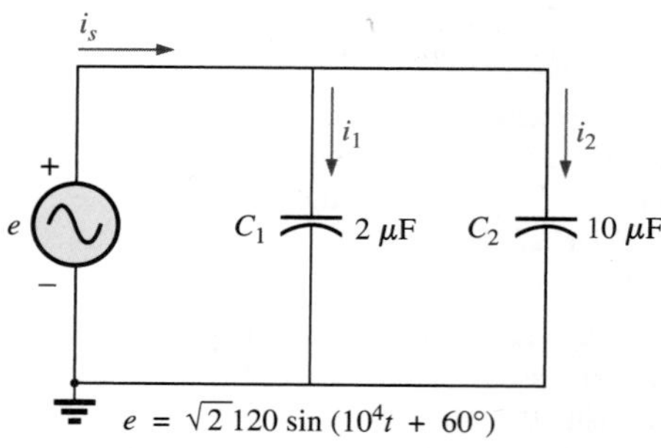

FIG. 14.85
Problem 37.

*38. For the network in Fig. 14.86 and the applied source:
   a. Determine the source voltage v_s.
   b. Find the currents i_1 and i_2.

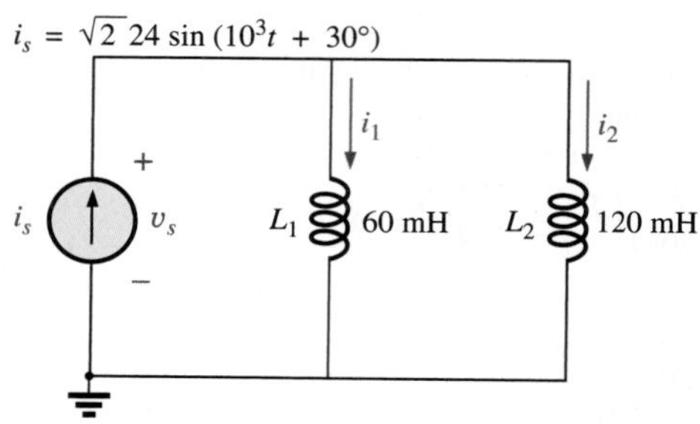

FIG. 14.86
Problem 38.

SECTION 14.9 Conversion between Forms

39. Convert the following from rectangular to polar form:
   a. $4 + j3$
   b. $2 + j2$
   c. $6 + j16$
   d. $100 + j1000$
   e. $1000 + j400$
   f. $0.001 + j0.0065$
   g. $7.6 - j9$
   h. $-8 - j4$
   i. $-15 - j60$
   j. $+78 - j65.3$
   k. $-2400 + j3600$
   l. $5 \times 10^{-3} - j25 \times 10^{-3}$

40. Convert the following from polar to rectangular form:
   a. $6 \angle 30°$
   b. $40 \angle 80°$
   c. $7400 \angle 70°$
   d. $4 \times 10^{-4} \angle 8°$
   e. $0.04 \angle 90°$
   f. $0.0093 \angle 42°$
   g. $65 \angle 150°$
   h. $1.2 \angle 135°$
   i. $500 \angle 200°$
   j. $6320 \angle -35°$
   k. $7.52 \angle -125°$
   l. $8 \times 10^{-3} \angle 210°$

41. Convert the following from rectangular to polar form:
   a. $1 + j15$
   b. $60 + j5$
   c. $0.01 + j0.3$
   d. $100 - j200$
   e. $-5.6 + j86$
   f. $-2.7 - j38.6$

42. Convert the following from polar to rectangular form:
   a. $13 \angle 5°$
   b. $160 \angle 87°$
   c. $7 \times 10^{-6} \angle 2°$
   d. $8.7 \angle 177°$
   e. $76 \angle -4°$
   f. $396 \angle +265°$

SECTION 14.10 Mathematical Operations with Complex Numbers

Perform the following operations.

43. Addition and subtraction (express your answers in rectangular form):
- **a.** $(4.2 + j6.8) + (7.6 + j0.2)$
- **b.** $(142 + j7) + (9.8 + j42) + (0.1 + j0.9)$
- **c.** $(4 \times 10^{-6} + j76) + (7.2 \times 10^{-7} - j5)$
- **d.** $(9.8 + j6.2) - (4.6 + j4.6)$
- **e.** $(167 + j243) - (-42.3 - j68)$
- **f.** $(-36.0 + j78) - (-4 - j6) + (10.8 - j72)$
- **g.** $6 \angle 20° + 8 \angle 80°$
- **h.** $42 \angle 45° + 62 \angle 60° - 70 \angle 120°$

44. Multiplication [express your answers in rectangular form for parts (a) through (d), and in polar form for parts (e) through (h)]:
- **a.** $(2 + j3)(6 + j8)$
- **b.** $(7.8 + j1)(4 + j2)(7 + j6)$
- **c.** $(0.002 + j0.006)(-4 + j8)$
- **d.** $(400 - j200)(-0.01 - j0.5)(-1 + j3)$
- **e.** $(2 \angle 60°)(4 \angle -40°)$
- **f.** $(6.9 \angle 8°)(7.2 \angle -72°)$
- **g.** $(0.002 \angle 120°)(0.5 \angle 200°)(40 \angle +80°)$
- **h.** $(540 \angle -20°)(-5 \angle 180°)(6.2 \angle 0°)$

45. Division (express your answer in polar form):
- **a.** $(42 \angle 10°)/(7 \angle 60°)$
- **b.** $(0.006 \angle 120°)/(30 \angle +60°)$
- **c.** $(4360 \angle -20°)/(40 \angle -210°)$
- **d.** $(650 \angle -80°)/(8.5 \angle 360°)$
- **e.** $(8 + j8)/(2 + j2)$
- **f.** $(8 + j42)/(-6 - j4)$
- **g.** $(0.05 + j0.25)/(8 - j60)$
- **h.** $(-4.5 - j6)/(0.1 - j0.8)$

***46.** Perform the following operations (express your answers in rectangular form):

a. $$\frac{(4 + j3) + (6 - j8)}{(3 + j3) - (2 + j3)}$$

b. $$\frac{8 \angle 60°}{(2 \angle 0°) + (100 + j400)}$$

c. $$\frac{(6 \angle 20°)(120 \angle -40°)(3 + j8)}{2 \angle -30°}$$

d. $$\frac{(0.4 \angle 60°)^2(300 \angle 40°)}{3 + j9}$$

e. $$\left(\frac{1}{(0.02 \angle 10°)^2}\right)\left(\frac{2}{j}\right)^3\left(\frac{1}{6^2 - j\sqrt{900}}\right)$$

***47. a.** Determine a solution for x and y if

$$(x + j4) + (3x + jy) - j7 = 16 \angle 0°$$

b. Determine x if

$$(10 \angle 20°)(x \angle -60°) = 30.64 - j25.72$$

c. Determine a solution for x and y if

$$(5x + j10)(2 - jy) = 90 - j70$$

d. Determine θ if

$$\frac{80 \angle 0°}{20 \angle \theta} = 3.464 - j2$$

SECTION 14.12 Phasors

48. Express the following in phasor form:
- **a.** $\sqrt{2}(160) \sin(\omega t + 30°)$
- **b.** $\sqrt{2}(25 \times 10^{-3}) \sin(157t - 40°)$
- **c.** $100 \sin(\omega t - 90°)$
- **d.** $20 \sin(377t + 0°)$
- **e.** $6 \times 10^{-6} \cos \omega t$
- **f.** $3.6 \times 10^{-6} \cos(754t - 20°)$

49. Express the following phasor currents and voltages as sine waves if the frequency is 60 Hz:
- **a.** $\mathbf{I} = 40 \text{ A} \angle 20°$
- **b.** $\mathbf{V} = 120 \text{ V} \angle 10°$
- **c.** $\mathbf{I} = 8 \times 10^{-3} \text{ A} \angle 120°$
- **d.** $\mathbf{V} = 5 \text{ V} \angle 90°$
- **e.** $\mathbf{I} = 1200 \text{ A} \angle -50°$
- **f.** $\mathbf{V} = \frac{6000}{\sqrt{2}} \text{ V} \angle -180°$

50. For the system in Fig. 14.87, find the sinusoidal expression for the unknown voltage v_a if

$$e_{in} = 60 \sin(377t + 20°)$$
$$v_b = 20 \sin(377t - 20°)$$

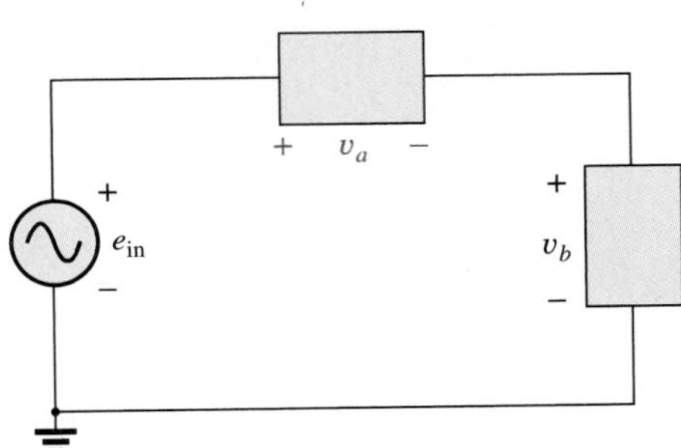

FIG. 14.87
Problem 50.

51. For the system in Fig. 14.88, find the sinusoidal expression for the unknown current i_1 if

$$i_s = 20 \times 10^{-6} \sin(\omega t + 60°)$$
$$i_2 = 6 \times 10^{-6} \sin(\omega t - 30°)$$

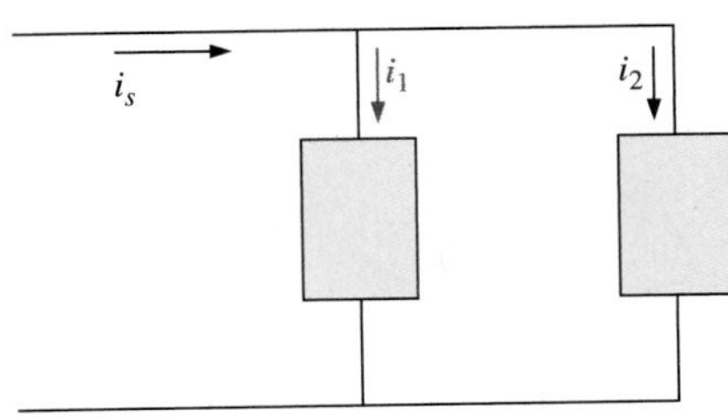

FIG. 14.88
Problem 51.

52. Find the sinusoidal expression for the applied voltage e for the system in Fig. 14.89 if

$$v_a = 60 \sin(\omega t + 30°)$$
$$v_b = 30 \sin(\omega t + 60°)$$
$$v_c = 40 \sin(\omega t + 120°)$$

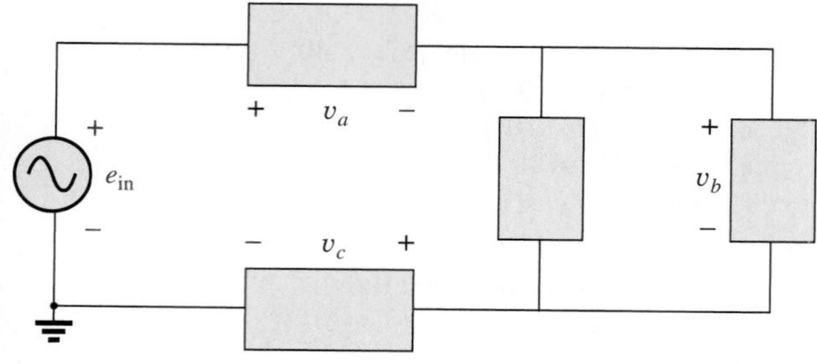

FIG. 14.89
Problem 52.

53. Find the sinusoidal expression for the current i_s for the system in Fig. 14.90 if

$$i_1 = 6 \times 10^{-3} \sin(377t + 180°)$$
$$i_2 = 8 \times 10^{-3} \sin(377t - 180°)$$
$$i_3 = 2i_2$$

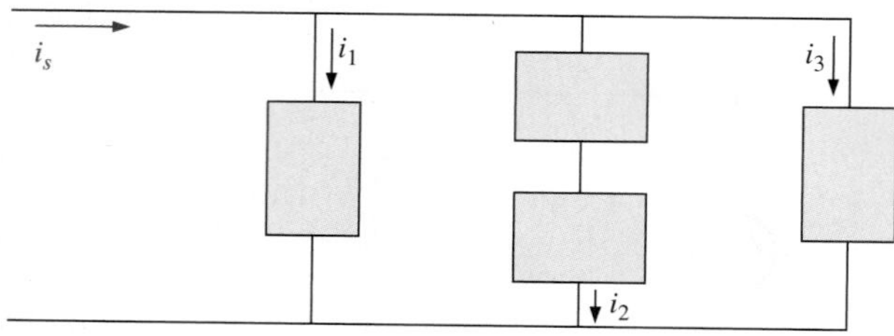

FIG. 14.90
Problem 53.

SECTION 14.13 Computer Analysis

PSpice or Multisim

54. Plot i_c and v_c versus time for the network in Fig. 14.79 for two cycles if the frequency is 0.2 kHz.

55. Plot the magnitude and phase angle of the current i_C versus frequency (100 Hz to 100 kHz) for the network in Fig. 14.79.

56. Plot the total impedance of the configuration in Fig. 14.27 versus frequency (100 kHz to 100 MHz) for the following parameter values: $C = 0.1\ \mu F$, $L_s = 0.2\ \mu H$, $R_s = 2\ M\Omega$, and $R_p = 100\ M\Omega$. For what frequency range is the capacitor "capacitive"?

GLOSSARY

Average or **real power** The power delivered to and dissipated by the load over a full cycle.

Complex conjugate A complex number defined by simply changing the sign of an imaginary component of a complex number in the rectangular form.

Complex number A number that represents a point in a two-dimensional plane located with reference to two distinct axes. It defines a vector drawn from the origin to that point.

Derivative The instantaneous rate of change of a function with respect to time or another variable.

Leading and **lagging power factors** An indication of whether a network is primarily capacitive or inductive in nature. Leading power factors are associated with capacitive networks, and lagging power factors with inductive networks.

Phasor A radius vector that has a constant magnitude at a fixed angle from the positive real axis and that represents a sinusoidal voltage or current in the vector domain.

Phasor diagram A "snapshot" of the phasors that represent a number of sinusoidal waveforms at $t = 0$.

Polar form A method of defining a point in a complex plane that includes a single magnitude to represent the distance from the origin, and an angle to reflect the counterclockwise distance from the positive real axis.

Power factor (F_p) An indication of how reactive or resistive an electrical system is. The higher the power factor, the greater the resistive component.

Reactance The opposition of an inductor or a capacitor to the flow of charge that results in the continual exchange of energy between the circuit and magnetic field of an inductor or the electric field of a capacitor.

Reciprocal A format defined by 1 divided by the complex number.

Rectangular form A method of defining a point in a complex plane that includes the magnitude of the real component and the magnitude of the imaginary component, the latter component being defined by an associated letter j.

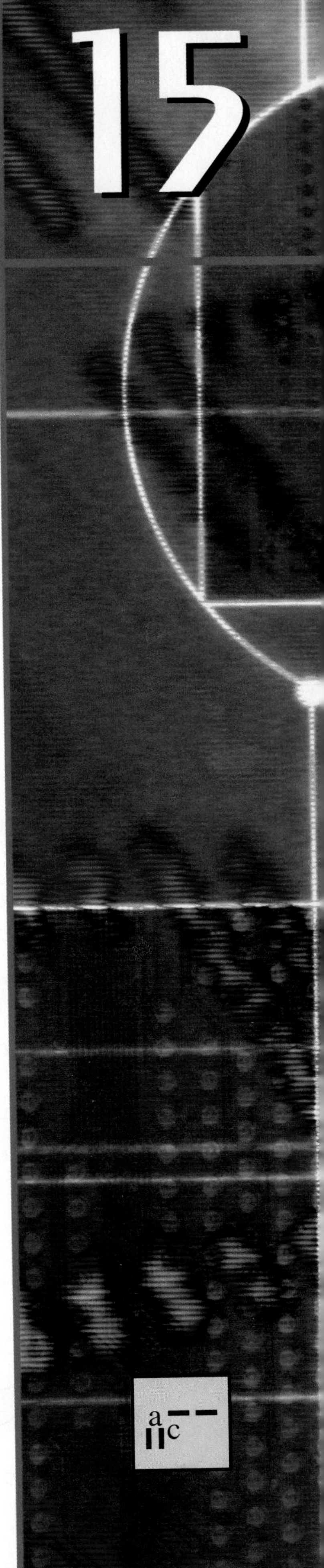

Series and Parallel ac Circuits

Objectives

- **Become familiar with the characteristics of series and parallel ac networks and be able to find current, voltage, and power levels for each element.**
- **Be able to find the total impedance of any series or parallel ac network and sketch the impedance and admittance diagram of each.**
- **Develop confidence in applying Kirchhoff's current and voltage law to any series or parallel configuration.**
- **Be able to apply the voltage divider rule or current divider rule to any ac network.**
- **Become adept at finding the frequency response of a series or parallel combination of elements.**

15.1 INTRODUCTION

In this chapter, phasor algebra is used to develop a quick, direct method for solving both series and parallel ac circuits. The close relationship that exists between this method for solving for unknown quantities and the approach used for dc circuits will become apparent after a few simple examples are considered. Once this association is established, many of the rules (current divider rule, voltage divider rule, and so on) for dc circuits can be readily applied to ac circuits.

SERIES ac CIRCUITS

15.2 IMPEDANCE AND THE PHASOR DIAGRAM

Resistive Elements

In Chapter 14, we found, for the purely resistive circuit in Fig. 15.1, that v and i were in phase, and the magnitude

$$I_m = \frac{V_m}{R} \quad \text{or} \quad V_m = I_m R$$

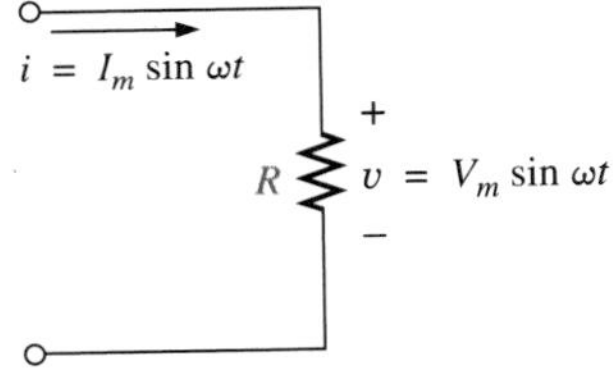

FIG. 15.1
Resistive ac circuit.

In phasor form,

$$v = V_m \sin \omega t \Rightarrow \mathbf{V} = V \angle 0°$$

where $V = 0.707V_m$.

Applying Ohm's law and using phasor algebra, we have

$$\mathbf{I} = \frac{V \angle 0°}{R \angle \theta_R} = \frac{V}{R} \angle 0° - \theta_R$$

Since i and v are in phase, the angle associated with i also must be 0°. To satisfy this condition, θ_R must equal 0°. Substituting $\theta_R = 0°$, we find

$$\mathbf{I} = \frac{V \angle 0°}{R \angle 0°} = \frac{V}{R} \angle 0° - 0° = \frac{V}{R} \angle 0°$$

so that in the time domain,

$$i = \sqrt{2}\left(\frac{V}{R}\right) \sin \omega t$$

We use the fact that $\theta_R = 0°$ in the following polar format to ensure the proper phase relationship between the voltage and current of a resistor:

$$\boxed{\mathbf{Z}_R = R \angle 0°} \qquad (15.1)$$

The boldface roman quantity $\mathbf{Z}_R$, having both magnitude and an associated angle, is referred to as the *impedance* of a resistive element. It is measured in ohms and is a measure of how much the element will "impede" the flow of charge through the network. The above format will prove to be a useful "tool" when the networks become more complex and phase relationships become less obvious. It is important to realize, however, that $\mathbf{Z}_R$ is *not a phasor,* even though the format $R \angle 0°$ is very similar to the phasor notation for sinusoidal currents and voltages. The term *phasor* is reserved for quantities that vary with time, and R and its associated angle of 0° are fixed, nonvarying quantities.

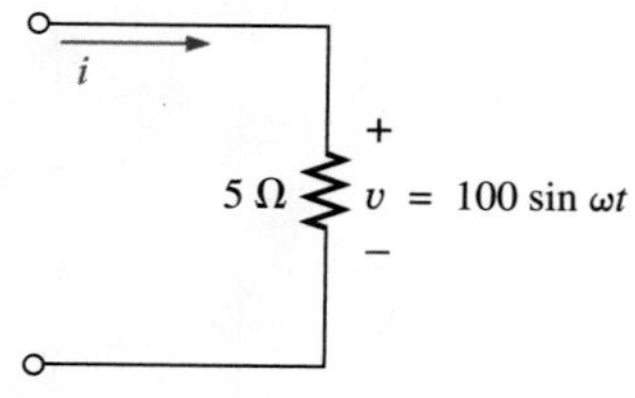

FIG. 15.2
Example 15.1.

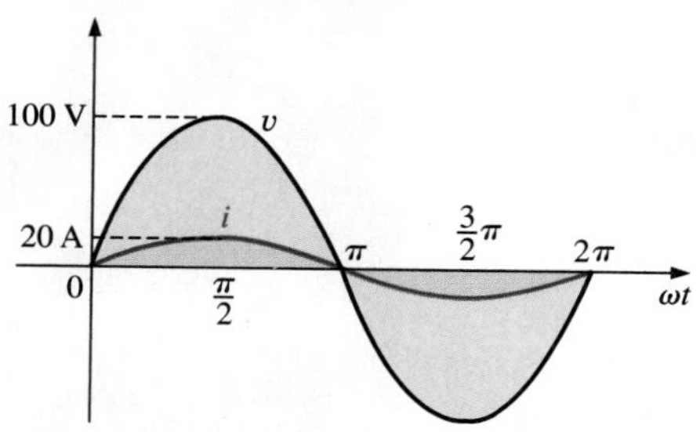

FIG. 15.3
Waveforms for Example 15.1.

EXAMPLE 15.1 Using complex algebra, find the current i for the circuit in Fig. 15.2. Sketch the waveforms of v and i.

Solution: Note Fig. 15.3:

$$v = 100 \sin \omega t \Rightarrow \text{phasor form } \mathbf{V} = 70.71 \text{ V} \angle 0°$$

$$\mathbf{I} = \frac{\mathbf{V}}{\mathbf{Z}_R} = \frac{V \angle \theta}{R \angle 0°} = \frac{70.71 \text{ V} \angle 0°}{5\ \Omega \angle 0°} = 14.14 \text{ A} \angle 0°$$

and $\quad i = \sqrt{2}(14.14) \sin \omega t = \mathbf{20 \sin \omega t}$

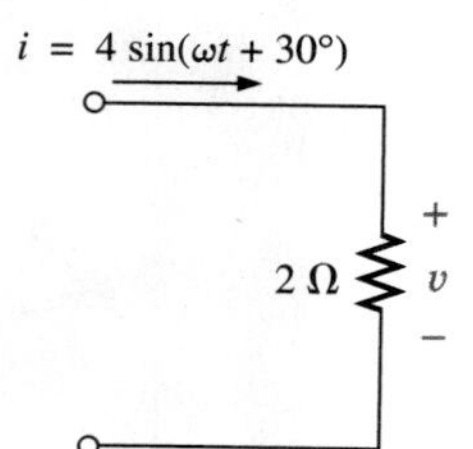

FIG. 15.4
Example 15.2.

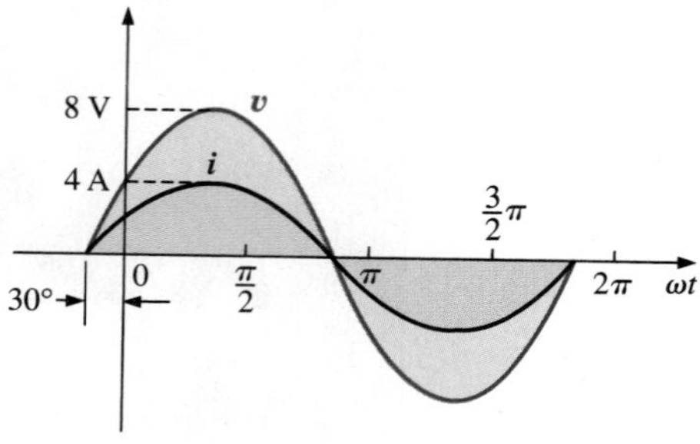

FIG. 15.5
Waveforms for Example 15.2.

EXAMPLE 15.2 Using complex algebra, find the voltage v for the circuit in Fig. 15.4. Sketch the waveforms of v and i.

Solution: Note Fig. 15.5:

$$i = 4 \sin(\omega t + 30°) \Rightarrow \text{phasor form } \mathbf{I} = 2.828 \text{ A} \angle 30°$$

$$\mathbf{V} = \mathbf{I}\mathbf{Z}_R = (I \angle \theta)(R \angle 0°) = (2.828 \text{ A} \angle 30°)(2\ \Omega \angle 0°) = 5.656 \text{ V} \angle 30°$$

and $\quad v = \sqrt{2}(5.656) \sin(\omega t + 30°) = \mathbf{8.0 \sin(\omega t + 30°)}$

It is often helpful in the analysis of networks to have a **phasor diagram,** which shows at a glance the *magnitudes* and *phase relations* among the various quantities within the network. For example, the phasor diagrams of the circuits considered in the two preceding examples would be as shown in Fig. 15.6. In both cases, it is immediately obvious that v and i are in phase since they both have the same phase angle.

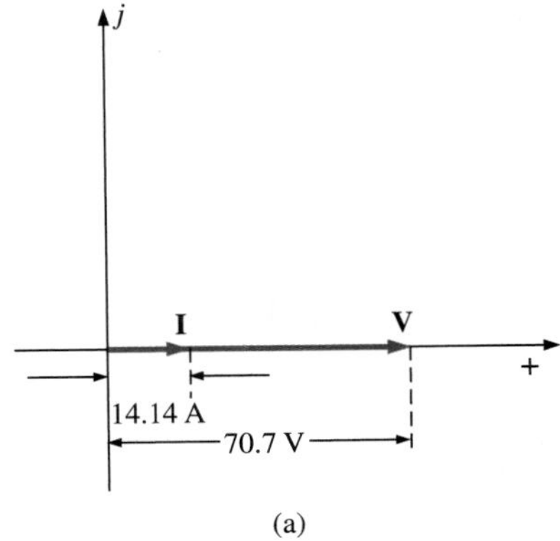

(a)

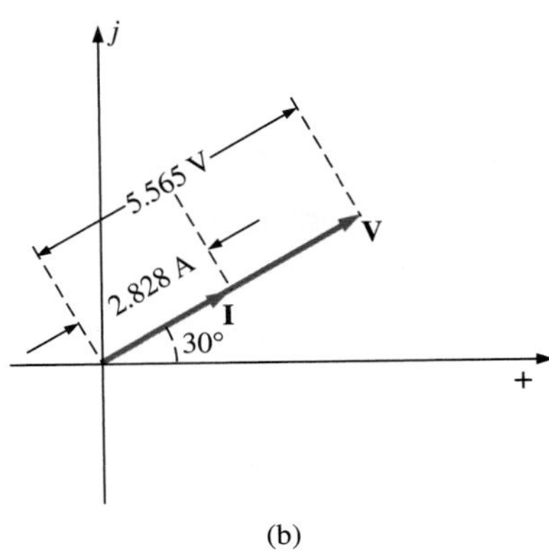

(b)

FIG. 15.6
Phasor diagrams for Examples 15.1 and 15.2.

Inductive Reactance

We learned in Chapter 13 that for the pure inductor in Fig. 15.7, the voltage leads the current by 90° and that the reactance of the coil X_L is determined by ωL.

$$v = V_m \sin \omega t \Rightarrow \text{phasor form } \mathbf{V} = V \angle 0^\circ$$

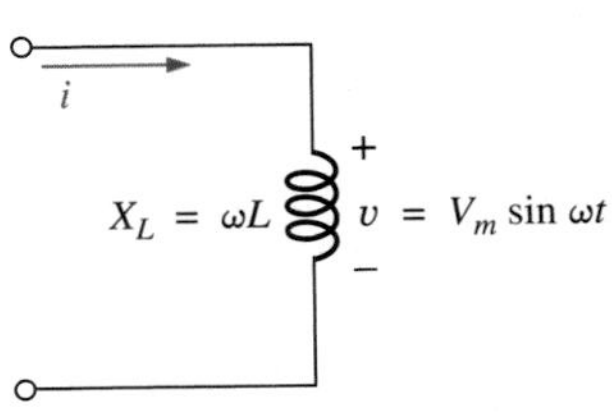

FIG. 15.7
Inductive ac circuit.

By Ohm's law,

$$\mathbf{I} = \frac{V \angle 0^\circ}{X_L \angle \theta_L} = \frac{V}{X_L} \angle 0^\circ - \theta_L$$

Since v leads i by 90°, i must have an angle of −90° associated with it. To satisfy this condition, θ_L must equal +90°. Substituting $\theta_L = 90°$, we obtain

$$\mathbf{I} = \frac{V \angle 0^\circ}{X_L \angle 90^\circ} = \frac{V}{X_L} \angle 0^\circ - 90^\circ = \frac{V}{X_L} \angle -90^\circ$$

so that in the time domain,

$$i = \sqrt{2}\left(\frac{V}{X_L}\right) \sin(\omega t - 90^\circ)$$

We use the fact that $\theta_L = 90°$ in the following polar format for inductive reactance to ensure the proper phase relationship between the voltage and current of an inductor:

$$\boxed{\mathbf{Z}_L = X_L \angle 90^\circ} \qquad \textbf{(15.2)}$$

The boldface roman quantity $\mathbf{Z}_L$, having both magnitude and an associated angle, is referred to as the *impedance* of an inductive element. It is measured in ohms and is a measure of how much the inductive element "controls or impedes" the level of current through the network (always keep in mind that inductive elements are storage devices and do not dissipate like resistors). The above format, like that defined for the resistive element, will prove to be a useful tool in the analysis of ac networks. Again, be aware that $\mathbf{Z}_L$ is not a phasor quantity, for the same reasons indicated for a resistive element.

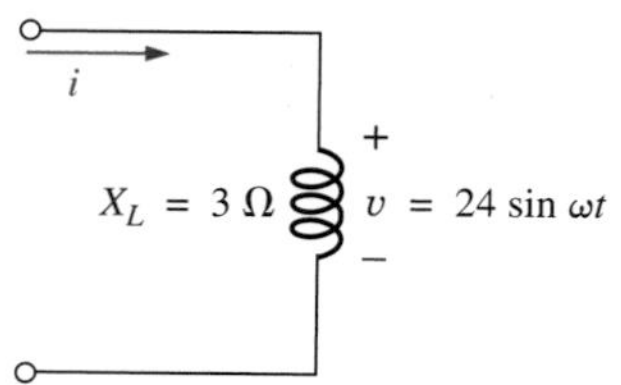

FIG. 15.8
Example 15.3.

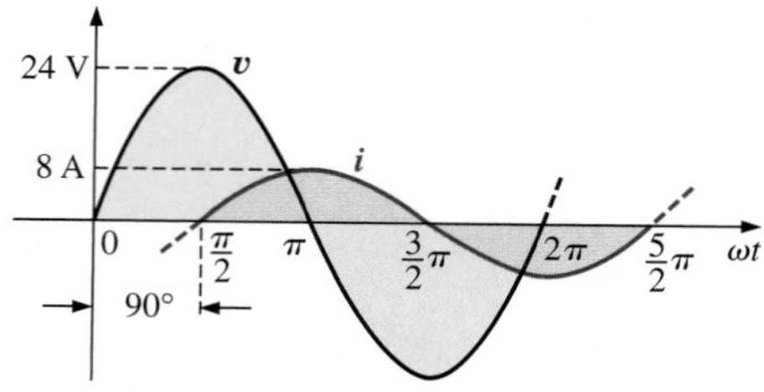

FIG. 15.9
Waveforms for Example 15.3.

EXAMPLE 15.3 Using complex algebra, find the current i for the circuit in Fig. 15.8. Sketch the v and i curves.

Solution: Note Fig. 15.9:

$$v = 24 \sin \omega t \Rightarrow \text{phasor form } \mathbf{V} = 16.968 \text{ V} \angle 0^\circ$$

$$\mathbf{I} = \frac{\mathbf{V}}{\mathbf{Z}_L} = \frac{V \angle \theta}{X_L \angle 90^\circ} = \frac{16.968 \text{ V} \angle 0^\circ}{3\ \Omega \angle 90^\circ} = 5.656 \text{ A} \angle -90^\circ$$

and $\quad i = \sqrt{2}(5.656) \sin(\omega t - 90^\circ) = \mathbf{8.0 \sin(\omega t - 90^\circ)}$

EXAMPLE 15.4 Using complex algebra, find the voltage v for the circuit in Fig. 15.10. Sketch the v and i curves.

Solution: Note Fig. 15.11:

$$i = 5 \sin(\omega t + 30^\circ) \Rightarrow \text{phasor form } \mathbf{I} = 3.535 \text{ A} \angle 30^\circ$$

$$\mathbf{V} = \mathbf{IZ}_L = (I \angle \theta)(X_L \angle 90^\circ) = (3.535 \text{ A} \angle 30^\circ)(4\ \Omega \angle +90^\circ)$$
$$= 14.140 \text{ V} \angle 120^\circ$$

and $\quad v = \sqrt{2}(14.140) \sin(\omega t + 120^\circ) = \mathbf{20 \sin(\omega t + 120^\circ)}$

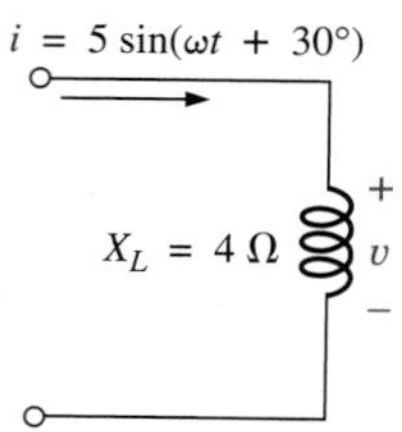

FIG. 15.10
Example 15.4.

The phasor diagrams for the two circuits of the two preceding examples are shown in Fig. 15.12. Both indicate quite clearly that the voltage leads the current by 90°.

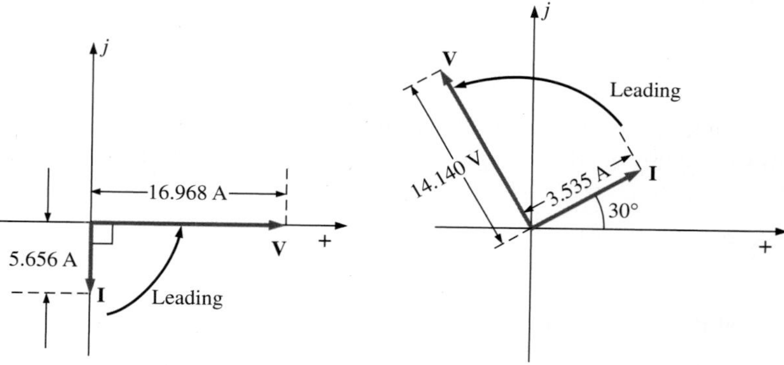

FIG. 15.12
Phasor diagrams for Examples 15.3 and 15.4.

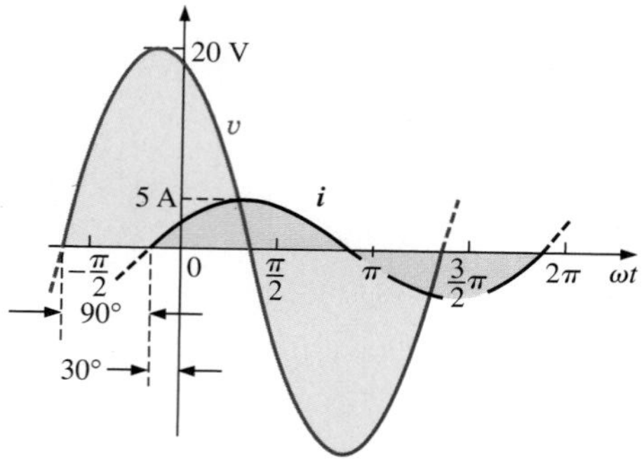

FIG. 15.11
Waveforms for Example 15.4.

Capacitive Reactance

We learned in Chapter 13 that for the pure capacitor in Fig. 15.13, the current leads the voltage by 90° and that the reactance of the capacitor X_C is determined by $1/\omega C$.

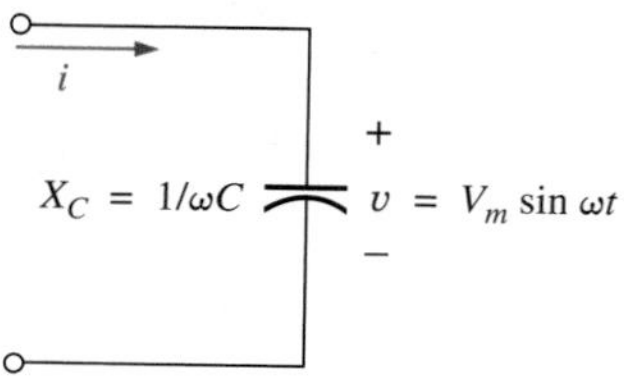

FIG. 15.13
Capacitive ac circuit.

$$v = V_m \sin \omega t \Rightarrow \text{phasor form } \mathbf{V} = V \angle 0^\circ$$

Applying Ohm's law and using phasor algebra, we find

$$\mathbf{I} = \frac{V \angle 0^\circ}{X_C \angle \theta_C} = \frac{V}{X_C} \underline{/0^\circ - \theta_C}$$

Since i leads v by 90°, i must have an angle of +90° associated with it. To satisfy this condition, θ_C must equal −90°. Substituting $\theta_C = -90^\circ$ yields

$$\mathbf{I} = \frac{V \angle 0^\circ}{X_C \angle -90^\circ} = \frac{V}{X_C} \underline{/0^\circ - (-90^\circ)} = \frac{V}{X_C} \angle 90^\circ$$

so, in the time domain,

$$i = \sqrt{2}\left(\frac{V}{X_C}\right) \sin(\omega t + 90^\circ)$$

We use the fact that $\theta_C = -90^\circ$ in the following polar format for capacitive reactance to ensure the proper phase relationship between the voltage and current of a capacitor:

$$\boxed{\mathbf{Z}_C = X_C \angle -90^\circ} \qquad \textbf{(15.3)}$$

The boldface roman quantity $\mathbf{Z}_C$, having both magnitude and an associated angle, is referred to as the *impedance* of a capacitive element. It is measured in ohms and is a measure of how much the capacitive element "controls or impedes" the level of current through the network (always keep in mind that capacitive elements are storage devices and do not dissipate like resistors). The above format, like that defined for the resistive element, will prove a very useful tool in the analysis of ac networks. Again, be aware that $\mathbf{Z}_C$ is not a phasor quantity, for the same reasons indicated for a resistive element.

EXAMPLE 15.5 Using complex algebra, find the current i for the circuit in Fig. 15.14. Sketch the v and i curves.

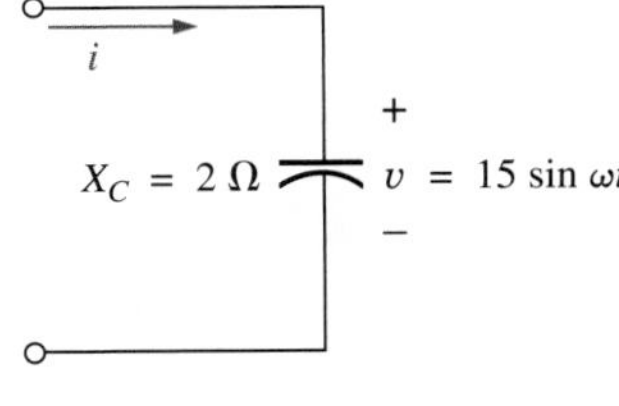

FIG. 15.14
Example 15.5.

Solution: Note Fig. 15.15:

$$v = 15 \sin \omega t \Rightarrow \text{phasor notation } \mathbf{V} = 10.605 \text{ V} \angle 0^\circ$$

$$\mathbf{I} = \frac{\mathbf{V}}{\mathbf{Z}_C} = \frac{V \angle \theta}{X_C \angle -90^\circ} = \frac{10.605 \text{ V} \angle 0^\circ}{2\ \Omega \angle -90^\circ} = 5.303 \text{ A} \angle 90^\circ$$

and $\quad i = \sqrt{2}(5.303) \sin(\omega t + 90^\circ) = \mathbf{7.5 \sin(\omega \mathit{t} + 90^\circ)}$

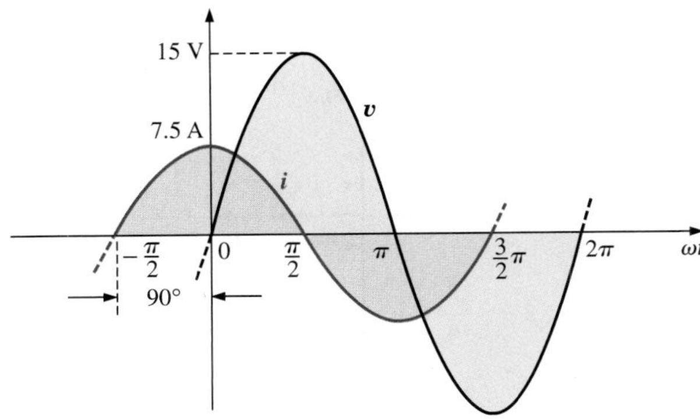

FIG. 15.15
Waveforms for Example 15.5.

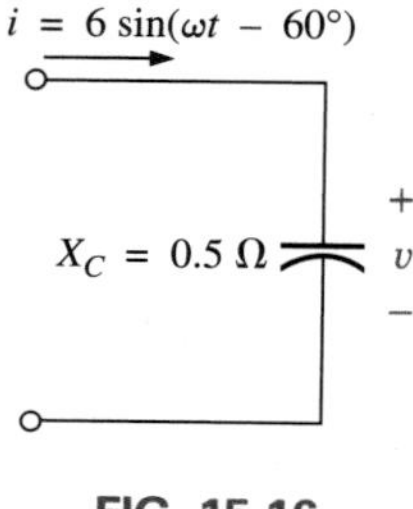

FIG. 15.16
Example 15.6.

EXAMPLE 15.6 Using complex algebra, find the voltage v for the circuit in Fig. 15.16. Sketch the v and i curves.

Solution: Note Fig. 15.17:

$$i = 6\sin(\omega t - 60^\circ) \Rightarrow \text{phasor notation } \mathbf{I} = 4.242\text{ A}\angle -60^\circ$$

$$\mathbf{V} = \mathbf{I}\mathbf{Z}_C = (I\angle\theta)(X_C\angle -90^\circ) = (4.242\text{ A}\angle -60^\circ)(0.5\ \Omega\angle -90^\circ)$$
$$= 2.121\text{ V}\angle -150^\circ$$

and

$$v = \sqrt{2}(2.121)\sin(\omega t - 150^\circ) = \mathbf{3.0\sin(\omega t - 150^\circ)}$$

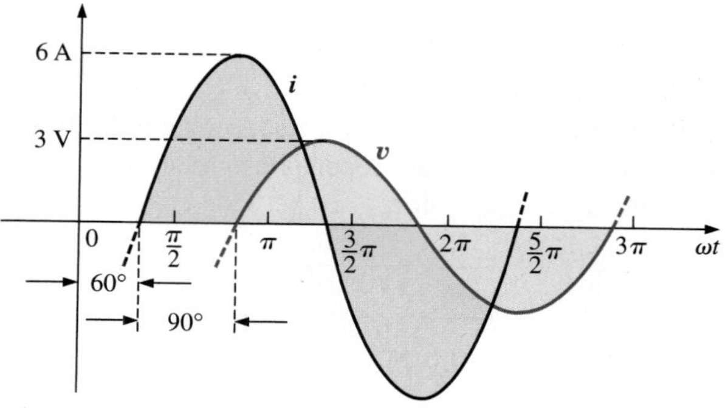

FIG. 15.17
Waveforms for Example 15.6.

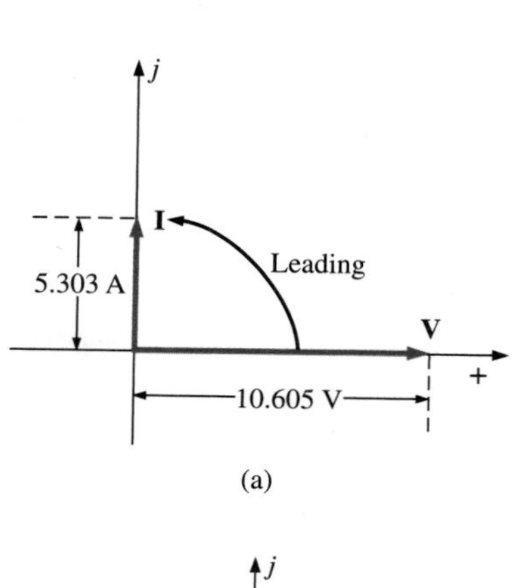

(a)

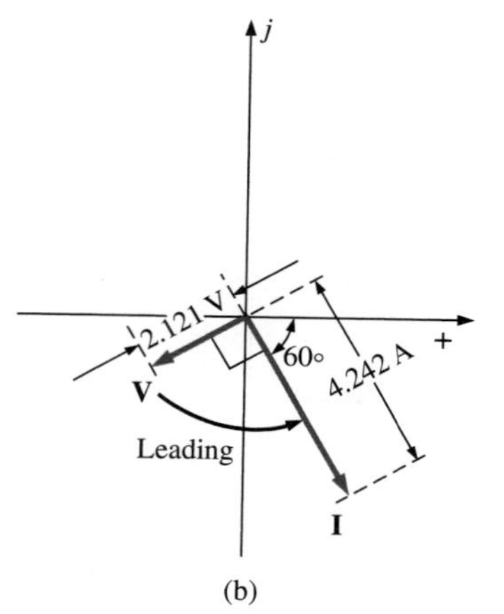

(b)

FIG. 15.18
Phasor diagrams for Examples 15.5 and 15.6.

The phasor diagrams for the two circuits of the two preceding examples are shown in Fig. 15.18. Both indicate quite clearly that the current i leads the voltage v by 90°.

Impedance Diagram

Now that an angle is associated with resistance, inductive reactance, and capacitive reactance, each can be placed on a complex plane diagram, as shown in Fig. 15.19. For any network, the resistance will *always* appear on the positive real axis, the inductive reactance on the positive imaginary axis, and the capacitive reactance on the negative imaginary axis.

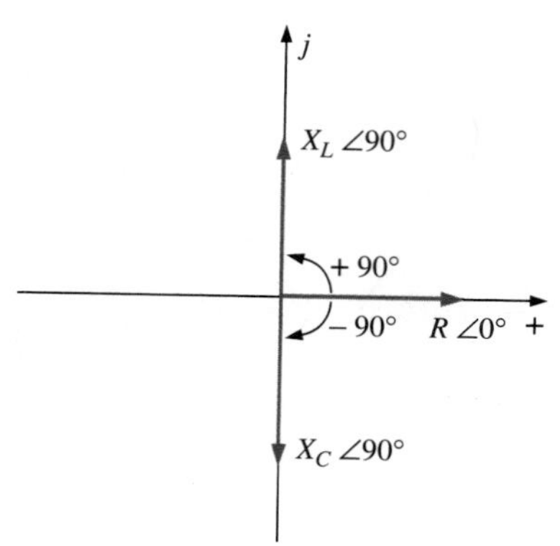

FIG. 15.19
Impedance diagram.

The result is an **impedance diagram** that can reflect the individual and total impedance levels of an ac network.

We will find in the rest of this text that networks combining different types of elements will have total impedances that extend from -90° to $+90^\circ$. If the total impedance has an angle of 0°, it is said to be resistive in nature. If it is closer to 90°, it is inductive in nature. If it is closer to -90°, it is capacitive in nature.

Of course, for single-element networks, the angle associated with the impedance will be the same as that of the resistive or reactive element, as revealed by Eqs. (15.1) through (15.3). It is important to remember that impedance, like resistance or reactance, is not a phasor quantity representing a time-varying function with a particular phase shift. It is simply an operating tool that is extremely useful in determining the magnitude and angle of quantities in a sinusoidal ac network.

Once the total impedance of a network is determined, its magnitude will define the resulting current level (through Ohm's law), whereas its angle will reveal whether the network is primarily inductive or capacitive or simply resistive.

For any configuration (series, parallel, series-parallel, and so on), the angle associated with the total impedance is the angle by which the applied voltage leads the source current. For inductive networks, θ_T will be positive, whereas for capacitive networks, θ_T will be negative.

15.3 SERIES CONFIGURATION

The overall properties of series ac circuits (Fig. 15.20) are the same as those for dc circuits. For instance, the total impedance of a system is the sum of the individual impedances:

$$\mathbf{Z}_T = \mathbf{Z}_1 + \mathbf{Z}_2 + \mathbf{Z}_3 + \cdots + \mathbf{Z}_N \tag{15.4}$$

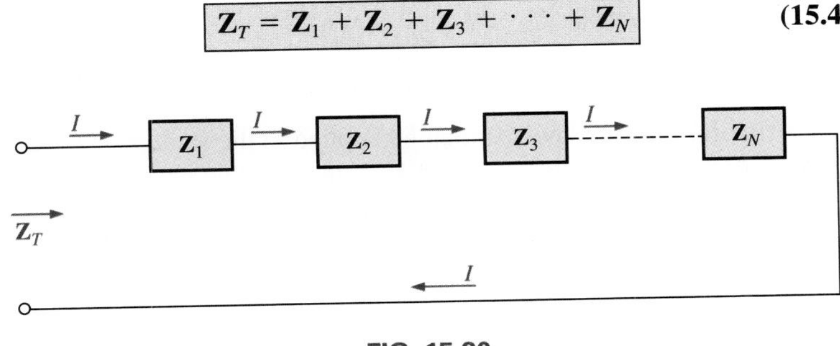

FIG. 15.20
Series impedances.

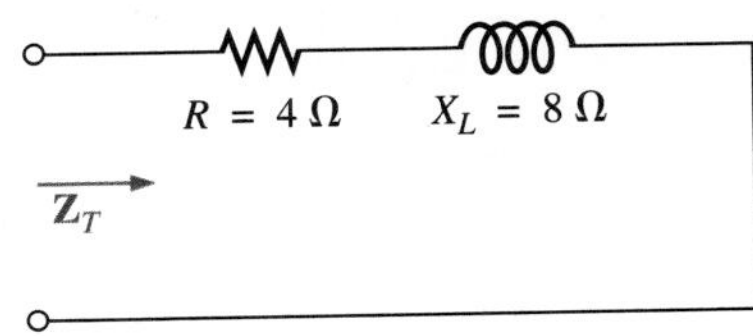

FIG. 15.21
Example 15.7.

EXAMPLE 15.7 Draw the impedance diagram for the circuit in Fig. 15.21, and find the total impedance.

Solution: As indicated by Fig. 15.22, the input impedance can be found graphically from the impedance diagram by properly scaling the real and imaginary axes and finding the length of the resultant vector Z_T and angle θ_T. Or, by using vector algebra, we obtain

$$\begin{aligned}\mathbf{Z}_T &= \mathbf{Z}_1 + \mathbf{Z}_2\\ &= R\,\angle 0^\circ + X_L\,\angle 90^\circ\\ &= R + jX_L = 4\ \Omega + j8\ \Omega\\ \mathbf{Z}_T &= \mathbf{8.94\ \Omega\,\angle 63.43^\circ}\end{aligned}$$

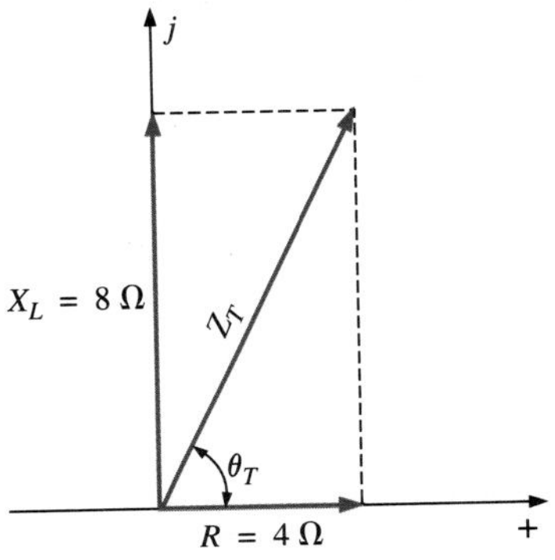

FIG. 15.22
Impedance diagram for Example 15.7.

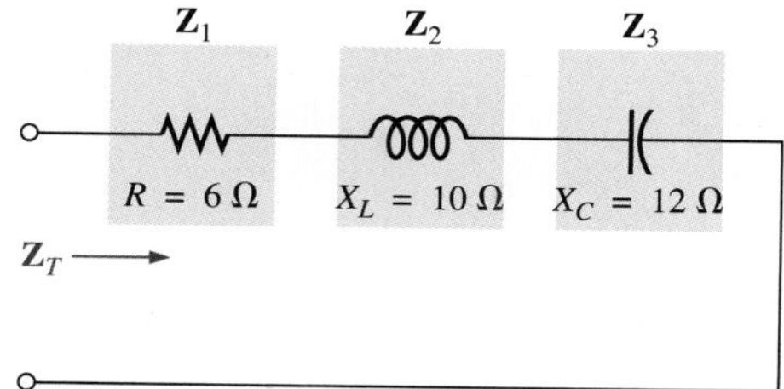

FIG. 15.23
Example 15.8

EXAMPLE 15.8 Determine the input impedance to the series network in Fig. 15.23. Draw the impedance diagram.

Solution:

$$\begin{aligned} \mathbf{Z}_T &= \mathbf{Z}_1 + \mathbf{Z}_2 + \mathbf{Z}_3 \\ &= R\angle 0^\circ + X_L\angle 90^\circ + X_C\angle -90^\circ \\ &= R + jX_L - jX_C \\ &= R + j(X_L - X_C) = 6\ \Omega + j(10\ \Omega - 12\ \Omega) = 6\ \Omega - j2\ \Omega \\ \mathbf{Z}_T &= \mathbf{6.32\ \Omega\ \angle -18.43^\circ} \end{aligned}$$

The impedance diagram appears in Fig. 15.24. Note that in this example, series inductive and capacitive reactances are in direct opposition. For the circuit in Fig. 15.23, if the inductive reactance were equal to the capacitive reactance, the input impedance would be purely resistive. We will have more to say about this particular condition in a later chapter.

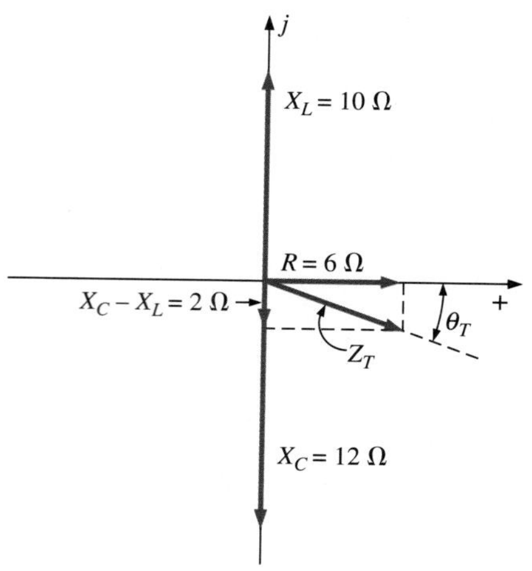

FIG. 15.24
Impedance diagram for Example 15.8.

For the representative **series ac configuration** in Fig. 15.25 having two impedances, *the current is the same through each element* (as it was for the series dc circuits) and is determined by Ohm's law:

$$\mathbf{Z}_T = \mathbf{Z}_1 + \mathbf{Z}_2$$

and

$$\mathbf{I} = \frac{\mathbf{E}}{\mathbf{Z}_T} \tag{15.5}$$

The voltage across each element can then be found by another application of Ohm's law:

$$\mathbf{V}_1 = \mathbf{IZ}_1 \tag{15.6a}$$

$$\mathbf{V}_2 = \mathbf{IZ}_2 \tag{15.6b}$$

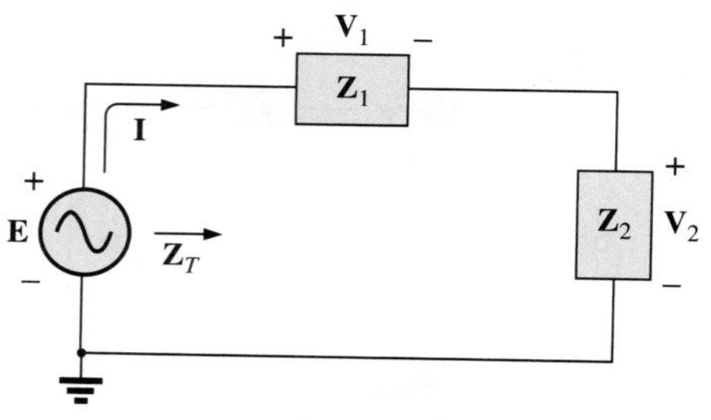

FIG. 15.25
Series ac circuit.

Kirchhoff's voltage law can then be applied in the same manner as it is employed for dc circuits. However, keep in mind that we are now dealing with the algebraic manipulation of quantities that have both magnitude and direction.

$$\mathbf{E} - \mathbf{V}_1 - \mathbf{V}_2 = 0$$

or

$$\mathbf{E} = \mathbf{V}_1 + \mathbf{V}_2 \tag{15.7}$$

The power to the circuit can be determined by

$$P = EI\cos\theta_T \tag{15.8}$$

where θ_T is the phase angle between **E** and **I.**

Now that a general approach has been introduced, the simplest of series configurations will be investigated in detail to further emphasize the similarities in the analysis of dc circuits. In many of the circuits to be considered, $3 + j4 = 5\angle 53.13^\circ$ and $4 + j3 = 5\angle 36.87^\circ$ are used quite frequently to ensure that the approach is as clear as possible and not lost in mathematical complexity. Of course, the problems at the end of the chapter will provide plenty of experience with random values.

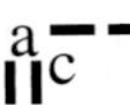

R-L

Refer to Fig. 15.26.

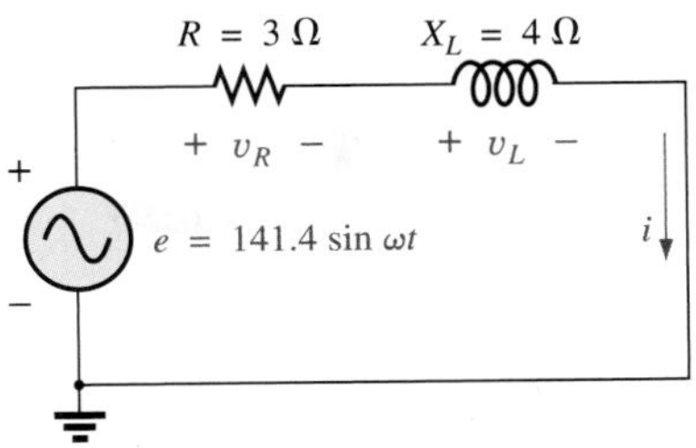

FIG. 15.26
Series R-L circuit.

Phasor Notation

$$e = 141.4 \sin \omega t \Rightarrow \mathbf{E} = 100\text{ V} \angle 0°$$

Note Fig. 15.27.

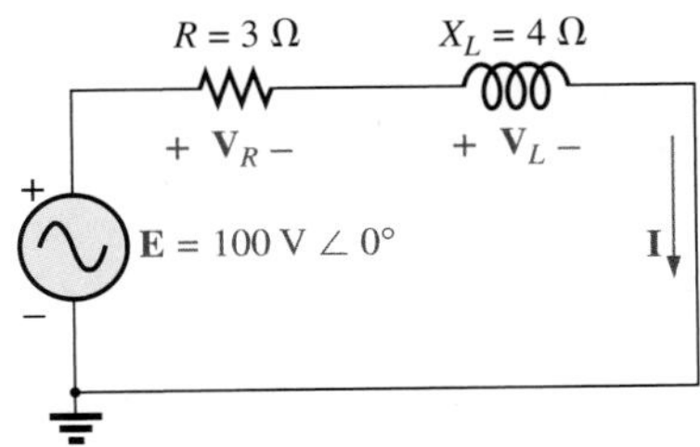

FIG. 15.27
Applying phasor notation to the network in Fig. 15.26.

$\mathbf{Z}_T$

$$\mathbf{Z}_T = \mathbf{Z}_1 + \mathbf{Z}_2 = 3\ \Omega \angle 0° + 4\ \Omega \angle 90° = 3\ \Omega + j4\ \Omega$$

and

$$\mathbf{Z}_T = \mathbf{5\ \Omega \angle 53.13°}$$

Impedance diagram: See Fig. 15.28.

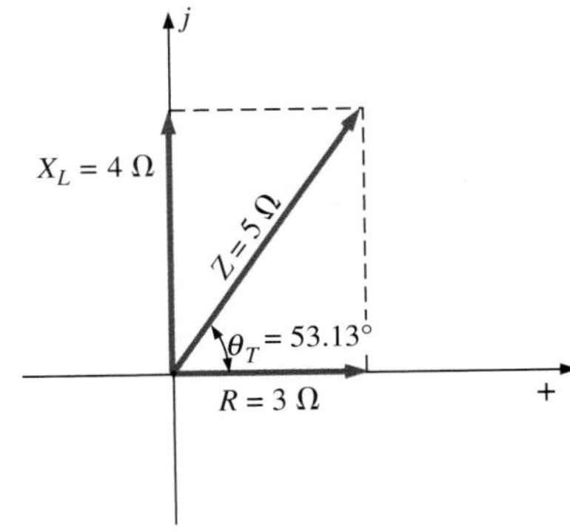

FIG. 15.28
Impedance diagram for the series R-L circuit in Fig. 15.26.

I

$$\mathbf{I} = \frac{\mathbf{E}}{\mathbf{Z}_T} = \frac{100\text{ V} \angle 0°}{5\ \Omega \angle 53.13°} = \mathbf{20\text{ A} \angle -53.13°}$$

$\mathbf{V}_R$ and $\mathbf{V}_L$

Ohm's law:

$$\begin{aligned}\mathbf{V}_R &= \mathbf{I}\mathbf{Z}_R = (20\text{ A} \angle -53.13°)(3\ \Omega \angle 0°) \\ &= \mathbf{60\text{ V} \angle -53.13°} \\ \mathbf{V}_L &= \mathbf{I}\mathbf{Z}_L = (20\text{ A} \angle -53.13°)(4\ \Omega \angle 90°) \\ &= \mathbf{80\text{ V} \angle 36.87°}\end{aligned}$$

Kirchhoff's voltage law:

$$\Sigma_{\circlearrowright} \mathbf{V} = \mathbf{E} - \mathbf{V}_R - \mathbf{V}_L = 0$$

or

$$\mathbf{E} = \mathbf{V}_R + \mathbf{V}_L$$

In rectangular form,

$$\begin{aligned}\mathbf{V}_R &= 60\text{ V} \angle -53.13° = 36\text{ V} - j48\text{ V} \\ \mathbf{V}_L &= 80\text{ V} \angle +36.87° = 64\text{ V} + j48\text{ V}\end{aligned}$$

and

$$\begin{aligned}\mathbf{E} = \mathbf{V}_R + \mathbf{V}_L &= (36\text{ V} - j48\text{ V}) + (64\text{ V} + j48\text{ V}) = 100\text{ V} + j0 \\ &= 100\text{ V} \angle 0°\end{aligned}$$

as applied.

Phasor diagram: Note that for the phasor diagram in Fig. 15.29, **I** is in phase with the voltage across the resistor and lags the voltage across the inductor by 90°.

Power: The total power in watts delivered to the circuit is

$$\begin{aligned}P_T &= EI \cos \theta_T \\ &= (100\text{ V})(20\text{ A}) \cos 53.13° = (2000\text{ W})(0.6) \\ &= \mathbf{1200\text{ W}}\end{aligned}$$

where E and I are effective values and θ_T is the phase angle between E and I, or

$$\begin{aligned}P_T &= I^2R \\ &= (20\text{ A})^2(3\ \Omega) = (400)(3) \\ &= \mathbf{1200\text{ W}}\end{aligned}$$

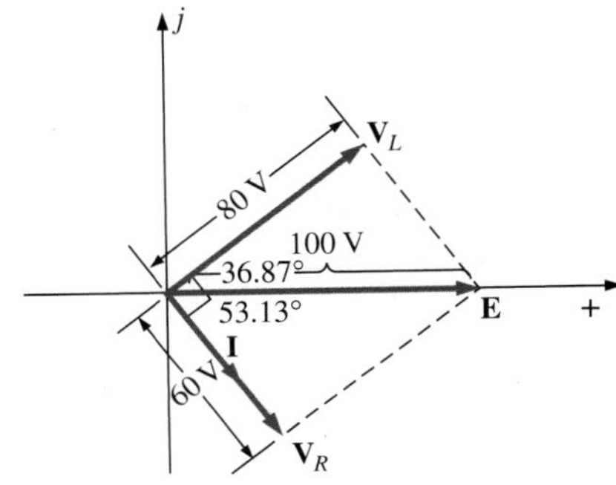

FIG. 15.29
Phasor diagram for the series R-L circuit in Fig. 15.26.

where I is the effective value, or, finally,

$$\begin{aligned} P_T = P_R + P_L &= V_R I \cos\theta_R + V_L I \cos\theta_L \\ &= (60\text{ V})(20\text{ A})\cos 0^\circ + (80\text{ V})(20\text{ A})\cos 90^\circ \\ &= 1200\text{ W} + 0 \\ &= \mathbf{1200\ W} \end{aligned}$$

where θ_R is the phase angle between $\mathbf{V}_R$ and $\mathbf{I}$, and θ_L is the phase angle between $\mathbf{V}_L$ and $\mathbf{I}$.

Power factor: The power factor F_p of the circuit is $\cos 53.13^\circ =$ **0.6 lagging,** where 53.13° is the phase angle between **E** and **I.**

If we write the basic power equation $P = EI\cos\theta$ as follows:

$$\cos\theta = \frac{P}{EI}$$

where E and I are the input quantities and P is the power delivered to the network, and then perform the following substitutions from the basic series ac circuit:

$$\cos\theta = \frac{P}{EI} = \frac{I^2R}{EI} = \frac{IR}{E} = \frac{R}{E/I} = \frac{R}{Z_T}$$

we find

$$\boxed{F_p = \cos\theta_T = \frac{R}{Z_T}} \qquad \textbf{(15.9)}$$

Reference to Fig. 15.28 also indicates that θ is the impedance angle θ_T as written in Eq. (15.9), further supporting the fact that the impedance angle θ_T is also the phase angle between the input voltage and current for a series ac circuit. To determine the power factor, it is necessary only to form the ratio of the total resistance to the magnitude of the input impedance. For the case at hand,

$$F_p = \cos\theta = \frac{R}{Z_T} = \frac{3\ \Omega}{5\ \Omega} = \mathbf{0.6\ lagging}$$

as found above.

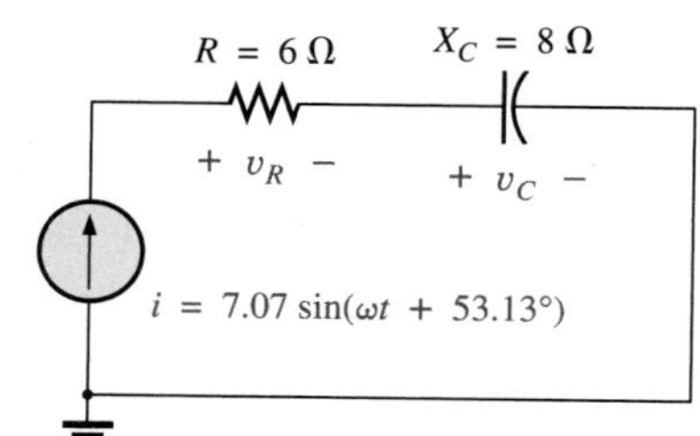

FIG. 15.30
Series R-C ac circuit.

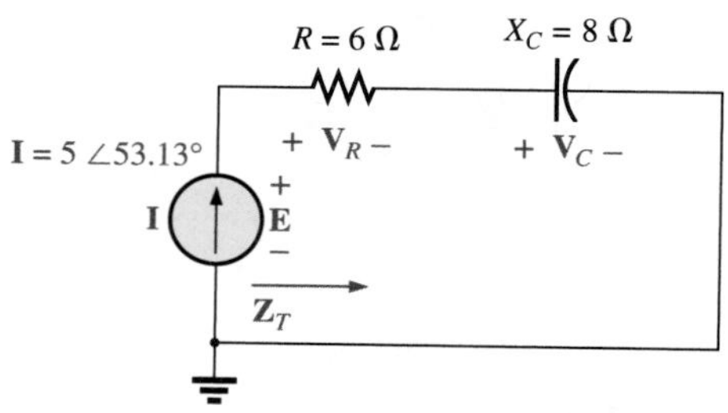

FIG. 15.31
Applying phasor notation to the circuit in Fig. 15.30.

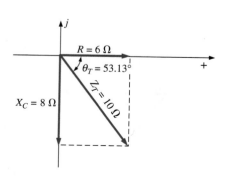

FIG. 15.32
Impedance diagram for the series R-C circuit in Fig. 15.30.

R-C

Refer to Fig. 15.30.

Phasor Notation

$$i = 7.07\sin(\omega t + 53.13^\circ) \Rightarrow \mathbf{I} = 5\text{ A}\angle 53.13^\circ$$

Note Fig. 15.31.

$\mathbf{Z}_T$

$$\mathbf{Z}_T = \mathbf{Z}_1 + \mathbf{Z}_2 = 6\ \Omega\angle 0^\circ + 8\ \Omega\angle -90^\circ = 6\ \Omega - j8\ \Omega$$

and

$$\mathbf{Z}_T = \mathbf{10\ \Omega\angle -53.13^\circ}$$

Impedance diagram: As shown in Fig. 15.32.

E

$$\mathbf{E} = \mathbf{I}\mathbf{Z}_T = (5\text{ A}\angle 53.13^\circ)(10\ \Omega\angle -53.13^\circ) = \mathbf{50\ V\angle 0^\circ}$$

V_R and V_C

$$\mathbf{V}_R = \mathbf{I}\mathbf{Z}_R = (I \angle\theta)(R \angle 0^\circ) = (5 \text{ A} \angle 53.13^\circ)(6 \ \Omega \angle 0^\circ) \\ = \mathbf{30 \ V \angle 53.13^\circ}$$

$$\mathbf{V}_C = \mathbf{I}\mathbf{Z}_C = (I \angle\theta)(X_C \angle -90^\circ) = (5 \text{ A} \angle 53.13^\circ)(8 \ \Omega \angle -90^\circ) \\ = \mathbf{40 \ V \angle -36.87^\circ}$$

Kirchhoff's voltage law:

$$\Sigma_{\circlearrowright} \mathbf{V} = \mathbf{E} - \mathbf{V}_R - \mathbf{V}_C = 0$$

or

$$\mathbf{E} = \mathbf{V}_R + \mathbf{V}_C$$

which can be verified by vector algebra as demonstrated for the *R-L* circuit.

Phasor diagram: Note on the phasor diagram in Fig. 15.33 that the current **I** is in phase with the voltage across the resistor and leads the voltage across the capacitor by 90°.

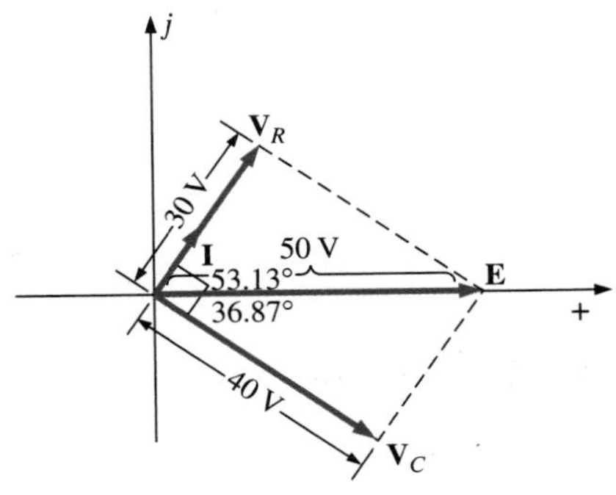

FIG. 15.33
Phasor diagram for the series R-C circuit in Fig. 15.30.

Time domain: In the time domain,

$$e = \sqrt{2}(50) \sin \omega t = \mathbf{70.70 \sin \omega t}$$
$$v_R = \sqrt{2}(30) \sin(\omega t + 53.13^\circ = \mathbf{42.42 \sin(\omega t + 53.13^\circ)}$$
$$v_C = \sqrt{2}(40) \sin(\omega t - 36.87^\circ = \mathbf{56.56 \ \sin(\omega t - 36.87^\circ)}$$

A plot of all of the voltages and the current of the circuit appears in Fig. 15.34. Note again that i and v_R are in phase and that v_C lags i by 90°.

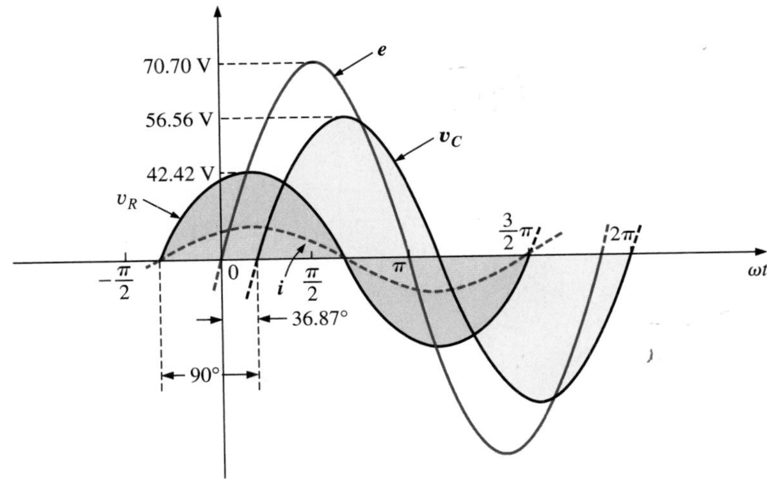

FIG. 15.34
Waveforms for the series R-C circuit in Fig. 15.30.

Power: The total power in watts delivered to the circuit is

$$P_T = EI \cos \theta_T = (50 \text{ V})(5 \text{ A}) \cos 53.13^\circ \\ = (250)(0.6) = \mathbf{150 \ W}$$

or

$$P_T = I^2R = (5 \text{ A})^2(6 \ \Omega) = (25)(6) \\ = \mathbf{150 \ W}$$

or, finally,

$$P_T = P_R + P_C = V_R I \cos \theta_R + V_C I \cos \theta_C \\ = (30 \text{ V})(5 \text{ A}) \cos 0^\circ + (40 \text{ V})(5 \text{ A}) \cos 90^\circ \\ = 150 \text{ W} + 0 \\ = \mathbf{150 \ W}$$

Power factor: The power factor of the circuit is

$$F_p = \cos\theta = \cos 53.13^\circ = \mathbf{0.6\ leading}$$

Using Eq. (15.9), we obtain

$$F_p = \cos\theta = \frac{R}{Z_T} = \frac{6\ \Omega}{10\ \Omega}$$
$$= \mathbf{0.6\ leading}$$

as determined above.

R-L-C

Refer to Fig. 15.35.

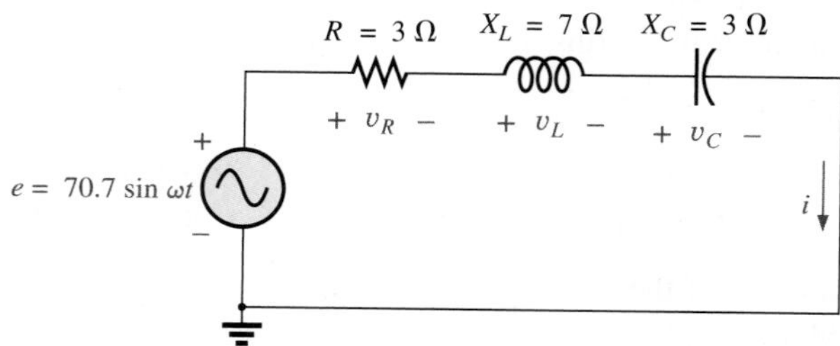

FIG. 15.35
Series R-L-C ac circuit.

Phasor Notation As shown in Fig. 15.36.

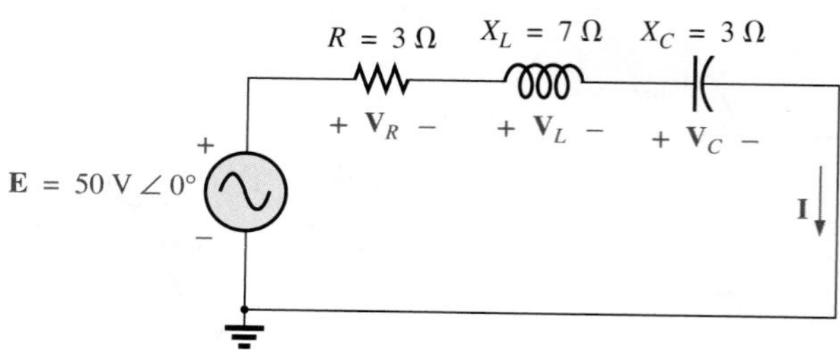

FIG. 15.36
Applying phasor notation to the circuit in Fig. 15.35.

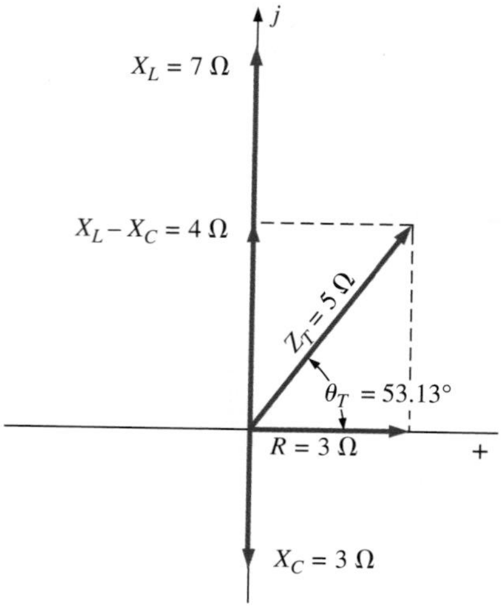

FIG. 15.37
Impedance diagram for the series R-L-C circuit in Fig. 15.35.

Z_T

$$\mathbf{Z}_T = \mathbf{Z}_1 + \mathbf{Z}_2 + \mathbf{Z}_3 = R\angle 0^\circ + X_L\angle 90^\circ + X_C\angle -90^\circ$$
$$= 3\ \Omega + j7\ \Omega - j3\ \Omega = 3\ \Omega + j4\ \Omega$$

and

$$\mathbf{Z}_T = \mathbf{5\ \Omega\ \angle 53.13^\circ}$$

Impedance diagram: As shown in Fig. 15.37.

I

$$\mathbf{I} = \frac{\mathbf{E}}{\mathbf{Z}_T} = \frac{50\ \text{V}\ \angle 0^\circ}{5\ \Omega\ \angle 53.13^\circ} = \mathbf{10\ A\ \angle\ -53.13^\circ}$$

$\mathbf{V}_R$, $\mathbf{V}_L$, and $\mathbf{V}_C$

$$\mathbf{V}_R = \mathbf{IZ}_R = (I \angle\theta)(R \angle 0^\circ) = (10 \text{ A} \angle -53.13^\circ)(3\ \Omega \angle 0^\circ) = \mathbf{30\ V \angle -53.13^\circ}$$

$$\mathbf{V}_L = \mathbf{IZ}_L = (I \angle\theta)(X_L \angle 90^\circ) = (10 \text{ A} \angle -53.13^\circ)(7\ \Omega \angle 90^\circ) = \mathbf{70\ V \angle 36.87^\circ}$$

$$\mathbf{V}_C = \mathbf{IZ}_C = (I \angle\theta)(X_C \angle -90^\circ) = (10 \text{ A} \angle -53.13^\circ)(3\ \Omega \angle -90^\circ) = \mathbf{30\ V \angle -143.13^\circ}$$

Kirchhoff's voltage law:

$$\Sigma_{\circlearrowright} \mathbf{V} = \mathbf{E} - \mathbf{V}_R - \mathbf{V}_L - \mathbf{V}_C = 0$$

or

$$\mathbf{E} = \mathbf{V}_R + \mathbf{V}_L + \mathbf{V}_C$$

which can also be verified through vector algebra.

Phasor diagram: The phasor diagram in Fig. 15.38 indicates that the current **I** is in phase with the voltage across the resistor, lags the voltage across the inductor by 90°, and leads the voltage across the capacitor by 90°.

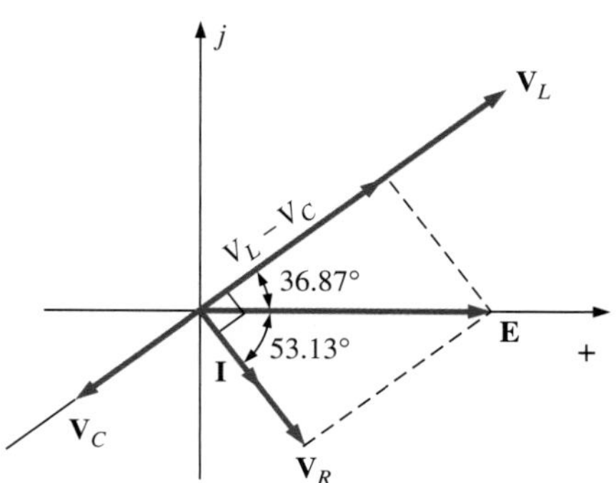

FIG. 15.38
Phasor diagram for the series R-L-C circuit in Fig. 15.35.

Time domain:

$$i = \sqrt{2}(10) \sin(\omega t - 53.13^\circ) = \mathbf{14.14 \sin(\omega t - 53.13^\circ)}$$
$$v_R = \sqrt{2}(30) \sin(\omega t - 53.13^\circ) = \mathbf{42.42 \sin(\omega t - 53.13^\circ)}$$
$$v_L = \sqrt{2}(70) \sin(\omega t + 36.87^\circ) = \mathbf{98.98 \sin(\omega t + 36.87^\circ)}$$
$$v_C = \sqrt{2}(30) \sin(\omega t - 143.13^\circ) = \mathbf{42.42 \sin(\omega t - 143.13^\circ)}$$

A plot of all the voltages and the current of the circuit appears in Fig. 15.39.

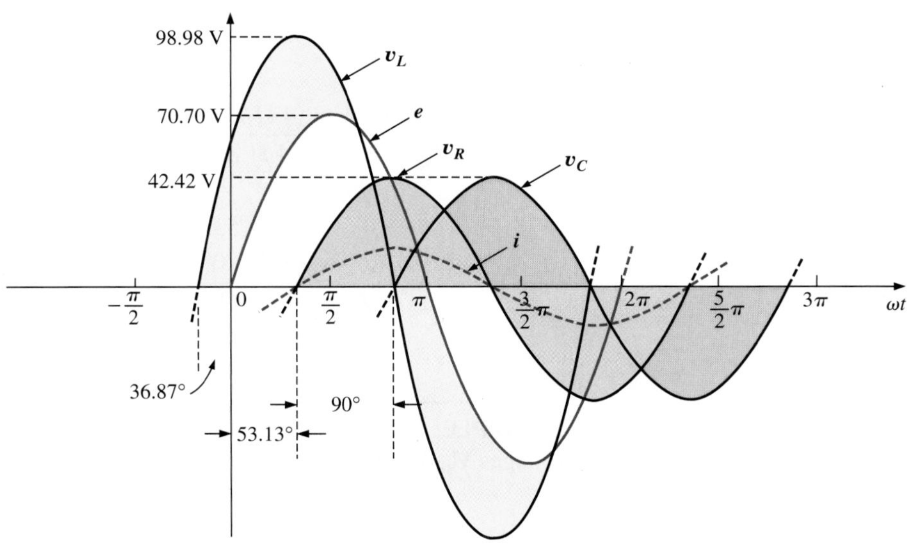

FIG. 15.39
Waveforms for the series R-L-C circuit in Fig. 15.35.

Power: The total power in watts delivered to the circuit is

$$P_T = EI \cos \theta_T = (50 \text{ V})(10 \text{ A}) \cos 53.13^\circ = (500)(0.6) = \mathbf{300\ W}$$

or

$$P_T = I^2R = (10 \text{ A})^2(3\ \Omega) = (100)(3) = \mathbf{300\ W}$$

or

$$\begin{aligned} P_T &= P_R + P_L + P_C \\ &= V_R I \cos\theta_R + V_L I \cos\theta_L + V_C I \cos\theta_C \\ &= (30\text{ V})(10\text{ A}) \cos 0° + (70\text{ V})(10\text{ A}) \cos 90° + (30\text{ V})(10\text{ A}) \cos 90° \\ &= (30\text{ V})(10\text{ A}) + 0 + 0 = \mathbf{300\text{ W}} \end{aligned}$$

Power factor: The power factor of the circuit is

$$F_p = \cos\theta_T = \cos 53.13° = \mathbf{0.6\ lagging}$$

Using Eq. (15.9), we obtain

$$F_p = \cos\theta = \frac{R}{Z_T} = \frac{3\ \Omega}{5\ \Omega} = \mathbf{0.6\ lagging}$$

15.4 VOLTAGE DIVIDER RULE

The basic format for the **voltage divider rule** in ac circuits is exactly the same as that for dc circuits:

$$\boxed{\mathbf{V}_x = \frac{\mathbf{Z}_x\mathbf{E}}{\mathbf{Z}_T}} \qquad \textbf{(15.10)}$$

where $\mathbf{V}_x$ is the voltage across one or more elements in a series that have total impedance $\mathbf{Z}_x$, $\mathbf{E}$ is the total voltage appearing across the series circuit, and $\mathbf{Z}_T$ is the total impedance of the series circuit.

EXAMPLE 15.9 Using the voltage divider rule, find the voltage across each element of the circuit in Fig. 15.40.

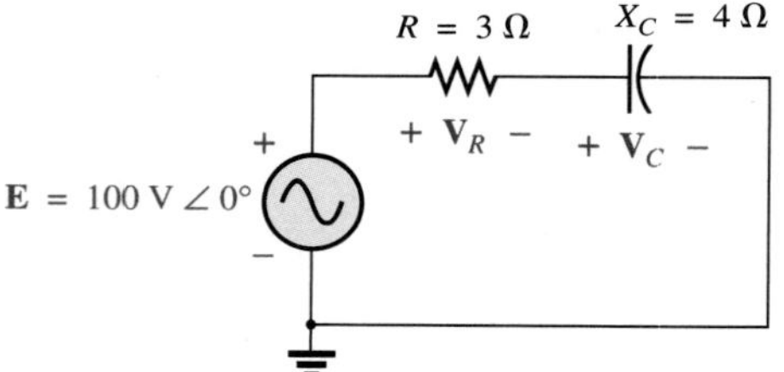

FIG. 15.40
Example 15.9.

Solution:

$$\mathbf{V}_C = \frac{\mathbf{Z}_C\mathbf{E}}{\mathbf{Z}_C + \mathbf{Z}_R} = \frac{(4\ \Omega \angle -90°)(100\text{ V} \angle 0°)}{4\ \Omega \angle -90° + 3\ \Omega \angle 0°} = \frac{400 \angle -90°}{3 - j4}$$

$$= \frac{400 \angle -90°}{5 \angle -53.13°} = \mathbf{80\text{ V} \angle -36.87°}$$

$$\mathbf{V}_R = \frac{\mathbf{Z}_R\mathbf{E}}{\mathbf{Z}_C + \mathbf{Z}_R} = \frac{(3\ \Omega \angle 0°)(100\text{ V} \angle 0°)}{5\ \Omega \angle -53.13°} = \frac{300 \angle 0°}{5 \angle -53.13°}$$

$$= \mathbf{60\text{ V} \angle +53.13°}$$

EXAMPLE 15.10 Using the voltage divider rule, find the unknown voltages $\mathbf{V}_R$, $\mathbf{V}_L$, $\mathbf{V}_C$, and $\mathbf{V}_1$ for the circuit in Fig. 15.41.

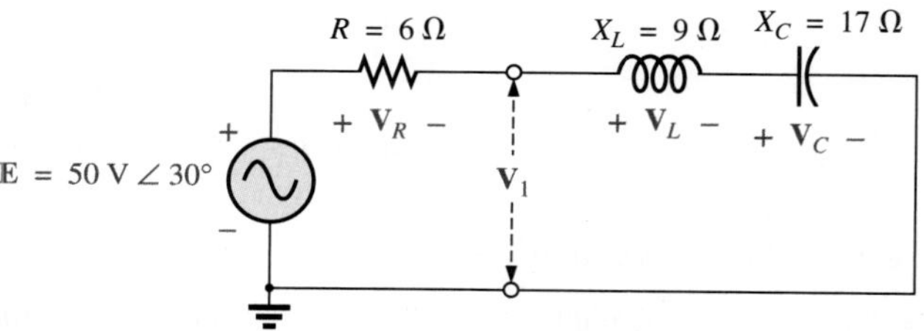

FIG. 15.41
Example 15.10.

Solution:

$$\mathbf{V}_R = \frac{\mathbf{Z}_R\mathbf{E}}{\mathbf{Z}_R + \mathbf{Z}_L + \mathbf{Z}_C} = \frac{(6\ \Omega\ \angle 0^\circ)(50\ \text{V}\ \angle 30^\circ)}{6\ \Omega\ \angle 0^\circ + 9\ \Omega\ \angle 90^\circ + 17\ \Omega\ \angle -90^\circ}$$

$$= \frac{300\ \angle 30^\circ}{6 + j\,9 - j\,17} = \frac{300\ \angle 30^\circ}{6 - j\,8}$$

$$= \frac{300\ \angle 30^\circ}{10\ \angle -53.13^\circ} = \mathbf{30\ V\ \angle 83.13^\circ}$$

Calculator The above calculation provides an excellent opportunity to demonstrate the power of today's calculators. Using the TI-89 calculator, the sequence of steps to calculate $\mathbf{V}_R$ are shown in Fig. 15.42.

(6 ∠ 0 °) × (5 0 ∠ 3 0 °) ÷ ((
6 ∠ 0 °) + (9 ∠ 9 0 °) + (1
7 ∠ (-) 9 0 °)) ▶ Polar ENTER ENTER **30.00E0 ∠ 83.13E0**

FIG. 15.42
Using the TI-89 calculator to determine $\mathbf{V}_R$ in Example 15.10.

$$\mathbf{V}_L = \frac{\mathbf{Z}_L\mathbf{E}}{\mathbf{Z}_T} = \frac{(9\ \Omega\ \angle 90^\circ)(50\ \text{V}\ \angle 30^\circ)}{10\ \Omega\ \angle -53.13^\circ} = \frac{450\ \text{V}\ \angle 120^\circ}{10\ \angle -53.13^\circ}$$

$$= \mathbf{45\ V \angle 173.13^\circ}$$

$$\mathbf{V}_C = \frac{\mathbf{Z}_C\mathbf{E}}{\mathbf{Z}_T} = \frac{(17\ \Omega\ \angle -90^\circ)(50\ \text{V}\ \angle 30^\circ)}{10\ \Omega\ \angle -53.13^\circ} = \frac{850\ \text{V}\ \angle -60^\circ}{10\ \angle -53^\circ}$$

$$= \mathbf{85\ V\ \angle\ -6.87^\circ}$$

$$\mathbf{V}_1 = \frac{(\mathbf{Z}_L + \mathbf{Z}_C)\mathbf{E}}{\mathbf{Z}_T} = \frac{(9\ \Omega\ \angle 90^\circ + 17\ \Omega\ \angle -90^\circ)(50\ \text{V}\ \angle 30^\circ)}{10\ \Omega\ \angle -53.13^\circ}$$

$$= \frac{(8\ \angle -90^\circ)(50\ \angle 30^\circ)}{10\ \angle -53.13^\circ}$$

$$= \frac{400\ \angle -60^\circ}{10\ \angle -53.13^\circ} = \mathbf{40\ V\ \angle -6.87^\circ}$$

EXAMPLE 15.11 For the circuit in Fig. 15.43:

a. Calculate **I,** $\mathbf{V}_R$, $\mathbf{V}_L$, and $\mathbf{V}_C$ in phasor form.
b. Calculate the total power factor.
c. Calculate the average power delivered to the circuit.

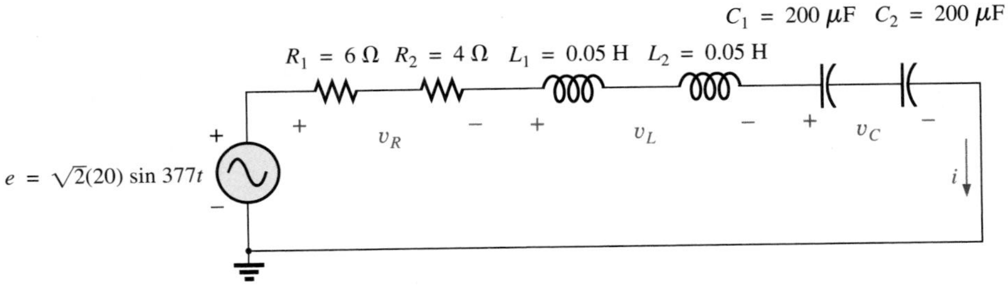

FIG. 15.43
Example 15.11.

d. Draw the phasor diagram.
e. Obtain the phasor sum of $\mathbf{V}_R$, $\mathbf{V}_L$, and $\mathbf{V}_C$, and show that it equals the input voltage **E.**
f. Find $\mathbf{V}_R$ and $\mathbf{V}_C$ using the voltage divider rule.

Solutions:

a. Combining common elements and finding the reactance of the inductor and capacitor, we obtain

$$R_T = 6\ \Omega + 4\ \Omega = 10\ \Omega$$
$$L_T = 0.05\ \text{H} + 0.05\ \text{H} = 0.1\ \text{H}$$
$$C_T = \frac{200\ \mu\text{F}}{2} = 100\ \mu\text{F}$$
$$X_L = \omega L = (377\ \text{rad/s})(0.1\ \text{H}) = 37.70\ \Omega$$
$$X_C = \frac{1}{\omega C} = \frac{1}{(377\ \text{rad/s})(100 \times 10^{-6}\ \text{F})} = \frac{10^6\ \Omega}{37{,}700} = 26.53\ \Omega$$

Redrawing the circuit using phasor notation results in Fig. 15.44.

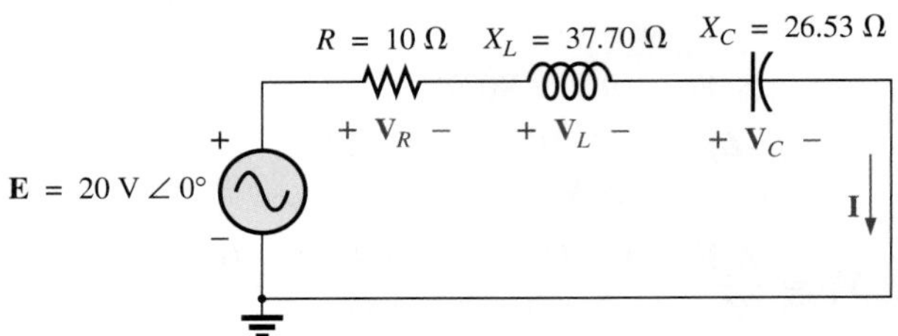

FIG. 15.44
Applying phasor notation to the circuit in Fig. 15.43.

For the circuit in Fig. 15.44,

$$\begin{aligned}\mathbf{Z}_T &= R\ \angle 0^\circ + X_L\ \angle 90^\circ + X_C\ \angle -90^\circ \\ &= 10\ \Omega + j37.70\ \Omega - j26.53\ \Omega \\ &= 10\ \Omega + j11.17\ \Omega = \mathbf{15\ \Omega\ \angle 48.16^\circ}\end{aligned}$$

The current **I** is

$$\mathbf{I} = \frac{\mathbf{E}}{\mathbf{Z}_T} = \frac{20\ \text{V}\angle 0^\circ}{15\ \Omega\ \angle 48.16^\circ} = \mathbf{1.33\ A\ \angle -48.16^\circ}$$

The voltage across the resistor, inductor, and capacitor can be found using Ohm's law:

$$\begin{aligned}\mathbf{V}_R = \mathbf{IZ}_R = (I\ \angle\theta)(R\ \angle 0^\circ) &= (1.33\ \text{A}\ \angle -48.16^\circ)(10\ \Omega\ \angle 0^\circ) \\ &= \mathbf{13.30\ V\ \angle -48.16^\circ}\end{aligned}$$
$$\begin{aligned}\mathbf{V}_L = \mathbf{IZ}_L = (I\ \angle\theta)(X_L\ \angle 90^\circ) &= (1.33\ \text{A}\ \angle -48.16^\circ)(37.70\ \Omega\ \angle 90^\circ) \\ &= \mathbf{50.14\ V\ \angle 41.84^\circ}\end{aligned}$$
$$\begin{aligned}\mathbf{V}_C = \mathbf{IZ}_C = (I\ \angle\theta)(X_C\ \angle -90^\circ) &= (1.33\ \text{A}\ \angle -48.16^\circ)(26.53\ \Omega\ \angle -90^\circ) \\ &= \mathbf{35.28\ V\ \angle -138.16^\circ}\end{aligned}$$

b. The total power factor, determined by the angle between the applied voltage **E** and the resulting current **I,** is 48.16°:

$$F_p = \cos\theta = \cos 48.16^\circ = \mathbf{0.667\ lagging}$$

or
$$F_p = \cos\theta = \frac{R}{Z_T} = \frac{10\ \Omega}{15\ \Omega} = \mathbf{0.667\ lagging}$$

c. The total power in watts delivered to the circuit is

$$P_T = EI\cos\theta = (20\text{ V})(1.33\text{ A})(0.667) = \mathbf{17.74\text{ W}}$$

d. The phasor diagram appears in Fig. 15.45.

e. The phasor sum of $\mathbf{V}_R$, $\mathbf{V}_L$, and $\mathbf{V}_C$ is

$$\begin{aligned}\mathbf{E} &= \mathbf{V}_R + \mathbf{V}_L + \mathbf{V}_C\\ &= 13.30\text{ V}\angle -48.16^\circ + 50.14\text{ V}\angle 41.84^\circ + 35.28\text{ V}\angle -138.16^\circ\\ \mathbf{E} &= 13.30\text{ V}\angle -48.16^\circ + 14.86\text{ V}\angle 41.84^\circ\end{aligned}$$

Therefore,

$$E = \sqrt{(13.30\text{ V})^2 + (14.86\text{ V})^2} = \mathbf{20\text{ V}}$$

and $\theta_E = \mathbf{0^\circ}$ (from phasor diagram)

and $\mathbf{E} = 20\angle 0^\circ$

f. $$\mathbf{V}_R = \frac{\mathbf{Z}_R\mathbf{E}}{\mathbf{Z}_T} = \frac{(10\ \Omega\angle 0^\circ)(20\text{ V}\angle 0^\circ)}{15\ \Omega\angle 48.16^\circ} = \frac{200\text{ V}\angle 0^\circ}{15\angle 48.16^\circ} = \mathbf{13.3\text{ V}\angle -48.16^\circ}$$

$$\mathbf{V}_C = \frac{\mathbf{Z}_C\mathbf{E}}{\mathbf{Z}_T} = \frac{(26.5\ \Omega\angle -90^\circ)(20\text{ V}\angle 0^\circ)}{15\ \Omega\angle 48.16^\circ} = \frac{530.6\text{ V}\angle -90^\circ}{15\angle 48.16^\circ} = \mathbf{35.37\text{ V}\angle -138.16^\circ}$$

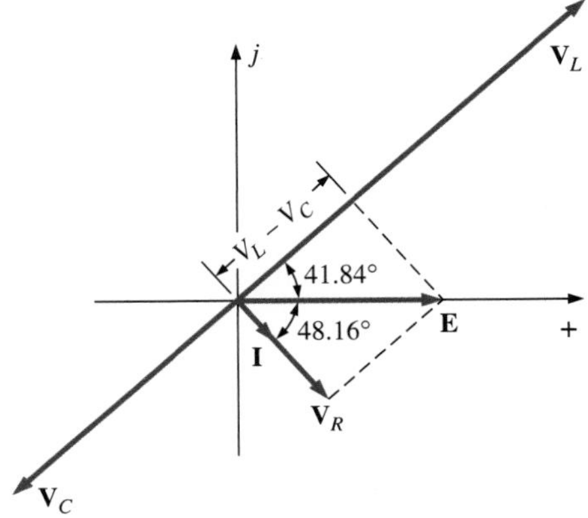

FIG. 15.45
Phasor diagram for the circuit in Fig. 15.43.

15.5 FREQUENCY RESPONSE FOR SERIES ac CIRCUITS

Thus far, the analysis has been for a fixed frequency, resulting in a fixed value for the reactance of an inductor or a capacitor. We now examine how the response of a series circuit changes as the frequency changes. We assume ideal elements throughout the discussion so that the response of each element will be as shown in Fig. 15.46. Each response in Fig. 15.46 was discussed in detail in Chapter 14.

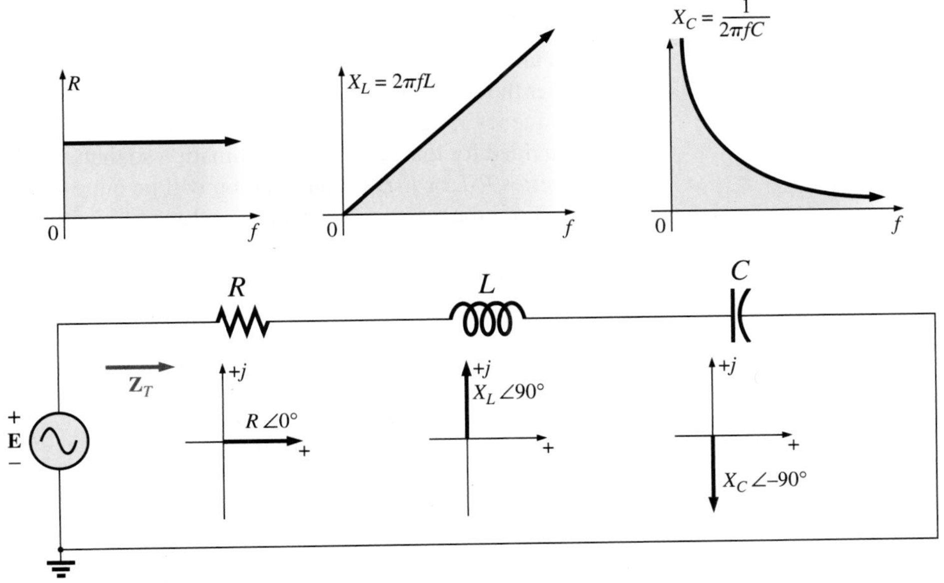

FIG. 15.46
Reviewing the frequency response of the basic elements.

When considering elements in series, remember that the total impedance is the sum of the individual elements and that the reactance of an inductor is in direct opposition to that of a capacitor. For Fig. 15.46, we are first aware that the resistance will remain fixed for the full range of frequencies: It will always be there, but, more importantly, its magnitude will not change. The inductor, however, will provide increasing levels of impedance as the frequency increases, while the capacitor will provide lower levels of impedance.

We are also aware from Chapter 14 that the inductor has a short-circuit equivalence at $f = 0$ Hz or very low frequencies, while the capacitor is nearly an open circuit for the same frequency range. For very high frequencies, the capacitor approaches the short-circuit equivalence, and the inductor approaches the open-circuit equivalence.

In general, therefore, if we encounter a series *R-L-C* circuit at very low frequencies, we can assume that the capacitor, with its very large impedance, will be the predominant factor. If the circuit is just an *R-L* series circuit, the impedance may be determined primarily by the resistive element since the reactance of the inductor is so small. As the frequency increases, the reactance of the coil increases to the point where it totally outshadows the impedance of the resistor. For an *R-L-C* combination, as the frequency increases, the reactance of the capacitor begins to approach a short-circuit equivalence, and the total impedance will be determined primarily by the inductive element. At very high frequencies, for an *R-C* series circuit, the total impedance eventually approaches that of the resistor since the impedance of the capacitor is dropping off so quickly.

In total, therefore,

when encountering a series ac circuit of any combination of elements, always use the idealized response of each element to establish some feeling for how the circuit will respond as the frequency changes.

Once you have a logical, overall sense for what the response will be, you can concentrate on working out the details.

Series *R-C* ac Circuit

As an example of establishing the frequency response of a circuit, consider the series *R-C* circuit in Fig. 15.47. As noted next to the source, the frequency range of interest is from 0 to 20 kHz. A great deal of detail is provided for this particular combination so that obtaining the response of a series *R-L* or *R-L-C* combination will be quite straightforward.

Since the resistance remains fixed at 5 kΩ for the full frequency range, and the total impedance is the sum of the impedances, it is immediately obvious that the lowest possible impedance is 5 kΩ. The highest imped-

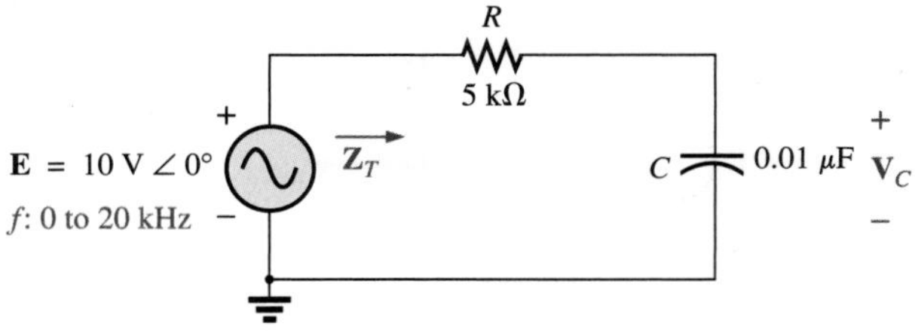

FIG. 15.47
Determining the frequency response of a series R-C circuit.

ance, however, is dependent on the capacitive element since its impedance at very low frequencies is extremely high. At very low frequencies we can conclude, without a single calculation, that the impedance is determined primarily by the impedance of the capacitor. At the highest frequencies, we can assume that the reactance of the capacitor has dropped to such low levels that the impedance of the combination will approach that of the resistance.

The frequency at which the reactance of the capacitor drops to that of the resistor can be determined by setting the reactance of the capacitor equal to that of the resistor as follows:

$$X_C = \frac{1}{2\pi f_1 C} = R$$

Solving for the frequency yields

$$\boxed{f_1 = \frac{1}{2\pi RC}} \tag{15.11}$$

This significant point appears in the frequency plots in Fig. 15.48. Substituting values, we find that it occurs at

$$f_1 = \frac{1}{2\pi RC} = \frac{1}{2\pi(5\text{ k}\Omega)(0.01\ \mu\text{F})} \cong 3.18\text{ kHz}$$

We now know that for frequencies greater than f_1, $R > X_C$ and that for frequencies less than f_1, $X_C > R$, as shown in Fig. 15.48.

Now for the details. The total impedance is determined by the following equation:

$$\mathbf{Z}_T = R - jX_C$$

and

$$\boxed{\mathbf{Z}_T = Z_T \angle \theta_T = \sqrt{R^2 + X_C^2} \angle -\tan^{-1}\frac{X_C}{R}} \tag{15.12}$$

The magnitude and angle of the total impedance can now be found at any frequency of interest by simply substituting into Eq. (15.12). The presence of the capacitor suggests that we start from a low frequency (100 Hz) and then open the spacing until we reach the upper limit of interest (20 kHz).

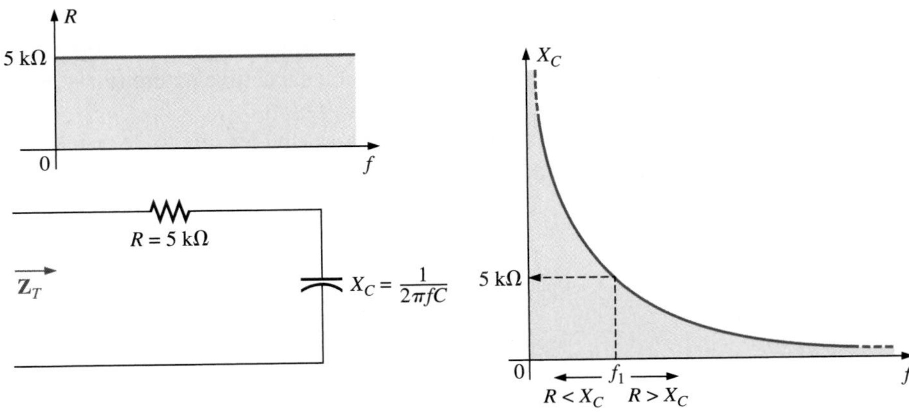

FIG. 15.48

The frequency response for the individual elements of a series R-C circuit.

f = 100 Hz

$$X_C = \frac{1}{2\pi fC} = \frac{1}{2\pi(100\text{ Hz})(0.01\ \mu\text{F})} = 159.16\text{ k}\Omega$$

and $$Z_T = \sqrt{R^2 + X_C^2} = \sqrt{(5\text{ k}\Omega)^2 + (159.16\text{ k}\Omega)^2} = 159.24\text{ k}\Omega$$

with $$\theta_T = -\tan^{-1}\frac{X_C}{R} = -\tan^{-1}\frac{159.16\text{ k}\Omega}{5\text{ k}\Omega} = -\tan^{-1}31.83$$
$$= -88.2^\circ$$

and $$\mathbf{Z}_T = \mathbf{159.24\ k\Omega \angle -88.2^\circ}$$

which compares very closely with $\mathbf{Z}_C = 159.16\text{ k}\Omega \angle -90^\circ$ if the circuit were purely capacitive ($R = 0\ \Omega$). Our assumption that the circuit is primarily capacitive at low frequencies is therefore confirmed.

f = 1 kHz

$$X_C = \frac{1}{2\pi fC} = \frac{1}{2\pi(1\text{ kHz})(0.01\ \mu\text{F})} = 15.92\text{ k}\Omega$$

and $$Z_T = \sqrt{R^2 + X_C^2} = \sqrt{(5\text{ k}\Omega)^2 + (15.92\text{ k}\Omega)^2} = 16.69\text{ k}\Omega$$

with $$\theta_T = -\tan^{-1}\frac{X_C}{R} = -\tan^{-1}\frac{15.92\text{ k}\Omega}{5\text{ k}\Omega}$$
$$= -\tan^{-1}3.18 = -72.54^\circ$$

and $$\mathbf{Z}_T = \mathbf{16.69\ k\Omega \angle -72.54^\circ}$$

A noticeable drop in the magnitude has occurred, and the impedance angle has dropped almost 17° from the purely capacitive level.

Continuing:

f = 5 kHz: $\mathbf{Z}_T = \mathbf{5.93\ k\Omega \angle -32.48^\circ}$
f = 10 kHz: $\mathbf{Z}_T = \mathbf{5.25\ k\Omega \angle -17.66^\circ}$
f = 15 kHz: $\mathbf{Z}_T = \mathbf{5.11\ k\Omega \angle -11.98^\circ}$
f = 20 kHz: $\mathbf{Z}_T = \mathbf{5.06\ k\Omega \angle -9.04^\circ}$

Note how close the magnitude of Z_T at f = 20 kHz is to the resistance level of 5 kΩ. In addition, note how the phase angle is approaching that associated with a pure resistive network (0°).

A plot of Z_T versus frequency in Fig. 15.49 completely supports our assumption based on the curves in Fig. 15.48. The plot of θ_T versus frequency in Fig. 15.50 further suggests that the total impedance made a transition from one of a capacitive nature ($\theta_T = -90^\circ$) to one with resistive characteristics ($\theta_T = 0^\circ$).

Applying the voltage divider rule to determine the voltage across the capacitor in phasor form yields

$$\mathbf{V}_C = \frac{\mathbf{Z}_C\mathbf{E}}{\mathbf{Z}_R + \mathbf{Z}_C}$$
$$= \frac{(X_C\angle -90^\circ)(E\angle 0^\circ)}{R - jX_C} = \frac{X_C E\angle -90^\circ}{R - jX_C}$$
$$= \frac{X_C E\angle -90^\circ}{\sqrt{R^2 + X_C^2}\ \angle -\tan^{-1}X_C/R}$$

or $$\mathbf{V}_C = V_C\angle\theta_C = \frac{X_C E}{\sqrt{R^2 + X_C^2}}\ \angle -90^\circ + \tan^{-1}(X_C/R)$$

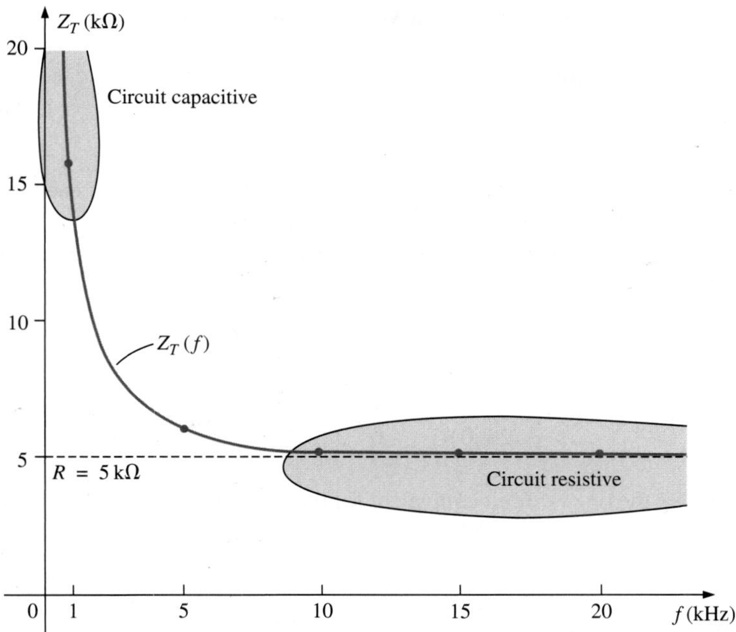

FIG. 15.49
The magnitude of the input impedance versus frequency for the circuit in Fig. 15.47.

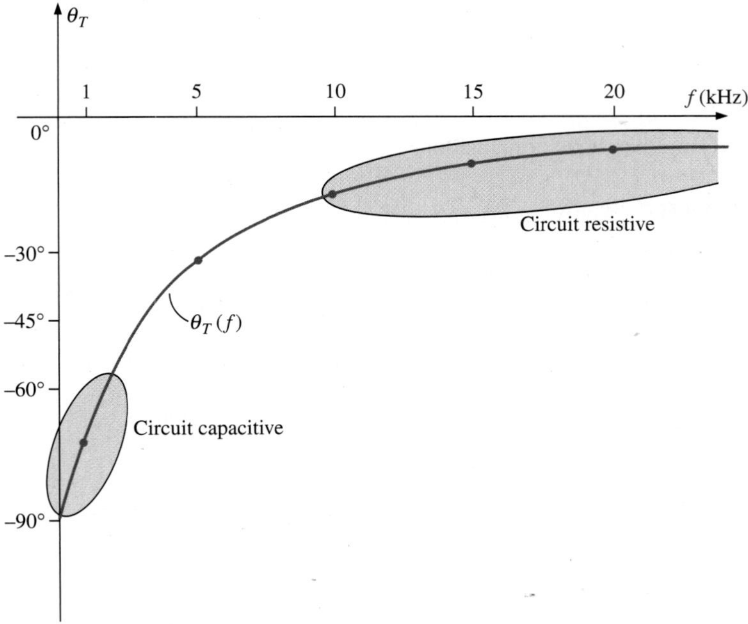

FIG. 15.50
The phase angle of the input impedance versus frequency for the circuit in Fig. 15.47.

The magnitude of $\mathbf{V}_C$ is therefore determined by

$$V_C = \frac{X_C E}{\sqrt{R^2 + X_C^2}} \tag{15.13}$$

and the phase angle θ_C by which $\mathbf{V}_C$ leads $\mathbf{E}$ is given by

$$\theta_C = -90^\circ + \tan^{-1}\frac{X_C}{R} = -\tan^{-1}\frac{R}{X_C} \qquad \textbf{(15.14)}$$

To determine the frequency response, X_C must be calculated for each frequency of interest and inserted into Eqs. (15.13) and (15.14).

To begin our analysis, we should consider the case of $f = 0$ Hz (dc conditions).

***f* = 0 Hz**

$$X_C = \frac{1}{2\pi(0)C} = \frac{1}{0} \Rightarrow \text{very large value}$$

Applying the open-circuit equivalent for the capacitor based on the above calculation results in the following:

$$\mathbf{V}_C = \mathbf{E} = 10\text{ V}\angle 0^\circ$$

If we apply Eq. (15.13), we find

$$X_C^2 \gg R^2$$

and $$\sqrt{R^2 + X_C^2} \cong \sqrt{X_C^2} = X_C$$

and $$V_C = \frac{X_C E}{\sqrt{R^2 + X_C^2}} = \frac{X_C E}{X_C} = E$$

with $$\theta_C = -\tan^{-1}\frac{R}{X_C} = -\tan^{-1} 0 = 0^\circ$$

verifying the above conclusions.

***f* = 1 kHz** Applying Eq. (15.13):

$$X_C = \frac{1}{2\pi f C} = \frac{1}{(2\pi)(1 \times 10^3\text{ Hz})(0.01 \times 10^{-6}\text{ F})} \cong \mathbf{15.92\ k\Omega}$$

$$\sqrt{R^2 + X_C^2} = \sqrt{(5\text{ k}\Omega)^2 + (15.92\text{ k}\Omega)^2} \cong 16.69\text{ k}\Omega$$

and $$V_C = \frac{X_C E}{\sqrt{R^2 + X_C^2}} = \frac{(15.92\text{ k}\Omega)(10)}{16.69\text{ k}\Omega} = \mathbf{9.54\ V}$$

Applying Eq. (15.14):

$$\theta_C = -\tan^{-1}\frac{R}{X_C} = -\tan^{-1}\frac{5\text{ k}\Omega}{15.9\text{ k}\Omega}$$
$$= -\tan^{-1} 0.314 = \mathbf{-17.46^\circ}$$

and $$\mathbf{V}_C = \mathbf{9.53\ V\angle -17.46^\circ}$$

As expected, the high reactance of the capacitor at low frequencies has resulted in the major part of the applied voltage appearing across the capacitor.

If we plot the phasor diagrams for $f = 0$ Hz and $f = 1$ kHz, as shown in Fig. 15.51, we find that $\mathbf{V}_C$ is beginning a clockwise rotation with an increase in frequency that will increase the angle θ_C and decrease the

FIG. 15.51
The phasor diagram for the circuit in Fig. 15.47 for $f = 0$ Hz and 1 kHz.

phase angle between **I** and **E**. Recall that for a purely capacitive network, **I** leads **E** by 90°. As the frequency increases, therefore, the capacitive reactance is decreasing, and eventually $R >> X_C$ with $\theta_C = -90°$, and the angle between **I** and **E** will approach 0°. Keep in mind as we proceed through the other frequencies that θ_C is the phase angle between $\mathbf{V}_C$ and **E** and that the magnitude of the angle by which **I** leads **E** is determined by

$$|\theta_I| = 90° - |\theta_C| \tag{15.15}$$

f = 5 kHz Applying Eq. (15.13):

$$X_C = \frac{1}{2\pi fC} = \frac{1}{(2\pi)(5 \times 10^3\ \text{Hz})(0.01 \times 10^{-6}\ \text{F})} \cong \mathbf{3.18\ k\Omega}$$

Note the dramatic drop in X_C from 1 kHz to 5 kHz. In fact, X_C is now less than the resistance R of the network, and the phase angle determined by $\tan^{-1}(X_C/R)$ must be less than 45°. Here,

$$V_C = \frac{X_C E}{\sqrt{R^2 + X_C^2}} = \frac{(3.18\ \text{k}\Omega)(10\ \text{V})}{\sqrt{(5\ \text{k}\Omega)^2 + (3.18\ \text{k}\Omega)^2}} = \mathbf{5.37\ V}$$

with

$$\theta_C = -\tan^{-1}\frac{R}{X_C} = -\tan^{-1}\frac{5\ \text{k}\Omega}{3.2\ \text{k}\Omega}$$
$$= -\tan^{-1} 1.56 = \mathbf{-57.38°}$$

f = 10 kHz

$$X_C \cong \mathbf{1.59\ k\Omega} \quad V_C = \mathbf{3.03\ V} \quad \theta_C = \mathbf{-72.34°}$$

f = 15 kHz

$$X_C \cong \mathbf{1.06\ k\Omega} \quad V_C = \mathbf{2.07\ V} \quad \theta_C = \mathbf{-78.02°}$$

f = 20 kHz

$$X_C \cong \mathbf{795.78\ \Omega} \quad V_C = \mathbf{1.57\ V} \quad \theta_C = \mathbf{-80.96°}$$

The phasor diagrams for f = 5 kHz and f = 20 kHz appear in Fig. 15.52 to show the continuing rotation of the $\mathbf{V}_C$ vector.

Note also from Figs. 15.51 and 15.52 that the vector $\mathbf{V}_R$ and the current **I** have grown in magnitude with the reduction in the capacitive reactance.

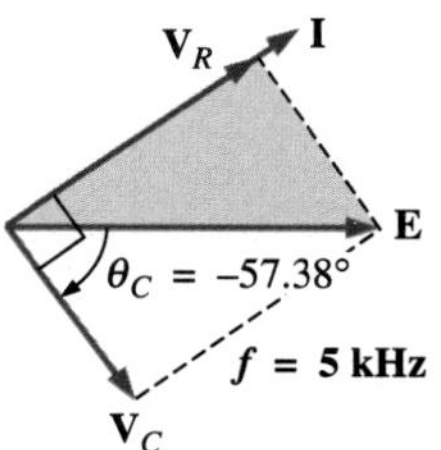

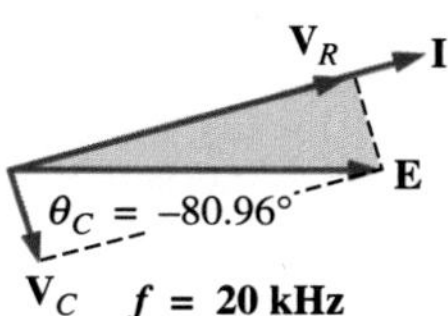

FIG. 15.52
The phasor diagram for the circuit in Fig. 15.47 for f = 5 kHz and 20 kHz.

Eventually, at very high frequencies X_C will approach zero ohms and the short-circuit equivalent can be applied, resulting in $V_C \cong 0$ V and $\theta_C \cong -90°$, and producing the phasor diagram in Fig. 15.53. The network is then resistive, the phase angle between **I** and **E** is essentially zero degrees, and V_R and I are their maximum values.

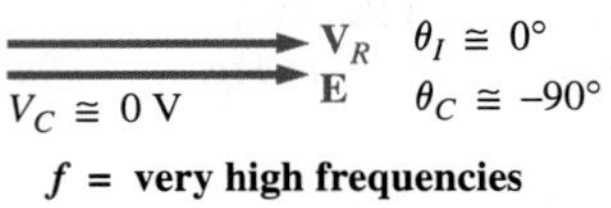

$\mathbf{V}_R$ $\theta_I \cong 0°$
$\mathbf{E}$ $\theta_C \cong -90°$
$V_C \cong 0$ V
f = **very high frequencies**

FIG. 15.53
The phasor diagram for the circuit in Fig. 15.47 at very high frequencies.

A plot of V_C versus frequency appears in Fig. 15.54. At low frequencies, $X_C >> R$, and V_C is very close to E in magnitude. As the applied frequency increases, X_C decreases in magnitude along with V_C as V_R captures more of the applied voltage. A plot of θ_C versus frequency is provided in Fig. 15.55. At low frequencies, the phase angle between $\mathbf{V}_C$ and **E** is very small since $\mathbf{V}_C \cong \mathbf{E}$. Recall that if two phasors are equal, they must have the same angle. As the applied frequency increases, the network becomes more resistive, and the phase angle between $\mathbf{V}_C$ and **E** approaches 90°. Keep in mind that, at high frequencies, **I** and **E** are approaching an in-phase situation and the angle between $\mathbf{V}_C$ and **E** will approach that between $\mathbf{V}_C$ and **I**, which we know must be 90° ($\mathbf{I}_C$ leading $\mathbf{V}_C$).

A plot of V_R versus frequency approaches E volts from zero volts with an increase in frequency, but remember $V_R \neq E - V_C$ due to the vector relationship. The phase angle between **I** and **E** could be plotted directly from Fig. 15.55 using Eq. (15.15).

In Chapter 21, the analysis of this section is extended to a much wider frequency range using a log axis for frequency. It will be demonstrated

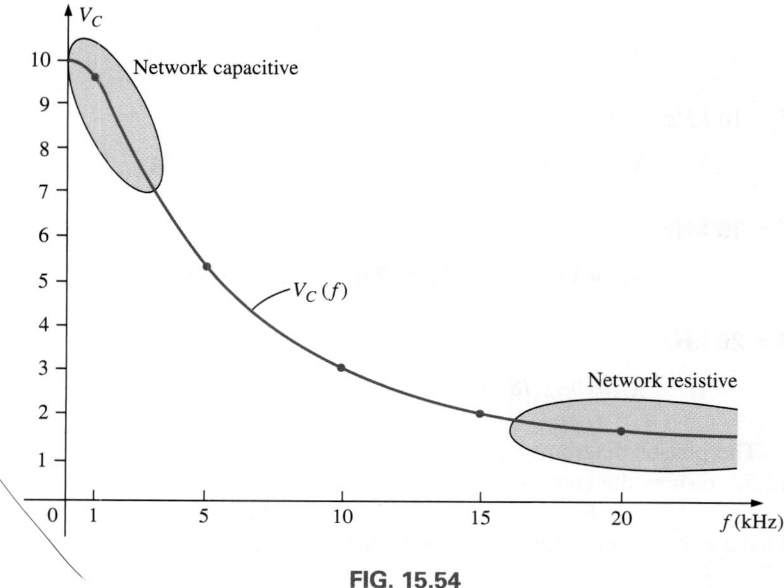

FIG. 15.54
The magnitude of the voltage V_C versus frequency for the circuit in Fig. 15.47.

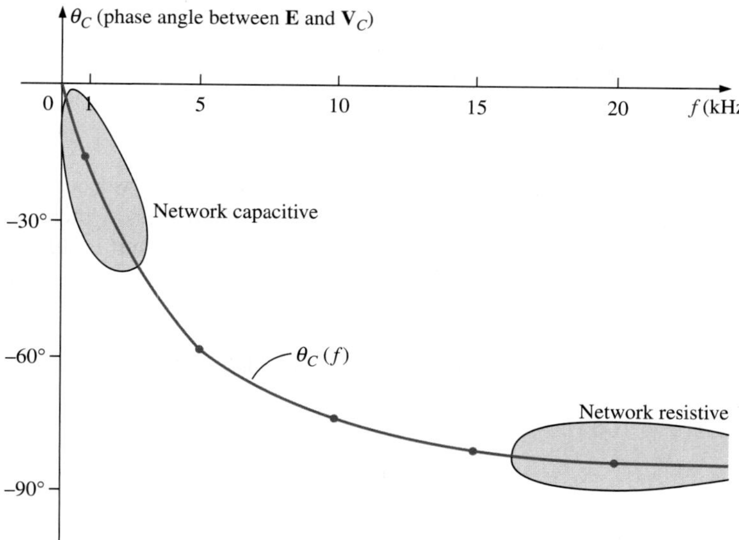

FIG. 15.55
The phase angle between **E** *and* $\mathbf{V}_C$ *versus frequency for the circuit in Fig. 15.47.*

that an R-C circuit such as that in Fig. 15.47 can be used as a filter to determine which frequencies will have the greatest impact on the stage to follow. From our current analysis, it is obvious that any network connected across the capacitor will receive the greatest potential level at low frequencies and be effectively "shorted out" at very high frequencies.

The analysis of a series R-L circuit proceeds in much the same manner, except that X_L and V_L increases with frequency and the angle between **I** and **E** approaches 90° (voltage leading the current) rather than 0°. If $\mathbf{V}_L$ is plotted versus frequency, $\mathbf{V}_L$ will approach **E,** as demonstrated in Example 15.12, and X_L will eventually attain a level at which the open-circuit equivalent is appropriate.

EXAMPLE 15.12 For the series R-L circuit in Fig. 15.56:

a. Determine the frequency at which $X_L = R$.
b. Develop a mental image of the change in total impedance with frequency without doing any calculations.
c. Find the total impedance at $f = 100$ Hz and 40 kHz, and compare your answer with the assumptions of part (b).
d. Plot the curve of V_L versus frequency.
e. Find the phase angle of the total impedance at $f = 40$ kHz. Can the circuit be considered inductive at this frequency? Why?

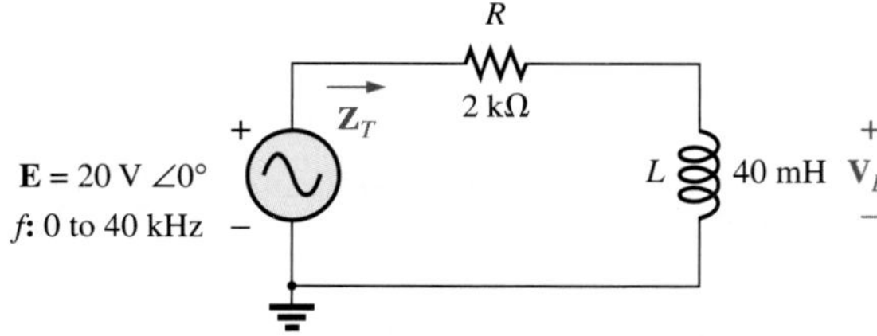

FIG. 15.56
Circuit for Example 15.12.

Solutions:

a. $X_L = 2\pi f_1 L = R$

and $$f_1 = \frac{R}{2\pi L} = \frac{2\text{ k}\Omega}{2\pi(40\text{ mH})} = \mathbf{7957.7\ Hz}$$

b. At low frequencies, $R > X_L$ and the impedance will be very close to that of the resistor, or 2 kΩ. As the frequency increases, X_L increases to a point where it is the predominant factor. The result is that the curve starts almost horizontal at 2 kΩ and then increases linearly to very high levels.

c. $Z_T = R + jX_L = Z_T\angle\theta_T = \sqrt{R^2 + X_L^2}\angle\tan^{-1}\frac{X_L}{R}$

At $f = 100$ Hz:

$$X_L = 2\pi fL = 2\pi(100\text{ Hz})(40\text{ mH}) = 25.13\ \Omega$$

and $$Z_T = \sqrt{R^2 + X_L^2} = \sqrt{(2\text{ k}\Omega)^2 + (25.13\ \Omega)^2} = 2000.16\ \Omega \cong R$$

At $f = 40$ kHz:

$$X_L = 2\pi fL = 2\pi(40\text{ kHz})(40\text{ mH}) \cong 10.05\text{ k}\Omega$$

and $$Z_T = \sqrt{R^2 + X_L^2} = \sqrt{(2\text{ k}\Omega)^2 + (10.05\text{ k}\Omega)^2} = 10.25\text{ k}\Omega \cong X_L$$

Both calculations support the conclusions of part (b).

d. Applying the voltage divider rule:

$$\mathbf{V}_L = \frac{\mathbf{Z}_L\mathbf{E}}{\mathbf{Z}_T}$$

From part (c), we know that at 100 Hz, $Z_T \cong R$ so that $V_R \cong 20$ V and $V_L \cong 0$ V. Part (c) revealed that at 40 kHz, $Z_T \cong X_L$ so that $V_L \cong 20$ V and $V_R \cong 0$ V. The result is two plot points for the curve in Fig. 15.57.

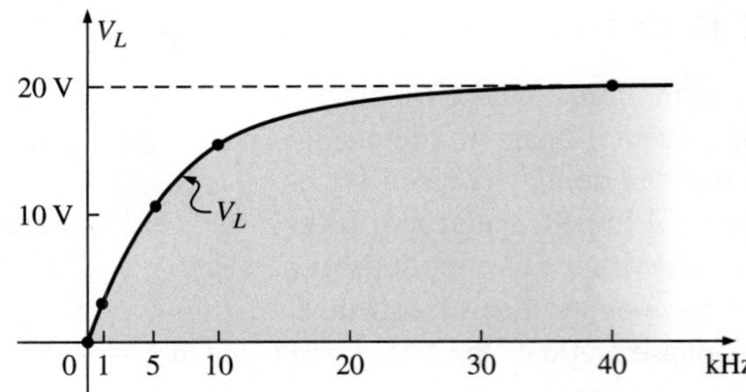

FIG. 15.57
Plotting V_L versus f for the series R-L circuit in Fig. 15.56.

At 1 kHz: $X_L = 2\pi fL \cong 0.25\text{ k}\Omega$

and $$\mathbf{V}_L = \frac{(0.25\text{ k}\Omega\angle 90°)(20\text{ V}\angle 0°)}{2\text{ k}\Omega + j\,0.25\text{ k}\Omega} = \mathbf{2.48\ V\angle 82.87°}$$

At 5 kHz: $X_L = 2\pi fL \cong 1.26\text{ k}\Omega$

and $$\mathbf{V}_L = \frac{(1.26\text{ k}\Omega\angle 90°)(20\text{ V}\angle 0°)}{2\text{ k}\Omega + j1.26\text{ k}\Omega} = \mathbf{10.68\ V\angle 57.79°}$$

At 10 kHz: $X_L = 2\pi fL \cong 2.5\text{ k}\Omega$

and $$\mathbf{V}_L = \frac{(2.5\text{ k}\Omega\ \angle 90°)(20\text{ V}\ \angle 0°)}{2.5\text{ k}\Omega + j\,2.5\text{ k}\Omega} = \mathbf{15.63\ V\angle 38.66°}$$

The complete plot appears in Fig. 15.57.

e. $$\theta_T = \tan^{-1}\frac{X_L}{R} = \tan^{-1}\frac{10.05\text{ k}\Omega}{2\text{ k}\Omega} = \mathbf{78.75°}$$

The angle θ_T is closing in on the 90° of a purely inductive network. Therefore, the network can be considered quite inductive at a frequency of 40 kHz.

15.6 SUMMARY: SERIES ac CIRCUITS

The following is a review of important conclusions that can be derived from the discussion and examples of the previous sections. The list is not all-inclusive, but it does emphasize some of the conclusions that should be carried forward in the future analysis of ac systems.

For series ac circuits with reactive elements:

1. ***The total impedance will be frequency dependent.***
2. ***The impedance of any one element can be greater than the total impedance of the network.***
3. ***The inductive and capacitive reactances are always in direct opposition on an impedance diagram.***
4. ***Depending on the frequency applied, the same circuit can be either predominantly inductive or predominantly capacitive.***
5. ***At lower frequencies, the capacitive elements will usually have the most impact on the total impedance, while at high frequencies the inductive elements will usually have the most impact.***
6. ***The magnitude of the voltage across any one element can be greater than the applied voltage.***
7. ***The magnitude of the voltage across an element compared to the other elements of the circuit is directly related to the magnitude of its impedance; that is, the larger the impedance of an element, the larger the magnitude of the voltage across the element.***
8. ***The voltages across a coil or capacitor are always in direct opposition on a phasor diagram.***
9. ***The current is always in phase with the voltage across the resistive elements, lags the voltage across all the inductive elements by 90°, and leads the voltage across all the capacitive elements by 90°.***
10. ***The larger the resistive element of a circuit compared to the net reactive impedance, the closer the power factor is to unity.***

PARALLEL ac CIRCUITS

15.7 ADMITTANCE AND SUSCEPTANCE

The discussion for **parallel ac circuits** is very similar to that for dc circuits. In dc circuits, *conductance* (G) was defined as being equal to $1/R$. The total conductance of a parallel circuit was then found by adding the conductance of each branch. The total resistance R_T is simply $1/G_T$.

In ac circuits, we define **admittance (Y)** as being equal to **1/Z.** The unit of measure for admittance as defined by the SI system is *siemens,*

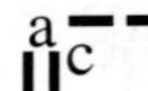

which has the symbol S. Admittance is a measure of how well an ac circuit will *admit,* or allow, current to flow in the circuit. The larger its value, therefore, the heavier the current flow for the same applied potential. The total admittance of a circuit can also be found by finding the sum of the parallel admittances. The total impedance $\mathbf{Z}_T$ of the circuit is then $1/\mathbf{Y}_T$; that is, for the network in Fig. 15.58:

$$\mathbf{Y}_T = \mathbf{Y}_1 + \mathbf{Y}_2 + \mathbf{Y}_3 + \cdots + \mathbf{Y}_N \tag{15.16}$$

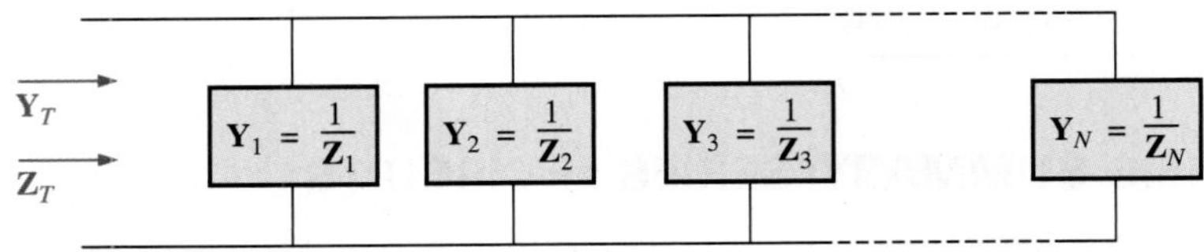

FIG. 15.58
Parallel ac network.

or, since $\mathbf{Z} = 1/\mathbf{Y}$,

$$\frac{1}{\mathbf{Z}_T} = \frac{1}{\mathbf{Z}_1} + \frac{1}{\mathbf{Z}_2} + \frac{1}{\mathbf{Z}_3} + \cdots + \frac{1}{\mathbf{Z}_N} \tag{15.17}$$

and

$$\mathbf{Z}_T = \frac{1}{\dfrac{1}{\mathbf{Z}_1} + \dfrac{1}{\mathbf{Z}_2} + \dfrac{1}{\mathbf{Z}_3} + \cdots + \dfrac{1}{\mathbf{Z}_N}} \tag{15.18}$$

matching Eq. (6.3) for dc networks.

For two impedances in parallel,

$$\frac{1}{\mathbf{Z}_T} = \frac{1}{\mathbf{Z}_1} + \frac{1}{\mathbf{Z}_2}$$

If the manipulations used in Chapter 6 to find the total resistance of two parallel resistors are now applied, the following similar equation results:

$$\mathbf{Z}_T = \frac{\mathbf{Z}_1\mathbf{Z}_2}{\mathbf{Z}_1 + \mathbf{Z}_2} \tag{15.19}$$

For N parallel equal impedances ($\mathbf{Z}_1$), the total impedance is determined by

$$\mathbf{Z}_T = \frac{\mathbf{Z}_1}{\mathbf{N}} \tag{15.20}$$

For three parallel impedances,

$$\mathbf{Z}_T = \frac{\mathbf{Z}_1\mathbf{Z}_2\mathbf{Z}_3}{\mathbf{Z}_1\mathbf{Z}_2 + \mathbf{Z}_2\mathbf{Z}_3 + \mathbf{Z}_1\mathbf{Z}_3} \tag{15.21}$$

As pointed out in the introduction to this section, conductance is the reciprocal of resistance, and

$$\mathbf{Y}_R = \frac{1}{\mathbf{Z}_R} = \frac{1}{R \angle 0^\circ} = G \angle 0^\circ \tag{15.22}$$

The reciprocal of reactance (1/*X*) is called **susceptance** and is a measure of how *susceptible* an element is to the passage of current through it. Susceptance is also measured in *siemens* and is represented by the capital letter *B*.

For the inductor,

$$\mathbf{Y}_L = \frac{1}{\mathbf{Z}_L} = \frac{1}{X_L \angle 90^\circ} = \frac{1}{X_L} \angle -90^\circ \tag{15.23}$$

Defining $$B_L = \frac{1}{X_L} \quad \text{(siemens, S)} \tag{15.24}$$

we have $$\mathbf{Y}_L = B_L \angle -90^\circ \tag{15.25}$$

Note that for inductance, an increase in frequency or inductance will result in a decrease in susceptance or, correspondingly, in admittance.

For the capacitor,

$$\mathbf{Y}_C = \frac{1}{\mathbf{Z}_C} = \frac{1}{X_C \angle -90^\circ} = \frac{1}{X_C} \angle 90^\circ \tag{15.26}$$

Defining $$B_C = \frac{1}{X_C} \quad \text{(siemens, S)} \tag{15.27}$$

we have $$\mathbf{Y}_C = B_C \angle 90^\circ \tag{15.28}$$

For the capacitor, therefore, an increase in frequency or capacitance will result in an increase in its susceptibility.

For parallel ac circuits, the **admittance diagram** is used with the three admittances, represented as shown in Fig. 15.59.

Note in Fig. 15.59 that the conductance (like resistance) is on the positive real axis, whereas inductive and capacitive susceptances are in direct opposition on the imaginary axis.

For any configuration (series, parallel, series-parallel, and so on), the angle associated with the total admittance is the angle by which the source current leads the applied voltage. For inductive networks, θ_T is negative, whereas for capacitive networks, θ_T is positive.

For parallel ac networks, the components of the configuration and the desired quantities determine whether to use an impedance or admittance approach. If the total impedance is requested, the most direct route may be to use impedance parameters. However, sometimes using admittance parameters can also be very efficient, as demonstrated in some of the examples in the rest of the text. In general, use the approach with which you are more comfortable. Naturally, if the format of the desired quantity is spelled out, it is usually best to work with those parameters.

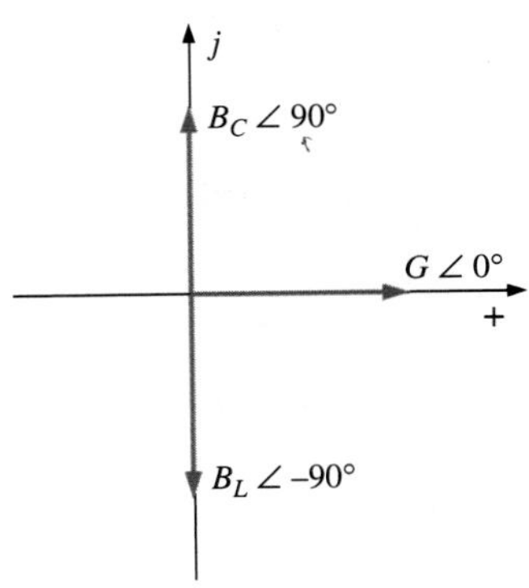

FIG. 15.59
Admittance diagram.

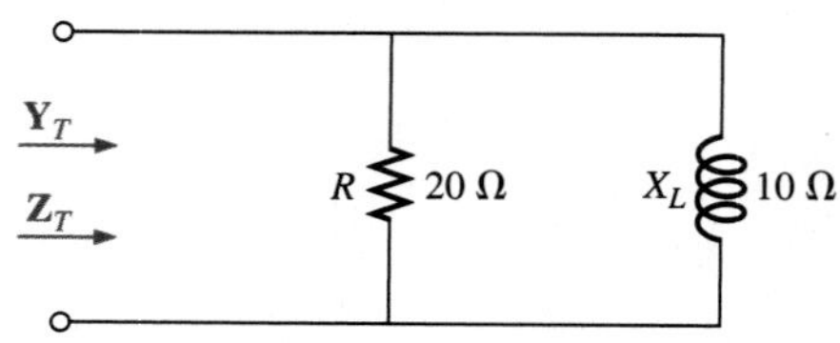

FIG. 15.60
Example 15.12.

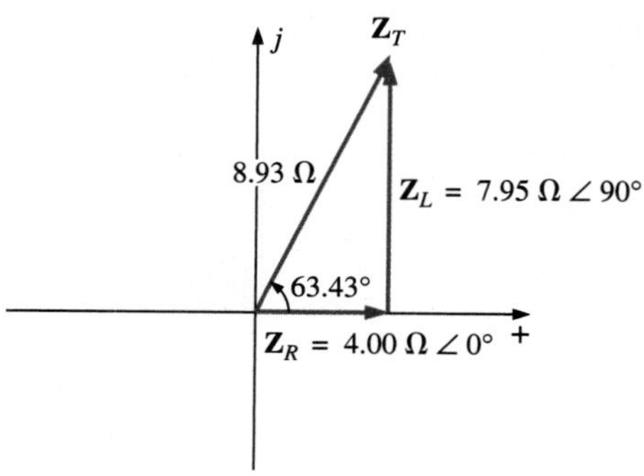

FIG. 15.61
Impedance diagram for the network in Fig. 15.60.

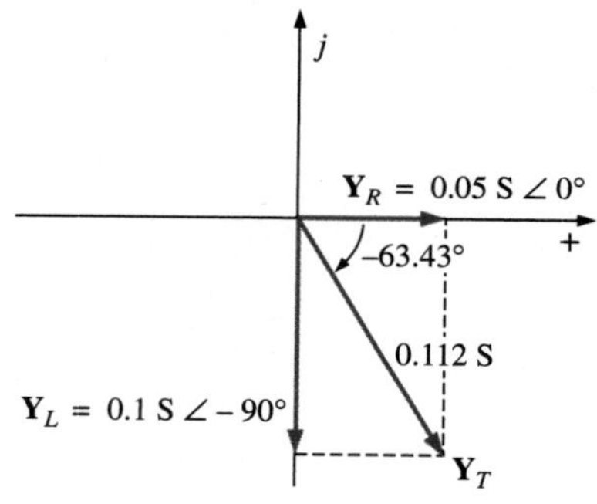

FIG. 15.62
Admittance diagram for the network in Fig. 15.60.

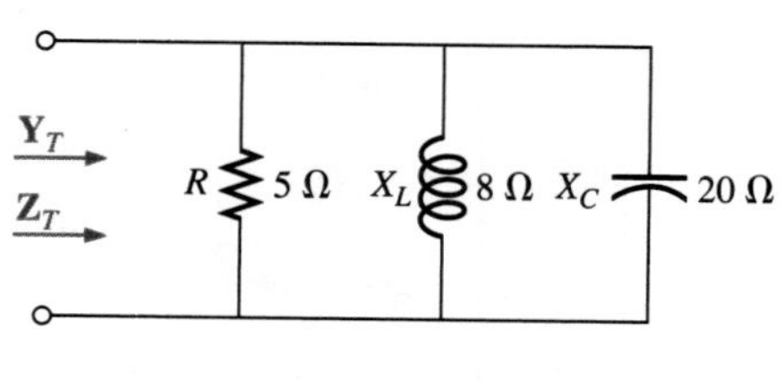

FIG. 15.63
Example 15.14.

EXAMPLE 15.13 For the network in Fig. 15.60:

a. Calculate the input impedance.
b. Draw the impedance diagram.
c. Find the admittance of each parallel branch.
d. Determine the input admittance and draw the admittance diagram.

Solutions:

a. $$\mathbf{Z}_T = \frac{\mathbf{Z}_R\mathbf{Z}_L}{\mathbf{Z}_R + \mathbf{Z}_L} = \frac{(20\ \Omega\ \angle 0^\circ)(10\ \Omega\ \angle 90^\circ)}{20\ \Omega + j\,10\ \Omega}$$
$$= \frac{200\ \Omega\ \angle 90^\circ}{22.361\ \angle 26.57^\circ} = \mathbf{8.93\ \Omega\ \angle 63.43^\circ}$$
$$= \mathbf{4.00\ \Omega + \mathit{j}\,7.95\ \Omega} = R_T + j\,X_{L_T}$$

b. The impedance diagram appears in Fig. 15.61.

c. $$\mathbf{Y}_R = G\angle 0^\circ = \frac{1}{R}\angle 0^\circ = \frac{1}{20\ \Omega}\angle 0^\circ = \mathbf{0.05\ S\ \angle 0^\circ}$$
$$= \mathbf{0.05\ S + \mathit{j}\,0}$$
$$\mathbf{Y}_L = B_L\angle -90^\circ = \frac{1}{X_L}\angle -90^\circ = \frac{1}{10\ \Omega}\angle -90^\circ$$
$$= \mathbf{0.1\ S\ \angle -90^\circ = 0 - \mathit{j}\,0.1\ S}$$

d. $$\mathbf{Y}_T = \mathbf{Y}_R + \mathbf{Y}_L = (0.05\ \text{S} + j\,0) + (0 - j\,0.1\ \text{S})$$
$$= \mathbf{0.05\ S - \mathit{j}\,0.1\ S} = G - jB_L$$

The admittance diagram appears in Fig. 15.62.

EXAMPLE 15.14 Repeat Example 15.13 for the parallel network in Fig. 15.63.

Solutions:

a. $$\mathbf{Z}_T = \frac{1}{\dfrac{1}{\mathbf{Z}_R} + \dfrac{1}{\mathbf{Z}_L} + \dfrac{1}{\mathbf{Z}_C}}$$
$$= \frac{1}{\dfrac{1}{5\ \Omega\ \angle 0^\circ} + \dfrac{1}{8\ \Omega\ \angle 90^\circ} + \dfrac{1}{20\ \Omega\ \angle -90^\circ}}$$
$$= \frac{1}{0.2\ \text{S}\ \angle 0^\circ + 0.125\ \text{S}\ \angle -90^\circ + 0.05\ \text{S}\ \angle 90^\circ}$$
$$= \frac{1}{0.2\ \text{S} - j\,0.075\ \text{S}} = \frac{1}{0.2136\ \text{S}\ \angle -20.56^\circ}$$
$$= \mathbf{4.68\ \Omega\ \angle 20.56^\circ}$$

or

$$\mathbf{Z}_T = \frac{\mathbf{Z}_R\mathbf{Z}_L\mathbf{Z}_C}{\mathbf{Z}_R\mathbf{Z}_L + \mathbf{Z}_L\mathbf{Z}_C + \mathbf{Z}_R\mathbf{Z}_C}$$
$$= \frac{(5\ \Omega\ \angle 0^\circ)(8\ \Omega\ \angle 90^\circ)(20\ \Omega\ \angle -90^\circ)}{(5\ \Omega\ \angle 0^\circ)(8\ \Omega\ \angle 90^\circ) + (8\ \Omega\ \angle 90^\circ)(20\ \Omega\ \angle -90^\circ) + (5\ \Omega\ \angle 0^\circ)(20\ \Omega\ \angle -90^\circ)}$$
$$= \frac{800\ \Omega\ \angle 0^\circ}{40\ \angle 90^\circ + 160\ \angle 0^\circ + 100\ \angle -90^\circ}$$

$$= \frac{800\ \Omega}{160 + j\,40 - j\,100} = \frac{800\ \Omega}{160 - j\,60}$$

$$= \frac{800\ \Omega}{170.88\ \angle -20.56°}$$

$$= \mathbf{4.68\ \Omega\ \angle 20.56° = 4.38\ \Omega + j\,1.64\ \Omega}$$

b. The impedance diagram appears in Fig. 15.64.

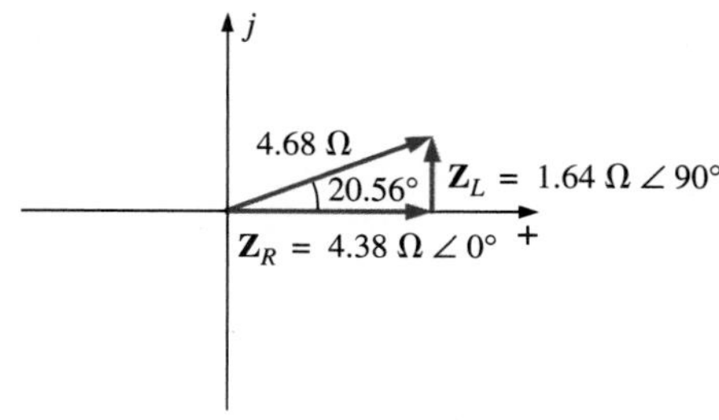

FIG. 15.64
Impedance diagram for the network in Fig. 15.63.

c. $\mathbf{Y}_R = G\ \angle 0° = \frac{1}{R}\ \angle 0° = \frac{1}{5\ \Omega}\ \angle 0°$

$= \mathbf{0.2\ S\ \angle 0° = 0.2\ S} + \boldsymbol{j}\,\mathbf{0}$

$\mathbf{Y}_L = B_L\ \angle -90° = \frac{1}{X_L}\ \angle -90° = \frac{1}{8\ \Omega}\ \angle -90°$

$= \mathbf{0.125\ S\ \angle -90° = 0} - \boldsymbol{j}\,\mathbf{0.125\ S}$

$\mathbf{Y}_C = B_C\ \angle 90° = \frac{1}{X_C}\ \angle 90° = \frac{1}{20\ \Omega}\ \angle 90°$

$= \mathbf{0.050\ S\ \angle +90° = 0} + \boldsymbol{j}\,\mathbf{0.050\ S}$

d. $\mathbf{Y}_T = \mathbf{Y}_R + \mathbf{Y}_L + \mathbf{Y}_C$

$= (0.2\ S + j\,0) + (0 - j\,0.125\ S) + (0 + j\,0.050\ S)$

$= 0.2\ S - j\,0.075\ S = \mathbf{0.214\ S\ \angle -20.56°}$

The admittance diagram appears in Fig. 15.65.

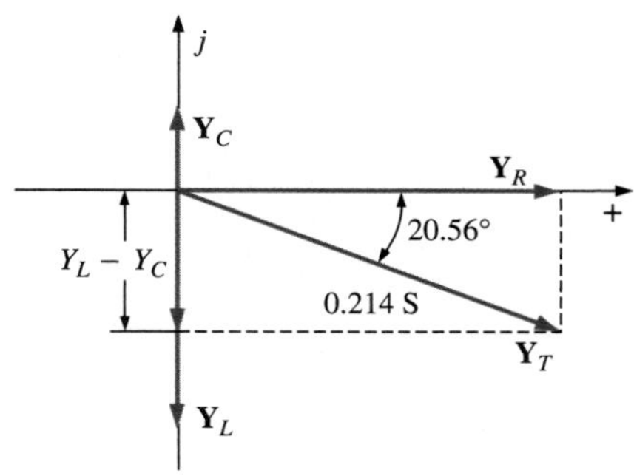

FIG. 15.65
Admittance diagram for the network in Fig. 15.63.

On many occasions, the inverse relationship $\mathbf{Y}_T = 1/\mathbf{Z}_T$ or $\mathbf{Z}_T = 1/\mathbf{Y}_T$ will require that we divide the number 1 by a complex number having a real and an imaginary part. This division, if not performed in the polar form, requires that we multiply the numerator and denominator by the conjugate of the denominator, as follows:

$$\mathbf{Y}_T = \frac{1}{\mathbf{Z}_T} = \frac{1}{4\ \Omega + j\,6\ \Omega} = \left(\frac{1}{4\ \Omega + j\,6\ \Omega}\right)\left(\frac{(4\ \Omega - j\,6\ \Omega)}{(4\ \Omega - j\,6\ \Omega)}\right)$$

$$= \frac{4 - j\,6}{4^2 + 6^2}$$

and

$$\mathbf{Y}_T = \frac{4}{52}\ S - j\,\frac{6}{52}\ S$$

To avoid this laborious task each time we want to find the reciprocal of a complex number in rectangular form, a format can be developed using the following complex number, which is symbolic of any impedance or admittance in the first or fourth quadrant:

$$\frac{1}{a_1 \pm j\,b_1} = \left(\frac{1}{a_1 \pm j\,b_1}\right)\left(\frac{a_1 \mp j\,b_1}{a_1 \mp j\,b_1}\right) = \frac{a_1 \mp j\,b_1}{a_1^2 + b_1^2}$$

or

$$\boxed{\frac{1}{a_1 \pm j\,b_1} = \frac{a_1}{a_1^2 + b_1^2} \mp j\,\frac{b_1}{a_1^2 + b_1^2}} \qquad \textbf{(15.29)}$$

Note that the denominator is simply the sum of the squares of each term. The sign is inverted between the real and imaginary parts. A few examples will develop some familiarity with the use of this equation.

EXAMPLE 15.15 Find the admittance of each set of series elements in Fig. 15.66.

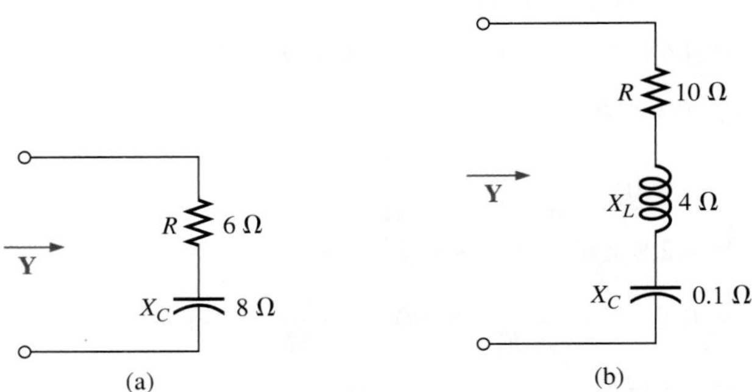

FIG. 15.66
Example 15.15.

Solutions:

a. $\mathbf{Z} = R - jX_C = 6\ \Omega - j\,8\ \Omega$

Eq. (15.29):

$$\mathbf{Y} = \frac{1}{6\ \Omega - j\,8\ \Omega} = \frac{6}{(6)^2 + (8)^2} + j\frac{8}{(6)^2 + (8)^2}$$
$$= \mathbf{\frac{6}{100}\ S + j\frac{8}{100}\ S}$$

b. $\mathbf{Z} = 10\ \Omega + j\,4\ \Omega + (-j\,0.1\ \Omega) = 10\ \Omega + j\,3.9\ \Omega$

Eq. (15.29):

$$\mathbf{Y} = \frac{1}{\mathbf{Z}} = \frac{1}{10\ \Omega + j\,3.9\ \Omega} = \frac{10}{(10)^2 + (3.9)^2} - j\frac{3.9}{(10)^2 + (3.9)^2}$$
$$= \frac{10}{115.21} - j\frac{3.9}{115.21} = \mathbf{0.087\ S - j\,0.034\ S}$$

15.8 PARALLEL ac NETWORKS

For the representative parallel ac network in Fig. 15.67, the total impedance or admittance is determined as described in the previous section, and the source current is determined by Ohm's law as follows:

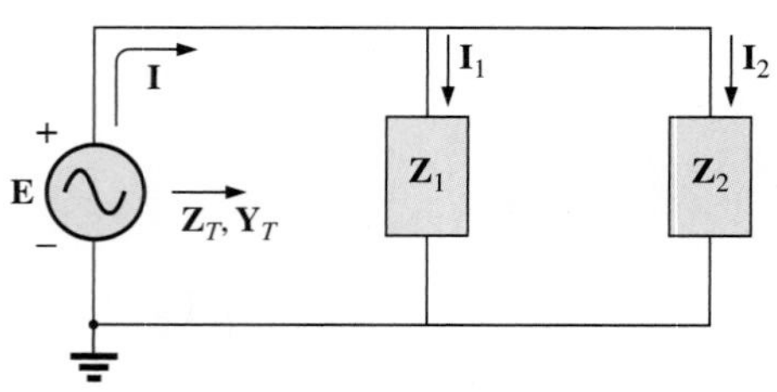

FIG. 15.67
Parallel ac network.

$$\mathbf{I} = \frac{\mathbf{E}}{\mathbf{Z}_T} = \mathbf{E}\mathbf{Y}_T \qquad \textbf{(15.30)}$$

Since the voltage is the same across parallel elements, the current through each branch can then be found through another application of Ohm's law:

$$\mathbf{I}_1 = \frac{\mathbf{E}}{\mathbf{Z}_1} = \mathbf{E}\mathbf{Y}_1 \qquad \textbf{(15.31a)}$$

$$\mathbf{I}_2 = \frac{\mathbf{E}}{\mathbf{Z}_2} = \mathbf{E}\mathbf{Y}_2 \qquad \textbf{(15.31b)}$$

Kirchhoff's current law can then be applied in the same manner as used for dc networks. However, keep in mind that we are now dealing with the algebraic manipulation of quantities that have both magnitude and direction.

$$\mathbf{I} - \mathbf{I}_1 - \mathbf{I}_2 = 0$$

or

$$\mathbf{I} = \mathbf{I}_1 + \mathbf{I}_2 \qquad \textbf{(15.32)}$$

The power to the network can be determined by

$$P = EI\cos\theta_T \qquad \textbf{(15.33)}$$

where θ_T is the phase angle between **E** and **I.**

Let us now look at a few examples carried out in great detail for the first exposure.

R-L

Refer to Fig. 15.68.

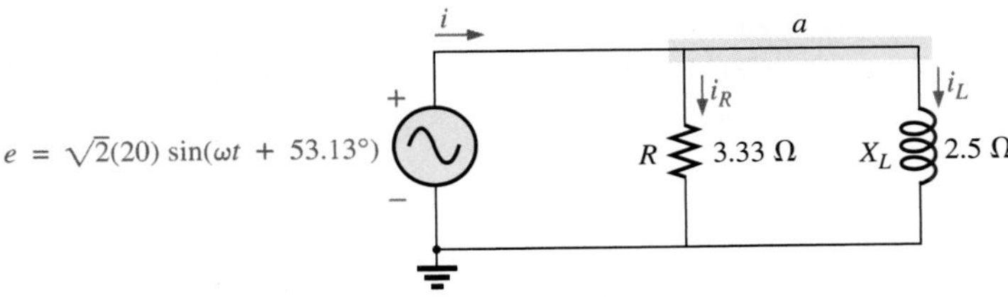

FIG. 15.68
Parallel R-L network.

Phasor Notation As shown in Fig. 15.69.

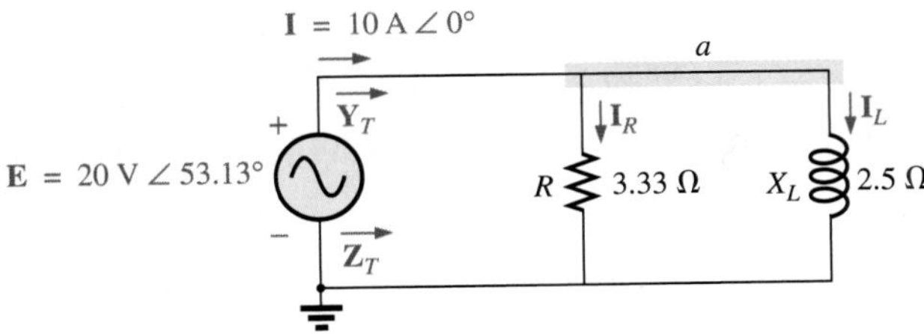

FIG. 15.69
Applying phasor notation to the network in Fig. 15.68.

$$\begin{aligned}
\mathbf{Y}_T &= \mathbf{Y}_R + \mathbf{Y}_L \\
&= G\angle 0° + B_L\angle -90° = \frac{1}{3.33\ \Omega}\angle 0° + \frac{1}{2.5\ \Omega}\angle -90° \\
&= 0.3\ \text{S}\ \angle 0° + 0.4\ \text{S}\ \angle -90° = 0.3\ \text{S} - j\,0.4\ \text{S} \\
&= \mathbf{0.5\ S\ \angle -53.13°}
\end{aligned}$$

$$\mathbf{Z}_T = \frac{1}{\mathbf{Y}_T} = \frac{1}{0.5\ \text{S}\ \angle -53.13°} = \mathbf{2\ \Omega\ \angle 53.13°}$$

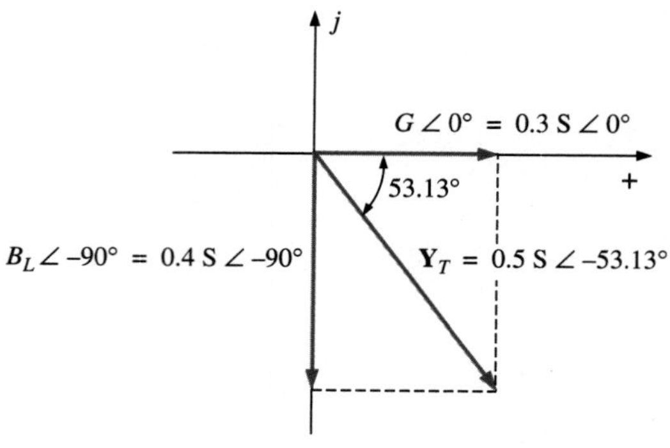

FIG. 15.70
Admittance diagram for the parallel R-L network in Fig. 15.68.

Admittance diagram: As shown in Fig. 15.70.

$$\mathbf{I} = \frac{\mathbf{E}}{\mathbf{Z}_T} = \mathbf{E}\mathbf{Y}_T = (20\text{ V}\angle 53.13°)(0.5\text{ S}\angle -53.13°) = \mathbf{10\ A\ \angle 0°}$$

$$\begin{aligned}\mathbf{I}_R &= \frac{E\angle\theta}{R\angle 0°} = (E\angle\theta)(G\angle 0°)\\ &= (20\text{ V}\angle 53.13°)(0.3\text{ S}\angle 0°) = \mathbf{6\ A\ \angle 53.13°}\end{aligned}$$

$$\begin{aligned}\mathbf{I}_L &= \frac{E\angle\theta}{X_L\angle 90°} = (E\angle\theta)(B_L\angle -90°)\\ &= (20\text{ V}\angle 53.13°)(0.4\text{ S}\angle -90°)\\ &= \mathbf{8\ A\ \angle -36.87°}\end{aligned}$$

Kirchhoff's current law: At node *a*,

$$\mathbf{I} - \mathbf{I}_R - \mathbf{I}_L = 0$$

or

$$\mathbf{I} = \mathbf{I}_R + \mathbf{I}_L$$

$$10\text{ A}\angle 0° = 6\text{ A}\angle 53.13° + 8\text{ A}\angle -36.87°$$
$$10\text{ A}\angle 0° = (3.60\text{ A} + j\,4.80\text{ A}) + (6.40\text{ A} - j\,4.80\text{ A}) = 10\text{ A} + j\,0$$

and

$$\mathbf{10\ A\ \angle 0° = 10\ A\ \angle 0°} \quad \text{(checks)}$$

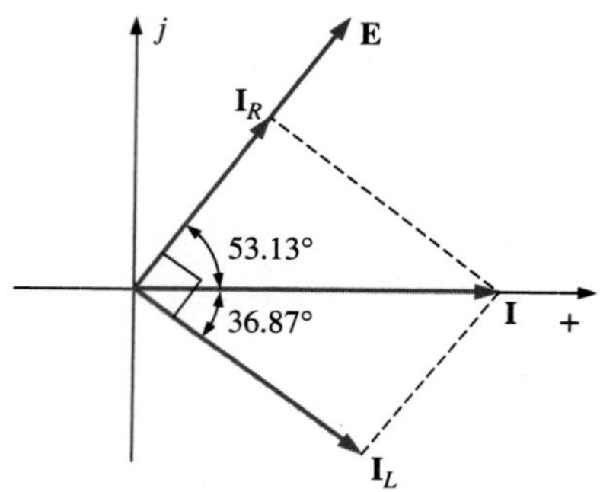

FIG. 15.71
Phasor diagram for the parallel R-L network in Fig. 15.68.

Phasor diagram: The phasor diagram in Fig. 15.71 indicates that the applied voltage **E** is in phase with the current $\mathbf{I}_R$ and leads the current $\mathbf{I}_L$ by 90°.

Power: The total power in watts delivered to the circuit is

$$\begin{aligned}P_T &= EI\cos\theta_T\\ &= (20\text{ V})(10\text{ A})\cos 53.13° = (200\text{ W})(0.6)\\ &= \mathbf{120\ W}\end{aligned}$$

or

$$P_T = I^2R = \frac{V_R^2}{R} = V_R^2\,G = (20\text{ V})^2(0.3\text{ S}) = \mathbf{120\ W}$$

or, finally,

$$\begin{aligned}P_T &= P_R + P_L = EI_R\cos\theta_R + EI_L\cos\theta_L\\ &= (20\text{ V})(6\text{ A})\cos 0° + (20\text{ V})(8\text{ A})\cos 90° = 120\text{ W} + 0\\ &= \mathbf{120\ W}\end{aligned}$$

Power factor: The power factor of the circuit is

$$F_p = \cos\theta_T = \cos 53.13° = \mathbf{0.6\ lagging}$$

or, through an analysis similar to that used for a series ac circuit,

$$\cos\theta_T = \frac{P}{EI} = \frac{E^2/R}{EI} = \frac{EG}{I} = \frac{G}{I/V} = \frac{G}{Y_T}$$

and

$$\boxed{F_p = \cos\theta_T = \frac{G}{Y_T}} \qquad \textbf{(15.34)}$$

where G and Y_T are the magnitudes of the total conductance and admittance of the parallel network. For this case,

$$F_p = \cos\theta_T = \frac{0.3\text{ S}}{0.5\text{ S}} = \mathbf{0.6\ lagging}$$

Impedance approach: The current **I** can also be found by first finding the total impedance of the network:

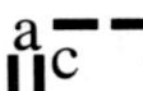

$$\mathbf{Z}_T = \frac{\mathbf{Z}_R\mathbf{Z}_L}{\mathbf{Z}_R + \mathbf{Z}_L} = \frac{(3.33\ \Omega\ \angle 0^\circ)(2.5\ \Omega\ \angle 90^\circ)}{3.33\ \Omega\ \angle 0^\circ + 2.5\ \Omega\ \angle 90^\circ}$$
$$= \frac{8.325\ \angle 90^\circ}{4.164\ \angle 36.87^\circ} = \mathbf{2\ \Omega\ \angle 53.13^\circ}$$

And then, using Ohm's law, we obtain

$$\mathbf{I} = \frac{\mathbf{E}}{\mathbf{Z}_T} = \frac{20\ \text{V}\ \angle 53.13^\circ}{2\ \Omega\ \angle 53.13^\circ} = \mathbf{10\ A\ \angle 0^\circ}$$

R-C

Refer to Fig. 15.72.

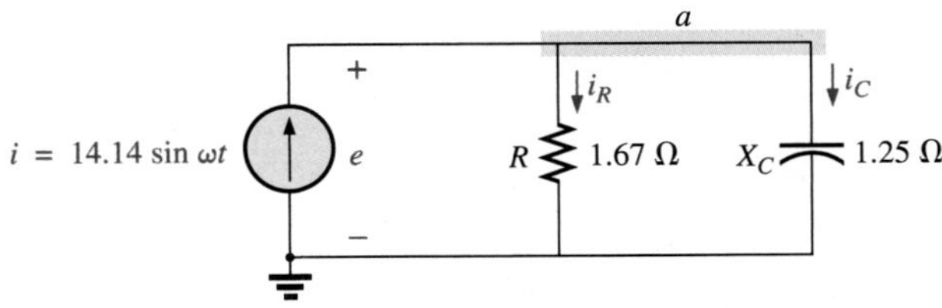

FIG. 15.72
Parallel R-C network.

Phasor Notation As shown in Fig. 15.73.

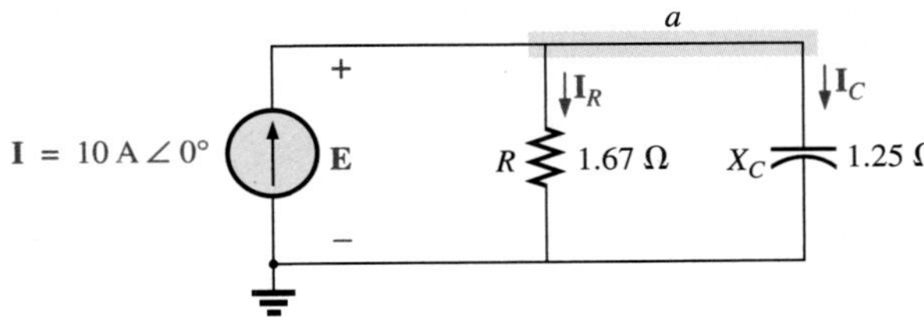

FIG. 15.73
Applying phasor notation to the network in Fig. 15.72.

$$\mathbf{Y}_T = \mathbf{Y}_R + \mathbf{Y}_C = G\ \angle 0^\circ + B_C \angle 90^\circ = \frac{1}{1.67\ \Omega} \angle 0^\circ + \frac{1}{1.25\ \Omega} \angle 90^\circ$$
$$= 0.6\ \text{S}\ \angle 0^\circ + 0.8\ \text{S}\ \angle 90^\circ = 0.6\ \text{S} + j\,0.8\ \text{S} = \mathbf{1.0\ S\ \angle 53.13^\circ}$$
$$\mathbf{Z}_T = \frac{1}{\mathbf{Y}_T} = \frac{1}{1.0\ \text{S}\ \angle 53.13^\circ} = \mathbf{1\ \Omega\ \angle -53.13^\circ}$$

Admittance diagram: As shown in Fig. 15.74.

$$\mathbf{E} = \mathbf{I}\mathbf{Z}_T = \frac{\mathbf{I}}{\mathbf{Y}_T} = \frac{10\ \text{A}\ \angle 0^\circ}{1\ \text{S}\ \angle 53.13^\circ} = \mathbf{10\ V\ \angle -53.13^\circ}$$
$$\mathbf{I}_R = (E\ \angle\theta)(G\ \angle 0^\circ)$$
$$= (10\ \text{V}\ \angle -53.13^\circ)(0.6\ \text{S}\ \angle 0^\circ) = \mathbf{6\ A\ \angle -53.13^\circ}$$
$$\mathbf{I}_C = (E\ \angle\theta)(B_C\ \angle 90^\circ)$$
$$= (10\ \text{V}\ \angle -53.13^\circ)(0.8\ \text{S}\ \angle 90^\circ) = \mathbf{8\ A\ \angle 36.87^\circ}$$

Kirchhoff's current law: At node *a,*

$$\mathbf{I} - \mathbf{I}_R - \mathbf{I}_C = 0$$

or

$$\mathbf{I} = \mathbf{I}_R + \mathbf{I}_C$$

which can also be verified (as for the *R-L* network) through vector algebra.

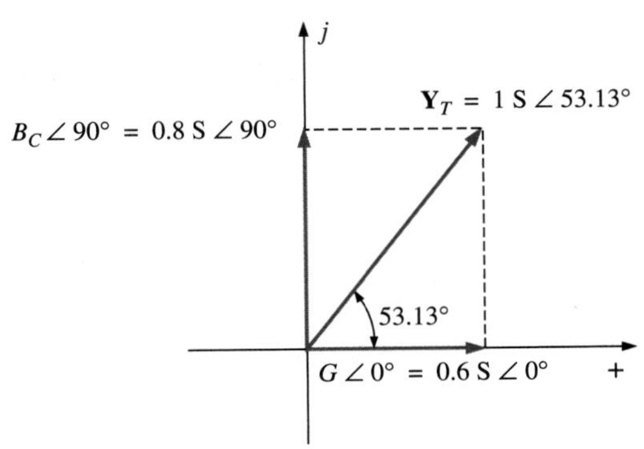

FIG. 15.74
Admittance diagram for the parallel R-C network in Fig. 15.72.

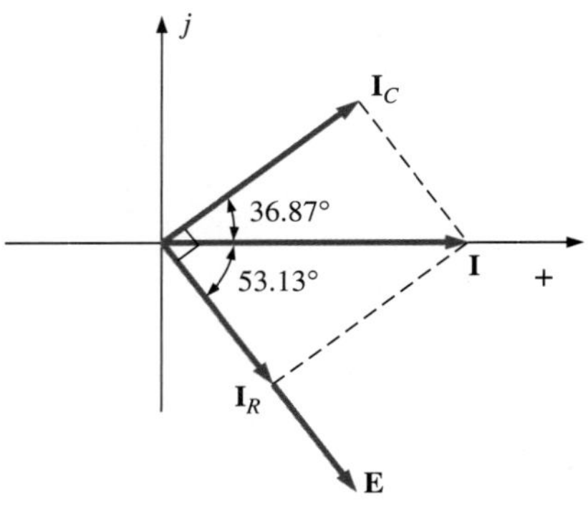

FIG. 15.75
Phasor diagram for the parallel R-C network in Fig. 15.74.

Phasor diagram: The phasor diagram in Fig. 15.75 indicates that **E** is in phase with the current through the resistor $\mathbf{I}_R$ and lags the capacitive current $\mathbf{I}_C$ by 90°.

Time domain:

$$e = \sqrt{2}(10)\sin(\omega t - 53.13°) = \mathbf{14.14\sin(\omega t - 53.13°)}$$
$$i_R = \sqrt{2}(6)\sin(\omega t - 53.13°) = \mathbf{8.48\sin(\omega t - 53.13°)}$$
$$i_C = \sqrt{2}(8)\sin(\omega t + 36.87°) = \mathbf{11.31\sin(\omega t + 36.87°)}$$

A plot of all of the currents and the voltage appears in Fig. 15.76. Note that e and i_R are in phase and e lags i_C by 90°.

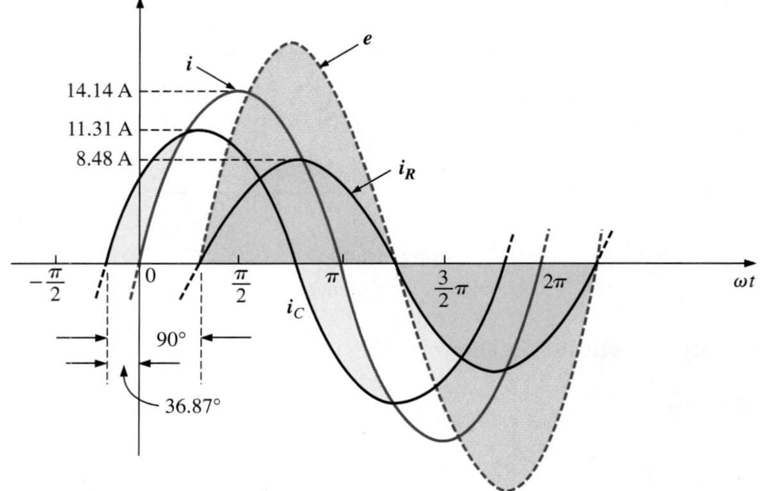

FIG. 15.76
Waveforms for the parallel R-C network in Fig. 15.72.

Power:

$$P_T = EI\cos\theta = (10\text{ V})(10\text{ A})\cos 53.13° = (10)^2(0.6)$$
$$= \mathbf{60\text{ W}}$$

or

$$P_T = E^2G = (10\text{ V})^2(0.6\text{ S}) = \mathbf{60\text{ W}}$$

or, finally,

$$P_T = P_R + P_C = EI_R\cos\theta_R + EI_C\cos\theta_C$$
$$= (10\text{ V})(6\text{ A})\cos 0° + (10\text{ V})(8\text{ A})\cos 90°$$
$$= \mathbf{60\text{ W}}$$

Power factor: The power factor of the circuit is

$$F_p = \cos 53.13° = \mathbf{0.6\ leading}$$

Using Eq. (15.34), we have

$$F_p = \cos\theta_T = \frac{G}{Y_T} = \frac{0.6\text{ S}}{1.0\text{ S}} = \mathbf{0.6\ leading}$$

Impedance approach: The voltage **E** can also be found by first finding the total impedance of the circuit:

$$\mathbf{Z}_T = \frac{\mathbf{Z}_R\mathbf{Z}_C}{\mathbf{Z}_R + \mathbf{Z}_C} = \frac{(1.67\ \Omega\ \angle 0°)(1.25\ \Omega\ \angle -90°)}{1.67\ \Omega\ \angle 0° + 1.25\ \Omega\ \angle -90°}$$
$$= \frac{2.09\ \angle -90°}{2.09\ \angle -36.81°} = \mathbf{1\ \Omega\ \angle -53.19°}$$

and then, using Ohm's law, we find

$$\mathbf{E} = \mathbf{IZ}_T = (10\text{ A}\angle 0°)(1\,\Omega\angle -53.19°) = \mathbf{10\text{ V}\angle -53.19°}$$

R-L-C

Refer to Fig. 15.77.

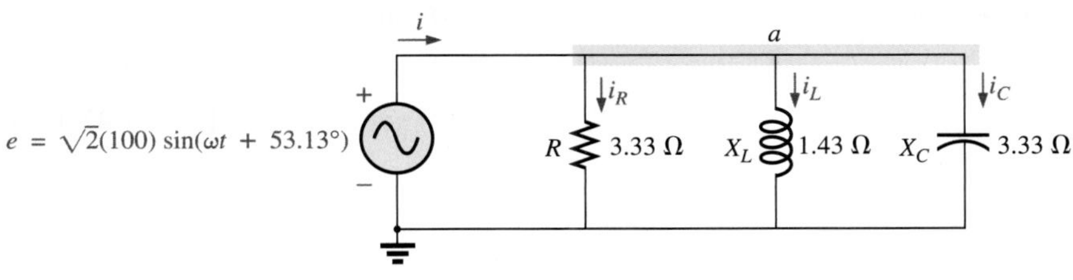

FIG. 15.77
Parallel R-L-C ac network.

Phasor notation: As shown in Fig. 15.78.

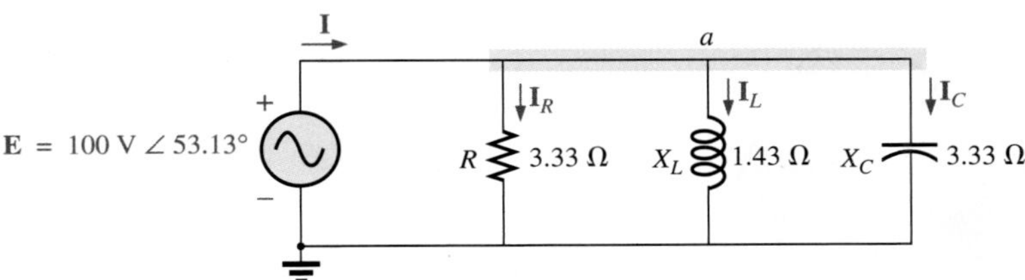

FIG. 15.78
Applying phasor notation to the network in Fig. 15.77.

$$\begin{aligned}
\mathbf{Y}_T &= \mathbf{Y}_R + \mathbf{Y}_L + \mathbf{Y}_C = G\angle 0° + B_L\angle -90° + B_C\angle 90° \\
&= \frac{1}{3.33\,\Omega}\angle 0° + \frac{1}{1.43\,\Omega}\angle -90° + \frac{1}{3.33\,\Omega}\angle 90° \\
&= 0.3\text{ S}\angle 0° + 0.7\text{ S}\angle -90° + 0.3\text{ S}\angle 90° \\
&= 0.3\text{ S} - j\,0.7\text{ S} + j\,0.3\text{ S} \\
&= 0.3\text{ S} - j\,0.4\text{ S} = \mathbf{0.5\text{ S}\angle -53.13°}
\end{aligned}$$

$$\mathbf{Z}_T = \frac{1}{\mathbf{Y}_T} = \frac{1}{0.5\text{ S}\angle -53.13°} = \mathbf{2\,\Omega\angle 53.13°}$$

Admittance diagram: As shown in Fig. 15.79.

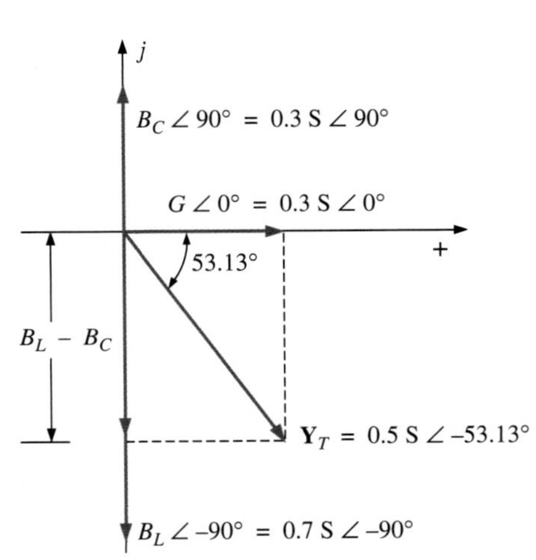

FIG. 15.79
Admittance diagram for the parallel R-L-C network in Fig. 15.77.

$$\mathbf{I} = \frac{\mathbf{E}}{\mathbf{Z}_T} = \mathbf{EY}_T = (100\text{ V}\angle 53.13°)(0.5\text{ S}\angle -53.13°) = \mathbf{50\text{ A}\angle 0°}$$

$$\begin{aligned}
\mathbf{I}_R &= (E\angle\theta(G\angle 0°) \\
&= (100\text{ V}\angle 53.13°)(0.3\text{ S}\angle 0°) = \mathbf{30\text{ A}\angle 53.13°} \\
\mathbf{I}_L &= (E\angle\theta)(B_L\angle -90°) \\
&= (100\text{ V}\angle 53.13°)(0.7\text{ S}\angle -90°) = \mathbf{70\text{ A}\angle -36.87°} \\
\mathbf{I}_C &= (E\angle\theta)(B_C\angle 90°) \\
&= (100\text{ V}\angle 53.13°)(0.3\text{ S}\angle +90°) = \mathbf{30\text{ A}\angle 143.13°}
\end{aligned}$$

Kirchhoff's current law: At node *a,*

$$\mathbf{I} - \mathbf{I}_R - \mathbf{I}_L - \mathbf{I}_C = 0$$

or

$$\mathbf{I} = \mathbf{I}_R + \mathbf{I}_L + \mathbf{I}_C$$

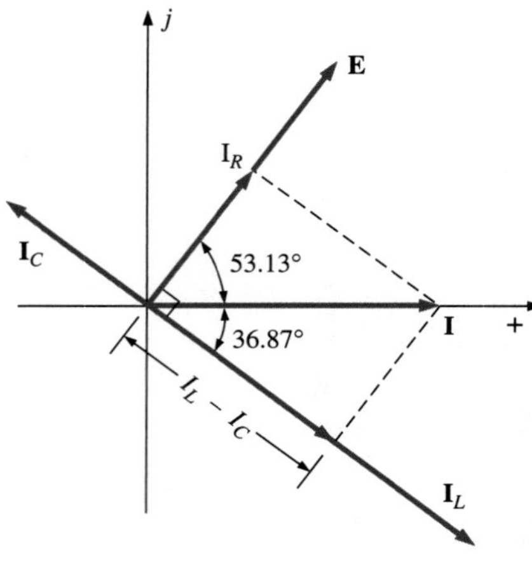

FIG. 15.80
Phasor diagram for the parallel R-L-C network in Fig. 15.77.

Phasor diagram: The phasor diagram in Fig. 15.80 indicates that the impressed voltage **E** is in phase with the current $\mathbf{I}_R$ through the resistor, leads the current $\mathbf{I}_L$ through the inductor by 90°, and lags the current $\mathbf{I}_C$ of the capacitor by 90°.

Time domain:

$$i = \sqrt{2}(50)\sin\omega t = \mathbf{70.70\sin\omega t}$$
$$i_R = \sqrt{2}(30)\sin(\omega t + 53.13°) = \mathbf{42.42\sin(\omega t + 53.13°)}$$
$$i_L = \sqrt{2}(70)\sin(\omega t - 36.87°) = \mathbf{98.98\sin(\omega t - 36.87°)}$$
$$i_C = \sqrt{2}(30)\sin(\omega t + 143.13°) = \mathbf{42.42\sin(\omega t + 143.13°)}$$

A plot of all of the currents and the impressed voltage appears in Fig. 15.81.

Power: The total power in watts delivered to the circuit is

$$P_T = EI\cos\theta = (100\text{ V})(50\text{ A})\cos 53.13° = (5000)(0.6) = \mathbf{3000\ W}$$

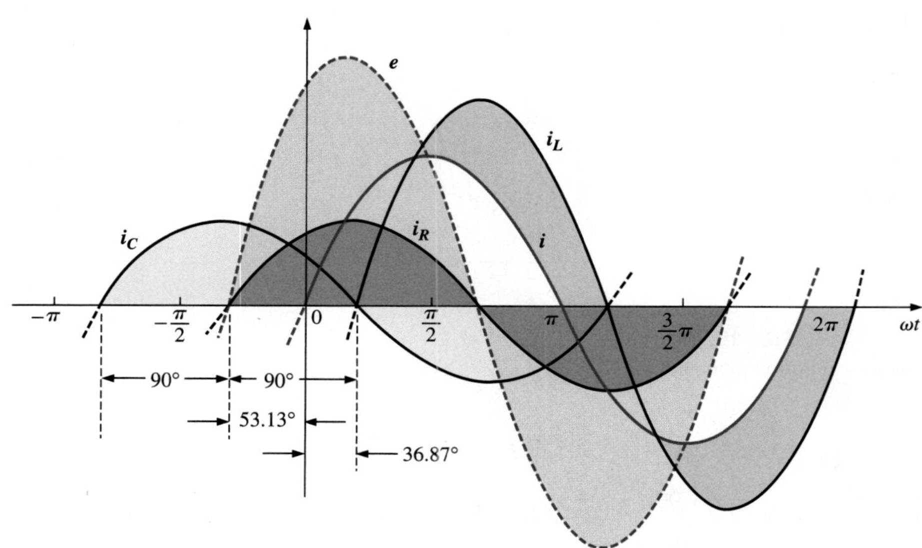

FIG. 15.81
Waveforms for the parallel R-L-C network in Fig. 15.77.

or
$$P_T = E^2G = (100\text{ V})^2(0.3\text{ S}) = \mathbf{3000\ W}$$

or, finally,

$$\begin{aligned}P_T &= P_R + P_L + P_C\\ &= EI_R\cos\theta_R + EI_L\cos\theta_L + EI_C\cos\theta_C\\ &= (100\text{ V})(30\text{ A})\cos 0° + (100\text{ V})(70\text{ A})\cos 90°\\ &\qquad + (100\text{ V})(30\text{ A})\cos 90°\\ &= 3000\text{ W} + 0 + 0\\ &= \mathbf{3000\ W}\end{aligned}$$

Power factor: The power factor of the circuit is

$$F_p = \cos\theta_T = \cos 53.13° = \mathbf{0.6\ lagging}$$

Using Eq. (15.34), we obtain

$$F_p = \cos\theta_T = \frac{G}{Y_T} = \frac{0.3\text{ S}}{0.5\text{ S}} = \mathbf{0.6\ lagging}$$

Impedance approach: The input current **I** can also be determined by first finding the total impedance in the following manner:

$$\mathbf{Z}_T = \frac{\mathbf{Z}_R\mathbf{Z}_L\mathbf{Z}_C}{\mathbf{Z}_R\mathbf{Z}_L + \mathbf{Z}_L\mathbf{Z}_C + \mathbf{Z}_R\mathbf{Z}_C} = \mathbf{2\ \Omega\ \angle 53.13^\circ}$$

and, applying Ohm's law, we obtain

$$\mathbf{I} = \frac{\mathbf{E}}{\mathbf{Z}_T} = \frac{100\text{ V}\ \angle 53.13^\circ}{2\ \Omega\ \angle 53.13^\circ} = \mathbf{50\ A\ \angle 0^\circ}$$

15.9 CURRENT DIVIDER RULE

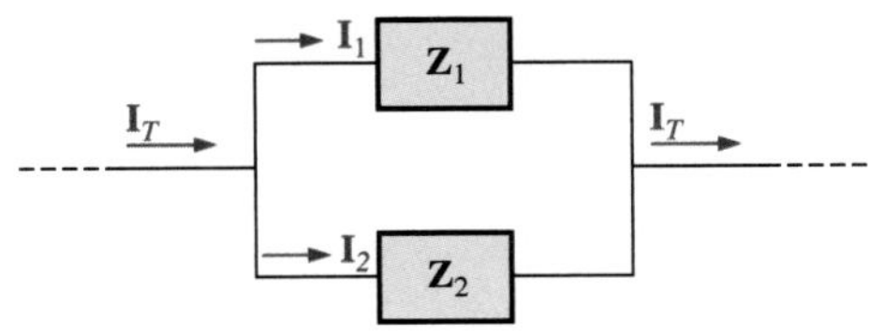

FIG. 15.82
Applying the current divider rule.

The basic format for the **current divider rule** in ac circuits is exactly the same as that for dc circuits; that is, for two parallel branches with impedances $\mathbf{Z}_1$ and $\mathbf{Z}_2$ as shown in Fig. 15.82,

$$\mathbf{I}_1 = \frac{\mathbf{Z}_2\mathbf{I}_T}{\mathbf{Z}_1 + \mathbf{Z}_2} \quad \text{or} \quad \mathbf{I}_2 = \frac{\mathbf{Z}_1\mathbf{I}_T}{\mathbf{Z}_1 + \mathbf{Z}_2} \qquad \mathbf{(15.35)}$$

EXAMPLE 15.16 Using the current divider rule, find the current through each impedance in Fig. 15.83.

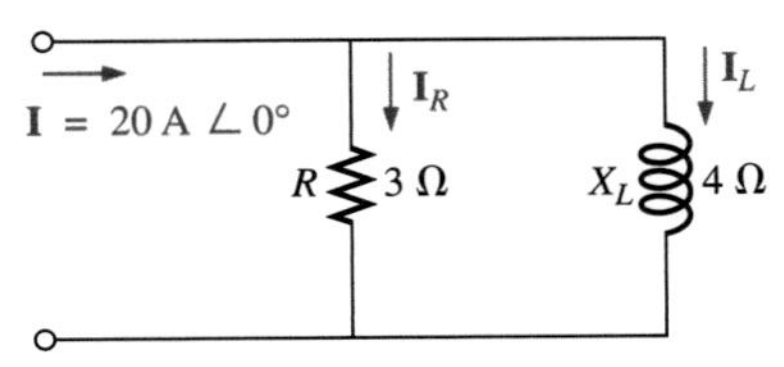

FIG. 15.83
Example 15.16.

Solution:

$$\mathbf{I}_R = \frac{\mathbf{Z}_L\mathbf{I}_T}{\mathbf{Z}_R + \mathbf{Z}_L} = \frac{(4\ \Omega\ \angle 90^\circ)(20\text{ A}\ \angle 0^\circ)}{3\ \Omega\ \angle 0^\circ + 4\ \Omega\ \angle 90^\circ} = \frac{80\text{ A}\ \angle 90^\circ}{5\ \angle 53.13^\circ}$$
$$= \mathbf{16\ A\ \angle 36.87^\circ}$$

$$\mathbf{I}_L = \frac{\mathbf{Z}_R\mathbf{I}_T}{\mathbf{Z}_R + \mathbf{Z}_L} = \frac{(3\ \Omega\ \angle 0^\circ)(20\text{ A}\ \angle 0^\circ)}{5\ \Omega\ \angle 53.13^\circ} = \frac{60\text{ A}\ \angle 0^\circ}{5\ \angle 53.13^\circ}$$
$$= \mathbf{12\ A\ \angle -53.13^\circ}$$

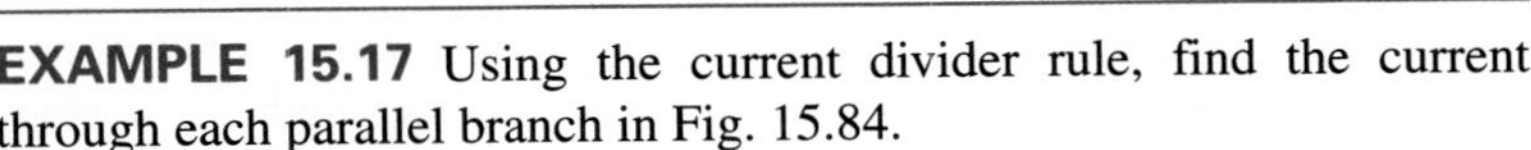

EXAMPLE 15.17 Using the current divider rule, find the current through each parallel branch in Fig. 15.84.

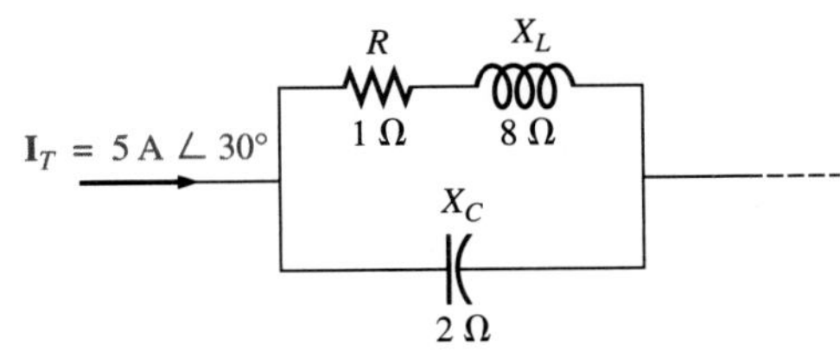

FIG. 15.84
Example 15.17.

Solution:

$$\mathbf{I}_{R\text{-}L} = \frac{\mathbf{Z}_C\mathbf{I}_T}{\mathbf{Z}_C + \mathbf{Z}_{R\text{-}L}} = \frac{(2\ \Omega\ \angle -90^\circ)(5\text{ A}\ \angle 30^\circ)}{-j\,2\ \Omega + 1\ \Omega + j\,8\ \Omega} = \frac{10\text{ A}\ \angle -60^\circ}{1 + j6}$$
$$= \frac{10\text{ A}\ \angle -60^\circ}{6.083\ \angle 80.54^\circ} \cong \mathbf{1.64\ A\ \angle -140.54^\circ}$$

$$\mathbf{I}_C = \frac{\mathbf{Z}_{R\text{-}L}\mathbf{I}_T}{\mathbf{Z}_{R\text{-}L} + \mathbf{Z}_C} = \frac{(1\ \Omega + j\,8\ \Omega)(5\text{ A}\ \angle 30^\circ)}{6.08\ \Omega\ \angle 80.54^\circ}$$
$$= \frac{(8.06\ \angle 82.87^\circ)(5\text{ A}\ \angle 30^\circ)}{6.08\ \angle 80.54^\circ} = \frac{40.30\text{ A}\ \angle 112.87^\circ}{6.083\ \angle 80.54^\circ}$$
$$= \mathbf{6.63\ A\ \angle 32.33^\circ}$$

15.10 FREQUENCY RESPONSE OF PARALLEL ELEMENTS

Recall that for elements in series, the total impedance is the direct sum of the impedances of each element, and the largest real or imaginary component has the most impact on the total impedance. For parallel elements, it is important to remember that *the smallest parallel resistor or the smallest parallel reactance will have the most impact on the real or imaginary component, respectively, of the total impedance.*

In Fig. 15.85, the frequency response has been included for each element of a parallel *R-L-C* combination. At very low frequencies, the impedance of the coil will be less than that of the resistor or capacitor, resulting in an inductive network in which the reactance of the inductor will have the most impact on the total impedance. As the frequency increases, the impedance of the inductor will increase while the impedance of the capacitor will decrease. Depending on the components chosen, it is possible that the reactance of the capacitor will drop to a point where it will equal the impedance of the coil before either one reaches the resistance level.

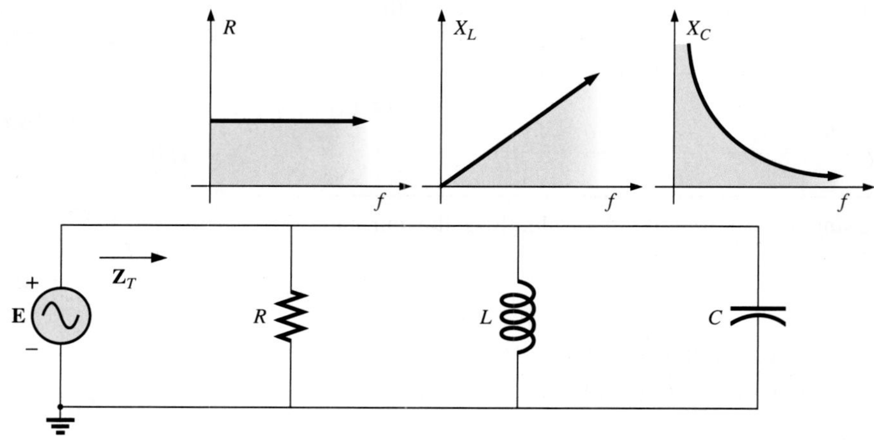

FIG. 15.85
Frequency response for parallel R-L-C elements.

Therefore, it is impossible to make too many broad statements about the effect of each element as the frequency increases. In general, however, for very low frequencies, we can assume that a parallel *R-L-C* network will be inductive as described above, and at very high frequencies it will be capacitive since X_C will drop to very low levels. In between, a point will result at which X_L will equal X_C and where X_L or X_C will equal *R*. The frequencies at which these events occur, however, depend on the elements chosen and the frequency range of interest. In general, however, keep in mind that the smaller the resistance or reactance, the greater its impact on the total impedance of a parallel system.

Let us now note the impact of frequency on the total impedance and inductive current for the parallel *R-L* network in Fig. 15.86 for a frequency range of zero through 40 kHz.

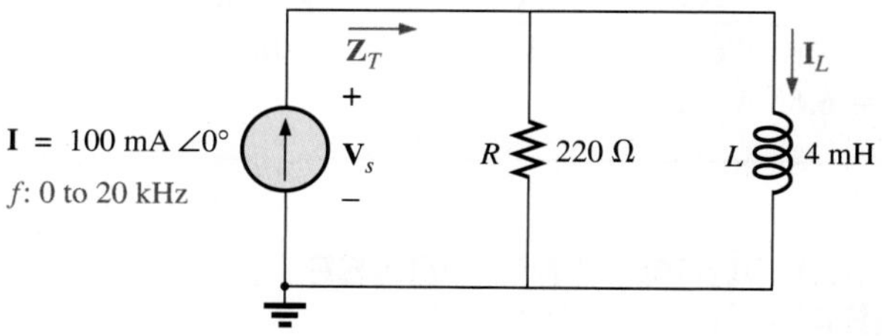

FIG. 15.86
Determining the frequency response of a parallel R-L network.

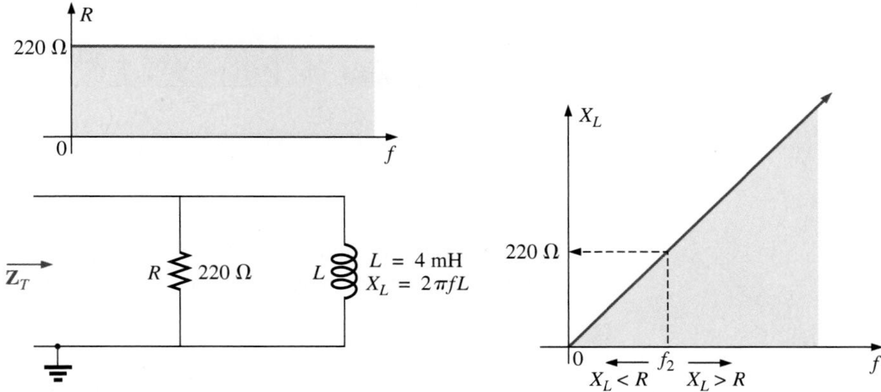

FIG. 15.87
The frequency response of the individual elements of a parallel R-L network.

$\mathbf{Z}_T$ Before getting into specifics, let us first develop a "sense" for the impact of frequency on the network in Fig. 15.86 by noting the impedance-versus-frequency curves of the individual elements, as shown in Fig. 15.87. The fact that the elements are now in parallel requires that we consider their characteristics in a different manner than occurred for the series *R-C* circuit in Section 15.5. Recall that for parallel elements, the element with the smallest impedance will have the greatest impact on the total impedance at that frequency. In Fig. 15.87, for example, X_L is very small at low frequencies compared to *R*, establishing X_L as the predominant factor in this frequency range. In other words, at low frequencies the network will be primarily inductive, and the angle associated with the total impedance will be close to 90°, as with a pure inductor. As the frequency increases, X_L increases until it equals the impedance of the resistor (220 Ω). The frequency at which this situation occurs can be determined in the following manner:

$$X_L = 2\pi f_2 L = R$$

and

$$\boxed{f_2 = \frac{R}{2\pi L}} \qquad \textbf{(15.36)}$$

which for the network in Fig. 15.86 is

$$f_2 = \frac{R}{2\pi L} = \frac{220\ \Omega}{2\pi(4 \times 10^{-3}\ \text{H})}$$
$$\cong \mathbf{8.75\ kHz}$$

which falls within the frequency range of interest.

For frequencies less than f_2, $X_L < R$, and for frequencies greater than f_2, $X_L > R$, as shown in Fig. 15.87. A general equation for the total impedance in vector form can be developed in the following manner:

$$\mathbf{Z}_T = \frac{\mathbf{Z}_R\mathbf{Z}_L}{\mathbf{Z}_R + \mathbf{Z}_L}$$
$$= \frac{(R\angle 0°)(X_L\angle 90°)}{R + jX_L} = \frac{RX_L\angle 90°}{\sqrt{R^2 + X_L^2}\angle \tan^{-1} X_L/R}$$

and

$$\mathbf{Z}_T = \frac{RX_L}{\sqrt{R^2 + X_L^2}}\underline{/90° - \tan^{-1} X_L/R}$$

so that
$$\mathbf{Z}_T = \frac{RX_L}{\sqrt{R^2 + X_L^2}} \tag{15.37}$$

and
$$\theta_T = 90° - \tan^{-1}\frac{X_L}{R} = \tan^{-1}\frac{R}{X_L} \tag{15.38}$$

The magnitude and angle of the total impedance can now be found at any frequency of interest simply by substituting Eqs. (15.37) and (15.38).

***f* = 1 kHz**

$$X_L = 2\pi fL = 2\pi(1\text{ kHz})(4 \times 10^{-3}\text{ H}) = 25.12\ \Omega$$

and

$$Z_T = \frac{RX_L}{\sqrt{R^2 + X_L^2}} = \frac{(220\ \Omega)(25.12\ \Omega)}{\sqrt{(220\ \Omega)^2 + (25.12\ \Omega)^2}} = \mathbf{24.96\ \Omega}$$

with
$$\theta_T = \tan^{-1}\frac{R}{X_L} = \tan^{-1}\frac{220\ \Omega}{25.12\ \Omega}$$
$$= \tan^{-1} 8.76 = 83.49°$$

and
$$\mathbf{Z}_T = \mathbf{24.96\ \Omega\ \angle 83.49°}$$

This value compares very closely with $X_L = 25.12\ \Omega\ \angle 90°$, which it would be if the network were purely inductive ($R = \infty\ \Omega$). Our assumption that the network is primarily inductive at low frequencies is therefore confirmed.

Continuing:

$f =$ 5 kHz: $\mathbf{Z}_T = \mathbf{109.1\ \Omega\ \angle 60.23°}$
$f =$ 10 kHz: $\mathbf{Z}_T = \mathbf{165.5\ \Omega\ \angle 41.21°}$
$f =$ 15 kHz: $\mathbf{Z}_T = \mathbf{189.99\ \Omega\ \angle 30.28°}$
$f =$ 20 kHz: $\mathbf{Z}_T = \mathbf{201.53\ \Omega\ \angle 23.65°}$
$f =$ 30 kHz: $\mathbf{Z}_T = \mathbf{211.19\ \Omega\ \angle 16.27°}$
$f =$ 40 kHz: $\mathbf{Z}_T = \mathbf{214.91\ \Omega\ \angle 12.35°}$

At $f = 40$ kHz, note how closely the magnitude of Z_T has approached the resistance level of 220 Ω and how the associated angle with the total impedance is approaching zero degrees. The result is a network with terminal characteristics that are becoming more and more resistive as the frequency increases, which further confirms the earlier conclusions developed by the curves in Fig. 15.87.

Plots of Z_T versus frequency in Fig. 15.88 and θ_T in Fig. 15.89 clearly reveal the transition from an inductive network to one that has resistive characteristics. Note that the transition frequency of 8.75 kHz occurs right in the middle of the "knee" of the curves for both Z_T and θ_T.

A review of Figs. 15.49 and 15.88 reveals that a series *R-C* and a parallel *R-L* network will have an impedance level that approaches the resistance of the network at high frequencies. The capacitive circuit approaches the level from above, whereas the inductive network does the same from below. For the series *R-L* circuit and the parallel *R-C* network, the total impedance will begin at the resistance level and then display the characteristics of the reactive elements at high frequencies.

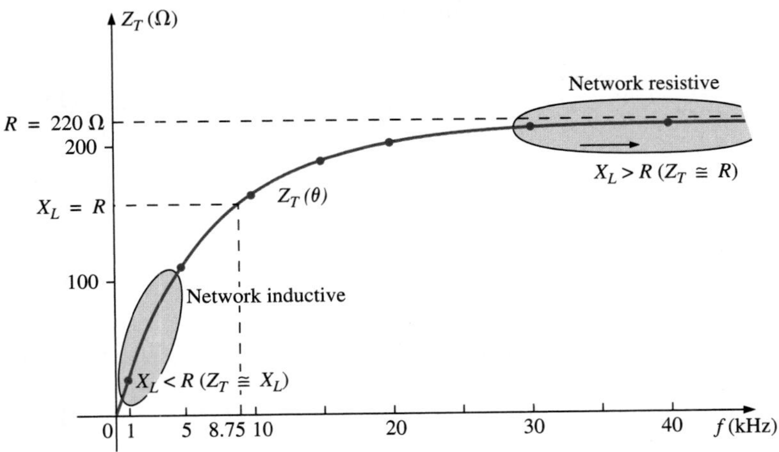

FIG. 15.88
The magnitude of the input impedance versus frequency for the network in Fig. 15.86.

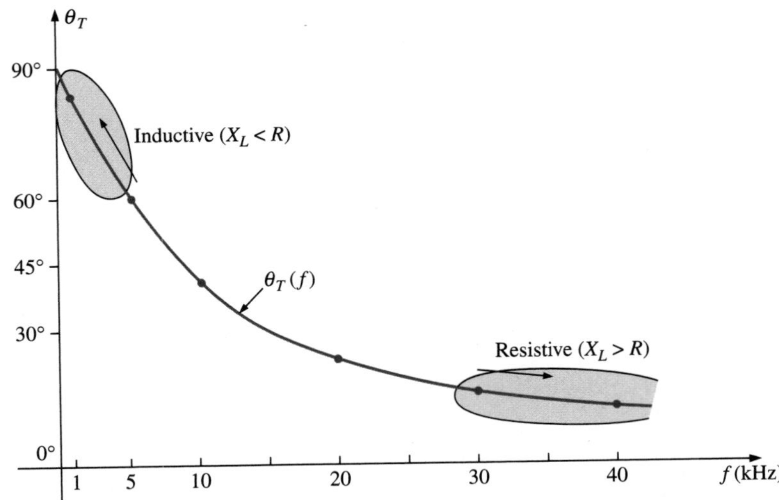

FIG. 15.89
The phase angle of the input impedance versus frequency for the network in Fig. 15.86.

$\mathbf{I}_L$ Applying the current divider rule to the network in Fig. 15.86 results in the following:

$$\mathbf{I}_L = \frac{\mathbf{Z}_R \mathbf{I}}{\mathbf{Z}_R + \mathbf{Z}_L}$$

$$= \frac{(R \angle 0°)(I \angle 0°)}{R + j X_L} = \frac{RI \angle 0°}{\sqrt{R^2 + X_L^2}\ \underline{/\tan^{-1} X_L/R}}$$

and
$$\mathbf{I}_L = I_L \angle \theta_L = \frac{RI}{\sqrt{R^2 + X_L^2}}\ \underline{/-\tan^{-1} X_L/R}$$

The magnitude of I_L is therefore determined by

$$I_L = \frac{RI}{\sqrt{R^2 + X_L^2}} \quad (15.39)$$

and the phase angle θ_L, by which $\mathbf{I}_L$ leads **I**, is given by

$$\theta_L = -\tan^{-1}\frac{X_L}{R} \quad (15.40)$$

Because θ_L is always negative, the magnitude of θ_L is, in actuality, the angle by which $\mathbf{I}_L$ lags **I**.

To begin our analysis, let us first consider the case of $f = 0$ Hz (dc conditions).

f = 0 Hz

$$X_L = 2\pi fL = 2\pi(0\text{ Hz})L = 0\ \Omega$$

Applying the short-circuit equivalent for the inductor in Fig. 15.86 results in

$$\mathbf{I}_L = \mathbf{I} = 100\text{ mA}\angle 0°$$

as appearing in Figs. 15.90 and 15.91.

f = 1 kHz Applying Eq. (15.39):

$$X_L = 2\pi fL = 2\pi(1\text{ kHz})(4\text{ mH}) = 25.12\ \Omega$$

and $$\sqrt{R^2 + X_L^2} = \sqrt{(220\ \Omega)^2 + (25.12\ \Omega)^2} = 221.43\ \Omega$$

and $$I_L = \frac{RI}{\sqrt{R^2 + X_L^2}} = \frac{(220\ \Omega)(100\text{ mA})}{221.43\ \Omega} = \mathbf{99.35\ mA}$$

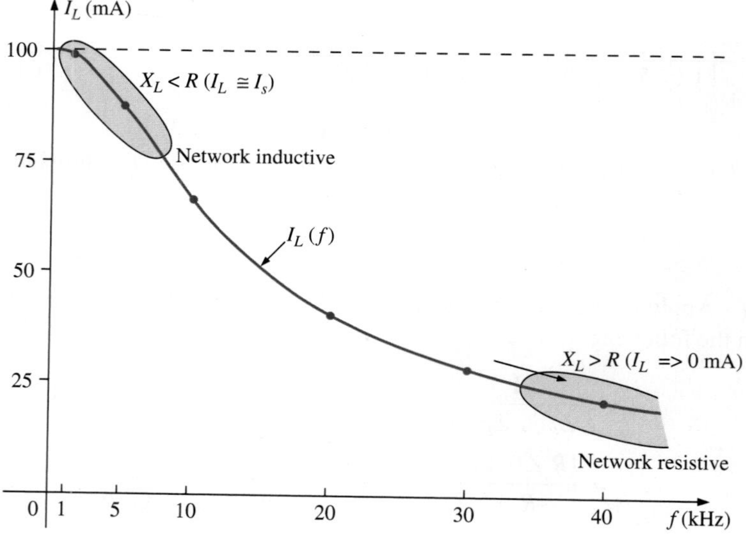

FIG. 15.90
The magnitude of the current $\mathbf{I}_L$ versus frequency for the parallel R-L network in Fig. 15.86.

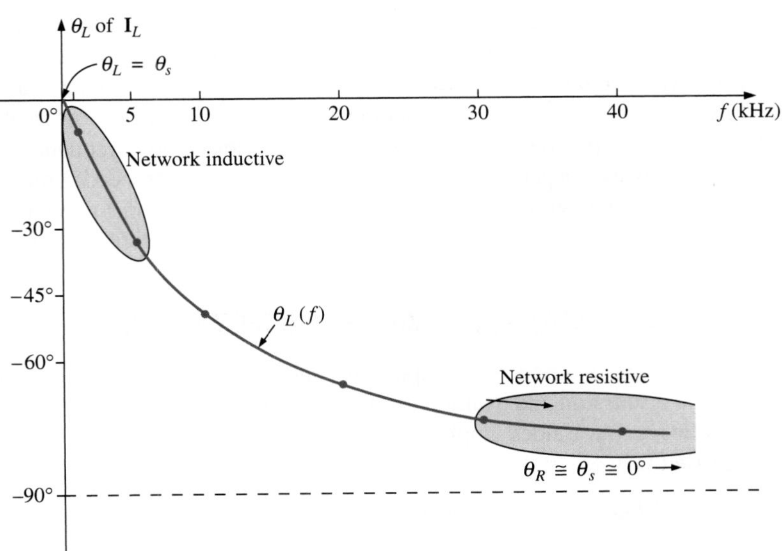

FIG. 15.91
The phase angle of the current $\mathbf{I}_L$ *versus frequency for the parallel R-L network in Fig. 15.86.*

with

$$\theta_L = \tan^{-1}\frac{X_L}{R} = -\tan^{-1}\frac{25.12\ \Omega}{220\ \Omega} = -\tan^{-1} 0.114 = \mathbf{-6.51°}$$

and $$\mathbf{I}_L = \mathbf{99.35\ mA} \angle \mathbf{-6.51°}$$

The result is a current $\mathbf{I}_L$ that is still very close to the source current $\mathbf{I}$ in both magnitude and phase.

Continuing:

$$
\begin{aligned}
f &= 5\text{ kHz:} & \mathbf{I}_L &= \mathbf{86.84\ mA} \angle \mathbf{-29.72°}\\
f &= 10\text{ kHz:} & \mathbf{I}_L &= \mathbf{65.88\ mA} \angle \mathbf{-48.79°}\\
f &= 15\text{ kHz:} & \mathbf{I}_L &= \mathbf{50.43\ mA} \angle \mathbf{-59.72°}\\
f &= 20\text{ kHz:} & \mathbf{I}_L &= \mathbf{40.11\ mA} \angle \mathbf{-66.35°}\\
f &= 30\text{ kHz:} & \mathbf{I}_L &= \mathbf{28.02\ mA} \angle \mathbf{-73.73°}\\
f &= 40\text{ kHz:} & \mathbf{I}_L &= \mathbf{21.38\ mA} \angle \mathbf{-77.65°}
\end{aligned}
$$

The plot of the magnitude of I_L versus frequency is provided in Fig. 15.90 and reveals that the current through the coil dropped from its maximum of 100 mA to almost 20 mA at 40 kHz. As the reactance of the coil increased with frequency, more of the source current chose the lower-resistance path of the resistor. The magnitude of the phase angle between $\mathbf{I}_L$ and $\mathbf{I}$ is approaching 90° with an increase in frequency, as shown in Fig. 15.91, leaving its initial value of zero degrees at $f = 0$ Hz far behind.

At $f = 1$ kHz, the phasor diagram of the network appears as shown in Fig. 15.92. First note that the magnitude and the phase angle of $\mathbf{I}_L$ are very close to those of $\mathbf{I}$. Since the voltage across a coil must lead the current through a coil by 90°, the voltage $\mathbf{V}_s$ appears as shown. The voltage across a resistor is in phase with the current through the resistor, resulting in the direction of $\mathbf{I}_R$ shown in Fig. 15.92. Of course, at this frequency $R > X_L$, and the current I_R is relatively small in magnitude.

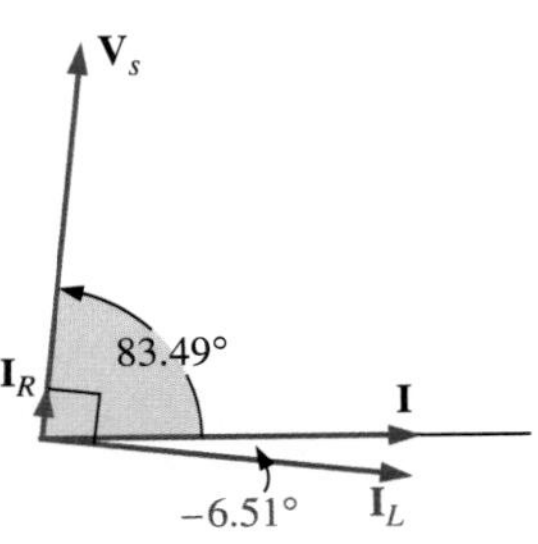

FIG. 15.92
The phasor diagram for the parallel R-L network in Fig. 15.86 at f = 1 kHz.

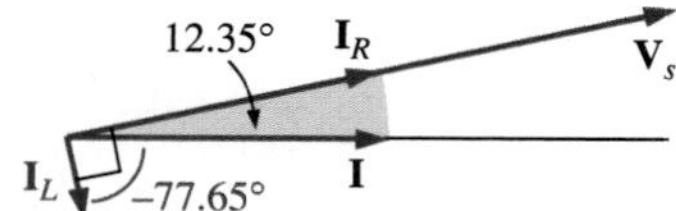

FIG. 15.93
The phasor diagram for the parallel R-L network in Fig. 15.86 at f = 40 kHz.

At $f = 40$ kHz, the phasor diagram changes to that appearing in Fig. 15.93. Note that now $\mathbf{I}_R$ and $\mathbf{I}$ are close in magnitude and phase because $X_L > R$. The magnitude of $\mathbf{I}_L$ has dropped to very low levels, and the phase angle associated with $\mathbf{I}_L$ is approaching $-90°$. The network is now more "resistive" compared to its "inductive" characteristics at low frequencies.

The analysis of a parallel *R-C* or *R-L-C* network would proceed in much the same manner, with the inductive impedance predominating at low frequencies and the capacitive reactance predominating at high frequencies.

15.11 SUMMARY: PARALLEL ac NETWORKS

The following is a review of important conclusions that can be derived from the discussion and examples of the previous sections. The list is not all-inclusive, but it does emphasize some of the conclusions that should be carried forward in the future analysis of ac systems.

For parallel ac networks with reactive elements:

1. ***The total admittance or impedance will be frequency dependent.***
2. ***Depending on the frequency applied, the same network can be either predominantly inductive or predominantly capacitive.***
3. ***The magnitude of the current through any one branch can be greater than the source current.***
4. ***The inductive and capacitive susceptances are in direct opposition on an admittance diagram.***
5. ***At lower frequencies, the inductive elements will usually have the most impact on the total impedance, while at high frequencies, the capacitive elements will usually have the most impact.***
6. ***The impedance of any one element can be less than the total impedance (recall that for dc circuits, the total resistance must always be less than the smallest parallel resistor).***
7. ***The magnitude of the current through an element, compared to the other elements of the network, is directly related to the magnitude of its impedance; that is, the smaller the impedance of an element, the larger the magnitude of the current through the element.***
8. ***The current through a coil is always in direct opposition with the current through a capacitor on a phasor diagram.***
9. ***The applied voltage is always in phase with the current through the resistive elements, leads the voltage across all the inductive elements by 90°, and lags the current through all capacitive elements by 90°.***
10. ***The smaller the resistive element of a network compared to the net reactive susceptance, the closer the power factor is to unity.***

15.12 EQUIVALENT CIRCUITS

In a series ac circuit, the total impedance of two or more elements in series is often equivalent to an impedance that can be achieved with fewer elements of different values, the elements and their values being determined by the frequency applied. This is also true for parallel circuits. For the circuit in Fig. 15.94(a),

$$\mathbf{Z}_T = \frac{\mathbf{Z}_C\mathbf{Z}_L}{\mathbf{Z}_C + \mathbf{Z}_L} = \frac{(5\ \Omega \angle -90°)(10\ \Omega \angle 90°)}{5\ \Omega \angle -90° + 10\ \Omega \angle 90°} = \frac{50 \angle 0°}{5 \angle 90°}$$
$$= 10\ \Omega \angle -90°$$

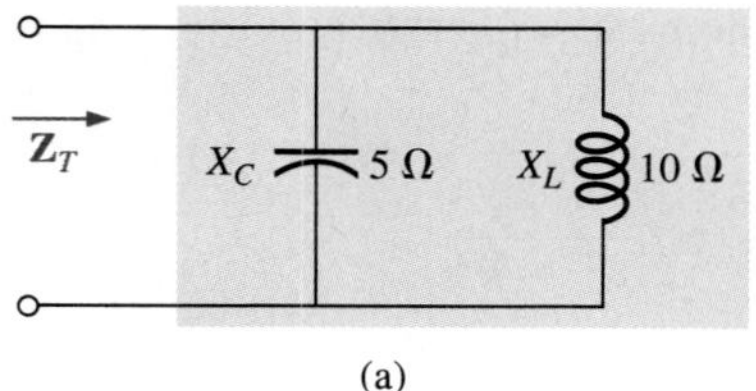

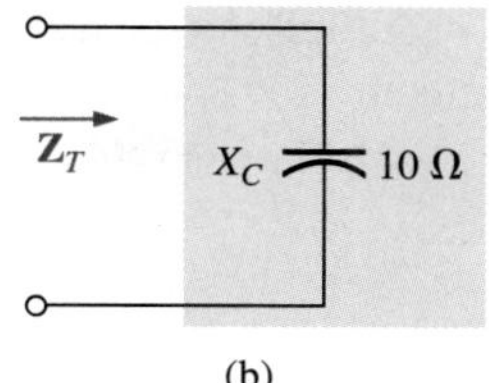

FIG. 15.94
Defining the equivalence between two networks at a specific frequency.

The total impedance at the frequency applied is equivalent to a capacitor with a reactance of 10 Ω, as shown in Fig. 15.94(b). Always keep in mind that this equivalence is true only at the applied frequency. If the frequency changes, the reactance of each element changes, and the equivalent circuit changes—perhaps from capacitive to inductive in the above example.

Another interesting development appears if the impedance of a parallel circuit, such as the one in Fig. 15.95(a), is found in rectangular form. In this case,

$$\begin{aligned}\mathbf{Z}_T &= \frac{\mathbf{Z}_L\mathbf{Z}_R}{\mathbf{Z}_L + \mathbf{Z}_R} = \frac{(4\ \Omega\ \angle 90^\circ)(3\ \Omega\ \angle 0^\circ)}{4\ \Omega\ \angle 90^\circ + 3\ \Omega\ \angle 0^\circ}\\ &= \frac{12\ \angle 90^\circ}{5\ \angle 53.13^\circ} = 2.40\ \Omega\ \angle 36.87^\circ\\ &= 1.92\ \Omega + j\,1.44\ \Omega\end{aligned}$$

which is the impedance of a series circuit with a resistor of 1.92 Ω and an inductive reactance of 1.44 Ω, as shown in Fig. 15.95(b).

The current **I** will be the same in each circuit in Fig. 15.94 or Fig. 15.95 if the same input voltage **E** is applied. For a parallel circuit of one resistive element and one reactive element, the series circuit with the same input impedance will always be composed of one resistive and one reactive element. The impedance of each element of the series circuit will be different from that of the parallel circuit, but the reactive elements will always be of the same type; that is, an *R-L* circuit and an *R-C* parallel circuit will have an equivalent *R-L* and *R-C* series circuit, respectively. The same is true when converting from a series to a parallel circuit. In the discussion to follow, keep in mind that

the term equivalent refers only to the fact that for the same applied potential, the same impedance and input current will result.

To formulate the equivalence between the series and parallel circuits, the equivalent series circuit for a resistor and reactance in parallel can be found by determining the total impedance of the circuit in rectangular form; that is, for the circuit in Fig. 15.96(a),

$$\mathbf{Y}_p = \frac{1}{R_p} + \frac{1}{\pm jX_p} = \frac{1}{R_p} \mp j\frac{1}{X_p}$$

and

$$\begin{aligned}\mathbf{Z}_p &= \frac{1}{\mathbf{Y}_p} = \frac{1}{(1/R_p) \mp j(1/X_p)}\\ &= \frac{1/R_p}{(1/R_p)^2 + (1/X_p)^2} \pm j\frac{1/X_p}{(1/R_p)^2 + (1/X_p)^2}\end{aligned}$$

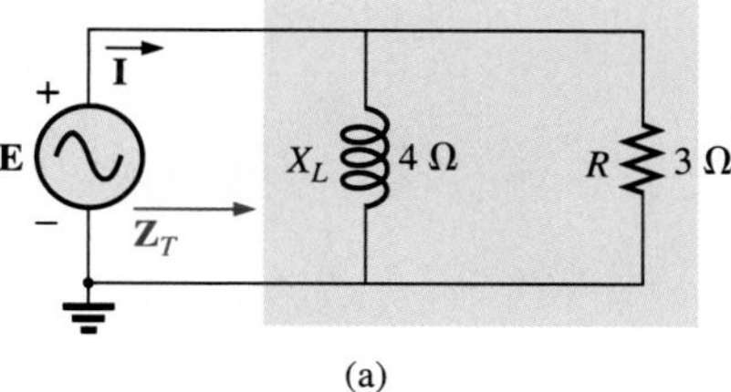

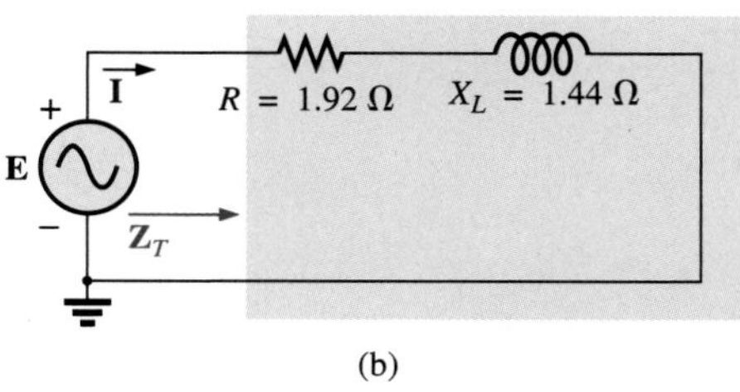

FIG. 15.95
Finding the series equivalent circuit for a parallel R-L network.

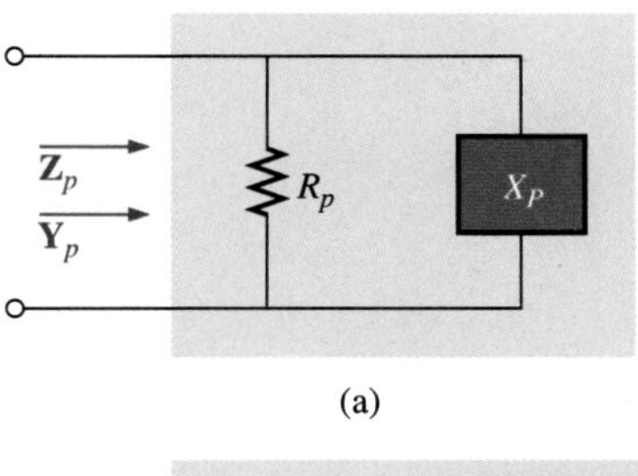

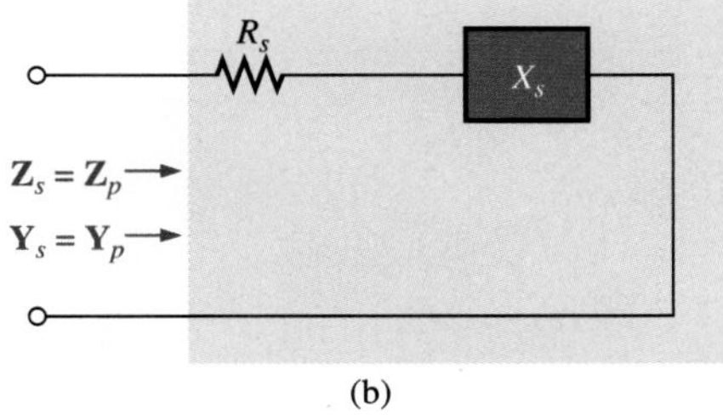

FIG. 15.96
Defining the parameters of equivalent series and parallel networks.

Multiplying the numerator and denominator of each term by $R_p^2 X_p^2$ results in

$$\mathbf{Z}_p = \frac{R_p X_p^2}{X_p^2 + R_p^2} \pm j \frac{R_p^2 X_p}{X_p^2 + R_p^2}$$
$$= R_s \pm j X_s \qquad \text{[Fig. 15.96(b)]}$$

and

$$R_s = \frac{R_p X_p^2}{X_p^2 + R_p^2} \tag{15.41}$$

with

$$X_s = \frac{R_p^2 X_p}{X_p^2 + R_p^2} \tag{15.42}$$

For the network in Fig. 15.95,

$$R_s = \frac{R_p X_p^2}{X_p^2 + R_p^2} = \frac{(3\ \Omega)(4\ \Omega)^2}{(4\ \Omega)^2 + (3\ \Omega)^2} = \frac{48\ \Omega}{25} = \mathbf{1.92\ \Omega}$$

and

$$X_s = \frac{R_p^2 X_p}{X_p^2 + R_p^2} = \frac{(3\ \Omega)^2(4\ \Omega)}{(4\ \Omega)^2 + (3\ \Omega)^2} = \frac{36\ \Omega}{25} = \mathbf{1.44\ \Omega}$$

which agrees with the previous result.

The equivalent parallel circuit for a circuit with a resistor and reactance in series can be found by finding the total admittance of the system in rectangular form; that is, for the circuit in Fig. 15.96(b),

$$\mathbf{Z}_s = R_s \pm jX_s$$
$$\mathbf{Y}_s = \frac{1}{\mathbf{Z}_s} = \frac{1}{R_s \pm j X_s} = \frac{R_s}{R_s^2 + X_s^2} \mp j \frac{X_s}{R_s^2 + X_s^2}$$
$$= G_p \mp j B_p = \frac{1}{R_p} \mp j \frac{1}{X_p} \qquad \text{[Fig. 15.96(a)]}$$

or

$$R_p = \frac{R_s^2 + X_s^2}{R_s} \tag{15.43}$$

with

$$X_p = \frac{R_s^2 + X_s^2}{X_s} \tag{15.44}$$

For the above example,

$$R_p = \frac{R_s^2 + X_s^2}{R_s} = \frac{(1.92\ \Omega)^2 + (1.44\ \Omega)^2}{1.92\ \Omega} = \frac{5.76\ \Omega}{1.92} = \mathbf{3.0\ \Omega}$$

and

$$X_p = \frac{R_s^2 + X_s^2}{X_s} = \frac{5.76\ \Omega}{1.44} = \mathbf{4.0\ \Omega}$$

as shown in Fig. 15.95(a).

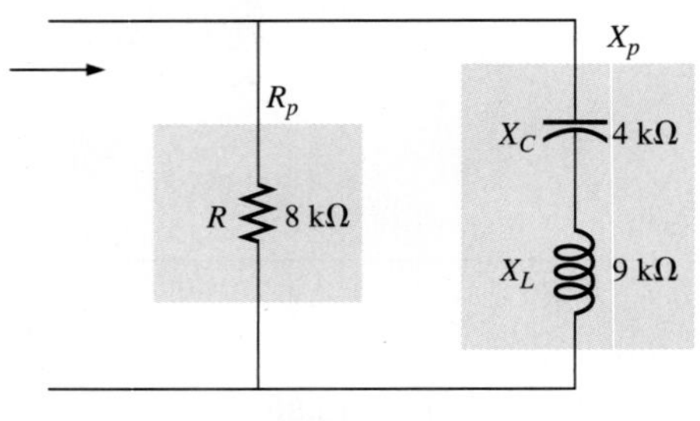

FIG. 15.97
Example 15.18.

EXAMPLE 15.18 Determine the series equivalent circuit for the network in Fig. 15.97.

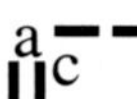

Solution:

$$R_p = 8\text{ k}\Omega$$
$$X_p\text{ (resultant)} = |X_L - X_C| = |9\text{ k}\Omega - 4\text{ k}\Omega|$$
$$= 5\text{ k}\Omega$$

and

$$R_s = \frac{R_p X_p^2}{X_p^2 + R_p^2} = \frac{(8\text{ k}\Omega)(5\text{ k}\Omega)^2}{(5\text{ k}\Omega)^2 + (8\text{ k}\Omega)^2} = \frac{200\text{ k}\Omega}{89} = \mathbf{2.25\text{ k}\Omega}$$

with

$$X_s = \frac{R_p^2 X_p}{X_p^2 + R_p^2} = \frac{(8\text{ k}\Omega)^2(5\text{ k}\Omega)}{(5\text{ k}\Omega)^2 + (8\text{ k}\Omega)^2} = \frac{320\text{ k}\Omega}{89}$$
$$= \mathbf{3.60\text{ k}\Omega} \quad \textbf{(inductive)}$$

The equivalent series circuit appears in Fig. 15.98.

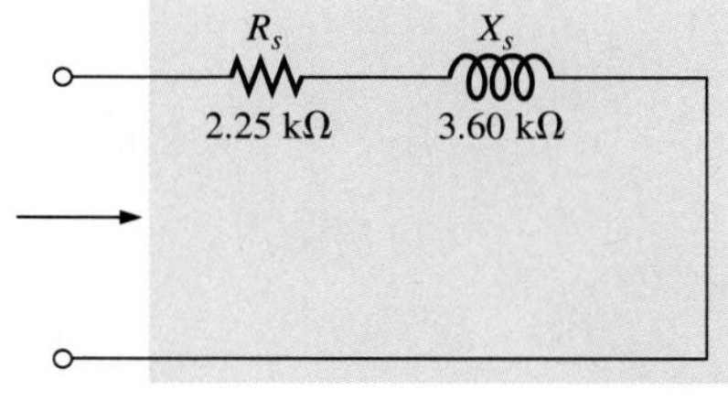

FIG. 15.98
The equivalent series circuit for the parallel network in Fig. 15.97.

EXAMPLE 15.19 For the network in Fig. 15.99:

a. Determine $\mathbf{Y}_T$ and $\mathbf{Z}_T$.
b. Sketch the admittance diagram.
c. Find $\mathbf{E}$ and $\mathbf{I}_L$.
d. Compute the power factor of the network and the power delivered to the network.
e. Determine the equivalent series circuit as far as the terminal characteristics of the network are concerned.
f. Using the equivalent circuit developed in part (e), calculate **E,** and compare it with the result of part (c).
g. Determine the power delivered to the network, and compare it with the solution of part (d).
h. Determine the equivalent parallel network from the equivalent series circuit, and calculate the total admittance $\mathbf{Y}_T$. Compare the result with the solution of part (a).

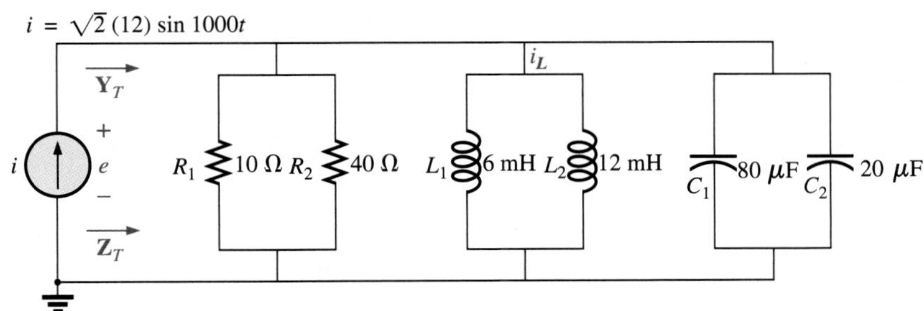

FIG. 15.99
Example 15.19.

Solutions:

a. Combining common elements and finding the reactance of the inductor and capacitor, we obtain

$$R_T = 10\ \Omega \parallel 40\ \Omega = 8\ \Omega$$
$$L_T = 6\text{ mH} \parallel 12\text{ mH} = 4\text{ mH}$$
$$C_T = 80\ \mu\text{F} + 20\ \mu\text{F} = 100\ \mu\text{F}$$
$$X_L = \omega L = (1000\text{ rad/s})(4\text{ mH}) = 4\ \Omega$$
$$X_C = \frac{1}{\omega C} = \frac{1}{(1000\text{ rad/s})(100\ \mu\text{F})} = 10\ \Omega$$

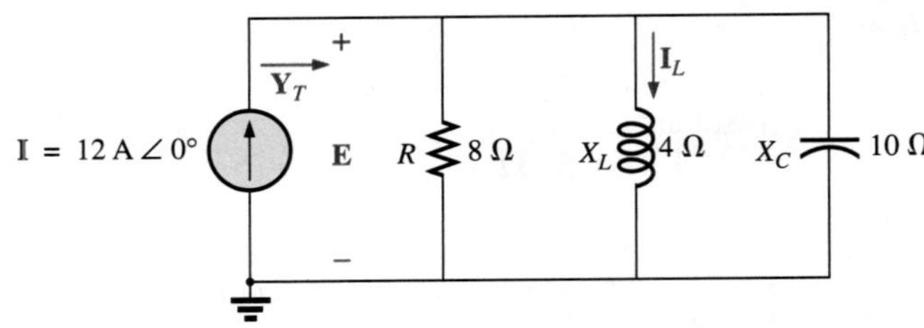

FIG. 15.100
Applying phasor notation to the network in Fig. 15.99.

The network is redrawn in Fig. 15.100 with phasor notation. The total admittance is

$$\begin{aligned}\mathbf{Y}_T &= \mathbf{Y}_R + \mathbf{Y}_L + \mathbf{Y}_C \\ &= G\angle 0° + B_L\angle -90° + B_C\angle +90° \\ &= \frac{1}{8\,\Omega}\angle 0° + \frac{1}{4\,\Omega}\angle -90° + \frac{1}{10\,\Omega}\angle +90° \\ &= 0.125\text{ S}\angle 0° + 0.25\text{ S}\angle -90° + 0.1\text{ S}\angle +90° \\ &= 0.125\text{ S} - j\,0.25\text{ S} + j\,0.1\text{ S} \\ &= 0.125\text{ S} - j\,0.15\text{ S} = \mathbf{0.195\text{ S}\angle -50.194°}\end{aligned}$$

$$\mathbf{Z}_T = \frac{1}{\mathbf{Y}_T} = \frac{1}{0.195\text{ S}}\angle -50.194° = \mathbf{5.13\ \Omega\angle 50.19°}$$

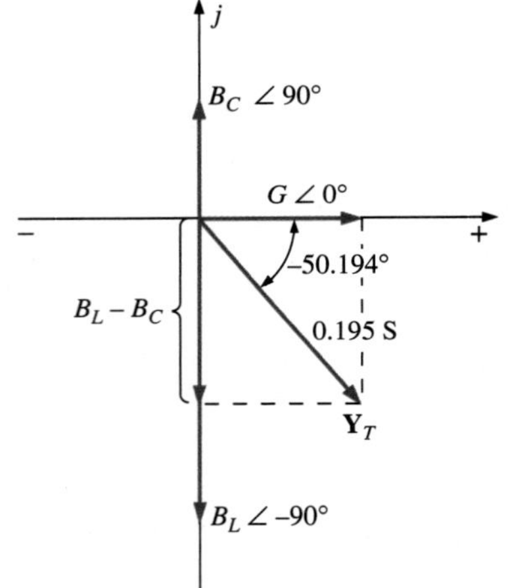

FIG. 15.101
Admittance diagram for the parallel R-L-C network in Fig. 15.99.

b. See Fig. 15.101.

c. $$\mathbf{E} = \mathbf{I}\mathbf{Z}_T = \frac{\mathbf{I}}{\mathbf{Y}_T} = \frac{12\text{ A}\angle 0°}{0.195\text{ S}\angle -50.194°} = \mathbf{61.54\text{ V}\angle 50.19°}$$

$$\mathbf{I}_L = \frac{\mathbf{V}_L}{\mathbf{Z}_L} = \frac{\mathbf{E}}{\mathbf{Z}_L} = \frac{61.538\text{ V}\angle 50.194°}{4\,\Omega\angle 90°} = \mathbf{15.39\text{ A}\angle -39.81°}$$

d. $$F_p = \cos\theta = \frac{G}{Y_T} = \frac{0.125\text{ S}}{0.195\text{ S}} = \mathbf{0.641\ lagging}\ (\mathbf{E}\text{ leads }\mathbf{I})$$

$$\begin{aligned}P = EI\cos\theta &= (61.538\text{ V})(12\text{ A})\cos 50.194° \\ &= \mathbf{472.75\ W}\end{aligned}$$

e. $$\begin{aligned}\mathbf{Z}_T = \frac{1}{\mathbf{Y}_T} = \frac{1}{0.195\text{ S}\angle -50.194°} &= 5.128\ \Omega\angle +50.194° \\ &= 3.28\ \Omega + j\,3.94\ \Omega \\ &= R + j\,X_L\end{aligned}$$

$$X_L = 3.94\ \Omega = \omega L$$

$$L = \frac{3.94\ \Omega}{\omega} = \frac{3.94\ \Omega}{1000\text{ rad/s}} = \mathbf{3.94\ mH}$$

The series equivalent circuit appears in Fig. 15.102.

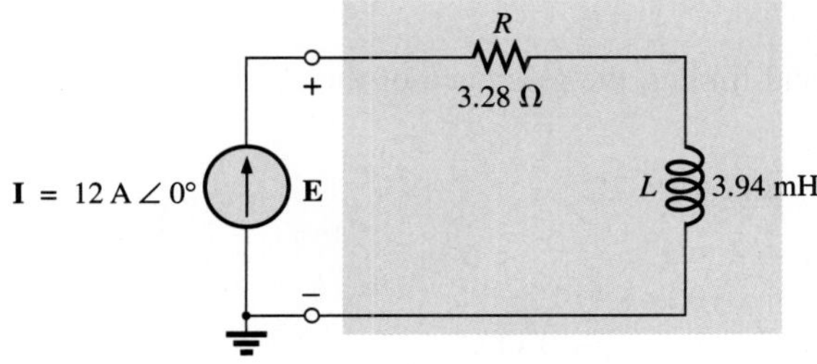

FIG. 15.102
Series equivalent circuit for the parallel R-L-C network in Fig. 15.99 with ω = 1000 rad/s.

f. $\mathbf{E} = \mathbf{I}\mathbf{Z}_T = (12 \text{ A} \angle 0°)(5.128 \ \Omega \angle 50.194°)$
$= \mathbf{61.54 \ V \angle 50.194°}$ (as above)

g. $P = I^2R = (12 \text{ A})^2(3.28 \ \Omega) = \mathbf{472.32 \ W}$ (as above)

h. $$R_p = \frac{R_s^2 + X_s^2}{R_s} = \frac{(3.28 \ \Omega)^2 + (3.94 \ \Omega)^2}{3.28 \ \Omega} = \mathbf{8 \ \Omega}$$

$$X_p = \frac{R_s^2 + X_s^2}{X_s} = \frac{(3.28 \ \Omega)^2 + (3.94 \ \Omega)^2}{3.94 \ \Omega} = \mathbf{6.67 \ \Omega}$$

The parallel equivalent circuit appears in Fig. 15.103.

$$\mathbf{Y}_T = G \angle 0° + B_L \angle -90° = \frac{1}{8 \ \Omega} \angle 0° + \frac{1}{6.675 \ \Omega} \angle -90°$$
$$= 0.125 \text{ S} \angle 0° + 0.15 \text{ S} \angle -90°$$
$$= 0.125 \text{ S} - j\,0.15 \text{ S} = \mathbf{0.195 \ S \angle -50.194°} \quad \text{(as above)}$$

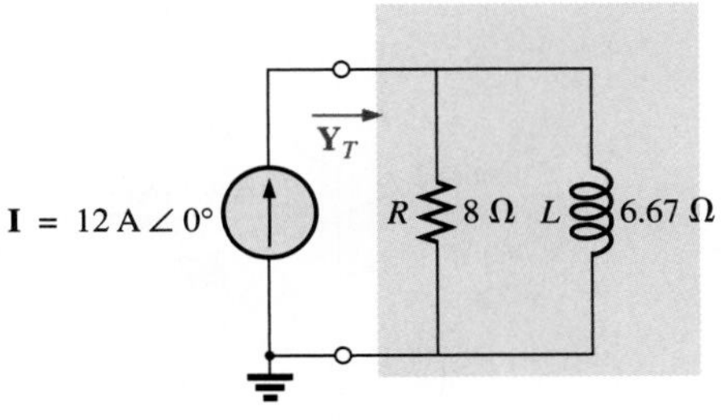

FIG. 15.103
Parallel equivalent of the circuit in Fig. 15.102.

15.13 PHASE MEASUREMENTS

Measuring the phase angle between quantities is one of the most important functions that an oscilloscope can perform. It is an operation that must be performed carefully, however, or you may obtain the incorrect result or damage the equipment. Whenever you are using the dual-trace capability of an oscilloscope, the most important thing to remember is that

both channels of a dual-trace oscilloscope must be connected to the same ground.

Measuring Z_T and θ_T

For ac parallel networks restricted to resistive loads, the total impedance can be found in the same manner as described for dc circuits: Simply remove the source and place an ohmmeter across the network terminals. However,

for parallel ac networks with reactive elements, the total impedance cannot be measured with an ohmmeter.

An experimental procedure must be defined that permits determining the magnitude and the angle of the terminal impedance.

The phase angle between the applied voltage and the resulting source current is one of the most important because (a) it is also the phase angle associated with the total impedance; (b) it provides an instant indication of whether a network is resistive or reactive; (c) it reveals whether a network is inductive or capacitive; and (d) it can be used to find the power delivered to the network.

In Fig. 15.104, a resistor has been added to the configuration between the source and the network to permit measuring the current and finding the phase angle between the applied voltage and the source current.

At the frequency of interest, the applied voltage establishes a voltage across the sensing resistor which can be displayed by one channel of the dual-trace oscilloscope. In Fig. 15.104, channel 1 is displaying the applied voltage, and channel 2 the voltage across the sensing resistor. Sensitivities for each channel are chosen to establish the waveforms appearing on the screen in Fig. 15.105. As emphasized above, note that both channels have the same ground connection. In fact, the need for a common ground connection is the only reason that the sensing resistor

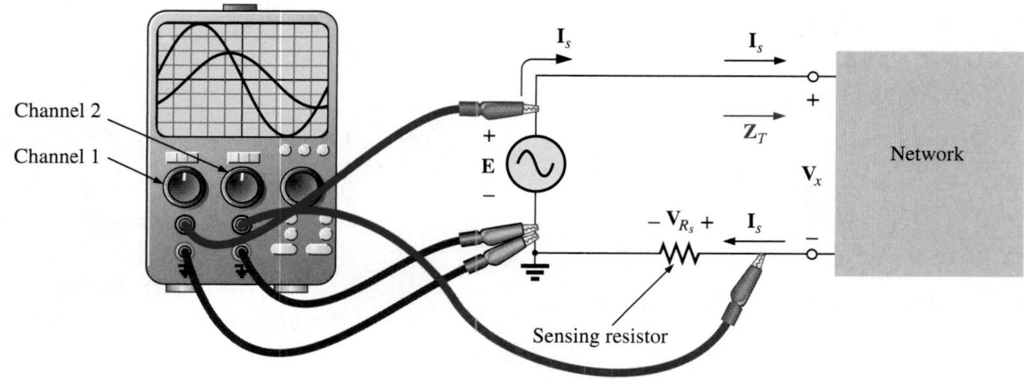

FIG. 15.104
Using an oscilloscope to measure Z_T and θ_T.

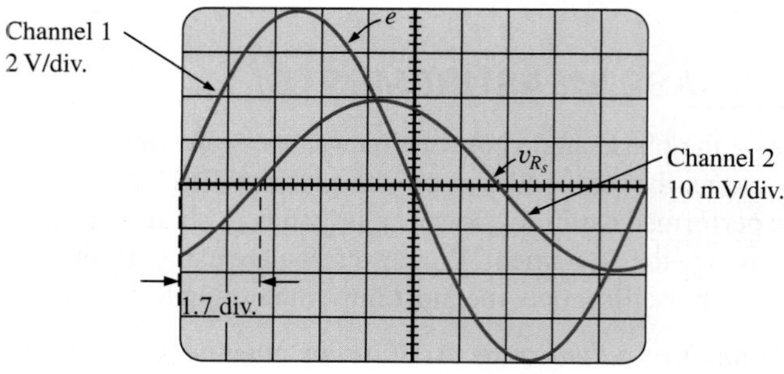

FIG. 15.105
e and v_{R_s}, for the configuration in Fig. 15.104.

was not connected to the positive side of the supply. Since oscilloscopes display only voltages versus time, the peak value of the source current must be found using Ohm's law. *Since the voltage across a resistor and the current through the resistor are in phase, the phase angle between the two voltages will be the same as that between the applied voltage and the resulting source current.*

Using the sensitivities, the peak value of the applied voltage is

$$E_m = (4 \text{ div.})(2 \text{ V/div.}) = 8 \text{ V}$$

while the peak value of the voltage across the sensing resistor is

$$V_{R_s(\text{peak})} = (2 \text{ div.})(10 \text{ mV/div.}) = 20 \text{ mV}$$

Using Ohm's law, the peak value of the current is

$$I_{s(\text{peak})} = \frac{V_{R_s(\text{peak})}}{R_s} = \frac{20 \text{ mV}}{10 \,\Omega} = 2 \text{ mA}$$

The sensing resistor is chosen small enough so that the voltage across the sensing resistor is small enough to permit the approximation $\mathbf{V}_x = \mathbf{E} - \mathbf{V}_{R_s} \cong \mathbf{E}$. The magnitude of the input impedance is then

$$Z_T = \frac{V_x}{I_s} \cong \frac{E}{I_s} = \frac{8 \text{ V}}{2 \text{ mA}} = \mathbf{4\ k\Omega}$$

For the chosen horizontal sensitivity, each waveform in Fig. 15.105 has a period T defined by ten horizontal divisions, and the phase angle between the two waveforms is 1.7 divisions. Using the fact that each pe-

riod of a sinusoidal waveform encompasses 360°, the following ratios can be set up to determine the phase angle θ:

$$\frac{10 \text{ div.}}{360^\circ} = \frac{1.7 \text{ div.}}{\theta}$$

and
$$\theta = \left(\frac{1.7}{10}\right)360^\circ = \mathbf{61.2^\circ}$$

In general,

$$\boxed{\theta = \frac{(\text{div. for } \theta)}{(\text{div. for } T)} \times 360^\circ} \qquad \textbf{(15.45)}$$

Therefore, the total impedance is

$$\mathbf{Z}_T = \mathbf{4\ k\Omega\ \angle 61.2^\circ} = \mathbf{1.93\ k\Omega + \mathit{j}\ 3.51\ k\Omega} = R + j\,X_L$$

which is equivalent to the series combination of a 1.93 kΩ resistor and an inductor with a reactance of 3.51 kΩ (at the frequency of interest).

Measuring the Phase Angle between Various Voltages

In Fig. 15.106, an oscilloscope is being used to find the phase relationship between the applied voltage and the voltage across the inductor. Note again that each channel shares the same ground connection. The resulting pattern appears in Fig. 15.107 with the chosen sensitivities. This time, both channels have the same sensitivity, resulting in the following peak values for the voltages:

$$E_m = (3 \text{ div.})(2 \text{ V/div.}) = \mathbf{6\ V}$$
$$V_{L(\text{peak})} = (1.6 \text{ div.})(2 \text{ V/div.}) = \mathbf{3.2\ V}$$

The phase angle is determined using Eq. 15.45:

$$\theta = \frac{(1 \text{ div.})}{(8 \text{ div.})} \times 360^\circ$$
$$\theta = \mathbf{45^\circ}$$

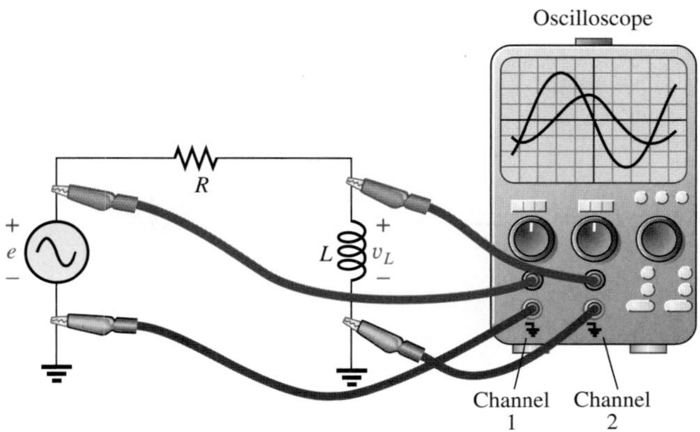

FIG. 15.106
Determining the phase relationship between e and v_L.

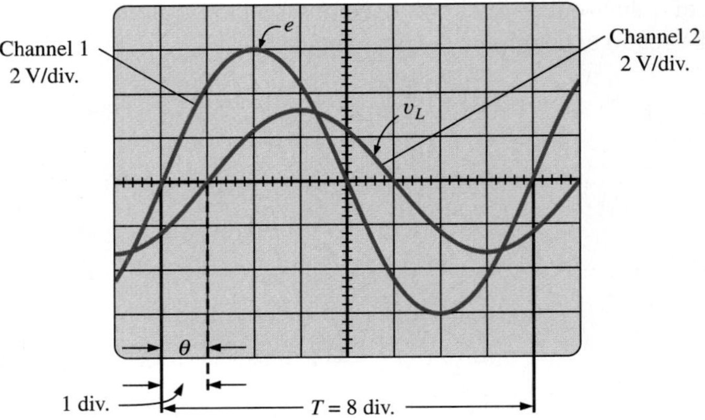

FIG. 15.107
Determining the phase angle between e and v_L for the configuration in Fig. 15.106.

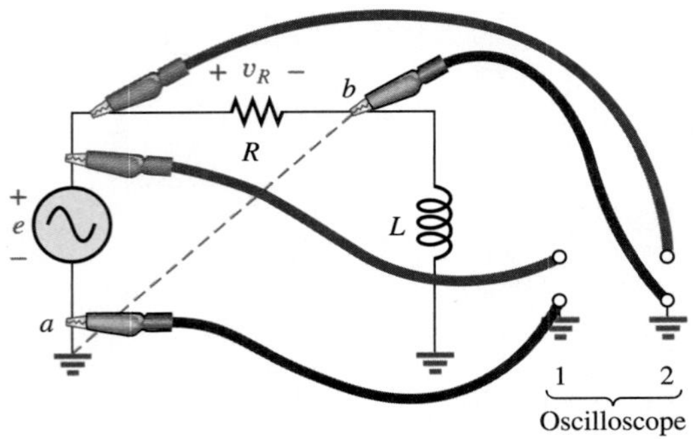

FIG. 15.108
An improper phase-measurement connection.

If the phase relationship between e and v_R is desired, the oscilloscope *cannot be connected* as shown in Fig. 15.108. *The grounds of each channel are internally connected in the oscilloscope,* forcing point b to have the same potential as point a. The result would be a direct connection between points a and b that would short out the inductive element. If the inductive element is the predominant factor in controlling the level of the current, the current in the circuit could rise to dangerous levels and damage the oscilloscope or supply. The easiest way to find the phase relationship between e and v_R would be to simply interchange the positions of the resistor and the inductor and proceed as before.

For the parallel network in Fig. 15.109, the phase relationship between two of the branch currents, i_R and i_L, can be determined using a sensing resistor, as shown in the figure. The value of the sensing resistor is chosen small enough in comparison to the value of the series inductive reactance to ensure that it will not affect the general response of the network. Channel 1 displays the voltage v_R, and channel 2 the voltage v_{R_s}. Since v_R is in phase with i_R, and v_{R_s} is in phase with i_L, the phase relationship between v_R and v_{R_s}, will be the same as between i_R and i_L. The peak value of each current can be found through a simple application of Ohm's law.

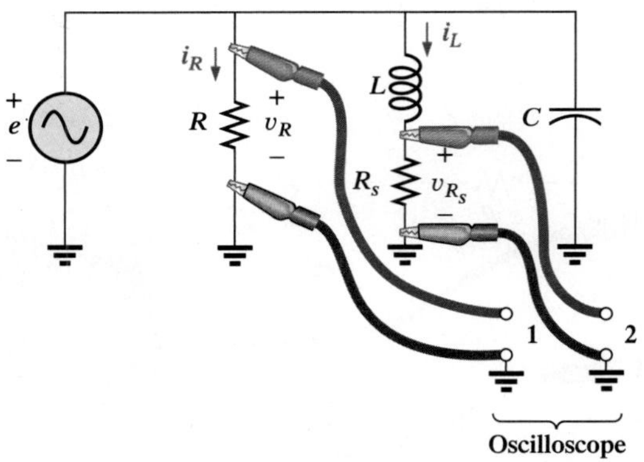

FIG. 15.109
Determining the phase relationship between i_R and i_L.

15.14 APPLICATIONS

Home Wiring

An expanded view of house wiring is provided in Fig. 15.110 to permit a discussion of the entire system. The house panel has been included with the "feed" and the important grounding mechanism. In addition, a number of typical circuits found in the home have been included to provide a sense for the manner in which the total power is distributed.

First note how the copper bars in the panel are laid out to provide both 120 V and 240 V. Between any one bar and ground is the single-phase 120 V supply. However, the bars have been arranged so that 240 V can be obtained between two vertical adjacent bars using a double-gang circuit breaker. When time permits, examine your own panel (but do not remove the cover), and note the dual circuit breaker arrangement for the 240 V supply.

For appliances such as fixtures and heaters that have a metal casing, the ground wire is connected to the metal casing to provide a direct path to ground path for a "shorting" or errant current as described in Section 6.8. For outlets that do not have a conductive casing, the ground lead is connected to a point on the outlet that distributes to all important points of the outlet.

Note the series arrangement between the thermostat and the heater but the parallel arrangement between heaters on the same circuit. In addition,

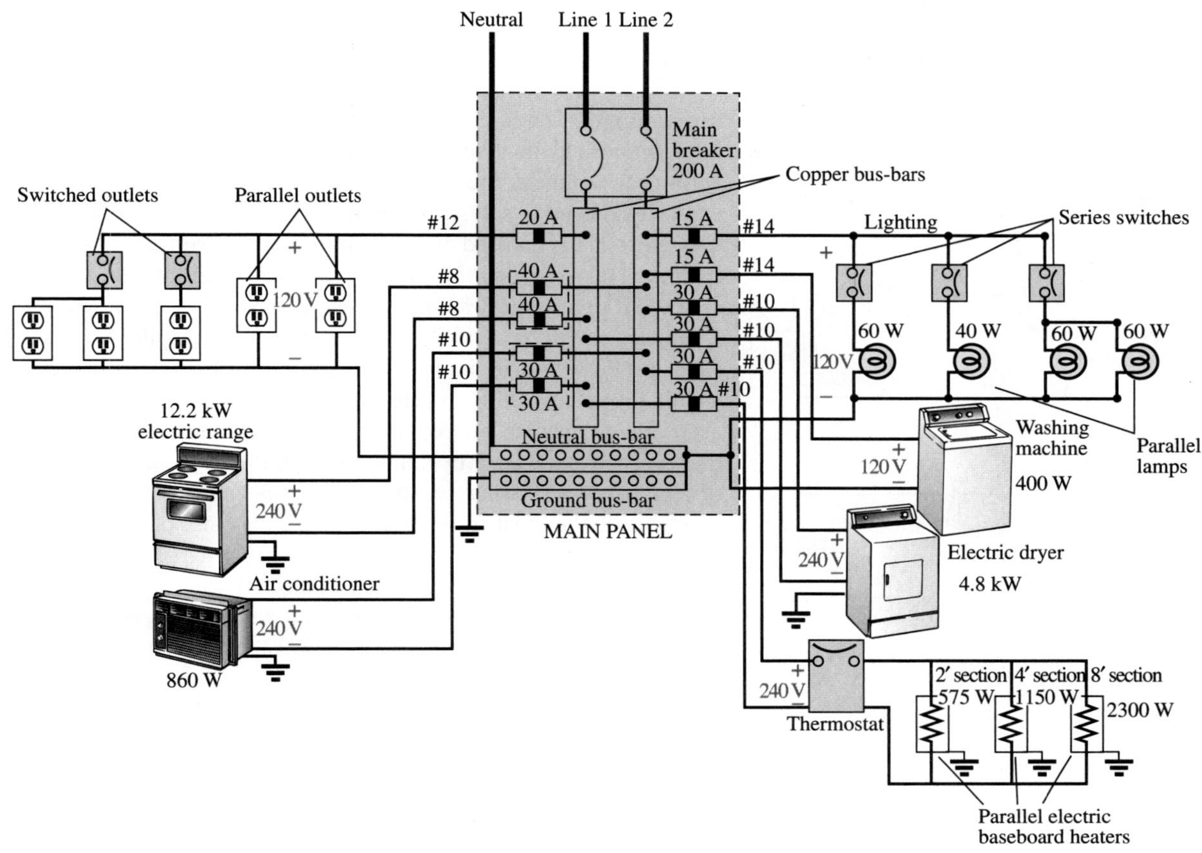

FIG. 15.110
Home wiring diagram.

note the series connection of switches to lights in the upper-right corner but the parallel connection of lights and outlets. Due to high current demand, the air conditioner, heaters, and electric stove have 30 A breakers. Keep in mind that the total current does not equal the product of the two (or 60 A) since each breaker is in a line and the same current will flow through each breaker.

In general, you now have a surface understanding of the general wiring in your home. You may not be a qualified, licensed electrician, but at least you should now be able to converse with some intelligence about the system.

Speaker Systems

The best reproduction of sound is obtained using a different speaker for the low-, mid-, and high-frequency regions. Although the typical audio range for the human ear is from about 100 Hz to 20 kHz, speakers are available from 20 Hz to 40 kHz. For the low-frequency range usually extending from about 20 Hz to 300 Hz, a speaker referred to as a *woofer* is used. Of the three speakers, it is normally the largest. The mid-range speaker is typically smaller in size and covers the range from about 100 Hz to 5 kHz. The *tweeter,* as it is normally called, is usually the smallest of the three speakers and typically covers the range from about 2 kHz to 25 kHz. There is an overlap of frequencies to ensure that frequencies aren't lost in those regions where the response of one drops off and the other takes over. A great deal more about the range of each speaker and their dB response (a term you may have heard when discussing speaker response) is covered in detail in Chapter 21.

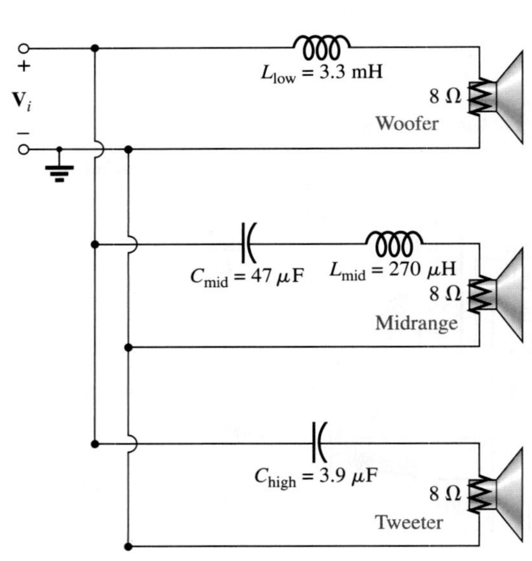

FIG. 15.111
Crossover speaker system.

One popular method for hooking up the three speakers is the *crossover* configuration in Fig. 15.111. Note that it is nothing more than a parallel network with a speaker in each branch and full applied voltage across each branch. The added elements (inductors and capacitors) were carefully chosen to set the range of response for each speaker. Note that each speaker is labeled with an impedance level and associated frequency. This type of information is typical when purchasing a quality speaker. It immediately identifies the type of speaker and reveals at which frequency it will have its maximum response. A detailed analysis of the same network will be included in Section 21.15. For now, however, it should prove interesting to determine the total impedance of each branch at specific frequencies to see if indeed the response of one will far outweigh the response of the other two. Since an amplifier with an output impedance of 8 Ω is to be used, maximum transfer of power (see Section 18.5 for ac networks) to the speaker results when the impedance of the branch is equal to or very close to 8 Ω.

Let us begin by examining the response of the frequencies to be carried primarily by the mid-range speaker since it represents the greatest portion of the human hearing range. Since the mid-range speaker branch is rated at 8 Ω at 1.4 kHz, let us test the effect of applying 1.4 kHz to all branches of the crossover network.

For the mid-range speaker:

$$X_C = \frac{1}{2\pi fC} = \frac{1}{2\pi(1.4\text{ kHz})(47\ \mu\text{F})} = 2.42\ \Omega$$
$$X_L = 2\pi fL = 2\pi(1.4\text{ kHz})(270\ \mu\text{H}) = 2.78\ \Omega$$
$$R = 8\ \Omega$$

$$\begin{aligned}\text{and} \quad \mathbf{Z}_{\text{mid-range}} &= R + j(X_L - X_C) = 8\ \Omega + j(2.78\ \Omega - 2.42\ \Omega) \\ &= 8\ \Omega + j\,0.36\ \Omega \\ &= 8.008\ \Omega \angle -2.58° \cong 8\ \Omega \angle 0° = R\end{aligned}$$

In Fig. 15.112(a), the amplifier with the output impedance of 8 Ω has been applied across the mid-range speaker at a frequency of 1.4 kHz. Since the total reactance offered by the two series reactive elements is so small compared to the 8 Ω resistance of the speaker, we can essentially replace the series combination of the coil and capacitor by a short circuit of 0 Ω. We are then left with a situation where the load impedance is an exact match with the output impedance of the amplifier, and maximum power will be delivered to the speaker. Because of the equal series impedances, each will capture half the applied voltage or 6 V. The power to the speaker is then $V^2/R = (6\text{ V})^2/8\ \Omega = 4.5$ W.

At a frequency of 1.4 kHz, we would expect the woofer and tweeter to have minimum impact on the generated sound. We will now test the validity of this statement by determining the impedance of each branch at 1.4 kHz.

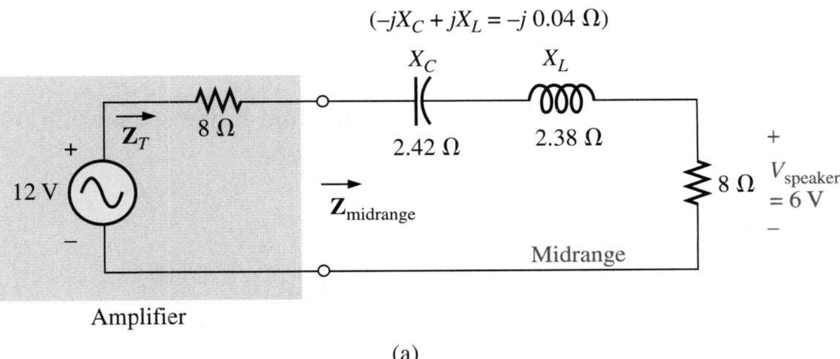

(a)

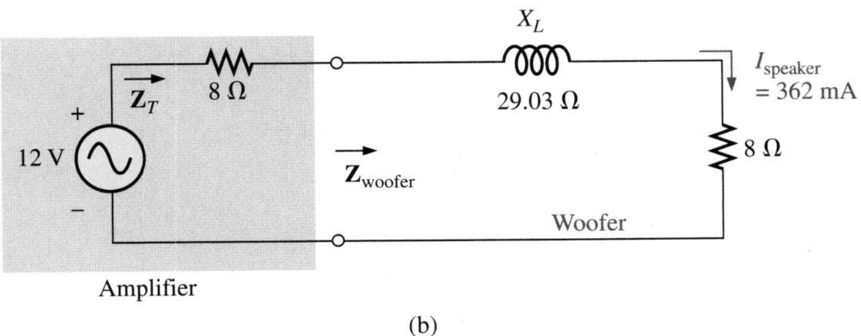

(b)

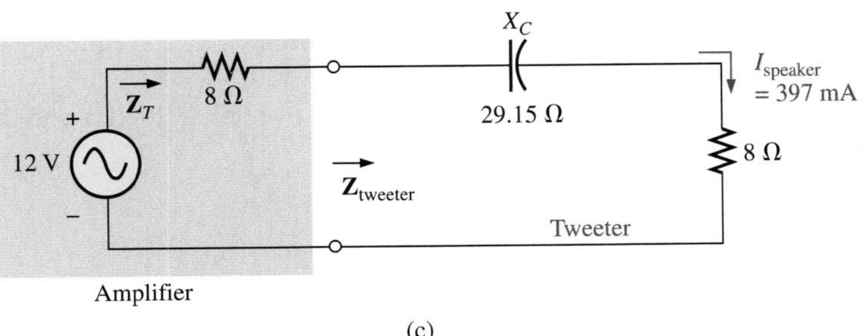

(c)

FIG. 15.112
Crossover network: (a) mid-range speaker at 1.4 kHz; (b) woofer at 1.4 kHz; (c) tweeter.

For the woofer:

$$X_L = 2\,\pi f L = 2\pi(1.4\text{ kHz})(3.3\text{ mH}) = 29.03\ \Omega$$

and

$$Z_{\text{woofer}} = R + jX_L = 8\ \Omega + j\,29.03\ \Omega = 30.11\ \Omega \angle 74.59°$$

which is a poor match with the output impedance of the amplifier. The resulting network is shown in Fig. 15.112(b).

The total load on the source of 12 V is

$$Z_T = 8\ \Omega + 8\ \Omega + j\,29.03\ \Omega = 16\ \Omega + j\,29.03\ \Omega = 33.15\ \Omega \angle 61.14°$$

and the current is

$$\mathbf{I} = \frac{\mathbf{E}}{\mathbf{Z}_T} = \frac{12\text{ V}\angle 0°}{33.15\ \Omega \angle 61.14°} = 362\text{ mA}\angle -61.14°$$

The power to the 8 Ω speaker is then

$$P_{\text{woofer}} = I^2R = (362\text{ mA})^2(8\ \Omega) = \mathbf{1.05\ W}$$

or about 1 W.

Consequently, the sound generated by the mid-range speaker far outweighs the response of the woofer (as it should).

For the tweeter in Fig. 15.112,

$$X_C = \frac{1}{2\pi fC} = \frac{1}{2\pi(1.4\text{ kHz})(3.9\ \mu\text{F})} = 29.15\ \Omega$$

and

$$Z_{\text{tweeter}} = R - jX_C = 8\ \Omega - j\,29.15\ \Omega = 30.23\ \Omega \angle -74.65°$$

which, as for the woofer, is a poor match with the output impedance of the amplifier. The current

$$\mathbf{I} = \frac{\mathbf{E}}{\mathbf{Z}_T} = \frac{12\text{ V}\angle 0°}{30.23\ \Omega \angle -74.65°} = 397\text{ mA}\angle 74.65°$$

The power to the 8 Ω speaker is then

$$P_{\text{tweeter}} = I^2R = (397\text{ mA})^2(8\ \Omega) = \mathbf{1.26\ W}$$

or about 1.3 W.

Consequently, the sound generated by the mid-range speaker far outweighs the response of the tweeter also.

All in all, the mid-range speaker predominates at a frequency of 1.4 kHz for the crossover network in Fig. 15.111.

Let us now determine the impedance of the tweeter at 20 kHz and the impact of the woofer at this frequency.

For the tweeter:

$$X_C = \frac{1}{2\pi fC} = \frac{1}{2\pi(20\text{ kHz})(3.9\ \mu\text{F})} = 2.04\ \Omega$$

with

$$Z_{\text{tweeter}} = 8\ \Omega - j\,2.04\ \Omega = 8.26\ \Omega \angle -14.31°$$

Even though the magnitude of the impedance of the branch is not exactly 8 Ω, it is very close, and the speaker will receive a high level of power (actually 4.43 W).

For the woofer:

$$X_L = 2\pi fL = 2\pi(20\text{ kHz})(3.3\text{ mH}) = 414.69\ \Omega$$

with $$Z_{woofer} = 8\ \Omega - j\,414.69\ \Omega = 414.77\ \Omega\ \angle 88.9°$$

which is a terrible match with the output impedance of the amplifier. Therefore, the speaker will receive a very low level of power (6.69 mW $\cong$ 0.007 W).

For all the calculations, note that the capacitive elements predominate at low frequencies, and the inductive elements at high frequencies. For the low frequencies, the reactance of the coil is quite small, permitting a full transfer of power to the speaker. For the high-frequency tweeter, the reactance of the capacitor is quite small, providing a direct path for power flow to the speaker.

Phase-Shift Power Control

In Chapter 11, the internal structure of a light dimmer was examined and its basic operation described. We can now turn our attention to how the power flow to the bulb is controlled.

If the dimmer were composed of simply resistive elements, all the voltages of the network would be in phase as shown in Fig. 15.113(a). If we assume that 20 V are required to turn on the triac in Fig. 11.68, the power will be distributed to the bulb for the period highlighted by the blue area of Fig. 15.113(a). For this situation, the bulb is close to full brightness since the applied voltage is available to the bulb for almost the entire cycle. To reduce the power to the bulb (and therefore reduce its brightness), the controlling voltage would need a lower peak voltage as shown in Fig. 15.113(b). In fact, the waveform in Fig. 15.113(b) is such that the turn-on voltage is not reached until the peak value occurs. In this case, power is delivered to the bulb for only half the cycle, and the brightness of the bulb is reduced. The problem with using only resistive elements in a dimmer now becomes apparent: The bulb can be made no dimmer than the situation depicted by Fig. 15.113(b). Any further reduction in the controlling voltage would reduce its peak value below the trigger level, and the bulb would never turn on.

This dilemma can be resolved by using a series combination of elements such as shown in Fig. 15.114(a) from the dimmer in Fig. 11.68. Note that the controlling voltage is the voltage across the capacitor, while the full line voltage of 120 V rms, 170 V peak, is across the entire branch. To describe the behavior of the network, let us examine the case defined by

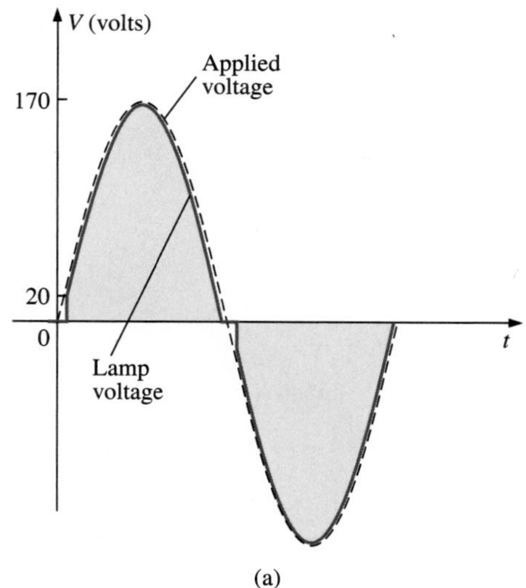

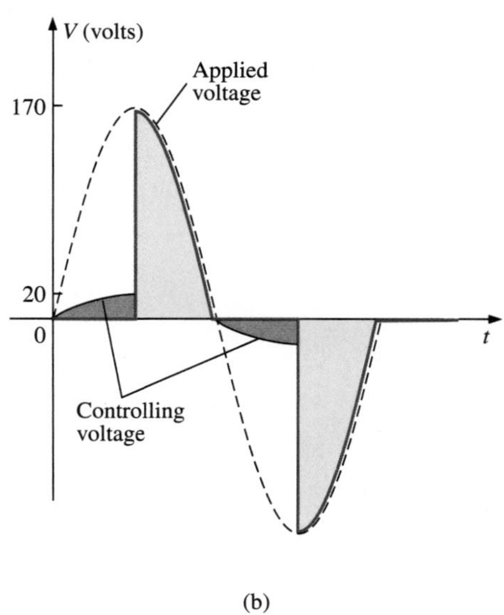

FIG. 15.113
Light dimmer: (a) with purely resistive elements; (b) half-cycle power distribution.

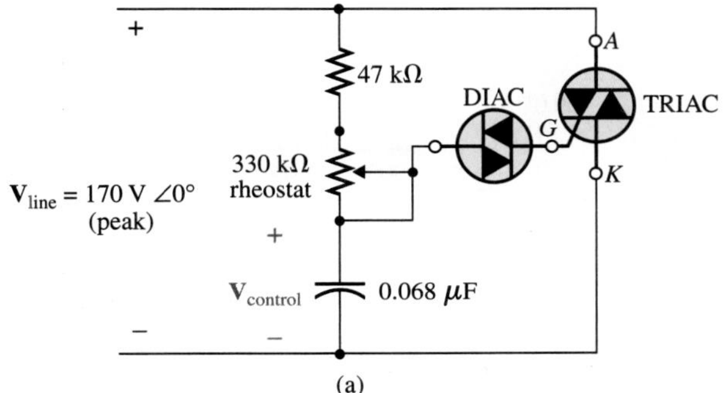

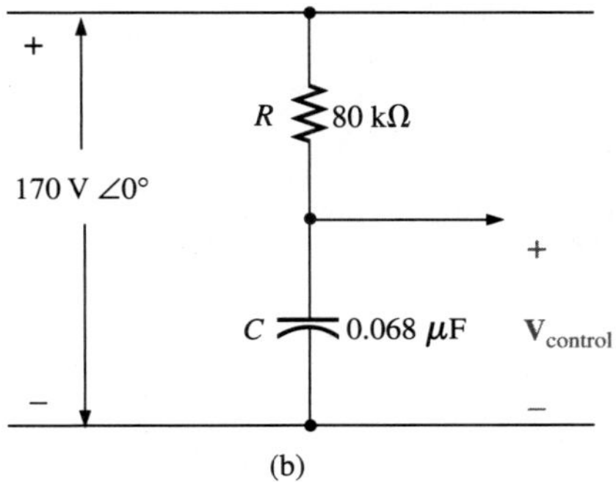

FIG. 15.114
Light dimmer: (a) from Fig. 11.68; (b) with rheostat set at 33 kΩ.

setting the potentiometer (used as a rheostat) to 1/10 its maximum value, or 33 kΩ. Combining the 33 kΩ with the fixed resistance of 47 kΩ results in a total resistance of 80 kΩ and the equivalent network in Fig. 15.114(b).

At 60 Hz, the reactance of the capacitor is

$$X_C = \frac{1}{2\pi fC} = \frac{1}{2\pi(60\text{ Hz})(62\ \mu\text{F})} = 42.78\text{ k}\Omega$$

Applying the voltage divider rule:

$$\begin{aligned}\mathbf{V}_{\text{control}} &= \frac{\mathbf{Z}_C\mathbf{V}_s}{\mathbf{Z}_R + \mathbf{Z}_C} \\ &= \frac{(42.78\text{ k}\Omega\angle -90^\circ)(V_s\angle 0^\circ)}{80\text{ k}\Omega - j\,42.78\text{ k}\Omega} = \frac{42.78\text{ k}\Omega\ V_s\angle -90^\circ}{90.72\text{ k}\Omega\angle -28.14^\circ} \\ &= 0.472\ V_s\angle -61.86^\circ\end{aligned}$$

Using a peak value of 170 V:

$$\begin{aligned}\mathbf{V}_{\text{control}} &= 0.472(170\text{ V})\angle -61.86^\circ \\ &= 80.24\text{ V}\angle -61.86^\circ\end{aligned}$$

producing the waveform in Fig. 15.115(a). The result is a waveform with a phase shift of 61.86° (lagging the applied line voltage) and a relatively high peak value. The high peak value results in a quick transition to the 20 V turn-on level, and power is distributed to the bulb for the major portion of the applied signal. Recall from the discussion in Chapter 11 that the response in the negative region is a replica of that achieved in the positive region. If we reduced the potentiometer resistance further, the phase angle would be reduced, and the bulb would burn brighter. The situation is now very similar to that described for the response in Fig. 15.113(a). In other words, nothing has been gained thus far by using the capacitive element in the control network. However, let us now increase the potentiometer resistance to 200 kΩ and note the effect on the controlling voltage.

That is,

$$R_T = 200\text{ k}\Omega + 47\text{ k}\Omega = 247\text{ k}\Omega$$

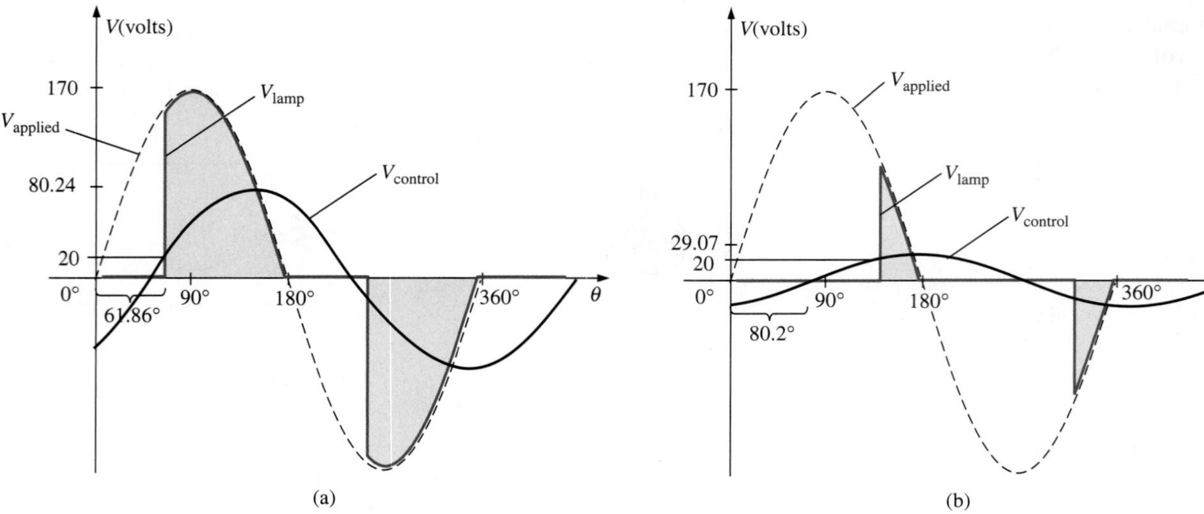

FIG. 15.115
Light dimmer in Fig. 11.68; (a) rheostat set at 33 kΩ; (b) rheostat set at 200 kΩ.

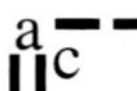

$$\begin{aligned}\mathbf{V}_{\text{control}} &= \frac{\mathbf{Z}_C\mathbf{V}_s}{\mathbf{Z}_R + \mathbf{Z}_C} \\ &= \frac{(42.78\ \text{k}\Omega\ \angle -90^\circ)(V_s\angle 0^\circ)}{247\ \text{k}\Omega - j42.78\ \text{k}\Omega} = \frac{42.78\ \text{k}\Omega\ V_s\angle -90^\circ}{250.78\ \text{k}\Omega\ \angle -9.8^\circ} \\ &= 0.171\ V_s\angle -80.2^\circ\end{aligned}$$

and using a peak value of 170 V, we have

$$\begin{aligned}\mathbf{V}_{\text{control}} &= 0.171(170\ \text{V})\ \angle -80.2^\circ \\ &= 29.07\ \text{V}\ \angle -80.2^\circ\end{aligned}$$

The peak value has been substantially reduced to only 29.07 V, and the phase-shift angle has increased to 80.2°. The result, as depicted by Fig. 15.115(b), is that the firing potential of 20 V is not reached until near the end of the positive region of the applied voltage. Power is delivered to the bulb for only a very short period of time, causing the bulb to be quite dim, significantly dimmer than obtained from the response in Fig. 15.113(b).

A conduction angle less than 90° is therefore possible due only to the phase shift introduced by the series *R-C* combination. Thus, it is possible to construct a network of some significance with a rather simple pair of elements.

15.15 COMPUTER ANALYSIS

PSpice

Series *R-L-C* Circuit The *R-L-C* network in Fig. 15.35 is now analyzed using OrCAD Capture. Since the inductive and capacitive reactances cannot be entered onto the screen, the associated inductive and capacitive levels were first determined as follows:

$$X_L = 2\pi fL \Rightarrow L = \frac{X_L}{2\pi f} = \frac{7\ \Omega}{2\pi(1\ \text{kHz})} = 1.114\ \text{mH}$$

$$X_C = \frac{1}{2\pi fC} \Rightarrow C = \frac{1}{2\pi fX_C} = \frac{1}{2\pi(1\ \text{kHz})3\ \Omega} = 53.05\ \mu\text{F}$$

Enter the values into the schematic as shown in Fig. 15.116. For the ac source, the sequence is **Place part** icon-**SOURCE-VSIN-OK** with **VOFF** set at 0 V, **VAMPL** set at 70.7 V (the peak value of the applied sinusoidal source in Fig. 15.35), and **FREQ** = 1 kHz. Double-click on the source symbol and the **Property Editor** appears, confirming the above choices and showing that **DF** = 0 s, **PHASE** = 0°, and **TD** = 0 s as set by the default levels. You are now ready to do an analysis of the circuit for the fixed frequency of 1 kHz.

The simulation process is initiated by first selecting the **New Simulation Profile** icon and inserting **PSpice 15-1** as the **Name** followed by **Create.** The **Simulation Settings** dialog appears and since you are continuing to plot the results against time, select the **Time Domain (Transient)** option under **Analysis type.** Since the period of each **cycle** of the applied source is 1 ms, set the **Run to time** at 5 ms so that five cycles appear. Leave the **Start saving data after** at 0 s even though there will be an oscillatory period for the reactive elements before the circuit settles down. Set the **Maximum step size** at 5 ms/1000 = 5 μs. Finally, select **OK** followed by the **Run PSpice** key. The result is a blank screen with an x-axis extending from 0 s to 5 ms.

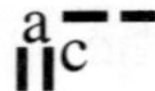

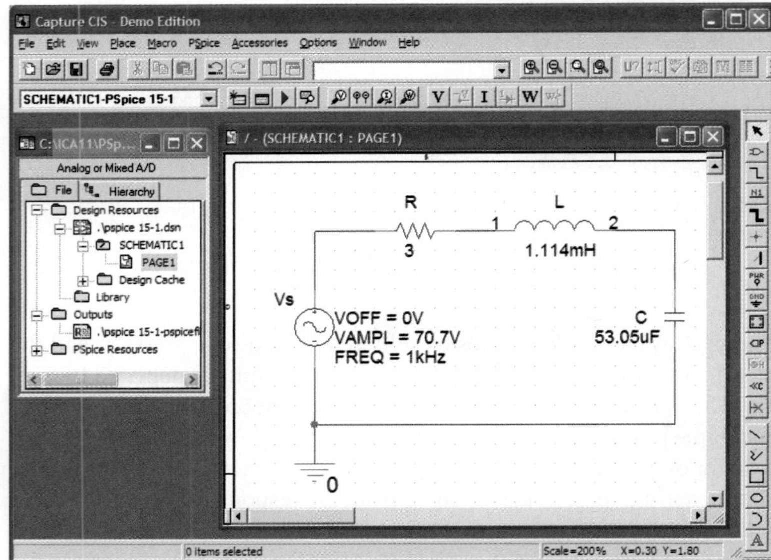

FIG. 15.116
Using PSpice to analyze a series R-L-C ac circuit.

The first quantity of interest is the current through the circuit, so select **Trace-Add-Trace** followed by **I(R)** and **OK.** The resulting plot in Fig. 15.117 clearly shows that there is a period of storing and discharging of the reactive elements before a steady-state level is established. It would appear that after 3 ms, steady-state conditions have been essentially established. Select the **Toggle cursor** key, and left-click; a cursor appears that can be moved along the axis near the maximum value around 1.4 ms.

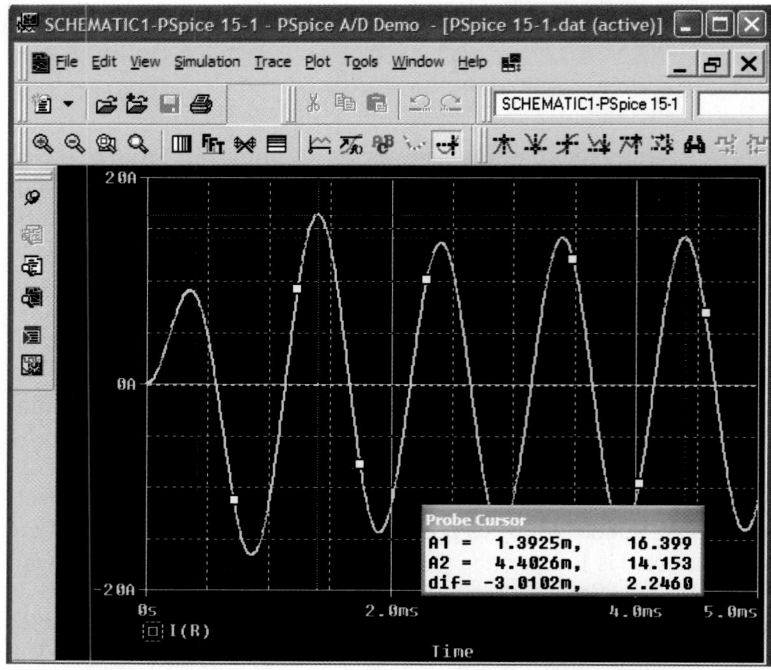

FIG. 15.117
A plot of the current for the circuit in Fig. 15.116 showing the transition from the transient state to the steady-state response.

In fact, the cursor reveals a maximum value of 16.4 A which exceeds the steady-state solution by over 2 A. Right-click to establish a second cursor on the screen that can be placed near the steady-state peak around 4.4 ms. The resulting peak value is about 14.15 A which is a match with the longhand solution for Fig. 15.35. We will therefore assume that steady-state conditions have been established for the circuit after 4 ms.

Now add the source voltage through **Trace-Add Trace-V(Vs:+)-OK** to obtain the multiple plot at the bottom of Fig. 15.118. For the voltage across the coil, the sequence **Plot-Add Plot to Window-Trace-Add Trace-V(L:1)-V(L:2)** results in the plot appearing at the top of Fig. 15.118. Take special note that the **Trace Expression** is **V(L:1)–V(L:2)** rather than just **V(L:1)** because **V(L:1)** would be the voltage from that point to ground which would include the voltage across the capacitor. In addition, the – sign between the two comes from the **Functions or Macros** list at the right of the **Add Traces** dialog box. Finally, since we know that the waveforms are fairly steady after 3 ms, cut away the waveforms before 3 ms with **Plot-Axis Settings-X axis-User Defined-3ms to 5ms-OK** to obtain the two cycles of Fig. 15.118. Now you can clearly see that the peak value of the voltage across the coil is 100 V to match the analysis of Fig. 15.35. It is also clear that the applied voltage leads the input current by an angle that can be determined using the cursors. First activate the cursor option by selecting the cursor key (a red plot through the origin) in the second toolbar down from the menu bar. Then select **V(Vs:+)** at the bottom left of the screen with a left click, and set it at that point where the applied voltage passes through the horizontal axis with a positive slope. The result is **A1** = 4 ms at $-4.243\ \mu V \cong 0$ V. Then select **I(R)** at the bottom left of the screen with a right click, and place it at the point where the current waveform passes through the horizontal axis with a positive slope. The result is **A2** = 4.15 ms at -55.15 mA = 0.55 A ≅

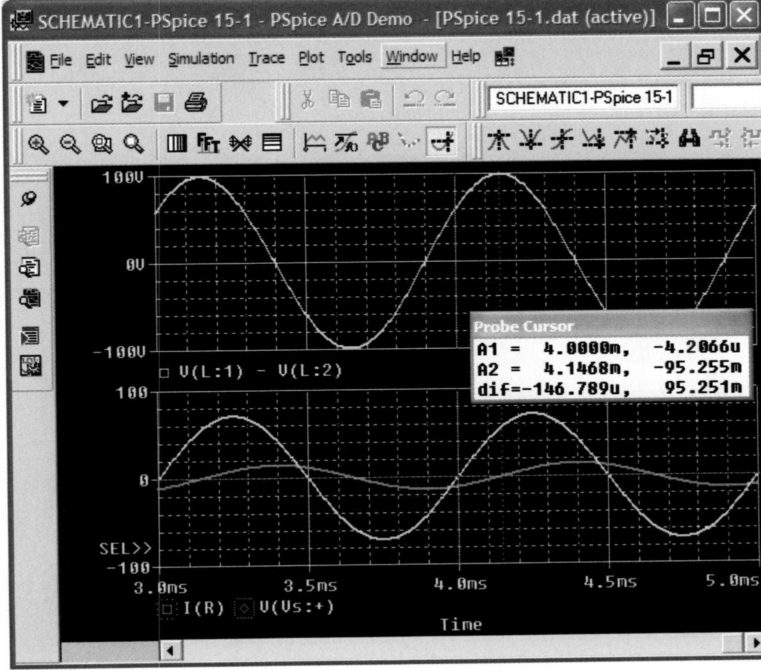

FIG. 15.118

A plot of the steady-state response (t > 3 ms) for v_L, v_s, and i for the circuit in Fig. 15.116.

0 A (compared to a peak value of 14.14 A). At the bottom of the **Probe Cursor** dialog box, the time difference is 147.24 μs.

Now set up the ratio

$$\frac{147.24\ \mu\text{s}}{1000\ \mu\text{s}} = \frac{\theta}{360^\circ}$$

$$\theta = 52.99^\circ$$

The phase angle by which the applied voltage leads the source is 52.99° which is very close to the theoretical solution of 53.13° obtained in Fig. 15.39. Increasing the number of data points for the plot would have increased the accuracy level and brought the results closer to 53.13°.

Multisim

We now examine the response of a network versus frequency rather than time using the network in Fig. 15.86 which now appears on the schematic in Fig. 15.119. The ac current source appears as **AC_CURRENT_SOURCE** under the **SIGNAL_CURRENT_SOURCES Family** listing. Note that the current source was given an amplitude of 1 A to establish a magnitude match between the response of the voltage across the network and the impedance of the network. That is,

$$|Z_T| = \left|\frac{V_s}{I_s}\right| = \left|\frac{V_s}{1\text{ A}}\right| = |V_s|$$

Before applying computer methods, we should develop a rough idea of what to expect so that we have something to which to compare the computer solution. At very high frequencies such as 1 MHz, the impedance of the inductive element will be about 25 kΩ which when placed in parallel with the 220 Ω will look like an open circuit. The result is that as the fre-

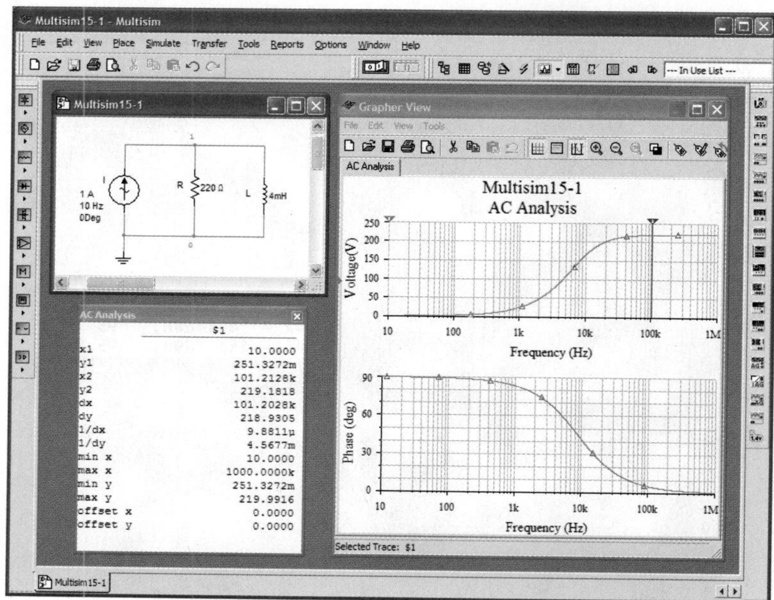

FIG. 15.119

Obtaining an impedance plot for a parallel R-L network using Multisim.

quency gets very high, we should expect the impedance of the network to approach the 220 Ω level of the resistor. In addition, since the network will take on resistive characteristics at very high frequencies, the angle associated with the input impedance should also approach 0 Ω. At very low frequencies the reactance of the inductive element will be much less than the 220 Ω of the resistor, and the network will take on inductive characteristics. In fact, at, say, 10 Hz, the reactance of the inductor is only about 0.25 Ω which is very close to a short-circuit equivalent compared to the parallel 220 Ω resistor. The result is that the impedance of the network is very close to 0 Ω at very low frequencies. Again, since the inductive effects are so strong at low frequencies, the phase angle associated with the input impedance should be very close to 90°.

Now for the computer analysis. The current source, the resistor element, and the inductor are all placed and connected using procedures described in detail in earlier chapters. However, there is one big difference this time: Since the output will be plotted versus frequency, the **AC Analysis Magnitude** in the **SIGNAL_CURRENT_SOURCES** dialog box for the source must be set to 1 A. In this case, the default level of **1A** matches that of the applied source, so you were set even if you failed to check the setting. In the future, however, a voltage or current source may be used that does not have an amplitude of 1, and proper entries must be made to this listing.

For the simulation, first apply the sequence **Simulate-Analyses-AC Analysis** to obtain the **AC Analysis** dialog box. Set the **Start frequency** at **10 Hz** so that you have entries at very low frequencies, and set the **Stop frequency** at **1MHz** so that you have data points at the other end of the spectrum. The **Sweep type** can remain **Decade,** but the number of points per decade will be 1000 so that you obtain a detailed plot. Set the **Vertical scale** on **Linear.** Within **Output variables,** only one node, **1,** is defined. Shifting it over to the **Selected variables for analysis** column using the **Add** keypad and then hitting the **Simulate** key results in the two plots in Fig. 15.119. Select the **Show/Hide Grid** key to place the grid on the graph, and select the **Show/Hide Cursors** key to place the **AC Analysis** dialog box appearing in Fig. 15.119. Since two graphs are present, define the one you are working on by clicking on the **Voltage** or **Phase** heading on the left side of each plot. A small red arrow appears when selected so you know which is the active plot. When setting up the cursors, be sure that you have activated the correct plot. When the red cursor is moved to 10 Hz (**x1**), you find that the voltage across the network is only 0.251 V (**y1**), resulting in an input impedance of only 0.25 Ω—quite small and matching your theoretical prediction. In addition, note that the phase angle is essentially at 90° in the other plot, confirming your other assumption above—a totally inductive network. If you set the blue cursor near 100 kHz (**x2** = 102.3 kHz), you find that the impedance at 219.2 Ω (**y2**) is closing in on the resistance of the parallel resistor of 220 Ω, again confirming the preliminary analysis above. As noted in the bottom of the **AC Analysis** box, the maximum value of the voltage is 219.99 Ω or essentially 220 Ω at 1 MHz. Before leaving the plot, note the advantages of using a log axis when you want a response over a wide frequency range.

PROBLEMS

SECTION 15.2 Impedance and the Phasor Diagram

1. Express the impedances in Fig. 15.120 in both polar and rectangular forms.

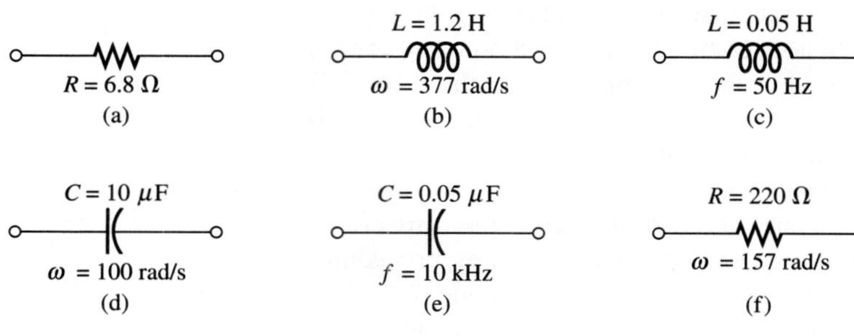

FIG. 15.120
Problem 1.

2. Find the current i for the elements in Fig. 15.121 using complex algebra. Sketch the waveforms for v and i on the same set of axes.

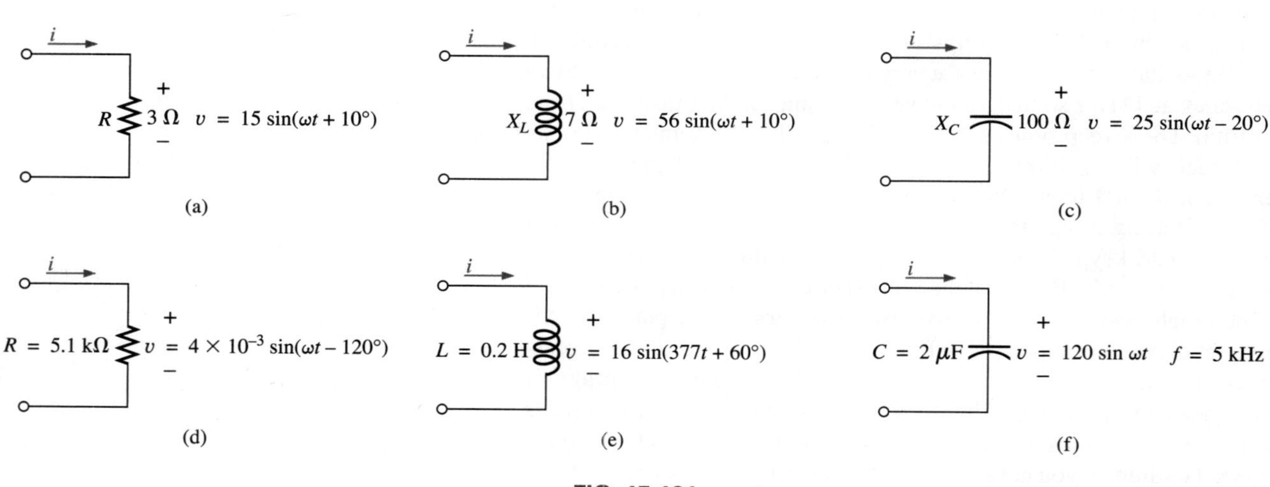

FIG. 15.121
Problem 2.

3. Find the voltage v for the elements in Fig. 15.122 using complex algebra. Sketch the waveforms of v and i on the same set of axes.

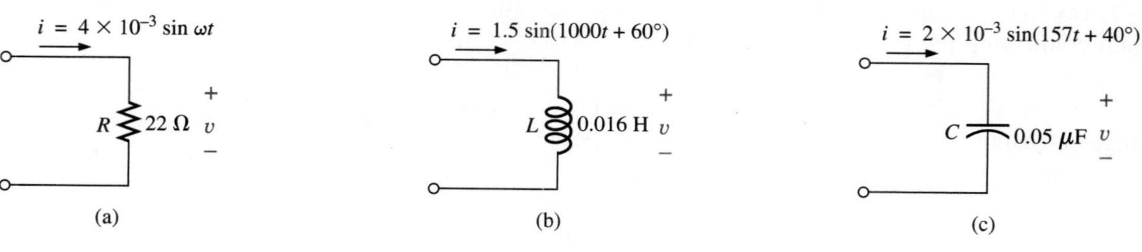

FIG. 15.122
Problem 3.

SECTION 15.3 Series Configuration

4. Calculate the total impedance of the circuits in Fig. 15.123. Express your answer in rectangular and polar forms, and draw the impedance diagram.

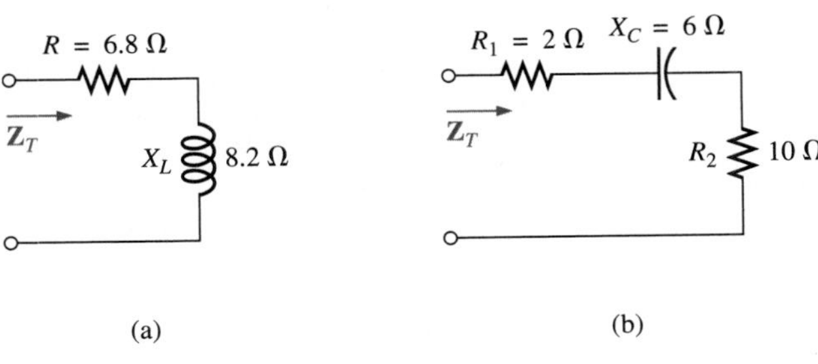

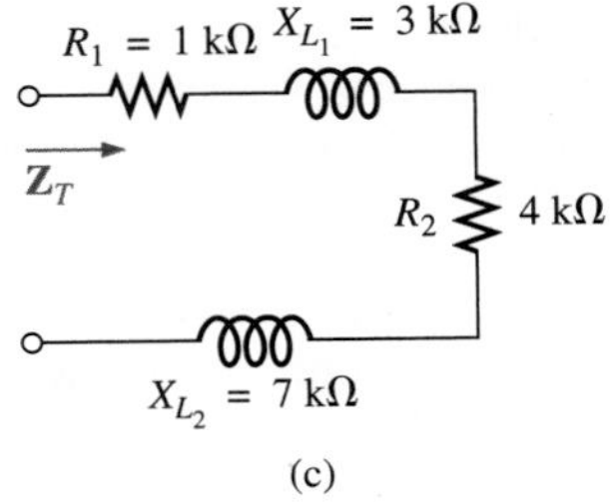

FIG. 15.123
Problem 4.

5. Calculate the total impedance of the circuits in Fig. 15.124. Express your answer in rectangular and polar forms, and draw the impedance diagram.

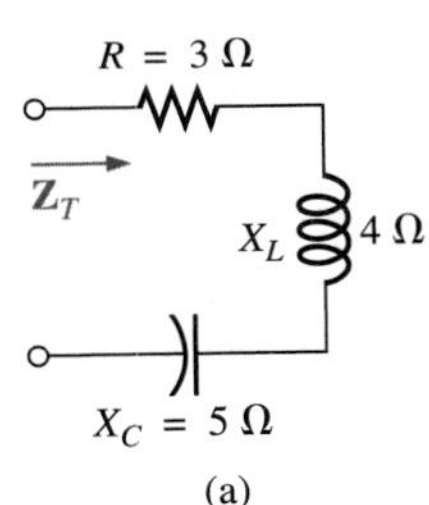

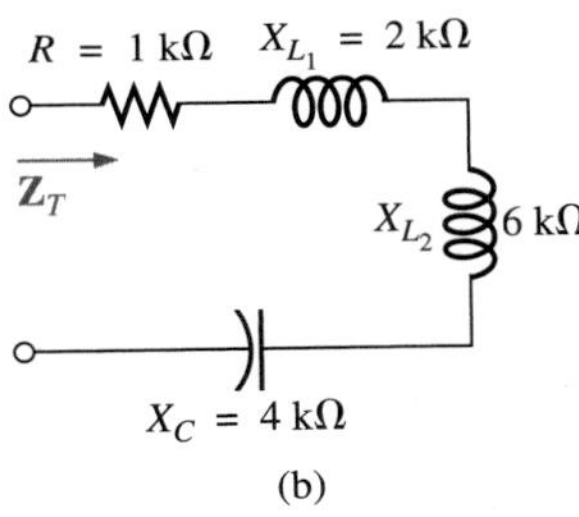

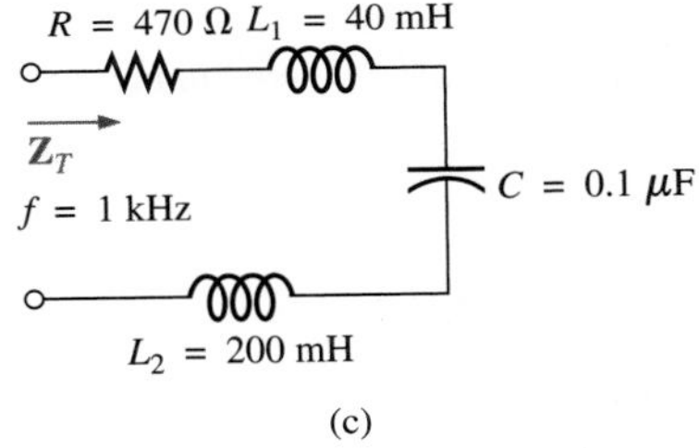

FIG. 15.124
Problem 5.

6. Find the type and impedance in ohms of the series circuit elements that must be in the closed container in Fig. 15.125 for the indicated voltages and currents to exist at the input terminals. (Find the simplest series circuit that will satisfy the indicated conditions.)

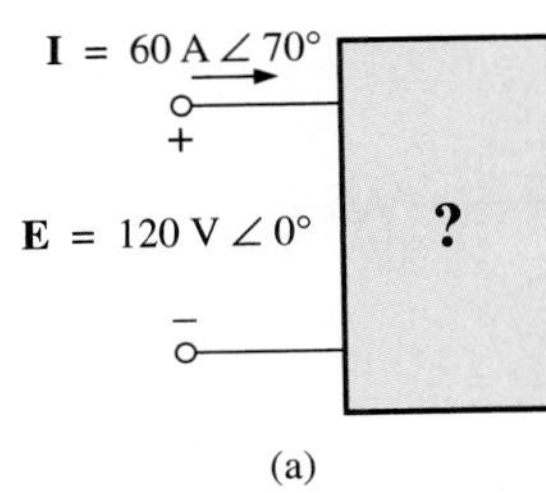

I = 20 mA ∠ 40°
+
E = 80 V ∠ 320°
?
–
(b)

I = 0.2 A ∠ –60°
+
E = 8 kV ∠ 0°
?
–
(c)

FIG. 15.125
Problems 6 and 27.

7. For the circuit in Fig. 15.126:
   a. Find the total impedance $\mathbf{Z}_T$ in polar form.
   b. Draw the impedance diagram.
   c. Find the current **I** and the voltages $\mathbf{V}_R$ and $\mathbf{V}_L$ in phasor form.
   d. Draw the phasor diagram of the voltages **E**, $\mathbf{V}_R$, and $\mathbf{V}_L$, and the current **I**.
   e. Verify Kirchhoff's voltage law around the closed loop.
   f. Find the average power delivered to the circuit.
   g. Find the power factor of the circuit, and indicate whether it is leading or lagging.
   h. Find the sinusoidal expressions for the voltages and current if the frequency is 60 Hz.
   i. Plot the waveforms for the voltages and current on the same set of axes.

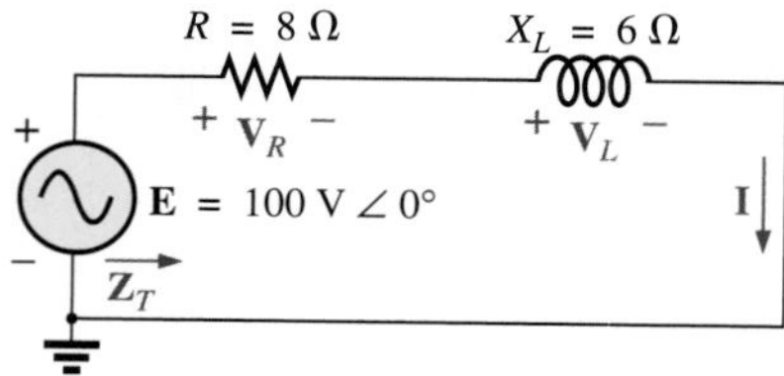

FIG. 15.126
Problems 7 and 48.

8. Repeat Problem 7 for the circuit in Fig. 15.127, replacing $\mathbf{V}_L$ with $\mathbf{V}_C$ in parts (c) and (d).

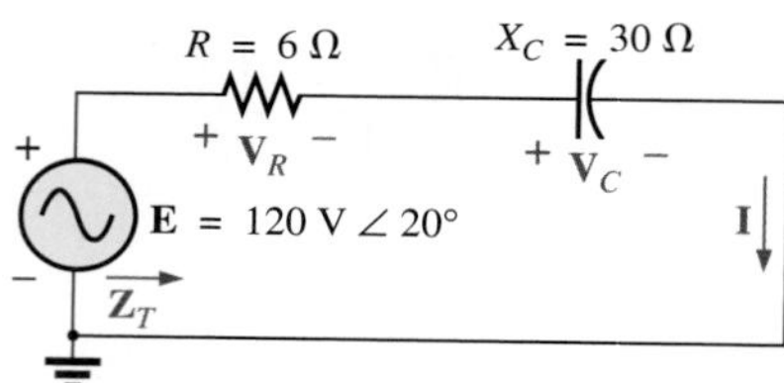

FIG. 15.127
Problem 8.

9. Given the network in Fig. 15.128:
   a. Determine $\mathbf{Z}_T$.
   b. Find **I**.
   c. Calculate $\mathbf{V}_R$ and $\mathbf{V}_C$.
   d. Find P and F_p.

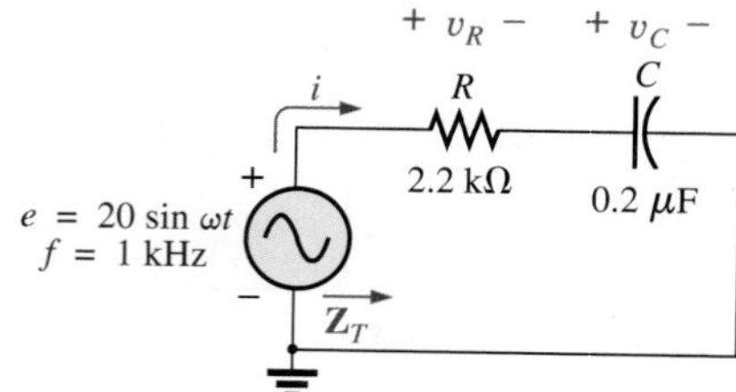

FIG. 15.128
Problems 9 and 50.

10. For the circuit in Fig. 15.129:
    a. Find the total impedance $\mathbf{Z}_T$ in polar form.
    b. Draw the impedance diagram.
    c. Find the value of C in microfarads and L in henries.
    d. Find the current **I** and the voltages $\mathbf{V}_R$, $\mathbf{V}_L$, and $\mathbf{V}_C$ in phasor form.
    e. Draw the phasor diagram of the voltages **E**, $\mathbf{V}_R$, $\mathbf{V}_L$, and $\mathbf{V}_C$, and the current **I**.
    f. Verify Kirchhoff's voltage law around the closed loop.
    g. Find the average power delivered to the circuit.
    h. Find the power factor of the circuit, and indicate whether it is leading or lagging.
    i. Find the sinusoidal expressions for the voltages and current.
    j. Plot the waveforms for the voltages and current on the same set of axes.

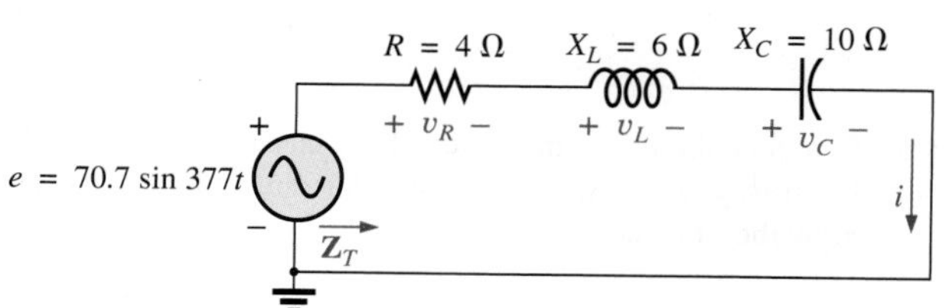

FIG. 15.129
Problem 10.

11. Repeat Problem 10 for the circuit in Fig. 15.130.

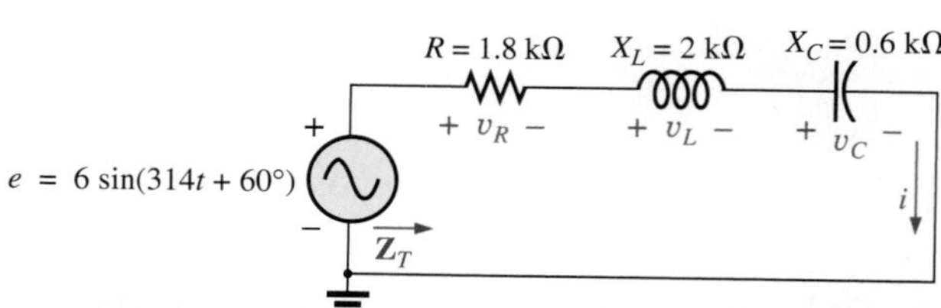

FIG. 15.130
Problem 11.

12. Using the oscilloscope reading in Fig. 15.131, determine the resistance R.

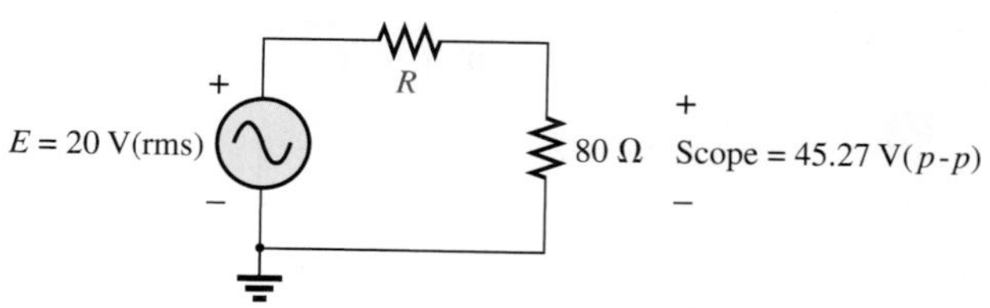

FIG. 15.131
Problem 12.

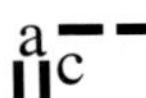

*13. Using the DMM current reading and the oscilloscope measurement in Fig. 15.132:

a. Determine the inductance L.

b. Find the resistance R.

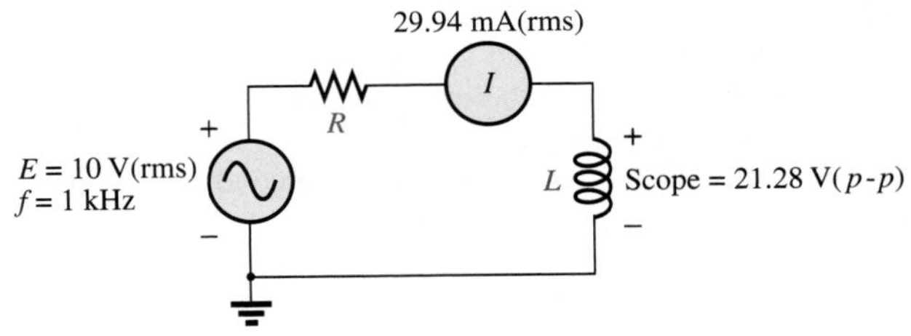

FIG. 15.132
Problem 13.

*14. Using the oscilloscope reading in Fig. 15.133, determine the capacitance C.

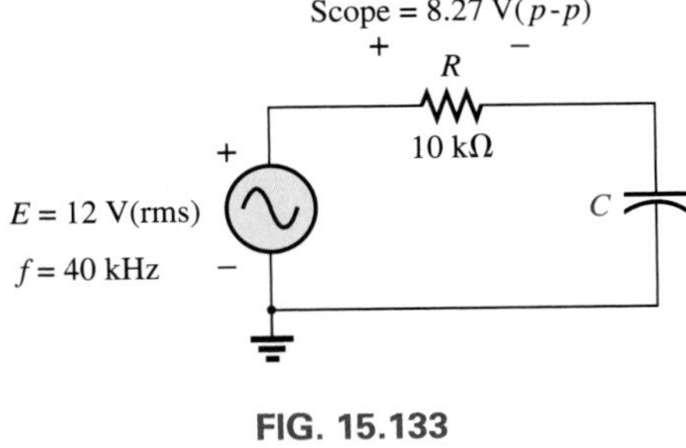

FIG. 15.133
Problem 14.

SECTION 15.4 Voltage Divider Rule

15. Calculate the voltages $\mathbf{V}_1$ and $\mathbf{V}_2$ for the circuits in Fig. 15.134 in phasor form using the voltage divider rule.

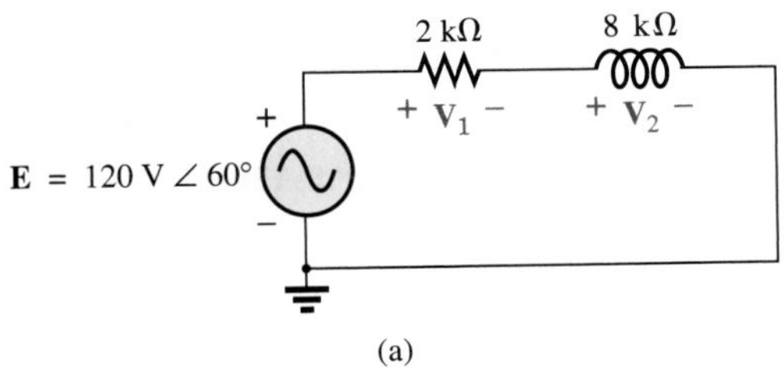

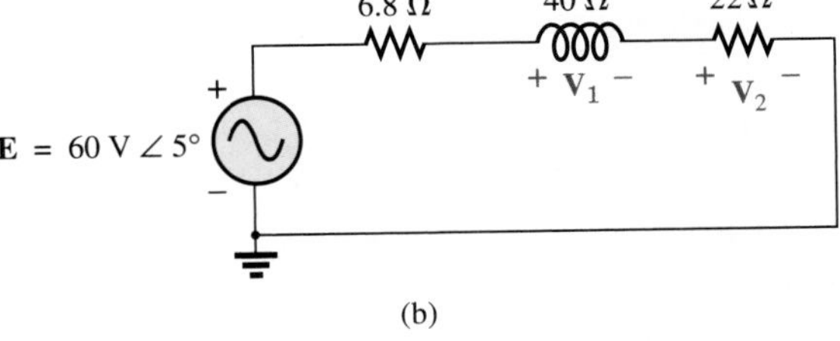

FIG. 15.134
Problem 15.

16. Calculate the voltages $\mathbf{V}_1$ and $\mathbf{V}_2$ for the circuits in Fig. 15.135 in phasor form using the voltage divider rule.

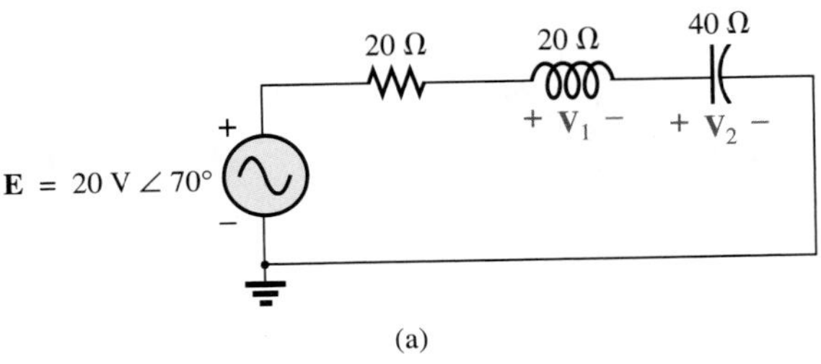

E = 120 V ∠ 0°

4.7 kΩ

30 kΩ

3.3 kΩ

V1

V2

10 kΩ

(b)

FIG. 15.135
Problem 16.

*17. For the circuit in Fig. 15.136:

a. Determine **I**, $\mathbf{V}_R$, and $\mathbf{V}_C$ in phasor form.

b. Calculate the total power factor, and indicate whether it is leading or lagging.

c. Calculate the average power delivered to the circuit.

d. Draw the impedance diagram.

e. Draw the phasor diagram of the voltages **E**, $\mathbf{V}_R$, and $\mathbf{V}_C$, and the current **I**.

f. Find the voltages $\mathbf{V}_R$ and $\mathbf{V}_C$ using the voltage divider rule, and compare them with the results of part (a) above.

g. Draw the equivalent series circuit of the above as far as the total impedance and the current i are concerned.

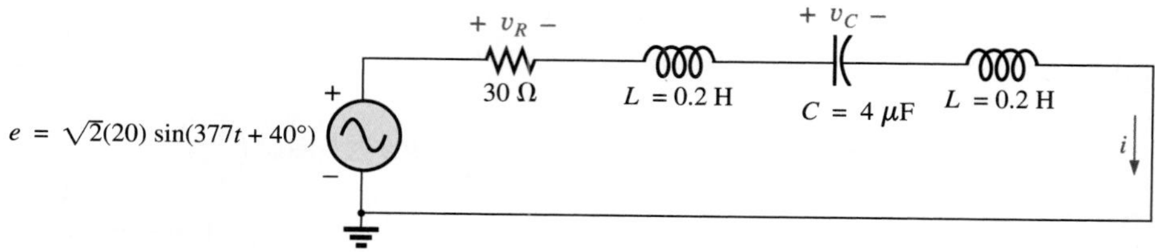

FIG. 15.136
Problems 17, 18, and 51.

*18. Repeat Problem 17 if the capacitance is changed to 220 μF.

19. An electrical load has a power factor of 0.8 lagging. It dissipates 8 kW at a voltage of 200 V. Calculate the impedance of this load in rectangular coordinates.

*20. Find the series element or elements that must be in the enclosed container in Fig. 15.137 to satisfy the following conditions:

a. Average power to circuit = 300 W.

b. Circuit has a lagging power factor.

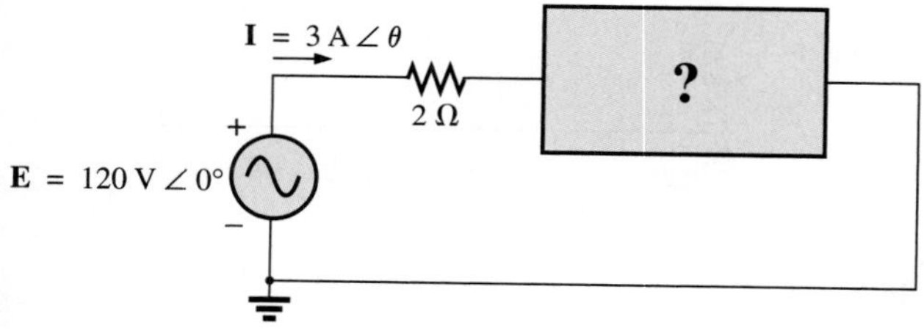

FIG. 15.137
Problem 20.

SECTION 15.5 Frequency Response for Series ac Circuits

*21. For the circuit in Fig. 15.138:

a. Plot Z_T and θ_T versus frequency for a frequency range of zero to 20 kHz.

b. Plot V_L versus frequency for the frequency range of part (a).

c. Plot θ_L versus frequency for the frequency range of part (a).

d. Plot V_R versus frequency for the frequency range of part (a).

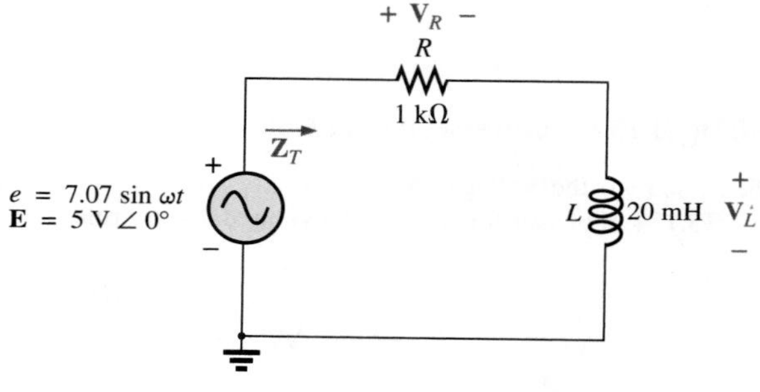

FIG. 15.138
Problem 21.

*22. For the circuit in Fig. 15.139:

a. Plot Z_T and θ_T versus frequency for a frequency range of zero to 10 kHz.

b. Plot V_C versus frequency for the frequency range of part (a).

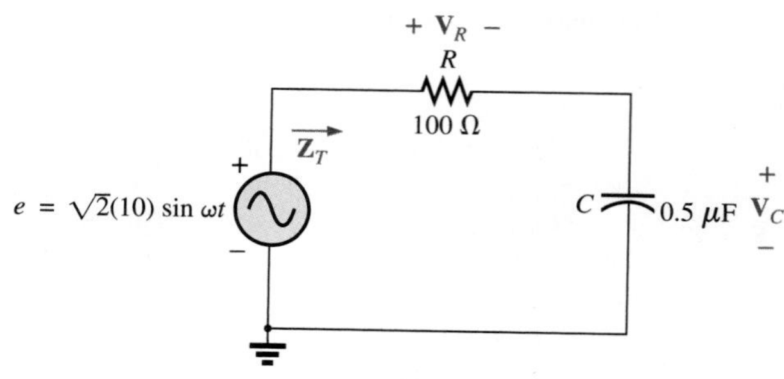

FIG. 15.139
Problem 22.

c. Plot θ_C versus frequency for the frequency range of part (a).

d. Plot V_R versus frequency for the frequency range of part (a).

*23. For the series *R-L-C* circuit in Fig. 15.140:

a. Plot Z_T and θ_T versus frequency for a frequency range of zero to 20 kHz in increments of 1 kHz.

b. Plot V_C (magnitude only) versus frequency for the same frequency range of part (a).

c. Plot I (magnitude only) versus frequency for the same frequency range of part (a).

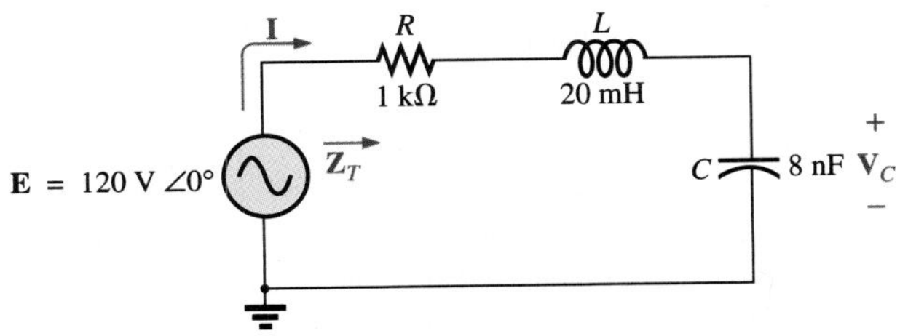

FIG. 15.140
Problem 23.

24. For the series *R-C* circuit in Fig. 15.141:

a. Determine the frequency at which $X_C = R$.

b. Develop a mental image of the change in total impedance with frequency without resorting to a single calculation.

c. Find the total impedance at 100 Hz and 10 kHz, and compare your answer with the assumptions of part (b).

d. Plot the curve of $\mathbf{V}_C$ versus frequency.

e. Find the phase angle of the total impedance at $f = 40$ kHz. Is the network resistive or capacitive at this frequency?

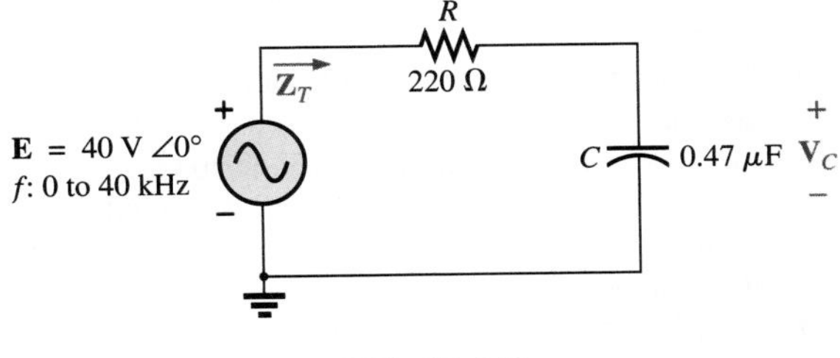

FIG. 15.141
Problem 24.

SECTION 15.7 Admittance and Susceptance

25. Find the total admittance and impedance of the circuits in Fig. 15.142. Identify the values of conductance and susceptance, and draw the admittance diagram.

26. Find the total admittance and impedance of the circuits in Fig. 15.143. Identify the values of conductance and susceptance, and draw the admittance diagram.

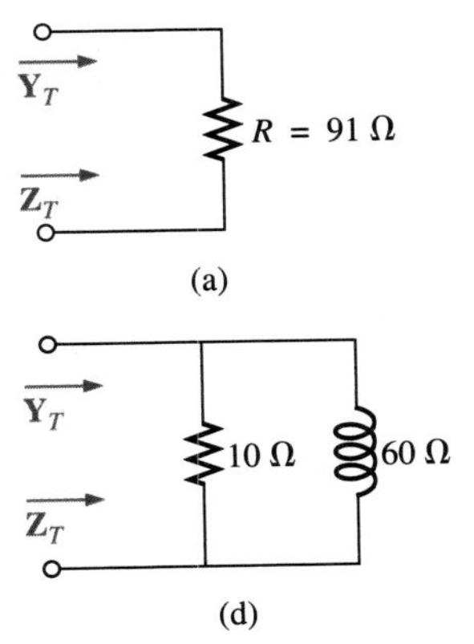

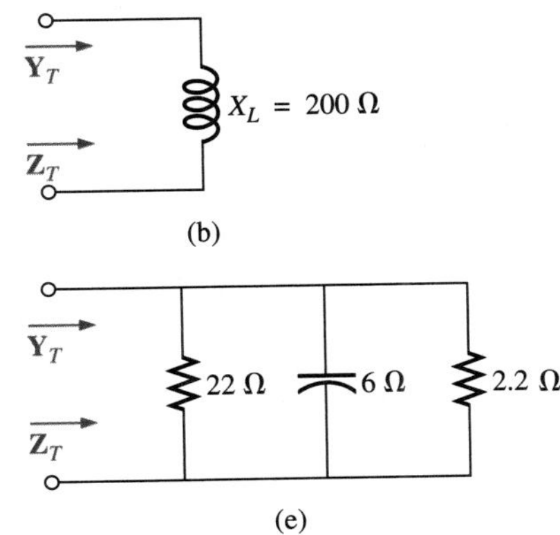

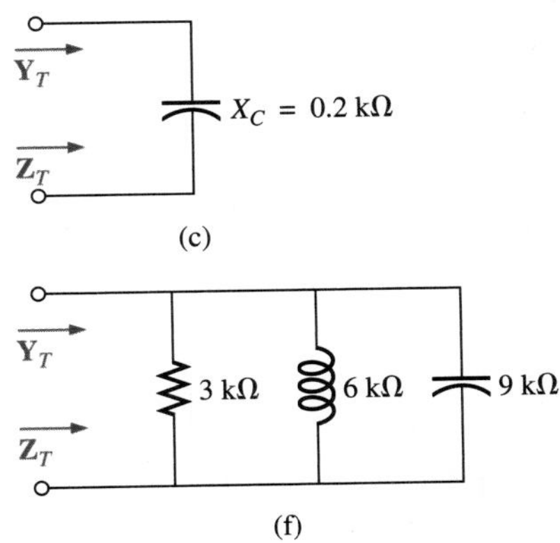

FIG. 15.142
Problem 25.

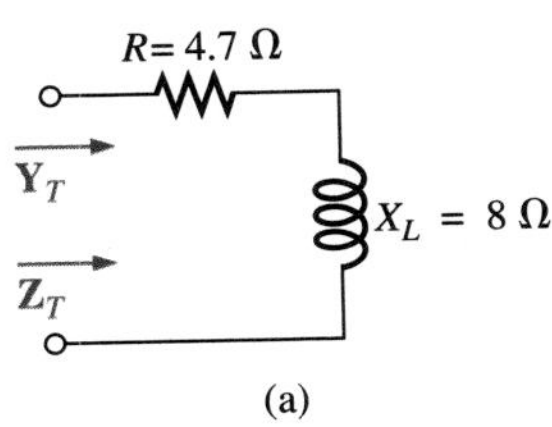

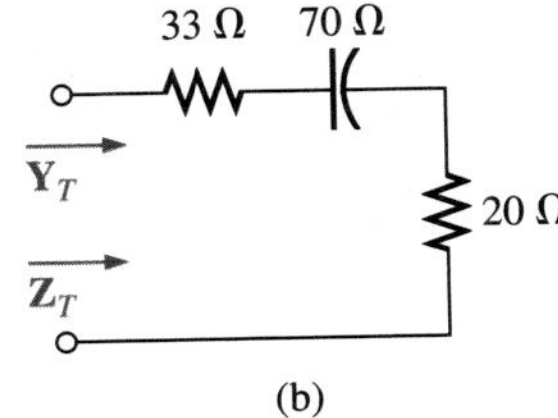

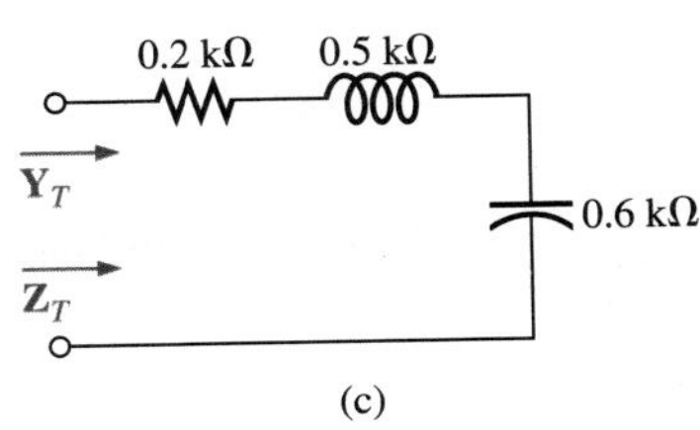

FIG. 15.143
Problem 26.

27. Repeat Problem 6 for the parallel circuit elements that must be in the closed container for the same voltage and current to exist at the input terminals. (Find the simplest parallel circuit that will satisfy the conditions indicated.)

SECTION 15.8 Parallel ac Networks

28. For the circuit in Fig. 15.144:

a. Find the total admittance $\mathbf{Y}_T$ in polar form.
b. Draw the admittance diagram.
c. Find the voltage **E** and the currents $\mathbf{I}_R$ and $\mathbf{I}_L$ in phasor form.
d. Draw the phasor diagram of the currents $\mathbf{I}_s$, $\mathbf{I}_R$, and $\mathbf{I}_L$, and the voltage **E**.
e. Verify Kirchhoff's current law at one node.
f. Find the average power delivered to the circuit.
g. Find the power factor of the circuit, and indicate whether it is leading or lagging.
h. Find the sinusoidal expressions for the currents and voltage if the frequency is 60 Hz.
i. Plot the waveforms for the currents and voltage on the same set of axes.

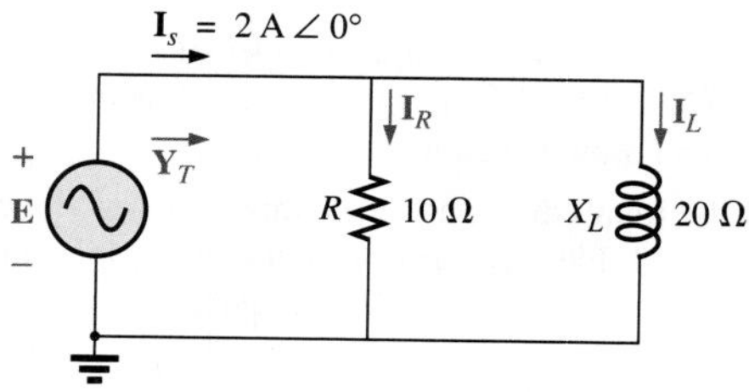

FIG. 15.144
Problem 28.

29. Repeat Problem 28 for the circuit in Fig. 15.145, replacing $\mathbf{I}_L$ with $\mathbf{I}_C$ in parts (c) and (d).

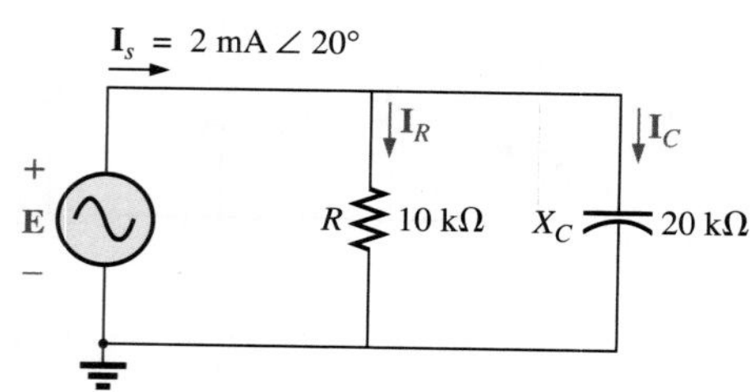

FIG. 15.145
Problem 29.

30. Repeat Problem 28 for the circuit in Fig. 15.146, replacing **E** with $\mathbf{I}_s$ in part (c).

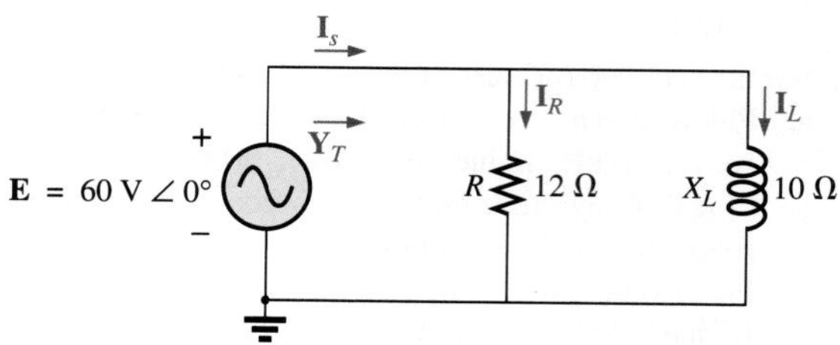

FIG. 15.146
Problems 30 and 49.

31. For the circuit in Fig. 15.147:

a. Find the total admittance and impedance in polar form.
b. Draw the admittance and impedance diagrams.
c. Find the value of C in microfarads and L in henries.
d. Find the voltage **E** and currents $\mathbf{I}_R$, $\mathbf{I}_L$, and $\mathbf{I}_C$ in phasor form.
e. Draw the phasor diagram of the currents $\mathbf{I}_s$, $\mathbf{I}_R$, $\mathbf{I}_L$, and $\mathbf{I}_C$, and the voltage **E**.
f. Verify Kirchhoff's current law at one node.
g. Find the average power delivered to the circuit.
h. Find the power factor of the circuit, and indicate whether it is leading or lagging.
i. Find the sinusoidal expressions for the currents and voltage.
j. Plot the waveforms for the currents and voltage on the same set of axes.

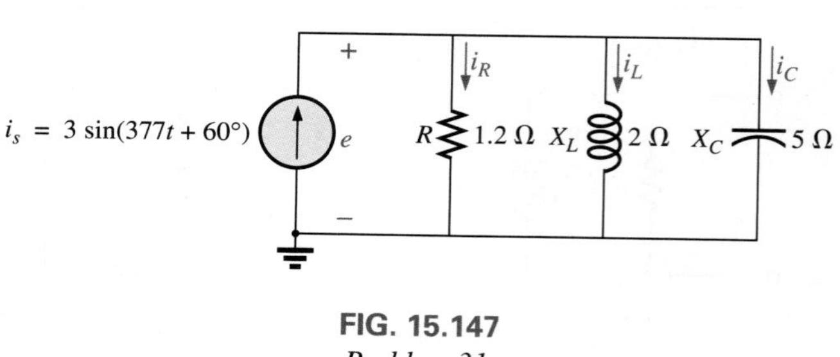

FIG. 15.147
Problem 31.

32. Repeat Problem 31 for the circuit in Fig. 15.148.

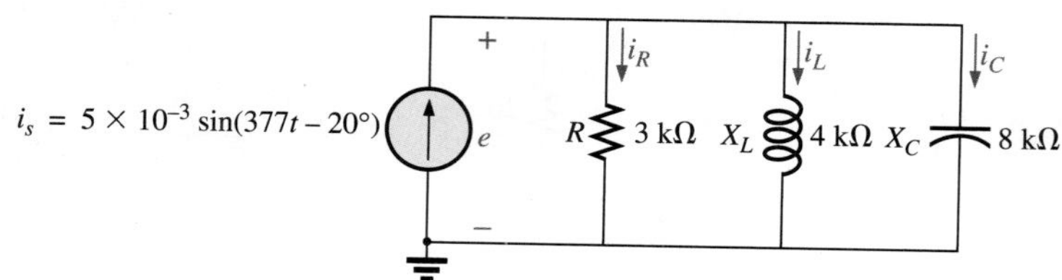

FIG. 15.148
Problem 32.

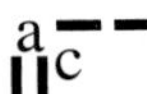

33. Repeat Problem 31 for the circuit in Fig. 15.149, replacing e with i_s in part (d).

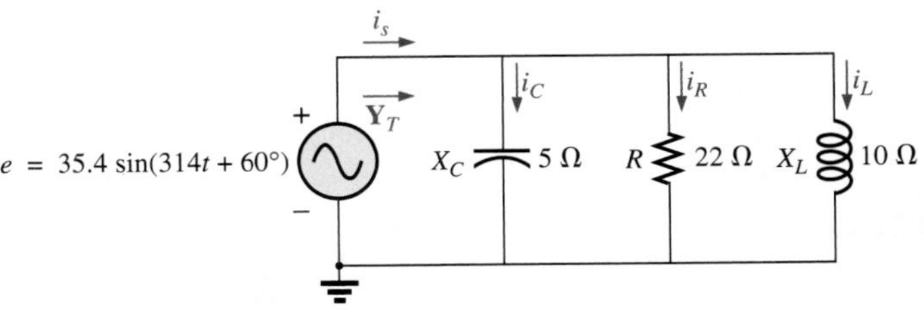

FIG. 15.149
Problem 33.

SECTION 15.9 Current Divider Rule

34. Calculate the currents $\mathbf{I}_1$ and $\mathbf{I}_2$ in Fig. 15.150 in phasor form using the current divider rule.

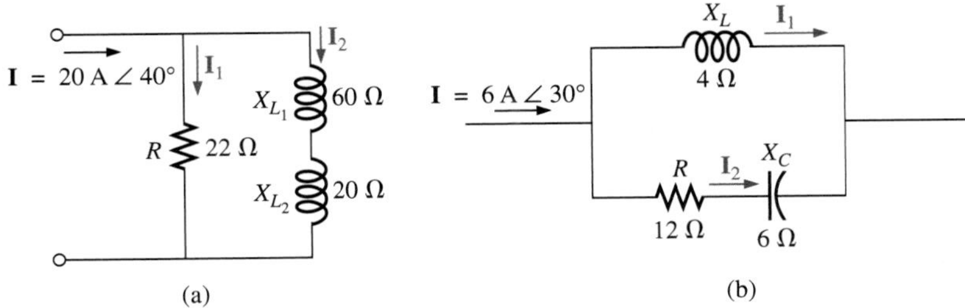

FIG. 15.150
Problem 34.

SECTION 15.10 Frequency Response of Parallel Elements

***35.** For the parallel *R-C* network in Fig. 15.151:

a. Plot Z_T and θ_T versus frequency for a frequency range of zero to 20 kHz.

b. Plot V_C versus frequency for the frequency range of part (a).

c. Plot I_R versus frequency for the frequency range of part (a).

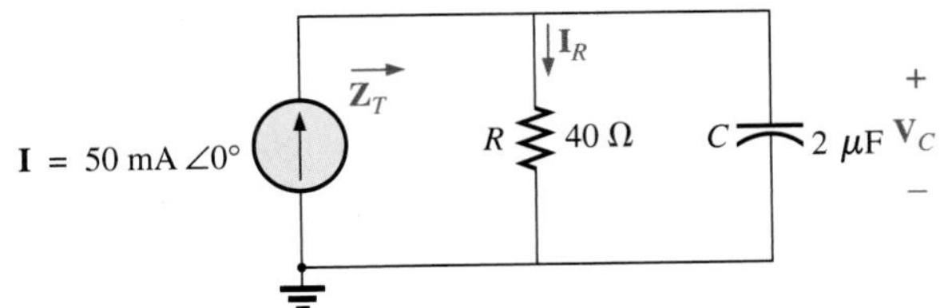

FIG. 15.151
Problems 35 and 37.

***36.** For the parallel *R-L* network in Fig. 15.152:

a. Plot Z_T and θ_T versus frequency for a frequency range of zero to 10 kHz.

b. Plot I_L versus frequency for the frequency range of part (a).

c. Plot I_R versus frequency for the frequency range of part (a).

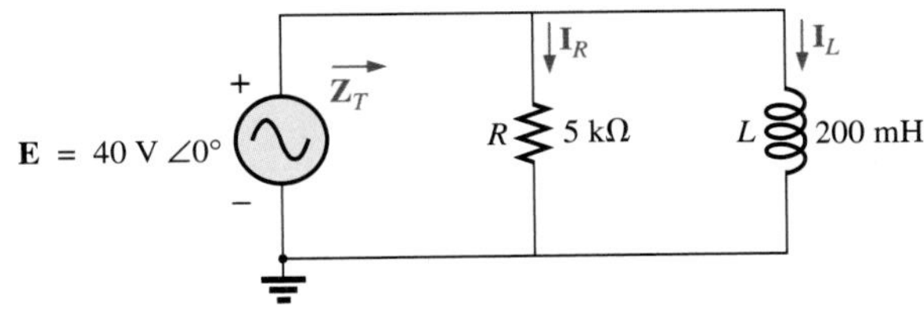

FIG. 15.152
Problems 36 and 38.

37. Plot Y_T and θ_T (of $\mathbf{Y}_T = Y_T \angle \theta_T$) for a frequency range of zero to 20 kHz for the network in Fig. 15.151.

38. Plot Y_T and θ_T (of $\mathbf{Y}_T = Y_T \angle \theta_T$) for a frequency range of zero to 10 kHz for the network in Fig. 15.152.

39. For the parallel *R-L-C* network in Fig. 15.153:

a. Plot Y_T and θ_T (of $\mathbf{Y}_T = Y_T \angle \theta_T$) for a frequency range of zero to 20 kHz.

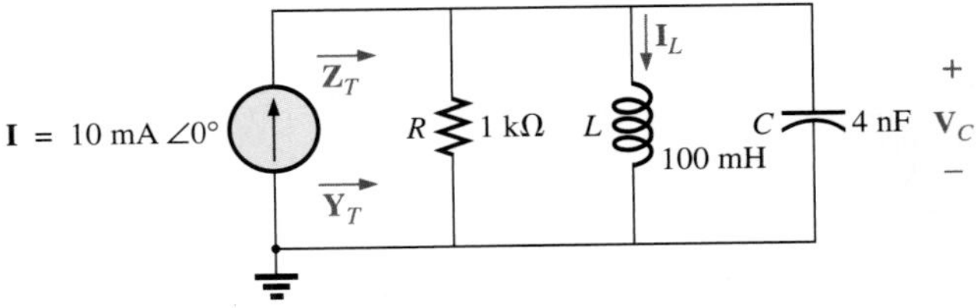

FIG. 15.153
Problem 39.

b. Repeat part (a) for Z_T and θ_T (of $\mathbf{Z}_T = Z_T \angle \theta_T$).
c. Plot V_C versus frequency for the frequency range of part (a).
d. Plot I_L versus frequency for the frequency range of part (a).

SECTION 15.12 Equivalent Circuits

40. For the series circuits in Fig. 15.154, find a parallel circuit that will have the same total impedance ($\mathbf{Z}_T$).

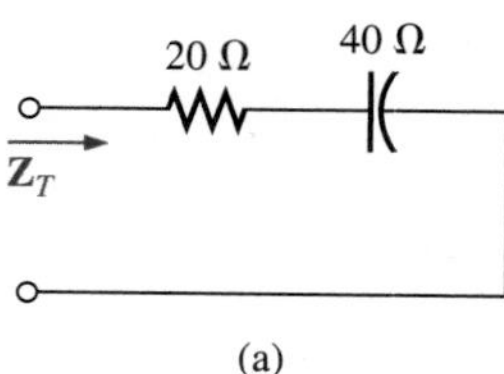

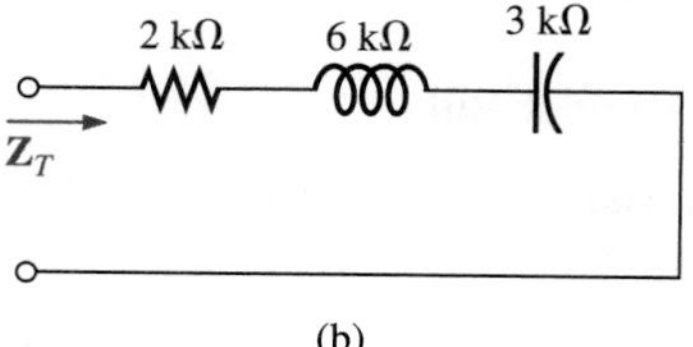

FIG. 15.154
Problem 40.

41. For the parallel circuits in Fig. 15.155, find a series circuit that will have the same total impedance.

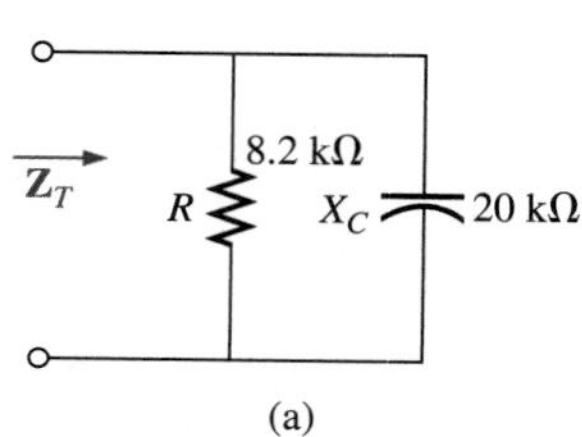

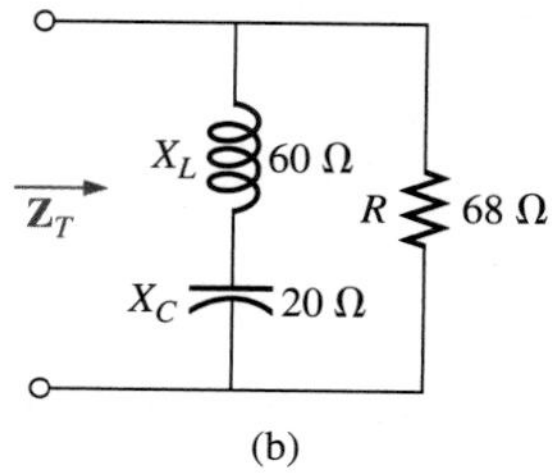

FIG. 15.155
Problem 41.

42. For the network in Fig. 15.156:
a. Calculate $\mathbf{E}$, $\mathbf{I}_R$, and $\mathbf{I}_L$ in phasor form.
b. Calculate the total power factor, and indicate whether it is leading or lagging.
c. Calculate the average power delivered to the circuit.
d. Draw the admittance diagram.
e. Draw the phasor diagram of the currents $\mathbf{I}_s$, $\mathbf{I}_R$, and $\mathbf{I}_L$, and the voltage $\mathbf{E}$.

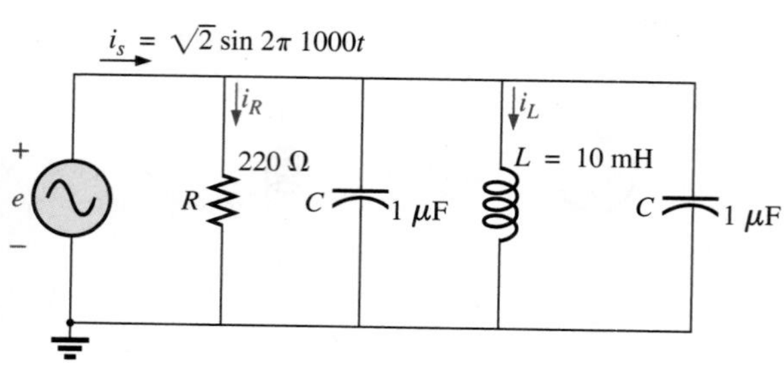

FIG. 15.156
Problems 42 and 43.

f. Find the current $\mathbf{I}_C$ for each capacitor using only Kirchhoff's current law.
g. Find the series circuit of one resistive and reactive element that will have the same impedance as the original circuit.

***43.** Repeat Problem 42 if the inductance is changed to 1 H.

44. Find the element or elements that must be in the closed container in Fig. 15.157 to satisfy the following conditions. (Find the simplest parallel circuit that will satisfy the indicated conditions.)
a. Average power to the circuit = 3000 W.
b. Circuit has a lagging power factor.

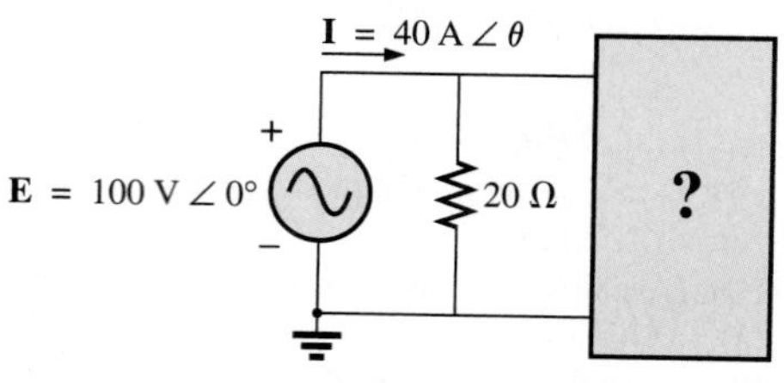

FIG. 15.157
Problem 44.

SECTION 15.13 Phase Measurements

45. For the circuit in Fig. 15.158, determine the phase relationship between the following using a dual-trace oscilloscope. The circuit can be reconstructed differently for each part, but do not use sensing resistors. Show all connections on a redrawn diagram.

a. e and v_C
b. e and i_s
c. e and v_L

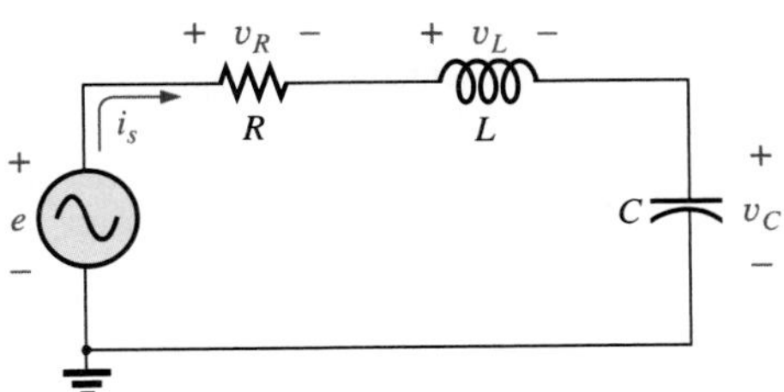

FIG. 15.158
Problem 45.

46. For the network in Fig. 15.159, determine the phase relationship between the following using a dual-trace oscilloscope. The network must remain as constructed in Fig. 15.159, but sensing resistors can be introduced. Show all connections on a redrawn diagram.

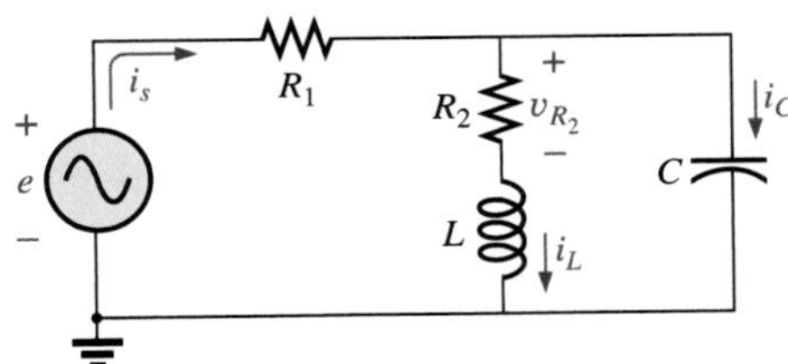

FIG. 15.159
Problem 46.

a. e and v_{R_2}
b. e and i_s
c. i_L and i_C

47. For the oscilloscope traces in Fig. 15.160:

a. Determine the phase relationship between the waveforms, and indicate which one leads or lags.
b. Determine the peak-to-peak and rms values of each waveform.
c. Find the frequency of each waveform.

SECTION 15.15 Computer Analysis

PSpice or Multisim

48. For the network in Fig. 15.126 (use $f = 1$ kHz):

a. Determine the rms values of the voltages $\mathbf{V}_R$ and $\mathbf{V}_L$ and the current $\mathbf{I}$.
b. Plot v_R, v_L, and i versus time on separate plots.
c. Place e, v_R, v_L, and i on the same plot, and label accordingly.

49. For the network in Fig. 15.146:

a. Determine the rms values of the currents $\mathbf{I}_s$, $\mathbf{I}_R$, and $\mathbf{I}_L$.
b. Plot i_s, i_R, and i_L versus time on separate plots.
c. Place e, i_s, i_R, and i_L on the same plot, and label accordingly.

50. For the network in Fig. 15.128:

a. Plot the impedance of the network versus frequency from 0 to 10 kHz.
b. Plot the current i versus frequency for the frequency range zero to 10 kHz.

***51.** For the network in Fig. 15.136:

a. Find the rms values of the voltages v_R and v_C at a frequency of 1 kHz.
b. Plot v_C versus frequency for the frequency range zero to 10 kHz.
c. Plot the phase angle between e and i for the frequency range zero to 10 kHz.

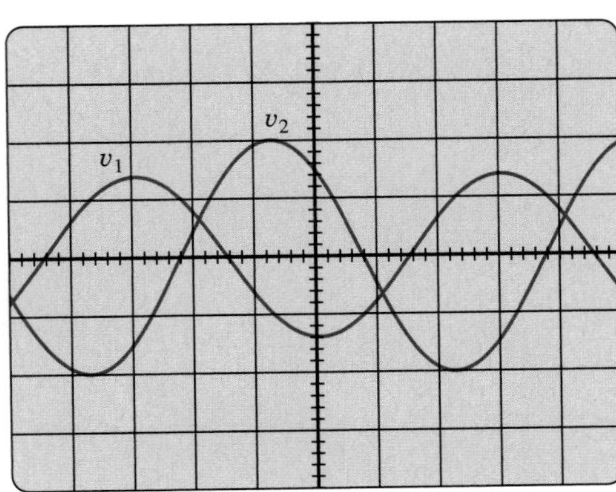

FIG. 15.160
Problem 47.

GLOSSARY

Admittance A measure of how easily a network will "admit" the passage of current through that system. It is measured in siemens, abbreviated S, and is represented by the capital letter *Y.*

Admittance diagram A vector display that clearly depicts the magnitude of the admittance of the conductance, capacitive susceptance, and inductive susceptance, and the magnitude and angle of the total admittance of the system.

Current divider rule A method by which the current through either of two parallel branches can be determined in an ac network without first finding the voltage across the parallel branches.

Equivalent circuits For every series ac network, there is a parallel ac network (and vice versa) that will be "equivalent" in the sense that the input current and impedance are the same.

Impedance diagram A vector display that clearly depicts the magnitude of the impedance of the resistive, reactive, and capacitive components of a network, and the magnitude and angle of the total impedance of the system.

Parallel ac circuits A connection of elements in an ac network in which all the elements have two points in common. The voltage is the same across each element.

Phasor diagram A vector display that provides at a glance the magnitude and phase relationships among the various voltages and currents of a network.

Series ac configuration A connection of elements in an ac network in which no two impedances have more than one terminal in common and the current is the same through each element.

Susceptance A measure of how "susceptible" an element is to the passage of current through it. It is measured in siemens, abbreviated S, and is represented by the capital letter *B*.

Voltage divider rule A method through which the voltage across one element of a series of elements in an ac network can be determined without first having to find the current through the elements.

Series-Parallel ac Networks

Objectives

- *Develop confidence in the analysis of series-parallel ac networks.*
- *Become proficient in the use of calculators and computer methods to support the analysis of ac series-parallel networks.*
- *Understand the importance of proper grounding in the operation of any electrical system.*

16.1 INTRODUCTION

In this chapter, we shall use the fundamental concepts of the previous chapter to develop a technique for solving **series-parallel ac networks.** A brief review of Chapter 7 may be helpful before considering these networks since the approach here is quite similar to that undertaken earlier. The circuits to be discussed have only one source of energy, either potential or current. Networks with two or more sources are considered in Chapters 17 and 18, using methods previously described for dc circuits.

In general, when working with series-parallel ac networks, consider the following approach:

1. *Redraw the network, using block impedances to combine obvious series and parallel elements, which will reduce the network to one that clearly reveals the fundamental structure of the system.*
2. *Study the problem and make a brief mental sketch of the overall approach you plan to use. Doing this may result in time- and energy-saving shortcuts. In some cases, a lengthy, drawn-out analysis may not be necessary. A single application of a fundamental law of circuit analysis may result in the desired solution.*
3. *After the overall approach has been determined, it is usually best to consider each branch involved in your method independently before tying them together in series-parallel combinations. In most cases, work back from the obvious series and parallel combinations to the source to determine the total impedance of the network. The source current can then be determined, and the path back to specific unknowns can be defined. As you progress back to the source, continually define those unknowns that have not been lost in the reduction process. It will save time when you have to work back through the network to find specific quantities.*
4. *When you have arrived at a solution, check to see that it is reasonable by considering the magnitudes of the energy source and the elements in the circuit. If not, either solve the network using another approach, or check over your work very carefully. At this point, a computer solution can be an invaluable asset in the validation process.*

16.2 ILLUSTRATIVE EXAMPLES

EXAMPLE 16.1 For the network in Fig. 16.1:

a. Calculate $\mathbf{Z}_T$.
b. Determine $\mathbf{I}_s$.
c. Calculate $\mathbf{V}_R$ and $\mathbf{V}_C$.

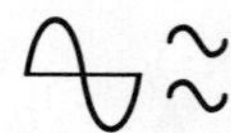

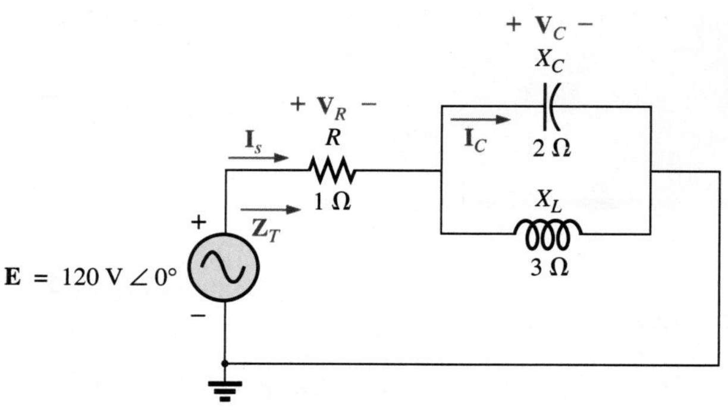

FIG. 16.1
Example 16.1.

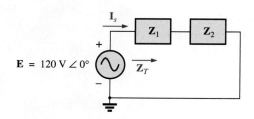

d. Find $\mathbf{I}_C$.
e. Compute the power delivered.
f. Find F_p of the network.

Solutions:

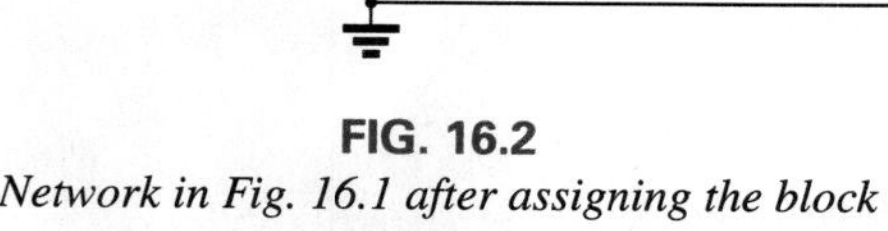

FIG. 16.2
Network in Fig. 16.1 after assigning the block impedances.

a. As suggested in the introduction, the network has been redrawn with block impedances, as shown in Fig. 16.2. The impedance $\mathbf{Z}_1$ is simply the resistor R of 1 Ω, and $\mathbf{Z}_2$ is the parallel combination of X_C and X_L. The network now clearly reveals that it is fundamentally a series circuit, suggesting a direct path toward the total impedance and the source current. For many such problems, you must work back to the source to find first the total impedance and then the source current. When the unknown quantities are found in terms of these subscripted impedances, the numerical values can then be substituted to find the magnitude and phase angle of the unknown. In other words, try to find the desired solution solely in terms of the subscripted impedances before substituting numbers. This approach will usually enhance the clarity of the chosen path toward a solution while saving time and preventing careless calculation errors. Note also in Fig. 16.2 that all the unknown quantities except $\mathbf{I}_C$ have been preserved, meaning that we can use Fig. 16.2 to determine these quantities rather than having to return to the more complex network in Fig. 16.1.

The total impedance is defined by

$$\mathbf{Z}_T = \mathbf{Z}_1 + \mathbf{Z}_2$$

with

$$\mathbf{Z}_1 = R\angle 0° = 1\ \Omega\angle 0°$$

$$\mathbf{Z}_2 = \mathbf{Z}_C \parallel \mathbf{Z}_L = \frac{(X_C\angle -90°)(X_L\angle 90°)}{-jX_C + jX_L} = \frac{(2\ \Omega\angle -90°)(3\ \Omega\angle 90°)}{-j2\ \Omega + j3\ \Omega}$$

$$= \frac{6\ \Omega\angle 0°}{j1} = \frac{6\ \Omega\angle 0°}{1\angle 90°} = 6\ \Omega\angle -90°$$

and

$$\mathbf{Z}_T = \mathbf{Z}_1 + \mathbf{Z}_2 = 1\ \Omega - j6\ \Omega = \mathbf{6.08\ \Omega\angle -80.54°}$$

b. $$\mathbf{I}_s = \frac{\mathbf{E}}{\mathbf{Z}_T} = \frac{120\text{ V}\angle 0°}{6.08\ \Omega\angle -80.54°} = \mathbf{19.74\text{ A}\angle 80.54°}$$

c. Referring to Fig. 16.2, we find that $\mathbf{V}_R$ and $\mathbf{V}_C$ can be found by a direct application of Ohm's law:

$$\mathbf{V}_R = \mathbf{I}_s\mathbf{Z}_1 = (19.74 \text{ A} \angle 80.54^\circ)(1\ \Omega \angle 0^\circ) = \mathbf{19.74\ V \angle 80.54^\circ}$$
$$\mathbf{V}_C = \mathbf{I}_s\mathbf{Z}_2 = (19.74 \text{ A} \angle 80.54^\circ)(6\ \Omega \angle -90^\circ) = \mathbf{118.44\ V \angle -9.46^\circ}$$

d. Now that $\mathbf{V}_C$ is known, the current $\mathbf{I}_C$ can also be found using Ohm's law.

$$\mathbf{I}_C = \frac{\mathbf{V}_C}{\mathbf{Z}_C} = \frac{118.44 \text{ V} \angle -9.46^\circ}{2\ \Omega \angle -90^\circ} = \mathbf{59.22\ A \angle 80.54^\circ}$$

e. $P_{\text{del}} = I_s^2R = (19.74 \text{ A})^2(1\ \Omega) = \mathbf{389.67\ W}$

f. $F_p = \cos\theta = \cos 80.54^\circ = \mathbf{0.164\ leading}$

The fact that the total impedance has a negative phase angle (revealing that $\mathbf{I}_s$ leads $\mathbf{E}$) is a clear indication that the network is capacitive in nature and therefore has a leading power factor. The fact that the network is capacitive can be determined from the original network by first realizing that, for the parallel L-C elements, the smaller impedance predominates and results in an R-C network.

EXAMPLE 16.2 For the network in Fig. 16.3:

a. If $\mathbf{I}$ is 50 A $\angle 30^\circ$, calculate $\mathbf{I}_1$ using the current divider rule.
b. Repeat part (a) for $\mathbf{I}_2$.
c. Verify Kirchhoff's current law at one node.

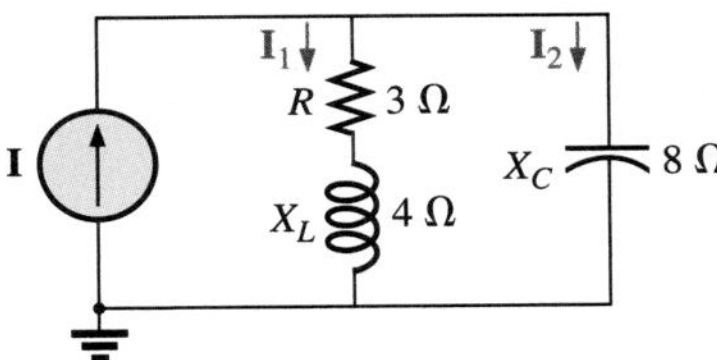

FIG. 16.3
Example 16.2.

Solutions:

a. Redrawing the circuit as in Fig. 16.4, we have

$$\mathbf{Z}_1 = R + jX_L = 3\ \Omega + j\,4\ \Omega = 5\ \Omega \angle 53.13^\circ$$
$$\mathbf{Z}_2 = -jX_C = -j\,8\ \Omega = 8\ \Omega \angle -90^\circ$$

Using the current divider rule yields

$$\mathbf{I}_1 = \frac{\mathbf{Z}_2\mathbf{I}}{\mathbf{Z}_2 + \mathbf{Z}_1} = \frac{(8\ \Omega \angle -90^\circ)(50 \text{ A} \angle 30^\circ)}{(-j\,8\ \Omega) + (3\ \Omega + j\,4\ \Omega)} = \frac{400 \angle -60^\circ}{3 - j\,4}$$
$$= \frac{400 \angle -60^\circ}{5 \angle -53.13^\circ} = \mathbf{80\ A \angle -6.87^\circ}$$

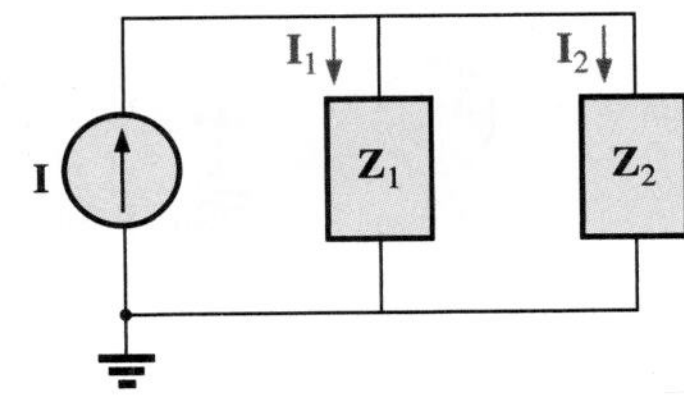

FIG. 16.4
Network in Fig. 16.3 after assigning the block impedances.

b. $$\mathbf{I}_2 = \frac{\mathbf{Z}_1\mathbf{I}}{\mathbf{Z}_2 + \mathbf{Z}_1} = \frac{(5\ \Omega \angle 53.13^\circ)(50 \text{ A} \angle 30^\circ)}{5\ \Omega \angle -53.13^\circ} = \frac{250 \angle 83.13^\circ}{5 \angle -53.13^\circ}$$
$$= \mathbf{50\ A \angle 136.26^\circ}$$

c.
$$\mathbf{I} = \mathbf{I}_1 + \mathbf{I}_2$$
$$50 \text{ A} \angle 30^\circ = 80 \text{ A} \angle -6.87^\circ + 50 \text{ A} \angle 136.26^\circ$$
$$= (79.43 - j\,9.57) + (-36.12 + j\,34.57)$$
$$= 43.31 + j\,25.0$$
$$50 \text{ A} \angle 30^\circ = 50 \text{ A} \angle 30^\circ \quad \text{(checks)}$$

EXAMPLE 16.3 For the network in Fig. 16.5:

a. Calculate the voltage $\mathbf{V}_C$ using the voltage divider rule.
b. Calculate the current $\mathbf{I}_s$.

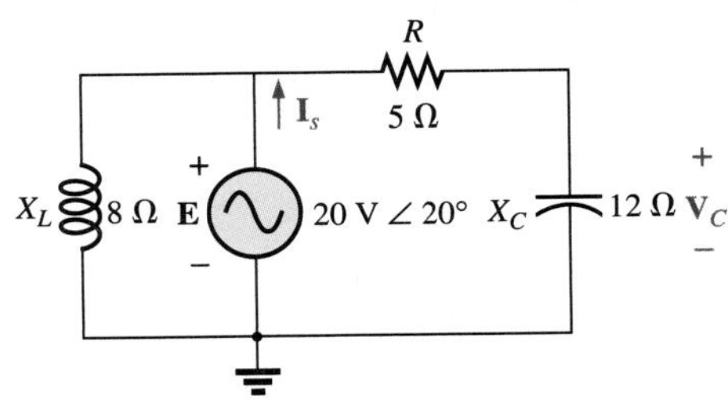

FIG. 16.5
Example 16.3.

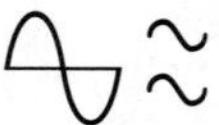

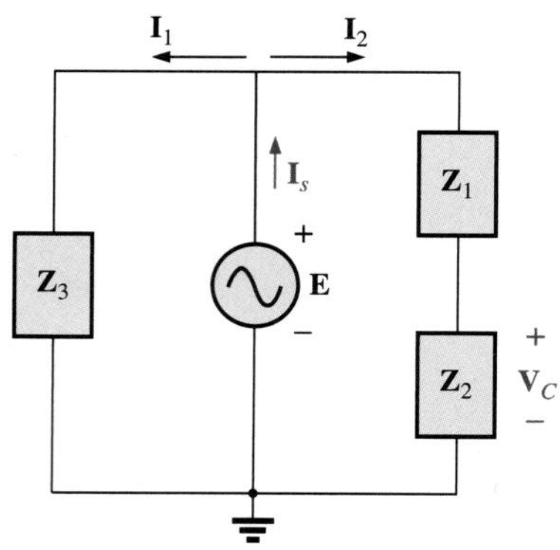

FIG. 16.6
Network in Fig. 16.5 after assigning the block impedances.

Solutions:

a. The network is redrawn as shown in Fig. 16.6, with

$$\mathbf{Z}_1 = 5\ \Omega = 5\ \Omega \angle 0^\circ$$
$$\mathbf{Z}_2 = -j\,12\ \Omega = 12\ \Omega \angle -90^\circ$$
$$\mathbf{Z}_3 = +j\,8\ \Omega = 8\ \Omega \angle 90^\circ$$

Since $\mathbf{V}_C$ is desired, we will not combine R and X_C into a single block impedance. Note also how Fig. 16.6 clearly reveals that $\mathbf{E}$ is the total voltage across the series combination of $\mathbf{Z}_1$ and $\mathbf{Z}_2$, permitting the use of the voltage divider rule to calculate $\mathbf{V}_C$. In addition, note that all the currents necessary to determine $\mathbf{I}_s$ have been preserved in Fig. 16.6, revealing that there is no need to ever return to the network of Fig. 16.5—everything is defined by Fig. 16.6.

$$\mathbf{V}_C = \frac{\mathbf{Z}_2\mathbf{E}}{\mathbf{Z}_1 + \mathbf{Z}_2} = \frac{(12\ \Omega \angle -90^\circ)(20\ \text{V} \angle 20^\circ)}{5\ \Omega - j\,12\ \Omega} = \frac{240\ \text{V} \angle -70^\circ}{13 \angle -67.38^\circ}$$
$$= \mathbf{18.46\ V \angle -2.62^\circ}$$

b. $$\mathbf{I}_1 = \frac{\mathbf{E}}{\mathbf{Z}_3} = \frac{20\ \text{V} \angle 20^\circ}{8\ \Omega \angle 90^\circ} = 2.5\ \text{A} \angle -70^\circ$$

$$\mathbf{I}_2 = \frac{\mathbf{E}}{\mathbf{Z}_1 + \mathbf{Z}_2} = \frac{20\ \text{V} \angle 20^\circ}{13\ \Omega \angle -67.38^\circ} = 1.54\ \text{A} \angle 87.38^\circ$$

and

$$\mathbf{I}_s = \mathbf{I}_1 + \mathbf{I}_2$$
$$= 2.5\ \text{A} \angle -70^\circ + 1.54\ \text{A} \angle 87.38^\circ$$
$$= (0.86 - j\,2.35) + (0.07 + j\,1.54)$$
$$\mathbf{I}_s = 0.93 - j\,0.81 = \mathbf{1.23\ A \angle -41.05^\circ}$$

EXAMPLE 16.4 For Fig. 16.7:

a. Calculate the current $\mathbf{I}_s$.
b. Find the voltage $\mathbf{V}_{ab}$.

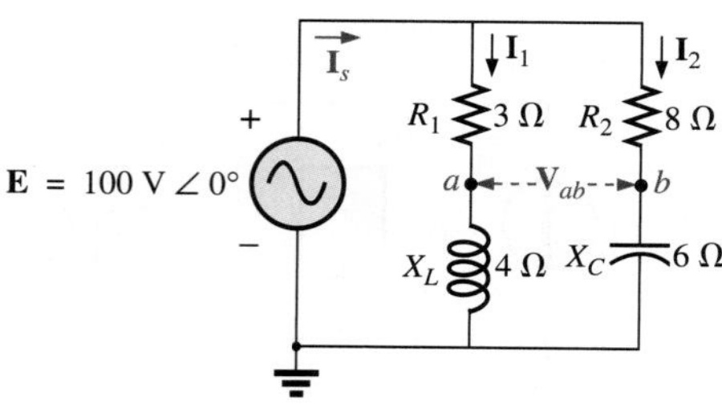

FIG. 16.7
Example 16.4.

Solutions:

a. Redrawing the circuit as in Fig. 16.8, we obtain

$$\mathbf{Z}_1 = R_1 + j\,X_L = 3\ \Omega + j\,4\ \Omega = 5\ \Omega \angle 53.13^\circ$$
$$\mathbf{Z}_2 = R_2 - j\,X_C = 8\ \Omega - j\,6\ \Omega = 10\ \Omega \angle -36.87^\circ$$

In this case the voltage $\mathbf{V}_{ab}$ is lost in the redrawn network, but the currents $\mathbf{I}_1$ and $\mathbf{I}_2$ remain defined for future calculations necessary to determine $\mathbf{V}_{ab}$. Fig. 16.8 clearly reveals that the total impedance can be found using the equation for two parallel impedances:

$$\mathbf{Z}_T = \frac{\mathbf{Z}_1\mathbf{Z}_2}{\mathbf{Z}_1 + \mathbf{Z}_2} = \frac{(5\ \Omega \angle 53.13^\circ)(10\ \Omega \angle -36.87^\circ)}{(3\ \Omega + j\,4\ \Omega) + (8\ \Omega - j\,6\ \Omega)}$$
$$= \frac{50\ \Omega \angle 16.26^\circ}{11 - j\,2} = \frac{50\ \Omega \angle 16.26^\circ}{11.18 \angle -10.30^\circ}$$
$$= \mathbf{4.472\ \Omega \angle 26.56^\circ}$$

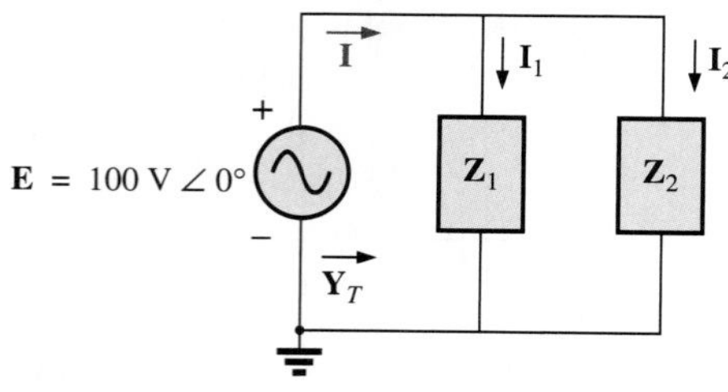

FIG. 16.8
Network in Fig. 16.7 after assigning the block impedances.

and

$$\mathbf{I}_s = \frac{\mathbf{E}}{\mathbf{Z}_T} = \frac{100\ \text{V} \angle 0^\circ}{4.472\ \Omega \angle 26.56^\circ} = \mathbf{22.36\ A \angle -26.56^\circ}$$

b. By Ohm's law,

$$\mathbf{I}_1 = \frac{\mathbf{E}}{\mathbf{Z}_1} = \frac{100 \text{ V} \angle 0^\circ}{5\ \Omega \angle 53.13^\circ} = \mathbf{20\ A \angle -53.13^\circ}$$

$$\mathbf{I}_2 = \frac{\mathbf{E}}{\mathbf{Z}_2} = \frac{100 \text{ V} \angle 0^\circ}{10\ \Omega \angle -36.87^\circ} = \mathbf{10\ A \angle 36.87^\circ}$$

Returning to Fig. 16.7, we have

$$\mathbf{V}_{R_1} = \mathbf{I}_1\mathbf{Z}_{R_1} = (20 \text{ A} \angle -53.13^\circ)(3\ \Omega \angle 0^\circ) = \mathbf{60\ V \angle -53.13^\circ}$$
$$\mathbf{V}_{R_2} = \mathbf{I}_1\mathbf{Z}_{R_2} = (10 \text{ A} \angle +36.87^\circ)(8\ \Omega \angle 0^\circ) = \mathbf{80\ V \angle +36.87^\circ}$$

Instead of using the two steps just shown, we could have determined $\mathbf{V}_{R_1}$ or $\mathbf{V}_{R_2}$ in one step using the voltage divider rule:

$$\mathbf{V}_{R_1} = \frac{(3\ \Omega \angle 0^\circ)(100 \text{ V} \angle 0^\circ)}{3\ \Omega \angle 0^\circ + 4\ \Omega \angle 90^\circ} = \frac{300 \text{ V} \angle 0^\circ}{5 \angle 53.13^\circ} = \mathbf{60\ V \angle -53.13^\circ}$$

To find $\mathbf{V}_{ab}$, Kirchhoff's voltage law must be applied around the loop (Fig. 16.9) consisting of the 3 Ω and 8 Ω resistors. By Kirchhoff's voltage law,

$$\mathbf{V}_{ab} + \mathbf{V}_{R_1} - \mathbf{V}_{R_2} = 0$$

or

$$\begin{aligned}\mathbf{V}_{ab} &= \mathbf{V}_{R_2} - \mathbf{V}_{R_1} \\ &= 80 \text{ V} \angle 36.87^\circ - 60 \text{ V} \angle -53.13^\circ \\ &= (64 + j\,48) - (36 - j\,48) \\ &= 28 + j\,96 \\ \mathbf{V}_{ab} &= \mathbf{100\ V \angle 73.74^\circ}\end{aligned}$$

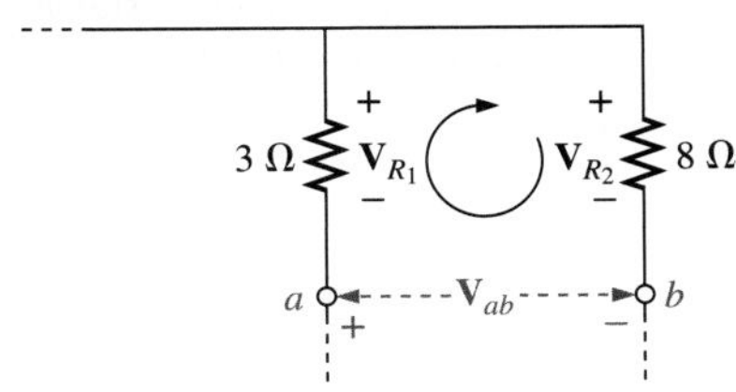

FIG. 16.9
Determining the voltage $\mathbf{V}_{ab}$ for the network in Fig. 16.7.

EXAMPLE 16.5 The network in Fig. 16.10 is frequently encountered in the analysis of transistor networks. The transistor equivalent circuit includes a current source **I** and an output impedance R_o. The resistor R_C is a biasing resistor to establish specific dc conditions, and the resistor R_i represents the loading of the next stage. The coupling capacitor is designed to be an open circuit for dc and to have as low an impedance as possible for the frequencies of interest to ensure that $\mathbf{V}_L$ is a maximum value. The frequency range of the example includes the entire audio (hearing) spectrum from 100 Hz to 20 kHz. The purpose of the example is to demonstrate that, for the full audio range, the effect of the capacitor can be ignored. It performs its function as a dc blocking agent but permits the ac to pass through with little disturbance.

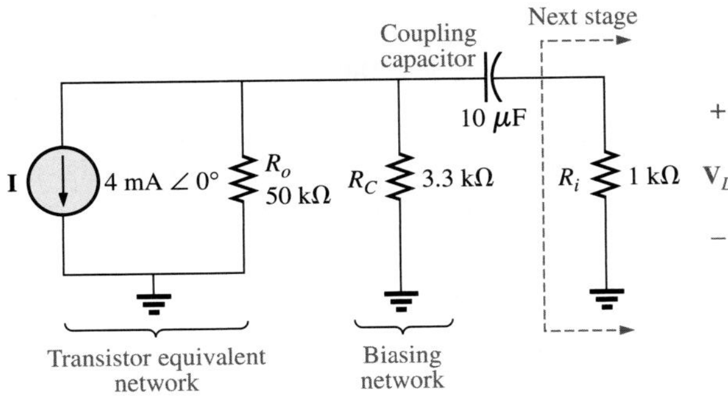

FIG. 16.10
Basic transistor amplifier.

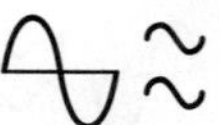

a. Determine $\mathbf{V}_L$ for the network in Fig. 16.10 at a frequency of 100 Hz.
b. Repeat part (a) at a frequency of 20 kHz.
c. Compare the results of parts (a) and (b).

Solutions:

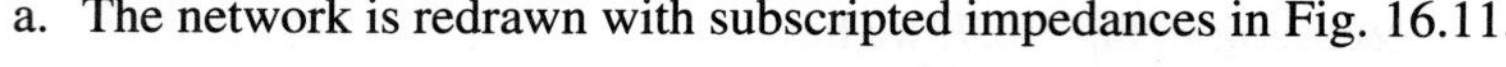

a. The network is redrawn with subscripted impedances in Fig. 16.11.

$$\mathbf{Z}_1 = 50 \text{ k}\Omega \angle 0° \parallel 3.3 \text{ k}\Omega \angle 0° = 3.096 \text{ k}\Omega \angle 0°$$
$$\mathbf{Z}_2 = R_i - j X_C$$

At $f = 100$ Hz: $X_C = \dfrac{1}{2\pi fC} = \dfrac{1}{2\pi(100 \text{ Hz})(10 \ \mu\text{F})} = 159.16 \ \Omega$

and $\qquad \mathbf{Z}_2 = 1 \text{ k}\Omega - j\,159.16 \ \Omega$

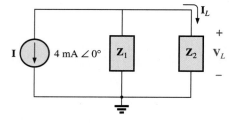

FIG. 16.11
Network in Fig. 16.10 following the assignment of the block impedances.

Current divider rule:

$$\begin{aligned}\mathbf{I}_L &= \frac{-\mathbf{Z}_1\mathbf{I}}{\mathbf{Z}_1 + \mathbf{Z}_2} = \frac{-(3.096 \text{ k}\Omega \angle 0°)(4 \text{ mA} \angle 0°)}{3.096 \text{ k}\Omega + 1 \text{ k}\Omega - j\,159.16 \ \Omega} \\ &= \frac{-12.384 \text{ A} \angle 0°}{4096 - j\,159.16} = \frac{-12.384 \text{ A} \angle 0°}{4099 \angle -2.225°} \\ &= -3.02 \text{ mA} \angle 2.23° = 3.02 \text{ mA} \angle\ 2.23° + 180° = 3.02 \text{ mA} \angle\ 182.23°\end{aligned}$$

and
$$\begin{aligned}\mathbf{V}_L &= \mathbf{I}_L\mathbf{Z}_R \\ &= (3.02 \text{ mA} \angle\ 182.23°)(1 \text{ k}\Omega \angle 0°) \\ &= \mathbf{3.02 \ V \angle\ 182.23°}\end{aligned}$$

b. At $f = 20$ kHz: $X_C = \dfrac{1}{2\pi fC} = \dfrac{1}{2\pi(20 \text{ kHz})(10 \ \mu\text{F})} = 0.796 \ \Omega$

Note the dramatic change in X_C with frequency. Obviously, the higher the frequency, the better the short-circuit approximation for X_C for ac conditions.

$$\mathbf{Z}_2 = 1 \text{ k}\Omega - j\,0.796 \ \Omega$$

Current divider rule:

$$\begin{aligned}\mathbf{I}_L &= \frac{-\mathbf{Z}_1\mathbf{I}}{\mathbf{Z}_1 + \mathbf{Z}_2} = \frac{-(3.096 \text{ k}\Omega \angle 0°)(4 \text{ mA} \angle 0°)}{3.096 \text{ k}\Omega + 1 \text{ k}\Omega - j\,0.796 \ \Omega} \\ &= \frac{-12.384 \text{ A} \angle 0°}{4096 - j\,0.796 \ \Omega} = \frac{-12.384 \text{ A} \angle 0°}{4096 \angle -0.011°} \\ &= -3.02 \text{ mA} \angle 0.01° = 3.02 \text{ mA} \angle 0.01° + 180° = 3.02 \text{ mA} \angle 180.01°\end{aligned}$$

and
$$\begin{aligned}\mathbf{V}_L &= \mathbf{I}_L\mathbf{Z}_R \\ &= (3.02 \text{ mA} \angle 180.01°)(1 \text{ k}\Omega \angle 0°) \\ &= \mathbf{3.02 \ V \angle 180.01°}\end{aligned}$$

c. The results clearly indicate that the capacitor had little effect on the frequencies of interest. In addition, note that most of the supply current reached the load for the typical parameters employed.

EXAMPLE 16.6 For the network in Fig. 16.12:

a. Determine the current **I**.
b. Find the voltage **V**.

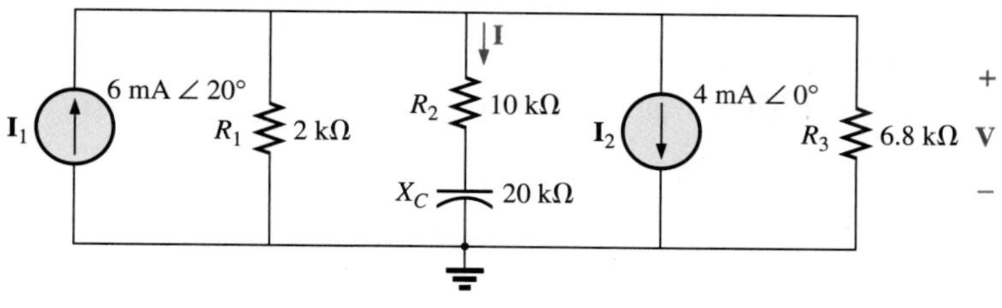

FIG. 16.12
Example 16.6.

Solutions:

a. The rules for parallel current sources are the same for dc and ac networks. That is, the equivalent current source is their sum or difference (as phasors). Therefore,

$$\begin{aligned}\mathbf{I}_T &= 6\text{ mA }\angle 20° - 4\text{ mA }\angle 0°\\ &= 5.638\text{ mA} + j\,2.052\text{ mA} - 4\text{ mA}\\ &= 1.638\text{ mA} + j\,2.052\text{ mA}\\ &= 2.626\text{ mA }\angle 51.402°\end{aligned}$$

Redrawing the network using block impedances results in the network in Fig. 16.13 where

$$\mathbf{Z}_1 = 2\text{ k}\Omega\angle 0° \parallel 6.8\text{ k}\Omega\angle 0° = 1.545\text{ k}\Omega\angle 0°$$

and $\quad \mathbf{Z}_2 = 10\text{ k}\Omega - j\,20\text{ k}\Omega = 22.\ 361\text{ k}\Omega\angle -63.435°$

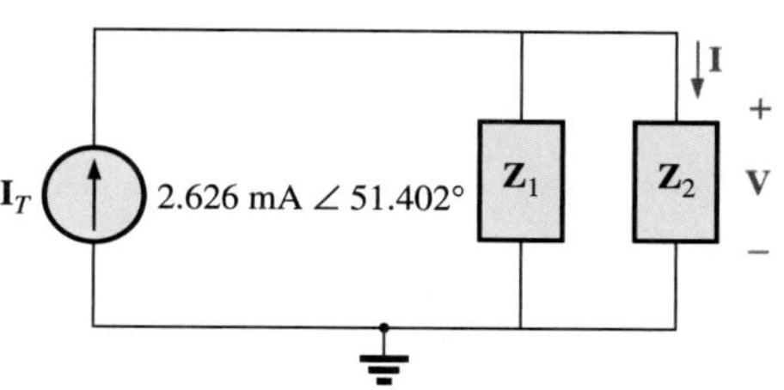

FIG. 16.13
Network in Fig. 16.12 following the assignment of the subscripted impedances.

Note that **I** and **V** are still defined in Fig. 16.13.

Current divider rule:

$$\begin{aligned}\mathbf{I} &= \frac{\mathbf{Z}_1\mathbf{I}_T}{\mathbf{Z}_1 + \mathbf{Z}_2} = \frac{(1.545\text{ k}\Omega\angle 0°)(2.626\text{ mA}\angle 51.402°)}{1.545\text{ k}\Omega + 10\text{ k}\Omega - j\,20\text{ k}\Omega}\\ &= \frac{4.057\text{ A}\angle 51.402°}{11.545\times 10^3 - j\,20\times 10^3} = \frac{4.057\text{ A}\angle 51.402°}{23.093\times 10^3\angle -60.004°}\\ &= \mathbf{0.18\ mA\ \angle 111.41°}\end{aligned}$$

b.
$$\begin{aligned}\mathbf{V} &= \mathbf{IZ}_2\\ &= (0.176\text{ mA}\angle 111.406°)(22.36\text{ k}\Omega\angle -63.435°)\\ &= \mathbf{3.94\ V\ \angle 47.97°}\end{aligned}$$

EXAMPLE 16.7 For the network in Fig. 16.14:

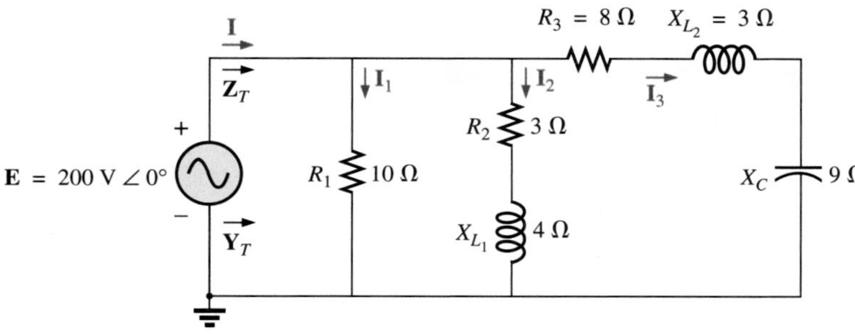

FIG. 16.14
Example 16.7.

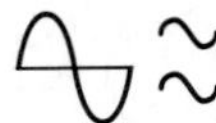

a. Compute **I**.
b. Find $\mathbf{I}_1$, $\mathbf{I}_2$, and $\mathbf{I}_3$.
c. Verify Kirchhoff's current law by showing that

$$\mathbf{I} = \mathbf{I}_1 + \mathbf{I}_2 + \mathbf{I}_3$$

d. Find the total impedance of the circuit.

Solutions:

a. Redrawing the circuit as in Fig. 16.15 reveals a strictly parallel network where

$$\mathbf{Z}_1 = R_1 = 10\ \Omega\ \angle 0°$$
$$\mathbf{Z}_2 = R_2 + j X_{L_1} = 3\ \Omega + j\,4\ \Omega$$
$$\mathbf{Z}_3 = R_3 + j X_{L_2} - j X_C = 8\ \Omega + j\,3\ \Omega - j\,9\ \Omega = 8\ \Omega - j\,6\ \Omega$$

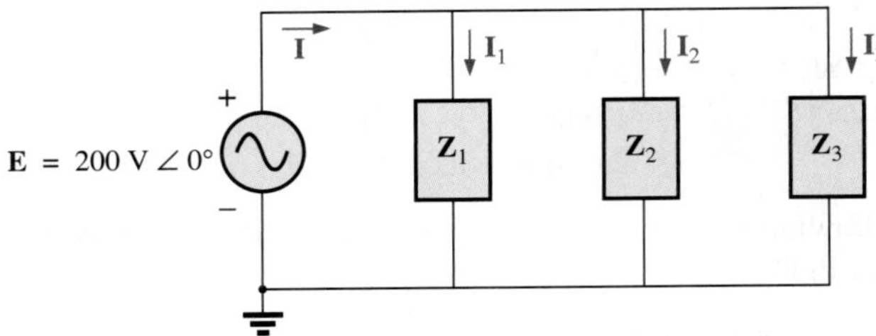

FIG. 16.15
Network in Fig. 16.14 following the assignment of the subscripted impedances.

The total admittance is

$$\begin{aligned}
\mathbf{Y}_T &= \mathbf{Y}_1 + \mathbf{Y}_2 + \mathbf{Y}_3 \\
&= \frac{1}{\mathbf{Z}_1} + \frac{1}{\mathbf{Z}_2} + \frac{1}{\mathbf{Z}_3} = \frac{1}{10\ \Omega} + \frac{1}{3\ \Omega + j\,4\ \Omega} + \frac{1}{8\ \Omega - j\,6\ \Omega} \\
&= 0.1\ \text{S} + \frac{1}{5\ \Omega\ \angle 53.13°} + \frac{1}{10\ \Omega\ \angle -36.87°} \\
&= 0.1\ \text{S} + 0.2\ \text{S}\ \angle -53.13° + 0.1\ \text{S}\ \angle 36.87° \\
&= 0.1\ \text{S} + 0.12\ \text{S} - j\,0.16\ \text{S} + 0.08\ \text{S} + j\,0.06\ \text{S} \\
&= 0.3\ \text{S} - j\,0.1\ \text{S} = 0.316\ \text{S}\ \angle -18.435°
\end{aligned}$$

Calculator The above mathematical exercise presents an excellent opportunity to demonstrate the power of some of today's calculators. Using the TI-89, the above operation is as shown in Fig. 16.16.

FIG. 16.16
Finding the total admittance for the network in Fig. 16.14 using the TI-89 calculator.

Be sure to use the negative sign for the complex number from the subtraction option and not the sign selection (−). The sign selection is used for negative angles in the polar form.

Converting to polar form requires the sequence shown in Fig. 16.17.

(• 3 − • 1 *i*) ▶ Polar ENTER ENTER **316.2E–3 ∠ –18.43E0**

FIG. 16.17
Converting the rectangular form in Fig. 16.16 to polar form.

Converting to polar form:
The current **I**:

$$\mathbf{I} = \mathbf{E}\mathbf{Y}_T = (200\text{ V}\angle 0°)(0.326\text{ S}\angle -18.435°)$$
$$= \mathbf{63.2\text{ A}\angle -18.44°}$$

b. Since the voltage is the same across parallel branches,

$$\mathbf{I}_1 = \frac{\mathbf{E}}{\mathbf{Z}_1} = \frac{200\text{ V}\angle 0°}{10\;\Omega\angle 0°} = \mathbf{20\text{ A}\angle 0°}$$

$$\mathbf{I}_2 = \frac{\mathbf{E}}{\mathbf{Z}_2} = \frac{200\text{ V}\angle 0°}{5\;\Omega\angle 53.13°} = \mathbf{40\text{ A}\angle -53.13°}$$

$$\mathbf{I}_3 = \frac{\mathbf{E}}{\mathbf{Z}_3} = \frac{200\text{ V}\angle 0°}{10\;\Omega\angle -36.87°} = \mathbf{20\text{ A}\angle +36.87°}$$

c.
$$\mathbf{I} = \mathbf{I}_1 + \mathbf{I}_2 + \mathbf{I}_3$$
$$60 - j\,20 = 20\angle 0° + 40\angle -53.13° + 20\angle +36.87°$$
$$= (20 + j\,0) + (24 - j\,32) + (16 + j\,12)$$
$$60 - j\,20 = 60 - j\,20 \qquad \text{(checks)}$$

d.
$$\mathbf{Z}_T = \frac{1}{\mathbf{Y}_T} = \frac{1}{0.316\text{ S}\angle -18.435°}$$
$$= \mathbf{3.17\;\Omega\angle 18.44°}$$

EXAMPLE 16.8 For the network in Fig. 16.18:

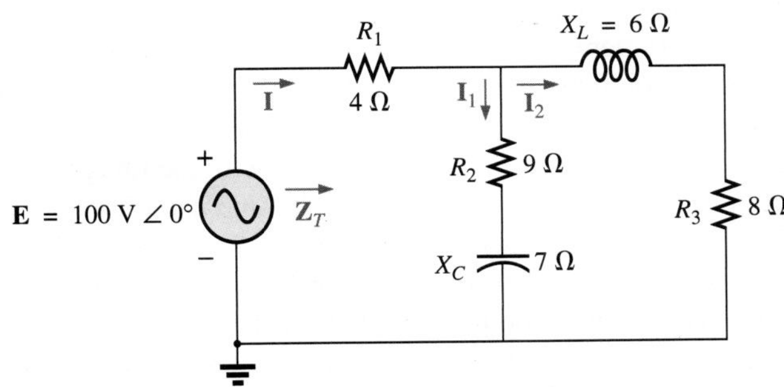

FIG. 16.18
Example 16.8.

a. Calculate the total impedance $\mathbf{Z}_T$.
b. Compute **I**.
c. Find the total power factor.
d. Calculate $\mathbf{I}_1$ and $\mathbf{I}_2$.
e. Find the average power delivered to the circuit.

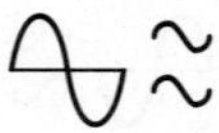

Solutions:

a. Redrawing the circuit as in Fig. 16.19, we have

$$\mathbf{Z}_1 = R_1 = 4\ \Omega \angle 0^\circ$$
$$\mathbf{Z}_2 = R_2 - j X_C = 9\ \Omega - j\,7\ \Omega = 11.40\ \Omega \angle -37.87^\circ$$
$$\mathbf{Z}_3 = R_3 + j X_L = 8\ \Omega + j\,6\ \Omega = 10\ \Omega \angle +36.87^\circ$$

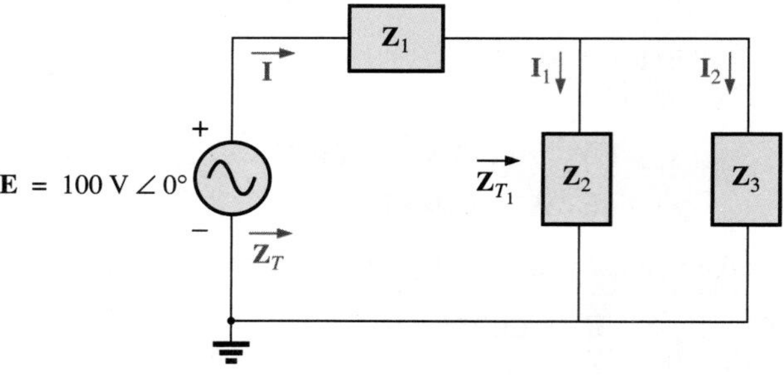

FIG. 16.19
Network in Fig. 16.18 following the assignment of the subscripted impedances.

Notice that all the desired quantities were conserved in the redrawn network. The total impedance:

$$\begin{aligned}
\mathbf{Z}_T &= \mathbf{Z}_1 + \mathbf{Z}_{T_1} \\
&= \mathbf{Z}_1 + \frac{\mathbf{Z}_2\mathbf{Z}_3}{\mathbf{Z}_2 + \mathbf{Z}_3} \\
&= 4\ \Omega + \frac{(11.4\ \Omega \angle -37.87^\circ)(10\ \Omega \angle 36.87^\circ)}{(9\ \Omega - j\,7\ \Omega) + (8\ \Omega + j\,6\ \Omega)} \\
&= 4\ \Omega + \frac{114\ \Omega \angle -1.00^\circ}{17.03 \angle -3.37^\circ} = 4\ \Omega + 6.69\ \Omega \angle 2.37^\circ \\
&= 4\ \Omega + 6.68\ \Omega + j\,0.28\ \Omega = 10.68\ \Omega + j\,0.28\ \Omega \\
\mathbf{Z}_T &= \mathbf{10.68\ \Omega \angle 1.5^\circ}
\end{aligned}$$

Mathcad Solution: The complex algebra just presented in detail provides an excellent opportunity to practice our Mathcad skills with complex numbers. Remember that the j must follow the numerical value of the imaginary part and **is not** multiplied by the numerical value. Simply type in the numerical value and then j. Also recall that unless you make a global change in the format, an i will appear with the imaginary part of the solution. As shown in Fig. 16.20, first define each impedance with **Shift:.** Then enter each impedance in sequence on the same line or succeeding lines. Next, define the equation for the total impedance using the brackets to ensure that the bottom summation is carried out before the division and also to provide the same format to the equation as appearing above. Then enter **ZT,** select the equal sign key, and the rectangular form for the total impedance appears as shown.

The polar form can be obtained by first going to the **Calculator** toolbar to obtain the magnitude operation and inserting **ZT** as shown in Fig. 16.20. Then selecting the equal sign results in the magnitude of 10.693 Ω. The angle is obtained by first going to the **Greek** toolbar and picking up theta, entering **T,** and defining the variable. The π comes from the **Calculator** toolbar, and the **arg**( ) from **Insert-*f(x)*-Function Name-arg.** Finally, enter the variable again and select the equal sign to obtain an angle of 1.478°. The computer solution of 10.693 Ω ∠1.478° is an excellent verification of the theoretical solution of 10.68 Ω ∠1.5°.

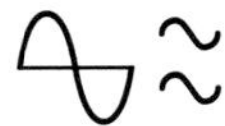

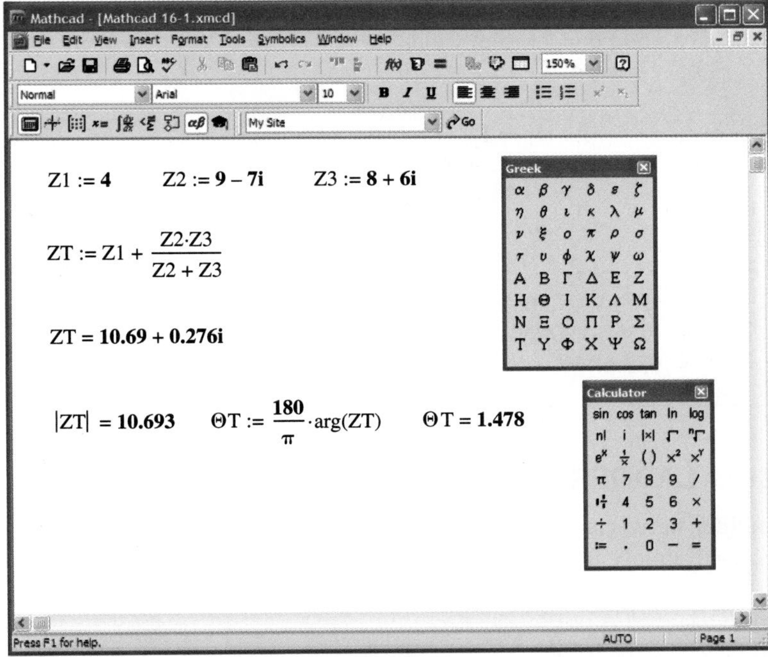

FIG. 16.20
Using Mathcad to determine the total impedance for the network in Fig. 16.18.

Calculator Another opportunity to demonstrate the versatility of the calculator! For the above operation, however, you must be aware of the priority of the mathematical operations, as demonstrated in the calculator display in Fig. 16.21. In most cases, the operations are performed in the same order they would be if you wrote them longhand.

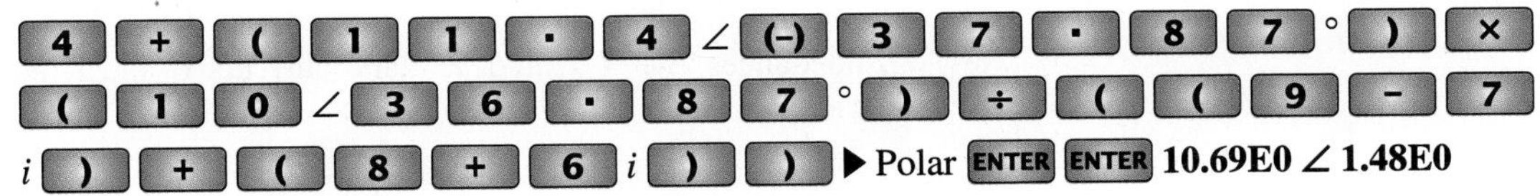

FIG. 16.21
Finding the total impedance for the network in Fig. 16.18 using the TI-89 calculator.

b. $\mathbf{I} = \dfrac{\mathbf{E}}{\mathbf{Z}_T} = \dfrac{100\text{ V}\ \angle 0^\circ}{10.684\ \Omega\ \angle 1.5^\circ} = \mathbf{9.36\ A\ \angle -1.5^\circ}$

c. $F_p = \cos\theta_T = \dfrac{R}{Z_T} = \dfrac{10.68\ \Omega}{10.684\ \Omega} \cong \mathbf{1}$

(essentially resistive, which is interesting, considering the complexity of the network)

d. Current divider rule:

$$\mathbf{I}_2 = \frac{\mathbf{Z}_2\mathbf{I}}{\mathbf{Z}_2 + \mathbf{Z}_3} = \frac{(11.40\ \Omega\ \angle -37.87^\circ)(9.36\text{ A}\ \angle -1.5^\circ)}{(9\ \Omega - j\,7\ \Omega) + (8\ \Omega + j\,6\ \Omega)}$$

$$= \frac{106.7\text{ A}\ \angle -39.37^\circ}{17 - j\,1} = \frac{106.7\text{ A}\ \angle -39.37^\circ}{17.03\ \angle -3.37^\circ}$$

$$\mathbf{I}_2 = \mathbf{6.27\ A\ \angle -36^\circ}$$

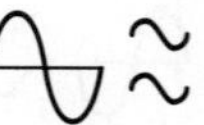

Applying Kirchhoff's current law (rather than another application of the current divider rule) yields

$$\mathbf{I}_1 = \mathbf{I} - \mathbf{I}_2$$

or

$$\begin{aligned}\mathbf{I} &= \mathbf{I}_1 - \mathbf{I}_2 \\ &= (9.36 \text{ A} \angle -1.5°) - (6.27 \text{ A} \angle -36°) \\ &= (9.36 \text{ A} - j\,0.25 \text{ A}) - (5.07 \text{ A} - j\,3.69 \text{ A}) \\ \mathbf{I}_1 &= 4.29 \text{ A} + j\,3.44 \text{ A} = \mathbf{5.5\ A \angle 38.72°}\end{aligned}$$

e.
$$\begin{aligned}P_T &= EI \cos \theta_T \\ &= (100 \text{ V})(9.36 \text{ A}) \cos 1.5° \\ &= (936)(0.99966) \\ P_T &= \mathbf{935.68\ W}\end{aligned}$$

16.3 LADDER NETWORKS

Ladder networks were discussed in some detail in Chapter 7. This section will simply apply the first method described in Section 7.6 to the general sinusoidal ac ladder network in Fig. 16.22. The current $\mathbf{I}_6$ is desired.

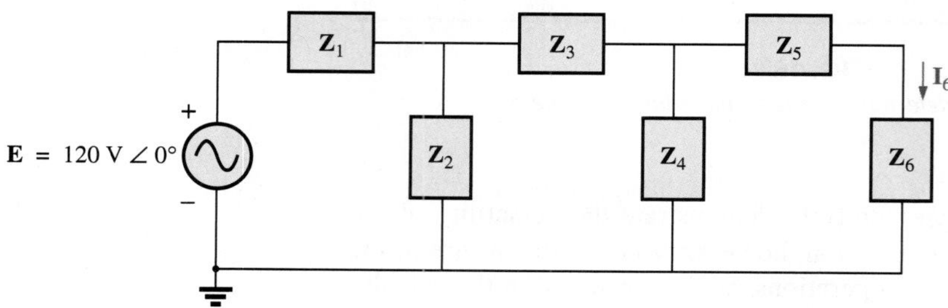

FIG. 16.22
Ladder network.

Impedances $\mathbf{Z}_T$, $\mathbf{Z}'_T$, and $\mathbf{Z}''_T$ and currents $\mathbf{I}_1$ and $\mathbf{I}_3$ are defined in Fig. 16.23.

$$\mathbf{Z}''_T = \mathbf{Z}_5 + \mathbf{Z}_6$$

and
$$\mathbf{Z}'_T = \mathbf{Z}_3 + \mathbf{Z}_4 \parallel \mathbf{Z}''_T$$

with
$$\mathbf{Z}_T = \mathbf{Z}_1 + \mathbf{Z}_2 \parallel \mathbf{Z}'_T$$

Then
$$\mathbf{I} = \frac{\mathbf{E}}{\mathbf{Z}_T}$$

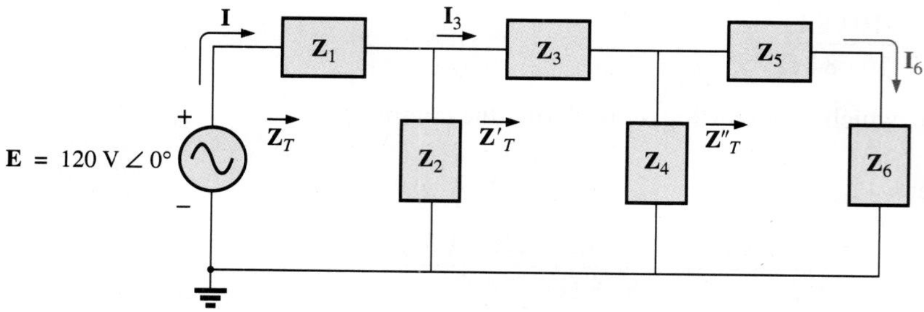

FIG. 16.23
Defining an approach to the analysis of ladder networks.

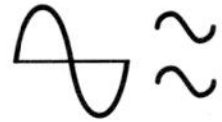

and $$\mathbf{I}_3 = \frac{\mathbf{Z}_2\mathbf{I}}{\mathbf{Z}_2 + \mathbf{Z}'_T}$$

with $$\mathbf{I}_6 = \frac{\mathbf{Z}_4\mathbf{I}_3}{\mathbf{Z}_4 + \mathbf{Z}''_T}$$

16.4 GROUNDING

Although usually treated too lightly in most introductory electrical or electronics texts, the impact of the ground connection and how it can provide a measure of safety to a design are very important topics. *Ground potential is 0 V at every point in a network that has a ground symbol.* Since all points are at the same potential, they can all be connected together, but for purposes of clarity most are left isolated on a large schematic. On a schematic, the voltage levels provided are always with respect to ground. A system can therefore be checked quite rapidly by simply connecting the black lead of the voltmeter to the ground connection and placing the red lead at the various points where the typical operating voltage is provided. A close match normally implies that that portion of the system is operating properly.

There are various types of grounds whose use depends on the application. An *earth ground* is one that is connected directly to the earth by a low-impedance connection. Under typical environmental conditions, *local* ground potentials are fairly uniform and can be defined as equal to zero volts. This local uniformity is due to sufficient conductive agents in the soil such as water and electrolytes to ensure that any difference in voltage on the surface is equalized by a flow of charge between the two points. However, between long distances on the earth's surface there can be significant changes in potential level. Every home has an earth ground, usually established by a long conductive rod driven into the ground and connected to the power panel. The electrical code requires a direct connection from earth ground to the cold-water pipes of a home for safety reasons. A "hot" wire touching a cold-water pipe draws sufficient current because of the low-impedance ground connection to throw the breaker. Otherwise, people in the bathroom could pick up the voltage when they touched the cold-water faucet, thereby risking bodily harm. Because water is a conductive agent, any area of the home with water, such as a bathroom or the kitchen, is of particular concern. Most electrical systems are connected to earth ground primarily for safety reasons. All the power lines in a laboratory, at industrial locations, or in the home are connected to earth ground.

A second type is referred to as a *chassis ground,* which may be *floating* or connected directly to an earth ground. A chassis ground simply stipulates that the chassis has a reference potential for all points of the network. If the chassis is not connected to earth potential (0 V), it is said to be *floating* and can have any other reference voltage for the other voltages to be compared to. For instance, if the chassis is sitting at 120 V, all measured voltages of the network will be referenced to this level. A reading of 32 V between a point in the network and the chassis ground will therefore actually be at 152 V with respect to earth potential. Most high-voltage systems are not left floating, however, because of loss of the safety factor. For instance, if someone should touch the chassis and be standing on a suitable ground, the full 120 V would fall across that individual.

Grounding can be particularly important when working with numerous pieces of measuring equipment in the laboratory. For instance, the supply and oscilloscope in Fig. 16.24(a) are each connected directly to

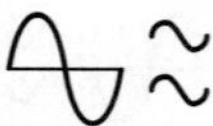

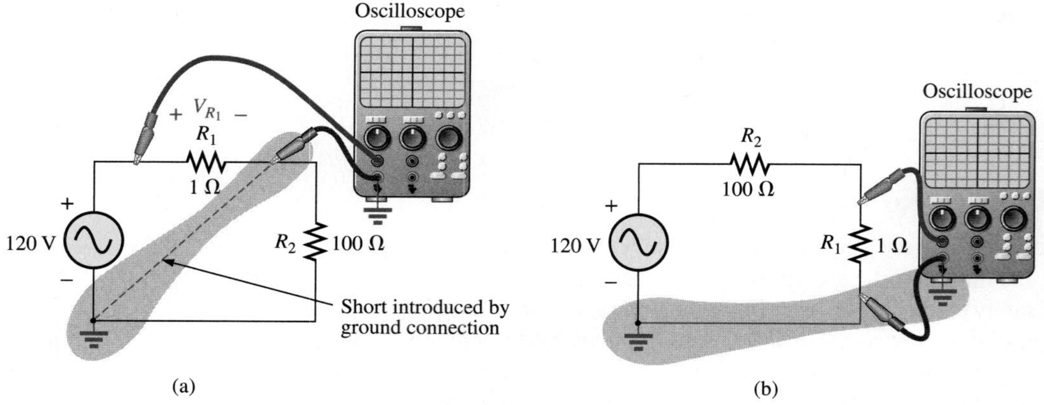

FIG. 16.24
Demonstrating the effect of the oscilloscope ground on the measurement of the voltage across resistor R_1.

an earth ground through the negative terminal of each. If the oscilloscope is connected as shown in Fig. 16.24(a) to measure the voltage V_{R_1}, a dangerous situation will develop. The grounds of each piece of equipment are connected together through the earth ground, and they effectively short out the resistor. Since the resistor is the primary current-controlling element in the network, the current will rise to a very high level and possibly damage the instruments or cause dangerous side effects. In this case, the supply or scope should be used in the floating mode, or the resistors interchanged as shown in Fig. 16.24(b). In Fig. 16.24(b), the grounds have a common point and do not affect the structure of the network.

The National Electrical Code requires that the "hot" (or *feeder*) line that carries current to a load be *black,* and the line (called the *neutral*) that carries the current back to the supply be *white.* Three-wire conductors have a ground wire that must be *green* or *bare,* which ensures a common ground but which is not designed to carry current. The components of a three-prong extension cord and wall outlet are shown in Fig. 16.25. Note that on both fixtures, the connection to the hot lead is smaller than the return leg and that the ground connection is partially circular.

The complete wiring diagram for a household outlet is shown in Fig. 16.26. Note that the current through the ground wire is zero and that both

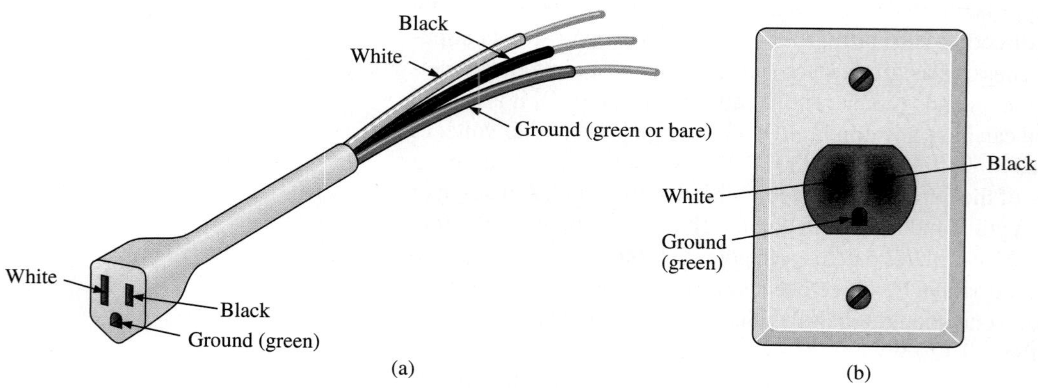

FIG. 16.25
Three-wire conductors: (a) extension cord; (b) home outlet.

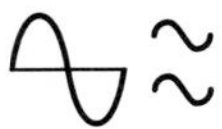

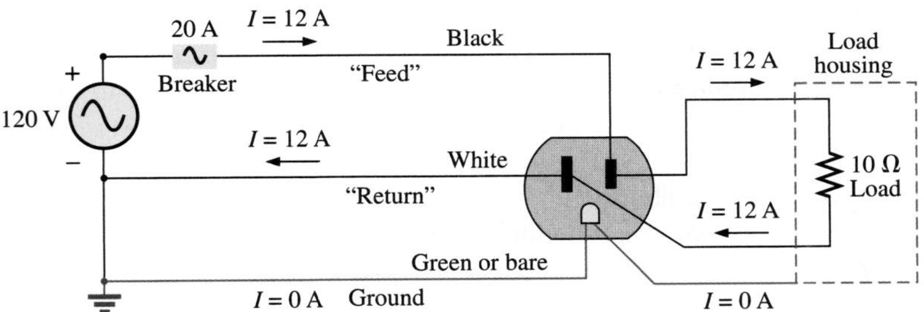

FIG. 16.26
Complete wiring diagram for a household outlet with a 10 Ω load.

the return wire and the ground wire are connected to an earth ground. The full current to the loads flows through the feeder and return lines.

The importance of the ground wire in a three-wire system can be demonstrated by the toaster in Fig. 16.27, rated 1200 W at 120 V. From the power equation $P = EI$, the current drawn under normal operating conditions is $I = P/E = 1200 \text{ W}/120 \text{ V} = 10 \text{ A}$. If a two-wire line were used as shown in Fig. 16.27(a), the 20 A breaker would be quite comfortable with the 10 A current, and the system would perform normally. However, if abuse to the feeder caused it to become frayed and to touch the metal housing of the toaster, the situation depicted in Fig. 16.27(b) would result. The housing would become "hot," yet the breaker would not "pop" because the current would still be the rated 10 A. A dangerous condition would exist because anyone touching the toaster would feel the

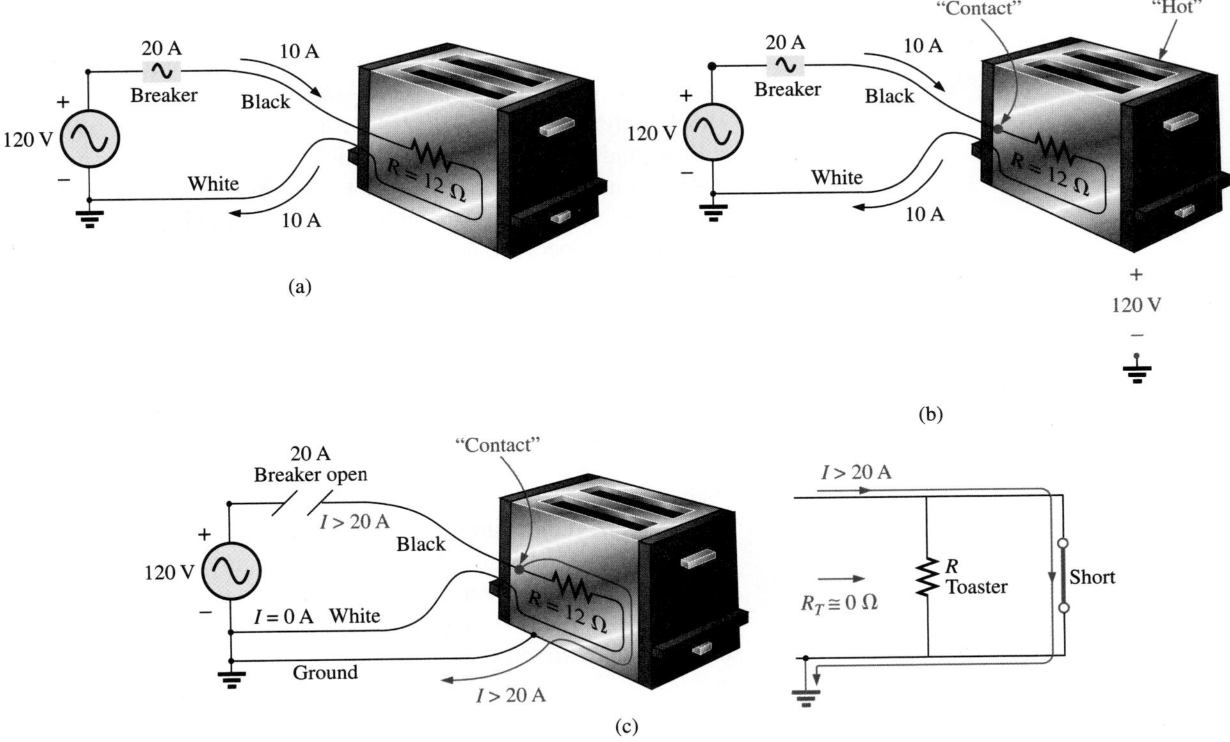

FIG. 16.27
Demonstrating the importance of a properly grounded appliance: (a) ungrounded; (b) ungrounded and undesirable contact; (c) grounded appliance with undesirable contact.

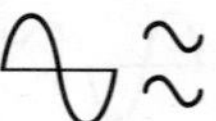

full 120 V to ground. If the ground wire were attached to the chassis as shown in Fig. 16.27(c), a low-resistance path would be created between the short-circuit point and ground, and the current would jump to very high levels. The breaker would "pop," and the user would be warned that a problem exists.

Although the above discussion does not cover all possible areas of concern with proper grounding or introduce all the nuances associated with the effect of grounds on a system's performance, you should understand the importance of its impact.

16.5 APPLICATIONS

The vast majority of the applications appearing throughout the text have been of the series-parallel variety. The following are series-parallel combinations of elements and systems used to perform important everyday tasks. The ground fault circuit interrupter (GFCI) outlet employs series protective switches and sensing coils and a parallel control system, while the ideal equivalent circuit for the coax cable employs a series-parallel combination of inductors and capacitors.

GFCI (Ground Fault Circuit Interrupter)

The National Electric Code, the "bible" for all electrical contractors, now requires that ground fault circuit interrupter (GFCI) outlets be used in any area where water and dampness could result in serious injury, such as in bathrooms, pools, marinas, and so on. The outlet looks like any other except that it has a reset button and a test button in the center of the unit as shown in Fig. 16.28(a). The primary difference between it and an ordinary outlet is that it will shut the power off much more quickly than the breaker all the way down in the basement could. You may still feel a shock with a GFCI outlet, but the current cuts off so quickly (in a few milliseconds) that a person in normal health should not receive a serious electrical injury. Whenever in doubt about its use, remember that its cost (relatively inexpensive) is well worth the increased measure of safety.

The basic operation is best described by the network in Fig. 16.28(b). The protection circuit separates the power source from the outlet itself. Note in Fig. 16.28(b) the importance of grounding the protection circuit to the central ground of the establishment (a water pipe, ground bar, and so on, connected to the main panel). In general, the outlet will be grounded to the same connection. Basically, the network shown in Fig. 16.28(b) senses both the current entering (I_i) and the current leaving (I_o) and provides a direct connection to the outlet when they are equal. If a fault should develop such as that caused by someone touching the hot leg while standing on a wet floor, the return current will be less than the feed current (just a few milliamperes is enough). The protection circuitry senses this difference, establishes an open circuit in the line, and cuts off the power to the outlet.

Fig. 16.29(a) shows the feed and return lines passing through the sensing coils. The two sensing coils are separately connected to the printed circuit board. There are two pulse control switches in the line and a return to establish an open circuit under errant conditions. The two contacts in Fig. 16.29(a) are the contacts that provide conduction to the outlet. When a fault develops, another set of similar contacts in the housing slides away, providing the desired open-circuit condition. The separation is created by the solenoid appearing in Fig. 16.29(b). When the solenoid is energized due to a fault condition, it pulls the plunger toward the solenoid, compressing the

(a)

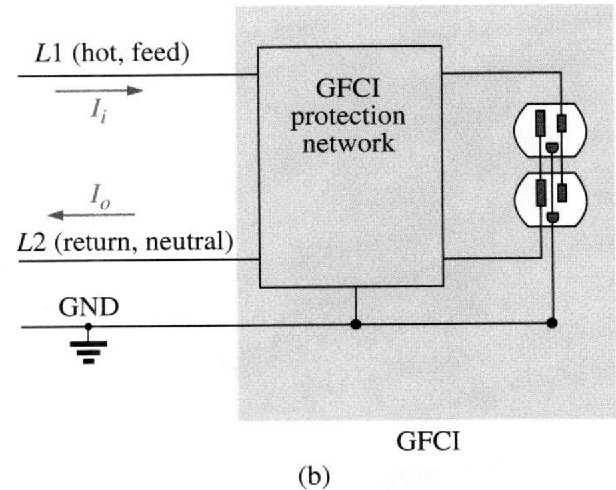

(b)

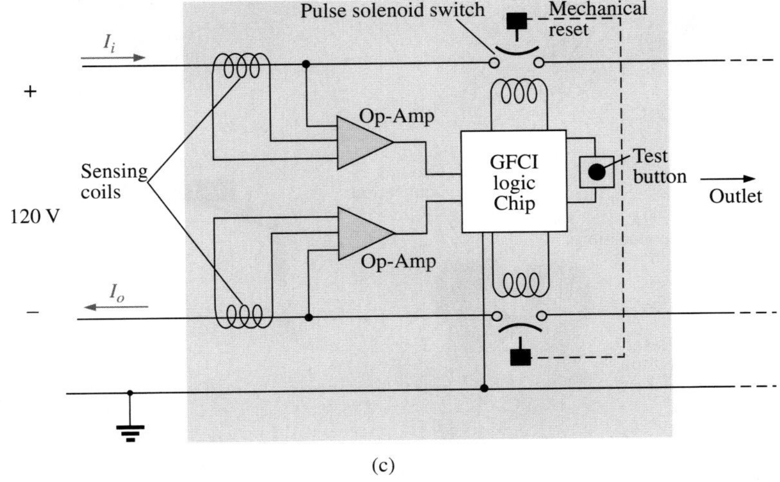

(c)

FIG. 16.28

GFCI outlet: (a) wall-mounted appearance; (b) basic operation; (c) schematic.

spring. At the same time, the slots in the lower plastic piece (connected directly to the plunger) shift down, causing a disconnect by moving the structure inserted in the slots. The test button is connected to the brass bar across the unit in Fig. 16.29(c) below the reset button. When pressed, it places a large resistor between the line and ground to "unbalance" the line and cause a fault condition. When the button is released, the resistor is separated from the line, and the unbalance condition is removed. The resistor is actually connected directly to one end of the bar and moves down with pressure on the bar as shown in Fig. 16.29(d). Note in Fig. 16.29(c) how the metal ground connection passes right through the entire unit and how it is connected to the ground terminal of an applied plug. Also note how it is separated from the rest of the network with the plastic housing. Although this unit appears simple on the outside and is relatively small in size, it is beautifully designed and contains a great deal of technology and innovation.

Before leaving the subject, note the logic chip in the center of Fig. 16.29(a) and the various sizes of capacitors and resistors. Note also the four diodes in the upper left region of the circuit board used as a bridge rectifier for the ac-to-dc conversion process. The transistor is the black element with the half-circle appearance. It is part of the driver circuit for

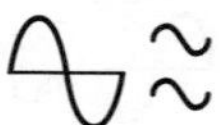

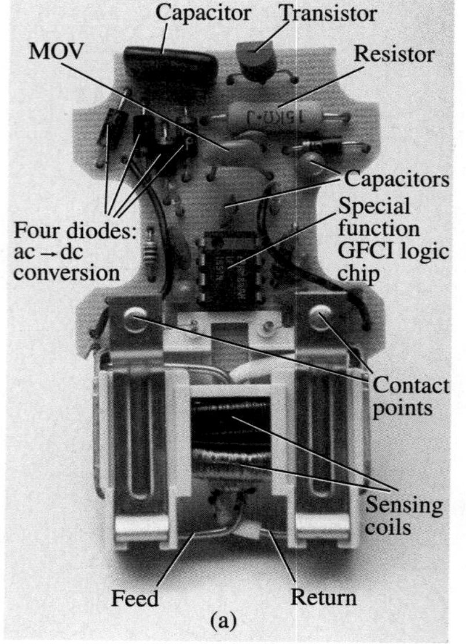

(a)

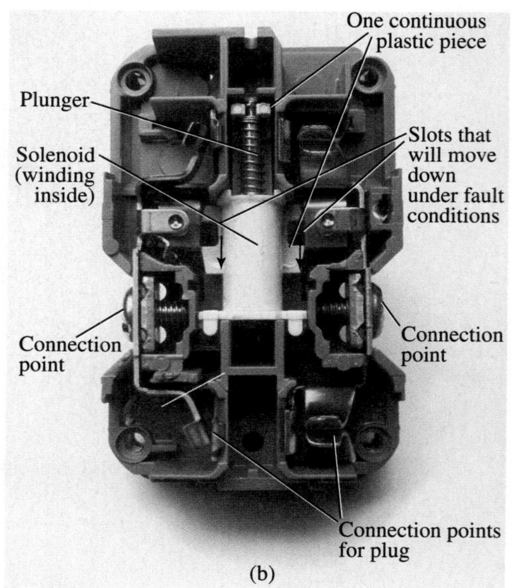

(b)

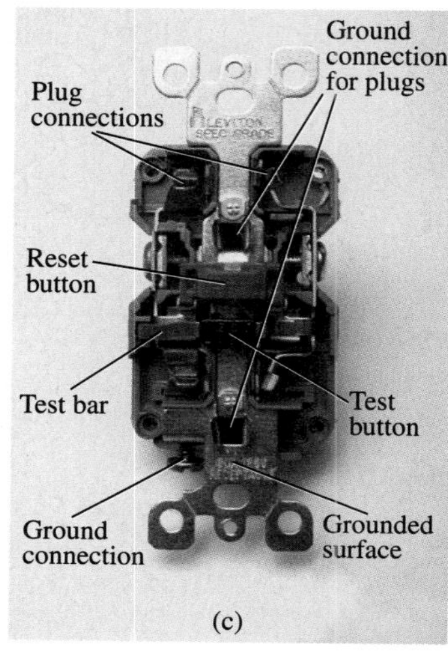

(c)

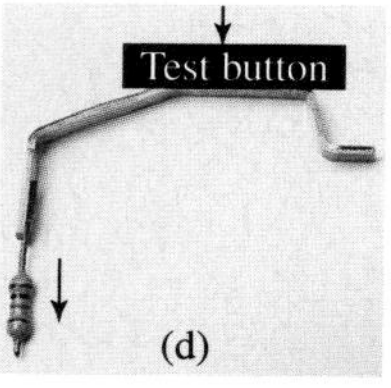

(d)

FIG. 16.29

GFCI construction: (a) sensing coils; (b) solenoid control (bottom view); (c) grounding (top view); (d) test bar.

the controlling solenoid. Because of the size of the unit, there wasn't a lot of room to provide the power to quickly open the circuit. The result is the use of a pulse circuit to control the motion of the controlling solenoid. In other words, the solenoid is pulsed for a short period of time to cause the required release. If the design used a system that would hold the circuit open on a continuing basis, the power requirement would be greater and the size of the coil larger. A small coil can handle the required power pulse for a short period of time without any long-term damage.

As mentioned earlier, if unsure, install a GFCI. It provides a measure of safety—at a very reasonable cost—that should not be ignored.

16.6 COMPUTER ANALYSIS

PSpice

ac Bridge Network We will use Example 16.4 to demonstrate the power of the **VPRINT** option in the **SPECIAL** library. It permits a di-

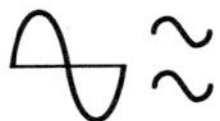

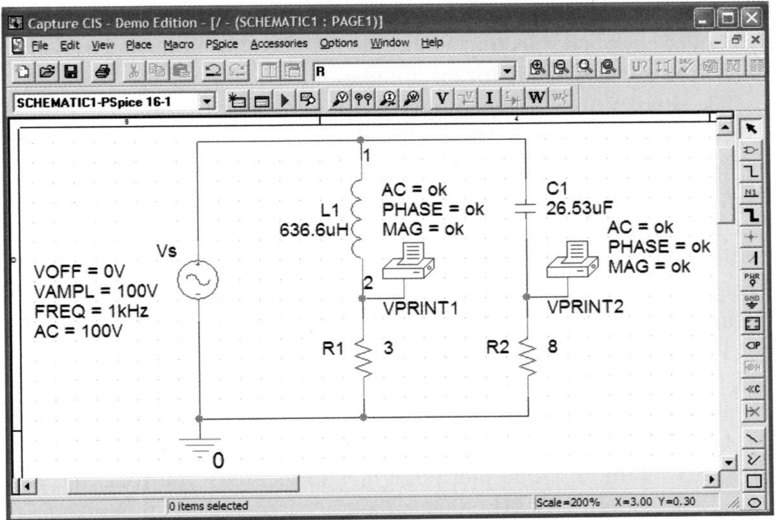

FIG. 16.30

*Determining the voltage across R_1 and R_2 using the **VPRINT** option of a PSpice analysis.*

rect determination of the magnitude and angle of any voltage in an ac network. Similarly, the **IPRINT** option does the same for ac currents. In Example 16.4, the ac voltages across R_1 and R_2 were first determined, and then Kirchhoff's voltage law was applied to determine the voltage between the two known points. Since PSpice is designed primarily to determine the voltage at a point with respect to ground, the network in Fig. 16.7 is entered as shown in Fig. 16.30 to permit a direct calculation of the voltages across R_1 and R_2.

The source and network elements are entered using a procedure that has been demonstrated several times in previous chapters, although for the **AC Sweep** analysis to be performed in this example, the source must carry an **AC** level also. Fortunately, it is the same as **VAMPL** as shown in Fig. 16.30. It is introduced into the source description by double-clicking on the source symbol to obtain the **Property Editor** dialog box. Select the **AC** column and enter 100 V in the box below. Then select **Display** and choose **Name and Value.** Click **OK** followed by **Apply,** and you can exit the dialog box. The result is **AC** = 100 V added to the source description on the diagram and in the system. Using the reactance values in Fig. 16.7, the values for L and C were determined using a frequency of 1 kHz. The voltage across R_1 and R_2 can be determined using the **Trace** command in the same manner as described in the previous chapter or by using the **VPRINT** option. Both approaches are discussed in this section because they have application to any ac network.

The **VPRINT** option is under the **SPECIAL** library in the **Place Part** dialog box. Once selected, the printer symbol appears on the screen next to the cursor, and it can be placed near the point of interest. Once the printer symbol is in place, double-click on it to display the **Property Editor** dialog box. Scrolling from left to right, type the word **ok** under **AC, MAG,** and **PHASE.** When each is active, select the **Display** key and choose the option **Name and Value** followed by **OK.** When all the entries have been made, choose **Apply** and exit the dialog box. The result appears in Fig. 16.30 for the two applications of the **VPRINT** option. If you prefer, **VPRINT1** and **VPRINT2** can be added to distinguish between the two when you review the output data. To do this, return to the

Property Editor dialog box for each by double-clicking on the printer symbol of each and selecting **Value** and then **Display** followed by **Value Only.** You are now ready for the simulation.

The simulation is initiated by selecting the **New Simulation Profile** icon and entering **PSpice 16-1** as the **Name.** Then select **Create** to bring up the **Simulation Settings** dialog box. This time, you want to analyze the network at 1 kHz but are not interested in plots against time. Thus, select the **AC Sweep/Noise** option under **Analysis type** in the **Analysis** section. An **AC Sweep Type** region then appears asking for the **Start Frequency.** Since you are interested in the response at only one frequency, the **Start** and **End Frequency** will be the same: 1 kHz. Since you need only one point of analysis, the **Points/Decade** will be 1. Click **OK,** and select the **Run PSpice** icon. The **SCHEMATIC1** screen appears, and the voltage across R_1 can be determined by selecting **Trace** followed by **Add Trace** and then **V(R1:1).** The result is the bottom display in Fig. 16.31 with only one plot point at 1 kHz. Since you fixed the frequency of interest at 1 kHz, this is the only frequency with a response. The magnitude of the voltage across R_1 is 60 V to match the longhand solution of Example 16.4. The phase angle associated with the voltage can be determined by the sequence **Plot-Add Plot to Window-Trace-Add Trace-P( )** from the **Functions or Macros** list and then **V(R1:1)** to obtain **P(V(R1:1))** in the **Trace Expression** box. Click **OK,** and the resulting plot shows that the phase angle is near just less than $-50°$ which is certainly a close match with the $-53.13°$ obtained in Example 16.4.

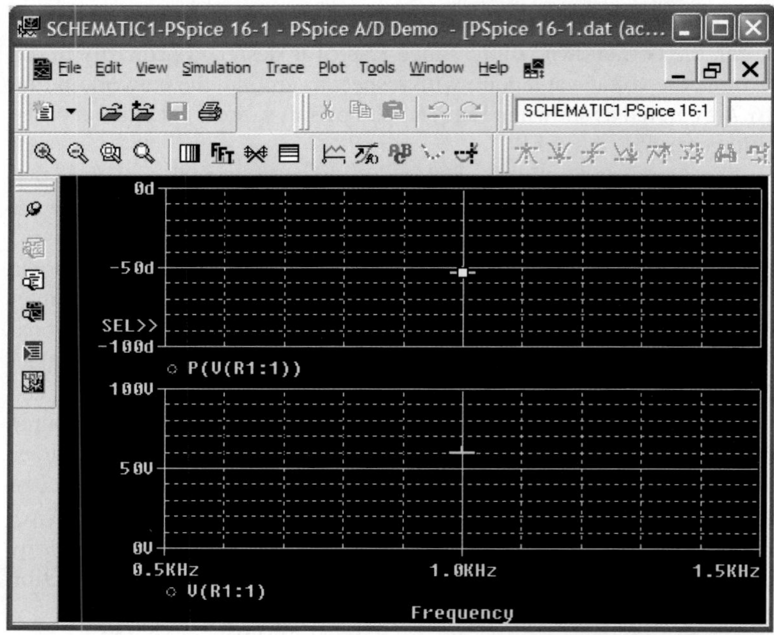

FIG. 16.31

The resulting magnitude and phase angle for the voltage V_{R_1} in Fig. 16.30.

The **VPRINT** option just introduced offers another method for analyzing voltage in a network. When the **SCHEMATIC1** window appears after the simulation, exit the window using the **X,** and select **PSpice** on the top menu bar of the resulting screen. Select **View Output File** from the list that appears. You will see a long list of data about the construction of the network and the results obtained from the simulation. In Fig. 16.32, the por-

```
** Profile: "SCHEMATIC1-PSpice 16-2"  [ C:\ICA11\PSpice\PSpice 16-1-
PSpiceFiles\SCHEMATIC1\PSpice 16-2.sim ]

****    AC ANALYSIS                    TEMPERATURE =  27.000 DEG C
******************************************************************************
 FREQ         VM(N00879)  VP(N00879)
 1.000E+03   6.000E+01    -5.313E+01

****    AC ANALYSIS                    TEMPERATURE =  27.000 DEG C
******************************************************************************
 FREQ         VM(N00875)  VP(N00875)
 1.000E+03   8.000E+01    3.687E+01
```

FIG. 16.32
*The **VPRINT1** (V_{R_1}) and **VPRINT2** (V_{R_2}) response for the network in Fig. 16.30.*

tion of the output file listing the resulting magnitude and phase angle for the voltages defined by **VPRINT1** and **VPRINT2** is provided. Note that the voltage across R_1 defined by **VPRINT1** is 60 V at an angle of 53.13°. The voltage across R_2 as defined by **VPRINT2** is 80 V at an angle of 36.87°. Both are exact matches of the solutions in Example 16.4. In the future, therefore, if the **VPRINT** option is used, the results will appear in the output file.

Now you can determine the voltage across the two branches from point *a* to point *b*. Return to **SCHEMATIC1,** and select **Trace** followed by **Add Trace** to obtain the list of **Simulation Output Variables.** Then, by applying Kirchhoff's voltage law around the closed loop, you find that the desired voltage is **V(R1:1)-V(R2:1)** which when followed by **OK** results in the plot point in the screen in the bottom of Fig. 16.33. Note that it is exactly 100 V as obtained in the longhand solution. Determine the phase angle through **Plot-Add Plot to Window-Trace-Add Trace,** creating the expression **P(V(R1:1)-V(R2:1)).** Remember that the expression can be generated using the lists of **Output variables** and **Functions,** but it can also be typed in from the keyboard. However, always remember that

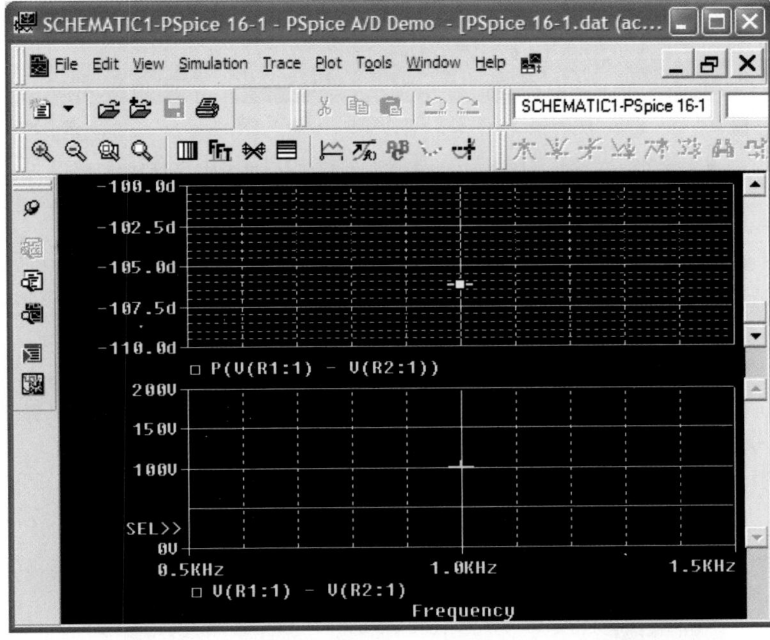

FIG. 16.33
The PSpice response for the voltage between the two points above resistors R_1 and R_2.

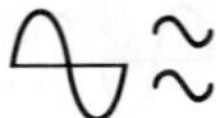

parentheses must be in sets—a left and a right. Click **OK,** and a solution near $-105°$ appears. A better reading can be obtained by using **Plot-Axis Settings-Y Axis-User Defined** and changing the scale to $-100°$ to $-110°$. The result is the top screen in Fig. 16.33 with an angle closer to $-106.5°$ or $73.5°$ which is very close to the theoretical solution of $73.74°$.

Finally, the last way to find the desired bridge voltage is to remove the **VPRINT2** option and place the ground at that point as shown in Fig. 16.34. Now the voltage generated from a point above R_1 to ground will be the desired voltage. Repeating a full simulation results in the plot in Fig. 16.35 with the same results as Fig. 16.33. Note, however, that even

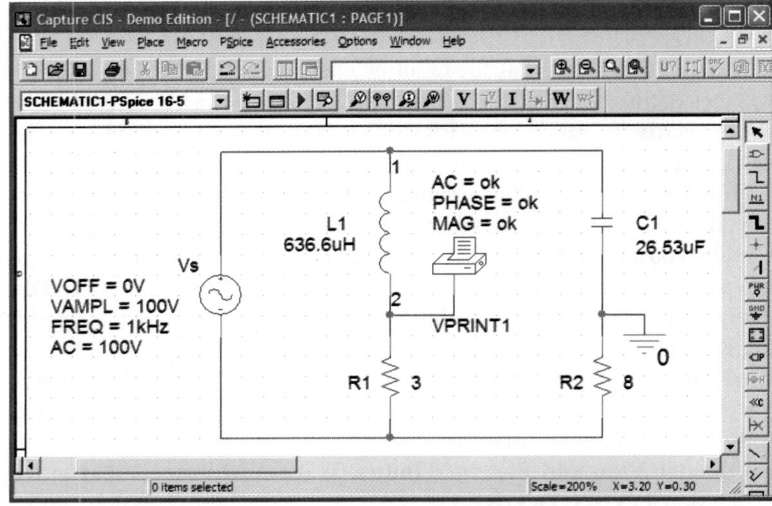

FIG. 16.34

Determining the voltage between the two points above resistors R_1 and R_2 by moving the ground connection in Fig. 16.30 to the position of ***VPRINT2.***

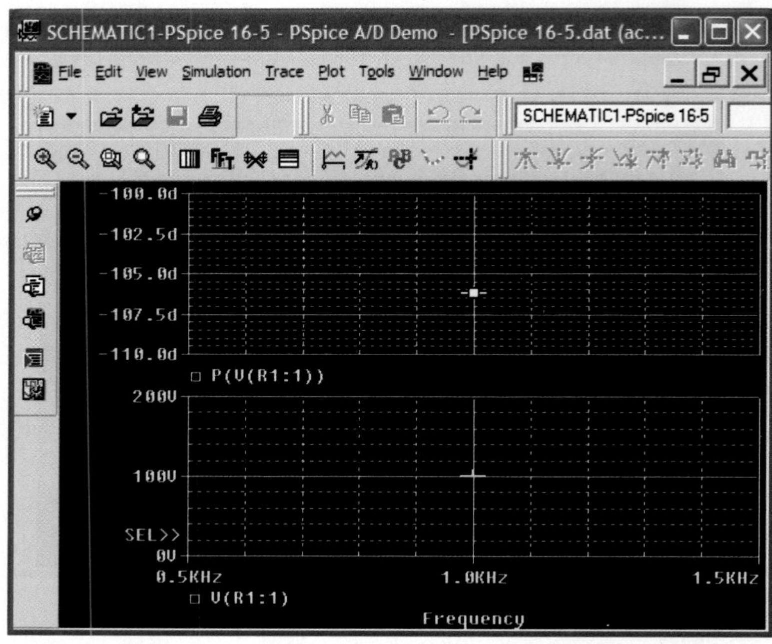

FIG. 16.35

PSpice response to the simulation of the network in Fig. 16.34.

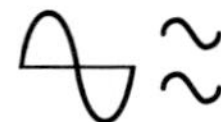

though the two figures look the same, the quantities listed in the bottom left of each plot are different.

Multisim

Multisim is now used to determine the voltage across the last element of the ladder network in Fig. 16.36. The mathematical content of this chapter suggests that this analysis would be a lengthy exercise in complex algebra, with one mistake (a single sign or an incorrect angle) enough to invalidate the results. However, it takes only a few minutes to "draw" the network on the screen and only a few seconds to generate the results—results you can usually assume are correct if all the parameters were entered correctly. The results are certainly an excellent check against a longhand solution.

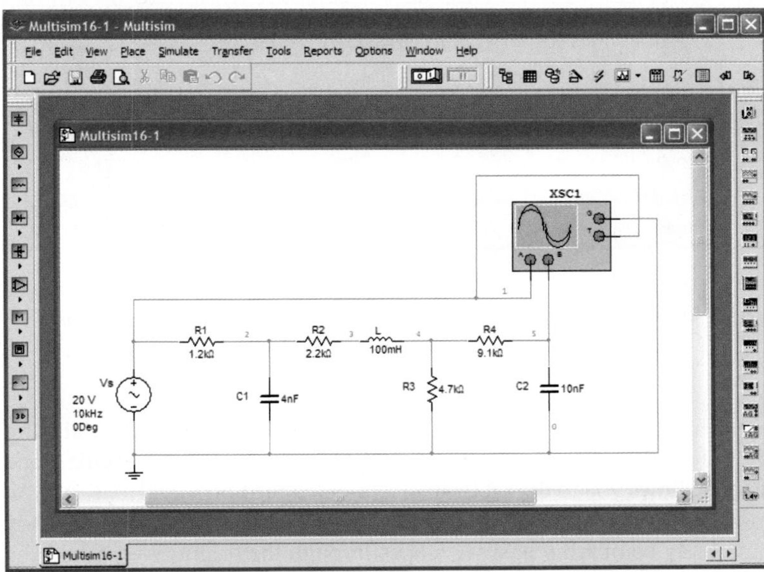

FIG. 16.36

Using the Multisim oscilloscope to determine the voltage across the capacitor C_2.

Our first approach is to use an oscilloscope to measure the amplitude and phase angle of the output voltage as shown in Fig. 16.36. Note that five nodes are defined, with node 5 the desired voltage. The oscilloscope settings include a **Time base** of 20 μs/div. since the period of the 10 kHz signal is 100 μs. Channel **A** is set on 10 V/div. so that the full 20 V of the applied signal will have a peak value encompassing two divisions. Note that **Channel A** in Fig. 16.36 is connected directly to the source **Vs** and to the **Trigger** input for synchronization. Expecting the output voltage to have a smaller amplitude resulted in a vertical sensitivity of 1 V/div. for **Channel B.** The analysis was initiated by placing the **Simulation** switch in the **1** position. It is important to realize that

when simulation is initiated, it will take time for networks with reactive elements to settle down and for the response to reach its steady-state condition. It is therefore wise to let a system run for a while after simulation before selecting Sing. (Single) on the oscilloscope to obtain a steady waveform for analysis.

The resulting plots in Fig. 16.37 clearly show that the applied voltage has an amplitude of 20 V and a period of 100 μs (5 div. at 20 μs/div). The

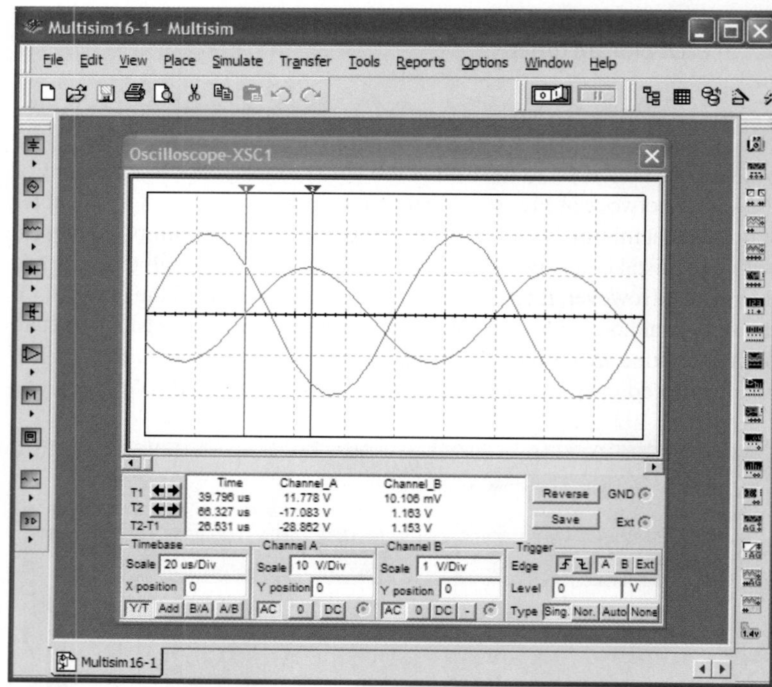

FIG. 16.37
Using Multisim to display the applied voltage and voltage across the capacitor C_2 for the network in Fig. 16.36.

cursors sit ready for use at the left and right edges of the screen. Clicking on the small red arrow (with number 1) at the top of the oscilloscope screen allows you to drag it to any location on the horizontal axis. As you move the cursor, the magnitude of each waveform appears in the **T1** box below. By comparing positive slopes through the origin, you should see that the applied voltage is leading the output voltage by an angle that is more than 90°. Setting the cursor at the point where the output voltage on channel B passes through the origin with a positive slope, you find that you cannot achieve exactly 0 V; but 0.01 V is certainly very close at 39.8 μs (**T1**).

Knowing that the applied voltage passed through the origin at 0 μs permits the following calculation for the phase angle:

$$\frac{39.8\ \mu\text{s}}{100\ \mu\text{s}} = \frac{\theta}{360°}$$

$$\theta = 143.28°$$

with the result that the output voltage has an angle of $-143.28°$ associated with it. The second cursor at the right edge of the screen is blue. Selecting it and moving it to the peak value of the output voltage results in 1.16 V at 66.33 μs (**T2**). The result of all the above is

$$\mathbf{V}_{C_2} = 1.16\ \text{V}\ \angle -143.28°$$

The second approach is to use the **AC Analysis** option under the **Simulate** heading. First, realize that when you use the oscilloscope as you just did, you did not need to pass through the sequence of dialog boxes to choose the desired analysis. All that was necessary was to simulate using either the switch or the **PSpice-Run** sequence—the oscilloscope was there

to measure the output voltage. Remember that the source defined the magnitude of the applied voltage, the frequency, and the phase shift. This time, use the sequence **Simulate-Analyses-AC Analysis** to obtain the **AC Analysis** dialog box in which the **Start** and **Stop frequencies** are 10 kHz and the **Selected variable for analysis** is node 5 only. Selecting **Simulate** results in a magnitude-phase plot with no apparent indicators at 10 kHz. However, this is easily corrected by first selecting one of the plots by clicking on the **Voltage** label at the left of the plot. Then select the **Show/Hide Grid, Show/Hide Legend,** and **Show/Hide Cursors** keys to obtain the cursors, legend, and **AC Analysis** dialog box. Hook on the red cursor and move it to 10 kHz. At that location, and that location only, **x1** appears as 10 kHz in the dialog box, and **y1** is ≅ 1.19 V as shown in Fig. 16.38. If you then select the **Phase** curve and repeat the procedure, you will find that at 10 kHz (**x1**) the angle is −142.15° (**y1**) which is very close to the −143.28° obtained above.

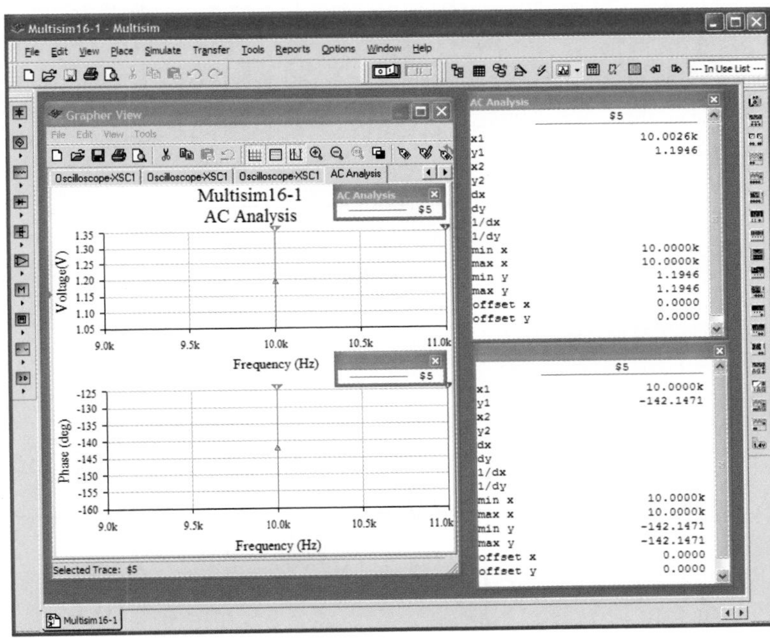

FIG. 16.38
*Using the **AC Analysis** option in Multisim to determine the magnitude and phase angle for the voltage V_{C_2} for the network in Fig. 16.36.*

In total, therefore, you have two methods to obtain an ac voltage in a network—one by instrumentation and the other through the computer methods. Both are valid, although, as expected, the computer approach has a higher level of accuracy.

PROBLEMS

SECTION 16.2 Illustrative Examples

1. For the series-parallel network in Fig. 16.39:
   a. Calculate $\mathbf{Z}_T$.
   b. Determine $\mathbf{I}$.
   c. Determine $\mathbf{I}_1$.
   d. Find $\mathbf{I}_2$ and $\mathbf{I}_3$.
   e. Find $\mathbf{V}_L$.

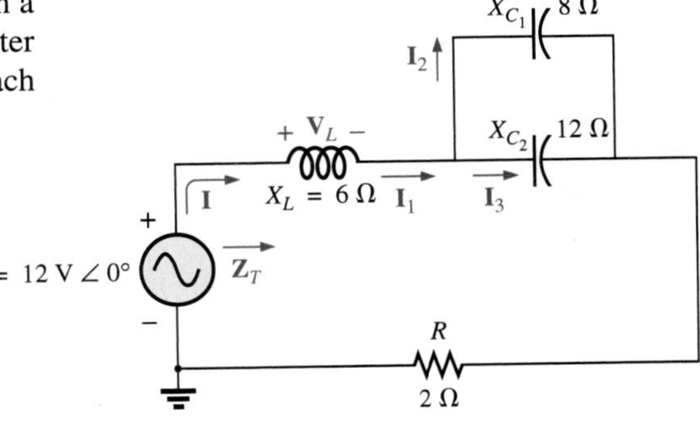

FIG. 16.39
Problem 1.

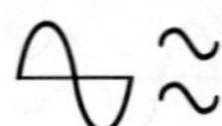

2. For the network in Fig. 16.40:
   a. Find the total impedance $\mathbf{Z}_T$.
   b. Determine the current $\mathbf{I}_s$.
   c. Calculate $\mathbf{I}_C$ using the current divider rule.
   d. Calculate $\mathbf{V}_L$ using the voltage divider rule.

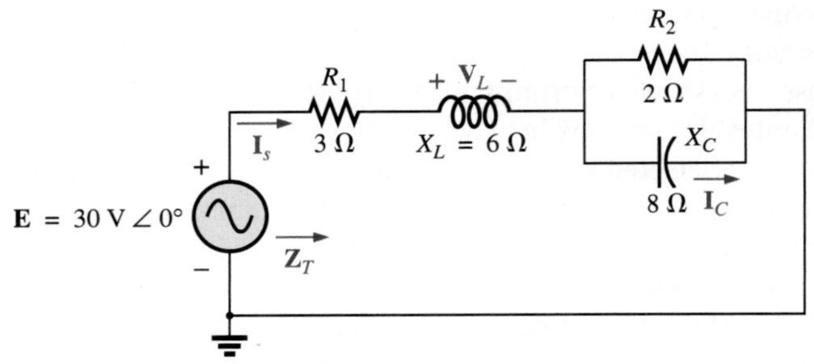

FIG. 16.40
Problems 2 and 15.

3. For the network in Fig. 16.41:
   a. Find the total impedance $\mathbf{Z}_T$ and the total admittance $\mathbf{Y}_T$.
   b. Find the current $\mathbf{I}_s$.
   c. Calculate $\mathbf{I}_2$ using the current divider rule.
   d. Calculate $\mathbf{V}_C$.
   e. Calculate the average power delivered to the network.

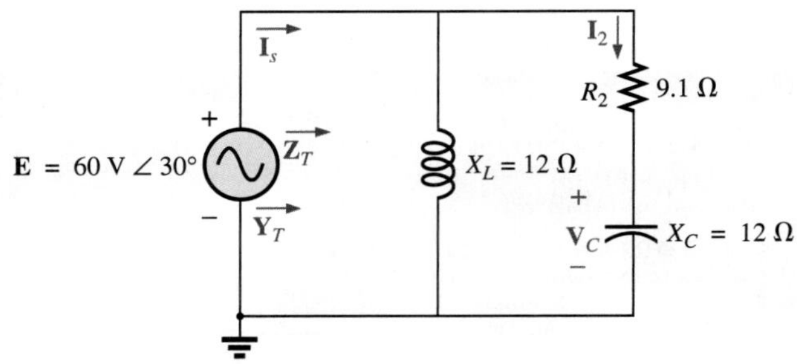

FIG. 16.41
Problem 3.

4. For the network in Fig. 16.42:
   a. Find the total impedance $\mathbf{Z}_T$.
   b. Calculate the voltage $\mathbf{V}_2$ and the current $\mathbf{I}_L$.
   c. Find the power factor of the network.

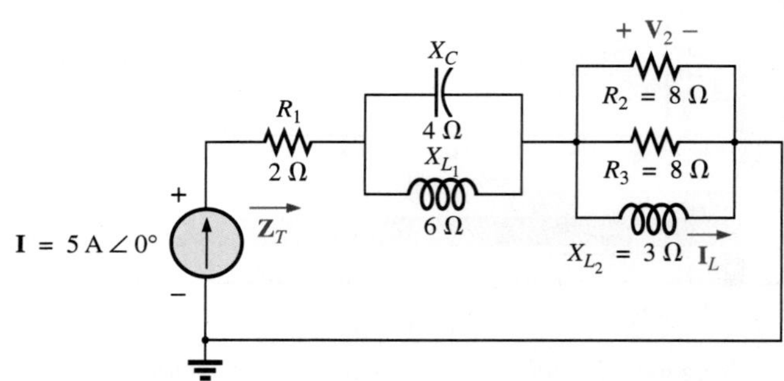

FIG. 16.42
Problem 4.

5. For the network in Fig. 16.43:
   a. Find the current $\mathbf{I}$.
   b. Find the voltage $\mathbf{V}_C$.
   c. Find the average power delivered to the network.

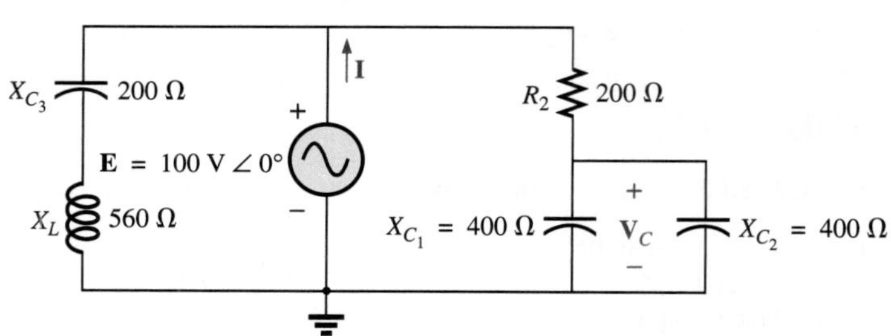

FIG. 16.43
Problem 5.

*6. For the network in Fig. 16.44:

a. Find the current $\mathbf{I}_1$.

b. Calculate the voltage $\mathbf{V}_C$ using the voltage divider rule.

c. Find the voltage $\mathbf{V}_{ab}$.

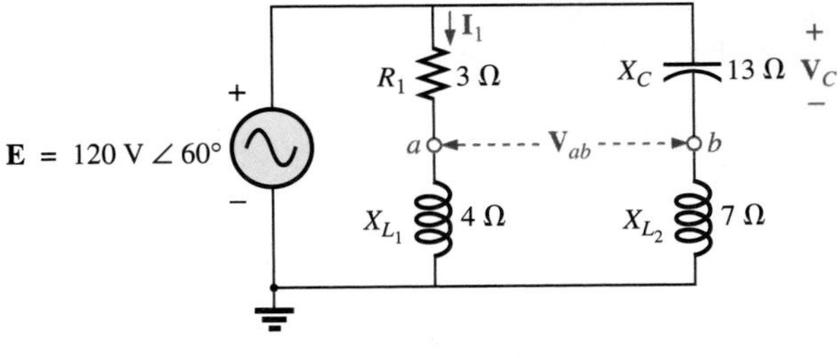

FIG. 16.44
Problem 6.

*7. For the network in Fig. 16.45:

a. Find the current $\mathbf{I}_1$.

b. Find the voltage $\mathbf{V}_1$.

c. Calculate the average power delivered to the network.

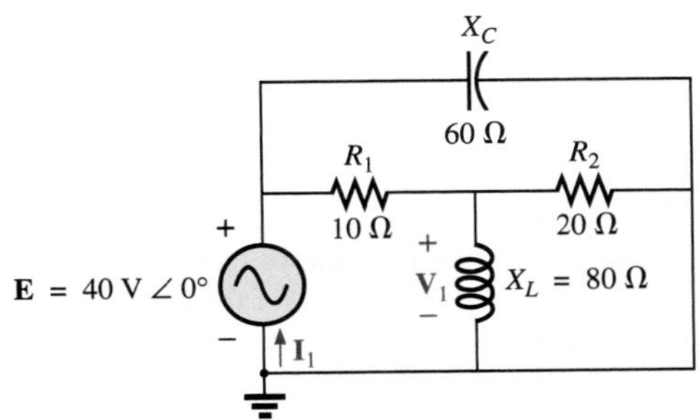

FIG. 16.45
Problems 7 and 16.

8. For the network in Fig. 16.46:

a. Find the total impedance $\mathbf{Z}_T$ and the admittance $\mathbf{Y}_T$.

b. Find the currents $\mathbf{I}_1$, $\mathbf{I}_2$, and $\mathbf{I}_3$.

c. Verify Kirchhoff's current law by showing that $\mathbf{I}_s = \mathbf{I}_1 + \mathbf{I}_2 + \mathbf{I}_3$.

d. Find the power factor of the network, and indicate whether it is leading or lagging.

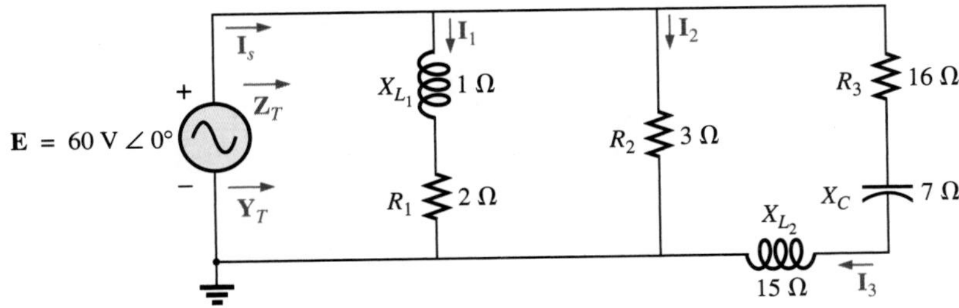

FIG. 16.46
Problem 8.

*9. For the network in Fig. 16.47:

a. Find the total admittance $\mathbf{Y}_T$.

b. Find the voltages $\mathbf{V}_1$ and $\mathbf{V}_2$.

c. Find the current $\mathbf{I}_3$.

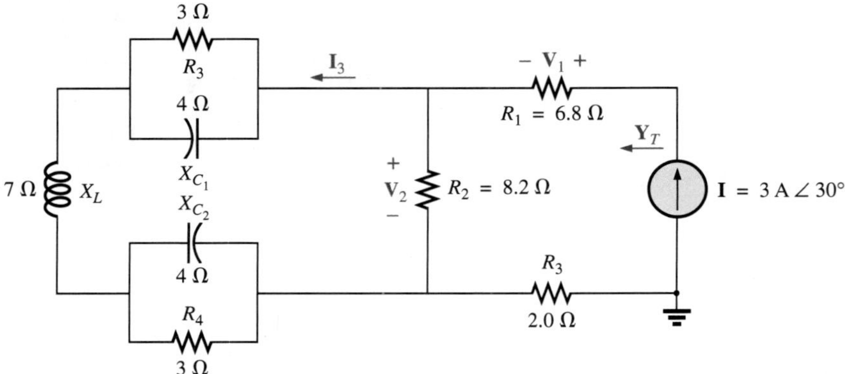

FIG. 16.47
Problem 9.

*10. For the network in Fig. 16.48:

a. Find the total impedance $\mathbf{Z}_T$ and the admittance $\mathbf{Y}_T$.

b. Find the source current $\mathbf{I}_s$ in phasor form.

c. Find the currents $\mathbf{I}_1$ and $\mathbf{I}_2$ in phasor form.

d. Find the voltages $\mathbf{V}_1$ and $\mathbf{V}_{ab}$ in phasor form.

e. Find the average power delivered to the network.

f. Find the power factor of the network, and indicate whether it is leading or lagging.

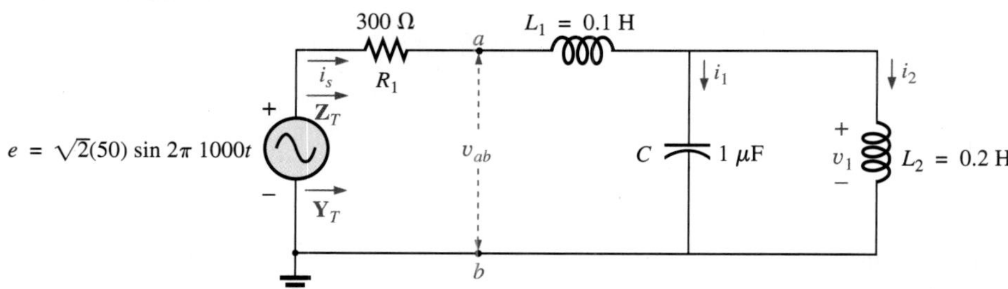

FIG. 16.48
Problem 10.

*11. Find the current **I** for the network in Fig. 16.49.

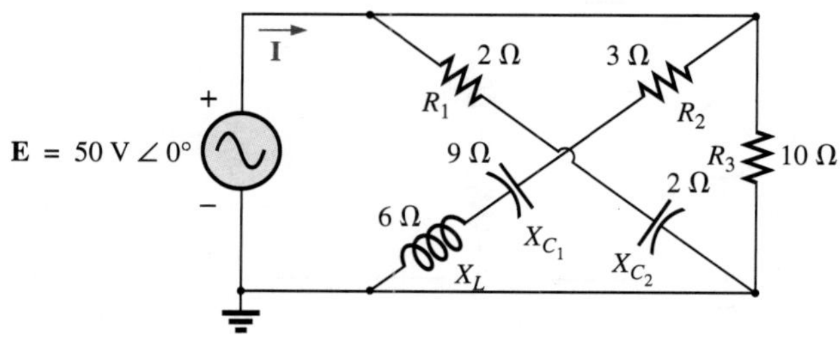

FIG. 16.49
Problems 11 and 17.

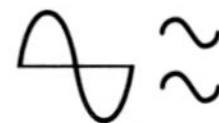

SECTION 16.3 Ladder Networks

12. Find the current $\mathbf{I}_5$ for the network in Fig. 16.50. Note the effect of one reactive element on the resulting calculations.

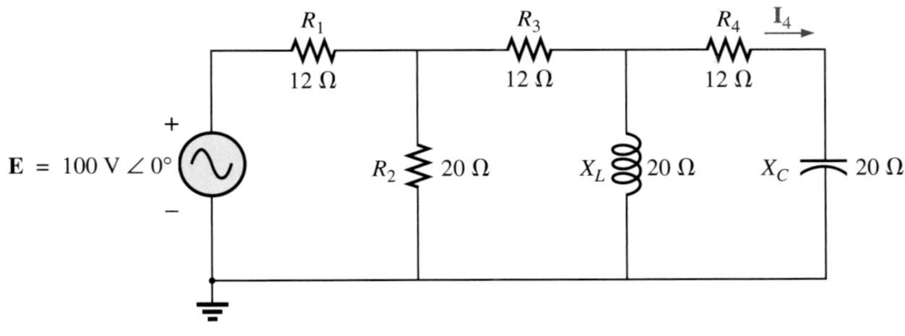

FIG. 16.50
Problem 12.

13. Find the average power delivered to R_4 in Fig. 16.51.

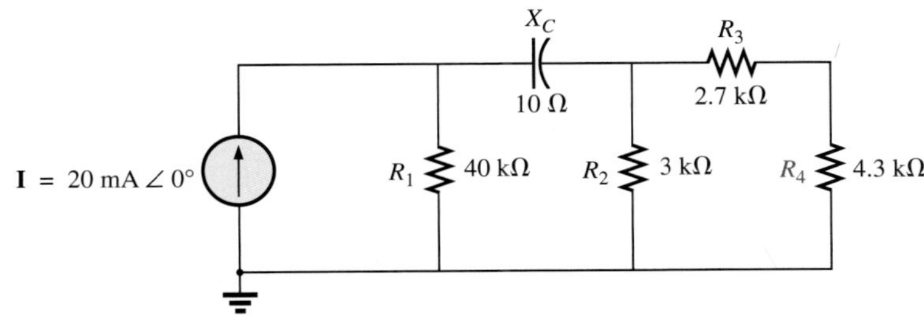

FIG. 16.51
Problem 13.

14. Find the current $\mathbf{I}_1$ for the network in Fig. 16.52.

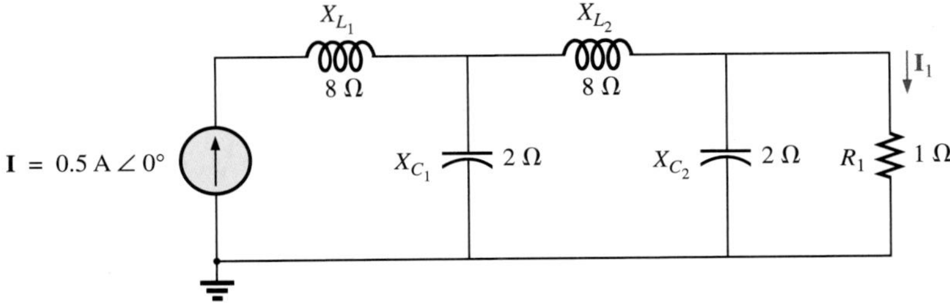

FIG. 16.52
Problems 14 and 18.

SECTION 16.6 Computer Analysis

PSpice or Multisim

For Problems 15 through 18, use a frequency of 1 kHz to determine the inductive and capacitive levels required for the input files. In each case, write the required input file.

*15. Repeat Problem 2 using PSpice or Multisim.

*16. Repeat Problem 7, parts (a) and (b), using PSpice or Multisim.

*17. Repeat Problem 11 using PSpice or Multisim.

*18. Repeat Problem 14 using PSpice or Multisim.

GLOSSARY

Ladder network A repetitive combination of series and parallel branches that has the appearance of a ladder.

Series-parallel ac network A combination of series and parallel branches in the same network configuration. Each branch may contain any number of elements whose impedance is dependent on the applied frequency.

Methods of Analysis and Selected Topics (ac)

Objectives

- ***Understand the differences between independent and dependent sources and how the magnitude and angle of a controlled source is determined by the dependent variable.***
- ***Be able to convert between voltage and current sources and vice versa in the ac domain.***
- ***Become proficient in the application of mesh and nodal analysis to ac networks with independent and controlled sources.***
- ***Be able to define the relationship between the elements of an ac bridge network that will establish a balance condition.***

17.1 INTRODUCTION

For networks with two or more sources that are not in series or parallel, the methods described in the last two chapters cannot be applied. Rather, methods such as mesh analysis or nodal analysis must be used. Since these methods were discussed in detail for dc circuits in Chapter 8, this chapter considers the variations required to apply these methods to ac circuits. Dependent sources are also introduced for both mesh and nodal analysis.

The branch-current method is not discussed again because it falls within the framework of mesh analysis. In addition to the methods mentioned above, the bridge network and Δ-Y, Y-Δ conversions are also discussed for ac circuits.

Before we examine these topics, however, we must consider the subject of independent and controlled sources.

17.2 INDEPENDENT VERSUS DEPENDENT (CONTROLLED) SOURCES

In the previous chapters, each source appearing in the analysis of dc or ac networks was an **independent source,** such as E and I (or $\mathbf{E}$ and $\mathbf{I}$) in Fig. 17.1.

The term independent specifies that the magnitude of the source is independent of the network to which it is applied and that the source displays its terminal characteristics even if completely isolated.

A dependent or controlled source is one whose magnitude is determined (or controlled) by a current or voltage of the system in which it appears.

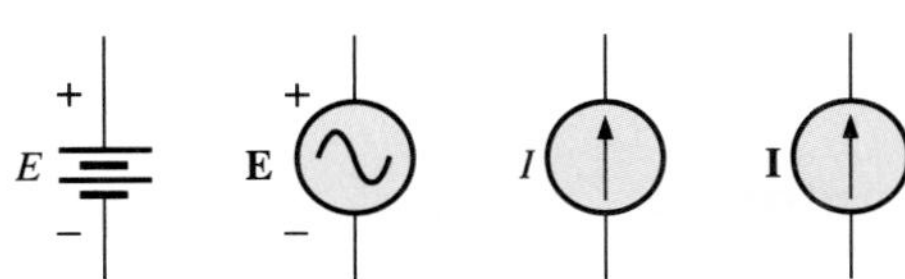

FIG. 17.1
Independent sources.

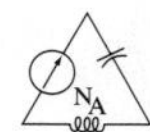

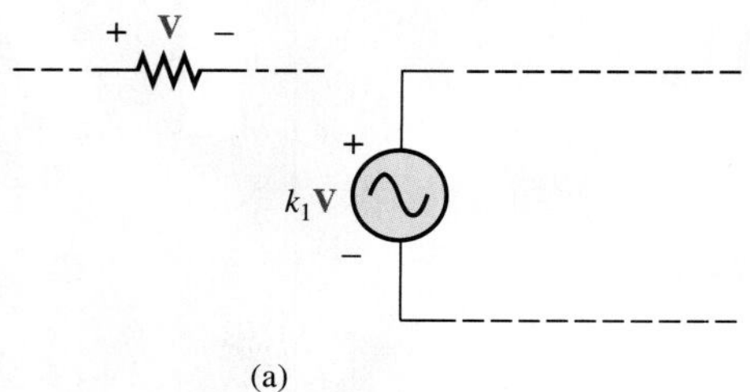

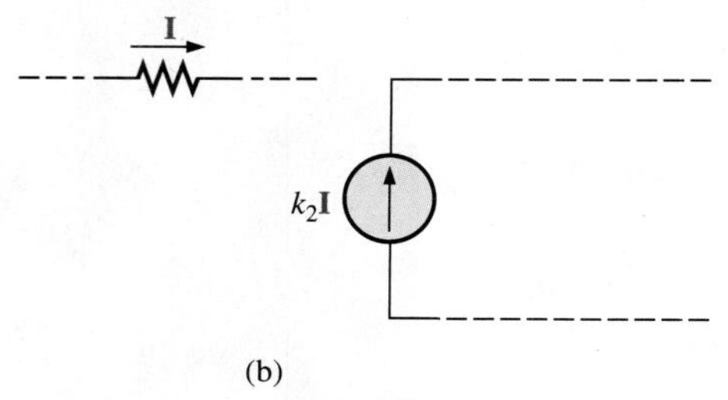

FIG. 17.2
Controlled or dependent sources.

Currently two symbols are used for controlled sources. One simply uses the independent symbol with an indication of the controlling element, as shown in Fig. 17.2. In Fig. 17.2(a), the magnitude and phase of the voltage are controlled by a voltage **V** elsewhere in the system, with the magnitude further controlled by the constant k_1. In Fig. 17.2(b), the magnitude and phase of the current source are controlled by a current **I** elsewhere in the system, with the magnitude further controlled by the constant k_2. To distinguish between the dependent and independent sources, the notation in Fig. 17.3 was introduced. In recent years, many respected publications on circuit analysis have accepted the notation in Fig. 17.3, although a number of excellent publications in the area of electronics continue to use the symbol in Fig. 17.2, especially in the circuit modeling for a variety of electronic devices such as the transistor and FET. This text uses the symbols in Fig. 17.3.

Possible combinations for controlled sources are indicated in Fig. 17.4. Note that the magnitude of current sources or voltage sources can be controlled by a voltage and a current, respectively. Unlike with the independent source, isolation such that **V** or **I** $= 0$ in Fig. 17.4(a) results in the short-circuit or open-circuit equivalent as indicated in Fig. 17.4(b). Note that the type of representation under these conditions is controlled by whether it is a current source or a voltage source, not by the controlling agent (**V** or **I**).

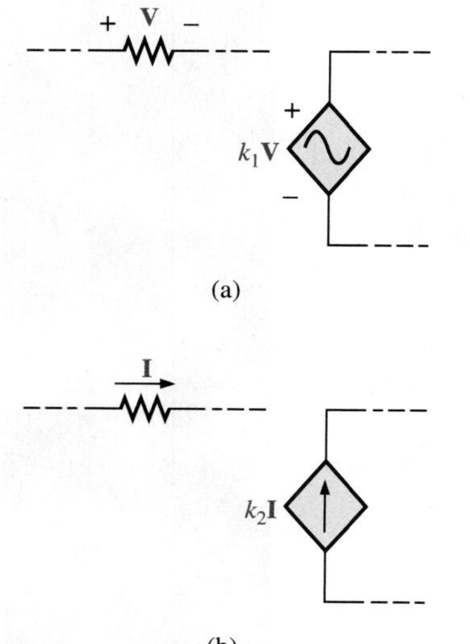

FIG. 17.3
Special notation for controlled or dependent sources.

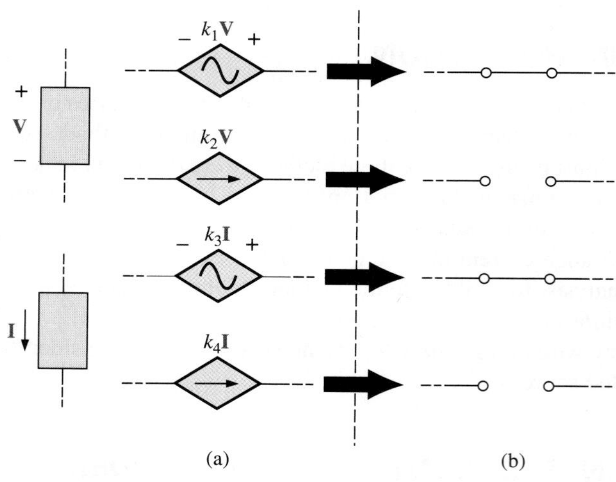

FIG. 17.4
Conditions of **V** $=$ *0 V and* **I** $=$ *0 A for a controlled source.*

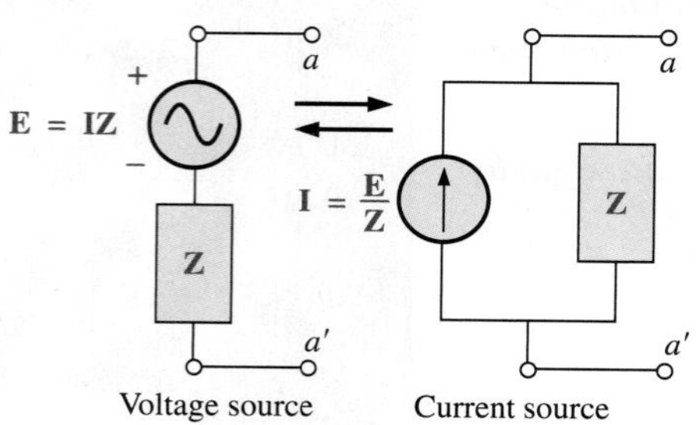

FIG. 17.5
Source conversion.

17.3 SOURCE CONVERSIONS

When applying the methods to be discussed, it may be necessary to convert a current source to a voltage source, or a voltage source to a current source. This **source conversion** can be accomplished in much the same manner as for dc circuits, except now we shall be dealing with phasors and impedances instead of just real numbers and resistors.

Independent Sources

In general, the format for converting one type of independent source to another is as shown in Fig. 17.5.

EXAMPLE 17.1 Convert the voltage source in Fig. 17.6(a) to a current source.

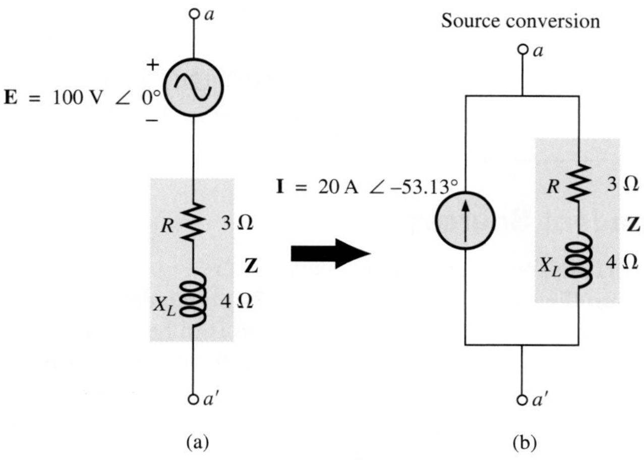

FIG. 17.6
Example 17.1.

Solution:

$$\mathbf{I} = \frac{\mathbf{E}}{\mathbf{Z}} = \frac{100\text{ V}\angle 0^\circ}{5\ \Omega\ \angle 53.13^\circ}$$
$$= \mathbf{20\ A\ \angle -53.13^\circ} \qquad \text{[Fig. 17.6(b)]}$$

EXAMPLE 17.2 Convert the current source in Fig. 17.7(a) to a voltage source.

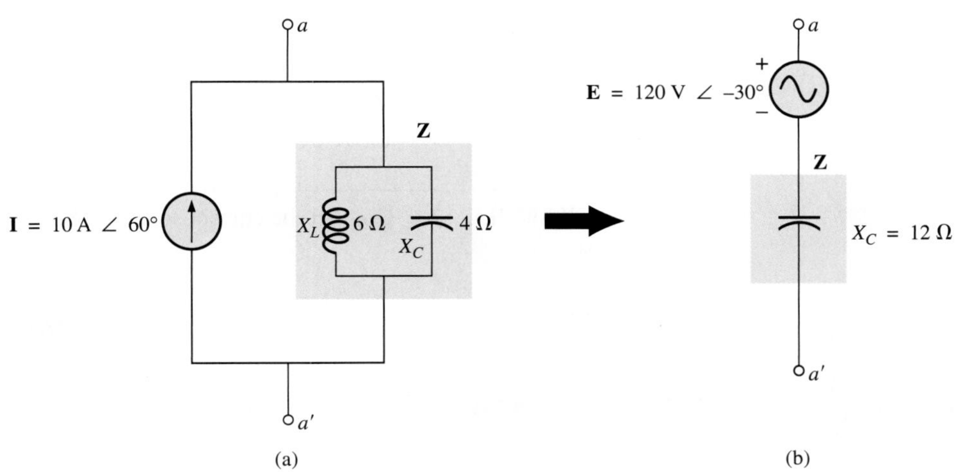

FIG. 17.7
Example 17.2.

Solution:

$$\mathbf{Z} = \frac{\mathbf{Z}_C\mathbf{Z}_L}{\mathbf{Z}_C + \mathbf{Z}_L} = \frac{(X_C \angle -90°)(X_L \angle 90°)}{-j X_C + j X_L}$$

$$= \frac{(4\ \Omega \angle -90°)(6\ \Omega \angle 90°)}{-j 4\ \Omega + j 6\ \Omega} = \frac{24\ \Omega \angle 0°}{2 \angle 90°}$$

$$= \mathbf{12\ \Omega \angle -90°}$$

$$\mathbf{E} = \mathbf{IZ} = (10\text{ A} \angle 60°)(12\ \Omega \angle -90°)$$

$$= \mathbf{120\ V \angle -30°} \qquad \text{[Fig. 17.7(b)]}$$

Dependent Sources

For dependent sources, the direct conversion in Fig. 17.5 can be applied if the controlling variable (**V** or **I** in Fig. 17.4) is not determined by a portion of the network to which the conversion is to be applied. For example, in Figs. 17.8 and 17.9, **V** and **I**, respectively, are controlled by an external portion of the network. Conversions of the other kind, where **V** and **I** are controlled by a portion of the network to be converted, are considered in Sections 18.3 and 18.4.

EXAMPLE 17.3 Convert the voltage source in Fig. 17.8(a) to a current source.

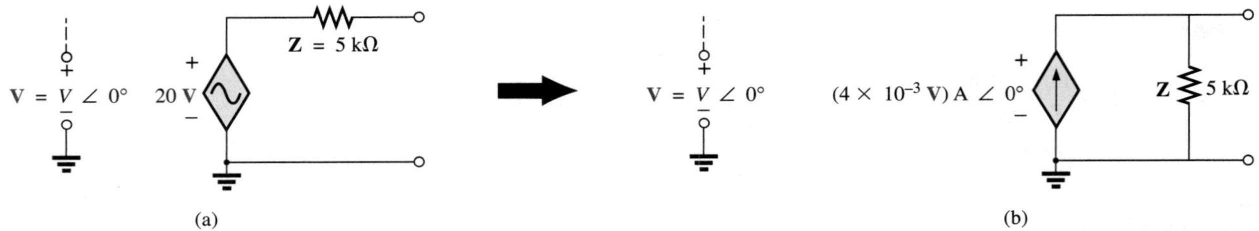

FIG. 17.8
Source conversion with a voltage-controlled voltage source.

Solution:

$$\mathbf{I} = \frac{\mathbf{E}}{\mathbf{Z}} = \frac{(20\mathbf{V})\ \text{V}\angle 0°}{5\text{ k}\Omega \angle 0°}$$

$$= \mathbf{(4 \times 10^{-3}\ V)A \angle 0°} \qquad \text{[Fig. 17.8(b)]}$$

EXAMPLE 17.4 Convert the current source in Fig. 17.9(a) to a voltage source.

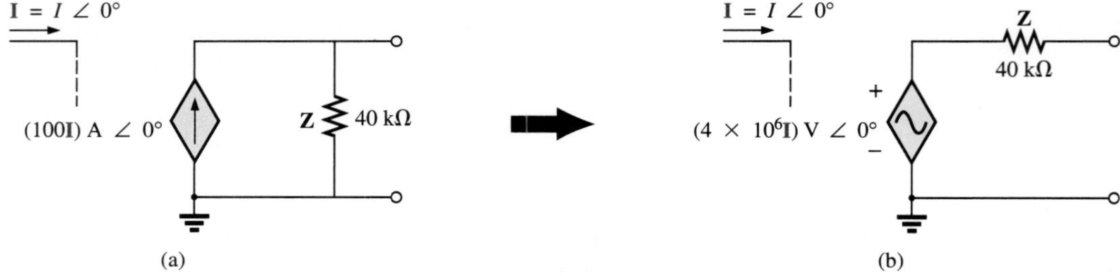

FIG. 17.9
Source conversion with a current-controlled current source.

Solution:

$$\mathbf{E} = \mathbf{IZ} = [(100\mathbf{I})\text{ A} \angle 0°][40\text{ k}\Omega \angle 0°]$$
$$= (\mathbf{4 \times 10^6 I})\textbf{ V} \angle \mathbf{0°} \qquad \text{[Fig. 17.9(b)]}$$

17.4 MESH ANALYSIS

General Approach

Independent Voltage Sources Before examining the application of the method to ac networks, the student should first review the appropriate sections on **mesh analysis** in Chapter 8 since the content of this section will be limited to the general conclusions of Chapter 8.

The general approach to mesh analysis for independent sources includes the same sequence of steps appearing in Chapter 8. In fact, throughout this section the only change from the dc coverage is to substitute impedance for resistance and admittance for conductance in the general procedure.

1. ***Assign a distinct current in the clockwise direction to each independent closed loop of the network. It is not absolutely necessary to choose the clockwise direction for each loop current. However, it eliminates the need to have to choose a direction for each application. Any direction can be chosen for each loop current with no loss in accuracy as long as the remaining steps are followed properly.***
2. ***Indicate the polarities within each loop for each impedance as determined by the assumed direction of loop current for that loop.***
3. ***Apply Kirchhoff's voltage law around each closed loop in the clockwise direction. Again, the clockwise direction was chosen to establish uniformity and to prepare us for the format approach to follow.***
   a. ***If an impedance has two or more assumed currents through it, the total current through the impedance is the assumed current of the loop in which Kirchhoff's voltage law is being applied, plus the assumed currents of the other loops passing through in the same direction, minus the assumed currents passing through in the opposite direction.***
   b. ***The polarity of a voltage source is unaffected by the direction of the assigned loop currents.***
4. ***Solve the resulting simultaneous linear equations for the assumed loop currents.***

The technique is applied as above for all networks with independent sources or for networks with *dependent sources where the controlling variable is not a part of the network under investigation.* If the controlling variable is part of the network being examined, a method to be described shortly must be applied.

EXAMPLE 17.5 Using the general approach to mesh analysis, find the current $\mathbf{I}_1$ in Fig. 17.10.

Solution: When applying these methods to ac circuits, it is good practice to represent the resistors and reactances (or combinations thereof) by subscripted impedances. When the total solution is found in terms of these subscripted impedances, the numerical values can be substituted to find the unknown quantities.

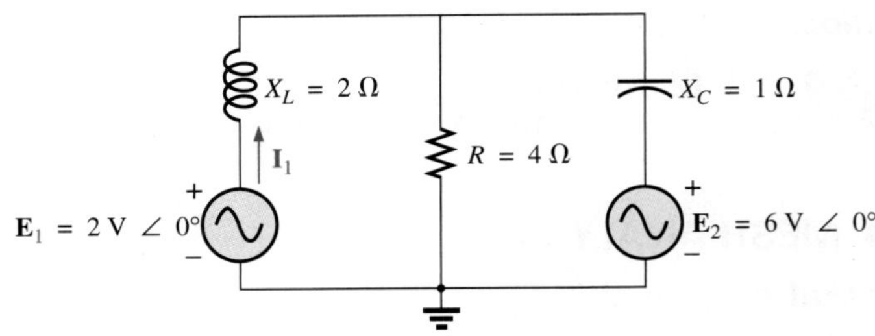

FIG. 17.10
Example 17.5.

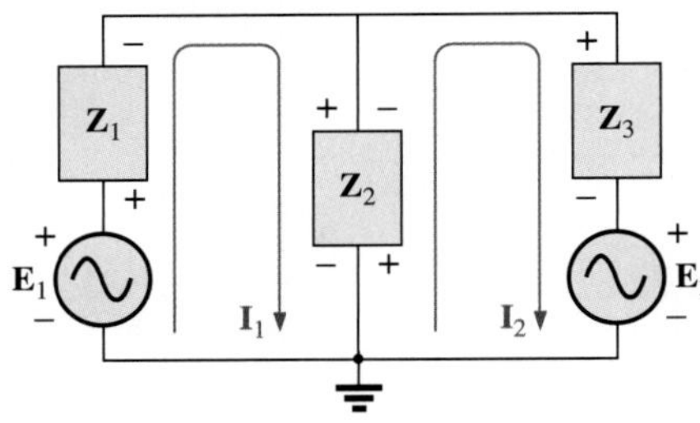

FIG. 17.11
Assigning the mesh currents and subscripted impedances for the network in Fig. 17.10.

The network is redrawn in Fig. 17.11 with subscripted impedances:

$$\mathbf{Z}_1 = +j\,X_L = +j\,2\ \Omega \qquad \mathbf{E}_1 = 2\ \text{V} \angle 0°$$
$$\mathbf{Z}_2 = R = 4\ \Omega \qquad \mathbf{E}_2 = 6\ \text{V} \angle 0°$$
$$\mathbf{Z}_3 = -j\,X_C = -j\,1\ \Omega$$

Steps 1 and 2 are as indicated in Fig. 17.11.

Step 3:

$$+\mathbf{E}_1 - \mathbf{I}_1\mathbf{Z}_1 - \mathbf{Z}_2(\mathbf{I}_1 - \mathbf{I}_2) = 0$$
$$-\mathbf{Z}_2(\mathbf{I}_2 - \mathbf{I}_1) - \mathbf{I}_2\mathbf{Z}_3 - \mathbf{E}_2 = 0$$

or

$$\mathbf{E}_1 - \mathbf{I}_1\mathbf{Z}_1 - \mathbf{I}_1\mathbf{Z}_2 + \mathbf{I}_2\mathbf{Z}_2 = 0$$
$$-\mathbf{I}_2\mathbf{Z}_2 + \mathbf{I}_1\mathbf{Z}_2 - \mathbf{I}_2\mathbf{Z}_3 - \mathbf{E}_2 = 0$$

so that

$$\mathbf{I}_1(\mathbf{Z}_1 + \mathbf{Z}_2) - \mathbf{I}_2\mathbf{Z}_2 = \mathbf{E}_1$$
$$\mathbf{I}_2(\mathbf{Z}_2 + \mathbf{Z}_3) - \mathbf{I}_1\mathbf{Z}_2 = -\mathbf{E}_2$$

which are rewritten as

$$\begin{aligned} \mathbf{I}_1(\mathbf{Z}_1 + \mathbf{Z}_2) - \mathbf{I}_2\mathbf{Z}_2 \qquad\qquad &= \mathbf{E}_1 \\ -\mathbf{I}_1\mathbf{Z}_2 \qquad\qquad + \mathbf{I}_2(\mathbf{Z}_2 + \mathbf{Z}_3) &= -\mathbf{E}_2 \end{aligned}$$

Step 4: Using determinants, we obtain

$$\mathbf{I}_1 = \frac{\begin{vmatrix} \mathbf{E}_1 & -\mathbf{Z}_2 \\ -\mathbf{E}_2 & \mathbf{Z}_2 + \mathbf{Z}_3 \end{vmatrix}}{\begin{vmatrix} \mathbf{Z}_1 + \mathbf{Z}_2 & -\mathbf{Z}_2 \\ -\mathbf{Z}_2 & \mathbf{Z}_2 + \mathbf{Z}_3 \end{vmatrix}} = \frac{\mathbf{E}_1(\mathbf{Z}_2 + \mathbf{Z}_3) - \mathbf{E}_2(\mathbf{Z}_2)}{(\mathbf{Z}_1 + \mathbf{Z}_2)(\mathbf{Z}_2 + \mathbf{Z}_3) - (\mathbf{Z}_2)^2} = \frac{(\mathbf{E}_1 - \mathbf{E}_2)\mathbf{Z}_2 + \mathbf{E}_1\mathbf{Z}_3}{\mathbf{Z}_1\mathbf{Z}_2 + \mathbf{Z}_1\mathbf{Z}_3 + \mathbf{Z}_2\mathbf{Z}_3}$$

Substituting numerical values yields

$$\mathbf{I}_1 = \frac{(2\ \text{V} - 6\ \text{V})(4\ \Omega) + (2\ \text{V})(-j\,1\ \Omega)}{(+j\,2\ \Omega)(4\ \Omega) + (+j\,2\ \Omega)(-j\,2\ \Omega) + (4\ \Omega)(-j\,2\ \Omega)}$$
$$= \frac{-16 - j\,2}{j\,8 - j^2\,2 - j\,4} = \frac{-16 - j\,2}{2 + j\,4} = \frac{16.12\ \text{A} \angle -172.87°}{4.47 \angle 63.43°}$$
$$= \mathbf{3.61\ A \angle -236.30°} \text{ or } \mathbf{3.61\ A \angle 123.70°}$$

Dependent Voltage Sources For dependent voltage sources, the procedure is modified as follows:

1. Steps 1 and 2 are the same as those applied for independent voltage sources.
2. Step 3 is modified as follows: Treat each dependent source like an independent source when Kirchhoff's voltage law is applied to each independent loop. However, once the equation is written, substitute the equation for the controlling quantity to ensure that the unknowns are limited solely to the chosen mesh currents.
3. Step 4 is as before.

EXAMPLE 17.6 Write the mesh currents for the network in Fig. 17.12 having a dependent voltage source.

Solution:

Steps 1 and 2 are defined in Fig. 17.12.

Step 3:
$$\mathbf{E}_1 - \mathbf{I}_1 R_1 - R_2(\mathbf{I}_1 - \mathbf{I}_2) = 0$$
$$R_2(\mathbf{I}_2 - \mathbf{I}_1) + \mu\mathbf{V}_x - \mathbf{I}_2 R_3 = 0$$

Then substitute $\mathbf{V}_x = (\mathbf{I}_1 - \mathbf{I}_2)R_2$

The result is two equations and two unknowns.

$$\mathbf{E}_1 - \mathbf{I}_1 R_1 - R_2(\mathbf{I} - \mathbf{I}_2) = 0$$
$$\underline{R_2(\mathbf{I}_2 - \mathbf{I}_1) + \mu R_2(\mathbf{I}_1 - \mathbf{I}_2) - \mathbf{I}_2 R_3 = 0}$$

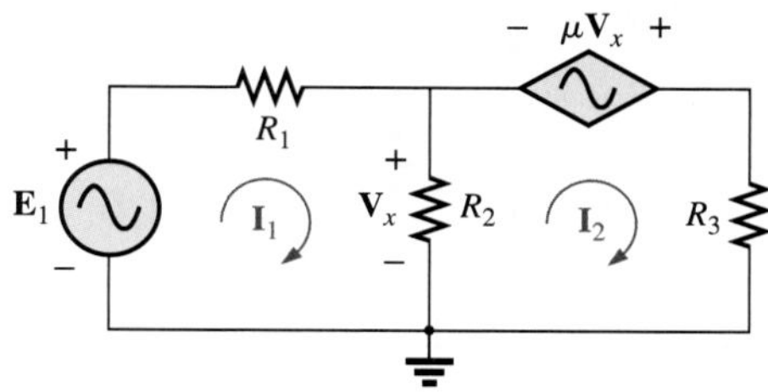

FIG. 17.12
Applying mesh analysis to a network with a voltage-controlled voltage source.

Independent Current Sources For independent current sources, the procedure is modified as follows:

1. Steps 1 and 2 are the same as those applied for independent sources.
2. Step 3 is modified as follows: Treat each current source as an open circuit (recall the *supermesh* designation in Chapter 8), and write the mesh equations for each remaining independent path. Then relate the chosen mesh currents to the dependent sources to ensure that the unknowns of the final equations are limited to the mesh currents.
3. Step 4 is as before.

EXAMPLE 17.7 Write the mesh currents for the network in Fig. 17.13 having an independent current source.

Solution:

Steps 1 and 2 are defined in Fig. 17.13.

Step 3: $\mathbf{E}_1 - \mathbf{I}_1\mathbf{Z}_1 + \mathbf{E}_2 - \mathbf{I}_2\mathbf{Z}_2 = 0$ (only remaining independent path)

with $\mathbf{I}_1 + \mathbf{I} = \mathbf{I}_2$

The result is two equations and two unknowns.

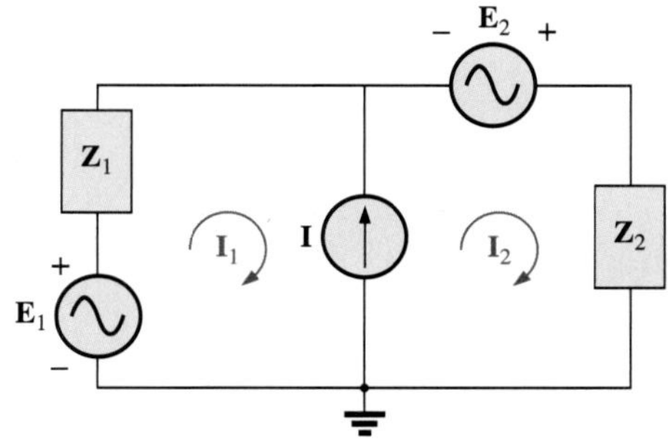

FIG. 17.13
Applying mesh analysis to a network with an independent current source.

Dependent Current Sources For dependent current sources, the procedure is modified as follows:

1. Steps 1 and 2 are the same as those applied for independent sources.

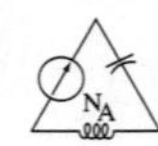

2. Step 3 is modified as follows: The procedure is essentially the same as that applied for independent current sources, except now the dependent sources have to be defined in terms of the chosen mesh currents to ensure that the final equations have only mesh currents as the unknown quantities.
3. Step 4 is as before.

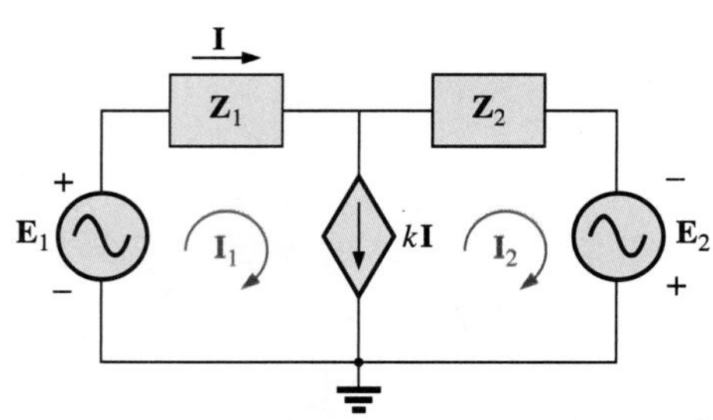

FIG. 17.14
Applying mesh analysis to a network with a current-controlled current source.

EXAMPLE 17.8 Write the mesh currents for the network in Fig. 17.14 having a dependent current source.

Solution:

Steps 1 and 2 are defined in Fig. 17.14.

Step 3: $\mathbf{E}_1 - \mathbf{I}_1\mathbf{Z}_1 - \mathbf{I}_2\mathbf{Z}_2 + \mathbf{E}_2 = 0$

and $k\mathbf{I} = \mathbf{I}_1 - \mathbf{I}_2$

Now $\mathbf{I} = \mathbf{I}_1$ so that $k\mathbf{I}_1 = \mathbf{I}_1 - \mathbf{I}_2$ or $\mathbf{I}_2 = \mathbf{I}_1(1 - k)$

The result is two equations and two unknowns.

Format Approach

The format approach was introduced in Section 8.8. The steps for applying this method are repeated here with changes for its use in ac circuits:

1. ***Assign a loop current to each independent closed loop (as in the previous section) in a clockwise direction.***
2. ***The number of required equations is equal to the number of chosen independent closed loops. Column 1 of each equation is formed by summing the impedance values of those impedances through which the loop current of interest passes and multiplying the result by that loop current.***
3. ***We must now consider the mutual terms that are always subtracted from the terms in the first column. It is possible to have more than one mutual term if the loop current of interest has an element in common with more than one other loop current. Each mutual term is the product of the mutual impedance and the other loop current passing through the same element.***
4. ***The column to the right of the equality sign is the algebraic sum of the voltage sources through which the loop current of interest passes. Positive signs are assigned to those sources of voltage having a polarity such that the loop current passes from the negative to the positive terminal. Negative signs are assigned to those potentials for which the reverse is true.***
5. ***Solve the resulting simultaneous equations for the desired loop currents.***

The technique is applied as above for all networks with independent sources or for networks with dependent sources where the controlling variable is not a part of the network under investigation. If the controlling variable is part of the network being examined, additional care must be taken when applying the above steps.

EXAMPLE 17.9 Using the format approach to mesh analysis, find the current $\mathbf{I}_2$ in Fig. 17.15.

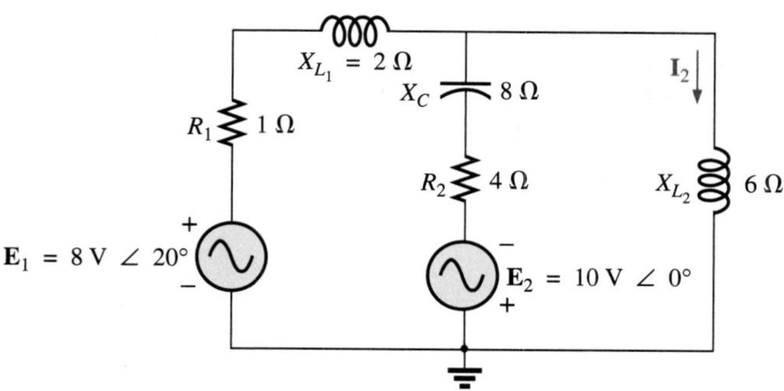

FIG. 17.15
Example 17.9.

Solution: The network is redrawn in Fig. 17.16:

$$\mathbf{Z}_1 = R_1 + jX_{L_1} = 1\ \Omega + j\,2\ \Omega \qquad \mathbf{E}_1 = 8\ \text{V}\ \angle 20^\circ$$
$$\mathbf{Z}_2 = R_2 - jX_C = 4\ \Omega - j\,8\ \Omega \qquad \mathbf{E}_2 = 10\ \text{V}\ \angle 0^\circ$$
$$\mathbf{Z}_3 = +jX_{L_2} = +j\,6\ \Omega$$

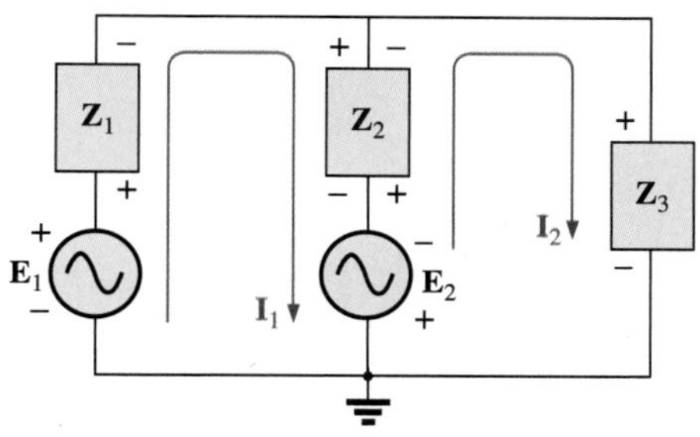

FIG. 17.16
Assigning the mesh currents and subscripted impedances for the network in Fig. 17.15.

Note the reduction in complexity of the problem with the substitution of the subscripted impedances.

Step 1 is as indicated in Fig. 17.16.

Steps 2 to 4:

$$\mathbf{I}_1(\mathbf{Z}_1 + \mathbf{Z}_2) - \mathbf{I}_2\mathbf{Z}_2 = \mathbf{E}_1 + \mathbf{E}_2$$
$$\mathbf{I}_2(\mathbf{Z}_2 + \mathbf{Z}_3) - \mathbf{I}_1\mathbf{Z}_2 = -\mathbf{E}_2$$

which are rewritten as

$$\begin{aligned} \mathbf{I}_1(\mathbf{Z}_1 + \mathbf{Z}_2) - \mathbf{I}_2\mathbf{Z}_2 \qquad &= \mathbf{E}_1 + \mathbf{E}_2 \\ -\mathbf{I}_1\mathbf{Z}_2 \qquad + \mathbf{I}_2(\mathbf{Z}_2 + \mathbf{Z}_3) &= -\mathbf{E}_2 \end{aligned}$$

Step 5: Using determinants, we have

$$\mathbf{I}_2 = \frac{\begin{vmatrix} \mathbf{Z}_1 + \mathbf{Z}_2 & \mathbf{E}_1 + \mathbf{E}_2 \\ -\mathbf{Z}_2 & -\mathbf{E}_2 \end{vmatrix}}{\begin{vmatrix} \mathbf{Z}_1 + \mathbf{Z}_2 & -\mathbf{Z}_2 \\ -\mathbf{Z}_2 & \mathbf{Z}_2 + \mathbf{Z}_3 \end{vmatrix}}$$
$$= \frac{-(\mathbf{Z}_1 + \mathbf{Z}_2)\mathbf{E}_2 + \mathbf{Z}_2(\mathbf{E}_1 + \mathbf{E}_2)}{(\mathbf{Z}_1 + \mathbf{Z}_2)(\mathbf{Z}_2 + \mathbf{Z}_3) - \mathbf{Z}_2^2}$$
$$= \frac{\mathbf{Z}_2\mathbf{E}_1 - \mathbf{Z}_1\mathbf{E}_2}{\mathbf{Z}_1\mathbf{Z}_2 + \mathbf{Z}_1\mathbf{Z}_3 + \mathbf{Z}_2\mathbf{Z}_3}$$

Substituting numerical values yields

$$\mathbf{I}_2 = \frac{(4\ \Omega - j\,8\ \Omega)(8\ \text{V}\ \angle 20^\circ) - (1\ \Omega + j\,2\ \Omega)(10\ \text{V}\ \angle 0^\circ)}{(1\ \Omega + j\,2\ \Omega)(4\ \Omega - j\,8\ \Omega) + (1\ \Omega + j\,2\ \Omega)(+j\,6\ \Omega) + (4\ \Omega - j\,8\ \Omega)(+j\,6\ \Omega)}$$
$$= \frac{(4 - j\,8)(7.52 + j\,2.74) - (10 + j\,20)}{20 + (j\,6 - 12) + (j\,24 + 48)}$$
$$= \frac{(52.0 - j\,49.20) - (10 + j\,20)}{56 + j\,30} = \frac{42.0 - j\,69.20}{56 + j\,30} = \frac{80.95\ \text{A}\ \angle -58.74^\circ}{63.53\ \angle 28.18^\circ}$$
$$= \mathbf{1.27\ A\ \angle -86.92^\circ}$$

Calculator Solution: The calculator can be an effective tool in performing the long, laborious calculations involved with the final equation appearing above. However, you must be very careful to use brackets to define the order of the arithmetic operations (remember that each open bracket must be followed by a close bracket). Using the TI-89 calculator, the sequence in Fig. 17.17 provides the solution for $\mathbf{I}_2$.

((4−8i)×(8∠20°)−(1+2i)×(10∠0°))÷((1+2i)×(4−8i)+(1+2i)×(6i)+(4−8i)×(6i)) ▶ Polar	1.27E0∠−86.95E0

FIG. 17.17
Determining $\mathbf{I}_2$ using the TI-89 calculator.

Mathcad Solution: This example provides an excellent opportunity to demonstrate the power of Mathcad. First, define the impedances and parameters for the equations to follow as shown in Fig. 17.18. Then enter the **guess** values of the mesh currents $\mathbf{I}_1$ and $\mathbf{I}_2$. Enter the label **Given** followed by the equations for the network. Note that in this example, you are not continuing with the analysis until the matrix is defined—you are working directly from the network equations. Once the equations have been properly entered, enter **Find(I1,I2)**. Then selecting the equal sign results in the single-column matrix with the results in rectangular form. Conversion to polar form requires defining a variable **A** and then calling for the magnitude and angle using the definitions entered earlier in the listing and both the **Calculator** and **Greek** toolbars. The result for $\mathbf{I}_2$ is 1.274 A ∠−86.94°, which is an excellent match with the theoretical solution.

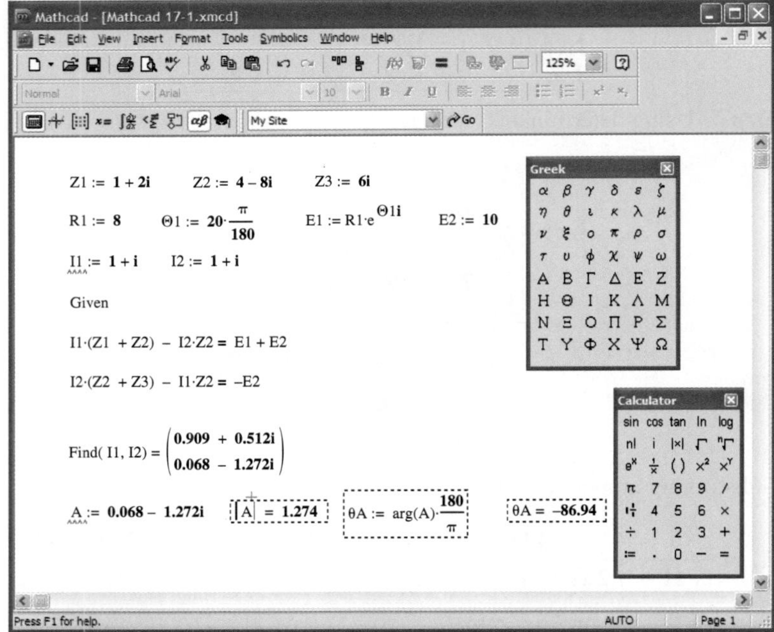

FIG. 17.18
Using Mathcad to verify the results in Example 17.9.

EXAMPLE 17.10 Write the mesh equations for the network in Fig. 17.19. Do not solve.

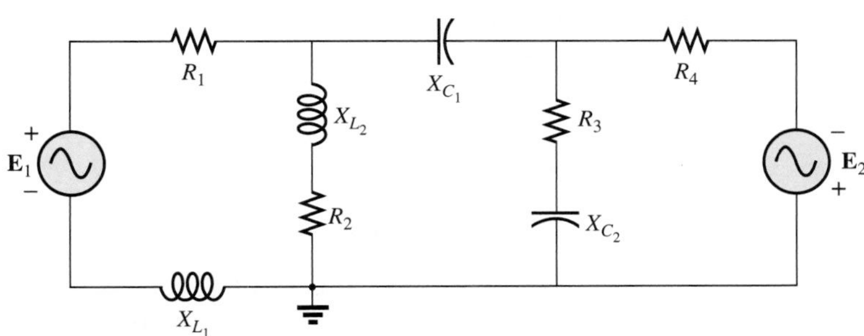

FIG. 17.19
Example 17.10.

Solution: The network is redrawn in Fig. 17.20. Again note the reduced complexity and increased clarity provided by the use of subscripted impedances:

$$\mathbf{Z}_1 = R_1 + j X_{L_1} \qquad \mathbf{Z}_4 = R_3 - j X_{C_2}$$
$$\mathbf{Z}_2 = R_2 + j X_{L_2} \qquad \mathbf{Z}_5 = R_4$$
$$\mathbf{Z}_3 = j X_{C_1}$$

and

$$\begin{aligned} \mathbf{I}_1(\mathbf{Z}_1 + \mathbf{Z}_2) - \mathbf{I}_2\mathbf{Z}_2 &= \mathbf{E}_1 \\ \mathbf{I}_2(\mathbf{Z}_2 + \mathbf{Z}_3 + \mathbf{Z}_4) - \mathbf{I}_1\mathbf{Z}_2 - \mathbf{I}_3\mathbf{Z}_4 &= 0 \\ \mathbf{I}_3(\mathbf{Z}_4 + \mathbf{Z}_5) - \mathbf{I}_2\mathbf{Z}_4 &= \mathbf{E}_2 \end{aligned}$$

or

$$\begin{array}{llll} \mathbf{I}_1(\mathbf{Z}_1 + \mathbf{Z}_2) & - \mathbf{I}_2(\mathbf{Z}_2) & + 0 & = \mathbf{E}_1 \\ \mathbf{I}_1\mathbf{Z}_2 & - \mathbf{I}_2(\mathbf{Z}_2 + \mathbf{Z}_3 + \mathbf{Z}_4) & - \mathbf{I}_3(\mathbf{Z}_4) & = 0 \\ 0 & - \mathbf{I}_2(\mathbf{Z}_4) & - \mathbf{I}_3(\mathbf{Z}_4 + \mathbf{Z}_5) & = \mathbf{E}_2 \end{array}$$

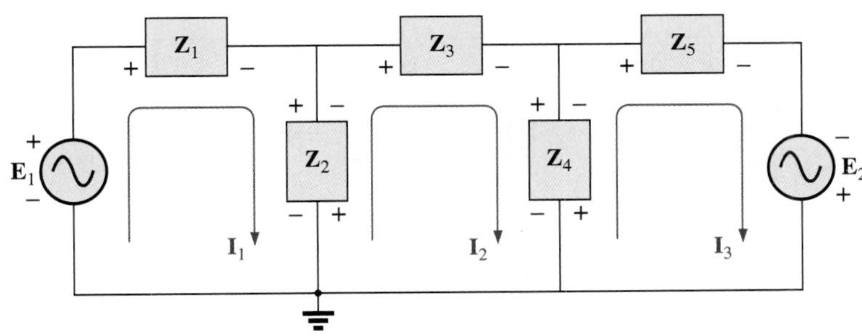

FIG. 17.20
Assigning the mesh currents and subscripted impedances for the network in Fig. 17.19.

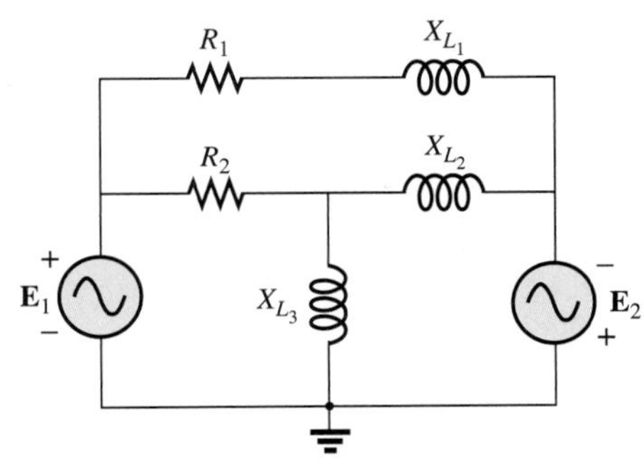

FIG. 17.21
Example 17.11.

EXAMPLE 17.11 Using the format approach, write the mesh equations for the network in Fig. 17.21.

Solution: The network is redrawn as shown in Fig. 17.22, where

$$\mathbf{Z}_1 = R_1 + j X_{L_1} \qquad \mathbf{Z}_3 = j X_{L_2}$$
$$\mathbf{Z}_2 = R_2 \qquad \mathbf{Z}_4 = j X_{L_3}$$

and

$$\begin{aligned} \mathbf{I}_1(\mathbf{Z}_2 + \mathbf{Z}_4) - \mathbf{I}_2\mathbf{Z}_2 - \mathbf{I}_3\mathbf{Z}_4 &= \mathbf{E}_1 \\ \mathbf{I}_2(\mathbf{Z}_1 + \mathbf{Z}_2 + \mathbf{Z}_3) - \mathbf{I}_1\mathbf{Z}_2 - \mathbf{I}_3\mathbf{Z}_3 &= 0 \\ \mathbf{I}_3(\mathbf{Z}_3 + \mathbf{Z}_4) - \mathbf{I}_2\mathbf{Z}_3 - \mathbf{I}_1\mathbf{Z}_4 &= \mathbf{E}_2 \end{aligned}$$

or

$$\begin{array}{llll} \mathbf{I}_1(\mathbf{Z}_2 + \mathbf{Z}_4) & - \mathbf{I}_2\mathbf{Z}_2 & - \mathbf{I}_3\mathbf{Z}_4 & = \mathbf{E}_1 \\ -\mathbf{I}_1\mathbf{Z}_2 & + \mathbf{I}_2(\mathbf{Z}_1 + \mathbf{Z}_2 + \mathbf{Z}_3) & - \mathbf{I}_3\mathbf{Z}_3 & = 0 \\ -\mathbf{I}_1\mathbf{Z}_4 & - \mathbf{I}_2\mathbf{Z}_3 & + \mathbf{I}_3(\mathbf{Z}_3 + \mathbf{Z}_4) & = \mathbf{E}_2 \end{array}$$

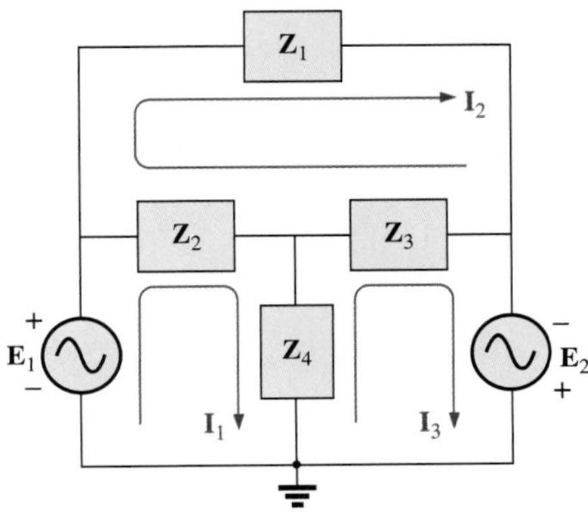

FIG. 17.22
Assigning the mesh currents and subscripted impedances for the network in Fig. 17.21.

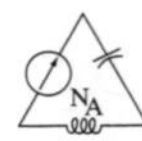

Note the symmetry *about* the diagonal axis; that is, note the location of $-\mathbf{Z}_2$, $-\mathbf{Z}_4$, and $-\mathbf{Z}_3$ off the diagonal.

17.5 NODAL ANALYSIS

General Approach

Independent Sources Before examining the application of the method to ac networks, a review of the appropriate sections on **nodal analysis** in Chapter 8 is suggested since the content of this section is limited to the general conclusions of Chapter 8.

The fundamental steps are the following:

1. ***Determine the number of nodes within the network.***
2. ***Pick a reference node and label each remaining node with a subscripted value of voltage:*** $\mathbf{V}_1$, $\mathbf{V}_2$, ***and so on.***
3. ***Apply Kirchhoff's current law at each node except the reference. Assume that all unknown currents leave the node for each application of Kirchhoff's current law.***
4. ***Solve the resulting equations for the nodal voltages.***

A few examples will refresh your memory about the content of Chapter 8 and the general approach to a nodal-analysis solution.

EXAMPLE 17.12 Determine the voltage across the inductor for the network in Fig. 17.23.

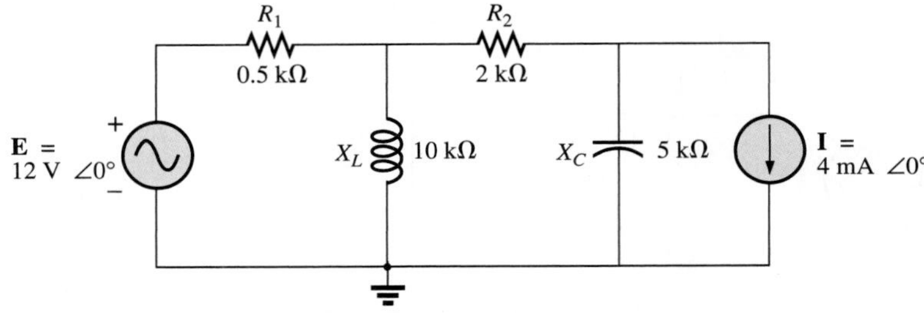

FIG. 17.23
Example 17.12.

Solution:

Steps 1 and 2 are as indicated in Fig. 17.24.

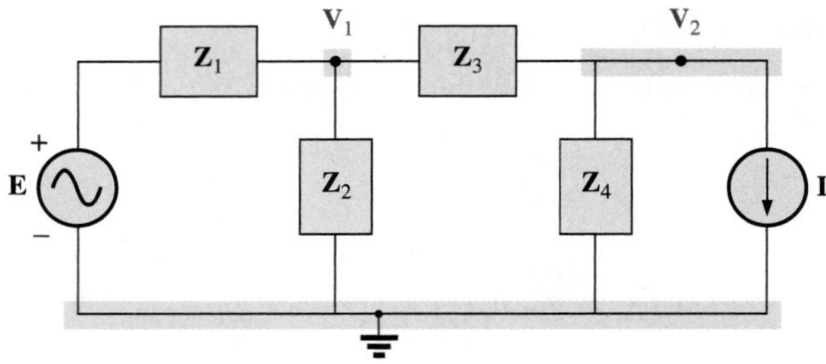

FIG. 17.24
Assigning the nodal voltages and subscripted impedances to the network in Fig. 17.23.

Step 3: Note Fig. 17.25 for the application of Kirchhoff's current law to node $\mathbf{V}_1$:

$$\Sigma \mathbf{I}_i = \Sigma \mathbf{I}_o$$
$$0 = \mathbf{I}_1 + \mathbf{I}_2 + \mathbf{I}_3$$
$$\frac{\mathbf{V}_1 - \mathbf{E}}{\mathbf{Z}_1} + \frac{\mathbf{V}_1}{\mathbf{Z}_2} + \frac{\mathbf{V}_1 - \mathbf{V}_2}{\mathbf{Z}_3} = 0$$

Rearranging terms:

$$\mathbf{V}_1\left[\frac{1}{\mathbf{Z}_1} + \frac{1}{\mathbf{Z}_2} + \frac{1}{\mathbf{Z}_3}\right] - \mathbf{V}_2\left[\frac{1}{\mathbf{Z}_3}\right] = \frac{\mathbf{E}_1}{\mathbf{Z}_1} \qquad \textbf{(17.1)}$$

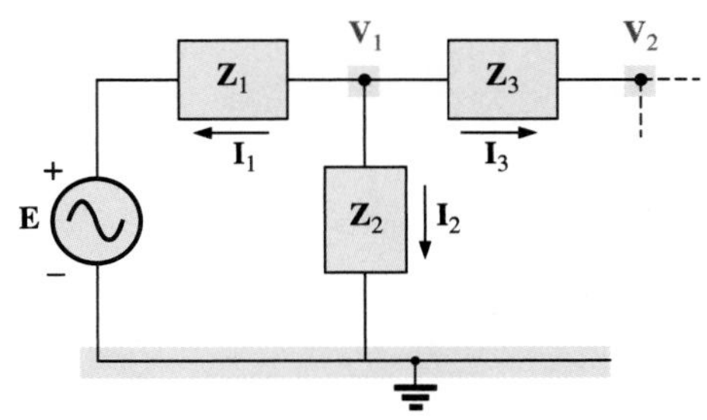

FIG. 17.25
Applying Kirchhoff's current law to the node $\mathbf{V}_1$ in Fig. 17.24.

Note Fig. 17.26 for the application of Kirchoff's current law to node $\mathbf{V}_2$.

$$0 = \mathbf{I}_3 + \mathbf{I}_4 + \mathbf{I}$$
$$\frac{\mathbf{V}_2 - \mathbf{V}_1}{\mathbf{Z}_3} + \frac{\mathbf{V}_2}{\mathbf{Z}_4} + \mathbf{I} = 0$$

Rearranging terms:

$$\mathbf{V}_2\left[\frac{1}{\mathbf{Z}_3} + \frac{1}{\mathbf{Z}_4}\right] - \mathbf{V}_1\left[\frac{1}{\mathbf{Z}_3}\right] = -\mathbf{I} \qquad \textbf{(17.2)}$$

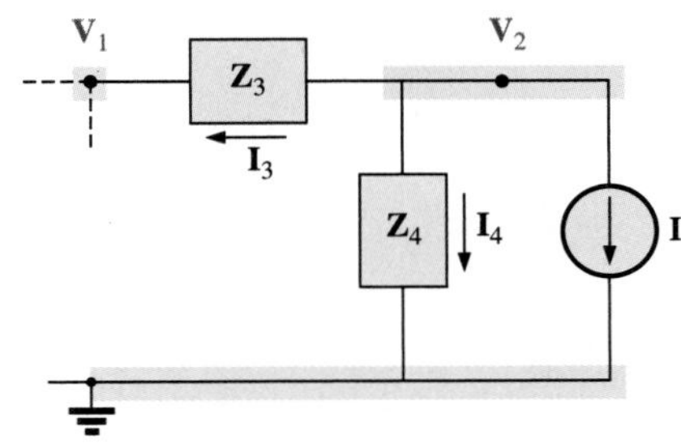

FIG. 17.26
Applying Kirchhoff's current law to the node $\mathbf{V}_2$ in Fig. 17.24.

Grouping equations:

$$\mathbf{V}_1\left[\frac{1}{\mathbf{Z}_1} + \frac{1}{\mathbf{Z}_2} + \frac{1}{\mathbf{Z}_3}\right] - \mathbf{V}_2\left[\frac{1}{\mathbf{Z}_3}\right] = \frac{\mathbf{E}}{\mathbf{Z}_1}$$
$$\mathbf{V}_1\left[\frac{1}{\mathbf{Z}_3}\right] - \mathbf{V}_2\left[\frac{1}{\mathbf{Z}_3} + \frac{1}{\mathbf{Z}_4}\right] = \mathbf{I}$$

$$\frac{1}{\mathbf{Z}_1} + \frac{1}{\mathbf{Z}_2} + \frac{1}{\mathbf{Z}_3} = \frac{1}{0.5\text{ k}\Omega} + \frac{1}{j\,10\text{ k}\Omega} + \frac{1}{2\text{ k}\Omega} = 2.5\text{ mS}\angle -2.29^\circ$$
$$\frac{1}{\mathbf{Z}_3} + \frac{1}{\mathbf{Z}_4} = \frac{1}{2\text{ k}\Omega} + \frac{1}{-j\,5\text{ k}\Omega} = 0.539\text{ mS}\angle 21.80^\circ$$

and

$$\mathbf{V}_1[2.5\text{ mS}\angle -2.29^\circ] - \mathbf{V}_2[0.5\text{ mS}\angle 0^\circ] = 24\text{ mA}\angle 0^\circ$$
$$\mathbf{V}_1[0.5\text{ mS}\angle 0^\circ] - \mathbf{V}_2[0.539\text{ mS}\angle 21.80^\circ] = 4\text{ mA}\angle 0^\circ$$

with

$$\mathbf{V}_1 = \frac{\begin{vmatrix} 24\text{ mA}\angle 0^\circ & -0.5\text{ mS}\angle 0^\circ \\ 4\text{ mA}\angle 0^\circ & -0.539\text{ mS}\angle 21.80^\circ \end{vmatrix}}{\begin{vmatrix} 2.5\text{ mS}\angle -2.29^\circ & -0.5\text{ mS}\angle 0^\circ \\ 0.5\text{ mS}\angle 0^\circ & -0.539\text{ mS}\angle 21.80^\circ \end{vmatrix}}$$

$$= \frac{(24\text{ mA}\angle 0^\circ)(-0.539\text{ mS}\angle 21.80^\circ) + (0.5\text{ mS}\angle 0^\circ)(4\text{ mA}\angle 0^\circ)}{(2.5\text{ mS}\angle -2.29^\circ)(-0.539\text{ mS}\angle 21.80^\circ) + (0.5\text{ mS}\angle 0^\circ)(0.5\text{ mS}\angle 0^\circ)}$$

$$= \frac{-12.94\times 10^{-6}\text{ V}\angle 21.80^\circ + 2\times 10^{-6}\text{ V}\angle 0^\circ}{-1.348\times 10^{-6}\angle 19.51^\circ + 0.25\times 10^{-6}\angle 0^\circ}$$

$$= \frac{-(12.01 + j\,4.81)\times 10^{-6}\text{ V} + 2\times 10^{-6}\text{ V}}{-(1.271 + j\,0.45)\times 10^{-6} + 0.25\times 10^{-6}}$$

$$= \frac{-10.01\text{ V} - j\,4.81\text{ V}}{-1.021 - j\,0.45} = \frac{11.106\text{ V}\angle -154.33^\circ}{1.116\angle -156.21^\circ}$$

$$\mathbf{V}_1 = \mathbf{9.95\ V\angle 1.88^\circ}$$

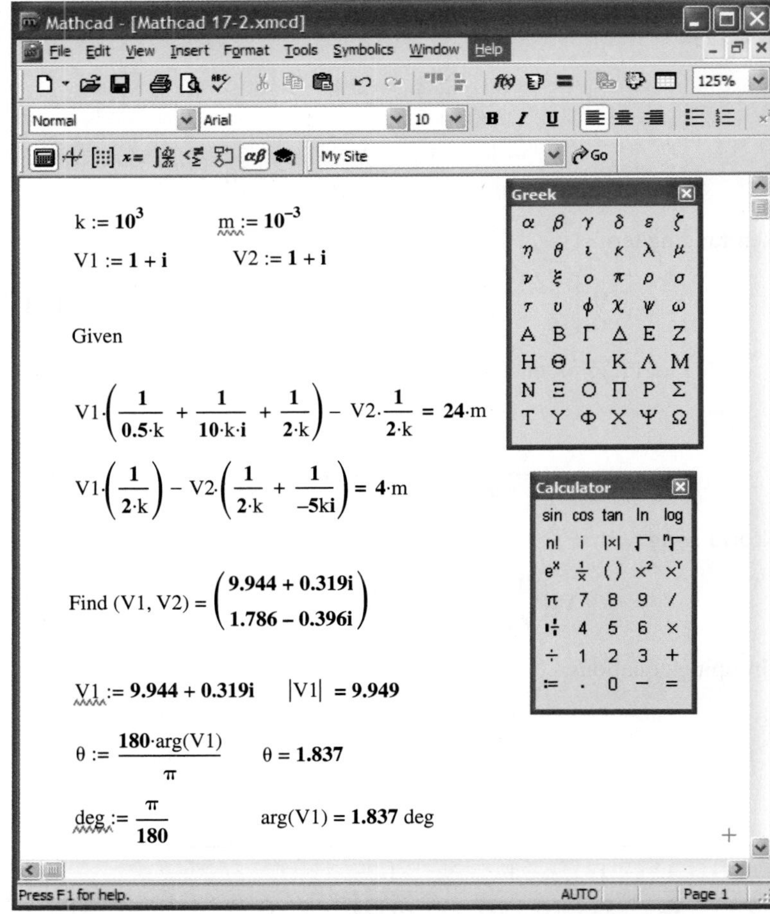

FIG. 17.27
Using Mathcad to verify the results of Example 17.12.

Mathcad Solution: The length and the complexity of the above mathematical development strongly suggest the use of an alternative approach such as Mathcad. The printout in Fig. 17.27 first defines the letters **k** and **m** to specific numerical values so that the power-of-ten format did not have to be included in the equations. Thus, the results are cleaner and easier to review. When entering the equations, remember that the *j* is entered as 1*j* *without* the multiplication sign between the 1 and the *j*. A multiplication sign between the two would define the *j* as another variable. Also be sure that the multiplication process is inserted between the nodal variables and the brackets. If an error signal continues to surface, it is often best to simply reenter the entire listing—errors are often not easy to spot simply by looking at the resulting equations.

Finally the results are obtained and converted to polar form for comparison with the theoretical solution. The solution of 9.949 A $\angle 1.837°$ is a very close confirmation of the longhand solution.

Before leaving this example, let's look at another method for obtaining the polar form of the solution. The method appears in the bottom of Fig. 17.27. First define **deg** as shown, and then pick up **arg** from the **Insert-*f(x)*-Insert Function-arg** sequence. Next enter **V1**; the result is in radian form but with a small black rectangle in the place where the units normally appear. Click on that black rectangle, and the bracket ap-

pears where you can type **deg.** When you select the equal sign, the angle in degrees appears.

Dependent Current Sources For dependent current sources, the procedure is modified as follows:

1. Steps 1 and 2 are the same as those applied for independent sources.
2. Step 3 is modified as follows: Treat each dependent current source like an independent source when Kirchhoff's current law is applied to each defined node. However, once the equations are established, substitute the equation for the controlling quantity to ensure that the unknowns are limited solely to the chosen nodal voltages.
3. Step 4 is as before.

EXAMPLE 17.13 Write the nodal equations for the network in Fig. 17.28 having a dependent current source.

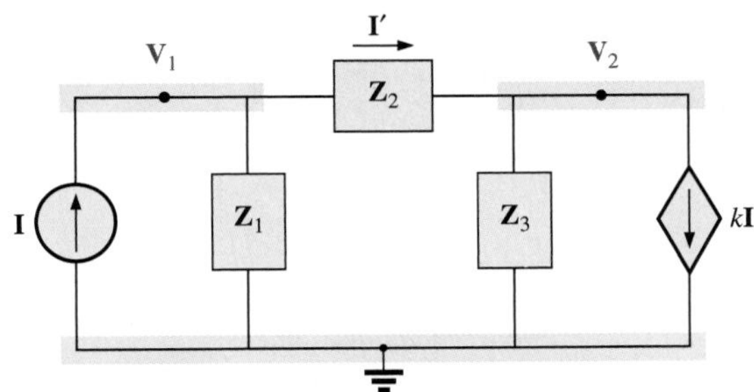

FIG. 17.28
Applying nodal analysis to a network with a current-controlled current source.

Solution:

Steps 1 and 2 are as defined in Fig. 17.28.

Step 3: At node $\mathbf{V}_1$,

$$\mathbf{I} = \mathbf{I}_1 + \mathbf{I}_2$$

$$\frac{\mathbf{V}_1}{\mathbf{Z}_1} + \frac{\mathbf{V}_1 - \mathbf{V}_2}{\mathbf{Z}_2} - \mathbf{I} = 0$$

and
$$\mathbf{V}_1\left[\frac{1}{\mathbf{Z}_1} + \frac{1}{\mathbf{Z}_2}\right] - \mathbf{V}_2\left[\frac{1}{\mathbf{Z}_2}\right] = \mathbf{I}$$

At node $\mathbf{V}_2$,

$$\mathbf{I}_2 + \mathbf{I}_3 + k\mathbf{I} = 0$$

$$\frac{\mathbf{V}_2 - \mathbf{V}_1}{\mathbf{Z}_2} + \frac{\mathbf{V}_2}{\mathbf{Z}_3} + k\left[\frac{\mathbf{V}_1 - \mathbf{V}_2}{\mathbf{Z}_2}\right] = 0$$

and
$$\mathbf{V}_1\left[\frac{1 - k}{\mathbf{Z}_2}\right] - \mathbf{V}_2\left[\frac{1 - k}{\mathbf{Z}_2} + \frac{1}{\mathbf{Z}_3}\right] = 0$$

resulting in two equations and two unknowns.

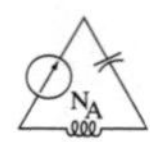

Independent Voltage Sources between Assigned Nodes For independent voltage sources between assigned nodes, the procedure is modified as follows:

1. Steps 1 and 2 are the same as those applied for independent sources.
2. Step 3 is modified as follows: Treat each source between defined nodes as a short circuit (recall the *supernode* classification in Chapter 8), and write the nodal equations for each remaining independent node. Then relate the chosen nodal voltages to the independent voltage source to ensure that the unknowns of the final equations are limited solely to the nodal voltages.
3. Step 4 is as before.

EXAMPLE 17.14 Write the nodal equations for the network in Fig. 17.29 having an independent source between two assigned nodes.

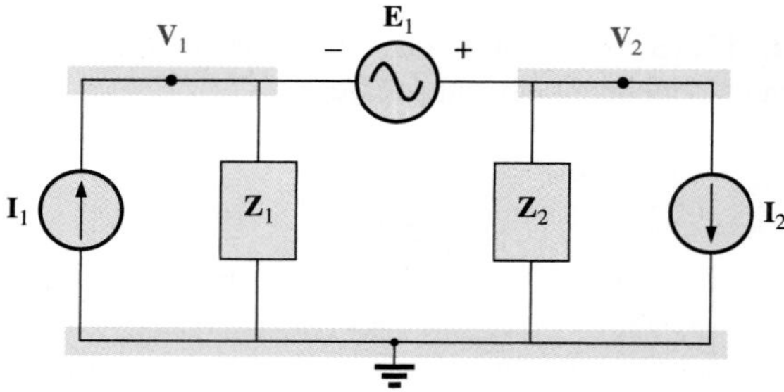

FIG. 17.29
Applying nodal analysis to a network with an independent voltage source between defined nodes.

Solution:

Steps 1 and 2 are defined in Fig. 17.29.

Step 3: Replacing the independent source $\mathbf{E}_1$ with a short-circuit equivalent results in a supernode that generates the following equation when Kirchhoff's current law is applied to node $\mathbf{V}_1$:

$$\mathbf{I}_1 = \frac{\mathbf{V}_1}{\mathbf{Z}_1} + \frac{\mathbf{V}_2}{\mathbf{Z}_2} + \mathbf{I}_2$$

with

$$\mathbf{V}_2 - \mathbf{V}_1 = \mathbf{E}_1$$

and we have two equations and two unknowns.

Dependent Voltage Sources between Defined Nodes For dependent voltage sources between defined nodes, the procedure is modified as follows:

1. Steps 1 and 2 are the same as those applied for independent voltage sources.
2. Step 3 is modified as follows: The procedure is essentially the same as that applied for independent voltage sources, except now the dependent sources have to be defined in terms of the chosen nodal voltages to ensure that the final equations have only nodal voltages as their unknown quantities.
3. Step 4 is as before.

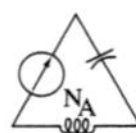

EXAMPLE 17.15 Write the nodal equations for the network in Fig. 17.30 having a dependent voltage source between two defined nodes.

Solution:

Steps 1 and 2 are defined in Fig. 17.30.

Step 3: Replacing the dependent source $\mu\mathbf{V}_x$ with a short-circuit equivalent results in the following equation when Kirchhoff's current law is applied at node $\mathbf{V}_1$:

$$\mathbf{I} = \mathbf{I}_1 + \mathbf{I}_2$$

$$\frac{\mathbf{V}_1}{\mathbf{Z}_1} + \frac{(\mathbf{V}_1 - \mathbf{V}_2)}{\mathbf{Z}_2} - \mathbf{I} = 0$$

and

$$\mathbf{V}_2 = \mu\mathbf{V}_x = \mu[\mathbf{V}_1 - \mathbf{V}_2]$$

or

$$\mathbf{V}_2 = \frac{\mu}{1 + \mu}\mathbf{V}_1$$

resulting in two equations and two unknowns. Note that because the impedance $\mathbf{Z}_3$ is in parallel with a voltage source, it does not appear in the analysis. It will, however, affect the current through the dependent voltage source.

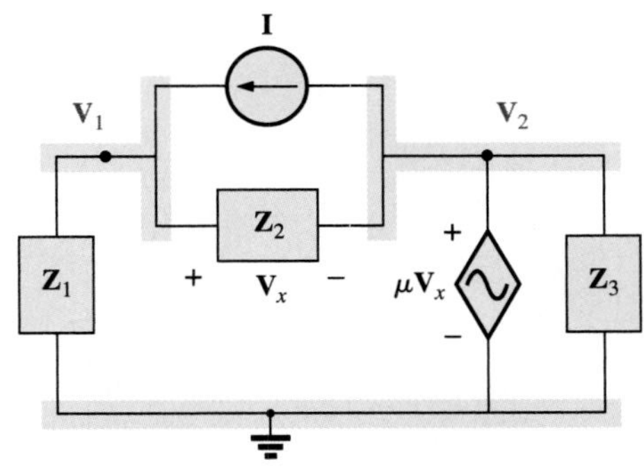

FIG. 17.30
Applying nodal analysis to a network with a voltage-controlled voltage source.

Format Approach

A close examination of Eqs. (17.1) and (17.2) in Example 17.12 reveals that they are the same equations that would have been obtained using the format approach introduced in Chapter 8. Recall that the approach required that the voltage source first be converted to a current source, but the writing of the equations was quite direct and minimized any chances of an error due to a lost sign or missing term.

The sequence of steps required to apply the format approach is the following:

1. ***Choose a reference node and assign a subscripted voltage label to the (N − 1) remaining independent nodes of the network.***
2. ***The number of equations required for a complete solution is equal to the number of subscripted voltages (N − 1). Column 1 of each equation is formed by summing the admittances tied to the node of interest and multiplying the result by that subscripted nodal voltage.***
3. ***The mutual terms are always subtracted from the terms of the first column. It is possible to have more than one mutual term if the nodal voltage of interest has an element in common with more than one other nodal voltage. Each mutual term is the product of the mutual admittance and the other nodal voltage tied to that admittance.***
4. ***The column to the right of the equality sign is the algebraic sum of the current sources tied to the node of interest. A current source is assigned a positive sign if it supplies current to a node, and a negative sign if it draws current from the node.***
5. ***Solve resulting simultaneous equations for the desired nodal voltages. The comments offered for mesh analysis regarding independent and dependent sources apply here also.***

EXAMPLE 17.16 Using the format approach to nodal analysis, find the voltage across the 4 Ω resistor in Fig. 17.31.

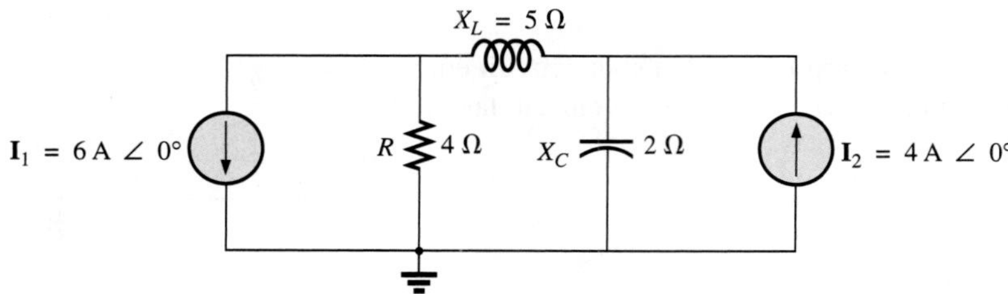

FIG. 17.31
Example 17.16.

Solution: Choosing nodes (Fig. 17.32) and writing the nodal equations, we have

$$\mathbf{Z}_1 = R = 4\ \Omega \quad \mathbf{Z}_2 = jX_L = j\,5\ \Omega \quad \mathbf{Z}_3 = -jX_C = -j\,2\ \Omega$$

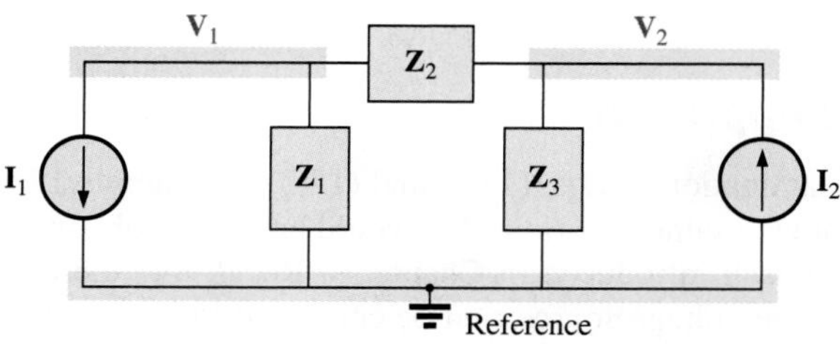

FIG. 17.32
Assigning the nodal voltages and subscripted impedances for the network in Fig. 17.31.

$$\mathbf{V}_1(\mathbf{Y}_1 + \mathbf{Y}_2) - \mathbf{V}_2(\mathbf{Y}_2) = -\mathbf{I}_1$$
$$\mathbf{V}_2(\mathbf{Y}_3 + \mathbf{Y}_2) - \mathbf{V}_1(\mathbf{Y}_2) = +\mathbf{I}_2$$

or

$$\begin{aligned} \mathbf{V}_1(\mathbf{Y}_1 + \mathbf{Y}_2) - \mathbf{V}_2(\mathbf{Y}_2) \qquad\qquad &= -\mathbf{I}_1 \\ -\mathbf{V}_1(\mathbf{Y}_2) \qquad + \mathbf{V}_2(\mathbf{Y}_3 + \mathbf{Y}_2) &= +\mathbf{I}_2 \end{aligned}$$

$$\mathbf{Y}_1 = \frac{1}{\mathbf{Z}_1} \quad \mathbf{Y}_2 = \frac{1}{\mathbf{Z}_2} \quad \mathbf{Y}_3 = \frac{1}{\mathbf{Z}_3}$$

Using determinants yields

$$\mathbf{V}_1 = \frac{\begin{vmatrix} -\mathbf{I}_1 & -\mathbf{Y}_2 \\ +\mathbf{I}_2 & \mathbf{Y}_3 + \mathbf{Y}_2 \end{vmatrix}}{\begin{vmatrix} \mathbf{Y}_1 + \mathbf{Y}_2 & -\mathbf{Y}_2 \\ -\mathbf{Y}_2 & \mathbf{Y}_3 + \mathbf{Y}_2 \end{vmatrix}} = \frac{\begin{vmatrix} -6\text{ A} + j\,0.2\text{ S} \\ +4\text{ A} + j\,0.3\text{ S} \end{vmatrix}}{\begin{vmatrix} 0.25\text{ S} - j\,0.2\text{ S} & +j\,0.2\text{ S} \\ +j\,0.2\text{ S} & +j\,0.3\text{ S} \end{vmatrix}}$$

$$= \frac{-(\mathbf{Y}_3 + \mathbf{Y}_2)\mathbf{I}_1 + \mathbf{I}_2\mathbf{Y}_2}{(\mathbf{Y}_1 + \mathbf{Y}_2)(\mathbf{Y}_3 + \mathbf{Y}_2) - \mathbf{Y}_2^2}$$

$$= \frac{-(\mathbf{Y}_3 + \mathbf{Y}_2)\mathbf{I}_1 + \mathbf{I}_2\mathbf{Y}_2}{\mathbf{Y}_1\mathbf{Y}_3 + \mathbf{Y}_2\mathbf{Y}_3 + \mathbf{Y}_1\mathbf{Y}_2}$$

Substituting numerical values, we have

$$\mathbf{V}_1 = \frac{-[(1/-j\,2\,\Omega) + (1/j\,5\,\Omega)]6\text{ A}\angle 0^\circ + 4\text{ A}\angle 0^\circ\,(1/j\,5\,\Omega)}{(1/4\,\Omega)(1/-j\,2\,\Omega) + (1/j\,5\,\Omega)(1/-j\,2\,\Omega) + (1/4\,\Omega)(1/j\,5\,\Omega)}$$

$$= \frac{-(+j\,0.5 - j\,0.2)6\angle 0^\circ + 4\angle 0^\circ\,(-j\,0.2)}{(1/-j\,8) + (1/10) + (1/j\,20)}$$

$$= \frac{(-0.3\angle 90^\circ)(6\angle 0^\circ) + (4\angle 0^\circ)(0.2\angle -90^\circ)}{j\,0.125 + 0.1 - j\,0.05}$$

$$= \frac{-1.8\angle 90^\circ + 0.8\angle -90^\circ}{0.1 + j\,0.075}$$

$$= \frac{2.6\text{ V}\angle -90^\circ}{0.125\angle 36.87^\circ}$$

$$\mathbf{V}_1 = \mathbf{20.80\text{ V}\angle -126.87^\circ}$$

Calculator Solution: Using the TI-89 calculator, enter the parameters for the determinant form for $\mathbf{V}_1$ in the rectangular form as shown by the sequence in Fig. 17.33. Note the different negative signs used to enter the data.

det ([(-) 6 , . 2 *i* ; 4 , . 3 *i* **]**) ÷
det ([. 2 5 - . 2 *i* , . 2 *i* ; . 2 *i* ,
. 3 *i* **]**) ▶ **Polar** ENTER **20.80E0 ∠ –126.8E0**

FIG. 17.33
Determining $\mathbf{V}_1$ *using the TI-89 calculator.*

EXAMPLE 17.17 Using the format approach, write the nodal equations for the network in Fig. 17.34.

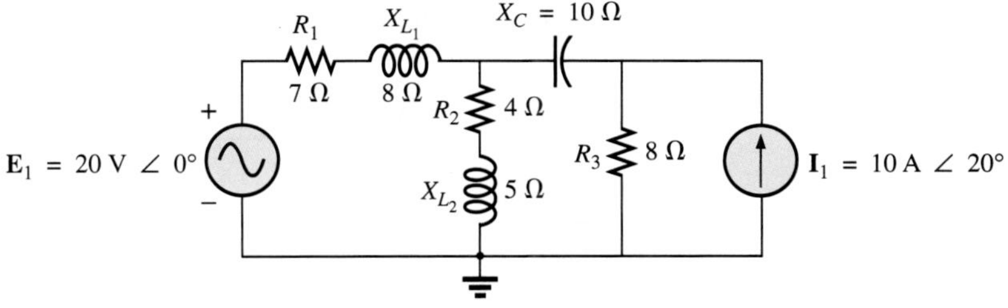

FIG. 17.34
Example 17.17.

Solution: The circuit is redrawn in Fig. 17.35, where

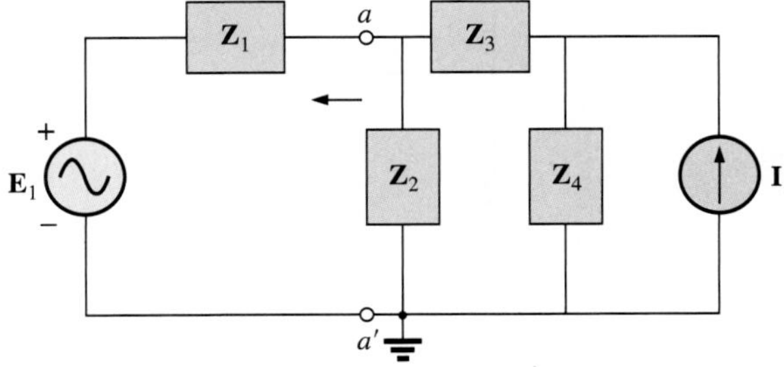

FIG. 17.35
Assigning the subscripted impedances for the network in Fig. 17.34.

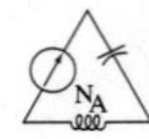

$$\mathbf{Z}_1 = R_1 + jX_{L_1} = 7\ \Omega + j\,8\ \Omega \qquad \mathbf{E}_1 = 20\ \text{V}\ \angle 0^\circ$$
$$\mathbf{Z}_2 = R_2 + jX_{L_2} = 4\ \Omega + j\,5\ \Omega \qquad \mathbf{I}_1 = 10\ \text{A}\ \angle 20^\circ$$
$$\mathbf{Z}_3 = -jX_C = -j\,10\ \Omega$$
$$\mathbf{Z}_4 = R_3 = 8\ \Omega$$

Converting the voltage source to a current source and choosing nodes, we obtain Fig. 17.36. Note the "neat" appearance of the network using the subscripted impedances. Working directly with Fig. 17.34 would be more difficult and could produce errors.

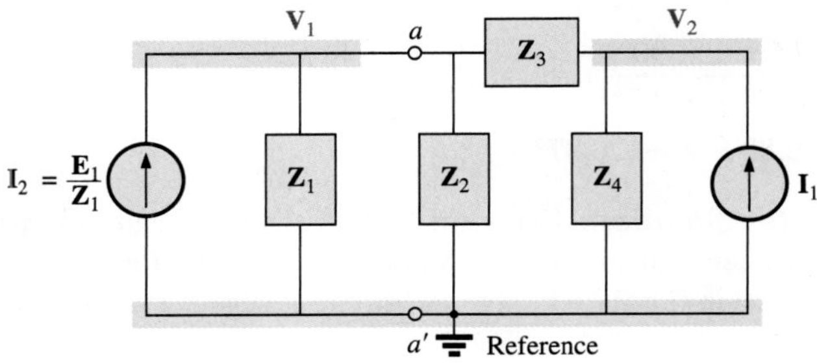

FIG. 17.36
Converting the voltage source in Fig. 17.35 to a current source and defining the nodal voltages.

Write the nodal equations:

$$\mathbf{V}_1(\mathbf{Y}_1 + \mathbf{Y}_2 + \mathbf{Y}_3) - \mathbf{V}_2(\mathbf{Y}_3) = +\mathbf{I}_2$$
$$\underline{\mathbf{V}_2(\mathbf{Y}_3 + \mathbf{Y}_4) - \mathbf{V}_1(\mathbf{Y}_3) = +\mathbf{I}_1}$$

$$\mathbf{Y}_1 = \frac{1}{\mathbf{Z}_1} \quad \mathbf{Y}_2 = \frac{1}{\mathbf{Z}_2} \quad \mathbf{Y}_3 = \frac{1}{\mathbf{Z}_3} \quad \mathbf{Y}_4 = \frac{1}{\mathbf{Z}_4}$$

which are rewritten as

$$\begin{aligned} \mathbf{V}_1(\mathbf{Y}_1 + \mathbf{Y}_2 + \mathbf{Y}_3) - \mathbf{V}_2(\mathbf{Y}_3) \qquad &= +\mathbf{I}_2 \\ -\mathbf{V}_1(\mathbf{Y}_3) \qquad + \mathbf{V}_2(\mathbf{Y}_3 + \mathbf{Y}_4) &= +\mathbf{I}_1 \end{aligned}$$

EXAMPLE 17.18 Write the nodal equations for the network in Fig. 17.37. Do not solve.

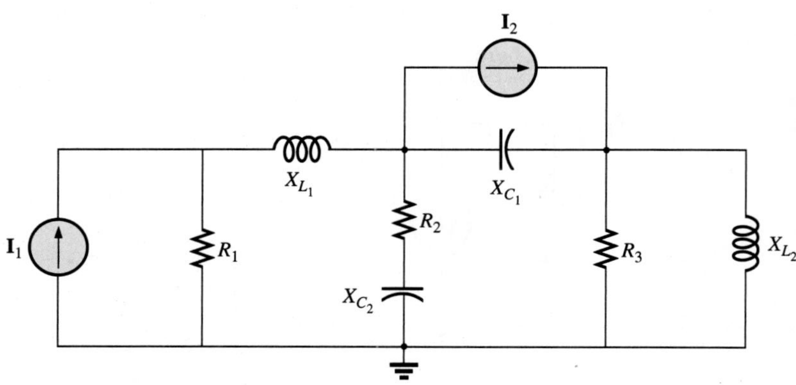

FIG. 17.37
Example 17.18.

Solution: Choose nodes (Fig. 17.38):

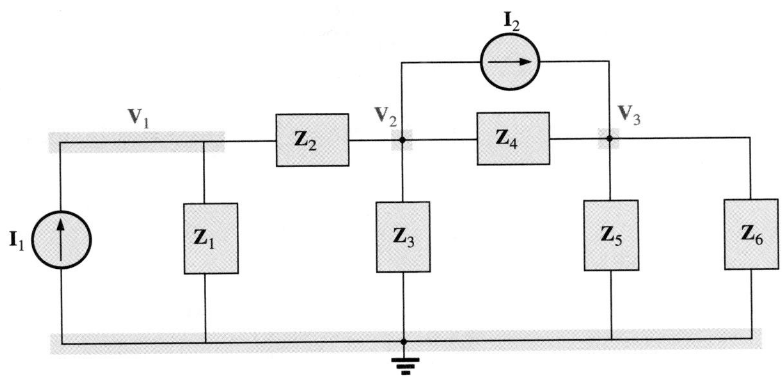

FIG. 17.38
Assigning the nodal voltages and subscripted impedances for the network in Fig. 17.35.

$$\mathbf{Z}_1 = R_1 \qquad \mathbf{Z}_2 = j\,X_{L_1} \quad \mathbf{Z}_3 = R_2 - j\,X_{C_2}$$
$$\mathbf{Z}_4 = -j\,X_{C_1} \quad \mathbf{Z}_5 = R_3 \qquad \mathbf{Z}_6 = j\,X_{L_2}$$

and write the nodal equations:

$$\begin{aligned}\mathbf{V}_1(\mathbf{Y}_1 + \mathbf{Y}_2) - \mathbf{V}_2(\mathbf{Y}_2) &= +\mathbf{I}_1\\ \mathbf{V}_2(\mathbf{Y}_2 + \mathbf{Y}_3 + \mathbf{Y}_4) - \mathbf{V}_1(\mathbf{Y}_2) - \mathbf{V}_3(\mathbf{Y}_4) &= -\mathbf{I}_2\\ \mathbf{V}_3(\mathbf{Y}_4 + \mathbf{Y}_5 + \mathbf{Y}_6) - \mathbf{V}_2(\mathbf{Y}_4) &= +\mathbf{I}_2\end{aligned}$$

which are rewritten as

$$\begin{array}{llll}\mathbf{V}_1(\mathbf{Y}_1 + \mathbf{Y}_2) - \mathbf{V}_2(\mathbf{Y}_2) & & + 0 & = +\mathbf{I}_1\\ -\mathbf{V}_1(\mathbf{Y}_2) & + \mathbf{V}_2(\mathbf{Y}_2 + \mathbf{Y}_3 + \mathbf{Y}_4) - \mathbf{V}_3(\mathbf{Y}_4) & & = -\mathbf{I}_2\\ 0 & - \mathbf{V}_2(\mathbf{Y}_4) & + \mathbf{V}_3(\mathbf{Y}_4 + \mathbf{Y}_5 + \mathbf{Y}_6) & = +\mathbf{I}_2\end{array}$$

$$\mathbf{Y}_1 = \frac{1}{R_1} \qquad \mathbf{Y}_2 = \frac{1}{j\,X_{L_1}} \quad \mathbf{Y}_3 = \frac{1}{R_2 - j\,X_{C_2}}$$
$$\mathbf{Y}_4 = \frac{1}{-j\,X_{C_1}} \quad \mathbf{Y}_5 = \frac{1}{R_3} \qquad \mathbf{Y}_6 = \frac{1}{j\,X_{L_2}}$$

Note the symmetry about the diagonal for this example and those preceding it in this section.

EXAMPLE 17.19 Apply nodal analysis to the network in Fig. 17.39. Determine the voltage $\mathbf{V}_L$.

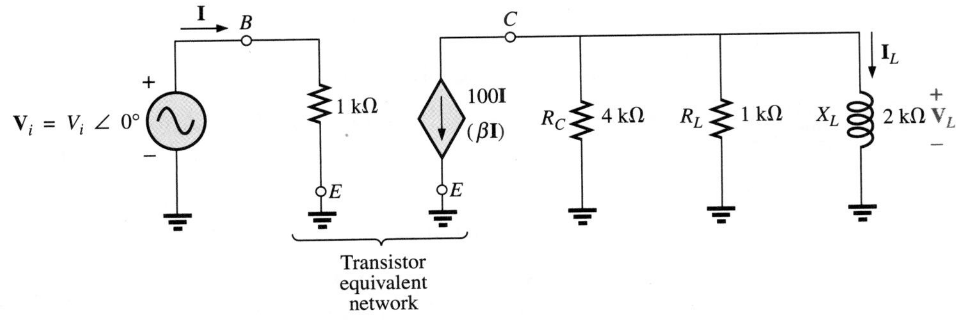

FIG. 17.39
Example 17.19.

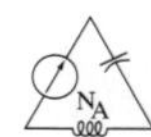

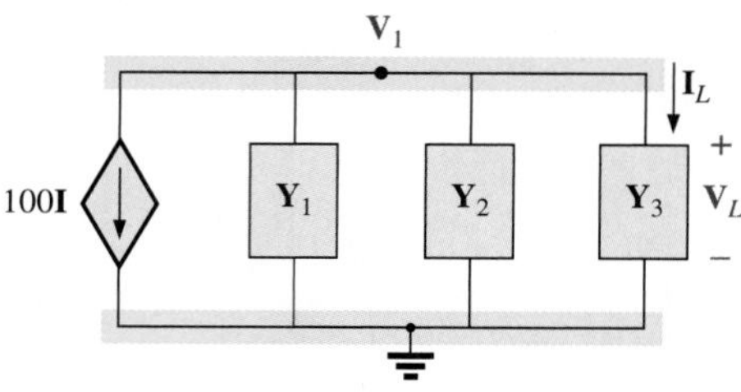

FIG. 17.40
Assigning the nodal voltage and subscripted impedances for the network in Fig. 17.39.

Solution: In this case, there is no need for a source conversion. The network is redrawn in Fig. 17.40 with the chosen nodal voltage and subscripted impedances.

Apply the format approach:

$$\mathbf{Y}_1 = \frac{1}{\mathbf{Z}_1} = \frac{1}{4\ \text{k}\Omega} = 0.25\ \text{mS}\ \angle 0° = G_1\ \angle 0°$$

$$\mathbf{Y}_2 = \frac{1}{\mathbf{Z}_2} = \frac{1}{1\ \text{k}\Omega} = 1\ \text{mS}\ \angle 0° = G_2\ \angle 0°$$

$$\mathbf{Y}_3 = \frac{1}{\mathbf{Z}_3} = \frac{1}{2\ \text{k}\Omega\ \angle 90°} = 0.5\ \text{mS}\ \angle -90°$$

$$= -j\,0.5\ \text{mS} = -j\,B_L$$

$$\mathbf{V}_1\text{: } (\mathbf{Y}_1 + \mathbf{Y}_2 + \mathbf{Y}_3)\mathbf{V}_1 = -100\mathbf{I}$$

and

$$\mathbf{V}_1 = \frac{-100\mathbf{I}}{\mathbf{Y}_1 + \mathbf{Y}_2 + \mathbf{Y}_3}$$

$$= \frac{-100\mathbf{I}}{0.25\ \text{mS} + 1\ \text{mS} - j\,0.5\ \text{mS}}$$

$$= \frac{-100 \times 10^3\mathbf{I}}{1.25 - j\,0.5} = \frac{-100 \times 10^3\mathbf{I}}{1.3463\ \angle -21.80°}$$

$$= -74.28 \times 10^3\mathbf{I}\ \angle 21.80°$$

$$= -74.28 \times 10^3 \left(\frac{\mathbf{V}_i}{1\ \text{k}\Omega}\right) \angle 21.80°$$

$$\mathbf{V}_1 = \mathbf{V}_L = \mathbf{-(74.28V_\mathit{i})V\ \angle 21.80°}$$

17.6 BRIDGE NETWORKS (ac)

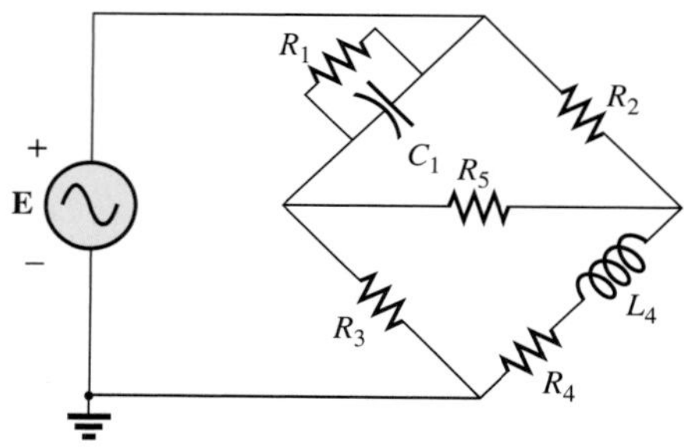

FIG. 17.41
Maxwell bridge.

The basic bridge configuration was discussed in some detail in Section 8.11 for dc networks. We now continue to examine **bridge networks** by considering those that have reactive components and a sinusoidal ac voltage or current applied.

We first analyze various familiar forms of the bridge network using mesh analysis and nodal analysis (the format approach). The balance conditions are investigated throughout the section.

Apply **mesh analysis** to the network in Fig. 17.41. The network is redrawn in Fig. 17.42, where

$$\mathbf{Z}_1 = \frac{1}{\mathbf{Y}_1} = \frac{1}{G_1 + j\,B_C} = \frac{G_1}{G_1^2 + B_C^2} - j\,\frac{B_C}{G_1^2 + B_C^2}$$

$$\mathbf{Z}_2 = R_2 \quad \mathbf{Z}_3 = R_3 \quad \mathbf{Z}_4 = R_4 + j\,X_L \quad \mathbf{Z}_5 = R_5$$

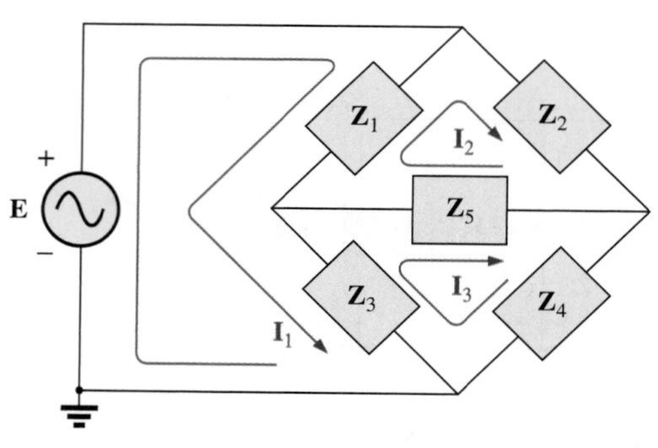

FIG. 17.42
Assigning the mesh currents and subscripted impedances for the network in Fig. 17.41.

Applying the format approach:

$$(\mathbf{Z}_1 + \mathbf{Z}_3)\mathbf{I}_1 - (\mathbf{Z}_1)\mathbf{I}_2 - (\mathbf{Z}_3)\mathbf{I}_3 = \mathbf{E}$$
$$(\mathbf{Z}_1 + \mathbf{Z}_2 + \mathbf{Z}_5)\mathbf{I}_2 - (\mathbf{Z}_1)\mathbf{I}_1 - (\mathbf{Z}_5)\mathbf{I}_3 = 0$$
$$(\mathbf{Z}_3 + \mathbf{Z}_4 + \mathbf{Z}_5)\mathbf{I}_3 - (\mathbf{Z}_3)\mathbf{I}_1 - (\mathbf{Z}_5)\mathbf{I}_2 = 0$$

which are rewritten as

$$\begin{aligned}
\mathbf{I}_1(\mathbf{Z}_1 + \mathbf{Z}_3) &- \mathbf{I}_2\mathbf{Z}_1 & &- \mathbf{I}_3\mathbf{Z}_3 &= \mathbf{E}\\
-\mathbf{I}_1\mathbf{Z}_1 &+ \mathbf{I}_2(\mathbf{Z}_1 + \mathbf{Z}_2 + \mathbf{Z}_5) & &- \mathbf{I}_3\mathbf{Z}_5 &= 0\\
-\mathbf{I}_1\mathbf{Z}_3 &- \mathbf{I}_2\mathbf{Z}_5 & &+ \mathbf{I}_3(\mathbf{Z}_3 + \mathbf{Z}_4 + \mathbf{Z}_5) &= 0
\end{aligned}$$

Note the symmetry about the diagonal of the above equations. For balance, $\mathbf{I}_{\mathbf{Z}_5} = 0\text{ A}$, and

$$\mathbf{I}_{\mathbf{Z}_5} = \mathbf{I}_2 - \mathbf{I}_3 = 0$$

From the above equations,

$$\mathbf{I}_2 = \frac{\begin{vmatrix} \mathbf{Z}_1 + \mathbf{Z}_3 & \mathbf{E} & -\mathbf{Z}_3 \\ -\mathbf{Z}_1 & 0 & -\mathbf{Z}_5 \\ -\mathbf{Z}_3 & 0 & (\mathbf{Z}_3 + \mathbf{Z}_4 + \mathbf{Z}_5) \end{vmatrix}}{\begin{vmatrix} \mathbf{Z}_1 + \mathbf{Z}_3 & -\mathbf{Z}_1 & -\mathbf{Z}_3 \\ -\mathbf{Z}_1 & (\mathbf{Z}_1 + \mathbf{Z}_2 + \mathbf{Z}_5) & -\mathbf{Z}_5 \\ -\mathbf{Z}_3 & -\mathbf{Z}_5 & (\mathbf{Z}_3 + \mathbf{Z}_4 + \mathbf{Z}_5) \end{vmatrix}}$$

$$= \frac{\mathbf{E}(\mathbf{Z}_1\mathbf{Z}_3 + \mathbf{Z}_1\mathbf{Z}_4 + \mathbf{Z}_1\mathbf{Z}_5 + \mathbf{Z}_3\mathbf{Z}_5)}{\Delta}$$

where Δ signifies the determinant of the denominator (or coefficients). Similarly,

$$\mathbf{I}_3 = \frac{\mathbf{E}(\mathbf{Z}_1\mathbf{Z}_3 + \mathbf{Z}_3\mathbf{Z}_2 + \mathbf{Z}_1\mathbf{Z}_5 + \mathbf{Z}_3\mathbf{Z}_5)}{\Delta}$$

and

$$\mathbf{I}_{\mathbf{Z}_5} = \mathbf{I}_2 - \mathbf{I}_3 = \frac{\mathbf{E}(\mathbf{Z}_1\mathbf{Z}_4 - \mathbf{Z}_3\mathbf{Z}_2)}{\Delta}$$

For $\mathbf{I}_{\mathbf{Z}_5} = 0$, the following must be satisfied (for a finite Δ not equal to zero):

$$\boxed{\mathbf{Z}_1\mathbf{Z}_4 = \mathbf{Z}_3\mathbf{Z}_2} \qquad \mathbf{I}_{\mathbf{Z}_5} = 0 \qquad \textbf{(17.3)}$$

This condition is analyzed in greater depth later in this section.

Applying **nodal analysis** to the network in Fig. 17.43 results in the configuration in Fig. 17.44, where

$$\mathbf{Y}_1 = \frac{1}{\mathbf{Z}_1} = \frac{1}{R_1 - jX_C} \qquad \mathbf{Y}_2 = \frac{1}{\mathbf{Z}_2} = \frac{1}{R_2}$$

$$\mathbf{Y}_3 = \frac{1}{\mathbf{Z}_3} = \frac{1}{R_3} \qquad \mathbf{Y}_4 = \frac{1}{\mathbf{Z}_4} = \frac{1}{R_4 + jX_L} \qquad \mathbf{Y}_5 = \frac{1}{R_5}$$

and

$$\begin{aligned} (\mathbf{Y}_1 + \mathbf{Y}_2)\mathbf{V}_1 - (\mathbf{Y}_1)\mathbf{V}_2 - (\mathbf{Y}_2)\mathbf{V}_3 &= \mathbf{I} \\ (\mathbf{Y}_1 + \mathbf{Y}_3 + \mathbf{Y}_5)\mathbf{V}_2 - (\mathbf{Y}_1)\mathbf{V}_1 - (\mathbf{Y}_5)\mathbf{V}_3 &= 0 \\ \underline{(\mathbf{Y}_2 + \mathbf{Y}_4 + \mathbf{Y}_5)\mathbf{V}_3 - (\mathbf{Y}_2)\mathbf{V}_1 - (\mathbf{Y}_5)\mathbf{V}_2} &= \underline{0} \end{aligned}$$

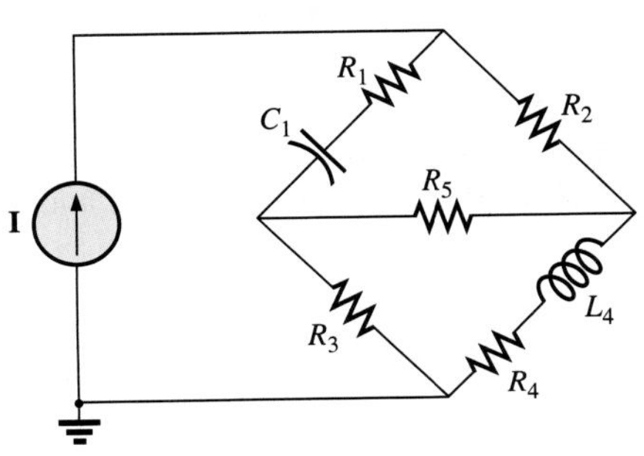

FIG. 17.43
Hay bridge.

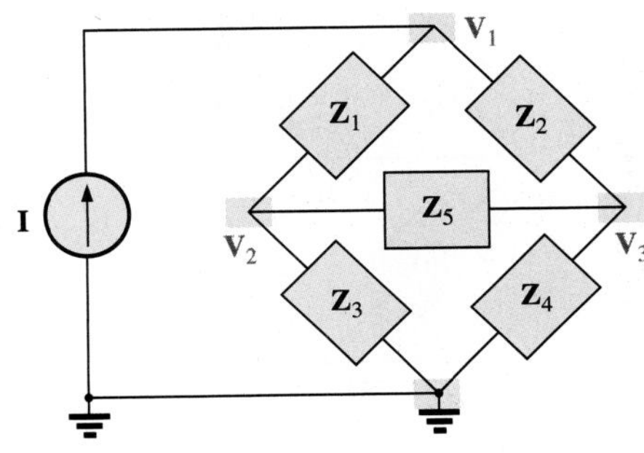

FIG. 17.44
Assigning the nodal voltages and subscripted impedances for the network in Fig. 17.43.

which are rewritten as

$$\begin{array}{llll} \mathbf{V}_1(\mathbf{Y}_1 + \mathbf{Y}_2) - \mathbf{V}_2\mathbf{Y}_1 & & - \mathbf{V}_3\mathbf{Y}_2 & = \mathbf{I} \\ -\mathbf{V}_1\mathbf{Y}_2 & + \mathbf{V}_2(\mathbf{Y}_1 + \mathbf{Y}_3 + \mathbf{Y}_5) & - \mathbf{V}_3\mathbf{Y}_5 & = 0 \\ -\mathbf{V}_1\mathbf{Y}_2 & - \mathbf{V}_2\mathbf{Y}_5 & + \mathbf{V}_3(\mathbf{Y}_2 + \mathbf{Y}_4 + \mathbf{Y}_5) & = 0 \end{array}$$

Again, note the symmetry about the diagonal axis. For balance, $\mathbf{V}_{\mathbf{Z}_5} = 0\text{ V}$, and

$$\mathbf{V}_{\mathbf{Z}_5} = \mathbf{V}_2 - \mathbf{V}_3 = 0$$

From the above equations,

$$\mathbf{V}_2 = \frac{\begin{vmatrix} \mathbf{Y}_1 + \mathbf{Y}_2 & \mathbf{I} & -\mathbf{Y}_2 \\ -\mathbf{Y}_1 & 0 & -\mathbf{Y}_5 \\ -\mathbf{Y}_2 & 0 & (\mathbf{Y}_2 + \mathbf{Y}_4 + \mathbf{Y}_5) \end{vmatrix}}{\begin{vmatrix} \mathbf{Y}_1 + \mathbf{Y}_2 & -\mathbf{Y}_1 & -\mathbf{Y}_2 \\ -\mathbf{Y}_1 & (\mathbf{Y}_1 + \mathbf{Y}_3 + \mathbf{Y}_5) & -\mathbf{Y}_5 \\ -\mathbf{Y}_2 & -\mathbf{Y}_5 & (\mathbf{Y}_2 + \mathbf{Y}_4 + \mathbf{Y}_5) \end{vmatrix}}$$

$$= \frac{\mathbf{I}(\mathbf{Y}_1\mathbf{Y}_3 + \mathbf{Y}_1\mathbf{Y}_4 + \mathbf{Y}_1\mathbf{Y}_5 + \mathbf{Y}_3\mathbf{Y}_5)}{\Delta}$$

Similarly,

$$\mathbf{V}_3 = \frac{\mathbf{I}(\mathbf{Y}_1\mathbf{Y}_3 + \mathbf{Y}_3\mathbf{Y}_2 + \mathbf{Y}_1\mathbf{Y}_5 + \mathbf{Y}_3\mathbf{Y}_5)}{\Delta}$$

Note the similarities between the above equations and those obtained for mesh analysis. Then

$$\mathbf{V}_{\mathbf{Z}_5} = \mathbf{V}_2 - \mathbf{V}_3 = \frac{\mathbf{I}(\mathbf{Y}_1\mathbf{Y}_4 - \mathbf{Y}_3\mathbf{Y}_2)}{\Delta}$$

For $\mathbf{V}_{\mathbf{Z}_5} = 0$, the following must be satisfied for a finite Δ not equal to zero:

$$\boxed{\mathbf{Y}_1\mathbf{Y}_4 = \mathbf{Y}_3\mathbf{Y}_2} \qquad \mathbf{V}_{\mathbf{Z}_5} = 0 \qquad \textbf{(17.4)}$$

However, substituting $\mathbf{Y}_1 = 1/\mathbf{Z}_1$, $\mathbf{Y}_2 = 1/\mathbf{Z}_2$, $\mathbf{Y}_3 = 1/\mathbf{Z}_3$, and $\mathbf{Y}_4 = 1/\mathbf{Z}_4$, we have

$$\frac{1}{\mathbf{Z}_1\mathbf{Z}_4} = \frac{1}{\mathbf{Z}_3\mathbf{Z}_2}$$

or

$$\boxed{\mathbf{Z}_1\mathbf{Z}_4 = \mathbf{Z}_3\mathbf{Z}_2} \qquad \mathbf{V}_{\mathbf{Z}_5} = 0$$

corresponding with Eq. (17.3) obtained earlier.

Let us now investigate the balance criteria in more detail by considering the network in Fig. 17.45, where it is specified that **I** and **V** = 0.

Since **I** = 0,

$$\boxed{\mathbf{I}_1 = \mathbf{I}_3} \qquad \textbf{(17.5a)}$$

and

$$\boxed{\mathbf{I}_2 = \mathbf{I}_4} \qquad \textbf{(17.5b)}$$

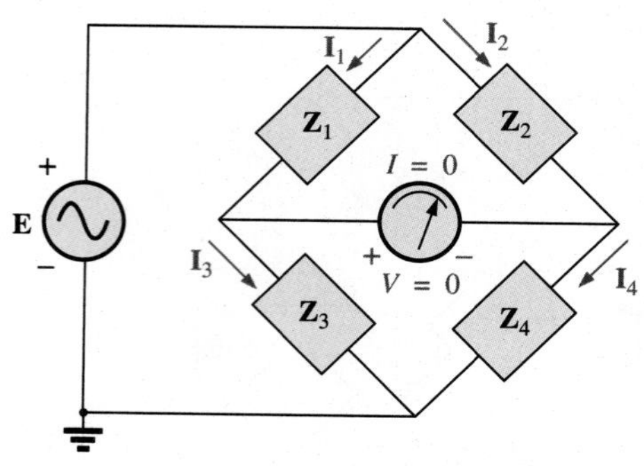

FIG. 17.45
Investigating the balance criteria for an ac bridge configuration.

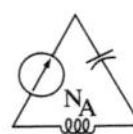

In addition, for $\mathbf{V} = 0$,

$$\boxed{\mathbf{I}_1\mathbf{Z}_1 = \mathbf{I}_2\mathbf{Z}_2} \qquad \textbf{(17.5c)}$$

and

$$\boxed{\mathbf{I}_3\mathbf{Z}_3 = \mathbf{I}_4\mathbf{Z}_4} \qquad \textbf{(17.5d)}$$

Substituting the preceding current relations into Eq. (17.5d), we have

$$\mathbf{I}_1\mathbf{Z}_3 = \mathbf{I}_2\mathbf{Z}_4$$

and

$$\mathbf{I}_2 = \frac{\mathbf{Z}_3}{\mathbf{Z}_4}\mathbf{I}_1$$

Substituting this relationship for $\mathbf{I}_2$ into Eq. (17.5c) yields

$$\mathbf{I}_1\mathbf{Z}_1 = \left(\frac{\mathbf{Z}_3}{\mathbf{Z}_4}\mathbf{I}_1\right)\mathbf{Z}_2$$

and

$$\mathbf{Z}_1\mathbf{Z}_4 = \mathbf{Z}_2\mathbf{Z}_3$$

as obtained earlier. Rearranging, we have

$$\boxed{\frac{\mathbf{Z}_1}{\mathbf{Z}_3} = \frac{\mathbf{Z}_2}{\mathbf{Z}_4}} \qquad \textbf{(17.6)}$$

corresponding with Eq. (8.2) for dc resistive networks.

For the network in Fig. 17.43, which is referred to as a **Hay bridge** when $\mathbf{Z}_5$ is replaced by a sensitive galvanometer,

$$\mathbf{Z}_1 = R_1 - j\,X_C$$
$$\mathbf{Z}_2 = R_2$$
$$\mathbf{Z}_3 = R_3$$
$$\mathbf{Z}_4 = R_4 + j\,X_L$$

This particular network is used for measuring the resistance and inductance of coils in which the resistance is a small fraction of the reactance X_L.

Substitute into Eq. (17.6) in the following form:

$$\mathbf{Z}_2\mathbf{Z}_3 = \mathbf{Z}_4\mathbf{Z}_1$$
$$R_2R_3 = (R_4 + j\,X_L)(R_1 - j\,X_C)$$

or

$$R_2R_3 = R_1R_4 + j\,(R_1X_L - R_4X_C) + X_C\,X_L$$

so that

$$R_2R_3 + j\,0 = (R_1R_4 + X_C\,X_L) + j\,(R_1X_L - R_4X_C)$$

For the equations to be equal, *the real and imaginary parts must be equal.* Therefore, for a balanced Hay bridge,

$$\boxed{R_2R_3 = R_1R_4 + X_CX_L} \qquad \textbf{(17.7a)}$$

and

$$\boxed{0 = R_1X_L - R_4X_C} \qquad \textbf{(17.7b)}$$

or substituting $\quad X_L = \omega L \quad$ and $\quad X_C = \dfrac{1}{\omega C}$

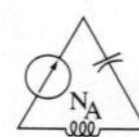

we have $$X_C X_L = \left(\frac{1}{\omega C}\right)(\omega L) = \frac{L}{C}$$

and $$R_2R_3 = R_1R_4 + \frac{L}{C}$$

with $$R_1\omega L = \frac{R_4}{\omega C}$$

Solving for R_4 in the last equation yields

$$R_4 = \omega^2 LCR_1$$

and substituting into the previous equation, we have

$$R_2R_3 = R_1(\omega^2 LCR_1) + \frac{L}{C}$$

Multiply through by C and factor:

$$CR_2R_3 = L(\omega^2 C^2 R_1^2 + 1)$$

and $$L = \frac{CR_2R_3}{1 + \omega^2 C^2 R_1^2} \qquad \textbf{(17.8a)}$$

With additional algebra this yields:

$$R_4 = \frac{\omega^2 C^2 R_1 R_2 R_3}{1 + \omega^2 C^2 R_1^2} \qquad \textbf{(17.8b)}$$

Eqs. (17.7) and (17.8) are the balance conditions for the Hay bridge. Note that each is frequency dependent. For different frequencies, the resistive and capacitive elements must vary for a particular coil to achieve balance. For a coil placed in the Hay bridge as shown in Fig. 17.43, the resistance and inductance of the coil can be determined by Eqs. (17.8a) and (17.8b) when balance is achieved.

The bridge in Fig. 17.41 is referred to as a **Maxwell bridge** when $\mathbf{Z}_5$ is replaced by a sensitive galvanometer. This setup is used for inductance measurements when the resistance of the coil is large enough not to require a Hay bridge.

Application of Eq. (17.6) in the form:

$$\mathbf{Z}_2\mathbf{Z}_3 = \mathbf{Z}_4\mathbf{Z}_1$$

and substituting

$$\mathbf{Z}_1 = R_1 \angle 0° \parallel X_{C_1} \angle -90° = \frac{(R_1 \angle 0°)(X_{C_1} \angle -90°)}{R_1 - j X_{C_1}}$$

$$= \frac{R_1 X_{C_1} \angle -90°}{R_1 - j X_{C_1}} = \frac{-j R_1 X_{C_1}}{R_1 - j X_{C_1}}$$

$$\mathbf{Z}_2 = R_2$$

$$\mathbf{Z}_3 = R_3$$

and $$\mathbf{Z}_4 = R_4 + j X_{L_4}$$

we have $$(R_2)(R_3) = (R_4 + j X_{L_4})\left(\frac{-j R_1 X_{C_1}}{R_1 - j X_{C_1}}\right)$$

$$R_2R_3 = \frac{-j R_1 R_4 X_{C_1} + R_1 X_{C_1} X_{L_4}}{R_1 - j X_{C_1}}$$

or $$(R_2R_3)(R_1 - jX_{C_1}) = R_1X_{C_1}X_{L_4} - jR_1R_4X_{C_1}$$

and $$R_1R_2R_3 - jR_2R_3X_{C_1} = R_1X_{C_1}X_{L_4} - jR_1R_4X_{C_1}$$

so that for balance

$$\cancel{R_1}R_2R_3 = \cancel{R_1}X_{C_1}X_{L_4}$$
$$R_2R_3 = \left(\frac{1}{\cancel{2\pi}fC_1}\right)(\cancel{2\pi}fL_4)$$

and $$\boxed{L_4 = C_1R_2R_3} \quad (17.9)$$

and $$R_2R_3\cancel{X_{C_1}} = R_1R_4\cancel{X_{C_1}}$$

so that $$\boxed{R_4 = \frac{R_2R_3}{R_1}} \quad (17.10)$$

Note the absence of frequency in Eqs. (17.9) and (17.10).

One remaining popular bridge is the **capacitance comparison bridge** of Fig. 17.46. An unknown capacitance and its associated resistance can be determined using this bridge. Application of Eq. (17.6) yields the following results:

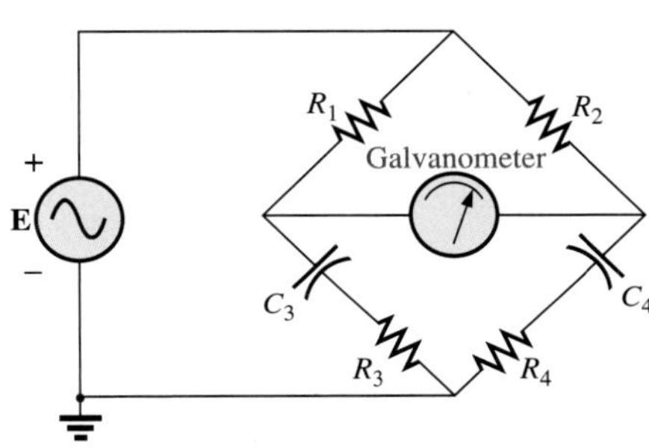

FIG. 17.46
Capacitance comparison bridge.

$$\boxed{C_4 = C_3\frac{R_1}{R_2}} \quad (17.11)$$

$$\boxed{R_4 = \frac{R_2R_3}{R_1}} \quad (17.12)$$

The derivation of these equations appears as a problem at the end of the chapter.

17.7 Δ-Y, Y-Δ CONVERSIONS

The Δ-Y, Y-Δ (or π-T, T-π as defined in Section 8.12) conversions for ac circuits are not derived here since the development corresponds exactly with that for dc circuits. Taking the **Δ-Y configuration** shown in Fig. 17.47, we find the general equations for the impedances of the Y in terms of those for the Δ:

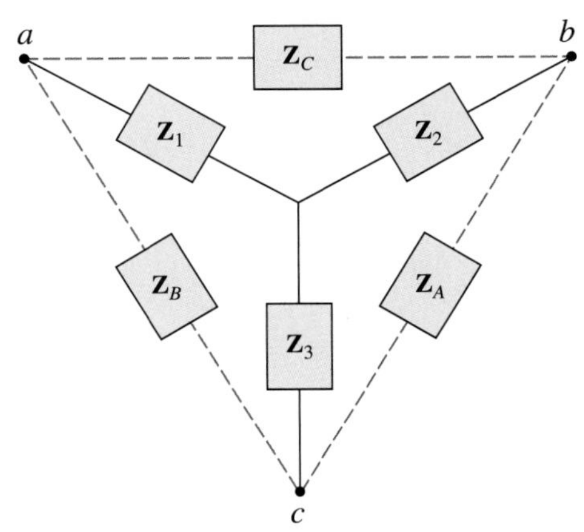

FIG. 17.47
Δ-Y configuration.

$$\boxed{\mathbf{Z}_1 = \frac{\mathbf{Z}_B\mathbf{Z}_C}{\mathbf{Z}_A + \mathbf{Z}_B + \mathbf{Z}_C}} \quad (17.13)$$

$$\boxed{\mathbf{Z}_2 = \frac{\mathbf{Z}_A\mathbf{Z}_C}{\mathbf{Z}_A + \mathbf{Z}_B + \mathbf{Z}_C}} \quad (17.14)$$

$$\boxed{\mathbf{Z}_3 = \frac{\mathbf{Z}_A\mathbf{Z}_B}{\mathbf{Z}_A + \mathbf{Z}_B + \mathbf{Z}_C}} \quad (17.15)$$

For the impedances of the Δ in terms of those for the Y, the equations are

$$\mathbf{Z}_B = \frac{\mathbf{Z}_1\mathbf{Z}_2 + \mathbf{Z}_1\mathbf{Z}_3 + \mathbf{Z}_2\mathbf{Z}_3}{\mathbf{Z}_2} \quad (17.16)$$

$$\mathbf{Z}_A = \frac{\mathbf{Z}_1\mathbf{Z}_2 + \mathbf{Z}_1\mathbf{Z}_3 + \mathbf{Z}_2\mathbf{Z}_3}{\mathbf{Z}_1} \quad (17.17)$$

$$\mathbf{Z}_C = \frac{\mathbf{Z}_1\mathbf{Z}_2 + \mathbf{Z}_1\mathbf{Z}_3 + \mathbf{Z}_2\mathbf{Z}_3}{\mathbf{Z}_3} \quad (17.18)$$

Note that each impedance of the Y is equal to the product of the impedances in the two closest branches of the Δ, divided by the sum of the impedances in the Δ.

Further, the value of each impedance of the Δ is equal to the sum of the possible product combinations of the impedances of the Y, divided by the impedances of the Y farthest from the impedance to be determined.

Drawn in different forms (Fig. 17.48), they are also referred to as the T and π configurations.

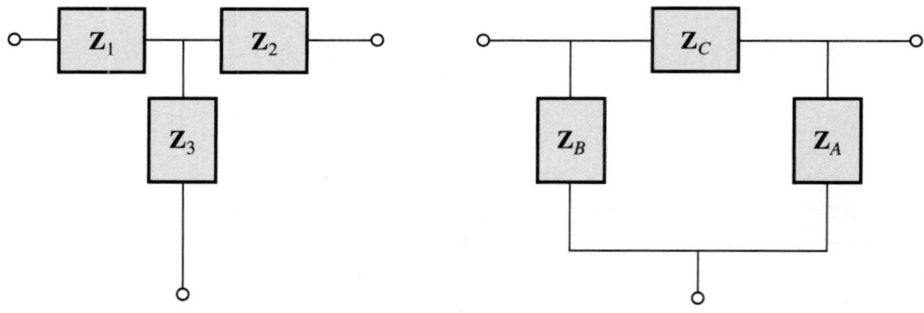

FIG. 17.48
The T and π configurations.

In the study of dc networks, we found that if all of the resistors of the Δ or Y were the same, the conversion from one to the other could be accomplished using the equation

$$R_\Delta = 3R_Y \quad \text{or} \quad R_Y = \frac{R_\Delta}{3}$$

For ac networks,

$$\mathbf{Z}_\Delta = 3\mathbf{Z}_Y \quad \text{or} \quad \mathbf{Z}_Y = \frac{\mathbf{Z}_\Delta}{3} \quad (17.19)$$

Be careful when using this simplified form. It is not sufficient for all the impedances of the Δ or Y to be of the same magnitude: *The angle associated with each must also be the same.*

EXAMPLE 17.20 Find the total impedance $\mathbf{Z}_T$ of the network in Fig. 17.49.

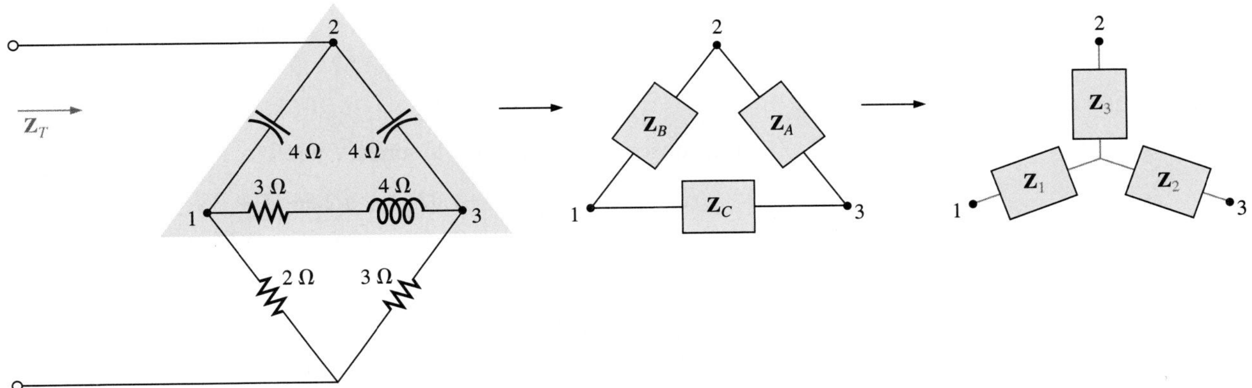

FIG. 17.49
Converting the upper Δ of a bridge configuration to a Y.

Solution:

$$\mathbf{Z}_B = -j\,4 \qquad \mathbf{Z}_A = -j\,4 \qquad \mathbf{Z}_C = 3 + j\,4$$

$$\begin{aligned}
\mathbf{Z}_1 &= \frac{\mathbf{Z}_B\mathbf{Z}_C}{\mathbf{Z}_A + \mathbf{Z}_B + \mathbf{Z}_C} = \frac{(-j\,4\,\Omega)(3\,\Omega + j\,4\,\Omega)}{(-j\,4\,\Omega) + (-j\,4\,\Omega) + (3\,\Omega + j\,4\,\Omega)} \\
&= \frac{(4\,\angle -90^\circ)(5\,\angle 53.13^\circ)}{3 - j\,4} = \frac{20\,\angle -36.87^\circ}{5\,\angle -53.13^\circ} \\
&= 4\,\Omega\,\angle 16.13^\circ = 3.84\,\Omega + j\,1.11\,\Omega
\end{aligned}$$

$$\begin{aligned}
\mathbf{Z}_2 &= \frac{\mathbf{Z}_A\mathbf{Z}_C}{\mathbf{Z}_A + \mathbf{Z}_B + \mathbf{Z}_C} = \frac{(-j\,4\,\Omega)(3\,\Omega + j\,4\,\Omega)}{5\,\Omega\,\angle -53.13^\circ} \\
&= 4\,\Omega\,\angle 16.13^\circ = 3.84\,\Omega + j\,1.11\,\Omega
\end{aligned}$$

Recall from the study of dc circuits that if two branches of the Y or Δ are the same, the corresponding Δ or Y, respectively, will also have two similar branches. In this example, $\mathbf{Z}_A = \mathbf{Z}_B$. Therefore, $\mathbf{Z}_1 = \mathbf{Z}_2$, and

$$\begin{aligned}
\mathbf{Z}_3 &= \frac{\mathbf{Z}_A\mathbf{Z}_B}{\mathbf{Z}_A + \mathbf{Z}_B + \mathbf{Z}_C} = \frac{(-j\,4\,\Omega)(-j\,4\,\Omega)}{5\,\Omega\,\angle -53.13^\circ} \\
&= \frac{16\,\Omega\,\angle -180^\circ}{5\,\angle -53.13^\circ} = 3.2\,\Omega\,\angle -126.87^\circ = -1.92\,\Omega - j\,2.56\,\Omega
\end{aligned}$$

Replace the Δ by the Y (Fig. 17.50):

$$\mathbf{Z}_1 = 3.84\,\Omega + j\,1.11\,\Omega \qquad \mathbf{Z}_2 = 3.84\,\Omega + j\,1.11\,\Omega$$
$$\mathbf{Z}_3 = -1.92\,\Omega - j\,2.56\,\Omega \qquad \mathbf{Z}_4 = 2\,\Omega$$
$$\mathbf{Z}_5 = 3\,\Omega$$

Impedances $\mathbf{Z}_1$ and $\mathbf{Z}_4$ are in series:

$$\begin{aligned}
\mathbf{Z}_{T_1} &= \mathbf{Z}_1 + \mathbf{Z}_4 = 3.84\,\Omega + j\,1.11\,\Omega + 2\,\Omega = 5.84\,\Omega + j\,1.11\,\Omega \\
&= 5.94\,\Omega\,\angle 10.76^\circ
\end{aligned}$$

Impedances $\mathbf{Z}_2$ and $\mathbf{Z}_5$ are in series:

$$\begin{aligned}
\mathbf{Z}_{T_2} &= \mathbf{Z}_2 + \mathbf{Z}_5 = 3.84\,\Omega + j\,1.11\,\Omega + 3\,\Omega = 6.84\,\Omega + j\,1.11\,\Omega \\
&= 6.93\,\Omega\,\angle 9.22^\circ
\end{aligned}$$

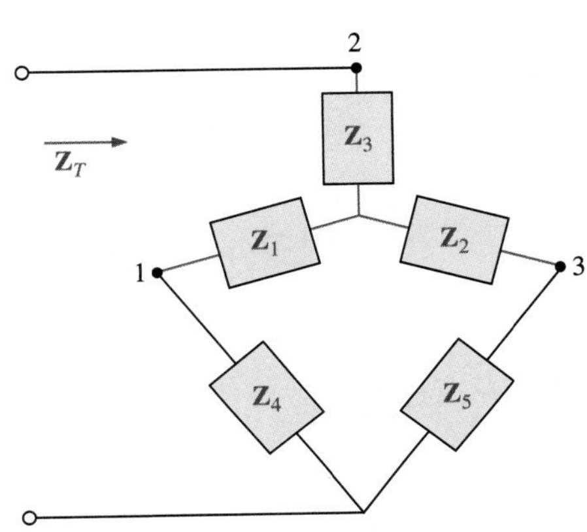

FIG. 17.50
The network in Fig. 17.49 following the substitution of the Y configuration.

Impedances $\mathbf{Z}_{T_1}$ and $\mathbf{Z}_{T_2}$ are in parallel:

$$\mathbf{Z}_{T_3} = \frac{\mathbf{Z}_{T_1}\mathbf{Z}_{T_2}}{\mathbf{Z}_{T_1} + \mathbf{Z}_{T_2}} = \frac{(5.94\ \Omega \angle 10.76^\circ)(6.93\ \Omega \angle 9.22^\circ)}{5.84\ \Omega + j\,1.11\ \Omega + 6.84\ \Omega + j\,1.11\ \Omega}$$

$$= \frac{41.16\ \Omega \angle 19.98^\circ}{12.68 + j\,2.22} = \frac{41.16\ \Omega \angle 19.98^\circ}{12.87 \angle 9.93^\circ} = 3.198\ \Omega \angle 10.05^\circ$$

$$= 3.15\ \Omega + j\,0.56\ \Omega$$

Impedances $\mathbf{Z}_3$ and $\mathbf{Z}_{T_3}$ are in series. Therefore,

$$\mathbf{Z}_T = \mathbf{Z}_3 + \mathbf{Z}_{T_3} = -1.92\ \Omega - j\,2.56\ \Omega + 3.15\ \Omega + j\,0.56\ \Omega$$

$$= 1.23\ \Omega - j\,2.0\ \Omega = \mathbf{2.35\ \Omega \angle -58.41^\circ}$$

EXAMPLE 17.21 Using both the Δ-Y and Y-Δ transformations, find the total impedance $\mathbf{Z}_T$ for the network in Fig. 17.51.

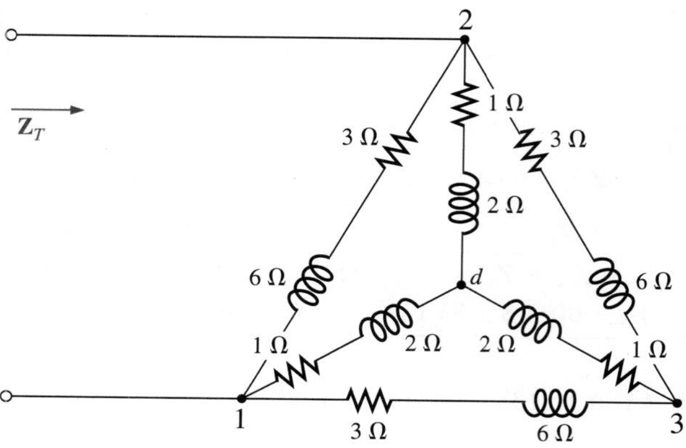

FIG. 17.51
Example 17.21.

Solution: *Using the Δ-Y transformation,* we obtain Fig. 17.52. In this case, since both systems are balanced (same impedance in each branch), the center point d' of the transformed Δ will be the same as point d of the original Y:

$$\mathbf{Z}_Y = \frac{\mathbf{Z}_\Delta}{3} = \frac{3\ \Omega + j\,6\ \Omega}{3} = 1\ \Omega + j\,2\ \Omega$$

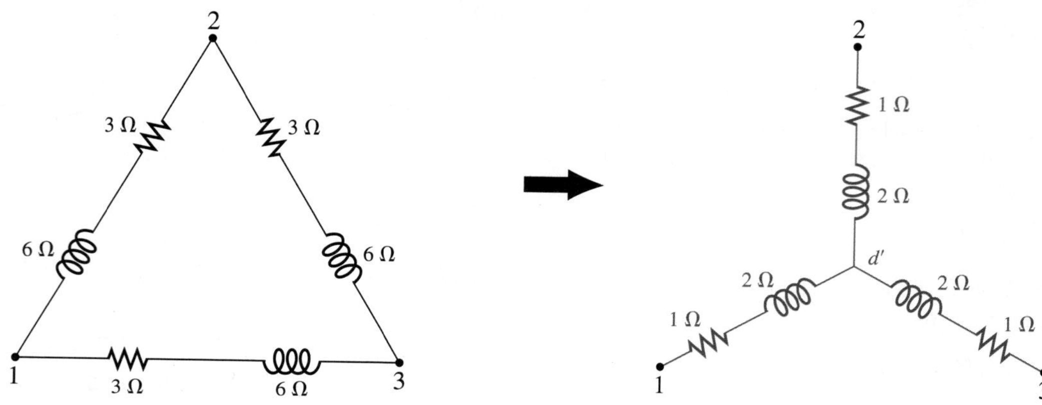

FIG. 17.52
Converting a Δ configuration to a Y configuration.

and (Fig. 17.53)

$$\mathbf{Z}_T = 2\left(\frac{1\ \Omega + j\,2\ \Omega}{2}\right) = \mathbf{1\ \Omega + j\,2\ \Omega}$$

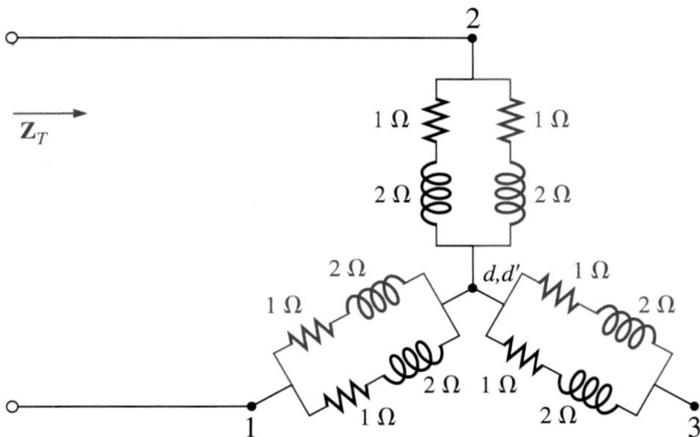

FIG. 17.53
Substituting the Y configuration in Fig. 17.52 into the network in Fig. 17.51.

Using the Y-Δ transformation (Fig. 17.54), we obtain

$$\mathbf{Z}_\Delta = 3\mathbf{Z}_Y = 3(1\ \Omega + j\,2\ \Omega) = 3\ \Omega + j\,6\ \Omega$$

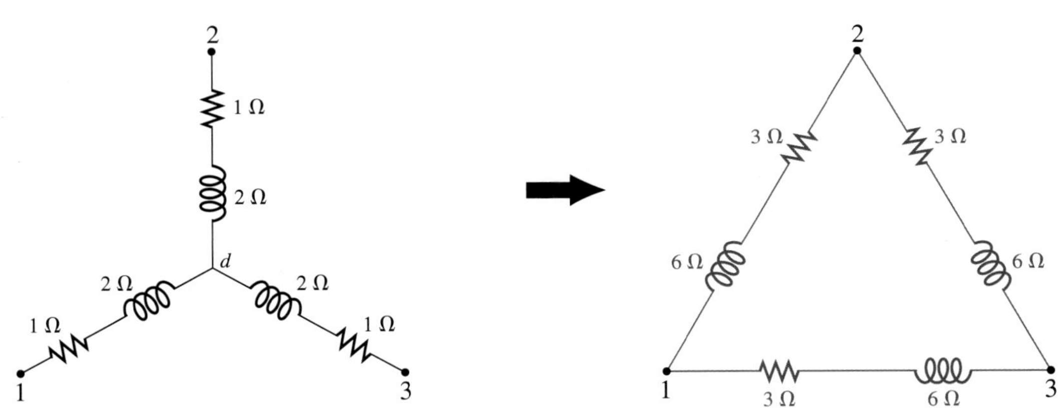

FIG. 17.54
Converting the Y configuration in Fig. 17.51 to a Δ.

Each resulting parallel combination in Fig. 17.55 will have the following impedance:

$$\mathbf{Z}' = \frac{3\ \Omega + j\,6\ \Omega}{2} = 1.5\ \Omega + j\,3\ \Omega$$

and

$$\mathbf{Z}_T = \frac{\mathbf{Z}'\,(2\mathbf{Z}')}{\mathbf{Z}' + 2\mathbf{Z}'} = \frac{2(\mathbf{Z}')^2}{3\mathbf{Z}'} = \frac{2\mathbf{Z}'}{3}$$

$$= \frac{2(1.5\ \Omega + j\,3\ \Omega)}{3} = \mathbf{1\ \Omega + j\,2\ \Omega}$$

which compares with the above result.

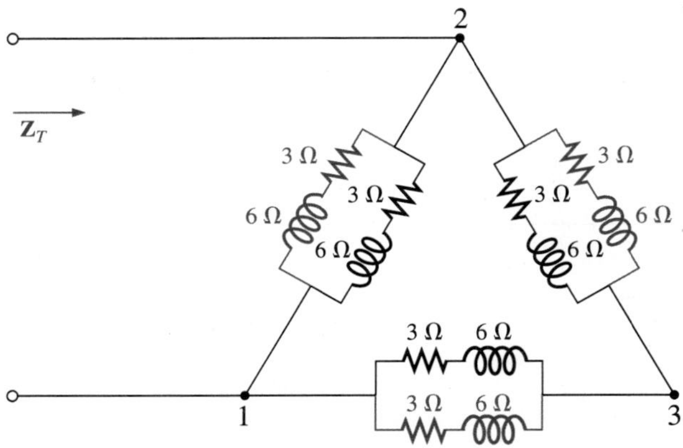

FIG. 17.55
Substituting the Δ configuration in Fig. 17.54 into the network in Fig. 17.51.

17.8 COMPUTER ANALYSIS

PSpice

Nodal Analysis The first application of PSpice is to determine the nodal voltages for the network in Example 17.16 and compare solutions. The network appears as shown in Fig. 17.56 using elements that were determined from the reactance level at a frequency of 1 kHz. There is no need to continually use 1 kHz. Any frequency will do, but remember to use the chosen frequency to find the network components and when setting up the simulation.

For the current sources, choose **ISIN** so that the phase angle can be specified (even though it is 0°), although the symbol does not have the arrow used in the text material. The direction must be recognized as pointing from the + to − sign of the source. That requires that the sources $\mathbf{I}_1$ and $\mathbf{I}_2$ be set as shown in Fig. 17.56. Reverse the source $\mathbf{I}_2$ by using the **Mirror Vertically** option obtained by right-clicking the source symbol on the screen. Setting up the **ISIN** source is the same as that used with the **VSIN** source. It can be found under the **SOURCE** library, and its attributes are the same as for the **VSIN** source. For each source, set **IOFF** to 0 A; the amplitude is the peak value of the source current. The frequency will be the same for each source. Then select **VPRINT1** from the

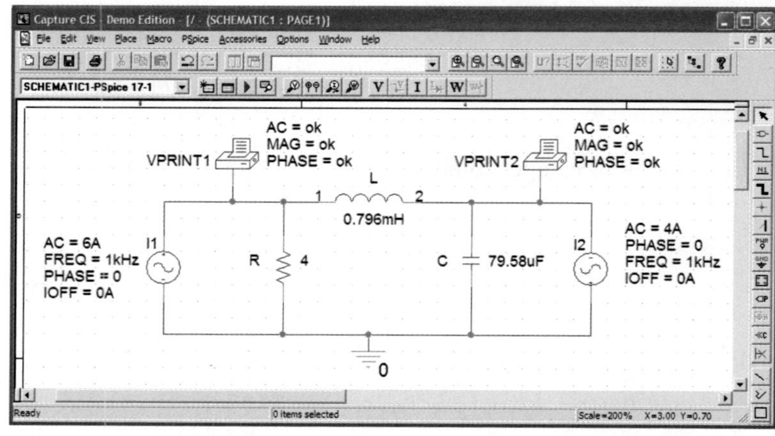

FIG. 17.56
Using PSpice to verify the results in Example 17.16.

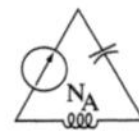

SPECIAL library and place it to generate the desired nodal voltages. Finally add the remaining elements to the network as shown in Fig. 17.56. For each source, double-click the symbol to generate the **Property Editor** dialog box. Set **AC** at the 6 A level for the $\mathbf{I}_1$ source and at 4 A for the $\mathbf{I}_2$ source, followed by **Display** and **Name and Value** for each. It appears as shown in Fig. 17.56. Double-clicking on each **VPRINT1** option also provides the **Property Editor,** so **OK** can be added under **AC, MAG,** and **PHASE.** For each quantity, select **Display** followed by **Name and Value** and **OK.** Then select **Value** and **VPRINT1** is displayed as **Value** only. Selecting **Apply** and leaving the dialog box results in the listing next to each source in Fig. 17.56. For **VPRINT2,** first change the listing on **Value** from **VPRINT1** to **VPRINT2** before selecting **Display** and **Apply.**

Now select the **New Simulation Profile** icon, and enter **PSpice 17-1** as the **Name** followed by **Create.** In the **Simulation Settings** dialog box, select **AC Sweep,** and set the **Start Frequency** and **End Frequency** at 1 kHz with 1 for the **Points/Decade.** Click **OK,** and select the **Run PSpice** icon; a **SCHEMATIC1** screen results. Exiting (**X**) brings you back to the **Orcad Capture** window. Selecting **PSpice** followed by **View Output File** results in the display in Fig. 17.57, providing exactly the same results as obtained in Example 17.16 with $\mathbf{V}_1 = 20.8\text{ V} \angle -126.9°$. The other nodal voltage is $8.617\text{ V} \angle -15.09°$.

```
** Profile: "SCHEMATIC1-PSpice 17-1"  [ C:\ICA11\PSpice\PSpice 17-1-
PSpiceFiles\SCHEMATIC1\PSpice 17-1.sim ]

****   AC ANALYSIS                TEMPERATURE =  27.000 DEG C
******************************************************************************

FREQ         VM(N01900)    VP(N01900)
1.000E+03    2.080E+01     -1.269E+02

FREQ         VM(N01908)    VP(N01908)
1.000E+03    8.617E+00     -1.509E+01
```

FIG. 17.57
Output file for the nodal voltages for the network in Fig. 17.56.

Current-Controlled Current Source (CCCS) Our interest now turns to controlled sources in the PSpice environment. Controlled sources are not particularly difficult to apply once a few important elements of their use are understood. The network in Fig. 17.14 has a current-controlled current source in the center leg of the configuration. The magnitude of the current source is k times the current through resistor R_1, where k can be greater or less than 1. The resulting schematic, appearing in Fig. 17.58, seems quite complex in the area of the controlled source, but once you understand the role of each component, the schematic is not that difficult to understand. First, since it is the only new element in the schematic, let us concentrate on the controlled source. Current-controlled current sources (CCCS) are called up under the **ANALOG** library as **F** and appear as shown in the center in Fig. 17.58. Pay attention to the direction of the current in each part of the symbol. In particular, note that the sensing current of **F** has the same direction as the defining controlling current in Fig. 17.14. In addition, note that the controlled current source also has the same direction as the source in Fig. 17.14. If you double-click on the CCCS symbol, the **Property Editor** dialog box appears with the **GAIN** (k as described above) set at 1. In this example, the gain must be set at **0.7,** so click on the region below the **GAIN** label and enter **0.7.** Then select **Display** followed by **Name and Value-OK.** Exit the **Property Editor,** and **GAIN = 0.7** appears with the CCCS as shown in Fig. 17.58.

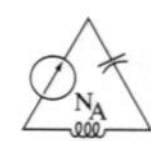

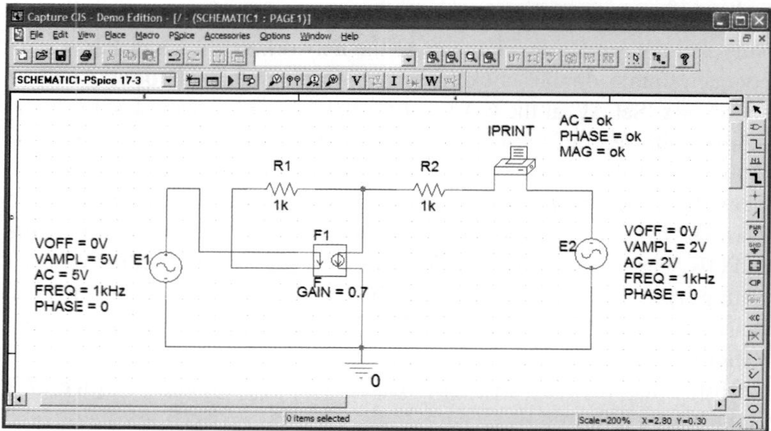

FIG. 17.58
Using PSpice to verify the results in Example 17.8.

The other new component in this schematic is **IPRINT;** it can be found in the **SPECIAL** library. It is used to tell the program to list the current in the branch of interest in the output file. If you fail to tell the program which output data you would like, it will simply run through the simulation and list specific features of the network but will not provide any voltages or currents. In this case, the current $\mathbf{I}_2$ through the resistor R_2 is desired. Double-clicking on the **IPRINT** component results in the **Property Editor** dialog box with a number of elements that need to be defined—much like that for **VPRINT.** First enter **OK** beneath **AC** and follow with **Display-Name and Value-OK.** Repeat for **MAG** and **PHASE,** and then select **Apply** before leaving the dialog box. The **OK** tells the software program that these are the quantities that it is "ok" to generate and provide. The purpose of the **Apply** at the end of each visit to the **Property Editor** dialog box is to "apply" the changes made to the network under investigation. When you exit the **Property Editor,** the three chosen parameters appear on the schematic with the **OK** directive. You may find that the labels will appear all over the **IPRINT** symbol. No problem—just click on each, and move to a more convenient location.

The remaining components of the network should be fairly familiar, but don't forget to **Mirror Vertically** the voltage source **E2.** In addition, do not forget to call up the **Property Editor** for each source and set the level of **AC, FREQ, VAMPL,** and **VOFF** and be sure that the **PHASE** is set on the default value of 0°. The value appears with each parameter in Fig. 17.58 for each source. Always be sure to select **Apply** before leaving the **Property Editor.** After placing all the components on the screen, you must connect them with a **Place wire** selection. Normally, this is pretty straightforward. However, with controlled sources, it is often necessary to cross over wires *without* making a connection. In general, when you're placing a wire over another wire and you don't want a connection to be made, click a spot on one side of the wire to be crossed to create the temporary red square. Then cross the wire, and click again to establish another red square. If the connection is done properly, the crossed wire should not show a connection point (a small red dot). In this example, the top of the controlling current was connected first from the **E1** source. Then a wire was connected from the lower end of the sensing current to the point where a 90° turn up the page was to be made. The wire was clicked in place at this point before crossing the original wire and clicked again before making the right turn to resistor R_1. You will not find a small red dot where the wires cross.

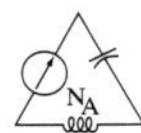

Now for the simulation. In the **Simulation Settings** dialog box, select **AC Sweep/Noise** with a **Start** and **End Frequency** of 1 kHz. There will be 1 **Point/Decade.** Click **OK,** and select the **Run Spice** key; a **SCHEMATIC1** results that should be exited to obtain the **Orcad Capture** screen. Select **PSpice** followed by **View Output File,** and scroll down until you read **AC ANALYSIS** (see Fig. 17.59). The magnitude of the desired current is 1.615 mA with a phase angle of 0°, a perfect match with the theoretical analysis to follow. One would expect a phase angle of 0° since the network is composed solely of resistive elements.

```
** Profile: "SCHEMATIC1-PSpice 17-3" [ C:\ICA11\PSpice\PSpice 17-3-
PSpiceFiles\SCHEMATIC1\PSpice 17-3.sim ]

****    AC ANALYSIS                    TEMPERATURE =  27.000 DEG C
******************************************************************************

  FREQ         IM(V_PRINT1)        IP(V_PRINT1)
  1.000E+03    1.615E-03           0.000E+00
```

FIG. 17.59
The output file for the mesh current $\mathbf{I}_2$ in Fig. 17.14.

The equations obtained earlier using the supermesh approach were

$$\mathbf{E} - \mathbf{I}_1\mathbf{Z}_1 - \mathbf{I}_2\mathbf{Z}_2 + \mathbf{E}_2 = 0 \quad \text{or} \quad \mathbf{I}_1\mathbf{Z}_1 + \mathbf{I}_2\mathbf{Z}_2 = \mathbf{E}_1 + \mathbf{E}_2$$

and

$$k\mathbf{I} = k\mathbf{I}_1 = \mathbf{I}_1 - \mathbf{I}_2$$

resulting in

$$\mathbf{I}_1 = \frac{\mathbf{I}_2}{1 - k} = \frac{\mathbf{I}_2}{1 - 0.7} = \frac{\mathbf{I}_2}{0.3} = 3.333\mathbf{I}_2$$

so that

$$\mathbf{I}_1(1\text{ k}\Omega) + \mathbf{I}_2(1\text{ k}\Omega) = 7\text{ V} \qquad \text{(from above)}$$

becomes

$$(3.333\mathbf{I}_2)1\text{ k}\Omega + \mathbf{I}_2(1\text{ k}\Omega) = 7\text{ V}$$

or

$$(4.333\text{ k}\Omega)\mathbf{I}_2 = 7\text{ V}$$

and

$$\mathbf{I}_2 = \frac{7\text{ V}}{4.333\text{ k}\Omega} = \mathbf{1.615\text{ mA} \angle 0°}$$

confirming the computer solution.

PROBLEMS

SECTION 17.2 Independent versus Dependent (Controlled) Sources

1. Discuss, in your own words, the difference between a controlled and an independent source.

SECTION 17.3 Source Conversions

2. Convert the voltage sources in Fig. 17.60 to current sources.

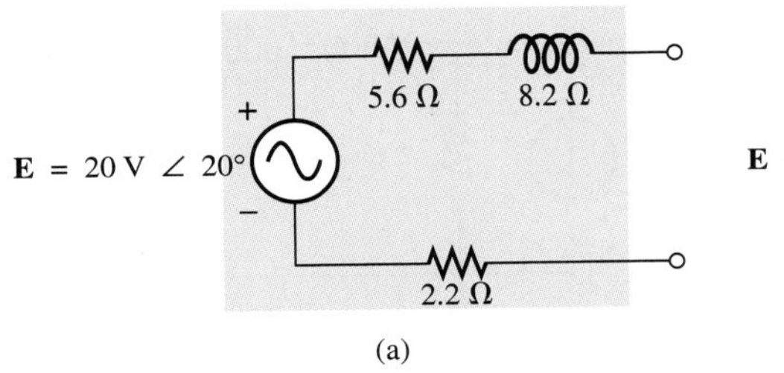

(a)

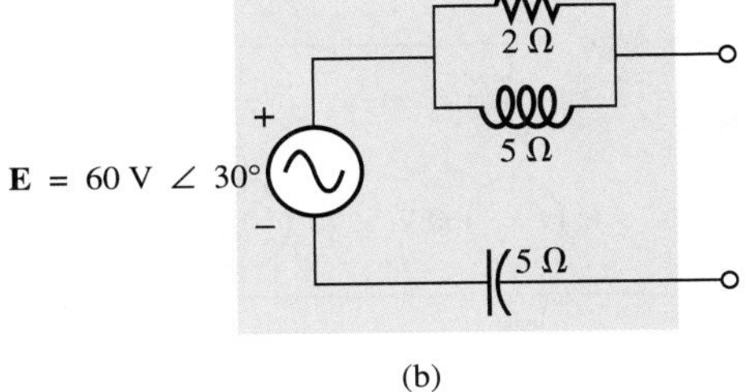

(b)

FIG. 17.60
Problem 2.

3. Convert the current sources in Fig. 17.61 to voltage sources.

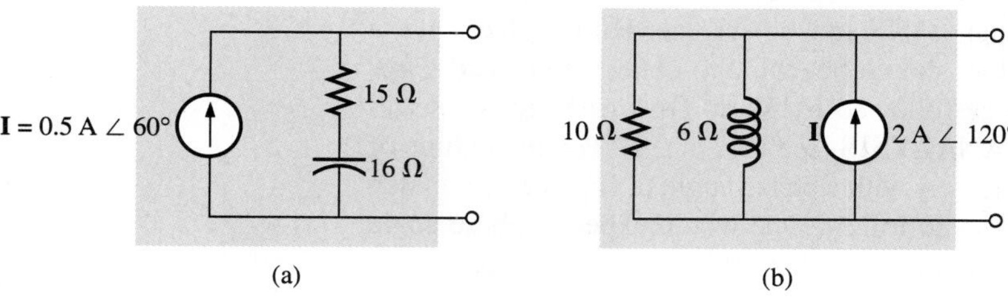

FIG. 17.61
Problem 3.

4. Convert the voltage source in Fig. 17.62(a) to a current source and the current source in Fig. 17.62(b) to a voltage source.

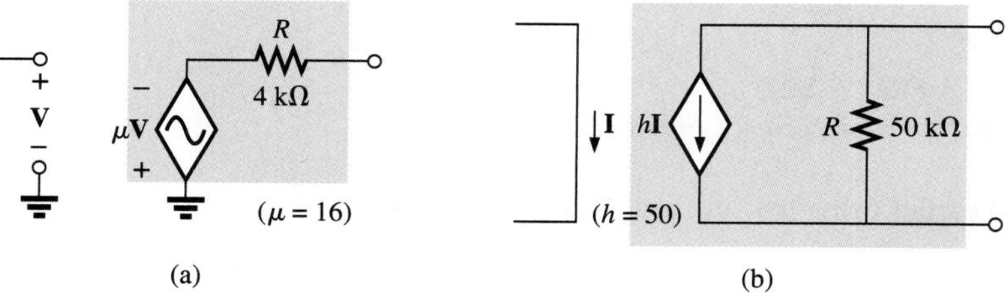

FIG. 17.62
Problem 4.

SECTION 17.4 Mesh Analysis

5. Write the mesh equations for the networks in Fig. 17.63. Determine the current through the resistor R.

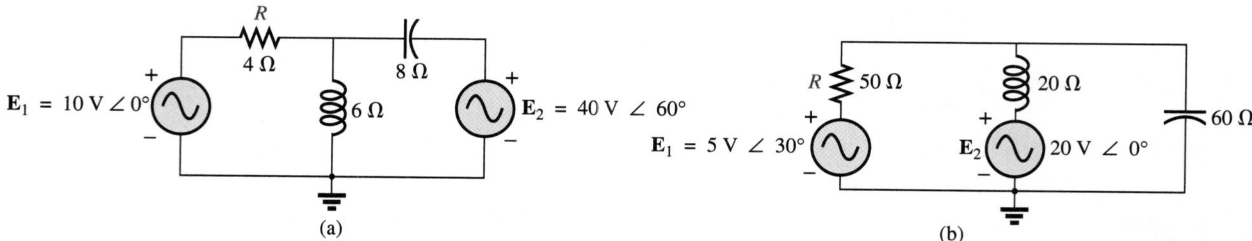

FIG. 17.63
Problems 5 and 34.

6. Write the mesh equations for the networks in Fig. 17.64. Determine the current through the resistor R_1.

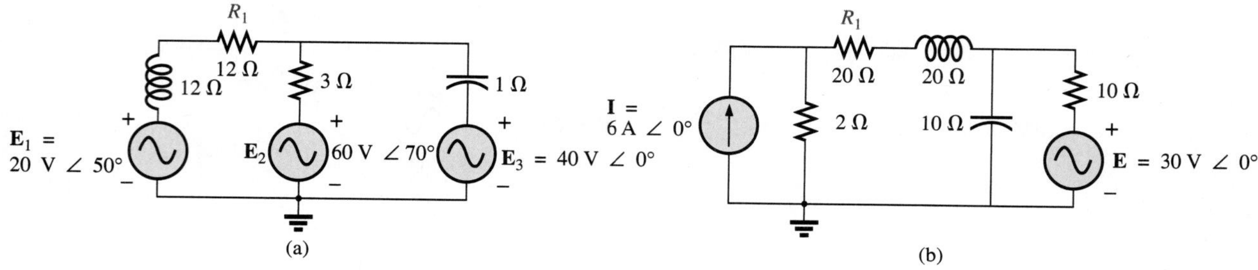

FIG. 17.64
Problems 6 and 16.

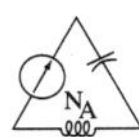

*7. Write the mesh equations for the networks in Fig. 17.65. Determine the current through the resistor R_1.

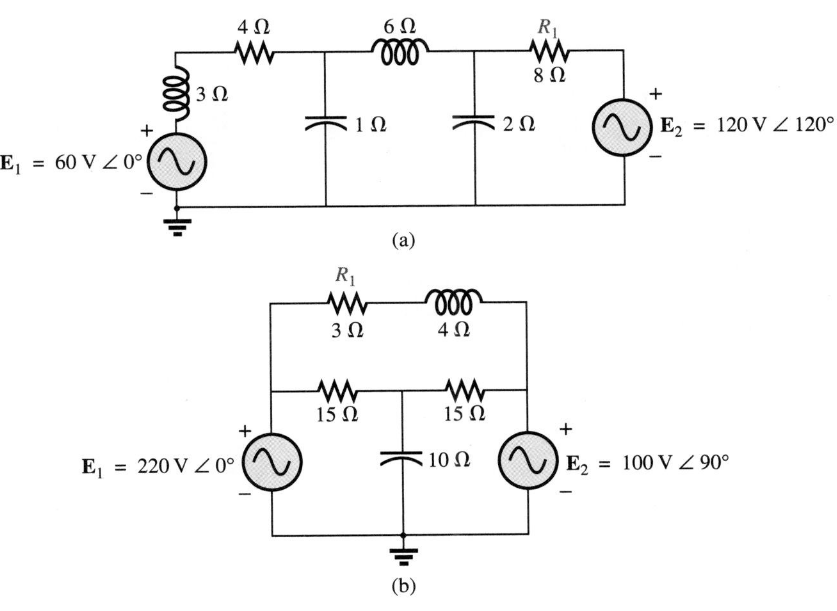

FIG. 17.65
Problems 7, 17, and 35.

*8. Write the mesh equations for the networks in Fig. 17.66. Determine the current through the resistor R_1.

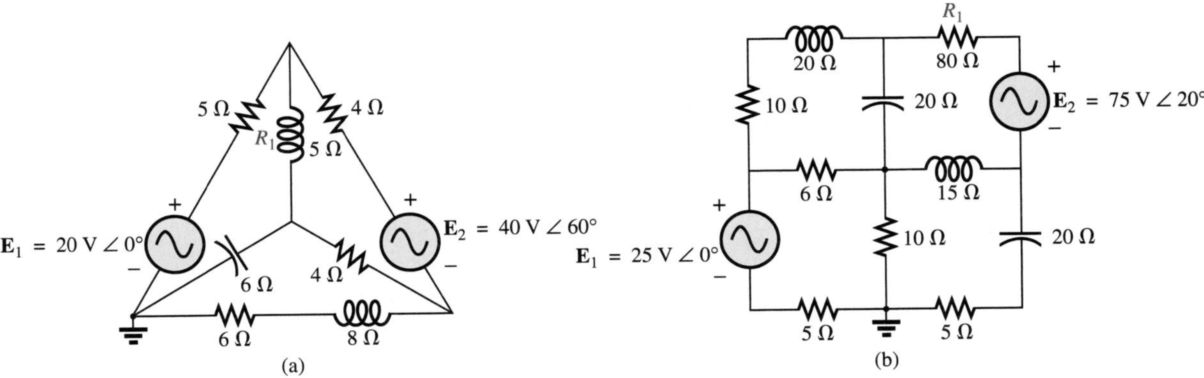

FIG. 17.66
Problems 8, 18, and 19.

9. Using mesh analysis, determine the current $\mathbf{I}_L$ (in terms of **V**) for the network in Fig. 17.67.

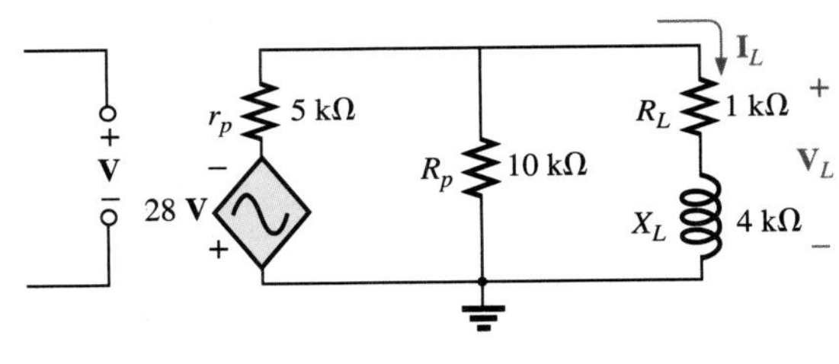

FIG. 17.67
Problem 9.

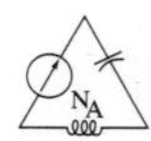

*10. Using mesh analysis, determine the current $\mathbf{I}_L$ (in terms of $\mathbf{I}$) for the network in Fig. 17.68.

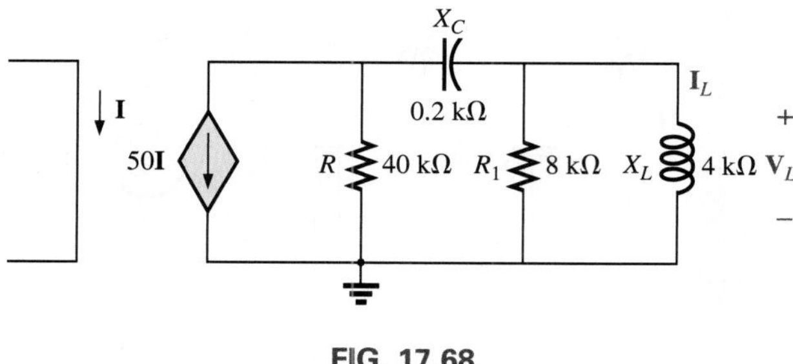

FIG. 17.68
Problem 10.

*11. Write the mesh equations for the network in Fig. 17.69, and determine the current through the 1 kΩ and 2 kΩ resistors.

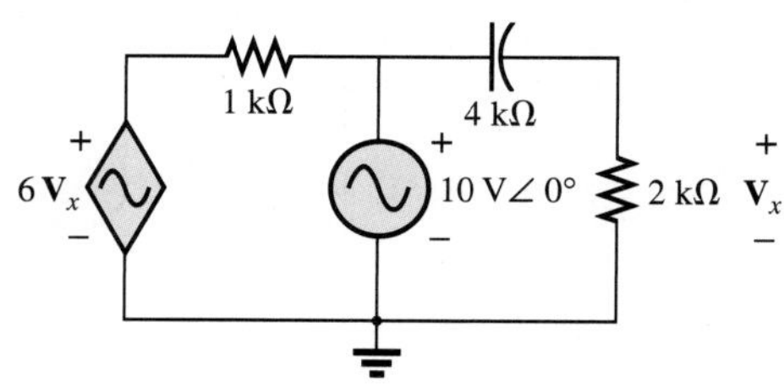

FIG. 17.69
Problems 11 and 36.

*12. Write the mesh equations for the network in Fig. 17.70, and determine the current through the 10 kΩ resistor.

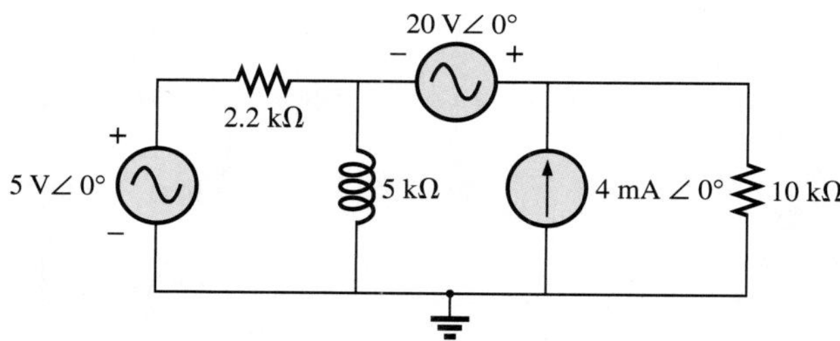

FIG. 17.70
Problems 12 and 37.

*13. Write the mesh equations for the network in Fig. 17.71, and determine the current through the inductive element.

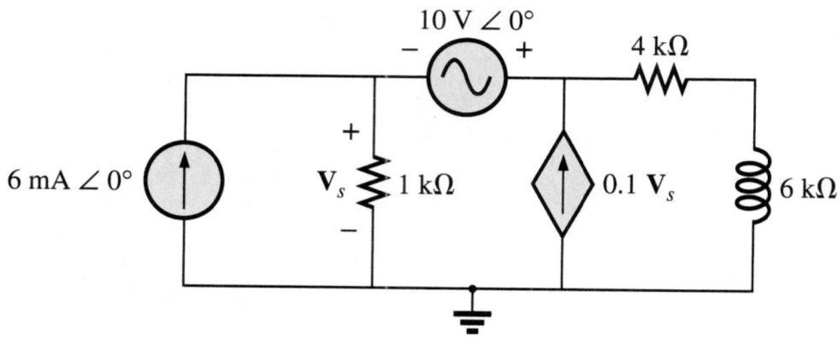

FIG. 17.71
Problems 13 and 38.

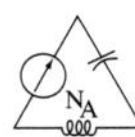

SECTION 17.5 Nodal Analysis

14. Determine the nodal voltages for the networks in Fig. 17.72.

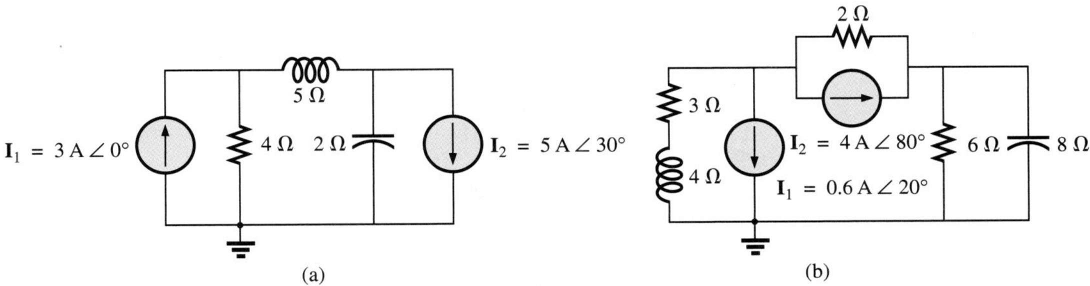

FIG. 17.72
Problems 14 and 39.

15. Determine the nodal voltages for the networks in Fig. 17.73.

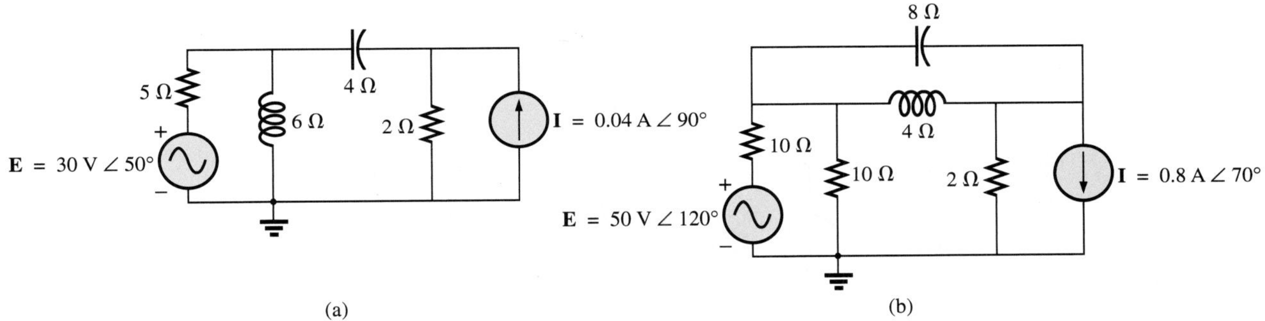

FIG. 17.73
Problem 15.

16. Determine the nodal voltages for the network in Fig. 17.64(b).

17. Determine the nodal voltages for the network in Fig. 17.65(b).

***18.** Determine the nodal voltages for the network in Fig. 17.66(a).

***19.** Determine the nodal voltages for the network in Fig. 17.66(b).

***20.** Determine the nodal voltages for the networks in Fig. 17.74.

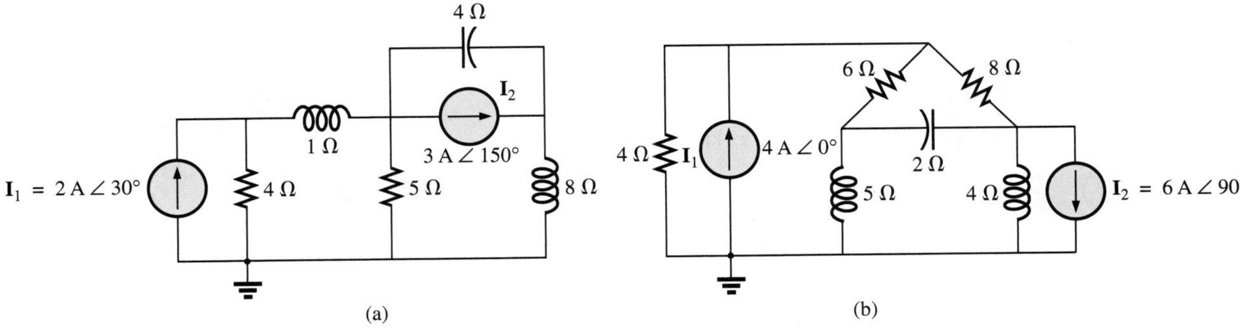

FIG. 17.74
Problem 20.

***21.** Write the nodal equations for the network in Fig. 17.75, and find the voltage across the 1 kΩ resistor.

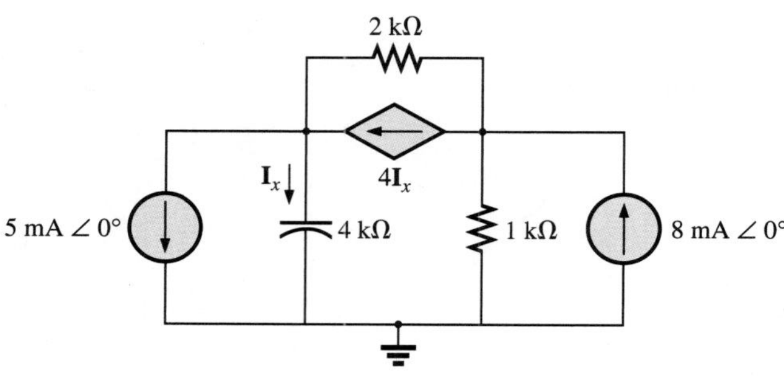

FIG. 17.75
Problems 21 and 40.

***22.** Write the nodal equations for the network in Fig. 17.76, and find the voltage across the capacitive element.

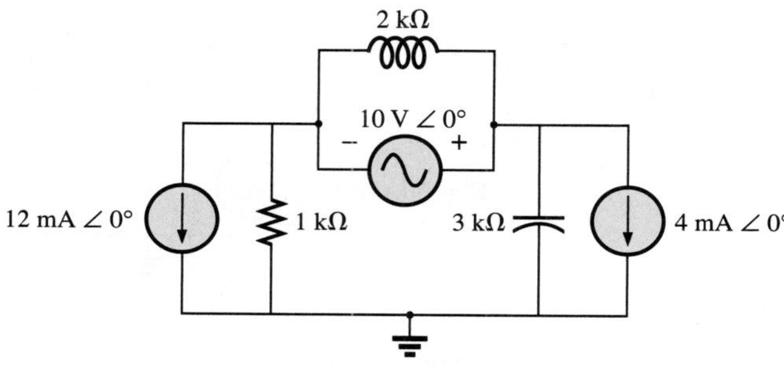

FIG. 17.76
Problems 22 and 41.

***23.** Write the nodal equations for the network in Fig. 17.77, and find the voltage across the 2 kΩ resistor.

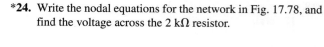

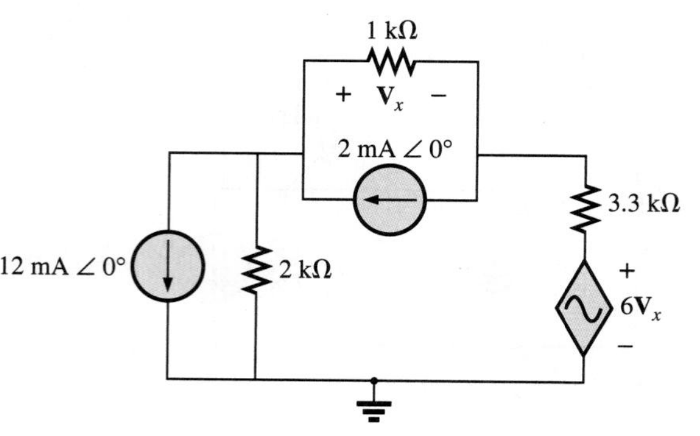

FIG. 17.77
Problems 23 and 42.

***24.** Write the nodal equations for the network in Fig. 17.78, and find the voltage across the 2 kΩ resistor.

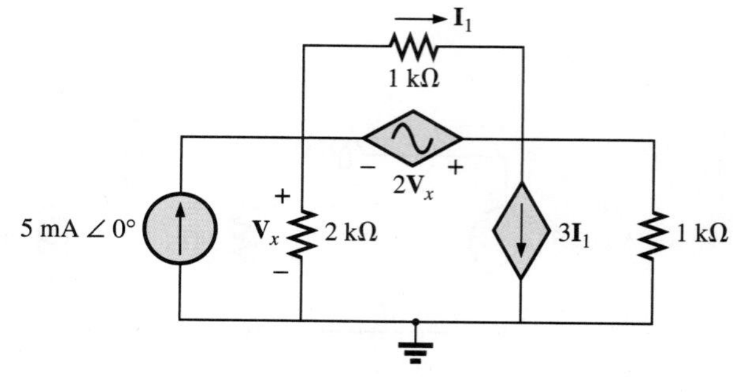

FIG. 17.78
Problems 24 and 43.

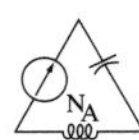

*25. For the network in Fig. 17.79, determine the voltage $\mathbf{V}_L$ in terms of the voltage $\mathbf{E}_i$.

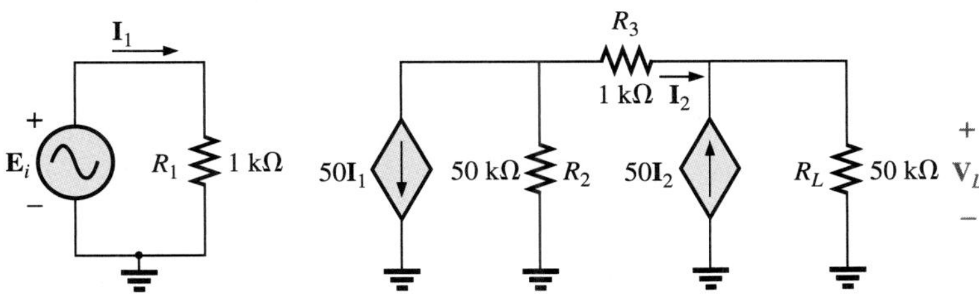

FIG. 17.79
Problem 25.

SECTION 17.6 Bridge Networks (ac)

26. For the bridge network in Fig. 17.80:
   a. Is the bridge balanced?
   b. Using mesh analysis, determine the current through the capacitive reactance.
   c. Using nodal analysis, determine the voltage across the capacitive reactance.

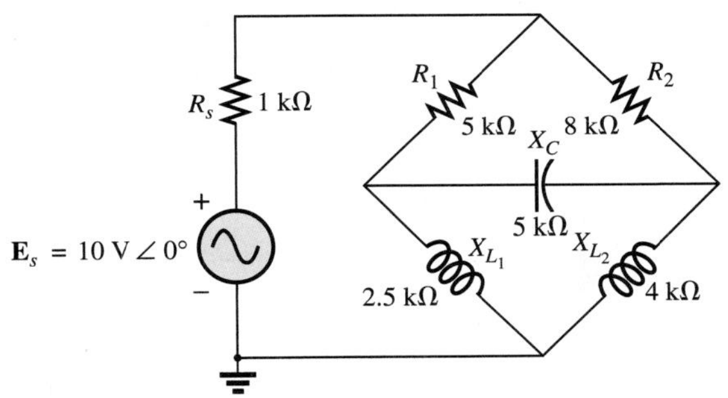

FIG. 17.80
Problem 26.

27. For the bridge network in Fig. 17.81:
   a. Is the bridge balanced?
   b. Using mesh analysis, determine the current through the capacitive reactance.
   c. Using nodal analysis, determine the voltage across the capacitive reactance.

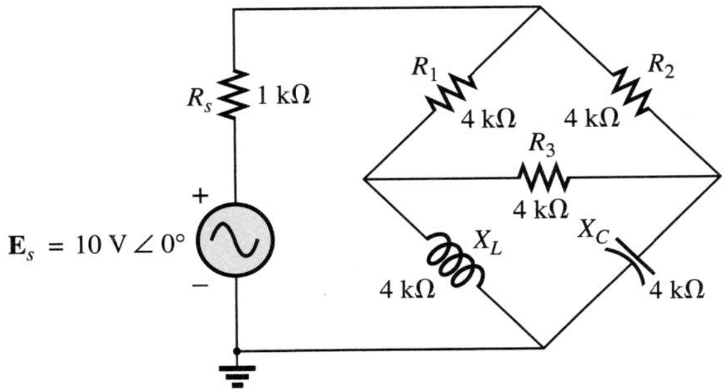

FIG. 17.81
Problem 27.

28. The Hay bridge in Fig. 17.82 is balanced. Using Eq. (17.3), determine the unknown inductance L_x and resistance R_x.

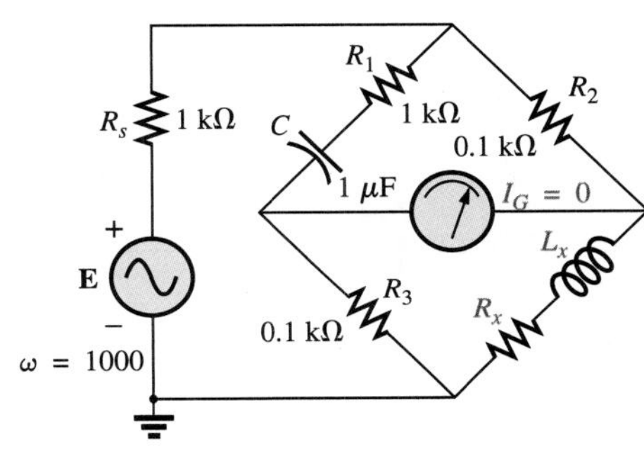

FIG. 17.82
Problem 28.

29. Determine whether the Maxwell bridge in Fig. 17.83 is balanced ($\omega = 1000$ rad/s).

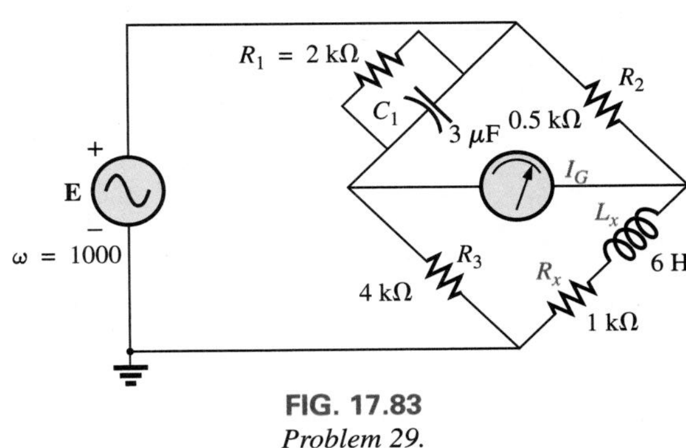

FIG. 17.83
Problem 29.

30. Derive the balance Eqs. (17.11) and (17.12) for the capacitance comparison bridge.

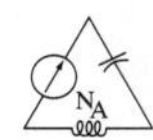

31. Determine the balance equations for the inductance bridge in Fig. 17.84.

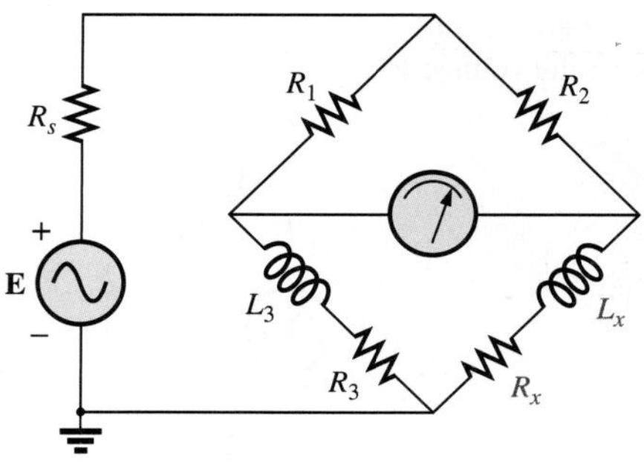

FIG. 17.84
Problem 31.

SECTION 17.7 Δ-Y, Y-Δ Conversions

32. Using the Δ-Y or Y-Δ conversion, determine the current **I** for the networks in Fig. 17.85.

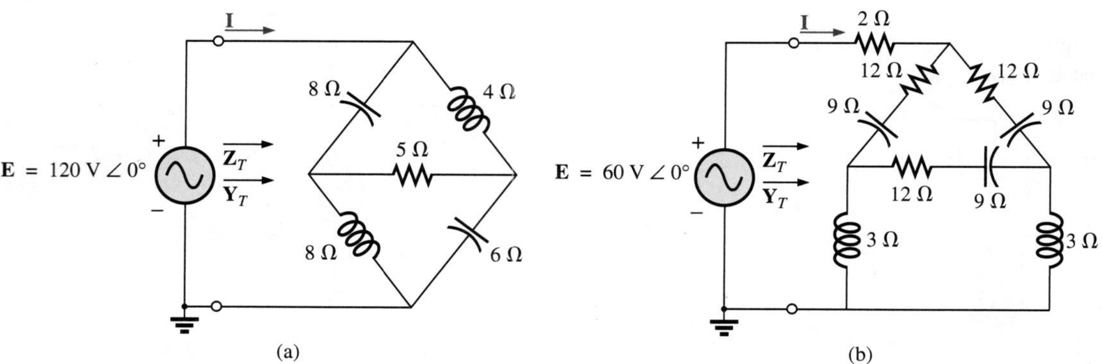

FIG. 17.85
Problem 32.

33. Using the Δ-Y or Y-Δ conversion, determine the current **I** for the networks in Fig. 17.86. (**E** = 100 V ∠0° in each case.)

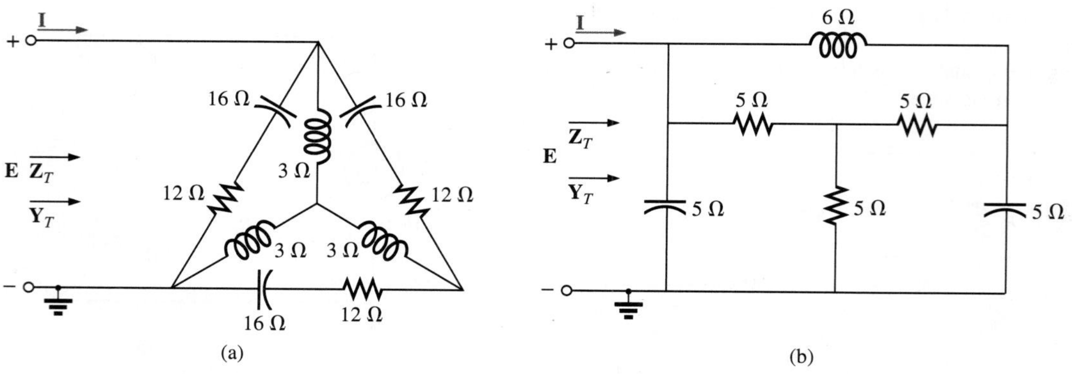

FIG. 17.86
Problem 33.

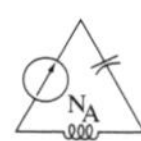

SECTION 17.8 Computer Analysis
PSpice or Multisim

34. Determine the mesh currents for the network in Fig. 17.63(a).

35. Determine the mesh currents for the network in Fig. 17.65(a).

***36.** Determine the mesh currents for the network in Fig. 17.69.

***37.** Determine the mesh currents for the network in Fig. 17.70.

***38.** Determine the mesh currents for the network in Fig. 17.71.

39. Determine the nodal voltages for the network in Fig. 17.72(b).

***40.** Determine the nodal voltages for the network in Fig. 17.75.

***41.** Determine the nodal voltages for the network in Fig. 17.76.

***42.** Determine the nodal voltages for the network in Fig. 17.77.

***43.** Determine the nodal voltages for the network in Fig. 17.78.

GLOSSARY

Bridge network A network configuration having the appearance of a diamond in which no two branches are in series or parallel.

Capacitance comparison bridge A bridge configuration having a galvanometer in the bridge arm that is used to determine an unknown capacitance and associated resistance.

Delta (Δ) configuration A network configuration having the appearance of the capital Greek letter *delta.*

Dependent (controlled) source A source whose magnitude and/or phase angle is determined (controlled) by a current or voltage of the system in which it appears.

Hay bridge A bridge configuration used for measuring the resistance and inductance of coils in those cases where the resistance is a small fraction of the reactance of the coil.

Independent source A source whose magnitude is independent of the network to which it is applied. It displays its terminal characteristics even if completely isolated.

Maxwell bridge A bridge configuration used for inductance measurements when the resistance of the coil is large enough not to require a Hay bridge.

Mesh analysis A method through which the loop (or mesh) currents of a network can be determined. The branch currents of the network can then be determined directly from the loop currents.

Nodal analysis A method through which the nodal voltages of a network can be determined. The voltage across each element can then be determined through application of Kirchhoff's voltage law.

Source conversion The changing of a voltage source to a current source, or vice versa, which will result in the same terminal behavior of the source. In other words, the external network is unaware of the change in sources.

Wye (Y) configuration A network configuration having the appearance of the capital letter *Y.*

Network Theorems (ac)

Objectives

- ***Be able to apply the superposition theorem to ac networks with independent and dependent sources.***
- ***Become proficient in applying Thévenin's theorem to ac networks with independent and dependent sources.***
- ***Be able to apply Norton's theorem to ac networks with independent and dependent sources.***
- ***Clearly understand the conditions that must be met for maximum power transfer to a load in an ac network with independent or dependent sources.***

18.1 INTRODUCTION

This chapter parallels Chapter 9, which dealt with network theorems as applied to dc networks. Reviewing each theorem in Chapter 9 before beginning this chapter is recommended because many of the comments offered there are not repeated here.

Due to the need for developing confidence in the application of the various theorems to networks with controlled (dependent) sources, some sections have been divided into two parts: independent sources and dependent sources.

Theorems to be considered in detail include the superposition theorem, Thévenin's and Norton's theorems, and the maximum power transfer theorem. The substitution and reciprocity theorems and Millman's theorem are not discussed in detail here because a review of Chapter 9 will enable you to apply them to sinusoidal ac networks with little difficulty.

18.2 SUPERPOSITION THEOREM

You will recall from Chapter 9 that the **superposition theorem** eliminated the need for solving simultaneous linear equations by considering the effects of each source independently. To consider the effects of each source, we had to remove the remaining sources. This was accomplished by setting voltage sources to zero (short-circuit representation) and current sources to zero (open-circuit representation). The current through, or voltage across, a portion of the network produced by each source was then added algebraically to find the total solution for the current or voltage.

The only variation in applying this method to ac networks with independent sources is that we are now working with impedances and phasors instead of just resistors and real numbers.

The superposition theorem is not applicable to power effects in ac networks since we are still dealing with a nonlinear relationship. It can be applied to networks with sources of different frequencies only if the total response for *each* frequency is found independently and the results are expanded in a nonsinusoidal expression, as appearing in Chapter 25.

One of the most frequent applications of the superposition theorem is to electronic systems in which the dc and ac analyses are treated separately and the total solution is the sum of the two. It is an important application of the theorem because the impact of the reactive elements

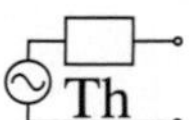

changes dramatically in response to the two types of independent sources. In addition, the dc analysis of an electronic system can often define important parameters for the ac analysis. Example 18.4 demonstrates the impact of the applied source on the general configuration of the network.

We first consider networks with only independent sources to provide a close association with the analysis of Chapter 9.

Independent Sources

EXAMPLE 18.1 Using the superposition theorem, find the current **I** through the 4 Ω reactance (X_{L_2}) in Fig. 18.1.

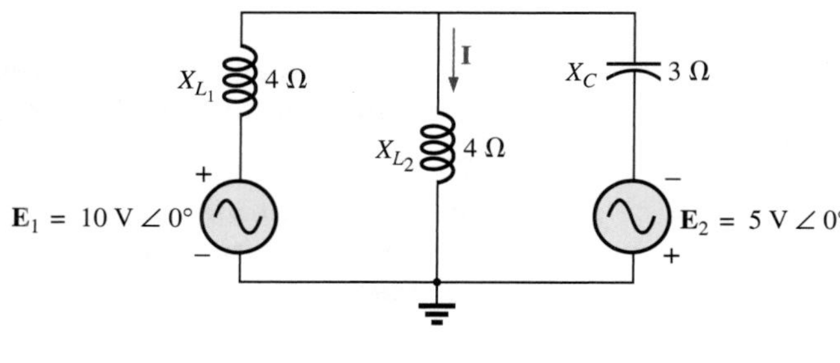

FIG. 18.1
Example 18.1.

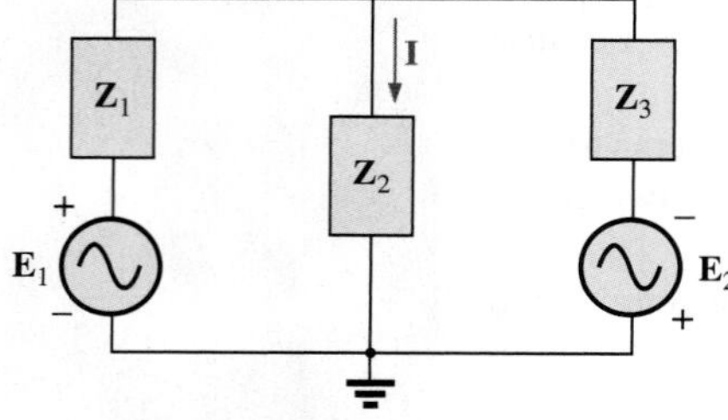

FIG. 18.2
Assigning the subscripted impedances to the network in Fig. 18.1.

Solution: For the redrawn circuit (Fig. 18.2),

$$\mathbf{Z}_1 = +jX_{L_1} = j\,4\,\Omega$$
$$\mathbf{Z}_2 = +jX_{L_2} = j\,4\,\Omega$$
$$\mathbf{Z}_3 = -jX_C = -j\,3\,\Omega$$

Considering the effects of the voltage source $\mathbf{E}_1$ (Fig. 18.3), we have

$$\mathbf{Z}_{2\|3} = \frac{\mathbf{Z}_2\mathbf{Z}_3}{\mathbf{Z}_2 + \mathbf{Z}_3} = \frac{(j\,4\,\Omega)(-j\,3\,\Omega)}{j\,4\,\Omega - j\,3\,\Omega} = \frac{12\,\Omega}{j} = -j\,12\,\Omega$$
$$= 12\,\Omega\,\angle -90^\circ$$

$$\mathbf{I}_{s_1} = \frac{\mathbf{E}_1}{\mathbf{Z}_{2\|3} + \mathbf{Z}_1} = \frac{10\text{ V}\,\angle 0^\circ}{-j\,12\,\Omega + j\,4\,\Omega} = \frac{10\text{ V}\,\angle 0^\circ}{8\,\Omega\,\angle -90^\circ}$$
$$= 1.25\text{ A}\,\angle 90^\circ$$

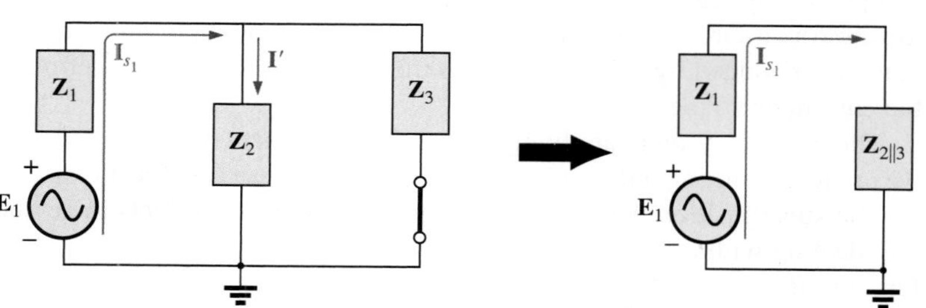

FIG. 18.3
Determining the effect of the voltage source $\mathbf{E}_1$ *on the current* **I** *of the network in Fig. 18.1.*

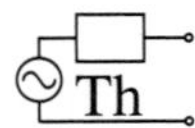

and

$$\mathbf{I}' = \frac{\mathbf{Z}_3\mathbf{I}_{s_1}}{\mathbf{Z}_2 + \mathbf{Z}_3} \qquad \text{(current divider rule)}$$

$$= \frac{(-j\,3\,\Omega)(j\,1.25\text{ A})}{j\,4\,\Omega - j\,3\,\Omega} = \frac{3.75\text{ A}}{j\,1} = 3.75\text{ A}\angle -90^\circ$$

Considering the effects of the voltage source $\mathbf{E}_2$ (Fig. 18.4), we have

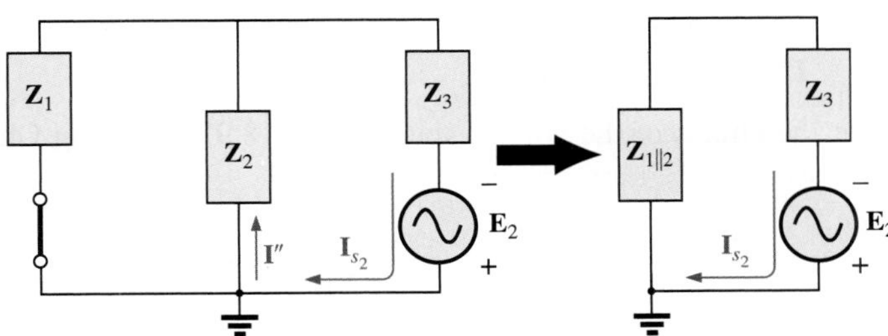

FIG. 18.4
Determining the effect of the voltage source $\mathbf{E}_2$ on the current $\mathbf{I}$ of the network in Fig. 18.1.

$$\mathbf{Z}_{1\|2} = \frac{\mathbf{Z}_1}{N} = \frac{j\,4\,\Omega}{2} = j\,2\,\Omega$$

$$\mathbf{I}_{s_2} = \frac{\mathbf{E}_2}{\mathbf{Z}_{1\|2} + \mathbf{Z}_3} = \frac{5\text{ V}\angle 0^\circ}{j\,2\,\Omega - j\,3\,\Omega} = \frac{5\text{ V}\angle 0^\circ}{1\,\Omega\angle -90^\circ} = 5\text{ A}\angle 90^\circ$$

and $$\mathbf{I}'' = \frac{\mathbf{I}_{s_2}}{2} = 2.5\text{ A}\angle 90^\circ$$

The resultant current through the 4 Ω reactance X_{L_2} (Fig. 18.5) is

$$\mathbf{I} = \mathbf{I}' - \mathbf{I}''$$
$$= 3.75\text{ A}\angle -90^\circ - 2.50\text{ A}\angle 90^\circ = -j\,3.75\text{ A} - j\,2.50\text{ A}$$
$$= -j\,6.25\text{ A}$$
$$\mathbf{I} = \mathbf{6.25\text{ A}\angle -90^\circ}$$

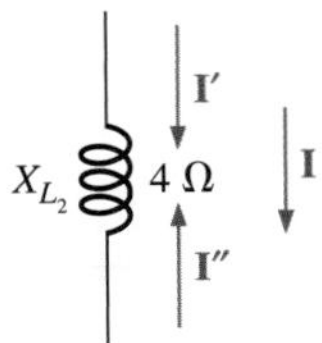

FIG. 18.5
Determining the resultant current for the network in Fig. 18.1.

EXAMPLE 18.2 Using superposition, find the current **I** through the 6 Ω resistor in Fig. 18.6.

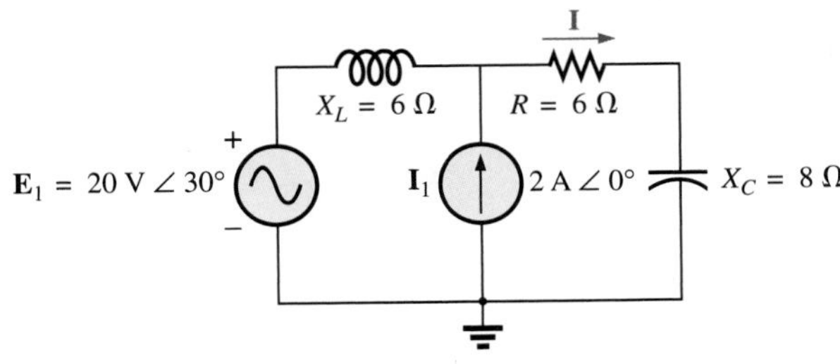

FIG. 18.6
Example 18.2.

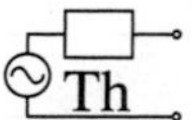

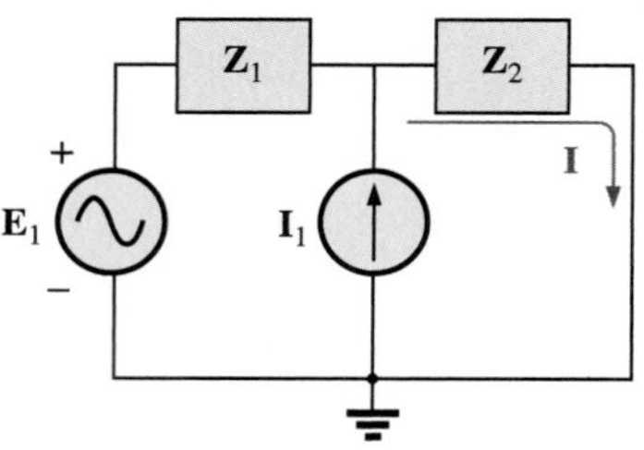

FIG. 18.7
Assigning the subscripted impedances to the network in Fig. 18.6.

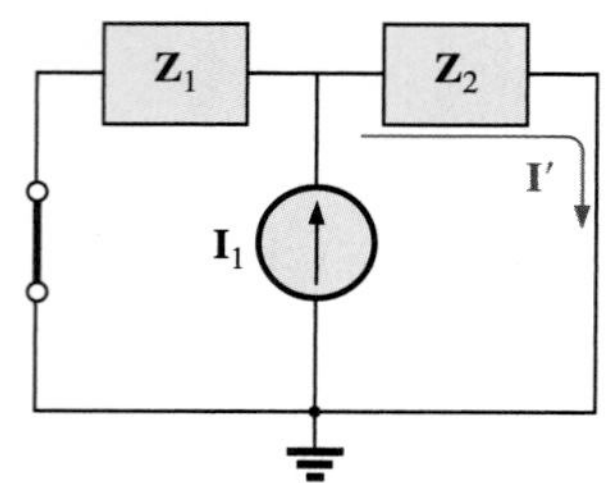

FIG. 18.8
Determining the effect of the current source $\mathbf{I}_1$ *on the current* **I** *of the network in Fig. 18.6.*

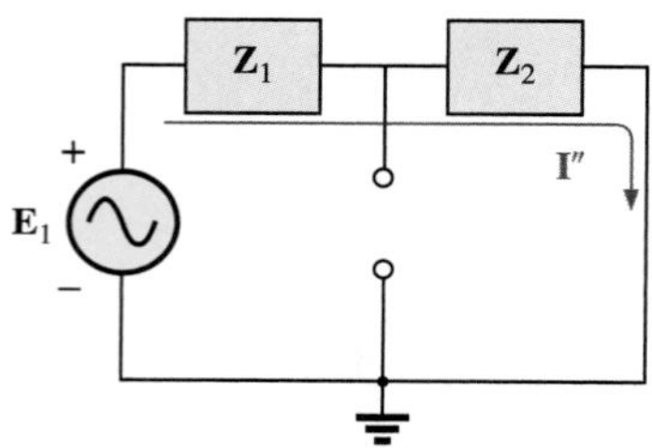

FIG. 18.9
Determining the effect of the voltage source $\mathbf{E}_1$ *on the current* **I** *of the network in Fig. 18.6.*

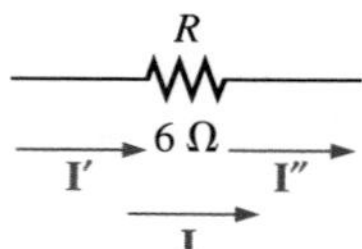

FIG. 18.10
Determining the resultant current **I** *for the network in Fig. 18.6.*

Solution: For the redrawn circuit (Fig. 18.7),

$$\mathbf{Z}_1 = j\,6\;\Omega \quad \mathbf{Z}_2 = 6\;\Omega - j\,8\;\Omega$$

Consider the effects of the current source (Fig. 18.8). Applying the current divider rule, we have

$$\mathbf{I}' = \frac{\mathbf{Z}_1\mathbf{I}_1}{\mathbf{Z}_1 + \mathbf{Z}_2} = \frac{(j\,6\;\Omega)(2\text{ A})}{j\,6\;\Omega + 6\;\Omega - j\,8\;\Omega} = \frac{j\,12\text{ A}}{6 - j\,2}$$

$$= \frac{12\text{ A}\;\angle 90^\circ}{6.32\;\angle -18.43^\circ}$$

$$\mathbf{I}' = 1.9\text{ A}\;\angle 108.43^\circ$$

Consider the effects of the voltage source (Fig. 18.9). Applying Ohm's law gives us

$$\mathbf{I}'' = \frac{\mathbf{E}_1}{\mathbf{Z}_T} = \frac{\mathbf{E}_1}{\mathbf{Z}_1 + \mathbf{Z}_2} = \frac{20\text{ V}\;\angle 30^\circ}{6.32\;\Omega\;\angle -18.43^\circ}$$

$$= 3.16\text{ A}\;\angle 48.43^\circ$$

The total current through the 6 Ω resistor (Fig. 18.10) is

$$\begin{aligned}
\mathbf{I} &= \mathbf{I}' + \mathbf{I}'' \\
&= 1.9\text{ A}\;\angle 108.43^\circ + 3.16\text{ A}\;\angle 48.43^\circ \\
&= (-0.60\text{ A} + j\,1.80\text{ A}) + (2.10\text{ A} + j\,2.36\text{ A}) \\
&= 1.50\text{ A} + j\,4.16\text{ A} \\
\mathbf{I} &= \mathbf{4.42\text{ A}\;\angle 70.2^\circ}
\end{aligned}$$

EXAMPLE 18.3 Using superposition, find the voltage across the 6 Ω resistor in Fig. 18.6. Check the results against $\mathbf{V}_{6\Omega} = \mathbf{I}(6\;\Omega)$, where **I** is the current found through the 6 Ω resistor in Example 18.2.

Solution: For the current source,

$$\mathbf{V}'_{6\Omega} = \mathbf{I}'(6\;\Omega) = (1.9\text{ A}\;\angle 108.43^\circ)(6\;\Omega) = 11.4\text{ V}\;\angle 108.43^\circ$$

For the voltage source,

$$\mathbf{V}''_{6\Omega} = \mathbf{I}''(6) = (3.16\text{ A}\;\angle 48.43^\circ)(6\;\Omega) = 18.96\text{ V}\;\angle 48.43^\circ$$

The total voltage across the 6 Ω resistor (Fig. 18.11) is

$$\begin{aligned}
\mathbf{V}_{6\Omega} &= \mathbf{V}'_{6\Omega} + \mathbf{V}''_{6\Omega} \\
&= 11.4\text{ V}\;\angle 108.43^\circ + 18.96\text{ V}\;\angle 48.43^\circ \\
&= (-3.60\text{ V} + j\,10.82\text{ V}) + (12.58\text{ V} + j\,14.18\text{ V}) \\
&= 8.98\text{ V} + j\,25.0\text{ V} \\
\mathbf{V}_{6\Omega} &= \mathbf{26.5\text{ V}\;\angle 70.2^\circ}
\end{aligned}$$

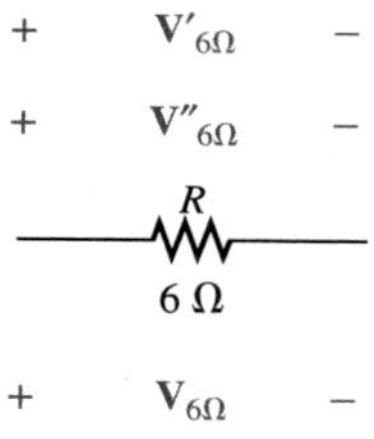

+ $\mathbf{V}'_{6\Omega}$ −

+ $\mathbf{V}''_{6\Omega}$ −

R

6 Ω

+ $\mathbf{V}_{6\Omega}$ −

FIG. 18.11
Determining the resultant voltage $\mathbf{V}_{6\Omega}$ *for the network in Fig. 18.6.*

Checking the result, we have

$$\mathbf{V}_{6\Omega} = \mathbf{I}(6\ \Omega) = (4.42\ \text{A}\ \angle 70.2°)(6\ \Omega)$$
$$= \mathbf{26.5\ V\ \angle 70.2°} \qquad \text{(checks)}$$

EXAMPLE 18.4 For the network in Fig. 18.12, determine the sinusoidal expression for the voltage v_3 using superposition.

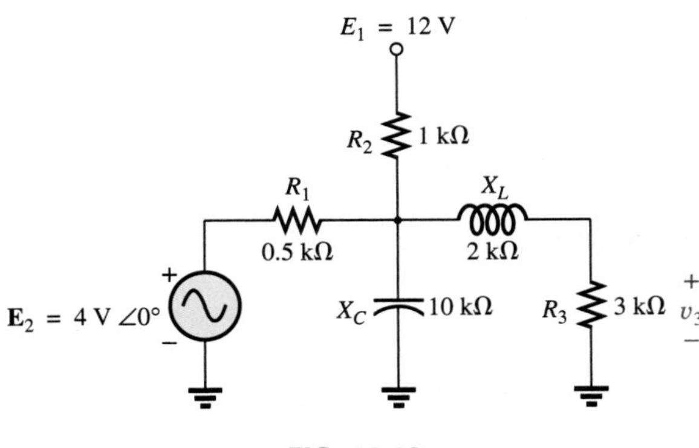

FIG. 18.12
Example 18.4.

Solution: For the dc analysis, the capacitor can be replaced by an open-circuit equivalent, and the inductor by a short-circuit equivalent. The result is the network in Fig. 18.13.

The resistors R_1 and R_3 are then in parallel, and the voltage V_3 can be determined using the voltage divider rule:

$$R' = R_1 \parallel R_3 = 0.5\ \text{k}\Omega \parallel 3\ \text{k}\Omega = 0.429\ \text{k}\Omega$$

and

$$V_3 = \frac{R'E_1}{R' + R_2}$$
$$= \frac{(0.429\ \text{k}\Omega)(12\ \text{V})}{0.429\ \text{k}\Omega + 1\ \text{k}\Omega} = \frac{5.148\ \text{V}}{1.429}$$
$$V_3 \cong \mathbf{3.6\ V}$$

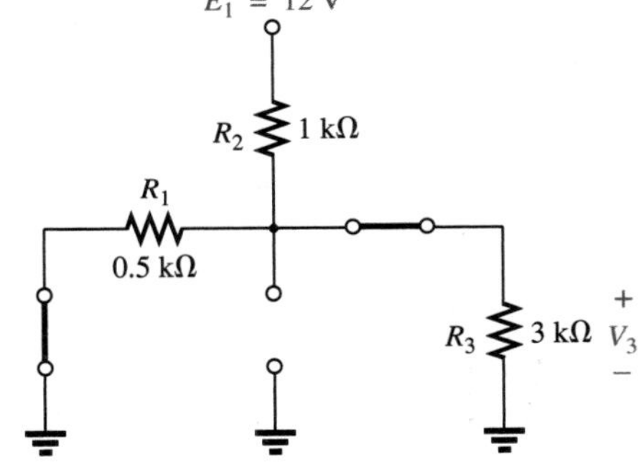

FIG. 18.13
Determining the effect of the dc voltage source E_1 on the voltage v_3 of the network in Fig. 18.12.

For the ac analysis, the dc source is set to zero and the network is redrawn, as shown in Fig. 18.14.

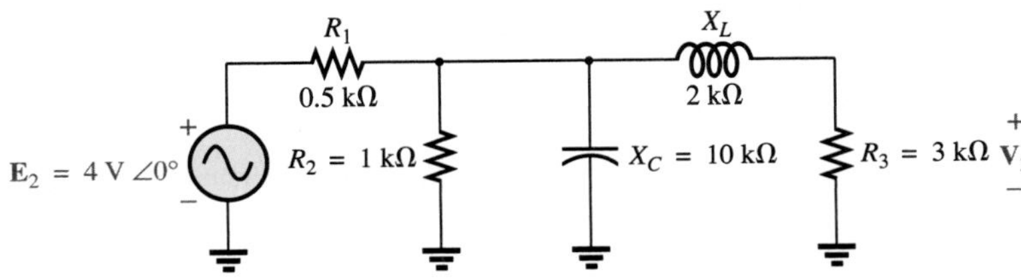

FIG. 18.14
Redrawing the network in Fig. 18.12 to determine the effect of the ac voltage source $\mathbf{E}_2$.

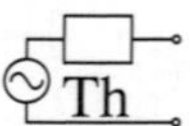

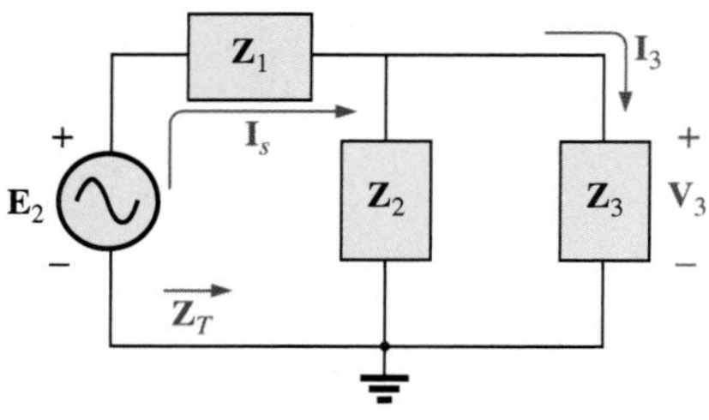

FIG. 18.15
Assigning the subscripted impedances to the network in Fig. 18.14.

The block impedances are then defined as in Fig. 18.15, and series-parallel techniques are applied as follows:

$$\mathbf{Z}_1 = 0.5\ \text{k}\Omega\ \angle 0°$$

$$\mathbf{Z}_2 = (R_2\ \angle 0°\ \|\ (X_C\ \angle -90°)$$

$$= \frac{(1\ \text{k}\Omega\ \angle 0°)(10\ \text{k}\Omega\ \angle -90°)}{1\ \text{k}\Omega - j\ 10\ \text{k}\Omega} = \frac{10\ \text{k}\Omega\ \angle -90°}{10.05\ \angle -84.29°}$$

$$= 0.995\ \text{k}\Omega\ \angle -5.71°$$

$$\mathbf{Z}_3 = R_3 + j\,X_L = 3\ \text{k}\Omega + j\ 2\ \text{k}\Omega = 3.61\ \text{k}\Omega\ \angle 33.69°$$

and

$$\mathbf{Z}_T = \mathbf{Z}_1 + \mathbf{Z}_2 \,\|\, \mathbf{Z}_3$$

$$= 0.5\ \text{k}\Omega + (0.995\ \text{k}\Omega\ \angle -5.71°)\ \|\ (3.61\ \text{k}\Omega\ \angle 33.69°)$$

$$= 1.312\ \text{k}\Omega\ \angle 1.57°$$

Calculator Solution: Performing the above on the TI-89 calculator requires the sequence of steps in Fig. 18.16.

.5+((0.995∠(−)5.71°)×(3.61∠33.69°))÷((0.995∠(−)5.71°)+(3.61∠33.69°)) ▶ Polar	1.31E0∠1.55E0

FIG. 18.16
Using the TI-89 calculator to determine $\mathbf{Z}_T$ for the network in Fig. 18.12.

$$\mathbf{I}_s = \frac{\mathbf{E}_2}{\mathbf{Z}_T} = \frac{4\ \text{V}\ \angle 0°}{1.312\ \text{k}\Omega\ \angle 1.57°} = 3.05\ \text{mA}\ \angle -1.57°$$

Current divider rule:

$$\mathbf{I}_3 = \frac{\mathbf{Z}_2\mathbf{I}_s}{\mathbf{Z}_2 + \mathbf{Z}_3} = \frac{(0.995\ \text{k}\Omega\ \angle -5.71°)(3.05\ \text{mA}\ \angle -1.57°)}{0.995\ \text{k}\Omega\ \angle -5.71° + 3.61\ \text{k}\Omega\ \angle 33.69°}$$

$$= 0.686\ \text{mA}\ \angle -32.74°$$

with

$$\mathbf{V}_3 = (I_3\ \angle \theta)(R_3\ \angle 0°)$$

$$= (0.686\ \text{mA}\ \angle -32.74°)(3\ \text{k}\Omega\ \angle 0°)$$

$$= \mathbf{2.06\ V\ \angle -32.74°}$$

The total solution:

$$v_3 = v_3\ (\text{dc}) + v_3\ (\text{ac})$$

$$= 3.6\ \text{V} + 2.06\ \text{V}\ \angle -32.74°$$

$$v_3 = \mathbf{3.6 + 2.91\ sin(\omega t - 32.74°)}$$

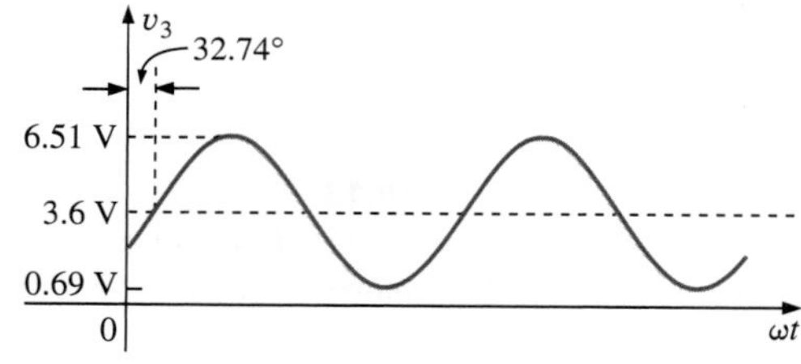

FIG. 18.17
The resultant voltage v_3 for the network in Fig. 18.12.

The result is a sinusoidal voltage having a peak value of 2.91 V riding on an average value of 3.6 V, as shown in Fig. 18.17.

Dependent Sources

For dependent sources in which *the controlling variable is not determined by the network to which the superposition theorem is to be applied,* the application of the theorem is basically the same as for independent sources. The solution obtained will simply be in terms of the controlling variables.

EXAMPLE 18.5 Using the superposition theorem, determine the current $\mathbf{I}_2$ for the network in Fig. 18.18. The quantities μ and h are constants.

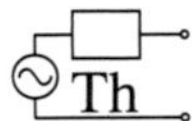

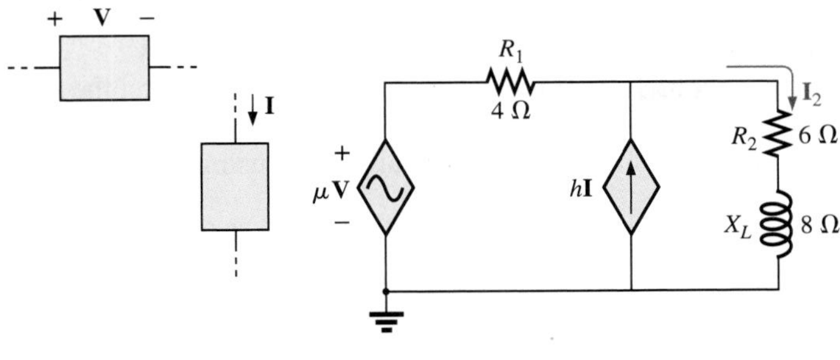

FIG. 18.18
Example 18.5.

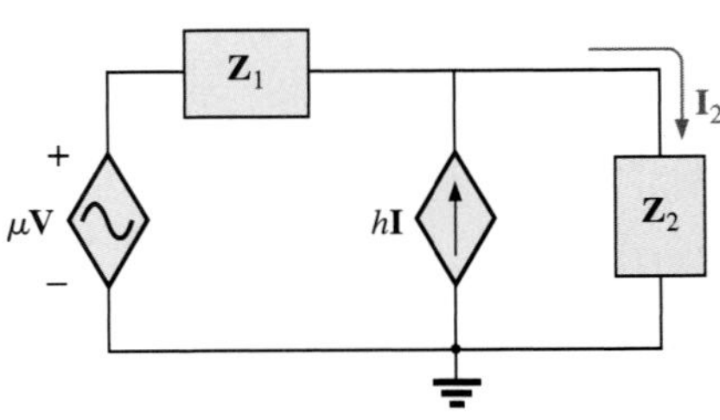

FIG. 18.19
Assigning the subscripted impedances to the network in Fig. 18.18.

Solution: With a portion of the system redrawn (Fig. 18.19),

$$\mathbf{Z}_1 = R_1 = 4\ \Omega \quad \mathbf{Z}_2 = R_2 + j\,X_L = 6\ \Omega + j\,8\ \Omega$$

For the voltage source (Fig. 18.20),

$$\mathbf{I}' = \frac{\mu\mathbf{V}}{\mathbf{Z}_1 + \mathbf{Z}_2} = \frac{\mu\mathbf{V}}{4\ \Omega + 6\ \Omega + j\,8\ \Omega} = \frac{\mu\mathbf{V}}{10\ \Omega + j\,8\ \Omega}$$
$$= \frac{\mu\mathbf{V}}{12.8\ \Omega\ \angle 38.66^\circ} = 0.078\ \mu\mathbf{V}/\Omega\ \angle{-38.66^\circ}$$

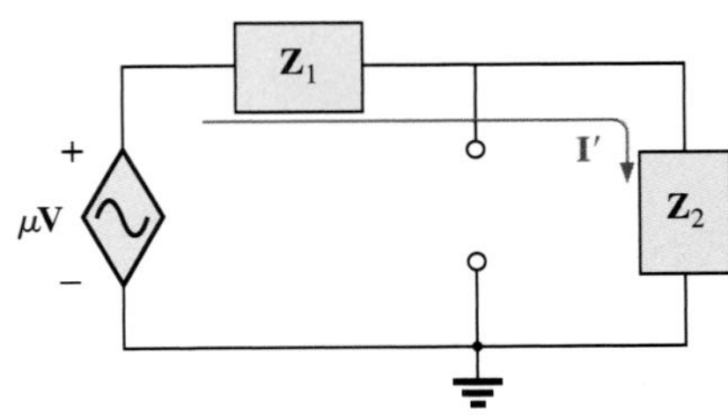

FIG. 18.20
Determining the effect of the voltage-controlled voltage source on the current $\mathbf{I}_2$ for the network in Fig. 18.18.

For the current source (Fig. 18.21),

$$\mathbf{I}'' = \frac{\mathbf{Z}_1(h\mathbf{I})}{\mathbf{Z}_1 + \mathbf{Z}_2} = \frac{(4\ \Omega)(h\mathbf{I})}{12.8\ \Omega\ \angle 38.66^\circ} = 4(0.078)h\mathbf{I}\ \angle{-38.66^\circ}$$
$$= 0.312h\mathbf{I}\ \angle{-38.66^\circ}$$

The current $\mathbf{I}_2$ is

$$\mathbf{I}_2 = \mathbf{I}' + \mathbf{I}''$$
$$= 0.078\ \mu\mathbf{V}/\Omega\ \angle{-38.66^\circ} + 0.312h\mathbf{I}\ \angle{-38.66^\circ}$$

For $\mathbf{V} = 10\ \text{V}\ \angle 0^\circ$, $\mathbf{I} = 20\ \text{mA}\ \angle 0^\circ$, $\mu = 20$, and $h = 100$,

$$\mathbf{I}_2 = 0.078(20)(10\ \text{V}\ \angle 0^\circ)/\Omega\ \angle{-38.66^\circ}$$
$$+ 0.312(100)(20\ \text{mA}\ \angle 0^\circ)\angle{-38.66^\circ}$$
$$= 15.60\ \text{A}\ \angle{-38.66^\circ} + 0.62\ \text{A}\ \angle{-38.66^\circ}$$
$$\mathbf{I}_2 = \mathbf{16.22\ A\ \angle{-38.66^\circ}}$$

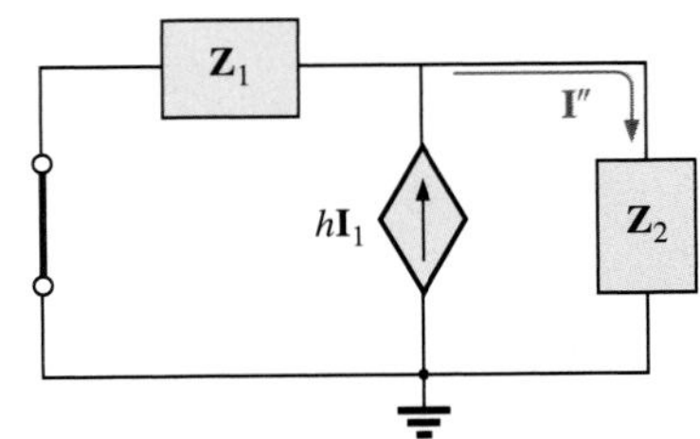

FIG. 18.21
Determining the effect of the current-controlled current source on the current $\mathbf{I}_2$ for the network in Fig. 18.18.

For dependent sources in which *the controlling variable is determined by the network to which the theorem is to be applied,* the dependent source cannot be set to zero unless the controlling variable is also zero. For networks containing dependent sources (as in Example 18.5) and dependent sources of the type just introduced above, the superposition theorem is applied for each independent source and each dependent source not having a controlling variable in the portions of the network under investigation. It must be reemphasized that dependent sources are not sources of energy in the sense that, if all independent sources are removed from a system, all currents and voltages must be zero.

EXAMPLE 18.6 Determine the current $\mathbf{I}_L$ through the resistor R_L in Fig. 18.22.

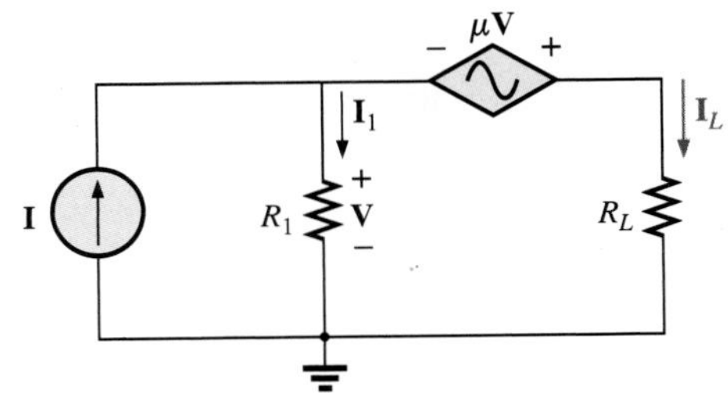

FIG. 18.22
Example 18.6.

Solution: Note that the controlling variable **V** is determined by the network to be analyzed. From the above discussions, it is understood that the

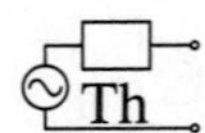

dependent source cannot be set to zero unless **V** is zero. If we set **I** to zero, the network lacks a source of voltage, and $\mathbf{V} = 0$ with $\mu\mathbf{V} = 0$. The resulting $\mathbf{I}_L$ under this condition is zero. Obviously, therefore, the network must be analyzed as it appears in Fig. 18.22, with the result that neither source can be eliminated, as is normally done using the superposition theorem.

Applying Kirchhoff's voltage law, we have

$$\mathbf{V}_L = \mathbf{V} + \mu\mathbf{V} = (1 + \mu)\mathbf{V}$$

and

$$\mathbf{I}_L = \frac{\mathbf{V}_L}{R_L} = \frac{(1 + \mu)\mathbf{V}}{R_L}$$

The result, however, must be found in terms of **I** since **V** and $\mu\mathbf{V}$ are only dependent variables.

Applying Kirchhoff's current law gives us

$$\mathbf{I} = \mathbf{I}_1 + \mathbf{I}_L = \frac{\mathbf{V}}{R_1} + \frac{(1 + \mu)\mathbf{V}}{R_L}$$

and

$$\mathbf{I} = \mathbf{V}\left(\frac{1}{R_1} + \frac{1 + \mu}{R_L}\right)$$

or

$$\mathbf{V} = \frac{\mathbf{I}}{(1/R_1) + [(1 + \mu)/R_L]}$$

Substituting into the above yields

$$\mathbf{I}_L = \frac{(1 + \mu)\mathbf{V}}{R_L} = \frac{(1 + \mu)}{R_L}\left(\frac{\mathbf{I}}{(1/R_1) + [(1 + \mu)/R_L]}\right)$$

Therefore,

$$\mathbf{I}_L = \frac{(1 + \mu)R_1\mathbf{I}}{R_L + (1 + \mu)R_1}$$

18.3 THÉVENIN'S THEOREM

Thévenin's theorem, as stated for sinusoidal ac circuits, is changed only to include the term *impedance* instead of *resistance;* that is,

any two-terminal linear ac network can be replaced with an equivalent circuit consisting of a voltage source and an impedance in series, as shown in Fig. 18.23.

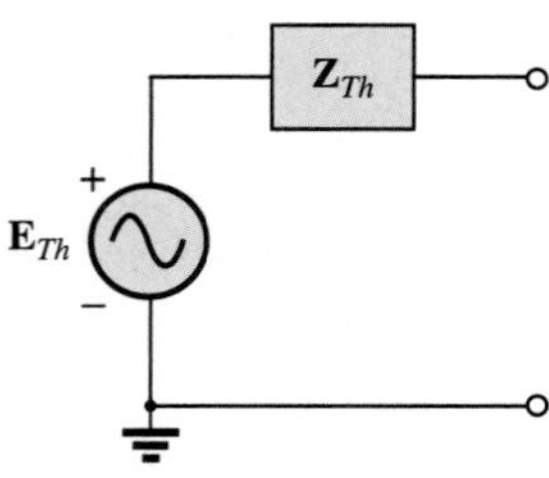

FIG. 18.23
Thévenin equivalent circuit for ac networks.

Since the reactances of a circuit are frequency dependent, the Thévenin circuit found for a particular network is applicable only at *one* frequency.

The steps required to apply this method to dc circuits are repeated here with changes for sinusoidal ac circuits. As before, the only change is the replacement of the term *resistance* with *impedance.* Again, dependent and independent sources are treated separately.

Example 18.9, the last example of the independent source section, includes a network with dc and ac sources to establish the groundwork for possible use in the electronics area.

Independent Sources

1. ***Remove that portion of the network across which the Thévenin equivalent circuit is to be found.***

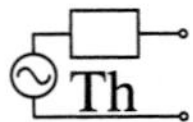

2. ***Mark (○, •, and so on) the terminals of the remaining two-terminal network.***
3. ***Calculate $\mathbf{Z}_{Th}$ by first setting all voltage and current sources to zero (short circuit and open circuit, respectively) and then finding the resulting impedance between the two marked terminals.***
4. ***Calculate $\mathbf{E}_{Th}$ by first replacing the voltage and current sources and then finding the open-circuit voltage between the marked terminals.***
5. ***Draw the Thévenin equivalent circuit with the portion of the circuit previously removed replaced between the terminals of the Thévenin equivalent circuit.***

EXAMPLE 18.7 Find the Thévenin equivalent circuit for the network external to resistor R in Fig. 18.24.

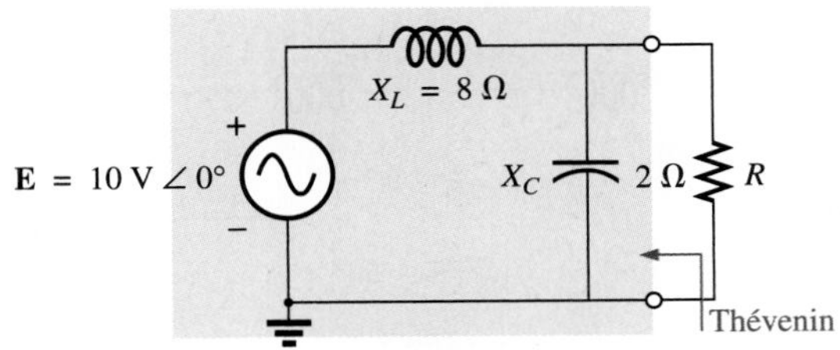

FIG. 18.24
Example 18.7.

Solution:

Steps 1 and 2 (Fig. 18.25):

$$\mathbf{Z}_1 = j\,X_L = j\,8\,\Omega \quad \mathbf{Z}_2 = -j\,X_C = -j\,2\,\Omega$$

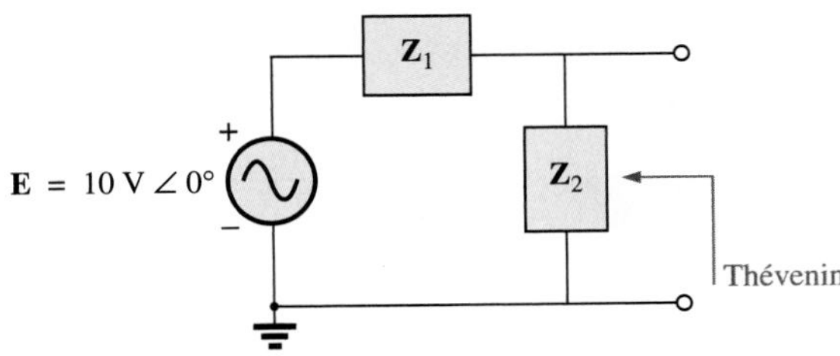

FIG. 18.25
Assigning the subscripted impedances to the network in Fig. 18.24.

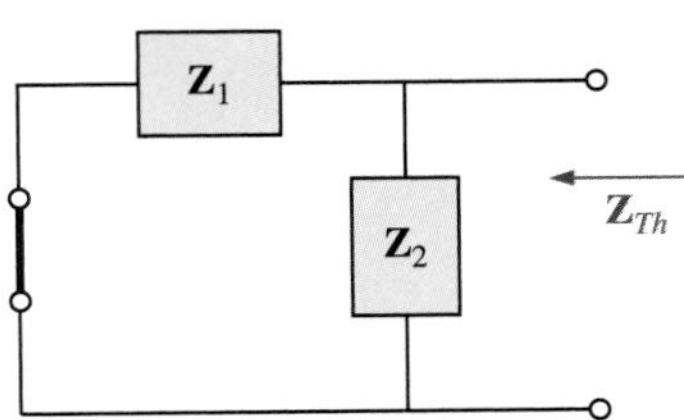

FIG. 18.26
Determining the Thévenin impedance for the network in Fig. 18.24.

Step 3 (Fig. 18.26):

$$\mathbf{Z}_{Th} = \frac{\mathbf{Z}_1\mathbf{Z}_2}{\mathbf{Z}_1 + \mathbf{Z}_2} = \frac{(j\,8\,\Omega)(-j\,2\,\Omega)}{j\,8\,\Omega - j\,2\,\Omega} = \frac{-j^2\,16\,\Omega}{j\,6} = \frac{16\,\Omega}{6\,\angle 90^\circ}$$
$$= \mathbf{2.67\,\Omega\,\angle -90^\circ}$$

Step 4 (Fig. 18.27):

$$\mathbf{E}_{Th} = \frac{\mathbf{Z}_2\mathbf{E}}{\mathbf{Z}_1 + \mathbf{Z}_2} \qquad \text{(voltage divider rule)}$$
$$= \frac{(-j\,2\,\Omega)(10\text{ V})}{j\,8\,\Omega - j\,2\,\Omega} = \frac{-j\,20\text{ V}}{j\,6} = \mathbf{3.33\text{ V}\,\angle -180^\circ}$$

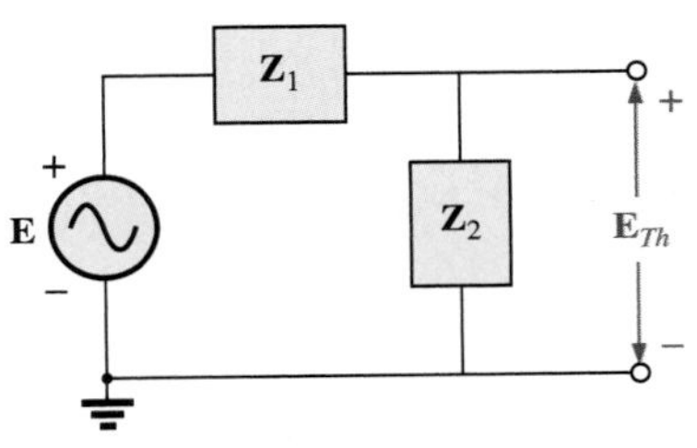

FIG. 18.27
Determining the open-circuit Thévenin voltage for the network in Fig. 18.24.

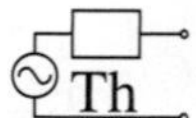

Step 5: The Thévenin equivalent circuit is shown in Fig. 18.28.

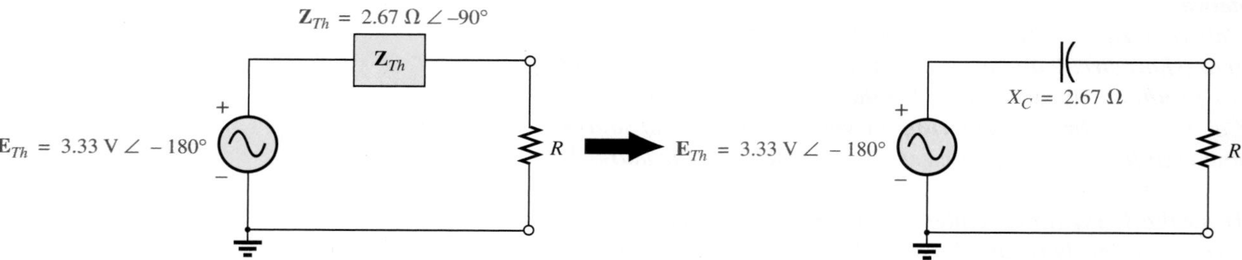

FIG. 18.28
The Thévenin equivalent circuit for the network in Fig. 18.24.

EXAMPLE 18.8 Find the Thévenin equivalent circuit for the network external to branch *a-a′* in Fig. 18.29.

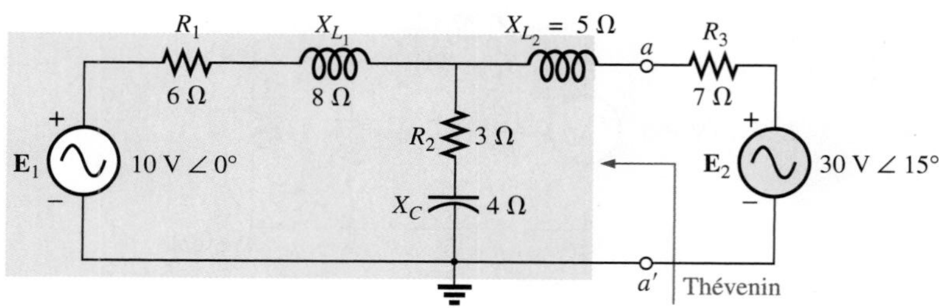

FIG. 18.29
Example 18.8.

Solution:

Steps 1 and 2 (Fig. 18.30): Note the reduced complexity with subscripted impedances:

$$\mathbf{Z}_1 = R_1 + j X_{L_1} = 6\ \Omega + j\,8\ \Omega$$
$$\mathbf{Z}_2 = R_2 - j X_C = 3\ \Omega - j\,4\ \Omega$$
$$\mathbf{Z}_3 = +j X_{L_2} = j\,5\ \Omega$$

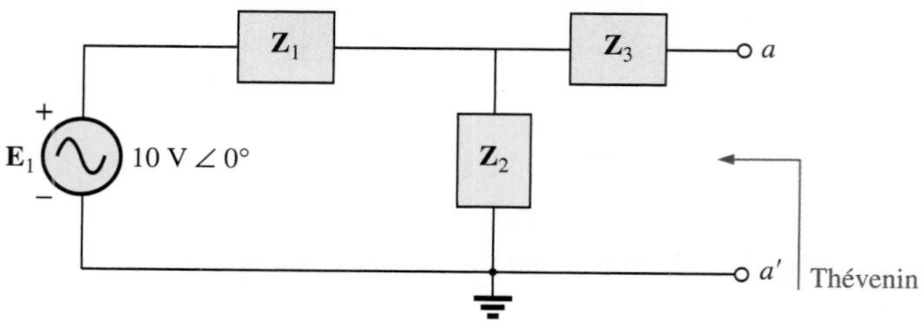

FIG. 18.30
Assigning the subscripted impedances for the network in Fig. 18.29.

Step 3 (Fig. 18.31):

$$\mathbf{Z}_{Th} = \mathbf{Z}_3 + \frac{\mathbf{Z}_1\mathbf{Z}_2}{\mathbf{Z}_1 + \mathbf{Z}_2} = j\,5\ \Omega + \frac{(10\ \Omega\ \angle 53.13^\circ)(5\ \Omega\ \angle -53.13^\circ)}{(6\ \Omega + j\,8\ \Omega) + (3\ \Omega - j\,4\ \Omega)}$$
$$= j\,5 + \frac{50\ \angle 0^\circ}{9 + j\,4} = j\,5 + \frac{50\ \angle 0^\circ}{9.85\ \angle 23.96^\circ}$$
$$= j\,5 + 5.08\ \angle -23.96^\circ = j\,5 + 4.64 - j\,2.06$$
$$\mathbf{Z}_{Th} = \mathbf{4.64\ \Omega + j\,2.94\ \Omega = 5.49\ \Omega\ \angle 32.36^\circ}$$

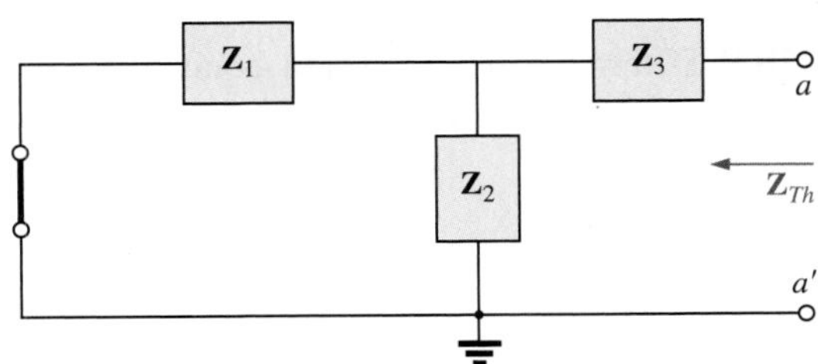

FIG. 18.31
Determining the Thévenin impedance for the network in Fig. 18.29.

Step 4 (Fig. 18.32): Since *a-a'* is an open circuit, $\mathbf{I}_{\mathbf{Z}_3} = 0$. Then $\mathbf{E}_{Th}$ is the voltage drop across $\mathbf{Z}_2$:

$$\mathbf{E}_{Th} = \frac{\mathbf{Z}_2\mathbf{E}}{\mathbf{Z}_2 + \mathbf{Z}_1} \qquad \text{(voltage divider rule)}$$

$$= \frac{(5\ \Omega\ \angle -53.13^\circ)(10\ \text{V}\ \angle 0^\circ)}{9.85\ \Omega\ \angle 23.96^\circ}$$

$$\mathbf{E}_{Th} = \frac{50\ \text{V}\ \angle -53.13^\circ}{9.85\ \angle 23.96^\circ} = \mathbf{5.08\ V\ \angle -77.09^\circ}$$

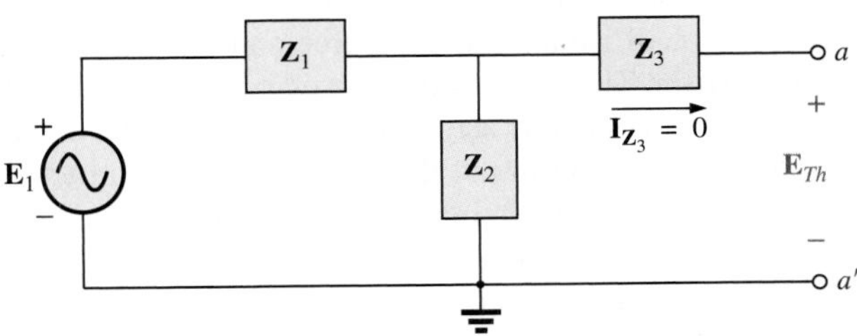

FIG. 18.32
Determining the open-circuit Thévenin voltage for the network in Fig. 18.29.

Step 5: The Thévenin equivalent circuit is shown in Fig. 18.33.

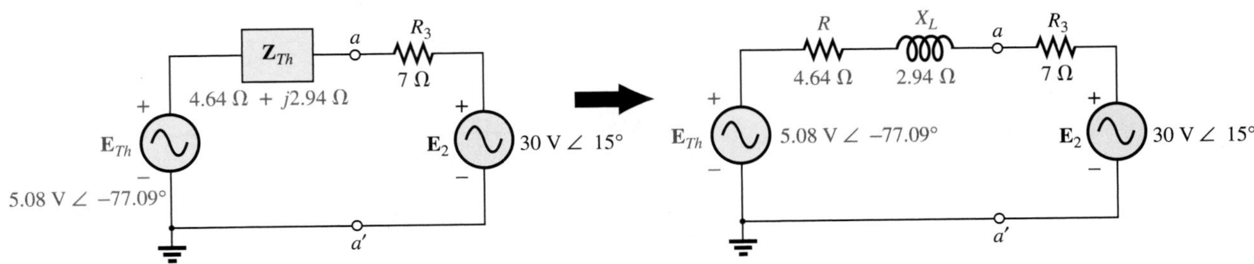

FIG. 18.33
The Thévenin equivalent circuit for the network in Fig. 18.29.

The next example demonstrates how superposition is applied to electronic circuits to permit *a separation of the dc and ac analyses.* The fact that the controlling variable in this analysis is not in the portion of the network connected directly to the terminals of interest permits an analysis of the network in the same manner as applied above for independent sources.

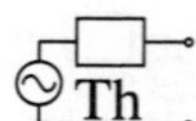

EXAMPLE 18.9 Determine the Thévenin equivalent circuit for the transistor network external to the resistor R_L in the network in Fig. 18.34. Then determine $\mathbf{V}_L$.

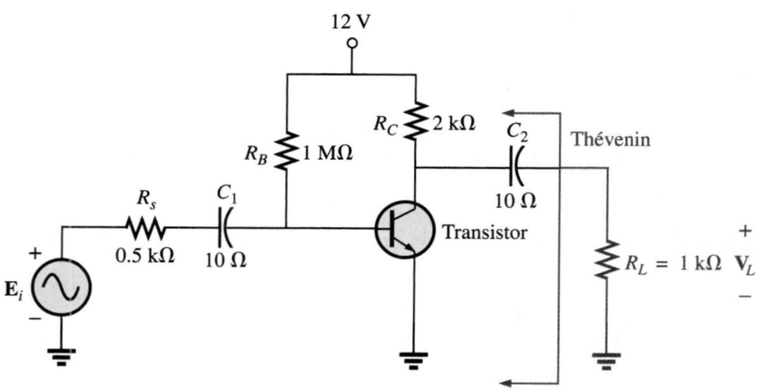

FIG. 18.34
Example 18.9.

Solution: Applying superposition.

dc Conditions Substituting the open-circuit equivalent for the coupling capacitor C_2 will isolate the dc source and the resulting currents from the load resistor. The result is that for dc conditions, $V_L = 0$ V. Although the output dc voltage is zero, the application of the dc voltage is important to the basic operation of the transistor in a number of important ways, one of which is to determine the parameters of the "equivalent circuit" to appear in the ac analysis to follow.

ac Conditions For the ac analysis, an equivalent circuit is substituted for the transistor, as established by the dc conditions above, that will behave like the actual transistor. A great deal more will be said about equivalent circuits and the operations performed to obtain the network in Fig. 18.35, but for now we limit our attention to the manner in which the Thévenin equivalent circuit is obtained. Note in Fig. 18.35 that the equivalent circuit includes a resistor of 2.3 kΩ and a controlled current source whose magnitude is determined by the product of a factor of 100 and the current I_1 in another part of the network.

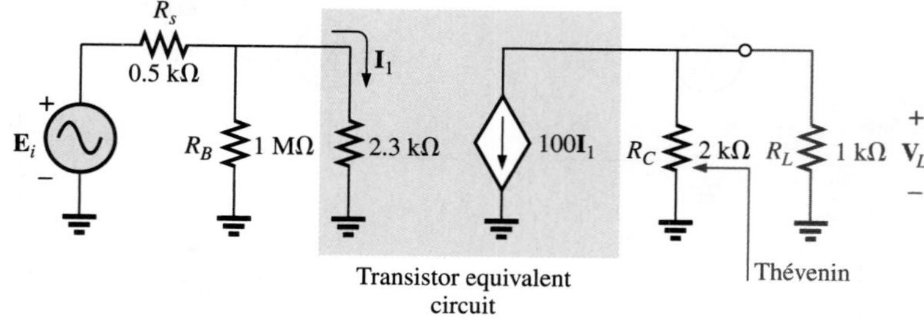

FIG. 18.35
The ac equivalent network for the transistor amplifier in Fig. 18.34.

Note in Fig. 18.35 the absence of the coupling capacitors for the ac analysis. In general, coupling capacitors are designed to be open circuits for dc analysis and short circuits for ac analysis. The short-circuit equivalent is valid because the other impedances in series with the coupling ca-

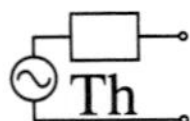

pacitors are so much larger in magnitude that the effect of the coupling capacitors can be ignored. Both R_B and R_C are now tied to ground because the dc source was set to zero volts (superposition) and replaced by a short-circuit equivalent to ground.

For the analysis to follow, the effect of the resistor R_B will be ignored since it is so much larger than the parallel 2.3 kΩ resistor.

$\mathbf{Z}_{Th}$ When $\mathbf{E}_i$ is set to zero volts, the current $\mathbf{I}_1$ will be zero amperes, and the controlled source $100\mathbf{I}_1$ will be zero amperes also. The result is an open-circuit equivalent for the source, as appearing in Fig. 18.36.

It is fairly obvious from Fig. 18.36 that

$$\mathbf{Z}_{Th} = \mathbf{2\ k\Omega}$$

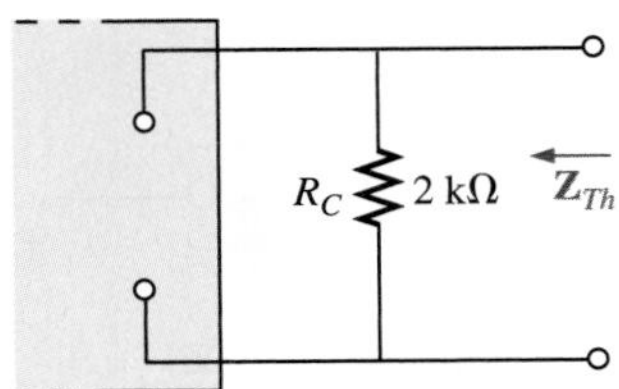

FIG. 18.36
Determining the Thévenin impedance for the network in Fig. 18.35.

$\mathbf{E}_{Th}$ For $\mathbf{E}_{Th}$, the current $\mathbf{I}_1$ in Fig. 18.35 will be

$$\mathbf{I}_1 = \frac{\mathbf{E}_i}{R_s + 2.3\ \text{k}\Omega} = \frac{\mathbf{E}_i}{0.5\ \text{k}\Omega + 2.3\ \text{k}\Omega} = \frac{\mathbf{E}_i}{2.8\ \text{k}\Omega}$$

and
$$100\mathbf{I}_1 = (100)\left(\frac{\mathbf{E}_i}{2.8\ \text{k}\Omega}\right) = 35.71 \times 10^{-3}/\Omega\ \mathbf{E}_i$$

Referring to Fig. 18.37, we find that

$$\begin{aligned}\mathbf{E}_{Th} &= -(100\mathbf{I}_1)R_C \\ &= -(35.71 \times 10^{-3}/\Omega\ \mathbf{E}_i)(2 \times 10^3\ \Omega) \\ \mathbf{E}_{Th} &= \mathbf{-71.42E}_i\end{aligned}$$

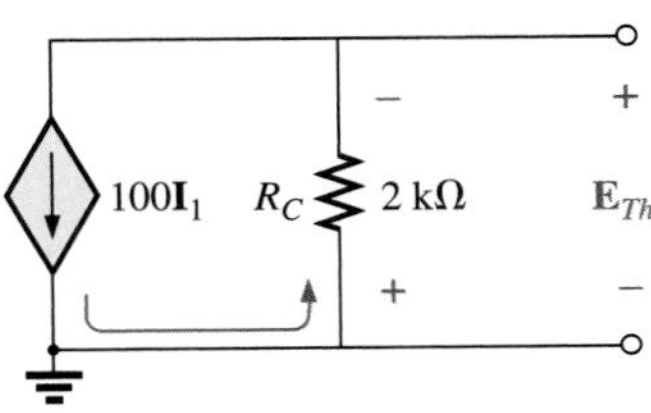

FIG. 18.37
Determining the Thévenin voltage for the network in Fig. 18.35.

The Thévenin equivalent circuit appears in Fig. 18.38 with the original load R_L.

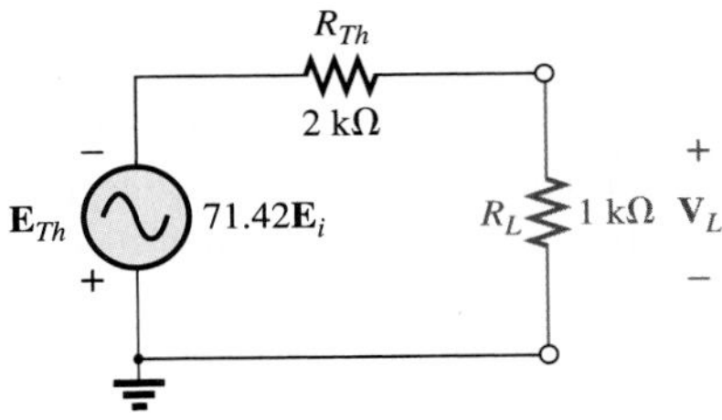

FIG. 18.38
The Thévenin equivalent circuit for the network in Fig. 18.35.

Output Voltage $\mathbf{V}_L$

$$\mathbf{V}_L = \frac{-R_L\mathbf{E}_{Th}}{R_L + R_{Th}} = \frac{-(1\ \text{k}\Omega)(71.42\mathbf{E}_i)}{1\ \text{k}\Omega + 2\ \text{k}\Omega}$$

and
$$\mathbf{V}_L = \mathbf{-23.81E}_i$$

revealing that the output voltage is 23.81 times the applied voltage with a phase shift of 180° due to the minus sign.

Dependent Sources

For dependent sources with a *controlling variable not in the network under investigation,* the procedure indicated above can be applied. However, for dependent sources of the other type, where the *controlling variable is part of the network to which the theorem is to be applied,* another approach must be used. The necessity for a different approach is demonstrated in an example to follow. The method is *not limited to dependent*

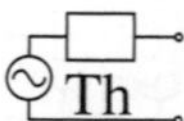

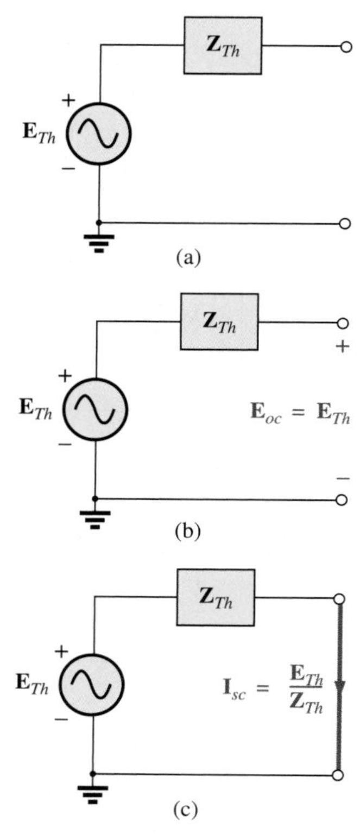

FIG. 18.39
Defining an alternative approach for determining the Thévenin impedance.

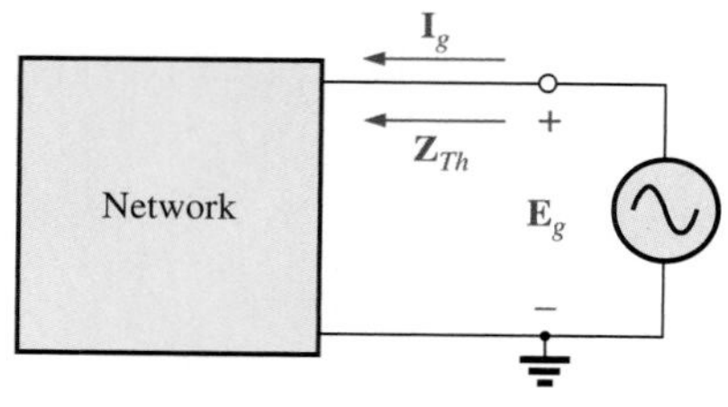

FIG. 18.40
Determining $\mathbf{Z}_{Th}$ using the approach $\mathbf{Z}_{Th} = \mathbf{E}_g/\mathbf{I}_g$.

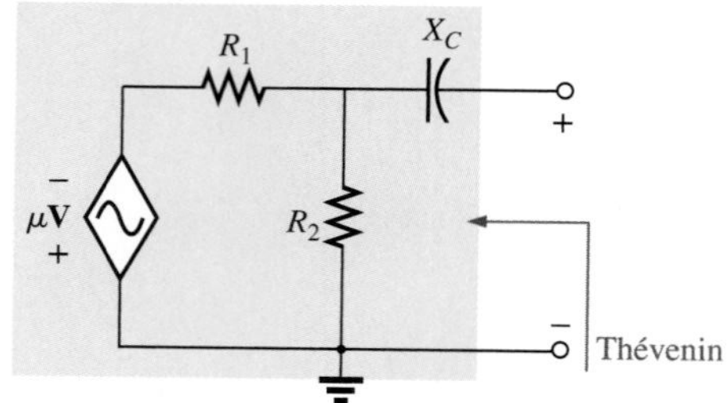

FIG. 18.41
Example 18.10.

sources of the latter type. It can also be applied to any dc or sinusoidal ac network. However, for networks of independent sources, the method of application used in Chapter 9 and presented in the first portion of this section is generally more direct, with the usual savings in time and errors.

The new approach to Thévenin's theorem can best be introduced at this stage in the development by considering the Thévenin equivalent circuit in Fig. 18.39(a). As indicated in Fig. 18.39(b), the open-circuit terminal voltage ($\mathbf{E}_{oc}$) of the Thévenin equivalent circuit is the Thévenin equivalent voltage; that is,

$$\boxed{\mathbf{E}_{oc} = \mathbf{E}_{Th}} \tag{18.1}$$

If the external terminals are short circuited as in Fig. 18.39(c), the resulting short-circuit current is determined by

$$\boxed{\mathbf{I}_{sc} = \frac{\mathbf{E}_{Th}}{\mathbf{Z}_{Th}}} \tag{18.2}$$

or, rearranged,

$$\mathbf{Z}_{Th} = \frac{\mathbf{E}_{Th}}{\mathbf{I}_{sc}}$$

and

$$\boxed{\mathbf{Z}_{Th} = \frac{\mathbf{E}_{oc}}{\mathbf{I}_{sc}}} \tag{18.3}$$

Eqs. (18.1) and (18.3) indicate that for any linear bilateral dc or ac network with or without dependent sources of any type, if the open-circuit terminal voltage of a portion of a network can be determined along with the short-circuit current between the same two terminals, the Thévenin equivalent circuit is effectively known. A few examples will make the method quite clear. The advantage of the method, which was stressed earlier in this section for independent sources, should now be more obvious. The current $\mathbf{I}_{sc}$, which is necessary to find $\mathbf{Z}_{Th}$, is in general more difficult to obtain since all of the sources are present.

There is a third approach to the Thévenin equivalent circuit that is also useful from a practical viewpoint. The Thévenin voltage is found as in the two previous methods. However, the Thévenin impedance is obtained by applying a source of voltage to the terminals of interest and determining the source current as indicated in Fig. 18.40. For this method, the source voltage of the original network is set to zero. The Thévenin impedance is then determined by the following equation:

$$\boxed{\mathbf{Z}_{Th} = \frac{\mathbf{E}_g}{\mathbf{I}_g}} \tag{18.4}$$

Note that for each technique, $\mathbf{E}_{Th} = \mathbf{E}_{oc}$, but the Thévenin impedance is found in different ways.

EXAMPLE 18.10 Using each of the three techniques described in this section, determine the Thévenin equivalent circuit for the network in Fig. 18.41.

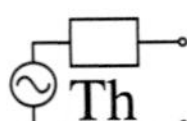

Solution: Since for each approach the Thévenin voltage is found in exactly the same manner, it is determined first. From Fig. 18.41, where $\mathbf{I}_{X_C} = 0$,

Due to the polarity for **V** and defined terminal polarities

$$\mathbf{V}_{R_1} = \mathbf{E}_{Th} = \mathbf{E}_{oc} = -\frac{R_2(\mu\mathbf{V})}{R_1 + R_2} = -\frac{\mu R_2 \mathbf{V}}{R_1 + R_2}$$

The following three methods for determining the Thévenin impedance appear in the order in which they were introduced in this section.

Method 1: See Fig. 18.42.

$$\mathbf{Z}_{Th} = \boldsymbol{R_1 \| R_2 - j\,X_C}$$

Method 2: See Fig. 18.43. Converting the voltage source to a current source (Fig. 18.44), we have (current divider rule)

$$\mathbf{I}_{sc} = \frac{-(R_1 \| R_2)\dfrac{\mu\mathbf{V}}{R_1}}{(R_1 \| R_2) - j\,X_C} = \frac{-\dfrac{R_1 R_2}{R_1 + R_2}\left(\dfrac{\mu\mathbf{V}}{R_1}\right)}{(R_1 \| R_2) - j\,X_C}$$

$$= \frac{\dfrac{-\mu R_2 \mathbf{V}}{R_1 + R_2}}{(R_1 \| R_2) - j\,X_C}$$

and

$$\mathbf{Z}_{Th} = \frac{\mathbf{E}_{oc}}{\mathbf{I}_{sc}} = \frac{\dfrac{-\mu R_2 \mathbf{V}}{R_1 + R_2}}{\dfrac{\dfrac{-\mu R_2 \mathbf{V}}{R_1 + R_2}}{(R_1 \| R_2) - j\,X_C}} = \frac{1}{\dfrac{1}{(R_1 \| R_2) - j\,X_C}}$$

$$= \boldsymbol{R_1 \| R_2 - j\,X_C}$$

Method 3: See Fig. 18.45.

$$\mathbf{I}_g = \frac{\mathbf{E}_g}{(R_1 \| R_2) - j\,X_C}$$

and

$$\mathbf{Z}_{Th} = \frac{\mathbf{E}_g}{\mathbf{I}_g} = \boldsymbol{R_1 \| R_2 - j\,X_C}$$

In each case, the Thévenin impedance is the same. The resulting Thévenin equivalent circuit is shown in Fig. 18.46.

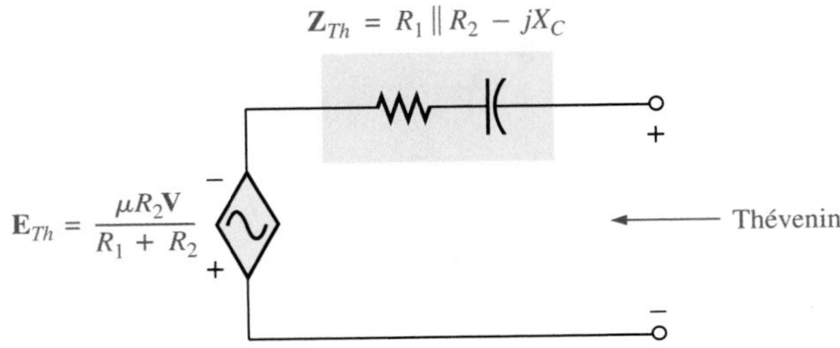

FIG. 18.46
The Thévenin equivalent circuit for the network in Fig. 18.41.

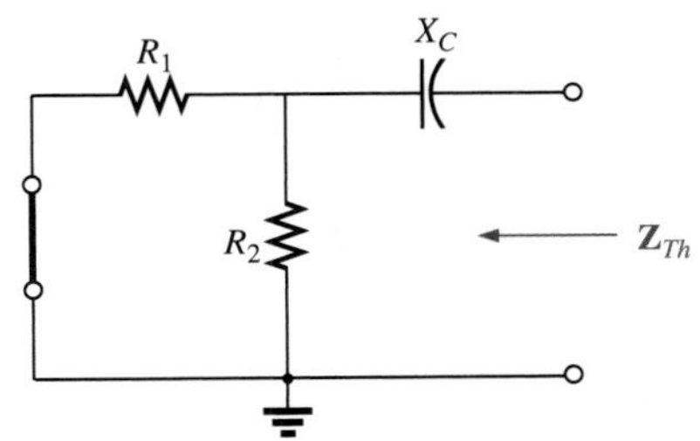

FIG. 18.42
Determining the Thévenin impedance for the network in Fig. 18.41.

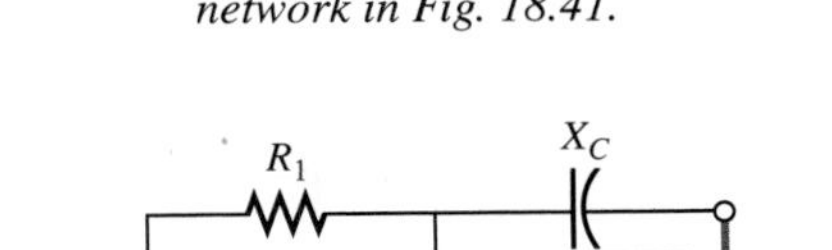

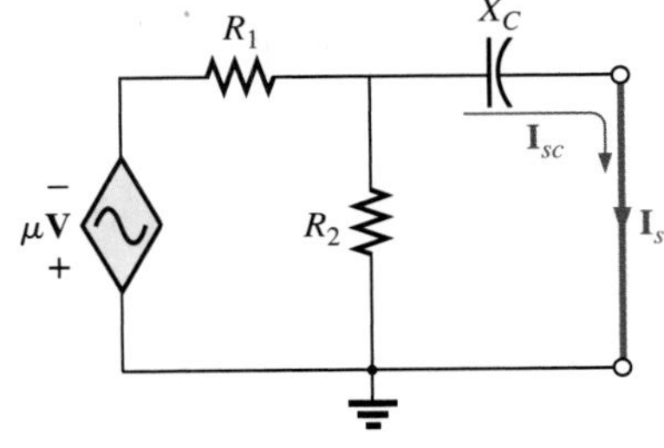

FIG. 18.43
Determining the short-circuit current for the network in Fig. 18.41.

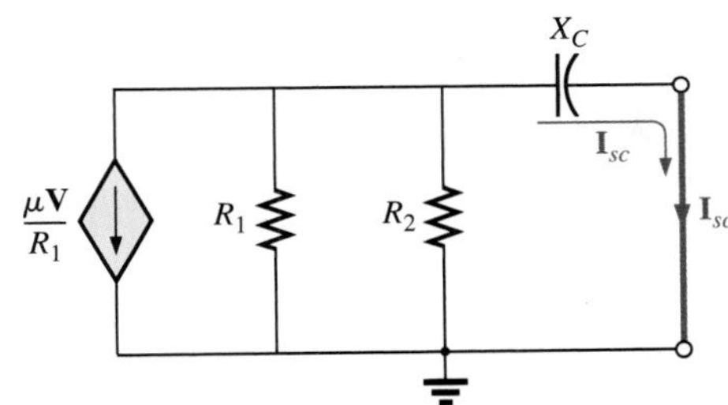

FIG. 18.44
Converting the voltage source in Fig. 18.43 to a current source.

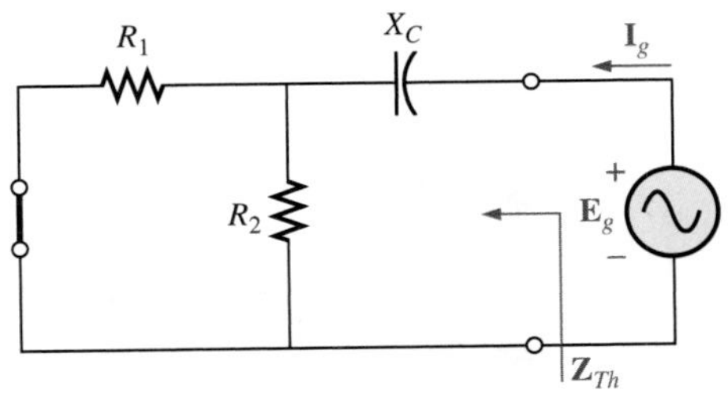

FIG. 18.45
Determining the Thévenin impedance for the network in Fig. 18.41 using the approach $\mathbf{Z}_{Th} = \mathbf{E}_g/\mathbf{I}_g$.

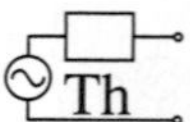

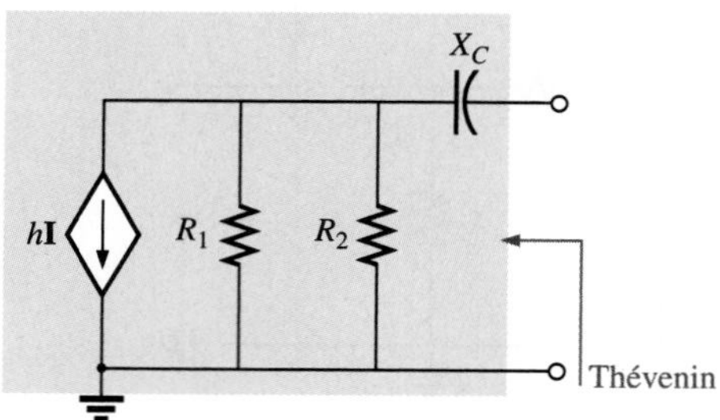

FIG. 18.47
Example 18.11.

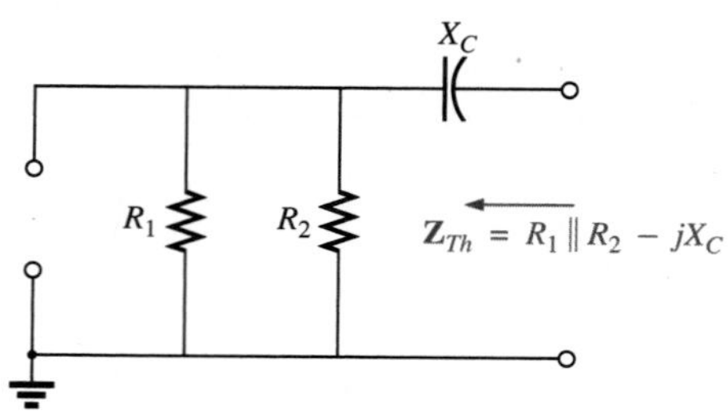

FIG. 18.48
Determining the Thévenin impedance for the network in Fig. 18.47.

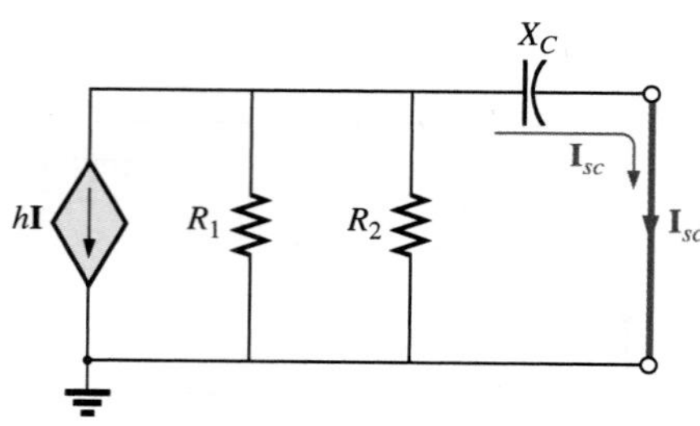

FIG. 18.49
Determining the short-circuit current for the network in Fig. 18.47.

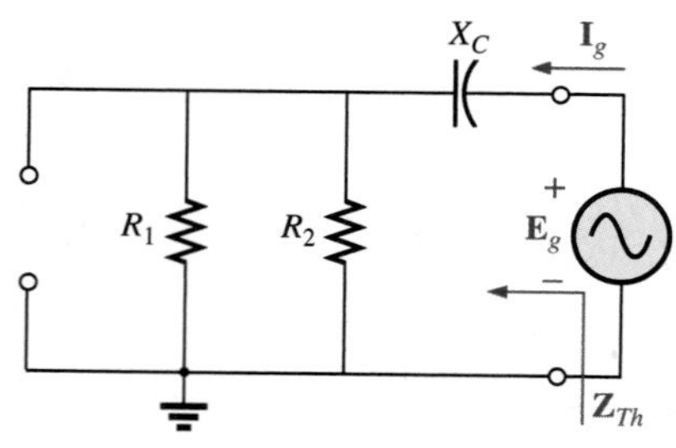

FIG. 18.50
Determining the Thévenin impedance using the approach $\mathbf{Z}_{Th} = \mathbf{E}_g/\mathbf{I}_g$.

EXAMPLE 18.11 Repeat Example 18.10 for the network in Fig. 18.47.

Solution: From Fig. 18.47, $\mathbf{E}_{Th}$ is

$$\mathbf{E}_{Th} = \mathbf{E}_{oc} = -h\mathbf{I}(R_1 \| R_2) = -\frac{\boldsymbol{hR_1R_2\mathrm{I}}}{\boldsymbol{R_1 + R_2}}$$

Method 1: See Fig. 18.48.

$$\mathbf{Z}_{Th} = \boldsymbol{R_1 \| R_2 - j\,X_C}$$

Note the similarity between this solution and that obtained for the previous example.

Method 2: See Fig. 18.49.

$$\mathbf{I}_{sc} = \frac{-(R_1 \| R_2)h\mathbf{I}}{(R_1 \| R_2) - j\,X_C}$$

and
$$\mathbf{Z}_{Th} = \frac{\mathbf{E}_{oc}}{\mathbf{I}_{sc}} = \frac{-h\mathbf{I}(R_1 \| R_2)}{\dfrac{-(R_1 \| R_2)h\mathbf{I}}{(R_1 \| R_2) - j\,X_C}} = \boldsymbol{R_1 \| R_2 - j\,X_C}$$

Method 3: See Fig. 18.50.

$$\mathbf{I}_g = \frac{\mathbf{E}_g}{(R_1 \| R_2) - j\,X_C}$$

and
$$\mathbf{Z}_{Th} = \frac{\mathbf{E}_g}{\mathbf{I}_g} = \boldsymbol{R_1 \| R_2 - j\,X_C}$$

The following example has a dependent source that will not permit the use of the method described at the beginning of this section for independent sources. All three methods will be applied, however, so that the results can be compared.

EXAMPLE 18.12 For the network in Fig. 18.51 (introduced in Example 18.6), determine the Thévenin equivalent circuit between the indicated terminals using each method described in this section. Compare your results.

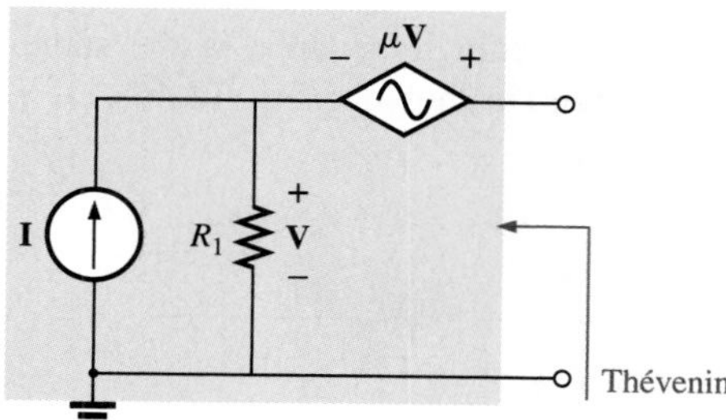

FIG. 18.51
Example 18.12.

Solution: First, using Kirchhoff's voltage law, $\mathbf{E}_{Th}$ (which is the same for each method) is written

$$\mathbf{E}_{Th} = \mathbf{V} + \mu\mathbf{V} = (1 + \mu)\mathbf{V}$$

However,
$$\mathbf{V} = \mathbf{I}R_1$$
so
$$\mathbf{E}_{Th} = (\mathbf{1} + \boldsymbol{\mu})\mathbf{I}\boldsymbol{R}_1$$

$\mathbf{Z}_{Th}$

Method 1: See Fig. 18.52. Since $\mathbf{I} = 0$, $\mathbf{V}$ and $\mu\mathbf{V} = 0$, and

$$\cancel{\mathbf{Z}_{Th} = R_1} \qquad \text{(incorrect)}$$

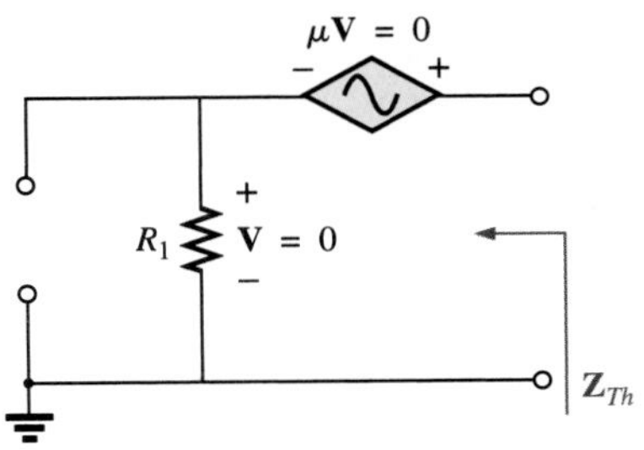

Method 2: See Fig. 18.53. Kirchhoff's voltage law around the indicated loop gives us

$$\mathbf{V} + \mu\mathbf{V} = 0$$

and
$$\mathbf{V}(1 + \mu) = 0$$

Since μ is a positive constant, the above equation can be satisfied only when $\mathbf{V} = 0$. Substitution of this result into Fig. 18.53 yields the configuration in Fig. 18.54, and

$$\mathbf{I}_{sc} = \mathbf{I}$$

with

$$\mathbf{Z}_{Th} = \frac{\mathbf{E}_{oc}}{\mathbf{I}_{sc}} = \frac{(1 + \mu)\mathbf{I}R_1}{\mathbf{I}} = (\mathbf{1} + \boldsymbol{\mu})\boldsymbol{R}_1 \qquad \text{(correct)}$$

Method 3: See Fig. 18.55.

$$\mathbf{E}_g = \mathbf{V} + \mu\mathbf{V} = (1 + \mu)\mathbf{V}$$

or
$$\mathbf{V} = \frac{\mathbf{E}_g}{1 + \mu}$$

and
$$\mathbf{I}_g = \frac{\mathbf{V}}{R_1} = \frac{\mathbf{E}_g}{(1 + \mu)R_1}$$

and
$$\mathbf{Z}_{Th} = \frac{\mathbf{E}_g}{\mathbf{I}_g} = (\mathbf{1} + \boldsymbol{\mu})\boldsymbol{R}_1 \qquad \text{(correct)}$$

The Thévenin equivalent circuit appears in Fig. 18.56, and

$$\mathbf{I}_L = \frac{(\mathbf{1} + \boldsymbol{\mu})\boldsymbol{R}_1\mathbf{I}}{\boldsymbol{R}_L + (\mathbf{1} + \boldsymbol{\mu})\boldsymbol{R}_1}$$

which compares with the result in Example 18.6.

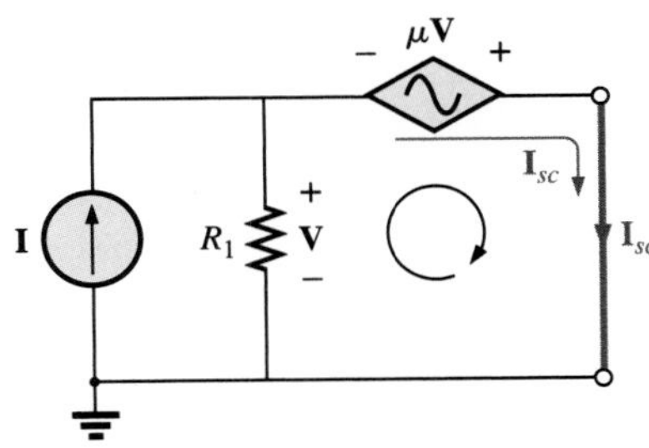

FIG. 18.52
Determining $\mathbf{Z}_{Th}$ incorrectly.

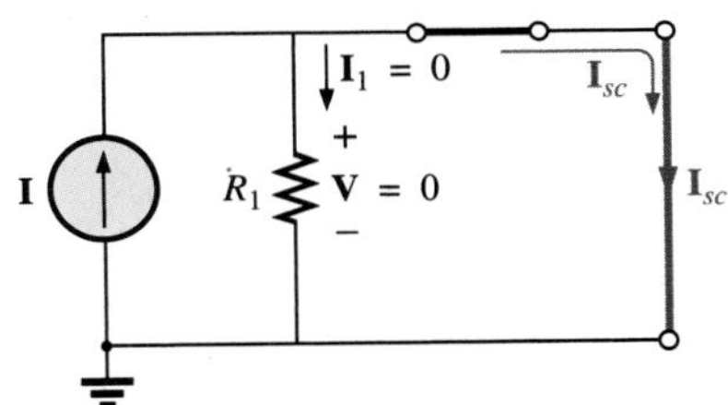

FIG. 18.53
Determining I_{sc} for the network in Fig. 18.51.

FIG. 18.54
Substituting $\mathbf{V} = 0$ into the network in Fig. 18.53.

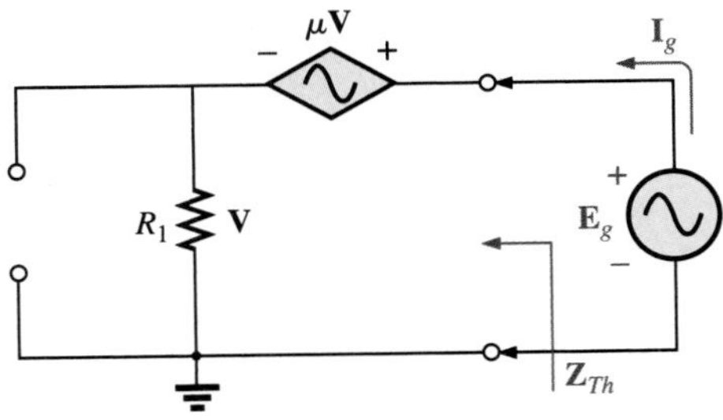

FIG. 18.55
Determining $\mathbf{Z}_{Th}$ using the approach $\mathbf{Z}_{Th} = \mathbf{E}_g/\mathbf{I}_g$.

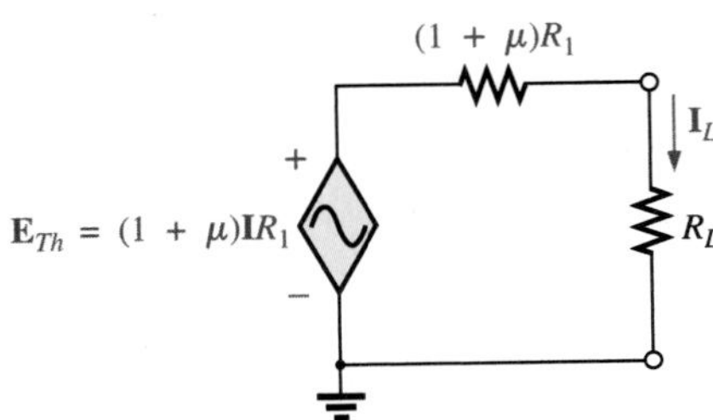

FIG. 18.56
The Thévenin equivalent circuit for the network in Fig. 18.51.

The network in Fig. 18.57 is the basic configuration of the transistor equivalent circuit applied most frequently today (although most texts in

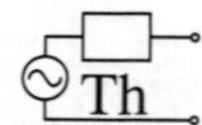

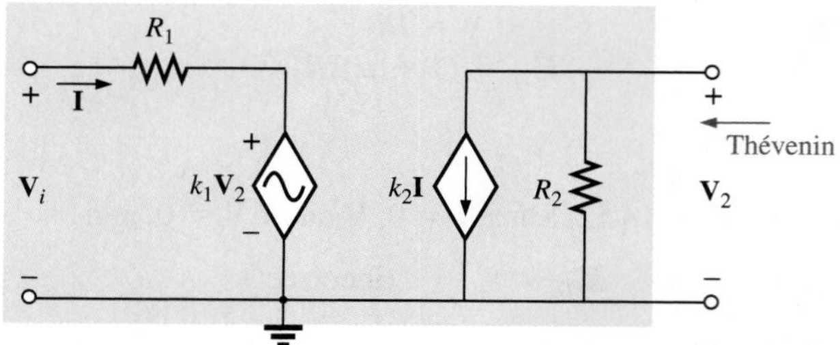

FIG. 18.57
Example 18.13: Transistor equivalent network.

electronics use the circle rather than the diamond outline for the source). Obviously, it is necessary to know its characteristics and to be adept in its use. Note that there are both a controlled voltage and a controlled current source, each controlled by variables in the configuration.

EXAMPLE 18.13 Determine the Thévenin equivalent circuit for the indicated terminals of the network in Fig. 18.57.

Solution: Apply the second method introduced in this section.

$\mathbf{E}_{Th}$

$$\mathbf{E}_{oc} = \mathbf{V}_2$$

$$\mathbf{I} = \frac{\mathbf{V}_i - k_1\mathbf{V}_2}{R_1} = \frac{\mathbf{V}_i - k_1\mathbf{E}_{oc}}{R_1}$$

and
$$\mathbf{E}_{oc} = -k_2\mathbf{I}R_2 = -k_2R_2\left(\frac{\mathbf{V}_i - k_1\mathbf{E}_{oc}}{R_1}\right)$$

$$= \frac{-k_2R_2\mathbf{V}_i}{R_1} + \frac{k_1k_2R_2\mathbf{E}_{oc}}{R_1}$$

or
$$\mathbf{E}_{oc}\left(1 - \frac{k_1k_2R_2}{R_1}\right) = \frac{-k_2R_2\mathbf{V}_i}{R_1}$$

and
$$\mathbf{E}_{oc}\left(\frac{R_1 - k_1k_2R_2}{\cancel{R}_1}\right) = \frac{-k_2R_2\mathbf{V}_i}{\cancel{R}_1}$$

so
$$\boxed{\mathbf{E}_{oc} = \frac{-k_2R_2\mathbf{V}_i}{R_1 - k_1k_2R_2} = \mathbf{E}_{Th}} \tag{18.5}$$

$\mathbf{I}_{sc}$ For the network in Fig. 18.58, where

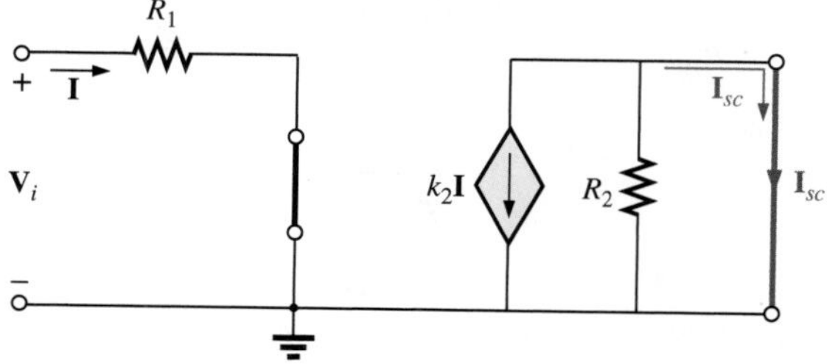

FIG. 18.58
Determining $\mathbf{I}_{sc}$ for the network in Fig. 18.57.

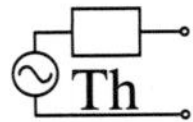

$$\mathbf{V}_2 = 0 \qquad k_1\mathbf{V}_2 = 0 \qquad \mathbf{I} = \frac{\mathbf{V}_i}{R_1}$$

and
$$\mathbf{I}_{sc} = -k_2\mathbf{I} = \frac{-k_2\mathbf{V}_i}{R_1}$$

so
$$\mathbf{Z}_{Th} = \frac{\mathbf{E}_{oc}}{\mathbf{I}_{sc}} = \frac{\dfrac{-k_2R_2\mathbf{V}_i}{R_1 - k_1k_2R_2}}{\dfrac{-k_2\mathbf{V}_i}{R_1}}$$

and
$$\boxed{\mathbf{Z}_{Th} = \frac{R_1R_2}{R_1 - k_1k_2R_2}} \qquad \textbf{(18.6)}$$

Frequently, the approximation $k_1 \cong 0$ is applied. Then the Thévenin voltage and impedance are

$$\boxed{\mathbf{E}_{Th} = \frac{-k_2R_2\mathbf{V}_i}{R_1}} \qquad k_1 = 0 \qquad \textbf{(18.7)}$$

$$\boxed{\mathbf{Z}_{Th} = R_2} \qquad k_1 = 0 \qquad \textbf{(18.8)}$$

Apply $\mathbf{Z}_{Th} = \mathbf{E}_g/\mathbf{I}_g$ to the network in Fig. 18.59, where

$$\mathbf{I} = \frac{-k_1\mathbf{V}_2}{R_1}$$

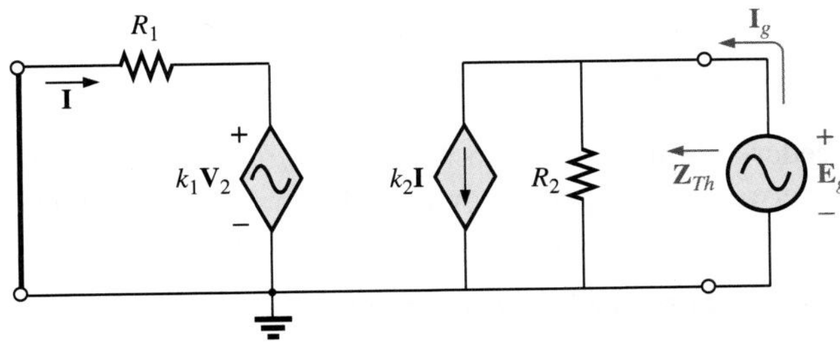

FIG. 18.59
Determining $\mathbf{Z}_{Th}$ using the procedure $\mathbf{Z}_{Th} = \mathbf{E}_g/\mathbf{I}_g$.

But
$$\mathbf{V}_2 = \mathbf{E}_g$$

so
$$\mathbf{I} = \frac{-k_1\mathbf{E}_g}{R_1}$$

Applying Kirchhoff's current law, we have

$$\mathbf{I}_g = k_2\mathbf{I} + \frac{\mathbf{E}_g}{R_2} = k_2\left(-\frac{k_1\mathbf{E}_g}{R_1}\right) + \frac{\mathbf{E}_g}{R_2}$$
$$= \mathbf{E}_g\left(\frac{1}{R_2} - \frac{k_1k_2}{R_1}\right)$$

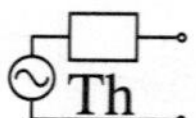

and
$$\frac{\mathbf{I}_g}{\mathbf{E}_g} = \frac{R_1 - k_1k_2R_2}{R_1R_2}$$

or
$$\mathbf{Z}_{Th} = \frac{\mathbf{E}_g}{\mathbf{I}_g} = \frac{\boldsymbol{R_1R_2}}{\boldsymbol{R_1 - k_1k_2R_2}}$$

as obtained above.

The last two methods presented in this section were applied only to networks in which the magnitudes of the controlled sources were dependent on a variable within the network for which the Thévenin equivalent circuit was to be obtained. Understand that both of these methods can also be applied to any dc or sinusoidal ac network containing only independent sources or dependent sources of the other kind.

18.4 NORTON'S THEOREM

The three methods described for Thévenin's theorem will each be altered to permit their use with **Norton's theorem.** Since the Thévenin and Norton impedances are the same for a particular network, certain portions of the discussion are quite similar to those encountered in the previous section. We first consider independent sources and the approach developed in Chapter 9, followed by dependent sources and the new techniques developed for Thévenin's theorem.

You will recall from Chapter 9 that Norton's theorem allows us to replace any two-terminal linear bilateral ac network with an equivalent circuit consisting of a current source and an impedance, as in Fig. 18.60.

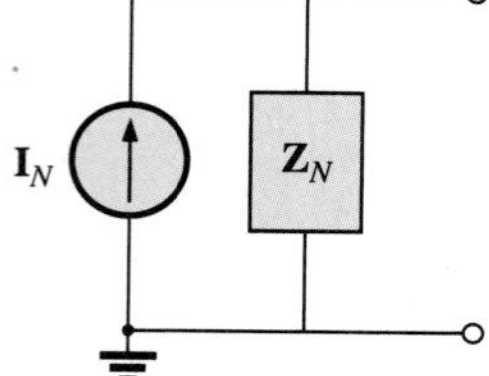

FIG. 18.60
The Norton equivalent circuit for ac networks.

The Norton equivalent circuit, like the Thévenin equivalent circuit, is applicable at only one frequency since the reactances are frequency dependent.

Independent Sources

The procedure outlined below to find the Norton equivalent of a sinusoidal ac network is changed (from that in Chapter 9) in only one respect: the replacement of the term *resistance* with the term *impedance.*

1. ***Remove that portion of the network across which the Norton equivalent circuit is to be found.***
2. ***Mark (○, •, and so on) the terminals of the remaining two-terminal network.***
3. ***Calculate $\mathbf{Z}_N$ by first setting all voltage and current sources to zero (short circuit and open circuit, respectively) and then finding the resulting impedance between the two marked terminals.***
4. ***Calculate $\mathbf{I}_N$ by first replacing the voltage and current sources and then finding the short-circuit current between the marked terminals.***
5. ***Draw the Norton equivalent circuit with the portion of the circuit previously removed replaced between the terminals of the Norton equivalent circuit.***

The Norton and Thévenin equivalent circuits can be found from each other by using the source transformation shown in Fig. 18.61. The source transformation is applicable for any Thévenin or Norton equivalent circuit determined from a network with any combination of independent or dependent sources.

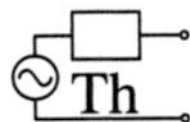

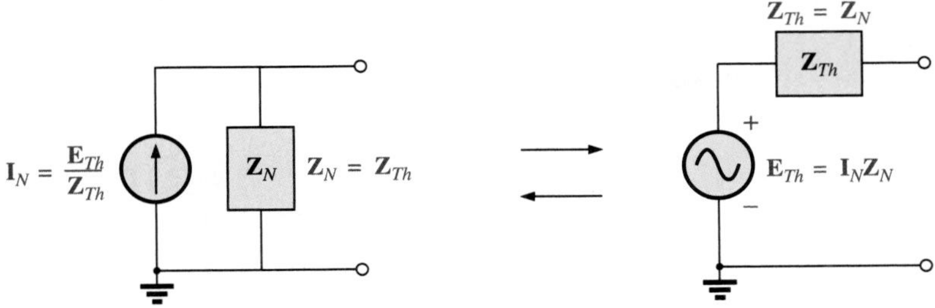

FIG. 18.61
Conversion between the Thévenin and Norton equivalent circuits.

EXAMPLE 18.14 Determine the Norton equivalent circuit for the network external to the 6 Ω resistor in Fig. 18.62.

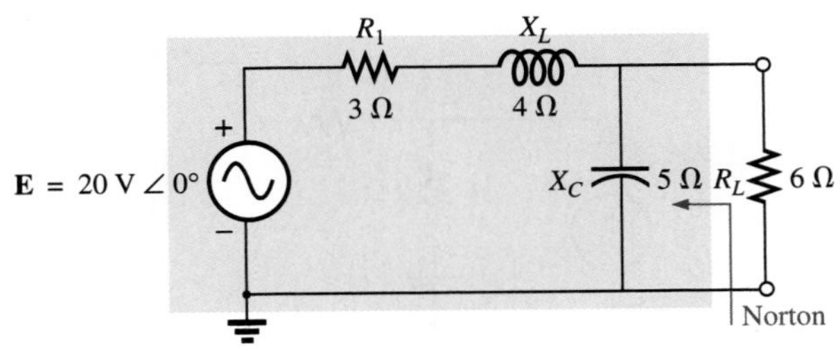

FIG. 18.62
Example 18.14.

Solution:

Steps 1 and 2 (Fig. 18.63):

$$\mathbf{Z}_1 = R_1 + j X_L = 3\ \Omega + j\,4\ \Omega = 5\ \Omega\ \angle 53.13^\circ$$
$$\mathbf{Z}_2 = -j X_C = -j\,5\ \Omega$$

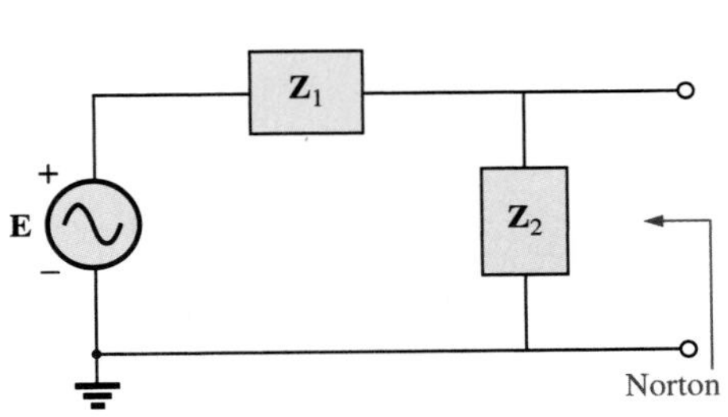

FIG. 18.63
Assigning the subscripted impedances to the network in Fig. 18.62.

Step 3 (Fig. 18.64):

$$\mathbf{Z}_N = \frac{\mathbf{Z}_1\mathbf{Z}_2}{\mathbf{Z}_1 + \mathbf{Z}_2} = \frac{(5\ \Omega\ \angle 53.13^\circ)(5\ \Omega\ \angle -90^\circ)}{3\ \Omega + j\,4\ \Omega - j\,5\ \Omega} = \frac{25\ \Omega\ \angle -36.87^\circ}{3 - j\,1}$$
$$= \frac{25\ \Omega\ \angle -36.87^\circ}{3.16\ \angle -18.43^\circ} = 7.91\ \Omega\ \angle -18.44^\circ = \mathbf{7.50\ \Omega - j\,2.50\ \Omega}$$

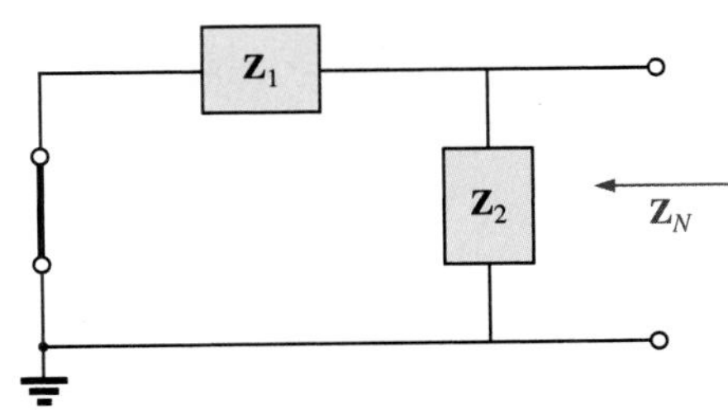

FIG. 18.64
Determining the Norton impedance for the network in Fig. 18.62.

Step 4 (Fig. 18.65):

$$\mathbf{I}_N = \mathbf{I}_1 = \frac{\mathbf{E}}{\mathbf{Z}_1} = \frac{20\ \text{V}\ \angle 0^\circ}{5\ \Omega\ \angle 53.13^\circ} = \mathbf{4\ A\ \angle -53.13^\circ}$$

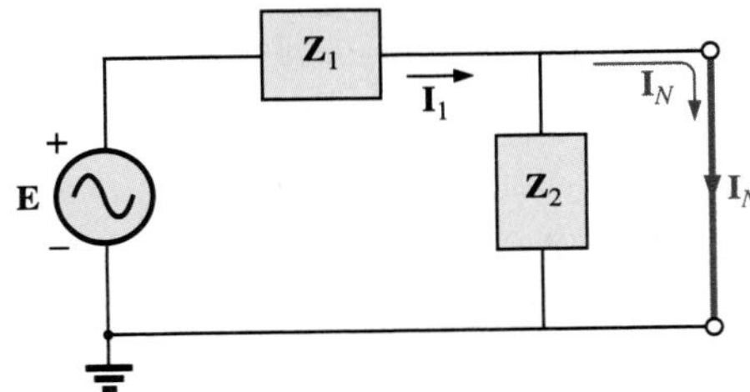

FIG. 18.65
Determining $\mathbf{I}_N$ for the network in Fig. 18.62.

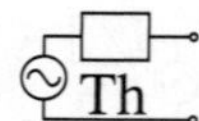

Step 5: The Norton equivalent circuit is shown in Fig. 18.66.

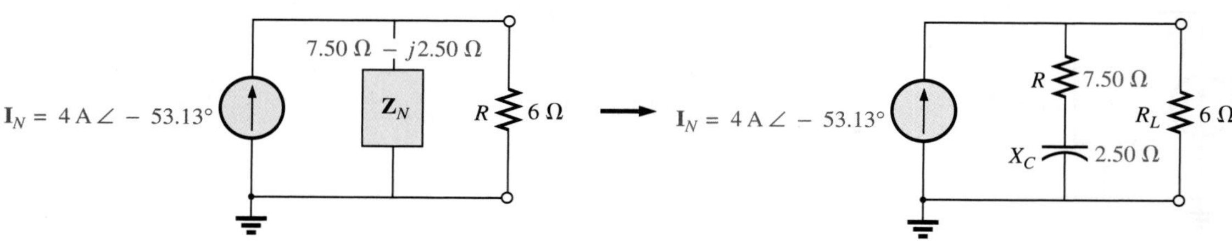

FIG. 18.66
The Norton equivalent circuit for the network in Fig. 18.62.

EXAMPLE 18.15 Find the Norton equivalent circuit for the network external to the 7 Ω capacitive reactance in Fig. 18.67.

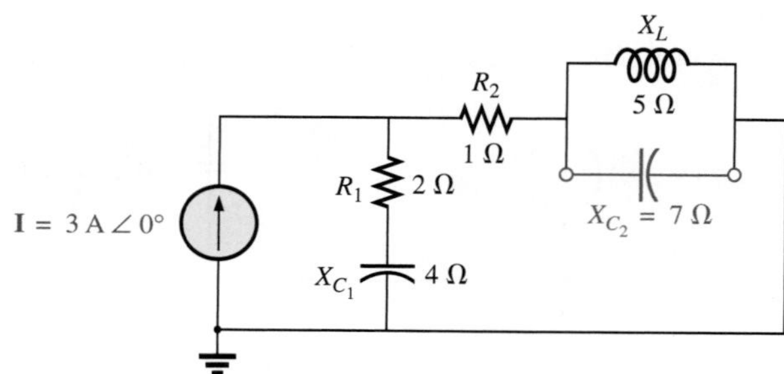

FIG. 18.67
Example 18.15.

Solution:

Steps 1 and 2 (Fig. 18.68):

$$\mathbf{Z}_1 = R_1 - j\,X_{C_1} = 2\,\Omega - j\,4\,\Omega$$
$$\mathbf{Z}_2 = R_2 = 1\,\Omega$$
$$\mathbf{Z}_3 = +j\,X_L = j\,5\,\Omega$$

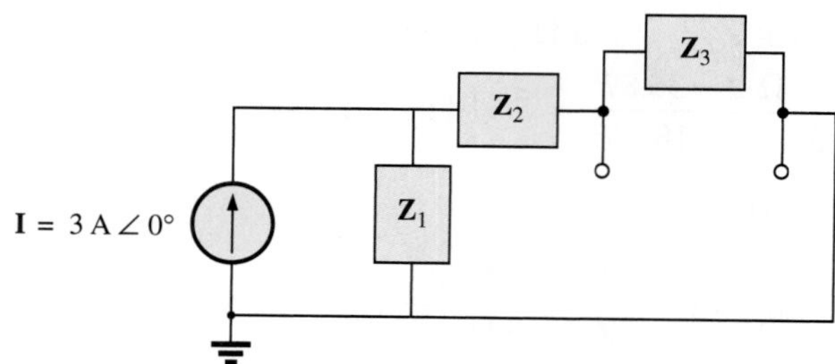

FIG. 18.68
Assigning the subscripted impedances to the network in Fig. 18.67.

Step 3 (Fig. 18.69):

$$\mathbf{Z}_N = \frac{\mathbf{Z}_3(\mathbf{Z}_1 + \mathbf{Z}_2)}{\mathbf{Z}_3 + (\mathbf{Z}_1 + \mathbf{Z}_2)}$$

$$\mathbf{Z}_1 + \mathbf{Z}_2 = 2\,\Omega - j\,4\,\Omega + 1\,\Omega = 3\,\Omega - j\,4\,\Omega = 5\,\Omega\,\angle{-53.13^\circ}$$

$$\mathbf{Z}_N = \frac{(5\,\Omega\,\angle 90^\circ)(5\,\Omega\,\angle{-53.13^\circ})}{j\,5\,\Omega + 3\,\Omega - j\,4\,\Omega} = \frac{25\,\Omega\,\angle 36.87^\circ}{3 + j\,1}$$

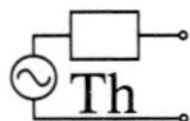

$$= \frac{25\ \Omega\ \angle 36.87^\circ}{3.16\ \angle +18.43^\circ}$$

$$\mathbf{Z}_N = 7.91\ \Omega\ \angle 18.44^\circ = \mathbf{7.50\ \Omega + j\,2.50\ \Omega}$$

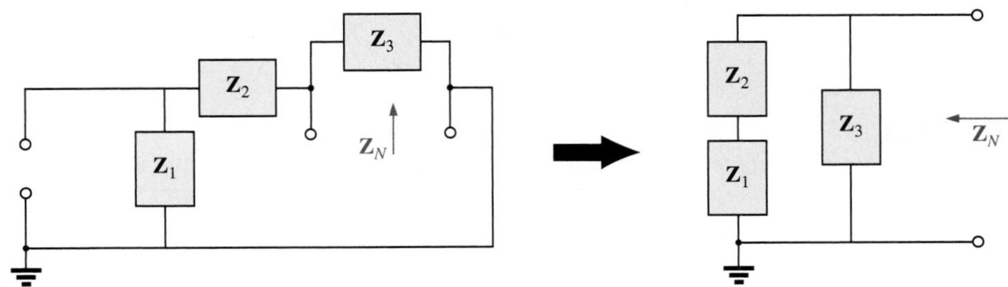

FIG. 18.69
Finding the Norton impedance for the network in Fig. 18.67.

Calculator Solution: Performing the above on the TI-89 calculator results in the sequence in Fig. 18.70:

((5i×(2−4i+1)))÷((5i+2−4i+1)) ▶ Polar	7.91E0∠18.43E0

FIG. 18.70

Step 4 (Fig. 18.71):

$$\mathbf{I}_N = \mathbf{I}_1 = \frac{\mathbf{Z}_1\mathbf{I}}{\mathbf{Z}_1 + \mathbf{Z}_2} \qquad \text{(current divider rule)}$$

$$= \frac{(2\ \Omega - j\,4\ \Omega)(3\ \text{A})}{3\ \Omega - j\,4\ \Omega} = \frac{6\ \text{A} - j\,12\ \text{A}}{5\ \angle -53.13^\circ} = \frac{13.4\ \text{A}\ \angle -63.43^\circ}{5\ \angle -53.13^\circ}$$

$$\mathbf{I}_N = \mathbf{2.68\ A\ \angle -10.3^\circ}$$

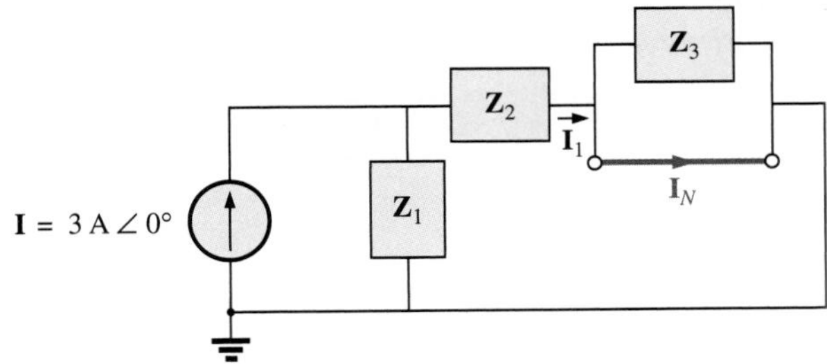

FIG. 18.71
Determining $\mathbf{I}_N$ for the network in Fig. 18.67.

Step 5: The Norton equivalent circuit is shown in Fig. 18.72.

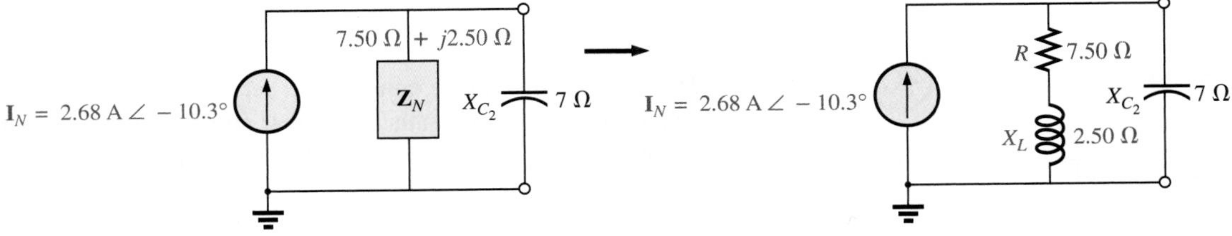

FIG. 18.72
The Norton equivalent circuit for the network in Fig. 18.67.

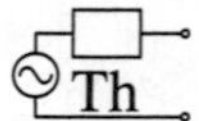

EXAMPLE 18.16 Find the Thévenin equivalent circuit for the network external to the 7 Ω capacitive reactance in Fig. 18.67.

Solution: Using the conversion between sources (Fig. 18.73), we obtain

$$\mathbf{Z}_{Th} = \mathbf{Z}_N = \mathbf{7.50\ \Omega + \mathit{j}\,2.50\ \Omega}$$
$$\mathbf{E}_{Th} = \mathbf{I}_N\mathbf{Z}_N = (2.68\text{ A}\ \angle -10.3°)(7.91\ \Omega\ \angle 18.44°)$$
$$= \mathbf{21.2\ V\ \angle 8.14°}$$

The Thévenin equivalent circuit is shown in Fig. 18.74.

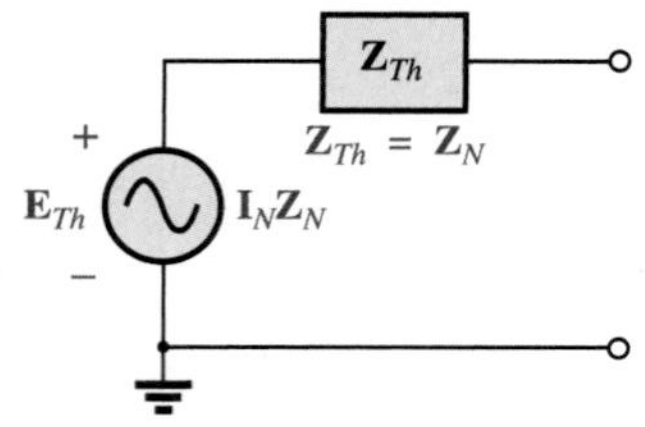

FIG. 18.73
Determining the Thévenin equivalent circuit for the Norton equivalent in Fig. 18.72.

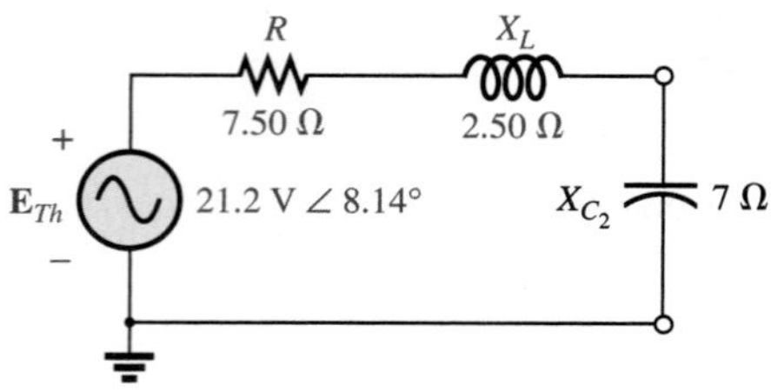

FIG. 18.74
The Thévenin equivalent circuit for the network in Fig. 18.67.

Dependent Sources

As stated for Thévenin's theorem, *dependent sources in which the controlling variable is not determined by the network* for which the Norton equivalent circuit is to be found do not alter the procedure outlined above.

For dependent sources of the other kind, one of the following procedures must be applied. Both of these procedures can also be applied to networks with any combination of independent sources and dependent sources not controlled by the network under investigation.

The Norton equivalent circuit appears in Fig. 18.75(a). In Fig. 18.75(b), we find that

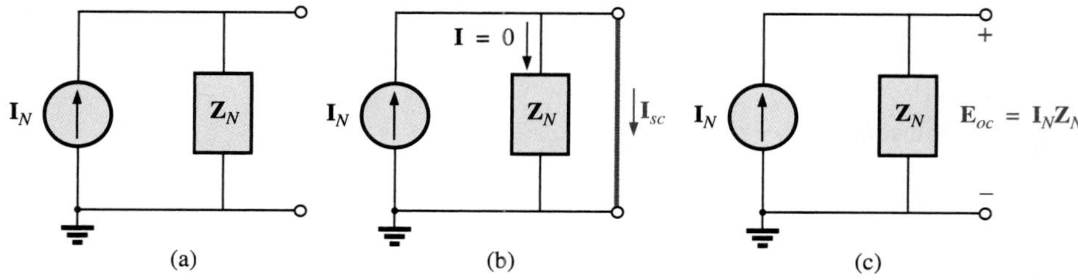

FIG. 18.75
Defining an alternative approach for determining $\mathbf{Z}_N$.

$$\boxed{\mathbf{I}_{sc} = \mathbf{I}_N} \qquad (18.9)$$

and in Fig. 18.75(c) that

$$\mathbf{E}_{oc} = \mathbf{I}_N\mathbf{Z}_N$$

Or, rearranging, we have

$$\mathbf{Z}_N = \frac{\mathbf{E}_{oc}}{\mathbf{I}_N}$$

and

$$\boxed{\mathbf{Z}_N = \frac{\mathbf{E}_{oc}}{\mathbf{I}_{sc}}} \qquad (18.10)$$

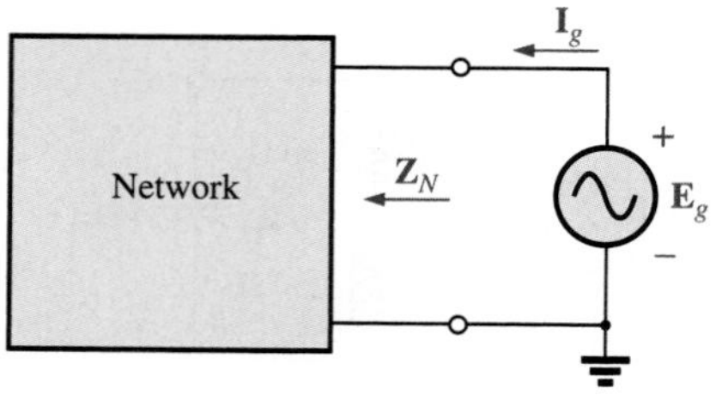

The Norton impedance can also be determined by applying a source of voltage $\mathbf{E}_g$ to the terminals of interest and finding the resulting $\mathbf{I}_g$, as shown in Fig. 18.76. All independent sources and dependent sources not controlled by a variable in the network of interest are set to zero, and

$$\boxed{\mathbf{Z}_N = \frac{\mathbf{E}_g}{\mathbf{I}_g}} \qquad (18.11)$$

FIG. 18.76
Determining the Norton impedance using the approach $\mathbf{Z}_N = \mathbf{E}_g/\mathbf{I}_g$.

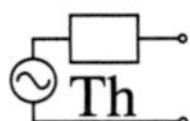

For this latter approach, the Norton current is still determined by the short-circuit current.

EXAMPLE 18.17 Using each method described for dependent sources, find the Norton equivalent circuit for the network in Fig. 18.77.

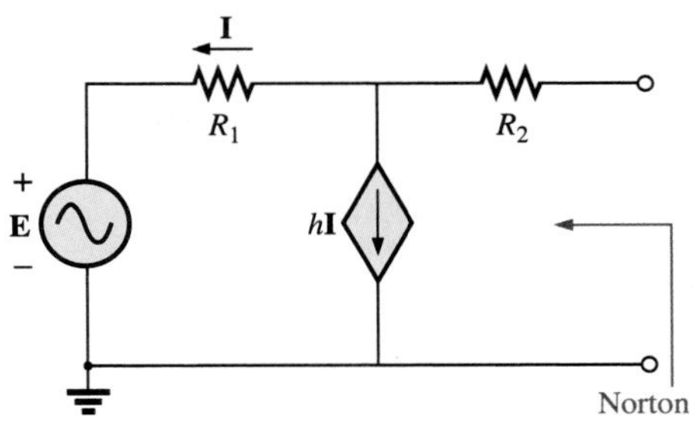

FIG. 18.77
Example 18.17.

Solution:

I_N For each method, $\mathbf{I}_N$ is determined in the same manner. From Fig. 18.78 using Kirchhoff's current law, we have

$$0 = \mathbf{I} + h\mathbf{I} + \mathbf{I}_{sc}$$

or
$$\mathbf{I}_{sc} = -(1 + h)\mathbf{I}$$

Applying Kirchhoff's voltage law gives us

$$\mathbf{E} + \mathbf{I}R_1 - \mathbf{I}_{sc}R_2 = 0$$

and
$$\mathbf{I}R_1 = \mathbf{I}_{sc}R_2 - \mathbf{E}$$

or
$$\mathbf{I} = \frac{\mathbf{I}_{sc}R_2 - \mathbf{E}}{R_1}$$

so
$$\mathbf{I}_{sc} = -(1 + h)\mathbf{I} = -(1 + h)\left(\frac{\mathbf{I}_{sc}R_2 - \mathbf{E}}{R_1}\right)$$

or
$$R_1\mathbf{I}_{sc} = -(1 + h)\mathbf{I}_{sc}R_2 + (1 + h)\mathbf{E}$$
$$\mathbf{I}_{sc}[R_1 + (1 + h)R_2] = (1 + h)\mathbf{E}$$
$$\mathbf{I}_{sc} = \frac{(1 + h)\mathbf{E}}{R_1 + (1 + h)R_2} = \mathbf{I}_N$$

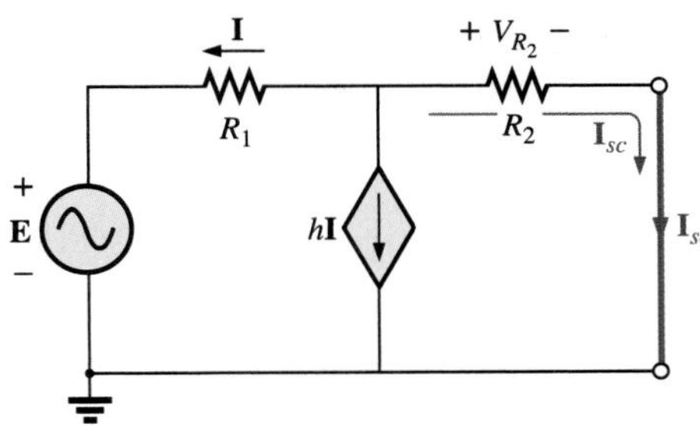

FIG. 18.78
Determining $\mathbf{I}_{sc}$ for the network in Fig. 18.77.

$\mathbf{Z}_N$

Method 1: $\mathbf{E}_{oc}$ is determined from the network in Fig. 18.79. By Kirchhoff's current law,

$$0 = \mathbf{I} + h\mathbf{I} \quad \text{or} \quad \mathbf{1}(h + 1) = 0$$

For h, a positive constant **I** must equal zero to satisfy the above. Therefore,

$$\mathbf{I} = 0 \quad \text{and} \quad h\mathbf{I} = 0$$

and
$$\mathbf{E}_{oc} = \mathbf{E}$$

with
$$\mathbf{Z}_N = \frac{\mathbf{E}_{oc}}{\mathbf{I}_{sc}} = \frac{\mathbf{E}}{\dfrac{(1 + h)\mathbf{E}}{R_1 + (1 + h)R_2}} = \frac{R_1 + (1 + h)R_2}{(1 + h)}$$

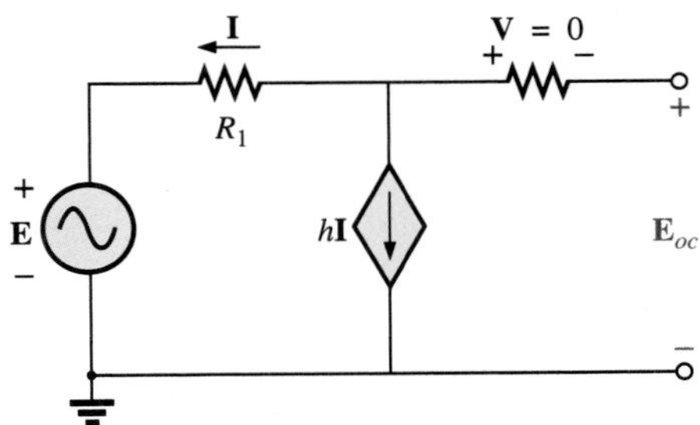

FIG. 18.79
Determining $\mathbf{E}_{oc}$ for the network in Fig. 18.77.

Method 2: Note Fig. 18.80. By Kirchhoff's current law,

$$\mathbf{I}_g = \mathbf{I} + h\mathbf{I} = (\mathrm{I} + h)\mathbf{I}$$

By Kirchhoff's voltage law,

$$\mathbf{E}_g - \mathbf{I}_gR_2 - \mathbf{I}R_1 = 0$$

or
$$\mathbf{I} = \frac{\mathbf{E}_g - \mathbf{I}_gR_2}{R_1}$$

Substituting, we have

$$\mathbf{I}_g = (1 + h)\mathbf{I} = (1 + h)\left(\frac{\mathbf{E}_g - \mathbf{I}_gR_2}{R_1}\right)$$

and
$$\mathbf{I}_gR_1 = (1 + h)\mathbf{E}_g - (1 + h)\mathbf{I}_gR_2$$

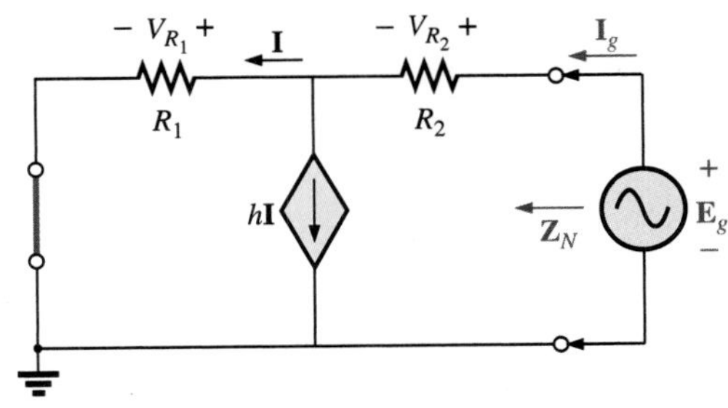

FIG. 18.80
Determining the Norton impedance using the approach $\mathbf{Z}_N = \mathbf{E}_g/\mathbf{E}_g$.

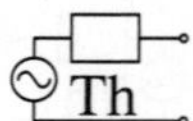

so $$\mathbf{E}_g(1 + h) = \mathbf{I}_g[R_1 + (1 + h)R_2]$$

or $$\mathbf{Z}_N = \frac{\mathbf{E}_g}{\mathbf{I}_g} = \frac{R_1 + (1 + h)R_2}{1 + h}$$

which agrees with the above.

EXAMPLE 18.18 Find the Norton equivalent circuit for the network configuration in Fig. 18.57.

Solution: By source conversion,

$$\mathbf{I}_N = \frac{\mathbf{E}_{Th}}{\mathbf{Z}_{Th}} = \frac{\dfrac{-k_2R_2\mathbf{V}_i}{R_1 - k_1k_2R_2}}{\dfrac{R_1R_2}{R_1 - k_1k_2R_2}}$$

and $$\boxed{\mathbf{I}_N = \frac{-k_2\mathbf{V}_i}{R_1}} \tag{18.12}$$

which is $\mathbf{I}_{sc}$ as determined in Example 18.13, and

$$\boxed{\mathbf{Z}_N = \mathbf{Z}_{Th} = \frac{R_2}{1 - \dfrac{k_1k_2R_2}{R_1}}} \tag{18.13}$$

For $k_1 \cong 0$, we have

$$\boxed{\mathbf{I}_N = \frac{-k_2\mathbf{V}_i}{R_1}} \qquad k_1 = 0 \tag{18.14}$$

$$\boxed{\mathbf{Z}_N = R_2} \qquad k_1 = 0 \tag{18.15}$$

18.5 MAXIMUM POWER TRANSFER THEOREM

When applied to ac circuits, the **maximum power transfer theorem** states that

maximum power will be delivered to a load when the load impedance is the conjugate of the Thévenin impedance across its terminals.

That is, for Fig. 18.81, for maximum power transfer to the load,

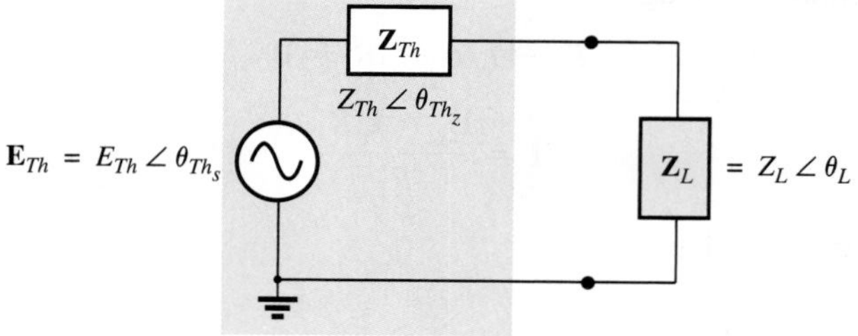

FIG. 18.81
Defining the conditions for maximum power transfer to a load.

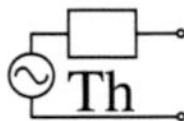

$$Z_L = Z_{Th} \text{ and } \theta_L = -\theta_{Th_Z} \qquad (18.16)$$

or, in rectangular form,

$$R_L = R_{Th} \quad \text{and} \quad \pm j\,X_{\text{load}} = \mp j\,X_{Th} \qquad (18.17)$$

The conditions just mentioned will make the total impedance of the circuit appear purely resistive, as indicated in Fig. 18.82:

$$\mathbf{Z}_T = (R \pm jX) + (R \mp j\,X)$$

and

$$\mathbf{Z}_T = 2R \qquad (18.18)$$

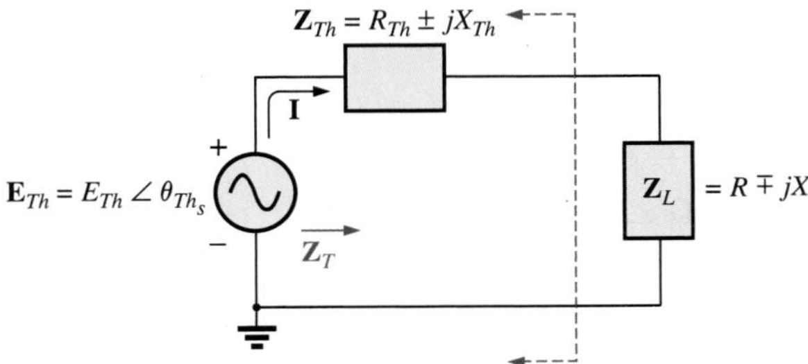

FIG. 18.82
Conditions for maximum power transfer to $\mathbf{Z}_L$.

Since the circuit is purely resistive, the power factor of the circuit under maximum power conditions is 1; that is,

$$F_p = 1 \qquad \text{(maximum power transfer)} \qquad (18.19)$$

The magnitude of the current **I** in Fig. 18.82 is

$$I = \frac{E_{Th}}{Z_T} = \frac{E_{Th}}{2R}$$

The maximum power to the load is

$$P_{\max} = I^2R = \left(\frac{E_{Th}}{2R}\right)^2 R$$

and

$$P_{\max} = \frac{E_{Th}^2}{4R} \qquad (18.20)$$

EXAMPLE 18.19 Find the load impedance in Fig. 18.83 for maximum power to the load, and find the maximum power.

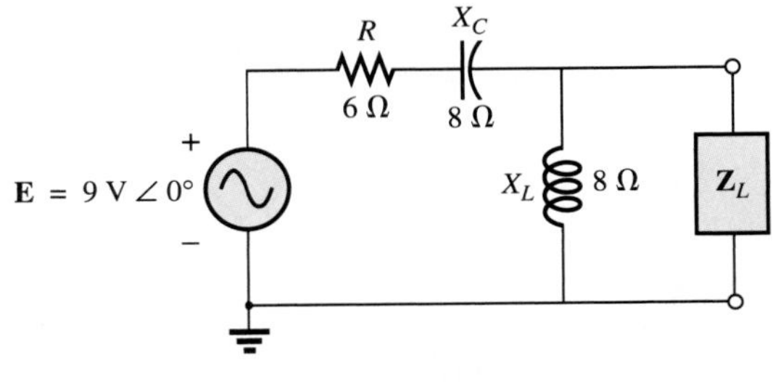

FIG. 18.83
Example 18.19.

Solution: Determine $\mathbf{Z}_{Th}$ [Fig. 18.84(a)]:

$$\mathbf{Z}_1 = R - j\,X_C = 6\ \Omega - j\,8\ \Omega = 10\ \Omega \angle -53.13°$$
$$\mathbf{Z}_2 = +j\,X_L = j\,8\ \Omega$$
$$\mathbf{Z}_{Th} = \frac{\mathbf{Z}_1\mathbf{Z}_2}{\mathbf{Z}_1 + \mathbf{Z}_2} = \frac{(10\ \Omega \angle -53.13°)(8\ \Omega \angle 90°)}{6\ \Omega - j\,8\ \Omega + j\,8\ \Omega} = \frac{80\ \Omega \angle 36.87°}{6 \angle 0°}$$
$$= 13.33\ \Omega \angle 36.87° = 10.66\ \Omega + j\,8\ \Omega$$

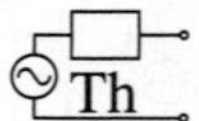

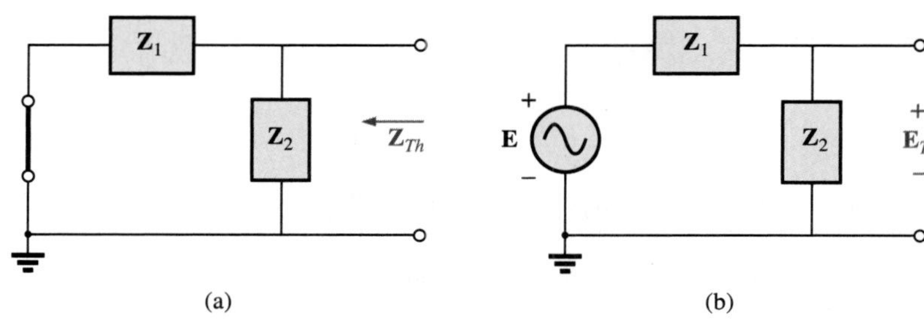

FIG. 18.84
Determining (a) $\mathbf{Z}_{Th}$ and (b) $\mathbf{E}_{Th}$ for the network external to the load in Fig. 18.83.

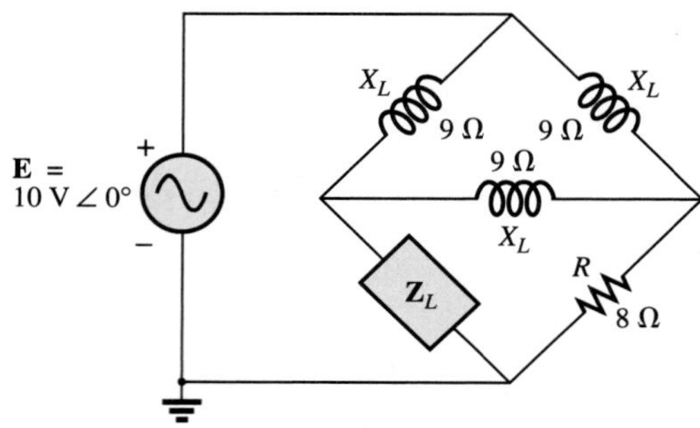

FIG. 18.85
Example 18.20.

and $$\mathbf{Z}_L = 13.3\ \Omega\ \angle -36.87° = \mathbf{10.66\ \Omega - j\,8\ \Omega}$$

To find the maximum power, we must first find $\mathbf{E}_{Th}$ [Fig. 18.84(b)], as follows:

$$\mathbf{E}_{Th} = \frac{\mathbf{Z}_2\mathbf{E}}{\mathbf{Z}_2 + \mathbf{Z}_1} \qquad \text{(voltage divider rule)}$$

$$= \frac{(8\ \Omega\ \angle 90°)(9\ \text{V}\ \angle 0°)}{j\,8\ \Omega + 6\ \Omega - j\,8\ \Omega} = \frac{72\ \text{V}\ \angle 90°}{6\ \angle 0°} = 12\ \text{V}\ \angle 90°$$

Then $$P_{\max} = \frac{E_{Th}^2}{4R} = \frac{(12\ \text{V})^2}{4(10.66\ \Omega)} = \frac{144}{42.64} = \mathbf{3.38\ W}$$

EXAMPLE 18.20 Find the load impedance in Fig. 18.85 for maximum power to the load, and find the maximum power.

Solution: First we must find $\mathbf{Z}_{Th}$ (Fig. 18.86).

$$\mathbf{Z}_1 = +j\,X_L = j\,9\ \Omega \quad \mathbf{Z}_2 = R = 8\ \Omega$$

Converting from a Δ to a Y (Fig. 18.87), we have

$$\mathbf{Z'}_1 = \frac{\mathbf{Z}_1}{3} = j\,3\ \Omega \quad \mathbf{Z}_2 = 8\ \Omega$$

The redrawn circuit (Fig. 18.88) shows

$$\mathbf{Z}_{Th} = \mathbf{Z'}_1 + \frac{\mathbf{Z'}_1(\mathbf{Z'}_1 + \mathbf{Z}_2)}{\mathbf{Z'}_1 + (\mathbf{Z'}_1 + \mathbf{Z}_2)}$$

$$= j\,3\ \Omega + \frac{3\ \Omega\ \angle 90°(j\,3\ \Omega + 8\ \Omega)}{j\,6\ \Omega + 8\ \Omega}$$

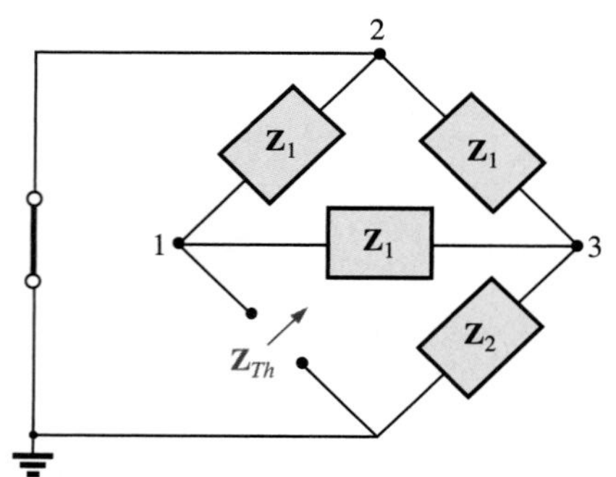

FIG. 18.86
Defining the subscripted impedances for the network in Fig. 18.85.

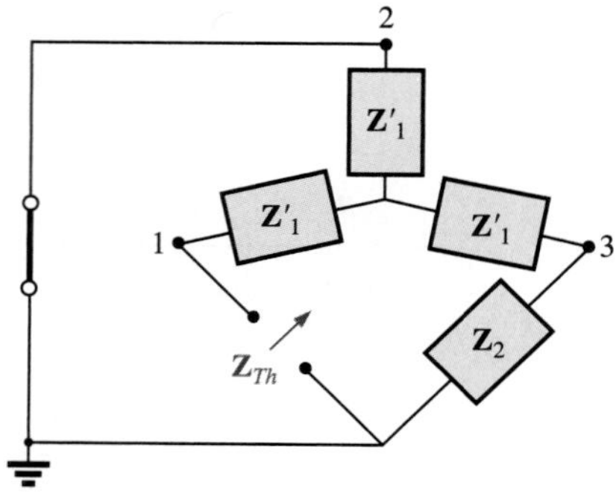

FIG. 18.87
Substituting the Y equivalent for the upper Δ configuration in Fig. 18.86.

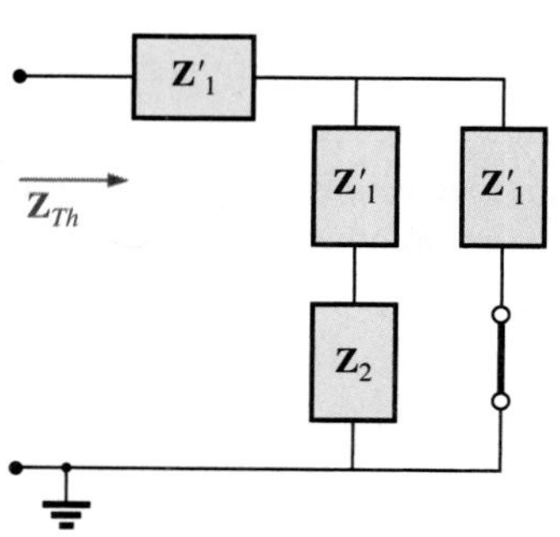

FIG. 18.88
Determining $\mathbf{Z}_{Th}$ for the network in Fig. 18.85.

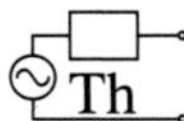

$$= j\,3 + \frac{(3\ \angle 90^\circ)(8.54\ \angle 20.56^\circ)}{10\ \angle 36.87^\circ}$$

$$= j\,3 + \frac{25.62\ \angle 110.56^\circ}{10\ \angle 36.87^\circ} = j\,3 + 2.56\ \angle 73.69^\circ$$

$$= j\,3 + 0.72 + j\,2.46$$

$$\mathbf{Z}_{Th} = 0.72\ \Omega + j\,5.46\ \Omega$$

and $$\mathbf{Z}_L = \mathbf{0.72\ \Omega - j\,5.46\ \Omega}$$

For $\mathbf{E}_{Th}$, use the modified circuit in Fig. 18.89 with the voltage source replaced in its original position. Since $I_1 = 0$, $\mathbf{E}_{Th}$ is the voltage across the series impedance of $\mathbf{Z}'_1$ and $\mathbf{Z}_2$. Using the voltage divider rule gives us

$$\mathbf{E}_{Th} = \frac{(\mathbf{Z}'_1 + \mathbf{Z}_2)\mathbf{E}}{\mathbf{Z}'_1 + \mathbf{Z}_2 + \mathbf{Z}'_1} = \frac{(j\,3\ \Omega + 8\ \Omega)(10\ \text{V}\ \angle 0^\circ)}{8\ \Omega + j\,6\ \Omega}$$

$$= \frac{(8.54\ \angle 20.56^\circ)(10\ \text{V}\ \angle 0^\circ)}{10\ \angle 36.87^\circ}$$

$$\mathbf{E}_{Th} = 8.54\ \text{V}\ \angle -16.31^\circ$$

and $$P_{\max} = \frac{E_{Th}^2}{4R} = \frac{(8.54\ \text{V})^2}{4(0.72\ \Omega)} = \frac{72.93}{2.88}\ \text{W}$$

$$= \mathbf{25.32\ W}$$

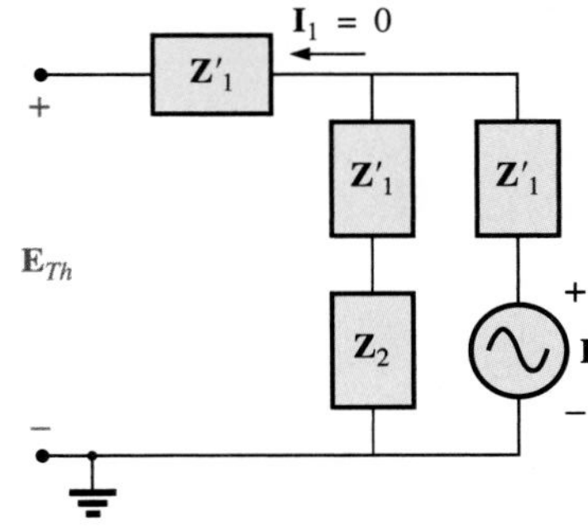

FIG. 18.89
Finding the Thévenin voltage for the network in Fig. 18.85.

If the load resistance is adjustable but the magnitude of the load reactance cannot be set equal to the magnitude of the Thévenin reactance, then the maximum power *that can be delivered* to the load will occur when the load reactance is made as close to the Thévenin reactance as possible and the load resistance is set to the following value:

$$R_L = \sqrt{R_{Th}^2 + (X_{Th} + X_{\text{load}})^2} \qquad \textbf{(18.21)}$$

where each reactance carries a positive sign if inductive and a negative sign if capacitive.

The power delivered will be determined by

$$P = E_{Th}^2/4R_{\text{av}} \qquad \textbf{(18.22)}$$

where $$R_{\text{av}} = \frac{R_{Th} + R_L}{2} \qquad \textbf{(18.23)}$$

The derivation of the above equations is given in Appendix G of the text. The following example demonstrates the use of the above.

EXAMPLE 18.21 For the network in Fig. 18.90:

a. Determine the value of R_L for maximum power to the load if the load reactance is fixed at 4 Ω.
b. Find the power delivered to the load under the conditions of part (a).
c. Find the maximum power to the load if the load reactance is made adjustable to any value, and compare the result to part (b) above.

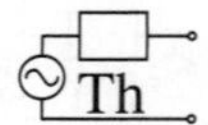

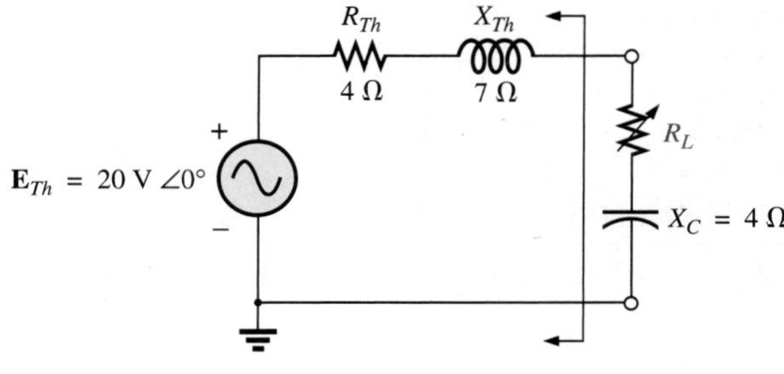

FIG. 18.90
Example 18.21.

Solutions:

a. Eq. (18.21):

$$R_L = \sqrt{R_{Th}^2 + (X_{Th} + X_{\text{load}})^2}$$
$$= \sqrt{(4\ \Omega)^2 + (7\ \Omega - 4\ \Omega)^2}$$
$$= \sqrt{16 + 9} = \sqrt{25}$$
$$R_L = \mathbf{5\ \Omega}$$

b. Eq. (18.23):

$$R_{\text{av}} = \frac{R_{Th} + R_L}{2} = \frac{4\ \Omega + 5\ \Omega}{2}$$
$$= \mathbf{4.5\ \Omega}$$

Eq. (18.22):

$$P = \frac{E_{Th}^2}{4R_{\text{av}}}$$
$$= \frac{(20\ \text{V})^2}{4(4.5\ \Omega)} = \frac{400}{18}\ \text{W}$$
$$\cong \mathbf{22.22\ W}$$

c. For $\mathbf{Z}_L = 4\ \Omega - j\,7\ \Omega$,

$$P_{\text{max}} = \frac{E_{Th}^2}{4R_{Th}} = \frac{(20\ \text{V})^2}{4(4\ \Omega)}$$
$$= \mathbf{25\ W}$$

exceeding the result of part (b) by 2.78 W.

18.6 SUBSTITUTION, RECIPROCITY, AND MILLMAN'S THEOREMS

As indicated in the introduction to this chapter, the **substitution** and **reciprocity theorems** and **Millman's theorem** will not be considered here in detail. A careful review of Chapter 9 will enable you to apply these theorems to sinusoidal ac networks with little difficulty. A number of problems in the use of these theorems appear in the Problems section at the end of the chapter.

18.7 APPLICATION

Electronic Systems

One of the blessings in the analysis of electronic systems is that the superposition theorem can be applied so that the dc analysis and ac analy-

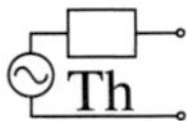

sis can be performed separately. The analysis of the dc system will affect the ac response, but the analysis of each is a distinct, separate process. Even though electronic systems have not been investigated in this text, a number of important points can be made in the description to follow that support some of the theory presented in this and recent chapters, so inclusion of this description is totally valid at this point. Consider the network in Fig. 18.91 with a transistor power amplifier, an 8 Ω speaker as the load, and a source with an internal resistance of 800 Ω. Note that each component of the design was isolated by a color box to emphasize the fact that each component must be carefully weighed in any good design.

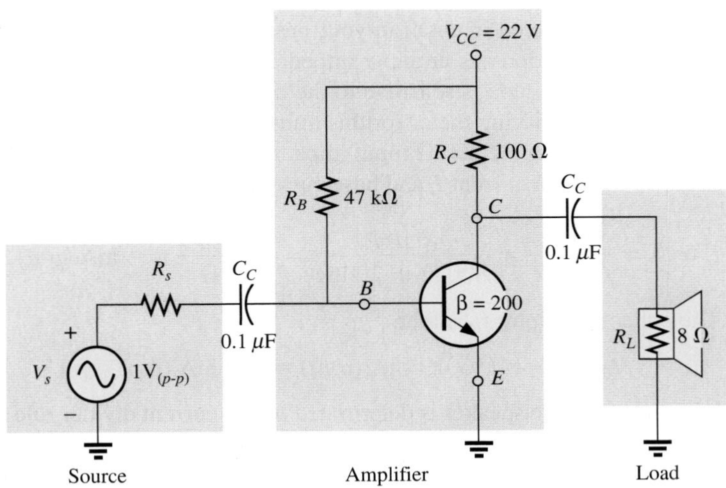

FIG. 18.91
Transistor amplifier.

As mentioned above, the analysis can be separated into a dc and an ac component. For the dc analysis, the two capacitors can be replaced by an open-circuit equivalent (Chapter 10), resulting in an isolation of the amplifier network as shown in Fig. 18.92. Given the fact that V_{BE} will be about 0.7 V dc for any operating transistor, the base current I_B can be found as follows using Kirchhoff's voltage law:

$$I_B = \frac{V_{R_B}}{R_B} = \frac{V_{CC} - V_{BE}}{R_B} = \frac{22\text{ V} - 0.7\text{ V}}{47\text{ k}\Omega} = \mathbf{453.2\ \mu A}$$

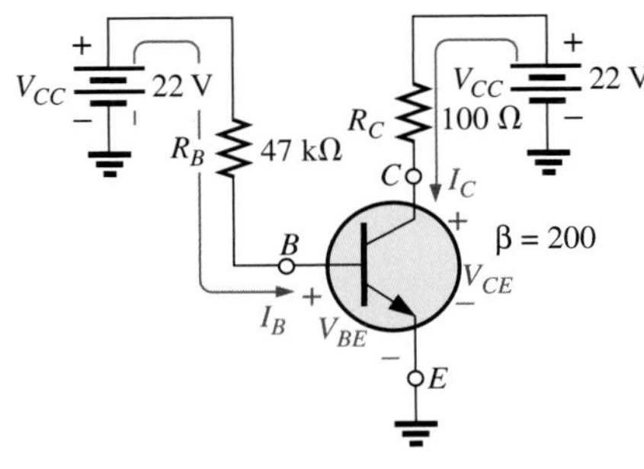

FIG. 18.92
dc equivalent of the transistor network in Fig. 18.91.

For transistors, the collector current I_C is related to the base current by $I_C = \beta I_B$, and

$$I_C = \beta I_B = (200)(453.2\ \mu\text{A}) = \mathbf{90.64\ mA}$$

Finally, through Kirchhoff's voltage law, the collector voltage (also the collector-to-emitter voltage since the emitter is grounded) can be determined as follows:

$$V_C = V_{CE} = V_{CC} - I_C R_C = 22\text{ V} - (90.64\text{ mA})(100\ \Omega) = \mathbf{12.94\ V}$$

For the dc analysis, therefore,

$$I_B = \mathbf{453.2\ \mu A} \quad I_C = \mathbf{90.64\ mA} \quad V_{CE} = \mathbf{12.94\ V}$$

which will define a point of dc operation for the transistor. This is an important aspect of electronic design since the dc operating point will have an effect on the ac gain of the network.

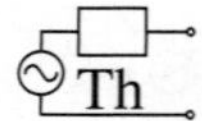

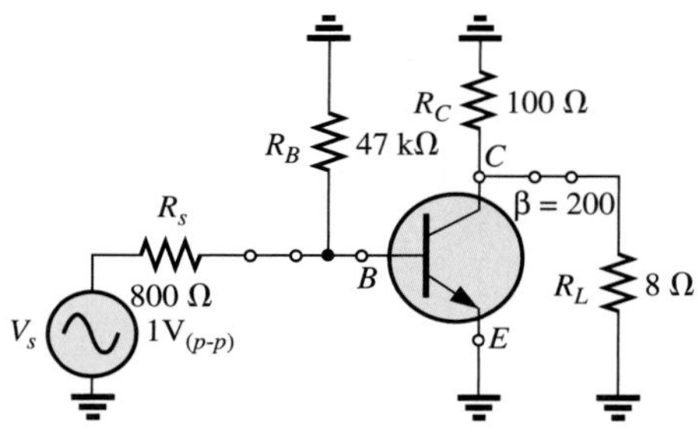

FIG. 18.93
ac equivalent of the transistor network in Fig. 18.91.

Now, using superposition, we can analyze the network from an ac viewpoint by setting all dc sources to zero (replaced by ground connections) and replacing both capacitors by short circuits as shown in Fig. 18.93. Substituting the short-circuit equivalent for the capacitors is valid because at 10 kHz (the midrange for human hearing response), the reactance of the capacitor is determined by $X_C = 1/2\pi fC = 15.92\ \Omega$ which can be ignored when compared to the series resistors at the source and load. In other words, the capacitor has played the important role of isolating the amplifier for the dc response and completing the network for the ac response.

Redrawing the network as shown in Fig. 18.94(a) permits an ac investigation of its response. The transistor has now been replaced by an equivalent network that represents the behavior of the device. This process will be covered in detail in your basic electronics courses. This transistor configuration has an input impedance of 200 Ω and a current source whose magnitude is sensitive to the base current in the input circuit and to the amplifying factor for this transistor of 200. The 47 kΩ resistor in parallel with the 200 Ω input impedance of the transistor can be ignored, so the input current I_i and base current I_b are determined by

$$I_i \cong I_b = \frac{V_s}{R_s + R_i} = \frac{1\text{ V}(p\text{-}p)}{800\ \Omega + 200\ \Omega} = \frac{1\text{ V}(p\text{-}p)}{1\text{ k}\Omega} = 1\text{ mA }(p\text{-}p)$$

The collector current I_C is then

$$I_C = \beta I_b = (200)(1\text{ mA }(p\text{-}p)) = 200\text{ mA }(p\text{-}p)$$

and the current to the speaker is determined by the current divider rule as follows:

$$I_L = \frac{100\ \Omega(I_C)}{100\ \Omega + 8\ \Omega} = 0.926I_C = 0.926(200\text{ mA }(p\text{-}p))$$
$$= 185.2\text{ mA }(p\text{-}p)$$

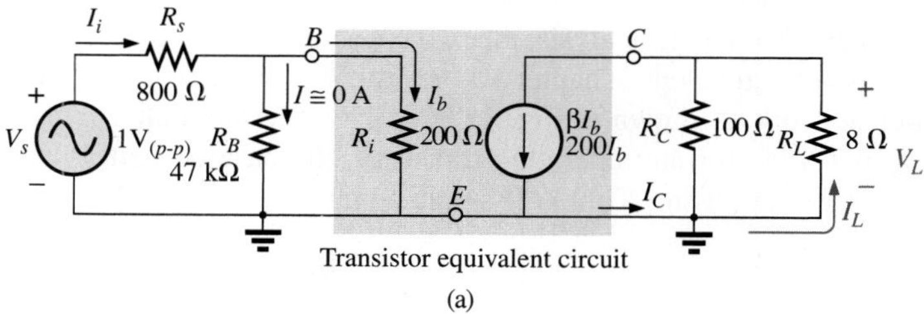

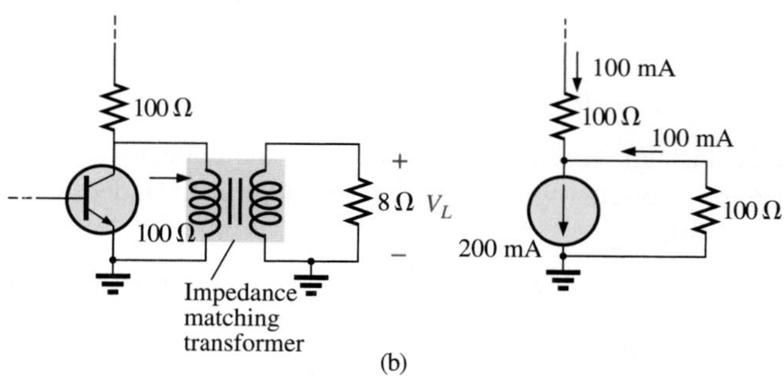

FIG. 18.94
(a) Network in Fig. 18.93 following the substitution of the transistor equivalent network; (b) effect of the matching transformer.

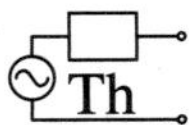

with the voltage across the speaker being

$$V_L = -I_L R_L = -(185.2 \text{ mA } (p\text{-}p))(8\ \Omega) = -1.48 \text{ V}$$

The power to the speaker is then determined as follows:

$$P_{\text{speaker}} = V_{L_{\text{rms}}} \cdot I_{L_{\text{rms}}} = \frac{(V_{L(p\text{-}p)})(I_{L(p\text{-}p)})}{8} = \frac{(1.48 \text{ V})(185.2 \text{ mA } (p\text{-}p))}{8}$$
$$= \mathbf{34.26\ mW}$$

which is relatively low. It initially appears that the above was a good design for distribution of power to the speaker because a majority of the collector current went to the speaker. However, you must always keep in mind that power is the product of voltage and current. A high current with a very low voltage results in a lower power level. In this case, the voltage level is too low. However, if we introduce a matching transformer that makes the 8 Ω resistive load "look like" 100 Ω as shown in Fig. 18.94(b), establishing maximum power conditions, the current to the load drops to half of the previous amount because current splits through equal resistors. But the voltage across the load increases to

$$V_L = I_L R_L = (100 \text{ mA } (p\text{-}p))(100\ \Omega) = 10 \text{ V } (p\text{-}p)$$

and the power level to

$$P_{\text{speaker}} = \frac{(V_{L(p\text{-}p)})(I_{L(p\text{-}p)})}{8} = \frac{(10 \text{ V})(100 \text{ mA})}{8} = \mathbf{125\ mW}$$

which is 3.6 times the gain without the matching transformer.

For the 100 Ω load, the dc conditions are unaffected due to the isolation of the capacitor C_C, and the voltage at the collector is 12.94 V as shown in Fig. 18.95(a). For the ac response with a 100 Ω load, the output voltage as determined above will be 10 V peak-to-peak (5 V peak) as shown in Fig. 18.95(b). Note the out-of-phase relationship with the input due to the opposite polarity of V_L. The full response at the collector terminal of the transistor can then be drawn by superimposing the ac response on the dc response as shown in Fig. 18.95(c) (another application of the superposition theorem). In other words, the dc level simply shifts the ac waveform up or down and does not disturb its shape. The peak-to-peak value remains the same, and the phase relationship is unaltered. The

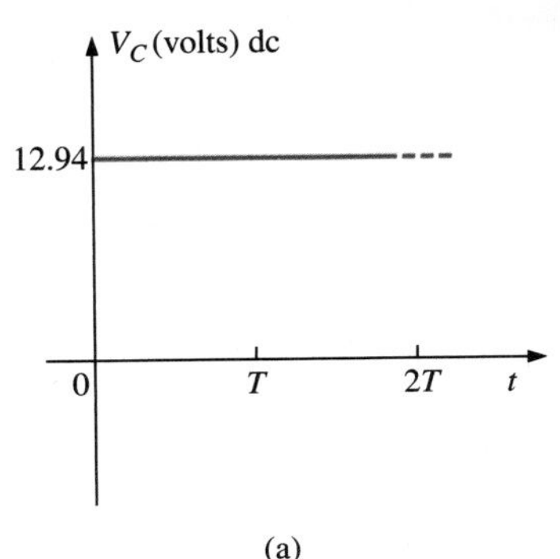

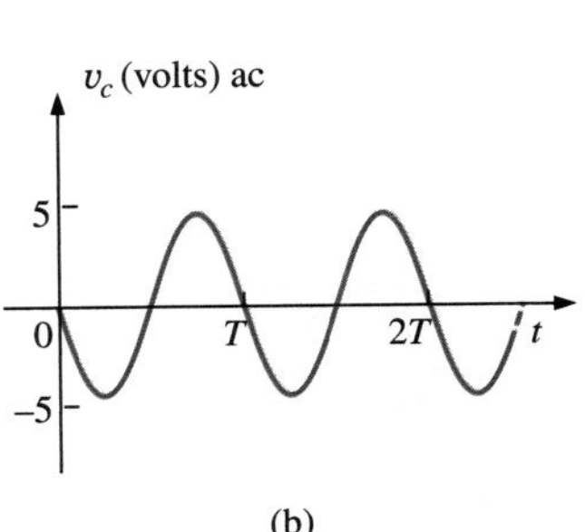

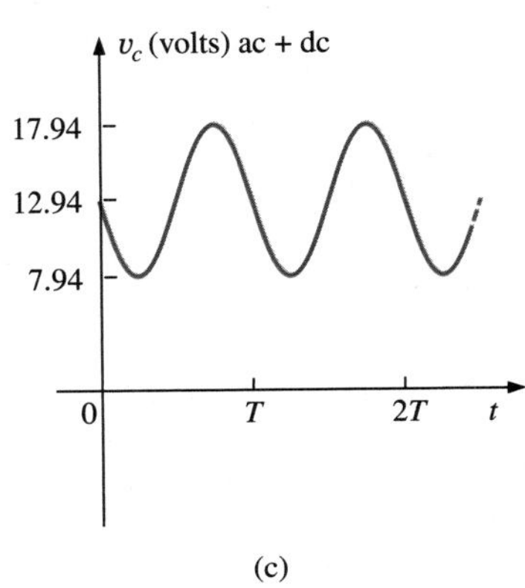

FIG. 18.95
Collector voltage for the network in Fig. 18.91: (a) dc; (b) ac; (c) dc and ac.

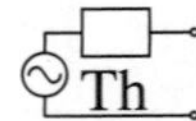

total waveform at the load will include only the ac response of Fig. 18.95(b) since the dc component has been blocked out by the capacitor.

The voltage at the source appears as shown in Fig. 18.96(a), while the voltage at the base of the transistor appears as shown in Fig. 18.96(b) because of the presence of the dc component.

A number of important concepts were presented in the above example, with some probably leaving a question or two because of your lack of experience with transistors. However, you should understand that the superposition theorem has the power to permit an isolation of the dc and ac responses and the ability to combine both if the total response is desired.

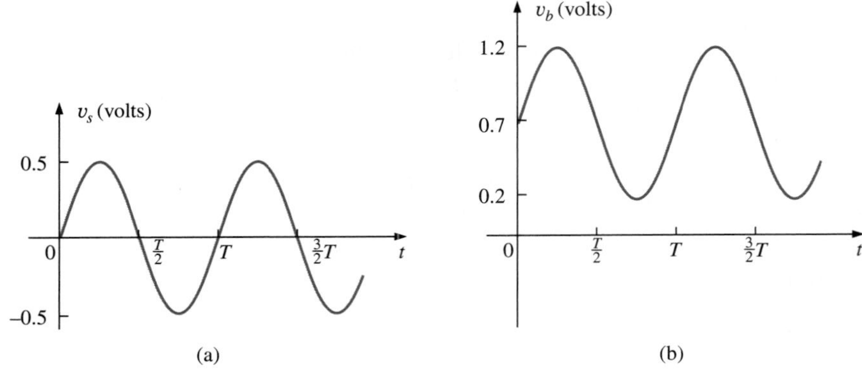

FIG. 18.96
Applied signal: (a) at the source; (b) at the base of the transistor.

18.8 COMPUTER ANALYSIS

PSpice

Thévenin's Theorem This application parallels the methods used to determine the Thévenin equivalent circuit for dc circuits. The network in Fig. 18.29 appears as shown in Fig. 18.97 when the open-circuit Thévenin voltage is to be determined. The open circuit is simulated by using a resistor of 1 T (1 million MΩ). The resistor is necessary to establish a connection between the right side of inductor L_2 and ground—nodes cannot be left floating for OrCAD simulations. Since the magnitude and the angle of the voltage are required, **VPRINT1** is introduced as shown in Fig 18.97. The simulation was an **AC Sweep** simulation at

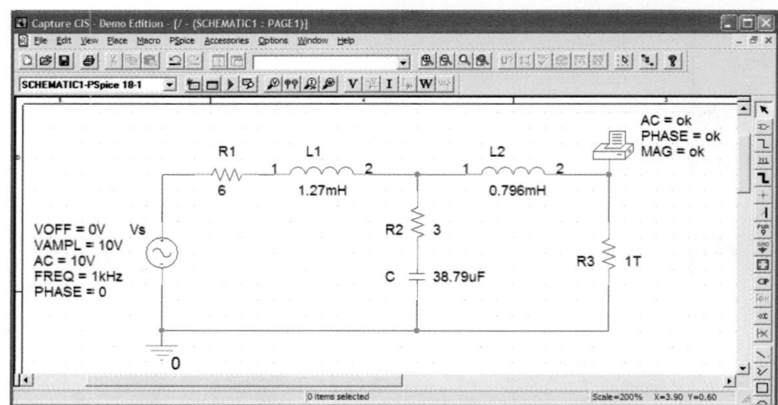

FIG. 18.97
Using PSpice to determine the open-circuit Thévenin voltage.

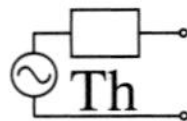

1 kHz, and when the **Orcad Capture** window was obtained, the results appearing in Fig. 18.98 were taken from the listing resulting from the **PSpice-View Output File.** The magnitude of the Thévenin voltage is 5.187 V to compare with the 5.08 V of Example 18.8, while the phase angle is $-77.13°$ to compare with the $-77.09°$ of the same example—excellent results.

```
** Profile: "SCHEMATIC1-PSpice 18-1"  [ C:\ICA11\PSpice\PSpice 18-1-
PSpiceFiles\SCHEMATIC1\PSpice 18-1.sim ]

****    AC ANALYSIS              TEMPERATURE =  27.000 DEG C
*****************************************************************************

 FREQ        VM(N01052)         VP(N01052)

 1.000E+03   5.187E+00          -7.713E+01
```

FIG. 18.98

The output file for the open-circuit Thévenin voltage for the network in Fig. 18.97.

Next, the short-circuit current is determined using **IPRINT** as shown in Fig. 18.99, to permit a determination of the Thévenin impedance. The resistance R_{coil} of 1 $\mu\Omega$ had to be introduced because inductors cannot be treated as ideal elements when using PSpice; they must all show some series internal resistance. Note that the short-circuit current will pass directly through the printer symbol for **IPRINT.** Incidentally, there is no need to exit the **SCHEMATIC1** developed above to determine the Thévenin voltage. Simply delete **VPRINT** and **R3,** and insert **IPRINT.** Then run a new simulation to obtain the results in Fig. 18.100. The

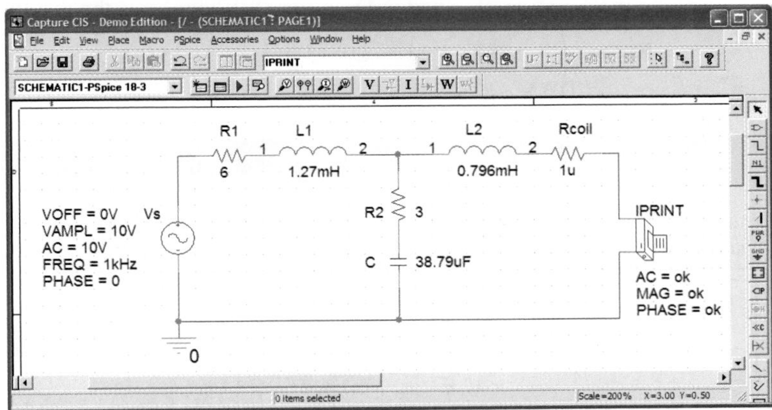

FIG. 18.99

Using PSpice to determine the short-circuit current.

```
** Profile: "SCHEMATIC1-PSpice 18-3"  [ C:\ICA11\PSpice\pspice 18-3-
pspicefiles\schematic1\pspice 18-3.sim ]

****    AC ANALYSIS              TEMPERATURE =  27.000 DEG C
*****************************************************************************
FREQ         IM(V_PRINT1)       IP(V_PRINT1)

1.000E+03    9.361E-01          -1.086E+02
```

FIG. 18.100

The output file for the short-circuit current for the network in Fig. 18.99.

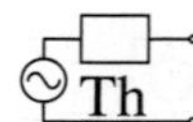

magnitude of the short-circuit current is 936.1 mA at an angle of −108.6°. The Thévenin impedance is then defined by

$$\mathbf{Z}_{Th} = \frac{\mathbf{E}_{Th}}{\mathbf{I}_{sc}} = \frac{5.187\text{ V }\angle -77.13^\circ}{936.1\text{ mA }\angle -108.6^\circ} = 5.54\ \Omega\ \angle 31.47^\circ$$

which is an excellent match with 5.49 Ω ∠32.36° obtained in Example 18.8.

VCVS The next application will verify the results in Example 18.12 and provide some practice using controlled (dependent) sources. The network in Fig. 18.51, with its **voltage-controlled voltage source** (VCVS), will have the schematic appearance in Fig. 18.101. The VCVS appears as **E** in the **ANALOG** library, with the voltage **E1** as the controlling voltage and **E** as the controlled voltage. In the **Property Editor** dialog box, change the **GAIN** to 20, but leave the rest of the columns as is. After **Display-Name and Value,** select **Apply** and exit the dialog box. This results in **GAIN = 20** near the controlled source. Take particular note of the second ground inserted near **E** to avoid a long wire to ground that may overlap other elements. For this exercise, the current source **ISRC** is used because it has an arrow in its symbol, and frequency is not important for this analysis since there are only resistive elements present. In the **Property Editor** dialog box, set the **AC** level to 5 mA and the **DC** level to 0 A; both are displayed using **Display-Name and value. VPRINT1** is set up as in past exercises. The resistor **Roc** (open circuit) was given a very large value so that it appears as an open circuit to the rest of the network. **VPRINT1** provides the open circuit Thévenin voltage between the points of interest. Running the simulation in the **AC Sweep** mode at 1 kHz results in the output file appearing in Fig. 18.102, revealing that the Thévenin voltage is 210 V ∠0°. Substituting the numerical values of this example into the equation obtained in Example 18.12 confirms the result:

$$\begin{aligned}\mathbf{E}_{Th} &= (1+\mu)\mathbf{I}R_1 = (1+20)\,(5\text{ mA }\angle 0^\circ)(2\text{ k}\Omega)\\ &= \mathbf{210\ V\angle 0^\circ}\end{aligned}$$

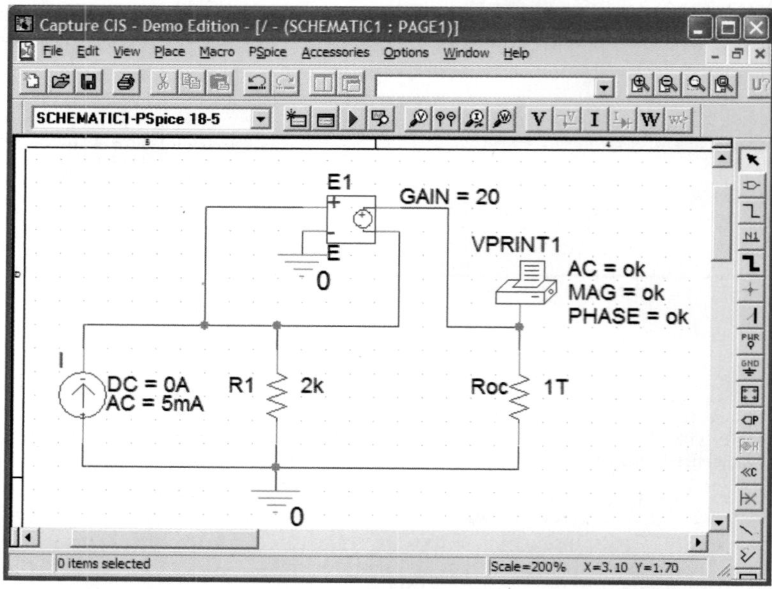

FIG. 18.101
Using PSpice to determine the open-circuit Thévenin voltage for the network in Fig. 18.51.

```
** Profile: "SCHEMATIC1-PSpice 18-5"  [ C:\ICA11\PSpice\pspice 18-5-
pspicefiles\schematic1\pspice 18-5.sim ]

****   AC ANALYSIS                TEMPERATURE =  27.000 DEG C
**************************************************************************
FREQ         VM(N01097)  VP(N01097)

1.000E+03    2.100E+02    0.000E+00
```

FIG. 18.102
The output file for the open-circuit Thévenin voltage for the network in Fig. 18.101.

Next, determine the short-circuit current using the **IPRINT** option. Note in Fig. 18.103 that the only difference between this network and that in Fig. 18.102 is the replacement of **Roc** with **IPRINT** and the removal of **VPRINT1.** Therefore, you do not need to completely "redraw" the network. Just make the changes and run a new simulation. The result of the new simulation as shown in Fig. 18.104 is a current of 5 mA at an angle of 0°.

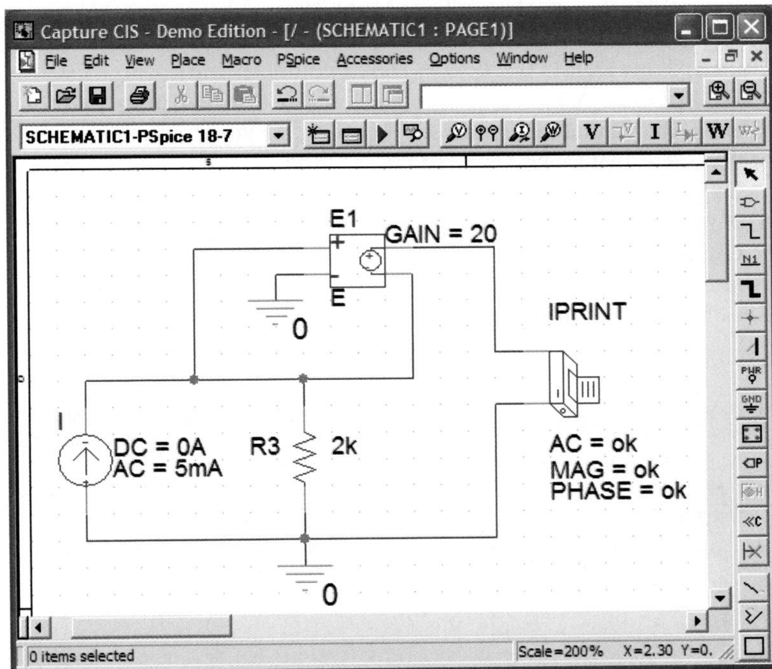

FIG. 18.103
Using PSpice to determine the short-circuit current for the network in Fig. 18.51.

```
** Profile: "SCHEMATIC1-PSpice 18-7"  [ C:\ICA11\PSpice\pspice 18-7-
pspicefiles\schematic1\PSpice 18-7.sim ]

****   AC ANALYSIS                TEMPERATURE =  27.000 DEG C

**************************************************************************

FREQ         IM(V_PRINT1)     IP(V_PRINT1)

1.000E+03    5.000E-03        0.000E+00
```

FIG. 18.104
The output file for the short-circuit current for the network in Fig. 18.103.

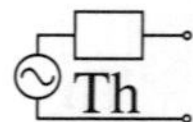

The ratio of the two measured quantities results in the Thévenin impedance:

$$\mathbf{Z}_{Th} = \frac{\mathbf{E}_{oc}}{\mathbf{I}_{sc}} = \frac{\mathbf{E}_{Th}}{\mathbf{I}_{sc}} = \frac{210\ \text{V} \angle 0^\circ}{5\ \text{mA} \angle 0^\circ} = \mathbf{42\ k\Omega}$$

which also matches the longhand solution in Example 18.12:

$$\mathbf{Z}_{Th} = (1 + \mu)R_1 = (1 + 20)2\ \text{k}\Omega = (21)2\ \text{k}\Omega = \mathbf{42\ k\Omega}$$

Multisim

Superposition This analysis begins with the network in Fig. 18.12 from Example 18.4 because it has both an ac and a dc source. You will find in the analysis to follow that it is not necessary to set up a separate network for each source. Once the network is set up, the dc levels will appear during simulation, and the ac response can be found from a **View** option.

The resulting schematic appears in Fig. 18.105. The construction is quite straightforward with the parameters of the ac source set as follows: **Voltage(RMS):** 4 V; **AC Analysis Magnitude:** 4 V; **Phase:** 0 Degrees; **AC Analysis Phase:** 0 Degrees; **Frequency (F):** 1 kHz; **Voltage Offset:** 0 V; and **Time Delay:** 0 Seconds. The dc voltmeter **Indicator** is listed as **VOLTMETER_V** under **Component** in the **Select a Component** dialog box. Recall that the indicators appear on the keypad on the left edge of the screen that looks like a red 8 LED display.

To perform the analysis, use the following sequence to obtain the **AC Analysis** dialog box: **Simulate-Analyses-AC Analysis.** In the dialog box, make the following settings under the **Frequency Parameters** heading: **Start frequency:** 1 kHz; **Stop frequency:** 1 kHz; **Sweep Type:** Decade; **Number of points:** 1000; **Vertical scale:** Linear. Then shift to

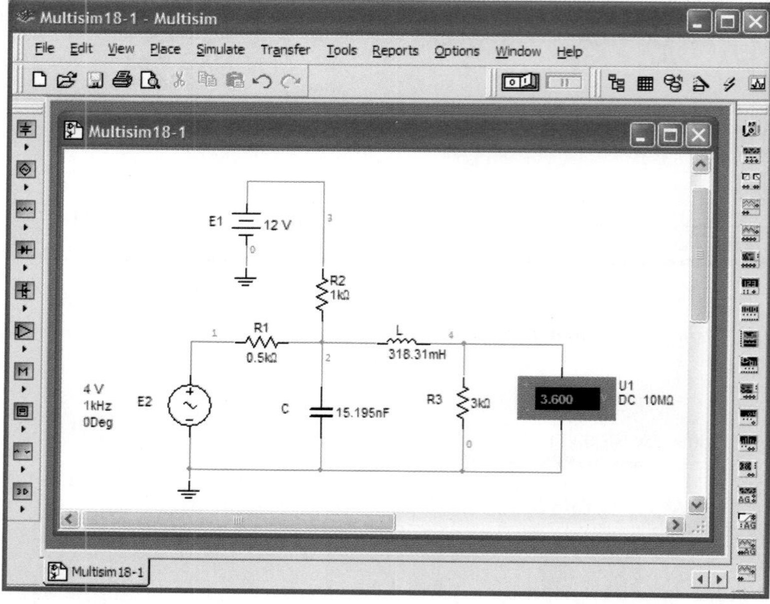

FIG. 18.105
Using Multisim to apply superposition to the network in Fig. 18.12.

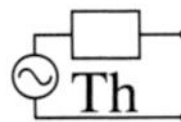

the **Output** option and select \$4 (note node 4 on the constructed network) under **Variables in circuit** followed by **Add** to place it in the **Selected variables for analysis** column. Move any other variables in the selected list back to the variable list using the **Remove** option. Then select **Simulate,** and the **Grapher View** response of Fig. 18.106 results. During the simulation process, the dc solution of 3.6 V appears on the voltmeter display (an exact match with the longhand solution). There are two plots in Fig. 18.106: one of magnitude versus frequency and the other of phase versus frequency. Left-click to select the upper graph, and a red arrow shows up along the left edge of the plot. The arrow reveals which plot is currently active. To change the label for the vertical axis from **Magnitude** to **Voltage (V)** as shown in Fig. 18.106, select the **Properties** key from the top toolbar and choose **Left Axis.** Then change the label to **Voltage (V)** followed by **OK,** and the label appears as shown in Fig. 18.106. Next, to read the levels indicated on each graph with a high degree of accuracy, select the **Show/Hide Cursor** keypad on the toolbar. The keypad has a small red sine wave with two vertical markers. The result is a set of markers at the left edge of each figure. By selecting a marker from the left edge of the voltage plot and moving it to 1 kHz, you can find the value of the voltage in the accompanying table. Note that at a frequency of 1.006 kHz or essentially 1 kHz, the voltage is 2.06 V which is an exact match with the longhand solution in Example 18.4. If you then drop down to the phase plot, you find at the same frequency that the phase angle is -32.65, which is very close to the -32.74 in the longhand solution.

In general, therefore, the results are an excellent match with the solutions in Example 18.4 using techniques that can be applied to a wide variety of networks that have both dc and ac sources.

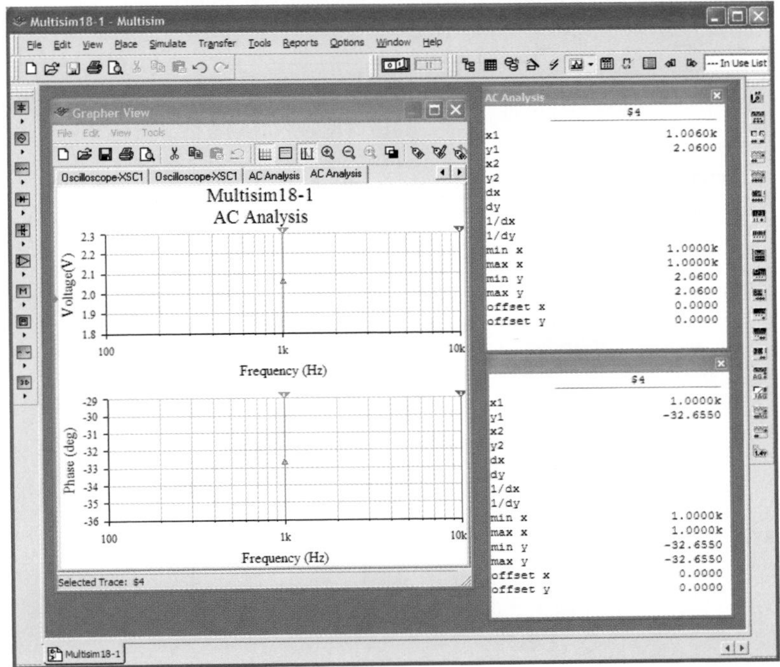

FIG. 18.106
The output results from the simulation of the network in Fig. 18.105.

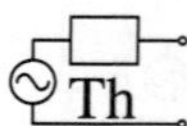

PROBLEMS

SECTION 18.2 Superposition Theorem

1. Using superposition, determine the current through the inductance X_L for each network in Fig. 18.107.

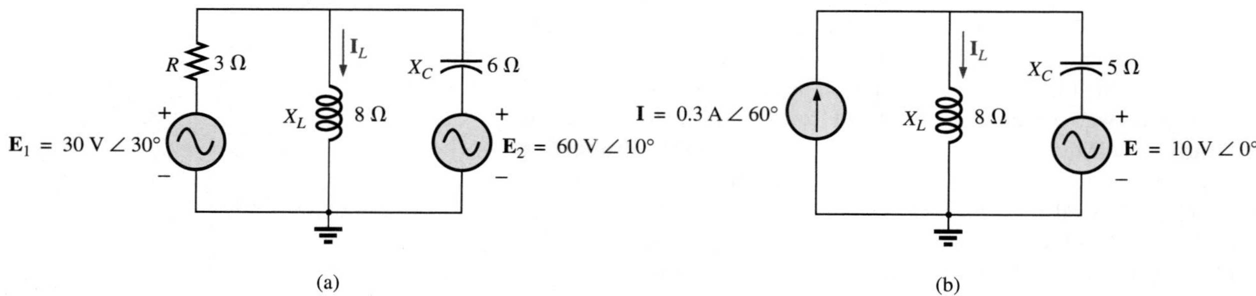

FIG. 18.107
Problem 1.

*2. Using superposition, determine the current $\mathbf{I}_L$ for each network in Fig. 18.108.

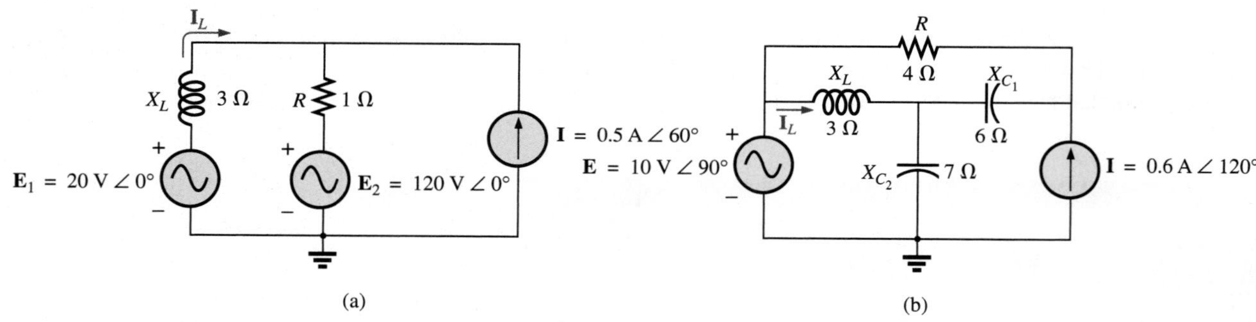

FIG. 18.108
Problem 2.

*3. Using superposition, find the sinusoidal expression for the current i for the network in Fig. 18.109.

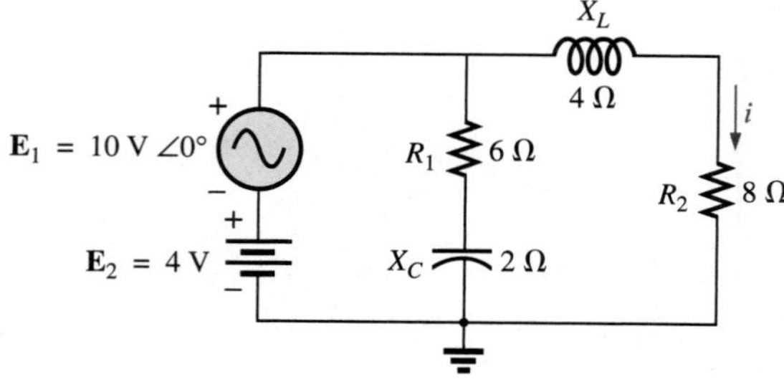

FIG. 18.109
Problems 3, 15, 30, and 42.

4. Using superposition, find the sinusoidal expression for the voltage v_C for the network in Fig. 18.110.

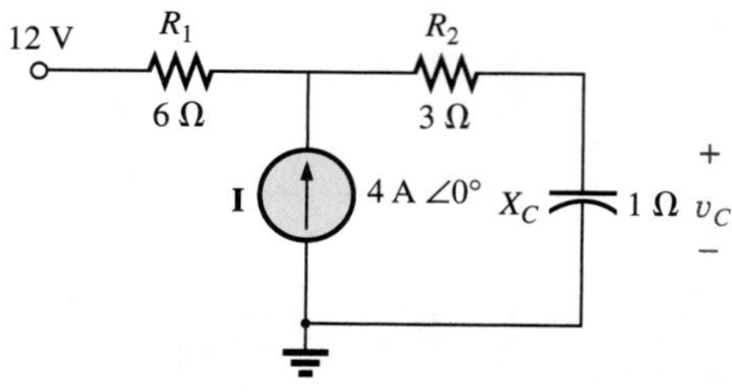

FIG. 18.110
Problems 4, 16, 31, and 43.

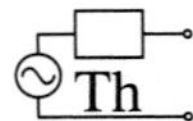

*5. Using superposition, find the current **I** for the network in Fig. 18.111.

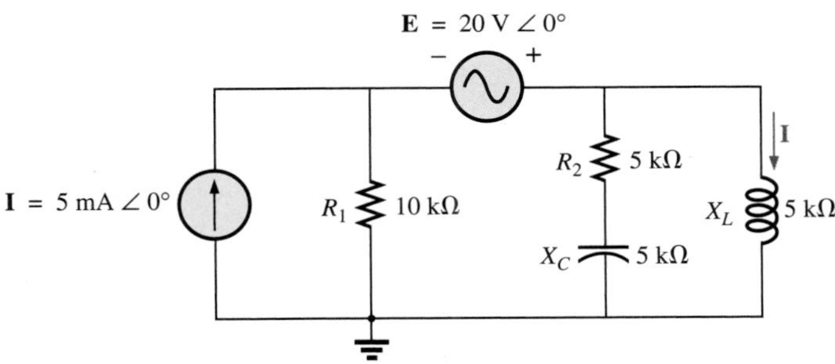

FIG. 18.111
Problems 5, 17, 32, and 44.

6. Using superposition, determine the current $\mathbf{I}_L$ ($h = 100$) for the network in Fig. 18.112.

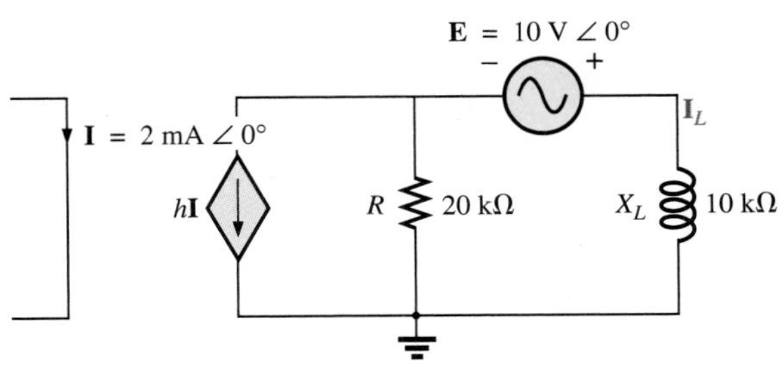

FIG. 18.112
Problems 6 and 20.

7. Using superposition, for the network in Fig. 18.113, determine the voltage $\mathbf{V}_L$ ($\mu = 20$).

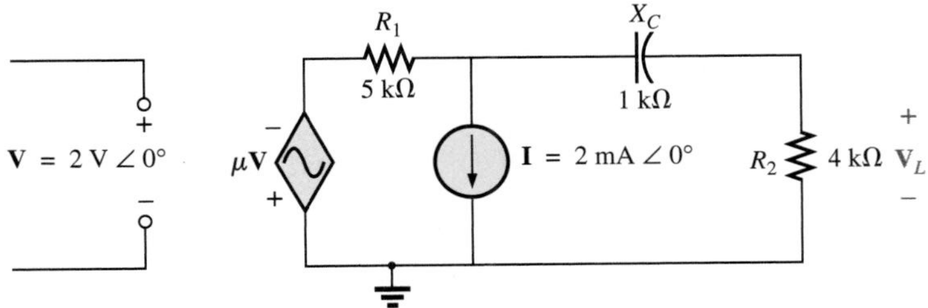

FIG. 18.113
Problems 7, 21, and 35.

*8. Using superposition, determine the current $\mathbf{I}_L$ for the network in Fig. 18.114 ($\mu = 20$; $h = 100$).

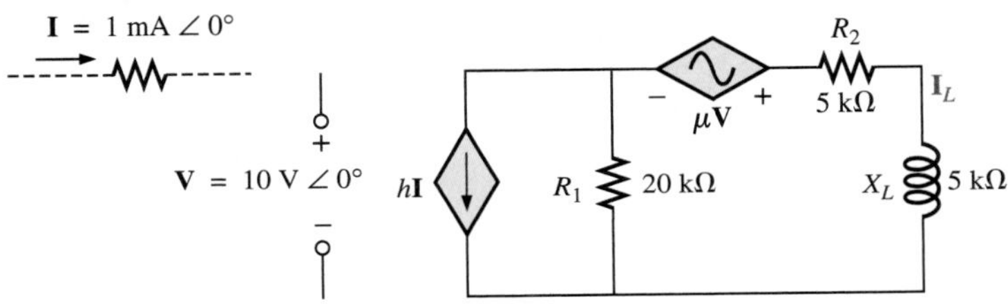

FIG. 18.114
Problems 8, 22, and 36.

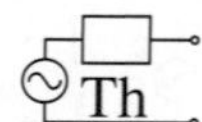

*9. Determine $\mathbf{V}_L$ for the network in Fig. 18.115 ($h = 50$).

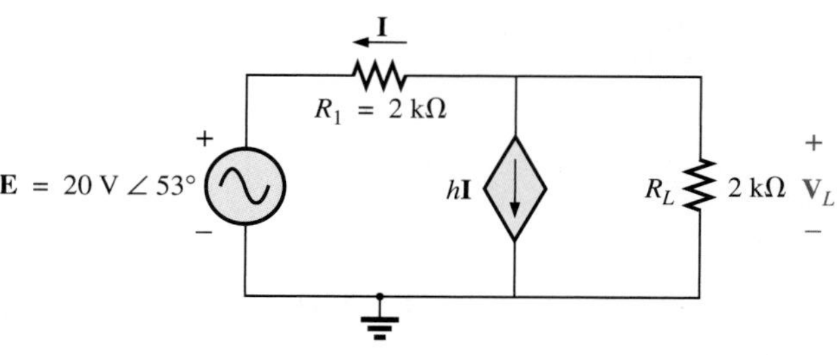

FIG. 18.115
Problems 9 and 23.

*10. Calculate the current **I** for the network in Fig. 18.116.

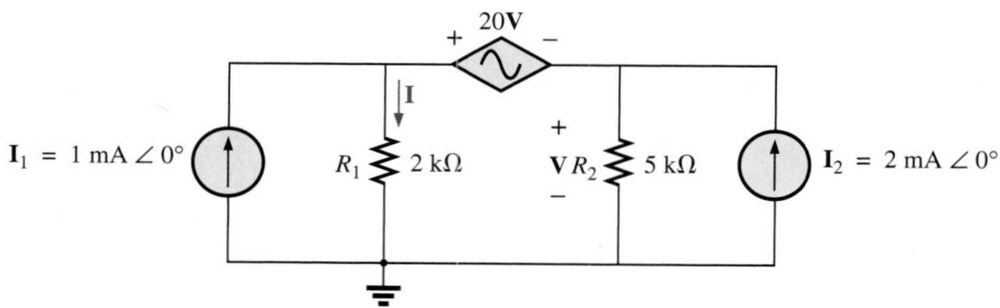

FIG. 18.116
Problems 10, 24, and 38.

11. Find the voltage $\mathbf{V}_s$ for the network in Fig. 18.117.

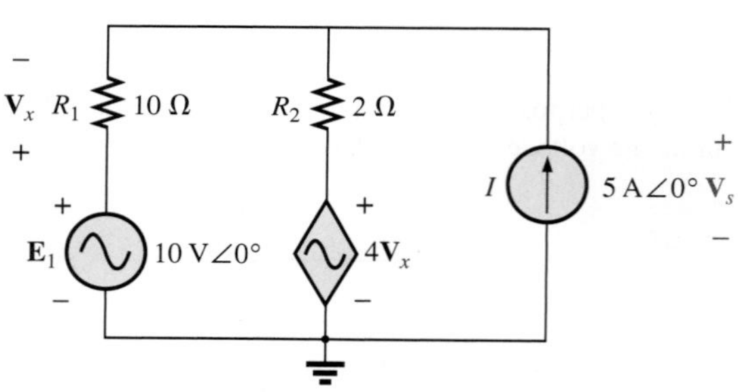

FIG. 18.117
Problem 11.

SECTION 18.3 Thévenin's Theorem

12. Find the Thévenin equivalent circuit for the portions of the networks in Fig. 18.118 external to the elements between points *a* and *b*.

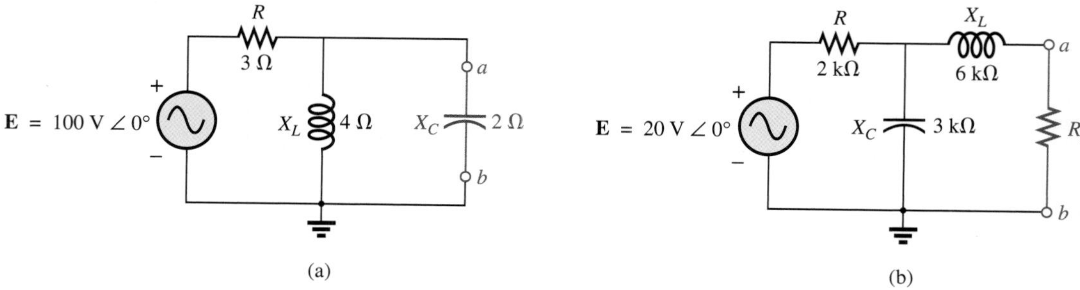

FIG. 18.118
Problems 12 and 26.

*13. Find the Thévenin equivalent circuit for the portions of the networks in Fig. 18.119 external to the elements between points a and b.

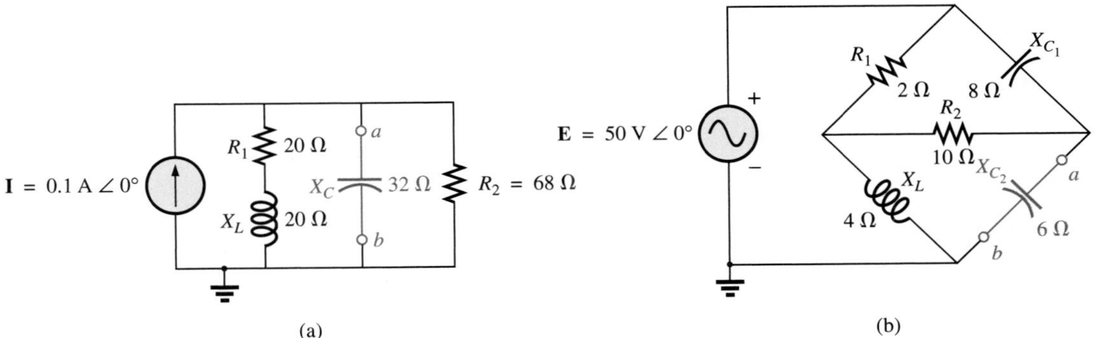

FIG. 18.119
Problems 13 and 27.

*14. Find the Thévenin equivalent circuit for the portions of the networks in Fig. 18.120 external to the elements between points a and b.

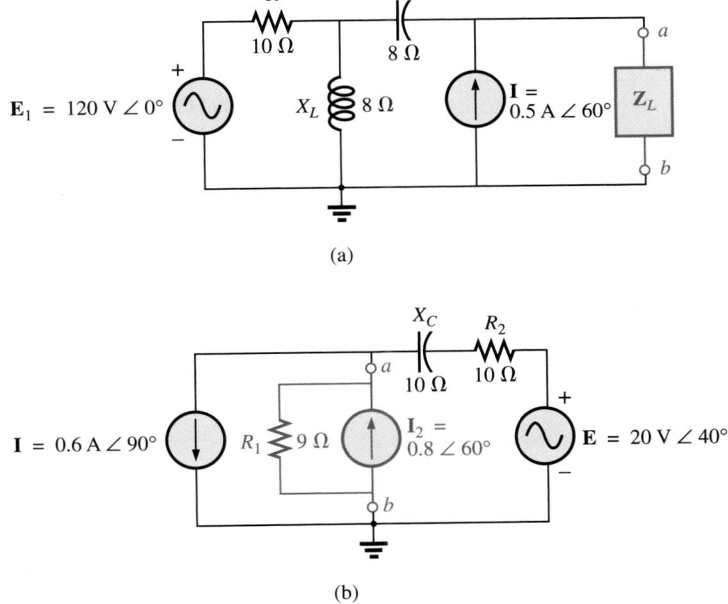

FIG. 18.120
Problems 14 and 28.

*15. **a.** Find the Thévenin equivalent circuit for the network external to the resistor R_2 in Fig. 18.109.
b. Using the results of part (a), determine the current i of the same figure.

16. **a.** Find the Thévenin equivalent circuit for the network external to the capacitor in Fig. 18.110.
b. Using the results of part (a), determine the voltage $\mathbf{V}_C$ for the same figure.

*17. **a.** Find the Thévenin equivalent circuit for the network external to the inductor in Fig. 18.111.
b. Using the results of part (a), determine the current **I** of the same figure.

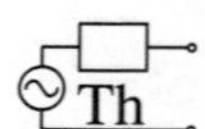

18. Determine the Thévenin equivalent circuit for the network external to the 5 kΩ inductive reactance in Fig. 18.121 (in terms of **V**).

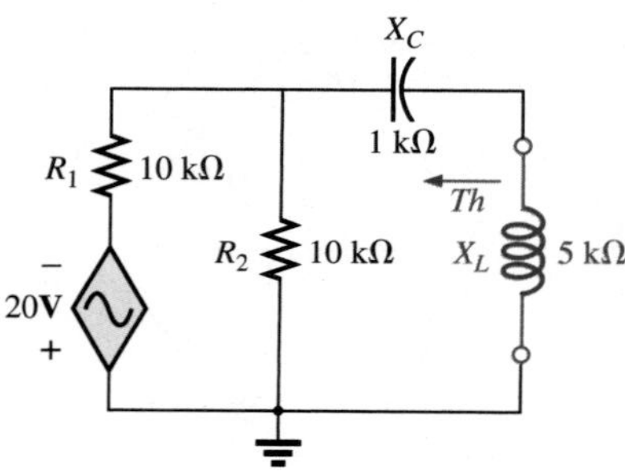

FIG. 18.121
Problems 18 and 33.

19. Determine the Thévenin equivalent circuit for the network external to the 4 kΩ inductive reactance in Fig. 18.122 (in terms of **I**).

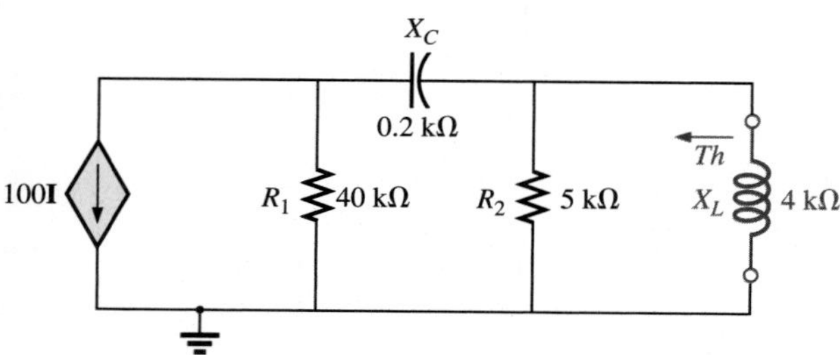

FIG. 18.122
Problems 19 and 34.

20. Find the Thévenin equivalent circuit for the network external to the 10 kΩ inductive reactance in Fig. 18.112.

21. Determine the Thévenin equivalent circuit for the network external to the 4 kΩ resistor in Fig. 18.113.

***22.** Find the Thévenin equivalent circuit for the network external to the 5 kΩ inductive reactance in Fig. 18.114.

***23.** Determine the Thévenin equivalent circuit for the network external to the 2 kΩ resistor in Fig. 18.115.

***24.** Find the Thévenin equivalent circuit for the network external to the resistor R_1 in Fig. 18.116.

***25.** Find the Thévenin equivalent circuit for the network to the left of terminals *a*-*a'* in Fig. 18.123.

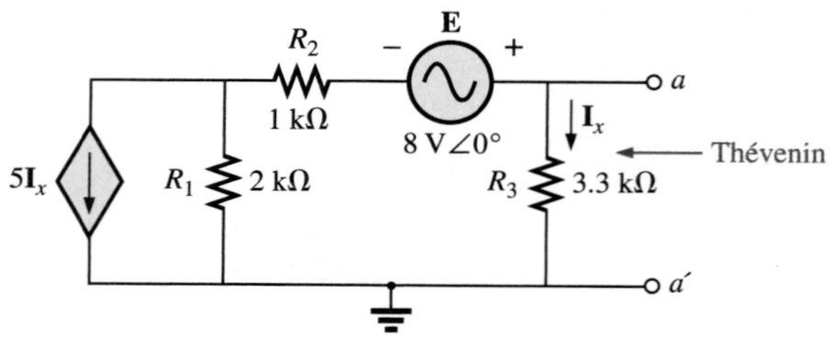

FIG. 18.123
Problem 25.

SECTION 18.4 Norton's Theorem

26. Find the Norton equivalent circuit for the network external to the elements between *a* and *b* for the networks in Fig. 18.118.

27. Find the Norton equivalent circuit for the network external to the elements between *a* and *b* for the networks in Fig. 18.119.

28. Find the Norton equivalent circuit for the network external to the elements between *a* and *b* for the networks in Fig. 18.120.

***29.** Find the Norton equivalent circuit for the portions of the networks in Fig. 18.124 external to the elements between points *a* and *b*.

***30.** **a.** Find the Norton equivalent circuit for the network external to the resistor R_2 in Fig. 18.109.
b. Using the results of part (a), determine the current **I** of the same figure.

***31.** **a.** Find the Norton equivalent circuit for the network external to the capacitor in Fig. 18.110.
b. Using the results of part (a), determine the voltage $\mathbf{V}_C$ for the same figure.

***32.** **a.** Find the Norton equivalent circuit for the network external to the inductor in Fig. 18.111.
b. Using the results of part (a), determine the current **I** of the same figure.

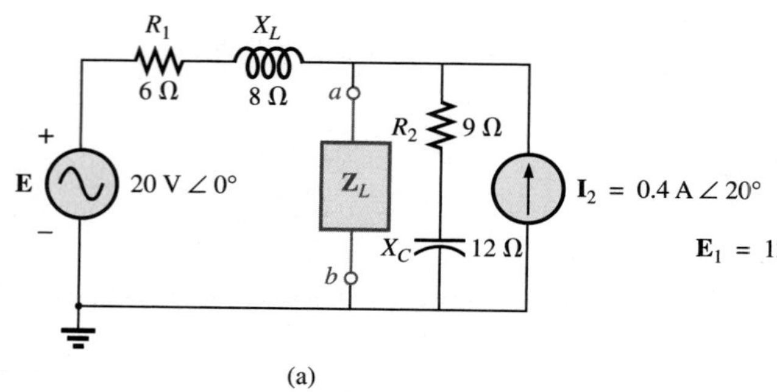

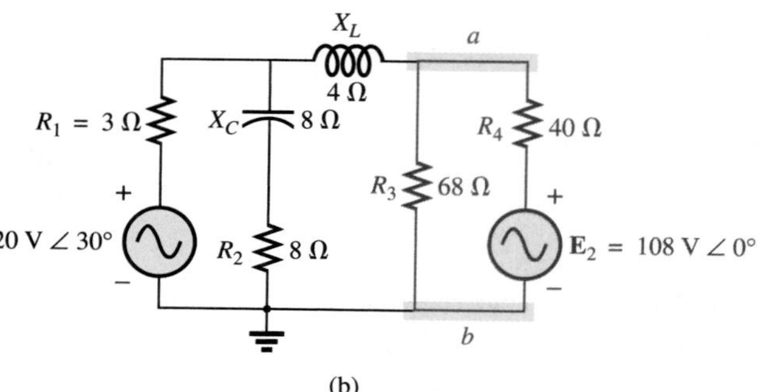

FIG. 18.124
Problem 29.

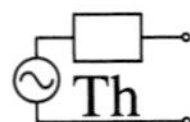

33. Determine the Norton equivalent circuit for the network external to the 5 kΩ inductive reactance in Fig. 18.121.

34. Determine the Norton equivalent circuit for the network external to the 4 kΩ inductive reactance in Fig. 18.122.

35. Find the Norton equivalent circuit for the network external to the 4 kΩ resistor in Fig. 18.113.

***36.** Find the Norton equivalent circuit for the network external to the 5 kΩ inductive reactance in Fig. 18.114.

***37.** For the network in Fig. 18.125, find the Norton equivalent circuit for the network external to the 2 kΩ resistor.

***38.** Find the Norton equivalent circuit for the network external to the $\mathbf{I}_1$ current source in Fig. 18.116.

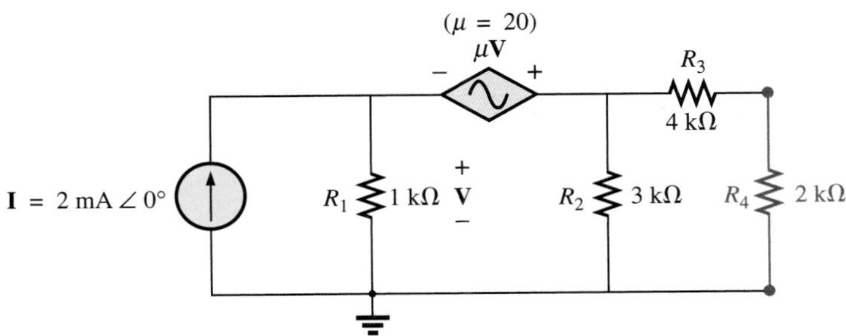

FIG. 18.125
Problem 37.

SECTION 18.5 Maximum Power Transfer Theorem

39. Find the load impedance $\mathbf{Z}_L$ for the networks in Fig. 18.126 for maximum power to the load, and find the maximum power to the load.

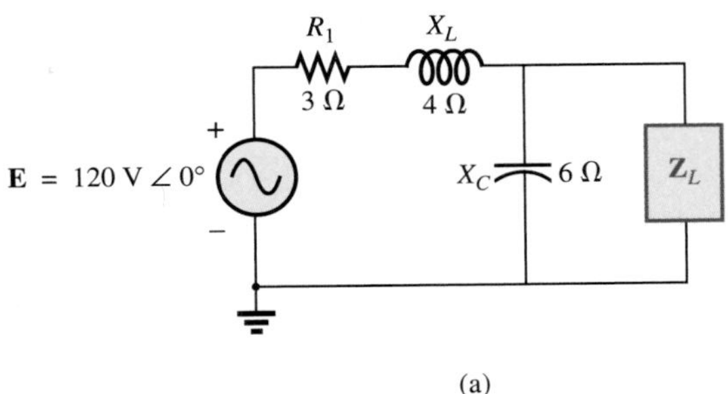

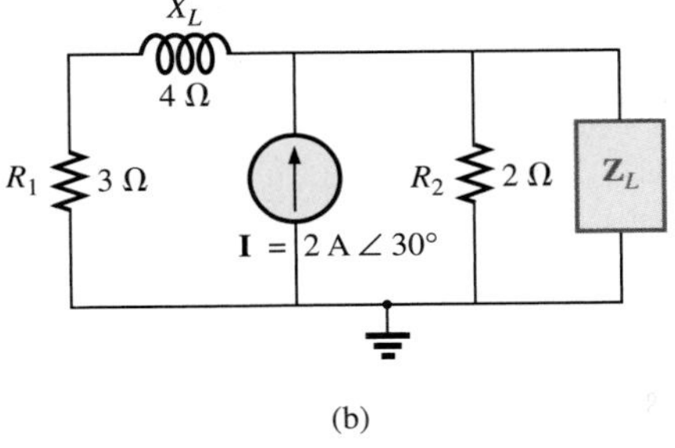

FIG. 18.126
Problem 39.

***40.** Find the load impedance $\mathbf{Z}_L$ for the networks in Fig. 18.127 for maximum power to the load, and find the maximum power to the load.

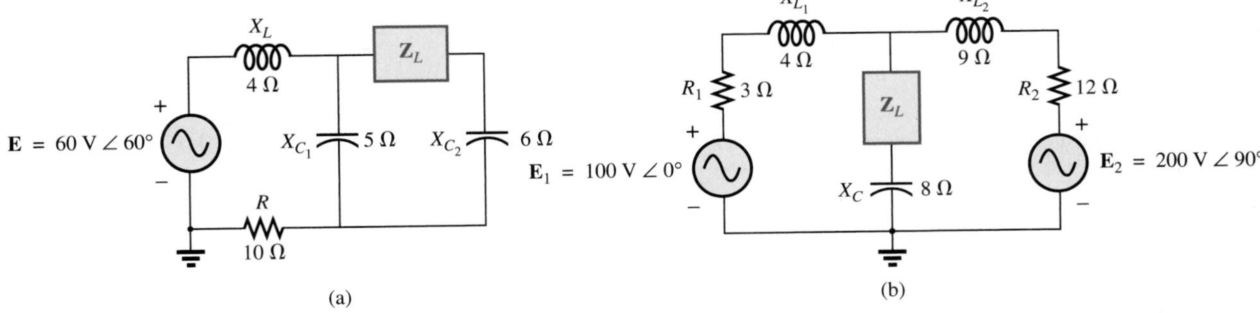

FIG. 18.127
Problem 40.

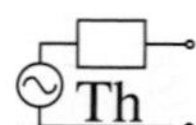

41. Find the load impedance R_L for the network in Fig. 18.128 for maximum power to the load, and find the maximum power to the load.

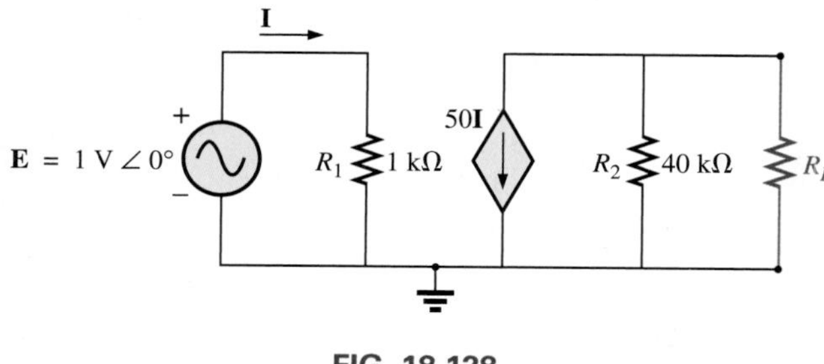

FIG. 18.128
Problem 41.

***42.** **a.** Determine the load impedance to replace the inductor X_L in Fig. 18.109 to ensure maximum power to the load.

b. Using the results of part (a), determine the maximum power to the load.

***43.** **a.** Determine the load impedance to replace the capacitor X_C in Fig. 18.110 to ensure maximum power to the load.

b. Using the results of part (a), determine the maximum power to the load.

***44.** **a.** Determine the load impedance to replace the inductor X_L in Fig. 18.111 to ensure maximum power to the load.

b. Using the results of part (a), determine the maximum power to the load.

45. **a.** For the network in Fig. 18.129, determine the value of R_L that will result in maximum power to the load.

b. Using the results of part (a), determine the maximum power delivered.

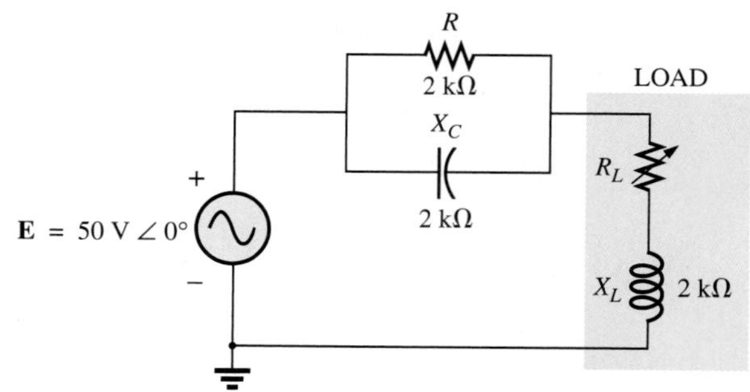

FIG. 18.129
Problem 45.

***46.** **a.** For the network in Fig. 18.130, determine the level of capacitance that will ensure maximum power to the load if the range of capacitance is limited to 1 nF to 10 nF.

b. Using the results of part (a), determine the value of R_L that will ensure maximum power to the load.

c. Using the results of parts (a) and (b), determine the maximum power to the load.

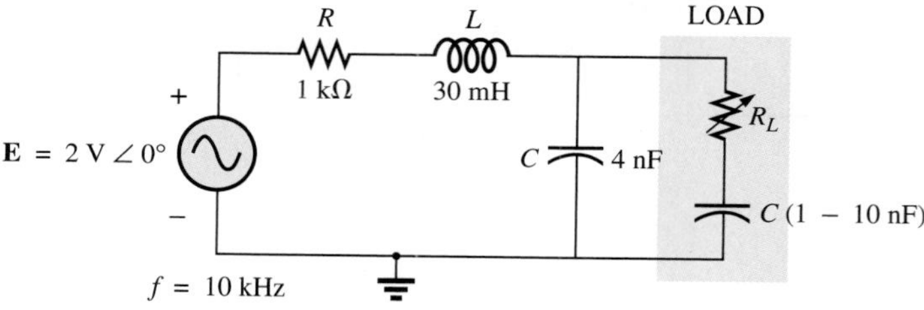

FIG. 18.130
Problem 46.

SECTION 18.6 Substitution, Reciprocity, and Millman's Theorems

47. For the network in Fig. 18.131, determine two equivalent branches through the substitution theorem for the branch *a-b*.

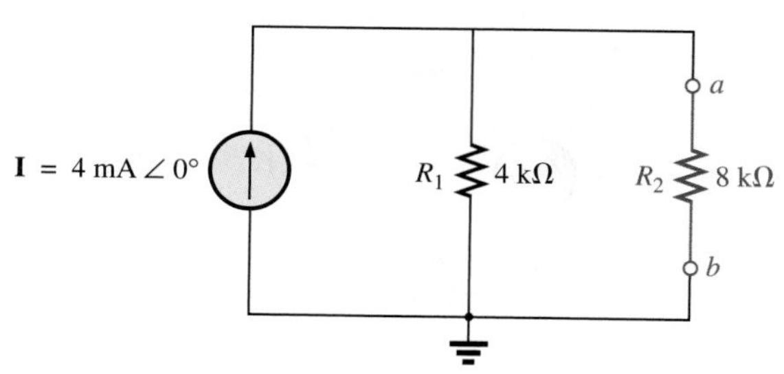

FIG. 18.131
Problem 47.

48. **a.** For the network in Fig. 18.132(a), find the current **I**.
b. Repeat part (a) for the network in Fig. 18.132(b).
c. Do the results of parts (a) and (b) compare?

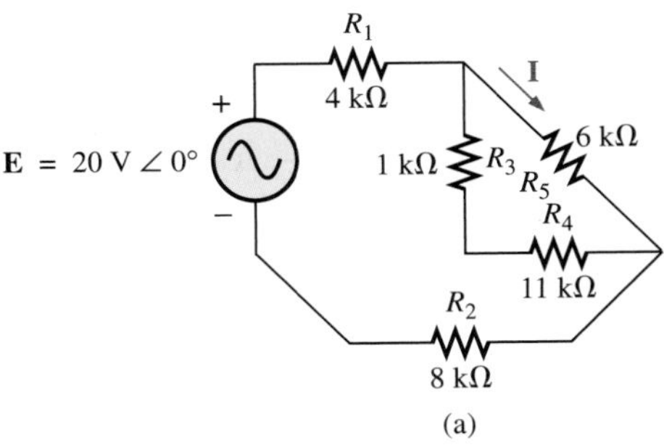

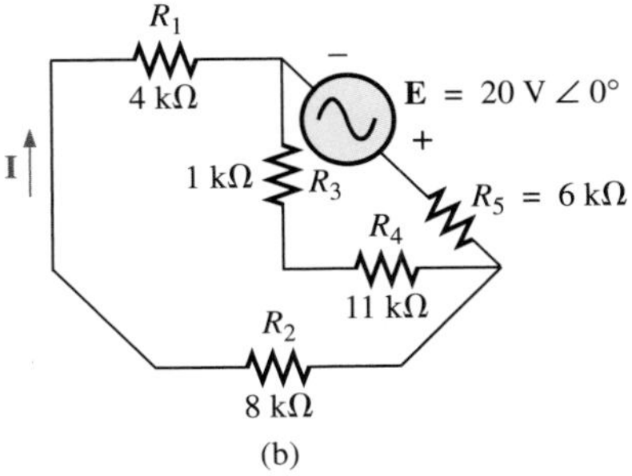

FIG. 18.132
Problem 48.

49. Using Millman's theorem, determine the current through the 4 kΩ capacitive reactance of Fig. 18.133.

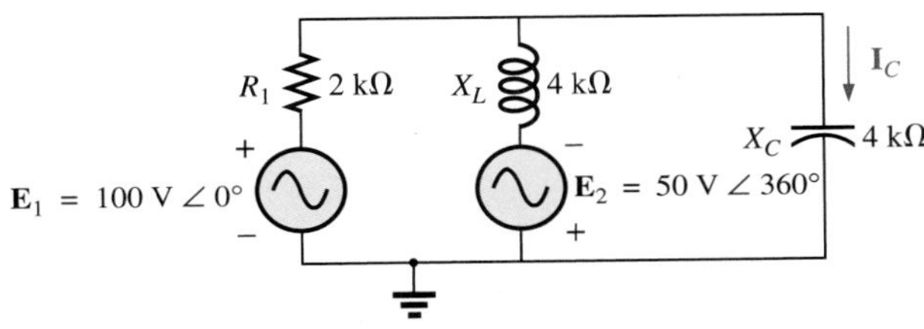

FIG. 18.133
Problem 49.

SECTION 18.8 Computer Analysis

PSpice or Multisim

50. Apply superposition to the network in Fig. 18.6. That is, determine the current **I** due to each source, and then find the resultant current.

***51.** Determine the current $\mathbf{I}_L$ for the network in Fig. 18.22 using schematics.

***52.** Using schematics, determine $\mathbf{V}_2$ for the network in Fig. 18.57 if $\mathbf{V}_i = 1\text{ V}\angle 0°$, $R_1 = 0.5\text{ k}\Omega$, $k_1 = 3 \times 10^{-4}$, $k_2 = 50$, and $R_2 = 20\text{ k}\Omega$.

***53.** Find the Norton equivalent circuit for the network in Fig. 18.77 using schematics.

***54.** Using schematics, plot the power to the *R-C* load in Fig. 18.90 for values of R_L from 1 Ω to 10 Ω.

GLOSSARY

Maximum power transfer theorem A theorem used to determine the load impedance necessary to ensure maximum power to the load.

Millman's theorem A method using voltage-to-current source conversions that will permit the determination of unknown variables in a multiloop network.

Norton's theorem A theorem that permits the reduction of any two-terminal linear ac network to one having a single current source and parallel impedance. The resulting configuration can then be used to determine a particular current or voltage in the original network or to examine the effects of a specific portion of the network on a particular variable.

Reciprocity theorem A theorem stating that for single-source networks, the magnitude of the current in any branch of a network, due to a single voltage source anywhere else in the network, will equal the magnitude of the current through the branch in which the source was originally located if the source is placed in the branch in which the current was originally measured.

Substitution theorem A theorem stating that if the voltage across and current through any branch of an ac bilateral network are known, the branch can be replaced by any combination of elements that will maintain the same voltage across and current through the chosen branch.

Superposition theorem A method of network analysis that permits considering the effects of each source independently. The resulting current and/or voltage is the phasor sum of the currents and/or voltages developed by each source independently.

Thévenin's theorem A theorem that permits the reduction of any two-terminal linear ac network to one having a single voltage source and series impedance. The resulting configuration can then be employed to determine a particular current or voltage in the original network or to examine the effects of a specific portion of the network on a particular variable.

Voltage-controlled voltage source (VCVS) A voltage source whose parameters are controlled by a voltage elsewhere in the system.

Power (ac)

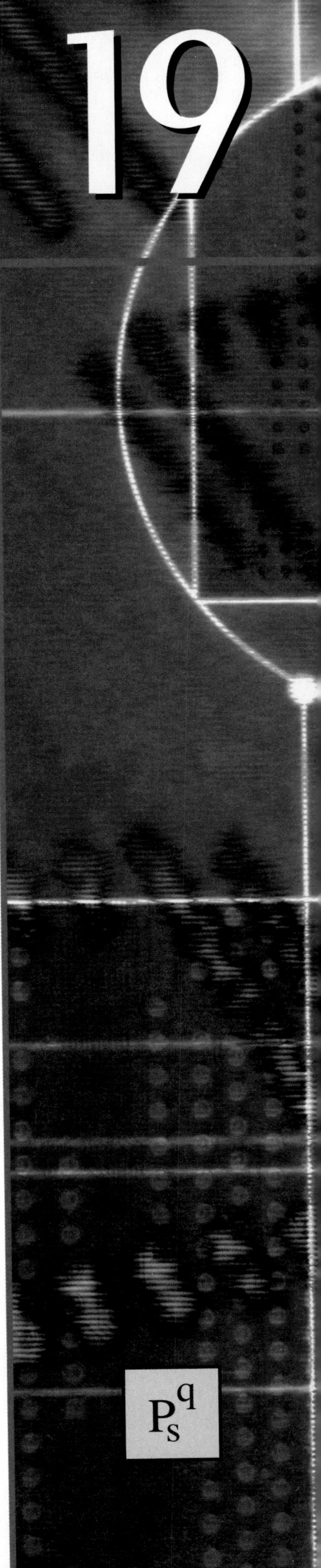

Objectives

- *Become familiar with the differences between average, apparent, and reactive power and how to calculate each for any combination of resistive and reactive elements.*
- *Understand that the energy dissipated by a load is the area under the power curve for the period of time of interest.*
- *Become aware of how the real, apparent, and reactive power are related in an ac network and how to find the total value of each for any configuration.*
- *Understand the concept of power-factor correction and how to apply it to improve the terminal characteristics of a load.*
- *Develop some understanding of energy losses in an ac system that are not present under dc conditions.*

19.1 INTRODUCTION

The discussion of power in Chapter 14 included only the average or real power delivered to an ac network. We now examine the total power equation in a slightly different form and introduce two additional types of power: **apparent** and **reactive.**

19.2 GENERAL EQUATION

For any system such as in Fig. 19.1, the power delivered to a load at any instant is defined by the product of the applied voltage and the resulting current; that is,

$$p = vi$$

In this case, since v and i are sinusoidal quantities, let us establish a general case where

$$v = V_m \sin(\omega t + \theta)$$

and

$$i = I_m \sin \omega t$$

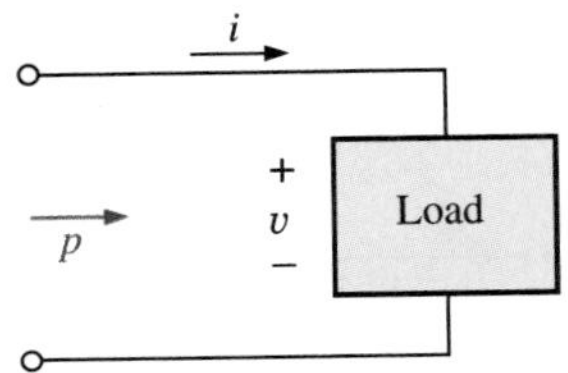

FIG. 19.1
Defining the power delivered to a load.

The chosen v and i include all possibilities because, if the load is purely resistive, $\theta = 0°$. If the load is purely inductive or capacitive, $\theta = 90°$ or $\theta = -90°$, respectively. For a network that is primarily inductive, θ is positive (v leads i). For a network that is primarily capacitive, θ is negative (i leads v).

Substituting the above equations for v and i into the power equation results in

$$p = V_m I_m \sin \omega t \sin(\omega t + \theta)$$

If we now apply a number of trigonometric identities, the following form for the power equation results:

$$p = VI \cos \theta(1 - \cos 2\omega t) + VI \sin \theta(\sin 2\omega t) \tag{19.1}$$

where V and I are the rms values. The conversion from peak values V_m and I_m to rms values resulted from the operations performed using the trigonometric identities.

It would appear initially that nothing has been gained by putting the equation in this form. However, the usefulness of the form of Eq. (19.1) is demonstrated in the following sections. The derivation of Eq. (19.1) from the initial form appears as an assignment at the end of the chapter.

If Eq. (19.1) is expanded to the form

$$p = \underbrace{VI \cos \theta}_{\text{Average}} - \underbrace{VI \cos \theta}_{\text{Peak}} \cos \underbrace{2\omega t}_{2x} + \underbrace{VI \sin \theta}_{\text{Peak}} \sin \underbrace{2\omega t}_{2x}$$

there are two obvious points that can be made. First, the average power still appears as an isolated term that is time independent. Second, both terms that follow vary at a frequency twice that of the applied voltage or current, with peak values having a very similar format.

In an effort to ensure completeness and order in presentation, each basic element (R, L, and C) is treated separately.

19.3 RESISTIVE CIRCUIT

For a purely resistive circuit (such as that in Fig. 19.2), v and i are in phase, and $\theta = 0°$, as appearing in Fig. 19.3. Substituting $\theta = 0°$ into Eq. (19.1), we obtain

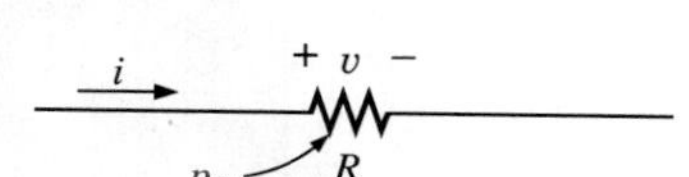

FIG. 19.2
Determining the power delivered to a purely resistive load.

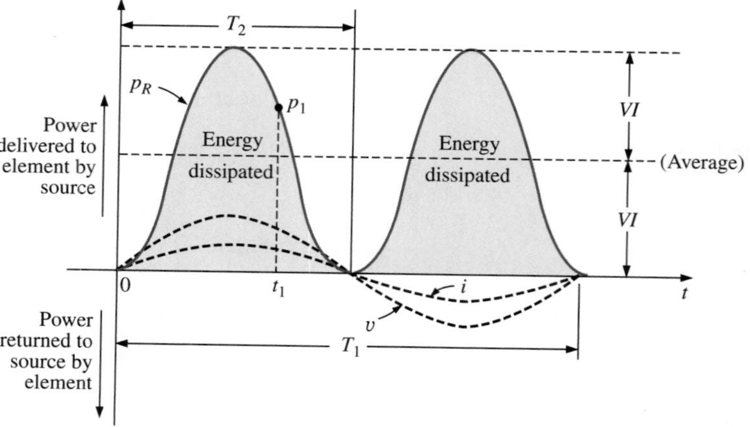

FIG. 19.3
Power versus time for a purely resistive load.

$$\begin{aligned} p_R &= VI\cos(0°)(1 - \cos 2\omega t) + VI\sin(0°)\sin 2\omega t \\ &= VI\,(1 - \cos 2\omega t) + 0 \end{aligned}$$

or

$$\boxed{p_R = VI - VI\cos 2\omega t} \qquad \textbf{(19.2)}$$

where VI is the average or dc term and $-VI\cos 2\omega t$ is a negative cosine wave with twice the frequency of either input quantity (v or i) and a peak value of VI. The plot in Fig. 19.3 has the same characteristics as obtained in Fig. 14.30.

Note that

$$T_1 = \text{period of input quantities}$$
$$T_2 = \text{period of power curve } p_R$$

In addition, the power curve passes through two cycles about its average value of VI for each cycle of either v or i ($T_1 = 2T_2$ or $f_2 = 2f_1$). Consider also that since the peak and average values of the power curve are the same, the curve is always above the horizontal axis. This indicates that

the total power delivered to a resistor will be dissipated in the form of heat.

The power returned to the source is represented by the portion of the curve below the axis, which is zero in this case. The power dissipated by the resistor at any instant of time t_1 can be found by simply substituting the time t_1 into Eq. (19.2) to find p_1, as indicated in Fig. 19.3. The **average (real) power** from Eq. (19.2), or Fig. 19.3, is VI; or, as a summary,

$$\boxed{P = VI = \frac{V_m I_m}{2} = I^2R = \frac{V^2}{R}} \qquad \text{(watts, W)} \qquad \textbf{(19.3)}$$

as derived in Chapter 14.

The energy dissipated by the resistor (W_R) over one full cycle of the applied voltage is the area under the power curve in Fig. 19.3. It can be found using the following equation:

$$W = Pt$$

where P is the average value and t is the period of the applied voltage; that is,

$$\boxed{W_R = VIT_1} \qquad \text{(joules, J)} \qquad \textbf{(19.4)}$$

or, since $T_1 = 1/f_1$,

$$\boxed{W_R = \frac{VI}{f_1}} \qquad \text{(joules, J)} \qquad \textbf{(19.5)}$$

EXAMPLE 19.1 For the resistive circuit in Fig. 19.4,

a. Find the instantaneous power delivered to the resistor at times t_1 through t_6.
b. Plot the results of part (a) for one full period of the applied voltage.
c. Find the average value of the curve of part (b) and compare the level to that determined by Eq. (19.3).

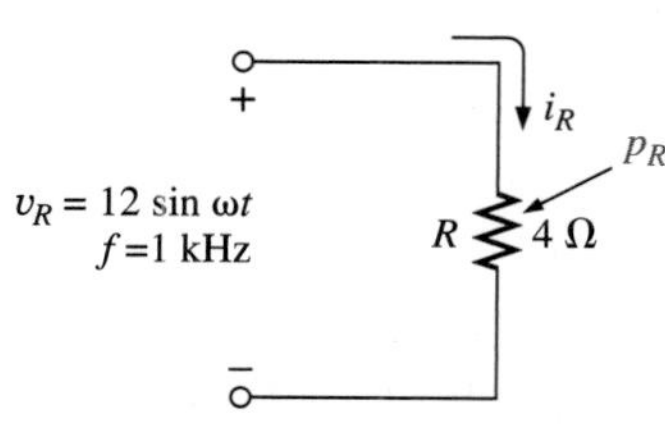

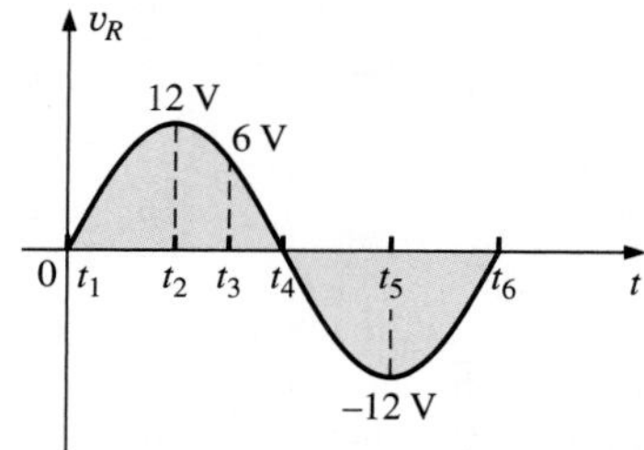

FIG. 19.4
Example 19.1.

d. Find the energy dissipated by the resistor over one full period of the applied voltage.

Solutions:

a. t_1: $v_R = 0\text{ V}$ and $p_R = v_R\, i_R = \mathbf{0\ W}$
t_2: $v_R = 12\text{ V}$ and $i_R = 12\text{ V}/4\ \Omega = \mathbf{3\ A}$
$p_R = v_R i_R = (12\text{ V})(3\text{ A}) = \mathbf{36\ W}$
t_3: $v_R = 6\text{ V}$ and $i_R = 6\text{ V}/4\ \Omega = \mathbf{1.5\ A}$
$p_R = v_R\, i_R = (6\text{ V})(1.5\text{ A}) = \mathbf{9\ W}$
t_4: $v_R = 0\text{ V}$ and $p_R = v_R\, i_R = \mathbf{0\ W}$
t_5: $v_R = -12\text{ V}$ and $i_R = -12\text{ V}/4\ \Omega = \mathbf{-3\ A}$
$p_R\ v_R\, i_R = (-12\text{ V})(-3\text{ A}) = \mathbf{36\ W}$
t_6: $v_R = 0\text{ V}$ and $p_R = v_R\, i_R = \mathbf{0\ W}$

b. The resulting plot of v_R, i_R, and p_R appears in Fig. 19.5.

c. The average value of the curve in Fig. 19.5 is 18 W, which is an exact match with that obtained using Eq. (19.3). That is,

$$P = \frac{V_m I_m}{2} = \frac{(12\text{ V})(3\text{ A})}{2} = \mathbf{18\ W}$$

d. The area under the curve is determined by Eq. (19.5):

$$W_R = \frac{VI}{f_1} = \frac{V_m I_m}{2 f_1} = \frac{(12\text{ V})(3\text{ A})}{2(1\text{ kHz})} = \mathbf{18\ mJ}$$

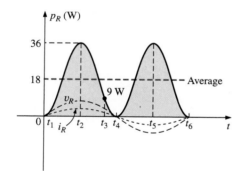

FIG. 19.5
Power curve for Example 19.1.

19.4 APPARENT POWER

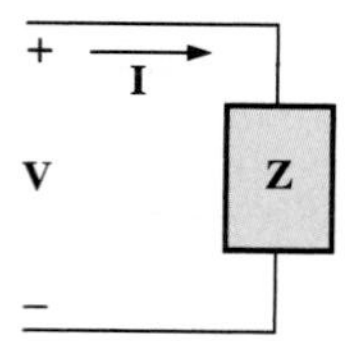

FIG. 19.6
Defining the apparent power to a load.

From our analysis of dc networks (and resistive elements above), it would seem *apparent* that the power delivered to the load in Fig. 19.6 is determined by the product of the applied voltage and current, with no concern for the components of the load; that is, $P = VI$. However, we found in Chapter 14 that the power factor ($\cos \theta$) of the load has a pronounced effect on the power dissipated, less pronounced for more reactive loads. Although the product of the voltage and current is not always the power delivered, it is a power rating of significant usefulness in the description and analysis of sinusoidal ac networks and in the maximum rating of a number of electrical components and systems. It is called the **apparent power** and is represented symbolically by S.* Since it is simply the product of voltage and current, its units are *volt-amperes* (VA). Its magnitude is determined by

*Prior to 1968, the symbol for apparent power was the more descriptive P_a.

$$\boxed{S = VI} \quad \text{(volt-amperes, VA)} \tag{19.6}$$

or, since
$$V = IZ \quad \text{and} \quad I = \frac{V}{Z}$$

then
$$\boxed{S = I^2Z} \quad \text{(VA)} \tag{19.7}$$

and
$$\boxed{S = \frac{V^2}{Z}} \quad \text{(VA)} \tag{19.8}$$

The average power to the load in Fig. 19.4 is

$$P = VI\cos\theta$$

However,
$$S = VI$$

Therefore,
$$\boxed{P = S\cos\theta} \quad \text{(W)} \tag{19.9}$$

and the power factor of a system F_p is

$$\boxed{F_p = \cos\theta = \frac{P}{S}} \quad \text{(unitless)} \tag{19.10}$$

The power factor of a circuit, therefore, is the ratio of the average power to the apparent power. For a purely resistive circuit, we have

$$P = VI = S$$

and
$$F_p = \cos\theta = \frac{P}{S} = 1$$

In general, power equipment is rated in volt-amperes (VA) or in kilovolt-amperes (kVA) and not in watts. By knowing the volt-ampere rating and the rated voltage of a device, we can readily determine the *maximum* current rating. For example, a device rated at 10 kVA at 200 V has a maximum current rating of $I = 10{,}000\text{ VA}/200\text{ V} = 50\text{ A}$ when operated under rated conditions. The volt-ampere rating of a piece of equipment is equal to the wattage rating only when the F_p is 1. It is therefore a maximum power dissipation rating. This condition exists only when the total impedance of a system $Z \angle\theta$ is such that $\theta = 0°$.

The exact current demand of a device, when used under normal operating conditions, can be determined if the wattage rating and power factor are given instead of the volt-ampere rating. However, the power factor is sometimes not available, or it may vary with the load.

The reason for rating some electrical equipment in kilovolt-amperes rather than in kilowatts can be described using the configuration in Fig. 19.7. The load has an apparent power rating of 10 kVA and a current rating of 50 A at the applied voltage, 200 V. As indicated, the current demand of 70 A is above the rated value and could damage the load element, yet the reading on the wattmeter is relatively low since the load is highly reactive. In other words, the wattmeter reading is an indication of the watts dissipated and may not reflect the magnitude of the current drawn. Theoretically, if the load were purely reactive, the wattmeter reading would be zero even if the load was being damaged by a high current level.

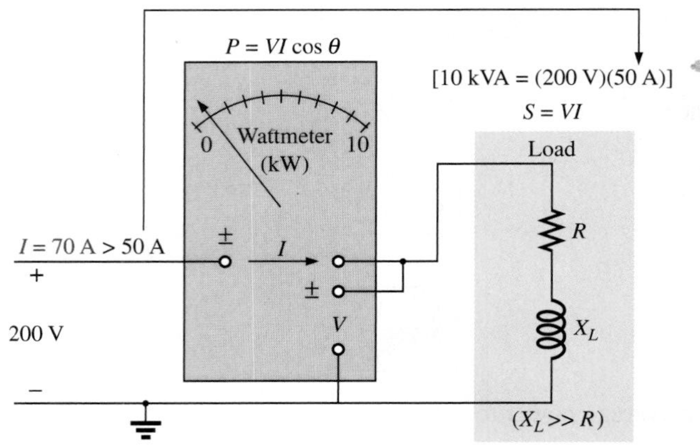

FIG. 19.7
Demonstrating the reason for rating a load in kVA rather than kW.

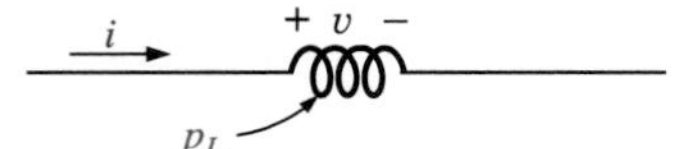

FIG. 19.8
Defining the power level for a purely inductive load.

19.5 INDUCTIVE CIRCUIT AND REACTIVE POWER

For a purely inductive circuit (such as that in Fig. 19.8), v leads i by 90°, as shown in Fig. 19.9. Therefore, in Eq. (19.1), $\theta = 90°$. Substituting $\theta = 90°$ into Eq. (19.1) yields

$$p_L = VI\cos(90°)(1 - \cos 2\omega t) + VI\sin(90°)(\sin 2\omega t)$$
$$= 0 + VI\sin 2\omega t$$

or

$$\boxed{p_L = VI\sin 2\omega t} \qquad \textbf{(19.11)}$$

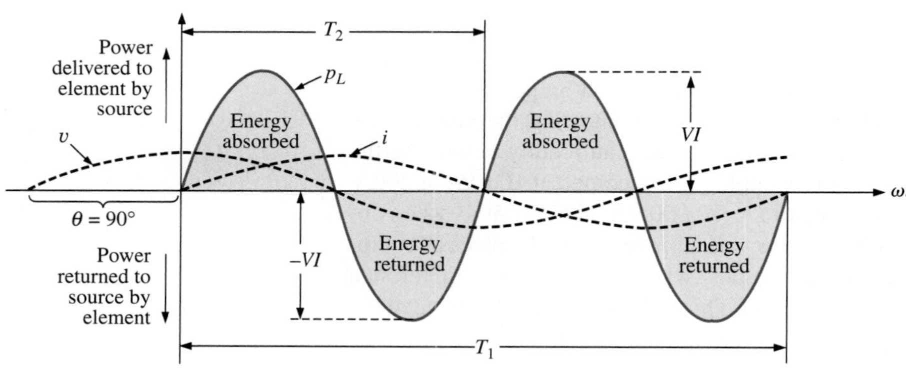

FIG. 19.9
The power curve for a purely inductive load.

where $VI\sin 2\omega t$ is a sine wave with twice the frequency of either input quantity (v or i) and a peak value of VI. Note the absence of an average or constant term in the equation.

Plotting the waveform for p_L (Fig. 19.9), we obtain

$$T_1 = \text{period of either input quantity}$$
$$T_2 = \text{period of } p_L \text{ curve}$$

Note that over one full cycle of p_L (T_2), the area above the horizontal axis in Fig. 19.9 is exactly equal to that below the axis. This indicates that over a full cycle of p_L, the power delivered by the source to the inductor is exactly equal to that returned to the source by the inductor.

The net flow of power to the pure (ideal) inductor is zero over a full cycle, and no energy is lost in the transaction.

The power absorbed or returned by the inductor at any instant of time t_1 can be found simply by substituting t_1 into Eq. (19.11). The peak value of the curve VI is defined as the **reactive power** associated with a pure inductor.

In general, the reactive power associated with any circuit is defined to be $VI \sin \theta$, a factor appearing in the second term of Eq. (19.1). Note that it is the peak value of that term of the total power equation that produces no net transfer of energy. The symbol for reactive power is Q, and its unit of measure is the *volt-ampere reactive* (VAR).* The Q is derived from the quadrature (90°) relationship between the various powers, to be discussed in detail in a later section. Therefore,

$$\boxed{Q_L = VI \sin \theta} \qquad \text{(volt-ampere reactive, VAR)} \qquad \textbf{(19.12)}$$

where θ is the phase angle between V and I.

For the inductor,

$$\boxed{Q_L = VI} \qquad \text{(VAR)} \qquad \textbf{(19.13)}$$

or, since $V = IX_L$ or $I = V/X_L$,

$$\boxed{Q_L = I^2 X_L} \qquad \text{(VAR)} \qquad \textbf{(19.14)}$$

or

$$\boxed{Q_L = \frac{V^2}{X_L}} \qquad \text{(VAR)} \qquad \textbf{(19.15)}$$

The apparent power associated with an inductor is $S = VI$, and the average power is $P = 0$, as noted in Fig. 19.9. The power factor is therefore

$$F_p = \cos \theta = \frac{P}{S} = \frac{0}{VI} = 0$$

If the average power is zero, and the energy supplied is returned within one cycle, why is reactive power of any significance? The reason is not obvious but can be explained using the curve in Fig. 19.9. At every instant of time along the power curve that the curve is above the axis (positive), energy must be supplied to the inductor, even though it will be returned during the negative portion of the cycle. This power requirement during the positive portion of the cycle requires that the generating plant provide this energy during that interval. Therefore, the effect of reactive elements such as the inductor can be to raise the power requirement of the generating plant, even though the reactive power is not dissipated but simply "borrowed." The increased power demand during these intervals is a cost factor that must be passed on to the industrial consumer. In fact, most larger users of electrical energy pay for the apparent power demand rather than the watts dissipated since the volt-amperes used are sensitive to the reactive power requirement (see Section 19.7). In other words, the closer the power factor of an industrial outfit is to 1, the more efficient the plant's operation since it is limiting its use of "borrowed" power.

---

*Prior to 1968, the symbol for reactive power was the more descriptive P_q.

The energy stored by the inductor during the positive portion of the cycle (Fig. 19.9) is equal to that returned during the negative portion and can be determined using the following equation:

$$W = Pt$$

where P is the average value for the interval and t is the associated interval of time.

Recall from Chapter 14 that the average value of the positive portion of a sinusoid equals 2(peak value/π) and $t = T_2/2$. Therefore,

$$W_L = \left(\frac{2VI}{\pi}\right) \times \left(\frac{T_2}{2}\right)$$

and

$$W_L = \frac{VIT_2}{\pi} \quad \text{(J)} \qquad \textbf{(19.16)}$$

or, since $T_2 = 1/f_2$, where f_2 is the frequency of the p_L curve, we have

$$W_L = \frac{VI}{\pi f_2} \quad \text{(J)} \qquad \textbf{(19.17)}$$

Since the frequency f_2 of the power curve is twice that of the input quantity, if we substitute the frequency f_1 of the input voltage or current, Eq. (19.17) becomes

$$W_L = \frac{VI}{\pi(2f_1)} = \frac{VI}{\omega_1}$$

However,

$$V = IX_L = I\omega_1 L$$

so that

$$W_L = \frac{(I\omega_1 L)I}{\omega_1}$$

and

$$W_L = LI^2 \quad \text{(J)} \qquad \textbf{(19.18)}$$

providing an equation for the energy stored or released by the inductor in one half-cycle of the applied voltage in terms of the inductance and rms value of the current squared.

EXAMPLE 19.2 For the inductive circuit in Fig. 19.10,

a. Find the instantaneous power level for the inductor at times t_1 through t_5.

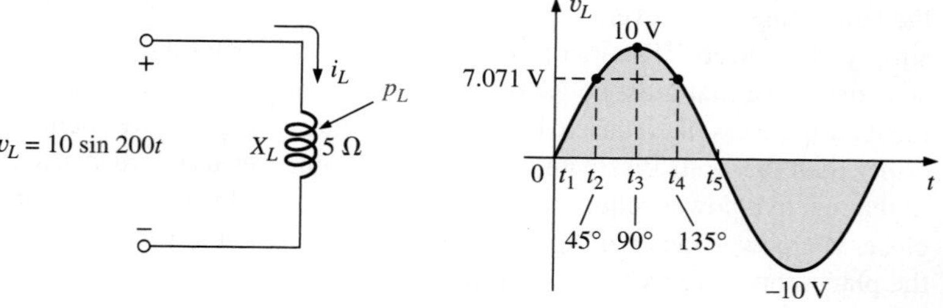

FIG. 19.10
Example 19.2.

b. Plot the results of part (a) for one full period of the applied voltage.
c. Find the average value of the curve of part (b) over one full cycle of the applied voltage and compare the peak value of each pulse with the value determined by Eq. (19.13).
d. Find the energy stored or released for any one pulse of the power curve.

Solutions:

a. t_1: $v_L = 0\text{ V}, p_L = v_L i_L = \mathbf{0\ W}$

$$t_2: v_L = 7.071\text{ V},\ i_L = \frac{V_m}{X_L}\sin(\alpha - 90°)$$

$$= \frac{10\text{ V}}{5\ \Omega}\sin(\alpha - 90°) = 2\sin(\alpha - 90°)$$

At $\alpha = 45°$, $i_L = 2\sin(45° - 90°) = 2\sin(-45°) = -1.414\text{ A}$

$$p_L = v_L i_L = (7.071\text{ V})(-1.414\text{ A}) = \mathbf{-10\ W}$$

t_3: $i_L = 0\text{ A}, p_L = v_L i_L = \mathbf{0\ W}$

t_4: $v_L = 7.071\text{ V}, i_L = 2\sin(\alpha - 90°) = 2\sin(135° - 90°)$

$$= 2\sin 45° = 1.414\text{ A}$$

$$p_L = v_L i_L = (7.071\text{ V})(1.414\text{ A}) = \mathbf{+10\ W}$$

t_5: $v_L = 0\text{ V}, p_L = v_L i_L = \mathbf{0\ W}$

b. The resulting plot of v_L, i_L, and p_L appears in Fig. 19.11.

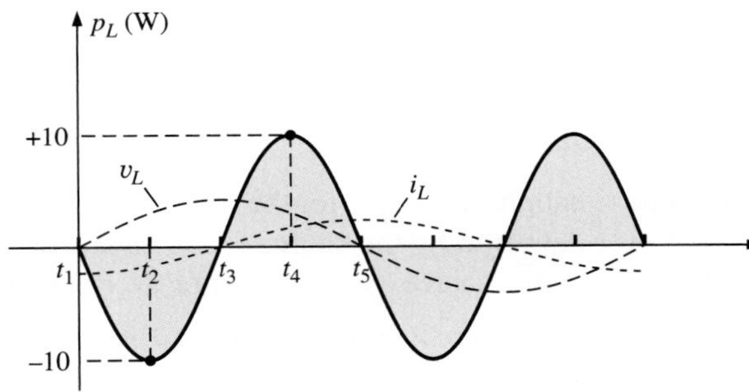

FIG. 19.11
Power curve for Example 19.2.

c. The average value for the curve in Fig. 19.11 is 0 W over one full cycle of the applied voltage. The peak value of the curve is 10 W which compares directly with that obtained from the product

$$VI = \frac{V_m I_m}{2} = \frac{(10\text{ V})(2\text{ A})}{2} = \mathbf{10\ W}$$

d. The energy stored or released during each pulse of the power curve is:

$$W_L = \frac{VI}{\omega_1} = \frac{V_m I_m}{2\,\omega_1} = \frac{(10\text{ V})(2\text{ A})}{2\,(200\text{ rad/s})} = \mathbf{50\ mJ}$$

19.6 CAPACITIVE CIRCUIT

For a purely capacitive circuit (such as that in Fig. 19.12), i leads v by 90°, as shown in Fig. 19.13. Therefore, in Eq. (19.1), $\theta = -90°$. Substituting $\theta = -90°$ into Eq. (19.1), we obtain

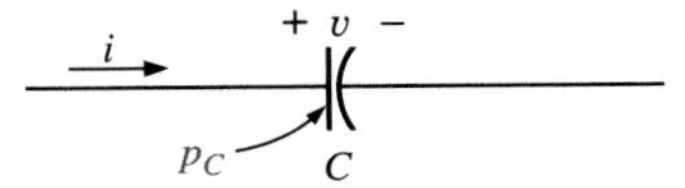

FIG. 19.12
Defining the power level for a purely capacitive load.

$$p_C = VI\cos(-90°)(1 - \cos 2\omega t) + VI\sin(-90°)(\sin 2\omega t)$$
$$= 0 - VI\sin 2\omega t$$

or

$$\boxed{p_C = -VI\sin 2\omega t} \qquad (19.19)$$

where $-VI\sin 2\omega t$ is a negative sine wave with twice the frequency of either input (v or i) and a peak value of VI. Again, note the absence of an average or constant term.

Plotting the waveform for p_C (Fig. 19.13) gives us

$$T_1 = \text{period of either input quantity}$$
$$T_2 = \text{period of } p_C \text{ curve}$$

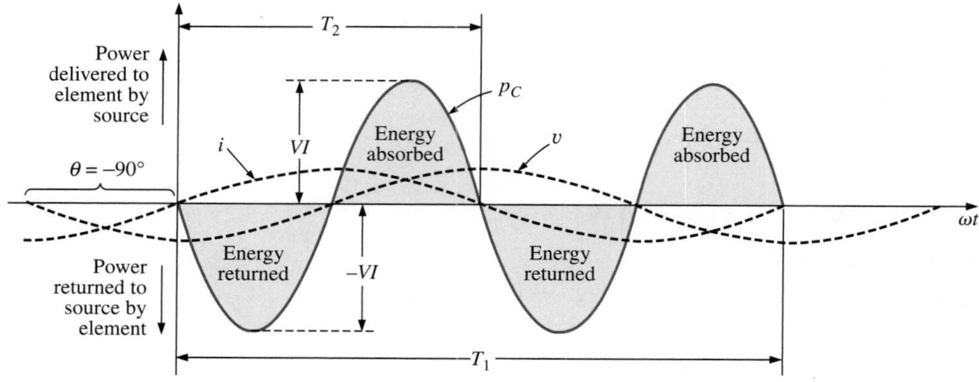

FIG. 19.13
The power curve for a purely capacitive load.

Note that the same situation exists here for the p_C curve as existed for the p_L curve. The power delivered by the source to the capacitor is exactly equal to that returned to the source by the capacitor over one full cycle.

The net flow of power to the pure (ideal) capacitor is zero over a full cycle,

and no energy is lost in the transaction. The power absorbed or returned by the capacitor at any instant of time t_1 can be found by substituting t_1 into Eq. (19.19).

The reactive power associated with the capacitor is equal to the peak value of the p_C curve, as follows:

$$\boxed{Q_C = VI} \qquad \text{(VAR)} \qquad (19.20)$$

But, since $V = IX_C$ and $I = V/X_C$, the reactive power to the capacitor can also be written

$$\boxed{Q_C = I^2X_C} \qquad \text{(VAR)} \qquad (19.21)$$

and

$$\boxed{Q_C = \frac{V^2}{X_C}} \qquad \text{(VAR)} \qquad (19.22)$$

The apparent power associated with the capacitor is

$$\boxed{S = VI} \qquad \text{(VA)} \qquad (19.23)$$

and the average power is $P = 0$, as noted from Eq. (19.19) or Fig. 19.13. The power factor is, therefore,

$$F_p = \cos\theta = \frac{P}{S} = \frac{0}{VI} = 0$$

The energy stored by the capacitor during the positive portion of the cycle (Fig. 19.13) is equal to that returned during the negative portion and can be determined using the equation $W = Pt$.

Proceeding in a manner similar to that used for the inductor, we can show that

$$\boxed{W_C = \frac{VIT_2}{\pi}} \quad \text{(J)} \qquad \textbf{(19.24)}$$

or, since $T_2 = 1/f_2$, where f_2 is the frequency of the p_C curve,

$$\boxed{W_C = \frac{VI}{\pi f_2}} \quad \text{(J)} \qquad \textbf{(19.25)}$$

In terms of the frequency f_1 of the input quantities υ and i,

$$W_C = \frac{VI}{\pi(2f_1)} = \frac{VI}{\omega_1} = \frac{V(V\omega_1 C)}{\omega_1}$$

and

$$\boxed{W_C = CV^2} \quad \text{(J)} \qquad \textbf{(19.26)}$$

providing an equation for the energy stored or released by the capacitor in one half-cycle of the applied voltage in terms of the capacitance and rms value of the voltage squared.

19.7 THE POWER TRIANGLE

The three quantities **average power, apparent power,** and **reactive power** can be related in the vector domain by

$$\boxed{\mathbf{S} = \mathbf{P} + \mathbf{Q}} \qquad \textbf{(19.27)}$$

with

$$\mathbf{P} = P\angle 0^\circ \quad \mathbf{Q}_L = Q_L\angle 90^\circ \quad \mathbf{Q}_C = Q_C\angle -90^\circ$$

For an inductive load, the *phasor power* **S,** as it is often called, is defined by

$$\mathbf{S} = P + j\,Q_L$$

as shown in Fig. 19.14.

The 90° shift in Q_L from P is the source of another term for reactive power: *quadrature power.*

For a capacitive load, the phasor power **S** is defined by

$$\mathbf{S} = P - j\,Q_C$$

as shown in Fig. 19.15.

If a network has both capacitive and inductive elements, the reactive component of the power triangle will be determined by the *difference* between the reactive power delivered to each. If $Q_L > Q_C$, the resultant

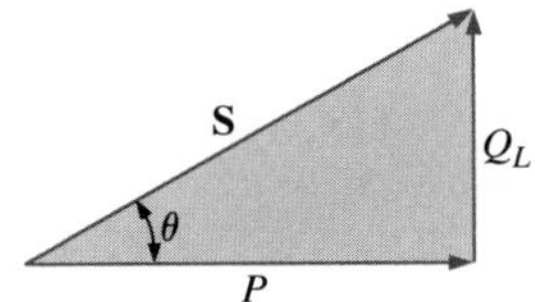

FIG. 19.14
Power diagram for inductive loads.

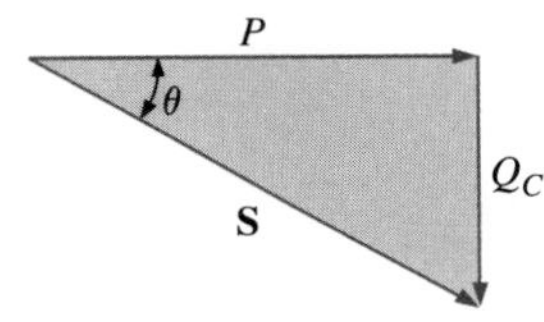

FIG. 19.15
Power diagram for capacitive loads.

power triangle will be similar to Fig. 19.14. If $Q_C > Q_L$, the resultant power triangle will be similar to Fig. 19.15.

That the total reactive power is the difference between the reactive powers of the inductive and capacitive elements can be demonstrated by considering Eqs. (19.11) and (19.19). Using these equations, the reactive power delivered to each reactive element has been plotted for a series *L-C* circuit on the same set of axes in Fig. 19.16. The reactive elements were chosen such that $X_L > X_C$. Note that the power curve for each is exactly 180° out of phase. The curve for the resultant reactive power is therefore determined by the algebraic resultant of the two at each instant of time. Since the reactive power is defined as the peak value, the reactive component of the power triangle is as indicated in Fig. 19.16: $I^2 (X_L - X_C)$.

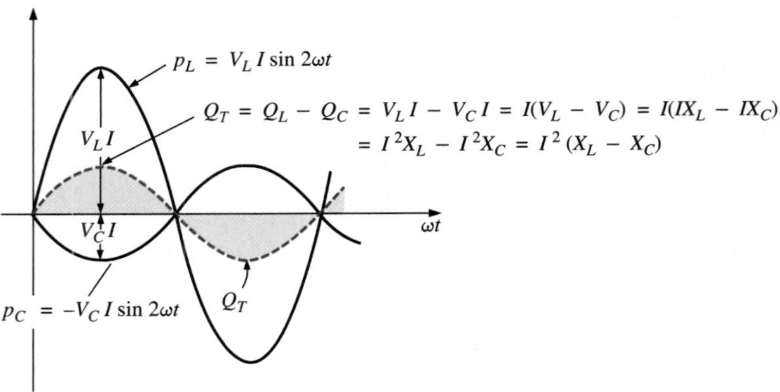

FIG. 19.16
Demonstrating why the net reactive power is the difference between that delivered to inductive and capacitive elements.

An additional verification can be derived by first considering the impedance diagram of a series *R-L-C* circuit (Fig. 19.17). If we multiply each radius vector by the current squared (I^2), we obtain the results shown in Fig. 19.18, which is the power triangle for a predominantly inductive circuit.

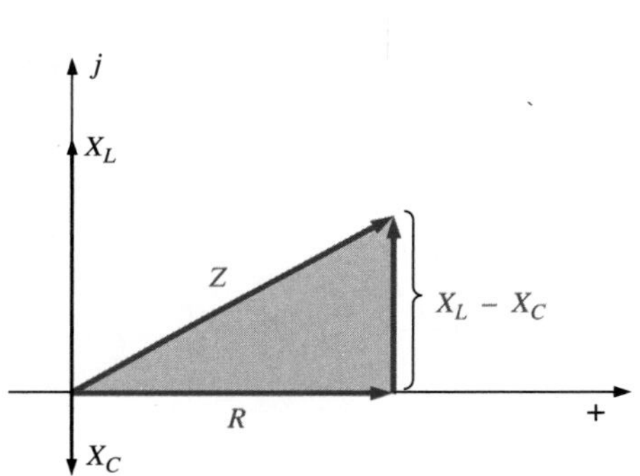

FIG. 19.17
Impedance diagram for a series R-L-C circuit.

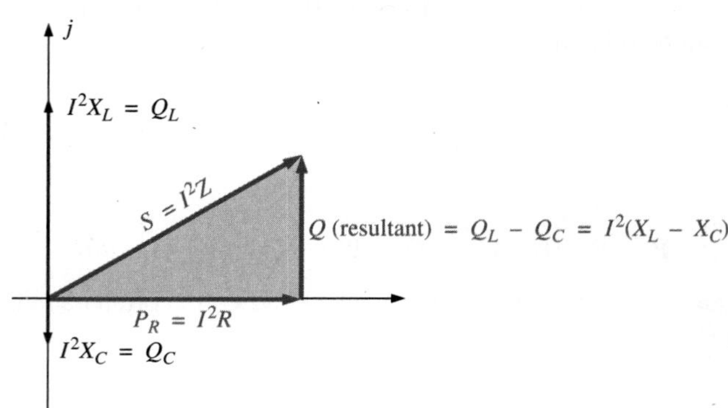

FIG. 19.18
The result of multiplying each vector in Fig. 19.17 by I^2 for a series R-L-C circuit.

Since the reactive power and average power are always angled 90° to each other, the three powers are related by the Pythagorean theorem; that is,

$$S^2 = P^2 + Q^2 \tag{19.28}$$

Therefore, the third power can always be found if the other two are known.

It is particularly interesting that the equation

$$\boxed{\mathbf{S} = \mathbf{VI}^*} \tag{19.29}$$

will provide the vector form of the apparent power of a system. Here, **V** is the voltage across the system, and $\mathbf{I}^*$ is the complex conjugate of the current.

Consider, for example, the simple *R-L* circuit in Fig. 19.19, where

$$\mathbf{I} = \frac{\mathbf{V}}{\mathbf{Z}_T} = \frac{10\ \text{V}\angle 0°}{3\ \Omega + j\,4\ \Omega} = \frac{10\ \text{V}\angle 0°}{5\ \Omega\ \angle 53.13°} = 2\ \text{A}\ \angle -53.13°$$

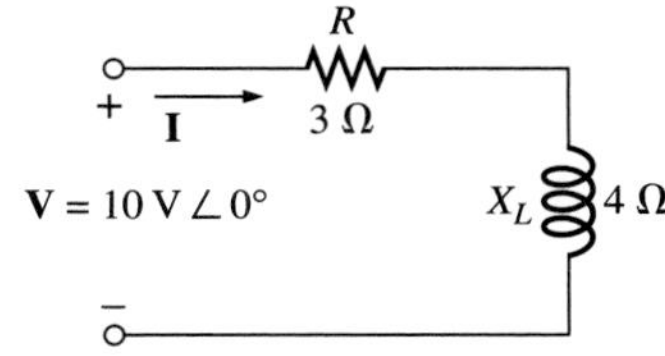

FIG. 19.19
Demonstrating the validity of Eq. (19.29).

The real power (the term *real* being derived from the positive real axis of the complex plane) is

$$P = I^2R = (2\ \text{A})^2(3\ \Omega) = 12\ \text{W}$$

and the reactive power is

$$Q_L = I^2X_L = (2\ \text{A})^2(4\ \Omega) = 16\ \text{VAR}\ (L)$$

with $\mathbf{S} = P + j\,Q_L = 12\ \text{W} + j\,16\ \text{VAR}\ (L) = 20\ \text{VA}\ \angle 53.13°$

as shown in Fig. 19.20. Applying Eq. (19.29) yields

$$\mathbf{S} = \mathbf{VI}^* = (10\ \text{V}\ \angle 0°)(2\text{A}\ \angle +53.13°) = 20\ \text{VA}\ \angle 53.13°$$

as obtained above.

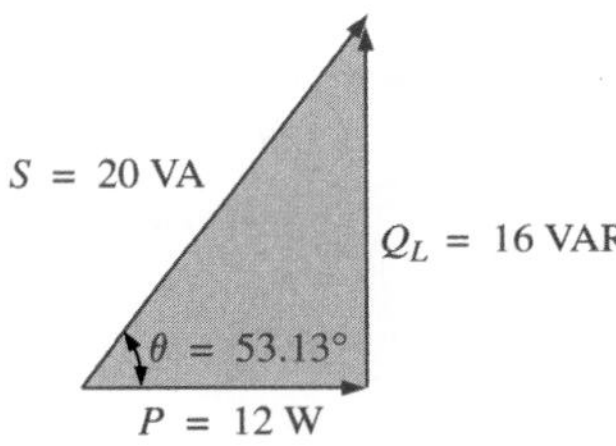

FIG. 19.20
The power triangle for the circuit in Fig. 19.19.

The angle θ associated with **S** and appearing in Figs. 19.14, 19.15, and 19.20 is the power-factor angle of the network. Since

$$P = VI\cos\theta$$

or

$$P = S\cos\theta$$

then

$$\boxed{F_p = \cos\theta = \frac{P}{S}} \tag{19.30}$$

19.8 THE TOTAL *P*, *Q*, AND *S*

The total number of watts, volt-amperes reactive, and volt-amperes, and the power factor of any system can be found using the following procedure:

1. ***Find the real power and reactive power for each branch of the circuit.***
2. ***The total real power of the system (P_T) is then the sum of the average power delivered to each branch.***
3. ***The total reactive power (Q_T) is the difference between the reactive power of the inductive loads and that of the capacitive loads.***
4. ***The total apparent power is*** $S_T = \sqrt{P_T^2 + Q_T^2}$.
5. ***The total power factor is*** P_T/S_T.

There are two important points in the above tabulation. First, the total apparent power must be determined from the total average and reactive powers and *cannot* be determined from the apparent powers of each branch. Second, and more important, it is *not necessary* to consider the series-parallel arrangement of branches. In other words, the total real, reactive, or apparent power is independent of whether the loads are in series, parallel, or series-parallel. The following examples demonstrate the relative ease with which all of the quantities of interest can be found.

EXAMPLE 19.3 Find the total number of watts, volt-amperes reactive, and volt-amperes, and the power factor F_p of the network in Fig. 19.21. Draw the power triangle and find the current in phasor form.

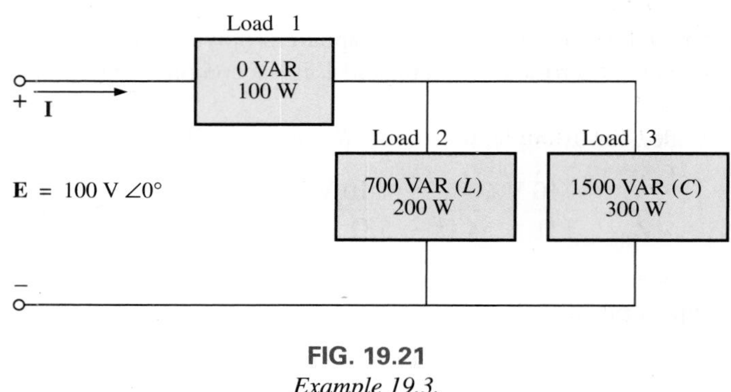

FIG. 19.21
Example 19.3.

Solution: Construct a table such as shown in Table 19.1.

TABLE 19.1

Load	W	VAR	VA
1	100	0	100
2	200	700 (L)	$\sqrt{(200)^2 + (700)^2} = 728.0$
3	300	1500 (C)	$\sqrt{(300)^2 + (1500)^2} = 1529.71$
	$P_T =$ **600** Total power dissipated	$Q_T =$ **800 (C)** Resultant reactive power of network	$S_T = \sqrt{(600)^2 + (800)^2} =$ **1000** (Note that $S_T \neq$ sum of each branch: $1000 \neq 100 + 728 + 1529.71$)

Thus,

$$F_p = \frac{P_T}{S_T} = \frac{600\text{ W}}{1000\text{ VA}} = \mathbf{0.6\ leading}\ (C)$$

The power triangle is shown in Fig. 19.22.

Since $S_T = VI = 1000$ VA, $I = 1000$ VA/100 V $= 10$ A; and since θ of $\cos\theta = F_p$ is the angle between the input voltage and current:

$$\mathbf{I} = \mathbf{10\ A\ \angle\ +53.13^\circ}$$

The plus sign is associated with the phase angle since the circuit is predominantly capacitive.

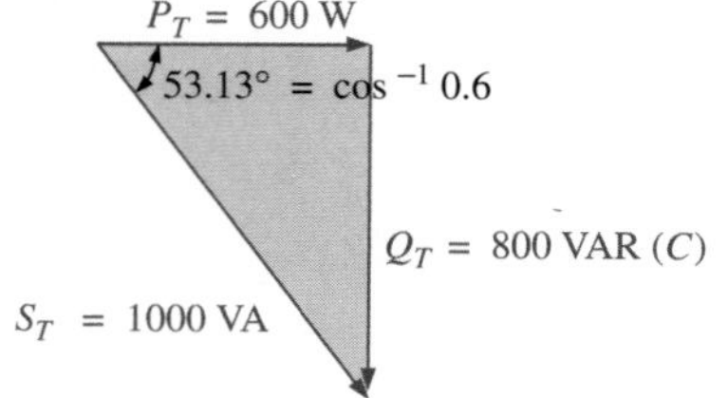

FIG. 19.22
Power triangle for Example 19.3.

EXAMPLE 19.4

a. Find the total number of watts, volt-amperes reactive, and volt-amperes, and the power factor F_p for the network in Fig. 19.23.
b. Sketch the power triangle.
c. Find the energy dissipated by the resistor over one full cycle of the input voltage if the frequency of the input quantities is 60 Hz.
d. Find the energy stored in, or returned by, the capacitor or inductor over one half-cycle of the power curve for each if the frequency of the input quantities is 60 Hz.

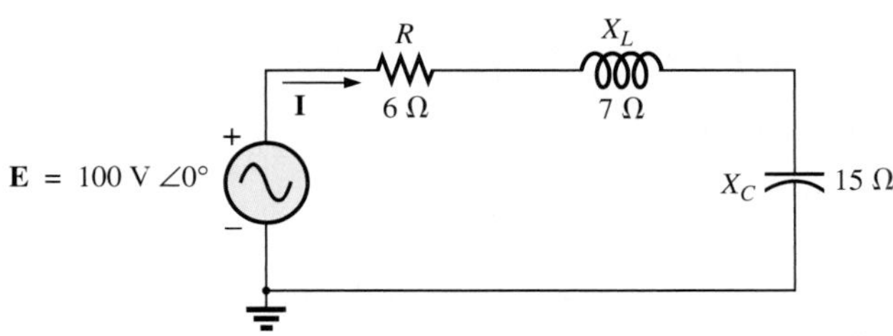

FIG. 19.23
Example 19.4.

Solutions:

a. $$\mathbf{I} = \frac{\mathbf{E}}{\mathbf{Z}_T} = \frac{100\text{ V}\angle 0°}{6\ \Omega + j\,7\ \Omega - j\,15\ \Omega} = \frac{100\text{ V}\angle 0°}{10\ \Omega\angle -53.13°} = 10\text{ A}\angle 53.13°$$

$$\mathbf{V}_R = (10\text{ A}\angle 53.13°)(6\ \Omega\angle 0°) = 60\text{ V}\angle 53.13°$$
$$\mathbf{V}_L = (10\text{ A}\angle 53.13°)(7\ \Omega\angle 90°) = 70\text{ V}\angle 143.13°$$
$$\mathbf{V}_C = (10\text{ A}\angle 53.13°)(15\ \Omega\angle -90°) = 150\text{ V}\angle -36.87°$$

$$P_T = EI\cos\theta = (100\text{ V})(10\text{ A})\cos 53.13° = \mathbf{600\text{ W}}$$
$$= I^2R = (10\text{ A})^2(6\ \Omega) = \mathbf{600\text{ W}}$$
$$= \frac{V_R^2}{R} = \frac{(60\text{ V})^2}{6} = \mathbf{600\text{ W}}$$

$$S_T = EI = (100\text{ V})(10\text{ A}) = \mathbf{1000\text{ VA}}$$
$$= I^2Z_T = (10\text{ A})^2(10\ \Omega) = \mathbf{1000\text{ VA}}$$
$$= \frac{E^2}{Z_T} = \frac{(100\text{ V})^2}{10\ \Omega} = \mathbf{1000\text{ VA}}$$

$$Q_T = EI\sin\theta = (100\text{ V})(10\text{ A})\sin 53.13° = \mathbf{800\text{ VAR}}$$
$$= Q_C - Q_L$$
$$= I^2(X_C - X_L) = (10\text{ A})^2(15\ \Omega - 7\ \Omega) = \mathbf{800\text{ VAR}}$$

$$Q_T = \frac{V_C^2}{X_C} - \frac{V_L^2}{X_L} = \frac{(150\text{ V})^2}{15\ \Omega} - \frac{(70\text{ V})^2}{7\ \Omega}$$
$$= 1500\text{ VAR} - 700\text{ VAR} = \mathbf{800\text{ VAR}}$$

$$F_p = \frac{P_T}{S_T} = \frac{600\text{ W}}{1000\text{ VA}} = \mathbf{0.6\ leading\ (C)}$$

b. The power triangle is as shown in Fig. 19.24.

c. $$W_R = \frac{V_RI}{f_1} = \frac{(60\text{ V})(10\text{ A})}{60\text{ Hz}} = \mathbf{10\text{ J}}$$

d. $$W_L = \frac{V_LI}{\omega_1} = \frac{(70\text{ V})(10\text{ A})}{(2\pi)(60\text{ Hz})} = \frac{700\text{ J}}{377} = \mathbf{1.86\text{ J}}$$

$$W_C = \frac{V_CI}{\omega_1} = \frac{(150\text{ V})(10\text{ A})}{377\text{ rad/s}} = \frac{1500\text{ J}}{377} = \mathbf{3.98\text{ J}}$$

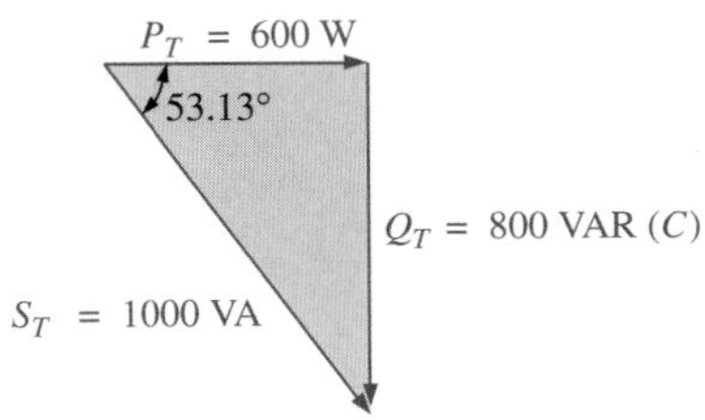

FIG. 19.24
Power triangle for Example 19.4.

EXAMPLE 19.5 For the system in Fig. 19.25,

a. Find the average power, apparent power, reactive power, and F_p for each branch.

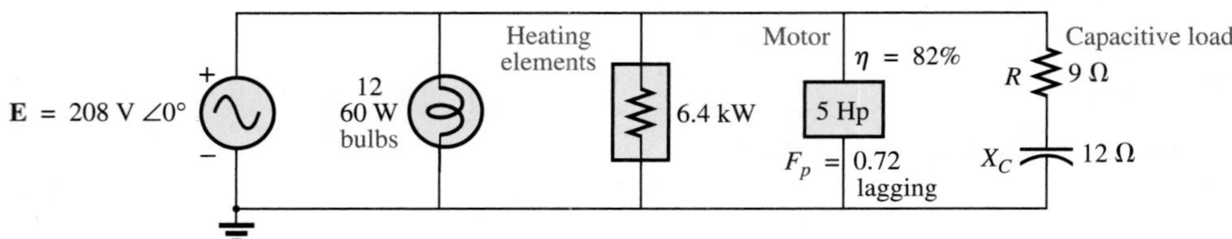

FIG. 19.25
Example 19.5.

b. Find the total number of watts, volt-amperes reactive, and volt-amperes, and the power factor of the system. Sketch the power triangle.
c. Find the source current I.

Solutions:

a. *Bulbs:*

Total dissipation of applied power

$$P_1 = 12(60\text{ W}) = \mathbf{720\text{ W}}$$
$$Q_1 = \mathbf{0\text{ VAR}}$$
$$S_1 = P_1 = \mathbf{720\text{ VA}}$$
$$F_{p_1} = \mathbf{1}$$

Heating elements:

Total dissipation of applied power

$$P_2 = \mathbf{6.4\text{ kW}}$$
$$Q_2 = \mathbf{0\text{ VAR}}$$
$$S_2 = P_2 = \mathbf{6.4\text{ kVA}}$$
$$F_{p_2} = \mathbf{1}$$

Motor:

$$\eta = \frac{P_o}{P_i} \rightarrow P_i = \frac{P_o}{\eta} = \frac{5(746\text{ W})}{0.82} = \mathbf{4548.78\text{ W}} = P_3$$
$$F_p = \mathbf{0.72\text{ lagging}}$$
$$P_3 = S_3\cos\theta \rightarrow S_3 = \frac{P_3}{\cos\theta} = \frac{4548.78\text{ W}}{0.72} = \mathbf{6317.75\text{ VA}}$$

Also, $\theta = \cos^{-1} 0.72 = 43.95°$, so that

$$Q_3 = S_3\sin\theta = (6317.75\text{ VA})(\sin 43.95°)$$
$$= (6317.75\text{ VA})(0.694) = \mathbf{4384.71\text{ VAR }(L)}$$

Capacitive load:

$$\mathbf{I} = \frac{\mathbf{E}}{\mathbf{Z}} = \frac{208\text{ V}\angle 0°}{9\ \Omega - j\,12\ \Omega} = \frac{208\text{ V}\angle 0°}{15\ \Omega\angle -53.13°} = 13.87\text{ A}\angle 53.13°$$
$$P_4 = I^2R = (13.87\text{ A})^2 \cdot 9\ \Omega = \mathbf{1731.39\text{ W}}$$
$$Q_4 = I^2X_C = (13.87\text{ A})^2 \cdot 12\ \Omega = \mathbf{2308.52\text{ VAR }(C)}$$
$$S_4 = \sqrt{P_4^2 + Q_4^2} = \sqrt{(1731.39\text{ W})^2 + (2308.52\text{ VAR})^2}$$
$$= \mathbf{2885.65\text{ VA}}$$
$$F_p = \frac{P_4}{S_4} = \frac{1731.39\text{ W}}{2885.65\text{ VA}} = \mathbf{0.6\text{ leading}}$$

b. $P_T = P_1 + P_2 + P_3 + P_4$
$= 720\text{ W} + 6400\text{ W} + 4548.78\text{ W} + 1731.39\text{ W}$
$= \mathbf{13{,}400.17\ W}$
$Q_T = \pm Q_1 \pm Q_2 \pm Q_3 \pm Q_4$
$= 0 + 0 + 4384.71\text{ VAR }(L) - 2308.52\text{ VAR }(C)$
$= \mathbf{2076.19\ VAR\ (L)}$
$S_T = \sqrt{P_T^2 + Q_T^2} = \sqrt{(13{,}400.17\text{ W})^2 + (2076.19\text{ VAR})^2}$
$= 13{,}560.06\text{ VA}$
$F_p = \dfrac{P_T}{S_T} = \dfrac{13.4\text{ kW}}{13{,}560.06\text{ VA}} = \mathbf{0.988\ lagging}$
$\theta = \cos^{-1} 0.988 = 8.89°$

Note Fig. 19.26.

c. $S_T = EI \rightarrow I = \dfrac{S_T}{E} = \dfrac{13{,}559.89\text{ VA}}{208\text{ V}} = 65.19\text{ A}$

Lagging power factor: **E** leads **I** by 8.89°, and

$$\mathbf{I} = \mathbf{65.19\ A\ \angle\ -8.89°}$$

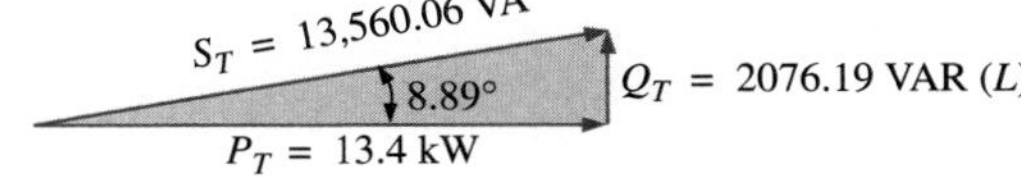

FIG. 19.26
Power triangle for Example 19.5.

EXAMPLE 19.6 An electrical device is rated 5 kVA, 100 V at a 0.6 power-factor lag. What is the impedance of the device in rectangular coordinates?

Solution:

$$S = EI = 5000\text{ VA}$$

Therefore,

$$I = \frac{5000\text{ VA}}{100\text{ V}} = 50\text{ A}$$

For $F_p = 0.6$, we have

$$\theta = \cos^{-1} 0.6 = 53.13°$$

Since the power factor is lagging, the circuit is predominantly inductive, and **I** lags **E**. Or, for $\mathbf{E} = 100\text{ V}\ \angle 0°$,

$$\mathbf{I} = 50\text{ A}\ \angle\ -53.13°$$

However,

$$\mathbf{Z}_T = \frac{\mathbf{E}}{\mathbf{I}} = \frac{100\text{ V}\ \angle 0°}{50\text{ A}\ \angle -53.13°} = 2\ \Omega\ \angle 53.13° = \mathbf{1.2\ \Omega + \mathit{j}\ 1.6\ \Omega}$$

which is the impedance of the circuit in Fig. 19.27.

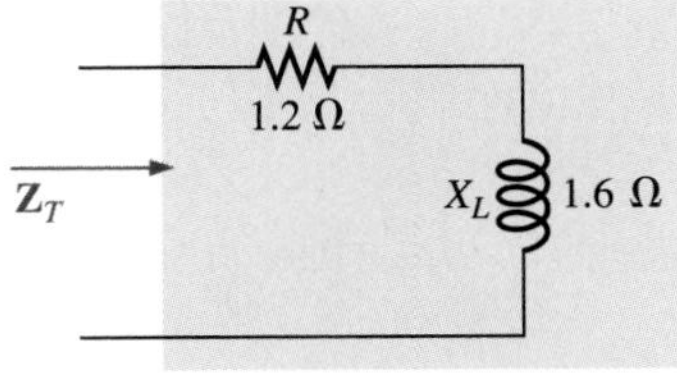

FIG. 19.27
Example 19.6.

19.9 POWER-FACTOR CORRECTION

The design of any power transmission system is very sensitive to the magnitude of the current in the lines as determined by the applied loads. Increased currents result in increased power losses (by a squared factor since $P = I^2R$) in the transmission lines due to the resistance of the lines. Heavier currents also require larger conductors, increasing the amount of copper needed for the system, and, quite obviously, they require increased generating capacities by the utility company.

Every effort must therefore be made to keep current levels at a minimum. Since the line voltage of a transmission system is fixed, the apparent power is directly related to the current level. In turn, the smaller the net apparent power, the smaller the current drawn from the supply. Minimum current is therefore drawn from a supply when $S = P$ and $Q_T = 0$. Note the effect of decreasing levels of Q_T on the length (and magnitude) of S in Fig. 19.28 for the same real power. Note also that the power-factor angle approaches zero degrees and F_p approaches 1, revealing that the network is appearing more and more resistive at the input terminals.

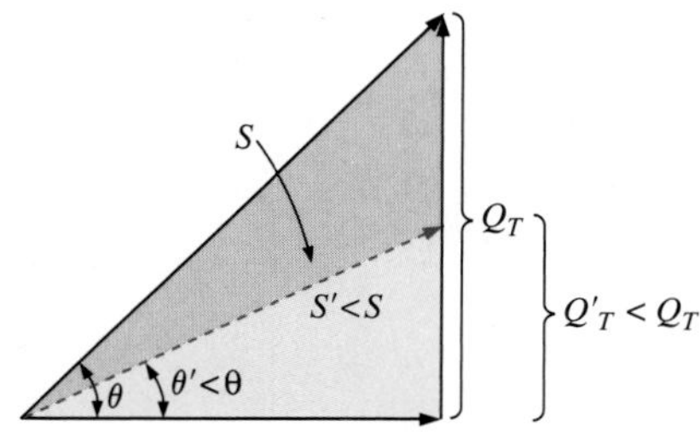

FIG. 19.28
Demonstrating the impact of power-factor correction on the power triangle of a network.

The process of introducing reactive elements to bring the power factor closer to unity is called **power-factor correction.** Since most loads are inductive, the process normally involves introducing elements with capacitive terminal characteristics having the sole purpose of improving the power factor.

In Fig. 19.29(a), for instance, an inductive load is drawing a current I_L that has a real and an imaginary component. In Fig. 19.29(b), a capacitive load was added in parallel with the original load to raise the power factor of the total system to the unity power-factor level. Note that by placing all the elements in parallel, the load still receives the same terminal voltage and draws the same current I_L. In other words, the load is unaware of and unconcerned about whether it is hooked up as shown in Fig. 19.29(a) or Fig. 19.29(b).

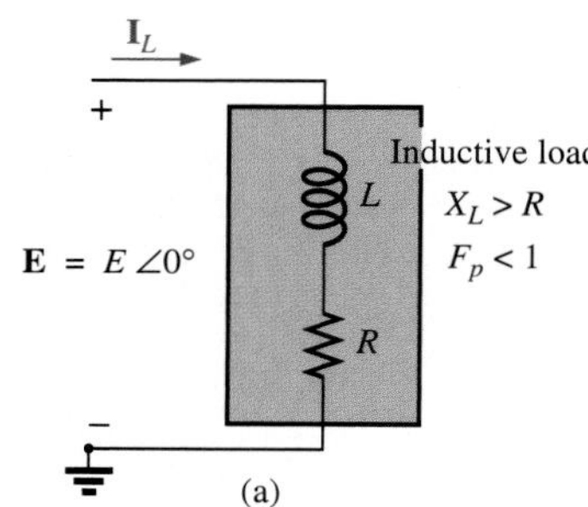

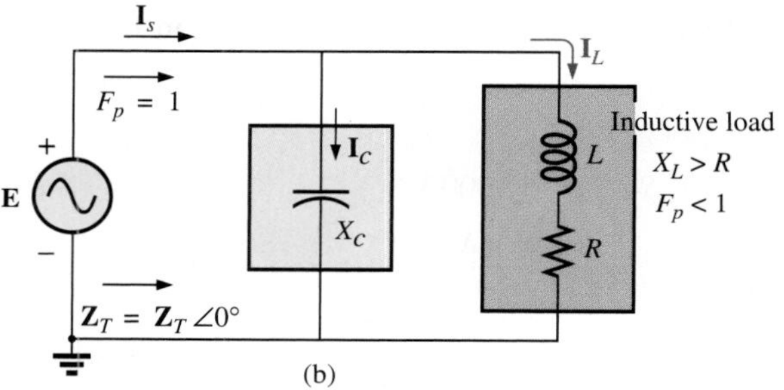

FIG. 19.29
Demonstrating the impact of a capacitive element on the power factor of a network.

Solving for the source current in Fig. 19.29(b):

$$\begin{aligned}\mathbf{I}_s &= \mathbf{I}_C + \mathbf{I}_L \\ &= j\,I_C(I_{\text{mag}}) + I_L(R_e) + j\,I_L(I_{\text{mag}}) = j\,I_C + I_L - j\,I_L \\ &= I_L(R_e) + j\,[I_L(I_{\text{mag}}) + I_C(I_{\text{mag}})] = I_L + j\,[I_C + I_L]\end{aligned}$$

If X_C is chosen such that $I_C = I_L$, then

$$\mathbf{I}_s = I_L + j\,(0) = I_L \angle 0°$$

The result is a source current whose magnitude is simply equal to the real part of the inductive load current, which can be considerably less than the magnitude of the load current in Fig. 19.29(a). In addition, since the phase angle associated with both the applied voltage and the source current is the same, the system appears "resistive" at the input terminals, and all of the power supplied is absorbed, creating maximum efficiency for a generating utility.

EXAMPLE 19.7 A 5 hp motor with a 0.6 lagging power factor and an efficiency of 92% is connected to a 208 V, 60 Hz supply.

a. Establish the power triangle for the load.
b. Determine the power-factor capacitor that must be placed in parallel with the load to raise the power factor to unity.
c. Determine the change in supply current from the uncompensated to the compensated system.
d. Find the network equivalent of the above, and verify the conclusions.

Solutions:

a. Since 1 hp = 746 W,

$$P_o = 5 \text{ hp} = 5(746 \text{ W}) = 3730 \text{ W}$$

and $\quad P_i \text{ (drawn from the line)} = \dfrac{P_o}{\eta} = \dfrac{3730 \text{ W}}{0.92} = 4054.35 \text{ W}$

Also, $\quad F_p = \cos\theta = 0.6$

and $\quad \theta = \cos^{-1} 0.6 = 53.13°$

Applying $\quad \tan\theta = \dfrac{Q_L}{P_i}$

we obtain $\quad Q_L = P_i \tan\theta = (4054.35 \text{ W}) \tan 53.13° = 5405.8 \text{ VAR } (L)$

and

$$S = \sqrt{P_i^2 + Q_L^2} = \sqrt{(4054.35 \text{ W})^2 + (5405.8 \text{ VAR})^2} = 6757.25 \text{ VA}$$

The power triangle appears in Fig. 19.30.

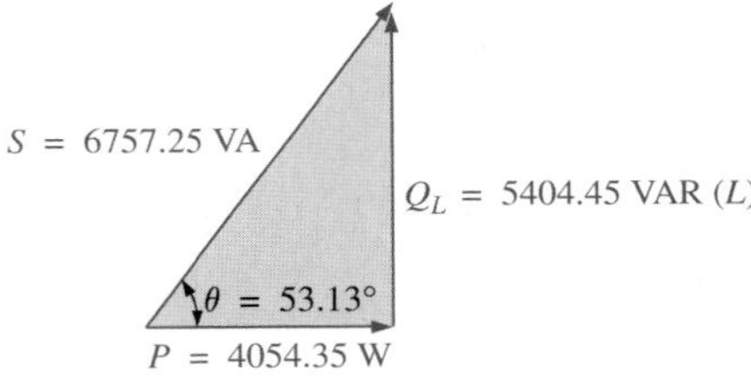

FIG. 19.30
Initial power triangle for the load in Example 19.7.

b. A net unity power-factor level is established by introducing a capacitive reactive power level of 5405.8 VAR to balance Q_L. Since

$$Q_C = \frac{V^2}{X_C}$$

then $\quad X_C = \dfrac{V^2}{Q_C} = \dfrac{(208 \text{ V})^2}{5405.8 \text{ VAR } (C)} = 8\ \Omega$

and $\quad C = \dfrac{1}{2\pi f X_C} = \dfrac{1}{(2\pi)(60 \text{ Hz})(8\ \Omega)} = \mathbf{331.6\ \mu F}$

c. **At 0.6F_p,**

$$S = VI = 6757.25 \text{ VA}$$

and $\quad I = \dfrac{S}{V} = \dfrac{6757.25 \text{ VA}}{208 \text{ V}} = \mathbf{32.49\ A}$

At unity F_p,

$$S = VI = 4054.35 \text{ VA}$$

and $\quad I = \dfrac{S}{V} = \dfrac{4054.35 \text{ VA}}{208 \text{ V}} = \mathbf{19.49\ A}$

producing a 40% reduction in supply current.

d. For the motor, the angle by which the applied voltage leads the current is

$$\theta = \cos^{-1} 0.6 = 53.13°$$

and $P = EI\cos\theta = 4054.35$ W, from above, so that

$$I = \frac{P}{E\cos\theta} = \frac{4054.35 \text{ W}}{(208 \text{ V})(0.6)} = \mathbf{32.49\ A} \qquad \text{(as above)}$$

resulting in

$$\mathbf{I} = 32.49 \text{ A } \angle -53.13^\circ$$

Therefore,

$$\mathbf{Z} = \frac{\mathbf{E}}{\mathbf{I}} = \frac{208 \text{ V } \angle 0^\circ}{32.49 \text{ A } \angle -53.13^\circ} = 6.4\ \Omega\ \angle 53.13^\circ$$
$$= 3.84\ \Omega + j\,5.12\ \Omega$$

as shown in Fig. 19.31(a).

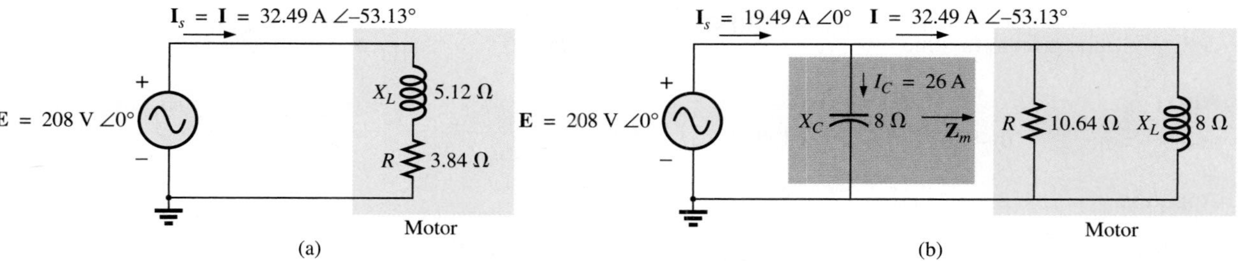

FIG. 19.31
Demonstrating the impact of power-factor corrections on the source current.

The equivalent parallel load is determined from

$$\mathbf{Y} = \frac{1}{\mathbf{Z}} = \frac{1}{6.4\ \Omega\ \angle 53.13^\circ}$$
$$= 0.156 \text{ S } \angle -53.13^\circ = 0.0936 \text{ S} - j\,0.125 \text{ S}$$
$$= \frac{1}{10.68\ \Omega} + \frac{1}{j\,8\ \Omega}$$

as shown in Fig. 19.31(b).

It is now clear that the effect of the 8 Ω inductive reactance can be compensated for by a parallel capacitive reactance of 8 Ω using a power-factor correction capacitor of 332 μF.

Since

$$\mathbf{Y}_T = \frac{1}{-j\,X_C} + \frac{1}{R} + \frac{1}{+j\,X_L} = \frac{1}{R}$$

$$I_s = EY_T = E\left(\frac{1}{R}\right) = (208 \text{ V})\left(\frac{1}{10.68\ \Omega}\right) = \mathbf{19.49\ A} \quad \text{as above}$$

In addition, the magnitude of the capacitive current can be determined as follows:

$$I_C = \frac{E}{X_C} = \frac{208 \text{ V}}{8\ \Omega} = \mathbf{26\ A}$$

EXAMPLE 19.8

a. A small industrial plant has a 10 kW heating load and a 20 kVA inductive load due to a bank of induction motors. The heating elements are considered purely resistive ($F_p = 1$), and the induction motors have a lagging power factor of 0.7. If the supply is 1000 V at 60 Hz, determine the capacitive element required to raise the power factor to 0.95.
b. Compare the levels of current drawn from the supply.

Solutions:

a. For the induction motors,

$$S = VI = 20 \text{ kVA}$$
$$P = S\cos\theta = (20 \times 10^3 \text{ VA})(0.7) = 14 \text{ kW}$$
$$\theta = \cos^{-1} 0.7 \cong 45.6^\circ$$

and

$$Q_L = VI\sin\theta = (20 \text{ kVA})(0.714) = 14.28 \text{ kVAR } (L)$$

The power triangle for the total system appears in Fig. 19.32.

Note the addition of real powers and the resulting S_T:

$$S_T = \sqrt{(24 \text{ kW})^2 + (14.28 \text{ kVAR})^2} = 27.93 \text{ kVA}$$

with
$$I_T = \frac{S_T}{E} = \frac{27.93 \text{ kVA}}{1000 \text{ V}} = \mathbf{27.93\ A}$$

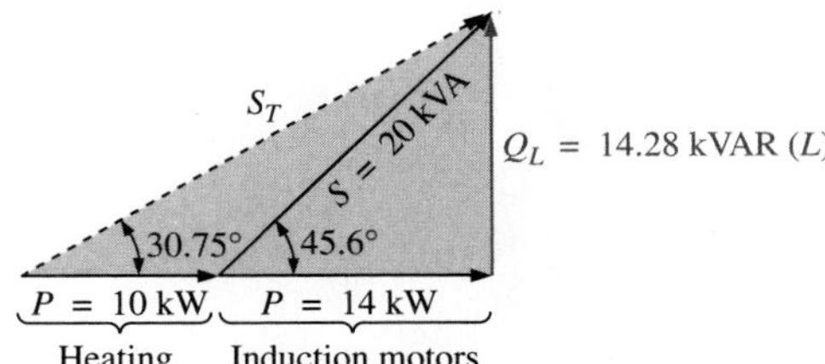

FIG. 19.32
Initial power triangle for the load in Example 19.8.

The desired power factor of 0.95 results in an angle between S and P of

$$\theta = \cos^{-1} 0.95 = 18.19^\circ$$

changing the power triangle to that in Fig. 19.33:

$$\text{with } \tan\theta = \frac{Q'_L}{P_T} \rightarrow Q'_L = P_T \tan\theta = (24 \text{ kW})(\tan 18.19^\circ)$$
$$= (24 \text{ kW})(0.329) = 7.9 \text{ kVAR } (L)$$

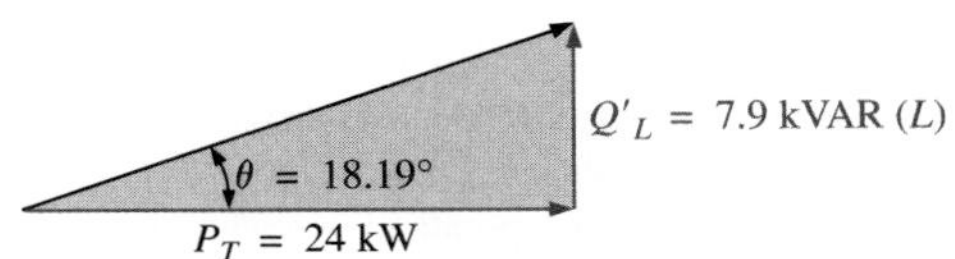

FIG. 19.33
Power triangle for the load in Example 19.8 after raising the power factor to 0.95.

The inductive reactive power must therefore be reduced by

$$Q_L - Q'_L = 14.28 \text{ kVAR } (L) - 7.9 \text{ kVAR } (L) = 6.38 \text{ kVAR } (L)$$

Therefore, $Q_C = 6.38$ kVAR, and using

$$Q_C = \frac{E^2}{X_C}$$

we obtain

$$X_C = \frac{E^2}{Q_C} = \frac{(10^3 \text{ V})^2}{6.38 \text{ kVAR}} = 156.74\ \Omega$$

and
$$C = \frac{1}{2\pi f X_C} = \frac{1}{(2\pi)(60 \text{ Hz})(156.74\ \Omega)} = \mathbf{16.93\ \mu F}$$

b.
$$S_T = \sqrt{(24 \text{ kW})^2 + [7.9 \text{ kVAR } (L)]^2}$$
$$= 25.27 \text{ kVA}$$
$$I_T = \frac{S_T}{E} = \frac{25.27 \text{ kVA}}{1000 \text{ V}} = \mathbf{25.27\ A}$$

The new I_T is

$$I_T = \mathbf{25.27\ A} \angle \mathbf{27.93\ A} \qquad \text{(original)}$$

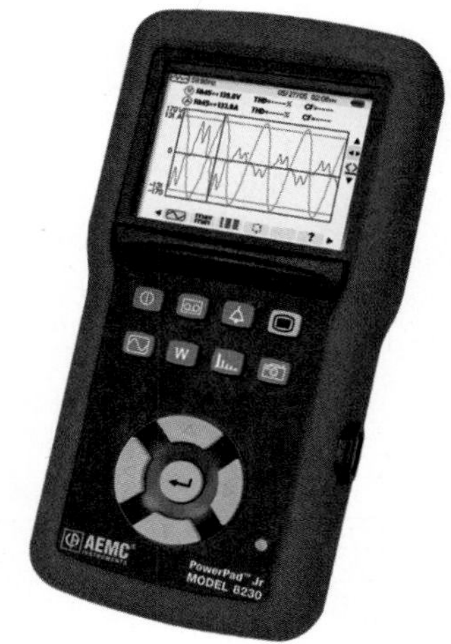

FIG. 19.34
Digital single-phase and three-phase power meter.
(Courtesy of AEMC® Instruments. Foxborough, MA.)

19.10 POWER METERS

The power meter in Fig. 19.34 uses a sophisticated electronic package to sense the voltage and current levels and has an analog-to-digital conversion unit that displays the levels in digital form. It is capable of providing a digital readout for distorted nonsinusoidal waveforms, and it can provide the phase power, total power, apparent power, reactive power, and power factor. It can also measure currents up to 500 A, voltages up to 600 V, and frequencies from 30 Hz to 1000 Hz.

The power quality analyzer in Fig. 19.35 can also display the real, reactive, and apparent power levels along with the power factor. However, it has a broad range of other options, including providing the harmonic content of up to 51 terms for the voltage, current, and power. The power range extends from 250 W to 2.5 MW, and the current can be read up to 1000 A. The meter can also be used to measure resistance levels from 500 Ω to 30 MΩ, capacitance levels from 50 nF to 500 μF, and temperature in both °C and °F.

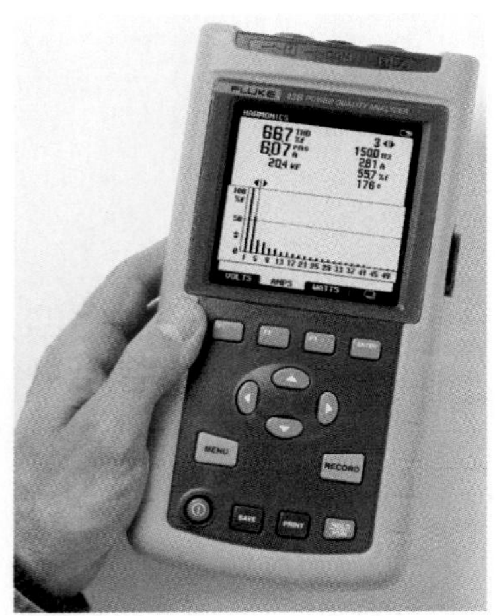

FIG. 19.35
Power quality analyzer capable of displaying the power in watts, the current in amperes, and the voltage in volts.
(Courtesy of Fluke Corporation. Reproduced with Permission.)

19.11 EFFECTIVE RESISTANCE

The resistance of a conductor as determined by the equation $R = \rho(l/A)$ is often called the *dc, ohmic,* or *geometric* resistance. It is a constant quantity determined only by the material used and its physical dimensions. In ac circuits, the actual resistance of a conductor (called the **effective resistance**) differs from the dc resistance because of the varying currents and voltages that introduce effects not present in dc circuits.

These effects include radiation losses, skin effect, eddy currents, and hysteresis losses. The first two effects apply to any network, while the latter two are concerned with the additional losses introduced by the presence of ferromagnetic materials in a changing magnetic field.

Experimental Procedure

The effective resistance of an ac circuit cannot be measured by the ratio V/I since this ratio is now the impedance of a circuit that may have both resistance and reactance. The effective resistance can be found, however, by using the power equation $P = I^2R$, where

$$\boxed{R_{\text{eff}} = \frac{P}{I^2}} \tag{19.31}$$

A wattmeter and an ammeter are therefore necessary for measuring the effective resistance of an ac circuit.

Radiation Losses

Let us now examine the various losses in greater detail. The **radiation loss** is the loss of energy in the form of electromagnetic waves during the transfer of energy from one element to another. This loss in energy requires that the input power be larger to establish the same current I, causing R to increase as determined by Eq. (19.31). At a frequency of 60 Hz, the effects of radiation losses can be completely ignored. However, at radio frequencies, this is an important effect and may in fact become the main effect in an electromagnetic device such as an antenna.

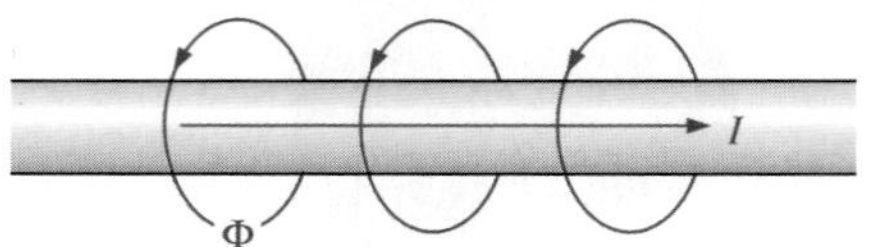

FIG. 19.36
Demonstrating the skin effect on the effective resistance of a conductor.

Skin Effect

The explanation of **skin effect** requires the use of some basic concepts previously described. Remember from Chapter 12 that a magnetic field exists around every current-carrying conductor (Fig. 19.36). Since the amount of charge flowing in ac circuits changes with time, the magnetic field surrounding the moving charge (current) also changes. Recall also that a wire placed in a changing magnetic field will have an induced voltage across its terminals as determined by Faraday's law, $e = N \times (d\phi/dt)$. The higher the frequency of the changing flux as determined by an alternating current, the greater the induced voltage.

For a conductor carrying alternating current, the changing magnetic field surrounding the wire links the wire itself, thus developing within the wire an induced voltage that opposes the original flow of charge or current. These effects are more pronounced at the center of the conductor than at the surface because the center is linked by the changing flux inside the wire as well as that outside the wire. As the frequency of the applied signal increases, the flux linking the wire changes at a greater rate. An increase in frequency therefore increases the counter-induced voltage at the center of the wire to the point where the current, for all practical purposes, flows on the surface of the conductor. At 60 Hz, the skin effect is almost noticeable. However, at radio frequencies, the skin effect is so pronounced that conductors are frequently made hollow because the center part is relatively ineffective. The skin effect, therefore, reduces the effective area through which the current can flow, and it causes the resistance of the conductor, given by the equation $R\uparrow = \rho(l/A\downarrow)$, to increase.

Hysteresis and Eddy Current Losses

As mentioned earlier, hysteresis and eddy current losses appear when a ferromagnetic material is placed in the region of a changing magnetic field. To describe eddy current losses in greater detail, we consider the effects of an alternating current passing through a coil wrapped around a ferromagnetic core. As the alternating current passes through the coil, it develops a changing magnetic flux Φ linking both the coil and the core that develops an induced voltage within the core as determined by Faraday's law. This induced voltage and the geometric resistance of the core $R_C = \rho(l/A)$ cause currents to be developed within the core, $i_{core} = (e_{ind}/R_C)$, called **eddy currents.** The currents flow in circular paths, as shown in Fig. 19.37, changing direction with the applied ac potential.

The eddy current losses are determined by

$$P_{eddy} = i^2_{eddy}R_{core}$$

The magnitude of these losses is determined primarily by the type of core used. If the core is nonferromagnetic—and has a high resistivity like wood or air—the eddy current losses can be neglected. In terms of the frequency of the applied signal and the magnetic field strength produced, the eddy current loss is proportional to the square of the frequency times the square of the magnetic field strength:

$$P_{eddy} \propto f^2B^2$$

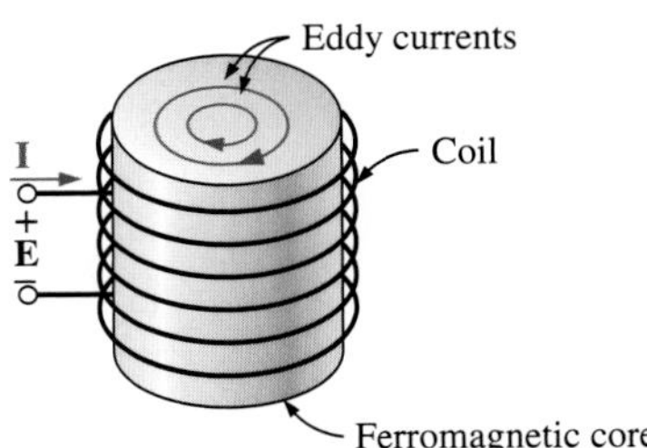

FIG. 19.37
Defining the eddy current losses of a ferromagnetic core.

Eddy current losses can be reduced if the core is constructed of thin, laminated sheets of ferromagnetic material insulated from one another and aligned parallel to the magnetic flux. Such construction reduces the magnitude of the eddy currents by placing more resistance in their path.

Hysteresis losses were described in Section 12.6. You will recall that in terms of the frequency of the applied signal and the magnetic field strength produced, the hysteresis loss is proportional to the frequency to the 1st power times the magnetic field strength to the nth power:

$$P_{hys} \propto f^1B^n$$

where n can vary from 1.4 to 2.6, depending on the material under consideration.

Hysteresis losses can be effectively reduced by the injection of small amounts of silicon into the magnetic core, constituting some 2% or 3% of the total composition of the core. This must be done carefully, however, because too much silicon makes the core brittle and difficult to machine into the shape desired.

EXAMPLE 19.9

a. An air-core coil is connected to a 120 V, 60 Hz source as shown in Fig. 19.38. The current is found to be 5 A, and a wattmeter reading of 75 W is observed. Find the effective resistance and the inductance of the coil.

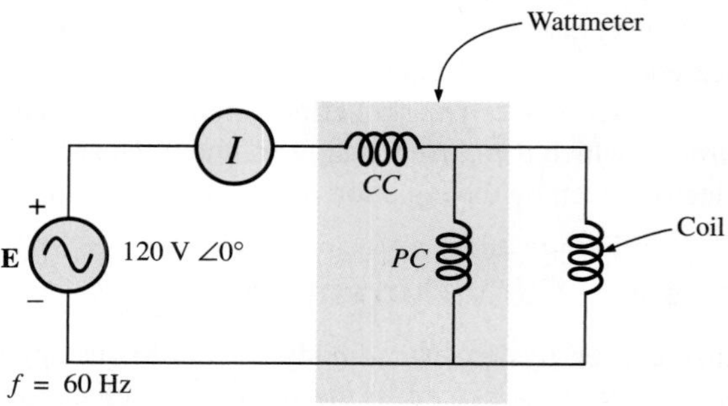

FIG. 19.38
The basic components required to determine the effective resistance and inductance of the coil.

b. A brass core is then inserted in the coil. The ammeter reads 4 A, and the wattmeter 80 W. Calculate the effective resistance of the core. To what do you attribute the increase in value over that in part (a)?

c. If a solid iron core is inserted in the coil, the current is found to be 2 A, and the wattmeter reads 52 W. Calculate the resistance and the inductance of the coil. Compare these values to those in part (a), and account for the changes.

Solutions:

a. $R = \dfrac{P}{I^2} = \dfrac{75 \text{ W}}{(5 \text{ A})^2} = \mathbf{3\ \Omega}$

$$Z_T = \frac{E}{I} = \frac{120 \text{ V}}{5 \text{ A}} = 24\ \Omega$$

$$X_L = \sqrt{Z_T^2 - R^2} = \sqrt{(24\ \Omega)^2 - (3\ \Omega)^2} = 23.81\ \Omega$$

and $X_L = 2\pi f L$

or $$L = \frac{X_L}{2\pi f} = \frac{23.81\ \Omega}{377 \text{ rad/s}} = \mathbf{63.16 \text{ mH}}$$

b. $R = \dfrac{P}{I^2} = \dfrac{80 \text{ W}}{(4 \text{ A})^2} = \dfrac{80\ \Omega}{16} = \mathbf{5\ \Omega}$

The brass core has less reluctance than the air core. Therefore, a greater magnetic flux density B will be created in it. Since $P_{eddy} \propto$

f^2B^2, and $P_{hys} \propto f^1B^n$, as the flux density increases, the core losses and the effective resistance increase.

c. $R = \dfrac{P}{I^2} = \dfrac{52\text{ W}}{(2\text{ A})^2} = \dfrac{52\ \Omega}{4} = \mathbf{13\ \Omega}$

$$Z_T = \frac{E}{I} = \frac{120\text{ V}}{2\text{ A}} = 60\ \Omega$$

$$X_L = \sqrt{Z_T^2 - R^2} = \sqrt{(60\ \Omega)^2 - (13\ \Omega)^2} = 58.57\ \Omega$$

$$L = \frac{X_L}{2\pi f} = \frac{58.57\ \Omega}{377\text{ rad/s}} = \mathbf{155.36\text{ mH}}$$

The iron core has less reluctance than the air or brass cores. Therefore, a greater magnetic flux density B will be developed in the core. Again, since $P_{eddy} \propto f^2B^2$, and $P_{hys} \propto f^1B^n$, the increased flux density will cause the core losses and the effective resistance to increase.

Since the inductance L is related to the change in flux by the equation $L = N\,(d\phi/di)$, the inductance will be greater for the iron core because the changing flux linking the core will increase.

19.12 APPLICATIONS

Portable Power Generators

Even though it may appear that 120 V ac are just an extension cord away, there are times—such as in a remote cabin, on a job site, or while camping—that we are reminded that not every corner of the globe is connected to an electric power source. As you travel farther away from large urban communities, gasoline generators such as shown in Fig. 19.39 appear in increasing numbers in hardware stores, lumber yards, and other retail establishments to meet the needs of the local community. Since ac generators are driven by a gasoline motor, they must be properly ventilated and cannot be run indoors. Usually, because of the noise and fumes that result, they are placed as far away as possible and are connected by a long, heavy-duty, weather-resistant extension cord. Any connection points must be properly protected and placed to ensure that the connections will not sit in a puddle of water or be sensitive to heavy rain or snow. Although there is some effort involved in setting up generators and constantly ensuring that they have enough gas, most users think that they are priceless.

FIG. 19.39
Single-phase portable generator. (Courtesy of Coleman Powermate, Inc.)

The vast majority of generators are built to provide between 1750 W and 5000 W of power, although larger units can provide up to 20,000 W. At first encounter, you may assume that 5000 W are more than adequate. However, keep in mind that the unit purchased should be rated at least 20% above your expected load because of surge currents that result when appliances, motors, tools, and so on, are turned on. Remember that even a light bulb develops a large turn-on current due to the cold, low-resistance state of the filament. If you work too closely to the rated capacity, experiences such as a severe drop in lighting can result when an electric saw is turned on—almost to the point where it appears that the lights go out altogether. Generators are like any other piece of equipment: If you apply a load that is too heavy, they will shut down. Most have protective fuses or circuit breakers to ensure that the excursions above rated conditions are monitored and not exceeded beyond reason. The 20% protective barrier drops the output power from a 5000 W unit to 4000 W, and already we begin to wonder about the load we can apply. Although 4000 W are sufficient to run a number of 60 W bulbs, a TV, an oil burner, and so on, troubles develop whenever a unit is hooked up for direct heating (such as heaters, hair dryers, and clothes dryers). Even microwaves at 1200 W command quite a power drain. Add a small electric heater at 1500 W with six 60 W bulbs (360 W), a 250 W TV, and a 250 W oil burner, and then turn on an electric hair dryer at 1500 W—suddenly you are very close to your maximum of 4000 W. It doesn't take long to push the limits when it comes to energy-consuming appliances.

Table 19.2 provides a list of specifications for the broad range of portable gasoline generators. Since the heaviest part of a generator is the gasoline motor, anything over 5 hp gets pretty heavy, especially when you add the weight of the gasoline. Most good units providing over 2400 W will have receptacles for 120 V and 220 V at various current levels, with an outlet for 12 V dc. They are also built so that they tolerate outdoor conditions of a reasonable nature and can run continuously for long periods of time. At 120 V, a 5000 W unit can provide a maximum current of about 42 A.

TABLE 19.2
Specifications for portable gasoline-driven ac generators.

Continuous output power	1750–3000 W	2000–5000 W	2250–7500 W
Horsepower of gas motor	4–11 hp	5–14 hp	5–16 hp
Continuous output current	At 120 V: 15–25 A At 220 V(3ϕ): 8–14 A	At 120 V: 17–42 A At 220 V(3ϕ): 9–23 A	At 120 V: 19–63 A At 220 V(3ϕ): 10–34 A
Output voltage	120 V or 3ϕ: 120 V/220 V	120 V or 3ϕ: 120 V/220 V	120 V or 3ϕ: 120 V/220 V
Receptacles	2	2–4	2–4
Fuel tank	½ to 2 gallons gasoline	½ to 3 gallons gasoline	1 to 5 gallons gasoline

Business Sense

Because of the costs involved, every large industrial plant must continuously review its electric utility bill to ensure its accuracy and to consider ways to conserve energy. As described in this chapter, the power factor associated with the plant as a whole can have a measurable effect on the drain current and therefore the kVA drain on the power line. Power companies are aware of this problem and actually add a surcharge if the power factor fades below about 0.9. In other words, to ensure that the load appears as resistive in nature as possible, the power company asks users to try to ensure that their power factor is between 0.9 and 1 so that the kW demand is very close to the kVA demand.

Consider the following monthly bill for a fairly large industrial plant:

kWh consumption	146.5 MWh
peak kW demand	241 kW
kW demand	233 kW
kVA demand	250 kVA

The rate schedule provided by the local power authority is the following:

Energy First 450 kWh @ 22.3¢/kWh Next 12 MWh @ 17.1¢/kWh
Additional kWh @ 8.9¢/kWh

Power First 240 kW @ free
Additional kW @ $12.05/kW

Note that this rate schedule has an energy cost breakdown and a power breakdown. This second fee is the one sensitive to the overall power factor of the plant.

The electric bill for the month is then calculated as follows:

$$\begin{aligned}\text{Cost} &= (450\text{ kWh})(22.3¢/\text{kWh}) + (12\text{ MWh})(17.1¢/\text{kWh}) \\ &\quad + [146.2\text{ MWh} - (12\text{ MWh} + 450\text{ kWh})](8.9¢/\text{kWh}) \\ &= \$100.35 + \$2052.00 + \$11{,}903.75 \\ &= \mathbf{\$14{,}056.10}\end{aligned}$$

Before examining the effect of the power fee structure, we can find the overall power factor of the load for the month with the following ratio taken from the monthly statement:

$$F_p = \frac{P}{P_a} = \frac{233\text{ kW}}{250\text{ kVA}} = \mathbf{0.932}$$

Since the power factor is larger than 0.9, the chances are that there will not be a surcharge or that the surcharge will be minimal.

When the power component of the bill is determined, the kVA demand is multiplied by the magic number of 0.9 to determine a kW level at this power factor. This kW level is compared to the metered level, and the consumer pays for the higher level.

In this case, if we multiply the 250 kVA by 0.9, we obtain 225 kW which is slightly less than the metered level of 233 kW. However, both levels are less than the free level of 240 kW, so there is no additional charge for the power component. The total bill remains at $14,056.10.

If the kVA demand of the bill were 388 kVA with the kW demand staying at 233 kW, the situation would change because 0.9 times 388 kVA would result in 349.2 kW which is much greater than the metered 233 kW. The 349.2 kW would then be used to determine the bill as follows:

$$349.2\text{ kW} - 240\text{ kW} = 109.2\text{ kW}$$

$$(109.2\text{ kW})(\$12.05/\text{kW}) = \mathbf{\$1315.86}$$

which is significant.

The total bill can then be determined as follows:

$$\begin{aligned}\text{Cost} &= \$14{,}056.10 + \$1{,}315.86 \\ &= \mathbf{\$15{,}371.96}\end{aligned}$$

Thus, the power factor of the load dropped to 233 kW/388 kVA = 0.6 which would put an unnecessary additional load on the power plant. It is certainly time to consider the power-factor-correction option as described in this text. It is not uncommon to see large capacitors sitting at the point where power enters a large industrial plant to perform a needed level of power-factor correction.

All in all, therefore, it is important to fully understand the impact of a poor power factor on a power plant—whether you someday work for the supplier or for the consumer.

19.13 COMPUTER ANALYSIS

PSpice

Power Curve: Resistor The computer analysis begins with a verification of the curves in Fig. 19.3 which show the in-phase relationship between the voltage and current of a resistor. The figure shows that the power curve is totally above the horizontal axis and that the curve has a

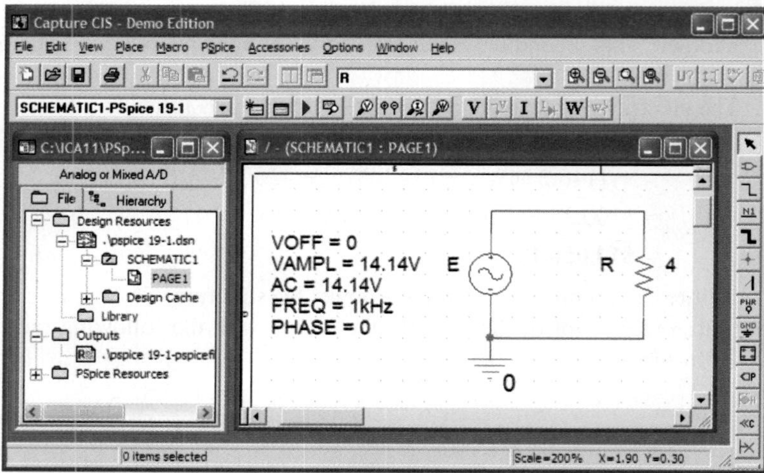

FIG. 19.40

Using PSpice to review the power curve for a resistive element in an ac circuit.

frequency twice the applied frequency and a peak value equal to twice the average value. First, set up the simple schematic of Fig. 19.40. Then, use the **Time Domain(Transient)** option to get a plot versus time, and set the **Run to time** to 1 ms and the **Maximum step size** to 1 ms/1000 = 1 μs. Select **OK** and then the **Run PSpice** icon to perform the simulation. Then **Trace-Add Trace-V1(R)** results in the curve appearing in Fig. 19.41. Next, **Trace-Add Trace-I(R)** results in the curve for the current as appearing in Fig. 19.41. Finally, plot the power curve using **Trace-Add Trace-V1(R)*I(R)** from the basic power equation, and the larger curve of Fig. 19.41 results. The original plot had a y-axis that extended from -50 to $+50$. Since all of the data points are from -20 to $+50$, the y-axis was changed to this new range through **Plot-Axis Settings-Y Axis-User Defined-($-$20 to +50)-OK** to obtain the plot in Fig. 19.41.

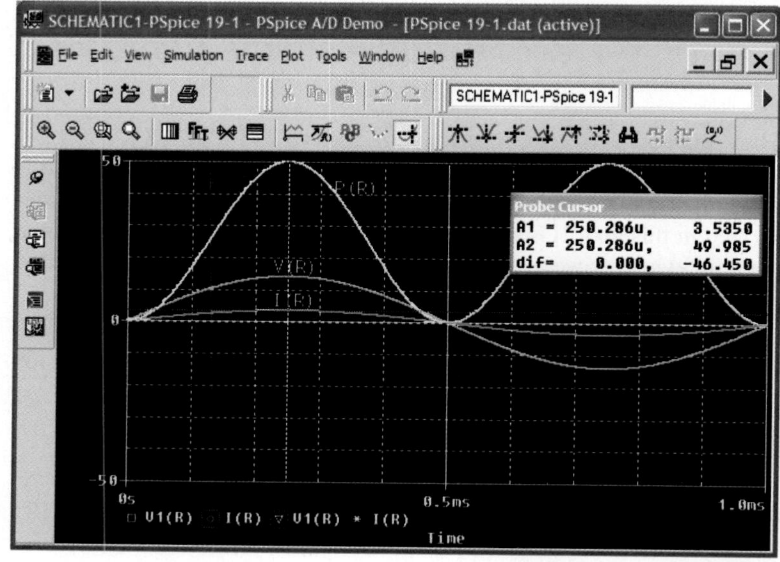

FIG. 19.41

The resulting plots for the power, voltage, and current for the resistor in Fig. 19.40.

You can distinguish between the curves by looking at the symbol next to each quantity at the bottom left of the plot. In this case, however, to make it even clearer, a different color was selected for each trace by right-clicking on each trace, selecting **Properties,** and choosing the color and width of each curve. However, you can also add text to the screen by selecting the **ABC** icon to obtain the **Text Label** dialog box, entering the label such as **P(R),** and clicking **OK.** The label can then be placed anywhere on the screen. By selecting the **Toggle cursor** key and then clicking on **I(R)** at the bottom of the screen, you can use the cursor to find the maximum value of the current. At **A1** = 250 μs or ¼ of the total period of the input voltage, the current is a peak at 3.54 A. The peak value of the power curve can then be found by right-clicking on **V1(R)*I(R),** clicking on the graph, and then finding the peak value (also available by clicking on the **Cursor Peak** icon to the right of the **Toggle cursor** key). It occurs at the same point as the maximum current at a level of 50 W. In particular, note that the power curve shows two cycles, while both v_R and i_R show only one cycle. Clearly, the power curve has twice the frequency of the applied signal. Also note that the power curve is totally above the zero line, indicating that power is being absorbed by the resistor through the entire displayed cycle. Further, the peak value of the power curve is twice the average value of the curve; that is, the peak value of 50 W is twice the average value of 25 W.

The results of the above simulation can be verified by performing the longhand calculation using the rms value of the applied voltage. That is,

$$P = \frac{V_R^2}{R} = \frac{(10\text{ V})^2}{4\ \Omega} = \mathbf{25\ W}$$

Power Curves: Series *R-L-C* Circuit The network in Fig. 19.42, with its combination of elements, is now used to demonstrate that, no matter what the physical makeup of the network, the average value of the power curve established by the product of the applied voltage and resulting source current is equal to that dissipated by the network. At a frequency of 1 kHz, the reactance of the 1.273 mH inductor will be 8, and the reactance of the capacitor will be 4 Ω, resulting in a lagging network. An analysis of the network results in

$$\mathbf{Z}_T = 4\ \Omega + j4\ \Omega = 5.657\ \Omega\ \angle 45°$$

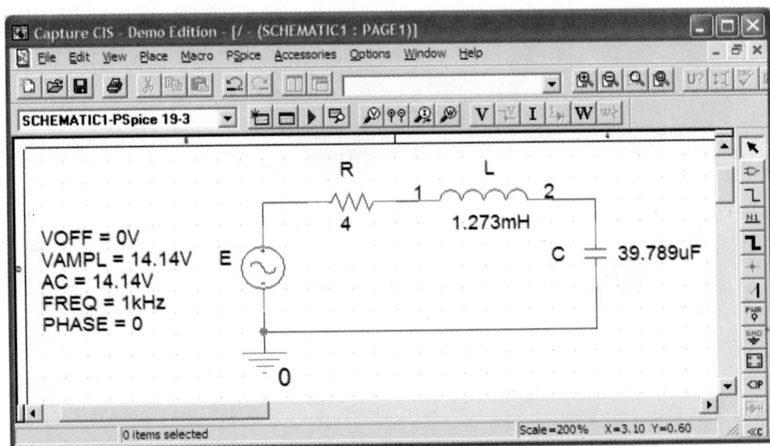

FIG. 19.42
Using PSpice to examine the power distribution in a series R-L-C circuit.

with
$$\mathbf{I} = \frac{\mathbf{E}}{\mathbf{Z}_T} = \frac{10\text{ V}\angle 0^\circ}{5.657\ \Omega\ \angle 45^\circ} = 1.768\text{ A}\ \angle -45^\circ$$

and
$$P = I^2R = (1.768\text{ A})^2\ 4\ \Omega = \mathbf{12.5\ W}$$

The three curves in Fig. 19.43 are obtained using the **Simulation Output Variables V(E:+), I(R),** and **V(E:+)*I(R).** The **Run to time** under the **Simulation Profile** listing was 20 ms, although 1 μs was chosen as the **Maximum step size** to ensure a good plot. In particular, note that the horizontal axis does not start until t = 18 ms to ensure that you are in a steady-state mode and not in a transient stage (where the peak values of the waveforms could change with time). Set the horizontal axis to extend from 18 ms to 20 ms by selecting **Plot-Axis Settings-X Axis-User Defined-18ms to 20ms-OK.** First note that the current lags the applied voltage as expected for the lagging network. The phase angle between the two is 45° as determined above. Second, be aware that the elements are chosen so that the same scale can be used for the current and voltage. The vertical axis does not have a unit of measurement, so the proper units must be mentally added for each plot. Using **Plot-Label-Line,** draw a line across the screen at the average power level of 12.5 W. A pencil appears that can be clicked in place at the left edge at the 12.5 W level. Drag the pencil across the page to draw the desired line. Once you are at the right edge, release the mouse, and the line is drawn. Obtain the different colors for the traces by right-clicking on a trace and selecting from the choices under **Properties.** Note that the 12.5 W level is indeed the average value of the power curve. It is interesting to note that the power curve dips below the axis for only a short period of time. In other words, during the two visible cycles, power is being absorbed by the circuit most of the time. The small region below the axis is the return of energy to the network by the reactive elements. In general, therefore, the source must supply power to the circuit most of the time, even though a good percentage of the power may be delivering energy to the reactive elements, not being dissipated.

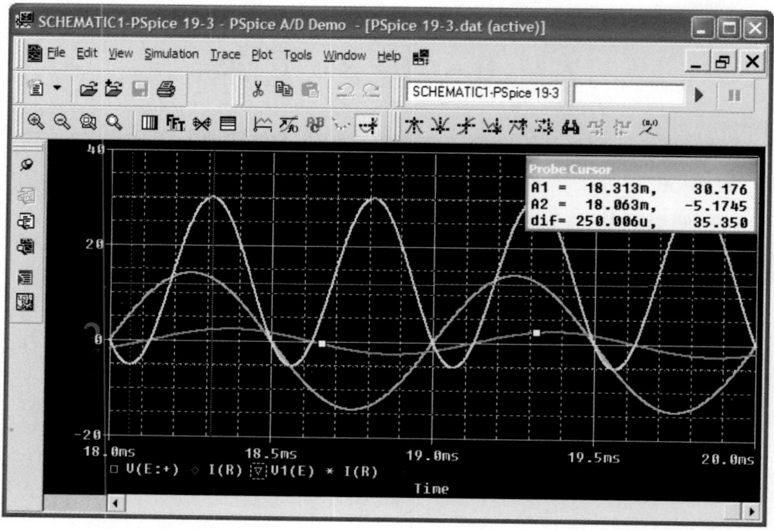

FIG. 19.43

Plots of the applied voltage e, current $i_R = i_s$, and power delivered $p_s = e \cdot i_s$ for the circuit in Fig. 19.42.

PROBLEMS

SECTIONS 19.1 THROUGH 19.8

1. For the battery of bulbs (purely resistive) appearing in Fig. 19.44:
   a. Determine the total power dissipation.
   b. Calculate the total reactive and apparent power.
   c. Find the source current I_s.
   d. Calculate the resistance of each bulb for the specified operating conditions.
   e. Determine the currents I_1 and I_2.

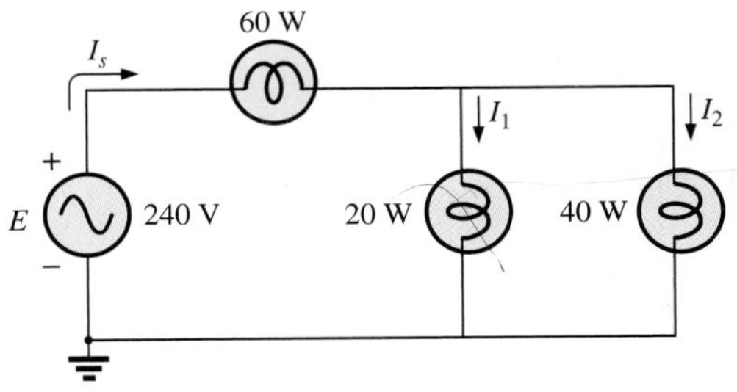

FIG. 19.44
Problem 1.

2. For the network in Fig. 19.45:
   a. Find the average power delivered to each element.
   b. Find the reactive power for each element.
   c. Find the apparent power for each element.
   d. Find the total number of watts, volt-amperes reactive, and volt-amperes, and the power factor F_p of the circuit.
   e. Sketch the power triangle.
   f. Find the energy dissipated by the resistor over one full cycle of the input voltage.
   g. Find the energy stored or returned by the capacitor and the inductor over one half-cycle of the power curve for each.

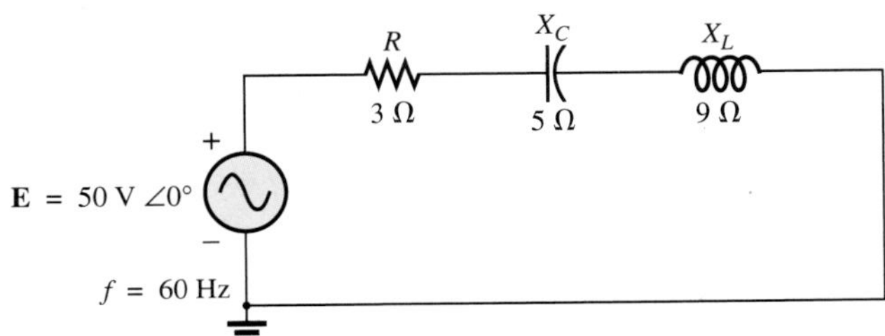

FIG. 19.45
Problem 2.

3. For the system in Fig. 19.46:
   a. Find the total number of watts, volt-amperes reactive, and volt-amperes, and the power factor F_p.
   b. Draw the power triangle.
   c. Find the current $\mathbf{I}_s$.
4. For the system in Fig. 19.47:
   a. Find P_T, Q_T, and S_T.
   b. Determine the power factor F_p.
   c. Draw the power triangle.
   d. Find $\mathbf{I}_s$.

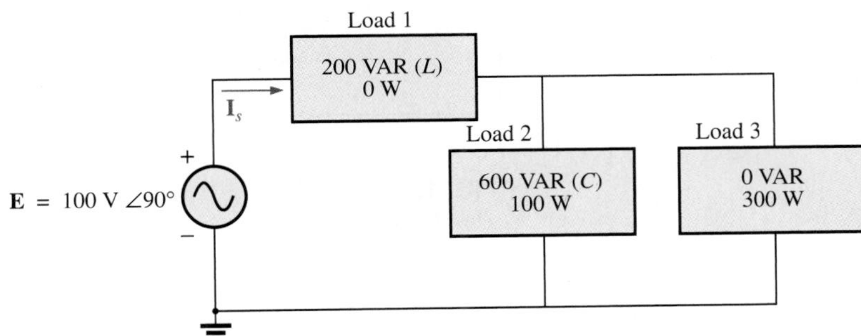

FIG. 19.46
Problem 3.

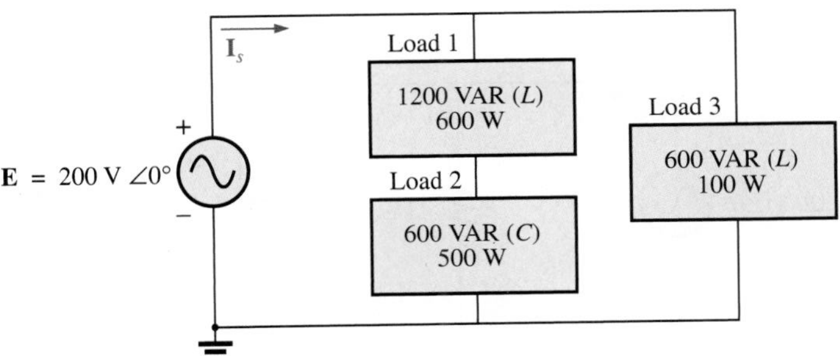

FIG. 19.47
Problem 4.

5. For the system in Fig. 19.48:
   a. Find P_T, Q_T, and S_T.
   b. Find the power factor F_p.
   c. Draw the power triangle.
   d. Find $\mathbf{I}_s$.

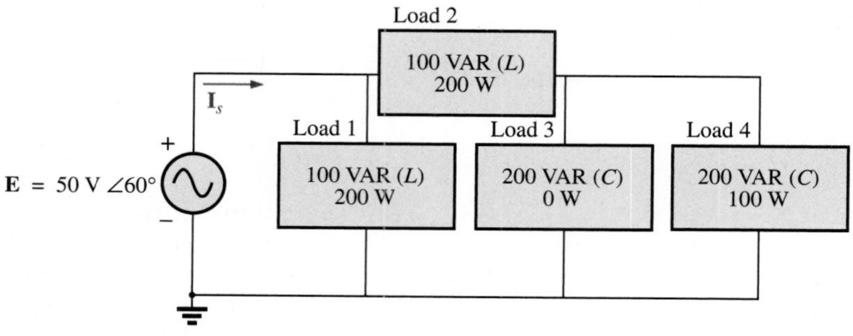

FIG. 19.48
Problem 5.

6. For the circuit in Fig. 19.49:
   a. Find the average, reactive, and apparent power for the 20 Ω resistor.
   b. Repeat part (a) for the 10 Ω inductive reactance.
   c. Find the total number of watts, volt-amperes reactive, and volt-amperes, and the power factor F_p.
   d. Find the current $\mathbf{I}_s$.

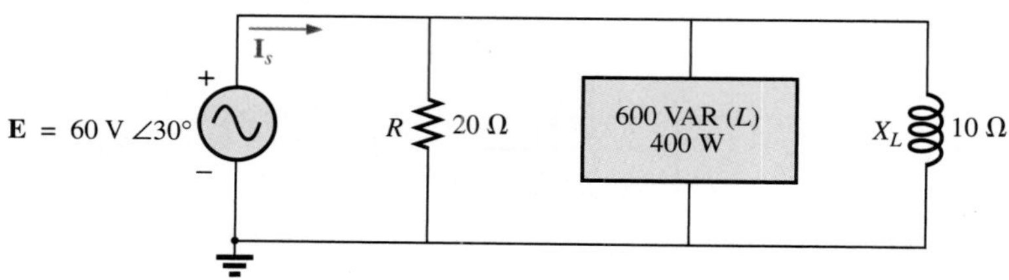

FIG. 19.49
Problem 6.

7. For the network in Fig. 19.50:
   a. Find the average power delivered to each element.
   b. Find the reactive power for each element.
   c. Find the apparent power for each element.
   d. Find P_T, Q_T, S_T, and F_p for the system.
   e. Sketch the power triangle.
   f. Find $\mathbf{I}_s$.

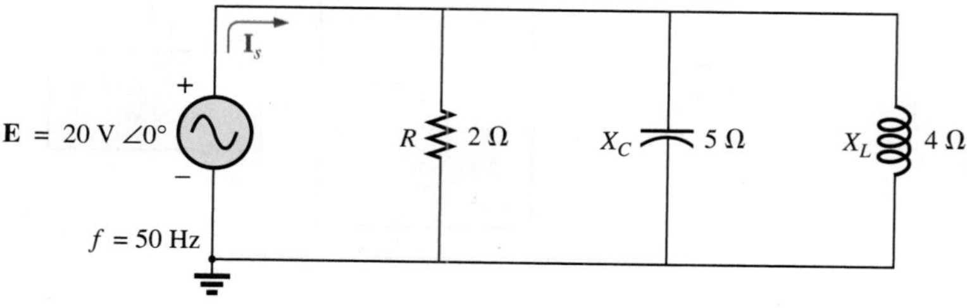

FIG. 19.50
Problem 7.

8. Repeat Problem 7 for the circuit in Fig. 19.51.

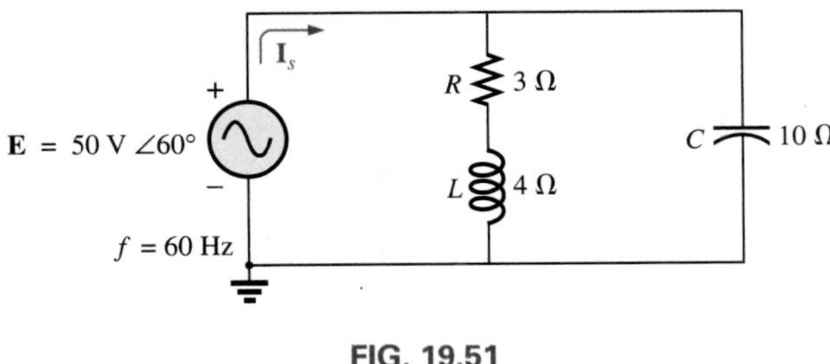

FIG. 19.51
Problem 8.

*9. For the network in Fig. 19.52:
   a. Find the average power delivered to each element.
   b. Find the reactive power for each element.
   c. Find the apparent power for each element.
   d. Find the total number of watts, volt-amperes reactive, and volt-amperes, and the power factor F_p of the circuit.
   e. Sketch the power triangle.
   f. Find the energy dissipated by the resistor over one full cycle of the input voltage.
   g. Find the energy stored or returned by the capacitor and the inductor over one half-cycle of the power curve for each.

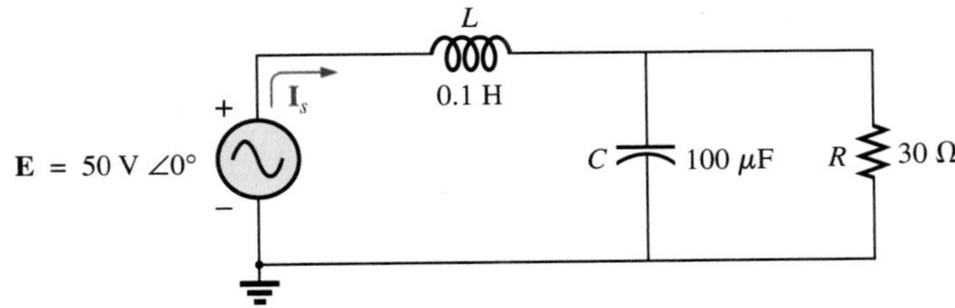

FIG. 19.52
Problem 9.

10. An electrical system is rated 10 kVA, 200 V at a 0.5 leading power factor.
   a. Determine the impedance of the system in rectangular coordinates.
   b. Find the average power delivered to the system.

11. An electrical system is rated 5 kVA, 120 V, at a 0.8 lagging power factor.
   a. Determine the impedance of the system in rectangular coordinates.
   b. Find the average power delivered to the system.

*12. For the system in Fig. 19.53.
   a. Find the total number of watts, volt-amperes reactive, and volt-amperes, and F_p.
   b. Find the current $\mathbf{I}_s$.
   c. Draw the power triangle.
   d. Find the type of elements and their impedance in ohms within each electrical box. (Assume that all elements of a load are in series.)
   e. Verify that the result of part (b) is correct by finding the current $\mathbf{I}_s$ using only the input voltage **E** and the results of part (d). Compare the value of $\mathbf{I}_s$ with that obtained for part (b).

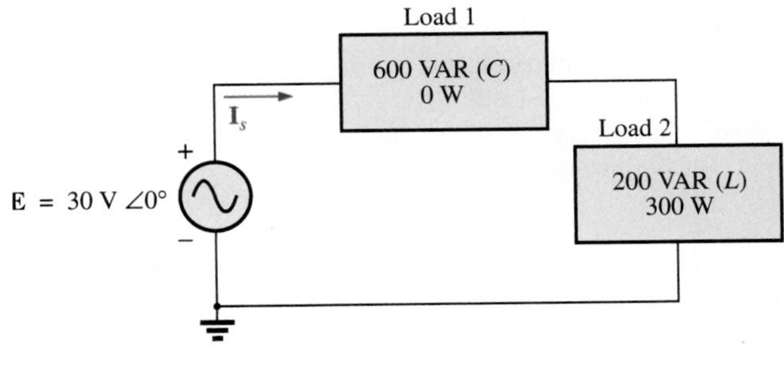

FIG. 19.53
Problem 12.

*13. Repeat Problem 12 for the system in Fig. 19.54.

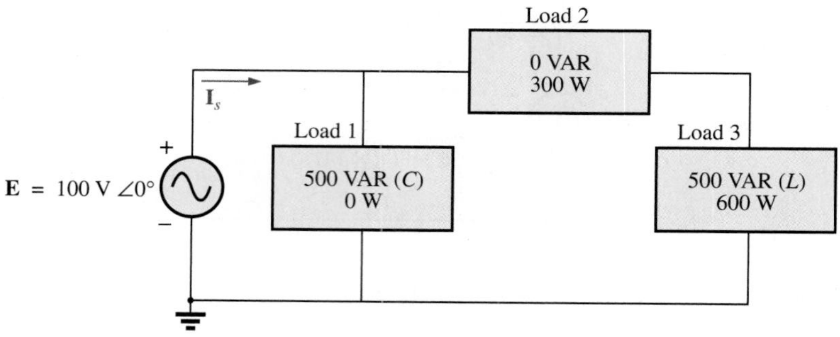

FIG. 19.54
Problem 13.

*14. For the circuit in Fig. 19.55:
   a. Find the total number of watts, volt-amperes reactive, and volt-amperes, and F_p.
   b. Find the current $\mathbf{I}_s$.
   c. Find the type of elements and their impedance in each box. (Assume that the elements within each box are in series.)

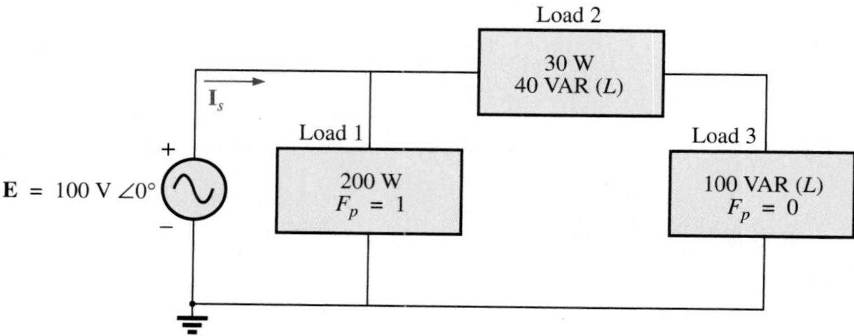

FIG. 19.55
Problem 14.

15. For the circuit in Fig. 19.56:
   a. Find the total number of watts, volt-amperes reactive, and volt-amperes, and F_p.
   b. Find the voltage **E**.
   c. Find the type of elements and their impedance in each box. (Assume that the elements within each box are in series.)

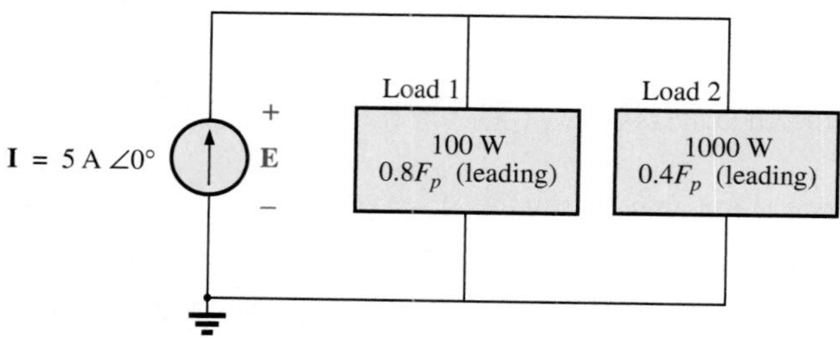

FIG. 19.56
Problem 15.

SECTION 19.9 Power-Factor Correction

*16. The lighting and motor loads of a small factory establish a 10 kVA power demand at a 0.7 lagging power factor on a 208 V, 60 Hz supply.
  a. Establish the power triangle for the load.
  b. Determine the power-factor capacitor that must be placed in parallel with the load to raise the power factor to unity.
  c. Determine the change in supply current from the uncompensated to the compensated system.
  d. Repeat parts (b) and (c) if the power factor is increased to 0.9.

17. The load on a 120 V, 60 Hz supply is 5 kW (resistive), 8 kVAR (inductive), and 2 kVAR (capacitive).
  a. Find the total kilovolt-amperes.
  b. Determine the F_p of the combined loads.
  c. Find the current drawn from the supply.
  d. Calculate the capacitance necessary to establish a unity power factor.
  e. Find the current drawn from the supply at unity power factor, and compare it to the uncompensated level.

18. The loading of a factory on a 1000 V, 60 Hz system includes:
  20 kW heating (unity power factor)
  10 kW (P_i) induction motors (0.7 lagging power factor)
  5 kW lighting (0.85 lagging power factor)
  a. Establish the power triangle for the total loading on the supply.
  b. Determine the power-factor capacitor required to raise the power factor to unity.
  c. Determine the change in supply current from the uncompensated to the compensated system.

SECTION 19.10 Power Meters

19. a. A wattmeter is connected with its current coil as shown in Fig. 19.57 and with the potential coil across points *f-g*. What does the wattmeter read?
  b. Repeat part (a) with the potential coil (*PC*) across *a-b*, *b-c*, *a-c*, *a-d*, *c-d*, *d-e*, and *f-e*.

20. The voltage source in Fig. 19.58 delivers 660 VA at 120 V, with a supply current that lags the voltage by a power factor of 0.6.
  a. Determine the voltmeter, ammeter, and wattmeter readings.
  b. Find the load impedance in rectangular form.

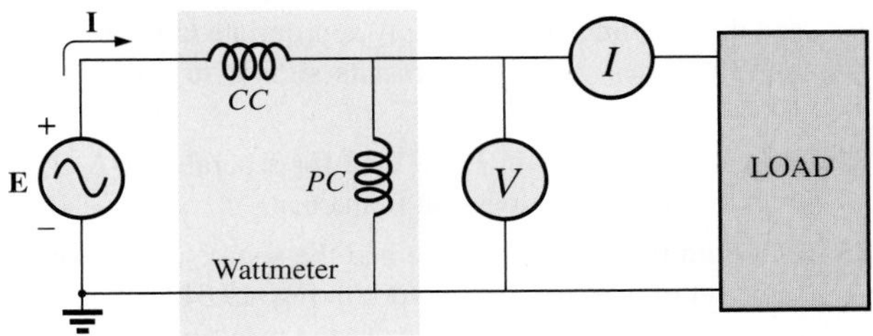

FIG. 19.58
Problem 20.

SECTION 19.11 Effective Resistance

21. a. An air-core coil is connected to a 200 V, 60 Hz source. The current is found to be 4 A, and a wattmeter reading of 80 W is observed. Find the effective resistance and the inductance of the coil.
  b. A brass core is inserted in the coil. The ammeter reads 3 A, and the wattmeter reads 90 W. Calculate the effective resistance of the core. Explain the increase over the value in part (a).
  c. If a solid iron core is inserted in the coil, the current is found to be 2 A, and the wattmeter reads 60 W. Calculate the resistance and inductance of the coil. Compare these values to the values in part (a), and account for the changes.

22. a. The inductance of an air-core coil is 0.08 H, and the effective resistance is 4 Ω when a 60 V, 50 Hz source is connected across the coil. Find the current passing through the coil and the reading of a wattmeter across the coil.
  b. If a brass core is inserted in the coil, the effective resistance increases to 7 Ω, and the wattmeter reads 30 W. Find the current passing through the coil and the inductance of the coil.
  c. If a solid iron core is inserted in the coil, the effective resistance of the coil increases to 10 Ω, and the current decreases to 1.7 A. Find the wattmeter reading and the inductance of the coil.

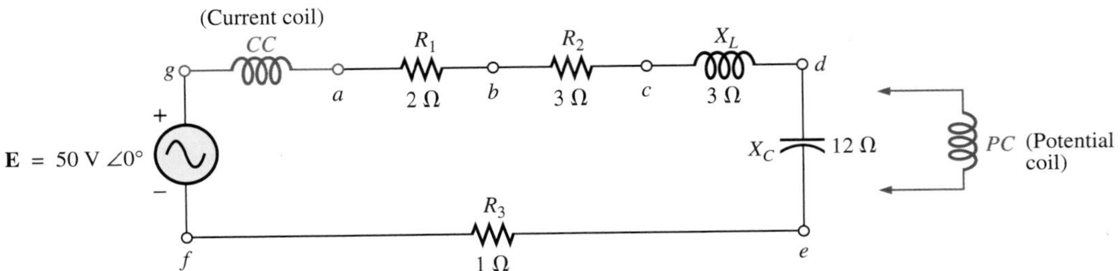

FIG. 19.57
Problem 19.

SECTION 19.13 Computer Analysis

PSpice or Multisim

23. Using PSpice or Multisim, obtain a plot of reactive power for a pure capacitor of 636.62 μF at a frequency of 1 kHz for one cycle of the input voltage using an applied voltage $\mathbf{E} = 10\text{ V}\angle 0°$. On the same graph, plot both the applied voltage and the resulting current. Apply appropriate labels to the resulting curves to generate results similar to those in Fig. 19.41.

24. Repeat the analysis in Fig. 19.42 for a parallel *R-L-C* network of the same values and frequency.

25. Plot both the applied voltage and the source current on the same set of axes for the network in Fig. 19.31(b), and show that they are both in phase due to the resulting unity power factor.

GLOSSARY

Apparent power The power delivered to a load without consideration of the effects of a power-factor angle of the load. It is determined solely by the product of the terminal voltage and current of the load.

Average (real) power The delivered power dissipated in the form of heat by a network or system.

Eddy currents Small, circular currents in a paramagnetic core causing an increase in the power losses and the effective resistance of the material.

Effective resistance The resistance value that includes the effects of radiation losses, skin effect, eddy currents, and hysteresis losses.

Hysteresis losses Losses in a magnetic material introduced by changes in the direction of the magnetic flux within the material.

Power-factor correction The addition of reactive components (typically capacitive) to establish a system power factor closer to unity.

Radiation losses The losses of energy in the form of electromagnetic waves during the transfer of energy from one element to another.

Reactive power The power associated with reactive elements that provides a measure of the energy associated with setting up the magnetic and electric fields of inductive and capacitive elements, respectively.

Skin effect At high frequencies, a counter-induced voltage builds up at the center of a conductor, resulting in an increased flow near the surface (skin) of the conductor and a sharp reduction near the center. As a result, the effective area of conduction decreases and the resistance increases as defined by the basic equation for the geometric resistance of a conductor.

RESONANCE

Objectives

- *Become familiar with the frequency response of a series resonant circuit and how to calculate the resonant and cutoff frequencies.*
- *Be able to calculate a tuned network's quality factor, bandwidth, and power levels at important frequency levels.*
- *Become familiar with the frequency response of a parallel resonant circuit and how to calculate the resonant and cutoff frequencies.*
- *Understand the impact of the quality factor on the frequency response of a series or parallel resonant network.*
- *Begin to appreciate the difference between defining parallel resonance at either the frequency where the input impedance is a maximum or where the network has a unity power factor.*

20.1 INTRODUCTION

This chapter introduces the very important *resonant* (or *tuned*) *circuit,* which is fundamental to the operation of a wide variety of electrical and electronic systems in use today. The resonant circuit is a combination of R, L, and C elements having a frequency response characteristic similar to the one appearing in Fig. 20.1. Note in the figure that the response is a maximum for the frequency f_r, decreasing to the right and left of this frequency. In other words, for a particular range of frequencies, the response will be near or equal to the maximum. The frequencies to the far left or right have very low voltage or current levels and, for all practical purposes, have little effect on the system's response. The radio or television receiver has a response curve for each broadcast station of the type indicated in Fig. 20.1. When the receiver is set (or tuned) to a particular station, it is set on or near the frequency f_r in Fig. 20.1. Stations transmitting at

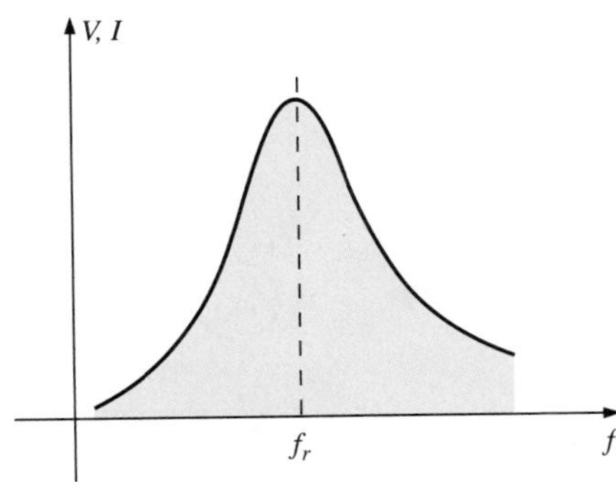

FIG. 20.1
Resonance curve.

frequencies to the far right or left of this resonant frequency are not carried through with significant power to affect the program of interest. The tuning process (setting the dial to f_r) as described above is the reason for the terminology *tuned circuit.* When the response is at or near the maximum, the circuit is said to be in a state of **resonance.**

The concept of resonance is not limited to electrical or electronic systems. If mechanical impulses are applied to a mechanical system at the proper frequency, the system will enter a state of resonance in which sustained vibrations of very large amplitude will develop. The frequency at which this occurs is called the *natural frequency* of the system. The classic example of this effect was the Tacoma Narrows Bridge built in 1940 over Puget Sound in Washington State. Four months after the bridge, with its suspended span of 2800 ft, was completed, a 42 mi/h pulsating gale set the bridge into oscillations at its natural frequency. The amplitude of the oscillations increased to the point where the main span broke up and fell into the water below. It has since been replaced by the new Tacoma Narrows Bridge, completed in 1950.

The resonant electrical circuit *must* have both inductance and capacitance. In addition, resistance will always be present due either to the lack of ideal elements or to the control offered on the shape of the resonance curve. When resonance occurs due to the application of the proper frequency (f_r), the energy absorbed by one reactive element is the same as that released by another reactive element within the system. In other words, energy pulsates from one reactive element to the other. Therefore, once an ideal (pure C, L) system has reached a state of resonance, it requires no further reactive power since it is self-sustaining. In a practical circuit, there is some resistance associated with the reactive elements that will result in the eventual "damping" of the oscillations between reactive elements.

There are two types of resonant circuits: *series* and *parallel.* Each will be considered in some detail in this chapter.

SERIES RESONANCE

20.2 SERIES RESONANT CIRCUIT

A resonant circuit (series or parallel) must have an inductive and a capacitive element. A resistive element is always present due to the internal resistance of the source (R_s), the internal resistance of the inductor (R_l), and any added resistance to control the shape of the response curve (R_{design}). The basic configuration for the series resonant circuit appears in Fig. 20.2(a) with the resistive elements listed above. The "cleaner" ap-

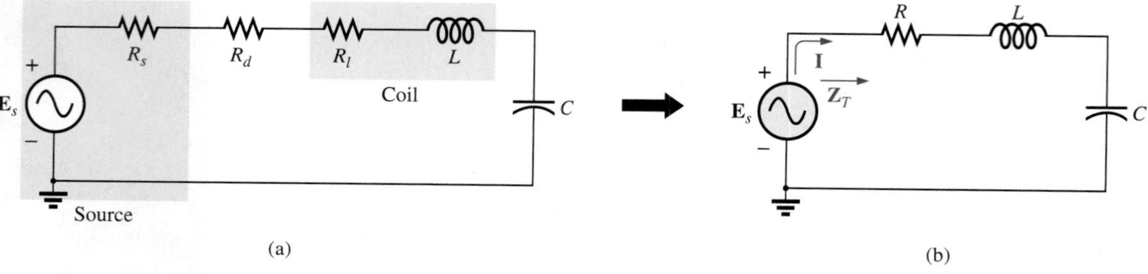

FIG. 20.2
Series resonant circuit.

pearance in Fig. 20.2(b) is a result of combining the series resistive elements into one total value. That is,

$$R = R_s + R_l + R_d \qquad \textbf{(20.1)}$$

The total impedance of this network at any frequency is determined by

$$\mathbf{Z}_T = R + j X_L - j X_C = R + j (X_L - X_C)$$

The resonant conditions described in the introduction occurs when

$$X_L = X_C \qquad \textbf{(20.2)}$$

removing the reactive component from the total impedance equation. The total impedance at resonance is then

$$\mathbf{Z}_{T_s} = R \qquad \textbf{(20.3)}$$

representing the minimum value of $\mathbf{Z}_T$ at any frequency. The subscript s is employed to indicate series resonant conditions.

The resonant frequency can be determined in terms of the inductance and capacitance by examining the defining equation for resonance [Eq. (20.2)]:

$$X_L = X_C$$

Substituting yields

$$\omega L = \frac{1}{\omega C} \quad \text{and} \quad \omega^2 = \frac{1}{LC}$$

and

$$\omega_s = \frac{1}{\sqrt{LC}} \qquad \textbf{(20.4)}$$

or

$$f_s = \frac{1}{2\pi\sqrt{LC}} \qquad \begin{aligned} f &= \text{hertz (Hz)} \\ L &= \text{henries (H)} \\ C &= \text{farads (F)} \end{aligned} \qquad \textbf{(20.5)}$$

The current through the circuit at resonance is

$$\mathbf{I} = \frac{E \angle 0°}{R \angle 0°} = \frac{E}{R} \angle 0°$$

which is the maximum current for the circuit in Fig. 20.2 for an applied voltage **E** since $\mathbf{Z}_T$ is a minimum value. Consider also that *the input voltage and current are in phase at resonance.*

Since the current is the same through the capacitor and inductor, the voltage across each is equal in magnitude but 180° out of phase at resonance:

$$\left.\begin{aligned} \mathbf{V}_L &= (I \angle 0°)(X_L \angle 90°) = IX_L \angle 90° \\ \mathbf{V}_C &= (I \angle 0°)(X_C \angle -90°) = IX_C \angle -90° \end{aligned}\right\} \begin{array}{l} 180° \\ \text{out of} \\ \text{phase} \end{array}$$

and, since $X_L = X_C$, the magnitude of V_L equals V_C at resonance; that is,

$$V_{L_s} = V_{C_s} \qquad \textbf{(20.6)}$$

Fig. 20.3, a phasor diagram of the voltages and current, clearly indicates that the voltage across the resistor at resonance is the input voltage, and **E, I,** and $\mathbf{V}_R$ are in phase at resonance.

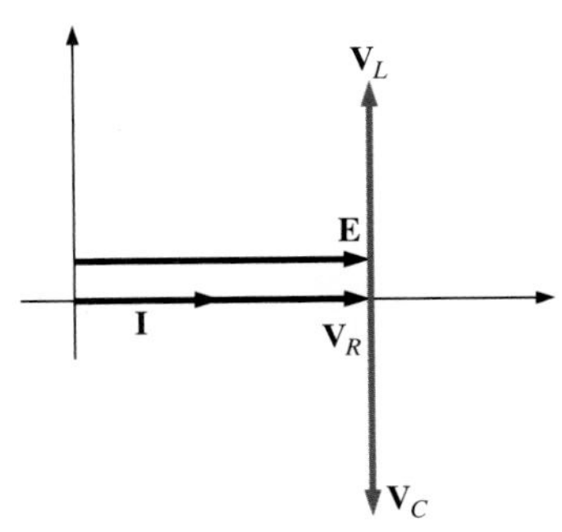

FIG. 20.3
Phasor diagram for the series resonant circuit at resonance.

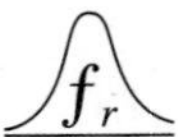

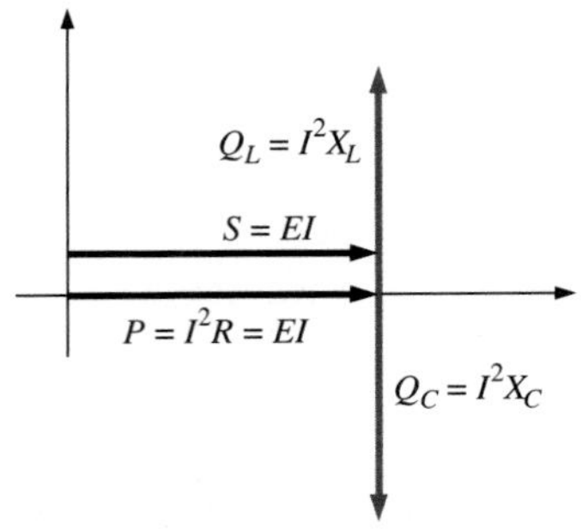

FIG. 20.4
Power triangle for the series resonant circuit at resonance.

The average power to the resistor at resonance is equal to I^2R, and the reactive power to the capacitor and inductor are I^2X_C and I^2X_L, respectively.

The power triangle at resonance (Fig. 20.4) shows that the total apparent power is equal to the average power dissipated by the resistor since $Q_L = Q_C$. The power factor of the circuit at resonance is

$$F_p = \cos\theta = \frac{P}{S}$$

and

$$F_{p_s} = 1 \tag{20.7}$$

Plotting the power curves of each element on the same set of axes (Fig. 20.5), we note that, even though the total reactive power at any instant is equal to zero (note that $t = t'$), energy is still being absorbed and released by the inductor and capacitor at resonance.

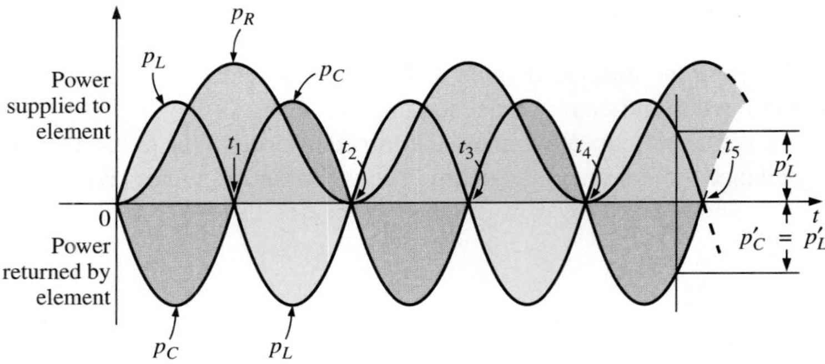

FIG. 20.5
Power curves at resonance for the series resonant circuit.

A closer examination reveals that the energy absorbed by the inductor from time 0 to t_1 is the same as the energy released by the capacitor from 0 to t_1. The reverse occurs from t_1 to t_2, and so on. Therefore, the total apparent power continues to be equal to the average power, even though the inductor and capacitor are absorbing and releasing energy. This condition occurs only at resonance. The slightest change in frequency introduces a reactive component into the power triangle, which increases the apparent power of the system above the average power dissipation, and resonance no longer exists.

20.3 THE QUALITY FACTOR (*Q*)

The **quality factor** Q of a series resonant circuit is defined as the ratio of the reactive power of either the inductor or the capacitor to the average power of the resistor at resonance; that is,

$$Q_s = \frac{\text{reactive power}}{\text{average power}} \tag{20.8}$$

The quality factor is also an indication of how much energy is placed in storage (continual transfer from one reactive element to the other) compared to that dissipated. The lower the level of dissipation for the same reactive power, the larger the Q_s factor and the more concentrated and intense the region of resonance.

Substituting for an inductive reactance in Eq. (20.8) at resonance gives us

$$Q_s = \frac{I^2 X_L}{I^2 R}$$

and

$$Q_s = \frac{X_L}{R} = \frac{\omega_s L}{R} \quad \textbf{(20.9)}$$

If the resistance R is just the resistance of the coil (R_l), we can speak of the Q of the coil, where

$$Q_{\text{coil}} = Q_l = \frac{X_L}{R_l} \quad \textbf{(20.10)}$$

Since the quality factor of a coil is typically the information provided by manufacturers of inductors, it is often given the symbol Q without an associated subscript. It appears from Eq. (20.10) that Q_l increases linearly with frequency since $X_L = 2\pi fL$. That is, if the frequency doubles, then Q_l also increases by a factor of 2. This is approximately true for the low range to the midrange of frequencies such as shown for the coils in Fig. 20.6. Unfortunately, however, as the frequency increases, the effective resistance of the coil also increases, due primarily to skin effect phenomena, and the resulting Q_l decreases. In addition, the capacitive effects between the windings increases, further reducing the Q_l of the coil. For this reason, Q_l must be specified for a particular frequency or frequency range. For wide frequency applications, a plot of Q_l versus frequency is often provided. The maximum Q_l for most commercially available coils is less than 200, with most having a maximum near 100. Note in Fig. 20.6 that for coils of the same type, Q_l drops off more quickly for higher levels of inductance.

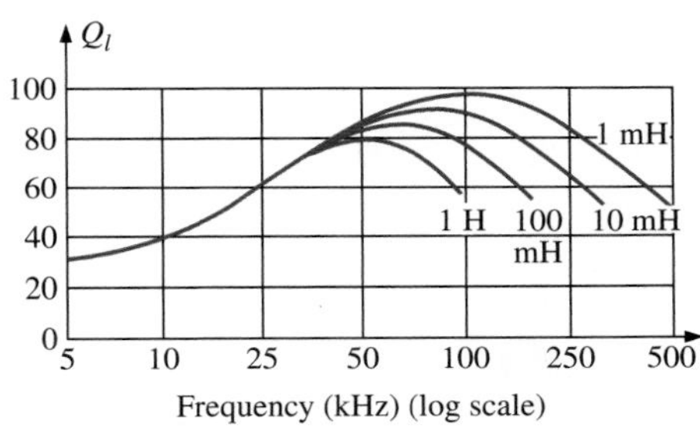

FIG. 20.6
Q_l versus frequency for a series of inductors of similar construction.

If we substitute

$$\omega_s = 2\pi f_s$$

and then

$$f_s = \frac{1}{2\pi\sqrt{LC}}$$

into Eq. (20.9), we have

$$Q_s = \frac{\omega_s L}{R} = \frac{2\pi f_s L}{R} = \frac{2\pi}{R}\left(\frac{1}{2\pi\sqrt{LC}}\right)L$$

$$= \frac{L}{R}\left(\frac{1}{\sqrt{LC}}\right) = \left(\frac{\sqrt{L}}{\sqrt{L}}\right)\frac{L}{R\sqrt{LC}}$$

and

$$Q_s = \frac{1}{R}\sqrt{\frac{L}{C}} \quad \textbf{(20.11)}$$

providing Q_s in terms of the circuit parameters.

For series resonant circuits used in communication systems, Q_s is usually greater than 1. By applying the voltage divider rule to the circuit in Fig. 20.2, we obtain

$$V_L = \frac{X_L E}{Z_T} = \frac{X_L E}{R} \quad \text{(at resonance)}$$

and $$\boxed{V_{L_s} = Q_s E} \tag{20.12}$$

or $$V_C = \frac{X_C E}{Z_T} = \frac{X_C E}{R}$$

and $$\boxed{V_{C_s} = Q_s E} \tag{20.13}$$

Since Q_s is usually greater than 1, the voltage across the capacitor or inductor of a series resonant circuit can be significantly greater than the input voltage. In fact, in many cases the Q_s is so high that careful design and handling (including adequate insulation) are mandatory with respect to the voltage across the capacitor and inductor.

In the circuit in Fig. 20.7, for example, which is in the state of resonance,

$$Q_s = \frac{X_L}{R} = \frac{480\ \Omega}{6\ \Omega} = 80$$

and $$V_L = V_C = Q_s E = (80)(10\ \text{V}) = \mathbf{800\ V}$$

which is certainly a potential of significant magnitude.

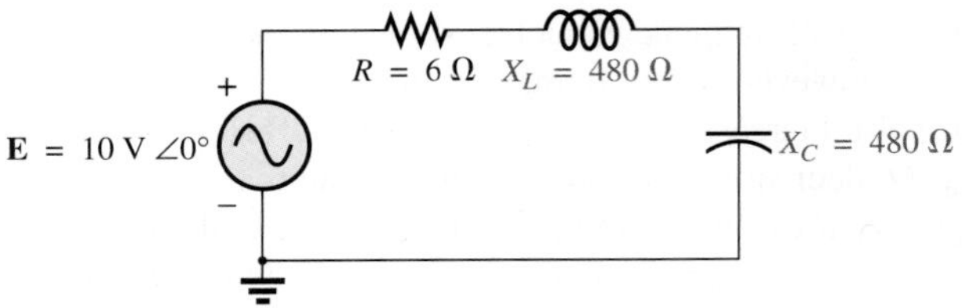

FIG. 20.7
High-Q series resonant circuit.

20.4 Z_T VERSUS FREQUENCY

The total impedance of the series *R-L-C* circuit in Fig. 20.2 at any frequency is determined by

$$\mathbf{Z}_T = R + jX_L - jX_C \quad \text{or} \quad \mathbf{Z}_T = R + j(X_L - X_C)$$

The magnitude of the impedance $\mathbf{Z}_T$ versus frequency is determined by

$$Z_T = \sqrt{R^2 + (X_L - X_C)^2}$$

The total-impedance-versus-frequency curve for the series resonant circuit in Fig. 20.2 can be found by applying the impedance-versus-frequency curve for each element of the equation just derived, written in the following form:

$$\boxed{Z_T(f) = \sqrt{[R(f)]^2 + [X_L(f) - X_C(f)]^2}} \tag{20.14}$$

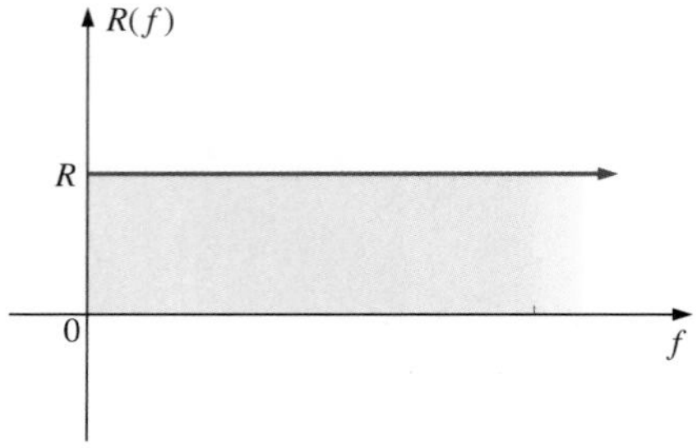

where $Z_T(f)$ "means" the total impedance as a *function* of frequency. For the frequency range of interest, we assume that the resistance R does not change with frequency, resulting in the plot in Fig. 20.8. The curve for the inductance, as determined by the reactance equation, is a straight line intersecting the origin with a slope equal to the inductance of the coil. The mathematical expression for any straight line in a two-dimensional plane is given by

$$y = mx + b$$

FIG. 20.8
Resistance versus frequency.

Thus, for the coil,

$$\begin{array}{cccccc} X_L = 2\pi fL + 0 = & (2\pi L) & (f) & + 0 \\ \downarrow & \downarrow & \downarrow & \downarrow \\ y = & m & \cdot\ x & +\ b \end{array}$$

(where $2\pi L$ is the slope), producing the results shown in Fig. 20.9.

For the capacitor,

$$X_C = \frac{1}{2\pi fC} \quad \text{or} \quad X_C f = \frac{1}{2\pi C}$$

which becomes $yx = k$, the equation for a hyperbola, where

$$\begin{aligned} y\,(\text{variable}) &= X_C \\ x\,(\text{variable}) &= f \\ k\,(\text{constant}) &= \frac{1}{2\pi C} \end{aligned}$$

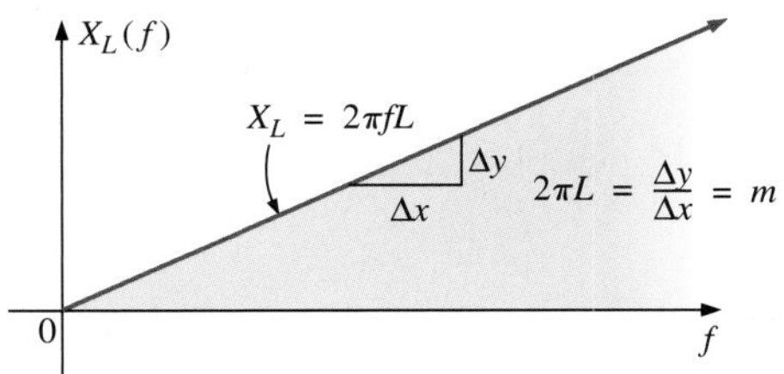

FIG. 20.9
Inductive reactance versus frequency.

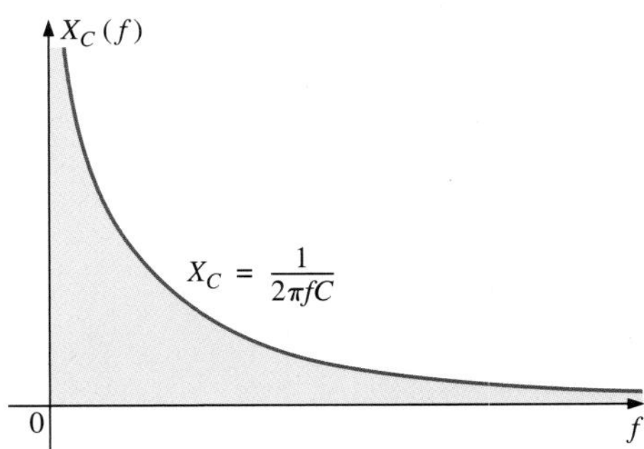

FIG. 20.10
Capacitive reactance versus frequency.

The hyperbolic curve for $X_C(f)$ is plotted in Fig. 20.10. In particular, note its very large magnitude at low frequencies and its rapid dropoff as the frequency increases.

If we place Figs. 20.9 and 20.10 on the same set of axes, we obtain the curves in Fig. 20.11. The condition of resonance is now clearly defined by the point of intersection, where $X_L = X_C$. For frequencies less than f_s, it is also quite clear that the network is primarily capacitive ($X_C > X_L$). For frequencies above the resonant condition, $X_L > X_C$, and the network is inductive.

Applying

$$\begin{aligned} Z_T(f) &= \sqrt{[R(f)]^2 + [X_L(f) - X_C(f)]^2} \\ &= \sqrt{[R(f)]^2 + [X(f)]^2} \end{aligned}$$

to the curves in Fig. 20.11, where $X(f) = X_L(f) - X_C(f)$, we obtain the curve for $Z_T(f)$ as shown in Fig. 20.12. The minimum impedance occurs at the resonant frequency and is equal to the resistance R. Note that the curve is not symmetrical about the resonant frequency (especially at higher values of Z_T).

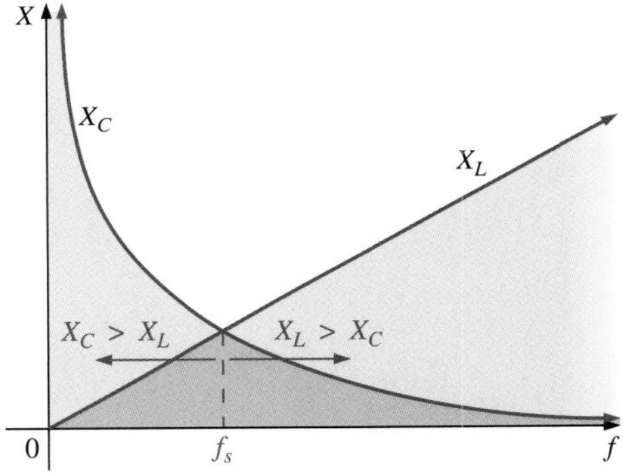

FIG. 20.11
Placing the frequency response of the inductive and capacitive reactance of a series R-L-C circuit on the same set of axes.

The phase angle associated with the total impedance is

$$\theta = \tan^{-1}\frac{(X_L - X_C)}{R} \tag{20.15}$$

For the $\tan^{-1}x$ function (resulting when $X_L > X_C$), the larger x is, the larger the angle θ (closer to 90°). However, for regions where $X_C > X_L$, one must also be aware that

$$\tan^{-1}(-x) = -\tan^{-1}x \tag{20.16}$$

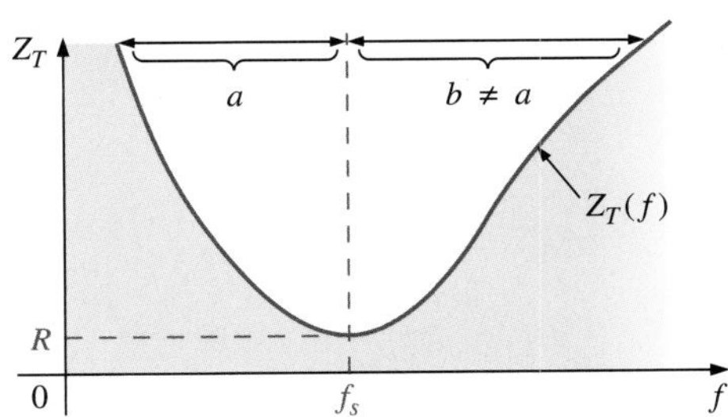

FIG. 20.12
Z_T versus frequency for the series resonant circuit.

At low frequencies, $X_C > X_L$, and θ approaches −90° (capacitive), as shown in Fig. 20.13, whereas at high frequencies, $X_L > X_C$, and θ approaches 90°. In general, therefore, for a series resonant circuit:

$f < f_s$:	network capacitive; **I** leads **E**
$f > f_s$:	network inductive; **E** leads **I**
$f = f_s$:	network resistive; **E** and **I** are in phase

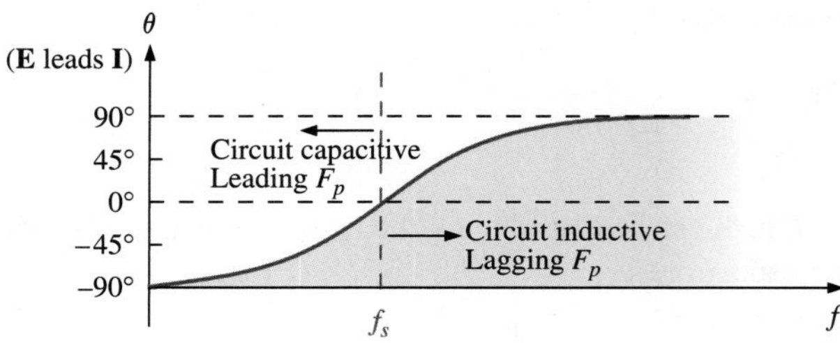

FIG. 20.13
Phase plot for the series resonant circuit.

20.5 SELECTIVITY

If we now plot the magnitude of the current $I = E/Z_T$ versus frequency for a *fixed* applied voltage E, we obtain the curve shown in Fig. 20.14, which rises from zero to a maximum value of E/R (where Z_T is a minimum) and then drops toward zero (as Z_T increases) at a slower rate than it rose to its peak value. The curve is actually the inverse of the impedance-versus-frequency curve. Since the Z_T curve is not absolutely symmetrical about the resonant frequency, the curve of the current versus frequency has the same property.

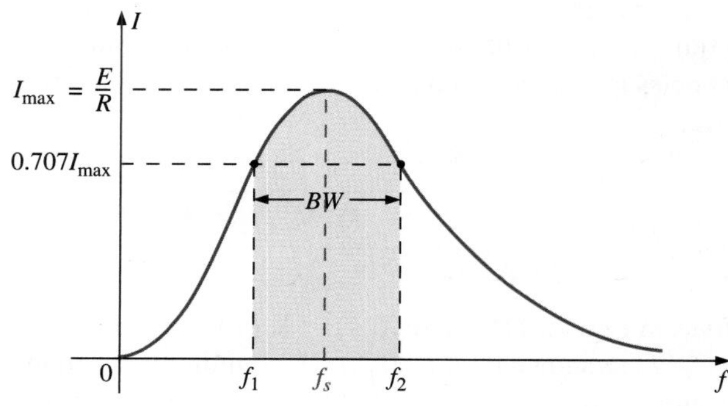

FIG. 20.14
I versus frequency for the series resonant circuit.

There is a definite range of frequencies at which the current is near its maximum value and the impedance is at a minimum. Those frequencies corresponding to 0.707 of the maximum current are called the **band frequencies, cutoff frequencies, half-power frequencies,** or **corner frequencies.** They are indicated by f_1 and f_2 in Fig. 20.14. The range of frequencies between the two is referred to as the **bandwidth** (abbreviated BW) of the resonant circuit.

Half-power frequencies are those frequencies at which the power delivered is one-half that delivered at the resonant frequency; that is,

$$P_{HPF} = \frac{1}{2} P_{max} \tag{20.17}$$

The above condition is derived using the fact that

$$P_{max} = I_{max}^2 R$$

and $$P_{HPF} = I^2 R = (0.707 I_{max})^2 R = (0.5)(I_{max}^2 R) = \frac{1}{2} P_{max}$$

Since the resonant circuit is adjusted to select a band of frequencies, the curve in Fig. 20.14 is called the **selectivity curve.** The term is derived from the fact that one must be *selective* in choosing the frequency to ensure that it is in the bandwidth. The smaller the bandwidth, the higher the selectivity. The shape of the curve, as shown in Fig. 20.15, depends on each element of the series *R-L-C* circuit. If the resistance is made smaller with a fixed inductance and capacitance, the bandwidth decreases and the selectivity increases. Similarly, if the ratio L/C increases with fixed resistance, the bandwidth again decreases with an increase in selectivity.

In terms of Q_s, if R is larger for the same X_L, then Q_s is less, as determined by the equation $Q_s = \omega_s L/R$.

A small Q_s, therefore, is associated with a resonant curve having a large bandwidth and a small selectivity, while a large Q_s indicates the opposite.

For circuits where $Q_s \geq 10$, a widely accepted approximation is that the resonant frequency bisects the bandwidth and that the resonant curve is symmetrical about the resonant frequency.

These conditions are shown in Fig. 20.16, indicating that the cutoff frequencies are then equidistant from the resonant frequency.

For any Q_s, the preceding is not true. The cutoff frequencies f_1 and f_2 can be found for the general case (any Q_s) by first using the fact that a drop in current to 0.707 of its resonant value corresponds to an increase in impedance equal to $1/0.707 = \sqrt{2}$ times the resonant value, which is R.

Substituting $\sqrt{2}R$ into the equation for the magnitude of Z_T, we find that

$$Z_T = \sqrt{R^2 + (X_L - X_C)^2}$$

becomes

$$\sqrt{2}R = \sqrt{R^2 + (X_L - X_C)^2}$$

or, squaring both sides, that

$$2R^2 = R^2 + (X_L - X_C)^2$$

and,

$$R^2 = (X_L - X_C)^2$$

Taking the square root of both sides gives us

$$R = X_L - X_C \quad \text{or} \quad R - X_L + X_C = 0$$

Let us first consider the case where $X_L > X_C$, which relates to f_2 or ω_2. Substituting $\omega_2 L$ for X_L and $1/\omega_2 C$ for X_C and bringing both quantities to the left of the equal sign, we have

$$R - \omega_2 L + \frac{1}{\omega_2 C} = 0 \quad \text{or} \quad R\omega_2 - \omega_2^2 L + \frac{1}{C} = 0$$

which can be written

$$\omega_2^2 - \frac{R}{L}\omega_2 - \frac{1}{LC} = 0$$

Solving the quadratic, we have

$$\omega_2 = \frac{-(-R/L) \pm \sqrt{[-(R/L)]^2 - [-(4/LC)]}}{2}$$

and

$$\omega_2 = +\frac{R}{2L} \pm \frac{1}{2}\sqrt{\frac{R^2}{L^2} + \frac{4}{LC}}$$

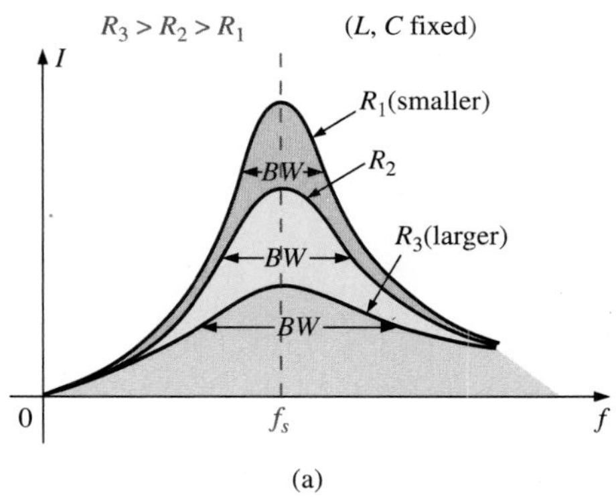

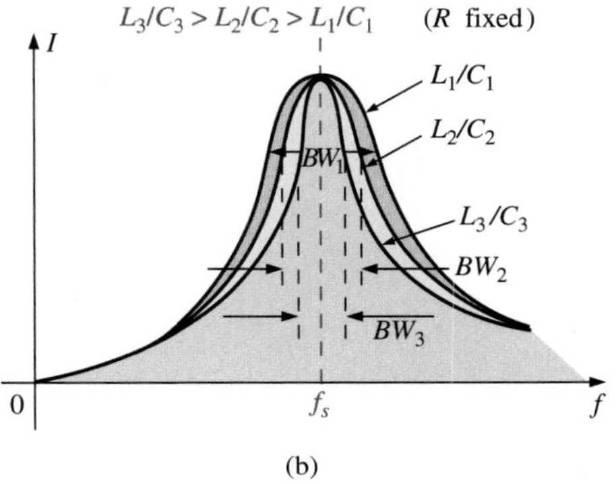

FIG. 20.15
Effect of R, L, and C on the selectivity curve for the series resonant circuit.

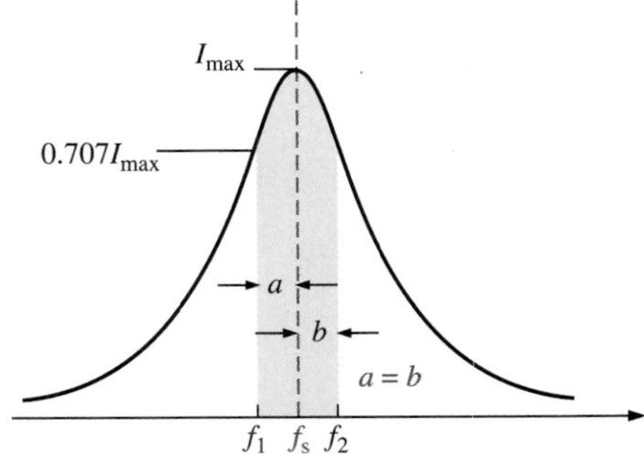

FIG. 20.16
Approximate series resonance curve for $Q_s \geq 10$.

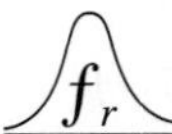

with
$$f_2 = \frac{1}{2\pi}\left[\frac{R}{2L} + \frac{1}{2}\sqrt{\left(\frac{R}{L}\right)^2 + \frac{4}{LC}}\right] \quad \text{(Hz)} \tag{20.18}$$

The negative sign in front of the second factor was dropped because $(1/2)\sqrt{(R/L)^2 + 4/LC}$ is always greater than $R/(2L)$. If it were not dropped, there would be a negative solution for the radian frequency ω.

If we repeat the same procedure for $X_C > X_L$, which relates to ω_1 or f_1 such that $Z_T = \sqrt{R^2 + (X_C - X_L)^2}$, the solution f_1 becomes

$$f_1 = \frac{1}{2\pi}\left[-\frac{R}{2L} + \frac{1}{2}\sqrt{\left(\frac{R}{L}\right)^2 + \frac{4}{LC}}\right] \quad \text{(Hz)} \tag{20.19}$$

The bandwidth (BW) is

$$BW = f_2 - f_1 = \text{Eq. (20.18)} - \text{Eq. (20.19)}$$

and
$$BW = f_2 - f_1 = \frac{R}{2\pi L} \tag{20.20}$$

Substituting $R/L = \omega_s/Q_s$ from $Q_s = \omega_s L/R$ and $1/2\pi = f_s/\omega_s$ from $\omega_s = 2\pi f_s$ gives us

$$BW = \frac{R}{2\pi L} = \left(\frac{1}{2\pi}\right)\left(\frac{R}{L}\right) = \left(\frac{f_s}{\omega_s}\right)\left(\frac{\omega_s}{Q_s}\right)$$

or
$$BW = \frac{f_s}{Q_s} \tag{20.21}$$

which is a very convenient form since it relates the bandwidth to the Q_s of the circuit. As mentioned earlier, Eq. (20.21) verifies that the larger the Q_s, the smaller the bandwidth, and vice versa.

Written in a slightly different form, Eq. (20.21) becomes

$$\frac{f_2 - f_1}{f_s} = \frac{1}{Q_s} \tag{20.22}$$

The ratio $(f_2 - f_1)/f_s$ is sometimes called the *fractional bandwidth,* providing an indication of the width of the bandwidth compared to the resonant frequency.

It can also be shown through mathematical manipulations of the pertinent equations that the resonant frequency is related to the geometric mean of the band frequencies; that is,

$$f_s = \sqrt{f_1 f_2} \tag{20.23}$$

20.6 V_R, V_L, AND V_C

Plotting the magnitude (effective value) of the voltages $\mathbf{V}_R$, $\mathbf{V}_L$, and $\mathbf{V}_C$ and the current $\mathbf{I}$ versus frequency for the series resonant circuit on the same set of axes, we obtain the curves shown in Fig. 20.17. Note that the V_R curve has the same shape as the I curve and a peak value equal to the magnitude of the input voltage E. The V_C curve builds up slowly at first from

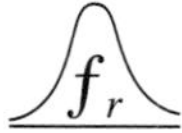

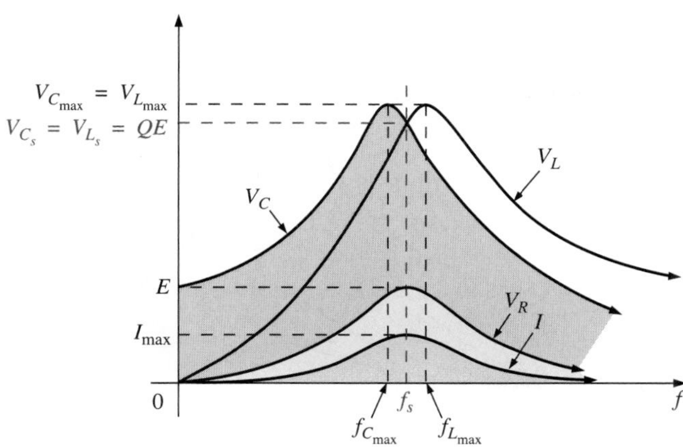

FIG. 20.17
V_R, V_L, V_C, and I versus frequency for a series resonant circuit.

a value equal to the input voltage since the reactance of the capacitor is infinite (open circuit) at zero frequency and the reactance of the inductor is zero (short circuit) at this frequency. As the frequency increases, $1/\omega C$ of the equation

$$V_C = IX_C = (I)\left(\frac{1}{\omega C}\right)$$

becomes smaller, but I increases at a rate faster than that at which $1/\omega C$ drops. Therefore, V_C rises and will continue to rise due to the quickly rising current, until the frequency nears resonance. As it approaches the resonant condition, the rate of change of I decreases. When this occurs, the factor $1/\omega C$, which decreased as the frequency rose, overcomes the rate of change of I, and V_C starts to drop. The peak value occurs at a frequency just before resonance. After resonance, both V_C and I drop in magnitude, and V_C approaches zero.

The higher the Q_s of the circuit, the closer $f_{C_{max}}$ will be to f_s, and the closer $V_{C_{max}}$ will be to Q_sE. For circuits with $Q_s \geq 10$, $f_{C_{max}} \cong f_s$, and $V_{C_{max}} \cong Q_sE$.

The curve for V_L increases steadily from zero to the resonant frequency since both quantities ωL and I of the equation $V_L = IX_L = (I)(\omega L)$ increase over this frequency range. At resonance, I has reached its maximum value, but ωL is still rising. Therefore, V_L reaches its maximum value after resonance. After reaching its peak value, the voltage V_L drops toward E since the drop in I overcomes the rise in ωL. It approaches E because X_L will eventually be infinite, and X_C will be zero.

As Q_s of the circuit increases, the frequency $f_{L_{max}}$ drops toward f_s, and $V_{L_{max}}$ approaches Q_sE. For circuits with $Q_s \geq 10$, $f_{L_{max}} \cong f_s$, and $V_{L_{max}} \cong Q_sE$.

The V_L curve has a greater magnitude than the V_C curve for any frequency above resonance, and the V_C curve has a greater magnitude than the V_L curve for any frequency below resonance. This again verifies that the series *R-L-C* circuit is predominantly capacitive from zero to the resonant frequency and predominantly inductive for any frequency above resonance.

For the condition $Q_s \geq 10$, the curves in Fig. 20.17 appear as shown in Fig. 20.18. Note that they each peak (on an approximate basis) at the resonant frequency and have a similar shape.

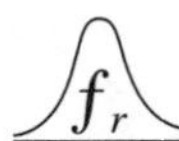

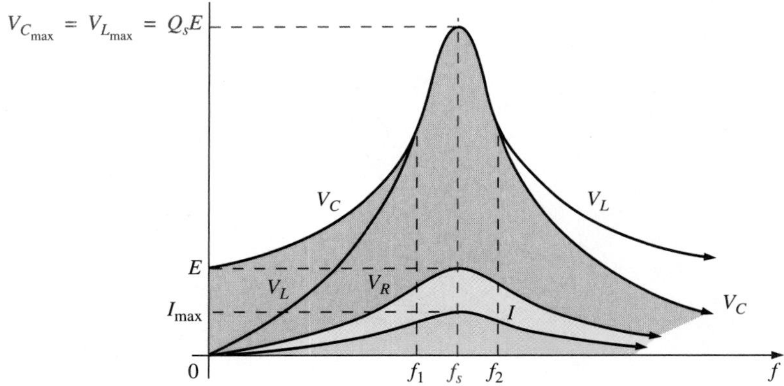

FIG. 20.18
V_R, V_L, V_C, and I for a series resonant circuit where $Q_s \geq 10$.

In review,

1. V_C and V_L are at their maximum values at or near resonance (depending on Q_s).
2. At very low frequencies, V_C is very close to the source voltage and V_L is very close to zero volts, whereas at very high frequencies, V_L approaches the source voltage and V_C approaches zero volts.
3. Both V_R and I peak at the resonant frequency and have the same shape.

20.7 EXAMPLES (SERIES RESONANCE)

EXAMPLE 20.1

a. For the series resonant circuit in Fig. 20.19, find **I**, $\mathbf{V}_R$, $\mathbf{V}_L$, and $\mathbf{V}_C$ at resonance.
b. What is the Q_s of the circuit?
c. If the resonant frequency is 5000 Hz, find the bandwidth.
d. What is the power dissipated in the circuit at the half-power frequencies?

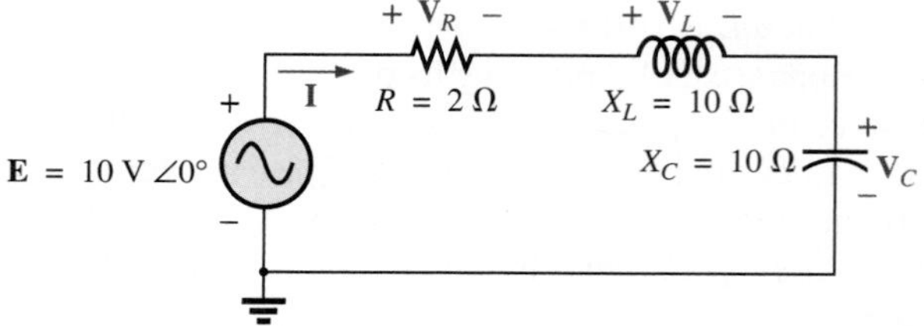

FIG. 20.19
Example 20.1.

Solutions:

a. $\mathbf{Z}_{T_s} = R = 2\ \Omega$

$$\mathbf{I} = \frac{\mathbf{E}}{\mathbf{Z}_{T_s}} = \frac{10\ \text{V} \angle 0°}{2\ \Omega \angle 0°} = \mathbf{5\ A \angle 0°}$$

$$\mathbf{V}_R = \mathbf{E} = \mathbf{10\ V \angle 0°}$$

$$\mathbf{V}_L = (I \angle 0°)(X_L \angle 90°) = (5\text{ A} \angle 0°)(10\ \Omega \angle 90°) = \mathbf{50\ V \angle 90°}$$

$$\mathbf{V}_C = (I \angle 0°)(X_C \angle -90°) = (5\text{ A} \angle 0°)(10\ \Omega \angle -90°) = \mathbf{50\ V \angle -90°}$$

b. $Q_s = \dfrac{X_L}{R} = \dfrac{10\ \Omega}{2\ \Omega} = \mathbf{5}$

c. $BW = f_2 - f_1 = \dfrac{f_s}{Q_s} = \dfrac{5000\text{ Hz}}{5} = \mathbf{1000\ Hz}$

d. $P_{\text{HPF}} = \dfrac{1}{2}P_{\max} = \dfrac{1}{2}I_{\max}^2 R = \left(\dfrac{1}{2}\right)(5\text{ A})^2(2\ \Omega) = \mathbf{25\ W}$

EXAMPLE 20.2 The bandwidth of a series resonant circuit is 400 Hz.

a. If the resonant frequency is 4000 Hz, what is the value of Q_s?
b. If $R = 10\ \Omega$, what is the value of X_L at resonance?
c. Find the inductance L and capacitance C of the circuit.

Solutions:

a. $BW = \dfrac{f_s}{Q_s}$ or $Q_s = \dfrac{f_s}{BW} = \dfrac{4000\text{ Hz}}{400\text{ Hz}} = \mathbf{10}$

b. $Q_s = \dfrac{X_L}{R}$ or $X_L = Q_s R = (10)(10\ \Omega) = \mathbf{100\ \Omega}$

c. $X_L = 2\pi f_s L$ or $L = \dfrac{X_L}{2\pi f_s} = \dfrac{100\ \Omega}{2\pi(4000\text{ Hz})} = \mathbf{3.98\ mH}$

$$X_C = \frac{1}{2\pi f_s C} \quad \text{or} \quad C = \frac{1}{2\pi f_s X_C} = \frac{1}{2\pi(4000\text{ Hz})(100\ \Omega)} = \mathbf{397.89\ nF}$$

EXAMPLE 20.3 A series *R-L-C* circuit has a series resonant frequency of 12,000 Hz.

a. If $R = 5\ \Omega$, and if X_L at resonance is 300 Ω, find the bandwidth.
b. Find the cutoff frequencies.

Solutions:

a. $Q_s = \dfrac{X_L}{R} = \dfrac{300\ \Omega}{5\ \Omega} = 60$

$$BW = \frac{f_s}{Q_s} = \frac{12{,}000\text{ Hz}}{60} = \mathbf{200\ Hz}$$

b. Since $Q_s \geq 10$, the bandwidth is bisected by f_s. Therefore,

$$f_2 = f_s + \frac{BW}{2} = 12{,}000\text{ Hz} + 100\text{ Hz} = \mathbf{12{,}100\ Hz}$$

and $f_1 = 12{,}000\text{ Hz} - 100\text{ Hz} = \mathbf{11{,}900\ Hz}$

EXAMPLE 20.4

a. Determine the Q_s and bandwidth for the response curve in Fig. 20.20.
b. For $C = 101.5$ nF, determine L and R for the series resonant circuit.
c. Determine the applied voltage.

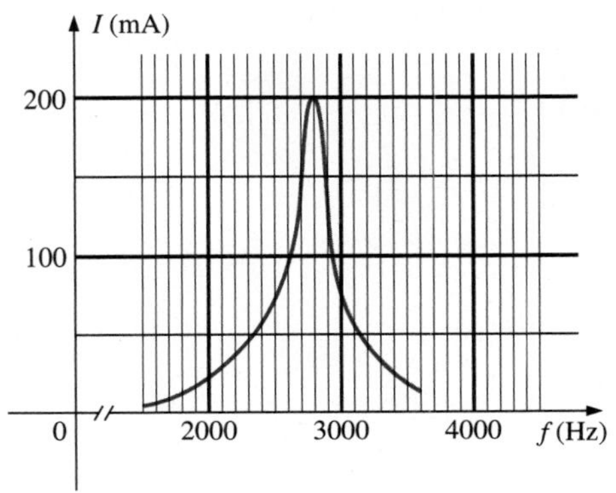

FIG. 20.20
Example 20.4

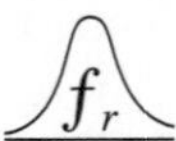

Solutions:

a. The resonant frequency is 2800 Hz. At 0.707 times the peak value,

$$BW = \mathbf{200\ Hz}$$

and
$$Q_s = \frac{f_s}{BW} = \frac{2800\text{ Hz}}{200\text{ Hz}} = \mathbf{14}$$

b. $$f_s = \frac{1}{2\pi\sqrt{LC}} \quad \text{or} \quad L = \frac{1}{4\pi^2 f_s^2 C}$$

$$= \frac{1}{4\pi^2\,(2.8\text{ kHz})^2(101.5\text{ nF})}$$

$$= \mathbf{31.83\ mH}$$

$$Q_s = \frac{X_L}{R} \quad \text{or} \quad R = \frac{X_L}{Q_s} = \frac{2\pi(2800\text{ Hz})(31.832\text{ mH})}{14}$$

$$= \mathbf{40\ \Omega}$$

c. $$I_{\max} = \frac{E}{R} \quad \text{or} \quad E = I_{\max}R$$

$$= (200\text{ mA})(40\ \Omega) = \mathbf{8\ V}$$

EXAMPLE 20.5 A series *R-L-C* circuit is designed to resonate at $\omega_s = 10^5$ rad/s, have a bandwidth of $0.15\omega_s$, and draw 16 W from a 120 V source at resonance.

a. Determine the value of *R*.
b. Find the bandwidth in hertz.
c. Find the nameplate values of *L* and *C*.
d. Determine the Q_s of the circuit.
e. Determine the fractional bandwidth.

Solutions:

a. $$P = \frac{E^2}{R} \quad \text{and} \quad R = \frac{E^2}{P} = \frac{(120\text{ V})^2}{16\text{ W}} = \mathbf{900\ \Omega}$$

b. $$f_s = \frac{\omega_s}{2\pi} = \frac{10^5\text{ rad/s}}{2\pi} = 15{,}915.49\text{ Hz}$$

$$BW = 0.15f_s = 0.15(15{,}915.49\text{ Hz}) = \mathbf{2387.32\ Hz}$$

c. Eq. (20.20):

$$BW = \frac{R}{2\pi L} \quad \text{and} \quad L = \frac{R}{2\pi BW} = \frac{900\ \Omega}{2\pi(2387.32\text{ Hz})} = \mathbf{60\ mH}$$

$$f_s = \frac{1}{2\pi\sqrt{LC}} \quad \text{and} \quad C = \frac{1}{4\pi^2 f_s^2\, L}$$

$$= \frac{1}{4\pi^2(15{,}915.49\text{ Hz})^2(60 \times 10^{-3})}$$

$$= \mathbf{1.67\ nF}$$

d. $$Q_s = \frac{X_L}{R} = \frac{2\pi f_s L}{R} = \frac{2\pi(15{,}915.49\text{ Hz})(60\text{ mH})}{900\ \Omega} = \mathbf{6.67}$$

e. $$\frac{f_2 - f_1}{f_s} = \frac{BW}{f_s} = \frac{1}{Q_s} = \frac{1}{6.67} = \mathbf{0.15}$$

PARALLEL RESONANCE

20.8 PARALLEL RESONANT CIRCUIT

The basic format of the series resonant circuit is a series *R-L-C* combination in series with an applied voltage source. The parallel resonant circuit has the basic configuration in Fig. 20.21, a parallel *R-L-C* combination in parallel with an applied current source.

For the series circuit, the impedance was a minimum at resonance, producing a significant current that resulted in a high output voltage for $\mathbf{V}_C$ and $\mathbf{V}_L$. For the parallel resonant circuit, the impedance is relatively high at resonance, producing a significant voltage for $\mathbf{V}_C$ and $\mathbf{V}_L$ through the Ohm's law relationship ($\mathbf{V}_C = \mathbf{IZ}_T$). For the network in Fig. 20.21, resonance occurs when $X_L = X_C$, and the resonant frequency has the same format obtained for series resonance.

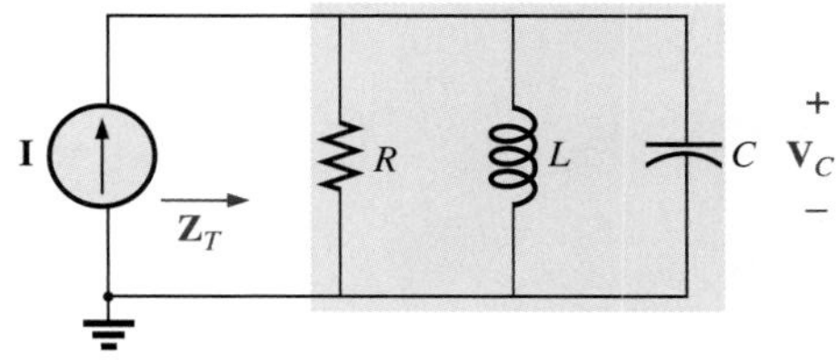

FIG. 20.21
Ideal parallel resonant network.

If the practical equivalent in Fig. 20.22 had the format in Fig. 20.21, the analysis would be as direct and lucid as that experienced for series resonance. However, in the practical world, the internal resistance of the coil must be placed in series with the inductor, as shown in Fig. 20.22. The resistance R_l can no longer be included in a simple series or parallel combination with the source resistance and any other resistance added for design purposes. Even though R_l is usually relatively small in magnitude compared with other resistance and reactance levels of the network, it does have an important impact on the parallel resonant condition, as demonstrated in the sections to follow. In other words, the network in Fig. 20.21 is an ideal situation that can be assumed only for specific network conditions.

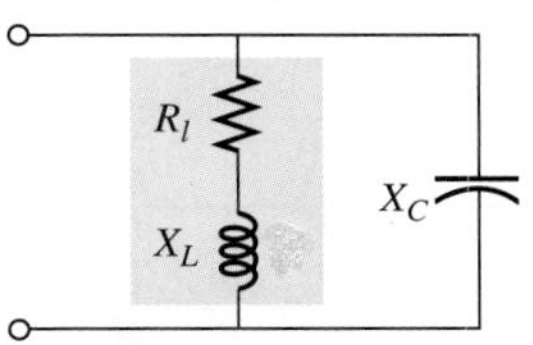

FIG. 20.22
Practical parallel L-C network.

Our first effort is to find a parallel network equivalent (at the terminals) for the series *R-L* branch in Fig. 20.22 using the technique introduced in Section 15.10. That is,

$$\mathbf{Z}_{R\text{-}L} = R_l + j\,X_L$$

and
$$\mathbf{Y}_{R\text{-}L} = \frac{1}{\mathbf{Z}_{R\text{-}L}} = \frac{1}{R_l + j\,X_L} = \frac{R_l}{R_l^2 + X_L^2} - j\frac{X_L}{R_l^2 + X_L^2}$$

$$= \frac{1}{\dfrac{R_l^2 + X_L^2}{R_l}} + \frac{1}{j\left(\dfrac{R_l^2 + X_L^2}{X_L}\right)} = \frac{1}{R_p} + \frac{1}{j\,X_{L_p}}$$

with
$$R_p = \frac{R_l^2 + X_L^2}{R_l} \qquad (20.24)$$

and
$$X_{L_p} = \frac{R_l^2 + X_L^2}{X_L} \qquad (20.25)$$

as shown in Fig. 20.23.

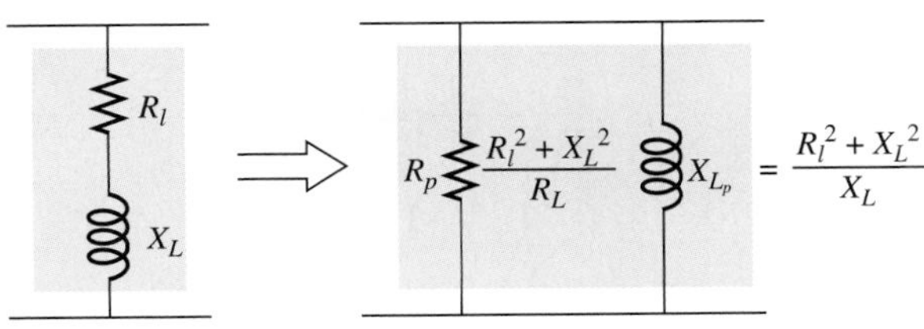

FIG. 20.23
Equivalent parallel network for a series R-L combination.

Redrawing the network in Fig. 20.22 with the equivalent in Fig. 20.23 and a practical current source having an internal resistance R_s results in the network in Fig. 20.24.

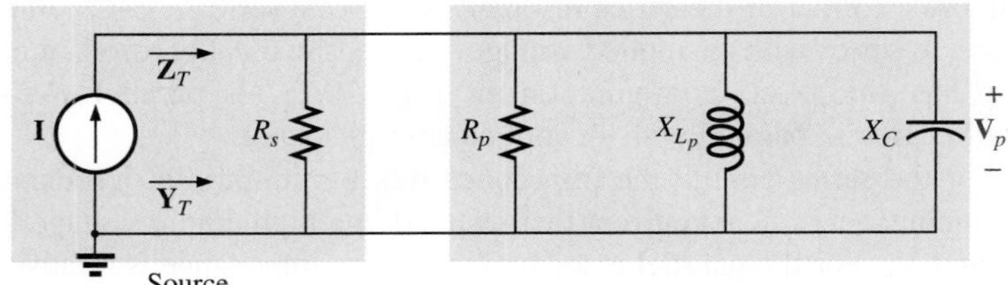

FIG. 20.24
Substituting the equivalent parallel network for the series R-L combination in Fig. 20.22.

If we define the parallel combination of R_s and R_p by the notation

$$\boxed{R = R_s \parallel R_p} \tag{20.26}$$

the network in Fig. 20.25 results. It has the same format as the ideal configuration in Fig. 20.21.

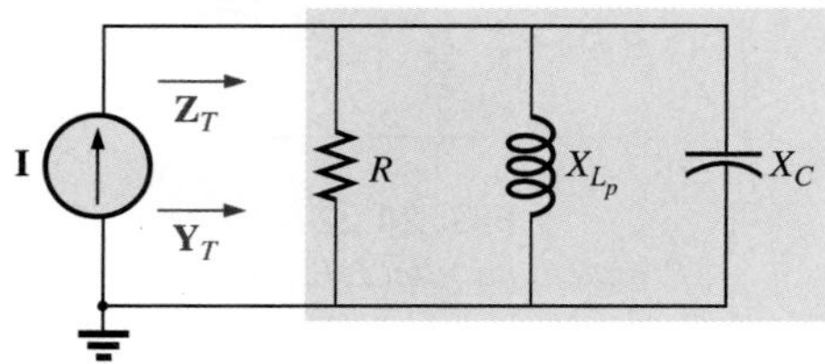

FIG. 20.25
Substituting $R = R_s \parallel R_p$ for the network in Fig. 20.24.

We are now at a point where we can define the resonance conditions for the practical parallel resonant configuration. Recall that for series resonance, the resonant frequency was the frequency at which the impedance was a minimum, the current a maximum, and the input impedance purely resistive, and the network had a unity power factor. For parallel networks, since the resistance R_p in our equivalent model is frequency dependent, the frequency at which maximum V_C is obtained is not the same as required for the unity-power-factor characteristic. Since both conditions are often used to define the resonant state, the frequency at which each occurs is designated by different subscripts.

Unity Power Factor, f_p

For the network in Fig. 20.25,

$$\mathbf{Y}_T = \frac{1}{\mathbf{Z}_1} + \frac{1}{\mathbf{Z}_2} + \frac{1}{\mathbf{Z}_3} = \frac{1}{R} + \frac{1}{j\,X_{L_p}} + \frac{1}{-j\,X_C}$$
$$= \frac{1}{R} - j\left(\frac{1}{X_{L_p}}\right) + j\left(\frac{1}{X_C}\right)$$

and

$$\boxed{\mathbf{Y}_T = \frac{1}{R} + j\left(\frac{1}{X_C} - \frac{1}{X_{L_p}}\right)} \tag{20.27}$$

For unity power factor, the reactive component must be zero as defined by

$$\frac{1}{X_C} - \frac{1}{X_{L_p}} = 0$$

Therefore,

$$\frac{1}{X_C} = \frac{1}{X_{L_p}}$$

and

$$\boxed{X_{L_p} = X_C} \tag{20.28}$$

Substituting for X_{L_p} yields

$$\frac{R_l^2 + X_L^2}{X_L} = X_C \tag{20.29}$$

The resonant frequency, f_p, can now be determined from Eq. (20.29) as follows:

$$R_l^2 + X_L^2 = X_C X_L = \left(\frac{1}{\omega C}\right)\omega L = \frac{L}{C}$$

or

$$X_L^2 = \frac{L}{C} - R_l^2$$

with

$$2\pi f_p L = \sqrt{\frac{L}{C} - R_l^2}$$

and

$$f_p = \frac{1}{2\pi L}\sqrt{\frac{L}{C} - R_l^2}$$

Multiplying the top and bottom of the factor within the square root sign by C/L produces

$$f_p = \frac{1}{2\pi L}\sqrt{\frac{1 - R_l^2(C/L)}{C/L}} = \frac{1}{2\pi L\sqrt{C/L}}\sqrt{1 - \frac{R_l^2 C}{L}}$$

and

$$f_p = \frac{1}{2\pi\sqrt{LC}}\sqrt{1 - \frac{R_l^2 C}{L}} \tag{20.30}$$

or

$$f_p = f_s\sqrt{1 - \frac{R_l^2 C}{L}} \tag{20.31}$$

where f_p is the resonant frequency of a parallel resonant circuit (for $F_p = 1$) and f_s is the resonant frequency as determined by $X_L = X_C$ for series resonance. Note that unlike a series resonant circuit, the resonant frequency f_p is a function of resistance (in this case R_l). Note also, however, the absence of the source resistance R_s in Eqs. (20.30) and (20.31). Since the factor $\sqrt{1 - (R_l^2 C/L)}$ is less than 1, f_p is less than f_s. Recognize also that as the magnitude of R_l approaches zero, f_p rapidly approaches f_s.

Maximum Impedance, f_m

At $f = f_p$ the input impedance of a parallel resonant circuit will be near its maximum value but not quite its maximum value due to the frequency dependence of R_p. The frequency at which maximum impedance occurs is defined by f_m and is slightly more than f_p, as demonstrated in Fig. 20.26. The frequency f_m is determined by differentiating (calculus) the general equation for Z_T with respect to frequency and then determining the frequency at which the resulting equation is equal to zero. The algebra is quite extensive and cumbersome and is not included here. The resulting equation, however, is the following:

$$f_m = f_s\sqrt{1 - \frac{1}{4}\left(\frac{R_l^2 C}{L}\right)} \tag{20.32}$$

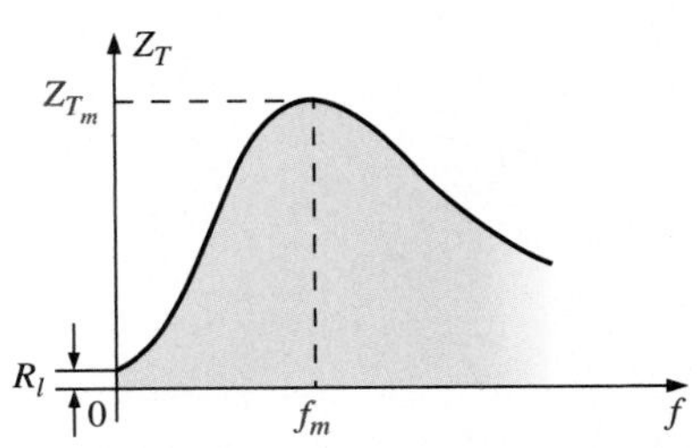

FIG. 20.26
Z_T versus frequency for the parallel resonant circuit.

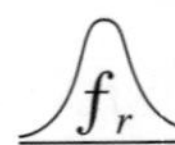

Note the similarities with Eq. (20.31). Since the square root factor of Eq. (20.32) is always more than the similar factor of Eq. (20.31), f_m is always closer to f_s and more than f_p. In general,

$$f_s > f_m > f_p \tag{20.33}$$

Once f_m is determined, the network in Fig. 20.25 can be used to determine the magnitude and phase angle of the total impedance at the resonance condition simply by substituting $f = f_m$ and performing the required calculations. That is,

$$Z_{T_m} = R \parallel X_{L_p} \parallel X_C \Big|_{f = f_m} \tag{20.34}$$

20.9 SELECTIVITY CURVE FOR PARALLEL RESONANT CIRCUITS

The Z_T-versus-frequency curve in Fig. 20.26 clearly reveals that a parallel resonant circuit exhibits maximum impedance at resonance (f_m), unlike the series resonant circuit, which experiences minimum resistance levels at resonance. Note also that Z_T is approximately R_l at $f = 0$ Hz since $Z_T = R_s \parallel R_l \cong R_l$.

Since the current I of the current source is constant for any value of Z_T or frequency, the voltage across the parallel circuit will have the same shape as the total impedance Z_T, as shown in Fig. 20.27.

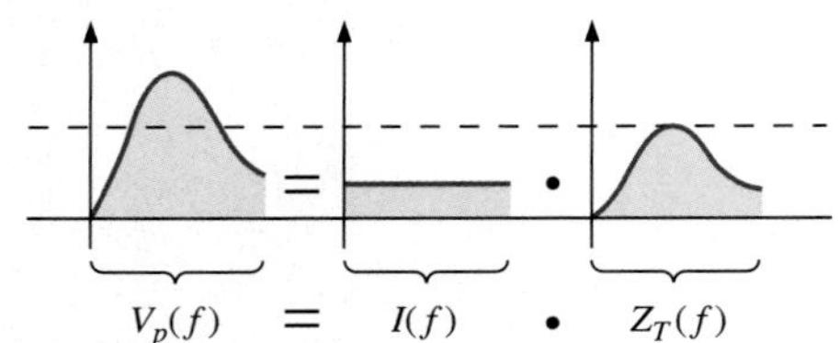

FIG. 20.27
Defining the shape of the $V_p(f)$ curve.

For the parallel circuit, the resonance curve of interest is that of the voltage V_C across the capacitor. The reason for this interest in V_C derives from electronic considerations that often place the capacitor at the input to another stage of a network.

Since the voltage across parallel elements is the same,

$$V_C = V_p = IZ_T \tag{20.35}$$

The resonant value of V_C is therefore determined by the value of Z_{T_m} and the magnitude of the current source I.

The quality factor of the parallel resonant circuit continues to be determined by the ratio of the reactive power to the real power. That is,

$$Q_p = \frac{V_p^2/X_{L_p}}{V_p^2/R}$$

where $R = R_s \parallel R_p$, and V_p is the voltage across the parallel branches. The result is

$$Q_p = \frac{R}{X_{L_p}} = \frac{R_s \parallel R_p}{X_{L_p}} \tag{20.36a}$$

or since $X_{L_p} = X_C$ at resonance,

$$Q_p = \frac{R_s \parallel R_p}{X_C} \tag{20.36b}$$

For the ideal current source ($R_s = \infty\ \Omega$) or when R_s is sufficiently large compared to R_p, we can make the following approximation:

$$R = R_s \| R_p \cong R_p$$

and
$$Q_p = \frac{R_s \| R_p}{X_{L_p}} = \frac{R_p}{X_{L_p}} = \frac{(R_l^2 + X_L^2)/R_l}{(R_l^2 + X_L^2)/X_L}$$

so that
$$Q_p = \frac{X_L}{R_l} = Q_l \Bigg|_{R_s >> R_p} \tag{20.37}$$

which is simply the quality factor Q_l of the coil.

In general, the bandwidth is still related to the resonant frequency and the quality factor by

$$BW = f_2 - f_1 = \frac{f_r}{Q_p} \tag{20.38}$$

The cutoff frequencies f_1 and f_2 can be determined using the equivalent network in Fig. 20.25 and the unity power condition for resonance. The half-power frequencies are defined by the condition that the output voltage is 0.707 times the maximum value. However, for parallel resonance with a current source driving the network, the frequency response for the driving point impedance is the same as that for the output voltage. This similarity permits defining each cutoff frequency as the frequency at which the input impedance is 0.707 times its maximum value. Since the maximum value is the equivalent resistance R in Fig. 20.25, the cutoff frequencies are associated with an impedance equal to $0.707R$ or $(1/\sqrt{2})R$.

Setting the input impedance for the network in Fig. 20.25 equal to this value results in the following relationship:

$$\mathbf{Z} = \frac{1}{\dfrac{1}{R} + j\left(\omega C - \dfrac{1}{\omega L}\right)} = 0.707R$$

which can be written as

$$\mathbf{Z} = \frac{1}{\dfrac{1}{R}\left[1 + jR\left(\omega C - \dfrac{1}{\omega L}\right)\right]} = \frac{R}{\sqrt{2}}$$

or
$$\frac{R}{1 + jR\left(\omega C - \dfrac{1}{\omega L}\right)} = \frac{R}{\sqrt{2}}$$

and finally
$$\frac{1}{1 + jR\left(\omega C - \dfrac{1}{\omega L}\right)} = \frac{1}{\sqrt{2}}$$

The only way the equality can be satisfied is if the magnitude of the imaginary term on the bottom left is equal to 1 because the magnitude of $1 + j\,1$ must be equal to $\sqrt{2}$.

The following relationship, therefore, defines the cutoff frequencies for the system:

$$R\left(\omega C - \frac{1}{\omega L}\right) = 1$$

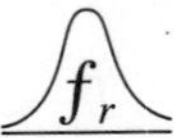

Substituting $\omega = 2\pi f$ and rearranging results in the following quadratic equation:

$$f^2 - \frac{f}{2\pi RC} - \frac{1}{4\pi^2 LC} = 0$$

having the form $$af^2 + bf + c = 0$$

with $$a = 1 \quad b = -\frac{1}{2\pi RC} \quad \text{and} \quad c = -\frac{1}{4\pi^2 LC}$$

Substituting into the equation:

$$f = \frac{-b \pm \sqrt{b^2 - 4ac}}{2a}$$

results in the following after a series of careful mathematical manipulations:

$$f_1 = \frac{1}{4\pi C}\left[\frac{1}{R} - \sqrt{\frac{1}{R^2} + \frac{4C}{L}}\right] \tag{20.39a}$$

$$f_2 = \frac{1}{4\pi C}\left[\frac{1}{R} + \sqrt{\frac{1}{R^2} + \frac{4C}{L}}\right] \tag{20.39b}$$

Since the term in the brackets of Eq. (20.39a) is always negative, simply associate f_1 with the magnitude of the result.

The effect of R_l, L, and C on the shape of the parallel resonance curve, as shown in Fig. 20.28 for the input impedance, is quite similar to their effect on the series resonance curve. Whether or not R_l is zero, the parallel resonant circuit frequently appears in a network schematic as shown in Fig. 20.28.

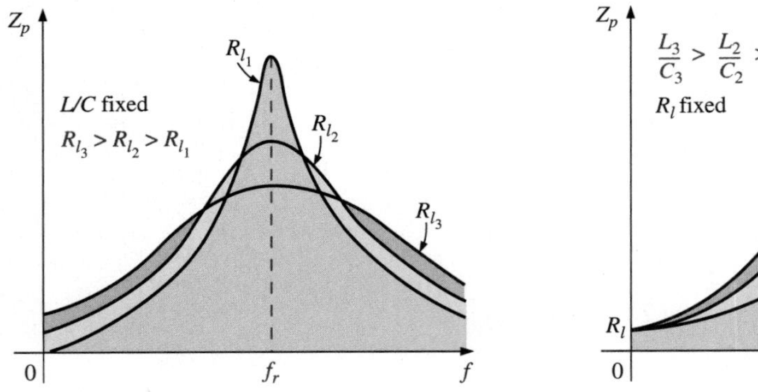

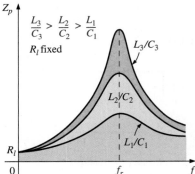

FIG. 20.28
Effect of R_l, L, and C on the parallel resonance curve.

At resonance, an increase in R_l or a decrease in the ratio L/C results in a decrease in the resonant impedance, with a corresponding increase in the current. The bandwidth of the resonance curves is given by Eq. (20.38). For increasing R_l or decreasing L (or L/C for constant C), the bandwidth increases as shown in Fig. 20.28.

At low frequencies, the capacitive reactance is quite high, and the inductive reactance is low. Since the elements are in parallel, the total im-

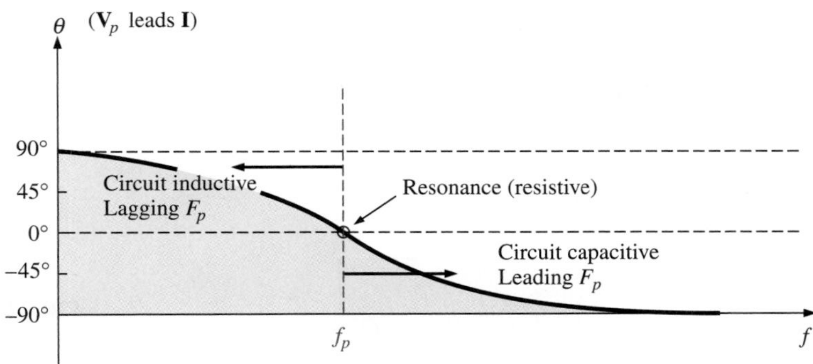

FIG. 20.29
Phase plot for the parallel resonant circuit.

pedance at low frequencies is therefore inductive. At high frequencies, the reverse is true, and the network is capacitive. At resonance (f_p), the network appears resistive. These facts lead to the phase plot in Fig. 20.29. Note that it is the inverse of that appearing for the series resonant circuit because at low frequencies the series resonant circuit was capacitive and at high frequencies it was inductive.

20.10 EFFECT OF $Q_L \geq 10$

The content of the previous section may suggest that the analysis of parallel resonant circuits is significantly more complex than encountered for series resonant circuits. Fortunately, however, this is not the case since, for the majority of parallel resonant circuits, the quality factor of the coil Q_l is sufficiently large to permit a number of approximations that simplify the required analysis.

Inductive Reactance, X_{L_p}

If we expand X_{L_p} as

$$X_{L_p} = \frac{R_l^2 + X_L^2}{X_L} = \frac{R_l^2(X_L)}{X_L(X_L)} + X_L = \frac{X_L}{Q_l^2} + X_L$$

then, for $Q_l \geq 10$, $X_L/Q_l^2 \cong 0$ compared to X_L, and

$$\boxed{X_{L_p} \cong X_L} \quad {}_{Q_l \geq 10} \tag{20.40}$$

and since resonance is defined by $X_{L_p} = X_C$, the resulting condition for resonance is reduced to:

$$\boxed{X_L \cong X_C} \quad {}_{Q_l \geq 10} \tag{20.41}$$

Resonant Frequency, f_p (Unity Power Factor)

We can rewrite the factor R_l^2C/L of Eq. (20.31) as

$$\frac{R_l^2C}{L} = \frac{1}{\dfrac{L}{R_l^2C}} = \frac{1}{\dfrac{(\omega)}{(\omega)}\dfrac{L}{R_l^2C}} = \frac{1}{\dfrac{\omega L}{R_l^2\omega C}} = \frac{1}{\dfrac{X_L X_C}{R_l^2}}$$

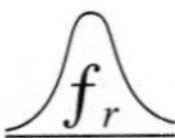

and substitute Eq. (20.41) ($X_L \cong X_C$):

$$\frac{1}{\dfrac{X_L X_C}{R_l^2}} = \frac{1}{\dfrac{X_L^2}{R_l^2}} = \frac{1}{Q_l^2}$$

Eq. (20.31) then becomes

$$\boxed{f_p = f_s\sqrt{1 - \frac{1}{Q_l^2}}} \quad Q_l \geq 10 \tag{20.42}$$

clearly revealing that as Q_l increases, f_p becomes closer and closer to f_s.

For $Q_l \geq 10$,

$$1 - \frac{1}{Q_l^2} \cong 1$$

and

$$\boxed{f_p \cong f_s = \frac{1}{2\pi\sqrt{LC}}} \quad Q_l \geq 10 \tag{20.43}$$

Resonant Frequency, f_m (Maximum V_C)

Using the equivalency $R_l^2C/L = 1/Q_l^2$ derived for Eq. (20.42), Eq. (20.32) takes on the following form:

$$\boxed{f_m \cong f_s\sqrt{1 - \frac{1}{4}\left(\frac{1}{Q_l^2}\right)}} \quad Q_l \geq 10 \tag{20.44}$$

The fact that the negative term under the square root will always be less than that appearing in the equation for f_p reveals that f_m will always be closer to f_s than f_p.

For $Q_l \geq 10$, the negative term becomes very small and can be dropped from consideration, leaving:

$$\boxed{f_m \cong f_s = \frac{1}{2\pi\sqrt{LC}}} \quad Q_l \geq 10 \tag{20.45}$$

In total, therefore, for $Q_l \geq 10$,

$$\boxed{f_p \cong f_m \cong f_s} \quad Q_l \geq 10 \tag{20.46}$$

R_p

$$R_p = \frac{R_l^2 + X_L^2}{R_l} = R_l + \frac{X_L^2}{R_l}\left(\frac{R_l}{R_l}\right) = R_l + \frac{X_L^2}{R_l^2}R_l$$
$$= R_l + Q_l^2R_l = (1 + Q_l^2)R_l$$

For $Q_l \geq 10$, $1 + Q_l^2 \cong Q_l^2$, and

$$\boxed{R_p \cong Q_l^2R_l} \quad Q_l \geq 10 \tag{20.47}$$

Applying the approximations just derived to the network in Fig. 20.24 results in the approximate equivalent network for $Q_l \geq 10$ in Fig. 20.30, which is certainly a lot "cleaner" in general appearance.

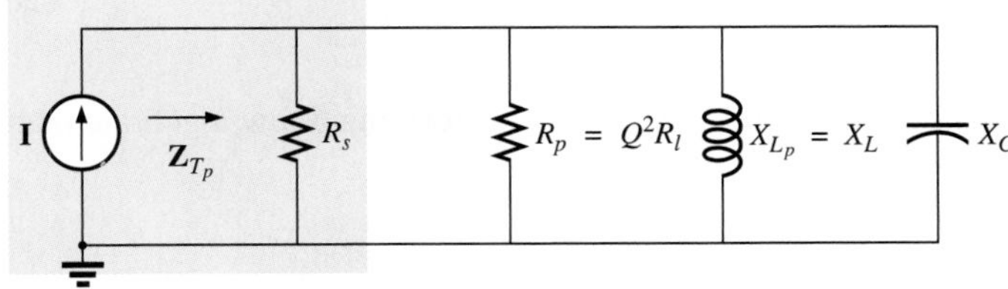

FIG. 20.30
Approximate equivalent circuit for $Q_l \geq 10$.

Substituting $Q_l = \dfrac{X_L}{R_l}$ into Eq. (20.47),

$$R_p \cong Q_l^2 R_l = \left(\frac{X_L}{R_l}\right)^2 R_l = \frac{X_L^2}{R_l} = \frac{X_L X_C}{R_l} = \frac{2\pi f L}{R_l(2\pi f C)}$$

and

$$\boxed{R_p \cong \frac{L}{R_l C}} \quad Q_l \geq 10 \qquad \textbf{(20.48)}$$

Z_{T_p}

The total impedance at resonance is now defined by

$$\boxed{Z_{T_p} \cong R_s \parallel R_p = R_s \parallel Q_l^2 R_l} \quad Q_l \geq 10 \qquad \textbf{(20.49)}$$

For an ideal current source ($R_s = \infty\ \Omega$), or if $R_s >> R_p$, the equation reduces to

$$\boxed{Z_{T_p} \cong Q_l^2 R_l} \quad Q_l \geq 10, R_s >> R_p \qquad \textbf{(20.50)}$$

Q_p

The quality factor is now defined by

$$\boxed{Q_p = \frac{R}{X_{L_p}} \cong \frac{R_s \parallel Q_l^2 R_l}{X_L}} \qquad \textbf{(20.51)}$$

Quite obviously, therefore, R_s does have an impact on the quality factor of the network and the shape of the resonant curves.

If an ideal current source ($R_s = \infty\ \Omega$) is used, or if $R_s >> R_p$,

$$Q_p \cong \frac{R_s \parallel Q_l^2 R_l}{X_L} = \frac{Q_l^2 R_l}{X_L} = \frac{Q_l^2}{X_L/R_l} = \frac{Q_l^2}{Q_l}$$

and

$$\boxed{Q_p \cong Q_l} \quad Q_l \geq 10, R_s >> R_p \qquad \textbf{(20.52)}$$

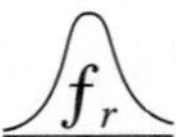

BW

The bandwidth defined by f_p is

$$BW = f_2 - f_1 = \frac{f_p}{Q_p} \tag{20.53}$$

By substituting Q_p from above and performing a few algebraic manipulations, we can show that

$$BW = f_2 - f_1 \cong \frac{1}{2\pi}\left[\frac{R_l}{L} + \frac{1}{R_s C}\right] \tag{20.54}$$

clearly revealing the impact of R_s on the resulting bandwidth. Of course, if $R_s = \infty\ \Omega$ (ideal current source):

$$BW = f_2 - f_1 \cong \frac{R_l}{2\pi L}\Bigg|_{R_s = \infty\ \Omega} \tag{20.55}$$

I_L and I_C

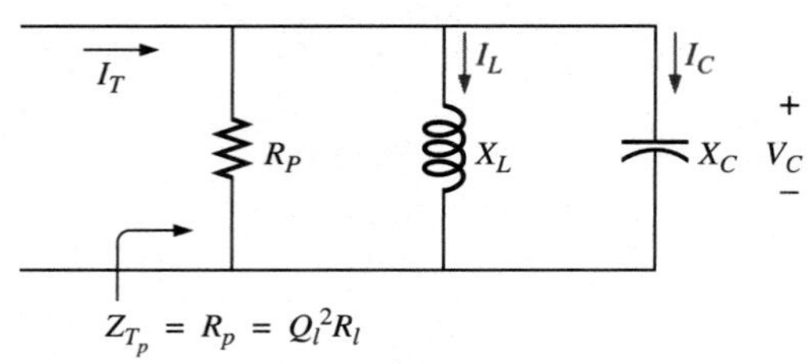

FIG. 20.31
Establishing the relationship between I_C and I_L and the current I_T.

A portion of Fig. 20.30 is reproduced in Fig. 20.31, with I_T defined as shown.

As indicated, Z_{T_p} at resonance is $Q_l^2 R_l$. The voltage across the parallel network is, therefore,

$$V_C = V_L = V_R = I_T Z_{T_p} = I_T Q_l^2 R_l$$

The magnitude of the current I_C can then be determined using Ohm's law, as follows:

$$I_C = \frac{V_C}{X_C} = \frac{I_T Q_l^2 R_l}{X_C}$$

Substituting $X_C = X_L$ when $Q_l \geq 10$,

$$I_C = \frac{I_T Q_l^2 R_l}{X_L} = I_T \frac{Q_l^2}{\dfrac{X_L}{R_l}} = I_T \frac{Q_l^2}{Q_l}$$

and

$$I_C \cong Q_l I_T \Big|_{Q_l \geq 10} \tag{20.56}$$

revealing that the capacitive current is Q_l times the magnitude of the current entering the parallel resonant circuit. For large Q_l, the current I_C can be significant.

A similar derivation results in

$$I_L \cong Q_l I_T \Big|_{Q_l \geq 10} \tag{20.57}$$

Conclusions

The equations resulting from the application of the condition $Q_l \geq 10$ are obviously a great deal easier to apply than those obtained earlier. It is, therefore, a condition that should be checked early in an analysis to determine which approach must be applied. Although the condition $Q_l \geq 10$ was applied throughout, many of the equations are still good approximations for $Q_l < 10$. For instance, if $Q_l = 5$, $X_{L_p} = (X_L/Q_l^2) + X_l = (X_L/25) + X_L = 1.04X_L$, which is very close to X_L. In fact, for $Q_l = 2$, $X_{L_p} = (X_L/4) + X_L = 1.25X_L$, which is not X_L, but it is only 25% off. In general, be aware that the approximate equations can be applied with good accuracy with $Q_l < 10$. The smaller the level of Q_l, however, the less valid the approximation. The approximate equations are certainly valid for a range of values of $Q_l < 10$ if a rough approximation to the actual response is desired rather than one accurate to the hundredths place.

20.11 SUMMARY TABLE

In an effort to limit any confusion resulting from the introduction of f_p and f_m and an approximate approach dependent on Q_l, the summary in Table 20.1 was developed. You can always use the equations for any Q_l, but a proficiency in applying the approximate equations defined by Q_l will pay dividends in the long run.

For the future, the analysis of a parallel resonant network may proceed as follows:

1. Determine f_s to obtain some idea of the resonant frequency. Recall that for most situations, f_s, f_m, and f_p will be relatively close to each other.
2. Calculate an approximate Q_l using f_s from below, and compare it to the condition $Q_l \geq 10$. If the condition is satisfied, the approximate approach should be the chosen path unless a high degree of accuracy is required.
3. If Q_l is less than 10, the approximate approach can be applied, but it must be understood that the smaller the level of Q_l, the less accurate the solution. However, considering the typical variations

TABLE 20.1
Parallel resonant circuit ($f_s = 1/(2\pi\sqrt{LC})$).

	Any Q_l	**$Q_l \geq 10$**	**$Q_l \geq 10, R_s >> Q_l^2 R_l$**
f_p	$f_s\sqrt{1 - \frac{R_l^2 C}{L}}$	f_s	f_s
f_m	$f_s\sqrt{1 - \frac{1}{4}\left[\frac{R_l^2 C}{L}\right]}$	f_s	f_s
Z_{T_p}	$R_s \parallel R_p = R_s \parallel \left(\frac{R_l^2 + X_L^2}{R_l}\right)$	$R_s \parallel Q_l^2 R_l$	$Q_l^2 R_l$
Z_{T_m}	$R_s \parallel \mathbf{Z}_{R-L} \parallel \mathbf{Z}_C$	$R_s \parallel Q_l^2 R_l$	$Q_l^2 R_l$
Q_p	$\frac{Z_{T_p}}{X_{L_p}} = \frac{Z_{T_p}}{X_C}$	$\frac{Z_{T_p}}{X_L} = \frac{Z_{T_p}}{X_C}$	Q_l
BW	$\frac{f_p}{Q_p}$ or $\frac{f_m}{Q_p}$	$\frac{f_p}{Q_p} = \frac{f_s}{Q_p}$	$\frac{f_p}{Q_l} = \frac{f_s}{Q_l}$
I_L, I_C	Network analysis	$I_L = I_C = Q_l I_T$	$I_L = I_C = Q_l I_T$

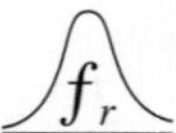

from nameplate values for many of our components and that a resonant frequency to the tenths place is seldom required, the use of the approximate approach for many practical situations is usually quite valid.

20.12 EXAMPLES (PARALLEL RESONANCE)

EXAMPLE 20.6 Given the parallel network in Fig. 20.32 composed of "ideal" elements:

a. Determine the resonant frequency f_p.
b. Find the total impedance at resonance.
c. Calculate the quality factor, bandwidth, and cutoff frequencies f_1 and f_2 of the system.
d. Find the voltage V_C at resonance.
e. Determine the currents I_L and I_C at resonance.

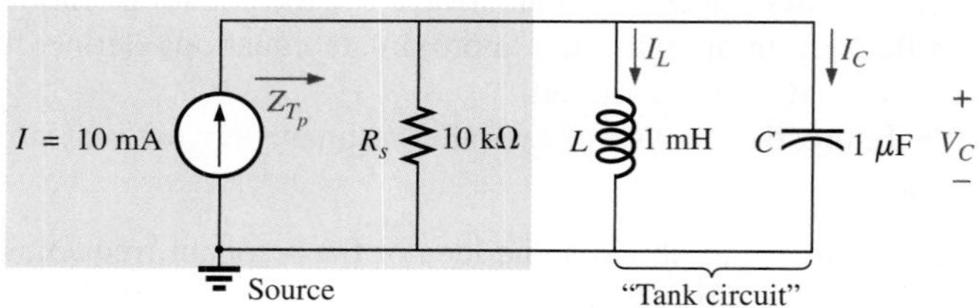

FIG. 20.32
Example 20.6.

Solutions:

a. The fact that R_l is zero ohms results in a very high $Q_l\ (= X_L/R_l)$, permitting the use of the following equation for f_p:

$$f_p = f_s = \frac{1}{2\pi\sqrt{LC}} = \frac{1}{2\pi\sqrt{(1\text{ mH})(1\ \mu\text{F})}} = \mathbf{5.03\ kHz}$$

b. For the parallel reactive elements:

$$\mathbf{Z}_L \parallel \mathbf{Z}_C = \frac{(X_L \angle 90^\circ)(X_C \angle -90^\circ)}{+j(X_L - X_C)}$$

but $X_L = X_C$ at resonance, resulting in a zero in the denominator of the equation and a very high impedance that can be approximated by an open circuit. Therefore,

$$Z_{T_p} = R_s \parallel \mathbf{Z}_L \parallel \mathbf{Z}_C = R_s = \mathbf{10\ k\Omega}$$

c. $$Q_p = \frac{R_s}{X_{L_p}} = \frac{R_s}{2\pi f_p L} = \frac{10\text{ k}\Omega}{2\pi(5.03\text{ kHz})(1\text{ mH})} = \mathbf{316.41}$$

$$BW = \frac{f_p}{Q_p} = \frac{5.03\text{ kHz}}{316.41} = \mathbf{15.90\ Hz}$$

Eq. (20.39a):

$$f_1 = \frac{1}{4\pi C}\left[\frac{1}{R} - \sqrt{\frac{1}{R^2} + \frac{4C}{L}}\right]$$

$$= \frac{1}{4\pi(1\ \mu\text{F})}\left[\frac{1}{10\ \text{k}\Omega} - \sqrt{\frac{1}{(10\ \text{k}\Omega)^2} + \frac{4(1\ \mu\text{F})}{1\ \text{mH}}}\right]$$
$$= \mathbf{5.03\ kHz}$$

Eq. (20.39b):

$$f_2 = \frac{1}{4\pi C}\left[\frac{1}{R} + \sqrt{\frac{1}{R^2} + \frac{4C}{L}}\right]$$
$$= \mathbf{5.04\ kHz}$$

d. $V_C = IZ_{T_p} = (10\ \text{mA})(10\ \text{k}\Omega) = \mathbf{100\ V}$

e. $I_L = \dfrac{V_L}{X_L} = \dfrac{V_C}{2\pi f_p L} = \dfrac{100\ \text{V}}{2\pi(5.03\ \text{kHz})(1\ \text{mH})} = \dfrac{100\ \text{V}}{31.6\ \Omega} = \mathbf{3.16\ A}$

$I_C = \dfrac{V_C}{X_C} = \dfrac{100\ \text{V}}{31.6\ \Omega} = \mathbf{3.16\ A}\ (= Q_p I)$

Example 20.6 demonstrates the impact of R_s on the calculations associated with parallel resonance. The source impedance is the only factor to limit the input impedance and the level of V_C.

EXAMPLE 20.7 For the parallel resonant circuit in Fig. 20.33 with $R_s = \infty\ \Omega$:

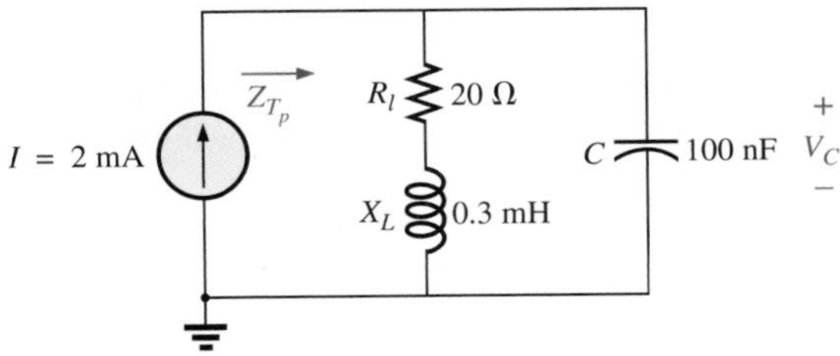

FIG. 20.33
Example 20.7.

a. Determine f_s, f_m, and f_p, and compare their levels.
b. Calculate the maximum impedance and the magnitude of the voltage V_C at f_m.
c. Determine the quality factor Q_p.
d. Calculate the bandwidth.
e. Compare the above results with those obtained using the equations associated with $Q_l \geq 10$.

Solutions:

a. $f_s = \dfrac{1}{2\pi\sqrt{LC}} = \dfrac{1}{2\pi\sqrt{(0.3\ \text{mH})(100\ \text{nF})}} = \mathbf{29.06\ kHz}$

$$f_m = f_s\sqrt{1 - \frac{1}{4}\left[\frac{R_l^2 C}{L}\right]}$$
$$= (29.06\ \text{kHz})\sqrt{1 - \frac{1}{4}\left[\frac{(20\ \Omega)^2\,(100\ \text{nF})}{0.3\ \text{mH}}\right]}$$
$$= \mathbf{25.58\ kHz}$$

$$f_p = f_s\sqrt{1 - \frac{R_l^2 C}{L}} = (29.06\text{ kHz})\sqrt{1 - \left[\frac{(20\ \Omega)^2(100\text{ nF})}{0.3\text{ mH}}\right]}$$
$$= \mathbf{27.06\ kHz}$$

Both f_m and f_p are less than f_s, as predicted. In addition, f_m is closer to f_s than f_p, as forecast. f_m is about 0.5 kHz less than f_s, whereas f_p is about 2 kHz less. The differences among f_s, f_m, and f_p suggest a low Q network.

b. $\mathbf{Z}_{T_m} = (R_l + jX_L) \parallel -jX_C$ at $f = f_m$

$$X_L = 2\pi f_m L = 2\pi(28.58\text{ kHz})(0.3\text{ mH}) = 53.87\ \Omega$$

$$X_C = \frac{1}{2\pi f_m C} = \frac{1}{2\pi(28.58\text{ kHz})(100\text{ nF})} = 55.69\ \Omega$$

$$R_l + jX_L = 20\ \Omega + j\,53.87\ \Omega = 57.46\ \Omega\ \angle 69.63^\circ$$

$$\mathbf{Z}_{T_m} = \frac{(57.46\ \Omega\ \angle 69.63^\circ)(55.69\ \Omega\ \angle -90^\circ)}{20\ \Omega + j\,53.87\ \Omega - j\,55.69\ \Omega}$$
$$= \mathbf{159.34\ \Omega\ \angle -15.17^\circ}$$

$$V_{C_{\max}} = IZ_{T_m} = (2\text{ mA})(159.34\ \Omega) = \mathbf{318.68\ mV}$$

c. $R_s = \infty\ \Omega$; therefore,

$$Q_p = \frac{R_s \parallel R_p}{X_{L_p}} = \frac{R_p}{X_{L_p}} = Q_l = \frac{X_L}{R_l}$$

$$= \frac{2\pi(27.06\text{ kHz})(0.3\text{ mH})}{20\ \Omega} = \frac{51\ \Omega}{20\ \Omega} = \mathbf{2.55}$$

The low Q confirms our conclusion of part (a). The differences among f_s, f_m, and f_p are significantly less for higher Q networks.

d. $$BW = \frac{f_p}{Q_p} = \frac{27.06\text{ kHz}}{2.55} = \mathbf{10.61\ kHz}$$

e. For $Q_l \geq 10$, $f_m = f_p = f_s = \mathbf{29.06\ kHz}$

$$Q_p = Q_l = \frac{2\pi f_s L}{R_l} = \frac{2\pi(29.06\text{ kHz})(0.3\text{ mH})}{20\ \Omega} = \mathbf{2.74}$$

(versus 2.55 above)

$$Z_{T_p} = Q_l^2 R_l = (2.74)^2 \cdot 20\ \Omega = \mathbf{150.15\ \Omega\ \angle 0^\circ}$$

(versus 159.34 Ω ∠−15.17° above)

$$V_{C_{\max}} = IZ_{T_p} = (2\text{ mA})(150.15\ \Omega) = \mathbf{300.3\ mV}$$

(versus 318.68 mV above)

$$BW = \frac{f_p}{Q_p} = \frac{29.06\text{ kHz}}{2.74} = \mathbf{10.61\ kHz}$$

(versus 10.61 kHz above)

The results reveal that, even for a relatively low Q system, the approximate solutions are still close compared to those obtained using the full equations. The primary difference is between f_s and f_p (about 7%), with the difference between f_s and f_m at less than 2%. For the future, using f_s to determine Q_l will certainly provide a measure of Q_l that can be used to determine whether the approximate approach is appropriate.

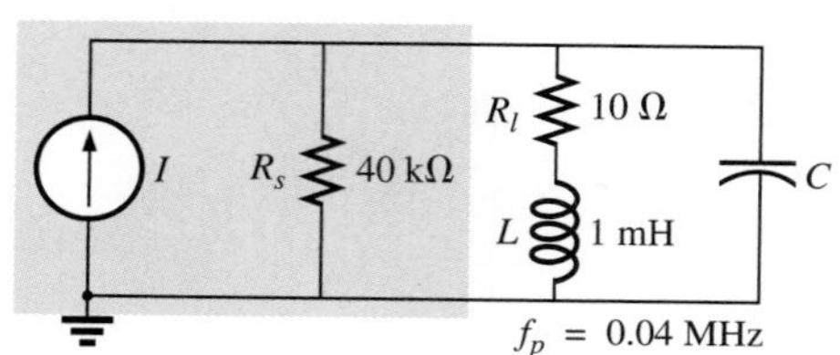

FIG. 20.34
Example 20.8.

EXAMPLE 20.8 For the network in Fig. 20.34 with f_p provided:

a. Determine Q_l.
b. Determine R_p.

c. Calculate Z_{T_p}.
d. Find C at resonance.
e. Find Q_p.
f. Calculate the BW and cutoff frequencies.

Solutions:

a. $Q_l = \dfrac{X_L}{R_l} = \dfrac{2\pi f_p L}{R_l} = \dfrac{2\pi(0.04 \text{ MHz})(1 \text{ mH})}{10 \ \Omega} = \mathbf{25.12}$

b. $Q_l \geq 10$. Therefore,

$$R_p \cong Q_l^2 R_l = (25.12)^2(10 \ \Omega) = \mathbf{6.31 \ k\Omega}$$

c. $Z_{T_p} = R_s \| R_p = 40 \text{ k}\Omega \| 6.31 \text{ k}\Omega = \mathbf{5.45 \ k\Omega}$

d. $Q_l \geq 10$. Therefore,

$$f_p \cong \frac{1}{2\pi\sqrt{LC}}$$

and $C = \dfrac{1}{4\pi^2 f^2 L} = \dfrac{1}{4\pi^2 (0.04 \text{ MHz})^2 (1 \text{ mH})} = \mathbf{15.83 \ nF}$

e. $Q_l \geq 10$. Therefore,

$$Q_p = \frac{Z_{T_p}}{X_L} = \frac{R_s \| Q_l^2 R_l}{2\pi f_p L} = \frac{5.45 \text{ k}\Omega}{2\pi(0.04 \text{ MHz})(1 \text{ mH})} = \mathbf{21.68}$$

f. $BW = \dfrac{f_p}{Q_p} = \dfrac{0.04 \text{ MHz}}{21.68} = \mathbf{1.85 \ kHz}$

$$\begin{aligned} f_1 &= \frac{1}{4\pi C}\left[\frac{1}{R} - \sqrt{\frac{1}{R^2} + \frac{4C}{L}}\right] \\ &= \frac{1}{4\pi(15.9 \text{ mF})}\left[\frac{1}{5.45 \text{ k}\Omega} - \sqrt{\frac{1}{(5.45 \text{ k}\Omega)^2} + \frac{4(15.9 \text{ mF})}{1 \text{ mH}}}\right] \\ &= 5.005 \times 10^6[183.486 \times 10^{-6} - 7.977 \times 10^{-3}] \\ &= 5.005 \times 10^6[-7.794 \times 10^{-3}] \\ &= \mathbf{39 \ kHz} \qquad \text{(ignoring the negative sign)} \end{aligned}$$

$$\begin{aligned} f_2 &= \frac{1}{4\pi C}\left[\frac{1}{R} + \sqrt{\frac{1}{R^2} + \frac{4C}{L}}\right] \\ &= 5.005 \times 10^6[183.486 \times 10^{-6} + 7.977 \times 10^{-3}] \\ &= 5.005 \times 10^6[8.160 \times 10^{-3}] \\ &= \mathbf{40.84 \ kHz} \end{aligned}$$

Note that $f_2 - f_1 = 40.84 \text{ kHz} - 39 \text{ kHz} = 1.84 \text{ kHz}$, confirming our solution for the bandwidth above. Note also that the bandwidth is not symmetrical about the resonant frequency, with 1 kHz below and 840 Hz above.

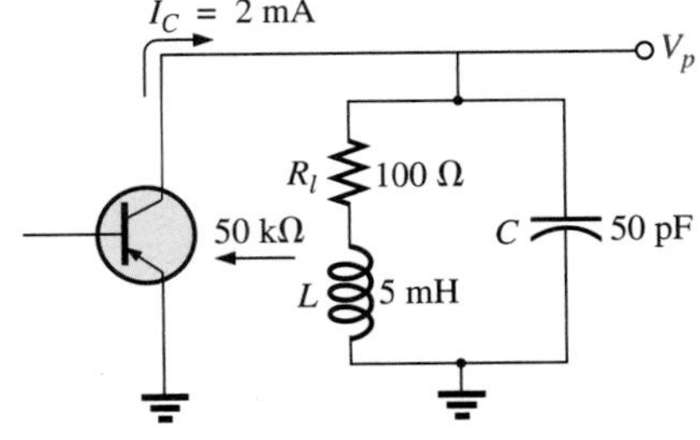

FIG. 20.35
Example 20.9.

EXAMPLE 20.9 The equivalent network for the transistor configuration in Fig. 20.35 is provided in Fig. 20.36.

a. Find f_p.
b. Determine Q_p.
c. Calculate the BW.
d. Determine V_p at resonance.
e. Sketch the curve of V_C versus frequency.

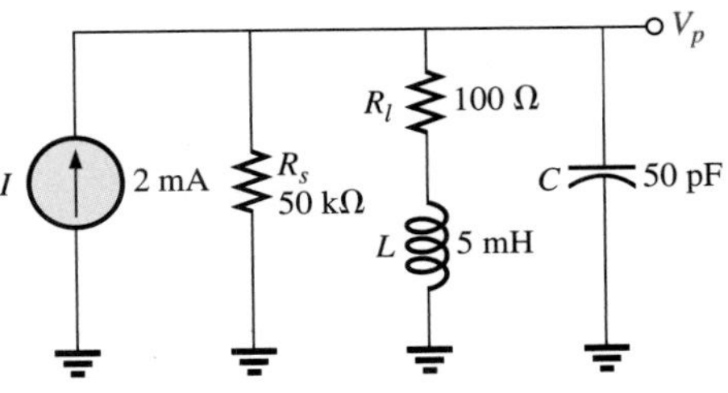

FIG. 20.36
Equivalent network for the transistor configuration in Fig. 20.35.

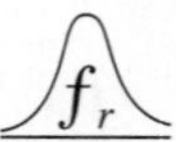

Solutions:

a. $$f_s = \frac{1}{2\pi\sqrt{LC}} = \frac{1}{2\pi\sqrt{(5\text{ mH})(50\text{ pF})}} = 318.31\text{ kHz}$$

$$X_L = 2\pi f_s L = 2\pi(318.31\text{ kHz})(5\text{ mH}) = 10\text{ k}\Omega$$

$$Q_l = \frac{X_L}{R_l} = \frac{10\text{ k}\Omega}{100\text{ k}\Omega} = 100 > 10$$

Therefore, $f_p = f_s =$ **318.31 kHz.** Using Eq. (20.31) results in $\cong$ 318.5 kHz.

b. $$Q_p = \frac{R_s \parallel R_p}{X_L}$$

$$R_p = Q_l^2 R_l = (100)^2 100\ \Omega = 1\text{ M}\Omega$$

$$Q_p = \frac{50\text{ k}\Omega \parallel 1\text{ M}\Omega}{10\text{ k}\Omega} = \frac{47.62\text{ k}\Omega}{10\text{ k}\Omega} = \mathbf{4.76}$$

Note the drop in Q from $Q_l = 100$ to $Q_p = 4.76$ due to R_s.

c. $$BW = \frac{f_p}{Q_p} = \frac{318.31\text{ kHz}}{4.76} = \mathbf{66.87\text{ kHz}}$$

On the other hand,

$$BW = \frac{1}{2\pi}\left(\frac{R_l}{L} + \frac{1}{R_s C}\right) = \frac{1}{2\pi}\left[\frac{100\ \Omega}{5\text{ mH}} + \frac{1}{(50\text{ k}\Omega)(50\text{ pF})}\right]$$
$$= \mathbf{66.85\text{ kHz}}$$

compares very favorably with the above solution.

d. $V_p = IZ_{T_p} = (2\text{ mA})(R_s \parallel R_p) = (2\text{ mA})(47.62\text{ k}\Omega) = \mathbf{95.24\text{ V}}$

e. See Fig. 20.37.

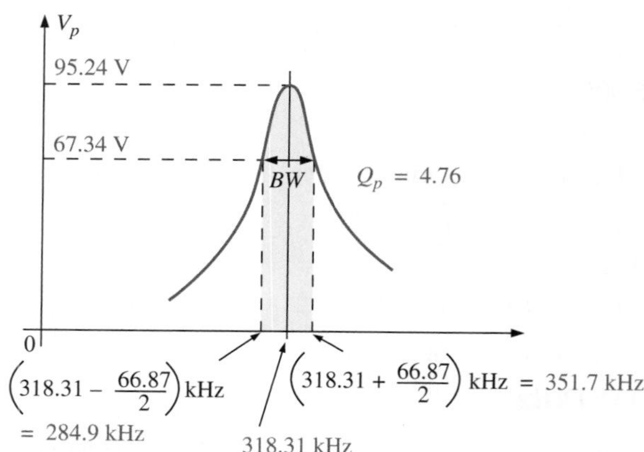

FIG. 20.37
Resonance curve for the network in Fig. 20.36.

EXAMPLE 20.10 Repeat Example 20.9, but ignore the effects of R_s, and compare results.

Solutions:

a. f_p is the same, **318.31 kHz.**

b. For $R_s = \infty\ \Omega$,

$$Q_p = Q_l = \mathbf{100} \qquad \text{(versus 4.76)}$$

c. $$BW = \frac{f_p}{Q_p} = \frac{318.31\text{ kHz}}{100} = \mathbf{3.18\text{ kHz}} \qquad \text{(versus 66.87 kHz)}$$

d. $Z_{T_p} = R_p = 1\text{ M}\Omega$ (versus $47.62\text{ k}\Omega$)

$V_p = IZ_{T_p} = (2\text{ mA})(1\text{ M}\Omega) = \mathbf{2000\text{ V}}$ (versus 95.24 V)

The results obtained clearly reveal that the source resistance can have a significant impact on the response characteristics of a parallel resonant circuit.

EXAMPLE 20.11 Design a parallel resonant circuit to have the response curve in Fig. 20.38 using a 1 mH, 10 Ω inductor and a current source with an internal resistance of 40 kΩ.

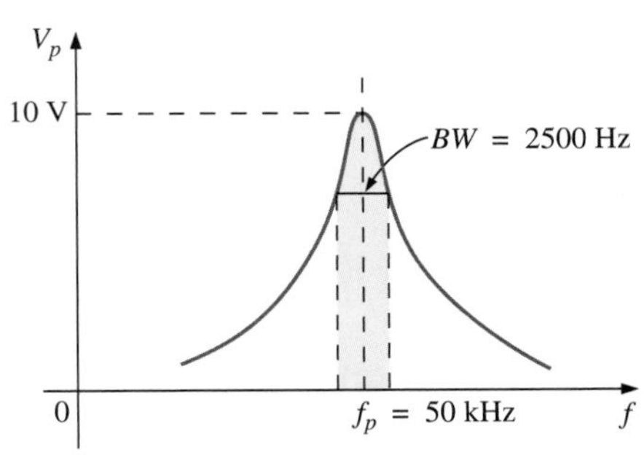

FIG. 20.38
Example 20.11.

Solution:

$$BW = \frac{f_p}{Q_p}$$

Therefore,

$$Q_p = \frac{f_p}{BW} = \frac{50{,}000\text{ Hz}}{2500\text{ Hz}} = \mathbf{20}$$

$$X_L = 2\pi f_p L = 2\pi(50\text{ kHz})(1\text{ mH}) = 314\ \Omega$$

and
$$Q_l = \frac{X_L}{R_l} = \frac{314\ \Omega}{10\ \Omega} = \mathbf{31.4}$$

$$R_p = Q_l^2 R = (31.4)^2(10\ \Omega) = \mathbf{9859.6\ \Omega}$$

$$Q_p = \frac{R}{X_L} = \frac{R_s \parallel 9859.6\ \Omega}{314\ \Omega} = 20 \qquad \text{(from above)}$$

so that
$$\frac{(R_s)(9859.6)}{R_s + 9859.6} = 6280$$

resulting in
$$R_s = 17.298\text{ k}\Omega$$

However, the source resistance was given as 40 kΩ. We must therefore add a parallel resistor (R') that will reduce the 40 kΩ to approximately 17.298 kΩ; that is,

$$\frac{(40\text{ k}\Omega)(R')}{40\text{ k}\Omega + R'} = 17.298\text{ k}\Omega$$

Solving for R':

$$R' = \mathbf{30.48\text{ k}\Omega}$$

The closest commercial value is **30 kΩ.** At resonance, $X_L = X_C$, and

$$X_C = \frac{1}{2\pi f_p C}$$

$$C = \frac{1}{2\pi f_p X_C} = \frac{1}{2\pi(50\text{ kHz})(314\ \Omega)}$$

and
$$C \cong \mathbf{0.01\ \mu F} \qquad \text{(commercially available)}$$

$$Z_{T_p} = R_s \parallel Q_l^2 R_l$$
$$= 17.298\text{ k}\Omega \parallel 9859.6\ \Omega$$
$$= 6.28\text{ k}\Omega$$

with
$$V_p = IZ_{T_p}$$

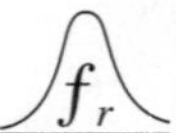

and
$$I = \frac{V_p}{Z_{T_p}} = \frac{10\text{ V}}{6.28\text{ k}\Omega} \cong \mathbf{1.6\text{ mA}}$$

The network appears in Fig. 20.39.

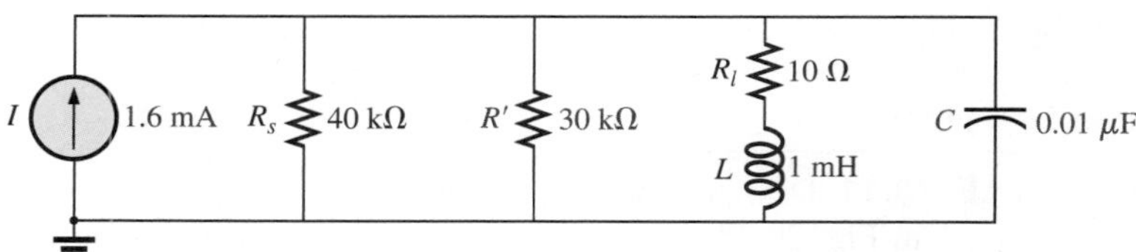

FIG. 20.39
Network designed to meet the criteria in Fig. 20.38.

20.13 APPLICATIONS

Stray Resonance

Stray resonance, like stray capacitance and inductance and unexpected resistance levels, can occur in totally unexpected situations and can severely affect the operation of a system. All that is required to produce stray resonance is, for example, a level of capacitance introduced by parallel wires or copper leads on a printed circuit board, or simply two parallel conductive surfaces with residual charge and inductance levels associated with any conductor or components such as tape recorder heads, transformers, and so on, that provide the elements necessary for a resonance effect. In fact, this resonance effect is a very common effect in a cassette tape recorder. The play/record head is a coil that can act like an inductor and an antenna. Combine this factor with the stray capacitance and real capacitance in the network to form the tuning network, and the tape recorder with the addition of a semiconductor diode can respond like an AM radio. As you plot the frequency response of any transformer, you normally find a region where the response has a peaking effect (look ahead at Fig. 25.21). This peaking is due solely to the inductance of the coils of the transformer and the stray capacitance between the wires.

In general, any time you see an unexpected peaking in the frequency response of an element or a system, it is normally caused by a resonance condition. If the response has a detrimental effect on the overall operation of the system, a redesign may be in order, or a filter can be added that will block the frequencies that result in the resonance condition. Of course, when you add a filter composed of inductors and/or capacitors, you must be careful that you don't add another unexpected resonance condition. It is a problem that can be properly weighed only by constructing the system and exposing it to the full range of tests.

Graphic and Parametric Equalizers

We have all noticed at one time or another that the music we hear in a concert hall doesn't quite sound the same when we play a recording of it on our home entertainment center. Even after we check the specifications of the speakers and amplifiers and find that both are nearly perfect (and the most expensive we can afford), the sound is still not what it should be. In general, we are experiencing the effects of the local environmental characteristics on the sound waves. Some typical problems are hard walls

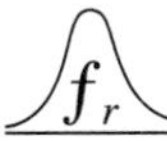

or floors (stone, cement) that make high frequencies sound louder. Curtains and rugs, on the other hand, absorb high frequencies. The shape of the room and the placement of the speakers and furniture also affect the sound that reaches our ears. Another criterion is the echo or reflection of sound that occurs in the room. Concert halls are designed very carefully with their vaulted ceilings and curved walls to allow a certain amount of echo. Even the temperature and humidity characteristics of the surrounding air affect the quality of the sound. It is certainly impossible, in most cases, to redesign your listening area to match a concert hall, but with the proper use of electronic systems you can develop a response that has all the qualities that you want from a home entertainment center.

For a quality system, a number of steps can be taken: *characterization and digital delay (surround sound)* and *proper speaker and amplifier selection and placement.* Characterization is a process whereby a thorough sound absorption check of the room is performed and the frequency response determined. A *graphic equalizer* such as appearing in Fig. 20.40(a) is then used to make the response "flat" for the full range of frequencies. In other words, the room is made to appear as though all the frequencies receive equal amplification in the listening area. For instance, if the room is fully carpeted with full draping curtains, there is a considerable amount of high-frequency absorption, requiring that the high frequencies have additional amplification to match the sound levels of the mid and low frequencies. To *characterize* the typical rectangular-shaped room, a setup such as shown in Fig. 20.40(b) may be used. The amplifier and speakers are placed in the center of one wall, with additional speakers in the corners of the room facing the reception area. A mike is then placed in the reception area about 10 ft from the amplifier and centered between the two other speakers. A *pink noise* is then sent out from a spectrum analyzer (often an integral part of the graphic equalizer) to the amplifier and speakers. Pink noise is actually a square-wave signal whose amplitude and frequency can be controlled. A square-wave signal was chosen because a Fourier breakdown of a square-wave signal results in a broad range of frequencies for the system to check. You will find in Chapter 24 that a square wave can be constructed of an infinite series of sine waves of different frequencies. Once the proper volume of pink noise is established, the spectrum analyzer can be used to set the response of each slide band to establish the desired flat response. The center frequencies for the slides of the graphic equalizer in Fig. 20.40(a) are provided in Fig. 20.40(c), along with the frequency response for a number of adjoining frequencies evenly spaced on a logarithmic scale. Note that each center frequency is actually the resonant frequency for that slide. The design is such that each slide can control the volume associated with that frequency, but the bandwidth and frequency response stay fairly constant. A good spectrum analyzer has each slide set against a decibel (dB) scale (decibels are discussed in detail in Chapter 21). The decibel scale simply establishes a scale for the comparison of audio levels. At a normal listening level, usually a change of about 3 dB is necessary for the audio change to be detectable by the human ear. At low levels of sound, a 2 dB change may be detectable, but at loud sounds probably a 4 dB change would be necessary for the change to be noticed. These are not strict laws but guidelines commonly used by audio technicians. For the room in question, the mix of settings may be as shown in Fig. 20.40(c). Once set, the slides are not touched again. A flat response has been established for the room for the full audio range so that every sound or type of music is covered.

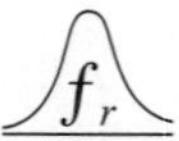

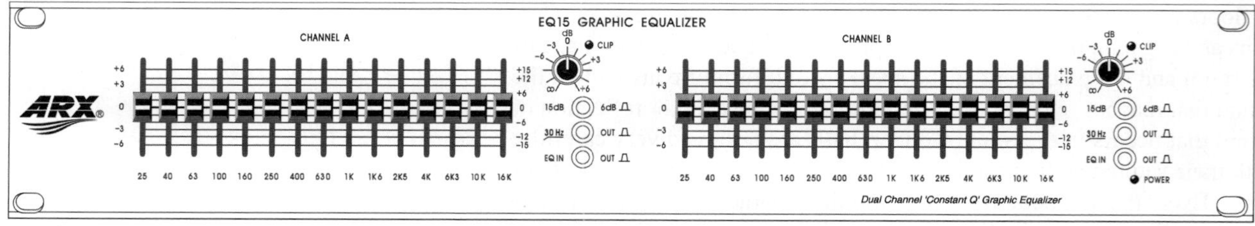

(a)

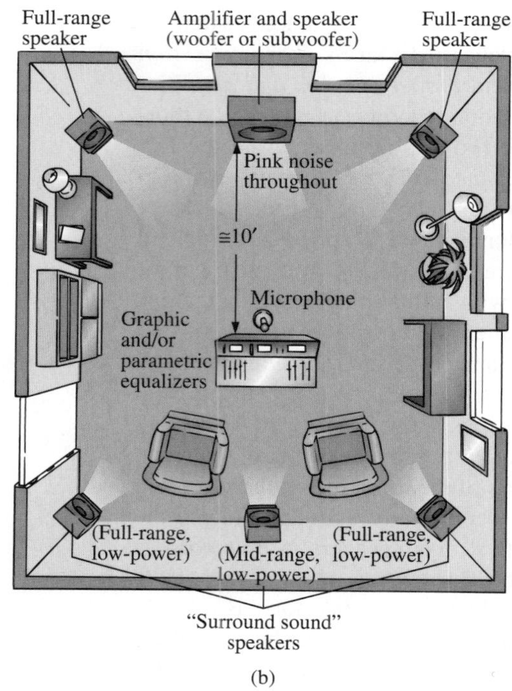

(b)

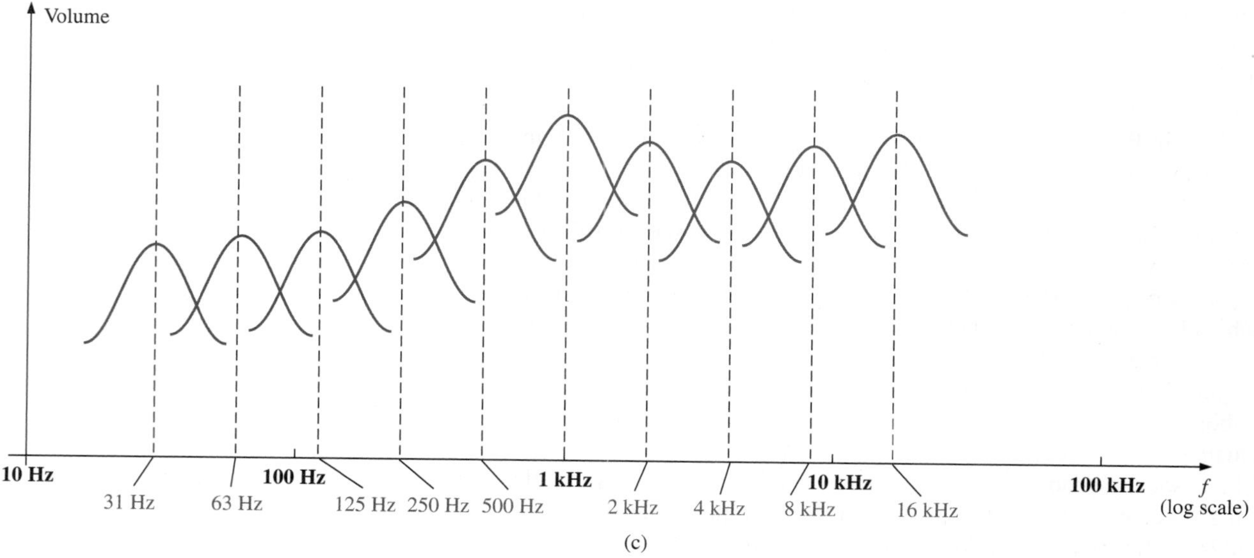

(c)

FIG. 20.40

(a) Dual-channel 15-band "Constant Q" graphic equalizer (Courtesy of ARX Systems.); *(b) setup; (c) frequency response.*

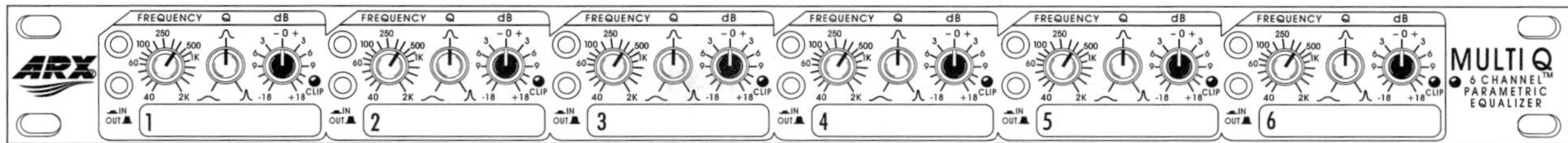

FIG. 20.41
Six-channel parametric equalizer.
(Courtesy of ARX Systems.)

A *parametric equalizer* such as appearing in Fig. 20.41 is similar to a *graphic equalizer,* but instead of separate controls for the individual frequency ranges, it uses three basic controls over three or four broader frequency ranges. The typical controls—the *gain, center frequency,* and *bandwidth*—are typically available for the *low-*, *mid-*, and *high-frequency* ranges. Each is fundamentally an independent control; that is, a change in one can be made without affecting the other two. For the parametric equalizer in Fig. 20.41, each of the six channels has a frequency control switch which, in conjunction with the $f \times 10$ switch, gives a range of center frequencies from 40 Hz through 16 kHz. It has controls for BW ("Q") from 3 octaves to $\frac{1}{20}$ octave, and ± 18 dB cut and boost. Some like to refer to the parametric equalizer as a *sophisticated tone control* and actually use them to enrich the sound after the flat response has been established by the graphic equalizer. The effect achieved with a standard tone control knob is sometimes referred to as "boring" compared to the effect established by a good parametric equalizer, primarily because the former can control only the volume and not the bandwidth or center frequency. In general, graphic equalizers establish the important *flat response* while parametric equalizers are adjusted to provide the *type* and *quality* of sound you like to hear. You can "notch out" the frequencies that bother you and remove tape "hiss" and the "sharpness" often associated with CDs.

One characteristic of concert halls that is more difficult to fake is the fullness of sound that concert halls are able to provide. In the concert hall, you have the direct sound from the instruments and the reflection of sound off the walls and the vaulted ceilings which were all carefully designed expressly for this purpose. Any reflection results in a delay in the sound waves reaching the ear, creating the fullness effect. Through digital delay, speakers can be placed to the back and side of a listener to establish the surround sound effect. In general, the delay speakers are much lower in wattage, with 20 W speakers typically used with a 100 W system. The echo response is one reason that people often like to play their stereos louder than they should for normal hearing. By playing the stereo louder, they create more echo and reflection off the walls, bringing into play some of the fullness heard at concert halls.

It is probably safe to say that any system composed of quality components, a graphic and parametric equalizer, and surround sound will have all the components necessary to have a quality reproduction of the concert hall effect.

20.14 COMPUTER ANALYSIS

PSpice

Series Resonance This chapter provides an excellent opportunity to demonstrate what computer software programs can do for us. Imagine having to plot a detailed resonance curve with all the calculations

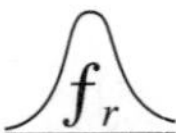

required for each frequency. At every frequency, the reactance of the inductive and capacitive elements changes, and the phasor operations would have to be repeated—a long and arduous task. However, with PSpice, taking a few moments to enter the circuit and establish the desired simulation results in a detailed plot in a few seconds that can have plot points every microsecond!

For the first time, the horizontal axis is in the frequency domain rather than in the time domain as in all the previous plots. For the series resonant circuit in Fig. 20.42, the magnitude of the source was chosen to produce a maximum current of $I = 400\text{ mV}/40\ \Omega = 10\text{ mA}$ at resonance, and the reactive elements establish a resonant frequency of

$$f_s = \frac{1}{2\pi\sqrt{LC}} = \frac{1}{2\pi\sqrt{(30\text{ mH})(0.1\ \mu\text{F})}} \cong \mathbf{2.91\ kHz}$$

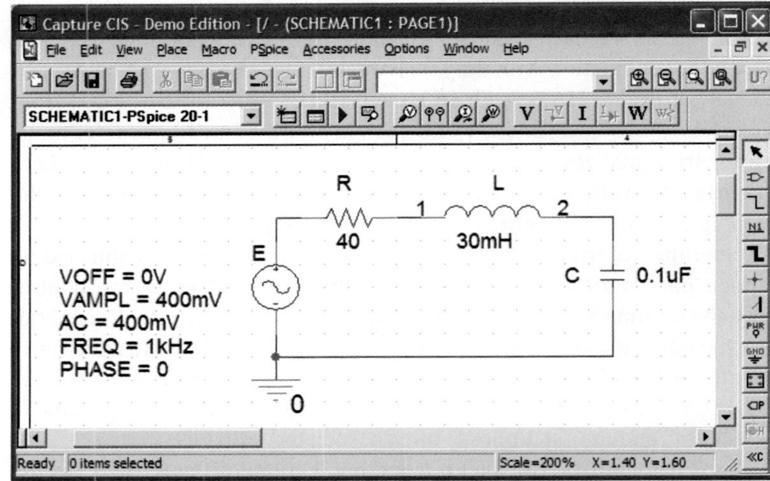

FIG. 20.42
Series resonant circuit to be analyzed using PSpice.

The quality factor is

$$Q_l = \frac{X_L}{R_l} = \frac{546.64\ \Omega}{40\ \Omega} \cong \mathbf{13.7}$$

which is relatively high and should give us a nice sharp response.

The bandwidth is

$$BW = \frac{f_s}{Q_l} = \frac{2.91\text{ kHz}}{13.7} \cong \mathbf{212\ Hz}$$

which will be verified using our cursor options.

For the ac source, choose **VSIN.** All the parameters are set by double-clicking on the source symbol and entering the values in the **Property Editor** dialog box. For each, select **Name and Value** under **Display** followed by **Apply** before leaving the dialog box.

In the **Simulation Settings** dialog box, select **AC Sweep/Noise** and set the **Start Frequency** at 1 kHz, the **End Frequency** at 10 kHz, and the **Points/Decade** at 10,000. The **Logarithmic scale** and **Decade** settings remain at their default values. Choose 10,000 for **Points/Decade** to ensure a number of data points near the peak value. When the **SCHEMATIC1**

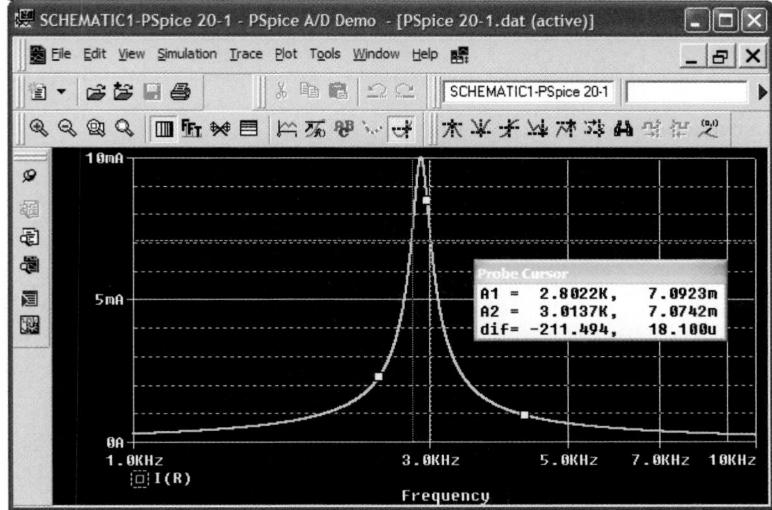

FIG. 20.43

Resonance curve for the current of the circuit in Fig. 20.42.

screen in Fig. 20.43 appears, **Trace-Add Trace-I(R)-OK** results in a logarithmic plot that peaks just to the left of 3 kHz. The spacing between grid lines on the **X-axis** should be increased, so select **Plot-Axis Settings-X Grid-unable Automatic-Spacing-Log-0.1-OK.** Next, select the **Toggle cursor** icon, and right-click to move the right cursor as close to 7.07 mA as possible (0.707 of the peak value to define the bandwidth) to obtain **A1** with a frequency of 2.80 kHz at a level of 7.09 mA—the best we can do with the 10,000 data points per decade. Now left-click, and place the left cursor as close to the same level as possible. The result is 3.01 kHz at a level of 7.07 mA for **A2.** The cursors were set in the order above to obtain a positive answer for the difference of the two as appearing in the third line of the **Probe Cursor** box. The resulting 211.49 Hz is an excellent match with the calculated value of 212 Hz.

Parallel Resonance Let us now investigate the parallel resonant circuit in Fig. 20.33 and compare the results with the longhand solution. The network appears in Fig. 20.44 using **ISRC** as the ac source voltage. Under the **Property Editor** heading, set the following values: **DC** = 0 A, **AC** = 2 mA, and **TRAN** = 0. Under **Display,** select **Do Not Display** for both **DC** and **TRAN** since they do not play a part in our analysis. In the **Simulation Settings** dialog box, select **AC Sweep/Noise,** and select the **Start Frequency** at 10 kHz since we know that it will resonate near 30 kHz. Choose the **End Frequency** as 100 kHz for a first run to see the results. Set the **Points/Decade** at 10,000 to ensure a good number of data points for the peaking region. After simulation, **Trace-Add Trace-V(C:1)-OK** results in the plot in Fig. 20.45 with a resonant frequency near 30 kHz. The selected range appears to be a good one, but the initial plot needed more grid lines on the x-axis, so use **Plot-Axis Settings-X-Grid-unenable Automatic-Spacing-Log-0.1-OK** to obtain a grid line at 10 kHz intervals. Next select the **Toggle cursor** pad and a left-click cursor is established on the screen. Choose the **Cursor Peak** pad to find the peak value of the curve. The result is **A1** = 319.45 mV at 28.94 kHz which is a very close match with the calculated value of 318.68 mV at 28.57 kHz for the maximum value of V_C. The bandwidth is defined at a level of 0.707(319.45 mV)

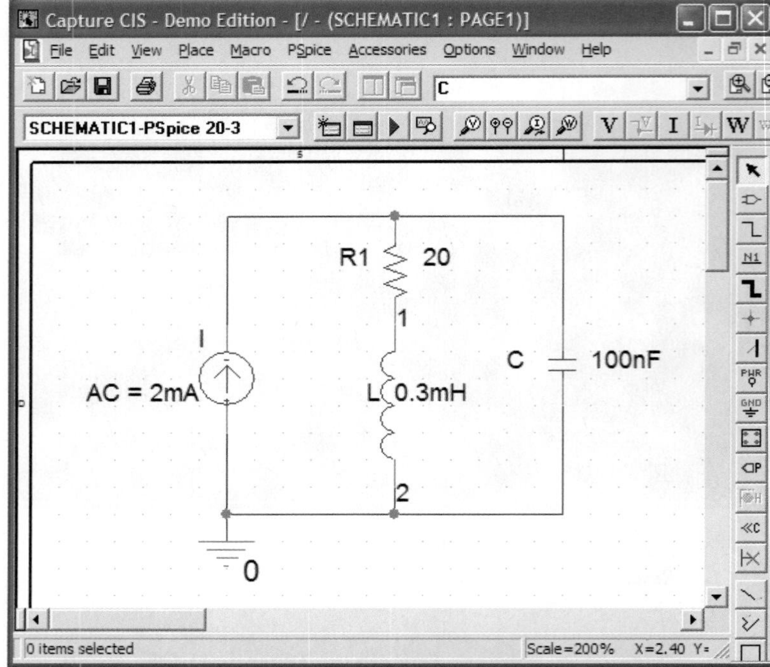

FIG. 20.44
Parallel resonant network to be analyzed using PSpice.

= 225.85 mV. Using the right-click cursor, you find that the closest value is 224.57 mV for the 10,000 points of data per decade. The resulting frequency is 34.69 kHz as shown in the **Probe Cursor** box in Fig. 20.45.

Now use the left-click cursor to find the same level to the left of the peak value so that you can determine the bandwidth. The closest that the left-click cursor can come to 225.85 mV is 224.96 mV at a frequency of 23.97 kHz. The bandwidth then appears as 10.72 kHz in the **Probe Cursor** box, comparing very well with the longhand solution of 10.68 kHz in Example 20.7.

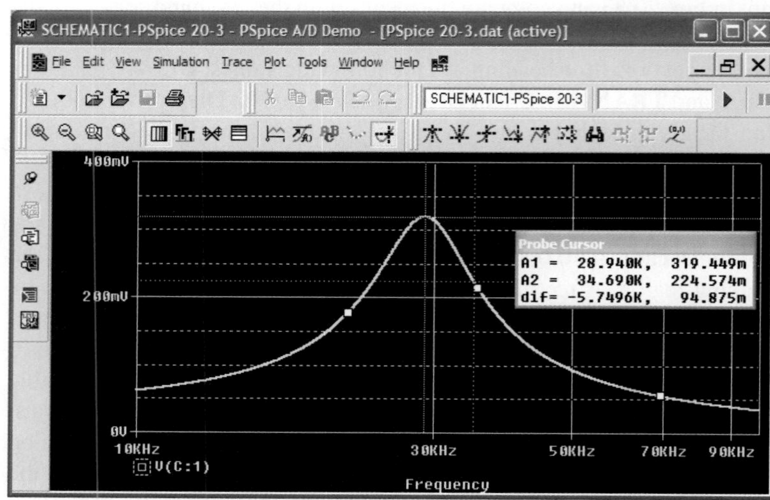

FIG. 20.45
Resonance curve for the voltage across the capacitor in Fig. 20.44.

You can now look at the phase angle of the voltage across the parallel network to find the frequency when the network appears resistive and the phase angle is 0°. First use **Trace-Delete All Traces,** and call up **P(V(C:1))** followed by **OK.** The result is the plot in Fig. 20.46, revealing that the phase angle is close to −90° at very high frequencies as the capacitive element with its decreasing reactance takes over the characteristics of the parallel network. At 10 kHz, the inductive element has a lower reactance than the capacitive element, and the network has a positive phase angle. Using the cursor option, move the left cursor along the horizontal axis until the phase angle is at its minimum value. As shown in Fig. 20.46, the smallest angle available with the determined data points is 49.86 mdegrees = 0.05° which is certainly very close to 0°. The corresponding frequency is 27.046 kHz which is essentially an exact match with the longhand solution of 27.051 kHz. Clearly, therefore, the frequency at which the phase angle is zero and the total impedance appears resistive is less than the frequency at which the output voltage is a maximum.

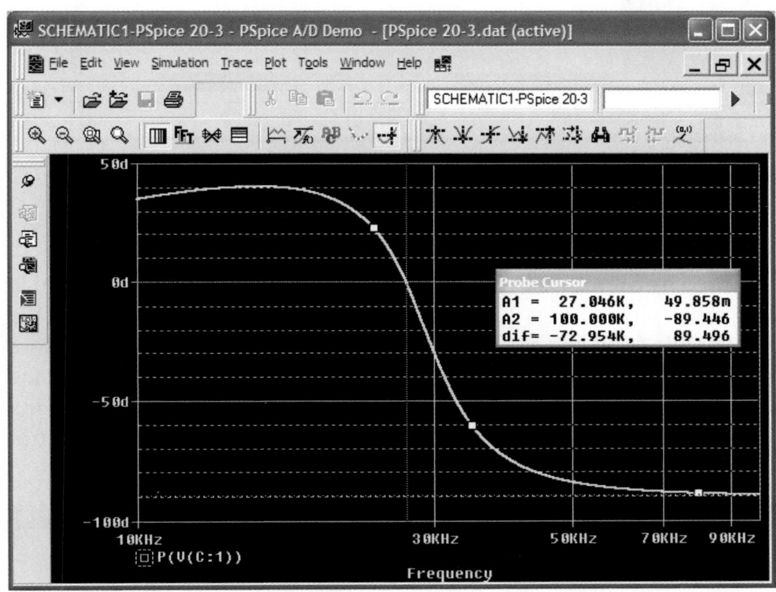

FIG. 20.46

Phase plot for the voltage v_C for the parallel resonant network in Fig. 20.44.

Multisim

The results of Example 20.9 are now confirmed using Multisim. The network in Fig. 20.36 appears as shown in Fig. 20.47 after all the elements have been placed as described in earlier chapters. In particular, note that the frequency assigned to the 2 mA ac current source is 100 kHz. Since we have some idea that the resonant frequency is a few hundred kilohertz, it seems appropriate that the starting frequency for the plot begins at 100 kHz and extends to 1 MHz. Also, be sure that the **AC Magnitude** is set to 2 mA in the **Analysis Setup** within the **AC Current** dialog box.

For simulation, first select the sequence **Simulate-Analyses-AC Analysis** to obtain the **AC Analysis** dialog box. Set the **Start frequency** at 100 kHz, the **Stop frequency** at 1 MHz, **Sweep type** at **Decade, Number of points per decade** at 1000, and the **Vertical scale** at **Linear.** Under **Output variables,** select node number 1 as a **Variable for analysis**

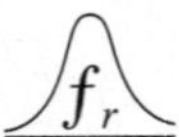

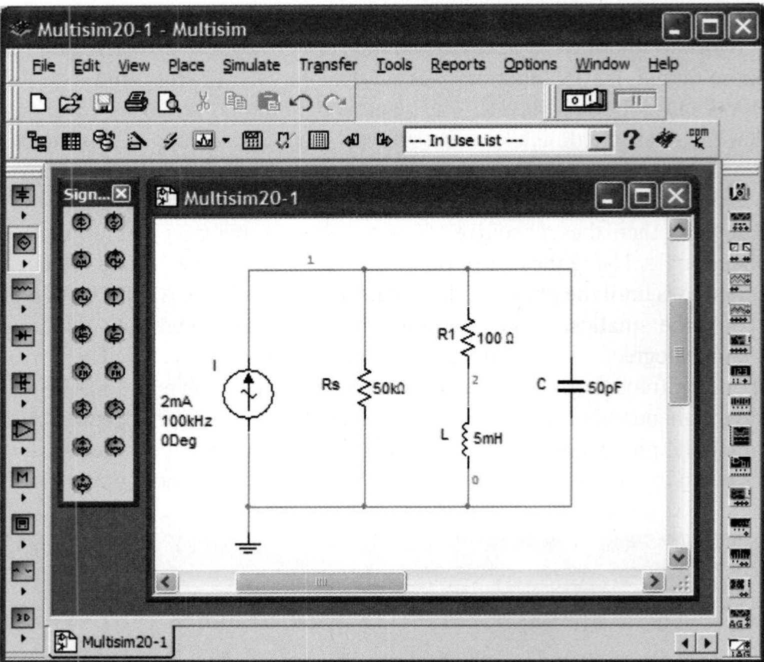

FIG. 20.47
Using Multisim to confirm the results of Example 20.9.

followed by **Simulate** to run the program. The results are the magnitude and phase plots in Fig. 20.48. Starting with the **Voltage** plot, select the **Show/Hide Grid** key, **Show/Hide Legend** key, and **Show/Hide Cursors** key. You will immediately note under the **AC Analysis** cursor box that the maximum value is 95.24 V and the minimum value is 6.94 V. By moving the cursor until you reach 95.24 V (**y1**), you can find the resonant fre-

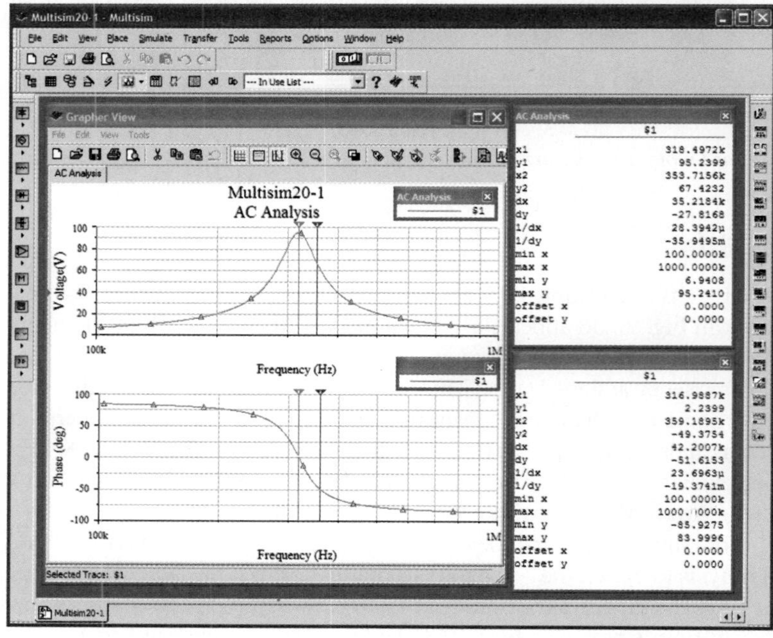

FIG. 20.48
Magnitude and phase plots for the voltage υ_C of the network in Fig. 20.47.

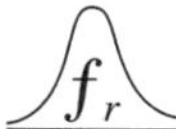

quency. As shown in the top cursor dialog box in Fig. 20.48, this is achieved at 318.97 kHz (**x1**). The other (blue) cursor can be used to define the high cutoff frequency for the bandwidth by first calculating the 0.707 level of the output voltage. The result is 0.707(95.24 V) = 67.33 V. The closest you can come to this level with the cursor is 67.42 V (**y2**) which defines a frequency of 353.71 kHz (**x2**). If you now use the red cursor to find the corresponding level below the resonant frequency, you find a level of 67.49 V (**y1**) at 287.08 kHz (**x1**). The resulting bandwidth is therefore 353.71 kHz − 287.08 kHz = 66.63 kHz.

You can now determine the resonant frequency if you define resonance as that frequency that results in a phase angle of 0° for the output voltage. By repeating the process described above for the phase plot, set the red cursor as close to 0° as possible. The result is 2.24° (**y1**) at 316.98 kHz (**x1**), clearly revealing that the resonant frequency defined by the phase angle is less than that defined by the peak voltage. However, with a Q_l of about 100, the difference of 1.99 kHz is not significant. Also note that when the second cursor was set on 359.19 kHz, the phase angle of −49.37° is close to the 45° expected at the cutoff frequency.

Again, the computer solution is a very close match with the longhand solution in Example 20.9 with a perfect match of 95.24 V for the peak value and only a small difference in bandwidth with 66.87 kHz in Example 20.9 and 66.63 kHz here. For the high cutoff frequency, the computer generated a result of 353.71 kHz, while the theoretical solution was 351.7 kHz. For the low cutoff frequency, the computer responded with 287.08 kHz compared to a theoretical solution of 284 kHz.

PROBLEMS

SECTIONS 20.2 THROUGH 20.7 Series Resonance

1. Find the resonant ω_s and f_s for the series circuit with the following parameters:
   a. $R = 10\ \Omega$, $L = 1$ H, $C = 16\ \mu$F
   b. $R = 300\ \Omega$, $L = 0.5$ H, $C = 0.16\ \mu$F
   c. $R = 20\ \Omega$, $L = 0.28$ mH, $C = 7.46\ \mu$F
2. For the series circuit in Fig. 20.49:
   a. Find the value of X_C for resonance.
   b. Determine the total impedance of the circuit at resonance.
   c. Find the magnitude of the current I.
   d. Calculate the voltages V_R, V_L, and V_C at resonance. How are V_L and V_C related? How does V_R compare to the applied voltage E?
   e. What is the quality factor of the circuit? Is it a high or low Q circuit?
   f. What is the power dissipated by the circuit at resonance?

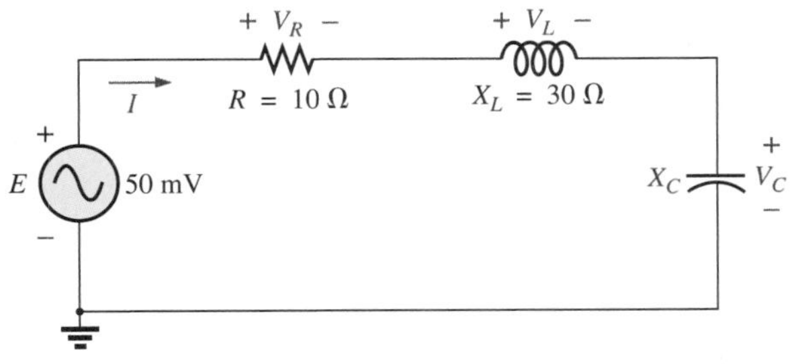

FIG. 20.49
Problem 2.

3. For the series circuit in Fig. 20.50:
   a. Find the value of X_L for resonance.
   b. Determine the magnitude of the current I at resonance.
   c. Find the voltages V_R, V_L, and V_C at resonance, and compare their magnitudes.
   d. Determine the quality factor of the circuit. Is it a high or low Q circuit?
   e. If the resonant frequency is 5 kHz, determine the value of L and C.
   f. Find the bandwidth of the response if the resonant frequency is 5 kHz.
   g. What are the low and high cutoff frequencies?

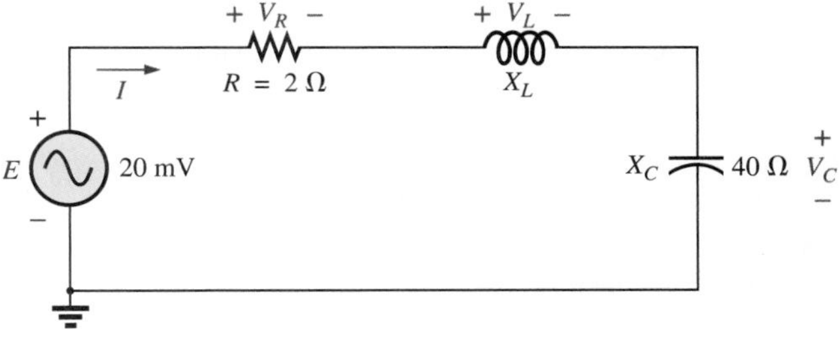

FIG. 20.50
Problem 3.

4. For the circuit in Fig. 20.51:
   a. Find the value of L in millihenries if the resonant frequency is 1800 Hz.
   b. Calculate X_L and X_C. How do they compare?
   c. Find the magnitude of the current I_{rms} at resonance.
   d. Find the power dissipated by the circuit at resonance.
   e. What is the apparent power delivered to the system at resonance?
   f. What is the power factor of the circuit at resonance?
   g. Calculate the Q of the circuit and the resulting bandwidth.
   h. Find the cutoff frequencies, and calculate the power dissipated by the circuit at these frequencies.

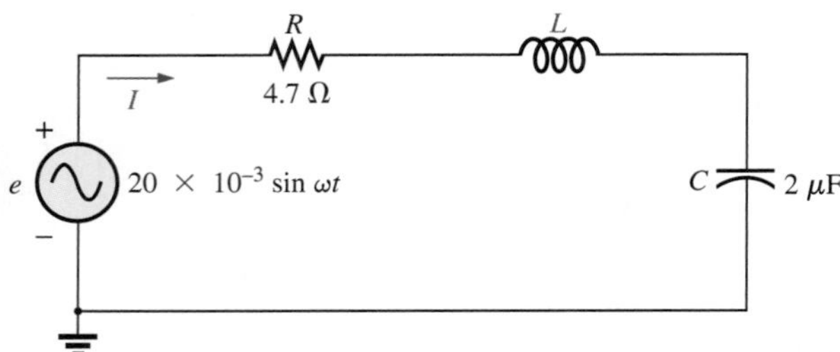

FIG. 20.51
Problem 4.

5. a. Find the bandwidth of a series resonant circuit having a resonant frequency of 6000 Hz and a Q_s of 15.
   b. Find the cutoff frequencies.
   c. If the resistance of the circuit at resonance is 3 Ω, what are the values of X_L and X_C in ohms?
   d. What is the power dissipated at the half-power frequencies if the maximum current flowing through the circuit is 0.5 A?
6. A series circuit has a resonant frequency of 10 kHz. The resistance of the circuit is 5 Ω, and X_C at resonance is 200 Ω.
   a. Find the bandwidth.
   b. Find the cutoff frequencies.
   c. Find Q_s.
   d. If the input voltage is 30 V $\angle 0°$, find the voltage across the coil and capacitor in phasor form.
   e. Find the power dissipated at resonance.
7. a. The bandwidth of a series resonant circuit is 200 Hz. If the resonant frequency is 2000 Hz, what is the value of Q_s for the circuit?
   b. If $R = 2\ \Omega$, what is the value of X_L at resonance?
   c. Find the value of L and C at resonance.
   d. Find the cutoff frequencies.
8. The cutoff frequencies of a series resonant circuit are 5400 Hz and 6000 Hz.
   a. Find the bandwidth of the circuit.
   b. If Q_s is 9.5, find the resonant frequency of the circuit.
   c. If the resistance of the circuit is 2 Ω, find the value of X_L and X_C at resonance.
   d. Find the value of L and C at resonance.

*9. Design a series resonant circuit with an input voltage of 5 V $\angle 0°$ to have the following specifications:
   a. A peak current of 500 mA at resonance
   b. A bandwidth of 120 Hz
   c. A resonant frequency of 8400 Hz
   d. Find the value of L and C and the cutoff frequencies.

*10. Design a series resonant circuit to have a bandwidth of 400 Hz using a coil with a Q_l of 20 and a resistance of 2 Ω. Find the values of L and C and the cutoff frequencies.

*11. A series resonant circuit is to resonate at $\omega_s = 2\pi \times 10^6$ rad/s and draw 20 W from a 120 V source at resonance. If the fractional bandwidth is 0.16:
   a. Determine the resonant frequency in hertz.
   b. Calculate the bandwidth in hertz.
   c. Determine the values of R, L, and C.
   d. Find the resistance of the coil if $Q_l = 80$.

*12. A series resonant circuit will resonate at a frequency of 1 MHz with a fractional bandwidth of 0.2. If the quality factor of the coil at resonance is 12.5 and its inductance is 100 μH, determine the following:
   a. The resistance of the coil.
   b. The additional resistance required to establish the indicated fractional bandwidth.
   c. The required value of capacitance.

SECTIONS 20.8 THROUGH 20.12 **Parallel Resonance**

13. For the "ideal" parallel resonant circuit in Fig. 20.52:
   a. Determine the resonant frequency (f_p).
   b. Find the voltage V_C at resonance.
   c. Determine the currents I_L and I_C at resonance.
   d. Find Q_p.

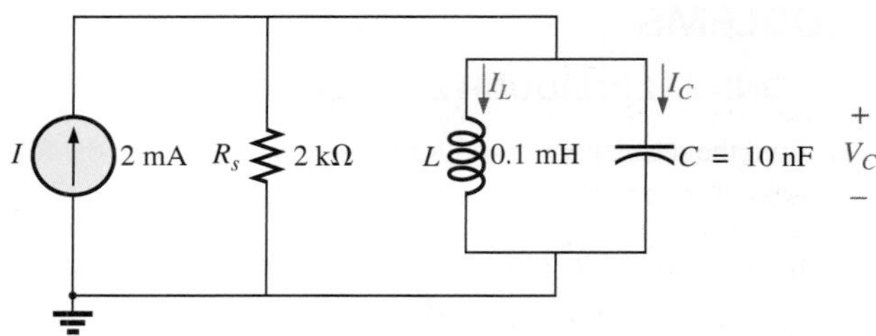

FIG. 20.52
Problem 13.

14. For the parallel resonant network in Fig. 20.53:
   a. Calculate f_s.
   b. Determine Q_l using $f = f_s$. Can the approximate approach be applied?
   c. Determine f_p and f_m.
   d. Calculate X_L and X_C using f_p. How do they compare?

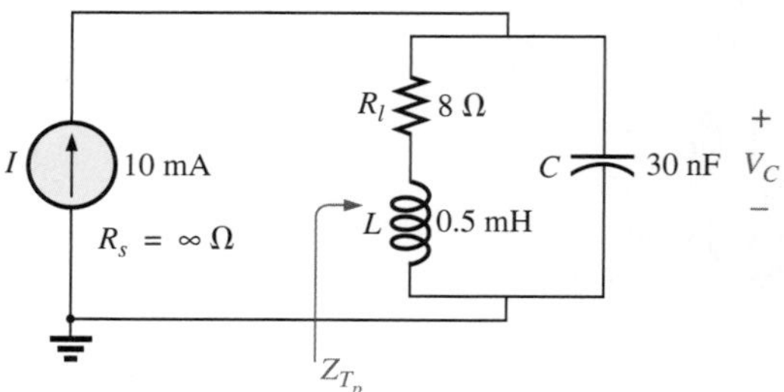

FIG. 20.53
Problem 14.

e. Find the total impedance at resonance (f_p).
f. Calculate V_C at resonance (f_p).
g. Determine Q_p and the BW using f_p.
h. Calculate I_L and I_C at f_p.

15. Repeat Problem 14 for the network in Fig. 20.54.

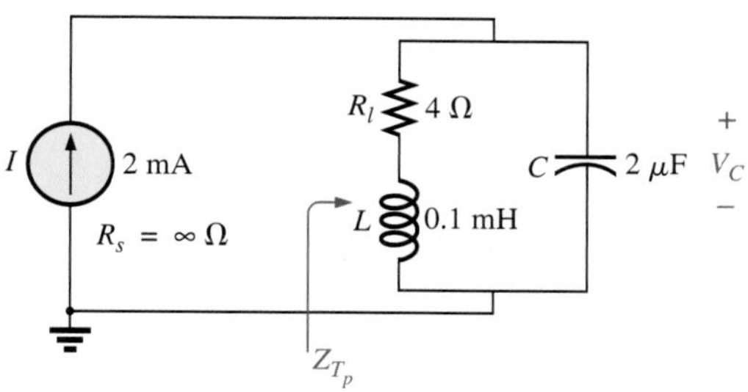

FIG. 20.54
Problem 15.

16. For the network in Fig. 20.55:
a. Find the value of X_C at resonance (f_p).
b. Find the total impedance Z_{T_p} at resonance (f_p).
c. Find the currents I_L and I_C at resonance (f_p).
d. If the resonant frequency is 20,000 Hz, find the value of L and C at resonance.
e. Find Q_p and the BW.

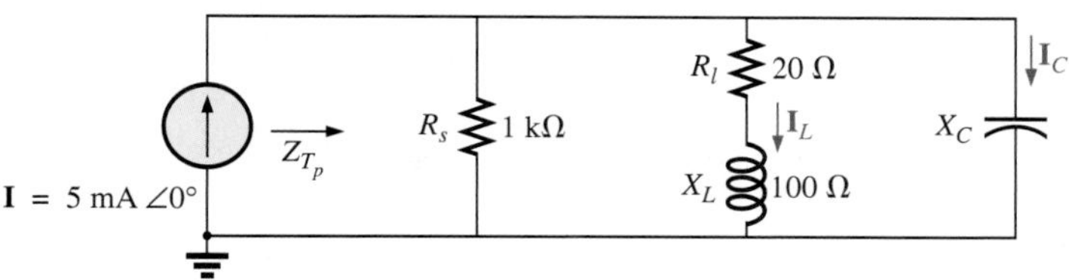

FIG. 20.55
Problem 16.

17. Repeat Problem 16 for the network in Fig. 20.56.

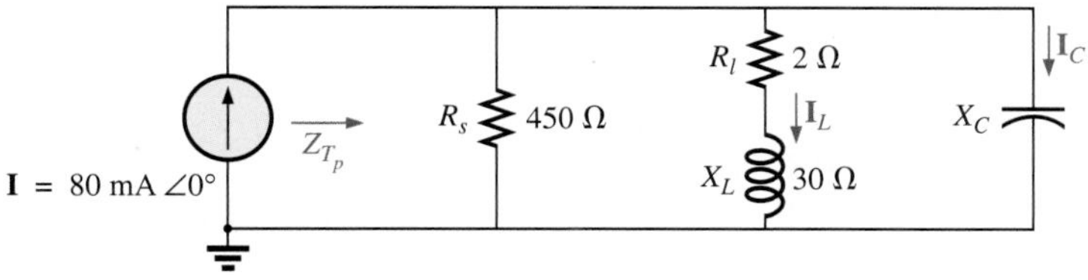

FIG. 20.56
Problem 17.

18. For the network in Fig. 20.57:

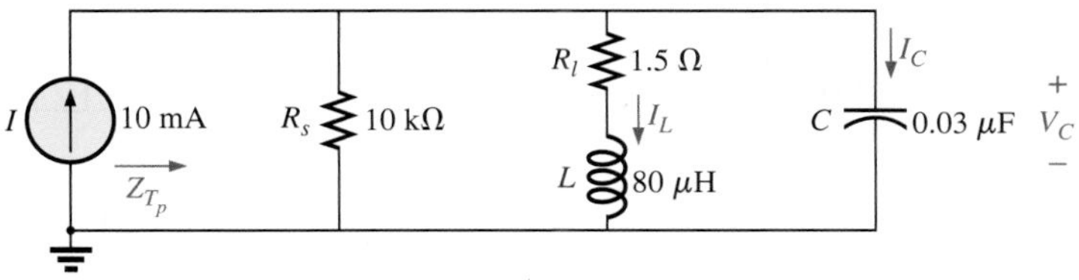

FIG. 20.57
Problem 18.

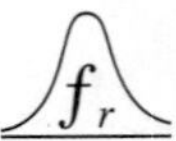

a. Find the resonant frequencies f_s, f_p, and f_m. What do the results suggest about the Q_p of the network?
b. Find the values of X_L and X_C at resonance (f_p). How do they compare?
c. Find the impedance Z_{T_p} at resonance (f_p).
d. Calculate Q_p and the BW.
e. Find the magnitude of currents I_L and I_C at resonance (f_p).
f. Calculate the voltage V_C at resonance (f_p).

***19.** Repeat Problem 18 for the network in Fig. 20.58.

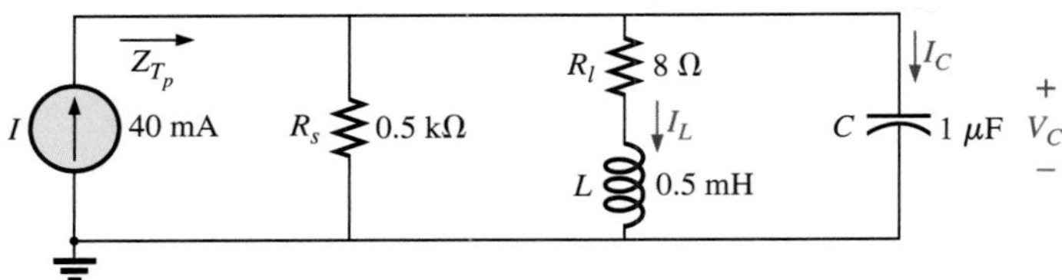

FIG. 20.58
Problems 19 and 29.

20. It is desired that the impedance $\mathbf{Z}_T$ of the high Q circuit in Fig. 20.59 be 50 kΩ ∠0° at resonance (f_p).
a. Find the value of X_L.
b. Compute X_C.
c. Find the resonant frequency (f_p) if L = 16 mH.
d. Find the value of C.

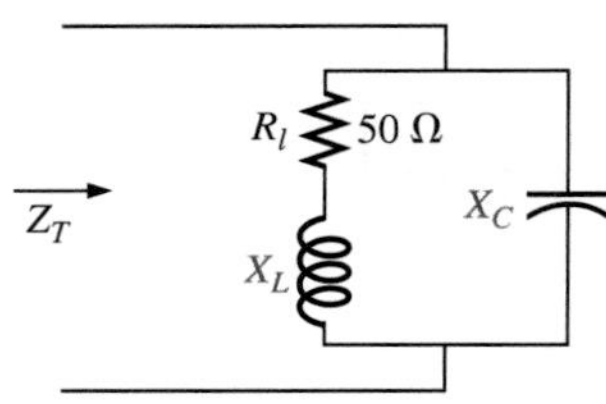

FIG. 20.59
Problem 20.

21. For the network in Fig. 20.60:
a. Find f_p.
b. Calculate the magnitude of V_C at resonance (f_p).
c. Determine the power absorbed at resonance.
d. Find the BW.

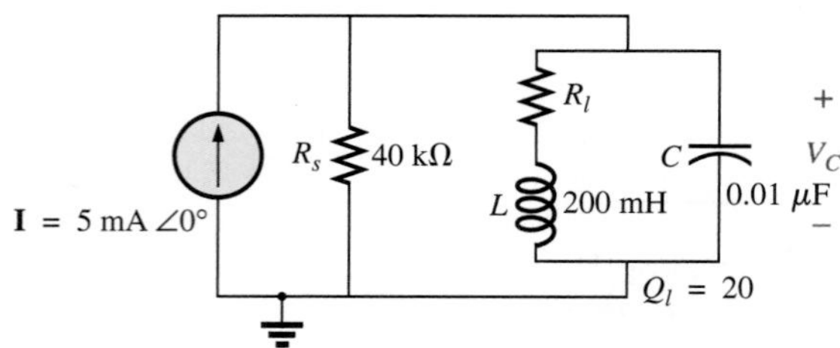

FIG. 20.60
Problem 21.

***22.** For the network in Fig. 20.61:
a. Find the value of X_L for resonance.
b. Find Q_l.
c. Find the resonant frequency (f_p) if the bandwidth is 1 kHz.
d. Find the maximum value of the voltage V_C.
e. Sketch the curve of V_C versus frequency. Indicate its peak value, resonant frequency, and band frequencies.

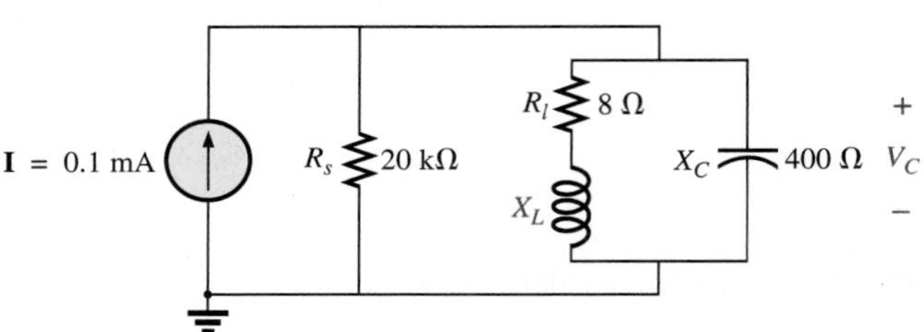

FIG. 20.61
Problem 22.

***23.** Repeat Problem 22 for the network in Fig. 20.62.

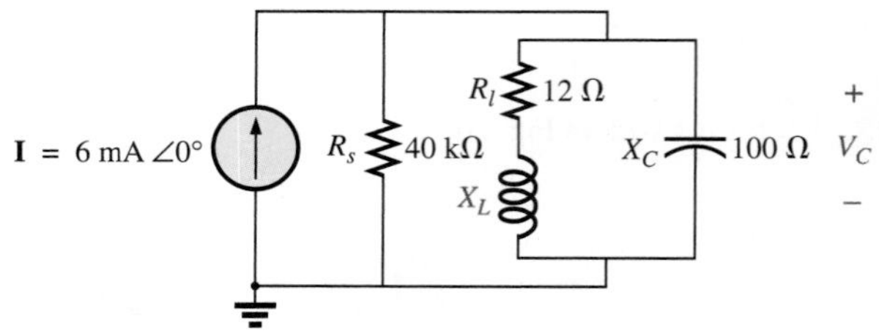

FIG. 20.62
Problem 23.

*24. For the network in Fig. 20.63:

a. Find f_s, f_p, and f_m.

b. Determine Q_l and Q_p at f_p after a source conversion is performed.

c. Find the input impedance Z_{T_p}.

d. Find the magnitude of the voltage V_C.

e. Calculate the bandwidth using f_p.

f. Determine the magnitude of the currents I_C and I_L.

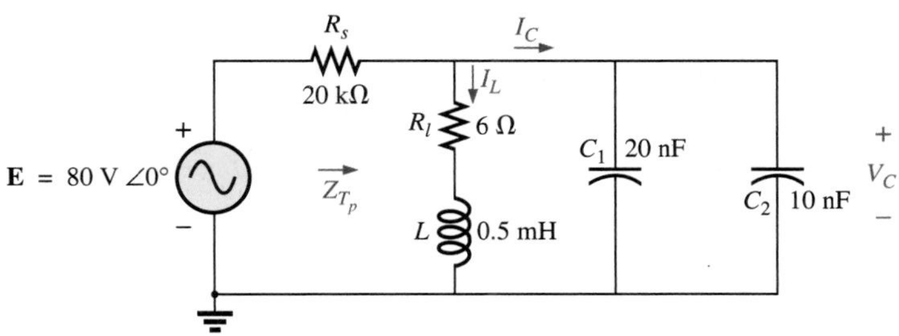

FIG. 20.63
Problem 24.

*25. For the network in Fig. 20.64, the following are specified:

$$f_p = 20 \text{ kHz}$$
$$BW = 1.8 \text{ kHz}$$
$$L = 2 \text{ mH}$$
$$Q_l = 80$$

Find R_s and C.

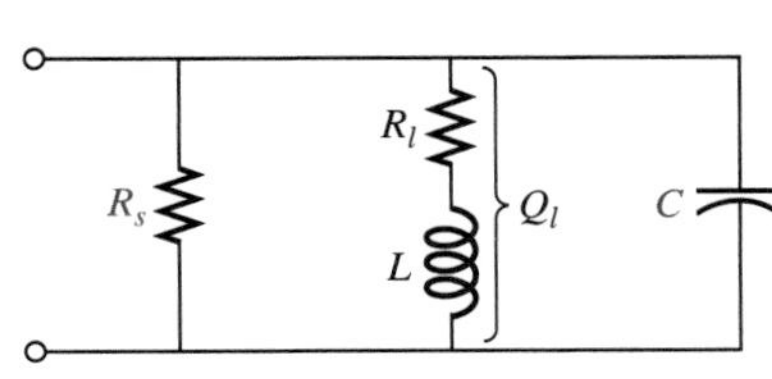

FIG. 20.64
Problem 25.

*26. Design the network in Fig. 20.65 to have the following characteristics:

a. $BW = 500$ Hz

b. $Q_p = 30$

c. $V_{C_{max}} = 1.8$ V

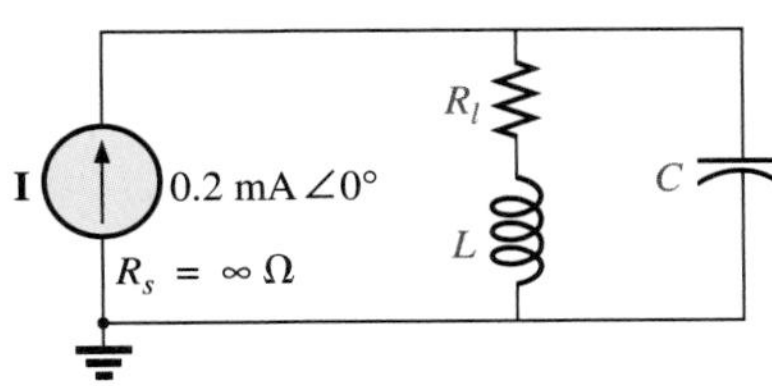

FIG. 20.65
Problem 26.

*27. For the parallel resonant circuit in Fig. 20.66:

a. Determine the resonant frequency.

b. Find the total impedance at resonance.

c. Find Q_p.

d. Calculate the BW.

e. Repeat parts (a) through (d) for $L = 20\ \mu$H and $C = 20$ nF.

f. Repeat parts (a) through (d) for $L = 0.4$ mH and $C = 1$ nF.

g. For the network in Fig. 20.66 and the parameters of parts (e) and (f), determine the ratio L/C.

h. Do your results confirm the conclusions in Fig. 20.28 for changes in the L/C ratio?

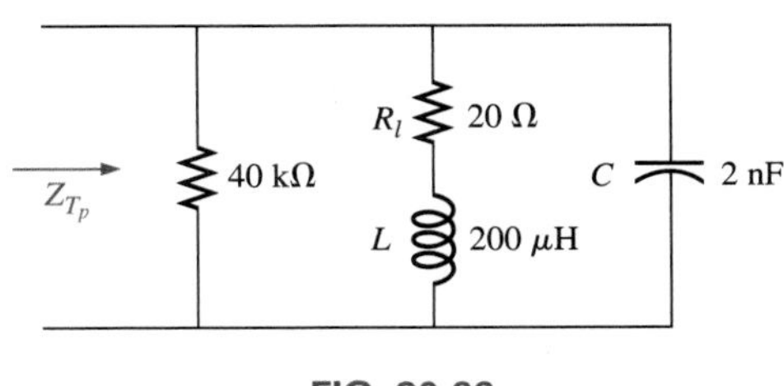

FIG. 20.66
Problem 27.

SECTION 20.14 Computer Analysis

PSpice or Multisim

28. Verify the results in Example 20.8. That is, show that the resonant frequency is 40 kHz, the cutoff frequencies are as calculated, and the bandwidth is 1.85 kHz.

29. Find f_p and f_m for the parallel resonant network in Fig. 20.58, and comment on the resulting bandwidth as it relates to the quality factor of the network.

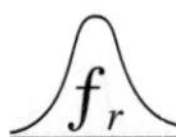

GLOSSARY

Band (cutoff, half-power, corner) frequencies Frequencies that define the points on the resonance curve that are 0.707 of the peak current or voltage value. In addition, they define the frequencies at which the power transfer to the resonant circuit will be half the maximum power level.

Bandwidth *(BW)* The range of frequencies between the band, cutoff, or half-power frequencies.

Quality factor *(Q)* A ratio that provides an immediate indication of the sharpness of the peak of a resonance curve. The higher the Q, the sharper the peak and the more quickly it drops off to the right and left of the resonant frequency.

Resonance A condition established by the application of a particular frequency (the resonant frequency) to a series or parallel R-L-C network. The transfer of power to the system is a maximum, and, for frequencies above and below, the power transfer drops off to significantly lower levels.

Selectivity A characteristic of resonant networks directly related to the bandwidth of the resonant system. High selectivity is associated with small bandwidth (high Q's), and low selectivity with larger bandwidths (low Q's).

Decibels, Filters, and Bode Plots

Objectives

- *Develop confidence in the use of logarithms and decibels in the description of power and voltage levels.*
- *Become familiar with the frequency response of high- and low-pass filters. Learn to calculate the cutoff frequency and describe the phase response.*
- *Be able to calculate the cutoff frequencies and sketch the frequency response of a pass-band, stop-band, or double-tuned filter.*
- *Develop skills in interpreting and establishing the Bode response of any filter.*
- *Become aware of the characteristics and operation of a crossover network.*

21.1 LOGARITHMS

The use of logarithms in industry is so extensive that a clear understanding of their purpose and use is an absolute necessity. At first exposure, logarithms often appear vague and mysterious due to the mathematical operations required to find the logarithm and antilogarithm using the longhand table approach that is typically taught in mathematics courses. However, almost all of today's scientific calculators have the common and natural log functions, eliminating the complexity of applying logarithms and allowing us to concentrate on the positive characteristics of the function.

Basic Relationships

Let us first examine the relationship between the variables of the logarithmic function. The mathematical expression

$$N = (b)^x$$

states that the number N is equal to the base b taken to the power x. A few examples:

$$100 = (10)^2$$
$$27 = (3)^3$$
$$54.6 = (e)^4 \qquad \text{where } e = 2.7183$$

To find the power x to satisfy the equation

$$1200 = (10)^x$$

you can determine the value of x using logarithms in the following manner:

$$x = \log_{10} 1200 = \mathbf{3.079}$$

revealing that

$$10^{3.079} = 1200$$

Note that the logarithm was taken to the base 10—the number to be taken to the power of x. There is no limitation to the numerical value of the base except that tables and calculators are designed to handle either a base of 10 (common logarithm, LOG) or base $e = 2.7183$ (natural logarithm, IN). In review, therefore,

$$\text{If } N = (b)^x, \text{ then } x = \log_b N. \tag{21.1}$$

The base to be used is a function of the area of application. If a conversion from one base to the other is required, the following equation can be applied:

$$\log_e x = 2.3 \log_{10} x \tag{21.2}$$

In this chapter, we concentrate solely on the common logarithm. However, a number of the conclusions are also applicable to natural logarithms.

Some Areas of Application

The following are some of the most common applications of the logarithmic function:

1. The response of a system can be plotted for a range of values that may otherwise be impossible or unwieldy with a linear scale.
2. Levels of power, voltage, and the like, can be compared without dealing with very large or very small numbers that often cloud the true impact of the difference in magnitudes.
3. A number of systems respond to outside stimuli in a nonlinear logarithmic manner. The result is a mathematical model that permits a direct calculation of the response of the system to a particular input signal.
4. The response of a cascaded or compound system can be rapidly determined using logarithms if the gain of each stage is known on a logarithmic basis. This characteristic is demonstrated in an example to follow.

Graphs

Graph paper is available in **semilog** and **log-log** varieties. Semilog paper has only one log scale, with the other a linear scale. Both scales of log-log paper are log scales. A section of semilog paper appears in Fig. 21.1. Note the linear (even-spaced-interval) vertical scaling and the repeating intervals of the log scale at multiples of 10.

The spacing of the log scale is determined by taking the common log (base 10) of the number. The scaling starts with 1, since $\log_{10} 1 = 0$. The distance between 1 and 2 is determined by $\log_{10} 2 = 0.3010$, or approximately 30% of the full distance of a log interval, as shown on the graph. The distance between 1 and 3 is determined by $\log_{10} 3 = 0.4771$, or about 48% of the full width. For future reference, keep in mind that almost 50% of the width of one log interval is represented by a 3 rather than by the 5 of a linear scale. In addition, note that the number 5 is about 70% of the full width, and 8 is about 90%. Remembering the percentage of full width of the lines 2, 3, 5, and 8 is particularly useful when the various lines of a log plot are left unnumbered.

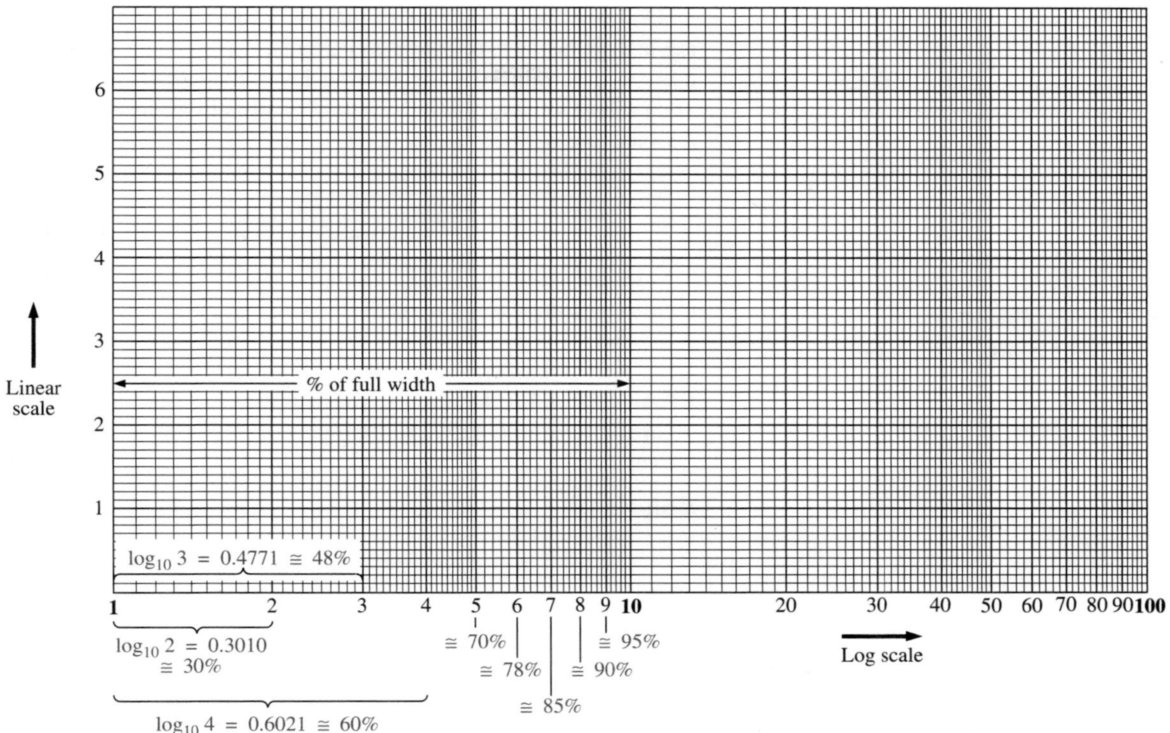

FIG. 21.1
Semilog graph paper.

Since

$$\begin{aligned}\log_{10} 1 &= 0 \\ \log_{10} 10 &= 1 \\ \log_{10} 100 &= 2 \\ \log_{10} 1000 &= 3 \\ &\vdots\end{aligned}$$

the spacing between 1 and 10, 10 and 100, 100 and 1000, and so on, is the same as shown in Figs. 21.1 and 21.2.

Note in Figs. 21.1 and 21.2 that the log scale becomes compressed at the high end of each interval. With increasing frequency levels assigned to each interval, a single graph can provide a frequency plot extending from 1 Hz to 1 MHz, as shown in Fig. 21.2, with particular reference to the 30%, 50%, 70%, and 90% levels of each interval.

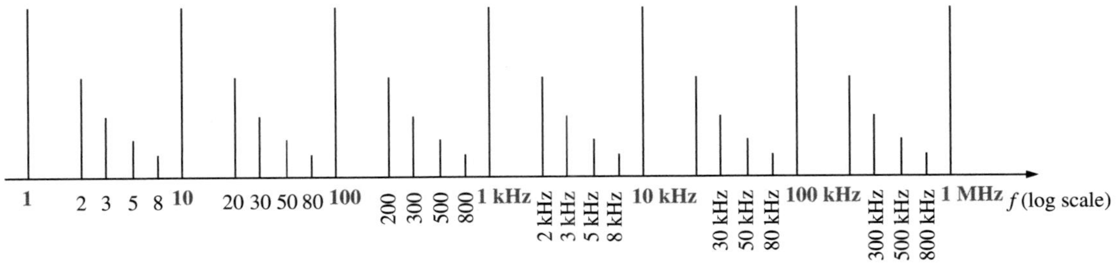

FIG. 21.2
Frequency log scale.

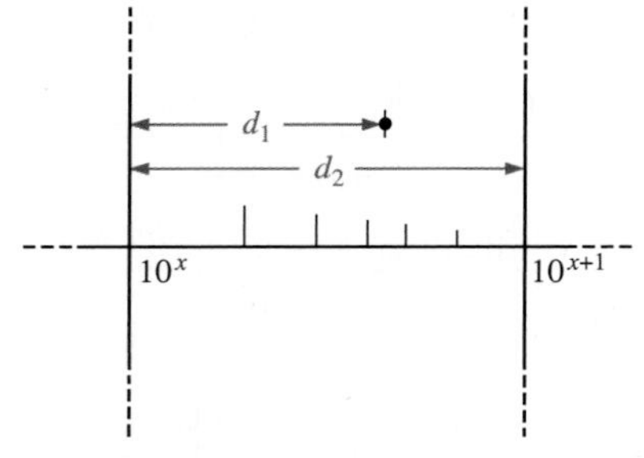

FIG. 21.3
Finding a value on a log plot.

On many log plots, the tick marks for most of the intermediate levels are left off because of space constraints. The following equation can be used to determine the logarithmic level at a particular point between known levels using a ruler or simply estimating the distances. The parameters are defined by Fig. 21.3.

$$\text{Value} = 10^x \times 10^{d_1/d_2} \tag{21.3}$$

The derivation of Eq. (21.3) is simply an extension of the details regarding distance appearing in Fig. 21.1.

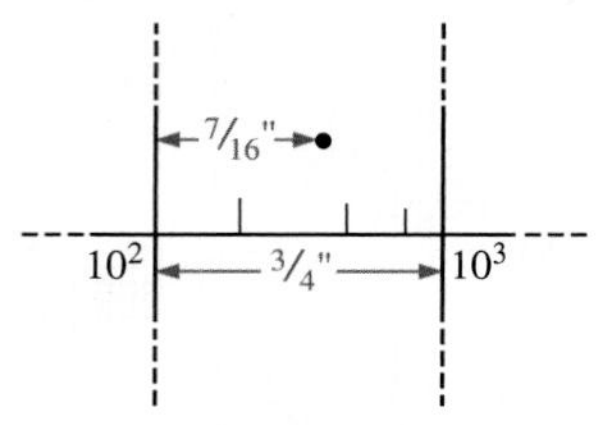

FIG. 21.4
Example 21.1.

EXAMPLE 21.1 Determine the value of the point appearing on the logarithmic plot in Fig. 21.4 using the measurements made by a ruler (linear).

Solution:

$$\frac{d_1}{d_2} = \frac{7/16''}{3/4''} = \frac{0.438''}{0.750''} = 0.584$$

Using a calculator:

$$10^{d_1/d_2} = 10^{0.584} = 3.837$$

Applying Eq. (21.3):

$$\begin{aligned}\text{Value} &= 10^x \times 10^{d_1/d_2} = 10^2 \times 3.837\\ &= \mathbf{383.7}\end{aligned}$$

21.2 PROPERTIES OF LOGARITHMS

There are a few characteristics of logarithms that should be emphasized:

1. ***The common or natural logarithm of the number 1 is 0.***

$$\log_{10} 1 = 0 \tag{21.4}$$

just as $10^x = 1$ requires that $x = 0$.

2. ***The log of any number less than 1 is a negative number.***

$$\log_{10} \tfrac{1}{2} = \log_{10} 0.5 = -0.3$$
$$\log_{10} \tfrac{1}{10} = \log_{10} 0.1 = -1$$

3. ***The log of the product of two numbers is the sum of the logs of the numbers.***

$$\log_{10} ab = \log_{10} a + \log_{10} b \tag{21.5}$$

4. ***The log of the quotient of two numbers is the log of the numerator minus the log of the denominator.***

$$\log_{10} \frac{a}{b} = \log_{10} a - \log_{10} b \tag{21.6}$$

5. The log of a number taken to a power is equal to the product of the power and the log of the number.

$$\log_{10} a^n = n \log_{10} a \qquad \textbf{(21.7)}$$

Calculator Functions

Using the TI-89 calculator, the common logarithm of a number is determined by first selecting the CATALOG key and then scrolling to find the common logarithm function. The time involved in scrolling through the options can be reduced by first selecting the key with the first letter of the desired function—in this case, *L* as shown below to find the common logarithm of the number 80.

CATALOG [4 (L)] ↓ LOG [(] [ENTER] [8] [0] [)] [ENTER] **1.903**

For the reverse process, to find N, or the antilogarithm, use the function 10. In this case, the function 10 is found after the letter *Z* in the catalog. It is easier if you first select *A* to start at the top of the listing and then go backwards to the power of ten function by using the upward moving scroll option. The antilogarithm of the number 0.6 is found as follows:

CATALOG [= (A)] ↑ 10^ [(] [ENTER] [0] [.] [6] [)] [ENTER] **3.981**

EXAMPLE 21.2 Evaluate each of the following logarithmic expressions:

a. $\log_{10} 0.004$
b. $\log_{10} 250{,}000$
c. $\log_{10}(0.08)(240)$
d. $\log_{10} \dfrac{1 \times 10^4}{1 \times 10^{-4}}$
e. $\log_{10}(10)^4$

Solutions:

a. $\mathbf{-2.398}$
b. $\mathbf{+5.398}$
c. $\log_{10}(0.08)(240) = \log_{10} 0.08 + \log_{10} 240 = -1.097 + 2.380 = \mathbf{1.283}$
d. $\log_{10} \dfrac{1 \times 10^4}{1 \times 10^{-4}} = \log_{10} 1 \times 10^4 - \log_{10} 1 \times 10^{-4} = 4 - (-4) = \mathbf{8}$
e. $\log_{10} 10^4 = 4 \log_{10} 10 = 4(1) = \mathbf{4}$

21.3 DECIBELS

Power Gain

Two levels of power can be compared using a unit of measure called the *bel*, which is defined by the following equation:

$$B = \log_{10} \frac{P_2}{P_1} \qquad \text{(bels)} \qquad \textbf{(21.8)}$$

However, to provide a unit of measure of *less* magnitude, a **decibel** is defined, where

$$\boxed{1 \text{ bel} = 10 \text{ decibels (dB)}} \tag{21.9}$$

The result is the following important equation, which compares power levels P_2 and P_1 in decibels:

$$\boxed{\text{dB} = 10 \log_{10} \frac{P_2}{P_1}} \quad \text{(decibels, dB)} \tag{21.10}$$

If the power levels are equal ($P_2 = P_1$), there is no change in power level, and dB $= 0$. If there is an increase in power level ($P_2 > P_1$), the resulting decibel level is positive. If there is a decrease in power level ($P_2 < P_1$), the resulting decibel level will be negative.

For the special case of $P_2 = 2P_1$, the gain in decibels is

$$\text{dB} = 10 \log_{10} \frac{P_2}{P_1} = 10 \log_{10} 2 = \mathbf{3\ dB}$$

Therefore, for a speaker system, a 3 dB increase in output requires that the power level be doubled. In the audio industry, it is a generally accepted rule that an increase in sound level is accomplished with 3 dB increments in the output level. In other words, a 1 dB increase is barely detectable, and a 2 dB increase just discernible. A 3 dB increase normally results in a readily detectable increase in sound level. An additional increase in the sound level is normally accomplished by simply increasing the output level another 3 dB. If an 8 W system were in use, a 3 dB increase would require a 16 W output, whereas an additional increase of 3 dB (a total of 6 dB) would require a 32 W system, as demonstrated by the calculations below:

$$\text{dB} = 10 \log_{10} \frac{P_2}{P_1} = 10 \log_{10} \frac{16}{8} = 10 \log_{10} 2 = \mathbf{3\ dB}$$

$$\text{dB} = 10 \log_{10} \frac{P_2}{P_1} = 10 \log_{10} \frac{32}{8} = 10 \log_{10} 4 = \mathbf{6\ dB}$$

For $P_2 = 10P_1$,

$$\text{dB} = 10 \log_{10} \frac{P_2}{P_1} = 10 \log_{10} 10 = 10(1) = \mathbf{10\ dB}$$

resulting in the unique situation where the power gain has the same magnitude as the decibel level.

For some applications, a reference level is established to permit a comparison of decibel levels from one situation to another. For communication systems, a commonly applied reference level is

$$P_{ref} = 1 \text{ mW} \quad \text{(across a 600 } \Omega \text{ load)}$$

Eq. (21.10) is then typically written as

$$\boxed{\text{dB}_m = 10 \log_{10} \frac{P}{1 \text{ mW}} \bigg|_{600\ \Omega}} \tag{21.11}$$

Note the subscript m to denote that the decibel level is determined with a reference level of 1 mW.

In particular, for $P = 40$ mW,

$$\text{dB}_m = 10 \log_{10} \frac{40 \text{ mW}}{1 \text{ mW}} = 10 \log_{10} 40 = 10(1.6) = \mathbf{16\ dB}_m$$

whereas for $P = 4$ W,

$$\text{dB}_m = 10 \log_{10} \frac{4000 \text{ mW}}{1 \text{ mW}} = 10 \log_{10} 4000 = 10(3.6) = \mathbf{36\ dB}_m$$

Even though the power level has increased by a factor of 4000 mW/ 40 mW = 100, the dB_m increase is limited to 20 dB_m. In time, the significance of dB_m levels of 16 dB_m and 36 dB_m will generate an immediate appreciation regarding the power levels involved. An increase of 20 dB_m is also associated with a significant gain in power levels.

Voltage Gain

Decibels are also used to provide a comparison between voltage levels. Substituting the basic power equations $P_2 = V_2^2/R_2$ and $P_1 = V_1^2/R_1$ into Eq. (21.10) results in

$$\begin{aligned}\text{dB} &= 10 \log_{10} \frac{P_2}{P_1} = 10 \log_{10} \frac{V_2^2/R_2}{V_1^2/R_1} \\ &= 10 \log_{10} \frac{V_2^2/V_1^2}{R_2/R_1} = 10 \log_{10}\left(\frac{V_2}{V_1}\right)^2 - 10 \log_{10}\left(\frac{R_2}{R_1}\right)\end{aligned}$$

and
$$\text{dB} = 20 \log_{10} \frac{V_2}{V_1} - 10 \log_{10} \frac{R_2}{R_1}$$

For the situation where $R_2 = R_1$, a condition normally assumed when comparing voltage levels on a decibel basis, the second term of the preceding equation drops out ($\log_{10} 1 = 0$), and

$$\boxed{\text{dB}_v = 20 \log_{10} \frac{V_2}{V_1}} \qquad \text{(dB)} \qquad \textbf{(21.12)}$$

Note the subscript v to define the decibel level obtained.

EXAMPLE 21.3 Find the voltage gain in dB of a system where the applied signal is 2 mV and the output voltage is 1.2 V.

Solution:

$$\text{dB}_v = 20 \log_{10} \frac{V_o}{V_i} = 20 \log_{10} \frac{1.2 \text{ V}}{2 \text{ mV}} = 20 \log_{10} 600 = \mathbf{55.56\ dB}$$

for a voltage gain $A_v = V_o/V_i$ of 600.

EXAMPLE 21.4 If a system has a voltage gain of 36 dB, find the applied voltage if the output voltage is 6.8 V.

Solution:

$$dB_v = 20 \log_{10} \frac{V_o}{V_i}$$

$$36 = 20 \log_{10} \frac{V_o}{V_i}$$

$$1.8 = \log_{10} \frac{V_o}{V_i}$$

From the antilogarithm:

$$\frac{V_o}{V_i} = 63.096$$

and $$V_i = \frac{V_o}{63.096} = \frac{6.8 \text{ V}}{63.096} = \mathbf{107.77 \text{ mV}}$$

TABLE 21.1

V_o/V_i	$dB = 20 \log_{10} (V_o/V_i)$
1	0 dB
2	6 dB
10	20 dB
20	26 dB
100	40 dB
1,000	60 dB
100,000	100 dB

Table 21.1 compares the magnitude of specific gains to the resulting decibel level. In particular, note that when voltage levels are compared, a doubling of the level results in a change of 6 dB rather than 3 dB as obtained for power levels.

In addition, note that an increase in gain from 1 to 100,000 results in a change in decibels that can easily be plotted on a single graph. Also note that doubling the gain (from 1 to 2 and 10 to 20) results in a 6 dB increase in the decibel level, while a change of 10 to 1 (from 1 to 10, 10 to 100, and so on) always results in a 20 dB decrease in the decibel level.

Human Auditory Response

One of the most frequent applications of the decibel scale is in the communication and entertainment industries. The human ear does not respond in a linear fashion to changes in source power level; that is, a doubling of the audio power level from 1/2 W to 1 W does not result in a doubling of the loudness level for the human ear. In addition, a change from 5 W to 10 W is received by the ear as the same change in sound intensity as experienced from 1/2 W to 1 W. In other words, the ratio between levels is the same in each case (1 W/0.5 W = 10 W/5 W = 2), resulting in the same decibel or logarithmic change defined by Eq. (21.7). The ear, therefore, responds in a logarithmic fashion to changes in audio power levels.

To establish a basis for comparison between audio levels, a reference level of 0.0002 **microbar** (μbar) was chosen, where 1 μbar is equal to the sound pressure of 1 dyne per square centimeter, or about 1 millionth of the normal atmospheric pressure at sea level. The 0.0002 μbar level is the threshold level of hearing. Using this reference level, the sound pressure level in decibels is defined by the following equation:

$$dB_s = 20 \log_{10} \frac{P}{0.0002 \ \mu\text{bar}} \qquad \textbf{(21.13)}$$

where P is the sound pressure in microbars.

The decibel levels in Fig. 21.5 are defined by Eq. (21.13). Meters designed to measure audio levels are calibrated to the levels defined by Eq. (21.13) and shown in Fig. 21.5.

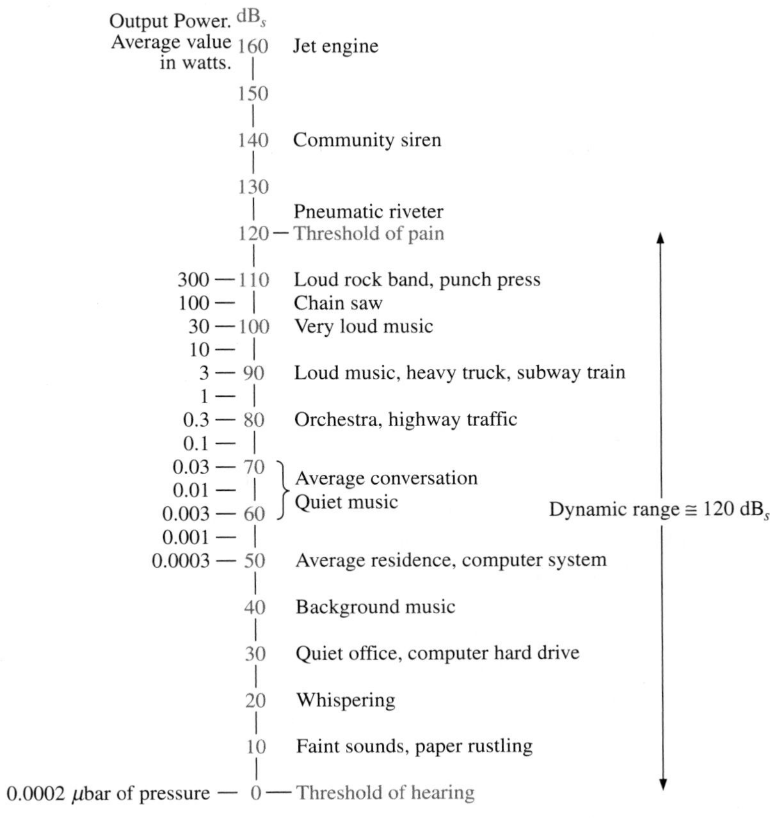

FIG. 21.5
Typical sound levels and their decibel levels.

A common question regarding audio levels is how much the power level of an acoustical source must be increased to double the sound level received by the human ear. The question is not as simple as it first seems due to considerations such as the frequency content of the sound, the acoustical conditions of the surrounding area, the physical characteristics of the surrounding medium, and—of course—the unique characteristics of the human ear. However, a general conclusion can be formulated that has practical value if we note the power levels of an acoustical source appearing to the left in Fig. 21.5. Each power level is associated with a particular decibel level, and a change of 10 dB in the scale corresponds with an increase or a decrease in power by a factor of 10. For instance, a change from 90 dB to 100 dB is associated with a change in wattage from 3 W to 30 W. Through experimentation, it has been found that on an average basis the loudness level doubles for every 10 dB change in audio level—a conclusion somewhat verified by the examples to the right in Fig. 21.5.

To double the sound level received by the human ear, the power rating of the acoustical source (in watts) must be increased by a factor of 10.

In other words, doubling the sound level available from a 1 W acoustical source requires moving up to a 10 W source.

Instrumentation

A number of modern VOMs and DMMs have a dB scale designed to provide an indication of power ratios referenced to a standard level of 1 mW at 600 Ω. Since the reading is accurate only if the load has a characteristic

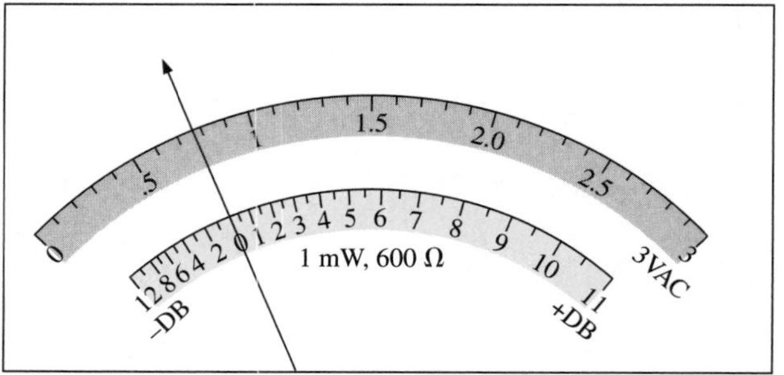

FIG. 21.6
Defining the relationship between a dB scale referenced to 1 mW, 600 Ω and a 3 V rms voltage scale.

impedance of 600 Ω, the 1 mW, 600 Ω reference level is normally printed somewhere on the face of the meter, as shown in Fig. 21.6. The dB scale is usually calibrated to the lowest ac scale of the meter. In other words, when making the dB measurement, choose the lowest ac voltage scale, but read the dB scale. If a higher voltage scale is chosen, a correction factor must be used, which is sometimes printed on the face of the meter but is always available in the meter manual. If the impedance is other than 600 Ω or not purely resistive, other correction factors must be used that are normally included in the meter manual. Using the basic power equation $P = V^2/R$ reveals that 1 mW across a 600 Ω load is the same as applying 0.775 V rms across a 600 Ω load; that is, $V = \sqrt{PR} = \sqrt{(1\text{ mW})(600\ \Omega)} = 0.775\text{ V}$. The result is that an analog display will have 0 dB [defining the reference point of 1 mW, $\text{dB} = 10 \log_{10} P_2/P_1 = 10 \log_{10} (1\text{ mW}/1\text{ mW(ref)}) = 0\text{ dB}$] and 0.775 V rms on the same pointer projection, as shown in Fig. 21.6. A voltage of 2.5 V across a 600 Ω load results in a dB level of $\text{dB} = 20 \log_{10} V_2/V_1 = 20 \log_{10} 2.5\text{ V}/0.775 = 10.17\text{ dB}$, resulting in 2.5 V and 10.17 dB appearing along the same pointer projection. A voltage of less than 0.775 V, such as 0.5 V, results in a dB level of $\text{dB} = 20 \log_{10} V_2/V_1 = 20 \log_{10} 0.5\text{ V}/0.775\text{ V} = -3.8\text{ dB}$, also shown on the scale in Fig. 21.6. Although a reading of 10 dB reveals that the power level is 10 times the reference, don't assume that a reading of 5 dB means that the output level is 5 mW. The 10 : 1 ratio is a special one in logarithmic circles. For the 5 dB level, the power level must be found using the antilogarithm (3.126), which reveals that the power level associated with 5 dB is about 3.1 times the reference or 3.1 mW. A conversion table is usually provided in the manual for such conversions.

21.4 FILTERS

Any combination of passive (*R, L,* and *C*) and/or active (transistors or operational amplifiers) elements designed to select or reject a band of frequencies is called a **filter.** In communication systems, filters are used to pass those frequencies containing the desired information and to reject the remaining frequencies. In stereo systems, filters can isolate particular bands of frequencies for increased or decreased emphasis by the output acoustical system (amplifier, speaker, and so on). Filters are used to filter out any unwanted frequencies, commonly called *noise,* due to the nonlinear characteristics of some electronic devices or signals picked up from the surrounding medium. In general, there are two classifications of filters:

1. **Passive filters** are those filters composed of series or parallel combinations of *R, L,* and *C* elements.

2. **Active filters** are filters that employ active devices such as transistors and operational amplifiers in combination with R, L, and C elements.

Since this text is limited to passive devices, the analysis of this chapter is limited to passive filters. In addition, only the most fundamental forms are examined in the next few sections. The subject of filters is a very broad one that continues to receive extensive research support from industry and the government as new communication systems are developed to meet the demands of increased volume and speed. There are courses and texts devoted solely to the analysis and design of filter systems that can become quite complex and sophisticated. In general, however, all filters belong to the four broad categories of **low-pass, high-pass, pass-band,** and **stop-band,** as depicted in Fig. 21.7. For each form, there are critical frequencies that define the regions of pass-bands and stop-bands (often called *reject* bands). Any frequency in the pass-band will pass through to the next

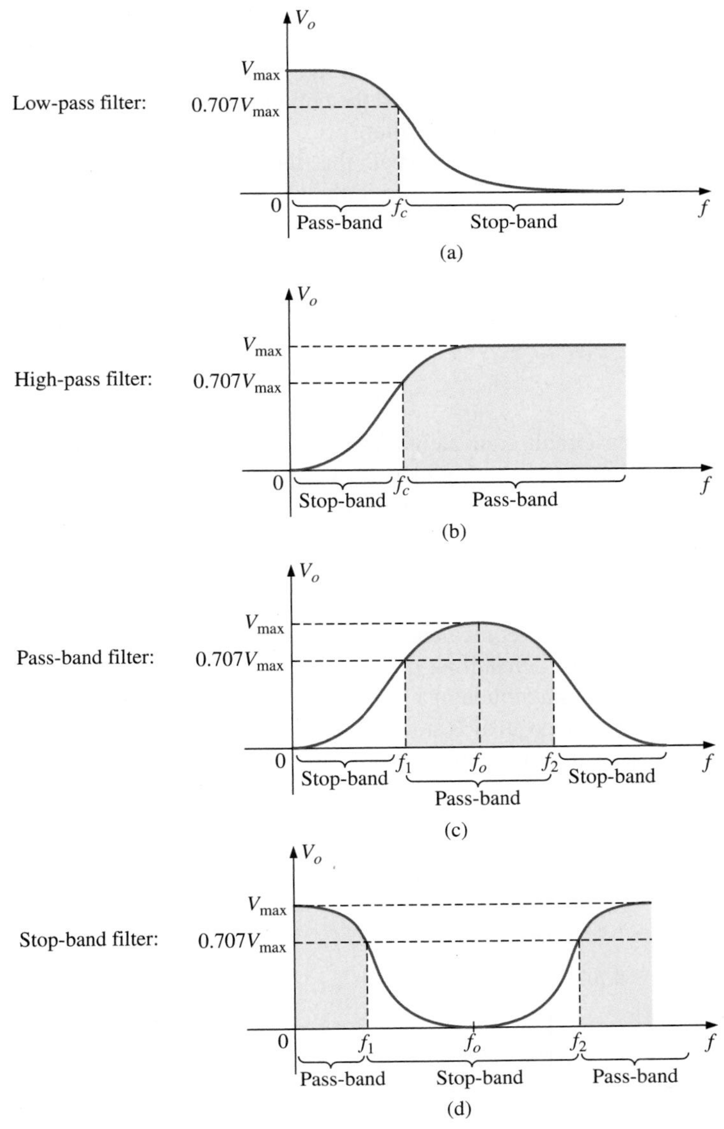

FIG. 21.7
Defining the four broad categories of filters.

stage with at least 70.7% of the maximum output voltage. Recall the use of the 0.707 level to define the bandwidth of a series or parallel resonant circuit (both with the general shape of the pass-band filter).

For some stop-band filters, the stop-band is defined by conditions other than the 0.707 level. In fact, for many stop-band filters, the condition that $V_o = 1/1000V_{max}$ (corresponding with -60 dB in the discussion to follow) is used to define the stop-band region, with the pass-band continuing to be defined by the 0.707 V level. The resulting frequencies between the two regions are then called the *transition frequencies* and establish the *transition region.*

At least one example of each filter in Fig. 21.7 is discussed in some detail in the sections to follow. Take particular note of the relative simplicity of some of the designs.

21.5 *R-C* LOW-PASS FILTER

The *R-C* filter, incredibly simple in design, can be used as a low-pass or a high-pass filter. If the output is taken off the capacitor, as shown in Fig. 21.8, it responds as a low-pass filter. If the positions of the resistor and capacitor are interchanged and the output is taken off the resistor, the response is that of a high-pass filter.

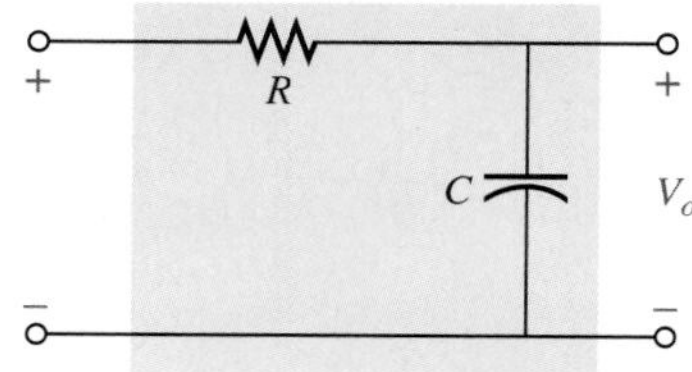

FIG. 21.8
Low-pass filter.

A glance at Fig. 21.7(a) reveals that the circuit should behave in a manner that results in a high-level output for low frequencies and a declining level for frequencies above the critical value. Let us first examine the network at the frequency extremes of $f = 0$ Hz and very high frequencies to test the response of the circuit.

At $f = 0$ Hz,

$$X_C = \frac{1}{2\pi fC} = \infty\ \Omega$$

and the open-circuit equivalent can be substituted for the capacitor, as shown in Fig. 21.9, resulting in $\mathbf{V}_o = \mathbf{V}_i$.

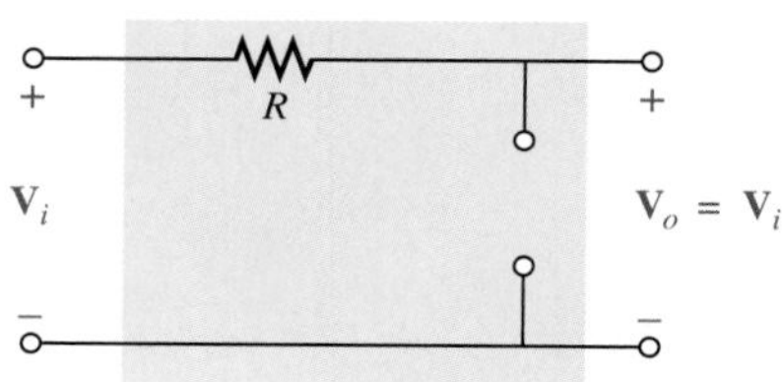

FIG. 21.9
R-C low-pass filter at low frequencies.

At very high frequencies, the reactance is

$$X_C = \frac{1}{2\pi fC} \cong 0\ \Omega$$

and the short-circuit equivalent can be substituted for the capacitor, as shown in Fig. 21.10, resulting in $\mathbf{V}_o = 0$ V.

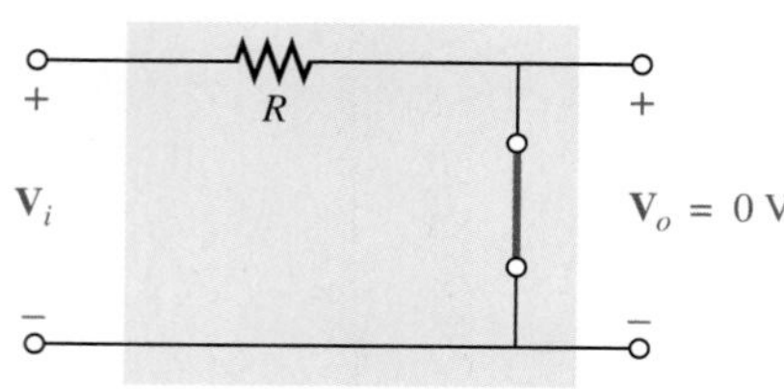

FIG. 21.10
R-C low-pass filter at high frequencies.

A plot of the magnitude of V_o versus frequency results in the curve in Fig. 21.11. Our next goal is now clearly defined: Find the frequency at which the transition takes place from a pass-band to a stop-band.

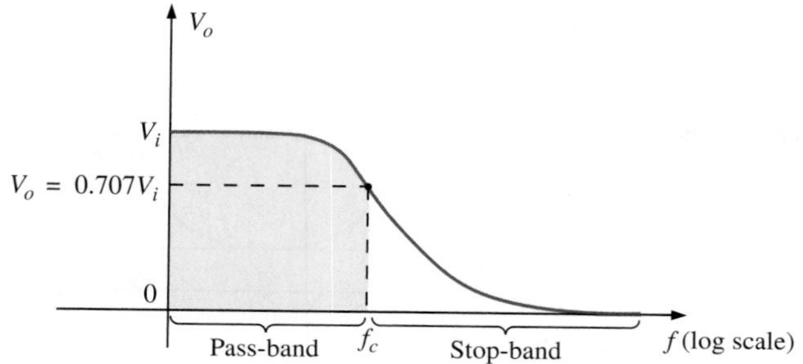

FIG. 21.11
V_o versus frequency for a low-pass R-C filter.

For filters, a normalized plot is used more often than the plot of V_o versus frequency in Fig. 21.11.

Normalization is a process whereby a quantity such as voltage, current, or impedance is divided by a quantity of the same unit of measure to establish a dimensionless level of a specific value or range.

A normalized plot in the filter domain can be obtained by dividing the plotted quantity such as V_o in Fig. 21.11 with the applied voltage V_i for the frequency range of interest. Since the maximum value of V_o for the low-pass filter in Fig. 21.8 is V_i, each level of V_o in Fig. 21.11 is divided by the level of V_i. The result is the plot of $A_v = V_o/V_i$ in Fig. 21.12. Note that the maximum value is 1 and the cutoff frequency is defined at the 0.707 level.

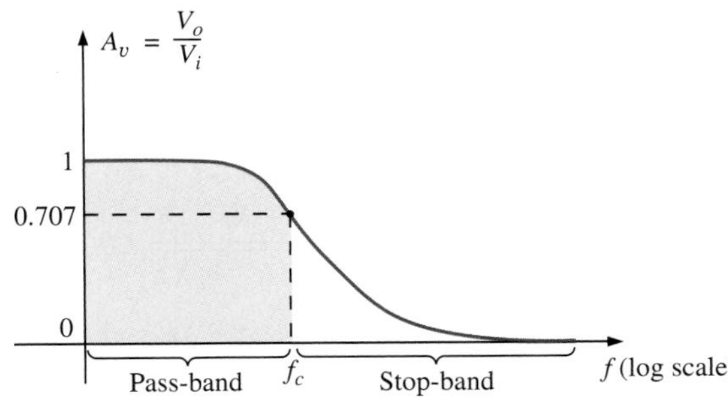

FIG. 21.12
Normalized plot of Fig. 21.11.

At any intermediate frequency, the output voltage $\mathbf{V}_o$ in Fig. 21.8 can be determined using the voltage divider rule:

$$\mathbf{V}_o = \frac{X_C \angle -90° \mathbf{V}_i}{R - j X_C}$$

or

$$\mathbf{A}_v = \frac{\mathbf{V}_o}{\mathbf{V}_i} = \frac{X_C \angle -90°}{R - j X_C} = \frac{X_C \angle -90°}{\sqrt{R^2 + X_C^2} \angle -\tan^{-1}(X_C/R)}$$

and

$$\mathbf{A}_v = \frac{\mathbf{V}_o}{\mathbf{V}_i} = \frac{X_C}{\sqrt{R^2 + X_C^2}} \angle -90° + \tan^{-1}\left(\frac{X_C}{R}\right)$$

The magnitude of the ratio V_o/V_i is therefore determined by

$$A_v = \frac{V_o}{V_i} = \frac{X_C}{\sqrt{R^2 + X_C^2}} = \frac{1}{\sqrt{\left(\frac{R}{X_C}\right)^2 + 1}} \tag{21.14}$$

and the phase angle is determined by

$$\theta = -90° + \tan^{-1}\frac{X_C}{R} = -\tan^{-1}\frac{R}{X_C} \tag{21.15}$$

For the special frequency at which $X_C = R$, the magnitude becomes

$$A_v = \frac{V_o}{V_i} = \frac{1}{\sqrt{\left(\frac{R}{X_C}\right)^2 + 1}} = \frac{1}{\sqrt{1+1}} = \frac{1}{\sqrt{2}} = 0.707$$

which defines the critical or cutoff frequency in Fig. 21.12.

The frequency at which $X_C = R$ is determined by

$$\frac{1}{2\pi f_c C} = R$$

and

$$\boxed{f_c = \frac{1}{2\pi RC}} \tag{21.16}$$

The impact of Eq. (21.16) extends beyond its relative simplicity. For any low-pass filter, the application of any frequency less than f_c results in an output voltage V_o that is at least 70.7% of the maximum. For any frequency above f_c, the output is less than 70.7% of the applied signal.

Solving for $\mathbf{V}_o$ and substituting $\mathbf{V}_i = V_i \angle 0°$ gives

$$\mathbf{V}_o = \left[\frac{X_C}{\sqrt{R^2 + X_C^2}} \angle\theta\right]\mathbf{V}_i = \left[\frac{X_C}{\sqrt{R^2 + X_C^2}} \angle\theta\right] V_i \angle 0°$$

and

$$\mathbf{V}_o = \frac{X_C V_i}{\sqrt{R^2 + X_C^2}} \angle\theta$$

The angle θ is, therefore, the angle by which $\mathbf{V}_o$ leads $\mathbf{V}_i$. Since $\theta = -\tan^{-1} R/X_C$ is always negative (except at $f = 0$ Hz), it is clear that $\mathbf{V}_o$ will always lag $\mathbf{V}_i$, leading to the label *lagging network* for the network in Fig. 21.8.

At high frequencies, X_C is very small and R/X_C is quite large, resulting in $\theta = -\tan^{-1} R/X_C$ approaching $-90°$.

At low frequencies, X_C is quite large and R/X_C is very small, resulting in θ approaching 0°.

At $X_C = R$, or $f = f_c$, $-\tan^{-1} R/X_C = -\tan^{-1} 1 = -45°$.

A plot of θ versus frequency results in the phase plot in Fig. 21.13.

The plot is of $\mathbf{V}_o$ leading $\mathbf{V}_i$, but since the phase angle is always negative, the phase plot in Fig. 21.14 ($\mathbf{V}_o$ lagging $\mathbf{V}_i$) is more appropriate. Note that a change in sign requires that the vertical axis be changed to the angle by which $\mathbf{V}_o$ lags $\mathbf{V}_i$. In particular, note that the phase angle between $\mathbf{V}_o$ and $\mathbf{V}_i$ is less than 45° in the pass-band and approaches 0° at lower frequencies.

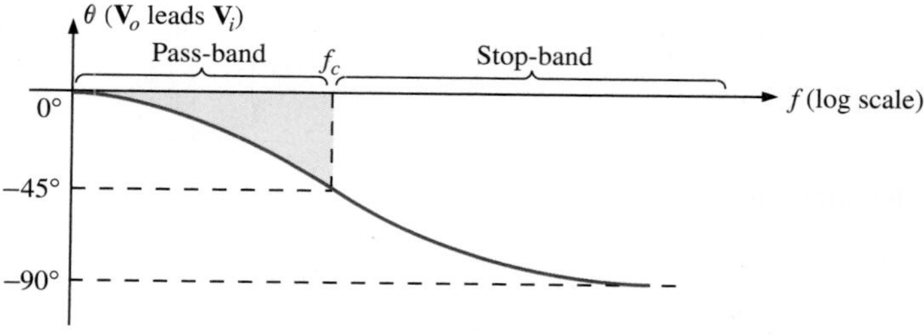

FIG. 21.13
Angle by which $\mathbf{V}_o$ leads $\mathbf{V}_i$.

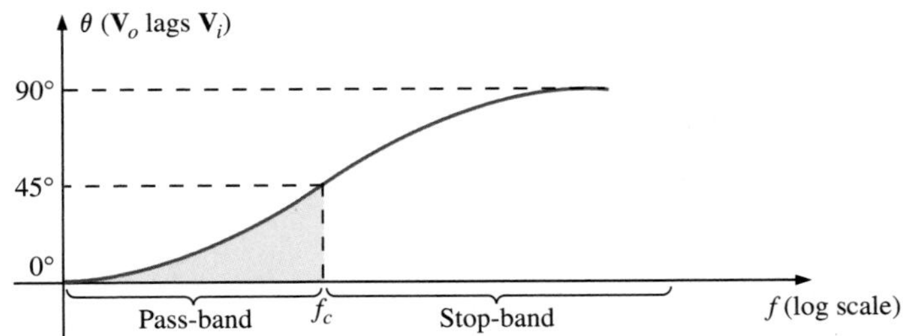

FIG. 21.14
Angle by which $\mathbf{V}_o$ lags $\mathbf{V}_i$.

In summary, for the low-pass *R-C* filter in Fig. 21.8:

$$f_c = \frac{1}{2\pi RC}$$

For	$f < f_c,$	$V_o > 0.707V_i$
whereas for	$f > f_c,$	$V_o < 0.707V_i$
At f_c,	$\mathbf{V}_o$ lags $\mathbf{V}_i$ by 45°	

The low-pass filter response in Fig. 21.7(a) can also be obtained using the *R-L* combination in Fig. 21.15 with

$$f_c = \frac{R}{2\pi L} \tag{21.17}$$

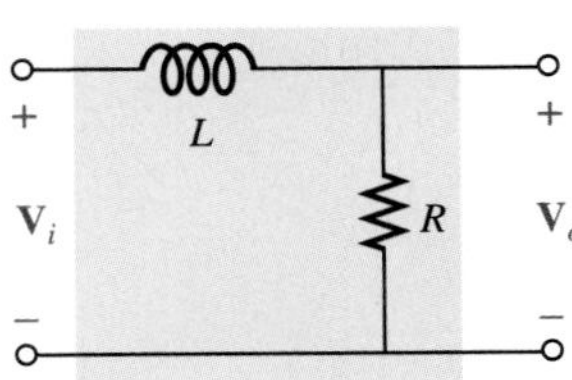

FIG. 21.15
Low-pass R-L filter.

In general, however, the *R-C* combination is more popular due to the smaller size of capacitive elements and the nonlinearities associated with inductive elements. The details of the analysis of the low-pass *R-L* can be an exercise for independent study.

EXAMPLE 21.5

a. Sketch the output voltage V_o versus frequency for the low-pass *R-C* filter in Fig. 21.16.
b. Determine the voltage V_o at $f = 100$ kHz and 1 MHz, and compare the results to the results obtained from the curve in part (a).
c. Sketch the normalized gain $A_v = V_o/V_i$.

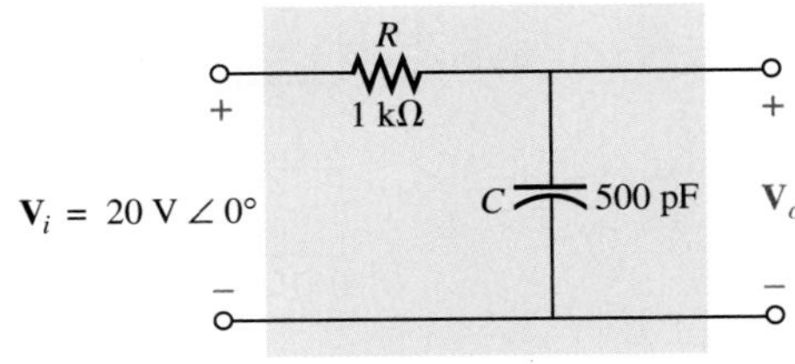

FIG. 21.16
Example 21.5.

Solutions:

a. Eq. (21.16):

$$f_c = \frac{1}{2\pi RC} = \frac{1}{2\pi(1\text{ k}\Omega)(500\text{ pF})} = \mathbf{318.31\ kHz}$$

At f_c, $V_o = 0.707(20\text{ V}) = 14.14$ V. See Fig. 21.17.

b. Eq. (21.14):

$$V_o = \frac{V_i}{\sqrt{\left(\frac{R}{X_C}\right)^2 + 1}}$$

At $f = 100$ kHz:

$$X_C = \frac{1}{2\pi fC} = \frac{1}{2\pi(100\text{ kHz})(500\text{ pF})} = 3.18\text{ k}\Omega$$

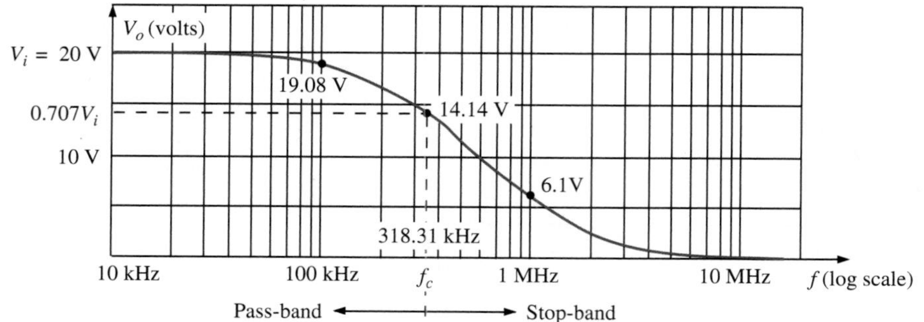

FIG. 21.17
Frequency response for the low-pass R-C network in Fig. 21.16.

and
$$V_o = \frac{20\text{ V}}{\sqrt{\left(\dfrac{1\text{ k}\Omega}{3.18\text{ k}\Omega}\right)^2 + 1}} = \mathbf{19.08\text{ V}}$$

At $f = 1$ MHz:

$$X_C = \frac{1}{2\pi fC} = \frac{1}{2\pi(1\text{ MHz})(500\text{ pF})} = 0.32\text{ k}\Omega$$

and
$$V_o = \frac{20\text{ V}}{\sqrt{\left(\dfrac{1\text{ k}\Omega}{0.32\text{ k}\Omega}\right)^2 + 1}} = \mathbf{6.1\text{ V}}$$

Both levels are verified by Fig. 21.17.

c. Dividing every level in Fig. 21.17 by $V_i = 20$ V results in the normalized plot in Fig. 21.18.

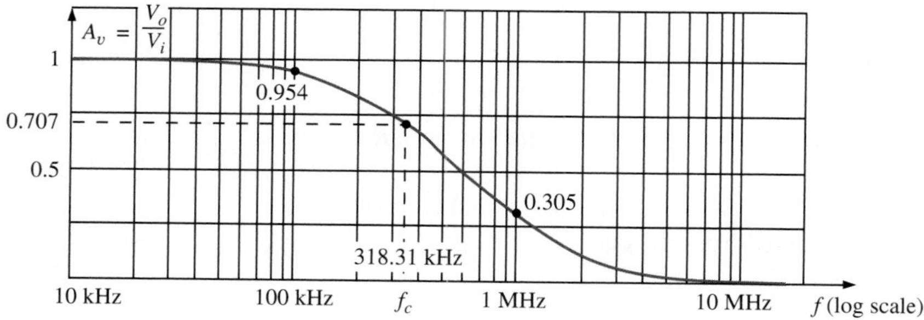

FIG. 21.18
Normalized plot of Fig. 21.17.

21.6 *R-C* HIGH-PASS FILTER

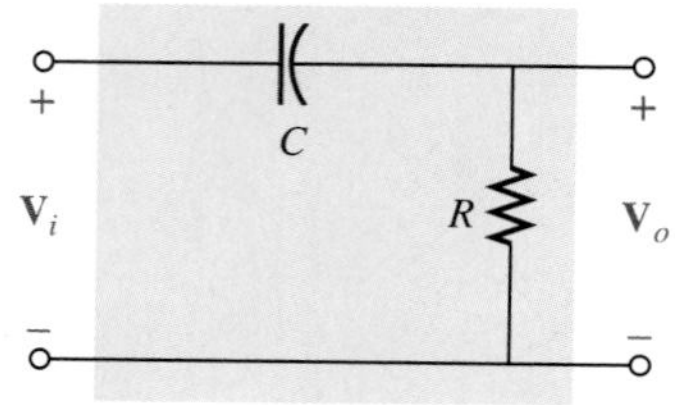

As noted in Section 21.5, a high-pass *R-C* filter can be constructed by simply reversing the positions of the capacitor and resistor, as shown in Fig. 21.19.

FIG. 21.19
High-pass filter.

At very high frequencies, the reactance of the capacitor is very small, and the short-circuit equivalent can be substituted, as shown in Fig. 21.20. The result is that $\mathbf{V}_o = \mathbf{V}_i$.

At $f = 0$ Hz, the reactance of the capacitor is quite high, and the open-circuit equivalent can be substituted, as shown in Fig. 21.21. In this case, $\mathbf{V}_o = 0$ V.

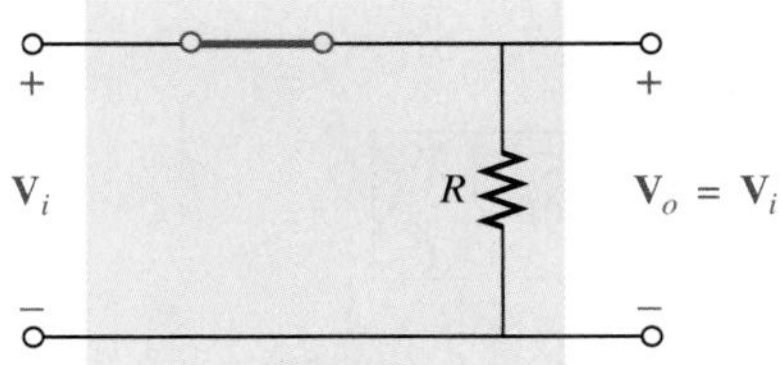

FIG. 21.20
R-C high-pass filter at very high frequencies.

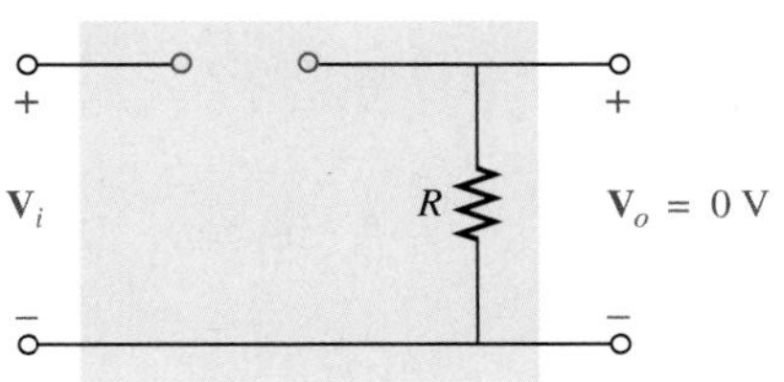

FIG. 21.21
R-C high-pass filter at $f = 0$ Hz.

A plot of the magnitude versus frequency is provided in Fig. 21.22, with the normalized plot in Fig. 21.23.

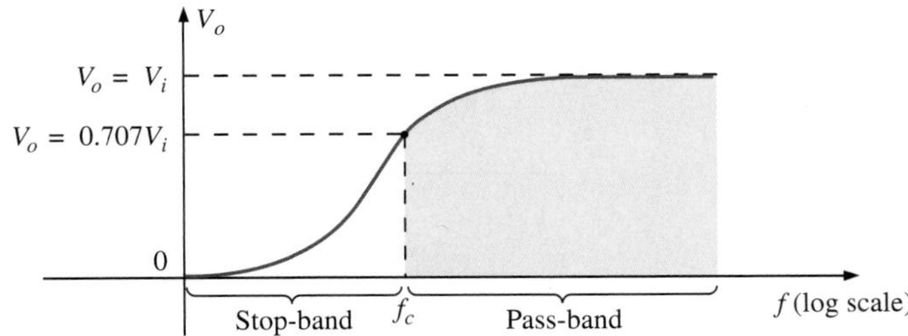

FIG. 21.22
V_o versus frequency for a high-pass R-C filter.

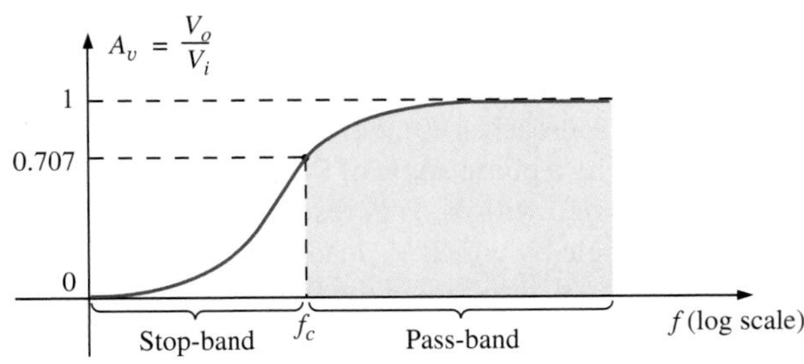

FIG. 21.23
Normalized plot of Fig. 21.22.

At any intermediate frequency, the output voltage can be determined using the voltage divider rule:

$$\mathbf{V}_o = \frac{R \angle 0^\circ \mathbf{V}_i}{R - j X_C}$$

or

$$\frac{\mathbf{V}_o}{\mathbf{V}_i} = \frac{R \angle 0^\circ}{R - j X_C} = \frac{R \angle 0^\circ}{\sqrt{R^2 + X_C^2} \angle -\tan^{-1}(X_C/R)}$$

and

$$\frac{\mathbf{V}_o}{\mathbf{V}_i} = \frac{R}{\sqrt{R^2 + X_C^2}} \angle \tan^{-1}(X_C/R)$$

The magnitude of the ratio $\mathbf{V}_o/\mathbf{V}_i$ is therefore determined by

$$A_v = \frac{V_o}{V_i} = \frac{R}{\sqrt{R^2 + X_C^2}} = \frac{1}{\sqrt{1 + \left(\frac{X_C}{R}\right)^2}} \qquad \textbf{(21.18)}$$

and the phase angle θ by

$$\theta = \tan^{-1}\frac{X_C}{R} \tag{21.19}$$

For the frequency at which $X_C = R$, the magnitude becomes

$$\frac{V_o}{V_i} = \frac{1}{\sqrt{1 + \left(\frac{X_C}{R}\right)^2}} = \frac{1}{\sqrt{1+1}} = \frac{1}{\sqrt{2}} = 0.707$$

as shown in Fig. 21.23.

The frequency at which $X_C = R$ is determined by

$$X_C = \frac{1}{2\pi f_c C} = R$$

and

$$f_c = \frac{1}{2\pi RC} \tag{21.20}$$

For the high-pass *R-C* filter, the application of any frequency greater than f_c results in an output voltage V_o that is at least 70.7% of the magnitude of the input signal. For any frequency below f_c, the output is less than 70.7% of the applied signal.

For the phase angle, high frequencies result in small values of X_C, and the ratio X_C/R approaches zero with $\tan^{-1}(X_C/R)$ approaching 0°, as shown in Fig. 21.24. At low frequencies, the ratio X_C/R becomes quite large, and $\tan^{-1}(X_C/R)$ approaches 90°. For the case $X_C = R$, $\tan^{-1}(X_C/R) = \tan^{-1} 1 = 45°$. Assigning a phase angle of 0° to $\mathbf{V}_i$ such that $\mathbf{V}_i = V_i \angle 0°$, the phase angle associated with $\mathbf{V}_o$ is θ, resulting in $\mathbf{V}_o = V_o \angle \theta$ and revealing that θ is the angle by which $\mathbf{V}_o$ leads $\mathbf{V}_i$. Since the angle θ is the angle by which $\mathbf{V}_o$ leads $\mathbf{V}_i$ throughout the frequency range in Fig. 21.24, the high-pass *R-C* filter is referred to as a *leading network.*

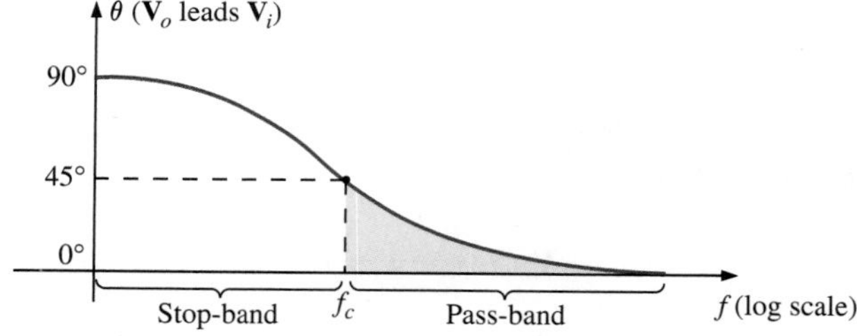

FIG. 21.24
Phase-angle response for the high-pass R-C filter.

In summary, for the high-pass *R-C* filter:

$$f_c = \frac{1}{2\pi RC}$$

For	$f < f_c$,	$V_o < 0.707V_i$
whereas for	$f > f_c$,	$V_o > 0.707V_i$
At f_c,	$\mathbf{V}_o$ leads $\mathbf{V}_i$ by 45°	

The high-pass filter response in Fig. 21.23 can also be obtained using the same elements in Fig. 21.15 but interchanging their positions, as shown in Fig. 21.25.

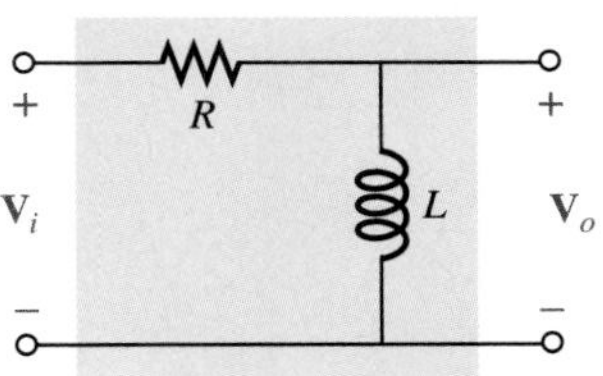

FIG. 21.25
High-pass R-L filter.

EXAMPLE 21.6 Given $R = 20\ \text{k}\Omega$ and $C = 1200\ \text{pF}$:

a. Sketch the normalized plot if the filter is used as both a high-pass and a low-pass filter.
b. Sketch the phase plot for both filters in part (a).
c. Determine the magnitude and phase of $\mathbf{A}_v = \mathbf{V}_o/\mathbf{V}_i$ at $f = \frac{1}{2}f_c$ for the high-pass filter.

Solutions:

a. $$f_c = \frac{1}{2\pi RC} = \frac{1}{(2\pi)(20\ \text{k}\Omega)(1200\ \text{pF})}$$
$$= \mathbf{6631.46\ Hz}$$

The normalized plots appear in Fig. 21.26.

b. The phase plots appear in Fig. 21.27.

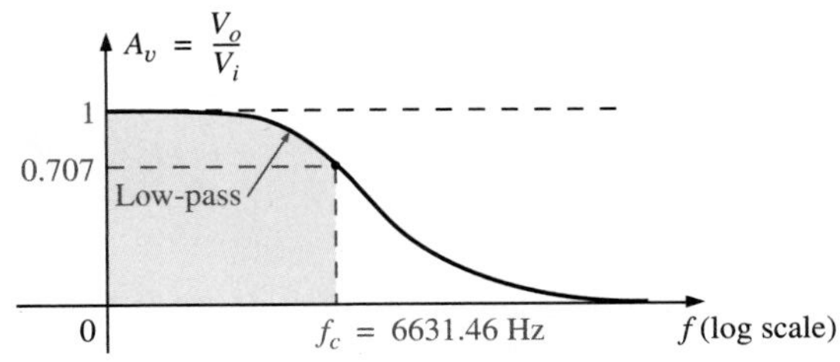

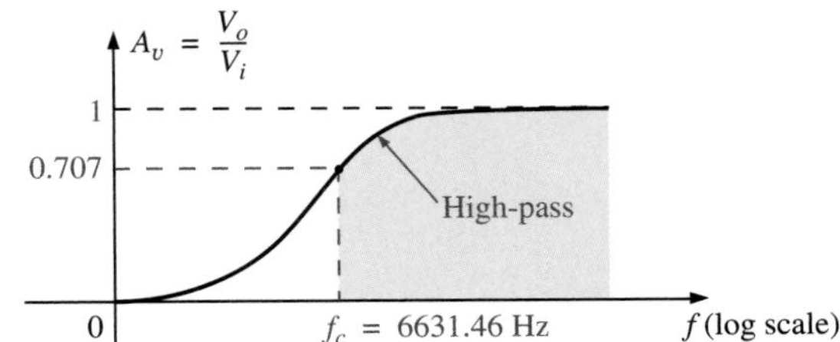

FIG. 21.26
Normalized plots for a low-pass and a high-pass filter using the same elements.

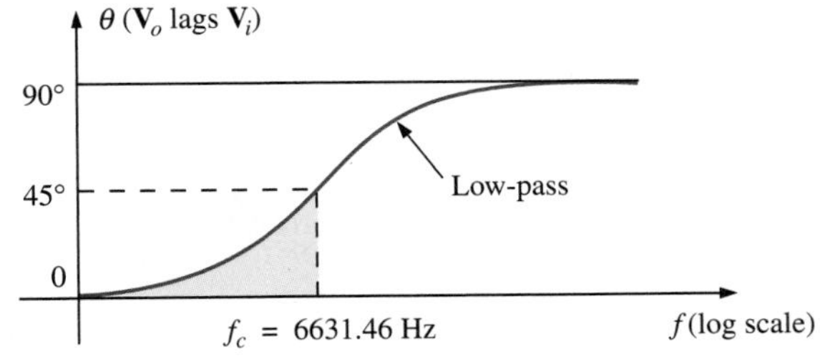

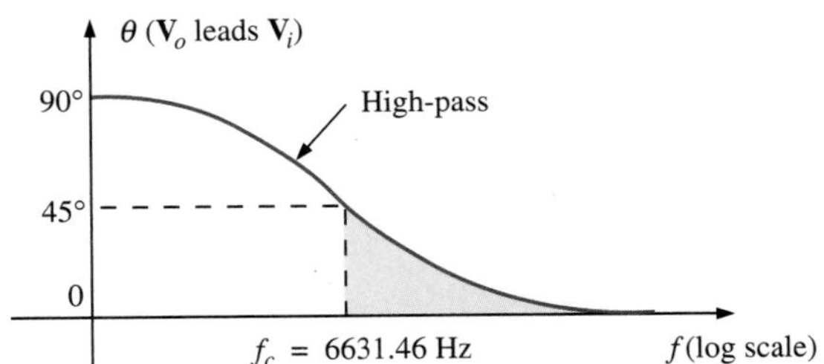

FIG. 21.27
Phase plots for a low-pass and a high-pass filter using the same elements.

c. $$f = \frac{1}{2}f_c = \frac{1}{2}(6631.46\ \text{Hz}) = 3315.73\ \text{Hz}$$

$$X_C = \frac{1}{2\pi fC} = \frac{1}{(2\pi)(3315.73\ \text{Hz})(1200\ \text{pF})}$$
$$\cong 40\ \text{k}\Omega$$

$$A_v = \frac{V_o}{V_i} = \frac{1}{\sqrt{1 + \left(\frac{X_C}{R}\right)^2}} = \frac{1}{\sqrt{1 + \left(\frac{40\ \text{k}\Omega}{20\ \text{k}\Omega}\right)^2}} = \frac{1}{\sqrt{1 + (2)^2}}$$

$$= \frac{1}{\sqrt{5}} = 0.4472$$

$$\theta = \tan^{-1}\frac{X_C}{R} = \tan^{-1}\frac{40\ \text{k}\Omega}{20\ \text{k}\Omega} = \tan^{-1} 2 = 63.43^\circ$$

and
$$\mathbf{A}_v = \frac{\mathbf{V}_o}{\mathbf{V}_i} = \mathbf{0.447 \angle 63.43^\circ}$$

21.7 PASS-BAND FILTERS

A number of methods are used to establish the pass-band characteristic in Fig. 21.7(c). One method uses both a low-pass and a high-pass filter in cascade, as shown in Fig. 21.28.

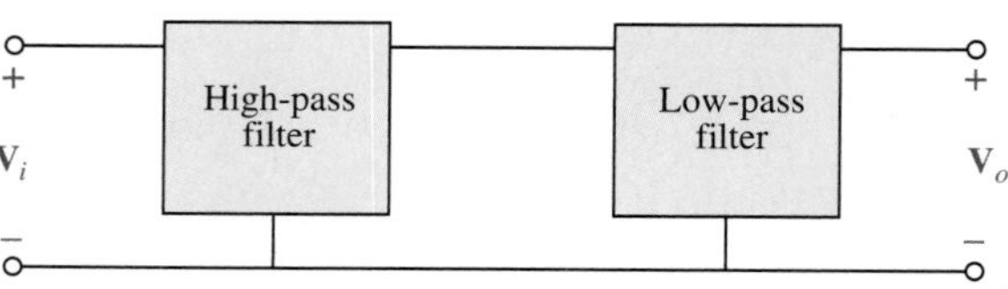

FIG. 21.28
Pass-band filter.

The components are chosen to establish a cutoff frequency for the high-pass filter that is lower than the critical frequency of the low-pass filter, as shown in Fig. 21.29. A frequency f_1 may pass through the low-pass filter but have little effect on V_o due to the reject characteristics of the high-pass filter. A frequency f_2 may pass through the high-pass filter unmolested but be prohibited from reaching the high-pass filter by the low-pass characteristics. A frequency f_o near the center of the pass-band passes through both filters with very little degeneration.

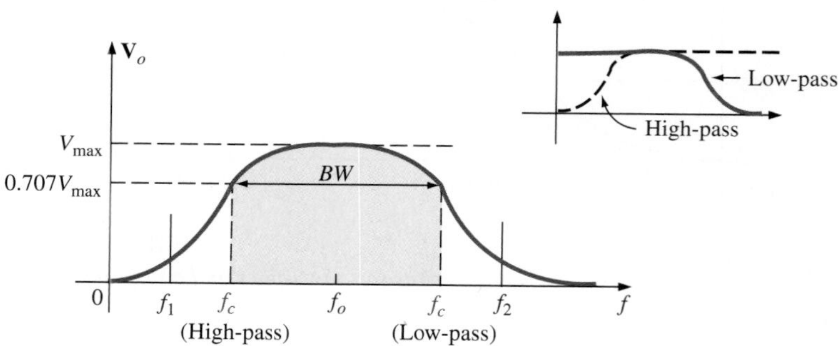

FIG. 21.29
Pass-band characteristics.

The network in Example 21.7 generates the characteristics of Fig. 21.29. However, for a circuit such as the one shown in Fig. 21.30, there is a loading between stages at each frequency that affects the level of $\mathbf{V}_o$. Through proper design, the level of $\mathbf{V}_o$ may be very near the level of $\mathbf{V}_i$

in the pass-band, but it will never equal it exactly. In addition, as the critical frequencies of each filter get closer and closer together to increase the quality factor of the response curve, the peak values within the pass-band continue to drop. For cases where $V_{o_{max}} \neq V_{i_{max}}$ the bandwidth is defined at 0.707 of the resulting $V_{o_{max}}$.

EXAMPLE 21.7 For the pass-band filter in Fig. 21.30:

a. Determine the critical frequencies for the low- and high-pass filters.
b. Using only the critical frequencies, sketch the response characteristics.
c. Determine the actual value of V_o at the high-pass critical frequency calculated in part (a), and compare it to the level that defines the upper frequency for the pass-band.

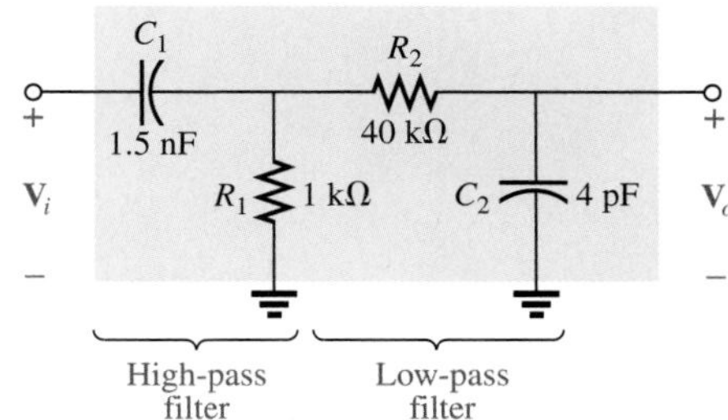

FIG. 21.30
Pass-band filter.

Solutions:

a. High-pass filter:

$$f_c = \frac{1}{2\pi R_1 C_1} = \frac{1}{2\pi(1\text{ k}\Omega)(1.5\text{ nF})} = \mathbf{106.1\text{ kHz}}$$

Low-pass filter:

$$f_c = \frac{1}{2\pi R_2 C_2} = \frac{1}{2\pi(40\text{ k}\Omega)(4\text{ pF})} = \mathbf{994.72\text{ kHz}}$$

b. In the mid-region of the pass-band at about 500 kHz, an analysis of the network reveals that $V_o \cong 0.9V_i$ as shown in Fig. 21.31. The bandwidth is therefore defined at a level of $0.707(0.9V_i) = 0.636V_i$ as also shown in Fig. 21.31.

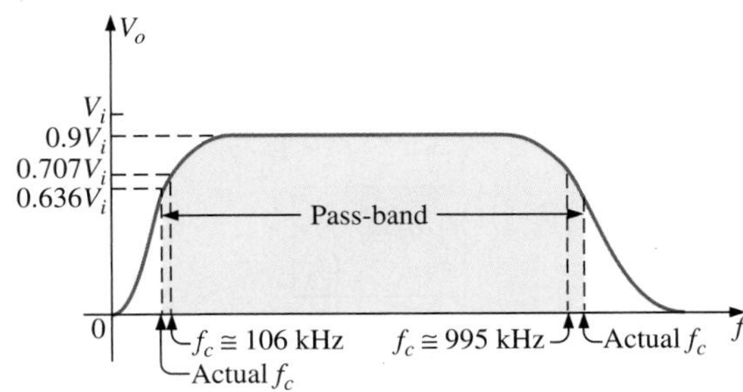

FIG. 21.31
Pass-band characteristics for the filter in Fig. 21.30.

c. At $f = 994.72$ kHz,

$$X_{C_1} = \frac{1}{2\pi f C_1} \cong 107\ \Omega$$

and

$$X_{C_2} = \frac{1}{2\pi f C_2} = R_2 = 40\text{ k}\Omega$$

resulting in the network in Fig. 21.32.

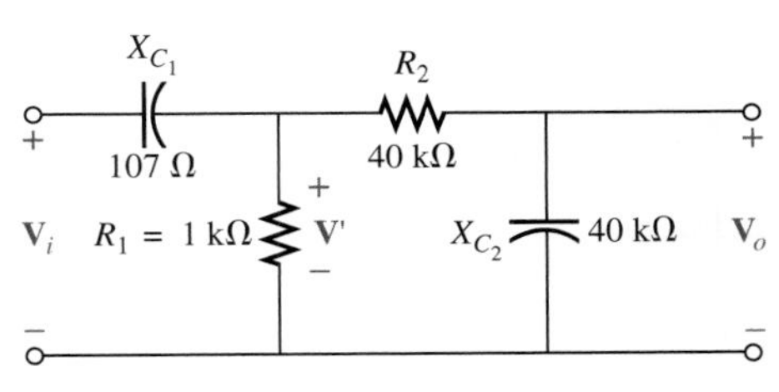

FIG. 21.32
Network of Fig. 21.30 at f = 994.72 kHz.

The parallel combination $R_1 \parallel (R_2 - jX_{C_2})$ is essentially 0.976 kΩ $\angle 0°$ because the $R_2 - X_{C_2}$ combination is so large compared to the parallel resistor R_1.

Then

$$\mathbf{V}' = \frac{0.976\text{ k}\Omega \angle 0°(\mathbf{V}_i)}{0.976\text{ k}\Omega - j0.107\text{ k}\Omega} \cong 0.994\mathbf{V}_i \angle 6.26°$$

with

$$\mathbf{V}_o = \frac{(40\text{ k}\Omega \angle -90°)(0.994\mathbf{V}_i \angle 6.26°)}{40\text{ k}\Omega - j40\text{ k}\Omega}$$

$$\mathbf{V}_o \cong 0.703\mathbf{V}_i \angle -39°$$

so that

$$V_o \cong \mathbf{0.703V}_i \quad \text{at } f = 994.72\text{ kHz}$$

Since the bandwidth is defined at $0.636V_i$ the upper cutoff frequency will be higher than 994.72 kHz as shown in Fig. 21.31.

The pass-band response can also be obtained using the series and parallel resonant circuits discussed in Chapter 20. In each case, however, V_o will not be equal to V_i in the pass-band, but a frequency range in which V_o will be equal to or greater than $0.707V_{max}$ can be defined.

For the series resonant circuit in Fig. 21.33, $X_L = X_C$ at resonance, and

$$V_{o_{max}} = \frac{R}{R + R_l} V_i \qquad f = f_s \tag{21.21}$$

and

$$f_s = \frac{1}{2\pi\sqrt{LC}} \tag{21.22}$$

with

$$Q_s = \frac{X_L}{R + R_l} \tag{21.23}$$

and

$$BW = \frac{f_s}{Q_s} \tag{21.24}$$

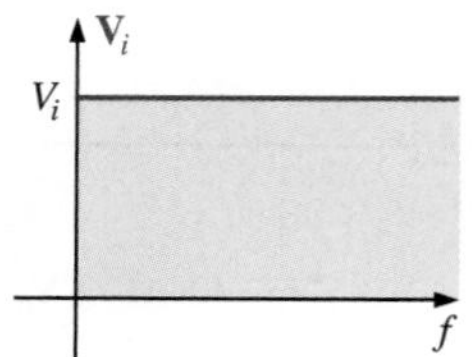

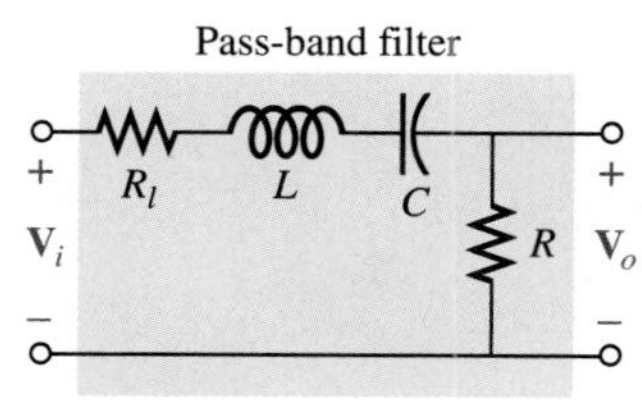

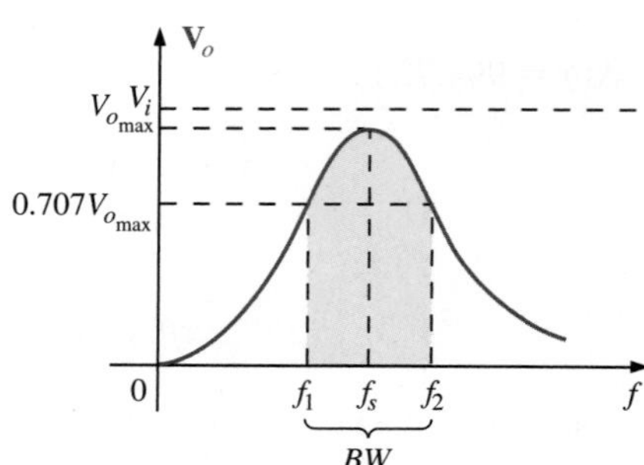

FIG. 21.33
Series resonant pass-band filter.

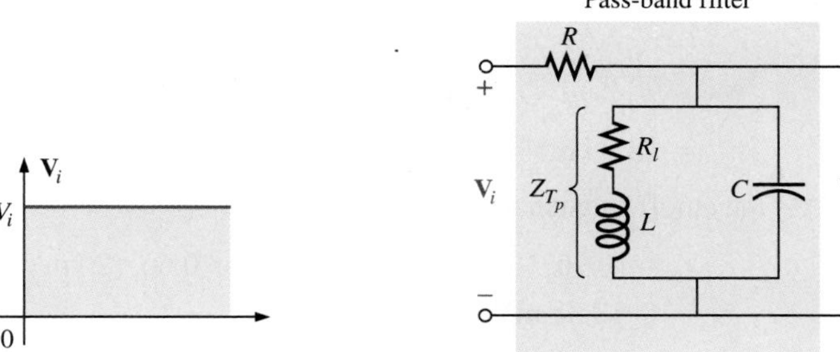

$V_o = V_C$
$V_{o_{max}}$ V_i
$0.707V_{o_{max}}$
0 f_1 f_p f_2 f
BW

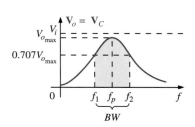

FIG. 21.34
Parallel resonant pass-band filter.

For the parallel resonant circuit in Fig. 21.34, Z_{T_p} is a maximum value at resonance, and

$$V_{o_{max}} = \frac{Z_{T_p} V_i}{Z_{T_p} + R} \Bigg|_{f = f_p} \qquad \textbf{(21.25)}$$

with

$$Z_{T_p} = Q_l^2 R_l \quad Q_l \geq 10 \qquad \textbf{(21.26)}$$

and

$$f_p = \frac{1}{2\pi\sqrt{LC}} \quad Q_l \geq 10 \qquad \textbf{(21.27)}$$

For the parallel resonant circuit

$$Q_p = \frac{X_L}{R_l} \qquad \textbf{(21.28)}$$

and

$$BW = \frac{f_p}{Q_p} \qquad \textbf{(21.29)}$$

As a first approximation that is acceptable for most practical applications, it can be assumed that the resonant frequency bisects the bandwidth.

EXAMPLE 21.8

a. Determine the frequency response for the voltage V_o for the series circuit in Fig. 21.35.
b. Plot the normalized response $A_v = V_o/V_i$.
c. Plot a normalized response defined by $A'_v = A_v/A_{v_{max}}$.

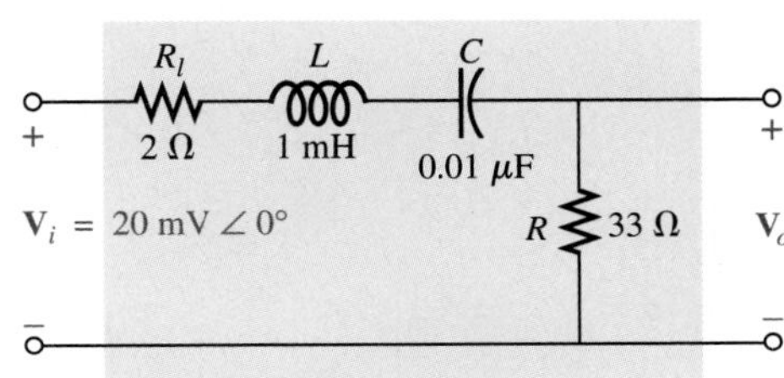

FIG. 21.35
Series resonant pass-band filter for Example 21.8.

Solutions:

a. $f_s = \dfrac{1}{2\pi\sqrt{LC}} = \dfrac{1}{2\pi\sqrt{(1\text{ mH})(0.01\ \mu\text{F})}} = \mathbf{50{,}329.21\ Hz}$

$$Q_s = \frac{X_L}{R + R_l} = \frac{2\pi(50{,}329.21\text{ Hz})(1\text{ mH})}{33\ \Omega + 2\ \Omega} = \mathbf{9.04}$$

$$BW = \frac{f_s}{Q_s} = \frac{50{,}329.21\text{ Hz}}{9.04} = \mathbf{5.57\ kHz}$$

At resonance:

$$V_{o_{max}} = \frac{RV_i}{R + R_l} = \frac{33\ \Omega(V_i)}{33\ \Omega + 2\ \Omega} = 0.943V_i = 0.943(20\text{ mV})$$
$$= \mathbf{18.86\ mV}$$

At the cutoff frequencies:

$$V_o = (0.707)(0.943V_i) = 0.667V_i = 0.667(20\text{ mV})$$
$$= \mathbf{13.34\ mV}$$

Note Fig. 21.36.

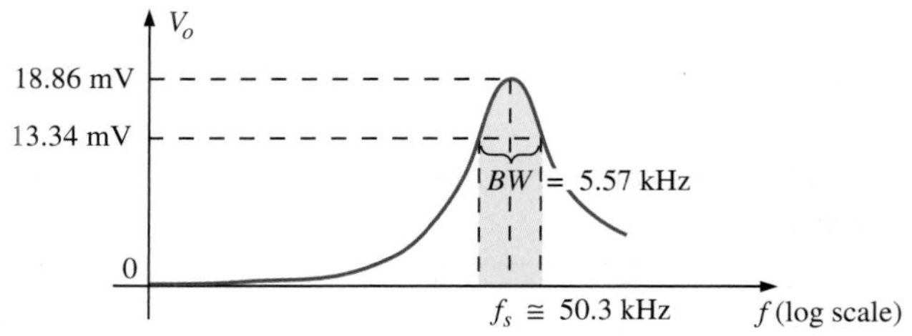

FIG. 21.36
Pass-band response for the network.

b. Dividing all levels in Fig. 21.36 by $V_i = 20$ mV results in the normalized plot in Fig. 21.37(a).
c. Dividing all levels in Fig. 21.37(a) by $A_{v_{max}} = 0.943$ results in the normalized plot in Fig. 21.37(b).

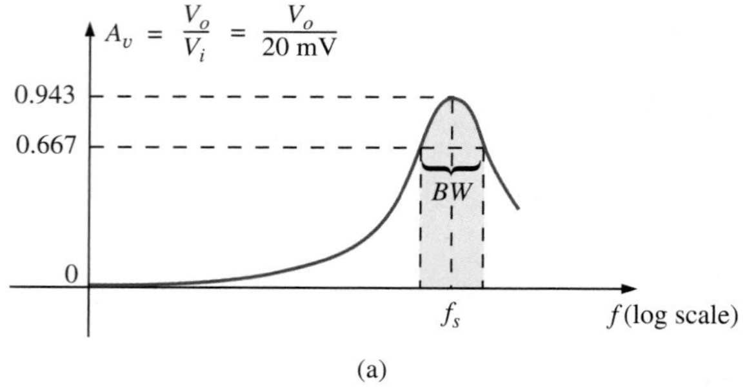

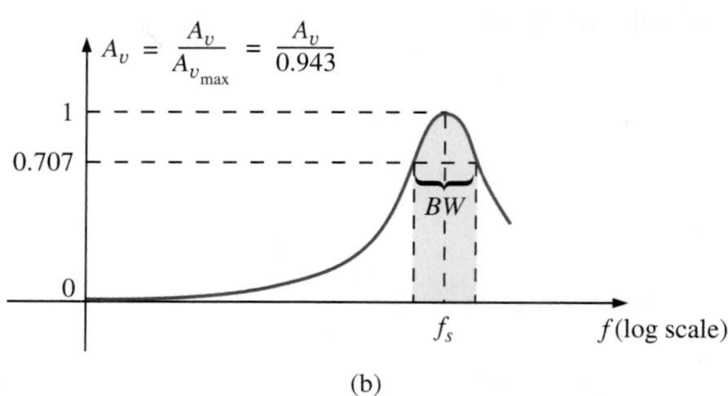

FIG. 21.37
Normalized plots for the pass-band filter in Fig. 21.35.

21.8 STOP-BAND FILTERS

Stop-band filters can also be constructed using a low-pass and a high-pass filter. However, rather than the cascaded configuration used for the pass-band filter, a parallel arrangement is required, as shown in Fig. 21.38. A low-frequency f_1 can pass through the low-pass filter, and a higher-frequency f_2 can use the parallel path, as shown in Figs. 21.38 and 21.39. However, a frequency such as f_o in the reject-band is higher than the low-pass critical frequency and lower than the high-pass critical frequency, and is therefore prevented from contributing to the levels of V_o above $0.707V_{max}$.

The characteristics of a stop-band filter are the inverse of the pattern obtained for the pass-band filters; therefore, at any frequency, the sum of

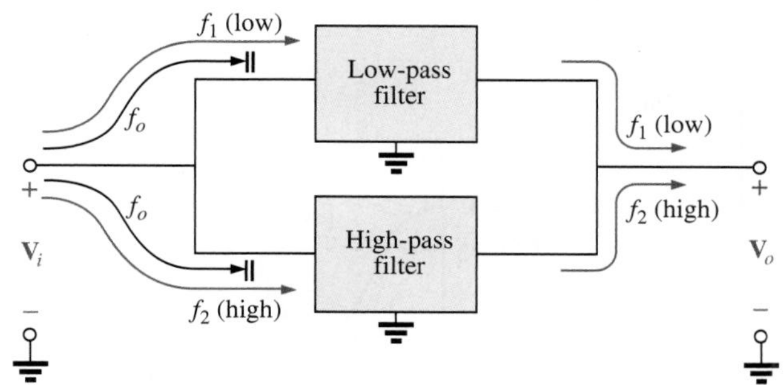

FIG. 21.38
Stop-band filter.

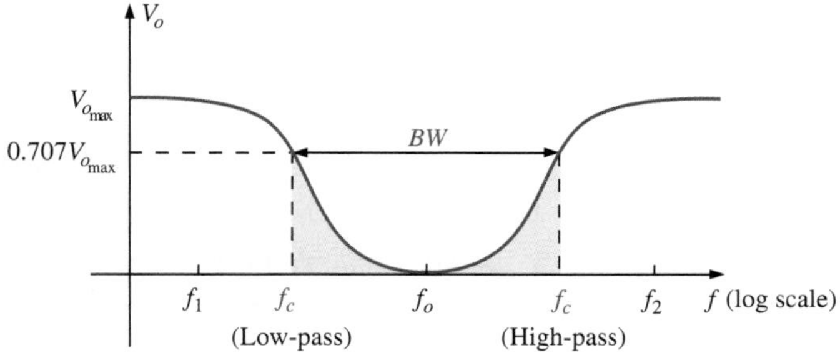

FIG. 21.39
Stop-band characteristics.

the magnitudes of the two waveforms to the right of the equals sign in Fig. 21.40 equals the applied voltage V_i.

For the pass-band filters in Figs. 21.33 and 21.34, therefore, if we take the output off the other series elements as shown in Figs. 21.41 and 21.42, a stop-band characteristic is obtained, as required by Kirchhoff's voltage law.

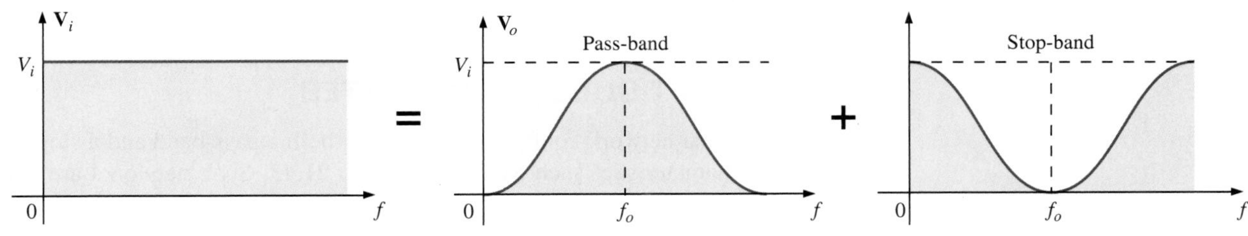

FIG. 21.40
Demonstrating how an applied signal of fixed magnitude can be broken down into a pass-band and stop-band response curve.

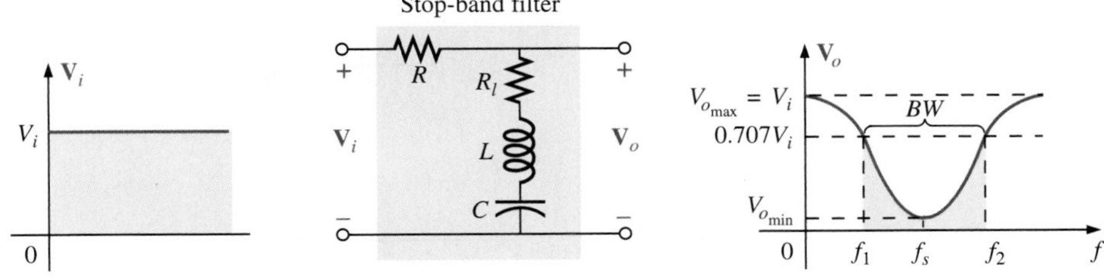

FIG. 21.41
Stop-band filter using a series resonant circuit.

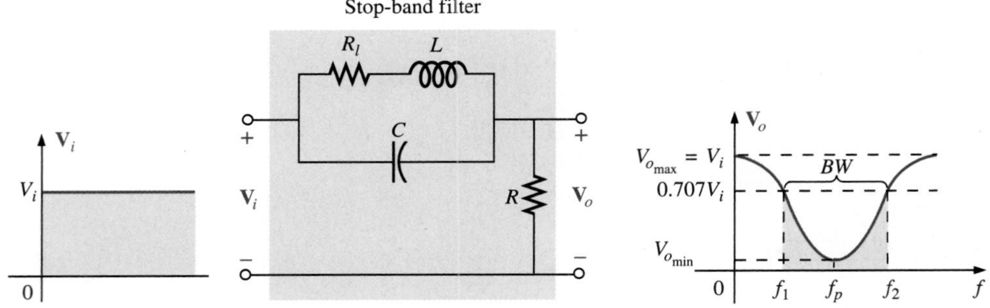

FIG. 21.42
Stop-band filter using a parallel resonant network.

For the series resonant circuit in Fig. 21.41, Eqs. (21.22) through (21.24) still apply, but now, at resonance,

$$V_{o_{min}} = \frac{R_l V_i}{R_l + R} \tag{21.30}$$

For the parallel resonant circuit in Fig. 21.42, Eqs. (21.26) through (21.29) are still applicable, but now, at resonance,

$$V_{o_{min}} = \frac{RV_i}{R + Z_{T_p}} \tag{21.31}$$

The maximum value of V_o for the series resonant circuit is V_i at the low end due to the open-circuit equivalent for the capacitor and V_i at the high end due to the high impedance of the inductive element.

For the parallel resonant circuit, at $f = 0$ Hz, the inductor can be replaced by a short-circuit equivalent, and the capacitor can be replaced by its open circuit and $V_o = RV_i/(R + R_l)$. At the high-frequency end, the capacitor approaches a short-circuit equivalent, and V_o increases toward V_i.

21.9 DOUBLE-TUNED FILTER

Some network configurations display both a pass-band and a stop-band characteristic, such as shown in Fig. 21.43. Such networks are called

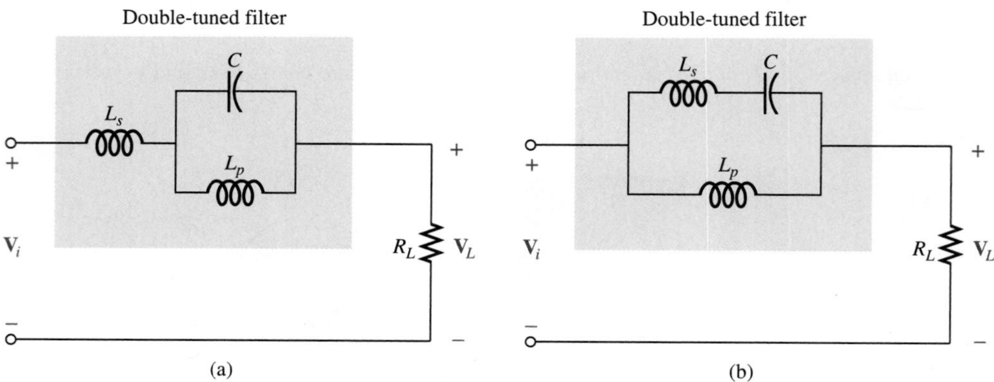

FIG. 21.43
Double-tuned networks.

double-tuned filters. For the network in Fig. 21.43(a), the parallel resonant circuit establishes a stop-band for the range of frequencies not permitted to establish a significant V_L. The greater part of the applied voltage appears across the parallel resonant circuit for this frequency range due to its very high impedance compared with R_L. For the pass-band, the parallel resonant circuit is designed to be capacitive (inductive if L_s is replaced by C_s). The inductance L_s is chosen to cancel the effects of the resulting net capacitive reactance at the resonant pass-band frequency of the tank circuit, thereby acting as a series resonant circuit. The applied voltage then appears across R_L at this frequency.

For the network in Fig. 21.43(b), the series resonant circuit still determines the pass-band, acting as a very low impedance across the parallel inductor at resonance. At the desired stop-band resonant frequency, the series resonant circuit is capacitive. The inductance L_p is chosen to establish parallel resonance at the resonant stop-band frequency. The high impedance of the parallel resonant circuit results in a very low load voltage V_L.

For rejected frequencies below the pass-band, the networks should appear as shown in Fig. 21.43. For the reverse situation, L_s in Fig. 21.43(a) and L_p in Fig. 21.43(b) are replaced by capacitors.

EXAMPLE 21.9 For the network in Fig. 21.43(b), determine L_s and L_p for a capacitance C of 500 pF if a frequency of 200 kHz is to be rejected and a frequency of 600 kHz accepted.

Solution: For series resonance, we have

$$f_s = \frac{1}{2\pi\sqrt{LC}}$$

and $$L_s = \frac{1}{4\pi^2 f_s^2 C} = \frac{1}{4\pi^2(600\text{ kHz})^2(500\text{ pF})} = \mathbf{140.7\ \mu H}$$

At 200 kHz,

$$X_{L_s} = \omega L = 2\pi f_s L_s = (2\pi)(200\text{ kHz})(140.7\ \mu\text{H}) = 176.8\ \Omega$$

and $$X_C = \frac{1}{\omega C} = \frac{1}{(2\pi)(200\text{ kHz})(500\text{ pF})} = 1591.5\ \Omega$$

For the series elements,

$$j\,(X_{L_s} - X_C) = j\,(176.8\ \Omega - 1591.5\ \Omega) = -j\,1414.7\ \Omega = -j\,X'_C$$

At parallel resonance ($Q_l \geq 10$ assumed),

$$X_{L_p} = X'_C$$

and $$L_p = \frac{X_{L_p}}{\omega} = \frac{1414.7\ \Omega}{(2\pi)(200\text{ kHz})} = \mathbf{1.13\ mH}$$

The frequency response for the preceding network appears as one of the examples of PSpice in the last section of the chapter.

21.10 BODE PLOTS

There is a technique for sketching the frequency response of such factors as filters, amplifiers, and systems on a decibel scale that can save a great deal of time and effort and provide an excellent way to compare decibel levels at different frequencies.

FIG. 21.44
Hendrik Wade Bode.
Courtesy of AT&T Archives.

American (Madison, WI; Summit, NJ; Cambridge, MA)
(1905–81)
V.P. at Bell Laboratories
Professor of Systems Engineering, Harvard University

In his early years at Bell Laboratories, Hendrik Bode was involved with *electric filter* and *equalizer design.* He then transferred to the Mathematics Research Group, where he specialized in research pertaining to electrical networks theory and its application to long-distance communication facilities. In 1948 he was awarded the Presidential Certificate of Merit for his work in electronic fire control devices. In addition to the publication of the book *Network Analysis and Feedback Amplifier Design* in 1945, which is considered a classic in its field, he has been granted 25 patents in electrical engineering and systems design. Upon retirement, Bode was elected Gordon McKay Professor of Systems Engineering at Harvard University. He was a fellow of the IEEE and American Academy of Arts and Sciences.

The curves obtained for the magnitude and/or phase angle versus frequency are called Bode plots (Fig. 21.44). Through the use of straight-line segments called idealized Bode plots, the frequency response of a system can be found efficiently and accurately.

To ensure that the derivation of the method is correctly and clearly understood, the first network to be analyzed is examined in some detail. The second network is treated in a shorthand manner, and finally a method for quickly determining the response is introduced.

High-Pass *R-C* Filter

Let us start by reexamining the high-pass filter in Fig. 21.45. The high-pass filter was chosen as our starting point because the frequencies of primary interest are at the low end of the frequency spectrum.

The voltage gain of the system is given by

$$\mathbf{A}_v = \frac{\mathbf{V}_o}{\mathbf{V}_i} = \frac{R}{R - jX_C} = \frac{1}{1 - j\dfrac{X_C}{R}} = \frac{1}{1 - j\dfrac{1}{2\pi fCR}}$$

$$= \frac{1}{1 - j\left(\dfrac{1}{2\pi RC}\right)\dfrac{1}{f}}$$

If we substitute

$$f_c = \frac{1}{2\pi RC} \tag{21.32}$$

which we recognize as the cutoff frequency of earlier sections, we obtain

$$\mathbf{A}_v = \frac{1}{1 - j\,(f_c/f)} \tag{21.33}$$

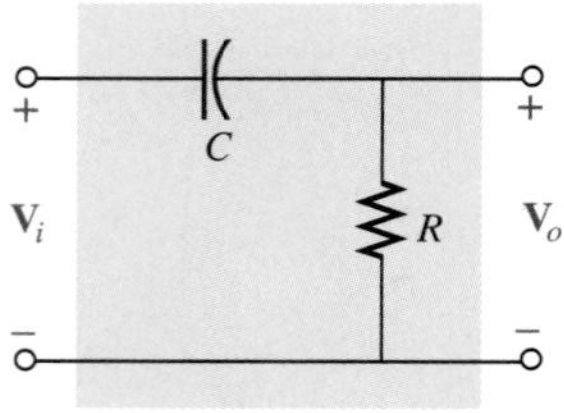

FIG. 21.45
High-pass filter.

We will find in the analysis to follow that the ability to reformat the gain to one having the general characteristics of Eq. (21.33) is critical to the application of the Bode technique. Different configurations result in variations of the format of Eq. (21.33), but the desired similarities become obvious as we progress through the material.

In magnitude and phase form:

$$\mathbf{A}_v = \frac{\mathbf{V}_o}{\mathbf{V}_i} = A_v \angle\theta = \frac{1}{\sqrt{1 + (f_c/f)^2}} \angle \tan^{-1}(f_c/f) \tag{21.34}$$

providing an equation for the magnitude and phase of the high-pass filter in terms of the frequency levels.

Using Eq. (21.12),

$$A_{v_{\text{dB}}} = 20 \log_{10} A_v$$

and, substituting the magnitude component of Eq. (21.34),

$$Av_{\text{dB}} = 20 \log_{10} \frac{1}{\sqrt{1 + (f_c/f)^2}} = \underbrace{20 \log_{10} 1}_{0} - 20 \log_{10} \sqrt{1 + (f_c/f)^2}$$

and $$\mathbf{A}_{v_{\text{dB}}} = -20 \log_{10} \sqrt{1 + \left(\frac{f_c}{f}\right)^2}$$

Recognizing that $\log_{10}\sqrt{x} = \log_{10} x^{1/2} = \frac{1}{2}\log_{10} x$, we have

$$A_{v_{\text{dB}}} = -\frac{1}{2}(20)\log_{10}\left[1 + \left(\frac{f_c}{f}\right)^2\right]$$

$$= -10\log_{10}\left[1 + \left(\frac{f_c}{f}\right)^2\right]$$

For frequencies where $f << f_c$ or $(f_c/f)^2 >> 1$,

$$1 + \left(\frac{f_c}{f}\right)^2 \cong \left(\frac{f_c}{f}\right)^2$$

and $$A_{v_{\text{dB}}} = -10\log_{10}\left(\frac{f_c}{f}\right)^2$$

but $$\log_{10} x^2 = 2\log_{10} x$$

resulting in $$A_{v_{\text{dB}}} = -20\log_{10}\frac{f_c}{f}$$

However, logarithms are such that

$$-\log_{10} b = +\log_{10}\frac{1}{b}$$

and substituting $b = f_c/f$, we have

$$\boxed{A_{v_{\text{dB}}} = +20\log_{10}\frac{f}{f_c}} \quad f << f_c \qquad \textbf{(21.35)}$$

First note the similarities between Eq. (21.35) and the basic equation for gain in decibels: $G_{\text{dB}} = 20\log_{10} V_o/V_i$. The comments regarding changes in decibel levels due to changes in V_o/V_i can therefore be applied here also, except now a change in frequency by a 2 : 1 ratio results in a 6 dB change in gain. A change in frequency by a 10 : 1 ratio results in a 20 dB change in gain.

Two frequencies separated by a 2 : 1 ratio are said to be an octave apart. For Bode plots, a change in frequency by one octave will result in a 6 dB change in gain.

Two frequencies separated by a 10 : 1 ratio are said to be a decade apart. For Bode plots, a change in frequency by one decade will result in a 20 dB change in gain.

One may wonder about all the mathematical development to obtain an equation that initially appears confusing and of limited value. As specified, Eq. (21.35) is accurate only for frequency levels much less than f_c.

First, realize that the mathematical development of Eq. (21.35) does not have to be repeated for each configuration encountered. Second, the equation itself is seldom applied but simply used to define a straight line on a log plot that permits a sketch of the frequency response of a system with a minimum of effort and a high degree of accuracy.

To plot Eq. (21.35), consider the following levels of increasing frequency:

For $f = f_c/10$, $f/f_c = 0.1$ and $+20\log_{10} 0.1 = -20$ dB
For $f = f_c/4$, $f/f_c = 0.25$ and $+20\log_{10} 0.25 = -12$ dB
For $f = f_c/2$, $f/f_c = 0.51$ and $+20\log_{10} 0.5 = -6$ dB
For $f = f_c$, $f/f_c = 1$ and $+20\log_{10} 1 = 0$ dB

Note from the above equations that as the frequency of interest approaches f_c, the dB gain becomes less negative and approaches the final normalized value of 0 dB. The positive sign in front of Eq. (21.35) can therefore be interpreted as an indication that the dB gain will have a positive slope with an increase in frequency. A plot of these points on a log scale results in the straight-line segment in Fig. 21.46 to the left of f_c.

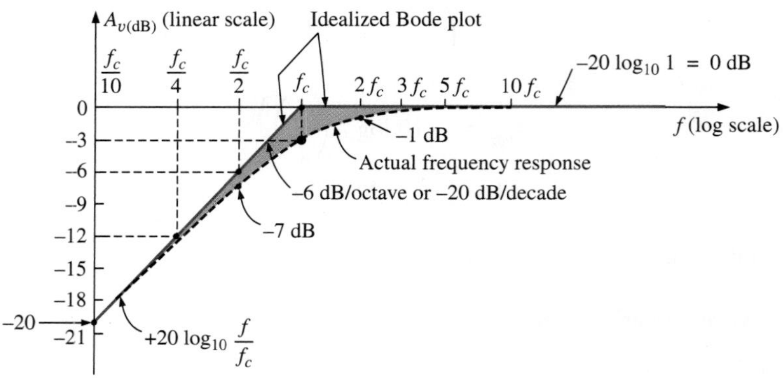

FIG. 21.46
Idealized Bode plot for the low-frequency region.

For the future, note that the resulting plot is a straight line intersecting the 0 dB line at f_c. It increases to the right at a rate of +6 dB per octave or +20 dB per decade. In other words, once f_c is determined, find $f_c/2$, and a plot point exists at −6 dB (or find $f_c/10$, and a plot point exists at −20 dB).

Bode plots are straight-line segments because the dB change per decade or octave is constant.

The actual response approaches an asymptote (straight-line segment) defined by $A_{v_{\text{dB}}} = 0$ dB since at high frequencies

$$f >> f_c \quad \text{and} \quad f_c/f \cong 0$$

with
$$A_{v_{\text{dB}}} = 20 \log_{10} \frac{1}{\sqrt{1 + (f_c/f)^2}} = 20 \log_{10} \frac{1}{\sqrt{1 + 0}}$$
$$= 20 \log_{10} 1 = 0 \text{ dB}$$

The two asymptotes defined above intersect at f_c, as shown in Fig. 21.46, forming an envelope for the actual frequency response.

At $f = f_c$, the cutoff frequency,

$$A_{v_{\text{dB}}} = 20 \log_{10} \frac{1}{\sqrt{1 + (f_c/f)^2}} = 20 \log_{10} \frac{1}{\sqrt{1 + 1}} = 20 \log_{10} \frac{1}{\sqrt{2}}$$
$$= \mathbf{-3\ dB}$$

At $f = 2f_c$,

$$A_{v_{\text{dB}}} = -20 \log_{10} \sqrt{1 + \left(\frac{f_c}{2f_c}\right)^2} = -20 \log_{10} \sqrt{1 + \left(\frac{1}{2}\right)^2}$$
$$= -20 \log_{10} \sqrt{1.25} = \mathbf{-1\ dB}$$

as shown in Fig. 21.46.

At $f = f_c/2$,

$$A_{v_{dB}} = -20 \log_{10} \sqrt{1 + \left(\frac{f_c}{f_c/2}\right)^2} = -20 \log_{10} \sqrt{1 + (2)^2}$$
$$= -20 \log_{10} \sqrt{5}$$
$$= \mathbf{-7\ dB}$$

separating the idealized Bode plot from the actual response by 7 dB − 6 dB = 1 dB, as shown in Fig. 21.46.

Reviewing the above,

at $f = f_c$, the actual response curve is 3 dB down from the idealized Bode plot, whereas at $f = 2f_c$ and $f_c/2$, the actual response curve is 1 dB down from the asymptotic response.

The phase response can also be sketched using straight-line asymptotes by considering a few critical points in the frequency spectrum.

Eq. (21.34) specifies the phase response (the angle by which $\mathbf{V}_o$ leads $\mathbf{V}_i$) by

$$\theta = \tan^{-1}\frac{f_c}{f} \qquad \textbf{(21.36)}$$

For frequencies well below f_c ($f << f_c$), $\theta = \tan^{-1}(f_c/f)$ approaches 90° and for frequencies well above f_c ($f >> f_c$), $\theta = \tan^{-1}(f_c/f)$ will approach 0°, as discovered in earlier sections of the chapter. At $f = f_c$, $\theta = \tan^{-1}(f_c/f) = \tan^{-1} 1 = 45°$.

Defining $f << f_c$ for $f = f_c/10$ (and less) and $f >> f_c$ for $f = 10f_c$ (and more), we can define

an asymptote at $\theta = 90°$ for $f << f_c/10$, an asymptote at $\theta = 0°$ for $f >> 10f_c$, and an asymptote from $f_c/10$ to $10f_c$ that passes through $\theta = 45°$ at $f = f_c$.

The asymptotes defined above all appear in Fig. 21.47. Again, the Bode plot for Eq. (21.36) is a straight line because the change in phase angle will be 45° for every tenfold change in frequency.

Substituting $f = f_c/10$ into Eq. (21.36),

$$\theta = \tan^{-1}\left(\frac{f_c}{f_c/10}\right) = \tan^{-1} 10 = \mathbf{84.29°}$$

for a difference of $90° - 84.29° \cong 5.7°$ from the idealized response.

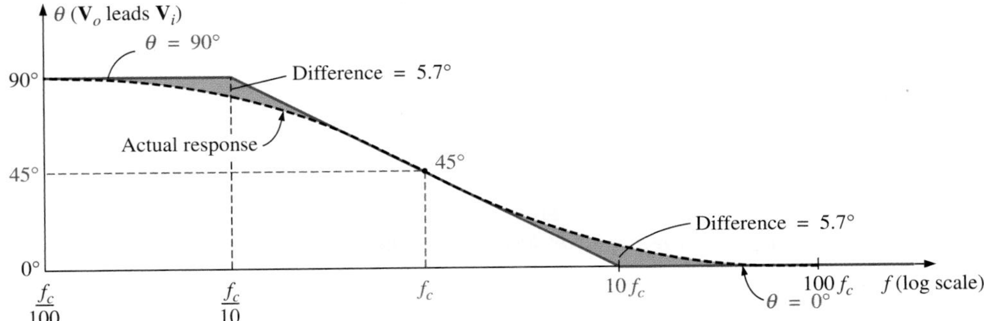

FIG. 21.47
Phase response for a high-pass R-C filter.

Substituting $f = 10f_c$,

$$\theta = \tan^{-1}\left(\frac{f_c}{10f_c}\right) = \tan^{-1}\frac{1}{10} \cong \mathbf{5.7^\circ}$$

In summary, therefore,

at $f = f_c$, $\theta = 45^\circ$, whereas at $f = f_c/10$ and $10f_c$, the difference between the actual phase response and the asymptotic plot is 5.7°.

EXAMPLE 21.10

a. Sketch $A_{v_{dB}}$ versus frequency for the high-pass *R-C* filter in Fig. 21.48.
b. Determine the decibel level at $f = 1$ kHz.
c. Sketch the phase response versus frequency on a log scale.

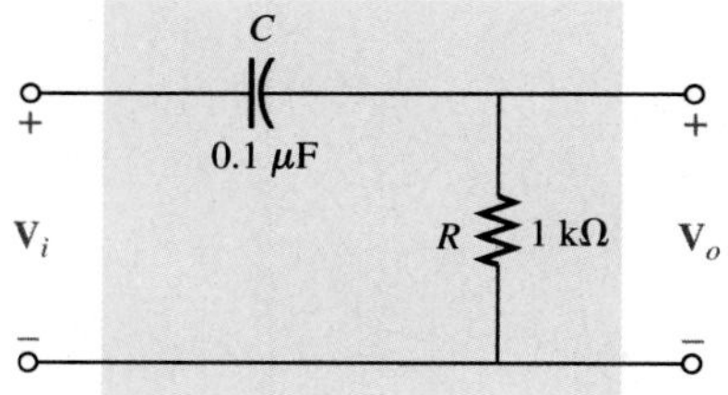

FIG. 21.48
Example 21.10.

Solutions:

a. $$f_c = \frac{1}{2\pi RC} = \frac{1}{(2\pi)(1\text{ k}\Omega)(0.1\ \mu\text{F})} = 1591.55\text{ Hz}$$

The frequency f_c is identified on the log scale as shown in Fig. 21.49. A straight line is then drawn from f_c with a slope that will intersect -20 dB at $f_c/10 = 159.15$ Hz or -6 dB at $f_c/2 = 795.77$ Hz. A second asymptote is drawn from f_c to higher frequencies at 0 dB. The actual response curve can then be drawn through the -3 dB level at f_c approaching the two asymptotes of Fig. 21.49. Note the 1 dB difference between the actual response and the idealized Bode plot at $f = 2f_c$ and $0.5f_c$.

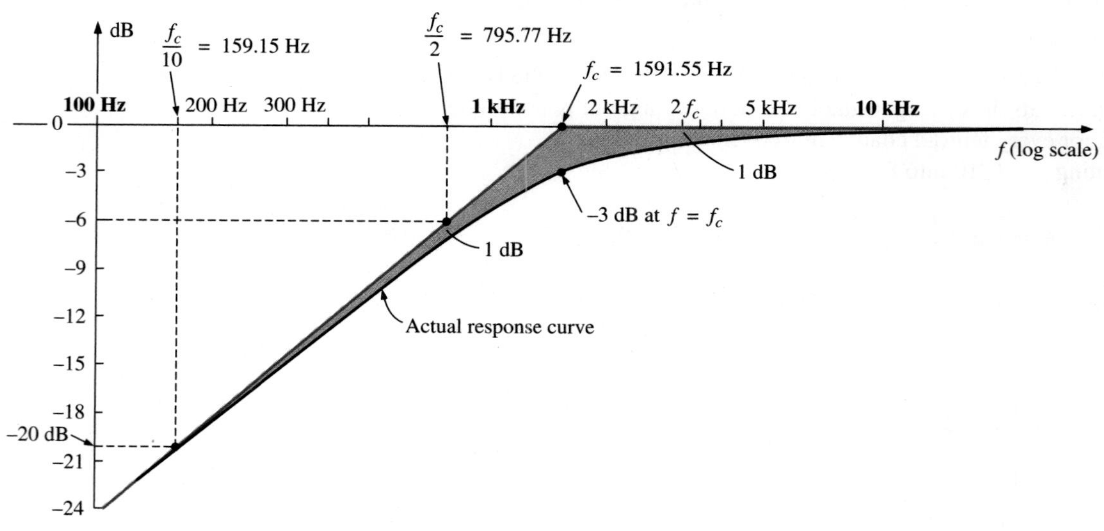

FIG. 21.49
Frequency response for the high-pass filter in Fig. 21.48.

Note that in the solution to part (a), there is no need to use Eq. (21.35) or to perform any extensive mathematical manipulations.

b. Eq. (21.33):

$$|A_{v\text{dB}}| = 20\log_{10}\frac{1}{\sqrt{1+\left(\frac{f_c}{f}\right)^2}} = 20\log_{10}\frac{1}{\sqrt{1+\left(\frac{1591.55\text{ Hz}}{1000}\right)^2}}$$

$$= 20\log_{10}\frac{1}{\sqrt{1+(1.592)^2}} = 20\log_{10}0.5318 = \mathbf{-5.49\ dB}$$

as verified by Fig. 21.49.

c. See Fig. 21.50. Note that $\theta = 45°$ at $f = f_c = 1591.55$ Hz, and the difference between the straight-line segment and the actual response is 5.7° at $f = f_c/10 = 159.2$ Hz and $f = 10f_c = 15{,}923.6$ Hz.

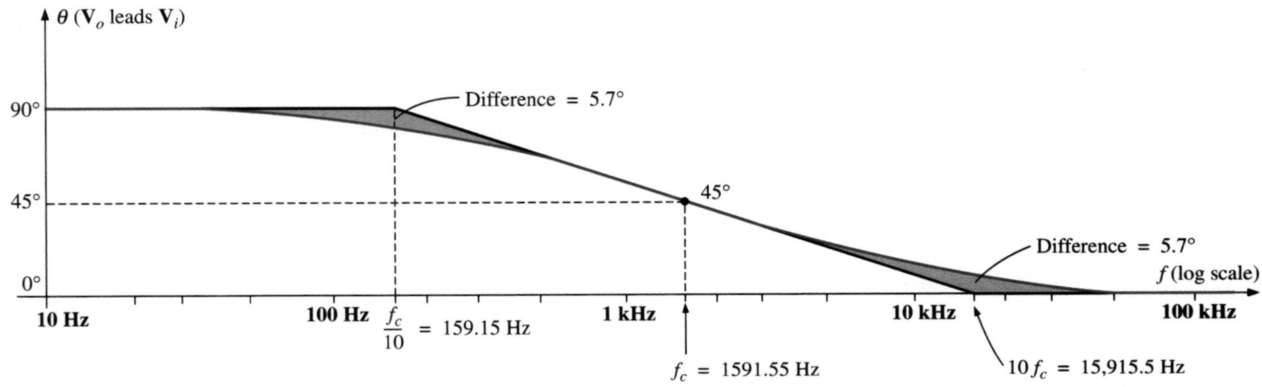

FIG. 21.50
Phase plot for the high-pass R-C filter.

Low-Pass *R-C* Filter

For the low-pass filter in Fig. 21.51,

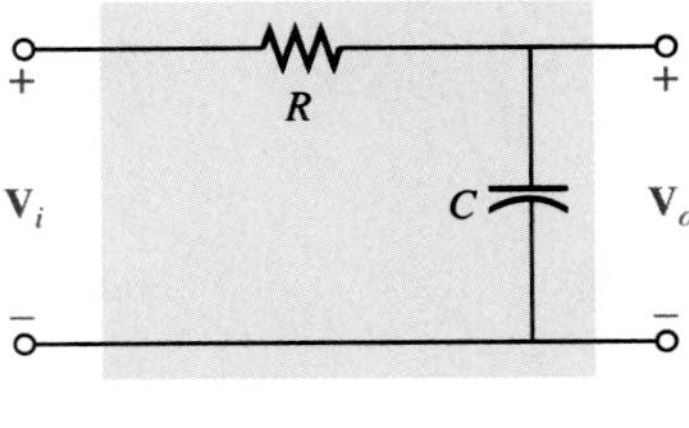

FIG. 21.51
Low-pass filter.

$$\mathbf{A}_v = \frac{\mathbf{V}_o}{\mathbf{V}_i} = \frac{-jX_C}{R - jX_C} = \frac{1}{\frac{R}{-jX_C}+1}$$

$$= \frac{1}{1 + j\frac{R}{X_C}} = \frac{1}{1 + j\frac{R}{\frac{1}{2\pi fC}}} = \frac{1}{1 + j\frac{f}{\frac{1}{2\pi RC}}}$$

and

$$\boxed{\mathbf{A}_v = \frac{1}{1 + j(f/f_c)}} \qquad \textbf{(21.37)}$$

with

$$\boxed{f_c = \frac{1}{2\pi RC}} \qquad \textbf{(21.38)}$$

as defined earlier.

Note that now the sign of the imaginary component in the denominator is positive and f_c appears in the denominator of the frequency ratio rather than in the numerator, as in the case of f_c for the high-pass filter.

In terms of magnitude and phase,

$$\mathbf{A}_v = \frac{\mathbf{V}_o}{\mathbf{V}_i} = \mathbf{A}_v \angle\theta = \frac{1}{\sqrt{1 + (f/f_c)^2}} \angle -\tan^{-1}(f/f_c) \tag{21.39}$$

An analysis similar to that performed for the high-pass filter results in

$$A_{v_{\text{dB}}} = -20 \log_{10} \frac{f}{f_c} \quad f >> f_c \tag{21.40}$$

Note in particular that the equation is exact only for frequencies much greater than f_c, but a plot of Eq. (21.40) does provide an asymptote that performs the same function as the asymptote derived for the high-pass filter. In addition, note that it is exactly the same as Eq. (21.35), except for the minus sign, which suggests that the resulting Bode plot will have a negative slope [recall the positive slope for Eq. (21.35)] for increasing frequencies beyond f_c.

A plot of Eq. (21.40) appears in Fig. 21.52 for $f_c = 1$ kHz. Note the 6 dB drop at $f = 2f_c$ and the 20 dB drop at $f = 10f_c$.

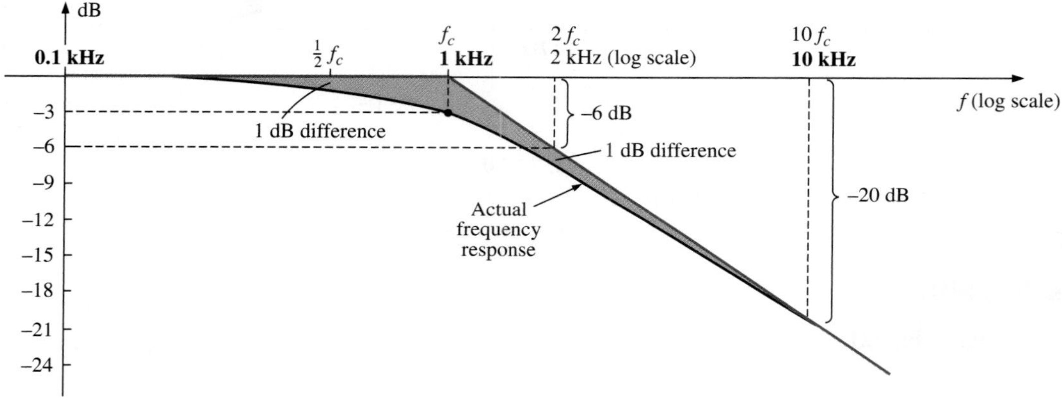

FIG. 21.52
Bode plot for the high-frequency region of a low-pass R-C filter.

At $f >> f_c$, the phase angle $\theta = -\tan^{-1}(f/f_c)$ approaches $-90°$, whereas at $f << f_c$, $\theta = -\tan^{-1}(f/fc)$ approaches 0°. At $f = f_c$, $\theta = -\tan^{-1} 1 = -45°$, establishing the plot in Fig. 21.53. Note again the 45° change in phase angle for each tenfold increase in frequency.

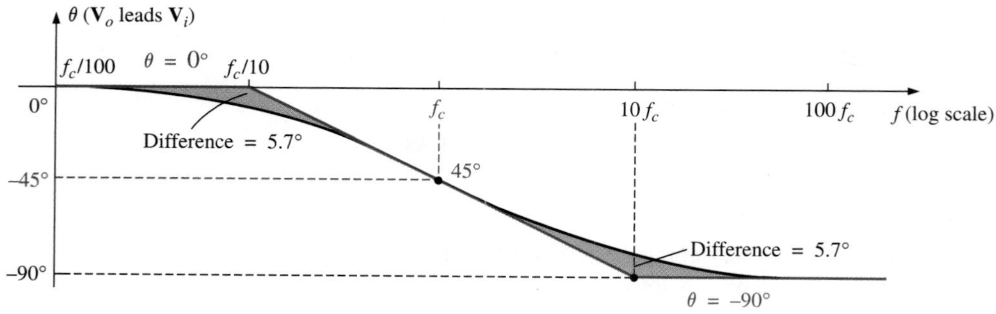

FIG. 21.53
Phase plot for a low-pass R-C filter.

Even though the preceding analysis has been limited solely to the *R-C* combination, the results obtained will have an impact on networks that are a great deal more complicated. One good example is the high- and low-frequency response of a standard transistor configuration. Some capacitive elements in a practical transistor network affect the low-frequency response, and others affect the high-frequency response. In the absence of the capacitive elements, the frequency response of a transistor ideally stays level at the midband value. However, the coupling capacitors at low frequencies and the bypass and parasitic capacitors at high frequencies define a bandwidth for numerous transistor configurations. In the low-frequency region, specific capacitors and resistors form an *R-C* combination that defines a low cutoff frequency. There are then other elements and capacitors forming a second *R-C* combination that define a high cutoff frequency. Once the cutoff frequencies are known, the −3 dB points are set, and the bandwidth of the system can be determined.

21.11 SKETCHING THE BODE RESPONSE

In the previous section, we found that normalized functions of the form appearing in Fig. 21.54 had the Bode envelope and the dB response indicated in the same figure. In this section, we introduce additional functions and their responses that can be used in conjunction with those in Fig. 21.54 to determine the dB response of more sophisticated systems in a systematic, time-saving, and accurate manner.

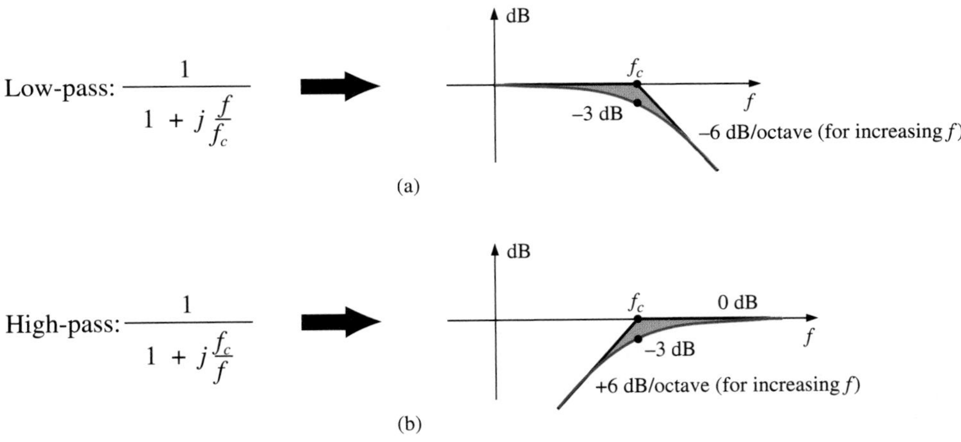

FIG. 21.54
dB response of (a) low-pass filter and (b) high-pass filter.

As an avenue toward introducing an additional function that appears quite frequently, let us examine the high-pass filter in Fig. 21.55 which has a high-frequency output less than the full applied voltage.

Before developing a mathematical expression for $\mathbf{A}_v = \mathbf{V}_o/\mathbf{V}_i$, let us first make a rough sketch of the expected response.

At $f = 0$ Hz, the capacitor assumes its open-circuit equivalence, and $V_o = 0$ V. At very high frequencies, the capacitor can assume its short-circuit equivalence, and

$$V_o = \frac{R_2}{R_1 + R_2} V_i = \frac{4\text{ k}\Omega}{1\text{ k}\Omega + 4\text{ k}\Omega} V_i = 0.8V_i$$

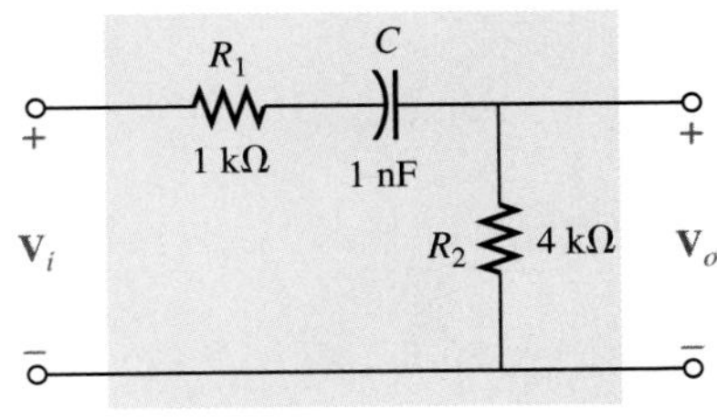

FIG. 21.55
High-pass filter with attenuated output.

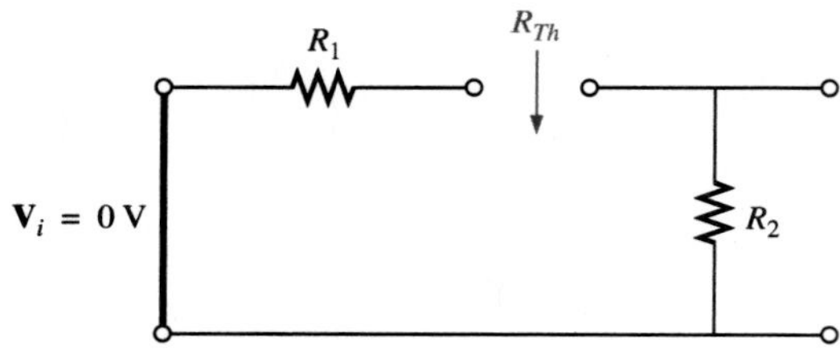

FIG. 21.56
Determining R_{Th} for the equation for cutoff frequency.

The resistance to be used in the equation for cutoff frequency can be determined by determining the Thévenin resistance "seen" by the capacitor. Setting $V_i = 0$ V and solving for R_{Th} (for the capacitor C) results in the network in Fig. 21.56, where it is quite clear that

$$R_{Th} = R_1 + R_2 = 1\text{ k}\Omega + 4\text{ k}\Omega = 5\text{ k}\Omega$$

Therefore,

$$f_c = \frac{1}{2\pi R_{Th}C} = \frac{1}{2\pi(5\text{ k}\Omega)(1\text{ nF})} = 31.83\text{ kHz}$$

A sketch of V_o versus frequency is provided in Fig. 21.57(a). A normalized plot using V_i as the normalizing quantity results in the response in Fig. 21.57(b). If the maximum value of A_v is used in the normalization process, the response in Fig. 21.57(c) is obtained. For all the plots obtained in the previous section, V_i was the maximum value, and the ratio V_o/V_i had a maximum value of 1. For many situations, this will not be the case, and we must be aware of which ratio is being plotted versus frequency. The dB response curves for the plots in Figs. 21.57(b) and 21.57(c) can both be obtained quite directly using the foundation established by the conclusions depicted in Fig. 21.54, but we must be aware of what to expect and how they will differ. In Fig. 21.57(b), we are comparing the output level to the input voltage. In Fig. 21.57(c) we are plotting A_v versus the maximum value of A_v. On most data sheets and for the majority of the investigative techniques commonly used, the normalized plot in Fig. 21.57(c) is used because it establishes 0 dB as an asymptote for the dB plot. To ensure that the impact of using either Fig. 21.57(b) or Fig. 21.57(c) in a frequency plot is understood, the analysis of the filter in Fig. 21.55 includes the resulting dB plot for both normalized curves.

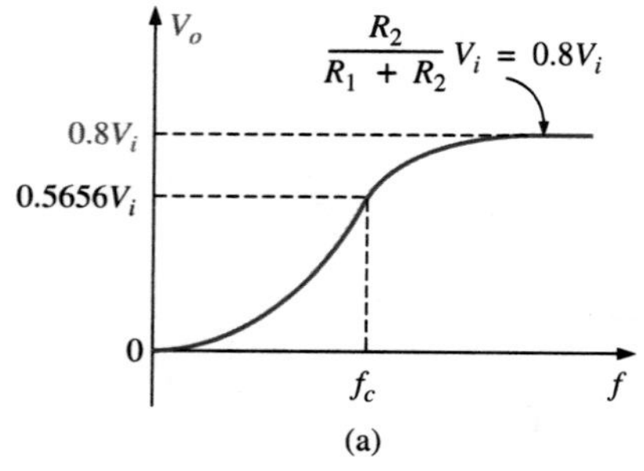

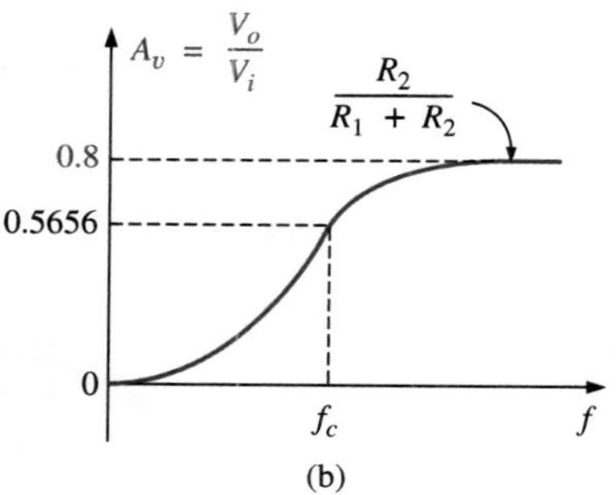

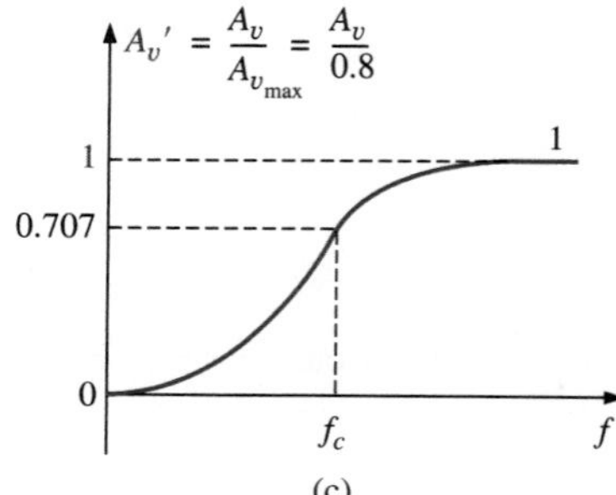

FIG. 21.57
Finding the normalized plot for the gain of the high-pass filter in Fig. 21.55 with attenuated output.

For the network in Fig. 21.55:

$$\mathbf{V}_o = \frac{R_2\mathbf{V}_i}{R_1 + R_2 - jX_C} = R_2\left[\frac{1}{R_1 + R_2 - jX_C}\right]\mathbf{V}_i$$

Dividing the top and bottom of the equation by $R_1 + R_2$ results in

$$\mathbf{V}_o = \frac{R_2}{R_1 + R_2}\left[\frac{1}{1 - j\dfrac{X_C}{R_1 + R_2}}\right]$$

but $$-j\frac{X_C}{R_1+R_2} = -j\frac{1}{\omega(R_1+R_2)C} = -j\frac{1}{2\pi f(R_1+R_2)C}$$

$$= -j\frac{f_c}{f} \quad \text{with} \quad f_c = \frac{1}{2\pi R_{Th}C} \quad \text{and} \quad R_{Th} = R_1 + R_2$$

so that $$\mathbf{V}_o = \frac{R_2}{R_1+R_2}\left[\frac{1}{1-j(f_c/f)}\right]\mathbf{V}_i$$

If we divide both sides by $\mathbf{V}_i$, we obtain

$$\mathbf{A}_v = \frac{\mathbf{V}_o}{\mathbf{V}_i} = \frac{R_2}{R_1+R_2}\left[\frac{1}{1-j(f_c/f)}\right] \tag{21.41}$$

from which the magnitude plot in Fig. 21.57(b) can be obtained. If we divide both sides by $\mathbf{A}_{v_{max}} = R_2/(R_1 + R_2)$, we have

$$\mathbf{A}'_v = \frac{\mathbf{A}_v}{\mathbf{A}_{v_{max}}} = \frac{1}{1-j\,(f_c/f)} \tag{21.42}$$

from which the magnitude plot in Fig. 21.57(c) can be obtained.

Based on the past section, a dB plot of the magnitude of $A'_v = A_v/A_{v_{max}}$ is now quite direct using Fig. 21.54(b). The plot appears in Fig. 21.58.

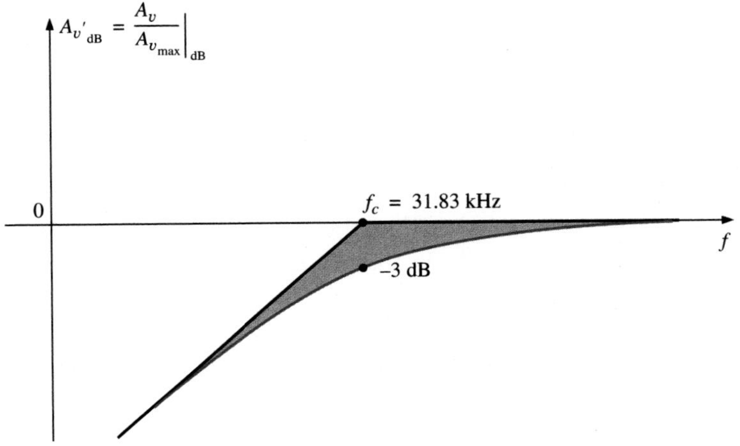

FIG. 21.58
dB plot for A'_v for the high-pass filter in Fig. 21.55.

For the gain $A_v = V_o/V_i$, we can apply Eq. (21.5):

$$20\log_{10}ab = 20\log_{10}a + 20\log_{10}b$$

where

$$20\log_{10}\left\{\frac{R_2}{R_1+R_2}\left[\frac{1}{1-j\,(f_c/f)}\right]\right\}$$
$$= 20\log_{10}\frac{R_2}{R_1+R_2} + 20\log_{10}\frac{1}{\sqrt{1+(f_c/f)^2}}$$

The second term results in the same plot in Fig. 21.58, but the first term must be added to the second to obtain the total dB response.

Since $R_2/(R_1 + R_2)$ must always be less than 1, we can rewrite the first term as

$$20 \log_{10} \frac{R_2}{R_1 + R_2} = 20 \log_{10} \frac{1}{\dfrac{R_1 + R_2}{R_2}} = \underbrace{20 \log_{10} 1}_{0} - 20 \log_{10} \frac{R_1 + R_2}{R_2}$$

and

$$20 \log_{10} \frac{R_2}{R_1 + R_2} = -20 \log_{10} \frac{R_1 + R_2}{R_2} \tag{21.43}$$

providing the drop in dB from the 0 dB level for the plot. Adding one log plot to the other *at each frequency,* as permitted by Eq. (21.5), results in the plot in Fig. 21.59.

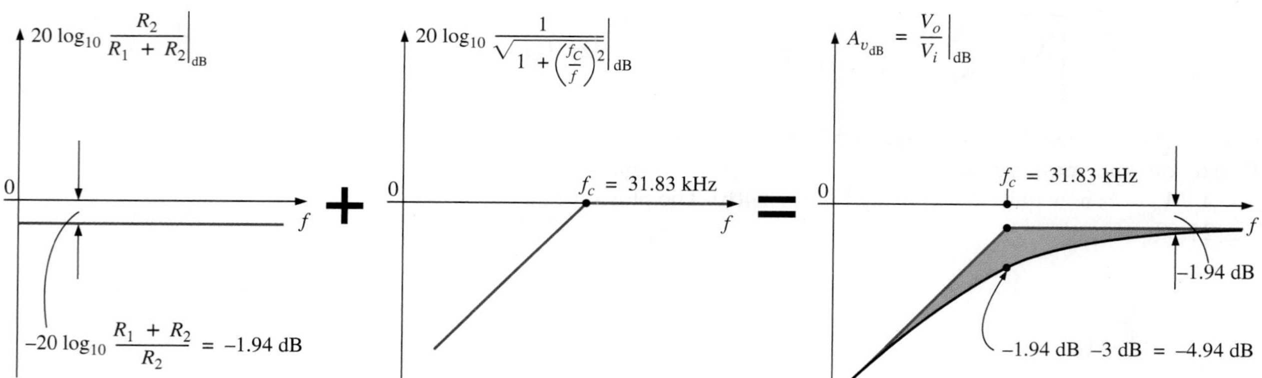

FIG. 21.59

Obtaining a dB plot of $A_{v_{dB}} = \left.\dfrac{V_o}{V_i}\right|_{dB}$.

For the network in Fig. 21.55, the gain $\mathbf{A}_v = \mathbf{V}_o/\mathbf{V}_i$ can also be found in the following manner:

$$\mathbf{V}_o = \frac{R_2 \mathbf{V}_i}{R_1 + R_2 - jX_C}$$

$$\mathbf{A}_v = \frac{\mathbf{V}_o}{\mathbf{V}_i} = \frac{R_2}{R_1 + R_2 - jX_C} = \frac{jR_2}{j(R_1 + R_2) + X_C}$$
$$= \frac{jR_2/X_C}{j(R_1 + R_2)/X_C + 1} = \frac{j\omega R_2 C}{1 + j\omega(R_1 + R_2)C}$$
$$= \frac{j2\pi f R_2 C}{1 + j2\pi f(R_1 + R_2)C}$$

and

$$\mathbf{A}_v = \frac{\mathbf{V}_o}{\mathbf{V}_i} = \frac{j(f/f_1)}{1 + j(f/f_c)} \tag{21.44}$$

with $\quad f_1 = \dfrac{1}{2\pi R_2 C} \quad$ and $\quad f_c = \dfrac{1}{2\pi(R_1 + R_2)C}$

The bottom of Eq. (21.44) is a match of the denominator of the low-pass function in Fig. 21.54(a). The numerator, however, is a new function that defines a unique Bode asymptote that will prove useful for a variety of network configurations.

Applying Eq. (21.5):

$$20 \log_{10} \frac{V_o}{V_i} = 20 \log_{10} \left[\frac{f}{f_1}\right]\left[\frac{1}{\sqrt{1 + (f/f_c)^2}}\right]$$

$$= 20 \log_{10}(f/f_1) + 20 \log_{10} \frac{1}{\sqrt{1 + (f/f_c)^2}}$$

Let us now consider specific frequencies for the first term.

At $f = f_1$:

$$20 \log_{10} \frac{f}{f_1} = 20 \log_{10} 1 = 0 \text{ dB}$$

At $f = 2f_1$:

$$20 \log_{10} \frac{f}{f_1} = 20 \log_{10} 2 = +6 \text{ dB}$$

At $f = \frac{1}{2}f_1$:

$$20 \log_{10} \frac{f}{f_1} = 20 \log_{10} 0.5 = -6 \text{ dB}$$

A dB plot of $20 \log_{10} (f/f_1)$ is provided in Fig. 21.60. Note that the asymptote passes through the 0 dB line at $f = f_1$ and has a positive slope of +6 dB/octave (or 20 dB/decade) for frequencies above and below f_1 for increasing values of f.

If we examine the original function A_v, we find that the phase angle associated with $j f/f_1 = f/f_1 \angle 90°$ is fixed at 90°, resulting in a phase angle for $\mathbf{A}_v$ of $90° - \tan^{-1}(f/f_c) = +\tan^{-1}(f_c/f)$.

Now that we have a plot of the dB response for the magnitude of the function f/f_1, we can plot the dB response of the magnitude of $\mathbf{A}_v$ using a procedure outlined by Fig. 21.61.

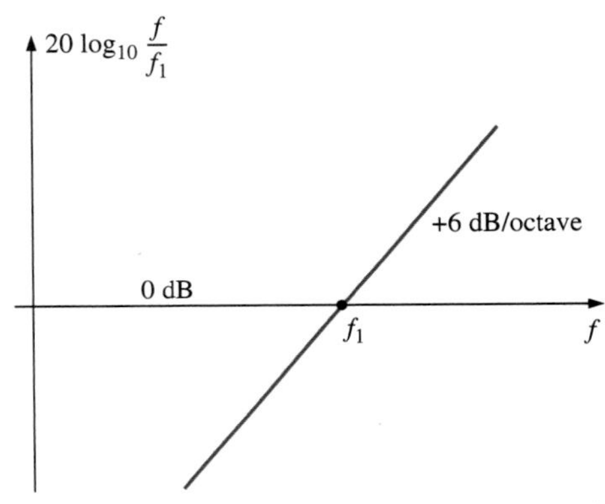

FIG. 21.60
dB plot of f/f_1.

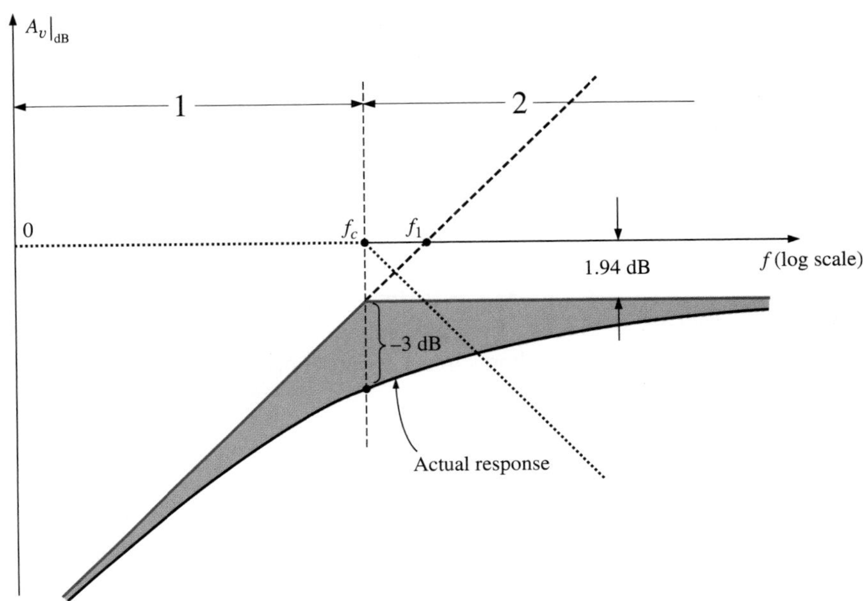

FIG. 21.61
Plot of $A_v|_{dB}$ for the network in Fig. 21.55.

Solving for f_1 and f_c:

$$f_1 = \frac{1}{2\pi R_2 C} = \frac{1}{2\pi(4\text{ k}\Omega)(1\text{ nF})} = 39.79\text{ kHz}$$

with $$f_c = \frac{1}{2\pi(R_1 + R_2)C} = \frac{1}{2\pi(5\text{ k}\Omega)(1\text{ nF})} = 31.83\text{ kHz}$$

For this development, the straight-line asymptotes for each term resulting from the application of Eq. (21.5) are drawn on the same frequency axis to permit an examination of the impact of one line section on the other. For clarity, the frequency spectrum in Fig. 21.61 has been divided into two regions.

In region 1, we have a 0 dB asymptote and one increasing at 6 dB/octave for increasing frequencies. The sum of the two as defined by Eq. (21.5) is simply the 6 dB/octave asymptote shown in the figure.

In region 2, one asymptote is increasing at 6 dB, and the other is decreasing at -6 dB/octave for increasing frequencies. The net effect is that one cancels the other for the region greater than $f = f_c$, leaving a horizontal asymptote beginning at $f = f_c$. A careful sketch of the asymptotes on a log scale reveals that the horizontal asymptote is at -1.94 dB, as obtained earlier for the same function. The horizontal level can also be determined by simply plugging $f = f_c$ into the Bode plot defined by f/f_1; that is,

$$20\log\frac{f}{f_1} = 20\log_{10}\frac{f_c}{f_1} = 20\log_{10}\frac{31.83\text{ kHz}}{39.79\text{ kHz}}$$
$$= 20\log_{10}0.799 = -1.94\text{ dB}$$

The actual response can then be drawn using the asymptotes and the known differences at $f = f_c$ (-3 dB) and at $f = 0.5f_c$ or $2f_c$ (-1 dB).

In summary, therefore, the same dB response for $\mathbf{A}_v = \mathbf{V}_o/\mathbf{V}_i$ can be obtained by isolating the maximum value or defining the gain in a different form. The latter approach permitted the introduction of a new function for our catalog of idealized Bode plots that will prove useful in the future.

21.12 LOW-PASS FILTER WITH LIMITED ATTENUATION

Our analysis now continues with the low-pass filter in Fig. 21.62, which has limited attenuation at the high-frequency end. That is, the output will not drop to zero as the frequency becomes relatively high. The filter is similar in construction to Fig. 21.55, but note that now $\mathbf{V}_o$ includes the capacitive element.

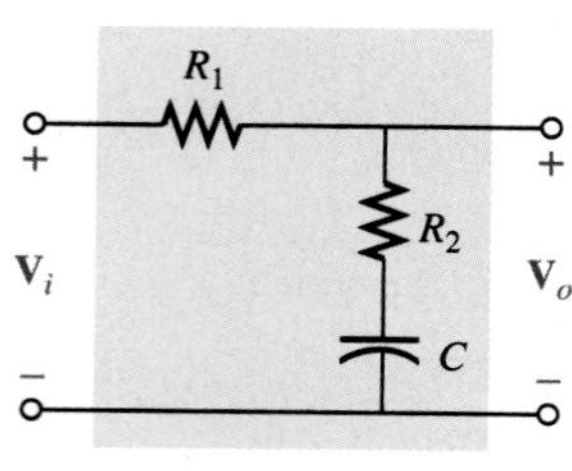

FIG. 21.62
Low-pass filter with limited attenuation.

At $f = 0$ Hz, the capacitor can assume its open-circuit equivalence, and $\mathbf{V}_o = \mathbf{V}_i$. At high frequencies, the capacitor can be approximated by a short-circuit equivalence, and

$$\mathbf{V}_o = \frac{R_2}{R_1 + R_2}\mathbf{V}_i$$

A plot of V_o versus frequency is provided in Fig. 21.63(a). A sketch of $A_v = V_o/V_i$ appears as shown in Fig. 21.63(b).

An equation for $\mathbf{V}_o$ in terms of $\mathbf{V}_i$ can be derived by first applying the voltage divider rule:

$$\mathbf{V}_o = \frac{(R_2 - jX_C)\mathbf{V}_i}{R_1 + R_2 - jX_C}$$

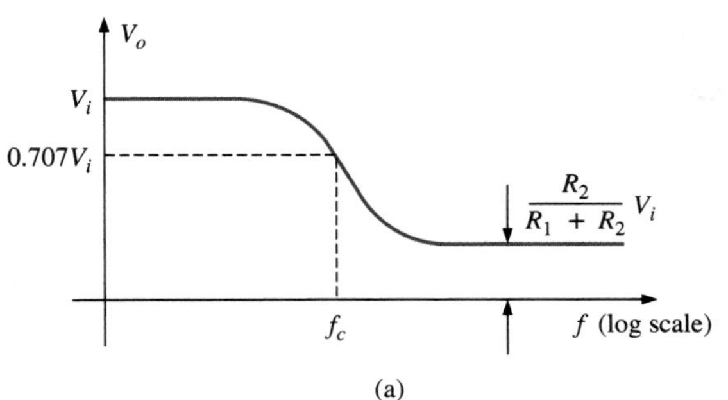

(a)

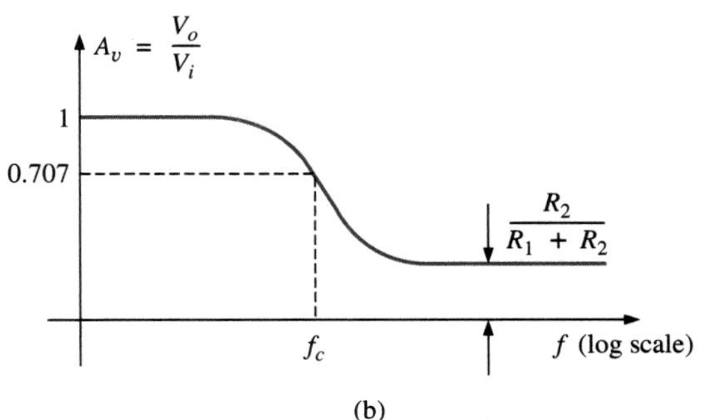

(b)

FIG. 21.63
Low-pass filter with limited attenuation.

and
$$\mathbf{A}_v = \frac{\mathbf{V}_o}{\mathbf{V}_i} = \frac{R_2 - jX_C}{R_1 + R_2 - jX_C} = \frac{R_2/X_C - j}{(R_1 + R_2)/X_C - j}$$
$$= \frac{(j)(R_2X_C - j)}{(j)((R_1 + R_2)/X_C - j)}$$
$$= \frac{j(R_2/X_C) + 1}{j((R_1 + R_2)/X_C) + 1} = \frac{1 + j\,2\pi f R_2 C}{1 + j\,2\pi f(R_1 + R_2)C}$$

so that
$$\mathbf{A}_v = \frac{\mathbf{V}_o}{\mathbf{V}_i} = \frac{1 + j\,(f/f_1)}{1 + j\,(f/f_c)} \qquad \textbf{(21.45)}$$

with
$$f_1 = \frac{1}{2\pi R_2 C} \quad \text{and} \quad f_c = \frac{1}{2\pi(R_1 + R_2)C}$$

The denominator of Eq. (21.45) is simply the denominator of the low-pass function in Fig. 21.54(a). The numerator, however, is new and must be investigated.

Applying Eq. (21.5):

$$A_{v_{\text{dB}}} = 20 \log_{10} \frac{V_o}{V_i} = 20 \log_{10} \sqrt{1 + (f/f_1)^2} + 20 \log_{10} \frac{1}{\sqrt{1 + (f/f_c)^2}}$$

For $f >> f_1$, $(f/f_1)^2 >> 1$, and the first term becomes

$$20 \log_{10} \sqrt{(f/f_1)^2} = 20 \log_{10}((f/f_1)^2)^{1/2} = 20 \log_{10}(f/f_1)|_{f \gg f_1}$$

which defines the idealized Bode asymptote for the numerator of Eq. (21.45).

At $f = f_1$, $20 \log_{10} 1 = 0$ dB, and at $f = 2f_1$, $20 \log_{10} 2 = 6$ dB. For frequencies much less than f_1, $(f/f_1)^2 << 1$, and the first term of the Eq. (21.5) expansion becomes $20 \log_{10} \sqrt{1} = 20 \log_{10} 1 = 0\text{ dB}$, which establishes the low-frequency asymptote.

The full idealized Bode response for the numerator of Eq. (21.45) is provided in Fig. 21.64.

We are now in a position to determine $A_v|_{\text{dB}}$ by plotting the asymptote for each function of Eq. (21.45) on the same frequency axis, as shown in Fig. 21.65. Note that f_c must be less than f_1 since the denominator of f_1 includes only R_2, whereas the denominator of f_c includes both R_2 and R_1.

Since $R_2/(R_1 + R_2)$ will always be less than 1, we can use an earlier development to obtain an equation for the drop in dB below the 0 dB axis at high frequencies. That is,

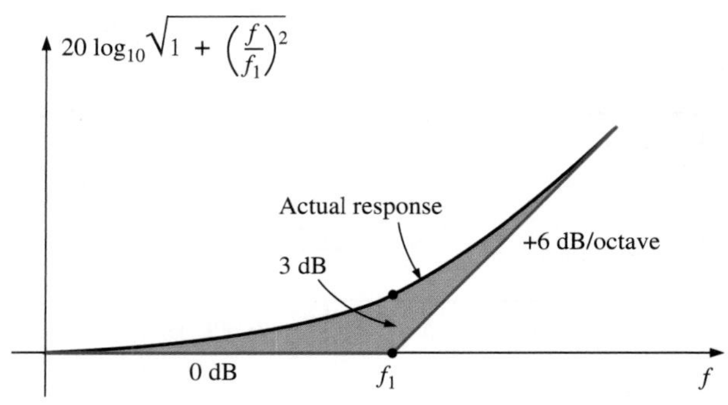

FIG. 21.64
Idealized and actual Bode response for the magnitude of $(1 + j(f/f_1))$.

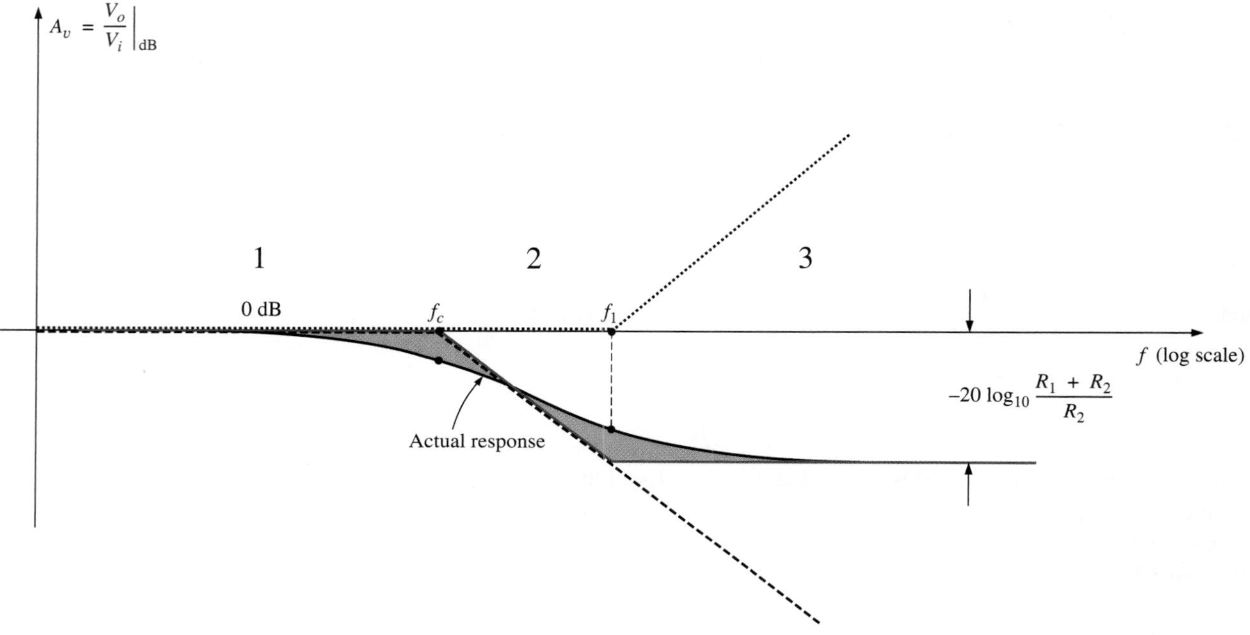

FIG. 21.65
$A_{v_{dB}}$ versus frequency for the low-pass filter with limited attenuation of Fig. 21.62.

$$\begin{aligned}20 \log_{10} R_2/(R_1 + R_2) &= 20 \log_{10} 1/((R_1 + R_2)/R_2)\\ &= \underbrace{20 \log_{10} 1}_{0} - 20 \log_{10}((R_1 + R_2)/R_2)\end{aligned}$$

and

$$\boxed{20 \log_{10} \frac{R_2}{R_1 + R_2} = -20 \log_{10} \frac{R_1 + R_2}{R_2}} \tag{21.46}$$

as shown in Fig. 21.65.

In region 1 in Fig. 21.65, both asymptotes are at 0 dB, resulting in a net Bode asymptote at 0 dB for the region. At $f = f_c$, one asymptote maintains its 0 dB level, whereas the other is dropping by 6 dB/octave. The sum of the two is the 6 dB drop per octave shown for the region. In region 3, the −6 dB/octave asymptote is balanced by the +6 dB/octave asymptote, establishing a level asymptote at the negative dB level attained by the f_c asymptote at $f = f_1$. The dB level of the horizontal asymptote in region 3 can be determined using Eq. (21.46) or by substituting $f = f_1$ into the asymptotic expression defined by f_c.

The full idealized Bode envelope is now defined, permitting a sketch of the actual response by shifting 3 dB in the right direction at each corner frequency, as shown in Fig. 21.65.

The phase angle associated with $\mathbf{A}_v$ can be determined directly from Eq. (21.45). That is,

$$\boxed{\theta = \tan^{-1} f/f_1 - \tan^{-1} f/f_c} \qquad \textbf{(21.47)}$$

A full plot of θ versus frequency can be obtained by substituting various key frequencies into Eq. (21.47) and plotting the result on a log scale.

The first term of Eq. (21.47) defines the phase angle established by the numerator of Eq. (21.45). The asymptotic plot established by the numerator is provided in Fig. 21.66. Note the phase angle of 45° at $f = f_1$ and the straight-line asymptote between $f_1/10$ and $10f_1$.

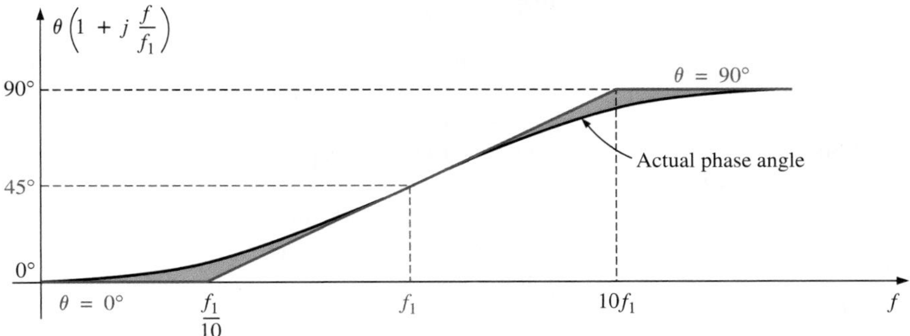

FIG. 21.66
Phase angle for $(1 + j\,(f/f_1))$.

Now that we have an asymptotic plot for the phase angle of the numerator, we can plot the full phase response by sketching the asymptotes for both functions of Eq. (21.45) on the same graph, as shown in Fig. 21.67.

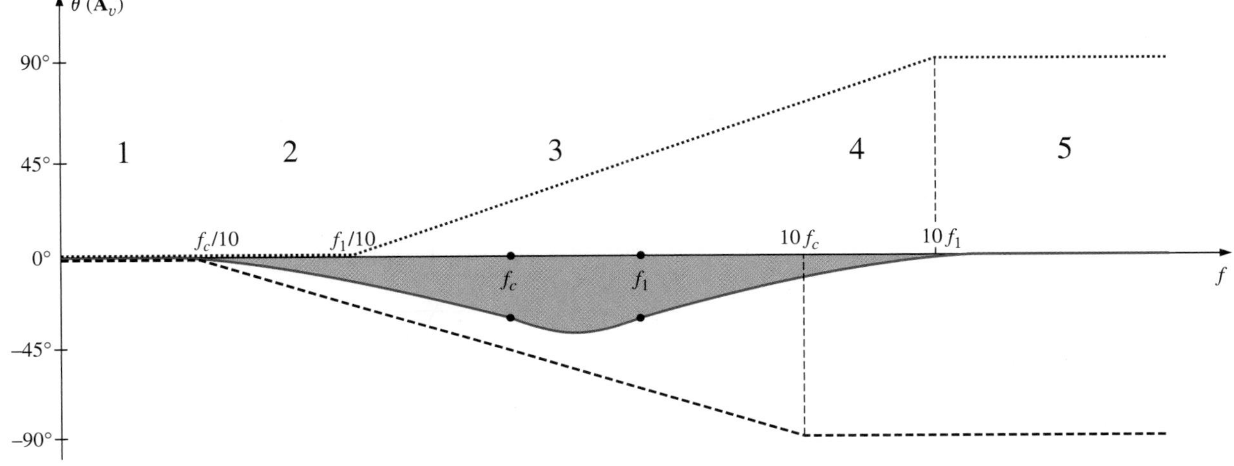

FIG. 21.67
Phase angle for the low-pass filter in Fig. 21.62.

The asymptotes in Fig. 21.67 clearly indicate that the phase angle will be 0° in the low-frequency range and 0° (90° − 90° = 0°) in the high-frequency range. In region 2, the phase plot drops below 0° due to the impact of the f_c asymptote. In region 4, the phase angle increases since the asymptote due to f_c remains fixed at −90°, whereas that due to f_1 is increasing. In the midrange, the plot due to f_1 is balancing the continued negative drop due to the f_c asymptote, resulting in the leveling response indicated. Due to the equal and opposite slopes of the asymptotes in the midregion, the angles of f_1 and f_c will be the same, but note that they are less than 45°. The maximum negative angle will occur between f_1 and f_c. The remaining points on the curve of Fig. 21.67 can be determined by simply substituting specific frequencies into Eq. (21.45). However, it is also useful to know that the most dramatic (the quickest) changes in the phase angle occur when the dB plot of the magnitude also goes through its greatest changes (such as at f_1 and f_c).

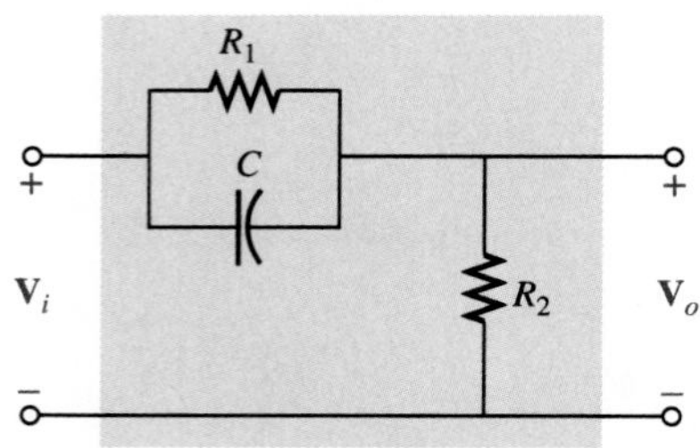

FIG. 21.68
High-pass filter with limited attenuation.

21.13 HIGH-PASS FILTER WITH LIMITED ATTENUATION

The filter in Fig. 21.68 is designed to limit the low-frequency attenuation in much the same manner as described for the low-pass filter of the previous section.

At $f = 0$ Hz, the capacitor can assume its open-circuit equivalence, and $\mathbf{V}_o = [R_2/(R_1 + R_2)]\mathbf{V}_i$. At high frequencies, the capacitor can be approximated by a short-circuit equivalence, and $\mathbf{V}_o = \mathbf{V}_i$.

The resistance to be used when determining f_c can be found by finding the Thévenin resistance for the capacitor C, as shown in Fig. 21.69. A careful examination of the resulting configuration reveals that $R_{Th} = R_1 \parallel R_2$ and $f_c = 1/2\pi(R_1 \parallel R_2)C$.

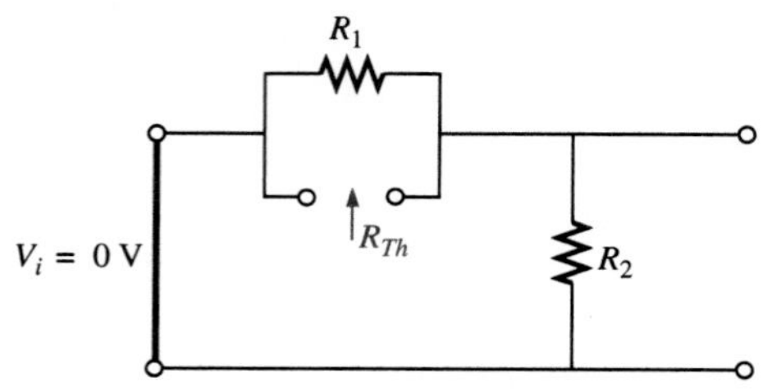

FIG. 21.69
Determining R for the f_c calculation for the filter in Fig. 21.68.

A plot of V_o versus frequency is provided in Fig. 21.70(a), and a sketch of $A_v = V_o/V_i$ appears in Fig. 21.70(b).

An equation for $\mathbf{A}_v = \mathbf{V}_o/\mathbf{V}_i$ can be derived by first applying the voltage divider rule:

$$\mathbf{V}_o = \frac{R_2\mathbf{V}_i}{R_2 + R_1 \parallel -j X_C}$$

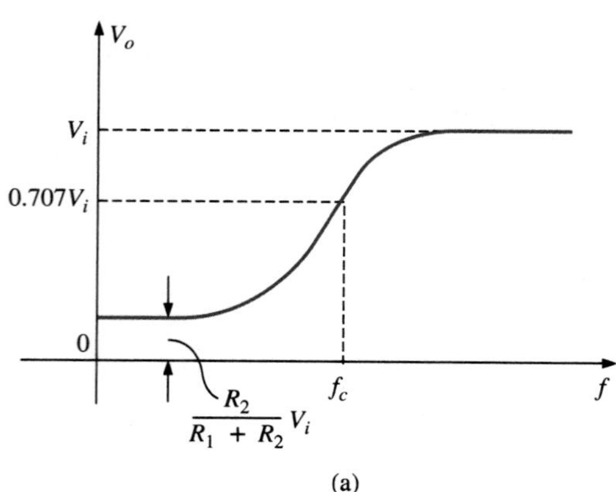

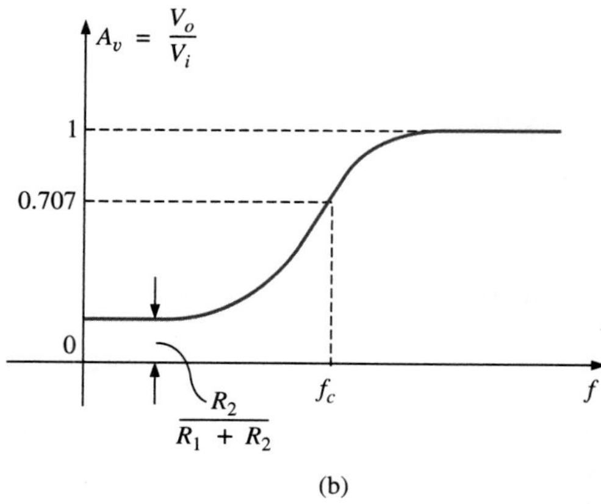

FIG. 21.70
High-pass filter with limited attenuation.

and $$\mathbf{A}_v = \frac{\mathbf{V}_o}{\mathbf{V}_i} = \frac{R_2}{R_2 + R_1 \parallel -jX_C} = \frac{R_2}{R_2 + \dfrac{R_1(-jX_C)}{R_1 - jX_C}}$$

$$= \frac{R_2(R_1 - jX_C)}{R_2(R_1 - jX_C) - jR_1X_C} = \frac{R_1R_2 - jR_2X_C}{R_1R_2 - jR_2X_C - jR_1X_C}$$

$$= \frac{R_1R_2 - jR_2X_C}{R_1R_2 - j(R_1 + R_2)X_C} = \frac{1 - j\dfrac{R_2X_C}{R_1R_2}}{1 - j\dfrac{(R_1 + R_2)}{R_1R_2}X_C}$$

$$= \frac{1 - j\dfrac{X_C}{R_1}}{1 - j\dfrac{X_C}{\dfrac{R_1R_2}{R_1 + R_2}}} = \frac{1 - j\dfrac{X_C}{R_1}}{1 - j\dfrac{X_C}{R_1 \parallel R_2}} = \frac{1 - j\dfrac{1}{2\pi fR_1C}}{1 - j\dfrac{1}{2\pi f(R_1 \parallel R_2)C}}$$

so that $$\boxed{\mathbf{A}_v = \frac{\mathbf{V}_o}{\mathbf{V}_i} = \frac{1 - j(f_1/f)}{1 - j(f_c/f)}} \tag{21.48}$$

with $$f_1 = \frac{1}{2\pi R_1C} \quad \text{and} \quad f_c = \frac{1}{2\pi(R_1 \parallel R_2)C}$$

The denominator of Eq. (21.48) is simply the denominator of the high-pass function in Fig. 21.54(b). The numerator, however, is new and must be investigated.

Applying Eq. (21.5):

$$A_{v_{dB}} = 20\log_{10}\frac{V_o}{V_i} = 20\log_{10}\sqrt{1 + (f_1/f)^2} + 20\log_{10}\frac{1}{\sqrt{1 + (f_c/f)^2}}$$

For $f << f_1$, $(f_1/f)^2 >> 1$, and the first term becomes

$$20\log_{10}\sqrt{(f_1/f)^2} = 20\log_{10}(f_1/f)\,|_{f \ll f_1}$$

which defines the idealized Bode asymptote for the numerator of Eq. (21.48).

At $f = f_1$, $20\log_{10} 1 = 0$ dB
At $f = 0.5f_1$, $20\log_{10} 2 = 6$ dB
At $f = 0.1f_1$, $20\log_{10} 10 = 20$ dB

For frequencies greater than f_1, $f_1/f << 1$ and $20\log_{10} 1 = 0$ dB, which establishes the high-frequency asymptote. The full idealized Bode plot for the numerator of Eq. (21.48) is provided in Fig. 21.71.

We are now in a position to determine $A_{v_{dB}}$ by plotting the asymptotes for each function of Eq. (21.48) on the same frequency axis, as shown in Fig. 21.72. Note that f_c must be more than f_1 since $R_1 \parallel R_2$ must be less than R_1.

When determining the linearized Bode response, let us first examine region 2, where one function is 0 dB and the other is dropping at 6 dB/octave for decreasing frequencies. The result is a decreasing asymptote from f_c to f_1. At the intersection of the resultant of region 2 with f_1, we enter region 1, where the asymptotes have opposite slopes and cancel the effect of each other. The resulting level at f_1 is determined by

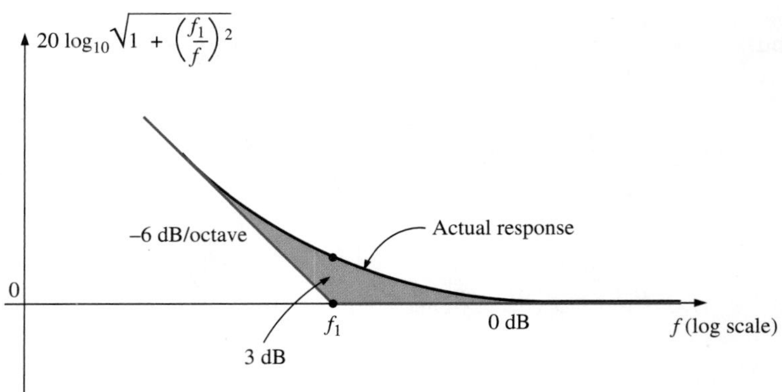

FIG. 21.71
Idealized and actual Bode response for the magnitude of $(1 - j(f_1/f))$.

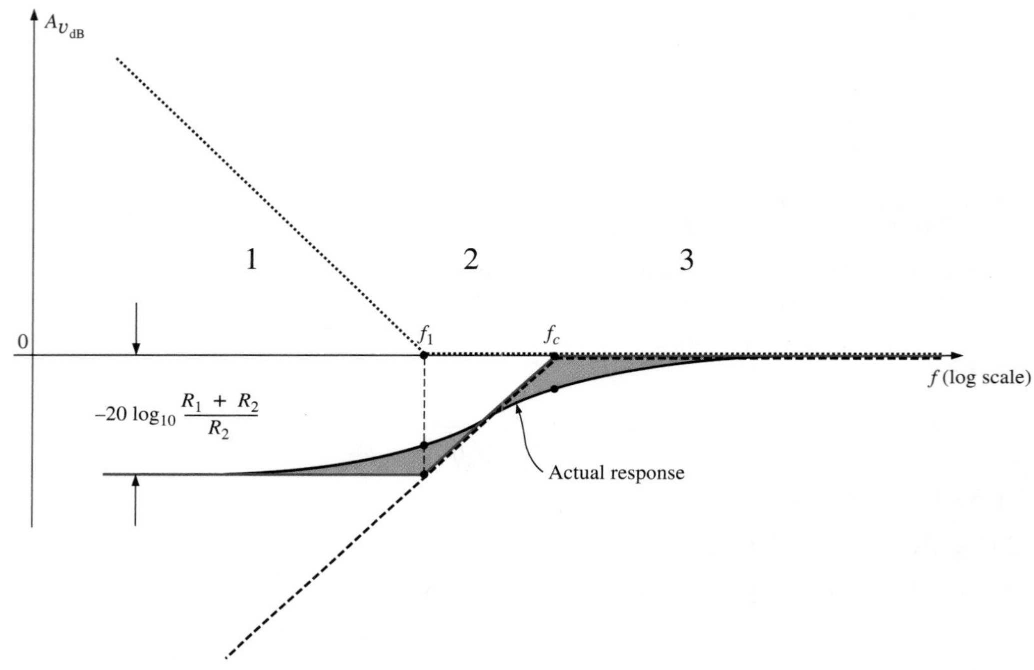

FIG. 21.72
$A_{v_{dB}}$ versus frequency for the high-pass filter with limited attenuation in Fig. 21.68.

$-20 \log_{10}(R_1 + R_2)/R_2$, as found in earlier sections. The drop can also be determined by substituting $f = f_1$ into the asymptotic equation defined for f_c. In region 3, both are at 0 dB, resulting in a 0 dB asymptote for the region. The resulting asymptotic and actual responses both appear in Fig. 21.72.

The phase angle associated with A_v can be determined directly from Eq. (21.48); that is,

$$\theta = -\tan^{-1}\frac{f_1}{f} + \tan^{-1}\frac{f_c}{f} \qquad \textbf{(21.49)}$$

A full plot of θ versus frequency can be obtained by substituting various key frequencies into Eq. (21.49) and plotting the result on a log scale.

The first term of Eq. (21.49) defines the phase angle established by the numerator of Eq. (21.48). The asymptotic plot resulting from the numer-

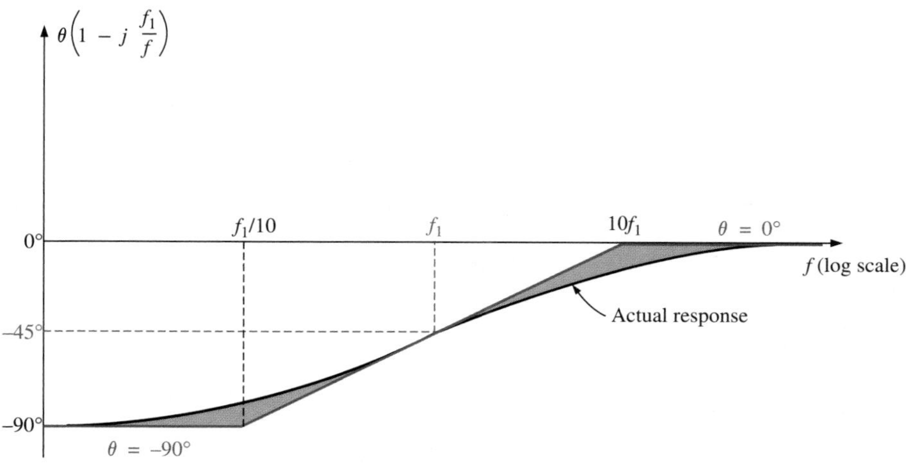

FIG. 21.73
Phase angle for $(1 - j\,(f_1/f))$.

ator is provided in Fig. 21.73. Note the leading phase angle of 45° at $f = f_1$ and the straight-line asymptote from $f_1/10$ to $10f_1$.

Now that we have an asymptotic plot for the phase angle of the numerator, we can plot the full phase response by sketching the asymptotes for both functions of Eq. (21.48) on the same graph, as shown in Fig. 21.74.

The asymptotes in Fig. 21.74 clearly indicate that the phase angle will be 90° in the low-frequency range and 0° (90° − 90° = 0°) in the high-frequency range. In region 2, the phase angle is increasing above 0° because one angle is fixed at 90° and the other is becoming less negative. In region 4, one is 0° and the other is decreasing, resulting in a decreasing θ for this region. In region 3, the positive angle is always greater than the negative angle, resulting in a positive angle for the entire region. Since the slopes of the asymptotes in region 3 are equal but opposite, the angles at f_c and f_1 are the same. Fig. 21.74 reveals that the angle at f_c and f_1 will be less than 45°. The maximum angle occurs between f_c and f_1, as shown in

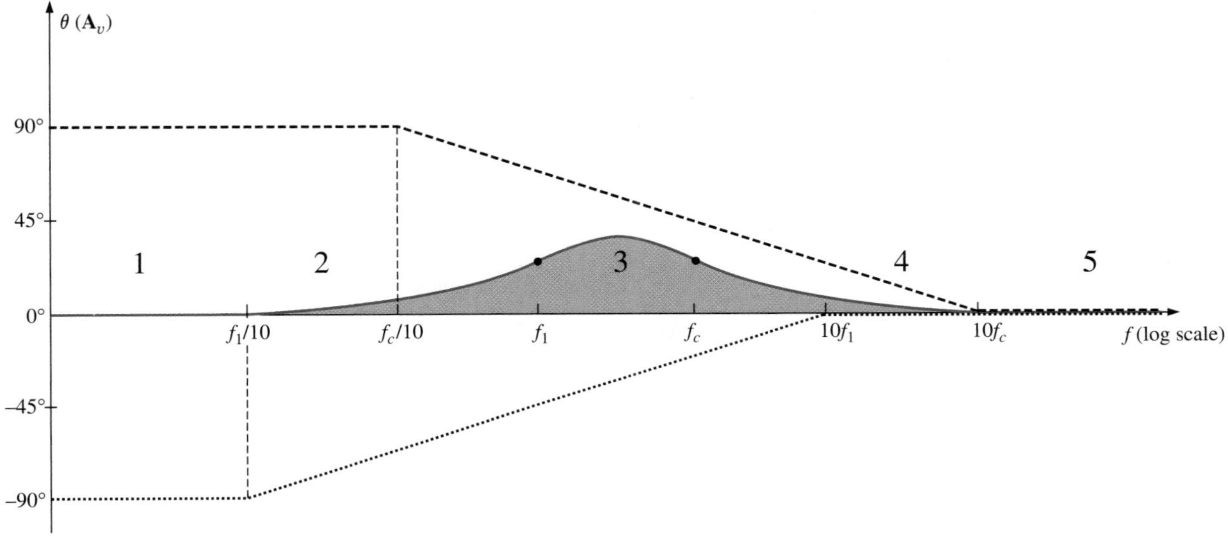

FIG. 21.74
Phase response for the high-pass filter in Fig. 21.68.

the figure. Note again that the greatest change in θ occurs at the corner frequencies, matching the regions of greatest change in the dB plot.

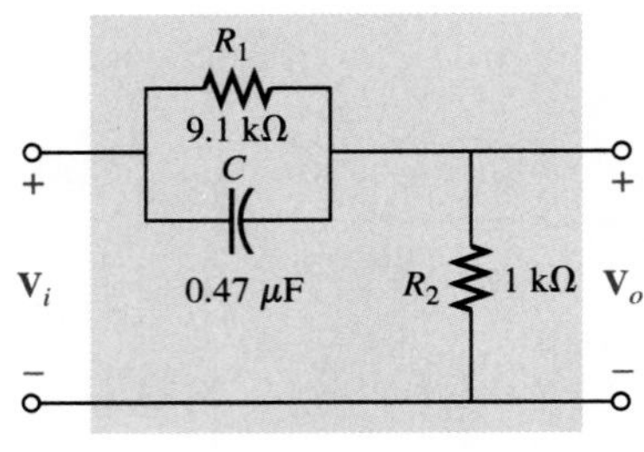

FIG. 21.75
Example 21.11.

EXAMPLE 21.11 For the filter in Fig. 21.75:

a. Sketch the curve of $A_{v_{dB}}$ versus frequency using a log scale.
b. Sketch the curve of θ versus frequency using a log scale.

Solutions:

a. For the break frequencies:

$$f_1 = \frac{1}{2\pi R_1 C} = \frac{1}{2\pi(9.1\text{ k}\Omega)(0.47\ \mu\text{F})} = \mathbf{37.2\ Hz}$$

$$f_c = \frac{1}{2\pi\left(\dfrac{R_1R_2}{R_1 + R_2}\right)C} = \frac{1}{2\pi(0.9\text{ k}\Omega)(0.47\ \mu\text{F})} = \mathbf{376.25\ Hz}$$

The maximum low-level attenuation is

$$-20\log_{10}\frac{R_1 + R_2}{R_2} = -20\log_{10}\frac{9.1\text{ k}\Omega + 1\text{ k}\Omega}{1\text{ k}\Omega}$$
$$= -20\log_{10}10.1 = \mathbf{-20.09\ dB}$$

The resulting plot appears in Fig. 21.76.

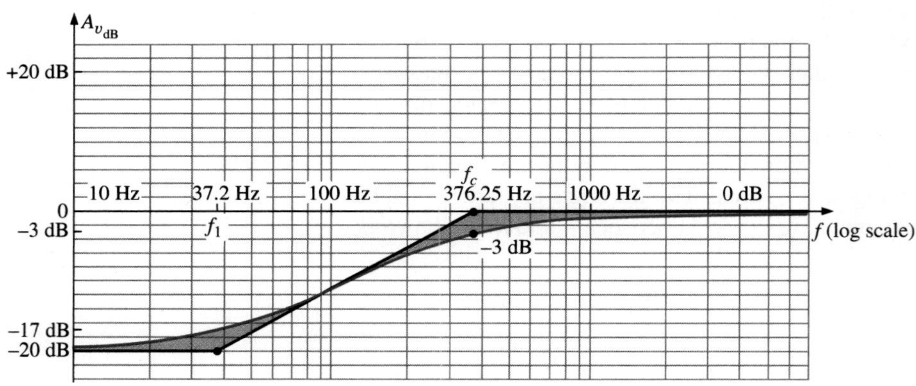

FIG. 21.76
$A_{v_{dB}}$ versus frequency for the filter in Fig. 21.75.

b. For the break frequencies:

At $f = f_1 = 37.2$ Hz,

$$\theta = -\tan^{-1}\frac{f_1}{f} + \tan^{-1}\frac{f_c}{f}$$
$$= -\tan^{-1}1 + \tan^{-1}\frac{376.25\text{ Hz}}{37.2\text{ Hz}}$$
$$= -45^\circ + 84.35^\circ$$
$$= \mathbf{39.35^\circ}$$

At $f = f_c = 376.26$ Hz,

$$\theta = -\tan^{-1}\frac{37.2 \text{ Hz}}{376.26 \text{ Hz}} + \tan^{-1} 1$$
$$= -5.65° + 45°$$
$$= \mathbf{39.35°}$$

At a frequency midway between f_c and f_1 on a log scale, for example, 120 Hz:

$$\theta = -\tan^{-1}\frac{37.2 \text{ Hz}}{120 \text{ Hz}} + \tan^{-1}\frac{376.26 \text{ Hz}}{120 \text{ Hz}}$$
$$= -17.22° + 72.31°$$
$$= \mathbf{55.09°}$$

The resulting phase plot appears in Fig. 21.77.

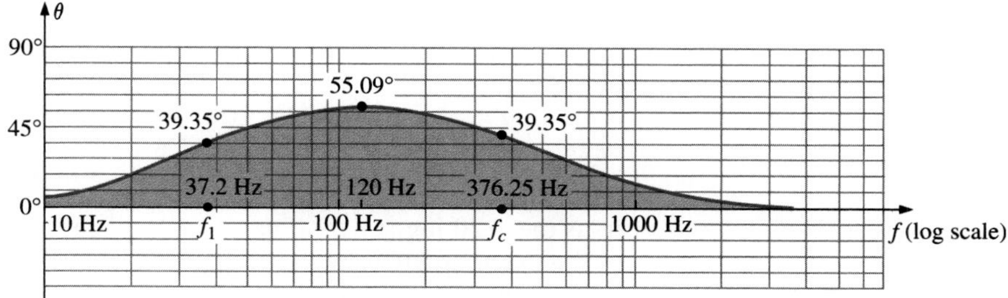

FIG. 21.77
θ (the phase angle associated with A_v) versus frequency for the filter in Fig. 21.75.

21.14 ADDITIONAL PROPERTIES OF BODE PLOTS

Bode plots are not limited to filters but can be applied to any system for which a dB-versus-frequency plot is desired. Although the previous sections did not cover all the functions that lend themselves to the idealized linear asymptotes, many of those most commonly encountered have been introduced.

We now examine some of the special situations that can develop that further demonstrate the adaptability and usefulness of the linear Bode approach to frequency analysis.

In all the situations described in this chapter, there was only one term in the numerator or denominator. For situations where there is more than one term, there will be an interaction between functions that must be examined and understood. In many cases, the use of Eq. (21.5) will prove useful. For example, if $\mathbf{A}_v$ has the format

$$\mathbf{A}_v = \frac{200(1 - jf_2/f)(jf/f_1)}{(1 - jf_1/f)(1 + jf/f_2)} = \frac{(a)(b)(c)}{(d)(e)} \qquad \textbf{(21.50)}$$

we can expand the function in the following manner:

$$A_{v_{\text{dB}}} = 20 \log_{10}\frac{(a)(b)(c)}{(d)(e)}$$
$$= 20 \log_{10} a + 20 \log_{10} b + 20 \log_{10} c - 20 \log_{10} d - 20 \log_{10} e$$

revealing that the net or resultant dB level is equal to the algebraic sum of the contributions from all the terms of the original function. We can, therefore, add algebraically the linearized Bode plots of all the terms in each frequency interval to determine the idealized Bode plot for the full function.

If two terms happen to have the same format and corner frequency, as in the function

$$\mathbf{A}_v = \frac{1}{(1 - jf_1/f)(1 - jf_1/f)}$$

the function can be rewritten as

$$\mathbf{A}_v = \frac{1}{(1 - jf_1/f)^2}$$

so that

$$A_{v_{dB}} = 20 \log_{10} \frac{1}{(\sqrt{1 + (f_1/f)^2})^2} = -20 \log_{10}(1 + (f_1/f)^2)$$

for $f << f_1$, $(f_1/f)^2 >> 1$, and

$$A_{v_{dB}} = -20 \log_{10}(f_1/f)^2 = -40 \log_{10} f_1/f$$

versus the $-20 \log_{10}(f_1/f)$ obtained for a single term in the denominator. The resulting dB asymptote will drop, therefore, at a rate of -12 dB/octave (-40 dB/decade) for decreasing frequencies rather than -6 dB/octave. The corner frequency is the same, and the high-frequency asymptote is still at 0 dB. The idealized Bode plot for the above function is provided in Fig. 21.78.

Note the steeper slope of the asymptote and the fact that the actual curve now passes -6 dB below the corner frequency rather than -3 dB, as for a single term.

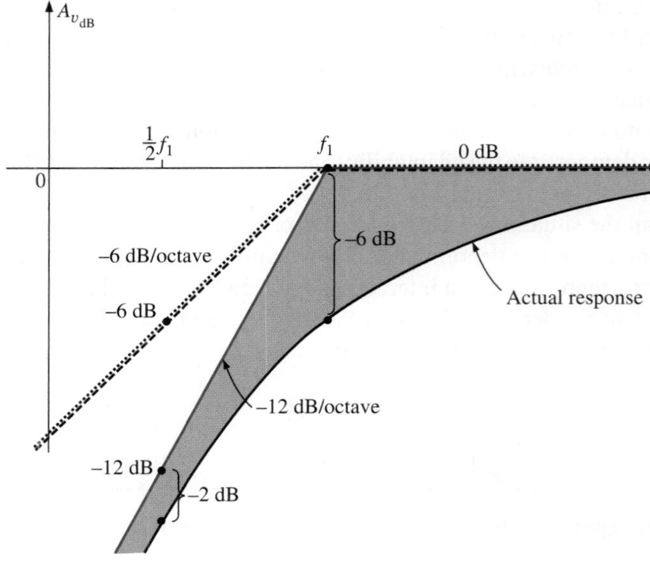

FIG. 21.78

Plotting the linearized Bode plot of $\dfrac{1}{(1 - j(f_1/f))^2}$.

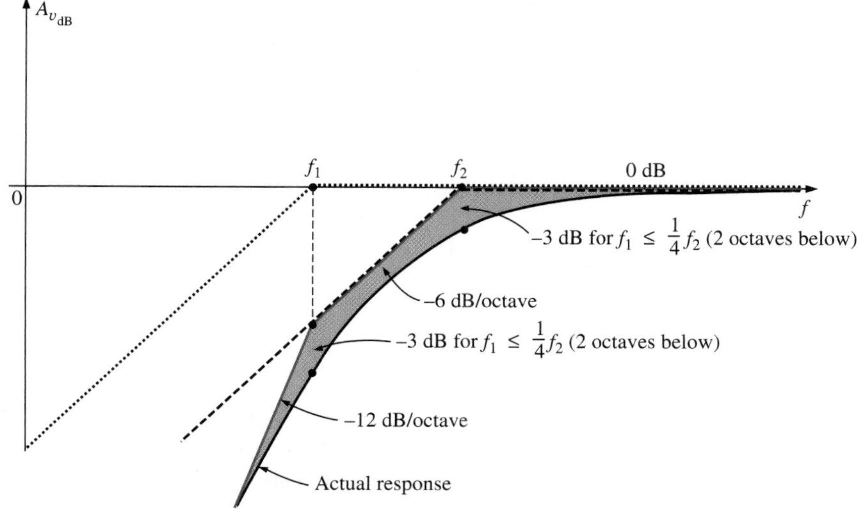

FIG. 21.79

Plot of $A_{v_{dB}}$ *for* $\dfrac{1}{(1 - j(f_1/f))(1 - j(f_2/f))}$ *with* $f_1 < f_2$.

Keep in mind that if the corner frequencies of the two terms in the numerator or denominator are close but not exactly equal, the total dB drop is the algebraic sum of the contributing terms of the expansion. For instance, consider the linearized Bode plot in Fig. 21.79 with corner frequencies f_1 and f_2.

In region 3, both asymptotes are 0 dB, resulting in an asymptote at 0 dB for frequencies greater than f_2. For region 2, one asymptote is at 0 dB, whereas the other drops at −6 dB/octave for decreasing frequencies. The net result for this region is an asymptote dropping at −6 dB, as shown in the same figure. At f_1, we find two asymptotes dropping off at −6 dB for decreasing frequencies. The result is an asymptote dropping off at −12 dB/octave for this region.

If f_1 and f_2 are at least two octaves apart, the effect of one on the plotting of the actual response for the other can almost be ignored. In other words, for this example, if $f_1 < \frac{1}{4}f_2$, the actual response is down −3 dB at $f = f_2$ and f_1.

The above discussion can be expanded for any number of terms at the same frequency or in the same region. For three equal terms in the denominator, the asymptote will drop at −18 dB/octave, and so on. Eventually, the procedure will become self-evident and relatively straightforward to apply. In many cases, the hardest part of finding a solution is to put the original function in the desired form.

EXAMPLE 21.12 A transistor amplifier has the following gain:

$$\mathbf{A}_v = \frac{100}{\left(1 - j\dfrac{50\text{ Hz}}{f}\right)\left(1 - j\dfrac{200\text{ Hz}}{f}\right)\left(1 + j\dfrac{f}{10\text{ kHz}}\right)\left(1 + j\dfrac{f}{20\text{ kHz}}\right)}$$

a. Sketch the normalized response $A'_v = A_v/A_{v_{max}}$, and determine the bandwidth of the amplifier.

b. Sketch the phase response, and determine a frequency where the phase angle is close to 0°.

Solutions:

a. $$A'_v = \frac{A_v}{A_{v_{max}}} = \frac{A_v}{100}$$

$$= \frac{1}{\left(1 - j\frac{50\text{ Hz}}{f}\right)\left(1 - j\frac{200\text{ Hz}}{f}\right)\left(1 + j\frac{f}{10\text{ kHz}}\right)\left(1 + j\frac{f}{20\text{ kHz}}\right)}$$

$$= \frac{1}{(a)(b)(c)(d)} = \left(\frac{1}{a}\right)\left(\frac{1}{b}\right)\left(\frac{1}{c}\right)\left(\frac{1}{d}\right)$$

and

$$A'_{v_{dB}} = -20\log_{10} a - 20\log_{10} b - 20\log_{10} c - 20\log_{10} d$$

clearly substantiating the fact that the total number of decibels is equal to the algebraic sum of the contributing terms.

A careful examination of the original function reveals that the first two terms in the denominator are high-pass filter functions, whereas the last two are low-pass functions. Fig. 21.80 demonstrates how the combination of the two types of functions defines a bandwidth for the amplifier. The high-frequency filter functions have defined the low cutoff frequency, and the low-frequency filter functions have defined the high cutoff frequency.

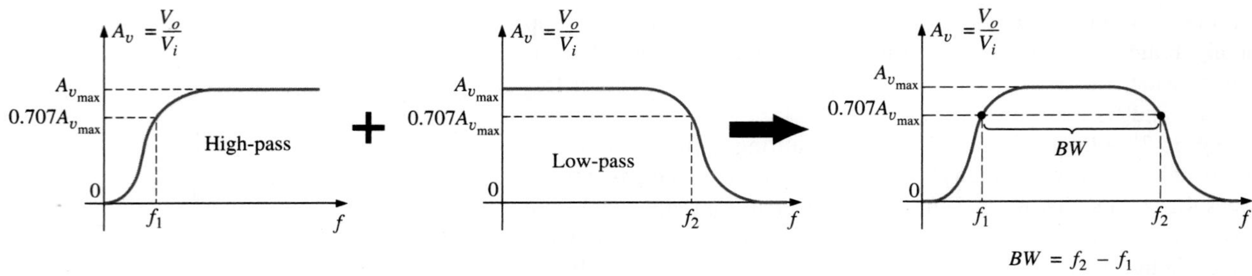

FIG. 21.80
Finding the overall gain versus frequency for Example 21.12.

Plotting all the idealized Bode plots on the same axis results in the plot in Fig. 21.81. Note for frequencies less than 50 Hz that the result-

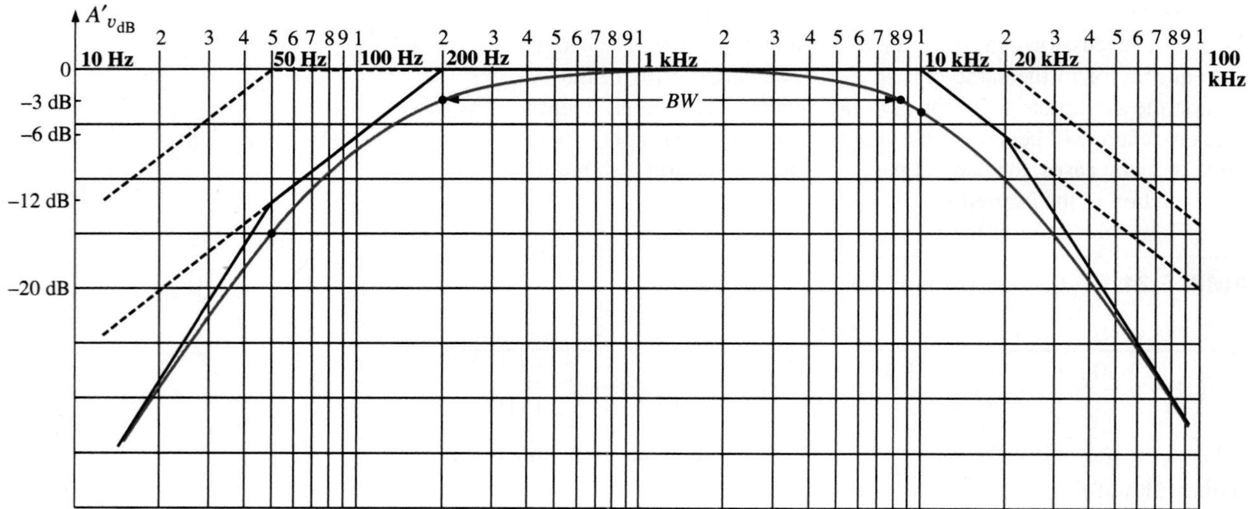

FIG. 21.81
$A'_{v_{dB}}$ versus frequency for Example 21.12.

ing asymptote drops off at −12 dB/octave. In addition, since 50 Hz and 200 Hz are separated by two octaves, the actual response will be down by only about −3 dB at the corner frequencies of 50 Hz and 200 Hz.

For the high-frequency region, the corner frequencies are not separated by two octaves, and the difference between the idealized plot and the actual Bode response must be examined more carefully. Since 10 kHz is one octave below 20 kHz, we can use the fact that the difference between the idealized response and the actual response for a single corner frequency is 1 dB. If we add an additional −1 dB drop due to the 20 kHz corner frequency to the −3 dB drop at $f = 10$ kHz, we can conclude that the drop at 10 kHz will be −4 dB, as shown on the plot. To check the conclusion, let us write the full expression for the dB level at 10 kHz and find the actual level for comparison purposes.

$$\begin{aligned} A'_{v_{dB}} &= -20 \log_{10}\sqrt{1+\left(\frac{50\text{ Hz}}{10\text{ kHz}}\right)^2} - 20\log_{10}\sqrt{1+\left(\frac{200\text{ Hz}}{10\text{ kHz}}\right)^2} \\ &\quad - 20\log_{10}\sqrt{1+\left(\frac{10\text{ kHz}}{10\text{ kHz}}\right)^2} - 20\log_{10}\sqrt{1+\left(\frac{10\text{ kHz}}{20\text{ kHz}}\right)^2} \\ &= -0.00011\text{ dB} - 0.0017\text{ dB} - 3.01\text{ dB} - 0.969\text{ dB} \\ &= -3.98\text{ dB} \cong \mathbf{-4\ dB} \qquad \text{as before} \end{aligned}$$

An examination of the above calculations reveals that the last two terms predominate in the high-frequency region and essentially eliminate the need to consider the first two terms in that region. For the low-frequency region, examining the first two terms is sufficient.

Proceeding in a similar fashion, we find a −4 dB difference at $f = 20$ kHz, resulting in the actual response appearing in Fig. 21.81. Since the bandwidth is defined at the −3 dB level, a judgment must be made as to where the actual response crosses the −3 dB level in the high-frequency region. A rough sketch suggests that it is near 8.5 kHz. Plugging this frequency into the high-frequency terms results in

$$\begin{aligned} A'_{v_{dB}} &= -20\log_{10}\sqrt{1+\left(\frac{8.5\text{ kHz}}{10\text{ kHz}}\right)^2} - 20\log_{10}\sqrt{1+\left(\frac{8.5\text{ kHz}}{20\text{ kHz}}\right)^2} \\ &= -2.148\text{ dB} - 0.645\text{ dB} \cong \mathbf{-2.8\ dB} \end{aligned}$$

which is relatively close to the −3 dB level, and

$$BW = f_{\text{high}} - f_{\text{low}} = 8.5\text{ kHz} - 200\text{ Hz} = \mathbf{8.3\ kHz}$$

In the midrange of the bandwidth, $A'_{v_{dB}}$ approaches 0 dB. At $f =$ 1 kHz:

$$\begin{aligned} A'_{v_{dB}} &= -20\log_{10}\sqrt{1+\left(\frac{50\text{ Hz}}{1\text{ kHz}}\right)^2} - 20\log_{10}\sqrt{1+\left(\frac{200\text{ Hz}}{1\text{ kHz}}\right)^2} \\ &\quad - 20\log_{10}\sqrt{1+\left(\frac{1\text{ kHz}}{10\text{ kHz}}\right)^2} - 20\log_{10}\sqrt{1+\left(\frac{1\text{ kHz}}{20\text{ kHz}}\right)^2} \\ &= -0.0108\text{ dB} - 0.1703\text{ dB} - 0.0432\text{ dB} - 0.0108\text{ dB} \\ &= \mathbf{-0.235\ dB} \cong \mathbf{-\frac{1}{5}\ dB} \end{aligned}$$

which is certainly close to the 0 dB level, as shown on the plot.

b. The phase response can be determined by substituting a number of key frequencies into the following equation, derived directly from the original function $\mathbf{A}_v$:

$$\theta = \tan^{-1}\frac{50\text{ Hz}}{f} + \tan^{-1}\frac{200\text{ Hz}}{f} - \tan^{-1}\frac{f}{10\text{ kHz}} - \tan^{-1}\frac{f}{20\text{ kHz}}$$

However, let us make full use of the asymptotes defined by each term of $\mathbf{A}_v$ and sketch the response by finding the resulting phase angle at critical points on the frequency axis. The resulting asymptotes and phase plot are provided in Fig. 21.82. Note that at $f = 50$ Hz, the sum of the two angles determined by the straight-line asymptotes is $45° + 75° = 120°$ (actual $= 121°$). At $f = 1$ kHz, if we subtract 5.7° for one corner frequency, we obtain a net angle of $14° - 5.7° \cong 8.3°$ (actual $= 5.6°$).

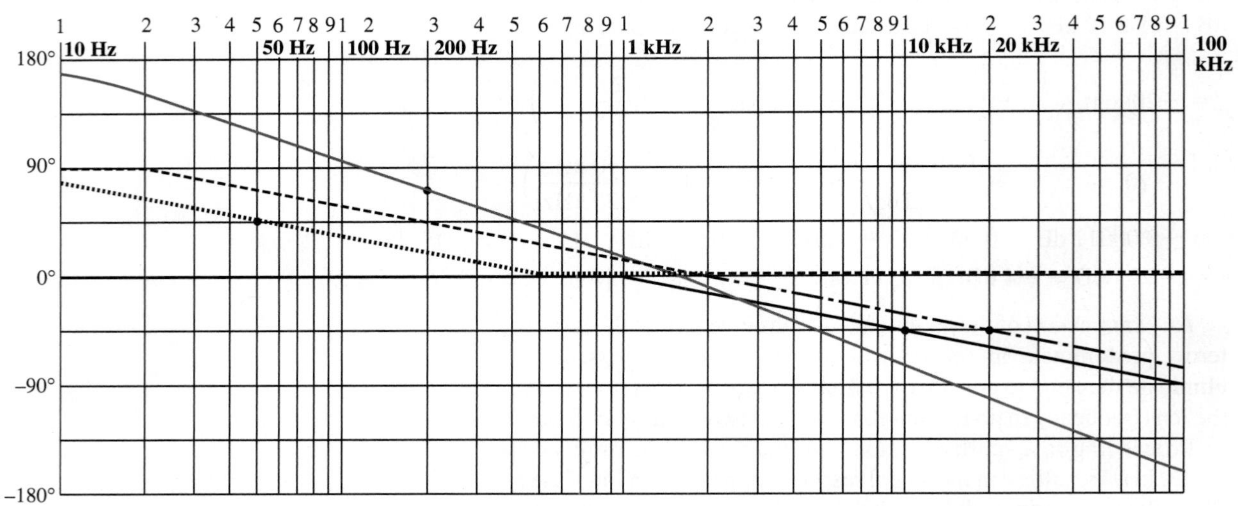

FIG. 21.82
Phase response for Example 21.12.

At 10 kHz, the asymptotes leave us with $\theta \cong -45° - 32° = -77°$ (actual $= -71.56°$). The net phase plot appears to be close to 0° at about 1300 Hz. To check on our assumptions and the use of the asymptotic approach, plug in $f = 1300$ Hz into the equation for θ:

$$\begin{aligned}\theta &= \tan^{-1}\frac{50\text{ Hz}}{1300\text{ Hz}} + \tan^{-1}\frac{200\text{ Hz}}{1300\text{ Hz}} - \tan^{-1}\frac{1300\text{ Hz}}{10\text{ kHz}} - \tan^{-1}\frac{1300\text{ Hz}}{20\text{ kHz}}\\ &= 2.2° + 8.75° - 7.41° - 3.72°\\ &= -0.18° \cong 0° \qquad \text{as predicted}\end{aligned}$$

In total, the phase plot appears to shift from a positive angle of 180° ($\mathbf{V}_o$ leading $\mathbf{V}_i$) to a negative angle of 180° as the frequency spectrum extends from very low frequencies to high frequencies. In the midregion, the phase plot is close to 0° ($\mathbf{V}_o$ in phase with $\mathbf{V}_i$), much like the response to a common-base transistor amplifier.

Table 21.2 consolidates some of the material introduced in this chapter and provides a reference for future investigations. It includes the linearized dB and phase plots for the functions appearing in the first column. There are many other functions, but these provide a foundation to which others can be added.

Reviewing the development of the filters in Sections 21.12 and 21.13 shows that establishing the function $\mathbf{A}_v$ in the proper form is the most

TABLE 21.2
Idealized Bode plots for various functions.

Function	dB Plot	Phase Plot
$\mathbf{A}_v = 1 - j\frac{f_1}{f}$	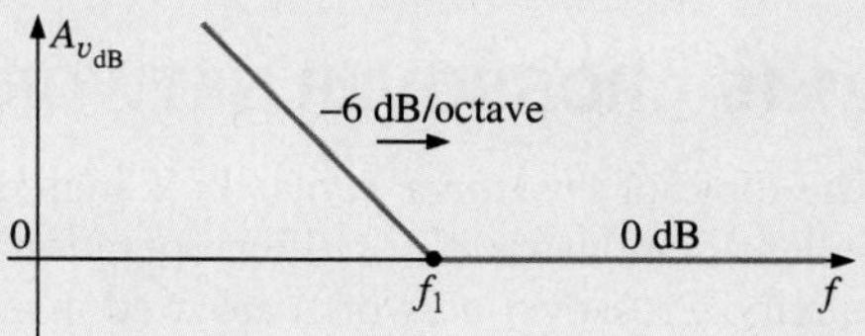 $A_{v_{dB}}$; 0; −6 dB/octave; 0 dB; f_1; f	θ ($\mathbf{V}_o$ leads $\mathbf{V}_i$); 90°; 45°; 0°; −45°; −90°; $f_1/10$; f_1; $10f_1$; 0°; −45°; −90°; f
$\mathbf{A}_v = 1 + j\frac{f}{f_1}$	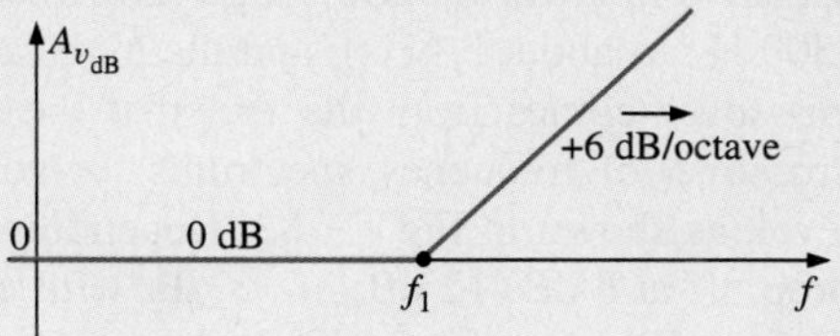 $A_{v_{dB}}$; 0; 0 dB; +6 dB/octave; f_1; f	θ ($\mathbf{V}_o$ leads $\mathbf{V}_i$); 90°; 45°; 0°; −90°; 0°; +45°; +90°; $\frac{f_1}{10}$; f_1; $10f_1$; f
$\mathbf{A}_v = j\frac{f}{f_1}$	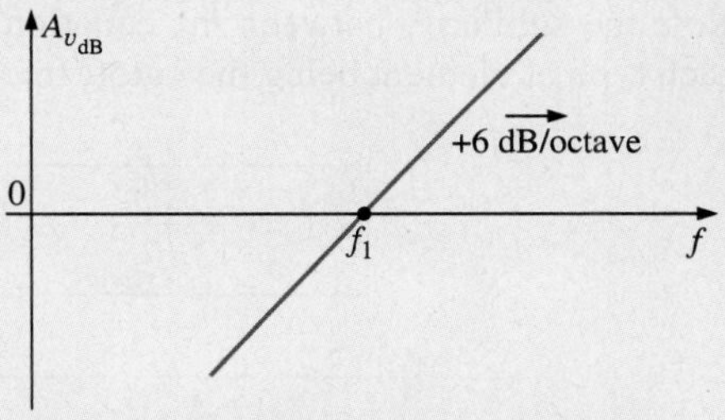 $A_{v_{dB}}$; 0; +6 dB/octave; f_1; f	θ ($\mathbf{V}_o$ leads $\mathbf{V}_i$); 90°; 45°; 0°; +90°; $\frac{f_1}{10}$; f_1; $10f_1$; f
$\mathbf{A}_v = \frac{1}{1 - j\frac{f_1}{f}}$	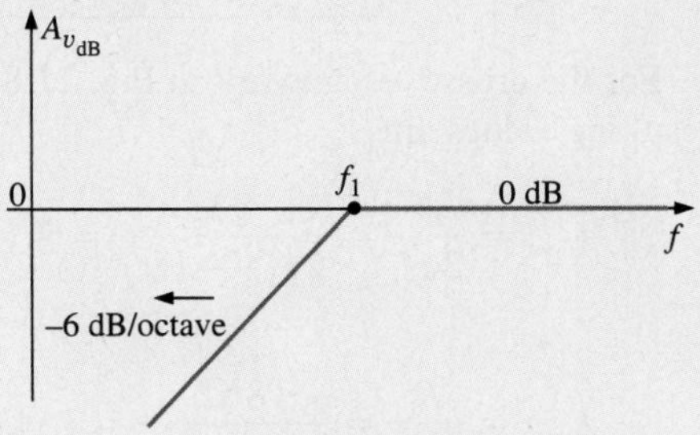 $A_{v_{dB}}$; 0; f_1; 0 dB; −6 dB/octave; f	θ ($\mathbf{V}_o$ leads $\mathbf{V}_i$); +90°; 90°; 45°; 0°; +45°; 0°; $\frac{f_1}{10}$; f_1; $10f_1$; f
$\mathbf{A}_v = \frac{1}{1 + j\frac{f}{f_1}}$	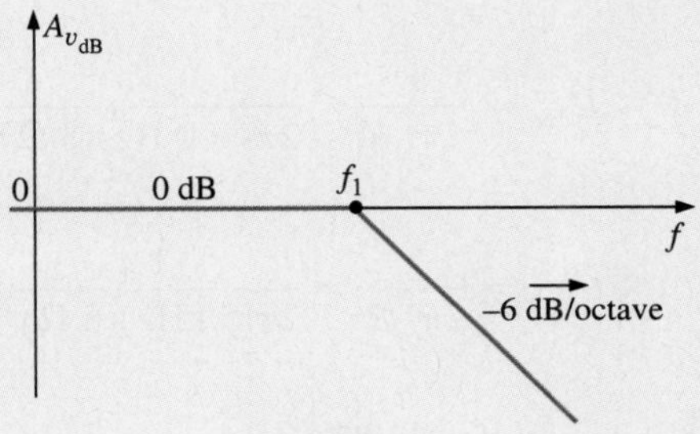 $A_{v_{dB}}$; 0; 0 dB; f_1; −6 dB/octave; f	θ ($\mathbf{V}_o$ leads $\mathbf{V}_i$); 90°; 45°; 0°; −45°; −90°; 0°; $f_1/10$; f_1; $10f_1$; −45°; −90°; f

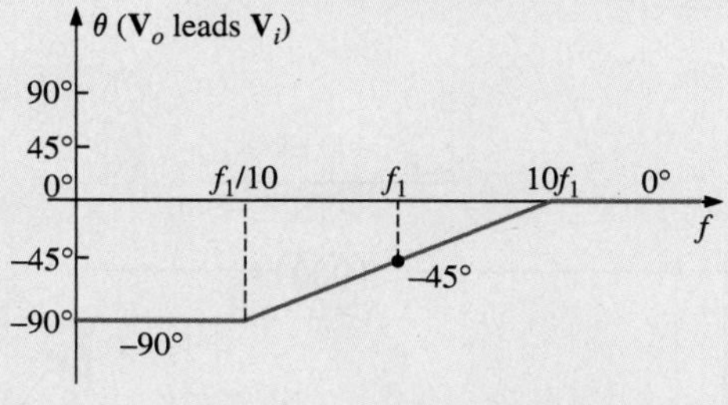

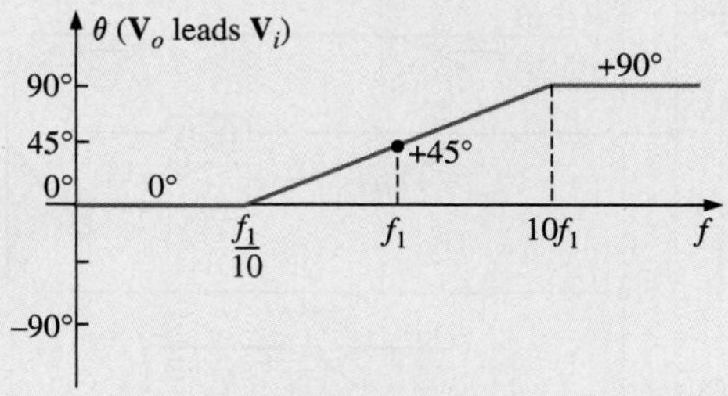

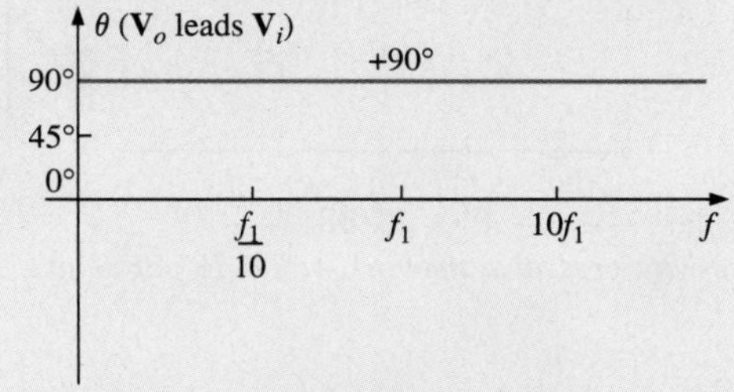

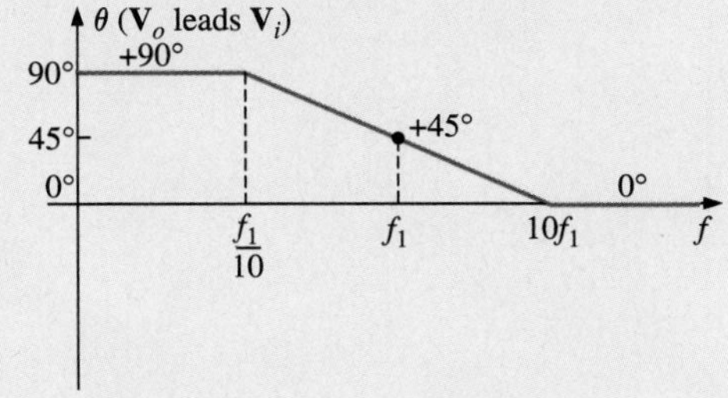

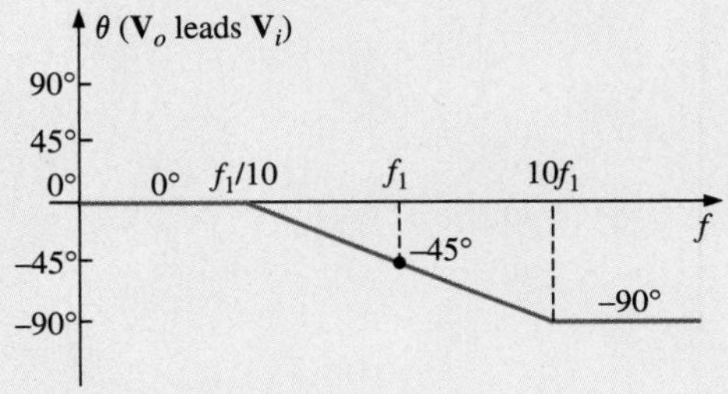

difficult part of the analysis. However, with practice and an awareness of the desired format, you will discover methods that will significantly reduce the effort involved.

21.15 CROSSOVER NETWORKS

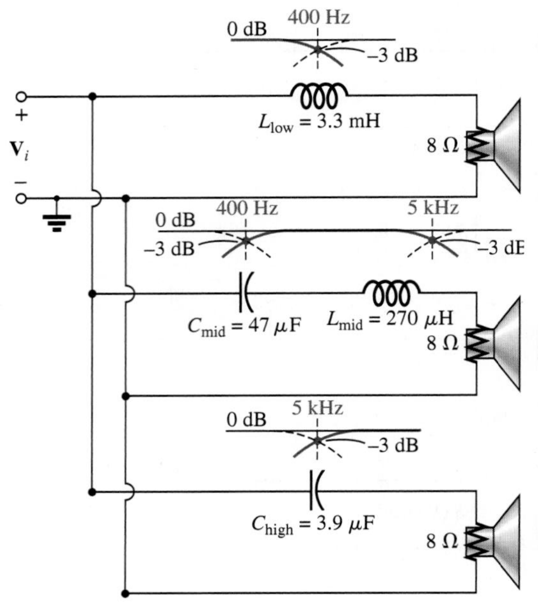

FIG. 21.83
Three-way, crossover network with 6 dB per octave.

The topic of *crossover networks* is included primarily to present an excellent demonstration of filter operation without a high level of complexity. Crossover networks are used in audio systems to ensure that the proper frequencies are channeled to the appropriate speaker. Although less expensive audio systems have only one speaker to cover the full audio range from about 20 Hz to 20 kHz, better systems have at least three speakers to cover the low range (20 Hz to about 500 Hz), the midrange (500 Hz to about 5 kHz), and the high range (5 kHz and up). The term *crossover* comes from the fact that the system is designed to have a crossover of frequency spectrums for adjacent speakers at the −3 dB level, as shown in Fig. 21.83. Depending on the design, each filter can drop off at 6 dB, 12 dB, or 18 dB, with complexity increasing with the desired dB drop-off rate. The three-way crossover network in Fig. 21.83 is quite simple in design, with a low-pass *R-L* filter for the *woofer,* an *R-L-C* pass-band filter for the midrange, and a high-pass *R-C* filter for the *tweeter.* The basic equations for the components are provided below. Note the similarity between the equations, with the only difference for each type of element being the cutoff frequency.

$$L_{low} = \frac{R}{2\pi f_1} \qquad L_{mid} = \frac{R}{2\pi f_2} \tag{21.51}$$

$$C_{mid} = \frac{1}{2\pi f_1 R} \qquad C_{high} = \frac{1}{2\pi f_2 R} \tag{21.52}$$

For the crossover network in Fig. 21.83 with three 8 Ω speakers, the resulting values are

$$L_{low} = \frac{R}{2\pi f_1} = \frac{8\ \Omega}{2\pi(400\text{ Hz})} = 3.183\text{ mH} \rightarrow 3.3\text{ mH}$$
(commercial value)

$$L_{mid} = \frac{R}{2\pi f_2} = \frac{8\ \Omega}{2\pi(5\text{ kHz})} = 254.65\ \mu\text{H} \rightarrow 270\ \mu\text{H}$$
(commercial value)

$$C_{mid} = \frac{1}{2\pi f_1 R} = \frac{1}{2\pi(400\text{ Hz})(8\ \Omega)} = 49.736\ \mu\text{F} \rightarrow 47\ \mu\text{F}$$
(commercial value)

$$C_{high} = \frac{1}{2\pi f_2 R} = \frac{1}{2\pi(5\text{ kHz})(8\ \Omega)} = 3.979\ \mu\text{F} \rightarrow 3.9\ \mu\text{F}$$
(commercial value)

as shown in Fig. 21.83.

For each filter, a rough sketch of the frequency response is included to show the crossover at the specific frequencies of interest. Because all three speakers are in parallel, the source voltage and impedance for each are the same. The total loading on the source is obviously a function of the frequency applied, but the total delivered is determined solely by the speakers since they are essentially resistive in nature.

To test the system, let us apply a 4 V signal at a frequency of 1 kHz (a predominant frequency of the typical human auditory response curve) and see which speaker has the highest power level.

At $f = 1$ kHz,

$$X_{L_{\text{low}}} = 2\pi f L_{\text{low}} = 2\pi(1\text{ kHz})(3.3\text{ mH}) = 20.74\ \Omega$$

$$\mathbf{V}_o = \frac{(\mathbf{Z}_R \angle 0^\circ)(V_i \angle 0^\circ)}{\mathbf{Z}_T} = \frac{(8\ \Omega \angle 0^\circ)(4\text{ V} \angle 0^\circ)}{8\ \Omega + j\,20.74\ \Omega}$$

$$= 1.44\text{ V} \angle -68.90^\circ$$

$$X_{L_{\text{mid}}} = 2\pi f L_{\text{mid}} = 2\pi(1\text{ kHz})(270\ \mu\text{H}) = 1.696\ \Omega$$

$$X_{C_{\text{mid}}} = \frac{1}{2\pi f C_{\text{mid}}} = \frac{1}{2\pi(1\text{ kHz})(47\ \mu\text{F})} = 3.386\ \Omega$$

$$\mathbf{V}_o = \frac{(\mathbf{Z}_R \angle 0^\circ)(V_i \angle 0^\circ)}{\mathbf{Z}_T} = \frac{(8\ \Omega \angle 0^\circ)(4\text{ V} \angle 0^\circ)}{8\ \Omega + j\,1.696\ \Omega - j\,3.386\ \Omega}$$

$$= 3.94\text{ V} \angle 11.93^\circ$$

$$X_{C_{\text{high}}} = \frac{1}{2\pi f C_{\text{high}}} = \frac{1}{2\pi(1\text{ kHz})(3.9\ \mu\text{F})} = 40.81\ \Omega$$

$$\mathbf{V}_o = \frac{(Z_R \angle 0^\circ)(V_i \angle 0^\circ)}{\mathbf{Z}_T} = \frac{(8\ \Omega \angle 0^\circ)(4\text{ V} \angle 0^\circ)}{8\ \Omega - j\,40.81\ \Omega}$$

$$= 0.77\text{ V} \angle 78.91^\circ$$

Using the basic power equation $P = V^2/R$, the power to the woofer is

$$P_{\text{low}} = \frac{V^2}{R} = \frac{(1.44\text{ V})^2}{8\ \Omega} = \mathbf{0.259\ W}$$

to the midrange speaker,

$$P_{\text{mid}} = \frac{V^2}{R} = \frac{(3.94\text{ V})^2}{8\ \Omega} = \mathbf{1.94\ W}$$

and to the tweeter,

$$P_{\text{high}} = \frac{V^2}{R} = \frac{(0.77\text{ V})^2}{8\ \Omega} = \mathbf{0.074\ W}$$

resulting in a power ratio of 7.5 : 1 between the midrange and the woofer and 26 : 1 between the midrange and the tweeter. Obviously, the response of the midrange speaker totally overshadows the other two.

21.16 APPLICATIONS

Attenuators

Attenuators are, by definition, any device or system that can reduce the power or voltage level of a signal while introducing little or no distortion. There are two general types: passive and active. The passive type uses only resistors, while the active type uses electronic devices such as transistors and integrated circuits. Since electronics is a subject for the courses to follow, only the resistive type is covered here. Attenuators are commonly used in audio equipment (such as the graphic and parametric equalizers introduced in Chapter 20), antenna systems, AM or FM systems where attenuation may be required before the signals are mixed, and any other application where a reduction in signal strength is required.

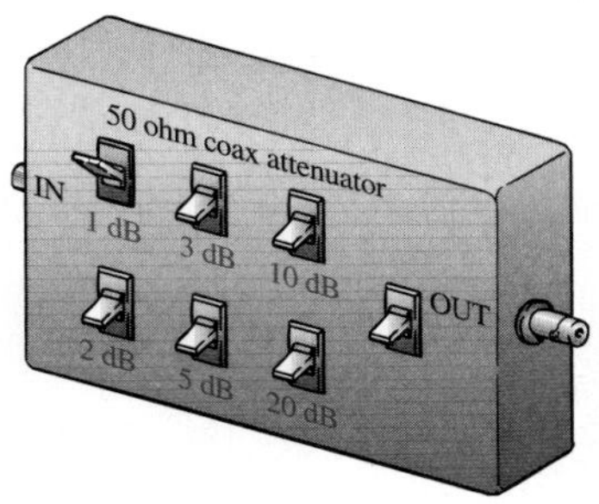

FIG. 21.84
Passive coax attenuator.

The unit in Fig. 21.84 has coaxial input and output terminals and switches to set the level of dB reduction. It has a flat response from dc to about 6 GHz, which essentially means that its introduction into the network will not affect the frequency response for this band of frequencies. The design is rather simple with resistors connected in either a *tee* (T) or a *wye* (Y) configuration as shown in Figs. 21.85 and 21.86, respectively, for a 50 Ω coaxial system. In each case, the resistors are chosen to ensure that the input impedance and output impedance match the line. That is, the input and output impedances of each configuration will be 50 Ω. For a number of dB attenuations, the resistor values for the T and Y are provided in Figs. 21.85 and 21.86. Note in each design that two of the resistors are the same, while the third is a much smaller or larger value.

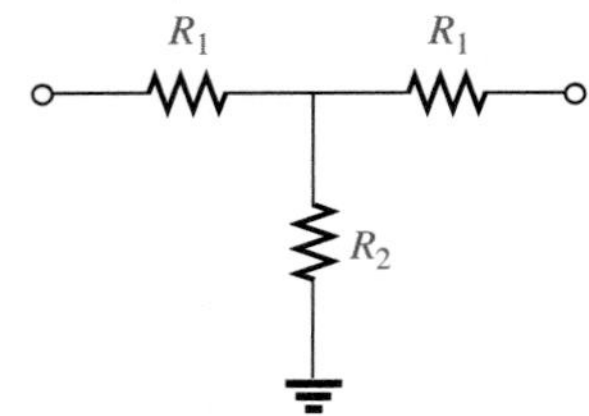

Attenuation	R_1	R_2
1 dB	2.9 Ω	433.3 Ω
2 dB	5.7 Ω	215.2 Ω
3 dB	8.5 Ω	141.9 Ω
5 dB	14.0 Ω	82.2 Ω
10 dB	26.0 Ω	35.0 Ω
20 dB	41.0 Ω	10.0 Ω

FIG. 21.85
Tee (T) configuration.

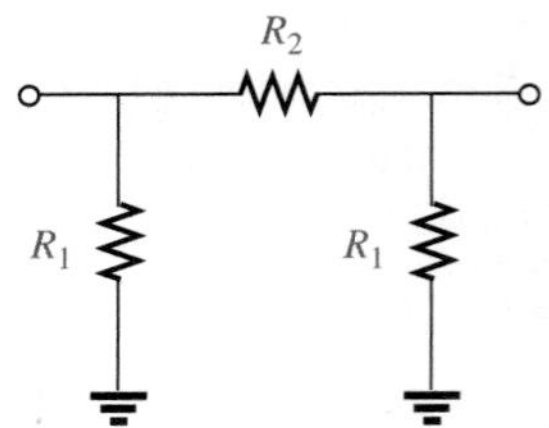

Attenuation	R_1	R_2
1 dB	870.0 Ω	5.8 Ω
2 dB	436.0 Ω	11.6 Ω
3 dB	292.0 Ω	17.6 Ω
5 dB	178.6 Ω	30.4 Ω
10 dB	96.2 Ω	71.2 Ω
20 dB	61.0 Ω	247.5 Ω

FIG. 21.86
Wye (Y) configuration.

For the 1 dB attenuation, the resistor values were inserted for the T configuration in Fig. 21.87(a). Terminating the configuration with a 50 Ω load, we find through the following calculations that the input impedance is, in fact, 50 Ω:

$$\begin{aligned} R_i &= R_1 + R_2 \,\|\, (R_1 + R_L) = 2.9\ \Omega + 433.3\ \Omega \,\|\, (2.9\ \Omega + 50\ \Omega) \\ &= 2.9\ \Omega + 47.14\ \Omega \\ &= \mathbf{50.04\ \Omega} \end{aligned}$$

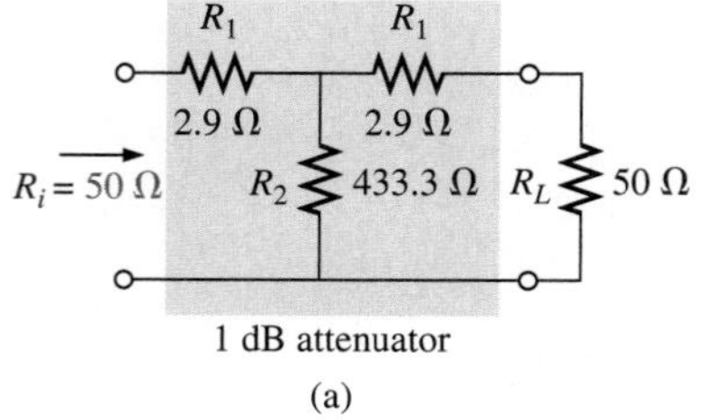

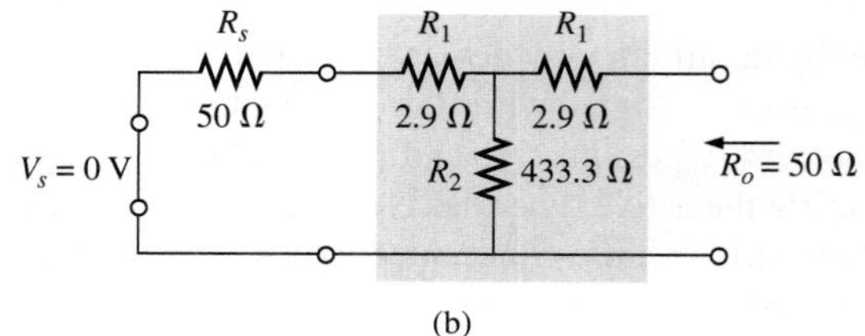

FIG. 21.87
1 dB attenuator: (a) loaded; (b) finding R_o.

Looking back from the load as shown in Fig. 21.87(b) with the source set to zero volts, we find through the following calculations that the output impedance is also 50 Ω:

$$R_o = R_1 + R_2 \parallel (R_1 + R_s) = 2.9\ \Omega + 433.3\ \Omega \parallel (2.9\ \Omega + 50\ \Omega)$$
$$= 2.9\ \Omega + 47.14\ \Omega$$
$$= \mathbf{50.04\ \Omega}$$

In Fig. 21.88, a 50 Ω load has been applied, and the output voltage is determined as follows:

$$R' = R_2 \parallel (R_1 + R_L) = 47.14\ \Omega \qquad \text{from above}$$

and
$$V_{R_2} = \frac{R'V_s}{R' + R_1} = \frac{47.14\ \Omega\ V_s}{47.14\ \Omega + 2.9\ \Omega} = 0.942V_s$$

with
$$V_L = \frac{R_L V_{R_2}}{R_L + R_1} = \frac{50\ \Omega\ (0.942V_s)}{50\ \Omega + 2.9\ \Omega} = 0.890V_s$$

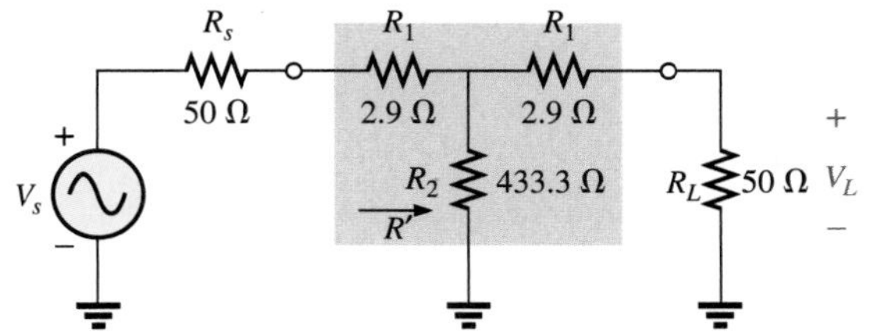

FIG. 21.88
Determining the voltage levels for the 1 dB attenuator in Fig. 21.87(a).

Calculating the drop in dB results in the following:

$$A_{v_{dB}} = 20 \log_{10} \frac{V_L}{V_s} = 20 \log_{10} \frac{0.890\ V_s}{V_s}$$
$$= 20 \log_{10} 0.890 = \mathbf{-1.01\ dB}$$

substantiating the fact that there is a 1 dB attenuation.

As mentioned earlier, there are other methods for attenuation that are more sophisticated in design and beyond the scope of the coverage of this text. However, the above designs are quite effective, relatively inexpensive, and perform quite well.

Noise Filters

Noise is a problem that can occur in any electronic system. In general, the presence of any unwanted signal can affect the overall operation of a system. It can come from a power source (60 Hz hum), from feedback networks, from mechanical systems connected to electrical systems, from stray capacitive and inductive effects, or possibly from a local signal source that is not properly shielded—the list is endless. To solve a noise problem, an analyst needs a broad practical background, a sense for the origin for the unwanted noise, and the ability to remove it in the simplest, most direct way. Generally noise problems arise during the testing phase, not during the original design phase. Although sophisticated methods may be needed, most situations are resolved simply by rearranging an element or two of a value sensitive to the problem.

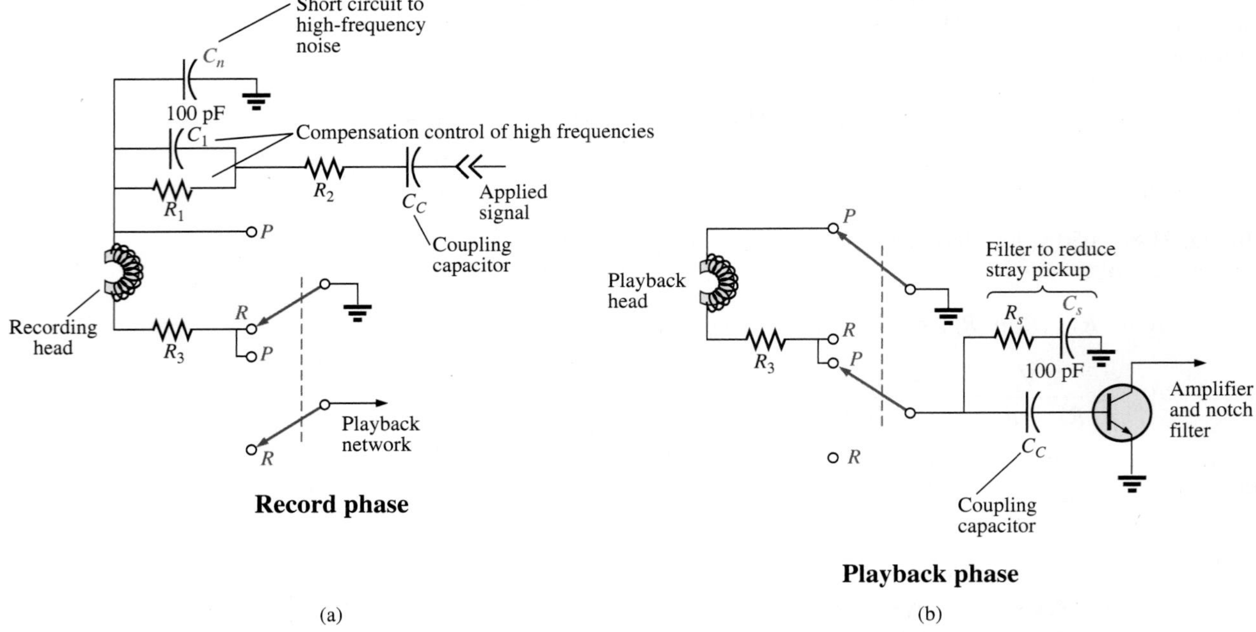

FIG. 21.89
Noise reduction in a tape recorder.

In Fig. 21.89, two capacitors have been strategically placed in the tape recording and playback sections of a tape recorder to remove the undesirable high-frequency noise (rushing sound) that can result from unexpected, randomly placed particles on a magnetic tape, noise coming down the line, or noise introduced from the local environment. During the record mode, with the switches in the positions shown (R), the 100 pF capacitor at the top of the schematic acts as a short circuit to the high-frequency noise. The capacitor C_1 is included to compensate for the fact that recording on a tape is not a linear process versus frequency. In other words, certain frequencies are recorded at higher amplitudes than others.

In Fig. 21.90 a sketch of recording level versus frequency has been provided, clearly indicating that the human audio range of about 40 Hz to 20 kHz is very poor for the tape recording process, starting to rise only after 20 kHz. Thus, tape recorders must include a fixed biasing frequency which when added to the actual audio signal brings the frequency range to be amplified to the region of high-amplitude recording. On some tapes,

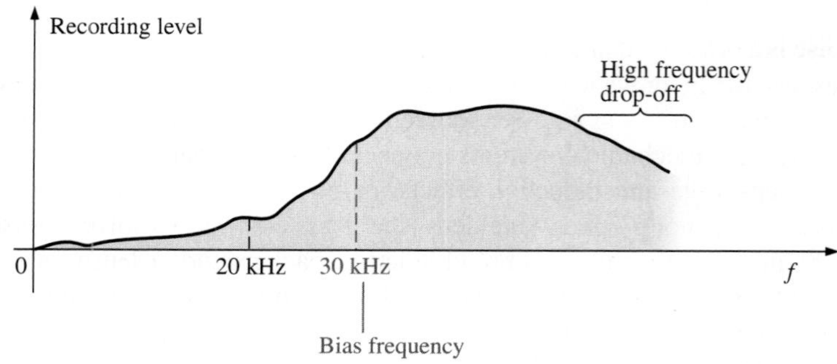

FIG. 21.90
Noise reduction in a tape recorder.

the *actual bias frequency* is provided, while on others the phrase *normal bias* is used. Even after you pass the bias frequency, there is a frequency range that follows that drops off considerably. Compensation for this drop-off is provided by the parallel combination of the resistor R_1 and the capacitor C_1 mentioned above. At frequencies near the bias frequency, the capacitor is designed to act essentially like an open circuit (high reactance), and the head current and voltage are limited by the resistors R_1 and R_2. At frequencies in the region where the tape gain drops off with frequency, the capacitor begins to take on a lower reactance level and reduce the net impedance across the parallel branch of R_1 and C_1. The result is an increase in head current and voltage due to the lower net impedance in the line, resulting in a leveling in the tape gain following the bias frequency. Eventually, the capacitor begins to take on the characteristics of a short circuit, effectively shorting out the resistance R_1, and the head current and voltage will be a maximum. During playback, this bias frequency is eliminated by a notch filter so that the original sound is not distorted by the high-frequency signal.

During playback (P), the upper circuit in Fig. 21.89 is set to ground by the upper switch, and the lower network comes into play. Again note the second 100 pF capacitor connected to the base of the transistor to short to ground any undesirable high-frequency noise. The resistor is there to absorb any power associated with the noise signal when the capacitor takes on its short-circuit equivalence. Keep in mind that the capacitor was chosen to act as a short-circuit equivalent for a particular frequency range and not for the audio range where it is essentially an open circuit.

Alternators in a car are notorious for developing high-frequency noise down the line to the radio, as shown in Fig. 21.91(a). This problem is usually alleviated by placing a high-frequency filter in the line as shown. The inductor of 1 H offers a high impedance for the range of noise frequencies, while the capacitor (1000 μF to 47,000 μF) acts as a short-circuit equivalent to any noise that happens to get through. For the speaker system in Fig. 21.91(b), the push-pull power arrangement of transistors in the output section can often develop a short period of time between pulses where

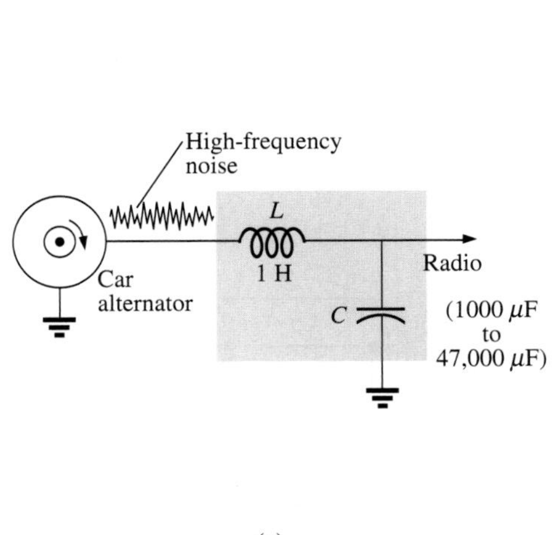

(a)

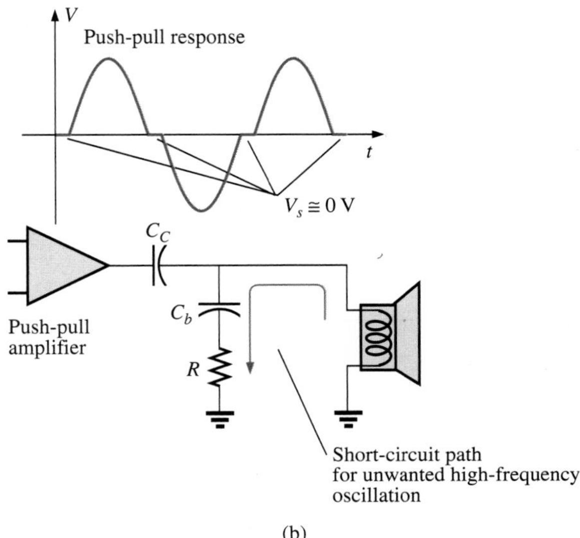

(b)

FIG. 21.91
Noise generation: (a) due to a car alternator; (b) from a push-pull amplifier.

the strong signal voltage is zero volts. During this short period, the coil of the speaker rears its inductive effects, sees an unexpected path to ground like a switch opening, and quickly cuts off the speaker current. Through the familiar relationship $v_L = L(di_L/dt)$, an unexpected voltage develops across the coil and sets a high-frequency oscillation on the line that finds its way back to the amplifier and causes further distortion. This effect can be subdued by placing an *R-C* path to ground that offers a low-resistance path from the speaker to ground for a range of frequencies typically generated by this signal distortion. Since the capacitor assumes a short-circuit equivalence for the range of noise disturbance, the resistor was added to limit the current and absorb the energy associated with the signal noise.

In regulators, such as the 5 V regulator in Fig. 21.92(a), when a spike in current comes down the line for any number of reasons, there is a voltage drop along the line, and the input voltage to the regulator drops. The regulator, performing its primary function, senses this drop in input voltage and increases its amplification level through a feedback loop to maintain a constant output. However, the spike is of such short duration that the output voltage has a spike of its own because the input voltage has quickly returned to its normal level, and with the increased amplification the output jumps to a higher level. Then the regulator senses its error and quickly cuts its gain. The sensitivity to changes in the input level has caused the output level to go through a number of quick oscillations that can be a real problem for the equipment to which the dc voltage is applied: A high-frequency noise signal has been developed. One way to subdue this reaction and, in fact, slow the system response down so that very short interval spikes have less impact is to add a capacitor across the output as shown in Fig. 21.92(b). Since the regulator is providing a fixed dc level, a large capacitor of 1 μF can be used to short-circuit a wide range of high-frequency disturbances. However, you don't want to make the capacitor too large or you'll get too much *damping,* and large overshoots and undershoots can develop. To maximize the input of the added capacitor, you must place it physically closer to the regulator to ensure that noise is not picked up between the regulator and capacitor and to avoid developing any delay time between output signal and capacitive reaction.

In general, as you examine the schematic of working systems and see elements that don't appear to be part of any standard design procedure, you can assume that they are either protective devices or due to noise on the line that is affecting the operation of the system. Noting their type, value, and location often reveals their purpose and modus operandi.

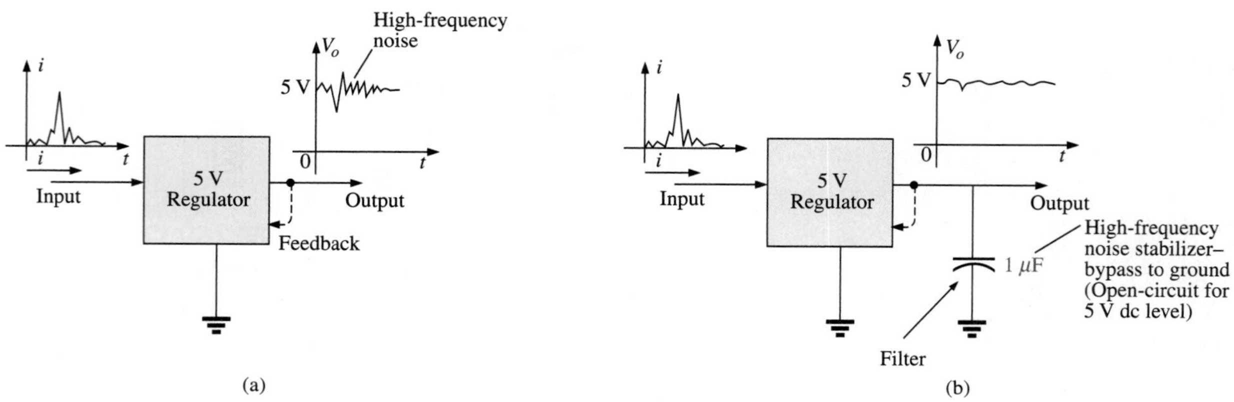

FIG. 21.92
Regulator: (a) effect of spike in current on the input side; (b) noise reduction.

21.17 COMPUTER ANALYSIS

PSpice

Double-Tuned Filter Our analysis now turns to a fairly complex-looking filter for which an enormous amount of time would be required to generate a detailed plot of gain versus frequency using a handheld calculator. It is the same filter examined in Example 21.9, so we have a chance to test our theoretical solution. The schematic appears in Fig. 21.93 with **VAC** again chosen since the frequency range of interest is set by the **Simulation Profile.** Again, the attributes for the source are set in the **Property Editor** box rather than by selecting the components from the screen. Note the need for the two resistors in series with the inductors since inductors cannot be considered ideal elements. The small value of the resistive elements, however, has no effect on the results obtained.

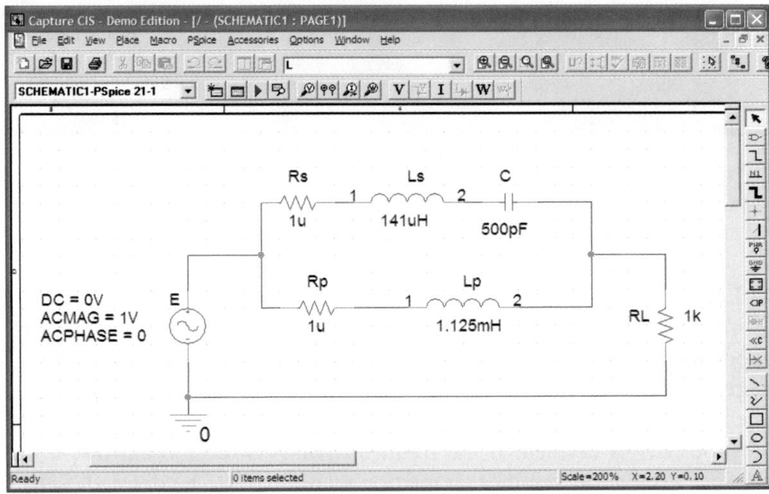

FIG. 21.93
Using PSpice to analyze a double-tuned filter.

In the **Simulation Settings** dialog box, select **AC Sweep** again with a **Start Frequency** of 100 Hz and an **End Frequency** of 10 MHz (be sure to enter this value as **10MEGHZ**) to ensure that the full-range effect is provided. Then use the axis controls to close in on the desired plot. The **Points/Decade** remains at 10k, although with this range of frequencies it may take a few seconds to simulate. Once the **SCHEMATIC1** appears, **Trace-Add Trace-V(RL:1)-OK** results in the plot in Fig. 21.94. Quite obviously, there is a reject-band around 200 kHz and a pass-band around 600 kHz. Note that up to 10 kHz, there is another pass-band as the inductor L_p provides an almost direct path of low impedance from input to output. At frequencies approaching 10 MHz, there is a continous stop-band due to the open-circuit equivalence of the L_p inductor. Using the cursor option, place the left-click cursor on the minimum point of the graph by using the **Cursor Trough** keypad (the second pad to the right of the **Toggle cursor** pad). Right-click to identify the frequency of the maximum point on the curve near 600 kHz. The results appearing in the **Probe Cursor** box clearly support our theoretical calculations of 200 kHz for the band-stop minimum (**A1** = 200.03 kHz with a magnitude of essentially 0 V) and 600 kHz for the pass-band maximum (**A2** = 604.51 kHz with a magnitude of 1 V).

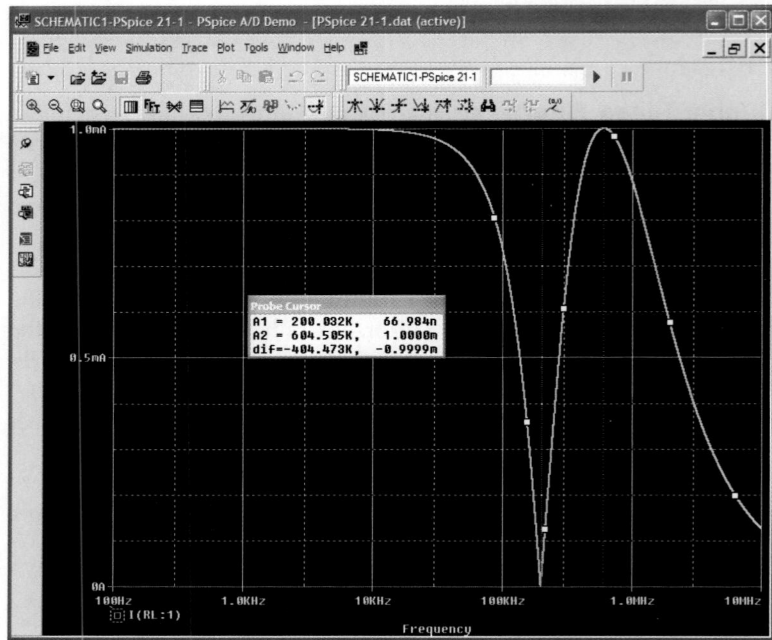

FIG. 21.94
Magnitude plot versus frequency for the voltage across R_L of the network in Fig. 21.93.

Let us now concentrate on the range from 10 kHz to 1 MHz where most of the filtering action is taking place. That was the advantage of choosing such a wide range of frequencies when the **Simulation Settings** were set up. The data have been established for the broad range of frequencies, and you can simply select a band of interest once the region of most activity is defined. If the frequency range were too narrow in the original simulation, another simulation would have to be defined. Select **Plot-Axis Settings-X Axis-User Defined-10kHz to 1MEGHz-OK** to obtain the plot at the bottom of Fig. 21.95. A dB plot of the results can

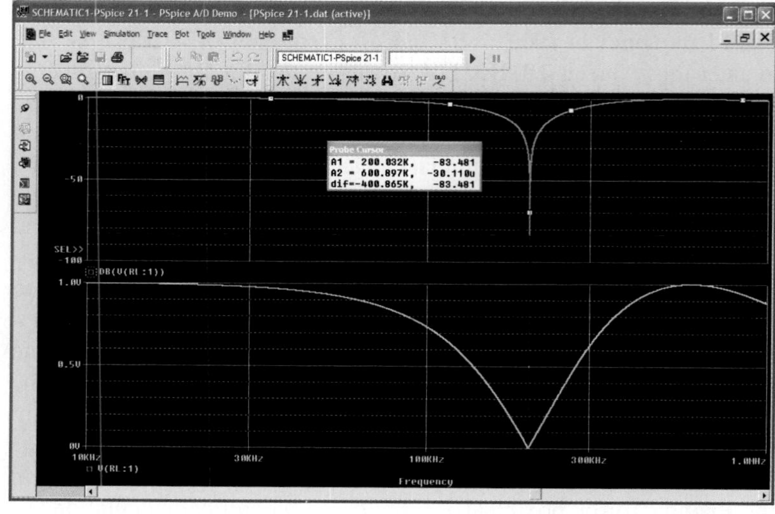

FIG. 21.95
dB and magnitude plot for the voltage across R_L of the network in Fig. 21.93.

also be displayed in the same figure by selecting **Plot-Add Plot to Window-Trace-Add Trace-DB(V(RL:1))-OK,** resulting in the plot at the top of the figure. Using the left-click cursor option and the **Cursor Trough** key, you find that the minimum is at -83.48 dB at a frequency of 200 kHz, which is an excellent characteristic for a band-stop filter. Using the right-click cursor and setting it on 600 kHz, you find that the drop is -30.11 μdB or essentially 0 dB, which is excellent for the pass-band region.

Multisim

High-Pass Filter The Multisim computer analysis is an investigation of the high-pass filter in Fig. 21.96. The cutoff frequency is determined by $f = 1/2\pi RC = 1.592$ kHz, with the voltage across the resistor approaching 1 V at high frequencies at a phase angle of 0°.

For this analysis, the **Component: AC_POWER** under **POWER_SOURCES** was chosen. The components of the source were set in the **POWER_SOURCES** dialog box as shown in Fig. 21.96. Use the sequence **Simulate-Analyses-AC Analysis** to obtain the **AC Analysis** dialog box. Select the following settings: **Start frequency:** 10 Hz, **Stop frequency:** 100 kHz, **Sweep type:** Decade (logarithmic), **Number of points per decade:** 1000, **Vertical scale:** Linear. Under the **Output** option, move \$2 to **Selected variables for analysis** using the **Add** option and remove the \$1 option using the **Remove** option. Select **Simulate** to obtain the response in Fig. 21.96. Add the grid option to each, and then select **Show/Hide Cursors** to permit a determination of the magnitude and phase angle at the cutoff frequency. As shown in Fig. 21.96, the magnitude is 0.707 at 1.589 kHz and the phase angle is 45.14 at 1.58 kHz—very close to the expected results.

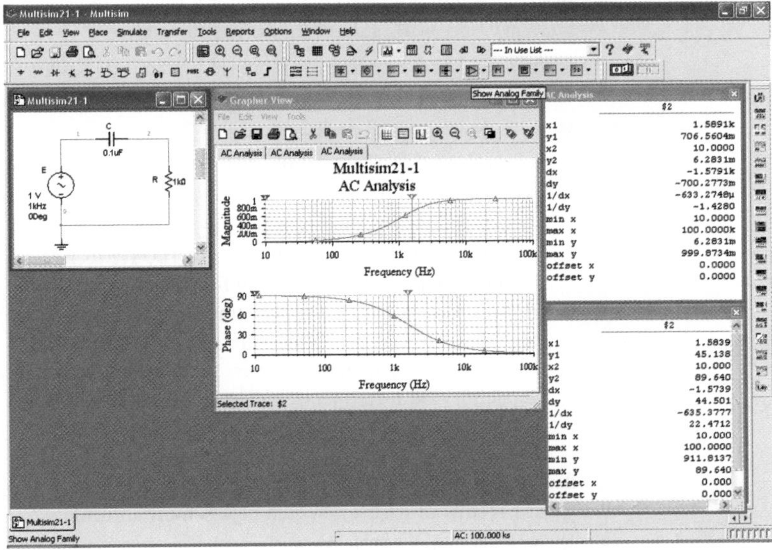

FIG. 21.96

High-pass R-C filter to be investigated using PSpice.

PROBLEMS

SECTION 21.1 Logarithms

1. a. Determine the frequencies (in kHz) at the points indicated on the plot in Fig. 21.97(a).
   b. Determine the voltages (in mV) at the points indicated on the plot in Fig. 21.97(b).

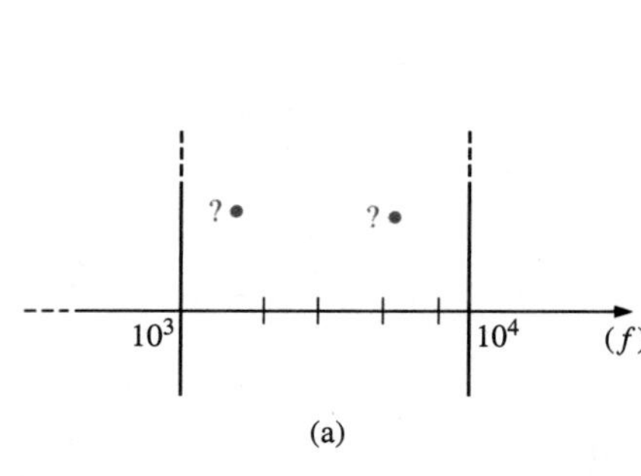

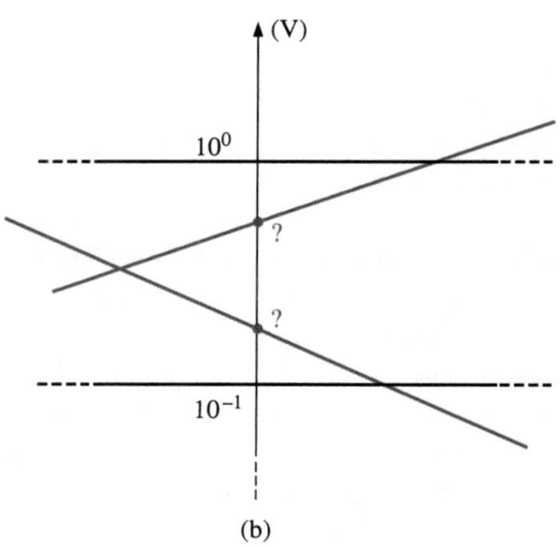

FIG. 21.97
Problem 1.

SECTION 21.2 Properties of Logarithms

2. Determine $\log_{10}x$ for each value of x.
   a. 100,000 b. 0.0001
   c. 10^8 d. 10^{-6}
   e. 20 f. 8643.4
   g. 56,000 h. 0.318
3. Given $N = \log_{10}x$, determine x for each value of N.
   a. 3 b. 12
   c. 0.2 d. 0.04
   e. 10 f. 3.18
   g. 1.001 h. 6.1
4. Determine $\log_e x$ for each value of x.
   a. 100,000 b. 0.0001
   c. 20 d. 8643.4

   Compare with the solutions to Problem 2.
5. Determine $\log_{10} 48 = \log_{10}(8)(6)$, and compare to $\log_{10} 8 + \log_{10} 6$.
6. Determine $\log_{10} 0.2 = \log_{10} 18/90$, and compare to $\log_{10} 18 - \log_{10} 90$.
7. Verify that $\log_{10} 0.5$ is equal to $-\log_{10} 1/0.5 = -\log_{10} 2$.
8. Find $\log_{10}(3)^3$, and compare with $3 \log_{10} 3$.

SECTION 21.3 Decibels

9. a. Determine the number of bels that relate power levels of $P_2 = 280$ mW and $P_1 = 4$ mW.
   b. Determine the number of decibels for the power levels of part (a), and compare results.
10. A power level of 100 W is 6 dB above what power level?
11. If a 2 W speaker is replaced by one with a 40 W output, what is the increase in decibel level?
12. Determine the dB_m level for an output power of 120 mW.
13. Find the dB_v gain of an amplifier that raises the voltage level from 0.1 mV to 8.4 mV.
14. Find the output voltage of an amplifier if the applied voltage is 20 mV and a dB_v gain of 22 dB is attained.
15. If the sound pressure level is increased from 0.001 μbar to 0.016 μbar, what is the increase in dB_s level?
16. What is the required increase in acoustical power to raise a sound level from that of quiet music to very loud music? Use Fig. 21.5.
17. a. Using semilog paper, plot X_L versus frequency for a 10 mH coil and a frequency range of 100 Hz to 1 MHz. Choose the best vertical scaling for the range of X_L.
    b. Repeat part (a) using log-log graph paper. Compare to the results of part (a). Which plot is more informative?
    c. Using semilog paper, plot X_C versus frequency for a 1 μF capacitor and a frequency range of 10 Hz to 100 kHz. Again choose the best vertical scaling for the range of X_C.
    d. Repeat part (a) using log-log graph paper. Compare to the results of part (c). Which plot is more informative?
18. a. For the meter of Fig. 21.6, find the power delivered to a load for an 8 dB reading.
    b. Repeat part (a) for a -5 dB reading.

SECTION 21.5 *R-C* Low-Pass Filter

19. For the *R-C* low-pass filter in Fig. 21.98:

a. Sketch $A_v = V_o/V_i$ versus frequency using a log scale for the frequency axis. Determine $A_v = V_o/V_i$ at $0.1f_c$, $0.5f_c$, f_c, $2f_c$, and $10f_c$.

b. Sketch the phase plot of θ versus frequency, where θ is the angle by which $\mathbf{V}_o$ leads $\mathbf{V}_i$. Determine θ at $f = 0.1f_c$, $0.5f_c$, f_c, $2f_c$, and $10f_c$.

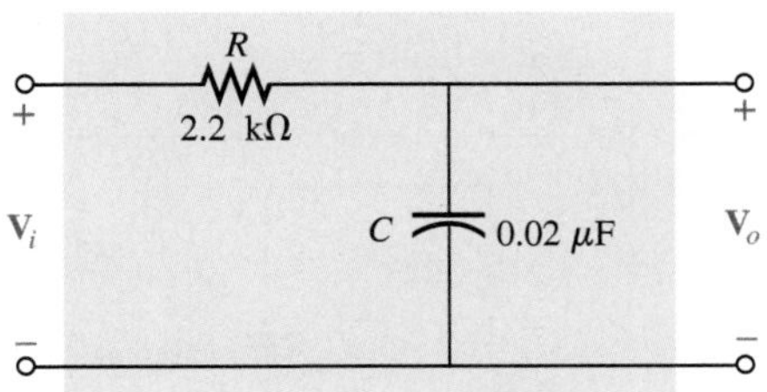

FIG. 21.98
Problem 19.

***20.** For the network in Fig. 21.99:

a. Determine V_o at a frequency one octave above the critical frequency.

b. Determine V_o at a frequency one decade below the critical frequency.

c. Do the levels of parts (a) and (b) verify the expected frequency plot of V_o versus frequency for the filter?

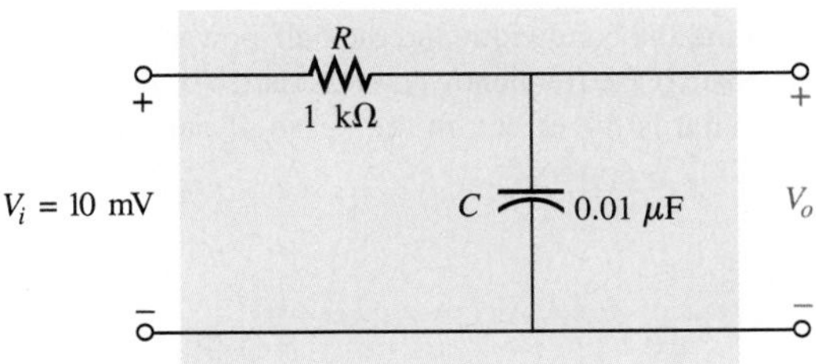

FIG. 21.99
Problem 20.

21. Design an *R-C* low-pass filter to have a cutoff frequency of 500 Hz using a resistor of 1.2 kΩ. Then sketch the resulting magnitude and phase plot for a frequency range of $0.1f_c$ to $10f_c$.

22. For the low-pass filter in Fig. 21.100:

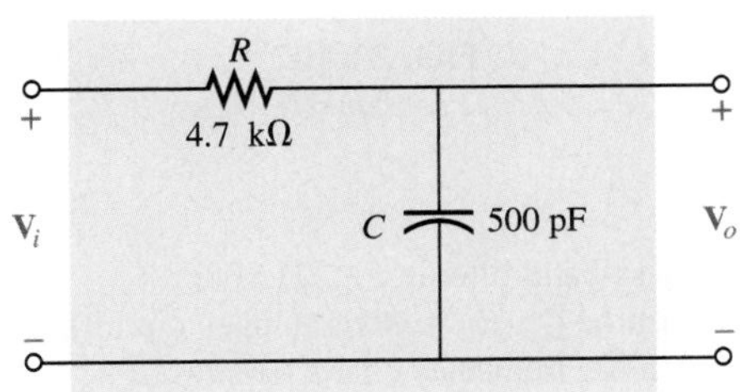

FIG. 21.100
Problem 22.

a. Determine f_c.

b. Find $A_v = V_o/V_i$ at $f = 0.1f_c$, and compare to the maximum value of 1 for the low-frequency range.

c. Find $A_v = V_o/V_i$ at $f = 10f_c$, and compare to the minimum value of 0 for the high-frequency range.

d. Determine the frequency at which $A_v = 0.01$ or $V_o = \frac{1}{100}V_i$.

SECTION 21.6 *R-C* High-Pass Filter

23. For the *R-C* high-pass filter in Fig. 21.101:

a. Sketch $A_v = V_o/V_i$ versus frequency using a log scale for the frequency axis. Determine $A_v = V_o/V_i$ at f_c, one octave above and below f_c, and one decade above and below f_c.

b. Sketch the phase plot of θ versus frequency, where θ is the angle by which $\mathbf{V}_o$ leads $\mathbf{V}_i$. Determine θ at the same frequencies noted in part (a).

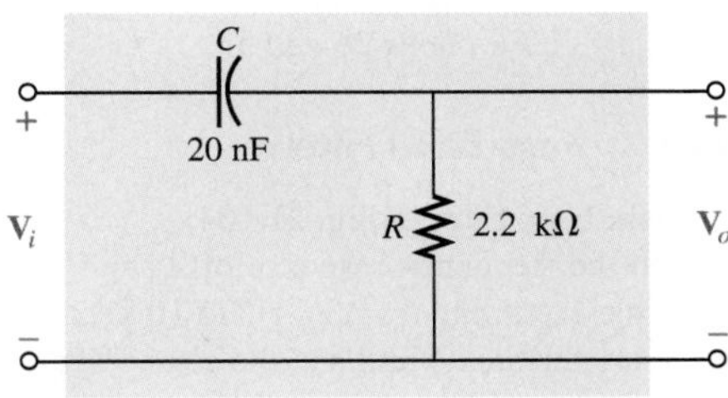

FIG. 21.101
Problem 23.

24. For the network in Fig. 21.102:

a. Determine $A_v = V_o/V_i$ at $f = f_c$ for the high-pass filter.

b. Determine $A_v = V_o/V_i$ at two octaves above f_c. Is the rise in V_o significant from the $f = f_c$ level?

c. Determine $A_v = V_o/V_i$ at two decades above f_c. Is the rise in V_o significant from the $f = f_c$ level?

d. If $V_i = 10$ mV, what is the power delivered to R at the critical frequency?

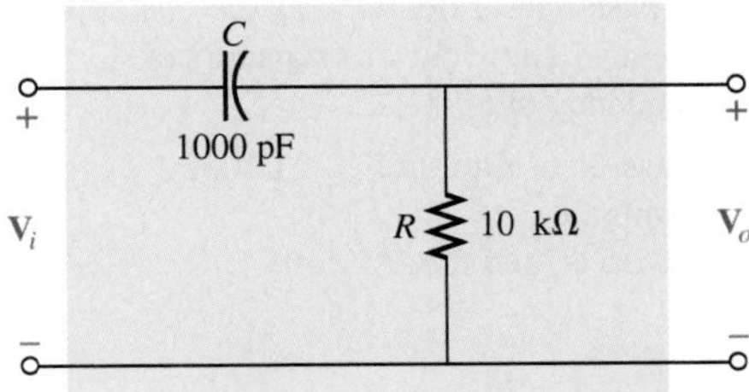

FIG. 21.102
Problem 24.

25. Design a high-pass *R-C* filter to have a cutoff or corner frequency of 2 kHz, given a capacitor of 0.1 μF. Choose the closest commercial value for *R*, and then recalculate the resulting corner frequency. Sketch the normalized gain $A_v = V_o/V_i$ for a frequency range of $0.1f_c$ to $10f_c$.

26. For the high-pass filter in Fig. 21.103:
 a. Determine f_c.
 b. Find $A_v = V_o/V_i$ at $f = 0.01f_c$, and compare to the minimum level of 0 for the low-frequency region.
 c. Find $A_v = V_o/V_i$ at $f = 100f_c$, and compare to the maximum level of 1 for the high-frequency region.
 d. Determine the frequency at which $V_o = \frac{1}{2}V_i$.

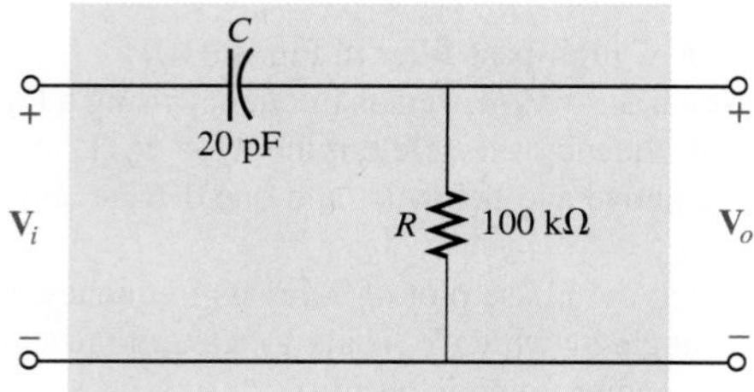

FIG. 21.103
Problems 26 and 54.

SECTION 21.7 Pass-Band Filters

27. For the pass-band filter in Fig. 21.104:
 a. Sketch the frequency response of $A_v = V_o/V_i$ against a log scale extending from 10 Hz to 10 kHz.
 b. What are the bandwidth and the center frequency?

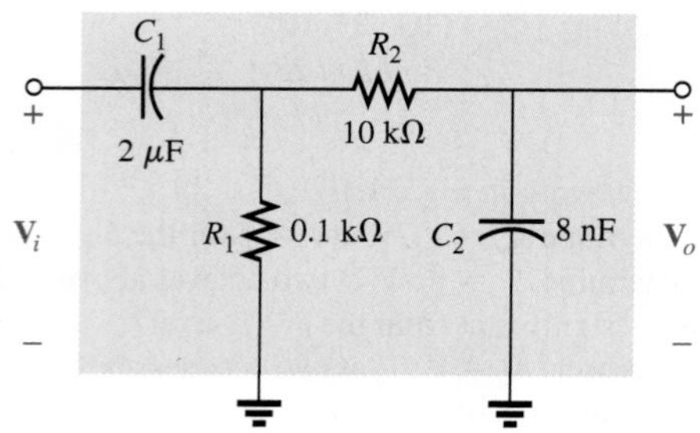

FIG. 21.104
Problems 27 and 28.

***28.** Design a pass-band filter such as the one appearing in Fig. 21.104 to have a low cutoff frequency of 4 kHz and a high cutoff frequency of 80 kHz.

29. For the pass-band filter in Fig. 21.105:
 a. Determine f_s.
 b. Calculate Q_s and the BW for $\mathbf{V}_o$.

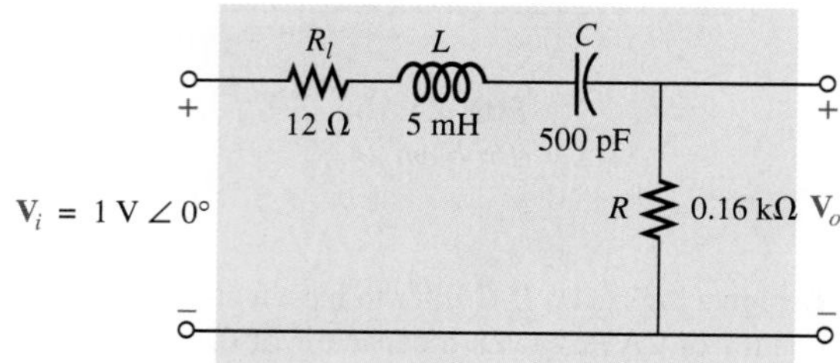

FIG. 21.105
Problem 29.

 c. Sketch $A_v = V_o/V_i$ for a frequency range of 1 kHz to 1 MHz.
 d. Find the magnitude of V_o at $f = f_s$ and the cutoff frequencies.

30. For the pass-band filter in Fig. 21.106:
 a. Determine the frequency response of $A_v = V_o/V_i$ for a frequency range of 100 Hz to 1 MHz.
 b. Find the quality factor Q_p and the BW of the response.

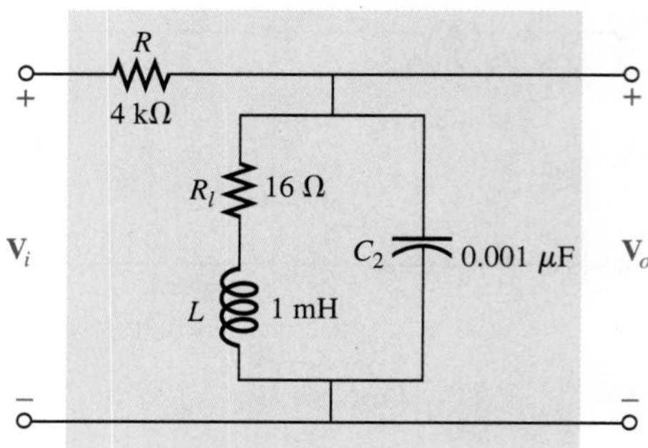

FIG. 21.106
Problems 30 and 55.

SECTION 21.8 Stop-Band Filters

***31.** For the stop-band filter in Fig. 21.107:
 a. Determine Q_s.
 b. Find the bandwidth and the half-power frequencies.
 c. Sketch the frequency characteristics of $A_v = V_o/V_i$.
 d. What is the effect on the curve of part (c) if a load of 2 kΩ is applied?

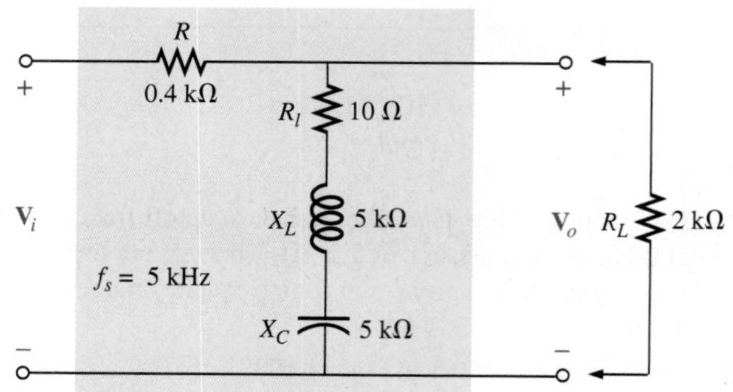

FIG. 21.107
Problem 31.

***32.** For the pass-band filter in Fig. 21.108:
 a. Determine Q_p ($R_L = \infty\ \Omega$, an open circuit).
 b. Sketch the frequency characteristics of $A_v = V_o/V_i$.
 c. Find Q_p (loaded) for $R_L = 100\ \text{k}\Omega$, and indicate the effect of R_L on the characteristics of part (b).
 d. Repeat part (c) for $R_L = 20\ \text{k}\Omega$.

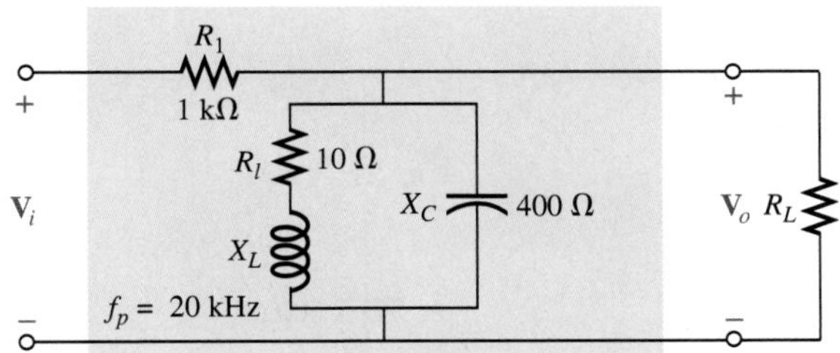

FIG. 21.108
Problem 32.

SECTION 21.9 Double-Tuned Filter

33. a. For the network in Fig. 21.43(a), if $L_p = 400\ \mu H$ ($Q > 10$), $L_s = 60\ \mu H$, and $C = 120$ pF, determine the rejected and accepted frequencies.
   b. Sketch the response curve for part (a).

34. a. For the network in Fig. 21.43(b), if the rejected frequency is 30 kHz and the accepted is 100 kHz, determine the values of L_s and L_p ($Q > 10$) for a capacitance of 200 pF.
   b. Sketch the response curve for part (a).

SECTION 21.10 Bode Plots

35. a. Sketch the idealized Bode plot for $A_v = V_o/V_i$ for the high-pass filter in Fig. 21.109.
   b. Using the results of part (a), sketch the actual frequency response for the same frequency range.
   c. Determine the decibel level at f_c, $\frac{1}{2}f_c$, $2f_c$, $\frac{1}{10}f_c$, and $10f_c$.
   d. Determine the gain $A_v = V_o/V_i$ as $f = f_c$, $\frac{1}{2}f_c$, and $2f_c$.
   e. Sketch the phase response for the same frequency range.

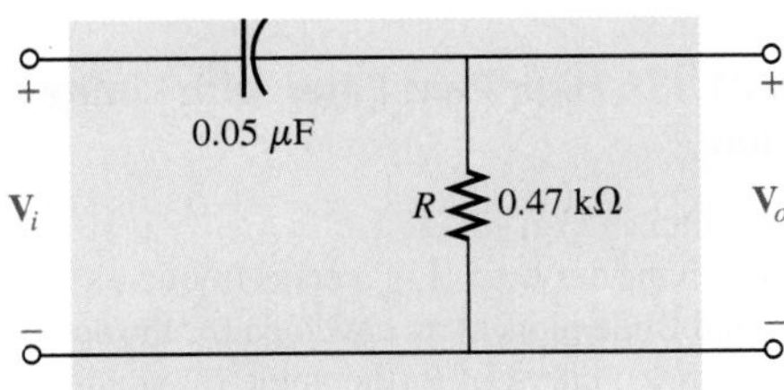

FIG. 21.109
Problem 35.

*36. a. Sketch the response of the magnitude of V_o (in terms of V_i) versus frequency for the high-pass filter in Fig. 21.110.
   b. Using the results of part (a), sketch the response $A_v = V_o/V_i$ for the same frequency range.
   c. Sketch the idealized Bode plot.
   d. Sketch the actual response, indicating the dB difference between the idealized and the actual response at $f = f_c$, $0.5f_c$, and $2f_c$.
   e. Determine $A_{v_{dB}}$ at $f = 1.5f_c$ from the plot of part (d), and then determine the corresponding magnitude of $A_v = V_o/V_i$.
   f. Sketch the phase response for the same frequency range (the angle by which $\mathbf{V}_o$ leads $\mathbf{V}_i$).

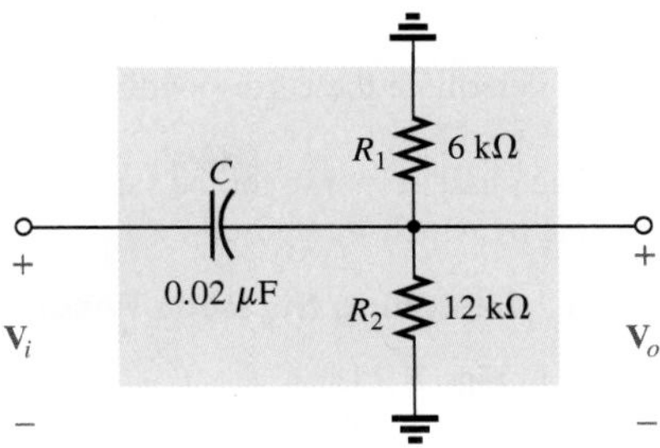

FIG. 21.110
Problem 36.

37. a. Sketch the idealized Bode plot for $A_v = V_o/V_i$ for the low-pass filter in Fig. 21.111.
   b. Using the results of part (a), sketch the actual frequency response for the same frequency range.
   c. Determine the decibel level at f_c, $\frac{1}{2}f_c$, $2f_c$, $\frac{1}{10}f_c$, and $10f_c$.
   d. Determine the gain $A_v = V_o/V_i$ at $f = f_c$, $\frac{1}{2}f_c$, and $2f_c$.
   e. Sketch the phase response for the same frequency range.

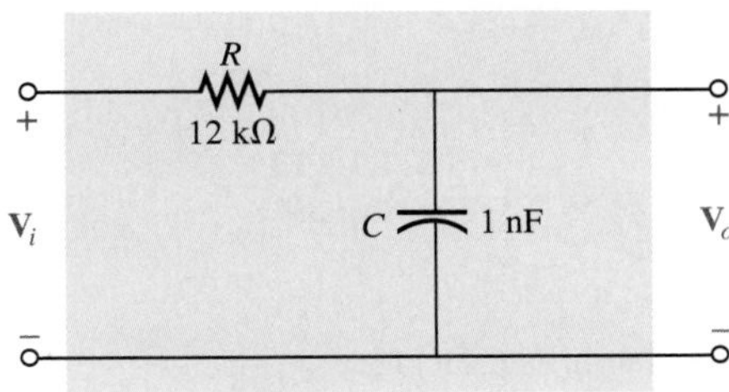

FIG. 21.111
Problem 37.

*38. a. Sketch the response of the magnitude of V_o (in terms of V_i) versus frequency for the low-pass filter in Fig. 21.112.
   b. Using the results of part (a), sketch the response $A_v = V_o/V_i$ for the same frequency range.
   c. Sketch the idealized Bode plot.
   d. Sketch the actual response indicating the dB difference between the idealized and the actual response at $f = f_c$, $0.5f_c$, and $2f_c$.

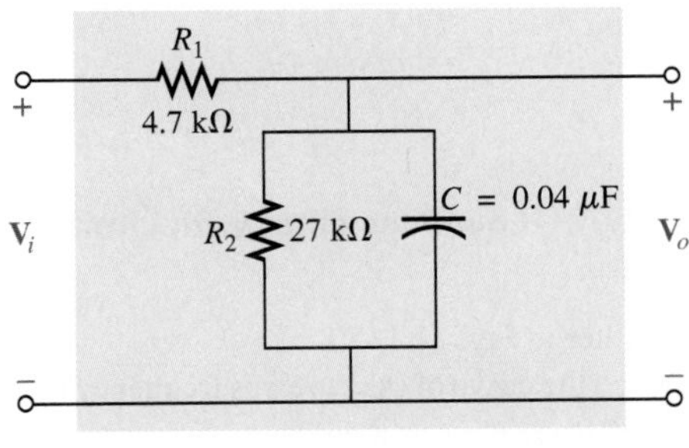

FIG. 21.112
Problem 38.

e. Determine $A_{v_{dB}}$ at $f = 0.25f_c$ from the plot of part (d), and then determine the corresponding magnitude of $A_v = V_o/V_i$.

f. Sketch the phase response for the same frequency range (the angle by which $\mathbf{V}_o$ leads $\mathbf{V}_i$).

SECTION 21.11 Sketching the Bode Response

39. For the filter in Fig. 21.113:

a. Sketch the curve of $A_{v_{dB}}$ versus frequency using a log scale.

b. Sketch the curve of θ versus frequency for the same frequency range as in part (a).

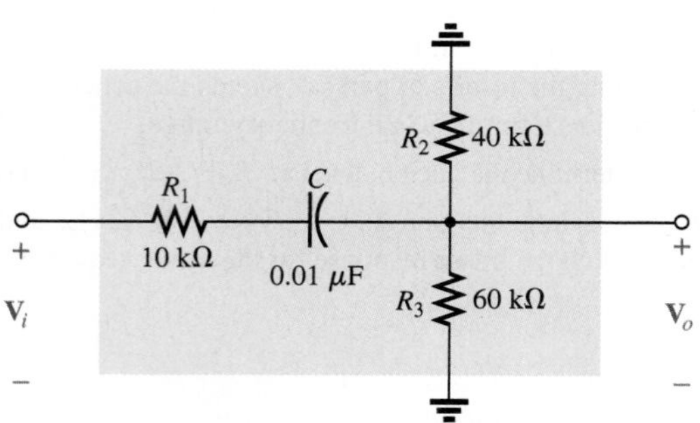

FIG. 21.113
Problem 39.

***40.** For the filter in Fig. 21.114:

a. Sketch the curve of $A_{v_{dB}}$ versus frequency using a log scale.

b. Sketch the curve of θ versus frequency for the same frequency range as in part (a).

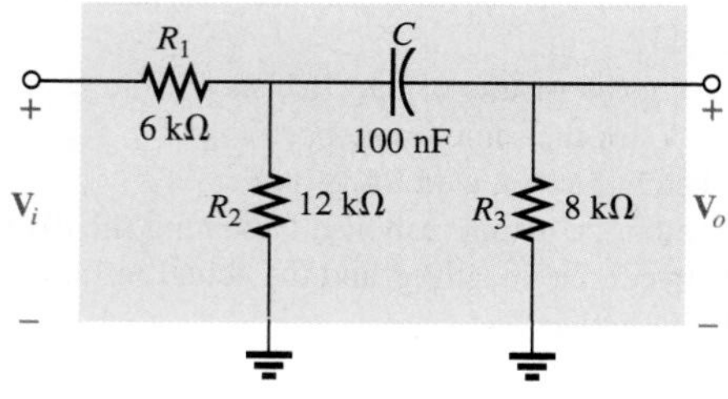

FIG. 21.114
Problem 40.

SECTION 21.12 Low-Pass Filter with Limited Attenuation

41. For the filter in Fig. 21.115:

a. Sketch the curve of $A_{v_{dB}}$ versus frequency using the idealized Bode plots as a guide.

b. Sketch the curve of θ versus frequency.

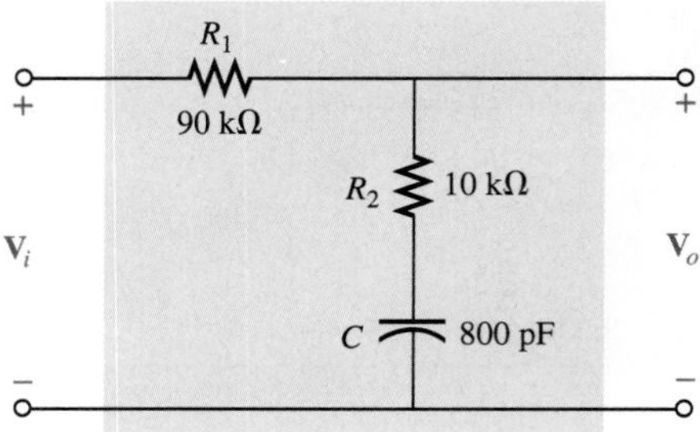

FIG. 21.115
Problem 41.

***42.** For the filter in Fig. 21.116:

a. Sketch the curve of $A_{v_{dB}}$ versus frequency using the idealized Bode plots as a guide.

b. Sketch the curve of θ versus frequency.

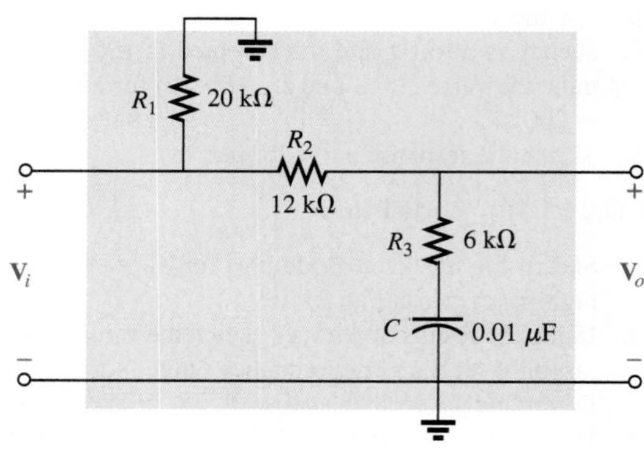

FIG. 21.116
Problem 42.

SECTION 21.13 High-Pass Filter with Limited Attenuation

43. For the filter in Fig. 21.117:

a. Sketch the curve of $A_{v_{dB}}$ versus frequency using the idealized Bode plots as an envelope for the actual response.

b. Sketch the curve of θ (the angle by which $\mathbf{V}_o$ leads $\mathbf{V}_i$) versus frequency.

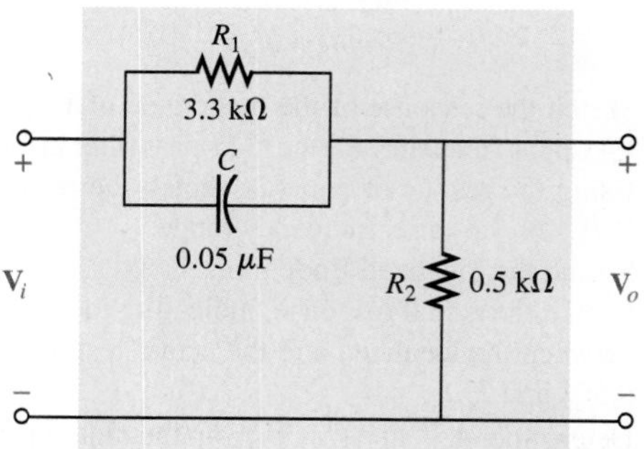

FIG. 21.117
Problem 43.

*44. For the filter in Fig. 21.118:

a. Sketch the curve of $A_{v_{dB}}$ versus frequency using the idealized Bode plots as an envelope for the actual response.

b. Sketch the curve of θ (the angle by which $\mathbf{V}_o$ leads $\mathbf{V}_i$) versus frequency.

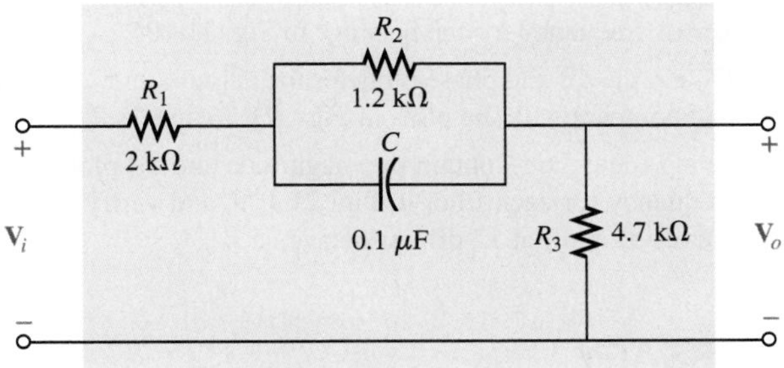

FIG. 21.118
Problem 44.

SECTION 21.14 Additional Properties of Bode Plots

45. A bipolar transistor amplifier has the following gain:

$$\mathbf{A}_v = \frac{160}{\left(1 - j\dfrac{100\text{ Hz}}{f}\right)\left(1 - j\dfrac{130\text{ Hz}}{f}\right)\left(1 + j\dfrac{f}{20\text{ kHz}}\right)\left(1 + j\dfrac{f}{50\text{ kHz}}\right)}$$

a. Sketch the normalized Bode response $A'_{v_{dB}} = (A_v/A_{v_{max}})|_{dB}$, and determine the bandwidth of the amplifier. Be sure to note the corner frequencies.

b. Sketch the phase response, and determine a frequency where the phase angle is relatively close to 45°.

46. A JFET transistor amplifier has the following gain:

$$\mathbf{A}_v = \frac{-5.6}{\left(1 - j\dfrac{10\text{ Hz}}{f}\right)\left(1 - j\dfrac{45\text{ Hz}}{f}\right)\left(1 - j\dfrac{68\text{ Hz}}{f}\right)\left(1 + j\dfrac{f}{23\text{ kHz}}\right)\left(1 + j\dfrac{f}{50\text{ kHz}}\right)}$$

a. Sketch the normalized Bode response $A'_{v_{dB}} = (A_v/A_{v_{max}}|_{dB})$, and determine the bandwidth of the amplifier. When you normalize, be sure that the maximum value of A'_v is $+1$. Clearly indicate the cutoff frequencies on the plot.

b. Sketch the phase response, and note the regions of greatest change in phase angle. How do the regions correspond to the frequencies appearing in the function $\mathbf{A}_v$?

47. A transistor amplifier has a midband gain of -120, a high cutoff frequency of 36 kHz, and a bandwidth of 35.8 kHz. In addition, the actual response is also about -15 dB at $f = 50$ Hz. Write the transfer function $\mathbf{A}_v$ for the amplifier.

48. Sketch the Bode plot of the following function:

$$\mathbf{A}_v = \frac{0.05}{0.05 - j\,100/f}$$

49. Sketch the Bode plot of the following function:

$$\mathbf{A}_v = \frac{200}{200 + j\,0.1f}$$

50. Sketch the Bode plot of the following function:

$$\mathbf{A}_v = \frac{jf/1000}{(1 + jf/1000)(1 + jf/10{,}000)}$$

***51.** Sketch the Bode plot of the following function:

$$\mathbf{A}_v = \frac{(1 + jf/1000)(1 + jf/2000)}{(1 + jf/3000)^2}$$

***52.** Sketch the Bode plot of the following function (note the presence of ω rather than f):

$$\mathbf{A}_v = \frac{40(1 + j\,0.001\omega)}{(j\,0.001\omega)(1 + j\,0.0002\omega)}$$

SECTION 21.15 Crossover Networks

***53.** The three-way crossover network in Fig. 21.119 has a 12 dB rolloff at the cutoff frequencies.

a. Determine the ratio V_o/V_i for the woofer and tweeter at the cutoff frequencies of 400 Hz and 5 kHz, respectively, and compare to the desired level of 0.707.

b. Calculate the ratio V_o/V_i for the woofer and tweeter at a frequency of 3 kHz, where the midrange speaker is designed to predominate.

c. Determine the ratio V_o/V_i for the midrange speaker at a frequency of 3 kHz, and compare to the desired level of 1.

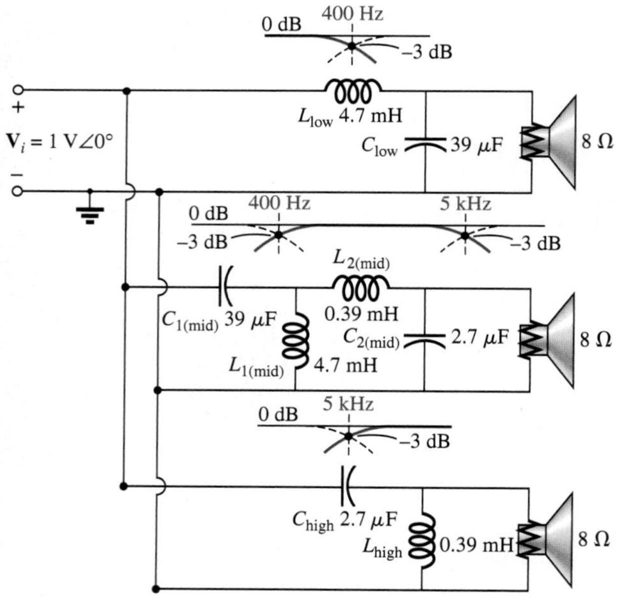

FIG. 21.119
Problems 53 and 57.

SECTION 21.17 Computer Analysis

PSpice or Multisim

54. Using schematics, obtain the magnitude and phase response versus frequency for the network in Fig. 21.103.

55. Using schematics, obtain the magnitude and phase response versus frequency for the network in Fig. 21.106.

***56.** Obtain the dB and phase plots for the network in Fig. 21.75, and compare with the plots in Figs. 21.76 and 21.77.

***57.** Using schematics, obtain the magnitude and dB plot versus frequency for each filter in Fig. 21.119, and verify that the curves drop off at 12 dB per octave.

GLOSSARY

Active filter A filter that uses active devices such as transistors or operational amplifiers in combination with *R*, *L*, and *C* elements.

Bode plot A plot of the frequency response of a system using straight-line segments called *asymptotes*.

Decibel A unit of measurement used to compare power levels.

Double-tuned filter A network having both a pass-band and a stop-band region.

Filter Networks designed to either pass or reject the transfer of signals at certain frequencies to a load.

High-pass filter A filter designed to pass high frequencies and reject low frequencies.

Log-log paper Graph paper with vertical and horizontal log scales.

Low-pass filter A filter designed to pass low frequencies and reject high frequencies.

Microbar (μbar) A unit of measurement for sound pressure levels that permits comparing audio levels on a dB scale.

Pass-band (band-pass) filter A network designed to pass signals within a particular frequency range.

Passive filter A filter constructed of series, parallel, or series-parallel *R*, *L*, and *C* elements.

Semilog paper Graph paper with one log scale and one linear scale.

Stop-band filter A network designed to reject (block) signals within a particular frequency range.

Transformers

Objectives

- *Become familiar with the flux linkages that exist between the coils of a transformer and how the voltages across the primary and secondary are established.*
- *Understand the operation of an iron-core and air-core transformer and how to calculate the currents and voltages of the primary and secondary circuits.*
- *Be aware of how the transformer is used for impedance matching purposes to ensure a high level of power transfer.*
- *Become aware of all the components that make up the equivalent circuit of a transformer and how they affect its performance and frequency response.*
- *Understand how to use and interpret the dot convention of mutually coupled coils in a network.*

22.1 INTRODUCTION

Chapter 11 discussed the *self-inductance* of a coil. We shall now examine the **mutual inductance** that exists between coils of the same or different dimensions. Mutual inductance is a phenomenon basic to the operation of the *transformer,* an electrical device used today in almost every field of electrical engineering. This device plays an integral part in power distribution systems and can be found in many electronic circuits and measuring instruments. In this chapter, we discuss three of the basic applications of a transformer: to build up or step down the voltage or current, to act as an impedance matching device, and to isolate (no physical connection) one portion of a circuit from another. In addition, we will introduce the **dot convention** and will consider the transformer equivalent circuit. The chapter concludes with a word about writing mesh equations for a network with mutual inductance.

22.2 MUTUAL INDUCTANCE

A transformer is constructed of two coils placed so that the changing flux developed by one links the other, as shown in Fig. 22.1. This results in an induced voltage across each coil. To distinguish between the coils, we will apply the transformer convention that

the coil to which the source is applied is called the primary, and the coil to which the load is applied is called the secondary.

For the primary of the transformer in Fig. 22.1, an application of Faraday's law [Eq. (11.9)] results in

$$e_p = N_p \frac{d\phi_p}{dt} \qquad \text{(volts, V)} \qquad \textbf{(22.1)}$$

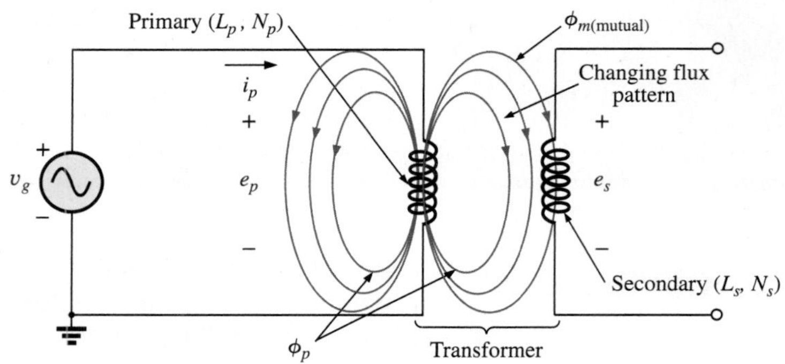

FIG. 22.1
Defining the components of a transformer.

revealing that the voltage induced across the primary is directly related to the number of turns in the primary and the rate of change of magnetic flux linking the primary coil. Or, from Eq. (11.11),

$$e_p = L_p \frac{di_p}{dt} \qquad \text{(volts, V)} \tag{22.2}$$

revealing that the induced voltage across the primary is also directly related to the self-inductance of the primary and the rate of change of current through the primary winding.

The magnitude of e_s, the voltage induced across the secondary, is determined by

$$e_s = N_s \frac{d\phi_m}{dt} \qquad \text{(volts, V)} \tag{22.3}$$

where N_s is the number of turns in the secondary winding and ϕ_m is the portion of the primary flux ϕ_p that links the secondary winding.

If all of the flux linking the primary links the secondary, then

$$\phi_m = \phi_p$$

and

$$e_s = N_s \frac{d\phi_p}{dt} \qquad \text{(volts, V)} \tag{22.4}$$

The **coefficient of coupling** (k) between two coils is determined by

$$k \text{ (coefficient of coupling)} = \frac{\phi_m}{\phi_p} \tag{22.5}$$

Since the maximum level of ϕ_m is ϕ_p, the coefficient of coupling between two coils can never be greater than 1.

The coefficient of coupling between various coils is indicated in Fig. 22.2. In Fig. 22.2(a), the ferromagnetic steel core ensures that most of the flux linking the primary also links the secondary, establishing a coupling coefficient very close to 1. In Fig. 22.2(b), the fact that both coils are overlapping results in the flux of one coil linking the other coil, with the result

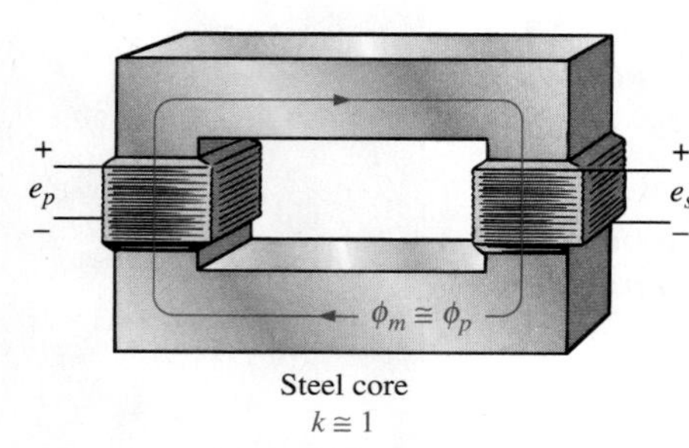

Steel core
$k \cong 1$
(a)

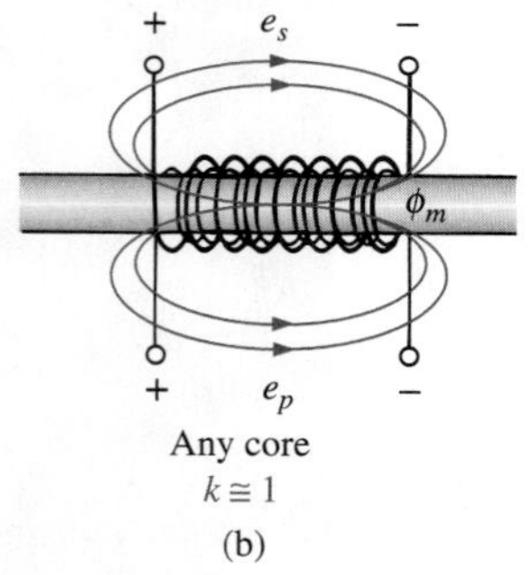

Any core
$k \cong 1$
(b)

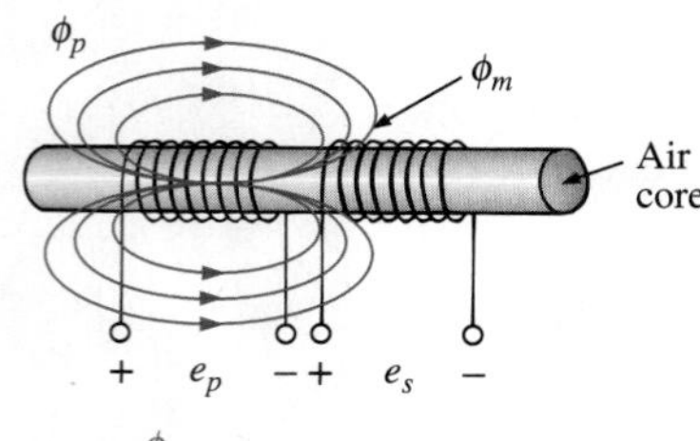

$k = \frac{\phi_m}{\phi_1} << 1\ (0.01 \rightarrow 0.3)$

(c)

FIG. 22.2
Windings having different coefficients of coupling.

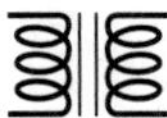

that the coefficient of coupling is again very close to 1. In Fig. 22.2(c), the absence of a ferromagnetic core results in low levels of flux linkage between the coils. The closer the two coils are, the greater the flux linkage, and the higher the value of *k,* although it will never approach a level of 1. Those coils with low coefficients of coupling are said to be **loosely coupled.**

For the secondary, we have

$$e_s = N_s \frac{d\phi_m}{dt} = N_s \frac{dk\phi_p}{dt}$$

and

$$e_s = kN_s \frac{d\phi_p}{dt} \qquad \text{(volts, V)} \qquad \textbf{(22.6)}$$

The mutual inductance between the two coils in Fig. 22.1 is determined by

$$M = N_s \frac{d\phi_m}{di_p} \qquad \text{(henries, H)} \qquad \textbf{(22.7)}$$

or

$$M = N_p \frac{d\phi_p}{di_s} \qquad \text{(henries, H)} \qquad \textbf{(22.8)}$$

Note in the above equations that the symbol for mutual inductance is the capital letter M and that its unit of measurement, like that of self-inductance, is the *henry.* In words, Eqs. (22.7) and (22.8) state that the

mutual inductance between two coils is proportional to the instantaneous change in flux linking one coil due to an instantaneous change in current through the other coil.

In terms of the inductance of each coil and the coefficient of coupling, the mutual inductance is determined by

$$M = k\sqrt{L_pL_s} \qquad \text{(henries, H)} \qquad \textbf{(22.9)}$$

The greater the coefficient of coupling (greater flux linkages), or the greater the inductance of either coil, the higher the mutual inductance between the coils. Relate this fact to the configurations in Fig. 22.2.

The secondary voltage e_s can also be found in terms of the mutual inductance if we rewrite Eq. (22.3) as

$$e_s = N_s\left(\frac{d\phi_m}{di_p}\right)\left(\frac{di_p}{dt}\right)$$

and, since $M = N_s\,(d\phi_m/di_p)$, it can also be written

$$e_s = M\frac{di_p}{dt} \qquad \text{(volts, V)} \qquad \textbf{(22.10)}$$

Similarly,

$$e_p = M\frac{di_s}{dt} \qquad \text{(volts, V)} \qquad \textbf{(22.11)}$$

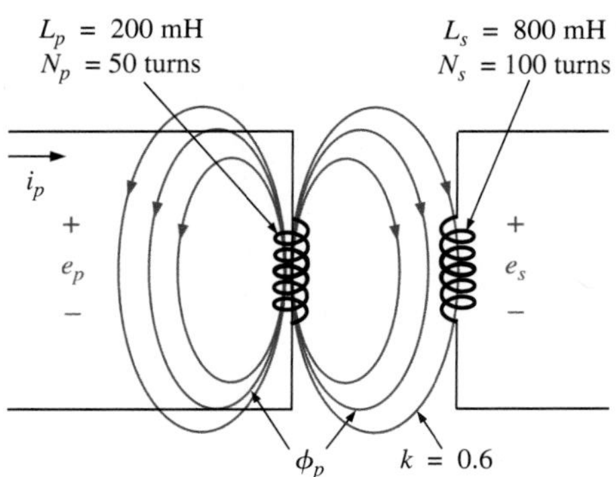

FIG. 22.3
Example 22.1

EXAMPLE 22.1 For the transformer in Fig. 22.3:

a. Find the mutual inductance M.
b. Find the induced voltage e_p if the flux ϕ_p changes at the rate of 450 mWb/s.
c. Find the induced voltage e_s for the same rate of change indicated in part (b).
d. Find the induced voltages e_p and e_s if the current i_p changes at the rate of 0.2 A/ms.

Solutions:

a. $M = k\sqrt{L_pL_s} = 0.6\sqrt{(200\text{ mH})(800\text{ mH})}$
$= 0.6\sqrt{16 \times 10^{-2}} = (0.6)(400 \times 10^{-3}) = \mathbf{240\ mH}$

b. $e_p = N_p \dfrac{d\phi_p}{dt} = (50)(450\text{ mWb/s}) = \mathbf{22.5\ V}$

c. $e_s = kN_s \dfrac{d\phi_p}{dt} = (0.6)(100)(450\text{ mWb/s}) = \mathbf{27\ V}$

d. $e_p = L_p \dfrac{di_p}{dt} = (200\text{ mH})(0.2\text{ A/ms})$
$= (200\text{ mH})(200\text{ A/s}) = \mathbf{40\ V}$

$e_s = M \dfrac{di_p}{dt} = (240\text{ mH})(200\text{ A/s}) = \mathbf{48\ V}$

22.3 THE IRON-CORE TRANSFORMER

An iron-core transformer under loaded conditions is shown in Fig. 22.4. The iron core will serve to increase the coefficient of coupling between the coils by increasing the mutual flux ϕ_m. Recall from Chapter 11 that magnetic flux lines always take the path of least reluctance, which in this case is the iron core.

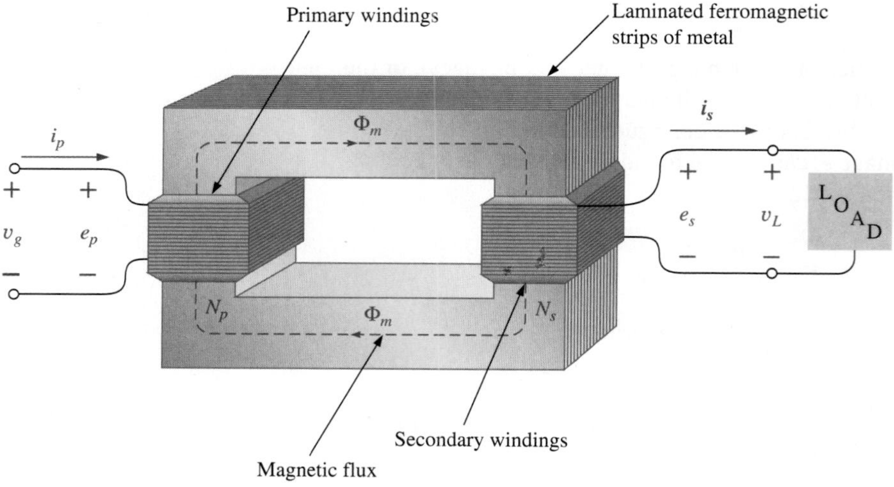

FIG. 22.4
Iron-core transformer.

In the analyses in this chapter, we assume that all of the flux linking coil 1 will link coil 2. In other words, the coefficient of coupling is its maximum value, 1, and $\phi_m = \phi_p = \phi_s$. In addition, we first analyze the transformer from an ideal viewpoint; that is, we neglect losses such as the geometric or dc resistance of the coils, the leakage reactance due to the flux linking either coil that forms no part of ϕ_m, and the hysteresis and eddy current losses. This is not to convey the impression, however, that we are far from the actual operation of a transformer. Most transformers manufactured today can be considered almost ideal. The equations we develop under ideal conditions are, in general, a first approximation to the actual response, which is never off by more than a few percentage points. The losses are considered in greater detail in Section 22.6.

When the current i_p through the primary circuit of the iron-core transformer is a maximum, the flux ϕ_m linking both coils is also a maximum. In fact, the magnitude of the flux is directly proportional to the current through the primary windings. Therefore, the two are in phase, and for sinusoidal inputs, the magnitude of the flux varies as a sinusoid also. That is, if

$$i_p = \sqrt{2}I_p \sin \omega t$$

then

$$\phi_m = \Phi_m \sin \omega t$$

The induced voltage across the primary due to a sinusoidal input can be determined by Faraday's law:

$$e_p = N_p \frac{d\phi_p}{dt} = N_p \frac{d\phi_m}{dt}$$

Substituting for ϕ_m gives us

$$e_p = N_p \frac{d}{dt}(\Phi_m \sin \omega t)$$

and differentiating, we obtain

$$e_p = \omega N_p \Phi_m \cos \omega t$$

or

$$e_p = \omega N_p \Phi_m \sin (\omega t + 90^\circ)$$

indicating that the induced voltage e_p leads the current through the primary coil by 90°.

The effective value of e_p is

$$E_p = \frac{\omega N_p \Phi_m}{\sqrt{2}} = \frac{2\pi f N_p \Phi_m}{\sqrt{2}}$$

and

$$\boxed{E_p = 4.44 f N_p \Phi_m} \qquad \textbf{(22.12)}$$

which is an equation for the rms value of the voltage across the primary coil in terms of the frequency of the input current or voltage, the number of turns of the primary, and the maximum value of the magnetic flux linking the primary.

For the case under discussion, where the flux linking the secondary equals that of the primary, if we repeat the procedure just described for the induced voltage across the secondary we get

$$\boxed{E_s = 4.44 f N_s \Phi_m} \qquad \textbf{(22.13)}$$

Dividing Eq. (22.12) by Eq. (22.13), as follows:

$$\frac{E_p}{E_s} = \frac{4.44fN_p\Phi_m}{4.44fN_s\Phi_m}$$

we obtain

$$\boxed{\frac{E_p}{E_s} = \frac{N_p}{N_s}} \tag{22.14}$$

revealing an important relationship for transformers:

The ratio of the magnitudes of the induced voltages is the same as the ratio of the corresponding turns.

If we consider that

$$e_p = N_p\frac{d\phi_m}{dt} \quad \text{and} \quad e_s = N_s\frac{d\phi_m}{dt}$$

and divide one by the other, that is,

$$\frac{e_p}{e_s} = \frac{N_p(d\phi_m/dt)}{N_s(d\phi_m/dt)}$$

then

$$\frac{e_p}{e_s} = \frac{N_p}{N_s}$$

The *instantaneous* values of e_1 and e_2 are therefore related by a constant determined by the turns ratio. Since their instantaneous magnitudes are related by a constant, the induced voltages are in phase, and Eq. (22.14) can be changed to include phasor notation; that is,

$$\boxed{\frac{\mathbf{E}_p}{\mathbf{E}_s} = \frac{N_p}{N_s}} \tag{22.15}$$

or, since $\mathbf{V}_g = \mathbf{E}_1$ and $\mathbf{V}_L = \mathbf{E}_2$ for the ideal situation,

$$\boxed{\frac{\mathbf{V}_g}{\mathbf{V}_L} = \frac{N_p}{N_s}} \tag{22.16}$$

The ratio N_p/N_s, usually represented by the lowercase letter a, is referred to as the **transformation ratio:**

$$\boxed{a = \frac{N_p}{N_s}} \tag{22.17}$$

If $a < 1$, the transformer is called a **step-up transformer** since the voltage $E_s > E_p$; that is,

$$\frac{E_p}{E_s} = \frac{N_p}{N_s} = a \quad \text{or} \quad E_s = \frac{E_p}{a}$$

and, if $a < 1$,

$$E_s > E_p$$

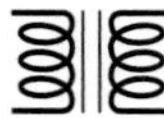

If $a > 1$, the transformer is called a **step-down transformer** since $E_s < E_p$; that is,

$$E_p = aE_s$$

and, if $a > 1$, then

$$E_p > E_s$$

EXAMPLE 22.2 For the iron-core transformer in Fig. 22.5:

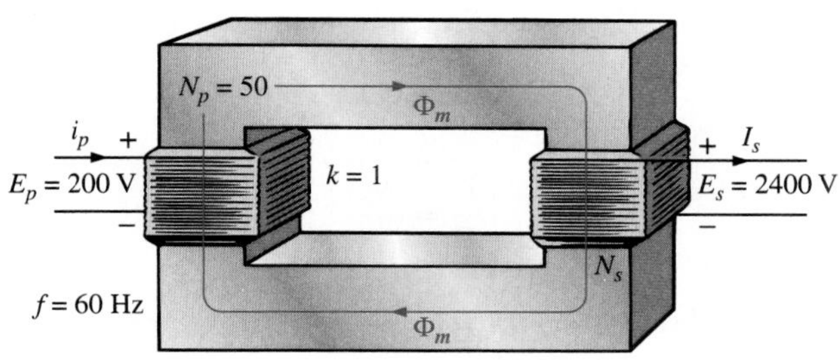

FIG. 22.5
Example 22.2.

a. Find the maximum flux Φ_m.
b. Find the secondary turns N_s.

Solutions:

a. $E_p = 4.44N_p f\Phi_m$

Therefore, $$\Phi_m = \frac{E_p}{4.44\,N_p f} = \frac{200\text{ V}}{(4.44)(50\text{ t})(60\text{ Hz})}$$

and $$\Phi_m = \mathbf{15.02\ mWb}$$

b. $$\frac{E_p}{E_s} = \frac{N_p}{N_s}$$

Therefore, $$N_s = \frac{N_p E_s}{E_p} = \frac{(50\text{ t})(2400\text{ V})}{200\text{ V}}$$

$$= \mathbf{600\ turns}$$

The induced voltage across the secondary of the transformer in Fig. 22.4 establishes a current i_s through the load Z_L and the secondary windings. This current and the turns N_s develop an mmf $N_s i_s$ that are not present under no-load conditions since $i_s = 0$ and $N_s i_s = 0$. Under loaded or unloaded conditions, however, the net ampere-turns on the core produced by both the primary and the secondary must remain unchanged for the same flux ϕ_m to be established in the core. The flux ϕ_m must remain the same to have the same induced voltage across the primary and to balance the voltage impressed across the primary. To counteract the mmf of the secondary, which is tending to change ϕ_m, an additional current must flow in the primary. This current is called the *load component of the primary current* and is represented by the notation i'_p.

For the balanced or equilibrium condition,

$$N_p i'_p = N_s i_s$$

The total current in the primary under loaded conditions is

$$i_p = i'_p + i_{\phi_m}$$

where i_{ϕ_m} is the current in the primary necessary to establish the flux ϕ_m. For most practical applications, $i'_p > i_{\phi_m}$. For our analysis, we assume $i_p \cong i'_p$, so

$$N_p i_p = N_s i_s$$

Since the instantaneous values of i_p and i_s are related by the turns ratio, the phasor quantities $\mathbf{I}_p$ and $\mathbf{I}_s$ are also related by the same ratio:

$$N_p \mathbf{I}_p = N_s \mathbf{I}_s$$

or

$$\boxed{\frac{\mathbf{I}_p}{\mathbf{I}_s} = \frac{N_s}{N_p}} \qquad \textbf{(22.18)}$$

The primary and secondary currents of a transformer are therefore related by the inverse ratios of the turns.

Keep in mind that Eq. (22.18) holds true only if we neglect the effects of i_{ϕ_m}. Otherwise, the magnitudes of $\mathbf{I}_p$ and $\mathbf{I}_s$ are not related by the turns ratio, and $\mathbf{I}_p$ and $\mathbf{I}_s$ are not in phase.

For the step-up transformer, $a < 1$, and the current in the secondary, $I_s = aI_p$, is less in magnitude than that in the primary. For a step-down transformer, the reverse is true.

22.4 REFLECTED IMPEDANCE AND POWER

In the previous section we found that

$$\frac{\mathbf{V}_g}{\mathbf{V}_L} = \frac{N_p}{N_s} = a \quad \text{and} \quad \frac{\mathbf{I}_p}{\mathbf{I}_s} = \frac{N_s}{N_p} = \frac{1}{a}$$

Dividing the first by the second, we have

$$\frac{\mathbf{V}_g/\mathbf{V}_L}{\mathbf{I}_p/\mathbf{I}_s} = \frac{a}{1/a}$$

or

$$\frac{\mathbf{V}_g/\mathbf{I}_p}{\mathbf{V}_L/\mathbf{I}_s} = a^2 \quad \text{and} \quad \frac{\mathbf{V}_g}{\mathbf{I}_p} = a^2\frac{\mathbf{V}_L}{\mathbf{I}_s}$$

However, since

$$\mathbf{Z}_p = \frac{\mathbf{V}_g}{\mathbf{I}_p} \quad \text{and} \quad \mathbf{Z}_L = \frac{\mathbf{V}_L}{\mathbf{I}_s}$$

then

$$\boxed{\mathbf{Z}_p = a^2\mathbf{Z}_L} \qquad \textbf{(22.19)}$$

That is, the impedance of the primary circuit of an ideal transformer is the transformation ratio squared times the impedance of the load. If a transformer is used, therefore, an impedance can be made to appear larger or smaller at the primary by placing it in the secondary of a step-down ($a > 1$) or step-up ($a < 1$) transformer, respectively. Note that if the load is capacitive or inductive, the **reflected impedance** is also capacitive or inductive.

For the ideal iron-core transformer,

$$\frac{E_p}{E_s} = a = \frac{I_s}{I_p}$$

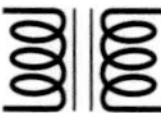

or

$$E_pI_p = E_sI_s \tag{22.20}$$

and

$$P_{in} = P_{out} \quad \text{(ideal conditions)} \tag{22.21}$$

EXAMPLE 22.3 For the iron-core transformer in Fig. 22.6:

a. Find the magnitude of the current in the primary and the impressed voltage across the primary.
b. Find the input resistance of the transformer.

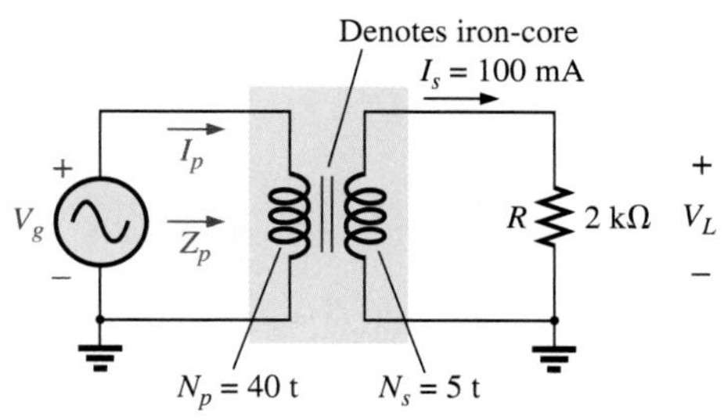

FIG. 22.6
Example 22.3.

Solutions:

a. $\dfrac{I_p}{I_s} = \dfrac{N_s}{N_p}$

$$I_p = \frac{N_s}{N_p}I_s = \left(\frac{5\text{ t}}{40\text{ t}}\right)(0.1\text{ A}) = \mathbf{12.5\text{ mA}}$$

$$V_L = I_sZ_L = (0.1\text{ A})(2\text{ k}\Omega) = 200\text{ V}$$

Also, $\dfrac{V_g}{V_L} = \dfrac{N_p}{N_s}$

$$V_g = \frac{N_p}{N_s}V_L = \left(\frac{40\text{ t}}{5\text{ t}}\right)(200\text{ V}) = \mathbf{1600\text{ V}}$$

b. $Z_p = a^2Z_L$

$$a = \frac{N_p}{N_s} = 8$$

$$Z_p = (8)^2(2\text{ k}\Omega) = R_p = \mathbf{128\text{ k}\Omega}$$

EXAMPLE 22.4 For the residential supply appearing in Fig. 22.7, determine (assuming a totally resistive load) the following:

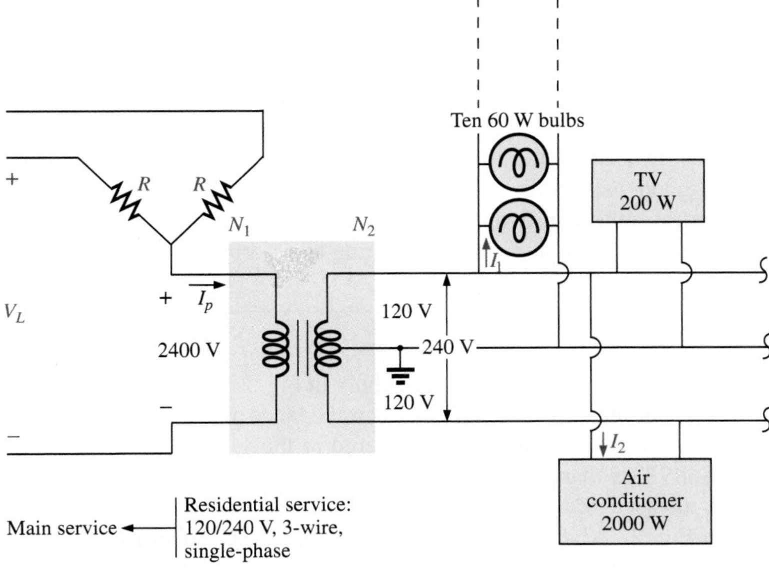

FIG. 22.7
Single-phase residential supply.

a. the value of R to ensure a balanced load
b. the magnitude of I_1 and I_2
c. the line voltage V_L
d. the total power delivered for a balanced three-phase load
e. the turns ratio $a = N_p/N_s$

Solutions:

a. $P_T = (10)(60\text{ W}) + 200\text{ W} + 2000\text{ W}$
$= 600\text{ W} + 200\text{ W} + 2000\text{ W} = 2800\text{ W}$
$P_{in} = P_{out}$
$V_pI_p = V_sI_s = 2800\text{ W}$ (purely resistive load)
$(2400\text{ V})I_p = 2800\text{ W}$ and $I_p = 1.17\text{ A}$

$$R = \frac{V_\phi}{I_p} = \frac{2400\text{ V}}{1.17\text{ A}} = \mathbf{2051.28\ \Omega}$$

b. $P_1 = 600\text{ W} = VI_1 = (120\text{ V})I_1$

and $I_1 = \mathbf{5\ A}$

$P_2 = 2000\text{ W} = VI_2 = (240\text{ V})I_2$

and $I_2 = \mathbf{8.33\ A}$

c. $V_L = \sqrt{3}V_\phi = 1.73(2400\text{ V}) = \mathbf{4152\ V}$

d. $P_T = 3P_\phi = 3(2800\text{ W}) = \mathbf{8.4\ kW}$

e. $$a = \frac{N_p}{N_s} = \frac{V_p}{V_s} = \frac{2400\text{ V}}{240\text{ V}} = \mathbf{10}$$

22.5 IMPEDANCE MATCHING, ISOLATION, AND DISPLACEMENT

Transformers can be particularly useful when you are trying to ensure that a load receives maximum power from a source. Recall that maximum power is transferred to a load when its impedance is a match with the internal resistance of the supply. Even if a perfect match is unattainable, the closer the load matches the internal resistance, the greater the power to the load and the more efficient the system. Unfortunately, unless it is planned as part of the design, most loads are not a close match with the internal impedance of the supply. However, transformers have a unique relationship between their primary and secondary impedances that can be put to good use in the impedance matching process. Example 22.5 demonstrates the significant difference in the power delivered to the load with and without an impedance matching transformer.

EXAMPLE 22.5

a. The source impedance for the supply in Fig. 22.8(a) is 512 Ω, which is a poor match with the 8 Ω input impedance of the speaker. You can expect only that the power delivered to the speaker will be significantly less than the maximum possible level. Determine the power to the speaker under the conditions in Fig. 22.8(a).
b. In Fig. 22.8(b), an audio impedance matching transformer was introduced between the speaker and the source, and it was designed to ensure maximum power to the 8 Ω speaker. Determine the input impedance of the transformer and the power delivered to the speaker.

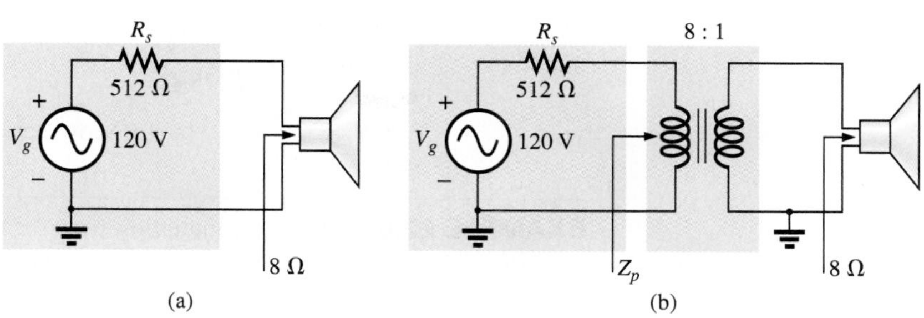

FIG. 22.8
Example 22.5.

c. Compare the power delivered to the speaker under the conditions of parts (a) and (b).

Solutions:

a. The source current:

$$I_s = \frac{E}{R_T} = \frac{120\text{ V}}{512\ \Omega + 8\ \Omega} = \frac{120\text{ V}}{520\ \Omega} = 230.8\text{ mA}$$

The power to the speaker:

$$P = I^2R = (230.8\text{ mA})^2 \cdot 8\ \Omega = \mathbf{426.15\ mW} \cong \mathbf{0.43\ W}$$

or less than half a watt.

b. $Z_p = a^2Z_L$

$$a = \frac{N_p}{N_s} = \frac{8}{1} = 8$$

and $$Z_p = (8)^2 8\ \Omega = \mathbf{512\ \Omega}$$

which matches that of the source. Maximum power transfer conditions have been established, and the source current is now determined by

$$I_s = \frac{E}{R_T} = \frac{120\text{ V}}{512\ \Omega + 512\ \Omega} = \frac{120\text{ V}}{1024\ \Omega} = 117.19\text{ mA}$$

The power to the primary (which equals that to the secondary for the ideal transformer) is

$$P = I^2R = (117.19\text{ mA})^2\, 512\ \Omega = \mathbf{7.032\ W}$$

The result is not in milliwatts, as obtained above, and exceeds 7 W, which is a significant improvement.

c. Comparing levels, 7.032 W/426.15 mW = 16.5, or more than 16 times the power delivered to the speaker using the impedance matching transformer.

Another important application of the impedance matching capabilities of a transformer is the matching of the 300 Ω twin line transmission line from a television antenna to the 75 Ω input impedance of a television (ready-made for the 75 Ω coaxial cable), as shown in Fig. 22.9. A match must be made to ensure the strongest signal to the television receiver.

Using the equation $Z_p = a^2Z_L$ we find

$$300\ \Omega = a^2 75\ \Omega$$

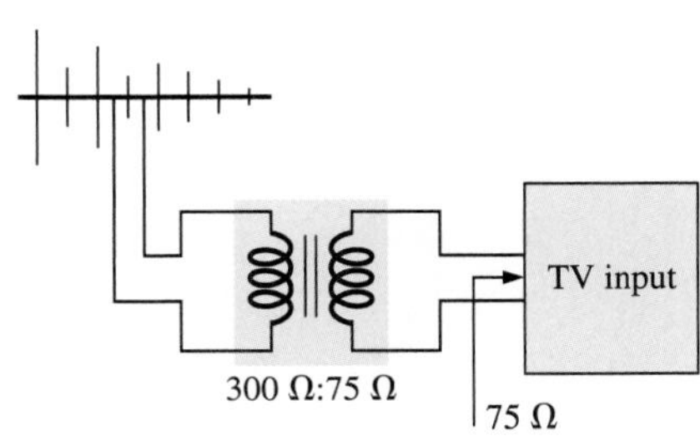

FIG. 22.9
Television impedance matching transformer.

and $$a = \sqrt{\frac{300\ \Omega}{75\ \Omega}} = \sqrt{4} = \mathbf{2}$$

with $N_p : N_s = 2 : 1$ (a step-down transformer)

EXAMPLE 22.6 Impedance matching transformers are also quite evident in public address systems, such as the one appearing in the 70.7 V system in Fig. 22.10. Although the system has only one set of output terminals, up to four speakers can be connected to this system (the number is a function of the chosen system). Each 8 Ω speaker is connected to the 70.7 V line through a 10 W audio-matching transformer (defining the frequency range of linear operation).

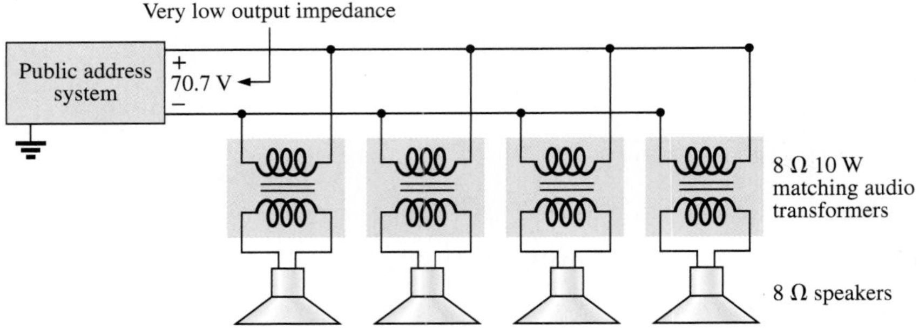

FIG. 22.10
Public address system.

a. If each speaker in Fig. 22.10 can receive 10 W of power, what is the maximum power drain on the source?
b. For each speaker, determine the impedance seen at the input side of the transformer if each is operating under its full 10 W of power.
c. Determine the turns ratio of the transformers.
d. At 10 W, what are the speaker voltage and current?
e. What is the load seen by the source with one, two, three, or four speakers connected?

Solutions:

a. Ideally, the primary power equals the power delivered to the load, resulting in a maximum of **40 W** from the supply.
b. The power at the primary:

$$P_p = V_p I_p = (70.7\ \text{V})\, I_p = 10\ \text{W}$$

and $$I_p = \frac{10\ \text{W}}{70.7\ \text{V}} = 141.4\ \text{mA}$$

so that $$Z_p = \frac{V_p}{I_p} = \frac{70.7\ \text{V}}{141.4\ \text{mA}} = \mathbf{500\ \Omega}$$

c. $$Z_p = a^2 Z_L \Rightarrow a = \sqrt{\frac{Z_p}{Z_L}} = \sqrt{\frac{500\ \Omega}{8\ \Omega}} = \sqrt{62.5} = \mathbf{7.91 \cong 8:1}$$

d. $$V_s = V_L = \frac{V_p}{a} = \frac{70.7\ \text{V}}{7.91} = \mathbf{8.94\ V \cong 9\ V}$$

e. All the speakers are in parallel. Therefore,

One speaker: $R_T = \mathbf{500\ \Omega}$

Two speakers: $R_T = \dfrac{500\ \Omega}{2} = \mathbf{250\ \Omega}$

Three speakers: $R_T = \dfrac{500\ \Omega}{3} = \mathbf{167\ \Omega}$

Four speakers: $R_T = \dfrac{500\ \Omega}{4} = \mathbf{125\ \Omega}$

Even though the load seen by the source varies with the number of speakers connected, the source impedance is so low (compared to the lowest load of 125 Ω) that the terminal voltage of 70.7 V is essentially constant. This is not the case where the desired result is to match the load to the input impedance; rather, it was to ensure 70.7 V at each primary, no matter how many speakers were connected, and to limit the current drawn from the supply.

The transformer is frequently used to isolate one portion of an electrical system from another. *Isolation* implies the absence of any direct physical connection. As a first example of its use as an isolation device, consider the measurement of line voltages on the order of 40,000 V (Fig. 22.11).

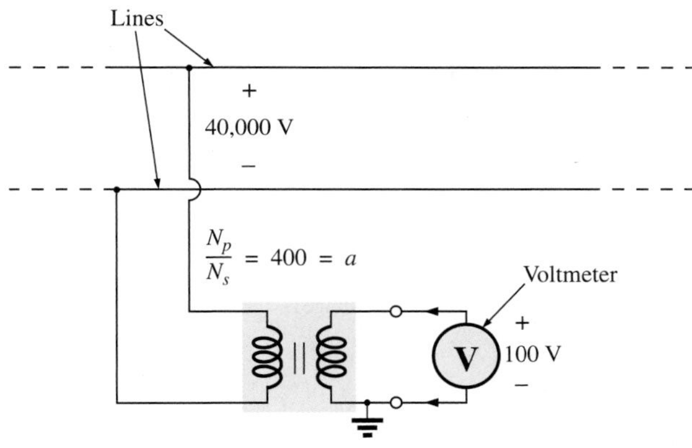

FIG. 22.11
Isolating a high-voltage line from the point of measurement.

To apply a voltmeter across 40,000 V would obviously be a dangerous task due to the possibility of physical contact with the lines when making the necessary connections. Including a transformer in the transmission system as original equipment can bring the potential down to a safe level for measurement purposes and can determine the line voltage using the turns ratio. Therefore, the transformer serves both to isolate and to step down the voltage.

As a second example, consider the application of the voltage v_x to the vertical input of the oscilloscope (a measuring instrument) in Fig. 22.12. If the connections are made as shown, and if the generator and oscilloscope have a common ground, the impedance $\mathbf{Z}_2$ has been effectively shorted out of the circuit by the ground connection of the oscilloscope. The input voltage to the oscilloscope is therefore meaningless as far as the

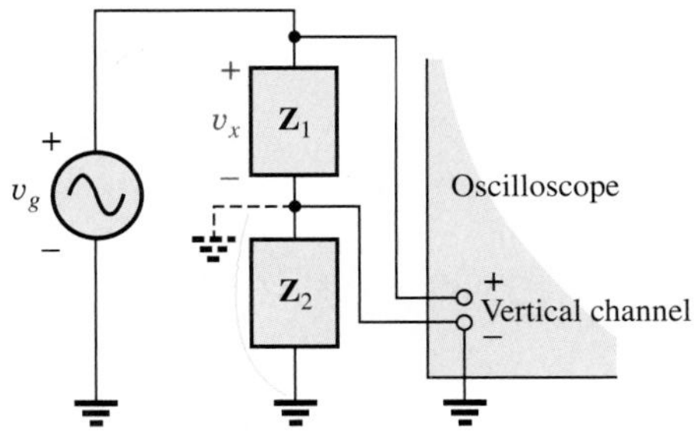

FIG. 22.12
Demonstrating the shorting effect introduced by the grounded side of the vertical channel of an oscilloscope.

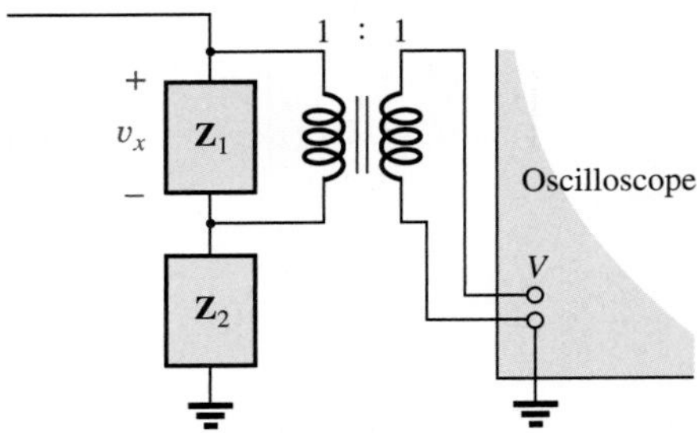

FIG. 22.13
Correcting the situation of Fig. 22.12 using an isolation transformer.

voltage v_x is concerned. In addition, if $\mathbf{Z}_2$ is the current-limiting impedance in the circuit, the current in the circuit may rise to a level that causes severe damage to the circuit. If a transformer is used as shown in Fig. 22.13, this problem is eliminated, and the input voltage to the oscilloscope will be v_x.

The linear variable differential transformer (LVDT) is a sensor that can reveal displacement using transformer effects. In its simplest form, the LVDT has a central winding and two secondary windings, as shown in Fig. 22.14(a). A ferromagnetic core inside the windings is free to move as dictated by some external force. A constant, low-level ac voltage is applied to the primary, and the output voltage is the difference between the voltages induced in the secondaries. If the core is in the position shown in Fig. 22.14(b), a relatively large voltage is induced across the secondary winding labeled coil 1, and a relatively small voltage is induced across the secondary winding labeled coil 2 (essentially an air-core transformer for this position). The result is a relatively large secondary output voltage. If the core is in the position shown in Fig. 22.14(c), the flux linking each coil is the same, and the output voltage (being the difference) will be quite small. In total, therefore, the position of the core can be related to the secondary voltage, and a position-versus-voltage graph can be developed as shown in Fig. 22.14(d). Due to the nonlinearity of the *B-H* curve, the curve becomes somewhat nonlinear if the core is moved too far out of the unit.

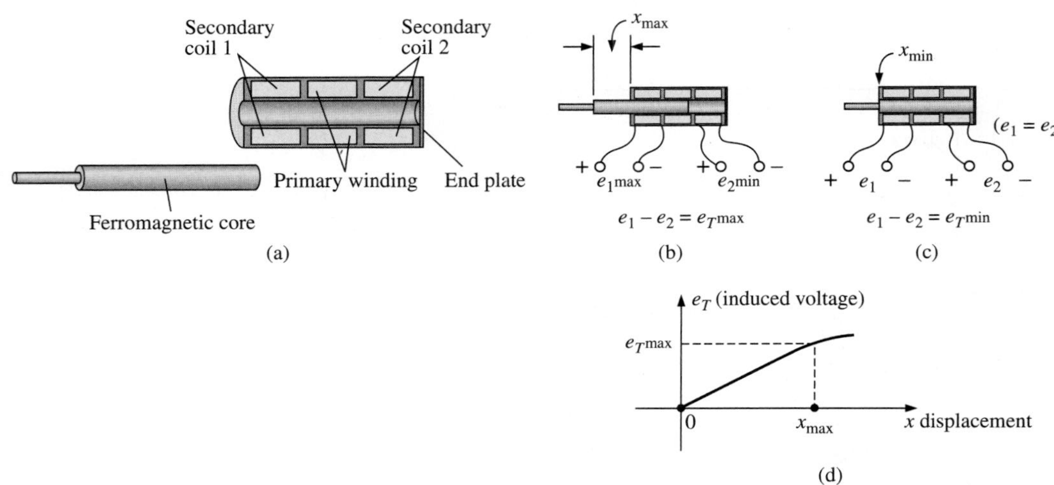

FIG. 22.14
LVDT transformer: (a) construction; (b) maximum displacement; (c) minimum displacement; (d) graph of induced voltage versus displacement.

22.6 EQUIVALENT CIRCUIT (IRON-CORE TRANSFORMER)

For the nonideal or practical iron-core transformer, the equivalent circuit appears as in Fig. 22.15. As indicated, part of this equivalent circuit includes an ideal transformer. The remaining elements of Fig. 22.15 are those elements that contribute to the nonideal characteristics of the device. The resistances R_p and R_s are simply the dc or geometric resistance of the primary and secondary windings, respectively. For the primary and secondary coils of a transformer, there is a small amount of flux that links

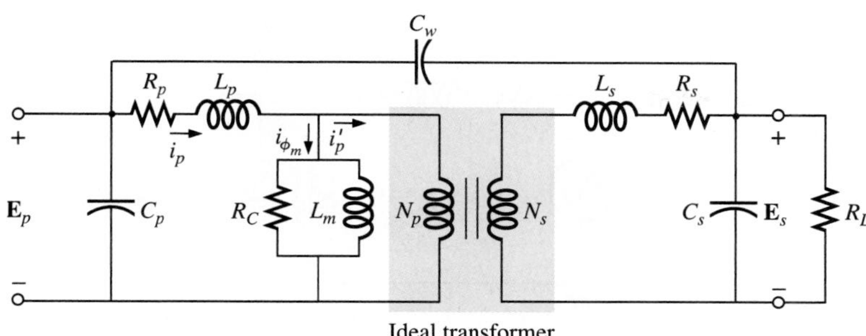

FIG. 22.15
Equivalent circuit for the practical iron-core transformer.

each coil but does not pass through the core, as shown in Fig. 22.16 for the primary winding. This **leakage flux,** representing a definite loss in the system, is represented by an inductance L_p in the primary circuit and an inductance L_s in the secondary.

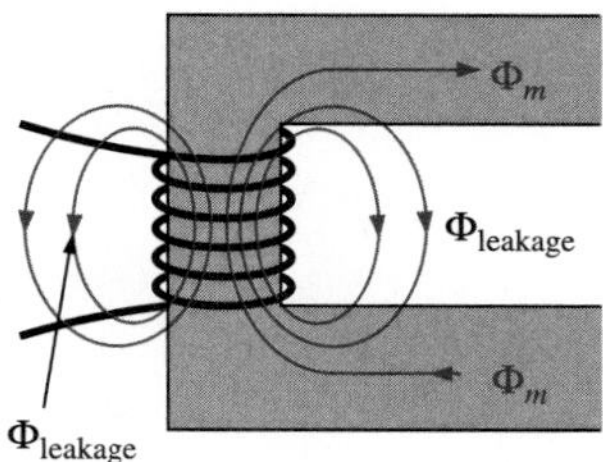

FIG. 22.16
Identifying the leakage flux of the primary.

The resistance R_c represents the hysteresis and eddy current losses (core losses) within the core due to an ac flux through the core. The inductance L_m (magnetizing inductance) is the inductance associated with the magnetization of the core, that is, the establishing of the flux Φ_m in the core. The capacitances C_p and C_s are the lumped capacitances of the primary and secondary circuits, respectively, and C_w represents the equivalent lumped capacitances between the windings of the transformer.

Since i'_p is normally considerably larger than i_{ϕ_m} (the magnetizing current), we will ignore i_{ϕ_m} for the moment (set it equal to zero), resulting in the absence of R_c and L_m in the reduced equivalent circuit in Fig. 22.17. The capacitances C_p, C_w, and C_s do not appear in the equivalent circuit in Fig. 22.17 since their reactance at typical operating frequencies do not appreciably affect the transfer characteristics of the transformer.

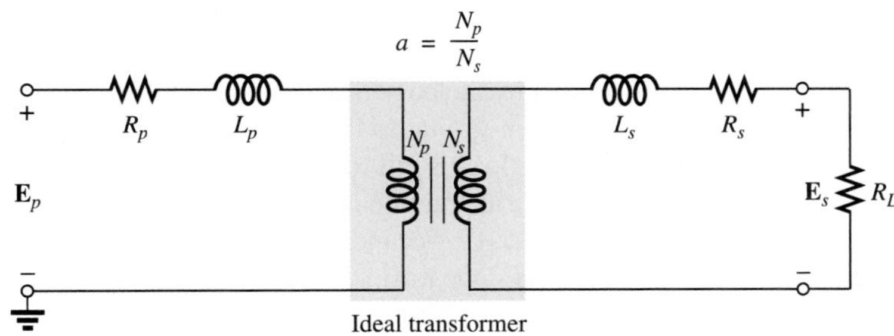

FIG. 22.17
Reduced equivalent circuit for the nonideal iron-core transformer.

If we now reflect the secondary circuit through the ideal transformer using Eq. (22.19), as shown in Fig. 22.18(a), we will have the load and generator voltage in the same continuous circuit. The total resistance and inductive reactance of the primary circuit are determined by

$$R_{\text{equivalent}} = R_e = R_p + a^2R_s \tag{22.22}$$

and

$$X_{\text{equivalent}} = X_e = X_p + a^2X_s \tag{22.23}$$

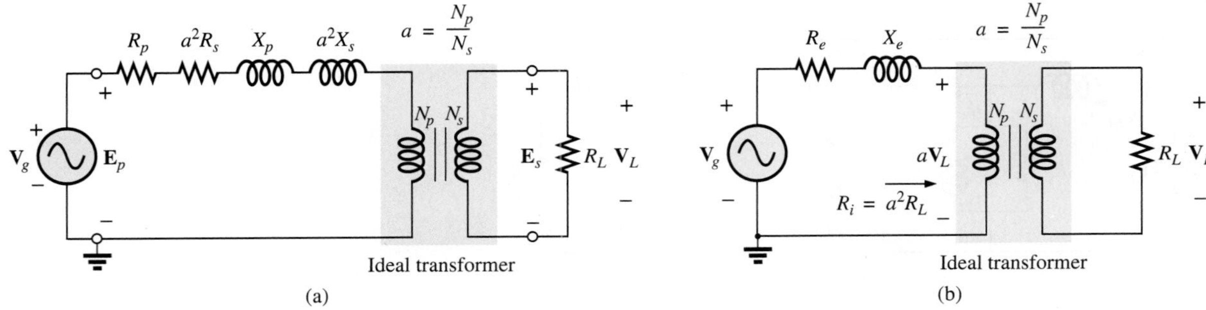

FIG. 22.18
Reflecting the secondary circuit into the primary side of the iron-core transformer.

which result in the useful equivalent circuit of Fig. 22.18(b). The load voltage can be obtained directly from the circuit in Fig. 22.18(b) through the voltage divider rule:

$$a\mathbf{V}_L = \frac{R_i\mathbf{V}_g}{(R_e + R_i) + jX_e}$$

and

$$\mathbf{V}_L = \frac{a^2R_L\mathbf{V}_g}{(R_e + a^2R_L) + jX_e} \qquad \textbf{(22.24)}$$

FIG. 22.19
Phasor diagram for the iron-core transformer with (a) unity power-factor load (resistive) and (b) lagging power-factor load (inductive).

The network in Fig. 22.18(b) also allows us to calculate the generator voltage necessary to establish a particular load voltage. The voltages across the elements in Fig. 22.18(b) have the phasor relationship indicated in Fig. 22.19(a). Note that the current is the reference phasor for drawing the phasor diagram. That is, the voltages across the resistive elements are *in phase* with the current phasor, while the voltage across the equivalent inductance leads the current by 90°. The primary voltage, by Kirchhoff's voltage law, is then the phasor sum of these voltages, as indicated in Fig. 22.19(a). For an inductive load, the phasor diagram appears in Fig. 22.19(b). Note that $a\mathbf{V}_L$ leads $\mathbf{I}$ by the power-factor angle of the load. The remainder of the diagram is then similar to that for a resistive load. (The phasor diagram for a capacitive load is left to the reader as an exercise.)

The effect of R_e and X_e on the magnitude of $\mathbf{V}_g$ for a particular $\mathbf{V}_L$ is obvious from Eq. (22.24) or Fig. 22.19. For increased values of R_e or X_e, an increase in V_g is required for the same load voltage. For R_e and $X_e = 0$, $\mathbf{V}_L$ and $\mathbf{V}_g$ are simply related by the turns ratio.

EXAMPLE 22.7 For a transformer having the equivalent circuit in Fig. 22.20:

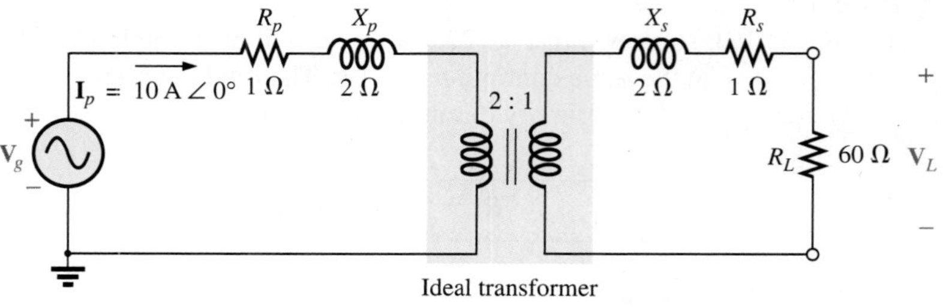

FIG. 22.20
Example 22.7

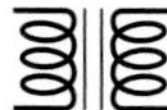

a. Determine R_e and X_e.
b. Determine the magnitude of the voltages V_L and V_g.
c. Determine the magnitude of the voltage V_g to establish the same load voltage in part (b) if R_e and $X_e = 0\ \Omega$. Compare with the result of part (b).

Solutions:

a. $R_e = R_p + a^2R_s = 1\ \Omega + (2)^2(1\ \Omega) = \mathbf{5\ \Omega}$
$X_e = X_p + a^2X_s = 2\ \Omega + (2)^2(2\ \Omega) = \mathbf{10\ \Omega}$

b. The transformed equivalent circuit appears in Fig. 22.21.

$$aV_L = (I_p)(a^2R_L) = 2400\ \text{V}$$

Thus,

$$V_L = \frac{2400\ \text{V}}{a} = \frac{2400\ \text{V}}{2} = \mathbf{1200\ V}$$

and

$$\begin{aligned}\mathbf{V}_g &= \mathbf{I}_p(R_e + a^2R_L + jX_e)\\ &= 10\ \text{A}(5\ \Omega + 240\ \Omega + j\,10\ \Omega) = 10\ \text{A}(245\ \Omega + j\,10\ \Omega)\\ \mathbf{V}_g &= 2450\ \text{V} + j\,100\ \text{V} = 2452.04\ \text{V}\ \angle 2.34^\circ\\ &= \mathbf{2452.04\ V\ \angle 2.34^\circ}\end{aligned}$$

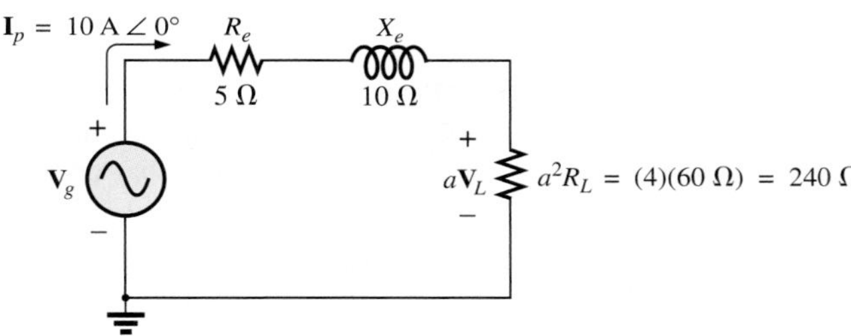

FIG. 22.21
Transformed equivalent circuit of Fig. 22.20.

c. For R_e and $X_e = 0$, $V_g = aV_L = (2)(1200\ \text{V}) = 2400\ \text{V}$.
Therefore, it is necessary to increase the generator voltage by 52.04 V (due to R_e and X_e) to obtain the same load voltage.

22.7 FREQUENCY CONSIDERATIONS

For certain frequency ranges, the effect of some parameters in the equivalent circuit of the iron-core transformer in Fig. 22.15 should not be ignored. Since it is convenient to consider a low-, mid-, and high-frequency region, the equivalent circuits for each are now introduced and briefly examined.

For the low-frequency region, the series reactance ($2\pi fL$) of the primary and secondary leakage reactances can be ignored since they are small in magnitude. The magnetizing inductance must be included, however, since it appears in parallel with the secondary reflected circuit, and small impedances in a parallel network can have a dramatic impact on the terminal characteristics. The resulting equivalent network for the low-frequency region is provided in Fig. 22.22(a). As the frequency decreases, the reactance of the magnetizing inductance reduces in magnitude, causing a reduction in the voltage across the secondary circuit. For $f = 0$ Hz, L_m is ideally a short

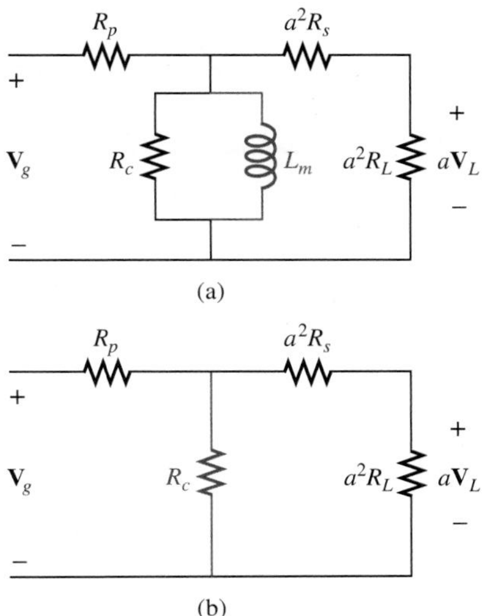

FIG. 22.22
(a) Low-frequency reflected equivalent circuit; (b) mid-frequency reflected circuit.

circuit, and $V_L = 0$. As the frequency increases, the reactance of L_m is eventually sufficiently large compared with the reflected secondary impedance to be neglected. The mid-frequency reflected equivalent circuit then appears as shown in Fig. 22.22(b). Note the absence of reactive elements, resulting in an *in-phase* relationship between load and generator voltages.

For higher frequencies, the capacitive elements and primary and secondary leakage reactances must be considered, as shown in Fig. 22.23. For discussion purposes, the effects of C_w and C_s appear as a lumped capacitor C in the reflected network in Fig. 22.23; C_p does not appear since the effect of C predominates. As the frequency of interest increases, the capacitive reactance ($X_C = 1/2\pi fC$) decreases to the point that it will have a shorting effect across the secondary circuit of the transformer, causing V_L to decrease in magnitude.

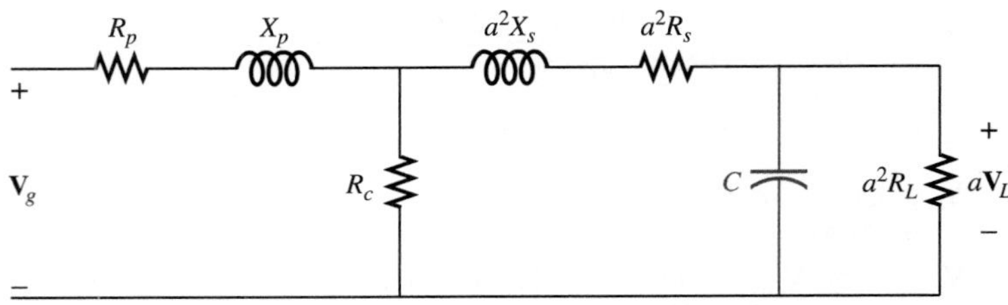

FIG. 22.23
High-frequency reflected equivalent circuit.

A typical iron-core transformer-frequency response curve appears in Fig. 22.24. For the low- and high-frequency regions, the primary element responsible for the drop-off is indicated. The peaking that occurs in the high-frequency region is due to the series resonant circuit established by the inductive and capacitive elements of the equivalent circuit. In the peaking region, the series resonant circuit is in, or near, its resonant or tuned state.

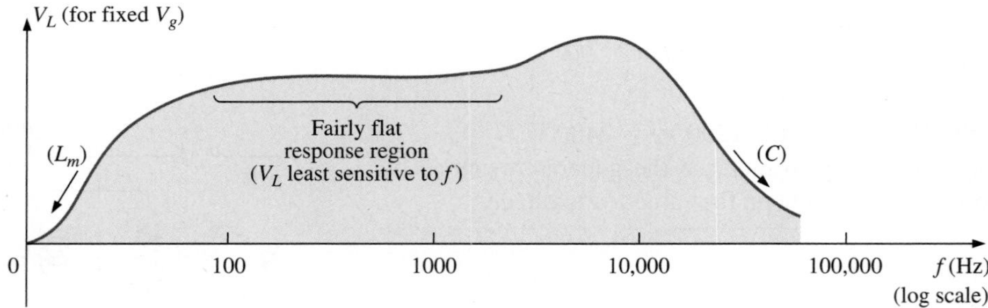

FIG. 22.24
Transformer-frequency response curve.

22.8 SERIES CONNECTION OF MUTUALLY COUPLED COILS

In Chapter 11, we found that the total inductance of series isolated coils was determined simply by the sum of the inductances. For two coils that are connected in series but also share the same flux linkages, such as those in Fig. 22.25(a), a mutual term is introduced that alters the total inductance of the series combination. The physical picture of how the coils are connected is indicated in Fig. 22.25(b). An iron core is included, although the equations to be developed are for any two mutually coupled coils with any value of coefficient of coupling k. When referring to the voltage induced across the inductance L_1 (or L_2) due to the change in flux linkages of the inductance L_2 (or L_1, respectively), the mutual inductance is represented by M_{12}. This type of subscript notation is particularly important when there are two or more mutual terms.

Due to the presence of the mutual term, the induced voltage e_1 is composed of that due to the self-inductance L_1 and that due to the mutual inductance M_{12}. That is,

$$e_1 = L_1\frac{di_1}{dt} + M_{12}\frac{di_2}{dt}$$

However, since $i_1 = i_2 = i$,

$$e_1 = L_1\frac{di}{dt} + M_{12}\frac{di}{dt}$$

or

$$e_1 = (L_1 + M_{12})\frac{di}{dt} \qquad \text{(volts, V)} \qquad \textbf{(22.25)}$$

and, similarly,

$$e_2 = (L_2 + M_{12})\frac{di}{dt} \qquad \text{(volts, V)} \qquad \textbf{(22.26)}$$

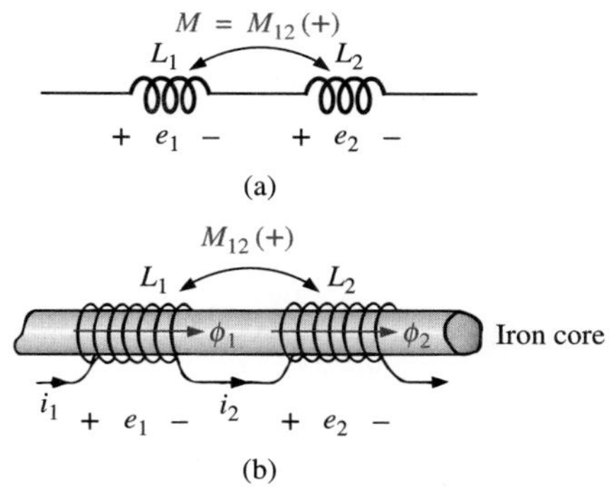

FIG. 22.25
Mutually coupled coils connected in series.

For the series connection, the total induced voltage across the series coils, represented by e_T, is

$$e_T = e_1 + e_2 = (L_1 + M_{12})\frac{di}{dt} + (L_2 + M_{12})\frac{di}{dt}$$

or

$$e_T = (L_1 + L_2 + M_{12} + M_{12})\frac{di}{dt}$$

and the total effective inductance is

$$L_{T(+)} = L_1 + L_2 + 2M_{12} \qquad \text{(henries, H)} \qquad \textbf{(22.27)}$$

The subscript (+) was included to indicate that the mutual terms have a positive sign and are added to the self-inductance values to determine the total inductance. If the coils are wound such as shown in Fig. 22.26, where ϕ_1 and ϕ_2 are in opposition, the induced voltages due to the mutual terms oppose that due to the self-inductance, and the total inductance is determined by

$$L_{T(-)} = L_1 + L_2 - 2M_{12} \qquad \text{(henries, H)} \qquad \textbf{(22.28)}$$

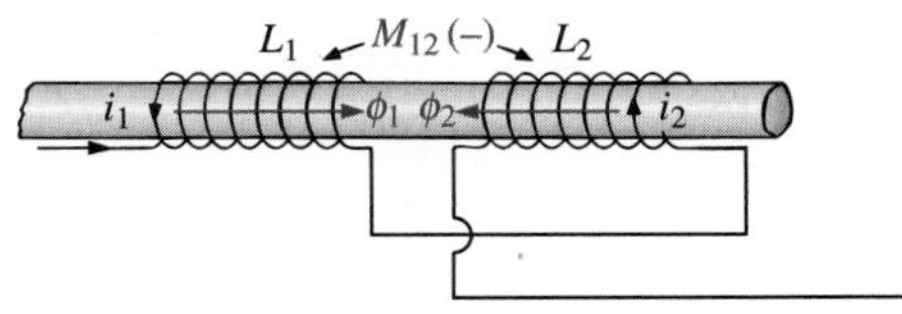

FIG. 22.26
Mutually coupled coils connected in series with negative mutual inductance.

Through Eqs. (22.27) and (22.28), the mutual inductance can be determined by

$$M_{12} = \frac{1}{4}(L_{T(+)} - L_{T(-)}) \qquad \textbf{(22.29)}$$

Eq. (22.29) is very effective in determining the mutual inductance between two coils. It states that the mutual inductance is equal to one-quarter the difference between the total inductance with a positive and negative mutual effect.

From the preceding, it should be clear that the mutual inductance directly affects the magnitude of the voltage induced across a coil since it determines the net inductance of the coil. Additional examination reveals that the sign of the mutual term for each coil of a coupled pair is the same. For $L_{T(+)}$ they are both positive, and for $L_{T(-)}$ they are both negative. On a network schematic where it is inconvenient to indicate the windings and the flux path, a system of dots is used that determines whether the mutual terms are to be positive or negative. The dot convention is shown in Fig. 22.27 for the series coils in Figs. 22.25 and 22.26.

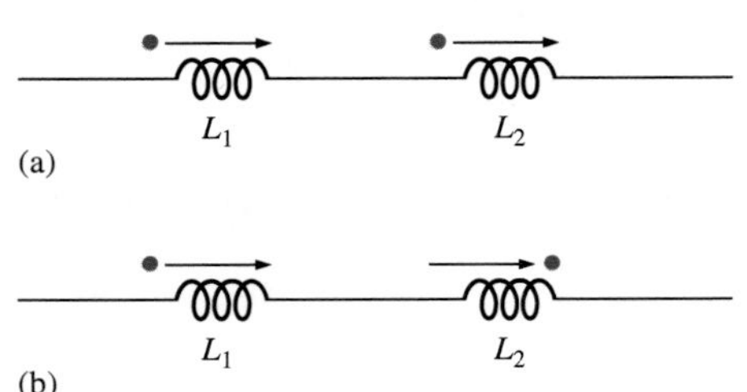

FIG. 22.27
Dot convention for the series coils in (a) Fig. 22.25 and (b) Fig. 22.26.

If the current through *each* of the mutually coupled coils is going away from (or toward) the dot as it *passes through the coil,* the mutual term will be positive, as shown for the case in Fig 22.27(a). If the arrow indicating current direction through the coil is leaving the dot for one coil and entering the dot for the other, the mutual term is negative.

A few possibilities for mutually coupled transformer coils are indicated in Fig. 22.28(a). The sign of M is indicated for each. When determining the sign, be sure to examine the current direction within the coil itself. In Fig. 22.28(b), one direction is indicated outside for one coil and through for the other. It initially may appear that the sign should be positive since both currents enter the dot, but the current *through* coil 1 is leaving the dot; hence a negative sign is in order.

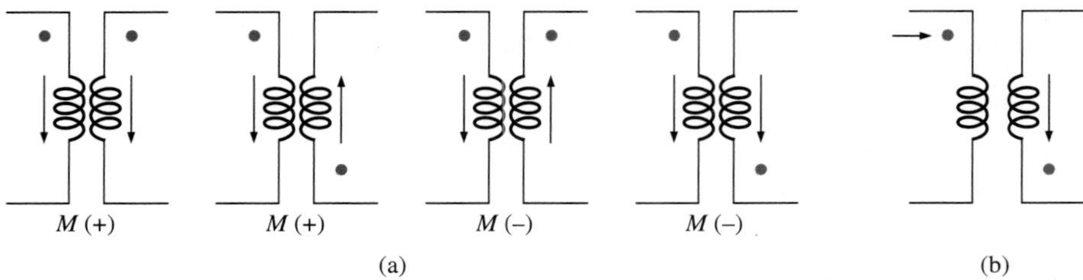

FIG. 22.28
Defining the sign of M for mutually coupled transformer coils.

The dot convention also reveals the polarity of the *induced* voltage across the mutually coupled coil. If the reference direction for the current *in* a coil leaves the dot, the polarity at the dot for the induced voltage of the mutually coupled coil is positive. In the first two figures in Fig. 22.28(a), the polarity at the dots of the induced voltages is positive. In the third figure in Fig. 22.28(a), the polarity at the dot of the right coil is negative, while the polarity at the dot of the left coil is positive, since the current enters the dot (within the coil) of the right coil. The comments for the third figure in Fig. 22.28(a) can also be applied to the last figure in Fig. 22.28(a).

EXAMPLE 22.8 Find the total inductance of the series coils in Fig. 22.29.

Solution:

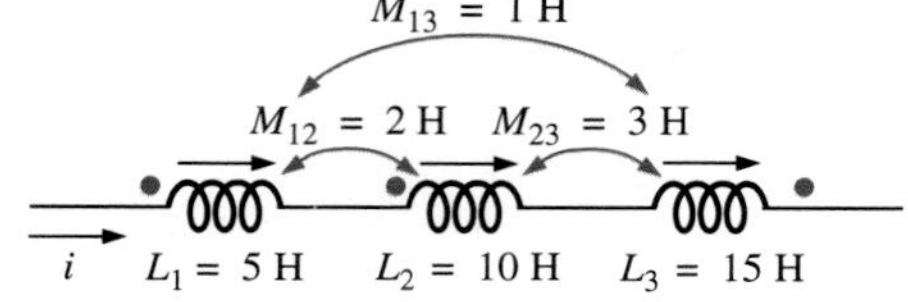

FIG. 22.29
Example 22.8.

Current vectors leave dot.

Coil 1: $L_1 + M_{12} - M_{13}$

One current vector enters dot, while one leaves.

Coil 2: $L_2 + M_{12} - M_{23}$

Coil 3: $L_3 - M_{23} - M_{13}$

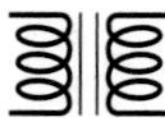

and

$$L_T = (L_1 + M_{12} - M_{13}) + (L_2 + M_{12} - M_{23}) + (L_3 - M_{23} - M_{13})$$
$$= L_1 + L_2 + L_3 + 2M_{12} - 2M_{23} - 2M_{13}$$

Substituting values, we find

$$L_T = 5\text{ H} + 10\text{ H} + 15\text{ H} + 2(2\text{ H}) - 2(3\text{ H}) - 2(1\text{ H})$$
$$= 34\text{ H} - 8\text{ H} = \mathbf{26\text{ H}}$$

EXAMPLE 22.9 Write the mesh equations for the transformer network in Fig. 22.30.

Solution: For each coil, the mutual term is positive, and the sign of M in $\mathbf{X}_m = \omega M \angle 90°$ is positive, as determined by the direction of $\mathbf{I}_1$ and $\mathbf{I}_2$. Thus,

$$\mathbf{E}_1 - \mathbf{I}_1 R_1 - \mathbf{I}_1 X_{L_1} \angle 90° - \mathbf{I}_2 X_m \angle 90° = 0$$

or

$$\mathbf{E}_1 - \mathbf{I}_1(R_1 + j X_{L_1}) - \mathbf{I}_2 X_m \angle 90° = 0$$

For the other loop,

$$-\mathbf{I}_2 X_{L_2} \angle 90° - \mathbf{I}_1 X_m \angle 90° - \mathbf{I}_2 R_L = 0$$

or

$$\mathbf{I}_2(R_L + j X_{L_2}) - \mathbf{I}_1 X_m \angle 90° = 0$$

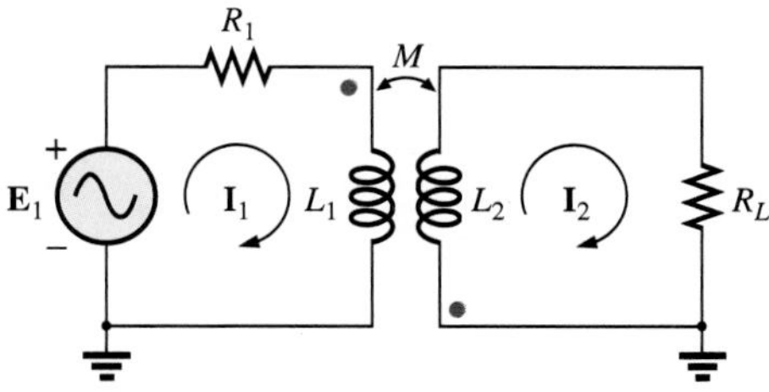

FIG. 22.30
Example 22.9.

22.9 AIR-CORE TRANSFORMER

As the name implies, the air-core transformer does not have a ferromagnetic core to link the primary and secondary coils. Rather, the coils are placed sufficiently close to have a mutual inductance that establishes the desired transformer action. In Fig. 22.31, current direction and polarities have been defined for the air-core transformer. Note the presence of a mutual inductance term M, which is positive in this case, as determined by the dot convention.

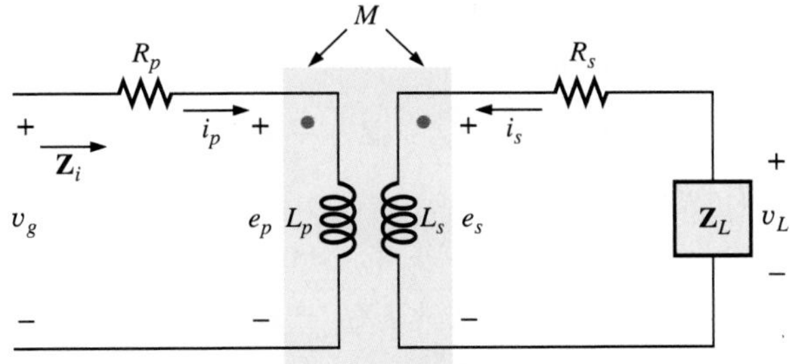

FIG. 22.31
Air-core transformer equivalent circuit.

From past analysis in this chapter, we now know that

$$e_p = L_p \frac{di_p}{dt} + M \frac{di_s}{dt} \qquad \textbf{(22.30)}$$

for the primary circuit.

We found in Chapter 11 that for the pure inductor, with no mutual inductance present, the mathematical relationship

$$v_1 = L\frac{di_1}{dt}$$

resulted in the following useful form of the voltage across an inductor:

$$\mathbf{V}_1 = \mathbf{I}_1 X_L \angle 90^\circ \qquad \text{where } X_L = \omega L$$

Similarly, it can be shown, for a mutual inductance, that

$$v_1 = M\frac{di_2}{dt}$$

results in

$$\boxed{\mathbf{V}_1 = \mathbf{I}_2 X_m \angle 90^\circ \quad \text{where } X_m = \omega M} \tag{22.31}$$

Eq. (22.30) can then be written (using phasor notation) as

$$\boxed{\mathbf{E}_p = \mathbf{I}_p X_{L_p} \angle 90^\circ + \mathbf{I}_s X_m \angle 90^\circ} \tag{22.32}$$

and $$\mathbf{V}_g = \mathbf{I}_p R_p \angle 0^\circ + \mathbf{I}_p X_{L_p} \angle 90^\circ + \mathbf{I}_s X_m \angle 90^\circ$$

or $$\boxed{\mathbf{V}_g = \mathbf{I}_p(R_p + j X_{L_p}) + \mathbf{I}_s X_m \angle 90^\circ} \tag{22.33}$$

For the secondary circuit,

$$\boxed{\mathbf{E}_s = \mathbf{I}_s X_{L_s} \angle 90^\circ + \mathbf{I}_p X_m \angle 90^\circ} \tag{22.34}$$

and $$\mathbf{V}_L = \mathbf{I}_s R_s \angle 0^\circ + \mathbf{I}_s X_{L_s} \angle 90^\circ + \mathbf{I}_p X_m \angle 90^\circ$$

or $$\boxed{\mathbf{V}_L = \mathbf{I}(R_s + j X_{L_s}) + \mathbf{I}_p X_m \angle 90^\circ} \tag{22.35}$$

Substituting $$\mathbf{V}_L = -\mathbf{I}_s \mathbf{Z}_L$$

into Eq. (22.35) results in

$$0 = \mathbf{I}_s(R_s + j X_{L_s} + \mathbf{Z}_L) + \mathbf{I}_p X_m \angle 90^\circ$$

Solving for $\mathbf{I}_s$, we have

$$\mathbf{I}_s = \frac{-\mathbf{I}_p X_m \angle 90^\circ}{R_s + j X_{L_s} + \mathbf{Z}_L}$$

and, substituting into Eq. (22.33), we obtain

$$\mathbf{V}_g = \mathbf{I}_p(R_p + j X_{L_p}) + \left(\frac{-\mathbf{I}_p X_m \angle 90^\circ}{R_s + j X_{L_s} + \mathbf{Z}_L}\right) X_m \angle 90^\circ$$

Thus, the input impedance is

$$\mathbf{Z}_i = \frac{\mathbf{V}_g}{\mathbf{I}_p} = R_p + j X_{L_p} - \frac{(X_m \angle 90^\circ)^2}{R_s + j X_{L_s} + \mathbf{Z}_L}$$

or, defining

$$\mathbf{Z}_p = R_p + j X_{L_p} \quad \mathbf{Z}_s = R_s + j X_{L_s} \quad \text{and} \quad X_m \angle 90^\circ = +j\,\omega M$$

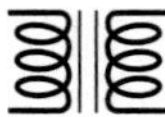

we have

$$\mathbf{Z}_i = \mathbf{Z}_p - \frac{(+j\,\omega M)^2}{\mathbf{Z}_s + \mathbf{Z}_L}$$

and

$$\mathbf{Z}_i = \mathbf{Z}_p - \frac{(\omega M)^2}{\mathbf{Z}_s + \mathbf{Z}_L} \qquad \textbf{(22.36)}$$

The term $(\omega M)^2/(\mathbf{Z}_s + \mathbf{Z}_L)$ is called the *coupled impedance,* and it is independent of the sign of M since it is squared in the equation. Consider also that since $(\omega M)^2$ is a constant with 0° phase angle, if the load $\mathbf{Z}_L$ is resistive, the resulting coupled impedance term appears capacitive due to division of $(\mathbf{Z}_s + R_L)$ into $(\omega M)^2$. This resulting capacitive reactance opposes the series primary inductance L_p, causing a reduction in $\mathbf{Z}_i$. Including the effect of the mutual term, the input impedance to the network appears as shown in Fig. 22.32.

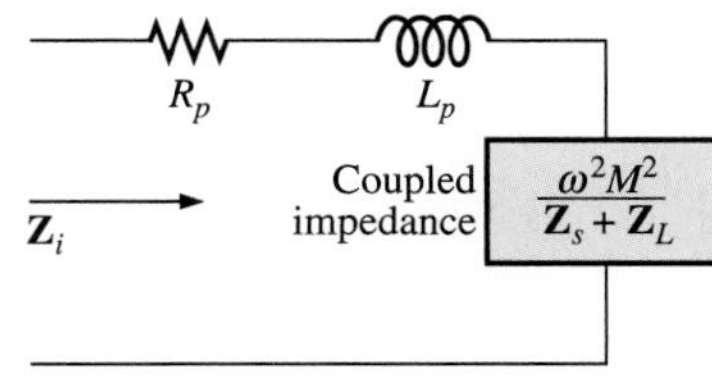

FIG. 22.32
Input characteristics for the air-core transformer.

EXAMPLE 22.10 Determine the input impedance to the air-core transformer in Fig. 22.33.

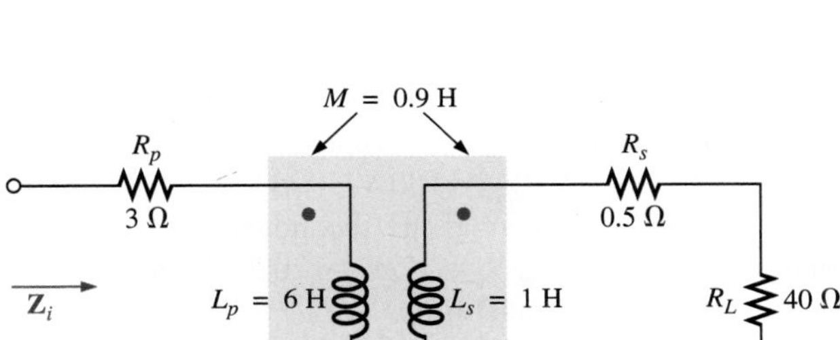

FIG. 22.33
Example 22.10.

Solution:

$$\begin{aligned}
\mathbf{Z}_i &= \mathbf{Z}_p + \frac{(\omega M)^2}{\mathbf{Z}_s + \mathbf{Z}_L} \\
&= R_p + j\,X_{L_p} + \frac{(\omega M)^2}{R_s + j\,X_{L_s} + R_L} \\
&= 3\ \Omega + j\,2.4\ \text{k}\Omega + \frac{((400\ \text{rad/s})(0.9\ \text{H}))^2}{0.5\ \Omega + j\,400\ \Omega + 40\ \Omega} \\
&\cong j\,2.4\ \text{k}\Omega + \frac{129.6 \times 10^3\ \Omega}{40.5 + j\,400} \\
&= j\,2.4\ \text{k}\Omega + \underbrace{322.4\ \Omega\ \angle -84.22^\circ}_{\text{capacitive}} \\
&= j\,2.4\ \text{k}\Omega + (0.0325\ \text{k}\Omega - j\,0.3208\ \text{k}\Omega) \\
&= 0.0325\ \text{k}\Omega + j\,(2.40 - 0.3208)\ \text{k}\Omega
\end{aligned}$$

and $\quad \mathbf{Z}_i = R_i + j\,X_{L_i} = \mathbf{32.5\ \Omega + \mathit{j}\,2079\ \Omega = 2079.25\ \Omega\ \angle 89.10^\circ}$

22.10 NAMEPLATE DATA

A typical iron-core power transformer rating, included in the **nameplate data** for the transformer, might be the following:

5 kVA 2000/100 V 60 Hz

The 2000 V or the 100 V can be either the primary or the secondary voltage; that is, if 2000 V is the primary voltage, then 100 V is the secondary voltage, and vice versa. The 5 kVA is the apparent power ($S = VI$) rating of the transformer. If the secondary voltage is 100 V, then the maximum load current is

$$I_L = \frac{S}{V_L} = \frac{5000 \text{ VA}}{100 \text{ V}} = 50 \text{ A}$$

and if the secondary voltage is 2000 V, then the maximum load current is

$$I_L = \frac{S}{V_L} = \frac{5000 \text{ VA}}{2000 \text{ V}} = 2.5 \text{ A}$$

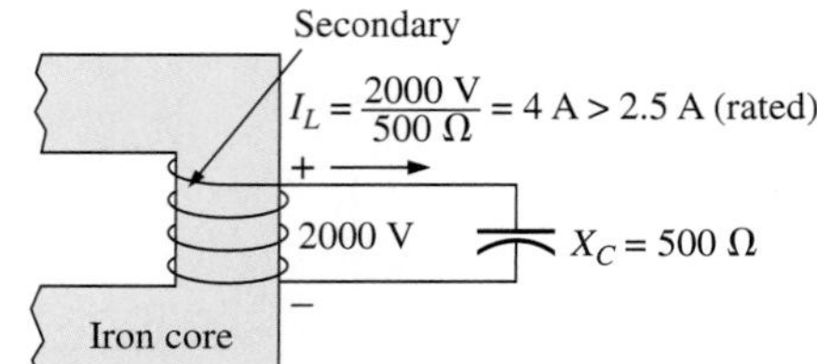

FIG. 22.34
Demonstrating why transformers are rated in kVA rather than kW.

The transformer is rated in terms of the apparent power rather than the average, or real, power for the reason demonstrated by the circuit in Fig. 22.34. Since the current through the load is greater than that determined by the apparent power rating, the transformer may be permanently damaged. Note, however, that since the load is purely capacitive, the average power to the load is zero. The wattage rating is therefore meaningless regarding the ability of this load to damage the transformer.

The transformation ratio of the transformer under discussion can be either of two values. If the secondary voltage is 2000 V, the transformation ratio is $a = N_p/N_s = V_g/V_L = 100 \text{ V}/2000 \text{ V} = 1/20$, and the transformer is a step-up transformer. If the secondary voltage is 100 V, the transformation ratio is $a = N_p/N_s = V_g/V_L = 2000 \text{ V}/100 \text{ V} = 20$, and the transformer is a step-down transformer.

The rated primary current can be determined by applying Eq. (22.18):

$$I_p = \frac{I_s}{a}$$

which is equal to $[2.5 \text{ A}/(1/20)] = 50$ A if the secondary voltage is 2000 V, and $(50 \text{ A}/20) = 2.5$ A if the secondary voltage is 100 V.

To explain the necessity for including the frequency in the nameplate data, consider Eq. (22.12):

$$E_p = 4.44 f_p N_p \Phi_m$$

and the *B-H* curve for the iron core of the transformer (Fig. 22.35).

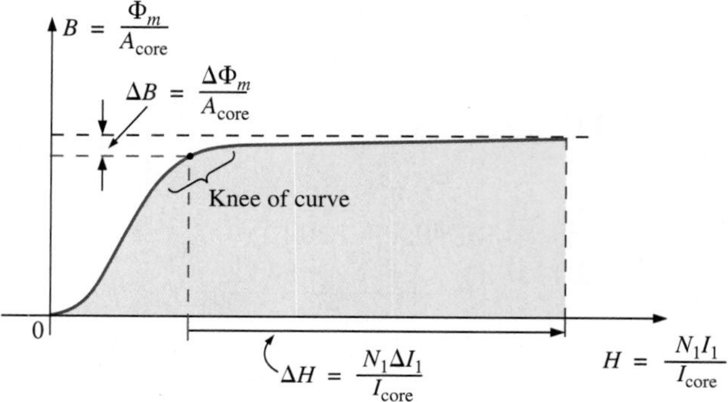

FIG. 22.35
Demonstrating why the frequency of application is important for transformers.

The point of operation on the *B-H* curve for most transformers is at the knee of the curve. If the frequency of the applied signal drops, and N_p and E_p remain the same, then Φ_m must increase in magnitude, as determined by Eq. (22.12):

$$\Phi_m \uparrow = \frac{E_p}{4.44 f_p \downarrow N_p}$$

The result is that B increases, as shown in Fig. 22.35, causing H to increase also. The resulting ΔI could cause a very high current in the primary, resulting in possible damage to the transformer.

22.11 TYPES OF TRANSFORMERS

Transformers are available in many different shapes and sizes. Some of the more common types include the power transformer, audio transformer, IF (intermediate frequency) transformer, and RF (radio frequency) transformer. Each is designed to fulfill a particular requirement in a specific area of application. The symbols for some of the basic types of transformers are shown in Fig. 22.36.

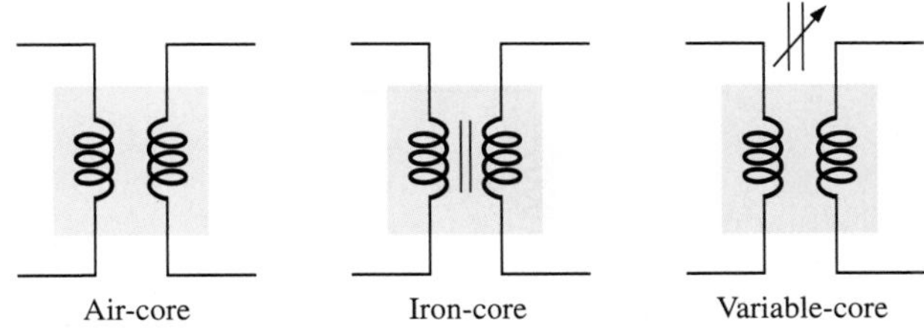

FIG. 22.36
Transformer symbols.

The method of construction varies from one transformer to another. Two of the many different ways in which the primary and secondary coils can be wound around an iron core are shown in Fig. 22.37. In either case, the core is made of laminated sheets of ferromagnetic material separated by an insulator to reduce the eddy current losses. The sheets themselves also contain a small percentage of silicon to increase the electrical resistivity of the material and further reduce the eddy current losses.

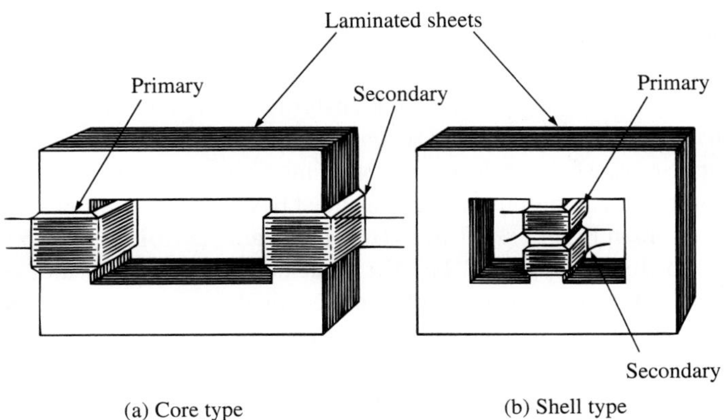

FIG. 22.37
Types of ferromagnetic core construction.

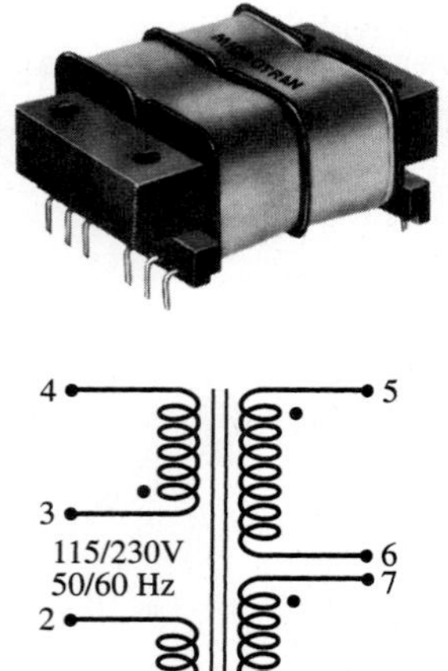

FIG. 22.38
Split bobbin, low-profile power transformer.
(Courtesy of Microtran Company, Inc.)

A variation of the core-type transformer appears in Fig. 22.38. This transformer is designed for low-profile (the 2.5 VA size has a maximum height of only 0.65 in.) applications in power, control, and instrumentation applications. There are actually two transformers on the same core, with the primary and secondary of each wound side by side. The schematic representation appears in the same figure. Each set of terminals on the left can accept 115 V at 50 or 60 Hz, whereas each side of the output provides 230 V at the same frequency. Note the dot convention, as described earlier in the chapter.

The **autotransformer** [Fig. 22.39(b)] is a type of power transformer that, instead of employing the two-circuit principle (complete isolation between coils), has one winding common to both the input and the output circuits. The induced voltages are related to the turns ratio in the same manner as that described for the two-circuit transformer. If the proper connection is used, a two-circuit power transformer can be used as an autotransformer. The advantage of using it as an autotransformer is that a larger apparent power can be transformed. This can be demonstrated by the two-circuit transformer of Fig. 22.39(a), shown in Fig. 22.39(b) as an autotransformer.

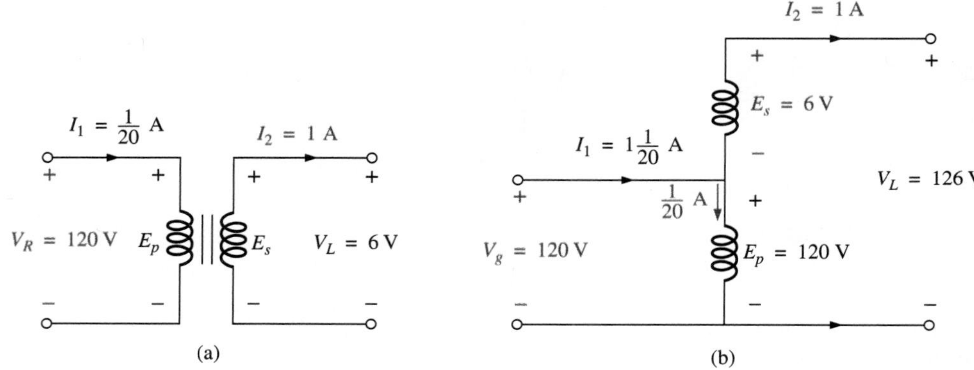

FIG. 22.39
(a) Two-circuit transformer; (b) autotransformer.

For the two-circuit transformer, note that $S = (\frac{1}{20}\text{ A})(120\text{ V}) = 6$ VA, whereas for the autotransformer, $S = (1\frac{1}{20}\text{ A})(120\text{ V}) = 126$ VA, which is many times that of the two-circuit transformer. Note also that the current and voltage of each coil are the same as those for the two-circuit configuration. The disadvantage of the autotransformer is obvious: loss of the isolation between the primary and secondary circuits.

A pulse transformer designed for printed-circuit applications where high-amplitude, long-duration pulses must be transferred without saturation appears in Fig. 22.40. Turns ratios are available from 1 : 1 to 5 : 1 at maximum line voltages of 240 V rms at 60 Hz. The upper unit is for printed-circuit applications with isolated dual primaries, whereas the lower unit is the bobbin variety with a single primary winding.

Two miniature (¼ in. by ¼ in.) transformers with plug-in or insulated leads appear in Fig. 22.41, along with their schematic representations. Power ratings of 100 mW or 125 mW are available with a variety of turns ratios, such as 1 : 1, 5 : 1, 9.68 : 1, and 25 : 1.

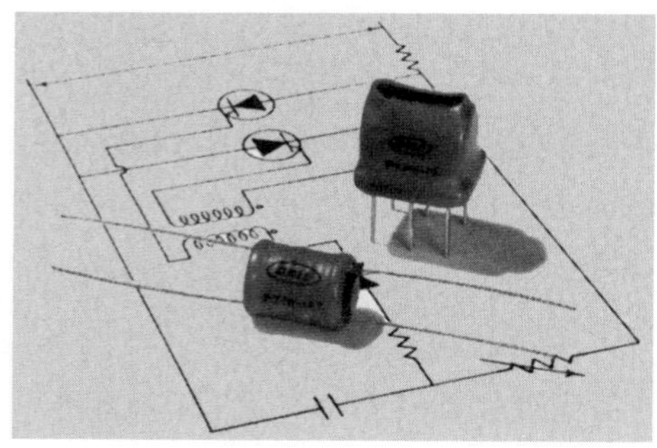

FIG. 22.40
Pulse transformers.
(Courtesy of DALE Electronics, Inc.)

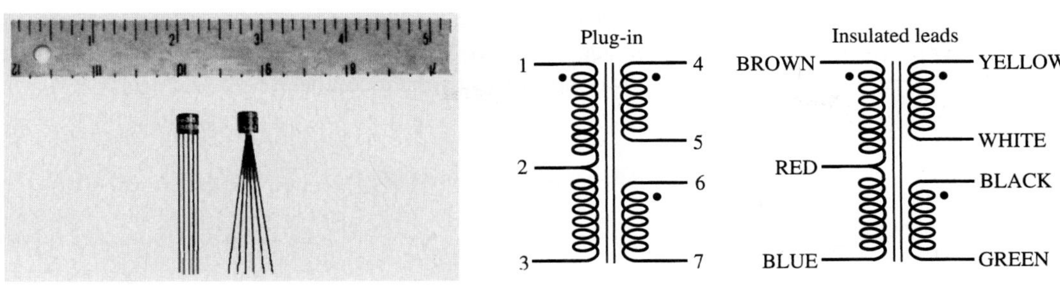

FIG. 22.41
Miniature transformers.
(Courtesy of PICO Electronics, Inc.)

22.12 TAPPED AND MULTIPLE-LOAD TRANSFORMERS

For the **center-tapped** (primary) **transformer** in Fig. 22.42, where the voltage from the center tap to either outside lead is defined as $E_p/2$, the relationship between E_p and E_s is

$$\frac{\mathbf{E}_p}{\mathbf{E}_s} = \frac{N_p}{N_s} \tag{22.37}$$

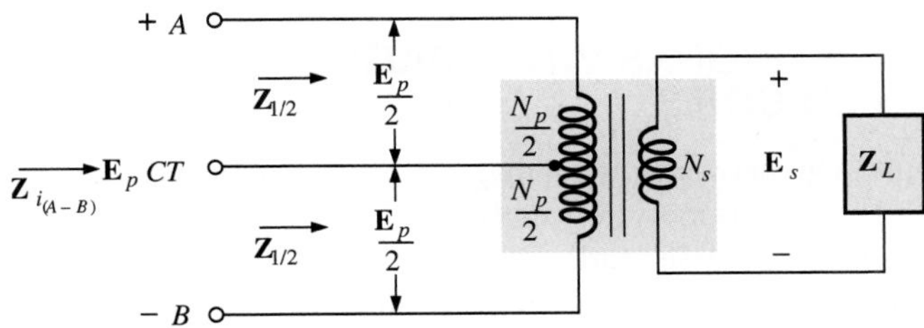

FIG. 22.42
Ideal transformer with a center-tapped primary.

For each half-section of the primary,

$$\mathbf{Z}_{1/2} = \left(\frac{N_p/2}{N_s}\right)^2 \mathbf{Z}_L = \frac{1}{4}\left(\frac{N_p}{N_s}\right)^2 \mathbf{Z}_L$$

with

$$\mathbf{Z}_{i_{(A-B)}} = \left(\frac{N_p}{N_s}\right)^2 \mathbf{Z}_L$$

Therefore,

$$\mathbf{Z}_{1/2} = \frac{1}{4}\mathbf{Z}_i \tag{22.38}$$

For the **multiple-load transformer** in Fig. 22.43, the following equations apply:

$$\frac{\mathbf{E}_i}{\mathbf{E}_2} = \frac{N_1}{N_2} \qquad \frac{\mathbf{E}_i}{\mathbf{E}_3} = \frac{N_1}{N_3} \qquad \frac{\mathbf{E}_2}{\mathbf{E}_3} = \frac{N_2}{N_3} \tag{22.39}$$

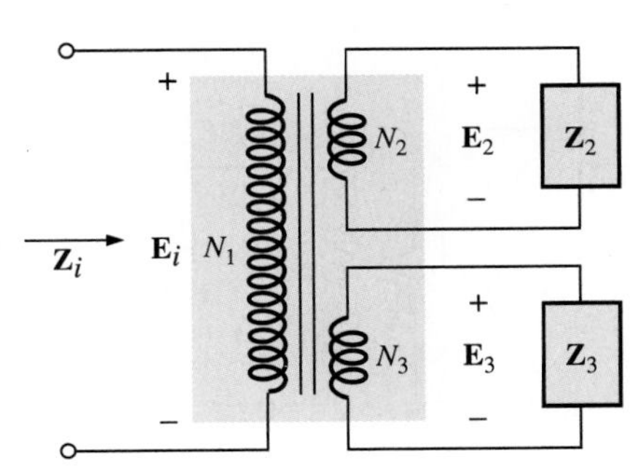

FIG. 22.43
Ideal transformer with multiple loads.

The total input impedance can be determined by first noting that, for the ideal transformer, the power delivered to the primary is equal to the power dissipated by the load; that is,

$$P_1 = P_{L_2} + P_{L_3}$$

and, for resistive loads ($\mathbf{Z}_i = R_i$, $\mathbf{Z}_2 = R_2$, and $\mathbf{Z}_3 = R_3$),

$$\frac{E_i^2}{R_i} = \frac{E_2^2}{R_2} = \frac{E_3^2}{R_3}$$

or, since $$E_2 = \frac{N_2}{N_1}E_i \quad \text{and} \quad E_3 = \frac{N_3}{N_1}E_1$$

then $$\frac{E_i^2}{R_i} = \frac{[(N_2/N_1)E_i]^2}{R_2} + \frac{[(N_3/N_1)E_i]^2}{R_3}$$

and $$\frac{E_i^2}{R_i} = \frac{E_i^2}{(N_1/N_2)^2R_2} + \frac{E_i^2}{(N_1/N_3)^2R_3}$$

Thus, $$\boxed{\frac{1}{R_i} = \frac{1}{(N_1/N_2)^2R_2} + \frac{1}{(N_1/N_3)^2R_3}} \tag{22.40}$$

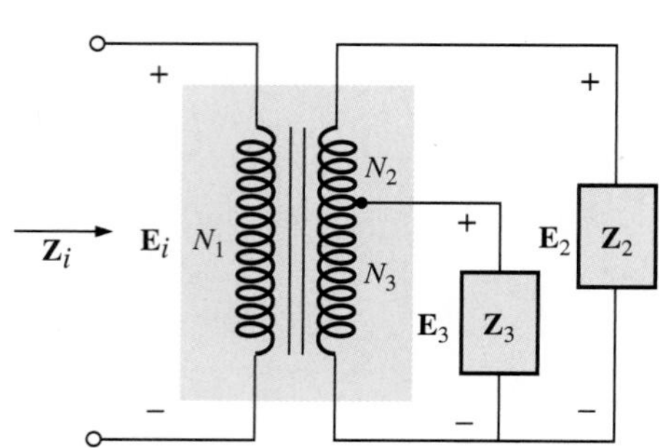

FIG. 22.44
Ideal transformer with a tapped secondary and multiple loads.

indicating that the load resistances are reflected in parallel.

For the configuration in Fig. 22.44, with E_2 and E_3 defined as shown, Eqs. (22.39) and (22.40) are applicable.

22.13 NETWORKS WITH MAGNETICALLY COUPLED COILS

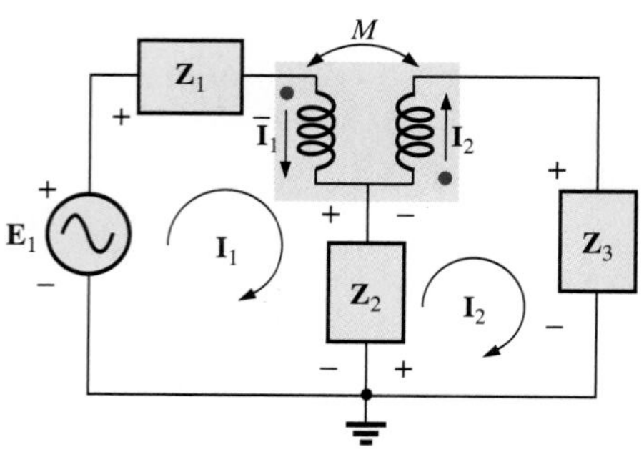

FIG. 22.45
Applying mesh analysis to magnetically coupled coils.

For multiloop networks with magnetically coupled coils, the mesh-analysis approach is most frequently applied. A firm understanding of the dot convention discussed earlier should make the writing of the equations quite direct and free of errors. Before writing the equations for any particular loop, first determine whether the mutual term is positive or negative, keeping in mind that it will have the same sign as that for the other magnetically coupled coil. For the two-loop network in Fig. 22.45, for example, the mutual term has a positive sign since the current through each coil leaves the dot. For the primary loop,

$$\mathbf{E}_1 - \mathbf{I}_1\mathbf{Z}_1 - \mathbf{I}_1\mathbf{Z}_{L_1} - \mathbf{I}_2\mathbf{Z}_m - \mathbf{Z}_2(\mathbf{I}_1 - \mathbf{I}_2) = 0$$

where M of $\mathbf{Z}_m = \omega M \angle 90°$ is positive, and

$$\mathbf{I}_1(\mathbf{Z}_1 + \mathbf{Z}_{L_1} + \mathbf{Z}_2) - \mathbf{I}_2(\mathbf{Z}_2 - \mathbf{Z}_m) = \mathbf{E}_1$$

Note in the above that the mutual impedance was treated as if it were an additional inductance in series with the inductance L_1 having a sign determined by the dot convention and the voltage across which is determined by the current in the magnetically coupled loop.

For the secondary loop,

$$-\mathbf{Z}_2(\mathbf{I}_2 - \mathbf{I}_1) - \mathbf{I}_2\mathbf{Z}_{L_2} - \mathbf{I}_1\mathbf{Z}_m - \mathbf{I}_2\mathbf{Z}_3 = 0$$

or $$\mathbf{I}_2(\mathbf{Z}_2 + \mathbf{Z}_{L_2} + \mathbf{Z}_3) - \mathbf{I}_1(\mathbf{Z}_2 - \mathbf{Z}_m) = 0$$

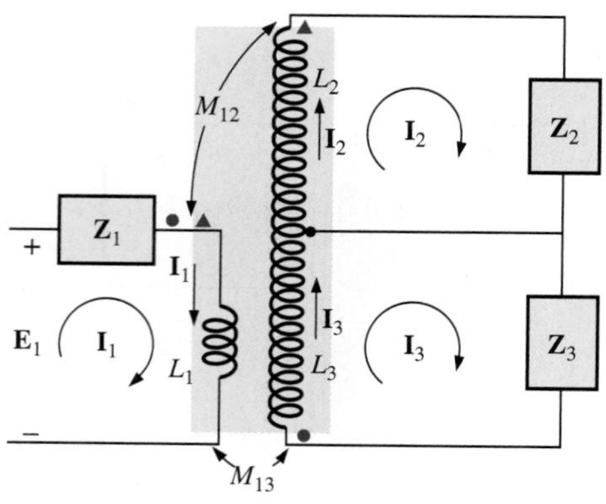

FIG. 22.46
Applying mesh analysis to a network with two magnetically coupled coils.

For the network in Fig. 22.46, we find a mutual term between L_1 and L_2 and L_1 and L_3, labeled M_{12} and M_{13}, respectively.

For the coils with the dots (L_1 and L_3), since each current through the coils leaves the dot, M_{13} is positive for the chosen direction of I_1 and I_3.

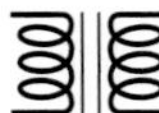

However, since the current I_1 leaves the dot through L_1, and I_2 enters the dot through coil L_2, M_{12} is negative. Consequently, for the input circuit,

$$\mathbf{E}_1 - \mathbf{I}_1\mathbf{Z}_1 - \mathbf{I}_1\mathbf{Z}_{L_1} - \mathbf{I}_2(-\mathbf{Z}_{m_{12}}) - \mathbf{I}_3\mathbf{Z}_{m_{13}} = 0$$

or

$$\mathbf{E}_1 - \mathbf{I}_1(\mathbf{Z}_1 + \mathbf{Z}_{L_1}) + \mathbf{I}_2\mathbf{Z}_{m_{12}} - \mathbf{I}_3\mathbf{Z}_{m_{13}} = 0$$

For loop 2,

$$-\mathbf{I}_2\mathbf{Z}_2 - \mathbf{I}_2\mathbf{Z}_{L_2} - \mathbf{I}_1(-\mathbf{Z}_{m_{12}}) = 0$$

$$-\mathbf{I}_1\mathbf{Z}_{m_{12}} + \mathbf{I}_2(\mathbf{Z}_2 + \mathbf{Z}_{L_2}) = 0$$

and for loop 3,

$$-\mathbf{I}_3\mathbf{Z}_3 - \mathbf{I}_3\mathbf{Z}_{L_3} - \mathbf{I}_1\mathbf{Z}_{m_{13}} = 0$$

or

$$\mathbf{I}_1\mathbf{Z}_{m_{13}} + \mathbf{I}_3(\mathbf{Z}_3 + \mathbf{Z}_{L_3}) = 0$$

In determinant form,

$$\begin{array}{llll}
\mathbf{I}_1(\mathbf{Z}_1 + \mathbf{Z}_{L_1}) & - \mathbf{I}_2\mathbf{Z}_{m_{12}} & + \mathbf{I}_3\mathbf{Z}_{m_{13}} & = \mathbf{E}_1 \\
-\mathbf{I}_1\mathbf{Z}_{m_{12}} & + \mathbf{I}_2(\mathbf{Z}_2 + \mathbf{Z}_{L_{12}}) & + 0 & = 0 \\
\mathbf{I}_1\mathbf{Z}_{m_{13}} & + 0 & + \mathbf{I}_3(\mathbf{Z}_3 + \mathbf{Z}_{13}) & = 0
\end{array}$$

22.14 APPLICATIONS

The transformer has appeared throughout the text in a number of described applications, from the basic dc supply to the soldering gun to the flyback transformer of a simple flash camera. Transformers were used to increase or decrease the voltage or current level, to act as an impedance matching device, or in some cases to play a dual role of transformer action and reactive element. They are so common in such a wide variety of systems that it is important to become very familiar with their general characteristics. For most applications, transformer design can be considered 100% efficient. That is, the power applied is the power delivered to the load. In general, however, transformers are frequently the largest element of a design and because of the nonlinearity of the *B-H* curve can cause some distortion of the transformed waveform. Therefore, they are useful only in situations where the applied voltage is changing with time. The application of a dc voltage to the primary results in 0 V at the secondary, but the application of a voltage that changes with time, no matter what its general appearance, results in a voltage on the secondary. Remember that even though it can provide isolation between the primary and secondary circuits, a transformer can transform the load impedance to the primary circuit at a level that can significantly impact on the behavior of the network. Even the smallest impedance in the secondary can be made to appear very large in the primary when a step-down transformer is used.

Transformers, like every other component you may use, have power ratings. The larger the power rating, the larger the resulting transformer, primarily because of the larger conductors in the windings to handle the current. The size of a transformer is also a function of the frequency involved. The lower the frequency, the larger the required transformer, as easily recognized by the size of large power transformers (also affected by the current levels as mentioned above). For the same power level, the higher the frequency of transformation, the smaller the transformer can be. Because of eddy current and hysteresis losses in a transformer, the design of the core is quite important. A solid core would introduce high levels of such losses, whereas a core constructed of sheets of high-permeability steel with the proper insulation between the sheets would reduce the losses significantly.

Although very fundamental in their basic structure, transformers are recognized as one of the major building blocks of electrical and electronic systems. There isn't a publication on new components published that does not include a new design for the variety of applications developing every day.

Low-Voltage Compensation

At times during the year, peak demands from the power company can result in a reduced voltage down the line. In midsummer, for example, the line voltage may drop from 120 V to 100 V because of the heavy load often due primarily to air conditioners. However, air conditioners do not run as well under low-voltage conditions, so the following option using an autotransformer may be the solution.

In Fig. 22.47(a), an air conditioner drawing 10 A at 120 V is connected through an autotransformer to the available supply that has dropped to 100 V. Assuming 100% efficiency, the current drawn from the line would have to be 12 A to ensure that $P_i = P_o = 1200$ W. Using the analysis introduced in Section 22.11, we find that the current in the primary winding is 2 A with 10 A in the secondary. The 12 A exist only in the line connecting the source to the primary. If the voltage level is increased using the traditional step-up transformer shown in Fig. 22.47(b), the same currents result at the source and load. However, note that the current through the primary is now 12 A which is 6 times that in the autotransformer. The result is that the winding in the autotransformer can be much thinner due to the significantly lower current level.

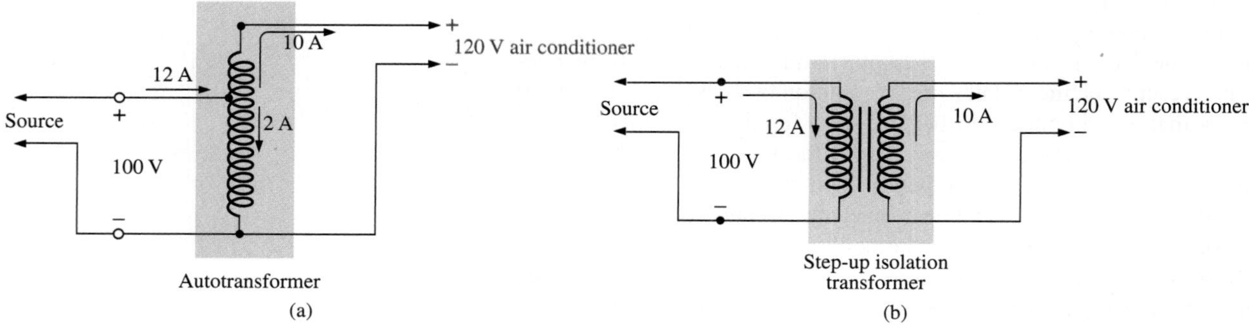

FIG. 22.47
Maintaining a 120 V supply for an air conditioner: (a) using an autotransformer; (b) using a traditional step-up transformer.

Let us now examine the turns ratio required and the number of turns involved for each setup (associating one turn with each volt of the primary and secondary).

For the autotransformer:

$$\frac{N_s}{N_p} = \frac{V_s}{V_p} = \frac{10\text{ V}}{100\text{ V}} \Rightarrow \frac{10\text{ t}}{100\text{ t}}$$

For the traditional transformer:

$$\frac{N_s}{N_p} = \frac{V_s}{V_p} = \frac{120\text{ V}}{100\text{ V}} \Rightarrow \frac{120\text{ t}}{100\text{ t}}$$

In total, therefore, the autotransformer has only 10 turns in the secondary, whereas the traditional has 120. For the autotransformer, we need only 10 turns of heavy wire to handle the current of 10 A, not the full 120 required for the traditional transformer. In addition, the total number of

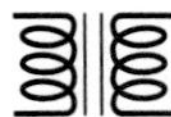

turns for the autotransformer is 110 compared to 220 for the traditional transformer.

The net result of all the above is that even though the protection offered by the isolation feature is lost, the autotransformer can be much smaller in size and weight and, therefore, less costly.

Ballast Transformer

Until just recently, all fluorescent lights like those in Fig. 22.48(a) had a ballast transformer as shown in Fig. 22.48(b). In many cases, its weight alone is almost equal to that of the fixture itself. In recent years, a solid-state equivalent transformer has been developed that in time will replace most of the ballast transformers.

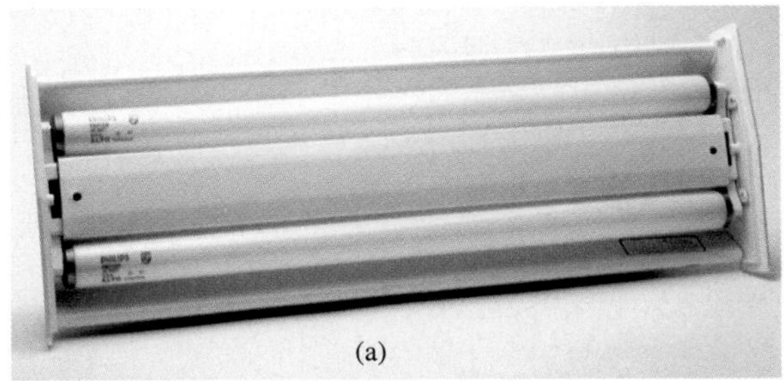

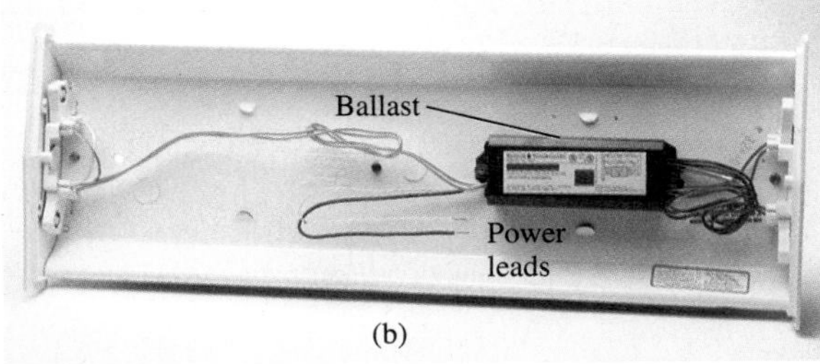

FIG. 22.48
Fluorescent lamp: (a) general appearance; (b) internal view with ballast.

The basic connections for a single-bulb fluorescent light are provided in Fig. 22.49(a). Note that the transformer is connected as an autotransformer with the full applied 120 V across the primary. When the switch is closed, the applied voltage and the voltage across the secondary will add and establish a current through the filaments of the fluorescent bulb. The starter is initially a short circuit to establish the continuous path through the two filaments. In older fluorescent bulbs, the starter was a cylinder with two contacts, as shown in Fig. 22.49(b), which had to be replaced on occasion. It sat right under the fluorescent bulb near one of the bulb connections. Now, as shown by the sketch of the inside of a ballast transformer in Fig. 22.49(c), the starter is now commonly built into the ballast and can no longer be replaced. The voltage established by the autotransformer action is sufficient to heat the filaments but not light the fluorescent bulb. The fluorescent lamp is a long tube with a coating of fluorescent paint on the inside. It is filled with an inert gas and a small amount of liquid mercury. The distance between the electrodes at the ends of the lamp is too much for the applied autotransformer voltage to establish conduction. To overcome this problem, the filaments are first heated as described above to convert the mercury (a good conductor)

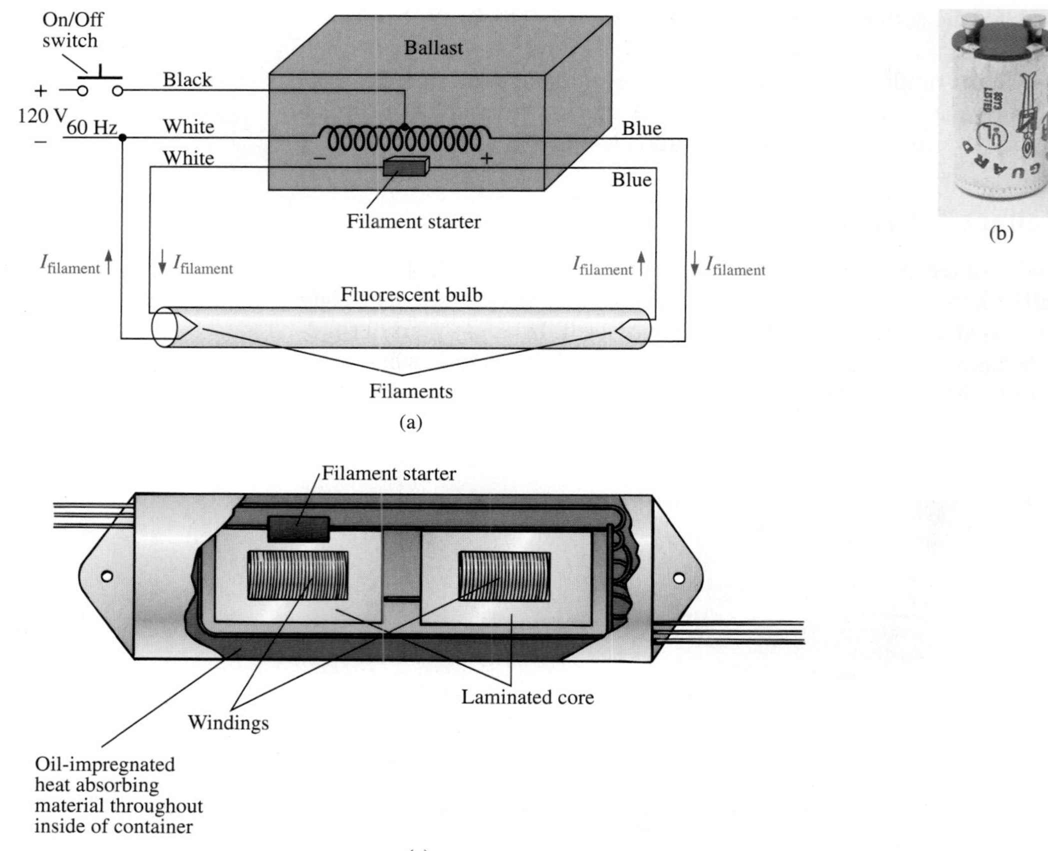

FIG. 22.49
(a) Schematic of single-bulb fluorescent lamp; (b) starter; (c) internal view of ballast transformer.

from a liquid to a gas. Conduction can then be established by the application of a large potential across the electrodes. This potential is established when the starter (a thermal switch that opens when it reaches a particular temperature) opens and causes the inductor current to drop from its established level to zero amperes. This quick drop in current establishes a very high spike in voltage across the coils of the autotransformer as determined by $v_L = L(di_L/dt)$. This significant spike in voltage also appears across the bulb and establishes current between the electrodes. Light is then given off as the electrons hit the fluorescent surface on the inside of the tube. It is the persistence of the coating that helps hide the oscillation in conduction level due to the low-frequency (60 Hz) power that can result in a flickering light. The starter remains open until the next time the bulb is turned on. The flow of charge between electrodes is then maintained solely by the voltage across the autotransformer. This current is relatively low in magnitude because of the reactance of the secondary winding in the resulting series circuit. In other words, the autotransformer has shifted to one that is now providing a reactance to the secondary circuit to limit the current through the bulb. Without this limiting factor, the current through the bulb would be too high, and the bulb would quickly burn out. This action of the coils of the transformer generating the required voltage and then acting as a coil to limit the current has resulted in the general terminology of *swinging choke.*

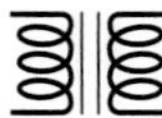

The fact that the light is not generated by an *IR* drop across a filament of a bulb is the reason fluorescent lights are so energy efficient. In fact, in an incandescent bulb, about 75% of the applied energy is lost in heat, with only 25% going to light emission. In a fluorescent bulb, more than 70% goes to light emission and 30% to heat losses. As a rule of thumb, the lighting from a 40 W fluorescent lamp [such as the unit in Fig. 22.48(a) with its two 20 W bulbs] is equivalent to that of a 100 W incandescent bulb.

One other interesting difference between incandescent and fluorescent bulbs is the method of determining whether they are good or bad. For the incandescent light, it is immediately obvious when it fails to give light at all. For the fluorescent bulb, however, assuming that the ballast is in good working order, the bulb begins to dim as its life wears on. The electrodes become coated and less efficient, and the coating on the inner surface begins to deteriorate.

Rapid-start fluorescent lamps are different in operation only in that the voltage generated by the transformer is sufficiently large to atomize the gas upon application and initiate conduction, thereby removing the need for a starter and eliminating the warm-up time of the filaments. In time, the solid-state ballast will probably be the unit of choice because of its quick response, higher efficiency, and lighter weight, but the transition will take some time. The basic operation will remain the same, however.

Because of the fluorine gas (hence the name *fluorescent* bulb) and the mercury in fluorescent lamps, they must be discarded with care. Ask your local disposal facility where to take bulbs. Breaking them for insertion in a plastic bag could be very dangerous. If you happen to break a bulb and get cut in the process, go immediately to a medical facility since you could sustain fluorine or mercury poisoning.

22.15 COMPUTER ANALYSIS

PSpice

Transformer (Controlled Sources) The simple transformer configuration in Fig. 22.50 is now investigated using controlled sources to mimic the behavior of the transformer as defined by its basic voltage and current relationships.

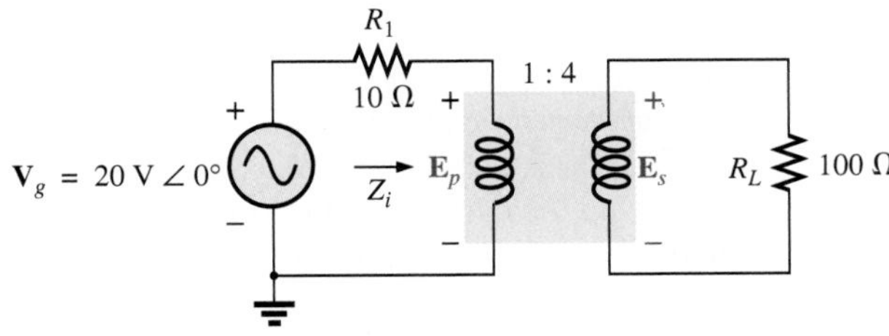

FIG. 22.50
Applying PSpice to a step-up transformer.

For comparison purposes, a theoretical solution of the network yields the following:

$$
\begin{aligned}
Z_i &= a^2 Z_L \\
&= \left(\frac{1}{4}\right)^2 100\ \Omega \\
&= 6.25\ \Omega
\end{aligned}
$$

and $$E_p = \frac{(6.25\ \Omega)(20\ \text{V})}{6.25\ \Omega + 10\ \Omega} = 7.692\ \text{V}$$

with $$E_s = \frac{1}{a}E_p = \frac{1}{(1/4)}(7.692\ \text{V}) = 4(7.692\ \text{V}) = 30.77\ \text{V}$$

and $$V_L = E_s = \mathbf{30.77\ V}$$

For the ideal transformer, the secondary voltage is defined by $E_s = N_s/N_p(E_p)$ which is $E_s = 4E_p$ for the network in Fig. 22.50. The fact that the magnitude of one voltage is controlled by another requires that we use the **Voltage-Controlled Voltage Source (VCVS)** source in the **ANALOG** library. It appears as **E** in the **Parts List** and has the format appearing in Fig. 22.51. The sensing voltage is **E1,** and the controlled voltage appears across the two terminals of the circular symbol for a voltage source. Double-click on the source symbol to set the **GAIN** to 4 for this example. Note in Fig. 22.51 that the sensing voltage is the primary voltage of the circuit in Fig. 22.50, and the output voltage is connected directly to the load resistor **RL.** There is no real problem making the necessary connections because of the format of the **E** source.

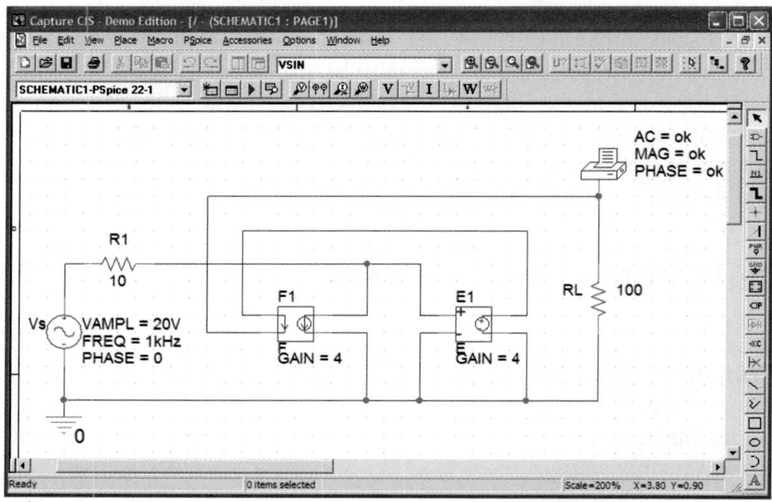

FIG. 22.51
Using PSpice to determine the magnitude and phase angle for the load voltage of the network in Fig. 22.50.

The next step is to set up the current relationship for the transformer. Since the magnitude of one current is controlled by the magnitude of another current in the same configuration, a **Current-Controlled Current Source (CCCS)** must be employed. It also appears in the **ANALOG** library under the **Parts List** as **F** and has the format appearing in Fig. 22.51. Note that both currents have a direction associated with them. For the ideal transformer, $I_p = N_s/N_p\ (I_s)$ which is $I_p = 4I_s$ for the network in Fig. 22.50. The gain for the part can be set using the same procedure defined for the **E** source. Since the secondary current is the controlling current, its level must be fed into the **F** source in the same direction as indicated in the controlled source. When making this connection, be sure to click the wire in location before crossing the wire of the primary circuit and then clicking it again after crossing the wire. If you do this properly, a connection point indicated by a small red dot will not appear. The controlled current I_{R_1} can be connected as shown because the connection

E1 is only sensing a voltage level, essentially has infinite impedance, and can be looked upon as an open circuit. In other words, the current through **R1** will be the same as through the controlled source of **F.**

A simulation was set up with **AC Sweep** and 1 kHz for the **Start** and **End Frequencies.** One data point per decade was selected, and the simulation was initiated. After the **SCHEMATIC1** screen appeared, the window was exited, and **PSpice-View Output File** was selected to result in the **AC ANALYSIS** solution of Fig. 22.52. Note that the voltage is 30.77 V, which is an exact match with the theoretical solution.

```
** Profile: "SCHEMATIC1-PSpice 22-1"  [ C:\ICA11\PSpice\pspice 22-1-
pspicefiles\schematic1\PSpice 22-1.sim ]

****    AC ANALYSIS                    TEMPERATURE =  27.000 DEG C

******************************************************************************

 FREQ         VM(N01850)          VP(N01850)

 1.000E+03    3.077E+01           0.000E+00
```

FIG. 22.52

The output file for the analysis indicated in Fig. 22.51.

Transformer (Library) The same network can be analyzed by choosing one of the transformers from the **EVAL** library as shown in Fig. 22.53. Choose the transformer labeled **K3019PL_3C8,** and place the proper attributes in the **Property Editor** dialog box. The only three required are **COUPLING** set at 1, **L1_TURNS** set at 1, and **L2_TURNS** set at 4. In the **Simulation Settings,** choose **AC Sweep** and use **1MEGHz** for both the **Start and End Frequency** because it was found that it acts as an almost ideal transformer at this frequency—a little bit of run and test. When the simulation is run, the results under **PSpice-View Output File** appear as shown in Fig. 22.54— almost an exact match with the theoretical solution of 30.77 V.

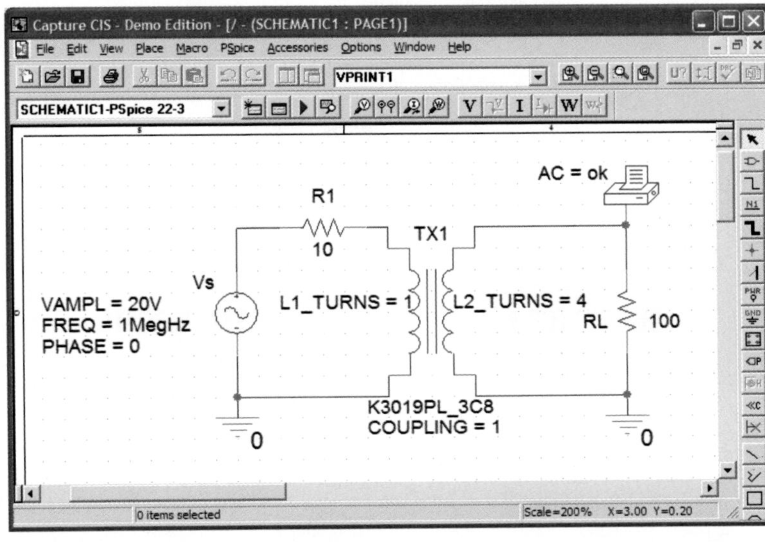

FIG. 22.53

Using a transformer provided in the ***EVAL*** *library to analyze the network in Fig. 22.50.*

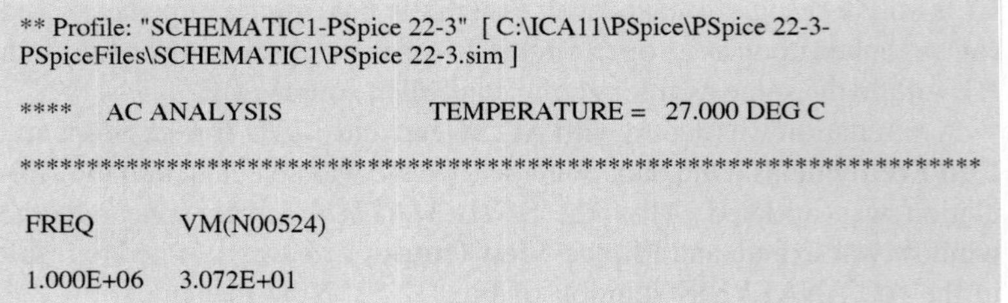
** Profile: "SCHEMATIC1-PSpice 22-3" [C:\ICA11\PSpice\PSpice 22-3-PSpiceFiles\SCHEMATIC1\PSpice 22-3.sim]

**** AC ANALYSIS TEMPERATURE = 27.000 DEG C

****

FREQ	VM(N00524)
1.000E+06	3.072E+01

FIG. 22.54
The output file for the analysis indicated in Fig. 22.53

Multisim

Transformer(Library) Multisim is now used to analyze the same transformer configuration just investigated using PSpice. In Fig. 22.55, obtain the source by first selecting **Place Source** to open the **Select a Component** dialog box. Select **SIGNAL_VOLTAGE** followed by **AC_VOLTAGE** and click **OK.** For the source, peak values are set, hence the difference in the set value in Fig. 22.55 and the rms multimeter reading. Obtain the transformer by selecting **Place Basic-Family-BASIC_VIRTUAL-TS_VIRTUAL.** Then select the turns ratio of 2 to open the **BASIC_VIRTUAL** dialog box. Change the **Primary-to-Secondary Turns Ratio:** to 0.25 for this example.

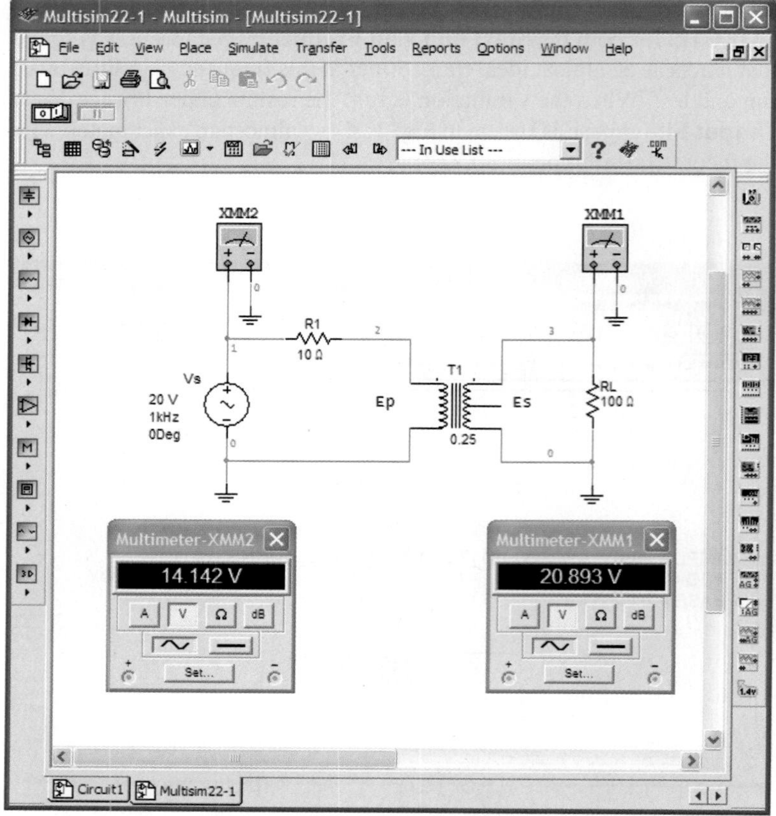

FIG. 22.55

The rest of the configuration is constructed using techniques described earlier. A simulation results in the meter displays in Fig. 22.55. Changing the rms reading of 20.893 V to a peak value results in 29.54 V which is a close match to that obtained using PSpice.

PROBLEMS

SECTION 22.2 Mutual Inductance

1. For the air-core transformer in Fig. 22.56:
   a. Find the value of L_s if the mutual inductance M is equal to 80 mH.
   b. Find the induced voltages e_p and e_s if the flux linking the primary coil changes at the rate of 0.08 Wb/s.
   c. Find the induced voltages e_p and e_s if the current i_p changes at the rate of 0.3 A/ms.

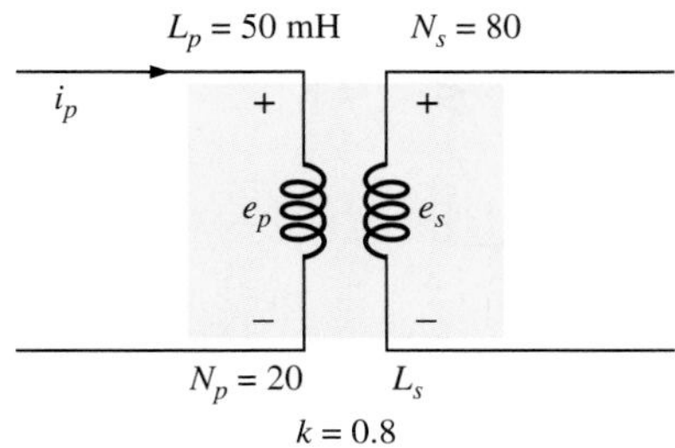

FIG. 22.56
Problems 1, 2, and 3.

2. **a.** Repeat Problem 1 if k is changed to 1.
   b. Repeat Problem 1 if k is changed to 0.2.
   c. Compare the results of parts (a) and (b).

3. Repeat Problem 1 for $k = 0.9$, $N_p = 300$ turns, and $N_s = 25$ turns.

SECTION 22.3 The Iron-Core Transformer

4. For the iron-core transformer ($k = 1$) in Fig. 22.57:
   a. Find the magnitude of the induced voltage E_s.
   b. Find the maximum flux Φ_m.

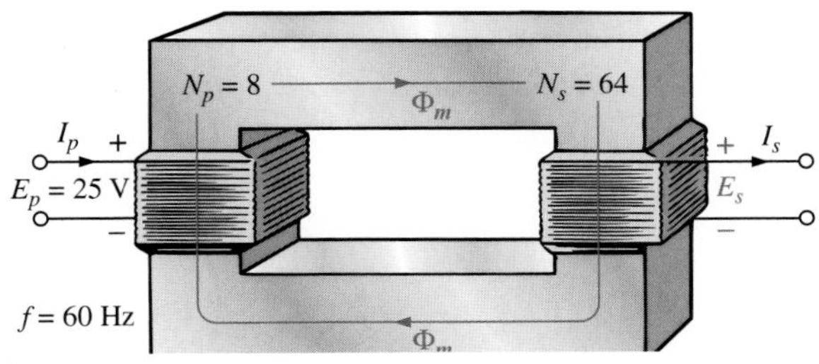

FIG. 22.57
Problems 4, 5, and 7.

5. Repeat Problem 4 for $N_p = 240$ and $N_s = 30$.

6. Find the applied voltage of an iron-core transformer if the secondary voltage is 240 V, and $N_p = 60$ with $N_s = 720$.

7. If the maximum flux passing through the core of Problem 4 is 12.5 mWb, find the frequency of the input voltage.

SECTION 22.4 Reflected Impedance and Power

8. For the iron-core transformer in Fig. 22.58:
   a. Find the magnitude of the current I_L and the voltage V_L if $a = 1/5$, $I_p = 2$ A, and $Z_L = 2\ \Omega$ resistor.
   b. Find the input resistance for the data specified in part (a).

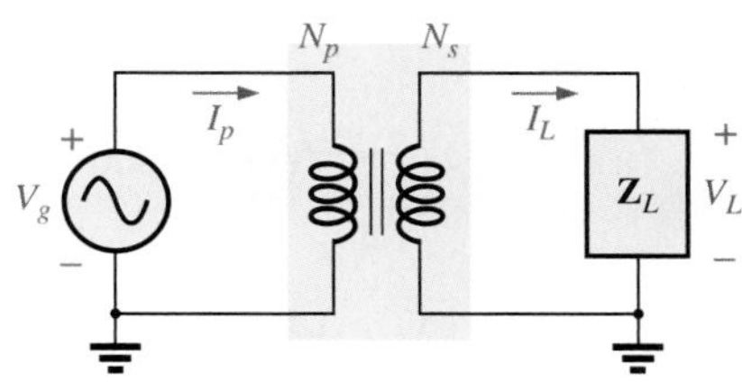

FIG. 22.58
Problems 8 through 12.

9. Find the input impedance for the iron-core transformer of Fig. 22.58 if $a = 2$, $I_p = 4$ A, and $V_g = 1600$ V.

10. Find the voltage V_g and the current I_p if the input impedance of the iron-core transformer in Fig. 22.58 is 4 Ω, and $V_L = 1200$ V and $a = 1/4$.

11. If $V_L = 240$ V, $Z_L = 20\ \Omega$ resistor, $I_p = 0.05$ A, and $N_s = 50$, find the number of turns in the primary circuit of the iron-core transformer in Fig. 22.58.

12. **a.** If $N_p = 400$, $N_s = 1200$, and $V_g = 100$ V, find the magnitude of I_p for the iron-core transformer in Fig. 22.58 if $Z_L = 9\ \Omega + j\ 12\ \Omega$.
   b. Find the magnitude of the voltage V_L and the current I_L for the conditions of part (a).

SECTION 22.5 Impedance Matching, Isolation, and Displacement

13. **a.** For the circuit in Fig. 22.59, find the transformation ratio required to deliver maximum power to the speaker.
   b. Find the maximum power delivered to the speaker.

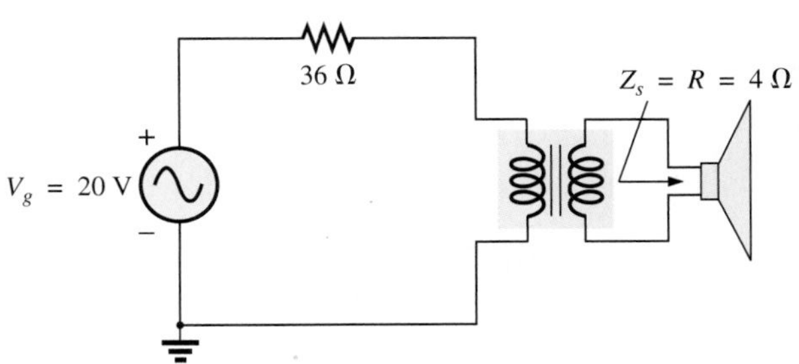

FIG. 22.59
Problem 13.

SECTION 22.6 Equivalent Circuit (Iron-Core Transformer)

14. For the transformer in Fig. 22.60, determine
  a. the equivalent resistance R_e.
  b. the equivalent reactance X_e.
  c. the equivalent circuit reflected to the primary.
  d. the primary current for $V_g = 50\text{ V}\angle 0°$.
  e. the load voltage V_L.
  f. the phasor diagram of the reflected primary circuit.
  g. the new load voltage if we assume the transformer to be ideal with a 4 : 1 turns ratio. Compare the result with that of part (e).

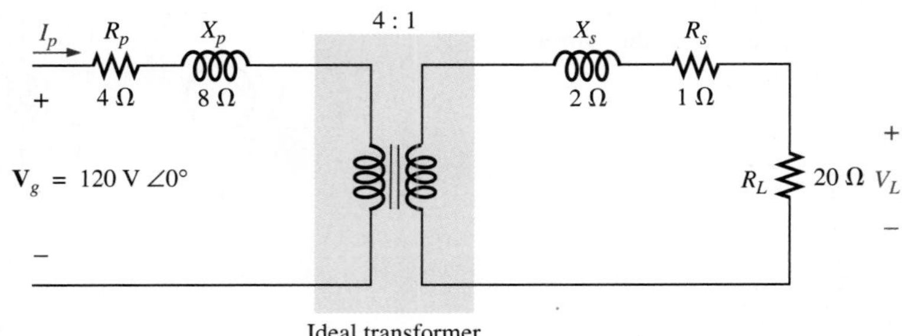

FIG. 22.60
Problems 14 through 16, 30, and 31.

15. For the transformer in Fig. 22.60, if the resistive load is replaced by an inductive reactance of 20 Ω:
  a. Determine the total reflected primary impedance.
  b. Calculate the primary current, $\mathbf{I}_p$.
  c. Determine the voltage across R_e and X_e, and find the reflected load.
  d. Draw the phasor diagram.

16. Repeat Problem 15 for a capacitive load having a reactance of 20 Ω.

SECTION 22.7 Frequency Considerations

17. Discuss in your own words the frequency characteristics of the transformer. Use the applicable equivalent circuit and frequency characteristics appearing in this chapter.

SECTION 22.8 Series Connection of Mutually Coupled Coils

18. Determine the total inductance of the series coils in Fig. 22.61.

19. Determine the total inductance of the series coils in Fig. 22.62.

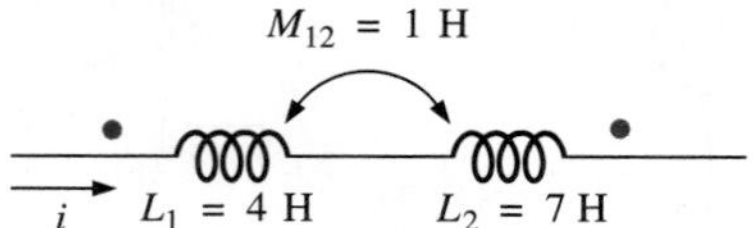

FIG. 22.61
Problem 18.

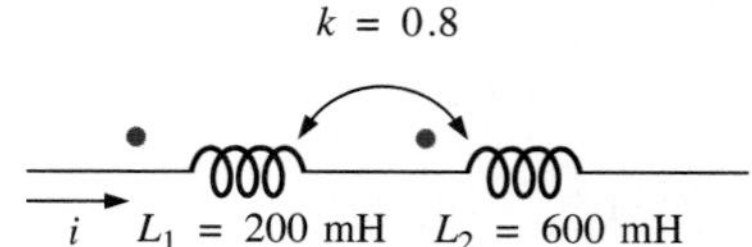

FIG. 22.62
Problem 19.

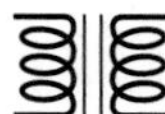

20. Determine the total inductance of the series coils in Fig. 22.63.

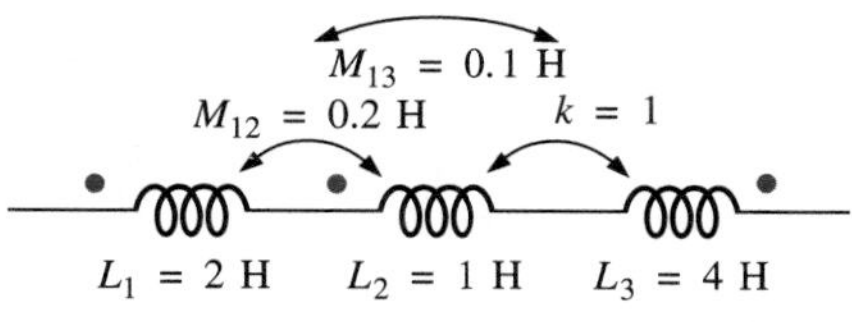

FIG. 22.63
Problem 20.

21. Write the mesh equations for the network in Fig. 22.64.

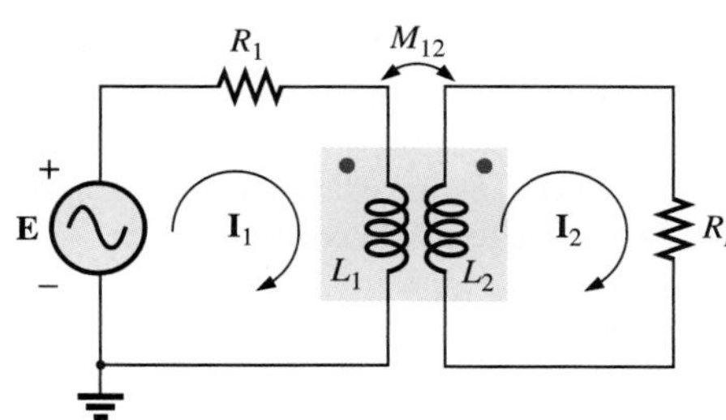

FIG. 22.64
Problem 21.

SECTION 22.9 Air-Core Transformer

22. Determine the input impedance to the air-core transformer in Fig. 22.65. Sketch the reflected primary network.

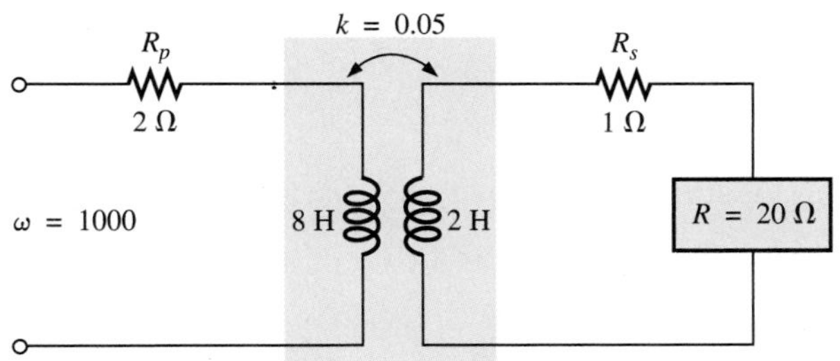

FIG. 22.65
Problems 22 and 32.

SECTION 22.10 Nameplate Data

23. An ideal transformer is rated 10 kVA, 2400/120 V, 60 Hz.
   a. Find the transformation ratio if the 120 V is the secondary voltage.
   b. Find the current rating of the secondary if the 120 V is the secondary voltage.
   c. Find the current rating of the primary if the 120 V is the secondary voltage.
   d. Repeat parts (a) through (c) if the 2400 V is the secondary voltage.

SECTION 22.11 Types of Transformers

24. Determine the primary and secondary voltages and currents for the autotransformer in Fig. 22.66.

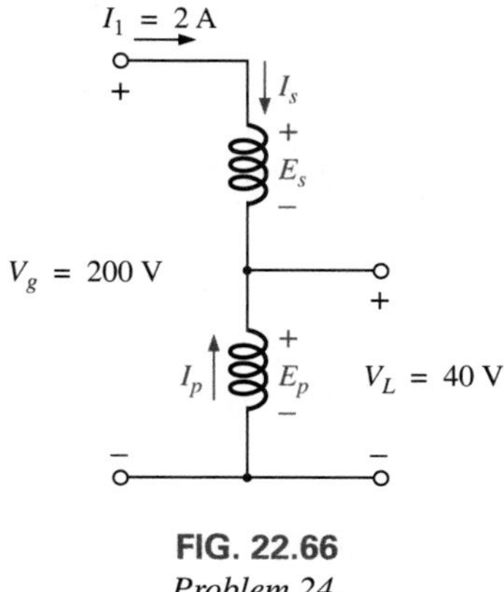

FIG. 22.66
Problem 24.

SECTION 22.12 Tapped and Multiple-Load Transformers

25. For the center-tapped transformer in Fig. 22.42 where N_p = 100, N_s = 25, $Z_L = R\,\angle 0° = 5\ \Omega\,\angle 0°$, and $\mathbf{E}_p = 100\text{ V}\,\angle 0°$.
   a. Determine the load voltage and current.
   b. Find the impedance $\mathbf{Z}_i$.
   c. Calculate the impedance $\mathbf{Z}_{1/2}$.

26. For the multiple-load transformer in Fig. 22.43 where N_1 = 90, N_2 = 15, N_3 = 45, $\mathbf{Z}_2 = R_2\,\angle 0° = 8\ \Omega\,\angle 0°$, $\mathbf{Z}_3 = R_L\,\angle 0° = 5\ \Omega\,\angle 0°$, and $\mathbf{E}_i = 60\text{ V}\,\angle 0°$:
   a. Determine the load voltages and currents.
   b. Calculate $\mathbf{Z}_1$.

27. For the multiple-load transformer in Fig. 22.44 where N_1 = 120, N_2 = 40, N_3 = 30, $\mathbf{Z}_2 = R_2\,\angle 0° = 12\ \Omega\,\angle 0°$, $\mathbf{Z}_3 = R_3\,\angle 0° = 10\ \Omega\,\angle 0°$, and $\mathbf{E}_1 = 120\text{ V}\,\angle 60°$:
   a. Determine the load voltages and currents.
   b. Calculate $\mathbf{Z}_1$.

SECTION 22.13 Networks with Magnetically Coupled Coils

28. Write the mesh equations for the network in Fig. 22.67.

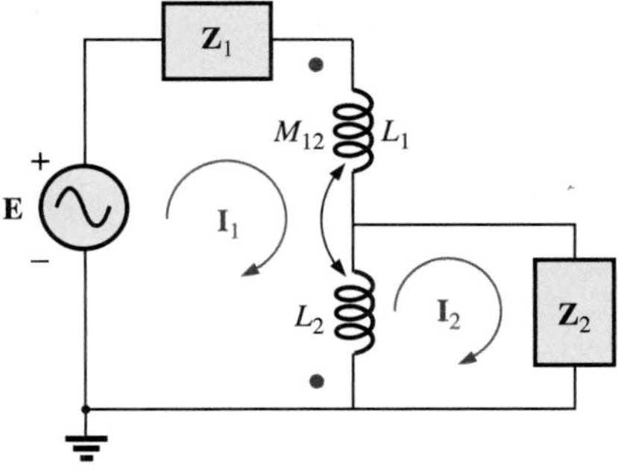

FIG. 22.67
Problem 28.

29. Write the mesh equations for the network in Fig. 22.68.

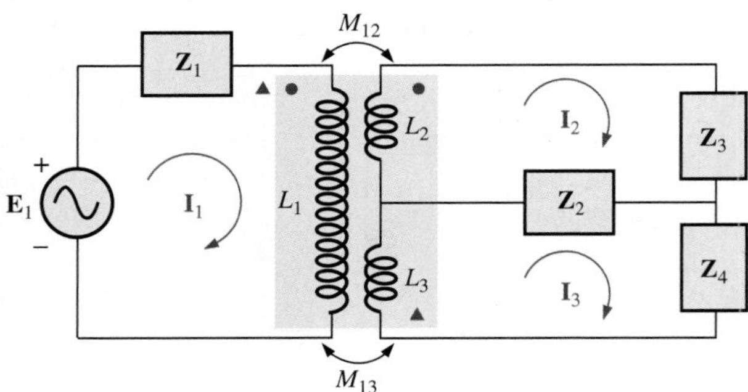

FIG. 22.68
Problem 29.

SECTION 22.15 Computer Analysis

PSpice or Multisim

*30. Generate the schematic for the network in Fig. 22.60, and find the voltage V_L.

*31. Develop a technique using PSpice or Multisim to find the input impedance at the source for the network in Fig. 22.60.

*32. Using a transformer from the library, find the load voltage for the network in Fig. 22.65 for an applied voltage of 40 V $\angle 0°$.

GLOSSARY

Autotransformer A transformer with one winding common to both the primary and the secondary circuits. A loss in isolation is balanced by the increase in its kilovolt-ampere rating.

Coefficient of coupling *(k)* A measure of the magnetic coupling of two coils that ranges from a minimum of 0 to a maximum of 1.

Dot convention A technique for labeling the effect of the mutual inductance on a net inductance of a network or system.

Leakage flux The flux linking the coil that does not pass through the ferromagnetic path of the magnetic circuit.

Loosely coupled A term applied to two coils that have a low coefficient of coupling.

Multiple-load transformers Transformers having more than a single load connected to the secondary winding or windings.

Mutual inductance The inductance that exists between magnetically coupled coils of the same or different dimensions.

Nameplate data Information such as the kilovolt-ampere rating, voltage transformation ratio, and frequency of application that is of primary importance in choosing the proper transformer for a particular application.

Primary The coil or winding to which the source of electrical energy is normally applied.

Reflected impedance The impedance appearing at the primary of a transformer due to a load connected to the secondary. Its magnitude is controlled directly by the transformation ratio.

Secondary The coil or winding to which the load is normally applied.

Step-down transformer A transformer whose secondary voltage is less than its primary voltage. The transformation ratio *a* is greater than 1.

Step-up transformer A transformer whose secondary voltage is greater than its primary voltage. The magnitude of the transformation ratio *a* is less than 1.

Tapped transformer A transformer having an additional connection between the terminals of the primary or secondary windings.

Transformation ratio *(a)* The ratio of primary to secondary turns of a transformer.

Polyphase Systems

Objectives

- *Become familiar with the operation of a three-phase generator and the magnitude and phase relationship between the three phase voltages.*
- *Be able to calculate the voltages and currents for a three-phase Y-connected generator and Y-connected load.*
- *Understand the significance of the phase sequence on the generated voltages of a three-phase Y-connected or Δ-connected generator.*
- *Be able to calculate the voltages and currents for a three-phase Δ-connected generator and Δ-connected load.*
- *Understand how to calculate the real, reactive, and apparent power to all the elements of a Y- or Δ-connected load and be able to measure the power to the load.*

23.1 INTRODUCTION

An ac generator designed to develop a single sinusoidal voltage for each rotation of the shaft (rotor) is referred to as a **single-phase ac generator.** If the number of coils on the rotor is increased in a specified manner, the result is a **polyphase ac generator,** which develops more than one ac phase voltage per rotation of the rotor. In this chapter, the three-phase system is discussed in detail since it is the most frequently used for power transmission.

In general, three-phase systems are preferred over single-phase systems for the transmission of power for many reasons, including the following:

1. Thinner conductors can be used to transmit the same kVA at the same voltage, which reduces the amount of copper required (typically about 25% less) and in turn reduces construction and maintenance costs.
2. The lighter lines are easier to install, and the supporting structures can be less massive and farther apart.
3. Three-phase equipment and motors have preferred running and starting characteristics compared to single-phase systems because of a more even flow of power to the transducer than can be delivered with a single-phase supply.
4. In general, most larger motors are three phase because they are essentially self-starting and do not require a special design or additional starting circuitry.

The frequency generated is determined by the number of poles on the *rotor* (the rotating part of the generator) and the speed with which the shaft is turned. In the United States, the line frequency is 60 Hz, whereas in Europe the chosen standard is 50 Hz. Both frequencies were chosen primarily because they can be generated by a relatively efficient and stable mechanical design that is sensitive to the size of the generating systems and the demand that must be met during peak periods. On aircraft and ships, the demand levels permit the use of a 400 Hz line frequency.

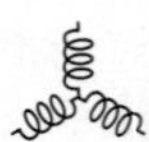

The three-phase system is used by almost all commercial electric generators. This does not mean that single-phase and two-phase generating systems are obsolete. Most small emergency generators, such as the gasoline type, are one-phase generating systems. The two-phase system is commonly used in servomechanisms, which are self-correcting control systems capable of detecting and adjusting their own operation. Servomechanisms are used in ships and aircraft to keep them on course automatically, or, in simpler devices such as a thermostatic circuit, to regulate heat output. In many cases, however, where single-phase and two-phase inputs are required, they are supplied by one and two phases of a three-phase generating system rather than generated independently.

The number of **phase voltages** that can be produced by a polyphase generator is not limited to three. Any number of phases can be obtained by spacing the windings for each phase at the proper angular position around the stator. Some electrical systems operate more efficiently if more than three phases are used. One such system involves the process of rectification, which is used to convert an alternating output to one having an average, or dc, value. The greater the number of phases, the smoother the dc output of the system.

23.2 THREE-PHASE GENERATOR

The three-phase generator in Fig. 23.1(a) has three induction coils placed 120° apart on the stator, as shown symbolically by Fig. 23.1(b). Since the three coils have an equal number of turns, and each coil rotates with the same angular velocity, the voltage induced across each coil has the same peak value, shape, and frequency. As the shaft of the generator is turned by some external means, the induced voltages e_{AN}, e_{BN}, and e_{CN} are generated simultaneously, as shown in Fig. 23.2. Note the 120° phase shift between waveforms and the similarities in appearance of the three sinusoidal functions.

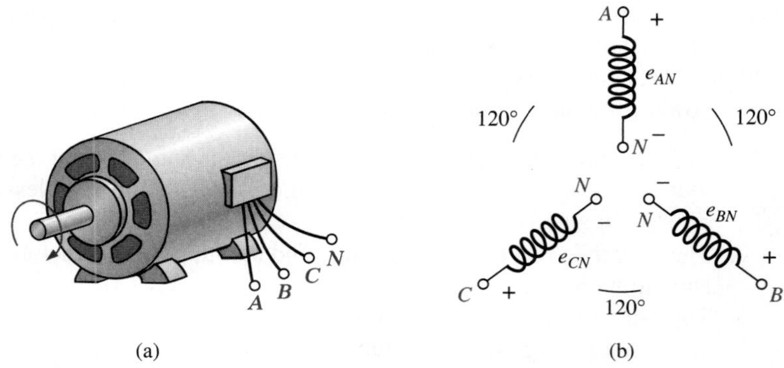

FIG. 23.1
(a) Three-phase generator; (b) induced voltages of a three-phase generator.

In particular, note that

at any instant of time, the algebraic sum of the three phase voltages of a three-phase generator is zero.

This is shown at $\omega t = 0$ in Fig. 23.2, where it is also evident that *when one induced voltage is zero, the other two are 86.6% of their positive or negative maximums. In addition, when any two are equal in magnitude*

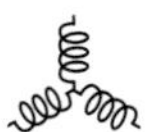

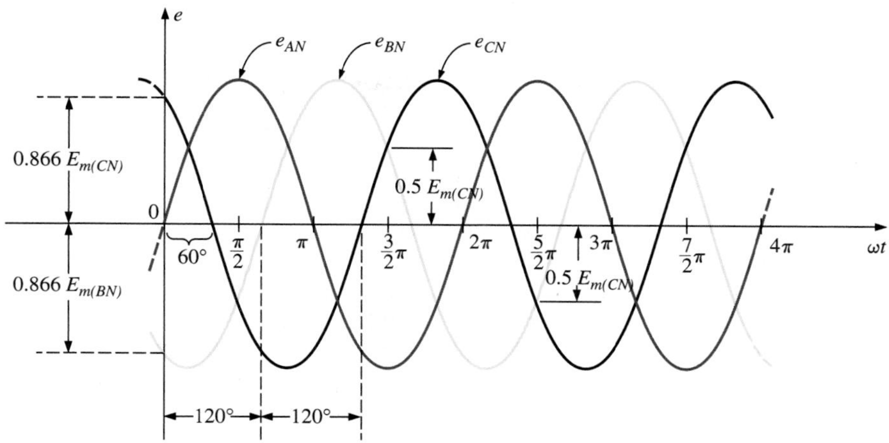

FIG. 23.2
Phase voltages of a three-phase generator.

and sign (at $0.5E_m$), the remaining induced voltage has the opposite polarity and a peak value.

The sinusoidal expression for each of the induced voltages in Fig. 23.2 is

$$\begin{aligned} e_{AN} &= E_{m(AN)} \sin \omega t \\ e_{BN} &= E_{m(BN)} \sin(\omega t - 120°) \\ e_{CN} &= E_{m(CN)} \sin(\omega t - 240°) = E_{m(CN)} \sin(\omega t + 120°) \end{aligned} \qquad \textbf{(23.1)}$$

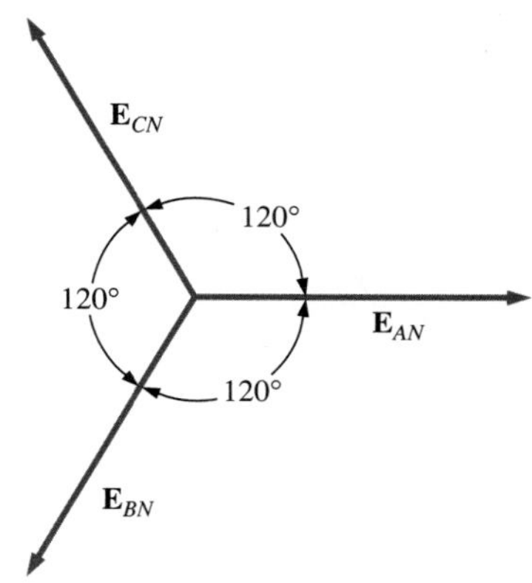

FIG. 23.3
Phasor diagram for the phase voltages of a three-phase generator.

The phasor diagram of the induced voltages is shown in Fig. 23.3, where the effective value of each is determined by

$$E_{AN} = 0.707E_{m(AN)}$$
$$E_{BN} = 0.707E_{m(BN)}$$
$$E_{CN} = 0.707E_{m(CN)}$$

and

$$\mathbf{E}_{AN} = E_{AN} \angle 0°$$
$$\mathbf{E}_{BN} = E_{BN} \angle -120°$$
$$\mathbf{E}_{CN} = E_{CN} \angle +120°$$

By rearranging the phasors as shown in Fig. 23.4 and applying a law of vectors which states that *the vector sum of any number of vectors drawn such that the "head" of one is connected to the "tail" of the next, and that the head of the last vector is connected to the tail of the first is zero,* we can conclude that the phasor sum of the phase voltages in a three-phase system is zero. That is,

$$\mathbf{E}_{AN} + \mathbf{E}_{BN} + \mathbf{E}_{CN} = 0 \qquad \textbf{(23.2)}$$

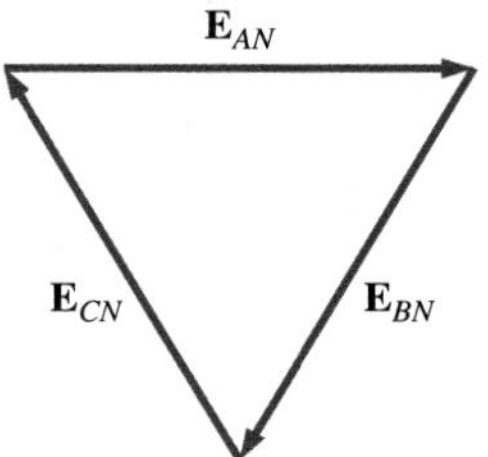

FIG. 23.4
Demonstrating that the vector sum of the phase voltages of a three-phase generator is zero.

23.3 Y-CONNECTED GENERATOR

If the three terminals denoted N in Fig. 23.1(b) are connected together, the generator is referred to as a **Y-connected three-phase generator**

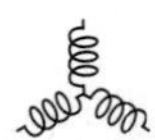

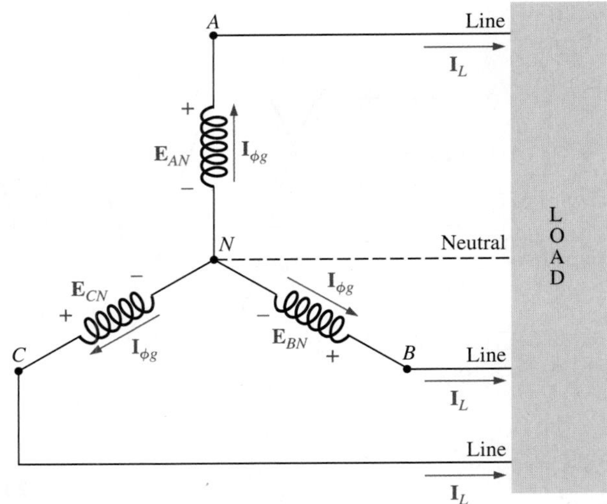

FIG. 23.5
Y-connected generator.

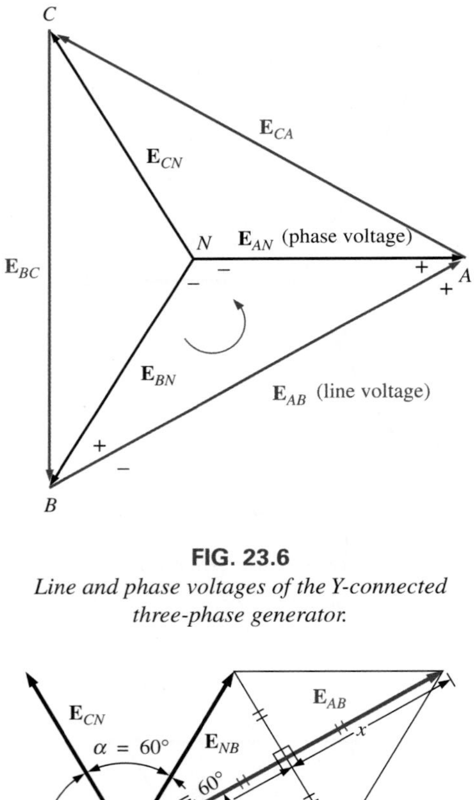

FIG. 23.6
Line and phase voltages of the Y-connected three-phase generator.

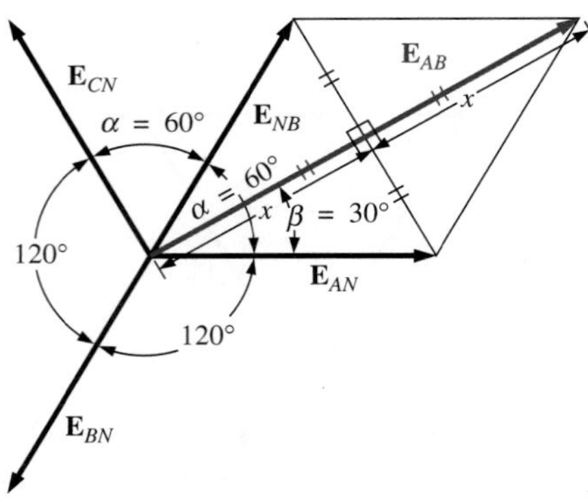

FIG. 23.7
Determining a line voltage for a three-phase generator.

(Fig. 23.5). As indicated in Fig. 23.5, the Y is inverted for ease of notation and for clarity. The point at which all the terminals are connected is called the *neutral point.* If a conductor is not attached from this point to the load, the system is called a *Y-connected, three-phase, three-wire generator.* If the neutral is connected, the system is a *Y-connected, three-phase, four-wire generator.* The function of the neutral will be discussed in detail when we consider the load circuit.

The three conductors connected from *A, B,* and *C* to the load are called *lines.* For the Y-connected system, it should be obvious from Fig. 23.5 that the **line current** equals the **phase current** for each phase; that is,

$$\boxed{\mathbf{I}_L = \mathbf{I}_{\phi g}} \tag{23.3}$$

where ϕ is used to denote a phase quantity and g is a generator parameter.

The voltage from one line to another is called a **line voltage.** On the phasor diagram (Fig. 23.6) it is the phasor drawn from the end of one phase to another in the counterclockwise direction.

Applying Kirchhoff's voltage law around the indicated loop in Fig. 23.6, we obtain

$$\mathbf{E}_{AB} - \mathbf{E}_{AN} + \mathbf{E}_{BN} = 0$$

or

$$\mathbf{E}_{AB} = \mathbf{E}_{AN} - \mathbf{E}_{BN} = \mathbf{E}_{AN} + \mathbf{E}_{NB}$$

The phasor diagram is redrawn to find $\mathbf{E}_{AB}$ as shown in Fig. 23.7. Since each phase voltage, when reversed ($\mathbf{E}_{NB}$), bisects the other two, $\alpha = 60°$. The angle β is 30° since a line drawn from opposite ends of a rhombus divides in half both the angle of origin and the opposite angle. Lines drawn between opposite corners of a rhombus also bisect each other at right angles.

The length x is

$$x = E_{AN} \cos 30° = \frac{\sqrt{3}}{2} E_{AN}$$

and

$$E_{AB} = 2x = (2)\frac{\sqrt{3}}{2}E_{AN} = \sqrt{3}E_{AN}$$

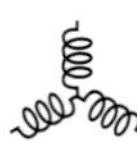

Noting from the phasor diagram that θ of $\mathbf{E}_{AB} = \beta = 30°$, the result is

$$\mathbf{E}_{AB} = E_{AB} \angle 30° = \sqrt{3}E_{AN} \angle 30°$$

and
$$\mathbf{E}_{CA} = \sqrt{3}E_{CN} \angle 150°$$
$$\mathbf{E}_{BC} = \sqrt{3}E_{BN} \angle 270°$$

In words, the magnitude of the line voltage of a Y-connected generator is $\sqrt{3}$ times the phase voltage:

$$\boxed{E_L = \sqrt{3}E_\phi} \qquad \textbf{(23.4)}$$

with the phase angle between any line voltage and the nearest phase voltage at 30°.

In sinusoidal notation,

$$e_{AB} = \sqrt{2}E_{AB}\sin(\omega t + 30°)$$
$$e_{CA} = \sqrt{2}E_{CA}\sin(\omega t + 150°)$$

and
$$e_{BC} = \sqrt{2}E_{BC}\sin(\omega t + 270°)$$

The phasor diagram of the line and phase voltages is shown in Fig. 23.8. If the phasors representing the line voltages in Fig. 23.8(a) are rearranged slightly, they will form a closed loop [Fig. 23.8(b)]. Therefore, we can conclude that the sum of the line voltages is also zero; that is,

$$\boxed{\mathbf{E}_{AB} + \mathbf{E}_{CA} + \mathbf{E}_{BC} + 0} \qquad \textbf{(23.5)}$$

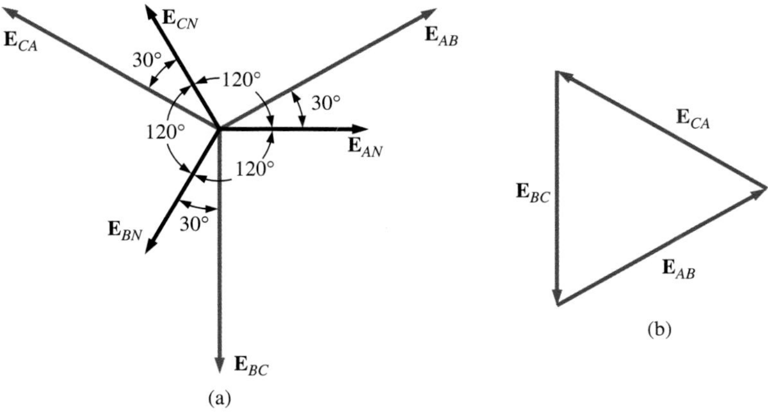

FIG. 23.8
(a) Phasor diagram of the line and phase voltages of a three-phase generator; (b) demonstrating that the vector sum of the line voltages of a three-phase system is zero.

23.4 PHASE SEQUENCE (Y-CONNECTED GENERATOR)

The **phase sequence** can be determined by the order in which the phasors representing the phase voltages pass through a fixed point on the phasor diagram if the phasors are rotated in a counterclockwise direction. For example, in Fig. 23.9 the phase sequence is *ABC*. However, since the fixed point can be chosen anywhere on the phasor diagram, the sequence can

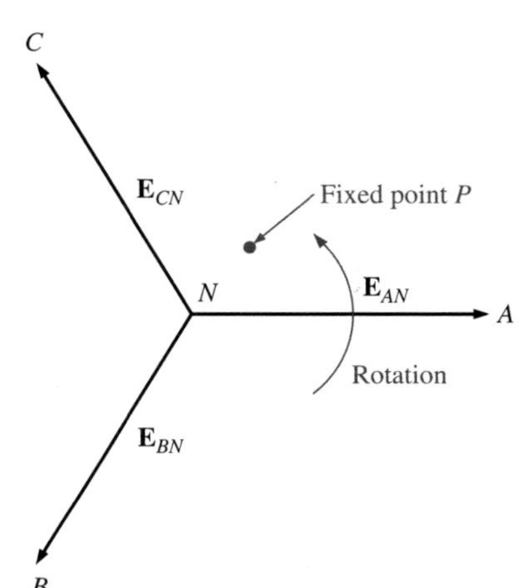

FIG. 23.9
Determining the phase sequence from the phase voltages of a three-phase generator.

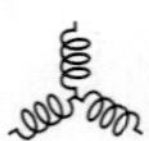

also be written as *BCA* or *CAB*. The phase sequence is quite important in the three-phase distribution of power. In a three-phase motor, for example, if two phase voltages are interchanged, the sequence will change, and the direction of rotation of the motor will be reversed. Other effects will be described when we consider the loaded three-phase system.

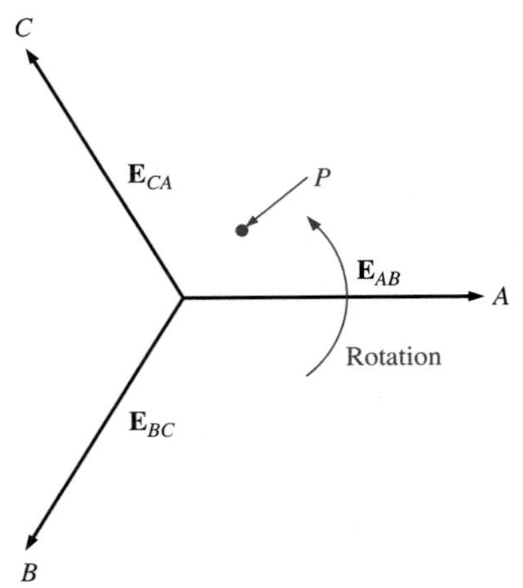

FIG. 23.10
Determining the phase sequence from the line voltages of a three-phase generator.

The phase sequence can also be described in terms of the line voltages. Drawing the line voltages on a phasor diagram in Fig. 23.10, we are able to determine the phase sequence by again rotating the phasors in the counterclockwise direction. In this case, however, the sequence can be determined by noting the order of the passing first or second subscripts. In the system in Fig. 23.10, for example, the phase sequence of the first subscripts passing point *P* is *ABC,* and the phase sequence of the second subscripts is *BCA*. But we know that *BCA* is equivalent to *ABC,* so the sequence is the same for each. Note that the phase sequence is the same as that of the phase voltages described in Fig. 23.9.

If the sequence is given, the phasor diagram can be drawn by simply picking a reference voltage, placing it on the reference axis, and then drawing the other voltages at the proper angular position. For a sequence of *ACB,* for example, we might choose E_{AB} to be the reference [Fig. 23.11(a)] if we wanted the phasor diagram of the line voltages, or E_{AN} for the phase voltages [Fig. 23.11(b)]. For the sequence indicated, the phasor diagrams would be as in Fig. 23.11. In phasor notation,

$$\text{Line voltages} \begin{cases} \mathbf{E}_{AB} = E_{AB} \angle 0^\circ & \text{(reference)} \\ \mathbf{E}_{CA} = E_{CA} \angle -120^\circ \\ \mathbf{E}_{BC} = E_{BC} \angle +120^\circ \end{cases}$$

$$\text{Phase voltages} \begin{cases} \mathbf{E}_{AN} = E_{AN} \angle 0^\circ & \text{(reference)} \\ \mathbf{E}_{CN} = E_{CN} \angle -120^\circ \\ \mathbf{E}_{BN} = E_{BN} \angle +120^\circ \end{cases}$$

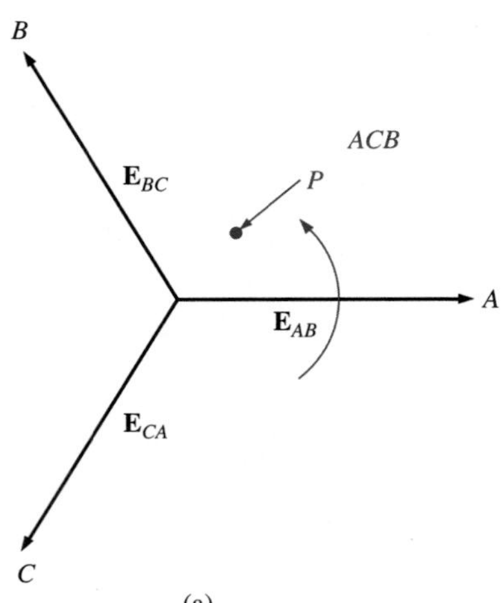

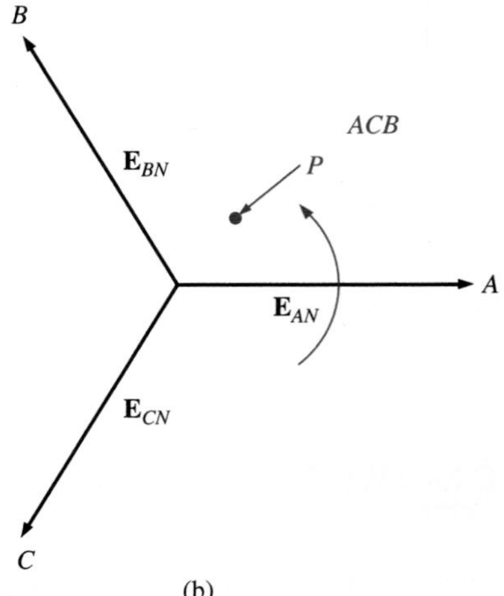

FIG. 23.11
Drawing the phasor diagram from the phase sequence.

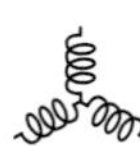

23.5 Y-CONNECTED GENERATOR WITH A Y-CONNECTED LOAD

Loads connected to three-phase supplies are of two types: the Y and the Δ. If a Y-connected load is connected to a Y-connected generator, the system is symbolically represented by Y-Y. The physical setup of such a system is shown in Fig. 23.12.

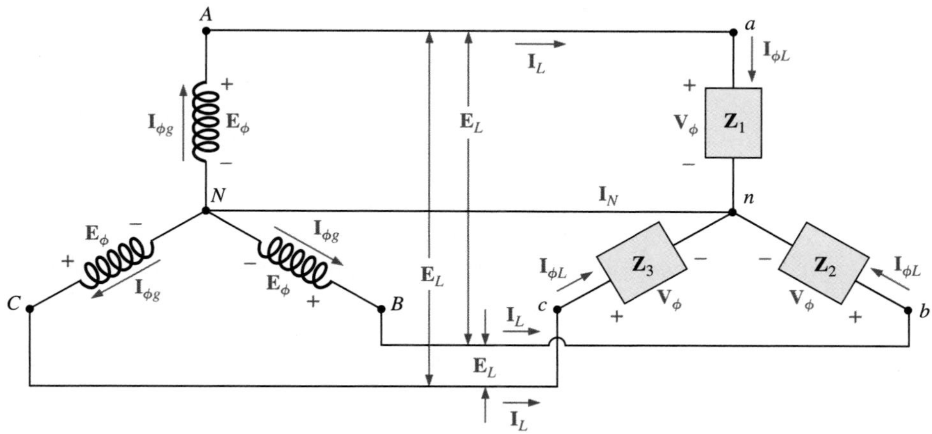

FIG. 23.12
Y-connected generator with a Y-connected load.

If the load is balanced, the **neutral connection** can be removed without affecting the circuit in any manner; that is, if

$$\mathbf{Z}_1 = \mathbf{Z}_2 = \mathbf{Z}_3$$

then I_N will be zero. (This will be demonstrated in Example 23.1.) Note that in order to have a balanced load, the phase angle must also be the same for each impedance—a condition that was unnecessary in dc circuits when we considered balanced systems.

In practice, if a factory, for example, had only balanced, three-phase loads, the absence of the neutral would have no effect since, ideally, the system would always be balanced. The cost would therefore be less since the number of required conductors would be reduced. However, lighting and most other electrical equipment use only one of the phase voltages, and even if the loading is designed to be balanced (as it should be), there is never perfect continuous balancing since lights and other electrical equipment are turned on and off, upsetting the balanced condition. The neutral is therefore necessary to carry the resulting current away from the load and back to the Y-connected generator. This is demonstrated when we consider unbalanced Y-connected systems.

We shall now examine the *four-wire Y-Y-connected system.* The current passing through each phase of the generator is the same as its corresponding line current, which in turn for a Y-connected load is equal to the current in the phase of the load to which it is attached:

$$\boxed{\mathbf{I}_{\phi g} = \mathbf{I}_L = \mathbf{I}_{\phi L}} \qquad \textbf{(23.6)}$$

For a balanced or an unbalanced load, since the generator and load have a common neutral point, then

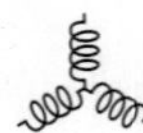

$$\mathbf{V}_\phi = \mathbf{E}_\phi \tag{23.7}$$

In addition, since $\mathbf{I}_{\phi L} = \mathbf{V}_\phi/\mathbf{Z}_\phi$, the magnitude of the current in each phase is equal for a balanced load and unequal for an unbalanced load. Recall that for the Y-connected generator, the magnitude of the line voltage is equal to $\sqrt{3}$ times the phase voltage. This same relationship can be applied to a balanced or an unbalanced four-wire Y-connected load:

$$E_L = \sqrt{3}V_\phi \tag{23.8}$$

For a voltage drop across a load element, the first subscript refers to that terminal through which the current enters the load element, and the second subscript refers to the terminal from which the current leaves. In other words, the first subscript is, by definition, positive with respect to the second for a voltage drop. Note Fig. 23.13, in which the standard double subscripts for a source of voltage and a voltage drop are indicated.

EXAMPLE 23.1 The phase sequence of the Y-connected generator in Fig. 23.13 is *ABC*.

a. Find the phase angles θ_2 and θ_3.
b. Find the magnitude of the line voltages.
c. Find the line currents.
d. Verify that, since the load is balanced, $\mathbf{I}_N = 0$.

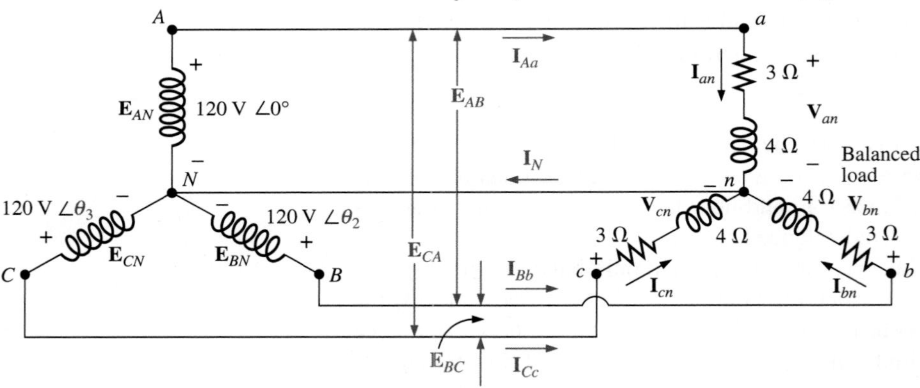

FIG. 23.13
Example 23.1.

Solutions:

a. For an *ABC* phase sequence,

$$\theta_2 = \mathbf{-120^\circ} \quad \text{and} \quad \theta_3 = \mathbf{+120^\circ}$$

b. $E_L = \sqrt{3}E_\phi = (1.73)(120 \text{ V}) = 208 \text{ V}$. Therefore,

$$E_{AB} = E_{BC} = E_{CA} = \mathbf{208\ V}$$

c. $\mathbf{V}_\phi = \mathbf{E}_\phi$. Therefore,

$$\mathbf{V}_{an} = \mathbf{E}_{AN} \qquad \mathbf{V}_{bn} = \mathbf{E}_{BN} \qquad \mathbf{V}_{cn} = \mathbf{E}_{CN}$$

$$\mathbf{I}_{\phi L} = \mathbf{I}_{an} = \frac{\mathbf{V}_{an}}{\mathbf{Z}_{an}} = \frac{120 \text{ V} \angle 0^\circ}{3\ \Omega + j\,4\ \Omega} = \frac{120 \text{ V} \angle 0^\circ}{5\ \Omega \angle 53.13^\circ}$$
$$= 24 \text{ A} \angle -53.13^\circ$$

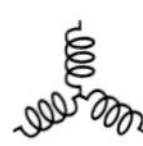

$$\mathbf{I}_{bn} = \frac{\mathbf{V}_{bn}}{\mathbf{Z}_{bn}} = \frac{120 \text{ V} \angle -120^\circ}{5 \ \Omega \angle 53.13^\circ} = 24 \text{ A} \angle -173.13^\circ$$

$$\mathbf{I}_{cn} = \frac{\mathbf{V}_{cn}}{\mathbf{Z}_{cn}} = \frac{120 \text{ V} \angle +120^\circ}{5 \ \Omega \angle 53.13^\circ} = 24 \text{ A} \angle 66.87^\circ$$

and, since $\mathbf{I}_L = \mathbf{I}_{\phi L}$,

$$\mathbf{I}_{Aa} = \mathbf{I}_{an} = \mathbf{24 \text{ A} \angle -53.13^\circ}$$
$$\mathbf{I}_{Bb} = \mathbf{I}_{bn} = \mathbf{24 \text{ A} \angle -173.13^\circ}$$
$$\mathbf{I}_{Cc} = \mathbf{I}_{cn} = \mathbf{24 \text{ A} \angle 66.87^\circ}$$

d. Applying Kirchhoff's current law, we have

$$\mathbf{I}_N = \mathbf{I}_{Aa} + \mathbf{I}_{Bb} + \mathbf{I}_{Cc}$$

In rectangular form,

$$\begin{aligned}
\mathbf{I}_{Aa} &= 24 \text{ A} \angle -53.13^\circ &&= 14.40 \text{ A} - j\,19.20 \text{ A} \\
\mathbf{I}_{Bb} &= 24 \text{ A} \angle -173.13^\circ &&= -22.83 \text{ A} - j\,2.87 \text{ A} \\
\mathbf{I}_{Cc} &= 24 \text{ A} \angle 66.87^\circ &&= 9.43 \text{ A} + j\,22.07 \text{ A} \\
\Sigma(\mathbf{I}_{Aa} &+ \mathbf{I}_{Bb} + \mathbf{I}_{Cc}) &&= 0 + j\,0
\end{aligned}$$

and $\mathbf{I}_N$ is in fact equals to **zero,** as required for a balanced load.

23.6 Y-Δ SYSTEM

There is no neutral connection for the Y-Δ system in Fig. 23.14. Any variation in the impedance of a phase that produces an unbalanced system simply varies the line and phase currents of the system.

For a balanced load,

$$\boxed{\mathbf{Z}_1 = \mathbf{Z}_2 = \mathbf{Z}_3} \qquad (23.9)$$

The voltage across each phase of the load is equal to the line voltage of the generator for a balanced or an unbalanced load:

$$\boxed{\mathbf{V}_\phi = \mathbf{E}_L} \qquad (23.10)$$

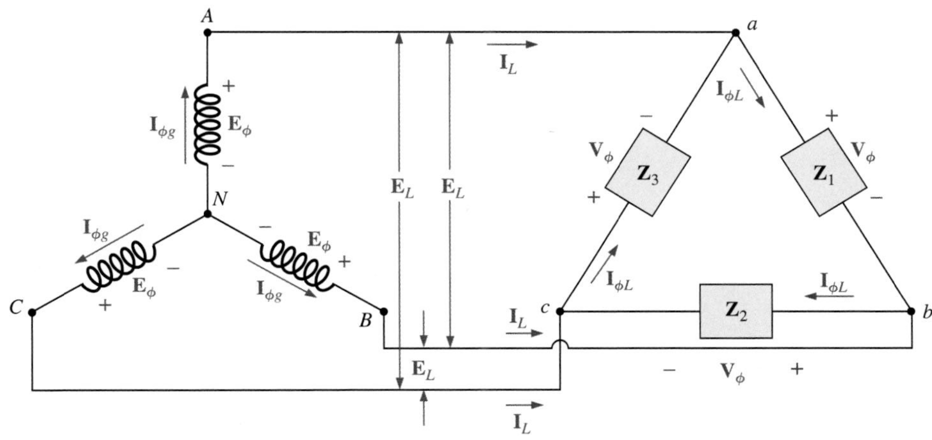

FIG. 23.14
Y-connected generator with a Δ-connected load.

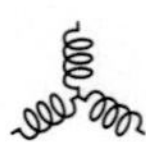

The relationship between the line currents and phase currents of a balanced Δ load can be found using an approach very similar to that used in Section 23.3 to find the relationship between the line voltages and phase voltages of a Y-connected generator. For this case, however, Kirchhoff's current law is used instead of Kirchhoff's voltage law.

The results obtained are

$$\boxed{I_L = \sqrt{3} I_\phi} \qquad \textbf{(23.11)}$$

and the phase angle between a line current and the nearest phase current is 30°. A more detailed discussion of this relationship between the line and phase currents of a Δ-connected system can be found in Section 23.7.

For a balanced load, the line currents will be equal in magnitude, as will the phase currents.

EXAMPLE 23.2 For the three-phase system in Fig. 23.15:

a. Find the phase angles θ_2 and θ_3.
b. Find the current in each phase of the load.
c. Find the magnitude of the line currents.

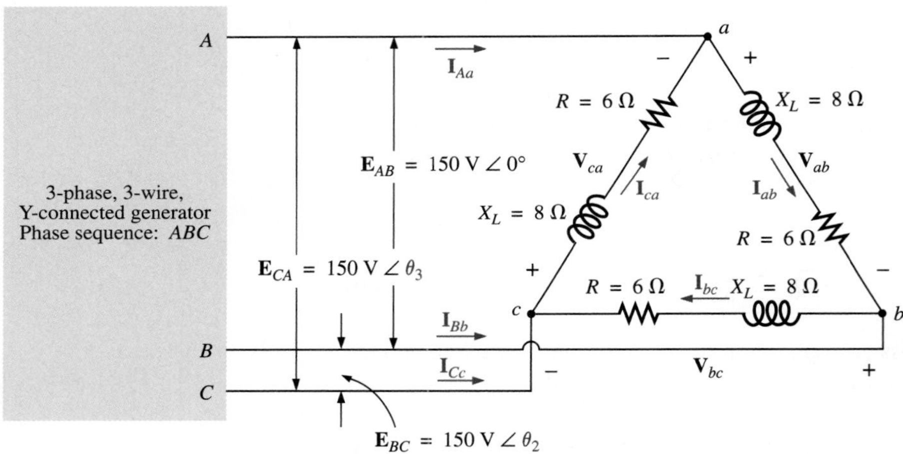

FIG. 23.15
Example 23.2.

Solutions:

a. For an *ABC* sequence,

$$\theta_2 = \mathbf{-120°} \quad \text{and} \quad \theta_3 = \mathbf{+120°}$$

b. $\mathbf{V}_\phi = \mathbf{E}_L$. Therefore,

$$\mathbf{V}_{ab} = \mathbf{E}_{AB} \qquad \mathbf{V}_{ca} = \mathbf{E}_{CA} \qquad \mathbf{V}_{bc} = \mathbf{E}_{BC}$$

The phase currents are

$$\mathbf{I}_{ab} = \frac{\mathbf{V}_{ab}}{\mathbf{Z}_{ab}} = \frac{150 \text{ V} \angle 0°}{6\ \Omega + j\,8\ \Omega} = \frac{150 \text{ V} \angle 0°}{10\ \Omega \angle 53.13°} = \mathbf{15\ A \angle -53.13°}$$

$$\mathbf{I}_{bc} = \frac{\mathbf{V}_{bc}}{\mathbf{Z}_{bc}} = \frac{150 \text{ V} \angle -120°}{10\ \Omega \angle 53.13°} = \mathbf{15\ A \angle -173.13°}$$

$$\mathbf{I}_{ca} = \frac{\mathbf{V}_{ca}}{\mathbf{Z}_{ca}} = \frac{150 \text{ V} \angle +120°}{10\ \Omega \angle 53.13°} = \mathbf{15\ A \angle 66.87°}$$

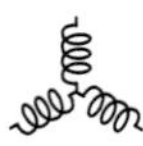

c. $I_L = \sqrt{3}I_\phi = (1.73)(15\text{ A}) = 25.95\text{ A}$. Therefore,

$$I_{Aa} = I_{Bb} = I_{Cc} = \mathbf{25.95\ A}$$

23.7 Δ-CONNECTED GENERATOR

If we rearrange the coils of the generator in Fig. 23.16(a) as shown in Fig. 23.16(b), the system is referred to as a *three-phase, three-wire,* **Δ-connected ac generator.** In this system, the phase and line voltages are equivalent and equal to the voltage induced across each coil of the generator; that is,

$$\left.\begin{aligned} \mathbf{E}_{AB} &= \mathbf{E}_{AN} \text{ and } e_{AN} = \sqrt{2}E_{AN}\sin\omega t \\ \mathbf{E}_{BC} &= \mathbf{E}_{BN} \text{ and } e_{BN} = \sqrt{2}E_{BN}\sin(\omega t - 120^\circ) \\ \mathbf{E}_{CA} &= \mathbf{E}_{CN} \text{ and } e_{CN} = \sqrt{2}E_{CN}\sin(\omega t + 120^\circ) \end{aligned}\right\} \begin{array}{l}\text{Phase}\\ \text{sequence}\\ ABC\end{array}$$

or

$$\mathbf{E}_L = \mathbf{E}_{\phi g} \tag{23.12}$$

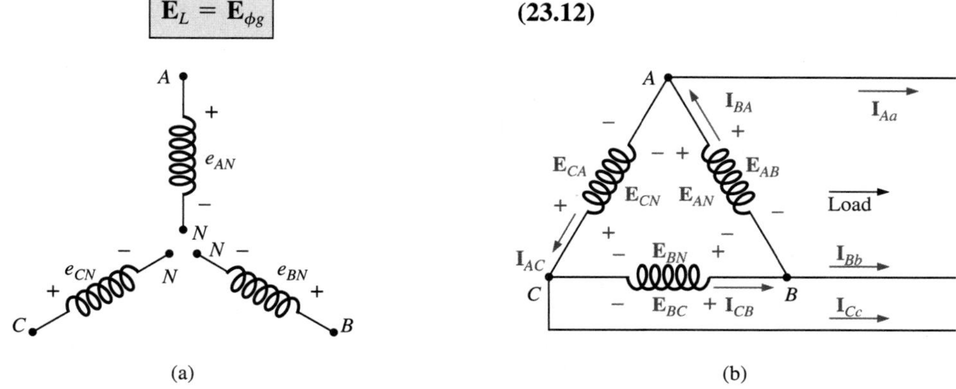

FIG. 23.16
Δ-connected generator.

Note that only one voltage (magnitude) is available instead of the two available in the Y-connected system.

Unlike the line current for the Y-connected generator, the line current for the Δ-connected system is not equal to the phase current. The relationship between the two can be found by applying Kirchhoff's current law at one of the nodes and solving for the line current in terms of the phase currents; that is, at node *A*,

$$\mathbf{I}_{BA} = \mathbf{I}_{Aa} + \mathbf{I}_{AC}$$

or

$$\mathbf{I}_{Aa} = \mathbf{I}_{BA} - \mathbf{I}_{AC} = \mathbf{I}_{BA} + \mathbf{I}_{CA}$$

The phasor diagram is shown in Fig. 23.17 for a balanced load.

Using the same procedure to find the line current as was used to find the line voltage of a Y-connected generator produces the following:

$$\begin{aligned} I_{Aa} &= \sqrt{3}I_{BA}\angle -30^\circ \\ I_{Bb} &= \sqrt{3}I_{CB}\angle -150^\circ \\ I_{Cc} &= \sqrt{3}I_{AC}\angle 90^\circ \end{aligned}$$

In general:

$$I_L = \sqrt{3}I_{\phi g} \tag{23.13}$$

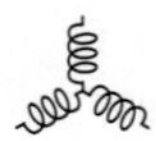

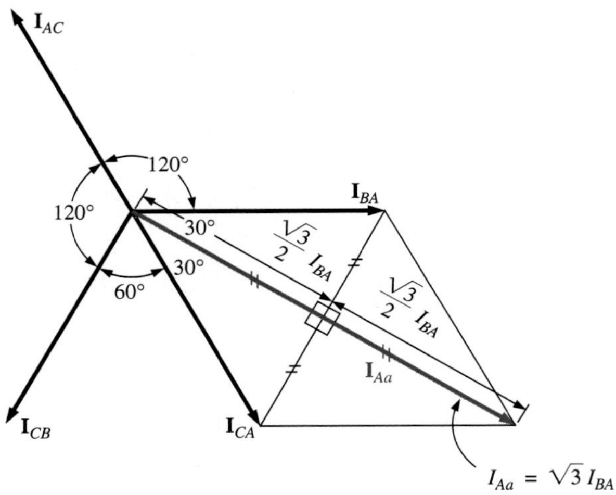

FIG. 23.17
Determining a line current from the phase currents of a Δ-connected, three-phase generator.

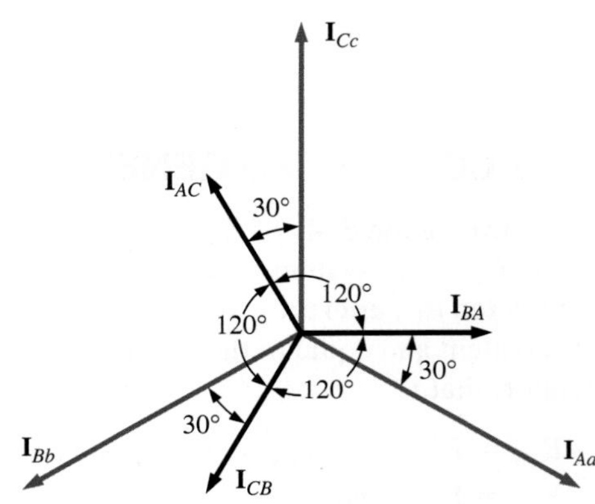

FIG. 23.18
The phasor diagram of the currents of a three-phase, Δ-connected generator.

with the phase angle between a line current and the nearest phase current at 30°. The phasor diagram of the currents is shown in Fig. 23.18.

Just as for the voltages of a Y-connected generator, the phasor sum of the line currents or phase currents for Δ-connected systems with balanced loads is zero.

23.8 PHASE SEQUENCE (Δ-CONNECTED GENERATOR)

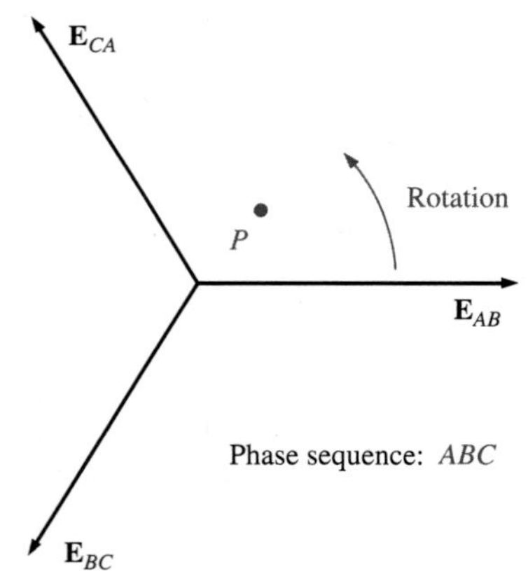

FIG. 23.19
Determining the phase sequence for a Δ-connected, three-phase generator.

Even though the line and phase voltages of a Δ-connected system are the same, it is standard practice to describe the phase sequence in terms of the line voltages. The method used is the same as that described for the line voltages of the Y-connected generator. For example, the phasor diagram of the line voltages for a phase sequence *ABC* is shown in Fig. 23.19. In drawing such a diagram, one must take care to have the sequence of the first and second subscripts the same. In phasor notation,

$$\mathbf{E}_{AB} = E_{AB} \angle 0^\circ$$
$$\mathbf{E}_{BC} = E_{BC} \angle -120^\circ$$
$$\mathbf{E}_{CA} = E_{CA} \angle 120^\circ$$

23.9 Δ-Δ, Δ-Y THREE-PHASE SYSTEMS

The basic equations necessary to analyze either of the two systems (Δ-Δ, Δ-Y) have been presented at least once in this chapter. Following are two descriptive examples, one with a Δ-connected load and one with a Y-connected load.

EXAMPLE 23.3 For the Δ-Δ system shown in Fig. 23.20:

a. Find the phase angles θ_2 and θ_3 for the specified phase sequence.
b. Find the current in each phase of the load.
c. Find the magnitude of the line currents.

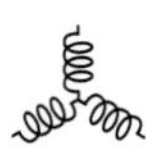

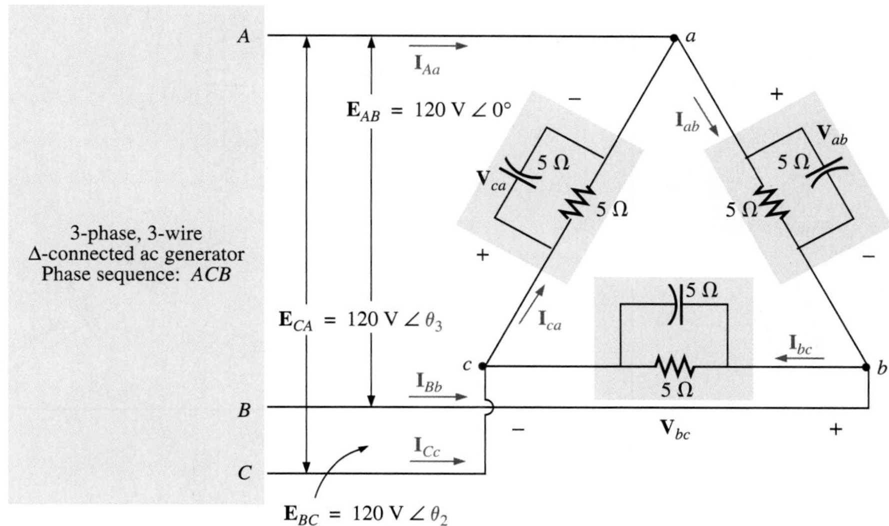

FIG. 23.20
Example 23.3: Δ-Δ system.

Solutions:

a. For an *ACB* phase sequence,

$$\theta_2 = \mathbf{120^\circ} \quad \text{and} \quad \theta_3 = \mathbf{-120^\circ}$$

b. $\mathbf{V}_\phi = \mathbf{E}_L$. Therefore,

$$\mathbf{V}_{ab} = \mathbf{E}_{AB} \quad \mathbf{V}_{ca} = \mathbf{E}_{CA} \quad \mathbf{V}_{bc} = \mathbf{E}_{BC}$$

The phase currents are

$$\mathbf{I}_{ab} = \frac{\mathbf{V}_{ab}}{\mathbf{Z}_{ab}} = \frac{120\text{ V}\angle 0^\circ}{\dfrac{(5\ \Omega\angle 0^\circ)(5\ \Omega\angle -90^\circ)}{5\ \Omega - j\,5\ \Omega}} = \frac{120\text{ V}\angle 0^\circ}{\dfrac{25\ \Omega\angle -90^\circ}{7.071\angle -45^\circ}}$$

$$= \frac{120\text{ V}\angle 0^\circ}{3.54\ \Omega\angle -45^\circ} = \mathbf{33.9\ A\angle 45^\circ}$$

$$\mathbf{I}_{bc} = \frac{\mathbf{V}_{bc}}{\mathbf{Z}_{bc}} = \frac{120\text{ V}\angle 120^\circ}{3.54\ \Omega\angle -45^\circ} = \mathbf{33.9\ A\angle 165^\circ}$$

$$\mathbf{I}_{ca} = \frac{\mathbf{V}_{ca}}{\mathbf{Z}_{ca}} = \frac{120\text{ V}\angle -120^\circ}{3.54\ \Omega\angle -45^\circ} = \mathbf{33.9\ A\angle -75^\circ}$$

c. $I_L = \sqrt{3}I_\phi = (1.73)(34\text{ A}) = 58.82\text{ A}$. Therefore,

$$I_{Aa} = I_{Bb} = I_{Cc} = \mathbf{58.82\ A}$$

EXAMPLE 23.4 For the Δ-Y system shown in Fig. 23.21:

a. Find the voltage across each phase of the load.
b. Find the magnitude of the line voltages.

Solutions:

a. $\mathbf{I}_{\phi L} = \mathbf{I}_L$. Therefore,

$$\mathbf{I}_{an} = \mathbf{I}_{Aa} = 2\text{ A}\angle 0^\circ$$
$$\mathbf{I}_{bn} = \mathbf{I}_{Bb} = 2\text{ A}\angle -120^\circ$$
$$\mathbf{I}_{cn} = \mathbf{I}_{Cc} = 2\text{ A}\angle 120^\circ$$

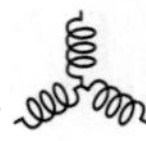

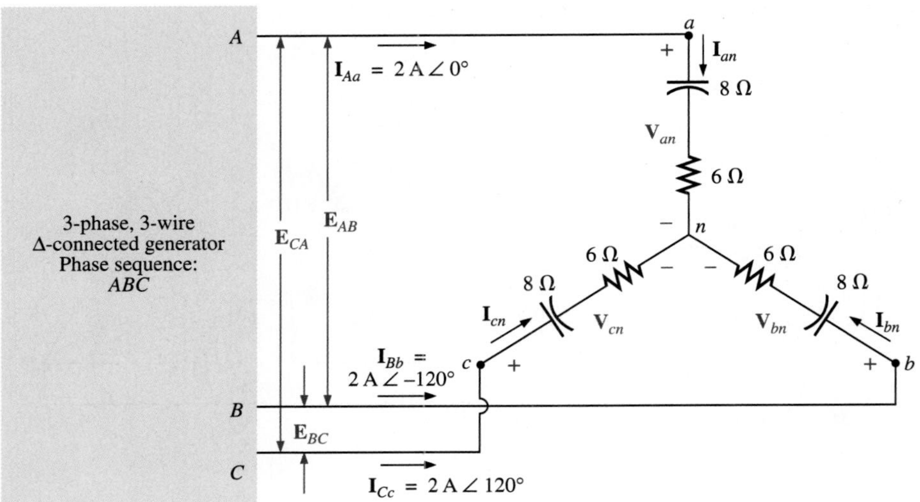

FIG. 23.21
Example 23.4: Δ-Y system.

The phase voltages are

$$\mathbf{V}_{an} = \mathbf{I}_{an}\mathbf{Z}_{an} = (2\text{ A }\angle 0°)(10\ \Omega\ \angle -53.13°) = \mathbf{20\ V\ \angle -53.13°}$$
$$\mathbf{V}_{bn} = \mathbf{I}_{bn}\mathbf{Z}_{bn} = (2\text{ A }\angle -120°)(10\ \Omega\ \angle -53.13°) = \mathbf{20\ V\ \angle -173.13°}$$
$$\mathbf{V}_{cn} = \mathbf{I}_{cn}\mathbf{Z}_{cn} = (2\text{ A }\angle 120°)(10\ \Omega\ \angle -53.13°) = \mathbf{20\ V\ \angle 66.87°}$$

b. $E_L = \sqrt{3}V_\phi = (1.73)(20\text{ V}) = 34.6\text{ V}$. Therefore,

$$E_{BA} = E_{CB} = E_{AC} = \mathbf{34.6\ V}$$

23.10 POWER

Y-Connected Balanced Load

Please refer to Fig. 23.22 for the following discussion.

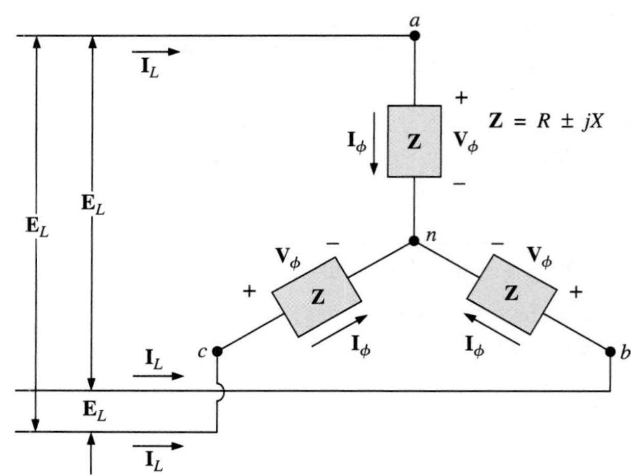

FIG. 23.22
Y-connected balanced load.

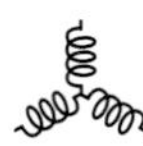

Average Power The average power delivered to each phase can be determined by Eq. (23.14).

$$P_\phi = V_\phi I_\phi \cos \theta_{I_\phi}^{V_\phi} = I_\phi^2 R_\phi = \frac{V_R^2}{R_\phi} \quad \text{(watts, W)} \qquad \textbf{(23.14)}$$

where $\theta_{I_\phi}^{V_\phi}$ indicates that θ is the phase angle between V_ϕ and I_ϕ.

The total power delivered can be determined by Eq. (23.15) or Eq. (23.16).

$$P_T = 3P_\phi \quad \text{(W)} \qquad \textbf{(23.15)}$$

or, since

$$V_\phi = \frac{E_L}{\sqrt{3}} \quad \text{and} \quad I_\phi = I_L$$

then

$$P_T = 3\frac{E_L}{\sqrt{3}} I_L \cos \theta_{I_\phi}^{V_\phi}$$

But

$$\left(\frac{3}{\sqrt{3}}\right)(1) = \left(\frac{3}{\sqrt{3}}\right)\left(\frac{\sqrt{3}}{\sqrt{3}}\right) = \frac{3\sqrt{3}}{3} = \sqrt{3}$$

Therefore,

$$P_T = \sqrt{3} E_L I_L \cos \theta_{I_\phi}^{V_\phi} = 3I_L^2 R_\phi \quad \text{(W)} \qquad \textbf{(23.16)}$$

Reactive Power The reactive power of each phase (in volt-amperes reactive) is

$$Q_\phi = V_\phi I_\phi \sin \theta_{I_\phi}^{V_\phi} = I_\phi^2 X_\phi = \frac{V_X^2}{X_\phi} \quad \text{(VAR)} \qquad \textbf{(23.17)}$$

The total reactive power of the load is

$$Q_T = 3Q_\phi \quad \text{(VAR)} \qquad \textbf{(23.18)}$$

or, proceeding in the same manner as above, we have

$$Q_T = \sqrt{3} E_L I_L \sin \theta_{I_\phi}^{V_\phi} = 3I_L^2 X_\phi \quad \text{(VAR)} \qquad \textbf{(23.19)}$$

Apparent Power The apparent power of each phase is

$$S_\phi = V_\phi I_\phi \quad \text{(VA)} \qquad \textbf{(23.20)}$$

The total apparent power of the load is

$$S_T = 3S_\phi \quad \text{(VA)} \qquad \textbf{(23.21)}$$

or, as before,

$$S_T = \sqrt{3} E_L I_L \quad \text{(VA)} \qquad \textbf{(23.22)}$$

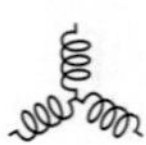

Power Factor The power factor of the system is given by

$$F_p = \frac{P_T}{S_T} = \cos \theta_{I_\phi}^{V_\phi} \quad \text{(leading or lagging)} \qquad \textbf{(23.23)}$$

EXAMPLE 23.5 For the Y-connected load in Fig. 23.23:

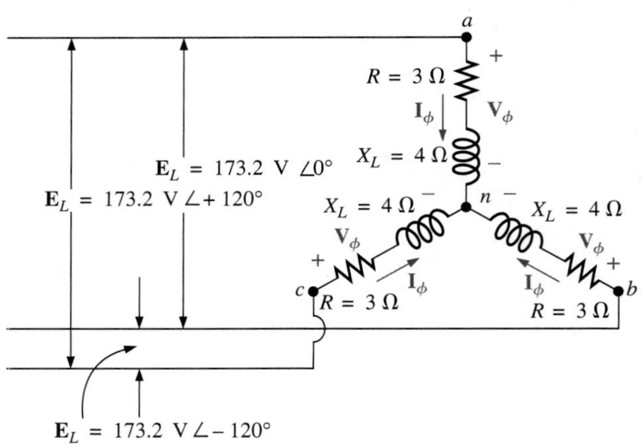

FIG. 23.23
Example 23.5.

a. Find the average power to each phase and the total load.
b. Determine the reactive power to each phase and the total reactive power.
c. Find the apparent power to each phase and the total apparent power.
d. Find the power factor of the load.

Solutions:

a. The *average power* is

$$P_\phi = V_\phi I_\phi \cos \theta_{I_\phi}^{V_\phi} = (100 \text{ V})(20 \text{ A}) \cos 53.13^\circ = (2000)(0.6) = \mathbf{1200\ W}$$

$$P_\phi = I_\phi^2 R_\phi = (20 \text{ A})^2(3\ \Omega) = (400)(3) = \mathbf{1200\ W}$$

$$P_\phi = \frac{V_R^2}{R_\phi} = \frac{(60 \text{ V})^2}{3\ \Omega} = \frac{3600}{3} = \mathbf{1200\ W}$$

$$P_T = 3P_\phi = (3)(1200 \text{ W}) = \mathbf{3600\ W}$$

or

$$P_T = \sqrt{3}E_L I_L \cos \theta_{I_\phi}^{V_\phi} = (1.732)(173.2 \text{ V})(20 \text{ A})(0.6) = \mathbf{3600\ W}$$

b. The *reactive power* is

$$Q_\phi = V_\phi I_\phi \sin \theta_{I_\phi}^{V_\phi} = (100 \text{ V})(20 \text{ A}) \sin 53.13^\circ = (2000)(0.8) = \mathbf{1600\ VAR}$$

or
$$Q_\phi = I_\phi^2 X_\phi = (20 \text{ A})^2(4\ \Omega) = (400)(4) = \mathbf{1600\ VAR}$$

$$Q_T = 3Q_\phi = (3)(1600 \text{ VAR}) = \mathbf{4800\ VAR}$$

or

$$Q_T = \sqrt{3}E_L I_L \sin \theta_{I_\phi}^{V_\phi} = (1.732)(173.2 \text{ V})(20 \text{ A})(0.8) = \mathbf{4800\ VAR}$$

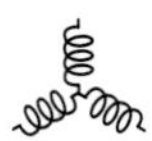

c. The *apparent power* is

$$S_\phi = V_\phi I_\phi = (100\text{ V})(20\text{ A}) = \mathbf{2000\text{ VA}}$$
$$S_T = 3S_\phi = (3)(2000\text{ VA}) = \mathbf{6000\text{ VA}}$$

or $S_T = \sqrt{3}E_L I_L = (1.732)(173.2\text{ V})(20\text{ A}) = \mathbf{6000\text{ VA}}$

d. The *power factor* is

$$F_p = \frac{P_T}{S_T} = \frac{3600\text{ W}}{6000\text{ VA}} = \mathbf{0.6\ lagging}$$

Δ-Connected Balanced Load

Please refer to Fig. 23.24 for the following discussion.

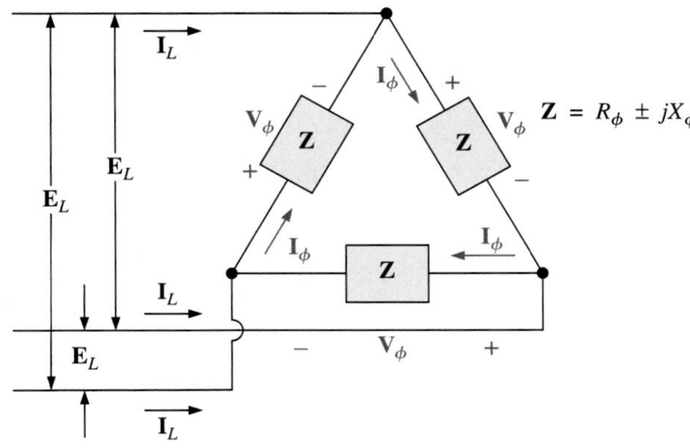

FIG. 23.24
Δ-connected balanced load.

Average Power

$$P_\phi = V_\phi I_\phi \cos\theta_{I_\phi}^{V_\phi} = I_\phi^2 R_\phi = \frac{V_R^2}{R_\phi} \quad \text{(W)} \qquad \textbf{(23.24)}$$

$$P_T = 3P_\phi \quad \text{(W)} \qquad \textbf{(23.25)}$$

Reactive Power

$$Q_\phi = V_\phi I_\phi \sin\theta_{I_\phi}^{V_\phi} = I_\phi^2 X_\phi = \frac{V_X^2}{X_\phi} \quad \text{(VAR)} \qquad \textbf{(23.26)}$$

$$Q_T = 3Q_\phi \quad \text{(VAR)} \qquad \textbf{(23.27)}$$

Apparent Power

$$S_\phi = V_\phi I_\phi \quad \text{(VA)} \qquad \textbf{(23.28)}$$

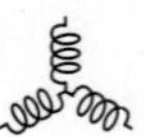

$$S_T = 3S_\phi = \sqrt{3}E_L I_L \qquad \text{(VA)} \tag{23.29}$$

Power Factor

$$F_p = \frac{P_T}{S_T} \tag{23.30}$$

EXAMPLE 23.6 For the Δ-Y connected load in Fig. 23.25, find the total average, reactive, and apparent power. In addition, find the power factor of the load.

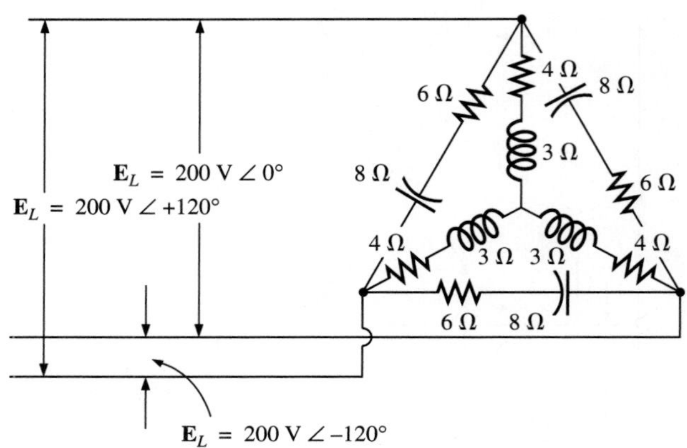

FIG. 23.25
Example 23.6.

Solution: Consider the Δ and Y separately.

For the Δ:

$$Z_\Delta = 6\ \Omega - j\,8\ \Omega = 10\ \Omega\ \angle -53.13^\circ$$

$$I_\phi = \frac{E_L}{Z_\Delta} = \frac{200\ \text{V}}{10\ \Omega} = 20\ \text{A}$$

$$P_{T_\Delta} = 3I_\phi^2 R_\phi = (3)(20\ \text{A})^2(6\ \Omega) = \mathbf{7200\ W}$$

$$Q_{T_\Delta} = 3I_\phi^2 X_\phi = (3)(20\ \text{A})^2(8\ \Omega) = \mathbf{9600\ VAR}\ (C)$$

$$S_{T_\Delta} = 3V_\phi I_\phi = (3)(200\ \text{V})(20\ \text{A}) = \mathbf{12{,}000\ VA}$$

For the Y:

$$Z_Y = 4\ \Omega + j\,3\ \Omega = 5\ \Omega\ \angle 36.87^\circ$$

$$I_\phi = \frac{E_L/\sqrt{3}}{Z_Y} = \frac{200\ \text{V}/\sqrt{3}}{5\ \Omega} = \frac{116\ \text{V}}{5\ \Omega} = 23.12\ \text{A}$$

$$P_{T_Y} = 3I_\phi^2 R_\phi = (3)(23.12\ \text{A})^2(4\ \Omega) = \mathbf{6414.41\ W}$$

$$Q_{T_Y} = 3I_\phi^2 X_\phi = (3)(23.12\ \text{A})^2(3\ \Omega) = \mathbf{4810.81\ VAR}\ (L)$$

$$S_{T_Y} = 3V_\phi I_\phi = (3)(116\ \text{V})(23.12\ \text{A}) = \mathbf{8045.76\ VA}$$

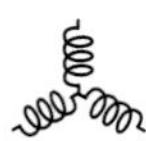

For the total load:

$$P_T = P_{T_\Delta} + P_{T_Y} = 7200\ \text{W} + 6414.41\ \text{W} = \mathbf{13{,}614.41\ W}$$

$$Q_T = Q_{T_\Delta} - Q_{T_Y} = 9600\ \text{VAR}\ (C) - 4810.81\ \text{VAR}\ (I)$$

$$= \mathbf{4789.19\ VAR}\ (\boldsymbol{C})$$

$$S_T = \sqrt{P_T^2 + Q_T^2} = \sqrt{(13{,}614.41\ \text{W})^2 + (4789.19\ \text{VAR})^2}$$

$$= \mathbf{14{,}432.2\ VA}$$

$$F_p = \frac{P_T}{S_T} = \frac{13{,}614.41\ \text{W}}{14{,}432.20\ \text{VA}} = \mathbf{0.943\ leading}$$

EXAMPLE 23.7 Each transmission line of the three-wire, three-phase system in Fig. 23.26 has an impedance of $15\ \Omega + j\,20\ \Omega$. The system delivers a total power of 160 kW at 12,000 V to a balanced three-phase load with a lagging power factor of 0.86.

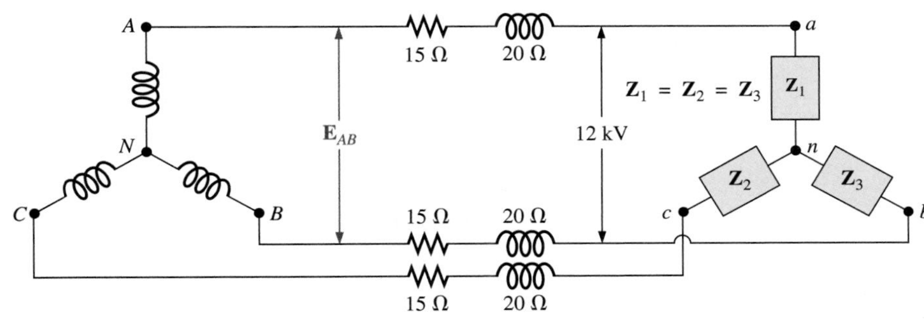

FIG. 23.26
Example 23.7.

a. Determine the magnitude of the line voltage E_{AB} of the generator.
b. Find the power factor of the total load applied to the generator.
c. What is the efficiency of the system?

Solutions:

a. $$V_\phi\ (\text{load}) = \frac{V_L}{\sqrt{3}} = \frac{12{,}000\ \text{V}}{1.73} = 6936.42\ \text{V}$$

$$P_T\ (\text{load}) = 3V_\phi I_\phi \cos\theta$$

and

$$I_\phi = \frac{P_T}{3V_\phi \cos\theta} = \frac{160{,}000\ \text{W}}{3(6936.42\ \text{V})(0.86)}$$

$$= \mathbf{8.94\ A}$$

Since $\theta = \cos^{-1} 0.86 = 30.68°$, assigning $\mathbf{V}_\phi$ an angle of 0° or $\mathbf{V}_\phi = V_\phi \angle 0°$, a lagging power factor results in

$$\mathbf{I}_\phi = 8.94\ \text{A} \angle -30.68°$$

For each phase, the system will appear as shown in Fig. 23.27, where

$$\mathbf{E}_{AN} - \mathbf{I}_\phi \mathbf{Z}_{\text{line}} - \mathbf{V}_\phi = 0$$

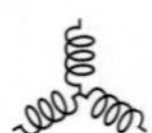

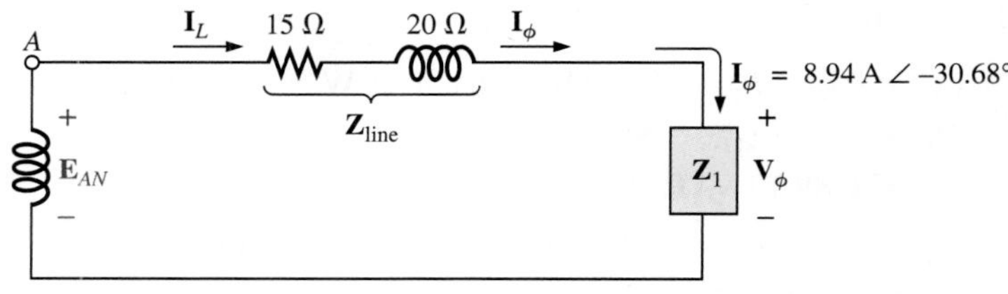

FIG. 23.27
The loading on each phase of the system in Fig. 23.26.

or

$$\begin{aligned}\mathbf{E}_{AN} &= \mathbf{I}_\phi \mathbf{Z}_{\text{line}} + \mathbf{V}_\phi \\ &= (8.94\text{ A}\angle -30.68^\circ)(25\ \Omega\ \angle 53.13^\circ) + 6936.42\text{ V}\angle 0^\circ \\ &= 223.5\text{ V}\angle 22.45^\circ + 6936.42\text{ V}\angle 0^\circ \\ &= 206.56\text{ V} + j\,85.35\text{ V} + 6936.42\text{ V} \\ &= 7142.98\text{ V} + j\,85.35\text{ V} \\ &= 7143.5\text{ V}\angle 0.68^\circ\end{aligned}$$

Then

$$\begin{aligned}E_{AB} &= \sqrt{3}E_{\phi g} = (1.73)(7143.5\text{ V}) \\ &= \mathbf{12{,}358.26\ V}\end{aligned}$$

b.
$$\begin{aligned}P_T &= P_{\text{load}} + P_{\text{lines}} \\ &= 160\text{ kW} + 3(I_L)^2 R_{\text{line}} \\ &= 160\text{ kW} + 3(8.94\text{ A})^2 15\ \Omega \\ &= 160{,}000\text{ W} + 3596.55\text{ W} \\ &= 163{,}596.55\text{ W}\end{aligned}$$

and

$$P_T = \sqrt{3}V_L I_L \cos\theta_T$$

or

$$\cos\theta_T = \frac{P_T}{\sqrt{3}V_L I_L} = \frac{163{,}596.55\text{ W}}{(1.73)(12{,}358.26\text{ V})(8.94\text{ A})}$$

and

$$F_p = \mathbf{0.856} < 0.86 \text{ of load}$$

c.
$$\eta = \frac{P_o}{P_i} = \frac{P_o}{P_o + P_{\text{losses}}} = \frac{160\text{ kW}}{160\text{ kW} + 3596.55\text{ W}} = 0.978$$
$$= \mathbf{97.8\%}$$

23.11 THREE-WATTMETER METHOD

The power delivered to a balanced or an unbalanced four-wire, Y-connected load can be found by the **three-wattmeter method,** that is, by using three wattmeters in the manner shown in Fig. 23.28. Each wattmeter measures the power delivered to each phase. The potential coil of each wattmeter is connected parallel with the load, while the current coil is in series with the load. The total average power of the system can be found by summing the three wattmeter readings; that is,

$$\boxed{P_{T_Y} = P_1 + P_2 + P_3} \qquad \textbf{(23.31)}$$

For the load (balanced or unbalanced), the wattmeters are connected as shown in Fig. 23.29. The total power is again the sum of the three wattmeter readings:

$$\boxed{P_{T_\Delta} = P_1 + P_2 + P_3} \qquad \textbf{(23.32)}$$

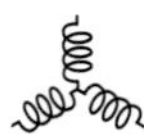

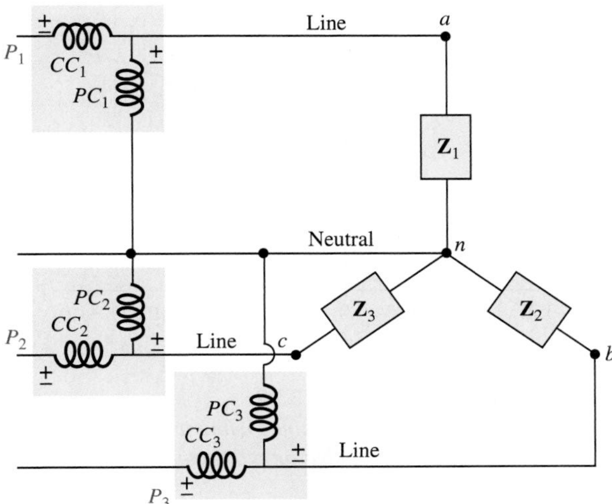

FIG. 23.28
Three-wattmeter method for a Y-connected load.

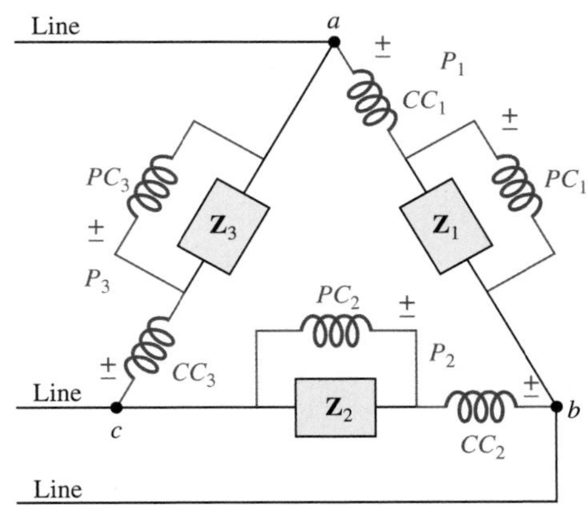

FIG. 23.29
Three-wattmeter method for a Δ-connected load.

If in either of the cases just described the load is balanced, the power delivered to each phase will be the same. The total power is then just three times any one wattmeter reading.

23.12 TWO-WATTMETER METHOD

The power delivered to a three-phase, three-wire, Δ- or Y-connected, balanced or unbalanced local can be found using only two wattmeters if the proper connection is employed and if the wattmeter readings are interpreted properly. The basic connections of this **two-wattmeter method** are shown in Fig. 23.30. One end of each potential coil is connected to the same line. The current coils are then placed in the remaining lines.

The connection shown in Fig. 23.31 also satisfies the requirements. A third hookup is also possible, but this is left to the reader as an exercise.

The total power delivered to the load is the algebraic sum of the two wattmeter readings. For a *balanced* load, we now consider two methods of determining whether the total power is the sum or the difference of the two wattmeter readings. The first method to be described requires that we know or are able to find the power factor (leading or lagging) of any one

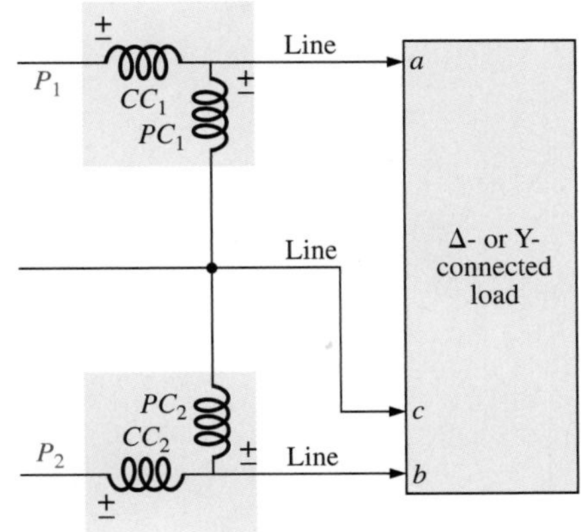

FIG. 23.30
Two-wattmeter method for a Δ- or a Y-connected load.

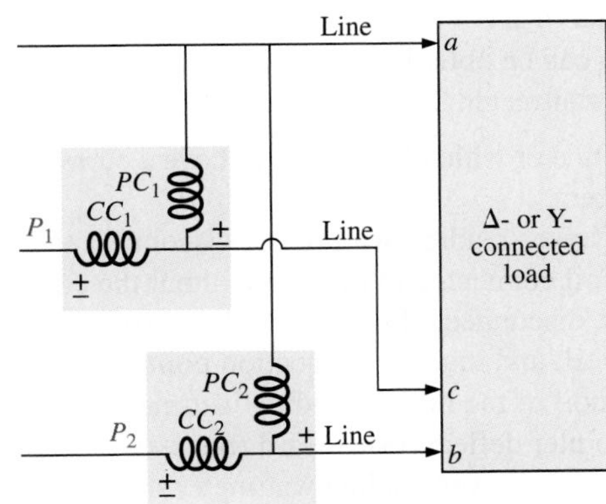

FIG. 23.31
Alternative hookup for the two-wattmeter method.

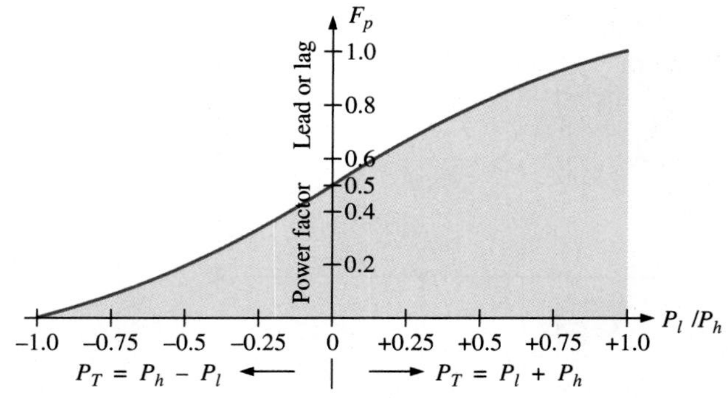

FIG. 23.32
Determining whether the readings obtained using the two-wattmeter method should be added or subtracted.

phase of the load. When this information has been obtained, it can be applied directly to the curve in Fig. 23.32.

The curve in Fig. 23.32 is a plot of the power factor of the load (phase) versus the ratio P_l/P_h, where P_l and P_h are the magnitudes of the lower- and higher-reading wattmeters, respectively. Note that for a power factor (leading or lagging) greater than 0.5, the ratio has a positive value. This indicates that both wattmeters are reading positive, and the total power is the sum of the two wattmeter readings; that is, $P_T = P_l + P_h$. For a power factor less than 0.5 (leading or lagging), the ratio has a negative value. This indicates that the smaller-reading wattmeter is reading negative, and the total power is the difference of the two wattmeter readings; that is, $P_T = P_h - P_l$.

A closer examination reveals that, when the power factor is 1 ($\cos 0° = 1$), corresponding to a purely resistive load, $P_l/P_h = 1$ or $P_l = P_h$, and both wattmeters have the same wattage indication. At a power factor equal to 0 ($\cos 90° = 0$), corresponding to a purely reactive load, $P_l/P_h = -1$ or $P_l = -P_h$, and both wattmeters again have the same wattage indication but with opposite signs. The transition from a negative to a positive ratio occurs when the power factor of the load is 0.5 or $\theta = \cos^{-1} 0.5 = 60°$. At this power factor, $P_l/P_h = 0$, so that $P_l = 0$, while P_h reads the total power delivered to the load.

The second method for determining whether the total power is the sum or difference of the two wattmeter readings involves a simple laboratory test. For the test to be applied, both wattmeters must first have an up-scale deflection. If one of the wattmeters has a below-zero indication, an up-scale deflection can be obtained by simply reversing the leads of the current coil of the wattmeter. To perform the test:

1. Take notice of which line does not have a current coil sensing the line current.
2. For the lower-reading wattmeter, disconnect the lead of the potential coil connected to the line without the current coil.
3. Take the disconnected lead of the lower-reading wattmeter's potential coil, and touch a connection point on the line that has the current coil of the higher-reading wattmeter.
4. If the pointer deflects downward (below zero watts), the wattage reading of the lower-reading wattmeter should be subtracted from that of the higher-reading wattmeter. Otherwise, the readings should be added.

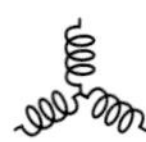

For a *balanced system,* since

$$P_T = P_h \pm P_l = \sqrt{3}E_L I_L \cos \theta_{I_\phi}^{V_\phi}$$

the power factor of the load (phase) can be found from the wattmeter readings and the magnitude of the line voltage and current:

$$F_p = \cos \theta_{I_\phi}^{V_\phi} = \frac{P_h \pm P_l}{\sqrt{3}E_L I_L} \tag{23.33}$$

EXAMPLE 23.8 For the unbalanced Δ-connected load in Fig. 23.33 with two properly connected wattmeters:

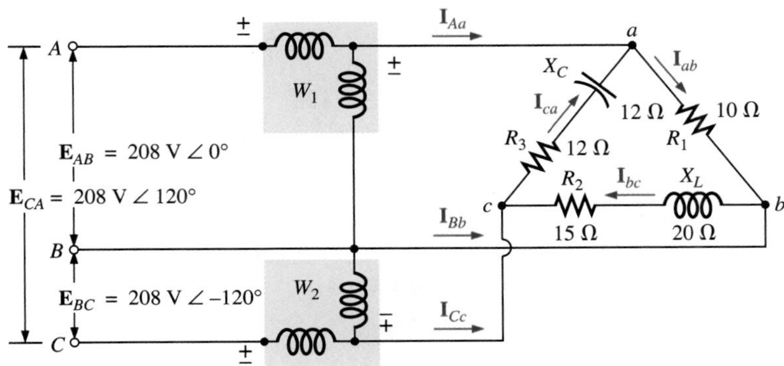

FIG. 23.33
Example 23.8.

a. Determine the magnitude and angle of the phase currents.
b. Calculate the magnitude and angle of the line currents.
c. Determine the power reading of each wattmeter.
d. Calculate the total power absorbed by the load.
e. Compare the result of part (d) with the total power calculated using the phase currents and the resistive elements.

Solutions:

a. $\mathbf{I}_{ab} = \dfrac{\mathbf{V}_{ab}}{\mathbf{Z}_{ab}} = \dfrac{\mathbf{E}_{AB}}{\mathbf{Z}_{ab}} = \dfrac{208 \text{ V} \angle 0^\circ}{10\ \Omega \angle 0^\circ} = \mathbf{20.8 \text{ A} \angle 0^\circ}$

$\mathbf{I}_{bc} = \dfrac{\mathbf{V}_{bc}}{\mathbf{Z}_{bc}} = \dfrac{\mathbf{E}_{BC}}{\mathbf{Z}_{bc}} = \dfrac{208 \text{ V} \angle -120^\circ}{15\ \Omega + j\,20\ \Omega} = \dfrac{208 \text{ V} \angle -120^\circ}{25\ \Omega \angle 53.13^\circ}$
$= \mathbf{8.32 \text{ A} \angle -173.13^\circ}$

$\mathbf{I}_{ca} = \dfrac{\mathbf{V}_{ca}}{\mathbf{Z}_{ca}} = \dfrac{\mathbf{E}_{CA}}{\mathbf{Z}_{ca}} = \dfrac{208 \text{ V} \angle +120^\circ}{12\ \Omega + j\,12\ \Omega} = \dfrac{208 \text{ V} \angle +120^\circ}{16.97\ \Omega \angle -45^\circ}$
$= \mathbf{12.26 \text{ A} \angle 165^\circ}$

b. $\mathbf{I}_{Aa} = \mathbf{I}_{ab} - \mathbf{I}_{ca}$
$= 20.8 \text{ A} \angle 0^\circ - 12.26 \text{ A} \angle 165^\circ$
$= 20.8 \text{ A} - (-11.84 \text{ A} + j\,3.17 \text{ A})$
$= 20.8 \text{ A} + 11.84 \text{ A} - j\,3.17 \text{ A} = 32.64 \text{ A} - j\,3.17 \text{ A}$
$= \mathbf{32.79 \text{ A} \angle -5.55^\circ}$

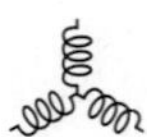

$$\begin{aligned}\mathbf{I}_{Bb} &= \mathbf{I}_{bc} - \mathbf{I}_{ab}\\ &= 8.32\text{ A}\angle{-173.13^\circ} - 20.8\text{ A}\angle 0^\circ\\ &= (-8.26\text{ A} - j\,1\text{ A}) - 20.8\text{ A}\\ &= -8.26\text{ A} - 20.8\text{ A} - j\,1\text{ A} = -29.06\text{ A} - j\,1\text{ A}\\ &= \mathbf{29.08\ A\ \angle{-178.03^\circ}}\end{aligned}$$

$$\begin{aligned}\mathbf{I}_{Cc} &= \mathbf{I}_{ca} - \mathbf{I}_{bc}\\ &= 12.26\text{ A}\angle 165^\circ - 8.32\text{ A}\angle{-173.13^\circ}\\ &= (-11.84\text{ A} + j\,3.17\text{ A}) - (-8.26\text{ A} - j\,1\text{ A})\\ &= -11.84\text{ A} + 8.26\text{ A} + j\,(3.17\text{ A} + 1\text{ A}) = -3.58\text{ A} + j\,4.17\text{ A}\\ &= \mathbf{5.5\ A\ \angle 130.65^\circ}\end{aligned}$$

c. $P_1 = V_{ab}I_{Aa}\cos\theta_{\mathbf{I}_{Aa}}^{\mathbf{V}_{ab}}$ $\quad \mathbf{V}_{ab} = 208\text{ V}\angle 0^\circ$

$$\begin{aligned}\mathbf{I}_{Aa} &= 32.79\text{ A}\angle{-5.55^\circ}\\ &= (208\text{ V})(32.79\text{ A})\cos 5.55^\circ\\ &= \mathbf{6788.35\ W}\end{aligned}$$

$$\mathbf{V}_{bc} = \mathbf{E}_{BC} = 208\text{ V}\angle{-120^\circ}$$

but $\quad \mathbf{V}_{cb} = \mathbf{E}_{CB} = 208\text{ V}\angle{-120^\circ} + 180^\circ = 208\text{ V}\angle 60^\circ$

with $\quad \mathbf{I}_{Cc} = 5.5\text{ A}\angle 130.65^\circ$

$$\begin{aligned}P_2 &= V_{cb}I_{Cc}\cos\theta_{\mathbf{I}_{Cc}}^{\mathbf{V}_{cb}}\\ &= (208\text{ V})(5.5\text{ A})\cos 70.65^\circ\\ &= \mathbf{379.1\ W}\end{aligned}$$

d. $P_T = P_1 + P_2 = 6788.35\text{ W} + 379.1\text{ W} = \mathbf{7167.45\ W}$

e.
$$\begin{aligned}P_T &= (I_{ab})^2R_1 + (I_{bc})^2R_2 + (I_{ca})^2R_3\\ &= (20.8\text{ A})^2 10\ \Omega + (8.32\text{ A})^2 15\ \Omega + (12.26\text{ A})^2 12\ \Omega\\ &= 4326.4\text{ W} + 1038.34\text{ W} + 1803.69\text{ W}\\ &= \mathbf{7168.43\ W}\end{aligned}$$

(The slight difference is due to the level of accuracy carried through the calculations.)

23.13 UNBALANCED, THREE-PHASE, FOUR-WIRE, Y-CONNECTED LOAD

For the three-phase, four-wire, Y-connected load in Fig. 23.34, conditions are such that *none* of the load impedances are equal—hence we have an **unbalanced polyphase load.** Since the neutral is a common point between the load and source, no matter what the impedance of each phase of the load and source, the voltage across each phase is the phase voltage of the generator:

$$\mathbf{V}_\phi = \mathbf{E}_\phi \tag{23.34}$$

The phase currents can therefore be determined by Ohm's law:

$$I_{\phi_1} = \frac{\mathbf{V}_{\phi_1}}{\mathbf{Z}_1} = \frac{\mathbf{E}_{\phi_1}}{\mathbf{Z}_1} \quad \text{and so on} \tag{23.35}$$

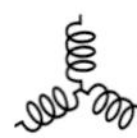

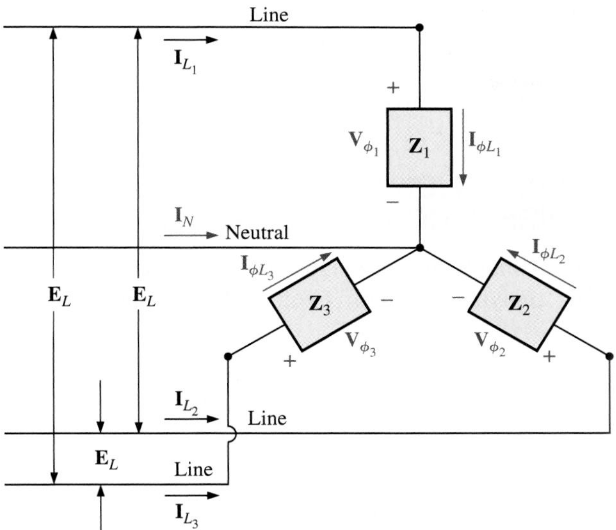

FIG. 23.34
Unbalanced Y-connected load.

The current in the neutral for any unbalanced system can then be found by applying Kirchhoff's current law at the common point *n:*

$$\mathbf{I}_N = \mathbf{I}_{\phi_1} + \mathbf{I}_{\phi_2} + \mathbf{I}_{\phi_3} = \mathbf{I}_{L_1} + \mathbf{I}_{L_2} + \mathbf{I}_{L_3} \quad \textbf{(23.36)}$$

Because of the variety of equipment found in an industrial environment, both three-phase power and single-phase power are usually provided with the single-phase obtained off the three-phase system. In addition, since the load on each phase is continually changing, a four-wire system (with a neutral) is normally used to ensure steady voltage levels and to provide a path for the current resulting from an unbalanced load. The system in Fig. 23.35 has a three-phase transformer dropping the line voltage from 13,800 V to 208 V. All the lower-power-demand loads such as lighting, wall outlets, security, etc., use the single-phase, 120 V line to neutral voltage. Higher power loads, such as air conditioners, electric ovens or dryers, etc., use the single-phase, 208 V available from line to line. For larger motors and special high-demand equipment, the full three-phase power can be taken directly off the system, as shown in Fig. 23.35. In the design and construction of a commercial establishment, the

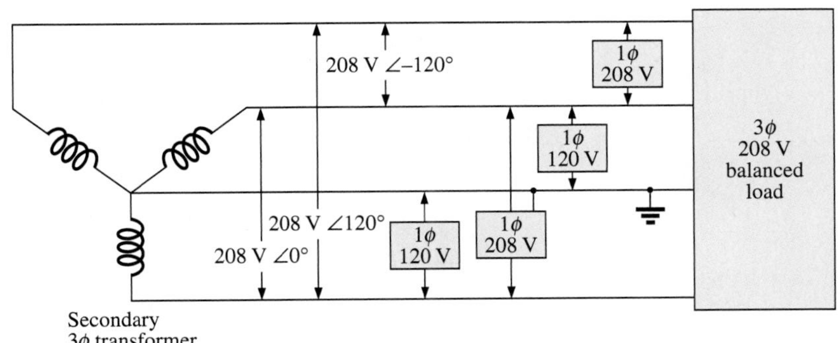

FIG. 23.35
3φ/1φ, 208 V/120 V industrial supply.

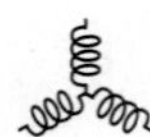

National Electric Code requires that every effort be made to ensure that the expected loads, whether they be single- or multiphase, result in a total load that is as balanced as possible between the phases, thus ensuring the highest level of transmission efficiency.

23.14 UNBALANCED, THREE-PHASE, THREE-WIRE, Y-CONNECTED LOAD

For the system shown in Fig. 23.36, the required equations can be derived by first applying Kirchhoff's voltage law around each closed loop to produce

$$\mathbf{E}_{AB} - \mathbf{V}_{an} + \mathbf{V}_{bn} = 0$$
$$\mathbf{E}_{BC} - \mathbf{V}_{bn} + \mathbf{V}_{cn} = 0$$
$$\mathbf{E}_{CA} - \mathbf{V}_{cn} + \mathbf{V}_{an} = 0$$

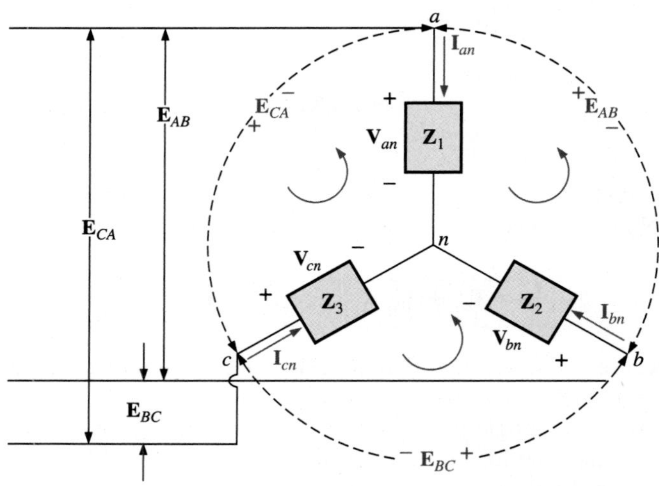

FIG. 23.36
Unbalanced, three-phase, three-wire, Y-connected load.

Substituting, we have

$$\mathbf{V}_{an} = \mathbf{I}_{an}\mathbf{Z}_1 \quad \mathbf{V}_{bn} = \mathbf{I}_{bn}\mathbf{Z}_2 \quad \mathbf{V}_{cn} = \mathbf{I}_{cn}\mathbf{Z}_3$$

$$\mathbf{E}_{AB} = \mathbf{I}_{an}\mathbf{Z}_1 - \mathbf{I}_{bn}\mathbf{Z}_2 \qquad \textbf{(23.37a)}$$
$$\mathbf{E}_{BC} = \mathbf{I}_{bn}\mathbf{Z}_2 - \mathbf{I}_{cn}\mathbf{Z}_3 \qquad \textbf{(23.37b)}$$
$$\mathbf{E}_{CA} = \mathbf{I}_{cn}\mathbf{Z}_3 - \mathbf{I}_{an}\mathbf{Z}_1 \qquad \textbf{(23.37c)}$$

Applying Kirchhoff's current law at node n results in

$$\mathbf{I}_{an} + \mathbf{I}_{bn} + \mathbf{I}_{cn} = 0 \quad \text{and} \quad \mathbf{I}_{bn} = -\mathbf{I}_{an} - \mathbf{I}_{cn}$$

Substituting for $\mathbf{I}_{bn}$ in Eqs. (23.37a) and (23.37b) yields

$$\mathbf{E}_{AB} = \mathbf{I}_{an}\mathbf{Z}_1 - [-(\mathbf{I}_{an} + \mathbf{I}_{cn})]\mathbf{Z}_2$$
$$\mathbf{E}_{BC} = -(\mathbf{I}_{an} + \mathbf{I}_{cn})\mathbf{Z}_2 - \mathbf{I}_{cn}\mathbf{Z}_3$$

which are rewritten as

$$\mathbf{E}_{AB} = \mathbf{I}_{an}(\mathbf{Z}_1 + \mathbf{Z}_2) + \mathbf{I}_{cn}\mathbf{Z}_2$$
$$\mathbf{E}_{BC} = \mathbf{I}_{an}(-\mathbf{Z}_2) + \mathbf{I}_{cn}[-(\mathbf{Z}_2 + \mathbf{Z}_3)]$$

Using determinants, we have

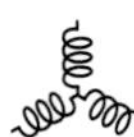

$$\mathbf{I}_{an} = \frac{\begin{vmatrix} \mathbf{E}_{AB} & \mathbf{Z}_2 \\ \mathbf{E}_{BC} & -(\mathbf{Z}_2 + \mathbf{Z}_3) \end{vmatrix}}{\begin{vmatrix} \mathbf{Z}_1 + \mathbf{Z}_2 & \mathbf{Z}_2 \\ -\mathbf{Z}_2 & -(\mathbf{Z}_2 + \mathbf{Z}_3) \end{vmatrix}}$$

$$= \frac{-(\mathbf{Z}_2 + \mathbf{Z}_3)\mathbf{E}_{AB} - \mathbf{E}_{BC}\mathbf{Z}_2}{-\mathbf{Z}_1\mathbf{Z}_2 - \mathbf{Z}_1\mathbf{Z}_3 - \mathbf{Z}_2\mathbf{Z}_3 - \mathbf{Z}_2^2 + \mathbf{Z}_2^2}$$

$$\mathbf{I}_{an} = \frac{-\mathbf{Z}_2(\mathbf{E}_{AB} + \mathbf{E}_{BC}) - \mathbf{Z}_3\mathbf{E}_{AB}}{-\mathbf{Z}_1\mathbf{Z}_2 - \mathbf{Z}_1\mathbf{Z}_3 - \mathbf{Z}_2\mathbf{Z}_3}$$

Applying Kirchhoff's voltage law to the line voltages:

$$\mathbf{E}_{AB} + \mathbf{E}_{CA} + \mathbf{E}_{BC} = 0 \quad \text{or} \quad \mathbf{E}_{AB} + \mathbf{E}_{BC} = -\mathbf{E}_{CA}$$

Substituting for ($\mathbf{E}_{AB} + \mathbf{E}_{CB}$) in the above equation for $\mathbf{I}_{an}$:

$$\mathbf{I}_{an} = \frac{-\mathbf{Z}_2(-\mathbf{E}_{CA}) - \mathbf{Z}_3\mathbf{E}_{AB}}{-\mathbf{Z}_1\mathbf{Z}_2 - \mathbf{Z}_1\mathbf{Z}_3 - \mathbf{Z}_2\mathbf{Z}_3}$$

and

$$\boxed{\mathbf{I}_{an} = \frac{\mathbf{E}_{AB}\mathbf{Z}_3 - \mathbf{E}_{CA}\mathbf{Z}_2}{\mathbf{Z}_1\mathbf{Z}_2 + \mathbf{Z}_1\mathbf{Z}_3 + \mathbf{Z}_2\mathbf{Z}_3}} \tag{23.38}$$

In the same manner, it can be shown that

$$\boxed{\mathbf{I}_{cn} = \frac{\mathbf{E}_{CA}\mathbf{Z}_2 - \mathbf{E}_{BC}\mathbf{Z}_1}{\mathbf{Z}_1\mathbf{Z}_2 + \mathbf{Z}_1\mathbf{Z}_3 + \mathbf{Z}_2\mathbf{Z}_3}} \tag{23.39}$$

Substituting Eq. (23.39) for $\mathbf{I}_{cn}$ in the right-hand side of Eq. (23.37b), we obtain

$$\boxed{\mathbf{I}_{bn} = \frac{\mathbf{E}_{BC}\mathbf{Z}_1 - \mathbf{E}_{AB}\mathbf{Z}_3}{\mathbf{Z}_1\mathbf{Z}_2 + \mathbf{Z}_1\mathbf{Z}_3 + \mathbf{Z}_2\mathbf{Z}_3}} \tag{23.40}$$

EXAMPLE 23.9 A *phase-sequence indicator* is an instrument that can display the phase sequence of a polyphase circuit. A network that performs this function appears in Fig. 23.37. The applied phase sequence is *ABC*. The bulb corresponding to this phase sequence burns more brightly than the bulb indicating the *ACB* sequence because a greater current is passing through the *ABC* bulb. Calculating the phase currents demonstrates that this situation does in fact exist:

$$Z_1 = X_C = \frac{1}{\omega C} = \frac{1}{(377\ \text{rad/s})(16 \times 10^{-6}\ \text{F})} = 166\ \Omega$$

By Eq. (23.39),

$$\mathbf{I}_{cn} = \frac{\mathbf{E}_{CA}\mathbf{Z}_2 - \mathbf{E}_{BC}\mathbf{Z}_1}{\mathbf{Z}_1\mathbf{Z}_2 + \mathbf{Z}_1\mathbf{Z}_3 + \mathbf{Z}_2\mathbf{Z}_3}$$

$$= \frac{(200\ \text{V} \angle 120^\circ)(200\ \Omega \angle 0^\circ) - (200\ \text{V} \angle -120^\circ)(166\ \Omega \angle -90^\circ)}{(166\ \Omega \angle -90^\circ)(200\ \Omega \angle 0^\circ) + (166\ \Omega \angle -90^\circ)(200\ \Omega \angle 0^\circ) + (200\ \Omega \angle 0^\circ)(200\ \Omega \angle 0^\circ)}$$

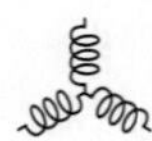

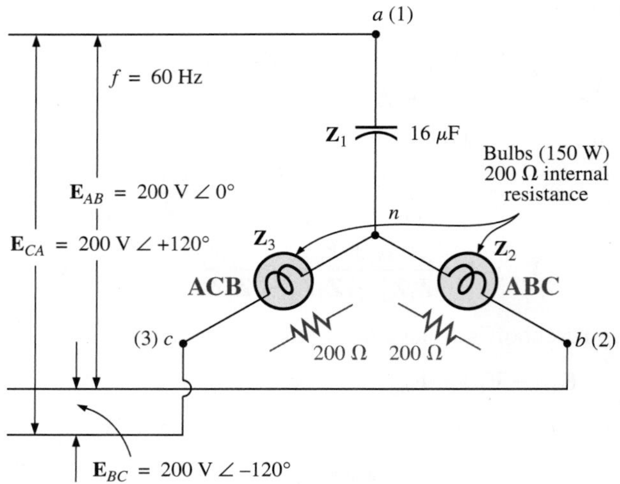

FIG. 23.37
Example 23.9.

$$\mathbf{I}_{cn} = \frac{40{,}000\text{ V}\angle 120° + 33{,}200\text{ V}\angle -30°}{33{,}200\ \Omega\angle -90° + 33{,}200\ \Omega\angle -90° + 40{,}000\ \Omega\angle 0°}$$

Dividing the numerator and denominator by 1000 and converting both to the rectangular domain yields

$$\mathbf{I}_{cn} = \frac{(-20 + j\,34.64) + (28.75 - j\,16.60)}{40 - j\,66.4}$$

$$= \frac{8.75 + j\,18.04}{77.52\angle -58.93°} = \frac{20.05\angle 64.13°}{77.52\angle -58.93°}$$

$$\mathbf{I}_{cn} = \mathbf{0.259\text{ A}\angle 123.06°}$$

By Eq. (23.40),

$$\mathbf{I}_{bn} = \frac{\mathbf{E}_{BC}\mathbf{Z}_1 - \mathbf{E}_{AB}\mathbf{Z}_3}{\mathbf{Z}_1\mathbf{Z}_2 + \mathbf{Z}_1\mathbf{Z}_3 + \mathbf{Z}_2\mathbf{Z}_3}$$

$$= \frac{(200\text{ V}\angle -120°)(166\angle -90°) - (200\text{ V}\angle 0°)(200\angle 0°)}{77.52 \times 10^3\ \Omega\angle -58.93°}$$

$$\mathbf{I}_{bn} = \frac{33{,}200\text{ V}\angle -210° - 40{,}000\text{ V}\angle 0°}{77.52 \times 10^3\ \Omega\angle -58.93°}$$

Dividing by 1000 and converting to the rectangular domain yields

$$\mathbf{I}_{bn} = \frac{-28.75 + j\,16.60 - 40.0}{77.52\angle -58.93°} = \frac{-68.75 + j\,16.60}{77.52\angle -58.93°}$$

$$= \frac{70.73\angle 166.43°}{77.52\angle -58.93°} = \mathbf{0.91\text{ A}\angle 225.36°}$$

and $I_{bn} > I_{cn}$ by a factor of more than 3 : 1. Therefore, the bulb indicating an *ABC* sequence will burn more brightly due to the greater current. If the phase sequence were *ACB,* the reverse would be true.

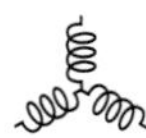

PROBLEMS

SECTION 23.5 Y-Connected Generator with a Y-Connected Load

1. A balanced Y load having a 10 Ω resistance in each leg is connected to a three-phase, four-wire, Y-connected generator having a line voltage of 208 V. Calculate the magnitude of
   - **a.** the phase voltage of the generator.
   - **b.** the phase voltage of the load.
   - **c.** the phase current of the load.
   - **d.** the line current.
2. Repeat Problem 1 if each phase impedance is changed to a 12 Ω resistor in series with a 16 Ω capacitive reactance.
3. Repeat Problem 1 if each phase impedance is changed to a 10 Ω resistor in parallel with a 10 Ω capacitive reactance.
4. The phase sequence for the Y-Y system in Fig. 23.38 is *ABC*.
   - **a.** Find the angles θ_2 and θ_3 for the specified phase sequence.
   - **b.** Find the voltage across each phase impedance in phasor form.
   - **c.** Find the current through each phase impedance in phasor form.
   - **d.** Draw the phasor diagram of the currents found in part (c), and show that their phasor sum is zero.
   - **e.** Find the magnitude of the line currents.
   - **f.** Find the magnitude of the line voltages.
5. Repeat Problem 4 if the phase impedances are changed to a 9 Ω resistor in series with a 12 Ω inductive reactance.
6. Repeat Problem 4 if the phase impedances are changed to a 6 Ω resistance in parallel with an 8 Ω capacitive reactance.
7. For the system in Fig. 23.39, find the magnitude of the unknown voltages and currents.

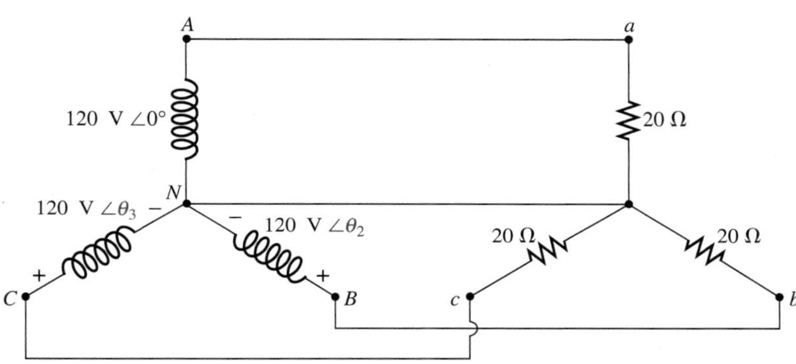

FIG. 23.38
Problems 4, 5, 6, and 31.

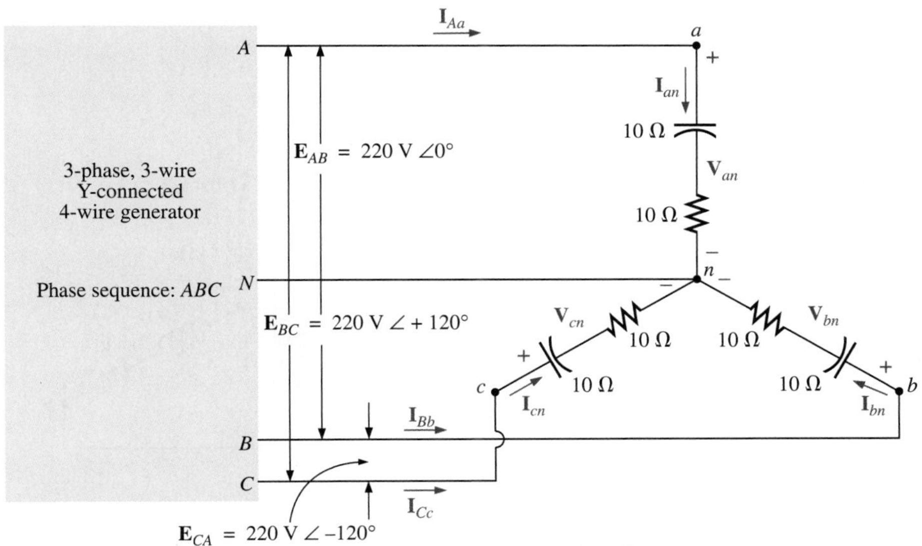

FIG. 23.39
Problems 7, 32, and 44.

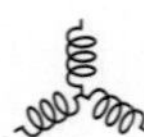

***8.** Compute the magnitude of the voltage E_{AB} for the balanced three-phase system in Fig. 23.40.

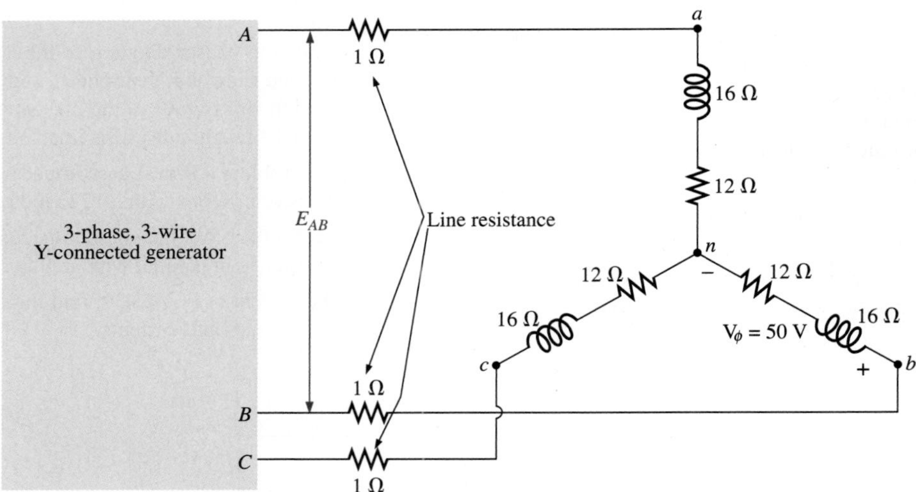

FIG. 23.40
Problem 8.

***9.** For the Y-Y system in Fig. 23.41:

a. Find the magnitude and angle associated with the voltages $\mathbf{E}_{AN}$, $\mathbf{E}_{BN}$, and $\mathbf{E}_{CN}$.

b. Determine the magnitude and angle associated with each phase current of the load: $\mathbf{I}_{an}$, $\mathbf{I}_{bn}$, and $\mathbf{I}_{cn}$.

c. Find the magnitude and phase angle of each line current: $\mathbf{I}_{Aa}$, $\mathbf{I}_{Bb}$, and $\mathbf{I}_{Cc}$.

d. Determine the magnitude and phase angle of the voltage across each phase of the load: $\mathbf{V}_{an}$, $\mathbf{V}_{bn}$, and $\mathbf{V}_{cn}$.

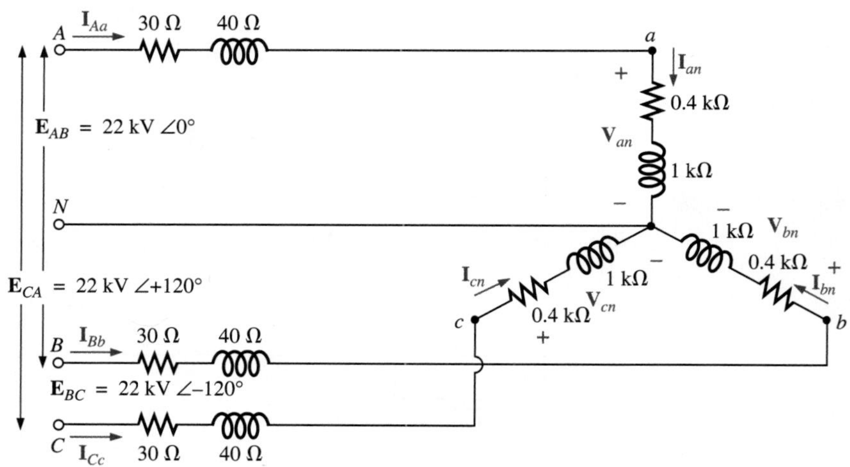

FIG. 23.41
Problem 9.

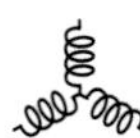

SECTION 23.6 Y-Δ System

10. A balanced Δ load having a 20 Ω resistance in each leg is connected to a three-phase, three-wire, Y-connected generator having a line voltage of 208 V. Calculate the magnitude of

a. the phase voltage of the generator.
b. the phase voltage of the load.
c. the phase current of the load.
d. the line current.

11. Repeat Problem 10 if each phase impedance is changed to a 6.8 Ω resistor in series with a 14 Ω inductive reactance.

12. Repeat Problem 10 if each phase impedance is changed to an 18 Ω resistance in parallel with an 18 Ω capacitive reactance.

13. The phase sequence for the Y-Δ system in Fig. 23.42 is *ABC*.

a. Find the angles θ_2 and θ_3 for the specified phase sequence.
b. Find the voltage across each phase impedance in phasor form.
c. Draw the phasor diagram of the voltages found in part (b), and show that their sum is zero around the closed loop of the Δ load.
d. Find the current through each phase impedance in phasor form.
e. Find the magnitude of the line currents.
f. Find the magnitude of the generator phase voltages.

14. Repeat Problem 13 if the phase impedances are changed to a 100 Ω resistor in series with a capacitive reactance of 100 Ω.

15. Repeat Problem 13 if the phase impedances are changed to a 3 Ω resistor in parallel with an inductive reactance of 4 Ω.

16. For the system in Fig. 23.43, find the magnitude of the unknown voltages and currents.

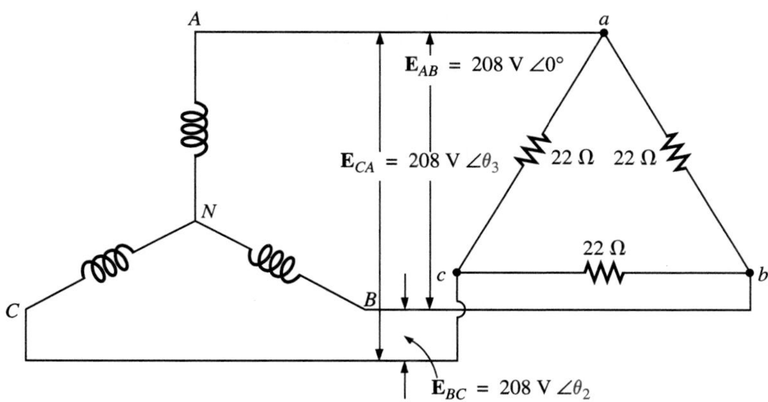

FIG. 23.42
Problems 13, 14, 15, 34, and 45.

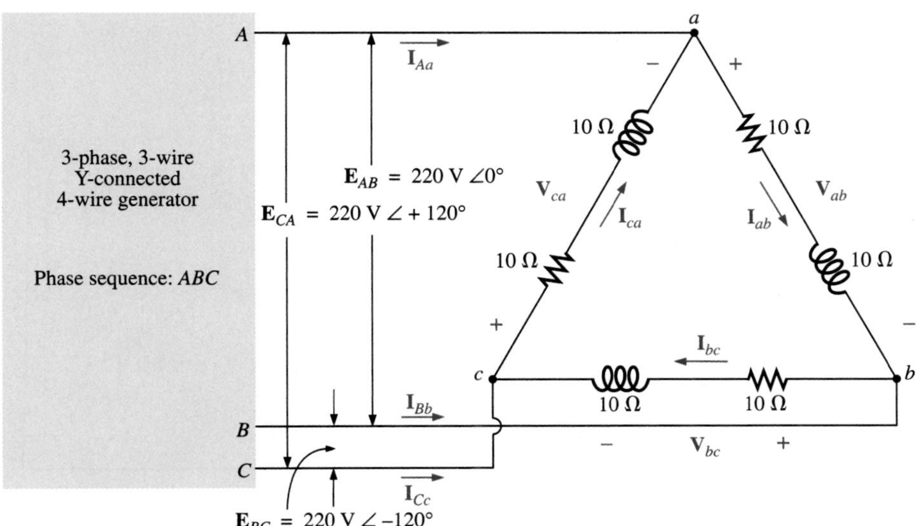

FIG. 23.43
Problems 16, 35, and 47.

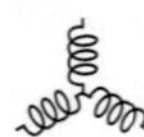

*17. For the Δ-connected load in Fig. 23.44:
  a. Find the magnitude and angle of each phase current $\mathbf{I}_{ab}$, $\mathbf{I}_{bc}$, and $\mathbf{I}_{ca}$.
  b. Calculate the magnitude and angle of each line current $\mathbf{I}_{Aa}$, $\mathbf{I}_{Bb}$, and $\mathbf{I}_{Cc}$.
  c. Determine the magnitude and angle of the voltages $\mathbf{E}_{AB}$, $\mathbf{E}_{BC}$, and $\mathbf{E}_{CA}$.

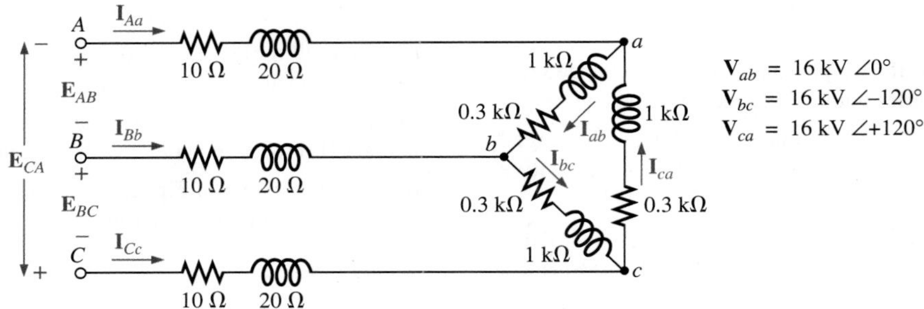

FIG. 23.44
Problem 17.

SECTION 23.9 Δ-Δ, Δ-Y Three-Phase Systems

18. A balanced Y load having a 30 Ω resistance in each leg is connected to a three-phase, Δ-connected generator having a line voltage of 208 V. Calculate the magnitude of
  a. the phase voltage of the generator.
  b. the phase voltage of the load.
  c. the phase current of the load.
  d. the line current.

19. Repeat Problem 18 if each phase impedance is changed to a 12 Ω resistor in series with a 12 Ω inductive reactance.

20. Repeat Problem 18 if each phase impedance is changed to a 15 Ω resistor in parallel with a 20 Ω capacitive reactance.

*21. For the system in Fig. 23.45, find the magnitude of the unknown voltages and currents.

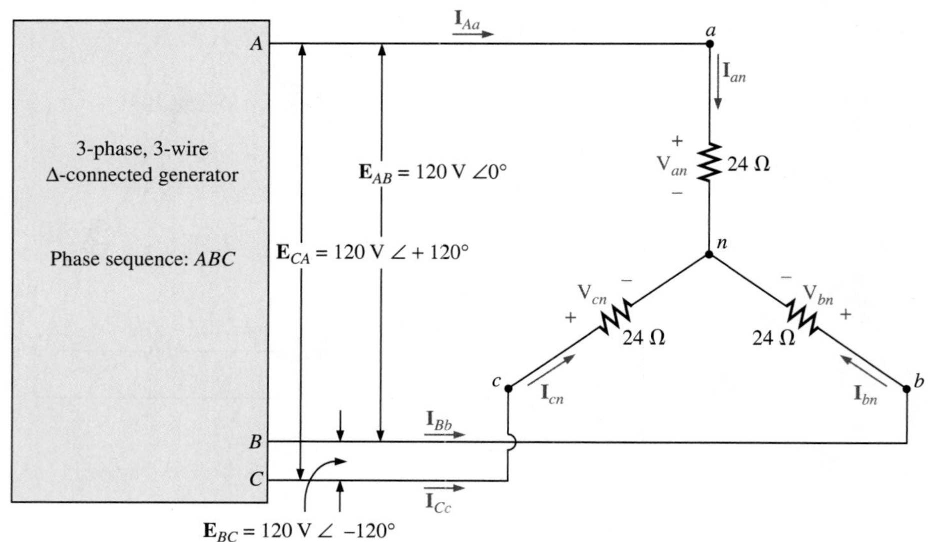

FIG. 23.45
Problems 21, 22, 23, and 37.

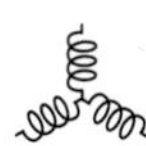

22. Repeat Problem 21 if each phase impedance is changed to a 10 Ω resistor in series with a 20 Ω inductive reactance.

23. Repeat Problem 21 if each phase impedance is changed to a 20 Ω resistor in parallel with a 15 Ω capacitive reactance.

24. A balanced Δ load having a 220 Ω resistance in each leg is connected to a three-phase, Δ-connected generator having a line voltage of 440 V. Calculate the magnitude of
 a. the phase voltage of the generator.
 b. the phase voltage of the load.
 c. the phase current of the load.
 d. the line current.

25. Repeat Problem 24 if each phase impedance is changed to a 12 Ω resistor in series with a 9 Ω capacitive reactance.

26. Repeat Problem 24 if each phase impedance is changed to a 22 Ω resistor in parallel with a 22 Ω inductive reactance.

27. The phase sequence for the Δ-Δ system in Fig. 23.46 is *ABC*.
 a. Find the angles θ_2 and θ_3 for the specified phase sequence.
 b. Find the voltage across each phase impedance in phasor form.
 c. Draw the phasor diagram of the voltages found in part (b), and show that their phasor sum is zero around the closed loop of the Δ load.
 d. Find the current through each phase impedance in phasor form.
 e. Find the magnitude of the line currents.

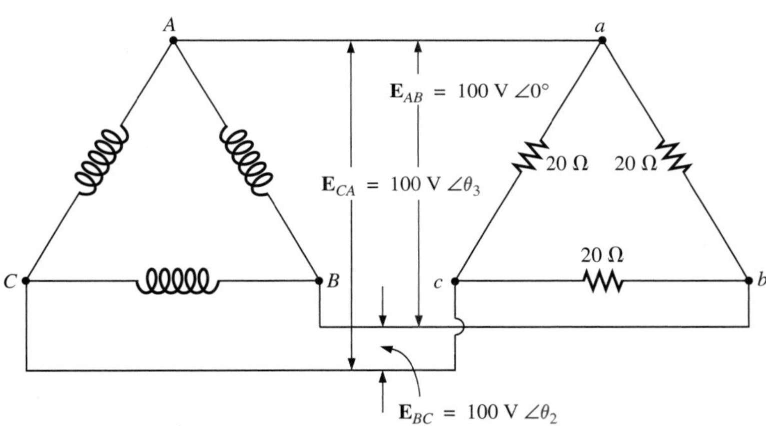

FIG. 23.46
Problem 27.

28. Repeat Problem 25 if each phase impedance is changed to a 12 Ω resistor in series with a 16 Ω inductive reactance.

29. Repeat Problem 25 if each phase impedance is changed to a 20 Ω resistor in parallel with a 20 Ω capacitive reactance.

SECTION 23.10 Power

30. Find the total watts, volt-amperes reactive, volt-amperes, and F_p of the three-phase system in Problem 2.

31. Find the total watts, volt-amperes reactive, volt-amperes, and F_p of the three-phase system in Problem 4.

32. Find the total watts, volt-amperes reactive, volt-amperes, and F_p of the three-phase system in Problem 7.

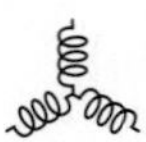

33. Find the total watts, volt-amperes reactive, volt-amperes, and F_p of the three-phase system in Problem 12.

34. Find the total watts, volt-amperes reactive, volt-amperes, and F_p of the three-phase system in Problem 14.

35. Find the total watts, volt-amperes reactive, volt-amperes, and F_p of the three-phase system in Problem 16.

36. Find the total watts, volt-amperes reactive, volt-amperes, and F_p of the three-phase system in Problem 20.

37. Find the total watts, volt-amperes reactive, volt-amperes, and F_p of the three-phase system in Problem 22.

38. Find the total watts, volt-amperes reactive, volt-amperes, and F_p of the three-phase system in Problem 26.

39. Find the total watts, volt-amperes reactive, volt-amperes, and F_p of the three-phase system in Problem 28.

40. A balanced, three-phase, Δ-connected load has a line voltage of 200 and a total power consumption of 4800 W at a lagging power factor of 0.8. Find the impedance of each phase in rectangular coordinates.

41. A balanced, three-phase, Y-connected load has a line voltage of 208 and a total power consumption of 1200 W at a leading power factor of 0.6. Find the impedance of each phase in rectangular coordinates.

***42.** Find the total watts, volt-amperes reactive, volt-amperes, and F_p of the system in Fig. 23.47.

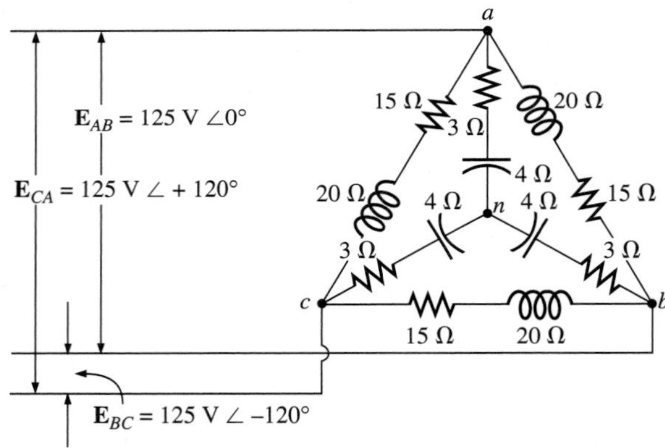

FIG. 23.47
Problem 42.

***43.** The Y-Y system in Fig. 23.48 has a balanced load and a line impedance $\mathbf{Z}_{\text{line}} = 4\ \Omega + j\ 20\ \Omega$. If the line voltage at the generator is 16,000 V and the total power delivered to the load is 1200 kW at 80 A, determine each of the following:

a. The magnitude of each phase voltage of the generator.
b. The magnitude of the line currents.
c. The total power delivered by the source.
d. The power factor angle of the entire load "seen" by the source.
e. The magnitude and angle of the current $\mathbf{I}_{Aa}$ if $\mathbf{E}_{AN} = E_{AN}\ \angle 0°$.
f. The magnitude and angle of the phase voltage $\mathbf{V}_{an}$.
g. The impedance of the load of each phase in rectangular coordinates.

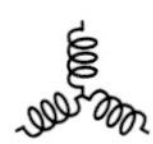

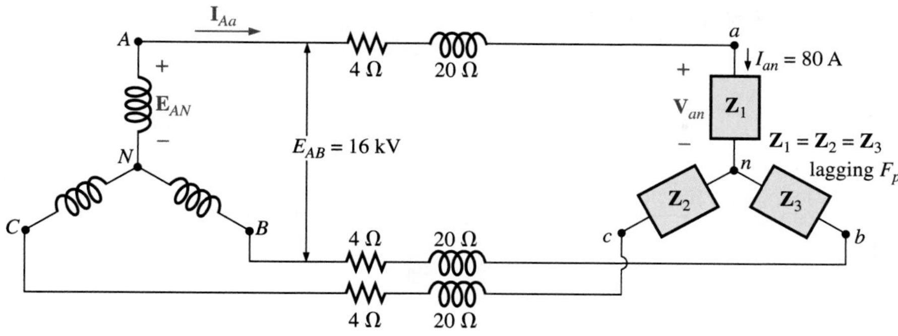

FIG. 23.48
Problem 43.

h. The difference between the power factor of the load and the power factor of the entire system (including $\mathbf{Z}_{\text{line}}$).

i. The efficiency of the system.

SECTION 23.11 Three-Wattmeter Method

44. **a.** Sketch the connections required to measure the total watts delivered to the load in Fig. 23.39 using three wattmeters.

b. Determine the total wattage dissipation and the reading of each wattmeter.

45. Repeat Problem 44 for the network in Fig. 23.42.

SECTION 23.12 Two-Wattmeter Method

46. **a.** For the three-wire system in Fig. 23.49, properly connect a second wattmeter so that the two measure the total power delivered to the load.

b. If one wattmeter has a reading of 200 W and the other a reading of 85 W, what is the total dissipation in watts if the total power factor is 0.8 leading?

c. Repeat part (b) if the total power factor is 0.2 lagging and $P_l = 100$ W.

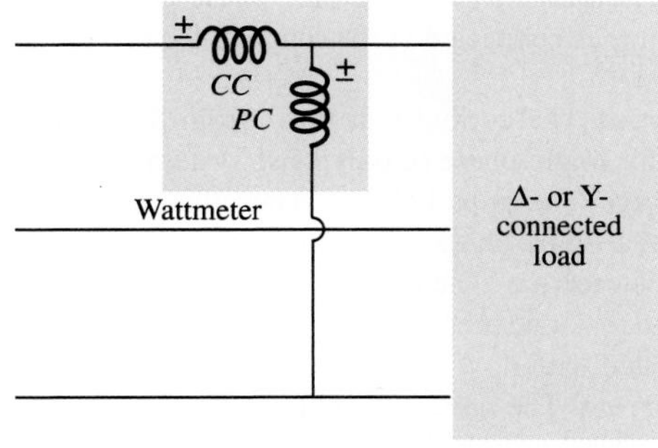

FIG. 23.49
Problem 46.

47. Sketch three different ways that two wattmeters can be connected to measure the total power delivered to the load in Problem 16.

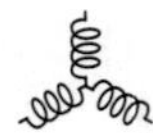

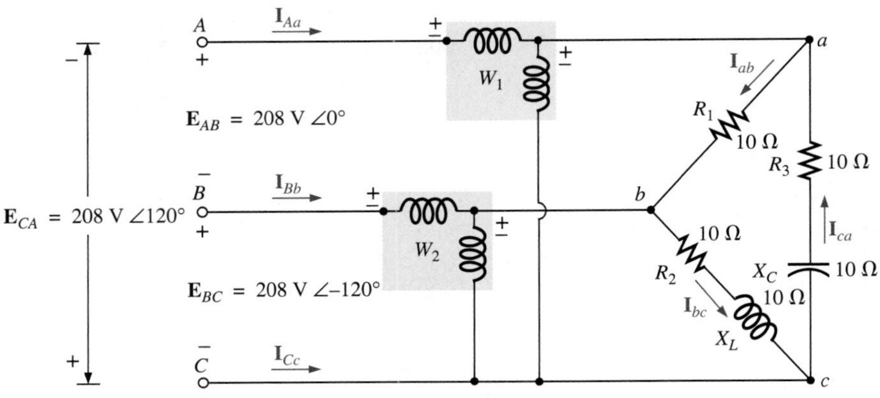

FIG. 23.50
Problem 48.

*48. For the Y-Δ system in Fig. 23.50:
- a. Determine the magnitude and angle of the phase currents.
- b. Find the magnitude and angle of the line currents.
- c. Determine the reading of each wattmeter.
- d. Find the total power delivered to the load.

SECTION 23.13 Unbalanced, Three-Phase, Four-Wire, Y-Connected Load

*49. For the system in Fig. 23.51:
- a. Calculate the magnitude of the voltage across each phase of the load.
- b. Find the magnitude of the current through each phase of the load.
- c. Find the total watts, volt-amperes reactive, volt-amperes, and F_p of the system.
- d. Find the phase currents in phasor form.
- e. Using the results of part (c), determine the current $\mathbf{I}_N$.

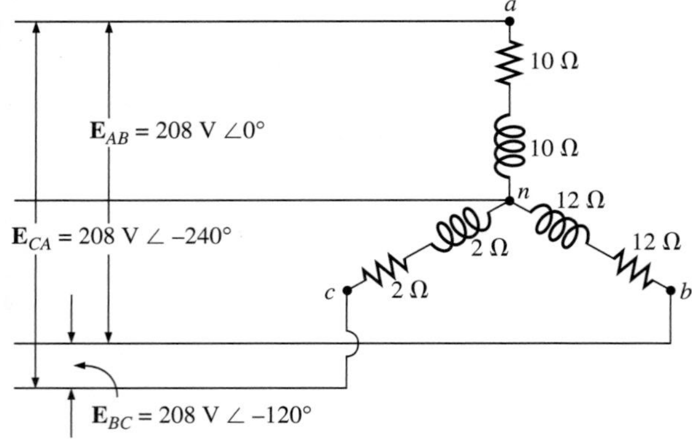

FIG. 23.51
Problem 49.

SECTION 23.14 Unbalanced, Three-Phase, Three-Wire, Y-Connected Load

*50. For the three-phase, three-wire system in Fig. 23.52, find the magnitude of the current through each phase of the load, and find the total watts, volt-amperes reactive, volt-amperes, and F_p of the load.

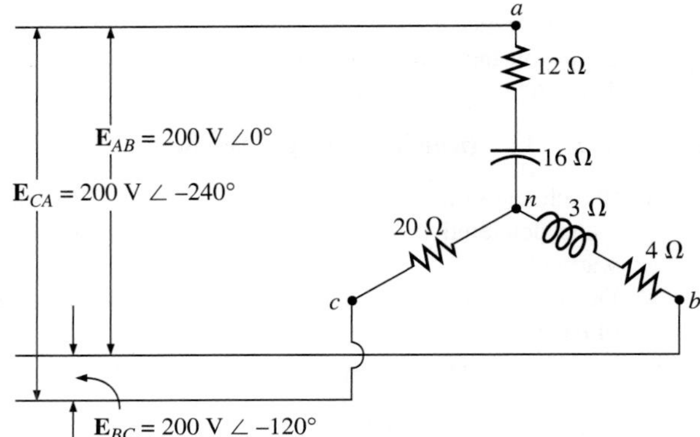

FIG. 23.52
Problem 50.

GLOSSARY

Δ-connected ac generator A three-phase generator having the three phases connected in the shape of the capital Greek letter *delta* (Δ).

Line current The current that flows from the generator to the load of a single-phase or polyphase system.

Line voltage The potential difference that exists between the lines of a single-phase or polyphase system.

Neutral connection The connection between the generator and the load that, under balanced conditions, will have zero current associated with it.

Phase current The current that flows through each phase of a single-phase (or polyphase) generator or load.

Phase sequence The order in which the generated sinusoidal voltages of a polyphase generator will affect the load to which they are applied.

Phase voltage The voltage that appears between the line and neutral of a Y-connected generator and from line to line in a Δ-connected generator.

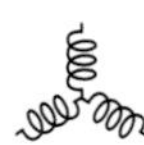

Polyphase ac generator An electromechanical source of ac power that generates more than one sinusoidal voltage per rotation of the rotor. The frequency generated is determined by the speed of rotation and the number of poles of the rotor.

Single-phase ac generator An electromechanical source of ac power that generates a single sinusoidal voltage having a frequency determined by the speed of rotation and the number of poles of the rotor.

Three-wattmeter method A method for determining the total power delivered to a three-phase load using three wattmeters.

Two-wattmeter method A method for determining the total power delivered to a Δ- or Y-connected three-phase load using only two wattmeters and considering the power factor of the load.

Unbalanced polyphase load A load not having the same impedance in each phase.

Y-connected three-phase generator A three-phase source of ac power in which the three phases are connected in the shape of the letter Y.

Pulse Waveforms and the R-C Response

Objectives

- *Become familiar with the specific terms that define a pulse waveform and how to calculate various parameters such as the pulse width, rise and fall times, and tilt.*
- *Be able to calculate the pulse repetition rate and the duty cycle of any pulse waveform.*
- *Become aware of the parameters that define the response of an R-C network to a square-wave input.*
- *Understand how a compensator probe of an oscilloscope is used to improve the appearance of an output pulse waveform.*

24.1 INTRODUCTION

Our analysis thus far has been limited to alternating waveforms that vary in a sinusoidal manner. This chapter introduces the basic terminology associated with the pulse waveform and examines the response of an *R-C* circuit to a square-wave input. The importance of the pulse waveform to the electrical/electronics industry cannot be overstated. A vast array of instrumentation, communication systems, computers, radar systems, and so on, all use pulse signals to control operation, transmit data, and display information in a variety of formats.

The response to a pulse signal of the networks described thus far is quite different from that obtained for sinusoidal signals. In fact, we must refer to the dc chapter on capacitors (Chapter 10) for a few fundamental concepts and equations that will help us in the analysis to follow. This chapter is just an introduction, designed to provide the fundamentals that will be helpful when the pulse waveform is encountered in specific areas of application.

24.2 IDEAL VERSUS ACTUAL

The **ideal pulse** in Fig. 24.1 has vertical sides, sharp corners, and a flat peak characteristic; it starts instantaneously at t_1 and ends just as abruptly at t_2.

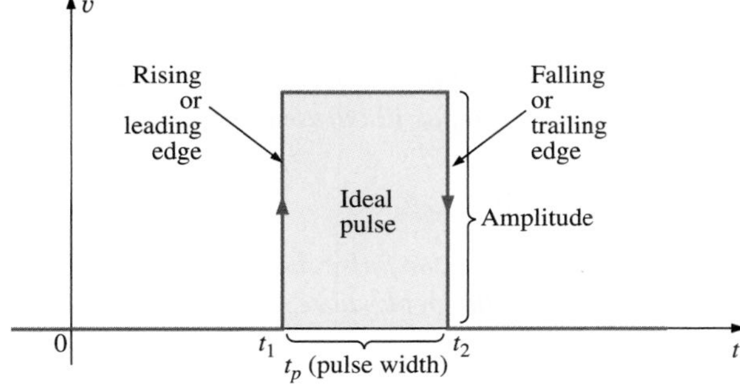

FIG. 24.1
Ideal pulse waveform.

The waveform in Fig. 24.1 is applied in the analysis in this chapter and probably in the initial investigation of areas of application beyond the scope of this text. Once the fundamental operation of a device, package, or system is clearly understood using ideal characteristics, the effect of an **actual** (or **true** or **practical**) **pulse** must be considered. If an attempt were made to introduce all the differences between an ideal and actual pulse in a single figure, the result would probably be complex and confusing. A number of waveforms are therefore used to define the critical parameters.

The reactive elements of a network, in their effort to prevent instantaneous changes in voltage (capacitor) and current (inductor), establish a slope to both edges of the pulse waveform, as shown in Fig. 24.2. The *rising edge* of the waveform in Fig. 24.2 is defined as the edge that increases from a lower to a higher level.

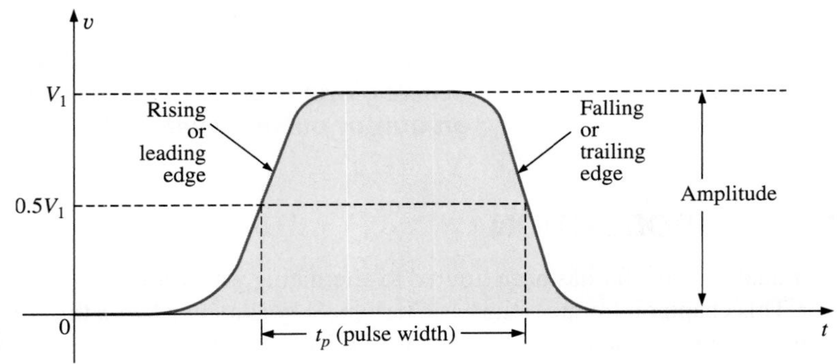

FIG. 24.2
Actual pulse waveform.

The falling edge is defined by the region or edge where the waveform decreases from a higher to a lower level. Since the rising edge is the first to be encountered (closest to t = 0 s), it is also called the leading edge. The falling edge always follows the leading edge and is therefore often called the trailing edge.

Both regions are defined in Figs. 24.1 and 24.2.

Amplitude

For most applications, the **amplitude of a pulse waveform** is defined as the peak-to-peak value. Of course, if the waveforms all start and return to the zero-volt level, then the peak and peak-to-peak values are synonymous.

For the purposes of this text, the amplitude of a pulse waveform is the peak-to-peak value, as illustrated in Figs. 24.1 and 24.2.

Pulse Width

The pulse width (t_p), or pulse duration, is defined by a pulse level equal to 50% of the peak value.

For the ideal pulse in Fig. 24.1, the pulse width is the same at any level, whereas t_p for the waveform in Fig. 24.2 is a very specific value.

Base-Line Voltage

The base-line voltage (V_b) is the voltage level from which the pulse is initiated.

The waveforms in Figs. 24.1 and 24.2 both have a 0 V base-line voltage. In Fig. 24.3(a) the base-line voltage is 1 V, whereas in Fig. 24.3(b) the base-line voltage is −4 V.

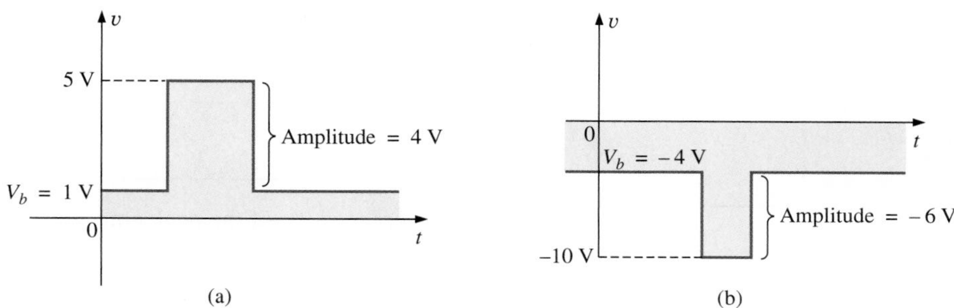

FIG. 24.3
Defining the base-line voltage.

Positive-Going and Negative-Going Pulses

A positive-going pulse increases positively from the base-line voltage, whereas a negative-going pulse increases in the negative direction from the base-line voltage.

The waveform in Fig. 24.3(a) is a positive-going pulse, whereas the waveform in Fig. 24.3(b) is a negative-going pulse.

Even though the base-line voltage in Fig. 24.4 is negative, the waveform is positive-going (with an amplitude of 10 V) since the voltage increased in the positive direction from the base-line voltage.

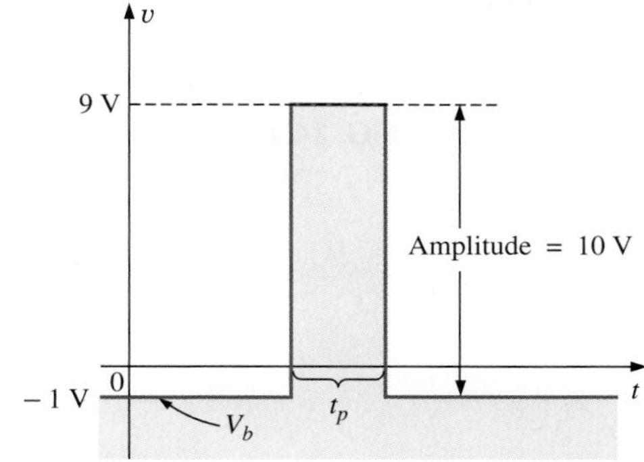

FIG. 24.4
Positive-going pulse.

Rise Time (t_r) and Fall Time (t_f)

The time required for the pulse to shift from one level to another is of particular importance. The *rounding* (defined in Fig. 24.5) that occurs at the beginning and end of each transition makes it difficult to define the exact point at which the rise time should be initiated and terminated. For this reason,

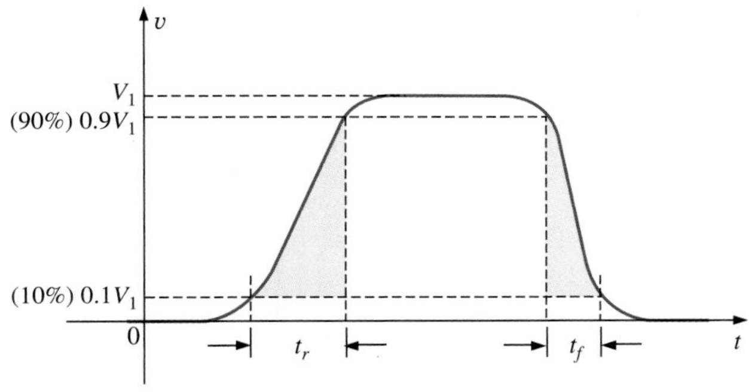

FIG. 24.5
Defining t_r and t_f.

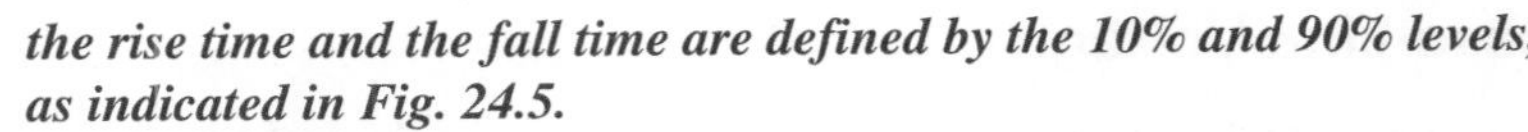

the rise time and the fall time are defined by the 10% and 90% levels, as indicated in Fig. 24.5.

Note that there is no requirement that t_r equal t_f.

Tilt

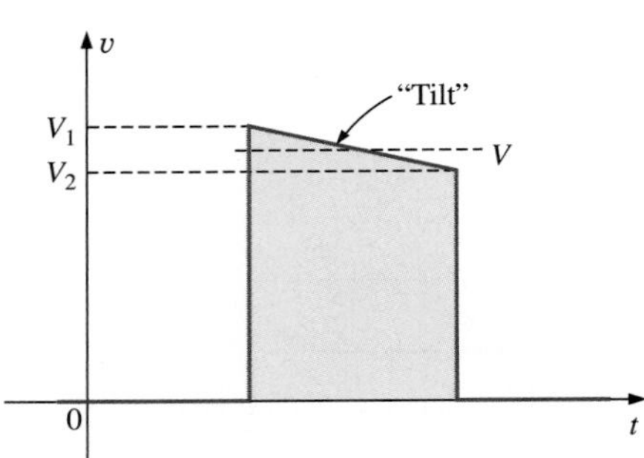

FIG. 24.6
Defining tilt.

Fig. 24.6 illustrates an undesirable but common distortion normally occurring due to a poor low-frequency response characteristic of the system through which a pulse has passed. The drop in peak value is called **tilt, droop,** or **sag.** The percentage tilt is defined by

$$\% \text{ tilt} = \frac{V_1 - V_2}{V} \times 100\% \tag{24.1}$$

where V is the average value of the peak amplitude as determined by

$$V = \frac{V_1 + V_2}{2} \tag{24.2}$$

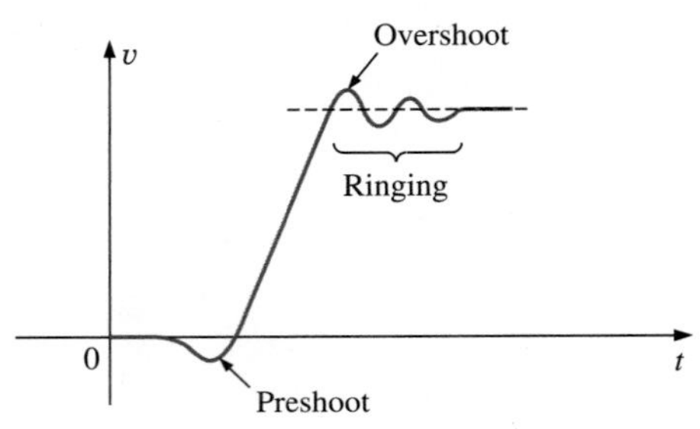

FIG. 24.7
Defining preshoot, overshoot, and ringing.

Naturally, the less the percentage tilt or sag, the more ideal the pulse. Due to rounding, it may be difficult to define the values of V_1 and V_2. It is then necessary to approximate the sloping region by a straight-line approximation and use the resulting values of V_1 and V_2.

Other distortions include the *preshoot* and *overshoot* appearing in Fig. 24.7, normally due to pronounced high-frequency effects of a system, and *ringing,* due to the interaction between the capacitive and inductive elements of a network at their natural or resonant frequency.

EXAMPLE 24.1 Determine the following for the pulse waveform in Fig. 24.8:

a. positive- or negative-going?
b. base-line voltage
c. pulse width
d. maximum amplitude
e. tilt

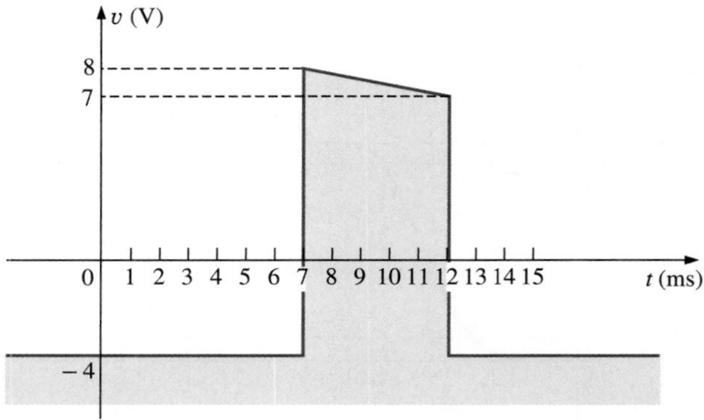

FIG. 24.8
Example 24.1.

Solutions:

a. **positive-going**
b. $V_b = \mathbf{-4\ V}$
c. $t_p = (12 - 7)\text{ ms} = \mathbf{5\ ms}$
d. $V_{max} = 8\text{ V} + 4\text{ V} = \mathbf{12\ V}$
e. $V = \dfrac{V_1 + V_2}{2} = \dfrac{12\text{ V} + 11\text{ V}}{2} = \dfrac{23\text{ V}}{2} = 11.5\text{ V}$

$$\%\text{ tilt} = \frac{V_1 - V_2}{V} \times 100\% = \frac{12\text{ V} - 11\text{ V}}{11.5\text{ V}} \times 100\% = \mathbf{8.7\%}$$

(Remember, V is defined by the average value of the peak amplitude.)

EXAMPLE 24.2 Determine the following for the pulse waveform in Fig. 24.9:

a. positive- or negative-going?
b. base-line voltage
c. tilt
d. amplitude
e. t_p
f. t_r and t_f

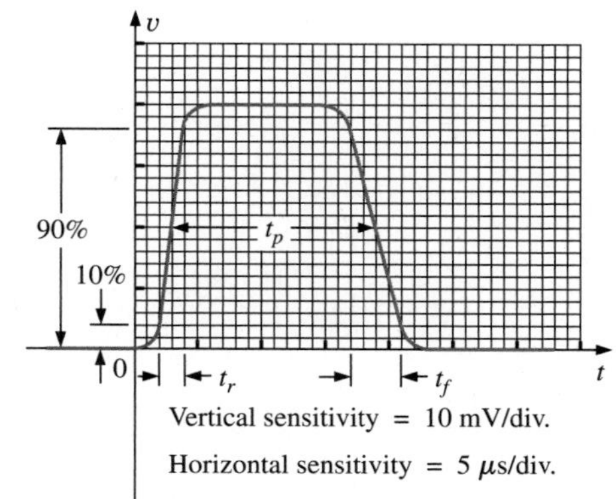

FIG. 24.9
Example 24.2.

Solutions:

a. **positive-going**
b. $V_b = \mathbf{0\ V}$
c. % tilt = **0%**
d. amplitude = (4 div.)(10 mV/div.) = **40 mV**
e. $t_p = (3.2\text{ div.})(5\ \mu\text{s/div.}) = \mathbf{16\ \mu s}$
f. $t_r = (0.4\text{ div.})(5\ \mu\text{s/div.}) = \mathbf{2\ \mu s}$
$t_f = (0.8\text{ div.})(5\ \mu\text{s/div.}) = \mathbf{4\ \mu s}$

24.3 PULSE REPETITION RATE AND DUTY CYCLE

A series of pulses such as those appearing in Fig. 24.10 is called a **pulse train.** The varying widths and heights may contain information that can be decoded at the receiving end.

If the pattern repeats itself in a periodic manner as shown in Fig. 24.11(a) and (b), the result is called a **periodic pulse train.**

The *period (T)* of the pulse train is defined as the time differential between any two similar points on the pulse train, as shown in Figs. 24.11(a) and (b).

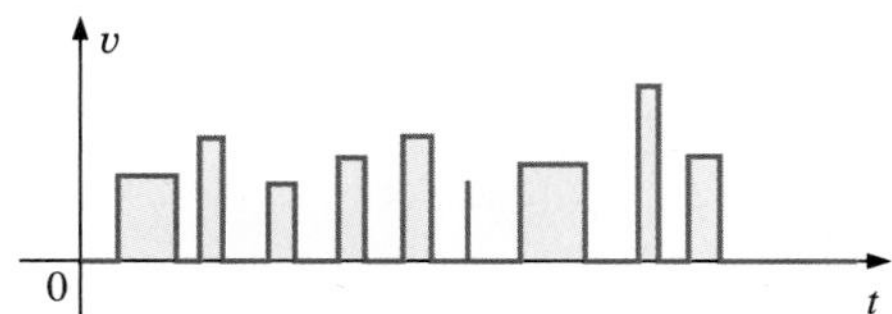

FIG. 24.10
Pulse train.

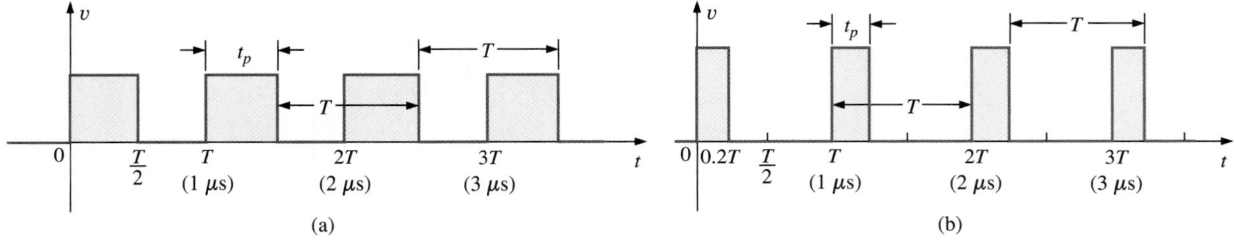

FIG. 24.11
Periodic pulse trains.

The **pulse repetition frequency** (prf), or **pulse repetition rate** (prr), is defined by

$$\text{prf (or prr)} = \frac{1}{T} \quad \text{Hz or pulses/s} \tag{24.3}$$

Applying Eq. (24.3) to each waveform in Fig. 24.11 results in the same pulse repetition frequency since the periods are the same. The result clearly reveals that

the shape of the periodic pulse does not affect the determination of the pulse repetition frequency.

The pulse repetition frequency is determined solely by the period of the repeating pulse. The factor that reveals how much of the period is encompassed by the pulse is called the **duty cycle,** defined as follows:

$$\text{Duty cycle} = \frac{\text{pulse width}}{\text{period}} \times 100\%$$

or

$$\text{Duty cycle} = \frac{t_p}{T} \times 100\% \tag{24.4}$$

For Fig. 24.11(a) (a square-wave pattern),

$$\text{Duty cycle} = \frac{0.5T}{T} \times 100\% = \mathbf{50\%}$$

and for Fig. 24.11(b),

$$\text{Duty cycle} = \frac{0.2T}{T} \times 100\% = \mathbf{20\%}$$

The above results clearly reveal that

the duty cycle provides a percentage indication of the portion of the total period encompassed by the pulse waveform.

EXAMPLE 24.3 Determine the pulse repetition frequency and the duty cycle for the periodic pulse waveform in Fig. 24.12.

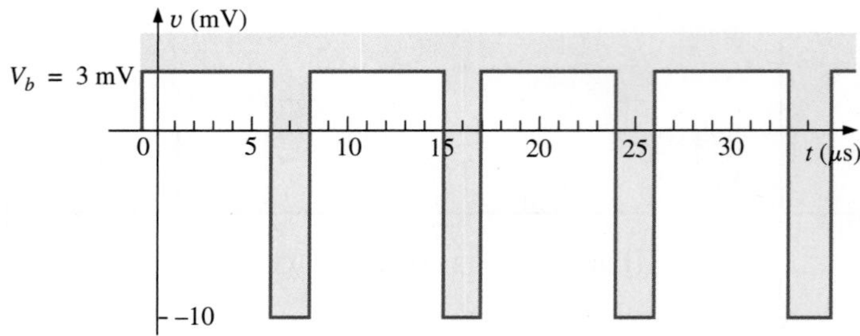

FIG. 24.12
Example 24.3.

Solution:

$$T = (15 - 6)\ \mu\text{s} = 9\ \mu\text{s}$$

$$\text{prf} = \frac{1}{T} = \frac{1}{9\ \mu\text{s}} \cong \mathbf{111.11\ kHz}$$

$$\text{Duty cycle} = \frac{t_p}{T} \times 100\% = \frac{(8 - 6)\ \mu\text{s}}{9\ \mu\text{s}} \times 100\%$$

$$= \frac{2}{9} \times 100\% \cong \mathbf{22.22\%}$$

EXAMPLE 24.4 Determine the pulse repetition frequency and the duty cycle for the oscilloscope pattern in Fig. 24.13 having the indicated sensitivities.

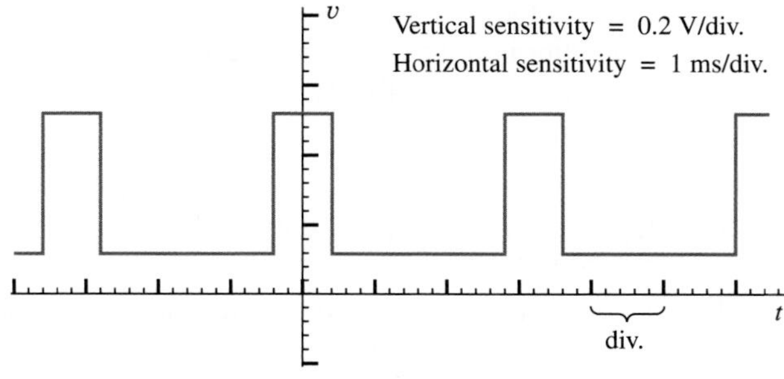

FIG. 24.13
Example 24.4.

Solution:

$$T = (3.2\ \text{div.})(1\ \text{ms/div.}) = 3.2\ \text{ms}$$

$$t_p = (0.8\ \text{div.})(1\ \text{ms/div.}) = 0.8\ \text{ms}$$

$$\text{prf} = \frac{1}{T} = \frac{1}{3.2\ \text{ms}} = \mathbf{312.5\ Hz}$$

$$\text{Duty cycle} = \frac{t_p}{T} \times 100\% = \frac{0.8\ \text{ms}}{3.2\ \text{ms}} \times 100\% = \mathbf{25\%}$$

EXAMPLE 24.5 Determine the pulse repetition rate and duty cycle for the trigger waveform in Fig. 24.14.

Solution:

$$T = (2.6\ \text{div.})(10\ \mu\text{s/div.}) = 26\ \mu\text{s}$$

$$\text{prf} = \frac{1}{T} = \frac{1}{26\ \mu\text{s}} = \mathbf{38{,}462\ kHz}$$

$$t_p \cong (0.2\ \text{div.})(10\ \mu\text{s/div.}) = 2\ \mu\text{s}$$

$$\text{Duty cycle} = \frac{t_p}{T} \times 100\% = \frac{2\ \mu\text{s}}{26\ \mu\text{s}} \times 100\% = \mathbf{7.69\%}$$

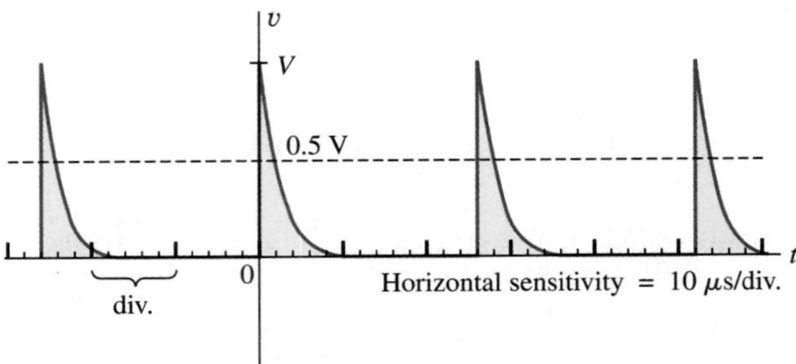

FIG. 24.14
Example 24.5.

24.4 AVERAGE VALUE

The average value of a pulse waveform can be determined using one of two methods. The first is the procedure outlined in Section 13.7, which can be applied to any alternating waveform. The second can be applied only to pulse waveforms since it utilizes terms specifically related to pulse waveforms; that is,

$$V_{av} = (\text{duty cycle})(\text{peak value}) + (1 - \text{duty cycle})(V_b) \tag{24.5}$$

In Eq. (24.5), the peak value is the maximum deviation from the reference or zero-volt level, and the duty cycle is in decimal form. Eq. (24.5) does not include the effect of any tilt pulse waveforms with sloping sides.

EXAMPLE 24.6 Determine the average value for the periodic pulse waveform in Fig. 24.15.

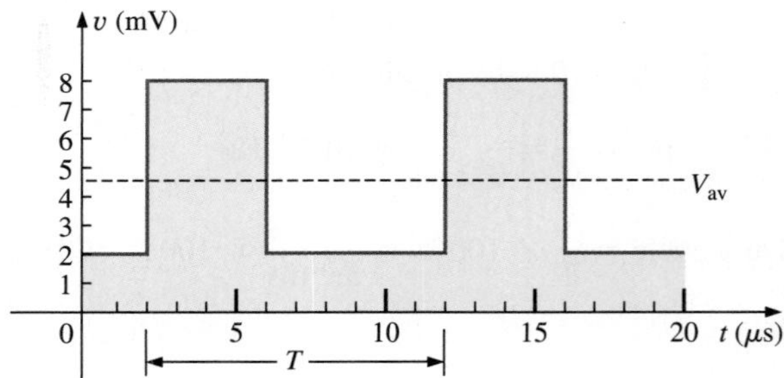

FIG. 24.15
Example 24.6.

Solution: By the method in Section 13.7,

$$G = \frac{\text{area under curve}}{T}$$

$$T = (12 - 2)\ \mu\text{s} = 10\ \mu\text{s}$$

$$G = \frac{(8\text{ mV})(4\ \mu\text{s}) + (2\text{ mV})(6\ \mu\text{s})}{10\ \mu\text{s}} = \frac{32 \times 10^{-9} + 22 \times 10^{-9}}{10 \times 10^{-6}}$$

$$= \frac{44 \times 10^{-9}}{10 \times 10^{-6}} = \mathbf{4.4\ mV}$$

By Eq. (24.5),

$$V_b = +2 \text{ mV}$$

$$\text{Duty cycle} = \frac{t_p}{T} = \frac{(6-2)\ \mu\text{s}}{10\ \mu\text{s}} = \frac{4}{10} = 0.4 \text{ (decimal form)}$$

$$\text{Peak value (from 0 V reference)} = 8 \text{ mV}$$

$$\begin{aligned} V_{av} &= (\text{duty cycle})(\text{peak value}) + (1 - \text{duty cycle})(V_b) \\ &= (0.4)(8 \text{ mV}) + (1 - 0.4)(2 \text{ mV}) \\ &= 3.2 \text{ mV} + 1.2 \text{ mV} = \mathbf{4.4\ mV} \end{aligned}$$

as obtained above.

EXAMPLE 24.7 Given a periodic pulse waveform with a duty cycle of 28%, a peak value of 7 V, and a base-line voltage of −3 V:

a. Determine the average value.
b. Sketch the waveform.
c. Verify the result of part (a) using the method of Section 13.7.

Solutions:

a. By Eq. (24.5),

$$\begin{aligned} V_{av} &= (\text{duty cycle})\ (\text{peak value}) + (1 - \text{duty cycle})\ (V_b) \\ &= (0.28)(7 \text{ V}) + (1 - 0.28)(-3 \text{ V}) = 1.96 \text{ V} + (-2.16 \text{ V}) \\ &= \mathbf{-0.2\ V} \end{aligned}$$

b. See Fig. 24.16.

c. $$G = \frac{(7 \text{ V})(0.28T) - (3 \text{ V})(0.72T)}{T} = 1.96 \text{ V} - 2.16 \text{ V}$$

$$= \mathbf{-0.2\ V}$$

as obtained above.

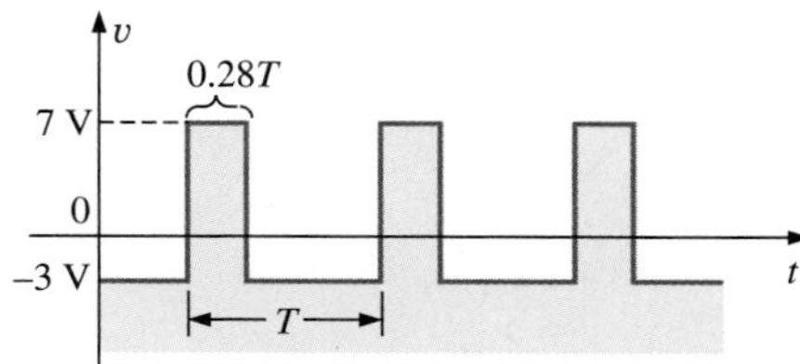

FIG. 24.16
Solution to part (b) of Example 24.7.

Instrumentation

The average value (dc value) of any waveform can be easily determined using the oscilloscope. If the mode switch of the scope is set in the ac position, the average or dc component of the applied waveform is blocked by an internal capacitor from reaching the screen. The pattern can be adjusted to establish the display in Fig. 24.17(a). If the mode switch is then placed in the dc position, the vertical shift (positive or negative) reveals the average or dc level of the input signal, as shown in Fig. 24.17(b).

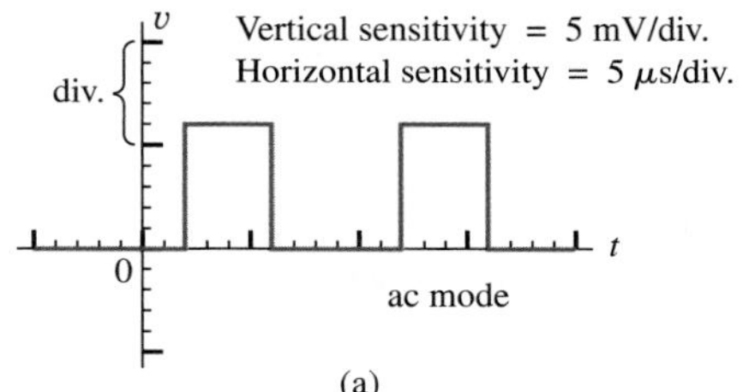

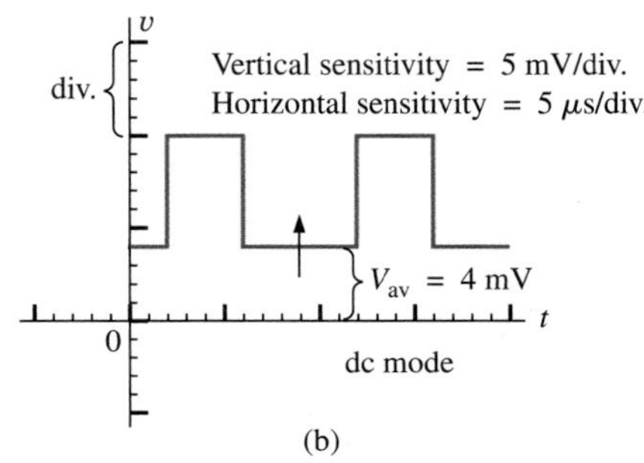

FIG. 24.17
Determining the average value of a pulse waveform using an oscilloscope.

24.5 TRANSIENT *R-C* NETWORKS

In Chapter 10, the general solution for the transient behavior of an *R-C* network with or without initial values was developed. The resulting equation for the voltage across a capacitor is repeated below for convenience.

$$\boxed{v_C = V_f + (V_i - V_f)e^{-t/RC}} \qquad \textbf{(24.6)}$$

Recall that V_i is the initial voltage across the capacitor when the transient phase is initiated as shown in Fig. 24.18. The voltage V_f is the steady-state (resting) value of the voltage across the capacitor when the transient phase has ended. The transient period is approximated as 5τ, where τ is the time constant of the network and is equal to the product *RC*.

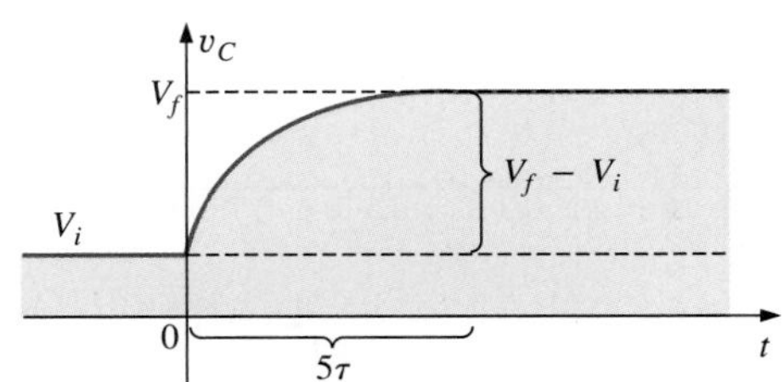

FIG. 24.18
Defining the parameters of Eq. (24.6).

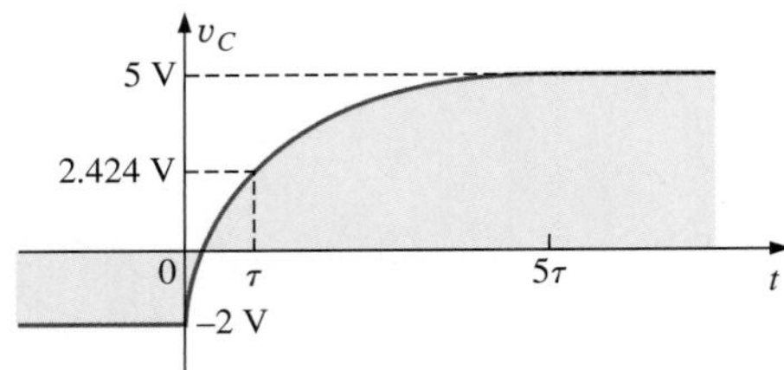

FIG. 24.19
Example of the use of Eq. (24.6).

For the situation where the initial voltage is zero volts, the equation reduces to the following familiar form, where V_f is often the applied voltage:

$$\boxed{v_C = V_f(1 - e^{-t/RC})}\Big|_{V_i = 0\text{ V}} \tag{24.7}$$

For the case in Fig. 24.19, $V_i = -2$ V, $V_f = +5$ V, and

$$\begin{aligned} v_C &= V_i + (V_f - V_i)(1 - e^{-t/RC}) \\ &= -2\text{ V} + [5\text{ V} - (-2\text{ V})](1 - e^{-t/RC}) \\ v_C &= -2\text{ V} + 7\text{ V}(1 - e^{-t/RC}) \end{aligned}$$

For the case where $t = \tau = RC$,

$$\begin{aligned} v_C &= -2\text{ V} + 7\text{ V}(1 - e^{-t/\tau}) = -2\text{ V} + 7\text{ V}(1 - e^{-1}) \\ &= -2\text{ V} + 7\text{ V}(1 - 0.368) = -2\text{ V} + 7\text{ V}(0.632) \\ v_C &= 2.424\text{ V} \end{aligned}$$

as verified by Fig. 24.19.

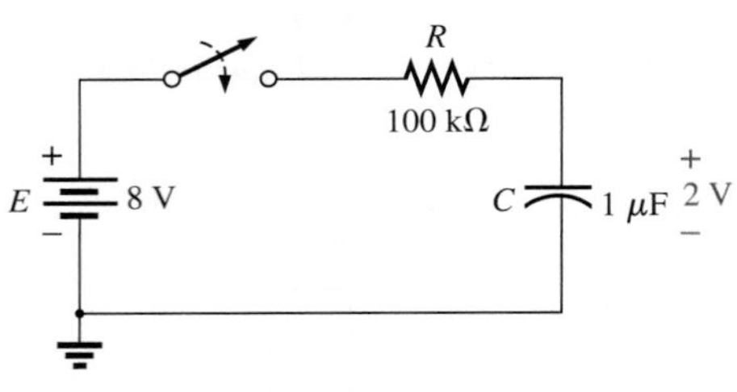

FIG. 24.20
Example 24.8.

EXAMPLE 24.8 The capacitor in Fig. 24.20 is initially charged to 2 V before the switch is closed. The switch is then closed.

a. Determine the mathematical expression for v_C.
b. Determine the mathematical expression for i_C.
c. Sketch the waveforms of v_C and i_C.

Solutions:

a. $V_i = 2$ V
V_f (after 5τ) $= E = 8$ V
$\tau = RC = (100\text{ k}\Omega)(1\ \mu\text{F}) = 100$ ms

By Eq. (24.6),

$$\begin{aligned} v_C &= V_f + (V_i - V_f)e^{-t/RC} \\ &= 8\text{ V} + (2\text{ V} - 8\text{ V})e^{-t/\tau} \end{aligned}$$

and $$v_C = \mathbf{8\ V - 6\ V}\ \boldsymbol{e^{-t/\tau}}$$

b. When the switch is first closed, the voltage across the capacitor cannot change instantaneously, and $V_R = E - V_i = 8\text{ V} - 2\text{ V} = 6$ V. The current therefore jumps to a level determined by Ohm's law:

$$I_{R_{max}} = \frac{V_R}{R} = \frac{6\text{ V}}{100\text{ k}\Omega} = 0.06\text{ mA}$$

The current then decays to zero amperes with the same time constant calculated in part (a), and

$$i_C = \mathbf{0.06\ mA}\boldsymbol{e^{-t/\tau}}$$

c. See Fig. 24.21.

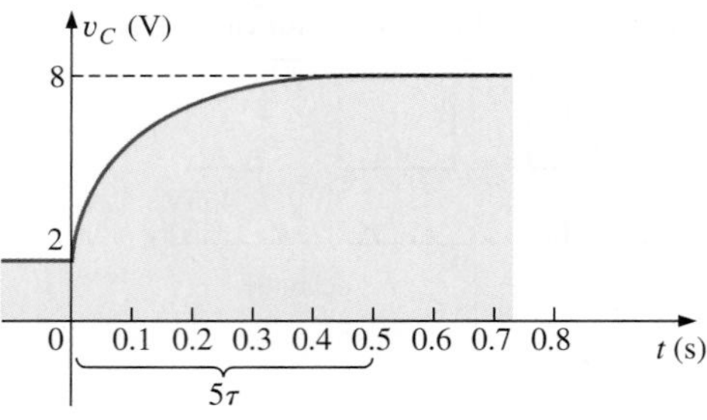

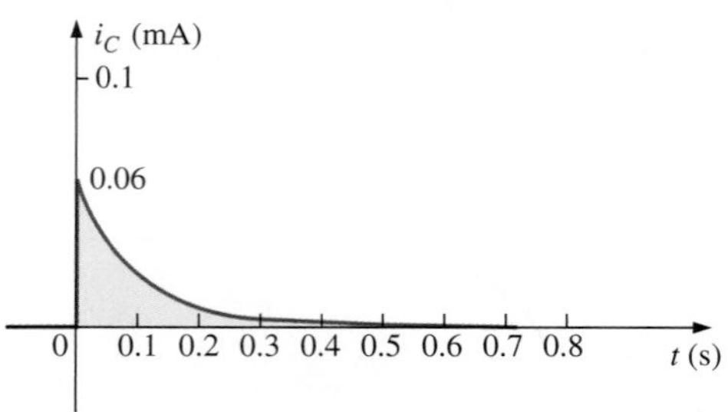

FIG. 24.21
v_C and i_C for the network in Fig. 24.20.

EXAMPLE 24.9 Sketch v_C for the step input shown in Fig. 24.22. Assume that the -4 mV has been present for a period of time in excess of five time constants of the network. Then determine when $v_C = 0$ V if the step changes levels at $t = 0$ s.

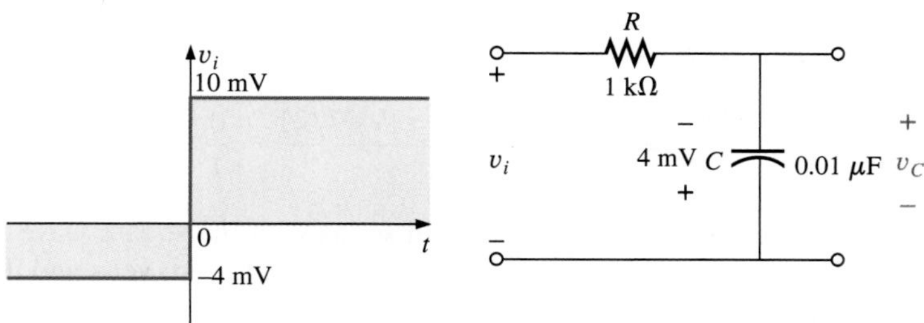

FIG. 24.22
Example 24.9.

Solution:

$$V_i = -4 \text{ mV} \qquad V_f = 10 \text{ mV}$$
$$\tau = RC = (1 \text{ k}\Omega)(0.01 \;\mu\text{F}) = 10 \;\mu\text{s}$$

By Eq. (24.6),

$$\begin{aligned} v_C &= V_f + (V_i - V_f)e^{-t/RC} \\ &= 10 \text{ mV} + (-4\text{mV} - 10 \text{ mV})e^{-t/10\mu\text{s}} \end{aligned}$$

and $\qquad \boldsymbol{v_C = 10 \textbf{ mV} - 14 \textbf{ mV } e^{-t/10\mu s}}$

The waveform appears in Fig. 24.23.

Substituting $v_C = 0$ V into the above equation yields

$$v_C = 0 \text{ V} = 10 \text{ mV} - 14 \text{ mV } e^{-t/10\mu\text{s}}$$

and $\qquad \dfrac{10 \text{ mV}}{14 \text{ mV}} = e^{-t/10\mu\text{s}}$

or $\qquad 0.714 = e^{-t/10\mu\text{s}}$

but $\qquad \log_e 0.714 = \log_e(e^{-t/10\mu\text{s}}) = \dfrac{-t}{10 \;\mu\text{s}}$

and $\qquad t = -(10 \;\mu\text{s})\log_e 0.714 = -(10 \;\mu\text{s})(-0.377) = \mathbf{3.37 \;\mu s}$

as indicated in Fig. 24.23.

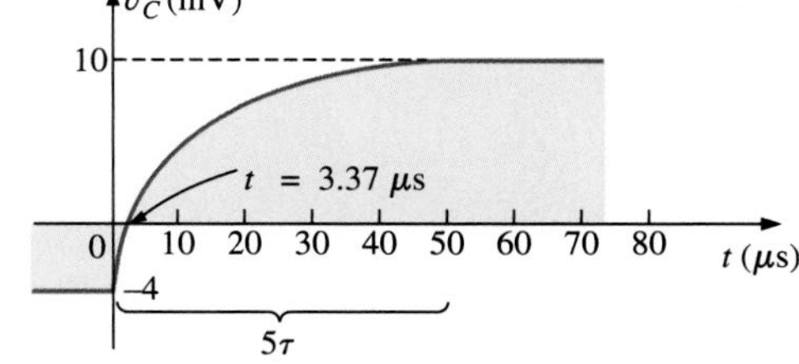

FIG. 24.23
v_C for the network in Fig. 24.22.

24.6 *R-C* RESPONSE TO SQUARE-WAVE INPUTS

The **square wave** in Fig. 24.24 is a particular form of pulse waveform. It has a duty cycle of 50% and an average value of zero volts, as calculated below:

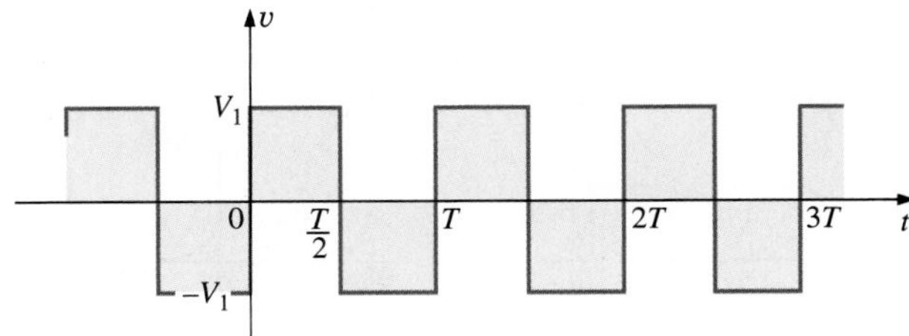

FIG. 24.24
Periodic square wave.

$$\text{Duty cycle} = \frac{t_p}{T} \times 100\% = \frac{T/2}{T} \times 100\% = \mathbf{50\%}$$

$$V_{av} = \frac{(V_1)(T/2) + (-V_1)(T/2)}{T} = \frac{0}{T} = \mathbf{0\ V}$$

The application of a dc voltage V_1 in series with the square wave in Fig. 24.24 can raise the base-line voltage from $-V_1$ to zero volts and the average value to V_1 volts.

If a square wave such as developed in Fig. 24.25 is applied to an *R-C* circuit as shown in Fig. 24.26, the period of the square wave can have a pronounced effect on the resulting waveform for v_C.

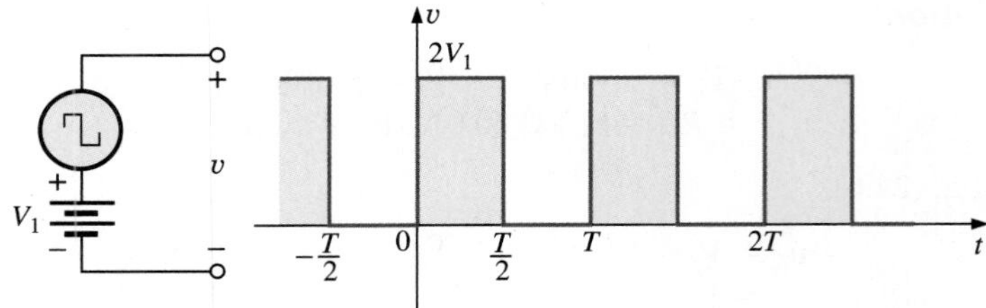

FIG. 24.25
Raising the base-line voltage of a square wave to zero volts.

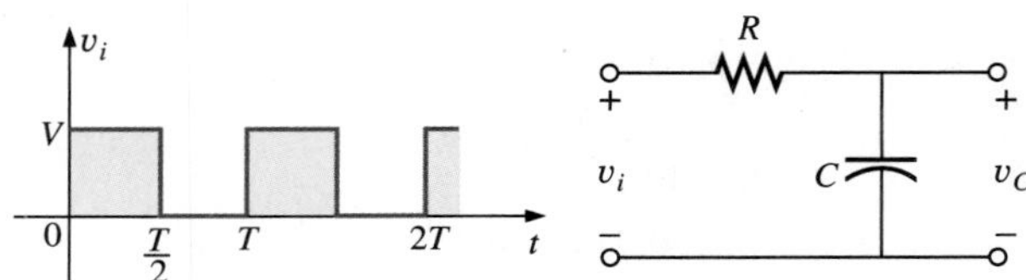

FIG. 24.26
Applying a periodic square-wave pulse train to an R-C network.

For the analysis to follow, we will assume that steady-state conditions will be established after a period of five time constants has passed. The types of waveforms developed across the capacitor can then be separated into three fundamental types: $T/2 > 5\tau$, $T/2 = 5\tau$, and $T/2 < 5\tau$.

$T/2 > 5\tau$

The condition $T/2 > 5\tau$ or $T > 10\tau$, establishes a situation where the capacitor can charge to its steady-state value in advance of $t = T/2$. The resulting waveforms for v_C and i_C appear as shown in Fig. 24.27. Note how

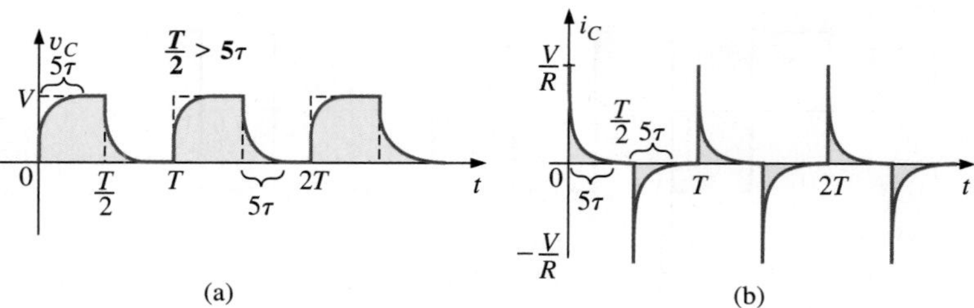

FIG. 24.27
v_C and i_C for $T/2 > 5\tau$.

closely the voltage v_C shadows the applied waveform and how i_C is nothing more than a series of very sharp spikes. Note also that the change of V_i from V to zero volts during the trailing edge results in a rapid discharge of v_C to zero volts. In essence, when $V_i = 0$, the capacitor and resistor are in parallel and the capacitor discharges through R with a time constant equal to that encountered during the charging phase but with a direction of charge flow (current) opposite to that established during the charging phase.

$T/2 = 5\tau$

If the frequency of the square wave is chosen such that $T/2 = 5\tau$ or $T = 10\tau$, the voltage v_C reaches its final value just before beginning its discharge phase, as shown in Fig. 24.28. The voltage v_C no longer resembles the square-wave input and, in fact, has some of the characteristics of a triangular waveform. The increased time constant has resulted in a more rounded v_C, and i_C has increased substantially in width to reveal the longer charging period.

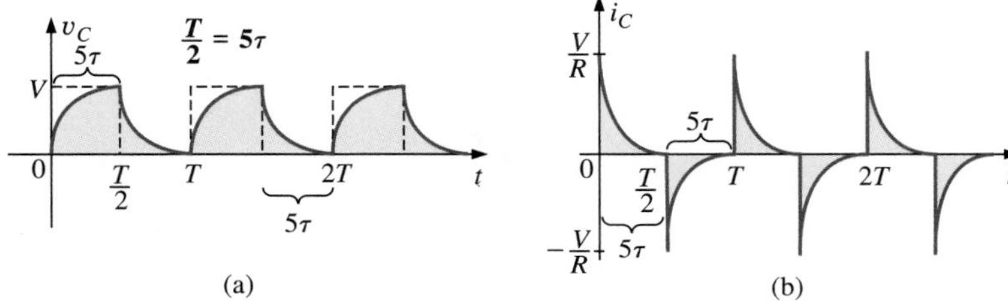

FIG. 24.28
v_C and i_C for $T/2 = 5\tau$.

$T/2 < 5\tau$

If $T/2 < 5\tau$ or $T < 10\tau$, the voltage v_C will not reach its final value during the first pulse (Fig. 24.29), and the discharge cycle will not return to zero volts. In fact, the initial value for each succeeding pulse changes until steady-state conditions are reached. In most instances, it is a good approximation to assume that steady-state conditions have been established in five cycles of the applied waveform.

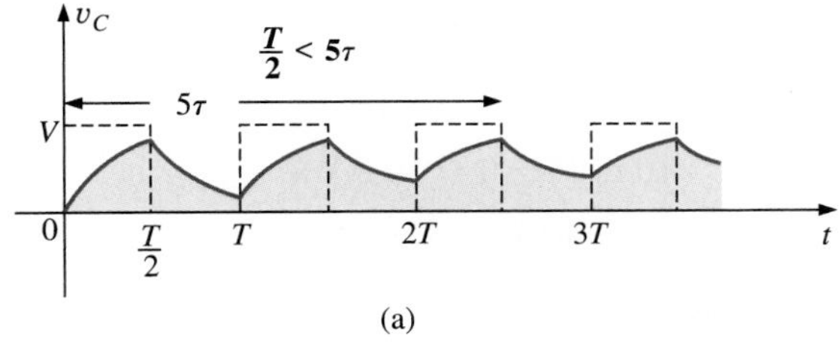

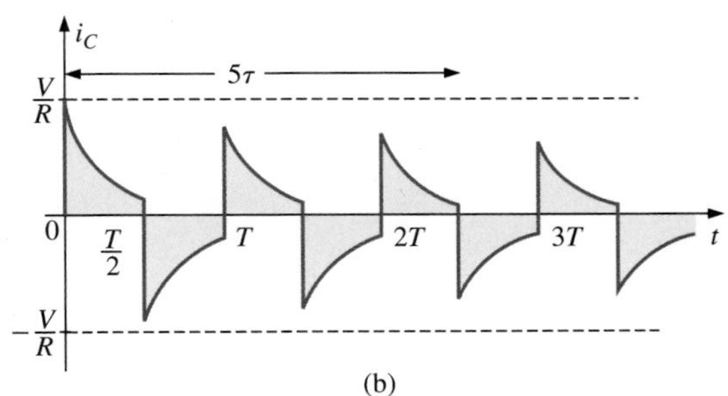

FIG. 24.29
v_C and i_C for $T/2 < 5\tau$.

As the frequency increases and the period decreases, there will be a flattening of the response for v_C until a pattern like that in Fig. 24.30 results. Fig. 24.30 begins to reveal an important conclusion regarding the response curve for v_C:

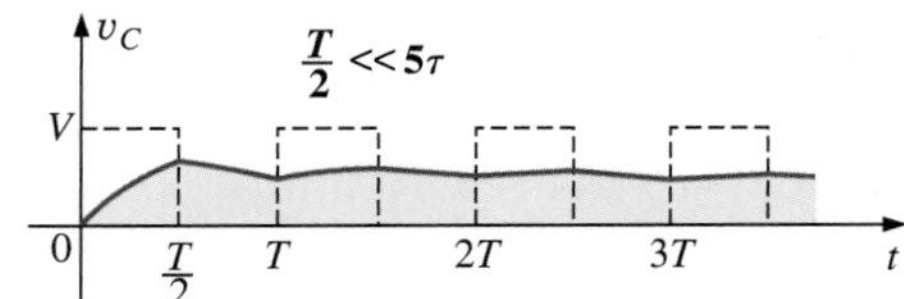

FIG. 24.30
v_C for $T/2 << 5\tau$ or $T << 10\tau$.

Under steady-state conditions, the average value of v_C will equal the average value of the applied square wave.

Note in Figs. 24.29 and 24.30 that the waveform for v_C approaches an average value of $V/2$.

EXAMPLE 24.10 The 1000 Hz square wave in Fig. 24.31 is applied to the *R-C* circuit of the same figure.

a. Compare the pulse width of the square wave to the time constant of the circuit.
b. Sketch v_C.
c. Sketch i_C.

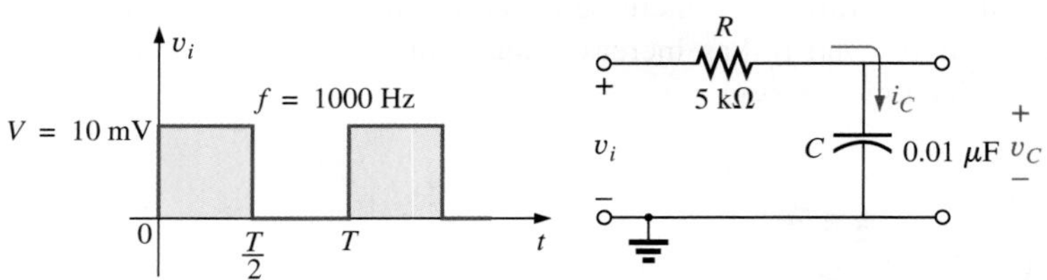

FIG. 24.31
Example 24.10.

Solutions:

a. $T = \dfrac{1}{f} = \dfrac{1}{1000} = 1\text{ ms}$

$$t_p = \frac{T}{2} = 0.5\text{ ms}$$

$$\tau = RC = (5 \times 10^3\ \Omega)(0.01 \times 10^{-6}\text{ F}) = 0.05\text{ ms}$$

$$\frac{t_p}{\tau} = \frac{0.5\text{ ms}}{0.05\text{ ms}} = 10 \text{ and}$$

$$t_p = \mathbf{10\tau} = \frac{\boldsymbol{T}}{\mathbf{2}}$$

The result reveals that v_C charges to its final value in half the pulse width.

b. For the charging phase, $V_i = 0$ V and $V_f = 10$ mV, and

$$v_C = V_f + (V_i - V_f)e^{-t/RC}$$
$$= 10\text{ mV} + (0 - 10\text{ mV})e^{-t/\tau}$$

and $$v_C = \mathbf{10\ mV(1 - e^{-t/\tau})}$$

For the discharge phase, $V_i = 10$ mV and $V_f = 0$ V, and

$$v_C = V_f + (V_i - V_f)e^{-t/\tau}$$
$$= 0\text{ V} + (10\text{ mV} - 0\text{ V})e^{-t/\tau}$$

and $$v_C = \mathbf{10\ mV}e^{-t/\tau}$$

The waveform for v_C appears in Fig. 24.32.

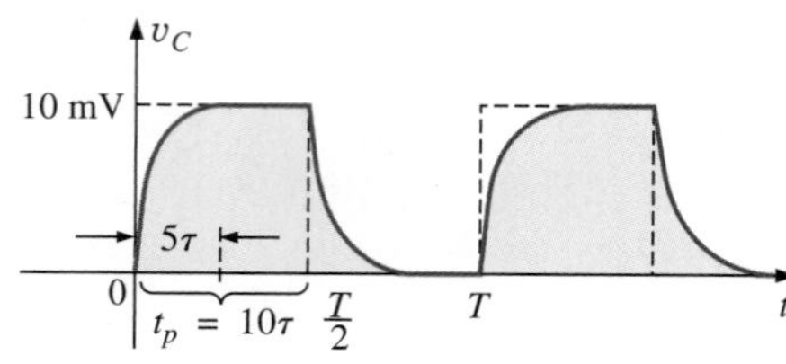

FIG. 24.32
v_C for the R-C network in Fig. 24.31.

c. For the charging phase at $t = 0$ s, $V_R = V$ and $I_{R_{max}} = V/R = 10\text{ mV}/5\text{ k}\Omega = 2\ \mu\text{A}$ and

$$i_C = I_{max}e^{-t/\tau} = \mathbf{2\ \mu A}e^{-t/\tau}$$

For the discharge phase, the current will have the same mathematical formulation but the opposite direction, as shown in Fig. 24.33.

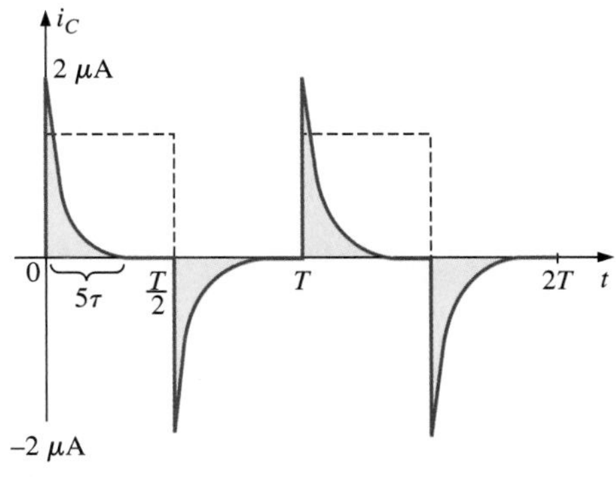

FIG. 24.33
i_C for the R-C network in Fig. 24.31.

EXAMPLE 24.11 Repeat Example 24.10 for $f = 10$ kHz.

Solution:

$$T = \frac{1}{f} = \frac{1}{10\text{ kHz}} = 0.1\text{ ms}$$

and
$$\frac{T}{2} = 0.05\text{ ms}$$

with
$$\tau = t_p = \frac{T}{2} = 0.05\text{ ms}$$

In other words, the pulse width is exactly equal to the time constant of the network. The voltage v_C will not reach the final value before the first pulse of the square-wave input returns to zero volts.

For t in the range $t = 0$ to $T/2$, $V_i = 0$ V and $V_f = 10$ mV, and

$$v_C = 10\text{ mV}(1 - e^{-t/\tau})$$

Recall from Chapter 10 that at $t = \tau$, $v_C = 63.2\%$ of the final value. Substituting $t = \tau$ into the equation above yields

$$\begin{aligned} v_C &= (10\text{ mV})(1 - e^{-1}) = (10\text{ mV})(1 - 0.368) \\ &= (10\text{ mV})(0.632) = 6.32\text{ mV} \end{aligned}$$

as shown in Fig. 24.34.

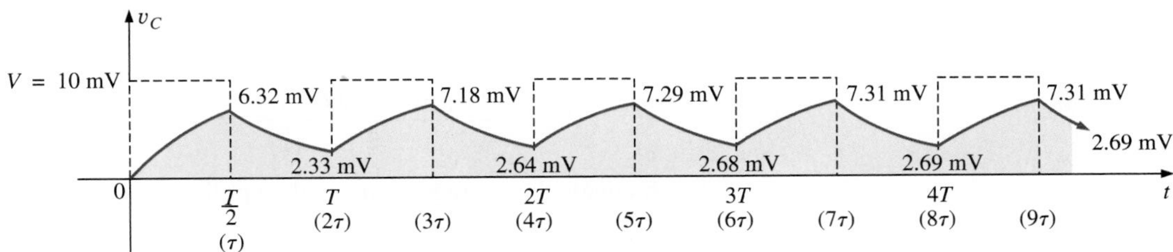

FIG. 24.34
v_C response for $t_p = \tau = T/2$.

For the discharge phase between $t = T/2$ and T, $V_i = 6.32$ mV and $V_f = 0$ V, and

$$\begin{aligned} v_C &= V_f + (V_i - V_f)e^{-t/\tau} \\ &= 0\text{ V} + (6.32\text{ mV} - 0\text{ V})e^{-t/\tau} \\ v_C &= 6.32\text{ mV}e^{-t/\tau} \end{aligned}$$

with t now being measured from $t = T/2$ in Fig. 24.34. In other words, for each interval in Fig. 24.34, the beginning of the transient waveform is defined as $t = 0$ s. The value of v_C at $t = T$ is therefore determined by substituting $t = \tau$ into the above equation, and not 2τ as defined by Fig. 24.34.

Substituting $t = \tau$,

$$\begin{aligned} v_C &= (6.32\text{ mV})(e^{-1}) = (6.32\text{ mV})(0.368) \\ &= 2.33\text{ mV} \end{aligned}$$

as shown in Fig. 24.34.

For the next interval, $V_i = 2.33$ mV and $V_f = 10$ mV, and

$$\begin{aligned} \upsilon_C &= V_f + (V_i - V_f)e^{-t/\tau} \\ &= 10 \text{ mV} + (2.33 \text{ mV} - 10 \text{ mV})e^{-t/\tau} \\ \upsilon_C &= 10 \text{ mV} - 7.67 \text{ mV}e^{-t/\tau} \end{aligned}$$

At $t = \tau$ (since $t = T = 2\tau$ is now $t = 0$ s for this interval),

$$\begin{aligned} \upsilon_C &= 10 \text{ mV} - 7.67 \text{ mV}e^{-1} \\ &= 10 \text{ mV} - 2.82 \text{ mV} \\ \upsilon_C &= 7.18 \text{ mV} \end{aligned}$$

as shown in Fig. 24.34.

For the discharge interval, $V_i = 7.18$ mV and $V_f = 0$ V, and

$$\begin{aligned} \upsilon_C &= V_f + (V_i - V_f)e^{-t/\tau} \\ &= 0 \text{ V} + (7.18 \text{ mV} - 0)e^{-t/\tau} \\ \upsilon_C &= 7.18 \text{ mV}e^{-t/\tau} \end{aligned}$$

At $t = \tau$ (measured from 3τ in Fig. 24.34),

$$\begin{aligned} \upsilon_C &= (7.18 \text{ mV})(e^{-1}) = (7.18 \text{ mV})(0.368) \\ &= 2.64 \text{ mV} \end{aligned}$$

as shown in Fig. 24.34.

Continuing in the same manner, the remaining waveform for υ_C is generated as depicted in Fig. 24.34. Note that repetition occurs after $t = 8\tau$, and the waveform has essentially reached steady-state conditions in a period of time less than 10τ, or five cycles of the applied square wave.

A closer look reveals that both the peak and the lower levels continued to increase until steady-state conditions were established. Since the exponential waveforms between $t = 4T$ and $t = 5T$ have the same time constant, the average value of υ_C can be determined from the steady-state 7.31 mV and 2.69 mV levels as follows:

$$V_{av} = \frac{7.31 \text{ mV} + 2.69 \text{ mV}}{2} = \frac{10 \text{ mV}}{2} = 5 \text{ mV}$$

which equals the average value of the applied signal as stated earlier in this section.

We can use the results in Fig. 24.34 to plot i_C. At any instant of time,

$$\upsilon_i = \upsilon_R + \upsilon_C \quad \text{or} \quad \upsilon_R = \upsilon_i - \upsilon_C$$

and

$$i_R = i_C = \frac{\upsilon_i - \upsilon_C}{R}$$

At $t = 0^+$, $\upsilon_C = 0$ V, and

$$i_R = \frac{\upsilon_i - \upsilon_C}{R} = \frac{10 \text{ mV} - 0}{5 \text{ k}\Omega} = 2\ \mu\text{A}$$

as shown in Fig. 24.35.

As the charging process proceeds, the current i_C decays at a rate determined by

$$i_C = 2\ \mu\text{A}e^{-t/\tau}$$

At $t = \tau$,

$$\begin{aligned} i_C &= (2\ \mu\text{A})(e^{-\tau/\tau}) = (2\ \mu\text{A})(e^{-1}) = (2\ \mu\text{A})(0.368) \\ &= 0.736\ \mu\text{A} \end{aligned}$$

as shown in Fig. 24.35.

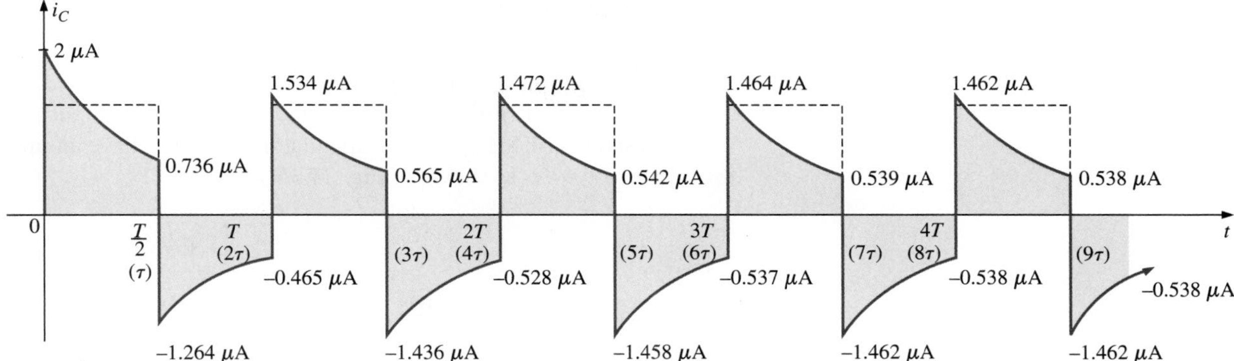

FIG. 24.35
i_C response for $t_p = \tau = T/2$.

For the trailing edge of the first pulse, the voltage across the capacitor cannot change instantaneously, resulting in the following when v_i drops to zero volts:

$$i_C = i_R = \frac{v_i - v_C}{R} = \frac{0 - 6.32\text{ mV}}{5\text{ k}\Omega} = -1.264\ \mu\text{A}$$

as illustrated in Fig. 24.35. The current then decays as determined by

$$i_C = -1.264\ \mu\text{A}e^{-t/\tau}$$

and at $t = \tau$ (actually $t = 2\tau$ in Fig. 24.35),

$$\begin{aligned} i_C &= (-1.264\ \mu\text{A})(e^{-\tau/\tau}) = (-1.264\ \mu\text{A})(e^{-1}) \\ &= (-1.264\ \mu\text{A})(0.368) = -0.465\ \mu\text{A} \end{aligned}$$

as shown in Fig. 24.35.

At $t = T\ (t = 2\tau)$, $v_C = 2.33$ mV, and v_i returns to 10 mV, resulting in

$$i_C = i_R = \frac{v_i - v_C}{R} = \frac{10\text{ mV} - 2.33\text{ mV}}{5\text{ k}\Omega} = 1.534\ \mu\text{A}$$

The equation for the decaying current is now

$$i_C = 1.534\ \mu\text{A}e^{-t/\tau}$$

and at $t = \tau$ (actually $t = 3\tau$ in Fig. 24.35),

$$i_C = (1.534\ \mu\text{A})(0.368) = 0.565\ \mu\text{A}$$

The process continues until steady-state conditions are reached at the same time they were attained for v_C. Note in Fig. 24.35 that the positive peak current decreased toward steady-state conditions while the negative peak became more negative. Note that the current waveform becomes symmetrical about the axis when steady-state conditions are established. The result is that the net average current over one cycle is zero, as it should be in a series *R-C* circuit. Recall from Chapter 10 that the capacitor under dc steady-state conditions can be replaced by an open-circuit equivalent, resulting in $I_C = 0$ A.

Although both examples provided above started with an uncharged capacitor, the same approach can be used effectively for initial conditions. Simply substitute the initial voltage on the capacitor as V_i in Eq. (24.6) and proceed as above.

24.7 OSCILLOSCOPE ATTENUATOR AND COMPENSATING PROBE

The $\times 10$ **attenuator probe** used with oscilloscopes is designed to reduce the magnitude of the input voltage by a factor of 10. If the input impedance to a scope is 1 MΩ, the $\times 10$ attenuator probe will have an internal resistance of 9 MΩ, as shown in Fig. 24.36.

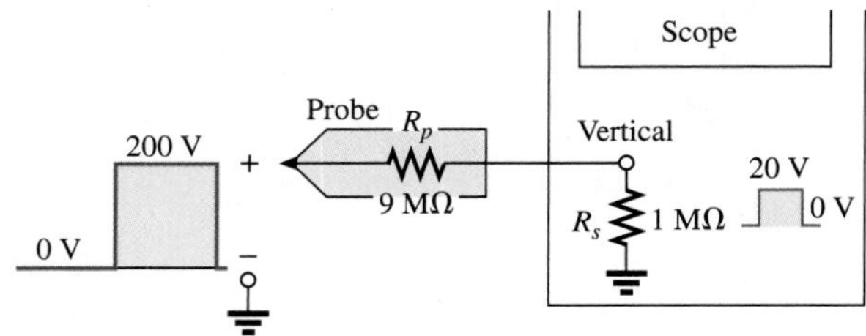

FIG. 24.36
$\times 10$ attenuator probe.

Applying the voltage divider rule,

$$V_{\text{scope}} = \frac{(1\ \text{M}\Omega)(V_i)}{1\ \text{M}\Omega + 9\ \text{M}\Omega} = \frac{1}{10}V_i$$

In addition to the input resistance, oscilloscopes have some internal input capacitance, and the probe adds an additional capacitance in parallel with the oscilloscope capacitance, as shown in Fig. 24.37. The probe

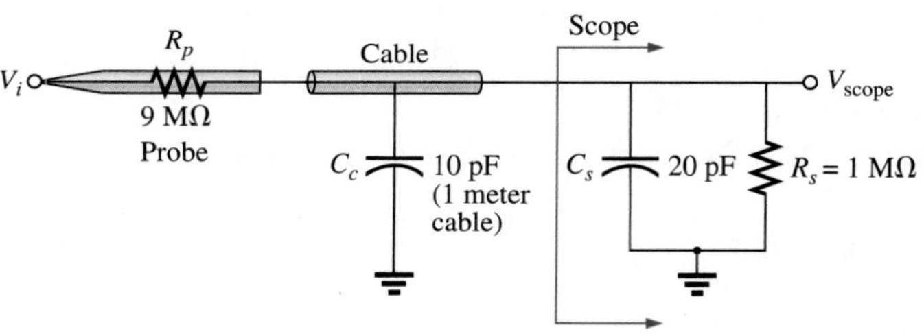

FIG. 24.37
Capacitive elements present in an attenuator probe arrangement.

R_p, 9 MΩ, V_i, V_{scope}, R_s 1 MΩ, $C_i = C_c + C_s = 30$ pF

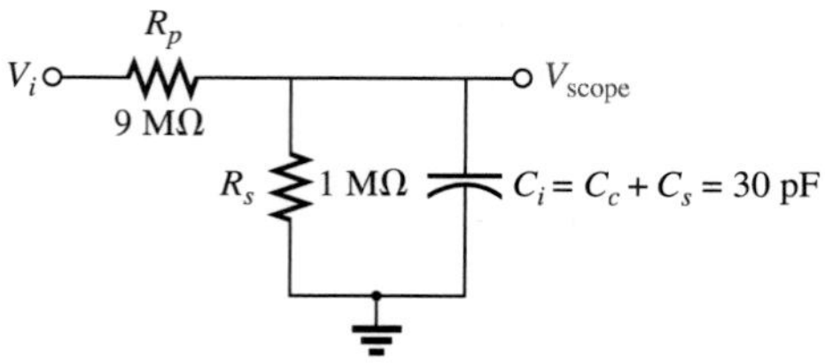

FIG. 24.38
Equivalent network in Fig. 24.37.

capacitance is typically about 10 pF for a 1 m (3.3 ft) cable, reaching about 15 pF for a 3 m (9.9 ft) cable. The total input capacitance is therefore the sum of the two capacitive elements, resulting in the equivalent network in Fig. 24.38.

For the analysis to follow, let us determine the Thévenin equivalent circuit for the capacitor C_i:

$$E_{Th} = \frac{(1\ \text{M}\Omega)(V_i)}{1\ \text{M}\Omega + 9\ \text{M}\Omega} = \frac{1}{10}V_i$$

and $$R_{Th} = 9\ \text{M}\Omega \parallel 1\ \text{M}\Omega = 0.9\ \text{M}\Omega$$

The Thévenin network is shown in Fig. 24.39.

For $v_i = 200$ V (peak),

$$E_{Th} = 0.1v_i = 20\ \text{V (peak)}$$

and for v_C, $V_f = 20$ V and $V_i = 0$ V, with

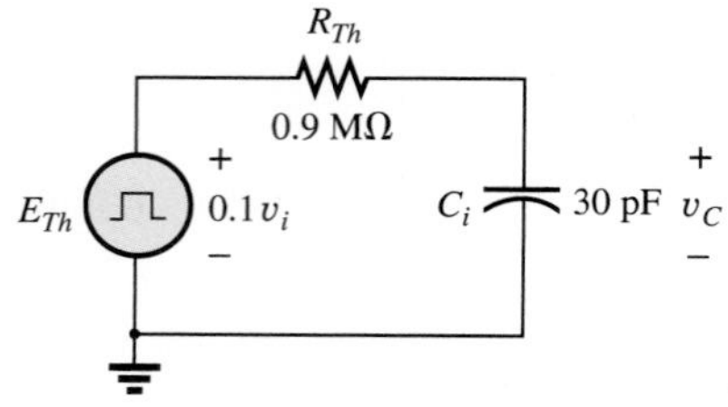

FIG. 24.39
Thévenin equivalent for C_i in Fig. 24.38.

$$\tau = RC = (0.9 \times 10^6\ \Omega)(30 \times 10^{-12}\ \text{F}) = 27\ \mu\text{s}$$

For an applied frequency of 5 kHz,

$$T = \frac{1}{f} = 0.2\ \text{ms} \quad \text{and} \quad \frac{T}{2} = 0.1\ \text{ms} = 100\ \mu\text{s}$$

with $5\tau = 135\ \mu\text{s} > 100\ \mu\text{s}$, as shown in Fig. 24.40, clearly producing a severe rounding distortion of the square wave and a poor representation of the applied signal.

To improve matters, a variable capacitor is often added in parallel with the resistance of the attenuator, resulting in a **compensated attenuator probe** such as the one shown in Fig. 24.41. In Chapter 22, it was demonstrated that a square wave can be generated by a summation of sinusoidal signals of particular frequency and amplitude. If we therefore design a network such as the one shown in Fig. 24.42 that ensures that V_{scope} is $0.1v_i$ for any frequency, then the rounding distortion is removed, and V_{scope} has the same appearance as v_i.

Applying the voltage divider rule to the network in Fig. 24.42,

$$\mathbf{V}_{\text{scope}} = \frac{\mathbf{Z}_s\mathbf{V}_i}{\mathbf{Z}_s + \mathbf{Z}_p} \tag{24.8}$$

If the parameters are chosen or adjusted such that

$$R_pC_p = R_sC_s \tag{24.9}$$

the phase angle of $\mathbf{Z}_s$ and $\mathbf{Z}_p$ will be the same, and Eq. (24.8) will reduce to

$$\mathbf{V}_{\text{scope}} = \frac{R_s\mathbf{V}_i}{R_s + R_p} \tag{24.10}$$

which is insensitive to frequency since the capacitive elements have dropped out of the relationship.

In the laboratory, simply adjust the probe capacitance using a standard or known square-wave signal until the desired sharp corners of the square wave are obtained. If you avoid the calibration step, you may make a rounded signal look square since you assumed a square wave at the point of measurement.

Too much capacitance results in an overshoot effect, whereas too little continues to show the rounding effect.

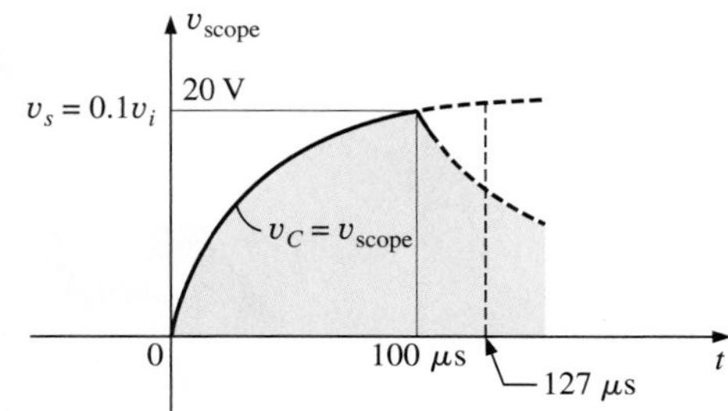

FIG. 24.40
The scope pattern for the conditions in Fig. 24.38 with v_i = 200 V peak.

FIG. 24.41
Commercial compensated 10 : 1 attenuator probe.
(Courtesy of Tektronix, Inc.)

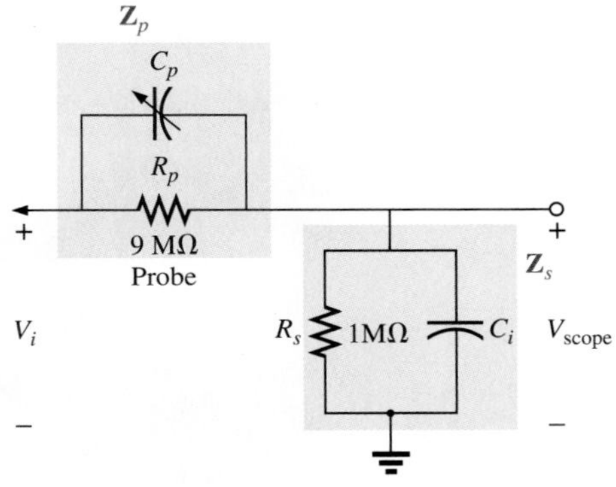

FIG. 24.42
Compensated attenuator and input impedance to a scope, including the cable capacitance.

24.8 APPLICATION

TV Remote

The TV remote works in many ways like a garage door opener or car alarm transmitter. There is no visible connection between the transmitter and the receiver, and each transmitter is linked to its receiver with a special code. The only major difference is that the TV remote uses an infrared frequency while the other two use a much lower radio frequency.

The TV remote in Fig. 24.43(a) has been opened to reveal the internal construction of its keypad and face in Fig. 24.43(b). The three components in Fig. 24.43(b) are lined up to show how the holes in the cover

(a)

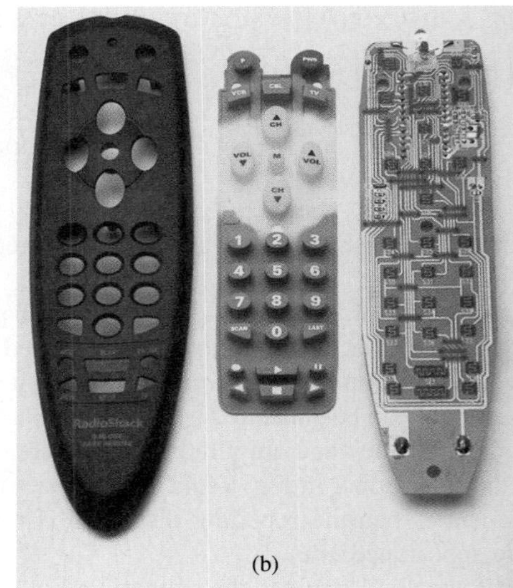

(b)

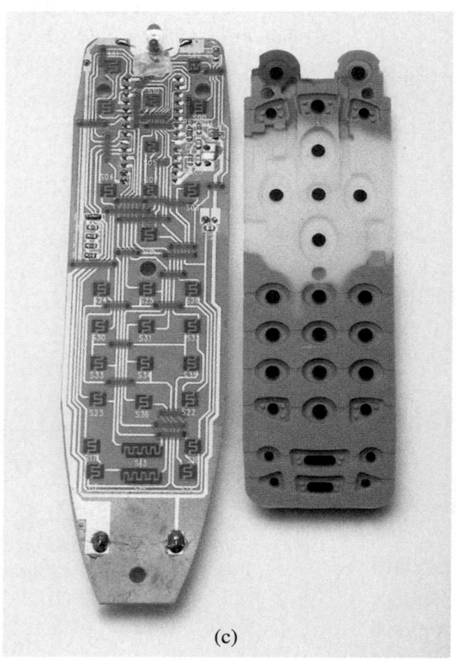

(c)

(d)

FIG. 24.43
TV remote: (a) external appearance; (b) internal construction; (c) carbon keypads; (d) enlarged view of S31 keypad.

match the actual keys in the switch membrane and where each button on the keypad hits on the face of the printed circuit board. Note on the printed circuit board that there is a black pad to match each key on the membrane. The back side of the switch membrane in Fig. 24.43(c) shows the soft carbon contacts that make contact with the carbon contacts on the printed board when the buttons are depressed. An enlarged view of one of the contacts (S31) in Fig. 24.43(c) is shown in Fig. 24.43(d) to illustrate the separation between circuits and the pattern used to ensure continuity when the solid round carbon pad at the bottom of the key is put in place.

All the connections established when a key is pressed are passed on to a relatively large switch-matrix-encoder IC chip appearing on the back side of the printed circuit board as shown in Fig. 24.44. For the pad (S31) in Fig. 24.43(d), three wires of the matrix appearing in Fig. 24.43(b) are connected when the corresponding key (number 5) is pressed. The encoder then reacts to this combination and sends out the appropriate signal as an infrared (IR) signal from the IR LED appearing at the end of the remote control, as shown in Fig. 24.43(b) and Fig. 24.44. The second smaller LED (red on actual unit) appearing at the top of Fig. 24.43(b) blinks during transmission. Once the batteries are inserted, the CMOS electronic circuitry that controls the operation of the remote is *always on.* This is possible only because of the very low power drain of CMOS circuitry. The power (PWR) button is used only to turn the TV on and activate the receiver.

The signal sent out by the majority of remotes is one of the two types appearing in Fig. 24.45. In each case there is a *key pulse* to initiate the signal sequence and to inform the receiver that the coded signal is about to arrive. In Fig. 24.45(a), a 4-bit binary-coded signal is transmitted using pulses in specific locations to represent the "ones" and using the absence of a pulse to represent the "zeros." That coded signal can then be interpreted by the receiver unit and the proper operation performed. In Fig. 24.45(b), the signal is frequency controlled. Each key has a different frequency associated with it. The result is that each key has a specific transmission frequency. Since each TV receiver responds to a different pulse train, a remote must be coded for the TV under control. There are *fixed program* remotes that can be used with only one TV. Then there are *smart* remotes that are preprogrammed internally with a number of remote control codes. You have to set up remotes of this type according to the TV you have, using a three-digit coding system accessed through the TV setup screen. *Learning* remotes are those that can use the old remote to learn the code and then store it for future use. In this case, one remote is set directly in front of the other, and the information is transferred from one to the other when both are energized. Remotes are also available that are a combination of the last two.

The remote in Fig. 24.43 uses four AAA batteries in series for a total of 6 V. It has its own local crystal oscillator separate from the IC as shown by the discrete elements to the top right and midleft of the printed circuit

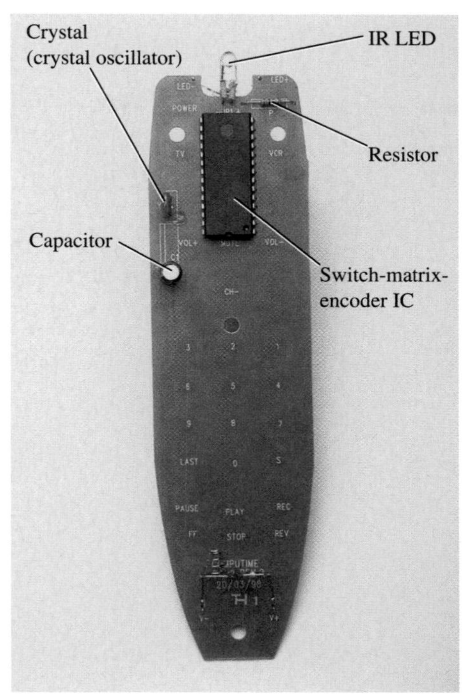

FIG. 24.44
Back side of TV remote in Fig. 24.43.

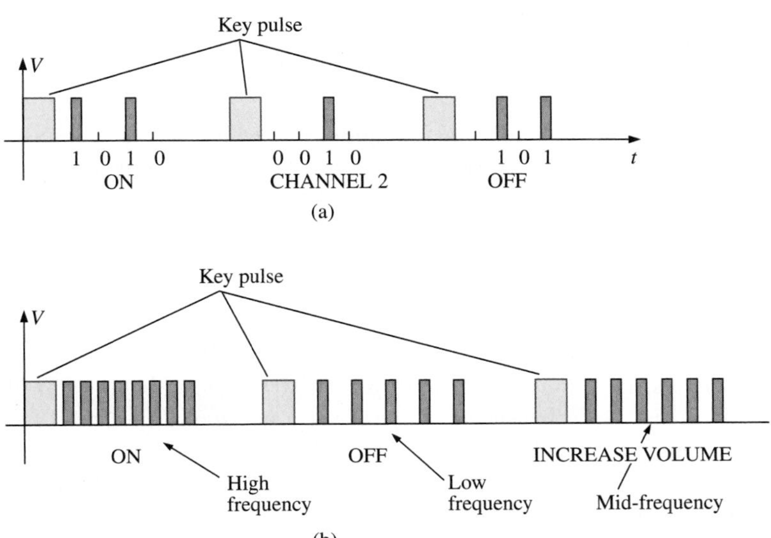

FIG. 24.45
Signal transmission: (a) pulse train; (b) variation.

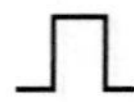

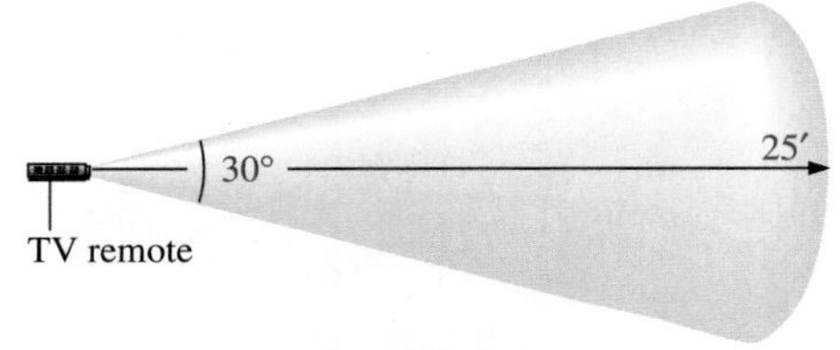

FIG. 24.46
Range and coverage arc for TV remote in Fig. 24.43.

board in Fig. 24.43(c). The crystal itself, which is relatively large compared to the other elements, appears on the other side of the board just above the electrolytic capacitor in Fig. 24.44. The oscillator generates the pulse signal required for proper IC operation. Note how flush most of the discrete elements are in Fig. 24.43(b) and the rather large electrolytic capacitor on the back of the printed circuit board in Fig. 24.44. The specifications on the unit give it a range control of 25 ft with a 30° coverage arc as shown in Fig. 24.46. The arc coverage of your unit can easily be tested by pointing it directly at the TV and then moving it in any direction until it no longer controls the TV.

24.9 COMPUTER ANALYSIS

PSpice

***R-C* Response** Our analysis begins with a verification of the results of Example 24.10 which examined the response of the series *R-C* circuit appearing on the schematic in Fig. 24.47. The source is one used in

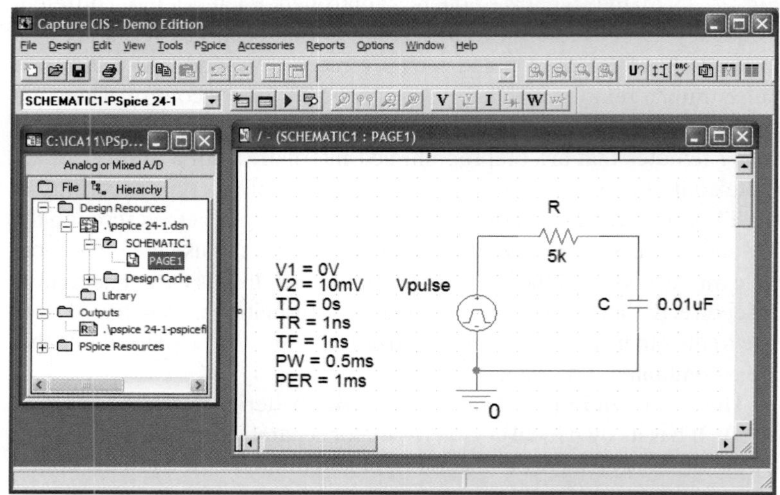

FIG. 24.47
Using PSpice to verify the results of Example 24.10.

Chapters 10 and 11 to replicate the action of a switch in series with a dc source. The defining attributes for the pulse waveform are repeated for convenience in Fig. 24.48. Recall that the **PW** was made long enough so that the full transient period could be examined. In this analysis, the pulse width is adjusted to permit viewing the transient behavior of an *R-C* network between changing levels of the applied pulse. Initially the **PW** is set at 10 times the time constant of the network so that the full transient response can occur between changes in voltage level. The time constant of the network is $\tau = RC = (5\text{ k}\Omega)(0.01\ \mu\text{F}) = 0.05$ ms, resulting in a **PW** of 0.5 ms in Fig. 24.47. To establish a square-wave appearance, the period was chosen as twice the pulse width or 1 ms as shown in the **VPulse** listing.

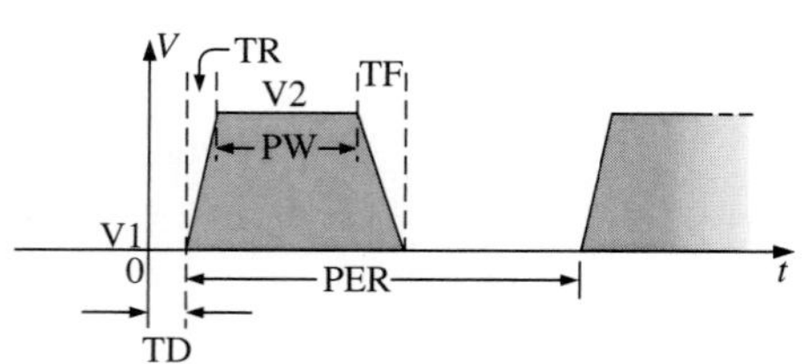

FIG. 24.48
Defining the PSpice ***Vpulse*** *parameters.*

In the **Simulation Settings** dialog box, select **Time Domain(Transient)** to get a response versus time. Select the **Run to time** at 2 ms so that two full cycles result. Leave the **Start saving data after** on the default value of 0 s, and set the **Maximum step size** at 2 ms/1000 = 2 μs. After simulation, **Trace-Add Trace-I(C)-OK,** the bottom plot in Fig. 24.49 is the result. Note that the maximum current is 2 μA as determined by

$I_{C_{max}} = 10\text{ mV}/5\text{ k}\Omega$, and the full transient response appears within each pulse. Note also that the current dropped below the axis to reveal a change in direction when the applied voltage dropped from the 10 mV level to 0 V. Through **Plot-Add Plot to Window-Trace-Add Trace-V(Vpulse:+)-OK-Trace-Add Trace-V(C:1)-OK,** the plots of the applied voltage and the voltage across the capacitor can be displayed in the upper graph in Fig. 24.49. First, select the upper graph in Fig. 24.49 so that you can move the **SEL>>,** and then select the **Toggle cursor** key. Now left-click on **V(C:1)** at the bottom right of the graph and left-click again to set a cursor on the graph. Setting the cursor at five time constants reveals that the transient voltage has reached 9.935 mV. Setting the right-click cursor at ten time constants reveals that V_C has essentially reached the 10 mV level.

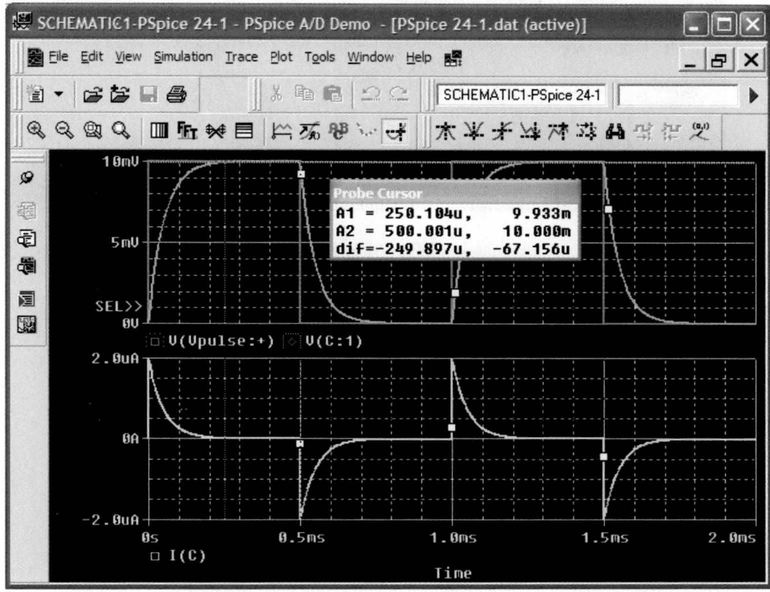

FIG. 24.49

Plot of v_{pulse}, v_C, and i_C for the circuit in Fig. 24.47.

Setting $t_p = \tau = T/2$ The parameters of the source will now be modified by changing the frequency of the pulse waveform to 10 kHz with a period of 0.1 ms and a pulse width of 0.05 ms. For **Vpulse,** the changes are **PW** = 0.05 ms and **PER** = 0.1 ms. The time constant of the network remains the same at 0.05 ms, so the pulse width equals the time constant of the circuit. The result is that it will take a number of pulses before the voltage across the capacitor reaches its final value of 10 mV. Under the **Simulation Settings,** change the **Run to time** to 0.5 ms = 500 μs or five cycles of the applied voltage. Change the **Maximum step size** to 500 μs/1000 = 500 ns = 0.5 μs. Under the **SCHEMATIC1** window, select **Trace-Add Trace-V(C:1)-OK** to obtain the transient voltage across the capacitor. Select **Trace-Add Trace-V(Vpulse:+)-OK** to place the applied voltage on the same screen. Note in the resulting plots in Fig. 24.50 that the voltage builds up from 0 V until it appears to reach a fairly steady state after 400 μs. At 400 μs, use a left cursor (**A1**) to find the minimum point with 2.71 mV resulting—a close match with the longhand calculation of Example 24.11 at 2.69 mV. At 450 μs, the right-click cursor **(A2)** provides a level of 7.29 mV which is again a close match with the calculated level of 7.31 mV.

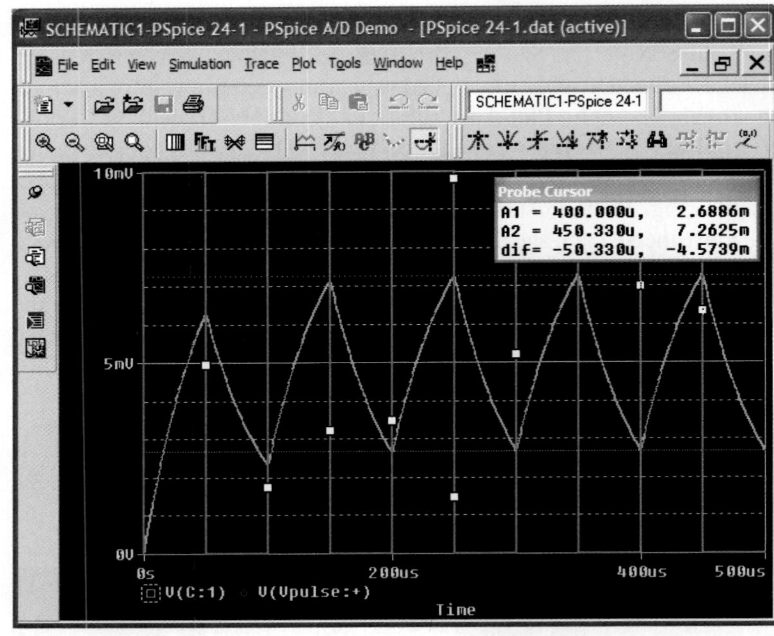

FIG. 24.50
Plot of v_C for the circuit in Fig. 24.47 with $t_p = \tau = T/2$.

PROBLEMS

SECTION 24.2 Ideal versus Actual

1. Determine the following for the pulse waveform in Fig. 24.51:
   a. positive- or negative-going?
   b. base-line voltage
   c. pulse width
   d. amplitude
   e. % tilt

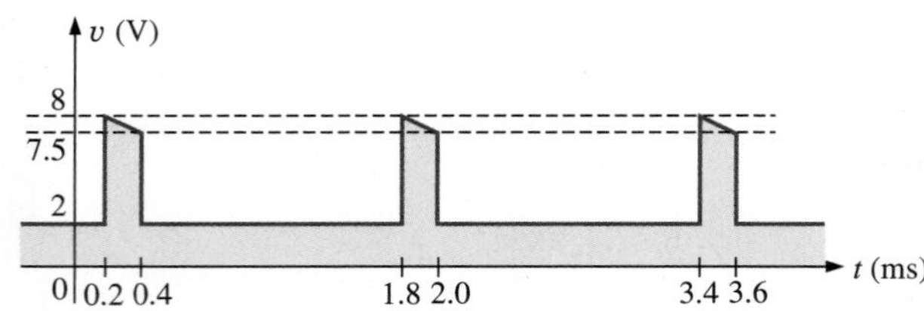

FIG. 24.51
Problems 1, 8, and 12.

2. Repeat Problem 1 for the pulse waveform in Fig. 24.52.

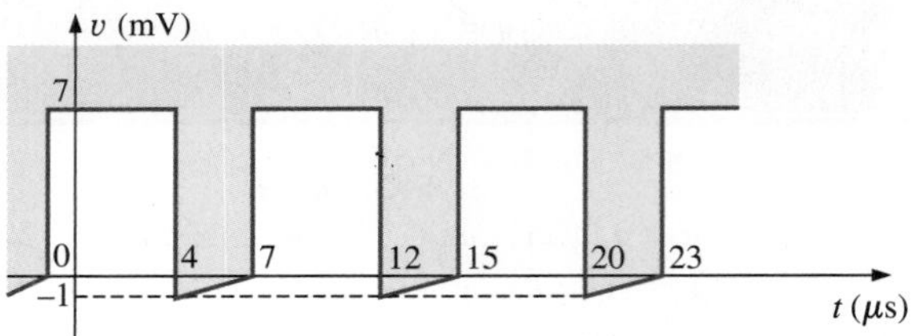

FIG. 24.52
Problems 2 and 9.

3. Repeat Problem 1 for the pulse waveform in Fig. 24.53.

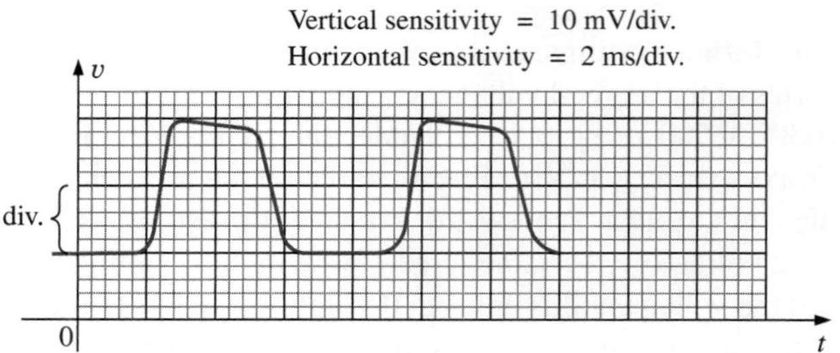

FIG. 24.53
Problems 3, 4, 10, and 13.

4. Determine the rise and fall times for the waveform in Fig. 24.53.

5. Sketch a pulse waveform that has a base-line voltage of -5 mV, a pulse width of 2 μs, an amplitude of 15 mV, a 10% tilt, a period of 10 μs, and vertical sides, and that is positive-going.

6. For the waveform in Fig. 24.54, established by straight-line approximations of the original waveform:
   a. Determine the rise time.
   b. Find the fall time.
   c. Find the pulse width.
   d. Calculate the frequency.

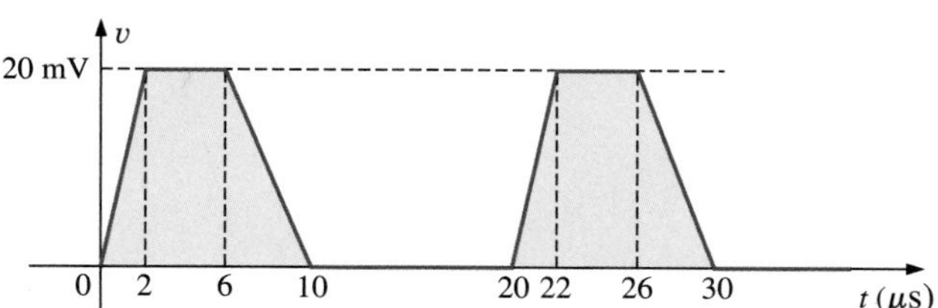

FIG. 24.54
Problems 6 and 14.

7. For the waveform in Fig. 24.55:
   a. Determine the period.
   b. Find the frequency.
   c. Find the maximum and minimum amplitudes.

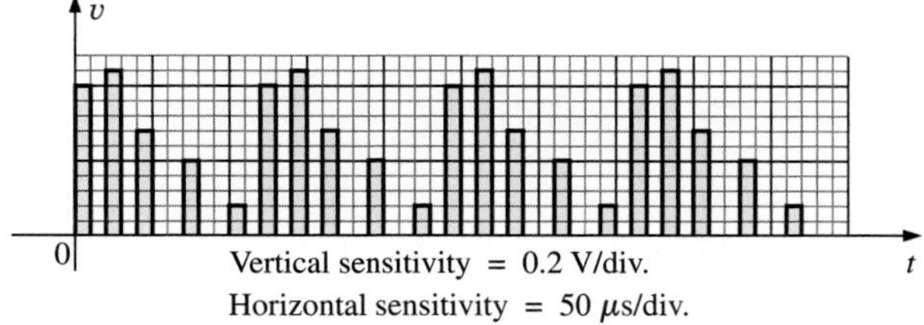

FIG. 24.55
Problems 7 and 15.

SECTION 24.3 Pulse Repetition Rate and Duty Cycle

8. Determine the pulse repetition frequency and duty cycle for the waveform in Fig. 24.51.

9. Determine the pulse repetition frequency and duty cycle for the waveform in Fig. 24.52.

10. Determine the pulse repetition frequency and duty cycle for the waveform in Fig. 24.53.

SECTION 24.4 Average Value

11. For the waveform in Fig. 24.56, determine the
    a. period.
    b. pulse width.
    c. pulse repetition frequency.
    d. average value.
    e. effective value.

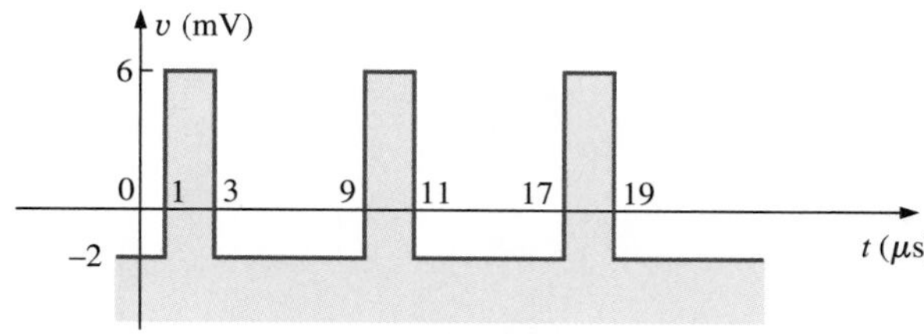

FIG. 24.56
Problem 11.

12. Determine the average value of the periodic pulse waveform in Fig. 24.51.

13. To the best accuracy possible, determine the average value of the waveform in Fig. 24.53.

14. Determine the average value of the waveform in Fig. 24.54.

15. Determine the average value of the periodic pulse train in Fig. 24.55.

SECTION 24.5 Transient *R-C* Networks

16. The capacitor in Fig. 24.57 is initially charged to 5 V, with the polarity indicated in the figure. The switch is then closed at $t = 0$ s.
    a. What is the mathematical expression for the voltage v_C?
    b. Sketch v_C versus t.
    c. What is the mathematical expression for the current i_C?
    d. Sketch i_C versus t.

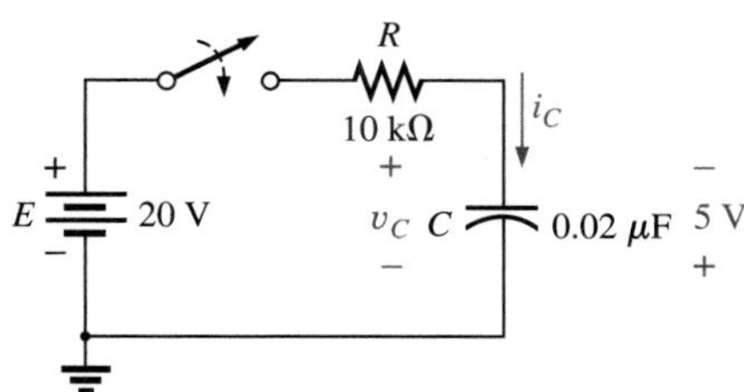

FIG. 24.57
Problem 16.

17. For the input voltage v_i appearing in Fig. 24.58, sketch the waveform for v_o. Assume that steady-state conditions were established with $v_i = 8$ V.

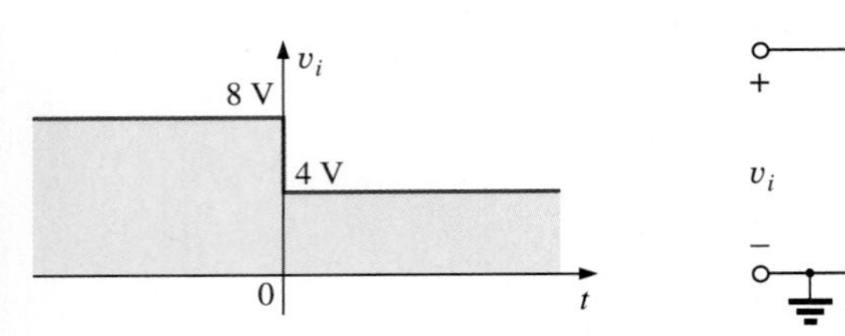

FIG. 24.58
Problem 17.

18. The switch in Fig. 24.59 is in position 1 until steady-state conditions are established. Then the switch is moved (at $t = 0$ s) to position 2. Sketch the waveform for the voltage v_C.

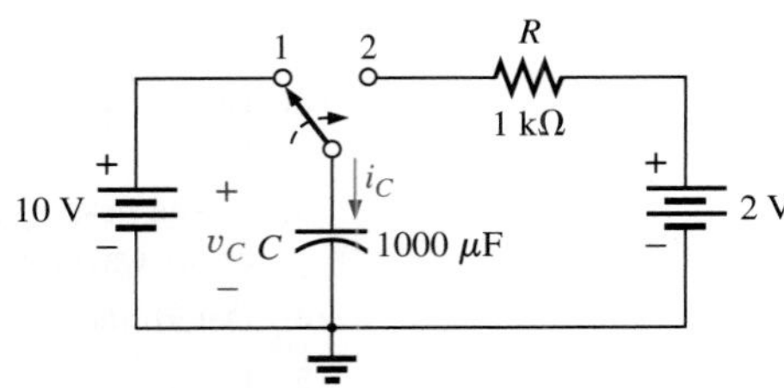

FIG. 24.59
Problems 18 and 19.

19. Sketch the waveform for i_C for Problem 18.

SECTION 24.6 *R-C* Response to Square-Wave Inputs

20. Sketch the voltage v_C for the network in Fig. 24.60 due to the square-wave input of the same figure with a frequency of

a. 500 Hz.
b. 100 Hz.
c. 5000 Hz.

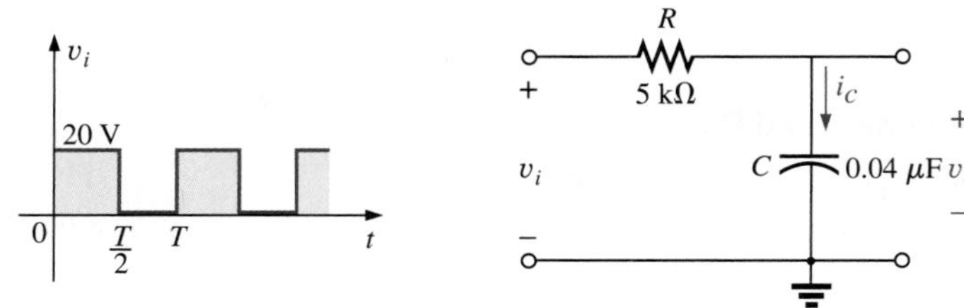

FIG. 24.60
Problems 20, 21, 23, 24, 27, and 28.

21. Sketch the current i_C for each frequency in Problem 20.

22. Sketch the response v_C of the network in Fig. 24.60 to the square-wave input in Fig. 24.61.

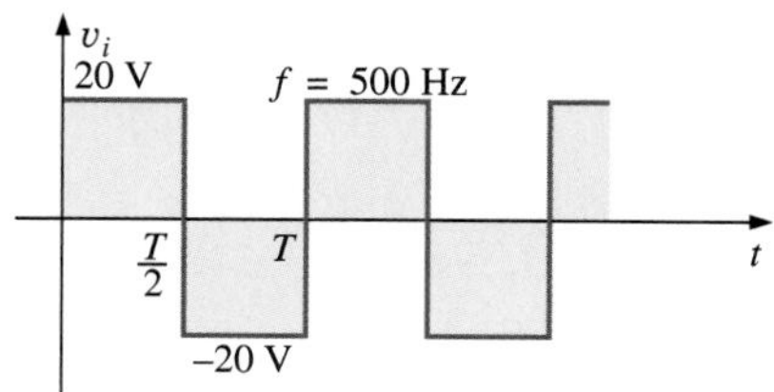

FIG. 24.61
Problem 22.

23. If the capacitor in Fig. 24.60 is initially charged to 20 V, sketch the response v_C to the same input signal (in Fig. 24.60) at a frequency of 500 Hz.

24. Repeat Problem 23 if the capacitor is initially charged to −10 V.

SECTION 24.7 Oscilloscope Attenuator and Compensating Probe

25. Given the network in Fig. 24.42 with $R_p = 9\ \text{M}\Omega$ and $R_s = 1\ \text{M}\Omega$, find $\mathbf{V}_{\text{scope}}$ in polar form if $C_p = 3$ pF, $C_s = 18$ pF, $C_c = 9$ pF, and $v_i = \sqrt{2}\,(100) \sin 2\pi 10{,}000t$. That is, determine $\mathbf{Z}_s$ and $\mathbf{Z}_p$, substitute into Eq. (24.8), and compare the results obtained with Eq. (24.10). Is it verified that the phase angle of $\mathbf{Z}_s$ and $\mathbf{Z}_p$ is the same under the condition $R_pC_p = R_sC_s$?

26. Repeat Problem 25 at $\omega = 10^5$ rad/s.

SECTION 24.9 Computer Analysis

PSpice

27. Using schematics, obtain the waveforms for v_C and i_C for the network in Fig. 24.60 for a frequency of 1 kHz.

*28. Using schematics, place the waveforms of v_i, v_C, and i_C on the same printout for the network in Fig. 24.60 at a frequency of 2 kHz.

*29. Using schematics, obtain the waveform appearing on the scope in Fig. 24.37 with a 20 V pulse input at a frequency of 5 kHz.

*30. Place a capacitor in parallel with R_p in Fig. 24.37 that will establish an in-phase relationship between v_{scope} and v_i. Using schematics, obtain the waveform appearing on the scope in Fig. 24.37 with a 20 V pulse input at a frequency of 5 kHz.

GLOSSARY

Actual (true, practical) pulse A pulse waveform having a leading edge and a trailing edge that are not vertical, along with other distortion effects such as tilt, ringing, or overshoot.

Amplitude of a pulse waveform The peak-to-peak value of a pulse waveform.

Attenuator probe A scope probe that will reduce the strength of the signal applied to the vertical channel of a scope.

Base-line voltage The voltage level from which a pulse is initiated.

Compensated attenuator probe A scope probe that can reduce the applied signal and balance the effects of the input capacitance of a scope on the signal to be displayed.

Duty cycle Factor that reveals how much of a period is encompassed by the pulse waveform.

Fall time (t_f) The time required for the trailing edge of a pulse waveform to drop from the 90% to the 10% level.

Ideal pulse A pulse waveform characterized as having vertical sides, sharp corners, and a flat peak response.

Negative-going pulse A pulse that increases in the negative direction from the base-line voltage.

Periodic pulse train A sequence of pulses that repeats itself after a specific period of time.

Positive-going pulse A pulse that increases in the positive direction from the base-line voltage.

Pulse repetition frequency (pulse repetition rate) The frequency of a periodic pulse train.

Pulse train A series of pulses that may have varying heights and widths.

Pulse width (t_p) The pulse width defined by the 50% voltage level.

Rise time (t_r) The time required for the leading edge of a pulse waveform to travel from the 10% to the 90% level.

Square wave A periodic pulse waveform with a 50% duty cycle.

Tilt (droop, sag) The drop in peak value across the pulse width of a pulse waveform.

NONSINUSOIDAL CIRCUITS

Objectives

- *Become familiar with the components of the Fourier series expansion for any sinusoidal or nonsinusoidal function.*
- *Understand how the appearance and time axis plot of a waveform can identify which terms of a Fourier series will be present.*
- *Be able to determine the response of a network to any input defined by a Fourier series expansion.*
- *Learn how to add two or more waveforms defined by Fourier series expansions.*

25.1 INTRODUCTION

Any waveform that differs from the basic description of the sinusoidal waveform is referred to as **nonsinusoidal.** The most obvious and familiar are the dc, square-wave, triangular, sawtooth, and rectified waveforms in Fig. 25.1.

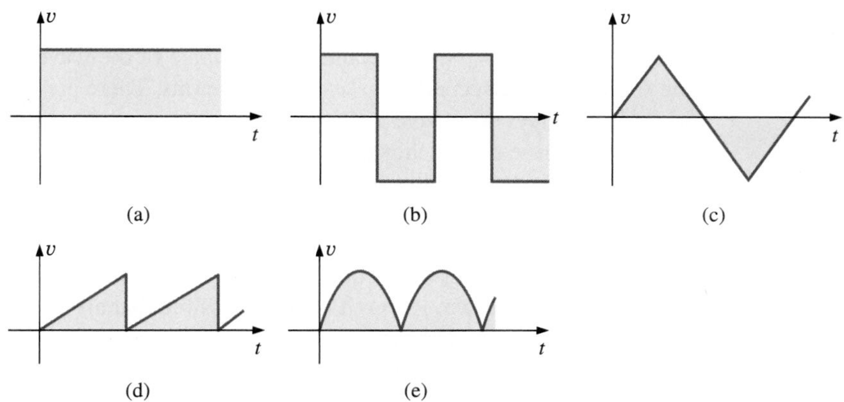

FIG. 25.1
Common nonsinusoidal waveforms: (a) dc; (b) square-wave; (c) triangular; (d) sawtooth; (e) rectified.

The output of many electrical and electronic devices are nonsinusoidal, even though the applied signal may be purely sinusoidal. For example, the network in Fig. 25.2 uses a diode to clip off the negative portion of the applied signal in a process called *half-wave rectification,* which is used in the development of dc levels from a sinusoidal input. You will find in your electronics courses that the diode is similar to a mechanical switch, but it is different because it can conduct current in only one direction. The output waveform is definitely nonsinusoidal, but note that it has the same period as the applied signal and matches the input for half the period.

This chapter demonstrates how a nonsinusoidal waveform like the output in Fig. 25.2 can be represented by a series of terms. It also explains how to determine the response of a network to such an input.

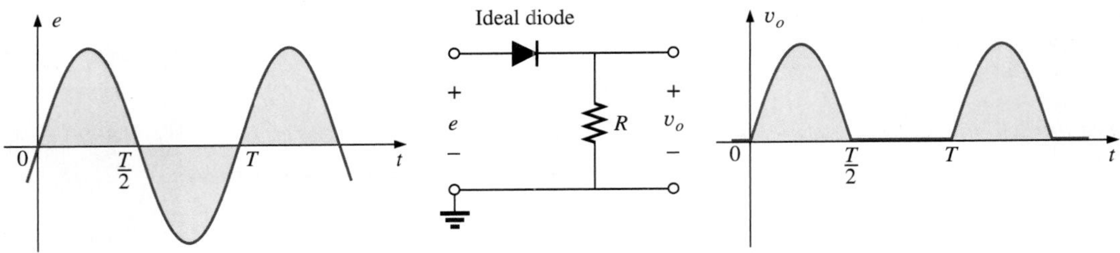

FIG. 25.2
Half-wave rectifier producing a nonsinusoidal waveform.

25.2 FOURIER SERIES

Fourier series refers to a series of terms, developed in 1822 by Baron Jean Fourier (Fig. 25.3), that can be used to represent a nonsinusoidal periodic waveform. In the analysis of these waveforms, we solve for each term in the Fourier series:

$$f(t) = \underbrace{A_0}_{\text{dc or average value}} + \underbrace{A_1 \sin \omega t + A_2 \sin 2\omega t + A_3 \sin 3\omega t + \cdots + A_n \sin n\omega t}_{\text{sine terms}} + \underbrace{B_1 \cos \omega t + B_2 \cos 2\omega t + B_3 \cos 3\omega t + \cdots + B_n \cos n\omega t}_{\text{cosine terms}} \tag{25.1}$$

FIG. 25.3
Baron Jean Fourier.
Courtesy of the Smithsonian Institution
Photo No. 56,822

French (Auxerre, Grenoble, Paris)
(1768–1830)
Mathematician, Egyptologist, and **Administrator**
Professor of Mathematics, École Polytechnique

Best known for an infinite mathematical series of sine and cosine terms called the *Fourier series* which he used to show how the conduction of heat in solids can be analyzed and defined. Although he was primarily a mathematician, a great deal of Fourier's work revolved around real-world physical occurrences such as heat transfer, sunspots, and the weather. He joined the École Polytechnique in Paris as a faculty member when the institute first opened. Napoleon requested his aid in the research of Egyptian antiquities, resulting in a three-year stay in Egypt as Secretary of the Institut d'Égypte. Napoleon made him a baron in 1809, and he was elected to the Académie des Sciences in 1817.

Depending on the waveform, a large number of these terms may be required to approximate the waveform closely for the purpose of circuit analysis.

As shown in Eq. (25.1), the Fourier series has three basic parts. The first is the dc term A_0, which is the average value of the waveform over one full cycle. The second is a series of sine terms. There are no restrictions on the values or relative values of the amplitudes of these sine terms, but each will have a frequency that is an integer multiple of the frequency of the first sine term of the series. The third part is a series of cosine terms. There are again *no* restrictions on the values or relative values of the amplitudes of these cosine terms, but each will have a frequency that is an integer multiple of the frequency of the first cosine term of the series. For a particular waveform, it is quite possible that all of the sine *or* cosine terms are zero. Characteristics of this type can be determined by simply examining the nonsinusoidal waveform and its position on the horizontal axis.

The first term of the sine and cosine series is called the **fundamental component.** It represents the minimum frequency term required to represent a particular waveform, and it also has the same frequency as the waveform being represented. A fundamental term, therefore, must be present in any Fourier series representation. The other terms with higher-order frequencies (integer multiples of the fundamental) are called the **harmonic terms.** A term that has a frequency equal to twice the fundamental is the second harmonic; three times, the third harmonic; and so on.

Average Value: A_0

The dc term of the Fourier series is the average value of the waveform over one full cycle. If the net area above the horizontal axis equals that

below in one full period, $A_0 = 0$, and the dc term does not appear in the expansion. If the area above the axis is greater than that below over one full cycle, A_0 is positive and will appear in the Fourier series representation. If the area below the axis is greater, A_0 is negative and will appear with the negative sign in the expansion.

Odd Function (Point Symmetry)

If a waveform is such that its value for $+t$ is the negative of that for $-t$, it is called an odd function or is said to have point symmetry.

Fig. 25.4(a) is an example of a waveform with point symmetry. Note that the waveform has a peak value at t_1 that matches the magnitude (with the opposite sign) of the peak value at $-t_1$. For waveforms of this type, all the parameters $B_{1\to\infty}$ of Eq. (25.1) will be zero. In fact,

waveforms with point symmetry can be fully described by just the dc and sine terms of the Fourier series.

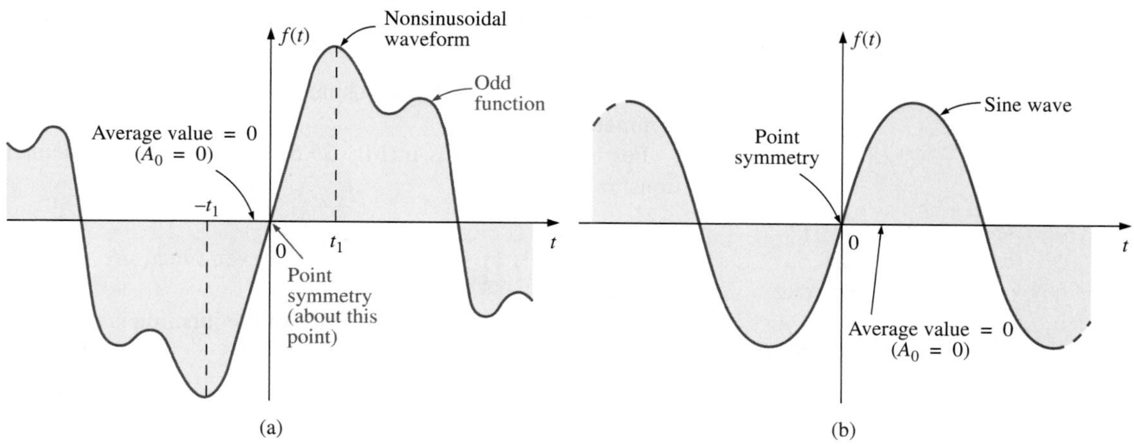

FIG. 25.4
Point symmetry.

Note in Fig. 25.4(b) that a sine wave is an odd function with point symmetry.

For both waveforms in Fig. 25.4, the following mathematical relationship is true:

$$f(t) = -f(-t) \quad \text{(odd function)} \qquad \textbf{(25.2)}$$

In words, it states that the magnitude of the function at $+t$ is equal to the negative of the magnitude at $-t$ [t_1 in Fig. 25.4(a)].

Even Function (Axis Symmetry)

If a waveform is symmetric about the vertical axis, it is called an even function or is said to have axis symmetry.

Fig. 25.5(a) is an example of such a waveform. Note that the value of the function at t_1 is equal to the value at $-t_1$. For waveforms of this type, all the parameters $A_{1\to\infty}$ will be zero. In fact,

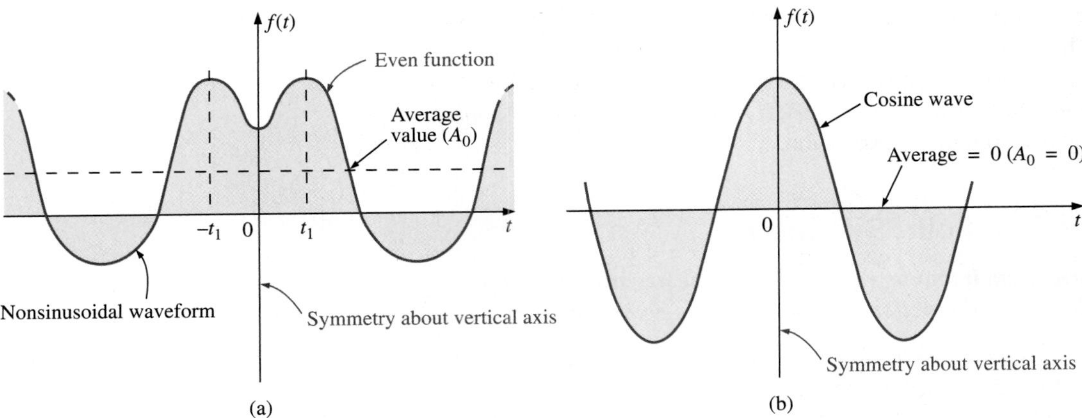

FIG. 25.5
Axis symmetry.

waveforms with axis symmetry can be fully described by just the dc and cosine terms of the Fourier series.

Note in Fig. 25.5(b) that a cosine wave is an even function with axis symmetry.

For both waveforms in Fig. 25.5, the following mathematical relationship is true:

$$\boxed{f(t) = f(-t)} \qquad \text{(even function)} \qquad \textbf{(25.3)}$$

In words, it states that the magnitude of the function is the same at $+t_1$ as at $-t$ [t_1 in Fig. 25.5(a)].

Mirror or Half-Wave Symmetry

If a waveform has half-wave or mirror symmetry as demonstrated by the waveform of Fig. 25.6, the even harmonics of the series of sine and cosine terms will be zero.

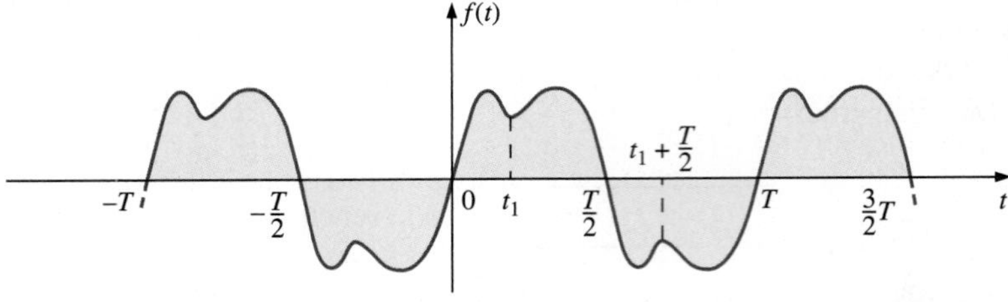

FIG. 25.6
Mirror symmetry.

In functional form, the waveform must satisfy the following relationship:

$$\boxed{f(t) = -f\left(t + \frac{T}{2}\right)} \qquad \textbf{(25.4)}$$

NON

Eq. (25.4) states that the waveform encompassed in one time interval $T/2$ will repeat itself in the next $T/2$ time interval, but in the negative sense (t_1 in Fig. 25.6). For example, the waveform in Fig. 25.6 from zero to $T/2$ will repeat itself in the time interval $T/2$ to T, but below the horizontal axis.

Repetitive on the Half-Cycle

The repetitive nature of a waveform can determine whether specific harmonics will be present in the Fourier series expansion. In particular,

if a waveform is repetitive on the half-cycle as demonstrated by the waveform in Fig. 25.7, the odd harmonics of the series of sine and cosine terms are zero.

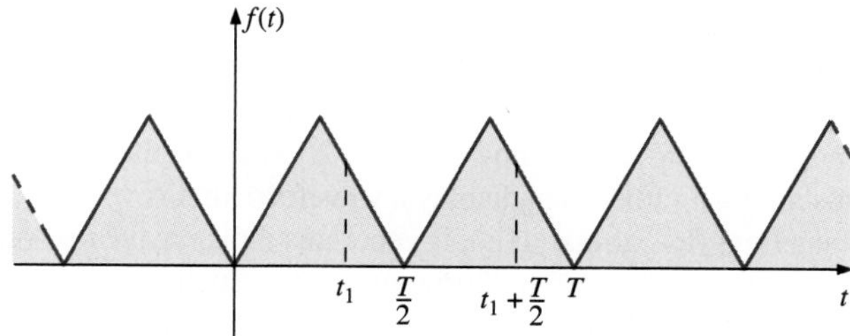

FIG. 25.7
A waveform repetitive on the half-cycle.

In functional form, the waveform must satisfy the following relationship:

$$f(t) = f\left(t + \frac{T}{2}\right) \qquad (25.5)$$

Eq. (25.5) states that the function repeats itself after each $T/2$ time interval (t_1 in Fig. 25.7). The waveform, however, will also repeat itself after each period T. In general, therefore, for a function of this type, if the period T of the waveform is chosen to be twice that of the minimum period ($T/2$), the odd harmonics will all be zero.

Mathematical Approach

The constants A_0, $A_{1\to n}$, and $B_{1\to n}$ can be determined by using the following integral formulas:

$$A_0 = \frac{1}{T}\int_0^T f(t)\,dt \qquad (25.6)$$

$$A_n = \frac{2}{T}\int_0^T f(t)\sin n\omega t\,dt \qquad (25.7)$$

$$B_n = \frac{2}{T}\int_0^T f(t)\cos n\omega t\,dt \qquad (25.8)$$

These equations have been presented for recognition purposes only; they are not used in the following analysis.

Instrumentation

Three types of instrumentation are available that reveal the dc, fundamental, and harmonic content of a waveform: *the spectrum analyzer, wave analyzer,* and *Fourier analyzer.* The purpose of such instrumentation is not solely to determine the composition of a particular waveform but also to reveal the level of distortion that may have been introduced by a system. For instance, an amplifier may be increasing the applied signal by a factor of 50, but in the process it may have distorted the waveform in a way that is quite unnoticeable from the oscilloscope display. The amount of distortion appears in the form of harmonics at frequencies that are multiples of the applied frequency. Each of the above instruments reveal which frequencies are having the most impact on the distortion, permitting their removal with properly designed filters.

The spectrum analyzer is shown in Fig. 25.8. It has the appearance of an oscilloscope but rather than display a waveform that is voltage (vertical axis) versus time (horizontal axis), it generates a display scaled off in dB (vertical axis) versus frequency (horizontal axis). Such a display is said to be in the *frequency domain* versus the *time domain* of the standard oscilloscope. The height of the vertical line in the display of Fig. 25.8 reveals the impact of that frequency on the shape of the waveform. Spectrum analyzers are unable to provide the phase angle associated with each component.

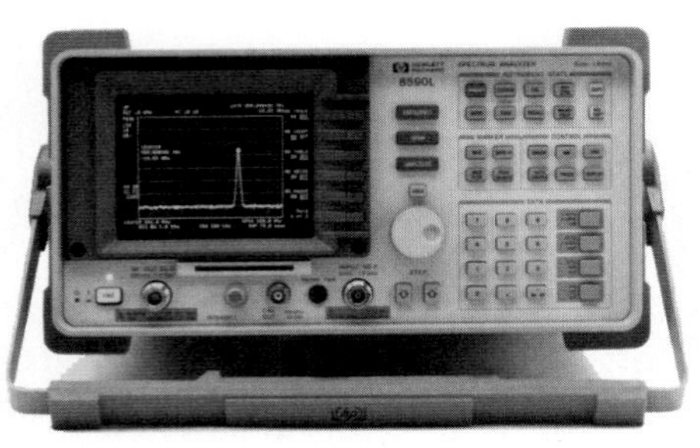

FIG. 25.8
Spectrum analyzer.
(Courtesy of Hewlett Packard)

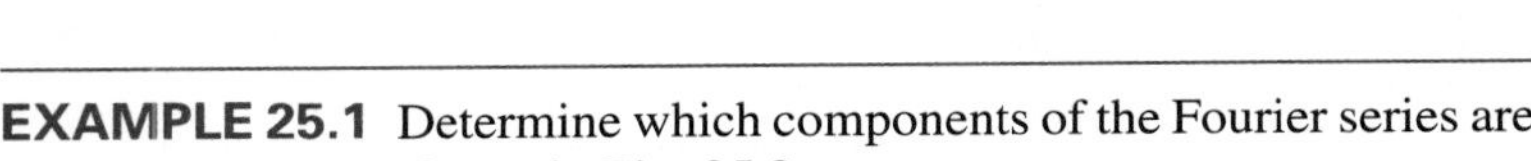

EXAMPLE 25.1 Determine which components of the Fourier series are present in the waveforms in Fig. 25.9.

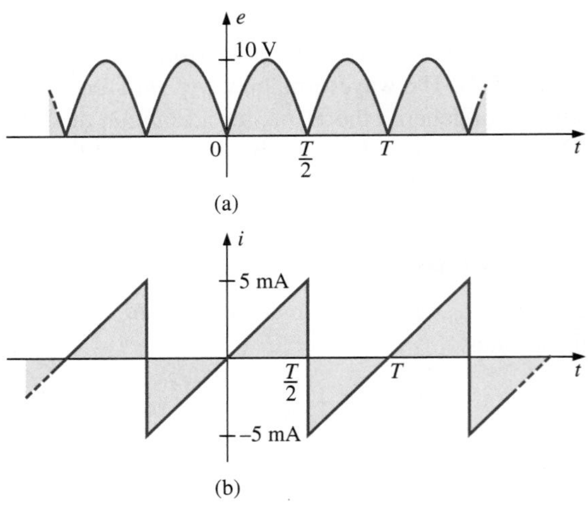

FIG. 25.9
Example 25.1.

Solutions:

a. The waveform has a net area above the horizontal axis and therefore will have **a positive dc term A_0.**

The waveform has axis symmetry, resulting in **only cosine terms** in the expansion.

The waveform has half-cycle symmetry, resulting in **only even terms** in the cosine series.

b. The waveform has the same area above and below the horizontal axis within each period, resulting in $\boldsymbol{A_0 = 0}$.

The waveform has point symmetry, resulting in **only sine terms** in the expansion.

EXAMPLE 25.2 Write the Fourier series expansion for the waveforms in Fig. 25.10.

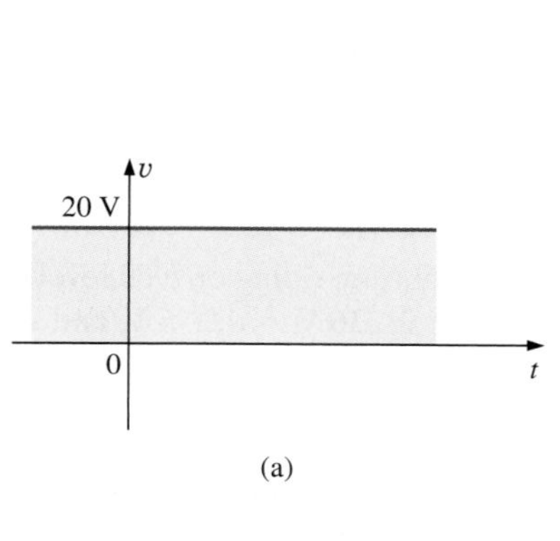

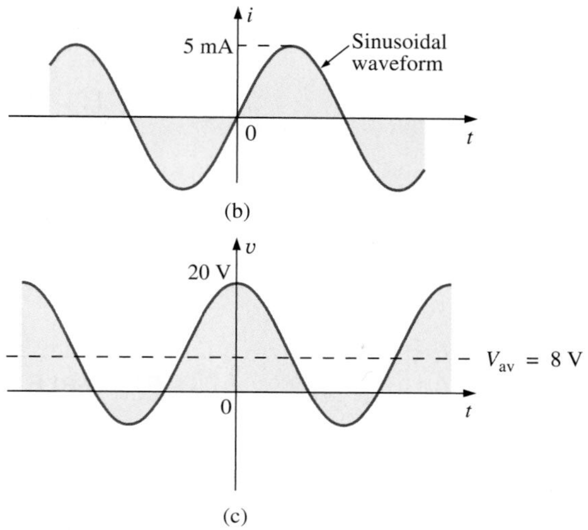

FIG. 25.10
Example 25.2.

Solutions:

a. $A_0 = 20 \qquad A_{1\to n} = 0 \qquad B_{1\to n} = 0$
$\boldsymbol{v = 20}$

b. $A_0 = 0 \qquad A_1 = 5 \times 10^{-3} \qquad A_{2\to n} = 0 \qquad B_{1\to n} = 0$
$\boldsymbol{i = 5 \times 10^{-3} \sin \omega t}$

c. $A_0 = 8 \qquad A_{1\to n} = 0 \qquad B_1 = 12 \qquad B_{2\to n} = 0$
$\boldsymbol{v = 8 + 12 \cos \omega t}$

EXAMPLE 25.3 Sketch the following Fourier series expansion:

$$v = 2 + 1 \cos \alpha + 2 \sin \alpha$$

Solution: Note Fig. 25.11.

The solution could be obtained graphically by first plotting all of the functions and then considering a sufficient number of points on the horizontal axis; or phasor algebra could be used as follows:

$$\begin{aligned} 1 \cos \alpha + 2 \sin \alpha &= 1\text{ V} \angle 90° + 2\text{ V} \angle 0° = j\,1\text{ V} + 2\text{ V} \\ &= 2\text{ V} + j\,1\text{ V} = 2.236\text{ V} \angle 26.57° \\ &= 2.236 \sin(\alpha + 26.57°) \end{aligned}$$

and $$\boldsymbol{v = 2 + 2.236 \sin (\alpha + 26.57°)}$$

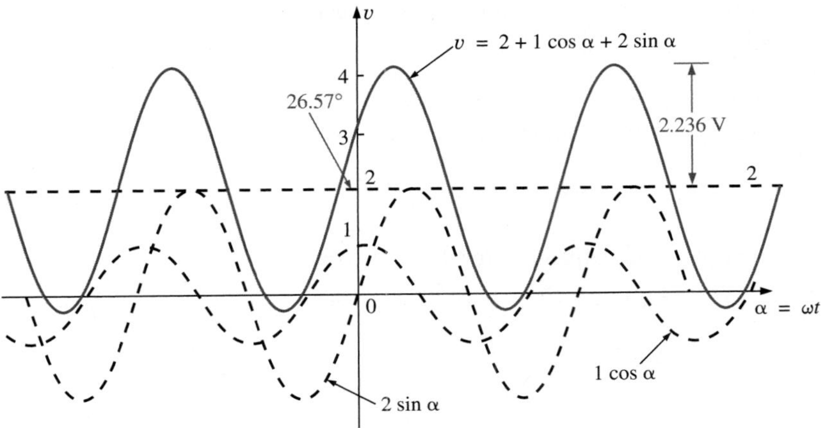

FIG. 25.11
Example 25.3.

which is simply the sine wave portion riding on a dc level of 2 V. That is, its positive maximum is 2 V + 2.236 V = 4.236 V, and its minimum is 2 V − 2.236 V = −0.236 V.

EXAMPLE 25.4 Sketch the following Fourier series expansion:

$$i = 1 \sin \omega t + 1 \sin 2\omega t$$

Solution: See Fig. 25.12. Note that in this case the sum of the two sinusoidal waveforms of different frequencies is *not* a sine wave. Recall that complex algebra can be applied only to waveforms having the *same* frequency. In this case, the solution is obtained graphically point by point, as shown for $t = t_1$.

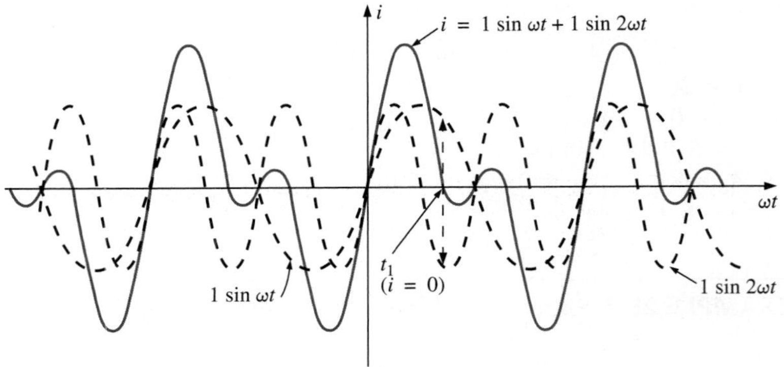

FIG. 25.12
Example 25.4.

As an additional example in the use of the Fourier series approach, consider the square wave shown in Fig. 25.13. The average value is zero, so $A_0 = 0$. It is an odd function, so all the constants $B_{1\to n}$ equal zero; only sine terms are present in the series expansion. Since the waveform satisfies the criteria for $f(t) = -f(t + T/2)$, the even harmonics are also zero.

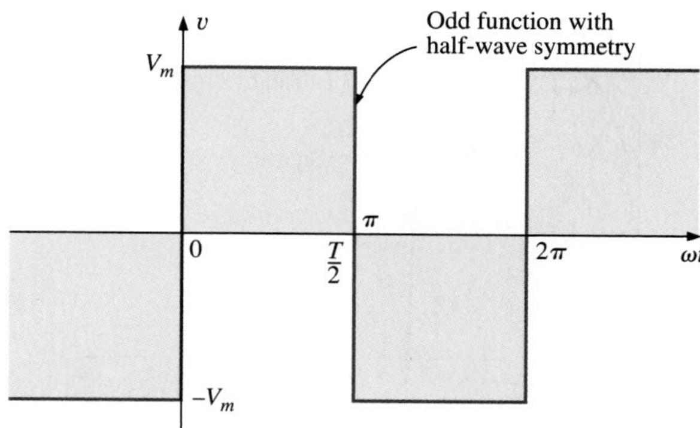

FIG. 25.13
Square wave.

The expression obtained after evaluating the various coefficients using Eq. (25.8) is

$$v = \frac{4}{\pi} V_m\left(\sin \omega t + \frac{1}{3}\sin 3\omega t + \frac{1}{5}\sin 5\omega t + \frac{1}{7}\sin 7\omega t + \cdots + \frac{1}{n}\sin n\omega t\right) \quad (25.9)$$

Note that the fundamental does indeed have the same frequency as that of the square wave. If we add the fundamental and third harmonics, we obtain the results shown in Fig. 25.14.

Even with only the first two terms, a few characteristics of the square wave are beginning to appear. If we add the next two terms (Fig. 25.15), the width of the pulse increases, and the number of peaks increases.

As we continue to add terms, the series better approximate the square wave. Note, however, that the amplitude of each succeeding term diminishes to the point at which it is negligible compared with those of the first few terms. A good approximation is to assume that the waveform is composed of the harmonics up to and including the ninth. Any higher

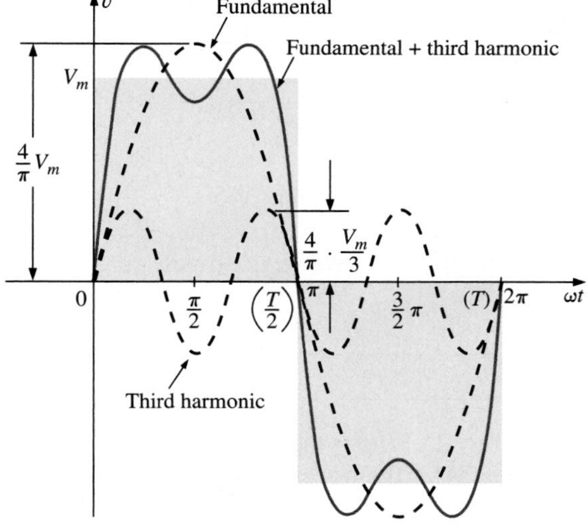

FIG. 25.14
Fundamental plus third harmonic.

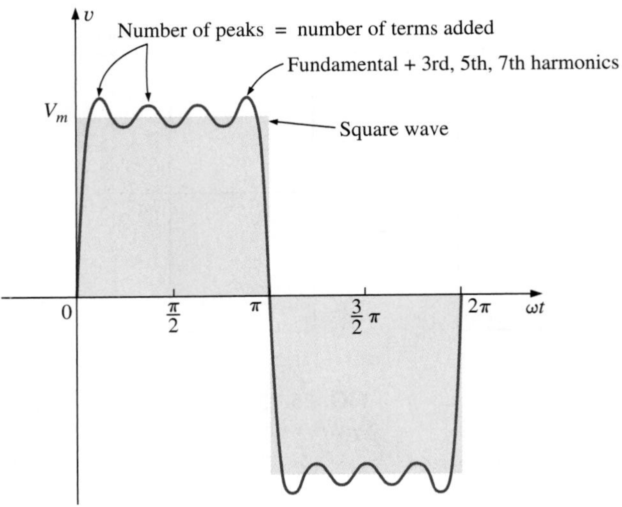

FIG. 25.15
Fundamental plus third, fifth, and seventh harmonics.

harmonics would be less than one-tenth the fundamental. If the waveform just described were shifted above or below the horizontal axis, the Fourier series would be altered only by a change in the dc term. Fig. 25.16(c), for example, is the sum of Fig. 25.16(a) and (b). The Fourier series for the complete waveform is, therefore,

$$\begin{aligned} v = v_1 + v_2 &= V_m + \text{Eq. (25.9)} \\ &= V_m + \frac{4}{\pi} V_m\left(\sin \omega t + \frac{1}{3} \sin 3\omega t + \frac{1}{5} \sin 5\omega t + \frac{1}{7} \sin 7\omega t + \cdots \right) \end{aligned}$$

and $$v = V_m\left[1 + \frac{4}{\pi}\left(\sin \omega t + \frac{1}{3} \sin 3\omega t + \frac{1}{5} \sin 5\omega t + \frac{1}{7} \sin 7\omega t + \cdots \right)\right]$$

FIG. 25.16
Shifting a waveform vertically with the addition of a dc term.

The equation for the half-wave rectified pulsating waveform in Fig. 25.17(b) is

$$v_2 = 0.318V_m + 0.500V_m \sin \alpha - 0.212V_m \cos 2\alpha - 0.0424V_m \cos 4\alpha - \cdots \qquad \textbf{(25.10)}$$

The waveform in Fig. 25.17(c) is the sum of the two in Fig. 25.17(a) and (b). The Fourier series for the waveform in Fig. 25.17(c) is, therefore,

$$\begin{aligned} v_T = v_1 + v_2 &= -\frac{V_m}{2} + \text{Eq. (25.10)} \\ &= -0.500V_m + 0.318V_m + 0.500V_m \sin \alpha - 0.212V_m \cos 2\alpha - 0.0424V_m \cos 4\alpha + \cdots \end{aligned}$$

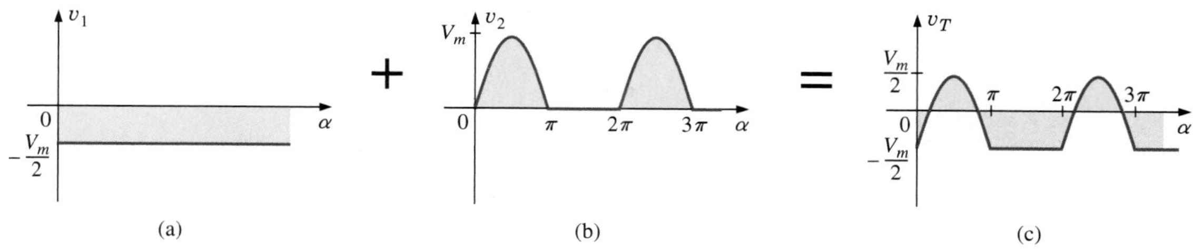

FIG. 25.17
Lowering a waveform with the addition of a negative dc component.

and $v_T = -0.182V_m + 0.5V_m \sin \alpha - 0.212V_m \cos 2\alpha - 0.0424V_m \cos 4\alpha + \cdots$

If either waveform were shifted to the right or left, the phase shift would be subtracted from or added to, respectively, the sine and cosine terms. The dc term would not change with a shift to the right or left.

If the half-wave rectified signal is shifted 90° to the left, as in Fig. 25.18, the Fourier series becomes

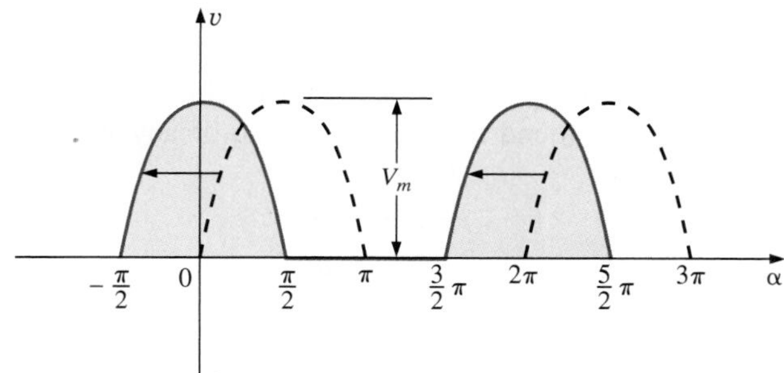

FIG. 25.18
Changing the phase angle of a waveform.

$$v = 0.318V_m + 0.500V_m \underbrace{\sin(\alpha + 90°)}_{\cos \alpha} - 0.212V_m \cos 2(\alpha + 90°) - 0.0424V_m \cos 4(\alpha + 90°) + \cdots$$

$$= 0.318V_m + 0.500V_m \cos \alpha - 0.212V_m \cos(2\alpha + 180°) - 0.0424V_m \cos(4\alpha + 360°) + \cdots$$

and $v = 0.318V_m + 0.500V_m \cos \alpha + 0.212V_m \cos 2\alpha - 0.0424V_m \cos 4\alpha + \cdots$

25.3 CIRCUIT RESPONSE TO A NONSINUSOIDAL INPUT

The Fourier series representation of a nonsinusoidal input can be applied to a linear network using the principle of superposition. Recall that this theorem allowed us to consider the effects of each source of a circuit independently. If we replace the nonsinusoidal input with the terms of the Fourier series deemed necessary for practical considerations, we can use superposition to find the response of the network to each term (Fig. 25.19).

The total response of the system is then the algebraic sum of the values obtained for each term. The major change between using this theorem for nonsinusoidal circuits and using it for the circuits previously described is that the frequency will be different for each term in the nonsinusoidal application. Therefore, the reactances

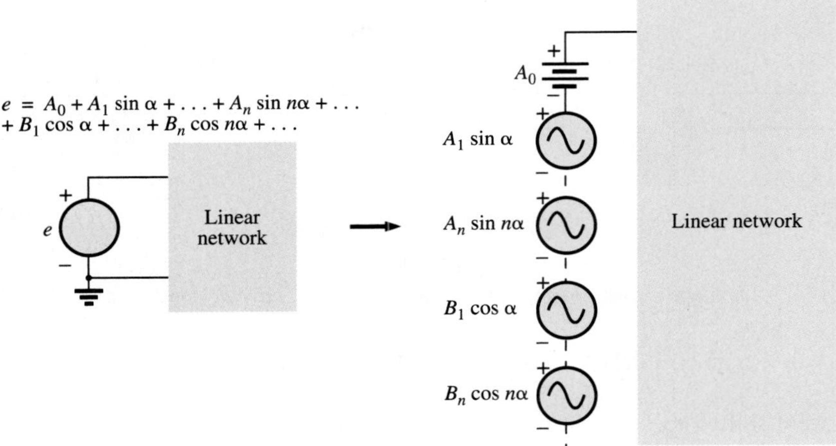

FIG. 25.19
Setting up the application of a Fourier series of terms to a linear network.

$$X_L = 2\pi fL \quad \text{and} \quad X_C = \frac{1}{2\pi fC}$$

will change for each term of the input voltage or current.

In Chapter 13, we found that the rms value of any waveform was given by

$$\sqrt{\frac{1}{T}\int_0^T f^2(t)\, dt}$$

If we apply this equation to the following Fourier series:

$$v(\alpha) = V_0 + V_{m_1} \sin \alpha + \cdots + V_{m_n} \sin n\alpha + V'_{m_1} \cos \alpha + \cdots + V'_{m_n} \cos n\alpha$$

then

$$V_{\text{rms}} = \sqrt{V_0^2 + \frac{V_{m_1}^2 + \cdots + V_{m_n}^2 + V'^2_{m_1} + \cdots + V'^2_{m_n}}{2}} \quad (25.11)$$

However, since

$$\frac{V_{m_1}^2}{2} = \left(\frac{V_{m_1}}{\sqrt{2}}\right)\left(\frac{V_{m_1}}{\sqrt{2}}\right) = (V_{1_{\text{rms}}})(V_{1_{\text{rms}}}) = V_{1_{\text{rms}}}^2$$

then

$$V_{\text{rms}} = \sqrt{V_0^2 + V_{1_{\text{rms}}}^2 + \cdots + V_{n_{\text{rms}}}^2 + V'^2_{1_{\text{rms}}} + \cdots + V'^2_{n_{\text{rms}}}} \quad (25.12)$$

Similarly, for

$$i(\alpha) = I_0 + I_{m_1} \sin \alpha + \cdots + I_{m_n} \sin n\alpha + I'_{m_1} \cos \alpha + \cdots + I'_{m_n} \cos n\alpha$$

we have

$$I_{\text{rms}} = \sqrt{I_0^2 + \frac{I_{m_1}^2 + \cdots + I_{m_n}^2 + I'^2_{m_1} + \cdots + I'^2_{m_n}}{2}} \quad (25.13)$$

and

$$I_{rms} = \sqrt{I_0^2 + I_{1_{rms}}^2 + \cdots + I_{n_{rms}}^2 + I'^2_{1_{rms}} + \cdots + I'^2_{n_{rms}}} \qquad (25.14)$$

The total power delivered is the sum of that delivered by the corresponding terms of the voltage and current. In the following equations, all voltages and currents are rms values:

$$P_T = V_0 I_0 + V_1 I_1 \cos\theta_1 + \cdots + V_n I_n \cos\theta_n + \cdots \qquad (25.15)$$

$$P_T = I_0^2 R + I_1^2 R + \cdots + I_n^2 R + \cdots \qquad (25.16)$$

or

$$P_T = I_{rms}^2 R \qquad (25.17)$$

with I_{rms} as defined by Eq. (25.13), and, similarly,

$$P_T = \frac{V_{rms}^2}{R} \qquad (25.18)$$

with V_{rms} as defined by Eq. (25.11).

EXAMPLE 25.5

a. Sketch the input resulting from the combination of sources in Fig. 25.20.
b. Determine the rms value of the input in Fig. 25.20.

Solutions:

a. Note Fig. 25.21.
b. Eq. (25.12):

$$V_{rms} = \sqrt{V_0^2 + \frac{V_m^2}{2}}$$

$$= \sqrt{(4\text{ V})^2 + \frac{(6\text{ V})^2}{2}} = \sqrt{16 + \frac{36}{2}}\text{ V} = \sqrt{34}\text{ V}$$

$$= \mathbf{5.831\ V}$$

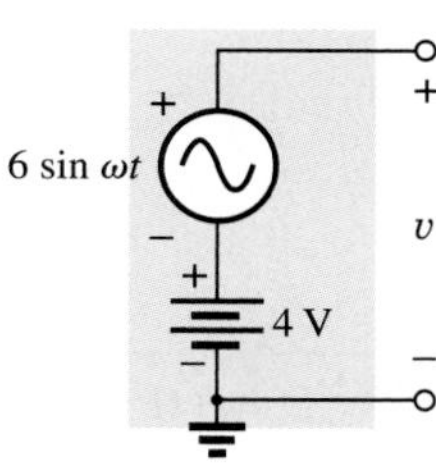

FIG. 25.20
Example 25.5.

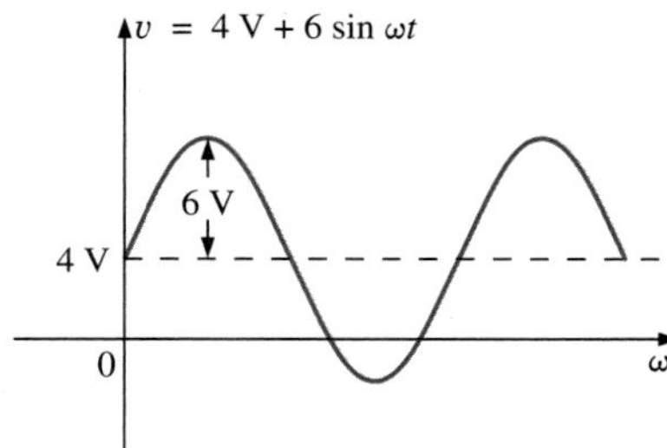

FIG. 25.21
Wave pattern generated by the source in Fig. 25.20.

It is particularly interesting to note from Example 25.5 that the rms value of a waveform having both dc and ac components is not simply the sum of the effective values of each. In other words, there is a temptation in the absence of Eq. (25.12) to state that V_{rms} = 4 V + 0.707 (6 V) = 8.242 V, which is incorrect and, in fact, exceeds the correct level by some 41%.

Instrumentation

It is important to realize that not every DMM will read the rms value of nonsinusoidal waveforms such as the one appearing in Fig. 25.21. Many are designed to read the rms value of sinusoidal waveforms only. It is

important to read the manual provided with the meter to see if it is a *true rms* meter that can read the rms value of any waveform.

We learned in Chapter 13 that the rms value of a square wave is the peak value of the waveform. Let us test this result using the Fourier expansion and Eq. (25.11).

EXAMPLE 25.6 Determine the rms value of the square wave of Fig. 25.13 with $V_m = 20$ V using the first six terms of the Fourier expansion, and compare the result to the actual rms value of 20 V.

Solution:

$$v = \frac{4}{\pi}(20\text{ V})\sin\omega t + \frac{4}{\pi}\left(\frac{1}{3}\right)(20\text{ V})\sin 3\omega t + \frac{4}{\pi}\left(\frac{1}{5}\right)(20\text{ V})\sin 5\omega t + \frac{4}{\pi}\left(\frac{1}{7}\right)(20\text{ V})\sin 7\omega t$$

$$+ \frac{4}{\pi}\left(\frac{1}{9}\right)(20\text{ V})\sin 9\omega t + \frac{4}{\pi}\left(\frac{1}{11}\right)(20\text{ V})\sin 11\omega t$$

$$v = 25.465\sin\omega t + 8.488\sin 3\omega t + 5.093\sin 5\omega t + 3.638\sin 7\omega t + 2.829\sin 9\omega t + 2.315\sin 11\omega t$$

Eq. (25.11):

$$V_{\text{rms}} = \sqrt{V_0^2 + \frac{V_{m_1}^2 + V_{m_2}^2 + V_{m_3}^2 + V_{m_4}^2 + V_{m_5}^2 + V_{m_6}^2}{2}}$$

$$= \sqrt{(0\text{ V})^2 + \frac{(25.465\text{ V})^2 + (8.488\text{ V})^2 + (5.093\text{ V})^2 + (3.638\text{ V})^2 + (2.829\text{ V})^2 + (2.315\text{ V})^2}{2}}$$

$$= \mathbf{19.66\text{ V}}$$

The solution differs less than 0.4 V from the correct answer of 20 V. However, each additional term in the Fourier series brings the result closer to the 20 V level. An infinite number results in an exact solution of 20 V.

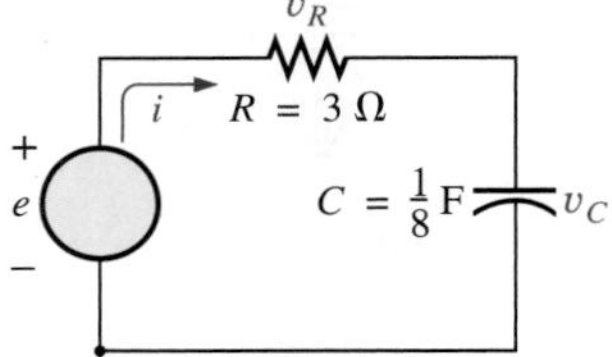

FIG. 25.22
Example 25.7.

EXAMPLE 25.7 The input to the circuit in Fig. 25.22 is the following:

$$e = 12 + 10\sin 2t$$

a. Find the current i and the voltages v_R and v_C.
b. Find the rms values of i, v_R, and v_C.
c. Find the power delivered to the circuit.

Solutions:

a. Redraw the original circuit as shown in Fig. 25.23. Then apply superposition:

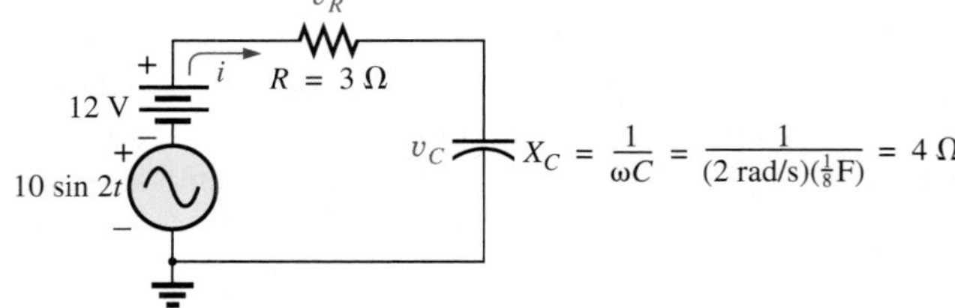

FIG. 25.23
Circuit in Fig. 25.22 with the components of the Fourier series input.

1. *For the 12 V dc supply portion of the input, $I = 0$* since the capacitor is an open circuit to dc when v_C has reached its final (steady-state) value. Therefore,

$$V_R = IR = 0\text{ V} \quad \text{and} \quad V_C = 12\text{ V}$$

2. *For the ac supply,*

$$\mathbf{Z} = 3\ \Omega - j\,4\ \Omega = 5\ \Omega \angle -53.13^\circ$$

and
$$\mathbf{I} = \frac{\mathbf{E}}{\mathbf{Z}} = \frac{\dfrac{10}{\sqrt{2}}\ \text{V} \angle 0^\circ}{5\ \Omega \angle -53.13^\circ} = \frac{2}{\sqrt{2}}\ \text{A} \angle +53.13^\circ$$

$$\mathbf{V}_R = (I \angle \theta)(R \angle 0^\circ) = \left(\frac{2}{\sqrt{2}}\ \text{A} \angle +53.13^\circ\right)(3\ \Omega \angle 0^\circ)$$
$$= \frac{6}{\sqrt{2}}\ \text{V} \angle +53.13^\circ$$

and

$$\mathbf{V}_C = (I \angle \theta)(X_C \angle -90^\circ) = \left(\frac{2}{\sqrt{2}}\ \text{A} \angle +53.13^\circ\right)(4\ \Omega \angle -90^\circ)$$
$$= \frac{8}{\sqrt{2}}\ \text{V} \angle -36.87^\circ$$

In the time domain,

$$i = \mathbf{0 + 2\ sin(2}t\ \mathbf{+ 53.13^\circ)}$$

Note that even though the dc term was present in the expression for the input voltage, the dc term for the current in this circuit is zero:

$$v_R = \mathbf{0 + 6\ sin(2}t\ \mathbf{+ 53.13^\circ)}$$

and
$$v_C = \mathbf{12 + 8\ sin(2}t\ \mathbf{- 36.87^\circ)}$$

b. Eq. (25.14): $I_{\text{rms}} = \sqrt{(0)^2 + \dfrac{(2\ \text{A})^2}{2}} = \sqrt{2}\ \text{A} = \mathbf{1.414\ A}$

Eq. (25.12): $V_{R_{\text{rms}}} = \sqrt{(0)^2 + \dfrac{(6\ \text{V})^2}{2}} = \sqrt{18}\ \text{V} = \mathbf{4.243\ V}$

Eq. (25.12): $V_{C_{\text{rms}}} = \sqrt{(12\ \text{V})^2 + \dfrac{(8\ \text{V})^2}{2}} = \sqrt{176}\ \text{V} = \mathbf{13.267\ V}$

c. $P = I_{\text{rms}}^2 R = \left(\dfrac{2}{\sqrt{2}}\ \text{A}\right)^2 (3\ \Omega) = \mathbf{6\ W}$

EXAMPLE 25.8 Find the response of the circuit in Fig. 25.24 to the input shown.

$$e = 0.318E_m + 0.500E_m \sin \omega t - 0.212E_m \cos 2\omega t - 0.0424E_m \cos 4\omega t + \ldots$$

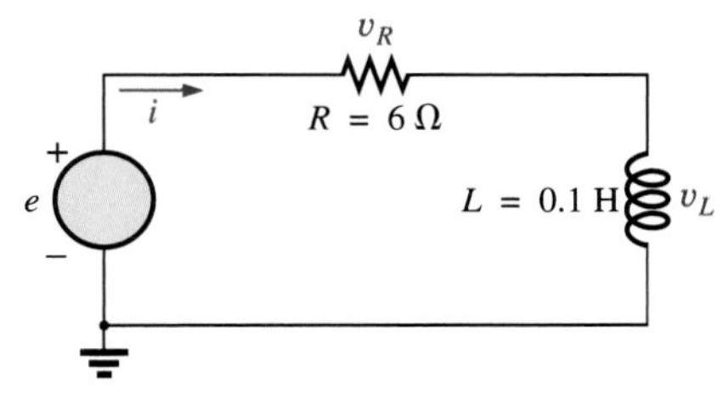

(a)

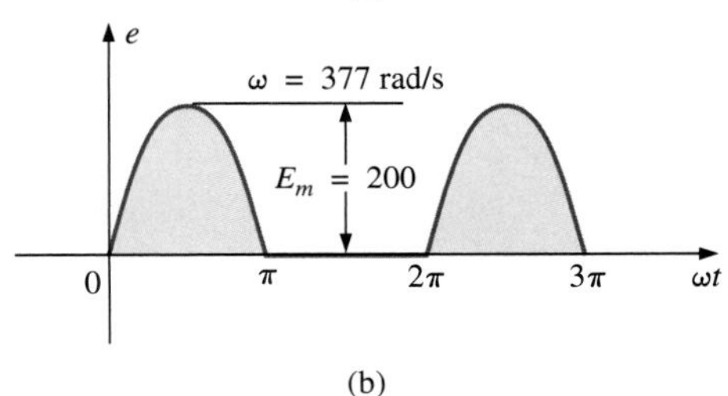

(b)

FIG. 25.24
Example 25.8.

Solution: For discussion purposes, only the first three terms are used to represent *e*. Converting the cosine terms to sine terms and substituting for E_m gives us

$$e = 63.60 + 100.0 \sin \omega t - 42.40 \sin(2\omega t + 90^\circ)$$

Using phasor notation, the original circuit becomes like the one shown in Fig. 25.25.

Applying Superposition *For the dc term* ($E_0 = 63.6$ V):

$$X_L = 0 \qquad \text{(short for dc)}$$
$$\mathbf{Z}_T = R \angle 0^\circ = 6\ \Omega \angle 0^\circ$$

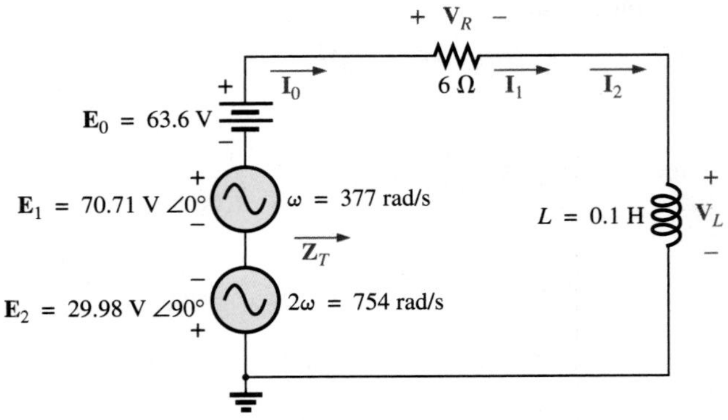

FIG. 25.25
Circuit in Fig. 25.24 with the components of the Fourier series input.

$$I_0 = \frac{E_0}{R} = \frac{63.6\text{ V}}{6\ \Omega} = 10.60\text{ A}$$
$$V_{R_0} = I_0 R = E_0 = 63.60\text{ V}$$
$$V_{L_0} = 0$$

The average power is

$$P_0 = I_0^2 R = (10.60\text{ A})^2(6\ \Omega) = 674.2\text{ W}$$

For the fundamental term ($\mathbf{E}_1 = 70.71\text{ V}\ \angle 0°$, $\omega = 377$):

$$X_{L_1} = \omega L = (377\text{ rad/s})(0.1\text{ H}) = 37.7\ \Omega$$
$$\mathbf{Z}_{T_1} = 6\ \Omega + j\,37.7\ \Omega = 38.17\ \Omega\ \angle 80.96°$$
$$\mathbf{I}_1 = \frac{\mathbf{E}_1}{\mathbf{Z}_{T_1}} = \frac{70.71\text{ V}\ \angle 0°}{38.17\ \Omega\ \angle 80.96°} = 1.85\text{ A}\ \angle -80.96°$$
$$\mathbf{V}_{R_1} = (I_1\ \angle\theta)(R\ \angle 0°) = (1.85\text{ A}\ \angle -80.96°)(6\ \Omega\ \angle 0°)$$
$$= 11.10\text{ V}\ \angle -80.96°$$
$$\mathbf{V}_{L_1} = (I_1\ \angle\theta)(X_{L_1}\ \angle 90°) = (1.85\text{ A}\ \angle -80.96°)(37.7\ \Omega\ \angle 90°)$$
$$= 69.75\text{ V}\ \angle 9.04°$$

The average power is

$$P_1 = I_1^2 R = (1.85\text{ A})^2(6\ \Omega) = 20.54\text{ W}$$

For the second harmonic ($\mathbf{E}_2 = 29.98\text{ V}\ \angle -90°$, $\omega = 754$): The phase angle of $\mathbf{E}_2$ was changed to $-90°$ to give it the same polarity as the input voltages $\mathbf{E}_0$ and $\mathbf{E}_1$.

$$X_{L_2} = \omega L = (754\text{ rad/s})(0.1\text{ H}) = 75.4\ \Omega$$
$$\mathbf{Z}_{T_2} = 6\ \Omega + j\,75.4\ \Omega = 75.64\ \Omega\ \angle 85.45°$$
$$\mathbf{I}_2 = \frac{\mathbf{E}_2}{\mathbf{Z}_{T_2}} = \frac{29.98\text{ V}\ \angle -90°}{75.64\ \Omega\ \angle 85.45°} = 0.396\text{ A}\ \angle -174.45°$$
$$\mathbf{V}_{R_2} = (I_2\ \angle\theta)(R\ \angle 0°) = (0.396\text{ A}\ \angle -174.45°)(6\ \Omega\ \angle 0°)$$
$$= 2.38\text{ V}\ \angle -174.45°$$
$$\mathbf{V}_{L_2} = (I_2\ \angle\theta)(X_{L_2}\ \angle 90°) = (0.396\text{ A}\ \angle -174.45°)(75.4\ \Omega\ \angle 90°)$$
$$= 29.9\text{ V}\ \angle -84.45°$$

The average power is

$$P_2 = I_2^2 R = (0.396\text{ A})^2(6\ \Omega) = 0.941\text{ W}$$

The Fourier series expansion for i is

$$i = \mathbf{10.6 + \sqrt{2}(1.85)\sin(377t - 80.96^\circ) + \sqrt{2}(0.396)\sin(754t - 174.45^\circ)}$$

and

$$I_{\text{rms}} = \sqrt{(10.6\text{ A})^2 + (1.85\text{ A})^2 + (0.396\text{ A})^2} = \mathbf{10.77\text{ A}}$$

The Fourier series expansion for v_R is

$$v_R = \mathbf{63.6 + \sqrt{2}(11.10)\sin(377t - 80.96^\circ) + \sqrt{2}(2.38)\sin(754t - 174.45^\circ)}$$

and

$$V_{R_{\text{rms}}} = \sqrt{(63.6\text{ V})^2 + (11.10\text{ V})^2 + (2.38\text{ V})^2} = \mathbf{64.61\text{ V}}$$

The Fourier series expansion for v_L is

$$v_L = \mathbf{\sqrt{2}(69.75)\sin(377t + 9.04^\circ) + \sqrt{2}(29.93)\sin(754t - 84.45^\circ)}$$

and $$V_{L_{\text{rms}}} = \sqrt{(69.75\text{ V})^2 + (29.93\text{ V})^2} = \mathbf{75.90\text{ V}}$$

The total average power is

$$P_T = I^2_{\text{rms}}R = (10.77\text{ A})^2(6\ \Omega) = \mathbf{695.96\text{ W}} = P_0 + P_1 + P_2$$

25.4 ADDITION AND SUBTRACTION OF NONSINUSOIDAL WAVEFORMS

The Fourier series expression for the waveform resulting from the addition or subtraction of two nonsinusoidal waveforms can be found using phasor algebra if the terms having the same frequency are considered separately.

For example, the sum of the following two nonsinusoidal waveforms is found using this method:

$$v_1 = 30 + 20\sin 20t + \cdots + 5\sin(60t + 30^\circ)$$
$$v_2 = 60 + 30\sin 20t + 20\sin 40t + 10\cos 60t$$

1. dc terms:

$$V_{T_0} = 30\text{ V} + 60\text{ V} = 90\text{ V}$$

2. $\omega = 20$:

$$V_{T_{1(\max)}} = 30\text{ V} + 20\text{ V} = 50\text{ V}$$

and $$v_{T_1} = 50\sin 20t$$

3. $\omega = 40$:

$$v_{T_2} = 20\sin 40t$$

4. $\omega = 60$:

$$5\sin(60t + 30^\circ) = (0.707)(5)\text{ V}\angle 30^\circ = 3.54\text{ V}\angle 30^\circ$$
$$10\cos 60t = 10\sin(60t + 90^\circ) \Rightarrow (0.707)(10)\text{ V}\angle 90^\circ = 7.07\text{ V}\angle 90^\circ$$

$$\mathbf{V}_{T_3} = 3.54\text{ V}\angle 30^\circ + 7.07\text{ V}\angle 90^\circ = 3.07\text{ V} + j\,1.77\text{ V} + j\,7.07\text{ V} = 3.07\text{ V} + j\,8.84\text{ V}$$
$$\mathbf{V}_{T_3} = 9.36\text{ V}\angle 70.85^\circ$$

and $$v_{T_3} = 13.24\sin(60t + 70.85^\circ)$$

with

$$v_T = v_1 + v_2 = \mathbf{90 + 50 \sin 20}\boldsymbol{t}\mathbf{ + 20 \sin 40}\boldsymbol{t}\mathbf{ + 13.24 \sin(60}\boldsymbol{t}\mathbf{ + 70.85°)}$$

25.5 COMPUTER ANALYSIS

PSpice

Fourier Series The computer analysis begins with a verification of the waveform in Fig. 25.15, demonstrating that only four terms of a Fourier series can generate a waveform that has a number of characteristics of a square wave. The square wave has a peak value of 10 V at a frequency of 1 kHz, resulting in the following Fourier series using Eq. (25.9) (and recognizing that $\omega = 2\pi f = 6283.19$ rad/s):

$$v = \frac{4}{\pi}(10\text{ V})\left(\sin \omega t + \frac{1}{3}\sin 3\omega t + \frac{1}{5}\sin 5\omega t + \frac{1}{7}\sin 7\omega t\right)$$
$$= 12.732 \sin \omega t + 4.244 \sin 3\omega t + 2.546 \sin 5\omega t + 1.819 \sin 7\omega t$$

Each term of the Fourier series is treated as an independent ac source as shown in Fig. 25.26 with its peak value and applicable frequency. The sum of the source voltages appears across the resistor R and generates the waveform in Fig. 25.27.

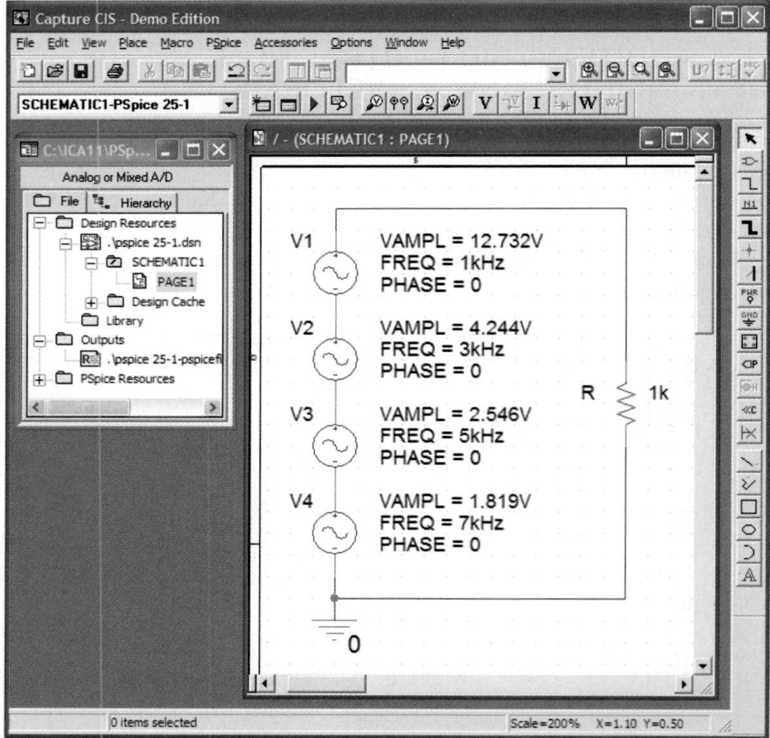

FIG. 25.26

Using PSpice to apply four terms of the Fourier expansion of a 10 V square wave to a load resistor of 1 kΩ.

Each source used **VSIN,** and since we want to display the result against time, choose **Time Domain(Transient)** in the **Simulation Settings.** For each source, select the **Property Editor** dialog box. Set **AC, FREQ, PHASE, VAMPL,** and **VOFF** (at 0 V). (Due to limited space, only **VAMPL, FREQ,** and **PHASE** are displayed in Fig. 25.26.) Under **Display,** set all of the remaining quantities on **Do Not Display.**

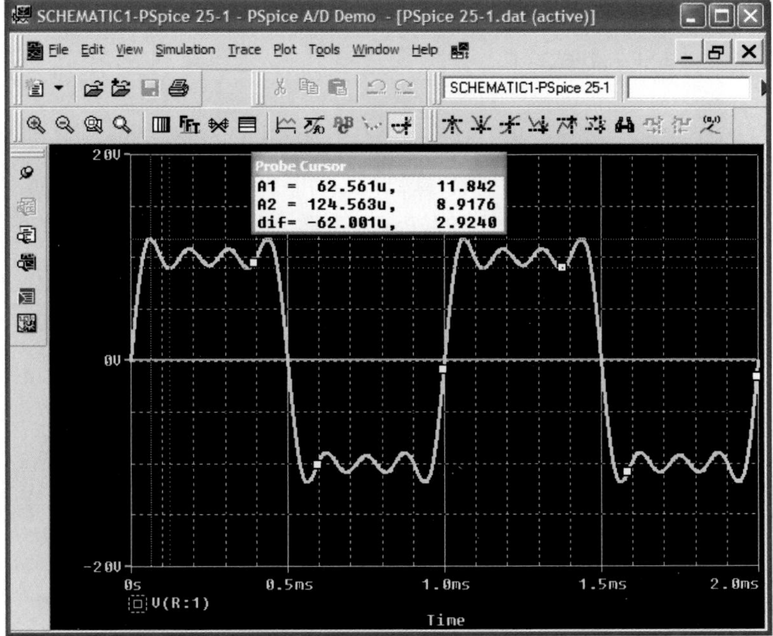

FIG. 25.27

The resulting waveform of the voltage across the resistor R in Fig. 25.26.

Set the **Run to time** at 2 ms so that two cycles of the fundamental frequency of 1 kHz appear. The **Start saving data after** remains at the default value of 0 s, and the **Maximum step size** at 1 μs, even though 2 ms/1000 = 2 μs, because we want to have additional plot points for the complex waveform. Once the **SCHEMATIC1** window appears, **Trace-Add Trace-V(R:1)-OK** results in the waveform in Fig. 25.27. To make the horizontal line at 0 V heavier, right-click on the line, select **Properties,** and then choose the green color and wider line. Click **OK,** and the wider line in Fig. 25.27 results, making it a great deal clearer where the 0 V line is located. Through the same process, make the curve yellow and wider as shown in the same figure. Using the cursors, you find that the first peak reaches 11.84 V and then drops to 8.920 V. The average value of the waveform is clearly +10 V in the positive region as shown by the dashed line entered using **Plot-Label-Line.** In every respect, the waveform is beginning to have the characteristics of a periodic square wave with a peak value of 10 V and a frequency of 1 kHz.

Fourier Components A frequency spectrum plot revealing the magnitude and frequency of each component of a Fourier series can be obtained by returning to **Plot** and selecting **Axis Settings** followed by **X Axis** and then **Fourier** under **Processing Options.** Click **OK,** and a number of spikes appear on the far left of the screen, with a frequency spectrum that extends from 0 Hz to 600 kHz. Select **Plot-Axis Settings** again, go to **Data Range,** and select **User Defined** to change the range to 0 Hz to 10 kHz since this is the range of interest for this waveform. Click **OK,** and the graph in Fig. 25.28 results, giving the magnitude and frequency of the components of the waveform. Using the left cursor, you find that the highest peak is 12.738 V at 1 kHz, comparing very well with the source **VI** having a peak value of 12.732 V at 1 kHz. Using the right-click cursor, you can move over to 3 kHz and find a magnitude of 4.246 V, again comparing very well with source **V2** with a peak value of 4.244 V.

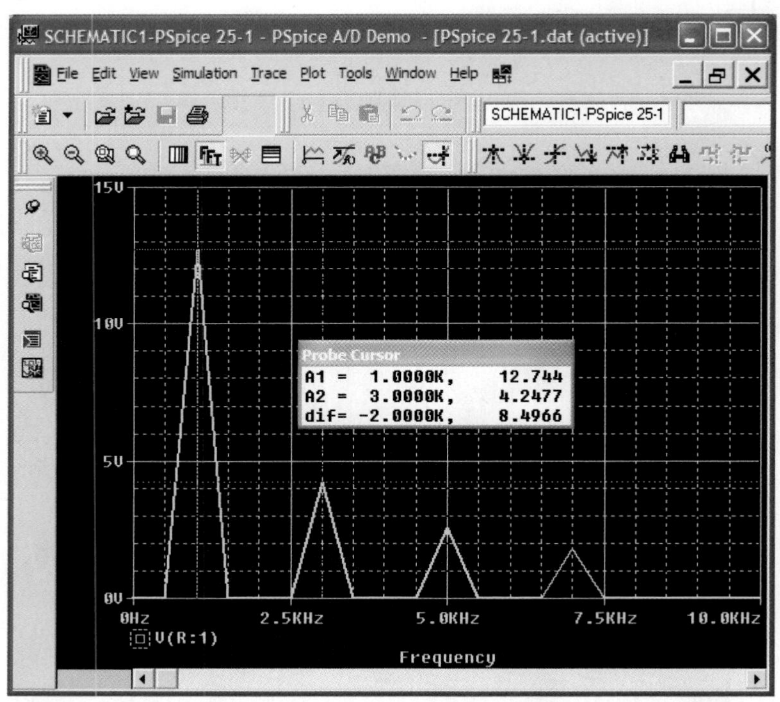

FIG. 25.28
The Fourier components of the waveform in Fig. 25.27.

PROBLEMS

SECTION 25.2 Fourier Series

1. For the waveforms in Fig. 25.29, determine whether the following will be present in the Fourier series representation:

   a. dc term
   b. cosine terms
   c. sine terms
   d. even-ordered harmonics
   e. odd-ordered harmonics

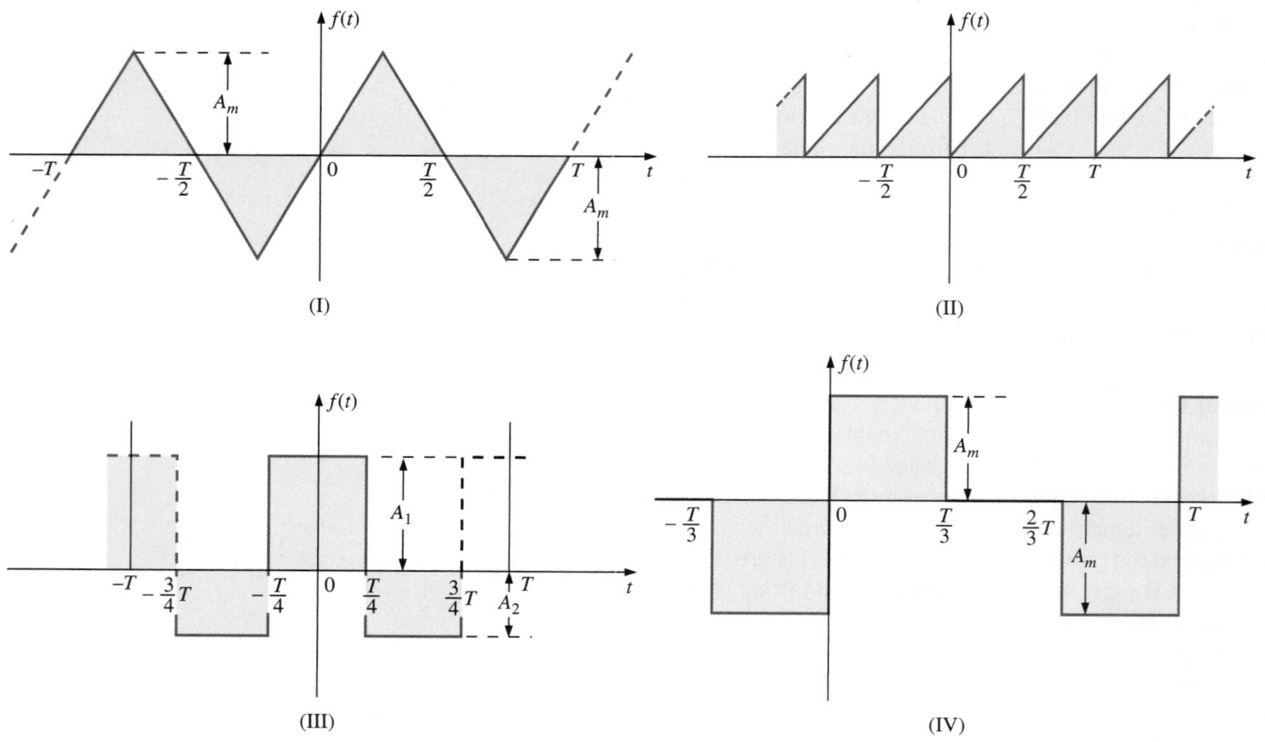

FIG. 25.29
Problem 1.

2. If the Fourier series for the waveform in Fig. 25.30(a) is

$$i = \frac{2I_m}{\pi}\left(1 + \frac{2}{3}\cos 2\omega t - \frac{2}{15}\cos 4\omega t + \frac{2}{35}\cos 6\omega t + \cdots\right.$$

find the Fourier series representation for waveforms (b) through (d).

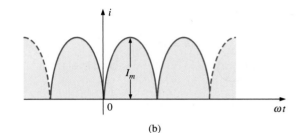

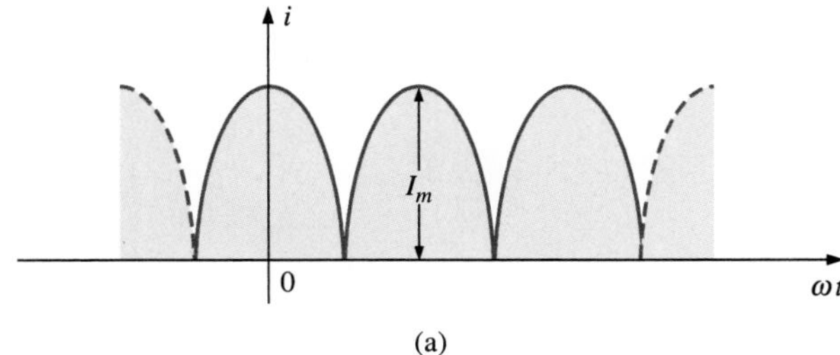

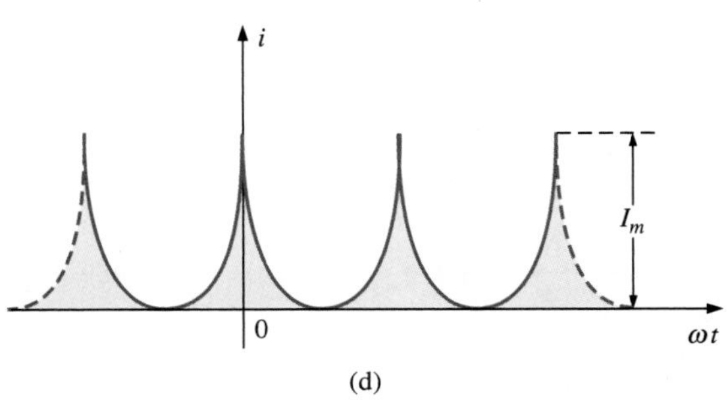

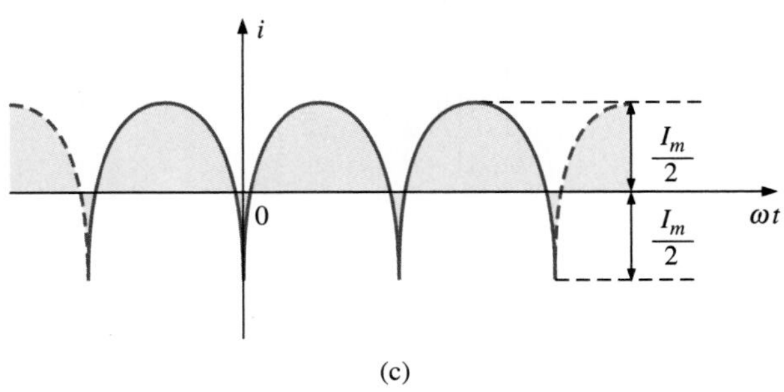

(d)

FIG. 25.30
Problem 2.

3. Sketch the following nonsinusoidal waveforms with $\alpha = \omega t$ as the abscissa:
   a. $v = -4 + 2 \sin \alpha$
   b. $v = (\sin \alpha)^2$
   c. $i = 2 - 2 \cos \alpha$
4. Sketch the following nonsinusoidal waveforms with α as the abscissa:
   a. $i = 3 \sin \alpha - 6 \sin 2\alpha$
   b. $v = 2 \cos 2\alpha + \sin \alpha$
5. Sketch the following nonsinusoidal waveforms with ωt as the abscissa:
   a. $i = 50 \sin \omega t + 25 \sin 3\omega t$
   b. $i = 50 \sin \alpha - 25 \sin 3\alpha$
   c. $i = 4 + 3 \sin \omega t + 2 \sin 2\omega t - 1 \sin 3\omega t$

SECTION 25.3 Circuit Response to a Nonsinusoidal Input

6. Find the average and effective values of the following nonsinusoidal waves:
   a. $v = 100 + 50 \sin \omega t + 25 \sin 2\omega t$
   b. $i = 3 + 2 \sin(\omega t - 53°) + 0.8 \sin(2\omega t - 70°)$
7. Find the rms value of the following nonsinusoidal waves:
   a. $v = 20 \sin \omega t + 15 \sin 2\omega t - 10 \sin 3\omega t$
   b. $i = 6 \sin(\omega t + 20°) + 2 \sin(2\omega t + 30°)$
      $-1 \sin(3\omega t + 60°)$
8. Find the total average power to a circuit whose voltage and current are as indicated in Problem 6.
9. Find the total average power to a circuit whose voltage and current are as indicated in Problem 7.
10. The Fourier series representation for the input voltage to the circuit in Fig. 25.31 is

$$e = 18 + 30 \sin 400t$$

   a. Find the nonsinusoidal expression for the current i.
   b. Calculate the rms value of the current.
   c. Find the expression for the voltage across the resistor.
   d. Calculate the rms value of the voltage across the resistor.
   e. Find the expression for the voltage across the reactive element.
   f. Calculate the rms value of the voltage across the reactive element.
   g. Find the average power delivered to the resistor.

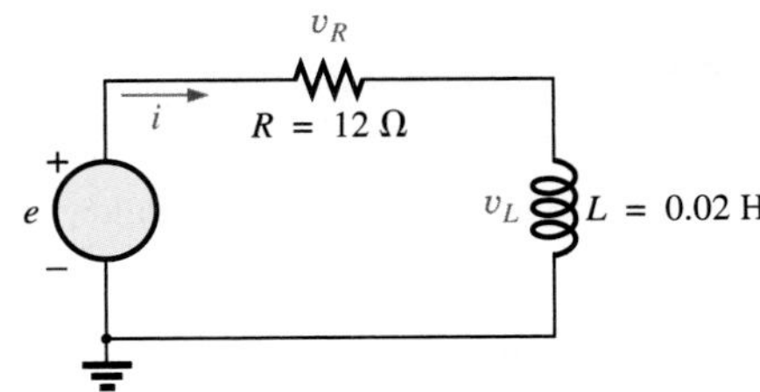

FIG. 25.31
Problems 10, 11, and 12.

11. Repeat Problem 10 for

$$e = 24 + 30 \sin 400t + 10 \sin 800t$$

12. Repeat Problem 10 for the following input voltage:

$$e = -60 + 20 \sin 300t - 10 \sin 600t$$

13. Repeat Problem 10 for the circuit in Fig. 25.32.

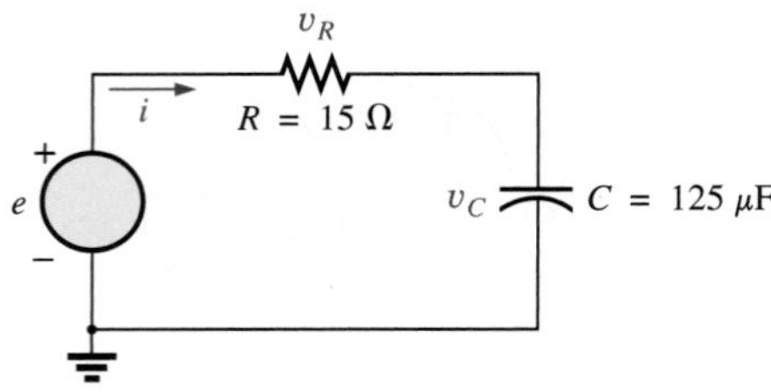

FIG. 25.32
Problem 13.

*14. The input voltage in Fig. 25.33(a) to the circuit in Fig. 25.33(b) is a full-wave rectified signal having the following Fourier series expansion:

$$e = \frac{(2)(100\text{ V})}{\pi}\left(1 + \frac{2}{3}\cos 2\omega t - \frac{2}{15}\cos 4\omega t + \frac{2}{53}\cos 6\omega t + \cdots\right)$$

where $\omega = 377$.

a. Find the Fourier series expression for the voltage v_o using only the first three terms of the expression.
b. Find the rms value of v_o.
c. Find the average power delivered to the 1 kΩ resistor.

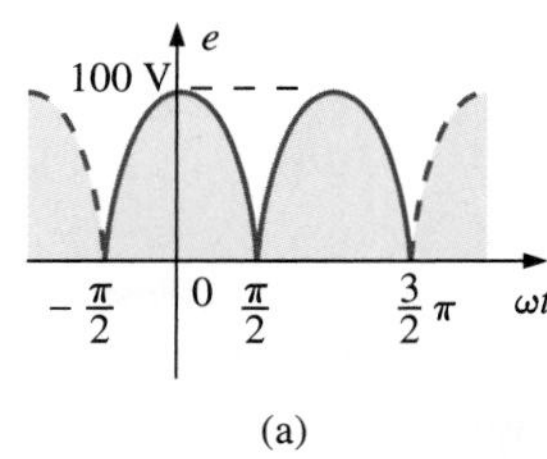

(a)

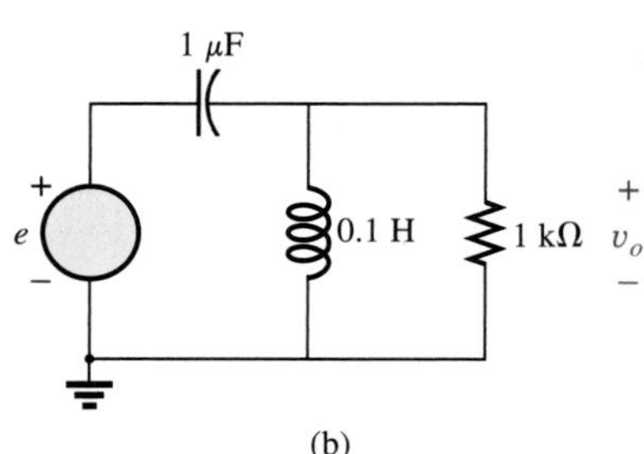

(b)

FIG. 25.33
Problem 14.

*15. Find the Fourier series expression for the voltage v_o in Fig. 25.34.

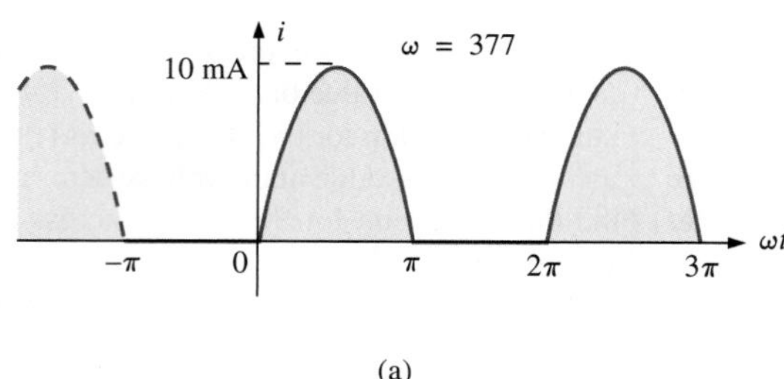

(a)

(b)

FIG. 25.34
Problem 15.

SECTION 25.4 Addition and Subtraction of Nonsinusoidal Waveforms

16. Perform the indicated operations on the following nonsinusoidal waveforms:
   a. $[60 + 70 \sin \omega t + 20 \sin(2\omega t + 90°) + 10 \sin(3\omega t + 60°)] + [20 + 30 \sin \omega t - 20 \cos 2\omega t + 5 \cos 3\omega t]$
   b. $[20 + 60 \sin \alpha + 10 \sin(2\alpha - 180°) + 5 \cos(3\alpha + 90°)] - [5 - 10 \sin \alpha + 4 \sin(3\alpha - 30°)]$

17. Find the nonsinusoidal expression for the current i_s of the diagram in Fig. 25.35.

$i_2 = 10 + 30 \sin 20t - 0.5 \sin(40t + 90°)$
$i_1 = 20 + 4 \sin(20t + 90°) + 0.5 \sin(40t + 30°)$

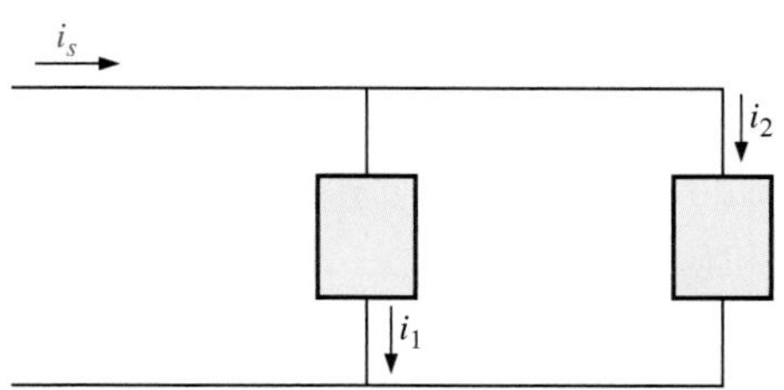

FIG. 25.35
Problem 17.

18. Find the nonsinusoidal expression for the voltage e of the diagram in Fig. 25.36.

$v_1 = 20 - 200 \sin 600t + 100 \cos 1200t + 75 \sin 1800t$
$v_2 = -10 + 150 \sin(600t + 30°) + 50 \sin(1800t + 60°)$

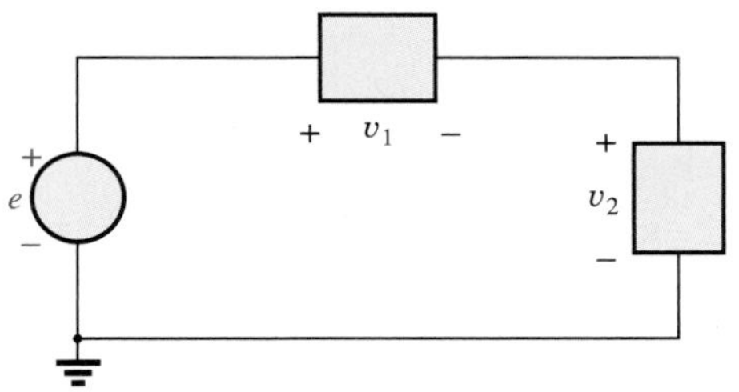

FIG. 25.36
Problem 18.

SECTION 25.5 Computer Analysis

PSpice

19. Plot the waveform in Fig. 25.11 for two or three cycles. Then obtain the Fourier components, and compare them to the applied signal.

20. Plot a half-rectified waveform with a peak value of 20 V using Eq. (25.10). Use the dc term, the fundamental term, and four harmonics. Compare the resulting waveform to the ideal half-rectified waveform.

21. Demonstrate the effect of adding two more terms to the waveform in Fig. 25.27, and generate the Fourier spectrum.

GLOSSARY

Axis symmetry A sinusoidal or nonsinusoidal function that has symmetry about the vertical axis.

Even harmonics The terms of the Fourier series expansion that have frequencies that are even multiples of the fundamental component.

Fourier series A series of terms, developed in 1826 by Baron Jean Fourier, that can be used to represent a nonsinusoidal function.

Fundamental component The minimum frequency term required to represent a particular waveform in the Fourier series expansion.

Half-wave (mirror) symmetry A sinusoidal or nonsinusoidal function that satisfies the relationship

$$f(t) = -f\left(t + \frac{T}{2}\right)$$

Harmonic terms The terms of the Fourier series expansion that have frequencies that are integer multiples of the fundamental component.

Nonsinusoidal waveform Any waveform that differs from the fundamental sinusoidal function.

Odd harmonics The terms of the Fourier series expansion that have frequencies that are odd multiples of the fundamental component.

Point symmetry A sinusoidal or nonsinusoidal function that satisfies the relationship $f(\alpha) = -f(-\alpha)$.

Appendixes

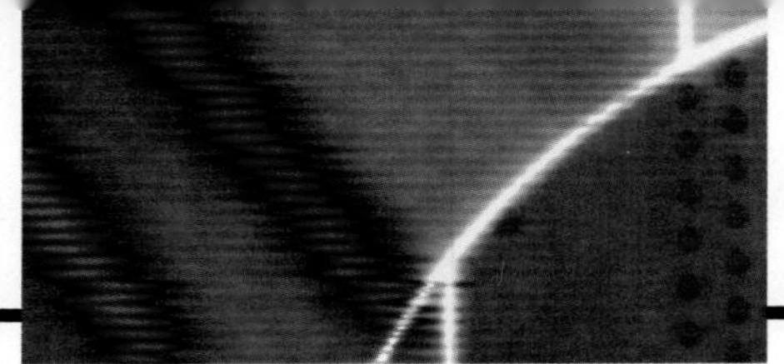

Appendix A

CONVERSION FACTORS

To Convert from	To	Multiply by
Btus	Calorie-grams	251.996
	Ergs	1.054×10^{10}
	Foot-pounds	777.649
	Hp-hours	0.000393
	Joules	1054,35
	Kilowatthours	0.000293
	Wattseconds	1054.35
Centimeters	Angstrom units	1×10^{8}
	Feet	0.0328
	Inches	0.3937
	Meters	0.01
	Miles (statute)	6.214×10^{-6}
	Millimeters	10
Circular mils	Square centimeters	5.067×10^{-6}
	Square inches	7.854×10^{-7}
Cubic inches	Cubic centimeters	16.387
	Gallons (U.S. liquid)	0.00433
Cubic meters	Cubic feet	35.315
Days	Hours	24
	Minutes	1440
	Seconds	86,400
Dynes	Gallons (U.S. liquid)	264.172
	Newtons	0.00001
	Pounds	2.248×10^{-6}
Electronvolts	Ergs	1.60209×10^{-12}
Ergs	Dyne-centimeters	1.0
	Electronvolts	6.242×10^{11}
	Foot-pounds	7.376×10^{-8}
	Joules	1×10^{-7}
	Kilowatthours	2.777×10^{-14}
Feet	Centimeters	30.48
	Meters	0.3048
Foot-candles	Lumens/square foot	1.0
	Lumens/square meter	10.764
Foot-pounds	Dyne-centimeters	1.3558×10^{7}
	Ergs	1.3558×10^{7}
	Horsepower-hours	5.050×10^{-7}
	Joules	1.3558
	Newton-meters	1.3558
Gallons (U.S. liquid)	Cubic inches	231
	Liters	3.785
	Ounces	128
	Pints	8

To Convert from	To	Multiply by
Gauss	Maxwells/square centimeter	1.0
	Lines/square centimeter	1.0
	Lines/square inch	6.4516
Gilberts	Ampere-turns	0.7958
Grams	Dynes	980.665
	Ounces	0.0353
	Pounds	0.0022
Horsepower	Btus/hour	2547.16
	Ergs/second	7.46×10^9
	Foot-pounds/second	550.221
	Joules/second	746
	Watts	746
Hours	Seconds	3600
Inches	Angstrom units	2.54×10^8
	Centimeters	2.54
	Feet	0.0833
	Meters	0.0254
Joules	Btus	0.000948
	Ergs	1×10^7
	Foot-pounds	0.7376
	Horsepower-hours	3.725×10^{-7}
	Kilowatthours	2.777×10^{-7}
	Wattseconds	1.0
Kilograms	Dynes	980,665
	Ounces	35.2
	Pounds	2.2
Lines	Maxwells	1.0
Lines/square centimeter	Gauss	1.0
Lines/square inch	Gauss	0.1550
	Webers/square inch	1×10^{-8}
Liters	Cubic centimeters	1000.028
	Cubic inches	61.025
	Gallons (U.S. liquid)	0.2642
	Ounces (U.S. liquid)	33.815
	Quarts (U.S. liquid)	1.0567
Lumens	Candle power (spher.)	0.0796
Lumens/square centimeter	Lamberts	1.0
Lumens/square foot	Foot-candles	1.0
Maxwells	Lines	1.0
	Webers	1×10^{-8}
Meters	Angstrom units	1×10^{10}
	Centimeters	100
	Feet	3.2808
	Inches	39.370
	Miles (statute)	0.000621

To Convert from	To	Multiply by
Miles (statute)	Feet	5280
	Kilometers	1.609
	Meters	1609.344
Miles/hour	Kilometers/hour	1.609344
Newton–meters	Dyne–centimeters	1×10^7
	Kilogram-meters	0.10197
Oersteds	Ampere-turns/inch	2.0212
	Ampere-turns/meter	79.577
	Gilberts/centimeter	1.0
Quarts (U.S. liquid)	Cubic centimeters	946.353
	Cubic inches	57.75
	Gallons (U.S. liquid)	0.25
	Liters	0.9463
	Pints (U.S. liquid)	2
	Ounces (U.S. liquid)	32
Radians	Degrees	57.2958
Slugs	Kilograms	14.5939
	Pounds	32.1740
Watts	Btus/hour	3.4144
	Ergs/second	1×10^7
	Horsepower	0.00134
	Joules/second	1.0
Webers	Lines	1×10^8
	Maxwells	1×10^8
Years	Days	365
	Hours	8760
	Minutes	525,600
	Seconds	3.1536×10^7

Appendix B

TI-86 CALCULATOR

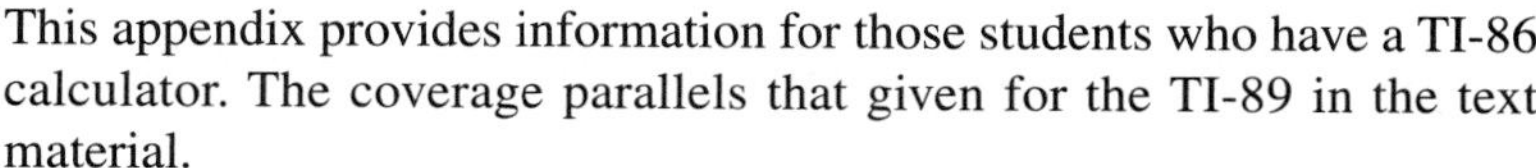

This appendix provides information for those students who have a TI-86 calculator. The coverage parallels that given for the TI-89 in the text material.

INITIAL SETTINGS

For the TI-86 calculator, pressing the 2nd function (yellow) key followed by the MODE key provides a list of options for the initial settings of the calculator. For each item in the MODE list, use the scroll keys to make a selection and then select the ENTER key.

ORDER OF OPERATIONS

The order of operations is the same as for the TI-89 calculator.

DETERMINANTS

The determinant operator is obtained by the sequence: **2nd FUNCTION-MATRX-MATH.** The **det** operator appears at the far left of the listing at the bottom of the screen. Select it by pressing the **F1** key. Once selected, enter all the parameters of the matrix within a set of brackets. Enter the first row of the determinant within a second set of brackets with a comma between each entry. Enter the second row in the same manner. After adding the closing bracket, select the **ENTER** key to provide the solution. Always be aware that the number of brackets forming a left enclosure must equal the number forming a right enclosure.

Following are the determinants for the current I_1 and the calculator input required:

$$I_1 = \frac{\begin{vmatrix} 2 & 4 \\ 6 & 5 \end{vmatrix}}{\begin{vmatrix} 6 & 4 \\ 4 & 5 \end{vmatrix}} = \frac{10 - 24}{30 - 16} = \frac{-14}{14} = \mathbf{-1\ A}$$

det[[2,4][6,5]]/det[[6,4][4,5]] ENTER	−1

Following is a third-order determinant and the calculator input required:

$$I_3 = I_{10\Omega} = \frac{\begin{vmatrix} 11 & -3 & 15 \\ -3 & 10 & 0 \\ -8 & -5 & 0 \end{vmatrix}}{\begin{vmatrix} 11 & -3 & -8 \\ -3 & 10 & -5 \\ -8 & -5 & 23 \end{vmatrix}} = \mathbf{1.22\ A}$$

det[[11,−3,15][−3, 10,0][−8,−5,0]]/det[[11,−3,−8][−3,10,−5][−8,−5,23]] ENTER	1.22

EXPONENTIAL FUNCTION

The exponential function e^x is obtained through the sequence 2nd Function e^x.

For the equation

$$v = 20\text{ V}(1 - e^{-2.5})$$

the calculator sequence would be the following:

20 (1 − 2nd e^x (−) 2.5) **ENTER**	18.3

LOGARITHMS

The natural logarithm ($\log_e$) is obtained using the LN key. That is, $\log_e 2$ is obtained using the following sequence:

LN 2 **ENTER**	693.15E−3

The common logarithm ($\log_{10}$) is obtained using the LOG key. That is, $\log_{10}2$ is obtained using the following sequence:

LOG 2 **ENTER**	301.03E−3

The antilogarithm of a natural logarithm permits working backwards on the above operation. That is, it provides the number we have to take the natural logarithm of to obtain the given number. For instance, the natural logarithm of what number will result in 693.15E-3? This is expressed in equation form as follows:

LN (?) = 693.15E−3

The result is obtained using the exponential function in the following manner:

$$e^{693.15\text{E}-3} = 2.00$$

The calculator sequence is:

2nd e^x 693.15EE(−)3 **ENTER**	2.00

The common logarithm,

LOG (?) = 301.03E−3

is processed with

2nd 10^x 301.03EE(−)3 **ENTER**	2.00

TRIGONOMETRIC FUNCTIONS

Degrees vs. Radians

The choice is made by selecting one or the other from the **MODE** listing. Don't forget the **ENTER** operation before leaving the screen!

Sin, Cos, and Tan

Select the appropriate key followed by the angle in degrees as in the following:

SIN 30 ENTER	0.50

Sin^{-1}, Cos^{-1}, and Tan^{-1}

The 2nd function must be used to obtain the desired function as shown below:

2nd SIN^{-1} 0.5 ENTER	30

COMPLEX NUMBERS

To convert complex numbers on the TI-86 calculator, you must first call up the 2nd function **CPLX** from the keyboard, which results in a menu at the bottom of the display including conj, real, imag, abs, and angle. If you choose the key **MORE, ▶ Rec** and **▶ Pol** appear as options (for the conversion process). To convert from one form to another, enter the current form in brackets with a comma between components for the rectangular form and an angle symbol for the polar form. Follow this form with the operation to be performed, and select **ENTER.** The result appears on the screen in the desired format.

To convert $3 - j4$ to polar form:

(3, −4) ▶ Pol ENTER	(5.00E0 ∠ −53.13E0)

To convert $0.006^{20.6}$ to rectangular form:

(0.006 ∠20.6) ▶ Rec ENTER	(5.62E−3, 2.11E−3)

To solve this equation:

$$\frac{(2\angle 20^\circ)^2(3 + j\,4)}{8 - j\,6} = ?$$

use this sequence:

((2 ∠20)^2 * (3,4))/(8,−6) ▶ Pol ENTER	(2.00E0 ∠130.00E0)

Appendix C

PSPICE, MULTISIM, AND MATHCAD

PSPICE 10.0

The PSpice software package used throughout this text is derived from programs developed at the University of California at Berkeley during the early 1970s. SPICE is an acronym for Simulation Program with Integrated Circuit Emphasis. Although a number of companies have customized SPICE for their particular use, Cadence Design Systems offers both a commercial and a demo version of OrCAD. The commercial or professional versions that engineering companies use can be quite expensive, so Cadence offers a free distribution of a demo version to provide an introduction to the power of this simulation package. This text uses the OrCAD 10.0 Demo version. An application for a free demo version is available online through the web site www.orcad.com. Select **OrCAD Capture** and then **OrCAD Demo CD Request**. Fill out the provided form to receive a copy of the demo version.

You can also contact EMA Mid Atlantic at 1-877-EMA-EDA1 or visit the EMA web site at www.ema-eda.com

Minimum system requirements:

Pentium 4 (32-bit) equivalent or faster
Edition, or Windows 2000 (SP4)
256 MB RAM (512 MB recommended)
300 MB swap space (or more)
CD-ROM drive
32,768 color Windows display

MULTISIM 8

Multisim is a product of Electronics Workbench, a subsidiary of National Instruments. Its web site is www.electronicsworkbench.com and phone number in the U.S. and Canada is 1 800-263-5552. In Europe, the web site is www.ewbeurope.com and phone number is 31-35-694-4444.

Copies can be purchased at the above web sites and through Prentice Hall at www.prenhall.com.

Minimum system requirements:

Windows 2000/XP (Windows XP Professional recommended)
Pentium III (Pentium 4 recommended)
128 MB RAM (256 MB recommended)
150 MB hard disk space (500 MB recommended)
CD-ROM drive
800×600 screen resolution (1024×768 recommended)

MATHCAD 12

Mathcad is a product of MathSoft Engineering and Education, Inc. located at 101 Main Street, Cambridge, MA 02142-1521. The web site is www.mathsoft.com. The academic version of Mathcad is available online for purchase at www.edu.com and www.journeyed.com. Professors can purchase the current Mathcad 12 Academic Edition by calling the customer service department at 800-628-4223 or 617-444-8000 or email sales-info@mathsoft.com.

Minimum system requirements for Mathcad 12:

PC with 300 MHz or higher processor clock speed (400 MHz or higher recommended)
Windows 2000, XP, or higher
128 MB RAM (256 MB or higher recommended)
Video card and monitor support for 16-bit or higher color depth and 800×600 or higher screen resolution
100 MB disk space
CD-ROM or DVD drive

Appendix D

DETERMINANTS

Determinants are used to find the mathematical solutions for the variables in two or more simultaneous equations. Once the procedure is properly understood, solutions can be obtained with a minimum of time and effort and usually with fewer errors than when using other methods.

Consider the following equations, where x and y are the unknown variables and a_1, a_2, b_1, b_2, c_1, and c_2 are constants:

$$\begin{array}{ccccc} \text{Col. 1} & & \text{Col. 2} & & \text{Col. 3} \\ \hline a_1x & + & b_1y & = & c_1 \\ a_2x & + & b_2y & = & c_2 \end{array} \qquad \begin{array}{r} \textbf{(D.1a)} \\ \textbf{(D.1b)} \end{array}$$

It is certainly possible to solve for one variable in Eq. (D.1a) and substitute into Eq. (D.1b). That is, solving for x in Eq. (D.1a),

$$x = \frac{c_1 - b_1y}{a_1}$$

and substituting the result in Eq. (D.1b),

$$a_2\left(\frac{c_1 - b_1y}{a_1}\right) + b_2y = c_2$$

It is now possible to solve for y, since it is the only variable remaining, and then substitute into either equation for x. This is acceptable for two equations, but it becomes a very tedious and lengthy process for three or more simultaneous equations.

Using determinants to solve for x and y requires that the following formats be established for each variable:

$$\begin{array}{cc} \begin{array}{cc} \text{Col.} & \text{Col.} \\ 1 & 2 \end{array} & \begin{array}{cc} \text{Col.} & \text{Col.} \\ 1 & 2 \end{array} \\ \hline x = \dfrac{\begin{vmatrix} c_1 & b_1 \\ c_2 & b_2 \end{vmatrix}}{\begin{vmatrix} a_1 & b_1 \\ a_2 & b_2 \end{vmatrix}} & y = \dfrac{\begin{vmatrix} a_1 & c_1 \\ a_2 & c_2 \end{vmatrix}}{\begin{vmatrix} a_1 & b_1 \\ a_2 & b_2 \end{vmatrix}} \end{array} \qquad \textbf{(D.2)}$$

First note that only constants appear within the vertical brackets and that the denominator of each is the same. In fact, the denominator is simply the coefficients of x and y in the same arrangement as in Eqs. (D.1a) and (D.1b). When solving for x, replace the coefficients of x in the numerator by the constants to the right of the equal sign in Eqs. (D.1a) and (D.1b), and repeat the coefficients of the y variable. When solving for y, replace the y coefficients in the numerator by the constants to the right of the equal sign, and repeat the coefficients of x.

Each configuration in the numerator and denominator of Eq. (D.2) is referred to as a *determinant (D)*, which can be evaluated numerically in the following manner:

$$\text{Determinant} = D = \begin{array}{c} \begin{array}{cc} \text{Col.} & \text{Col.} \\ \underline{1} & \underline{2} \end{array} \\ \begin{vmatrix} a_1 & b_1 \\ a_2 & b_2 \end{vmatrix} \end{array} = a_1b_2 - a_2b_1 \qquad \textbf{(D.3)}$$

The expanded value is obtained by first multiplying the top left element by the bottom right and then subtracting the product of the lower left and upper right elements. This particular determinant is referred to as a *second-order* determinant, since it contains two rows and two columns.

It is important to remember when using determinants that the columns of the equations, as indicated in Eqs. (D.1a) and (D.1b), must be placed in the same order within the determinant configuration. That is, since a_1 and a_2 are in column 1 of Eqs. (D.1a) and (D.1b), they must be in column 1 of the determinant. (The same is true for b_1 and b_2.)

Expanding the entire expression for x and y, we have the following:

$$x = \frac{\begin{vmatrix} c_1 & b_1 \\ c_2 & b_2 \end{vmatrix}}{\begin{vmatrix} a_1 & b_1 \\ a_2 & b_2 \end{vmatrix}} = \frac{c_1b_2 - c_2b_1}{a_1b_2 - a_2b_1} \qquad \textbf{(D.4a)}$$

$$y = \frac{\begin{vmatrix} a_1 & c_1 \\ a_2 & c_2 \end{vmatrix}}{\begin{vmatrix} a_1 & b_1 \\ a_2 & b_2 \end{vmatrix}} = \frac{a_1c_2 - a_2c_1}{a_1b_2 - a_2b_1} \qquad \textbf{(D.4b)}$$

EXAMPLE D.1 Evaluate the following determinants:

a. $\begin{vmatrix} 2 & 2 \\ 3 & 4 \end{vmatrix} = (2)(4) - (3)(2) = 8 - 6 = \mathbf{2}$

b. $\begin{vmatrix} 4 & -1 \\ 6 & 2 \end{vmatrix} = (4)(2) - (6)(-1) = 8 + 6 = \mathbf{14}$

c. $\begin{vmatrix} 0 & -2 \\ -2 & 4 \end{vmatrix} = (0)(4) - (-2)(-2) = 0 - 4 = \mathbf{-4}$

d. $\begin{vmatrix} 0 & 0 \\ 3 & 10 \end{vmatrix} = (0)(10) - (3)(0) = \mathbf{0}$

EXAMPLE D.2 Solve for x and y:

$$\begin{aligned} 2x + y &= 3 \\ 3x + 4y &= 2 \end{aligned}$$

Solution:

$$x = \frac{\begin{vmatrix} 3 & 1 \\ 2 & 4 \end{vmatrix}}{\begin{vmatrix} 2 & 1 \\ 3 & 4 \end{vmatrix}} = \frac{(3)(4) - (2)(1)}{(2)(4) - (3)(1)} = \frac{12 - 2}{8 - 3} = \frac{10}{5} = \mathbf{2}$$

$$y = \frac{\begin{vmatrix} 2 & 3 \\ 3 & 2 \end{vmatrix}}{5} = \frac{(2)(2) - (3)(3)}{5} = \frac{4 - 9}{5} = \frac{-5}{5} = \mathbf{-1}$$

Check:

$$\begin{aligned} 2x + y &= (2)(2) + (-1) \\ &= 4 - 1 = 3 \qquad \text{(checks)} \\ 3x + 4y &= (3)(2) + (4)(-1) \\ &= 6 - 4 = 2 \qquad \text{(checks)} \end{aligned}$$

EXAMPLE D.3 Solve for x and y:

$$\begin{aligned} -x + 2y &= 3 \\ 3x - 2y &= -2 \end{aligned}$$

Solution: In this example, note the effect of the minus sign and the use of parentheses to ensure that the proper sign is obtained for each product:

$$x = \frac{\begin{vmatrix} 3 & 2 \\ -2 & -2 \end{vmatrix}}{\begin{vmatrix} -1 & 2 \\ 3 & -2 \end{vmatrix}} = \frac{(3)(-2) - (-2)(2)}{(-1)(-2) - (3)(2)}$$

$$= \frac{-6 + 4}{2 - 6} = \frac{-2}{-4} = \mathbf{\frac{1}{2}}$$

$$y = \frac{\begin{vmatrix} -1 & 3 \\ 3 & -2 \end{vmatrix}}{-4} = \frac{(-1)(-2) - (3)(3)}{-4}$$

$$= \frac{2 - 9}{-4} = \frac{-7}{-4} = \mathbf{\frac{7}{4}}$$

EXAMPLE D.4 Solve for x and y:

$$\begin{aligned} x &= 3 - 4y \\ 20y &= -1 + 3x \end{aligned}$$

Solution: In this case, the equations must first be placed in the format of Eqs. (D.1a) and (D.1b):

$$\begin{aligned} x + 4y &= 3 \\ -3x + 20y &= -1 \end{aligned}$$

$$x = \frac{\begin{vmatrix} 3 & 4 \\ -1 & 20 \end{vmatrix}}{\begin{vmatrix} 1 & 4 \\ -3 & 20 \end{vmatrix}} = \frac{(3)(20) - (-1)(4)}{(1)(20) - (-3)(4)}$$

$$= \frac{60 + 4}{20 + 12} = \frac{64}{32} = \mathbf{2}$$

$$y = \frac{\begin{vmatrix} 1 & 3 \\ -3 & -1 \end{vmatrix}}{32} = \frac{(1)(-1) - (-3)(3)}{32}$$

$$= \frac{-1 + 9}{32} = \frac{8}{32} = \mathbf{\frac{1}{4}}$$

The use of determinants is not limited to the solution of two simultaneous equations; determinants can be applied to any number of simultaneous linear equations. First we examine a shorthand method that is applicable to third-order determinants only, since most of the problems in the text are limited to this level of difficulty. We then investigate the general procedure for solving any number of simultaneous equations.

Consider the three following simultaneous equations:

Col. 1		Col. 2		Col. 3		Col. 4
a_1x	$+$	b_1y	$+$	c_1z	$=$	d_1
a_2x	$+$	b_2y	$+$	c_2z	$=$	d_2
a_3x	$+$	b_3y	$+$	c_3z	$=$	d_3

in which x, y, and z are the variables, and $a_{1,\,2,\,3}$, $b_{1,\,2,\,3}$, $c_{1,\,2,\,3}$, and $d_{1,\,2,\,3}$ are constants.

The determinant configuration for x, y, and z can be found in a manner similar to that for two simultaneous equations. That is, to solve for x, find the determinant in the numerator by replacing column 1 with the elements to the right of the equal sign. The denominator is the determinant of the coefficients of the variables (the same applies to y and z). Again, the denominator is the same for each variable.

$$x = \frac{\begin{vmatrix} d_1 & b_1 & c_1 \\ d_2 & b_2 & c_2 \\ d_3 & b_3 & c_3 \end{vmatrix}}{D}, \quad y = \frac{\begin{vmatrix} a_1 & d_1 & c_1 \\ a_2 & d_2 & c_2 \\ a_3 & d_3 & c_3 \end{vmatrix}}{D}, \quad z = \frac{\begin{vmatrix} a_1 & b_1 & d_1 \\ a_2 & b_2 & d_2 \\ a_3 & b_3 & d_3 \end{vmatrix}}{D}$$

where

$$D = \begin{vmatrix} a_1 & b_1 & c_1 \\ a_2 & b_2 & c_2 \\ a_3 & b_3 & c_3 \end{vmatrix}$$

A shorthand method for evaluating the third-order determinant consists of repeating the first two columns of the determinant to the right of the determinant and then summing the products along specific diagonals as shown below:

$$D = \begin{vmatrix} a_1 & b_1 & c_1 \\ a_2 & b_2 & c_2 \\ a_3 & b_3 & c_3 \end{vmatrix} \begin{matrix} a_1 & b_1 \\ a_2 & b_2 \\ a_3 & b_3 \end{matrix}$$

4(−) 5(−) 6(−)

1(+) 2(+) 3(+)

The products of the diagonals 1, 2, and 3 are positive and have the following magnitudes:

$$+a_1b_2c_3 + b_1c_2a_3 + c_1a_2b_3$$

The products of the diagonals 4, 5, and 6 are negative and have the following magnitudes:

$$-a_3b_2c_1 - b_3c_2a_1 - c_3a_2b_1$$

The total solution is the sum of the diagonals 1, 2, and 3 minus the sum of the diagonals 4, 5, and 6:

$$+ (a_1b_2c_3 + b_1c_2a_3 + c_1a_2b_3) - (a_3b_2c_1 + b_3c_2c_1 + c_3a_2b_1) \qquad \textbf{(D.5)}$$

Warning: This method of expansion is good only for third-order determinants! It cannot be applied to fourth- and higher-order systems.

EXAMPLE D.5 Evaluate the following determinant:

$$\begin{vmatrix} 1 & 2 & 3 \\ -2 & 1 & 0 \\ 0 & 4 & 2 \end{vmatrix} \rightarrow \left|\begin{array}{ccc|cc} 1 & 2 & 3 & 1 & 2 \\ -2 & 1 & 0 & -2 & 1 \\ 0 & 4 & 2 & 0 & 4 \end{array}\right. \quad \begin{matrix} (-) & (-) & (-) \\ (+) & (+) & (+) \end{matrix}$$

Solution:

$$\begin{aligned} &[(1)(1)(2) + (2)(0)(0) + (3)(-2)(4)] \\ &\qquad - [(0)(1)(3) + (4)(0)(1) + (2)(-2)(2)] \\ &\qquad = (2 + 0 - 24) - (0 + 0 - 8) = (-22) - (-8) \\ &\qquad = -22 + 8 = \mathbf{-14} \end{aligned}$$

EXAMPLE D.6 Solve for x, y, and z:

$$\begin{aligned} 1x + 0y - 2z &= -1 \\ 0x + 3y + 1z &= +2 \\ 1x + 2y + 3z &= 0 \end{aligned}$$

Solution:

$$x = \frac{\left|\begin{array}{ccc|cc} -1 & 0 & -2 & -1 & 0 \\ 2 & 3 & 1 & 2 & 3 \\ 0 & 2 & 3 & 0 & 2 \end{array}\right.}{\left|\begin{array}{ccc|cc} 1 & 0 & -2 & 1 & 0 \\ 0 & 3 & 1 & 0 & 3 \\ 1 & 2 & 3 & 1 & 2 \end{array}\right.}$$

$$= \frac{[(-1)(3)(3) + (0)(1)(0) + (-2)(2)(2)] - [(0)(3)(-2) + (2)(1)(-1) + (3)(2)(0)]}{[(1)(3)(3) + (0)(1)(1) + (-2)(0)(2)] - [(1)(3)(-2) + (2)(1)(1) + (3)(0)(0)]}$$

$$= \frac{(-9 + 0 - 8) - (0 - 2 + 0)}{(9 + 0 + 0) - (-6 + 2 + 0)}$$

$$= \frac{-17 + 2}{9 + 4} = -\frac{\mathbf{15}}{\mathbf{13}}$$

$$y = \frac{\left|\begin{array}{ccc|cc} 1 & -1 & -2 & 1 & -1 \\ 0 & 2 & 1 & 0 & 2 \\ 1 & 0 & 3 & 1 & 0 \end{array}\right.}{13}$$

$$= \frac{[(1)(2)(3) + (-1)(1)(1) + (-2)(0)(0)] - [(1)(2)(-2) + (0)(1)(1) + (3)(0)(-1)]}{13}$$

$$= \frac{(6 - 1 + 0) - (-4 + 0 + 0)}{13}$$

$$= \frac{5 + 4}{13} = \frac{\mathbf{9}}{\mathbf{13}}$$

$$z = \frac{\left|\begin{array}{ccc} 1 & 0 & -1 \\ 0 & 3 & 2 \\ 1 & 2 & 0 \end{array}\right|\begin{array}{cc} 1 & 0 \\ 0 & 3 \\ 1 & 2 \end{array}}{13}$$

$$= \frac{[(1)(3)(0) + (0)(2)(1) + (-1)(0)(2)] - [(1)(3)(-1) + (2)(2)(1) + (0)(0)(0)]}{13}$$

$$= \frac{(0 + 0 + 0) - (-3 + 4 + 0)}{13}$$

$$= \frac{0 - 1}{13} = \mathbf{-\frac{1}{13}}$$

or from $0x + 3y + 1z = +2$,

$$z = 2 - 3y = 2 - 3\left(\frac{9}{13}\right) = \frac{26}{13} - \frac{27}{13} = \mathbf{-\frac{1}{13}}$$

Check:

$$\begin{array}{l} 1x + 0y - 2z = -1 \\[2ex] 0x + 3y + 1z = +2 \\[2ex] 1x + 2y + 3z = 0 \end{array} \left\{ \begin{array}{l} -\frac{15}{13} + 0 + \frac{2}{13} = -1 \\[2ex] 0 + \frac{27}{13} + \frac{-1}{13} = +2 \\[2ex] -\frac{15}{13} + \frac{18}{13} + \frac{-3}{13} = 0 \end{array} \right\} \begin{array}{l} -\frac{13}{13} = -1 \checkmark \\[2ex] \frac{26}{13} = +2 \checkmark \\[2ex] -\frac{18}{13} + \frac{18}{13} = 0 \checkmark \end{array}$$

The general approach to third-order or higher determinants requires that the determinant be expanded in the following form. There is more than one expansion that will generate the correct result, but this form is typically used when the material is first introduced.

$$D = \begin{vmatrix} a_1 & b_1 & c_1 \\ a_2 & b_2 & c_2 \\ a_3 & b_3 & c_3 \end{vmatrix} = \underset{\substack{\uparrow \\ \text{Multiplying} \\ \text{factor}}}{a_1} \underbrace{\left(+\underbrace{\begin{vmatrix} b_2 & c_2 \\ b_3 & c_3 \end{vmatrix}}_{\text{Minor}}\right)}_{\text{Cofactor}} + \underset{\substack{\uparrow \\ \text{Multiplying} \\ \text{factor}}}{b_1} \underbrace{\left(-\underbrace{\begin{vmatrix} a_2 & c_2 \\ a_3 & c_3 \end{vmatrix}}_{\text{Minor}}\right)}_{\text{Cofactor}} + \underset{\substack{\uparrow \\ \text{Multiplying} \\ \text{factor}}}{c_1} \underbrace{\left(+\underbrace{\begin{vmatrix} a_2 & b_2 \\ a_3 & b_3 \end{vmatrix}}_{\text{Minor}}\right)}_{\text{Cofactor}}$$

This expansion was obtained by multiplying the elements of the first row of D by their corresponding cofactors. It is not a requirement that the first row be used as the multiplying factors. In fact, any *row* or *column* (not diagonals) may be used to expand a third-order determinant.

The sign of each cofactor is dictated by the position of the multiplying factors (a_1, b_1, and c_1 in this case) as in the following standard format:

$$\left|\begin{array}{ccc} + \rightarrow & - & + \\ \downarrow & & \\ - & + & - \\ + & - & + \end{array}\right|$$

Note that the proper sign for each element can be obtained by assigning the upper left element a positive sign and then changing signs as you move horizontally or vertically to the neighboring position.

For the determinant D, the elements would have the following signs:

$$\begin{vmatrix} a_1^{(+)} & b_1^{(-)} & c_1^{(+)} \\ a_2^{(-)} & b_2^{(+)} & c_2^{(-)} \\ a_3^{(+)} & b_3^{(-)} & c_3^{(+)} \end{vmatrix}$$

The minors associated with each multiplying factor are obtained by covering up the row and column in which the multiplying factor is located and writing a second-order determinant to include the remaining elements in the same relative positions that they have in the third-order determinant.

Consider the cofactors associated with a_1 and b_1 in the expansion of D. The sign is positive for a_1 and negative for b_1 as determined by the standard format. Following the procedure outlined above, we can find the minors of a_1 and b_1 as follows:

$$a_{1(\text{minor})} = \begin{vmatrix} a_1 & b_1 & c_1 \\ a_2 & b_2 & c_2 \\ a_3 & b_3 & c_3 \end{vmatrix} = \begin{vmatrix} b_2 & c_2 \\ b_3 & c_3 \end{vmatrix}$$

$$b_{1(\text{minor})} = \begin{vmatrix} a_1 & b_1 & c_1 \\ a_2 & b_2 & c_2 \\ a_3 & b_3 & c_3 \end{vmatrix} = \begin{vmatrix} a_2 & c_2 \\ a_3 & c_3 \end{vmatrix}$$

It was pointed out that any row or column may be used to expand the third-order determinant, and the same result will still be obtained. Using the first column of D, we obtain the expansion

$$D = \begin{vmatrix} a_1 & b_1 & c_1 \\ a_2 & b_2 & c_2 \\ a_3 & b_3 & c_3 \end{vmatrix} = a_1\left(+\begin{vmatrix} b_2 & c_2 \\ b_3 & c_3 \end{vmatrix}\right) + a_2\left(-\begin{vmatrix} b_1 & c_1 \\ b_3 & c_3 \end{vmatrix}\right) + a_3\left(+\begin{vmatrix} b_1 & c_1 \\ b_2 & c_2 \end{vmatrix}\right)$$

The proper choice of row or column can often effectively reduce the amount of work required to expand the third-order determinant. For example, in the following determinants, the first column and third row, respectively, would reduce the number of cofactors in the expansion:

$$D = \begin{vmatrix} 2 & 3 & -2 \\ 0 & 4 & 5 \\ 0 & 6 & 7 \end{vmatrix} = 2\left(+\begin{vmatrix} 4 & 5 \\ 6 & 7 \end{vmatrix}\right) + 0 + 0 = 2(28 - 30)$$
$$= \mathbf{-4}$$

$$D = \begin{vmatrix} 1 & 4 & 7 \\ 2 & 6 & 8 \\ 2 & 0 & 3 \end{vmatrix} = 2\left(+\begin{vmatrix} 4 & 7 \\ 6 & 8 \end{vmatrix}\right) + 0 + 3\left(+\begin{vmatrix} 1 & 4 \\ 2 & 6 \end{vmatrix}\right)$$
$$= 2(32 - 42) + 3(6 - 8) = 2(-10) + 3(-2)$$
$$= \mathbf{-26}$$

EXAMPLE D.7 Expand the following third-order determinants:

a. $$D = \begin{vmatrix} 1 & 2 & 3 \\ 3 & 2 & 1 \\ 2 & 1 & 3 \end{vmatrix} = 1\left(+\begin{vmatrix} 2 & 1 \\ 1 & 3 \end{vmatrix}\right) + 3\left(-\begin{vmatrix} 2 & 3 \\ 1 & 3 \end{vmatrix}\right) + 2\left(+\begin{vmatrix} 2 & 3 \\ 2 & 1 \end{vmatrix}\right)$$

$$= 1[6 - 1] + 3[-(6 - 3)] + 2[2 - 6]$$
$$= 5 + 3(-3) + 2(-4)$$
$$= 5 - 9 - 8$$
$$= \mathbf{-12}$$

b. $$D = \begin{vmatrix} 0 & 4 & 6 \\ 2 & 0 & 5 \\ 8 & 4 & 0 \end{vmatrix} = 0 + 2\left(-\begin{vmatrix} 4 & 6 \\ 4 & 0 \end{vmatrix}\right) + 8\left(+\begin{vmatrix} 4 & 6 \\ 0 & 5 \end{vmatrix}\right)$$

$$= 0 + 2[-(0 - 24)] + 8[(20 - 0)]$$
$$= 0 + 2(24) + 8(20)$$
$$= 48 + 160$$
$$= \mathbf{208}$$

Appendix E

GREEK ALPHABET

Letter	Capital	Lowercase	Some Applications
Alpha	A	α	Area, angles, coefficients
Beta	B	β	Angles, coefficients, flux density
Gamma	Γ	γ	Specific gravity, conductivity
Delta	Δ	δ	Density, variation
Epsilon	E	ε	Base of natural logarithms
Zeta	Z	ζ	Coefficients, coordinates, impedance
Eta	H	η	Efficiency, hysteresis coefficient
Theta	Θ	θ	Phase angle, temperature
Iota	I	ι	
Kappa	K	κ	Dielectric constant, susceptibility
Lambda	Λ	λ	Wavelength
Mu	M	μ	Amplification factor, micro, permeability
Nu	N	ν	Reluctivity
Xi	Ξ	ξ	
Omicron	O	o	
Pi	Π	π	3.1416
Rho	P	ρ	Resistivity
Sigma	Σ	σ	Summation
Tau	T	τ	Time constant
Upsilon	Υ	υ	
Phi	Φ	ϕ	Angles, magnetic flux
Chi	X	χ	
Psi	Ψ	ψ	Dielectric flux, phase difference
Omega	Ω	ω	Ohms, angular velocity

Appendix F

MAGNETIC PARAMETER CONVERSIONS

	SI (MKS)	CGS	English
Φ	webers (Wb) 1 Wb	maxwells $= 10^8$ maxwells	lines $= 10^8$ lines
B	T $1\text{T} = 1\ \text{Wb/m}^2$	gauss (maxwells/cm^2) $= 10^4$ gauss	lines/in.^{2} $= 6.452 \times 10^4$ lines/in.^{2}
A	$1\ \text{m}^2$	$= 10^4\ \text{cm}^2$	$= 1550\ \text{in.}^2$
μ_o	$4\pi \times 10^{-7}$ Wb/Am	= 1 gauss/oersted	= 3.20 lines/Am
$\mathscr{F}$	NI (ampere-turns, At) 1 At	$0.4\pi NI$ (gilberts) = 1.257 gilberts	NI (At) 1 gilbert = 0.7958 At
H	NI/l (At/m) 1At/m	$0.4\pi NI/l$ (oersteds) $= 1.26 \times 10^{-2}$ oersted	NI/l(At/in.) $= 2.54 \times 10^{-2}$ At/in.
H_g	$7.97 \times 10^5 B_g$ (At/m)	B_g (oersteds)	$0.313B_g$ (At/in.)

Appendix G

MAXIMUM POWER TRANSFER CONDITIONS

Derivation of maximum power transfer conditions for the situation where the resistive component of the load is adjustable, but the load reactance is set in magnitude.*

For the circuit in Fig. G.1, the power delivered to the load is determined by

$$P = \frac{V_{R_L}^2}{R_L}$$

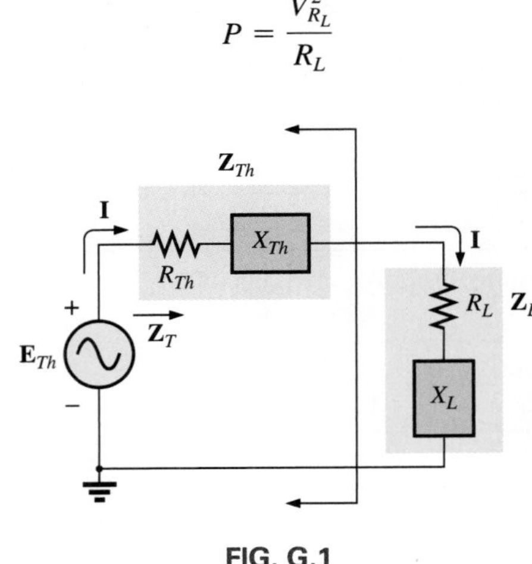

FIG. G.1

Applying the voltage divider rule:

$$\mathbf{V}_{R_L} = \frac{R_L\mathbf{E}_{Th}}{R_L + R_{Th} + X_{Th}\angle 90^\circ + X_L\angle 90^\circ}$$

The magnitude of $\mathbf{V}_{R_L}$ is determined by

$$V_{R_L} = \frac{R_L E_{Th}}{\sqrt{(R_L + R_{Th})^2 + (X_{Th} + X_L)^2}}$$

and
$$V_{R_L}^2 = \frac{R_L^2 E_{Th}^2}{(R_L + R_{Th})^2 + (X_{Th} + X_L)^2}$$

with
$$P = \frac{V_{R_L}^2}{R_L} = \frac{R_L E_{Th}^2}{(R_L + R_{Th})^2 + (X_{Th} + X_L)^2}$$

Using differentiation (calculus), maximum power will be transferred when $dP/dR_L = 0$. The result of the preceding operation is that

$$R_L = \sqrt{R_{Th}^2 + (X_{Th} + X_L)^2} \qquad \text{[Eq. (18.21)]}$$

The magnitude of the total impedance of the circuit is

$$Z_T = \sqrt{(R_{Th} + R_L)^2 + (X_{Th} + X_L)^2}$$

*With sincerest thanks for the input of Professor Harry J. Franz of the Beaver Campus of Pennsylvania State University.

Substituting this equation for R_L and applying a few algebraic maneuvers will result in

$$Z_T = 2R_L(R_L + R_{Th})$$

and the power to the load R_L will be

$$P = I^2R_L = \frac{E_{Th}^2}{Z_T^2}R_L = \frac{E_{Th}^2R_L}{2R_L(R_L + R_{Th})}$$

$$= \frac{E_{Th}^2}{4\left(\dfrac{R_L + R_{Th}}{2}\right)}$$

$$= \frac{E_{Th}^2}{4R_{av}}$$

with $$R_{av} = \frac{R_L + R_{Th}}{2}$$

Appendix H

ANSWERS TO SELECTED ODD-NUMBERED PROBLEMS

Chapter 1

5. 29.05 mph
11. 737.8 ft-lbs
13. **(a)** 10^4 **(b)** 10^6 **(c)** 10^3 **(d)** 10^{-3} **(e)** 10^0 **(f)** 10^{-1}
15. **(a)** 5.22×10^4 **(b)** 4.50×10^5 **(c)** 4.4×10^{-4} **(d)** 6.5×10^2
17. **(a)** 1.5×10^1 **(b)** 4.4 **(c)** 2.296×10^2 **(d)** 8.40×10^3
19. **(a)** 2.50×10^7 **(b)** 6.67×10^{-8} **(c)** 4.4 **(d)** 1.95×10^{25}
21. **(a)** 1.6×10^5 **(b)** 2.16×10^{-7} **(c)** 1.44×10^3 **(d)** 1.11×10^{11}
23. **(a)** 300 **(b)** 200.0×10^3 **(c)** 9.0×10^{12} **(d)** 150.0×10^{-9} **(e)** 24.0×10^{12} **(f)** 800.0×10^{18} **(g)** 56.4×10^3
25. **(a)** 50 ms **(b)** 2 ms **(c)** 40 μs **(d)** 0.0084 μs **(e)** 4000 mm **(f)** 0.26 km
27. **(a)** 10^5 pF **(b)** 8 cm **(c)** 60×10^{-5} km **(d)** 11.52×10^6 ms **(e)** 16 μm **(f)** 60×10^{-4} m^2
29. 5,280 ft, 1760 yds, 1609.35 m, 1.61 km
31. 3.40 s
33. 73.33 days
35. 3600 quarters
37. 345.6 m
39. 47.30 min/mi
41. **(a)** 4.74×10^{-3} Btu **(b)** 7.1×10^{-4} m^3 **(c)** 1.21×10^5 s **(d)** 2113.38 pints
43. 13
45. 0.64
47. 2.95
49. 1.20×10^{12}

Chapter 2

3. **(a)** 1.11 μN **(b)** 0.31 N **(c)** 1138.34 kN
5. 10 mm
7. 3 kV
9. 6 C
11. 4.29 mA
13. 1.92 C
15. 3 s
17. 2.25×10^{18} electrons
19. 22.43 mA
21. 6.67 V
23. 3.34 A
25. 60.0 Ah
35. 600 C

Chapter 3

1. **(a)** 500 mils **(b)** 20 mils **(c)** 250 mils **(d)** 1000 mils **(e)** 240 mils **(f)** 39.37 mils
3. **(a)** 0.04 in. **(b)** 0.029 in. **(c)** 0.2 in. **(d)** 0.025 in. **(e)** 0.0025 in. **(f)** 0.01 in.
5. 92.81 Ω
7. 3.58 ft
9. **(a)** silver **(b)** silver: 99 Ω; copper: 5.19 Ω; aluminum: 1.36 Ω
11. **(a)** 21.71 $\mu\Omega$ **(b)** 35.59 $\mu\Omega$
13. 942.28 mΩ
15. **(a)** #8: 1.13 Ω; #18: 11.49 Ω **(b)** #18: #8 = 10 : 1 **(c)** #18: #8 = 1 : 10
17. **(a)** 1.09 mA/CM **(b)** 1.39 kA/in.^{2} **(c)** 3.6 in.^{2}
19. **(a)** 21.71 $\mu\Omega$ **(b)** 35.59 $\mu\Omega$
21. 0.15 in.
23. 2.57 Ω
25. 3.67 Ω
27. 46 mΩ
29. **(a)** 27.85°C **(b)** −210.65°C
31. **(a)** 0.00393 **(b)** 83.61°C
33. 1.751 Ω
35. 100.30 Ω
41. 6.5 kΩ
45. **(a)** Brown, red, brown, silver **(b)** Gray, red, gold, silver **(c)** Blue, gray, red, silver **(d)** Orange, orange, green, silver
47. no overlap
49. **(a)** 8.33 mS **(b)** 0.25 mS **(c)** 0.46 μS
51. 500 S
55. **(a)** −50°C: 10^5 Ω-cm 50°C: 500 Ω-cm 200°C: 7 Ω-cm **(b)** negative **(c)** No **(d)** 3.6 Ω-cm/°C
57. **(a)** 0.5 mA: 195 V; 1 mA: 200 V; 5 mA: 215 V **(b)** 20 V **(c)** 10 : 1 vs 1.08 : 1

Chapter 4

1. 117.5 V
3. 4 kΩ
5. 72 mV
7. 54.55 Ω
9. 28.57 Ω
11. 1.2 kΩ
13. **(a)** 12.63 Ω **(b)** 4.1×10^6 J
21. 16 s
23. 250 W
25. 4.8 W
27. 10.44 mA
29. 2.14 mA
31. 461.27 μA, no
33. 405 mW
35. 32 Ω, 120 V
37. 70.71 mA, 1.42 kV
39. 59.80 kWh
41. 0.65¢
43. 44.44 h
45. **(a)** 4.1 W **(b)** 19.78 Ω **(c)** 88.56 kJ
47. 74.21¢
49. 94.43%
51. 84.77%
53. 16.06 A
55. 56.52 A
57. 65.25%
59. 80%
61. $\eta_1 = 40\%$, $\eta_2 = 80\%$

Chapter 5

1. **(a)** E and R_1 **(b)** R_1 and R_2 **(c)** E and R_1 **(d)** E and R_1, R_3 and R_4
3. **(a)** 7.7 kΩ **(b)** 17.5 kΩ
5. **(a)** 62 Ω **(b)** 1.8 kΩ **(c)** 27 kΩ **(d)** $R_1 = 8$ kΩ, $R_2 = 16$ kΩ
7. **(a)** 40 Ω **(b)** 3 A **(c)** $V_1 = 30$ V; $V_2 = 36$ V; $V_3 = 54$ V
9. **(a)** 88 V **(b)** 20 V
11. **(a)** 8.18 mA, 18 V **(b)** 2.5 mA, 20 V **(c)** 9.94 μA, 99.35 V
13. **(a)** 82.0 Ω, 250 mA; $V_{R_1} = 5.50$ V; $V_{R_2} = 2.50$ V; $V_{R_3} = 11.75$ V; $V_{R_4} = 0.75$ V **(b)** $P_{R_1} = 1.38$ W; $P_{R_2} = 625.00$ mW; $P_{R_3} = 2.94$ W; $P_{R_4} = 187.50$ mW **(c)** 5.13 W **(d)** 5.13 W **(e)** same **(f)** 47 Ω **(g)** dissipated **(h)** R_1: 2 W; R_2: $\frac{1}{2}$ W; R_3: 5 W; R_4: 1/2 W

15. **(a)** 0.53 A **(b)** 8 W **(c)** 15 V **(d)** all out
17. **(a)** $V_{ab} = 0$ V **(b)** $V_{ab} = -6$ V **(c)** $V_{ab} = 14$ V
19. **(a)** 10 V, 2 kΩ **(b)** 42 V, 1.5 kΩ
21. **(a)** 28 V **(b)** 4 V
23. **(a)** $V_1 = 9$ V; $V_2 = 8$ V **(b)** $V_1 = 11$ V; $V_2 = 7$ V
25. **(a)** 8.2 kΩ **(b)** V_3: $V_2 = 8.2 : 1$ V_3: $V_1 = 82 : 1$ **(c)** 52.90 V **(d)** 59.35 V
27. **(a)** $V_1 = 60$ V; $V_2 = 40$ V; $E = 120$ V **(b)** $V_1 = 40$ V; $V_3 = 70$ V **(c)** $V_2 = 20$ V; $V_1 = 10$ V; $E = 1030$ V **(d)** $V_1 = 10$ V; $V_2 = 3$ V
29. **(a)** 1.6 kΩ **(b)** 1.5 Ω
31. **(a)** $R_x = 80$ Ω **(b)** $\frac{1}{4}$ W
33. $V_{R_1} = 12$ V; $V_{R_2} = 42$ V; $V_{R_3} = 6$ V
35. **(a)** $V_a = 4$ V; $V_b = -8$ V; $V_{ab} = 12$ V **(b)** $V_a = 14$ V; $V_b = 4$ V; $V_{ab} = 10$ V **(c)** $V_a = 13$ V; $V_b = 6$ V; $V_{ab} = 7$ V
37. **(a)** $V_a = 12$ V; $V_1 = 8$ V **(b)** $V_a = 10$ V; $V_1 = 12$ V
39. $R_1 = 8$ Ω; $R_3 = 4$ Ω
41. **(a)** $V_a = 44$ V; $V_b = 40$ V; $V_c = 32$ V; $V_d = 20$ V **(b)** $V_{ab} = 4$ V; $V_{cb} = -8$ V; $V_{cd} = 12$ V **(c)** $V_{ad} = 24$ V; $V_{ca} = -12$ V
43. $V_0 = 0$ V; $V_{03} = 0$ V; $V_2 = 8$ V; $V_{23} = 8$ V; $V_{12} = 12$ V; $I_i = 17$ mA
45. **(a)** 11.82 V **(b)** 1.52% **(c)** $P_s = 42.96$ W; $P_{int} = 0.64$ W

Chapter 6

1. **(a)** R_2 and R_3 **(b)** E and R_3 **(c)** E and R_1 **(d)** R_2, R_3, and R_4 **(e)** E, R_1, R_2, R_3, and R_4 **(f)** E, R_1, R_2, and R_3 **(g)** R_2 and R_3
3. **(a)** 6.04 Ω **(b)** 545.55 Ω **(c)** 90.09 Ω **(d)** 5.99 kΩ **(e)** 2.62 Ω **(f)** 0.99 Ω
5. **(a)** 8 Ω **(b)** 18 kΩ **(c)** 20 kΩ **(d)** 3.3 kΩ **(e)** $R_1 = R_2 =$ 6.4 kΩ; $R_3 = 3.2$ kΩ
7. **(a)** 1.6 Ω **(b)** ∞ Ω **(c)** ∞ Ω **(d)** 1.18 Ω
9. 120 Ω
11. **(a)** 2.12 Ω **(b)** 18 V **(c)** $I_s = 8.5$ A; $I_1 = 6$ A; $I_2 = 2$ A; $I_3 = 0.5$ A **(d)** $I_s = 8.5$ A $= I_T$
13. **(a)** 1 kΩ **(b)** 1.003 kΩ **(c)** I_3 the most; I_4 the least **(d)** $I_{R_1} = 4.4$ mA; $I_{R_2} = 2$ mA; $I_{R_3} = 36.67$ mA; $I_{R_4} = 0.79$ mA **(e)** $I_s = 43.87$ mA $= I_T$ **(f)** always greater
15. $V = 12$ V; $I' = 12$ A; $I'' = 8$ A
17. **(a)** $I_s = 7.5$ A; $I_1 = 1.5$ A **(b)** $I_s = 9.6$ mA; $I_1 = 0.8$ mA
19. **(a)** $R_T = 867.86$ Ω; $I_{R_1} = 100$ mA; $I_{R_2} = 3.03$ mA; $I_{R_3} = 12.2$ mA **(b)** $P_{R_1} = 10$ W; $P_{R_2} = 0.30$ W; $P_{R_3} = 1.22$ W **(c)** 115.2 W **(d)** $P_s = 11.52$ W $= P_T$ **(e)** R_1—smallest resistor
21. 1.26 kW
23. **(a)** 14.67 A **(b)** 256 W **(c)** 14.67 A
25. **(a)** $I_1 = 1$ A; $I_2 = 3$ A **(b)** $I_1 = 4$ A; $I_2 = 9$ A; $I_3 = 6$ A; $I_4 = 13$ A
27. $R_1 = 3$ kΩ; $R_3 = 6$ kΩ; $R_T = 1.33$ kΩ; $E = 12$ V
29. $I_2 = 2$ A; $I_3 = 12$ A; $I_4 = 0.6$ A; $I_T = 20.6$ A
31. **(a)** 9 A **(b)** 10, 0.9 A **(c)** 1000, 9 mA **(d)** 100,000, 90 μA **(e)** little effect **(f)** 9.1 A **(g)** 0.91 A **(h)** 9.1 mA **(i)** 91 μA
33. **(a)** 6 kΩ **(b)** $I_1 = 24$ mA; $I_2 = 8$ mA
35. **(a)** $I_1 = I_2 = 3$ A **(b)** 36 W **(c)** 72 W **(d)** 6 A
37. 3 A, 2 Ω
39. **(a)** 6.13 V **(b)** 9 V **(c)** 9 V
41. **(a)** 16.48 V **(b)** 16.47 V **(c)** 16.32 V **(d)** *a:* 13.33 V *b:* 13.25 V *c:* 11.43 V
43. No, 6 kΩ not connected
45. **(a)** 1 kΩ not connected **(b)** Used +4 V source

Chapter 7

1. **(a)** E and R_1 in series; R_2, R_3, and R_4 in parallel **(b)** E and R_1 in series; R_2, R_3, and R_4 in parallel **(c)** R_1 and R_2 in series; E, R_3, and R_4 in parallel **(d)** E and R_1 in series; R_4 and R_5 in series; R_2 and R_3 in parallel **(e)** E and R_1 in series; R_2 and R_3 in parallel **(f)** E, R_1, and R_4 in parallel; R_6 and R_7 in series; R_2 and R_3 in parallel.
3. **(a)** yes **(b)** 6 A **(c)** yes **(d)** 6 V **(e)** 3.73 Ω **(f)** 1 A **(g)** 20 W
5. **(a)** 4 Ω **(b)** $I_s = 9$ A; $I_1 = 6$ A; $I_2 = 3$ A **(c)** 6 V
7. **(a)** $I_s = 16$ mA; $I_2 = 2.33$ mA; $I_6 = 2$ mA **(b)** $V_1 = 28$ V; $V_5 = 7.2$ V **(c)** 261.33 mW
9. $I_1 = 4$ A; $I_2 = 0.72$ A
11. **(a)** $I_s = 5$ A; $I_1 = 1$ A; $I_3 = 4$ A; $I_4 = 0.5$ A **(b)** $V_a = 17$ V; $V_{bc} = 10$ V
13. **(a)** $I_E = 2$ mA $= I_C$ **(b)** 24 μA **(c)** $V_B = 2.7$ V; $V_C = 3.6$ V **(d)** $V_{CE} = 1.6$ V; $V_{BC} = -0.9$ V
15. **(a)** 174.12 Ω **(b)** 11.89 V **(c)** 20.11 V **(d)** 11.89 V **(e)** 20.54 mA
17. **(a)** $I_2 = 1.67$ A; $I_6 = 1.11$ A; $I_8 = 0$ A **(b)** $V_4 = 10$ V; $V_8 = 0$ V
19. **(a)** 1.88 Ω **(b)** $V_1 = V_4 = 32$ V **(c)** $I_3 = 8$ A **(d)** 1.88 Ω
21. **(a)** 6.75 A **(b)** 32 V
23. 8.33 Ω
25. **(a)** 24 A **(b)** 8 A **(c)** $V_3 = 48$ V; $V_5 = 24$ V; $V_7 = 16$ V **(d)** $P_{R_7} = 128$ W; $P_s = 5760$ W
27. 4.44 W
29. **(a)** 64 V **(b)** $R_{L_2} = 4$ kΩ $R_{L_3} = 3$ kΩ **(c)** $R_1 = 0.5$ kΩ $R_2 = 1.2$ kΩ $R_3 = 2$ kΩ
31. **(a)** yes **(b)** $R_1 = 750$ Ω; $R_2 = 250$ Ω **(c)** $R_1 = 745$ Ω; $R_2 = 255$ Ω
33. **(a)** 1 mA **(b)** $R_{shunt} = 5$ mΩ
35. **(a)** $R_s = 300$ kΩ **(b)** 20,000 Ω/V
37. 0.05 μA

Chapter 8

1. **(a)** $I_2 = I_3 = 10$ mA **(b)** 10 V **(c)** 37.6 V
3. 28 V
5. 1.6 V, 0.1 A
7. **(a)** 3 A, 6 Ω **(b)** 4.09 mA, 2.2 kΩ
9. **(a)** 11.76 A **(b)** 1.2 kV, 100 Ω
11. **(a)** 2 A **(b)** 8 V
13. **(b)** $V_{ab} = -6.44$ V **(c)** 1.07 A
15. **(a)** $I_{R_1} = -\frac{1}{7}$ A; $I_{R_2} = \frac{5}{7}$ A; $I_{R_3} = \frac{4}{7}$ A **(b)** $I_{R_1} = 3.06$ A; $I_{R_2} = 3.25$ A; $I_{R_3} = 0.19$ A
17. **(I):** 8.55 A, −22.75 V **(II):** 1.27 A, −0.92 V
19. **(a)** $I_B = 63.02$ μA; $I_C = 4.42$ mA; $I_E = 4.48$ mA **(b)** $V_B = 2.98$ V;

$V_E = 2.28$ V; $V_C = 10.28$ V
(c) 70.14
21. **(I):** $I_{R_1} = 1.45$ mA; $I_{R_2} = 8.51$ mA; $I_{R_3} = 7.06$ mA **(II):** $I_{R_1} = 2.03$ mA; $I_{R_2} = 0.80$ mA; $I_{R_3} = I_{R_4} = 1.23$ mA
23. **(a)** 63.69 mA
25. **(I):** $I_1 = 1.21$ mA; $I_2 = -0.48$ mA; $I_3 = -0.62$ mA (all CW) **(II):** $I_1 = -0.24$ A; $I_2 = -0.52$ A; $I_3 = -1.28$ A (all CW)
27. **(a)** $I_{24\,V} = I_{6\,\Omega} = I_{10\,\Omega} = I_{12\,V} = 3$ A (CW); $I_{4\,\Omega} = 3$ A (CCW) **(b)** $I_{20\,V} = I_{4\,\Omega} = 5.53$ A, $I_{6\,\Omega} = 2.47$ A, $I_{8\,\Omega} = 0.53$ A, $I_{1\,\Omega} = 8.53$ A
29. **(I):** **(b)** $I_1 = 1.45$ mA; $I_2 = -8.51$ mA (all CW) **(c)** $I_{R_1} = 1.45$ mA; $I_{R_2} = 8.51$ mA; $I_{R_3} = 9.96$ mA **(II):** **(b)** $I_1 = 2.03$ mA; $I_2 = 1.23$ mA (all CW) **(c)** $I_{R_1} = 2.03$ mA; $I_{R_2} = 0.80$ mA; $I_{R_3} = I_{R_4} = 1.23$ mA
31. 63.69 mA
33. **(I):** $I_1 = 1.21$ mA; $I_2 = -0.48$ mA; $I_3 = -0.62$ mA (all CW) **(II):** $I_1 = -0.24$ A; $I_2 = -0.52$ A; $I_3 = -1.28$ A (all CW)
35. **(a)** $V_1 = 8.08$ V; $V_2 = 9.39$ V **(b)** $V_1 = 4.80$ V; $V_2 = 6.40$ V
37. **(I):** **(b)** $V_1 = -2.65$ V; $V_2 = 0.95$ V **(c)** $V_{R_1} = 15.35$ V; $V_{R_2} = 2.05$ V; $V_{R_3} = 2.65$ V; $V_{R_4} = 3.60$ V; $V_{R_5} = 0.95$ V **(II):** **(b)** $V_1 = 8.88$ V; $V_2 = 9.83$ V; $V_3 = -3.01$ V **(c)** $V_{R_1} = 4.12$ V; $V_{R_2} = 0.84$ V; $V_{R_3} = 5.17$ V; $V_{R_4} = 3.01$ V; $V_{R_5} = 0.95$ V
39. **(I):** $V_1 = -5.31$ V; $V_2 = -0.62$ V; $V_3 = 3.75$ V **(II):** $V_1 = -6.92$ V; $V_2 = 12$ V; $V_3 = 2.3$ V
41. **(a)** $V_1 = 8.08$ V; $V_2 = 9.39$ V **(b)** $V_1 = 4.8$ V; $V_2 = 6.4$ V
43. **(I):** **(b)** $V_1 = 7.24$ V; $V_2 = -2.45$ V; $V_3 = 1.41$ V **(c)** $V_{R_1} = 7.76$ V; $V_{R_2} = 2.45$ V; $V_{R_3} = 1.41$ V; $V_{R_4} = 3.86$ V; $V_{R_5} = 9.69$ V; $V_{R_6} = 5.83$ V **(II):** **(b)** $V_1 = -6.64$ V; $V_2 = 1.29$ V; $V_3 = 10.66$ V **(c)** $V_{R_1} = 6.64$ V; $V_{R_2} = 5.34$ V; $V_{R_3} = 1.29$ V; $V_{R_4} = 7.93$ V; $V_{R_5} = 9.37$ V; $V_{R_6} = 17.30$ V
45. **(b)** 40 mA **(c, d)** no
47. **(b)** 0 mA **(c, d)** yes
51. **(a)** 7.36 A **(b)** 1.76 mA
53. 2.14 A
55. **(a)** $R_T = 1$ kΩ **(b)** 5.71 mA

Chapter 9

1. **(a)** $I_{R_1} = 4$ A; $I_{R_2} = 3$ A; $I_{R_3} = 7$ A **(b)** E_1: 48 W; E_2: 48 W **(c)** 192 W **(d)** no
3. 3.17 A
5. 10.66 V
7. **(a)** $R_{Th} = 6\ \Omega$; $E_{Th} = 6$ V **(b)** 2 Ω: 0.75 A; 30 Ω: 166.67 mA; 100 Ω: 56.60 mA
9. **(a)** $R_{Th} = 7.5\ \Omega$; $E_{Th} = 10$ V **(b)** 2 Ω: 2.22 W; 100 Ω: 0.87 W
11. $R_{Th} = 1.58$ kΩ; $E_{Th} = -1.15$ V
13. **(I):** $R_{Th} = 45\ \Omega$; $E_{Th} = -5$ V **(II):** $R_{Th} = 2.06$ kΩ; $E_{Th} = 16.77$ V
15. $R_{Th} = 4.04$ kΩ; $E_{Th} = 9.74$ V
17. **(a)** $R_{Th} = 12.5$ kΩ; $E_{Th} = 20$ V **(b)** $R_{Th} = 2.72$ kΩ; $E_{Th} = 60$ mV **(c)** $R_{Th} = 2.2$ kΩ; $E_{Th} = 16$ V
19. **(a)** $R_N = 7.5\ \Omega$; $I_N = 1.34$ A
21. **(I):** **(a)** $R_N = 9.76\ \Omega$; $I_N = 0.95$ A **(II):** **(b)** $R_N = 2\ \Omega$; $I_N = 30$ A
23. **(a)** $R_N = 3\ \Omega$; $I_N = 5$ A **(b)** $R_N = 2\ \Omega$; $I_N = 0.75$ A
25. **(I):** **(a)** 10 Ω **(b)** 100 mW **(II):** **(a)** 4.03 kΩ **(b)** 8.93 mW
27. **(a)** 2 Ω **(b)** 60.5 W
29. **(a)** ∞ Ω **(b)** no
31. 6.12 A, 18.37 V
33. 0.38 A, 76.52 V
35. 2.32 mA, 15.78 V
39. **(a)** 0.5 mA **(b)** 0.5 mA **(c)** yes
41. **(a)** 4 V **(b)** 4 V **(c)** yes

Chapter 10

1. **(a)** 9×10^3 N/C **(b)** 36×10^9 N/C
3. 120 μF
5. 50 V/m
7. 8×10^3 V/m
9. 1.11 nF
11. mica
13. **(a)** 10^6 V/m **(b)** 4.96 μC **(c)** 24.80 nF
15. 29.04 kV
17. 0.35 μF
19. 176 μF → 264 μF
21. **(a)** 0.51 s **(b)** $\upsilon_C = 20\text{ V}(1 - e^{-t/0.51\text{ s}})$ **(c)** 1τ: 12.64 V; 3τ: 19 V; 5τ: 19.87 V **(d)** $i_C = 0.2\text{ mA } e^{-t/0.51\text{ s}}$, $\upsilon_R = 20\text{ V}e^{-t/0.51\text{ s}}$
23. **(a)** 5.5 ms **(b)** $\upsilon_C = 100\text{ V}(1 - e^{-t/5.5\text{ ms}})$ **(c)** 1τ: 63.21 V; 3τ: 95.02 V; 5τ: 99.33 V **(d)** $i_C = 18.18\text{ mA}e^{-t/5.5\text{ ms}}$, $\upsilon_R = 60\text{ V}e^{-t/5.5\text{ ms}}$
25. **(a)** 5 ms **(b)** 19.8 mV **(c)** 60 mV
27. **(a)** 200 ms **(b)** $\upsilon_C = 8\text{ V}(1 - e^{-t/200\text{ ms}})$; $i_C = 4\text{ mA}e^{-t/200\text{ ms}}$ **(c)** $\upsilon_C = 7.95$ V; $i_C = 26.95$ μA **(d)** $\upsilon_C = 7.95\text{ V}e^{-t/200\text{ ms}}$; $i_C = 3.98\text{ mA}e^{-t/200\text{ ms}}$
29. **(a)** $\upsilon_C = 50\text{ V}(1 - e^{-t/100\text{ ms}})$; $i_C = 10\text{ mA}e^{-t/100\text{ ms}}$; $\upsilon_{R_1} = 30\text{ V}e^{-t/100\text{ ms}}$ **(b)** $\upsilon_C = 31.6$ V; $i_C = 3.68$ mA; $\upsilon_{R_1} = 11.04$ V **(c)** $\upsilon_C = 31.6\text{ V}e^{-t/40\text{ ms}}$; $i_C = -15.8\text{ mA}e^{-t/40\text{ ms}}$; $\upsilon_{R_2} = -31.6\text{ V}e^{-t/40\text{ ms}}$
31. **(a)** 10 μs **(b)** 3 kA **(c)** yes
33. $\upsilon_C = 40\text{ V}e^{-t/4.4\text{ s}}$; $i_C = 18.18\text{ mA}e^{-t/4.4\text{ s}}$; $\upsilon_R = 40\text{ V}e^{-t/4.4\text{ s}}$
35. **(a)** $\upsilon_C = 52\text{ V} - 40\text{ V}e^{-t/123.8\text{ ms}}$; $i_C = 2.20\text{ mA}e^{-t/123.8\text{ ms}}$
37. 0.73 s
39. **(a)** 166.80 ms **(b)** 1 mA **(c)** 43.20 mW
41. **(a)** 22.07 V **(b)** 0.81 μA **(c)** 3.58 s
43. **(a)** $\upsilon_C = 27.2\text{ V} - 25.2\text{ V}e^{-t/18.26\text{ ms}}$; $i_C = 3.04\text{ mA}e^{-t/18.26\text{ ms}}$
45. **(a)** $\upsilon_C = 3.27\text{ V}(1 - e^{-t/53.80\text{ ms}})$; $i_C = 1.22\text{ mA}e^{-t/53.80\text{ ms}}$
47. **(a)** 19.63 V **(b)** 2.32 s **(c)** 1.15 s
49. 0 → 20 μs: $i_C = -1.18$ A; 20 → 30 μs: $i_C = 0$ A; 30 → 50 μs: $i_C = 2.35$ A; 50 → 80 μs: $i_C = -2.35$ A; 80 → 90 μs: $i_C = 4.7$ A; 90 → 100 μs: $i_C = 0$ A
51. 12 μF
53. $V_1 = 10$ V; $Q_1 = 60$ μC; $V_2 = 6.67$ V; $Q_2 = 40$ μC; $V_3 = 3.33$ V; $Q_3 = 40$ μC
55. $V_1 = 16$ V; $Q_1 = 144$ μC; $V_2 = 8$ V; $Q_2 = 80$ μC; $V_3 = 7.11$ V; $Q_3 = 64$ μC; $V_4 = 0.89$ V; $Q_4 = 64$ μC
57. 8,640 pJ
59. **(a)** $W_{6\,\mu F} = 1.19$ mJ; $W_{12\,\mu F} = 0.38$ mJ

(b) $W_{6\,\mu F} = 85.23\ \mu J$; $W_{12\,\mu F} = 42.77\ \mu J$

Chapter 11

1. (a) $0.04\ Wb/m^2$ (b) 0.04 T (c) 88 At (d) 0.4×10^3 gauss
3. 12.54 mH
5. (a) 45 mH (b) 1.67 mH (c) 80 mH (d) 1875 mH
7. 6.0 V
9. 14 turns
11. 5 V
13. (a) 2.27 μs (b) $i_L = 5.45\ mA\ (1 - e^{-t/2.27\ \mu s})$ (c) $v_L = 12\ Ve^{-t/2.27\ \mu s}$; $v_R = 12\ V\ (1 - e^{-t/2.27\ \mu s})$ (d) i_L: $1\tau = 3.45$ mA; $3\tau = 5.18$ mA; $5\tau = 5.41$ mA v_L: $1\tau = 4.42$ V; $3\tau = 0.60$ V; $5\tau = 0.08$V
15. (a) $i_L = 9.23\ mA - 1.23\ mAe^{-t/30.77\ \mu s}$ $v_L = 4.8\ Ve^{-t/30.77\ \mu s}$
17. (a) $i_L = 1.76\ mA - 4.76\ mAe^{-t/588.2\ \mu s}$ $v_L = 16.2\ Ve^{-t/588.2\ \mu s}$
19. (a) $i_L = 2\ mA\ (1 - e^{-t/1\ \mu s})$; $v_L = 20\ Ve^{-t/1\ \mu s}$ (b) $i_L = 2\ mAe^{-t/1\ \mu s}$; $v_L = -40\ Ve^{-t/1\ \mu s}$
21. (a) $i_L = 0.88\ mA\ (1 - e^{-t/0.74\ \mu s})$; $v_L = 6\ Ve^{-t/0.74\ \mu s}$ (b) $i_L = 0.88\ mAe^{-t/0.33\ \mu s}$; $v_L = -13.23\ Ve^{-t/0.33\ \mu s}$
23. (a) $i_L = 1.33\ mA\ (1 - e^{-t/55.56\ ns})$; $v_L = 48\ Ve^{-t/55.56\ ns}$ (b) $i_L = 1.11$ mA; $v_L = 7.93$ V
25. (a) 0.24 V (b) 29.47 V (c) 18.96 V (d) 2.03 ms
27. (a) $i_L = 3\ mA\ (1 - e^{-t/6.67\ \mu s})$; $v_L = 2.25\ Ve^{-t/6.67\ \mu s}$ (b) $i_L = 2.60$ mA; $v_L = 0.30$ V (c) $i_L = 2.60\ mAe^{-t/3.33\ \mu s}$; $v_L = -3.90\ Ve^{-t/3.33\ \mu s}$
29. (a) $i_L = -3.48\ mA - 7.43\ mAe^{-t/173.9\ \mu s}$ $v_L = 51.28\ Ve^{-t/173.9\ \mu s}$
31. (a) 20 V (b) 12 μA (c) 5.38 μs (d) 0.37 V
33. 0 → 2 ms: −50 mV; 2 → 5 ms: 0 V; 5 → 11 ms: 25 mV; 11 → 18 ms: −7.14 V; 18 ms →: 0 V
35. (a) 10 H (b) 2.4 H
37. (a) 16 mH in series with 18 μF
39. (a) $i_L = 3.56\ mA\ (1 - e^{-t/8.33\ \mu s})$; $v_L = 12.8\ Ve^{-t/8.33\ \mu s}$
41. $I_1 = 7$ A; $I_2 = 2$ A
43. $I_1 = 3$ A; $I_2 = 0$ A; $V_1 = 12$ V; $V_2 = 0$ V

Chapter 12

1. (Φ) CGS: 5×10^4 Maxwells; English: 5×10^4 lines (*B*) CGS: 8 Gauss; English: 51.62 lines/in.^{2}
3. (a) 0.04 T
5. 952.4×10^3 At/Wb
7. 2,624.67 At/m
9. 2.13 A
11. (a) 60 t (b) 13.34×10^{-4} Wb/Am
13. 2.70 A
15. 1.35 N
17. (a) 2.02 A (b) 2 N
19. 6.12 mWb
21. (a) $B = 1.5\ T\ (1 - e^{-H/700\ At/m})$ (b) 900 At/m: 1.09 T; 1800 At/m: 1.39 T; 2700 At/m: 1.47 T (c) $H = -700 \log_e \left[1 - \frac{B}{1.5}\right]$ (d) 1 T: 769.03 At/m; 1.4 T: 1895 At/m (e) 40.1 mA

Chapter 13

1. (a) 20 mA (b) 15 ms: −20 mA, 20 ms: 0 mA (c) 40 mA (d) 20 ms (e) 2.5 cycles
3. (a) 8 mV (b) 3 μs: −8 mV, 9 μs: 0 mV (c) 16 mV (d) 4.5 μs (e) 2.22 cycles
5. (a) 60 Hz (b) 100 Hz (c) 25 Hz (d) 40 kHz
7. 0.3 ms
9. (a) 150 mV (b) 40 μs (c) 25 kHz
11. (a) 45° (b) 30° (c) 18° (d) 108°
13. (a) 314.16 rad/s (b) 3769.91 rad/s (c) 12.56×10^3 rad/s (d) 25.13×10^3 rad/s
15. 2.08 ms
17. (a) 20, 60 Hz (b) 5, 120 Hz (c) 10^6, 1591.55 Hz (d) −6.4, 149.92 Hz
21. 0.48 A
23. 11.54°, 168.46°
27. (a) $v = 25 \sin(\omega t + 30°)$ (b) $i = 3 \times 10^{-3} \sin(6.28 \times 10^3 t - 60°)$
29. v leads i by 10°
31. i leads v by 80°
33. i leads v by 190°
35. 13.95 μs
37. 2 V
39. 3.87 mA
41. (a) 40 μs (b) 25 kHz (c) 17.13 mV
43. (a) $v = 14.14 \sin 377t$ (b) $i = 70.7 \times 10^{-3} \sin 377t$ (c) $v = 2.83 \times 10^3 \sin 377t$
45. 2.16 V
47. (a) $T = 40\ \mu s$, $f = 25$ kHz, Average = 20 mV, Peak = 40 mV, rms = 34.64 mV (b) $T = 100\ \mu s$, $f = 10$ kHz, Average = −0.3 V, Peak = 0.3 mV, rms = 367.42 mV

Chapter 14

3. (a) $3770 \cos 377t$ (b) $452.4 \cos(754t + 20°)$ (c) $4440.63 \cos(157t - 20°)$ (d) $200 \cos t$
5. (a) $v = 700 \sin 1000t$ (b) $v = 14.8 \sin(400t - 120°)$ (c) $v = 42 \times 10^{-3} \sin(\omega t + 88°)$ (d) $v = 28 \sin \omega t$
7. (a) 1.59 H (b) 2.65 H (c) 1.68 H
9. (a) $v = 100 \sin(\omega t + 90°)$ (b) $v = 0.8 \sin(\omega t + 150°)$ (c) $v = 120 \sin(\omega t - 120°)$ (d) $v = 60 \sin(\omega t + 190°)$
11. (a) $i = 2.4 \sin(\omega t - 90°)$ (b) $i = 0.6 \sin(\omega t - 70°)$ (c) $i = 0.8 \sin(\omega t + 10°)$ (d) $i = 1.6 \sin(377t + 130°)$
13. (a) ∞ Ω (b) 530.79 Ω (c) 265.39 Ω (d) 15.92 Ω (e) 62.83 Ω
15. (a) 31.83 Hz (b) 4.66 Hz (c) 9.31 Hz (d) 1.59 Hz
17. (a) $i = 6 \times 10^{-3} \sin(200t + 90°)$ (b) $i = 22.64 \times 10^{-6} \sin(377t + 90°)$ (c) $i = 44.94 \times 10^{-3} \sin(374t + 300°)$ (d) $i = 56 \times 10^{-3} \sin(\omega t + 160°)$
19. (a) $v = 1334 \sin(300t - 90°)$ (b) $v = 42.48 \sin(377t - 90°)$ (c) $v = 159 \sin 754t$ (d) $v = 100 \sin(1600t - 170°)$
21. (a) *C* (b) $L = 254.78$ mH (c) $R = 5\ \Omega$
25. 318.47 mH
27. 5.07 nF
29. 0 W
31. 192 W
33. $i = 40 \sin(\omega t - 50°)$
35. (a) $i = 4 \sin(314t - 30°)$ (b) 79.62 mH (c) 0 W
37. (a) $i_1 = 3.39 \sin(10^4 t + 150°)$, $i_2 = 16.97 \sin(10^4 t + 150°)$ (b) $i_s = 20.36 \sin(10^4 t + 150°)$
39. (a) $5.0 \angle 36.87°$ (b) $2.83 \angle 45°$ (c) $17.09 \angle 69.44°$ (d) $1.0 \times 10^3 \angle 84.29°$ (e) $1077.03 \angle 21.80°$ (f) $6.58 \times 10^{-3} \angle 81.25°$ (g) $11.78 \angle -49.82°$

(h) $8.94 \angle -153.43°$
(i) $61.85 \angle -104.04°$
(j) $101.73 \angle -39.94°$
(k) $4{,}326.66 \angle 123.69°$
(l) $25.5 \times 10^{-3} \angle -78.69°$

41. (a) $15.03 \angle 86.19°$
(b) $60.21 \angle 4.76°$
(c) $0.30 \angle 88.09°$
(d) $223.61 \angle -63.43°$
(e) $86.18 \angle 93.73°$
(f) $38.69 \angle -94.0°$

43. (a) $11.8 + j\,7.0$
(b) $151.90 + j\,49.90$
(c) $4.72 \times 10^{-6} + j\,71$
(d) $5.20 + j\,1.60$
(e) $209.30 + j\,311.0$
(f) $-21.20 + j\,12.0$
(g) $7.03 + j\,9.93$
(h) $95.7 + j\,22.77$

45. (a) $6.0 \angle -50°$
(b) $200 \times 10^{-6} \angle 60°$
(c) $109 \angle -170°$
(d) $76.47 \angle -80°$ (e) $4 \angle 0°$
(f) $5.93 \angle -134.47°$
(g) $4.21 \times 10^{-3} \angle 161.10°$
(h) $9.30 \angle -43.99°$

47. (a) $x = 4, y = 3$ (b) 4
(c) $x = 3, y = 6$ or $x = 6, y = 3$
(d) 30°

49. (a) $56.57 \sin(377t + 20°)$
(b) $169.68 \sin(377t + 10°)$
(c) $11.31 \times 10^{-3} \sin(377t + 120°)$
(d) $7.07 \sin(377t + 90°)$
(e) $1696.8 \sin(377t - 50°)$
(f) $6000 \sin(377t - 180°)$

51. $i_1 = 20.88 \times 10^{-6} \sin(\omega t + 76.70°)$

53. $i_s = -21.21 \times 10^{-3} \sin 377t$

Chapter 15

1. (a) $6.8\ \Omega \angle 0° = 6.8$
(b) $452.4\ \Omega \angle 90° = j\,452.4\ \Omega$
(c) $15.7\ \Omega \angle 90° = j\,15.7\ \Omega$
(d) $1\ k\Omega \angle -90° = -j\,1\ k\Omega$
(e) $318.47\ \Omega \angle -90° = -j\,318.47\ \Omega$ (f) $220\ \Omega \angle 0° = 220\ \Omega$

3. (a) $v = 88 \times 10^{-3} \sin \omega t$
(b) $v = 16.98 \sin(1000t + 150°)$
(c) $v = 254.7 \sin(157t - 50°)$

5. (a) $\mathbf{Z}_T = 3\ \Omega - j\,1\ \Omega = 3.16\ \Omega \angle -18.43°$ (b) $\mathbf{Z}_T = 1\ k\Omega + j\,4\ k\Omega = 4.12\ k\Omega \angle 75.96°$
(c) $\mathbf{Z}_T = 470\ \Omega - j\,80\ \Omega = 476.76\ \Omega \angle -9.66°$

7. (a) $10\ \Omega \angle 36.87°$
(c) $\mathbf{I} = 10\ A \angle -36.87°$, $\mathbf{V}_R = 80\ V \angle -36.87°$, $\mathbf{V}_L = 60\ V \angle 53.13°$
(f) 800 W (g) 0.8 lagging
(h) $v_R = 113.12 \sin(\omega t - 36.87°)$
$v_L = 84.84 \sin(\omega t + 53.13°)$
$i = 14.14 \sin(\omega t - 36.87°)$

9. (a) $2.34\ k\Omega \angle -19.89°$
(b) $6.04\ mA \angle 19.89°$
(c) $\mathbf{V}_R = 13.29\ V \angle 19.89°$, $\mathbf{V}_C = 4.81\ V \angle -70.11°$
(d) 80.26 mW, 0.94 leading

11. (a) $2.16\ k\Omega \angle 33.69°$ (c) $5.31\ \mu F$, 6.37 H
(d) $\mathbf{I} = 1.96\ mA \angle 26.31°$, $\mathbf{V}_R = 3.53\ V \angle 26.31°$, $\mathbf{V}_L = 2.68\ V \angle 116.31°$, $\mathbf{V}_C = 1.18\ V \angle -63.69°$
(g) 6.91 mW (h) 0.832 lagging
(i) $i = 2.77 \times 10^{-3} \sin(\omega t + 26.31°)$
$v_R = 4.99 \sin(\omega t + 26.31°)$
$v_L = 3.79 \sin(\omega t + 116.31°)$
$v_C = 1.67 \sin(\omega t - 63.69°)$

13. (a) 40 mH (b) $220\ \Omega$

15. (a) $\mathbf{V}_1 = 29.09\ V \angle -15.96°$, $\mathbf{V}_2 = 116.36\ V \angle 74.04°$
(b) $\mathbf{V}_1 = 48.69\ V \angle 40.75°$, $\mathbf{V}_2 = 26.78\ V \angle -49.25°$

17. (a) $\mathbf{I} = 39\ mA \angle 126.65°$, $\mathbf{V}_R = 1.17\ V \angle 126.65°$, $\mathbf{V}_C = 25.86\ V \angle 36.65°$
(b) 0.058 leading (c) 45.63 mW
(f) $\mathbf{V}_R = 1.17\ V \angle 126.65°$, $\mathbf{V}_C = 25.84\ V \angle 36.65°$
(g) $\mathbf{Z}_T = 30\ \Omega - j\,512.2\ \Omega$

19. $3.2\ \Omega + j\,2.4\ \Omega$

25. (a) $\mathbf{Z}_T = 91\ \Omega \angle 0°$, $\mathbf{Y}_T = 10.99\ mS \angle 0°$
(b) $\mathbf{Z}_T = 200\ \Omega \angle 90°$, $\mathbf{Y}_T = 5\ mS \angle -90°$ (c) $\mathbf{Z}_T = 0.2\ k\Omega \angle -90°$, $\mathbf{Y}_T = 5\ mS \angle 90°$
(d) $\mathbf{Z}_T = 9.86\ \Omega \angle 9.46°$, $\mathbf{Y}_T = 0.10\ S \angle -9.46°$
(e) $\mathbf{Z}_T = 1.90\ \Omega \angle -18.43°$, $\mathbf{Y}_T = 0.53\ S \angle 18.43°$
(f) $\mathbf{Z}_T = 2.94\ k\Omega \angle 9.55°$, $\mathbf{Y}_T = 0.34\ mS \angle -9.55°$

27. (a) $R = 5.85\ \Omega$, $X_C = 2.13\ \Omega$
(b) $R = 23.26\ k\Omega$, $X_C = 4.07\ k\Omega$
(c) $R = 80\ k\Omega$, $X_L = 46.19\ k\Omega$

29. (a) $\mathbf{Y}_T = 0.112\ mS \angle 26.57°$
(c) $\mathbf{E} = 17.89\ V \angle -6.57°$, $\mathbf{I}_R = 1.79\ mA \angle -6.57°$, $\mathbf{I}_C = 0.90\ mA \angle 83.44°$ (f) 32.04 mW
(g) 0.894 leading (h) $i_s = 2.83 \times 10^{-3} \sin(\omega t + 20°)$
$i_R = 2.53 \times 10^{-3} \sin(\omega t - 6.57°)$
$i_C = 1.27 \times 10^{-3} \sin(\omega t + 83.44°)$
$e = 25.3 \sin(\omega t - 6.57°)$

31. (a) $\mathbf{Y}_T = 0.89\ S \angle -19.81°$, $\mathbf{Z}_T = 1.12\ \Omega \angle 19.81°$
(c) $531\ \mu F$, 5.31 mH
(d) $\mathbf{E} = 2.40\ V \angle 79.81°$, $\mathbf{I}_R = 2.00\ A \angle 79.81°$, $\mathbf{I}_L = 1.20\ A \angle -10.19°$, $\mathbf{I}_C = 0.48\ A \angle 169.81°$ (g) 4.8 W
(h) 0.941 lagging
(i) $e = 3.39 \sin(377t + 79.81°)$
$i_R = 2.83 \sin(377t + 79.81°)$
$i_L = 1.70 \sin(377t - 10.19°)$
$i_C = 0.68 \sin(377t + 169.81°)$

33. (a) $\mathbf{Y}_T = 0.11\ S \angle 65.77°$, $\mathbf{Z}_T = 9.09\ \Omega \angle -65.77°$
(c) $636.9\ \mu F$, 31.8 mH
(d) $\mathbf{E} = 25.03\ V \angle 60°$, $\mathbf{I}_s = 2.75\ A \angle 125.77°$, $\mathbf{I}_C = 5\ A \angle 150°$, $\mathbf{I}_R = 1.14\ A \angle 60°$, $\mathbf{I}_L = 2.50\ A \angle -30°$
(g) 28.59 W (h) 0.409 leading
(i) $e = 35.4 \sin(314t + 60°)$
$i_s = 3.89 \sin(314t + 125.77°)$
$i_C = 7.07 \sin(314t + 150°)$
$i_R = 1.61 \sin(314t + 60°)$
$i_L = 3.54 \sin(314t - 30°)$

41. (a) $7.02\ k\Omega - j\,2.88\ k\Omega$
(b) $17.48\ \Omega + j\,29.72\ \Omega$

43. (a) $\mathbf{E} = 75.6\ V \angle -70.11°$, $\mathbf{I}_R = 0.34\ A \angle -70.11°$, $\mathbf{I}_L = 12.04\ mA \angle -160.11°$
(b) 0.340 leading (c) 25.97 W
(f) $0.47\ A \angle 19.63°$
(g) $25.72\ \Omega - j\,71.08\ \Omega$

47. (I): (a) v_1 leads v_2 by 72°
(b) v_1: p-p = 2.5 V, rms = 0.88 V
v_2: p-p = 1.2 V, rms = 0.42 V
(c) 1.25 kHz
(II): (a) v_1 leads v_2 by 132°
(b) v_1: p-p = 5.6 V, rms = 1.98 V
v_2: p-p = 8 V, rms = 2.83 V
(c) 16.67 kHz

Chapter 16

1. (a) $2.33\ \Omega \angle 30.96°$
(b) $5.15\ A \angle -30.96°$
(c) $5.15\ A \angle -30.96°$
(d) $\mathbf{I}_2 = 3.09\ A \angle -30.96°$, $\mathbf{I}_3 = 2.06\ A \angle -30.96°$
(e) $30.9\ V \angle 59.04°$

3. (a) $\mathbf{Z}_T = 19.86\ \Omega \angle 37.17°$, $\mathbf{Y}_T = 50.35\ mS \angle -37.17°$
(b) $3.02\ A \angle -7.17°$
(c) $3.98\ A \angle 82.83°$
(d) $47.81\ V \angle -7.17°$
(e) 167.07 W

5. (a) $0.25\ A \angle -6.34°$
(b) $70.71\ V \angle -45°$
(c) 24.85 W

7. (a) $1.42\ A \angle 18.26°$
(b) $26.57\ V \angle 4.76°$
(c) 54.07 W

9. (a) $82.51\ mS \angle -8.11°$
(b) $\mathbf{V}_1 = 20.4\ V \angle 30°$, $\mathbf{V}_2 = 10.89\ V \angle 58.13°$
(c) $1.93\ A \angle 11.11°$

11. 32.89 A $\angle 38.89°$
13. 139.71 mW

Chapter 17

3. (a) $\mathbf{Z} = 21.93\ \Omega\ \angle -46.85°$, $\mathbf{E} = 10.97\ \text{V}\ \angle 13.15°$ (b) $\mathbf{Z} = 5.15\ \Omega\ \angle 59.04°$, $\mathbf{E} = 10.30\ \text{V}\ \angle 179.04°$
5. (a) 5.15 A $\angle -24.5°$ (b) 0.44 A $\angle 143.48°$
7. (a) 13.07 A $\angle -33.71°$ (b) 48.33 A $\angle -77.57°$
9. $\mathbf{I}_L = -3.17 \times 10^{-3}\ \mathbf{V}\ \angle 137.29°$
11. $\mathbf{I}_{1k\Omega} = 10\ \text{mA}\ \angle 0°$, $\mathbf{I}_{2k\Omega} = 1.67\ \text{mA}\ \angle 0°$
13. 1.38 mA $\angle -56.31°$
15. (a) $\mathbf{V}_1 = 19.86\ \text{V}\ \angle 43.8°$, $\mathbf{V}_2 = 8.94\ \text{V}\ \angle 106.9°$ (b) $\mathbf{V}_1 = 19.78\ \text{V}\ \angle 132.48°$, $\mathbf{V}_2 = 13.37\ \text{V}\ \angle 98.78°$
17. $\mathbf{V}_1 = 100\ \text{V}\ \angle 90°$, $\mathbf{V}_2 = 96.66\ \text{V}\ \angle -12.43°$, $\mathbf{V}_3 = 0\ \text{V}\ \angle 0°$
19. $\mathbf{V}_1 = 14.62\ \text{V}\ \angle -5.86°$, $\mathbf{V}_2 = 35.03\ \text{V}\ \angle -37.69°$ $\mathbf{V}_3 = 32.4\ \text{V}\ \angle -73.34°$, $\mathbf{V}_4 = 5.67\ \text{V}\ \angle 23.53°$
21. $\mathbf{V}_{1k\Omega} = 2.25\ \text{V}\ \angle 17.63°$
23. $\mathbf{V}_{2k\Omega} = 10.67\ \text{V}\ \angle 180°$
25. $\mathbf{V}_L = -2451.92\ \mathbf{E}_i$
27. (a) no (b) 1.76 mA $\angle -71.54°$ (c) 7.03 V $\angle -18.46°$
29. yes
31. $L_x = R_2 L_3/R_1$
33. (a) 11.57 A $\angle -67.13°$ (b) 36.9 A $\angle 23.87°$

Chapter 18

1. (a) 6.09 A $\angle -32.12°$ (b) 3.77 A $\angle -93.8°$
3. $0.5 + 1.58 \sin(\omega t - 26.57°)$
5. 6.26 mA $\angle -63.43°$
7. -22.09 V $\angle 6.34°$
9. 19.62 V $\angle 53°$
11. 10 V $\angle 0°$
13. (a) $\mathbf{E}_{Th} = 2.13\ \text{V}\ \angle 32.2°$, $\mathbf{Z}_{Th} = 21.31\ \Omega\ \angle 32.2°$ (b) $\mathbf{E}_{Th} = 57.95\ \text{V}\ \angle 11.10°$, $\mathbf{Z}_{Th} = 6.81\ \Omega\ \angle -54.23°$
15. (a) $\mathbf{E}_{Th} = 4\ \text{V} + 10\ \text{V}\ \angle 0°$, $\mathbf{Z}_{Th} = 4\ \Omega\ \angle 90°$ (b) $0.5 + 1.58 \sin(\omega t - 26.57°)$
17. (a) $\mathbf{E}_{Th} = 31.31\ \text{V}\ \angle -26.57°$, $\mathbf{Z}_{Th} = 4.47\ \text{k}\Omega\ \angle -26.57°$ (b) 6.26 mA $\angle 63.44°$
19. $\mathbf{E}_{Th} = -444.45 \times 10^3\ \mathbf{I}\ \angle 0.26°$, $\mathbf{Z}_{Th} = 4.44\ \text{k}\Omega\ \angle -0.03°$
21. $\mathbf{E}_{Th} = -50\ \text{V}\ \angle 0°$, $\mathbf{Z}_{Th} = 5.10\ \text{k}\Omega\ \angle -11.31°$
23. $\mathbf{E}_{Th} = 20\ \text{V}\ \angle 53°$, $\mathbf{Z}_{Th} = -39.22\ \Omega\ \angle 0°$
25. $\mathbf{E}_{Th} = 1.62\ \text{V}\ \angle 0°$, $\mathbf{Z}_{Th} = 607.42\ \Omega\ \angle 0°$
27. (a) $\mathbf{I}_N = 0.1\ \text{A}\ \angle 0°$, $\mathbf{Z}_N = 21.31\ \Omega\ \angle 32.2°$ (b) $\mathbf{I}_N = 8.51\ \text{A}\ \angle 65.32°$, $\mathbf{Z}_N = 6.81\ \Omega\ \angle -54.23°$
29. (a) $\mathbf{I}_N = 2.15\ \text{A}\ \angle -42.87°$, $\mathbf{Z}_N = 9.66\ \Omega\ \angle 14.93°$ (b) $\mathbf{I}_N = 22.83\ \text{A}\ \angle -34.65°$, $\mathbf{Z}_N = 4.37\ \Omega\ \angle 55.67°$
31. (a) $\mathbf{I}_N = 1.33\ \text{A} + 2.67\ \text{A}\ \angle 0°$, $\mathbf{Z}_N = 9\ \Omega\ \angle 0°$ (b) $12\ \text{V} + 2.65\ \text{V}\ \angle -83.66°$
33. $\mathbf{I}_N = -1.96 \times 10^{-3}\ \mathbf{V}\ \angle 11.31°$, $\mathbf{Z}_N = 5.1\ \text{k}\Omega\ \angle -11.31°$
35. $\mathbf{I}_N = 9.81\ \text{mA}\ \angle 11.31°$, $\mathbf{Z}_N = 5.1\ \text{k}\Omega\ \angle -11.31°$
37. $\mathbf{I}_N = 0.79\ \text{mA}\ \angle 0°$, $\mathbf{Z}_N = 6.63\ \text{k}\Omega\ \angle 0°$
39. (a) $\mathbf{Z}_L = 8.32\ \Omega\ \angle 3.18°$, 1198.2 W (b) $\mathbf{Z}_L = 1.56\ \Omega\ \angle 14.47°$, 1.61 W
41. 40 kΩ, 25 W
43. (a) 9 Ω (b) 20 W
45. (a) 1.41 kΩ (b) 516.53 mW
49. 25.77 mA $\angle 104.4°$

Chapter 19

1. (a) 120 W (b) $Q_T = 0$ VAR, $S_T = 120$ VA (c) 0.5 A (d) 20 W: 720 Ω, 40 W: 360 Ω (e) $I_1 = 0.17$ A, $I_2 = 0.33$ A
3. (a) $P_T = 400$ W, $Q_T = -400$ VAR (C), $S_T = 565.69$ VA, $F_P = 0.707$ leading (c) 5.66 A $\angle 135°$
5. (a) $P_T = 500$ W, $Q_T = -200$ VAR (C), $S_T = 538.52$ VA (b) $F_P = 0.928$ leading (d) 10.78 A $\angle 21.88°$
7. (a) R: 200 W, L: 0 W, C: 0 W (b) R: 0 VAR, L: 100 VAR, C: 80 VAR (c) R: 200 VA, L: 100 VAR, C: 80 VA (d) $P_T = 200$ W, $Q_T = 20$ VAR (L), $S_T = 200$ VA, $F_P = 0.995$ lagging (f) 10.05 A $\angle -5.73°$
9. (a–c) L: 0 W, 126.74 VAR, 126.74 VA, C: 0 W, 46.92 VAR, 46.92 VA, R: 38.99 W, 0 VAR, 38.99 VA (d) $P_T = 38.99$ W, $Q_T = 79.82$ VAR (L), $S_T = 88.83$ VA, $F_p = 0.439$ lagging (f) $W_R = 0.31$ J (g) $W_L = 0.32$ J, $W_C = 0.12$ J
11. (a) $\mathbf{Z} = 2.30\ \Omega + j\,1.73\ \Omega$ (b) 4000 W
13. (a) $P_T = 900$ W, $Q_T = 0$ VAR, $S_T = 900$ VA, $F_p = 1$ (b) 9 A $\angle 0°$ (d) $\mathbf{Z}_1$: $X_C = 20\ \Omega$, $\mathbf{Z}_2$: $R = 2.83\ \Omega$, $\mathbf{Z}_3$: $R = 5.66\ \Omega$, $X_L = 4.72\ \Omega$
15. (a) $P_T = 1100$ W, $Q_T = 2366.26$ VAR, $S_T = 2609.44$ VA, $F_P = 0.422$ leading (b) 521.89 V $\angle -65.07°$ (c) $\mathbf{Z}_1$: $R = 1743.38\ \Omega$, $X_C = 1307.53\ \Omega$, $\mathbf{Z}_2$: $R = 43.59\ \Omega$, $X_C = 99.88\ \Omega$
17. (a) 7.81 kVA (b) 0.640 lagging (c) 65.08 A (d) 1105 μF (e) 41.67 A
19. (a) 128.14 W (b) *a–b:* 42.69 W, *b–c:* 64.03 W, *a–c:* 106.72 W, *a–d:* 106.72 W, *c–d:* 0 W, *d–e:* 0 W, *f–e:* 21.34 W
21. (a) $R = 5\ \Omega$, 132.03 mH (b) 10 Ω (c) 15 Ω, 262.39 mH

Chapter 20

1. (a) $\omega_s = 250$ rad/s, $f_s = 39.79$ Hz (b) $\omega_s = 3535.53$ rad/s, $f_s = 562.7$ Hz (c) $\omega_s = 21{,}880$ rad/s, $f_s = 3482.31$ Hz
3. (a) 40 Ω (b) 10 mA (c) $V_R = 20$ mV, $V_L = 400$ mV, $V_C = 400$ mV (d) 20 (high) (e) 1.27 mH, 795.77 nF (f) 250 Hz (g) 4.88 kHz, 5.13 kHz
5. (a) 400 Hz (b) 5800 Hz, 6200 Hz (c) 45 Ω (d) 375 mW
7. (a) 10 (b) 20 Ω (c) 1.59 mH, 3.98 μF (d) 1900 Hz, 2100 Hz
9. $R = 10\ \Omega$, $L = 13.26$ mH, $C = 27.07$ nF, 8340 Hz, 8460 Hz
11. (a) 1 MHz (b) 160 kHz (c) $R = 720\ \Omega$, $L = 0.716$ mH, $C = 35.37$ pF (d) 56.25 Ω
13. (a) 159.16 kHz (b) 4 V (c) 40 mA (d) 20
15. (a) 11,253.95 Hz (b) 1.77 (no) (c) $f_p = 9{,}280.24$ Hz, $f_m = 10{,}794.41$ Hz (d) $X_L = 5.83\ \Omega$, $X_C = 8.57\ \Omega$, $X_C > X_L$ (e) 12.5 Ω (f) 25 mV (g) $Q_p = 1.46$, $BW = 6.36$ kHz (h) $I_C = 2.92$ mA, $I_L = 3.54$ mA
17. (a) 30 Ω (b) 225 Ω (c) $\mathbf{I}_C = 0.6\ \text{A}\ \angle 90°$, $\mathbf{I}_L \cong 0.6\ \text{A}\ \angle -86.19°$ (d) $L = 0.239$ mH, $C = 265.26$ nF (e) $Q_p = 7.5$, $BW = 2.67$ kHz
19. (a) $f_s = 7.12$ kHz, $f_p = 6.65$ kHz, $f_m = 7$ kHz, low Q_p (b) $X_L = 20.88\ \Omega$, $X_C = 23.94\ \Omega$ (c) 55.56 Ω (d) $Q_p = 2.32$, $BW = 2.87$ kHz

(e) $I_C = 92.73$ mA, $I_L = 99.28$ mA
(f) 2.22 V

21. (a) 3558.81 Hz (b) 138.2 V (c) 691 mW (d) 575.86 Hz

23. (a) 98.54 Ω (b) 8.21 (c) 8.05 kHz (d) 4.83 V (e) $f_1 = 7.55$ kHz, $f_2 = 8.55$ kHz

25. $R_s = 3.24$ kΩ, $C = 31.66$ nF

27. (a) 251.65 kHz, (b) 4.44 kΩ (c) 14.05, (d) 17.91 kHz (e) *a:* 251.65 kHz, *b:* 49.94 Ω, *c:* 2.04, *d:* 95.55 kHz (f) *a:* 251.65 kHz, *b:* 13.33 kΩ, *c:* 21.08, *d:* 11.94 kHz (g) *Network:* 100×10^3, *part (e):* 1×10^3, *part (f):* 400×10^3 (h) As *L/C* increased, *BW* decreased

Chapter 21

1. (a) 1.54 kHz, 5.62 kHz (b) 0.22 V, 0.52 V

3. (a) 1000 (b) 10^{12} (c) 1.59 (d) 1.1 (e) 10^{10} (f) 1513.56 (g) 10.02 (h) 1,258,925.41

5. 1.68

7. −0.30

9. (a) 1.85 bels (b) 18.45 bels

11. 13.01 dB

13. 38.49 dB_v

15. 24.08 dB_s

19. (a) f_c: 0.707, $0.1f_c$: 0.995, $0.5f_c$: 0.894, $2f_c$: 0.447, $10f_c$: 0.0995 (b) f_c: −45°, $0.1f_c$: −5.71°, $0.5f_c$: −26.57°, $2f_c$: −63.43°, $10f_c$: −84.29°

21. $C = 0.265\ \mu$F

23. (a) f_c: 0.707, $2f_c$: 0.894, $0.5f_c$: 0.447, $10f_c$: 0.995, $0.1f_c$: 0.0995 (b) f_c: 45°, $2f_c$: 26.57°, $0.5f_c$: 63.43°, $10f_c$: 5.71°, $0.1f_c$: 84.29°

25. $R = 797$ Ω

27. (a) low-pass: $f_{c_1} = 795.77$ Hz, high-pass: $f_{c_2} = 1989.44$ Hz (b) $f_c = 1940$ Hz, $BW \cong 2.9$ kHz

29. (a) 100.66 kHz (b) $Q_s = 18.39$, $BW = 5{,}473.52$ Hz (d) f_s: 0.93 V, f_1: 0.66 V, f_2: 0.66 V

31. (a) 12.2 (b) $BW = 409.84$ Hz, $f_1 = 4.80$ kHz, $f_2 = 5.20$ kHz (d) little change

33. (a) $f_p = 726.44$ kHz (band-stop), $f = 2.01$ MHz (pass-band)

35. (a–b) $f_c = 772.55$ Hz (c) f_c: −3 dB, $0.5f_c$: −7 dB, $2f_c$: −0.969 dB, $0.1f_c$: −20.04 dB, $10f_c$: −0.043 dB (d) f_c: 0.707, $0.5f_c$: 0.447, $2f_c$: 0.894 (e) f_c: 45°, $0.5f_c$: 63.43°, $2f_c$: 26.57°

37. (a–b) $f_c = 13.26$ kHz (c) f_c: −3 dB, $0.5f_c$: −0.97 dB, $2f_c$: −6.99 dB, $0.1f_c$: −0.04 dB, $10f_c$: −20.04 dB (d) f_c: 0.707, $0.5f_c$: 0.894, $2f_c$: 0.447 (e) f_c: −45°, $0.5f_c$: −26.57°, $2f_c$: −63.43°

39. (a) $f_1 = 663.15$ Hz, $f_c = 468.1$ Hz (b) f_1: 45°, f_c: 54.78°, $2f_1$: 26.57°, $10f_1$: 5.71°

41. (a) $f_1 = 19{,}894.37$ Hz, $f_c = 1{,}989.44$ Hz (b) f_1: −39.29°, f_c: −39.29°, 10 kHz: −52.06°

43. (a) $f_1 = 964.58$ Hz, $f_c = 7{,}334.33$ Hz (b) f_1: 37.51°, f_c: 37.51°, 1.3 kHz: 43.38°

45. (a) $f = 180$ Hz: $A_{v_{dB}} \cong -3$ dB, $f = 18$ kHz: $A_{v_{dB}} = -3.105$ dB (b) $f = 180$ Hz: ≅ 90°, $f = 1.8$ kHz: ≅ 0°, $f = 18$ kHz: ≅ −90°

47. $A_v = \dfrac{-120}{[1 - j\frac{50}{f}][1 - j\frac{200}{f}][1 + j\frac{f}{36\text{ kHz}}]}$

49. $f_c = 2$ kHz

51. $f_1 = 1$ kHz: $A_{v_{dB}} = 3.06$ dB, $f_2 = 2$ kHz: $A_{v_{dB}} = 6.81$ dB, $f_3 = 3$ kHz: $A_{v_{dB}} = 9.1$ dB

53. (a) Woofer: 400 Hz: $A_v = 0.673$ Tweeter: 5 kHz: $A_v = 0.678$ (b) Woofer: 3 kHz: $A_v = 0.015$ Tweeter: 3 kHz: $A_v = 0.337$ (c) mid-range: 3 kHz: $A_v = 0.998$

Chapter 22

1. (a) 0.2 H (b) $e_p = 1.6$ V, $e_s = 5.12$ V (c) $e_p = 15$ V, $e_s = 24$ V

3. (a) 158.02 mH (b) $e_p = 24$ V, $e_s = 1.8$ V (c) $e_p = 15$ V, $e_s = 24$ V

5. (a) 3.13 V (b) 391.02 μWb

7. 56.31 Hz

9. 400 Ω

11. 12,000 turns

13. (a) 3 (b) 2.78 W

15. (a) 360.56 Ω ∠86.82° (b) 332.82 mA ∠−86.82° (c) $\mathbf{V}_{R_e} = 6.66$ V ∠−36.82°, $\mathbf{V}_{X_e} = 13.31$ V ∠3.18°, $\mathbf{V}_{X_L} = 106.50$ V ∠3.18°

19. 1.35 H

21. $\mathbf{I}_1(\mathbf{Z}_{R_1} + \mathbf{Z}_{L_1}) + \mathbf{I}_2(\mathbf{Z}_m) = \mathbf{E}_1$
$\mathbf{I}_1(\mathbf{Z}_m) + \mathbf{I}_2(\mathbf{Z}_{L_2} + \mathbf{Z}_{R_1}) = 0$
$\mathbf{X}_m = -\omega M \angle 90°$

23. (a) 20 (b) 83.33 A (c) 4.17 A (d) *a:* 0.05, *b:* 4.17 A, *c:* 83.33 A

25. (a) $\mathbf{V}_L = 25$ V ∠0°, $\mathbf{I}_s = 5$ A ∠0° (b) 80 Ω ∠0° (c) 20 Ω ∠0°

27. (a) $\mathbf{E}_2 = 40$ V ∠60°, $\mathbf{I}_2 = 3.33$ A ∠60°, $\mathbf{E}_3 = 30$ V ∠60°, $\mathbf{I}_3 = 3$ A ∠60° (b) 64.52 Ω

Chapter 23

1. (a) 120.1 V (b) 120.1 V (c) 12.01 A (d) 12.01 A

3. (a) 120.1 V (b) 120.1 V (c) 16.98 A (d) 16.98 A

5. (a) $\theta_2 = -120°$, $\theta_3 = +120°$ (b) $\mathbf{V}_{an} = 120$ V∠0°, $\mathbf{V}_{bn} = 120$ V∠−120°, $\mathbf{V}_{cn} = 120$ V∠120° (c) $\mathbf{I}_{an} = 8$ A ∠−53.13°, $\mathbf{I}_{bn} = 8$ A ∠−173.13°, $\mathbf{I}_{cn} = 8$ A ∠66.87° (e) 8 A (f) 207.85 V

7. $V_\phi = 127.0$ V, $I_\phi = 8.98$ A, $I_L = 8.98$ A

9. (a) $\mathbf{E}_{AN} = 12.7$ kV ∠−30°, $\mathbf{E}_{BN} = 12.7$ kV ∠−150°, $\mathbf{E}_{CN} = 12.7$ kV ∠90° (b–c) $\mathbf{I}_{an} = \mathbf{I}_{Aa} = 11.29$ A ∠−97.54°, $\mathbf{I}_{bn} = \mathbf{I}_{Bb} = 11.29$ A ∠−217.54°, $\mathbf{I}_{cn} = \mathbf{I}_{Cc} = 11.29$ A ∠22.46° (d) $\mathbf{V}_{an} = 12.16$ kV ∠−29.34°, $\mathbf{V}_{bn} = 12.16$ kV ∠−149.34°, $\mathbf{V}_{cn} = 12.16$ kV ∠90.66°

11. (a) 120.1 V (b) 208 V (c) 13.36 A (d) 23.15 A

13. (a) $\theta_2 = -120°$, $\theta_3 = +120°$ (b) $\mathbf{V}_{ab} = 208$ V ∠0°, $\mathbf{V}_{bc} = 208$ V ∠−120°, $\mathbf{V}_{ca} = 208$ V ∠120° (d) $\mathbf{I}_{ab} = 9.46$ A ∠0°, $\mathbf{I}_{bc} = 9.46$ A ∠−120°, $\mathbf{I}_{ca} = 9.46$ A ∠120° (e) 16.38 A (f) 120.1 V

15. (a) $\theta_2 = -120°$, $\theta_3 = +120°$ (b) $\mathbf{V}_{ab} = 208$ V ∠0°, $\mathbf{V}_{bc} = 208$ V ∠−120°, $\mathbf{V}_{ca} = 208$ V ∠120° (d) $\mathbf{I}_{ab} = 86.67$ A ∠−36.87°, $\mathbf{I}_{bc} = 16.67$ A ∠−156.87° $\mathbf{I}_{ca} = 86.67$ A ∠83.13° (e) 150.11 A (f) 120.1 V

17. (a) $\mathbf{I}_{ab} = 15.33$ A ∠−73.30°, $\mathbf{I}_{bc} = 15.33$ A ∠−193.30°, $\mathbf{I}_{ca} = 15.33$ A ∠46.7° (b) $\mathbf{I}_{Aa} = 26.55$ A ∠−103.30°, $\mathbf{I}_{Bb} = 26.55$ A ∠136.70°, $\mathbf{I}_{Cc} = 26.55$ A ∠16.70° (c) $\mathbf{E}_{AB} = 17.01$ kV ∠−0.59°, $\mathbf{E}_{BC} = 17.01$ kV ∠−120.59°, $\mathbf{E}_{CA} = 17.01$ kV ∠119.41°

19. **(a)** 208 V **(b)** 120.09 V **(c)** 7.08 A **(d)** 7.08 A

21. $V_\phi = 69.28$ V, $I_\phi = 2.89$ A, $I_L = 2.89$ A

23. $V_\phi = 69.28$ V, $I_\phi = 5.77$ A, $I_L = 5.77$ A

25. **(a)** 440 V **(b)** 440 V **(c)** 29.33 A **(d)** 50.8 A

27. **(a)** $\theta_2 = -120°$, $\theta_3 = +120°$ **(b)** $\mathbf{V}_{ab} = 100\text{ V} \angle 0°$, $\mathbf{V}_{bc} = 100\text{ V} \angle -120°$, $\mathbf{V}_{ca} = 100\text{ V} \angle 120°$ **(d)** $\mathbf{I}_{ab} = 5\text{ A} \angle 0°$, $\mathbf{I}_{bc} = 5\text{ A} \angle -120°$, $\mathbf{I}_{ca} = 5\text{ A} \angle 120°$ **(e)** 8.66 A

29. **(a)** $\theta_2 = -120°$, $\theta_3 = +120°$ **(b)** $\mathbf{V}_{ab} = 100\text{ V} \angle 0°$, $\mathbf{V}_{bc} = 100\text{ V} \angle -120°$, $\mathbf{V}_{ca} = 100\text{ V} \angle 120°$ **(d)** $\mathbf{I}_{ab} = 7.07\text{ A} \angle 45°$, $\mathbf{I}_{bc} = 7.07\text{ A} \angle -75°$, $\mathbf{I}_{ca} = 7.07\text{ A} \angle 165°$ **(e)** 12.25 A

31. $P_T = 2160$ W, $Q_T = 0$ VAR, $S_T = 2160$ VA, $F_p = 1$

33. $P_T = 7210.67$ W, $Q_T = 7210.67$ VAR (C), $S_T = 10{,}197.42$ VA, $F_P = 0.707$ (leading)

35. $P_T = 7.26$ kW, $Q_T = 7.26$ kVAR (L), $S_T = 10.27$ kVA, $F_P = 0.707$ (lagging)

37. $P_T = 287.93$ W, $Q_T = 575.86$ VAR (L), $S_T = 643.83$ VA, $F_p = 0.447$ (lagging)

39. $P_T = 900$ W, $Q_T = 1200$ VAR (L), $S_T = 1500$ VA, $F_p = 0.6$ (lagging)

41. $12.98\ \Omega - j\,7.31\ \Omega$

43. **(a)** 9,237.6 V **(b)** 80 A **(c)** 1276.8 kW **(d)** 0.576 lagging **(e)** $\mathbf{I}_{Aa} = 80\text{ A} \angle -54.83°$ **(f)** $\mathbf{V}_{an} = 7773.45\text{ V} \angle -4.87°$ **(g)** $62.52\ \Omega + j\,74.38\ \Omega$ **(h)** System: 0.576 lagging; Load: 0.643 lagging **(i)** 93.98%

45. **(b)** $P_T = 5899.64$ W, $P_{\text{meter}} = 1966.55$ W

49. **(a)** 120.09 V **(b)** $\mathbf{I}_{an} = 8.49$ A, $\mathbf{I}_{bn} = 7.08$ A, $\mathbf{I}_{cn} = 42.47$ A **(c)** $P_T = 4.93$ kW, $Q_T = 4.93$ kVAR (L), $S_T = 6.97$ kVA, $F_p = 0.707$ (lagging) **(d)** $\mathbf{I}_{an} = 8.49\text{ A} \angle -75°$, $\mathbf{I}_{bn} = 7.08\text{ A} \angle -195°$, $\mathbf{I}_{cn} = 42.47\text{ A} \angle 45°$ **(e)** $35.09\text{ A} \angle -43.00°$

Chapter 24

1. **(a)** positive-going **(b)** 2 V **(c)** 0.2 ms **(d)** 6 V **(e)** 6.5%

3. **(a)** positive-going **(b)** 10 mV **(c)** 3.2 ms **(d)** 20 mV **(e)** 6.9%

5. $V_2 = 13.58$ mV

7. **(a)** 120 μs **(b)** 8.33 kHz **(c)** maximum: 440 mV minimum: 80 mV

9. prf = 125 kHz Duty cycle = 62.5%

11. **(a)** 8 μs **(b)** 2 μs **(c)** 125 kHz **(d)** 0 V **(e)** 3.46 mV

13. 18.88 mV

15. 117 mV

17. $v_C = 4\text{ V}(1 - e^{-t/20\text{ms}})$

19. $i_C = -8\text{ mA}e^{-t}$

21. $i_C = 4\text{ mA}\ e^{-t/0.2\text{ ms}}$

23. $0 \rightarrow T/2$: $v_C = 20$ V, $T/2 \rightarrow T$: $v_C = 20\text{ V}e^{-t/02.\text{ ms}}$, $T \rightarrow 3/2\,T$: $v_C = 20\text{ V}(1 - e^{-t/0.2\text{ ms}})$ $3/2\,T \rightarrow 2T$: $v_C = 20\text{ V}e^{-t/02.\text{ ms}}$,

25. $\mathbf{V}_{\text{scope}} = 10\text{ V} \angle 0°$, $\theta_{\mathbf{Z}_S} = \theta_{\mathbf{Z}_P} = -59.5°$

Chapter 25

1. **(I):** **(a)** no **(b)** no **(c)** yes **(d)** no **(e)** yes **(II):** **(a)** yes **(b)** yes **(c)** yes **(d)** yes **(e)** no **(III):** **(a)** yes **(b)** yes **(c)** no **(d)** yes **(e)** yes **(IV):** **(a)** no **(b)** no **(c)** yes **(d)** yes **(e)** yes

7. **(a)** 19.04 V **(b)** 4.53 A

9. 71.87 W

11. **(a)** $2 + 2.08 \sin(400t - 33.69°) + 0.5 \sin(800t - 53.13°)$ **(b)** 2.51 A **(c)** $24 + 24.96 \sin(400t - 33.69°) + 6 \sin(800t - 53.13°)$ **(d)** 30.09 V **(e)** $16.64 \sin(400t + 56.31°) + 8 \sin(800t + 36.87°)$ **(f)** 13.06 V **(g)** 75.48 W

13. **(a)** $1.2 \sin(400t + 53.13°)$ **(b)** 0.85 A **(c)** $18 \sin(400t + 53.13°)$ **(d)** 12.73 V **(e)** $18 + 23.98 \sin(400t - 36.87°)$ **(f)** 24.73 V **(g)** 10.79 W

15. $2.26 \times 10^{-3} \sin(377t + 93.66°) + 1.92 \times 10^{-3} \sin(754t + 1.64°)$

17. $30 + 30.27 \sin(20t + 7.59°) + 0.5 \sin(40t - 30°)$

Index

Q

R

S